Osvaldo Gervasi   Marina L. Gavrilova
Vipin Kumar   Antonio Laganà
Heow Pueh Lee   Youngsong Mun
David Taniar   Chih Jeng Kenneth Tan (Eds.)

Computational Science and Its Applications – ICCSA 2005

International Conference
Singapore, May 9-12, 2005
Proceedings, Part IV

Volume Editors

Osvaldo Gervasi
University of Perugia
E-mail: ogervasi@computer.org

Marina L. Gavrilova
University of Calgary
E-mail: marina@cpsc.ucalgary.ca

Vipin Kumar
University of Minnesota
E-mail: kumar@cs.umn.edu

Antonio Laganà
University of Perugia
E-mail: lag@dyn.unipg.it

Heow Pueh Lee
Institute of High Performance Computing, IHPC
E-mail: hplee@ihpc.a-star.edu.sg

Youngsong Mun
Soongsil University
E-mail: mun@computing.soongsil.ac.kr

David Taniar
Monash University
E-mail: David.Taniar@infotech.monash.edu.au

Chih Jeng Kenneth Tan
Queen's University Belfast
E-mail: cjtan@optimanumerics.com

Library of Congress Control Number: Applied for

CR Subject Classification (1998): D, F, G, H, I, J, C.2.3

ISSN   0302-9743
ISBN-10   3-540-25863-9 Springer Berlin Heidelberg New York
ISBN-13   978-3-540-25863-6 Springer Berlin Heidelberg New York

This work is subject to copyright. All rights are reserved, whether the whole or part of the material is concerned, specifically the rights of translation, reprinting, re-use of illustrations, recitation, broadcasting, reproduction on microfilms or in any other way, and storage in data banks. Duplication of this publication or parts thereof is permitted only under the provisions of the German Copyright Law of September 9, 1965, in its current version, and permission for use must always be obtained from Springer. Violations are liable to prosecution under the German Copyright Law.

Springer is a part of Springer Science+Business Media

springeronline.com

© Springer-Verlag Berlin Heidelberg 2005
Printed in Germany

Typesetting: Camera-ready by author, data conversion by Scientific Publishing Services, Chennai, India
Printed on acid-free paper   SPIN: 11424925   06/3142   5 4 3 2 1 0

Preface

The four-volume set assembled following *the 2005 International Conference on Computational Science and Its Applications*, ICCSA 2005, held in Suntec International Convention and Exhibition Centre, Singapore from 9 May 2005 till 12 May 2005, represents the fine collection of 540 refereed papers selected from nearly 2700 submissions.

Computational science has firmly established itself as a vital part of many scientific investigations, affecting researchers and practitioners in areas ranging from applications such as aerospace and automotive, to emerging technologies such as bioinformatics and nanotechnologies, to core disciplines such as mathematics, physics, and chemistry. Due to the sheer size of many challenges in computational science, the use of supercomputing, parallel processing, and sophisticated algorithms is inevitable and becomes a part of fundamental theoretical research as well as endeavors in emerging fields. Together, these far-reaching scientific areas contribute to shape this conference in the realms of state-of-the-art computational science research and applications, encompassing the facilitating theoretical foundations and the innovative applications of such results in other areas.

The topics of the refereed papers span all the traditional as well as emerging computational science realms, and are structured according to six main conference themes:

- Computational Methods and Applications
- High-Performance Computing, Networks and Optimization
- Information Systems and Information Technologies
- Scientific Visualization, Graphics and Image Processing
- Computational Science Education
- Advanced and Emerging Applications

In addition, papers from 27 workshops and technical sessions on specific topics of interest, including information security, mobile communication, grid computing, modeling, optimization, computational geometry, virtual reality, symbolic computations, molecular structures, Web systems and intelligence, spatial analysis, bioinformatics and geocomputation, to name a few, complete this comprehensive collection.

The warm response of the great number of researchers to the offer to present high-quality papers in ICCSA 2005 took the conference to record heights. The continuous support of computational science researchers has helped build ICCSA into a firmly established forum in this area. We look forward to building on this symbiotic relationship together to grow ICCSA further.

We recognize the contribution of the International Steering Committee and we deeply thank the International Program Committee for their tremendous support in putting this conference together, nearly 900 referees for their diligent work, and the Institute of High Performance Computing, Singapore for its generous assistance in hosting the event.

We also thank our sponsors for their continuous support without which this conference would not have been possible.

Finally, we thank all authors for their submissions and all invited speakers and conference attendees for making the ICCSA conference truly one of the premium events in the scientific community, facilitating the exchange of ideas, fostering new collaborations, and shaping the future of computational science.

May 2005

Marina L. Gavrilova
Osvaldo Gervasi

on behalf of the co-editors

Vipin Kumar
Antonio Laganà
Heow Pueh Lee
Youngsong Mun
David Taniar
Chih Jeng Kenneth Tan

Organization

ICCSA 2005 was organized by the Institute of High Performance Computing (Singapore), the University of Minnesota (Minneapolis, MN, USA), the University of Calgary (Canada) and the University of Perugia (Italy).

Conference Chairs

Vipin Kumar (Army High Performance Computing Center and University of Minnesota, USA), Honorary Chair
Marina L. Gavrilova (University of Calgary, Canada), Conference Co-chair, Scientific
Osvaldo Gervasi (University of Perugia, Italy), Conference Co-chair, Program
Jerry Lim (Institute of High Performance Computing, Singapore), Conference Co-chair, Organization

International Steering Committee

Alexander V. Bogdanov (Institute for High Performance Computing and Information Systems, Russia)
Marina L. Gavrilova (University of Calgary, Canada)
Osvaldo Gervasi (University of Perugia, Italy)
Kurichi Kumar (Institute of High Performance Computing, Singapore)
Vipin Kumar (Army High Performance Computing Center and University of Minnesota, USA)
Andres Iglesias (University de Cantabria, Spain)
Antonio Laganà (University of Perugia, Italy)
Heow Pueh Lee (Institute of High Performance Computing, Singapore)
Youngsong Mun (Soongsil University, Korea)
Chih Jeng Kenneth Tan (OptimaNumerics Ltd., and Queen's University Belfast, UK)
David Taniar (Monash University, Australia)

Local Organizing Committee

Kurichi Kumar (Institute of High Performance Computing, Singapore)
Heow Pueh Lee (Institute of High Performance Computing, Singapore)

Workshop Organizers

Approaches or Methods of Security Engineering
Haeng Kon Kim (Catholic University of Daegu, Korea)
Tai-hoon Kim (Korea Information Security Agency, Korea)

Authentication, Authorization and Accounting
Eui-Nam John Huh (Seoul Women's University, Korea)

Component-Based Software Engineering and Software Process Models
Haeng Kon Kim (Catholic University of Daegu, Korea)

Computational Geometry and Applications (CGA 2005)
Marina Gavrilova (University of Calgary, Canada)

Computer Graphics and Geometric Modeling (TSCG 2005)
Andres Iglesias (University of Cantabria, Spain)
Deok-Soo Kim (Hanyang University, Korea)

Computer Graphics and Rendering
Jiawan Zhang (Tianjin University, China)

Data Mining and Bioinformatics
Xiaohua Hu (Drexel University, USA)
David Taniar (Monash University, Australia)

Digital Device for Ubiquitous Computing
Hong Joo Lee (Daewoo Electronics Corp, Korea)

Grid Computing and Peer-to-Peer (P2P) Systems
Jemal H. Abawajy (Deakin University, Australia)
Maria S. Perez (Universitad Politecnica de Madrid, Spain)

Information and Communication Technology (ICT) Education
Woochun Jun (Seoul National University, Korea)

Information Security and Hiding, ISH 2005
Raphael C.W. Phan (Swinburne University of Technology, Sarawak, Malaysia)

Intelligent Multimedia Services and Synchronization in Mobile Multimedia Networks

Dong Chun Lee (Howon University, Korea)
Kuinam J. Kim (Kyonggi University, Korea)

Information Systems Information Technologies (ISIT)

Youngsong Mun (Soongsil University, Korea)

Internet Comunications Security (WICS)

Josè Sierra-Camara (University Carlos III of Madrid, Spain)
Julio Hernandez-Castro (University Carlos III of Madrid, Spain)
Antonio Izquierdo (University Carlos III of Madrid, Spain)
Joaquin Torres (University Carlos III of Madrid, Spain)

Methodology of Information Engineering

Sangkyun Kim (Somansa Ltd., Korea)

Mobile Communications

Hyunseung Choo (Sungkyunkwan University, Korea)

Modeling Complex Systems

Heather J. Ruskin (Dublin City University, Ireland)
Ruili Wang (Massey University, New Zealand)

Modeling of Location Management in Mobile Information Systems

Dong Chun Lee (Howon University, Korea)

Molecular Structures and Processes

Antonio Laganà (University of Perugia, Italy)

Optimization: Theories and Applications (OTA 2005)

In-Jae Jeong (Hanyang University, Korea)
Dong-Ho Lee (Hanyang University, Korea)
Deok-Soo Kim (Hanyang University, Korea)

Parallel and Distributed Computing

Jiawan Zhang (Tianjin University, China)

Pattern Recognition and Ubiquitous Computing

Woongjae Lee (Seoul Women's University, Korea)

Spatial Analysis and GIS: Local or Global?

Stefania Bertazzon (University of Calgary, Canada)
Borruso Giuseppe (University of Trieste, Italy)
Falk Huettmann (Institute of Arctic Biology, USA)

Specific Aspects of Computational Physics for Modeling Suddenly Emerging Phenomena

Paul E. Sterian (Politehnica University, Romania)
Cristian Toma (Titu Maiorescu University, Romania)

Symbolic Computation (SC 2005)

Andres Iglesias (University of Cantabria, Spain)
Akemi Galvez (University of Cantabria, Spain)

Ubiquitous Web Systems and Intelligence

David Taniar (Monash University, Australia)
Wenny Rahayu (La Trobe University, Australia)

Virtual Reality in Scientific Applications and Learning (VRSAL 2005)

Osvaldo Gervasi (University of Perugia, Italy)
Antonio Riganelli (University of Perugia, Italy)

Program Committee

Jemal Abawajy (Deakin University, Australia)
Kenny Adamson (EZ-DSP, UK)
Srinivas Aluru (Iowa State University, USA)
Frank Baetke (Hewlett-Packard, USA)
Mark Baker (Portsmouth University, UK)
Young-Cheol Bang (Korea Polytechnic University, Korea)
David Bell (Queen's University Belfast, UK)
Stefania Bertazzon (University of Calgary, Canada)
Sergei Bespamyatnikh (Duke University, USA)
J.A. Rod Blais (University of Calgary, Canada)
Alexander V. Bogdanov (Institute for High Performance Computing and
    Information Systems, Russia)
Richard P. Brent (University of Oxford, UK)
Peter Brezany (University of Vienna, Austria)
Herve Bronnimann (Polytechnic University, NY, USA)
John Brooke (University of Manchester, UK)
Martin Buecker (Aachen University, Germany)
Rajkumar Buyya (University of Melbourne, Australia)
YoungSik Choi (University of Missouri, USA)

Hyunseung Choo (Sungkyunkwan University, Korea)
Bastien Chopard (University of Geneva, Switzerland)
Min Young Chung (Sungkyunkwan University, Korea)
Toni Cortes (Universidad de Catalunya, Spain)
Yiannis Cotronis (University of Athens, Greece)
Danny Crookes (Queen's University Belfast, UK)
Josè C. Cunha (New University of Lisbon, Portugal)
Brian J. d'Auriol (University of Texas at El Paso, USA)
Alexander Degtyarev (Institute for High Performance Computing and Data Bases, Russia)
Frédéric Desprez (INRIA, France)
Tom Dhaene (University of Antwerp, Belgium)
Beniamino Di Martino (Second University of Naples, Italy)
Hassan Diab (American University of Beirut, Lebanon)
Ivan Dimov (Bulgarian Academy of Sciences, Bulgaria)
Iain Duff (Rutherford Appleton Laboratory, UK and CERFACS, France)
Thom Dunning (NCSA, USA)
Fabrizio Gagliardi (CERN, Switzerland)
Marina L. Gavrilova (University of Calgary, Canada)
Michael Gerndt (Technical University of Munich, Germany)
Osvaldo Gervasi (University of Perugia, Italy)
Bob Gingold (Australian National University, Australia)
James Glimm (SUNY Stony Brook, USA)
Christopher Gold (Hong Kong Polytechnic University, China)
Yuriy Gorbachev (Institute of High Performance Computing and Information Systems, Russia)
Andrzej Goscinski (Deakin University, Australia)
Jin Hai (Huazhong University of Science and Technology, China)
Ladislav Hlucky (Slovak Academy of Science, Slovakia)
Shen Hong (Japan Advanced Institute of Science and Technology, Japan)
Paul Hovland (Argonne National Laboratory, USA)
Xiaohua Hu (Drexel University, USA)
Eui-Nam John Huh (Seoul Women's University, Korea)
Terence Hung (Institute of High Performance Computing, Singapore)
Andres Iglesias (University of Cantabria, Spain)
In-Jae Jeong (Hanyang University, Korea)
Elisabeth Jessup (University of Colorado, USA)
Peter K. Jimack (University of Leeds, UK)
Christopher Johnson (University of Utah, USA)
Benjoe A. Juliano (California State University at Chico, USA)
Peter Kacsuk (MTA SZTAKI Research Institute, Hungary)
Kyung Woo Kang (KAIST, Korea)
Carl Kesselman (University of Southern California, USA)
Daniel Kidger (Quadrics, UK)
Deok-Soo Kim (Hanyang University, Korea)

Haeng Kon Kim (Catholic University of Daegu, Korea)
Jin Suk Kim (KAIST, Korea)
Tai-hoon Kim (Korea Information Security Agency, Korea)
Yoonhee Kim (Syracuse University, USA)
Mike Kirby (University of Utah, USA)
Jacek Kitowski (AGH University of Science and Technology, Poland)
Dieter Kranzlmueller (Johannes Kepler University Linz, Austria)
Kurichi Kumar (Institute of High Performance Computing, Singapore)
Vipin Kumar (Army High Performance Computing Center and University of Minnesota, USA)
Domenico Laforenza (Italian National Research Council, Italy)
Antonio Laganà (University of Perugia, Italy)
Joseph Landman (Scalable Informatics LLC, USA)
Francis Lau (University of Hong Kong, Hong Kong, China)
Bong Hwan Lee (Texas A&M University, USA)
Dong Chun Lee (Howon University, Korea)
Dong-Ho Lee (Hanyang University, Korea)
Heow Pueh Lee (Institute of High Performance Computing, Singapore)
Sang Yoon Lee (Georgia Institute of Technology, USA)
Tae Jin Lee (Sungkyunkwan University, Korea)
Bogdan Lesyng (ICM Warszawa, Poland)
Zhongze Li (Chinese Academy of Sciences, China)
Laurence Liew (Scalable Systems Pte., Singapore)
David Lombard (Intel Corporation, USA)
Emilio Luque (Universitat Autonoma of Barcelona, Spain)
Michael Mascagni (Florida State University, USA)
Graham Megson (University of Reading, UK)
John G. Michopoulos (US Naval Research Laboratory, USA)
Edward Moreno (Euripides Foundation of Marilia, Brazil)
Youngsong Mun (Soongsil University, Korea)
Jiri Nedoma (Academy of Sciences of the Czech Republic, Czech Republic)
Genri Norman (Russian Academy of Sciences, Russia)
Stephan Olariu (Old Dominion University, USA)
Salvatore Orlando (University of Venice, Italy)
Robert Panoff (Shodor Education Foundation, USA)
Marcin Paprzycki (Oklahoma State University, USA)
Gyung-Leen Park (University of Texas, USA)
Ron Perrott (Queen's University Belfast, UK)
Dimitri Plemenos (University of Limoges, France)
Richard Ramaroson (ONERA, France)
Rosemary Renaut (Arizona State University, USA)
Alexey S. Rodionov (Russian Academy of Science, Russia)
Paul Roe (Queensland University of Technology, Australia)
Reneé S. Renner (California State University at Chico, USA)
Heather J. Ruskin (Dublin City University, Ireland)

Ole Saastad (Scali, Norway)
Muhammad Sarfraz (King Fahd University of Petroleum and Minerals, Saudi Arabia)
Edward Seidel (Louisiana State University, USA and Albert Einstein Institute, Germany)
Josè Sierra-Camara (University Carlos III of Madrid, Spain)
Dale Shires (US Army Research Laboratory, USA)
Vaclav Skala (University of West Bohemia, Czech Republic)
Burton Smith (Cray, USA)
Masha Sosonkina (University of Minnesota, USA)
Alexei Sourin (Nanyang Technological University, Singapore)
Elena Stankova (Institute for High Performance Computing and Data Bases, Russia)
Gunther Stuer (University of Antwerp, Belgium)
Kokichi Sugihara (University of Tokyo, Japan)
Boleslaw Szymanski (Rensselaer Polytechnic Institute, USA)
Ryszard Tadeusiewicz (AGH University of Science and Technology, Poland)
Chih Jeng Kenneth Tan (OptimaNumerics, UK and Queen's University Belfast, UK)
David Taniar (Monash University, Australia)
John Taylor (Quadrics, UK)
Ruppa K. Thulasiram (University of Manitoba, Canada)
Pavel Tvrdik (Czech Technical University, Czech Republic)
Putchong Uthayopas (Kasetsart University, Thailand)
Mario Valle (Visualization Group, Swiss National Supercomputing Centre, Switzerland)
Marco Vanneschi (University of Pisa, Italy)
Piero Giorgio Verdini (University of Pisa and Istituto Nazionale di Fisica Nucleare, Italy)
Jesus Vigo-Aguiar (University of Salamanca, Spain)
Jens Volkert (University of Linz, Austria)
Koichi Wada (University of Tsukuba, Japan)
Kevin Wadleigh (Hewlett-Packard, USA)
Jerzy Wasniewski (Technical University of Denmark, Denmark)
Paul Watson (University of Newcastle upon Tyne)
Jan Weglarz (Poznan University of Technology, Poland)
Tim Wilkens (Advanced Micro Devices, USA)
Roman Wyrzykowski (Technical University of Czestochowa, Poland)
Jinchao Xu (Pennsylvania State University, USA)
Chee Yap (New York University, USA)
Osman Yasar (SUNY at Brockport, USA)
George Yee (National Research Council and Carleton University, Canada)
Yong Xue (Chinese Academy of Sciences, China)
Igor Zacharov (SGI Europe, Switzerland)
Xiaodong Zhang (College of William and Mary, USA)

Aledander Zhmakin (SoftImpact, Russia)
Krzysztof Zielinski (ICS UST/CYFRONET, Poland)
Albert Zomaya (University of Sydney, Australia)

Sponsoring Organizations

The Institute of High Performance Computing, Singapore
University of Perugia, Perugia, Italy
University of Calgary, Calgary, Canada
University of Minnesota, Minneapolis, USA
Queen's University Belfast, UK
Society for Industrial and Applied Mathematics, USA
The Institution of Electrical Engineers, UK
OptimaNumerics Ltd., UK
MASTER-UP, Italy

Table of Contents – Part IV

Information and Communication Technology (ICT) Education Workshop

Exploring Constructivist Learning Theory and Course Visualization on Computer Graphics
  Yiming Zhao, Mingming Zhang, Shu Wang, Yefang Chen 1

A Program Plagiarism Evaluation System
  Young-Chul Kim, Jaeyoung Choi 10

Integrated Development Environment for Digital Image Computing and Configuration Management
  Jeongheon Lee, YoungTak Cho, Hoon Heo, Oksam Chae 20

E-Learning Environment Based on Intelligent Synthetic Characters
  Lu Ye, Jiejie Zhu, Mingming Zhang, Ruth Aylett, Lifeng Ren, Guilin Xu 30

SCO Control Net for the Process-Driven SCORM Content Aggregation Model
  Kwang-Hoon Kim, Hyun-Ah Kim, Chang-Min Kim 38

Design and Implementation of a Web-Based Information Communication Ethics Education System for the Gifted Students in Computer
  Woochun Jun, Sung-Keun Cho, Byeong Heui Kwak 48

International Standards Based Information Technology Courses: A Case Study from Turkey
  Mustafa Murat Inceoglu 56

Design and Implementation of the KORI: Intelligent Teachable Agent and Its Application to Education
  Sung-il Kim, Sung-Hyun Yun, Mi-sun Yoon, Yeon-hee So, Won-sik Kim, Myung-jin Lee, Dong-seong Choi, Hyung-Woo Lee ... 62

Digital Device for Ubiquitous Computing Workshop

A Space-Efficient Flash Memory Software for Mobile Devices
  Yeonseung Ryu, Tae-sun Chung, Myungho Lee 72

Security Threats and Their Countermeasures of Mobile Portable
Computing Devices in Ubiquitous Computing Environments
   Sang ho Kim, Choon Seong Leem 79

A Business Model (BM) Development Methodology in Ubiquitous
Computing Environment
   Choon Seong Leem, Nam Joo Jeon, Jong Hwa Choi,
   Hyoun Gyu Shin 86

Developing Business Models in Ubiquitous Era: Exploring
Contradictions in Demand and Supply Perspectives
   Jungwoo Lee, Sunghwan Lee 96

Semantic Web Based Intelligent Product and Service Search Framework
for Location-Based Services
   Wooju Kim, SungKyu Lee, DeaWoo Choi 103

A Study on Value Chain in a Ubiquitous Computing Environment
   Hong Joo Lee, Choon Seong Leem 113

A Study on Authentication Mechanism Using Robot Vacuum Cleaner
   Hong Joo Lee, Hee Jun Park, Sangkyun Kim 122

Design of Inside Information Leakage Prevention System in Ubiquitous
Computing Environment
   Hangbae Chang, Kyung-kyu Kim 128

Design and Implementation of Home Media Server Using TV-Anytime
for Personalized Broadcasting Service
   Changho Hong, Jongtae Lim 138

Optimization: Theories and Applications (OTA) 2005 Workshop

Optimal Signal Control Using Adaptive Dynamic Programming
   Chang Ouk Kim, Yunsun Park, Jun-Geol Baek 148

Inverse Constrained Bottleneck Problems on Networks
   Xiucui Guan, Jianzhong Zhang 161

Dynamic Scheduling Problem of Batch Processing Machine in
Semiconductor Burn-in Operations
   Pei-Chann Chang, Yun-Shiow Chen, Hui-Mei Wang 172

Polynomial Algorithm for Parallel Machine Mean Flow Time Scheduling
Problem with Release Dates
 Peter Brucker, Svetlana A. Kravchenko 182

Differential Approximation of MIN SAT, MAX SAT and Related Problems
 Bruno Escoffier, Vangelis Th. Paschos 192

Probabilistic Coloring of Bipartite and Split Graphs
 *Federico Della Croce, Bruno Escoffier, Cécile Murat,
 Vangelis Th. Paschos* 202

Design Optimization Modeling for Customer-Driven Concurrent
Tolerance Allocation
 *Young Jin Kim, Byung Rae Cho, Min Koo Lee,
 Hyuck Moo Kwon* 212

Application of Data Mining for Improving Yield in Wafer Fabrication
System
 Dong-Hyun Baek, In-Jae Jeong, Chang-Hee Han 222

Determination of Optimum Target Values for a Production Process
Based on Two Surrogate Variables
 Min Koo Lee, Hyuck Moo Kwon, Young Jin Kim, Jongho Bae 232

An Evolution Algorithm for the Rectilinear Steiner Tree Problem
 Byounghak Yang 241

A Two-Stage Recourse Model for Production Planning with Stochastic
Demand
 K.K. Lai, Stephen C.H. Leung, Yue Wu 250

A Hybrid Primal-Dual Algorithm with Application to the Dual
Transportation Problems
 Gyunghyun Choi, Chulyeon Kim 261

Real-Coded Genetic Algorithms for Optimal Static Load Balancing in
Distributed Computing System with Communication Delays
 Venkataraman Mani, Sundaram Suresh, HyoungJoong Kim 269

Heterogeneity in and Determinants of Technical Efficiency in the Use
of Polluting Inputs
 Taeho Kim, Jae-Gon Kim 280

A Continuation Method for the Linear Second-Order Cone
Complementarity Problem
 Yu Xia, Jiming Peng 290

Fuzzy Multi-criteria Decision Making Approach for Transport Projects
Evaluation in Istanbul
  E. Ertugrul Karsak, S. Sebnem Ahiska 301

An Improved Group Setup Strategy for PCB Assembly
  V. Jorge Leon, In-Jae Jeong 312

A Mixed Integer Programming Model for Modifying a Block Layout to
Facilitate Smooth Material Flows
  Jae-Gon Kim, Marc Goetschalckx 322

An Economic Capacity Planning Model Considering Inventory and
Capital Time Value
  S.M. Wang, K.J. Wang, H.M. Wee, J.C. Chen 333

A Quantity-Time-Based Dispatching Policy for a VMI System
  Wai-Ki Ching, Allen H. Tai 342

An Exact Algorithm for Multi Depot and Multi Period Vehicle
Scheduling Problem
  Kyung Hwan Kang, Young Hoon Lee, Byung Ki Lee 350

Determining Multiple Attribute Weights Consistent with Pairwise
Preference Orders
  Byeong Seok Ahn, Chang Hee Han 360

A Pricing Model for a Service Inventory System When Demand Is Price
and Waiting Time Sensitive
  Peng-Sheng You 368

A Bi-population Based Genetic Algorithm for the Resource-Constrained
Project Scheduling Problem
  Dieter Debels, Mario Vanhoucke 378

Optimizing Product Mix in a Multi-bottleneck Environment Using
Group Decision-Making Approach
  Alireza Rashidi Komijan, Seyed Jafar Sadjadi 388

Using Bipartite and Multidimensional Matching to Select the Roots of
a System of Polynomial Equations
  Henk Bekker, Eelco P. Braad, Boris Goldengorin 397

Principles, Models, Methods, and Algorithms for the Structure
Dynamics Control in Complex Technical Systems
  B.V. Sokolov, R.M. Yusupov, E.M. Zaychik 407

Applying a Hybrid Ant Colony System to the Vehicle Routing Problem
Chia-Ho Chen, Ching-Jung Ting, Pei-Chann Chang 417

A Coevolutionary Approach to Optimize Class Boundaries for
Multidimensional Classification Problems
Ki-Kwang Lee 427

Analytical Modeling of Closed-Loop Conveyors with Load Recirculation
Ying-Jiun Hsieh, Yavuz A. Bozer 437

A Multi-items Ordering Model with Mixed Parts Transportation
Problem in a Supply Chain
Beumjun Ahn, Kwang-Kyu Seo 448

Artificial Neural Network Based Life Cycle Assessment Model for
Product Concepts Using Product Classification Method
Kwang-Kyu Seo, Sung-Hwan Min, Hun-Woo Yoo 458

New Heuristics for No-Wait Flowshop Scheduling with Precedence
Constraints and Sequence Dependent Setup Time
Young Hae Lee, Jung Woo Jung 467

Efficient Dual Methods for Nonlinearly Constrained Networks
Eugenio Mijangos 477

A First-Order ε-Approximation Algorithm for Linear Programs
and a Second-Order Implementation
*Ana Maria A.C. Rocha, Edite M.G.P. Fernandes,
João L.C. Soares* 488

Inventory Allocation with Multi-echelon Service Level Considerations
Jenn-Rong Lin, Linda K. Nozick, Mark A. Turnquist 499

A Queueing Model for Multi-product Production System
Ho Woo Lee, Tae Hoon Kim 509

Discretization Approach and Nonparametric Modeling for Long-Term
HIV Dynamic Model
Jianwei Chen, Jin-Ting Zhang, Hulin Wu 519

Performance Analysis and Optimization of an Improved Dynamic
Movement-Based Location Update Scheme in Mobile Cellular Networks
Jang Hyun Baek, Jae Young Seo, Douglas C. Sicker 528

Capacitated Disassembly Scheduling: Minimizing the Number of
Products Disassembled
   *Jun-Gyu Kim, Hyong-Bae Jeon, Hwa-Joong Kim, Dong-Ho Lee,
   Paul Xirouchakis* 538

Ascent Phase Trajectory Optimization for a Hypersonic Vehicle Using
Nonlinear Programming
   *H.M. Prasanna, Debasish Ghose, M.S. Bhat,
   Chiranjib Bhattacharyya, J. Umakant* 548

Estimating Parameters in Repairable Systems Under Accelerated Stress
   Won Young Yun, Eun Suk Kim 558

Optimization Model for Remanufacturing System at Strategic and
Operational Level
   Kibum Kim, Bongju Jeong, Seung-Ju Jeong 566

A Novel Procedure to Identify the Minimized Overlap Boundary of
Two Groups by DEA Model
   Dong Shang Chang, Yi Chun Kuo 577

A Parallel Tabu Search Algorithm for Optimizing Multiobjective VLSI
Placement
   Mahmood R. Minhas, Sadiq M. Sait 587

A Coupled Gradient Network Approach for the Multi-machine Earliness
and Tardiness Scheduling Problem
   Derya Eren Akyol, G. Mirac Bayhan 596

An Analytic Model for Correlated Traffics in Computer-Communication
Networks
   Si-Yeong Lim, Sun Hur 606

Product Mix Decisions in the Process Industry
   Seung J. Noh, Suk-Chul Rim 615

On the Optimal Workloads Allocation of an FMS with Finite In-process
Buffers
   Soo-Tae Kwon 624

NEOS Server Usage in Wastewater Treatment Cost Minimization
   *Isabel A.C.P. Espoírito-Santo, Edite M.G.P Fernandes,
   Madalena M. Araújo, Eugenio C. Ferreira* 632

Branch and Price Algorithm for Content Allocation Problem in VOD
Network
    Jungman Hong, Seungkil Lim 642

Regrouping Service Sites: A Genetic Approach Using a Voronoi Diagram
    Jeong-Yeon Seo, Sang-Min Park, Seoung Soo Lee, Deok-Soo Kim ... 652

Profile Association Rule Mining Using Tests of Hypotheses Without
Support Threshold
    Kwang-Il Ahn, Jae-Yearn Kim 662

The Capacitated max-k-cut Problem
    Daya Ram Gaur, Ramesh Krishnamurti 670

A Cooperative Multi-Colony Ant Optimization Based Approach to
Efficiently Allocate Customers to Multiple Distribution Centers in a
Supply Chain Network
    Srinivas, Yogesh Dashora, Alok Kumar Choudhary,
    Jenny A. Harding, Manoj Kumar Tiwari 680

Experimentation System for Efficient Job Performing in Veterinary
Medicine Area
    Leszek Koszalka, Piotr Skworcow............................... 692

An Anti-collision Algorithm Using Two-Functioned Estimation for
RFID Tags
    Jia Zhai, Gi-Nam Wang 702

A Proximal Solution for a Class of Extended Minimax Location Problem
    Oscar Cornejo, Christian Michelot 712

A Lagrangean Relaxation Approach for Capacitated Disassembly
Scheduling
    Hwa-Joong Kim, Dong-Ho Lee, Paul Xirouchakis 722

General Tracks

DNA-Based Algorithm for 0-1 Planning Problem
    Lei Wang, Zhiping P. Chen, Xinhua H. Jiang 733

Clustering for Image Retrieval via Improved Fuzzy-ART
    Sang-Sung Park, Hun-Woo Yoo, Man-Hee Lee, Jae-Yeon Kim,
    Dong-Sik Jang... 743

Mining Schemas in Semi-structured Data Using Fuzzy Decision Trees
  Sun Wei, Liu Da-xin 753

Parallel Seismic Propagation Simulation in Anisotropic Media by
Irregular Grids Finite Difference Method on PC Cluster
  Weitao Sun, Jiwu Shu, Weimin Zheng 762

The Web Replica Allocation and Topology Assignment Problem in
Wide Area Networks: Algorithms and Computational Results
  Marcin Markowski, Andrzej Kasprzak 772

Optimal Walking Pattern Generation for a Quadruped Robot Using
Genetic-Fuzzy Algorithm
  Bo-Hee Lee, Jung-Shik Kong, Jin-Geol Kim 782

Modelling of Process of Electronic Signature with Petri Nets and
(Max, Plus) Algebra
  Ahmed Nait-Sidi-Moh, Maxime Wack 792

Evolutionary Algorithm for Congestion Problem in Connection-Oriented
Networks
  Michał Przewoźniczek, Krzysztof Walkowiak 802

Design and Development of File System for Storage Area Networks
  Gyoung-Bae Kim, Myung-Joon Kim, Hae-Young Bae 812

Transaction Reordering for Epidemic Quorum in Replicated Databases
  Huaizhong Lin, Zengwei Zheng, Chun Chen 826

Automatic Boundary Tumor Segmentation of a Liver
  Kyung-Sik Seo, Tae-Woong Chung 836

Fast Algorithms for l1 Norm/Mixed l1 and l2 Norms for Image
Restoration
  *Haoying Fu, Michael Kwok Ng, Mila Nikolova, Jesse Barlow,
  Wai-Ki Ching* 843

Intelligent Semantic Information Retrieval in Medical Pattern Cognitive
Analysis
  Marek R. Ogiela, Ryszard Tadeusiewicz, Lidia Ogiela 852

FSPN-Based Genetically Optimized Fuzzy Polynomial Neural Networks
  Sung-Kwun Oh, Seok-Beom Roh, Daehee Park, Yong-Kab Kim 858

Unsupervised Color Image Segmentation Using Mean Shift and
Deterministic Annealing EM
    Wanhyun Cho, Jonghyun Park, Myungeun Lee, Soonyoung Park   867

Identity-Based Key Agreement Protocols in a Multiple PKG
Environment
    Hoonjung Lee, Donghyun Kim, Sangjin Kim, Heekuck Oh   877

Evolutionally Optimized Fuzzy Neural Networks Based on Evolutionary
Fuzzy Granulation
    Sung-Kwun Oh, Byoung-Jun Park, Witold Pedrycz,
    Hyun-Ki Kim   887

Multi-stage Detailed Placement Algorithm for Large-Scale Mixed-Mode
Layout Design
    Lijuan Luo, Qiang Zhou, Xianlong Hong, Hanbin Zhou   896

Adaptive Mesh Smoothing for Feature Preservation
    Weishi Li, Li Ping Goh, Terence Hung, Shuhong Xu   906

A Fuzzy Grouping-Based Load Balancing for Distributed Object
Computing Systems
    Hyo Cheol Ahn, Hee Yong Youn   916

DSP-Based ADI-PML Formulations for Truncating Linear Debye and
Lorentz Dispersive FDTD Domains
    Omar Ramadan   926

Mobile Agent Based Adaptive Scheduling Mechanism in Peer to Peer
Grid Computing
    SungJin Choi, MaengSoon Baik, ChongSun Hwang, JoonMin Gil,
    HeonChang Yu   936

Comparison of Global Optimization Methods for Drag Reduction in
the Automotive Industry
    Laurent Dumas, Vincent Herbert, Frédérique Muyl   948

Multiple Intervals Versus Smoothing of Boundaries in the Discretization
of Performance Indicators Used for Diagnosis in Cellular Networks
    Raquel Barco, Pedro Lázaro, Luis Díez, Volker Wille   958

Visual Interactive Clustering and Querying of Spatio-Temporal Data
    Olga Sourina, Dongquan Liu   968

Breakdown-Free ML(k)BiCGStab Algorithm for Non-Hermitian Linear Systems
  Kentaro Moriya, Takashi Nodera 978

On Algorithm for Efficiently Combining Two Independent Measures in Routing Paths
  Moonseong Kim, Young-Cheol Bang, Hyunseung Choo 989

Real Time Hand Tracking Based on Active Contour Model
  Jae Sik Chang, Eun Yi Kim, KeeChul Jung, Hang Joon Kim 999

Hardware Accelerator for Vector Quantization by Using Pruned Look-Up Table
  Pi-Chung Wang, Chun-Liang Lee, Hung-Yi Chang, Tung-Shou Chen 1007

Optimizations of Data Distribution Localities in Cluster Grid Environments
  Ching-Hsien Hsu, Shih-Chang Chen, Chao-Tung Yang, Kuan-Ching Li 1017

Abuse-Free Item Exchange
  Hao Wang, Heqing Guo, Jianfei Yin, Qi He, Manshan Lin, Jun Zhang 1028

Transcoding Pattern Generation for Adaptation of Digital Items Containing Multiple Media Streams in Ubiquitous Environment
  Maria Hong, DaeHyuck Park, YoungHwan Lim, YoungSong Mun, Seongjin Ahn 1036

Identity-Based Aggregate and Verifiably Encrypted Signatures from Bilinear Pairing
  Xiangguo Cheng, Jingmei Liu, Xinmei Wang 1046

Element-Size Independent Analysis of Elasto-Plastic Damage Behaviors of Framed Structures
  Yutaka Toi, Jeoung-Gwen Lee 1055

On the Rila-Mitchell Security Protocols for Biometrics-Based Cardholder Authentication in Smartcards
  Raphael C.-W. Phan, Bok-Min Goi 1065

On-line Fabric-Defects Detection Based on Wavelet Analysis
  Sungshin Kim, Hyeon Bae, Seong-Pyo Cheon, Kwang-Baek Kim 1075

Application of Time-Series Data Mining for Fault Diagnosis of
Induction Motors
  Hyeon Bae, Sungshin Kim, Yon Tae Kim, Sang-Hyuk Lee 1085

Distortion Measure for Binary Document Image Using Distance and
Stroke
  Guiyue Jin, Ki Dong Lee 1095

Region and Shape Prior Based Geodesic Active Contour and
Application in Cardiac Valve Segmentation
  Yanfeng Shang, Xin Yang, Ming Zhu, Biao Jin, Ming Liu 1102

Interactive Fluid Animation Using Particle Dynamics Simulation and
Pre-integrated Volume Rendering
  Jeongjin Lee, Helen Hong, Yeong Gil Shin 1111

Performance of Linear Algebra Code: Intel Xeon EM64T and ItaniumII
Case Examples
  Terry Moreland, Chih Jeng Kenneth Tan 1120

Dataset Filtering Based Association Rule Updating in Small-Sized
Temporal Databases
  Jason J. Jung, Geun-Sik Jo 1131

A Comparison of Model Selection Methods for Multi-class Support
Vector Machines
  Huaqing Li, Feihu Qi, Shaoyu Wang 1140

Fuzzy Category and Fuzzy Interest for Web User Understanding
  SiHun Lee, Jee-Hyong Lee, Keon-Myung Lee, Hee Yong Youn 1149

Automatic License Plate Recognition System Based on Color Image
Processing
  Xifan Shi, Weizhong Zhao, Yonghang Shen 1159

Exploiting Locality Characteristics for Reducing Signaling Load in
Hierarchical Mobile IPv6 Networks
  Ki-Sik Kong, Sung-Ju Roh, Chong-Sun Hwang 1169

Parallel Feature-Preserving Mesh Smoothing
  Xiangmin Jiao, Phillip J. Alexander 1180

On Multiparametric Sensitivity Analysis in Minimum Cost Network
Flow Problem
  Sanjeet Singh, Pankaj Gupta, Davinder Bhatia 1190

Mining Patterns of Mobile Users Through Mobile Devices and the
Musics They Listen
John Goh, David Taniar 1203

Scheduling the Interactions of Multiple Parallel Jobs and Sequential
Jobs on a Non-dedicated Cluster
Adel Ben Mnaouer 1212

Feature-Correlation Based Multi-view Detection
Kuo Zhang, Jie Tang, JuanZi Li, KeHong Wang 1222

BEST: Buffer-Driven Efficient Streaming Protocol
*Sunhun Lee, Jungmin Lee, Kwangsue Chung, WoongChul Choi,
Seung Hyong Rhee* 1231

A New Neuro-Dominance Rule for Single Machine Tardiness Problem
Tarık Çakar 1241

Sinogram Denoising of Cryo-Electron Microscopy Images
Taneli Mielikäinen, Janne Ravantti 1251

Study of a Cluster-Based Parallel System Through Analytical Modeling
and Simulation
Bahman Javadi, Siavash Khorsandi, Mohammad K. Akbari 1262

Robust Parallel Job Scheduling Infrastructure for Service-Oriented
Grid Computing Systems
J.H. Abawajy 1272

SLA Management in a Service Oriented Architecture
James Padgett, Mohammed Haji, Karim Djemame 1282

Attacks on Port Knocking Authentication Mechanism
*Antonio Izquierdo Manzanares, Joaquín Torres Márquez,
Juan M. Estevez-Tapiador, Julio César Hernández Castro* 1292

Marketing on Internet Communications Security for Online Bank
Transactions
José M. Sierra, Julio C. Hernández, Eva Ponce, Jaime Manera 1301

A Formal Analysis of Fairness and Non-repudiation in the RSA-CEGD
Protocol
*Almudena Alcaide, Juan M. Estévez-Tapiador, Antonio Izquierdo,
José M. Sierra* 1309

Distribution Data Security System Based on Web Based Active Database
  *Sang-Yule Choi, Myong-Chul Shin, Nam-Young Hur,
  Jong-Boo Kim, Tai-Hoon Kim, Jae-Sang Cha* 1319

Data Protection Based on Physical Separation: Concepts and Application Scenarios
  Stefan Lindskog, Karl-Johan Grinnemo, Anna Brunstrom 1331

Some Results on a Class of Optimization Spaces
  K.C. Sivakumar, J. Mercy Swarna 1341

Author Index 1349

Table of Contents – Part I

Information Systems and Information Technologies (ISIT) Workshop

The Technique of Test Case Design Based on the UML Sequence Diagram for the Development of Web Applications
  Yongsun Cho, Woojin Lee, Kiwon Chong 1

Flexible Background-Texture Analysis for Coronary Artery Extraction Based on Digital Subtraction Angiography
  Sung-Ho Park, Jeong-Hee Cha, Joong-Jae Lee, Gye-Young Kim 11

New Size-Reduced Visual Secret Sharing Schemes with Half Reduction of Shadow Size
  Ching-Nung Yang, Tse-Shih Chen 19

An Automatic Resource Selection Scheme for Grid Computing Systems
  Kyung-Woo Kang, Gyun Woo 29

Matching Colors with KANSEI Vocabulary Using Similarity Measure Based on WordNet
  Sunkyoung Baek, Miyoung Cho, Pankoo Kim 37

A Systematic Process to Design Product Line Architecture
  Soo Dong Kim, Soo Ho Chang, Hyun Jung La 46

Variability Design and Customization Mechanisms for COTS Components
  Soo Dong Kim, Hyun Gi Min, Sung Yul Rhew 57

A Fast Lossless Multi-resolution Motion Estimation Algorithm Using Selective Matching Units
  Jong-Nam Kim 67

Developing an XML Document Retrieval System for a Digital Museum
  Jae-Woo Chang 77

WiCTP: A Token-Based Access Control Protocol for Wireless Networks
  Raal Goff, Amitava Datta 87

An Optimized Internetworking Strategy of MANET and WLAN
  Hyewon K. Lee, Youngsong Mun 97

An Internetworking Scheme for UMTS/WLAN Mobile Networks
Sangjoon Park, Youngchul Kim, Jongchan Lee 107

A Handover Scheme Based on HMIPv6 for B3G Networks
*Eunjoo Jeong, Sangjoon Park, Hyewon K. Lee, Kwan-Joong Kim,
Youngsong Mun, Byunggi Kim* 118

Collaborative Filtering for Recommendation Using Neural Networks
Myung Won Kim, Eun Ju Kim, Joung Woo Ryu 127

Dynamic Access Control Scheme for Service-Based Multi-netted
Asymmetric Virtual LAN
Wonwoo Choi, Hyuncheol Kim, Seongjin Ahn, Jinwook Chung 137

New Binding Update Method Using GDMHA in Hierarchical Mobile
IPv6
*Jong-Hyouk Lee, Young-Ju Han, Hyung-Jin Lim,
Tai-Myung Chung* 146

Security in Sensor Networks for Medical Systems Torso Architecture
Chaitanya Penubarthi, Myuhng-Joo Kim, Insup Lee 156

Multimedia: An SIMD – Based Efficient 4x4 2 D Transform Method
*Sang-Jun Yu, Chae-Bong Sohn, Seoung-Jun Oh,
Chang-Beom Ahn* 166

A Real-Time Cooperative Swim-Lane Business Process Modeler
Kwang-Hoon Kim, Jung-Hoon Lee, Chang-Min Kim 176

A Focused Crawling for the Web Resource Discovery Using a Modified
Proximal Support Vector Machines
YoungSik Choi, KiJoo Kim, MunSu Kang 186

A Performance Improvement Scheme of Stream Control Transmission
Protocol over Wireless Networks
*Kiwon Hong, Kugsang Jeong, Deokjai Choi,
Choongseon Hong* 195

Cache Management Protocols Based on Re-ordering for Distributed
Systems
SungHo Cho, Kyoung Yul Bae 204

DRC-BK: Mining Classification Rules by Using Boolean Kernels
Yang Zhang, Zhanhuai Li, Kebin Cui 214

General-Purpose Text Entry Rules for Devices with 4x3 Configurations of Buttons
Jaewoo Ahn, Myung Ho Kim 223

Dynamic Load Redistribution Approach Using Genetic Information in Distributed Computing
Seonghoon Lee, Dongwoo Lee, Donghee Shim, Dongyoung Cho 232

A Guided Search Method for Real Time Transcoding a MPEG2 P Frame into H.263 P Frame in a Compressed Domain
Euisun Kang, Maria Hong, Younghwan Lim, Youngsong Mun, Seongjin Ahn 242

Cooperative Security Management Enhancing Survivability Against DDoS Attacks
Sung Ki Kim, Byoung Joon Min, Jin Chul Jung, Seung Hwan Yoo 252

Marking Mechanism for Enhanced End-to-End QoS Guarantees in Multiple DiffServ Environment
Woojin Park, Kyuho Han, Sinam Woo, Sunshin An 261

An Efficient Handoff Mechanism with Web Proxy MAP in Hierarchical Mobile IPv6
Jonghyoun Choi, Youngsong Mun 271

A New Carried-Dependence Self-scheduling Algorithm
Hyun Cheol Kim 281

Improved Location Management Scheme Based on Autoconfigured Logical Topology in HMIPv6
Jongpil Jeong, Hyunsang Youn, Hyunseung Choo, Eunseok Lee 291

Ontological Model of Event for Integration of Inter-organization Applications
Wang Wenjun, Luo Yingwei, Liu Xinpeng, Wang Xiaolin, Xu Zhuoqun 301

Secure XML Aware Network Design and Performance Analysis
Eui-Nam Huh, Jong-Youl Jeong, Young-Shin Kim, Ki-Young Mun 311

A Probe Detection Model Using the Analysis of the Fuzzy Cognitive Maps
Se-Yul Lee, Yong-Soo Kim, Bong-Hwan Lee, Suk-Hoon Kang, Chan-Hyun Youn 320

Mobile Communications (Mobicomm) Workshop

QoS Provisioning in an Enhanced FMIPv6 Architecture
  Zheng Wan, Xuezeng Pan, Lingdi Ping 329

A Novel Hierarchical Routing Protocol for Wireless Sensor Networks
  Trong Thua Huynh, Choong Seon Hong 339

A Vertical Handoff Algorithm Based on Context Information in CDMA-WLAN Integrated Networks
  Jang-Sub Kim, Min-Young Chung, Dong-Ryeol Shin 348

Scalable Hash Chain Traversal for Mobile Device
  Sung-Ryul Kim 359

A Rate Separation Mechanism for Performance Improvements of Multi-rate WLANs
  Chae-Tae Im, Dong-Hee Kwon, Young-Joo Suh 368

Improved Handoff Scheme for Supporting Network Mobility in Nested Mobile Networks
  *Han-Kyu Ryu, Do-Hyeon Kim, You-Ze Cho, Kang-Won Lee,
  Hee-Dong Park* 378

A *Prompt Retransmit* Technique to Improve TCP Performance for Mobile Ad Hoc Networks
  Dongkyun Kim, Hanseok Bae 388

Enhanced Fast Handover for Mobile IPv6 Based on IEEE 802.11 Network
  *Seonggeun Ryu, Younghwan Lim, Seongjin Ahn,
  Youngsong Mun* 398

An Efficient Macro Mobility Scheme Supporting Fast Handover in Hierarchical Mobile IPv6
  Kyunghye Lee, Youngsong Mun 408

Study on the Advanced MAC Scheduling Algorithm for the Infrared Dedicated Short Range Communication
  Sujin Kwag, Jesang Park, Sangsun Lee 418

Design and Evaluation of a New Micro-mobility Protocol in Large Mobile and Wireless Networks
  Young-Chul Shim, Hyun-Ah Kim, Ju-Il Lee 427

Performance Analysis of Transmission Probability Control Scheme in
Slotted ALOHA CDMA Networks
  In-Taek Lim 438

RWA Based on Approximated Path Conflict Graphs in Optical Networks
  *Zhanna Olmes, Kun Myon Choi, Min Young Chung, Tae-Jin Lee,
  Hyunseung Choo* 448

Secure Routing in Sensor Networks: Security Problem Analysis and
Countermeasures
  Youngsong Mun, Chungsoo Shin 459

Policy Based Handoff in MIPv6 Networks
  *Jong-Hyouk Lee, Byungchul Park, Hyunseung Choo,
  Tai-Myung Chung* 468

An Effective Location Management Strategy for Cellular Mobile
Networks
  In-Hye Shin, Gyung-Leen Park, Kang Soo Tae 478

Authentication Authorization Accounting (AAA) Workshop

On the Rila-Mitchell Security Protocols for Biometrics-Based
Cardholder Authentication in Smartcards
  Raphael C.-W. Phan, Bok-Min Goi 488

An Efficient Dynamic Group Key Agreement for Low-Power Mobile
Devices
  Seokhyang Cho, Junghyun Nam, Seungjoo Kim, Dongho Won 498

Compact Linear Systolic Arrays for Multiplication Using a Trinomial
Basis in $GF(2^m)$ for High Speed Cryptographic Processors
  Soonhak Kwon, Chang Hoon Kim, Chun Pyo Hong 508

A Secure User Authentication Protocol Based on One-Time-Password
for Home Network
  Hea Suk Jo, Hee Yong Youn 519

On AAA with Extended IDK in Mobile IP Networks
  Hoseong Jeon, Min Young Chung, Hyunseung Choo 529

Secure Forwarding Scheme Based on Session Key Reuse Mechanism in
HMIPv6 with AAA
  Kwang Chul Jeong, Hyunseung Choo, Sungchang Lee 540

A Hierarchical Authentication Scheme for MIPv6 Node with Local Movement Property
  Miyoung Kim, Misun Kim, Youngsong Mun 550

An Effective Authentication Scheme for Mobile Node with Fast Roaming Property
  Miyoung Kim, Misun Kim, Youngsong Mun 559

A Study on the Performance Improvement to AAA Authentication in Mobile IPv4 Using Low Latency Handoff
  Youngsong Mun, Sehoon Jang 569

Authenticated Key Agreement Without Subgroup Element Verification
  Taekyoung Kwon 577

Multi-modal Biometrics with PKIs for Border Control Applications
  Taekyoung Kwon, Hyeonjoon Moon 584

A Scalable Mutual Authentication and Key Distribution Mechanism in a NEMO Environment
  Mihui Kim, Eunah Kim, Kijoon Chae 591

Service-Oriented Home Network Middleware Based on OGSA
  Tae Dong Lee, Chang-Sung Jeong 601

Implementation of Streamlining PKI System for Web Services
  Namje Park, Kiyoung Moon, Jongsu Jang, Sungwon Sohn, Dongho Won 609

Efficient Authentication for Low-Cost RFID Systems
  Su Mi Lee, Young Ju Hwang, Dong Hoon Lee, Jong In Lim 619

An Efficient Performance Enhancement Scheme for Fast Mobility Service in MIPv6
  Seung-Yeon Lee, Eui-Nam Huh, Sang-Bok Kim, Young-Song Mun ... 628

Face Recognition by the LDA-Based Algorithm for a Video Surveillance System on DSP
  Jin Ok Kim, Jin Soo Kim, Chin Hyun Chung 638

Computational Geometry and Applications (CGA'05) Workshop

Weakly Cooperative Guards in Grids
  Michał Małafiejski, Paweł Żyliński 647

Mesh Generation for Symmetrical Geometries
  Krister Åhlander 657

A Certified Delaunay Graph Conflict Locator for Semi-algebraic Sets
  François Anton 669

The Offset to an Algebraic Curve and an Application to Conics
  *François Anton, Ioannis Emiris, Bernard Mourrain,
  Monique Teillaud* 683

Computing the Least Median of Squares Estimator in Time $O(n^d)$
  Thorsten Bernholt 697

Pocket Recognition on a Protein Using Euclidean Voronoi Diagram of Atoms
  *Deok-Soo Kim, Cheol-Hyung Cho, Youngsong Cho, Chung In Won,
  Dounguk Kim* 707

Region Expansion by Flipping Edges for Euclidean Voronoi Diagrams of 3D Spheres Based on a Radial Data Structure
  Donguk Kim, Youngsong Cho, Deok-Soo Kim 716

Analysis of the Nicholl-Lee-Nicholl Algorithm
  Frank Dévai 726

Flipping to Robustly Delete a Vertex in a Delaunay Tetrahedralization
  Hugo Ledoux, Christopher M. Gold, George Baciu 737

A Novel Topology-Based Matching Algorithm for Fingerprint Recognition in the Presence of Elastic Distortions
  Chengfeng Wang, Marina L. Gavrilova 748

Bilateral Estimation of Vertex Normal for Point-Sampled Models
  Guofei Hu, Jie Xu, Lanfang Miao, Qunsheng Peng 758

A Point Inclusion Test Algorithm for Simple Polygons
  Weishi Li, Eng Teo Ong, Shuhong Xu, Terence Hung 769

A Modified Nielson's Side-Vertex Triangular Mesh Interpolation Scheme
  Zhihong Mao, Lizhuang Ma, Wuzheng Tan 776

An Acceleration Technique for the Computation of Voronoi Diagrams Using Graphics Hardware
  Osami Yamamoto 786

On the Rectangular Subset Closure of Point Sets
  Stefan Porschen 796

Computing Optimized Curves with NURBS Using Evolutionary
Intelligence
  Muhammad Sarfraz, Syed Arshad Raza, M. Humayun Baig......... 806

A Novel Delaunay Simplex Technique for Detection of Crystalline
Nuclei in Dense Packings of Spheres
  A.V. Anikeenko, M.L. Gavrilova, N.N. Medvedev 816

Recognition of Minimum Width Color-Spanning Corridor and
Minimum Area Color-Spanning Rectangle
  Sandip Das, Partha P. Goswami, Subhas C. Nandy 827

Volumetric Reconstruction of Unorganized Set of Points with Implicit
Surfaces
  Vincent Bénédet, Loïc Lamarque, Dominique Faudot 838

Virtual Reality in Scientific Applications and Learning (VRSAL 2005) Workshop

Guided Navigation Techniques for 3D Virtual Environment Based on
Topic Map
  *Hak-Keun Kim, Teuk-Seob Song, Yoon-Chu Choy,
  Soon-Bum Lim* 847

Image Sequence Augmentation Using Planar Structures
  Juwan Kim, Dongkeun Kim 857

MultiPro: A Platform for PC Cluster Based Active Stereo Display
System
  Qingshu Yuan, Dongming Lu, Weidong Chen, Yunhe Pan.......... 865

Two-Level 2D Projection Maps Based Horizontal Collision Detection
Scheme for Avatar in Collaborative Virtual Environment
  Yu Chunyan, Ye Dongyi, Wu Minghui, Pan Yunhe 875

A Molecular Modeling System Based on Dynamic Gestures
  Sungjun Park, Jun Lee, Jee-In Kim 886

Face Modeling Using Grid Light and Feature Point Extraction
  Lei Shi, Xin Yang, Hailang Pan 896

Virtual Chemical Laboratories and Their Management on the Web
*Antonio Riganelli, Osvaldo Gervasi, Antonio Laganà,
Johannes Froehlich* 905

Tangible Tele-meeting System with DV-ARPN (Augmented Reality Peripheral Network)
Yong-Moo Kwon, Jin-Woo Park 913

Integrating Learning and Assessment Using the Semantic Web
*Osvaldo Gervasi, Riccardo Catanzani, Antonio Riganelli,
Antonio Laganà* 921

The Implementation of Web-Based Score Processing System for WBI
Young-Jun Seo, Hwa-Young Jeong, Young-Jae Song 928

ELCHEM: A Metalaboratory to Develop Grid e-Learning Technologies and Services for Chemistry
*A. Laganà, A. Riganelli, O. Gervasi, P. Yates, K. Wahala,
R. Salzer, E. Varella, J. Froeklich* 938

Client Allocation for Enhancing Interactivity in Distributed Virtual Environments
Duong Nguyen Binh Ta, Suiping Zhou 947

IMNET: An Experimental Testbed for Extensible Multi-user Virtual Environment Systems
Tsai-Yen Li, Mao-Yung Liao, Pai-Cheng Tao 957

Application of MPEG-4 in Distributed Virtual Environment
Qiong Zhang, Taiyi Chen, Jianzhong Mo 967

A New Approach to Area of Interest Management with Layered-Structures in 2D Grid
Yu Chunyan, Ye Dongyi, Wu Minghui, Pan Yunhe 974

Awareness Scheduling and Algorithm Implementation for Collaborative Virtual Environment
Yu Sheng, Dongming Lu, Yifeng Hu, Qingshu Yuan 985

M of N Features vs. Intrusion Detection
Zhuowei Li, Amitabha Das 994

Molecular Structures and Processes Workshop

High-Level Quantum Chemical Methods for the Study of Photochemical Processes
   Hans Lischka, Adélia J.A. Aquino, Mario Barbatti, Mohammad Solimannejad 1004

Study of Predictive Abilities of the Kinetic Models of Multistep Chemical Reactions by the Method of Value Analysis
   Levon A. Tavadyan, Avet A. Khachoyan, Gagik A. Martoyan, Seyran H. Minasyan 1012

Lateral Interactions in O/Pt(111): Density-Functional Theory and Kinetic Monte Carlo
   A.P.J. Jansen, W.K. Offermans 1020

Intelligent Predictive Control with Locally Linear Based Model Identification and Evolutionary Programming Optimization with Application to Fossil Power Plants
   Mahdi Jalili-Kharaajoo 1030

Determination of Methanol and Ethanol Synchronously in Ternary Mixture by NIRS and PLS Regression
   Q.F. Meng, L.R. Teng, J.H. Lu, C.J. Jiang, C.H. Gao, T.B. Du, C.G. Wu, X.C. Guo, Y.C. Liang 1040

Ab Initio and Empirical Atom Bond Formulation of the Interaction of the Dimethylether-Ar System
   Alessandro Costantini, Antonio Laganà, Fernando Pirani, Assimo Maris, Walther Caminati 1046

A Parallel Framework for the Simulation of Emission, Transport, Transformation and Deposition of Atmospheric Mercury on a Regional Scale
   Giuseppe A. Trunfio, Ian M. Hedgecock, Nicola Pirrone 1054

A Cognitive Perspective for Choosing Groupware Tools and Elicitation Techniques in Virtual Teams
   Gabriela N. Aranda, Aurora Vizcaíno, Alejandra Cechich, Mario Piattini 1064

A Fast Method for Determination of Solvent-Exposed Atoms and Its Possible Applications for Implicit Solvent Models
   Anna Shumilina 1075

Thermal Rate Coefficients for the $N + N_2$ Reaction: Quasiclassical, Semiclassical and Quantum Calculations
Noelia Faginas Lago, Antonio Laganà, Ernesto Garcia, X. Gimenez 1083

A Molecular Dynamics Study of Ion Permeability Through Molecular Pores
Leonardo Arteconi, Antonio Laganà 1093

Theoretical Investigations of Atmospheric Species Relevant for the Search of High-Energy Density Materials
Marzio Rosi 1101

Pattern Recognition and Ubiquitous Computing Workshop

ID Face Detection Robust to Color Degradation and Facial Veiling
Dae Sung Kim, Nam Chul Kim 1111

Detection of Multiple Vehicles in Image Sequences for Driving Assistance System
SangHoon Han, EunYoung Ahn, NoYoon Kwak 1122

A Computational Model of Korean Mental Lexicon
Heui Seok Lim, Kichun Nam, Yumi Hwang 1129

A Realistic Human Face Modeling from Photographs by Use of Skin Color and Model Deformation
Kyongpil Min, Junchul Chun 1135

An Optimal and Dynamic Monitoring Interval for Grid Resource Information System
Angela Song-Ie Noh, Eui-Nam Huh, Ji-Yeun Sung, Pill-Woo Lee 1144

Real Time Face Detection and Recognition System Using Haar-Like Feature/HMM in Ubiquitous Network Environments
Kicheon Hong, Jihong Min, Wonchan Lee, Jungchul Kim 1154

A Hybrid Network Model for Intrusion Detection Based on Session Patterns and Rate of False Errors
Se-Yul Lee, Yong-Soo Kim, Woongjae Lee 1162

Energy-Efficiency Method for Cluster-Based Sensor Networks
Kyung-Won Nam, Jun Hwang, Cheol-Min Park, Young-Chan Kim 1170

A Study on an Efficient Sign Recognition Algorithm for a Ubiquitous
Traffic System on DSP
  Jong Woo Kim, Kwang Hoon Jung, Chung Chin Hyun 1177

Real-Time Implementation of Face Detection for a Ubiquitous
Computing
  Jin Ok Kim, Jin Soo Kim 1187

On Optimizing Feature Vectors from Efficient Iris Region Normalization
for a Ubiquitous Computing
  Bong Jo Joung, Woongjae Lee 1196

On the Face Detection with Adaptive Template Matching and Cascaded
Object Detection for Ubiquitous Computing Environment
  Chun Young Chang, Jun Hwang 1204

On Improvement for Normalizing Iris Region for a Ubiquitous
Computing
  *Bong Jo Joung, Chin Hyun Chung, Key Seo Lee, Wha Young Yim,
  Sang Hyo Lee* 1213

Author Index 1221

Table of Contents – Part II

Approaches or Methods of Security Engineering Workshop

Implementation of Short Message Service System to Be Based Mobile Wireless Internet
Hae-Sool Yang, Jung-Hun Hong, Seok-Hyung Hwang, Haeng-Kon Kim 1

Fuzzy Clustering for Documents Based on Optimization of Classifier Using the Genetic Algorithm
Ju-In Youn, He-Jue Eun, Yong-Sung Kim 10

P2P Protocol Analysis and Blocking Algorithm
Sun-Myung Hwang 21

Object Modeling of RDF Schema for Converting UML Class Diagram
Jin-Sung Kim, Chun-Sik Yoo, Mi-Kyung Lee, Yong-Sung Kim 31

A Framework for Security Assurance in Component Based Development
Gu-Beom Jeong, Guk-Boh Kim 42

Security Framework to Verify the Low Level Implementation Codes
Haeng-Kon Kim, Hae-Sool Yang 52

A Study on Evaluation of Component Metric Suites
Haeng-Kon Kim 62

The K-Means Clustering Architecture in the Multi-stage Data Mining Process
Bobby D. Gerardo, Jae-Wan Lee, Yeon-Sung Choi, Malrey Lee 71

A Privacy Protection Model in ID Management Using Access Control
Hyang-Chang Choi, Yong-Hoon Yi, Jae-Hyun Seo, Bong-Nam Noh, Hyung-Hyo Lee 82

A Time-Variant Risk Analysis and Damage Estimation for Large-Scale Network Systems
InJung Kim, YoonJung Chung, YoungGyo Lee, Dongho Won 92

Efficient Multi-bit Shifting Algorithm in Multiplicative Inversion
Problems
    Injoo Jang, Hyeong Seon Yoo 102

Modified Token-Update Scheme for Site Authentication
    Joungho Lee, Injoo Jang, Hyeong Seon Yoo 111

A Study on Secure SDP of RFID Using Bluetooth Communication
    Dae-Hee Seo, Im-Yeong Lee, Hee-Un Park 117

The Semantic Web Approach in Location Based Services
    Jong-Woo Kim, Ju-Yeon Kim, Hyun-Suk Hwang, Sung-Seok Park,
    Chang-Soo Kim, Sung-gi Park 127

SCTE: Software Component Testing Environments
    Haeng-Kon Kim, Oh-Hyun Kwon 137

Computer Security Management Model Using MAUT and SNMP
    Jongwoo Chae, Jungkyu Kwon, Mokdong Chung 147

Session and Connection Management for QoS-Guaranteed Multimedia
Service Provisioning on IP/MPLS Networks
    Young-Tak Kim, Hae-Sun Kim, Hyun-Ho Shin 157

A GQS-Based Adaptive Mobility Management Scheme Considering the
Gravity of Locality in Ad-Hoc Networks
    Ihn-Han Bae, Sun-Jin Oh 169

A Study on the E-Cash System with Anonymity and Divisibility
    Seo-Il Kang, Im-Yeong Lee 177

An Authenticated Key Exchange Mechanism Using One-Time Shared
Key
    Yonghwan Lee, Eunmi Choi, Dugki Min 187

Creation of Soccer Video Highlight Using the Caption Information
    Oh-Hyung Kang, Seong-Yoon Shin 195

The Information Search System Using Neural Network and Fuzzy
Clustering Based on Mobile Agent
    Jaeseon Ko, Bobby D. Gerardo, Jaewan Lee, Jae-Jeong Hwang 205

A Security Evaluation and Testing Methodology for Open Source
Software Embedded Information Security System
    Sung-ja Choi, Yeon-hee Kang, Gang-soo Lee 215

An Effective Method for Analyzing Intrusion Situation Through
IP-Based Classification
    *Minsoo Kim, Jae-Hyun Seo, Seung-Yong Lee, Bong-Nam Noh,
    Jung-Taek Seo, Eung-Ki Park, Choon-Sik Park* 225

A New Stream Cipher Using Two Nonlinear Functions
    Mi-Og Park, Dea-Woo Park 235

New Key Management Systems for Multilevel Security
    *Hwankoo Kim, Bongjoo Park, JaeCheol Ha, Byoungcheon Lee,
    DongGook Park* 245

Neural Network Techniques for Host Anomaly Intrusion Detection
Using Fixed Pattern Transformation
    ByungRae Cha, KyungWoo Park, JaeHyun Seo 254

The Role of Secret Sharing in the Distributed MARE Protocols
    Kyeongmo Park 264

Security Risk Vector for Quantitative Asset Assessment
    *Yoon Jung Chung, Injung Kim, NamHoon Lee, Taek Lee,
    Hoh Peter In* 274

A Remote Video Study Evaluation System Using a User Profile
    Seong-Yoon Shin, Oh-Hyung Kang 284

Performance Enhancement of Wireless LAN Based on Infrared
Communications Using Multiple-Subcarrier Modulation
    Hae Geun Kim 295

Modeling Virtual Network Collaboration in Supply Chain Management
    Ha Jin Hwang 304

SPA-Resistant Simultaneous Scalar Multiplication
    Mun-Kyu Lee 314

HSEP Design Using F2mHECC and ThreeB Symmetric Key Under
e-Commerce Environment
    Byung-kwan Lee, Am-Sok Oh, Eun-Hee Jeong 322

A Fault Distance Estimation Method Based on an Adaptive Data
Window for Power Network Security
    *Chang-Dae Yoon, Seung-Yeon Lee, Myong-Chul Shin,
    Ho-Sung Jung, Jae-Sang Cha* 332

Distribution Data Security System Based on Web Based Active Database
Sang-Yule Choi, Myong-Chul Shin, Nam-Young Hur, Jong-Boo Kim, Tai-hoon Kim, Jae-Sang Cha 341

Efficient DoS Resistant Multicast Authentication Schemes
JaeYong Jeong, Yongsu Park, Yookun Cho 353

Development System Security Process of ISO/IEC TR 15504 and Security Considerations for Software Process Improvement
Eun-ser Lee, Malrey Lee 363

Flexible ZCD-UWB with High QoS or High Capacity Using Variable ZCD Factor Code Sets
Jaesang Cha, Kyungsup Kwak, Changdae Yoon, Chonghyun Lee 373

Fine Grained Control of Security Capability and Forward Security in a Pairing Based Signature Scheme
Hak Soo Ju, Dae Youb Kim, Dong Hoon Lee, Jongin Lim, Kilsoo Chun 381

The Large Scale Electronic Voting Scheme Based on Undeniable Multi-signature Scheme
Sung-Hyun Yun, Hyung-Woo Lee 391

IPv6/IPsec Conformance Test Management System with Formal Description Technique
Hyung-Woo Lee, Sung-Hyun Yun, Jae-Sung Kim, Nam-Ho Oh, Do-Hyung Kim 401

Interference Cancellation Algorithm Development and Implementation for Digital Television
Chong Hyun Lee, Jae Sang Cha 411

Algorithm for ABR Traffic Control and Formation Feedback Information
Malrey Lee, Dong-Ju Im, Young Keun Lee, Jae-deuk Lee, Suwon Lee, Keun Kwang Lee, HeeJo Kang 420

Interference-Free ZCD-UWB for Wireless Home Network Applications
Jaesang Cha, Kyungsup Kwak, Sangyule Choi, Taihoon Kim, Changdae Yoon, Chonghyun Lee 429

Safe Authentication Method for Security Communication in Ubiquitous
Hoon Ko, Bangyong Sohn, Hayoung Park, Yongtae Shin 442

Pre/Post Rake Receiver Design for Maximum SINR in MIMO
Communication System
  Chong Hyun Lee, Jae Sang Cha 449

SRS-Tool: A Security Functional Requirement Specification
Development Tool for Application Information System of Organization
  Sang-soo Choi, Soo-young Chae, Gang-soo Lee 458

Design Procedure of IT Systems Security Countermeasures
  Tai-hoon Kim, Seung-youn Lee 468

Similarity Retrieval Based on Self-organizing Maps
  *Dong-Ju Im, Malrey Lee, Young Keun Lee, Tae-Eun Kim,
  SuWon Lee, Jaewan Lee, Keun Kwang Lee, Kyung Dal Cho* 474

An Expert System Development for Operating Procedure Monitoring
of PWR Plants
  Malrey Lee, Eun-ser Lee, HeeJo Kang, HeeSook Kim 483

Security Evaluation Targets for Enhancement of IT Systems Assurance
  Tai-hoon Kim, Seung-youn Lee 491

Protection Profile for Software Development Site
  Seung-youn Lee, Myong-chul Shin 499

Information Security and Hiding (ISH 2005) Workshop

Improved RS Method for Detection of LSB Steganography
  Xiangyang Luo, Bin Liu, Fenlin Liu 508

Robust Undetectable Interference Watermarks
  *Ryszard Grząślewicz, Jarosław Kutyłowski, Mirosław Kutyłowski,
  Wojciech Pietkiewicz* 517

Equidistant Binary Fingerprinting Codes. Existence and Identification
Algorithms
  Marcel Fernandez, Miguel Soriano, Josep Cotrina 527

Color Cube Analysis for Detection of LSB Steganography in RGB
Color Images
  Kwangsoo Lee, Changho Jung, Sangjin Lee, Jongin Lim 537

Compact and Robust Image Hashing
  Sheng Tang, Jin-Tao Li, Yong-Dong Zhang 547

Watermarking for 3D Mesh Model Using Patch CEGIs
  Suk-Hwan Lee, Ki-Ryong Kwon 557

Related-Key and Meet-in-the-Middle Attacks on Triple-DES and DES-EXE
  Jaemin Choi, Jongsung Kim, Jaechul Sung, Sangjin Lee, Jongin Lim 567

Fault Attack on the DVB Common Scrambling Algorithm
  Kai Wirt 577

HSEP Design Using F2mHECC and ThreeB Symmetric Key Under e-Commerce Envrionment
  Byung-kwan Lee, Am-Sok Oh, Eun-Hee Jeong 585

Perturbed Hidden Matrix Cryptosystems
  Zhiping Wu, Jintai Ding, Jason E. Gower, Dingfeng Ye 595

Identity-Based Identification Without Random Oracles
  Kaoru Kurosawa, Swee-Huay Heng 603

Linkable Ring Signatures: Security Models and New Schemes
  Joseph K. Liu, Duncan S. Wong 614

Practical Scenarios for the Van Trung-Martirosyan Codes
  Marcel Fernandez, Miguel Soriano, Josep Cotrina 624

Obtaining True-Random Binary Numbers from a Weak Radioactive Source
  Ammar Alkassar, Thomas Nicolay, Markus Rohe 634

Modified Sequential Normal Basis Multipliers for Type II Optimal Normal Bases
  Dong Jin Yang, Chang Han Kim, Youngho Park, Yongtae Kim, Jongin Lim 647

A New Method of Building More Non-supersingular Elliptic Curves
  Shi Cui, Pu Duan, Choong Wah Chan 657

Accelerating AES Using Instruction Set Extensions for Elliptic Curve Cryptography
  Stefan Tillich, Johann Großschädl 665

Modeling of Location Management in Mobile Information Systems Workshop

Access Control Capable Integrated Network Management System for TCP/IP Networks
Hyuncheol Kim, Seongjin Ahn, Younghwan Lim,
Youngsong Mun 676

A Directional-Antenna Based MAC Protocol for Wireless Sensor Networks
Shen Zhang, Amitava Datta 686

An Extended Framework for Proportional Differentiation: Performance Metrics and Evaluation Considerations
Jahwan Koo, Seongjin Ahn 696

QoS Provisioning in an Enhanced FMIPv6 Architecture
Zheng Wan, Xuezeng Pan, Lingdi Ping 704

Delay of the Slotted ALOHA Protocol with Binary Exponential Backoff Algorithm
Sun Hur, Jeong Kee Kim, Dong Chun Lee 714

Design and Implementation of Frequency Offset Estimation, Symbol Timing and Sampling Clock Offset Control for an IEEE 802.11a Physical Layer
Kwang-ho Chun, Seung-hyun Min, Myoung-ho Seong,
Myoung-seob Lim 723

Automatic Subtraction Radiography Algorithm for Detection of Periodontal Disease in Internet Environment
Yonghak Ahn, Oksam Chae 732

Improved Authentication Scheme in W-CDMA Networks
Dong Chun Lee, Hyo Young Shin, Joung Chul Ahn,
Jae Young Koh 741

Memory Reused Multiplication Implementation for Cryptography System
Gi Yean Hwang, Jia Hou, Kwang Ho Chun, Moon Ho Lee 749

Scheme for the Information Sharing Between IDSs Using JXTA
Jin Soh, Sung Man Jang, Geuk Lee 754

Workflow System Modeling in the Mobile Healthcare B2B Using
Semantic Information
*Sang-Young Lee, Yung-Hyeon Lee, Jeom-Goo Kim,
Dong Chun Lee* 762

Detecting Water Area During Flood Event from SAR Image
Hong-Gyoo Sohn, Yeong-Sun Song, Gi-Hong Kim 771

Position Based Handover Control Method
Jong chan Lee, Sok-Pal Cho, Hong-jin Kim 781

Improving Yellow Time Method of Left-Turning Traffic Flow at
Signalized Intersection Networks by ITS
Hyung Jin Kim, Bongsoo Son, Soobeom Lee, Joowon Park 789

Intelligent Multimedia Services and Synchronization in Mobile Multimedia Networks Workshop

A Multimedia Database System Using Dependence Weight Values for a
Mobile Environment
Kwang Hyoung Lee, Hee Sook Kim, Keun Wang Lee 798

A General Framework for Analyzing the Optimal Call Admission
Control in DS-CDMA Cellular Network
Wen Chen, Feiyu Lei, Weinong Wang 806

Heuristic Algorithm for Traffic Condition Classification with Loop
Detector Data
Sangsoo Lee, Sei-Chang Oh, Bongsoo Son 816

Spatial Data Channel in a Mobile Navigation System
Yingwei Luo, Guomin Xiong, Xiaolin Wang, Zhuoqun Xu 822

A Video Retrieval System for Electrical Safety Education Based on a
Mobile Agent
Hyeon Seob Cho, Keun Wang Lee 832

Fuzzy Multi-criteria Decision Making-Based Mobile Tracking
Gi-Sung Lee 839

Evaluation of Network Blocking Algorithm based on ARP Spoofing
and Its Application
Jahwan Koo, Seongjin Ahn, Younghwan Lim, Youngsong Mun 848

Design and Implementation of Mobile-Learning System for Environment Education
Keun Wang Lee, Jong Hee Lee 856

A Simulation Model of Congested Traffic in the Waiting Line
Bongsoo Son, Taewan Kim, Yongjae Lee 863

Core Technology Analysis and Development for the Virus and Hacking Prevention
Seung-Jae Yoo 870

Development of Traffic Accidents Prediction Model with Intelligent System Theory
SooBeom Lee, TaiSik Lee, Hyung Jin Kim, YoungKyun Lee 880

Prefetching Scheme Considering Mobile User's Preference in Mobile Networks
Jin Ah Yoo, In Seon Choi, Dong Chun Lee 889

System Development of Security Vulnerability Diagnosis in Wireless Internet Networks
Byoung-Muk Min, Sok-Pal Cho, Hong-jin Kim, Dong Chun Lee 896

An Active Node Management System for Secure Active Networks
Jin-Mook Kim, In-sung Han, Hwang-bin Ryou 904

Ubiquitous Web Systems and Intelligence Workshop

A Systematic Design Approach for XML-View Driven Web Document Warehouses
Vicky Nassis, Rajugan R., Tharam S. Dillon, Wenny Rahayu 914

Clustering and Retrieval of XML Documents by Structure
Jeong Hee Hwang, Keun Ho Ryu 925

A New Method for Mining Association Rules from a Collection of XML Documents
Juryon Paik, Hee Yong Youn, Ungmo Kim 936

Content-Based Recommendation in E-Commerce
Bing Xu, Mingmin Zhang, Zhigeng Pan, Hongwei Yang 946

A Personalized Multilingual Web Content Miner: *PMWebMiner*
Rowena Chau, Chung-Hsing Yeh, Kate A. Smith 956

Context-Based Recommendation Service in Ubiquitous Commerce
  Jeong Hee Hwang, Mi Sug Gu, Keun Ho Ryu 966

A New Continuous Nearest Neighbor Technique for Query Processing
on Mobile Environments
  Jeong Hee Chi, Sang Ho Kim, Keun Ho Ryu 977

Semantic Web Enabled Information Systems: Personalized Views on
Web Data
  *Robert Baumgartner, Christian Enzi, Nicola Henze, Marc Herrlich,
  Marcus Herzog, Matthias Kriesell, Kai Tomaschewski* 988

Design of Vehicle Information Management System for Effective
Retrieving of Vehicle Location
  Eung Jae Lee, Keun Ho Ryu 998

Context-Aware Workflow Language Based on Web Services for
Ubiquitous Computing
  Joohyun Han, Yongyun Cho, Jaeyoung Choi 1008

A Ubiquitous Approach for Visualizing Back Pain Data
  T. Serif, G. Ghinea, A.O. Frank 1018

Prototype Design of Mobile Emergency Telemedicine System
  *Sun K. Yoo, S.M. Jung, B.S. Kim, H.Y. Yun, S.R. Kim,
  D.K. Kim* 1028

An Intermediate Target for Quick-Relay of Remote Storage to Mobile
Devices
  Daegeun Kim, MinHwan Ok, Myong-soon Park 1035

Reflective Middleware for Location-Aware Application Adaptation
  *Uzair Ahmad, S.Y. Lee, Mahrin Iqbal, Uzma Nasir, A. Ali,
  Mudeem Iqbal* 1045

Efficient Approach for Interactively Mining Web Traversal Patterns
  Yue-Shi Lee, Min-Chi Hsieh, Show-Jane Yen 1055

Query Decomposition Using the XML Declarative Description Language
  Le Thi Thu Thuy, Doan Dai Duong 1066

On URL Normalization
  Sang Ho Lee, Sung Jin Kim, Seok Hoo Hong 1076

Clustering-Based Schema Matching of Web Data for Constructing
Digital Library
  Hui Song, Fanyuan Ma, Chen Wang 1086

Bringing Handhelds to the Grid Resourcefully: A Surrogate Middleware
Approach
  *Maria Riaz, Saad Liaquat Kiani, Anjum Shehzad,
  Sungyoung Lee* 1096

Mobile Mini-payment Scheme Using SMS-Credit
  Simon Fong, Edison Lai 1106

Context Summarization and Garbage Collecting Context
  Faraz Rasheed, Yong-Koo Lee, Sungyoung Lee 1115

EXtensible Web (xWeb): An XML-View Based Web Engineering
Methodology
  *Rajugan R., William Gardner, Elizabeth Chang,
  Tharam S. Dillon* 1125

A Web Services Framework for Integrated Geospatial Coverage Data
  Eunkyu Lee, Minsoo Kim, Mijeong Kim, Inhak Joo 1136

Open Location-Based Service Using Secure Middleware Infrastructure
in Web Services
  Namje Park, Howon Kim, Seungjoo Kim, Dongho Won 1146

Ubiquitous Systems and Petri Nets
  *David de Frutos Escrig, Olga Marroquín Alonso,
  Fernando Rosa Velardo* 1156

Virtual Lab Dashboard: Ubiquitous Monitoring and Control in a Smart
Bio-laboratory
  *XiaoMing Bao, See-Kiong Ng, Eng-Huat Chua,
  Wei-Khing For* 1167

On Discovering Concept Entities from Web Sites
  Ming Yin, Dion Hoe-Lian Goh, Ee-Peng Lim 1177

Modelling Complex Systems Workshop

Towards a Realistic Microscopic Traffic Simulation at an Unsignalised
Interscetion
  Mingzhe Liu, Ruili Wang, Ray Kemp 1187

Complex Systems: Particles, Chains, and Sheets
  R.B Pandey 1197

Discretization of Delayed Multi-input Nonlinear System via Taylor Series and Scaling and Squaring Technique
  Yuanliang Zhang, Hyung Jo Choi, Kil To Chong 1207

On the Scale-Free Intersection Graphs
  Xin Yao, Changshui Zhang, Jinwen Chen, Yanda Li 1217

A Stochastic Viewpoint on the Generation of Spatiotemporal Datasets
  MoonBae Song, KwangJin Park, Ki-Sik Kong, SangKeun Lee 1225

A Formal Approach to the Design of Distributed Data Warehouses
  Jane Zhao 1235

A Mathematical Model for Genetic Regulation of the Lactose Operon
  Tianhai Tian, Kevin Burrage 1245

Network Emergence in Immune System Shape Space
  Heather J. Ruskin, John Burns 1254

A Multi-agent System for Modelling Carbohydrate Oxidation in Cell
  Flavio Corradini, Emanuela Merelli, Marco Vita 1264

Characterizing Complex Behavior in (Self-organizing) Multi-agent Systems
  Bingcheng Hu, Jiming Liu 1274

Protein Structure Abstraction and Automatic Clustering Using Secondary Structure Element Sequences
  Sung Hee Park, Chan Yong Park, Dae Hee Kim, Seon Hee Park, Jeong Seop Sim 1284

A Neural Network Method for Induction Machine Fault Detection with Vibration Signal
  Hua Su, Kil To Chong, A.G. Parlos 1293

Author Index 1303

Table of Contents – Part III

Grid Computing and Peer-to-Peer (P2P) Systems Workshop

Resource and Service Discovery in the iGrid Information Service
Giovanni Aloisio, Massimo Cafaro, Italo Epicoco, Sandro Fiore, Daniele Lezzi, Maria Mirto, Silvia Mocavero 1

A Comparison of Spread Methods in Unstructured P2P Networks
Zhaoqing Jia, Bingzhen Pei, Minglu Li, Jinyuan You 10

A New Service Discovery Scheme Adapting to User Behavior for Ubiquitous Computing
Yeo Bong Yoon, Hee Yong Youn 19

The Design and Prototype of RUDA, a Distributed Grid Accounting System
M.L. Chen, A. Geist, D.E. Bernholdt, K. Chanchio, D.L. Million ... 29

An Adaptive Routing Mechanism for Efficient Resource Discovery in Unstructured P2P Networks
Luca Gatani, Giuseppe Lo Re, Salvatore Gaglio 39

Enhancing UDDI for Grid Service Discovery by Using Dynamic Parameters
Brett Sinclair, Andrzej Goscinski, Robert Dew 49

A New Approach for Efficiently Achieving High Availability in Mobile Computing
M. Mat Deris, J.H. Abawajy, M. Omar 60

A Flexible Communication Scheme to Support Grid Service Emergence
Lei Gao, Yongsheng Ding 69

A Kernel-Level RTP for Efficient Support of Multimedia Service on Embedded Systems
Dong Guk Sun, Sung Jo Kim 79

Group-Based Scheduling Scheme for Result Checking in Global Computing Systems
HongSoo Kim, SungJin Choi, MaengSoon Baik, KwonWoo Yang, HeonChang Yu, Chong-Sun Hwang 89

Service Discovery Supporting Open Scalability Using FIPA-Compliant
Agent Platform for Ubiquitous Networks
Kee-Hyun Choi, Ho-Jin Shin, Dong-Ryeol Shin 99

A Mathematical Predictive Model for an Autonomic System to Grid
Environments
Alberto Sánchez, María S. Pérez 109

Spatial Analysis and GIS: Local or Global? Workshop

Spatial Analysis: Science or Art?
Stefania Bertazzon 118

Network Density Estimation: Analysis of Point Patterns over a Network
Giuseppe Borruso 126

Linking Global Climate Grid Surfaces with Local Long-Term Migration
Monitoring Data: Spatial Computations for the Pied Flycatcher to
Assess Climate-Related Population Dynamics on a Continental Scale
Nikita Chernetsov, Falk Huettmann 133

Classifying Internet Traffic Using Linear Regression
Troy D. Mackay, Robert G.V. Baker 143

Modeling Sage Grouse: Progressive Computational Methods for Linking
a Complex Set of Local, Digital Biodiversity and Habitat Data Towards
Global Conservation Statements and Decision-Making Systems
Anthonia Onyeahialam, Falk Huettmann, Stefania Bertazzon 152

Local Analysis of Spatial Relationships: A Comparison of GWR and
the Expansion Method
Antonio Páez 162

Middleware Development for Remote Sensing Data Sharing and Image
Processing on HIT-SIP System
*Jianqin Wang, Yong Xue, Chaolin Wu, Yanguang Wang,
Yincui Hu, Ying Luo, Yanning Guan, Shaobo Zhong, Jiakui Tang,
Guoyin Cai* 173

A New and Efficient K-Medoid Algorithm for Spatial Clustering
Qiaoping Zhang, Isabelle Couloigner 181

Computer Graphics and Rendering Workshop

Security Management for Internet-Based Virtual Presentation of Home Textile Product
    Lie Shi, Mingmin Zhang, Li Li, Lu Ye, Zhigeng Pan 190

An Efficient Approach for Surface Creation
    L.H. You, Jian J. Zhang 197

Interactive Visualization for OLAP
    Kesaraporn Techapichetvanich, Amitava Datta 206

Interactive 3D Editing on Tiled Display Wall
    Xiuhui Wang, Wei Hua, Hujun Bao 215

A Toolkit for Automatically Modeling and Simulating 3D Multi-articulation Entity in Distributed Virtual Environment
    Xiaohui Liang, Chuanpeng Wang, Yinghui Che, Jiangying Yu, Na Qu 225

Footprint Analysis and Motion Synthesis
    Qinping Zhao, Xiaoyan Hu 235

An Adaptive and Efficient Algorithm for Polygonization of Implicit Surfaces
    Mingyong Pang, Zhigeng Pan, Mingmin Zhang, Fuyan Zhang 245

A Framework of Web GIS Based Unified Public Health Information Visualization Platform
    Xiaolin Lu 256

An Improved Colored-Marker Based Registration Method for AR Applications
    Xiaowei Li, Yue Liu, Yongtian Wang, Dayuan Yan, Dongdong Weng, Tao Yang 266

Non-photorealistic Tour into Panorama
    Yang Zhao, Ya-Ping Zhang, Dan Xu 274

Image Space Silhouette Extraction Using Graphics Hardware
    Jiening Wang, Jizhou Sun, Ming Che, Qi Zhai, Weifang Nie 284

Adaptive Fuzzy Weighted Average Filter for Synthesized Image
    Qing Xu, Liang Ma, Weifang Nie, Peng Li, Jiawan Zhang, Jizhou Sun 292

Data Mining and Bioinformatics Workshop

The Binary Multi-SVM Voting System for Protein Subcellular Localization Prediction
  Bo Jin, Yuchun Tang, Yan-Qing Zhang, Chung-Dar Lu, Irene Weber 299

Gene Network Prediction from Microarray Data by Association Rule and Dynamic Bayesian Network
  Hei-Chia Wang, Yi-Shiun Lee 309

Protein Interaction Prediction Using Inferred Domain Interactions and Biologically-Significant Negative Dataset
  Xiao-Li Li, Soon-Heng Tan, See-Kiong Ng 318

Semantic Annotation of Biomedical Literature Using Google
  Rune Sætre, Amund Tveit, Tonje Stroemmen Steigedal, Astrid Lægreid 327

Fast Parallel Algorithms for the Longest Common Subsequence Problem Using an Optical Bus
  Xiaohua Xu, Ling Chen, Yi Pan, Ping He 338

Estimating Gene Networks from Expression Data and Binding Location Data via Boolean Networks
  Osamu Hirose, Naoki Nariai, Yoshinori Tamada, Hideo Bannai, Seiya Imoto, Satoru Miyano 349

Efficient Matching and Retrieval of Gene Expression Time Series Data Based on Spectral Information
  Hong Yan 357

SVM Classification to Predict Two Stranded Anti-parallel Coiled Coils Based on Protein Sequence Data
  Zhong Huang, Yun Li, Xiaohua Hu 374

Estimating Gene Networks with cDNA Microarray Data Using State-Space Models
  Rui Yamaguchi, Satoru Yamashita, Tomoyuki Higuchi 381

A Penalized Likelihood Estimation on Transcriptional Module-Based Clustering
  Ryo Yoshida, Seiya Imoto, Tomoyuki Higuchi 389

Conceptual Modeling of Genetic Studies and Pharmacogenetics
  Xiaohua Zhou, Il-Yeol Song 402

Parallel and Distributed Computing Workshop

A Dynamic Parallel Volume Rendering Computation Mode Based on Cluster
Weifang Nie, Jizhou Sun, Jing Jin, Xiaotu Li, Jie Yang, Jiawan Zhang 416

Dynamic Replication of Web Servers Using Rent-a-Servers
Young-Chul Shim, Jun-Won Lee, Hyun-Ah Kim 426

Survey of Parallel and Distributed Volume Rendering: Revisited
Jiawan Zhang, Jizhou Sun, Zhou Jin, Yi Zhang, Qi Zhai 435

Scheduling Pipelined Multiprocessor Tasks: An Experimental Study with Vision Architecture
M. Fikret Ercan 445

Universal Properties Verification of Parameterized Parallel Systems
Cecilia E. Nugraheni 453

Symbolic Computation, SC 2005 Workshop

2d Polynomial Interpolation: A Symbolic Approach with Mathematica
Ali Yazici, Irfan Altas, Tanil Ergenc 463

Analyzing the Synchronization of Chaotic Dynamical Systems with Mathematica: Part I
Andres Iglesias, Akemi Gálvez 472

Analyzing the Synchronization of Chaotic Dynamical Systems with Mathematica: Part II
Andres Iglesias, Akemi Gálvez 482

A Mathematica Package for Computing and Visualizing the Gauss Map of Surfaces
Ruben Ipanaqué, Andres Iglesias 492

Numerical-Symbolic *Matlab* Toolbox for Computer Graphics and Differential Geometry
Akemi Gálvez, Andrés Iglesias 502

A LiE Subroutine for Computing Prehomogeneous Spaces Associated with Real Nilpotent Orbits
Steven Glenn Jackson, Alfred G. Noël 512

Applications of Graph Coloring
  Ünal Ufuktepe, Goksen Bacak 522

Mathematica Applications on Time Scales
  Ahmet Yantır, Ünal Ufuktepe 529

A Discrete Mathematics Package for Computer Science and Engineering Students
  Mustafa Murat Inceoglu 538

Circle Inversion of Two-Dimensional Objects with Mathematica
  Ruben T. Urbina, Andres Iglesias 547

Specific Aspects of Computational Physics for Modeling Suddenly-Emerging Phenomena Workshop

Specific Aspects of Training IT Students for Modeling Pulses in Physics
  Adrian Podoleanu, Cristian Toma, Cristian Morarescu,
  Alexandru Toma, Theodora Toma 556

Filtering Aspects of Practical Test-Functions and the Ergodic Hypothesis
  Flavia Doboga, Ghiocel Toma, Stefan Pusca, Mihaela Ghelmez,
  Cristian Morarescu 563

Definition of Wave-Corpuscle Interaction Suitable for Simulating Sequences of Physical Pulses
  Minas Simeonidis, Stefan Pusca, Ghiocel Toma, Alexandru Toma,
  Theodora Toma 569

Practical Test-Functions Generated by Computer Algorithms
  Ghiocel Toma 576

Possibilities for Obtaining the Derivative of a Received Signal Using Computer-Driven Second Order Oscillators
  Andreea Sterian, Ghiocel Toma 585

Simulating Laser Pulses by Practical Test Functions and Progressive Waves
  Rodica Sterian, Cristian Toma 592

Statistical Aspects of Acausal Pulses in Physics and Wavelets Applications
  Cristian Toma, Rodica Sterian 598

Wavelet Analysis of Solitary Wave Equation
Carlo Cattani 604

Numerical Analysis of Some Typical Finite Differences Simulations of the Waves Propagation Through Different Media
Dan Iordache, Stefan Pusca, Ghiocel Toma 614

B–Splines and Nonorthogonal Wavelets
Nikolay Strelkov 621

Optimal Wavelets
Nikolay Strelkov, Vladimir Dol'nikov 628

Dynamics of a Two-Level Medium Under the Action of Short Optical Pulses
Valerică Ninulescu, Andreea-Rodica Sterian 635

Nonlinear Phenomena in Erbium-Doped Lasers
Andreea Sterian, Valerică Ninulescu 643

Internet Communications Security (WICS) Workshop

An e-Lottery Scheme Using Verifiable Random Function
Sherman S.M. Chow, Lucas C.K. Hui, S.M. Yiu, K.P. Chow 651

Related-Mode Attacks on Block Cipher Modes of Operation
Raphael C.-W. Phan, Mohammad Umar Siddiqi 661

A Digital Cash Protocol Based on Additive Zero Knowledge
Amitabh Saxena, Ben Soh, Dimitri Zantidis 672

On the Security of Wireless Sensor Networks
Rodrigo Roman, Jianying Zhou, Javier Lopez 681

Dependable Transaction for Electronic Commerce
Hao Wang, Heqing Guo, Manshan Lin, Jianfei Yin, Qi He, Jun Zhang 691

On the Security of a Certified E-Mail Scheme with Temporal Authentication
Min-Hua Shao, Jianying Zhou, Guilin Wang 701

Security Flaws in Several Group Signatures Proposed by Popescu
Guilin Wang, Sihan Qing 711

A Simple Acceptance/Rejection Criterium for Sequence Generators in
Symmetric Cryptography
Amparo Fúster-Sabater, Pino Caballero-Gil 719

Secure Electronic Payments in Heterogeneous Networking: New
Authentication Protocols Approach
*Joaquin Torres, Antonio Izquierdo, Arturo Ribagorda,
Almudena Alcaide*... 729

Component Based Software Engineering and Software Process Model Workshop

Software Reliability Measurement Use Software Reliability Growth
Model in Testing
Hye-Jung Jung, Hae-Sool Yang 739

Thesaurus Construction Using Class Inheritance
Gui-Jung Kim, Jung-Soo Han 748

An Object Structure Extraction Technique for Object Reusability
Improvement Based on Legacy System Interface
Chang-Mog Lee, Cheol-Jung Yoo, Ok-Bae Chang 758

Automatic Translation Form Requirements Model into Use Cases
Modeling on UML
Haeng-Kon Kim, Youn-Ky Chung................................ 769

A Component Identification Technique from Object-Oriented Model
Mi-Sook Choi, Eun-Sook Cho 778

Retrieving and Exploring Ontology-Based Human Motion Sequences
*Hyun-Sook Chung, Jung-Min Kim, Yung-Cheol Byun,
Sang-Yong Byun* 788

An Integrated Data Mining Model for Customer Credit Evaluation
Kap Sik Kim, Ha Jin Hwang 798

A Study on the Component Based Architecture for Workflow Rule
Engine and Tool
Ho-Jun Shin, Kwang-Ki Kim, Bo-Yeon Shim 806

A Fragment-Driven Process Modeling Methodology
Kwang-Hoon Kim, Jae-Kang Won, Chang-Min Kim................. 817

A FCA-Based Ontology Construction for the Design of Class Hierarchy
   Suk-Hyung Hwang, Hong-Gee Kim, Hae-Sool Yang 827

Component Contract-Based Formal Specification Technique
   Ji-Hyun Lee, Hye-Min Noh, Cheol-Jung Yoo, Ok-Bae Chang 836

A Business Component Approach for Supporting the Variability of the Business Strategies and Rules
   Jeong Ah Kim, YoungTaek Jin, SunMyung Hwang 846

A CBD Application Integration Framework for High Productivity and Maintainability
   Yonghwan Lee, Eunmi Choi, Dugki Min 858

Integrated Meta-model Approach for Reengineering from Legacy into CBD
   Eun Sook Cho 868

Behavior Modeling Technique Based on EFSM for Interoperability Testing
   Hye-Min Noh, Ji-Hyen Lee, Cheol-Jung Yoo, Ok-Bae Chang 878

Automatic Connector Creation for Component Assembly
   Jung-Soo Han, Gui-Jung Kim, Young-Jae Song 886

MaRMI-RE: Systematic Componentization Process for Reengineering Legacy System
   Jung-Eun Cha, Chul-Hong Kim 896

A Study on the Mechanism for Mobile Embedded Agent Development Based on Product Line
   Haeng-Kon Kim 906

Frameworks for Model-Driven Software Architecture
   Soung Won Kim, Myoung Soo Kim, Haeng Kon Kim 916

Parallel and Distributed Components with Java
   Chang-Moon Hyun 927

CEB: Class Quality Evaluator for BlueJ
   Yu-Kyung Kang, Suk-Hyung Hwang, Hae-Sool Yang, Jung-Bae Lee, Hee-Chul Choi, Hyun-Wook Wee, Dong-Soon Kim 938

Workflow Modeling Based on Extended Activity Diagram Using ASM Semantics

Eun-Jung Ko, Sang-Young Lee, Hye-Min Noh, Cheol-Jung Yoo, Ok-Bae Chang 945

Unification of XML DTD for XML Documents with Similar Structure

Chun-Sik Yoo, Seon-Mi Woo, Yong-Sung Kim 954

Secure Payment Protocol for Healthcare Using USIM in Ubiquitous

Jang-Mi Baek, In-Sik Hong 964

Verification of UML-Based Security Policy Model

Sachoun Park, Gihwon Kwon 973

Computer Graphics and Geometric Modeling (TSCG 2005) Workshop

From a Small Formula to Cyberworlds

Alexei Sourin 983

Visualization and Analysis of Protein Structures Using Euclidean Voronoi Diagram of Atoms

Deok-Soo Kim, Donguk Kim, Youngsong Cho, Joonghyun Ryu, Cheol-Hyung Cho, Joon Young Park, Hyun Chan Lee 993

C^2 Continuous Spline Surfaces over Catmull-Clark Meshes

Jin Jin Zheng, Jian J. Zhang, Hong Jun Zhou, Lianguan G. Shen .. 1003

Constructing Detailed Solid and Smooth Surfaces from Voxel Data for Neurosurgical Simulation

Mayumi Shimizu, Yasuaki Nakamura 1013

Curvature Estimation of Point-Sampled Surfaces and Its Applications

Yongwei Miao, Jieqing Feng, Qunsheng Peng 1023

The Delaunay Triangulation by Grid Subdivision

Si Hyung Park, Seoung Soo Lee, Jong Hwa Kim 1033

Feature-Based Texture Synthesis

Tong-Yee Lee, Chung-Ren Yan 1043

A Fast 2D Shape Interpolation Technique

Ping-Hsien Lin, Tong-Yee Lee 1050

Triangular Prism Generation Algorithm for Polyhedron Decomposition
  Jaeho Lee, JoonYoung Park, Deok-Soo Kim, HyunChan Lee 1060

Tweek: A Framework for Cross-Display Graphical User Interfaces
  Patrick Hartling, Carolina Cruz-Neira 1070

Surface Simplification with Semantic Features Using Texture and Curvature Maps
  Soo-Kyun Kim, Jung Lee, Cheol-Su Lim, Chang-Hun Kim 1080

Development of a Machining Simulation System Using the Octree Algorithm
  Y.H. Kim, S.L. Ko 1089

A Spherical Point Location Algorithm Based on Barycentric Coordinates
  Yong Wu, Yuanjun He, Haishan Tian 1099

Realistic Skeleton Driven Skin Deformation
  X.S. Yang, Jian J. Zhang 1109

Implementing Immersive Clustering with VR Juggler
  Aron Bierbaum, Patrick Hartling, Pedro Morillo, Carolina Cruz-Neira 1119

Adaptive Space Carving with Texture Mapping
  Yoo-Kil Yang, Jung Lee, Soo-Kyun Kim, Chang-Hun Kim 1129

User-Guided 3D Su-Muk Painting
  Jung Lee, Joon-Yong Ji, Soo-Kyun Kim, Chang-Hun Kim 1139

Sports Equipment Based Motion Deformation
  Jong-In Choi, Chang-Hun Kim, Cheol-Su Lim 1148

Designing an Action Selection Engine for Behavioral Animation of Intelligent Virtual Agents
  Francisco Luengo, Andres Iglesias 1157

Interactive Transmission of Highly Detailed Surfaces
  Junfeng Ji, Sheng Li, Enhua Wu, Xuehui Liu 1167

Contour-Based Terrain Model Reconstruction Using Distance Information
  Byeong-Seok Shin, Hoe Sang Jung 1177

An Efficient Point Rendering Using Octree and Texture Lookup
  Yun-Mo Koo, Byeong-Seok Shin 1187

Faces Alive: Reconstruction of Animated 3D Human Faces
  Yu Zhang, Terence Sim, Chew Lim Tan 1197

Quasi-interpolants Based Multilevel B-Spline Surface Reconstruction
from Scattered Data
  Byung-Gook Lee, Joon-Jae Lee, Ki-Ryoung Kwon 1209

Methodology of Information Engineering Workshop

Efficient Mapping Rule of IDEF for UMM Application
  Kitae Shin, Chankwon Park, Hyoung-Gon Lee, Jinwoo Park 1219

A Case Study on the Development of Employee Internet Management
System
  Sangkyun Kim, Ilhoon Choi 1229

Cost-Benefit Analysis of Security Investments: Methodology and Case
Study
  Sangkyun Kim, Hong Joo Lee 1239

A Modeling Framework of Business Transactions for Enterprise
Integration
  Minsoo Kim, Dongsoo Kim, Yong Gu Ji, Hoontae Kim 1249

Process-Oriented Development of Job Manual System
  Seung-Hyun Rhee, Hoseong Song, Hyung Jun Won, Jaeyoung Ju,
  Minsoo Kim, Hyerim Bae 1259

An Information System Approach and Methodology for Enterprise
Credit Rating
  Hakjoo Lee, Choon Seong Leem, Kyungup Cha 1269

Privacy Engineering in ubiComp
  Tae Joong Kim, Sang Won Lee, Eung Young Lee 1279

Development of a BSC-Based Evaluation Framework for
e-Manufacturing Project
  Yongju Cho, Wooju Kim, Choon Seong Leem, Honzong Choi 1289

Design of a BPR-Based Information Strategy Planning (ISP) Framework
  Chiwoon Cho, Nam Wook Cho 1297

An Integrated Evaluation System for Personal Informatization Levels
and Their Maturity Measurement: Korean Motors Company Case
   *Eun Jung Yu, Choon Seong Leem, Seoung Kyu Park,
   Byung Wan Kim* 1306

Critical Attributes of Organizational Culture Promoting Successful KM
Implementation
   Heejun Park 1316

Author Index 1327

Exploring Constructivist Learning Theory and Course Visualization on Computer Graphics

Yiming Zhao[1], Mingmin Zhang[2], Shu Wang[3], and Yefang Chen[1]

[1] Department of Computer Science and Technology, Ningbo University,
Ningbo, 315211, P.R.China
[2] College of Computer Science, Zhejiang University,
Hangzhou, 310027, P.R.China
[3] Transaction newsroom of Ningbo University, Ningbo, 315211, P.R.China

Abstract. Constructivist learning theory has shown its advantages on learning system by improving students' interests and ability. In this paper, we introduce some research works of constructivist learning theory and some successful methods from ACM/IEEE-CS CC2001 course. Application for GV (Graphics and Visual Computing) shows the power that integrating constructivist view of learning with the real learning environment. Based on *PLATFORM AND MODULE* theory, we analyze visualization methods of Computer Graphics course material, represent a new four-lay division of GV curriculum. And we introduce scaffolding learning method, which helps students improve knowledge exploring ability. CVCS1.1 and GraphVCF1.0 are two efficient virtual platforms used in our virtual learning process. Experiments show that constructivist view of learning and its application in GV bring more efficiency and convenience to build a virtual learning environment.

1 Introduction

The newly established course of ACM/IEEE-CS for computer science and technology (CC2001[1]) build a knowledge GV(Graphics and Visual Computing). And they divide GV curriculum into two kernel modules and nine additional modules. The former includes basic graphics technology and graphics system, the latter is composed of graphics communication, geometry modeling, basic rending, custom rending, advanced technology, computer animation, virtual reality and compute vision. Results from CC1991 and CC2001 show that great changes have taken place in GV field since 1990's.

In fact, high development of computer hardware technology requires a much more harmonious interactive environment. Computer graphic application users show great interest in friend and convenient interface. Real time render system is becoming the main trend of in graphic area. Thus, computer graphics and relative graphical technologies become one of the most activist branches. Technology improvements requiring teaching methods on computer courses in high education part change the traditional way. Actually, many universities and colleges in China [11,12,13,14,15] have recognized the urgent situation and started computer graphics course and other related courses. However, content of GV is very complex. Present courses are hard to cover all ranges. And traditional teaching method have been outdated: bald content in textbooks, conflicts between few available teaching time and expanded teaching material, theory and practice, research developments and application usability become much deeper and harder. We believe new GV course system based on constructivist learning theory meet the common demands. And updating teaching material and innovating teaching methods are effective ways to reflect the newest achievement in GV field.

Discussion on computer pedagogy guided by CC2001 is still hot, but seldom applications in constructivist learning theory have been taken into practice. This paper presents some detailed pedagogical contents and pedagogical methods executed in our system which is running on local department at present. This paper is organized as follows: first part is the concept and teaching method of constructivism. Second part provides the design principle of GV course. CVCS1.1 and GraphVCF1.0, two efficient virtual platforms will be introduced in the third part. At the third part, we will present teaching control methods. Finally, the conclusion is drawn.

2 Constructivist Learning Theory

2.1 Concept of Constructivism

J.Piaget defines constructivism as a cognitive theory of objectivism[2, 3]. Cognitive theory believes that knowledge is separated from the learner, and study is the result of stimulation and response. Constructivists stress on the process of construction, which include the construction of the new incoming information and reconstruction of the existing experience. However, educator's destination is to pass the external information to students. Thus, they always emphasize on the content and the structure of knowledge, trying to set up the precise donation of knowledge. Under such circumstance, constructivists believe that the knowledge of learner can be gained from the help of others through communication with others.

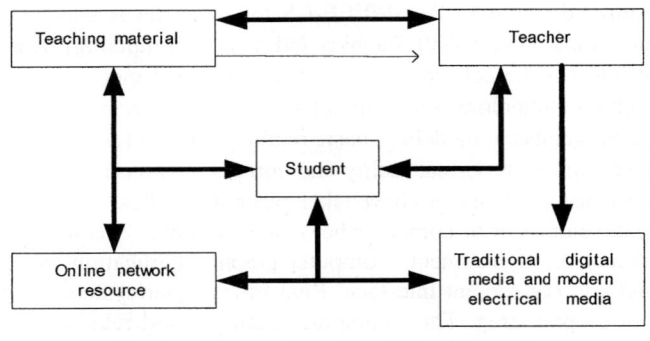

Remark: real line deligate strong interaction
broken line deligate weak interaction

Fig. 1. Four factors in new teaching model

Constructivism not only emphasizes the mutual-direction of pedagogical activity, but also focuses on the role of learning environment and learning information. Therefore, students, teachers, teaching information and learning environment are becoming the four key factors in new teaching pattern. (See Figure 1.)

Learning Environment includes several parts, context, cooperation, communication and meaning construction. Cooperation runs through the whole activity. Communication is the most basic mode during cooperation [16,17,18]. For instance, in order to finish the task, learning group members have to communicate with each other. The meaning construction is the destination of teaching process that finishes after students master it. From this point, students can understand the inner relationship between objects.

2.2 Scaffolding Instruction

From constructivism learning theory, there exists several ripe teaching methods, such as the scaffolding instruction, anchored instruction, random access instruction etc. Scaffolding is a process that requires direct teaching and monitoring. To be successful with this teaching strategy, the teacher must provide all information that a student needs to complete a given task. Detailed advice, direction of the student's attention, and alerts to the student of the sequence of activities are essential to a student's ability to perform within the scaffolded teaching environment. From this way, Children are able to direct their own attention, plan, and control their activities. Scaffolding is often compared to the scaffolding of a building. The scaffolding of a building is gradually moved until the structure is able to support its own weight. The support that is given in educational scaffolding comes in the form of modeling, giving students cues, prompts, hints, and partial solutions. If students have trouble to learn, teachers need model a new learning process. However, this also gives student another opportunity to observe the thinking and behavior. Those students who are not having difficulty are receiving reinforcement. Five different instructional scaffolding techniques can be divided:

1) Modeling of desired behaviors. Modeling is generally the first step in instructional scaffolding. It is defined as "teaching behavior that shows how one should feel, think or act within a given situation".
2) Offering explanations. Explicit statements adjusted to fit the learners' emerging understandings about what is being learned, why and when it is used and how it is used.
3) Inviting student participation. This practice engages the student in learning and provides her with ownership of the learning experience. When students contribute their ideas about a topic or skill, the instructor can add her own ideas to guide the discussion. If the students' understandings are incorrect or only partially correct, the teacher can correct them and tailor her explanations based upon whatever the students have brought to the discussion.

4) Verifying and clarifying student understanding. When the students gain experience with new material, it is important for the instructor to continuously assess their understanding and offer feedback. Verifying and clarifying student understanding is essentially offering affirmative feedback to reasonable understandings, corrective feedback to unreasonable understandings.
5) Inviting students to contribute clues. Teachers should encourage students to contribute information just as learning in a real classroom. The imagination of students will be extended in this phase. By and by, they can get the rule of the new material.

3 Design Principles of GV Curriculum

Based on GV core knowledge ideas and scaffolding instruction supported by constructivist learning theory, we provide a new four-lay based division of GV course material and course arrangement (See Table 1).

Arrangement methods showed in table 1 of above courses meets the requirements of PLATFORM plus MOJULE curriculum system. Without adding too many classes, it successfully contains all the core content provided in CC2001. Content arrangements consider basic skills, advanced skills, algorithm theory and application. Traditional paper examination pattern is changed to result evaluation in order to improve students' self-learning and knowledge exploring ability.

We also set up a systematic practice learning process for students. It consists of concept validation, project design and project training. Basic skills in GV are introduced to students from rich speeches and training actions. Knowing how to operate modeling software like 2D plot and 3Dmax, students first build up their own graphical concept framework. And then, they begin application construction. Basic skill training starts at the 2nd semester, and continues until students finish their graduate thesis. In the 6th semester, data structure course and program course like VC++ launched in order to deep students understanding of graphical view and visualization. Among them, computer graphics basics course contains 6 algorithm virtual experiments and one algorithm exercise system. Several topics are detailed discussed such as how to create common 3D graphics, how to design friendly user interface. 3D model design provides several virtual experiments considering with rendering of sketch, basic character model. Software tool exercises include AUTOCAD and SOLIDEDGE. GV algorithm and system design in 7th semester use the OPENGL to process geometry transformation, simulate the real world and handling images conversion. Computer animation course help students understand the main principles of animation process and how to model it such as model rendering, light effects and modeling etc. Virtual reality course focuses on how to use 3Dmax model various models and export them into VRML format, then represent them on the Web.

Table 1. Teaching arrangement and curriculum content design of GV

Layer	Course name	Teaching content	Corresponding to CC2001	Timing (Hour)	Teaching pattern	Evaluation
First layer: Conception construction skills	CV application elements	Simple color model; Graphics system and input & output equipment; light source; material; environment (Photoshop; AutoCAD; 3Dmax; Solid Edged)	GV1 GV2 GV5 GV8	2nd semester Short semester (12)	Cathedra: from the technology application, gives the conception framework	Evaluation: Picture process; Engineering draw; 3D model;
Second layer: basic graphics skills	Basic computer graphics & OpenGL	Summarize of graphics; Graphics interface and interaction technology with computer; The generation of the 2D graphics; The generation of 3D graphics and its transformation.	GV4 GV5 GV10	6th Semester (54)	Lesson: build the concept of framework;	Exercises; Group discussion; Appraisal
Third layer: High graphics skills	CV algorithm and its system	Vivid graphics generating technology (Simple light model; shadow generating; light tracing algorithm; texture mapping technology); Visual(Conception;3D interaction technology);Introduction of computer animation; Virtual reality	GV3 GV4GV5 GV6GV7 GV8GV9	7th semester (54)	Lesson	Finish a middle-large exercise (a practical system or optimize an algorithm)
Fourth layer: Advanced graphical skill	The application of Graphics and its newly development	Newly technology development	GV10 GV11	The 8th semester	Cathedra	The articles of new technology

4 Multi-dimension Teaching Environment

Constructivist learning theory emphasizes the process of building an effective teaching environment. Teachers are responsible of setting up such learning environment and help students establish GV concept. To implement new curriculum system, we develop two network-based virtual classrooms, named CVCS1.1 and GraphVCF1.0.

CVCS1.1 consists of virtual classroom subsystem [see Figure 2], virtual lab managing subsystem [see Figure 3], virtual tutorship subsystem [see Figure 4,5], homework submit subsystem, BBS and students learning states managing subsystem. CVCS1.1 supports 3D user interface. Exploring virtual 3D classroom, students have chance to select different courses. By using this system, students gain rich electrical media documents, such as course outline, course progress, course target and task arrangement. Material on the net will be added and updated by teachers at anytime. CVCS1.1 is also a shared resource database where teachers and students can communicate with each other. Other functions are also supported, such as online examination etc.

Fig. 2. CVSC1.0 main interface   **Fig. 3.** 3D virtual classroom

 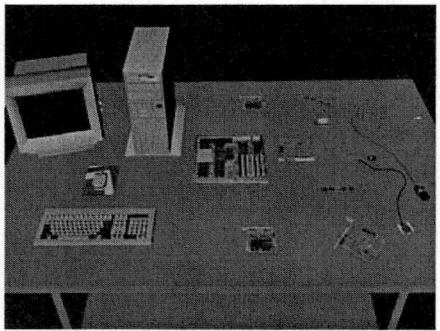

Fig. 4. 3D virtual classroom   **Fig. 5.** Virtual equipment

GraphVCF1.0 is a net-based framework for VC++ programming. At present, window system is becoming more and more complex. Although VC++ is a power program tools, it is difficult to understand all the principles. GraphVCF1.0 is designed to help students grasp basic concepts of DOCUMENT/VIEW in short time. Students put the algorithm with parameters in the prompt place, and without debug, the algorithm can run.

In virtual laboratory, virtual assemble system are build for graphic hardware construction. Students can realistic carry their experiment with video and picture supported. From doing this, students can correct their wrong actions in the real environment. Instructional resources and advising material are located on the net to provide useful reference for students. Online resource system is developing to include as many resources to guide students' virtual experiments. Students can access electrical teaching resource of computer graphics and visual field freely, which includes digital library, electrical reading room.

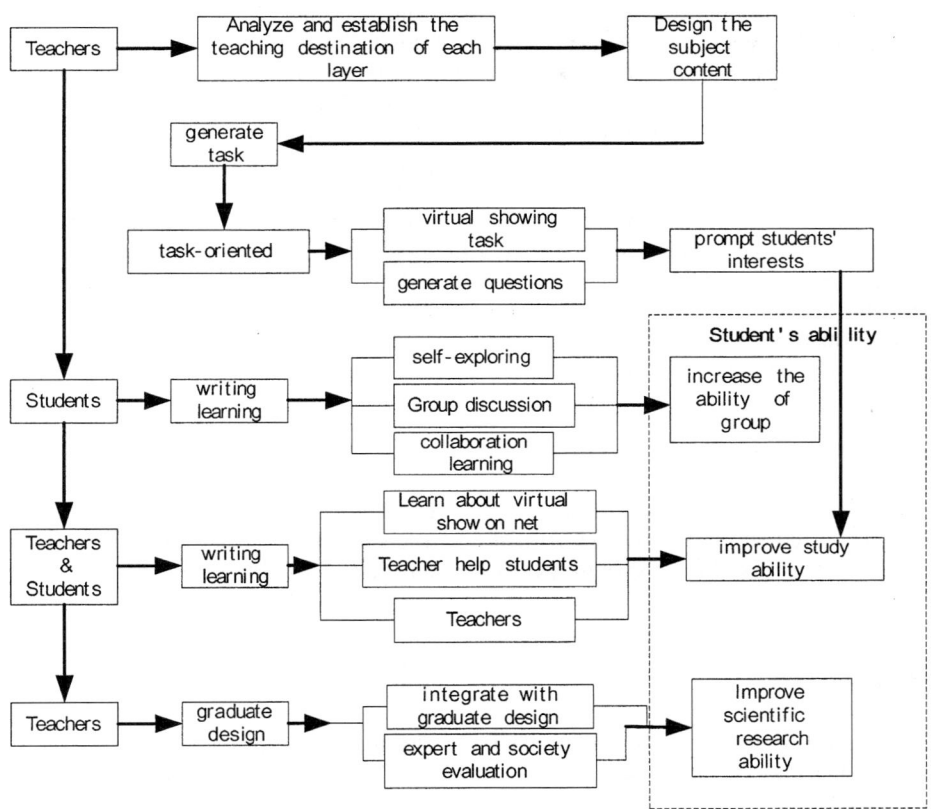

Fig. 6. The teaching progress

Constructivist view of teaching process requires the free transferring of resource distribution. And there is lots of factor influent the process. Teachers must make a good preparation for each teaching process. Students in the learning environment activated by the teacher should do their best to use their knowledge to answer questions and solve the practical problem. But instructional control and regulation are very important. If students are out of control, the concept construction will fail. To avoid this situation, we have designed a teaching-progress follow chart (Figure 6.), which based on Scaffolding teaching method. Teachers can follow the chart to implement their teaching method. This chart also provides a practicable reference to estimate the teaching quality.

5 Conclusion

Constructivist learning theory has shown prosperous since its first introduction on 1990 to China. Years of teaching experience in computer science[3], especially in computer graphics proves that constructivist view of teaching and learning theory meets the requirement of expanded teaching and learning group. Three advantages can be outlined briefly to support this opinion. Firstly, it can improve student's learning interest. There are plenty of algorithm and programming in GV teaching course, with the support of new teaching method, the students can easily get the idea of abstract concept, which will boost their learning passion and enthusiasm. Secondly, this teaching method improves the efficiency of algorithm what they are learning, for this new teaching method not only the relationship between teachers and students, but also the convenient way of knowledge acquirements for students can be gained. Thus, it gives the students the changes of consider the problem much more freely and deeply. Finally, by using network and virtual reality technology, students communicate with teachers and other students more frequently without regional limitation. This teaching method also provides students with chances of join in the scientific research and project, which can improve their research ability.

Acknowledgment

This project is under support of Scientific Research Fund of Zhejiang Provincial Education and co-supported by Department&Kuancheng Wang Education Fund. The authors would also like to give thanks to Zhaohui Zhu, Jiejie Zhu who help us to build the entire architecture of synthetic characters.

References

1. The Computer Society of the Institute for Electrical and Electronic Engineers (IEEE-CS) and the Association for Computing Machinery (ACM). Computing Curricula Final Draft–December15,2001.2001-12-15,
   http:// www.acm.org / sigcse / cc2001 [Z].
2. El-Hindi A., Donald J. "Beyond classroom boundaries: constructivist teaching with the Internet" Reading Teacher (1998) p.694-701
3. Kekang H. "Constructivism – The basic theory of innovate traditional teaching" Subject education (1998) (In Chinese)
4. Tang Z.S., Zhou J.Y, Li X.Y. "Computer graphics basis" TingHua publishing company (2000) (In Chinese)
5. Shi J.Y, Cai W.L. "Algorithm of visualization" Scientific publishing company (1996) (In Chinese)
6. Peng Q.S., Bao H.J, Jin X.G. "The basic algorithm of computer realistic graphics" Science publishing company. (1999) (In Chinese)
7. Zhang W.H, Zhou Y. "Computer Graphics in education" Computer Science (2003) 30 (6) p.104-105 (In Chinese)
8. Zhao Y.M. "The exploration of several questions in computer engineer education" Journal of NingBo University, (2003) 25(5) (In Chinese)

9. Von Glasserfeld E. "An introduction to radical constructivism in P. W. Watzlawick" The Invented Reality. W. Norton and Company, New York. (1984). p.17-40
10. MichealD.W. "Integrating Technology Into Teachingand Learning" Pearson Education Asia (2000)
11. Pan Z.G., Zhu C.H., Rui P., etc. "Emotion modeling and interaction in intelligent virtual environment" In the Proceedings of 3IA (2004) p.1-6
12. Pan Z.G., Xu W.W., Huang J., Zhang M.M., etc. "EasyBowling: A Virtual Bowling System based on Virtual Simulation" Computers&Graphics (2003) 27(2) p.231-238
13. Pan Z.G., Shi J.Y. "Virtual Reality and Application" Computers&Graphics (2003) 27(2)
14. Pan Z.G., Shi J.Y. "Virtual Reality Technology Development in China: An Overview" The International Journal of Virtual Reality (2000) 4(3) p.2-10
15. Shi J.Y., Pan Z.G. "China: Computer Graphics Education Available at Universities, Institutes and Training Centers" Computer Graphics (1999) 31(3) p.7-9
16. Zhu J.J., Hu W.H., Pan Z.G. "Design and implementation of Virtual Multimedia Classroom" Journal of CAD&CG, 2004. 16(1) 2004 p.73-79 (in Chinese)
17. Pan, Z.G., Zhu, J.J., Hu, W.H. "Interactive Learning of CG in Networked Virtual Environments" Computers & Graphics (2005) 29(2)
18. Hu. W.H., Zhu,J.J., Pan, Z.G. "Learning By Doing: A Case for Constructivist Virtual Learning Environment" In the Proceedings of Eurographics Education Program (2003) p.6-15
19. Pan Z.G., Zhu C.H., Rui P., etc. "Emotion modeling and interaction in intelligent virtual environment" In the Proceedings of 3IA (2004) p.1-6

A Program Plagiarism Evaluation System

Young-Chul Kim and Jaeyoung Choi

School of Computing, Soongsil University,
1-1 Sangdo-dong, Dongjak-gu, Seoul 156-743, Korea
yckim@ss.ssu.ac.kr, choi@ssu.ac.kr

Abstract. In this paper, we introduce an evaluation system for identifying program similarity by comparing syntax-trees for the given programs. By using syntax-trees, this system is capable of distinguishing plagiarism in spite of changes in program styles such as indent, white space and comments. It can also recognize plagiarism patterns resulting from changes in program structure such as statement exchanges, code block and function. Syntax-trees are created after program parsing, so they have the advantage of performing syntax and semantic analysis automatically. We also introduce an evaluation algorithm for program similarity and a grouping algorithm for the sake of reducing the count of comparisons. The experiment and estimation proves that a grouping algorithm can reduce a lot of counts of comparison.

1 Introduction

Today, program developers and students who study a program language can easily find and use examples of a desired program due to many books on programming and the development of the Internet as a medium. The development of these media helps many users who learn programming, on the other hand there are negative effects in that the study of programming is neglected and homework is easily done using shared programs. These similar assignments which students hand in make it difficult for the marker to accurately compare and evaluate the assignments with one another. Especially, assignments with slightly modified code and assignments with a changed style are very difficult to evaluate, and the results of evaluation can differ according to the marker [1, 2, 3]. This research introduces a system which evaluates similarity among different programs using the AST (Abstract Syntax Tree) of the programs submitted. The system evaluates similarity between two programs using AST produced in the process of parsing. If a similarity is found using AST, this system can check whether or not it is structurally similar or not structurally without regard to modification of the program's source code, and can perform a syntax error check. The main object of this research is to overcome difficulties of evaluating a program's source code that students submitted.

Plagiarism of a program becomes more varied as time goes by. An item which attracts public attention goes into newspapers and some students spend their time editing program's source rather than doing their homework by themselves. It is important to be able to evaluate whether a program's source code is similar or not to another's.

Especially, there is a keen need for a system that automatically evaluates whether two programs are similar or not. Fig. 1 shows a program similarity evaluation system model. In Fig. 1, the AST pool is a collection of AST generated in the process of parsing. The program similarity evaluation engine, a similarity table, and a similarity grouping are later described in Section 3.

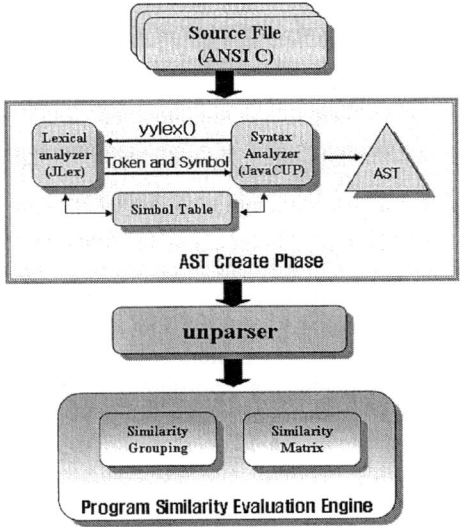

Fig. 1. Program similarity system model

This paper is as follows. Section 2 discusses related research. Section 3 describes a program similarity evaluation algorithm and grouping. Section 4 describes testing and evaluation. Finally, Section 5 contains the conclusion and suggests additional research tasks.

2 Related Research

Researches related to program similarity have been ongoing for the last 20 ~ 30 years ago. Researches related to similarity of documents were performed in the early stage of our research, with the aim of performing program similarity evaluation with these researches as a basis.

2.1 Similarity Evaluation System

In writing, plagiarism of documents means to borrow from original research and use it as one's own opinion without identifying the source [6]. Through documents have a linear structure, similarity is evaluated using statistical features rather than structural features because the length of the linear structure varies according to the document. Applying a statistical structure to the criteria for evaluating the similarity of documents is termed fingerprint [7]. Fingerprint is a method which examines the similarity between two documents for a selected word, and compares the frequency of the selected word. The similarity method for the initial program evaluates program similarity using a fingerprint. Namely, similarity is evaluated between two programs using the statistical parts of the program. The initial system is based on fingerprint using Halsted matrix [8]. Ottenstein [9], Donaldson [10], Berghel [11], among others made the system by using or extending the matrix which Halstead suggested.

Donaldson suggests a plagiarism method which is worth noticing. It mixes the program structure like the number of loop statements and procedure statements and uses it with the Halstead matrix. Recently evaluation systems were developed which evaluate program similarity using "token" generated in the process of lexical analysis. By using a program "token," these systems become insensitive to elements regardless of the syntax of a program such as the program style, statement, indentation, and so on being different from existing systems. The systems which evaluate a similarity program token include YAP3 [12], MOSS [13], Jplag [14], Sim [15], SID [16]. There is research using bioinformatics which is one way of using token string. That is, it is a method which uses sequence analysis in order to search for similar parts by comparing genomic sequences. Research in genomic sequence analysis searches for similar parts by comparing DNA sequences or protein sequences, and this is done in order to analyze the function of homogeneous or heterogeneous genomic sequence. It is based on the assumption that genes of identical sequences have the same function.

2.2 A Method of Evaluating Program Similarity

YAP3 [12] which was developed by M.J. Wise is a similarity evaluation system using the structural matrix method. YAP3 resets programs in order to evaluate their similarity. Namely, removing statements and string constants, changing letters from capital to lower case, changing from a source program to the same or similar synonym, rearranging an order of function call and so forth is performed. Similarity evaluation in YAP is evaluated as follows:

$$Match = \frac{(same - diff)}{\min file} - \frac{\max file - \min file}{\max file} \quad (1)$$

In (1), "max *file*" and "min *file*" represent the larger and the smaller of the two files respectively. Also, "*same*" means the number of tokens which is found in both of the files and "*diff*" means the number of other lines that are inside the block. Sim [15] also uses the value of a real number in a range from 0 to 1 in order to evaluate the similarity between two token strings. Sim uses the formula of similarity evaluation below:

$$S = 2 * \frac{score(s,t)}{score(s,s) + score(t,t)} \quad (2)$$

In (2), "*score(s, t)*" indicates whether the tokens(s and t) of two programs are equal or not. Namely, because of using the gene acids sequence technique, the weight of this sequence expresses a score in SIM, too. SID [16] is based on Kolmogorov complexity. SID expresses the value of similarity in a range from 0 to 1 like Sim, and similarity is evaluated as shown in the following (3):

$$R(s,t) = \frac{K(s) - K(s \mid t)}{K(st)} \quad (3)$$

In (3), "$K(s|t)$" expresses the Kolmogorov complexity, and if "t" is empty, the Kolmogorov complexity becomes "$K(s)$". Also, "$K(s)$-$K(s|t)$" is mutual information that expresses the difference between "$K(s)$" and "$K(t)$". On the contrary, "$K(st)$" expresses common information. D. Baxter [18] and others use a dummy code fragment to search between two programs. The dummy code is called "clone" in this research, and similarity was represented through a dummy code. Their research uses the numerical formula of similarity evaluation as follows:

$$Similarity = 2 * \frac{S}{2*S + L + R} \quad (4)$$

In (4), "S" is the amount of shared node, "L" is the amount of node which doesn't exist in the target source but is present among trees in the original source.

3 Similarity Evaluation and Grouping

AST is generated in the process of parsing, and it is arranged in sequential order of nodes linearly by an "unparser." Therefore, two programs are finally converted to a node string, and compared in order to perform a plagiarism check. A node string is a string that enumerates nodes of AST linearly. The value of similarity between two programs is as follows:

$$0 \leq Similarity(program1, program2) < 1 \quad (5)$$

A full plagiarism is a value of a similarity as 1 exactly. A strong similarity defines a value of a similarity as a real number within a range from 0.9 to 1. Also, a middle similarity defines a value of similarity as a value within a range from 0.7 to 0.9. A weak similarity defines a value of a similarity as 0.7 or less. This paper shows the following similarity check algorithm in order to test a lot of programs.

```
double Sim(NodeString A, NodeString B, long int minlength) {
        String matchstring, totalmatchstring;      /* Match string */
        int maxmatch = 0;         /* Initialize the number of match string */
        long int matchlength = 0; /* Initialize the number of all match string */
        Set(totalmatchstring) = { }; /* A set of all match string*/
        do {    matchstring = ""; /* Match string */
                matchstring = MatchString(A, B);
                Set(totalmatchstring) = Set(totalmatchstring) + matchstring;
        } while (maxmatch > minlength);
        for each matchstring in Set(totalmatchstring)
                matchlength = matchlength + Length(matchstring);
        end for
return ( S ) }
```

S's equation is as like,

$$S = 2 * \frac{matchlength}{Length(p1nodestring) + Length(p2nodestring)} \quad (6)$$

The *minlength*, which is used as inputs of program similarity evaluation algorithm, is defined by minimum numbers of substrings that are consistent among node strings. The *maxmatch* represents the number of consistent strings. To compare strings continues until *maxmatch* is bigger than *minlength*. Set(totalmatchstring) is defined as a function that stores all substring which is found in two node string. Set(totalmatchstring) is defined as a function that stores all match substring. Length(X) is defined as a function showing the length of node string X. Length(X) function which is used for a similarity evaluation is a function which calculates the length of node string. That is, this function find a length of matched substring and inputted the length of node string, and it's used in a numerical formula of a similarity finally.

In this paper, grouping is performed on assignments which have a high similarity among their programs. Hence the similarity of two different programs must be evaluated before performing a grouping. This paper shows the following grouping algorithm in order to test a lot of programs and bind similar groups.

```
file *P;     /* input programs to compare */
int g;       /* input a value of a global similarity */
boolean addgroup = false;  /* the flag about adding to a group */
add P to G(1); /* add the first program to a group in order to compare by force */
i =1; /* a counter as the number of a group */
Set G(1) = ∅     /* initialize a group */
while( not eof) {
   input P;       /* input program as the subject of comparison */
   for each i in G(i)    /* execute with regard to all groups */
        if Sim(P, G(i)) > g then   /* the case included in a group */ {
            add P to G(i); /* add to a group */
            addgroup = true;  }
   end for
   /* create a group if it isn't added to a group */
   if (not addgroup) then {
        i = i + 1; /* add a counter of a group */
        Set G(i) = ∅   /* initialize a group */
        add P to G(i); /* the case all is not included */    }
}
```

4 Testing and Evaluation

This system is implemented using Java language and jdk1.3.1. Therefore, it can be applied to windows and UNIX environments and programs written by ANSI C established in 1989 to check for plagiarism. Also, the JLex [4] and JavaCUP [5] utility is used in this system in order to execute syntax analysis and lexical analysis of C language.

Testing 1. Similarity Evaluation of Two Programs. Let us suppose that there are two programs which find the least common multiple (L.C.M.) like in the following Table 1 and 2 for testing. In table 2 the style is changed from that of Table 1. It is changed into a similar control statement.

Table 1. The original program (source.C)

```
#include <stdio.h>
void main(void) {
  long result; int index=0;
  int n1, n2, m1, m2, divs[100], lnum, i, flag;
  printf("Input 2 numbers for calculating GCM...\n");
  scanf("%d %d", &n1, &n2); m1= n1; m2 = n2;
  while(1) {  lnum = (m1>=m2?m1:m2); flag = 0;
     for(i=2; i<=lnum; i++) {  if(m1%i==0 && m2%i==0) {
        m1/=i; m2/=i; divs[index++]=i; flag=1;  break; }    }
     if(flag==0) break;   }
  result = m1 * m2;
  for(i=0; i<index; i++) result *= divs[i];
  printf("LCM of %d and %d is %ld.\n", n1, n2, result);   }
```

Table 2. A similar program (sim.C)

```
#include <stdio.h>
void main(void) {  /* Style Change */
long result; int index=0; /* Statement Position Change */
int n1, n2, m1, m2, divs[100], lnum, i, flag;
printf("Input 2 numbers for calculating GCM...\n");
scanf("%d %d", &n1, &n2); m1= n1; m2 = n2;
for(;;) {   /* Control Structure Change : while ==> for */
flag = 0; lnum = (m1>=m2?m1:m2); /* Statement Position Change */
for(i=2; i<=lnum; i++) {  /* Style Change */
if(m1%i==0 && m2%i==0) {
m1/=i; m2/=i; divs[index++]=i; flag=1; break; }} if(flag==0) break; }
result = m2 * m1; /* Operand Change */ result = m2 * m1; /* Dummy Code Addition */
result = m2 * m1; result = m2 * m1; /* Dummy Code Addition */
result = m2 * m1; /* Dummy Code Addition */ for(i=0; i<index; i++) result *= divs[i];
printf("LCM of %d and %d is %ld.\n", n1, n2, result); }
```

The result of comparison of the two programs is shown in Fig. 2. In Fig. 2, the value of similarity of the two programs is 92.98934. In testing, the value of similarity is decreased in spite of them being the same program, because many dummy codes are inserted as shown in Table 2. Hence, if there are many dummy codes, in this system, the value of similarity will decrease remarkably.

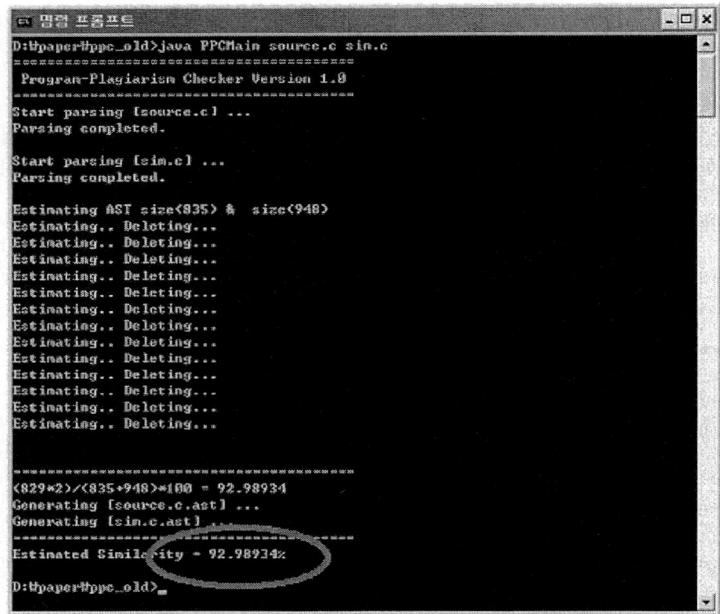

Fig. 2. A program similarity evaluation of source.C and sim.C

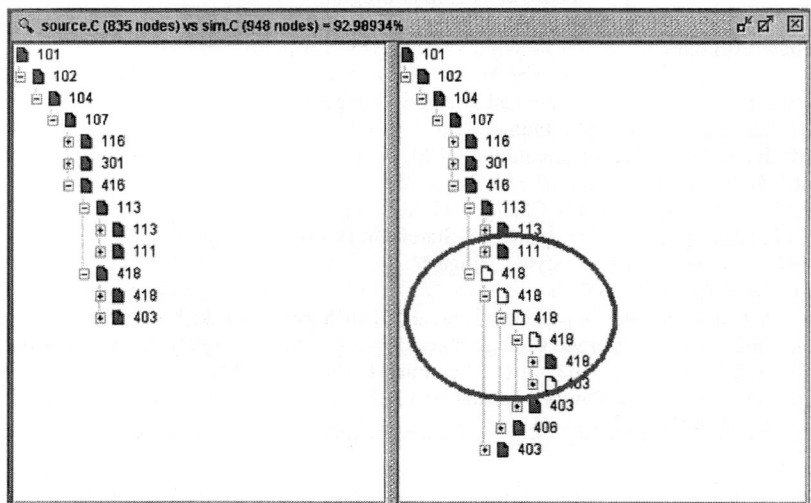

Fig. 3. A program syntax tree of source.C and sim.C

The following Fig. 3 shows the AST of comparing source.c with sim.c. A positive number in the figure means a node of AST. Fig. 3 shows the different parts of the two programs clearly. A match code is represented by a red mark (■) between two programs, and no match code is represented by a white mark (□). Therefore, this system shows that the more dummy codes that are inserted, the more the value of similarity is decreased.

Testing 2. Similarity Table. A similarity evaluation was performed and analyzed with regard to assignment S which university students majoring in computer submitted. The assignments were an example of a sort program written with C language, and about ten students' assignments were extracted and a similarity evaluation performed.

Table 3. The results of the entire program similarity

File Name	129.c	035.c	039.c	078.c	180.c	469.c	486.c	123.c	456.c
160.c	0.6669	0.85556	0.92987	0.89830	0.75231	0.82222	-1	0.84886	0.87282
129.c		0.78691	0.70613	0.66399	0.52589	0.80889	-1	0.60466	0.64314
035.c			0.89058	0.84	0.69112	0.91719	-1	0.76459	0.81351
039.c				0.90946	0.76179	0.82220	-1	0.83590	0.86497
078.c					0.81253	0.79964	-1	0.89170	0.90638
180.c						0.65712	-1	0.85480	0.84421
469.c							-1	0.73117	0.77869
486.c								-1	-1
123.c									0.91350

Table 3 shows the results. It represents a similarity table which compares all of the programs. A value of -1 value in the table means that a comparison was not performed yet; the other values represent a value of similarity. Therefore, because "486.c" file has syntax errors, it was excluded from the other programs. As you can see from Table 3, all programs have a similarity value within a range of 0.5 to 1. There was no full plagiarism among these programs, but a strong similarity appeared among many. If an evaluation similarity was performed on these programs using the clustering discussed before, the results would be similar to Testing 3 below. The following Table 4 is the results of a grouping test which fixes the similarity value at 0.9.

Testing 3. Perform a Grouping-set Global Similarity at 0.9. In Table 4, the point to observe is that there is less comparison when the global similarity value is 0.9 (strong similarity) than with the comparison of all programs. In order to determine the reliability of this system, clustering was checked in person and the results gave a similarity value close 0.9. That is, there was no full plagiarism found as the result of examining one group which formed part of the cluster. However, there were many parts that were considerably similar. This proves that the program which formed grouping did indeed have a similarity value of 0.9.

Table 4. The results of performing a grouping with global similarity set at 0.9 (Group0 : 160.c, 039.c, Group1 : 129.c, Group2 : 035.c, 469.c, Group3 : 078.c, 456.c, Group4 : 180.c, Group5: 486.c, Group6 : 123.c)

Group Name	Group1	Group2	Group3	Group4	Group5	Group6
Group0	0.66690	0.85556	0.89830	0.75231	-1	0.84886
Group1		0.78691	0.66399	0.52589	-1	0.60466
Group2			0.84	0.69112	-1	0.76459
Group3				0.81253	-1	0.89170
Group4					-1	0.85480
Group5						-1

5 Conclusion

This paper introduced a system to evaluate the similarity between two different programs, and showed how to implement it. And, it performed a clustering in order to reduce the number of comparisons of all programs. Evaluating the similarity of all programs was done with regard to the number of n programs. This showed that the number of comparisons can be reduced from the existing "n(n-1)/2" times to the minimum "n-1" times. In the testing and evaluation part of this system, program similarity evaluation and clustering are tested using AST. This system revealed problems with dummy code such as an unnecessary variables and statements and so on. Therefore, these problems can be solved by executing code optimization and using the dummy code checker which has been researched within the software engineering field. However, further research is needed regarding the execution rate. From the results of testing, this system showed the execution rate decreases, considerably with regard to a program which has about 500 lines or less. Therefore, study must be done to improve the execution rate of algorithms. In addition, similarity evaluation of an XML document, a similarity check using code generation and optimization and the use of a web-based program plagiarism check system are also needed. Various program plagiarism and documents plagiarism checks can be carried out using the algorithm which is used in this research.

Acknowledgements

This work was supported by Korea Research Foundation Grant. (KRF-2004-005-D00172)

References

1. Joy, M., Luck, M.: Plagiarism in Programming Assignments, IEEE Transaction in Education, Vol. 42 (1999) 129-133
2. Cunningham, P., Mikpyan, A. N.: Using CBR Techniques to Detect Plagiarism in Programming Assignments, available at Department of Computer Science, Trinity College, Dublin (1993)

3. Faidhi, J. A., Robinson, S. k.: An Empirical Approach for Detecting Program Similarity within a University Programming Environment, Computers and Education, Vol. 11 (1987) 11-19
4. Lin, J.: JLex Tutorial, available at http://bmrc.berkeley.edu/courseware/cs164/spring98/proj/jlex/tutorial.html
5. Hudson, S. E.: CUP Parser Generator for Java, available at http://www.cs.princeton.edy/~appel/modern/java/CUP/
6. Edlun, J. R.: What is "Plagiarism" and why do people do it?, available at http://www.calstatela.edu/centers/write_cn/plagiarism.htm, University Writing Centre Director, California State University, LA (1998)
7. Johnson, J. H.: Identifying Redundancy in Source Code using Fingerprints, In Proc. of CASCON 93 (1993) 171-183
8. Howard, M., Halstead: Elements of Software Science, Elsevier (1977)
9. Ottenstein, K. J.: an Algorithmic Approach to the Detection and Prevention of Plagiarism, ACM SIGSCE Bulletin, Vol. 8 (1976) 30-41
10. Donaldson, J. L., Lancaster, A. M.: A Plagiarism Detection System, ASM SIGSCE Bulletin (Proc. of 12th SIGSCE Technical Symp.), Vol. 13 (1981) 21-25
11. Berghel, H. L., Sallach, D. L.: Measurements of Program Similarity in Identical Task Environments, ACM SIGPLAN Notices, Vol. 19 (1984) 65-76
12. Wise, M. J.: Detection of Similarities in Student Programs: YAP'ing may be Preferable to Plague'ing, ACM SIGSCE Bulletin (Proc. of 23rd SIGCSE Technical Symp.), Vol. 24 (1992) 268-271
13. Aiken, A.: MOSS (Measure of Software Similarity) Plagiarism detection system, available at http://www.cs.berkeley.edu/~moss/, University of Berkeley, CA (2000)
14. Prechelt, L., Malpohl, G., Philppsen, M.: JPlag: Finding Plagiarism Among a Set of Programs, available at http://wwwipd.ira.uka.de/EIR/ D-76128 Karlsruhe, Germany, Technical Report 2000-1 (2000)
15. Gitchell, D., Tran, N.: Sim: A Utility For Detecting Similarity in Computer Programs, available at ftp://ftp.cs.vu.nl/pub/dick/similarity_tester/, In Proc. of 30th SCGCSE Technical Symp., New Orleans, USA (1998) 266-270
16. Chen, X., Li, M., Mckinnon, B., Seker, A.: A Theory of Uncheatable Program Plagiarism Detection and Its Practical Implementation, University of California, Santa Barbara (2002)
17. Chen, X., Kwong, S., Li, M.: A Compression Algorithm for DNA Sequence and its Applications in Genome Comparison, In Proc. of the 20th Workshop on Genome Information (1999) 52-61
18. Baxter, I., D., Yahin, A., Moura, L., Sant'Anna, M., Bier, L.: Clone Detection using Abstract Syntax Trees, In Proc. of the International Conference on Software Maintenance, Bethesda, Maryland (1998) 368-378

Integrated Development Environment for Digital Image Computing and Configuration Management

Jeongheon Lee, YoungTak Cho, Hoon Heo, and Oksam Chae

Department of Computer Engineering, Kyung Hee University,
1 Seochun-ri, Kiheung-eup, Yongin-si, Kyunggi-do, South Korea, 449-701
opendori@paran.com, greizen@hanmail.net,
hhoon@naver.com, oschae@khu.ac.kr

Abstract. In this paper, we present a system referred to as "Hello-Vision." Hello-Vision is a software development environment that can be used in conjunction with research and application development in the area of image processing. It is an integrated environment that supports reusable image processing algorithm developments, systematic management of algorithms and related information. Hello-Vision's function is to simplify image processing applications using the algorithms and to provide an ideal environment for the education of practical image processing engineers. Hello-Vision adopts a true object-oriented approach supporting well-separated data classes similar to IUE (Image Understanding Environment) classes. This process allows algorithm to be written by their functions without user interface programming by separating the interface layer from data processing functions. The user-defined functions are easily registered in the online algorithm management systems and treated as a system command. Hello-Vision is equipped with a visual programming environment where a user can easily create image- processing applications by manipulating visual command icons and user-defined function icons that are created automatically when they are registered. Hello-Vision also provides an ideal environment where instructors can register their lecture materials for image processing theory, which allows hands-on experimentation with Hello-Vision through user interaction.

1 Introduction

The image processing development environment should be an integrated environment that supports the development and management of user-defined functions and application development based on the functions. This development environment allows for more efficient research and application development. There are no general solutions that work reliably under various conditions in the field of image processing. Thus, most of the application systems are developed as special purpose systems that work only for the given conditions. This results in high development cost and long development time. An integrated development environment increases the efficiency of the application system development by demanding the user to create standardized algorithms and reuses them for other applications. An integrated environment is also necessary for training application developers with practical sense. In the integrated environment, students can analyze various existing solutions, which allows for a more simplified implementation and testing of their ideas. The integrated environment is an ideal environment for promoting related technology transfers.

Much research has been done to create an environment for easy algorithm and application development. In the 70's, researchers tried to standardize image data and created few basic image processing algorithm libraries [1][2]. In the 80's, many image processing packages were developed. The image processing packages consist of a basic image processing library and a simple test or programming environment. In the late 80's, some researchers developed special purpose programming languages to generate reusable image processing algorithms. These special languages were not widely accepted because ordinary developers did not want to learn new special purpose languages [3][4][6][7]. In the 90's, several integrated development environments including Khoros [8] and Wit [9] were introduced. Khoros has a visual programming environment, which maintains all algorithms in an executable file format and transfers data between algorithms through a file. It provides a good environment for idea verification in research. Wit is a commercial system with a better user interface and visual programming tool which allow it to be faster and better suited for algorithm development than Khoros. From the late 90s, JAVA based development environments have been developed by targeting machine vision industry [10]. Unfortunately, these systems have some drawbacks. They do not provide a true object-oriented class hierarchy although they support similar concepts including abstract data types. They do not support online information management to supply information for online help and a parameter check in visual programming. Those systems require users to use platform dependent utility functions to create user-defined function.

In this paper, we propose an integrated development environment called " Hello-Vision ". Hello-Vision is the result of our long efforts to develop user-friendly integrated systems for both researchers and application developers. We have two goals in this research. The first goal is to crate an ideal image processing educational environment where students learn theories in the effective multimedia environment and apply them to solve real problems as they learn to build practical sense. The second goal is to develop an expert system with which any field engineer with a little knowledge about computer vision can generate his own application system. As a part of the effort, we have completed an integrated environment called " Hello-Vision ". Hello-Vision is equipped with an online algorithm or function information manager. The algorithm information manager (AIM) manages registered user-defined functions and related knowledge and provides necessary information for various operations such as; function selection, visual programming, and automatic user interface generation.

With the help of the AIM, Hello-Vision's visual programming environment recognizes user-defined functions as soon as they are registered and checks the parameter compatibility among icons representing the functions before execution to reduce the number of runtime errors. By separating the user interface for parameter input from user-defined functions, Hello-Vision allows the algorithm developer to create the functions without worrying about GUI programming.

2 Overall Description of Hello-Vision

Hello-Vision consists of two tools; Development Tool (DT) and Configuration Manager (CM). Hello-Vision DT consists of seven modules; standard data classes, algorithm information management, algorithm execution, plug-in image analysis function management, image processing education, standalone application system generation, resource management, automatic user interface generation and an operating system dependent kernel library. And Hello-Vision CM consists of seven modules; local pc management, user management, delta engine, statistics management, file converter, configuration management, repository RDBMS. Fig. 1 shows a block diagram of the proposed system.

The standardized data class module consists of various 2D image buffers and image feature data classes defined to produce reusable functions. The algorithm information management module (AIM) is an online function information manager that manages the information necessary for function execution and user interface generation. The function execution module provides two ways of executing user-defined functions registered in the AIM; an icon based visual programming tool and a menu. By choosing one, the user can test individual function or create their own application by combining registered functions. The interface generation system automatically generates user interface based on information in the AIM to read in the parameters interactively. This system feature will reduce tedious GUI programming from many algorithm developers. The resource manager keeps track of the resources, mainly memory resources, for reliable execution of Hello-Vision DT. The kernel library is a collection of hardware and operating system dependent functions. We have isolated the system dependent functions from the rest of the systems to increase portability. The standalone application generation module provides users with an interactive output screen creation tool for an application program, which generates the execution file for the application program constructed by the visual programming environment.

The plug-in function management module provides various functions to analyze image- processing results interactively. It is designed based on plug-in functions for easy expansion. It automatically searches plug-in image analysis functions in the designated file directory and registers them in the image analysis menu. Thus, users can add their own image analysis function to the system by storing it in the designated plug-in directory. The education manager accepts multimedia lecture materials in various file formats including **.htm** and **.ppt** and presents them with hands-on experiments. In this environment, we can design education programs whose lectures and experiments are tightly integrated and present them in the same environment. This is a very important feature for the field of image processing where comparative experience, with many existing solutions, is important to build practical sense.

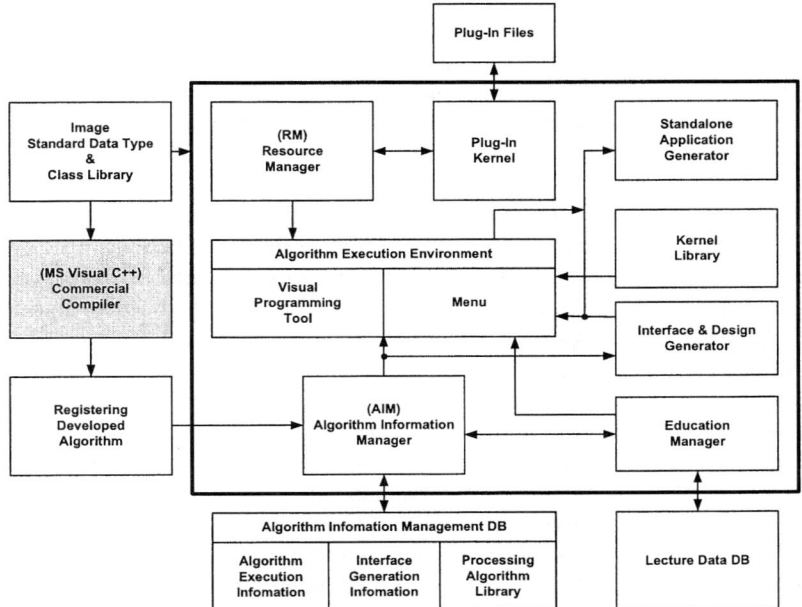

(a) Hello-Vision DT (Development Tool)

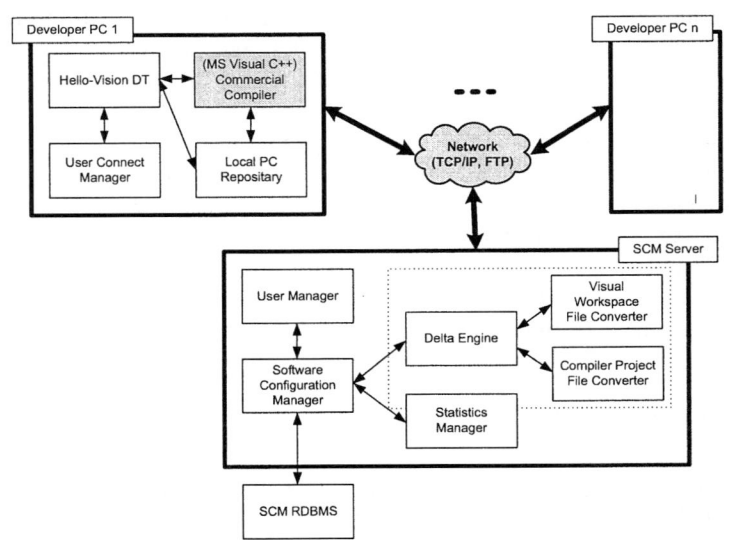

(b) Hello-Vision CM (Configuration Manager)

Fig. 1. Block diagram of Hello-Vision

Hello-Vision DT allows users to create their own functions by using a commercial programming environment similar to Microsoft Visual C++. The data classes in the standard data class module must be used to represent the data that will be transferred to other functions to make the user defined function reusable. All the user-defined functions are compiled and maintained in DLL form. Once a function is created, it can be registered into the AIM together with the information necessary for the selection and execution of the function. The function is treated like any other command or function in Hello-Vision DT's programming environment. The user can easily create a new application program by putting registered functions together using the visual programming tool. The modification of a user-defined program is simple. Firstly edit and then compile the function solely. There is no need to link it with Hello-Vision DT. To simplify user-defined function generation and modification, Hello-Vision DT creates a template C++ source file and workspace for the function, and then loads them in commercial C compiler as soon as the function is registered. Simultaneously, a new icon representing the function is created in the function management window. The icon is then ready to be used when users insert algorithm in the template file and compile. A modification can be made by reloading the source code, pushing the right mouse button and recompiling it. It is much easier than that of existing development environments which go through several steps; code generation, DLL file generation and DLL loading by the loading program.

Hello-Vision CM is the proven, full-featured solution for version control and revision management in team software development by using Hello-Vision DT. Hello-Vision CM organizes, manages and protects assets automatically to reduce common team development errors and speed project completion. More than simply storing file versions, Hello-Vision CM enables and automates complex team development tasks such as parallel development, visual differencing, branching and merging, identification of merge conflicts, promotion levels and team workflow. Hello-Vision DT and command line clients assure permissions protected yet flexible access to Hello-Vision CM by all team members, whenever and wherever they work. Layered client-server protection includes ability to assign and control user rights, standards-based encryption for secure access. The robust archive database and optimized file server architecture provide protection against data corruption as projects grow, enable centralized control of assets across the team, and assure high performance for key operations, including check-in, check-out and labeling.

3 Implementation and Results

Hello-Vision DT is implemented for IBM PC compatible machines working under the Windows 95/98/NT/ME/2000/XP operating system. Hell-Vision CM is implemented under the Windows NT/2000/2003 server and MS SQL. For the extensibility and maintenance of the systems, Hello-Vision is designed and implemented using the object-oriented programming technique. It is divided into several independent modules, which are constructed by well-defined objects. For portability, the functions depending on the operating system and hardware are grouped into a separate module. This is one of the important features of Hello-Vision.

Fig. 2 shows the layout of the Hello-Vision DT screen. The Hello-Vision DT screen consists of object window, function window, trace window and visual programming window. The object window shows the list of the data classes supported by the Hello-Vision DT and objects allocated during function execution as shown in Fig. 2. A User can display an allocated object on the output window by double-clicking the object.

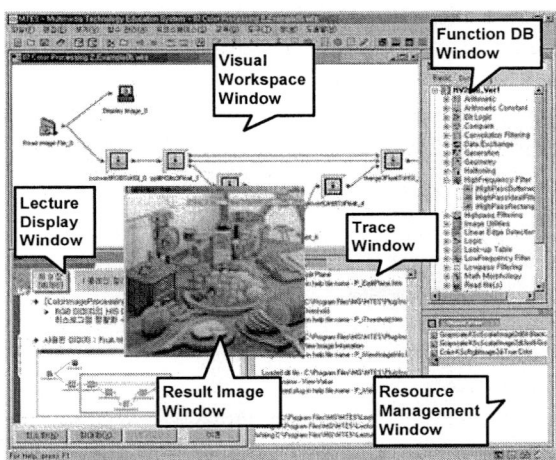

Fig. 2. The layout of Hello-Vision DT

The function DB window consists of windows of user-defined function and visual program command. The DB window of user-defined functions shows the list of user-defined functions registered in the AIM as shown in Fig. 2. The function window is served as a menu. The user can execute one of user-defined functions shown in the window by moving the mouse on the desired function and double-clicking it. Online help information of the selected user-defined function can be displayed by using the commands on the menu popping up with right button press as shown in Fig. 3. The "VP Ex file" button in the bottom of Fig. 3 retrieves the visual program that can test the selected function. The source code of the selected function can be edited and recompiled in the same manner by using the command in the popup menu if the source code is in the designated direction.

Hello-Vision DT provides a simple registration as shown in Fig. 4 for easy creation and registration of user-defined functions. When user defines the function to be created, it automatically generates the template source file and projection file for the function and then loads them on the commercial C compiler like MS Visual C++ as shown in Fig. 4. Users can create their function by inserting the algorithm in the template source file and compiling it. For the modification of any user-defined function, the source code of the function on the compiler can be reloaded by using the right button popup menu. The modification is effective as soon as the source code is compiled.

The visual program command window displays the list of basic commands and I/O commands for visual programming as shown in Fig. 2. The visual programming window is the workspace that allows users to create image processing applications by connecting visual command icons and user-defined function icons in the function window. Unlike most existing visual programming environments, which troubleshoot parameter compatibility in run time, the proposed system troubleshoots the parameter compatibility while connecting icons for programming. This allows less chance of error in run time. Fig. 5 shows the highlighted input nodes compatible for the output node to be connected.

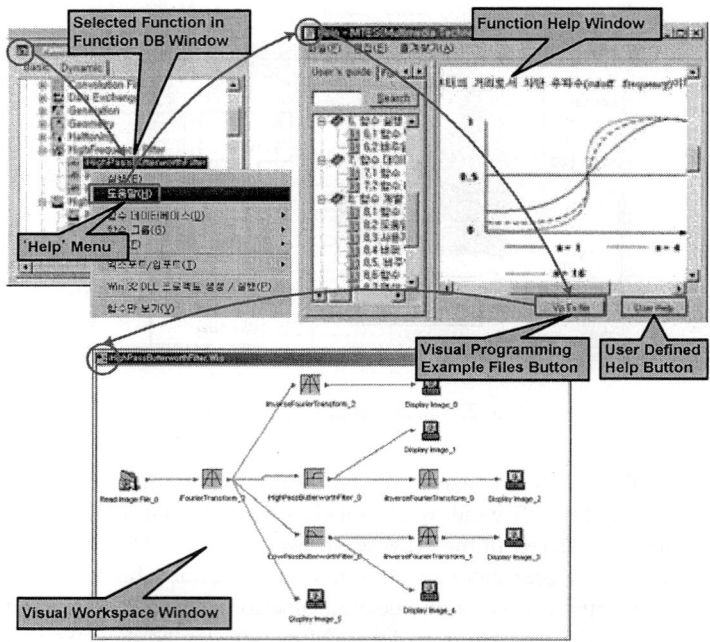

Fig. 3. Function Help Mechanism

The trace window tracks the command executions as shown in Fig. 2. The user can see the current status and the past record of the program execution from this window. The property window displays information for the selected icon or data object. After moving the mouse on an allocated object in the object window, a single mouse click will display property of the object. The property of a function icon in the visual workspace is displayed in the same manner.

Fig. 6 shows some of the tools provided by Hello-Vision DT for image analysis. They are grayscale-value display, histogram display for a gray level image, 2D and 3D color

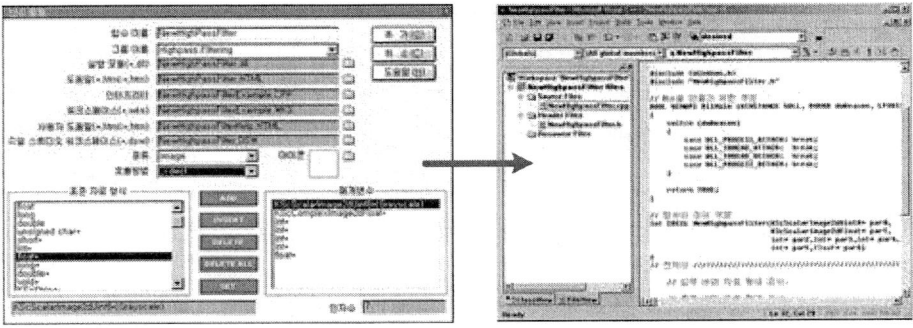

Fig. 4. Function Registration and automatic generation of template code

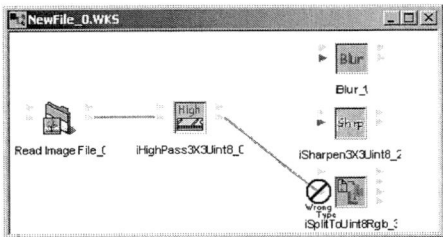

Fig. 5. Compatibility Checks

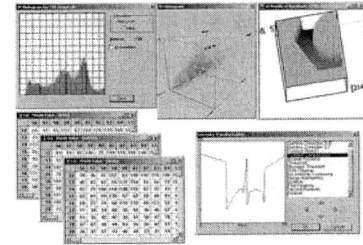

Fig. 6. image analysis tools

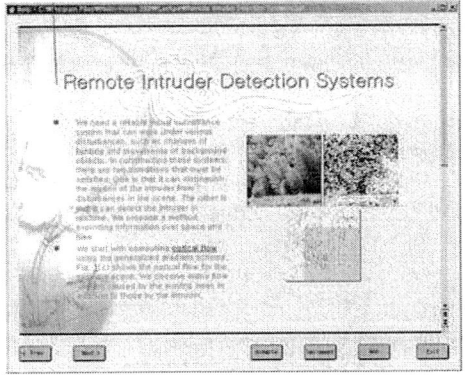

(a) Lecture Screen

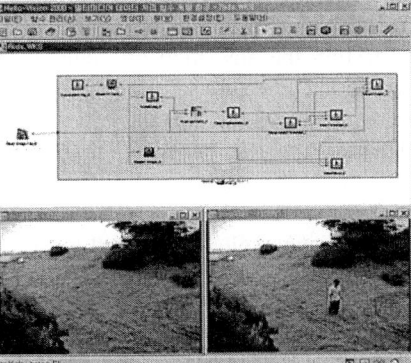

(b) Visual program for algorithm testing

Fig. 7. Education Environment

image histogram, and 3D display of a 2D image. New image analysis functions can be added as plug-in functions by inserting the functions in designated plug-in directories.

The educational environment of Hello-Vision DT supports the lecture, interactive analysis of existing solutions, and new algorithm programming and testing. Each lecture

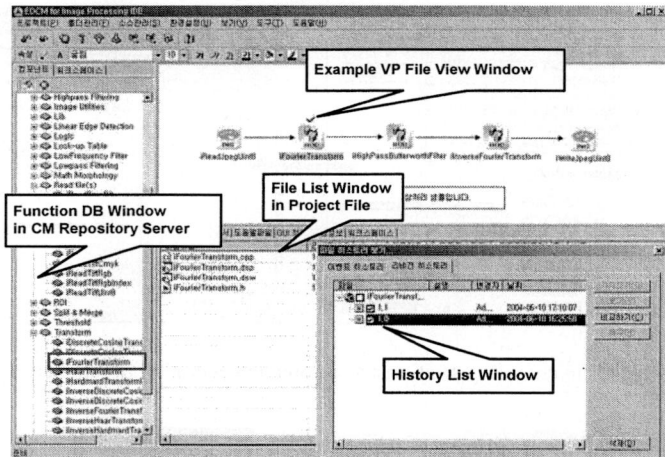

Fig. 8. The layout of Hello-Vision CM

set consists of lecture notes or viewable graphs in a different format, visual program that shows examples related to the lecture, and template source code for the algorithm to be implemented as an exercise. Fig. 7 shows a lecture screen displaying a viewgraph in Power Point file format. The buttons on the bottom right of the screen initiate experiments related to the lecture. When the 'example' button is pressed, a related visual program will be loaded and activated. Students can compare various related solutions registered in Hello-Vision by executing this visual program. The 'practice' button will load the template C source file for the algorithm to be implemented and a visual program for testing the algorithm on Hello-Vision DT. Students can implement the algorithm defined in the lecture screen by simply inserting proper code in the template file, which is similar to Fig. 5 but contains more detailed information. The algorithm can be tested by executing the test program loaded on the system. Fig. 7 shows an example visual program for test. This educational environment will help students to understand new image processing theory which will help to build up their knowledge about how to make use of existing solutions related to the theory and how to create new algorithm to solve their problems. Hello-Vision CM supports the software configuration management (SCM) method for a Hello-Vision DT's configuration data and for the properties in the visual programming environment. And Hello-Vision CM provides SCM method that based on visual workspace. For the project manager, Hello-Vision CM makes it easy to assign tasks across distributed teams and remote developers, regardless of Hello-Vision DT. Version Manager's quick implementation and ease use out of the box enable project teams to rapidly implement secure version management processes. Managers provide assurance and control over every stage of the development process by setting permissions of their choosing to control access to any file, project or sub-project. Common, unifying SCM practices can be established across distributed teams, resulting in higher team productivity. Automated and consistent branching techniques can be defined to support parallel development, promotion groups and labeling strategies. These steps help with reuse of code, which in turn delivers savings of cost in future projects and reduces rework in ongoing projects. Developers would rather be coding than keeping track of complex project processes, the status of revisions and what the developer in the next cubicle is doing that might affect their code. Hello-Vision CM offloads development task management and revision control, so developers can maintain focus on creativity in their own programming. Fig. 8 shows the layout of the Hello-Vision CM screen.

4 Conclusions

In this paper, we introduced the integrated development environment called "Hello-Vision" for image processing application development. It provides a simple and fast way of generating user-defined functions for both C and C++ users and supports an object-oriented approach for sharable function generation. It also offers a systematic function management scheme based on specially designed on-line function management system. The proposed system has already been implemented and distributed among researchers and educators whose initial test results show good promise of success in developing image processing applications. It also demonstrates that it can be a powerful training tool for image processing engineers.

References

1. Bertrand Zavidovique, Veronique Serfaty, Christian Fortunel, Mechanism to Capture and Communicate Image-Processing Expertise, in IEEE Software 11, p37-46, 1991
2. Hideyuki Tamura, Shigeyuki Sakane, Fumiaki Tomita, Naokazu Yokoya, Design and Implementation of SPIDER - A Transportable Image Processing Software Package, in Computer Vision, Graphics, and Image Processing Proc., Vol.23, No 3, p 273-294, 1982.
3. R. Taniguchi, M. Amamiya, E. Kawaguchi, Knowledge-based Image Processing System : IPSSENS-II, Proc. Int'l Conf Image Processing and Its Applications, p462-466, 1989.
4. H. Sato, H. Okazaki, T. Kawai, H. Yamamoto, H. Tamura, The VIEW-Station Environment : Tools and Architecture for a Platform - Independent Image-Processing Workstation, in Proc. 10th International Conference on Pattern Recognition, p576-583, 1990.
5. IUE Documentation version 4.1, Amerinex A.I., 1999.
6. JeongHun Lee, OkSam Chae, The Integrated Development Environment for the Management and Reuse of the Computer Vision and Image Processing Algorithm, in Journal of KISS(The Korea Information Science Society), Vol.24, No.3, 1997.
7. V. Cappellini, A Del Bimbo and P. Nesi, Integrating Object-Oriented Programming Paradigm Concepts in Designing a Vision and Pattern Recognition System Architecture, IEEE, p 572-575, 1990.
8. Mark Young, Danielle Argiro, and Steven Kubica, "Cantata: Visual Programming Environment for the Khoros System", Computer Graphics, Volume 29 No 2, May 1995, pp 22-24.
9. Wit, http://www.logicalvision.com
10. P.F. Whelan and D. Molloy, Machine Vision Algorithms in Java: Techniques and Implementation, Springer (London), 1-30, 2000.

E-Learning Environment Based on Intelligent Synthetic Characters

Lu Ye[1], Jiejie Zhu[1], Mingmin Zhang[1], Ruth Aylett[2], Lifeng Ren[1], and Guilin Xu[1]

[1] College of Computer Science, Zhejiang University, Hangzhou, 310037, P.R.China
ly@zust.edu.cn
[2] School of Maths and Computer Science, Montbatten Building, Heriot-Watt University, UK
ruth@macs.hw.ac.uk

Abstract. This paper introduces our approach to apply synthetic characters to behavior training in primary schools. Project groups include research centers from UK, Portugal, and China. Research members include artificial intelligent group, graphical group, educational group and psychological group. 24 months co-operation work and 4 meetings exchanging idea proof plentiful and substantial results. A closer link between China and EU countries in use of 3D interactive graphic environments in e-Learning has been established. Investigate results of synthetic characters in Potential Virtual Environments (PSE) is achieved from students' bullying phenomenon, empathy change theory, In addition, the open source movement of designing synthetic characters has been launched. Both Asia IT&C committee and project groups are looking forward to new results.

1 Introduction

Virtual Environment (VE) has been explored new development in recent years. Synthetic characters with its significant feature of interactive and intelligent behavior shows remarkable usage and advantages in VEs[1,2,3,4]. In virtual learning environment, such characters attract learners' interest in a great sense. However, in spite of all technological and artistic advances, synthetic characters lacking of intelligence and human like behavior are still far from perfect.

E-Learning with Virtual Interactive Synthetic characters (ELVIS) project aims to explore the potential of synthetic characters' behavior training in primary school. It helps create a liaison between associated schools in China, and the members and associates (including education authorities and schools) of two e-Learning projects in the EU, who form the direct target groups. Chinese education bodies including schools in Zhejiang province, Macau and the rest of China form later indirect targets. Two years project activities will include visits by project participants to schools in China and the EU, assessment of cultural and social differences, production of open source software, dissemination and adaptation to Chinese conditions of outputs.

A number of worthy systems and architectures for synthetic character behavior control have been proposed. Bruce Blumberg[4] etc. believe the ability to learn is a potentially compelling and important quality for interactive synthetic characters. And an autonomous animated dog was build that can be trained with a technique used to train real dogs. Mazin Assanie[5] builds synthetic characters that can accept and enact direction in real-time from the narrative manager. These characters must adhere to both the character design as intended by the author, as well as the performance requirements of their autonomous behavior, such as military doctrine or social roles. John E. Laird[6] etc. develop and evaluate synthetic characters with multiple skill levels and with human-like behavior. Through empirical and human judgments they evaluated a bat skill level and humanness of the variations. Their results suggest that both decision time and aiming skill are critical parameters when attempting to create human-like behavior. Song-Yee Yoon[7] believes compelling synthetic characters must behave in ways that reflect their past experience and thus allow for individual personalization. Rui[8],etc., built BeLife a multi-agent system for up-level behavior control for synthetic character. This system is a simulation tool with teaching objectives that include educating core concepts behind the management of greenhouses, fostering the understanding of experimentation in this particular context. Burke[9] focus on a particular cognitive ability, and found it instructive to build complete systems that creatures can interact with each other and with human participants.

This paper constructs in a way that first gives an overview of our project, which aims to apply synthetic characters to behavior training in primary schools. And in the second part, two researches result will be introduced to illustrate a clear idea of using synthetic character software to make psychology investigation results and open source movement program for building a mapping interface for high-level AI model and low-level animation engine. A possible architecture of entire synthetic character model is also represented.

2 Project Overview

2.1 Constraints

China has the largest education system in the world and educates 25 percent of the world's students. Technology is seen there as a key to development and to moving the country into a leading international position, and applying Information Communication Technology (ICT) in the classroom is being actively pursued in many regions under a central government national IT&C education policy [19,20]. Thus there are schools in some eastern provinces where ICT presentation resources are available in every classroom and a visualiser display is for interactive teaching & learning. An 'e-learning for life' initiative funded by Coca-Cola is leading to the setting up of e-learning centers. However there is a danger that this activity will concentrate on lower-level ICT to the exclusion of new approaches developed in EU programmers such as the i3 network ('Classrooms of the Future'), IST programmers in e-learning, and MINERVA-funded classroom activity. Many of these new approaches are based on 3D interactive graphics (virtual reality - VR) and the use of synthetic characters (sometimes known as Intelligent Virtual Agents, Virtual Humans, or, incorrectly, Avatars).

2.2 Tasks

The aim of the project is to produce liaison between organizations in China and Europe to facilitate and improve contacts with and the participation of Chinese partners in existing European IT&C initiatives. The project will focus particularly on the European Commission's Research and Technological Development Program, Framework V Information Society Technologies projects in e-Learning with synthetic characters, in particular the i3 project, NIMIS, VICTEC (Virtual ICT with Empathic Characters) and with the European Commission's Education and Training Socrates Minerva projects in ICT in education, especially ELVES. The specific objectives to be addressed in order to achieve the liaison and contacts will be:

1. To investigate the potential of Virtual Environments (VEs) in general, and Synthetic Characters in VEs in particular, to enhance the school curriculum in China
2. To transfer synthetic character technology into Open Source form so that it can easily be taken up in Chinese schools.
3. To identify the strengths and weaknesses of using VR and synthetic character solutions for personal and social education program in Chinese schools with particular reference to social, organizational and cultural differences.
4. To integrate and transfer Chinese graphics programming skills into e-Learning with synthetic characters in Europe.
5. To strengthen the world Open Source movement in the 3D graphics area by providing library support for synthetic characters.
6. To disseminate project outcomes through the Open Source movement, through the web, and through educational channels in China and Europe.

3 Research Results

3.1 E-Teatrix

E-Teatrix[10] is a tool for designing virtual interactive characters with emotions to help them study in traditional classroom. It can not only enrich the free time of children, but also improve the creative ability and design ability of children. In E-Teatrix, every child controls a character, and each character supports five emotional states. Figure 1 shows the Teatrix architecture.

The character that every user controls are an agent, each agent has some sensors. When these sensors get information, they turn this information into perception. Having filtered useless information, these perceptions are sent to the mind. The mind accepts those perceptions and analyzes the whole scene.

E-Teatrix helps educational researchers to explore following investigation in China(see Figure 2):

- Review of Potential Virtual Environments' application in General
- Analysis of bullying research and results in elementary schools in China

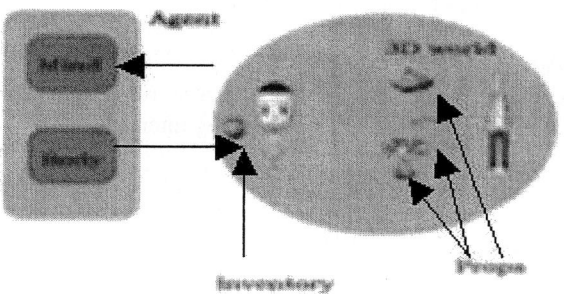

Fig. 1. E-Teatrix architecture

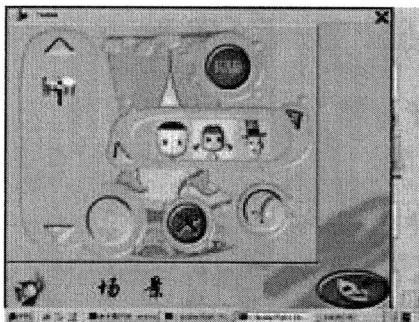 

(a) Select Characters in scene   (b) Exchanging ideas controlling characters

Fig. 2. Example of using E-Teatrix

- Analysis of empathy research and results in elementary schools in China
- Analysis story creating skills of elementary students in China
- Evaluation and testing this software in a different culture country

3.2 Open Source Movement

Standards like VRML, OpenSG contains no support for the development of synthetic characters, leading EU projects such as VICTEC (and others) to interface to proprietary systems such as games engines which do offer such support. Thus to make ICT using synthetic characters widely accessible in countries like China, some technical work is required to develop Open Source support for synthetic characters. The sub-project name titled AgentLib which means linking agent mind to bottom 3D engine. The graphics expertise in Chinese Universities is such that this work could be carried out there if a strong graphics group such as that at Zhejiang University was able to work alongside EU program in the area. This work would then feed back into the Open Source movement in Europe and internationally.

AgentLib Ontology

Figure 3 shows our designed architecture for AgentLib from high level to bottom render division. The important part is the mapping interface.

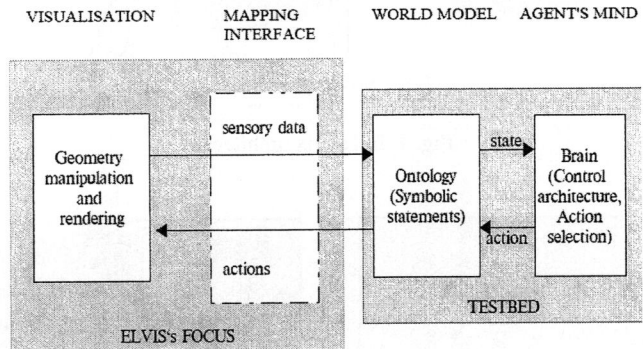

Fig. 3. 3-Layer Design for AgentLib

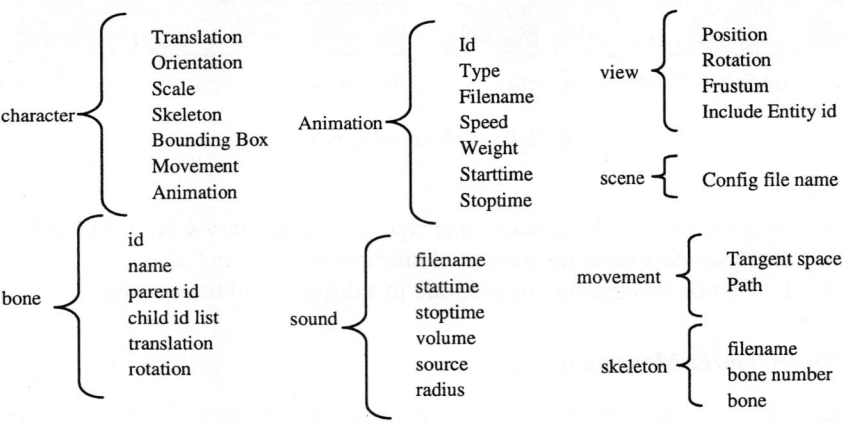

Fig. 4. Visualization world components

Visualization World Components

The visualization world is not equal the real rendering world. It concerns only the components that the abstract world is interested. Figure 5 shows a detailed components of visualization world.

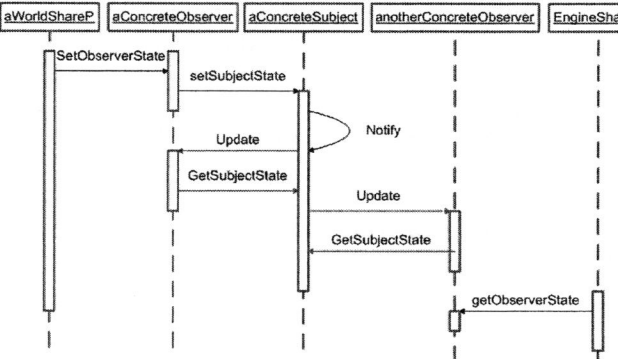

Fig. 5. Working process of observer and subject

Mapping Interface
The mapping interface should keep data consistent between abstract world and visualization world. We use observer design pattern[11] to implement it. Abstract world doesn't necessary know how to call the basic rendering functions to execute a specific character action. And the engine doesn't know which goal is the next target, either. We provide observer box as a black box to be used to encapsulate all concrete observers and a subject box is used to encapsulate all concrete subjects. Each concrete subject is independent to other ones, and they are linked with each concrete observer. For example, skeleton subject is linked with skeleton observer. If abstract world needs access skeleton subject, it set up the skeleton observer and the observer will update all changed data. Figure 6 shows the process of subject and observer working process.

Fig. 6. "YY" imaginary and implementation scenery

On-going Testbed Example
A testbed example called *"YY" goes mushroom picking* is designing to explore the whole ideas. It is a story about a Chinese little girl called *YY* picks mushrooms for her parents to cook her favorite meal (mushroom soup). The *YY* model is composed of body, basket and two pigtails. The model supports a set of animation concerning pick-

ing mushrooms. Different sizes of mushrooms are designed to be one factor to change *YY* picking path. Besides, *YY* sings specially when finds good mushrooms and she find mushrooms by *"smell"*.Figure 7 shows the imaginary scene and implemented scene. Mushrooms are randomly generated according to *YY*'s location. Which mushroom should be picked first was the goal generated by the mind.

Expand considerations including adding a small dog character which represents the animal character and its behavior. This little dog will change the example a little bit, since two characters have to communicate with each other, which will cause a series of complex state change.

4 Conclusion

Computer graphics and VR technology are widely used in Chinese education field[16]. More and more people, especially students, need virtual interactive characters with emotions to help them study in traditional classroom. This paper investigates the potential of VEs in general and Synthetic Character in particular to enhance the school curriculum in China and transfers synthetic character technology into open source form so that it can easily be taken up in Chinese school. E-Teatrix is a tool for solving this problem. It can not only enrich the free time of children, but also improve the creative ability and design ability of children. AgentLib supports the creation of 3D synthetic characters and merge together 3D and agent approaches which providing 3D representation for agent systems and means for building intelligent behavior for animated characters. Expected future results include analysis of impact of social and cultural factors on architecture, release of final version of Open Source library, and a final workshop will be held to exchange ideas of the related results. Besides, investigate impact on target groups is also an important sum up for further research.

Acknowledgments

This project is under support of 973 project (2002CB312103) and co-supported by China-EU project (ELVIS). The authors would also like to give thanks to Rui Parad., Marco Vala., Steve Ho., and Carlos Delgada., Zhaohui Zhu, who help us to build the entire architecture of synthetic characters.

References

1. John F., Xiaoyuan T., Demetri T. "Cognitive Modeling: Knowledge, Reasoning and Planning for Intelligent Characters". In the Proceedings of SIGGRAPH (1999) p.29-38
2. Martinho, A.P., "Pathematic Agents: Rapid development of Believable Emotional Agents in Intelligent Virtual Environments", In the Proceedinsg of Autonomous Agents (1999) p.1-8
3. Paiva A., Prada R., Machado I. "Heroes, villains, magicians: Dramatis Personae in a virtual story creation environment" In the Proceedings of Intelligent User Interfaces (2001)

4. Bruce B., Marc D., Yuri A. I., etc. "Integrated Learning for Interactive Synthetic Characters" In the Proceedings of SIGGRAPH (2002) p.417-426
5. Magerko, B., Laird, J.E., Assanie, M., Kerfoot, A., and Stokes, D. "AI Characters and Directors for Interactive Computer Games" In the Proceedings of 16th Innovative Applications of Artificial Intelligence Conference (2004) p. 877-883.
6. Laird J.E, Duchi, J.C. "Creating Human-like Synthetic Characters with Multiple Skill Levels:A Case Study using the Soar Quakebot". In the Proceedings of the AAAI Fall Symposium (2000)
7. Song Y.Y., Robert C. B., Bruce B, etc. "Interactive Training for Synthetic Characters Interactive training for synthetic characters" In the Proceedings of AAAI (2002) p.249--254
8. Prada R., Otero N., Vala A., Paiva A. "BeLife: Teaching Greenhouse Management Using an Agent Based Simulator" In the Proceedings of Agent Based Simulation Workshop (2004)
9. Burke, R., Isla, D., Downie, M., etc. "Creature smarts: The art and architecture of avirtual brain" In the Proceedings of the Computer Game developers Conference (2001)
10. Pan Z.G., Zhu C.H., Rui P., etc. "Emotion modeling and interaction in intelligent virtual environment" In the Proceedings of 3IA (2004) p.1-6
11. Erich G., Richard H., Ralph J., etc. "Design Patterns, Elements of Reusable Object-Oriented Software" Addison Wesley ISBN 0-201-63361-2.
12. Tracy J.M., James R. "The struture of emotion: A nonlinear dynamic systems approach" In Tracy J. Mayne & George A. Bonanno(Eds.), Emotions: Current Issues and Future Directions. New York: The Guilford Press (2001) p.1-36
13. Pan Z.G., Xu W.W., Huang J., Zhang M.M., etc. "EasyBowling: A Virtual Bowling System based on Virtual Simulation" Computers&Graphics (2003) 27(2) p.231-238
14. Wu J.H, "The research way to emotion", In the Proceedings of ACII (2003) p.6-12
15. Roussos M., Johnson A., Leigh J., etc. "Constructing Collaborative Stories Within Virtual Learning Landscapes", In the Proceedings of AIED (1996) p.129-135
16. Pan Z.G., Shi J.Y. "Virtual Reality and Application" Computers&Graphics (2003) 27(2)
17. Downie, M. "behavior, animation, music: the music and movement of synthetic characters" Master.s thesis, The Media Lab, MIT. 2000.
18. Yoon S., Blumburg, B., Schneider, G. "Motivation-driven learning for interactive synthetic characters" In the Proceedings of the Fourth International Conference on Autonomous Agents. 2000.
19. Pan Z.G., Shi J.Y. "Virtual Reality Technology Development in China: An Overview" The International Journal of Virtual Reality (2000) 4(3) p.2-10
20. Shi J.Y., Pan Z.G. "China: Computer Graphics Education Available at Universities, Institutes and Training Centers" Computer Graphics (1999) 31(3) p.7-9

SCO Control Net for the Process-Driven SCORM Content Aggregation Model

Kwang-Hoon Kim[1], Hyun-Ah Kim[1], and Chang-Min Kim[2]

[1] Collaboration Technology Research Lab.,
Department of Computer Science,
KYONGGI UNIVERSITY,
San 94-6 Yiuidong Youngtonggu Suwonsi Kyonggido, South Korea 442-760
{kwang, haKim}@kyonggi.ac.kr
http://ctrl.kyonggi.ac.kr
[2] Division of Computer Science,
SUNGKYUL UNIVERSITY,
147-2 Anyang-8dong, Manangu, Anyangsi, Kyonggido, South Korea 430-742
kimcm@sungkyul.ac.kr

Abstract. In this paper, we propose a new learning process (or experience) model, which is called SCO Control Net, to be possibly applicable to the SCORM[1] content aggregation model for e-Learning management systems. In terms of the representation of the learning process model, the current SCORM content aggregation model supports a rule-based model for defining a set of rules that describe the intended sequence and ordering of learning activities in a tree-structured content organization. On the other hand, the SCO Control Net aims to support a process-based model for explicitly defining the rules. That is, it is intended to realize the concept of process-driven content organization by separating the learning process from the SCORM's content organization. We would call that the former is content-driven aggregation model, and the later is process-driven aggregation model. This paper formalizes a theoretical basis of the process-driven aggregation model by representing the sequence of learning activities through the SCO control net. We strongly believe that the process-driven aggregation model delivers a way of much more convenient content aggregating work and system, in terms of not only defining the intended sequence and ordering of learning activities, but also building the runtime environment for sequencing and navigation of learning activities and experiences.

Keywords: SCORM Content Aggregation Model, Sharable Content Object, e-Learning Management System, SCO Control Net, Process-driven Content Aggregation Model, Sequencing and Navigation of Learning Process and Experience.

---

[1] Sharable Content Object Reference Model by the Advance Distributed Learning (ADL) Initiative that aims to foster the e-Learning specifications and standards.

1 Introduction

The e-Learning technology have been enhanced for many years to effectively improve efficiency and reduce costs as well. Particularly, based upon the outcomes of a lot of empirical studies, it is quite generally acceptable that individually tailored instruction using information technology sometimes offers ideal learning outcomes, because, in contrast to classroom learning, information technologies can adjust the pace, sequence, content and method of instruction to better fit each student's learning style, interests and goals [1]. However, that realizing the promise of improved learning efficiency even though the use of the most current and cutting-edge instructional technologies - such as Web-based instruction, interactive multimedia instruction, Intelligent Tutoring Systems and SCORM-based e-Learning Management Systems - still depends on the ability of those technologies to sufficiently and easily tailor quality and appropriate learning process and experience to the needs of individuals.

Also, the one-on-one individualization capabilities (or tailoring capabilities of the learning process) of technology-based instruction, in contrast to one-on-many classroom-based instruction, may approximate and perhaps exceed the effectiveness of one-on-one tutoring [2]. This adaptability to individual learners and their needs can be seen in several categories of e-Learning products. The typical standardized e-Learning product is the ADL[2]'s SCORM that aims to foster creations of reusable learning contents for computer and web-based learning. Especially, the SCORM Evolution Book at Version 1.3[3], that has been recently released, newly announces the sequencing and navigation standards (SN 1.3 Edition) so as to define a set of the rules and behaviors for the sequencing and ordering of learning activities. The sequencing rules and presentation strategies for learning activities are encoded in XML, defined as part of the items of the tree-structured content organization and finally composed into an activity tree. We would call the SCORM's strategy the activity tree-driven content organizing approach, in which the sequencing information is embedded onto the inside of the activities (items) represented in the tree-structured content organization.

In this paper, we propose a process-driven content organizing approach, in which, in contrast to the SCORM's approach, the sequencing information is explicitly define in not the tree-structured but the process-structured content organization. That is, we conceive a learning process model, which is called SCO control net (SCOCN), and formalize the model. The model aims to represent a way for composing the process-driven content organization applicable into the SCORM content aggregation model. So, we don't need to keep the hierarchical structure (or the activity tree) anymore in aggregating contents. At the same time, we are able to easily represent the learning process without any further restrictions by using not only the sequential-style of learning activity flow control but also the conjunctive-style (parallel-style) and disjunctive-style (alternative-

---

[2] The Advanced Distributed Learning Initiative launched by the Department of Defense and the White House Office of Science and Technology Policy.

style) of learning activity flow control. The sequencing information represented by the SCO control net is encoded in XML, too.

The remainder of this paper is organized as follows. The next section briefly describes about the motivation and related work of this paper. Section 3 works out the concept of the SCO control net by defining its graphical and formal notations. And, we describe how to apply the SCO control net in realizing the process-driven content aggregation model. Finally, the paper finalizes with an operational architecture of the e-Learning management system and some future works.

2 Motivation and Related Work

The primary motivation of this paper is on the structure of the Content Organization of the SCORM content aggregation model. Fig. 1 shows the structural differences between the SCORM's activity tree-driven content organizing approach and the SCOCN's process-driven content organizing approach proposed in the paper.

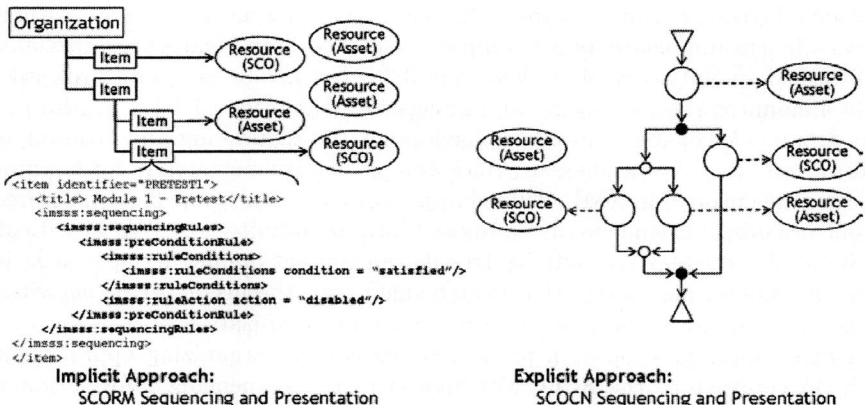

Fig. 1. Comparison of the Content Organization and Aggregation Approaches - Implicit sequencing vs. Explicit sequencing

In the current SCORM, the content organization is a map that represents the intended use of the content through tree-structured units of instruction (Activities). The map shows how Activities relate to one another as shown in the left-hand side of Fig. 1. The Activities, that are represented as items in the figure, may consist of other Activities (subActivities), which may themselves consist of other Activities, too. There is no set limit to the number of levels of nesting for Activities. Only the leaf Activities have an associated learning resource (SCO or Asset) that is used to perform the activity. In terms of the representation of learning process, sequencing only applies to Activities in the SCORM approach. The sequencing information is embedded as part of each Activities, as presented in the lower part of the left-hand side of Fig. 1. That's why we call it implicit sequencing, which means that the SCORM content aggregation model is based upon the activity tree and its learning process is implicitly scattered into the inside of each activity. The problem of this approach is on the fact that it is hard to define and visualize the learning process. At the same time, it is extremely inconvenient to adapt to dynamic changes and alternative/parallel flows, and to reflect the external factors, that can affect on the learning flows at rum time, such as learners' status, SCO's ownership and so on.

As shown in the right-hand side of Fig. 1, we propose a new way to resolve the problems caused from the current SCORM content aggregation model. We call it the process-driven content organizing approach with explicit sequencing and presentation of the learning activities. This approach is based on the strong belief that the learning process is the most effective factor in the e-Learning instruction. Also, through this approach, we are able to realize the interactive learning flow control and management at run time. The main contribution of this approach is on that the learning process modeled by the SCO control net represents not only the structure of the Content Organization but also the sequencing and navigation information of the learning activities, in contrast to that the SCORM's approach separately represents the structure of content organization and the sequencing information. The details of this approach is described in the next section.

In summary, before the advent of the SCORM and the shift toward an interoperable development strategy, it was extremely difficult to share content between different authoring environments and equally difficult to reuse content in other contexts that involved different sequencing requirements. Within SCORM, those difficulties were gotten rid of by doing that let sequencing information being defined on the Activities represented in the Content Organization and being external to the learning resources associated with those Activities. However, the SCORM has still some limitations and difficulties on representing the sequencing information. Therefore, our approach, the SCOCN's approach, is being proposed in order to be a remedy for getting rid of the SCORM's limitations simply pointed out in the previous.

3 SCO Control Net

The SCO Control Net is a mathematical formalism designed to model graphically learning processes eventually being activated by an e-Learning management system. The SCO Control Net aims to specify and analyze the learning flow and process for e-Learning instructions. It can be used within actual as well as hypothetical automated e-Learning content aggregation model to yield a comprehensive description of learning activities, to test the underlying the process-driven content aggregation description for certain flaws and inconsistencies, to quantify certain aspects of e-Learning content control flow, and to suggest possible e-Learning process restructuring permutations. In this section, we introduce the nomenclature used in modeling the SCO control net, and formally define the SCO control net and its implications. After that, we propose a graphical notation for the SCO control net and also we encode the SCO control net language (SCOCNL) in XML.

3.1 Nomenclature of the SCO Control Net

The SCORM-based e-Learning arena has a vast published literature, and sometimes a confusing array of terms and meanings. In describing the paper, it uses the basic SCOCN terminology of the learning process, learning activity, learning resources consisting of SCO (Sharable Content Object) and Asset, learning-case, actor including group, and role. These terms become the primitive components used for modeling learning processes through the SCOCN, and also they have some relationships with each other as shown in Fig. 2.

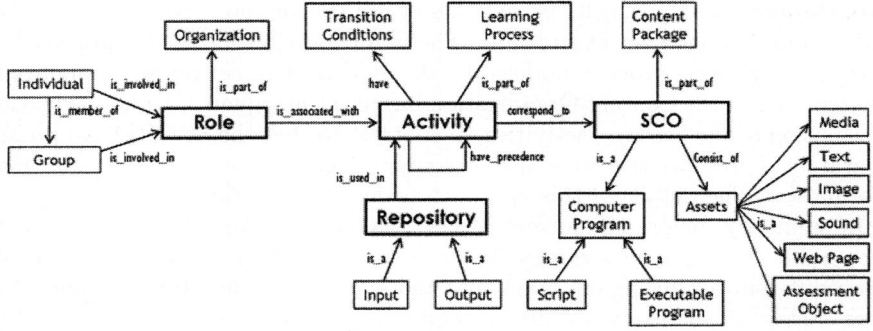

Fig. 2. Components and their Relationships of the SCO Control Net

Let's define them as follows: A **learning process** is a predefined or intended set of learning steps, called activities, and a partial ordering of these activities. A e-Learning management system helps to organize, control, and execute these learning activities associated with the learning process. Activities can be related to each other by conjunctive logic (after activity A, do activities B and C) or by disjunctive logic (after activity A, do activity B or C) with predicates attached. A **learning activity** is either a compound activity containing another learning process, or a basic unit of learn called an elementary activity. Only the elementary activity have an associated learning resource (SCO or Asset) that is used to perform the activity in one of three modes - manual, automatic, or hybrid. Typically one or more participants (including learners as well as teachers) are associated with each activity via roles. A **role** is a named designator for one or more participants, which conveniently acts as the basis for partitioning of teach, learn, skills, access controls, execution controls, and authority / responsibility. An **actor** is a person, program, or entity that can fulfill roles to execute, to be responsible for, or to be associated in some way with activities and learning processes. Multiple copies of the same learning process may be in various stages of execution. Thus, the learning process can be considered as a class (in object oriented terminology), and each execution, called a learning-case, can be considered an instance. A learning-case is thus defined as the locus of control for a particular execution of a learning process. If a learning activity is executed in automatic or hybrid mode, this means that whole/part of the SCO (or Asset) resource associated with the activity is automatically launched by an e-Learning enactment service. Computer programs that automatically perform activities or provide automated assistance within hybrid activities are called scripts. Finally, a **repository** is a set of input and output parameters of a learning activity.

3.2 Graphical Notation of the SCO Control Net

A e-Learning management system supporting SCOCN will have to provide a graphical SCOCN modeling tool to design, analyze, and evolve learning processes' specifications. In Fig. 3, we present a possible set of the graphical notations for SCOCN. As a result, an SCOCN control flow graph is composed of a set of learning activities represented by large circles, OR control flow nodes represented by small-open circles, AND control flow nodes represented by small-filled circles, and edges to interconnect these nodes. An arc represents precedence among nodes: if activity A leads to activity B (i.e., (A, B) is an edge in the graph), then activity A must be applied to an individual learning activity before activity B can be applied to it.

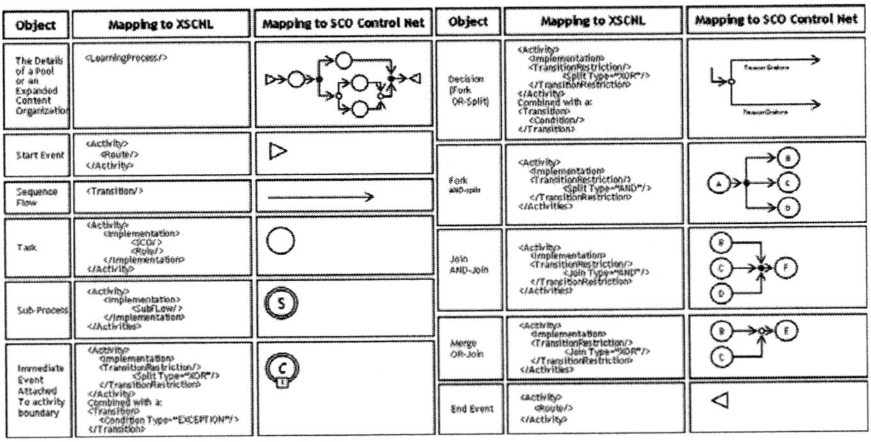

Fig. 3. SCOCN Graphical Notations and SCOCN XML Language

Also, the SCOCN can be encoded by the XML-based SCOCN Language (XSCNL) as shown in Fig 3. The XSCNL is a collaboration language that allows administrators to dynamically define, describe, schedule, modify their learning processes, their inter-relationships, and any organizational information associated with them. We can straightforwardly transform from the graphical notation of SCOCN to the XSCNL without further modifications.

As an example, let's consider an e-Learning classroom for the reading and comprehension lecture of TOEFL. This learning process is presented in Fig. 4. It consists of five learning activities from the placement test activity to the student evaluation activity. The detailed learning activities are following:

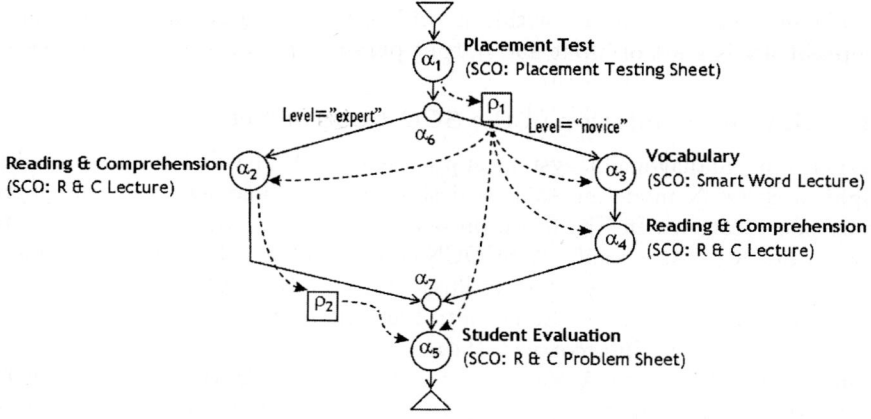

Fig. 4. Graphical Representation of A SCO Control Net Model

- Placement Test (α_1) (Learners take the placement test before joining the lecture. SCO is a web-based exam sheet.)
- Reading & Comprehension Lecture (α_2) (Learners being classified into an expert group take this learning path. SCO is a R/C lecture.)
- Vocabulary Lecture (α_3) (Learners being classified into a novice group take a vocabulary lecture prior to the R/C lecture. SCO is a vocabulary lecture.)
- Reading & Comprehension Lecture (α_4) (After taking the vocabulary lecture, learners take the R/C lecture. SCO is exactly same to the previous one.)
- Student Evaluation (α_5) (Learners are evaluated after the R/C lecture. SCO is a web-based R/C problem sheet.)

3.3 Formal Notation of the SCO Control Net

A basic SCOCN is 8-tuple $\Gamma = (\delta, \vartheta, \gamma, \varepsilon, \pi, \kappa, I, O)$ over a set of A activities (including a set of group activities), a set of S of SCOs (or Assets), a set T of transition conditions, a set R of repositories, a set P of roles, and a set C of actors (including a set of actor groups), where

- I is a finite set of initial input repositories, assumed to be loaded with information by some external learning process before execution of the SCOCN;

Table 1. Formal Representation of the SCO Control Net Model

$\Gamma = (\delta, \gamma, \varepsilon, \pi, \kappa, I, O)$ over A, R, P, C, T $The\,Order\,Procedure\,in\,ICN$
$A = \{\alpha_1, \alpha_2, \alpha_3, \alpha_4, \alpha_5, \alpha_6, \alpha_7, \alpha_I, \alpha_F\}$ $Activities$
$R = \{\rho_1, \rho_2\}$ $Repositories$
$S = \{\varpi_1, \varpi_2, \varpi_3, \varpi_4, \varpi_5\}$ $SCOs/Assets$
$T = \{d(default), or_1('novice'), or_2('expert')\}$ $Transition\,Conditions$
$I = \{\emptyset\}$ $Initial\,Input\,Repositories$
$O = \{\emptyset\}$ $Final\,Output\,Repositories$

$\delta = \delta_i \cup \delta_o$

$\delta_i(\alpha_I) = \{\emptyset\};$ $\delta_o(\alpha_I) = \{\alpha_1\};$
$\delta_i(\alpha_1) = \{\alpha_I\};$ $\delta_o(\alpha_1) = \{\alpha_6\};$
$\delta_i(\alpha_2) = \{\alpha_6\};$ $\delta_o(\alpha_2) = \{\alpha_7\};$
$\delta_i(\alpha_3) = \{\alpha_6\};$ $\delta_o(\alpha_3) = \{\alpha_4\};$
$\delta_i(\alpha_4) = \{\alpha_3\};$ $\delta_o(\alpha_4) = \{\alpha_7\};$
$\delta_i(\alpha_5) = \{\alpha_7\};$ $\delta_o(\alpha_5) = \{\alpha_F\};$
$\delta_i(\alpha_6) = \{\alpha_1\};$ $\delta_o(\alpha_6) = \{\alpha_2, \alpha_3\};$
$\delta_i(\alpha_7) = \{\alpha_2, \alpha_4\};$ $\delta_o(\alpha_7) = \{\alpha_5\};$
$\delta_i(\alpha_F) = \{\alpha_5\};$ $\delta_o(\alpha_F) = \{\emptyset\};$

$\vartheta = \vartheta_{sco} \cup \vartheta_a$

$\vartheta_{sco}(\alpha_I) = \{\emptyset\};$ $\vartheta_a(\varpi_1) = \{\alpha_1\};$
$\vartheta_{sco}(\alpha_1) = \{\varpi_1\};$ $\vartheta_a(\varpi_2) = \{\alpha_2\};$
$\vartheta_{sco}(\alpha_2) = \{\varpi_2\};$ $\vartheta_a(\varpi_3) = \{\alpha_3\};$
$\vartheta_{sco}(\alpha_3) = \{\varpi_3\};$ $\vartheta_a(\varpi_4) = \{\alpha_4\};$
$\vartheta_{sco}(\alpha_4) = \{\varpi_4\};$ $\vartheta_a(\varpi_5) = \{\alpha_5\};$
$\vartheta_{sco}(\alpha_5) = \{\varpi_5\};$
$\vartheta_{sco}(\alpha_6) = \{\emptyset\};$
$\vartheta_{sco}(\alpha_7) = \{\emptyset\};$
$\vartheta_{sco}(\alpha_F) = \{\emptyset\};$

$\gamma = \gamma_i \cup \gamma_o$

$\gamma_i(\alpha_I) = \{\emptyset\};$ $\gamma_o(\alpha_I) = \{\emptyset\};$
$\gamma_i(\alpha_1) = \{\emptyset\};$ $\gamma_o(\alpha_1) = \{\rho_1\};$
$\gamma_i(\alpha_2) = \{\rho_1\};$ $\gamma_o(\alpha_2) = \{\rho_2\};$
$\gamma_i(\alpha_3) = \{\rho_1\};$ $\gamma_o(\alpha_3) = \{\emptyset\};$
$\gamma_i(\alpha_4) = \{\rho_1\};$ $\gamma_o(\alpha_4) = \{\emptyset\};$
$\gamma_i(\alpha_5) = \{\rho_1, \rho_2\};$ $\gamma_o(\alpha_5) = \{\emptyset\};$
$\gamma_i(\alpha_6) = \{\emptyset\};$ $\gamma_o(\alpha_6) = \{\emptyset\};$
$\gamma_i(\alpha_7) = \{\emptyset\};$ $\gamma_o(\alpha_7) = \{\emptyset\};$
$\gamma_i(\alpha_F) = \{\emptyset\};$ $\gamma_o(\alpha_1) = \{\emptyset\};$

$\kappa = \kappa_i \cup \kappa_o$

$\kappa_i(\alpha_I) = \{\emptyset\};$ $\kappa_o(\alpha_I) = \{d\};$
$\kappa_i(\alpha_1) = \{d\};$ $\kappa_o(\alpha_1) = \{d\};$
$\kappa_i(\alpha_2) = \{or_1\};$ $\kappa_o(\alpha_2) = \{d\};$
$\kappa_i(\alpha_3) = \{or_2\};$ $\kappa_o(\alpha_3) = \{d\};$
$\kappa_i(\alpha_4) = \{d\};$ $\kappa_o(\alpha_4) = \{d\};$
$\kappa_i(\alpha_5) = \{d\};$ $\kappa_o(\alpha_5) = \{d\};$
$\kappa_i(\alpha_6) = \{d\};$ $\kappa_o(\alpha_6) = \{or_1, or_2\};$
$\kappa_i(\alpha_7) = \{d\};$ $\kappa_o(\alpha_7) = \{d\};$
$\kappa_i(\alpha_F) = \{d\};$ $\kappa_o(\alpha_F) = \{\emptyset\};$

- O is a finite set of final output repositories, perhaps containing information used by some external learning process after execution of the SCOCN;
- $\delta = \delta_i \cup \delta_o$
  where, $\delta_o : A \longrightarrow \wp(A)$ is a multi-valued mapping of an activity to its sets of (immediate) successors, and $\delta_i : A \longrightarrow \wp(A)$ is a multi-valued mapping of an activity to its sets of (immediate) predecessors; (For any given set S, $\wp(S)$ denotes the power set of S.)
- $\vartheta = \vartheta_{sco} \cup \vartheta_a$
  where $\vartheta_{sco} : S \longrightarrow \wp(A)$ is a single-valued mapping of an activity to one of the SCOs, and $\vartheta_a : A \longrightarrow \wp(S)$ is a multi-valued mapping of a SCO to its sets of associated activities;
- $\gamma = \gamma_i \cup \gamma_o$
  where $\gamma_o : A \longrightarrow \wp(R)$ is a multi-valued mapping (function) of an activity to its set of output repositories, and $\gamma_i : A \longrightarrow \wp(R)$ is a multi-valued mapping (function) of an activity to its set of input repositories;
- $\varepsilon = \varepsilon_a \cup \varepsilon_p$
  where $\varepsilon_p : A \longrightarrow P$ is a single-valued mapping of an activity to one of the roles, and $\varepsilon_a : P \longrightarrow \wp(A)$ is a multi-valued mapping of a role to its sets of associated activities;
- $\pi = \pi_p \cup \pi_c$
  where, $\pi_c : P \longrightarrow \wp(C)$ is a multi-valued mapping of a role to its sets of associated actors, and $\pi_p : C \longrightarrow \wp(P)$ is a multi-valued mapping of an actor to its sets of associated roles;
- $\kappa = \kappa_i \cup \kappa_o$
  where κ_i: sets of control-transition conditions, T, on each arc, $(\delta_i(\alpha), \alpha)$, $\alpha \in A$; and κ_o: sets of control-transition conditions, T, on each arc, $(\alpha, \delta_o(\alpha))$, $\alpha \in A$; where the set $T = \{default, or(conditions), and(conditions)\}$.

Table 1 is to represent the learning process model of Fig. 4 by using the formal notation of SCOCN. Note that we do not represent the parts of roles and actors assignments in the table for the sake of simplification.

4   Conclusions

So far, this paper has presented the sharable content object control net model for realizing process-driven e-Learning management systems. Particularly, in this paper we pointed out the limitations of the current SCORM aggregation model and proposed a solution for the limitations, as well. The solution is just the SCO control net that eventually gives us higher-level of efficiency in designing and launching learning activities and processes. That is, we newly formalize a theoretical basis for implementing the process-driven SCO aggregation model and system. The SCO control net will affect to not only the SCORM content aggregation model but also the SCORM runtime environment in order for SCORM to adapt the process-driven content aggregating approach. Also, the SCORM sequencing and navigation under SCOCN can support a much more sophisticated and convenient way in defining the intended sequencing and ordering of the learning activities. Finally, it will be necessary for a typical process-driven SCO aggregation model and system to be implemented in the near future by the collaboration technology research lab of Kyonggi University. Conclusively, we are strongly believe that the SCO control net might be the first footprint for pioneering the process-driven e-Learning technologies and systems.

Notes and Comments. The research was supported by the research fund of KOSEF (Korea Science and Engineering Foundation), No. R05-2002-000-01431-0.

References

1. "SCORM 2004 Overview", Advanced Distributed Learning, (2004)
2. "SCORM Content Aggregation Model, Version 1.3", Advanced Distributed Learning, (2004)
3. "SCORM Run-Time Environment, Version 1.3", Advanced Distributed Learning, (2004)
4. Clarence A. Ellis, Gary J. Nutt, "Office Information Systems and Computer Science", ACM Computing Surveys, Vol. 12, No. 1, (1980)
5. Clarence A. Ellis, Gary J. Nutt, "The Modeling and Analysis of Coordination Systems", University of Colorado/Dept. of Computer Science Technical Report, CU-CS-639-93, (1993)
6. Clarence A. Ellis, "Formal and Informal Models of Office Activity", Proceedings of the 1983 Would Computer Congress, Paris, France, (1983)
7. James H. Bair, "Contrasting Workflow Models: Getting to the Roots of Three Vendors", Proceedings of International CSCW Conference, (1990)
8. Kwang-Hoon Kim, "Practical Experience on Workflow: Hiring Process Automation by FlowMark", IBM Internship Report, IBM/ISSC Boulder Colorado, (1996)
9. Kwang-Hoon Kim and Su-Ki Paik, "Practical Experiences and Requirements on Workflow", Lecture Notes Asian '96 Post-Conference Workshop: Coordination Technology for Col-laborative Applications, The 2nd Asian Computer Science Conference, Singapore, (1996)

Design and Implementation of a Web-Based Information Communication Ethics Education System for the Gifted Students in Computer

Woochun Jun[1], Sung-Keun Cho[2], and Byeong Heui Kwak[3]

[1] Dept. of Computer Education, Seoul National University of Education, Seoul, Korea
{wocjun, kwak}@snue.ac.kr
[2] Seoul Ujang Elementary School, Seoul, Korea
vision1970@hanmail.net
[3] University Library, Seoul National University of Education, Seoul, Korea

Abstract. While current school education on computers and communication for the gifted students has mainly focused on knowledge and skills on computer, ethical issues on computer knowledge and technologies have not been dealt with sufficiently. In this paper, we introduce a Web-based information communication ethics education system for the gifted elementary school students in computer. The proposed system has the following characteristics: First, our system provides comprehensive areas of information communication ethics issues for the gifted elementary school students in computer. Second, for the sake of active participation and arousing interests, avatars are used to explain contents and guide students with voice message. Third, for each subject, there are three steps, *Introduction, Development* and *Discussion* to follow so that students can learn ethics by discussion rather than learning by heart. Fourth, for encouraging active communication between a teacher and students, various tools are supported. Finally, our system supports online survey or vote to check students' changes in information communication ethics.

1 Introduction

Recent advances in computer and communication technologies have rapidly changed our life in various ways. These advances provide new ways for people to communicate on a global scale and assess vast amounts of information. However, along with positive effects, side effects such as hacking, reckless use of cyber communication language, spreading harmful computer virus which might collapse entire computer system, overflow of lewd materials, copyright violation come to the front. In fact, more various sorts of those harmful effects are added everyday. In additions, the harmful effects on students become more serious [3].

In order to reduce the copious reverse effects raised in current information-oriented society, people used to set up strict rules or develop programs as a part of technical steps. However, those actions taken have turned out to be effective within a boundary. Experts are now moving into developing educational countermeasures against such harmful effects. They suggest that people can eradicate contrary effects and casualties by instructing students the essential information communication ethics. Therefore, the information communication ethics education will be the fundamental way of rooting up serious side effects [2].

Especially for elementary school students, information communication ethics education is more important since they usually do not have firm morality and are they're about to start learning computer knowledge and skills. Teachers must not to force them to learn the information communication ethics as a kind of cramming class subject. Instead, teachers are supposed to arrange their learning circumstances so that elementary school students can realize the proper code of conduct on the Web, and then internalize it naturally by themselves. That's why the information communication ethics education stresses the learning of one's free will as the groundwork. The information communication ethics education will let students constitute their own standard of ethics, hold it inside their mind, then finally translate it into action in real situation.

For gifted students, teaching information communication ethics is more important since they usually have higher sensibility and bigger knowledge and skills in computer than ordinary students [4]. Thus, it is necessary to select appropriate contents and teach those contents at their early stage. In this paper, a Web-based information communication ethics education system for gifted elementary school students in computer is introduced. The proposed system has the following characteristics: First, extensive contents for gifted elementary school students are organized. Second, the system adopts various types of avatars for delivering hard contents in an interesting manner. Third, the system supports various types of interactions by providing interactive environments. Students can exchange their ideas with other students, teachers, as well as experts in this field.

This paper is organized as follows. In Section 2, the basic concepts and principles of information communication ethics are presented. In Section 3, definition and characteristics of gifted students in computer are discussed. In Section 4, the design and implementation of a Web-based information communication ethics education system is explained. Finally conclusions and further research issues are provided in Section 5.

2 Definition and Principles of Information Communication Ethics

Information communication ethics is the theory analyzing social roles and effectiveness of computer technology, drawing up and justifying regulations for it's ethical usage [1]. In [7], four principles are proposed for the information communication ethics. Those principles are as follows. 1) Have respect for intellectual property rights 2) Have respect for one's privacy 3) Use legal language 4) Try not to cause harmful effects. Four steps for application of these principles are as follows [7].

1) Collect facts accurately.
2) Confirm the moral dilemma.
3) Evaluate the moral dilemma using the principles of the information communication ethics to decide which side guarantees major ethical supports.
4) Verify own solution based on the rule of generalization possibility.

From above 4 principles for the information communication ethics, we can realize that modern society clearly requires for upgraded information communication ethics than ever before. Four steps for application of above principles suggest good guideline.

Even though those principles and steps can give way in terms of educational context, still they are lack of illustrating conformity, selecting and organizing details for instruction. Particularly for elementary school students who never have used computer before, personal experience must be emphasized on just rather than forcing study.

3 Definition and Characteristics of the Gifted Students in Computer

There have been some works for defining gifted student in computer [5,6,8,9]. Based on those definitions, we define the gifted student as follows. First, gifted students have above-average ability in general intelligence, strong curiosity in computer, high creativeness, high mathematical and linguistic talent and tenacity. Second, gifted students are interested in application software, programming, game and multimedia and excellent in intuitiveness, generalization, inference, and adaptability for computer. Third, gifted students have talent in expressing ideas using computer and applying computers to other real-world problems.

Recently, in [9], characteristics and conditions of gifted students in computer are presented. Their works is summarized in Table 1, Table 2, Table 3 and Table 4, respectively.

Table 1. Characteristics and Conditions for gifted students: Logic area

Area	Subject	Contents
Logic	Application of computer knowledge	Ability to apply computer ability to real life
	Accomplishment of computer subjects	Scholastic achievement for computer-related subjects
	IQ	General intellectual ability
	Logical thinking	Ability to take steps based on logical thinking
	Algorithmic thinking	Ability to use computer for problem solving
	Inference ability	Ability to infer based on some evidences

Table 2. Characteristics and Conditions for gifted students: Intelligence area

Area	Subject	Contents
Intelligence	Programming ability	Ability to write a problem for problem solving
	Software use ability	Ability to use software at will
	Knowledge on software	Theoretical knowledge on software
	Mathematical ability	Knowledge on mathematics
	Multimedia use ability	Ability to use multimedia
	Aptitude to computer	Aptitude to acquire skills and knowledge on computer
	Confidence in computer	Confidence in computer-related fields

Table 3. Characteristics and Conditions for gifted students: Emotion area

Area	Subject	Contents
Emotion	Motive induction	Objectives and direction establishment for problem solving using computer
	Curiosity	Interest to computer-related fields
	Concentration	Curiosity to computer-related fields
	Perception on computer	Concentrating ability to computer-related fields
	Tenacity on assignment	Ability to identify principles for computer-related fields
	Potential development	Possibility to potential development for computer-related subjects
	Desire to accomplish in compute subjects	Desire to accomplish in computer-related subjects

Table 4. Characteristics and Conditions for gifted students: Creativeness area

Area	Subject	Contents
Creativeness	Will to computer study	Strong will for computer-related subjects
	Computer problem solving	Ability to solve problems on computer-related subjects
	Infinite imagination	Infinite imaginative power to computer-related subjects
	Originality in thinking	Original thinking to computer-related subjects
	Generalization of computer theories	Ability to generalize general facts and relationships among elements
	Intuitiveness to computer	Intuitiveness in computer problem solving
	Divergent thinking	Ability to diffuse computer-related theories

4 Design and Implementation of a Web-Based Information Communication Ethics Education System for the Gifted Students in Computer

4.1 Design of the Proposed System

Various materials and contents have been investigated to provide contents for gifted elementary school students in computer. Since gifted students learn or are likely to learn advanced techniques quicker than ordinary students, more emphasis is focused on contents leading gifted students to understand dysfunction of computer and harmfulness of various types of addictions, resulted from misuse of computer. The contents can be classified into 4 subjects, *Information communication ethics Pros and cons of cyber world, Netiquette and Cyber symptoms*, respectively. Table 5 shows those arrears and their corresponding subjects.

Table 5. Areas and subjects of contents

Area	Subject
Information communication ethics	-Meaning of information communication ethics -Ethics code of netizen -Terminologies of information communication -Chatting language
Pros and cons of cyber world	-Pros and cons of Internet -Illegal copy -Computer Virus -Hacking -Cyber crime
Netiquette	-Meaning of netiquette -Rules of netiquette -Observance of netiquette -Spam mail
Cyber symptoms	-Game addiction -Internet addiction -Unsound site -Chatting addiction

Table 6. Subject and study points of *Information communication ethics* category

Area	Subject	Study point
Information communication ethics	Meaning of information communication ethics	Meaning and necessity of information communication ethics education
	Ethics code of netizen	10 principles of ethics code
	Terminologies of Information communication	Definition of information communication terminology
	Chatting language	Manners for chatting

Table 7. Subject and study points of *Pros and cons of cyber world* category

Area	Subject	Study point
Pros and cons of cyber world	Pros and cons of Internet	Advantages and disadvantages of Internet use
	Illegal copy	Cases of copyright violation
	Computer virus	Harmfulness and prevention methods of virus
	Hacking	Types, problems and prevention methods of hacking
	Cyber crime	Types and prevention methods of cyber crime

Table 8. Subject and study points of *Netiquette* category

Area	Subject	Study point
Netiquette	Meaning of netiquette	Definition and necessity of netiquette
	Observance of netiquette	Observance of netiquette for different cases
	Spam mail	Harmfulness and prevention methods of Spam mail

Table 9. Subject and study points of *Cyber symptoms* category

Area	Subject	Study point
Cyber symptoms	Game addiction	Symptoms and self-diagnosis of game addiction
	Internet addiction	Symptoms, causes and self-diagnosis of Internet addiction
	Unsound site	Types, effects and countermeasures to unsound site
	Chatting addiction	Symptoms and self-diagnosis of chatting addiction

Subjects and their study points of the *Information communication ethics, Pros and cons of cyber world, Netiquette and Cyber symptoms* category are defined further as in Table 6, 7, 8 and 9, respectively.

The proposed system has the following characteristics. First, our system provides comprehensive areas of information communication ethics issues for gifted elementary school students in computer. Second, for the sake of active participation and arousing interests, avatars are used to explain contents and guide students with voice message. Third, for each subject, there are three steps, *Introduction, Development* and *Discussion* to follow. Students identify ethical issues introduced by Flash video (*Introduction* stage), grasp and try to solve problems based video data (*Development* stage) and discuss and exchange their ideas for themselves (*Discussion* stage). Finally, students are supposed to take a quiz for their study. Fourth, for encouraging active communication between a teacher and students, various tools such as Q & A, FAQ on BBS (Bulletin Board System) are supported. Finally, our system supports online survey to check students' changes in information communication ethics.

4.2 Implementation of the Proposed System

Our system has been implemented using ordinary hardware and software. For our implementation, we adopt PHP 4.0, Flash MX, and JavaScript for script language, Mysql for database, and Apache for Web server, respectively. Web site address of our system is http://comedu.snue.ac.kr/~m20011513. The followings are some screens captured to illustrate our system.

Figure 1 shows *Initial* Screen. In *Initial* Screen, students can choose diverse menus. Figure 2 shows *Study Guide* Screen. This screen provides general introduction of the proposed system to students.

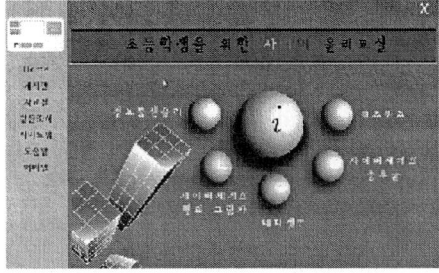

Fig. 1. *Initial* Screen

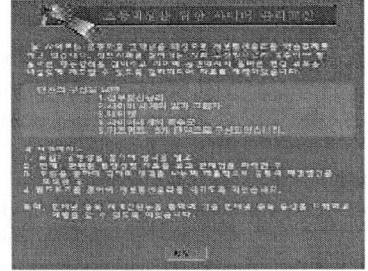

Fig. 2. *Study Guide* Screen

Figure 3 and 4 show *Netiquette* Screen and *Information Communication Ethics* Screen, respectively. Those screens provide contents for dealing with definitions of netiquette and information communication ethics, respectively.

Fig. 3. *Netiquette* Screen

Fig. 4. *Information Communication Ethics* Screen

Figure 5 and 6 represent data storage room and BBS, respectively. In data storage room, various contents containing multimedia data are provided. Also, students can ask questions or exchange ideas on BBS.

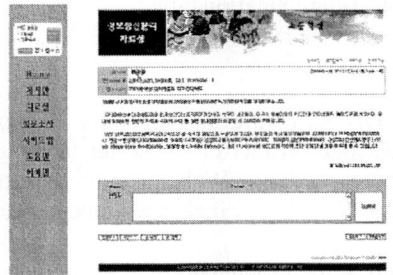

Fig. 5. *Data Storage Room* Screen

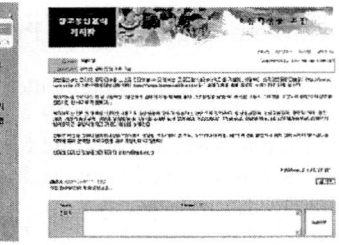

Fig. 6. *BBS* Screen

5 Conclusions and Further Work

In this paper we introduce a Web-based information communication ethics education system for gifted elementary school students in computer. The proposed system has the following characteristics: First, our system provides comprehensive areas of information communication ethics issues for gifted elementary school students in computer. Second, avatars are used to explain contents and guide students with voice message in interesting manners. Third, the proposed system supports students to learn ethics by discussion rather than memorization. Fourth, the system supports various types of interactions for encouraging active communication between a teacher and students. Finally, our system supports online survey to check students' changes in information communication ethics.

For our research, we need to extend our work as follows. First, evaluating effectiveness of our system is an immediate research issue. Second, we need to develop various delivering media to arouse interests.

References

[1] S. L. Edgar, Morality and Machines. Johns and Bartlett, 1997.
[2] http://www.kedi.re.kr/Exec/Pds/Cnt/128-16.htm
[3] Korea Education and Research Information Service, Ministry of Education & Human Resources Development, "Study on Preventing the Reverse Effects of Information-oriented Educational Institution", 2000.
[4] D. Na, "Development of a Curriculum of Education for the Elementary Gifted Children of Information Science, Elementary Computer Education major, Graduate School, Inchon National University of Education, 2003.
[5] S. Oh, "Definition and Recognition System for Gifted Students in Computer", Master Thesis, Computer Education Major, Graduate School of Education, Sungkyunkwan University, Korea, 2002.
[6] K. Ryu and J. Lee, "A Study on Information & Communication Ethics Education for the Gifted Elementary School Students in Information Science", Proceedings of the 8th Korean Association of Information Education Summer Conference, pp. 199-206, 2003.
[7] R. J. Severson, The principles of information ethics. ME Sharpe, 1997.
[8] S. Shin, "A Study on Development of a Program for the Identification of Information Gifted Children based on the Ability of a Discrete Thinking", Master Thesis, Elementary Computer Education major, Graduate School, Korea National University of Education, 2004.
[9] K. Yu, "Study on Recognition of Computer Teachers for Gifted Students in Information Science", Master Thesis, Graduate School of Education, Hanyang University, 2002.

International Standards Based Information Technology Courses: A Case Study from Turkey

Mustafa Murat Inceoglu

Ege Univ. Dept. of Computer Eng.,
Bornova, 35100, Izmir, Turkey
inceoglu@bornova.ege.edu.tr

Abstract. In this study, two programs which are being implemented in Ege University [13] since 2002 are presented. These programs are based on international standards. One of the programs is for the students of the university, the other one is for the people who are unemployed or who want to be employed in Information Technology (IT) business. The first education program that covers all university students is based on the standard of international European Computer Driving License (ECDL). The second program is IT employment program that is based on the curriculum of American Computing Association (ACM). The design, implementation and results of both of these programs are discussed in this study.

1 Introduction

Nowadays, almost every industry, especially finance, is becoming more and more dependent on information and communication technologies. In developing countries like Turkey, whatever profession (s)he has, every university graduate must have an acceptable knowledge level on information and communication technologies.

For this purpose, Ege University has started the "The University of Information Society Project" since 2003. A total of more than 3100 students enrolled from 11 different faculties (faculty of dentistry, pharmacy, letters, education, science, communication, economic and administrative sciences, engineering, fisheries, medicine and agriculture) and 8 different university high schools and vocational training schools (physical education and sports school, music conservatory, nursing school, Ataturk health school, Odemis medical school, Bayindir vocational training school, Tire-Kutsan vocational training school, Ege vocational training school) (approximately 18% of total number of students) in fall semester of 2003 and more than 3300 students in fall semester in 2004 (approximately 19% of total number of students) have taken the two-semester core course "Information Systems". As a result of this course, the students who graduate 3 years later will have an acceptable knowledge level on computers. In order to standardize the contents of the information systems course, the syllabus has been based on European Computer Driving Licence (ECDL) syllabus version 4.0 [2].

ECDL content consists of seven different modules and covers the required topics for an acceptable knowledge level on computers [1]. These modules are, concepts of information technology, using the computer and managing files, word processing, spreadsheets, database, presentation, information and communication.

Hardware, software, information networks, the use of information technology, in everyday life, health and safety, environment, security, copyright and the law topics are covered in concepts of information technology module. In the second module, using the computer and managing files, basic usage and basic file operations of an operating system (for example Microsoft Windows 2000/XP or Linux) are introduced. In word processing, spreadsheets, database ve presentation modules; basic operations on consecutively Microsoft Word, Excel, Access and Powerpoint (or Linux compatible) are introduced. In the seventh and the last module, information and communication module covers surfing on the web, using search engines on the web, e-mail operations etc.

2 Methodology

Different disciplines are investigated about how the acceptable knowledge level on computers are introduced and defined before the planning [3, 4, 5, 6, 14]. The first module, concepts of information technology is introduced theoretically and the other six modules are presented while every student is using a computer. Information systems course is a campus-wide core course of 3 credits. The students who possess an ECDL certificate or succeed in an exam based on ECDL before applying for university are not obliged to take this course.

3 Evaluation

A commission including representatives from eleven faculties and eight university high schools make a project planning before the start of academic year for common examination dates. In order to discover the success levels of teaching staff and students, the same questions are encouraged to be asked at each department of the faculties. The same commission makes a meeting at the end of academic year and evaluate the campus-wide success level of information systems course. The students can also take examinations in the ECDL test center of the university to achieve an ECDL certificate.

4 A Programming Course Based on Standards

Every year, more than 1,5 million high school graduates of Turkey enter the central examination for university education [11] and only 20% of these students can be placed to departments they desire. The people who cannot have the chance for university education or the students who cannot be placed in desired IT departments must be educated in this different and developing area.

The second study in Ege University with the university of information society project is the course for unemployed people to be educated as programmers. The aim of the course is not to educate people as software or computer engineers, but to educate them with basic required information for programming including web programming. This course program has been started in fall semester of 2002 by the department of computer engineering.

Table 1. Course curriculum

---

1. Semester (week 1 to 8)
Introduction to Algorithms, (12 hours/week):
Algorithms, procedural programming concepts, recursion, introduction to object oriented programming.
Information Systems, (8 hours/week):
Concepts of information technology, using the computer and managing files (e.g. Microsoft Windows/Linux), word processing (e.g. Microsoft Word/Wordperfect), spreadsheets (e.g. Microsoft Excel), database (e.g. Microsoft Access), presentation (e.g. Microsoft Powerpoint), information and communication (e.g. Microsoft Internet Explorer/Netscape, Outlook) (all of the ECDL modules).

2. Semester (week 9 to 16)
Java Programming-I, (12 hours/week):
Classes and objects in Java, encapsulation, control structures, arrays, strings, inheritance, interfaces.
Database, (8 hours/week):
Theory of Databases, Structured Query Language (SQL), Java Data Base Connection (JDBC).

3. Semester (week 17 to 24)
Java Programming-II, (8 hours/week):
Data types, elementary data structures, trees, storage management, sorting and searching programs, search trees, file operations.
Java Server Pages, (12 hours/week):
Java beans, cookies, sessions, Java Server Pages (JSP), data base applications with JSP, extensible markup language (XML) and Java, JSP Servlets.

4. Semester (week 25 to 32)
Project Management, (5 hours/week):
Project planning and implementation of computer program plans using Microsoft Project.
Microsoft .NET Programming, (15 hours/week):
Microsoft .NET technology, Microsoft C#.NET programming, ASP.NET, ADO.NET (It has been thought that educating people about Microsoft technologies could be useful for employment because they are widely being used in Turkey).

4.1 Course Contents and Implementation

Before defining the course contents, the "Computing Curricula" [7] developed by Institute of Electrical & Electronics Engineers (IEEE) and Association for Computing Machinery (ACM) has been studied in 2001. Considering the international and sector requirements, the curriculum has been modified for introduction to programming because the aim of this course is not to educate people as software or computer engineers. However successful participants can continue their education on software engineering because the course plan includes the basics of [7].

Recent similar studies are mostly for software engineering education, assisting software engineering education [9, 10] and web programming education [8]. Each participant are required to take an exam to be evaluated on analytic and arithmetic skills. People who pass this exam are invited to take the course. The course itself consists of 5 semesters where each semester is 2 months. The course is given 5 days per week and 4 hours per day in because of the intensive training schedule. As a result, 640 hours of training is given in four semesters. The fifth semester of the course is project semester. The curriculum of the course is shown in Table-1.

Participants are required to develop a programming project under the supervision of a teacher at the project semester. The participants are also asked to work in groups of at least two people and prepare an 8-week project plan. Before starting the project, the participants are given seminers by professional programmers for 1 week about program development, software medias being used and software methodologies. Following the end of seminers and project planning, the projects are expected to be concluded in 8 weeks. When the projects are completed, they are evaluated by a commission of 3 teachers and certificates are given to successful participants.

4.2 Evaluation

The yearly information including the age, sex, education level and employment (not in IT sector) about participants are shown in Table-2.

The information about the participants who successfully completed their education in 2002 and 2003 and employed in IT field after the course are shown in Table-3.

Table 2. Some information about participiants

		2002	2003	2004
Gender	Male	20	17	14
	Female	14	6	7
Age Group	18-25	23	8	12
	26-35	8	14	9
	36+	3	1	0
Graduation	High School	17	12	10
	Univ.	17	11	11
Number of Workers (non IT sector)		8	9	7
TOTAL (1)		34	23	21

Table 3. Participants who successfully completed their education in 2002 and 2003 and employed in IT field after the course

YEAR	GENDER		TOTAL (2)	RATIO
	M	F		TOTAL(2) / TOTAL(1)
2002	5	3	8	23%
2003	5	4	9	39%
TOTAL	10	7	17	

5 Conclusion

Information systems course has been started for creating employment advantage for graduating students against their counterparts from other universities, making students successful in business life and incresing the chance of students to be employed in international fields.

It has also been an important study for companies about increasing employment efficiency, satisfying the qualified employer requirements and decreasing IT budget for personnel education.

Starting from the fall semester of 2005, e-citizen [12] program, which has been introduced by ECDL foundation and bearing small differences from the contents given in university, is decided to be added to the information systems course. e-Citizen is a new end-user computer skills certification program developed by ECDL. It is an improver-level certification specifically developed to cater for those with a limited knowledge of computers and the Internet. e-Citizen is intended to help candidates bridge the digital divide by giving them the necessary skills to interact on-line across a broad range of environments, from dealing with government departments to communicating on-line with family and friends

In the second course, the programming course, it is aimed to increase the IT employment and to be used in continuous education. Although, this course has been open for society for 3 years, most of the participants are people who fail in the central examination for university and the university graduates who want career on IT field. It can be seen from the employment results of the course (Table-3) that, although the total number of participants are decreasing, the IT employment rate in increasing.

References

1. ECDL Standards, available at: http://www.ecdl.com
2. ECDL Syllabus version 4.0, available at: http://ww.ecdl.com/main/syll4.php
3. Logan, JR, Price, SL: Computer science education for medical informaticians, International Journal of Medical Informatics, 73 (2004), 139-144
4. Sanchez, G, Gardey, M: Computer literacy and undergraduate dental studies, Journal of Dental Research, 82 (2003), 13
5. Smith, RD: The application of information technology in the teaching of veterinary epidemiology and public health, Journal of Veterinary Medical Education, 30 (2003), 344-350

6. Childers, S: Computer literacy: Necessity or buzzword?, Information Technology and Libraries, 22 (2003), 100-104
7. ACM&IEEE Computing Curricula 2001 Computer Science. IEEE Press, available at: http://www.acm.org/sigcse/cc2001
8. Sridharan K: A Course on Web Languages and Web-Based Applications, IEEE Transactions on Education, 47 (2004), 254-260
9. Becker-Pechau P., Bleek WG, Lilienthal C, Schmolitzky A: Educating Non-Programmers to Flexible, Communicative Software Engineers in a 10 Month Training Program, Proceedings of the 17th Conference on Software Engineering Education and Training (CSEET'04)
10. Wegmann A: Theory and Practice Behind the Course Designing Enterprisewide IT Systems, IEEE Transactions on Education, 47 (2004), 490-496
11. Selection and Placement of Students in Higher Education Institutions in Turkey, available at: http://www.osym.gov.tr
12. e-citizen, available at: http://www.ecdl.com/main/ecit.php
13. Ege University web site, available at: http://www.ege.edu.tr
14. Wu A, Leung CH: IT in Education: What is Really Needed of Teachers' IT Competencies?, Proceedings of the International Conference on Computers in Education (ICCE'02)

Design and Implementation of the KORI: Intelligent Teachable Agent and Its Application to Education

Sung-il Kim[1], Sung-Hyun Yun[2], Mi-sun Yoon[3], Yeon-hee So[1], Won-sik Kim[1], Myung-jin Lee[1], Dong-seong Choi[4], and Hyung-Woo Lee[5]

[1] Dept. of Education, Korea University, Seoul, Korea
sungkim@korea.ac.kr
[2] Div. of Information and Communication Engineering, Cheonan University, Cheonan, Korea
shyoon@cheonan.ac.kr
[3] Dept. of Teacher Education, Jeonju University, Jeonju, Korea
msyoon@jj.ac.kr
[4] Div. of Design and Image, Cheonan University, Cheonan, Korea
hcilab@cheonan.ac.kr
[5] Dept. of Software, Hanshin University, Osan, Korea
hwlee@hs.ac.kr

Abstract. Most of the intelligent tutoring system (ITS) is used to deliver knowledge and train skills based on the expert model. According to the theories of leaning, learning by teaching method is more efficient for enhancing motivation to learn and cognitive ability than learning by listening or learning by reading. For the purpose of developing an adaptive intelligent agent to enhance the motivation to learn, the new type of teachable agent were designed and implemented, the KORI (KORea university Intelligent agent), in which the user plays a role of a tutor by teaching the agent using a concept map, posing questions, and providing feedbacks. KORI consists of four independent modules: teach module, Q&A module, test module, and resource module. In teach module, the KORI's knowledge is structured and organized through the concept map and the KORI makes new knowledge from the inference engine. In Q&A module, the KORI can provide answers to the users' questions through an interactive window. It is expected that providing the user with the active role of teaching the agent enhance the motivation to learn and the positive attitude toward the subject matter as well as cognition.

Keywords: TA (Teachable Agent), intelligent tutoring system, learning by teaching, concept map, motivation to learn, inference engine, knowledge representation.

1 Introduction

The traditional intelligent tutoring system (ITS) provides the learning materials and practice drills repetitiously to train the students and the level of student's learning is evaluated by the computer. The ITS has received the criticism that the iterative nature of learning and the passive role of the learner does not enhance the learner's motivation and cognition[1,2,3]. To overcome this limitation of ITS, the system should provide the learners with a chance for deep learning and allow them to play an active role in the process of learning. One way of providing an active role for the learners is to give them an opportunity to teach. The researchers in the field of cognitive science and learning science suggest that the teaching activity induce the elaborative and meaningful learning. A lot of research findings have shown the beneficial effect of learning by teaching[1,4,5].

This study introduces the design of the new type of intelligent teachable agent system, called KORI (KORea university Intelligent agent), for maximizing the motivation to learn and enhancing cognitive ability based on concept of 'learning by teaching'. For the KORI as a student, it is necessary to use AI (Artificial Intelligence) algorithm to construct the knowledge-based of the KORI. The concept map is used to teach the learning material and the inference engine is used to create the new knowledge.

The KORI consists of four independent modules: teach module, Q&A module, test module, and resource module. In teach module, the users teach KORI by drawing a concept map. In Q&A module, both the users and KORI ask questions and answer each other through an interactive window. To assess KORI's knowledge and provide feedback to the users, the test module consists of a set of predetermined questions that KORI should pass. In the resource module, the users can refer to the basic concepts to teach or explore the additional expanded information whenever they want in order to teach, ask questions, and provide feedbacks.

The KORI's brain can represent the knowledge based on the concept map and make new knowledge through the inference engine and it becomes more intelligent by the student tutor.

In section 2, the Teachable Agent (TA) and our motivation to implement KORI are introduced. In section 3, the structure of the KORI's brain and the implementation details are discussed. In section 4, the conclusion and the implications for the future work are described.

2 Teachable Agent

The fact that the active and meaningful learning occurs through the teaching process is reported consistently in the research area of learning science[1]. [1] showed that students who studied for the purpose of teaching other students understood the article better than those who studied for the qualifying examination. In addition, there are many variations of the instructional methods based on the concept of 'learning by teaching', such as reciprocal teaching, peer tutoring, or self-explanation[2,3]. TA is a computer program where students teach the computer agent to maximize their motivation to learn and to enhance the cognitive ability[5,6]. TA provides the student tutors with an active role so that they can have a positive attitude toward the subject matter. Teaching activity consisted of sub-activities such as memorization and comprehension, knowledge reorganization, explanation, demonstration, questioning, answering, and evaluation, and so on. These sub-activities lead to elaboration, organization, inference, and metacognition. In terms of motivational aspects, allowing the learner to play a tutor role can enhance the learners' motivation, which instills a sense of responsibility and increases a feeling of engagement, self-relevance, and situational interests to persist in learning[5].

3 The Architecture of the KORI's Brain and Implementation Details

Students teach the KORI by drawing a concept map and then the KORI creates its own knowledge through the inference engine, and updates it through the feedback mechanism. Figure 1 shows the schematic representation of the interface between KORI and the user to construct the KORI knowledge.

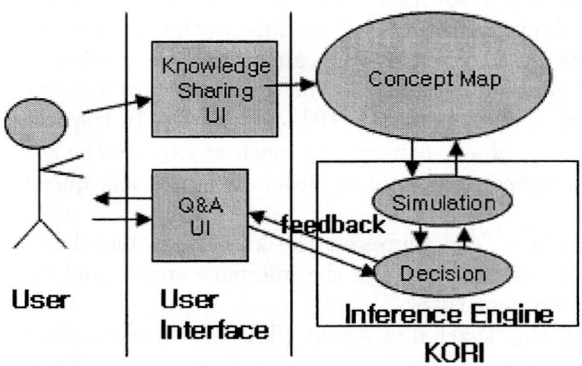

Fig. 1. Knowledge construction interface between the user and KORI

The following modules are used to design and implement the KORI's brain.

Concept map - to construct the KORI's knowledge
Knowledge inference engine - to make inferences and decisions
Search engine - to make, modify and search link paths between objects in the concept map

3.1 Knowledge Representation Based on the Concept Map

Concept map is used to structure and organize knowledge into objects and relations [7,8]. In KORI system, the objects are represented by the boxes and the relations between objects are represented by the arrows with the specific description of the relation. KORI also can represent the inheritance between objects as in semantic network.

The user can put the concepts whatever he/she want and draw arrows between concepts to indicate their relations. In the main window of the screen, the user types the name of rock in the box and represents the transformation between rocks with the arrow. The process of transformation was represented by mathematical symbol. The plus symbol (+) means the increase of the weathering factors while the minus symbol (-) means decrease. Below the concept mapping window, there are four taps, each of which has a different function, including KORI's talk to enhance the perceived interactivity, KORI's interpretation of the concept or the relation, learning resource, KORI's quiz score, and automatic feedback on KORI's performance by the system. The user can interact with KORI through these windows.

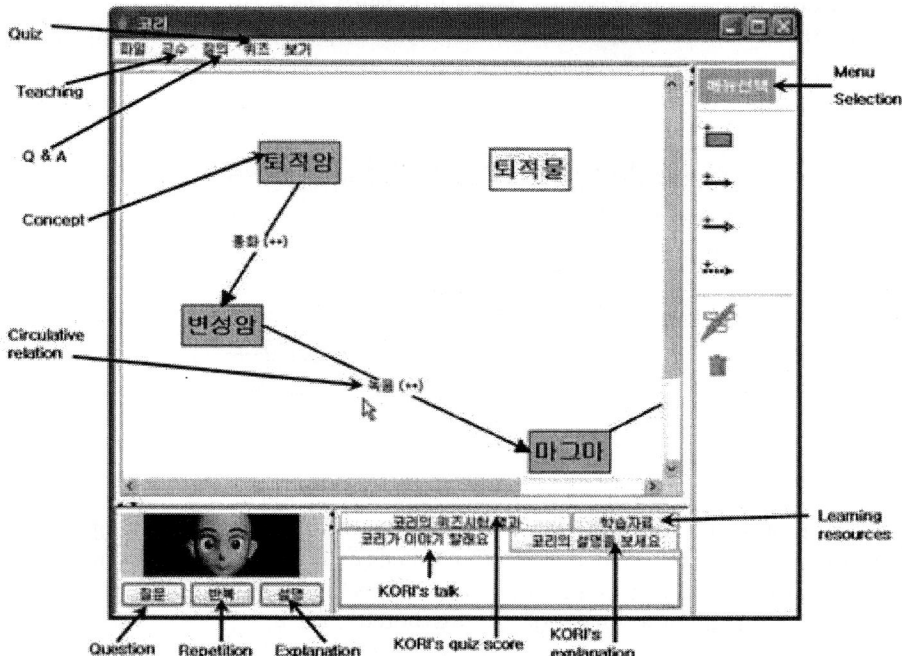

Fig. 2. The interface of KORI

Figure 3 shows an example of the concept map construction. In this study, the learning material is about the 'rock cycle'. The object 'sedimentary rock' has a relation with the object 'deposit'. To represent the relation between these two objects, an arrow is used with the specific description of 'be weathered'. In this example, each object has a transitive relation. The users can deliver their knowledge on the rock cycle through the concept map.

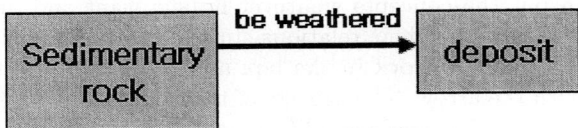

Fig. 3. Representation of objects and their relations

3.2 Knowledge Inference Engine

Using the forward and backward reasoning on objects, the KORI's knowledge inference engine makes new relations between objects that are not represented in the concept map. In KORI, the reasoning process occurs when the user asks the specific type of questions to the KORI. With the structured input interface where all objects and relations are displayed in pull-down menu, and the user can generate questions by selecting them.

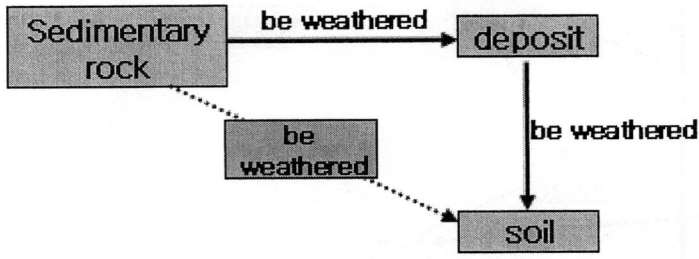

Fig. 4. New knowledge construction through the inference engine

Figure 4 shows an example of reasoning process through the knowledge inference engine. The user defines the 'sedimentary rock' object and the 'deposit' object. He/she specifies the relationship between these two objects by drawing an arrow with the description of 'be weathered'. Then, the user also defines the 'soil' object and draws an arrow with the 'be weathered' label from the 'deposit' object to the 'soil' object. After the user has finished drawing the concept map, the KORI activates inference engine with the predetermined rules that find a new relations between the 'sedimentary rock' and the 'soil'. The dotted arrow with 'be weathered' label shows the result of this reasoning processing. Using these two diagrams, the KORI represents its own knowledge and generate answers to students' questions.

3.3 User Interface

KORI consists of four independent modules. Figure 5 shows the interaction between them. In teach module, student teaches KORI through the concept map input interface. In Q&A module, KORI and the student communicate each other through dialogue box which supports structured input only. In test module, KORI is evaluated by built in expert module. The resource module consists of web site links or documents related to learning.

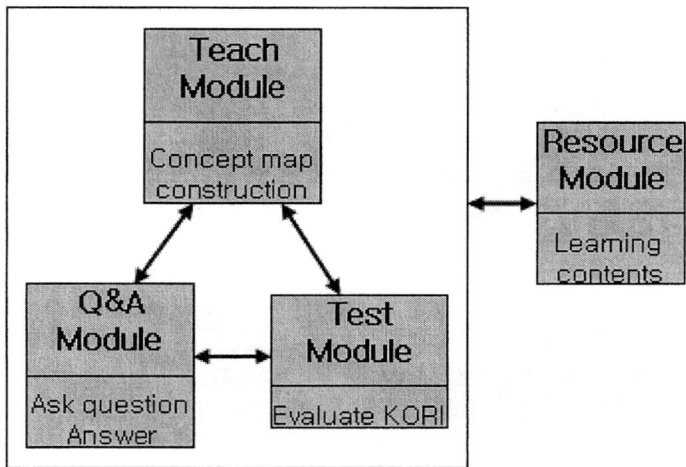

Fig. 5. The KORI module

The KORI is implemented based on JAVA platform. For GUI user interface, JAVA Swing and Jgraph components are used. KORI's brain consists of two modules: Concept map class and inference engine class, which are coded to represent and organize KORI's knowledge.

(1) Teach Module. Teach module consists of two stages: map selection and concept map teaching. In the first stage, map selection, depending on the developmental stage and the level of perceived competence of each user, one of the following maps is presented.

partial map - parts of the complete map is shown
null map - no map is shown
incorrect map - intentionally modified incorrect map is shown

Figure 6 shows an example of partial map. If students select partial map, they should update it to make complete map.

Before starting to teach KORI, each user's level of perceived competence is measured. [9] found that the partial map is more efficient to enhance their interests than any other types of concept maps for 4th graders who have high level of competence whereas the null map enhances the interest only for 5th graders who are highly competent. Based on the previous findings, the appropriate type of the concept map is automatically selected and presented, depending on their level of perceived competence.

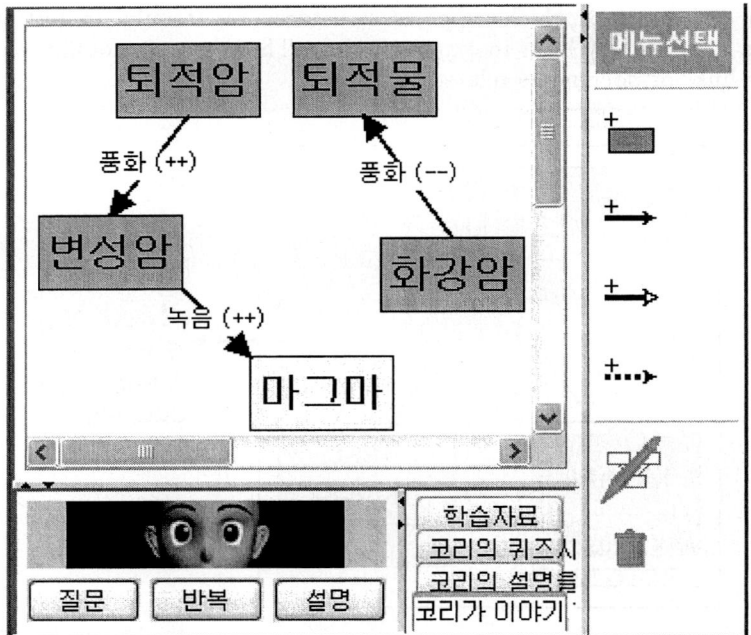

Fig. 6. An example of partial map

In the second stage, the concept map teaching, the user can use teaching tool box to draw the objects and the relations among them into the concept map to build the KORI's knowledge base. The rectangle tool is used to draw the objects and the arrow tool to represent relations between objects. Then the user types the specific descriptions in the rectangle and above the arrow to represent the meaning of objects and the relations.

(2) Q&A Module. If the user completes the construction of concept map, he/she moves to the Q&A module to update the KORI's knowledge base. Q&A module consists of various menus such as *Dialogue, Ask to KORI, Answer to KORI,* and *Ask to tutor.* Built-in *Ask to tutor* module is automatically activated when KORI reaches decision state during the reasoning process.

Figure 7 shows an example of Q&A interface. The user can answer to the KORI's question by choosing *Right* or *False* option. When KORI asks a question, the corresponding information of the concept map is highlighted with a red frame so that the user can easily notice the misconception of KORI and update concept map appropriately. The user also can either edit the concept map directly through the Modify menu or give a command through the structured input interface if KORI asks a wrong question.

(3) Test & Resource Module. KORI's knowledge is evaluated through the test menu in which KORI should answer to the predetermined built-in test ques-

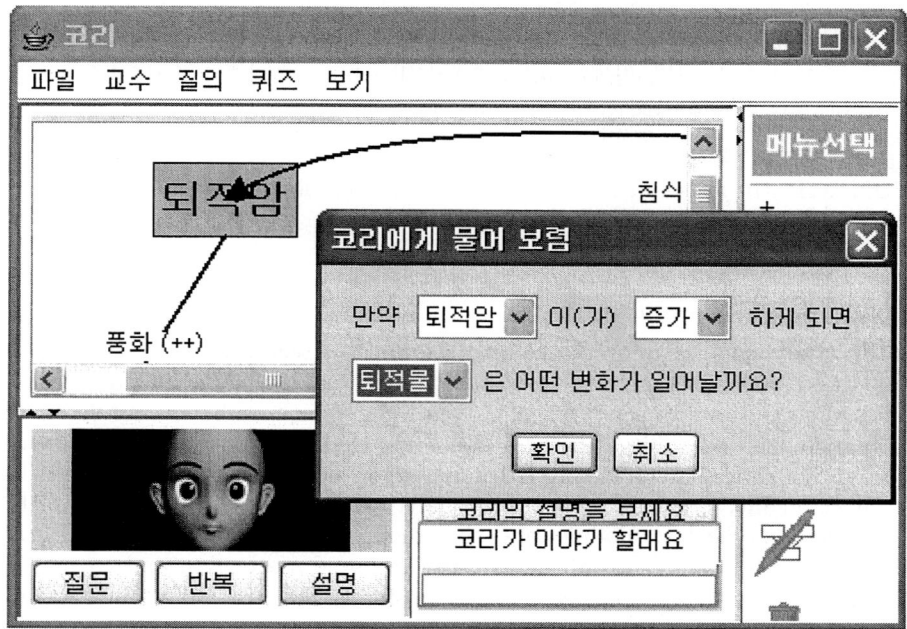

Fig. 7. An example of Q & A interface

tions, based on what they have been taught. The results of the test are showed in a separate window. The resource module provides the basic and expanded knowledge about rocks and their transformation. The user can access to this module by clicking the icons whenever they want to know more about rocks while teaching KORI. The resource is made of hypertext that is linked the basic concepts to the concrete images and examples.

3.4 Implementation of Concept Map and Inference Engine

To simulate KORI's knowledge, the following three classes are implemented.

Frame class - keeps objects and relations in the concept map window
ConceptMap class - manages objects, relations, links/paths between objects
InferenceEngine class - inherits inference rules and reasoning process from
 the ConceptMap class

In the InferenceEngine class, the following three methods are used to proceed to generate KORI's reasoning.

get_reasoning() - acquires new knowledge through reasoning process
generate_reasoning() - makes reasoning result to users' question
ask_question() - asks question to users in case that there exist several paths
 to go to the next state during reasoning process.

The ConceptMap class is used to construct and organize KORI's knowledge. It gets knowledge from both users and the inference engine. During the reasoning process, the InferenceEngine class uses the following methods inherited from the ConceptMap class to search objects and relations.

buildLinkPath() - sets the link between objects
addNextPath() - sets the next object and its relationship

4 Conclusion

In this study, the design and implementation of the teachable agent KORI was introduced, which is a modified version of traditional intelligent tutoring system that assigns the user for the tutor role to teach the agent to enhance motivation to learn and cognition ability. The KORI's knowledge is constructed and organized through the concept map and KORI makes new knowledge from it through the inference engine. The user interface of KORI consists of the teach module, the Q&A module, the test module, and the resource module. It is expected that teaching KORI would not only maximize the users' motivation and cognition, but also increase their self-efficacy and responsibility through various interactions and an immediate feedback.

Future research in the field of the intelligent tutoring agent should focus on the learners' motivation. The new generation of teachable agent should be able to reflect the individual differences in cognitive abilities, interest and motivation, and ongoing changes of the interest level. In addition, the dialog based query input interface to increase the interactivity and the learner control would make the teachable agent more interesting and efficient. To enhance the intelligence of teachable agent, various ways of knowledge construction should be developed and the collaborative learning system, in which multi-user can teach the agent by sharing knowledge, should be implemented.

Acknowledgments

This research was supported by Brain informatics Research Program sponsored by Korean Ministry of Science and Technology.

References

1. Bargh, J. A. & Schul, Y, "On the cognitive benefits of teaching," Journal of Educational Psychology, 72, pp.593-604, 1980.
2. Chi, M. T.H., Siler, S. A., Jeong, H., Yamauchi, T., & Hausmann, R. G, "Learning from human tutoring," Cognitive Science, 25(4), pp.471-533, 2001.
3. Graesser, A. C., Person, N., & Magliano, J, "Collaborative dialog patterns in naturalistic one-on-one tutoring," Applied Cognitive Psychologist, 9, pp.359-387, 1995.

4. Biswas, G., Schwartz, D., Bransford, J. & TAG-V, "Technology support for complex problem solving: From SAD environments to AI," In Forbus and Feltovich, (Eds.), Smart Machines in Education (pp.71-98). Menlo Park, CA: AAAI Press, 2001.
5. Kim. S., Kim, W.S., Yoon, M.S., So, Y.H., Kwon, E.J., Choi, J.S., Kim, M.S. Lee, M.J., & Park, T.J., "Understanding and Designing Teachable Agent," Journal of Korea Cognitive Science, 14(3), pp.13-21, 2003.
6. Brophy, S., Biswas, G., Kaltzberger, T., Bransford, J.,schwartz,d., "Teachable agent: Combining insights from learning Theory and Computer Science," Vandervielt, LTC, 1999.
7. Novak, J. D, "Concept mapping as a tool for improving science teaching and learning," In:D. F. Treagust; R, Duit; and B.J.Fraser (Eds), Improving Teachable and Learning in Sceince and Mathematics (pp. 32-43). London: Teachers College Press, 1996.
8. Stoyanov, S., & Kommers, P, "Agent-support for problem solving through conceptmapping," Journal of Interative Learning Research, 10 (3/4), pp.401-442, 1999.
9. Kim. S., Kwon, E.J., Yoon, M.S., So, Y.H., Kim, W.S., Lee, S., "The effects of types of concept map and science self-efficacy on interest and comprehension: A comparison of 4th and 5th graders", Korean Journal of Educational Psychology, 18(4), 17-31, 2004.

A Space-Efficient Flash Memory Software for Mobile Devices

Yeonseung Ryu*, Tae-sun Chung, and Myungho Lee

Department of Computer Software,
Myongji University,
Nam-dong, Yongin, Gyeonggi-do, 449-728, Korea
ysryu@mju.ac.kr

Abstract. Flash memory is becoming popular storage media for mobile computing devices. In this paper, we study a new block management scheme in Flash Translation Layer (FTL) for flash memory storages which considers the space utilization. Proposed scheme classifies data blocks according to their write access frequencies and improves the space utilization by managing the blocks according to their hotness degree. To evaluate the proposed scheme, we developed a simulator and performed trace-driven simulations.

1 Introduction

Flash memory is becoming important as nonvolatile storages for mobile devices because of its superiority in fast access speeds, low power consumption, shock resistance, high reliability, small size, and light weight [10, 15, 18, 11, 8, 5]. Because of these attractive features, and the decreasing of price and the increasing of capacity, flash memory will be widely used in consumer electronics, embedded systems, and mobile computers [1, 12]. Though flash memory has many advantages, its special hardware characteristics impose design challenges on storage systems. First, flash memory cannot be written over existing data unless erased in advance. Second, erase operations can be performed in a larger unit than the write operation. For an update of even a single byte, an erase operation of a large amount of data would be required. Besides it takes an order of magnitude longer than a write operation. Third, the number of times an erasure unit can be erased is limited (e.g., 10,000 to 1,000,000 times).

To overcome these problems, an software called a *Flash Translation Layer* (FTL) has been employed between host system and flash memory [7, 2, 9, 14, 13, 19, 16]. Figure 1 shows a typical software organization for NAND-type flash memory. The FTL is the driver that works in conjunction with file system to make flash memory appear to the system like a disk drive. Applications use system calls to access files on the flash memory. The file system then issues

---

* This work was supported by grant No. R08-2004-000-10391-0 from Ministry of Science and Technology.

Table 1. Characteristics of different storage media

Media	Access time		
	Read	Write	Erase
DRAM	60ns (2B)	60ns (2B)	
	2.56μs (1512B)	2.56μs (512B)	
NAND Flash	10.2μs (1B)	201μs (1B)	2-3ms (16KB)
	35.9μs (512B)	226μs (512B)	
Disk	avg. 12.4ms (512B)	avg. 12.4ms (512B)	

read/write commands along with logical block address and the request size to the FTL. Upon receipt of a command, address, and the size, the FTL translates them into a sequence of flash memory intrinsic commands (read/write/erase) and physical addresses. The FTL usually maintains the address translation table in order to map logical address of I/O requests to physical address in flash memory. The address translation table is indexed by logical block address (LBA), and each entry of the table contains the physical address of the corresponding LBA. The FTL usually uses *non-in-place update* mechanism to avoid having to erase on every data update. Under this mechanism, the FTL remaps each update request to different location (i.e., data updates are written to empty space) and set obsolete data as garbage, which a software cleaning process later reclaims [5].

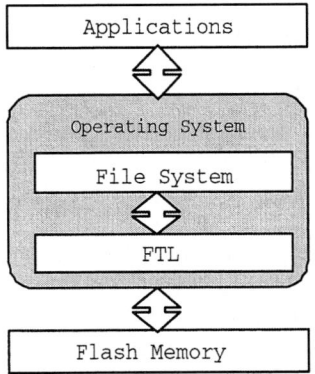

Fig. 1. Software organization for flash memory

In this paper, we propose a new FTL scheme which shows better space utilization than legacy schemes. Previous FTL schemes do not consider skewed access pattern and waste the space of flash memory. Proposed scheme exploits the hotness of data and increases the space utilization by managing the blocks according to their hotness degree. To evaluate the proposed scheme, we developed a simulator and performed trace-driven simulations. Proposed scheme performs substantially better than previous schemes with regard to the space utilization.

The rest of this paper is organized as follows. In Section 2, we review previous works that are relevant for this paper. In Section 3, we present a new flash translation layer scheme. Section 4 presents the experimental results to show the performance of proposed scheme. The conclusions of this paper are given in Section 5.

2 Background

A NAND flash memory is organized in terms of *blocks*, where each block is of a fixed number of *pages* [11, 8]. A block is the smallest unit of erase operation, while reads and writes are handled by pages. The size of page is fixed from 512B to 2KB and the size of block is somewhere between 4KB and 128KB depending on the product. There is a *spare area* appended to every page, where out-of-band data could be written. The typical size of spare area is 16B or 64B depending on the product. When the free space on flash memory is written, the space cannot be updated unless it is erased.

The mapping between the logical address and the physical address can be maintained either at the page level or at the block level [2, 6, 3, 4]. In the page-level address mapping, a logical page can be mapped to any physical page in flash memory. However, this mapping requires a large amount of space to store the needed mapping table. In the block-level address mapping, the logical address is divided into a logical block address and a block offset, and only the logical block address is translated into a physical block address in flash memory. This block address mapping has a restriction that the block offset in the mapped physical block be the same as that in the logical block. When there is an update request to a single page in a block, the physical block that contains the requested page is remapped to a free physical block, the write operation is performed to the page in the new physical block with the same block offset, and all the other page in the same block are copied from the original physical block to the new physical block.

To eliminate expensive copy operation in the basic block scheme, a technique called *log block scheme* was proposed [13]. The log block scheme manages most of the blocks at the block level, while a small fixed number of blocks are managed at the page level. The former holds ordinary data and are called data blocks. The latter are called log blocks. When an update to a page in a data block is requested, a log block is allocated and the update is performed to the log block incrementally from the first page. For the log blocks, a page-level mapping table is maintained. Once a log block is allocated for a data block, update requests to the data block can be performed in the log block until all the pages in the log block are consumed.

The problem of the log block scheme is that it does not consider the space utilization of the log blocks. Ruemmler and Wilkes at Hewlett-Packard collected disk-level traces of an HP-UX workstation which was used primarily for document preparation and electronic mail [17]. They reported that access locations are highly skewed. Roughly one third of all accesses go to the ten most frequently accessed blocks. Since usage patterns of mobile computers are likely to be similar to the personal computer, we can assume the access patterns to the flash memory are also likely to be highly skewed. The skewed access pattern implies that most of the spaces are not often accessed. In the log block scheme, it is possible that the space of the dedicated log block would be wasted when update requests in a data block are not frequent.

3   New Block Management Scheme

Our scheme is based on the log block scheme since it exhibits good performance for the write and the erase operations. The space of flash memory are managed as two types of blocks, data blocks and log blocks. The data blocks hold ordinary data and the log blocks are used as temporary space for update writes to data blocks. When an update to a page in a data block arrives, the update is performed to the log block incrementally from the first page. The pages in the log block that have the same logical page number can be written multiple times without any special handling. The FTL can identify the up-to-date copy by scanning the log block backward from the last page. The block mapping table manages the corresponding log block number for each data block. For a read request, the requested page is serviced either from the data block or from the log block depending on where the up-to-date copy is present.

Each data block is associated with a state indicating the hotness. Initially all data blocks is defined as the cold block. The degree of hotness of each block is determined by the number of times the block has been updated within the time window. When a data block is updated frequently, its state is changed to 'hot'.

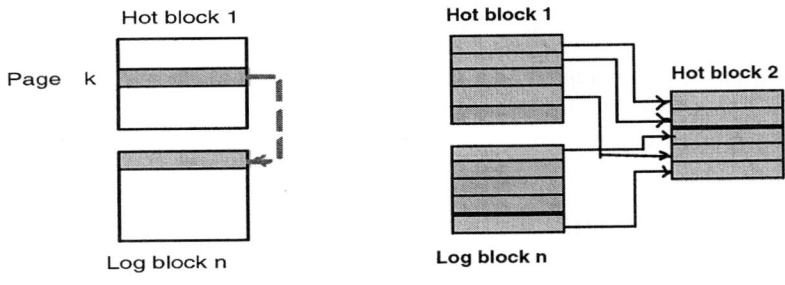

(a) A page k is updated   (b) Merge: When a log block n is full

Fig. 2. Operations for Hot block

Because hot blocks are likely to be updated soon and filled up fast, a dedicated log block is allocated to the hot block. If the hot log block becomes full, it is reclaimed by the *merge* operation. The merge operation allocates a free block and then fills each page with the up-to-date page, either from the log block if the

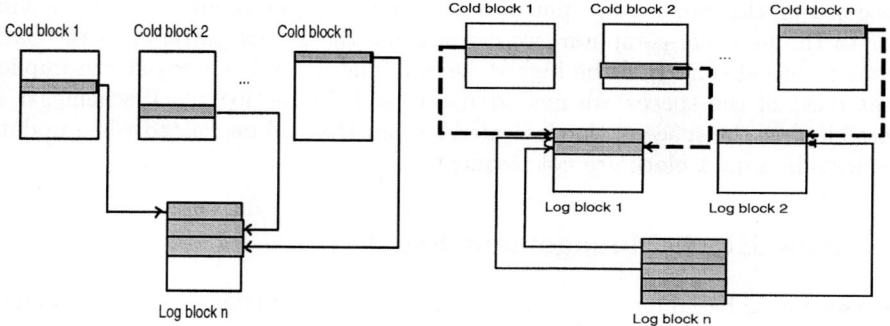

(a) Cold blocks share a log block n (b) Split: When a log block n is full

Fig. 3. Operations for cold block

corresponding page is present, or from the data block otherwise. After copying all the pages, the new block now becomes the data block, and the former data block and the log block are returned to the pool of free blocks, waiting to be erased.

On the other hand, cold blocks *share* the log blocks. When an update to a cold block is newly requested, the FTL allocates either a log block used by other cold blocks if it exists, or a new log block from the pool of free blocks otherwise. By sharing the log block, it avoids wasting the log block space. When the log block becomes full, it is reclaimed by the *split* operation. The split operation checks the number of data blocks that shares the log block. If two or more data blocks share the log block, the FTL allocates two log blocks from the free blocks and then distributes the up-to-date pages from the former log block into the new two log blocks. If the log block is used by only one data block, the FTL allocates a new log block and copies up-to-data pages from old log block to the new log block. In either case, the former log block is returned to the free blocks, waiting to be erased.

4 Simulation Studies

To evaluate the proposed scheme, we developed a simulator for the log block scheme and the proposed scheme, and performed trace-driven simulations. We used the traces of a digital camera[1] (WL1 in Fig. 4). We also generated synthetic trace long enough to simulate high locality of update accesses (WL2 in Fig. 4). The generated traces have very skewed access patterns like disk access patterns in HP personal workstation [17].

We define the number of extra erase operations as the number of erase operations minus the number of erase operations from an ideal scheme. The ideal

---

[1] We obtained trace data from an author of [13].

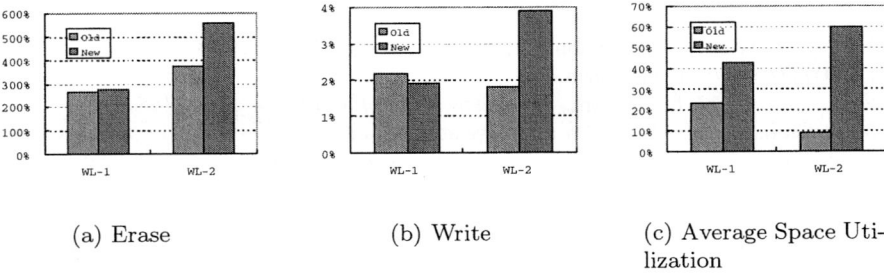

(a) Erase   (b) Write   (c) Average Space Utilization

Fig. 4. Performance Comparison

scheme is defined as a scheme that performs one erase operation for every n-page write requests, where n is the number of pages per block. Similarly, the number of extra write operations is defined as the number of write operations minus the number of writes requested. Performance metrics are the ratio of the number of extra erase operations to the number of erase operations from ideal scheme, the ratio of the number of extra write operations to the number of write requests and the average space utilization of the log blocks.

In fig. 4, 'Old' denotes the log block scheme and 'New' denotes the proposed scheme. The proposed scheme significantly improves the space utilization of the log blocks. For example, it increases the average space utilization from 10% to 60% in case of WL2. However, since the shared log blocks will be used frequently like the log blocks for the hot blocks, this incurs more reclamation process and thus increases the number of erase operations and write(i.e., extra data copy) operations. In the log block scheme, on the contrary, since a log block is dedicated to only one data block and a large portion of it remains unused for a long time in case of the cold blocks, it wastes the space but needs fewer erase operations. Simulation results show that there is a tradeoff between the number of erase/write operations and the space utilization.

5 Concluding Remarks

The primary concern in implementing the flash translation layer has been to improve the write performance by minimizing the number of erase operations and data copy operations. Previous log block scheme exhibits good performance for the write and the erase operations, but does not consider the space usage of the log blocks. Our approach is to classify data blocks according to their write access frequencies and to share the log blocks in order to improve the space utilization. Simulation results show that the proposed scheme improves the space utilization and there is a tradeoff between the space utilization and the number of erase/write operations. For the future works, we plan to study the allocation scheme of the shared log blocks to improve overall performance. We also plan to implement proposed scheme in the real system.

References

1. CompactFlash Association. Information about compactflash. http://www.compactflash.org.
2. A. Ban. Flash file system. In *United States Patent, no. 5,404,485*, 1995.
3. L. Chang and T. Kuo. An adaptive striping architecture for flash memory storage systems of embedded systems. In *Proceedings of the 8th IEEE Real-Time and Embedded Technology and Applications Symposium*, 2002.
4. L. Chang and T. Kuo. An efficient management scheme for large-scale flash memory storage systems. In *Proceedings of ACM Symposium on Applied Computing*, 2004.
5. M. Chiang and R. Chang. Cleaning policies in mobile computers using flash memory. *Journal of Systems and Software*, 48(3):213–231, 1999.
6. M. Chiang, P. Lee, and R. Chang. Using data clustering to improve cleaning performace for flash memory. *Software: Practice and Experience*, 29(3):267–290, 1999.
7. T. Chung, D. Park, Y. Ryu, and S. Hong. Lstaff: System software for large block flash memory. *Lecture Notes in Computer Science*, 3398:704–710, 2005.
8. Intel Corporation. Intel strataflash memory product overview. http://www.intel.com.
9. Intel Corporation. Understanding the flash translation layer (ftl) specification. http://developer.intel.com, December 1998.
10. F. Douglis, R. Caceres, F. Kaashoek, K. Li, B. Marsh, and J. Tauber. Storage alternatives for mobile computers. In *Proceedings of the 1st Symposium on Operating Systems Design and Implementation*, 1994.
11. Samsung Electronics. 256m x 8bit / 128m x 16bit nand flash memory. http://www.samsungelectronics.com.
12. SSFDC Forum. Features and specfications of smartmedia. http://www.ssfdc.or.jp.
13. J. Kim, S. Noh, S. Min, and Y. Cho. A space-efficient flash translation layer for compactflash systems. *IEEE Trans. on Consumer Electronics*, 48(2):366–375, 2002.
14. M-Systems. Trueffs. http://www.m-systems.com/.
15. B. Marsh, F. Douglis, and P. Krishnan. Flash memory file caching for mobile computers. In *Proceedings of the 27th Hawaii International Conference on Systems Sciences*, 1994.
16. MTD. Memory technology device (mtd) sub-system for linux. http://www.linux-mtd.infradead.org.
17. C. Ruemmler and J. Wilkes. Unix disk access patterns. In *Proceedings of 1993 Winter USENIX Conference*, pages 405 – 420, 1993.
18. M. Wu and W. Zwanepoel. envy: A non-volatile, main memory storage system. In *Proceedings of the 6th Internation Conference on Architectural Support for Programming Languages and Operating Systems*, 1994.
19. K. Yim, H. Bahn, and K. Koh. A flash compression layer for smartmedia card systems. *IEEE Trans. on Consumer Electronics*, 50(1):192–197, 2004.

Security Threats and Their Countermeasures of Mobile Portable Computing Devices in Ubiquitous Computing Environments

Sang ho Kim[1] and Choon Seong Leem[2]

[1] KISA, 78, Garak-Dong, Songpa-Gu, Seoul, Korea
shkim@kisa.or.kr
http://www.kisa.or.kr
[2] Yonsei University, 134, Shinchon-Dong, Seodaemun-Gu, Seoul, Korea
leem@yonsei.ac.kr
http://ebiz.yonsei.ac.kr

Abstract. Security has been more crucial issue as targets of attacker are wide ranges from personal information to various things in ubiquitous computing environments. Portable computing devices such as cellular phone, PDA, and smart phone in such conditions have been exposed under security threats, which cause attacker to exploit malicious code and modification of storied data. In this paper we present security threats and their countermeasures in technical, manageable, and physical aspect of mobile portable computing devices. We believe that this paper will contributes on initiating a research on security issues of mobile portable computing devices and provide users with guidance to keep their information safe in mobile communication environments.

1 Introduction

Recent advances in hardware and software technologies have created a plethora of mobile devices with a wide range of communication, computing, and storage capabilities. As wireless communication has especially been advanced, demands of portable personal computing device has been increased and services of wireless internet like downloading of a variety of contents, mobile banking, and information searches have been commercialized rapidly

However, mobile portable computing devices which are important components in ubiquitous computing environments have been exposed in security threats such as denial service attack exploiting low information processing capability of low-powered CPU, malicious code attacks exploiting vulnerabilities of mobile platforms and application programs, and exposure of information by unauthorized users. Recently, incident cases caused by attack exploiting vulnerabilities of mobile portable computing devices have been occurred all over the world including Sweden, Finland and Japan. It has been expected that damages of such attacks is serious as Hacking Group like 29A has developed and announced worm and virus exploiting vulnerabilities of mobile portable computing devices, number of users of mobile portable computing devices and services provided by them have been increased .Therefore, study on security threats and their countermeasures in of mobile portable computing devices is needed. From now on, previous works for above subjects have been initial step and the results are insufficient [1-3].

In this paper, we analyze security threats of mobile portable computing devices and then we suggest their countermeasures in technical, manageable, and physical aspects. This paper consists of four sections. In section two, we analyze security threats of mobile portable computing devices, which include examining well-known incident cases and vulnerabilities of them. In section three, we present countermeasures against security threats, which are the results form experiences of incident handling for worm and virus in PC environment, analyzing security threats we found in the aspects of technology, management and physical. In Section 4, we end with a conclusion and suggest future works.

2 Security Threats

Security threats of mobile portable computing devices comprising confidentiality, integrity and availability are malicious code, vulnerabilities of mobile platform and its application, attack on communication path from wired network to wireless network, and data robbery & damages.

2.1 Malicious Code

Malicious codes include worm, virus and Trojan horses. As examples, there are 'Cabir' in June, 2004 and 'WinCE.Dust.A' in July, 2004. Various services such as web search, e-mail and File translation provided by mobile portable computing devices in wireless communication environment have brought high possibilities of malicious code incidents. Figure 1 shows major infection rout of malicious codes.

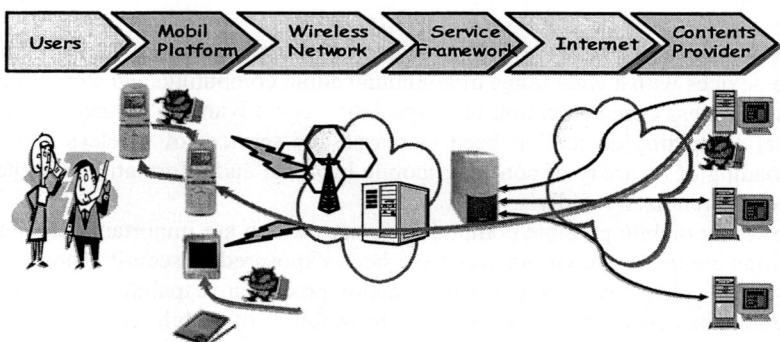

Fig. 1. Infection routs of malicious codes

Table 1 shows known malicious codes for mobile portable computing devices from now on.

Table 1. Known malicious code for mobile portable computing devices

Malicious code name	Date	Descriptions of characteristics and damages
Mosquito Trojan horses	2004. 7	·When illegal copy of Mosquito game was executed, great amounts of SMS was produced and distributed ·It was proven that Ojom, developing company of Mosquito game, implemented malicious codes in contents of its products and distributed.
WinCE.Dust A	2004. 7	·First Virus to infect files of Windows CE of Pocket PC · EXE files more than 4,096 byte were infected in directory execution of Virus.
Cabir	2004. 6	·Made by '29A', Hacker Group. ·It was executed in mobile phone supported by Symbian Operating System and it made 'Carbire' messages to screen. ·Propagation through Bluetooth communication with masquerading to security utility file, Caribe.sis
I-Mode Malicious code	2004. 6	·Police telephone number in Japan, 110 was connected automatically when SMS was checked
Phage	2004. 9	·A PDA virus infected by share of files. ·Programs of PDA was not operated when it was infected
Timofonica	2004. 6	·Transmission of great amount of SMS containing slander of specific communication Company to arbitrary Phone numbers.
SMS Malicious code	2004. 1	·Operations of Nokia cellular phone were stopped when specific SMS was received

Results of analyzing infection routs of malicious codes are as follows;

2.2.1 Infection by Applications

Malicious codes are included as a part of normal application and they can be included in platform itself and downloaded application. Malicious codes are activated and transmit a great amount of SMS, stop of other application or stop of operation of mobile portable computing devices when users execute application. Propagation speed of these kinds of malicious codes is relatively slow. As an example, there is 'Mosquito Trojan horses'.

2.2.2 Infection by Contents

Malicious contents through SMS or E-mail ca infect malicious code in mobile portable computing devices. In case of I-Mode malicious code, Police telephone number in Japan, 110 was connected automatically when received SMS was checked. Propagation speed of these kinds of malicious codes is very fast. As examples, there are 'Timofonica', 'SMS malicious code', and 'I-Mode malicious code'.

2.2.3 Infection by Communication Routs Between Mobile Portable Computing Devices

Malicious codes can be infected in case of sharing files using SYN cable, USB, Infrared communication or short length communication like Bluetooth. As examples, there are 'Phage', 'Cabir', and 'WinCE Dust.A'.

2.2 Vulnerabilities of Mobile Platform and Its Applications

Like normal PC environments, mobile portable computing devices can be attacked to exploit vulnerabilities such as Buffer overflow, Format string, Parsing error. Table 2 shows known vulnerabilities of mobile platform and its applications from now on.

Malicious code or Virus exploiting vulnerabilities of mobile platform and its applications is not found but it is high possibility to appear. It is expected that damages are high if worms exploiting such vulnerabilities appear, which act destructively like DoS attack and exposure of data. Especially, economic loss will be high in case of vulnerabilities of Smart phone which include sensitive data used in mobile banking, and mobile electronic commerce and personal information such as social identification.

2.3 Attacks on Communication Path from Wired Network to Wireless Network

Data can be eavesdropped and unauthorized users can access to mobile portable computing devices on wireless communication. As security mechanisms of IEEE 802.11 are not strong to prevent eavesdropping, it is easy for attackers to eavesdrop sensitive information by using SNIFFER tools. Attacker also can intrude network nodes between wireless and wired network, and eavesdrop sensitive data, exploiting vulnerabilities of wireless network which is applied to vulnerabilities and security threats of wired network as wired and wireless network have been integrated into single network. Therefore, worm and virus can be propagated widely through wired and wireless network. These cause to be denial of service, stop services themselves and bring economic damages in ubiquitous computing environments.

2.4 Data Robbery and Damages

Attached software to mobile portable computing devices which are connected to Personnel Computer or annexed software purchased separately can cause to exposure of data to attacker. Mobile portable computing devices normally provide logging functions to prevent these kinds of data robbery but these functions can't solve all of these problems.

Table 2. Known malicious code for mobile portable computing devices

Vulnerability name	Date	Descriptions
Anonymous Bluetooth access vulnerability	2004. 5	Users with Bluetooth can access to mobile phone without authentication when Bluetooth of some mobile phone is executed.
PalmOS MemoPad Memo Hiding Vulnerability	2003. 7	Bypass vulnerability exploiting edit applications is executed when security application installed in Palm OS is set to be low-level functions.
SIMENS mobile phone %IMG_NAME DoS Vulnerability	2003. 5	Denial of Service attack in Simens mobile phone is occurred when SMS message attached to modified images process.
Nikia SGSN SNMP Vulnerability	2003. 3	Vulnerability able to read SNMP options which have Any community with SNMP Deamon of DX 200 based Network elements in Nokia SGSN phone exists.
Nikia 6210 SMS DoS Vulnerability	2003. 2	Attacker can send malicious vCard to mobile phone used in exchanging address lists and supported by Nokia 6210.
SIMENCE Mobile phone SMS DoS Vulnerability	2002. 1	Mobile phone received by SMS messages including specific character can not show these messages and cannot delete them.
PalmOS TCP Scan remote DoS Vulnerability	2002. 1	Vulnerability that PDA is unstable exists when TCP connect() requests PDA installed on Palm OS 3.5
PalmOS Debugger Password bypass Vulnerability	2001. 3	Vulnerability that users able to access physically in PDA installed on Palm OS debugging mode bypass

3 Countermeasures

3.1 Technical Countermeasures

3.1.1 Installation and Operation of Anti-virus Products and Security Software

Antivirus products for mobile portable computing devices have been developed as worm and virus is emerged Security software to provide user authentication, access control or vulnerability scan. This is most easy and effective countermeasures for users to remove security threats. Update of antivirus products reflecting recent attack information is also important.

3.1.2 Applying Strong Encryption Algorithm and Authentication Methods

Sensitive data should be encrypted because access of mobile portable computing devices is easier than PC or Server. Capability of mobile portable computing devices and battery length are also considered in addition to its strength when encryption algorithm is used because discrepancy of processing overhead can be found for encryption algorithm. When Virtual Private Network is used, it provides confidentiality and integrity to support secure remote access.

Strong Authentication of mobile portable computing devices themselves such as CHAP (Challenge-Handshake Authentication Protocol), Mobile Access Number, authentication information management integrating central directories should be needed.

3.1.3 Enhancement of Security for Mobile Platform and Contents Server

Enhancement of security like security API for mobile platform and security checks of contents server is most important to prevent security threats. Applying recent security patches to mobile platform and contents server is also curtail. Filtering abnormal traffics on communication nodes should be needed by mobile service provider.

3.2 Manageable Countermeasures

3.2.1 Implementation of Security Policy and Periodic Training

Security policy for secure mobile communication such as monitoring and filtering policy for abnormal traffics and secure operation procedure for systems should be considered in aspect of mobile service provider. Security policy like sorts of storage contents allowed, network connection policy, prevention for use of extended hardware should be implemented in aspects of users. Update management is important to reflect modification of existed system or set-up of new system environment.

3.2.2 Periodic Data Back-Up

Data stored in mobile portable computing devices should be periodically backs up to data server located inside of Firewall or other mobile portable computing devices. It can minimize damages causing loss or destruction of sensitive data when incidents are occurred.

3.3 Physical Countermeasures

Firstly, mobile portable computing devices which are not used should be locked, keep them in the case to be not identical to unauthorized users and keep exterior memory devices separately removing from mobile portable computing devices. Secondly, Sensitive data should be encrypted if the data are stored in exterior memory devices like memory sticks and USB flash memory. Thirdly, ID should be removed or inactivated immediately if mobile portable computing devices are stolen or lost.

4 Conclusions and Future Works

In this paper, we analyze security threats of mobile portable computing devices as malicious code, vulnerabilities of mobile platform and its application, attack on communication path from wired network to wireless network, and data robbery & damages.

And then, we suggest their countermeasures in technical, manageable, and physical aspects. This paper contributes to initiate research on security issues of mobile portable computing devices and provide users with guidance to keep their information safe in ubiquitous computing environments.

Case study on analyzing chrematistics of worm and virus sample, and research on security enhanced mobile platform including security API are to future works.

References

1. Didi Barnes, " Portable Computing Device Security", September 2003.
2. Symentec, "Wireless Handheld and Smart phone Security", 2003.
3. Palm, "Handheld Security for the Mobile Enterprise:, September 2002.
4. Tom Karygiannis et al., NIST, "Wireless Network Security", September 2002.
5. R.Ramjee et al., "IP-based Access Network Infrastructure for Next-Generation Wireless Data Networks," IEEE Personal Communications, Aug, 2000.
6. F.Stajano et al., "The Resurrecting Duckling: Security Issues for Ubiquitous Computing", IEEE security and Privacy, 2002.
7. J.M Seigneur, S Farrell et al., "Secure ubiquitous computing based on entity recognition", Workshop on Security in ubiquitous computing, 4th International UBICOMP, 2002.

A Business Model (BM) Development Methodology in Ubiquitous Computing Environments

Choon Seong Leem, Nam Joo Jeon, Jong Hwa Choi, and Hyoun Gyu Shin

School of Computer and Industrial Engineering, Yonsei University,
134 Shinchon-dong Sudaemoon-gu, Seoul 120-749, South Korea
{leem, jeonnj, jhchoe, coolshg}@yonsei.ac.kr
http://ebiz.yonsei.ac.kr

Abstract. Even though the importance of a Business Model (BM) in ubiquitous computing environments has been growing, current research mainly focuses only on technology. Without a proper evaluation method for a BM, a promising BM has not been verified. In this research, we suggested a BM development methodology for ubiquitous computing, which contains the Business Model Analysis (BMA) framework for analyzing a BM and the Business Model Feasibility Analysis (BMFA) framework for evaluating a BM. The methodology is composed of phases, activities, and tasks in detail and depth, having systematic relations to each other. A prospective BM of ubiquitous computing environments could be extracted by this methodology.

1 Introduction

The development of Information Technology has been carried on through the technological revolution of information and communication. The means for the revolution are the changes in the computing environment, which consists of four stages [13]: The first step is a main frame computing revolution, and the second is a personal computing revolution. The third is a breakup computing revolution and the final is a ubiquitous computing revolution.

The ubiquitous computing revolution, which is considered as an Information Technology to fuse real space and cyber space, based on the networks among existing things in the real world, has been conducted over the whole of society. Even more, the future oriented technology has changed the axis of the paradigm of Information Technology.

Accordingly as the ubiquitous computing technology appears, the importance of a suitable Business Model (BM) has been increasing more and more in these new Information Technology environments. Nowadays, many studies about suggesting BMs of ubiquitous computing environments and the implementation of these have been tried by many researchers. However, most of these researchers focus only on the technological views of the development of BMs, and its procedure has a short logical process and systematic design. Besides, it is not easy to measure the BMs' value because a suitable estimation method for a BM does not exist.

From this point, this research will present a methodology, which can be used for developing a BM of ubiquitous environments, according to a systematic and logical approach in a BM field.

This research is composed of five chapters: Chapter two discusses the concept of ubiquitous computing and a BM, Chapter three discusses a BM development methodology which reflects technological and environmental characteristics of ubiquitous computing, Chapter four is a case which is applied to the methodology, and Chapter five discusses the conclusions and directions for further research.

2 Previous Research

2.1 Ubiquitous Computing Definitions and Characteristics

Even though there are many common factors about the definition of ubiquitous computing, these vary and are slightly different according to the scholar, time and organization. Table1 arranges the definitions of ubiquitous computing [4], [11], [12], [19].

Table 1. Concept comparison of ubiquitous computing by scholars and research institutes

Scholars and research institutes	Definition
Friedemann Mattern (2001)	Tomorrow everyday objects will become smart and they will all be interconnected.
K. Sakamura (1987)	Ubiquitous computing is making us to be able to use computers anywhere and anytime.
Mark Weiser (1993)	Ubiquitous computing has as its goal the enhancing computer use by making many computers available throughout the physical environment, but making them effectively invisible to the user.
IBM (2004)	Pervasive computing delivers mobile access to business information without limits from any device, over any network, using any style of interaction. It gives people control over the time and the place, on demand.

With many changes of the definition of ubiquitous computing according to technological progress, we have refined the definition of ubiquitous computing in this research. We defined that ubiquitous computing is a technology, in which invisible computers are embedded and connected with all things so that anyone can communicate, exchange and share information anywhere anytime.

Based on these characteristics of ubiquitous computing, the United States of America, Europe, Japan and Korea recently have chosen their own concepts of ubiquitous computing, and have been trying to benefit from the highly focused R&D. Table 2 compares the concepts of ubiquitous computing of each country [18].

To sum up these various concepts of a BM, a BM is to identify diverse components such as the products and services, business strategies and processes, and stakeholders of a BM, to express the value created between the players by combining the components, and to set up long-term business strategy for an operating company.

Table 2. Ubiquitous computing concept comparison of U.S, Europe, Japan, Korea

Country	U.S	Europe	Japan	Korea
Concept	Ubiquitous computing, Pervasive computing	Disappearing computing, Ambient computing	Ubiquitous network	Ubiquitous Appliance
Value	Service by smart devices	Intelligent cooperation by information artifacts	Anywhere connection by small chip, smart card, context roaming	Single function appliance using short range wireless Interface
Research field	Computer devices	Every objects	Network	Appliance
Core Technology	Short-distance radio communication, Sensor, MEMS, Small size object chip			

Even the researches of each country for ubiquitous computing have slightly different strategies, they pursue the three common characters: Smart, Seamless networks, and Mobility.

2.2 Business Model (BM) Concept

The concepts for a BM are different according to the researcher, transaction channel and so on. Table 3 shows representative concepts of a BM [1], [2], [15], [17].

Table 3. Business Model (BM) definition by scholars

Scholars	Definition
Timmers (1998)	An architecture for the product, service and information flows, including a description of the various business actors and their roles; and A description of the potential benefits for the various business actors; and A description of the sources of revenues.
Rappar (1999)	A business model is the method of doing business by which a company can sustain itself - that is, generate revenue.
Leem (2000)	Establishment and management strategy of enterprise for revenue model, business operation and cooperation for providing customer services.
Amit & Zott (2001)	The business model depicts the design of transaction content, structure, and governance so as to create value through the exploitation of business opportunities.

To sum up these various concepts of a BM, a BM is to identify diverse components such as the products and services, business strategies and processes, and stakeholders of a BM, to express the value created between the players by combining the components, and to set up long-term business strategy for an operating company.

Having dealt with the development of IT, most researchers have been concerned about classifying a BM or analyzing success factors for a BM. Despite these researches to define structural forms of a BM and establish a suitable strategy for a BM, they do not present a specific method to develop a BM.

The demanded for research for developing a BM are the analysis methodology for a BM based on the BM's component, the design methodology to define the flow of goods, services, and values etc., and the measurement methodology for the value of a BM. Table 4 shows research of existing development methods concerned with a BM [1], [5], [6], [7], [9], [10], [14], [16].

Table 4. Business model analysis, modeling and evaluation method by scholars

Category	Scholar	Research model
Analysis method	Gordjin (2000)	e3-valueTM ontology
	Barnes (2002)	The m-commerce value chain
	An, et al (2003)	Business model analysis framework
Design method	Gulla & Brasethvik (2000)	Business model process design
	Kang, et al (2002)	Multi-layer design framework
Evaluation method	Reilly (1996)	The feasibility analysis
	Amit & Zott (2001)	Value creation model
	Kim, et al (2003)	Value creation indices

2.3 Limitations of Previous Research

While the present ubiquitous computing related research is mainly technology oriented, recent research tends to show that the interest of developing a BM is increasing gradually [3]. However, there is still not enough research to suggest the necessary processes and specific contents for a BM development, beside some limits in applying the prior research related to developing for a BM in a ubiquitous computing field are exist. They are as follows:

- It is difficult to understand the factors such as analysis, design, and evaluation for a BM configurationally because of the separation of the related research for each area.
- They do not present a synthetic process and roles for BM development.
- When they come to the evaluation of a BM, they do not deal with all viewpoints but mainly with the viewpoint of cost and profit.
- Because most of them are the studies based on the limited channel circumstances such as e-business and m-business, they need to be modified and corrected so as to adjust for a ubiquitous computing field.

To develop a suitable BM in ubiquitous computing environments, a logical and systematic development method reflecting the characteristics of ubiquitous computing is needed. In this research we defined a suitable standard BM development methodology, which is based on ubiquitous computing environments, has 4 phases. They are planning, design, implementation, and management. Each area has its own specific tasks.

3 A Business Model (BM) Development Methodology in Ubiquitous Computing Environments

A BM development methodology in ubiquitous computing environments consists of 4 phases which are a BM planning, design, implementation, and management, and each phase has 14 activities composed of 26 detailed tasks. Each step of the methodology is related as are the results from each step. Fig. 1 shows the whole structure of the methodology.

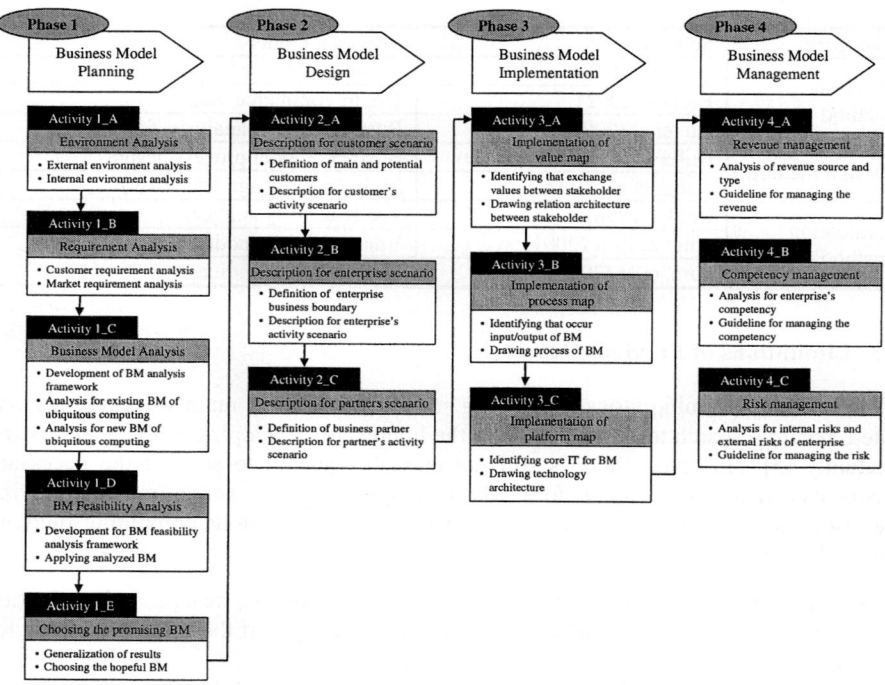

Fig. 1. A BM development methodology in ubiquitous computing environments

3.1 Business Model Planning (Phase 1)

The first step of the ubiquitous business model development methodology is BM planning (Phase 1). This step defines the necessary activities and detailed tasks to draw out a promising BM in ubiquitous business environments.

Analyzing the forms of a BM is conducted through environment analysis and requirement analysis in the BM planning phase. Subsequently, a series of activities and tasks for promising BMs are defined, which includes the development of the business feasibility analysis framework, applying and extracting a promising BM.

3.2 Business Model Design (Phase 2)

The BM design phase consists of activities and tasks so as to select a specific scenario, based on the events which might occur in the BM, with the perspective of major players such as operator, customer, and partner. This stage suggests a scenario as an operating plan for the promising BM which was selected in the earlier step, then uses it as a designing plan.

Table 5. Definition for detailed activities and tasks of BM planning phase

Phase	Activity	Description and detailed tasks
Phase 1 BM planning	Activity 1_A Environment analysis	Environment analysis for a company : competitive circumstances analysis, value chain analysis, and etc. for external and SWOT analysis, resource analysis, and etc. for internal
	Activity 1_B Requirement analysis	Requirement analysis about standardization, policy supporting and etc. for market and analysis about purpose of use, satisfaction, and etc. for customer, which are influenced by emerging ubiquitous computing environments
	Activity 1_C Business Model Analysis (BMA)	Development of Business Model Analysis (BMA) framework to analyze the BM of ubiquitous computing environments, and applying existing and new BMs of ubiquitous computing to the framework
	Activity 1_D Business Model Feasibility Analysis (BMFA)	Development for Business Model Feasibility Analysis (BMFA) framework based on ubiquitous computing environments, and applying the analyzed BMs to it
	Activity 1_E Choosing the promising BM	Summarization of results from BMFA, and choosing a promising BM in ubiquitous computing environments

Table 6. Definition for detailed activities and tasks of BM design phase

Phase	Activity	Description and detailed tasks
Phase 2 BM design	Activity 2_A Description for customer scenario	Definition of main and potential customers for the BM and description of the customers' activity scenario
	Activity 2_B Description for enterprise scenario	Definition of enterprise business boundary, and description of the enterprise's activity scenario including delivery ways of goods and service, and relationship of the business partners
	Activity 2_C Description for partner scenario	Definition of the business partner to be need for the BM operation, and description of the partners' roles and activity scenario

3.3 Business Model Implementation (Phase 3)

Based on the scenarios of customers, operators, and partners, the BM implementation phase consists of activities and tasks which describe factors needed for implementation of the BM and the relations of them. This phase classifies the factors such as value, process, platform layers, and suggests maps to structure the key values of each layer as architecture for commercialization of the BM.

Table 7. Definition for detailed activities and tasks of BM implementation phase

Phase	Activity	Description and detailed tasks
Phase 3 BM implementation	Activity 3_A Implementation of value map	Identifying the exchanged values between stakeholders, and drawing the relations between them
	Activity 3_B Implementation of process map	Identifying the flow of the inputs/outputs of BM from the start to end, and drawing the flow as a map
	Activity 3_C Implementation of platform map	Drawing technological architecture for implementing the BM

Table 8. Definition for detailed activities and tasks of BM management phase

Phase	Activity	Description and detailed tasks
Phase 4 BM management	Activity 4_A Revenue management	Analysis on resources and types of revenue of the BM, and proposal for revenue management
	Activity 4_B Competency management	Proposal for competency management to be needed for commercialization of the BM
	Activity 4_C Risk management	Proposal for risk management to overcome the risk factors which might be existing among the commercialization processes

3.4 Business Model Management (Phase 4)

The BM management phase consists of activities and tasks to propose guidelines for revenue, competency, and risk management. This step suggests management guides for commercialization of the BM, according to the results from the analysis of each BM carried out in the earlier planning phase.

4 Business Model Feasibility Analysis (BMFA) Framework

The ubiquitous computing BM development methodology suggested in this research includes a framework of Business Model Feasibility Analysis (BMFA). Not only does the framework play a crucial role in the planning phase but also the results of the analysis are significant factors as inputs for the activities of design, implementation and management phases. Fig. 2 illustrates the overall structure of the ubiquitous BMFA framework.

The ubiquitous BMFA framework is a filtering system for suitable BMs of ubiquitous computing environments, and is composed of pre BMFA framework and post BMFA framework; The pre BMFA framework, which consists of technical characteristics, requirements for uses, and enterprise strategies, is a system to evaluate the feasibility of execution of BMs, while the post BMFA framework being composed of competencies of enterprise, return on investment, and risk factors is a system to measure the feasibility of achievement of the BMs.

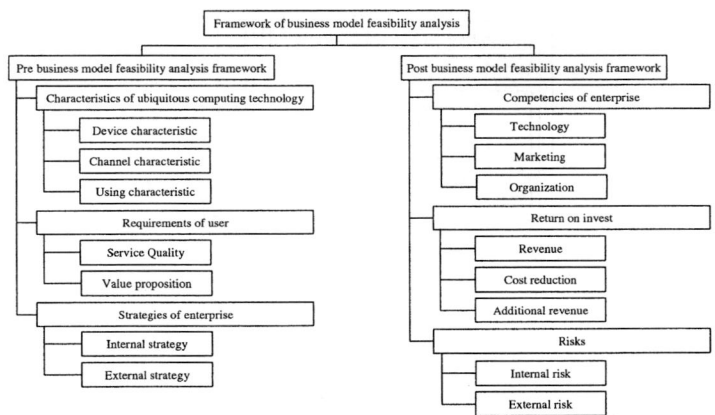

Fig. 2. The structure of Business Model Feasibility Analysis (BMFA) framework

The procedures of the BMFA are as follows: Extracting the highly executable BMs through the result of the pre BMFA, evaluating the extracted BMs based upon the post BMFA framework, and selecting a promising BM of ubiquitous computing environments according to the results of the evaluation. In addition, the components of ubiquitous BMFA framework are systematically related to each other so that the outputs of former activities are used as inputs for following activities.

The BMFA framework is much more meaningful in terms of the fact that various BMs could be evaluated objectively. Closely looking at the relations of the activities, the outputs of the environment analysis (activity 1_A) and requirement analysis (activity 1_B) are used as inputs for establishing the detailed evaluating indices of the pre/post BMFA framework. Likewise, the output of BMA (activity 1_C) is used for the detailed evaluation indices of the BMFA and used as basis inputs for performing the activity 2_A, activity 2_B, and activity 2_C of phase 2 as well. The output of BMFA (activity 1_D) is also used as inputs of implementation for platform map (activity 3_C) and phase 4 which comprise of activity 4_A, activity 4_B, and activity 4_C. In addition, the overall outputs of phase 2 serve as inputs for implementing the process map (activity 2_A) and value map (activity 2_B). Fig. 3 shows the relations between activities consisting of the ubiquitous BM development explained earlier.

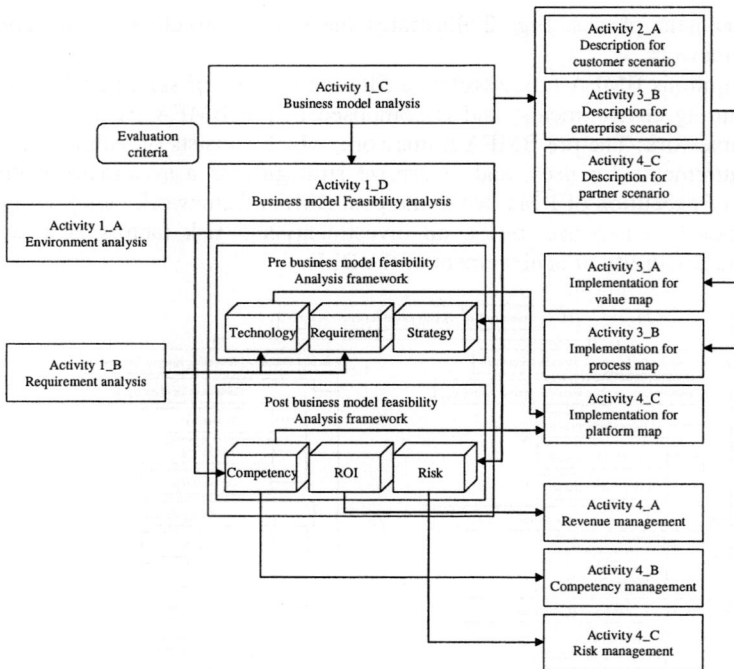

Fig. 3. Relation between BM feasibility and other activities

5 Conclusions and Future Research

Recently there has been research about ubiquitous computing and efforts for applying that technology to various fields such as the government, education, culture, industry, etc. more than ever. It is forecasted that these efforts will gradually become reality and make improvements in the effectiveness for individuals and society. However, the actual situation is that no promising BM has been suggested despite of the many applicable areas of ubiquitous computing.

In this research, we have suggested a suitable development methodology for ubiquitous computing environments, which has logical and systematic procedures, and defined the phases, activities, and tasks of the methodology. In addition, we have addressed that each activity of the methodology is not separated independently but has systematic relations to each other. The analysis framework for evaluating a BM and the feasibility analysis framework are also a vital result of our study.

However, there are certain limits that the methodology is not verified practically and does not include the case of developing an actual ubiquitous BM. Nevertheless, the limits could be removed by further researches for applying the methodology to various applicable areas along with establishing a repository.

References

1. Amit, R, Zott, C.: Value Creation in E-Business. Strategic Management Journal, vol. 22 (2001) 493-520
2. Choon Seong Leem: e-Business file. Yonjin.com (2000)
3. Fano, A., Gershman, A.: The Future of Business Services in the Age of Ubiquitous Computing. Communications of the ACM, vol.45, no.12 (2002) 83-87
4. Friedemann Mattern: The Vision and Technical Foundations of Ubiquitous Computing. UPGRADE, vol. II, no. 5 (2001) 3-6
5. Gordijn, J., Akkermans, H., Van Vliet, H.: What's in an electronic business. Lecture Notes in Computer Science (2000) 257-273
6. Gulla, A., Brasethvik, T.: On the challenges of business modeling in large-scale reengineering projects, 4th International Conference on Requirements engineering, USA (2000) 17-26
7. Hyo Keun Kim, Yung Ah Lee, Min Seon Kim: A Study on Value Creation Evaluation Indices of Internet business model. Spring Semiannual Conferences of The Korea Society of Management Information Systems, vol. 2003, no. 0 (2003) 449-457
8. Hyoun Gyu Shin, Choon Seong Leem, Hyung Sik Seo: A Taxonomy of Ubiquitous Computing Applications. Fall Semiannual Conferences of The Korea Society of Management Information Systems, vol. 2003, no. 0 (2003) 243-249
9. In Tae Kang, Yong Ho Lee, Jong Seo Yang, Yong Tae Park: On the framework to design Multi-layer business model. Spring Semiannual Conferences of Korea Institute of Industrial Engineers/Korea Operations Research and Management Science Society, vol. 0, no. 0. (2002) 70-73
10. Ji Hang An, Sang Hun Choe, Seog Gwon Jang, Yong Ho Kim: Development and Application of Business Model Analysis Framework. Information Systems Review, Vol.5, no.1 (2003) 19-14
11. K. Sakamura: The TRON Project. IEEE Micro, vol. 7, no. 2 (1987) 8-14
12. Mark Weiser: Ubiquitous Computing. IEEE Computer (1993)
13. Mark Weiser, John Seely Brown: The Coming Age of Calm Technology. PowerGrid Journal, vol.1.01 (1996)
14. Michael D. Reilly, Norma L. Millkin: Starting a Small Business: The Feasibility Analysis. MONTGUIDE (1996)
15. Rappa, M.: Business model on the Web. http://digitalenterprise.org/models/models.html. (1999)
16. Stuart J. Barnes: The mobile commerce value chain: analysis and future developments. International Journal of Information Management, vol. 22 (2002) 91-108.
17. Timmers, P.: Business models for Electronic Markets. Electronic Market, vol. 8, no. 2 (1998)
18. Wan Seok Kim, Jeong Kook Kim, Hyo Kee Kim, Chang Seok Kim, Heung Seo Koo, Sang Beom Lee, Tae Woong Park, Seong Kook Kim: The Technology, Infrastructure and Trend of Ubiquitous Computing. Korea Information Processing Society Review, vol. 10, no. 4 (2003)
19. http://www-306.ibm.com/software/pervasive/module/index.shtml (2004)

Developing Business Models in Ubiquitous Era: Exploring Contradictions in Demand and Supply Perspectives

Jungwoo Lee[1] and Sunghwan Lee[2]

[1] Associate Professor of Information Systems,
Graduate School of Information, Yonsei University,
Sudaemun Gu, Shinchon Dong 134, Seoul, Korea
[2] Graduate School of Information, Yonsei University,
Sudaemun Gu, Shinchon Dong 134, Seoul, Korea
{jlee, sunghwan}@yonsei.ac.kr,

Abstract. This study reviews and analyzes contradictions embedded in two different perspectives looking at ubiquitous computing: demand-side and supply-side. It is argued here that these differences in perspective may contain contractions in terms of assumptions which may deter developers from properly conceiving future applications and services in ubiquitous computing. Five distinct aspects – anybody, anytime, anywhere, any service, and any device – were used as an analysis framework against 'for me,' 'right now,' 'right here,' 'what I need,' and 'what I have.' Underlying factors that makes differences are suggested and discussions are made. Implications are discussed and further research is suggested.

1 Introduction

Since Mark Weiser termed 'ubiquitous' as a new paradigm for computing in 1988, efforts have been concerted to develop and advance information technologies towards 'connecting, invisible calm and silent, and real' ubiquitous computing [1]. Despite huge investment and good progresses on the technology side following dream applications, practitioners are still agonizing over what are appropriate business models which may provide incredible value as a new ubiquitous business or as converged with traditional businesses.

Most traditional business models focuses on supply-side view in which business processes are controlled by suppliers mostly based on predictions derived from preceding research outcomes. Researches are conducted much earlier than the time of demand or consumption, and resulting products or services are pushed onto the market, wishing these to be sold. In ubiquitous environment, this supply side view would not prevail any more. Paradigm is shifting from supplier-centric to customer-centric, from cost-reduction to value creation, mostly because information can be gathered and disseminated at the speed of light.

In other words, perspective on ubiquitous computing taken by supply-side are focused on such issues as anybody, anytime, anywhere, any service, and any device [2]. Emphasis is on to reach anybody, on the network, any time, anywhere, through any device for any service. However, when you take the demand-side view, these requirements needs to be translated as 'to me,' 'right now,' 'right here,' 'whatever I need,' and 'whatever I have' (Refer to figure 1). These differences in perspective may play an important role in determining requirements for ubiquitous computing business model. For example, 'anybody' requirement can be translated into service universally standardized to different users from the supply-side view, but from the demand-side, it would better be translated as personally customizable service that can be accessed universally. In other words, the contradictions embedded in assumptions of these different perspectives may prevent developers of ubiquitous computing from directing their attention to appropriate 'newly emerging' business models. Developers of ubiquitous computing and application need to reconcile differences in views in order to develop successful technological artifacts which fit with the business model in mind.

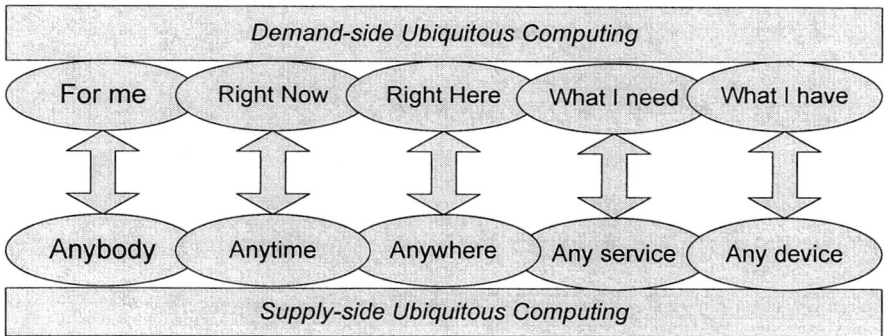

Fig. 1. Different Views on Ubiquitous Computing

It is suggested here that these contradictions are essential problems to be analyzed and solved for successful business model development for ubiquitous computing. This paper focuses on revealing the contradictions embedded in these differences of perspectives and related assumptions, and on deriving implications for business models.

2 Analysis

In this section, arguments are developed and presented for contradictions in these five aspects in detail with anecdotal evidence and ethnographic observations.

3 Human Dimension: Anybody Versus 'for me'

Ubiquitous concept proposes anybody can be connected to the network anytime anywhere. The human dimension of 'anybody' seems to connote collective treatment of public service. For example, mobile environments present many challenges for ubiquitous computing as it differs from fixed indoor contexts such as offices or houses in many important ways.

For example, when a person walks on the street, a device on the wall, such as an advertising board, identifies him/her and presents a notification that he has an emergency email on the device. It is a very common picture the ubiquitous computing draws as a future. However, this kind of application presents a challenge to developers as it contains contradictory requirements. Any body can walk down the street, but the application should be able to identify a specific person and provide a personal service among the crowd. Collective treatment has a characteristic of current broadcasting station where the same contents are provided to multiple audiences at the same time. Audiences can only choose to accept it or not. Compared to this broadcasting model, most Internet applications are very personal. People may not want to read their emails on an advertising board, unless the application converts the public device into very personal one. Internalized factors such as tasks and goals are different and externalized factors such s social resources are dynamic and unpredictable. This kind of contradictions presents a challenge as well as a problem.

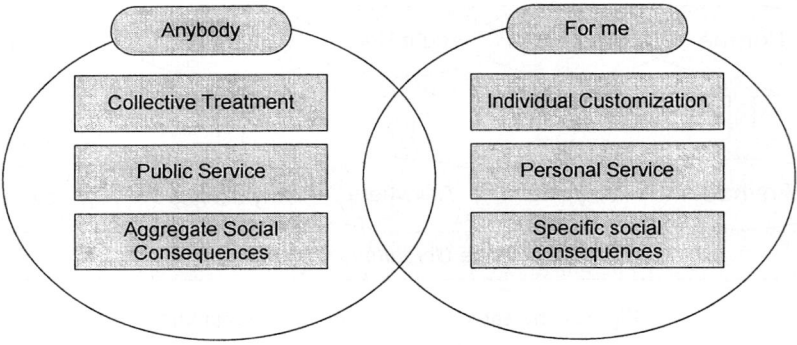

4 Temporal Dimension: Anytime Versus 'right now'

Temporal dimension focuses on time. Current computing applications assume constant and permanent services to users and users expect the application to be available any time. When it comes to ubiquitous computing where mobility and flexibility is the focal point, time plays a critical role and defines the context. As human activities governed by time, applications need to realize the context in which a human exist and ideally help him/her to proceed in time. It has been argued that mobile devices free people from the limitations of time and place. However, it is not yet clear how the temporal tension that we have in the normal daily lives be resolved as much as science fiction depicts[3].

Anytime perspective connotes constant services with historical data while 'right now' perspective requires the services to be reconfigurable and modifiable depending upon context in time. Ubiquitous service needs to be not only constantly available but also adaptable to contextual changes in time. Also, data contained should be updated in a timely fashion and to be time-stamped so that these can be referenced whenever needed. TimeSpace research focuses on integrating these perspectives [4].

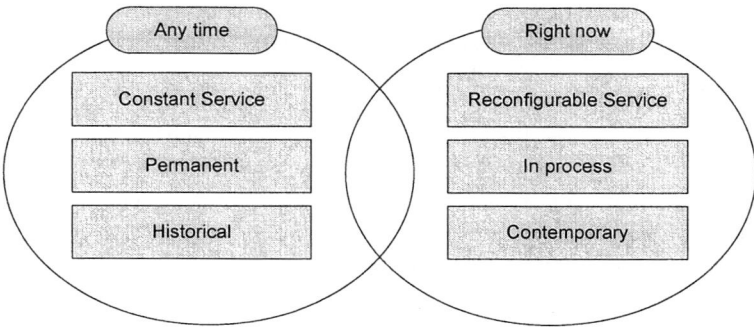

5 Locale Dimension: Anywhere Versus 'right here'

Locale dimension relates to location. Anywhere perspective connotes mobility, portability and reliability across different locations. However, mobility and portability presents a challenge for context awareness from the users' side. People gather, process, and share information in many ways, in and across many places. It is difficult to create appropriate ubiquitous computational support for this constant processing of information. Moreover, despite the mobility, users demand coherence throughout different places relating to different people, colleagues and family members.

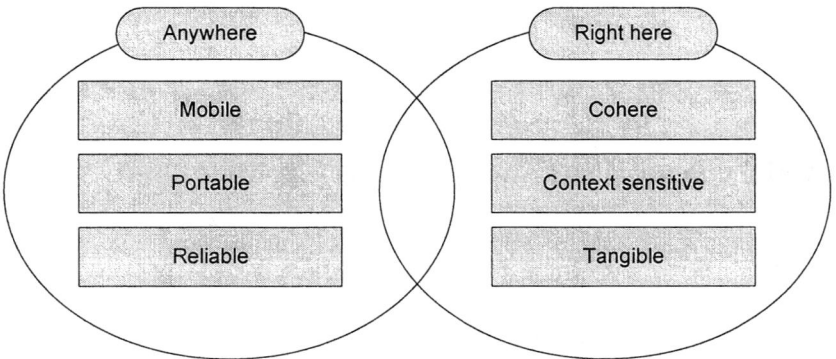

For this purpose, a range of information appliances, services and media need to be able to work together in support of increasingly mobile users. Mobile coherence should include collaborative working space as well as personal space. Also, portability may contradict context-sensitiveness. Portability should not limit itself into limited portability in which the same interface ports to different situation without considering context. Last but not the least; 'right here' perspective implies tangibility on the spot. Development of tangible user interface focuses on solving these contradictions between tangibility and reliability [5].

6 Service Dimension: Any Service Versus 'what I need'

Any service perspective connotes variety, consistency and dedication of applications and services. Services are diverse in ubiquitous environment so that users may have freedom in interacting with ubiquitous artifacts and these services are consistent in the sense that users may identify and perceive without confusion wherever they are and whenever the time of usage.

However, the variety needs to be reconciled with context dependency requirements. Context dependent users would demand various services, applications and devices to be combined and provided as necessary in a flexible and time-shared manner. Services needs to be integrated and support context-dependency, flexibility and time-share while maintaining variety and consistency with perceived dedication in ubiquitous environment.

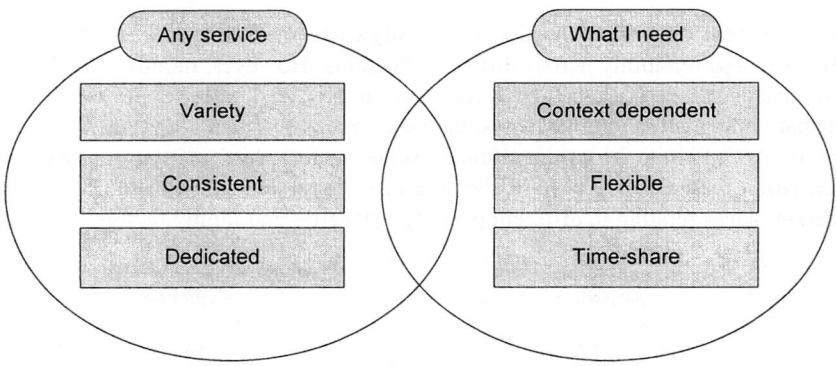

7 Artifact Dimension: Any Device Versus 'what I have'

One of the leading proposal applications of ubiquitous computing is based on universal appliance interactions. Let users use arbitrary devices to interact with arbitrary services and appliances. Home networking suggests to use cell phones to turn on the lights, extinguishing gas range, fill up bathtubs with warm waters at the right time before going home, etc. Most of services described are personal services supporting personal activities. However, appliances described in these cases are mostly public. Using public device for personal service requires modifications and adjustments. It is not just one time modifications but constant modifications adjusting for context and users. The device should be able to predict the task that the user wants to engage in, and produce appropriate interface for each context.

The flip side of this coin is whether each individual should carry multiple devices for different types of interactions or a single device can handle different interfaces that can handle multiple services. Historically, multi-purpose devices didn't get much attention from users. Watches with data processing capability turned out to be a fad that fascinates a small number of early adopters and multi-purpose remote control also seems to run down the similar path. However, in ubiquitous environment, it seems that multi-purpose device is necessary, but may not be sufficient. Developers of ubiquitous computing needs to take these contradictions into consideration in designing ubiquitous devices and services [6].

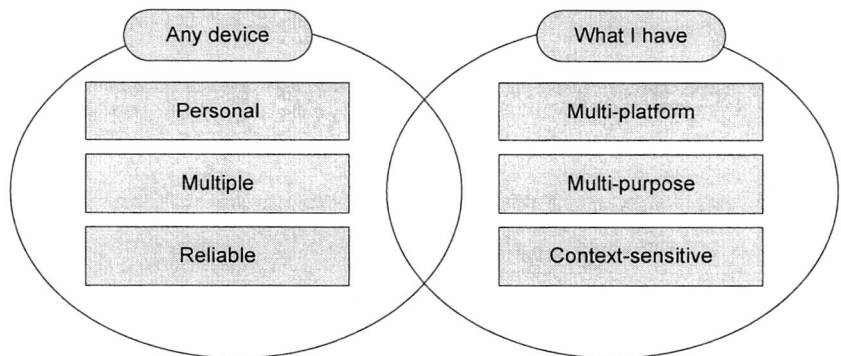

8 Discussion

This study is still in progress. Contradiction underlying the assumptions and belief may deter killer applications to be developed for future ubiquitous computing. It seems important to articulate these contradictions and reconcile them as ubiquitous computing advances. Contradiction dimensions needs further development with supporting theory as well as practical examples. Case studies and scenario analyses would be the next step in revealing implications of these contradictions and interdisciplinary theoretical development should accompany phenomenological observations and experiments.

References

[1] Mark Weiser. 'Hot topic: Ubiquitous Computing', *IEEE Computer*, pages 71 ~ 72, October 1993.
[2] Weiser & Brown, 'Designing Calm Technology', *PowerGrid Journal*, v 1.01, July 1996.
[3] Sakari Tamminen, Antti Oulasvirta, Kalle Toiskallio and Anu Kankainen, 'Understanding Mobile Context,' *Personal Ubiquitous Computing*, 8, pages 135-143, 2004

[4] Aparna Krishnan and Steve Jones, 'TimeSpace: activity-based temporal visualization of personal information spaces,' *Personal Ubiquitous Computing*, 9, pages 45-65, 2005
[5] Kenneth P. Fishkin, 'A taxonomy for and analysis of tangible interfaces,' *Personal Ubiquitous Computing*, 8, pages 347-358, 2004
[6] Andrew J. May, Tracy Ross, Steven H. Bayer and Mikko J. Tarkiainen, 'Pedestrian navigation aids: information requirements and design implications,' *Personal Ubiquitous Computing*, 7, pages 331-338, 2003

Semantic Web Based Intelligent Product and Service Search Framework for Location-Based Services

Wooju Kim[1], SungKyu Lee[2], and DeaWoo Choi[3]

[1,2] Department of Computer and Industrial Systems Engineering College of Engineering,
Yonsei University, Seoul, South Korea
{wkim, oop760}@yonsei.ac.kr
[3] e-Business Research Center, Yonsei University, Seoul, South Korea
qorwkr@nate.com

Abstract. As the potential market of L-commerce grows quickly, the necessity of location aware product and service search and recommendation methodology is also growing rapidly. However, existing related researches still suffer from problems such as rigidity of location aware service platform since their approaches are strongly dependent on their own proprietary ontologies, which make them closed systems to the world. To resolve this problem, we propose a semantic web based intelligent product and service search framework for L-commerce. The proposed framework makes a more flexible and effective product and service search available by providing a real-time ontology mapping mechanism between heterogeneous ontologies and taxonomies for products and services. As result, this makes accessible product and service information more abundant and at the same time, improves our search mechanism's precision while achieving user satisfaction of the search service. Our approach recommends the most relevant products or services by adopting multi-attribute based location aware search and evaluation methodologies in mobile environment. Our proposed framework is expected to contribute to realization of L-commerce by maximizing user satisfaction of the search for products and services in the future.

1 Introduction

We are now entering a ubiquitous era [1] based on mobile and location-based technology, where customers can search and utilize various products and services more than at any time and place. One of the state-of-the art technologies leading this ubiquitous era is the "Location-Based Services (LBS) [2]." LBS consists of services closely connected with real life such as navigation services, logistics shipping information services, and telematics. The so-called L-commerce based on location information is a combination of LBS and e-commerce, and it is becoming more significant. Therefore, even though the related researches and developments still remain in the initial stage, this area has a high potential power because the advantages of mobile can be applied to full capacity.

Before achieving these location-based services, a few problems of product and service searches in mobile environment should be resolved. From the viewpoint of information collection, existing location-based search engines end up neglecting varieties of information types because they collect information by hand or information providers input their information to the fixed format of search agents by themselves. Also, real time information problems can occur because existing search engines cannot react to the changes of information sensitively. From the viewpoint of information search, users cannot represent their real search intent and appropriate preference on the evaluation of the retrieved products and services because most existing location-based search engines [3,4,5] adopt only keyword based search methodologies or limited property based search methodologies. Lastly, search agents cannot provide semantic interoperability between service providers and users.

To cope with the problems raised above, we propose a semantic web based service search framework for L-commerce that utilizes web service and semantic web technologies. The proposed framework makes it possible to represent various search intents by devising the representation methodology of products and services search including users' location information, and to search relevant products and services from various information providers who use different ontologies by generating queries appropriate to each ontology. In addition, we develop a recommendation mechanism based on relevancy evaluation methodology based multiple attributes (e.g. users' location, properties of each service).

In the next section, we explain related works. In section 3, we describe our proposed framework, and in section 4, we briefly introduce the required methodologies to implement the proposed framework. In the final section, we present our conclusions.

2 Related Works

Recently, LBS has become popular with the development of mobile related technologies. LBS must often consider various context information such as location information, time related events, and schedules into its core mechanism, and such context-awareness capability is being emphasized more and more. Researches on subjects such as Cyberguide[6], Tourist guide[7], exhibition guidance[8], Shopper's eye[9], and Context-aware Office Assistant[10] that use various context information are already being done actively. However, because they have focused only on the navigation problems for users' preferences and used only simple location-based information, the researches still have many limitations. They especially cannot easily deal with contemporaneousness and varieties of information search because they assume that all the information including the store and product information to be processed is stored in the mobile device.

There are now some different types of researches such as Impulse[11] and Wherehoo server[3] that provide location-based information search service which enables users to search products or services based on users' current location. However, they still do not address the issues of not only the syntactic but also semantic interoperability among service requestors, service providers, and informediaries. Without an appropriate solution to such interoperability problem, it is almost impossible to access rich enough location related information, and so it is also unlikely that a practical and commercial L-commerce business model will be realized.

So far, we have discussed the related works and their pros and cons. Now, we propose a framework to cope with the problems mentioned in section 1 and to achieve our goals.

3 Intelligent Product and Service Search Framework

In this section, we introduce a semantic web based intelligent product and framework for L-commerce as shown in Figure 1, and explain the core components in the framework and external entities. Two main technologies adopted in the framework are web service and semantic web. Web service allows syntactic interoperability between service client and search agent, while semantic web admits semantic interoperability among all participating entities. Our framework is designed to fully utilize both technologies to perform a more accurate and intelligent product and service search than existing methodologies.

Since semantic web is not popular enough at the commercial level, we assume that service providers describe their products and services with metadata documents such as RDF and then publish them on the Web. Under this assumption, the proposed framework consists of three parts: service client, service providers, and intelligent product and service search agent. Let us discuss each of them one by one.

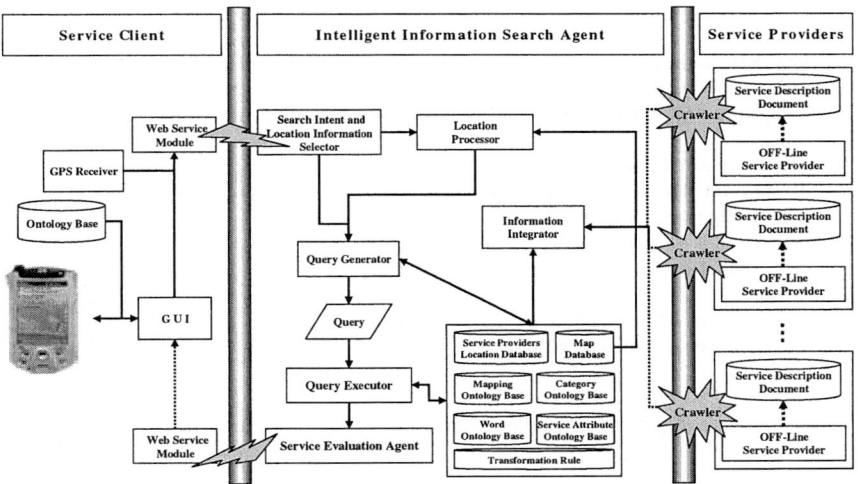

Fig. 1. Intelligent Product and Service Information Search Framework

3.1 Service Client

First external entity is service client, which is a module for mobile devices and allows users to describe their search intention. Its primary task is to provide a means for users to represent their product and service search intention with the provided search intention representation scheme, which will be discussed later in the paper. The users' current location information, captured by embedded GPS, is of course also automatically considered in the resulting representation.

The second task of service client is to communicate the obtained search request from users and corresponding results with the intelligent product service and search agent via web service technology.

The service client consists of four components: GUI, user specific ontology base, location information module, and web service module. The GUI allows users to describe their product and service search intention in a graphical form and shows the resulting list of the extracted services from the intelligent information search agent. The ontology base stores users' specific categories and attribute related ontologies. The location information module extracts current location information (e.g. latitude, longitude, speed) of users through a GPS receiver. Finally, the web service module is in charge of data communication between the intelligent information search agent and service client.

3.2 Service Providers

As a second external entity, service providers first describe their products and services using their own ontologies. Furthermore, they are required to publish the written metadata documents on the web. For example, a "Happy Restaurant" which serves Seafood and Steak cuisines makes a service description document by using the ontology available in restaurant.com. An example service description document is shown in Fig. 2.

```
<rdf:RDF xmlns:rdf="http://www.w3.org/1999/02/22-rdf-syntax-ns#"
xmlns:rdfs="http://www.w3.org/2000/01/rdf-schema#"
xmlns:owl="http://www.w3.org/2002/07/owl#" xmlns="http://www.restaurant.com#" >
    ...
<Restaurants rdf:ID="Happy_ Restaurant">
<restaurant_name>Happy Trails Restaurant</ restaurant_name >
<cuisine_type>Seafood</cuisine_type>
<cuisine_type>Steak House</cuisine_type>
<address>32498 US 431 Roanoke, AL xxxx</address>
<city>Roanoke</city>
<special_needs>Non Smoking Area</special_needs>
  <special_needs>Children's Menu</special_needs>
    ...
```

Fig. 2. An Example of Service Description Document

3.3 Intelligent Product and Service Search Agent

As a main part of the proposed framework, the intelligent product and service search agent first harvests service providers' information using a semantic web robot. Then, it integrates the collected information written in various ontologies into a unified product and service ontology and instance repository by rewriting it in its ontology. Based on this established integrated product and service repository, the search agent receives service client's search requests, retrieves relevant service providers, evaluates retrieved results, and recommends the most relevant services to users' requests.

To perform the tasks mentioned above, the intelligent product and service search agent uses of four major components: information integrator, location processor, query generator, and product and service evaluation agent.

First, the information integrator collects the category and attribute information about products and services from various service providers on the Web using a semantic web robot and then transforms it into a unified repository written in its ontology. Once a user search request arrives from a service client module, the location processor decides the range of regions to be searched by capturing the user's location included in the request message.

Then, the query generator generates a query to product and service ontology instance base by interpreting the user's request written in his private ontology and also considering the regional constraints generated by the location processor. Finally, the service evaluation agent measures the retrieved results by considering the user's search intent and recommends the most relevant products and services to the user.

To develop the whole framework for intelligent product and service search, three key issues need to be addressed. First, we have to provide a methodology to integrate product and service information written in different ontologies into a unified repository. Second, we have to provide an appropriate representation mechanism for user's search intent and also a transformation method to interpret the search request written in the user's private ontology to make it available for searching against a unified ontology instance repository located in the search agent. Finally, an evaluation mechanism has to be prepared to measure relevancies of products and services while considering both user's search intent and user's location. The following section addresses our approaches and methodologies for all these issues but briefly due to lack of space.

4 Search Procedure and Its Implementation

4.1 Product and Service Repository Construction Using Information Integrator

To establish a product and service ontology instance repository, the information integrator periodically collects service description documents that service providers publish on the Web and constructs an integrated repository for them based on a given ontology.

To achieve these goals, we first devise an ontology-matching algorithm [12] and create an ontology mapping rule base by applying the algorithm to each pair of search agent's ontology and service provider's ontology. Then, the information integrator transforms collected service description documents from various service providers into a unified form in search agent's specific ontology and stores them in the repository. The reader can refer to [12] regarding the ontology-matching algorithm in more detail.

4.2 Representation of User's Search Intent

Now, let us go over how we represent user's search intent for products and services. For more accurate search results than keyword based approach, we need to provide a more elaborate representation scheme for user's search intent. To do this, we essentially adopt a multi-attributes based representation paradigm. In addition, we also allow users to describe the search context of the object that they want to find by selecting a category from their own product and service taxonomy. We call this portion of user's search representation 'search domain'. Finally, we append user's location information captured by GPS embedded in user's mobile device to the final representation scheme for user's search intent as shown in Fig. 3.

For example, let us suppose a user intends to find an interesting, low-priced pizza restaurant reachable within 30 minutes. Then, with the GUI provided in the service client module embedded in the user's mobile device, the user can express his or her search intent as illustrated in Fig. 3. As depicted in Fig. 3, the user first selects a category such as 'Restaurant' and describes attributes and their related information that the user might want to consider. If the user wants, he or she might also assign relative weights between the attributes. In the example, we can see the user prefers closest restaurants over others. Finally, such search intent representation obtained from user is transferred to the intelligent product and service search agent via web service.

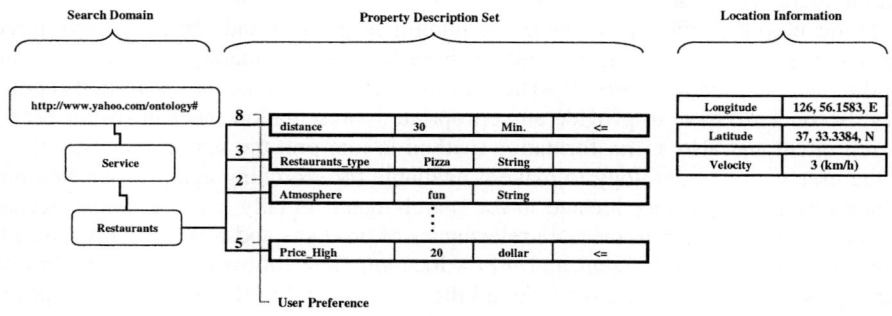

Fig. 3. An Example of User Configuration

4.3 Location Processor

The search intent and location information selector first receives a user's search representation and picks only the location information. The extracted location information is then sent to the location processor and the remaining part is sent directly to the query generator. The location processor selects geographical blocks that satisfy the user's location related preference by comparing them with the user's current location by GPS in terms of travel time. For example, as shown in Fig. 4, Blocks 6, 7, 10, 11, 14, and 15 are selected in this case. The location processor then passes them to query generator.

4.4 Query Generation Process

The query generator creates a query to the integrated product and service ontology instance repository based on not only the user's search intent but also the location information obtained from the location processor. In this paper, we adopt RDQL [13] as a query representation method, and so, the original search request from the user is transmitted as a form of RDQL. However, we still need to transform the original query for the search agent to understand it from the agent's own ontological perspective. Furthermore, we have to modify it to take into account the information from location processor such as the selected blocks.

To conduct this transformation, we develop three consecutive processes for query generation, which are respectively service category mapping algorithm, service property mapping algorithm, and location-based query conversion algorithm. Details of each algorithm can also be found in [12]. Figure 5 illustrates an example of re-generated query as a result of applying these three algorithms to the original query as shown in Fig. 3.

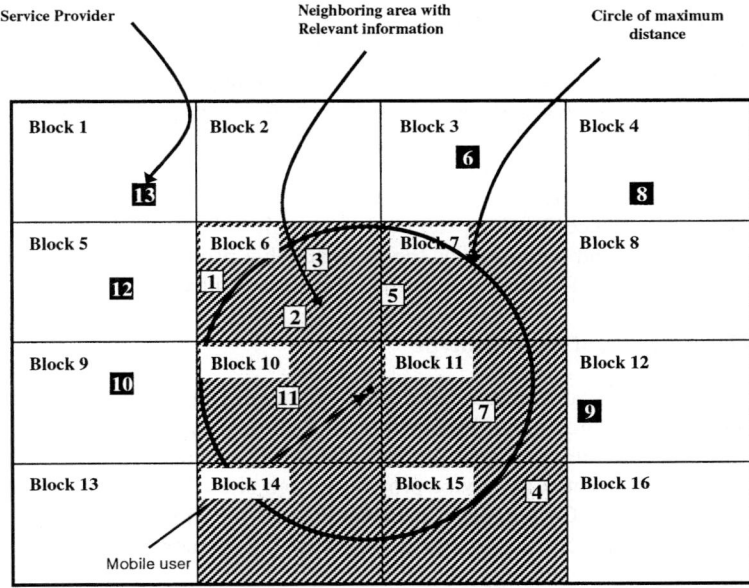

Fig. 4. Location Process Procedure

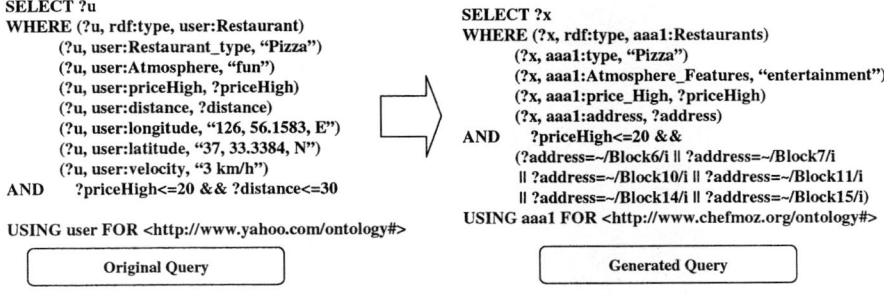

Fig. 5. An Example of Query Conversion

Then, the re-generated query is handed to the query executor and the query executor retrieves the product and service providers that satisfy all conditions represented on the right side in Fig. 5. For a more efficient filtering, we first filter out the irrelevant service providers according to location related constraints. In our example, the restaurants 1, 2, 3, 4, 5, 7, 11 survive this filtering and they are re-filtered with non-location related conditions. Finally, the re-filtered results are sent to the service evaluation agent and, in the end, {2, 3, 5, 7, 11} are passed in the example case.

4.6 Evaluation of Relevancies of the Extracted Products and Services

As a last step of the framework, the service evaluation agent performs three consecutive steps. The first step evaluates relevancies for each attribute. In this step, we compute relevancy based on two aspects, semantic similarity and syntactic similarity. Syntactic similarity measures how close the words are to each other lexicographically, while semantic similarity measures structural similarities such as similarity between the taxonomies depicted on the left side in Fig. 3. The second step computes a unified measurement by synthesizing those relevancies for each attribute, while considering their relative importance level represented as weights. Finally, the third step sorts them by relevancies and recommends the most suitable product or service provider to the user's search intent. Detailed evaluation procedure can be found in [12].

4.7 Implementation of the Prototype

To validate our framework empirically, we developed a prototype of the intelligent product and service search agent Framework, and it consists of three different parts, a service client module, an intelligent product and service search agent, and service description documents located at service providers' web sites. The service client module was implemented using C# language in PocketPC and the intelligent product and service search agent was developed based on Java. For demonstrational purpose, we adopted the 'restaurant' ontology available from Yahoo[1] as a service client's private

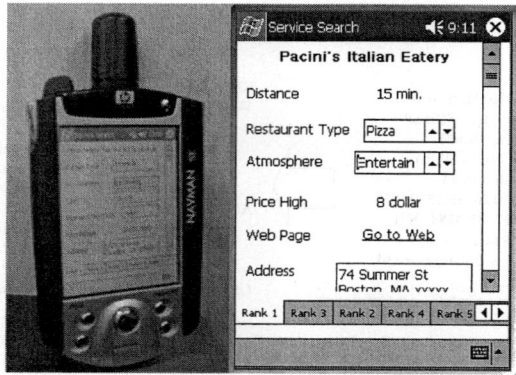

Fig. 6. The screenshot for the Intelligent Information Search Agent

---

[1] http://local.yahoo.com/

ontology and we also adopted the ontology from chefmoz.org as ontology for intelligent product and service search agent. Furthermore, we prepared various service providers' information using two different ontologies from restaurant.com and restaurantrow.com. Fig. 6 shows a screenshot of the service client recommending the most relevant restaurants to the user's search intent represented in Fig. 3. Currently, we are validating our work in a real world situation by cooperating with a leading portal in Korea.

5 Conclusions

In this paper, we discussed two major bottlenecks to the proliferation of L-commerce. First, there is no appropriate product and service search methodology yet for users to locate the most relevant services in the mobile environment. Second, the product and service information is still too poor to be ready for service for L-commerce because most of the existing frameworks for L-commerce are too independent of each other, which means there are serious interoperability problems.

To cope with these two obstacles, we propose a semantic web based intelligent product and service search framework for L-commerce. Our proposed framework allows service providers to easily publish information about their products and services independent of informediaries that provide information search services. The informediaries can also easily construct their own L-commerce ready information repository by using a real-time ontology mapping mechanism between heterogeneous ontologies and taxonomies. Moreover, our framework provides advanced semantic web based search mechanism to improve its precision while recommending the most relevant products or services by adopting multi-attribute based location aware search and evaluation methodologies, and therefore, maximizing user satisfaction.

We implemented a prototype system for our framework and are currently validating it in real world situation. We expect that our proposed framework will contribute to proliferation of L-commerce in the near future.

Acknowledgements. This research was supported by University IT Research Center Project.

References

1. Mark Weiser, The Computer for the Twenty-First Century. *Scientific American*, Sep., PP. 94-104 (1991)
2. Eija Kaasinen, User Needs for Location-aware Mobile Services, In *Personal and ubiquitous Computing*, Issue: Volume 7, Number 1/May 2003, published by Springer-Verlag London Ltd.
3. J. Youll, R. Krikorian. Wherehoo Server: An interactive location service for software agents and intelligent systems, In *Workshop on Infrastructure for Smart Devices*. Sept. 2000.
4. Yahoo! Local (http://local.yahoo.com/)
5. SuperPages.com (http://yellowpages.superpages.com/)
6. G. Abowed et al., Cyberguide: a Mobile Context-Aware Tour Guide, *ACM Wireless Networks 3*, pp. 421-433 (1997)

7. Cheverst K, Davies N, Mitchell K, Friday A, Efstratiou C, Developing a context-aware electronic tourist guide: some issues and experiences. In: *CHI 2000 Conference Proceedings*. ACM pp 17-24 (2000)
8. Bieber G, Giershich M, Personal mobile navigation systems – design considerations experiences. *Computers & Graphics 25*, pp 563-570 (2001)
9. Fano, A Shopper's Eye: Using Location-based Filtering for a Shopping Agent in the Physical World. In *Proceedings of the second International Conference on Autonomous Agents*, Minneapolis, MN, pp. 416-421 (1998)
10. H. Yan and T. Selker. Context aware office assistant. In P*roceedings of the 2000 International Conference on Intelligent User Interfaces*, New Orleans, LA USA. pp. 276-279 (2000)
11. J. Youll, J. Morris, R.C. Krikorian, P. Maes, Impulse: location-based agent assistance, In *Proceedings of the 4th International Conference on Autonomous Agents*, 2000.
12. W. Kim, D. Choi, Semantic Web Service Based Intelligent Product and Information Search Framework for Agents and Shopping Mall, In *KMIS Conference on Digital Convergence & IT Management*, Seoul, South Korea, pp. 39-46 (2004)
13. Seaborne, A.: RDQL – A Query Language for RDF, W3C Member Submission. (http://www.w3.org/Submission/2004/SUBM-RDQL-20040109/) (2004).

A Study on Value Chain in a Ubiquitous Computing Environment

Hong Joo Lee[1] and Choon Seong Leem[2]

[1] Quality & Reliability Lab. DAEWOO Electronics Corp.,
412-2 Cheungcheun2-dong, Bupyeong-gu, Incheon, Korea
phileo21@empas.com
[2] Department of Computer and Industrial Engineering, Yonsei University,
134, Shinchondong, Seodaemoongu, Seoul 129-749, Korea
leem@yonsei.ac.kr

Abstract. As we enter the 21st century, the term "ubiquitous computing" is emerging. The emerging benefits of "ubiquitous computing" have the potential of fundamentally altering the way we live and work. Most companies are in business to win, and outperform their competitors. They adopt new technologies to fend off new competitors, reinforce an exciting competitive advantage, leapfrog competitors, or just to make money in new markets. Performance is critical. Only recently have companies been in a better position to comprehend how to use new technology such as ubiquitous computing. So now, too many companies are interested in using ubiquitous computing. Our research includes the following: First, we used previous research about the definition of business models and characteristics of ubiquitous computing. Second, we analyzed value chains for ubiquitous computing. Finally, we suggest a new value chain considering ubiquitous computing environment.

1 Introduction

During the last 20 years, the business environment and the new technology embedded within it have experienced tremendous change. Also, new technologies have provided opportunities for value creation, which underpins intelligent and proactive strategy. Companies are in a better position to comprehend how to use new technology such as ubiquitous computing.

2 Previous Research

2.1 Characteristic of Ubiquitous Computing

In a ubiquitous computing environment, services through computers are provided without the user's direct intervention through the connected network.[26]

Ubiquitous computing is a calm technology that is defined by *MARK WISER*.[26] In ubiquitous computing environment, computers can intelligently recognize the problem context as well as proactively provide the suitable decision.[25]

Table 1. Characteristics of Ubiquitous Computing

Characteristic	Definition
Any one	Can be connected at any one
Any device	Can be connected at any device.
Any where	Can be connected at any where
Any application	Can be connected at any application
Any time	Can be connected at any time

2.2 Benefits of Ubiquitous Computing

The major benefits of ubiquitous computing can be summarized as enabling us to utilize information in several ways. The purpose of a software infrastructure for ubiquitous computing is to retrieve information from our real world that could not be made available before, and to control various everyday objects that could not be controlled before by embedding computers.[10] Some of the most important issues in ubiquitous computing are to provide context-awareness, to integrate physical and cyber spaces, to personalize our real world and reduce the complexities in our daily lives [8]. Many researchers are working on similar topics, such as sentient computing [12], pervasive computing [7], tangible bits [14], affective computing [22], ensemble computing [24] and proactive computing [23]. Such research shows that a software infrastructure to support ubiquitous computing is a key to realising its vision. The infrastructure makes it possible to share various devices and sensors, and to build ubiquitous computing applications easily.

Table 2. Benefits Ubiquitous Computing

Benefits of Ubiquitous Computing	Definition
Anywhere	Ubiquitous computing removes location dependency. This is embodied in the remote control of networked home appliances and information accessibility of mobile terminals.
Without conscious effort	Ubiquitous computing technology performs a variety of tasks without the beneficiary's being aware of it.
With a sense of super-realism	Ubiquitous computing does not block access to any information, and work can done remotely as if doing it on-site.
From the user's point of view	Ubiquitous computing technology implements functions from the user's point of view, rather than from that of the information and operations providers.

2.3 Business Model

A company's business model is management's model of how the strategies they pursue will allow the company to gain a competitive advantage and achieve superior profitability.[2]

Business models are usually based on financial projections of the pricing that the company can attain if it successfully implements its strategies and meets its goals.[3]

2.4 Business Model Concept

The business model provides a coherent framework that takes technological characteristics and potentials as inputs and converts them through customers and markets into economic outputs. The business model is thus conceived as a focusing device that mediates between technology development and economic value creation.

The Following is previous research of business models:

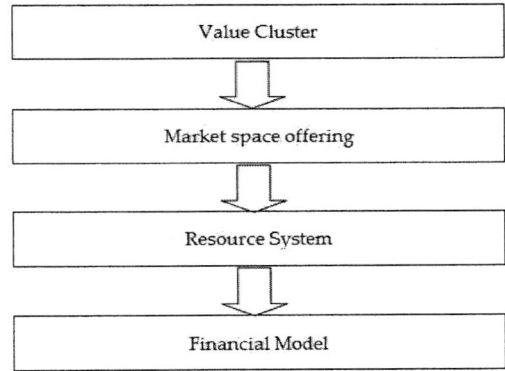

Fig. 1. Components of business model
(Source: Jeffrey F.Rayport et al, E-COMMERCE, McGraw-Hill,2001)

2.5 Differences Between the Business Model and Strategy

A Strategy is an action that a manager takes to execute a business model,[3] but, our concept of the business model differs from the focus of strategy in at least three ways.[10] First, the business model starts by creating value for the customer, and constructs the model around delivering that value. There is some attention to capturing a portion of the value created, but the emphasis upon value captured and sustainability is much stronger in the realm of strategy. There, the competitive threats to returns posed by current and potential entrants take center stage, whereas these are less central in the business model. A second difference lies in the creation of value for the business, versus creation of value for the shareholder. Oftentimes, the financial dimensions of a business are left out of the business model. The model is assumed to be financed out of internal corporate resources, so that financing issues do not figure prominently in the business model. A final difference, which we will explain below, lies in the assumptions made about the state of knowledge held by the company, its customers and third parties. The business model construct consciously assumes that this knowledge is cognitively limited, and biased by the earlier successes of the company.[13]

Table 3. Previous research of business model

Researcher	Previous Research
Alfred Chandler's seminal Strategy and Structure (1962)	Systematic and comparative account of growth and change in the modern industrial corporation
Chandler (1990)	Scale and Scope economies provided new growth opportunities for the corporation during the second industrial revolution.
Ansoff (1965)	Ideas from Strategy and Structure and applied them to emerging concepts of corporate strategy. Strategy came to be seen as a conscious plan to align the company with opportunities and threats posed by its environment.
Andrew (1987)	Corporate strategy was a superset of business strategy. A company's current business influenced its choice of likely future business as well
Christensen (1997)	A company must avoid internal resource allocation processes, and manage disruptive technologies outside the main business

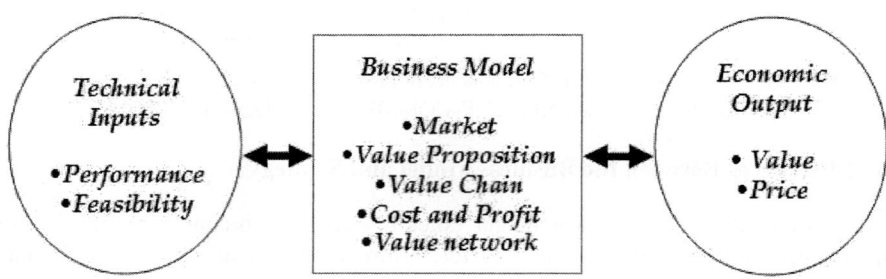

Fig. 2. The business model mediates between the technical and economic domains

2.6 Business Process for Ubiquitous Computing

This paper; business process for ubiquitous computing, considering value chain, suggests a new value chain.

This is previous research about the ubiquitous computing business model.

Table 4. Business model for ubiquitous computing

Researcher	Previous Research
NRI (2003)	3 types innovation business models using ubiquitous computing -Management business model for generic consumer -Asset management for company -Business model for public
Oh, Jae In (2004)	Business model framework using The u-Matrix -u-Trade -u-Hub -u-Care -u-Support
Kim, Jae Yoon (2003)	5 types Business Model considering supplier perspectives. -Technology Enabler -Component Provider -Set -System Integrator -Value Added Service

Table 5. Critical Success Factors in Business Model for Ubiquitous Computing

Factor		Variable
Application	Characteristic	Freshness Localization Variety
	Management	Scalability Timeliness
Technology		Accuracy/ Reliability Resolution Utility Personalization Useful
Cost		Transmission Cost Device Cost Application Cost
Control		Trust Privacy

(Source: Oh, J.I, Service@ Ubiquitous Space, The Electronic Times, 2004)

2.7 Business Value Chain

In every business, there are sequences of activities that companies perform to produce goods and services, deliver them to customers, and make or lose money. This sequence of activities is called a business system or value chain.[1] The Value chain of the business unit is only one part of a larger set of value-adding activities in an industry. The value chain of any company therefore needs to be understood as part of the larger 'system' of related value chains.

3 Ubiquitous Computing and the Value Chain

Obviously, business performance is dependent on the processes that gather and disseminate information. Normal business transactions (invoices, orders, payments, etc.) could be addressed by a company with most of its customers and suppliers who have computers, simply by connection via the Internet. This has indeed already happened in some industries especially those dominated by large retailers, where the majority of basic business transactions with suppliers are now electronic. This basic use of e-commerce is spreading through different industries at varying rates.

Therefore, business process is influenced by new technology such as ubiquitous computing technology. In many companies a more effective value chain is created using ubiquitous computing technology.

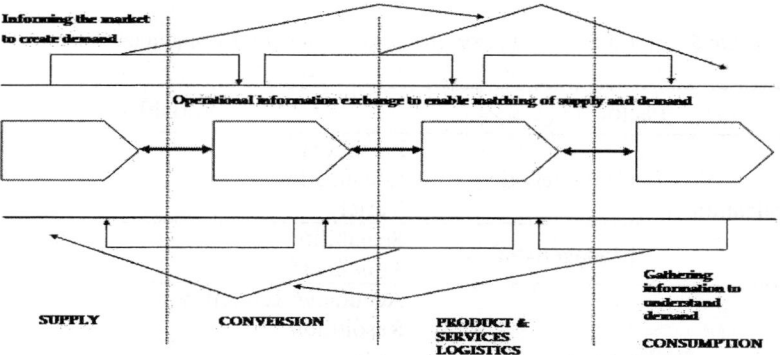

Fig. 3. Understanding the information issues in the value chain (Source : after J.F.Rayport and J.J. Sviokla)

It not only improves the economics of transaction processing but also enables the whole chain to respond more effectively to real-time supply and demand change, provided transaction information is shared. It considers the impact of ubiquitous computing in relation to:

Suppliers– Anyone supplying essential resources. It may be necessary to organized them either by the nature of what they supply or their strength, or their ability to exert pressure on you and other customers.

***Customers*–** This could include the consumers as well as direct customers if the latter are essentially distributors. The customers should be segmented in terms of what they buy or how much leverage they exert.

***Competitors*–** Obvious competitors who sell very similar products or services should be supplemented by actual or potential new entrants into the market and 'threatening' substitute products and services should be included as competition.

4 Value Chain for Ubiquitous Computing

The connections created by a network might allow a company to learn more about end users; that is, a denser social web may allow marketing and sales functions to be in more direct contact with downstream and end users. This will allow a two-way flow of information corresponding with the dual role of marketing. Ubiquitous computing, by enabling a wider geographic scope, represents another medium in which to market and sell to customers. The ability to cover a larger geographic area and to shift in time also affects the earlier stages of the value chain; specifically, it may allow a wider choice of inputs, distributed manufacturing, and remote testing. The main and most celebrated effect of ubiquitous computing on the value chain is a company's ability to carry lower amounts of inventory by ordering directly from a manufacturer and shipping directly to a customer. The activities of the value chain have been influenced by the development of ubiquitous computing.

Table 6. Ubiquitous computing affects these activities in four main ways

No	Main ways
1	It enables a larger scale of operations.
2	It widens the geographic scope of the area company represents
3	It allows more information to be collected and processed by the service provider
4	It enables a new delivery medium or mechanism

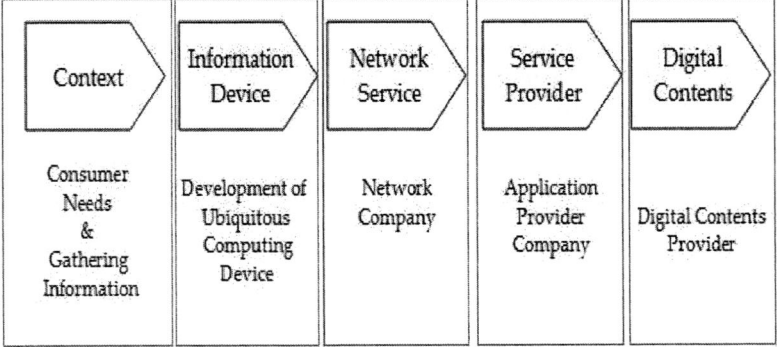

Fig. 4. Value chains in a ubiquitous computing environment

Value creation has a role in lowering the cost structure and increasing the perceived value of products through differentiation. As the first step in examining this concept, consider the value chain, which is illustrated in Figure 4. The process of transformation is composed of a number of primary activities and support activities that add value to the product/service.

Figure 4 is characteristics of value chain. Each stage definition is provided as follows:

Table 7. Each stage definition of value chain for ubiquitous computing

Stage	Definition
Context	Analyze consumer needs, provide service for consumer satisfaction.
Information Device	A company makes information devices for consumer. They want easy computing and easy data gathering. This service is a way to make a money for an information device production company.
Network Service	It is a very important part for wire/wireless network connections. It is a way to make money at network companies.
Service Provider	In ubiquitous computing environment, information device is operated by system software and application software.
Digital Contents	Digital contents are concerned with the creation of a service. (ex : cyber community) The role of the service function is to provide digital contents service and support. This function can create a perception of superior value in the minds of customers by solving customer problems and supporting customers

5 Conclusion

Ubiquitous computing technology enables a large number of new applications to have an impact on business processes. This paper focused on ubiquitous computing characteristics and the development of value chain for value creation. Our research suggests new 5 stage value chain considering ubiquitous computing environment and analyzed each-stage in the value chain. Further researches will be done on business models based on company characteristics.

References

1. Allan afuah.: Business Model A strategic management approach, McGraw Hill(2004)
2. Allan Afuah: Business Models, McGraw-Hill(2003)
3. Allan Afuah & Christopher L. Tucci.: Internet Business Models and Strategies, McGraw-Hill (2001)
4. Alfred Chandler, A.D, Strategy and Structure.: Chapters in the History of American Industrial Enterprise, MIT Press:Cambridge,MA (1962)

5. Ansoff.I.: Corporate Strategy,McGraw-Hill NewYork (1965)
6. Andrews,K.: The Concept of Corporate Strategy,Irwin:Homewood,IL(1987)
7. Banavar G, Beck J, Gluzberg E, Munson J, Sussman J, Zukowski D. Challenges An application model for pervasive computing. Proceedings Mobicom(2000)
8. Buxton. W.: Less is more (more or less), In: Denning PJ (ed) Invisible Computing(2002)
9. Chandler,A.D.:Sclae and Scope The Dynamics of Industrial Capitalism, Harvard University Press:Cambridge,MA (1990)
10. Charles W.L Hill & Gareth R.Jones.: Strategic management theory an integrated approach, Houghton Mifflin(2004)
11. Cristensen,C.:The Innovator's Dilemma,Harvard Business School Press:Bostro,MA (1997)
12. Harter A, Hopper A, Steggles P, Ward A, Webster P.: The anatomy of a context-aware Application, Proceedings 5th Annual ACM/IEEE International conference on Mobile Computing and Networking(1999)
13. Henry C. Lucas.: Strategies for electronic commerce and the internet, The MIT Press(2002)
14. Ishii H, Ullmer B.: Tangible bits towards seamless interfaces between people, bits and atoms. Proceedings Conference on Human Factors in Computing Systems(1997)
15. Jeffrey F.Rayport et al. : E-COMMERCE, McGraw-Hill(2001)
16. 16.J.F.Rayport and J.J. Sviokla.:Exploiting the virtual value chain, Harvard Business Review, November-December(1995),75-78
17. John ward and Joe Peppard.: Strategic planning for information systems, WILEY(2003)
18. Kim,Jae Yoon.: Ubiquitous Computing Business Model, SERI(2003)
19. Kim H.R., and Kim, H.G.: Semantic web technology for ubiquitous services, proceedings of Korean management Information system, Korea Society of Management Information System, (2003)
20. Lee, H.J and Lee, J.W.: Ubiquitous Innovation, e-co book(2004)
21. Oh, J.I. : Service@ Ubiquitous Space, The Electronic Times(2004)
22. Picard R.: Affective Computing. MIT Press(1997)
23. Tennenhouse.: Proactive computing. Communication ACM 43(5), (2000)
24. Thomas P.: Ensemble computing. Int'J Human-Compute Interact 13(2),(2001)
25. TOSHITADA NAGUMO.: Innovative Business Model in the Era of Ubiquitous Networks, , , NRI (2002)
26. Weiser M.: Hot Topics: Ubiquitous Computing, IEEE Computer, Vol.26,(1993).71-72

A Study on Authentication Mechanism Using Robot Vacuum Cleaner

Hong Joo Lee[1], Hee Jun Park[2], and Sangkyun Kim[3]

[1] Quality & Reliability Lab. DAEWOO Electronics Corp.,
412-2 Cheungcheun2-dong, Bupyeong-gu, Incheon, Korea
phileo21@empas.com
[2] Department of Computer and Industrial Engineering, Yonsei University,
134, Shinchondong, Seodaemoongu, Seoul 129-749, Korea
h.park@yonsei.ac.kr
[3] Somansa, Woorim e-Biz Center, Yangpyeongdong 3-ga,
Yeongdeungpogu, Seoul, Korea
saviour@somansa.com

Abstract. Robot is not a new field. It has been around for decades. In fact, most people have robots in their own home, even if they don't recognize the robots as such. For example, a dishwasher automatically washes and dries your dishes, then grinds up the rinsed-off food so the organic matter doesn't clog your drains. A washing machine soaks, soaps, agitates, and rinses your clothes. Down the street, the car wash-n-wax cleans, brushes, washes, and waxes your car, all for a few dollars. One of the better known home-oriented robots is robot vacuum cleaner which has already good cleaner for assistant house keeper. In this paper, we suggest that home security system using robot vacuum cleaner. Our robot vacuum cleaner makes an additional security function using camera. But, these technologies lack in provision of the authentication architecture of customers for the accountability, confidentiality. This paper presents the architecture of authentication mechanism to deal with the specific challenges arising from these applications and to overcome the weakness of traditional architecture for the secured infrastructure of home network environment.

1 Introduction

With the convergence of digital communication networks and the Internet, the usage of homenetwork applications utilizing digital devices such as cellular phone, robot vacuum cleaner been increased. Vacuum cleaner robot, is a particular case of house keeping using the home application considering home network environment. In home network environment, customer wants to be easy housework and robot help it.

Robot vacuum cleaner that finds its own way around a carpet heralds the beginning of the end for housework. The robot navigates around corners and large objects using an acoustic radar system similar to that of a bat and is fitted with a sensitive bumper to stop it damaging the furniture.

In a network environment, the customer possesses a public and private key pair of any encryption algorithm using public key mechanism, and maintains a certificate issued by a certification authority. If an authentication is required, the customer engages in an authentication protocol which is relying on a public-key infrastructure. In this environment, researcher makes a study robot vacuum cleaner adding security function. This paper, we propose the mechanism of authentication for robot vacuum cleaner in a homenetwork environment.

This paper proposed the mechanism of authentication using robot vacuum cleaner includes:

First, we have previous research about mechanism of security authentication.

Second, in a homenetwork environment, we suggest home security system using robot vacuum cleaner. Finally, we suggest the authentication of private using the personal mobile device.

2 Robotics and Homenetwork

2.1 Definition of Robotics

Two possible definitions are:

The Definition Supplied by the *Concise Oxford Dictionary*:
'Apparently human automaton, intelligent and obedient but impersonal machine'.
 This definition cannot, however, be entirely accurate since no existing robot in use resembles a human being, nor is intended to. [2]

The Definition Supplied by the Robot Institute of America :
'A reprogrammable and multifunctional manipulator, devised for the transport of materials, parts, tools or specialized systems, with varied and programmed movements, with the aim of carrying out varied tasks'. [2]

2.2 Characteristics of Robotics

Versatilely :
The structural/mechanical potential for performing varied tasks and/or performing the same task in different ways. This implies a mechanical structure with variable geometry. All existing robots have this quality.[1][2]

Auto-adaptability to the Environment:
This complex expression simply means that a robot must be designed to achieve its objective (the performance of a task) by itself, despite unforeseen, but limited, change in the environment during the performance of the task. Robot sense so far developed are modest compared with the human capacity for interpreting the environment, but intensive research is being carried out in this area. [2][3]

2.3 Homenetwork

Home Network is the collection of intelligent appliances and the connection of them in home.

First, Intelligent appliances includes computers, digitalized white appliances and AV appliances, controls, sensors, devices, and so on. Second, Services which make life convenient and various contents are supplied through the network. Finally, home network market is divided into appliances market and services market. [6]

3 Robot Vacuum Cleaner Authentication Mechanism

3.1 Introduction to Robot Vacuum Cleaner

Robot vacuum cleaner that cleans customers home all by itself.

The autonomous, cordless machine uses ultrasound to avoid obstacles and to work out the most efficient route around a room it has been set to clean. Left to its own devices, the compact, rechargeable appliance vacuums the perimeter of the room, and then cleans the remaining floor space in random zig-zag movements.

The robot navigates around corners and large objects using an acoustic radar system.

In these days, customer wants house to keep comfortably and observation from going out. Thus, some company develop robot vacuum cleaner with home security function.

The robot has observation function using camera and it controlled by customer using mobile device. Within this situation, several problems of information security are arisen apparently. To process a home security of customer according to the legal liabilities, the authentication between the customers and security service providers must be established. Furthermore, the control messages sent to all parties must be confidential, and no party should be able to alter identification information during transmission without any means of detection [5].

Both the customers and security service providers should not be able to deny their agreement unfaithfully on the purchasing conditions in later. Customers will not fully accept services that erode their privacy. In homenetwork environment, customer want to solve problem of home security using robot vacuum cleaner.

3.2 Security Requirements of Robot Vacuum Cleaner

Key Factors of Robot Vacuum Cleaner Protocols:
The design of robot vacuum cleaner protocol is strongly influenced by the availability of hardware and its functions. For example, the hardware in robot vacuum cleaner might contain special units which process electronics waves, but lack in the provision of general purpose units which support a public key cryptosystem.

The existence or absence of a traditional public-key infrastructure and the organizational constraints of security service providers and robot vacuum cleaner such as the requirements of accounting between the security service provider and the robot vacuum cleaner are also key factors of the design of home security protocol.

Basis of Security Design:
The home security applications partially operating with hardware-based security solutions are usually feasible in security using mobile device. It may be assumed that

customer is equipped with at least one information device that contains a unique string of identification called Hardware-ID. Furthermore, customer has a robot vacuum cleaner and unique code of customer ID. [7]

In these aspects, the problem of home security using robot must be resolved on the basis of decentralized authentication mechanism. It is one of the major axioms to design information security that the system should provide security features in a transparent way so that a minimum amount of customer interaction is required.

3.2.1 Security Requirements of Home Security Using Robot

The following security requirements for home security using robot.

Confidentiality:
Persons not involved in an electronic transaction should not be able to gain any information about home security (observation home); thus, all messages sent in the protocol must be encrypted.

Authentication:
As contracts must be legally binding, an authentication mechanism for both customers and home security service providers is required.

Customer Authentication:
To prevent theft of mobile devices, the customer has to authenticate himself or herself to the device before starting any transaction.

However, the overhead of authentication must be as low as possible; otherwise the usability of the mobile device might be affected.

Integrity:
Messages sent in the home security protocol using robot vacuum cleaner should not be altered by any party not involved in the transaction.

Partial Traceability:
In case of the single enforcement of illegal actions, it must be possible to retrieve an identity of that person.

This is complicated by the partial anonymity requirement, but possible in a similar fashion such as tracing of email: every party knows "someone" who is closer to knowing the real identity of a given party ; by following these links, a party with authoritative information on a customer's identity will be eventually found.

3.3 Authentication Mechanism of Home Security Using Robot Vacuum Cleaner

This paper suggests new authentication mechanism using mobile devices to overcome the limitation of traditional authentication mechanism using robot vacuum cleaner. This mechanism improves the level of security of customer authentication. This model consists of thirteen steps which are described below.

Step 1. The customer connected security service provider using mobile device.
Step 2. The security service provider sends an one time authentication code to customer's mobile device to execute second authentication.

Step 3. The customer inputs an authentication code provided by the security service provider into robot vacuum cleaner.

Step 4. Vacuum cleaner robot asks a authentication agency with the result of an authentication code.

Step 5. If customer input wrong code, send message to registration customer at security service provider.

The mutual relationships and data processes are illustrated in figure 1.

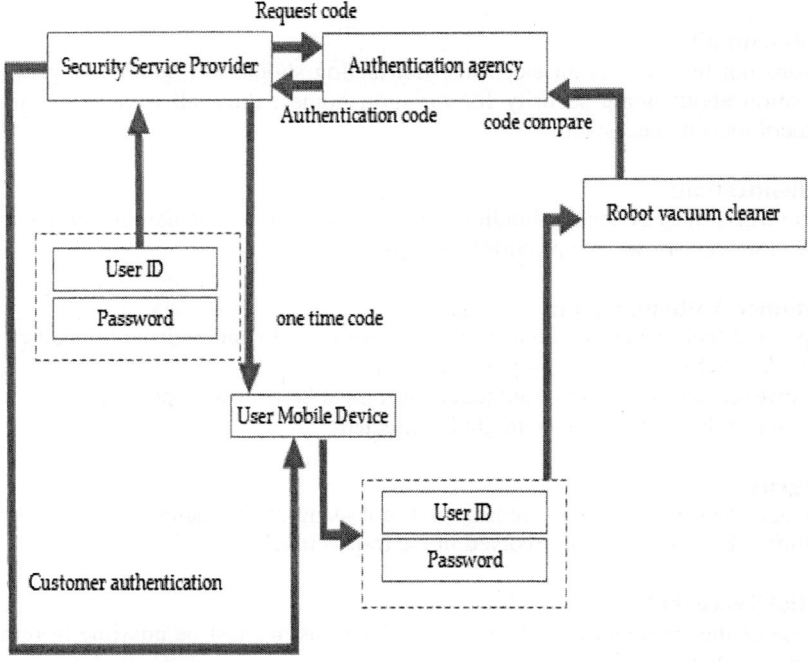

Fig. 1. Authentication mechanism for homenetwork environment using robot vacuum cleaner

3.4 Significance of This Mechanism

The customers are authenticated by authentication agency using mobile device.

In addition, if customer input wrong code, send message to registration customer at security service provider.

In this paper, we suggest enhanced mechanism of two steps authentication which provides personal authentication code using personal mobile device such as cellular phone and receives an authentication code that the customer input by himself to solve these problems.

4 Conclusion

This paper provides new authentication mechanism using mobile devices to overcome the limitation of existing mechanism. To protect the risks of breaches of private information or home security. We suggest an authentication mechanism which has the characteristics:

First step of authentication is processed using mobile device, Second After the successful processing of first step of authentication using mobile device, the authentication agency issues second authentication code via the mobile device that the customer already has. The customer inputs an authentication code by himself with mobile device such as cellular phone. Finally, The security service provider is authenticated by authentication code. New authentication mechanism suggested in this paper may protect the security risks that the existing mechanism exposes, but it requires one more step using mobile device which makes the home security service using robot vacuum cleaner customer to suffer from difficulties. So, further researches must provide: The easy of usage of secured authentication, and the protection mechanism.

References

1. ANDERS OREBACK.: A Component Framework for Autonomous Mobile Robots, Doctoral Thesis (2004)
2. Coiffet et al.: An introduction to robot technology, Hermes (1982)
3. Jim Butler.: Robotics and Microelectronics Mobile Robots as Gateways into Wireless Sensor Networks, Technology@Intel Magazine (2003)
4. Katzenbeisser, S. and Tomsich, P.: Applied Information Security for m-Commerce and Digital Television Environment,Lecture Notes In Computer Science 2115,(2001),165-175
5. Lee, H.J.: A Study on Method Using Effectiveness Interaction TV. HCI Conference (2003)
6. Lee, H.J.: Ubiquitous Innovation, e-co book (2004)
7. Tanh, D.V.: Security Issues in Mobile eCommerce. Lecture Notes in Computer Science (2000), 467-476

Design of Inside Information Leakage Prevention System in Ubiquitous Computing Environment

Hangbae Chang[1] and Kyung-kyu Kim[2]

[1] SoftCamp Co., Ltd,
5F Handong Bldg. 828-7 Yeoksam Dong Ganamgu ,Seoul, Korea
hbchang@softcamp.co.kr
[2] Graduate School of Information Yonsei University,
134 Sinchon-Dong, Seodaemun-Gu, Seoul, 120-749, Korea.
kyu.kim@yonsei.ac.kr

Abstract. This study designed a system that prevents the leakage of information flowing through physical communication networks in ubiquitous computing environment. The information security system was designed from the viewpoint of information life cycle based on u-Office' service scenario and, as a result, security holes were identified including indiscreet information access, malicious information leakage, unauthorized information alteration, illegal information retrieval, information leakage using various portable storage devices and printed-out information leakage. In addition, this study defined and designed essential security technology required to resolve these security holes.

1 Introduction

Existing business administration has emphasized the specialization of each staff in performing tasks but the introduction of information technology has changed the traditional business pattern. That is, information technology brought integrated information systems to support information sharing and task performance in companies. Examples of such systems are intranet, groupware, knowledge management system, electronic document management system, management information system, etc.

Particularly with the rapid development of relevant technologies, the scope of information sharing is expanding as ubiquitous computing environment makes it possible for subminiature computer devices connected to networks to be imbedded in actual physical spaces and to utilize information while the user does not recognize it.

In such computing environment, anyone, as long as he/she is authenticated as "I", can access a specific place within the company and inside information at any time and in any place. From a different viewpoint, however, the accessibility of all shared information at any time and in any place means that information can be easily exposed, altered and leaked out. Thus there are more security holes than before and the safe management of information becomes more difficult.

The present study designed a system that minimize information leakage, which may happen in intelligent information environment, using mobile terminals in order to promote the realization of real ubiquitous services.

2 Related Research

(1) Ubiquitous Computing Environment

"Ubiquitous' is a Latin word meaning 'anytime and anywhere' or 'exist simultaneously.' The term is currently used to describe computing environment in which users can communicate information using any device on any network (portable) and information is transmitted in the optimal method as the context of users' requirements are autonomously recognized while the users are not aware of it.

In ubiquitous environment, information service is highly mobile and embedded that information users become dynamic and computer devices become diversified. Moreover, different from existing information service that only provides information required by users, in ubiquitous environment the computer not only serves information but also performs necessary actions as it intelligently recognizes the concerned situation.

However, because computer equipment and networks composing ubiquitous computing environment have characteristics such as heterogeneity, openness, mobility and dynamicity, we need to apply distributed access control while guaranteeing the identification of each component rather than centralized one.

(2) Inside Information Leakage

The trend of informatization resulting from the development of information technology is spreading infinitely including managing and improving resources and processes in enterprises, collecting and processing data and producing meaning information, and communicating and utilizing information and creating knowledge.

The development of information technology, however, brings forth not only such eufunctions but also fatal dysfunctions. Basic human rights are infringed as personal information and life are disclosed and communication confidentiality is not guaranteed. In organizations as well, information management is getting harder as unauthorized persons can access information in the environment of integrated system and important information is computerized. Furthermore, the development of communication technology makes easier hacking by outsiders and information leakage by insiders.

Particularly in case of information leakage by insiders, because information users have to access important information inevitably for their works, control over the leakage of information accessed is mostly in user's hand because of work efficiency and technological limitations. Information leakage by insiders is more problematic when the asset value of information is higher. The leakage of technology information, customer information, national secrets, information related to rights and interests, etc. may threaten the competitiveness of enterprises and public institutions and drain national wealth. According to reports on information security accidents, 14.5% of companies surveyed experienced inside information leakage, and 82.5% of information leakage accidents were by insiders. This shows how serious inside information leakage is.

In ubiquitous computing environment where information sharing and information accessibility are heightened, the problem is even more serious. Thus it is urgently necessary to develop security technology that applies more strict control to inside information leakage while enabling staffs inside the company to access inside information at any time and in any place and supporting high work efficiency.

(3) Trend of Relevant Researches in Korea and Other Countries

① Trend of ubiquitous computing security technology

Foundational technology for ubiquitous computing security has been developed neither in Korea nor in other countries yet. Thus, rather than developing actual security technology, this study planed expected services and designed a new security platform, dividing required security technologies from the viewpoint of network and those from the viewpoint of application service.

In order to cope with security holes from the viewpoint of network, it is essential to develop security technologies for right management, denial prevention, availability and anonymity in addition to existing security services in wireless network environment such as authentication, confidentiality and integrity. In addition, problems such as lost and stolen mobile terminals, signal interrupting attacks, battery draining attack, unidentified connection and the invasion of privacy should be solved.

Technologies necessary from the viewpoint of application service are those for authentication to minimize users' intervention, distributed reliability management, access control to information and resources, the provision of digital identity information composed of user identities and properties in distributed environment, etc.

In Korea, the development of ubiquitous computing security technologies is led by the Team of Ubiquitous Computing Project under the Ministry of Science and Technology. In other countries, Carnegie Melon University in the U.S., Cambridge University in the U.K., etc. are developing next generation security systems in ubiquitous computing environment (particularly technologies for hiding the location of users to protect individuals' privacy) and technologies for the security of data transmitted through networks.

② Trend of technologies for preventing inside information leakage

Technologies for preventing inside information leakage are mainly to control and manage rights to access information and to observe the contents of inside communication. The latter may violate the Protection of Communication Secret Act and the Act on Promotion of Information and Communication Network Utilization and Information Protection, which guarantee individuals' privacy, in the course of monitoring indiscreet communication contents, say, the contents of mails and their efficiency may be low when communication is encrypted. Thus it is more effective to maximize security effect by introducing technologies and management procedure that control information flow (including information flow through communication and mobile storage devices) inside the organization according to integrated security policies and procedure of the organization rather than to monitor communication that may result in unnecessary troubles without security effect.

A number of research institutes and companies in Korea and other countries are developing information leakage prevention technologies with their own technological abilities (personal computer management technology, digital copyright management technology, user authentication technology, etc.), but these technologies cannot perfectly provide functions such as information encryption and authorization among mobile computers (Tablet PC, PDA, Smart Phone, etc.). In addition, because they have to use different programs according to the form of inside information crated, encrypted information cannot be used in other incompatible programs. In these ways, convenient information flow among various devices and programs, which is the biggest characteristic of ubiquitous computing environment, is impeded.

3 Research Methods

The present study purposes to observe and model the flow of users' behavior in performing tasks in ubiquitous computing environment, in which mobility is emphasized, and to design a security technology that solves security holes discovered in the flow.

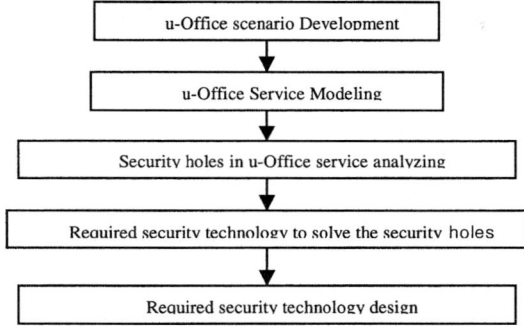

Fig. 1. Research Procedure

① Develop a scenario of task performance (u-Office Service) feasible in ubiquitous computing environment

Organization-level ubiquitous service 'u - Office' provides convenient office environment and, at the same time, supplies various information services in accord with users' location and the change of environment.

※ Examples of ubiquitous service (u-Office Service)

A company staff enters the office. As a sensor on the entrance door recognizes his fingerprints the door opens automatically. As he gets into the office, a sensor senses his position and turn on the light. In order to do works in arrears, he booted his computer in the office remotely from home. He connects to the company information system and finishes his works using information with limited rights. Then a mobile terminal, which tells schedule information, informs automatically that a repair and maintenance work is scheduled with a customer. As information on the location of the repair and maintenance work is forwarded to the vehicle, the optimal route to the location is decided according to traffic condition. Arriving at the customer's site, the staff meets a situation that requires important information on products kept in the company information system. He passes through the procedure of authentication and uses information within the limit allowed to him according to his rights. In addition, if he wants to continue at the company his work that has not been finished outside due to the shortage of time and information, his computer restores the state at the moment he stopped the work outside. The web browser and the spreadsheet program, with which he worked, are shown on the computer as they were, and are downloaded video clips as well. All the information may be displayed using a wide wall that can be seen at a glance and all these procedures may be controlled by voice.

② Model users' various task-performing behaviors in ubiquitous computing environment (Agent-based Modeling) and analyze security holes related to the behaviors

This study designed External AOR Model and Internal AOR Model using AORML (Agent Object Relationship Modeling Language) suggested by Wagner. Fig. 2 shows agent diagrams and an interaction frame diagram as parts of External AOR Model, which were modified based on the scenario presented above.

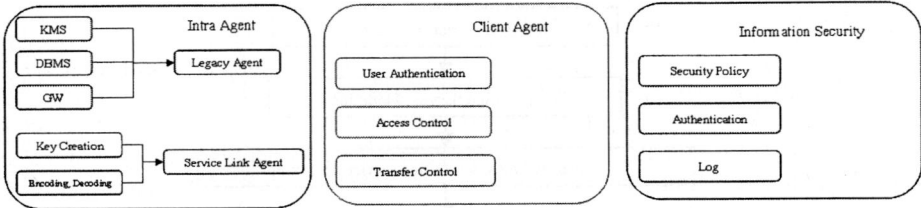

Fig. 2. Agent Diagrams(External AOR Model)

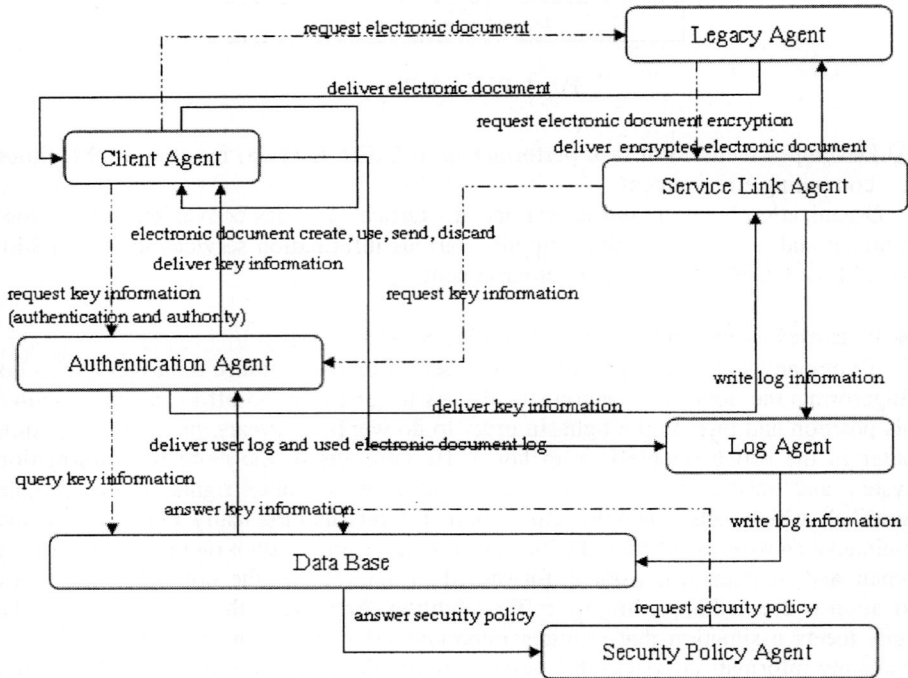

Fig. 3. Interaction Frame Diagram (External AOR Model)

③ Analyze security holes along the life cycle of information based on the model of expected u-Office' service

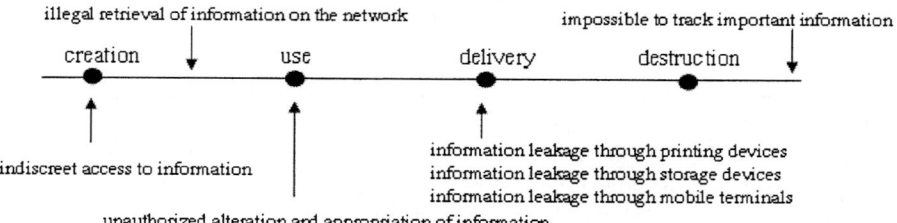

Fig. 4. Analysis of security holes throughout the life cycle of information

- Indiscreet access to information: If there is no appropriate management and access control system for newly created information, unspecified people (who should not see the information) may access the information indiscreetly. Then the value of information is diluted and information leakage is highly possible.
- Unauthorized alteration and appropriation of information: It is highly possible for information to be altered, misappropriated and misused.
- Indiscreet leakage of information: As devices for information storage and communication are diversified with the development of online and mobile storage devices, information leakage is getting easier and faster.
- Impossible to track important information: There is no means to monitor the use and flow of important information due to lack of devices to track those involved in information leakage and to call to account those who are responsible.

④ The contents of security technologies and the characteristics of information security required to solve security holes that are expected in the life cycle of information in u-Office service are as follows.

Table 1. Security technologies for preventing inside information leakage

Information life cycle	Required security technology	Security		
		Confidentiality	Integrity	Availability
Creation	Real-time encryption of user files and folders	O	O	
	Encryption of information kept in information system	O	O	
Use	Control over the use of information through authorization	O		O
	Just-in-time user authentication	O		
	Limiting the edition of confidential information	O		O
	Addition of watermarking to print-outs		O	
Delivery	Control over mobile terminals taken in and out	O		O
	Control over mobile storage devices	O		O
	Creation of security files for outside transmission	O	O	O
Destruction	Automatic destruction of confidential information	O		

⑤ Define major security technologies that need to develop and design for implementing the technologies

- Real-time encryption of user files and folders: Information created by users must be encrypted selectively or compulsorily according to the information security policy.

    If a separate security folder is designated and the access right policy is defined, all information stored or moved to the security folder must be encrypted automatically according to access right. In addition, information in the subfolders of the security folder must be encrypted according to access right in the same way. Information copied or moved to other folders must be kept in the encrypted state.

- Encryption of information kept in information system: In case a user retrieves important information stored in a system linked to the company information system, the document (including attached files) is encrypted automatically. For compatibility with other systems in the linkage of search engines and virus diagnosis, however, all documents in the information system are stored as plain texts so that they do not affect existing work flows.
- Control over the use of information through authorization: A malicious inside user may use important company information at a level exceeding his right. Thus, rights to use information should be defined specifically by individuals and groups so that they use information within defined limits (items to define rights: reading, editing, printing, releasing, taking-out, valid term, automatic destruction, etc.).
- Real-time user authentication: When a user uses confidential information inside or outside the company, he can use it only when he is allowed to access by real-time comparison of information on his access right with information on access right contained in the information.
- Limiting the edition of confidential information: In a joint work, a number of people share the same information and it is hard to distinguish information editors from information users. In this case, not only the whole information but also important data contained in the information must be protected.
- Addition of watermarking to print-outs: When confidential information is printed out, all forms of print-outs must contain watermarking so that printing actions can be monitored. When confidential information is printed out, the information and image of the output are sent to the management server and then information number, the person who prints, his staff number, his department, the time of printing, etc. are included in the print-out of the information.
- Control over mobile storage devices: Important information in the company may be taken out not only through the network by mail and folder sharing but also through mobile storage devices (floppy disks, hard disks, CD-RW, PDA, etc.). If the information security policy of the organization requires only selective encryption of information crated by users, there is no way to prevent information creators from taking out important information, which is not encrypted, using mobile storage devices. Thus strict security must be applied to the routes of information leakage by controlling rights to use mobile storage devices in individuals' and groups' computers.

- Creation of security files for outside transmission: It is possible to control the delivery of confidential information through setting access rights by document. When performing a joint task together with external organizations, however, it is impossible to deliver encrypted information because it cannot be retrieved or used. On the other hand, if important information is sent in the form of plain text, it cannot be protected. Thus, user authentication information + access control information + encrypted information are sent in the form of an executable file so that only specific designated users can retrieve the content of the information. The receiver of the file can retrieve and use the information by executing the file without installing a separate program in his terminal.
- Automatic destruction of confidential information: In case the number of retrievals, the number of printings, valid term, etc. have exceeded their limits as set in access right information, the confidential information is automatically destroyed and the destroyed information cannot be recovered with general recovery program.

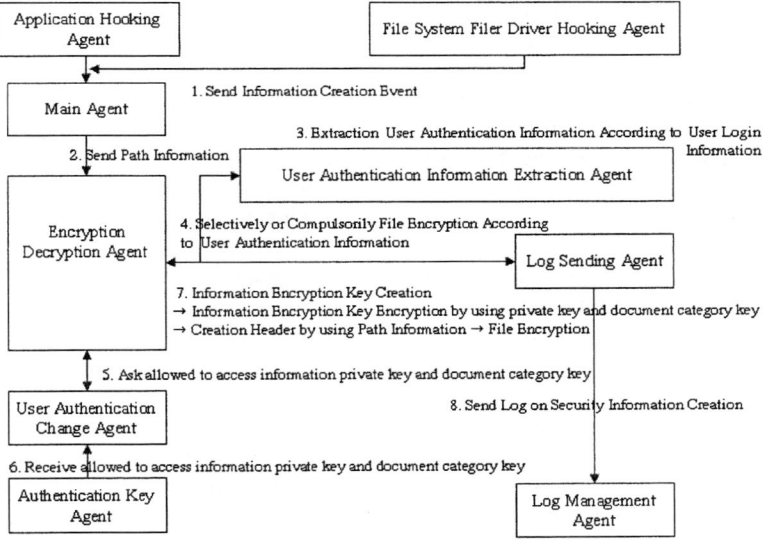

Fig. 5. Process of real-type encryption of user files and folders

4 Conclusions and Future Studies

Ubiquitous computing environment, in which any information can be communicated at any time and in any place and the optimal information service is provided to users based on automatically recognized user environment and requirement, is important for convenient and ceaseless task performance but, on the other hand, worsens the problem of inside information leakage.

The present study designed a system the prevent information leakage out of physical communication networks in ubiquitous computing environment. In particular, the information security system was designed from the viewpoint of information life cycle based on an assumed u-Office' service scenario. As a result, we identified security holes such as indiscreet information access, malicious information leakage, unauthorized information alteration, illegal information retrieval, information leakage through various portable storage devices, and print-out information.

In addition, this study defined and designed essential security technologies required to solve these security holes from the viewpoints of three steps of information security, which are user authentication, information access control and information use control. Defined technologies include those for real-time encryption of user files and folders, encryption of information kept in information system, control over the use of information through authorization, just-in-time user authentication, limiting the edition of confidential information, addition of watermarking to print-outs, control over mobile terminals taken in and out, control over mobile storage devices, creation of security files for outside transmission and automatic destruction of confidential information.

This research established policies and systems to protect important information, which is a valuable asset for the competitiveness of companies, securely even in mobile environment and constructed information system monitoring environment to track information leakage accidents caused by insiders, which are usually kept in secrecy. As the information security problem, which is an obstacle to the implementation of ubiquitous services, has been solved, we expect that the results of this study may accelerate ubiquitous services and promote the development and standardization of information security technologies in ubiquitous computing environment, which is still in the early stage.

Although technologies to be developed can protect general information in a file through file encryption and access control, however, it is impossible to protect design and drawing documents and source codes, which include various input files, temporary files created in each stage, target files, etc., just through the encryption of some files.

Thus it is necessary in the future to develop a virtual file system that encrypts and manages information and protects it from any access except that by authenticated users and application programs in order to cut off the possibility of leakage of information on designs and developments while preserving users' work environment.

References

[1] G. Wagner. "Agent Oriented Analysis and Design of Organizational Information Systems". Proc. of Fourth IEEE International Baltic Workshop on Databases and Information Systems. Vilnius (Lithuania), 2000. 5.
[2] Mark Weiser, "The Computer for the 21st Century," Scientific American Volume 265, Number 3, 1991. 9.
[3] Y. Suematsu, K. Takadama, N. Nawa, K. Shimohara, O. Katai, "Analyzing levels of the microapproach and its implications in the agent-based simulation", in: Proceedings of the 6th International Conference on Complex Systems, Chuo University, Tokyo, Japan, pp. 44–51, 9–11 2002. 9.

[4] G. Wagner. "The Agent Object Relationship metamodel: Towards a uni-fied conceptual view of state and behavior. Technical report", Eindhoven Univ. of Technology, Fac. of Technology Management, Available from http://AOR.rezearch.info, To appear in Information Systems, 2002. 5.
[5] F. Bergenti and A. Poggi. "Ubiquitous Information Agents.", International Journal of Cooperative Information Systems, 11(3–4):231–244, 2002.
[6] Lalana K, Tim F, Anupam J. "Developing secure agent systems using delegation based trust management", In Proceedings of Security of Mobile Multi-Agent Systems Workshop (AAMAS 2002) 2002;.

Design and Implementation of Home Media Server Using TV-Anytime for Personalized Broadcasting Service

Changho Hong[1] and Jongtae Lim[2]

[1] Digital Research Center, DAEWOO Electronics Corp. ASPD B/D.,
254-8 Gongdok-dong, Mapo-gu, Seoul, Korea
chhong@dwe.co.kr
[2] School of Electronics, Telecommunication and Computer Engineering,
HANKUK Aviation University
lim@hau.ac.kr

Abstract. In this paper, we introduce the design and implementation of a Home Media Server for personalized broadcasting service in ubiquitous environment. A implemented Home Media Server support multimedia contents like audio, video and metadata using wired/wireless digital interface. In bi-directional network environment, the special request of the HMS is sent to TV-Anytime metadata service providers and then the metadata server provides the personalized metadata back to HMS. We describe the actual system configuration and usage example of metadata for contents management.

1 Introduction

Personalized broadcasting service provides contents demanded by individual users at time when the users want to use. The expansion of digital broadcasting enabled personalized broadcasting and there are also increasing demands for personalized broadcasting. Particularly with the popularization of the Internet, users became able to access broadcasting service servers through return-channel and, as a result, people are getting more interested in bi-directional personalized broadcasting service.

Users can obtain detailed information about programs such as title, synopsis, schedule and review, and search specific programs using metadata received in unidirectional broadcasting environment. Metadata expressing such information were standardized in TV-Anytime Forum [1]. In unidirectional broadcasting environment, however, the volume of metadata transmitted to users is limited due to limited bandwidth allocated to the transmission of metadata and, accordingly, services provided to individual users are also restricted. Moreover, for the provision of services well-customized to individual demands, it is necessary to deliver information and requests of individual users to personalized service providers. If, for the delivery, a communication function is added to broadcasting HMS through return-channel, bi-directional personalized service can be provided. Representative examples of bi-directional personalized service are a person's retrieval of personalized programs based on the individual user's preference or choice, recommendation of personalized programs based on individual users' history of service uses. The standard for metadata communication between user HMS and metadata servers for bi-directional personalized service was also established in TV-Anytime Forum [2].

The implemented personalized broadcasting HMS transmits through the return-channel metadata about individual users' choice or preference and their pattern and history of watching to metadata servers providing personalized services. Particularly when there is no information about metadata servers providing personalized service, the system provides the function of metadata service retrieval through UDDI [3] standard in order to find metadata servers.

As shown in Figure 1, the general structure of bi-directional personalized service including the implemented bi-directional personalized broadcasting HMS is composed of digital terrestrial publisher, metadata server, broadcasting contents server, bi-directional personalized broadcasting HMS and UDDI service registry server. The metadata server manages various types of information about contents, and the UDDI registry server manages different kinds of services provided by metadata servers and information about how to access the services. A metadata service provider, who is separated from digital broadcasters, may run an independent metadata server or a digital broadcaster can run a metadata server directly. Bi-directional personalized broadcasting HMS receives and records digital programs broadcasted by digital terrestrial publishers and, in response to users' requests for personalized broadcasting service, it accesses the metadata server and exchanges metadata necessary for bi-directional personalized service.

The structure of the present thesis is as follows. First, Chapter 2 briefs the outline of metadata service in bi-directional broadcasting, and Chapter 3 describes the design and components of the hardware and software of bi-directional personalized broadcasting HMS and modules for implementing bi-directional metadata service. Chapter 4 explains cases of broadcasting services implemented through the developed broadcasting HMS, and Chapter 5 draws conclusions.

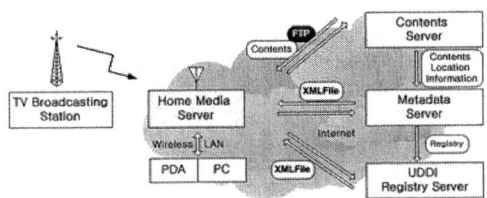

Fig. 1. The overall structure for bi-directional broadcasting service

2 Metadata Service in Bi-directional Broadcasting Environment

Metadata can be transmitted through unidirectional broadcasting or a bi-directional network. The transmission of metadata through a bi-directional network is more advantageous in several points than that through unidirectional broadcasting. For example, metadata can be provided according to users' personal demands. A larger volume of metadata can be transmitted in a bi-directional network than in unidirectional broadcasting, which has a limitation in bandwidth. Moreover, users can be provided with metadata without receiving broadcasting programs. Bi-directional metadata service means the exchange of metadata between broadcasting HMS and metadata servers using the return-channel of user-personalized broadcasting HMS, and it is divided into metadata retrieval and user-centric metadata transmission [2].

2.1 Metadata Retrieval

As shown in Figure 2 (left), metadata retrieval means a user's request for specific metadata service through bi-directional personalized broadcasting HMS and, in response to the request, the provision of metadata by the metadata server. For example, in order to retrieve specific programs, a user requests the metadata server to provide information about CRID (Content Reference Identifier), title, genre, keyword and publisher of contents. The request is made through the operation of TV-Anytime called 'get_Data' and the user is provided by the server with information.

2.2 Transmission of User-Centric Metadata

User-centric metadata is transmitted through the operation of TV-Anytime 'submit_Data.' Metadata related to the user such as UsageHistory and UserPreferenc is sent to the metadata server. On receiving the data, the metadata server sends back 'Acknowledgement' for the user-centric metadata as in Figure 2 (left). UserPreference is metadata about the user's personal preference, and UsageHistory about the user's usage history. Using UsageHistory metadata, which show how the user has consumed contents, the metadata server understand the user's pattern of consumption and, based on the understanding, recommend suitable contents or provides targeting services. Figure 2 (right) shows the general structure of UsageHistory information.

In order to create UsageHistory metadata, broadcasting HMS must update the user's UsageHistory information such as contents watched by the user and time when the user watched the programs. More efficient UsageHistory metadata may include the user's methods of consuming contents, e.g. the operation of Play, Preset Recording, Pause and FF/REW, the time of operation and the frequency of function.

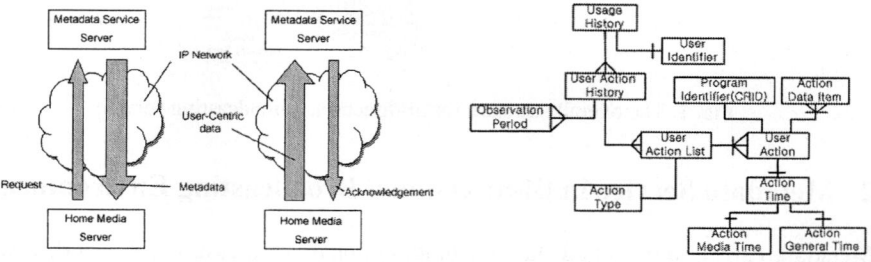

Fig. 2. Bi-directional Metadata Services(left) and UsageHistory Data Structure(right)

2.3 Transport Protocol

For the exchange of XML documents containing metadata between the server and personalized HMS, there should be network protocol that can be shared by the two sides. TV-Anytime adopted SOAP (Simple Object Access Protocol) [4] and HTTP [5], and standardized the transmission of TV-Anytime XML documents according to the protocols. The XML document requested to the server is contained in SOAP <Envelop>. HTTP header is added to the document and the document is sent to TCP/IP. HTTP includes connecting host IP, the type of contents, text coding method, the length of message and the type of SOAPAction.

3 Design of Bi-directional Personalized Broadcasting HMS

Broadcasting HMS for bi-directional personalized service is composed of a part of receiving and storing digital terrestrial broadcasts, hardware including network interface to exchange TV-Anytime metadata with metadata servers, and software including operating system, device drivers, TV-Anytime middleware and application programs.

3.1 Hardware

Hardware designed and implemented for bi-directional personalized broadcasting HMS in this study is composed of CPU module, system control module, MPEG-2 decoder & graphic processor module, RF tuber module and I/O module. The composition and function of each module are as follows.

3.1.1 CPU Module and System Control Module
This module drives the entire system to execute commands from the operating system and applications. Because of the characteristics of broadcasting HMS different from conventional PC, the module is composed of CPU, memory and I/O devices fit for embedded environment. HMS implemented in this study used 350MHz QED 5231 MIPS CPU. The CPU used 128Mbyte SDRAM system memory, 512Kbyte boot flash memory and Serial EEPROM. System control module plays the role of linking CPU, memories, PCI bus, local bus, peripherals and transport streams and, as in Figure 3, it is composed of memory controller, local bus controller, PCI bridge, serial I/O controller, I2C controller. In addition, it provides IDE interface for high-speed data transmission with hard disk, which is to store essential contents in personalized broadcasting HMS. In addition, it provides Ethernet interface through PCI bus for connection to IP network for bi-directional personalized broadcasting. I2C controller controls RF tuner module and RTC (Real Time Clock). The present system used TL811 IC of Zoran as the system controller.

3.1.2 MPEG-2 Decoder and Graphic Processor Module
The module creates video output and audio output by decoding MPEG-2 transport stream received from the system controller, and also plays the role of graphic processor. The present system used TL855 IC of Zoran. As shown in Figure 3, into TL855 are integrated MPEG-2 video decoder, transport demux, 2D graphic accelerator, video processor, audio processor. A separate memory is required for the demux of MPEG-2 and for audio/video decoding. The present system used 64Mbyte SDRAM. For audio decoding, both software decoding and hardware decoding using an audio process are available. Tuner module receives and demodulates a specific channel selected among

MPEG-2 HD stream of multiple channels transmitted through terrestrial broadcasting, and provide the result in the form of transport stream.

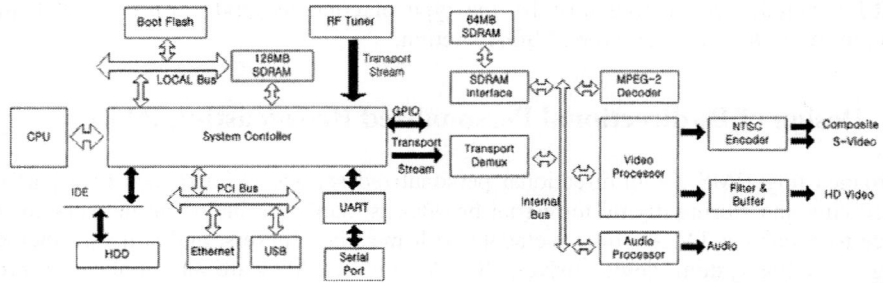

Fig. 3. Overall Hardware Structure of a Broadcasting Home Media Server

3.2 Software

As in Figure 4 (left), the software of bi-directional personalized broadcasting HMS is composed of operating system and device driver module, PVR (Personal Video Recorder) module performing functions such as time shift, PresetRecording and Trick-Play, parser that parses XML documents, database manager module that manages databases, network module for exchanging XML documents through the network, GUI module for providing user interface on the screen.

3.2.1 Operating System and Device Driver Module

Due to its characteristics, broadcasting HMS uses embedded operating system. Because basic audio and video decoding should not be interrupted by other tasks, there must be real-time functions. The present system used embedded Linux. The device driver module controls input-related hardware including tuner, transport stream demux, MPEG-2 decoder and graphic processor, which are for receiving digital broadcasts, and Ethernet controller and remote controller.

3.2.2 PVR Module

As in Figure 4 (right), PVR module decodes MPEG-2 transport streams from NIM (Network Interface Module) composed of tuner and demodulator and shows them on a display device, stores transport streams in storages such as hard disk and replays them, and performs the time shift function of storing and replaying simultaneously. The present system enables preset recording using PSIP (Program and System Information Protocol)[6] information like ordinary digital broadcasting recorders and, at the same time, enables users to select records based on detailed information about programs by providing them with information, which is more detailed than that provided by PSIP, through metadata retrieval as in Figure 2.

3.2.3 XML Parser

XML parser is a module that interprets metadata standardized based on XML [7]. In the personalized broadcasting HMS of this study, a parser called libXML [8] implemented in C was used, which runs smoothly with limited CPU and memory resource in embedded system environment. The process that XML documents are parsed is as follows. First, a XML file received through the network is stored in hard disk, its being well-formed is determined, and the validity of the document is checked. In case there is an abnormality in network environment or in the XML document sent by the metadata server, HMS requests re-transmission three times and, if transmission is still abnormal, it display an error message on the screen. Then, the header and tail related to SOAP are removed from the document and necessary data are extracted. Extracted data are reconstructed into information for the implementation of personalized service and provided to the user.

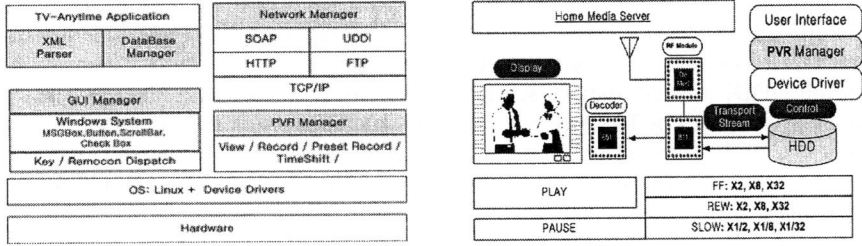

Fig. 4. Structure of Software Layer (left) and Basic Function of PVR (right)

3.2.4 Database Manager Module

Database manager module performs managing functions such as selecting, inserting, updating and deleting user information, recorded list, TV-Anytime metadata. In existing broadcast storage system such as PVR, only fragmentary information needs to be stored in addition to audio and video data, so information is stored in low-capacity non-volatile memory such as EEPROM or recorded in a file system embedded in PVR. However, in order to implement personalized information like TV-Anytime, a large volume of metadata about contents and user information such as users' contents preference and Usage-History should be stored, and this requires the use of DBMS (Database Management System). The present personalized broadcasting HMS used MySQL [9],which is suitable for an embedded system with insufficient resources. In order to provide personalized service, it manages each viewer's personal account and each individual's UserPreference data including genres, keywords, actors and publishers. UserPreference must be entered directly by each user when the account is created. Preference is managed by account, converted into UserPreference metadata, and sent to servers who provide TV-Anytime service. On receiving a request from a user, the metadata server retrieves programs based on UserPreference and provides a list of programs consistent with the preference and detailed information on the programs to the user in the form of TV-Anytime metadata [10]. In this way, the metadata server must have the function of retrieval based on User-Preference. On the other hand, a user's preference can be automatically generated and updated based on the user's UsageHistory as in TV3P [11], rather than being entered directly by the user. However, TV3P is implemented on PC, and it is not easy to implement and operate the functions in embedded system environment with limited resource. Thus, it is necessary make a separate research on optimization for the environment.

The present HMS created UsageHistory metadata by applying relatively simple algorithm based on the accumulated frequency of watching by program genre but, for more efficient UsageHistory metadata, it is necessary to develop an algorithm that utilizes the details of consumption including all metadata about programs, the type and time of program operation, frequencies. Algorithm development is one of areas studied intensively for personalized broadcasting service [11].

3.2.5 Network Module

Network module uses a transport stack explained in Section 3 of Chapter 2 in order to exchange metadata through bi-directional communication. In addition, it receives transport streams through FTP, and delivers the state of data transmission and errors.

Figure 5 shows the process of metadata service retrieval. Bi-directional personalized broadcasting HMS sends a query to the UDDI service registry server using <find_business>, and receives metadata containing <businessKey> and <serviceKey> from the UDDI service registry server. Broadcasting HMS parses the metadata, extracts <serviceKey> of TV-Anytime from the data, and sends a query to the server in the form of <get_ serviceDetail> in order to get an access point to the metadata server for the service. Then, from the replied document, broadcasting HMS extracts the IP of the metadata server, which is the access point of metadata service, and connects to the metadata server. If connection is accepted, HMS sends the server a XML document corresponding to 'get_Data' to retrieve information from the server. Then the metadata server sends a reply to the request to the personalized broadcasting HMS.

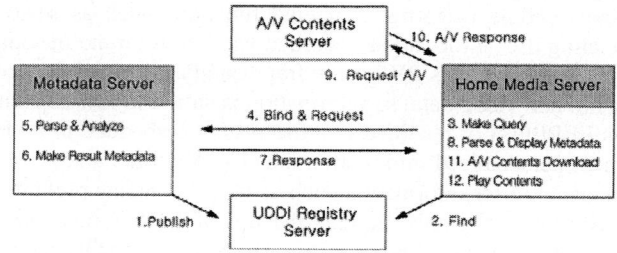

Fig. 5. Process of TV-Anytime Service discovery using UDDI

4 Implementation of Bi-directional Personalized Broadcasting HMS and an Example of Service

Bi-directional personalized broadcasting HMS can receive and record digital terrestrial broadcasts, and provides various bi-directional personalized service through network module. If the user wants to obtain information about programs selected from contents that are currently being broadcasted, the system receives data on the programs from the user and requests information to the metadata server. It again receives a broadcasting list and detailed information meeting the user's request and provides them to the user so that the user can select and watch preferred broadcasts. In addition, the system provides bi-directional personalized services by sending user-centric metadata UserPreference or UsageHistory to the server and receiving the server's recommendation of programs fit for the user.

Bi-directional personalized services implemented in the present broadcasting HMS are explained below. We explains the service of sending a user's preference to the metadata server and receiving from the server recommendation of programs fit for the user's preference, the service of sending a user's UsageHistory and receiving from the server recommendation of programs based on the user's tendency, and an example of metadata service retrieval.

4.1 Broadcasting Program Retrieval Using User Preference

Metadata on UserPreference contains preferred genres, program times, publishers, keywords. These items are directly entered by the user when the user account is opened in broadcasting HMS, and may be changed when the user retrieves programs. Figure 6 (left) shows a screen for entering preference data. In the figure, three preferred genres Fiction, Sports and Leisure were selected. Users who want to select keywords or publishers can use the Detail Information ('More') menu on the right bottom.

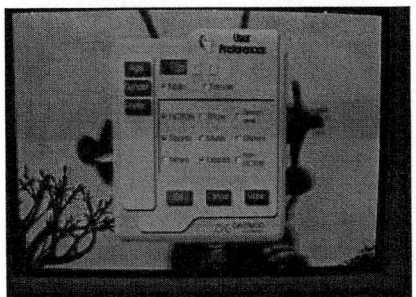 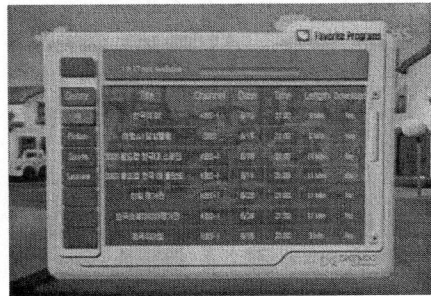

Fig. 6. Snapshot of setting the user preference (left) and Snapshot of metadata service server response for user preference (right)

Based on preference stored or entered as above, a program retrieval request is prepared as a XML document and sent to the metadata server. Then, in reply to the request, the server sends a reply XML document to HMS. Figure 6 (right) shows the server's reply to the preference. It shows a list of programs of the preferred genres, namely, fiction, sports and leisure, broadcasting time, broadcasters, download ability. In the figure, recommended programs are shown together in a list. If the Recommended Genre menu on the left side of the screen is selected, recommended programs only for the selected genre can be viewed. Download ability tells whether or not the contents can be downloaded from the contents server at <ProgramURL> as specified in ProgramLocationTable of transmitted metadata. In addition, if the user wants detailed information about each of recommended programs, the HMS may request the information to the metadata server. Figure 7 (left) shows detailed information about program 'Travel Show! Escape from Routine' such as its synopsis and review sent by the server in reply to a request for detailed information.

4.2 Broadcasting Program Recommendation Using UsageHistory Metadata

In this example of service, UsageHistory metadata created based on an individual user's UsageHistory is sent to the metadata server and programs are recommended based on the result of analyzing the UsageHistory. User-centric metadata are sent to the server and used in identifying users' tendency in watching programs and recommending appropriate programs. As shown in Figure 7 (right), a list of recommended programs personalized to each user's consumption pattern is provided to the user together with information such as program title, publisher, broadcasting time, length and download ability, and the user can select a program from the list and watch it or do preset recording. The figure shows that mainly show programs and leisure programs are recommended to users who have a tendency to watch programs of these genres (Show and Leisure). Recommended programs can be viewed by genre or all together. The example in the figure shows all programs recommended together. Listing by genre is available by selecting the Recommended Genre menu on the left side of the screen.

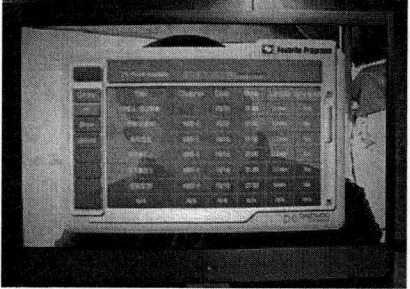

Fig. 7. Detailed Information of selected program (left) and snapshot of program list recommended by metadata service server (right)

5 Conclusion

In order to utilize bi-directional personalized broadcasting service, the present study presented an example of bi-directional personalized service by designing and implementing bi-directional personalized broadcasting HMS that satisfies TV-Anytime standard. It showed that broadcasting HMS composed of hardware and software for bi-directional personalized service can provide bi-directional personalized broadcasting services effectively in addition to the basic functions of broadcast receiving and recording under limited resources of the embedded system of digital broadcasting HMS. The implemented bi-directional personalized broadcasting HMS sends metadata about individual users' choice or preference and their UsageHistory to metadata servers through the return-channel, receives replies from the servers, and provides personalized services to the users based on the information from the metadata servers. Bi-directional environment enables services personalized to each individual's demand, which are not available in unidirectional environment. The activation of bi-directional personalized services must be preceded by the directional personalized services must be preceded by the expansion of TV-Anytime metadata service providers who provide services diversified and specialized for different groups of users.

References

1. SP003v1.3, TV-Anytime Specification-Metadata, 2003, "http://www.tv-anytime.org"
2. SP006v1.0, TV-Anytime Specification - Metadata Services over a Bi-directional Network, 2003, "http://www.tv-anytime.org"
3. Universal Description Discovery & Integration, Version 3.0, "http://uddi.org/pugs/uddi-v3.00-published-20020719.htm"
4. Simple Object Access Protocol (SOAP) 1.1, W3C Note, 8 May 2001, "http://www.w3.org/TR/2000/NOTE-SOAP-20000508"
5. RFC1945- Hypertext Transfer Protocol, HTTP/1.0, "http:// www.ietf.org/rfc/rfc1945.txt"
6. ATSC Standard A/65. Program and System Information Protocol for Terrestrial Broadcast and Cable, http://www.atsc.org"
7. W3C, Extensible Markup Language (XML) Version 1.0 Recommendation, February 1998, "http://www.w3.org/TR/2004/ REC-xml-20040204"
8. "http://www.xmlsoft.org"
9. "http://www.mysql.com"
10. Lee.j.s. Lee.suk.phil "Design and Implementation of bi-directional TV-Anytime system for personalized broadcasting service" The Korean Internet Broadcasting / TV Institute Vol 3, No.1, 2003.
11. Z. Yu and X. Zhou, "TV3P: An Adaptive Assistant for Personalized TV," IEEE transaction on Consumer Electronics, Vol. 50, No. 1, Feb. 2004

Optimal Signal Control Using Adaptive Dynamic Programming

Chang Ouk Kim[1], Yunsun Park[2], and Jun-Geol Baek[3]

[1] Department of Information and Industrial Engineering Yonsei University,
Sinchon-dong, Seodaemun-gu,Seoul, 120-749, Republic of Korea
[2] Department of Industrial and Systems Engineering, Myong Ji University, Yongin,
Kyungki-Do, 449-728, Republic of Korea
[3] Department of Industrial Systems Engineering, Induk Institute of Technology,
Seoul, 139-749, Republic of Korea

Abstract. This paper develops an adaptive, optimal planning algorithm for signal control at a single intersection using an efficient dynamic programming technique. It is called ADPAS (Adaptive Dynamic Programming Algorithm for Signaling). The objective of ADPAS is to minimize the total delay experienced by vehicles passing through an intersection. ADPAS can generate any sequence of green phases to optimize signal control without restriction to fixed cycles of green phases. The algorithm employs *reaching* as the method to solve the forward DP functional equation, which does not require any prior knowledge of the states of the DP network. The efficiency of the algorithm results from two acceleration techniques that adaptively eliminate inferior states as the algorithm progresses. We verify computational efficiency of ADPAS with several test cases.

1 Introduction

ITS (Intelligent Transportation Systems) has evolved to address the worsening traffic problems of congestion, inefficiency, danger, and pollution. One of the most important issues for successful ITS implementation is to devise a real time, adaptive signal planning scheme for efficient signal control at an intersection. In this paper, we address this issue by developing an optimal signal planning algorithm that gains its efficiency via computational acceleration schemes. Our objective is to minimize the total delay experienced by the vehicles passing through an intersection. However, other performance measures can be used by an appropriate replacement of the cost function.

We can classify traffic control schemes into two major categories. One is composed of fixed cycle schemes and the other consists of adaptive control schemes. The first method developed based on the fixed cycle scheme is Webster's Method [9]. This method and its variants have been widely used. They provide formulas for the optimal partition of green time at a fixed cycle intersection. They also find efficient cycle times. However, fixed cycle signal planning is generally based on historically recorded or observed data. Thus, to accommodate the actual real-time traffic and to take advantage of the potential for adaptive control, more sophisticated types of traffic control algorithms are required.

SCOOT [4] and SCATS [8] are two notable real-time, adaptive traffic control systems that have been more recently developed and widely used. However, signal plans they generate are centralized, i.e., they are selecting predetermined signal plans. SCOOT [4] is an off-line optimization procedure used in a real time mode, incrementally adjusting the signal timing plan employed as traffic conditions change. OPAC [2], PRODYN [3], UTOPIA [5], SPORT [10], and ALLONS-D [6] are also adaptive signal control methods, which are all heuristic and decentralized. They collect real-time data, find a signal plan and implement the signal plan found all on-line. They employ the rolling horizon concept of breaking up the time period into stages and implementing only the first decision of each stage.

More recently, a general purpose algorithm for real time traffic control at an intersection, called Controlled Optimization of Phases (COP), was developed by Sen and Head [7]. COP uses a Dynamic Programming (DP) approach, as we will do, but it may not produce an optimal signal policy when the objective is to minimize total delay at an intersection since it aggregates states for computational simplicity. The state in COP is composed of a single variable, i.e. time. The number of vehicles at the intersection, which should be considered to obtain an optimal signal plan to minimize total delay, is sacrificed for computational feasibility. In COP, lengths of vehicles are updated for accounting purposes but are not part of the state information.

The other schemes mentioned above fail to result in optimal solutions as well. OPAC [2] uses a restricted search heuristic which enumerates limited feasible solutions. PRODYN [3] discuss DP formulation, but fails to utilize the resulting opportunity to optimally solve the resulting DP. ALLONS-D [6] could be used as an optimal scheme, but its computational inefficiency cannot allow this since its main idea is a tree search for an optimal solution.

In this study, we will present an adaptive, optimal, and efficient signal planning algorithm to minimize total delay at an intersection. The algorithm is based on dynamic programming techniques complemented by two acceleration methods to adaptively eliminate, or prune (without loss of optimality) unnecessary states, thus eliminating these states and their successor states from further consideration. It is called ADPAS (Adaptive Dynamic Programming Algorithm for Signaling). Compared with other schemes, it generates an optimal signal plan. Despite its ability to solve to optimality, because of the two acceleration methods, we believe that it can be successfully applied in (near) on-line mode within a rolling horizon environment. We make decisions by selecting the phase to be green for an (additional) predetermined amount of time at each decision epoch. Hence, our method is not restricted to fixed cycle times and moreover can generate any sequence of phases.

In general, there are two approaches to DP formulations, backward and forward. Backward DP, which is usually solved by a recursive fixing method, would require complete prior knowledge of the DP network states for computation. In backward recursion, it is not possible in this case to know the boundary states in advance to start the recursion. However, if we adopt a forward DP formulation, we can generate relevant DP network states as the algorithm proceeds, removing the computational burden of searching the entire DP network prior to algorithmic implementation. Thus, in this paper, we develop a forward dynamic programming algorithm to perform an unconstrained, optimal search for traffic signal plans. The algorithm will be accelerated in efficiency by maintaining an *incumbent* value (the current best feasible solution) throughout the algorithm to eliminate states whose cost (delay) already exceeds the incumbent. Additional acceleration is achieved by comparing two states at the same time and green phase. As we will show later, if each phase on one state has more vehicles than those of the other state, the state with more vehicles can be eliminated without loss of optimality, thus reducing the number of states that need to be explored. In summary, this paper presents an efficient forward DP algorithm based on reaching together with acceleration techniques using incumbent and dominated state comparisons.

Section 2 will present our DP formulation of the signal control problem at an intersection. Based on the formulation in Section II, Section 3 will discuss our proposed acceleration schemes as well as the corresponding DP algorithm for the problem. In Section 4, we present a small example to illustrate the operational mechanism of the algorithm. In addition, numerical solutions to large examples are presented in Section IV to explore the computational efficiency of the algorithm. Finally, Section 5 will conclude this study with several remarks and suggestions for future research.

2 Dynamic Programming Formulation of the Signal Planning Problem

As we have discussed in the previous section, most signal control schemes employ heuristic approaches or approximations. This section formulates the signal control problem using dynamic programming to solve the problem optimally. It is the basis for the algorithm developed in Section III.

2.1 Problem Definition

General knowledge on procedures for dynamic programming formulations is well documented in [1]. To formulate the signal control problem of minimizing total delay experienced by the vehicles at an intersection during time 0 to H (the problem horizon) using dynamic programming, one of the most important aspects is to define an appropriate *state* for the problem. *State* in dynamic programming is so fundamental that if we can discover the proper state for the problem, we can readily complete the dynamic programming formulation by providing a dynamic recursive functional equation based on the state defined.

Many different DP state definitions for a problem are possible, depending on the underlying decisions used to formulate the problem, ultimately resulting in different DP formulations. In our study, the decision we will make at each epoch is which phase will be green until the next decision epoch. A *phase* is defined as a combination of non-conflicting movements that allow safe passage of vehicles through the intersection. Thus, the set of all phases means the set of such possible vehicle movements at the intersection. Another possible decision is how long the current phase in a cycle should be kept, as in COP (our decision can generate any signal plan COP does by simple adjustment of Δ below). Thus, if we decide to maintain the current phase as green until the next decision epoch, we maintain the current phase as green for the pre-specified amount of green time (Δ). On the contrary, if we decide to make another phase green, we impose the predefined amount of a yellow-red time (t_{YR}) to clear the intersection followed by the predefined minimum green time (t_{MG}) for the new green phase.

Each decision leads to a new state. The state should incorporate sufficient information about preceding decisions made to complete the decision tree arising from this point. Then, by adopting decisions of selecting the next phase to be green at each decision epoch, the corresponding state should have the following information: current time, current green phase, and current number of vehicles waiting at each phase. We assume that the traffic condition at the intersection is not heavy enough to consider a maximum green duration. Green time is to some extent limited by the optimality criterion. Its inclusion as a constraint would increase computation time significantly, although it is possible to do.

If we assume a two-phase intersection, the state variable at time t, $S(t)$ is

$$S(t) = (t, p(t), m(t), n(t))$$
$$= \text{(current time, current green phase,}$$
$$\text{current number of vehicles on phase 1,}$$
$$\text{current number of vehicles on phase 2)}$$

From the current state $S(t)$ at time t, we make a decision, $x(t)$, incurring cost of delay, $\sum_{phases,l} \sum_{vehicles,i} D_{i,l,t}(x(t))$, and the decision leads to a new state after the corresponding time, t_{step}, where

$$x(t) = \begin{cases} 0 \text{ if the current phase is maintained as green} \\ 1 \text{ otherwise} \end{cases}$$

$$t_{step} = \begin{cases} \Delta & \text{if } x(t) = 0 \\ t_c = t_{YR} + t_{MG} & \text{if } x(t) = 1 \end{cases}$$

$D_{i,l,t}(x(t))$ = delay incurred by vehicle i at phase l by the decision $x(t)$ at time t

$\sum_l \sum_i D_{i,l,t}(x(t))$ = total delay incurred at an intersection by the decision $x(t)$ at time t

Figure 1 shows the dynamics of a decision at time t.

Basically, the delay incurred by the vehicle is the difference between the departure time and arrival time of the vehicle, and it is affected by the phase

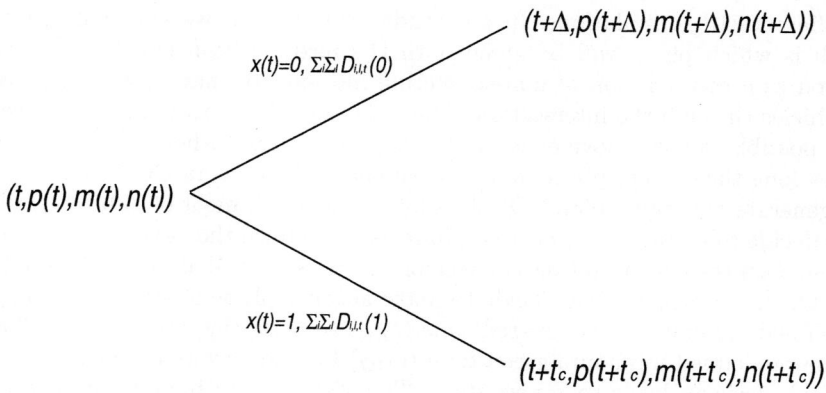

Fig. 1. Dynamics of Decisions from Current State at time t

experienced by the vehicle at the intersection as well as by the saturation flow rate of each phase. The way $D_{i,l,t}(x(t))$ is calculated is elaborated in detail in [6].

Since we have defined our decision, the corresponding state, and its dynamics, the remaining task to complete the DP formulation is to define the DP recursion; that is, to define the DP functional equation based on the optimal value function with an appropriate boundary condition to start the recursion. The value function is the total cost "from" or "to" a state according to the recursion used: backward or forward.

As mentioned, there are two approaches to DP formulation yielding the same result, forward recursion and backward recursion. Generally, they are equivalent in computational complexity. However, in some cases, one can be more efficient than the other as discussed before. The problem with the backward recursion is that the method of recursive fixing to solve this backward recursion requires that we should know all possible states of DP network even before we start the recursion, making it computationally inefficient. However, forward recursion of this problem does not require all the states to be known priori. Thus, we formulate the problem based on forward recursion to develop a correspondingly efficient computational algorithm with some acceleration techniques.

2.2 Problem Formulation-Forward Recursion

As discussed previously, backward recursion is not efficient computationally. Now, we present the forward recursion which can exploit an efficient computational methodology called *reaching*. Since forward recursion starts from the beginning of the DP network and moves forward until the horizon is reached, we define the optimal value function of the forward recursion $f(t, p(t), m(t), n(t))$ as the total minimum delay incurred from time 0 to time t of the current state, $(t, p(t), m(t), n(t))$. Moreover, the decisions among which we select to minimize are those "to" the current state, $(t, p(t), m(t), n(t))$, not "from" the current state

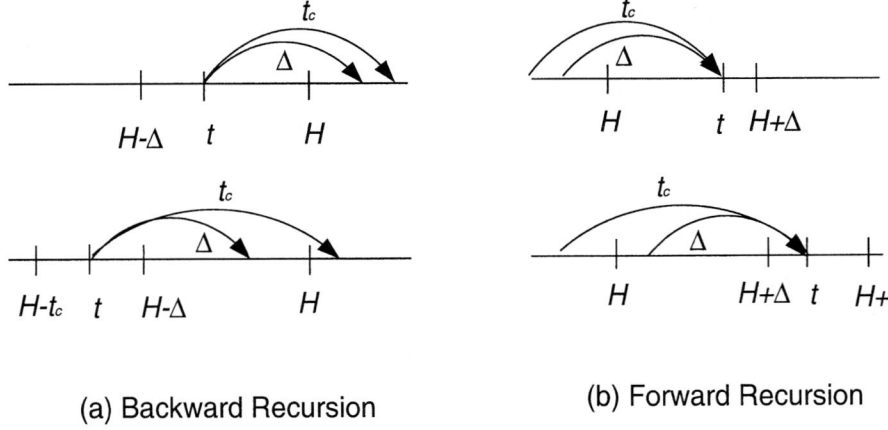

(a) Backward Recursion (b) Forward Recursion

Fig. 2. Decisions "to" Current State

as in the backward recursion. Let $Y(t)$ be the set of decisions leading to the current state, $(t, p(t), m(t), n(t))$ (refer to Figure 2). Note that $Y(t)$ can have more than two decisions to the current state by the state aggregation from the decision tree.

Then the forward functional equation is:

$f(t, p(t), m(t), n(t)) =$
$\min_{y(t) \in Y(t)} \{ f(t - t_y, p(t - t_y), m(t - t_y), n(t - t_y)) + \sum_l \sum_i D_{l,i,t-t_y}(y(t)) \}$
for $t = \Delta, 2\Delta, \ldots, t_c + \Delta, t_c + 2\Delta, \ldots, 2t_c + \Delta, 2t_c + 2\Delta, \ldots$
and appropriate $p(t), m(t), n(t)$,

where

$$t_y = \begin{cases} \Delta \text{ if } y(t) = 0 \\ t_c \text{ if } y(t) = 1 \end{cases}$$

Since forward recursion starts from the beginning of the DP network, the boundary condition is simple: $f(0, p(0), m(0), n(0)) = 0$. However, because of the end of horizon effect, the minimum total delay we will obtain at the end of the recursion is a little complicated, as in the following (refer to Figure 3):

$$\min \begin{cases} \min_{H \leq t \leq H+\Delta} \{ \min_{y(t)=0,1} \{ f(t - t_y, p(t - t_y), m(t - t_y), n(t - t_y)) + \sum_l \sum_i D_{l,i,t-t_y}(y(t)/H) \} \} \\ \min_{H+\Delta < t \leq H+t_c} \{ \min_{y(t) \neq 0} \{ f(t - t_y, p(t - t_y), m(t - t_y), n(t - t_y)) + \sum_l \sum_i D_{l,i,t-t_y}(y(t)/H) \} \} \end{cases}$$

Generally, there are two solution methodologies to solve DP forward recursion: recursive fixing and reaching [1]. As in the backward case, recursive fixing would require knowledge of some possible states in advance. However, if we employ the reaching methodology, we can have an efficient algorithm through which

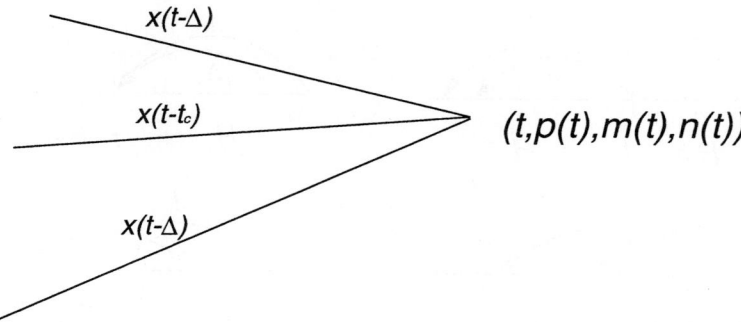

Fig. 3. End-of-Horizon Effects

we can perform the DP recursion without prior knowledge of all possible states of the DP network. The next section will discuss an efficient DP algorithm using reaching with acceleration by dominated state pruning.

3   Dynamic Programming Algorithm: Reaching and Pruning

Although the forward recursion does not require full prior knowledge of the state space as in the case of backward recursion, the forward recursive fixing method still requires partial knowledge of next states from the current state to solve each functional equation. However, if we employ the reaching methodology for the forward recursion, we do not need any prior knowledge on the state space of the DP network: we can evaluate the functional equation recursively as we create states. Thus, we will choose reaching over recursive fixing as the algorithm to solve the forward recursion.

To accelerate the algorithm, we will maintain an *incumbent* throughout the algorithm, the best feasible solution so far. The incumbent will be used to eliminate any state with value function greater than or equal to the incumbent value. Another acceleration is possible by observing two states of the same phase at the same time. For example, consider two states $(t, p(t), m(t), n(t))$ and $(t, p(t), m'(t), n'(t))$ with respective value functions f and f'. If $m(t) \leq m'(t)$, $n(t) \leq n'(t)$, and $f \leq f'$, we can eliminate the state, $(t, p(t), m'(t), n'(t))$, without loss of optimality, since under the same circumstances, more vehicles on one phase mean more delays in the future. This way of eliminating states will accelerate the algorithm by pruning inferior states before we investigate them. The following lemma and theorem formalize the above intuition.

Lemma 1. *Consider two states $(t, p(t), m(t), n(t))$ and $(t, p(t), m'(t), n'(t))$. If $(m(t), n(t)) \leq (m'(t), n'(t))$, then vehicles of the state, $(t, p(t), m(t), n(t))$ are a subset of the vehicles of the state, $(t, p(t), m'(t), n'(t))$.*

Proof. Since the arrival pattern at time t which has led to each state is the same, and two states are at the same time and the same green phase, $(m(t), n(t)) \le (m'(t), n'(t))$ means that among the vehicles which have arrived at the intersection by time t, the sequence of decisions leading to state, $(t, p(t), m(t), n(t))$, has cleared more vehicles out of the intersection. Thus, the vehicles of the state, $(t, p(t), m'(t), n'(t))$ include the vehicles of the state $(t, p(t), m(t), n(t))$.

From the lemma, we can show that we can eliminate inferior states from consideration without loss of optimality.

Theorem 1. *Consider two states, $(t, p(t), m(t), n(t))$ and $(t, p(t), m'(t), n'(t))$, with respective value functions f and f'. If $(m(t), n(t)) \le (m'(t), n'(t))$ and $f \le f'$, then the state, $(t, p(t), m'(t), n'(t))$, can be eliminated without loss of optimality.*

Proof. The future process from each state only depends on the number of vehicles on each phase since the two states are on the same green phase at the same time. From the lemma $(m(t), n(t)) \le (m'(t), n'(t))$ means that there may be additional vehicles on each phase of state, $(t, p(t), m'(t), n'(t))$. Moreover, since $f \le f'$, the future process from state, $(t, p(t), m'(t), n'(t))$, will incur more delay than the process from state, $(t, p(t), m(t), n(t))$, for any sequence of decisions.

Now, we are ready to show a reaching algorithm for optimal signal control with acceleration of state pruning through incumbent and state comparison.

Throughout the algorithm, we will maintain:

$Incumb$: incumbent, the current best feasible solution
P : set of permanent states
T : set of temporary states

P is the set of states whose permanent *optimal* value functions have been found, whereas T is the set of states whose optimal value functions are not found yet. However, temporary states have (temporary) value functions. Also, we will include the value function in the state for convenience, so the modified state consists of $(t, p(t), m(t), n(t), f)$.

ALGORITHM

Step 0: Initialization
  - $t = 0$, $Incumb = \infty$, $P = \emptyset$, $T = \{(0, p(0), m(0), n(0), 0)\}$

Step 1: Stopping Criterion
  - eliminate any state with $f \ge Incumb$ from T.
  - if $T = \emptyset$, stop.

Step 2: State Selection
  - select a state from T with smallest t (if tie, smallest value function; if tie, smallest total number of vehicles).
  - call it current state, $(t, p(t), m(t), n(t), f)$.

Step 3: Pruning by State Comparison
- if $(m(t), n(t), f)$ of the current state is greater than or equal to that of any state in P with the same time and green phase, eliminate the current state and go to Step 1.

Step 4: Reaching
- create new states $(t', p(t'), m(t'), n(t'), f')$ from each decision of the current state, where

$$t' = \begin{cases} \min\{H, t+\Delta\} & \text{if } x(t) = 0 \\ \min\{H, t+t_c\} & \text{if } x(t) = 1 \end{cases}$$

$$p(t') = \begin{cases} p(t) & \text{if } x(t) = 0 \\ p^c(t) \text{ (complement of } p(t)) & \text{if } x(t) = 1 \end{cases}$$

$$m(t') = m(t) + \#Arrv1(t, t') - \#Dept1(t, t')$$
$$n(t') = n(t) + \#Arrv2(t, t') - \#Dept2(t, t')$$

$$f' = f + \begin{cases} \sum_l \sum_i D_{l,i,t}(x(t)) & \text{if } t' < H \\ \sum_l \sum_i D_{l,i,t}(x(t)/H) & \text{if } t' = H \end{cases}$$

where $Arrvi(t, t')$ ($Depti(t, t')$) is the number of arrived (departed) vehicles during (t, t') at phase i.
- include the current state into P.

Step 5: Pruning by Incumbent for each state generated by Step 4
- if $t' = H$ and $f' < Incumb$, $Incumb \leftarrow f'$.
- if $f' \geq Incumb$, eliminate the new state.

Step 6: State Aggregation for each state generated by Step 4
- if there exists a state in T, $(t', p(t'), m(t'), n(t'), f'')$, with the same time, the same green phase, and the same vehicles as the new state, $(t', p(t'), m(t'), n(t'), f')$, created in Step 4, then aggregate two states into a single state, $(t', p(t'), m(t'), n(t'), \min(f', f''))$.
- if $t' < H$, include the aggregated state into T
- if $t' = H$, include the aggregated state into P
- go to Step 1

The algorithm stops when the temporary state set, T, is empty, that is, when every state in the network is either assigned its optimal value function or eliminated. Then, the optimal solution can be found by selecting a state from the set P with the smallest value function at the horizon. Step 3 and Step 5 are acceleration procedures which eliminate unnecessary states by comparing states and comparing value functions with the incumbent, respectively. Note that Step 1 also has incumbent comparison to remove some of the temporary states which were included by the previous (worse) incumbent. Step 6 is the aggregation process for states from the tree produced by the reaching process of Step 4. It is this aggregation process which gives the name of dynamic programming to this algorithm.

4 Experiments

In this section, we will at first solve an example problem using the reaching algorithm developed in the previous section. The example is taken from [7]. The arrival pattern during the next 10 unit times is depicted in Figure 4.

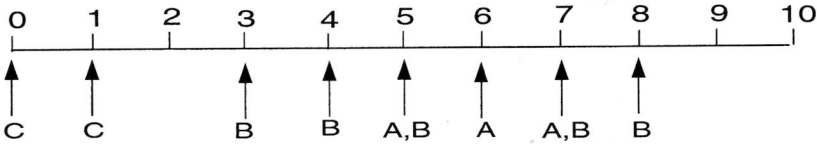

Fig. 4. Vehicle Arrival Pattern for Example

The intersection is composed of 3 phases, A, B, and C. $\Delta = 2$, $t_{MG} = 2$, $t_{YG} = 1$, and the saturation flow rate is assumed to be infinity for this example. We assume that the last phase to be green is not required to meet the minimum green time. We want to minimize the total delay incurred by the vehicles arriving at the intersection during next ten unit times. For this example, a vehicle's delay will be simply calculated as the difference between the departure time from the intersection and the arrival time to the intersection, ignoring any headway constraints. That is, vehicles leave the intersection as soon as the green light is turned on regardless of the vehicles in front of them.

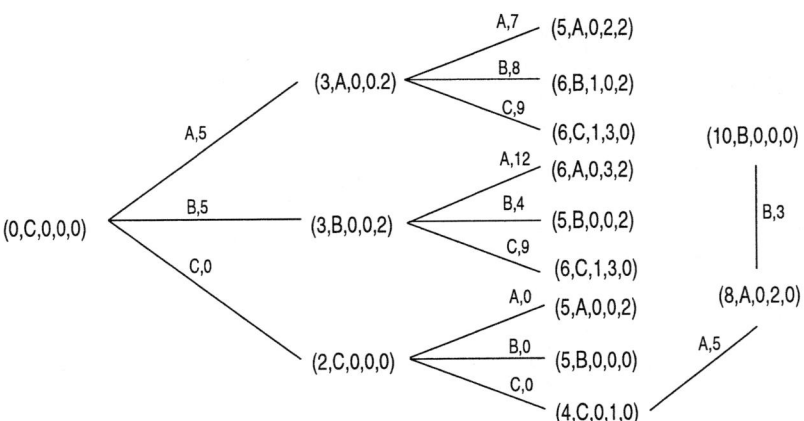

Fig. 5. Part of Decision Tree for Example

Figure 5 shows a part of the whole decision tree for the example. Note that by aggregating the two states $(6, C, 1, 3, 0)$ on the decision tree, we can transform this decision tree into a DP network (Step 6). The optimal solution is imposing 2 unit times of green time to Phase C, 2 unit times of green time to Phase B,

2 unit times of green time to Phase A, and 1 unit time of green time to Phase B with yellow-red times between phase transitions as depicted in Figure 6. The total delay caused by the sequence of green phases is 8. Note that the last green phase does not meet the minimum green time requirement, as we assumed.

Fig. 6. Optimal Signal Policy for Example

To verify the computational efficiency in terms of CPU time, we compared it with standard Reaching-based Forward DP (RFDP) algorithm where the two acceleration schemes are not employed. A personal computer and the C++ language were used to implement both algorithms. Table 2 shows the comparison result. Three-phased intersection is assumed for all test cases. From this result, it is observed that, as planning horizon increases, the computational time of RFDP exponentially grows because of the curse of dimensionality. On the contrary, ADPAS has been shown to greatly reduce computation burden and to outperform RFDP, since it is capable of eliminating many inferior states during DP procedure (refer to the last column of Table 1).

Table 1. Comparison of Computational Efficiency

Planning horizon	No. of vehicles arrived	Optimal delay	CPU time RFDP	CPU time ADPAS	Aggregation	Acceleration
10	16	33	0.05	0.0001	7	15
11	18	43	0.511	0.0003	6	23
12	17	40	0.831	0.0007	12	52
13	20	47	1.722	0.01	24	81
14	20	42	4.086	0.02	34	87
15	21	53	13.61	0.07	72	226
16	23	67	64.362	0.14	112	484
17	25	67	123.376	0.26	158	631
18	26	73	434.535	0.39	243	1004
19	26	69	N/A	0.641	298	1300
20	28	91	N/A	1.352	446	2380
21	30	80	N/A	2.464	509	3617
22	32	100	N/A	3.113	646	4119
23	34	119	N/A	4.786	866	6088
24	36	123	N/A	6.62	1044	7875
25	38	132	N/A	10.114	1261	11392
26	40	141	N/A	14.12	1523	14548
27	41	161	N/A	19.468	1862	17937

5 Conclusion

In this paper, we developed an efficient adaptive dynamic programming algorithm for the optimal signal control of a single intersection to minimize the total delay incurred by the vehicles passing through the intersection. We called it ADPAS (Adaptive Dynamic Programming Algorithm for Signaling). Compared to other algorithms, including COP which also adopts a DP approach, it is the first algorithm which gives an exact optimal signal policy to minimize total delay at an intersection with any phase sequencing. COP's decision is "how many green times in each fixed sequence of phase" from each state, whereas ADPAS decides "which phase to be green" for a predetermined amount of time. ADPAS can be easily modified to give any of the signal plans under COP's decision making policy by selecting an appropriate amount of green time.

ADPAS assumes a fixed horizon for the signal plan by ignoring the effect of uncleared vehicles at the intersection when the horizon is reached. However, the end-of-horizon effect should be studied further since the sequence of decisions leading to vehicle-remaining by the horizon could be penalized by its incompleteness. Finding a reasonably good (not time consuming) starting incumbent might be a future research topic, since the acceleration can be increased with this good incumbent. An important future research topic is to develop a coordinated signal control system for a traffic network with an adaptive signal control scheme at each intersection. Without the coordination of signal plans at each intersection, even the best signal policy at each intersection could fail to capture savings possible through system wide coordination.

References

1. E.V. Denardo, *Dynamic Programming: Models and Applications*, Prentice Hall, 1982.
2. N.H. Gartner, "OPAC: A demand-responsive strategy for traffic signal control," *Transportation Research Record*, vol. 906, pp. 75-81, 1983.
3. J.J. Henry, J.L. Farges, and J. Tuffal, "The PRODYN real time traffic algorithm," *4th IFAC-IFIP-IFORS Conference on Control in Transportation System*, Baden Baden, Germany, 1983.
4. P.B. Hunt, D.I. Robertson, R.D. Bretherton, and R.I. Winton, "SCOOT - a traffic responsive method for coordinating signals," *Laboratory Report no. LR 1014, Transportation and Road Research*, Crowthorne, Berkshire, England, 1981.
5. V. Mauro and D. DiTaronto, "UTOPIA," *Proceedings of the 6th IFAC/IFIP/IFORS Symposium on Control Computers and Communication in Transportation*, no. 12, pp. 245-252, 1990.
6. I. Porche and S. Lafortune, "Adaptive look-ahead optimization of traffic signals," *Technical Report no. CGR 97-11, Department of Electric Engineering and Computer Science, The University of Michigan*, 1997.
7. S. Sen and K.L. Head, "Controlled optimization at an intersection," *Transportation Science*, vol. 31, pp. 5-17, 1997.

8. A.G. Sims, "The Sydney coordinated adaptive traffic system," *Urban Transport Division of ASCE Proceedings, Engineering Foundation Conference on Research Directions in Computer Control of Urban Traffic Systems*, New York, NY, pp. 12-27, 1979.
9. F.V. Webster, "Traffic Signal Settings," *Road Research Technical Paper, no. 39*, HMSO, London, 1958.
10. S. Yargar and B. Han, "A procedure for real-time signal control that considers transit interference and priority," *Transportation Research B*, vol. 28, pp. 315-331, 1994.

Inverse Constrained Bottleneck Problems on Networks

Xiucui Guan and Jianzhong Zhang

Department of Mathematics,
City University of Hong Kong,
Hong Kong
xcguan@163.com

Abstract. In this paper, we consider a class of Inverse Constrained Bottleneck Problems (ICBP). Given two weight vectors w^1 and w^2, the constrained bottleneck problem is to find a solution that minimizes w^1-bottleneck weight subject to a budget constraint on w^2-sum weight. Whereas, in the inverse problem (ICBP), a candidate solution is given, we wish to modify the two weight vectors under bound restrictions so that the given solution becomes an optimal one to the constrained bottleneck problem and the deviation of the weights under some norm is minimum. When the modifications of two weight vectors are proportioned, we present a general method to solve the problem (ICBP) under weighted l_∞ norm and give two examples of the method including the inverse constrained bottleneck spanning tree problem and the inverse constrained bottleneck assignment problem.

1 Introduction

Inverse optimization problem has become a favorite topic recently, which was motivated by its background in traffic planning, and then was developed into more and more application fields, such as high speed communication, computerized tomography, isotonic regression, conjoint analysis in marketing, etc [4]. In an inverse optimization problem, a candidate solution is given and the goal is to modify parameters of the original problem so that the given solution becomes an optimal one under the new parameters and simultaneously minimize the costs incurred by the modification of parameters [5]. The modification cost function is typically a convex function of the magnitude of the deviation, where the norms l_1, l_2, l_∞ are generally used. When the objective function of the original problem is in the form of $\min\max\{c_ix_i|i \in I\}$ (or $\max\min\{c_ix_i|i \in I\}$), for example, Bottleneck Spanning Tree problem (BST), only few results on the corresponding inverse optimization problems are given, where Cai, Yang and Zhang [3] proved that the inverse center location problem is strongly NP-hard and Zhang et al. [8,10] presented polynomial algorithms for some inverse max-min (or min-max) optimization problems.

Before we introduce the Inverse Constrained Bottleneck Problem (ICBP), we first present the Constrained Bottleneck Problem (CBP) on a network $G =$

(V, E), which is a special case of the problem discussed in [1]. For each edge e, define two weights $w^1(e)$ and $w^2(e)$. Let F be a set of edges that have some required properties. We refer to F as a feasible solution and let $\mathcal{F}$ be the set of all feasible solutions. For any $F \in \mathcal{F}$, let $g(F, w^1) := \max_{e \in F} w^1(e)$ be w^1-weight bottleneck function and $f(F, w^2) := \sum_{e \in F} w^2(e)$ be w^2-weight sum function. The problem (CBP) is to find a feasible solution that minimizes the bottleneck weight subject to a budget constraint on the sum weight, which is mathematically formulated as follows:

$$\min_{F \in \mathcal{F}} g(F, w^1)$$
$$(CBP) \text{ s.t. } f(F, w^2) \leq B.$$

There are a variety of problems suitable to the model, such as $\mathcal{F}$ is the set of all spanning trees and the set of all paths between a pair of nodes. Now we define the inverse problem (ICBP) we focus on in this paper. Given a network $G = (V, E)$ and a feasible solution $\widetilde{F} \in \mathcal{F}$, we wish to modify the two weight vectors w^1 and w^2 such that $\widetilde{F}$ becomes an optimal solution to the constrained bottleneck problem under new weight vectors $\widetilde{w}^1$ and $\widetilde{w}^2$. Moreover, let $b^1_+, b^1_-, b^2_+, b^2_- \geq 0$ be four bound functions and $c^1, c^2 > 0$ be two cost functions defined on E, we require that $w^1 - b^1_- \leq \widetilde{w}^1 \leq w^1 + b^1_+$, $w^2 - b^2_- \leq \widetilde{w}^2 \leq w^2 + b^2_+$, and $\max\{\max_{e \in E}\{c^1(e)|\widetilde{w}^1(e) - w^1(e)|\}, \max_{e \in E}\{c^2(e)|\widetilde{w}^2(e) - w^2(e)|\}\}$ is minimum. We call it *the inverse problem under weighted l_∞ norm*, denoted by (ICBP$_b$) for simplicity.

The inverse problem (ICBP) has many potential applications in real world. An interpretation is given in the context of a transportation network G as follows [6]: Let $w^1(e)$ be the time for constructing the edge e in G and $w^2(e)$ be its construction cost. Assume that all edges are constructed simultaneously. Given a spanning tree $\widetilde{T}$ of G, we want to modify the two weight vectors, such that $\widetilde{T}$ can be constructed in the least amount of time satisfying a prescribed budget B on the construction cost under the new parameters, and the modification is as small as possible. Obviously, it is just the Inverse Constrained Bottleneck Spanning Tree Problem (ICBST), in which no restrictions on the modifications of weights are required.

Note that the inverse constrained bottleneck problem is to modify the parameters both in the objective function and in the constraints. Most published literatures on inverse optimization problems have concerned the modification of coefficients only in the objective function. One of the few except is the inverse maximal flow problem considered in [9].

The remainder of this paper is organized as follows. When the modifications of two weight vectors are proportioned, we present a general method to solve the problem (ICBP) under weighted l_∞ norm in Section 2. Conclusion remarks and further research are given in Section 3.

2   The Inverse Problem Under Weighted l_∞ Norm

Before discussing the inverse constrained bottleneck problem, let us first analyze the Bottleneck Problem (BP), which is to find a feasible solution $F^* \in \mathcal{F}$ in an edge weighted network $G = (V, E, w^1)$ such that $g(F^*, w^1) = \min_{F \in \mathcal{F}} g(F, w^1)$. In fact, we can easily prove that

Lemma 1. *For any feasible solution F^+, F^+ is an optimal solution to the bottleneck problem under a weight vector w^1 if and only if $E^+ := \{e \in E | w^1(e) \geq g(F^+, w^1)\}$ is a cut of $\mathcal{F}$. Here a cut of a family $\mathcal{F}$ is a subset of E which intersects all $F \in \mathcal{F}$.*

For the inverse problem (ICBP), we assume that the modifications of weight vectors w^1 and w^2 are proportioned throughout the paper in the form of $\widetilde{w}^2(e) - w^2(e) = \Delta c(e)(w^1(e) - \widetilde{w}^1(e))$, where $\Delta c(e) > 0$ is the modification ratio. The proportioned condition means that if $w^1(e)$ decreases (increases) by one unit, then $w^2(e)$ increases (decreases) by $\Delta c(e)$ units. We take the problem (ICBST) in a transportation network as an example. If we want to reduce the construction time, then we need to employ more workers and use more machines, which will incur more construction costs. Taking the bounds and proportioned conditions on the modifications of weights into consideration, let $\alpha(e) := \max\{w^1(e) - b_-^1(e), w^1(e) - b_+^2(e)/\Delta c(e)\}$ and $\beta(e) := \min\{w^1(e) + b_+^1(e), w^1(e) + b_-^2(e)/\Delta c(e)\}$, then the possible value of $g(\widetilde{F}, \widetilde{w}^1)$ is in the interval $[g_l(\widetilde{F}), g_u(\widetilde{F})]$, where $g_l(\widetilde{F}) := \max_{e \in \widetilde{F}} \alpha(e)$ and $g_u(\widetilde{F}) := \max_{e \in \widetilde{F}} \beta(e)$.

To solve the inverse problem (ICBP$_b$), we first consider *the restricted version of problem (ICBP$_b$) with value $p \in [g_l(\widetilde{F}), g_u(\widetilde{F})]$, denoted by (ICBP$_b(p)$), in which we require that $\widetilde{F}$ be an optimal solution to the problem (CBP) under the weight vector $(\widetilde{w}_p^1, \widetilde{w}_p^2)$ such that $g(\widetilde{F}, \widetilde{w}_p^1) = p$, $\widetilde{w}_p^1$ and $\widetilde{w}_p^2$ satisfy the bound restrictions and the modification is minimum.*

$$
\begin{aligned}
&\min \max_{e \in E} \rho^1(e) |\widetilde{w}_p^1(e) - w^1(e)| \\
&\text{s.t. } \max_{e \in \widetilde{F}} \widetilde{w}_p^1(e) = p \leq \max_{e \in F} \widetilde{w}_p^1(e), \ \forall F \in \mathcal{F}, \\
&\qquad \sum_{e \in \widetilde{F}} \widetilde{w}_p^2(e) \leq B, \\
&\qquad \widetilde{w}_p^2(e) - w^2(e) = \Delta c(e)(w^1(e) - \widetilde{w}_p^1(e)), \ \forall e \in E, \\
&\qquad \alpha(e) \leq \widetilde{w}_p^1(e) \leq \beta(e), \ \forall e \in E,
\end{aligned}
$$

(ICBP$_b(p)$)

where $\rho^1(e) := \max\{c^1(e), \Delta c(e) c^2(e)\}$ for any $e \in E$. Denote by $C(p)$ the optimal objective value of problem (ICBP$_b(p)$). Then we will determine the optimal solution to the inverse problem (ICBP$_b$) by the properties of function $C(p)$ on the interval $[g_l(\widetilde{F}), g_u(\widetilde{F})]$.

2.1 Solve the Restricted Inverse Problem (ICBP$_b(p)$)

In this subsection, we propose a general method to solve the restricted inverse problem (ICBP$_b(p)$) by considering two cases of the problem according to whether the edge subset $E^+(p) := \{e \in E | w^1(e) \geq p\}$ is a cut of $\mathcal{F}$ or not.

Let $\widetilde{F}(p) := \{e \in \widetilde{F} | w^1(e) \geq p\}$. Clearly, for each edge $e \in \widetilde{F}(p)$, we need to reduce the weight $w^1(e)$ to p in order to make the maximum weight on $\widetilde{F}$ equal to p. Let

$$\begin{cases} w_p^1(e) := p, \quad w_p^2(e) := w^2(e) + \Delta c(e)(w^1(e) - p), \forall e \in \widetilde{F}(p), \\ w_p^1(e) := w^1(e), \; w_p^2(e) := w^2(e), \hfill \forall e \in E \setminus \widetilde{F}(p). \end{cases} \quad (1)$$

Then the cost incurred by modifications on $\widetilde{F}(p)$ is

$$\tilde{c}_p(\widetilde{F}(p)) := \max\nolimits_{e \in \widetilde{F}(p)} \rho^1(e)(w^1(e) - p).$$

Case 1: If $E^+(p)$ is a cut of $\mathcal{F}$, then $\widetilde{F}$ is already an optimal solution to the bottleneck problem with respect to the weight vector w_p^1 and $g(\widetilde{F}, w_p^1) = p$. If $\sum_{e \in \widetilde{F}} w_p^2(e) \leq B$ also holds, then (w_p^1, w_p^2) is obviously the optimal solution to problem (ICBP$_b(p)$). Now we assume that $\sum_{e \in \widetilde{F}} w_p^2(e) > B$, let $\widehat{F}(p) := \{e \in \widetilde{F} | w^1(e) < p\}$, $B_p := B - \sum_{e \in \widetilde{F}(p)} w_p^2(e)$, $\beta_p(e) := \min\{p, \beta(e)\}$, $\mu_p(e) := w^2(e) - \Delta c(e)(\beta_p(e) - w^1(e))$, then we need to decrease the weights $w^2(e)$ to $\widehat{w}_p^2(e)$ for edges in $\widehat{F}(p)$ such that $\sum_{e \in \widehat{F}(p)} \widehat{w}_p^2(e) \leq B_p$ and $\widehat{w}_p^1(e) := w^1(e) + \frac{1}{\Delta c(e)}(w^2(e) - \widehat{w}_p^2(e)) \leq \beta_p(e)$, which is equivalent to $\widehat{w}_p^2(e) \geq \mu_p(e)$. Mathematically, it can be formulated as the following problem:

$$(ICP_b(p)) \quad \text{s.t.} \quad \begin{aligned} & \min \max_{e \in \widehat{F}(p)} \rho^2(e)(w^2(e) - \widehat{w}_p^2(e)) \\ & \sum_{e \in \widehat{F}(p)} \widehat{w}_p^2(e) \leq B_p, \\ & \widehat{w}_p^2(e) \geq \mu_p(e), \; \forall e \in \widehat{F}(p), \end{aligned} \quad (2)$$

where $\rho^2(e) := \max\{c^2(e), \frac{c^1(e)}{\Delta c(e)}\}$. We call it the Inverse Constraint Problem under weighted l_∞ norm with value p, denoted by (ICP$_b(p)$). Note that if $\sum_{e \in \widehat{F}(p)} \mu_p(e) > B_p$, then the instance is infeasible. Suppose that $\sum_{e \in \widehat{F}(p)} \mu_p(e) \leq B_p$. Sort the costs $\{\rho^2(e)(w^2(e) - \mu_p(e)) | e \in \widehat{F}(p)\}$ in a strictly increasing order and denote the ordered set of costs by R$(p) := \{r_1, \cdots, r_i, \cdots, r_l\}$, where $l \leq |\widehat{F}(p)|$. For any value x, let $\widehat{F}_p(x) := \{e \in \widehat{F}(p) | \rho^2(e)(w^2(e) - \mu_p(e)) \leq x\}$. Then $\widehat{w}_p^2(e) \geq \max\{\mu_p(e), w^2(e) - \frac{x}{\rho^2(e)}\} = \mu_p(e)$ for any edge $e \in \widehat{F}_p(x)$ and $\widehat{w}_p^2(e) \geq w^2(e) - \frac{x}{\rho^2(e)}$ for any edge $e \in \widehat{F}(p) \setminus \widehat{F}_p(x)$. Define a function $h_p(x) := \min \sum_{e \in \widehat{F}(p)} \widehat{w}_p^2(e)$, then

$$h_p(x) = \sum\nolimits_{e \in \widehat{F}_p(x)} \mu_p(e) + \sum\nolimits_{e \in \widehat{F}(p) \setminus \widehat{F}_p(x)} (w^2(e) - x/\rho^2(e)).$$

Obviously, $h_p(x)$ is a decreasing, piecewise linear function of x in the interval $[r_1, r_l]$. Denote by $\hat{c}_p(\widehat{F}(p))$ the optimal objective value of problem (ICP$_b(p)$).

To solve the problem (ICP$_b(p)$), we conduct a binary search in R(p) to find a subinterval $[r_i, r_{i+1})$ such that $h_p(r_i) > B_p$ and $h_p(r_{i+1}) \leq B_p$. Then we have

$$\hat{c}_p(\widehat{F}(p)) := \begin{cases} (\sum_{e \in \widehat{F}(p)} w^2(e) - B_p)/\sum_{e \in \widehat{F}(p)} \frac{1}{\rho^2(e)}, & \text{if } h_p(r_1) < B_p, \\ r_1, & \text{if } h_p(r_1) = B_p, \\ r_{i+1}, & \text{if } h_p(r_{i+1}) = B_p, \\ \frac{\sum_{e \in \widehat{F}_p(r_i)} \mu_p(e) + \sum_{e \in \widehat{F}(p) \setminus \widehat{F}_p(r_i)} w^2(e) - B_p}{\sum_{e \in \widehat{F}(p) \setminus \widehat{F}_p(r_i)} \frac{1}{\rho^2(e)}}, & \text{if } h_p(r_i) > B_p > h_p(r_{i+1}), \end{cases}$$

and

$$\widehat{w}_p^2(e) := \max\{\mu_p(e), w^2(e) - \hat{c}_p(\widehat{F}(p))/\rho^2(e)\}, \forall e \in \widehat{F}(p). \tag{3}$$

Note that the algorithm can find the optimal value $\hat{c}_p(\widehat{F}(p))$ in $O(|F| \log |F|)$ operations. Furthermore, the optimal solution to problem (ICBP$_b(p)$) is

$$\begin{cases} \widetilde{w}_p^2(e) := \widehat{w}_p^2(e), \widetilde{w}_p^1(e) := w^1(e) + \frac{1}{\Delta c(e)}(w^2(e) - \widehat{w}_p^2(e)), \forall e \in \widehat{F}(p) \\ \widetilde{w}_p^2(e) := w_p^2(e), \widetilde{w}_p^1(e) := w_p^1(e), & \text{otherwise,} \end{cases} \tag{4}$$

and the modification cost is $C(p) := \max\{\hat{c}_p(\widehat{F}(p)), \tilde{c}_p(\widetilde{F}(p))\}$.

Case 2: Now we consider the case that $E^+(p)$ is not a cut of $\mathcal{F}$. We first solve two subproblems of the restricted version (ICBP$_b(p)$), then show that the optimal solution to problem (ICBP$_b(p)$) can be obtained by the optimal solutions to the two subproblems.

Let $E(p) := \{e \in E | w^1(e) < p\}$ and $\mathcal{F}(p) := \{F \in \mathcal{F} | F \subseteq E(p)\}$, then $\mathcal{F}(p) \neq \emptyset$ since $E^+(p)$ is not a cut of $\mathcal{F}$. Hence we need to increase the weight $w^1(e)$ of at least one edge to p for each $F \in \mathcal{F}(p)$. Mathematically, it can be formulated as solving $\bar{w}_p^1(e)$ for all $e \in E(p)$ from the following problem:

$$
\text{(IBP}_b(p)) \quad \text{s.t.} \quad
\begin{aligned}
&\min \max_{e \in E(p)} \rho^1(e)(\bar{w}_p^1(e) - w^1(e)) \\
&\max_{e \in F} \bar{w}_p^1(e) = p, \forall F \in \mathcal{F}(p), \\
&\bar{w}_p^1(e) \leq \beta_p(e), \forall e \in E(p).
\end{aligned}
\tag{5}
$$

We call it the Inverse Bottleneck Problem under weighted l_∞ norm with value p, denoted by (IBP$_b(p)$). Next we solve the problem (IBP$_b(p)$) based on the following property.

Lemma 2. If the problem (IBP$_b(p)$) is feasible, then there is an optimal solution $\breve{w}_p^1$ to problem (IBP$_b(p)$) such that for any $e \in E(p)$,

$$\breve{w}_p^1(e) = \begin{cases} p, & \text{if } \breve{w}_p^1(e) > w^1(e), \\ w^1(e), & \text{otherwise.} \end{cases} \tag{6}$$

Proof: Let $\bar{w}_p^1$ be any optimal solution to problem $(IBP_b(p))$, then we show that

$$\breve{w}_p^1(e) := \begin{cases} w^1(e), & if\ e \in E^1 := \{e \in E(p)|p > \bar{w}_p^1(e) > w^1(e)\}, \\ p, & if\ e \in E^2 := \{e \in E(p)|\bar{w}_p^1(e) = p\}, \\ w^1(e), & if\ e \in E^3 := \{e \in E(p)|\bar{w}_p^1(e) = w^1(e)\}, \end{cases}$$

is also an optimal solution, which satisfies the condition (6). Obviously, $E(p) = E^1 \cup E^2 \cup E^3$. For any $F \in \mathcal{F}(p)$, we claim that $F' := F \cap (E^2 \cup E^3) \neq \emptyset$. Otherwise, $\max_{e \in F} \bar{w}_p^1(e) = \max_{e \in E^1} \bar{w}_p^1(e) < p$, which contradicts the feasibility of $\bar{w}_p^1$. Then $\max_{e \in F} \breve{w}_p^1(e) = \max_{e \in F'} \breve{w}_p^1(e) = \max_{e \in F'} \bar{w}_p^1(e) = p$. Note that $\breve{w}_p^1(e) \leq \bar{w}_p^1(e) \leq \beta_p(e)$ for any $e \in E(p)$. Therefore, $\breve{w}_p^1$ is a feasible solution. For the optimality of $\breve{w}_p^1$, it is easy to see that

$$\max_{e \in E(p)} \rho^1(e)(\bar{w}_p^1(e) - w^1(e)) \geq \max_{e \in E^2} \rho^1(e)(\bar{w}_p^1(e) - w^1(e))$$
$$= \max_{e \in E^2} \rho^1(e)(\breve{w}_p^1(e) - w^1(e)) = \max_{e \in E(p)} \rho^1(e)(\breve{w}_p^1(e) - w^1(e)). \quad \square$$

Next we only consider the optimal solution to problem $(IBP_b(p))$ satisfying the condition (6). Let $\overline{E}(p) := \{e \in E(p)|p \leq \beta(e)\}$ be the set of edges whose w^1-weight can be increased to p. In order to solve the problem $(IBP_b(p))$, we define a capacity vector $c_p : E(p) \to R$ as follows:

$$c_p(e) := \begin{cases} \rho^1(e)(p - w^1(e)), & \forall e \in \overline{E}(p), \\ +\infty, & \forall e \in E(p) \setminus \overline{E}(p). \end{cases} \quad (7)$$

Then we consider an associated subproblem $(BC(p))$ defined as finding the Bottleneck Cut of family $\mathcal{F}(p)$ under the capacity vector c_p.

$$\min_K \max_{e \in K} c_p(e)$$
$$(BC(p))\ s.t.\ K \subseteq E(p)\ is\ a\ cut\ of\ family\ \mathcal{F}(p). \quad (8)$$

Let $K(p)$ be the optimal solution to problem $(BC(p))$ and let $c_p(K(p)) := \max_{e \in K(p)} c_p(e)$ be the corresponding objective value. If $c_p(K(p)) = +\infty$, then the instance is infeasible; otherwise we define $\bar{w}_p^1$ as follows:

$$\bar{w}_p^1(e) := \begin{cases} p, & \forall e \in K(p), \\ w^1(e), & \forall e \in E(p) \setminus K(p). \end{cases} \quad (9)$$

Theorem 1. The weight vector $\bar{w}_p^1$ defined as (9) is an optimal solution to problem $(IBP_b(p))$, and the objective value is $c_p(K(p)) := \max_{e \in K(p)} c_p(e)$.

Proof: It is obvious that $\max_{e \in F} \bar{w}_p^1(e) = p$ for any $F \in \mathcal{F}(p)$ since $K(p)$ is a cut of $\mathcal{F}(p)$. Furthermore, $\bar{w}_p^1(e) \leq \beta_p(e)$ for any $e \in E(p)$. Hence $\bar{w}_p^1$ is a feasible solution of problem $(IBP_b(p))$. To prove the optimality of $\bar{w}_p^1$, we suppose $\breve{w}_p^1$ is an optimal solution to problem $(IBP_b(p))$ satisfying the condition (6). Define $\breve{E}(p) := \{e \in E(p)|\breve{w}_p^1(e) = p\}$, then $\breve{E}(p)$ is a cut of $\mathcal{F}(p)$. Otherwise, there is a feasible solution $F' \in \mathcal{F}(p)$ such that $F' \cap \breve{E}(p) = \emptyset$, then $\breve{w}_p^1(e) = w^1(e) < p$ for any $e \in F'$, and $\max_{e \in F'} \breve{w}_p^1(e) < p$, which contradicts the feasibility of $\breve{w}_p^1$. Moreover,

$$\max_{e\in E(p)} \rho^1(e)(\breve{w}_p^1(e) - w^1(e)) = \max_{e\in \breve{E}(p)} \rho^1(e)(p - w^1(e))$$
$$= \max_{e\in \breve{E}(p)} c_p(e) \geq \max_{e\in K(p)} c_p(e) = c_p(K(p))$$

and hence $\bar{w}_p^1$ is an optimal solution to problem (IBP$_b(p)$). □

After solving the problem (IBP$_b(p)$), let us consider the restricted inverse problem (ICBP$_b(p)$). Let $\widetilde{C}(p) := \max\{\tilde{c}_p(\widehat{F}(p)), c_p(K(p))\}$ and

$$\bar{w}_p^2(e) := \begin{cases} w^2(e) - \Delta c(e)(p - w^1(e)), & \forall e \in K(p), \\ \max\{\mu_p(e), w^2(e) - \frac{\widetilde{C}(p)}{\rho^2(e)}\}, & \forall e \in \widehat{F}(p)\setminus K(p). \end{cases} \quad (10)$$

If $\sum_{e\in \widehat{F}(p)} \bar{w}_p^2(e) \leq B_p$, then $(\widetilde{w}_p^1, \widetilde{w}_p^2)$ is an optimal solution to problem (ICBP$_b(p)$), where $\widetilde{w}_p^2(e) := \begin{cases} \bar{w}_p^2(e), \forall e \in \widehat{F}(p) \cup K(p) \\ w_p^2(e), for\ other\ edges \end{cases}$ and $\widetilde{w}_p^1(e) := w^1(e) + \frac{w^2(e) - \widetilde{w}_p^2(e)}{\Delta c(e)}$. Furthermore, the modification cost incurred is $C(p) := \widetilde{C}(p)$.

Otherwise, $C(p) > \widetilde{C}(p)$, we need to solve an inverse constraint problem (ICP$_b(p)$) for $\widehat{F}(p)\setminus K(p)$ in the form of problem (2) except for replacing $\widehat{F}(p)$ with $\widehat{F}(p)\setminus K(p)$ and replacing B_p with $B_p' := B_p - \sum_{e\in \widehat{F}(p)\cap K(p)} \bar{w}_p^2(e)$. Let the optimal solution obtained by (3) be $\widehat{w}_p^2(e), \forall e \in \widehat{F}(p)\setminus K(p)$, then we define

$$\begin{cases} \widetilde{w}_p^1(e) := p, & \widetilde{w}_p^2(e) := w^2(e) - \Delta c(e)(p - w^1(e)), \forall e \in K(p), \\ \widetilde{w}_p^2(e) := \widehat{w}_p^2(e), \widetilde{w}_p^1(e) := w^1(e) + \frac{w^2(e) - \widehat{w}_p^2(e)}{\Delta c(e)}, & \forall e \in \widehat{F}(p)\setminus K(p), \\ \widetilde{w}_p^1(e) := p, & \widetilde{w}_p^2(e) := w^2(e) + \Delta c(e)(w^1(e) - p), \forall e \in \widetilde{F}(p), \\ \widetilde{w}_p^1(e) := w^1(e), & \widetilde{w}_p^2(e) := w^2(e), & for\ other\ edges. \end{cases} \quad (11)$$

Remark 1. Note that in this case (i.e., $C(p) > \widetilde{C}(p)$)

$$\sum_{e\in \widetilde{F}} \widetilde{w}_p^2(e) = \sum_{e\in \widetilde{F}(p)} w_p^2(e) + \sum_{e\in \widehat{F}(p)\cap K(p)} \widetilde{w}_p^2(e) + \sum_{e\in \widehat{F}(p)\setminus K(p)} \widehat{w}_p^2(e) = B.$$

Furthermore, $\bar{w}_p^2(e) = \mu_p(e), \forall e \in \widehat{F}(p) \cap K(p)$, then $\widetilde{w}_p^2(e), \forall e \in \widehat{F}(p)$, is also an optimal solution to the inverse constrained problem ($\widetilde{ICP}_b(p)$) defined as (2).

Since the optimal modification cost $C(p)$ is at least the modification cost on $\widehat{F}(p)$ in order to make $\sum_{e\in \widehat{F}(p)} \widetilde{w}_p^2(e) \leq B_p$ hold, that is, $C(p) \geq \max_{e\in \widehat{F}(p)} \rho^1(e) (\widetilde{w}_p^1(e) - w^1(e))$, then we can conclude that

Theorem 2. If $\sum_{e\in \widehat{F}(p)} \bar{w}_p^2(e) > B_p$, then $(\widetilde{w}_p^1, \widetilde{w}_p^2)$ defined as (11) is an optimal solution to problem ($\widetilde{ICBP}_b(p)$). Furthermore, the modification cost is $C(p) := \max_{e\in \widehat{F}(p)\setminus K(p)} \rho^1(e)(\widetilde{w}_p^1(e) - w^1(e))$.

2.2 Solve the Original Inverse Problem (ICBP$_b$)

Now we turn to the original inverse problem (ICBP$_b$) under weighted l_∞ norm. Consider the collection of distinctive values of the weights w^1 and the upper bounds on weights within the interval $[g_l(\widetilde{F}), g_u(\widetilde{F})]$, that is, the set

$(\{w^1(e), w^1(e) + b_+^1(e), w^1(e) + \frac{b_-^2(e)}{\Delta c(e)} | e \in E\} \cup \{g_l(\widetilde{F})\}) \cap [g_l(\widetilde{F}), g_u(\widetilde{F})]$,
sort these values in a strictly increasing order, say $g_l(\widetilde{F}) = q_1 < q_2 < \cdots < q_t = g_u(\widetilde{F})$, where $t = O(|E|)$. Next we analyze the properties of the modification cost $C(p)$ in the subinterval $(q_k, q_{k+1}]$, then find the optimal solution to problem (ICBP$_b$) based on the properties.

Note that the possible value of the modification cost $C(p)$ is in the form of

$$\tilde{c}_p(\widetilde{F}(p)) := \max_{e \in \widetilde{F}(p)} \rho^1(e)(w^1(e) - p),$$

$$\hat{c}_p(\widehat{F}(p)) := (\sum_{e \in \widehat{F}_p(r_i)} \mu_p(e) + \sum_{e \in \widehat{F}(p) \setminus \widehat{F}_p(r_i)} w^2(e) - B_p) / \sum_{e \in \widehat{F}(p) \setminus \widehat{F}_p(r_i)} \frac{1}{\rho^2(e)},$$

and

$$c_p(K(p)) := \min_{K \in \mathcal{K}(p)} \max_{e \in K} \rho^1(e)(p - w^1(e)),$$

where $\mathcal{K}(p)$ is the collection of feasible cuts of $\mathcal{F}(p)$, that is, $K \in \mathcal{K}(p)$ if and only if K is a cut of $\mathcal{F}(p)$ and $\beta(e) \geq p$ for all $e \in K$. It is known from [10] that if $c_{p'}(K(p')) < +\infty$ for some $p' \in (q_k, q_{k+1})$, then $c_p(K(p)) < +\infty$ for all $p \in (q_k, q_{k+1}]$. Moreover $\mathcal{K}(p) = \mathcal{K}(q_{k+1})$, $c_p(K(p)) = \min_{K \in \mathcal{K}(q_{k+1})} \max_{e \in K} \rho^1(e)(p - w^1(e))$ and $\widetilde{F}(p) = \widetilde{F}(q_{k+1})$, $\tilde{c}_p(\widetilde{F}(p)) = \max_{e \in \widetilde{F}(q_{k+1})} \rho^1(e)(w^1(e) - p)$ for all $p \in (q_k, q_{k+1}]$. Similarly, $\widehat{F}(p) = \widehat{F}(q_{k+1})$ and $\widehat{F}_p(x) = \widehat{F}_{q_{k+1}}(x)$. Note that both

$$B_p := B - \sum_{e \in \widetilde{F}(q_{k+1})} [w^2(e) + \Delta c(e)(w^1(e) - p)]$$

and

$$\mu_p(e) := w^2(e) - \Delta c(e) \min\{p - w^1(e), b_+^1(e), b_-^2(e)/\Delta c(e)\}$$

are linear functions of p in the subinterval $(q_k, q_{k+1}]$, so is $\hat{c}_p(\widehat{F}(p))$.

Therefore, $C(p)$ is a piecewise linear zigzag function in the interval and the minimizer must be reached at some intersection point of linear functions, including $\{\rho^1(e)(w^1(e) - p) | e \in \widetilde{F}\} \cup \{\rho^1(e)(p - w^1(e)) | e \in E\}$, or an endpoint of the interval. Note that the total number of checking points is at most $3m + 1 + \frac{1}{2}(m + |\widetilde{F}|)(m + |\widetilde{F}| - 1)$, where $m = |E|$. Moreover, at each point, the main computation is to find a bottleneck cut. If $\overline{E}(p)$ is not a cut of $\mathcal{F}(p)$, then $\overline{E}(q)$ is not a cut of $\mathcal{F}(q)$ for all $q \geq p$. Hence, we can conclude that

Theorem 3. Under the requirement that the modifications of weight vectors w^1 and w^2 are proportioned in the form of $\widetilde{w}^2(e) - w^2(e) = \Delta c(e)(w^1(e) - \widetilde{w}^1(e))$, where $\Delta c(e) > 0$, the inverse constrained bottleneck problem under weighted l_∞ norm can be solved by the method presented in this paper. And the complexity of the method is not more than $O(m^2(|F| \log |F| + T_b))$, where $O(|F| \log |F|)$ is the complexity for the inverse constrained problem (ICP$_b$) and T_b is the complexity of the corresponding bottleneck cut problem. Moreover, if the bottleneck cut problem can be solved in strongly polynomial time, then the inverse problem can be solved in strongly polynomial time, too.

2.3 Examples of the Inverse Constrained Bottleneck Problems

We have shown that the inverse problem (ICBP$_b$) can be solved by a series of restricted inverse problems and the main computation in the restricted problem is to find a bottleneck cut. In fact, for many families $\mathcal{F}$, there are efficient algorithms to find their bottleneck cut. Here we give two examples.

Inverse Constrained Bottleneck Spanning Tree Problem (ICBST$_b$): To solve the bottleneck cut problem when $\mathcal{F}(p)$ is a family of all spanning trees, it is sufficient to find a max-sum spanning tree $T(p)$ of $G(p) = (V, E(p))$ with respect to $c_p(e)$ [10]. Let $e(p) := \arg\min_{e \in T(p)} c_p(e)$, S be the node set of one connected component of $T(p) - e(p)$ and $\bar{S} := V \backslash S$, then $K(p) := [S, \bar{S}]$ is the bottleneck cut with respect to $c_p(K)$ among all cut K. Note that the max-sum spanning tree can be obtained in $O(m + n \log n)$ [7] and the inverse constrained problem (ICP$_b(p)$) can be done in $O(n \log n)$ since $|\widetilde{F}| = n - 1$. Hence, the complexity of the problem (ICBST$_b$) is $O(m^2(m + n \log n))$.

Inverse Constrained Bottleneck Assignment Problem (ICBAP$_b$):
Given a complete bipartite graph $G = (X, Y; E)$ with nodes sets X and Y satisfying $|X| = |Y| = n$, where each pair (edge) (i, j) takes time $w^1(i, j)$ and incurs cost $w^2(i, j)$. An assignment S is a set of n pairs, where each node in X is assigned to exactly one node in Y and each node in Y is assigned to exactly one node in X. The function $g(S, w^1)$ denotes the maximum time that an assigned job requires and $f(S, w^2)$ denotes the total cost that an assignment incurs. Let $\widetilde{S}$ be a given assignment, then the problem (ICBAP$_b$) is to modify the weight vectors w^1 and w^2 satisfying the bound restrictions as little as possible under weighted l_∞ norm so that $\widetilde{S}$ becomes the assignment that minimizes the maximum time $g(\widetilde{S}, \widetilde{w}^1)$ and satisfies the constraint $f(\widetilde{S}, \widetilde{w}^2) \leq B$. Recall that the original problem (CBAP) can be solved in $O(n^3 \log m)$ operations [1].

Let $\mathcal{F}$ be a system of all assignment solutions, then a minimal cut K of $\mathcal{F}$ is the edge set of a subgraph of G induced by $V_X \cup V_Y$ with $V_X \subseteq X$, $V_Y \subseteq Y$ and $|V_X| + |V_Y| = n + 1$ [2]. Here a minimal cut K of $\mathcal{F}$ means that K is a cut of $\mathcal{F}$, but $K \backslash e$ is not a cut of $\mathcal{F}$ for any edge $e \in K$. Next we present the main idea of finding a bottleneck cut of $\mathcal{F}(p)$ by performing a binary search on the capacities $c_p(e)$ of edges $e \in \overline{E}(p)$. In each iteration, we consider the edge subset E_k including the edges in $\overline{E}(p)$ whose capacity is no more than the current considered capacity, then we are mainly to check if the set E_k is a cut of $\mathcal{F}(p)$ or not, which is equivalent to checking if $\bar{G}_k = (V, E(p) \backslash E_k)$ contains or not an assignment solution of $\mathcal{F}(p)$. Thus the bottleneck cut can be found in $O(n^3 \log m)$ operations and accordingly the complexity of the problem (ICBAP$_b$) is $O(m^2 n^3 \log m)$. A numerical example is available, but due to space limitation, please see the complete version of the paper from the web site http://www6.cityu.edu.hk/ma/people/jason.html.

3 Conclusion and Further Research

In this paper, we consider a class of inverse constrained bottleneck problems. When the modifications of two weight vectors are proportioned in the form of $\widetilde{w}^2(e) -$

$w^2(e) = \Delta c(e)(w^1(e) - \widetilde{w}^1(e))$, where $\Delta c(e) > 0$, we present a general method to solve the problem (ICBP) under weighted l_∞ norm. We can also consider the problem under weighted l_1 norm, denoted by (ICBP$_s$), where the objective function is min $\sum_{e \in E}[c^1(e)\,|\widetilde{w}^1(e) - w^1(e)| + c^2(e)|\widetilde{w}^2(e) - w^2(e)|]$. Similar to the case of l_∞ norm, the main problem to consider is the restricted problem (ICBP$_s(p)$) when $E^+(p)$ is not a cut of $\mathcal{F}$, which is equivalent to solving a minimum constrained cut problem defined as follows:

min $\sum_{e \in E(p)} (c^2(e) + c^1(e)/\Delta c(e))(w^2(e) - \widetilde{w}_p^2(e))$

s.t. $\widetilde{E}(p) := \{e \in \overline{E}(p) | \widetilde{w}_p^2(e) = w^2(e) - \Delta c(e)(p - w^1(e))\}$ is a cut of $\mathcal{F}(p)$,

$\sum_{e \in \widehat{F}(p)} \widetilde{w}_p^2(e) \leq B_p,$ (12)

$\widetilde{w}_p^2(e) \geq \mu_p(e), \forall e \in E(p).$

Let $K(p) \subseteq \overline{E}(p)$ be the minimum cut of $\mathcal{F}(p)$ under the capacity function c_p, which is defined as (7) by replacing $\rho^1(e)$ with $c^1(e) + \Delta c(e)c^2(e)$. Note that when we reduce the weights $w^2(e)$ on edges $e \in \widehat{F}(p)\setminus K(p)$, $(\widehat{F}(p) \cup K(p)) \cap \widetilde{E}(p)$ may contain other cuts of $\mathcal{F}(p)$ except $K(p)$, thus we may obtain less modification costs to satisfy all the constraints in (12). This is the main dilemma to solve such a problem. Therefore, it is difficult to present a general method and we hope to obtain some specific method based on its combinatorial properties.

Note that another proportioned relationship between the modifications of two vectors is $w^2(e) - \widetilde{w}^2(e) = \Delta c(e)(w^1(e) - \widetilde{w}^1(e))$, where $\Delta c(e) > 0$, which means that if the weight $w^1(e)$ decreases (increases) by one unit, then $w^2(e)$ decreases (increases) by $\Delta c(e)$ units. Now we briefly analysis the difficulties to solve the problem (ICBP) in this case. Note that when we reduce the weights $w^1(e)$ to $w_p^1(e) := p$ for edges in $\widetilde{F}(p)$, the wights $w^2(e)$ decrease correspondingly. Furthermore, to satisfy the budget constraint on the sum of w^2-weights on $\widetilde{F}$, the weights $w^2(e)$ of all the edges in $\widetilde{F}$, even of the edges whose w_p^1-weights are equal to p, can be reduced. On the other hand, when we increase the weights $w^1(e)$ to p to obtain a cut of $\mathcal{F}$, the weights $w^2(e)$ increase, too. Thus, no matter whether $E^+(p)$ is a cut of $\mathcal{F}$ or not, the inverse problem can not be transformed into two sub-problems to solve as in Section 2. We have to consider the modifications of the two weights simultaneously, that is, to satisfy the bottleneck restriction on w^1-weights and the budget constraint on the sum of w^2-weights simultaneously. Therefore, the inverse problem under weighted l_∞ (or l_1) norm is hard to solve, and we hope to solve the problem for some specific system $\mathcal{F}$ based on its combinatorial structure.

References

1. Berman, O., Einay, D., Handler, G.: The constrained bottleneck problem in networks. Oper. Res. **38** (1988) 178–181
2. Burkard, R.E., Lin, Y., Zhang, J.: Weight reduction problems with certain bottleneck objectives. European J. Oper. Res. **153** (2004) 191–199

3. Cai, M., Yang, X., Zhang, J.: The complexity analysis of the inverse center location problem. J. Global Optim. **5** (1999) 213–218
4. Heuberger, C.: Inverse optimization, a survey on problems, methods, and results. J. Comb. Optim. **329** (2004) 329–361
5. Hochbaum, D.S.: Efficient algorithms for the inverse spanning tree problem. Oper. Res. **51** (2003) 785–797
6. Punnen, A., Nair, K.: An improved algorithm for the constrained bottleneck spanning tree problem. INFORMS J. Comp. **8** (1996) 41–44
7. Schrijver, A.: Combinatorial Optimization: Polyhedra and Efficiency. Springer-Verlag Berlin (2003)
8. Yang, C., Zhang, J.: Inverse maximum capacity problems. OR Spektrum. **20** (1998) 97–100
9. Yang, C., Zhang, J., Ma, Z.: Inverse maximum flow and minimum cut problems. Optimization. **40** (1997) 147–170
10. Zhang, J., Yang, X., Cai, M.: Some inverse min-max network problems under weighted l_1 and l_∞ norms. Techniqual report. (2004)

Dynamic Scheduling Problem of Batch Processing Machine in Semiconductor Burn-in Operations

Pei-Chann Chang[1], Yun-Shiow Chen[1], and Hui-Mei Wang[1,2]

[1]Department of Industrial Engineering and Management,
Yuan Ze University,
135 Yuan-Tung Road, Chungli, Taoyuan, Taiwan 32026, R.O.C.
iepchang@saturn.yzu.edu.tw
[2]Department of Industrial Management,
Van Nung University,
amywang@cc.vit.edu.tw

Abstract. The dynamic scheduling problem of batch processing machine in semiconductor burn-in operations is studied and the objective is to minimize the total flow time of all jobs. This paper presents a constraint programming approach and develops different bounds, i.e., job-based bound and batch-based bound which can be embedded in branch and bound procedure to further improve the efficiency of the procedure. Also, some properties of the problem are also developed to further reduce the complexity of the problem.

1 Introduction

In the final testing stage of semiconductor manufacturing, burn-in oven is modeled as a batch-processing machine. Different Integrated Circuits (IC) require different minimum baking time and IC of different jobs (lots) are inserted onto suitable boards and then transferred to the oven for processing. After a long processing time of the thermal stress heating process, IC with latent defects due to infant mortality can be brought out. To ensure the quality of the product, the processing time of a batch is assigned to the longest processing time among all jobs in the batch, and once the processing of the batch starts, it cannot be interrupted and no job can be removed from the oven until the processing of the batch is complete.

Lee et al. (1992) first introduced the semiconductor burn-in scheduling problem and they have fully described the production characteristics of the burn-in operation and provided an efficient algorithm to minimize the number of tardy jobs and maximum tardiness under $1/B/agr.p_i, d_i$, $1/r_i, p_i = p, B/agr.r_i, d_i$ assumptions. Li and Lee (1997) not only extended the above assumption to $1/r_i, B/agr.r_i, d_i, p_i$, that is, jobs can be arranged in such a way that $r_1 \leq r_2 \leq \cdots \leq r_n$, $d_1 \leq d_2 \leq \cdots \leq d_n$ and $p_1 \leq p_2 \leq \cdots \leq p_n$, but also proved that those two problems, including minimizing maximum tardiness and minimizing total number of tardy jobs are strongly NP-hard. Besides, they also extended Lee et al. (1992) algorithms to solve $1/r_i, B/T_{\max}(\sum U_i)agr.r_i, d_i, p_i$, respectively. Chandru et al. (1993a) proposed a branch-and-bound algorithm and heuristics to minimize total completion time on a batch processing machine. Dupont and Jolai Ghazvini(1997) proposed some properties for the same problem but also including properties and elimination rule to solve the problems with up to 100 jobs. Lee and Uzsoy (1999) proposed several heuristics to minimize makespan for the problem where jobs' arrival times are not equal to zero. Sung and Choung (2000) provide a branch-and-bound algorithm to obtain optimal solution for the same problem, while they also proposed a heuristic to obtain near-optimal solutions.

From the literature surveyed above, they all assume that a board is referred as a single job such that the capacity of a machine is equal to the accommodated number of jobs. Uzsoy (1994) studied the case where jobs have different capacity requirements and compatible job families, in which he has proved that the problems of minimizing makespan and total completion time are NP-hard and he provided a number of heuristics to obtain near-optimal solutions. Jolai Ghazvini and Dupont (1998) examined the same problem and proposed a number of heuristics compared to Uzsoy's. Kempf et al. (1998) examined the problem where boards are considered as a secondary resource constraint and incompatible job families, under which assumption, they developed heuristics to obtain near-optimal solutions for minimizing makespan and total completion time. Chang and Wang (2004b) also provided a heuristic for a batch processing machine scheduling problem with the objective to minimize total completion time under non-identical job sizes assumption. Recently, soft-computing methods such as simulated annealing and genetic algorithm are also applied to solve this kind of problems as shown in Melouk et al. (2004), Uzoy and Wang (2002), Chang and Wang (2004a).

In practice, during the burn-in operation of Semiconductor manufacturing factory job sizes may be greater or smaller than the capacity of the oven, and each job can not be transferred to the next processing station unless all sub-jobs, i.e. a job split into sub-jobs to be processed, are finished. These two characteristics indicate that the above assumption by Kempf et al. (1998) is not appropriate for the flow time objective function. In this research we take these two characteristics into consideration and new assumptions are listed as follows:

1. There are n jobs, i.e., $J_1, J_2, ..., J_n$, to be processed on a batch processing machine and the release time of job i is denoted by r_i.
2. All jobs are compatible, i.e., in a same recipe and same temperature profile, and job processing time p_j and sizes s_i are deterministic and known *a priori*. According to realistic environment, s_i herein is without any restriction.
3. Burn-in oven has a capacity M and the total number of jobs in a batch must be less than M.
4. The number of batches needed for job i is equal to $\lceil \frac{s_i}{M} \rceil$, where $\lceil \frac{s_i}{M} \rceil$ denotes the smallest integer greater than $\frac{s_i}{M}$.
5. Machine breakdown is not allowed.

6. Once processing of a batch begins, other jobs cannot be added or removed from the machine until the processing is finished. The processing time of a batch is given by the longest processing time of the jobs in the batch.
7. The objective is to minimize the total flow times of jobs, $\sum F_i$

The problem we studied here is strongly NP-hard (Uzsoy, 1994) and the problem can be denoted as $1/r_i, B, s_i/\sum F_i$ according to Lee. et al. (1992).

2 A Constrain Programming Approach

Chang et al. (2000) developed an integer-programming model to obtain optimal solution for the same problem. However, dominant rules or expert knowledge are difficult to be constructed within traditional programming procedure, so it only can solve small size of problems. This paper proposes a constraint programming approach that includes if-then-else rule capability to solve the problem. Notations and definitions used to describe the scheduling problem are listed as follows:

2.1 Known Variables

N : a set of jobs to be processed in the burn-in oven and $N = \{1, 2, ..., n\}$.
M : Oven capacity.
r_i : the available time of job i, $i = 1, 2, \cdots, n$.
p_i : the processing time of job i, $i = 1, 2, \cdots, n$.
s_i : the processing quantity (sizes) of job i, $i = 1, 2, \cdots, n$.
Q_i : $Q_i = \lceil \frac{s_i}{M} \rceil$ is the total number of batches needed for job i in the burn-in oven.
K : the total number of batches needed for the set of jobs N in the burn-in oven and $K = \sum_{i=1}^{n} \lceil \frac{s_i}{M} \rceil = \sum_{i=1}^{n} Q_i$.

2.2 Auxiliary Variables

R^k : the available time of batch B^k.
P^k : the processing time of batch B^k.
C^k : the completion time of batch B^k.
F^k : the flow time of batch B^k.
c_i : the completion time of job i.
f_i : the flow time of job i and $f_i = c_i - r_i$.
x_i^k : denoting whether job i is processed on batch B^k,

$$\begin{cases} x_i^k = 1 \text{ if } q_i^k > 0 \\ x_i^k = 0 \text{ if } q_i^k = 0 \end{cases}$$

2.3 Decision Variables

q_i^k : the quantity of job i to be handled on the kth batch B^k

2.4 The Constraint Programming Model

The constraint programming model is formulated as follows:

Min $\sum_{i=1}^{n} f_i = \sum_{i=1}^{n} (c_i - r_i)$

s.t.

$$\sum_{i=1}^{n} q_i^k \leq M, \quad k = 1, 2, \cdots, K \tag{1}$$

$$\sum_{k=1}^{k} q_i^k = s_i, \quad i = 1, 2, \cdots, n \tag{2}$$

$$\sum_{k=1}^{K} x_i^k = Q_i, \quad i = 1, 2, \cdots, n \tag{3}$$

$$\begin{cases} q_i^k > 0 & \Rightarrow \quad x_i^k = 1 \\ q_i^k = 0 & \Rightarrow \quad x_i^k = 0 \end{cases}$$
$$i = 1, 2, ..., n; k = 1, 2, ..., K \tag{4}$$

$$\begin{cases} x_i^k = 0 & \Rightarrow \quad q_i^k = 0 \\ x_i^k = 1 & \Rightarrow \quad (q_i^k = M) \vee [q_i^k = \text{mod}\left(\frac{s_i}{M}\right)] \end{cases}$$
$$i = 1, 2, ..., n; k = 1, 2, ..., K \tag{5}$$

$$R^k \geq r_i \times x_i^k \quad i = 1, 2, ..., n; k = 1, 2, ..., K \tag{6}$$

$$P^k \geq p_i \times x_i^k \quad i = 1, 2, ..., n; k = 1, 2, ..., K \tag{7}$$

$$C^k \geq R^k + P^k \quad k = 1, 2, \cdots, K \tag{8}$$

$$c_i \geq C^k \times x_i^k \quad i = 1, 2, \cdots, n, \quad k = 1, 2, \cdots, K \tag{9}$$

$$R^{k+1} \geq C^k \quad k = 1, 2, \cdots, K-1 \tag{10}$$

Constraint (1) defines that the total flow time is equal to the sum of job's completion time minus job's available time. Constraint (1) ensures that the quantity of each batch does not exceed the capacity of a machine. Constraint (2) ensures that the sizes (quantity) of each job. Constraint (3) specifies that the total number of batches that each job is handled. In constraint (4), if job i is processed in batch B^k then x_i^k will be assigned as 1 otherwise x_i^k will be 0. In constraint (5), the processing quantity of job i in batch B^k, i.e., q_i^k, will be equal to M or mod $\left(\frac{s_i}{M}\right)$. (If $s_i^k > M$, then we will split the job into two sub-lots, i.e., M and mod $\left(\frac{s_i}{M}\right)$; otherwise $q_i^k = s_i^k$.). Constraint (6) defines the available time of the kth batch. Constraint (7) specifies the processing time of the kth batch. Constraint (8) defines the completion time of the kth batch. Constraint (9) defines the completion time of each job. Constraint (10) ensures that the starting time of the kth batch is greater than or equal to the previous batch's completion time.

The solution procedure of constraint programming is by constraint propagation that will eliminate the solution space from the set of variable's domain and infer new constraint from that. Therefore, the software package ILOG OPL is applied to solve the problem.

3 Properties of the Batch Processing Machine

3.1 Definition of Variables

σ : a set of sequenced jobs and $\sigma = \{J_{[1]}, J_{[2]}, ..., J_{[h]}\}$.
σ' : a set of un-sequenced jobs.
$C(\sigma)$: the completion time of the set of sequenced jobs σ.
$C(\sigma')$: the completion time of the set of un-sequenced jobs σ'.
$F(\sigma)$: the total flow time of the set of sequenced jobs σ, i.e.,

$$F(\sigma) = \sum_{i \in \sigma} f_i.$$

$F(\sigma')$: the total flow time of the set of sequenced jobs σ', i.e.,

$$F(\sigma') = \sum_{i \in \sigma'} f_i.$$

I^k : the idle time before batch B^k.
$|B^k|$: the processing quantity in batch B^k.
L^k : number of jobs in batch B^k.
w : the indicator of batch B^k.
w_i : the new indicator after assigning job i into batch B^k.
t : the ready time of burn-in oven.

3.2 Properties

Property 1: If there exits a job α with $r_\alpha = \min_i r_i$ and $p_\alpha = \min_i p_i$, and there is a job β with $r_\beta \geq \max(t, r_\alpha) + p_\alpha$ and $s_\alpha + s_\beta \leq M$ for $\alpha, \beta \in \sigma'$, then sequence $\{\sigma, (\alpha, \beta)\}$ is dominated by sequence $\{\sigma, \alpha\}$.

Proof: According to Dessouky and Deogun (1981), sequence $\{\sigma, \beta\}$ is dominated by sequence $\{\sigma, \alpha\}$ when $r_\beta \geq \max(t, r_\alpha) + p_\alpha$. Next, we need to prove that sequence $\{\sigma, \alpha\}$ dominates over sequence $\{\sigma, (\alpha, \beta)\}$.

Given a sequence $\{\sigma, \alpha, \beta\}$ and a set of un-sequenced jobs σ', the total flow times (TF) of this schedule, i.e., $\{\sigma, \alpha, \beta, \sigma'\}$ is equal to $TF = F(\sigma) + \max(t, r_\alpha) + p_\alpha + \max\{\max(t, r_\alpha) + p_\alpha, r_\beta\} + p_\beta + F(\sigma')$

Since $r_\beta \geq \max(t, r_\alpha) + p_\alpha$, the above formula can be simplified as follows:

$$TF = F(\sigma) + \max(t, r_\alpha) + p_\alpha + r_\beta + p_\beta + F(\sigma')$$

For a new sequence $\{\sigma(\alpha, \beta)\}$ and a set of un-sequenced jobs σ', when job α and job β are processed in the same batch, the total flow times (TF') of this schedule, i.e., $\{\sigma, (\alpha, \beta), \sigma'\}$ is equal to

$$TF' = F(\sigma) + 2\{\max(t, r_\beta) + \max(p_\alpha, p_\beta)\} + F(\sigma')$$

Since $r_\beta \geq t$ and $p_\beta \geq p_\alpha$, the above formula can be simplified as follows:

$$TF' = F(\sigma) + 2(r_\beta + p_\beta) + F(\sigma') \quad \ldots \quad (11)$$

Formula (1) − Formula (1), we can derive the following:

$$TF' - TF = r_\beta + p_\beta - \max(t, r_\alpha) - p_\alpha \geq 0 \quad \ldots \quad (12)$$

Again, $r_\beta \geq \max(t, r_\alpha) + p_\alpha$, thus $\{\sigma, (\alpha, \beta)\}$ is dominated by $\{\sigma, \alpha\}$

Property 2: Let LB_1 be one of the lower bound of $1/r_i, B, s_i/\sum F_i$, then $LB_1 \geq \sum \lceil \frac{s_i}{M} \rceil \times p_i$.

Proof: Intuitively, the flow time of each job is greater than or equal to its processing time. Therefore, the total flow time of each job to be processed in the burn-in operation will be greater than the sum of batch processing time on the burn-in oven, i.e., $LB_1 \geq \sum \lceil \frac{s_i}{M} \rceil \times p_i$. Since LB_1 is calculated based on job's attributes, we name it as job-based bound.

Property 3: Let LB_2 be another lower bound of $1/r_i, B, s_i/\sum F_i$, then $LB_2 \geq LB_{BWSPT}$.

Proof: According to Chandru et al. (1993b), an optimal schedule with **BWSPT** (Batch weighted short processing time) order is a sequence ranked as follows:

$$\frac{P^1}{|B^1|} \leq \frac{P^2}{|B^2|} \leq \ldots \leq \frac{P^k}{|B^k|}$$

and $LB_{BWSPT} = \sum_{i=1}^{k} P^i L^i$, however for a dynamic scheduling problem, i.e., $r_i \geq 0$, there may exit idle time before each batch (Oven is ready however job is not available, i.e., $r_i \geq t$.) Therefore,

$$LB_2 = \sum_{i=1}^{k}(I^i + P^i)L^i \geq \sum_{i=1}^{k} P^i L^i = LB_{BWSPT}.$$ Since LB_2 is calculated based on batches' attributes, we name it as batch-based bound.

Property 4: Let LB_3 is another lower bound of $1/r_i, B, s_i/\sum F_i$, then

$$LB_3 = n\left(I_1 + P^1\right) + \sum_{j=1}^{k-1}\left[(k-j+1)P^{j'}\right] + P^{LPT} + \sum_{i=1}^{n}\left\{\left[\frac{s_i - \text{mod}\left(\frac{s_i}{M}\right)}{M}\right]P^i\right\}$$

Proof:

1. Assume $s_i \leq M \ \forall \ i = 1, 2, 3, ..., n$ and each job is only processed once in one batch, then
    $F^1 = L^1 \times \left(I_{[1]} + P^{[1]}\right)$
    $F^2 = L^2 \times \left(I_{[1]} + P^{[1]} + I_{[2]} + P^{[2]}\right)$
    $\vdots$
    $F^k = L^k \times \left(I_{[1]} + P^{[1]} + I_{[2]} + P^{[2]} + \cdots + I_{[k]} + P^{[k]}\right)$
    Sum up the above formula in the right hand side and left hand side, then we can derive the following:

$$TF = \sum_{i=1}^{k} F^k = \sum_{i=1}^{k}\left\{\left(\sum_{j=i}^{k} L^j\right)(I_i + P^i)\right\}$$

$$= \left[\sum_{j=1}^{k} L^j\right](I_1 + P^1) + \sum_{i=2}^{k}\left\{\left(\sum_{j=i}^{k} j^j\right)(I_i + P^i)\right\}$$

$$\geq n\left(I_1 + P^1\right) + \sum_{i=2}^{k}\left\{\left(\sum_{j=i}^{k} j^j\right)(I_i + P^i)\right\}$$

$$\left(\text{since } \left[\sum_{j=1}^{k} L^j\right] \geq n\right)$$

$$\geq n\left(I_1 + P^1\right) + \sum_{j=1}^{k-1}\left[(k-j+1)P^{j'}\right] + P^{LPT}$$

(where $P^{j'}$ is in SPT order and $(k - j + 1)$ is the weight for job p'_j; P^{LPT} is the largest job among the set of unscheduled jobs.)

In Property 2, by relaxing the dynamic problem into static problem and applying SPT rule to derive the lower bound, we can arrange all batches except first batch in SPT order and the corresponding processing time for each batch is p'_j. Here, p_{LPT} is the largest batch processing time among all batches except first batch. And notice that we do not have idle time before or after each batch.

2. For jobs with $s_i > M$, then we can partition the problem into two sub-problems, one problem is $s_i < M$ and another problem is $s_i \geq M$. Then combining LB1 and condition (1), we can obtain the following result.

$$LB_3 = n(I_1 + p_1) + \sum_{j=1}^{k-1}\left[(k-j+1)p'_j\right] + p_{LPT} + \sum_{i=1}^{n}\left\{\left[\frac{s_i - \text{mod}\left(\frac{s_i}{M}\right)}{M}\right]p_i\right\}$$

Property 5: Given a sequence σ, if there are two successive jobs that can be combined into a new batch to form a new sequence σ'. If the completion time of a new batch is greater than that of the original batch in σ, then the total flow time TF' of σ' will be greater than the total flow time TF of σ.

Proof: Suppose a successive job list contains two jobs i and j and they can be combined into a single batch since $s_i + s_j \leq M$. However, Job j arrives later than the completion time of Job i, therefore Job i can only be combined with job j and forms a new batch. Figure 1 shows the difference between the schedule of σ and σ'. Since job 1 is delayed, the increment in flow time is θ_1. As for job j, the increment in flow time will be θ_2 and it is the difference between batch j processing time and batch (i,j) processing time. $\theta_2 > 0$ if job i has a larger processing time; otherwise $\theta_2 = 0$. This result is true for two more jobs in successive processing orders.

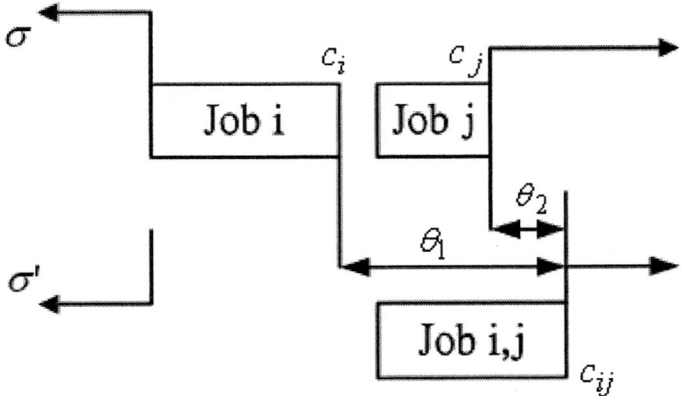

Fig. 1. The effect of the total flow time between sequences σ and σ'

According to BWSPT, weights for job i and job j before and after the merge are as follows:

$$w_i = \frac{p_i}{1} \text{ and } w_j = \frac{p_j}{1} \text{ before the merge}$$

$$\text{and } w_{ij} = \frac{p_i + p_j + \theta_1 + \theta_2}{|B^k|} \text{ after the merge.}$$

There is only one batch after the merge but $w' \geq w$ since $\theta_1 \geq p_i$ and $\theta_2 \geq p_j$. Therefore, we will not combine job i and job j into the same batch

4 Conclusion

The problem we considered, i.e., $1/r_j, s_j, B/\sum F$, where the job sizes can be greater or smaller than the capacity of oven, distinguishes this paper from past work in this area. For small size of problem, we constructed a constraint-programming model to obtain an optimal solution. In addition, we also provided some properties, which have been applied in developing the heuristic algorithm, to further reduce the size of the solution space. Computational results show that the heuristic is capable of obtaining a good solution with very reasonable computation times.

There are several directions for future research in our study, such as developing a tighter lower bound and more efficient heuristics. In addition, extensions of this study to include other soft-computing approaches such as evolutional approach, tabu search and simulated annealing, etc., are very applicable in the future research. Also, different performance measures and parallel batch processing machines can be further considered as an extension of this research.

References

1. Chandru, V., Lee, C.-Y., and Uzsoy, R.: Minimizing total completion time on batch processing machines. International Journal of Production Research. **31** (1993a) 2097-2121.
2. Chandru, V., Lee, C.-Y., and Uzsoy, R.: Minimizing total completion time on a batch processing machine with job families. Operations Research letter. **13** (1993b) 61-65.
3. Chang, P.C., Wang, H.M., and Su, L.H.: Scheduling job with burn in board constraint on Semiconductor Burn-In operations to minimize the mean flow time, Proceeding of the 5th Annual Conference on Industrial Engineering-Theory, Applications, and Practice December 13-15, Hsin-Chu Taiwan, R.O.C. (2000).
4. Chang, P.C. and Wang, H.M.: A Genetic Algorithm to minimize makespan for dynamic single batch processor scheduling, Proceeding of the 33rd International Conference on Computers and Industrial Engineering, (2004a) 834-840.
5. Chang, P.C., and Wang, H.M.: A heuristic for a batch processing machine scheduled to minimize total completion time with non-identical job sizes. International Journal of Advanced Manufacturing Technology. **24** (2004b) 615-620.
6. Dessouky, M. I. and J. S. Deogun: Sequencing jobs with unequal ready times to minimize mean flow time. SIAM Journal of Computing. **10** (1981) 192-202.
7. DuPont, L., and Jolai Ghazvini, F.: A branch and bound algorithm for minimizing mean flow time on a single batch processing machine. International Journal of Industrial Engineering. **4** (1997) 197-203.
8. DuPont, L., and Dhaenens-Flipo, C.: Minimizing the makespan on a batch machine with non-identical job sizes: an exact procedure. Computers & Operations Research. **29** (2002) 807-819.
9. Kempf, K.G., Uzsoy, R., and Wang, C.-S.: Scheduling a Single Batch Processing Machine with Secondary Resource Constraints. Journal of Manufacturing Systems. **17** (1998) 37-51.
10. Lee, C.-Y., Uzsoy, R., and Martin-Vega, L.A.: Efficient algorithms for scheduling semiconductor burn-in operations. Operations Research. **40** (1992) 764-775.
11. Lee C.-Y. and Uzsoy, R.: Minimizing makespan on a single batch processing machine. International Journal of Production Research. **37** (1999) 219-236.
12. Li, C.-L. and Lee, C.-Y.: Scheduling with agreeable release times and due dates on a batch processing machine European Journal of Operational Research. **96** (1997) 564-569.

13. ILOG, 2001, ILOG OPL Studio 3.5 User's Manual, ILOG, France.
14. Jolai Ghazvini, F. and Dupont, L.: Minimizing mean flow times criteria on a single batch processing machine with non-identical jobs sizes, International Journal of Production Economics. **55** (1998) 273-280.
15. Melouk, S., Damodaran P., and Chang P.Y.: Minimizing makespan for single machine batch processing with non-identical job sizes using simulated annealing. International Journal of Production Economics. **87** (2004) 141-147.
16. Sung, C.S. and Choung, Y.I.: Minimizing makespan on a single burn-in oven in semiconductor manufacturing. European Journal of Operational Research. **120** (2000) 559-574.
17. Uzsoy, R.: Scheduling a single batch processing machine with non-identical job sizes. International Journal of Production Research. **32** (199e4) 1615-1635.
18. Uzsoy, R. and Wang, C. S.: A Genetic Algorithm to Minimize Maximum Lateness on a Batch Processing Machine. Computers and Operations Research. **29** (2002) 1621-1640.

Polynomial Algorithm for Parallel Machine Mean Flow Time Scheduling Problem with Release Dates

Peter Brucker[1] and Svetlana A. Kravchenko[2]

[1] Universität Osnabrück, Fachbereich Mathematik/Informatik,
49069 Osnabrück, Germany
peter@mathematik.uni-osnabrueck.de

[2] United Institute of Informatics Problems, Surganova St. 6,
220012 Minsk, Belarus
kravch@newman.bas-net.by

Abstract. In this paper we give a polynomial algorithm for the problem $P \mid r_i, p_i = p, \text{pmtn} \mid \sum C_i$. This result is applied to derive a polynomial algorithm for the problem $O \mid r_i, p_{ij} = 1 \mid \sum C_i$.

1 Introduction

The problem considered can be stated as follows. There are n independent jobs and m identical parallel machines. For each job $i = 1, \ldots, n$ we know its processing time p_i, and its release time r_i. Each machine can process only one job at a time, and at any moment any job can be processed only by one but arbitrary machine. Preemptions of processing are allowed, i.e. the processing of any job may be interrupted at any time and resumed later, possibly on a different machine, the total time of processing job i being equal to p_i. For a schedule s, let $C_i(s)$ denote the time at which job i completes its processing. If no ambiguity arises, we drop the reference to schedule s and write C_i. The problem is to schedule all the jobs so as to minimize the optimality criterion $\sum_{i=1}^{n} C_i$. Following the well known classification scheme [7] suggested by Graham et al. the described problem can be denoted as $P \mid r_i, \text{pmtn} \mid \sum C_i$.

For the problem $P \mid \text{pmtn} \mid \sum w_i C_i$ with the more general objective function $\sum w_i C_i$ and release times $r_i = 0$ for all jobs i McNaughton [12] has shown that preemption is redundant, i.e. $P \mid \text{pmtn} \mid \sum w_i C_i$ has the same solution value as the non-preemptive problem $P \mid\mid \sum w_i C_i$. Thus, $P \mid p_i = p, \text{pmtn} \mid \sum w_i C_i$ can be solved by solving $P \mid p_i = p \mid \sum w_i C_i$, which can be formulated as a network flow problem. If we have release times $r_i \neq 0$ we have a different situation even for problems with objective function $\sum C_i$. For $P \mid r_i, p_i = p \mid \sum C_i$ an optimal schedule can be constructed by scheduling jobs j in an order of nondecreasing release times at the earliest time $t \geq r_i$ at which some machine is available. However, the preemptive version $P \mid r_i, p_i = p, \text{pmtn} \mid \sum C_i$ usually provides a schedule with smaller objective value than the optimal objective value for an instance of $P \mid r_i, p_i = p \mid \sum C_i$, as demonstrated in Figure 1. The main result in this paper is a polynomial algorithm for $P \mid r_i, p_i = p, \text{pmtn} \mid \sum C_i$. This algorithm has two steps. In a first step the job finishing times for an optimal schedule are calculated by solving a linear program with $O(n^3)$ variables and constraints. Using the corresponding finishing times an optimal schedule can be constructed by solving a maximum flow problem in a second step. We also

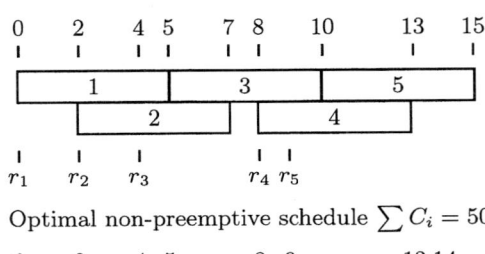

(a) Optimal non-preemptive schedule $\sum C_i = 50$

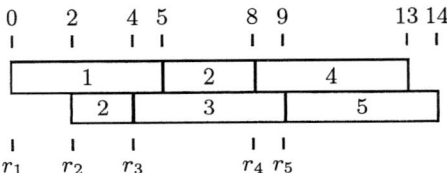

(b) Preemptive schedule with better $\sum C_i = 49$

Fig. 1. Preemption provides a better schedule for $P \mid r_i, p_i = p \mid \sum C_i$

show that our result can be applied to obtain a polynomial algorithm for the open shop problem $O \mid r_i, p_{ij} = 1 \mid \sum C_i$. Finally, we prove that the problem $P \mid r_i, \text{pmtn} \mid \sum C_i$ with arbitrary processing times is unary NP-hard.

Previous results which are related to our results are listed below.

$P2 \mid r_i, p_i = p, \text{pmtn} \mid \sum C_i$	$O(n \log n)$ algorithm	[8]
$Om \mid r_i, p_{ij} = 1 \mid \sum C_i$	$O(n^2 m^{3m})$ algorithm	[15]
$P2 \mid r_i, \text{pmtn} \mid \sum C_i$	binary NP-hardness	[5]
$P \mid r_i, p_i = p, \text{pmtn} \mid \sum w_i C_i$	unary NP-hardness	[10]

We refer the reader desiring to know more about the listed problems to [3], [13].

The paper is organized as follows. In Section 2 we describe a polynomial algorithm for $P \mid r_i, p_i = p, \text{pmtn} \mid \sum C_i$. In Section 3 we give a polynomial algorithm for $O \mid r_i, p_{ij} = 1 \mid \sum C_i$. Throughout the paper we suppose that all the jobs are enumerated in such a way that $r_1 \leq \ldots \leq r_n$ holds. Furthermore we assume that $r_1 = 0$.

2 A Polynomial Algorithm for $P \mid r_i, p_i = p, \text{pmtn} \mid \sum C_i$

In this section we derive a polynomial algorithm for problem $P \mid r_i, p_i = p, \text{pmtn} \mid \sum C_i$.

Statement 1. *For problem $P \mid r_i, p_i = p, \text{pmtn} \mid \sum C_i$ an optimal schedule can be found in the class of schedules, where $C_1 \leq \ldots \leq C_n$ holds.*

We set $r_{n+1} = r_n + p \cdot n$, i.e. r_{n+1} is a time point after which no job will be processed. Let s^* be an optimal schedule for $P \mid r_i, p_i = p, \text{pmtn} \mid \sum C_i$ with $C_1(s^*) \leq \ldots \leq C_n(s^*)$. We define $C_0^i = r_i$, $C_{n+1}^i = r_{i+1}$ for all $i = 1, \ldots, n$ and set for job $j = 1, \ldots, n$

$$C_j^i = \begin{cases} C_j(s^*) & \text{if } r_i < C_j(s^*) < r_{i+1} \\ r_i & \text{if } C_j(s^*) \leq r_i \\ r_{i+1} & \text{if } C_j(s^*) \geq r_{i+1} \end{cases}$$

Thus, for each $i = 1, \ldots, n$ the values

$$r_i = C_0^i \leq C_1^i \leq \ldots \leq C_{n+1}^i = r_{i+1} \qquad (1)$$

define a partition of the interval $[r_i, r_{i+1}]$. Note that C_0^i and C_{n+1}^i are added for technical reasons. Furthermore, we denote by $v_j^{(i,k)}$ the processing time of job j in the interval $[C_k^i, C_{k+1}^i]$. Then the values C_j^i ($j = 0, \ldots, n+1$; $i = 1, \ldots, n$) and $v_j^{(i,k)}$ ($i, j = 1, \ldots, n$; $k = 0, \ldots, n$) define a feasible solution of the following linear program which we denote by SLP. SLP has $O(n^3)$ variables and constraints, i.e. it can be solved polynomially (see [4] for computational aspects of linear programming problems).

Minimize

$$\sum_{i=1}^{n} ((C_1^i - r_i) + \ldots + (C_n^i - r_i)) \qquad (2)$$

subject to

$$C_2^1 = \ldots = C_n^1 = r_2; \ldots; C_{k+1}^k = \ldots = C_n^k = r_{k+1}; \ldots; C_n^{n-1} = r_n; \qquad (3)$$

$$v_j^{(i,k)} \leq C_{k+1}^i - C_k^i, \qquad i, j = 1, \ldots, n, \qquad k = 0, \ldots, n; \qquad (4)$$

$$\sum_{j=1}^{n} v_j^{(i,k)} \leq m(C_{k+1}^i - C_k^i), \qquad i = 1, \ldots, n, \qquad k = 0, \ldots, n; \qquad (5)$$

$$\sum_{i=j}^{n} \sum_{k=0}^{n} v_j^{(i,k)} = p, \qquad j = 1, \ldots, n; \qquad (6)$$

$$v_j^{(i,j)} = \ldots = v_j^{(i,n)} = 0, \qquad i, j = 1, \ldots, n; \qquad (7)$$

$$r_i = C_0^i \leq C_1^i \leq \ldots \leq C_n^i \leq C_{n+1}^i = r_{i+1}, \qquad i = 1, \ldots, n; \qquad (8)$$

$$C_j^i \geq 0, \quad i, j = 1, \ldots, n; \quad v_j^{(i,k)} \geq 0, \quad i, j = 1, \ldots, n, \quad k = 0, \ldots, n. \qquad (9)$$

Clearly restrictions (3) and (8) hold for s^*. Restrictions (4) and (5) are capacity constraints for m machines in the interval (C_k^i, C_{k+1}^i). (6) must hold because the job must be totally processed.

Let $i(j)$ be the index such that $r_{i(j)} < C_j(s^*) = C_j^{i(j)} \leq r_{i(j)+1}$. Restriction (7) must hold because if $i \geq i(j)$ then $C_j(s^*) = C_j^i \leq C_{j+1}^i \leq \ldots \leq C_{n+1}^i$. Thus, job j is not processed in $(C_j^i, C_{j+1}^i), \ldots, (C_n^i, C_{n+1}^i)$.

If $i < i(j)$ then $r_{i+1} \leq C_j(s^*) \leq C_v(s^*)$ for all $v \geq j$. Thus $C_v^i = r_{i+1}$ for all $v \geq j$. Finally $C_j^i \geq 0$ because $0 \leq r_1 \leq \ldots \leq r_{n+1}$.

Furthermore, $\sum_{i=1}^n ((C_1^i - r_i) + \ldots + (C_n^i - r_i)) = \sum_{j=1}^n (\sum_{i=1}^n (C_j^i - r_i)) = \sum_{j=1}^n C_j(s^*)$ because $\sum_{i=1}^n (C_j^i - r_i) = \sum_{i<i(j)} (r_{i+1} - r_i) + C_j(s^*) - r_{i(j)} + \sum_{i>i(j)} (r_i - r_i) = \sum_{j=1}^n C_j(s^*)$.

Summarizing, we have proven

Theorem 1. *For any optimal schedule s^* for the problem $P \mid r_i, p_i = p, pmtn \mid \sum C_i$ with $C_1(s^*) \leq \ldots \leq C_n(s^*)$ there is a corresponding feasible solution of SLP such that*

$$\sum_{i=1}^n C_i(s^*) = \sum_{i=1}^n ((C_1^i - r_i) + \ldots + (C_n^i - r_i)) \tag{10}$$

holds.

The next theorem shows that conversely any feasible solution of SLP also provides a feasible schedule.

Theorem 2. *Any feasible solution of problem SLP provides a feasible schedule for the scheduling problem $P \mid r_i, p_i = p, pmtn \mid \sum C_i$. This schedule can be constructed in $O(n^3)$ time and the total number of preemptions is bounded by $O(n^3)$.*

Note, that C_j is a real completion time of the job j, i.e. there is some $\delta > 0$, such that job j is processed in $(C_j - \delta, C_j)$, and job j is not processed in the interval (C_j, ∞). Whereas for C_j^i it may happen that there is some $\delta > 0$, such that job j is not processed in $(C_j^i - \delta, C_j^i)$. We now study the relationships between the $\sum_{i=1}^n C_i$-value and the optimal solution value of SLP.

Unfortunately, it is possible that for a feasible schedule the values C_j^i can be chosen in such a way that together with the corresponding $v_j^{(i,k)}$-values they are feasible for SLP and

$$\sum_{i=1}^n ((C_1^i - r_i) + \ldots + (C_n^i - r_i)) < \sum_{i=1}^n C_i \tag{11}$$

holds. This is illustrated by the following example.

Example 1. Consider the following schedule, see Figure 2. Here we have $C_1 = C_2 = 10$, $C_3 = C_4 = C_5 = 23$, $C_6 = 24$, $C_7 = 26$, $C_8 = C_9 = 32$, $C_{10} = 36$. Thus, $\sum_{i=1}^{10} C_i = 239$. However, the values $C_1^5 = 10$, $C_2^5 = 10$, $C_3^5 = C_4^5 = C_5^5 = 11$, $C_1^7 = C_2^7 = r_7$, $C_3^7 = 15$, $C_4^7 = 17$, $C_5^7 = C_6^7 = C_7^7 = 18$, $C_1^9 = C_2^9 = C_3^9 = C_4^9 = r_9$, $C_5^9 = 19$, $C_6^9 = C_7^9 = C_8^9 = C_9^9 = 22$, $C_1^{10} = C_2^{10} = r_{10}$, $C_3^{10} = C_4^{10} = C_5^{10} = 23$, $C_6^{10} = 24$, $C_7^{10} = 26$, $C_8^{10} = C_9^{10} = C_{10}^{10} = 27$ provide a feasible solution for SLP with $\sum_{i=1}^{10} ((C_1^i - r_i) + \ldots + (C_{10}^i - r_i)) = 224$.

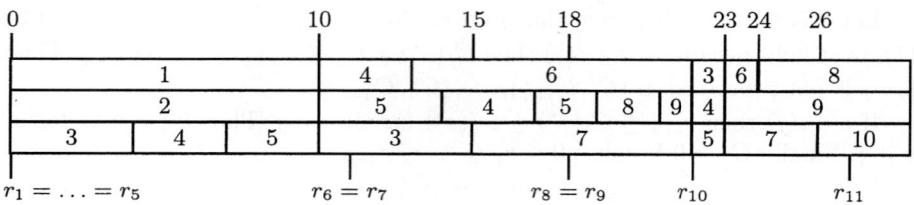

Fig. 2. Here $m = 3$, $p = 10$, $r_1 = \ldots = r_5 = 0$, $r_6 = r_7 = 11$, $r_8 = r_9 = 18$, $r_{10} = 22$, $r_{11} = 27$ hold

Next we show, that a schedule provided by an optimal solution of problem SLP can be transformed into a schedule such that

$$\sum_{i=1}^{n} C_i = \sum_{i=1}^{n} ((C_1^i - r_i) + \ldots + (C_n^i - r_i)) \qquad (12)$$

holds. To derive this result we need some preparations.

Theorem 3. *Let s be a schedule corresponding to an optimal solution of our linear program SLP. Then for each job j there is some $i \leq n + 1$ such that $C_j = C_j^i$ and $C_j^k = r_k$ holds for all $n + 1 \geq k > i$.*

Suppose that for some instance of $P \mid r_i, p_i = p, \text{pmtn} \mid \sum C_i$ we obtain a schedule s by solving the corresponding problem SLP. Then we have the following

Theorem 4. *If s is a schedule such that for any job j with $C_j = C_j^i$, we have $C_j^q = r_{q+1}$ for each $q < i$, then $\sum_{i=1}^{n} C_i = \sum_{i=1}^{n}((C_1^i - r_i) + \ldots + (C_n^i - r_i))$ holds.*

Thus, if the solution of the corresponding problem SLP is such that the condition of Theorem 4 holds, then we obtain an optimal schedule for the problem $P \mid r_i, p_i = p, \text{pmtn} \mid \sum C_i$. However, the condition of Theorem 4 does not necessarily hold for any solution of SLP. Now we show that if the condition of Theorem 4 does not hold then there is a procedure, such that any solution of SLP can be transformed into a new schedule s without increasing the $\sum_{i=1}^{n}((C_1^i - r_i) + \ldots + (C_n^i - r_i))$ - value, and such that the condition of Theorem 4 holds for s.

In the remainder of this chapter we will prove

Theorem 5. *Consider an optimal solution of problem SLP with values C_j^i for $i, j = 1, \ldots, n$ and let s be a corresponding schedule for $P \mid r_i, p_i = p, \text{pmtn} \mid \sum C_i$. Then s can be transformed into a schedule s' such that*

$$C_i(s') = (C_i^1 - r_1) + \ldots + (C_i^n - r_n)$$

holds for each $i = 1, \ldots, n$.

Let s be the solution of problem SLP, such that not for all jobs j with $C_j = C_j^i$, $C_j^q = r_{q+1}$ holds for $q < i$. Say that job j is incorrect if $C_j^q = r_{q+1}$ does not hold for all $q < i$, where $C_j^i = C_j$.

Let $t_1 < t_2 < \ldots < t_N$ be all the preemption points, i.e. all values t_i such that there is a job j and some $\delta > 0$ such that j is processed in $(t_i - \delta, t_i)$ and j is not processed in $(t_i, t_i + \delta)$.

An iteration of our transformation can be described in 5 steps as follows:

Step 1. Take the maximal point t_p such that at least one incorrect job is completing at time t_p, $t_p \in]r_i, r_{i+1}]$, see Figure 3. It is possible that there are several jobs completing at t_p and some of them are incorrect. Then by Lemma 1 we can denote all incorrect jobs completing at t_p by $j + 1, j+2, \ldots, j+z$, and the conditions $C_j^i < C_{j+1}^i = C_{j+2}^i = \ldots = C_{j+z}^i \leq C_{j+z+1}^i$ hold.

Step 2. Take the minimal value of g, $g < i$, such that $C_{j+1}^g \neq r_{g+1}$ holds for (r_g, r_{g+1}), see Figure 3. Let y jobs from the set $\{j+1, \ldots, n\}$ finish at time C_{j+1}^g, i.e. $C_j^g \leq C_{j+1}^g = C_{j+2}^g = \ldots = C_{j+y}^g < C_{j+y+1}^g$. By Lemma 2, $y \leq z$ holds.

Step 3. Take $\delta = \min\{\delta_1, \delta_2\}$, where
- $\delta_1 > 0$ is the greatest value such that in $(C_{j+1}^g, C_{j+1}^g + \delta_1)$ at most m jobs are processed;
- $\delta_2 > 0$ is the greatest value such that in $(C_{j+1}^i - \delta_2, C_{j+1}^i)$ jobs $j+1, \ldots, j+y$ are processed without interruptions.

Step 4. Take y (possibly empty[1]) jobs from $(C_{j+1}^g, C_{j+1}^g + \delta)$, such that they are not processed in $(C_{j+1}^i - \delta, C_{j+1}^i)$. By Lemma 3 this is always possible.

Step 5. Switch these y jobs taken from $(C_{j+1}^g, C_{j+1}^g + \delta)$ with jobs $j+1, \ldots, j+y$ from $(C_{j+1}^i - \delta, C_{j+1}^i)$. Add to the set $t_1, \ldots, t_N$ the points $C_{j+1}^i - \delta$ and $C_{j+1}^g + \delta$, see Figure 3.

By Lemma 4 any iteration of our transformation does not change $\sum_{i=1}^n ((C_1^i - r_i) + \ldots + (C_n^i - r_i))$ value.

Example 2. Now we apply our transformation to the schedule from Example 1. When we make the first iteration then in Step 1 $t_p = 23$, $\{j+1, j+2, \ldots, j+z\} = \{3, 4, 5\}$ hold, see Figure 2. In Step 2 $g = 7, y = 1$ hold. In Step 3 $\delta = 1$. In Step 4 we take job 7 from $(C_{j+1}^g, C_{j+1}^g + \delta) = (C_3^7, C_3^7 + 1)$. In Step 5 we switch job 7 from $(C_3^7, C_3^7 + 1)$ and job 3 from $(r_{10}, r_{10} + 1)$. Further iterations switch job 4 from the interval $(22, 23)$ with job 6 from the interval $(17, 18)$, and job 5 from the interval $(22, 23)$ with job 8 from the interval $(19, 20)$. Now $\sum_{i=1}^n C_i = 224$ holds, see Figure 4.

We will apply Step 1 - Step 5 to our schedule until all incorrect jobs are eliminated. In Lemma 4, and 5 we prove that in a finite number of steps we will obtain an optimal schedule.

---

[1] If on a machine no job is scheduled in $(C_{j+1}^g, C_{j+1}^g + \delta)$ this defines an empty job.

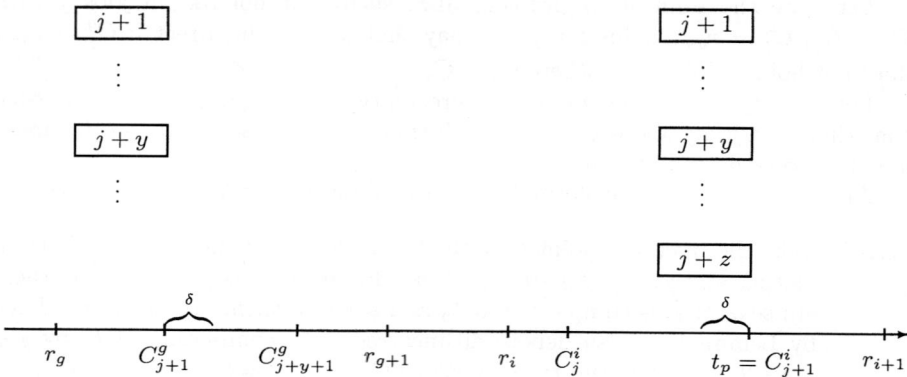

Fig. 3. Here $y \leq z$, $C^g_{j+1} = \ldots = C^g_{j+y} < C^g_{j+y+1}$, and $C^i_j < t_p = C^i_{j+1} = \ldots = C^i_{j+z}$ hold

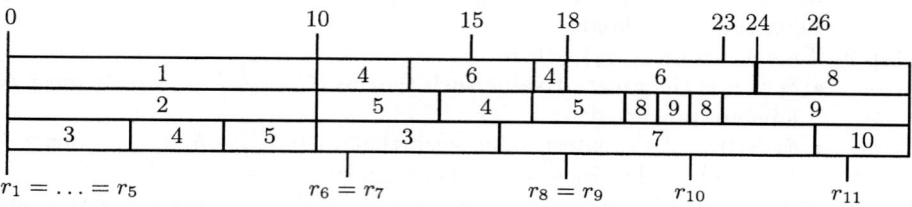

Fig. 4. The schedule obtained after the transformation

Lemma 1. *If in Step 1 there are z incorrect jobs finishing at t_p, then we can denote them by $j+1, \ldots, j+z$, and $C^i_j < C^i_{j+1} = C^i_{j+2} = \ldots = C^i_{j+z} \leq C^i_{j+z+1}$ holds.*

Lemma 2. *In Step 2 the condition $y \leq z$ holds.*

Lemma 3. *In Step 4 we can always choose the necessary number of jobs from $(C^g_{j+1}, C^g_{j+1} + \delta)$ such that they are not processed within $(C^i_{j+1} - \delta, C^i_{j+1})$.*

Lemma 4. *Applying Steps 1 to 5 to schedule s does not change the*

$$\sum_{i=1}^{n}((C^i_1 - r_i) + \ldots + (C^i_n - r_i)) \tag{13}$$

value.

We will apply Steps 1-5 to s as long as we find incorrect job. When all jobs in our schedule are not incorrect, the obtained schedule is an optimal one.

Lemma 5. *The described transformation takes the finite number of steps.*

Thus, we see that if the optimal solution of problem SLP has the objective value

$$\sum_{i=1}^{n}((C_1^i - r_i) + \ldots + (C_n^i - r_i)) \tag{14}$$

then for the optimal schedule

$$\sum_{i=1}^{n} C_i = \sum_{i=1}^{n}((C_1^i - r_i) + \ldots + (C_n^i - r_i)) \tag{15}$$

and

$$C_i = (C_i^1 - r_1) + (C_i^2 - r_2) + \ldots + (C_i^n - r_n) \text{ for each } i = 1, \ldots, n \tag{16}$$

hold.

Therefore, to find an optimal schedule we do not need to apply our transformation. We can solve the following network flow problem, see [11]:

Suppose that all points $r_1, \ldots, r_n, C_1, \ldots, C_n$ are enumerated in increasing order of their values: $t_1 < \ldots < t_k$. We construct a network which has the following vertices:

- a source u and a sink w,
- job vertices $\{j \mid j = 1, \ldots, n\}$,
- time interval vertices $\{(t_i, t_{i+1}) \mid i = 1, \ldots, 2n - 1\}$.

The arcs in our network are

- for each $j, j = 1, \ldots, n$, there is an arc (u, j) with capacity p, and there is an arc $(j, (t_i, t_{i+1}))$ if $r_j \leq t_i$ and $t_{i+1} \leq C_j$ hold,
- for each vertices $(t_i, t_{i+1}), i = 1, \ldots, 2n - 1$ there is an arc $((t_i, t_{i+1}), t)$ with capacity $m \cdot (t_{i+1} - t_i)$.

Any maximal flow will define an optimal schedule for the instance of problem $P \mid r_i, p_i = p, \text{pmtn} \mid \sum C_i$.

3   A Polynomial Algorithm for $O \mid r_i, p_{ij} = 1 \mid \sum C_i$

It is known [2] that problem $O \mid r_i, p_{ij} = 1 \mid \sum C_i$ is equivalent to the problem $P \mid r_i, p_i = m, \text{pmtn}^+ \mid \sum C_i$, where m is the number of machines. The meaning of pmtn$^+$ is that jobs can preempt at integer points only. Thus, to show that $O \mid r_i, p_{ij} = 1 \mid \sum C_i$ is polynomially solvable, we have to show

Theorem 6. *Any schedule for $P \mid r_i, p_i = m, \text{pmtn} \mid \sum C_i$, where r_i and m are integers, can be transformed in a polynomial time into a new schedule with preemptions and finishing times at integer points only, in such a way that the $\sum C_i$-value will not increase.*

4 Concluding Remarks

We want to note that the algorithm for $P \mid r_i, p_i = p, \text{pmtn} \mid \sum C_i$ can be generalized to solve problem $P \mid r_i, \text{pmtn} \mid \sum w_i C_i$ with arbitrary processing times p_i if the order of completion times in an optimal schedule is known. This is the case if the processing times, the due dates, and the release times are agreeable, i.e. if $p_i \leq p_j$ and $w_i \geq w_j$ hold for all i, j with $r_i \leq r_j$. Then again an optimal schedule can be found in the class of schedules where $C_i \leq C_j$ if $r_i \leq r_j$.

In this generalized algorithm one has to replace p by p_j in equation (6) and $((C_1^i - r_i) + \ldots + (C_n^i - r_i))$ by $w_i((C_1^i - r_i) + \ldots + (C_n^i - r_i))$ in (2). To achieve this general result one has to modify the proof of Theorem 3 slightly. The other arguments in Section 2 are the same.

Due to the fact that the modified problem SLP provides a feasible schedule for any ordering of the completion times one may apply local search procedures to find an optimal ordering of completion time for problem $P \mid r_i, \text{pmtn} \mid \sum w_i C_i$.

The most interesting related open problem is problem $P \mid r_i, p_i = p \mid \sum U_i$. It has been shown in [9] that the corresponding preemptive problem $P \mid r_i, p_i = p$, $\text{pmtn} \mid \sum U_i$ is unary NP-hard and it is possible that the nonpreemptive version of this problem is polynomially solvable.

Acknowledgments. The results described in this paper were attained during a visit of the second author at the University of Osnabrück which was supported by a fellowship from the Alexander von Humboldt Foundation the second author was supported also by Belarussian Republican Foundation for Fundamental Research, Project F03MS-039.

The first author was supported by INTAS Project 00-217

References

1. Baptiste, Ph., Brucker, P.: Scheduling Equal Processing Time Jobs: A Survey. In: Leung, Y. T.(ed.): Handbook of Scheduling: Algorithms, Models, and Performance Analysis, CRC Press LLC, Boca Raton, FR, (2004) 78-96
2. Brucker, P., Jurisch, B., Jurisch, M.: Open shop problems with unit time operations. Z. Oper. Res. **37** (1993) 59-73
3. Brucker, P.: Scheduling Algorithms. 3rd edn. Springer, Berlin (2001)
4. Chvatal, V.: Linear programming. Freeman, New York (1983)
5. Du, J., Leung, J. Y.-T., Young, G.H.: Minimizing mean flow time with release time constraint. Theoretical Computer Science **75** (1990) 347-355
6. Garey, M. R., Johnson, D.S.: Computers and intractability. A guide to the theory of NP-completeness. W.H. Freeman and Company, San Francisco, (1979)
7. Graham, R. L., Lawler, E.L., Lenstra, J.K., Rinnooy Kan, A.H.G.: Optimization and approximation in deterministic sequencing and scheduling: A survey. Ann Discrete Math **5** (1979) 287-326
8. Herrbach, L.A., Leung, J. Y.-T.: Preemptive scheduling of equal length jobs on two machines to minimize mean flow time. Operations Research **38** (1990) 487-494
9. Kravchenko, S.A.: On the complexity of minimizing the number of late jobs in unit time open shop. Discrete Applied Mathematics **100** (2000) 127-132

10. Leung, J.Y.-T., Young, G.H.: Preemptive scheduling to minimize mean weighted flow time. Information Processing Letters **34** (1990) 47-50
11. Lawler, E.L.: Combinatorial Optimization: Networks and Matroids. Holt, Rinehart and Winston, New York (1976).
12. McNaughton, R.: Scheduling with deadlines and loss functions. Management Science **6** (1959/1960) 1-12
13. Mokotoff, E.: Parallel machine scheduling problem: survey. Asia-Pacific Journal of Operational Research **18** (2001) 193-242
14. Schrijver, A.: Combinatorial Optimization. Springer, Berlin (2003)
15. Tautenhahn, T., Woeginger, G. J.: Minimizing the total completion time in a unit-time open shop with release times. Operations Research Letters **20** (1997) 207-212

Differential Approximation of MIN SAT, MAX SAT and Related Problems

(Extended Abstract)

Bruno Escoffier and Vangelis Th. Paschos

LAMSADE, Université Paris-Dauphine and CNRS UMR 7024 Place du Maréchal De, Lattre de Tassigny, 75775 Paris Cedex 16, France
{escoffier, paschos}@lamsade.dauphine.fr

Abstract. We present differential approximation results (both positive and negative) for optimal satisfiability, optimal constraint satisfaction, and some of the most popular restrictive versions of them. As an important corollary, we exhibit an interesting structural difference between the landscapes of approximability classes in standard and differential paradigms.

1 Introduction and Preliminaries

In this paper we deal with the approximation of some of the most famous and classical problems in the domain of the polynomial time approximation theory, the MIN and MAX SAT as well as the MIN and MAX DNF and some of their restricted versions, namely MAX and MIN k and EkSAT and MAX and MIN k and EkDNF. We study their approximability using the so-called *differential approximation ratio* which, informally, for an instance x of a combinatorial optimization problem Π, *measures the relative position of the value of an approximated solution in the interval between the worst-value of x, i.e., the value of a worst feasible solution of x, and the optimal-value of x, i.e., the value of a best solution of x.*

Given a set of clauses (i.e., disjunctions) $C_1, \ldots, C_m$ on n variables $x_1, \ldots, x_n$, MAX SAT (resp., MIN SAT) consists of determining a truth assignment to the variables that maximizes (minimizes) the number of clauses satisfied. On the other hand, given a set of cubes (i.e., conjunctions) $C_1, \ldots, C_m$ on n variables $x_1, \ldots, x_n$, MAX DNF (resp., MIN DNF) consists of determining a truth assignment to the variables that maximizes (minimizes) the number of conjunctions satisfied. For an integer $k \geqslant 2$, MAX kSAT, MAX kDNF, MIN kSAT, MIN kDNF (resp., MAX EkSAT, MAX EkDNF, MIN EkSAT, MIN EkDNF) are the versions of MAX SAT, MAX DNF, MIN SAT, MIN DNF where each clause or conjunction has size at most (resp., exactly) k. Finally, let us quote two particular weighted satisfiability versions, namely, MAX WSAT and MIN WSAT. In the former, given a set of clauses $C_1, \ldots, C_m$ on n variables $x_1, \ldots, x_n$, with non-negative integer weights $w(x)$ on any variable x, we wish to compute a truth assignment to the variables that both satisfies all the clauses and maximizes the sum of the weights of the variables set to 1. We consider that the assignment setting all the variables to 0 (even if it does not satisfy all the clauses) is feasible and represents the worst-value solution for the problem. The latter problem is similar to the former one, up to the fact that we wish to minimize the sum of the weights of the variables set to 1 and that feasible is now considered the assignment setting all the variables to 1.

A problem Π in **NPO** is a quadruple $(\mathcal{I}_\Pi, \text{Sol}_\Pi, m_\Pi, \text{opt}(\Pi))$ where: $\mathcal{I}_\Pi$ is the set of instances (and can be recognized in polynomial time); given $x \in \mathcal{I}_\Pi$, $\text{Sol}_\Pi(x)$ is the set of feasible solutions of x; the size of a feasible solution of x is polynomial in the size $|x|$ of the instance; moreover, one can determine in polynomial time if a solution is feasible or not; given $x \in \mathcal{I}_\Pi$ and $y \in \text{Sol}_\Pi(x)$, $m_\Pi(x, y)$ denotes the value of the solution y of the instance x; m_Π is called the objective function, and is computable in polynomial time; we suppose here that $m_\Pi(x, y) \in \mathbb{N}$; $\text{opt}(\Pi) \in \{\min, \max\}$.

Given an instance x of an optimization problem Π and a feasible solution $y \in \text{Sol}_\Pi(x)$, we denote by $\text{opt}_\Pi(x)$ the value of an optimal solution of x, and by $\omega_\Pi(x)$ the value of a worst solution of x. The *standard approximation ratio* of y is defined as $r_\Pi(x,y) = m_\Pi(x,y)/\text{opt}_\Pi(x)$, while the *differential approximation ratio* of y is defined as $\delta_\Pi(x,y) = |m_\Pi(x,y) - \omega_\Pi(x)|/|\text{opt}_\Pi(x) - \omega_\Pi(x)|$.

For a function f of $|x|$, an algorithm is a *standard f-approximation algorithm* (resp., *differential f-approximation algorithm*) for a problem Π if, for any instance x of Π, it returns a solution y such that $r(x, y) \leqslant f(|x|)$, if $\text{opt}(\Pi) = \min$, or $r(x,y) \geqslant f(|x|)$, if $\text{opt}(\Pi) = \max$ (resp., $\delta(x,y) \geqslant f(|x|)$).

With respect to the best approximation ratios known for them, **NPO** problems can be classified into approximability classes. One of the most notorious such classes is the class **APX** (or **DAPX** when dealing with the differential paradigm) including the problems for which there exists a polynomial algorithm achieving standard or differential approximation ratio $f(|x|)$ where function f is constant (it does not depend on any parameter of the instance).

We now define a kind of reduction, called *affine reduction* and denoted by AF, which, as we will see, is very natural in the differential approximation paradigm.

Definition 1. *Let Π and Π' be two **NPO** problems. Then, Π AF-reduces to Π' ($\Pi \leqslant_{\mathsf{AF}} \Pi'$), if there exist two functions f and g such that: (i) for any $x \in \mathcal{I}_\Pi$, $f(x) \in \mathcal{I}_\Pi$; (ii) for any $y \in \text{Sol}_{\Pi'}(x)$, $g(x,y) \in \text{Sol}_\Pi(x)$; moreover, $\text{Sol}_\Pi(x) = g(x, \text{Sol}_{\Pi'}(f(x)))$; (iii) for any $x \in \mathcal{I}_\Pi$, there exist $K \in \mathbb{R}$ and $k \in \mathbb{R}^*$ ($k > 0$ if $\text{opt}(\Pi) = \text{opt}(\Pi')$, $k < 0$, otherwise) such that, for any $y \in \text{Sol}_{\Pi'}(f(x))$, $m_{\Pi'}(f(x), y) = km_\Pi(x, g(x,y)) + K$. If $\Pi \leqslant_{\mathsf{AF}} \Pi'$ and $\Pi' \leqslant_{\mathsf{AF}} \Pi$, then Π and Π' are called* affine equivalent. *This equivalence will be denoted by $\Pi \equiv_{\mathsf{AF}} \Pi'$.*

It is easy to see that differential approximation ratio is stable under affine reduction. Formally, *if, for $\Pi, \Pi' \in \mathbf{NPO}$, $\mathsf{R} = (f, g)$ is an AF-reduction from Π to Π', then for any $x \in \mathcal{I}_\Pi$ and for any $y \in \text{Sol}_{\Pi'}(f(x))$, $\delta_\Pi(x, g(x, y)) = \delta_{\Pi'}(f(x), y)$.* Indeed, by Condition (ii) of Definition 1, worst and optimal solutions in x and $f(x)$ coincide. Since the value of any feasible solution of Π' is an affine transformation of the same solution seen as a solution of Π, the differential ratios for y and $g(x, y)$ coincide also. Hence, the following holds.

Proposition 1. *If $\Pi \equiv_{\mathsf{AF}} \Pi'$, then, for any constant r, any r-differential approximation algorithm for one of them is an r-differential approximation algorithm for the other one.*

Optimization satisfiability problems as MIN SAT and MAX SAT are of great interest from both theoretical and practical points of view. On the one hand, the satisfiability problem (SAT) is the first complete problem for **NP** and MAX SAT, MIN SAT have generalizations or restrictions that are the first problems proved complete for numerous approximation classes under various approximability preserving reductions ([1, 2]). For instance, MAX 3SAT is **APX**-complete under the AP-reduction and **Max-SNP**-complete under the L-reduction ([3]), MAX WSAT and MIN WSAT are **NPO**-complete under the AP-reduction ([4]), etc. In general, many optimal satisfiability problems have for the polynomial approximation theory the same status as SAT for **NP**-completeness theory. On the other hand, many problems in mathematical logic and in artificial intelligence can be expressed in terms of versions of SAT; constraints satisfaction is one such version. Also problems in database integrity constraints, query optimization, or in knowledge bases can be seen as optimization satisfiability problems. Finally, some approaches to inductive inference can be modeled as MAX SAT problems ([5, 6]).

Let us note that differential approximability of the problems dealt here, has already been studied in [7]. There, among other results, it was shown that MAX SAT and MIN DNF, as well as MIN SAT and MAX DNF are equivalent for the differential approximation, that all these problems are not solvable by polynomial time differential approximation schemata, unless **P** = **NP**, and, finally, that MIN SAT cannot be approximately solved within differential approximation ratio $1/m^{1-\epsilon}$, for any $\epsilon > 0$ (where m is the number of the clauses in its instance), unless **NP** = **co-RP**. Finally, let us mention here that both MAX WSAT and MIN WSAT belong to **0-DAPX**, the class of the problems for which no algorithm can guarantee differential approximation ratio strictly greater than 0, unless **P** = **NP** ([8]). This class has been also introduced in [7] and represents the worst possible configuration for differential approximation since it includes the problems for which no polynomial time approximation algorithm can guarantee differential ratio greater than 0. Inclusion of the problems dealt here in **0-DAPX** or not, was a major question we handled since [7].

In this paper, we further study differential approximability of MAX SAT, MIN SAT, MIN DNF and MAX DNF, and give approximation results and inapproximability bounds for several versions of these problems. For instance, we show that MAX SAT is not approximable within a constant approximation ratio, unless **P** = **NP**. This result is very interesting since it indicates that **Max-NP** ([3]) is not included in **DAPX**. This is an important difference with the standard approximability classes landscape where **Max-NP** $\subset$ **APX**. Another assessment with respect to our results is that the gap between lower and upper approximation bounds for the problems dealt is still large. However, this paper undertakes a systematic study of satisfiability problems in the differential paradigm, it extends the results of [7] and shows that none of the most classical satisfiability problems is in **0-DAPX**.

Results here are given without detailed proofs that can be found in [9].

2 Affine Reductions Between Optimal Satisfiability Problems

Let us first note that there does not exist general technique in order to transfer approximation results from differential (resp., standard) paradigm to standard (resp., differential) one, except for the case of maximization problems and for transfers between differential and standard paradigms. Proposition 2 just below deals with this last case.

Proposition 2. *If a maximization problem Π can be solved within differential approximation ratio δ, then it can be solved within standard approximation ratio δ, also.*

Corollary 1. *Any standard inapproximability bound for a maximization problem Π is also a differential inapproximability bound for Π.*

We give in this section some affine reductions and equivalences between the problems dealt in the paper. These results will allow us to focus ourselves only in the study of MAX SAT, MIN SAT and their restrictions without studying explicitly MAX and MIN DNF. We first recall a result already proved in [7].

Proposition 3. *([7])* MAX SAT $\equiv_{\mathsf{AF}}$ MIN DNF *and* MIN SAT $\equiv_{\mathsf{AF}}$ MAX DNF.

The following proposition shows that one can affinely pass from MAX EkSAT to MAX E$(k+1)$SAT. This, allows us to transfer inapproximability bounds from MAX E3SAT to MAX EkSAT, for any $k \geqslant 4$.

Proposition 4. MAX EkSAT $\leq_{\mathsf{AF}}$ MAX E$(k + 1)$SAT.

Proof (Sketch). For an instance ϕ of MAX EkSAT on n variables $x_1, \ldots, x_n$ and m clauses $C_1, \ldots, C_m$, consider a new variable y and build formula ϕ', instance of MAX E$(k + 1)$SAT as follows: for any clause $C_i = (\ell_{i_1}, \ldots, \ell_{i_k})$ of ϕ, where, for $j = 1, \ldots, k$, ℓ_{i_j} is a literal associated with x_{i_j}, ϕ' contains two new clauses $(\ell_{i_1}, \ldots, \ell_{i_k}, y)$ and $(\ell_{i_1}, \ldots, \ell_{i_k}, \bar{y})$. □

We now show that, for k fixed, problems kSAT and kDNF are affine equivalent.

Proposition 5. *For any fixed k,* MAX kSAT, MIN kSAT, MAX kDNF, MIN kDNF, MAX EkSAT, MIN EkSAT, MAX EkDNF *and* MIN EkDNF *are all affine equivalent.*

Proof (Sketch). For affine equivalence between MAX kSAT and MIN kSAT, given n variables $x_1, \ldots, x_n$, denote by $\mathcal{C}_k$ the set of clauses of size k and by $\mathcal{C}_{\leqslant k}$ the set of clauses of size at most k on the set $\{x_1, \ldots, x_n\}$. Remark that any truth assignment verifies the same number v_k of clauses on $\mathcal{C}_k$ and the same number $v_{\leqslant k}$ of clauses on $\mathcal{C}_{\leqslant k}$. Note also that, since k is assumed fixed, sets $\mathcal{C}_k$ and $\mathcal{C}_{\leqslant k}$ are of polynomial size. If ϕ is an instance of MAX EkSAT (resp. MIN EkSAT) on variable-set $\{x_1, \ldots, x_n\}$ and on a set $\mathcal{C} = \{C_1, \ldots, C_m\}$ of m clauses, consider for MIN EkSAT (resp., MAX EkSAT) the instance ϕ' on the clause-set $\mathcal{C}' = \mathcal{C}_k \setminus \mathcal{C}$. For the case of MAX and MIN kSAT, $\mathcal{C}' = \mathcal{C}_{\leqslant k} \setminus \mathcal{C}$

For equivalence between versions of SAT and the corresponding versions of DNF, given a clause $C = (\ell_{i_1} \vee \ldots \vee \ell_{i_k})$ on k literals, build the cube (conjunction) $D = (\bar{\ell}_{i_1} \wedge \ldots \wedge \bar{\ell}_{i_k})$.

Finally, in order to show that MAX kSAT $\leq_{\mathsf{AF}}$ MAX EkSAT proceed as in Proposition 4. □

It is shown in [10] (see also [1]), that MAX E3SAT is inapproximable within standard approximation ratio $(7/8) + \epsilon$, for any $\epsilon > 0$, and MAX E2SAT is inapproximable within standard approximation ratio $(21/22) + \epsilon$, for any $\epsilon > 0$ (in what follows for such results we will use, for simplicity, expression "within better than"). Discussion above, together with these bounds leads to the following result.

Proposition 6. MAX 2SAT, MAX E2SAT, MIN 2SAT, MIN E2SAT, MAX 2DNF, MAX E2DNF, MIN 2DNF and MIN E2DNF are inapproximable within differential approximation ratio better than 21/22. Furthermore, for any $k \geqslant 3$, MAX kSAT, MAX EkSAT, MIN kSAT and MIN EkSAT, MAX kDNF, MAX EkDNF, MIN kDNF and MIN EkDNF, are inapproximable within differential approximation ratio better than 7/8.

Since the satisfiability problems stated in Proposition 6 are particular cases either of MAX SAT, or of MIN SAT, or of MAX DNF, or, finally, of MIN DNF, application of Proposition 6 and of Proposition 3 concludes the following corollary.

Corollary 2. MAX SAT, MIN SAT, MAX DNF and MIN DNF are inapproximable within differential approximation 7/8.

Results of Corollary 2 are not the best ones. In Section 4, we strengthen the one for MAX SAT. On the other hand, as it is proved in [7], MIN SAT is inapproximable within differential ratio better than $m^{\epsilon-1}$, for any $\epsilon > 0$. Dealing with approximability of the problems stated in Proposition 5, the following remarks hold:

- if one of these problems is in **DAPX**, then all the other ones are so;
- problems MAX kSAT, MAX EkSAT, MIN k DNF and MIN EkDNF are approximable within differential ratios of $O(f(m))$ for a function f strictly decreasing with m if and only if one of them is $O(f(m))$ differentially approximable for $f(m) = O(m^\alpha)$, for some $\alpha > 0$, or $f(m) = O(\log m)$; the same holds for the quadruple MIN kSAT, MIN EkSAT, MAX k DNF and MAX EkDNF.

3 Positive Results

3.1 Maximum Satisfiability

Consider an instance ϕ of an optimal satisfiability problem, defined on n variables $x_1, \ldots, x_n$ and m clauses $C_1, \ldots, C_m$; consider also the very classical algorithm RSAT assigning at any variable value 1 with probability 1/2 and, obviously, value 0 with probability 1/2. Then, denoting by Sol(ϕ), the set of the 2^n

possible truth assignments for ϕ, and by $E(\text{RSAT}(\phi))$ the expectation of a solution computed by RSAT when running on ϕ, the following holds: $E(\text{RSAT}(\phi)) = \sum_{T \in \text{Sol}(\phi)} m(\phi, T)/2^n$.

Algorithm RSAT can be derandomized by the following technique denoted by DSAT. For $i = 1, \ldots, n$:

- compute $E'_i = E(m(\phi, T)|x_i = 1)$ and $E''_i = E(m(\phi, T)|x_i = 0)$, where T is a random assignment and the values of the $i - 1$ first variables have already been fixed in iterations $1, \ldots i - 1$;
- set $x_i = 1$, if $E'_i \geqslant E''_i$; otherwise, set $x_i = 0$.

Lemma 1. $m(\phi, \text{DSAT}(\phi)) \geqslant E(\text{RSAT}(\phi))$.

Note that DSAT is polynomial since, for any $i = 1, \ldots, n$, computation of E'_i and E''_i is performed in polynomial time. Indeed, for any such computation it suffices to determine with what probability any clause of ϕ is satisfied and to sum these probabilities over all the clauses of ϕ.

Proposition 7. *Algorithm DSAT achieves for* MAX EkSAT *differential approximation ratio* $2^k/(\text{opt}(\phi) + 2^k)$. *This ratio is bounded below by* $2^k/(m + 2^k)$.

We now propose a reduction transferring approximation results for MAX SAT problems from standard to differential paradigm. It will be used in order to achieve differential approximation results for MAX SAT, MAX 3SAT and MAX 2SAT.

Proposition 8. *If a maximum satisfiability problem is approximable on an instance* ϕ, *within standard approximation ratio* ρ, *then it is approximable in* ϕ *within differential approximation ratio* $\rho/((1 - \rho)\omega(\phi) + 1)$.

From the result of Proposition 8, we can deduce several corollaries by specifying values for $\omega(\phi)$ and ρ. The main such corollaries are stated in the proposition that follows.

Proposition 9. MAX SAT *is approximable within differential approximation ratio* $4.34/(m+4.34)$. MAX E2SAT *is approximable within differential approximation ratio* $17.9/(m + 19.3)$ *and* MAX 3SAT *within* $4.57/(m + 5.73)$.

3.2 Minimum Satisfiability

We finish this section by studying MIN SAT and some of its versions. Before stating our results, we note that algorithm RSAT can be derandomized in an exactly symmetric way, in order to provide a solution for MIN kSAT with value smaller than expectation's value.

Proposition 10. *If a minimum satisfiability problem is approximable on an instance* ϕ, *within standard approximation ratio* ρ, *then it is approximable in* ϕ *within differential approximation ratio*

$$\frac{\rho}{(\rho - 1)\left(1 - \frac{1}{2^k}\right) m + \rho}$$

The best standard approximation ratios known for MIN kSAT and MIN SAT are $2(1 - 2^{-k})$ and 2, respectively ([11]). With the ratio just mentioned for MIN kSAT, the result of Proposition 10 can be simplified as indicated in the following corollary.

Corollary 3. MIN kSAT *is approximable within differential ratio* $2^k/((2^{k-1} - 1)m + 2^k)$.

Proposition 11. MIN SAT *is approximable within differential ratio* $2/(m + 2)$.

Also, using Corollary 3 with $k = 2$ and $k = 3$, the following corollary holds and concludes the section.

Corollary 4. MIN 2SAT *and* MIN 3SAT *are approximable within differential ratios* $4/(m + 4)$ *and* $8/(3m + 8)$, *respectively.*

4 Inapproximability

As it is proved in [10], for any $p \geqslant 2$ and for any $\epsilon > 0$, MAX E3LINp[1] cannot be approximated within standard approximation ratio $(1/p) + \epsilon$, even if coefficients in the left-hand sides of the equations are all equal to 1. Note that, due to Corollary 1, this bound is immediately transferred to the differential paradigm.

Finally, let us recall the following result of [10], that will be used in this section.

Proposition 12. *([10]) Given a problem* $\Pi \in \mathbf{NP}$ *and a real* $\delta > 0$, *there exists a polynomial transformation g from any instance I of Π into an instance of* MAX E3LIN2 *such that: if I is a yes-instance of Π (we use here classical terminology from [12]), then* $\mathrm{opt}(g(I)) \geqslant (1 - \delta)m$, *otherwise*, $\mathrm{opt}(g(I)) \leqslant (1 + \delta)m/2$.

Proposition 12 shows, in fact, that MAX E3LIN2 is not approximable within standard ratio $1/2 + \epsilon$, for any $\epsilon > 0$.

We prove a result analogous to the one of Proposition 12 from any problem $\Pi \in \mathbf{NP}$ to MAX E3SAT.

Proposition 13. *Given a problem* $\Pi \in \mathbf{NP}$ *and a real* $\delta > 0$, *there exists a polynomial transformation f from any instance I of Π into an instance of* MAX E3SAT *such that:*

- *if I is a yes-instance of Π, then* $\mathrm{opt}(f(I)) - \omega(f(I)) \geqslant (1 - 2\delta)m/4$;
- *if I is a no-instance of Π, then* $\mathrm{opt}(f(I)) - \omega(f(I)) \leqslant \delta m/4$.

---

[1] In MAX E3LINp we are given a positive prime p, n variables $x_1, \ldots, x_n$ in $\mathbb{Z}/p\mathbb{Z}$, m linear equations of type $\alpha_{i_\ell} x_{i_\ell} + \alpha_{j_\ell} x_{j_\ell} + \alpha_{k_\ell} x_{k_\ell} = \beta_\ell$ and our objective is to determine an assignment on $x_1, \ldots, x_n$, in such a way that a maximum number among the m equations is satisfied.

Proof (Sketch). We first prove that the reduction of Proposition 12 can be translated into the differential paradigm also. Consider an instance $I' = g(I)$ of MAX E3LIN2 and a feasible solution $x = (x_1, x_2, \ldots, x_n)$ for I (we will use the same notation for both variables and their assignment) verifying k among the m equations of I'. Then, vector $\bar{x} = (1 - x_1, \ldots, 1 - x_n)$, verifies the $m - k$ equations not verified by x. In other words, $\mathrm{opt}(I) + \omega(I) = m$; hence, function g claimed by Proposition 12 is such that: if I is a *yes*-instance of Π, then $\mathrm{opt}(I') - \omega(I') \geqslant (1 - 2\delta)m$, otherwise $\mathrm{opt}(I') - \omega(I') \leqslant \delta m$.

Consider now an instance I of MAX E3LIN2 on n variables x_i, $i = 1, \ldots, n$ and m equations of type $x_i + x_j + x_k = \beta$ in $\mathbb{Z}/2\mathbb{Z}$, i.e., where variables and second members equal 0, or 1. We transform I into an instance $\phi = h(I)$ of MAX E3SAT in the following way: for any equation $x_i + x_j + x_k = 0$, we add in $h(I)$ the following four clauses: $(\bar{x}_i \vee x_j \vee x_k)$, $(x_i \vee \bar{x}_j \vee x_k)$, $(x_i \vee x_j \vee \bar{x}_k)$ and $(\bar{x}_i \vee \bar{x}_j \vee \bar{x}_k)$; for any equation $x_i + x_j + x_k = 1$, we add in $h(I)$ the following four clauses: $(x_i \vee x_j \vee x_k)$, $(\bar{x}_i \vee \bar{x}_j \vee x_k)$, $(\bar{x}_i \vee x_j \vee \bar{x}_k)$ and $(x_i \vee \bar{x}_j \vee \bar{x}_k)$.

Given a solution y for MAX E3SAT on $h(I)$, we construct a solution y' for I by setting $x_i = 1$ if $x_i = 1$ in $h(I)$ also; otherwise, we set $x_i = 0$.

It suffices then to remark that the composition $f = h \circ g$ (where g is as in Proposition 12) verifies the statement of the proposition. □

Proposition 13 has a very interesting corollary, expressed in the Proposition 14 just below, that exhibits another point of dissymmetry between standard and differential paradigms.

Proposition 14. *Unless $\mathbf{P = NP}$, no polynomial algorithm can compute, on an instance ϕ of MAX E3SAT a value that is a constant approximation of the quantity* $\mathrm{opt}(\phi) - \omega(\phi)$.

In view of Proposition 14, what is different between standard and differential paradigms with respect to the GAP-reduction is that in the former such a reduction immediately concludes the impossibility for a problem (assume that it is a maximization one) to be approximable within some ratio, by showing the impossibility for the optimal value to be approximated within this ratio. For that, it suffices that one reads the value of the solution returned by the approximation algorithm. In the latter paradigm such a conclusion is not always immediate. In fact, a reasoning similar to the one of the standard approximation is possible when computation of the worst solution can be done in polynomial time (this is, for instance, the case of maximum independent set and of many other **NP**-hard problems). In this case a simple reading of the value of the approximate solution is sufficient to give an approximation of $\mathrm{opt}(x) - \omega(x)$. A contrario, when it is **NP**-hard to compute $\omega(x)$ (this is the case of the problems dealt here –simply think that the worst solution for MAX SAT is the optimal one for MIN SAT and that both of them are **NP**-hard –, of traveling salesman, etc.), then reading the value $m(x, y)$ of the approximate solution does not provide us with knowledge about $m(x, y) - \omega(x)$ and, consequently no approximation of $\mathrm{opt}(x) - \omega(x)$ can be immediately estimated. So, use of GAP-reduction for achieving inapproximability results is different from the one paradigm to the other.

However, for the case we deal with, we will take advantage of a combination of Propositions 5 and 14 in order to achieve the inapproximability bound for MAX E3SAT given in Proposition 15 that follows.

Proposition 15. *Unless* $P = NP$, MAX E3SAT *is inapproximable within differential approximation ratio greater than* $1/2$.

Proof (Sketch). Assume that an approximation achieves differential ratio $\delta > 1/2$, for MAX E3SAT. Then, by Proposition 5, there exists an algorithm achieving the same differential ratio for MIN E3SAT. Denote by T_1 and T_2, respectively, the solutions computed by these algorithms on an instance ϕ of these problems. Then, one can prove that $m(\phi, T_1) - m(\phi, T_2) \geqslant (2\delta - 1)(\text{opt}(\phi) - \omega(\phi))$, and a simple reading of the values of T_1 and T_2, can provide us a constant approximation (since δ has been assumed to be a fixed constant greater than $1/2$) of the quantity $\text{opt}(\phi) - \omega(\phi)$, impossible by Proposition 14. □

Corollary 5. *For any* $k \geqslant 3$, MAX EkSAT, MIN EkSAT, MAX kSAT *and* MIN kSAT *are differentially inapproximable within ratios better than* $1/2$.

We now generalize the result of Proposition 13 in order to further strengthen inapproximability results of Corollary 5.

Proposition 16. *For any prime* $p > 0$, MAX E3LIN$p \leqslant_{\text{AF}}$ MAS E3$(p-1)$SAT.

The result of Proposition 16 together with the result of [10] stated in the beginning of the section and Proposition 1, lead to the following corollary.

Corollary 6. *For any prime* p, MAX E3$(p-1)$SAT *is inapproximable within differential ratio greater than* $1/p$.

Proposition 17. *For any* $k \geqslant 3$, *neither* MAX EkSAT, *nor* MIN EkSAT *can be approximately solved within differential ratio greater than* $1/p$, *where* p *is the largest positive prime such that* $3(p-1) \leqslant k$.

Corollary 7. *The following differential inapproximability bounds hold:* $1/2$ *for* MAX *and* MIN 3SAT 4SAT *and* 5SAT; $1/3$ *for* MAX *and* MIN 6SAT, ..., 11SAT; $1/5$ *for* MAX *and* MIN 12SAT, ..., 17SAT ...

Finally, MAX SAT being harder to approximate than any MAX kSAT problem, the following result holds and concludes the section.

Proposition 18. MAX SAT $\notin$ **DAPX**.

In [3] is defined a logical class of **NPO** maximization problems called **MAX-NP**. A maximization problem $\Pi \in$ **NPO** belongs to **Max-NP** if and only if there exist two finite structures $(U, \mathcal{I})$ and $(U, \mathcal{S})$, a quantifier-free first order formula ϕ and two constants k and ℓ such that, the optima of Π can be logically expressed as: $\max_{S \in \mathcal{S}} |\{x \in U^k : \exists y \in U^\ell, \phi(\mathcal{I}, S, x, y)\}|$. The predicate-set $\mathcal{I}$ draws the set of instances of Π, set $\mathcal{S}$ the solutions on $\mathcal{I}$ and ϕ the feasibility conditions for the solutions of Π. In the same article is proved that MAX SAT $\in$ **Max-NP** and that **MAX-NP** $\subset$ **APX**.

It is easy to see that the above definition of **Max-NP** identically holds in both standard and differential paradigms. So, Proposition 18 draws an important structural difference in the landscape of approximation classes in the two paradigms, since an immediate corollary of this proposition is that **MAX-NP** $\not\subset$ **DAPX**. We conjecture that the same holds for the other one of the celebrated logical classes of [3], the class **MAX-SNP**, i.e., we conjecture that **MAX-SNP** $\not\subset$ **DAPX**

We conclude the paper by strengthening the 21/22-inapproximability bound for MAX E2SAT.

Proposition 19. MAX E2LIN2 $\leq_{AF}$ MAX E2SAT. *Consequently,* MAX E2SAT *is differentially inapproximable within ratio greater than 11/12.*

References

1. Ausiello, G., Crescenzi, P., Gambosi, G., Kann, V., Marchetti-Spaccamela, A., Protasi, M.: Complexity and approximation. Combinatorial optimization problems and their approximability properties. Springer, Berlin (1999)
2. Vazirani, V.: Approximation algorithms. Springer, Berlin (2001)
3. Papadimitriou, C.H., Yannakakis, M.: Optimization, approximation and complexity classes. J. Comput. System Sci. **43** (1991) 425–440
4. Crescenzi, P., Kann, V., Silvestri, R., Trevisan, L.: Structure in approximation classes. SIAM J. Comput. **28** (1999) 1759–1782
5. Hooker, J.N.: Resolution vs. cutting plane solution of inference problems: some computational experience. Oper. Res. Lett. **7** (1988) 1–7
6. Kamath, A.P., Karmarkar, N.K., Ramakrishnan, K.G., Resende, M.G.: Computational experience with an interior point algorithm on the satisfiability problem. Ann. Oper. Res. **25** (1990) 43–58
7. Bazgan, C., Paschos, V.Th.: Differential approximation for optimal satisfiability and related problems. European J. Oper. Res. **147** (2003) 397–404
8. Monnot, J., Paschos, V. Th., Toulouse, S.: Approximation polynomiale des problèmes NP-difficiles : optima locaux et rapport différentiel. Hermès, Paris (2003)
9. Escoffier, B., Paschos, V. Th.: Differential approximation of MIN SAT, MAX SAT and related problems. Cahier du LAMSADE 220, LAMSADE, Université Paris-Dauphine (2004)
10. Håstad, J.: Some optimal inapproximability results. J. Assoc. Comput. Mach. **48** (2001) 798–859
11. Bertsimas, D., Teo, C.P., Vohra, R.: On dependent randomized rounding algorithms. In: Proc. International Conference on Integer Programming and Combinatorial Optimization, IPCO. Volume 1084 of Lecture Notes in Computer Science., Springer-Verlag (1996) 330–344
12. Garey, M.R., Johnson, D.S.: Computers and intractability. A guide to the theory of NP-completeness. W. H. Freeman, San Francisco (1979)

Probabilistic Coloring of Bipartite and Split Graphs

(Extended Abstract)

F. Della Croce[1], B. Escoffier[2], C. Murat[2], and V. Th. Paschos[2]

[1] D.A.I., Politecnico di Torino, Italy
federico.dellacroce@polito.it
[2] LAMSADE, Université Paris-Dauphine, France
{escoffier, murat, paschos}@lamsade.dauphine.fr

Abstract. We revisit in this paper the probabilistic coloring problem (PROBABILISTIC COLORING) and focus ourselves on bipartite and split graphs. We first give some general properties dealing with the optimal solution. We then show that the unique 2-coloring achieves approximation ratio 2 in bipartite graphs under any system of vertex-probabilities and propose a polynomial algorithm achieving tight approximation ratio 8/7 under identical vertex-probabilities. Then we deal with restricted cases of bipartite graphs. Main results for these cases are the following. Under non-identical vertex-probabilities PROBABILISTIC COLORING is polynomial for stars, for trees with bounded degree and a fixed number of distinct vertex-probabilities, and, consequently, also for paths with a fixed number of distinct vertex-probabilities. Under identical vertex-probabilities, PROBABILISTIC COLORING is polynomial for paths, for even and odd cycles and for trees whose leaves are either at even or at odd levels. Next, we deal with split graphs and show that PROBABILISTIC COLORING is **NP**-hard, even under identical vertex-probabilities. Finally, we study approximation in split graphs and provide a 2-approximation algorithm for the case of distinct probabilities and a polynomial time approximation schema under identical vertex-probabilities.

1 Preliminaries

In *minimum coloring problem*, the objective is to color the vertex-set V of a graph $G(V, E)$ with as few colors as possible so that no two adjacent vertices receive the same color. Since adjacent vertices are forbidden to be colored with the same color, a feasible coloring can be seen as a partition of V into *independent sets*. So, the optimal solution of minimum coloring is a *minimum-cardinality partition into independent sets*.

In the probabilistic version of minimum coloring, denoted by PROBABILISTIC COLORING, we are given: a graph $G(V, E)$ of order n, an n-vector $\mathbf{Pr} = (p_1, \ldots, p_n)$ of vertex-probabilities and a *modification strategy* M, i.e., an algorithm that when receiving a coloring $C = (S_1, \ldots, S_k)$ for V, called *a priori solution*, and a subgraph $G' = G[V']$ of G induced by a sub-set $V' \subseteq V$ as inputs, it modifies C in order to produce a coloring C' for G'. The objective is to determine a coloring C^* (a priori solution) of G minimizing the quantity (functional) $E(G, C, \mathtt{M}) = \sum_{V' \subseteq V} \Pr[V'] |C(V', \mathtt{M})|$ where $C(V', \mathtt{M})$ is the solution computed by $\mathtt{M}(C, V')$ (i.e., by $\mathtt{M}$ when executed with inputs the a priori solution C and the subgraph of G induced by V') and $\Pr[V'] = \prod_{i \in V'} p_i \prod_{i \in V \setminus V'} (1 - p_i)$ (there exist 2^n distinct sets V'; therefore, explicit computation of $E(G, C, \mathtt{M})$ is, a priori, not polynomial).

In this paper, we study PROBABILISTIC COLORING under the following simple but intuitive modification strategy $\mathtt{M}$: given an a priori solution C, take the set C ∩ V' as solution for G[V'], i.e., remove the absent vertices from C. Let us note that motivation of PROBABILISTIC COLORING by two real-world applications, the former dealing with timetabling and the latter with planning, is given in [1]. Since the modification strategy $\mathtt{M}$ is fixed for the rest of the paper we will simplify notations by using $E(G, C)$ instead of $E(G, C, \mathtt{M})$ and $C(V')$ instead of $C(V', \mathtt{M})$. In [1] it is shown that

$$E(G, C) = \sum_{j=1}^{k} \left(1 - \prod_{v_i \in S_j} (1 - p_i)\right) \qquad (1)$$

It is easy to see that computation of $E(G, C)$ can be performed in at most $O(n^2)$ steps, consequently, PROBABILISTIC COLORING $\in$ **NP**. Also, from (1), we can easily characterize the optimal a priori solution C^* for PROBABILISTIC COLORING: if the value of an independent set S_j of G is $1 - \prod_{v_i \in S_j}(1 - p_i)$ then *the optimal a priori coloring for G is the partition into independent sets for which the sum of their values is the smallest over all such partitions.*

PROBABILISTIC COLORING has been originally studied in [2, 1], where complexity and approximation issues have been considered for general graphs and several special configuration graphs such as bipartite graphs, complements of bipartite graphs and others.

Besides PROBABILISTIC COLORING, restricted versions of routing and network-design probabilistic minimization problems defined on complete graphs have been studied in [3, 4, 5, 6, 7, 8, 9, 10]. In [11], the minimum vertex covering problem in general and in bipartite graphs is studied, while in [12, 13] the longest path and the maximum independent set, respectively, are tackled.

In [2] it is shown that, for any $k \geqslant 2$, it is **NP**-hard to determine the best probabilistic k-coloring in bipartite graphs, even if the input has only four distinct vertex-probabilities with one of them being equal to 0. The **NP**-hardness result of [2] left, however, several open questions. For instance, "what is the complexity of PROBABILISTIC COLORING in paths, or trees, or cycles, or stars, ...?", etc. In this paper, we prove that, under non-identical vertex-probabilities, PROBABILISTIC COLORING is polynomial for stars and for trees with bounded degree and a fixed number of distinct vertex-probabilities and we deduce as a corollary that it is polynomial also for paths with a fixed number of distinct vertex-probabilities. Then, we show that, assuming identical vertex-probabilities, the problem is polynomial for paths, for cycles and for trees all leaves of which are either at even or at odd levels. We finally focus ourselves on split graphs and show that, in such graphs, PROBABILISTIC COLORING is **NP**-hard, even assuming identical vertex probabilities.

Let A be a polynomial time approximation algorithm for an **NP**-hard minimization graph-problem Π, let $m(G, S)$ be the value of the solution S provided by A on an instance G of Π, and opt(G) be the value of the optimal solution for G (following our notation for PROBABILISTIC COLORING, opt$(G) = E(G, C^*)$). The approximation ratio $\rho_\mathtt{A}(G)$ of the algorithm A on G is defined as $\rho_\mathtt{A}(G) = m(G, S)/\mathrm{opt}(G)$. An approximation algorithm achieving ratio, at most, ρ on any instance G of Π will be called ρ-approximation algorithm. A polynomial time approximation schema is a sequence $\mathtt{A}_\epsilon$ of polynomial time approximation algorithms which when they run with inputs a graph G (instance of Π) and any fixed constant $\epsilon > 0$, they produce a solution S such that $\rho_{\mathtt{A}_\epsilon}(G) \leqslant 1 + \epsilon$.

Dealing with approximation issues, we show that the unique 2-coloring (where all nodes of each set of the initial partition share the same color) achieves approximation ratio 2 in bipartite graphs under any system of vertex-probabilities. Furthermore, we propose a polynomial algorithm achieving approximation ratio 8/7 under identical vertex-probabilities. We also provide a 2-approximation polynomial time algorithm for split graphs under distinct vertex-probabilities and a polynomial time approximation schema when vertex-probabilities are identical.

Obviously, some of the results presented have several important corollaries. For instance, since PROBABILISTIC COLORING is approximable within ratio 2 in general (i.e., under any system of vertex-probabilities) bipartite graphs, it is so in general trees, paths and even cycles, also. Results here are given without detailed proofs which can be found in [14].

2 Properties

We give in this section some general properties about probabilistic colorings, upon which we will be based later in order to achieve our results. In what follows, given an a priori k-coloring $C = (S_1, \ldots, S_k)$ we will set: $f(C) = E(G, C)$, where $E(G, C)$ is given by (1), and, for $i = 1, \ldots, k$, $f(S_i) = 1 - \prod_{v_j \in S_i}(1 - p_j)$. By interchange arguments, it is possible to prove the following properties.

Property 1. Let $C = (S_1, \ldots, S_k)$ be a k-coloring and assume that colors are numbered so that $f(S_i) \leqslant f(S_{i+1})$, $i = 1, \ldots, k-1$. Consider a vertex x (of probability p_x) colored with S_i and a vertex y (of probability p_y) colored with S_j, $j > i$, such that $p_x \geqslant p_y$. If swapping colors of x and y leads to a new feasible coloring C', then $f(C') \leqslant f(C)$.

Property 1 has the following very natural particular case which will also be used in the sequel.

Property 2. Let $C = (S_1, \ldots, S_k)$ be a k-coloring and assume that colors are numbered so that $f(S_i) \leqslant f(S_{i+1})$, $i = 1, \ldots, k-1$. Consider a vertex x colored with color S_i. If it is feasible to color x with another color S_j, $j > i$, (by keeping colors of the other vertices unchanged), then the new feasible coloring C' verifies $f(C') \leq f(C)$.

Property 3. In any graph of maximum degree Δ, the optimal solution of PROBABILISTIC COLORING contains at most $\Delta + 1$ colors.

Properties seen until now in this section work for any graph and for any vertex-probability system. Let us now focus ourselves on the case of identical vertex-probabilities. Remark first that, for this case, Property 2 has a natural counterpart expressed as follows.

Property 4. Let $C = (S_1, \ldots, S_k)$ be a k-coloring and assume that colors are numbered so that $|S_i| \leq |S_{i+1}|$, $i = 1, \ldots, k-1$. If it is feasible to inflate a color S_j by "emptying" another color S_i with $i < j$, then the new coloring C', so created, verifies $f(C') \leq f(C)$.

Since, in Property 4, only the cardinalities of the colors intervene, the following corollary-property consequently holds.

Property 5. Let $C = (S_1, \ldots, S_k)$ be a k-coloring and assume that colors are numbered so that $|S_i| \leq |S_{i+1}|$, $i = 1, \ldots, k-1$. Consider two colors S_i and S_j, $i < j$, and a vertex-set $X \subset S_j$ such that, $|S_i| + |X| \geq |S_j|$. Consider (possibly unfeasible) coloring $C' = (S_1, \ldots, S_i \cup X, \ldots, S_j \setminus X, \ldots, S_k)$. Then, $f(C') \leq f(C)$.

From now on we define those colorings C such that Properties 1, or 2, or 4 hold, as "balanced colorings". In other words, for a balanced coloring C, there exists a coloring C', better than C, obtained as described in Properties 1, or 2, or 4. On the other hand, colorings for which transformations of the properties above cannot apply will be called "unbalanced colorings". From the above definitions, the following proposition immediately holds.

Proposition 1. *For any balanced coloring, there exist an unbalanced one dominating it.*

Let us further restrict ourselves to bipartite graphs. Remark first that the cases of vertex-probability 0 or 1 are trivial: for the former any a priori solution has value 0; for the latter, PROBABILISTIC COLORING coincides with the classical (deterministic) coloring problem where the (unique) 2-coloring is the best one.

Consider a bipartite graph $B(U, D, E)$ and, without loss of generality, assume $|U| \geq |D|$. Also, denote by $\alpha(B)$ the cardinality of a maximum independent set of B. Then it is possible to prove, by means of Proposition 1, the following.

Property 6. If $\alpha(B) = |U|$, then 2-coloring $C = (U, D)$ is optimal.

3 General Bipartite Graphs

We first give an easy result showing that the hard cases for PROBABILISTIC COLORING are the ones where vertex-probabilities are "small". Consider a bipartite graph $B(U, D, E)$ and denote by $p_{\min}$ its smallest vertex-probability.

Proposition 2. If $p_{\min} \geqslant 0.5$, then the unique 2-coloring $C = (U, D)$ is optimal for B.

When vertex-probabilities are generally and typically smaller than 0.5, it is possible to provide instances, even with identical vertex-probabilities, where the 2-coloring does not provide the optimal solution. For instance, consider a tree T on $2n$ vertices, where vertex 1 (the root) is linked to vertices $n+1, \ldots, 2n$ and vertex $2n$ is linked to vertices $1, \ldots, n$. Assume that vertex-probabilities on T are all equal to $p \ll 0.5$. Then, the 2-coloring $\{\{1, \ldots, n\}, \{n+1, \ldots, 2n\}\}$ has value $f_2 = 2(1 - (1-p)^n)$, while the 3-coloring $\{\{1\}, \{2, \ldots, 2n-1\}, \{2n\}\}$ has value $f_3 = 2(1 - (1-p)) + (1 - (1-p)^{2n-2})$. For p small enough and n large enough, we have $f_2 \approx 2$ and $f_3 \approx 1$.

Proposition 3. In any bipartite graph $B(U,D,E)$, its unique 2-coloring $C = (U, D)$ achieves approximation ratio bounded by 2. This bound is tight, even for paths under distinct vertex-probabilities.

We now restrict ourselves to the case of identical vertex-probabilities and consider the following algorithm, denoted by 3-COLOR in what follows:

1. compute and store the natural 2-coloring $C_0 = (U, D)$;
2. compute a maximum independent set S of B;
3. output the best coloring among C_0 and $C_1 = (S, U \setminus S, D \setminus S)$.

Obviously, 3-COLOR is polynomial, since computation of a maximum independent set can be performed in polynomial time in bipartite graphs ([15]).

Proposition 4. Algorithm 3-COLOR achieves approximation ratio 8/7 in bipartite graphs with identical vertex-probabilities. This bound is asymptotically tight.

Proof (Sketch). Consider an optimal solution $C^* = (S_1^*, S_2^*, \ldots S_k^*)$, and assume that $|S_1^*| \geqslant |S_2^*| \geqslant \ldots \geqslant |S_k^*|$. Set $n = |U \cup D|$, $n_1 = |S|$ and $n_2 = n - |S| = n - n_1$. Obviously, $n_1 \geqslant n_2$.

Based upon Property 4, the worst case for C_0 is reached when it is completely balanced, i.e., when $|U| = |D|$. In other words, $f(C_0) = f(U) + f(D) \leqslant 2(1-(1-p)^{(n_1+n_2)/2})$. By exactly the same reasoning, $f(C_1) = f(S) + f(U \setminus S) + f(D \setminus S) \leqslant 1 - (1-p)^{n_1} + 2(1-(1-p)^{n_2/2})$. Remark also that $|S_1^*| \leqslant |S_1| = n_1$. If this inequality is strict, then, applying Property 4, one, by emptying some colors S_j^*, $j > 1$, can obtain a (probably unfeasible) coloring C' such that $f(C') \leqslant f(C^*)$ and the largest color of C' is of size n_1; in other words, $f(C^*) \geqslant f(C') \geqslant 1 - (1-p)^{n_1} + 1 - (1-p)^{n_2}$. Setting $\beta = (1-p)^{n_1/2}$, $\alpha = (1-p)^{n_2/2}$ and using expressions for $f(C_0)$, $f(C_1)$ and $f(C^*)$, we get (omitting, for simplicity, to index ρ by 3-COLOR):

$$\rho(B) = \min\left\{\frac{f(C_0)}{f(C^*)}, \frac{f(C_1)}{f(C^*)}\right\} \leqslant \min\left\{\frac{2(1-\alpha\beta)}{2-\alpha^2-\beta^2}, \frac{3-\beta^2-2\alpha}{2-\alpha^2-\beta^2}\right\} \quad (2)$$

Then, with some algebra we get that equality of ratios in (2) holds when $\alpha = (1+\beta)/2$ and, in this case, the ratio is bounded above by 8/7.

For tightness, fix an $n \in \mathbb{N}$ and consider a bipartite graph $B(U, D, E)$ consisting of: (i) an independent set S_1 on $2n^2$ vertices; n^2 of them, denoted by $v_U^1, \ldots, v_U^{n^2}$ belong to U and the n^2 remaining ones, denoted by $v_D^1, \ldots, v_D^{n^2}$ belong to D; (ii) n paths $P_1, \ldots, P_n$ of size 4 (i.e. on 3 edges); set, for $i = 1, \ldots, n$, $P_i = (p_i^1, p_i^2, p_i^3, p_i^4)$; S_1 and the n paths P_i are completely disjoint; (iii) two vertices $u \in U$ and $v \in D$; u is linked to all the vertices of D and v to all the vertices of U; (iv) for any $v_i \in U \cup D$, $p_i = p = \ln 2/n$.

The graph so-constructed is balanced (i.e., $|U| = |D|$) and has size $2n^2+4n+2$.

Apply algorithm 3-COLOR to the so-constructed graph B. A maximum independent set of B consists of the $2n^2$ vertices of S_1 plus two vertices per any of the n paths P_i, $i = 1, \ldots, n$. Assume without loss of generality that the maximum independent set computed in Step 2 of algorithm 3-COLOR is $S = S_1 \cup_{i=1,\ldots,n} \{p_i^1, p_i^4\}$. Consider colorings: $C_0 = (U, D)$, $C_1 = (S, U \setminus S, D \setminus S)$, examined in Step 3, and $\hat{C} = (\hat{S}_1, \hat{S}_2, \hat{S}_3)$ of B with $\hat{S}_1 = S_1 \cup_{i=1,\ldots,n} \{p_i^1, p_i^3\}$, $\hat{S}_2 = \{v\} \cup_{i=1,\ldots,n} \{p_i^2, p_i^4\}$ and $\hat{S}_3 = \{u\}$. With elementary algebraic calculations, one gets that, for $n \to \infty$ and for $p = \ln 2/n$, $f(C_0) \to 2$, $f(C_1) \to 2$ and $f(C^*) \leqslant f(\hat{C}) \to 7/4$. $\square$

Let us note that in [14] we prove that the ratio 7/8 is tight even for trees.

4 Particular Families of Bipartite and "Almost" Bipartite Graphs: Trees and Cycles

Let us first note that for "trivial" families of bipartite graphs, as graphs isomorphic to a perfect matching, or to an independent set (i.e., collection of isolated vertices), PROBABILISTIC COLORING is polynomial, under any system of vertex-probabilities.

Recall that the counter-example of Proposition 3 shows that the natural 2-coloring is not always optimal in paths under distinct vertex-probabilities. In what follows, we first exhibit classes of trees where PROBABILISTIC COLORING is polynomial. As previously, we assume that $|U| \geqslant |D|$. Remark first that immediate application of Property 6 leads to the following result.

Proposition 5. PROBABILISTIC COLORING *is polynomial for paths under identical vertex-probabilities.*

Proposition 6. PROBABILISTIC COLORING *can be optimally solved in trees with complexity bounded above by* $(n + 1)^{\Delta(k\Delta+k+1)+1}$ *where k denotes the number of distinct vertex-probabilities. Hence the problem is polynomial in trees with bounded degree and with bounded number of distinct vertex-probabilities.*

Proof (Sketch). Consider a tree $T(N, E)$ of order n and denote by Δ its maximum degree. Let $p_1, \ldots, p_k$ be the k distinct vertex-probabilities in T, n_i be the number of vertices of T with probability p_i and set $M = \prod_{i=1}^{k} \{0, \ldots, n_i\}$. Recall finally that, from Property 3, any optimal solution of PROBABILISTIC COLORING in T uses at most $\Delta + 1$ colors.

Consider a vertex $v \in N$ with δ children and denote them by $v_1, \ldots, v_\delta$. Let $c \in \{1, \ldots, \Delta+1\}$ and $Q = \{q_1, \ldots, q_{\Delta+1}\} \in M^{\Delta+1}$ where, for any $j \in \{1, \ldots, \Delta+1\}$, $q_j = (q_{j_1}, \ldots, q_{j_k}) \in M$. We search if there exists a coloring of $T[v]$, i.e., of the sub-tree of T rooted at v verifying both of the following properties: (i)v is colored with color c and (ii) q_{i_j} vertices with probability p_i are colored with color j. For this, let us define predicate $P_v(c, Q)$ with value **true** if such a coloring exists. In other words, we consider any possible configuration (in terms of number of vertices of any probability in any of the possible colors) for all the feasible colorings for $T[v]$.

One can determine value of P_v if one can determine values of P_{v_i}, $i = 1, \ldots, \delta$. Indeed, it suffices that one looks-up all the alternatives, distributing the q_{i_j} vertices (of probability p_i colored with color j) over the δ children of v (q_{i_j} may be $q_{i_j}-1$ if $p(v) = p_i$ and $c = j$). Observe now that $|M| \leq (n+1)^k$ and, consequently, $|M^{\Delta+1}| \leq (n+1)^{k(\Delta+1)}$. For any vertex v, there exist at most $n|M^{\Delta+1}|$ values of P_v to be computed and for any of these computations, at most $(n|M^{\Delta+1}|)^\delta$ conjunctions, or disjunctions, have to be evaluated. Hence, the total complexity of this algorithm is bounded above by $n(n|M^{\Delta+1}|)^{\delta+1} \leq (n+1)^{\Delta(k\Delta+k+1)+1}$. To conclude it suffices to output the coloring corresponding to the best of the values of predicate $P_r(c, Q)$, where r is the root of T. □

Consider now two particular classes of trees, denoted by T_E and T_O, where all leaves lie exclusively either at even or at odd levels, respectively (root been considered at level 0). Obviously trees in both classes can be polynomially checked. The following proposition claims that, under identical vertex-probabilities, PROBABILISTIC COLORING is polynomial for both T_E and T_O.

Proposition 7. *Consider $T \in T_O$ (resp. in T_E). Then N_O (resp., N_E) is a maximum independent set of T. Henceforth, under identical vertex-probabilities,* PROBABILISTIC COLORING *is polynomial in T_O and T_E.*

Proposition 8. *Under any vertex-probability system 2-coloring is optimal for stars.*

We now deal with cycles with identical vertex-probabilities. In this case, the following proposition holds.

Proposition 9. PROBABILISTIC COLORING *is polynomial in cycles with identical vertex-probabilities.*

Proof (Sketch). The case of even cycles is immediate by Proposition 6. Consider an odd cycle $C_{2k+1} = \{v_1, \ldots, v_{2k+1}\}$. One can prove that 3-coloring (S_1, S_2, S_3) with $S_1 = \{v_1, v_3, \ldots, v_{2k-1}\}$, $S_2 = \{v_2, v_4, \ldots, v_{2k}\}$ and $S_3 = \{v_{2k+1}\}$ is optimal for C_{2k+1}. □

5 Split Graphs

We deal now with split graphs. This class of graphs is quite close to bipartite ones, since any split graph of order n is composed by a clique K_{n_1}, on n_1 vertices, an independent set S of size $n_2 = n - n_1$ and some edges linking vertices of $V(K_{n_1})$ to vertices of S. These graphs are, in some sense, on the midway between bipartite graphs and complements of bipartite graphs. In what follows, we first show that PROBABILISTIC COLORING is **NP**-hard in split graphs even under identical vertex-probabilities. For this, we prove that the decision counterpart of PROBABILISTIC COLORING in split graphs is **NP**-complete. This counterpart, denoted by PROBABILISTIC COLORING (K) is defined as follows: "given a split graph $G(V, E)$ a system of identical vertex-probabilities for G and a constant $K \leqslant |V|$, does there exist a coloring the functional of which is at most K?".

Proposition 10. *Even assuming identical vertex-probabilities,* PROBABILISTIC COLORING (K) *is **NP**-complete in split graphs.*

Proof (Sketch). The reduction is from 3-EXACT COVER ([15]). Consider a family $\mathcal{S} = \{S_1, S_2, \ldots, S_m\}$ of subsets of a ground set $\Gamma = \{\gamma_1, \gamma_2, \ldots, \gamma_n\}$ ($\cup_{S_i \in \mathcal{S}} S_i = \Gamma$) such that $|S_i| = 3$, $i = 1, \ldots, m$; assume that n is a multiple of 3 and set $q = n/3$. The split graph $G(V, E)$ for PROBABILISTIC COLORING will be constructed as follows: family $\mathcal{S}$ is replaced by a clique K_m (i.e., we take a vertex per set of $\mathcal{S}$); denote by $s_1, \ldots, s_m$ its vertices; ground set Γ is replaced by an independent set $X = \{v_1, \ldots, v_n\}$; $(s_i, v_j) \in E$ iff $\gamma_j \notin S_i$; $p > 1 - (1/q)$; $K = mp + q(1-p) - q(1-p)^4$.

One can prove that Γ can be partioned into q subsets of $\mathcal{S}$, if and only if G admits a coloring $C = (\{s_i, v_{i_1}, v_{i_2}, v_{i_3}\}_{i=1,\ldots,q}, \{s_{q+1}\}, \ldots, \{s_m\})$, where $S_i = \{\gamma_{i_1}, \gamma_{i_2}, \gamma_{i_3}\}$ with value $f(C) = K$. □

For the rest of this section we deal with approximation of PROBABILISTIC COLORING in split graphs. Let $G(K, S, E)$ be such a graph, where K is the vertex set of the clique ($|K| = m$) and S is the independent set ($|S| = n$). Fix an optimal PROBABILISTIC COLORING -solution $C^* = (S_1^*, S_2^*, \ldots, S_k^*)$ in $G(K, S, E)$. The following lemma straightforwardly holds.

Lemma 1. $m \leqslant k \leqslant m+1$.

Consider now the natural coloring, denoted by C, consisting of taking an unused color for any vertex of K and a color for the whole set S (in other words C uses $m+1$ colors for G).

Proposition 11. *Coloring C is a 2-approximation for split graphs under any system of vertex-probabilities.*

Proof (Sketch). Denote by $C^* = (S_1^*, S_2^*, \ldots, S_k^*)$, an optimal solution in G and assume that colors are ranged in decreasing-value order, i.e., $f(S_i^*) \geqslant f(S_{i+1}^*)$, $i = 1, \ldots, k-1$. From Lemma 1, $m \leqslant k \leqslant m+1$. If $k = m+1$ and S_1^* is the color that is a subset of S, then unbalancing arguments of Property 2 conclude that C is optimal. Hence, assume that S_1^* is a color including a vertex of K and vertices of S. Without loss of generality, assume also that, upon a reordering of vertices, vertex $v_i \in K$ is included in color S_i^*; also denote by p_i, the probability of vertex $v_i \in K$ and by q_i the probability of a vertex $v_i \in S$. Then, $f(C) = \sum_{i=1}^m p_i + (1 - \prod_{i=1}^n (1 - q_i))$ and $f(C^*) \geqslant \sum_{i=2}^m p_i + (1 - (1-p_1)\prod_{i=1}^n (1-q_i))$, where the last inequality holds thanks to the unbalancing arguments leading to Property 2, when we charge color S_1^* with all vertices of S. Observe also that: $1 - \prod_{i=1}^{n}(1 - q_i) \leqslant 1 - (1 - p_1)\prod_{i=1}^{n}(1 - q_i)$ and $1 - (1 - p_1)\prod_{i=1}^{n}(1 - q_i) \geqslant p_1$.

Combination of all the above leads, after some easy algebra to the claimed ratio. □

We now restrict ourselves to the case of identical graph probabilities. We will devise a polynomial time approximation schema for PROBABILISTIC COLORING in split graphs. For this we first need the following lemma.

Lemma 2. *Given a split graph $G(K, S, E)$, if there exists a vertex in S with degree m, then coloring C using $m + 1$ colors, one color per vertex of K and one color for the whole of vertices of S is optimal. On the other hand, if any vertex in S has degree strictly smaller than m, then any subset of S the vertices of which have all the same neighbors in K, will be colored with the same color in any optimal coloring of G.*

The following proposition affirms that if the size of the clique in G is fixed, then PROBABILISTIC COLORING can be solved in polynomial time.

Proposition 12. *If m, the size of K in $G(K, S, E)$, is fixed, then PROBABILISTIC COLORING can be solved in linear time.*

Proof (Sketch). Recall that we deal with the case where vertices of S have degree at most $m - 1$. One can prove that the number of all the possible m-colorings of the vertices of G satisfying the second part of Lemma 2 is bounded by m^{m+2^m}; this bound is a constant if m is so. So, one can choose the best among the $m + 1$-coloring of Lemma 2 and the m-colorings discussed just above, in order to produce an optimal solution for PROBABILISTIC COLORING. □

Consider now the following algorithm for PROBABILISTIC COLORING, denoted by SCHEMA:

1. fix an $\epsilon > 0$;
2. if $m \leqslant 1/\epsilon$, then optimally solve PROBABILISTIC COLORING by exhaustive look-up of all the feasible m-colorings as well as of coloring C of Proposition 12;
3. if $m \geqslant 1/\epsilon$, then output coloring C of Proposition 11.

Proposition 13. *SCHEMA is a polynomial time approximation schema for PROBABILISTIC COLORING in split graphs, under identical vertex-probabilities.*

Proof (Sketch). By Proposition 12, if Step 2 is executed, the solution computed, in polynomial time since ϵ is a fixed constant, is optimal for PROBABILISTIC COLORING. We deal now with Step 3 and the coloring C produced at this step. Denote by C^* an optimal coloring of G. Taking into account Property 4, the following expressions hold: $f(C) = m \times p + (1 - (1 - p)^n)$, $f(C^*) \geqslant (m - 1)p + (1 - (1 - p)^{n+1})$. Combination of them leads to the claimed result. □

References

1. Murat, C., Paschos, V. Th.: The probabilistic minimum coloring problem. Annales du LAMSADE 1, LAMSADE, Université Paris-Dauphine (2003) Available on http://l1.lamsade.dauphine.fr/~paschos/documents/a1pc.pdf.
2. Murat, C., Paschos, V. Th.: The probabilistic minimum coloring problem. In Bodlaender, H.L., ed.: Proc. 29th International Workshop on Graph Theoretical Concepts in Computer Science, WG'03. Volume 2880 of Lecture Notes in Computer Science., Springer-Verlag (2003) 346–357
3. Averbakh, I., Berman, O., Simchi-Levi, D.: Probabilistic a priori routing-location problems. Naval Res. Logistics **41** (1994) 973–989
4. Bertsimas, D.J.: On probabilistic traveling salesman facility location problems. Transportation Sci. **3** (1989) 184–191
5. Bertsimas, D.J.: The probabilistic minimum spanning tree problem. Networks **20** (1990) 245–275
6. Bertsimas, D.J., Jaillet, P., Odoni, A.: A priori optimization. Oper. Res. **38** (1990) 1019–1033
7. Jaillet, P.: Probabilistic traveling salesman problem. Technical Report 185, Operations Research Center, MIT, Cambridge Mass., USA (1985)
8. Jaillet, P.: A priori solution of a traveling salesman problem in which a random subset of the customers are visited. Oper. Res. **36** (1988) 929–936
9. Jaillet, P.: Shortest path problems with node failures. Networks **22** (1992) 589–605
10. Jaillet, P., Odoni, A.: The probabilistic vehicle routing problem. In Golden, B.L., Assad, A.A., eds.: Vehicle routing: methods and studies. North Holland, Amsterdam (1988)
11. Murat, C., Paschos, V. Th.: The probabilistic minimum vertex-covering problem. Int. Trans. Opl Res. **9** (2002) 19–32
12. Murat, C., Paschos, V. Th.: The probabilistic longest path problem. Networks **33** (1999) 207–219
13. Murat, C., Paschos, V. Th.: A priori optimization for the probabilistic maximum independent set problem. Theoret. Comput. Sci. **270** (2002) 561–590
14. Della Croce, F., Escoffier, B., Murat, C., Paschos, V. Th.: Probabilistic coloring of bipartite and split graphs. Cahier du LAMSADE 218, LAMSADE, Université Paris-Dauphine (2004)
15. Garey, M.R., Johnson, D.S.: Computers and intractability. A guide to the theory of NP-completeness. W. H. Freeman, San Francisco (1979)

Design Optimization Modeling for Customer-Driven Concurrent Tolerance Allocation

Young Jin Kim[1], Byung Rae Cho[2], Min Koo Lee[3], and Hyuck Moo Kwon[1]

[1] Department of Systems Management and Engineering,
Pukyong National University, Pusan 608-739, Korea
{youngk, iehmkwon}@pknu.ac.kr
[2] Department of Industrial Engineering, Clemson University, Clemson, SC 29634, US
bcho@clemson.edu
[3] Department of Information and Statistics,
Chungnam National University, Daejeon 305-764, Korea
sixsigma@cnu.ac.kr

Abstract. The majority of previous studies on tolerance allocation have viewed the issue as a design methodology to determine optimal component tolerances on behalf of a manufacturer, while meeting given assembly tolerance requirements. Although a considerable amount of research has been done on this issue, a couple of important questions still remain unanswered. First, how can a design engineer quantitatively incorporate a customer's perception on a product quality into a tolerance allocation scheme at the early design stage? Second, how can component tolerances and assembly tolerance be optimized in a simultaneous way? To answer these questions, this paper presents the so-called customer-driven concurrent tolerance allocation which is facilitated by the notion of truncated distribution and the use of mathematical programming techniques, while adopting the major principles of Taguchi philosophy. The procedures are demonstrated in a numerical example. The work presented in the paper is an effort to gain insights, which can be useful in practice when setting up guidelines for an overall tolerance allocation.

1 Introduction

In engineering design, tolerances are intended to capture variations from an idealized nominal dimension as part of a process of realization in many manufacturing processes. As an integral part of engineering design, tolerance optimization has become the focus of increased activity as manufacturing industries strive to increase productivity and improve the quality of their products. The effects of tolerance optimization are far-reaching. Not only do the tolerances affect the ability to assemble a final product in terms of functional performance, but they also affect customer satisfaction, product quality, inspection, and manufacturing costs of the designed product. Consequently, the importance of the notion of tolerance has generated a strong demand for enhanced tolerances at a competitive cost by stimulating remarkable changes in the upstream product design and development processes. Yet in spite of its importance, tolerance optimization is one of the least understood engineering tasks.

There have been two parallel developments in tolerance optimization. The first development, the so-called 'screening procedures', determines cost-effective tolerance limits for a product or assembled product, assuming that all products are subject to screening (i.e., 100% inspection). Screening may be an attractive industrial practice due to the rapid advancement of computerized inspection systems. Previous related works on this problem are well summarized in Tang and Tang [1], which provides an excellent discussion of the overall concept for the design of screening procedures within the framework of the various screening inspection environments. In an assembly-manufacturing environment, considerable research has been done on the second development known as 'tolerance synthesis'. In practice, an assembly is composed of a number of component parts and tolerances of individual components are stacked up to the tolerance of the assembly. A 'tolerance stack-up' is the term to describe how individual processes or component tolerances can combine to affect a final assembly dimension. Tolerance synthesis is an optimization procedure that determines the resulting component tolerance allocations, given a required assembly tolerance. Currently, tolerance allocation is still performed largely on a trial-and-error basis using assigned default assembly tolerances. Although this ad hoc approach might be easy to implement, it does not lead to optimal tolerances due to the lack of systematic evaluation and optimization. A survey of state-of-the-art tolerancing techniques can be found in Evans [2] and Zhang and Huq [3]. Readers are also referred to Zhang [4] for recent development in tolerance synthesis.

This article attempts to provide a new perspective on tolerance synthesis by rectifying existing problems in the current literature. First, the majority of previous studies on tolerance synthesis have been carried out based on an implicit relationship between tolerance and standard deviation, i.e., $t = 3\sigma$, where t and σ represent tolerance and standard deviation, respectively. The objective is to determine optimal tolerances on component dimensions, while meeting the stack-up tolerance so that the manufacturing cost would be minimized from a manufacturer's viewpoint. On the basis of the three-sigma relationship, the determination of component tolerances is basically equivalent to the selection of process precision (i.e., process variability). As pointed out in Vasseur et al. [5], however, there is no account for justifying the three-sigma relationship. From this perspective, the process precision levels as well as tolerances need to be determined by a systematic approach through quantitative modeling and design optimization. Secondly, the current practice of tolerance synthesis aims to optimize component tolerances in such a way that the overall assembly cost function can be minimized, while meeting the assembly tolerance within design specifications. Although optimizing the component tolerances with the fixed assembly tolerance is a current practice, there is room for improvement. As pointed out by Zhang [4], ideal tolerance designs should take account of the determination of assembly tolerance as well as tolerance allocation among components, which requires a concurrent tolerance optimization scheme in assembly and components. Furthermore, tolerance optimization by minimizing manufacturing costs solely on behalf of a manufacturer is unconvincing in the sense that customers must be foremost in a design engineer's mind from the beginning of tolerancing process. Thus, customer's perception on product quality needs to be taken into account in the process design stages. It will be discussed how the customer's perception on quality as well as the manufacturer's manufacturing costs can be quantitatively incorporated into the tolerance design process by adapting the concept of Taguchi's quality loss function. With these two aims, this article lays out a modeling foundation and presents an integrated optimization scheme for the simultaneous tolerance synthesis by combining mathematical programming techniques and Taguchi philosophy.

2 Tolerance Cost Modeling

In a general design practice, a final assembly tolerance is usually given based on a three-sigma limit, or calculated in such a way that manufacturing costs are minimized from a manufacturer's point of view. Since the assembly is often viewed as a final product to the customer, this perception must be translated into the selection of tolerance design parameters such as the mean and standard deviation of the assembly. With this aim, Taguchi loss function may be a good alternative in modeling the quality cost associated with customer's perception on the assembled product. The assembly tolerance must be defined and optimized mathematically to avoid ambiguous interpretation and to provide a sound basis for an overall tolerance design process.

One of the most important issues currently encountered in quality engineering is the selection of a proper quality loss function to relate the key characteristics of a product to its performance perceived by the customer. The quality loss function is a means of quantifying the quality loss of a product on a monetary scale, and is incurred when the product or its production process deviates from the customer-identified nominal dimension (target value) of the assembly. The functional relationship between the customer's dissatisfaction and assembly performance needs to be carefully defined since the choice of paths to performance enhancement for assembly depends heavily upon the type of quality loss function used at the tolerance design stage.

Among a number of quality loss functions proposed in the literature, Taguchi loss function may provide a good approximation to the quality loss, particularly over some range in the neighborhood of the nominal dimension of product performance. Taguchi loss function is represented by $L(y) = k(y-\tau)^2$, where k and τ are a positive loss coefficient and nominal target dimension of the assembly, respectively. This loss function basically says that a loss is always incurred when a product performance deviates from its nominal dimension regardless of how small the deviation is. A look at the loss function reveals a very desirable characteristic. It can easily be shown that $E[L(y)] = k[(\mu-\tau)^2 + \sigma^2]$, where $E[L(y)]$ represents the expected value of $L(y)$. Thus, in order to minimize customer's loss, both the bias ($|\mu - \tau|$) and the standard deviation (σ) must be reduced. Hence, one may employ Taguchi loss function as an approximation in situations where there is little or no information about the functional relationship between the assembly performance and the associated loss to the customer, or where there is no direct evidence to refute a quadratic representation. It is not our intention to provide detailed discussion on the loss function in this paper. Implementing a tight tolerance on an assembly may provide high outgoing quality (i.e., low loss to the customer), but it usually causes high manufacturing costs to the manufacturer. In contrast, implementing a loose tolerance may reduce the manufacturing costs to the manufacturer, but it may result in low outgoing quality (i.e., high loss to the customer). An immediate problem is then how to tradeoff these conflicting cost criteria to determine the most economical tolerance on the assembly.

Let Y be the functional dimension of an assembly, which can be represented by a function of n component dimensions X_i's, i.e., $Y = f(X_1, X_2, \cdots, X_n)$. It can be shown through Taylor series expansion that the mean and variance of Y, which are denoted by μ and σ^2, respectively, can be approximated by

$$\mu = f(\mu_1, \mu_2, \cdots, \mu_n) \quad \text{and} \quad \sigma^2 = \sum_{i=1}^{n} \left(\frac{\partial f}{\partial X_i}\right)^2 \sigma_i^2, \quad (1)$$

where μ_i and σ_i represent the mean and standard deviation of component dimension X_i, respectively. The loss $L(y)$ is incurred to the customer when $\tau - T \leq y \leq \tau + T$, where T denotes the assembly tolerance. Furthermore, if the deviation from the nominal target dimension is greater than T (i.e., $y < \tau - T$ or $y > \tau + T$), the assembly is assumed to be rejected (e.g., reworked or scrapped in a practical sense, depending upon the manufacturing environment) and excluded from shipment. These rejection costs would be incurred to the manufacturer. Let R_U and R_L represent rejection costs when $y > \tau + T$ and $y < \tau - T$, respectively. The expected total cost associated with the assembly manufacturing, which is denoted by $E[TC_{ASM}]$, is then expressed as

$$E[TC_{ASM}] = \int_{\tau-T}^{\tau+T} k(y-\tau)^2 f_Y(y)\, dy + R_U \cdot P(Y \geq \tau+T) + R_L \cdot P(Y \leq \tau-T), \quad (2)$$

where $f_Y(\cdot)$ represents the density function of the assembly dimension Y.

In addition to the assembly tolerance, tolerances on component dimensions need to be optimized so that the associated costs would be minimized. Two cost factors are considered in this article: operating cost and rejection cost. First, requiring better operating devices and more trained personnel, a manufacturing process with a higher precision (or lower variability) usually results in an increased operating cost. On the other hand, a manufacturing process with a lower precision may require a less operating cost, but the outgoing products may reveal a poor product quality due to high variability. Secondly, nonconforming products falling outside the tolerance limits need to be reworked or scrapped. With a given process precision, a tight tolerance limit results in a high outgoing quality accompanied by high rejection costs. By the same token, implementing a loose tolerance limit may reduce rejection costs at the expense of product quality. Thus, there is a need to find the optimum tolerance allocation scheme minimizing the sum of operating cost and rejection cost while meeting given design constraints.

Suppose that there are m alternative manufacturing processes for a component which have different levels of process precision (or variability). Let u_{ij} be the indicator variable, which takes 1 if the jth process is selected for producing the component dimension X_i. The operating cost and standard deviation of the jth process for dimension X_i are denoted by c_{ij} and σ_{ij}, respectively. Denoting the operating cost and standard deviation of dimension X_i by OC_i and σ_i, respectively, $OC_i = \sum_{j=1}^{m} c_{ij} u_{ij}$ and $\sigma_i = \sum_{j=1}^{m} \sigma_{ij} u_{ij}$. It is assumed that rejection costs are incurred when X_i falls outside the tolerance limits. Letting r_{U_i} and r_{L_i} denote the unit rejection costs when X_i falls above the upper tolerance limit and below the lower tolerance limit, respectively, the expected rejection cost for component dimension X_i, denoted by $E[RC_i]$, can then be written as

$$E[RC_i] = P(X_i \geq \mu_i + t_i) \cdot r_{U_i} + P(X_i \leq \mu_i - t_i) \cdot r_{L_i}, \qquad (3)$$

where t_i represents the tolerance of component dimension X_i, and the expected total cost associated with the tolerance allocation, denoted by $E[TC]$, can then be written as the sum of $E[TC_{ASM}]$, operating costs, and rejection costs as follows:

$$E[TC] = E[TC_{ASM}] + \sum_{i=1}^{n} (OC_i + E[RC_i])$$

$$= \int_{\tau-T}^{\tau+T} k(y-\tau)^2 f_Y(y)\,dy + R_U \cdot [1 - F_Y(\tau+T)] + R_L \cdot F_Y(\tau-T)$$

$$+ \sum_{i=1}^{n} \sum_{j=1}^{m} c_{ij} u_{ij} + \sum_{i=1}^{n} [(1 - F_i(\mu_i + t_i)) \cdot r_{U_i} + F_i(\mu_i - t_i) \cdot r_{L_i}], \quad (4)$$

where $F_Y(\cdot)$ and $F_i(\cdot)$ denote the cumulative distribution functions of assembly dimension Y and component dimension X_i, respectively. Note that $E[TC]$ is a function of process parameters, such as means, variances, and rejection costs, as well as component tolerances.

3 Optimization Model

To establish an optimization model for tolerance synthesis, one first needs to consider the stack-up of component tolerances to the assembly dimension. Suppose that components falling outside the tolerance limits are rejected. In this case, only the fraction of components within the tolerance limits is passed into the assembly process, which results in a truncated distribution. That is, the components are passed into the assembly process only when $\mu_i - t_i \leq X_i \leq \mu_i + t_i$. Thus, the corresponding truncated distribution represents the actual population of components passed into the assembly manufacturing and consequently its associated parameters, such as mean and variance of the truncated distribution, need to be obtained to describe the component dimensions realized in the assembly process. Let $f_i(x_i)$ and $\tilde{f}_i(x_i)$ denote the original and truncated probability density functions of X_i, respectively. It is well known that

$$\tilde{f}_i(x_i) = \frac{f_i(x_i)}{\int_{\mu_i-t_i}^{\mu_i+t_i} f_i(x_i)\,dx_i}.$$

The mean of the truncated distribution corresponds to μ_i since the original distribution is symmetrically truncated about the mean μ_i. Let $\tilde{\sigma}_i^2$ be the variance of the truncated distribution, which can be obtained by

$$\tilde{\sigma}_i^2 = \frac{\int_{\mu_i-t_i}^{\mu_i+t_i}(x_i-\mu_i)^2 f_i(x_i)\,dx_i}{\int_{\mu_i-t_i}^{\mu_i+t_i} f_i(x_i)\,dx_i}. \tag{5}$$

Since only the fraction of components within the tolerance limits is passed into the assembly process (see Fig. 1), the variance of assembly dimension is now

$$\sigma^2 = \sum_{i=1}^{n}\left(\frac{\partial f}{\partial X_i}\right)^2 \tilde{\sigma}_i^2. \tag{6}$$

Furthermore, it is a common practice to specify the desirable level of process capability for the assembly dimension which is denoted by C_p. The standard deviation of the assembly can then be written as $\sigma = T/(3C_p)$ to achieve the process capability of C_p for a given assembly tolerance T. From equation (6), the tolerances on component dimensions consequently need to be selected to satisfy

$$\sum_{i=1}^{n}\left(\frac{\partial f}{\partial X_i}\right)^2 [\tilde{\sigma}_i(t_i;\sigma_i)]^2 = \left(\frac{T}{3C_p}\right)^2.$$

In addition, the process means of component dimensions need to be set at the target to minimize quality loss, i.e., $\mu = f(\mu_1,\mu_2,\cdots,\mu_n) = \tau$.

Summing up these observations, the optimization model can then be written as shown in equation (7), where $t_i^{\min}$ and $t_i^{\max}$ represent the minimum and maximum of the component tolerance t_i, respectively, and $T^{\max}$ is the maximum allowable assembly tolerance. Examining a traditional tolerance synthesis problem, two observations are drawn. First, only the component tolerances are to be determined to minimize the manufacturing costs for a given fixed assembly tolerance. Second, the precision levels of component dimensions are accordingly adjusted on the basis of the three-sigma relationship. Note that decision variables in the proposed optimization model include the assembly tolerance, process means, and precision levels (variability) of component dimensions as well as the component tolerances. Thus, tolerancing processes in assembly and component domains are concurrently determined in a single optimization model.

$$\text{Minimize} \quad E[TC] = E[TC_{ASM}] + \sum_{i=1}^{n}\left(E[RC_i] + \sum_{j=1}^{m} c_{ij}u_{ij}\right)$$

Subject to $f(\mu_1, \mu_2, \cdots, \mu_n) = \tau$

$$\sum_{i=1}^{n}\left(\left.\frac{\partial f}{\partial X_i}\right|_{X_i=\mu_i}\right)^2 \tilde{\sigma}_i^2 = \left(\frac{T}{3C_p}\right)^2$$

$$t_i^{\min} \leq t_i \leq t_i^{\max}, \quad \text{for } i = 1, 2, \cdots, n \qquad (7)$$

$$\sum_{j=1}^{m} \sigma_{ij} u_{ij} = \sigma_i, \quad \text{for } i = 1, 2, \cdots, n$$

$$\sum_{j=1}^{m} u_{ij} = 1, \quad \text{for } i = 1, 2, \cdots, n$$

$$T \leq T^{\max}.$$

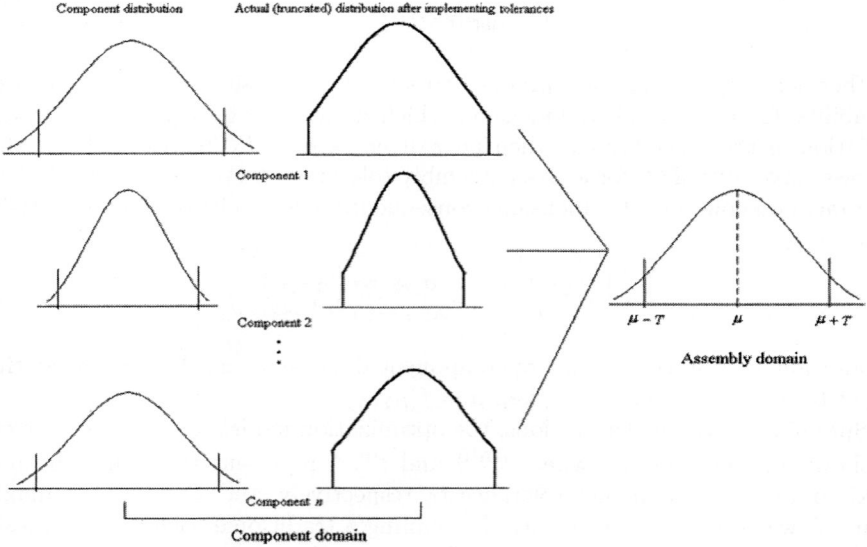

Fig. 1. Graphical Representation of Component and Assembly Domains

4 A Numerical Example

A glass bead example from Morrison [6] is adapted with some modifications to demonstrate the proposed model. In the early days of semiconductors, a glass bead has been used to seal the electrical connections to the transistor. As a major performance measure of glass beads, the glass volume is perceived to be critical to the sealing processes. More detailed descriptions can be found in Morrison [6]. The nominal dimensions of glass bead are set at 1.7mm in diameter (D), 0.6mm in bore (B), and 1.9mm in length (L). The glass volume, denoted by V, can be calculated by

$$V = \frac{\pi}{4}\left(D^2 - B^2\right)L.$$

The variances in individual dimensions contribute to the variations in glass volume as follows:

$$\sigma_V^2 = \left(\frac{\pi DL}{2}\right)^2 \sigma_D^2 + \left(\frac{\pi BL}{2}\right)^2 \sigma_B^2 + \left(\frac{\pi}{4}(D^2 - B^2)\right)^2 \sigma_L^2,$$

where σ_i^2's represent the variances of their respective dimension. Thus, the objective is to find the optimal precision level and tolerance for each dimension and the tolerance on the glass volume, while meeting the pre-specified process capability of glass bead manufacturing process.

The process precision levels and their operating cost data are given in Table 1. Further, the rejections costs associated with each dimension and glass volume are provided in Table 2. Suppose that the maximum allowable tolerance on the glass volume is specified at 0.5, i.e., $T \leq T^{\max} = 0.5$, to meet the functional requirement. The minimum and maximum tolerances on each dimension are given by

$$0.015 = t_D^{\min} \leq t_D \leq t_D^{\max} = 0.06, \quad 0.045 = t_B^{\min} \leq t_B \leq t_B^{\max} = 0.18,$$

$$\text{and} \quad 0.036 = t_L^{\min} \leq t_L \leq t_L^{\max} = 0.15.$$

Table 1. Manufacturing Precision Levels and Operating Cost Data

Dimension	j	σ_{ij}(mm)	c_{ij}($)
Diameter	1	0.025	0.43
	2	0.017	0.48
Bore	1	0.062	0.17
	2	0.039	0.22
Length	1	0.052	0.23
	2	0.026	0.28

Assuming that the manufacturing process for each dimension follows a normal distribution, one may establish the optimization model for the concurrent tolerance allocation as shown in equation (7). The proposed optimization model can be solved by using popular software such as Microsoft Excel and Matlab. We used the Solver option in Microsoft Excel to solve the optimization model. Suppose that the process capability of 1.0, i.e., $C_p = 1.0$, needs to be ensured. The optimal solutions are then found to be $\sigma_D = 0.025$, $\sigma_B = 0.062$, $\sigma_L = 0.026$, $t_D = 0.0552$, $t_B = 0.165$, and $t_L = 0.150$ with the expected total cost of $0.953 per bead.

Table 2. Rejection Cost Data

	Diameter	Bore	Length	Volume
$r_{U_i}/R_U(\$)$	0.8	0.9	0.6	2.9
$r_{L_i}/R_L(\$)$	1.2	1.4	1.3	3.8

Finally, a sensitivity analysis with respect to process capability is conducted for the case where $\sigma_D = 0.025$, $\sigma_B = 0.062$, and $\sigma_L = 0.026$. The numerical results are summarized in Fig. 2 and 3. Note that only a small portion of components is passed into the assembly dimension as the process capability becomes higher. The fractions falling within tolerance limits in terms of diameter, bore, and length are only 70.5%, 86.5%, and 93.8%, respectively, when $C_p = 2.0$ (see Fig. 2). This is because a tighter tolerance on each dimension needs to be imposed to ensure a highly capable manufacturing process of glass bead. This also explains the increase in the expected rejection cost for each component to achieve a higher process capability. It is meaningful to examine the effects of process capability on the optimal tolerances of assembly and components. The behavior of optimal assembly and component tolerances is depicted in Fig. 3. For smaller values of process capability, component tolerances are allowed to the maximum level of $t_i^{\max}$. On the other hand, tighter tolerances are imposed on individual components to keep the assembly tolerance at the maximum allowable level of $T^{\max}$ for a larger value of process capability.

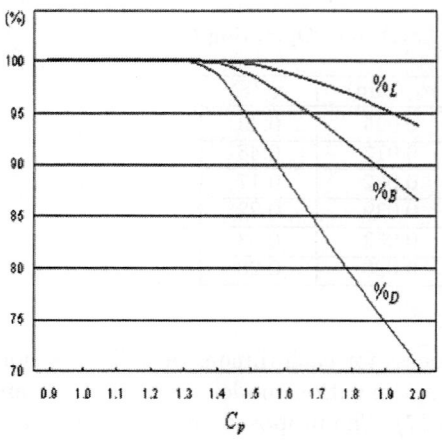

Fig. 2. Fractions of components passing into the assembly domain ($\sigma_D = 0.025$, $\sigma_B = 0.062$, and $\sigma_L = 0.026$)

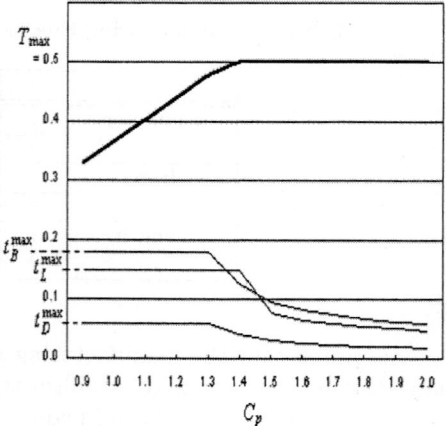

Fig. 3. Behavior of optimal assembly and component tolerances ($\sigma_D = 0.025$, $\sigma_B = 0.062$, and $\sigma_L = 0.026$)

5 Concluding Remarks

This article proposes a new method for tolerance allocation in which assembly tolerance and component tolerances are concurrently optimized within an integrated framework. In the traditional tolerance allocation problem, the component tolerances are determined to minimize the manufacturing costs for a given fixed assembly tolerance. In the proposed model, the assembly tolerance is viewed as a decision variable to minimize the customer's quality cost as well as manufacturing cost while meeting functional requirements. In addition, the process precision levels of component dimensions as well as component tolerances are to be determined in the optimization model. Consequently, the determination of assembly tolerance and its allocation among component dimensions are integrated in a single optimization model while incorporating the customer's perception on product quality. The proposed method may help a design engineer incorporate the customer's voice in a quantitative mechanism early at the design stage, and may provide practitioners with a more visible set of tools in an assembly manufacturing environment who seek cost effective high quality products.

References

1. Tang, K., Tang, J.: Design of Screening Procedures: A Review. Journal of Quality Technology. 26 (1994) 209–226
2. Evans, E.H.: Statistical Tolerancing: The State of The Art. Journal of Quality Technology. 6 (1974) 188–195
3. Zhang, H., Huq, M.: Tolerance Techniques: The State-of-The-Art. International Journal of Production Research. 30 (1992) 2111–2135
4. Zhang, H.: Advanced Tolerancing Techniques. John Wiley & Sons Inc. New York (1997)
5. Vasseur, H., Kurfess, T.R., Cagan, J.: Use of a Quality Loss Function to Select Statistical Tolerances. Journal of Manufacturing Science and Engineering. 119 (1997) 410–416
6. Morrison, S.J.: Variance Synthesis Revisited. Quality Engineering. 11 (1998) 149–155

Application of Data Mining for Improving Yield in Wafer Fabrication System

Dong-Hyun Baek[1], In-Jae Jeong[2], and Chang Hee Han[1]

[1] Department of Business Administration, Hanyang University, 1271 Sa 1-dong,
Sangrok-gu, Ansan-si, Gyeonggi-do 426-791, Korea
[2] Department of Industrial Engineering, Hanyang University,
17 Haengdang-dong, Seongdong-gu, Seoul 133-391, Korea
{estarbaek, chan, ijeong}@hanyang.ac.kr

Abstract. This paper presents a comprehensive and successful application of data mining methodologies to improve wafer yield in a semiconductor wafer fabrication system. To begin with, this paper applies a clustering method to automatically identify AUF (Area Uniform Failure) phenomenon from data instead of visual inspection that bad chips occurs in a specific area of wafer. Next, sequential pattern analysis and classification methods are applied to find out machines and parameters that are cause of low yield, respectively. Finally, this paper demonstrates an information system, Y2R-PLUS (Yield Rapid Ramp-up, Prediction, analysis & Up Support) that is developed in order to analyze wafer yield in a Korea semiconductor manufacturer.

1 Introduction

Yield is the most important factor in semiconductor industries as is often expressed "War on yields." Yield is defined as the ratio of the number of good products to the number of input products. As the memory chips become denser, maintaining the high level of yield becomes more difficult. In addition, yield can be affected by the lack of maintenances for processes and machines, human errors and parameter settings for designs and processes. In order to maintain competitiveness in market, it is essential to ramp-up the yield to the satisfactory level for mass productions.

Traditional semiconductor manufacturing process consists of 4 steps: wafer fabrication (FAB), probe test, assembly and final test. In case of 64M DRAM, FAB process consists of 350 individual steps, 500 manufacturing machines and 250 measuring machines. The amount of technical data gathered from FAB ranges more than 150 giga bytes in average in a month with 500 different types of parameters. This necessitates the development of more intelligent and effective decision supporting systems that can handle the mass production data to improve yield.

One of the most commonly used yield management system is the one that uses statistical quality control techniques. In statistical quality control, product qualities and the history of machine failures are managed using x control chart,

$\bar{x} - R$ control chart and process capability index. Also, statistical analysis uses techniques such as correlation analysis, regression and significance analysis. In case of inspection data, MAP analysis using graph is commonly used. Other types of yield management activity include 6-sigma and knowledge management activity which utilizes engineer's know-how.

In spite of the various activities to improve yields, there are many factors that cannot be recognized using traditional statistical methods, MAP analysis or the experimental study of engineers. Main reasons are as follows

1. It is time consuming process to analyze mass data and enormous correlated parameters
2. There exist many unstable processes that cannot be recognized by engineers
3. Integrated approaches are necessary in order to improve yield

This paper deals with the application of data mining in wafer fabrication system for yield improvement. Clustering analysis has been proposed to automatically determine AUF (Area Uniform Failure) in which bad chips occur frequently in specific location of the wafer when FAB process has been finished. In addition, we propose a method to determine machines and parameters that are the causes of low yield using data mining techniques such as sequential pattern analysis, classification analysis. We will introduce Y2R-PLUS (Yield Rapid Ramp-up, Prediction, analysis & Up Support) system, which implements the proposed methodology.

This paper consists of following sections. In section 2, we demonstrate data mining applications in quality control and a semiconductor manufacturing process Section 3 explains the problem under consideration, the proposed methodology and Y2R-PLUS system. Finally, section 4 concludes the paper.

2 Backgrounds

2.1 Data Mining Applications in Quality Control

Frawley (1991) defined data mining as a process to induce potentially useful information that is embedded in mass data that has not been recognized before. Rapid development in information technology has enabled the electronic gathering and storing of mass data. As a result, the amount of data has been increased exponentially, i.e., POS data, stock trading data in stock market, patient data in hospital, transaction data in bank, quality control related data in manufacturing system. Experts could analyze the data using statistical analysis and query when data size was not a significant problem. As the amount of data has been increased exponentially, it has become almost impossible to analyze mass data manually. Therefore this has initiated the need for the development of new techniques and intelligent methodologies to analyze mass data in order to assist decision maker (Fayyad et. al., 1996). Table 1 summarizes the applications of data mining techniques in the area of quality control.

Table 1. Applications of data mining in quality controls

Authors	Applications	Techniques	Objectives
Baek et.al (2000)	Etching process	decision tree	Detection of etching failures
Ahn et. al (1999)	Wheel casting process in automobile	classification and clustering technique for the mixed type of class data and continuous data	Finding optimal process condition
Hancock et. al (1996)	Valve casting process in automobile	ridge regression	Setting process parameters to reduce failures
Braha et.al (2002)	Cleansing process in semiconductor manufacturing	decision tree, neural network, composite classifiers, SOM	On-line monitoring of cleansing process
Kusiak et. al (2001)	PCB assembly	rough set theory	Detection of cause of fails
Yamashita (2000)	Simulator in Debutanizer factory	decision tree, Naive-Bayes classifier	Extraction of knowledge for factory operations
Lian et. al (2002)	sheet-metal frame for automobile industries	correlation analysis and maximal tree technique	Detection of dimensional variation

2.2 Semiconductor Manufacturing Process

Semiconductor manufacturing process is composed of wafer fabrication, probe test, assembly and final test. Hundreds of chips are mounted on silicon wafer in wafer fabrication process. Probe test is a process to test whether the chip possesses designed electrical properties. In assembly process, chips are removed from the wafer and electrical/physical properties are improved. In addition, chips are shaped to be protected from mechanical and physical impacts. Finally, final test examines electrical properties, functions and reliabilities of chips to classify into bad or good chips. It usually takes 3~4 months from wafer fabrication to final test.

Wafer fabrication (FAB) is the most expensive and complex process in the semiconductor manufacturing processes. In addition, FAB is an important process in terms of yields since the production life cycle is relatively long (2~3 months). Usually 25 wafers consist of a lot and wafers are processed and moved to next process as lots. FAB process consists of deposition, photolithography, etching and diffusion. Usually, it takes 10~30 layers to make a chip and one layer of chip requires deposition, photolithography, etching, and diffusion process. Therefore one chip must re-enter the same process consecutively, namely re-entrant flow. However not all chips follow the same process sequences. One or more processes can be skipped or be added in addition, the process sequences can be changed.

Usually, yield represents the ratio of good products to input products. For example, if a process produces 80 good products with 100 input products, yield of the process is 80%. Yield is especially important in semiconductor industries since it becomes more difficult to maintain high level of yields as chips become denser. In addition, reducing ramp-up period is also important for maintaining competitiveness in market share and profits. Yield can be deteriorated even in clean room environments because of human errors or the maintenance problems of processes and machines. Therefore it is important to figure out and eliminate the causes of the yield degradation.

3 Yield Management in FAB Process Using Data Mining

This section introduces a yield management system, Y2R-PLUS that has been developed for a large semiconductor company in Korea. Y2R-PLUS has been designed to analyze the cause of yield degradations in FAB. Also, we will introduce the problem statement and explain data mining techniques used in Y2R-PLUS.

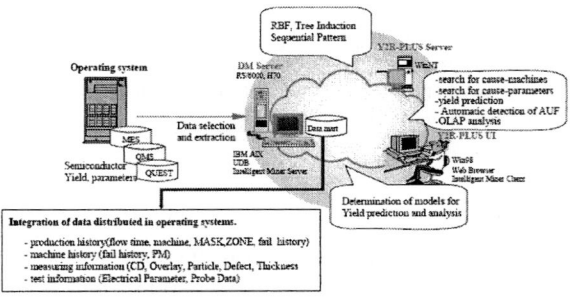

Fig. 1. Y2R-PLUS hierarchy

3.1 Introduction of Y2R-PLUS System

Fig. 1 shows the hierarchy of Y2R-PLUS system. Y2R-PLUS consists of 3 tiers structure such as data mining server, Y2R-PLUS server installed with Intelligent Miner developed by IBM and Y2R-PLUS UI (User Interface). Also, Y2R-PLUS system constructs data mart to extract/transform/transfer production history data, machine history data, measuring data and test data from MES (Manufacturing Execution System), QMS (Quality Management System) and QUEST. Users can access to data mining server to construct models for yield analysis using data mining techniques. Also, users can request to run RBF, tree induction and sequential pattern analysis that are already constructed by Web browsers. Y2R-PLUS system can provide a variety of reports to decision makers using OLAP (On-Line Analysis Processing).

Y2R-PLUS system provides functions such as

1. finding the process parameters and machines that are the cause of yield degradation

2. analyzing AUF/MAP using clustering techniques instead of visual inspections
3. yield predictions
4. reporting with OLAP function

3.2 Application of Data Mining Techniques

In this section, we explain the problem considered in this paper and the data mining techniques that are implemented in Y2R-PLUS system as shown in Table 2.

Table 2. Application of data mining techniques

Techniques	Applications
Clustering	Automatic detection of AUF
	-To find fails of specific type in wafers using clustering
Sequential Pattern	Search of cause-machine
	-To find machines or sequence of machines that are used to process common fails of wafers.
Classification	Search of cause-parameter
	-To find parameters which cause the difference between two groups of wafer using decision tree

Automatic detection of AUF. As explained in earlier section of the paper, probe test checks whether chips possess the desired electrical properties after FAB process. Chips in wafer are classified into good or bad products after probe test. In case of bad products, probe machine classifies products into several groups of fails according to the types of fails. For example, wiring fails can be grouped into around 10 categories such as 3-fail, 2-fail, W-fail, R-fail and so on. These are called as "Fail category" in industries.

In most cases, particular fail category occurs in a specific location of wafer. These fails are called AUF (area uniform failure) in industries. The left side of Fig. 2 shows 9 subregions of a wafer, AUF type can be classified into 12 categories according to the location of occurrences such as middle, top, bottom, left, right, top-left, top-right, bottom-left, bottom-right, edge, around and random. For example, if 3-fail occurs in middle area of a wafer, the type of AUF is 3-fail-middle. Dark dots in Fig. 2 represent the bad chips.

There are a variety of causes of AUF. For example, a particular machine generates pollutant in a specific location of wafers or the difference of temperature of wafer may cause fails in etching process. AUF is important in industries since the same AUF occurs by the same cause. That is, we can easily find the cause of AUF by tracking the common machines and common parameters on which AUF wafers are processed.

In visual inspection, engineers find AUF by inspecting wafer maps displayed on computer screen. Visual inspection is a time consuming process in addition, the quality of inspection heavily depends on the experience of inspectors.

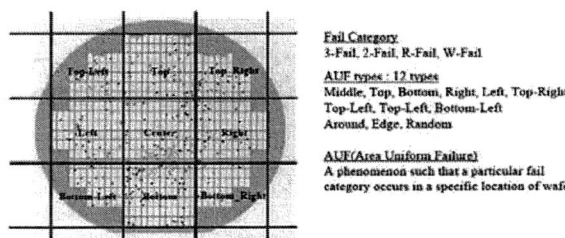

Fig. 2. AUF failure

The classification problem of AUF can be considered as a clustering problem. A clustering problem is to find subset of similar records given a set of records. Using a clustering technique, we can cluster fail categories into groups that are close to each other geometrically. Then we can determine AUF types according to the location of clusters. In this paper, we apply the modified single linkage method of hierarchical clustering techniques. Hierarchical clustering methods start with groups of individual objects. As the iteration goes on, clusters with close distances are grouped together. Hierarchical clustering can be classified into single linkage method, complete linkage method and average linkage method. Single linkage method uses minimum distance and complete linkage method utilizes maximum distance and average linkage method uses average distance between two clusters.

Fig. 3 shows the algorithm to determine the existence of AUF and AUF types in wafer according to fail categories (e.g., 3-fail and W-fail) after probe test. First, the set $S = \{C_1, C_1, ..., C_n\}$ of chips that have specific fail category (3-fail in Fig. 3) is generated. Chip C_i can be represented as relative location on chip, (X_i, Y_i) where X_i, Y_i is integers. (0, 0) is the top left location of the chip.

Second, distance between all pairs of C_i, C_j are calculated using the distance $d_{ij} = max(|X_i - X_j|, |Y_i - Y_j|)$. Two chips of minimum distance C_i, C_j are included in cluster $CLST$ and eliminated from set S. Next step puts all chips into $CLST$ when the distance between the chip and $CLST$ is less than 1 and eliminate the chip from set S. The distance between chip C_k and $CLST$ defined as minimum distance, that is the distance $= min\{d_{ik} : C_i \in CLST\}$.

If there are no chips with distance less than 1 from $CLST$, stop the step and save $CLST$. If S is null set, go to the next step, otherwise go to the previous step and form the next $CLST$. Let's suppose that the algorithm generates m clusters, $CLST_1, CLST_2, ... CLST_m$. We eliminate clusters containing less than a particular number of chips in the cluster since those do not have significant meaning as clusters.

Next we calculate the center coordinate of each cluster and determine the AUF type according to the center location. The center location can be calculated as an average location of chips in the cluster. As a result, multiple clusters of the same fail category can be formed in a wafer. For example, 3-fail occurs in top-left and middle forming 2 AUF types.

Finally, the last step of the algorithm is to determine clusters which satisfy the density constraint. The density constraint means that the density of AUF must exceed certain multiple times (e.g., 2 times) of the total density of fail category. Total density of a fail category is the number of chips containing the fail category divided by the number of total chips (i.e., net die). The density of AUF is the number of AUF type chips divided by the number of chips belonging to the AUF type area. For example, if the number of chips in $CLST_k$ is 10 and the central location of the cluster is in top-left (suppose that the number of chips in top-left area is 50), then the density of AUF of cluster $CLST_k$ is 10/50=0.2

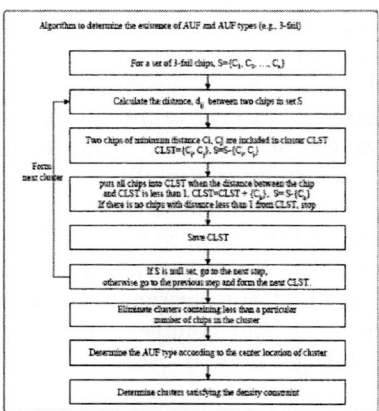

Fig. 3. Automatic failure detection algorithm for AUF

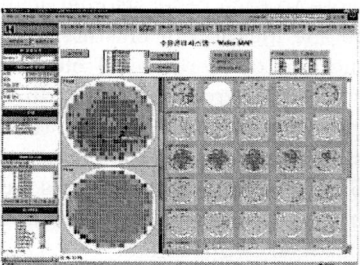

Fig. 4. Automatic detection of AUF

Fig. 4 shows the wafer of R-fail-edge AUF. Dark part of the wafer represents the chips with R-fail.

Clustering algorithm enables rapid cause analysis of yield degradation by defining AUF. The cause of yield degradation can be found by tracking the common process factors of wafers containing the same fail category. In visual inspection, engineers must determine AUF manually, which is time consuming process since tens of thousands of wafers are manufactured in industries. In addition, the performance of inspection is totally dependent on the engineers' experiences. Reducing time from the detection of AUF to the determination of the cause of the fail is very important in semiconductor manufacturing since the manufacturing consists of complicated and time consuming processes. We could reduce the inspection time from 1~2 days to 1~2 hours using the proposed clustering algorithm, the cause-machine search method and the cause-parameter search method. These search methods will be explained in the following sections. As a result, there was a significant improvement in yields.

Cause-machine search. Analysis groups of wafers are the collection of wafers in which we want to find the cause of AUF. The main reason to form analysis groups is to find the cause of fail using the set of wafers not individuals. Analysis groups can be formed using 1) the set of wafers containing the same AUF type 2) the set of low yield wafers 3) the set of wafers determined by engineers.

After forming analysis groups, the cause-machine search method finds the cause of fails in terms of production machines. That is, the method finds machines and machine sequences on which the wafers in the analysis groups are processed.

Y2R-PLUS system uses the sequential pattern mining function of Intelligent miner for the cause-machine search. Input data is the list of machines on which the wafers in the analysis group are processed in FAB process.

Table 3 shows the result of the cause-machine search. Support in the table means the ratio of wafers in the group that are processed on the corresponding machine. 90% of wafers in the analysis group are processed on E-9A-B machine of 1430 L-ASHER process as shown in the second column of Table 3.

Table 3. Results for cause-machine analysis

Machine	Support1 (Analysis)	Support2 (Comparing)	Difference
1430 L-ASHER E-9A-B	90%	89%	1%
1415 L-DRY-ET E-2B-05B	89%	25%	64%
1290 L-SOX-CL D-03-04	50%	45%	5%
1330 L-NIT-DE D-5A-06	30%	40%	-10%
2230 LNW ASHER E-9B-06	20%	22%	-2%

In this analysis, we consider not only analysis groups but also comparing groups. Comparing groups are the set of wafers that contains no fails or different types of fails in the similar production periods. The reason why we introduce the comparing groups is to prevent from the wrong interpretation of analysis when we consider only analysis groups. For example, we may wrongly conclude that the cause of fail of the analysis group in Table 3 is the E-9A-B machine of 1430 L-ASHER process since 90% of the wafers in the analysis group are processed on this machine. However 89% of the wafers in the comparing groups are also processed on the same machine. The reasons may be either the machine utilization of E-9A-B machine is originally high or many wafers are processed on this machine inevitably due to the failure of alternative machines. In case of 1415 L-DRY-ET process, 89% of wafers in the analysis group are processed on E-2B-05B machine. On the contrary, only 25% of wafers in comparing group are processed on this machine. Therefore we may conclude that the main cause of the fail of the analysis group is E-2B-05B machine.

Cause-parameter search. Usually, process parameters are controlled to be located within the predetermined specifications in every process. Each process parameters are monitored on-line using statistical quality control after each process is finished. The problem is that even if the parameters are under control, the parameters may be the cause of low yields when the specification of parameters is wrongly predetermined. This phenomenon frequently occurs especially in production lines producing new products since the production conditions of research/development lines are different from production lines. In addition, it is difficult to find cause-parameters of low yield due to a variety number of FAB processes and parameters.

The cause-parameter search problem can be classified into a classification problem. The cause-parameters of analysis group can be found by the comparison of the corresponding comparing group using decision tree technique. Suppose that the yield of analysis group is less than 50% and that of comparing group is more than 90%. Then we can find the process parameters that are the cause of the difference between two groups.

Table 4. Data type for cause-parameter analysis

Wafer ID	Input Parameter					Target Parameter
	CD_3716_ MP01_AVG	CD_4400_ MP02_Max	Overlay_22 50_Max_X _AVG	...	Overlay_23 50_Rotoffy _AVG	
PW4144	0.2354	0.4323	0.7863		0.0159	Analysis
PW4298	0.4746	0.5632	0.3243		0.0023	Comparing
PW4436	0.3120	0.4432	0.6545		0.1632	Analysis

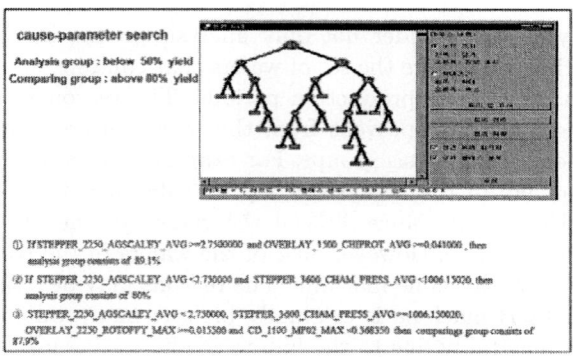

Fig. 5. Search results of cause parameters

Table 4 shows input data for classification problem. Input parameters are the measured parameters after processes are finished and the target parameter is either the parameter of analysis group or that of comparing group. The number of input parameters can be reduced if engineers can predict the set of parameters that are the possible causes of low yields. Fig. 5 shows the result of the cause-parameter search using the tree classification mining function of Intelligent miner.

4 Conclusion

This paper deals with the yield management methodology using data mining techniques for FAB process in semiconductor industries and its implementation in Y2R-PLUS (Yield Rapid Ramp-up, Prediction, analysis & Up Support) system for a Korean semiconductor company. In this yield management system, AUF can be found automatically using clustering techniques instead of visual inspection. Also, the cause-machines and cause-parameters of low yield wafers can be found using continuous pattern analysis and classification technique. Y2R-PLUS system has significantly improved yield in practice and we are currently working on the extension of its usage.

References

1. Ahn, J. S., Goh, Y. M., Jang, J. S.: Searching optimal process conditions using data mining techniques. Journal of the Korean institute of plant engineering, Vol. 4, No. 2 (1999) 129-144,
2. Baek, J. G., Kim, K. H., Kim, S. S., Kim, C. O.: Adaptive decision tree algorithm for data mining in real-time machine status database. Journal of the Korean institute of industrial engineers, Vol. 26, No. 2 (June 2000) 171-182.
3. Braha, D., and Shmilovici, A.: Data mining for improving a cleansing process in the semiconductor industry. IEEE transactions on semiconductor manufacturing, Vol. 15, No. 1 (2002) 91-101.
4. Fayyad, U., Piatetsky-Shapiro, G. and Smyth, P.(ed.): From data mining to knowledge discovery in databases. Advances in knowledge discovery and data mining, AAAI Press/MIT Press (1996) 1-34.
5. Frawley, W. J., Piatetsky-Shapiro, G., Matheus, C. J. (ed.): Knowledge discovery in databases: An overview, Knowledge discovery in databases. AAAI Press/MIT Press (1991) 1-27.
6. Hancock, W. M, Yoon, J. W., and Plont, R.: Use of ridge regression in the improved control of casting process. Quality engineering, Vol. 8, No. 3 (1996) 395-403.
7. Kusiak, A. and Kurasek, C.: Data mining of printed-circuit board defects. IEEE transactions on robotics and automation, Vol. 17, No. 2 (2001) 191-196.
8. Lian, J., Lai, X. M., Lin, Z. Q., and Yao, F. S.: Application of data mining and process knowledge discovering in sheet metal assembly dimensional variation diagnosis. Journal of materials processing technology, Vol. 129 (2002) 315-320.
9. Yamashita, Y.: Supervised learning for the analysis of process operational data. Computer and chemical engineering, Vol. 24, No. 2 (2000) 471-474.

Determination of Optimum Target Values for a Production Process Based on Two Surrogate Variables

Min Koo Lee[1], Hyuck Moo Kwon[2], Young Jin Kim[2], and Jongho Bae[1]

[1] Department of Information and Statistics, Chungnam National University,
Daejeon, 305-764, Korea
{sixsigma, bae-jongho}@cnu.ac.kr

[2] Department of Systems and Management Engineering,
Pukyong National University, Pusan 608-739, Korea
{iehmkwon, youngk}@pknu.ac.kr

Abstract. In this paper, we consider the problem of determining the optimum target values of the process mean and screening limits for a production process under single screening procedure. Two surrogate variables are observed simultaneously in single screening procedure. It is assumed that two surrogate variables are correlated with the quality characteristic of interest. A model is constructed that involve selling price and production, inspection and penalty costs. A method for finding the optimum target values of the process mean and screening limits is presented when the quality characteristic of interest and surrogate variables are assumed to be jointly normally distributed. A numerical example is presented.

1 Introduction

As a result of advances in automated manufacturing systems, censoring technology and automatic inspection equipment, complete inspections are increasingly used to improve the outgoing quality of the product. Every product is inspected to determine whether its quality characteristic satisfies the specification limits. Conforming products are sold at regular price, whereas nonconforming products are scraped, reprocessed or sold at a discounted price. Typical quality characteristics under consideration are weights, volume, and geometric dimensions. The quality characteristic may deviate from the target because of variations in raw material, labor and operation conditions. Thus, the process mean may be targeted to a higher value than necessary to reduce the fraction nonconforming. A higher process mean, however, may induce a higher production cost. Therefore, the process mean is better to be selected considering the tradeoff among production cost, payoff of conforming items, and the cost incurred due to nonconforming items.

Several researchers have studied this problem. Bettes [1], Golhar [2], and Golhar and Pollack [3] consider a filling process in which underfilled or overfilled containers are reprocessed at a fixed cost. Hunter and Kartha [4], Bisgaard et al. [5], Lee and Jang [6], and Hong et al. [7] study several sales condition for products in which the quality characteristic is smaller than the specification limit. Boucher and Jafari [8], and Al-Sultan [9] discussed situations in which the items are subjected to lot-by-lot acceptance sampling rather than complete inspections. Elsayed and Chen [10] determined optimum levels of process parameters for products with multiple characteristics, and Arcelus and Rahim [11] developed a model for simultaneously selecting optimum target means for both variable and attribute quality characteristics. Chen and Chung [12] considered an economic model for determining the most profitable target value and the optimum measuring precision level for a production process. Hong and Elsayed [13] studied the effects of measurement errors on process target, Pfeifer [14] showed the use of an electronic spreadsheet program as a solution method. Most recently, Rahim and Shaibu [15] applied the Taguchi loss function to determine the optimum process target and variance. Kim et al. [16] proposed a model for determining the optimal process target with the consideration of variance reduction and process capability. Teeravaraprug and Cho [17] designed the optimum process target levels for multiple quality characteristics, and Rahim et al. [18] considered the problem of selecting the most economical target mean and variance for a continuous production process. Finally, Duffuaa and Siddiqui [19] considered process targeting with multi-class screening and measurement error.

In all of these studies, inspection is performed on the quality characteristic of interest (performance variable). In some situations, it is impossible or not economical to directly inspect the performance variable. In such cases, the use of a surrogate variable which is highly correlated with performance variable is an attractive alternative, especially when inspecting the surrogate variable is relatively less expensive than inspecting the performance variable. In cement plants, for example, a performance measure of interest may be the weight of a cement bag, which is difficult to measure directly due to the high-speed of the packing line. The mil-ampere (mA) of the load cell is strongly correlated with the weight of a cement bag and does not require special effort to measure. Hence, it can be used as the surrogate variable (Bai and Lee [20]). The idea of selecting the cutoff value on surrogate variable has been studied by many researchers. Bai and Lee [20], Lee et al. [21] and Tang and Lo [22] present economic models that determine the process mean and the screening limit on the surrogate variable when inspection is based on surrogate variable instead of performance variable. In applications where quality assurance is critical, the outgoing quality improvement may be more important than the reduction in the inspection cost. Since a surrogate variable may not perfectly correlated with performance variable, some conforming items may be rejected and excluded from shipment while some nonconforming items may be accepted for shipment. These decision errors are likely to occur when the value of the surrogate variable is close to the screening limits. Consequently, in this situation, there may be an economic advantage to reduce the errors by observing the performance variable even though the inspection may be expensive. Of course, this can only be done when inspection of the performance variable is nondestructive.

This paper considers the problem of determination of the optimum process mean and the screening limits of two surrogate variables for a production process under single screening procedure. Two surrogate variables are observed simultaneously in single screening procedure. It is assumed that two surrogate variable are correlated with performance variable. A profit model is constructed involves selling price, production, inspection and penalty costs. Assuming that the performance and two surrogate variables are normally distributed, a method of finding the optimum process mean and the screening limits of two surrogate variables is presented. The proposed model is demonstrated with a numerical example and sensitivity analyzes are performed.

2 The Model

Consider a production process where items are produced continuously. Let Y be a performance variable representing the quality characteristic of interest. All items are inspected prior to shipment to determine whether they meet the lower and upper specification limits L and U on Y, respectively. All items are nonconforming if their y values are smaller than L or larger than U. Suppose that Y is normally distributed with an unknown process mean μ_y and known variance σ_y^2. The production cost per item is linearly related to Y, that is, $b + cy$ where b and c are constants. The expected production cost per item is given by

$$EPC = \int_{-\infty}^{\infty} (b + cy)f(y)dy = b + c\mu_y, \qquad (1)$$

Complete inspection is an another alternative for preventing the cost incurred by delivering nonconforming items to customers. In some situations, it is impossible or not economical to directly inspect the performance variable. In such cases, the use of a surrogate variable which is highly correlated with the performance variable is an attractive alternative, especially when inspecting the surrogate variable is relatively less expensive than inspecting the performance variable. Let X_1 and X_2 be the two surrogate variables. Assume that (Y, X_1, X_2) are jointly normally distributed with means $(\mu_y, \mu_{x_1}, \mu_{x_2})$ and variances $(\sigma_y^2, \sigma_{x_1}^2, \sigma_{x_2}^2)$ and correlation coefficients $(\rho_{0i}, \rho_{ij}, i,j = 1,2)$, where ρ_{0i} and ρ_{ij} denote the correlation between Y and X_i and X_j, $i,j = 1,2$, respectively. The relationship between Y and X_i, $i = 1,2$, are defined by the conditional distribution $g_i(x_i|y)$, which is assumed to be a normal density function with a mean $\alpha_{x_i} + \beta_{x_i}y$ and variance σ_i^2. Here, β_{x_i} is assumed to be positive so that Y and X_i have a positive relationship. If X_i is negatively correlated with Y, we then use $-X_i$ as the surrogate variable rather than X_i. The joint probability density function of Y and X_i is bivariate normal function with mean $(\mu_y, \mu_{x_1} = \alpha_i + \beta_{x_i}\mu_y)$ and variance $(\sigma_y^2, \sigma_{x_i}^2 = \beta_{x_i}^2\sigma_y^2 + \sigma_i^2)$ and correlation coefficient $\rho_{0i} = (\beta_{x_i}^2\sigma_y^2/(\beta_{x_i}^2\sigma_y^2 + \sigma_i^2))^{1/2}$, see Tang and Lo [22] for detailed derivation. In this paper, inspection is performed based on the linear combination of two surrogate variables, $Z = \lambda_1 X_1 + \lambda_2 X_2$, where λ_1 and λ_2 are selected to maximize the variance of $(Y - Z)$. We obtain

$$\lambda_1 = (\rho_{01} - \rho_{02}\rho_{12})/(1 - \rho_{12}^2)$$
$$\lambda_2 = (\rho_{02} - \rho_{01}\rho_{12})/(1 - \rho_{12}^2), \qquad (2)$$

See Anderson [23]. The standardized variables $Y' = (Y - \mu_y)/\sigma_y$ and $X'_i = (X_i - \mu_{x_i})/\sigma_{x_i}$, $i = 1,2$, jointly have a standard trivariate normal distribution with density function form

$$h(y', x'_1, x'_2) = |\Sigma|^{-1/2} \exp\left[-(y', x'_1, x'_2)\Sigma^{-1}(y', x'_1, x'_2)^T/2\right]/(2\pi)^{3/2}, \qquad (3)$$

where $(y', x'_1, x'_2)^T$ denotes the transpose of (y', x'_1, x'_2), and

$$\Sigma = \begin{bmatrix} 1 & \rho_{01} & \rho_{02} \\ \rho_{01} & 1 & \rho_{12} \\ \rho_{02} & \rho_{12} & 1 \end{bmatrix}.$$

Let δ_1 and δ_2 denote the screening limits on $Z = \lambda_1 X_1 + \lambda_2 X_2$. An accepted item with $\delta_1 \leq Z \leq \delta_2$ is sold at a fixed price a, the expected revenue per item is

$$ER = a \int_{\omega_1}^{\omega_2} \phi(z'/\theta)/\theta \, dz' = a\left(\Phi(\omega_2/\theta) - \Phi(\omega_1/\theta)\right), \qquad (4)$$

where $z' = \lambda_1 x'_1 + \lambda_2 x'_2$, $\omega_i = (\delta_i - \mu_{z'})/\sigma_{z'}$, $i = 1, 2$, $\mu_{z'} = \lambda_1 \mu_{x_1} + \lambda_2 \mu_{x_2}$, it can be shown that $\theta = \sigma_{z'} = \rho_{y'z'} = \left((\rho_{01}^2 + \rho_{02}^2 - 2\rho_{01}\rho_{02}\rho_{12})/(1 - \rho_{12}^2)\right)^{1/2}$, $\phi(\cdot)$ and $\Phi(\cdot)$ are the density function and distribution of the standard normal distribution, respectively. Since Y and X_i are not perfectly correlated, some nonconforming items may be accepted. If a nonconforming item is shipped to customers, penalty cost $r_l(y' < \eta_1)$ or $r_u(y' > \eta_2)$ is incurred including the cost of identifying and handling the nonconforming item, and service and replacement cost. The expected per item cost of acceptance is

$$EAC = r_l \int_{\omega_1}^{\omega_2} \int_{-\infty}^{\eta_1} f(y', z') \, dy' \, dz' + r_u \int_{\omega_1}^{\omega_2} \int_{\eta_2}^{\infty} f(y', z') \, dy' \, dz', \qquad (5)$$

where $\eta_1 = (L - \mu_y)/\sigma_y$, $\eta_2 = (U - \mu_y)/\sigma_y$, and $f(y', z')$ is the joint density function of Y' and Z'. Using the relationships

$$\int_{-\infty}^{\omega_2} \int_{-\infty}^{\eta_1} f(y', z') \, dy' \, dz' = \Psi(\eta_1, \omega_2/\theta; \theta), \qquad (6)$$

$$\int_{-\infty}^{\omega_2} \int_{\eta_2}^{\infty} f(y', z') \, dy' \, dz' = \Psi(-\eta_1 - \zeta, \omega_2/\theta; -\theta), \qquad (7)$$

where $\zeta = (U - L)/\sigma_y$, $\Psi(\cdot, \cdot; \theta)$ is the standardized bivariate normal distribution function with correlation coefficient θ, formula (5) can be rewritten as

$$EAC = r_l \left[\Psi(\eta_1, \omega_2/\theta; \theta) - \Psi(\eta_1, \omega_1/\theta; \theta)\right]$$
$$+ r_u \left[\Psi(-\eta_1 - \zeta, \omega_2/\theta; -\theta) - \Psi(-\eta_1 - \zeta, \omega_1/\theta; -\theta)\right]. \qquad (8)$$

Let c_{x_1} and c_{x_2} denote the surrogate variables inspection costs per item, then the expected profit per item is

$$EP = ER - EPC - EAC - c_{x_1} - c_{x_2}. \qquad (9)$$

The optimum values η_1^*, ω_1^* and ω_2^* can then be obtained by maximizing EP.

3 The Optimum Solution

In this Section, a method of finding the optimum process mean mean μ_y^* and screening limits δ_1^* and δ_2^* on $Z = \lambda_1 X_1 + \lambda_2 X_2$ is presented. The following identities will be used for the derivations.

$$\partial \Psi(\eta_1, \omega_i/\theta; \theta)/\partial \eta_1 = \Phi\left((\omega_i - \eta_1\theta^2)/((1-\theta^2)^{1/2}\theta)\right)\phi(\eta_1), \qquad (10)$$

$$\partial \Psi(-\eta_1 - \zeta, \omega_i/\theta; \theta)/\partial \eta_1 = -\Phi\left((\omega_i - (\eta_1+\zeta)\theta^2)/((1-\theta^2)^{1/2}\theta)\right)\phi(-\eta_1 - \zeta), \qquad (11)$$

$$\partial \Psi(\eta_1, \omega_i/\theta; \theta)/\partial \omega_i = \Phi\left((\eta_1 - \omega_i)/(1-\theta^2)^{1/2}\right)\phi(\omega_i/\theta)/\theta. \qquad (12)$$

Equating the first derivatives of EP with respect to η_1, ω_1 and ω_2 zero, respectively, and using formulas (10)-(12) yield

$$\begin{aligned} & r_l\left[\Phi\left((\omega_2^* - \eta_1^*\theta^2)/\Omega\right) - \Phi\left((\omega_1^* - \eta_1^*\theta^2)/\Omega\right)\right]\phi(\eta_1^*) \\ & -r_u\left[\Phi\left((\omega_2^* - (\eta_1^* + \zeta)\theta^2)/\Omega\right) - \Phi\left((\omega_1^* - (\eta_1^* + \zeta)\theta^2)/\Omega\right)\right]\phi(-\eta_1^* - \zeta) \\ & = c\sigma_y, \end{aligned} \qquad (13)$$

$$\Phi\left((-\eta_1^* - \zeta + \omega_1^*)\theta/\Omega\right) + \Phi\left((\eta_1^* - \omega_1^*)\theta/\Omega\right) = \frac{a}{r_l}, \qquad (14)$$

$$\Phi\left((-\eta_1^* - \zeta + \omega_2^*)\theta/\Omega\right) + \Phi\left((\eta_1^* - \omega_2^*)\theta/\Omega\right) = \frac{a}{r_u}, \qquad (15)$$

where $\Omega = (1-\theta^2)^{1/2}\theta$. If EP is a unimodal function of η_1, ω_1 and ω_2, the optimum values η_1^*, ω_1^* and ω_2^* are the values η_1, ω_1 and ω_2 simultaneously satisfying equations (13)-(15). Since these equations have $\phi(\cdot)$ and $\Phi(\cdot)$ and EP have $\Phi(\cdot)$ and $\Psi(\cdot, \cdot; \theta)$, it is difficult to find a closed form solution or show analytically that equations (13)-(15) have a unique solution or EP is a unimodal function of η_1, ω_1 and ω_2. Numerical studies over wide range of parameter values, however, indicate that the Hessian matrix at η_1^*, ω_1^* and ω_2^* is negative definite and it represents a maximum point. Therefore, the optimum values η_1^*, ω_1^* and ω_2^* can be obtained by solving equations (13)-(15) simultaneously, and a numerical search method such as Gauss-Seideel's iterative method can obtain η_1^*, ω_1^* and ω_2^*. IMSL [24] subroutines such as DNORDF and DBNRDF are used to evaluate the standard univariate and bivariate normal distribution functions, respectively. In most cases, the optimum values η_1^*, ω_1^* and ω_2^* can be obtained within a few seconds on a 586 PC. The optimum process mean μ_y^* and screening limits δ_1^* and δ_2^* on $Z = \lambda_1 X_1 + \lambda_2 X_2$ are then obtained by

$$\mu_y^* = L - \eta_1^*\sigma_y, \qquad (16)$$
$$\delta_1^* = \mu_{z'} + \omega_1^*\sigma_{z'}, \qquad (17)$$
$$\delta_2^* = \mu_{z'} + \omega_2^*\sigma_{z'}. \qquad (18)$$

4 A Numerical Example

In this section, an illustrative example is provided to demonstrate the proposed model and perform numerical analyzes to understand its properties.

Consider a production process of chemical factory. The packing operation consists two processes: a filling process and an inspection process. In the filling process of food-additives, chemicals and pharmaceuticals etc., it is very important to satisfy the specification limits. Each chemical container processed by the filling machine is moved to the loading and dispatching phases on a conveyor belt. Continuous weighing devices perform inspection based on the linear combination $Z = \lambda_1 X_1 + \lambda_2 X_2$ of two surrogate variables, which are correlated with the weight Y of the chemical container. From theoretical considerations and past experience, it is known that the variance of Y, $\sigma_y^2 = (0.75 kg)^2$, and that $X_i, i = 1, 2$, for given $Y = y$ is normally distributed

$$g_1(x_1|y) \sim N\left(4.0 + 0.08y, \sigma_1^2 = 0.050^2\right),$$

$$g_2(x_2|y) \sim N\left(4.0 + 0.085y, \sigma_2^2 = 0.055^2\right),$$

and $\rho_{12} = 0.6$. Suppose that the cost components and the specification limits for Y are $a = \$3.0$, $c_0 = \$0.1$, $c = \$0.06$, $r_l = \$6.5$, $r_u = \$6.0$, $c_{x_1} = \$0.004$, $c_{x_2} = \$0.0045$, $L = 40.0 kg$, and $U = 44.5 kg$. We obtain $\eta_1^* = -2.744$, $w_1^* = -2.693$ and $w_2^* = 3.256$ from equations (13)-(15). The optimum process mean μ_y^* and screening limits δ_1^* and δ_2^* on $Z = \lambda_1 X_1 + \lambda_2 X_2$ are obtained from equations (16)-(18), and are summarized in Table 1.

Table 1. Optimum Solutions

μ_{x_1}	7.365	μ_{x_2}	5.575	μ_z	6.915
η_1^*	-2.744	w_1^*	-2.693	w_2^*	3.256
μ_y^*	42.058	δ_1^*	6.903	δ_2^*	6.929
EP	0.3458	ρ_{01}	0.768	ρ_{02}	0.757

(i) *Effects of screening procedures.* For the above example, the solutions and their associated costs for single screening procedure and the screening procedures based on one surrogate variable are shown in Table 2. EP's of single screening screening procedure is larger than those of the screening procedures based on one surrogate variable.

(ii) *Effects of σ_y.* The screening limits δ_1 and δ_2 and the process mean μ_y for the above example are shown in Fig. 1 and 2 for selected values of the standard

Table 2. Optimum Solutions and Expected Profits of Three Different Screening Procedure

Use X_1 only	$\mu_y^* = 42.029$	$\delta_1^* = 7.156$	$\delta_2^* = 7.620$	$EP = 0.3428$
Use X_2 only	$\mu_y^* = 42.030$	$\delta_1^* = 7.350$	$\delta_2^* = 7.850$	$EP = 0.3420$
Use X_1 and X_2	$\mu_y^* = 42.058$	$\delta_1^* = 6.903$	$\delta_2^* = 6.929$	$EP = 0.3458$

deviation of Y from 0.6 to 1.0. Fig. 1 shows that the screening limits δ_1 and δ_2 increase as σ_y increases and the difference of screening limits δ_1 and δ_2 increases as σ_y increases. Fig. 2 indicates that the process mean μ_y tends to increase as σ_y increases.

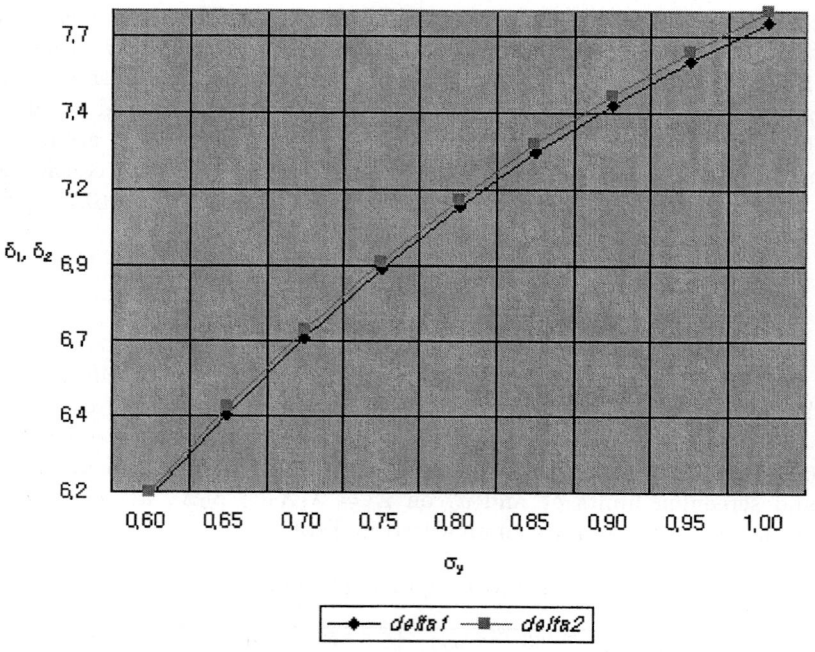

Fig. 1. Screening limits δ_1 and δ_2 as a function of σ_y

5 Concluding Remarks

This paper considered the problem of determining the optimum target values of the process mean and the screening limits for a production process under single screening procedure based on the linear combination of two surrogate variables in place of the quality characteristic of interest. A model is constructed under the assumption that the two surrogate variables and the quality characteristic of interest are jointly normally distributed. The optimum process mean and screening limits are jointly obtained by maximizing the expected profit function, which includes selling price, in addition to the production, inspection and penalty costs. It is difficult to find closed form solutions, or to show analytically that the solutions are optimum. Numerical studies over wide range of parameter values, however, indicate that the expected profit function is indeed unimodal. A numerical search method such as Gauss-Seideel's iterative method is used. IMSL subroutines such as DNORDF and DBNRDF are used to evaluate the standard univariate and bivariate normal distribution functions, respectively. In

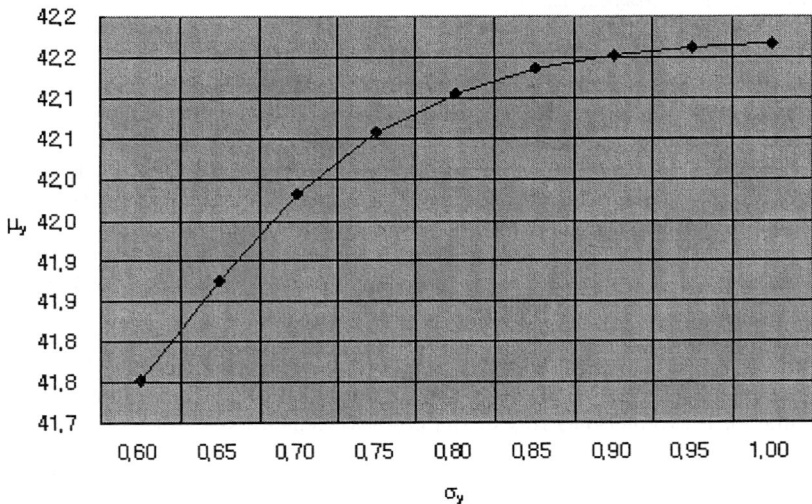

Fig. 2. Process mean μ_y as a function of σ_y

most cases, the optimum solutions are obtained within a few seconds on a 586 PC. Numerical results show that the expected profit of single screening screening procedure is larger than those of the screening procedures based on one surrogate variable. The screening limits increase as the standard deviation of quality characteristic of interest increases, and the process mean tends to increase as increases the standard deviation of quality characteristic of interest increases.

It will be of interest to consider the double screening procedure where one variable is used first to make one of three decisions - accept, reject, or undecided - and after the first screening, the second variable is employed to screen the undecided items.

References

1. Bettes, D.C.: Finding an Optimum Target Value in Relation to a Fixed Lower Limit and Arbitary Upper Limit. Applied Statistics. 11 (1962) 202–210
2. Golhar, D. Y.: Determination of the Best Mean Contents for a Canning Problem. Journal of Quality Technology. 19 (1987) 82–84
3. Golhar, D.Y., Pollock, S.M.: The Determination of the Best Mean and the Upper Limit for a Canning Problem. Journal of Quality Technology. 20 (1988) 188–192
4. Hunter, W.G., Kartha, C.D.: Determining the Most Profitable Target Value for a Production Process. Journal of Quality Technology. 9 (1977) 176–180
5. Bisgaard, S., Hunter, W.G., Pallesen, L.: Economic Selection of Quality of Manufactured Product. Technometrics. 26 (1984) 9–18
6. Lee, M. K. and Jang, J. S.: The Optimum Target Values for a Production Process with Three-class Screening. International Journal of Production Economics. 49 (1997) 91–99

7. Hong, S. H., Elsayed, E. A., Lee, M. K.: Optimum Mean Value and Screening Limits for Production Processes with Multi-Class Screening. International Journal of Production Research. 37 (1999) 155–163
8. Boucher, T.O., Jafari, M.A.: The Optimum Target Value for Single Filling Operations with Quality Sampling Plans. Journal of Quality Technology. 23 (1991) 44–47
9. Al-Sultan, K.S.: An Algorithm for the Determination of the Optimal Target Values for Two Machines in Series with Quality Sampling Plans. International Journal of Production Research. 12 (1994) 37–45
10. Elsayed, E.A., and Chen, A.: Optimal Levels of Process Parameters for Products with Multiple Characteristics. International Journal of Production Research. 31 (1993) 1117–1132
11. Arcelus, F. J., RAHIM, M. A.: Simultaneous Economic Selection of a Variables and an Attribute Target Mean. Journal of Quality Technology. 26 (1994) 125–133
12. Chen, S. L., Chung, K. J.: Selection of the Optimal Precision Level and Target Value for a Production Process: the Lower Specification Limit Case. IIE Transactions. 28 (1996) 979–985
13. Hong, S. H., Elsayed, E. A.: The Optimum Mean for Processes with Normally Distributed Measurement Error. Journal of Quality Technology. 31 (1999) 338–344
14. Pfeifer, P.E.: A General Piecewise Linear Canning Problem Model. Journal of Quality Technology. 31(1999) 326–337
15. Rahim, M.A., Shaibu, A.B.: Economic Selection of Optimal Target Values. Process Control and Quality. 11(2000) 369–381.
16. Kim, Y.J., Cho, B.R., Phillips, M.D.: Determination of the Optimal Process Mean with the Consideration of variance Reduction and Process Capability. Quality Engineering. 13 (2000) 251–260
17. Teeravaraprug, J. and Cho, B. R.: Designing the Optimal Process Target Levels for Multiple Quality Characteristics. International Journal of Production Research. 40(2002) 37–54
18. Rahim, M.A., Bhadury, J., Al-Sultan, K.S.: Joint Economic Selection of Target Mean and Variance. Engineering Optimization. 34(2002) 1–14.
19. Duffuaa, S., Siddiqui, A. W.: Process Targeting with Multi-Class Screening and Measurement Error. International Journal of Production Research. 41 (2003) 1373–1391
20. Bai, D.S., Lee, M.K.: Optimal target values for a Filling Process When Inspection is Based on a Correlaed Vraiable. International Journal of Production Economics. 32 (1993) 327–334
21. Lee, M. K., Hong, S. H., Elsayed, E. A.: The Optimum Target Value under Single and Two-Stage Screenings. Journal of Quality Technology. 33 (2001) 506–514
22. Tang, K., Lo, J.: Determination of the Process Mean When Inspection is Based on a Correlated Variable. IIE Transactions. 25 (1993) 66–72
23. Anderson T.W.: An Introduction to Multivariate Statistical Analysis. John Wiley & Sons Inc New York (1984)
24. Reference Manual: International Mathematical and Statistical Libraries. IMSL Library, Houston (1987)

An Evolution Algorithm for the Rectilinear Steiner Tree Problem

Byounghak Yang

Kyungwon University, Bockjung-dong San 65,
Sujung-gu, Seongnam-si, Kyunggi-do, Korea
byang@kyungwon.ac.kr
http://mhl.kyungwon.ac.kr

Abstract. The rectilinear Steiner tree problem (RSTP) is to find a minimum-length rectilinear interconnection of a set of terminals in the plane. It is well known that the solution to this problem will be the minimal spanning tree (MST) on some set Steiner points. The RSTP is known to be NP-complete. The RSTP has received a lot of attention in the literature and heuristic and optimal algorithms have been proposed. A key performance measure of the algorithm for the RSTP is the reduction rate that is achieved by the difference between the objective value of the RSTP and that of the MST without Steiner points. An evolution algorithm for RSTP based upon the Prim algorithm was presented. The computational results show that the evolution algorithm is better than the previously proposed other heuristics. The average reduction rate of solutions from the evolution algorithm was about 11%, which is almost similar to that of optimal solutions.

1 Introduction

Given a set V of n terminals in the plane, the rectilinear Steiner tree problem(RSTP) in the plane is to find a shortest network, a Steiner minimum tree, interconnecting S. The points in S are called Steiner points. We should find the optimal number of Steiner points and their location on rectilinear plane. It is well-known that the solution to RSTP will be the minimal spanning tree (MST) on some set of points V∪S.

The RSTP is known to be NP-complete [6]. Polynomial-time algorithm for the optimal solution is unlikely to be known [5]. Lee, Bose and Hwang [9] presented algorithm similar to the Prim algorithm for the Minimum Spanning tree. Smith and Liebman [11] and Smith, Lee and Liebman [10] presented heuristics algorithms for the RSTP.

Let MST(V) be the cost of MST on set of terminals V. Then the reduction rate R(V) = {MST(V∪S) - MST(V) } / MST(V) is used as performance measure for algorithms on the Steiner tree problem. Beasley [3] surveyed heuristics for RST problem and introduced his own heuristics. In his research, Beasley's heuristics gave the best reduction rate as about 10%. The only Genetic algorithm for Euclidian Steiner tree problem was presented by Hesser et al. [8]. The performance of the genetic algorithm was not so good and they didn't introduce algorithm for RSTP. Warme, Winter and Zachariasen [13] presented exact algorithm for RSTP and showed the average reduction rate is about 11% for the optimal solution. Main motivation of this research is to develop an evolution algorithm to meet the reduction rate to 11%.

Hanan showed that for any instance, an optimal RST exists in which every Steiner point lies at the intersection of two orthogonal lines that contain terminals [7]. Hanan's theorem implies that a graph G called the Hanan grid graph is guaranteed to contain an optimal RST. Hanan's grid graph is constructed as follows: draw a horizontal and vertical line through each terminal. The vertices in graph correspond to the intersections of the lines. In Fig. 1.(a), the black dots are the terminals and white dots are the intersection of two lines. The optimal Steiner point should be white dots in Hanan's gird.

For 3 terminals RST problem, we can get optimal Steiner point easily. Let (x_i, y_i) be the coordinates of the given terminal Ti; France [4] proved that the Steiner point S is located at (x_m, y_m), where x_m and y_m are the medians of $\{x_i\}$ and $\{y_i\}$, as like Fig. 1(b).

Fig. 1. Hanan's grid graph and Steiner point for 3 Terminals

2 Evolution Algorithm

Evolutionary algorithms are based on models of organic evolution. They model the collective the collective learning process within a population of individuals. The starting population is initialized by randomization or some heuristics method, and evolves toward successively better regions of search space by means of randomized process of recombination, mutation and selection. Each individual is evaluated as the fitness value, and the selection process favors those individuals of higher quality to reproduce more often than worse individuals. The recombination mechanism allows for mixing of parental information while passing it to their dependents, and mutation introduces innovation into the population. Although simplistic from a biologist's viewpoint, these algorithms are sufficiently complex to provide robust and powerful adaptive search mechanisms [1]. In this research, some Steiner points in Hanan's grid may become an individual, and the individual is evaluated by minimum spanning tree with terminals and Steiner points in the individual.

2.1 Individual and Evaluation

By the Hanan's theorem, the optimal Steiner points should be on the Hanan's grid. An individual are introduced as represented the location of Steiner point on Hanan's grid. Each vertical line and horizontal line is named as increasing number. Most left vertical line is the first vertical line, and most bottom horizontal line is the first horizontal line. Each individual has multiple (v_i, h_i) which is the location of i-th Steiner point S_i on Hanan's grid, which v_i is the index of vertical line on Steiner point S_i and h_i is the index of horizontal line on Steiner point S_i. And (x_i, y_i) is the real location on plane for Steiner point S_i, and is calculated from (v_i, h_i). In this research, individual is an assembly of (v_i, h_i) of a non-fixed number of Steiner points and number of Steiner points as follows ; {m, (v_1, h_1) , (v_2, h_2) ,......, (v_m, h_m) } where m is the number of Steiner points.

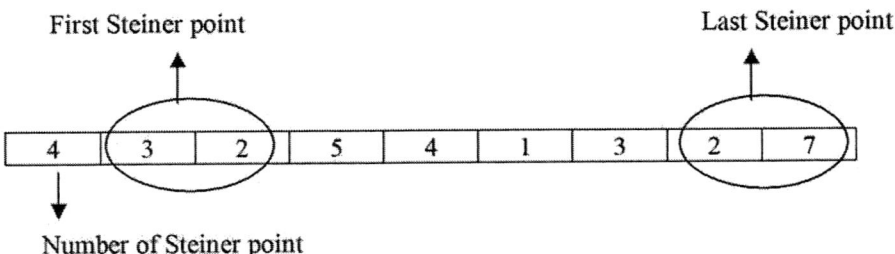

Fig. 2. Individual for Steiner points

Each individual has different length depending on its number of Steiner points, and represents one Steiner Tree. The fitness of the individual corresponds to the length of the MST than can be constructed by using the original terminals and the Steiner points in individual by Prim algorithm.

2.2 Recombination

The recombination or crossover operator exchanges a part of an individual between two individual. We choose some Steiner points in each individual and exchange with other Steiner points in the other individual. We have two parents to crossover as P1={3, (1,3), (2,4), (3,2)} and P2{2, (6,4), (7,8)}. By the random number, we choose the number of exchanging which is less than the number of Steiner points in both parents. Then by the random number, some Steiner points are choose and exchanged. In this case, we choose the 2 as exchanging number. First and third Steiner points are selected in P1, and two Steiner points are selected in P2. Therefore we have two children by performing crossover as follows: O1={3, (6,4), (2,4), (7,8)} and O2{2, (1,3), (3,2)}.

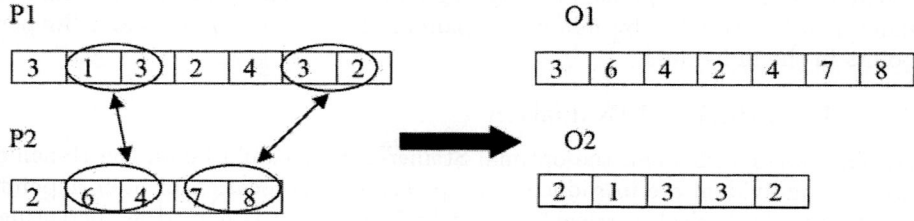

Fig. 3. Crossover Operator

2.3 Mutation

The mutation operator changes the locations of some selected Steiner points. Let (v_i, h_i) be the selected location of Steiner point. Then we can have new location $(v_i, +v, h_i + h)$ where v and h are some random integer number.

Fig. 4. Mutation Operator

2.4 Selection

We use the tournament selection. The k-tournament selection method select a single individual by choosing some number k of individuals randomly from the population and selecting the best individual from this group to survive to the next generation. The process is repeated as often as necessary to fill the new population. A common tournament size is k=2.

2.5 Initial population

For the 3-terminal RST problem, the median of terminals is the optimal Steiner point. We assume that each median point of every adjacent 3-terminal has high probability to be survived in the optimal Steiner tree. For the convenience to find adjacent 3-terminal, we make a minimum spanning tree for the V and find every directly connected 3 terminals in the minimum spanning tree as like Fig. 5(b). And a Steiner point for these connected 3 terminals is calculated. We make Steiner point for every connected 3 terminals in minimum spanning tree, and put in Steiner points pool as like Fig. 5(c) and use one of candidate Steiner point. We make initial population for the evolution algorithm as following procedure:

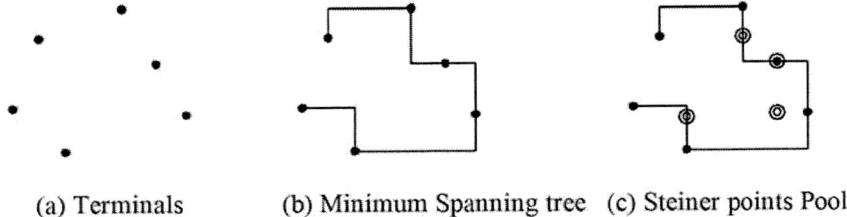

(a) Terminals   (b) Minimum Spanning tree   (c) Steiner points Pool

Fig. 5. A Steiner pool from minimum spanning tree for given terminals

Step 1: Make a minimum spanning tree (MST) for the terminals.
Step 2: Choose every connected three terminals in MST. Let them be 3-neighbor terminals (3NT).
Step 3: Make a Steiner point for each 3NT and add it in the Steiner points pool.
Step 4: Choose some Steiner points from Steiner points pool by random function and make one individual. Repeat Step 4 until all individuals are decided.

We make an initial population with 200 individuals.

2.6 Hybrid Operator

For searching new solution in evolution algorithm, the local search can be used. In this research, an insertion operator, a deletion operator and a moving opera-

Table 1. Computation Results

n	Reduction Rate			Number of Steiner points		
	min	average	max	min	average	max
10	4.377	10.611	19.299	2	3.933	6
20	7.462	11.721	14.734	7	9.067	12
30	8.496	11.356	16.542	12	13.867	16
40	8.367	10.807	13.884	15	18.533	22
50	8.628	10.705	12.457	19	22.933	27
60	9.292	11.755	13.940	24	27.933	31
70	9.733	11.163	12.775	29	32.133	34
80	8.957	11.081	13.212	34	37.467	41
90	9.453	11.203	13.000	37	40.600	45
100	9.777	11.501	13.478	41	45.400	51
250	10.016	11.090	11.845	102	111.133	117
500	10.231	10.811	12.054	209	217.000	230

Table 2. Algorithm comparison for RST problem

n	EA^a	Beasley[3]	Lee[9]	Smith 79[11]	Smith 80[9]
10	10.611	9.947	7- 8	7.1	8.316
20	11.721	10.59	8-10	7.621	7.65
30	11.356	10.25	8-10	7.978	8.306
40	10.807	9.956	*	5.887	8.641
50	10.705	9.522	*	*	*
60	11.755	10.146	*	*	*
70	11.163	9.779	*	*	*
80	11.081	9.831	*	*	*
90	11.203	10.128	*	*	*
100	11.501	10.139	*	*	*
250	11.090	9.964	*	*	*
500	10.811	9.879	*	*	*

a : Evolution Algorithm
* : They have not shown the results

tor are introduced. Mutation operator and crossover have the location of Steiner point moved to search a solution, but they don't search new Steiner point. We need to introduce a new Steiner point in current individual to enforce the variability of the solution. For some selected individual, randomly generated Steiner point in Hanan's grid is inserted in individual and increase the number of Steiner point as an insertion operator. On the other hand, some Steiner points in tree are connected with only one or two other node. Those kinds of Steiner point in tree are obviously useless to reduce the tree cost. We introduce deletion operator to delete some Steiner points which are connected less than 2 other node in Steiner tree. For the Steiner point connected with 3 other node, we know the optimal location of Steiner point as using the results of 3-terminal case. We trace the Steiner point (S) which is connected with exactly 3 other node(A,B,C) and calculate the optimal location(S*) of Steiner point for those 3 node(A,B,C) and move the location of S to S* as a moving operator.

2.7 Evolution Algorithm

Our described evolution algorithm is as follows:

Evolution Algorithm Procedure

```
begin
t ← 0
Initialize population P(t)
Fitness evaluation P(t)
while (do not satisfy termination criteria) do
        begin
        t ← t +1
        Selection P(t) from P(t-1)
```

        Recombination on P(t)
        Mutation on P(t)
        Local search on P(t)
        Fitness evaluation P(t)
      end
end

Table 3. Comparison with the optimal solutions

Instances	n	EA[a]	Beasley[3]	Optimal solution[3]
1	5	1.87	1.87	1.87
2	6	1.64	1.68	1.64
3	7	2.36	2.36	2.36
4	8	2.54	2.54	2.54
5	6	2.26	2.29	2.26
6	12	2.42	2.48	2.42
7	12	2.48	2.54	2.48
8	12	2.36	2.42	2.36
9	7	1.64	1.72	1.64
10	6	1.77	1.84	1.77
11	6	1.44	1.44	1.44
12	9	1.8	1.8	1.8
13	9	1.5	1.5	1.5
14	12	2.6	2.6	2.6
15	14	1.48	1.48	*
16	3	1.6	1.6	1.6
17	10	2	2.01	2
18	62	4.04	4.06	*
19	14	1.88	1.9	*
20	3	1.12	1.12	1.12
21	5	1.92	2.16	1.92
22	4	0.63	0.63	0.63
23	4	0.65	0.65	0.65
24	4	0.3	0.3	0.3
25	3	0.23	0.23	0.23
26	3	0.15	0.15	0.15
27	4	1.33	1.33	1.33
28	4	0.24	0.24	0.24
29	3	2	2	2
30	12	1.1	1.1	1.1
31	14	2.59	2.6	*
32	19	3.13	3.23	*
33	18	2.68	2.69	*
34	19	2.43	2.24	*
35	18	1.51	1.54	*
36	4	0.9	0.9	0.9
37	8	0.9	0.9	0.9
38	14	1.66	1.66	*
39	14	1.66	1.66	*
40	10	1.55	1.62	1.55
41	20	2.24	2.24	*
42	15	1.53	1.53	1.53
43	16	2.57	2.66	*
44	17	2.54	2.61	*
45	19	2.2	2.26	*
46	16	1.5	1.5	1.5

a : Evolution Algorithm
* : The optimal solution is unknown

3 Computational Experience

The computational study was made on Pentium IV processor. The evolution algorithm was programmed in Visual Studio. Problem instances are from OR-Library [2], 15 instances for each problem size 10, 20... 100,250,500. By our parameter setting, the crossover rate was 0.09, the mutation rate was 0.01 and the insertion rate was 0.31. The population size was 200. Each test problem was solved our evolution algorithm. The results are shown in Table 1. In Table 1, for each value of n, the minimum, the average, the maximum for the reduction rate and the number of Steiner points. For the RST problem, Beasley has solved same instances as our ones. The other research's results are from the survey of Beasley's research [3]. Those results are in Table2. In Table2, the evolution algorithm introduced in this paper gives larger percentage reduction (about 11%) in most instances than any other heuristics.

In order to compare the evolution algorithm with optimal solutions, the 46 test problem were solved by the evolution algorithm. The 46 test problems were given by Soukup and Chow [12]. Beasley solved this problem and showed some optimal solutions. Table 3 shows the Steiner tree cost of optimal solutions, and those of the evolution algorithm and Beasley's heuristics. In Table 3, the evolution algorithm found every optimal solution which is known. For the unknown optimal solution cases, our solutions are better than Beasley's solution.

4 Conclusion

The rectilinear Steiner tree problem is one of well known optimization problem. In this paper we have introduced an Evolution algorithm for Rectilinear Steiner tree problem as one of meta heurisitcs. Computational results showed that this Evolution algorithm gives better Steiner tree solution than the other heuristics. The average reduction rate of solutions which we found by the Evolution algorithm was 11%, which is almost similar to that of optimal solutions. For the optimal case problems, our Evolution algorithm found every known optimal solution. In class of Stener tree problem, there are Euclidean Steiner tree problem and graph Steiner tree problem. We will modify our Evolution algorithm for the Euclidean and graph Steiner tree problem.

References

1. Bäck, Thomas : Evolutionary Algorithm in Theory and Practice, Oxford University Press (1996)
2. Beasley, J. E.: OR-Library: distributing test problems by electronic mail. *Journal of the Operational Research Society* 41 (1990) 1069-1072
3. Beasley, J. E: A heuristic for Euclidean and rectilinear Steiner problems. *European Journal of Operational Research 58* (1992) 284-292
4. France, R.L.: A note on the optimum location of new machines in existing plant layouts, *J. Industrial Engineering* 14(1963) 57-59

5. Ganley, Joseph L.: Computing optimal rectilinear Steiner trees: A survey and experimental evaluation. *Discrete Applied Mathematics 90* (1999) 161-171
6. Garey, M. R., Johnson, D. S.: The rectilinear Steiner tree problem is NP-complete. *SIAM Journal on Applied Mathematics 32*(1977a) 826-834
7. Hanan, M.: On Steiner's problem with rectilinear distance. *SLAM Journal on Applied Mathematics* 14. (1966) 255-265
8. Hesser, J., Manner R., Stucky,O. : Optimization of Steiner Trees using Genetic Algorithms, *Proceedings of the Third International Conference on Genetic Algorithm* (1989) 231-236
9. Lee, J. L., Bose, N. K., Hwang, F. K.: Use of Steiner's problem in suboptimal routing in rectilinear metric. *IEEE Transaction son Circuits and Systems* 23(1976) 470-476
10. Smith, J. M., Lee. D. T., Liebman, J. S.: An O(n log n) heuristic algorithm for the rectilinear Steiner minimal tree problem. *Engineering Optimization 4* (1980) 172-192
11. Smith, J. M., Liebman, J. S.: Steiner trees, Steiner circuits and the interface problem in building design. *Engineering Optimization 2*(1979)15-36
12. Soukup, J., Chow, W.F.: Set of test problems for the minimum length connection networks. *ACM/SIGMAP Newsletter* 15(1973) 48-51
13. Warme, D.M, Winter, P., Zachariasen, M. : Exact Algorithms for Plane Steiner Tree Problems : A Computational Study In: D.Z. Du, J.M. Smith and J.H. Rubinstein (eds.): Advances in Steiner Tree, Kluser Academic Publishers (1998)

A Two-Stage Recourse Model for Production Planning with Stochastic Demand

K.K. Lai [1,2], Stephen C.H. Leung [1], and Yue Wu [3]

[1] Department of Management Sciences,
City University of Hong Kong, Hong Kong
{mssleung, mskklai}@cityu.edu.hk
[2] College of Business Administration, Hunan University, China
[3] School of Management, University of Southampton, Southampton, UK
Y.Wu@soton.ac.uk

Abstract. Production planning problems play a vital role in the supply chain management. The methodology of production planning problem can provide the quantity of production and the workforce level at each production plant to fulfil market demand. This paper develops a stochastic programming model with additional constraints. A set of data from a multinational lingerie company in Hong Kong is used to demonstrate the robustness and effectiveness of the proposed model.

1 Introduction

Medium-term production planning over a two- to 18-month planning horizon is classified as aggregate production planning (APP) [1]. Nam and Logendran [6] stated that APP is intended to translate forecast sales demand and production capacity into future production loading plans. As the planning takes place at the aggregate level, there is no need for APP to provide detailed material and capacity resource requirements for individual products and detailed schedules for facilities and personnel [1]. Masud and Hwang [5] compared three multiple criteria decision making models for the APP problem with maximizing contribution to profit, while minimizing changes in workforce level, inventory investment and backorders simultaneously using a set of data consisting of two products, a single production plant and eight planning periods. Baykasoglu [1] extended Masud and Hwang's model by adding subcontractor selection and set-up decisions. Hung and Hu [3] formulated a mixed integer programming model for production planning problems with set-up decisions. Due to the NP-hardness of the problems, a series of artificial intelligent approaches such as evolutionary algorithms [4], genetic algorithms [8] and decision-support systems [7] are also widely used to solve production planning problems with additional constraints and limitations.

This study is particularly motivated by the problems faced by a multinational lingerie company whose headquarters is in Hong Kong and sales branch offices, R&D, and customer services are spread across North America and Europe. Production plants are located in China, the Philippines, Thailand and other Southeast Asian countries to take advantage of lower wages and lower rental costs. The headquarters collects sales orders – which consist of the type of products, quantity, delivery date and location preference – through the sales branch offices. The company's products are mainly divided into two groups: cotton products, which contain at least 90% cotton; and silk products, which are mainly made of silk. Normally, labour costs vary for different products and for different production plants. It is not surprising to find that the labour costs incurred when manufacturing silk products are higher than those incurred in manufacturing cotton products because of the greater skill levels involved. In addition, manufacturing location preference plays an important role in the production planning problem. Sales orders from North America must be processed in Chinese production plants because the quality of materials required, such as silk, comes from China. European customers also favour Chinese production plants, but do not object to production plants in the Philippines or Sri Lanka being used as substitutes. In many circumstances, the production manager needs to negotiate between plants and customers to ensure that the manufacturing location preference can be fulfilled. Decision makers have to develop a production loading plan every three months with minimum total cost and stock-outs by considering the manufacturing capacity, workforce level, and other factors. According to the loading plan, the production plants are assigned a list of products with quantities to be produced over each period. Choosing the right level of the right production strategy involves a highly complex set of decisions. Solving production planning problems with uncertain demand data has therefore become a critical management task for the company.

The purpose of this paper is to formulate a two-stage recourse model for production planning problem with stochastic demand, in which the production cost, subcontracting cost, labour cost, inventory cost, hiring cost and lay-off cost, and penalty cost associated with under-fulfilment of realized demand under different economic growth scenarios are minimized. This paper is organized as follows. After this introductory section, the background to the two-stage recourse model is described. A two-stage recourse model is formulated for production planning problem in section 3, and a set of data from a Hong Kong company is used to test the effectiveness and efficiency of the proposed model in section 4. Our conclusions are given in the final section.

2   Framework of Two-Stage Recourse Model

In the following, the framework of two-stage stochastic programming model is briefly described. For detail, the reader is referred to Dantzig [2]. The stochastic linear programming model is expressed as follows:

$$\min c_1^T x_1 + \sum_{s=1}^{S} p_s(q^T y_2^s) \tag{1}$$

$$\text{s.t. } Ax_1 = b \tag{2}$$

$$T^s x_1 + W y_2^s = h^s \quad s = 1, ..., S \tag{3}$$

$$x_1, y_2^s \geq 0 \quad s = 1, ..., S \tag{4}$$

Equations (2) represent the first-stage model and equations (3) represent the second-stage model. x_1 is the vector of first-stage decision variables whish is scenario-independent. The optimal value of x_1 is not conditional on the realization of the uncertain parameters. c_1 is the vector of cost coefficient at the first stage. A is the first-stage coefficient matrix and b is the corresponding right-hand-side vectors. y_2 is the vector of second-stage (recourse) decision variables. q is the vector of cost (recourse) coefficient vectors at the second stage. W is second-stage (recourse) coefficient matrix and h^s is the corresponding right-hand-side vector and T^s is the matrix that ties the two stages together where $s \in \Omega$ represents scenarios in future and p_s is the probability that scenario s occurs. In the second-stage model, the random constraint defined in (3), $h^s - T^s x_1$, is the goal constraint: violations of this constraint are allowed, but the associated penalty cost, $q^T y_2$, will influence the choice of x_1. $q^T y_2$ is the recourse penalty cost or second-stage value function and $\sum_{s=1}^{S} p_s(q^T y_2^s)$ denotes the expected value of recourse penalty cost (second-stage value function).

3 A Two-Stage Recourse Model for Production Planning Problem

The multinational lingerie company in Hong Kong under investigation has to determine the quantity of product i, $i = 1, 2, ..., n$, manufactured from plant j, $j = 1, 2, ..., m$, to fulfil market demand over each period of time t, $t = 1, 2, ..., T$. In addition, the company is required to manufacture higher-quality products from Chinese plants, CF to fulfil some customers' orders. Model parameters and decision variables used throughout the paper are defined as follows.

Parameters:

Deterministic parameters:
C_{Wj} : the regular-time labour cost in plant j (\$/man-period)
C_{Oj} : the overtime labour cost of worker in plant j (\$/man-hour)
C_{Hj} : the cost of hiring one worker in plant j (\$/man)
C_{Lj} : the cost of laying off one worker in plant j(\$/man)
μ_i : the labour time for product i (man-hour/unit)
λ_i : the machine time for product i (machine-hour/unit)
δ : the working hours for each period (man-hour/man-period)
W_{j0} : the initial work force level in plant j (man-period)
W_{jt}^{min} : the minimum work force level available in plant j at period t (man-period)
W_{jt}^{max} : the maximum work force level available in plant j at period t (man-period)
α : the fraction of the work force available for overtime in each period
ε : the fraction of variation in the work force in each period
M_{jt} : the machine time capacity in plant j over period t (machine-hour)
β_t : the fraction of machine time capacity available for overtime use in period t

Recourse parameters:

C_{Pij}^s : the regular-time unit production cost for product i manufactured in plant j under scenario s(\$/unit)

C_{Yij}^s : the overtime unit production cost for product i manufactured in plant j under scenario s(\$/unit)

C_{Zij}^s : the unit production cost to contract product i under scenario s(\$/unit)

C_{Iij}^s : the unit inventory cost to hold product i in plant j at the end of each period under scenario s(\$/unit)

C_{Bi}^s : the unit shortage cost associated with the under-fulfilment of product i at the end of each period under scenario s(\$/unit)

D_{it}^s : the sales volume for product i in period t under scenario s (units)

I_{ij0} : the initial inventory of product i at the start of planning horizon in plant j(units)

θ_i : the fraction of product i that must be manufactured in Chinese plants

Decision variables:

First-stage decision variables:

P_{ijt}: the quantity of product i manufactured in plant j during regular time in period t (units)

Y_{ijt} : the quantity of product i manufactured in plant j during overtime in period t (units)

Z_{it} : the quantity of product i subcontracted during period t (units)

W_{jt} : the number of workers required in plant j during period t (man-period)

H_{jt} : the number of workers hired in plant j during period t (man-period)

L_{jt} : the number of workers laid-off in plant j during period t (man-period)

Second-stage decision variables:

I_{ijt}^s : the inventory of product i in plant j at the end of period t under scenario s (units)

B_{it}^s : the under-fulfilment of product i in period t under scenario s (units)

First-stage model:

$$\text{Min} \sum_{j \in J} \sum_{t \in T} \left(C_{Wj} \cdot W_{jt} + C_{Oj} \cdot \sum_{i \in I} \mu_i Y_{ijt} \right) + \sum_{j \in J} \sum_{t \in T} (C_{Hj} \cdot H_{jt} + C_{Lj} \cdot L_{jt}) \tag{5}$$

s.t.

$$W_{jt} = W_{jt-1} + H_{jt} - L_{jt} \quad \forall j, t \tag{6}$$

$$W_{jt}^{\min} \leq W_{jt} \leq W_{jt}^{\max} \quad \forall j, t \tag{7}$$

$$H_{jt} + L_{jt} \leq \varepsilon W_{jt-1} \quad \forall j, t \tag{8}$$

$$\sum_{i \in I} \mu_i \cdot P_{ijt} \leq \delta W_{jt} \quad \forall j, t \tag{9}$$

$$\sum_{i \in I} \mu_i \cdot Y_{ijt} \leq \delta \alpha W_{jt} \quad \forall j, t \tag{10}$$

$$\sum_{i \in I} \lambda_i \cdot P_{ijt} \leq M_{jt} \quad \forall j, t \tag{11}$$

$$\sum_{i \in I} \lambda_i \cdot Y_{ijt} \leq \beta_t M_{jt} \quad \forall j, t \tag{12}$$

$$P_{ijt}, Y_{ijt}, Z_{ijt}, W_{jt}, H_{jt}, L_{jt} \geq 0 \quad \forall i, j, t \tag{13}$$

The first component in expression (5) is the labour cost, which is associated with regular-time and overtime workers respectively. The last component is total hiring and laying off cost associated with changes in the workforce level. Constraint (6) ensures that the available workforce in any period equals the workforce from the previous period plus any change in workforce level during the current period. The change in workforce level may be due to either hiring extra workers or laying off redundant workers. It is noted that $H_{jt} \cdot L_{jt} = 0$ because either the net hiring or the net laying-off of workers takes place over a period, but not both. Constraint (7) ensures the upper- and lower-bound of change in workforce level over a period are provided. Constraint (8) ensures that the change in workforce level cannot exceed the proportion of workers employed during the previous period. Constraint (9) limits the regular-time production to the workers available. Constraint (10) limits the overtime hours of the available workers. Total production during each period by regular-time workers and overtime workers is limited by the available machine capacity, as shown by constraints (11) and (12) respectively. Constraint (13) ensures that all decision variables are non-negative.

Second-stage model:

$$Min \sum_{s \in S} p_s \left(\sum_{i \in I} \sum_{j \in J} \sum_{t \in T} (C^s_{Pij} \cdot P_{ijt} + C^s_{Yij} \cdot Y_{ijt} + C^s_{Zij} \cdot Z_{ijt}) \right)$$

$$+ \sum_{s \in S} p_s \left(\sum_{i \in I} \sum_{j \in J} \sum_{t \in T} C^s_{Iij} \cdot I^s_{ijt} \right) \tag{14}$$

$$+ \sum_{s \in S} p_s \left(\sum_{i \in I} \sum_{t \in T} C^s_{Bi} B^s_{it} \right)$$

s.t.

$$\sum_{j \in J} \left(I_{ijt-1}^s + P_{ijt} + Y_{ijt} + Z_{ijt} - I_{ijt}^s \right) = D_{it}^s - B_{it}^s \quad \forall i, t, s \quad (15)$$

$$\sum_{j \in CF} \left(P_{ijt} + Y_{ijt} + Z_{ijt} \right) = \theta_i \left(D_{it}^s - B_{it}^s \right) \quad \forall i, t, s \quad (16)$$

$$I_{ijt}^s, B_{it}^s \geq 0 \quad \forall i, j, t, s \quad (17)$$

Table 1. Unit production cost and unit inventory cost (in HK\$, 1US\$ = 7.8HK\$)

Plant, j	Situation, s	Production cost of regular-time worker, C_{Pij}^s, \$/unit						Production cost by overtime worker, C_{Yij}^s, \$/unit					
		Product, i											
		1	2	3	4	5	6	1	2	3	4	5	6
1	Boom	70	80	100	110	130	140	160	170	190	200	220	230
	Good	50	60	80	90	110	120	140	150	170	180	200	210
	Fair	40	50	70	80	100	110	130	140	160	170	190	200
	Poor	30	40	60	70	90	100	120	130	150	160	180	190
2	Boom	75	85	105	115	135	145	165	175	195	205	225	235
	Good	55	65	85	95	115	125	145	155	175	185	205	215
	Fair	45	55	75	85	105	115	135	145	165	175	195	205
	Poor	35	45	65	75	95	105	125	135	155	165	185	195
3	Boom	80	90	110	120	140	150	170	180	200	210	230	240
	Good	60	70	90	110	120	130	150	160	180	190	210	220
	Fair	50	60	80	90	110	120	140	150	170	180	200	210
	Poor	40	50	70	80	100	110	130	140	160	170	190	200

Plants, j	Situation, s	Production cost of subcontracting, C_{Zij}^s, \$/unit						Inventory cost, C_{Iij}^s, \$/unit					
		Product, i											
		1	2	3	4	5	6	1	2	3	4	5	6
1	Boom	190	200	220	230	250	260	20	25	35	40	50	55
	Good	170	180	200	210	230	240	15	20	30	35	45	50
	Fair	160	170	190	200	220	230	12	17	27	32	42	47
	Poor	150	160	180	190	210	220	10	15	25	30	40	45
2	Boom	190	200	220	230	250	260	18	23	33	38	48	53
	Good	170	180	200	210	230	240	13	18	28	33	43	48
	Fair	160	170	190	200	220	230	10	15	25	30	40	45
	Poor	150	160	180	190	210	220	8	13	23	28	38	43
3	Boom	190	200	220	230	250	260	15	20	30	35	45	50
	Good	170	180	200	210	230	240	10	15	25	30	40	45
	Fair	160	170	190	200	220	230	7	12	22	27	37	42
	Poor	150	160	180	190	210	220	5	10	20	25	35	40

Plant, j	Situation, s	Shortage cost, C_{Bi}^s, \$/unit					
		Product, i					
		1	2	3	4	5	6
1	Boom	350	400	500	550	650	700
	Good	300	350	450	500	600	650
	Fair	250	300	400	450	550	600
	Poor	200	250	350	400	500	550

The first component in expression (14) is the production cost, which is associated with the regular-time production, overtime production and subcontracting cost. The second component is the inventory cost associated with the storage of units of products in the warehouses for a period of time. The last component is the penalty cost associated with under-fulfilment of demand. Constraint (15) determines either the quantity of products stored in the warehouse

Table 2. Labour cost, hiring cost and lay-off cost

Plant, j	Labour cost of worker at regular time, C_{Wj}, $/man-period	Labour cost of worker at overtime, C_O, $/hour	Hiring cost, C_{Hj}, $/man	Lay-off cost, C_{Lj}, $/man	Minimum workforce	Change rate, ε	Initial workforce
1	250	10	100	120	300	0.40	300
2	225	9	90	110	300	0.45	300
3	200	8	80	100	300	0.50	300

Table 3. Market demand data ('000)

Product, i	Situation, s	Period, t											
		1	2	3	4	5	6	7	8	9	10	11	12
1	Boom	1.5	2.0	1.6	1.7	2.0	1.8	1.7	1.9	2.0	2.3	2.4	2.5
	Good	1.1	1.6	1.2	1.3	1.6	1.4	1.3	1.5	1.6	1.9	2.0	2.1
	Fair	0.7	1.2	0.8	0.9	1.2	1.0	0.9	1.1	1.2	1.5	1.6	1.7
	Poor	0.3	0.8	0.4	0.5	0.8	0.6	0.5	0.7	0.8	1.1	1.2	1.3
2	Boom	1.5	2.0	1.6	1.7	2.0	1.8	1.7	1.9	2.0	2.3	2.4	2.5
	Good	1.1	1.6	1.2	1.3	1.6	1.4	1.3	1.5	1.6	1.9	2.0	2.1
	Fair	0.7	1.2	0.8	0.9	1.2	1.0	0.9	1.1	1.2	1.5	1.6	1.7
	Poor	0.3	0.8	0.4	0.5	0.8	0.6	0.5	0.7	0.8	1.1	1.2	1.3
3	Boom	1.4	1.9	1.6	1.5	1.9	1.7	1.6	1.7	1.9	2.3	2.3	2.3
	Good	1.1	1.6	1.3	1.2	1.6	1.4	1.3	1.4	1.6	2.0	2.0	2.0
	Fair	0.8	1.3	1.0	0.9	1.3	1.1	1.0	1.1	1.3	1.7	1.7	1.7
	Poor	0.5	1.0	0.7	0.6	1.0	0.8	0.7	0.8	1.0	1.4	1.4	1.4
4	Boom	1.4	1.9	1.6	1.5	1.9	1.7	1.6	1.7	1.9	2.3	2.3	2.3
	Good	1.1	1.6	1.3	1.2	1.6	1.4	1.3	1.4	1.6	2.0	2.0	2.0
	Fair	0.8	1.3	1.0	0.9	1.3	1.1	1.0	1.1	1.3	1.7	1.7	1.7
	Poor	0.5	1.0	0.7	0.6	1.0	0.8	0.7	0.8	1.0	1.4	1.4	1.4
5	Boom	1.3	1.8	1.6	1.3	1.8	1.6	1.5	1.5	1.8	2.3	2.2	2.1
	Good	1.1	1.6	1.4	1.1	1.6	1.4	1.3	1.3	1.6	2.1	2.0	1.9
	Fair	0.9	1.4	1.2	0.9	1.4	1.2	1.1	1.1	1.4	1.9	1.8	1.7
	Poor	0.7	1.2	1.0	0.7	1.2	1.0	0.9	0.9	1.2	1.7	1.6	1.5
6	Boom	1.3	1.8	1.6	1.3	1.8	1.6	1.5	1.5	1.8	2.3	2.2	2.1
	Good	1.1	1.6	1.4	1.1	1.6	1.4	1.3	1.3	1.6	2.1	2.0	1.9
	Fair	0.9	1.4	1.2	0.9	1.4	1.2	1.1	1.1	1.4	1.9	1.8	1.7
	Poor	0.7	1.2	1.0	0.7	1.2	1.0	0.9	0.9	1.2	1.7	1.6	1.5

or the shortfall in meeting market demand. If the total quantity of products produced at the company's plants and supplied from subcontractors during period t plus previous stock at period $t-1$ (i.e. $I^s_{ijt-1} + P_{ijt} + Y_{ijt} + Z_{ijt}$) is greater than market demand D^s_{it}, then the stock at period t will be equal to $I^s_{ijt} = I^s_{ijt-1} + P_{ijt} + Y_{ijt} + Z_{ijt} - D^s_{it}$ and, under minimization, the deviation $B^s_{it} = 0$; whereas if $I^s_{ijt-1} + P_{ijt} + Y_{ijt} + Z_{ijt}$ is less than market demand, then $I^s_{ijt} = 0$ and $B^s_{it} = D^s_{it} - I^s_{ijt-1} - P_{ijt} - Y_{ijt} - Z_{ijt}$, indicating that market demand is not satisfied. Thus a penalty cost, $C^s_{Bi} \cdot B^s_{it}$, is incurred. Constraint (16) ensures that actual sales of products (i.e. $D^s_{it} - B^s_{it}$) are manufactured from Chinese plants. Constraint (17) ensures that the second-stage decision variables are non-negative.

4 Computational Results

The multinational lingerie company, under the study, in Hong Kong receives sales orders from its sales branches covering America and Europe. Each order may require one or more of six products, $i = 1, 2, \ldots, 6$. The products are manufactured in three main plants, $j = 1, 2, 3$, located in China, the Philippines and Thailand. The planning horizon covers 12 weeks, $t = 1, 2, \ldots, 12$. It is assumed that future economic conditions will fit into one of four possible situations – boom, good, fair and poor, with associated probabilities of 0.40, 0.25, 0.20 and 0.15 respectively.

Table 4. Machine capacity data ('000)

Plant, j	Period, t					
	1	2	3	4	5	6
1	500	400	450	550	400	450
2	400	350	350	500	300	350
3	300	300	250	450	200	250
Plant, j	Period, t					
	7	8	9	10	11	12
1	500	400	500	550	400	400
2	400	350	400	450	300	350
3	300	300	300	350	200	300

Table 5. Machine time data

Product, i	Labour production time, μ_i hour/unit, of worker	Machine time, λ_i hour/unit, of worker
1	1	0.75
2	1	0.75
3	1.5	1
4	1.5	1
5	2	1.5
6	2	1.5

Table 6. Production loading plan

	Plant, j	Product, i	Period, t											
			1	2	3	4	5	6	7	8	9	10	11	12
Regular-time production	1	1	1180	1520	1200	860	1040	1800	1280	900	1220	1900	2000	0
		2	740	1120	1200	1300	1600	840	880	1500	940	980	640	2160
		3	0	0	0	0	0	0	0	0	0	0	0	0
		4	0	0	0	0	0	0	0	0	0	0	0	0
		5	0	0	0	0	0	0	0	0	0	0	0	0
		6	0	0	0	0	0	0	0	0	0	0	0	0
	2	1	0	0	0	520	480	0	0	600	0	0	0	1820
		2	360	480	0	0	0	560	420	0	660	920	1760	340
		3	1040	1440	1300	1093	1440	1387	1160	0	0	1307	0	0
		4	0	0	300	0	0	0	0	1200	1000	0	587	0
		5	0	0	0	0	0	0	0	0	0	0	0	0
		6	0	0	0	0	0	0	0	0	0	0	0	0
	3	1	0	0	0	0	0	0	0	0	0	0	0	280
		2	0	0	0	0	0	0	0	0	0	0	0	0
		3	60	160	0	107	160	13	140	1400	1272	117	1760	1253
		4	1100	1600	1000	1200	1600	1400	1300	200	168	1803	0	0
		5	0	0	450	100	0	260	0	0	0	0	0	0
		6	90	0	0	0	0	0	0	0	0	0	0	0
Overtime production	1	1	0	0	0	0	0	0	0	0	0	0	0	0
		2	0	0	0	0	0	0	0	0	0	0	0	0
		3	0	0	0	0	0	0	0	0	0	576	0	0
		4	0	0	0	0	0	0	0	0	432	0	528	432
		5	0	396	230	324	396	0	324	360	0	0	0	0
		6	288	0	130	0	0	396	0	0	0	0	0	0
	2	1	0	0	0	0	0	0	0	0	0	0	0	0
		2	0	0	0	0	0	0	0	0	0	0	0	0
		3	0	0	0	0	0	0	0	0	328	0	240	0
		4	0	0	0	0	0	0	0	0	197	288	432	
		5	0	396	360	324	396	396	324	360	78	284	0	0
		6	288	0	0	0	0	0	0	0	0	0	0	0
	3	1	0	0	0	0	0	0	0	0	0	0	0	0
		2	0	0	0	0	0	0	0	0	0	0	0	0
		3	0	0	0	0	0	0	0	0	0	0	0	0
		4	0	0	0	0	0	0	0	0	0	0	528	432
		5	288	0	360	0	396	396	324	360	0	432	0	0
		6	0	396	0	324	0	0	0	0	324	0	0	0
Subcontract		1	0	0	0	0	0	0	0	0	0	0	0	0
		2	0	0	0	0	0	0	0	0	0	0	0	0
		3	0	0	0	0	0	0	0	0	0	0	0	747
		4	0	0	0	0	0	0	0	0	0	0	69	704
		5	812	808	0	352	412	348	328	220	1522	1384	2000	2100
		6	434	1204	1270	776	1600	1004	1300	1300	1276	2100	2000	2100
Hiring			30	45	0	0	30	0	0	15	0	45	0	0
Lay-off			0	0	15	15	0	0	30	0	15	0	15	30
Workforce			120	165	150	135	165	165	135	150	135	180	165	135

Table 7. Breakdown of costs

Production cost	Labour cost	Inventory cost	Hiring and lay-off cost	Operational cost	Penalty cost	Total cost
14,869,557	638,280	1,426,638	28,050	16,962,525	4,012,000	20,974,525

Table 1 shows the production cost and inventory cost for different products in each plant and the shortage costs with regard to different economic situations. Table 2 show the labour cost, hiring cost and lay-off costs, and the workforce data with regard to different plants. The estimated product demands under different economic conditions are shown in Table 3. Lastly, Table 4 shows the machine capacity of the three plants, while Table 5 shows the labour production time and machine time.

Table 6 reports the optimal production loading plan. It can be seen that the majority of products are produced by regular-time labour. In particular, all products 1 and 2 are come from regular-time production. It can also be noted that plant 1 is mainly used to produce products 1 and 2; plant 2 products 2 and 3; and plant 3 products 3 and 4. Products 5 and 6 are produced by overtime production and supplied by subcontractors. Lastly, the workforce level in each period attains the upper-bound limit. The corresponding number of workers hired and laid off can also be found in Table 6.

The breakdown of costs incurred is listed in Table 7. The operational cost, which is the sum of production cost, labour cost, inventory cost and hiring cost and lay-off cost, is $16,962,525. Clearly, when the demand requirements are smaller than the available products (from previous inventory, current production and subcontractor), the stock will be kept at the end of particular period t under scenario s, and the corresponding inventory cost will be incurred. On the other hand, when the demand requirements are not satisfied, the company's service level and goodwill will be damaged. Compensation may be considered to cover the excess demand. In this study, the penalty cost is considered. Under the optimal production loading plan, the penalty cost is $4,012,000. Overall, the total cost, which is the sum of operational and penalty cost, is $20,974,525.

5 Conclusions

In this paper, we developed a two-stage recourse model for production planning problem with stochastic demand. A set of data from a multinational lingerie company in Hong Kong is used to demonstrate the robustness and effectiveness of the proposed model. It is observed that the model can provide a credible and effective methodology for real-world production planning problems in an uncertain environment. However, there is still room for improvement and investigation. First, sensitivity analysis on the cost parameters in the objective function may be conducted to test the trade-off between total cost and stockouts. Second, the selection of scenarios of economic conditions could be further investigated.

Acknowledgement

The work described in this paper was supported by an Annual Grant of University of Southampton, UK.

References

1. Baykasoglu, A.: MOAPPS 1.0: Aggregate Production Planning Using the Multiple-Objective Tabu Search. International Journal of Production Research 39 (2001) 3685-3702.
2. Dantzig, G.B.: Linear Programming Under Uncertainty. Management Science 1 (1955) 197–206.
3. Hung, Y.F., Hu, Y.C.: Solving Mixed Integer Programming Production Planning Problems with Setups by Shadow Price Information. Computers and Operations Research 25 (1998) 1027-1042.
4. Hung, Y.F., Shih, C.C., Chen, C.P.: Evolutionary Algorithms for Production Planning Problems with Setup Decisions. Journal of the Operational Research Society 50 (1999) 857-866.
5. Masud, A.S.M., Hwang, C.L.: An Aggregate Production Planning Model and Application of Three Multiple Objective Decision Methods. International Journal of Production Research 18 (1980) 741-752.
6. Mazzola, J.B., Neebe, A.W., Rump, C.M.: Multiproduct Production Planning in the Presence of Work-Force Learning. European Journal of Operational Research 10 (1998) 336-356.
7. Nam, S.J., Logendran, R.: Aggregate Production Planning – A Survey of Models and Methodologies. European Journal of Operational Research 61 (1992) 255-272.
8. Ozdamar, L., Bozyel, M.A., Birbil, S.I.: A Hierarchical Decision Support System for Production Planning (with Case Study). European Journal of Operational Research 104 (1998) 403-422.
9. Wang, D., Fang, S.C.: A Genetics-Based Approach for Aggregated Production Planning in a Fuzzy Environment. IEEE Transactions on Systems, Man, and Cybernetics – Part A: Systems and Human 27 (1997) 636-645.

A Hybrid Primal-Dual Algorithm with Application to the Dual Transportation Problems

Gyunghyun Choi* and Chulyeon Kim

Dept. of Industrial Engineering, Hanyang University,
17 Haengdang-dong Seongdong-gu, Seoul, Korea
ghchoi@hanyang.ac.kr

Abstract. Subgradient optimization methods provide a valuable tool for obtaining a lower bound of specially structured linear programming or linear programming relaxation of discrete optimization problems. However, there is no practical rule for obtaining primal optimal solutions from subgradient-based approach other than the lower bounds. This paper presents a class of procedures to recover primal solutions directly from the information generated in the process of using subgradient optimization methods to solve such Lagrangian dual formulations. We also present a hybrid primal dual algorithm based on these methods and some computational results.

1 Introduction

Lagrangian dual(LD) methods are widely used approaches for solving specially structured constrained problems through the process of dualizing complicated constraints in order to obtain easier subproblems. These subproblems give a lower bound on the original objective function value. There are several issues that we need to consider in Lagrangian dual approaches. Among them, dualization strategies, solution methods, and primary recovery techniques are come of the most important ones.

Consider a linear programming problem stated as follow:

$$LP: \quad \min\{\mathbf{cx} : \mathbf{Ax} \leq \mathbf{b}, \mathbf{x} \in \mathbf{X}\} \tag{1}$$

where A is $m \times n$, $c \in R^n$, $b \in R^n$ and X is a nonempty polytope in R^n. We assume that X is so specially structured that it is relatively easy to solve linear programming problems over simply the set X.

While the subgradient optimization approach can be quite powerful in providing a quick lower bound on LP via the solution of LD, the disadvantage is that a primal optimal solution to LP is not usually available via this scheme. Such a primal solution is of importance not only when LP itself is the primary

---

* Corresponding Author.

problem of interest, but also in the context of branch and bound approaches where a solution to the linear programming relaxation LP might be required.

In this paper, we propose some practical primal recovery schemes, which use a specific convex combination weight strategy to recover primal feasible and optimal solutions under the fixed target value method to guarantee a rapid dual convergence as a suitable step length. If we need to further polish the primal solution to achieve near feasibility and optimality, we adopt the penalty function method.

2 Recovering Primal Solutions

Consider Problem LP of the form (1) and its Lagrangian dual and suppose that LD is solved using a subgradient approach under some suitable rule for selecting the step lengths $\lambda_k > 0$. At the k-th iteration of the dual subgradient method, let us define

$$X_k = \sum_{j=1}^{k} \mu_j^k X_{\pi_j} \quad where \quad \mu_j^k \geq 0 \quad for \quad 1 \leq j \leq k \quad and \quad \sum_{j=1}^{k} \mu_j^k = 1. \quad (2)$$

Hence, each X_k is a convex combination of optimal solutions X_{π_j} to the Lagrangian subproblems. Shor[5] presents a primal convergence theorem using a step-length rule given by the divergent series and the convex combination weight given by

$$\mu_j^k = \frac{\lambda_j}{\sum_{j=1}^{k} \lambda_j} \quad for \ all \ \ 1 \leq j \leq k \ . \quad (3)$$

Following this, Larsson and Liu[1] provided a similar results to that of Shor's by using the average weighting rule

$$\mu_j^k = \frac{1}{k} \quad for \ all \ \ 1 \leq j \leq k, \quad (4)$$

under a special step-length choice $\lambda_k = a/(b+ck)$ for the dual subgradient method, where and are some numbers. The average weighting rule (4) seems more reasonable than that of (3). However, Larsson and Liu's step-length rule is unpopular and untested for general problems. In other words, no computational performance for general Lagrangian dual problem has been reported. Their step-length rule has been used in their primal-dual heuristic algorithm for multicommodity flow problems. Considering the weakness of Larsson and Liu's procedure, Sherali and Choi[3] provides a primal convergence theorem for a wider class of step-length rules and convex combination weighting rules. To this end, let us define for each k,

$$\gamma_{jk} = \frac{\mu_j^k}{\lambda_j} \quad for \ \ j=1,...,k, \quad (5)$$

where λ_j and μ_j^k are the step-lengths and the convex combination weights, respectively, and let $\triangle\gamma_k^{max} = maximum\{\gamma_{jk} - \gamma_{(j-1)k} : 2 \leq j \leq k\}$.

Now, consider the following results.

Theorem 1 (Sherali and Choi[3]). *Suppose that the subgradient method operated under a suitable step-length rule attains dual convergence to some feasible solution, that is, $\pi_k \to \bar{\pi}$ as $k \to \infty$ for some $\bar{\pi} \geq 0$. If the step lengths λ_k and the convex combination weights μ_j^k, for all j and k are chosen to satisfy:*

(i) $\gamma_{jk} \geq \gamma_{(j-1)k}$ for all $j = 2, ..., k$, for each k
(ii) $\triangle\gamma_k^{max} \to 0$ as $k \to \infty$, and
(iii) $\gamma_{jk} \to 0$ as $k \to \infty$ and $\gamma_{kk} \leq \delta$ for all k, for some $\delta > 0$.

Then any accumulation point $\bar{x}$ of the sequence $\{x_k\}$ is feasible and $\bar{x}$ and $\bar{\pi}$ are optimal solutions to the primal and dual problems LP and LD, respectively.

The convergence theorem asserts that so long as the dual iterates are convergent, and the step lengths and the convex combination weights satisfy some conditions, the corresponding sequence of primal iterates will produce a feasible and optimal solution. Now, we present similar convergence results with using the most promising dual convergence step-length rule (6) and some other convex weighting rule (7). The proof is very straightforward and is omitted.

Theorem 2. *Let the step lengths λ_k and the convex combination weights μ_j^k for all j and k, are given by the follows.*

$$\lambda_k = \beta_k \frac{w - \theta(\pi_k)}{\|g_k\|^2} \quad for \ all \ k, \tag{6}$$

where w is a target value and $0 \leq \varepsilon_1 \leq \beta_k \leq \varepsilon_2 \leq 2$, and

$$\mu_j^k = \frac{\lambda_j}{\sum_{t=1}^{k} t\lambda_t} \quad for \ all \ j = 1, ..., k, \ for \ all \ k. \tag{7}$$

Then the primal-dual convergence property of Theorem 1 is attained.

Theorem 2 shows that Shor's rule can be extended by changing the divergent series step-length rule into fixed target value method(6) as a suitable step length. This method has a disadvantage that it assigns larger weights to earlier subproblem solutions and provides a similar result to that of Shor's. To solve this problem, we propose a specific primal recovery scheme in the following Theorem 3.

Theorem 3. *Let the step lengths λ_k and the convex combination weights μ_j^k for all j and k, are given by the follows.*

$$\lambda_k = \beta_k \frac{w - \theta(\pi_k)}{\|g_k\|^2} \quad for \ all \ k,$$

where w is a target value and $0 \leq \varepsilon_1 \leq \beta_k \leq \varepsilon_2 \leq 2$, and

$$\mu_j^k = \frac{j\lambda_j}{\sum_{t=1}^{k} t\lambda_t} \quad for\ all\ \ j=1,...,k,\ \ for\ all\ \ k. \tag{8}$$

Then the primal-dual convergence property of Theorem 1 is attained.

Note that every accumulation point of sequence the primal iterates generated at this scheme is an optimal solution to the primal problem. Moreover, it is easy to implement, using only the step length used in the subgradient algorithm itself. However, although this scheme satisfies primal convergence theorem, a primal solution obtained at some maximum iteration limits could be neither optimal nor even feasible to the primal problem. Hence, in next, we show that it is possible to achieve near feasible and optimal solutions by primal penalty function method.

3   A Hybrid Primal-Dual Algorithm

Note that our primal recovery schemes in Section2 guarantee only that every accumulation point of the sequence of the primal iterates $\{x_k\}$ is an optimal solution to the primal problem. Hence, a primal solution obtained at some maximum allowed iteration limit could be neither optimal nor even feasible to the primal problem. Therefore, in addition to the previous primal recovery scheme, we might need to further polish the primal solution to achieve near feasibility and optimality. Toward this end, we adopted the penalty function method of Sen and Sherail[2] and Sherali and Ulular [4]at the final stage of the proposed algorithm.

For a given dual solution $\bar{\pi} \geq 0$ that is obtained throughout the primal recovery scheme presented in Section 2, let us define the penalty function

$$h(\mathbf{x}) = \mathbf{c}^t \mathbf{x} + \sum_{i=1}^{m} (\bar{\pi}_i + \mathtt{w})\mathtt{max}\{0, (a_i^t \mathbf{x} - b_i)\} \tag{9}$$

where $\bar{\pi}_i$ denotes the i-th component of the vector $\bar{\pi}$ and $w > 0$ is a penalty parameter, and a_i4 and $b_i$are, respectively, the i-th row of the matrix $\mathbf{A}$ and the i-th component of $\mathbf{b}$.

The following Theorem 4 shows the convergence results of the penalty function method.

Theorem 4. *Suppose that the penalty parameter w is selected as $w = 2M + \triangle$ where $M = maximum\ \{(\bar{\pi})_i : i = 1,...,m\}$ and $\triangle$ is some constant.*

If $\theta(\bar{\pi}) \geq h(\mathbf{x}_k) - \varepsilon$ for some $\mathbf{x}_k \in \mathbf{X}$ and for some $\varepsilon > 0$, then we have

$$\mathtt{a}_j \mathbf{x}_k - \mathtt{b}_j \leq \frac{\varepsilon}{M} \quad for\ all\ \ 1 \leq j \leq m\ \ and\ \ |\bar{\pi}^t(\mathbf{A}\mathbf{x}_k - \mathtt{b})| \leq m\varepsilon\ .$$

Now, based on the contents in Section 2 and 3, let us formally design the hybrid primal dual algorithm. Algorithm for Recovering of Primal Solutions proposed in this paper consists of three stages. First, in the stage I, the algorithm attempts to solve the Lagrangian dual problem using a subgradient-based algorithm in conjunction with the fixed target value method (6). After a certain number of iterations, or when some other stopping criterion is satisfied, the algorithm turns to its second stage. In the stage II, the algorithm not only continues to perform the dual procedure as in the first stage, but also generates a sequence of updated primal solutions using some convex combination of the Lagrangian subproblem solutions. At the end of this stage, the algorithm evaluates the extent of feasibility and optimality of the incumbent primal solution using the available dual object function value. Since the objective of this research is on finding primal optimal solution, we may need any convergent subgradient-based algorithm during Stage I. If necessary, the algorithm then applies a penalty function method to the primal problem in the stage III to further improve the primal solution toward a near feasible and optimal solution.

4 Implementation and Computational Experience

For computational test runs, we have attempted to solve some transportation problems without exploiting any special structures in the Lagrangian dual scheme. This problem is to determine a feasible "shipping pattern" from origin to destinations that minimizes the total transportation cost. Let us assume that there are m supply nodes having respective supplies S_i for each $1 \leq j \leq m$, and n sink nodes having respective demands D_j for each $1 \leq j \leq n$. Denote $I(j)$ and $J(i)$ to be, respectively, the set of sink nodes that are linked to supply node i. Also, let us denote x_{ij} to be the flow from supply nodes i to the sink node j, at a corresponding cost c_{ij} units. Then, the dual transportation problem becomes to maximize $\{\theta(\pi)\}$, where

$$\theta(\pi) = Minimum \sum_{i \in I(j)} \sum_{j \in J(i)} (c_{ij} - \pi_j) x_{ij} + \sum_{j \in J(i)} \pi_{ij} D_j$$

$$\text{subject to} \sum_{j \in J(i)} x_{ij} = S_i \quad for \ all \ 1 \leq i \leq m$$

$$0 \leq x_{ij} \leq U_{ij} \quad for \ all \ i, j.$$

The test problems are randomly generated using a standard linear programming generation scheme so that the 10 problems have arcs between 5,000 and 20,000. For the computational test, the algorithm is coded in Visual C++ and is executed on Pentium 300MHz / 128 M DRAM IBM Compatable PC. Let $D(\%)$ and $P(\%)$, respectively, represent the dual and primal optimalities in percentage that were achieved during the procedure, and are computed as follows.

$$D = \left[1 - \frac{|f^* - \theta(\bar{\pi})|}{f^*}\right] \times 100^{(\%)}, \quad P = \left[1 - \frac{|\bar{f} - f(*)|}{f^*}\right] \times 100^{(\%)}$$

Moreover, we also present the magnitude of constraint violations at termination, by specifying the average violation magnitude, denote as AVG-VIO and the feasibility of primal solution recovered by some primal recovery schemes as presented in Section2 denoted as $F(\%)$. These are computed as follows.

$$AVG-VIO = \frac{\sum_{j \in V}\left|\sum_{i \in I(j)} x_{ij} - D_j\right|}{|V|} \text{ and } F = \left[1 - \frac{AVG-VIO}{AVG-DEM}\right] \times 100(\%)$$

where $|V|$ denotes the cardinality of the set, V, of the violated constraints. Table 1 presents the computational results obtained from Stage I-II using the primal recovery schemes proposed in Theorem 3.

Table 1. Computational Results for Stage I-II

PROB	f^*	$\bar{z}$	$D(\%)$	$\bar{f}$	$P(\%)$	$AVG-VIO$	$F(\%)$	CPU
TR1	113900	113855	99.96	113581	99.72	16.50	93.35	0.681
TR2	171000	170993	99.99	172202	99.30	42.82	87.52	1.052
TR3	227000	226743	99.89	231316	98.10	92.23	79.58	1.342
TR4	244200	243696	99.79	242976	99.50	99.74	80.15	1.512
TR5	336700	306568	99.93	305172	99.47	154.75	74.15	1.873
TR6	238300	237779	99.78	236397	99.20	91.32	81.83	1.493
TR7	262900	262044	99.67	263487	99.78	64.70	87.14	1.522
TR8	304100	303275	99.73	302020	99.32	188.85	68.57	1.823
TR9	366000	364860	99.69	361139	98.67	101.64	79.75	2.634
TR10	526300	524155	99.59	530823	99.14	151.96	69.63	3.775

The computational results show that the dual solution obtained at Stage I-II lies within 99.80 % of average optimality. Moreover, the primal recovery schemes presented in Theorem 2 and 3 produce primal solutions that lie on near optimal objective contours. The average optimality obtained is 99.20 % and 99.22%, respectively. This result indicates that schemes used in Theorem 2 and 3 can be quite powerful in providing a lower bound on LP. In case of scheme presented in Theorem 2, the mean value of F(%), the average constraint violation AVG-VIO as a percent of the average demand AVG-DEM, is 70.10 %. In case of scheme presented in Theorem 3, the mean value of F(%) have improved from 70.10% to 80.17%, compared with results of Theorem 2. However, we note that none of the 10 problems has satisfied the near-feasibility and near-optimality termination rules during the iterations of Stage II.

Table 2 presents the results for Stage III. Here, $\hat{f}$ represents the primal objective function value at termination and T(%) represent the primal optimalities in percentage that were achieved during the Stage III, and are computed similarly as before. ITR(CPU) indicates the number of iterations and the corresponding CPU time (in second) at termination.

Throughout these test runs, we have observed that the performance of the primal penalty function method depends very much on the quality of the dual

Table 2. Computational Results for Stage III

PROB	f^*	h(x)	$\hat{f}$	$T(\%)$	AVG-VIO	$F(\%)$	ITR	CPU
TR1	113900	114441	113600	99.74	0.26	99.89	27	0.771
TR2	171000	171938	169335	99.03	0.81	99.76	75	1.342
TR3	227000	227937	223215	98.33	1.52	99.66	68	1.682
TR4	244200	245130	238877	97.82	1.95	99.61	58	1.843
TR5	336700	308328	302402	98.57	1.93	99.68	61	2.293
TR6	238300	239178	234006	98.20	1.66	99.67	81	1.923
TR7	262900	263600	258160	98.20	1.67	99.67	71	1.903
TR8	304100	305008	298758	98.24	1.94	99.68	35	2.073
TR9	366000	366787	358549	97.96	1.70	99.66	55	3.094
TR10	526300	526992	519228	98.66	1.26	99.75	45	4.396

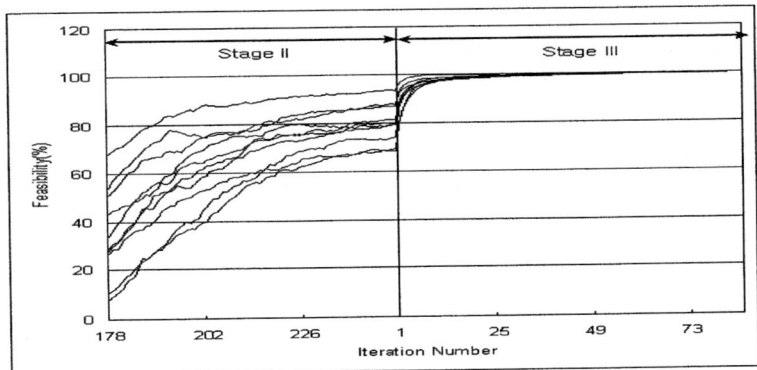

Fig. 1. Feasibility for Stage II and Stage III of Hybrid Primal Dual Algorithm

incumbent solution, as one might expect, since this influences the definition of the penalty function as well as the determination of the step-length. In an overall examination, the penalty function method seems to perform well over Stage II in the sense that the constraint violation magnitudes are reduced significantly. This fact indicates that feasibility is improved significantly at Stage III satisfying termination criterion for the relative duality gap between penalty function value and dual objective value.

Figure 1 shows that sequence of primal solution that a hybrid primal dual algorithm produces convergence to feasible point in this context.

5 Summary and Conclusions

In this research, we have proposed some practical primal recovery rules that satisfy primal convergence theorem in the Lagrangian dual subgradient-based method. We also have proposed a hybrid primal dual algorithm that meets the global convergence property and is based on practical primal recovery scheme. Throughout convergence proof and computational implementation, we verify that a hybrid primal dual algorithm produces a near-feasible and near-optimal solutions in a reasonably efficient manner. This can serve as a valuable tool in the context of Lagrangian dual/relaxation optimization. As for further research, we remark that there still needs to many experiments with different classes of important, special problems in order to understand the advantage or disadvantage of this algorithm.

References

1. Larsson, T and Z. Liu, "A primal convergence result for dual subgradient optimization with application to multi-commodity network flows", Research Report, Dept. of Mathematics Linkoping Instituted of Technology, S-581 83 Linkoping, Sweden, 1989.
2. Sen, S. and H. D. Sherali, "A class of convergent primal-dual subgradient algorithms for decomposable convex programs", Math. Programming, vol. **35**(1986), .279-297.
3. Sherali, H.D. and G. Choi, "Recovery of primal solutions when using subgradient optimization methods to solve Lagrangian duals of linear programs", OR Letters, Vol. **19**(1996), 105-113.
4. Sherali, H. D. and O. Ulular, "A primal-dual conjugate subgradient algorithm for specially structured linear and convex programming problems", Appl. Math. Optim, vol. **20**(1989), 193-221.
5. Shor, N. Z. Minimization Methods for Nondiffrentiable Functions, Translated from Russian by K. C. Kiwiel and A. Ruszczynski, Springer, Berlin, 1985.

Real-Coded Genetic Algorithms for Optimal Static Load Balancing in Distributed Computing System with Communication Delays

V. Mani[1], S. Suresh[1], and H. J. Kim[2]

[1] Department of Aerospace Engineering,
Indian Institute of Science, Bangalore, India
[2] Department of Control and Instrumentation Engineering,
Kangwon National University, Chunchon 200-701, Korea

Abstract. We consider the problem of static load balancing with the objective of minimizing the job response times. The jobs that arrive at a central scheduler are allocated to various processors in the system with certain probabilities. This optimization problem is solved using real-coded genetic algorithms. A comparison of this approach with the standard optimization methods are presented.

1 Introduction

The objective in load balancing is to schedule the jobs that arrive at a central scheduler to various processors in the system, to minimize the mean response time of a job. We consider a centralized model, in which jobs arrive at the central scheduler and are allocated to various processors in the system. In this situation, p_j is the probability that an arrived job is allocated to processor j and $\sum_{j=1}^{M} p_j = 1$, where M is the number of processors in the system [1, 2, 3, 4, 5]. This problem is studied in the context of optimally balancing the total load on a set of devices in memory hierarchy in [1], and in the context of optimally distributing the I/O request to disk in [2] and obtained a condition on the arrival rate of jobs to the system. For a similar problem, a dropout rule is analyzed in [3]. Dropout rule states that for some values of arrival rate, the slower processors in the system are never used. This problem is studied for different job classes in [4], and is also studied in the context of applying Bayesian decision theory to decentralized control of job scheduling in [5]. In all these studies [1, 2, 3, 4, 5], the mean communication delay for a job to processor j in the system is not considered.

The problem of optimal static load balancing in a distributed system incorporating the communication delay is first considered in [6]. In this study, two algorithms namely parametric-study algorithm and a single point algorithm are presented. The parametric-study algorithm generates the optimal solution as a function of communication time and the single point algorithm gives the optimal solution for a given system parameters. This study is further analyzed and another single point algorithm is presented in [7]. This load balancing problem in a tree hierarchy network configuration is studied in [8] and for the case of multi-class jobs is considered in [9]. In these studies, the mean communication delay is considered as a differentiable, nondecreasing and convex function.

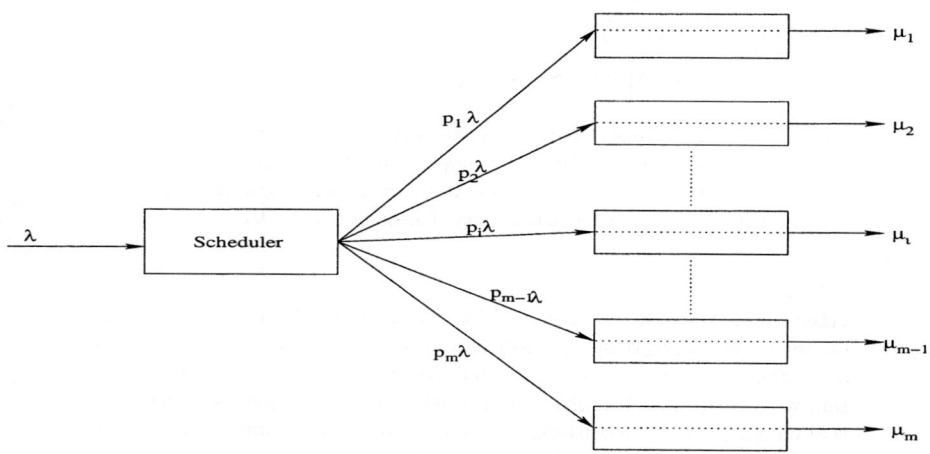

Fig. 1. Probabilistic Load Scheduling Model

In this paper, we consider the centralized probabilistic load scheduling problem with communication delay. Jobs arrive at the central scheduler in a Poisson arrival stream of rate λ. The scheduler allocates the jobs to the processors $j = 1, 2, \cdots, M$ according to probability distribution p_j, $j = 1, 2, \cdots, M$. Each processors in the network is modelled by an $M/M/1$ queue. For an M processor system, in each processor we have an independent $M/M/1$ queue. The j-th processor has a Poisson arrival rate with mean λ_j and has an exponentially distributed service rate with mean μ_j. In this paper, the mean communication delay incurred as a result of sending a job to processor j is denoted as g_j [6]. Thus, the mean response time for jobs at processor j is given by

$$T_j = \frac{1}{\mu_j - \lambda_j} + g_j \qquad (1)$$

where $\lambda_j = p_j \lambda$. Thus the problem of minimizing the mean response time T of the system can be formulated as:

$$Minimize \quad T = \sum_{j=1}^{M} \frac{p_j}{\mu_j - \lambda_j} + g_j$$

$$subject\ to \quad \sum_{j=1}^{M} p_j = 1$$

$$p_j \leq \frac{\mu_j}{\lambda}, \quad j = 1, 2, \cdots, M$$

$$p_j \geq 0, \quad j = 1, 2, \cdots, M \quad and \quad g_j = 0, \quad if\ p_j = 0 \qquad (2)$$

We use real-coded genetic algorithms for the above optimization problem to obtain the solution. Genetic algorithms have been used to solve difficult optimization problems with objective functions that do not possess 'nice' properties such as continuity, differentiability [10, 11, 12, 13]. We also present a study comparing the real-coded genetic algorithm with other optimization algorithms for this load balancing problem in terms of function evaluations.

2 Real-Coded Genetic Algorithms

Genetic algorithm (GA) is an intelligent search algorithm based on the mechanisms of evolution in nature. GA starts with an initial population of solutions (chromosomes) to the problem. Each solution is evaluated using a fitness function (objective function) and is assigned a fitness value. A selection method based on fitness value is used to decide which solutions are to be used for producing new solutions for the next generation. Genetic operators such as crossover, mutation and reproduction are applied on selected solutions to produce new population of solutions for the next generation. This process is repeated until the algorithm converges. Genetic algorithms use the concept "survival of the fittest" by passing good solutions to the next generation of solutions and combining different solutions to explore new search solutions. Details about how genetic algorithms work for a given problem is available in [10, 11, 12, 13].

In the initial studies on GAs [10, 11], the solutions are coded using binary strings. It is shown in [13] that for real valued numerical optimization problems, floating point representation of solutions (chromosomes) perform better than binary representation of solutions, because of consistency, precision and faster execution. GAs using real-number representation of solutions are called real-coded genetic algorithms and is used in our study. The real-coded genetic algorithm for our load balancing problem needs to address the following important issues: solution representation as a string; initial population; selection function; design of genetic operators; determination of fitness function; probabilities controlling the genetic operators.

2.1 Solution Representation

The solution for our problem is represented as a string of real numbers. Each element in the string represents the probability (p_j) that an arrived job is allocated to processor j in the system. In general for an M processor system, the length of the string is M. A valid string (solution) in our problem should satisfy the following constraints:

- (i) The sum of the elements of the string is equal to one. ($\sum_{j=1}^{M} p_j = 1$).
- (ii) Each element in the string is in the range $0 \leq p_j \leq \frac{\mu_j}{\lambda}$.

We must keep in mind the above constraints when we generate initial population of solutions and design genetic operators.

2.2 Population Initialization

One of the advantages of GAs is that it searches many solutions in the search space in parallel. This is due to the fact that the GAs search from a population of solution points instead of a single solution point. GA starts with an initial population of solutions to the given problem. The method of population initialization will affect the rate of convergence of the problem. The population size is problem-dependent. For our problem the initial population of solutions is selected in the following manner.

- Equal allocation: The value of p_j in the solution is $\frac{1}{M}$, for $j = 1, 2, \cdots, M$.
- Random allocation: Generate M random numbers. These random numbers are normalized such that the sum is equal to one.
- Zero allocation: Select a solution using equal or random allocation. Any one element in the selected solution is assigned zero and its value is equally allocated to other elements.
- Proportional allocation: The value of p_j in the solution is $p_j = \mu_j / \sum_{i=1}^{M} \mu_i$.

Some of the above allocation methods are based on the knowledge from the problem domain. For example, zero allocation is included to take care of the dropout condition. Drop-out condition states that for some values of arrival rate (λ), the slower processors are never used. We also know from the problem that the load fraction assigned to a faster processor is more than the load fraction assigned to a slower processor. This information is used in proportional allocation. It is possible in the above method of generating some of the solutions in initial population may not be a valid solution to the problem. In other words, it is possible in the above allocations that the constraint (ii) on the values of p_j may be violated. So we need some sort of a "repair algorithm". Repair algorithm would "repair" the solution making it a valid solution [13]. We can see that constraint (i) is always satisfied; i.e., the sum of the elements (probabilities) of the string is equal to one ($\sum_{j=1}^{M} p_j = 1$).

Repair Algorithm: If any of the p_i violates the constraint(ii), then that particular p_i is made 0.95 times $\frac{\mu_i}{\lambda}$ and the remaining load is equally or proportionately allocated to other processors in the system. This repair algorithm will always produce a valid solution because $\sum_{i=1}^{M} \mu_i > \lambda$. Here proportionately we mean that proportional to the service rates of the processors.

2.3 Selection Function

In GAs, the selection of solutions from the existing population (solutions) to produce new solutions for the next generation plays an important role. In our study, we have used the normalized geometric ranking method given in [14] for the selection process. In this method, the solutions in the current population are arranged in the decreasing order of their fitness values, and a rank is assigned to each of the solutions. If r_i is the rank of the solution i, then the probability of selection of solution i is

$$S_i = q^{'}(1-q)^{r_i - 1} \tag{3}$$

where q is the selection probability and $q' = \frac{q}{1-(1-q)^N}$ and N is the population size. We can see that the better solution has a better chance of being selected for producing new solutions using genetic operators.

2.4 Genetic Operators

Genetic operators such as crossover and mutation provide basic search mechanism in GAs. The operators used in our study are the following.

Crossover Operator: The crossover operator is regarded as the main search operator in genetic algorithms. The role of crossover operator is to use the information from existing solutions to produce better solutions. The crossover operator takes two solutions (parents) from the existing population and produce two new solutions (off springs). The crossover operators used in our study are the following:

Two Point Crossover (TPX): Let C_1 and C_2 are the two solutions selected for crossover operations. This operator first select two crossover points randomly i, j and $i < j$.

$$C_1 = \{c_1^1, c_2^1, \cdots, c_i^1, c_{i+1}^1, \cdots, c_j^1, c_{j+1}^1, \cdots, c_M^1\} \tag{4}$$

$$C_2 = \{c_1^2, c_2^2, \cdots, c_i^2, c_{i+1}^2, \cdots, c_j^2, c_{j+1}^2, \cdots, c_M^2\} \tag{5}$$

Let $K_1 = c_i^1 + c_{i+1}^1 + \cdots + c_j^1$ and $K_2 = c_i^2 + c_{i+1}^2 + \cdots + c_j^2$. Also let $x_1 = K_1/K_2$ and $x_2 = K_2/K_1$. Two new solutions H_1 and H_2 are obtained as

$$H_1 = \{c_1^1, c_2^1, \cdots, x_1 c_i^2, x_1 c_{i+1}^2, \cdots, x_1 c_j^2, c_{j+1}^1, \cdots, c_M^1\} \tag{6}$$

$$H_2 = \{c_1^2, c_2^2, \cdots, x_2 c_i^1, x_2 c_{i+1}^1, \cdots, x_2 c_j^1, c_{j+1}^2, \cdots, c_M^2\} \tag{7}$$

Simple Crossover (SCX): Here only one crossover point i is selected randomly and the second crossover point is M. Here Let $K_1 = c_i^1 + c_{i+1}^1 + \cdots + c_M^1$ and $K_2 = c_i^2 + c_{i+1}^2 + \cdots + c_M^2$. Also let $x_1 = K_1/K_2$ and $x_2 = K_2/K_1$. Two new solutions H_1 and H_2 are obtained as

$$H_1 = \{c_1^1, c_2^1, \cdots, x_1 c_i^2, x_1 c_{i+1}^2, \cdots, x_1 c_M^2\} \tag{8}$$

$$H_2 = \{c_1^2, c_2^2, \cdots, x_2 c_i^1, x_2 c_{i+1}^1, \cdots, x_2 c_M^1\} \tag{9}$$

Uniform Crossover (UCX): This operator first select two crossover points randomly i and j. Here Let $K_1 = c_i^1 + c_j^1$ and $K_2 = c_i^2 + c_j^2$. Also let $x_1 = K_1/K_2$ and $x_2 = K_2/K_1$. Two new solutions H_1 and H_2 are obtained as

$$H_1 = \{c_1^1, c_2^1, \cdots, x_1 c_i^2, c_{i+1}^1, \cdots, x_1 c_j^2, c_{j+1}^1, \cdots, c_M^1\} \tag{10}$$

$$H_2 = \{c_1^2, c_2^2, \cdots, x_2 c_i^1, c_{i+1}^2, \cdots, x_2 c_j^1, c_{j+1}^2, \cdots, c_M^2\} \tag{11}$$

Averaging Crossover (ACX): Two new solutions H_1 and H_2 are obtained by averaging C_1 and C_2.

$$H_1 = C_1 + \alpha(C_1 - C_2) \tag{12}$$

$$H_2 = C_2 + \alpha(C_2 - C_1) \tag{13}$$

where α is a scalar value in the range $0 \leq \alpha \leq 1$.

It is possible in the above crossover methods that some of the solutions obtained may not be a valid solution to the problem. For such invalid solutions we can use the repair algorithm presented earlier and obtain valid solutions.

Mutation Operator: The mutation operator use one solution to produce a new solution. Mutation operator is needed to ensure diversity in the population and to avoid premature convergence and local minima problems. Two different mutation operators used in our study are the following:

Swap Mutation (SM): Let C_1 be the solution selected for mutation operation. This operator first select two mutation points randomly i and j. The new solution (H_1) is generated by swapping the values at these mutation points.

$$C_1 = \{c_1^1, c_2^1, \cdots, c_i^1, c_{i+1}^1, \cdots, c_j^1, c_{j+1}^1, \cdots, c_M^1\} \tag{14}$$

$$H_1 = \{c_1^1, c_2^1, \cdots, c_j^1, c_{i+1}^1, \cdots, c_i^1, c_{j+1}^1, \cdots, c_M^1\} \tag{15}$$

Random Zero Mutation (RZM): This operator first select one mutation point randomly i. The new solution (H_1) is generated by making the value at this mutation point zero and distributes this value to other elements in the solution equally or proportionately.

$$C_1 = \{c_1^1, \cdots, c_i^1, c_{i+1}^1, \cdots, c_j^1, c_{j+1}^1, \cdots, c_M^1\} \tag{16}$$

$$H_1 = \{x_1 + c_1^1, \cdots, 0, x_{i+1} + c_{i+1}^1, \cdots, x_M + c_M^1\} \tag{17}$$

where $x_i = \frac{c_i^1}{M-1}$ in equal allocation, and $x_i = c_i^1 \frac{\mu_i}{\sum_{j=1, j\neq i}^{M} \mu_j}$ in proportional allocation. The random zero mutation operator is useful because one element is made zero implies the drop-out condition mentioned in [3, 4] for our problem. Also this random zero mutation operator increases the rate of convergence.

We have presented four different crossover operators and two mutation operators. Each crossover operator finds two new solutions and mutation operators produce one new solution. The type of genetic operator to be used depends on a particular problem. In fact one type of crossover operator may perform well for a problem may not perform well for a different problem. Even in the same problem one type of crossover may perform well in the earlier stages of the problem and may not perform well in the later stages of the same problem. This fact has been brought out in *no-free-lunch-theorem* [15]. Hence it is better to apply different crossovers simultaneously on the population for practical situations [16, 17]. This is true with mutation operators also. Hence, in this paper, we have used a hybrid operator methodology. In the hybrid method, the solutions selected for crossover and mutation are used to generate new solutions using different crossover and mutation operators. The best solution from these generated solutions are used for next generation. This is same as elitist model described in [12], and also the best solution from one generation is carried to the next generation. The advantage is that the search becomes faster and computation time is reduced.

2.5 Fitness Function

Fitness is the driving force in GAs. The only information used in the execution of genetic algorithms is the observed values of fitness of the solutions present in the population. The fitness function is the objective function of minimizing the response time for our problem. The calculation of the fitness function is simple. We get the vales of λ_j from the solutions and these values are used in the objective function of minimizing the response time for our problem. The GA will try to maximize the fitness and hence for our problem fitness function is

$$F = -T \qquad (18)$$

2.6 Termination Function

The most frequently used termination criterion are population convergence criteria and a specified maximum number of generations. In this study, a specified maximum number of generations (G) is used as termination criterion.

3 Simulation Results

The real-coded genetic algorithm for optimal static load balancing problem is implemented in MATLAB on a Pentium-IV machine. The genetic algorithm used the following parameters: S_m-mutation probability 0.05; S_c-crossover probability 0.6; q-selection probability 0.08; and $G = 500$. The following steps are carried out in real-coded genetic algorithm for optimal static load balancing problem.

- **STEP 1.** Initialization: An initial population of solutions of size (N) is generated using population initialization.
- **STEP 2.** Evaluation: The fitness value of each solution in the population is calculated according to the fitness function. The fitness is the negative of the mean response time for the given solution.
- **STEP 3.** Perform selection function using normalized geometric ranking, to select sequences for genetic operations.
- **STEP 4.** Genetic Operations: Perform crossover and mutation operations based on probability of crossover and mutation. Here we may get more solutions than the population size (N). Perform reproduction operation using elitist model [11] to obtain N best solutions.
- **STEP 5.** Repeat the steps 2, 3 and 4, until the algorithm converges.

We can see that GA use the Elitism strategy by passing the 'good' solutions to the next generation of solutions, and combining different solutions to explore new solutions. In this way, the GA converges to the optimal solution for the problem. In [18], a theory of convergence for real-coded genetic algorithm is analyzed. The advantage with genetic algorithms is that it starts with a random solutions and modify the solutions in successive generations, and the optimal solution is obtained. So we have used the genetic algorithm 5 times (runs) for each of the numerical examples in our simulation for verification.

We now present some numerical examples and also compare the performance of GA results with Nelder-Mead Simplex (NMS) method [19] and gradient descent (quasi-Newton) (QNM) technique [20]. The constrained optimization problem given in equation (2) is converted into a Lagrange function and solved using NMS and QNM techniques. The NMS method does not need the derivatives of the objective function. This method starts with an initial simplex of $(N+1)$ points; At each point of the simplex, the objective function is evaluated. Based on the objective function values at these points, elementary geometric transformation is done. In this manner, the worst point is replaced by a better point. The QNM need the derivative information. The conditions in this method are the objective function should be twice differentiable and the gradient vector and the Hessian matrix can be calculated at all points.

Table 1. Results For Numerical Example 1

Type	λ	p_1	p_2	p_3	p_4	T	NFC
	50	0	0.0972	0.3266	0.5762	0.01499	219
RGA	75	0.0097	0.1587	0.3262	0.5054	0.01756	263
	100	0.1006	0.1967	0.2985	0.4042	0.02923	274
	100	0	0.0983	0.3512	0.5505	0.01501	909
NMS	125	0.0097	0.1587	0.3262	0.5054	0.01756	449
	150	0.1006	0.1967	0.2985	0.4042	0.02923	438
	50	0	0.0972	0.3266	0.5762	0.01499	601
QNM	75	0.0097	0.1587	0.3262	0.5054	0.01756	342
	100	0.1006	0.1967	0.2985	0.4042	0.02923	374

Numerical Example 1. Consider a distributed computing system with four ($M = 4$) processors without communication delay; i.e., $g_j = 0, j = 1, 2, 3, 4$. The service rate of the four processors are $\mu_1 = 30$, $\mu_2 = 40$, $\mu_3 = 50$, and $\mu_4 = 60$. The solution obtained for different values of arrival rate (λ) are shown in Table 1. In this table, T is the response time and NFC is the number of function call. The number of function call (NFC) is a good indicator of the computation time of different methods. NFC is the number of times the objective function is evaluated. In real-coded genetic algorithm (RGA), it is the number of times the fitness function is evaluated. From the above table, we can observe that the RGA, NMS and QNM converges to optimal solutions but NMS and QNM techniques requires more function evaluations than the RGA. When any of the p_j are zero then the RGA is more efficient than the other methods.

The advantage of real-coded genetic algorithms is that it is easy to incorporate the communication delay. The reason is that only in the fitness function (the objective function) will change and all other parts of the algorithm are the same. We present numerical example 2. for four processor ($m = 4$) system with communication delays. For case (i), from the above Table 2 we observe the following: When the communication delay is 0.005 ($g_i = 0.005$, $i = 1,2,3,4$), the load fraction assigned to the processors are equal and is 0.25. The response time (T) is 0.03333, for arrival rate λ equals to 60. When the communication delay is increased by 100% for all the processors, one may think that the optimal load fraction assigned to the processors is equal distribution, because both the service rate and the communication delay is same for all the processors. But this is not the optimal load distribution. The optimal load distribution is obtained by removing one of the processors from the network. Or in other words one of the processors is dropped out from the load distribution process. With the communication delays, when all the four processors are used, the response time is 0.1 whereas when only three processors are used the response time is 0.08. So one of the processors is dropped out from the load distribution process in the real-coded genetic algorithm. Though in the Table 1 (case 2) we have shown the load fraction assigned to processor p_1 is zero, in our simulation we obtained one of the processors is assigned a zero load fraction and the load fraction assigned to the others are equal. Similarly, when the communication delay is 0.05 for all the processors, the optimal load distribution is obtained by removing any two processors from the network.

Table 2. Numerical Example 2

Case	λ	g_1	g_2	g_3	g_4	p_1	p_2	p_3	p_4	T
		Homogeneous Network: $\mu_1=\mu_2=\mu_3=\mu_4=40$								
	60	0.005	0.005	0.005	0.005	0.25	0.25	0.25	0.25	0.06
i	60	0.010	0.010	0.010	0.010	0.3333	0	0.3333	0.3333	0.08
	60	0.050	0.050	0.050	0.050	0.5	0	0	0.5	0.2
		Heterogeneous Network: $\mu_1=\mu_2=40$ and $\mu_3=\mu_4=80$								
	60	0.010	0.010	0.01	0.01	0	0	0.5	0.5	0.04
	60	0.050	0.050	0.05	0.05	0	0	0	1	0.1
ii	60	0.005	0.005	0.01	0.01	0	0	0.5	0.5	0.04
	60	0.005	0.005	0.02	0.02	0	0.25	0	0.75	0.5643
	60	0.005	0.005	0.10	0.10	0.5	0.5	0	0	0.11
		Heterogeneous Network: $\mu_1=40$, $\mu_2=60$, $\mu_3=80$, $\mu_4=100$								
	120	0.005	0.01	0.015	0.020	0	0	0.42893	0.57107	0.068177
iii	120	0.005	0.01	0.020	0.025	0.16456	0.28208	0	0.55335	0.075383
	120	0.005	0.01	0.015	0.030	0	0	0.4305	0.5695	0.078179

4 Conclusions

Load balancing is an important issue in distributed computing systems. In this paper, we proposed a real-coded genetic algorithm method for the load balancing problem in a centralized scheduler. The various issues involved in genetic algorithm such as solution representation, genetic operators are discussed. It is shown that how the knowledge from the problem are incorporated in solution representation and genetic operators. We also show that incorporating the communication delay is easy in our approach. Numerical results are presented to show the computation time performance of our method with standard optimization methods.

Acknowledgement. This work was in part supported by MSRC-ITRC, Kangwon National University, under the auspices of IITA and MIC, Korea.

References

1. Buzen, J. P., and Chen, P. P. S.: Optimal load balancing in memory hierarchies. Information Processing 74, The Netherlands; North-Holland. (1974) 271-275.
2. Piepmeier, W. F.: Optimal balancing of I/O requests to disks. Communications of the ACM. **18**(9) (1975) 524-527.
3. Agrawala, A. K., Tripathi, S. K. and Ricart, G.: Adaptive routing using a virtual waiting time technique. IEEE Trans. on Software Engineering. **8**(1) (1981) 76-81.
4. Ni, L. M. and Hwang, K.: Optimal load balancing in a multiple processor system with many job classes. IEEE Trans. on Software Engineering. **11**(5) (1985) 491-496.
5. Stankovic, J. A.: An application of bayesian decision theory to decentralized control of job scheduling. IEEE Trans. on Computers. **34**(2) (1985) 117-130.
6. Tantawi, A. N. and Towsley, D.: Optimal static load balancing in distributed computer systems. Journal of the Asso. for Computing Machinery. **32**(2) (1985) 445-465.
7. Kim, C. and Kameda, H.: An algorithm for optimal static load balancing in distributed computer systems. IEEE Trans. on Computers. **41**(3) (1992) 381-384.
8. Li, J. and Kameda, H.: A decomposition algorithm for optimal static load balancing in tree hierarchy network configurations. IEEE Trans. on Parallel and Distributed Systems. **5**(5) (1994) 540-548.
9. Li, J. and Kameda, H.: Load balancing problems for multiclass jobs in distributed/parallel computer systems. IEEE Trans. on Computers. **47**(3) (1998) 322-332.
10. Holland, H. J.: Adaptation in natural and artificial systems. University of Michigan Press, Ann Arbor, (1975).
11. Goldberg, D. E.: Genetic algorithms in search, optimization and machine learning Addison-Wesley, New York, (1989).
12. David, L.: Handbook of Genetic Algorithms, New York, (1991).
13. Michalewicz, Z.: Genetic algorithms + Data structures = Evolution programs, AI Series, Springer-Verlag, New York, (1994).
14. Houck, C. R., Joines, J. A. and Kay, M. G.: A genetic algorithm for function optimization: A Matlab implementation. ACM Trans. on Mathematical Software. **22** (1996) 1-14.

15. Wolpert, D. H. and Macready, W. G.: No free lunch theorems for optimization. IEEE Trans. on Evolutionary Computation. **1**(1) (1997) 67-82.
16. Herrera, F., Lozano, M. and Sanchez, A. M.: Hybrid crossover operators for real-coded genetic algorithms: An experimental study. Soft Computing - A Fusion of Foundations, Methodologies and Applications, 2002.
17. Herrera, F., Lozano, M. and Verdegay, J. L.: Tackling real-coded genetic algorithms: Operators and tools for behavioural analysis. Artificial Intelligence Review. **12**(4) (1998) 265-319.
18. Goldberg, D. E.: Real-coded genetic algorithms virtual alphabets, and blocking. Complex Systems. **5** (1991) 139-167.
19. Lagarias, J.C., Reeds, J. A., Wright, M. H. and Wright, P. E.: Convergence properties of the Nelder-Mead simplex method in low dimensions, SIAM Journal of Optimization. **9**(1) (1998) 112-147.
20. Shanno, D.F.: Conditioning of Quasi-Newton methods for function minimization. Mathematics of Computing. **24** (1970) 647-656.

Heterogeneity in and Determinants of Technical Efficiency in the Use of Polluting Inputs

Taeho Kim[1] and Jae-Gon Kim[2]

[1] College of Northeast Asian Studies, University of Incheon,
49-4 Dowha 3-dong, Nam-gu, Incheon, Korea (ROK)
{thkim, jaegkim}@incheon.ac.kr
http://benice.conasian.com

[2] Department of Industrial Engineering, University of Incheon,
177 Dowha 2-dong, Nam-gu, Incheon, Korea (ROK)

Abstract. Many industries are using damage control inputs to control the unpredictable damages to yield. For example, chemical production processes are using catalysts to promote reactions. However, use of these inputs can generate the pollutions and the efficient use of them can contribute to the protection of environment. This paper uses a famous nonparametric method (Data Envelopment Analysis: DEA) to estimate the efficiency in the use of polluting inputs such as damage control inputs and uses a regression analysis to identify factors which affect the efficiency. The results from application to an agricultural field show the existence of substantial inefficiency and heterogeneity across farms. This implies that uniform standards or incentives such as Pigouvian taxes are not useful to regulate the use of polluting inputs. Moreover, technical efficiency was determined by some farm-level characteristics.

1 Introduction

Traditional Pigouvian strategies (e.g. taxes or standards) have not been applied to manage the environmental impacts of industries, specially, agriculture. First, elasticities of use of polluting inputs with respect to their price appear to be small suggesting that substantial tax rates would be required to induce necessary changes in use, see e.g. [3]. Second, the impact of changes in use on environmental pollution is also estimated to be small [18]. Use of standards side steps these problems, however, they are most attractive when the regulated population is homogeneous. [3] has shown that heterogeneity across farm producers can be substantial and important to consider in estimation of the productivity of polluting inputs. Nonparametric results further extend evidence of this heterogeneity to technical efficiency in use of polluting inputs, [8], [15]. In both cases, substantial and heterogeneous inefficiency in the use of polluting inputs such as pesticides and fertilizers were reported.

Within this context, use of incentive-based regulation strategies becomes attractive, however, their design and effectiveness also depends upon the extent and nature of heterogeneity across the regulated population. In order to explain the extent and nature of the heterogeneity in inefficiency in the use of polluting inputs, it is possible to think about firm-level characteristics. In the case of agriculture, they include IPM[1] practices, BMP[2], and production scale.

This paper examines the hypothesis that substantial reduction in use of polluting inputs can be achieved through improvement in the technical efficiency of their use. Further, it evaluates the usefulness of information describing current use of specific IPM practices, other environmentally beneficial BMP, and operation scale to classify farm-level environmental performance as measured by technical efficiency in polluting input use. Though this paper is based on the data from US agriculture which has quite different characteristics from other industries such as computer industry and refinery industry and agriculture of other countries, it is certain that the analysis approach and model specifications of this paper can be applied to other industries and agriculture of other countries only with some modifications in characteristics.

2 Theoretical Backgrounds (Approaches)

The production frontier is defined as follows [13], [5].

$$f(x) = \max\{y : y \in P(x)\} = \min\{x : x \in L(y)\}. \tag{1}$$

where $P(x)$ describes the set of all output vectors that can be produced with each input vector x, and $L(y)$ describes the set of all input vectors that can produce each output vector, y. This production frontier provides the upper bound of production possibilities with a given input vector and the lower bound of inputs to produce a given output vector.

Technical efficiency is defined as follows: if the current technique is replaced with the one producing the most in its particular environment, how much less inputs could be used (input-oriented definition), or if only the current technique is replaced with the one producing the most in its particular environment, how much more could be produced (output-oriented definition) [11]. The input-oriented definition can be expressed as a function,

$$TE_I = \min\{\lambda : \lambda x \in L(y)\}. \tag{2}$$

where λ can be a scalar or a vector. If the efficiency score, λ, is a scalar, it means that firms adjust all inputs equi-proportionally (Radial measure). If λ is a vector, it means that firms adjust each input differently (Nonradial measure) [13]. The details on radial measure and nonradial measure are below. The output-oriented definition can be expressed as a function,

$$TE_O = \frac{1}{\max\{\eta : \eta y \in P(x)\}} \tag{3}$$

---

[1] Integrated Pest management such as pest management activities before, on, and after the pest emergence.
[2] Best Management Practices such as alternative land use practices, soil test, etc.

where η can be a scalar and a vector. Fig. 1 illustrates input-based technical efficiency (TE_I) and output-based technical efficiency (TE_O) from the perspective of an input-input space, an input-output space, and an output-output space.

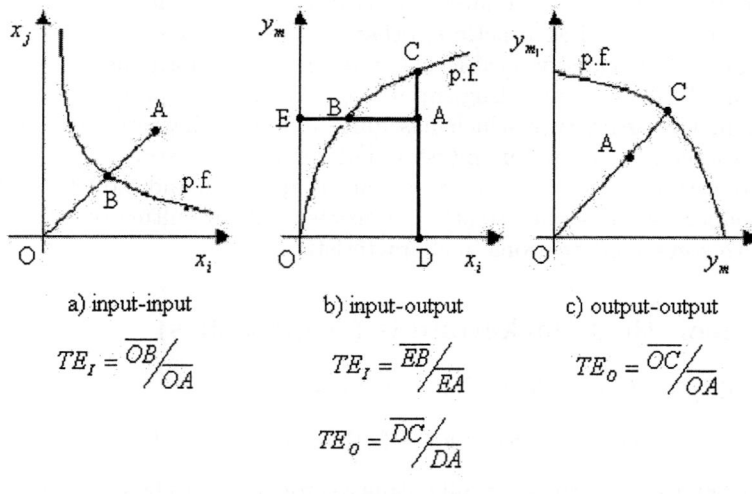

Fig. 1. Production frontiers and technical efficiencies

Input-based radial technical efficiency (RTE) model and input-based nonradial technical efficiency (NRTE) model are for the evaluated firm n_0 as follows [1], [5].

Objective function (RTE): $\quad\min\limits_{\lambda_{n_0},z} \lambda_{n_0}$

constraints:

$$\sum_{n=1}^{N} z_n x_{n,i} \leq x_{n_0,i}\lambda_{n_0}, \quad i=1,2,...,I : \text{Variable inputs}, \tag{4}$$

$$y_{n_0,j} \leq \sum_{n=1}^{N} z_n y_{n,j}, \quad j=1,2,...,J : \text{Outputs},$$

$$\sum_{n=1}^{N} z_n \leq 1, \text{Variable returns to scale}$$

Non-negative constraints

Object function (NRTE): $\min\limits_{\lambda_{n_0},z} \sum\limits_{n \in \{k | x_{n_0,k} \neq 0\}} \dfrac{\lambda_{n_0}}{N^+}$,

constraints:

$$\sum_{n=1}^{N} z_n x_{n,i} \leq x_{n_0,i} \lambda_{n_0}, \quad i = 1, 2, ..., I : \text{Variable inputs}, \qquad (5)$$

$$y_{n_0,j} \leq \sum_{n=1}^{N} z_n y_{n,j}, \quad j = 1, 2, ..., J : \text{Outputs},$$

$$\sum_{n=1}^{N} z_n \leq 1, \text{Variable returns to scale}$$

Non-negative constraints

As noted in equation (4) and (5), constraints set estimates the piecewise linear production frontier determining the optimal weights for each firm, and the objective function calculates the minimum ratio of the piecewise linear production frontier to current position of the evaluated firm.

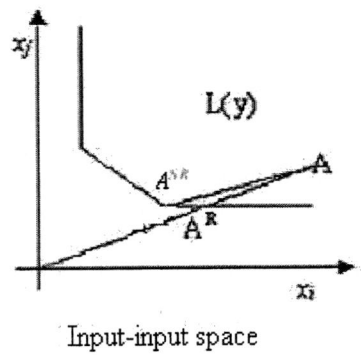

Input-input space

Fig. 2. Radial measure and nonradial measure in DEA

In DEA, as noted, technical efficiency can be measured in two ways: radially and nonradially. Farrell's measures of technical efficiency provide insights for total factor employment and propose equi-proportional reduction of all factors necessary to attain technical efficiency [7]. While this type of measure of technical efficiency may be useful for some questions, the differential impacts of inputs on the performance would seem to beg for a measure of input specific potential for the adjustment of input use. The former is called radial technical efficiency and the latter is called nonradial technical efficiency or Russell's measure [16].

As noted, radial measure adjusts the input vector back toward the origin. This radial efficient point (A^R) of Fig. 2 belongs to the weak efficient point in that it may have slacks [7]. In Fig. 2, although A^{NR} does not have any slack, A^R has the slack by $A^{NR} - A^R$. Because of this existence of slacks in RTE, it may seem to look incomplete. However, RTE is meaningful in that it can give us the overall technical efficiency of firms. In order to obtain a more complete (or at least not weak) measure, it is possible to think about a nonradial measure of technical efficiency. Each input factor is reduced in a different proportion. In Fig. 2, the adjustment from A to A^{NR} represents the nonradial adjustment. These nonradial technically efficient points (A^{NR}) belong to the efficiency points in that they do not have any slack. $NRTE$ is also meaningful in that it can give us insight into the source of the technical inefficiency of firms. This nonradial technical efficiency measure collapses to the radial measure when all technical efficiencies for all inputs are equal to the radial technical efficiency. Since nonradial measures can shrink an input vector at least as the radial measure can, the relationship between them is $0 \leq NRTE \leq RTE \leq 1$. The models for both radial measures and nonradial measures are in equation (4) and (5). This distinction between radial measure and nonradial measure comes from the piecewise linear property of the production frontier estimated by DEA.

Moreover, in order to understand what firm-level characteristics may be associated with technical efficiency, we use regression of technical efficiency scores on firm-level characteristics. The ordinary least square regression method is used in this paper $(y = \beta x + \epsilon)$. Many papers ([6], [4], [2], [8]) used double limited Tobit method to do the same thing. However it might be irrelevant because there is no truncation and censoring in this context. Though censored data come from sampling from a subset of a large population, for example, sampling data for income studies based on incomes above or below some income level [9], there is no censoring because our sample was drawn by random sampling. From the books of [9] and [10], truncation in our context means that though values greater than 1 can exist, the values are expressed by just 1. For example, though 1.2, 1.5, and even 2.5 can exist in data set, if all of them were expressed by 1 in regression analysis the data is said to be truncated by 1. However, in our context, because technical efficiency scores can never be over 1 (thus 1 exactly means 1), there is no truncation. In this paper, since dependent variable (technical efficiency) is not truncated or censored at all, there does not exist any convincing reason for utilizing Tobit regression instead of ordinary least square regression [9]. The model specification for winter wheat in this paper is in Table 1.

The specification of Table 1 is motivated as follows. [12] examines the relationship between farm size and technical efficiency of Brazilian farms. Technical efficiency is regressed on farm size, type of land tenure, technology indicators, and input usage. [14] regresses technical efficiency on education of farmer, land use practices, fertilizer use, and pesticide use for Nigerian farmers. [17] examines whether tillage, cropping practices, and farm characteristics such as farm size affect technical efficiency of farms growing wheat, sorghum, and soybean or not by using stochastic frontier method. Integrating the specifications of these

Table 1. Specification for regression analysis

Variable	Winter wheat
Dependent variables	Technical efficiency (Radial, Average nonradial, Nonradial of each input)
Independent variables	Land use practice: Strip cropping (BMP)
	Soil test (IPM)
	Land use practice: Contour farming (BMP)
	Land use practice: Using water way (BMP)
	Acres harvested ((0eratiion scale)

papers and concentrating on the title of this paper, our specification focuses on the relationship between technical efficiency, farm size, and IPM practices and BMP including tillage system, soil test (cropping practice), pesticide application practice, land use practices, and farm operation scale.

3 Data and Empirical Applications

Estimates are based on data collected in the 2000 Pennsylvania survey of winter wheat Production Practices. Extensive data on production practices were collected on selected fields across a sample of 90 producers selected as representative based on acreage planted. Data include farm level data describing acreage and production systems, as well as field specific data including seeding, fertilization, land use, pest management, field operation, and harvest practices. From them, we take an output factor, yield per planted acre, two general variable input factors, acres planted and seeding rate, and five damage control (Polluting) inputs factors. Thus, the production function can be expressed as $y = f(x, x^N)$ where x is the vector of variable inputs except damage control inputs such as pesticides and fertilizers, x^N is the vector of damage control inputs, and y is the output factor [3]. For wheat, we view yield per acre as outputs. Acres planted and seeding rate are regarded as variable inputs and applications of lime, nitrogen, phosphate, and potash, and acres treated with fertilizers are specified as damage control and polluting inputs. Yield varied from 17.0 to 100.0 bushels/acre. The descriptive statistics and specifications for DEA are in Table 2. An important issue with regard to DEA is the dimensionality which is generated by the numbers of outputs and inputs [8]. It suggests a rule for good dimensionality where if the ratio of the number of firms to the sum of the numbers of inputs and outputs is large enough (about 5), the model has a good dimensionality. In this paper, since the ratio is 11.25 ($\frac{90}{7+1}$), the DEA seems to have a good dimensionality.

Moreover, the summary of data for regression analysis is in Table 3 following the specification of Table 1.

Table 2. Specification for regression analysis

variable	N	Min	Max	Mean	Standard Deviation
Arcres planted	90	0.3	20	2.95	3.71
Seeding rate(lbs.)	90	60	180	134.25	29.71
Lime applied(tons)	90	0	30	10.47	9.74
Nitrogen applied(lbs.)	90	0	140	10.68	25.42
Phosphate applied(lbs.)	90	0	175	23.40	40.90
Potash applied(lbs.)	90	0	122.5	9.30	21.76
Acres treated of fertilizers	90	0	35	6.06	7.034
Yield/acre(bushedl)	90	17	100	61.27	17.10

Table 3. Summary of data for regression analysis

Binary variable	Adoption rate		Explanation	
Land use practice: Strip cropping	0.273		Adoption: 1, No: 0	
Soil test	0.545		Adoption: 1, No: 0	
Land use practice: Contour farming	0.318		Adoption: 1, No: 0	
Land use practice: Water way	0.227		Adoption: 1, No: 0	
Continuous variable	Min	Max	Mean	St Dev.
Acres harvested	1	35	6.925	6.59

4 Results and Discussions

First, the summary of efficiency estimates reported in Table 4 clarifies the existence of a substantial opportunity for reduction of polluting input use and a high degree of heterogeneity in technical efficiency among firms. Radial efficiency estimates range from 0.3250 to unity, while nonradial estimates (Average of input specific estimates) range from 0.1162 to unity. Only 34 firms are technically efficient and average radial technical efficiency is about 74%, which implies that there is an opportunity of radially average 26% reduction in polluting input use and average nonradial technical efficiency is about 64%, which implies that there is an opportunity of nonradially average 36% reduction in polluting input use. As shown, radial technical efficiency is always equal to or greater than nonradial technical efficiency (Average of input specific estimates).

In addition to estimating technical efficiency of variable inputs, we examine the impacts of farm management practices including IPM practices and BMP on technical efficiency to identify the factors affecting the efficiency. In other words, we examine what factors explain this heterogeneity on technical efficiency cross firms. Results of estimation of ordinary least square regression of the efficiency scores are reported in Table 5. Results reported indicate that efficiency is conditional on strip cropping and contour farming of land use practices. Use of strip cropping increases radial efficiency and nonradial efficiency by 0.246 and 0.308 respectively and use of contour farming increases radial efficiency, nonradial efficiency, and nonradial efficiency for lime use by 0.335, 0.229, and 0.293 respectively. It is also conditional on whether the individual producer do soil test or not. Use of soil test does affect technical efficiency most. It increases radial efficiency, nonradial efficiency, nonradial efficiency for lime use, and nonradial efficiency for fertilizer use by 0.502, 0.441, 0.198, and 0.301 respectively. Effect of production scale on technical efficiency is considerably high. 10 acre expansion increases radial efficiency, nonradial efficiency, nonradial efficiency for lime use,

Table 4. Summary of techincal efficiency results

Items	N	#of efficient farms	Min	Max	Mean
Radial efficiency	90	34	0.3250	1.0000	0.7773
Nonradial efficiency	90	34	0.1162	1.0000	0.6416
Acres planted	90	39	0.0126	1.0000	0.6522
Seed rate(lbs.)	90	49	0.3000	1.0000	0.8481
Lime applied(tons)	90	20	0.0144	1.0000	0.4282
Nitrogen applied(lbs.)	90	18	0.0253	1.0000	0.2995
Phosphate applied(lbs.)	90	30	0.0134	1.0000	0.4990
Potash applied(lbs.)	90	17	0.0253	1.0000	0.3007
Acres treated(fertilizer)	90	45	0.0709	1.0000	0.7121

Table 5. Summary of regression analysis

Factors	Radial efficiency	Average nonradial efficiency	Nonradial efficiency of Lime	Nonradial efficiency of Nitrogen	Nonradial efficiency of acres treated by Fertilizers
Land use practice: Strip cropping	0.246* (1.83)	0.308** (2.24)	-0.162 (-1.12)	0.069 (0.49)	0.182 (1.08)
Soil test	0.502*** (5.15)	0.441*** (4.42)	0.198* (1.89)	0.274*** (2.72)	0.301** (2.47)
Land use practice: Contour farming	0.335*** 2.71	0.229* (1.81)	0.293** (2.20)	-0.091 (-0.71)	0.108 (0.70)
Land use practice: Water way	-0.045 (-0.30)	0.024 (0.16)	0.05 (0.31)	-0.109 (-0.71)	0.202 (1.08)
Acres harvested	0.036*** (4.74)	0.025*** (3.17)	0.027*** (3.26)	0.004 (0.53)	0.021** (2.22)
F-value for regression model	34.42***	23.37***	8.46***	2.21*	9.48***
R^2	0.815	0.75	0.52	0.221	0.549

* Significant at the 10% level.
** Significant at the 5% level.
*** Significant at the 1% level.

and nonradial efficiency for fertilizer use by 0.36, 0.25, 0.27, and 0.21 respectively. Also, acres harvested are used to see the impact of production scale.

The above results imply that winter wheat farms should focus on strip cropping and contour farming in land use practice, and soil test in order to improve their technical efficiency in production. Additionally, winter wheat farms should also pay attention to their production scale. The approaches and model specifications can be applied to technical efficiency in the use of damage control inputs for other industries and agriculture of other industries with some modifications in variables.

5 Conclusion

This paper focuses on evaluation of the extent of heterogeneity across farm level use of potentially polluting inputs and the factors determining the heterogeneity across farms in Fertilizers and pesticides were considered. For samples of Pennsylvania winter wheat producers, farm level data were used to estimate nonparametric measures of the technical efficiency with which these farms utilize these inputs. Substantial heterogeneity was found in technical efficiency. Secondly, substantial opportunity was found for improving the efficiency through reduction in the use of these inputs on many farms. Both the observed heterogeneity across the farm population and the inefficiency in current use implies that traditional uniform Pigouvian approaches to public management of the use of these inputs would be inefficient.

Potential for targeting of farm specific instruments is affirmed by evidence that farm characteristics can be used to predict difficult to observe technical efficiency levels. For winter wheat farms, use of soil test increases efficiency most and strip cropping and contour farming in land use practice increase efficiency considerably. Moreover, efficiency for winter wheat farms is affected considerably more by acres harvested. The methods and results of this research can be applied to other industries and agriculture of other countries without any problem.

References

1. Banker, R.D., Charnes, A., Cooper, C.C.: Some Models for Estimating Technical and Scale Inefficiencies in Data Envelopment Analysis. Management Science, Vol. 30, No. 9 (1984) 1078-1092
2. Binam J.N., Sylla, K., Diarra, I., Nyambi, G..: Factors Affecting Technical Efficiency among Coffee Farmers in Cote d'Ivoire: Evidence from the Centre West Region. African Development Review, Vol. 15, No. 1 (2002) 66-76
3. Carpentier, A.C., Weaver, R.D.: Damage Control Productivity: Why Econometrics Matters. American Journal of Agricultural Economics, Vol.79 (1997) 47-61
4. Casu, B., Molyneux, P.: A Comparative Study of Efficiency in European Banking. Applied Economics, Vol. 35, No. 17 (2003) 1865-1876
5. Cooper, W.W., Seiford, L.M., Tone, T.: Data Envelopment Analysis: A Comprehensive Text with Models, Applications, References and DEA-solver Software. Kluwer Academic Publishers, Boston (2000)

6. Dhungana, B.R., Nuthall, P.L., Nartea, G..V.: Measuring the Economic Inefficiency of Nepalese Rice Farms Using Data Envelopment Analysis. The Australian Journal of Agricultural and Resource Economics, Vol. 48, No. 2 (2004) 347-369
7. Fare, R., Grosskopf, S., Lovell, K.: Production Frontiers. Cambridge University Press, Cambridge (1994)
8. Fernandez-Cornejo, J.: Nonradial Technical Efficiency and Chemical Input Use in Agriculture. Agricultural and Resource Economics Review (1994) 11-21
9. Greene, W.H.: Econometric Analysis. 4th edn. Prentice-Hall International, Upper Saddle River (2000)
10. Gujarati, D.N.: Basic Econometrics. 4th edn. McGraw Hill, Boston (2003)
11. Hall, M., Winsten, C.: The Ambiguous Notion of Efficiency. The Economic Journal, Vol. 69 (1959) 71-86
12. Helfand, S.M.: Farm Size and the Determinants of Productive Efficiency in the Brazilian Center-West. Contributed paper selected for presentation at the 25th International Conference of Agricultural Economists, Durban, South Africa (2003)
13. Kumbhakar, S.C., Lovell, K.: Stochastic Frontier Analysis. Cambridge: Cambridge University Press, Cambridge (2000)
14. Ogunyinka, E.O., Ajibefun, I.A.: Determinants of Technical Inefficiency in Farm Production: The Case of NDE Farmers in Ondo State, Nigeria. Selected Paper prepared for presentation at the Western Agricultural Economics Association Annual Meeting at The Denver Adam's Mark Hotel, Denver, Colorado (2003)
15. Piot-Lepetit, I., Vermersch, D., Weaver, R.D.: Agriculture's Environmental Externalities: DEA Evidence for French Agriculture. Applied Economics, Vol. 29, No. 3 (1997) 331-342
16. Russell, R.R.: Measures of Technical Efficiency. Journal of Economic Theory, Vol. 35, No. 1 (1985) 109-126
17. Sankranti, S., Langemeier, M.R.: Tillage Systems, Cropping Practices, Farm Characteristics and Efficiency. Selected paper prepared for presentation at the Southern Agricultural Economics Association Annual Meeting, Tulsa, (2004)
18. Weaver, R.D., Harper, J.K., Gillmeister, W.J.: Efficacy of Standards vs. Incentives for Managing the Environmental Impacts of Agriculture. Journal of Environmental Management, Vol. 2 (1996) 173-188

A Continuation Method for the Linear Second-Order Cone Complementarity Problem

Yu Xia[1],* and Jiming Peng[2],**

[1] The Institute of Statistical Mathematics,
4-6-7 Minami-Azabu, Minato-ku, Tokyo 106-8569, Japan
yuxia@ism.ac.jp
[2] Advanced optimization Lab, Department of Computing and
Software McMaster University, Hamilton, Ontario L8S 4K1, Canada
pengj@mcmaster.ca

Abstract. We reformulate the linear second-order cone complementarity problem into a system of nonlinear equations. Our reformulation is different from others. Our algorithm for the reformulation can start from an arbitrary point. We prove that our algorithm approximates an optimum of the linear second-order cone complementarity problem in finite steps under certain conditions. Finally, we show that the system of nonlinear equations of our reformulation is nonsingular at optimum under certain conditions.

1 Introduction

The linear second-order cone complementarity problem includes the linear complementarity problem and the second-order cone program[1] as special cases, and is the optimal condition to the convex quadratic second-order cone program. Especially, some problems in economics and engineering, such as facility location and Nash equilibrium, can be formulated as it. Second-order cone complementarity problems have been studied before; however, our algorithm is different from others. Our formulation is simpler, which means less computation; our method can be extended to P cone complementarity problems while others may not. In addition, our algorithm can start from an arbitrary point.

Next, we describe some of our notations. Column vectors are denoted by bold lower case letters and matrices by upper case letters. We use ";" to concatenate column vectors. Primal-dual variables are indexed from 0. Subscripts are used

---

* Research supported in part by a postdoctoral fellowship from JSPS (Japan Society for the Promotion of Science).
** Research partially supported by the grant # RPG 249635-02 of the National Sciences and Engineering Research Council of Canada (NSERC) and a PREA award. This research was also supported by the MITACS project "New Interior Point Methods and Software for Convex Conic-Linear Optimization and Their Application to Solve VLSI Circuit Layout Problems".

for indices of a vector or a matrix; superscripts are for iteration numbers. Thus, $(\mathbf{x}_i)_j$ is the $(j+1)th$ element of the vector $\mathbf{x}_i$. We use $\bar{\mathbf{x}}$ to represent the subverter of $\mathbf{x}$ excluding the first entry. As convention, I denotes the identity matrix.

The matrix R [2] is defined to be

$$R \stackrel{def}{=} \begin{pmatrix} 1 & & & \\ & -1 & & \\ & & \ddots & \\ & & & -1 \end{pmatrix}.$$

We use $\mathbf{0}$ to represent the zero vector and $\mathbf{1}$ to denote the vector of all one's. Let Q_n denote an $(n+1)$-dimensional second-order cone, i.e.,

$$Q_n \stackrel{def}{=} \{\mathbf{x} \in \mathbb{R}^{n+1} : x_0 \geq \|\bar{\mathbf{x}}\|_2\}.$$

We use Q to represent the Cartesian product of N second-order cones.

$$Q \stackrel{def}{=} Q_{n_1} \times Q_{n_2} \ldots Q_{n_N}.$$

Here, n_i are not necessarily the same for all the cones. We omit the subscripts when dimensionalities are clear from the context. The second-order cone induces a partial ordering. So we use $\mathbf{x} \geq_Q \mathbf{0}$ to denote $\mathbf{x} \in Q$ sometimes.

The linear second-order cone complementarity problem is generally written as the following.

$$\begin{aligned} \text{Find } & \mathbf{x} \in Q, \quad \mathbf{s} \in Q \\ \text{s.t. } & M\mathbf{x} + \mathbf{q} = \mathbf{s} \\ & \mathbf{x}^T \mathbf{s} = 0. \end{aligned} \quad (1)$$

Here, M is a real square matrix and $\mathbf{q}$ is a vector of proper order.

The remaining of the paper is organized as follows. In § 2, we give a new reformulation of the second-order cone complementarity problem; A globally convergent algorithm for the reformulation is presented in § 3. In § 4, we analyze the Jacobian of the reformulation at optima.

2 The Reformulation to a System of Nonlinear Equations

The constraints

$$\mathbf{x} \in Q, \quad \mathbf{s} \in Q, \quad \mathbf{x}^T \mathbf{s} = 0$$

are equivalent to (see [2]): for each $i = 1, \ldots N$,

$$(x_i)_0 (s_i)_j + (s_i)_0 (x_i)_j = 0 \quad (j = 1, \ldots, n_i) \quad (2a)$$
$$(x_i)_0 \geq \|\bar{\mathbf{x}}_i\|_2, \quad (s_i)_0 \geq \|\bar{\mathbf{s}}_i\|_2 \quad (2b)$$
$$(x_i)_0 = \|\bar{\mathbf{x}}_i\|_2 \text{ or } (s_i)_0 = \|\bar{\mathbf{s}}_i\|_2. \quad (2c)$$

For $j = 1, \ldots n_i$, we replace (2a) by

$$\sqrt{[(x_i)_0 + (s_i)_j]^2 + [(x_i)_j + (s_i)_0]^2} - \sqrt{(x_i)_0^2 + (x_i)_j^2 + (s_i)_0^2 + (s_i)_j^2} = 0. \quad (3)$$

(2b) is equivalent to (see[3]):

$$\sqrt{2}(x_i)_0 - \|\mathbf{x}_i\|_2 \geq 0, \quad \sqrt{2}(s_i)_0 - \|\mathbf{s}_i\|_2 \geq 0. \tag{4}$$

Compare to (2b), (4) has fewer nonsmooth points. We further reformulate (4) and (2c) via the Fischer-Burmeister function [4]:

$$\sqrt{(\sqrt{2}(x_i)_0 - \|\mathbf{x}_i\|_2)^2 + (\sqrt{2}(s_i)_0 - \|\mathbf{s}_i\|_2)^2}$$
$$- (\sqrt{2}(x_i)_0 - \|\mathbf{x}_i\|_2) - (\sqrt{2}(s_i)_0 - \|\mathbf{s}_i\|_2) = 0. \tag{5}$$

For $i = 1, \ldots, N$, let $\varphi_i(\mathbf{x}_i, \mathbf{s}_i)$ represent the system of equations including (5) and (3). Then, solving (1) is equivalent to finding a solution to the following system of nonlinear equations:

$$\begin{aligned} M\mathbf{x} + \mathbf{q} &= \mathbf{s} \\ \varphi_i(\mathbf{x}_i, \mathbf{s}_i) &= 0 \quad (i = 1, \ldots, N). \end{aligned} \tag{6}$$

3 Global Convergence

In this section, we give an algorithm for (6) and prove its finite convergence under certain conditions.

3.1 The Algorithm

Neither (3) nor (4) is Fréchet-differentiable everywhere. To overcome this problem and give global convergence results, we use the techniques of non-interior continuation/smoothing methods originally introduced by [5] and [6] for scalars. We add parameter μ to $\varphi_i(\mathbf{x}_i, \mathbf{s}_i)$.

$$\sqrt{(\sqrt{2}(x_i)_0 - \|\mathbf{x}_i\|_2)^2 + (\sqrt{2}(s_i)_0 - \|\mathbf{s}_i\|_2)^2 + 2\mu}$$
$$- (\sqrt{2}(x_i)_0 - \|\mathbf{x}_i\|_2) - (\sqrt{2}(s_i)_0 - \|\mathbf{s}_i\|_2) = 0,$$

$$\sqrt{[(x_i)_0 + (s_i)_j]^2 + [(x_i)_j + (s_i)_0]^2 + \mu} - \sqrt{(x_i)_0^2 + (x_i)_j^2 + (s_i)_0^2 + (s_i)_j^2 + 3\mu} = 0.$$

Denote the resulting system as $\varphi_i(\mathbf{x}_i, \mathbf{s}_i; \mu)$. Let

$$\varphi(\mathbf{x}, \mathbf{s}; \mu) \stackrel{\text{def}}{=} \begin{bmatrix} \varphi_1(\mathbf{x}_1, \mathbf{s}_1; \mu) \\ \varphi_2(\mathbf{x}_2, \mathbf{s}_2; \mu) \\ \vdots \\ \varphi_N(\mathbf{x}_N, \mathbf{s}_N; \mu) \end{bmatrix}.$$

Then $\mathbf{x}$ and $\mathbf{s}$ are said to be on the central path iff for some $\mu \geq 0$,

$$\begin{aligned} M\mathbf{x} + \mathbf{q} &= \mathbf{s} \\ \varphi(\mathbf{x}, \mathbf{s}; \mu) &= \mathbf{0}. \end{aligned} \tag{7}$$

Given $\beta > 0$, for $\mu > 0$, define the β-neighborhood of $(\mathbf{x}; \mathbf{s})$ as

$$\mathcal{N}_\beta(\mu) \stackrel{\text{def}}{=} \{(\mathbf{x}; \mathbf{s}) : \|M\mathbf{x} + \mathbf{q} - \mathbf{s}\| = 0,\ \|\varphi(\mathbf{x}, \mathbf{s}; \mu)\| \leq \beta\mu\}.$$

Note that we do not specify which norm $\|\cdot\|$ is used in $\mathcal{N}_\beta\mu$.
As $\mu \to 0$, $\mathcal{N}_\beta(\mu)$ converges to the solution set of (1).
Starting from any point $(\mathbf{x}, \mathbf{s})$, by solving the linear equation

$$M\,\Delta\mathbf{x} - \Delta\mathbf{s} = \mathbf{s} - \mathbf{q} - M\mathbf{x},$$

one can always get $(\mathbf{x}_0, \mathbf{s}_0) = (\mathbf{x} + \Delta\mathbf{x}, \mathbf{s} + \Delta\mathbf{s})$ satisfying

$$M\mathbf{x}_0 + \mathbf{q} - \mathbf{s}_0 = \mathbf{0}.$$

Therefor, we assume our initial point is feasible to the linear constraints.

We want to find an $(\epsilon_\mu, \epsilon_c)$-optimal solution $(\mathbf{x}^*; \mathbf{s}^*)$ such that $\mu \leq \epsilon_\mu$ and $\|\varphi(\mathbf{x}^*, \mathbf{s}^*; \mu)\| \leq \epsilon_c$. To achieve that, we reduce the parameter μ at each iteration, while keeping each iterate in some $\mathcal{N}_\beta(\mu)$ and updating the iterates by damped Newton's method, until the designed accuracy is achieved or the iterate is too large.

The Newton's direction $(\Delta\mathbf{x}, \Delta\mathbf{s})$ at the kth iteration is the solution to the following system.

$$\begin{aligned} M\,\Delta\mathbf{x} - \Delta\mathbf{s} &= \mathbf{0} \\ \frac{d\varphi}{d\mathbf{x}}\Delta\mathbf{x} + \frac{d\varphi}{d\mathbf{s}}\Delta\mathbf{s} &= -\varphi(\mathbf{x}^k, \mathbf{s}^k). \end{aligned} \quad (8)$$

Our algorithm is an extension of that in [7], which is designed for linear complementarity problem, while our problem is nonlinear. However, because of the special structure of our problem, our convergence analysis is more involved. Given $\Upsilon > 0$, $\epsilon_\mu > 0$, $\epsilon_c > 0$, our algorithm stops at an $(\epsilon_\mu, \epsilon_c)$ optimal solution or the iterates are unbounded. Below is the algorithm.

The Algorithm

1. For a given feasible starting point $(\mathbf{x}_0; \mathbf{s}_0)$, find positive scalars μ_0 and β, such that $(\mathbf{x}_0; \mathbf{s}_0) \in \mathcal{N}_\beta(\mu_0)$. Find μ_ϵ such that $\mu_\epsilon \leq \epsilon_\mu$ and $\epsilon_c \leq \beta\mu_\epsilon$. Set $k = 0$.
2. Do while $\mu_k > \mu_\epsilon$ and $\|(\mathbf{x}^k; \mathbf{s}^k)\| \leq \Upsilon$.
   (a) Let $(\Delta\mathbf{x}; \Delta\mathbf{s})$ be the solution to (8). Find the largest $0 < \hat{\alpha} \leq 1$ such that for any $\alpha \in [0, \hat{\alpha}]$,

   $$(\mathbf{x}^k + \alpha\,\Delta\mathbf{x};\ \mathbf{s}^k + \alpha\,\Delta\mathbf{s}) \in \mathcal{N}_\beta(\mu_k).$$

   Set

   $$\begin{aligned} \mathbf{x}^{k+1} &= \mathbf{x}^k + \hat{\alpha}\,\Delta\mathbf{x}, \\ \mathbf{s}^{k+1} &= \mathbf{s}^k + \hat{\alpha}\,\Delta\mathbf{s}. \end{aligned}$$

(b) Find the largest $0 < \hat{\mu} \le \mu_k$, such that for any $\mu \in [\hat{\mu}, \mu_k]$,
$$(\mathbf{x}^{k+1}; \mathbf{s}^{k+1}) \in \mathcal{N}_\beta(\mu) .$$

Set
$$\mu^{k+1} = \hat{\mu}^k ,$$
$$k \longleftarrow k+1 .$$

3.2 Global Convergence Results

In this part, we prove that our algorithm converges in finite steps under certain conditions.

Theorem 1. *Assume (8) is solvable at each iteration and $\|(\Delta\mathbf{x}^k; \Delta\mathbf{s}^k)\|$ doesn't converge to infinity. Then the algorithm terminates in finite iterations.*

Proof. We use contradiction.

Suppose the algorithm doesn't terminate after finite iterations. Then the Newton's direction $\|(\Delta\mathbf{x}^k; \Delta\mathbf{s}^k)\|$ doesn't converge to infinity, there is an infinite subsequence of $\{1, 2, \ldots\}$, denoted by $\{m_k\}_{k=1}^\infty$, such that $\{\|(\Delta\mathbf{x}^{m_k}; \Delta\mathbf{s}^{m_k})\|\}_{k=1}^\infty$ is upper bounded. Since the algorithm doesn't stop, $\|(\mathbf{x}^{m_k}; \mathbf{s}^{m_k})\|$ is also upper bounded and $\mu_k > \mu_\epsilon > 0$ for $k = 1, \ldots$. We use $\frac{1}{2}\Xi(\mathbf{x}, \mathbf{s}; \Delta\mathbf{x}, \Delta\mathbf{s})$ to represent the second-order term residue in the Taylor series expansion of $\varphi(\mathbf{x} + \Delta\mathbf{x}, \mathbf{s} + \Delta\mathbf{s})$ at $(\mathbf{x}; \mathbf{s})$. Then for all m_k and $0 \le \theta \le 1$, $\|\Xi(\mathbf{x}^{m_k}, \mathbf{s}^{m_k}; \theta\Delta\mathbf{x}, \theta\Delta\mathbf{s})\|$ is upper bounded, since $\varphi(\mathbf{x}, \mathbf{s})$, $\varphi(\mathbf{x} + \Delta\mathbf{x}, \mathbf{s} + \Delta\mathbf{s})$, and its first-order term is upper bounded.

By the mean value theorem, for any $0 \le \rho \le 1$ and any point $(\mathbf{x}; \mathbf{s})$, there exists a vector $\mathbf{0} \le \boldsymbol{\theta} \le \rho\mathbf{1}$, such that
$$\|\varphi(\mathbf{x}, \mathbf{s}; \mu) - \varphi(\mathbf{x}, \mathbf{s}; \rho\mu)\| = \left\|\frac{d\varphi(\mathbf{x}, \mathbf{s}; \boldsymbol{\theta}\mu)}{d\mu}\right\|(1 - \rho)\mu .$$

Since $\left\|\frac{d\varphi(\mathbf{x},\mathbf{s};\boldsymbol{\theta}\mu)}{d\mu}\right\|$ is upper bounded, there exists a scalar $\varpi > 0$ such that
$$\|\varphi(\mathbf{x}, \mathbf{s}; \mu^k) - \varphi(\mathbf{x}, \mathbf{s}; \rho\mu^k)\| \le \varpi(1 - \rho)\mu^k$$
for all $k = 1, \ldots$, and
$$\left\|\Xi(\mathbf{x}^{m_k}, \mathbf{s}^{m_k}; \tilde{\theta}\Delta\mathbf{x}, \tilde{\theta}\Delta\mathbf{s})\right\| \le \varpi$$
for all m_k and $0 \le \tilde{\theta} \le 1$.

By the triangular inequality, we have for all $(\mathbf{x}; \mathbf{s})$ and $0 \le \rho \le 1$,
$$\|\varphi(\mathbf{x}, \mathbf{s}; \rho\mu^k)\| \le \|\varphi(\mathbf{x}, \mathbf{s}; \mu^k)\| + \varpi(1 - \rho)\mu^k . \tag{9}$$

Since $(\Delta\mathbf{x}; \Delta\mathbf{s})$ solves (8), by Taylor series expansion, $\exists\, 0 \le \theta \le 1$, such that
$$\|\varphi(\mathbf{x}^{m_k} + \alpha\Delta\mathbf{x}^{m_k}, \mathbf{s}^{m_k} + \alpha\Delta\mathbf{s}^{m_k}; \mu^{m_k})\| \le (1 - \alpha)\|\varphi(\mathbf{x}^{m_k}, \mathbf{s}^{m_k}; \mu^{m_k})\|$$
$$+ \frac{\alpha^2}{2}\|\Xi(\mathbf{x}^{m_k}, \mathbf{s}^{m_k}; \theta\Delta\mathbf{x}^{m_k}, \theta\Delta\mathbf{s}^{m_k})\| . \tag{10}$$

Next we will show that μ^{m_k} can be reduced by a factor not less than a constant.

1. Suppose $\|\varphi(\mathbf{x}^{m_k}, \mathbf{s}^{m_k}; \mu^{m_k})\| < \|\Xi(\mathbf{x}^{m_k}, \mathbf{s}^{m_k}; \theta \Delta \mathbf{x}^{m_k}, \theta \Delta \mathbf{s}^{m_k})\|$.
   The minimum of the right-hand side of (10) as a function of α, which is convex, is obtained at a point where its derivative is zero, i.e., at

   $$\tilde{\alpha} = \|\varphi(\mathbf{x}^{m_k}, \mathbf{s}^{m_k}; \mu^{m_k})\| / \|\Xi(\mathbf{x}^{m_k}, \mathbf{s}^{m_k}; \theta \Delta \mathbf{x}^{m_k}, \theta \Delta \mathbf{s}^{m_k})\|.$$

   Therefore,

   $$\|\varphi(\mathbf{x}^{m_k+1}, \mathbf{s}^{m_k+1}; \mu^{m_k})\| \leq \|\varphi(\mathbf{x}^{m_k} + \tilde{\alpha}\Delta\mathbf{x}^{m_k}, \mathbf{s}^{m_k} + \tilde{\alpha}\Delta\mathbf{s}^{m_k}; \mu^{m_k})\|$$
   $$\leq \|\varphi(\mathbf{x}^{m_k}, \mathbf{s}^{m_k}; \mu^{m_k})\| - \frac{1}{2}\|\varphi(\mathbf{x}^{m_k}, \mathbf{s}^{m_k}; \mu^{m_k})\|^2/\varpi. \quad (11)$$

   Consider the function $f(a)$ defined on $a \in [0, \beta\mu^{m_k}]$:

   $$f(a) \stackrel{\text{def}}{=} a - \frac{1}{2}a^2/\varpi.$$

   (a) Assume $\varpi \leq \beta\mu^{m_k}$.
   The maximum of $f(a)$ is achieved at $\tilde{a} = \varpi$, with value

   $$f^{\max} = \frac{1}{2}\varpi \leq \frac{1}{2}\beta\mu^{m_k}.$$

   Combine (9) and (11),

   $$\|\varphi(\mathbf{x}^{m_k+1}, \mathbf{s}^{m_k+1}; \rho\mu^{m_k})\| \leq \frac{1}{2}\beta\mu^{m_k} + \varpi(1-\rho)\mu^{m_k}.$$

   When

   $$\rho \geq 1 - \frac{\beta}{2(\beta + \varpi)},$$

   $$\|\varphi(\mathbf{x}^{m_k+1}, \mathbf{s}^{m_k+1}; \rho\mu^{m_k})\| \leq \beta\rho\mu^{m_k}.$$

   (b) Assume $\varpi > \beta\mu^{m_k}$.
   The maximum of $f(a)$ is attained at $\tilde{a} = \beta\mu^{m_k}$, and

   $$f^{\max} = \beta\mu^{m_k} - \frac{1}{2}\frac{\beta^2(\mu^{m_k})^2}{\varpi}.$$

   Similar to (1a), when

   $$\rho \geq 1 - \frac{\beta^2\mu^{m_k}}{2\varpi(\beta + \varpi)},$$

   $$\|\varphi(\mathbf{x}^{m_k+1}, \mathbf{s}^{m_k+1}; \rho\mu^{m_k})\| \leq \beta\rho\mu^{m_k}.$$

2. Consider the case $\|\varphi(\mathbf{x}^{m_k},\mathbf{s}^{m_k};\mu^{m_k})\| \geq \|\Xi(\mathbf{x}^{m_k},\mathbf{s}^{m_k};\theta\Delta\theta\mathbf{x}^{m_k},\Delta\mathbf{s}^{m_k})\|$. Let
$$\tilde{\alpha} = 1.$$
Then
$$\|\varphi(\mathbf{x}^{m_k+1},\mathbf{s}^{m_k+1};\mu^{m_k})\| \leq \frac{1}{2}\|\Xi\| \leq \frac{1}{2}\varphi(\mathbf{x}^{m_k},\mathbf{s}^{m_k};\mu^{m_k}).$$
Since $(\mathbf{x}^{m_k};\mathbf{s}^{m_k}) \in \mathcal{N}_\beta(\mu^{m_k})$, it is not difficult to see that when
$$\rho \geq 1 - \frac{\beta}{2\varpi(\beta+\varpi)},$$
$$\|\varphi(\mathbf{x}^{m_k+1},\mathbf{s}^{m_k+1};\rho\mu^{m_k})\| \leq \beta\rho\mu^{m_k}.$$
Obviously, $\hat{\alpha} \geq \tilde{\alpha}$. Define
$$\rho_\epsilon \stackrel{def}{=} \max\left(1 - \frac{\beta}{2(\beta+\varpi)}, 1 - \frac{\beta}{2\varpi(\beta+\varpi)}, 1 - \frac{\beta^2\mu_\epsilon}{2\varpi(\beta+\varpi)}\right).$$
Then $\rho_\epsilon \in [0,1)$ is a constant and for all m_k,
$$\hat{\mu}^{m_k+1} \leq \rho_\epsilon \mu^{m_k}.$$

Since $\{\mu^k\}_{k=1}^\infty$ is a nonincreasing sequence, the above inequality implies that the sub-sequence $\{\mu^{m_k+1}\}_{k=1}^\infty$ converges to zero, which implies the whole sequence $\{\mu^k\}_{k=1}^\infty$ is less than μ_ϵ after finite iterations. That contradicts to the assumption $\mu^k \geq \mu_\epsilon$ for all $k = 1, \ldots$. □

Remark 1. The above results also show the global linear convergence rate of the algorithm.

4 Nonsingularity of the Jacobian at Optimum

In this part, we will show that the generalized Jacobian of (6) is nonsingular under certain conditions at optimum $(\mathbf{x}^*;\mathbf{s}^*)$. Then by [8], there is a neighborhood of $(\mathbf{x}^*;\mathbf{s}^*,0)$, such that for $\mu = 0$, the generalized Jacobian of (7) evaluated at each element in this neighborhood is nonsingular and is lower bounded. If our initial point is in that neighborhood, the search direction $(\Delta\mathbf{x}^k;\Delta\mathbf{s}^k)$ is upper bounded for each iteration k. Therefore, our algorithm terminates in finite iterations.

Assume strict complementarity is satisfied. That means we can divide the index set $Ind \stackrel{def}{=} \{1,\ldots,N\}$ into three parts

$$O \stackrel{def}{=} \{i \in Ind: \mathbf{x}_i = \mathbf{0}, \quad \mathbf{s}_i \in \text{int}\, Q_{n_i}\},$$
$$I \stackrel{def}{=} \{i \in Ind: \mathbf{x}_i \in \text{int}\, Q_{n_i}, \quad \mathbf{s}_i = \mathbf{0}\},$$
$$B \stackrel{def}{=} \{i \in Ind: \mathbf{x}_i \in \text{bd}\, Q_{n_i}, \quad \mathbf{s}_i \in \text{bd}\, Q_{n_i}\}.$$

Here, $\text{bd}\, Q_{n_i} \stackrel{def}{=} \{\mathbf{x} \in Q_{ni}: x_0 \geq \|\bar{\mathbf{x}}\|_2, x_0 \neq 0\}$. See [2].

To analyze the generalized Jacobian, we first consider $\partial_{\mathbf{x}_i}(\varphi_i)$ and $\partial_{\mathbf{s}_i}(\varphi_i)$ for $i \in O, I, B$ separately.

Since (2a) is satisfied at optimum, for $i \in O$, $\frac{d\varphi_i}{ds_i} = 0$ and any element in the subdifferential of φ_i at $\mathbf{x}_i$, denoted as $\partial_{\mathbf{x}_i}(\varphi_i)$, is the following arrow-shape matrix:

$$\begin{pmatrix} \xi_0 - \sqrt{2} & \xi_1 & \cdots & \xi_{n_i} \\ \frac{(s_i)_1}{\sqrt{(s_i)_1^2+(s_i)_0^2}} & \frac{(s_i)_0}{\sqrt{(s_i)_1^2+(s_i)_0^2}} & & \\ \vdots & & \ddots & \\ \frac{(s_i)_{n_i}}{\sqrt{(s_i)_{n_i}^2+(s_i)_0^2}} & & & \frac{(s_i)_0}{\sqrt{(s_i)_{n_i}^2+(s_i)_0^2}} \end{pmatrix}, \quad (12)$$

where $\boldsymbol{\xi}$, satisfying $\xi_0^2 + \cdots + \xi_{n_i}^2 = 1$, is a subgradient of the (n_i+1) dimensional vector $\mathbf{0}$. Next, we will show that any element in $\partial_{\mathbf{x}_i}(\varphi_i)$ is nonsingular. It is sufficient to prove that the determinant of (12) is negative. It is easy to verify that the determinant of (12) is

$$\frac{(s_i)_0^{n_i-1}}{\prod_{j=1}^{n_i} \sqrt{(s_i)_0^2 + (s_i)_j^2}} \left[(\xi_0 - \sqrt{2})(s_i)_0 - \sum_{j=1}^{n_i} \xi_j(s_i)_j \right]. \quad (13)$$

When $\bar{\boldsymbol{\xi}} = \mathbf{0}$, $|\xi_0|$ must be 1. Thus (13) < 0.

Now we consider the case $\bar{\boldsymbol{\xi}} \neq \mathbf{0}$. By Cauchy-Schwartz-Boniakovsky inequality and $\mathbf{s}_i \in \text{int } Q_{n_i}$,

$$-\sum_{j=1}^{n_i} \xi_j(s_i)_j \leq \|\bar{\boldsymbol{\xi}}\|_2 \|\bar{\mathbf{s}}_i\|_2 < \|\bar{\boldsymbol{\xi}}\|_2 (s_i)_0.$$

Note that

$$\xi_0 + \|\bar{\boldsymbol{\xi}}\|_2 \leq \sqrt{2\left(\xi_0^2 + \|\bar{\boldsymbol{\xi}}\|_2^2\right)} = \sqrt{2}.$$

Therefore, (13) < 0.

Hence, any element in $\partial_{\mathbf{x}_i}(\varphi_i)$ is nonsingular.

Similarly, one can verify that for $i \in I$, $\frac{d\varphi_i}{dx_i} = 0$ and the determinant of any element in $\partial_{\mathbf{s}_i}(\varphi_i)$ is negative; hence nonsingular.

For $i \in B$, any element in $\partial_{x_i}(\varphi_i)$ is

$$\partial_{\mathbf{x}_i}(\varphi_i, \zeta) = \begin{pmatrix} \zeta\left[\frac{(x_i)_0}{\|\mathbf{x}\|_2} - \sqrt{2}\right] & \zeta\frac{(x_i)_1}{\|\mathbf{x}\|_2} & \cdots & \zeta\frac{(x_i)_{n_i}}{\|\mathbf{x}\|_2} \\ \frac{(s_i)_1}{\sqrt{(s_i)_1^2+(s_i)_0^2}} & \frac{(s_i)_0}{\sqrt{(s_i)_1^2+(s_i)_0^2}} & & \\ \vdots & & \ddots & \\ \frac{(s_i)_{n_i}}{\sqrt{(s_i)_{n_i}^2+(s_i)_0^2}} & & & \frac{(s_i)_0}{\sqrt{(s_i)_{n_i}^2+(s_i)_0^2}} \end{pmatrix},$$

and its corresponding part in $\partial_{s_i}(\varphi_i)$ is

$$\partial_{\mathbf{s}_i}(\varphi_i,\zeta) = \begin{pmatrix} \eta\left[\frac{(s_i)_0}{\|\mathbf{s}\|_2} - \sqrt{2}\right] & \eta\frac{(s_i)_1}{\|\mathbf{s}\|_2} & \cdots & \eta\frac{(s_i)_{n_i}}{\|\mathbf{s}\|_2} \\ \frac{(x_i)_1}{\sqrt{(x_i)_1^2+(x_i)_0^2}} & \frac{(x_i)_0}{\sqrt{(x_i)_1^2+(x_i)_0^2}} & & \\ \vdots & & \ddots & \\ \frac{(x_i)_{n_i}}{\sqrt{(x_i)_{n_i}^2+(x_i)_0^2}} & & & \frac{(x_i)_0}{\sqrt{(x_i)_{n_i}^2+(x_i)_0^2}} \end{pmatrix},$$

where $(1-\zeta)^2 + (1-\eta)^2 = 1$.

The above equality implies $\zeta + \eta \neq 0$. Note that for any $i \in B$, $\frac{(x_i)_0}{\|\mathbf{x}_i\|_2} = \frac{(s_i)_0}{\|\mathbf{s}_i\|_2} = \frac{\sqrt{2}}{2}$. In addition, by (2a),

$$\frac{(x_i)_j}{\|\mathbf{x}_i\|_2} = -\frac{(s_i)_j}{\|\mathbf{s}_i\|_2} \ (j=1,\ldots,n_i) \,, \quad \frac{(x_i)_0}{\|\mathbf{x}_i\|_2} = \frac{(s_i)_0}{\|\mathbf{s}_i\|_2}.$$

Therefore,

$$\begin{pmatrix} -\eta & & & \\ & 1 & & \\ & & \ddots & \\ & & & 1 \end{pmatrix} \partial_{x_i}(\varphi_i,\zeta) \begin{pmatrix} -1 & & & \\ & 1 & & \\ & & \ddots & \\ & & & 1 \end{pmatrix} = \begin{pmatrix} \zeta & & & \\ & 1 & & \\ & & \ddots & \\ & & & 1 \end{pmatrix} \partial_{s_i}(\varphi_i,\zeta).$$

Next, we consider each element in the generalized Jacobian of (6), which is

$$J(\mathbf{x},\mathbf{s},\zeta) \stackrel{\text{def}}{=} \begin{pmatrix} M & -I \\ \partial_{\mathbf{x}}(\varphi_i,\zeta) & \partial_{\mathbf{s}}(\varphi_i,\zeta) \end{pmatrix}.$$

Below, we will give conditions under which each element in $J(\mathbf{x},\mathbf{s},\zeta)$ is nonsingular. We break up the columns of M into three parts according to the partition of the index set and divide the rows of M accordingly. We use $M_{(E,F)}$ to represent the submatrix of M with its rows from index set E and its columns from index set F. Thus $M_{(IB,B)}$ is the submatrix of M consisting of all columns corresponding to the boundary blocks and all rows corresponding to the interior and boundary blocks. Similar notations apply to the identity matrix I. Since $\partial_{\mathbf{x}_O}(\varphi_O)$ and $\partial_{\mathbf{s}_I}(\varphi_I)$ are nonsingular, nonsingularity of $J(\mathbf{x},\mathbf{s},\zeta)$ is equivalent to the following matrix being nonsingular:

$$\begin{pmatrix} M_{(I,I)} & M_{(I,B)} & \\ M_{(B,I)} & M_{(B,B)} & -I_B \\ \partial_{\mathbf{x}_B}(\varphi_B,\zeta) & \partial_{\mathbf{s}_B}(\varphi_B,\zeta) \end{pmatrix}.$$

First, we add the result of R left multiplying the columns corresponding to $(-I_B)$ to columns corresponding to $M_{(\cdot,B)}$. Note that due to (2a), for all $i \in B$, $j = 1,\ldots,n_i$:

$$\frac{(x_i)_j}{\sqrt{(x_i)_j^2 + (x_i)_0^2}} + \frac{(s_i)_j}{\sqrt{(s_i)_j^2 + (s_i)_0^2}} = 0, \quad \frac{(x_i)_j}{\|\mathbf{x}_i\|_2} + \frac{(s_i)_j}{\|\mathbf{s}_i\|_2} = 0.$$

Next, for all $i \in B$, we subtract the result of right multiplying $\frac{\bar{\mathbf{x}}_i}{(x_i)_0}$ to the columns corresponding to $(-I_{\bar{B}})$ from the column corresponding to $(I_i)_0$. Here, we use $\bar{B}$ to represent the columns (rows) corresponding to $\bar{\mathbf{x}}_i$, for all $i \in B$.

Without loss of generality, assume $B = \{1, 2, \ldots, m\}$. Because $\frac{(x_i)_0}{\|\mathbf{x}\|_2} = \frac{(s_i)_0}{\|\mathbf{s}\|_2} = \frac{\sqrt{2}}{2}$, $\eta + \zeta \neq 0$, the nonsingularity of $J(\mathbf{x}, \mathbf{s}, \zeta)$ is equivalent to the nonsingularity of the following matrix. Here, $A \stackrel{\text{def}}{=} \text{Diag}(R\mathbf{x}_1, \ldots, R\mathbf{x}_m)$.

$$\begin{bmatrix} M_{(I,I)} & M_{(I,B)} \\ M_{(B,I)} & (M_{(B,B)} - R) \; A \\ & A^T \end{bmatrix}$$

Remark 2. It is easy to verify that under the special case of second-order cone program, the above assumptions are the same as the primal-dual nondegeneracy and strict complementarity conditions of [9].

Remark 3. The above analysis is still applicable, if (3) is replaced by (2a) in (6), and in (5) the Fischer-Burmeister function is replaced by any other nonlinear complimentary function $\psi(a, b)$ with gradient satisfying

$$\begin{cases} \frac{d\psi(a,b)}{da} = 0, & \frac{d\psi(a,b)}{db} \neq 0 \quad (a \neq 0, b = 0) \\ \frac{d\psi(a,b)}{da} \neq 0, & \frac{d\psi(a,b)}{db} = 0 \quad (a = 0, b \neq 0). \end{cases}$$

Remark 4. For a boundary block $\mathbf{x}_i$, $\partial_{\mathbf{x}_i}(\varphi_i, \zeta)$ and $\partial_{\mathbf{s}_i}(\varphi_i, \zeta)$ are singular, since $\left[\text{Diag}\left(1, \frac{\sqrt{(s_i)_0^2 + (s_i)_j^2}}{\|\mathbf{s}_i\|_2}\right) \partial_{\mathbf{s}_i}(\varphi_i, \zeta) \right]$ and $\left[\text{Diag}\left(1, \frac{\sqrt{(x_i)_0^2 + (x_i)_j^2}}{\|\mathbf{x}_i\|_2}\right) \partial_{\mathbf{s}_i}(\varphi_i, \zeta) \right]$ have zero eigenvalues.

References

1. Alizadeh, F., Goldfarb, D.: Second-order cone programming. Math. Program. **95** (2003) 3–51
2. Adler, I., Alizadeh, F.: Primal-dual interior point algorithms for convex quadratically constrained and semidefinite optimization problems. Technical Report RRR 46-95, RUTCOR, Rutgers University (1995)
3. Fischer, A., Peng, J., Terlaky, T.: A new complementarity function for P cones. Manuscript (2003)
4. Fischer, A.: A special Newton-type optimization method. Optimization **24** (1992) 269–284
5. Chen, B., Harker, P.T.: A non-interior-point continuation method for linear complementarity problems. SIAM J. Matrix Anal. Appl. **14** (1993) 1168–1190
6. Kanzow, C.: Some noninterior continuation methods for linear complementarity problems. SIAM J. Matrix Anal. Appl. **17** (1996) 851–868

7. Burke, J.V., Xu, S.: The global linear convergence of a noninterior path-following algorithm for linear complementarity problems. Math. Oper. Res. **23** (1998) 719–734
8. Clarke, F.H.: Optimization and nonsmooth analysis. Canadian Mathematical Society Series of Monographs and Advanced Texts. John Wiley & Sons Inc., New York (1983)
9. Alizadeh, F., Schmieta, S.H.: Optimization with semidefinite, quadratic and linear constraints. Technical Report RRR 23-97, RUTCOR, Rutgers University (1997)

Fuzzy Multi-criteria Decision Making Approach for Transport Projects Evaluation in Istanbul [*]

E. Ertugrul Karsak and S. Sebnem Ahiska

Industrial Engineering Department, Galatasaray University,
Ortakoy, Istanbul 80840, Turkey
ekarsak@gsu.edu.tr, ssahiska@ncsu.edu

Abstract. Istanbul's current transport systems' capacity that stays inadequate faced with the uncontrolled growth of population has required the evaluation of several transport projects among which the construction of a new bridge, the construction of an under-water railway tunnel and the improvement of the current sea transport are cited as the most popular ones. This paper presents a robust two-phase fuzzy decision framework, which integrates the fuzzy Delphi method and a hierarchical distance-based fuzzy multi-criteria decision making (MCDM) approach, for evaluating these transport alternatives based on a comprehensive list of quantitative and qualitative performance attributes.

1 Introduction

Urban congestion is a serious worldwide problem that can adversely affect several objectives of transportation policy such as economic efficiency, environmental protection, accessibility, sustainability, economic regeneration, and equity [10]. Istanbul, the biggest metropolitan city of Turkey and one of the largest in Europe, has been encountering urban transportation problems for many years. Especially, the Bosphorus and Fatih bridges that connect the Asian and European regions of Istanbul suffer seriously from traffic congestion due to the fact that many people are obliged to make daily crossings from one side to the other for several purposes such as education, work or leisure activities.

Numerous alternative transport projects for the Bosphorus crossing, which includes the improvement of the Bosphorus and Fatih bridges and their connecting roads, the creation of new roads (e.g. bridges for wheeled or railway vehicles, tunnels for wheeled or railway vehicles), and the improvement of the current public transport system (e.g. buses or ferry system), have been suggested in the past decade [5, 13].

Transportation project selection is a complex decision problem since it usually involves the consideration of multiple conflicting objectives among which some may be qualitative or intangible that are very difficult if not impossible to express

---

[*] This research has been financially supported by Galatasaray University Research Fund.

in monetary or quantitative terms. The most frequently used techniques in transportation project appraisal can be classified as cost-benefit analysis (CBA) and multi-criteria analysis (MCA). CBA is criticized for generally excluding qualitative criteria, which are very difficult or impossible to transform into monetary values, from the analysis. MCA is suggested as a remedy to these problems for being capable of considering both monetary and non-monetary criteria simultaneously, and thus, deal with different dimensions of project evaluation such as social, environmental and financial aspects. A number of recent articles suggest the use of MCA for transportation projects evaluation [2, 11, 14, 15].

Several authors have employed fuzzy set theory to deal with imprecision faced while evaluating transportation projects. Teng and Tzeng [12] have proposed the use of a binary fuzzy multi-objective programming model for the selection of transportation investment projects. Avineri et al. [1] developed a fuzzy expert system, which enables the evaluation and selection of the interurban road projects in Israel.

This paper proposes a two-phase fuzzy multi-criteria decision making (MCDM) approach to address the transportation problem between the Asian and European sides of Istanbul. The proposed framework possesses a number of merits compared to previous studies. First, it enables the decision-makers to use linguistic terms when making qualitative assessments and thus reduces their cognitive burden in the evaluation process. Second, owing to the use of fuzzy set theory, it can handle data that are in forms of linguistic terms or fuzzy numbers as well as crisp numbers, which enables both quantitative and qualitative criteria to be incorporated into the decision process. Third, unlike most of the alternative approaches employed in evaluating transport alternatives, the proposed approach can handle evaluation criteria that are structured in multi-level hierarchies. Fourth, the proposed approach considers both ideal and anti-ideal solutions simultaneously, considering that people prefer to be as close as possible to the ideal and as distant as possible from the anti-ideal [17]. Fifth, unlike alternative fuzzy decision making tools, the developed methodology does not require the use of fuzzy number ranking methods that are reported to yield inconsistent results. Finally, the required computations are straightforward and can be easily programmed.

2 Two-Phase Hierarchical Fuzzy MCDM Approach

Real-world decision problems such as the selection of the best transport alternative for the Bosphorus crossing in Istanbul often involve the consideration of numerous performance attributes, yielding in general a multi-level hierarchical structure. Further, decision-makers often fail to assess precise importance weights or performance ratings to attributes that are qualitative or unpredictable. Thus, the decision tool should also possess the capability of quantifying the imprecision inherent in decision-makers' subjective assessments. This paper proposes a robust two-phase fuzzy MCDM approach, which can address decision problems having multi-level hierarchical structure where a number of qualitative as well as quantitative performance attributes are present.

Fuzzy set theory, which was introduced by Zadeh [16] to deal with problems in which a source of vagueness is involved, has been utilized for incorporating imprecise data into the decision framework. Recently, Karsak [7] has introduced a distance-based fuzzy MCDM approach for technology selection that is based on the proximity to ideal solution concept and has the capability of incorporating both crisp and fuzzy data. The origins of Karsak's approach can be found in the multi-criteria decision aid named TOPSIS (technique for order preference by similarity to ideal solution) [6].

When a large number of performance attributes are to be considered in the evaluation process, it may be preferred to structure them in a multi-level hierarchy in order to conduct a more effective decision analysis. This paper extends the distance-based fuzzy MCDM algorithm proposed by Karsak [7] to address decision problems having multi-level hierarchical structure. The decision framework set forth in here for the evaluation of transport alternatives in Istanbul consists of two phases. In the first phase, the fuzzy Delphi method, which is an effective group decision making tool, is applied to obtain the qualitative assessments of decision-makers regarding the importance of criteria and sub-criteria as well as the performance of alternatives with respect to qualitative sub-criteria. Instead of being merely constrained to using crisp values as with the traditional Delphi method, the fuzzy Delphi method enables the experts to express subjective information and opinions in terms of linguistic variables. A number of successful applications of the fuzzy Delphi method to achieve a consensus of experts' opinions which are denoted by fuzzy numbers have been presented [3, 4, 8]. In the second phase, the hierarchical distance-based fuzzy MCDM algorithm is employed for determining the best transport alternative.

The proposed two-phase fuzzy MCDM approach can be described as follows:

Phase 1. Obtaining consensus information to be used in evaluating alternatives employing the fuzzy Delphi method.

Step 1. The experts provide their qualitative assessments regarding the importance weights of criteria and sub-criteria, and the performance of alternatives with respect to qualitative sub-criteria using linguistic terms. Then, the importance weights and performance ratings assigned by each expert are expressed in the form of a triangular fuzzy number as

$$\tilde{w}_{jp} = (w_{ajp}, w_{bjp}, w_{cjp}), j = 1, 2, \ldots, n; p = 1, 2, \ldots, q$$

$$\tilde{w}_{jkp} = (w_{ajkp}, w_{bjkp}, w_{cjkp}), \forall j, p; k = 1, 2, \ldots, z_j$$

$$\tilde{y}_{ijkp} = (y_{aijkp}, y_{bijkp}, y_{cijkp}), \forall j, k, p; i = 1, 2, \ldots, m$$

where $\tilde{w}_{jp}$ represents the importance weight assigned to criterion j by expert p, $\tilde{w}_{jkp}$ represents the importance weight assigned to the sub-criterion k of criterion j by expert p, $\tilde{y}_{ijkp}$ denotes the performance rating assigned to alternative i with respect to the sub-criterion k of criterion j by expert p, m indicates the number of available alternatives, n indicates the number of criteria under consideration, z_j indicates the number of sub-criteria of criterion j, and q denotes the number of experts.

Step 2. First, the average importance weights for both criteria and sub-criteria, and the average performance ratings of the alternatives for qualitative sub-criteria are calculated using the arithmetic mean operator as follows:

$$\tilde{w}_j = (w_{aj}, w_{bj}, w_{cj}) = (\frac{1}{q}\sum_{p=1}^{q} w_{ajp}, \frac{1}{q}\sum_{p=1}^{q} w_{bjp}, \frac{1}{q}\sum_{p=1}^{q} w_{cjp}), \forall j \quad (1)$$

$$\tilde{w}_{jk} = (w_{ajk}, w_{bjk}, w_{cjk}) = (\frac{1}{q}\sum_{p=1}^{q} w_{ajkp}, \frac{1}{q}\sum_{p=1}^{q} w_{bjkp}, \frac{1}{q}\sum_{p=1}^{q} w_{cjkp}), \forall j, k \quad (2)$$

$$\tilde{y}_{ijk} = (\frac{1}{q}\sum_{p=1}^{q} y_{aijkp}, \frac{1}{q}\sum_{p=1}^{q} y_{bijkp}, \frac{1}{q}\sum_{p=1}^{q} y_{cijkp}), \forall i, j, k \quad (3)$$

where $\tilde{w}_j$ denotes the average importance weight of criterion j, $\tilde{w}_{jk}$ indicates the average importance weight of sub-criterion k of criterion j, and $\tilde{y}_{ijk}$ indicates the average performance rating of alternative i with respect to sub-criterion k of criterion j. Then, for each expert (E_p) the deviations of the individual assessments from their corresponding averages, i.e. the differences between $\tilde{w}_{jp}$ and $\tilde{w}_j$, the differences between $\tilde{w}_{jkp}$ and $\tilde{w}_{jk}$, and the differences between $\tilde{y}_{ijkp}$ and $\tilde{y}_{ijk}$ are computed, and sent back to the experts for re-evaluation.

Step 3. Each expert presents her revised assessments as $\tilde{w}_{jp}^1$, $\tilde{w}_{jkp}^1$, and $\tilde{y}_{ijkp}^1$. The averages of these revised assessments are calculated in an analogous way employing formulas (1), (2) and (3) leading to $\tilde{w}_j^1$, $\tilde{w}_{jk}^1$, and $\tilde{y}_{ijk}^1$, respectively. Then, the closeness of $\tilde{w}_j^1$ to $\tilde{w}_j$, $\tilde{w}_{jk}^1$ to $\tilde{w}_{jk}$, and $\tilde{y}_{ijk}^1$ to $\tilde{y}_{ijk}$, respectively, is calculated using a distance measure [3] as follows:

$$d(\tilde{w}_j^1, \tilde{w}_j) = \frac{1}{2}\left\{max\left(|w_{aj}^1 - w_{aj}|, |w_{cj}^1 - w_{cj}|\right) + |w_{bj}^1 - w_{bj}|\right\}, \forall j \quad (4)$$

$$d(\tilde{w}_{jk}^1, \tilde{w}_{jk}) = \frac{1}{2}\left\{max\left(|w_{ajk}^1 - w_{ajk}|, |w_{cjk}^1 - w_{cjk}|\right) + |w_{bjk}^1 - w_{bjk}|\right\}, \forall j, k \quad (5)$$

$$d(\tilde{y}_{ijk}^1, \tilde{y}_{ijk}) = \frac{1}{2}\left\{max\left(|y_{aijk}^1 - y_{aijk}|, |y_{cijk}^1 - y_{cijk}|\right) + |y_{bijk}^1 - y_{bijk}|\right\}, \forall i, j, k \quad (6)$$

This process starting with Step 2 is repeated until two successive averages for importance weights as well as performance ratings are reasonably close. It is assumed that the distance less than or equal to 0.2 corresponds to two reasonably close fuzzy numbers [4].

Step 4. If there is new important information available, the fuzzy estimates can be re-examined employing the same process at a later time.

Phase 2. Identifying the best alternative employing the hierarchical distance-based fuzzy MCDM approach.

Step 1. Construct the decision matrix that denotes the average fuzzy assessments corresponding to qualitative sub-criteria and the crisp values corresponding to quantitative sub-criteria for the considered alternatives.

Step 2. Normalize the crisp data to obtain unit-free and comparable sub-criteria values. The normalized values for crisp data regarding benefit-related as well as cost-related quantitative sub-criteria are calculated via a linear scale transformation as

$$\acute{y}_{ijk} = \begin{cases} \frac{y_{ijk} - y_{jk}^-}{y_{jk}^* - y_{jk}^-}, & k \in CB_j; i = 1, 2, \ldots, m; j = 1, 2, \ldots, n \\ \frac{y_{jk}^* - y_{ijk}}{y_{jk}^* - y_{jk}^-}, & k \in CC_j; i = 1, 2, \ldots, m; j = 1, 2, \ldots, n \end{cases} \quad (7)$$

where $\acute{y}_{ijk}$ denotes the normalized value of y_{ijk}, which is the crisp value assigned to alternative i with respect to the sub-criterion k of criterion j, m is the number of alternatives, n is the number of criteria, CBj is the set of benefit-related crisp sub-criteria of criterion j and CCj is the set of cost-related crisp sub-criteria of criterion j, $y_{jk}^* = max_i y_{ijk}$ and $y_{jk}^- = min_i y_{ijk}$. The normalized values for crisp data can be represented as $\tilde{y}_{ijk} = (\acute{y}_{aijk}, \acute{y}_{bijk}, \acute{y}_{cijk})$ in triangular fuzzy number format, where $\acute{y}_{aijk} = \acute{y}_{bijk} = \acute{y}_{cijk} = \acute{y}_{ijk}$.

Step 3. Aggregate the performance ratings of alternatives at the sub-criteria level to criteria level as follows:

$$\tilde{x}_{ij} = (x_{aij}, x_{bij}, x_{cij}) = \frac{\sum_k \tilde{w}_{jk}^1 \otimes \tilde{y}_{ijk}}{\sum_k \tilde{w}_{jk}^1}, \forall i, j \quad (8)$$

where $\tilde{x}_{ij}$ represents the aggregate performance rating of alternative i with respect to criterion j, $\tilde{w}_{jk}^1$ indicates the average importance weight assigned to sub-criterion k of criterion j at the end of Phase 1, and $\otimes$ is the fuzzy multiplication operator.

Step 4. Normalize the aggregate performance ratings at criteria level using a linear normalization procedure, which results in the best value to be equal to 1 and the worst one to be equal to 0, as follows:

$$\tilde{r}_{ij} = (r_{aij}, r_{bij}, r_{cij}) = \left(\frac{x_{aij} - x_{aj}^-}{x_{cj}^* - x_{aj}^-}, \frac{x_{bij} - x_{aj}^-}{x_{cj}^* - x_{aj}^-}, \frac{x_{cij} - x_{aj}^-}{x_{cj}^* - x_{aj}^-} \right), \forall i, j \quad (9)$$

where $x_{cj}^* = max_i x_{cij}$, $x_{aj}^- = min_i x_{aij}$, and $\tilde{r}_{ij}$ denotes the normalized aggregate performance rating of alternative i with respect to criterion j.

Step 5. Define the ideal solution $A^* = (r_1^*, r_2^*, \ldots, r_n^*)$ and the anti-ideal solution $A^- = (r_1^-, r_2^-, \ldots, r_n^-)$, where $r_j^* = (1, 1, 1)$ and $r_j^- = (0, 0, 0)$ for $j = 1, 2, \ldots, n$.

Step 6. Calculate the weighted distances from ideal solution and anti-ideal solution (D_i^* and D_i^-, respectively) for each alternative as follows:

$$D_i^* = \sum_j \frac{1}{2} \left\{ max \left(w_{aj}^1 |r_{aij} - 1|, w_{cj}^1 |r_{cij} - 1| \right) + w_{bj}^1 |r_{bij} - 1| \right\}, \forall i \quad (10)$$

$$D_i^- = \sum_j \frac{1}{2} \left\{ max \left(w_{aj}^1 |r_{aij} - 0|, w_{cj}^1 |r_{cij} - 0| \right) + w_{bj}^1 |r_{bij} - 0| \right\}, \forall i \quad (11)$$

Step 7. Calculate the proximity of the alternatives to the ideal solution, P_i^*, by considering the distances from ideal and anti-ideal solutions as

$$P_i^* = D_i^- / (D_i^* + D_i^-), \forall i \quad (12)$$

Step 8. Rank the alternatives according to P_i^* values in descending order. Identify the alternative with the highest P_i^* as the best alternative.

3 Application of the Framework to Istanbul's Transport Problem

In here, the proposed two-phase fuzzy MCDM approach is used for selecting the best transport alternative for the Bosphorus water-crossing among three mutually exclusive alternatives, namely the construction of a new bridge for wheeled vehicles, the construction of an under-water tube tunnel for railway vehicles, and the improvement of the current sea (ferry) transport. Benefiting from the literature on the evaluation of transport alternatives, economic criteria, environmental criteria, social and cultural criteria, criteria related to transportation policy, and transportation system criteria are identified as the selection criteria. Then, several relevant sub-criteria corresponding to these criteria are determined with the aim of conducting a comprehensive evaluation of transport alternatives. The elements of Istanbul's transport problem, i.e., the goal, the selection criteria and their associated sub-criteria as well as the transport alternatives to be evaluated, are defined as follows:

Goal (G): Selection of the best transport alternative for the Bosphorus water-crossing in Istanbul
Criteria (C_j) and their associated sub-criteria (SC_{jk}):
    C_1: Economic criteria
        SC_{11}: Construction cost
        SC_{12}: Operating and maintenance cost
        SC_{13}: Impact on the economic development of the region and the country
        SC_{14}: Possibility of increasing employment through the creation of new jobs
    C_2: Environmental criteria
        SC_{21}: Visual intrusion

SC_{22}: Damage on the ecosystem during construction
SC_{23}: Damage on the ecosystem during operation
C_3: Social and cultural criteria
SC_{31}: Impact on the population movement to the region
SC_{32}: Comfort offered to the passengers
SC_{33}: Risk of accidents resulting in death or injury during operation
SC_{34}: Accessibility to the transport alternative (in terms of existence of the connecting roads)
C_4: Criteria related to transportation policy
SC_{41}: Suitability to transportation master plan
SC_{42}: Dependence on the foreign countries during construction (in terms of finance, technology and energy)
C_5: Transportation system criteria
SC_{51}: Trip price (in terms of passenger cost)
SC_{52}: Time scheduling
SC_{53}: Capacity created (in terms of passenger/hour/destination)
SC_{54}: Vehicle traffic (in terms of vehicle/hour/destination)

Transport alternatives (A_i):
A_1: Construction of a new highway bridge
A_2: Construction of an under-water railway tunnel
A_3: Improvement of the current sea transport

A committee of 29 experts including university professors and specialists from the Ministry of Transportation and State Planning Organization is constructed.[1] A questionnaire that enables the experts to use linguistic terms in making their assessments is formed and distributed to each expert in the committee. Then, the fuzzy Delphi method is applied in order to obtain the assessments of experts regarding the importance of criteria and their associated sub-criteria as well as the performance of transport alternatives with respect to sub-criteria that are imprecise in nature.

Two different sets of linguistic terms are used in the questionnaire to express the subjectivity and vagueness inherent in the experts' assessments. In this study, the membership functions that were previously prescribed by Liang and Wang [9] for these linguistic terms are used. One set includes the linguistic terms "very low (VL)", "low (L)", "medium (M)", "high (H)" and "very high (VH)", which enable the experts to assess the importance of criteria and sub-criteria. The other set includes the linguistic terms "very poor (VP)", "poor (P)", "fair (F)", "good (G)" and "very good (VG)", which are used for rating the performance of the transport alternatives with respect to qualitative sub-criteria.

Once, all questionnaires are returned, linguistic assessments of experts are expressed in triangular fuzzy number format employing the membership functions, and the averages of the 29 experts' fuzzy assessments are calculated using equations (1), (2) and (3). Then, the deviations of the individual assessments from their corresponding averages are computed and sent back to experts for

---

[1] The authors are indebted to Asli Gul Oncel for her assistance in gathering data.

that they re-evaluate their responses. When the experts present their revised assessments, the distances between the averages of the revised assessments and the averages of the initial assessments are calculated using equations (4), (5) and (6). Following Cheng and Lin [4], the successive averages are assumed to be reasonably close when the distance between two fuzzy numbers is less than or equal to 0.2. Since the maximum distance between the successive averages is 0.059, we can conclude that the consensus condition is achieved.

The values of quantitative sub-criteria, namely construction cost, operating and maintenance cost, trip price and capacity created, and the fuzzy data, which include the averages of the revised fuzzy assessments of experts regarding the importance weights of criteria, the importance weights of sub-criteria and the performance ratings of alternatives with respect to qualitative sub-criteria are used as inputs to the hierarchical distance-based fuzzy MCDM algorithm to identify the best transport alternative for the Bosphorus crossing in Istanbul. The complete data set, which is not provided due to space limitations, can be obtained from the authors upon request.

The crisp data is normalized using equation (7). Then, sub-criteria values are aggregated to criteria level using equation (8), leading to the aggregate performance ratings of alternatives with respect to criteria. The normalized values of these aggregate performance ratings are computed using equation (9) and are represented in Table 1, where 0 indicates the worst value and 1 indicates the best value. From Table 1, we observe that A_1 has the highest performance with respect to economic criteria; A_2 has the highest performance with respect to environmental criteria, social and cultural criteria, and transportation system criteria; and A_3 has the highest performance with respect to criteria related to transportation policy.

Table 1. Normalized values of the aggregate performance ratings of transport alternatives with respect to criteria

	$\tilde{r}_{ij} = (r_{aij}, r_{bij}, r_{cij})$		
C_j	A_1	A_2	A_3
C_1	(0.083, 0.397, 1.000)	(0.017, 0.282, 0.755)	(0.000, 0.250, 0.739)
C_2	(0.000, 0.173, 0.636)	(0.112, 0.400, 1.000)	(0.086, 0.353, 0.939)
C_3	(0.000, 0.227, 0.744)	(0.106, 0.423, 1.000)	(0.062, 0.345, 0.921)
C_4	(0.000, 0.203, 0.656)	(0.082, 0.328, 0.783)	(0.131, 0.439, 1.000)
C_5	(0.000, 0.134, 0.432)	(0.203, 0.492, 1.000)	(0.087, 0.299, 0.704)

In order to determine the transport alternatives' overall performance, the ideal and anti-ideal solutions are obtained as defined in step 5 of the hierarchical distance-based fuzzy MCDM algorithm. Then, the weighted distances of each transport alternative from ideal and anti-ideal solutions, D_i^* and D_i^-, are calculated using equations (10) and (11), respectively. Finally, the proximity to the ideal solution for each transport alternative is computed employing equation (12). From Table 2, construction of an under-water railway tunnel, A_2, with the highest P_i^* value appears to be the best transport alternative, while A_3 and A_1 rank second and third, respectively.

Table 2. Ranking of the transport alternatives

A_i	P_i^*	Ranking
A_1	0.426	3
A_2	0.546	1
A_3	0.515	2

Sensitivity analyses are conducted to investigate the stability of the rankings with respect to the changes in the criteria sets. Economic criteria, with the highest importance weight assigned by the experts, are considered *sine qua non* for evaluating the transport alternatives, and thus, taken into account in all of the alternative scenarios. First, the delineated fuzzy MCDM procedure is applied to sets of four criteria, i.e. disregarding transportation system criteria (C_5), criteria related to transportation policy (C_4), social and cultural criteria (C_3), and environmental criteria (C_2), one at a time respectively. For four different combinations of criteria (i.e. $\{C_1, C_2, C_3, C_4\}$, $\{C_1, C_2, C_3, C_5\}$, $\{C_1, C_2, C_4, C_5\}$ and $\{C_1, C_3, C_4, C_5\}$), construction of an under-water railway tunnel, A_2, ranks first while A_3 and A_1 rank second and third, respectively. In other words, the rankings for four different criteria sets are the same as the one given in Table 2. Second, the decision analysis is conducted considering criteria sets of three criteria each, i.e. overlooking two criteria at a time. The results indicate that in four out of six scenarios, construction of an under-water railway tunnel ranks first with the highest P_i^* value, whereas for the criteria sets $\{C_1, C_2, C_4\}$ and $\{C_1, C_3, C_4\}$ improvement of the current sea transport is the best transport alternative. Construction of a new highway bridge ranks third in all of the alternative scenarios.

The Greater Municipality of Istanbul has also been favoring the construction of the railway tunnel since it contributes positively to public transport while preserving Istanbul's historical and natural structure, which is among the aims of the transportation master plan. On the other hand, the construction of the third bridge is not likely to be a long-term remedy to traffic congestion, since many studies reveal that creating new highway capacity leads to an increase in private car use, which in turn will create demand for more road space [5].

4 Concluding Remarks

The current transportation system of Istanbul that has remained insufficient faced with the high demand for daily crossings between the Asian and European sides, has led the local authorities to consider several transportation infrastructure projects such as constructing a new bridge or an under-water tunnel, which require large amounts of capital investments and thus careful and comprehensive examination before making a decision. This paper proposes a decision framework, which integrates the use of the fuzzy Delphi method and a hierarchical distance-based fuzzy MCDM algorithm, for determining the best transport alternative in Istanbul.

Besides having the capability of considering numerous factors that are structured in a multi-level hierarchy, the proposed decision framework enables the decision-makers to use linguistic terms when making their assessments regarding the importance of factors as well as the performance of transport alternatives with respect to qualitative factors, which reduces the cognitive burden of the evaluation process. In conclusion, due to its sound logic, effectiveness of quantifying the imprecision inherent in decision-maker's assessments and easily programmable structure, the decision framework presented in this paper appears to be a robust decision tool for comprehensive analysis of multi-level hierarchical transportation problems such as the Bosphorus crossing in Istanbul.

References

1. Avineri, E., Prashker, J., Ceder, A.: Transportation Projects Selection Process Using Fuzzy Sets Theory. Fuzzy Sets and Systems 116 (2000) 35-47
2. Bielli, M.: A DSS Approach to Urban Traffic Management. European Journal of Operational Research 61 (1992) 106-113
3. Bojadziev, G., Bojadziev, M.: Fuzzy Sets, Fuzzy Logic, Applications, Advances in Fuzzy Systems-Applications and Theory. Volume 5. World Scientific, Singapore (1995)
4. Cheng, C.H., Lin, Y.: Evaluating the Best Main Battle Tank Using Fuzzy Decision Theory with Linguistic Criteria Evaluation. European Journal of Operational Research 142 (2002) 174-186
5. Gercek, H., Yayla, N.: Crossing the Bosphorus Strait: A Railway Tunnel or Another Highway Bridge. Fourth Symposium on Strait Crossing, Norway (2001)
6. Hwang, C.-L., Yoon, K.: Multiple Attribute Decision Making: Methods and Applications. Springer, Heidelberg (1981)
7. Karsak, E.E.: Distance-Based Fuzzy MCDM Approach for Evaluating FMS Alternatives. International Journal of Production Research 40 (2002) 3167-3181
8. Karsak, E.E.: Fuzzy Multiple Objective Decision Making Approach to Prioritize Design Requirements in Quality Function Deployment. International Journal of Production Research 42 (2004) 3957-3974
9. Liang, G.S., Wang M.J.J.: A Fuzzy Multi-Criteria Decision-Making Approach for Robot Selection. Robotics & Computer Integrated Manufacturing 10 (1993) 267-274
10. May, A.D., Nash, C.A.: Urban Congestion: A European Perspective on Theory and Practice. Annual Review of Energy & the Environment 21 (1996) 239-260
11. Sayers, T.M., Jessop, A.T., Hills, P.J.: Multi-Criteria Evaluation of Transport Options–Flexible, Transparent and User-friendly?. Transport Policy 10 (2003) 95-105
12. Teng, J.Y., Tzeng, G.H.: Transportation Investment Project Selection Using Fuzzy Multiobjective Programming. Fuzzy Sets and Systems 96 (1998) 259-280
13. Ulengin, F., Topcu, Y.I., Sahin, S.O.: An Integrated Decision Aid System for Bosphorus Water-Crossing Problem. European Journal of Operational Research 134 (2001) 179-192

14. Vreeker, R., Nijkamp, P., Welle, C.T.: A Multicriteria Decision Support Methodology for Evaluating Airport Expansion Plans. Transportation Research Part D: Transport and Environment 7 (2002) 27-47
15. Yedla, S., Shrestha, R.M.: Multi-Criteria Approach for the Selection of Alternative Options for Environmentally Sustainable Transport System in Delhi. Transportation Research Part A: Policy and Practice 37 (2003) 717-729
16. Zadeh, L.A.: Fuzzy Sets. Information and Control 8 (1965) 338-353
17. Zeleny, M.: Multiple Criteria Decision Making. McGraw-Hill, New York (1982)

An Improved Group Setup Strategy for PCB Assembly

V. Jorge Leon[1] and In-Jae Jeong[2]

[1] Department of Engineering Technology,
Department of Industrial Engineering,Texas A&M University,
College Station, TX 77843-3367,USA
jleon@tamu.edu

[2] Department of Industrial Engineering, Hanyang University,
Seoul, 133-791, South Korea
ijeong@hanyang.ac.kr

Abstract. This paper considers a PCB group setup strategy in a single placement machine producing multiple products. The grouping objective is to minimize makespan that is composed of two attributes, the feeder change time and the placement time. Current group setup strategy considers only component similarity in grouping boards. In this paper, we propose an improved group setup strategy which considers both component similarity and geometric similarity using minimum metamorphic distance measures. The proposed group setup strategy is tested on a variety of production environments with existing group setup strategies and board's similarity coefficients. Test results show that the improved group setup strategy outperforms existing group setup strategies.

Keywords: PCB assembly, group setup, similarity coefficient.

1 Introduction

This paper considers the same group setup problem that is proposed by Leon and Peters (1998). We consider a single surface mount technology(SMT) machine that consists of three components: a table that holds the PCB, a feeder carriage that holds the components, and a head that picks up components from the feeders and places them on the PCB. The placement process begins from the movement of placement head. The head starts from a given home position, moves between feeders (on the machine) and placement location (on the PCB) until all components have been mounted. The head returns to the home position once all placements required for a board have been completed. The problem of determining a group setup strategy can be characterized by the grouping of similar boards. Feeders are loaded with the component types required by all of the members in a family. It is assumed that any board from a family can be processed on the machine without incurring changeover time. The conceptual formulation of the problem is as follows:

Notation:
f: family index $f = 1, \ldots K$
N_f: number of boards in family f
$N = \sum_{f=1}^{K} N_f$: total number of boards
i: board index $i = 1 \ldots N$
F: number of feeder slots on the machine
c_f: number of different component types required by boards in family f
σ: time to remove or install a feeder
n_f: number of feeder changes required from family f-1 to f
v: placement head velocity (mm/sec)
b_i: batch size of board i
d_i: length of tour followed by the head to assemble board i

Minimize: Makespan $= \sum_{f=1}^{K}(\sigma n_f + \sum_{i=1}^{N_f} \frac{b_i}{v} d_i)$
Subject to: Feeder capacity constraints
            Component placement constraints
            Component-feeder constraints

The objective is to minimize the makespan for producing multiple types of boards. The first term of the makespan is the setup time for a family and the second term is the placement time to produce boards in the family. We assume that the head velocity is constant for all types of boards. Also the time to remove or install a feeder is constant. The decision variables are the number of family, K, the types of boards in family f, N_f and the placement sequence of locations in board i and the component-feeder assignment for family f to determine $d_i, \forall f, i = 1 \ldots N_f$. Boards must be grouped such that within the family, boards share as many component types as possible in order to reduce setup time. Also the placement locations of boards must be similar to each others in order to reduce placement time without incurring additional setups.

The first constraints represent the feeder capacity constraints. Boards can be added to a family as long as the feeder capacity is not exceeded (i.e., $c_f \leq F, \forall f$). The second constraints, component placement constraints form a traveling salesman problem (TSP) constraints. That is, given a component-feeder assignment for each family, minimizing the time required to place all the components on the boards is equivalent to find a tour that visits all the placement locations. Component-feeder constraints means that each component needed for boards in a family must be assigned to a feeder. Only a single component type can reside in each feeder slot. If placement sequencing solution is given, the resulting problem is a linear assignment problem (LAP). As a result, a common approximate solution procedure solves TSP and LAP in cycles, using the solution of one problem as input to the next.

The contributions of this research is the enhancement of the traditional group setup strategy (Leon and Peter 1998). In the traditional group setup strategy, PCB families are formed with only respect to the component similarity. In the proposed improved group setup strategy, we consider both the component similarity and the geometric similarity of boards in determining the families of boards. The traditional group setup may be effective in cases where the feeder change time is high. However, it is conjectured when the feeder change time is relatively small and the placement time is large with respect to the makespan, the proposed improved group setup strategy may perform better than traditional group setup. Also this paper presents a comprehensive comparison among existing different setup strategies and board's similarity coefficients.

2 Literature Survey

2.1 Existing Setup Strategies

The traditional group setup procedure(Askin et al.,1994, Leon and Peters, 1996, 1998) is summarized in the following steps.
    Procedure group setup strategy.
Phase 1. Clustering (Form K families of boards with similar boards. Family sizes can not exceed the maximum number of feeder slots.)

    Step 1: Put each board-type in a single-member family
    Step 2: Compute similarity coefficient, s_{ij} for all pairs of family i and j
    Step 3: Compute clustering objective values
    Step 4: Set $T = max(s_{ij})$
    Step 5: Merge the pair of board i^* and j^*, if $s_{i^*j^*} = T$. Repeat until no more pairs can be merged at similarity level T.
    Step 6: Compute clustering objective and save the clustering solution if an improvement was achieved.
    Step 7: Repeat Step 2 through 6 while merging is possible.
Phase 2. Component-feeder assignment and placement sequence
    Step 8: Form a composite-board $H(f)$, f=1,..., K, this board consists of the superposition of all the placement locations with their corresponding components of the boards in family f.
    Step 9: Determine a feasible component-feeder assignment $C(H_f)$
    Step 10: For all $i \in N_f$, find a placement sequence $P(i)$, given $C(H_f)$
    Step 11: For all $i \in N_f$, find a component-feeder assignment $C(H_f)$ given $P(i)$
    Step 12: Repeat Step 10 and Step 11 for a predetermined number of iterations.

In Phase 1, the hierarchical clustering algorithm merges similar boards into a family. The clustering procedure continues until all boards form a single family. To form good families of boards, it is essential to develop a similarity coefficient which considers both the component similarity and the geometric similarity of any two boards. The clustering objective might be a minimization of the similarity coefficient between families or maximization of the similarity coefficient within families.

Component-feeder assignment and placement sequences are determined by solving iterative LAP and TSP proposed by Drezner and Nof(1984). For a given component-feeder assignment, $C(H_f)$, the placement sequencing problem can be solved as TSP problems. In this paper, we use the nearest-neighbor heuristic to solve the TSP. For a given placement sequences, $P(i)$, the component-feeder assignment problem is a LAP. The cost matrix of assignment is the component delivery time from feeder slots to placement locations. In this implementation, the LAP is solved using the shortest augmenting path algorithm proposed by Jonker and Vogenant(1987).

There are a number of other PCB setup strategies to reduce makespan in the literature (i.e., unique setup, minimum setup, group setup, partial setup). Refer to Drezner and Nof(1984), Ball and Magazine(1988), Crama et al.(1990) for unique setup, Lofgren and McGinnis(1986) for minimum setup and Leon and Peters(1996) for partial setup. Leon and Peters (1998) tested these setup strategies under a variety of production conditions and the results showed that partial setup performed better in average. Given a sequence of boards, partial setup specifies component-feeder assignment such that the placement time for the board is minimized if the increment of setup time is less than the reduction of placement time. If the reduction of place time is less than the increment of setup time, partial setup determines the sequence of boards in order to minimize the setup time. However, group setup might be a good alternative of partial setup strategy because the implementation of strategy is relatively easier than partial setup in industry. In group setup, the operator's responsibility is simple because all that he/she need to do is replacing a used feeders with new feeder which is already loaded with components. Contrarily, in partial setup, whenever a decision concerning to component-feeder assignment is made, operator put some components in a slot and changes feeder slots which might be a complicated operation. Therefore, we focus on the improvement of the traditional group setup strategy as an alternative of the partial setup strategy.

2.2 Existing Board's Similarity Coefficients for Group setups

This section introduces existing board's similarity coefficients that are used in the traditional group setup as shown in step 2 in section 2.1.

Component similarity based similarity coefficient

Leon and Peters(1998) proposed the component similarity based coefficient of board i and j as

$$s1_{ij} = \frac{\text{Number of common component types}}{\text{Number of total component types}} \quad (1)$$

The limitation of component similarity based coefficient is that it does not consider the geometric similarity of boards

Placement Location Matrix(PLM) based similarity coefficient

Quintana and Leon(1997) proposed PLM based similarity coefficient of board i and j which considers both component commonality and geometric similarity as follows;

Let point magnitude of coordinate (X, Y) be $\sqrt{X^2 + Y^2}$ and p_{ki} be the point magnitude of kth sorted placement location in ascending order for board i. n^* is the $\min(n_i, n_j)$ where n_i is the number of placement location of board i. NP_j is the number of placement locations of board j. Xrange and Yrange are the Cartesian distance of the largest board.

$$s_{ij} = 0.5((1 - D_{ij})F_{ij}) + 0.5 s1_{ij} \qquad (2)$$

where $D_{ij} = \dfrac{\sqrt{\sum_k^{n^*}(p_{ki}-p_{kj})^2}}{n^*\sqrt{Xrange^2+Yrange^2}}$, $F_{ij} = \dfrac{min(NP_i,NP_j)}{max(NP_i,NP_j)}$, $s1_{ij}$ is equivalent to equation (1).

The first term in equation (2) associates for geometric similarity and the second term is equivalent to the component commonality based similarity coefficient.

The limitation of the PLM measure is the assignment of equal weight for both component commonality and geometric similarity. As the authors mentioned, relative weights may vary depending on the problem under consideration.

3 Minimum Metamorphic Distance Based Similarity

In this section, we propose a geometric similarity coefficient for two boards based on the Euclidean distance between the respective placement locations of any two boards. Using the geometric similarity coefficient, we enhance the existing hierarchical clustering algorithm for group setup that is shown in section 2.1.

Consider the following unidimensional and single component boards in Figure 1. It is possible to transform the geometry of any one board to another one through a series of translations. In the example, these translations are represented as arrows in Figure 1. It can be observed that in order to transform board i into board j, the series of translations A, to A', B to B', C to C' and D to D' are required. Also, to transform board i into board k, the series of translations B to B'' and C to C'' are required. Hence, it is conjectured that the total translation required to transform the geometry of one board to that of another one is a surrogate measure of the dissimilarity, d, between these two boards. This distance d is defined as a metamorphic distance. In the example, $d_{ij} = a' + b' + c' + d'$, and $d_{ik} = b'' + c''$. The problem of finding the minimum transformation between two boards can be formulated as a linear assignment problem (LAP). Let the corresponding minimum metamorphic distance be MMD_{ij}. When boards with different number of locations is used, the issue of what to do with the excess locations on the board with more locations must be resolved. In Figure 1, boards j and k are identical except that the k does not have the first and last locations on board j. The LAP approach from above will yield $MMD_{ij} > MMD_{ik}$ because only two distances are considered in the assignment between i and k as a result, this could be wrongly interpreted as dissimilarity between i and j is greater than the dissimilarity between i and k.

One solution to the above problem is to transform all the locations on the board with more locations into the locations on the board with less number of locations. As a result, a dissimilarity in number of locations yields a larger total metamorphic distance (i.e., more dissimilarity). Let $|M| \geq |N|$ be the number of positions on each board, the total MMD_{ij} can be computed by applying the LAP method $\lceil \frac{|M|}{|N|} \rceil$ times, as follows:

M; Set of locations for board i.
N; Set of locations for board j.
M_k; Set of locations remaining in M at kth transformation.
MMD_{ijk}; MMD_{ij} after kth transformation.
$\{i\}_k$; Set of locations in board i that are transformed to board j at kth transformation

Initialize: $MMD_{ij} = 0$, $M_0 = M$, $\{i\}_0 = \emptyset$
For($k = 1$; $k < \lceil \frac{|M|}{|N|} \rceil$)
  Step 1: $M_k = M_{k-1} \setminus \{i\}_{k-1}$
  Step 2: Apply the LAP-method to transform N into $\{i\}_k \subset M_k$; get MMD_{ijk}
  Step 3: $MMD_{ij} = MMD_{ij} + MMD_{ijk}$
  Step 4: $k = k+1$

In step 1, the set of locations in board i is eliminated which is transformed from board j at the previous iteration. Step 2 calculates MMD between two boards by transforming all the locations in board j to board i. In step 3, MMD is cumulated until all the placement location in board j is transformed to board i. A simple extension for two dimension is to consider the Euclidean distance between positions. Let's suppose that different types of components are required on two different boards i and j, then total MMD_{ij} is defined as follows : $MMD_{ij} = \sum_{\forall c} MMD_{ij}^c$ where MMD_{ij}^c is MMD of component type c. Let p be the location of the component type c which belongs to board i and q be the location of the component of type c which belongs to board j. d_{pq}^c is the Euclidean distance between location p and q for component type c. The similarity coefficient based on MMD is defined as follows:

$$s_{ij} = 1 - \frac{\sum_{\forall c} MMD_{ij}^c}{\sum_{\forall c} \sum_{\forall p} max_{\forall q}(d_{pq}^c)} \qquad (3)$$

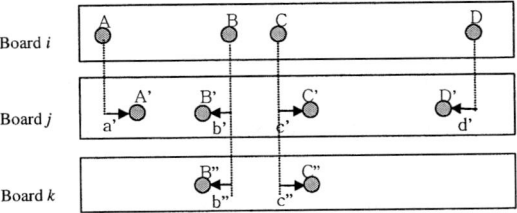

Fig. 1. Geometry transformation for the unequal number of placement locations

The nominator of the coefficient is the total sum of MMD for all types of components and the denominator is the normalizing factor. In the proposed improved group setup, we use the similarity coefficient in equation (3) and the clustering objective as the maximization of the average similarity within groups per unit feeder change. Therefore, the clustering objective is maximized when all boards in families are geometrically similar (i.e., placement time is minimized) and the number of feeder change is minimized (i.e., setup time is minimized).

4 Experiments

We consider a generic machine which has 70 feeder slots with 20mm between the slots. The board dimensions are maximum 320mm×245mm and the coordinates for each board were randomly generated from uniform distributions as follows: X=635+U(0,245), Y=254+U(0,320). The home position coordinate is (0,0) and the first feeder slot location is (457,0). The number of component types required per board were generated from U(6,20) from 70 different component types. We considered the time to install or remove feeder in cases of 30(sec) and 60(sec). The head velocity was tested for 100(mm/sec) and 300(mm/sec).

The batch size of boards were generated from U[50,100]. Also the total number of boards were generated from U[5,15]. The placement locations and the corresponding component types were generated from a seed board. A seed board is created with location $(L_{sx}(i), L_{sy}(i))$ and corresponding component type $C_s(i)$ for ith placement location. We fixed the number of placement location to 50 for the seed board. Then based on the component similarity (C) and geometric similarity (G), another board (i.e., a child board) is created using the following formula;

$$L_{cx}(i) = L_{sx}(i) + (1-G) \times 0.5 \times 245 \times U(-1,1) \qquad (4)$$

$$L_{cy}(i) = L_{sy}(i) + (1-G) \times 0.5 \times 320 \times U(-1,1) \qquad (5)$$

$$C_c(i) = \begin{cases} C_s(i) \text{ with probability } C \\ U(1, NC_c), \text{ otherwise} \end{cases} \qquad (6)$$

Where NC_c is the number of component type of the child board c. Equation (4), (5) means the coordinate of a child board's location is generated by the perturbation of the seed board's location. Equation (6) implies that the component type of a child board is the same as the seed board with probability C. Otherwise, the component types are randomly generated from the possible component types that are pre-assigned to the child board. 16 types of experiments were conducted using different problems sets as shown in Table 1. Each problem set consists of 20 random problems. For each problem instance, we tested 6 different levels of feeder capacity (20, 30, 40, 50, 60, 70). To measure the performance of the different setup strategies, the deviation from partial setup is computed as follows:

Percent deviation from partial setup=$\frac{M^{setupstrategy} - M^{PS}}{M^{PS}}$
Percent M^{PS} represents the makespan of partial setup (PS) strategy.

Table 1. Problem types

Problem type	Head velocity (mm/sec)	Feeder change time (sec)	Component similarity (C)	Geometric similarity (G)	Problem type	Head velocity (mm/sec)	Feeder change time (sec)	Component similarity (C)	Geometric similarity (G)
1	100	30	0.2	0.75	9	100	30	0.2	0.2
2	100	30	0.75	0.75	10	100	30	0.75	0.2
3	100	60	0.2	0.75	11	100	60	0.2	0.2
4	100	60	0.75	0.75	12	100	60	0.75	0.2
5	300	30	0.2	0.75	13	300	30	0.2	0.2
6	300	30	0.75	0.75	14	300	30	0.75	0.2
7	300	60	0.2	0.75	15	300	60	0.2	0.2
8	300	60	0.75	0.75	16	300	60	0.75	0.2

$M^{setup\ strategy}$ corresponds to the makespan of traditional group setup (GS), MMD based improved group setup (IGS) and PLM based group setup (PLM).

Figure 2-a),2-b) and Figure 3 shows the average setup time, placement time and makespan of PS, GS, IGS and PLM for problem type 1 (Head velocity 100mm/sec, feeder change time 30sec, component similarity 20%, geometric similarity 75%). For each random problem, we apply different level of the feeder capacity constraints (20, 30, 40, 50, 60, 70). Figure 2-a) shows that GS tends to perform better as the feeder capacity increases in terms of setup time. GS

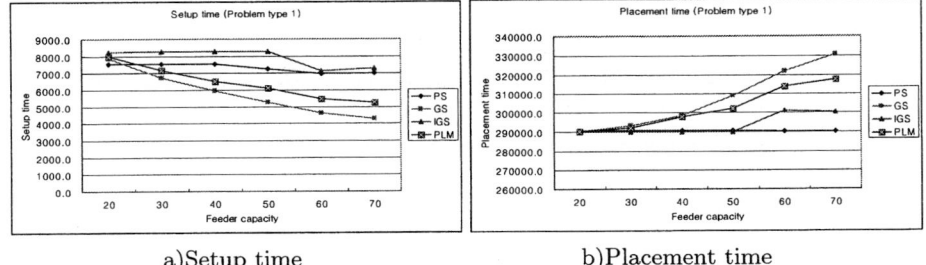

a) Setup time b) Placement time

Fig. 2. Result of setup time and placement time for problem type 1

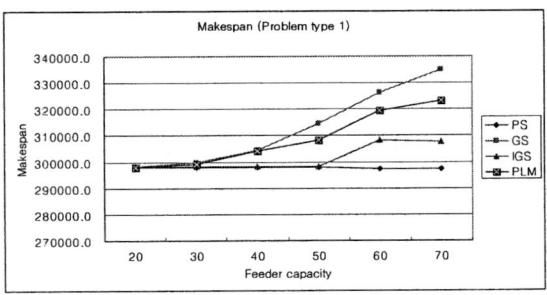

Fig. 3. Result of makespan for problem type 1

forms families of boards in order to reduce the setup time as long as the feeder capacity constraints are not violated. PLM behaves similar to GS. However, PS and IGS do not reduce the setup time much even if the feeder capacity does not restrict the solution space. This is because PS and IGS achieve an improvement in the reduction of the placement time instead of setup time as shown in Figure 2-b). The figure indicates that the placement time of GS increases as the feeder capacity increases since GS does not consider the geometry of boards. As shown in Figure 3, in case of maximum feeder capacity, IGS reduces about 8% of makespan compared to GS and deviates less than 3% from the partial setup strategy. PLM performs better than GS but the reduction of makespan is not significant (2.4% reduction from GS and 8.5% deviation from PS).

Table 2 summarizes the average setup time, placement time and makespan of different setup strategies with the feeder capacity 70 . The table shows that PS outperforms all other setup strategies. IGS outperforms GS and PLM in cases when the placement time becomes more important than the setup time (problem type 1, 3, 5, 9, 11, 13). The maximum percent deviation from PS of IGS is 5.6% while GS is 12.66% and PLM is 9.35%. In cases where the setup time is important (problem type 2, 4, 6, 8, 10, 12, 14, 16), the maximum percent deviation from PS of IGS is 1.21% while GS is 1.04% and PLM is 1.43%. This result implies that IGS balances the tradeoff between the setup time and the placement time and finds the solution that minimizes the makespan.

Table 2. Summary of experimental results for feeder capacity 70

Problem type	(GS-PS)/PS*100			(IGS-PS)/PS*100			(PLM-PS)/PS*100		
	Setup time Average	Placement time Average	Makespan Average	Setup time Average	Placement time Average	Makespan Average	Setup time Average	Placement time Average	Makespan Average
1	-38.77	13.91	12.66	3.93	3.45	3.46	-25.15	9.51	8.69
2	-29.54	1.19	0.73	-23.51	1.38	1.00	-21.58	1.21	0.86
3	-26.70	10.11	8.65	11.91	2.97	3.32	-16.64	8.37	7.37
4	-19.42	1.18	0.67	-10.99	1.52	1.21	-3.97	0.95	0.83
5	-28.13	11.16	8.97	12.25	4.89	5.30	-15.67	8.61	7.25
6	-6.17	0.66	0.43	-3.86	0.97	0.81	6.26	1.26	1.43
7	2.68	0.37	0.50	8.94	0.61	1.09	14.70	0.19	1.02
8	9.35	0.06	0.59	15.94	0.38	1.25	21.71	-0.03	1.19
9	-36.23	13.04	11.90	2.69	4.46	4.42	-22.17	8.96	8.24
10	-32.26	1.60	1.04	-26.87	1.66	1.19	-23.50	1.14	0.73
11	-29.85	11.28	9.60	4.97	4.93	4.94	-24.37	10.79	9.35
12	-15.78	1.01	0.58	-8.26	1.25	1.01	-2.62	0.78	0.70
13	-26.78	10.98	8.82	7.01	5.52	5.60	-12.87	7.93	6.74
14	-11.73	0.63	0.19	0.67	0.86	0.85	-5.38	1.36	1.12
15	4.87	-0.23	0.06	12.99	0.10	0.83	12.55	0.04	0.75
16	-25.14	0.90	0.27	-21.54	1.44	0.89	36.38	0.11	0.99
Overall									
Average	-19.35	4.87	4.10	-0.86	2.27	2.32	-5.14	3.82	3.58
Min	-38.77	-0.23	0.06	-26.87	0.10	0.81	-25.15	-0.03	0.70

5 Conclusions

This paper has presented an improved group setup strategy based on both component similarity and geometric similarity. It has demonstrated how the improved group setup strategy adapts to a variety of production conditions. The improved group setup strategy dominated the traditional group setup or PLM based group set up strategy. Overall, improved group setup strategy deviated from partial setup, maximum 5.6% and average 2.32%.

Acknowledgments

We thank Dr. R. Quintana for providing PLM code for the experimental comparison of the proposed methodology.

References

1. Askin, R. G., Dror, M., and Vakharia, A. J.: Printed circuit board scheduling and component loading in a multimachine, openshop manufacutring cell. Naval Research Logistics **415** (1994) 587–608
2. Ball, M. O. and Magazine, M. J.: Sequencing of insertions in printed circuit board assemblies. Opearations Research **362** (1988) 192–201
3. Crama, Y., Kolen, A. W. J., Oerlemans, A. G., and Spieksma, F. C. R.: Throughput rate optimization in the automated assembly of printed circuit boards. Annals of Operations Research **26** (1990) 455–480
4. Drenzner, Z and Nof, S. Y.: On optimizing bin picking and insertion plans for assembly robots. IIE Transactions **163** (1984) 262–270
5. Jonker, R. and Volgenant, A.: A shortest augmenting path algorithm for dense and sparse linear assignment problems. Computing **59** (1987) 231–340
6. Leon, V. J. and Peters B. A.: Re-planning and analysis of partial setup strategies in printed circuit board assembly systems. International Journal of Flexible Manufacturing Systems Special Issue in Electronics Manufacturing **84** 1996 389–412
7. Leon, V. J. and Peters B. A.: A comparison of setup strategies for printed circuit board assembly. Computers in Industrial Engineering **341** (1998) 219–234
8. Lofgren, C. B. and McGinnis, L. F.: Dynamic scheduling for flexible printed circuit card assembly. Proceedings of the IEEE Systems, Man, and Cybernetics Conference, Atlanta, GA, (1986)
9. Quintana, R., Leon, V. J.: An improved group setup management strategy for pick and place SMD assembly. Working paper (1997)

A Mixed Integer Programming Model for Modifying a Block Layout to Facilitate Smooth Material Flows

Jae-Gon Kim[1] and Marc Goetschalckx[2]

[1] Department of Industrial Engineering, University of Incheon,
177 Dowha-dong, Nam-gu, Incheon, Korea
jaegkim@incheon.ac.kr
[2] School of Industrial Systems and Engineering,
Georgia Institute of Technology, Atlanta, GA 30332, USA
marc.goetschalckx@isye.gatech.edu

Abstract. The conceptual block layouts generated by layout software packages very often yield detailed layouts with non-smooth material flow paths that have an excessive number of turns. Such non-smooth flow paths are not acceptable or feasible for actual construction. This paper presents a mixed integer programming model to modify a given block layout into a layout that allows the construction of smooth flow paths, while preserving the relative positions of the departments in the original block layout. Results of the computational experiments show that much smoother flow paths are generated in the modified block layouts than those in the original block layouts without significantly increasing the travel-based quality of final layouts.

1 Introduction

The facility layout problem has been traditionally solved in two stages because the complete problem is so complex and it cannot be solved in a reasonable amount of time for realistic instances. In the first step, a conceptual block is generated while ignoring the locations of the departmental input and output (I/O) points and the material flow paths. In the second step, a detailed layout is constructed by determining the locations of the I/O points of departments and constructing the material flow paths. When the block layout in the first stage is generated automatically by layout software, the material flow paths determined in the second phase are non-smooth, i.e. they have an excessive number of turns. Such flow paths cause many blind corners, excessive turning by the material handling carriers, increased construction costs of the facility, and possible violation of local fire and safety ordinances. We will develop a mathematical model to the block layout so that in the second phase a smooth material flow path can be constructed while at the same time preserving the basic structure of the block layout.

Few research results have been published on the problem of adjusting the block layout before generating the detailed layout. Usually layout designers have relied on their experience and skill rather than on automated methods to modify the block layout. Alagoz et al. [1] present a nonlinear programming model to adjust the areas and shapes of departments to avoid complicated flow paths with turns. However, their method can only be applied to layouts with a layer (or column) structure and the special flow path network. Lee and Kim [6] considered the problem of modifying a block layout with irregular shaped departments into one with rectangularly-shaped departments without significant changes in the relative positions of the departments. However, they did not consider the smoothness of material flow paths. In this paper, we will present a formulation for the problem of modifying a given block layout to allow construction of smooth flow paths.

2 Partial Edges

In this paper, it is assumed that all the departments in the initial block layout have a rectangular shape and that no empty space exists between any departments. The initial block layout is converted into an undirected graph. The vertices in the graph correspond to the points at which corner points of different departments meet and the edges correspond to the boundary segments of the departments between those points. In this graph, an edge is called a *partial edge* if its two vertices do not correspond to corner points of the same department. Fig. 1 shows a block layout and a corresponding graph for the block layout. In this graph, there exist 11 departments, 20 vertices, and 30 edges, including 7 partial edges. The edge between vertices 4 and 5 is a partial edge since vertices 4 and 5 correspond to corner points of different departments.

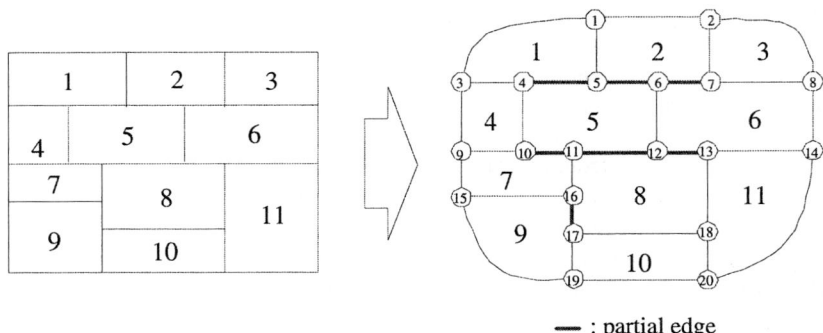

Fig. 1. A block layout and the corresponding undirected graph

Partial edges are the main obstacle in constructing smooth flow paths since they exist when departments are not aligned both horizontally and vertically. Any material-handling carrier traveling in an aisle perpendicular to the partial edge must make a 90-degree turn to continue its path. However, many partial edges can be removed by modifying the block layout, which includes changing the departmental areas but while preserving the relative positions of the departments in the block layout. If a partial edge exists between vertices u and v, we can remove it by modifying the block layout in such a way that the corner points of the departments which correspond to vertices u and v meet at the same point. In Fig. 1, the partial edge between vertices 4 and 5 will disappear if the block layout is modified by either moving the boundary between departments 1 and 2 leftward or the boundary between departments 4 and 5 rightward or both until both boundaries are aligned.

Partial edges have the following geometric characteristics: 1) a partial edge is adjacent to four different departments and 2) the two end vertices of a partial edge correspond to bottom-left, bottom-right, top-left and top-right corner points of the departments, respectively. In Fig. 1, the partial edge between vertices 4 and 5 is adjacent with departments 1, 2, 4 and 5, and vertex 4 corresponds to top-right and top-left corner points of departments 4 and 5, respectively, and vertex 5 corresponds to bottom-right and bottom-left corner points of departments 1 and 2, respectively. Let (u,v) be a partial edge between vertices u and v, where $u < v$, and let $d^{bl}_{(u,v)}$, $d^{br}_{(u,v)}$, $d^{tl}_{(u,v)}$, and $d^{tr}_{(u,v)}$ be the (indexes of) departments which are adjacent with (u,v) and each of which has vertex u or v at its bottom-left, bottom-right, top-left, or top-right corner point, respectively. A partial edge (u,v) disappears if the bottom-right corner point of $d^{br}_{(u,v)}$ (or the bottom-left corner point of $d^{bl}_{(u,v)}$) and the top-left corner point of $d^{tl}_{(u,v)}$ (or the top-right corner point of $d^{tr}_{(u,v)}$) meet at the same point in the modified block layout.

3 Mathematical Formulation

In this section we will present a mathematical formulation for modifying a given block layout while preserving the relative positions of departments with the objective of minimizing the number of partial edges in the modified block layout. The constraints of the model are as follows.

- The area of a department in the modified block layout should be no less than $100 \cdot \alpha\%$ of the area of that department in the initial block layout. In the computational study in this paper the value of α is set to 0.90 and 0.95.
- Departments should not overlap.
- The aspect or shape ratio of each department should be within a given range.
- The relative positions of the departments should remain unchanged from those in the initial layout.
- All departments should be placed within the given building restrictions and this rectangular building area is called the floor.
- There should be no empty space in the floor. This implies that the area of some departments may be increased.

We say that departments i and j overlap with each other in the x-dimension (y-dimension) if their corresponding line segments overlap when they are projected on x-axis (y-axis). The line segments overlap if they have a common segment of non-zero length. The relative position of two departments in the layout is said to remain the same if at least one of the non-overlap conditions remains valid in the modified layout. For example, if in the initial layout department A and B non-overlap only in the x-dimension, their relative position is said to remain the same if they also non-overlap in x-dimension in the modified layout. On the other hand, if in the initial layout department A and B non-overlap in both dimensions, it is sufficient for them to non-overlap in either the x or y-dimension in order for their relative position to remain unchanged.

The following notation will be used in the mathematical model.

Parameters

N	the number of departments
W	the width of the floor
L	the length of the floor
a_1	the area of department i in the given block layout
$\underline{\tau}_i, \bar{\tau}_i$	the lower and upper limits of the aspect ratio of department i
P	the set of partial edges in the given block layout
L_i	the set of departments that are adjacent with department i and placed on the left side of department i in the given block layout. (Two departments are considered to be adjacent if they have a common boundary segment or a common boundary corner point.)
R_i	the set of departments that are adjacent with department i and placed on the right side of department i in the given block layout
B_i	the set of departments that are adjacent with department i and placed below department i in the given block layout
A_i	the set of departments that are adjacent with department i and placed above department i in the given block layout
M	$=\max(W,L)$

Decision variables

x_i^l, x_i^r	x-coordinates of the left and right boundaries of department i in the modified block layout
y_i^b, y_i^t	y-coordinates of the bottom and top boundaries of department i in the modified block layout
$e_{u,v}$	the length of a partial edge (u,v) in the modified block layout
$z_{u,v}$	$= 1$ if a partial edge (u,v) in the given block layout still appear in the modified block layout, 0 otherwise

Now, the problem is formulated as a nonlinear mixed integer program.

[P1]   Minimize $\sum_{(u,v)\in P} z_{u,v}$

s.t. $e_{u,v} = |x^r_{d^{br}_{u,v}} - x^l_{d^{tl}_{u,v}}| + |y^b_{d^{br}_{u,v}} - y^t_{d^{tl}_{u,v}}|$ $\forall (u,v) \in P$ (1)

$e_{u,v} \leq M z_{u,v}$ $\forall (u,v) \in P$ (2)

$$(x_i^r - x_i^l)(y_i^t - y_i^b) \geq \alpha \cdot a_i \quad \forall i \tag{3}$$

$$\underline{\tau}_i(x_i^r - x_i^l) \leq y_i^t - y_i^b \leq \bar{\tau}_i(x_i^r - x_i^l) \quad \forall i \tag{4}$$

$$x_i^r = x_j^l \quad \forall i,j; i \in L_j \tag{5}$$

$$y_i^t = y_j^b \quad \forall i,j; i \in B_j \tag{6}$$

$$x_i^l = 0 \quad \forall i; L_i = \emptyset \tag{7}$$

$$x_i^r = W \quad \forall i; R_i = \emptyset \tag{8}$$

$$y_i^b = 0 \quad \forall i; B_i = \emptyset \tag{9}$$

$$y_i^t = L \quad \forall i; A_i = \emptyset \tag{10}$$

$$x_i^l, x_i^r, y_i^b, y_i^t \geq 0 \quad \forall i \tag{11}$$

$$e_{u,v} \geq 0, z_{u,v} \in \{0,1\} \quad \forall (u,v) \in P \tag{12}$$

The objective is to minimize the number of partial edges in the modified block layout. Constraint (1) specifies the length of partial edges and constraint (2) ensures that if the length of a partial edge is greater than zero, the partial edge exists in the modified block layout. Constraint (3) ensures that the departments satisfy their area requirements and constraint (4) ensure that the aspect ratios of the departments are within their given ranges. Constraints (5)-(10) ensure that the relative positions of the departments are remain equal to those in the given block layout and that all departments are placed within the floor, which will be proved in the next section.

The area requirement constraint (3) is nonlinear (quadratic), which makes the model harder to solve than if only linear constraints were present. In this study, we use the polyhedral outer approximation method presented by Kim and Goetschalckx [3] to linearize the department area constraint (3) with tangential supports. In this study, the number of tangential supports was set to 20, which is large enough to satisfy the area requirement almost exactly.

Now we present a mixed integer programming (MIP) model [P2], which is the same with [P1] except that the area requirement constraint is linearized.

[P2] $$\text{Minimize} \sum_{(u,v) \in P} z_{u,v}$$

s.t. (1), (2), (3), (4)-(12) and

$$y_i^{(k)}(x_i^r - x_i^l) + x_i^{(k)}(y_i^t - y_i^b) \geq 2\alpha \cdot a_i \quad \forall i \text{ and } k = 1,...,20 \quad (13)$$

[P2] can be solved using standard mixed integer programming packages. The only binary variables correspond to existing partial edges. In our numerical experiment, the number of partial edges grew approximately linear with the number of departments when the block layouts are randomly generated. Therefore, [P2] can be solved to optimality in a reasonable time even if the block layout has a large number of departments.

4 Validation of the MIP Model

Assume that department i is placed above (or to the left of) department j in the block layout. Then, two departments are said to be vertically (horizontally) coupled if the bottom (right) boundary of department i is linearly connected with the top (left) boundary of department j by bottom (right) or top (left) boundaries of other departments. In Fig. 1, departments 6 and 7 and departments 7 and 10 are vertically and horizontally coupled, respectively. Note that coupled departments are not adjacent to each other.

Proposition 1. *In the modified block layout, the relative positions of the departments remain identical to those in the initial block layout.*

Proof. To prove this proposition, we consider the three possible cases of the positional relationship between two departments in the given block layout.

Case 1) The two departments are adjacent to each other.
It is clear that the relative positions of the departments are kept the same in the modified block layout due to constraint (5) or (6). However, the departments may no longer be adjacent in the modified block layout but become vertically or horizontally coupled.

Case 2) The two departments are vertically or horizontally coupled.
We assume that departments i and j are vertically aligned and department i is above department j in the given block layout. Then geometrically there always exists a sequence of $2(m+1)$ departments $< i, c_1, c_2, ..., c_{2m}, j >$ denoted by $C_{i,j}^v$, where $m \geq 0$. In $C_{i,j}^v$, the department $2k - 1^{th}$ in the $2k^{th}$ position is above that in the $2k^{th}$ position while being adjacent to each other, where $1 \leq k \leq (m+1)$. According to constraint (6), the department in the $2k - 1^{th}$ position is placed above that in the $2k^{th}$ position in the modified block layout. Therefore, the department in the $2k - 1^{th}$ position is placed above that in the $2l^{th}$ position, where $k \leq l \leq (m+1)$, which means department i at the first position is placed above department j at the $2(m+1)^{th}$ position. Similarly, we can prove for horizontally-coupled departments.

Case 3) Two departments (departments i and j) are neither adjacent nor vertically or horizontally coupled.

Without loss of generality, we assume that department i is below or to the left of department j in the given block layout. Then geometrically there always exists a sequence of departments $<i, v_1, v_2, ..., v_p, j>$ denoted by $S_{i,j}^v$ or a sequence of departments $<i, h_1, h_2, ..., h_q, j>$ denoted by $S_{i,j}^h$, where $p \geq 1$ and $q \geq 1$. In $S_{i,j}^v$, the department in the kth position is below that in the $k+1^{th}$ position and they are adjacent or vertically coupled, where $1 \leq k \leq p+1$. In $S_{i,j}^h$, the department in the kth position is to the left of that in the $k+1^{th}$ position and they are adjacent or horizontally coupled, where $1 \leq k \leq q+1$. Note that $S_{i,j}^v$ and $S_{i,j}^h$ may not be unique and may have several alternatives. For an example, $S_{7,11}^v = <8,4,2>$ or $<8,5,2>$ or $<8,6,2>$ and $S_{7,11}^h = <7,8,11>$ or $<7,10,11>$ in Fig. 1. If $S_{i,j}^v$ exists, the department in the kth position in $S_{i,j}^v$ is placed below that in the $k+1^{th}$ position in the modified block layout by cases 1 and 2. Similarly, if $S_{i,j}^h$ exists, the department in the kth position in $S_{i,j}^h$ is placed to the left of that in the $k+1^{th}$ position in the modified block layout. Therefore, the relative position between departments i and j is preserved in the modified block layout. □

5 Computational Experiments

To test the suggested method, initial block layouts for the layout problems are required. We use the simulated annealing algorithm of Kim and Kim [4] to generate block layouts in which all departments have rectangular shapes and there is no empty space. Four block layouts were generated for the 10-department problem of Van Camp et al. [7], the 20-department problem of Armour and Buffa [2] and randomly generated 30- and 40-department problems, respectively. In the problems, the lower and upper limits of the aspect ratios of departments are set to 0.4 and 2.5, respectively, except for the problem in Van Camp et al. [7] for which the aspect ratios are specified. Note that Van Camp et al. [7] and Armour and Buffa [2] used upper triangular flow matrixes to represent flows between the departments in their test problems. Their flow matrixes were transformed by copying the flow value above the main diagonal to the corresponding cell below the main diagonal. The resulting matrix is symmetric and the total transportation distance (TTD) is doubled from the original problem. In the two randomly generated problems, the areas of the departments were randomly generated from the discrete uniform distribution with range [1, 10] and the width and length of the floor area were set to the square root of the sum of the department areas, which implies that the floor had a square shape. The inter-department flows were randomly generated while satisfying a flow intensity of 20% and the non-zero flow elements were sampled from a discrete uniform distribution with range [1, 100].

The four block layouts were modified by solving the MIP model. The tests were executed on a personal computer with a 700MHz Pentium-III processor and CPLEX 6.0 was used to solve the MIP in the method. Table 1 shows the results of modifying the block layouts using the method described above. In Table 1, the percentage area gap (PAG) represents the relative amount of the decreased area in the modified block layout compared to the prescribed area in the original layout. PAG of department i is calculated as $\max(a_i - a'_i)/a_i \times 100$, where a'_i is area of department i in the modified block layout. The amount of the increased area is not considered in calculating PAG since it is not necessary in checking the accuracy of the linearly approximated area requirement constraint, constraint (13) in [P2]. Note that departments with increased areas satisfy both the surrogate constraint and the original one but ones with decreased areas do not satisfy the original constraint while they satisfy the surrogate constraint.

Table 1. Results of Modifying the Block Layouts

α	Block layout	Num. of partial edges	Num. of total edges	Max. and avg. PAG(%)	CPU time (sec.)
0.90	10-Van Camp et al.	3 → 0	27 → 25	(10.05, 6.89)	0.12
	20-Armour&Buffa	14 → 1	57 → 48	(10.05, 5.52)	0.87
	30-random	24 → 5	87 → 74	(10.06, 6.12)	0.93
	40-random	37 → 7	117 → 99	(10.04, 5.57)	201.20
0.95	10-Van Camp et al.	3 → 1	27 → 26	(5.17, 3.94)	0.10
	20-Armour&Buffa	14 → 5	57 → 52	(5.04, 3.49)	0.32
	30-random	24 → 9	87 → 77	(5.04, 3.30)	2.81
	40-random	37 → 13	117 → 104	(5.07, 3.57)	107.52

As can be seen in the Table 1, the modified block layouts have a much smaller number of partial edges than the original block layouts with no new partial edge. In addition, the total number of edges also decreased by modifying the block layout as described above. Since the edges form the candidate set for the construction of the flow paths, a smaller number of edges simplifies that phase of the design problem. Based on our numerical experiment, which compares allowable shrinkage factors of 0.10 and 0.05 percent, the decrease in the number of partial edges is smaller when the allowable shrinkage is smaller. This is because that the closer the value of α is to 1, the more restricted the feasible domain is and the fewer partial edges can be eliminated. Note that if $\alpha = 1.0$, the given block layout cannot be modified at all. The maximum PAG is very close to $100(1-\alpha)$ for all the modified block layouts in our experiment; on the average, there was only 1% difference between two figures. This means that the non-linear area requirement constraint, constraint (3) is almost satisfied although it is replaced with the linear surrogate constraints by the polyhedral outer approximation method. The CPU times were very small (less than one second on the average) except to those for the 40-random block layout although they are sufficiently short also considering that the layout problem is a design problem which needs

Table 2. Effects of the Block layout Modification on the Flow Paths and TTD

Block layout		Total number of turns	Total lenth of flow paths	TTD
Initial	10-Van Camp et al.	8	45.84	13251.30
	20-Armour&Buffa	116	93.12	3347.59
	30-random	219	70.34	36118.82
	40-random	541	97.65	78043.47
Modified with $\alpha = 0.9$	10-Van Camp et al.	2	52.48	13715.41
	20-Armour&Buffa	62	91.21	3294.69
	30-random	108	70.60	35925.67
	40-random	240	97.51	78208.97
Modified with $\alpha = 0.95$	10-Van Camp et al.	2	44.83	14114.96
	20-Armour&Buffa	70	92.82	3284.87
	30-random	111	70.86	35750.15
	40-random	275	100.59	81631.27

not to be solved frequently. This means that the suggested MIP model can be solved easily within a short time even for the large-sized block layouts.

We determined the locations of I/O points of the departments and constructed flow paths on the modified block layouts to test the impact of the block layout modification on the total transportation distance (TTD). The branch and bound algorithm of Kim and Kim [5] is used to determine the optimal locations of I/O points with the objective of minimizing the TTD. Here, it is assumed that materials are transported along the shortest path between the I/O points. Once the locations of I/O points are determined, for each pair of I/O points which has a non-zero material flow between them, a flow path is constructed along the shortest path located on the boundaries of the departments between them. If there are more than one alternative shortest paths between the two I/O points, the shortest path with the smallest number of turns is selected for the flow path construction. Note that the cost of constructing flow paths is not considered here.

Table 2 shows the total number of turns in the flow paths, the total length of the flow paths, and the TTD in the given and modified block layouts when the flow paths are constructed with the algorithm described above. The total number of turns in the flow paths is the sum of the number of turns in all flow paths between I/O points with non-zero material flows. By modifying the block layouts, the total number of turns decreased more than about 50%, which means the flow paths became much smoother. There is not much difference between the total length of the flow paths and TTD when the initial and modified block layouts are compared although they increased about 1.2% and 1.7% on the average, respectively. However, the total material transportation time may decrease since average vehicle speed increases by eliminating turns since vehicles decreases their speed to make a turn in intersections or corners. Fig. 2 shows the locations of I/O points and the flow paths in the initial and modified block layouts for the

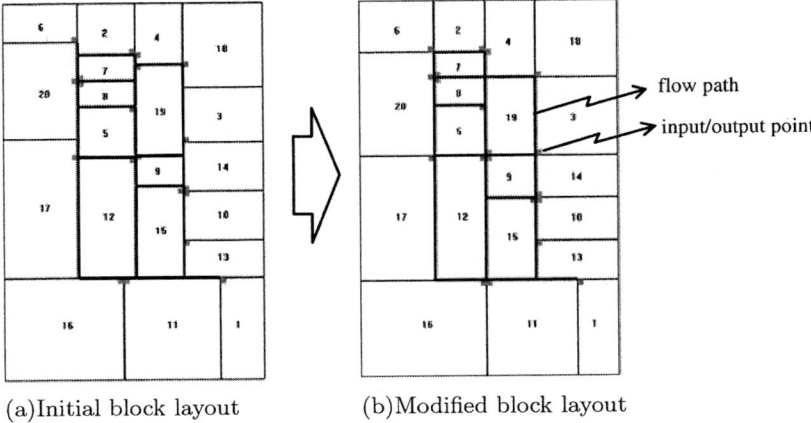

Fig. 2. I/O points and Flow Paths on the Initial and Modified Block Layouts for the Armour and Buffa Problem with $\alpha = 0.90$

20-department problem by Armour and Buffa with $\alpha = 0.90$. Note that in Fig. 2, the input and output points of each department are located at the same location since the flow matrix is symmetric in the 20-department problem by Armour and Buffa.

6 Conclusion

In this paper, we considered the problem of modifying an initial block layout into a block layout which will yield a flow path with fewer turns. It was shown that turns correspond to partial edges in a graph representation of the layout. An MIP model was developed that retained the minimum number of partial edges. The total number of partial edges was reduced by solving this model. The effectiveness of the MIP model was tested on several industrial-sized problems. The computational experiments showed that the suggested model could modify the block layouts into ones with more than 50% less flow-path turns within a few minute for well-known and large-sized random test problems.

References

1. Alagoz, O., Norman, B.A., Smith, A.E.: Designing Aisle Networks to Facilitate Material Flow. Proc. the 2002 Material Handling Research Colloquium (MHRC), Portland, Maine (2002)
2. Armour, G.C., Buffa, E.S: A Heuristic Algorithm and Simulation Approach to Relative Location of Facilities. Management Science, Vol. 9 (1963) 294-309
3. Kim, J-G., Goetschalckx, M.: An Integrated Approach for the Concurrent Determination of the Block Layout and I/O Point Location Based on the Contour Distance. International Journal of Production Research, to appear (2004)

4. Kim, J-G., Kim, Y-D.: A Space Partitioning Method for Facility Layout Problems with Shape Constraints. IIE Transactions, Vol. 30 (1998) 947-957
5. Kim, J-G. Kim, Y-D.: A Branch and Bound Algorithm for Locating I/O Points of Departments on the Block Layout. Journal of the Operational Research Society, Vol. 50 (1999) 517-525
6. Lee, G-C., Kim, Y-D.: Algorithms for Adjusting Shapes of Departments in Block Layouts on the Grid-Based Plane, Omega, The International Journal of Management Science, Vol. 28 (2000) 111-122
7. Van Camp, D.J., Carter, M.W., Vannelli, A.: A Nonlinear Optimization Approach for Solving Facility Layout Problems, European Journal of Operational Research, Vol. 57 (1991) 174-189

An Economic Capacity Planning Model Considering Inventory and Capital Time Value

S.M. Wang[1], K.J. Wang[2], H.M. Wee[1], and J.C. Chen[1]

[1] Department of Industrial Engineering,
Chung-Yuan Christian University, Chung Li, 320, Taiwan, R.O.C
[2] Department of Business Administration,
National Dong Hwa University, Hualien, 974, Taiwan, R.O.C
kjwang@mail.ndhu.edu.tw

Abstract. A company needs to implement several make-to-stock policies apart from a regular make-to-order production, so that the capacity of expensive resources can be fully utilized. The constraints to be considered in such capacity planning problem include finite budget for investing resources, lump demands of customers, decline of products price with time, different product mix for simultaneous manufacturing, time value of capital asset, technology levels of resources, and limited capacity of resources. We focus on the issues of resources acquisition and allocation decision in each production period. The goal is to maximize the long-term profit. This study formulates the problem as a non-linear mixed integer mathematical programming model. A constraint programming based genetic algorithm is developed to solve the problem efficiently.

1 Introduction

The investment on facilities for manufacturing many products requires a high amount of capital. Industries such as semiconductor, electronic products and TFT-LCD need to decide its make-to-stock and make-to-order policies, so that the capacity of expensive resources can be fully utilized. Decisions regarding resource planning/acquisition and allocation coupled together are very difficult and risky to company profitability. Moreover, the problem is very complex because many constraints such as finite budget, lumpy demands, uncertain product price, multiple production horizon, different products mix, time value of capital, technology innovation of resources, usage of multiple-function resources, and constrained capacity of resources need to be considered. In addition to revenue gained from products manufactured and resource assets, a decision maker needs to consider costs regarding inventory and resource acquisition costs such as procurement, renting, transfer and phase-out.

Therefore, we focuses on the following issues:

- How to plan on resources portfolio regarding purchasing, renting, transfer and phase-out alternatives?

- How to allocate resources to fulfill all orders of products in each production period while considering limited resource capability and capacity?
- How to decide on inventory levels on each production horizon such that holding cost and back-order cost can be reduced?

The rest of the paper is organized as follows. Section 2 reviews the related work. The problem is formally described in Section 3. Section 4 proposes a constrained programming based genetic algorithm to solve the problem. We illustrate the algorithm through sensitivity analysis in Section 5. Section 6 concludes the study.

2 Literature Review

Mixed integer programming and linear programming are the exact methods used for IC chip testing capacity planning. Leachman and Carmon (1992) addressed the production planning problem with alternative equipment. Rajagopalan (1994) presented a mix integer linear programming model applying to a situation in which the market demand for products and the number of available technologies could not decrease with time increases. Bashyam (1996) discussed the capacity expansion under demand uncertainty. Hung and Wang (1997) provided a linear programming model to solve material planning and resource allocation problems. Swaminathan (2000) addressed the demand uncertainty and focused on a make-to-order situation in aggregate capacity planning and equipment procurement.

Only a few studies have offered strategic concepts to support a resource replacement policy (Hsu 1998). Rajagopalan (1999) proposed a model useful in making timing decisions about adopting or replacing new equipment or processes. Li and Tirupati (1995) addressed the technology choice under stochastic demands for dynamic capacity allocation of two products.

Soft computing methods have emerged rapidly to tackle the resource allocation and expansion problem, which solved by balancing between solution efficiency and quality. For instance, Bard et al. (1999) considered the capacity planning of semiconductor manufacturing facilities using queuing and simulated annealing models. Wang and Hou (2003) solved the problem of capacity expansion and allocation in the semiconductor testing industry, considering multiple resources and limited budget.

From our literature survey, we can see that resource planning and allocation are highly correlated, but most academic studies have solved these problems separately. Some significant factors that affect the resource planning and allocation decisions in high-tech industries are the stocking policy used, the consideration of time value and the influence of deteriorating values of items. Decisions regarding resource acquisition and phase-out alternatives have not been thoughtfully considered. A more generalized mathematical model is required to describe precisely the problem facing the industries considering the follow factors.

3 Problem Formulation

The inherent characteristics of the problem include finite budget, lumpy demands, declined product price, multiple production horizon, products mix, time value of capital, technology innovation level of resources, operational ability of resources to per-form multiple functions, and capacity constraints of resources. In addition to the handling and acquisition of resources (such as procurement, renting, transfer and phase-out), a decision maker also needs to consider the balance between inventory and back-order, as well as an efficient task allocation.

A non-linear mixed integer programming model is developed to solve this problem. Further considerations about the model are explained as follows.

1. The profit gained (through the interest of remaining capital, the revenue of products manufactured, and from phasing out resources) in a preceding production period is reusable for increasing resources in its succeeding periods.
2. The proposed model can handle both make-to-order and make-to-stock products. Note that make-to-order decisions can also be degenerated by simply setting a high penalty cost for back-orders.
3. The asset and salvage of resources can be estimated. Besides, the technology level of a resource is reflected by its processing speed and salvage value.
4. The target utilization and throughput rate of resources are known.
5. Resources have the capabilities to process products of several types.
6. Inventory related cost is counted at the end of each period.

The objective to maximize the profit is formulated below:

$$Maximize \ \frac{1}{\prod_{t \in T}(1+I_t)} [\sum_{m \in M}(f_{t^{end},m} N_{t^{end},m})] + \frac{V_{t^{end}}}{\prod_{t \in T}(1+I_t)} \quad (1)$$

The first term of the objective function is the asset/salvage value of resources in its present value and the second one is for the remaining currency. I_t is the interest rate in period t. $f_{t,m}$ is the unit salvage of phasing-out the resource type m in period t. $N_{t,m}$ is the quantity of resource type m in period t, $N_{t,m} \in Z^+$. V_t is the balance of asset and capital in the end of period t, $V_t \in R^+$.

Inventory Balance Equations: The net inventory at the end of period t is computed by the data of the inventory in period $t\text{-}1$, quantity produced in period t and demand. Let $K_{t,d}$ be the inventory level of product type d in the end of period t, $K_{t,d} \in Z$; $a_{m,d}$ the production capability of resource type m to manufacture product type d, $m \in M$, which is a set of resource types and $a_{m,d}$ is a Boolean parameter. If resource type m can do product type $d \in D_m$, $a_{m,d} = 1$, otherwise $a_{m,d} = 0$. $X_{t,m,d}$ be the production quantity of resource type m producing product type d in period t, $X_{t,m,d} \in R^+$; $O_{t,d}$ the demand quantity of product type d in period t, $d \in D$, which is a set of product types, $t \in T$ which is a set of production periods 1, 2,...,t^{end}. Thus, $\forall t \in T, \forall d \in D$

$$K_{t,d} = K_{t-1,d} + \sum_{m \in M} a_{m,d} X_{t,m,d} - O_{t,d} \quad (2)$$

Inventory Cost Equations: Inventory cost occurs if the net inventory quantity of product type d is positive in period t. Otherwise, a back-order cost occurs. $Y_{t,d}$ is the inventory and back-order cost of product type d in period t, $Y_{t,d} \in R$; $h_{t,d}$ the unit holding cost of product type d in period t; $l_{t,d}$ the unit back-order cost of product type d in period t. $\forall t \in T, \forall d \in D$

$$Y_{t,d} = \begin{cases} \frac{h_{t,d}(K_{t-1,d})}{(1+I_t)^t} &, \text{ if } K_{t-1,d} \geq 0 \\ \frac{l_{t,d}|K_{t-1,d}|}{(1+I_t)^t} &, \text{ otherwise} \end{cases} \quad (3)$$

Capacity Balance Equations: The quantity of resource type m must be equal or larger than the quantity of allocated capacity. $Q_{t,m,r}$ is the renting/transfer quantity of the resource type m in period t, $Q_{t,m,r} \in Z^+$. $e_{m,d}$ the throughput of resource type m producing product type d; $w_{t,m}$ the available working hour of resource type m in period t; $z_{t,m}$ the target utilization of resource type m in period t. $\forall t \in T, \forall d \in D, \forall m \in M, \forall r \in R$

$$N_{t,m} + Q_{t,m,r} \geq \sum_{d \in D} \frac{a_{m,d} X_{t,m,d}}{e_{m,d} w_{t,m} z_{t,m}} \quad (4)$$

Net Asset and Capital Equations: When the number of resource type m in period t is greater than that in period t-1, then the asset change $G_{t,m}$ is positive, otherwise it is negative. Let $s_{t,m}$ be the unit procurement cost of resource type m in period t. $\forall t \in T, \forall m \in M$

$$G_{t,m} = \begin{cases} \frac{s_{t,m}(N_{t,m} - N_{t-1,m})}{(1+I_t)^t} &, \text{ if } N_{t,m} - N_{t-1,m} \geq 0 \\ \frac{f_{t,m}(N_{t,m} - N_{t-1,m})}{(1+I_t)^t} &, \text{ otherwise} \end{cases} \quad (5)$$

Thus, the net capital and asset in period t decrease by purchasing, renting or/and transferring resources, and inventory and back-order costs; and increase due to selling products. $\forall t \in T$

$$V_t = V_{t-1}(1+I_{t-1}) - \sum_{r \in P, m \in M} b_{t,m,r} Q_{t,m,r} - \sum_{m \in M} G_{t,m} + \sum_{m \in M, d \in D_m} p_{t,d} o_{t,m,d} - \sum_{d \in D} Y_{t,d} \quad (6)$$

where $b_{t,m,r}$ is the unit renting/transfer cost of resource type m in period t. $r \in R$, which is a set of resource acquisition approaches(by renting and transferring); $p_{t,d}$ the unit profit of product type d.

Note that constraints (4) and (5) the quantity of the resource in each period be an integer; however, the workload of the resource on the right hand side of constraint (4) can be a real number. Besides, constraint (2) is replaced by the $K_{t,d} = \sum_{m \in M} a_{m,d} X_{t,m,d} - \frac{K_{0,d} + \sum_{t \in T} o_{t,d}}{t_{end}}$ and $K_{t,d} = \sum_{m \in M} a_{m,d} X_{t,m,d} - o_{t,d}$ to realize the heuristics for leveling production and chasing production in each period, respectively.

The complexity (in terms of the size of variables and the number of constraints) increases exponentially upon increasing the resources type, the product type, the periods of production horizon, the amount of the capital and initial budget and the interest rate. So, we design a genetic algorithm to solve the problem effectively.

4 The Constraint Programming Based Genetic Algorithm

A distinguished property of the proposed algorithm as compared to the other genetic algorithm family is that it contains a chromosome repair mechanism to reduce unnecessary computational efforts of infeasible solutions which are unavoidably produced upon using a regular genetic algorithm in dealing with such non-linear integer programming problems.

This study develops a constraint programming based genetic algorithm (CPGA) and the chromosome structure for solving the problem addressed. The algorithm has been implemented by C++. We describe the algorithm as follows, in which $F(g)$ and $S(g)$ represent parents and offsprings of a generation (g), respectively.

```
Constraint Programming Based Genetic Algorithm (CPGA)
g ← 0;
Initialize and Evaluate F(g);
While (not termination condition) do
  Crossover: recombine chromosomes with leading fitness (by
Equation 1) in F(g) to yield S '(g);
  Mutation: alter the genes of randomly selected chromosomes
in F(g) to yield S "(g), and let S(g) = S '(g) + S "(g);
  Evaluate chromosomes in S(g): repair infeasible chromosomes
to be feasible (using the Constrain (4));
  Select F(g+1) from F(g) and S(g) using a roulette wheel
method;
  g ← g + 1;
End.
```

A multiple dimension structure of chromosome consists of the gene variables of $Q_{t,m,r}$, $X_{t,m,d}$ and $N_{t,m}$. Besides, in order to enhance searching diversity, this algorithm applies a cocktail crossover procedure blending the simple point operator, two-point operator, uniform operator, mathematical computing operator, uniform mathematical computing operator, and mix operator randomly. The mutation procedure of the algorithm uses the uniform operator (Gen and Cheng 2000).

The proposed CPGA is modified for the two new models; namely, the leveling based genetic algorithm (called Level-GA) and the chasing based genetic algorithm (called Chase-GA). Note that although demands are somewhat leveled or firmed in the two heuristics, a resource allocation decision problem still exists to be resolved.

5 An Illustration

An example modified from industry is to illustrate the proposed model and algorithm. The problem contains the data: (1) three types of resources, (2) 12 periods of production horizon are considered, (3) three types of products (items 1,2,and 3), (4) a 0.02 interest rate and an 0.80 target resource utilization, and (5) 1,800 working-hours in each period. Product unit profits and resource salvage decrease among time to reflect the change in technology level.

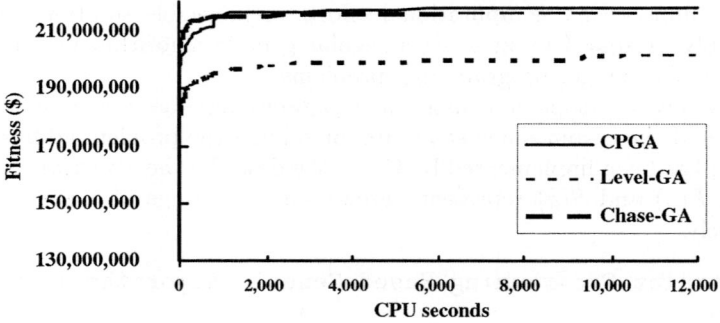

Fig. 1. Fitness evolution of CPGA, Level-GA and Chase-GA

Table 1. Capacity portfolio plan of CPGA

	Resource type (Units)		
	Resource 1	Resource 2	Resource 3
Period	On hand*(Renting) [Transfer]**Total**	On hand (Renting) [Transfer]**Total**	On hand (Renting) [Transfer]**Total**
1	2 (2) [0] **4**	1 (3) [0] **4**	1 (1) [0] **2**
2	2 (2) [0] **4**	1 (4) [3] **8**	1 (2) [0] **3**
3	2 (0) [0] **2**	1 (1) [0] **2**	1 (0) [0] **1**
4	2 (0) [0] **2**	1 (2) [0] **3**	1 (1) [0] **2**
5	2 (0) [0] **2**	1 (3) [3] **7**	1 (2) [0] **3**
6	2 (2) [0] **4**	1 (4) [1] **6**	1 (1) [0] **2**
7	2 (2) [0] **4**	1 (2) [0] **3**	1 (1) [0] **2**
8	2 (3) [0] **5**	1 (4) [0] **5**	0 (2) [0] **2**
9	2 (1) [0] **3**	1 (2) [0] **3**	0 (2) [0] **2**
10	2 (1) [0] **3**	1 (3) [0] **4**	0 (2) [0] **2**
11	2 (0) [0] **2**	1 (3) [0] **4**	0 (3) [0] **3**
12	2 (0) [0] **2**	1 (0) [0] **1**	0 (1) [0] **1**

*±procurement/phase-out

Figure 1 reveals that the proposed algorithms with a rapid convergence speed solve all the three models. The best result is achieved by CPGA among the three.

Table 2. Capacity allocation plan of CPGA

Period	Capacity type (Units)								
	Resource 1			Resource 2			Resource 3		
	Item1	Item2	Item3	Item1	Item2	Item3	Item1	Item2	Item3
1	6,680	39,400	-	11,435	-	37,750	-	16,869	12,905
2	36,088	8,717	-	29,493	-	63,942	-	20,174	22,424
3	7,132	15,895	-	11,106	-	11,274	-	8,094	6,400
4	17,079	5,961	-	20,597	-	11,482	-	30,305	4,856
5	5,573	17,467	-	20,530	-	65,334	-	36,007	11,697
6	37,548	3,887	-	19,996	-	50,775	-	27,361	4,941
7	27,335	18,741	-	9,732	-	26,695	-	24,906	8,096
8	41,165	16,148	-	13,342	-	48,480	-	36,986	831
9	31,686	2,874	-	14,027	-	20,674	-	21,740	9,996
10	28,529	6,031	-	15,479	-	32,089	-	36,744	991
11	21,261	1,779	-	11,725	-	37,344	-	49,620	4,788
12	21,945	1,091	-	4,000	-	7,840	-	15,826	2,024

Table 3. Demand and the resulting production and inventory plans

	Product type		
	Item1	Item2	Item3
Period	Demand (Production) [Inventory]	Demand (Production) [Inventory]	Demand (Production) [Inventory]
1	20,111 (18,115) [-1,996]	54,770 (56,269) [1,499]	56,956 (50,655) [-6,301]
2	59,392 (65,581) [4,193]	22,545 (28,891) [7,845]	55,796 (86,366) [24,269]
3	20,707 (18,238) [1,724]	26,406 (23,989) [5,428]	32,605 (17,674) [9,338]
4	37,579 (37,676) [1,821]	35,902 (36,266) [5,792]	39,885 (16,338) [-14,209]
5	28,473 (26,103) [-549]	54,852 (53,474) [4,414]	58,488 (77,031) [4,334]
6	53,266 (57,544) [3,729]	27,774 (31,248) [7,888]	55,695 (55,716) [4,355]
7	39,072 (37,067) [1,724]	46,104 (43,647) [5,431]	34,705 (34,791) [4,441]
8	53,965 (54,507) [2,266]	53,132 (53,134) [5,433]	46,810 (49,311) [6,942]
9	44,496 (45,713) [3,483]	21,370 (24,614) [8,677]	37,411 (30,670) [201]
10	45,770 (44,008) [1,721]	45,910 (42,775) [5,542]	26,197 (33,080) [7,084]
11	52,204 (32,986) [-17,497]	56,304 (51,399) [637]	50,541 (42,132) [-1,325]
12	54,243 (25,945) [-45,795]	46,277 (16,917) [-28,723]	42,224 (9,864) [-33,685]

Tables I and II present the resulting resource portfolio and allocation plans by CPGA. The capacity portfolio plan is represented by the quantities of resources on hand, by rent and by transfer. Table III presents the production and inventory plans.

6 Comparison of CPGA, Chase-GA and Level-GA Models

The performances of the three models (CPGA, Chase-GA and Level-GA) are examined under different setting of cost parameters. Two variations of market

Table 4. Demand and the resulting production and inventory plans

		Under high demand variation			
Total Profits CPGA		$h_{t,d}$(HIGH)		$h_{t,d}$(LOW)	
(deviation of Level-GA*) [deviation of Chase-GA+]		$l_{t,d}$ (HIGH)	$l_{t,d}$ (LOW)	$l_{t,d}$ (HIGH)	$l_{t,d}$ (LOW)
$s_{t,m}, f_{t,m}$ (HIGH)	$b_{t,m,r}$ (HIGH)	195,248,659 (-.1361)[-.0824]	220,531,693 (-.2565)[-.1875]	201,730,004 (-.1398)[-.1118]	220,963,773 (-.2334)[-.1891]
	$b_{t,m,r}$ (LOW)	217,184,776 (-.0795)[-.0067]	223,362,060 (-.0926)[-0.0342]	215,687,171 (-.0453)[-.0012]	227,732,470 (-.0913)[-.0527]
$s_{t,m}, f_{t,m}$ (LOW)	$b_{t,m,r}$ (HIGH)	205,373,333 (-.1252)[-.0806]	217,885,840 (-.1742)[-.1334]	201,818,477 (-.0961)[-.0644]	218,040,988 (-.1473)[-.1340]
	$b_{t,m,r}$ (LOW)	210,344,737 (-.0550)[-0.0016]	224,188,085 (-.1003)[-.0579]	219,423,739 (-.0842)[-.0374]	224,836,301 (-.0854)[-.0606]

*(Total Profits gained by Level-GA-Total Profits gained by CPGA)/Total Profits gained by CPGA
+(Total Profits gained by Chase-GA-Total Profits gained by CPGA)/Total Profits gained by CPGA

demand data are used. The high (low) level parameter of unit holding costs is set as 0.15 (0.05) of the product unit profit. The high (low) level parameter of unit back-order costs is set as 0.20 (0.10)of the product unit profit. The high (low) level parameter of unit renting cost of resources is set as 80,000 (40,000) times of average unit holding cost. The high (low) parameter of resource procurement costs is set as 6 (3) times of renting cost of the corresponding resource. Phase-out cost of a resource is 0.30 of its procurement cost.

The parameters of the CPGA (evolution time, crossover rate, mutation rate and population size) under different demands are also examined. Experimental results showed that the proposed algorithm is robust to the parameters. In order to achieve high performance, it is suggested to apply the best parameters under investigation of crossover rate (0.7), mutation rate (0.02), population size (15) and longer run time (4,000 CPU sec). This sensitivity analysis concludes that, as shown in Table IV for high demand variation, the performance of the CPGA outperforms the others two models.

7 Conclusions

This study has addressed the problem of resource acquisition and allocation in a make-to-stock production that requires a large amount of capital investment. Finite budget for investing resources, lumpy demands, long production horizon, many types of products to mix, time value of capital asset, technology level of resources, efficient usage of multiple-function resources, and limited capacity of resources are the constraints included in the problem. In addition to inventory plans, and resource acquisition, renting, transfer and phase-out have also been considered. This study has formulated the problem as a non-linear mixed integer mathematical programming model in which the goal is to maximize its overall profit. A constraint genetic algorithm based programming has been developed.

Examples have demonstrated that the proposed algorithm can solve the problem efficiently. Some applications of this model are to semiconductor testing industries.

References

1. Bard, J.F., K. Srinivasan, and D. Tirupati. An optimization approach to capacity expansion in semiconductor manufacturing facilities. *International Journal of Production Research*, 3359-3382, 1999.
2. Bashyam, T.C.A. Competitive capacity expansion under demand uncertainty. *European Journal of Operational Research*, 89-114, 1996.
3. Gen, M. and R. Cheng. *Genetic Algorithms and Engineering Optimization*, John Wiley and Sons, Inc, 2000.
4. Hsu, J.S. Equipment replacement policy- a survey. *Journal of Production and Inventory Management*, 23-27,1998.
5. Leachman, R.C. and T. Carmon, On capacity modeling for production planning with alternative machines. *IIE Transactions*, 62-72, 1992.
6. Li, S. and D.Tirupati, Technology choice with stochastic demands and dynamic capacity allocation: a two-product analysis. *Journal of Operations Management*, 239-258, 1995.
7. Rajagopalan, S. Capacity expansion with alternative technology choices. *European Journal of Operational Research*, 392-402, 1994.
8. Rajagopalan, S. Adoption timing of new equipment with another innovation anticipated. *IEEE Transactions on Engineering Management*, 14-25, 1999.
9. Swaminathan, J.M. Tool capacity planning for semiconductor fabrication facilities under demand uncertainty. *European Journal of Operational Research*, 545-558, 2000.
10. Wang, K.J. and T.C. Hou, Modeling and resolving the joint problem of capacity expansion and allocation with multiple resources and limited budget in semiconductor testing industry. *International Journal of Production Research*, 3217-3235, 2003.

A Quantity-Time-Based Dispatching Policy for a VMI System

Wai-Ki Ching and Allen H. Tai

Department of Mathematics,
The University of Hong Kong,
Pokfulam Road, Hong Kong, China

Abstract. In a Vendor-Managed Inventory (VMI) system, the supplier or the distributor is authorized to coordinate and consolidate the inventories at the retailers. The advantage of VMI is that the bullwhip effect can be minimized and the stock-out situations can also be reduced. Moreover, it provides a framework for synchronizing transportation decisions and hence reduce the transportation cost significantly. In this paper, we present an analytic model for quantity-time-based dispatching policy. The model discussed here takes into the account of the inventory cost, the transportation cost, the dispatching cost and the re-order cost. Since a new inventory cycle begins whenever there is a dispatching of products, the long-run average costs of the model can be obtained by using the renewal theory. We also derive a closed form solution of the optimal dispatching policy.

1 Introduction

In this paper, we consider a Vendor-Managed Inventory (VMI) system consisting of a vendor, a manufacturer and groups of retailers at different regions, see Figure 1. An analytic model of similar framework focusing on the Emergency Lateral Transshipment (ELT) has been studied by Ching [3]. Recent development in supply chain management focus on the coordination of different functional specialties and the integration of inventory control and transportation logistics, see Thomas [12] for instance. VMI is a supply chain initiative where the supplier or the distributor is responsible for all decisions regarding inventories at the retailers. Usually demands should be shipped immediately, but the vendor has the right of not delivering small orders to a region until an accumulated amount or an agreeable dispatching time. VMI requires the retailers to share the demands information with the supplier so as to allow making inventory replenishment decisions. This is usually achieved by using online data-retrieval systems and Electronic Data Interchange (EDI) technology, see for instance Chopra and Meindl [5] and Dyer and Nobeoka [6]. As a result, through the sharing of demands information the bullwhip effect can be reduced [5,9]. The bullwhip effect is the distortion of demands information transferred from the downstream retailers to the upstream suppliers, see Lee and Padmanabhan [8]. The current focus of

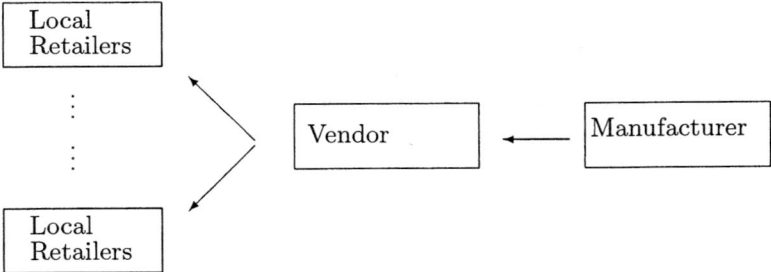

Fig. 1. The Supply Chain

VMI is the value of information sharing within a supply chain. Significant savings can also be achieved by carefully incorporating shipment consolidation and dispatching with stock replenishment decision in a VMI system, Higginson and Bookbinder [7]. Here shipment consolidation refers to the management of combining small shipments together in order to take the advantage of the decreased per unit transportation cost. Simulation is a useful tool for studying freight consolidation, Masters [10]. Other analytical approaches such as queueing theory and Markov decision process have been proposed to solve the consolidation models, see Higginson and Bookbinder [7] and Minkoff [11].

There are two types of dispatching policies: the quantity-based dispatching policy and the time-based dispatching policy, see for instance Higginson and Bookbinder [7]. A quantity-based policy dispatches whenever there is an accumulated load of size q. In this model, one has to determine the optimal dispatching size q and the optimal number of dispatches in each replenishment cycle. On the other hand, a time-based dispatching policy dispatches an accumulated load in every period of T. In this model, one has to determine the optimal quantity of replenishment Q and the optimal dispatching period T in each replenishment cycle. The time-based shipment consolidation have became a part of the transportation contract among the members of a supply chain and Delivery Time Guarantee (DTG) is a common marketing strategy in the competition of marketplaces, see Ching [4]. A VMI model based on time-based dispatching policy has been proposed and studied by Cetinkaya and Lee [2], they also discussed both advantages and disadvantages of the time-based and quantity-based dispatching policies. They remark that it is interesting to consider a model for the case of quantity-time-based dispatching policy. Here we propose an analytic model based on the simplified framework of [2] for the quantity-time-based dispatching policy. Our model takes into the account of the inventory cost, the transportation cost, the dispatching cost and the re-order cost. We remark that in modern E-business supply chain, inventory handling and transportation of products are the major costs, see Chopra and Meindl [5]. The dispatching cost is associated with the consolidation of shipment and the re-order cost corresponds to the inventory replenishment. In our model, for simplicity of discussion we assume that the demands of the retailers at a region is a simple Poisson process, the vendor applies a (q, Q, T) policy for replenishing the inventory and the lead time of the replenishment is assumed be negligible. The definition of a (q, Q, T) policy will be introduced shortly in Section 2. Since a new inventory cycle begins whenever there is a dispatching of products, the long-run average costs of the model can be obtained by using the renewal theory [1]. Moreover, closed form solution of optimal dispatching policy is also obtained.

The rest of the paper is organized as follows. In Section 2, we present the model for the quantity-time-based dispatching policy. In Section 3, we give a cost analysis of the model and derive the optimal dispatching policy with a numerical example. Finally, concluding remarks are given in Section 4 to conclude the paper and address further research issues.

2 The Quantity-Time-Based Dispatching Model

In this section, we give a model for quantity-time-based dispatching policy. In order to keep the models mathematically tractable, we consider models based on the simplified model discussed in [2]. Let us first define the following notations.

(i) λ^{-1}, the mean inter-arrival time of one unit of demand
(ii) I, the unit inventory cost per unit of time
(iii) D, the dispatching cost
(iv) F, the unit transportation cost
(v) C, the re-order cost
(vi) q, the size of a dispatching (quantity-based model)
(vii) r, the number of dispatches in a cycle (quantity-based model)
(viii) Q, the replenishment quantity (time-based model)
(ix) T, the dispatching period (time-based model)

Under this policy, a (q, Q, T) inventory replenishment is assumed. This means that the size of the replenishment is such that to clear the shortage and bring the inventory level back to Q. Moreover, a dispatching decision is made at the time $\min\{T_q, T\}$ where T_q is the time when a demands of size q is reached. The objective of this problem is to find the optimal values of q, Q and T such that the average long-run cost is minimized. The followings are the assumptions of the model.

(A1) The inventory level is under continuous review.
(A2) The vendor dispatches a load regularly for every period of T. If a size of demands q is accumulated before the planned dispatching time T, the vendor dispatches a load immediately.
(A3) The lead time of inventory replenishment is assumed to be negligible.
(A4) At the time of a dispatch, if the available inventory is not enough to clear the demand, we assume that the vendor can immediate replenish its stock from the manufacturer.

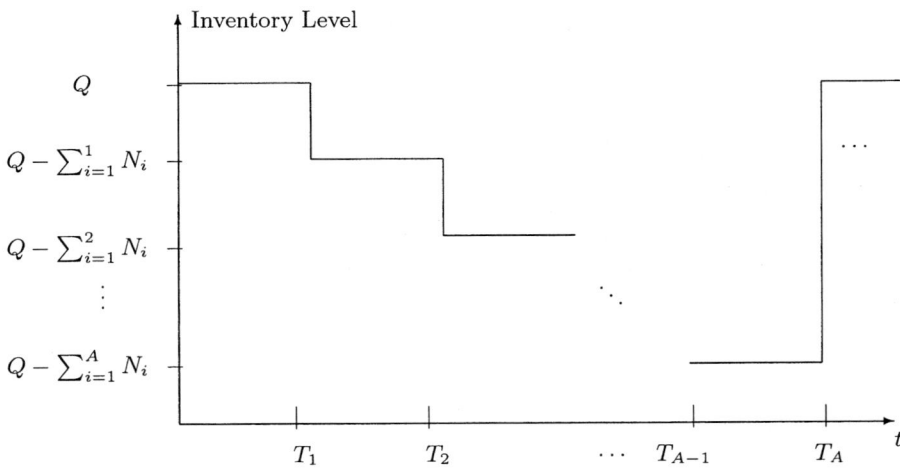

Fig. 2. The Inventory Level of a Cycle

A realization of the inventory levels in a replenishment cycle is shown in Figure 2. Here $N_i = N(T_i - T_{i-1})$ is the number of demands in the time interval $[T_{i-1}, T_i]$ and $T_i(i = 1, 2, \ldots, A)$ are the instants where a dispatch takes place. We note that $T_i = \min\{S_q, T\}$ where S_q is the time instant when the size of the demands is q. We remark that all T_i and S_q are random variables. At the time instant T_A (for certain A), the system is out of stock and an order is placed and arrived at once (as we assume zero lead time). Here

$$A = \inf\left\{a : \sum_{i=1}^{a} N(T_i - T_{i-1}) > Q\right\}$$

and A is a random variable representing the number of dispatch in a replenishment cycle. Moreover, the random variable $N(T)$ follows the Poisson distribution with mean λT. We aim at obtaining the optimal values of q, Q and T such that the average long-run cost of the system is minimized.

3 A Cost Analysis

In this section, we derive the expected long-run cost of the system by using renewal theory [1]. We note that a new inventory cycle begins whenever there is a dispatching of products, therefore the long-run average costs can be obtained by using the renewal theory. We will first derive the expected size of a dispatch $E(q_d)$ and an approximation for the expected number of dispatches $E(A)$ in each replenishment cycle. We then derive an approximate average cost.

We let d_T be the probability of dispatching a load at a planned dispatching time T, then

$$d_T = P(N(T) < q) = \sum_{i=0}^{q-1} \frac{(\lambda T)^i e^{-\lambda T}}{i!}.$$

Therefore the expected quantity of a dispatch is

$$E(q_d) = q(1 - d_T) + E(N(T))d_T = q(1 - d_T) + \lambda T d_T.$$

The expected time of a single dispatch is given by

$$E(T_q)(1 - d_T) + T d_T = \frac{q(1 - d_T)}{\lambda} + T d_T.$$

Meanwhile, since all stocks will be used up in a replenishment cycle, we have

$$E(\text{Number of dispatches}) \times E(\text{dispatching quantity}) > Q.$$

Also, since the stock is sufficient for the demand before a replenishment cycle ends, we have

$$Q > [E(\text{Number of dispatches}) - 1] \times E(\text{dispatching quantity}).$$

Therefore an upper bound and a lower bound of the expected number of dispatches are given by

$$N_{max} = \frac{Q}{q(1 - d_T) + \lambda T d_T} + 1$$

and

$$N_{min} = \frac{Q}{q(1 - d_T) + \lambda T d_T}$$

respectively. In view of the above bounds, we can approximate $E(A)$ by using N_{min}.

We then derive an approximate average long-run cost for the quantity-time-based model. Again we apply the renewal reward theorem, the average long-run cost is given by

$$C(q, Q, T) = \frac{\text{Replenishment Cycle Cost}}{\text{Replenishment Cycle Length}}.$$

(i) The expected inventory cost per cycle is given by

$$I \times \sum_{i=1}^{E(A)} [i \times E(T_i) \times E(N([T_i - T_{i-1}]))]$$

$$= I \times \sum_{i=1}^{E(A)} \left\{ i \times \left[\frac{q(1 - d_T)}{\lambda} + T d_T \right] \times [q(1 - d_T) + \lambda T d_T] \right\} \quad (1)$$

$$= I \times \frac{E(A)[E(A) + 1]}{2} \times \left[\frac{q(1 - d_T)}{\lambda} + T d_T \right] \times [q(1 - d_T) + \lambda T d_T]$$

$$= \frac{IQ}{2\lambda} \{Q + q(1 - d_T) + \lambda T d_T\}.$$

(ii) The expected dispatching cost per cycle is given by

$$D \times E(A) = \frac{DQ}{q(1-d_T) + \lambda T d_T}.$$

(iii) The expected transportation cost per cycle is given by

$$F \times E(A) \times E(N(T_i - T_{i-1})) = F \times \frac{Q[q(1-d_T) + \lambda T d_T]}{q(1-d_T) + \lambda T d_T}$$
$$= FQ.$$

(iv) The expected re-order cost per cycle is given by C.
(v) The expected length of a replenishment cycle is Q/λ.

Hence the expected cost is

$$C(q,Q,T) = \frac{IQ + Iq(1-d_T) + I\lambda T d_T}{2} + \frac{D\lambda}{q(1-d_T) + \lambda T d_T} + F\lambda + \frac{C\lambda}{Q}. \quad (2)$$

If we denote

$$V = q(1-d_T) + \lambda T d_T > 0$$

then (2) can be rewritten as

$$C(Q,V) = \frac{IQ + IV}{2} + \frac{D\lambda}{V} + \lambda F + \frac{C\lambda}{Q}. \quad (3)$$

From (3) we have

$$\begin{cases} \frac{\partial C(Q,V)}{\partial Q} = \frac{I}{2} - \frac{C\lambda}{Q^2} \\ \frac{\partial C(Q,V)}{\partial V} = \frac{I}{2} - \frac{D\lambda}{V^2} \\ \frac{\partial^2 C(Q,V)}{\partial Q^2} = \frac{2C\lambda}{Q^3} \\ \frac{\partial^2 C(Q,V)}{\partial V^2} = \frac{2D\lambda}{V^3}. \end{cases} \quad (4)$$

We note that the cost function $C(Q,V)$ is strictly convex for positive Q and V. Thus the unique global minimum for positive Q and V can be obtained by solving

$$\begin{cases} \frac{\partial C(Q,V)}{\partial Q} = \frac{I}{2} - \frac{C\lambda}{Q^2} = 0 \\ \frac{\partial C(Q,V)}{\partial V} = \frac{I}{2} - \frac{D\lambda}{V^2} = 0. \end{cases} \quad (5)$$

The optimal pair is then given by

$$(Q^*, V^*) = \left(\sqrt{\frac{2C\lambda}{I}}, \sqrt{\frac{2D\lambda}{I}}\right).$$

Therefore the optimal solution for minimizing $C(q,Q,T)$ is given by (q^*, Q^*, T^*), where

$$Q^* = \sqrt{\frac{2C\lambda}{I}}$$

and q^*, T^* satisfy the equation

$$q(1 - d_T) + \lambda T d_T = \sqrt{\frac{2D\lambda}{I}}$$

where $q \in N$ and $T \in (0, \infty)$. One possible choice of the optimal solution is the following:

$$(q^*, Q^*, T^*) = (\sqrt{\frac{2C\lambda}{I}}, \sqrt{\frac{2D\lambda}{I}}, \sqrt{\frac{2D}{\lambda I}}).$$

We note that if we set q^* to be large enough, then d_T will tend to 1 and

$$T^* = \sqrt{\frac{2D}{\lambda I}}.$$

Similarly if we set T^* to be large enough, d_T will tend to zero, then

$$q^* = \sqrt{\frac{2D\lambda}{I}}.$$

Example 1. Suppose that $\lambda = 10, D = 50$ and $I = 5$ then we have

$$V^* = \sqrt{\frac{2D\lambda}{I}} \approx 14.14.$$

In Table 1, we give some possible values of q and T such that $q(1 - d_T) + \lambda T d_T$ is close to 14.14.

Table 1. Solutions for q and T

q	T	$q(1 - d_T) + \lambda T d_T$
15	1.20	12.684
15	1.25	13.187
16	1.30	13.709
18	1.35	14.126
23	1.40	14.150
14	1.45	14.206
14	1.50	14.363

4 Concluding Remarks

In this paper, we discuss a Vendor-Managed Inventory (VMI) system where the vendor is authorized to coordinate and consolidate the inventory at the retailers. We present an analytic model for the quantity-time-based dispatching policy. Moreover, closed form solution of optimal dispatching policy is also obtained. For ease of discussion, the effect of the lead time in the inventory replenishment was not included in our model. It will be interesting to extend our model to include the lead time.

References

1. Barlow, R. and Proschan, F.:Mathematical Theory of Reliability, Classics in Applied Mathematics. SIAM, Phildaphia (1996)
2. Cetinkaya, S. and Lee, C.: Stock Replenishment and Shipment Scheduling for Vendor-Managed Inventory Systems. Manag. Sci., **46** (2000) 217–232
3. Ching, W.: Markov Modulated Poisson Processes for Multi-location Inventory Problems, Inter. J. Prod. Econ., **53** (1997) 217–223
4. Ching, W.: An Inventory Model for Manufacturing Systems with Delivery Time Guarantees, Comput. Operat. Res., **25** (1998) 367–377
5. Chopra, S. and Meindl, P.: Supply Chain Management, Strategy, Planning and Operation, Prentice Hall, New Jersey (2001)
6. Dyer, J. and Nobeoka, N.: Creating and Managing a High-performance Knowledge Sharing Network: The Toyota Case, Strategic Management Journal, **21** (2000) 345–367
7. Higginson, J. and Bookbinder, J.: Markovian Decision Processes in Shipment Consolidation, Transportation Sci., **29** (1995) 242–255
8. Lee, H. and Padmanabhan, L.: Information Distortion in a Supply Chain: The Bullwhip Effect, Manag. Sci., **43** (1997) 546–558
9. Lee, H., So, K. and Tang, C.: The Value of Information Sharing in a Two-level Supply Chain, Manag. Sci., **46** (2000) 626–643
10. Masters, J.: The Effect of Freight Consolidation on Customer Service, J. Busi. Logist., **2** (1980) 55–74
11. Minkoff, A.: A Markov Decision Model and Decomposition Heuristic for Dynamic Vehicle Dispatching, Oper. Res., **41** (1993) 77–90
12. Thomas, D.: Coordinated Supply Chain Management, Euro. J. Operat. Res., **94** (1996) 1–15

An Exact Algorithm for Multi Depot and Multi Period Vehicle Scheduling Problem

Kyung Hwan Kang, Young Hoon Lee, and Byung Ki Lee

Dept. of Industrial and Information Engineering, Yonsei University,
134 Shinchon-Dong, Seodaemoon-Gu, SEOUL 120-749 Korea
{optimal, youngh, leebk94}@yonsei.ac.kr

Abstract. This study is on the multi period vehicle scheduling problem in a supply chain where a fleet of vehicle delivers single type product from multi depots to multi retailers. The purpose of this model is to design the least costly schedule of vehicles in each depot to minimize transportation costs for product delivery and inventory holding costs at retailers over the planning period. A mixed integer programming formulation and an exact algorithm are suggested. In the exact algorithm, all feasible schedules are generated from each depot to each retailer and set of vehicle schedules are selected optimally by solving the shortest path problem. The effectiveness of the proposed procedure is evaluated by computational experiment.

1 Introduction

The purpose of Vehicle Routing Problem(VRP) and Vehicle Scheduling Problem(VSP) is to design the least costly(distance, time) routes for a fleet of capacitated vehicles to serve geographically scattered customers. There may be some restrictions such as the capacity for each vehicle, total traveling distance allowed for each vehicle, time window to visit the specific customers, and so forth. The decision which takes inventory holding cost and vehicle operating planning during the multi-period into consideration is required in the real supply chain. But, an objective function of general VRP/VSP is to minimize the transportation cost from a given depot to several customers within a single period. Our study is on multi period vehicle scheduling problem in a supply chain where a fleet of vehicle delivers single type product from multi depots to multi retailers. The purpose of this model is to design the least costly schedule of vehicles in each depot to minimize transportation costs in product delivery and inventory holding costs at retailers over the planning period. The example feasible solution of this model is shown in Figure 1.

In a supply chain which is composed of two depots and four retailers, three vehicles are available for each depot respectively. The loading capacity of vehicles of each depot is 30 and 60 respectively. The available number of vehicle for each depot is 3. The time spent in one way from each depot to retailers are (1, 2, 1, 1) and (2, 1, 2, 1). The demand of each retailer 1, 2, 3 and 4 is set at (0, 30, 10, 10, 0), (0, 10, 20, 20, 20), (0, 20, 10, 10, 20), (0, 10, 10, 20, 20) for the time

	1	2	3	4	5
Dep. 1 → R 1		(30)			
Dep. 1 → R 2		(30)			
Dep. 1 → R 3		(30)			
Dep. 1 → R 4					
Dep. 2 → R 1			(20)		
Dep. 2 → R 2				(40)	
Dep. 2 → R 3				(30)	
Dep. 2 → R 4		(60)			

Fig. 1. A feasible schedule of example data

period 5 respectively. The length of solid line and the dotted line represent the time spent in one way and returning trip. The number in the circle constitutes delivery quantity. The number of used vehicles of each depot is (1, 3, 3, 1, 0), (0, 2, 3, 3, 3) for the each time period. The delivery quantity is less than or equal to the loading capacity. Costs associated with vehicle set up cost, variable cost and inventory holding cost are considered. For example, from depot 2 to retailer 4, there is one vehicle set up cost, variable cost for 60(delivery quantity). The inventory of retailer 4 is (0, 50, 40, 20, 0) for the time period so that there is inventory holding cost from day 2 to day 4.

Although many researches concerning transportation planning and inventory/distribution have been studied extensively, much less is available on the combined problem. First, the research regarding the transportation planning was approached through the VRP/VSP. The VRP/VSP is considered an identical conception in the most recent research. The VRP/VSP have been extensively studied in Operations Research since Danzig and Ramser[1] introduced the traditional VRP. Diverse models that take into account realistic constraints have been studied. One of them is the VRP with time window(VRPTW) in which the customer service has to be performed within a certain time range, from the earliest time to the latest time window(Malandraki and Dial[2], Taillard et al.[3], Liu and Shen[4], Berger and Barkaoui[5]). Another one is the multi-depot VRP(MDVRP). Laporte[6] introduced the integer programming formulations for MDVRP. Renaud et al.[7] used a tabu search algorithm for this problem with capacity and route length restrictions. Salhi and Sari[8] extended the problem of simultaneously allocating customers to depots addressing the findings of the delivery routes and determining the vehicle fleet composition. Another model is the heterogeneous VRP(HVRP) which has a different vehicle capacity. Salhi and Rand[9] applied a perturbation procedure that uses reduction, reassignment, combination and relaxation within the existing or constructed routes to reduce the total cost of routing and acquisition by improving the utilization of the vehicles. Ochi[10] used the genetic algorithm, Gendreau et al.[11] applied tabu search respectively. Another example is the pick-up and delivery VRP which allows pick-up as well as delivery at each customer(the so-called delivery and backhaul problem). Mosheiov[12] presented tour partitioning heuristics in solving this problem which are based on breaking a basic tour into disjointed segments served by different vehicles.

Secondly, the research regarding the inventory/distribution has been executed a different point of view. The performance measures of an inventory and distribution model are generally expressed as functions of optimal purchasing volume, production, and shipping volume at each node, the optimal amount of every raw material, part, work-in-process and finished product to be stored at each supply chain stage and so on(Min and Zhou[13]). Similarly, the vehicle constrains such as vehicle capacity, the number of vehicle and operating cost of vehicle, did not attract attention in inventory/distribution problem. In other words, decisions concerning the inventory/distribution model and VRP/VSP are often studied separately. Kim and Kim[14] suggested an integration model which combined the vehicle operating planning and inventory/distribution model. In this problem, a multi-period vehicle scheduling problem(MPVSP) was suggested in a transportation system where a fleet of homogeneous vehicles delivers products of a single type from a single depot to multiple retailers. The objective is to minimize transportation costs for product delivery and inventory holding costs at retailers over the planning period. To solve this problem of large scale instances in a reasonable computation time, a two-phase heuristic algorithm was suggested based on a $k-th$ shortest path algorithm(Yen[15]).

Our research effort is designed to present integrated vehicle operating and inventory/distribution model as we extend the existing algorithm which suggests a near optimal solution in a single depot environment into an exact algorithm which would help guaranteeing an optimal solution in multi depot environment. The organization of this paper is as follows: In section 2, assumption and the mixed integer programming formulation are given, while section 3 introduces the relevant algorithm. The experimental result is suggested in section 4. The conclusion and the final discussion are in section 5.

2 Mathematical Formulation

There are I depots which supply a single product for the retailers J. No limitation is placed in supplying capacity for each depot during the planning period T. Additionally, each depot has K vehicles of the same capacity. Each vehicle can serve only one retailer per trip. The trip time of each vehicle is composed of the time spent in one way and returning trip. The time spent in one way is defined as the time taken from the depot departure to the retailer to be served and the time spent in returning trip is defined as the time taken from the retailer to the depot. For each period, the demand of each retailer is known to be deterministic but may vary by dates and retailers. It does not exceed the loading capacity of a single vehicle and has to be met before the end of the due date. Each vehicle is able to serve only one retailer for each time period, therefore, split delivery for demand quantity in each period is not permitted. There is no limit on the storage capacity for inventory at retailers and shortages are not allowed.

Logistic costs such as vehicle set up cost, variable cost in each depot and inventory holding cost in each retailer are considered. Vehicle set up cost is incurred when the vehicle departs to serve a delivery quantity from depot to retailer in time period. Variable cost is proportional to the total quantity of products transported. Inventory holding cost is also proportional to the quantity and the time length in retailers. The objective is to determine a vehicle operating planning, distributing quantities and inventory quantities for a given planning period to minimize a total cost. The mathematical formulation is as follows;

Notation

i: the index for the depot ($i = 1, 2, \ldots, I$)
j: the index for the retailer ($j = 1, 2, \ldots, J$)
k: the index for the vehicle ($k = 1, 2, \ldots, K$)
t: the index for the time period ($t = 1, 2, \ldots, T$)
Q_i: the capacity of the each vehicles in depot i
K: the number of vehicle in each depot
$D_{j,t}$: the demand of the retailer j in time period t
$V_{i,j,t}$: variable cost from depot i to retailer j in time period t
$S_{i,j,t}$: vehicle set up cost from depot i to retailer j in time period t
$\tau_{i,j}$: trip time from depot i to retailer j (the time spent in one way = the time spent in returning trip)
$H_{j,t}$: inventory holding cost of retailer j in time period t
M : big number
$X_{i,j,k,t}$: transportation quantity from depot i to retailer j by vehicle k in time period t
$R_{j,t}$: inventory level of retailer j in time period t
$Y_{i,j,k,t}=$ 1, if vehicle k departs from depot i to retailer j in time period t 0, otherwise

$$Min \sum_{j,t} H_{j,t}R_{j,t} + \sum_{i,j,k,t} V_{i,j,t}X_{i,j,k,t} + \sum_{i,j,k,t} S_{i,j,t}Y_{i,j,k,t} \qquad (1)$$

subject to

$$R_{j,t-1} + \sum_{i,k} X_{i,j,k,t} = D_{j,t} + R_{j,t} \qquad \forall j,t \qquad (2)$$

$$\sum_{k,j} Y_{i,j,k,t} \leq K \qquad \forall i,t \qquad (3)$$

$$\sum_j Y_{i,j,k,t} + \sum_j \sum_{u=t+1}^{t+2\tau_{i,j}-1} Y_{i,j,k,u} \leq 1 \qquad \forall i,j,k,t \qquad (4)$$

$$X_{i,j,k,t} \leq Q_i \qquad \forall i,j,k,t \qquad (5)$$

$$X_{i,j,k,t} \leq M \times Y_{i,j,k,t+1-\tau_{i,j}} \quad \forall i,j,k,t \tag{6}$$

$$R_{j,t}, X_{i,j,k,t} \geq 0 \quad \forall i,j,k,t \tag{7}$$

$$Y_{i,j,k,t} = 0 \text{ or } 1 \quad \forall i,j,k,t \tag{8}$$

The objective function of this model includes inventory holding cost at retailer, variable cost and vehicle set up cost. Balance equations on the product inventory and transportation for each retailer are shown in constraint (2). The constraints for the number of vehicle in each depot are shown in constraint (3). It is implied in constraint (4) that if a vehicle leaves the depot for a retailer, it would not be available for others during its returning trip and each vehicle is able to serve only one retailer for each time period also. The constraints for the vehicle loading capacity are shown in constraint (5). It is indicated that vehicle set up cost is incurred only when a delivery quantity is placed in constraint (6). The mixed integer programming provides an optimum solution within a reasonable time frame only for small scale problems.

3 Exact Algorithm

An exact algorithm which helps guaranteeing an optimal solution is suggested for this model. In the exact algorithm, all feasible schedules are generated from each depot to each retailer and set of vehicle schedules are selected optimally by solving the shortest path problem.

3.1 All Feasible Schedule Generation

All feasible schedules are generated by modifying the idea of Kim and Kim[14] who described the single depot vehicle scheduling problem as the shortest path problem. Each node represents a time period and each link represents a cost. The length of link between node m and node n represents the sum of vehicle set up cost, variable cost and inventory holding cost when vehicle serves the demand of retailer between $m+1$ and n day($n > m$). The arrival time of vehicle is the end of day $m+1$. The departure time of vehicle can be interpreted by calculating the trip time, that is to say, the departure time of vehicle is $m+2-\tau_{i,j}$ day. Some links do not appear in the graphs for real problems, since a link can be eliminated if delivery quantity by a trip corresponding to the link exceeds the sum of loading capacity of available vehicles. The total cost of a link is as followings; $C_{i,j,m,n}$ means the sum of vehicle set up cost, variable cost and inventory holding cost when vehicles of depot i serve retailer j's demand between $m+1$ and n day. It is implied in Figure 2 that the example of a generated feasible schedules from depot 1 to retailer 1. It is assumed that the trip time from depot 1 to retailer 1 is 2. $C_{1,1,1,4}$ represents the total cost when vehicles of depot 1 serve the demand of retailer 1 between day 2 and day 4. The time spent in one way and returning trip is (1, 2) and (3, 4) respectively. The number of used vehicle per each period

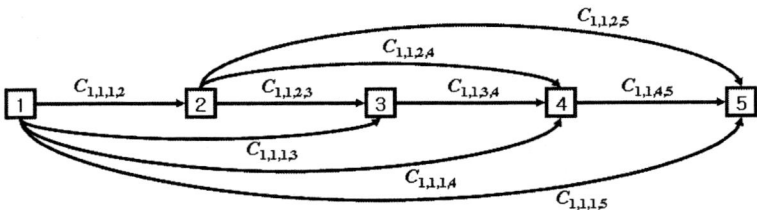

Fig. 2. An example of feasible schedule generation

is (1, 1, 1, 1, 0). If this method is applied in all depot and retailer during the time period, all feasible schedules less than $IJT^2/2$ are generated. In section 3.2, some schedules are selected optimally by solving the shortest path problem.

3.2 Optimal Selection of Generated Schedules

In the section 3.2, a set of feasible schedule is selected from the generated all feasible schedules in the section 3.1. If there is infinity number of vehicles in each depot, the shortest path from depot to each retailer is optimal solution. To find the optimal delivery schedule with the minimal value of objective function, we should find the shortest path from node 1 to node T per each retailer considering the constraints of available vehicles. The problem of selecting the best set from the all feasible schedules is formulated as a binary integer programming.

Notation

i: the index for the depot $(i = 1, 2, \ldots, I)$
j: the index for the retailer $(j = 1, 2, \ldots, J)$
t: the index for the time period $(t = 1, 2, \ldots, T)$
m: the index for the start time $(m = 1, 2, \ldots, T-1)$
n: the index for the end time $(n = 2, \ldots, T)$
K: the number of vehicle in each depot
$C_{i,j,m,n}$: the total cost of link(m, n) when depot i serves retailer j
$A_{i,j,m,n,t}$: the number of used vehicle in time period t for $C_{i,j,m,n}$
$Z_{i,j,m,n} = 1$, if the link $C_{i,j,m,n}$ is selected 0, otherwise

$$Min \sum_{i,j,m,n} C_{i,j,m,n} Z_{i,j,m,n} \qquad (9)$$

subject to

$$\sum_{i,m} Z_{i,j,m,p} - \sum_{i,n} Z_{i,j,p,n} = 0 \qquad \forall j, p = 2, \ldots, T-1 \qquad (10)$$

$$\sum_{i,n} Z_{i,j,1,n} = 1 \qquad \forall j \qquad (11)$$

$$\sum_{i,m} Z_{i,j,m,T} = 1 \quad \forall j \tag{12}$$

$$\sum_{j,m,n} A_{i,j,m,n,t} Z_{i,j,m,n} \leq K \quad \forall i, t \tag{13}$$

$$Z_{i,j,m,n} = 0 \text{ or } 1 \quad \forall i, j, m, n \tag{14}$$

Consequently, the objective function of equation (9) is equivalent to the original objective function of equation (1), i.e. minimizing the total cost. It is shown in constraint (10) that the link of network has to be connected continually. It is implied in constraint (11) that the start time of the link is time period 1. Constraint (12) makes that the end time of the link is time period T. The constraint for the number of vehicle in each depot is shown in constraint (13). Notice that the maximal number of routes generated is $IJT^2/2$ and the number of constraints is $JT + 2J + IT$.

4 Computational Experiments

The performance of the suggested exact algorithm was evaluated into two types of experiments. For small scale problems, solutions obtained by exact algorithm are compared with optimal solutions. Meanwhile, for large scale problem with single depot, performance is compared with the solution by Kim and Kim[14] method which presents a good performance in a single depot vehicle scheduling problem. For large scale problem with multi depot, computation time of an exact algorithm was presented.

For the small scale problems, 10 sets of instances were generated with respect to the total number of depots, retailers, and time periods. Each set also consists of 10 examples(total 100 examples). Optimal solutions with more than 3 depots, 8 retailers and 11 time periods cannot be found within a reasonable computation time. Thus, the number of depot, retailers and time periods are set at 2~6, 4~32, 11 and 16 respectively. The time spent in one way from each depot to retailers is assumed to be 1~2 day for each set. Inventory holding cost per unit for each retailer is randomly chosen from values of 0.1~0.2. Vehicle set up cost at the depot is also randomly chosen from values of 2~5. Transportation cost per unit is randomly chosen from values of 0.1~0.2. The demand quantity of retailer is uniformly distributed on 15~30 for every period. The number of vehicle is set at 3~25 for each set.

It is shown in Table 1 that the results of performances of exact algorithm is compared to optimal value. The optimal solutions were obtained using CPLEX 8.1, a commercial software package for mixed integer programming problems. Exact algorithm was coded in C++ language and CPLEX 8.1 and run on a personal computer with a Pentium IV 2.6Ghz CPU. The lower bound(LB) and upper bound(UB) have been computed by CPLEX under time limit of 30 minutes. The Gap* is defined as (Upper bound value - Lower bound value)100/Upperbound value. The upper bound value and lower bound value approach to the optimal value through the CPLEX iterative computation, in addition, the optimal value

Table 1. Performances of exact algorithm for a small scale instances

Set	No. of depot	No. of retailer	No. of period	Optimal value using the CPLEX 8.1					Exact algorithm	
				Upper Bound	Lower Bound	Gap^*	Com. time (sec.)	No. of success	Value	Com. time (sec.)
1	2	4	11	104.76	104.76	0.00	23.36	10/10	104.76	0.07
2	2	4	16	214.49	192.15	0.10	1800.27	0/10	214.49	0.22
3	2	8	11	215.97	215.95	0.00	34.00	10/10	215.97	0.11
4	2	8	16	422.81	395.81	0.06	1800.09	0/10	422.81	0.28
5	3	8	11	213.16	208.56	0.02	1586.62	4/10	213.16	0.10
6	3	8	16	423.19	358.90	0.15	1800.05	0/10	423.17	0.27
7	3	16	11	468.51	427.28	0.09	1800.07	0/10	467.70	0.27
8	3	16	16	852.27	715.23	0.16	1800.26	0/10	852.29	1.05
9	6	32	11	865.44	798.18	0.08	1800.18	0/10	865.44	0.97
10	6	32	16	1706.55	1434.43	0.16	1800.17	0/10	1704.17	3.31

Table 2. Performances of exact algorithm for a large scale instances with single depot

Set	No. of depot	No. of retailer	No. of period	Computation Time by Kim and Kim[14]	Computation Time by Exact algorithm	Gap^{**}
1	1	30	21	3.06	1.42	0.0213
2	1	50	21	5.00	2.15	0.0219
3	1	70	21	6.95	2.91	0.0222
1	1	90	21	9.43	3.98	0.0216

exists between the upper bound value and lower bound value. If the Gap^* is 0, the optimal value is found by solving the mixed integer programming using the CPLEX 8.1. It can be observed that for the networks with the more depots, retailers and time periods, the Gap^* gets larger. It comes from the fact that this mixed integer programming requires a considerable amount of computation time. The number of finding optimal solution for each set is shown in the column of "No. of success". Optimal solutions for the examples in the set 1, 3 could be found within a reasonable computation time. However, it is hard to find an optimal solution in most of the sets. On the other hand, the exact algorithm performs well compared to the optimal solution within significantly less computational time(0.67 second in average).

The result of large scale instances with single depot solved by exact algorithm and Kim and Kim[14] method is shown in Table 2. For these cases, 4 sets of instances were generated with respect to the total number of retailers, and time periods. Each set also consists of 30 examples(total 120 examples). The number of retailers and time periods are set at 30~90, 21 respectively. The Gap^{**} is defined as (Kim and Kim[14] value - exact algorithm value)×100/Kim and Kim[14] value).

For the computation time, suggested exact algorithm solved the cases more quickly than Kim and Kim[14] method. Even though the exact algorithm goes

Table 3. Performances of exact algorithm for a large scale instances with multi depot

Set	No. of depot	No. of retailer	No. of period	Computation Time by Exact algorithm
1	5	30	21	6.06
2	5	50	21	10.39
3	5	90	21	21.47
4	10	30	21	13.87
5	10	50	21	23.59
6	10	90	21	92.54
7	15	30	21	28.14
8	15	50	21	42.23
9	15	90	21	106.81

through an optimization procedure, the optimum solution can be obtained in a short while, as the number of constraints is considerably small. On the contrary, the calculation time takes longer in the Kim and Kim[14] method as it repeats iterative procedures in solving the $k-th$ shortest path problem. The performance of the exact algorithm proves to be better than that of the Kim and Kim[14] method. Because limited alternative schedules were generated and the vehicle schedule selection problem was solved by a heuristic based on the $k-th$ shortest path algorithm in Kim and Kim[14] method.

Table 3 shows the result of large scale instances with multi depot. We have set the number of depot at 5, 10, 15 and the number of retailers at 30, 50, 90. Each set also consists of 10 examples(total 90 examples). Exact algorithm guarantees the optimal value so that the criteria for measuring the performances of the algorithm can be depicted using the computation time. Exact algorithm finds the optimal solutions within a relatively short computation time.

5 Conclusion and Discussion

In this paper, we presented multi period vehicle scheduling problem in a supply chain where a fleet of vehicle delivers single type product from multi depots to multi retailers. The purpose of this model is to design the least costly schedule of vehicles in each depot to minimize transportation costs for product delivery and inventory holding costs at retailers over the planning period. Generally, the vehicle operating planning and the inventory/distribution model have been made separately. In this model, we considered integrated model with vehicle operating planning and the inventory/distribution. In an exact algorithm, all feasible schedules are generated from each depot to each retailer and set of vehicle schedules are selected optimally by solving the shortest path problem. We extended the existing algorithm which suggests a near optimal solution in a single depot environment into an exact algorithm which would help guaranteeing an optimal solution in multi depot environment. In the study, capacities for facilities are set loose, split delivery from depot to retailer is prohibited. Therefore, the investigation of these cases would attract many in the related fields. Further, the supply chain which is composed of multi stage and heterogeneous vehicle would also draw great interest.

References

1. Danzig, G.B., Ramser, J.H.: The Truck Dispatching Problem. Management Science, Vol. 6. (1959) 80–91
2. Malandraki, C., Dial, R.B.: A restricted dynamic programming heuristic algorithm for the traveling salesman problem. European Journal of Operational Research, Vol. 90. (1996) 45–55
3. Taillard, E., Badeau, P., Gendreau, M., Guertin, F., Potvin, J.Y.: A parallel tabu search heuristic for the vehicle routing problem with time windows. Transportation Research Part C: Emerging Technologies, Vol. 5. (1997) 109–122
4. Liu, F., Shen, S.: A route-neighborhood-based metaheuristic for vehicle routing problem with time windows. European Journal of Operational Research, Vol. 118. (1999) 485–504
5. Berger, J., Barkaoui, M.: A parallel hybrid genetic algorithm for the vehicle routing problem with time windows. Article in press, Computers and Operations Research, Vol. 31. (2004) 2037-2053
6. Laporte, G.: Integer programming formulations for the multi-depot vehicle routing problem. Comments on a paper by Kulkarni and Bhave, European Journal of Operational Research, Vol. 38. (1989) 228-237
7. Renaud, J., Laporte, G., Boctor, F.F.: A tabu search heuristic for the multi-depot vehicle routing problem. Computers and Operations Research, Vol. 23. (1996) 229-235
8. Salhi, S., Sari, M.: A multi-level composite heuristic for the multi-depot vehicle fleet mix problem. European Journal of Operational Research, Vol. 103. (1997) 95-112
9. Salhi, S., Rand, G.K.: Incorporating vehicle routing into the vehicle fleet composition problem. European Journal of Operational Research, Vol. 66. (1993) 313-330
10. Ochi, L.S.: A parallel evolutionary algorithm for the vehicle routing problem with heterogeneous fleet. Future Generation Computer Systems, Vol. 14. (1998) 285-292
11. Gendreau, M., Laporte, G., Musaraganyi, C., Taillard, E.D.: A tabu search heuristics for the heterogeneous fleet vehicle routing problem. Computers and Operations Research, Vol. 26. (1999) 1153-1173
12. Mosheiov, G.: Vehicle Routing with Pick-up and Delivery : Tour-Partitioning Heuristics. Computers and Industrial Engineering, Vol. 34. (1998) 669-684
13. Min, H., Zhou, G.: Supply chain modeling : past, present and future. Computers and Industrial Engineering, Vol. 43. (2002) 231-249
14. Kim, J., Kim, Y.: A decomposition approach to a multi-period vehicle scheduling problem. The International Journal of Management Science, Vol. 27. (1999) 421-430
15. Yen, J.Y.: Finding the K shortest loopless paths in a network. Management Science, Vol. 17. (1971) 712-716

Determining Multiple Attribute Weights Consistent with Pairwise Preference Orders

Byeong Seok Ahn[1] and Chang Hee Han[2]

[1] Associate Professor, Department of Business Administration, Hansung University,
389 Samsun 3, Sungbuk, Seoul 136-792, Korea
Tel.: 82-2-760-4412; Fax: 82-2-760-4482
bsahn@hansung.ac.kr

[2] Assistant Professor, Department of Business Administration, Hanyang University,
1271 Sa-1, Sangrok, Ansan, Gyeonggi 426-791, Korea
Tel.: 82-31-400-5634; Fax: 82-31-400-5591
chan@hanyang.ac.kr

Abstract. This paper presents a method for determining multiple attribute weights when pairwise comparison judgments on alternatives are specified and attribute consequences are captured in imprecise ways. A decision-maker or expert can express holistic pairwise comparisons on alternatives from his/her domain knowledge and decision alternatives are characterized by some tangible or possibly some intangible multiple attributes of which consequences can be represented by imprecise information. In this paper, attribute weights are to be estimated in the direction of minimizing the amount of violations and thus to be as consistent as possible with a decision-maker's ordered pairs. Multiple attribute weights that were determined with pairwise judgments on a subset of alternatives can be used to prioritize the other remaining alternatives.

Keywords: Multiattribute decision making; Weight estimation; Preference order; Imprecise information

1 Introduction

In multiple attribute decision making (MADM), one usually considers a finite discrete set of alternatives, each of which is valued by a finite discrete set of attributes. A classical evaluation of alternatives leads to the aggregation of multiple attributes into a unique criterion called a value function under certainty. A key component in the development of an additive multiple attribute model for selecting the best alternative is the method of obtaining the attribute weights. Various schemes aimed at eliciting exact weights from the decision-maker via interactive questions and answers may suffer on several counts. First of all, the weights are highly dependent on the elicitation method adopted [1, 2]. In addition, there is no agreement as to which method produces more accurate results since the true target weights remain unknown, or perhaps from a different perspective, are defined by the methodology. In order to avoid the difficulties associated with detailed weight elicitation, various surrogate weight approaches are presented in the decision analysis literature [3, 4].

A few studies which utilize a decision-maker's *a priori* pairwise comparison judgments on the set (or subset) of alternatives are devoted to estimating the unknown attribute weights while remaining consistent with pairwise preference orders. LINMAP, which is not based on mulitiattribute utility theory, estimates unknown attribute weights which minimize the Euclidean distance of each attribute consequence from the most preferred stimuli locations, called the ideal points [5]. Horsky and Rao [6] suggest an extended version of LINMAP in which they include comparisons of pairs of pairs, in addition to the collection of paired preference comparisons and more detailed statistical analysis of estimated attribute weights. Pekelman and Sen [7] propose a mathematical model which determines a set of weights for each decision-maker and present real world application. The UTA method, originally devoted to assessing a set of piecewise linear utility functions, can determine attribute weights which correspond to attribute-wise maximum utility value [8]. White et al. [9] present procedures for ranking alternatives and determining trade-off weights if the preference information gleaned from imprecise utility score assessment, trade-off weight assessment, and directly expressed preferences among alternatives is consistent.

In this paper, we use a decision-maker's paired comparison judgments on the set (or subset) of alternatives. The motivation for including ordered pairs as input is due to the facts that many decision-makers are willing and able to provide such data. For instance, many applications of multiple criteria decision aids have been published in the area of financial analysis, where a financial expert (or sometimes experts) articulates holistic paired comparisons between firms [10]. We consider, on the other hand, the imprecise preference data in the model since the need for handling imprecise preference data occurs in situations such as time pressure or lack of data, intrinsically intangible or non-monetary nature of attribute, decision-maker's limited attention and information processing capability and the like [11]. The types of imprecise information on attribute consequences considered in this paper cover a broad range of imprecision inherent in some attribute characteristics [12-14].

2 Conventional Mathematical Programming Formulations (Srinivasan and Shocker's Method)

This method uses the simple additive model and minimizes the total amount of violation of the pairwise preference constraints to obtain the attribute weights. Here, $A=\{x^1, x^2, \ldots, x^j, \ldots, x^m\}$ denotes the set of m alternatives on which pairwise preference judgments are to be made. Each of the m alternatives is described in terms of n attributes, $K=\{1, 2, \ldots, k, \ldots, n\}$. Also $Y=\{y_1^j, y_2^j, \ldots, y_k^j, \ldots, y_n^j\}$ denotes the jth alternative; i.e. Y_k^j is the attribute value of the kth attribute for the

jth alternative. Denoting the set of attribute weights $W = \{w_1, w_2, \ldots, w_k, \ldots, w_n\}$ for the n attributes which are the only parameters to be estimated, then the overall utility U^j, for the jth alternative, using simple additive model, is given by $U^j = \sum_{k \in K} w_k y_k^j$. This implies that the attribute weights w_ks are assumed to be not dependent on the attribute values Y_k^js. The constraints expressing the pairwise preference judgments of the decision-maker are of the form $U^i - U^j \geq 0$ for all $(i, j) \in \Omega$, in which $\Omega = [(i, j): x^i, x^j \in A]$ denote the set of pairwise preference judgments such that the alternative x^i is preferred to alternative x^j in a forced-choice pair comparison from the decision-maker. Then, according to the Srinivasan and Shocker [5], the set of weights W should be determined from the following linear programming problem:

$$\text{minimize} \sum_{(i,j) \in \Omega} z_{ij}$$

subject to
$$\sum_k w_k(y_k^i - y_k^j) + z_{ij} \geq 0, \text{ for all } (i,j) \in \Omega,$$
$$\sum_{(i,j) \in \Omega} \sum_k w_k(y_k^i - y_k^j) = 1,$$
$$z_{ij} \geq 0 \text{ for all } (i,j) \in \Omega,$$
$$w_k \geq 0 \text{ for all } k \in K,$$

where z_{ij} is a slack variable indicating the degree of deviation of the model from the true responses regarding the preference of alternative x^i to alternative x^j. The constraint $\sum_{(i,j) \in \Omega} \sum_k (y_k^i - y_k^j) = 1$ is simply added to preclude the trivial solution $w_k = 0$, $\forall k \in K$. Here the idea is that for every pair (i,j), U^i has to be greater than U^j for alternative x^i to be preferred to alternative x^j, and hence the sum of all $(U^i - U^j)$s must to be positive, even when the model is approximately valid. The right-hand side of the equation, which is added to preclude the trivial solution, can be any positive constant, which serves merely as a scaling factor for the set of weights W, and can, without loss of generality, be set at unity (please refer to the related works in [15, 16])

3 Attribute Weights Estimation Under Multiattribute Value Model

In addition to the notations defined above, let $v_k(x^i)$ be the value of alternative $x^i \in A$ on attribute $k \in K$ and w_k a scaling factor to represent the relative importance of the kth attribute. A classical means of evaluating an alternative leads to the aggregation of all criteria into a unique criterion called a value function under certainty and a utility function under uncertainty. In this paper, we assume that there exist additive value functions under preferential independence [17] and thus the underlying multiattribute value (MAV) function $v(x^i)$ of alternative x^i is denoted as follows:

$$v(x^i) = \sum_{k \in K} w_k v_k(x^i).$$

Suppose that the value of alternative x^i, $v_k(x^i)$ on attribute k is known, for example, only to lie within prescribed bounds such as $v_k^-(x^i) \leq v_k(x^i) \leq v_k^+(x^i)$. This may arise since the decision-maker can only specify a range $x^i \in [x^-, x^+]$ for the possible outcomes of alternative x^i on attribute k. The desired bounds for $v_k(x^i)$ can then be obtained from the known value function v_k in a manner that, for an increasing value function v_k, these bounds are given by $v_k^-(x^i) = v_k(x^-)$ and $v_k^+(x^i) = v_k(x^+)$. On the other hand, the decision-maker may sometimes prefer to specify preference as a ratio comparison between alternatives on a certain attribute and a verbal interpretation of the numerical value (i.e. α_{ij} in (1c)) used in the analytic hierarchy process literature can also be adopted in the model [18]. For example, the inequality, $v_k(x^i) \geq 3 \cdot v_k(x^j)$ is used to intend the value of alternative x^i is equal to or more than "slightly more important" than that of x^j on criterion k. As an extension, we take into account five forms of decision-maker's preferences on the attribute consequences;

- weak preference of some paired alternatives: $v_k(x^i) \geq v_k(x^j)$ (1a)
- strict preference of some paired alternatives: $v_k(x^i) - v_k(x^j) \geq \varepsilon$, where ε is a small positive number (1b)
- preference with ratio comparisons: $v_k(x^i) \geq \alpha_{ij} v_k(x^j)$ (1c)
- bounded preference: $v_k^-(x^i) \leq v_k(x^i) \leq v_k^+(x^i)$ (1d)
- preference differences of paired alternatives: $v_k(x^i) - v_k(x^j) \geq v_k(x^l) - v_k(x^m)$, for $i \neq j \neq l \neq m$ (1e)

Let Ω denote the set of *ordered* pairs (i, j) where i designates the preferred alternative from a paired comparison involving i and j. The set of ordered pairs could be comprised of 1) a set of past decision alternatives, 2) a subset of decision alternatives, especially when a set of alternatives is large, or 3) a set of fictitious alternatives, consisting of performances on the attributes which can be easily judged by the decision-maker to express his/her global comparisons [19].

Definition 1: Alternative x^i is said to be preferred to alternative x^j, based on the decision-maker's paired comparison judgment $(i, j) \in \Omega$ if and only if $\min[v(x^i)] > \max[v(x^j)]$, i.e., $\min \sum_k w_k v_k(x^i) > \max \sum_k w_k v_k(x^j)$.

Definition 2: Alternative x^i is pairwise dominant over alternative x^j if and only if the minimized value of the sum of weighted differences between alternative x^i and x^j is positive, i.e., $\min\{\sum_{k \in K} w_k [v_k(x^i) - v_k(x^j)]\} > 0$ subject to the constraints V and W.

Remark 1: Under imprecise attribute consequences, the aggregated value $v(x^i)$ of alternative x^i lies between an upper bound, $v^U(x^i)$ and a lower bound, $v^L(x^i)$, where $v^U(x^i) = \max_{V,W} \sum_k w_k v_k(x^i)$ and $v^L(x^i) = \min_{V,W} \sum_k w_k v_k(x^i)$ respectively. If a dominance relation based on Definition 1 holds between alternatives x^i and x^j, then a dominance relation based on Definition 2 always holds. This is easily proved. If alternatives x^i is preferred to x^j based on Definition 1, then it implies that $v^L(x^i) > v^U(x^j)$, i.e., $\min_{V,W} \sum_k w_k v_k(x^i) > \max_{V,W} \sum_k w_k v_k(x^j)$ and in turn, $\min_{V,W} \sum_k w_k v_k(x^i) + \min_{V,W} \sum_k -w_k v_k(x^j) > 0$. Hence,

$\min_{V,W} \sum_k w_k[v_k(x^i)-v_k(x^j)]$ in Definition 2 is greater than or equal to $\min_{V,W} \sum_k w_k v_k(x^i) + \min_{V,W} \sum_k -w_k v_k(x^j) > 0$.

Among the decision-maker's natural language statements assumed to be elicited, we notice the following general form: "The knowledge source indicates the relative importance of one alternative to another alternative, in a holistic manner that encompasses all the attributes." Let us define an optimal solution W^* to be a set of weights $\{W_k^*\}$ for $k \in K$. The solution would be consistent with the decision-maker's preferences if $v(x^i) - v(x^j) > 0$ for every a priori ordered pair $(i, j) \in \Omega$ and for all feasible values of W and V. We can state this as

$$\min\{\sum_k w_k[v_k(x^i) - v_k(x^j)] : w_k \in W, v_k \in V\} > 0 \text{ for } (i,j) \in \Omega.$$

Thus, the goal of analysis is determining the solution W^* for which the conditions such as $\sum_k w_k \cdot \min[v_k(x^i) - v_k(x^j)] \geq \varepsilon$ (ε is a small arbitrary positive number) for every a priori ordered pair $(i, j) \in \Omega$ are violated as minimally as possible. To attain the objective "as minimally as possible", we use auxiliary variables z_{ij} in $\sum_k w_k \cdot \min\{v_k(x^i) - v_k(x^j)\} + z_{ij} \geq \varepsilon$ for every ordered pair $(i, j) \in \Omega$ and minimize sum of auxiliary variables in the objective as shown below:

$$\text{minimize} \sum_{(i,j) \in \Omega} z_{ij} \quad (2a)$$

subject to

$$\sum_{k \in K} w_k \cdot \delta_k(x^i, x^j) + z_{ij} \geq \varepsilon \quad (2b)$$

$$w_k \in W, z_{ij} \geq 0, \varepsilon > 0 \text{ for } k \in K \text{ and } (i,j) \in \Omega \quad (2c)$$

in which $\delta_k(x^i, x^j) = \min\{v_k(x^i) - v_k(x^j) : v_k(x^i), v_k(x^j) \in V_k, k \in K\}$.

Theorem 1: The attribute weights $\{w_k^*\}_{k \in K}$ are totally consistent with the decision-maker's a priori ordered pairs consisting of strict preferences if and only if the optimal solution is $z_{ij}^* = 0$ for every pair $(i, j) \in \Omega$.

Proof: If the optimal solution, $z_{ij}^* = 0$ for every pair $(i, j) \in \Omega$, then this implies that from constraint (2b), $\min\{\sum_k w_k[v_k(x^i) - v_k(x^j)]\} > 0$ which, in turn, implies $v(x^i) > v(x^j)$ for every pair $(i, j) \in \Omega$. Thus ordered preferences specified by the decision-maker are fulfilled. If the attribute weights $\{w_k^*\}$ for $k \in K$ are consistent for every pair $(i, j) \in \Omega$, then it implies that $\sum_k w_k \cdot \min[v_k(x^i) - v_k(x^j)] \geq \varepsilon$ and thus $z_{ij}^* = 0$.

Remark 2: If the model allows for indifference preference between alternatives x^i and x^j, and the solution is totally consistent with the decision-maker's a priori ordered pairs consisting of strict preferences, then both z_{ij} and z_{ji} at the optimality take positive values and all the other z corresponding to the strict preferences takes zero values. For $z_{ij}^* > 0$ and $z_{ji}^* > 0$ implying $w^* \cdot \min[v(x^i) - v(x^j)] < 0$ and $w^* \cdot \min[v(x^j) - v(x^i)] < 0$, we can interpret that x^i is not preferred to x^j (not $x^i \succ x^j$) and x^j is not preferred to x^i (not $x^j \succ x^i$). By preference theory, this means that there exists indifference preference between alternatives x^i and x^j.

The minimization of potential violations is the simplest and most natural of several possible objectives. However, positive error terms for the strict preferences might produce contrary results to the original paired preference judgments even though the model is trying to attaining as minimum a value as possible. When this case occurs, further interactive modifications with the decision-maker can be investigated for reducing estimation errors; 1) checking and possible modification of imprecise value information and 2) checking and possible rearrangement of preference orders between alternatives specified by the decision-maker. In doing so, we identify the pairs which contribute violating terms from most to least and then proceed interactive modification with cooperative decision-maker.

Instead of the objective of minimization of potential violations, we can consider the objective which minimizes the number of violations because the number of violations might increase although the sum of the z_{ij} is attained at a minimum. Of course, with regard to the objectives there is no guarantee which objective is better for predicting attribute weights. One interesting point to be noted is that two objectives were tested in a real world application and the objective of minimizing the number of violation appeared to have an edge over the objective of minimizing the amount of violation [7].

Remark 3: If a decision-maker specifies her/his preference on a subset of alternatives set A, then the rest of the alternatives can be prioritized using the weights derived from Theorem 1. Suppose that we want to determine a dominance relation between alternatives x^g and x^h, which are not included in the set Ω, then x^g is preferred to x^h if $\sum_k w_k^* \cdot \min[v_k(x^g) - v_k(x^h)] > 0$.

4 An Illustrative Example

A numerical example is illustrated to show the method carried out in the Section 3. Suppose five non-dominated alternatives defined in terms of three attributes ($K = 3$) are as shown in Table 1. Here, all attribute outcomes are described in

Table 1. An example with three attributes and five alternatives

	Attributes		
Alternatives	1	2	3
x^1	1	0	[0.4, 0.5]
x^2	[0.5, 0.7]	[0.7, 0.8]	[0.8, 0.9]
x^3	[0.5, 0.6]	[0.6, 0.8]	1
x^4	[0.3, 0.4]	1	0
x^5	0	[0.8, 0.9]	[0.7, 0.8]

terms of bounded values to emphasize the decision-maker's imprecise information. The other forms such as (1a)-(1c) and (1e) can be generated depending on the decision situation considered.

Further, we assume that decision-maker indicates pairwise judgments on the alternatives such as $\Omega=\{(2, 3), (2, 4), (3, 2), (3, 4), (4, 1), (1, 5)\}$, i.e., alternative x^2 is at least preferred to alternative x^3, alternative x^2 to alternative x^4, etc. Note that we allow for indifference preference between alternatives x^2 and x^3. A mathematical program for estimating the attribute weights (i.e., w_1, w_2 and w_3) as consistent as the paired orders can be formulated as follows:

minimize $z_{23} + z_{24} + z_{32} + z_{34} + z_{41} + z_{15}$
subject to
$\sum_{k=1,2,3} w_k \cdot \min[v_k(x^i) - v_k(x^j)] + z_{ij} \geq \varepsilon$, for $(i, j) \in \Omega$
$W=\{w_1 + w_2 + w_3 = 1, w_1, w_2, w_3 \geq 0\}$, $\varepsilon > 0$.

Solving this program with a small arbitrary positive number ε replaced with 0.001, we obtain the optimal trade-off weights, $w_1^* = 0.4260$, $w_2^* = 0.3908$ and $w_3^* = 0.1832$ with objective value $z_{ij}^* = 0$ for every pair $(i, j) \in \Omega$ except $z_{23}^* = 0.119$ and $z_{32}^* = 0.146$ corresponding to the indifferent alternatives x^2 and x^3. This solution implies that in the viewpoint of a certain decision-maker, s/he considers that attribute 1 and 2 are almost equally important and they are about twice as important as attribute 3. For a reverse decision aiding purpose, suppose that there exist two alternatives x^6 and x^7 which are not evaluated *a priori* by the decision-maker and their value scores are also specified in bounded values such as

$V_1=\{v_1(x^6) \in [0.75, 0.8], v_1(x^7) \in [0.7, 0.8]\}$,
$V_2=\{v_2(x^6) \in [0.4, 0.5], v_2(x^7) \in [0.1, 0.2]\}$,
$V_3=\{v_3(x^6) \in [0.7, 0.8], v_3(x^7) \in [0.8, 0.9]\}$.

By Remark 3, $\sum_{k=1,2,3} w_k^* \cdot \min[v_k(x^6) - v_k(x^7)] = 0.02$, which indicates that alternative x^6 is preferred to x^7, utilizing the derived weights.

5 Concluding Remark

In a situation where the pairwise comparison judgments among the alternatives are gathered from a decision-maker and imprecise preference judgments on some attributes are provided in the model, a weight estimation methodology is presented. In a more realistic occasion, it is natural to extend the single decision-maker's weight estimation into representative multiple decision-makers' case. In usual cases, the quality of weight estimation can be enhanced by adopting group members' weight estimates and thus reducing the errors of a single decision-maker's estimates. Thus, various statistical analyses can be utilized to infer group attitude toward competing alternatives, which is left for further research.

References

1. K. Borcherding, D. von Winterfeldt, The effect of varying value trees on multiattribute evaluations, *Acta Psychologica* 68 (1988) 153-170.
2. P.J.H. Shoemaker, C.D. Waid, An experimental comparison of different approaches to determining weights in additive utility models, *Management Sci.* 28 (1982) 182-196.
3. W.G.. Stillwell, D.A. Seaver, W. Edwards, A comparison of weight approximation techniques in multiattribute utility decision making, *Org. Behav. Hum. Dec. Proc.* 28 (1981) 62-77.
4. F.H. Barron, B.E. Barrett, Decision quality using ranked attribute weights, *Management Sci.* 42 (1996) 1515-1523.
5. V. Srinivasan, A.D. Shocker, Linear programming techniques for multidimensional analysis of preferences, *Psychometrika* 38 (1973) 337-369.
6. D. Horsky, M.R. Rao, Estimation of attribute weights from preference comparisons, *Management Sci.* 30 (1984) 801-822.
7. D. Pekelman, S.K. Sen, Mathematical programming models for the determination of attributes weights. *Management Sci.* 20 (1974) 1217-1229.
8. E. Jacquet-Lagreze, J. Siskos, Assessing a set of additive utility functions for multicriteria decision-making, the UTA method, *Eur. J. Oper. Res.* 10 (1982) 151-164.
9. C.C. White, A.P. Sage, S. Dozono, A model of multiattribute decisionmaking and trade-off weight determination under uncertainty, *IEEE Trans. Syst. Man Cybernet.* 14 (1984) 223-229.
10. C. Zopounidis, A.I. Dimitras, Multicriteria Decision Aid Methods for the Prediction of Business Failure, Kluwer Academic Publishers, Dordrecht, 1998.
11. M. Weber, Decision making with incomplete information, *Eur. J. Oper. Res.* 28 (1987) 44-57.
12. A.P. Sage, C.C. White, ARIADNE: A Knowledge-based interactive system for planning and decision support, *IEEE Trans. Syst. Man Cybernet.* 14 (1984) 35-47.
13. K.S. Park, S.H. Kim, Tools for interactive multiattribute decision making with incompletely identified information, *Eur. J. Oper. Res.* 98 (1997) 111-123.
14. B.S. Ahn, K.S. Park, C.H. Han, J.K. Kim, Multi-attribute decision aid under hierarchical structure and incomplete information, *Eur. J. Oper. Res.* 125 (2000) 431-439.
15. C.K. Mustafi, M.J. Xavier, Mixed-integer linear programming formulation of a multi-attribute threshold model of choice, . J. Opl Res Soc. 36, (1985) 935-942.
16. M. Oral, O. Kettani, Modelling the process of multiattribute choice, J. Opl Res Soc. 40 (1989), 281-291.
17. R.L. Keeney, H. Raiffa, Decisions with Multiple Objectives: Preferences and Value Tradeoffs, Wiley, New York, 1976.
18. T.L. Saaty, The Analytic Hierarchy Process, McGraw-Hill, New York, 1980.
19. E. Jacquet-Lagreze, Y. Siskos, Invited Review: Preference disaggregation: 20 years of MCDA experience, *Eur. J. Oper. Res.* 130 (2001) 233-245.

A Pricing Model for a Service Inventory System When Demand Is Price and Waiting Time Sensitive

Peng-Sheng You

Graduate Institute of Transportation & Logistics,
National ChiaYi University,
300 Shiue-Fu Road, Chia-Yi 600, Taiwan

Abstract. This paper deals with the simultaneous determination of initial sales price, secondary sales price and resource capacity needed to be rented in a service rental system. It is assumed that demands are price and waiting time dependent. A deterministic model is developed for establishing the above-mentioned decisions. Analytical results show that optimal solutions can be found as a closed form formulation when the times to change sales prices are exogenous variables. In addition, we found sufficient conditions for optimization in the case where the times to change sales price are decision variables.

1 Introduction

Service providers seek a feasible and desirable solution for coping with puzzles in marketing environments all the time. The business challenges may be derived from a variety of possibilities such as how much of the capacity should be rented to satisfy future demands, how to adjust the capacity effectively by having the needed number of employees on duty, how to smooth the demand to match the service capacity, and so on. Unsuitable decisions on the above-mentioned problems may influence a company's profit.

For example, suppose a forwarder provides delivery service within a certain district. The forwarder does not have its own cargo and must rent its cargo capacity from cargo suppliers. The capacity renting decision is not easy to make. If the forwarder rents too much, it will have a higher probability of many remaining at vacant capacity by departure time and will result in a loss of the idle capacity contribution since cargo capacity is a perishable inventory with time. On the other hand, if the forwarder rents too few, the forwarder may lose a lot of sales. Thus, it is a challenge for forwarders to establish the suitable strategies to cope with the problems of capacity planning and demand management.

Likewise, additional business situations also emerge across many spheres such as transportation and internet service industries. Suppose a rental company provides a transportation service among a variety of cities which are connected by a number of rail legs. For a target market composed of n cities, the rental company can provide $0.5n(n+1)$ types of delivery service in the target market. To provide one unit of delivery service between any two cities requires one unit of capacity on each rail leg connecting the two cities.

Therefore, capacities on a certain train leg are used by several types of delivery service. To establish the delivery service on the whole network, the rental company should rent from the railways a set of capacities for all the legs of the whole network. Whenever the demand for the delivery service falls short of the rented capacity, the idle capacity is a loss due to the perishability of service capacity with time. On the contrary, if demand exceeds the capacity, the unsatisfied demand is considered a loss in sales. In order to maximize profit, the rental company must deal with the problem of how much capacity is needed for each leg and what is the suitable price for each type of delivery service.

Several studies on revenue management have been conducted on the problem of advanced sale issues. Researches on this area mostly focus on perishable inventories or service products. Readers are referred to the bibliography of [1, 3, 12, 19] for further information.

Liberman and Yechiali [9] developed a dynamic programming model for determining overbooking decision for a hotel reservation problem. Gale and Holmes [5] examined an airline model in which a monopoly firm employs an advance purchase discount to satisfy the demand for two flights with different departure times. They showed why a restriction on advance purchase may be a profit maximizing policy for a monopoly airline during a peak time period. James and Dana [7] proposed an airline model in which an airline faces two types of customers with different valuations and demand uncertainties. Luo [10] suggested an integrated inventory model for perishable goods in which backorder is allowed.

Moreover, Gilbert and Ballou [6] discussed a supply chain system in which advanced purchased information from downstream customers is employed to reduce operating costs. Their paper aimed to illustrate how advance purchase information can be used to reduce costs for a make-to-order supplier. A model for quantifying the benefits from advance commitments from downstream customers is developed. An example in steel industry is used to illustrate the analytic result of the developed model. Biyalogorsky et. al., [2] dealt with a resource allocation problem in which an overselling rule is used to reduce the potential yield and spoilage losses. A simple decision rule was developed to determine the resources allocation problem. They conclude that the overselling strategy can improve resource allocation efficiency by reducing the potential yield and spoilage losses. Ng, Wirtz, and Lee [14] discussed the problem of what strategy can be used to deal with the unused service capacity.

Subramanian, Stidham, and Lautenbacher [17] investigated an airline seat inventory control problem in which overbooking, cancellations and no-shows are allowed. You [21] proposed a multi-flight leg nested airline model for finding the optimal pricing decisions. Shugan and Xie [16] studied the separation of purchase and consumers on an advance sales system and proposed several remarkable implications. Lee and Ng [8] proposed a model for a monopolistic service firm for investigating the impact of market price sensitivity on the optimal price and capacity in an advance sale system. They showed that a policy with advance sales is superior to a policy without advance sales.

Meanwhile, Xie and Shugan [20] examined a selling model in which the sale interval is divided into an advance sale period and a spot sale period. They demonstrated that the advance sales can be profitable regardless of industry factors. Moe and Fader [13]) addressed a forecasting model in which advance purchase orders is used to predict new product sales. Tang, Rajaram, and Alptekinoglu [18] considered a perishable season product model in which advance booking discount is offered to customers who commit their orders prior to the selling season. They analyzed the benefits of the advance booking system and characterized the optimal discount parameters. McCardle, Rajaram and Tang [11] proposed a two firm competitive model for determining the optimal discount price and analyzing the benefits of the advance booking discount program.

Ringbom and Shy [15] dealt with an advance booking system in which customers' cancellation events and partial refunds are taking into consideration. Two cases are analyzed. In the first case, this paper aims to maximize the profit by determining the rate of partial refund offered to those customers who do not show up or cancel their reservations. In the second case, this paper determines the refund rate that maximizes the social welfare. Cachon [4] discussed the impact of the allocation of inventory risk between a risk neutral retailer and a supplier on supply chain efficiency. They found that if advance purchase discount is considered, the supply chain coordination is achievable by means of arbitrary profit allocation.

The aforementioned literature has discussed the problem of advance sales system on price setting and the price adjusting time. From the point of view of service firms which do not have their own service capacity, it is important to simultaneously establish their capacity renting and price setting policies to improve their revenues.

This paper deals with a service pricing problem in which a service firm rents a number of resources with distinct services over consecutive but non-overlappling time intervals from a service capacity supplier, assembles the rented service resources into a variety of service products, and employs an advanced sales system to sell the products to customers. Demand for products are assumed to be price and waiting time dependent. The purpose of this paper is to maximize the total profit over a planning period by determining the optimal initial and secondary prices, and required capacity size.

2 Assumptions and Formulation

This section develops a deterministic model of service rental organizations that investigate the joint service resource renting and pricing setting problem for a service firm which does not have its own service capacity. The problem is defined in terms of service resources and products. Assume that there exists L types of service resources and that products are assembled by one or several distinct service resources with consecutive but non-overlapping service time.

Postulate that the company requires customers to make reservations for their purchase. Let product type be defined by index pair-(i, j), $1 \leq i < j \leq L+1$. To provide a unit of product type-(i, j) requires one unit of resource types i, $i+1$, $\cdots$ and $j-1$, respectively. We define the ending time of the service of resource type i to be immediately prior to the starting time of the service of resource type $i+1$ for all i. Let T_i denotes the time to start the service of resource type i with the relationship $T_1 < T_2 \cdots < T_L < T_{L+1}$ where T_{L+1} is the stopping time of the service of resource type L. Then, it is noted that T_i also represents the stopping time to make reservation for product type-(i, k), $k = i+1, \cdots, L+1$. Assume that demand rate $d_i^j(p, t)$ for product type-(i, j) at time t is linearly decline with sales price p and waiting time $(T_i - t)$ where t is the time to make reservation. That is,

$$d_i^j(p, t) = \alpha_i^j - \beta_i^j p - \theta_i^j(T_i - t) \qquad (1)$$

where α_i^j, β_i^j and θ_i^j are positive constants. Although the assumption that demand function is linearly dependent on price and waiting time is restrictive. it has the advantage of providing theoretical solution and simple analysis.

We assume that the service firm rents each unit of the service resource type-i at the cost c_i from the service capacity provider. Since it is not possible that the demand rate is negative if the sales price is larger than the cost, we make a reasonable assumption that $\alpha_i^j - \beta_i^j \sum_{m=i}^{j-1} c_i - \theta_i^j T_i > 0$.

Moreover, suppose the service firm makes a single price change for product type-(i, j) at time point $g_i^j T_i$ where g_i^j represents a ratio of the time interval between the time to start reservation and the time to price change to the time interval between the time to start reservation and the starting time of the service of resource type-i, T_i. The purpose of the service firm is to maximize the total profit by determining the initial sales prices, secondary price and the amount of service resource needed to be rented.

Now, we will develop our formulation. Suppose the service firm initially sets the sales price of the product type-(i, j) at x_i^j and resets it at y_i^j at time point $g_i^j T_i$. Let matrices $\mathbf{x} = [x_i^j]$ and $\mathbf{y} = [y_i^j]$. Let $F(\mathbf{x}, \mathbf{y})$ be the net profit from all products. Then, we have

$$F(\mathbf{x}, \mathbf{y}) = \sum_{i=1}^{L} \sum_{j=i+1}^{L+1} R_i^j(x_i^j, y_i^j) - \sum_{i=1}^{L} s_i c_i \qquad (2)$$

where $R_i^j(x_i^j, y_i^j)$ and s_i respectively represent the total sales revenue from product type-(i, j) and the amount of service resource type-i needed to be rented, and are respectively given by

$$R_i^j(x_i^j, y_i^j) = \int_0^{g_i^j T_i} x_i^j d_i^j(x_i^j, t) dt + \int_{g_i^j T_i}^{T_i} y_i^j d_i^j(y_i^j, t) dt, \qquad (3)$$

$$s_i = \sum_{m=1}^{i} \sum_{j=i+1}^{L+1} \left(\int_0^{g_i^j T_i} d_i^j(x_i^j, t) dt + \int_{g_i^j T_i}^{T_i} d_i^j(y_i^j, t) dt \right). \qquad (4)$$

Our problem is to maximize $F(\mathbf{x}, \mathbf{r})$ through determining $\mathbf{x}$ and $\mathbf{y}$ subject to $d_i^j(x_i^j, t) \geq 0$ and $d_i^j(y_i^j, t) \geq 0$ for all (i, j). Integrating (3) and (4), and substituting the results into (2) gives

$$F(\mathbf{x}, \mathbf{y}) = \sum_{i=1}^{L} \sum_{j=i+1}^{L+1} \Big(-\beta_i^j T_i (g_i^j (x_i^j)^2 + (1 - g_i^j)(y_i^j)^2)$$
$$+ 0.5 T_i g_i^j (2\alpha_i^j + \theta_i^j T_i g_i^j - 2\theta_i^j T_i) x_i^j$$
$$+ 0.5 T_i (1 - g_i^j)(2\alpha_i^j - \theta_i^j T_i (1 - g_i^j)) y_i^j \Big)$$
$$- 0.5 \sum_{i=1}^{L} \sum_{m=1}^{i} T_m (2\alpha_m^j - \theta_m^j T_m - 2\beta_m^j (g_m^j x_m^j + (1 - g_m^j) y_m^j)) c_i. \quad (5)$$

After a little algebra, we can rewrite (5) as follows:

$$F(\mathbf{x}, \mathbf{y}) = \sum_{i=1}^{L} \sum_{j=i+1}^{L+1} \Big(-\beta_i^j T_i (g_i^j (x_i^j)^2 + (1 - g_i^j)(y_i^j)^2)$$
$$+ 0.5 T_i g_i^j (2\alpha_i^j + \theta_i^j T_i g_i^j - 2\theta_i^j T_i) x_i^j$$
$$+ 0.5 T_i (1 - g_i^j)(2\alpha_i^j - \theta_i^j T_i (1 - g_i^j)) y_i^j$$
$$- 0.5 T_i (2\alpha_i^j - \theta_i^j T_i - 2\beta_i^j (g_i^j x_i^j + (1 - g_i^j) y_i^j))(\sum_{m=i}^{j-1} c_m) \Big). \quad (6)$$

3 Decision Analysis

In this section, we provide solution procedure to find decisions for the proposed problem.

Lemma 1. *For fixed $\mathbf{g}$, $F(\mathbf{x}, \mathbf{y})$ is concave function of $\mathbf{x}$ and $\mathbf{y}$.*

Proof. First, we have $\frac{\partial^2 F(\mathbf{x}, \mathbf{y})}{\partial (x_i^j)^2} = -2\beta_i^j T_i g_i^j$, $\frac{\partial^2 F(\mathbf{x}, \mathbf{y})}{\partial (y_i^j)^2} = -2\beta_i^j T_i (1 - g_i^j)$ and $\frac{\partial^2 F(\mathbf{x}, \mathbf{y})}{\partial x_i^j r_i^j} = 0$ from which we can show that the Hessian matrix is negative-definite, thus we have completed the proof. □

The Lagrangian function is given by

$$L(\mathbf{x}, \mathbf{y}, \lambda, \eta, \mathbf{u}, \mathbf{v}) = F(\mathbf{x}, \mathbf{y}) - \sum_{i=1}^{L} \sum_{j=i+1}^{L+1} \lambda_i^j (-d_i^j (x_i^j, T_i) + (u_i^j)^2)$$
$$- \sum_{i=1}^{L} \sum_{j=i+1}^{L+1} \eta_i^j (-d_i^j (y_i^j, T_i) + (v_i^j)^2). \quad (7)$$

Taking the partial derivatives of L with respect to $\mathbf{x}, \mathbf{y}, \lambda, \eta, \mathbf{u}, \mathbf{v}$, we have the following KKT conditions:

$$\frac{\partial L}{\partial x_i^j} = -2\beta_i^j T_i g_i^j x_i^j + 0.5 T_i g_i^j (2\alpha_i^j + \theta_i^j T_i g_i^j - 2\theta_i^j T_i) + T_i \beta_i^j g_i^j (\sum_{m=i}^{j-1} c_m) - \lambda_i^j \beta_i^j$$
$$= 0, \tag{8}$$

$$\frac{\partial L}{\partial y_i^j} = -2\beta_i^j T_i (1 - g_i^j) y_i^j + 0.5 T_i (1 - g_i^j)(2\alpha_i^j - \theta_i^j T_i (1 - g_i^j))$$
$$+ T_i \beta_i^j (1 - g_i^j)(\sum_{m=i}^{j-1} c_m) - \eta_i^j \beta_i^j = 0, \tag{9}$$

$$\frac{\partial L}{\partial \lambda_i^j} = \alpha_i^j - \beta_i^j x_i^j - \theta_i^j T_i - (u_i^j)^2 = 0, \tag{10}$$

$$\frac{\partial L}{\partial \eta_i^j} = \alpha_i^j - \beta_i^j y_i^j - \theta_i^j T_i - (u_i^j)^2 = 0, \tag{11}$$

$$\frac{\partial L}{\partial u_i^j} = -2\lambda_i^j u_i^j = 0, \tag{12}$$

$$\frac{\partial L}{\partial u_i^j} = -2\eta_i^j v_i^j = 0. \tag{13}$$

Now, we will develop the optimal decisions. First, we will define the following function.

$$k_i^j = (\alpha_i^j - \beta_i^j \sum_{m=i}^{j-1} c_m))/(\theta_i^j T_i). \tag{14}$$

Theorem 1. *Suppose $g_i^j \leq 2k_i^j - 3$. Then the optimal initial sales price and secondary sales price are respectively given by $\bar{x}_i^j$ and $\bar{y}_i^j$ where*

$$\bar{x}_i^j = 0.25(2\alpha_i^j - \theta_i^j T_i (2 - g_i^j))/\beta_i^j + 0.5 \sum_{m=i}^{j-1} c_m, \tag{15}$$

$$\bar{y}_i^j = 0.25(2\alpha_i^j - \theta_i^j T_i (1 - g_i^j))/\beta_i^j + 0.5 \sum_{m=i}^{j-1} c_m. \tag{16}$$

Proof. Letting $\lambda_i^j = 0, \eta_i^j = 0$ and solving the equations (8) to (13) gives $x_i^j = \bar{x}_i^j$, $y_i^j = \bar{y}_i^j$ and $u_i^j = 0.5\alpha_i^j - 0.5\theta_i^j T_i - 0.5\beta_i^j \sum_{m=i}^{j-1} c_m - 0.25\theta_i^j T_i g_i^j \geq v_i^j = 0.5\alpha_i^j - 0.5\theta_i^j T_i - 0.5\beta_i^j \sum_{m=i}^{j-1} c_m - 0.25\theta_i^j T_i (1 + g_i^j) \geq 0.5\alpha_i^j - 0.5\theta_i^j T_i - 0.5\beta_i^j \sum_{m=i}^{j-1} c_m - 0.25\theta_i^j T_i (1 + (2k_i^j - 3)) = 0$ since v_i^j is decreasing in g_i^j and $g_i^j \leq 2k_i^j - 3$. In addition, since $\bar{x}_i^j \geq 0$ and $\bar{y}_i^j \geq 0$ by the assumption $\alpha_i^j - \beta_i^j \sum_{m=i}^{j-1} c_m - \theta_i^j T_i \geq 0$. Thus, we have completed the proof. □

Theorem 2. *Suppose $g_i^j > 2k_i^j - 3$. Then the optimal initial sales price and secondary sales price are respectively given by $\bar{x}_i^j$ and $\hat{y}_i^j$ where*

$$\hat{y}_i^j = (\alpha_i^j - \theta_i^j T_i)/\beta_i^j. \tag{17}$$

Proof. Letting $\lambda_i^j = 0$, $u_i^j = 0$ and solving the equations (8) to (13) gives $x_i^j = \bar{x}_i^j$, $y_i^j = \hat{y}_i^j > 0$ by modeling assumption, $u_i^j \geq 0$ and $\eta_i^j = 0.5T_i(1-g_i^j)(\theta_i^j T_i g_i^j + 3\theta_i^j T_i - 2\alpha_i^j + 2\beta_i^j \sum_{m=i}^{j-1} c_m)/\beta_i^j \geq 0$ due to $g_i^j > 2k_i^j - 3$. Thus, the proof is completed. □

The optimal initial and secondary prices have been developed under fixed values of $[g_i^j]$. Below, we allow **g** as decision variables and introduce the following sufficient conditions for optimality.

Theorem 3. *Suppose $k_i^j \geq 1.75$ for all i, j. Then the optimal initial sales price, secondary sales price and time ratio are respectively given by $\tilde{x}_i^j$, $\tilde{y}_i^j$ and $\tilde{g}_i^j$ where*

$$\tilde{x}_i^j = 0.125(2\alpha_i^j - 3\theta_i^j T_i)/\beta_i^j + 0.5 \sum_{m=i}^{j-1} c_m, \tag{18}$$

$$\tilde{y}_i^j = 0.5(\alpha_i^j - \theta_i^j T_i)/\beta_i^j + 0.5 \sum_{m=i}^{j-1} c_m, \tag{19}$$

$$\tilde{g}_i^j = 0.5. \tag{20}$$

Proof. For fixed $\mathbf{y}, \mathbf{g}$, from Theorems 2 and 3 we have F is concave in x and $F(\mathbf{x}, \mathbf{y}, \mathbf{g}) = F(\bar{\mathbf{x}}, \mathbf{y}, \mathbf{g})$. Substituting $\bar{\mathbf{x}}$ into $F(\mathbf{x}, \mathbf{y}, \mathbf{g})$ gives

$$F(\bar{\mathbf{x}}, \mathbf{y}, \mathbf{g}) = \sum_{i=1}^{L} \sum_{j=i+1}^{L+1} \Big(1/16 T_i^3 (\theta_i^j)^2 (g_i^j)^3 / \beta_i^j$$
$$+ 0.25(T_i)^2 \theta_i^j (\alpha_i^j - 2\beta_i^j y_i^j - \theta_i^j T_i + \beta_i^j \sum_{m=i}^{j-1} c_m)(g_i^j)^2/\beta_i^j$$
$$+ 0.25 T_i (\alpha_i^j - 2\beta_i^j y_i^j - \theta_i^j T_i + \beta_i^j \sum_{m=i}^{j-1} c_m)^2 g_i^j/\beta_i^j$$
$$+ 0.5 T_i (y_i^j - \sum_{m=i}^{j-1} c_m)(2\alpha_i^j - 2\beta_i^j y_i^j - \theta_i^j T_i) \Big). \tag{21}$$

Taking the first order derivatives with respect to g_i^j and y_i^j gives

$$\frac{\partial F(\bar{\mathbf{x}}, \bar{\mathbf{r}}, \mathbf{g})}{\partial g_i^j} = 3/16 T_i^3 (\theta_i^j)^2 (g_i^j)^2/\beta_i^j$$
$$+ 0.5(T_i)^2 \theta_i^j (\alpha_i^j - 2\beta_i^j y_i^j - \theta_i^j T_i + \beta_i^j \sum_{m=i}^{j-1} c_m) g_i^j/\beta_i^j$$
$$+ 0.25 T_i (\alpha_i^j - 2\beta_i^j y_i^j - \theta_i^j T_i + \beta_i^j \sum_{m=i}^{j-1} c_m)^2/\beta_i^j, \tag{22}$$

$$\frac{\partial^2 F(\bar{\mathbf{x}}, \bar{\mathbf{r}}, \mathbf{g})}{\partial y_i^j} = -0.5(T_i)^2 \theta_i^j (g_i^j)^2 - T_i(\alpha_i^j - 2\beta_i^j y_i^j - \theta_i^j T_i - \beta_i^j \sum_{m=i}^{j-1} c_m) g_i^j$$
$$+ 0.5 T_i (2\alpha_i^j - 2\beta_i^j y_i^j - \theta_i^j T_i) - \beta_i^j T_i (y_i^j - \sum_{m=i}^{j-1} c_m). \tag{23}$$

Equating the (22) and (23) to zero and solving them gives $y_i^j = \tilde{y}_i^j$ and $g_i^j = \tilde{g}_i^j$. Substituting $y_i^j = \tilde{y}_i^j$ into $\bar{x}_i^j$, we obtain the optimal initial sales price is $\tilde{x}_i^j$. Since $0 \leq \tilde{g}_i^j < 1$, $d_i^j(\tilde{x}_i^j, T_i) \geq 0$ and $d_i^j(\tilde{y}_i^j, T_i) \geq 0$, the values of $\tilde{\mathbf{x}}, \tilde{\mathbf{y}}, \tilde{\mathbf{g}}$ are optimal decisions if the Hessian matrix is negative definite at the points of $\tilde{\mathbf{x}}, \tilde{\mathbf{y}}, \tilde{\mathbf{g}}$. After a little algebra, we can show that $\frac{\partial^2 F}{\partial (y_i^j)^2}\big|_{y=\tilde{y}, g=\tilde{g}} = -T_i \beta_i^j$, $\frac{\partial^2 F}{\partial (g_i^j)^2}\big|_{y=\tilde{y}, g=\tilde{g}} = -3/16(T_i)^3(\theta_i^j)^2/\beta_i^j$ and $\frac{\partial^2 F}{\partial y_i^j g_i^j}\big|_{y=\tilde{y}, g=\tilde{g}} = 0.25(T_i)^2 \theta_i^j$ from which we obtain that the Hessian matrix is negative definite, thus we have completed the proof. □

4 Numerical Examples

To illustrate Theorems 2 to 4, we construct a numerical example with the following parameters: $L = 6$, $c_i = 5$ for all i and $\mathbf{T} = (30, 32, 34, 36, 38, 40)$. In addition, suppose the parameter of α, β, θ and $\mathbf{g}$ are assumed and listed in Table 1.

Table 1. Parameters for numerical examples

(i,j)	α_i^j	β_i^j	θ_i^j	g_i^j	k_i^j	(i,j)	α_i^j	β_i^j	θ_i^j	g_i^j	k_i^j	(i,j)	α_i^j	β_i^j	θ_i^j	g_i^j	k_i^j
(1,2)	17.2	1.84	0.09	0.1	2.97	(2,4)	16.4	1.06	0.01	0.4	18.19	(3,7)	14.8	0.40	0.06	0.3	3.32
(1,3)	13.2	0.78	0.03	0.2	5.96	(2,5)	10.0	0.31	0.08	0.4	2.07	(4,5)	10.0	1.12	0.04	0.4	3.06
(1,4)	11.6	0.40	0.07	0.4	2.68	(2,6)	17.2	0.53	0.03	0.8	6.79	(4,6)	10.0	0.45	0.08	0.1	1.90
(1,5)	12.4	0.39	0.02	0.2	7.73	(2,7)	16.4	0.36	0.07	0.1	3.32	(4,7)	12.4	0.48	0.03	0.7	4.85
(1,6)	13.2	0.32	0.04	0.8	4.33	(3,4)	12.4	1.20	0.09	0.5	2.10	(5,6)	17.2	1.99	0.06	0.3	3.18
(1,7)	17.2	0.31	0.09	0.1	2.88	(3,5)	13.2	0.75	0.05	0.7	3.34	(5,7)	17.2	0.85	0.10	0.5	2.28
(2,3)	14.0	1.45	0.08	0.2	2.64	(3,6)	16.4	0.58	0.09	0.2	2.51	(6,7)	15.6	1.71	0.07	0.7	2.52

For all cases, we have $g_i^j \leq 2k_i^j - 3$, thus we concluded from Theorem 2 that $\bar{\mathbf{x}}$ and $\bar{\mathbf{y}}$ are optimal initial and secondary sales prices. Table 2 reveals the optimal prices for all trips. The application of Table 2 can be interpreted as follows. For example, for trip-(3, 5), this table displays that $\bar{x}_3^5 = 13.06$ and $\bar{y}_3^5 = 13.63$ are optimal initial and secondary sales prices. Since the time of price change should be set at time $g_3^5 T_3 = 34 \times 0.7 = 23.8$, we derive that the sales prices \$13.06 and \$13.63 should be set during time $[0, 23.8]$ and $[23.8, 32]$, respectively. The illustration of the pricing strategies for other trips are similar to that of trip-(3, 5).

Substituting $\bar{\mathbf{x}}$ and $\bar{\mathbf{y}}$ in Table 2 into (6), we obtain that the optimal profit is equal to \$8,459.592. In addition, from (3) the optimal amount of capacity needed to be rented is $\mathbf{s} = (474.00, 815.53, 1022.15, 1004.07, 971.05, 623.07)$.

Suppose times to change price are decision variables. Then, since $k_i^j > 1.75$ for all (i,j), from Theorem 4 we have $\tilde{g}_i^j = 0.5$ for all (i,j). In addition, we concluded that the optimal initial and secondary sales prices are $\tilde{\mathbf{x}}$ and $\tilde{\mathbf{y}}$, respectively (See Table 2). The illustration of this case is similar to that without price change decision. The optimal profit in this case is \$8,487.048.

Table 2. Computational results for numerical examples

(i,j)	$\bar{x}_i^j$	$\bar{y}_i^j$	$\tilde{x}_i^j$	$\tilde{y}_i^j$	(i,j)	$\bar{x}_i^j$	$\bar{y}_i^j$	$\tilde{x}_i^j$	$\tilde{y}_i^j$	(i,j)	$\bar{x}_i^j$	$\bar{y}_i^j$	$\tilde{x}_i^j$	$\tilde{y}_i^j$
(1,2)	6.48	6.84	7.51	6.99	(2,4)	12.62	12.69	16.56	12.62	(3,7)	26.33	27.61	30.13	26.59
(1,3)	12.94	13.23	13.48	13.03	(2,5)	20.33	22.39	22.66	20.53	(4,5)	6.45	6.77	6.79	6.48
(1,4)	19.90	21.21	24.22	20.03	(2,6)	25.68	26.14	28.47	25.55	(4,6)	13.07	14.67	15.35	13.71
(1,5)	25.21	25.59	25.77	25.32	(2,7)	32.32	33.88	39.50	32.94	(4,7)	19.69	20.25	22.87	19.57
(1,6)	32.00	32.94	37.75	31.72	(3,4)	6.71	7.35	8.12	6.71	(5,6)	6.33	6.62	7.16	6.39
(1,7)	38.60	40.78	43.48	39.48	(3,5)	13.06	13.63	16.33	12.95	(5,7)	13.44	14.56	13.01	13.44
(2,3)	6.53	6.97	5.94	6.67	(3,6)	19.26	20.58	23.56	19.66	(6,7)	6.53	6.94	7.55	6.45

5 Conclusion

A mathematical model for simultaneously determining the discount amount and the regular sales prices for a service renting system is developed. This paper showed the optimal decision when the time of price changes are prescribed. In addition, we also found some sufficient conditions for optimality when the times of price changes are controllable variables.

We have analyzed the proposed problem under some restrictive assumption. For the model to be more realistic, it may be possible to extend the proposed model to the case with cancellations. Extension of the proposed model to a case with cancellation and other applications will be a focus of our future work.

Acknowledgment. The author would like to thank the National Science Council of the Republic of China for financially supporting this research under Contract No. 92-2416-H-150-001-.

References

1. Barut, M., Sridharan, V.: Design and evaluation of a dynamic capacity apportionment procedure. Eur J Oper Res. **155** (2004) 112-133
2. Biyalogorsky, E., Carmon, Z., Fruchter, E.G., Gerstner, E.: Research note: Overselling with opportunistic cancellations. Mark. Sci. **18** (1999) 605-610
3. Boer, S.V., Freling, R., Piersma, N: Mathematical programming for network revenue management revisited. Eur J Oper Res. **137** (2002) 72-92
4. Cachon, G.P.: The allocation of inventory risk in a supply chain: Push, pull, and advance-purchase discount contracts. Manage. Sci. **50** (2004) 222-238

5. Gale, I.L., Holmes, T.J.: Advance-purchase discounts and monopoly allocation of capacity, Amer. Econ. Rev. **83** (1993) 135-146
6. Gilbert, S.M., Ballou, R.H.: Supply chain benefits from advanced customer commitments. J. Oper. Manag. **78** 1999 61-73
7. James, D., Dana, Jr.: Advance purchase discounts and price discrimination in competitive markets, J. polit. Econ. **2** (1998) 395-422
8. Lee, K.S., Ng, I.C.L.: Advanced sale of service capacities: a theoretical analysis of the impact of price sensitivity on pricing and capacity allocations. J. Bus. Res. **54** (2001) 219-225
9. Liberman, V., Yechiali, U.: On the hotel overbooking problem - An inventory system with stochastic cancellations. Manage. Sci. **24** (1978) 1117-1126
10. Luo, W.: An integrated inventory system for perishable goods with backordering. Comput. Ind. Eng. **34** (1998) 685-693
11. McCardle, K., Rajaram, K., Tang, C.S.: Advance booking discount programs under retail competition. Manage. Sci. **50** (2004) 701-708
12. McGill J.I., Ryzin G.J.V.: Yield Management: Research Overview and Prospects. Transport. Sci. **33** (1999) 233-256
13. Moe, W.W., Fader, P.S.: Using advance purchase orders to forecast new product sales. Mark. Sci. **21** (2002) 347-364
14. Ng, I.C.L., Wirtz, J., Lee, K.S.: The strategic role of unused service capacity, Int. J. Serv. Ind. Manage. **10** (1999) 211-238
15. Ringbom, S., Shy, O.: Advance booking, cancellations, and partial refunds, Econ. Bulletin. **13** (2004) 1-7
16. Shugan, S.M., Xie, J.: Advance pricing of services and other implications of separating purchase and consumption. J. Ser. Res. **2** (2000) 227-239
17. Subramanian, J., Stidham, S.Jr., Lautenbacher, C.J.: Airline yield management with overbooking, cancellations, and no-shows, Transport. Sci. **33** (1999) 147-167.
18. Tang, C.S., Rajaram, K., Alptekinoglu, A.: The benefits of advance booking discount programs: Model and analysis. Manage. Sci. **50** (2004) 465-478
19. Weatherford, L.R., Bodily, S.E.: A taxonomy and research overview of perishable-asset revenue management: Yield management, overbooking, and pricing. Oper. Res. **40** (1992) 831-844
20. Xie, J., Shugan, S.M.: Electronic ticksts, smart cards, and online prepayments: When and how to advance sell. Mark. Sci. **20** (2001) 219-243
21. You, P.S.: Dynamic pricing in airline seat management for flights with multiple flight legs. Transport. Sci. **33** (1999) 192-206

A Bi-population Based Genetic Algorithm for the Resource-Constrained Project Scheduling Problem

Dieter Debels[1] and Mario Vanhoucke[1,2]

[1] Ghent University, Faculty of Economics and Business Administration,
Hoveniersberg 24, 9000 Ghent, Belgium
[2] Vlerick Leuven Gent Management School, Operations & Technology Management,
Centre, Reep 1, 9000 Ghent, Belgium
{dieter.debels, mario.vanhoucke}@ugent.be

Abstract. The resource-constrained project scheduling problem (RCPSP) is one of the most challenging problems in project scheduling. During the last couple of years many heuristic procedures have been developed for this problem, but still these procedures often fail in finding near-optimal solutions for more challenging problem instances. In this paper, we present a new genetic algorithm (GA) that, in contrast of a conventional GA, makes use of two separate populations. This bi-population genetic algorithm (BPGA) operates on both a population of left-justified schedules and a population of right-justified schedules in order to fully exploit the features of the iterative forward/backward scheduling technique. Comparative computational results reveal that this procedure can be considered as today's best performing RCPSP heuristic.

1 Introduction

We study the resource-constrained project scheduling problem (RCPSP), denoted as $m,1|cpm|C_{max}$ using the classification scheme of [9]. The RCPSP can be stated as follows. In a project network in AoN format G(N,A), we have a set of nodes N and a set of pairs A, representing the direct precedence relations. The set N contains n activities, numbered from 0 to $n-1$ ($|N| = n$). Furthermore, we have a set of resources R, and for each resource type $k \in R$, there is a constant availability a_k throughout the project horizon. Each activity $i \in N$ has a deterministic duration $d_i \in \mathbb{N}$ and requires $r_{ik} \in \mathbb{N}$ units of resource type k. We assume that $r_{ik} \leq a_k$ for $i \in N$ and $k \in R$. The dummy start and end activities 0 and n - 1 have zero duration and zero resource usage. A schedule S is defined by an n-vector of start times $s(S) = (s_0, \ldots, s_{n-1})$, which implies an n-vector of finish times $f(S)$ where $f_i = s_i + d_i, \forall i \in N$. A schedule S is said to be feasible if it is nonpreemptive and if the precedence and resource constraints are satisfied. If none of the activities can be scheduled forwards (backwards) due to precedence or resource constraints, then the schedule is said to be left-justified (right-justified). The objective of the RCPSP is to find a feasible schedule that minimizes the project makespan f_{n-1}.

2 Representation and Generation of Left- and Right-Justified Schedules

Each RCPSP meta-heuristic relies on a *schedule representation* to encode a schedule and a *schedule generation scheme* (SGS) to decode the schedule representation into a schedule S. For both the representation and generation of a schedule various approaches exist.

Table 1. Incorporation of TO condition

Justification of schedule	TO condition	implementation of TO in AL
Right-justified schedule	$s_i < s_j \Rightarrow p < q$	sort activities in increasing order of their start times
Left-justified schedule	$f_i > f_j \Rightarrow p < q$	sort activities in decreasing order of their finish times

Although five different methods are given in the literature [13], a schedule representation is simply a representation of a priority-structure between the activities. For our procedure we use the most frequently used [13] *activity list* (AL) representation where a sequence of non-dummy activities $\lambda = [\lambda_1, \ldots, \lambda_{n-2}]$ is used to determine the priority of each activity. When $\lambda_p = i$, we say that activity i is at position p in the AL. An activity i has a lower priority than all preceding activities in the sequence and a higher priority than all succeeding activities. An AL is said to be *precedence-feasible* if an activity never comes after the position of one of its successors (predecessors) in the list used for the generation of a left-justified (right-justified) schedule. In the current paper, we rely on the topological ordering (TO) condition [5, 28] for our AL representation. Our version of the TO condition and its implementation in the AL is described in Table 1, with p and q the positions of activity i and j in an AL. The table illustrates that the TO condition and the implementation depends on the justification of the schedule (left or right). Since the TO condition is based on start and finish times, and hence uses information from the corresponding schedule, we can only incorporate the TO condition after the schedule generation. In Sect. 3.3, the advantages of the TO condition will be illustrated on a project example.

Besides various schedule representations, there exist also two different types of SGSs in the literature; the serial SGS and the parallel SGS. As it is sometimes impossible to reach an optimal solution with the parallel SGS [17], we opt for the serial SGS where all activities are scheduled one-in-a-time and in the sequence of the AL. Each activity is scheduled as soon (as late) as possible within the precedence and resource constraints to construct a left-justified (right-justified) schedule. We introduce the example project depicted in Fig. 1, with a single renewable resource type with availability $a_1 = 2$. The problem is represented by an activity-on-the-node network. Corresponding to each activity we depicted the duration on top of the node and the resource demand below the node. Figure 2 represents a left-justified schedule 1, obtained by applying a serial SGS on the activity list [1, 2, 8, 5, 3, 4, 6, 7, 9]. The incorporation of the TO-condition on this schedule, leads to the activity list AL_1, depicted at the bottom of Fig. 2.

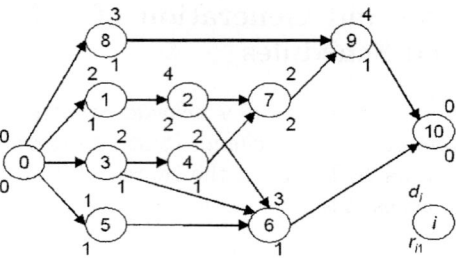

Fig. 1. Example project

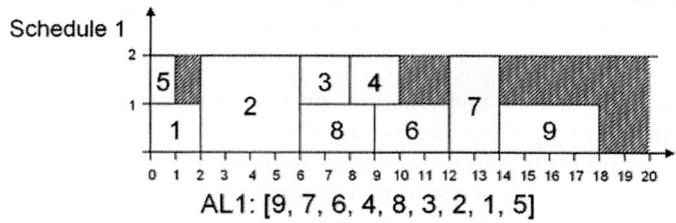

Fig. 2. A left-justified schedule and the corresponding AL after incorporation of the TO-condition

A well-known local search technique for RCPSP meta-heuristics is the iterative forward/backward scheduling technique. This technique is introduced by Li and Willis [20] and successfully implemented in various meta-heuristic procedures [1, 5, 11, 24, 25, 26, 27, 29]. The technique is based on the serial SGS and uses schedule information to determine the AL. Starting from a left-justified schedule, the procedure creates an AL by sorting the activities in decreasing order of their finish times (i.e. the TO condition for left-justified schedules of Table 1). Then, the serial SGS is used to build a right-justified schedule. In a following iteration, the activities are sorted in increasing order of the start times in the right-justified schedule (i.e. the TO condition for right-justified schedules of Table 1) and the serial SGS is used to generate a left-justified schedule. In doing so, only improvements can occur for each iteration. The procedure stops when no further improvements can be obtained. Assume that schedule 1 of Fig. 2 is our start left-justified schedule with an activity list AL_1 in which the activities are sorted in decreasing order of the finish times. The iterative forward/backward procedure uses this list to construct a right-justified schedule with corresponding activity list AL_2. In this list, the activities have been sorted in increasing order of their start times. This iteration (see schedule 2 of Fig. 3) leads to a makespan improvement of 2 time units. In a next iteration, the procedure constructs the left-justified schedule 3 with a corresponding activity list AL_3. The procedure could continue this process by using the activity list AL_3 to construct a right-justified schedule, but it is easy to see that no further makespan improvement can be achieved.

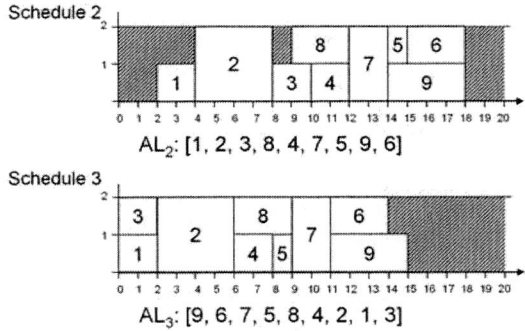

Fig. 3. The iterative forward/backward scheduling technique

3 The Bi-population Genetic Algorithm

The evolution of living beings motivated Holland [10] to solve complex optimization problems by using algorithms that simulate biological evolution. This approach gave rise to the technique known as a genetic algorithm (GA). In a GA, processes loosely based on natural selection, crossover and mutation are repeatedly applied to one population that represents potential solutions. In contrast to a regular GA, we use the bi-population genetic algorithm (BPGA) that makes use of two different populations: a population **LJS** that only contains left-justified schedules and a population **RJS** that only contains right-justified schedules. Both populations have the same population size. The procedure starts with the generation of an initial **LJS**, followed by an iterative process that continues until the stop criterion is satisfied. The iterative process consecutively adapts the population elements of **RJS** and **LJS**. **RJS** (**LJS**) is updated by feeding it with combinations of population elements taken from **LJS** (**RJS**) that are scheduled backwards (forwards) with the serial SGS. The remainder of this section reveals some further algorithmic details about the construction of the initial population, parent-selection, crossover-operator, diversification and selection mechanism of the BPGA.

3.1 Construction of the Initial Population

We start the genetic algorithm by building an initial population **LJS** of left-justified schedules. Each population element is created by randomly generating an AL, constructing the corresponding left-justified schedule and finally incorporating the TO condition of Table 1.

3.2 Parent Selection

For each population element a of **LJS** (**RJS**) we create a set of nrc right-justified (left-justified) children that are candidates to enter **RJS** (**LJS**). To create a child out of a, we select another parent b from **LJS** (**RJS**) by using the 2-tournament selection procedure. In this selection procedure two population-elements are chosen randomly, and the element with the lowest makespan is selected. Afterwards, we determine randomly whether a or b represents the father S_f. The other parent represents the mother S_m.

3.3 Generation of a Child

A right-justified (left-justified) child is created from two parents from **LJS** (**RJS**) in two phases. In both phases, the advantages of our TO-condition implementation are fully exploited.

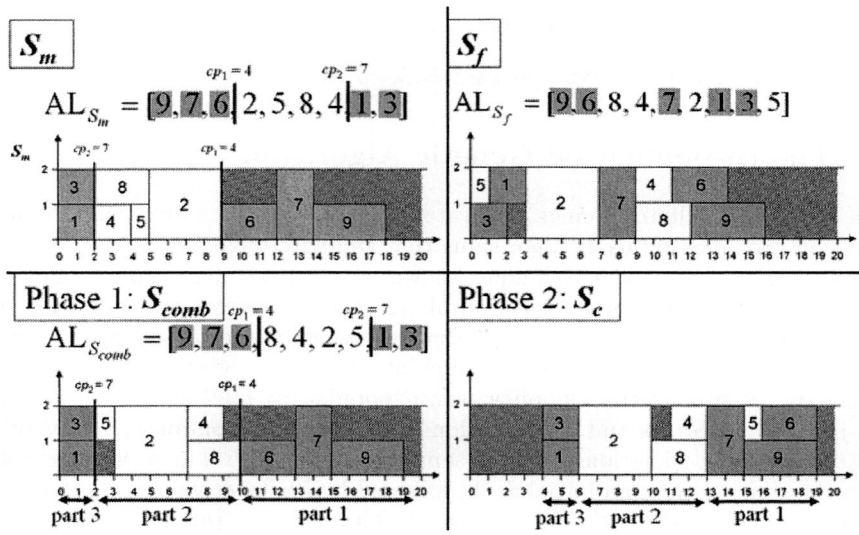

Fig. 4. Crossover operator

Phase 1: The Construction of a Combined Activity List $AL_{S_{comb}}$. Based on two parents from **LJS** (**RJS**), we use a 2-point crossover operator to generate the combined activity list $AL_{S_{comb}}$ which is used in phase 2 to construct a right-justified (left-justified) child S_c. To that purpose we select two crossover points cp_1 and cp_2 as follows. First, we randomly generate a crossover interval Δcp from $[1/4.f_{n-1}(S_m), 3/4.f_{n-1}(S_m)]$, where $f_{n-1}(S_m)$ denotes the makespan of the mother schedule. Then, we randomly generate cp_1 from $[0, f_{n-1}(S_m) - \Delta cp]$ and set $cp_2 = cp_1 + \Delta cp$. The TO condition allows the construction of $AL_{S_{comb}}$ and the combined schedule S_{comb} by simply copying parts from the AL of the mother and the father. More precisely, we copy all activity positions from the mother from the intervals $[1, cp_1[$ and $]cp_2, n]$. The remaining activities from the interval $[cp_1, cp_2]$ are copied in $AL_{S_{comb}}$ according to the AL ranking of the father. In Fig. 4, we have set cp_1 and cp_2 to 4 and 7, and AL_{S_f} and AL_{S_m} represent the activity lists of the parents in TO-format. The dark-colored activities from the interval $[1, 4[$ (i.e. 9, 7 and 6) and $]7, 9]$ (i.e. 1 and 3) are copied from AL_{S_m} to

$AL_{S_{comb}}$ and correspond to part 1 and part 3 in S_{comb}. The remaining activities (i.e. 2, 5, 8 and 4), displayed in white, are copied into $AL_{S_{comb}}$ according to the sequence of AL_{S_f}, i.e. 8, 4, 2 and 5 and represent part 2 in S_{comb}.

Phase 2: The Construction of a Right-Justified (Left-Justified) Child.
The combined schedule S_{comb} is often neither a left- or a right-justified schedule. Therefore, we transform this combined schedule into a left-justified (right-justified) schedule, when the parents belong to **RJS** (**LJS**), using the SGS. This can be done by running the iterative forward/backward scheduling procedure on the combined schedule S_{comb}. In doing so, only improvements can occur for each part of S_{comb}. In our example of Fig. 4, we transform S_{comb} in a right-justified schedule S_c, resulting in a makespan improvement of 3 time-units for part 1 and 1 time-unit for part 2.

3.4 Diversification

Diversification is necessary in every genetic algorithm to avoid the creation of a homogeneous population. We use a reactive method that only applies diversification to a child when it comes from two not mutually diverse parents. To define whether the parents are sufficiently diverse, we need a threshold τ and a distance measure. Our distance measure simply takes the sum of absolute deviations between the positions in the activity list of the father and the activity list of the mother for each activity and divides the obtained value by the number of non-dummy activities as defined in (1). Diversification is desirable if the distance exceeds the threshold τ and is exerted on $AL_{S_{comb}}$ by randomly swapping the positions of two activities for nrd times. In our example we calculate a distance of 2.0 between S_f and S_m as the sum of position differences for all activities is 18 and the number of non-dummy activities is 9.

$$\text{dist}(S_f, S_m) = \frac{1}{n-2} \sum_{i=1}^{n-2} |\text{position of } i \text{ in } AL_{S_f} - \text{position of } i \text{ in } AL_{S_m}| \quad (1)$$

3.5 Selection Mechanism

The selection mechanism determines the way in which the new generation replaces the old generation. In order to make the genetic algorithm successful, the 'survival of the fittest'-principle should be embedded. Good children should have a higher chance to enter the new generation than inferior ones in order to improve the quality of the population. The population **RJS** (**LJS**) is fed by children generated from **LJS** (**RJS**). In the following we will explain how we update **RJS**. The way in which we update **LJS** is analogue. As mentioned previously, we generate nrc children for each element of **LJS**. From the set that is created by the xth population-element, we select the child with the lowest makespan. This child will replace the xth element of **RJS**, even if this leads to a deterioration of the makespan. But, in order to prevent that we loose high-quality schedules, we do not automatically replace the xth element if this corresponds with the best-found schedule so far. In this case, we only perform replacement when the child represents a new best-found solution.

4 Comparative Computational Results

We have coded the procedure in Visual C++ 6.0 and performed computational tests on an Acer Travelmate 634LC with a Pentium IV 1.8 GHz processor using the well-known PSPLIB dataset [15] which we use to compare our procedure with other existing procedures from the literature. This dataset contains the subdatasets J30, J60 and J120 that contain problem-instances of 30, 60 and 120 activities respectively. We predefined the settings of the parameters as follows. The number of children nrc is fixed at 2, the diversification-parameter nrd is fixed at the number of non-dummy activities divided by 10 and the threshold τ for applying diversification is set equal to 2. The population size has been fine-tuned to an optimal value.

Table 2. Comparative results for J30, J60 and J120

J30	max. #schedules			J60	max. #schedules			J120	max. #schedules		
	1,000	5,000	50,000		1,000	5,000	50,000		1,000	5,000	50,000
[11]	0.10	0.04	0.00	BPGA	11.45	11.00	10.69	BPGA	34.29	32.34	30.75
[5]	0.27	0.11	0.01	[5]	11.73	11.10	10.71	[27]	34.07	32.54	31.24
BPGA	0.17	0.06	0.02	[27]	11.56	11.10	10.73	[5]	35.22	33.10	31.57
[27]	0.27	0.06	0.02	[11]	11.71	11.17	10.74	[30]	35.39	33.24	31.58
[1]	0.33	0.12	-	[30]	12.21	11.27	10.74	[11]	34.74	33.36	32.06
[30]	0.34	0.20	0.02	[7]	12.21	11.70	11.21	[30]	35.18	34.02	32.81
[26]	0.25	0.13	0.05	[6]	12.68	11.89	11.23	[7]	37.19	35.39	33.21
[22]	0.46	0.16	0.05	[26]	11.88	11.62	11.36	[26]	35.01	34.41	33.71
[24]	0.30	0.16	0.07	[25]	12.14	11.82	11.47	[21]	-	35.43	-
[7]	0.38	0.22	0.08	[1]	12.57	11.86	-	[6]	39.37	36.74	34.03
[6]	0.54	0.25	0.08	[24]	12.18	11.87	11.54	[25]	36.24	35.56	34.77
[25]	0.30	0.17	0.09	[3]	12.75	11.90	-	[24]	36.49	35.81	35.01
[30]	0.46	0.28	0.11	[22]	12.97	12.18	11.58	[1]	39.36	36.57	-
[3]	0.38	0.23	-	[30]	12.73	12.35	11.94	[22]	40.86	37.88	35.85
[4]	0.74	0.33	0.16	[23]	12.94	12.58	-	[4]	39.97	38.41	36.48
[23]	0.65	0.44	-	[4]	13.28	12.63	11.94	[30]	38.21	37.47	36.46
[2]	0.86	0.44	-	[6]	14.68	13.32	12.25	[3]	42.81	37.68	-
[12]	0.74	0.52	-	[6]	13.30	12.74	12.26	[6]	39.93	38.49	36.51
[6]	1.03	0.56	0.23	[12]	13.51	13.06	-	[23]	39.85	38.70	-
[17]	0.83	0.53	0.27	[17, 18]	13.66	13.21	-	[17]	39.60	38.75	37.74
[4]	0.81	0.54	0.28	[4]	13.80	13.31	12.83	[17, 18]	39.65	38.77	-
[16]	1.44	1.00	0.51	[17]	13.75	13.34	12.84	[6]	45.82	42.25	38.83
[17]	1.05	0.78	0.56	[17]	13.59	13.23	12.85	[17]	41.27	40.38	39.34
[6]	1.38	1.12	0.88	[2]	13.80	13.48	-	[12]	41.37	40.45	-
[17, 18]	1.40	1.28	-	[19]	14.33	13.49	-	[4]	41.36	40.46	39.41
[17]	1.40	1.29	1.13	[17]	13.96	13.53	12.97	[19]	42.91	40.69	-
[16]	1.77	1.48	1.22	[16]	14.89	14.30	13.66	[17]	42.84	41.84	40.63
[19]	2.08	1.59	-	[16]	15.94	15.17	14.22	[16]	44.46	43.05	41.44
								[16]	49.25	47.61	45.60

To be able to compare procedures for the RCPSP, Hartmann and Kolisch [8] presented a methodology in which all procedures can be tested on the PSPLIB-datasets by using 1,000 and 5,000 generated schedules as a stop condition. In [14] Hartmann and Kolisch give an update of the results, and also report on 50,000 schedules as a schedule limit. In Table 2 we compare our algorithm with these results for the datasets J30, J60 and J120 respectively. The average deviation from the optimal solution is used as a measure of quality for J30 and the average deviation from the critical path based lower bound for J60 and J120. For each dataset the heuristics are ranked by the corresponding deviation for 50,000 schedules. The results for 5,000 and 1,000 schedules are used as a tie-breaker, if necessary. The table reveals that our procedure is capable to report consistently good results. For the datasets J60 and J120 it outperforms all other procedures.

Table 3. Optimal values of the population size

Dataset	1,000 schedules		5,000 schedules		50,000 schedules	
	popsize	Avg.CPU	popsize	Avg.CPU	popsize	Avg.CPU
J30	55	0.02s	112	0.04s	416	0.39s
J60	30	0.03s	71	0.09s	390	0.89s
J120	20	0.09s	60	0.22s	290	2.19s

Only for J30, [11] and [5] report a slightly better result. Furthermore, our procedure often generates better solutions for the PSPLIB problem instances than the best solutions found so far (based on PSPLIB results on December 3, 2004, see http://www.bwl.uni-kiel.de/bwlinstitute/Prod/psplib/datasm.html). As an example, we obtained 15 improvements for J120 and with a stop condition of 50,000 schedules. In general we conclude that the more challenging the problem-instances are, the better our procedure performs compared to other procedures.

The optimal values of the population size used for the results of Table 2 and the average CPU-time needed to solve one problem-instance of the dataset are depicted in Table 3. This table reveals that the population size is positively related to the schedule limit and negatively related to the number of activities. The use of a large population avoids, similar to diversification, a homogeneous population, and this becomes more important for small problem instances and high values for the stop criterion.

5 Conclusion

In this paper we presented a genetic algorithm for the resource-constrained project scheduling problem (RCPSP) that operates on two separate populations. The first population only contains left-justified schedules and the second population only contains right-justified schedules. Our bi-population genetic algorithm (BPGA) combines schedules of the first population to create children that are candidate to enter the second population and vice versa. In this way the procedure is able to exploit the advantages of a local search technique denoted as the iterative forward/backward scheduling technique. The comparative computational results on the well-known PSPLIB dataset illustrated that the BPGA is currently the best meta-heuristic procedure for the RCPSP.

References

1. Alcaraz, J., Maroto, C.: A robust genetic algorithm for resource allocation in project scheduling, Annals of Operations Research, 102, 83-109 (2001).
2. Baar, T., Brucker, P., Knust, S.: Tabu-search algorithms and lower bounds for the resource-constrained project scheduling problem, Meta-heuristics: Advances and trends in local search paradigms for optimization, 1-8 (1998).
3. Bouleimen, K., Lecocq, H.: A new efficient simulated annealing algorithm for the resource-constrained project scheduling problem and its multiple mode version, European Journal of Operational Research, 149, 268-281 (2003).

4. Coelho, J., Tavares, L.: Comparative analysis of meta-heuricstics for the resource constrained project scheduling problem, Technical report, Department of Civil Engineering, Instituto Superior Tecnico, Portugal (2003).
5. Debels, D., De Reyck, B., Leus, R., Vanhoucke, M.: A scatter-search meta-heuristic for the resource-constrained project scheduling problem, European Journal of Operational Research, forthcoming.
6. Hartmann, S.: A competitive genetic algorithm for the resource-constrained project scheduling, Naval Research Logistics, 45, 733-750 (1998).
7. Hartmann, S.: A self-adapting genetic algorithm for project scheduling under resource constraints, Naval Research Logistics, 49, 433-448 (2002).
8. Hartmann, S., Kolisch, R.: Experimental evaluation of state-of-the-art heuristics for the resource-constrained project scheduling problem, European Journal of Operational Research, 127, 394-407 (2000).
9. Herroelen, W., Demeulemeester, E., De Reyck, B.: A classification scheme for project scheduling. In: Weglarz, J. (Ed.), Project Scheduling - Recent Models, Algorithms and Applications, International Series in Operations Research and Management Science, Kluwer Academic Publishers, Boston, 14, pp. 77-106 (1998).
10. Holland, J.H., 1975. Adaptation in natural and artificial systems. The University of Michigan Press, Ann Arbor.
11. Kochetov, Y., and Stolyar, A.: Evolutionary local search with variable neighbourhood for the resource constrained project scheduling problem, Proceedings of the 3rd International Workshop of Computer Science and Information Technologies (2003).
12. Kolisch, R., Drexl, A.: Adaptive search for solving hard project scheduling problems, Naval Research Logistics, 43, 23-40 (1996).
13. Kolisch, R., Hartmann, S.; Heuristic algorithms for solving the resource-constrained project scheduling problem: classification and computational analysis. In: Weglarz, J. (Ed.), Project Scheduling - Recent Models, Algorithms and Applications, Kluwer Academic Publishers, Boston, pp. 147-178 (1999).
14. Kolisch, R., Hartmann, S.: Experimental investigation of Heuristics for resource-constrained project scheduling: an update, working paper, Technical University of Munich (2004).
15. Kolisch, R., Sprecher, A.: PSPLIB - A project scheduling library, European Journal of Operational Research, 96, 205-216 (1996).
16. Kolisch, R.: Project scheduling under resource constraints - Efficient heuristics for several problem classes, Physica (1995).
17. Kolisch, R.: Serial and parallel resource-constrained project scheduling methods revisited: theory and computation, European Journal of Operational Research, 43, 23-40 (1996).
18. Kolisch, R.: Efficient priority rules for the resource-constrained project scheduling problem, Journal of Operations Management, 14, 179-192 (1996).
19. Leon V. J., Ramamoorthy, B.: Strength and adaptability of problem-space based neighbourhoods for resource-constrained scheduling, OR Spektrum, 17, 173-182 (1995).
20. Li, K.Y., Willis, R.J.: An iterative scheduling technique for resource-constrained project scheduling, European Journal of Operational Research, 56, 370-379 (1992).
21. Merkle, D., Middendorf, M., Schmeck, H.: Ant colony optimization for resource constrained project scheduling, IEEE Transaction on Evolutionary Computation, 6(4), 333-346 (2002).

22. Nonobe, K., Ibaraki, T.: Formulation and tabu search algorithm for the resource constrained project scheduling problem (RCPSP). In: Ribeiro, C.C., Hansen, P. (Eds.), Essays and Surveys in Meta-heuristics, Kluwer Academic Publishers, Boston, pp. 557-588 (2002).
23. Schirmer, A.: Case-based reasoning and improved adaptive search for project scheduling, Naval Research Logistics, 47, 201-222 (2000).
24. Tormos, P., Lova, A.: A competitive heuristic solution technique for resource-constrained project scheduling, Annals of Operations Research, 102, 65-81 (2001).
25. Tormos, P., Lova, A.: An efficient multi-pass heuristic for project scheduling with constrained resources, International Journal of Production Research, 41, 1071-1086 (2003).
26. Tormos, P., and Lova, A.: Integrating heuristics for resource constrained project scheduling: One step forward, Technical report, Department of Statistics and Operations Research, Universidad Politecnica de Valencia (2003).
27. Valls, V., Ballestin, F., Quintanilla, S.: A hybrid genetic algorithm for the Resource-constrained project scheduling problem with the peak crossover operator, Eighth International Workshop on Project Management and Scheduling, 368-371 (2002).
28. Valls, V., Quintanilla, S., Ballestin, F.: Resource-constrained project scheduling: a critical activity reordering heuristic, European Journal of Operational Research, 149, 282-301 (2003).
29. Valls, V., Ballestin, F., Quintanilla, S.: A population-based approach to the resource-constrained project scheduling problem, Annals of Operations Research, 131, 305-324 (2004).
30. Valls, V., Ballestin, F.: Quintanilla, S.: Justification and RCPSP: A technique that pays, European Journal of Operational Research, Forthcoming.

Optimizing Product Mix in a Multi-bottleneck Environment Using Group Decision-Making Approach

Alireza Rashidi Komijan[1] and S.J. Sadjadi[2]

[1] Islamic Azad University, Firoozkoh unit, Iran
alireza_rashidi@yahoo.com,
[2] Iran University of Science and Technology, Tehran, Iran
sjsadjadi@iust.ac.ir

Abstract. Determining product mix under the theory of constraints (TOC) has been widely considered by many researchers. The product mix problem in a multi-bottleneck environment is much more complicated than a single-bottleneck system. In all researches, which have been done in this field, only one criterion has been considered and it is the profit resulted from allocating one unit of bottleneck capacity to each product. In this paper, another criterion is considered together with the first one. The new criterion, which is very important in decision-making, is late delivery cost. Another contribution of the paper is that it assumes each bottleneck as a decision-maker and applies group decision-making techniques. The new approach benefits from the advantage of reaching optimum solution. This will be shown through an example and the results of the new approach will be used in the objective function of an integer linear programming (ILP) model in order to reach optimum solution.

1 Introduction

Theory of constraints is a production planning philosophy that focuses on the constraints (bottlenecks) of the system and tries to improve the throughput of the system by effective management of bottlenecks. This philosophy was first introduced in The Goal [1]. Then it was developed in other books [2,3]. Being a bottleneck in a system means that the system is unable to meet the demands of all products. So, the product mix should be determined in such a way that maximum throughput is obtained. When there is more than one bottleneck in the system, the problem becomes much more complicated. Although many algorithms have been developed in determining the product mix under TOC, all of them consider only one criterion in decision-making process. The criterion is resulted from dividing contribution margin of each product in to its processing time in the bottleneck station. In this paper, late delivery cost is proposed as a supplement of the above-mentioned criterion. This is a very important criterion and neglecting it may distort the final solution. Furthermore, this paper uses the approach of group decision-making in solving product mix problems. It will be discussed that there is an analogy between a multi-bottleneck environment and a group decision-making problem. This approach solves the product mix problem easily and benefits from the advantage of reaching the optimum solution.

2 Literature Review

In the early 1990s, the first algorithm on determining product mix under the theory of constraints was developed and supported by many researchers [4,5]. The algorithm determined the production priority using the ratio of contribution margin of each product to the processing time in the bottleneck station. At that time, it was stated that the algorithm could reach the optimum solution as ILP [4,5]. However, it was revealed that the algorithm was inefficient in handling two types of problems. The first type included problems associated with adding new product alternatives to an existing production line [6]. The second type included multi-bottleneck problems in which the algorithm could not reach the optimum solution. As a remedy for this problem, the concept of dominant bottleneck was proposed. The dominant bottleneck is the resource that has the largest difference between its available and required capacity. Using the term "dominant" causes the scheduler to think that it is the most important bottleneck and he should just focus on it. This approach is similar to focusing on local optimization instead of the global one. In 1997, the first algorithm was revised [7]. The revised algorithm started with an initial solution and tried to reach the optimum solution using neighborhood search. But the revised algorithm faced some disadvantages. Firstly, it examined many alternatives in the neighborhood search. This caused the algorithm to be time consuming in solving large-scale problems. Secondly, it was shown that the revised algorithm did not reach the optimum solution in all cases [8]. Thirdly, the revised algorithm considered the dominant bottleneck in all decision-making processes and considered other bottlenecks only in order to avoid infeasible solutions. Due to these disadvantages, the improved algorithm was developed [8]. The first advantage of the improved algorithm was that the speed of solving problems was independent from the dimension of the problem. The improved algorithm started with an initial solution and then, found the best path to reach the optimum solution with a logical procedure. This advantage caused the improved algorithm to reach the optimum solution without wasting time for searching in non-optimal paths. Secondly, the improved algorithm could reach the optimum solution as ILP. Thirdly, the improved algorithm, avoiding the concept of dominant bottleneck, considered all bottlenecks in decision-making process. This feature ensured that the final solution was optimal for the system, not just for the dominant bottleneck. All the previous algorithms considered only one criterion in determining product mix. This paper considers a new criterion together with the previous one. The new criterion is late delivery cost, which is very important in determining product mix. Also, this paper applies the approach of group decision-making in determining product mix. Using this approach is unique in the literature of TOC and product mix determination. It will be shown that the new approach can easily reach the optimum solution.

3 New Approach

In product mix problems, the scheduler should first determine the priority of each product. Then, the product mix will be developed with regard to the priorities. If there is one bottleneck in the system, the production priority will be easily obtained. It is calculated by dividing contribution margin of each product (the difference between the selling price and the raw material cost) in to its processing time in the bottleneck station. As the resulted product mix is optimal in view of the bottleneck, it is optimal for the whole system. In a multi-bottleneck environment, calculating priority becomes complicated. The question is that the processing time in which bottleneck should be considered. In some of the previous algorithms, the bottleneck which has the largest difference between its available and required capacity is assumed as the dominant bottleneck [4,5,7]. This causes the scheduler to consider only the dominant bottleneck and to ignore the importance of other bottlenecks. The product mix obtained regarding this philosophy will be optimal in view of the dominant bottleneck, not necessarily the whole system. The philosophy of this paper is that all bottlenecks have significant importance and should be considered in decision-making. Each bottleneck may lead the scheduler to a different priority sequence. So, the scheduler may have several priority sequences that all of them are important. There is an analogy between this problem and a group decision-making problem. Each bottleneck in the former corresponds to a decision-maker (DM) in the latter. So, the product mix problem can be solved using a suitable technique of group decision-making. In this paper, TOPSIS[1] method is applied. Another important aspect of the paper is that it considers two criteria in decision-making. All of the previous algorithms considered production priority as the sole criterion. Considering this criterion is necessary but it is not sufficient. Another criterion which has a high impact on product mix determination is late delivery cost. These two criteria should be considered simultaneously. In order to develop the algorithm and mathematical model, assume a system with n products and m bottlenecks. Products and bottlenecks are treated as alternatives and decision-makers respectively. So, the product mix problem in a multi-bottleneck environment is a group decision-making problem consisted of n alternatives, m decision-makers and two criteria: production priority (X) and late delivery cost (Y). In the algorithm, i, j and k denote alternative, criterion and decision-maker respectively.

Step 1. The production priority in view of bottleneck k is calculated by dividing contribution margin of each product in to the processing time in bottleneck k.

$$X_{ik} = \frac{CM_i}{t_{ik}}, \quad i = 1, \ldots, n, \text{ and } k = 1, \ldots, m, \tag{1}$$

where X_{ik} is the priority of product i in view of bottleneck k, CM_i is the contribution margin of product i and t_{ik} is processing time of product i in bottleneck k.

---

[1] Technique for order preference by similarity to ideal solution.

Step 2. For each decision-maker, a decision matrix is set as follows:

$$DM(k) = \begin{bmatrix} a_{11}^k & a_{12}^k \\ \vdots & \vdots \\ a_{i1}^k & a_{i2}^k \\ \vdots & \vdots \\ a_{n1}^k & a_{n2}^k \end{bmatrix} \quad k = 1, \ldots, m,$$

where a_{i1}^k is X_{ik} and a_{i2}^k is late delivery cost of product i. It is clear that late delivery cost of a product is constant and is not dependent to the bottlenecks. For simplicity, each value of decision matrix is shown as a_{ij}^k.

Step 3. The dimensionless decision matrix value r_{ij}^k is calculated as follows:

$$r_{ij}^k = \frac{a_{ij}^k}{\sqrt{\sum_{i=1}^n (a_{ij}^k)^2}}, \quad i = 1, \ldots, n; \; j = 1, 2, \text{ and } k = 1, \ldots, m. \tag{2}$$

Step 4. The weights vector of DMs is calculated by normalizing the absolute values of the difference between required and available capacity of each bottleneck. The vector is shown as $(t_1, \ldots, t_k, \ldots, t_m)$.

Step 5. Given the individual dimensionless decision matrices and the weights vector, group decision matrix is calculated. Each value of the group decision matrix (g_{ij}) is the weighted average of corresponding values in individual matrices.

Step 6. The weighted group decision matrix value v_{ij} is calculated as follows:

$$v_{ij} = g_{ij} w_j, \quad i = 1, \ldots, n, \text{ and } j = 1, 2, \tag{3}$$

where w_j is the weight of criterion j.

Step 7. In this step, the ideal and negative-ideal solutions are developed. As production priority and late delivery cost are benefit criteria, the ideal and negative-ideal solutions are calculated as follows:

$$A^+ = \left(\max_i v_{i1}, \max_i v_{i2} \right) = (v_1^+, v_2^+), \tag{4}$$

$$A^- = \left(\min_i v_{i1}, \min_i v_{i2} \right) = (v_1^-, v_2^-). \tag{5}$$

Note that late delivery cost is a profit criterion as the product with the highest late delivery cost is the most important one for production.

Step 8. The distances of alternatives from the ideal and the negative-ideal solutions are calculated using Euclidean distance:

$$S_i^+ = \sqrt{\sum_j (v_{ij} - v_j^+)^2} \quad i = 1, \ldots, n \text{ and } j = 1, 2, \tag{6}$$

$$S_i^- = \sqrt{\sum_j (v_{ij} - v_j^-)^2} \quad i = 1, \ldots, n \text{ and } j = 1, 2. \tag{7}$$

Step 9. The relative closeness of each alternative to the ideal solution is calculated as follows:

$$C_i^+ = \frac{S_i^-}{(S_i^+ + S_i^-)} \quad i = 1, \ldots, n. \tag{8}$$

Step 10. The weights vector of products is calculated by normalizing the relative closeness of all products to the ideal solution:

$$P_i = \frac{C_i^+}{\sum_{i=1}^n C_i^+} \quad i = 1, \ldots, n. \tag{9}$$

Step 11. In this step, the mathematical model of determining product mix is developed. In the model, decision variable is defined as the produced units of product i. The products weights are used as coefficients of variables in objective function. There are capacity and demand constraints in the model. Also, Z_i, P_i, t_{ik}, CP_k and D_i are defined as produced units of product i, weight or importance of product i, processing time of product i in bottleneck k, available capacity of bottleneck k and market demand of product i respectively.

$$\begin{aligned} \max \quad & \sum_{i=1}^n P_i Z_i \\ \text{subject to} \quad & \sum_{i=1}^n t_{ik} Z_i \leq CP_k \quad k = 1, \ldots, m \\ & Z_i \leq D_i \quad i = 1, \ldots, n \\ & Z_i \geq 0 \text{ and integer } i = 1, \ldots, n. \end{aligned} \tag{10}$$

4 Example

The new approach is explained through an example. Assume that a factory produces five products: A, B, C, D and E. Weekly demand, selling price and raw material cost of the products are shown in table 1. The contribution margin (CM) is the difference between the selling price and the raw material cost of each product. The factory uses four resources: shear, notch, pierce and bend. The processing time of each product in each station is presented in table 2. Assume that late delivery costs of products A to E are 3, 5, 3, 4 and 5 dollars respectively. For example, if the factory can not deliver 10 units of product A, 3×10 dollars should be paid. In other words, for each unit of A that can not be produced, 3 dollars should be paid.

Step 1. The production priority should be developed in view of each bottleneck. As shown in table 2, shear, notch and pierce are bottlenecks. So, there are three

Table 1. Weekly demand, selling price and raw material cost of each product

Product	Demand (unit)	Selling Price (dollar)	Raw material cost (dollar)	CM (dollar)
A	50	100	20	80
B	40	120	50	70
C	70	150	60	90
D	10	100	16	84
E	30	100	30	70

Table 2. Processing time in minute, available and required capacity

Product	Shear	Notch	Pierce	Bend
A	20	10	20	20
B	20	40	10	5
C	20	10	5	10
D	20	30	22	40
E	0	40	40	0
Available capacity	2400	2400	2400	2400
Required capacity	3400	4300	3170	2300
Difference	-1000	-1900	-770	100

decision-makers in the system. For shear, the production priority is calculated as follows:

Product	A	B	C	D	E
X_{i1}	4	3.5	4.5	4.2	-

In fact, the first DM assigns score 4.5 to the product C. In other words, product C has the highest priority in view of the first DM when criterion X is considered. In the same way, the production priority in view of notch and pierce are developed.

Product	A	B	C	D	E		Product	A	B	C	D	E
X_{i2}	8	1.75	9	2.8	1.75		X_{i3}	4	7	18	3.82	1.75

Step 2. The individual decision matrices are developed as follows:

$$DM(1) = \begin{bmatrix} 4 & 3 \\ 3.5 & 5 \\ 4.5 & 3 \\ 4.2 & 4 \\ 0 & 5 \end{bmatrix} \quad DM(2) = \begin{bmatrix} 8 & 3 \\ 1.75 & 5 \\ 9 & 3 \\ 2.8 & 4 \\ 1.75 & 5 \end{bmatrix} \quad DM(3) = \begin{bmatrix} 4 & 3 \\ 7 & 5 \\ 18 & 3 \\ 3.82 & 4 \\ 1.75 & 5 \end{bmatrix},$$

where each row represents the product $A, \ldots, E$, the first and the second columns are associated with the production priority and the late delivery cost, respectively.

Step 3. The dimensionless decision matrices are calculated as follows:

$$DM(1) = \begin{bmatrix} 0.491845 & 0.327327 \\ 0.430364 & 0.545545 \\ 0.553325 & 0.327327 \\ 0.516437 & 0.436436 \\ 0 & 0.545545 \end{bmatrix} \quad DM(2) = \begin{bmatrix} 0.634511 & 0.327327 \\ 0.138799 & 0.545545 \\ 0.713825 & 0.327327 \\ 0.222079 & 0.436436 \\ 0.138799 & 0.545545 \end{bmatrix}$$

$$DM(3) = \begin{bmatrix} 0.198357 & 0.327327 \\ 0.347124 & 0.545545 \\ 0.892605 & 0.327327 \\ 0.189431 & 0.436436 \\ 0.086781 & 0.545545 \end{bmatrix}$$

Step 4. As shown in table 2, the difference between required and available capacity of bottlenecks are 1000,1900 and 770 minutes. After normalizing these values, the weights vector of DMs is developed as (0.2725, 0.5177, 0.2098).

Step 5. In order to combine the individual opinions and to develop group decision matrix, weighted average technique is applied. Each value of the group decision matrix is the weighted average of corresponding values in individual matrices.

$$G = \begin{bmatrix} 0.504129 & 0.327327 \\ 0.261957 & 0.545545 \\ 0.707597 & 0.327327 \\ 0.295442 & 0.436436 \\ 0.090063 & 0.545545 \end{bmatrix}$$

Step 6. In this problem, it is assumed that X and Y have the same importance. Therefore, the weights vector of criteria is $(0.5, 0.5)$. The weighted group decision matrix is calculated as follows:

$$V = GW \begin{bmatrix} 0.504129 & 0.327327 \\ 0.261957 & 0.545545 \\ 0.707597 & 0.327327 \\ 0.295442 & 0.436436 \\ 0.090063 & 0.545545 \end{bmatrix} \begin{bmatrix} 0.5 & 0 \\ 0 & 0.5 \end{bmatrix} = \begin{bmatrix} 0.252065 & 0.163663 \\ 0.130979 & 0.272772 \\ 0.353798 & 0.163663 \\ 0.147721 & 0.218218 \\ 0.045032 & 0.272772 \end{bmatrix}$$

Step 7. As X and Y are benefit criteria, ideal and negative-ideal solutions are developed as follows:

$$A^+ = \{0.353798, 0.272772\} \text{ and } A^- = \{0.045032, 0.163663\}.$$

Step 8. The distances of alternatives from the ideal and the negative-ideal solutions are calculated as follows:

$S_A^+ = 0.14918 \quad S_B^+ = 0.22282 \quad S_C^+ = 0.109109 \quad S_D^+ = 0.213176 \quad S_E^+ = 0.308767$

$S_A^- = 0.207033 \quad S_B^- = 0.138894 \quad S_C^- = 0.308767 \quad S_D^- = 0.116281 \quad S_E^- = 0.109109$

Step 9. The relative closeness to the ideal solution is calculated as follows:

$$C_A^+ = 0.581207 \quad C_B^+ = 0.383989 \quad C_C^+ = 0.738896$$
$$C_D^+ = 0.352947 \quad C_E^+ = 0.261104$$

Step 10. The relative closeness values show the importance of alternatives (products). These values can be used in developing the objective function of an integer linear programming model. It is better to normalize them before using in objective function. The normalized values are the real priorities of products, which have been obtained considering all bottlenecks and criteria. The normalized relative closeness values are as follows:

$$P_A = 0.250721 \quad P_B = 0.165645 \quad P_C = 0.318745 \quad P_D = 0.152254 \quad P_E = 0.112635$$

Step 11. The mathematical model can be developed as follows:

max $\quad 0.250721A + 0.165645B + 0.318745C + 0.152254D + 0.112635E$
subject to
$$\begin{aligned}
20A + 20B + 20C + 20D &\leq 2400 \\
10A + 40B + 10C + 30D + 40E &\leq 2400 \\
20A + 10B + 5C + 22D + 40E &\leq 2400 \\
20A + 5B + 10C + 40D &\leq 2400 \\
A &\leq 50 \\
B &\leq 40 \\
C &\leq 70 \\
D &\leq 10 \\
E &\leq 30
\end{aligned}$$

A, B, C, D and E are integer and non-negative.

The optimum solution resulted from the above model is 50A, 70C and 26E. This is the product mix proposed by the new approach.

5 Conclusion

There have been developed several algorithms on determining product mix under TOC. The first algorithm was inefficient in handling multi-bottleneck problems. It could not reach the optimum solution in all cases. Due to the inefficiency of the first algorithm, the revised algorithm was developed. Its logic was based on reaching an initial solution and trying to improve it using neighborhood search. The main disadvantage of these algorithms was in defining dominant bottleneck. These algorithms focused on dominant bottleneck and ignored the importance of other bottlenecks. The product mix resulted from these algorithms was optimal in view of the dominant bottleneck, not necessarily the whole system. In late 2004, an improved algorithm was developed. The improved algorithm had the advantage of reaching optimum solution. Another advantage of the improved algorithm was that it did not waste time in non-optimal paths. All the algorithms discussed, considered the profitability of bottleneck as the only criterion of decision-making. This paper considers late delivery cost as a new criterion together with the previous one. Considering these two criteria will lead to more reliable results. Also, this paper benefits from the advantage of using the approach of group decision-making in determining product mix. This is the result of focus shift from the concept of dominant bottleneck to considering all bottlenecks in decision-making. Using the new approach can easily lead to the optimum solution and does not need neighborhood search.

References

1. Goldratt, E. M.: The Goal. Croton-on-Hudson, NY:North River Press (1984)
2. Goldratt, E. M.: Theory of Constraints. Croton-on-Hudson, NY:North River Press (1990)
3. Goldratt, E. M.: The Haystack Syndrome. Croton-on-Hudson, NY:North River Press (1990)
4. Luebbe, R., Finch, B.: Theory of Constraints and Linear Programming: A Comparison. International Journal of Production Research, **30** (1992) 1471–1478
5. Patterson, M.C.: The Product-Mix Decision: A Comparison of Theory of Constraints and Labor-Based Management Accounting. Production and Inventory Management Journal, **33** (1992) 80–85
6. Lee, T.N., Plenert, G.: Optimizing Theory of Constraints when New Product Alternatives Exist. Production and Inventory Management Journal **34** (1993) 51–57
7. Fredendall, L.D., Lea, B.R.: Improving the product Mix Heuristic in the Theory of Constraints. International Journal of Production Research, **35** (1997) 1535–1544
8. Aryanezhad, M.B., Komijan, A.R.: An Improved Algorithm for Optimizing Product Mix under the Theory of Constraints, International Journal of Production Research **42** (2004) 4221–4233

Using Bipartite and Multidimensional Matching to Select the Roots of a System of Polynomial Equations

H. Bekker*, E. P. Braad*, and B. Goldengorin**

* Department of Mathematics and Computing Science,
** Faculty of Economic Sciences, University of Groningen,
P.O.Box 800, 9700 AV Groningen, The Netherlands
bekker@cs.rug.nl, e.p.braad@wing.rug.nl, b.goldengorin@eco.rug.nl

Abstract. Assume that the system of two polynomial equations $f(x,y) = 0$ and $g(x,y) = 0$ has a finite number of solutions. Then the solution consists of pairs of an x-value and an y-value. In some cases conventional methods to calculate these solutions give incorrect results and are complicated to implement due to possible degeneracies and multiple roots in intermediate results. We propose and test a two-step method to avoid these complications. First all x-roots and all y-roots are calculated independently. Taking the multiplicity of the roots into account, the number of x-roots equals the number of y-roots. In the second step the x-roots and y-roots are matched by constructing a weighted bipartite graph, where the x-roots and the y-roots are the nodes of the graph, and the errors are the weights. Of this graph the minimum weight perfect matching is computed. By using a multidimensional matching method this principle may be generalized to more than two equations.

Keywords: combinatorial optimization, bipartite matching, system of polynomial equations.

1 Introduction

Consider a system of two polynomial equations

$$f(x,y) = 0 \quad g(x,y) = 0 \tag{1}$$

with numerical constants, and of low degree say ≤ 6. Assume that of this system it is known that the number of solutions in $\mathbb{C}$ is finite. Solving this system, using floating point arithmetic, is no problem for a computer algebra system, say MAPLE©. In general the calculations consist of two steps. First, with symbolic computations, a Groebner basis or resultants are calculated, and as a second step numerical calculations take place. For testing and prototyping this approach is very useful.

In this article a different situation is considered, that is, we assume that some of the constants in (1) are symbolic and we assume that (1) has to be solved for many thousands of problem instances per second for different values of the symbolic constants. Then, for the sake of speed, one is forced to program the algorithms in a conventional programming language, say C [1]. It is not acceptable, however, to copy the complete process as performed by the computer algebra system into C because the symbolic computations may be very time consuming, even in C. Therefore, the symbolic computations should be performed in a preprocessing phase by the computer algebra system, and the resulting equations should, for every problem instance, be solved by numerical methods implemented in C.

1.1 The Conventional Approach and Its Problems in More Detail

In more detail, the conventional approach means that from (1) a univariate polynomial, say $p(x)$, is derived by MAPLE. Obviously, $p(x)$ will contain symbolic constants. Now $p(x) = 0$, $f(x,y) = 0$ and $g(x,y) = 0$ are copied into the C program and for every problem instance the constants in these equations are replaced by numerical values. In general, solving $p(x) = 0$ is straightforward, giving the roots $x_1...x_n$. Subsequently, for every root x_i the corresponding root y_i has to be determined. To that end, x_i is backsubstituted in $f(x,y) = 0$ and $g(x,y) = 0$, giving the univariate polynomial equations $f(x_i, y) = 0$ and $g(x_i, y) = 0$. Solving $f(x_i, y) = 0$ for y gives the solutions $y_{f1}...y_{fm}$, and solving $g(x_i, y) = 0$ for y gives the solutions $y_{g1}...y_{gk}$. The value y_i occurring both in $y_{f1}...y_{fm}$ and $y_{g1}...y_{gk}$ is the desired value, i.e., the pair x_i, y_i is a root of (1). During this process, a number of complications may occur:

1. The equation $f(x_i, y) = 0$ may be degenerate, i.e. may be $0 = 0$, or even worse, may be near degenerate within the noise margin. The case of exact degeneracy is easily detected but it is not trivial to detect near degeneracy. In both cases every solution of the other equation, that is, of $g(x_i, y) = 0$ is a correct root. Analogously, $g(x_i, y) = 0$ may be degenerate, giving the same problems. As we know that (1) has a finite number of solutions the situation that $f(x_i, y) = 0$ and $g(x_i, y) = 0$ are both degenerate will not occur.
2. It is sometimes hard to select from $y_{f1}...y_{fm}$ and $y_{g1}...y_{gk}$ the collective value y_i because, by numerical errors, the actual value of y_i will be different in the two sets.
3. $p(x) = 0$ may have multiple roots, that is, the roots $x_1...x_n$ may contain (near) identical values. Let us assume that there is a double root, given by the the identical values x_j and x_{j+1}. Then there will be two matching roots y_j and y_{j+1}, not necessarily with the same value. When y_j is matched to x_j, in a later stage not y_j should be matched to x_{j+1} but y_{j+1}.

The problems mentioned in 1 are the hardest to detect and to handle. When this is not done properly, solutions may be lost or may be completely wrong. None of the problems in 1,2,3 can be foreseen, prevented or solved during the

---

[1] Of course, instead of MAPLE and C every other computer algebra system and programming language may be used.

preprocessing stage, they have to be dealt with during the numerical calculations. On the other hand, none of these problems is insurmountable. For example, when the problem in 1 occurs, in general the substitution $u = x+y$, $v = x-y$ gives non-degenerate equations. Also, by using exact computations, interval arithmetic or floating point filters the problem in item 2 can be avoided, and by bookkeeping properly the problem in item 3 is solved. However, apart from the speed penalty introduced by these techniques, the complexity of the implementation increases significantly by these measures. Moreover, every new set of equations (1) should be analysed, possible degeneracies should be signalled, and appropriate measures should be devised and implemented to handle these degeneracies. The method we propose avoids these complications. It is simple and robust, and most of all it is universal, i.e. no problem specific programming is required. It only assumes that two univariate polynomials $p(x)$ and $q(y)$ can be derived from (1), and it uses the minimum weight perfect bipartite matching algorithm.

For more than two equations a more general combinatorial optimization technique is required namely the multidimensional assignment method. Recently a new algorithm for this problem has been proposed [1,2] and in the last section it is discussed to which extent it may be used for more than two equations. Because the crucial part of our method is a combinatorial optimization technique we call it the CORS method, which stands for Combinatorial Optimization Root Selection method.

Literature. The CORS method is not mentioned in the literature. In [3] a method is discussed that avoids part of the problems in 1,2, and 3, but it only works for exact arithmetic. Essence of that method is that the root y_j, matching the root x_i, is found by substituting x_i in (1), and to calculate of these equations the greatest common devisor (GCD) with for example Euclid's algorithm. The root of the GCD is the desired value y_j.

Packages. Some numerical packages in C exists to solve (1) as for example SYNAPS [4]. The disadvantage of these packages is however that the symbolic computations are not handled in a preprocessing phase, which results in a significant decrease of the performance, but most of all, sometimes solutions are missed due to the problems mentioned earlier.

Not in this article. This article is not about a numerical method to obtain more precise results. It is about robustness. Solutions that are missed by other methods are found with the CORS method.

Article structure. In section 2 the minimum weight perfect bipartite matching is introduced and it is shown how this may be used to select roots. In section 3 the implementation and the results of CORS are discussed. In section 4 it is discussed how the CORS method may be generalized to more than two equations.

1.2 An Example Problem

To see how (1) is solved using the conventional approach let us look at the following small example.

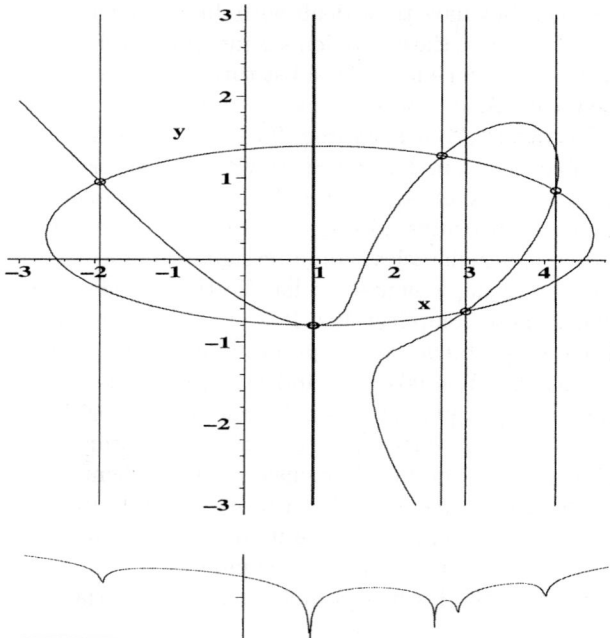

Fig. 1. The conventional method to solve (1). Lower part: the univariate polynomial $p(x)$ plotted logarithmically (in order to limit the dimension in the y direction). The local minima represent the roots of $p(x)$. Upper part: the curves $f(x, y)$ and $g(x, y)$, and vertical lines at positions corresponding with the roots of $p(x)$. To find the intersections of the curves, for every vertical line the intersections with the curves are calculated and identical y-values are selected. For more details see the main text

$$f(x, y) = (x - a)^3 + (y + b)^3 - 5(x - a)(y + b) - c = 0$$
$$g(x, y) = (0.3x - d)^2 + (y - e)^2 - h = 0 \qquad (2)$$

with $a = 1.5$, $b = 1$, $c = 0.4$, $d = 0.3$, $e = 0.29756$ and $h = 1.2$. For these constants the system has four single real roots and a double real root, see figure 1.

As an example, let us determine the root y_i, matching the root $x = 4.1481$ which is the rightmost x-root in this figure. The line $x = 4.1481$ intersects $f(x, y)$ three times, giving $y_{f1} = -5.192317423$, $y_{f2} = 0.8525335317$ and $y_{f3} = 1.339783891$ The line $x = 4.1481$ intersects $g(x, y)$ twice, giving $y_{g1} = 0.8525335293$ and $y_{g2} = -0.2574135293$. Hence, $y = 0.8525335$ is the root matching $x = 4.1481$. Clearly, $y_{f2} = 0.8525335317$ does not equal $y_{g1} = 0.8525335293$, which demonstrates point 3.

The CORS method will be tested on these equations and on the equations that were the starting point of this research, that is, the equations resulting from a computational geometry problem [5], given by

$$a_1 u^2 w^2 + b_1 u^2 w + c_1 uw^2 + d_1 u^2 + e_1 w^2 + f_1 u + g_1 w + h_1 = 0$$
$$a_2 u^2 w^2 + b_2 u^2 w + c_2 uw^2 + d_2 u^2 + e_2 w^2 + f_2 u + g_2 w + h_2 = 0. \quad (3)$$

Here, $a_1..h_1$ and $a_2..h_2$ are real constants, as they occur in the computational geometry problem.

2 Selecting Roots by Constructing a Minimal-Weight Matching

2.1 The Bipartite Weighted Graph and Matching

A complete weighted bipartite graph $G = (V, E, w)$ consists of a set of vertices $V = V_1 \cup V_2$, $|V_1| = |V_2| = n$, and of a set of n^2 arcs $(i,j) \in E \subseteq V_1 \times V_2$ with weights $w(i,j)$ for all $(i,j) \in E$. We define a feasible matching π as a permutation which maps V_1 onto V_2 and the weight of permutation π is $w(\pi) = \sum_{(i,j)\in\pi} w(i,j)$. The Minimum Weighted Bipartite Matching Problem is the problem of finding $\pi_0 \in \arg\min\{w(\pi) : \pi \in \mathbb{P}\}$. Here $\mathbb{P}$ is the set of all permutations, and π is called *feasible* if $w(\pi) < \infty$. This problem is well-known in the field of combinatorial optimization and is called the *Linear Assignment Problem* (LAP). Many efficient algorithms exist for solving the LAP with time complexity $O(n^3)$, see [8] and references within. All these algorithms are based on the shortest augmenting path method, the implementation of which, for example in the Hungarian algorithm, is based on the Koning-Egervary theorem [8].

In figure 2 part of a complete bipartite weighted graph is shown with four vertices in V_1 and four in V_2, and two matchings derived from this graph are shown. The rightmost matching has minimum weight.

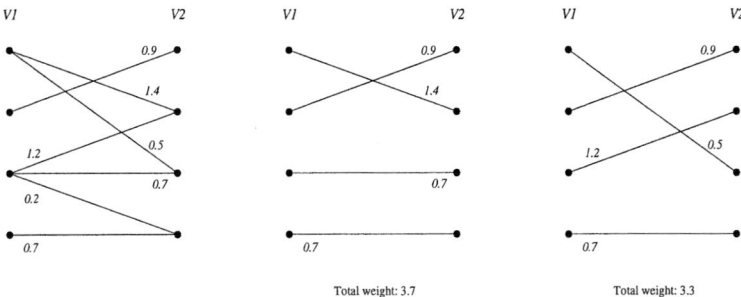

Fig. 2. Left: a weighted bipartite graph G with four vertices in V1 and in V2, and with seven arcs. Middle: a possible matching of G. Right: the minimum weight matching of G

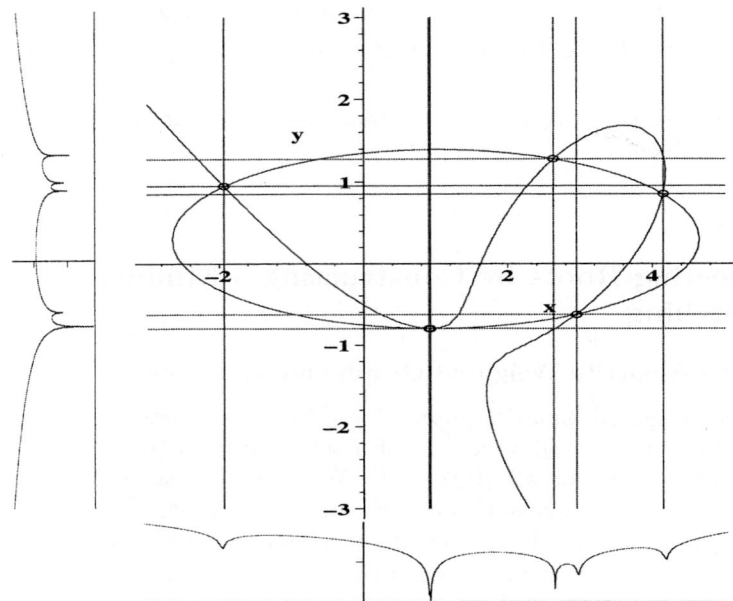

Fig. 3. The CORS method explained graphically. The figure consists of three parts. Lower part: the univariate polynomial $p(x)$ plotted logarithmically. The five local minima represent the roots of $p(x)$. The root at $x = 0.912$ is a double root. Left part: the univariate polynomial $q(y)$ plotted logarithmically. The five local minima represent the roots of $q(y)$. The root at $y = -0.797$ is a double root. Main part: the curves $f(x, y)$ and $g(x, y)$. The vertical lines are at positions corresponding with the roots of $p(x)$ and the horizontal lines are at positions corresponding with the roots of $q(y)$. The intersections of the curves are found by intersecting the horizontal lines with the vertical lines and selecting from all 36 intersections six intersections. The selection process is performed by calculating the minimum weight perfect bipartite matching. For more details see the main text

2.2 Selecting the Roots by Using a Bipartite Matching

Let us now see how a bipartite weighted graph and a minimum weight matching of this graph is used to select the roots of two polynomial equation. We consider again the polynomials with symbolic constants in (1). A computer algebra system is used to eliminate from (1) y and x, giving two univariate polynomials $p(x)$ and $q(y)$ respectively. Whether this done by calculating two Groebner basis or by calculating two resultants is irrelevant. Now $p(x)$, $q(y)$, $f(x, y)$ and $g(x, y)$ are copied into the C code. In these four equations, for every problem instance, the symbolic constants are replaced by numerical values and the roots in $\mathbb{C}$ of $p(x)$ and $q(y)$ are calculated numerically. Both $p(x)$ and $q(y)$ have n roots represented by $x_1..x_n$ and $y_1..y_n$ respectively. These roots are used to calculate n^2 weights, where $w_{i,j}$ is defined as

$$w_{i,j} = \sqrt{(|(f(x_i,y_j))^2| + |(g(x_i,y_j))^2|)}. \tag{4}$$

Subsequently a complete weighted bipartite graph G is constructed. The nodes in V_1 consist of the values $x_1..x_n$, and the nodes in V_2 consist of the values $y_1..y_n$. The arc between node x_i and y_j is assigned the weight $w_{i,j}$. Now of G the minimum-weight perfect matching π_0 is calculated. The n arcs in π_0 represent the optimal solutions of (1). Here, optimal means that the sum of the errors is minimal. In the following this method of root selection is called CORS1.

Instead of minimizing the sum of the errors it is also possible to minimize the maximum error. This is done by adjusting the weights in G as follows. All n^2 arcs and their weights are stored in a linear list L. Subsequently, L is sorted in increasing order of the weight. Now the weight in the first entry in L is set to 1, and the weight of item i is set to the sum of the weights in previous items plus one, i.e. $weight[i] = (\sum_{j=1}^{i-1} weight[j]) + 1$. Table (5) is generated with this method. In this way the matching is forced to consist of a subset of the first M items in L, with M as small as possible. We call this method CORS2.

3 Implementation, Tests and Results

Implementation. From (3) the univariate polynomials $p(u)$ and $q(w)$ are derived with MAPLE. The numerical calculations are implemented in C++ in double precission. Laguerre's method [6] is used to compute the roots of the polynomials $p(u)$ and $q(w)$. The LEDA [7] implementation of the minimum weight bipartite matching algorithm is used. Problem instances are obtained by randomly generating instances of the computational geometry problem [5], giving the constants $a_1..h_1$ and $a_2..h_2$.

Tests. We tested the CORS1 and CORS2 method on (2) and on (3). Only the results of the latter problem are presented here, the results of (2) are similar. Every problem instance is solved in two ways: with the CORS method and with SYNAPS, a C++ package for solving polynomial equations. We solved 10^5 problem instances with CORS and SYNAPS, and ≈ 400 with MAPLE. The latter problem instances were solved correctly by CORS and were missed by SYNAPS, i.e. we use MAPLE to decide whether CORS or SYNAPS gave the correct result.

Results. In general the results of CORS1 and CORS2 are identical. In the tests approximately 2.4% of the solutions is missed by SYNAPS and are found by CORS. This is confirmed by solving these problem instances with MAPLE. No solutions were missed by CORS. The average error of the solutions found by SYNAPS is $1.3 \; 10^{-10}$ and of CORS $6.5 \; 10^{-11}$. Running 10^5 problem instances with CORS takes 14 *sec.* and with SYNAPS 475 *sec.*

4 Applying the CORS Method to More Than Two Equations

The CORS method can easily be implemented for a system of two polynomial equations because many efficient implementations of the minimum weight bipartite matching algorithm exist. However, multidimensional matching algorithms are less abundant. Therefore, we now propose the Tolerance Based Algorithm (TBA) for solving the d-Dimensional Matching Problem (DMP). The idea of this algorithm is similar to the idea of a recently designed algorithm for solving the ATSP [2]. For sake of simplicity we first outline the TBA for the 2-DMP.

The algorithm is based on the tolerances for the Relaxed LAP (RLAP). A *feasible solution* θ to the RLAP is a mapping θ of V_1 into V_2 with $w(\theta) < \infty$. A feasible solution π to the LAP can be represented by a set of n arcs (i,j) such that the out-degree $d(i) = 1$ for all $i \in V_1$ and the in-degree $d(j) = 1$ for all $j \in V_2$, and a feasible solution θ to the RLAP can be represented by a set of n arcs (i,j) with the out-degree $d(i) = 1$ for all $i \in V_1$ and $\sum_{j \in V_2} d(j) = n$. We denote the set of all feasible solutions to the RLAP by $\mathbb{Q}$. It is clear that $\mathbb{P} \subset \mathbb{Q}$. Note that θ is feasible to the LAP if the in-degree $d(j) = 1$ for all $j \in V_2$. The RLAP is the problem of finding $\min\{w(\theta) : \theta \in \mathbb{Q}\} = w(\theta_0) \leq w(\pi_0)$ which is a lower bound $w(\theta_0)$ of $w(\pi_0)$.

The *tolerance problem* for the RLAP is the problem of finding, for each $e \in E$, the maximum decrease $l(e)$ and the maximum increase $u(e)$ of the arc weight $w(e)$ preserving the optimality of θ_0 under the assumption that the weights of all other arcs remain unchanged. The values $l(e)$ and $u(e)$ are called the *upper* and the *lower tolerances*, respectively, of edge e with respect to the optimal solution θ_0 and the function of arc weights w. For a selected arc, i.e an arc $[i_1, j_1(i_1)] \in \theta_0$ we define the *upper tolerance* $u[i, j_1(i)] = w[i, j_2(i)] - w[i, j_1(i)]$ with $w[i, j_1(i)] \leq w[i, j_2(i)] \leq ... \leq w[i, j_n(i)]$, and the *lower tolerance* $l[i, j_1(i)] = \infty$. Similarly, for an unselected arc, i.e an arc $[i^1(j), j] \notin \theta_0$ the *lower tolerance* $l[i^1(j), j] = w[i^1(j), j] - w[i^1(j), j_1(i^1(j))]$ with $w[i^1(j), j] \leq w[i^2(j), j] \leq ... \leq w[i^n(j), j]$, and the *upper tolerance* $l(i, j) = \infty$. The *bottleneck tolerance value* $b(\theta) = \max\{u(\theta), l(\theta)\}$ is defined as follows. For each $j \in V_2$, $u(j) = \min\{u(i, j) :$ for all (i, j) with $\deg(j) > 1\}$, and $u(\theta) = \max\{u(j) :$ for all j with $\deg(j) > 1\}$. Similarly, for each $j \in V_2$, $l(j) = \min\{l(i, j) :$ for all (i, j) with $\deg(j) = 0\}$, $l(\theta) = \max\{l(j) :$ for all j with $\deg(j) = 0\}$. Further we treat each π, and each θ as the sets of corresponding arcs such that $|\pi| = |\theta| = n$.

The TBA is based on the following results [1,2]:

(i) if $\theta_0 \notin \mathbb{P}$, then $w(\theta_0) + b(\theta_0) \leq w(\pi_0)$.
(ii) if $\mathbb{Q}_0 = \{\theta : w(\theta) = w(\theta_0)\}$ and $u(i, j) > 0$ for all $(i, j) \in \theta_0$ then $|\mathbb{Q}_0| = 1$.
(iii) if $u(i, j) > 0$ for $(i, j) \in \alpha \subset \theta_0$ and $u(i, j) = 0$ for $(i, j) \in \theta_0 \setminus \alpha$, then $|\mathbb{Q}_0| > 1$ and $|\cap \mathbb{Q}_0| = |\alpha|$. and can be outlined as follows.

Based on these properties of the RLAP, the LAP may be solved using the following tolerance based algorithm

1. Find θ_0.
2. If $\theta_0 \in \mathbb{P}$, then $\pi_0 = \theta_0$, and output π_0 with $w(\pi_0)$, stop.
3. Find $b(\theta)$.
4. If $b(\theta) = u[i, j_1(i)]$, then replace in θ_0 the arc $[i, j_1(i)]$ by the arc $[i, j_2(i)]$; otherwise (if $b(\theta) = l[i^1(j), j]$) replace in θ_0 the arc $[i^1(j), j_1(i^1(j))]$ by the arc $[i^1(j), j]$.
5. Return to step 2.

Note that for CORS2, after each iteration of the TBA for solving the LAP, the current number of vertices $j \in V_2$ with $d(j) = 1$ will be increased by at least one vertex. Hence, the time complexity of the TBA is $O(n^3)$.

In the following example we illustrate the TBA for the complete weighted bipartite graph below

$$\begin{array}{|c|c|c|c|} \hline 8 & 16384 & 256 & 4096 \\ \hline 2 & 16 & 4 & 64 \\ \hline 512 & 32768 & 128 & 32 \\ \hline 1 & 1024 & 8192 & 2048 \\ \hline \end{array} \tag{5}$$

$1.\theta_0 = \{(1,1), (2,1), (3,4), (4,1)\}; 2.\theta_0 \notin \mathbb{P}; 3. b(\theta) = 14; 4. b(\theta) = l(2,2)$, replace in in θ_0 the arc $(2,1)$ by $(2,2). 5$. Return to step 2. 2. $\theta_0 = \{(1,1), (2,2), (3,4), (4,1)\} \notin \mathbb{P}$; 3. $b(\theta) = 248$; 4. $b(\theta) = u(1,1)$, replace in in θ_0 the arc $(1,1)$ by $(1,3)$; 5. Return to step 2. 2. $\theta_0 = \{(1,3), (2,2), (3,4), (4,1)\} \in \mathbb{P}$, then $\pi_0 = \theta_0$, and output π_0 with $w(\pi_0) = 305$, stop.

This tolerance based algorithm may be generalized to more than two dimensions as follows. In case of d-DMP $V = \cup_{i=1}^d V_i$, $|V_i| = n$, the set of arcs $e \in E \subseteq \otimes_{i=1}^d V_i$ with weights $w(e)$ for all $e \in E$, and the in-degree $d(j)$ for all arcs (i,j) will be defined similarly to the 2-DMP with replacing V_2 by $\overline{V_2}$ which is a projection of the $\otimes_{i=2}^d V_i$ on V_2. If the objective function of d-DMP is to minimize the maximim error of all n matched roots, then TBA has $O((d-1)n^3)$ time complexity.

Currently we are implementing the multidimensional tolerance based matching algorithm, and we will test it on a system of three polynomial equations. For the 3D case we can not use Sylvester resultants to calculate the univariate polynomials $p(x)$, $q(y)$ and $r(z)$, we have to use a multiresultant method. These are widely available. No other algorithms are required to test the CORS method on three polynomial equations.

5 Discussion and Conclusion

As mentioned before, the CORS method is not meant to refine numerical results but to find solutions that are missed by other methods. The numerical quality of the solutions found by CORS depends on the quality of the univariate polynomial solver. We used Laguerre's method, but for multiple roots of even degree it is probably better to use Broydens method [6]. In spite of using a non-optimal univariate polynomial solver, our experiments show that the CORS method outperforms current methods in the sense that it does not lose roots. Moreover, it is faster and more acurate.

The CORS method is only suitable for systems of d equations with N solutions with small d and small N. That is because the number of arcs in the graph G is N^d. We think that the CORS method is efficient for $d \leq 3$ and $N \leq 20$, and may be used to analyze a system of equations with $d \leq 5$ and $N \leq 30$. Probably, for larger systems the overhead will be unacceptable.

Many problems in engineering and technology boil down to solving a small system of polynomial equations. Using the CORS method, only simple univariate equations have to be solved. Selecting the correct roots with a combinatorial optimization method proves to work well. In this way the user does not have to worry about degeneracies and other complications.

References

[1] Goldengorin, B., Sierksma, G. *Combinatorial optimization tolerances calculated in linear time*. SOM Research Report 03A30, University of Groningen, Groningen, The Netherlands, 2003(http://www.ub.rug.nl/eldoc/som/a/03A30/03a30.pdf).

[2] Goldengorin, B., Sierksma, G., and Turkensteen, M. *Tolerance Based Algorithms for the ATSP. Graph-Theoretic Concepts in Computer Science.* 30th International Workshop, WG 2004, Bad Honnef, Germany, June 21-23, 2004. Hromkovic J., Nagl M., Westfechtel B. (Eds.). Lecture Notes in Computer Science Vol. 3353, pp. 222-234, 2004.

[3] Sederberg, T., W., Zheng, J. *Algebraic methods for computer aided geometric design* in *Handbook of computer aided geometric design* Farin, G., Hoschek, J., Kim, M., S., (Eds.), North-Holland Elsevier (2002) p. 378.

[4] Synaps. Available at: http://www.inria.fr/galaad/logiciels/synaps/inex.html

[5] Bekker, H., Roerdink, J. B. T. M. *Calculating critical rotations of polyhedra for similarity measure evaluation*. Proceedings of IASTED International conference on Computer Graphics and Imaging, Palm springs, October 1999.

[6] Press, W. H., Flannery, B. P., Teukolsky, S. A., and Vetterling, W. T. *Numerical Recipes in C++, the Art of Scientific Computing*. Cambr. Univ. Press, New York.

[7] K. Melhorn, Näher, S. *LEDA A Platform for Combinatorial and Geometric Computing.* Cambridge University press, Cambridge. 1999

[8] Burkard, R.E. *Slected topics on assignment problems*. Discrete Applied Mathematics 123 (2002) 257–302.

Principles, Models, Methods, and Algorithms for the Structure Dynamics Control in Complex Technical Systems

B.V. Sokolov, R.M. Yusupov[1], and E.M. Zaychik[2]

[1] St.-Petersburg Institute for Informatics and Automation,
39, 14th Line, SPIIRAN, St.-Petersburg, 199178, Russia
sokol@iias.spb.su

[2] Public Corporation Vympel-Communications-R, 3, Konnogvardeiski Bulvar,
St.-Petersburg, 190000, Russia
EZaychik@beeline.ru

Abstract. One of the main features of modern complex technical systems (CTS) is the variability of their parameters and structures as caused by objective and subjective factors at different phases of the CTS life cycle. In other words we always come across the CTS structure dynamics in practice. Under the existing conditions the increment (stabilization) of CTS potentialities or reducing of degradation necessitate the control of CTS structures (including the control of structures reconfiguration). The aim of this investigation is to develop principles, models, methods and algorithms for the CTS structure dynamics control.

1 Introduction

The main subject of our investigation is complex systems. By complex systems we mean systems that should be studied through polytypic models and combined methods. In some instances investigations of complex systems require multiple methodological approaches, many theories and disciplines, and carrying out interdisciplinary researches. Different aspects of complexity can be considered to distinguish between a complex system and a simple one, for example: structure complexity, operational complexity, complexity of behavior choice, complexity of development [1,4,5,6].

Classic examples of complex systems are: control systems for various classes of moving objects such as surface and air transport, ships, space and launch vehicles, etc, geographically distributed heterogeneous networks, flexible computerized manufacturing.

One of the main features of modern complex technical systems (CTS) is the changeability of their parameters and structures due to objective and subjective causes at different stages of the CTS life cycle. In other words we always come across the CTS structure dynamics in practice. Under these conditions to increase (stabilize) CTS potentialities and capacity for work a structure control is to be performed. Reconfiguration is a widely used variant of the CTS structure control. Reconfiguration is a process of the CTS structure alteration with a view to increase, to keep, or to restore the level of CTS operability, or with a view to compensate the loss of CTS efficiency as caused by the degradation of its functions [11,12].

Typical technology of CTS reconfiguration under the condition of a single-resource failure includes the following main steps: **Step1**- fixing and analysis time and place of a resource failure, interruption of the task that used the defective resource, passing the task to another resource with or without retention of intermediate results; **Step2**- removal of the defective resources from the CTS configuration, making an attempt to use reserve resource of the same type or of another type with similar functionality; **Step3**- removal of connections with the faulty resource, prohibition on its use, as for the faulty resource itself, making an attempt of its recovery. If a task of a high priority uses the faulty resource than it can conflict with the tasks of the resource it is passed to, so it can be needed to preempt or to abort tasks of lower priority according to the service procedure. The described technology is usually implemented in modern CTS at a micro-level, which is at the level of CTS elements and blocks [7,8]. Special hardware and software modules are used. This reconfiguration sometimes is named blind reconfiguration, as the following operations are not fulfilled:

- Accounting and analysis of tasks, their characteristics, and functions,
- Analysis and estimation of the current state of CTS in whole,
- Real-time calculation, estimation and analysis of system's goal abilities for reasonable reallocation of CTS functions among its runnable elements and subsystems.

In real situation a single-resource failure can cause failures of some other resources or can reduce their efficiency. At that a substitution for a faulty resource may necessitate completely new efficient configurations of CTS.

The following intermediate conclusions can be considered now [11-13,15]:

- Firstly, besides the compensation of failures, the reconfiguration can be used for rise of operating efficiency of modern CTS,
- Secondly, to implement the proposed concept it is necessary to construct such formal tools that can join together the processes of CTS reconfiguration and the processes of CTS use at different phases of systems life cycle.

The presented considerations led us from a narrow traditional interpretation of CTS reconfiguration to a wide interpretation within a new applied theory of CTS structure-dynamics control. Developing of this theory is one the main aims of our investigations.

2 Approach and Results

As applied to CTS we distinguish the following main types of structures: the structure of CTS goals, functions and tasks; the organization structure; the technical structure; the topological structure; the structure of special software and mathematical tools; the technology structure (the structure of CTS control technology). By structure dynamics control we mean a process of control inputs producing and implementation for the CTS transition from the current macro-state to a given one.

The problem of CTS structure-dynamics control consists of the following groups of tasks: the tasks of structure dynamics analysis of CTS; the tasks of evaluation (observation) of structural states and CTS structural dynamics; the problems of optimal program synthesis for structure dynamics control in different situations.

From our point of view, the theory of structure-dynamics control will be interdisciplinary and will accumulate the results of classical control theory, operations research, artificial intelligence, systems theory, and systems analysis. The two last scientific directions will provide a structured definition of the structure-dynamics control problem instead of a weakly structured definition [2,3,9,10,16,17].

Today different methods and models are used for solving the problems of CTS structure dynamic control. The known approaches to these problems are based on the PERT description of scheduling and control problems and traditional dynamic interpretation. The realization of these dynamic approaches produces algorithmic and computational difficulties caused by high dimensionality, non-linearity, non-stationary, and uncertainty of the appropriate models [11,15].

We proposed to modify dynamic interpretation of operations control processes. The main idea of model simplification is to implement non-linear technological constraints in sets of allowable control inputs rather than in the right parts of differential equations. In this case, Lagrangian coefficients, keeping the information about technical and technological constraints, are defined via the local-sections method. Furthermore, we proposed to use interval constraints instead of relay ones [11]. Nevertheless the control inputs take on Boolean values as caused by the linearity of differential equations and convexity of the set of alternatives. The proposed substitution lets use fundamental scientific results of the modern control theory in various CTS control problems (including scheduling theory problems).

As provided by the concept of CTS multiple-model description the proposed general model includes particular dynamic models: the dynamic model of CTS motion control; the dynamic model of CTS channel control; the dynamic model of CTS operations control; the dynamic model of CTS flows control; the dynamic model of CTS resource control; the dynamic model of CTS operation parameters control; the dynamic model of CTS structure dynamic control; the dynamic model of CTS auxiliary operation control.

Procedures of structure-dynamics problem solving depend on the variants of transition and output functions (operators) implementation. Various approaches, methods, algorithms and procedures of coordinated choice through complexes of heterogeneous models are developed by now.

CTS structure-dynamic control problem has some specific features in comparison with classic optimal control problems [11-13]. **The first feature** is that the right parts of the differential equations undergo discontinuity at the beginning of interaction zones. The considered problems can be regarded as control problems with intermediate conditions. **The second feature** is the multi-criteria nature of the problems. **The third feature** is concerned with the influence of uncertainty factors. **The fourth feature** is the form of time-spatial, technical, and technological non-linear conditions that are mainly considered in control constraints and boundary conditions. On the whole the constructed model is a non-linear non-stationary finite-dimensional differential system with a re-configurable structure. Different variants of model aggregation were proposed. These variants produce a task of model quality selection that is the task of model complexity reduction. Decision-makers can select an appropriate level of model thoroughness in the interactive mode. The level of thoroughness depends on: input data, on external conditions, on required level of solution validity. The proposed interpretation of CTS structure dynamics control processes provides advantages of modern optimal control theory for CTS analysis and synthesis.

The preliminary investigations confirm that the most convenient concept for the formalization of CTS control processes is the concept of an active mobile object (AMO). In general case, it is an artificial object (a complex of devices) moving in space and interacting (by means of information, energy, or material flows) with other AMO and objects-in service (OS). The AMO consists of four subsystems relating to four processes (functioning forms): moving, interaction with OS and other AMO, functioning of the main (goal-oriented) and auxiliary facilities, resources consumption (replenishment).

The four functions of AMO are quite different, though the joint execution of these functions, the interaction being the main one, provide for AMO new characteristics. Thus, it becomes a specific object of investigation, and AMO control problems are strictly different than classical problems of mechanical-motion control. In general, AMO functioning includes informational, material, and energy interaction with OS, with other AMO, and with the environment. Along with the interaction, the facilities functioning, resource consumption (replenishment), and AMO motion are to be considered via functioning models.

The notion "Active Mobile Object" generalizes features of mobile elements dealing with different CTS types. Depending on the type of CTS Active Mobile Objects can move and interact in space, in air, on the ground, in water, or on water surface. Active Mobile Object can be also regarded as multi-agent system [11].

During the investigations we described the main classes of CTS structure dynamics problems. These problems include: AMO structure dynamics analysis problems; AMO structure dynamics diagnosis, observation, multi-layer control problems; problems of AMO generalized structural states synthesis and the choice problems of optimal transition programs providing the transition from a given CTS structural state to an allowable (optimal) structural state. Methodological and methodical basics for the theory of structure-dynamics control were developed. Methodological basics include: the methodologies of the generalized system analysis and the modern optimal control theory for CTS with re-configurable structures. The methodologies find their concrete reflection in the corresponding principles. The main principles are: *the principle of goal programmed control; the principle of external complement; the principle of necessary variety; the principles of multiple-*

model and multi-criteria approaches; the principle of new problems. The dynamic interpretation of structure-dynamics control processes lets apply the results, previously received in the theory of dynamic systems stability and sensitivity, for CTS analysis problems.

The multiple-model description of CTS structure-dynamics control processes is the base of comprehensive simulation technologies and of simulation systems. We assume the simulation system to be a specially organized complex. This complex consists of the following elements: *simulation models (the hierarchy of models); analytical models (the hierarchy of models) for a simplified (aggregated) description of objects being studied; informational subsystem that is a system of data bases (known ledge bases); control-and-coordination system for interrelation and joint use of previous elements and interaction with the user (decision-maker).*

The components of the simulation system were the main parts of the developed program prototypes during our investigation. The processes of AMO structure-dynamics control are hierarchical, multi-stage and multi-task ones. The structure of simulation system (SIS) models conforms the features of control processes.

- models of AMO CS and OS functioning (subsystem I of SIS);
- models of evaluation (observation) and analysis of structural states and AMO CS structure-dynamics (subsystem II of SIS);
- decision-making models for control processes in AMO CS (subsystem III of SIS).

The subsystem of models for AMO CS and OS functioning includes:

- models of AMO functioning, models of AMO classes functioning, and models of AMO system functioning;
- models of AMO interacting station (IS) functioning, models of functioning for control center (CC), central control station (CCS), and control station (CST);
- models of AMO CS subsystems interaction and models of interaction between AMO CS and OS;
- models of objects-in-service (OS) functioning;
- models of environmental impacts on AMO CS;
- simulation models of AMO CS goal directed applications under conditions of environmental impact.

The simulation system also includes: system of control, coordination and interpretation containing user interface and general control subsystem, local systems of control and coordination, subsystem of data processing, analysis, and interpretation for planning, control and modeling, subsystem of modeling scenarios formalization, subsystem of software parametric and structural adaptation, subsystem of recommendations producing for decision-making and modeling. The data-ware includes databases for AMO states, for AMO CS states and general situation, for SO states and data bases for analytical and simulation models of decision-making and of AMO CS functioning.

Existence of various alternative descriptions for CTS elements and control subsystems gives an opportunity of adaptive models selection (synthesis) for program control under changing environment. During our investigations the main phases and steps of a program-construction procedure for optimal structure-dynamics control in CTS were proposed.

At the first phase forming (generation) of allowable multi-structural macro-states is being performed. In other words a structure-functional synthesis of a new CTS make-up should be fulfilled in accordance with an actual or forecasted situation. Here the first-phase problems come to CTS structure-functional synthesis.

The general algorithm of the CTS structure-functional synthesis includes the following main steps.

Step 1. Gathering, analysis, and interrelation of input data for the synthesis of CTS multi-structural macro-states. Construction or correction of the appropriate models.
Step 2. Planning of a solving process for the problem of the CTS macro-states synthesis. Estimation of time and other resources needed for the problem.
Step 3. Construction and approximation of an attainability set (AS) for dynamic control system. This set contains indirect description of different variants of CTS make-up (variants of CTS multi-structural macro-states).
Step 4. Orthogonal projection of a set defining macro-state requirements to AS.
Step 5. Interpretation of output results and their transformation to a convenient form for future use (for example, the output data can be used for construction of adaptive plans of CTS development).

At *the second phase* a single multi-structural macro-state is being selected, and adaptive plans (programs) of CTS transition to the selected macro-state are constructed. These plans should specify transition programs, as well as programs of stable CTS operation in intermediate multi-structural macro-states. *The second phase* of program construction is aimed at a solution of multi-level multi-stage optimization problems. The general algorithm of problem solving should include the following steps.

Step 1. Input data for the problem are being prepared and analyzed in an interactive mode. During this step a structural and parametric adaptation of models, algorithms, and special software tools of simulation system (SS) (see the structure of SIS in second interim report) is being fulfilled to the past and to the current states of the environment, of object-in-service, of control subsystems embodied in existing and developing CTS. For missed data simulation experiments with SIS models or expert inquest can be used.
Step 2. Planning of comprehensive modeling of adaptive CTS control and development for the current and forecasted situation; planning of simulation experiments in SS; selection of models, selection of model structure; determination of methods and algorithms for particular modeling problems, selection of models and model structure for this problems; estimation of necessary time.
Step 3. Generating, via comprehensive modeling, of feasible variants of CTS functioning in initial, intermediate, and required multi-structural macro-states;

introducing of the results to a decision-maker; preliminary interactive structure-functional analysis of modeling results; producing of equivalent classes of CTS multi-structural macro-states.

Step 4. Automatic putting into operation of data of CTS functioning variants; analysis of constraints correctness; final selection of aggregation level for CTS SDC models, and for computation experiments aimed at CTS SDC program construction.

Step 5. Search for optimal CTS SDC programs for transition from a given multi-structural macro-state to a synthesized one and for stable CTS operation in intermediate multi-structural macro-states.

Step 6. Simulation of program execution under perturbation impacts for different variants of compensation control inputs received via methods and algorithms of real-time control.

Step 7. Structural and parametric adaptation of the plan and of SIS software to possible (forecasted through simulation models) states of SO, CS, and of the environment.

Here CTS structural redundancy should be provided for compensation of extra perturbation impacts. After reiterative computation experiments the stability of constructed CTS SDC plan is being estimated.

Step 8. Introducing of comprehensive adaptive planning results to a decision-maker; interpretation and correction of these results.

One of the main opportunities of the proposed method of CTS SDC program construction is that besides the vector of program control we receive a preferable multi-structural macro-state of CTS at final time. This is the state of CTS reliable operation in the current (forecasted) situation.

The combined methods and algorithms of optimal program construction for structure-dynamics control in centralized and non-centralized modes of CTS operation were developed too. The main combined method was based on joint use of the successive approximations method and the "branch and bounds" method. A theorem characterizing properties of the relaxed problem of CTS SDC optimal program construction was proved for a theoretical approval of the proposed method. An example was used to illustrate the main aspects of realization of the proposed combined method.

Classification and analysis of perturbation factors having an influence upon operation of a complex technical system were performed. Variants of perturbation-factors descriptions were considered for CTS SDC models. In our opinion, a comprehensive simulation of uncertainty factors with all adequate models and forms of description should be used during investigation of CTS SDC. Moreover, the abilities of CTS management should be estimated both in normal mode of operation and in emergency situations. It is important to estimate destruction "abilities" of perturbation impacts. In this case the investigation of CTS functioning should include the following phases:

- Determining of scenarios for CTS environment, particularly determining of extremely situations and impacts that can have catastrophic results.

- Analysis of CTS operation in a normal mode on the basis of a priori probability information (if any), simulation, and processing of expert information through theory of subjective probabilities and theory of fuzzy sets.
- Repetition of item b for the main extremely situations and estimation of guaranteed results of CTS operation in these situations.
- Computing of general (integral) efficiency measures of CTS structure-dynamics control.

Algorithms of parametric and structural adaptation for CTS SDC models were proposed. The algorithms were based on the methods of fuzzy clusterization, on the methods of hierarchy analysis, and on the methods of a joint use of analytical and simulation models.

The SDC application software for structure-dynamics control in complex technical systems was developed.

Operability of the software was shown for a space navigation system. The software was applied to construction of control programs for ground-based elements of the system and for navigation spacecrafts structure-dynamics. The considered system is a part of a space-facilities control system that is a classic example of CTS. A multiple-model description of the system, scheduling algorithms for ground-based technical facilities, and algorithms of structure-dynamics control for orbital system of navigation spacecrafts were developed.

The main distinguishing feature of the models and algorithms is the dynamic interpretation of operations within technological control loop of navigation spacecrafts. The dynamic interpretation of operations resulted in essential reduction of scheduling problem dimensionality and in advantages of the proposed algorithm because of its connectivity decrease. The problem dimensionality is determined by the number of independent paths in a network diagram of navigational ground-based-control-complex (GCC) operations and by current space-time, technical and technological constraints. In its turn, the degree of algorithmic connectivity depends on a dimensionality of the main and the conjugate state vectors [11]. If the vectors are known then the schedule calculation may be resumed after removal of appropriate constraints. Analysis of the proposed models, and algorithms and computational experiments with the developed application software showed that rational (optimal) scheduling of navigational GCC operations raises the total capacity of the control system, reduces delays in control loops, and prevents peak information loads caused by alteration of system's structure. Besides, the dynamic interpretation of GCC operation establishes a relationship between the control technology and the practical results of system's use (characteristics of navigational field). This relationship gives an opportunity of efficient control technology development for space navigation systems.

3 Conclusion

Methodological and methodical basis of the theory of CTS structure-dynamics control is developed by now. This theory can be widely used in practice. It has interdisciplinary basis provided by classic control theory, operations research, artificial intelligence, systems theory and systems analysis. The dynamic interpretation of CTS reconfiguration process provides strict mathematical base for complex technical-organizational problems that were never formalized before and have high practical importance [11-14]. The proposed approach to the problem of **CTS structure reconfiguration control** in the terms of general context of **CTS structural dynamics control** enables: common goals of CTS functioning to be directly linked with those implemented (realized) in CTS control process, a reasonable decision and selection (choice) of adequate consequence of problems solved and operations fulfilled related to structural dynamics to be made (in other words to synthesize and develop the CTS control method), a compromise distribution (trade-off) of a restricted resources appropriated for a structural dynamics control to be found voluntary.

This work was supported by European Office of Aerospace Research and Development (Project 1992p), Russian Foundation for Basic Research (grant 02-07-90463), Institute for System Analysis RAS (project 2.5).

References

1. Casti, J.L.: Connectivity, Complexity and Catastrophe in Large-Scale Systems. Wiley-Interscience, New York and London. (1979)
2. Doganovskii, S.A., Oseranii, N.A.: Automatic Restructurable Control Systems . Measurement, Check, Automation **90** (1990), N 1, 62–80 (in Russian).
3. Intelligent Control Systems: Theory and Applications. / Eds. M.M. Gupta, N.K. Sinka. N.Y.: IEEE Press. (1996).
4. Klir, G.J.: Architecture of Systems Problem Solving. Plenum Press, New York. (1985).
5. Mesarovic, M.D., Takahara, Y.: General Systems Theory: Mathematical Foundations. Academic Press, New York, Can Francisco, London. (1975)
6. Moiseev, N.N.: Elements of the Optimal Systems Theory. Nauka, Moscow. (1974) (in Russian).
7. Napolitano, M.R., Swaim, R.L.: An Aircraft Flight Control Reconfiguration Algorithm // Proc. AIAA Guidance, Navigation and Control Conf. Pt 1. (1989) 323–332
8. Napolitano, M.R., Swaim, R.L.: A New Technique for Aircraft Flight Control Reconfiguration // Proc. AIAA Guidance, Navigation and Control Conf. (1989) 323–332
9. Russell, S.J., Norvig, P.: Artificial Intelligence: A Modern Approach. Prentice-Hall Inc., a Simon and Schuster Company, Upper Saddle River, New Jersey (1995)
10. Shannon, R.E.: Systems Simulation. Prentice-Hall Inc., Englewood Cliffs, New Jersey. (1975)
11. Sokolov, B.V., Kalinin V.N.: Multi-model Approach to the Description of the Air-Space Facilities Control Process. Journal of Computer and System Sciences Internat. (1995), N 1, 149–156. (USA).
12. Sokolov, B.V., Yusupov, R.M.: Complex Simulation of Automated Control System of Navigation Spacecraft Operation. Journal of Automation and Information Sciences. **34** (2002), N 10, 19–30. (USA).
13. Sokolov, B.V. Yusupov, R.M.: Conceptual Foundations of Quality Estimation and Analysis for Models and Multiple-Model Systems. Journal of Computer and System Sciences Internat. (2004), N 6 (accepted for publication in USA).

14. Sokolov, B.V., Arkhipov, A.V., Ivanov, D.A.: Intelligent Supply Chain Planning in 'Virtual Organization' // Proc. PRO-VE'04 5th IFIP Working Conference on Virtual Enterprises, France, Toulouse, August, 22-27, 2004. **8** (2004), Part 8, 215–224.
15. Sokolov, B.V.: Dynamic models of comprehensive scheduling for ground-based facilities communication with navigation spacecrafts // Preprints 16th IFAC Symposium on Automatic Control in Aerospace, Saint-Petersburg, Russia. June 14-18, 2004. **1** (2004), 262–267.
16. Tsurlov, V.I.: Dynamic Problems of large Dimension. Nauka, Moscow (1989) (in Russian). Yusupov, R., Rozenwasser E.: Sensitivity of Automatic Control Systems. CRS. Press, London, New York (1999)

Applying a Hybrid Ant Colony System to the Vehicle Routing Problem

Chia-Ho Chen[1], Ching-Jung Ting[1], and Pei-Chann Chang[1]

[1]Department of Industrial Engineering and Management,
Yuan Ze University,
135 Yuan-Tung Road, Chungli, Taoyuan, Taiwan 32003, R.O.C.

Abstract. The vehicle routing problem (VRP) has been extensively studied because of the interest in its application in logistics and supply chain management. In this paper, we develop a hybrid algorithm (IACS-SA) that combines the strengths of improved ant colony system (IACS) and simulated annealing (SA) algorithm. The results of computational experiments on fourteen VRP benchmark problems show that our IACS-SA produces better solutions than those of other ACS in the literature. The results also indicate that such a hybrid algorithm is comparable with other meta-heuristic algorithms.

1 Introduction

Vehicle routing problem (VRP) has been largely studied because of the interest in its application in logistic and supply chains management. In the VRP, a fleet of K identical vehicles delivers goods to N customers whose demands are known. Each vehicle route starts and ends at the central depot. All vehicles have the same capacity and the maximum service time constraints. Each customer must be visited once by exactly one vehicle. The objective of the VRP is to minimize the total distance and/or the number of vehicles. The VRP has been proved to a NP-hard problem. Many researchers have used heuristic approaches to solve the VRP according to artificial intelligence, biological evolution and/or physics phenomenon, such as Simulated Annealing (SA) (Robuste et al., 1990; Alfa 1991; Osman, 1993; Breedam, 1995), Genetic Algorithms (GA) (Baker and Ayechew, 2003; Jaszkiewicz and Kominek, 2003), Tabu Search (TS) (Osman, 1993; Taillard, 1993; Rochat and Taillard, 1995; Rego and Roucairol, 1996; Xu and Kelly, 1996; Rego, 1998; Barbarrosoglu and Ozgur, 1999), and Ant System (AS) (Bullnheimer et al., 1998; Bullnheimer et al., 1999).

Ant algorithm is a new distributed meta-heuristic first introduced by Colorni et al. (1991). Dorigo et al. (1996) reported the ant system (AS) to solve the traveling salesman problem (Dorigo and Gambardella, 1997), the quadratic assignment problem (Maniezzo et al., 1994) and the job-shop scheduling (Colorni et al. 1994). Dorigo and Gambardella (1997) developed the ant colony system (ACS) to improve the performance of AS. They used a different state transition rule and added a local pheromone updating rule. Although ant algorithm has been applied to many combinatorial problems successfully, few researchers solved VRP using ACS. Bullnheimer et al. (1998) are the first researchers that used AS to solve VRP. Bullnheimer et al. (1999) developed an improved AS for VRP and improved the performance significantly. However, solutions produced by AS are not as good as other meta-heuristic approaches, such as TS and SA. Thus, Ting and Chen (2004) proposed an improved ant colony system (IACS) with a new route construction rule, a new pheromone updating rule and diverse local search approaches (2-opt and swap).

In recent years, many researchers have considered the topic of hybrid meta-heuristics in combinatorial optimization problems. The best solutions found for many practical and academic optimization problems are gained by hybrid algorithms. Combinations of algorithms such as local search, greedy heuristics, meta-heuristics are very powerful search algorithms that have been reported. Therefore, the objective of this research is to combine the advantages of population search in IACS and the single solution search in SA to develop a hybrid algorithm IACS-SA. To begin with, a solution is obtained by our IACS-SA. Then, we improve the solution using SA. Finally, we use our IACS-SA to solve 14 benchmark VRP problems to compare performance with other meta-heuristic approaches, including SA, GA, TS, and AS.

2 Improved Ant Colony System

The Improved Ant Colony System (IACS) was first proposed by Ting and Chen (2004). The IACS, which is based on the ACS proposed by Dorigo and Gambardella (1997), includes four steps as follows:

Step 1: Set parameters and initialize the pheromone trails.
Step 2: Each ant builds the solution by the state transition rule and carries out local pheromone update according to each ant's solution.
Step 3: Apply the local search to improve the ants' solution.
Step 4: Update the global pheromone information based on the best routes among all ants' solution.

2.1 Solution Construction

In the original ACS, each ant moves from present node i to the next node v according to the rule given by (1).

$$v = \begin{cases} \arg\max_{j \in U} [(\tau_{ij})^\alpha (\eta_{ij})^\beta] & q \leq q_0, \\ V & \text{otherwise.} \end{cases} \quad (1)$$

$$V : P_{ij} = \frac{(\tau_{ij})^\alpha (\eta_{ij})^\beta}{\sum_{j \in U} (\tau_{ij})^\alpha (\eta_{ij})^\beta} \quad (2)$$

where U is the set of nodes which are not visited yet, τ_{ij} is the pheromone level of edge (i, j), η_{ij} denotes the savings of combining two nodes i and j on one tour as opposed to serving them on two different tours. Thus, the η_{ij} is calculated as follows:

$$\eta_{ij} = d_{i0} + d_{0j} - d_{ij} \qquad (3)$$

where d_{ij} denotes the distance between nodes i and j, and the index 0 is the depot; and α, β are the parameters that determine the relative influence of pheromone versus distance ($\alpha, \beta > 0$). Moreover, q is a random number following uniform distribution in $[0, 1]$, and q_0 is a pre-defined parameter ($0 \leq q_0 \leq 1$). If $q \leq q_0$ then the best next node v is chosen according to eq. (1). On the contrary, the next node v is chosen according to V which is a random variable selected according to the probability distribution given in eq. (2). Thus, the parameter q_0 determines the relative importance of exploitation (v) versus exploration (V).

In our IACS, the best solution so far will be preserved and becomes the 1st solution in the next generation. We only reconstruct b-1 solutions (b is the number of ants) in each generation. The tour construction in IACS algorithm is as follows:

Step 1: Generate the ant's starting node randomly.
Step 2: Chose next node according to the tour construction rule by eq. (1).
Step 3: Repeat steps 2 until the ant visits all nodes.
Step 4: Divide the tour according to the vehicle capacity and/or maximum traveling time restrictions.

2.2 Local Search

In the original ACS, after constructing all ants' tours, local search is applied to improve each solution. However, local search is a time-consuming procedure of ACS. In our IACS, we apply local search to only one solution (the best among the b-1new constructed solutions) with two different types of local search: 2-opt and swap. Firstly, we apply 2-opt to the solution. Then, the swap move is carried out. The procedures of local search in IACS are as follows:

Step 1: Sort the solutions 2~b in ascending orders according to the objective function value.
Step 2: Carry out local search on the 2nd solution.

2.3 Pheromone Updating Rule

The pheromone updating of ACS includes global and local updating rules. Following Dorigo and Gambardella (1997), local updating rule in eq. (4) is applied to change the pheromone level of edges once an ant constructs a solution.

$$\tau_{ij}^{new} = \tau_{ij}^{old} + \rho\tau_0 \qquad (4)$$

where $0 \leq \rho \leq 1$ is the pheromone decay parameter, and $(\tau_0 = n * L_{nn})^{-1}$ is initial pheromone level of edges, n is the number of nodes and L_{nn} is the tour length evaluated by the Nearest Neighbor heuristic.

After local search, our global updating rule is applied to the best u solutions. The rule is described as follow:

$$\tau_{ij}^{new} = (1-\rho)\tau_{ij}^{old} + \rho \sum_{k=1}^{u} \Delta\tau_{ij}^{k} \qquad (5)$$

where

$$\Delta\tau_{ij}^{k} = \begin{cases} \frac{L_{u+1} - l_k}{L_{u+1}} & (i,j) \in \text{tour done by ant } k \\ 0 & \text{otherwise.} \end{cases} \qquad (6)$$

where L_k is the length of the tour done by ant k, u is the number of solutions whose pheromone will be updated.

2.4 Overall Procedure of IACS

The procedures of our IACS are described as follows

Step 1: Set parameters.
Step 2: Generate an initial solution using Nearest Neighbor heuristic.
Step 3: Apply the local search (2-opt and Swap) to the initial solution and let it to be the solution 1 of population. Set $g = 1, h = 2$.
Step 4: Construct solutions base on the route construction rule and progress local pheromone update. $h = h + 1$.
Step 5: If $h < b$, go to Step 4. Otherwise, go to Step 6 and $h = 2$.
Step 6: Sort the solutions $2\sim b$ in ascending order and apply local search (2-opt and Swap) to 2nd solution.
Step 7: Apply the global pheromone update for solutions $1\sim u$.
Step 8: Record the best solution so far and let it to be the solution 1 in the next generation. $g = g + 1$.
Step 9: If the stopping criterion (maximum number of iteration in this paper) is met, stop, then output the best solution. Otherwise, go to Step 4.

3 Simulated Annealing Algorithm

Simulated Annealing (SA) was first used to search the feasible solutions of an optimization problem by Kirkpatrick et al. (1983). SA is inspired by an analogy between the physical annealing of solid and combinatorial problems. In the physical process a solid is first melted at high temperature and then cooled very slowly, spending a long time at low temperatures, to obtain a more ordered structure corresponding to a minimum energy state. SA transfers this process to local search algorithms for combinatorial optimization problems. It does so by associating the set of solutions of the problem attacked with the states of the physical system, the objective function with the physical energy of the solid and the optimal solutions with the minimum energy states.

SA is a local search strategy which tries to avoid local minima by jumping out of them early in the computation. A worse variation is accepted as the new solution with a probability that decreases as the computation proceeds. In our research, the procedure is described as follows:

Step 1: Set parameters: T (initial temperature), γ (cooling parameter), m (maximum number of move operator), M (maximum number of iterations).

Step 2: Generate an initial solution x^0 using Nearest Neighbor heuristic. Set $x = x^0$.

Step 3: Apply the local search (2-opt and Swap) to the initial solution. $i = 1, j = 1$.

Step 4: Compute the objective function value of current solution $f(x)$.

Step 5: a. If $i \leq m$, apply the move operator (2-opt exchange and Swap move) to current solution to generate new solution x', and $i = i + 1$, then go to step 5b. Otherwise, go to step 6.
b. Evaluate $\Delta E = f(x') - f(x)$. If $\Delta E \leq 0$, go to step 5d; otherwise, go to step 5c.
c. Select a random variable $u \sim U(0, 1)$. If

$$u < P(\Delta E) = \exp(-\Delta E/T) \qquad (7)$$

then go to step 5d; otherwise, go to step 5a.
d. Accept the exchange, set $x = x'$ and $f(x) = f(x')$, then go to step 5a.

Step 6: If $j \leq M$ evaluate $T = \gamma T$, and $j = j + 1$, then go to step 3. Otherwise, stop.

4 Hybrid Algorithm IACS-SA

The IACS has advantage of searching for a variety of solutions, while SA can jump away from the local optimum by accepting worse solutions. Thus, we combine IACS and SA to develop a hybrid algorithm (IACS-SA) to take advantages of both heuristic algorithms. The framework of IACS-SA is shown in Figure 1. We first use Nearest Neighbor (NN) heuristic to generate an initial solution for both IACS and SA, and then the initial solution is improved by local search in advanced. Both IACS and SA will solve the VRP iteratively and independently. On one hand IACS generates new solutions according to tour construction rule and improves the solutions using local search in each iteration. On the other hand SA apply the 2-opt exchange to improve the solution m times and then the temperature cools down in each iteration. In each iteration IACS and SA will communicate their best solution so far. If the best solutions obtained by both algorithms at current iteration are different, the inferior one will be replaced by the other one. To communicate the solutions between IACS and SA, IACS-SA is able to obtain good local optimal solution. After a pre-specified number of iterations, IACS-SA outputs the best solution. Based on our experiment, IACS can find good solution quickly. We will only run IACS for small number of iterations to save computation times. Finally, the best solution will be further improved by SA for another pre-specified number of iterations.

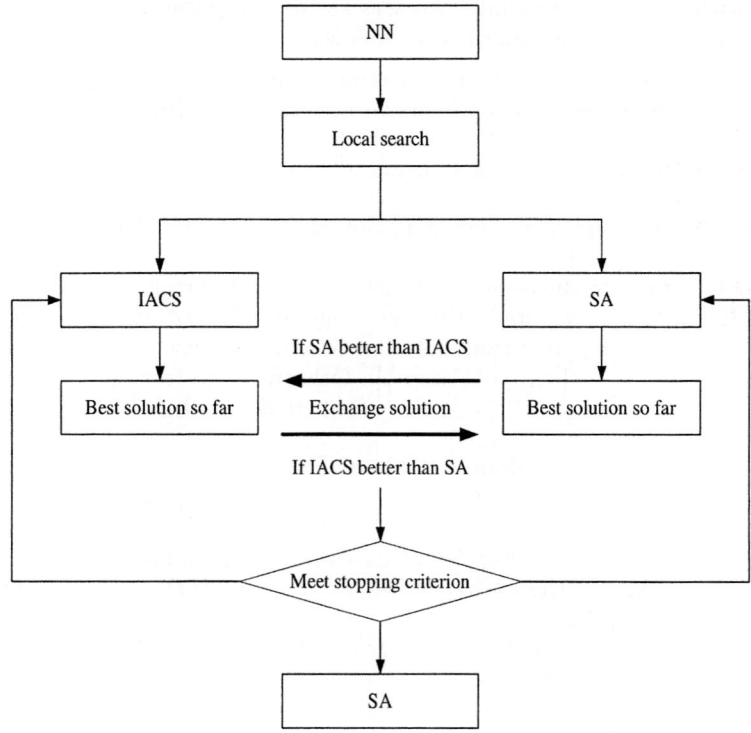

Fig. 1. The framework of IACS-SA

5 Numerical Analysis

In this section, 14 VRP benchmark problems described in Christofides et al. (1979) are tested by our IACS-SA. The IACS-SA is coded in Borland C++ Builder 5.0 and executed on a PC equipped with 128 MB of RAM and an Pentium processor running at 1000 MHz. The IACS-SA parameters used for VRP instances are $b = n/10, \alpha = 1, \beta = 4, \rho = 0.5, q_0 = 0.5, u = 2, G = 50, T = 1, \gamma = 0.999, m = n*n$ and $M = 3n$. The information of these test problems and computational results are summarized in Table 1. Columns 2-5 in table 1 show the problem size n, the vehicle capacity Q, the service time s, the maximum route length L, and the best-known solutions. The number of customers in these problems ranges from 50 to 199. The customers in problems 1-10 are generated by a random uniform distribution, while they are clustered in problems 11-14. Problems 1-5 and 6-10 are identical, except that the total length of each vehicle route is limited for the latter problems. Problems 13-14 are the counterparts of problems 11-12 with additional route length restriction. Columns 6-10 show the results that include the best solution, the worst solution, and the average solution, standard deviation and average run time (second) over 10 runs for each problem.

Table 1. The computational results of VRP instances. ^{a}: the solutions is same as the best-known solution

N0.	n	Q	s/L	Best know	Best	Worst	Average	S.D.	time
C1	50	160	0/∞	524.61	**524.61**^{a}	533.00	528.07	3.97	2
C2	75	140	0/∞	835.26	835.32	861.31	850.80	7.42	13
C3	100	200	0/∞	826.14	829.98	854.28	844.94	8.42	33
C4	150	200	0/∞	1028.42	1036.36	1062.71	1049.95	8.90	195
C5	199	200	0/∞	1291.45	1309.90	1337.72	1325.48	8.24	658
C6	50	160	10/200	555.43	**555.43**	573.47	565.03	5.85	4
C7	75	140	10/160	909.68	**909.67**	943.31	921.80	11.35	23
C8	100	200	10/230	865.94	**865.94**	891.95	879.88	8.69	62
C9	150	200	10/200	1162.55	1164.66	1228.70	1193.93	19.95	322
C10	199	200	10/200	1395.85	1417.57	1443.62	1433.00	7.74	1050
C11	120	200	0/∞	1042.11	**1042.11**	1044.95	1043.75	0.87	61
C12	100	200	0/∞	819.56	827.98	856.35	839.49	8.38	34
C13	120	200	50/720	1541.14	1543.85	1576.14	1557.74	9.77	124
C14	100	200	90/1040	866.37	**866.37**	868.79	867.26	0.70	49
Avg.					980.70	1005.45	992.94		188

Table 2. Comparison of IACS-SA with three ant algorithms. ^{a}: the best among four ant algorithms. ^{1}: AS by Bullnheimer et al. (1998), seconds on a Pentium 100 MHz PC. ^{2}: IAS by Bullnheimer et al. (1999), seconds on a Pentium 100 MHz PC. ^{3}: IACS by Ting and Chen (2004), seconds on a Pentium 1000 MHz PC

No.	AS^1		IAS^2		IACS^3		IACS-SA	
	RPD	Time	RPD	Time	RPD	Time	RPD	Time
C1	**0.00**	36	**0.00**	6	**0.00**	3	**0.00**	2
C2	4.23	144	1.08	78	0.11	26	**0.01**	13
C3	6.45	678	0.75	228	1.15	101	**0.47**	33
C4	11.57	1710	3.22	1104	0.95	617	**0.77**	195
C5	14.09	4932	4.03	5256	2.76	3080	**1.43**	658
C6	1.35	12	0.87	6	**0.00**	5	**0.00**	4
C7	4.23	210	0.72	102	**0.00**	41	**0.00**	23
C8	2.34	438	0.09	288	**0.00**	115	**0.00**	62
C9	3.39	1596	2.88	1650	0.96	853	**0.18**	322
C10	7.80	3438	4.00	4908	**1.29**	4223	1.56	1050
C11	2.91	972	2.22	552	**0.00**	204	**0.00**	61
C12	0.05	606	**0.00**	300	1.6	88	1.03	34
C13	3.20	258	1.22	660	0.38	428	**0.18**	124
C14	0.40	186	0.08	348	**0.00**	125	**0.00**	49
Avg.	4.43	1086	1.51	1106	0.66	708	**0.40**	188

Table 3. Comparison of IACS-SA with other meta-heuristics. a: the best among different algorithms. [1]: SA and TS by Osman (1993), seconds on a VAX 8600 computer. [2]: Hybrid GA by Baker and Ayechew (2003), seconds on a Pentium 266 MHz PC

No.	SA[1]		TS[1]		HGA[2]		IACS-SA	
	RPD	Time	RPD	Time	RPD	Time	RPD	Time
C1	0.65	167	**0.00**^{a}	67	**0.00**	23	**0.00**	2
C2	0.40	6434	1.05	71	0.43	617	**0.01**	13
C3	**0.37**	9334	1.07	675	0.40	717	0.47	33
C4	2.88	5012	2.29	3075	**0.62**	1961	0.77	195
C5	6.55	2318	4.84	1972	2.23	5261	**1.43**	658
C6	**0.00**	3410	**0.00**	140	**0.00**	429	**0.00**	4
C7	**0.00**	626	0.36	203	**0.00**	449	**0.00**	23
C8	0.09	957	0.01	1200	0.20	1904	**0.00**	62
C9	**0.14**	84301	2.19	2444	0.32	2242	0.18	322
C10	1.58	5708	1.87	3310	1.96	6433	**1.56**	1050
C11	12.85	315	**0.00**	1398	0.46	1483	**0.00**	61
C12	0.79	632	**0.00**	407	**0.00**	1285	1.03	34
C13	0.31	7622	0.38	1343	0.34	1063	**0.18**	124
C14	2.73	305	**0.00**	5579	0.09	585	**0.00**	49
Avg.	2.09	9081	1.00	1563	0.50	1746	**0.40**	188

In Table 2, results by IACS-SA are compared with two previous AS by Bullnheimer et al. (1998, 1999) and IACS by Ting and Chen (2004) in terms of relative percentage of deviation (RPD) over the best-known solutions as well as their computational times. The numbers in bold are the best solutions among four ant system algorithms. The average RPD from the best-known solution obtained by IACS-SA is the lowest among four ant system algorithms. IACS-SA yields the best solutions among four algorithms in 12 out of 14 problems and finds the optimal solutions in 6 problems. The run time not only depends on the CPU of the machines but also on the operation system, the compiler, the programming language and the precision used during the execution of the run. Therefore, a fair comparison of computational effort is difficult to establish. However, it is pretty obvious that the computational time of IACS-SA is much less than that of IACS which was implemented on the same PC. Thus, combining IACS with SA can improve the solution quality of IACS and save computational effort.

In Table 3 we compare the RPD from the best-known solution as well as the computation time of IACS-SA and three other meta-heuristic algorithms, which include the best-admissible variant of Osman's (1993) tabu search (denoted by TS), his simulated annealing algorithm (denoted by SA), and the hybrid genetic algorithm by Baker and Ayechew (2003) (denoted by Hybrid GA). Our hybrid algorithm seems to be superior in terms of solution quality with an average deviation of 0.4%. Osman's SA lacks the accuracy as compared to other approaches and cannot compete at the speed level. In addition, the IACS-SA yields the best solutions among four algorithms in 10 out of 14 problems. Thus, these results show that our hybrid algorithm is competitive with other meta-heuristic algorithms in terms of solution quality.

6 Conclusion

In this paper, we present the improvement of an ant colony system algorithm (IACS) and propose a hybrid algorithm combining IACS and simulated annealing algorithm (IACS-SA). The computational results of 14 benchmark problems reveal that our IACS-SA is much better than three previous ant system algorithms in the literature. Compared with other meta-heuristic algorithms, our IACS-SA also yields lower relative percentage of deviation than the other three approaches. It is believed that our IACS-SA is competitive and provides among the best results in solving VRP problems. Further research on additional modifications of the hybrid algorithm to extensions of the vehicle routing problem with time windows are of interest.

References

1. Alfa, A.S., Heragu S.S. and Chen, M.: A 3-opt Based Simulated Annealing Algorithm for Vehicle Routing Problems. Computers & Industrial Engineering, Vol. 21. (1991) 635-639
2. Baker, B.M., and Ayechew, M.A.: A Genetic Algorithm for the Vehicle Routing Problem. Computers and Operations Research, Vol. 30. (2003) 787-800
3. Barbarosoglu, G. and Ozgur, D.: A Tabu Search Algorithm for the Vehicle Routing Problem. Computers and Operations Research, Vol. 26. (1999) 255-270
4. Breedam, A.V.: Improvement Heuristics for the Vehicle Routing Problem Based on Simulated Annealing. European Journal of Operational Research, Vol. 86. (1995) 480-490
5. Bullnheimer, B., Hartl, R.F., and Strauss, C.: An Improved Ant System for the Vehicle Routing Problem. Annals of Operations Research, Vol. 89. (1999) 319-328
6. Bullnheimer, B., Hartl, R.F., and Strauss, C.: Applying the Ant System to the Vehicle Routing Problem. In S. Voss, S. Martello, I.H. Osman, and C. Roucairol, editors, Meta-Heuristics: Advances and Trends in Local Search Paradigms for Optimization, Kluwer, Boston, MA, (1998) 109-120
7. Christofides, N. Mingozzi, A., and Toth, P.: The Vehicle Routing Problem. In: Combinatorial Optimization, ed. N. Christofides, A. Mingozzi, P. Toth, and C. Sandi. Wiley, Chichester (1979)
8. Clarke, G. and Wright, J.W.: Scheduling of Vehicles from a Central Depot to a Number of Delivery Points. Operations Research, Vol. 12. (1964) 568-581
9. Colorni, A., Dorigo, M., and Maniezzo, V.: Distributed Optimization by Ant Colonies. In: F. Varela and P. Bourgine, editors, Proceeding of the European Conference on Artificial Life, Elsevier, Amsterdam (1991)

10. Colorni, A., Dorigo, M., Maniezzo, V., and Trubian, M.: Ant System for Job-Shop Scheduling. Belgian Journal of Operations Research, Statistics and Computer Science, Vol. 34. (1994) 39-53
11. Dorigo, M., and Gambardella, L.M. Ant Colony System: A Cooperative Learning Approach for the Traveling Salesman Problem. IEEE Transactions on Evolutionary Computation, Vol. 1. (1997) 53-66
12. Dorigo, M., Maniezzo, V., and Colorni, A.: Ant System: Optimization by a Colony of Cooperating Agents. IEEE Transactions on Systems, Man and Cybernetics Part B, Vol. 26. (1996) 29-41
13. Jaszkiewicz, A. and Kominek, P.: Genetic Local Search with Distance Preserving Recombination Operator for a Vehicle Routing Problem. European Journal of Operational Research, Vol. 151. (2003) 352-364
14. Maniezzo V., Colorni, A., and Dorigo, M.: The Ant System Applied to the Quadratic Assignment Problem. Tech. Rep. IRIDIA/94-28, Universite Libre de Bruxelles, Belgium (1994)
15. Osman, I.H.: Metastrategy Simulated Annealing and Tabu Search Algorithms for the Vehicle Routing Problem. Annals of Operations Research, Vol. 41. (1993) 421-451
16. Rego, C. and Roucairol, C.: A Parallel Tabu Search Algorithm Using Ejection Chains for the Vehicle Routing Problem. In I.H. Osman and J.P. Kelly, editors, Meta-heuristic: Theoy and Applications, Kluwer, Boston, MA, (1996) 667-675
17. Rego, C.: A Subpath Ejection Method for the Vehicle Routing Problem. Management Science, Vol. 44. (1998) 1447-1459
18. Rochat, Y., and Taillard, E.: Probabilistic and Intensification in Local Search for Vehicle Routing. Journal of Heuristics, Vol. 1. (1995) 147-167
19. Taillard, E.D.: Parallel Iterative Search Methods for Vehicle Routing Problems. Networks, Vol. 23. (1993) 661-673
20. Ting, C.J., and Chen, C.H.: An Improved Ant Colony System Algorithm for the Vehicle Routing Problem. Working Paper 2004-001, Department of Industrial Engineering and Management, Yuan Ze University, Taiwan, R.O.C. (2004)
21. Xu, J. and Kelly, J.P.: A Network Flow-based Tabu Search Heuristic for the Vehicle Routing Problem. Transportation Science, Vol. 30. (1996) 379-393

A Coevolutionary Approach to Optimize Class Boundaries for Multidimensional Classification Problems

Ki-Kwang Lee

School of Management, INJE University,
607 Kimhae, Kyungnam, South Korea
kiklee@inje.ac.kr

Abstract. This paper proposes a coevolutionary classification method to discover classifiers for multidimensional pattern classification problems with continuous features. The classification problems may be decomposed into two sub-problems, which are feature selection and classifier adaptation. A coevolutionary classification method is designed by coordinating the two sub-problems, whose performances are affected by each other. The proposed method establishes a group of partial subregions, defined by regional feature set, and then fits a finite number of classifiers to the data pattern by combining a genetic algorithm and a local adaptation algorithm in every sub-region. A cycle of the cooperation loop is completed by evolving the sub-regions based on the evaluation results of the fitted classifiers located in the corresponding sub-regions. The classifier system has been tested with well-known data sets from the UCI machine-learning database, showing superior performance to other methods such as the nearest neighbor, decision tree, and neural networks.

1 Introduction

Classification learning systems are useful for decision-making tasks in many application domains, such as financing, manufacturing, control, diagnostic applications, and prediction systems where classifying expertise is necessary [12]. This wide range of applicability motivated many researchers to further refine classification methods in various domains [5, 11]. The major objective of classification is to assign a new data object represented as features (sometimes referred to as attributes or input variables) to one of the possible classes with a minimal rate of misclassification. Solutions to a classification problem have been characterized in terms of parameterized or non-parameterized separation boundaries that could successfully distinguish the various classes in the feature space [9]. A primary focus of study to build the separation boundaries has been on learning from examples, where a classifier system accepts case descriptions that are pre-classified and then the system learns a set of separation surfaces that can classify new cases based on the pre-classified cases [8]. Various learning techniques have been contrived to design the separation surfaces, employing a variety of representation methods, such as mathematical functions, neural networks, fuzzy if-then rules, and decision trees. The method proposed in this paper to construct a classifier system consists of two levels, i.e. determining the feature space and searching the separation boundaries. The number and diversity of possible classifying features would easily dominate the amount of available decision data. When the number of features and the possible patterns are huge, a method of feature selection should be devised to find the most relevant features before automatic classification or decision learning [7]. We represent a feature set as a set of pairs of a feature and its operational range, which actually represents a hyper-rectangular sub-region in the dimensional space. The feature sets are obtained by iteratively adding a feature and its available interval to a current feature set in a sequential increasing manner, so that the sub-region expanded from the added feature can include as many positive examples as possible. In every sub-region, the classifiers, which are delineated by geometrical ellipsoids, adjust their parameters to search the separation boundaries by using a hybrid method of a genetic algorithm (GA) and a heuristic local search algorithm. Abe and Thawonmas [1] showed that a classifier with ellipsoidal regions had the generalization ability comparable or superior to those of classifiers with the other shapes. Motivated by the result of [1], the ellipsoids are adopted to fit the usual non-linear boundaries which hyper-rectangular sub-region cannot represent accurately. After the evolution stabilizes for the ellipsoidal regions in every feature sets represented by sub-regions in the dimensional space, the feature sets themselves are subject to evolution based on the evaluation results of the fitted ellipsoids located in the corresponding sub-regions. The two-level coevolution process is iterated until the termination condition is satisfied. The rest of this paper is organized as follows. Section 2 describes the details about a proposed classifier system for the pattern classification problem defined in this paper. Section 3 shows the experimental results from evaluating the performance of the proposed classifier system. Finally, conclusions are stated in section 4.

2   A Coevolutionary Classifier System

Let us assume that a pattern classification problem has c classes in an n-dimensional pattern space $[0, 1]^n$ with continuous input features. It is also supposed that a finite set of points $X = \{x_p, p = 1, 2, \ldots, m\}$ is given as the training data. Suppose that each point of X, $x_p = (x_{p1}, x_{p2}, \ldots, x_{pn})$, is assigned to one of the c classes, and let the corresponding subsets of X, having $N_1, N_2, \ldots, N_c$ points, respectively, be denoted by $X_1, X_2, \ldots, X_c$. Because the pattern space is $[0, 1]^n$, the feature values are $x_{pj} \in [0,1]$ for $p = 1, 2, \ldots, m$ and $j = 1, 2, \ldots, n$. It is desired that the subset $X_i(i = 1, 2, \ldots, c)$ are isolated by classifying regions labeled $L_{ij}(j = 1, \ldots)$, so that the new points can be assigned to one of the c classes. For the classification problem defined above, a coevolutionary method is proposed to construct separating boundaries of classes in feature subsets on the basis of the training data. The procedure for establishing a classifier system, which produces class boundaries, consists of two phases as follows.

Phase 1. Determine the feature subspaces of hyper-rectangles. In each subspace, a subset of selected features is used.

Phase 2. Search the separation boundaries by evolving ellipsoids in the determined feature subspaces through a hybrid genetic algorithm (GA).

Each phase of the procedure will be explained in detail in the following sections.

2.1 Initial Determination of the Feature Subspace

We propose in this subsection a spatial feature selection method by constructing sub-spaces, in each of which a specific subset of relevant features is to be considered. As a result, each subspace has different dimensions than the others. The spatial feature set constitutes pairs of a feature and its valid interval, which actually represents a hyper-rectangular subspace in the dimensional space. The feature subspaces are initially established so as to include as many positive examples as possible and to exclude negative examples according to its initial default class. A spatial feature set is obtained by iteratively adding a feature and its available interval to the current feature set in a sequential manner, so that the subspace expanded by the added feature can include as many positive examples as possible. In later phases, the feature spaces will be subject to an evolutionary process to maximize the performance of ellipsoids in them. A feature selection method to obtain the hyper-rectangles, which was similar to the initial establishment method of ours, is proposed in [6]. However, their approach differs from ours in the following points. First, their feature selection approach is performed in a backward manner, removing redundant features from the initiated maxi mum hyper-rectangles. It requires much computation cost to search the maximum hyper-rectangles in a large training dataset, while our method does not require the additional search for the maximum hyper-rectangles due to the forward sequential feature selection. Second, a binarization procedure is needed to apply their method to data with continuous features, while ours can be directly employed to the data with continuous features. The proposed method to build the feature subspaces is presented below and also illustrated by Fig. 1.

Step 1. An initial subspace T is constructed by defining the interval in the one-dimensional space of the reference feature v_0. The lower and upper limits of the interval are established on the basis of the nearest negative examples from some consecutive positive examples selected among training data set on the axis v_0.

Step 2. The initial subspace T is split into subspaces, $T_i, I = 1, 2, \ldots$, on the expanded dimensional space of the existing features and the new feature v'. Each T_i is created by defining lower and upper limits of its interval on the new feature v' in the same way of the Step 1.

Step 3. Each of the divided subspaces expands its interval along the existing features to the nearest negative example while keeping the interval of the new feature v'.

Step 4. If the largest one T^* among subspaces expanded from the initial subspace T includes more positive examples than those of the initial subspace T

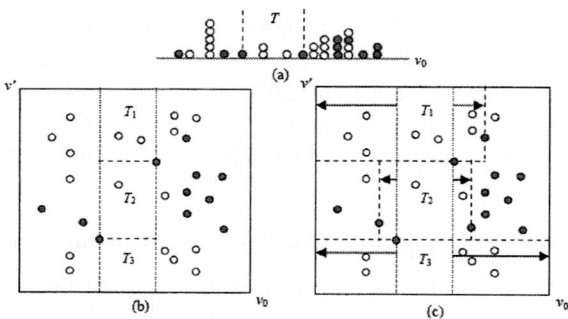

Fig. 1. An example of feature subspaces generation

then repeat Step 2 to Step 3 by considering the largest subspace T^* in Step 3 as an initial subspace T in Step 2. Otherwise the initial subspace T in Step 2 is inserted to the set of final sub-space set G on condition that the initial subspace in Step 2 has at least one positive example that is not included in the set of final subspaces.

Step 5. For every example in the training data set, the above procedure from Step 1 to Step 4 repeated to get the set of final subspaces G.

2.2 Evolution of the Ellipsoidal Classifiers

In every feature subspace, the classifiers represented by ellipsoids adjust their parameters to search the optimal separation boundaries through a hybrid GA. A set of adaptive operations is devised and used for the local search in the hybrid GA.

Classifier Representation With Ellipsoids. Assume that the data subset X_i for class C_i, where $i = 1, \ldots, c$, is covered by several ellipsoidal regions $L_{ij}(j = 1, \ldots)$, where L_{ij} denotes the jth region for class C_i. The ellipsoidal region L_{ij} is defined by two foci, $f_{ij}^{(1)}$ and $f_{ij}^{(2)}$ and a constant, i.e., size factor, D_{ij} as follows:

$$L_{ij} : dist\left(x, f_{ij}^{(1)}\right) + dist\left(x, f_{ij}^{(2)}\right) \leq D_{ij}, \text{where } dist(x,y) = \sqrt{(x-y)^t(x-y)}. \tag{1}$$

For each ellipsoidal region L_{ij}, we define the following classification rule where R_{ij} denotes the label of the jth rule for class C_i.

$$R_{ij} : \text{If } x \text{ is in } L_{ij} \text{ then } x \text{ belongs to class } C_i. \tag{2}$$

Classifier Strength and Determination of Class. For the pattern classification, it is reasonable to assume that the degree of membership of x for classification rule (2) increases as x moves toward the center of the ellipsoid L_{ij}, and decreases as x moves away from the center. To realize this characteristic, the degree of membership of x for a rule R_{ij} is defined as follows.

$$d_{ij}(x) = \frac{D_{ij}}{dist\left(x, f_{ij}^{(1)}\right) + dist\left(x, f_{ij}^{(2)}\right)}. \qquad (3)$$

If the value of $d_{ij}(x)$ in (3) is larger than 1, it indicates that point x is located within the ellipsoid L_{ij}. The value of (3) is less than 1 when x lies out of the boundary of the ellipsoid. Now the degree of membership of x for class C_i, denoted as $d_i(x)$, is given by $d_i(x) = \max_{j=1,...} \{d_{ij}(x)\}$. The class of input x is then determined as class C_{i*} such that $d_{i*}(x)$ is the maximum among $d_i(x), i = 1, \ldots, c$.

Chromosome Representation and Population Initiation. The chromosomes for ellipsoids are represented by strings of a floating-point value in $[0, 1]$, encoding the parameters of ellipsoids. Fig. 2 shows the structure of a chromosome in a three-dimensional feature subspace obtained in subsection 2.1 as an example.

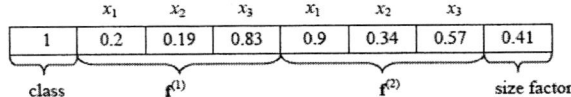

Fig. 2. An example of a chromosome of an ellipsoid in three-dimensional feature subspace

An initial population is generated in such a way that each individual assigned to one of the classes is encoded in terms of two foci, $f^{(1)}$ and $f^{(2)}$, and a size factor D, which are randomly allocated in the pattern space $[0, 1]^n$ to ensure sufficient diversity in the population. For each individual in half of the population, then, one of the foci is seeded with randomly selected training sample point for providing a good starting solution. The number of individuals with a certain class C_i in the population, denoted by $Pop(i)$, is determined in proportion to the number of training data with the same class. Consequently, the size of the population, denoted by Pop_size, is defined as the sum of $Pop(i), i = 1, \ldots, c$.

Fitness Computation. Two measures are considered to evaluate an ellipsoid based on the classification result of the corresponding classifier: generalization ability and classification rate. In order to obtain good generalization ability of an ellipsoid, the region that the ellipsoid covers needs to grow as large as possible. Therefore, when we divide the training data by ellipsoids, the number of data belonging to an ellipsoid should not be too small. For the high classification rate of an ellipsoid, the number of correctly classified data should be large relative to the number of incorrectly classified data among data belonging to the ellipsoid.

Considering the two measures, the fitness value of each ellipsoid is defined as follows, where $fitness(L_{ij})$ is the fitness value of the ellipsoid L_{ij}, $NC(L_{ij})$ is the number of training data that are correctly classified by L_{ij}, $NI(L_{ij})$ is the number of training data that are incorrectly classified by L_{ij}, and $weight(L_{ij})$ is the weight value that multiplies $NI(L_{ij})$.

$$fitness(L_{ij}) = NC(L_{ij}) - weight(L_{ij}) \times NI(L_{ij}). \qquad (4)$$

The weight value for an ellipsoid L_{ij} is used to determine the tradeoff between the generalization ability and the expected classification rate of the ellipsoid on the basis of the ratio of the number of data with same class C_i to the total number of remaining data. Based on the trade-off relation, the weight value of each ellipsoid is calculated as follows, where N_i is the number of data of which class is C_i among remaining training data, N_{remain} is the number of total remaining training data, and $\alpha(\alpha > 1)$ and $\beta(0 < \beta < 1)$ is constant.

$$weight(L_{ij}) = \exp\left\{\alpha \times \left(\frac{N_i}{N_{remain}} - \beta\right)\right\} \qquad (5)$$

Genetic Operations. In order to generate new offspring for class C_i, a pair of individuals with the same class C_i is selected from the current population. Each individual is selected by the following selection probability based on the roulette wheel selection with the linear scaling, where $fitness_{min}(S_i)$ is the minimum fitness value of the individuals in the current set S_i.

$$P(L_{ij}) = \frac{fitness(L_{ij}) - fitness_{min}(S_i)}{\sum_{L_{ik} \in S_i}\{fitness(L_{ik}) - fitness_{min}(S_i)\}}. \qquad (6)$$

From the selected pair of ellipsoids, the arithmetic crossover for randomly taken genes generates two offspring. For an example of the ith genes, a_i and b_i of the selected pair of ellipsoids are replaced by $\lambda a_i + (1-\lambda)b_i$ and $(1-\lambda)a_i + \lambda b_i$ respectively, where $0 < \lambda < 1$. Note that the size factor is determined by a random number drawn from a uniform distribution $U(dist(f^{(1)\prime}, f^{(2)\prime}), 1)$ in order to keep the size of the ellipsoid greater than distance between its two modified foci.

Each parameter of ellipsoids generated by the crossover operation is randomly replaced using a random number from $U(0, 1)$ at a pre-specified mutation probability. As in the crossover operation, the size factor is recomputed with the modified distance between the two altered foci.

Adaptive Operations. The adaptive operations consist of three operations, i.e., expansion, avoidance, and move. The most probable one of the three operations is selected for each ellipsoid based on its fitness value. The ellipsoid with a positive value of fitness is expanded to have a chance to contain more data patterns. If the fitness value of an ellipsoid is less than zero (i.e., an ellipsoid contains at least one misclassified data), the ellipsoid rotates or contracts to avoid the misclassified examples. Finally if an ellipsoid has a zero value of fitness (i.e., an ellipsoid does not contain any training data), the ellipsoid moves to other location in the pattern space. The fitness value of an ellipsoid can be zero even though the ellipsoid contains data from equation (4). However, there is a bare possibility that the number of correctly classified data is same as the number of misclassified data multiplied by weight value because the value of weight in equation (5) is a real number calculated by an exponential function with a real number of parameter. Nevertheless, the overall performance does not take a sudden turn for the worse. The following three subsections describe these operations in detail.

Avoidance. We propose three methods to avoid the misclassified examples considering the locations of the misclassified examples. Fig. 3 illustrates the three methods in two-dimensional pattern space. The first one is to avoid the misclassified examples by rotating the ellipsoid as shown in Fig. 3 (a). The rotation method is selected when the misclassified examples are located in near to boundary of the ellipsoid like the shading region in Fig. 3 (a). We rotate the ellipsoid by moving the focus nearby the misclassified examples in parallel to one variable axis while fixing the other focus and its size factor. The other avoidance method is to shrink the ellipsoid as shown in Fig. 3 (b). The contraction method is applied when the misclassified examples are located in the shading area of Fig. 3 (b). To avoid the misclassified examples, the ellipsoid is shrunk by moving two foci to the opposite directions of one another and its size factor fixed.

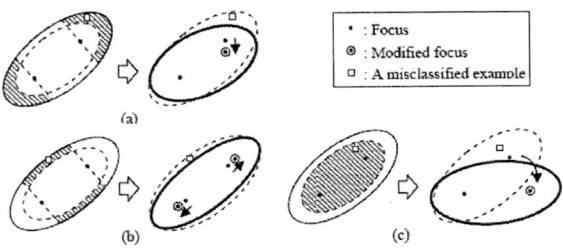

Fig. 3. Three methods of avoidance operation

Another avoidance method is carried out when the misclassified examples are located in the shading part of Figure 3 (c). The misclassified examples are hard to avoid by the previous two methods because they are located around the center of the ellipsoid. Therefore we propose the third method that randomly modifies the location of a randomly selected focus to avoid the misclassified examples around the center of the ellipsoid.

Expansion. An ellipsoid is expanded by one of following two methods (i.e., directed and undirected expansion) as shown in Fig. 4, which assumes the ellipsoid in a two-dimensional pattern space. In the directed expansion as expressed in

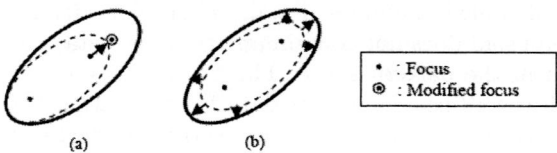

Fig. 4. Two methods of expansion operation

Fig. 4 (a), the ellipsoid extends its area to the opposite direction of a location-fixed focus, which is randomly selected among two foci. The directed expansion method is performed by modifying the locations of the foci and the size factor. In the undirected expansion, the ellipsoid is enlarged toward all directions as depicted in Fig. 4 (b). The undirected expansion of the ellipsoid is carried out by increasing only its size factor.

Move. If the ellipsoid does not contain any examples, it randomly moves to other location in the pattern space. For the fast adaptation of the ellipsoid, the ellipsoid needs to be moved to the area where no ellipsoids exist and training examples are densely distributed. Thus if there exist training examples which are not included in any ellipsoids, we place a focus to randomly selected one among the training examples.

Update of the Population. The proposed hybrid GA procedure applies genetic operations and adaptive operations after population elitist selection [3]. With the population elitist selection, pre-defined *Pop_size* individuals are selected from the current population and a set of the newly generated offspring. This updating method guarantees that the best *Pop_size* individuals seen so far always survived.Adaptive operations.

Termination Test. The termination criterion used in this study is to terminate the iteration of the GA operations and local adaptive operations when either all the training samples are covered by the ellipsoids in the population or the specified maximum number of iterations is exceeded. The final solution obtained by the hybrid GA procedure is not the final population itself but the best ellipsoids in the final population, which cover all the training samples contained by the final population. The selection of the best ellipsoids in the final population for the final output of the algorithm can eliminate the redundant ellipsoids whose removal does not change the recognition capability of the classifier.

2.3 Evolution of the Feature Subspaces

The feature subspaces evolve on the basis of the performance of the regional ellipsoid classifiers they contain in their regions. The evolution procedure for the feature sub-space population is composed of two phases, i.e., creating new individual feature subspaces and updating the current population. A feature subspace is considered to be evolved when the regional agents in the subspace have one of more positive examples included in other feature subspaces. This means that the feature subspace seeks for opportunities to expand to include the corresponding examples by adopting the features of the targeted examples that are not yet included in the subspace.

The newly obtained feature subset is inserted into the feature subspace population if any of its positive examples are not included in other feature subspaces. The insertion of new subspaces activates a deletion process that finds and deletes subspaces that are enclosed by those new subspaces.

3 Experimental Results

We applied the proposed methods to three data sets both to introduce a simple example of ellipsoids adaptation result and to verify the effectiveness of our methods. The data sets are available from the UCI Machine Learning Repository [2]. Each example of the data sets has continuous features. As a preprocessing of the data for our classifier system, the value of each feature is normalized as having the maximum value of one and the minimum value of zero.

Here we present the results achieved by the proposed coevolutionary approach and compare them with the performances of existing well-known classifiers, i.e., a k nearest neighbor [12], a decision tree with C4.5 [10], and a neural network with back-propagation [4]. We tried to guarantee the proper prediction power of the classifiers even under insufficient training data for scarce classes by adopting rather simple structures such as $k = 3$ and 1 hidden layer. Table 1 shows the comparison results of the average classification rate for the three data sets mentioned in the previous subsection. We can see that the proposed hybrid GA classification method with ellipsoidal regions achieved a superior classification rate of test data in comparison with other popular methods such as the k nearest neighbor, the decision tree, and the neural network.

Table 1. Comparison results in terms of classification rate

Classifier	Classification rate(%)	
	Glass	Ionsphere
k Nearest Neighbor ($k = 3$)	63.74	84.90
C4.5	65.70	88.89
Neural Network	63.36	90.03
This study	67.10	90.88

4 Conclusion

This paper proposes a coevolution-based classification method for multidimensional pattern classification problems. The method consists of two layers. Feature sets, pairs of a feature and its range, determine the sub-regions where they apply.

For each sub-region, a pool of ellipsoids is developed to fit the data patterns in the training examples. The ellipsoids are subject to the inner loop of adaptation process whereas the evolution of the feature sets forms the outer loop.

The proposed representation of ellipsoids, whose parameters are two foci and a size factor, has the advantage of interpretability, tractability and robust generalization ability. The GA procedure to fit the ellipsoids to the data patterns is expedited by a few common adaptive operations: expansion, avoidance, and move. The feature-ellipsoid coevolution allows robust performance in problems with a large number of features. The proposed coevolutionary classification method was applied to well-known data sets. The performance results showed the superiority of the proposed method to the existing methods regardless of the number of the features.

References

1. Abe, S., Thawonmas, R.: A fuzzy classifier with ellipsoidal regions. IEEE Trans. on Fuzzy Systems 5 (1997) 358-368
2. Blake, C.L., Merz, C.J.: UCI Repository of machine learning databases. Available online via ⟨http://www.ics.uci.edu/ ∼ mlearn/MLRepository.html⟩ (1998)
3. Eshelman, L.J.: The CHC adaptive search algorithm: How to have safe search when engaging in nontraditional genetic recombination. In: Rawlins, G.J.E.(eds.): Foundations of genetic algorithms, Morgan Kaufman, San Mateo, CA (1991) 265-283
4. Haykin, S.: Neural networks: a comprehensive foundation. Prentice-Hall, New Jersey (1994)
5. Jain, A. K., Duin, R.P.W., Mao, J.: Statistical pattern recognition: a review. IEEE Trans. on Pattern Analysis and Machine Intelligence 22 (2000) 4-37
6. Kudo, M., Shimbo, M.: Feature Selection Based on the Structural Indices of Categories. Pattern Recognition 26 (1993) 891-901
7. Liu, H., Setiono, R.: Incremental feature selection. Applied Intelligence 9 (1998) 217-230
8. Nolan, J.R.: Computer systems that learn: an empirical study of the effect of noise on the performance of three classification methods. Expert Systems with Applications 23 (2002) 39-47
9. Pal, S.K., Bandyopadhyay, S., Murthy, C.A.: Genetic algorithms for generation of class boundaries. IEEE Trans. on Systems, Man, and Cybernetics-Part B: Cybernetics 28 (1998) 816-828
10. Quinlan, J. R.: C4.5: Programs for machine learning. Morgan Kaufmann, San Mateo, CA (1993)
11. Simpson, P.K., Fuzzy min-max neural networks-part1: classification. IEEE Trans. on Neu-ral Networks 3 (1992) 776-786
12. Weiss, S.M., Kulikowski, C.A.: Computer systems that learn. Morgan Kaufmann, San Francisco, CA (1991)

Analytical Modeling of Closed-Loop Conveyors with Load Recirculation

Ying-Jiun Hsieh[1] and Yavuz A. Bozer[2]

[1] Graduate Institute of Technology and Innovation Management,
National Chung Hsing University, Taichung, Taiwan, R.O.C
arborfish@nchu.edu.tw

[2] Department of Industrial and Operations Engineering,
University of Michigan, Ann Arbor, MI, U.S.A
yabozer@umich.edu

Abstract. We present closed form analytical results to show the throughput performance of a discrete-window closed-loop conveyors system serving a user-specified set of stations with intermixed load/unload stations. The buffer capacity at the unloading stations is finite; loads that encounter a full buffer (i.e., blocked loads) are assumed to recirculate around the loop to try again. Given the job flow and routing data as well as the configuration of the conveyor loop, we present an analytical approach to approximate the expected overflow of loads on the conveyor (due to blocked loads). Given the expected overflow, we also show the stability condition for the conveyor system.

Keywords: Material handling; Manufacturing systems; Conveyor systems.

1 Introduction

For the past three decades, flexible manufacturing systems (FMS) and similar automated manufacturing systems that produce discrete units have been widely used in fabrication and machining. Such systems can be described as a set of workstations (machining centers with loading and unloading stations) linked by a material handling system (MHS), generally operating under computer control. In FMS and other medium to large-volume discrete-parts fabrication and machining operations, loop-based conveyors (with discrete-time windows or fixtures) are used quite extensively.

In this study we address a loop-based conveyor system where load recirculation may occur due to finite buffer capacity at unloading stations. A load is blocked when it reaches its destination unloading station and finds a full buffer at that station. A blocked load must recirculate around the loop and try again when it reaches its destination unloading station.

We are interested in estimating the expected load overflow on the conveyor due to blocked loads. Once the expected overflow is estimated, it is straightforward to verify whether or not the conveyor meets throughput (i.e., whether or not the conveyor system is "stable") using our previous results.

2 Basic Definitions and Literature Review

Consider a conveyor loop consisting of M loading and N unloading stations as shown in Figure 1. There is an input queue for each loading station and a finite-capacity output queue for each unloading station. We assume that all input queues are of infinite capacity so that loads arriving from outside are not rejected.

A load enters the system from outside through one of the loading stations and leaves the system through one of the unloading stations. No machining takes place at a loading station. However, at unloading station i there is a processing machine, and the departure rate is assumed to be Λ_i. Before leaving the system each load arriving at an unloading station must be processed through the corresponding processing machine. We also assume that loads arriving through the same loading station may exit the conveyor system at different unloading stations.

Loads arriving at loading station i are assumed to follow an independent Poisson process with rate Λ_i. The processing machine service time at unloading station j is exponentially distributed with rate μ_j. When a load arrives at a loading station, it joins the input queue. The load is then "served" on a First-Come-First-Serve (FCFS) basis. Each load is considered being "served" when it is waiting at the first, or Head-Of-Line (HOL) position, of the input queue [1].

Let P_{ij} denote the probability that a load arriving at loading station i needs to be delivered to unloading station j. (The P_{ij} values must be specified by the user.) Let f_{ij} denote the flow rate from station i to j. By definition, $f_{ij} = \lambda_i P_{ij}$. The conveyor, which provides continuous clock-wise or counterclockwise movement, is assumed to have a specified, constant velocity, length, and number of windows of fixed size; the velocity (V) is expressed in number of windows moved per time unit. The conveyor cycle time, $1/V$, is defined as the time required for the conveyor to move by one window. Each window on the conveyor can hold at most one load.

At the end of each conveyor cycle, the HOL load checks the status of the arriving window. If the arriving window is open, the load is automatically transferred on to that window through a transfer mechanism; otherwise, the (HOL) load waits until the next window arrives and again checks its status. The transfer time for each load is assumed to be constant and less than $1/V$.

After the load is placed on the conveyor, it travels towards its destination on the conveyor loop. When the load reaches its destination unloading station, it checks the status of the output queue. If the output queue is not full, the load is automatically transferred off the conveyor at the end of the conveyor cycle. Otherwise, the load must recirculate on the conveyor and try again later when it reaches its destination unloading station (since no accumulation is allowed on the loop). Once the load finds available space in the output queue, it is automatically

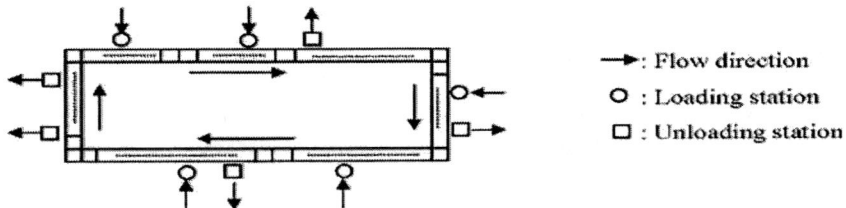

Fig. 1. Conveyor system with five loading and five unloading stations

diverted off the conveyor through a transfer mechanism. The transfer time is assumed to be constant and less than $1/V$. The diverted load joins the output queue and waits for processing, which is performed similarly on a FCFS basis at each unloading station. After the machining process, the load leaves the system. Throughout the paper, the processing machine at each unloading station will be referred to as the "server" at that station.

The analysis of overflow processes due to blocked loads in queueing systems has been the subject of research since Palm [6]. The general focus has been the form of the merge (or superposition) of the two stochastic processes, i.e. the overflow and the original input. Exact solutions are difficult to obtain and, therefore, approximations play an important role in the analysis of the above superposition.

Very few studies have considered conveyor loops. One of them is done by Sonderman [9], who studied the "traffic" behavior on the conveyor with load recirculation. His focus is on a conveyor loop with one loading and one unloading station. The arrivals are assumed to follow a general distribution and the service times at the unloading station are exponentially distributed. Furthermore, no waiting room is available at the unloading station; which is a special case of our study. He approximated the overflow process using an $M/M/1/1$ or $H_2/M/1/1$ loss system. However, due to our different assumptions concerning the conveyor characteristics, we utilized an approach based on conditional probabilities instead. In addition, Sonderman's approach is difficult to extend to the case with multiple and inter-mixed loading and unloading stations.

Pourbabai [8] extended Sonderman's research to include multiple unloading stations but with the "ordered entry" rule. That is, each load on the conveyor, as it travels, searches for an available unloading station in a sequential fashion. Pourbabai applied a similar approach to Soderman called the "two parameter" method to analyze the flow on the conveyor. Pourbabai and Sonderman [7] also studied the same system with "random access" instead of the "ordered entry" rule using a similar approach. The above studies are based on a conveyor loop with only one loading station, which reduces the complexity of the problem.

There have been several studies that address queueing systems with retrials. For example, Greenberg [4] studied an $M/G/1$ system with returning customers and no waiting line. A customer who finds the system full (i.e., the server busy) returns with probability α. The return times for returning customers are assumed to be exponentially distributed, and returning customers are assumed to see time averages. For a user-specified α value, Greenberg derives an upper bound on the server utilization.

Another study with returning customers is presented by Greenberg and Wolff [5], who generalize the results presented in [4]. The authors study an M/M/c/K queue where $K \geq 1$. An arriving customer who finds the system full returns with probability α_1. All returning customers who find the system full, return with probability α_2. It is again assumed that the return times are exponentially distributed and that returning customers see time averages. For user-specified α_1 and α_2 values, the authors derive an approximation for P_j, the proportion of time there are j customers $(j = 1, 2, \ldots, K)$ in the system, and present an upper bound on the server utilization.

An unloading station in the conveyor system we study can be modeled by setting $\alpha_1 = \alpha_2 = 1$ in [5], i.e., all blocked loads return to the unloading station for another attempt; no loads are lost. However, the return times on the conveyor are not exponentially distributed; in fact, the return times are constant since no accumulation or other delays occur on the conveyor loop. Furthermore, if no customers are lost, the effective arrival rate is equal to the original arrival rate and the server utilization is known. Instead of the server utilization, we are interested in the number of attempts a load makes before it is successfully unloaded at a particular station. That is, we are interested in the probability of blocking at an unloading station and the overflow it creates on the conveyor loop. We found that the P_K value obtained from [5] does not provide a good estimate for the probability of blocking. We believe this is due to non-exponential return times on the conveyor loop, and the fact that the conveyor speed (i.e., the mean time to return) is not captured in the approximation presented in [5].

Fredericks and Reisner [3] also developed an approximation that explicitly depends on the return rate. However, they studied a system with only one loading and one unloading station, and their approach is not applicable to our problem.

Hence, most of the existing studies represent a special case of our problem since they assume one loading station and one unloading station. Although a few studies considered the case with multiple unloading stations, their approach and results cannot be applied or generalized to our case. In next section, we will derive an approximation for the expected overflow and check the stability condition for the conveyor system.

3 The Expected Overflow and System Stability

The notation in this section is based on [1] except for those instances that relate to the derivation of the expected overflow. As an extension to [1], the total flow in each conveyor segment can be defined as the sum of the "base flow" (the flow without any load recirculation) and the overflow (the flow created by blocked loads.)

As shown in Figure 2, let Δ_i^{base} denote the base flow rate in conveyor segment S_i. By definition, $\Delta_i^{base} = \sum_k \sum_l f_{kl} x_{kl}^i$, where f_{kl} is the flow rate from loading station k to unloading station l (shown earlier), $x_{kl}^i = 1$ if the load from station k to l travels through segment S_i, zero otherwise. Note that x_{kl}^i is *not* a decision variable. Rather, it is simply obtained by determining whether or not the flow from station k to l goes through segment S_i.

Likewise, let $\Delta^{overflow}$ denote the expected total overflow rate on the conveyor. (Note that the expected total overflow rate will be the same in each segment of the conveyor.) The expected total flow and the window utilization in each conveyor segment S_i are given by,

$$\Delta_i = \Delta_i^{base} + \Delta^{overflow}. \tag{1}$$

and

$$q_i = \Delta_i/V. \tag{2}$$

In steady-state, provided the conveyor meets throughput, Δ_i^{base} can be directly and easily derived(shown earlier) from the load flow matrix. However, $\Delta^{overflow}$ is the sum of the expected unsuccessful departure rates at the unloading stations. In other words, if Ω denotes the set of the unloading stations, we have

$$\Delta^{overflow} = \sum_{i \in \Omega} \widetilde{\Lambda}_i. \tag{3}$$

where $\widetilde{\Lambda}_i$ is defined as the expected unsuccessful departure rate at unloading station i. Also, by definition, the probability of blocking at unloading station i, $P(B_i)$, is equal to $\widetilde{\Lambda}_i/(\Lambda_i + \widetilde{\Lambda}_i)$; or equivalently, $\widetilde{\Lambda}_i$ can be expressed in terms of $P(B_i)$ as follows:

$$\widetilde{\Lambda}_i = \Lambda_i P(B_i)/[1 - P(B_i)]. \tag{4}$$

When a window is arrives at unloading station i, let γ_i denote the conditional probability that a departure attempt will occur given that the arriving window is occupied. By definition, γ_i can be expressed as follows:

$$\gamma_i = (\Lambda_i + \widetilde{\Lambda}_i)/(\Delta_i^{base} + \Delta^{overflow}). \tag{5}$$

We use the "tagged load" approach to derive our model; we first treat the simplest case where no waiting room is available at the unloading stations. Consider unloading station i and the corresponding conveyor segment S_i shown in Figure 2.

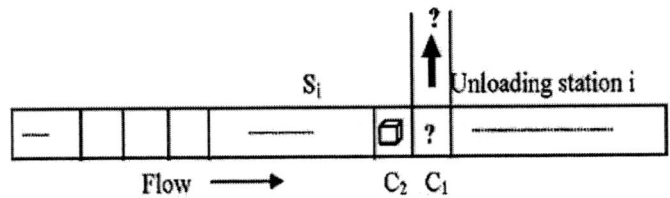

Fig. 2. Conveyor segment S_i and corresponding unloading station i

The conveyor speed is given as V windows per time unit. Suppose the tagged load is in window C_2 and it is destined to unloading station i. When the tagged load arrives at unloading station i and inspects it, by conditioning on the status of unloading station i when the preceding window (window C_1) inspected it one conveyor cycle ago, we are able to derive the probability of blocking for the tagged load. Note that the preceding window (window C_1) may or may not be full when it arrives at unloading station i.

Conditioning on $\beta_i(\bar{\beta}_i)$ which represents the event that, upon inspection, window C_1 observes an idle (busy) server at unloading station i, we have:

$$P(B_i) = P(B_i|\beta_i)P(\beta_i) + P(B_i|\bar{\beta}_i)P(\bar{\beta}_i). \tag{6}$$

The first conditional expression in Equation 6, $P(B_i|\beta_i)P(\beta_i)$, represents the conditional probability of blocking for the tagged load given that window C_1 observed an idle server. In other words, $P(B_i|\beta_i)P(\beta_i)$ is the probability that there is a load in window C_1 which is destined to station i and that specific load is still being served after one conveyor cycle time, i.e., $1/V$; plus the probability that there is no load in window C_1 which is destined to station i, but window C_2 finds a busy server upon arrival. Note that the second case occurs with zero probability. Now let $X_i \sim F_i$ be the service time (exponentially distributed with rate μ_i of the server at unloading station i. Furthermore, let p_i denote the probability that a departure attempt occurs when window C_2 arrives at unloading station i. We have,

$$P(B_i|\beta_i) = p_i[1 - F_i(1/V)]. \tag{7}$$

The second conditional expression in Equation 6, $P(B_i|\bar{\beta}_i)$, represents the conditional probability of blocking for the tagged load given that window C_1 observed a busy server. In other words, regardless of the content of window C_1, $P(B_i|\bar{\beta}_i)$ is the probability that the residual service time of the load in service is greater than the conveyor cycle time $1/V$. Since the residual service time is also exponentially distributed [5], we have:

$$P(B_i|\bar{\beta}_i) = [1 - F_i(1/V)]. \tag{8}$$

Assuming that window C_1 and C_2 have the same probability of observing a busy server upon arrival at unloading station i, we have $P(\bar{\beta}_i) = P(B_i)$ and $P(\beta_i) = 1 - P(B_i)$. Under this assumption, we substitute Equations 7 and 8 into 6 and rewrite the expression for $P(B_i)$ as follows:

$$P(B_i) = p_i[1 - F_i(1/V)](1 - P(B_i)) + [1 - F_i(1/V)]P(B_i). \tag{9}$$

Equation 9 can be further re-arranged as:

$$p_i = F_i(1/V)P(B_i)/\{[1 - F_i(1/V)][1 - P(B_i)]\}. \tag{10}$$

Recall that $P(B_i) = \widetilde{\Lambda}_i/(\Lambda_i + \widetilde{\Lambda}_i)$ Therefore, we can rewrite the right-hand side of Equation 10 and obtain p_i as follows:

$$p_i = F_i(1/V)\widetilde{\Lambda}_i/\{[1 - F_i(1/V)]\Lambda_i\}. \tag{11}$$

Momentarily setting Equation 11 aside, we now derive another expression for p_i. Let E_{1f} (E_{2f}) denote the event that window C_1 (C_2) is occupied. Furthermore, let $E_{1 \to i}$ ($E_{2 \to i}$) represent the event that there is a load in window C_1 (C_2) which is destined to unloading station i. By definition, we have

$$p_i = P(E_{1f} \text{ and } E_{1 \to i} | E_{2f} \text{ and } E_{2 \to i})$$
$$= P(E_{1f}| E_{2f} \text{ and } E_{2 \to i}) \, P(E_{1 \to i}| E_{2f}, E_{2 \to i}, \text{ and } E_{1f}).$$

Assuming that event E_{1f} is independent of events E_{2f} and $E_{2 \to i}$, we have:

$$P(E_{1f}|E_{2f} and E_{2 \to i}) = P(E_{1f}) = q_i. \tag{12}$$

Furthermore, assuming that event $E_{1 \to i}$ is independent of events E_{2f} and $E_{2 \to i}$, we have:

$$P(E_{1 \to i}|E_{2f}, E_{2 \to i}, and E_{1f}) = P(E_{1 \to i}|E_{1f}) = \gamma_i. \tag{13}$$

We substitute Equations 12 and 13 into 11 to obtain

$$p_i = q_i \gamma_i. \tag{14}$$

Substituting Equations 1, 2, and 5 into Equation 14, p_i can be expressed as a function of $\tilde{\Lambda}_i$. That is,

$$p_i = (\Lambda_i + \tilde{\Lambda}_i)/V. \tag{15}$$

Now $\tilde{\Lambda}_i$ can be solved through Equations 11 and 15; that is,

$$\tilde{\Lambda}_i = [1 - F_i(1/V)]\Lambda_i^2 / \{V F_i(1/V) - [1 - F_i(1/V)]\Lambda_i\}. \tag{16}$$

Let us define $\omega \equiv \tilde{\Lambda}_i / \Lambda_i$ as the *overflow ratio* at unloading i. Along with the definition that $P(B_i) = \tilde{\Lambda}_i/(\Lambda_i + \tilde{\Lambda}_i)$, $P(B_i)$ can be obtained from Equation 16 as:

$$P(B_i) = \omega_i/(1 + \omega_i). \tag{17}$$

We next derive $P(B_i)$ and $\tilde{\Lambda}_i$ for the more general case where the buffer capacity (including the load in service) at unloading station i, denoted as k, can be any positive integer.

Let $N_{i,j}^1$ ($N_{i,j}^2$) represent the event that, upon inspection, window $C_1(C_2)$ finds j loads present at unloading station i. The term $P(N_{i,j}^2)$ is obtained by conditioning on the possible events that can occur when window C_1 inspects unloading station i. In other words,

$$P(N_{i,j}^2) = \sum_{n=0}^{k} P(N_{i,j}^2|N_{i,n}^1) P(N_{i,n}^1). \tag{18}$$

There exist five possible events encountered by window C_1 upon arrival at unloading station i, if window C_2 observes j loads upon arrival.

Event 1. Since it is impossible for window C_2 to find two or more loads than what window C_1 found, we have:

$$P(N_{i,j}^2|N_{i,n}^1) = 0, \text{ if } j > n + 1. \tag{19}$$

Before discussing Events 2, 3, 4, and 5, we first present the following definitions: Let D_i represent the event that there is a load on window C_1 which is destined to unloading station i. Furthermore, let $Q_{i,n}$ denote the event that n load(s) are served within a conveyor cycle given that sufficient loads are available at unloading station i. We now discuss the following events:

Event 2. If window C_2 observes exactly one more load than window C_1, there must be a load in window C_1 which is destined to unloading station i. Moreover, no loads should finish service before window C_2 arrives. In other words,

$$P(N^2_{i,j}|N^1_{i,n}) = P(D_i|N^1_{i,n})P(Q_{i,0}), \text{ if } j = n+1. \tag{20}$$

Event 3. If window C_2 finds exactly the same number of loads (greater than zero but less than k loads) as window C_1, we have:

$$P(N^2_{i,j}|N^1_{i,n}) = P(D_i|N^1_{i,n})P(Q_{i,1}) + (1 - P(D_i|N^1_{i,n}))P(Q_{i,0}), \tag{21}$$

if $j = n$ and $0 < n < k$.

The first expression on the right-hand side of Equation 21, $P(D_i|N^1_{i,n})P(Q_{i,1})$ represents the probability that a departure attempt occurred as window C_1 arrived at unloading station i where n loads were observed; and exactly one load was served before window C_2 arrived. The second expression on the right-hand side of Equation 21,$(1 - P(D_i|N^1_{i,n}))P(Q_{i,0})$, represents the probability that no departure attempt occurred as window C_1 arrived at unloading station i where n loads were observed; and no loads were served before window C_2 arrived.

Event 4. If window C_2 finds a smaller number of loads than window C_1, and window C_1 does not find a full queue upon arrival at unloading station i, we have:

$$P(N^2_{i,j}|N^1_{i,n}) = P(D_i|N^1_{i,n})P(Q_{i,n-j+1}) + (1 - P(D_i|N^1_{i,n}))P(Q_{i,n-j}), \tag{22}$$

if $j < n < k$.

The first expression on the right-hand side of Equation 22, $P(D_i|N^1_{i,n})P(Q_{i,n-j+1})$,represents the probability that a departure attempt occurred as window C_1 arrived at unloading station i where n loads were observed; and exactly $(n - j + 1)$ loads were served before window C_2 arrived. The second expression on the right-hand side of Equation 22,$(1 - P(D_i|N^1_{i,n}))P(Q_{i,n-j})$, represents the probability that no departure attempt occurred as window C_1 arrived at unloading station i where n loads were observed; and $(n - j)$ loads were served before window C_2 arrived.

Event 5. Last, if window C_1 finds a full queue upon arrival at unloading station i, there must be exactly $(k - j)$ loads served before window C_2 arrived. In other words,

$$P(N^2_{i,j}|N^1_{i,n}) = P(Q_{i,n-j}), \text{ if } n = k. \tag{23}$$

Since we assumed that the process times of the server at unloading station i follows an independent and identical exponential distribution with mean $1/\mu_i$, the residual service times are exponentially distributed as well. Given that sufficient loads are available at unloading station i, the number of loads which are served within a conveyor cycle time, i.e., $1/V$, follows a Poisson distribution with mean μ_i/V. That is,

$$P(Q_{i,n}) = e^{-\mu_i/V}(\mu_i/V)^n/n!, n \geq 0. \tag{24}$$

If we assume that whether a departure attempt occurs or not does not depend on the number of loads observed at the corresponding unloading station, we have $P(D_i|N^1_{i,n}) = P(D_i)$. Note that $P(D_i) = p_i$ by definition. Therefore, from Equation 15 we have:

$$P(D_i|N^1_{i,n}) = (\Lambda_i + \widetilde{\Lambda}_i)/V, 0 \leq n \leq k. \tag{25}$$

Similar to the case of $k = 1$, we assume that windows C_1 and C_2 have the same probability of observing j loads at unloading station i upon arrival, i.e., $P(N^2_{i,j}) = P(N^1_{i,j})$. To obtain $P(B_i) = P(N^2_{i,k})$, the probability of blocking experienced by the tagged load when attempting to depart at unloading station i, we first substitute Equation 4 into 25, and then substitute Equations 19-25 into 18. The above substitutions yield a non-linear equation with k+1 unknowns: $P(N^2_{i,n}), n < k$. Note that the non-linearity is a consequence of the substitution of Equation 4 into 25. In a similar manner, the unknowns $P(N^2_{i,n}), n < k$, can be obtained through the same substitutions. Note that we also have the following boundary condition:

$$\sum_{n=0}^{k} P(N^2_{i,n}) = 1. \tag{26}$$

The term $P(B_i)$ can now be determined by solving the above set of non-linear system of equations. Solving such a set is not straightforward. Therefore, we propose an iterative method to compute $P(B_i)$ numerically.

Results we obtained earlier in [2] suggest that the status (empty or full) of two adjacent windows on the conveyor are correlated. Hence, we relax the assumption we made in Equation 12 and assume that the status of adjacent windows are correlated; we also utilize previous results obtained in [2] to estimate $\widetilde{a}_i$, the probability of having two adjacent full windows in conveyor segment S_i. In other words, we will replace the term q_i with $\widetilde{a}_i/q_i$ in Equation 12 to obtain a more accurate estimate of $P(E_{1f}|E_{2f} \text{ and } E_{2\to i})$. The iterative scheme (Algorithm 1) to compute $P(B_i)$ is presented as follows:

Algorithm 1:

Step 1:
Set the initial value of $P(B_i) = 0$, i.e., let $n = 0$, and $\widetilde{\Lambda}_i^{(n)} = 0 \ \forall \ i \in \Omega$.
Step 2:
Let $q_i^{(n)} = (\Delta_i^{base} + \sum_{j \in \Omega} \widetilde{\Lambda}_j^{(n)})/V, \forall i \in \{\theta, \Omega\}$,

$\gamma_i^{(n)} = (\Lambda_i + \widetilde{\Lambda}_i^{(n)})/(\Delta_i^{base} + \sum_{j \in \Omega} \widetilde{\Lambda}_j^{(n)})$, $\forall i \in \Omega$, and
$\bar{\rho}_i^{(n)} = \lambda_i/[V - (\Delta_i^{base} + \sum_{j \in \Omega} \tilde{\lambda}_j^{(n)})]$, $\forall i \in \theta$. Compute $\tilde{a}_i^{(n)}$, the probability of two adjacent full windows for each conveyor segment S_i, by solving the linear system of equations given by

$$\tilde{a}_{i+1}^{(n)} = \tilde{a}_i^{(n)} + (q_i^{(n)} - \tilde{a}_i^{(n)}) \left\{ \bar{\rho}_i^{(n)} + [F_{\lambda_i}(1/V)](1 - \bar{\rho}_i^{(n)}) \right\} + (q_i^{(n)} - \tilde{a}_i^{(n)})\bar{\rho}_i^{(n)} +$$
$$(\tilde{a}_i^{(n)} - 2q_i^{(n)} + 1) \left\{ \bar{\rho}_i^{(n)} - [F_{\lambda_i}(1/V)](1 - \bar{\rho}_i^{(n)}) \right\}, \forall i \in \theta; \text{ and } \tilde{a}_{i+1}^{(n)} = \tilde{a}_i^{(n)}(1 - \gamma_i^{(n)})^2, \forall i \in \Omega.$$

Step 3:
Compute the probability that a departure attempt occurs as the window preceding the tagged load arrives at unloading station i, i.e., $p_i^{(n)} = \tilde{a}_i^{(n)} \gamma_i^{(n)}/q_i^{(n)}$, $\forall i \in \Omega$. Use $P_i^{(n)}$ to solve for $P^{(n+1)}(N_{i,k}^2) \forall i \in \Omega$, i.e., $P(B_i)$ at the nth iteration, from Equation 18.

Step 4:
Update $\widetilde{\Lambda}_i^{(n+1)}$ with $P^{(n+1)}(N_{i,k}^2)\Lambda_i/[1 - P^{(n+1)}(N_{i,k}^2)]$, $\forall i \in \Omega$.

Step 5:
If $\left|\widetilde{\Lambda}_i^{(n)} - \widetilde{\Lambda}_i^{(n+1)}\right| < \epsilon, \forall i \in \Omega$ then
$P^{(n+1)}(N_{i,k}^2)$ obtained in *Step 3* is the probability of blocking for a
departure attempt at unloading station i, i.e., $P(B_i) = P^{(n+1)}(N_{i,k}^2)$, stop;
else
set $n = n + 1$, and go to *Step 2*.

Proving that Algorithm 1 will always return a (unique) solution for all k values is not straightforward. In order to test the conditions under which the above algorithm converges, we randomly generated the problems with various flow data. From these problems we observed empirically that the above algorithm always converges if the utilization of the processing machine at each unloading station is less than 1. We also observed that the above algorithm, if it converges, always returns the same solution for $P(B_i)$, regardless of the initial values selected for $P(B_i)$.

Once we compute the approximate values of expected load overflow $\widetilde{\Lambda}_i, \forall i \in \Omega$, we can substitute them into Equations 3 and 1 to obtain the expected total flow, Δ_i, in each conveyor segment S_i. Given the Δ_i value for each segment S_i, whether the conveyor system is stable or not can be easily verified through the stability condition we derived earlier in [1].

Furthermore, various first-moment performance measures may be approximated. For example, as a window arrives at unloading station i, we can obtain the observed expected WIP which is equal to $\sum_{n=0}^{k} nP(N_{i,n}^2)$.

4 Numerical Results and Conclusions

In order to test the performance of the analytical models, we simulated the conveyor loop with four loading and four unloading stations. In the simulation model, loads arrive at the loading stations in a Poisson fashion. The process times at each unloading stations are exponentially distributed with server utilization equal to 75%.

For various buffer size K, we examined three different conveyor speeds, i.e., slow, medium, and fast. We measured the probability of blocking experienced by the departure attempts, i.e., $P(B)$, at the unloading station. The results showed that the percentage differences between simulated $P(B)$ and calculated $P(B)$ values are generally within 2%. When we increase the conveyor speed or reduce the buffer capacity at the unloading stations, the expected overflow and the probability of blocking increase. Moreover, the change in expected overflow is less sensitive to changes in conveyor speed than changes in the buffer capacity at the unloading stations.

Similar to the conveyor loop without load recirculation we studied in [2], we observed a positive correlation between adjacent windows in each conveyor segment. Since the above correlation decreases as the conveyor speed is increased, the model gives more accurate estimates at higher conveyor speeds.

References

1. Bozer, Y.A., Hsieh, Y.J.: Throughput Performance Analysis and Machine Layout for Discrete-Space Closed-Loop Conveyors. IIE Transactions. **37**(2005) 77-89
2. Bozer, Y.A., Hsieh, Y.J.: Expected Waiting Times at Loading Stations in Discrete-space Closed-Loop Conveyors. European Journal of Operation Research. **155** (2004) 516-532
3. Fredericks, A.A., Reisner, G.A.: Approximation to Stochastic Service Systems with an Application to a Retrial Model. The Bell System Technical Journal. **58** (1979) 557-576
4. Greenberg, B.S.: M/G/1 Queueing Systems with Returning Customers. Journal of Applied Probability. **26**(1) (1989) 152-163
5. Greenberg, B.S., Wolff, R.W.: An Upper Bound on the Performance of Queues with Returning Customers. Journal of Applied Probability. **24**(2)(1987) 466-475
6. Palm, C.: Intensitatsschwankungen im Fernsprecheverkehr. Ericsson Technics. **44** (1943) 1-8
7. Pourbabai, B., Sonderman, D.: A Stochastic Recirculation System with Random Access. European Journal of Operational Research. **21** (1985) 367-378
8. Pourbabai, B.: Markovian Queueing Systems with Retrials and Heterogeneous Servers. Computational Mathematics Application. **13** (1987) 917-923
9. Sonderman, D.: An Analytical Model for Recirculating Conveyors with Stochastic Inputs and Outputs. International Journal of Production Research. **20** (1982) 591-605

A Multi-items Ordering Model with Mixed Parts Transportation Problem in a Supply Chain

Beumjun Ahn[1], and Kwang-Kyu Seo[1, *]

[1] Department of Industrial Information and Systems Engineering,
Sangmyung University, San 98-20, Anso-Dong,
Chonan, Chungnam 330-720, Korea
`{bjahn, kwangkyu}@smu.ac.kr`

Abstract. This paper explores the problem of making ordering decisions for multi-items with various constraints on the supply chain. With the advent of supply chain management, depots or warehouses fulfill a strategic role of achieving the logistics objectives of shorter cycle times, lower inventories, lower costs and better customer service. Many companies consider both their cost effectiveness and market proficiency to depend primarily on efficient logistics management. Depot management presently is considered a key to strengthening company logistics. In this paper, we propose a novel multi-items ordering model with the mixed parts transportation problem based on the depot system in a supply chain. Order range is especially introduced and used as decision parameters instead of order point to order multi-items. Finally, we test the model with a numerical example and show computational results that verify the effectiveness of the proposed model.

1 Introduction

To attain the customer service objectives in the overall supply chain, warehouses serve several value adding roles, which include transportation consolidation, product mixing, customer service, contingency protection and smoothing. In the past, depots or warehouses mainly focused on putting raw materials, in-process products and finished goods in storage. With the arrival of supply chain management, depots or warehouses serve a strategic role of achieving the logistics objectives of reduced cycle times, inventories and costs, and increased customer service levels [1].

Many companies consider both their cost effectiveness and market proficiency to depend primarily on efficient logistics management. Depot or warehouse management currently is considered a key to strengthening company logistics. Logistics activities mainly consist of transportation, inventory management, order fulfillment and warehousing. These activities establish an essential connection among suppliers, manufactures, distributors and customers in supply chains. Depots occupy an important position such as storing items, and retrieving items from storage to fulfill customer orders

---
* Corresponding author.

in the supply chain. In order to satisfy customer demands such as shorter order cycle times, products may stay in depot for just a few days or even a few hours. The depot expenditures which companies sustain involve considerable dollars amounts, hurrying the movement of parts in depots, therefore, has continued to become an essential issue for depot managers.

The parts transportation problem from a supplier to a manufacturer has been one of the major issues on the supply chain management [2]. In this paper, we propose a new ordering model for multi-items based on the depot system in a supply chain. The depot system is being utilized to adjust the inventory and the timely supply of the parts demanded by a manufacturer.

There have been several models about the multi-items ordering system. Shu proposed an ordering model for multi-items firstly [3]. He determined the optimal ordering cycle with respect to the rate of demands for two items. Nocturne formulated a model for joint replenishment of two items with his proposed original logic [4]. Doll and Whybark determined near optimal frequencies of production and the associated cycle time for production schedules [5]. They determined optimal ordering cycle as decision parameters without considering transportation condition. Tersine and Barman constructed a model considering transportation condition (freight rate) to determine ordering size in a deterministic EOQ (economic order quantity) system [6]. Buffa and Munn proposed an algorithm that could formulate an inbound consolidation strategy [7]. The model is available to the case of replenishing multi-items with some groups and the cost of transportation and inventory being affected by the strategy.

But these models did not simultaneously consider the transportation condition and multi-items. Therefore, we propose a new ordering model on the fixed order method with order range in order to overcome the mentioned shortcomings. The proposed model considers both the multi-items and mixed parts transportation condition in a supply chain.

2 Model Formulation

2.1 Notation

The following notations are used in the formulation of the proposed model.

k : time periods on transportation lead time ($k = 1, 2, ..., L1$)

D_t^i : demands for item i in period t

S^i : safety inventory level for item i

C_0^i : holding cost in the depot for item i

C_1 : transportation cost

Z^i : container size for item i

$F_{\min}, F_{\max}$: minimum and maximum available transportation size considering the mixed parts transportation and freight rate

W^i : value of minimum of order range for item i (this value is equal to order point in order point method)

I_t^i : inventory in the depot for item i in period t

Y_t : variable of transportation in period t

$Q_{t:tL1+1}^i$: transportation quantity ordered in period t for item i, this quantity is transported at the beginning of period $t + L1 + 1$

V^i : value of maximum in order range for item i (decision variable)

G^i: set of transportation quantity for item i, $G^i = \{g^i \mid g^i = \left\lceil (\sum_{t=1}^{T} D_t^i)/t \right\rceil, t \in H\}$

$\lceil X \rceil$: minimum integer larger than X

2.2 Concept of the Proposed Model

In the fixed order point method, an order is placed when the inventory on hand reaches a predetermined inventory level, to satisfy demand during the order cycle. The economic order quantity will be ordered whenever demand drops the inventory level to the reorder point.

This method is based on the following assumptions:

(1) A continuous, constant and known rate of demand.
(2) A constant and known replenishment or lead-time.
(3) Only one item in inventory.
(4) An infinite planning horizon, etc.

It is difficult to consider the impact of both multi-items and mixed parts transportation condition in the typical fixed order method. For this reason, the proposed model in this paper introduces and uses order range as decision parameters instead of order point.

Therefore, the order policy is changed as follows. We show the order policy of the proposed model for two items such as a and b.

Case 1: When the inventory level of a and b is within the order range of a and b, and the sum of transportation order quantities of a and b is in the limits of the available transportation quantity, order a and b. Otherwise do not order a and b.

Case 2: When the inventory level of a and b is greater than the upper order range of a and b, do not order a and b.

Case 3: When the inventory level of a is greater than the order range of a and that of b is within the order range of b, do not order a. If the transportation order quantity of b is in the limits of the available transportation quantity, then order b. Otherwise do not order b.

Case 4: When the inventory level of a is greater than the order range of a and that of b is less than the order range of b, do not order a but b.

Case 5: When the inventory level of a is within the order range of a and that of b is less than the order range of b and the transportation order quantity of a is in the limits of the available transportation quantity, order a. Otherwise do not order a but b.

Case 6: When the inventory level of a is less than the order range of a and that of less than the order range of b, order a and b.

Figure 1 shows the behavior of inventory level in the proposed model.

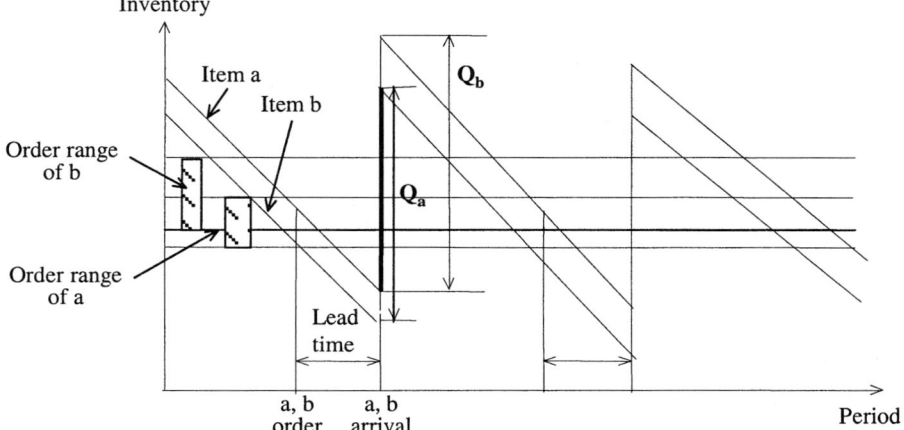

Fig. 1. Behavior of inventory level in the proposed model

In figure 1, the transportation order for the item a and b is placed on the order range predetermined respectively. The transportation quantity of the item a and b is Q_a and Q_b respectively. The decision factors of proposed model are determination of the order range and transportation order quantity.

The model is formulated to determine the maximum point of order range and transportation order quantity in such a way that total cost is minimized. The minimum point of order range is determined by adapting the order point in the typical fixed order method.

2.3 Assumption

The material and information flows in the proposed model are shown in figure 2. We focus on determining the best strategy for multi-items ordering from the supplier to the depot and from the depot to the manufacturer. We assume that transportation leadtime is L1 and demands are supplied to assembly line from the depot timely. Supplier produces M types of items and let $i \in \{1, 2, \cdots, M\}$ be index of items. Let $t \in \{1, 2, \cdots, T\}$ be an index of the time periods with the condition that the planning horizon starts at the beginning of period 1 and the end of period T. The set of time period is H.

The assumptions used in the paper are as follows.
(1) Demands for depot are deterministic process and suggested by assembly line.
(2) Inventory quantities in supplier's inventory point are always sufficient.
(3) During the planning horizon, inventory cost and transportation cost are constant.

(4) A minimum and maximum shipping quantity on the transportation is predetermined.
(5) Transportation for one period occurs once or not.

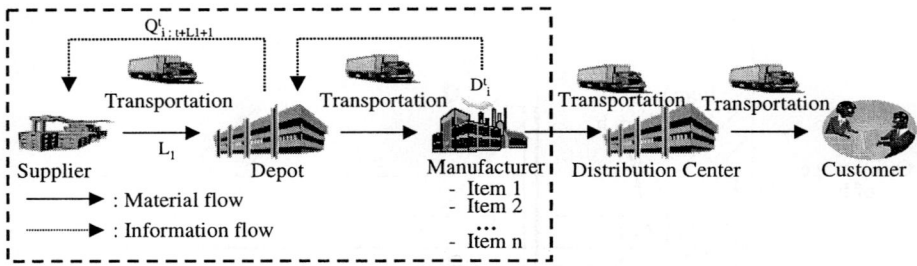

Fig. 2. Material and information flows of the proposed model in a supply chain

2.4 Formulation

We formulate the equations for the proposed model as follows.

(1) The inventory quantity in the depot is:

$$I_t^i = I_{t-1}^i + Y_{t-L1-1} \times Q_{t-L1-1:t}^i - D_t \quad (t=1,2,\ldots,T; i=1,2,\ldots,M) \tag{1}$$

I_0^i : initial inventory level for items i in the depot

$Y_{-L1} \times Q_{-L1:1}^i, \cdots, Y_0 \times Q_{0:1+L1}^i$: preordered transportation quantity

(2) The transportation order quantity is:

$$Q_{t:t+L1+1}^i = \begin{cases} 0 & \text{if } I_t^i + \sum_{k=1}^{L1} Y_{t-k} \times Q_{t-k:t+L1+1-k}^i > V^i \\ g^i & \text{if } I_t^i + \sum_{k=1}^{L1} Y_{t-k} \times Q_{t-k:t+L1+1-k}^i \leq V^i \end{cases} \tag{2}$$

$(t=1,2,\ldots,T; i=1,2,\ldots,M)$

(3) The criterion of transportation is:

$$Y_t = \begin{cases} 1 & \text{if } F_{\min} \leq \sum_{i=1}^{M} Z^i \times Q_{t:t+L1+1}^i \leq F_{\max} \quad \text{or} \\ & \exists i \in \{1,2,\cdots,M\}, I_t^i + \sum_{k=1}^{L1} Y_{t-k} \times Q_{t-k:t+L1+1-k}^i < W^i \\ 0 & \text{otherwise} \end{cases} \tag{3}$$

$(t=1,2,\ldots,T)$

(4) The condition for next planning horizon is:

$$\sum_{t=1}^{T} Y_t \times Q^i_{t:t+L1+1} \geq \sum_{t=1}^{T} D^i_t \qquad (i=1,2,\ldots,M) \qquad (4)$$

(5) The condition of safety inventory quantity is:

$$I^i_t \geq S^i \qquad (t=1,2,\ldots,T;\ i=1,2,\ldots,M) \qquad (5)$$

(6) The set of transportation order quantity in our model is:

$$G^i = \{g^i \mid g^i = \left\lceil (\sum_{t=1}^{T} D^i_t)/t \right\rceil, g^i \leq F_{max}/Z^i, t \in H\} \qquad (6)$$

$$(i=1,2,\ldots,M)$$

(7) The range of maximum of order range is:

$$V^i_{min} \leq V^i \leq V^i_{max} \qquad (i=1,2,\ldots,M) \qquad (7)$$

(8) The minimum of maximum of order range is:

$$V^i_{min} = \left\lceil L1 \times \overline{D^i} + \alpha \times L1 \times \delta^i_D \right\rceil \qquad (i=1,2,\ldots,M) \qquad (8)$$

(9) The maximum of maximum of order range is:

$$V^i_{max} = g^i_{max} + S^i \qquad (i=1,2,\ldots,M) \qquad (9)$$

(10) The minimum of transportation order quantity is:

$$g^i_{min} \leftarrow \min G^i \qquad (i=1,2,\ldots,M) \qquad (10)$$

(11) The maximum of transportation order quantity is:

$$g^i_{max} \leftarrow \max G^i \qquad (i=1,2,\ldots,M) \qquad (11)$$

where,

$\overline{D^i}$: average demands of item i
$L1$: transportation lead time
α : safety coefficient ($\alpha=1.65$ in this model)
δ^i_D : standard deviation of demands of item i

From the above formulation, we define an optimization equation as follows. This equation means the minimization of the sum of holding cost in the depot, and transportation cost between depot and supplier.

Minimize

$$TC_T = \sum_{t=1}^{T} \sum_{i=1}^{M} C^i_0 \times I^i_t + C_1 \times \sum_{t=1}^{T} Y_t \qquad (12)$$

subject to constraints (1) - (7)

The computational procedures to determine the transportation order quantities and order ranges are shown as the flow chart in figure 3.

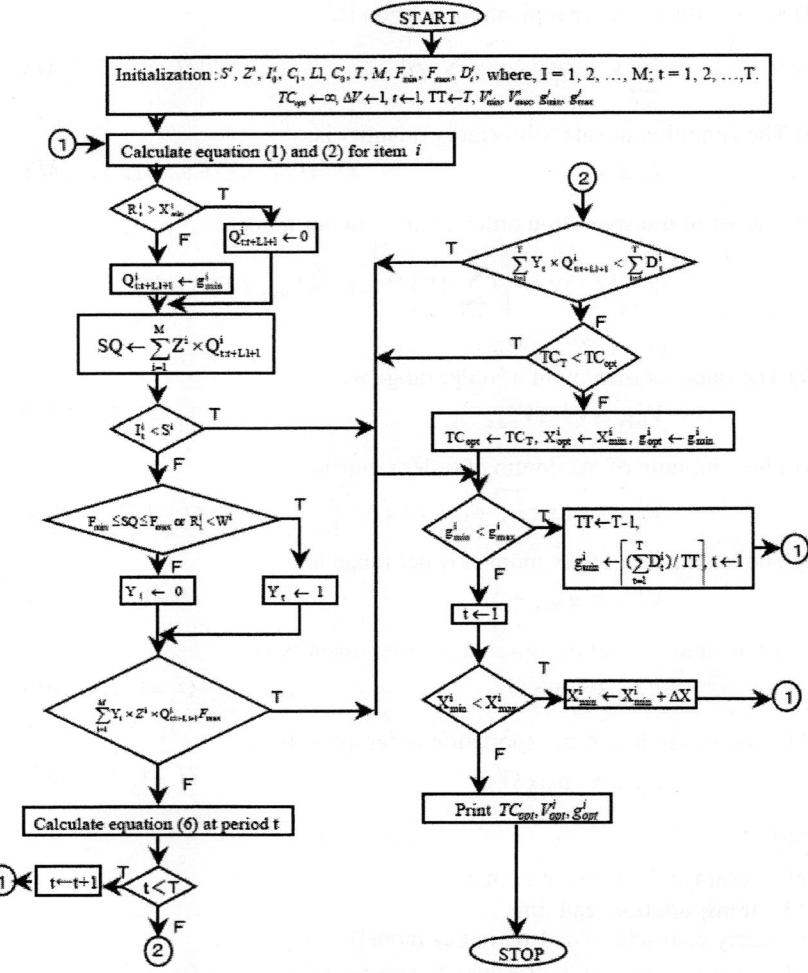

Fig. 3. Computation algorithm to determine the transportation order quantities and order ranges

3 Case Study

We experiment a numerical example in order to demonstrate the verification and effectiveness of the proposed model. We assume that there are three types of items and the planning horizon starts at the beginning of period 1 and finishes at the end of period T=10. And transportation lead-time is 1 period. Following data are used in the paper.

3.1 Input Data

We use the following input date to show the verification of the developed model.

(1) demands (container) (see table 1)
(2) size for each items (m^3)
$$Z^1 = 3 \quad Z^2 = 2 \quad Z^3 = 1$$
(3) minimum and maximum available transportation size (m^3)
If $L1 = 2$, then $F_{min} = 240$ and $F_{max} = 280$.
(4) holding and transportation cost ($)
$$C_0^1 = 1 \quad C_0^2 = 2 \quad C_0^3 = 1 \quad C_1 = 100$$
(5) initial inventory level, safety inventory level (container)
$$I_0^1 = 10 \quad I_0^2 = 29 \quad I_0^3 = 33 \quad S^i = 1 \quad (i=1, 2, 3)$$
(6) preordered transportation quantity (container)
If $L1 = 2$, then $Q^1_{-2:1} = 23 \quad Q^2_{-2:1} = 40 \quad Q^3_{-2:1} = 61 \quad Q^i_{-1:2} = 0$ (i=1, 2, 3)
$Q^1_{0:3} = 23 \quad Q^2_{0:3} = 40 \quad Q^i_{0:3} = 61 \quad Y_{-2} = 1 \quad Y_{-1} = 0 \quad Y_0 = 1$.

Table 1. The demand values of each period

$t =$	1	2	3	4	5	6	7	8	9	10
D_t^1	9	11	11	10	9	8	9	11	9	11
D_t^2	18	19	22	24	24	20	21	18	20	22
D_t^3	28	32	32	30	31	30	33	28	30	30

3.2 Computational Results

As a result of solving the developed computational procedures in case of $L1 = 2$, we obtained the table 2, 3, 4 and figure 3. Table 2 shows the computational results. The transportation on the planning horizon occurs 4 times for each item. Total cost is $1,709. The transportation orders (25 of item 1, 52 of item 2 and 77 of item 3) are occured at end of period of 3, 5, 8, 10 (see table 6). Table 7 shows the inventory level in the depot.

Table 2. Computational results ($L1 = 2$)

Item $i =$	1	2	3
Transportation order quantity	25	52	77
Occurrence of order	4	4	4
Order range	23-33	47-57	65-82
Total inventory quantity	175	283	368

Table 3. Tansportation order quantities ($L1 = 2$)

$t =$	1	2	3	4	5	6	7	8	9	10
$Q^1_{r,t+L1+1}$	0	0	25	0	25	0	0	25	0	25
$Q^2_{r,t+L1+1}$	0	0	52	0	52	0	0	52	0	52
$Q^3_{r,t+L1+1}$	0	0	77	0	77	0	0	77	0	77

Table 4. Inventory level in the depot ($L1 = 2$)

$t =$	1	2	3	4	5	6	7	8	9	10
I^1_t	24	13	25	15	6	23	14	28	19	8
I^2_t	49	30	50	26	2	34	13	47	27	5
I^3_t	66	34	63	33	2	49	16	65	35	5

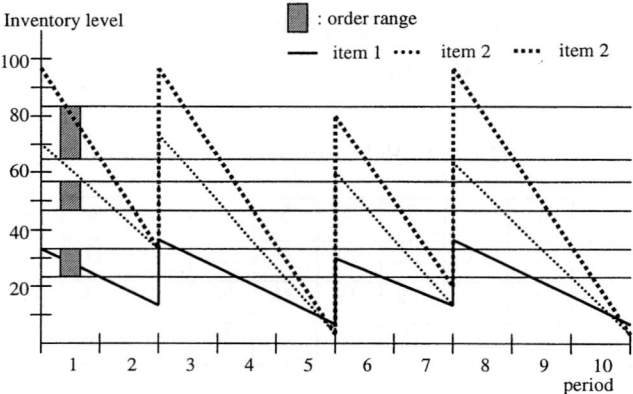

Fig. 4. Behavior of inventory level in the depot ($L1 = 2$)

Figure 4 illustrates the behavior of inventory level in the depot. The inventory level at period of 2 is less than the order range, but order was not placed on this period because there were preordered transportation quantity. The order was placed on the period 3 because the inventory levels on hand were within the order range.

From this example, we can conclude that the mixed parts transportation for multi-items occurs and the order is minimized by the proposed model under minimum total cost.

4 Conclusions

In order to improve the manufacturer satisfaction level, parts may wait in depot for just a short time. Therefore, reducing cycle times, inventories and costs in depots have continued to become an essential issue for depot mamagers. They are interested in discovering the most economical way for customer orders which minimizes the total costs occurred. The problem of ordering with constraints is usaually considered in a supply chain. This is largely an effect of the ordering policy and is regarded as an important issue on supply chain management.

In this paper, we proposed the new ordering model for multi-items in consideration of mixed parts tranportation condition. The computational algorithm was developed and the effectiviess of the proposed model was demonstrated through a case study.

In future we demonstrate the suitablity of the proposed model in the various supply chain models.

References

1. Coyle, J. J., Bardi, E. J., Langley Jr., C. J.: The Management of Business Logistics, 6th ed. West Publishing Company, New York, (1996)
2. Van den Berg, J. P.: A literature survey on planning and control of warehousing systems, IIE Transactions 31. (1999) 751–762
3. Doll, C. L., Whybark, C.: An iterative procedure for the single-machine multi-product lot scheduling problem, Managemnet Science Vol. 20(1). (1973) 50-55
4. Pirkul, H., Aras, O. A.: Capacitated multiple item ordering problem with quantity discounts. IIE Trans. 17. (1985) 206-211
5. Rosenblatt, M. J., Rothblum, U. G.: On the single resource capacity problem for multi-item inventory systems. Opns. Res. 38. (1990) 686-693
6. Shu, F. T.: Economic ordering frequency for two items jointly replenished. Management Science Vol. 17(6). (1971) B406-B410
7. Tersine, R. J., Barman, S.: Economic/Transport Lot Sizing with Quantity and Freight Rate Discounts. Decision Science 22. (1991) 1171-1179

Artificial Neural Network Based Life Cycle Assessment Model for Product Concepts Using Product Classification Method

Kwang-Kyu Seo[1]*, Sung-Hwan Min[2], and Hun-Woo Yoo[3]

[1] Department of Industrial Information and Systems Engineering, Sangmyung University,
San 98-20, Anso-Dong, Chonan, Chungnam 330-720, Korea
{kwangkyu}@smu.ac.kr
[2] Graduate School of Management, Korea Advanced Institute of Science and Technology,
207-43 Cheongryangri-Dong, Dongdaemun-Gu, Seoul 130-012, Korea
{shmin}@kgsm.kaist.ac.kr
[3] Center for Cognitive Science, Yonsei University,
134, Shinchon-Dong, Seodaemun-Ku, Seoul 120-749, Korea
{paulyhw}@yonsei.ac.kr

Abstract. Many companies are beginning to change the way they develop products due to increasing awareness of environmentally conscious product development. To copy with these trends, designers are being asked to incorporate environmental criteria into the design process. Recently Life Cycle Assessment (LCA) is used to support the decision-making for product design and the best alternative can be selected based on its estimated environmental impacts and benefits. Both the lack of detailed information and time for a full LCA for a various range of design concepts need the new approach for the environmental analysis. This paper presents an artificial neural network (ANN) based approximate LCA model of product concepts for product groups using a product classification method. A product classification method is developed to support the specialization of ANN based LCA model for different classes of products. Hierarchical clustering is used to guide a systematic identification of product groups based upon environmental categories using the C4.5 decision tree algorithm. Then, an artificial neural network approach is used to estimate an approximate LCA for classified products with product attributes and environmental impact drivers identified in this paper.

1 Introduction

The ability of a company to compete effectively on the increasingly competitive global market is influenced to a large extent by the cost as well as the quality of its products and the ability to bring products into the market in a timely manner. Traditionally, manufacturers focus on how to reduce the cost the company spends for materials acquisition, production, and logistics, but due to widespread consciousness of global environment problems and environmental legislative measures, companies also

---

* Corresponding author

should take environmental considerations into their decision making process of product development. These trends are driving many companies to consider environmental impacts during product development. Product designers are challenged with questions of what and how to consider environmental issues in relation to the products they are developing. In particular, it is quite relevant to understand how design changes can affect the environmental performance of product concepts in the early design process.

Life Cycle Assessment (LCA) is now the most sophisticated tools to consider and quantify the consumption of resources and the environmental impacts associated with a product or process [1]. By considering the entire life cycle and the associated environmental burdens, LCA identifies opportunities to improve environmental performance. Conceptually, a detailed LCA is an extremely useful method, but it may be rather costly, time consuming and sometimes difficult to communicate with non-environmental experts. Further the use of LCA poses some barriers at the conceptual stage of product development, where ideas are diverse and numerous, details are very scarce, and the environmental data for the assessment is short. This is unfortunate because the early phases of the design process are widely believed to be the most influential in defining the LCA of products. Therefore, a new methodology for estimating the environmental impacts of products is required in early design phase.

This paper presents an ANN based LCA model for products in early conceptual design phase by classifying products into groups according to their environmental and product characteristics. In order to specialize the proposed model for different classes of products, it is necessary to develop a method to classify products into general categories that can lead to more specific relationships between product attributes and the environmental performance of product concepts. ANNs and statistical analysis are applied to the proposed approach. The statistical analysis is used to check the correlation between product attributes and environmental impact drivers (EIDs) derived from environmental impact categories [2-3]. An ANN is trained on product attributes typically known in the conceptual phase and the LCA data from pre-existing detailed LCA studies.

2 An ANN Based Life Cycle Assessment Model

In this section, the reasonable EIDs and the meaningful product attributes are introduced and identified for the proposed model. The EIDs stand for environmental impact categories and the product attributes are meaningful to designers during conceptual design. The proposed model predicts the LCA results of product concepts using artificial neural networks with product attributes as inputs and EIDs as outputs. The concept of an ANN-based LCA model is also described.

2.1 Environmental Impact Drivers (EIDs) and Product Attributes

In order to estimate the environmental impacts of products for the entire life cycle, environmental impact drivers (EIDs) are introduced in this section. EIDs represent the key environmental characteristics that determine the environmental impacts of products and are strongly correlated with the key factors in life cycle inventory. EIDs eventually mean the environmental impact categories such as greenhouse effect, acidification, winter/summer smog, eutrophication, ozone depletion, solid material and energy, etc. which capture the environmental performance of product concepts and statistically tested for its ability to predict impact categories (see table 1) [2-3].

Table 1. The proposed environmental impact driver set

Impact category	Key factors	EID
life-cycle energy consumption	Energy	EID_{energy}
solid material waste	Material	$EID_{material}$
greenhouse effect	Carbon dioxide(CO_2)	$EID_{green.}$
ozone layer depletion	Chlorofluorocarbons (CFC)	EID_{ozone}
acidification	Sulfur dioxide(SO_2) & Nitrous oxides (NO_x)	$EID_{acid.}$
eutrophication	carbon, nitrogen, phosphorus	$EID_{eutro.}$
winter/summer smog	Nitrous oxides (NO_x), Hydrocarbons, SPM	EID_{smog}
......	......	......

Product attributes are envisioned to provide a learning interface between environmental experts and designers and can express the design changes. A list of meaningful general product attributes for use in training and querying the proposed model is thus defined (see table 2) [4-5]. The product attributes are selected such that they utilize only product information readily available during conceptual design; be compact to reduce demands on the proposed model, and be related to elements of the EID set. An identified set of product attributes is tested for first order relationships with elements in the EID set.

Table 2. Identified product attribute set

Q – Mass (kg)	Q – Lifetime (hours)
Q – Ceramics (%mass)	Q – Use time (hours)
Q – Fibers (%mass)	D – Mode of operation
Q – Ferrous metals (%mass)	B – Additional consumable
Q – Non-ferrous metals (%mass)	D – In use energy source
Q – Plastics (%mass)	Q – In use power consumption (Watt)
Q – Paper/Cardboard (%mass)	B – Modularity
Q – Chemicals (%mass)	B – Serviceability
Q – Wood (%mass)	B – Disassemblability
Q – Other materials (%mass)	(*Q: Quantitative,
B – Assemblability	D: Dimensionless,
D – Process	B: Binary)

2.2 An ANN Based Approximate Life Cycle Assessment Model

An ANN based LCA model is a different approach to other methods. Learning algorithms are trained to generalize on characteristics of product concepts using product attributes and corresponding EIDs from pre-existing LCA studies. The product designers query this ANN based LCA model with product attributes to quickly obtain an approximate LCA for a new product concept (see figure 1). Designers need to simply provide product attributes of new product concepts to gain LCA predictions. It has the flexibility to learn and grow as new information becomes available, but it does not require the creation of a new model to make as LCA prediction for a new product concept. Also, by supporting the extremely fast comparison of the environmental performance of product concepts, it does not delay product development.

However, insight gained about the effect of product groupings suggested it might be necessary to specialize the proposed models for different classes of products. This paper explores the method needed to classify products into general categories that can lead to more specific relationships between product attributes and the environmental performance of product concepts.

3 Product Classification Method for the ANN Based LCA Model

The goal is to develop a systematic product classification method that supports the development of the appropriately specialized ANN based LCA model.

The method may learn faster and more effectively if the learning space is narrowed into general but coherent product categories. The categorization should be based on properties that potentially create common dominant environmental impacts or similar scaling trends so that the proposed model is better able to emulate impacts of specific products within the group.

Classification criteria should be based upon the product attributes that are used to train and query the proposed model. The structure of the proposed model is shown in figure 1. In figure 1, products are classified by decision tree and then the approximate LCA of the new product concepts are predicted by ANN models. Product attributes and LCA data were gathered from 150 different products such as household appliances.

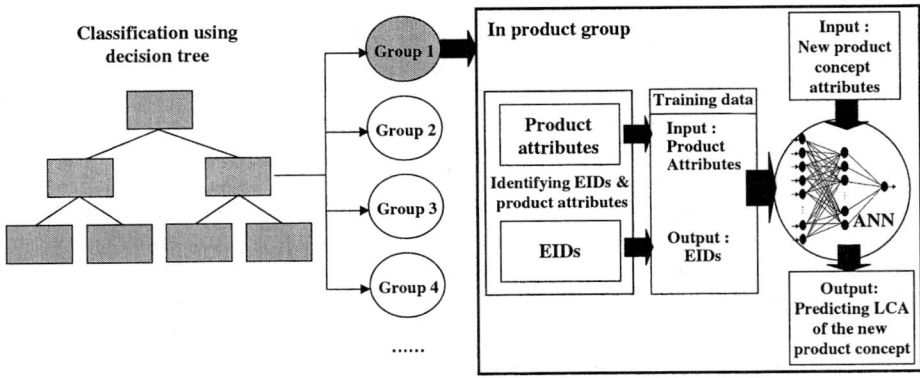

Fig. 1. The structure of the proposed model

3.1 Product Classification Method Using Decision Tree Algorithm

Product attributes data and environmentally driven categories were then used to develop a product classification system based on C4.5 decision tree algorithms [6].

C4.5 is a system for generating decision trees and classifying a set of instances. This algorithm employs the top down and recursive splitting technique to produce decision trees. A decision tree contains a root node, internal nodes representing the product attributes, branches characterizing the product attribute values and leaves expressing the binary decision.

The data flow of the decision tree is computed through the following steps.

Firstly, the raw datasets were pre-processed with the statistical packages and controlled for plausibility and missing values of the attributes.

Secondly, the model is constructed after pre-processing dataset. C4.5 performs the top down induction of the decision tree on the basis of a training set, which is used to train the algorithm and build the structure of the decision tree. The training set is split recursively into two sub-trees. This procedure is done at each node by calculating the influence of the target attribute. C4.5 calculates the descending order of the attributes within the tree with the gain ratio criterion. An information gain ration criterion is used to consistently choose the best possible test to decide which attribute will be tested. The information gain criterion measures the ratio of information relevant to classification that is provided by the division to the information produced by the division itself. C4.5 also contains heuristic methods for pruning decision trees to create more comprehensible structures without compromising accuracy on unseen cases. If the expected error rate in the sub-tree is greater than in a single leaf, pruning is done by replacing a whole sub-tree by a leaf node. The following parameters were also considered for analysis of the average pruned tree such as grouping attribute values, minimum cases, size, observed error rate for unseen cases and expected error rate for unseen cases.

Finally, the generated model is tested with respect to its explanatory power, with the stratified 10-fold cross-validation method. The number of correct classified instances determines whether the model can be applied to the datasets, or whether further preparation of the model will be necessary. The 10 cross-validation procedure was rerun with a pruning confidence level of 25% for different combinations of parameter options and product descriptors to explore distinct results on error estimations associated with different decision trees. The different combinations produce the decision trees differently structured classification systems as expected. Figure 2 displays the decision trees corresponding to the highlighted combinations and exhibits the smallest expected error rate relatively to the other combinations.

Figure 2 shows that the products are classified into 7 groups and the results of the classification analysis guided the definition of environmentally driven categories of products: (1) low-mass, passive products; (2) low-mass, active products; (3) durable, active, household appliances; (4) durable, mobile, active (external energy based) products; (5) durable, low-mass, active (external energy based) products; (6) low-mass, fiber-based, active (external energy based) products; (7) durable, mobile, active products.

The purpose of this section is to explore the viability of a product classification system that could be used to support the specialized ANN based LCA model. The approximate LCA of product concepts would be estimated by the selection of the suitable classification scheme.

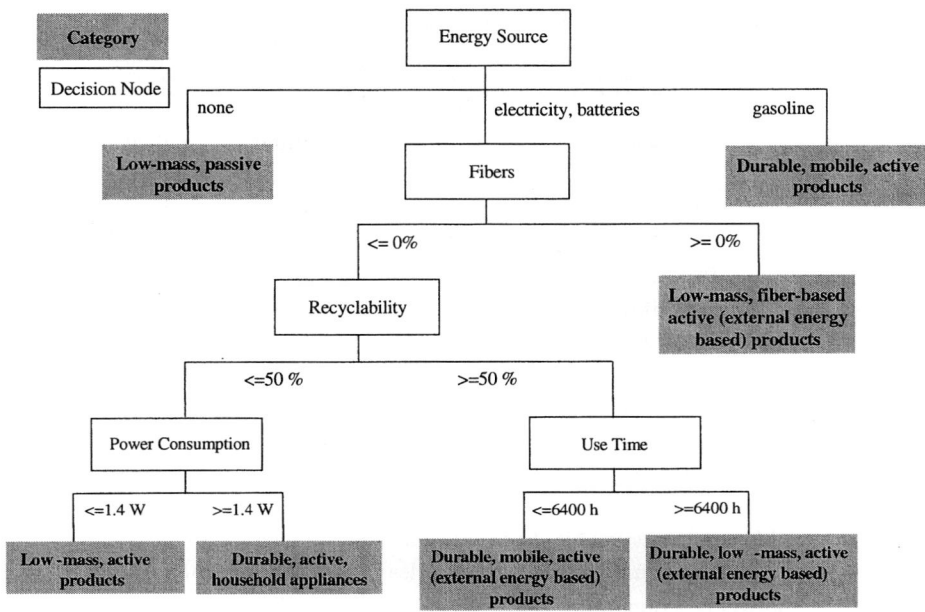

Fig. 2. The classification results by the decision tree algorithm

3.3 Approximate LCA of Classified Products Using an ANN Model

With product attributes and EIDs defined, ANN-based LCA models were trained in an effort to validate the concept. The feasibility test was conducted focusing on EID_{energy}. Training data with product attributes and corresponding life cycle energy consumption from true past studies were collected for 40 different products in the group (3) - durable, active, household appliances.

The identified products attributes are used to input of an ANN model and EIDs represented impact categories are used to output. The architecture for a backpropagation (BP) neural network is developed to estimate an approximate LCA of product concepts. The inputs in the ANN experiments consist of 19 product attributes strongly correlated with EID_{energy} and the output consists of one node (EID_{energy}) (see table 3).

More than 30 experiments were performed to determine the best combination of the learning rates (α), momentum (ρ), number of hidden layers, number of neurons in hidden layers, learning rules and transfer functions. The resulting network has a hidden layer with 16 neurons. The most popular learning rules, generalized delta rules and a sigmoid transfer function are used for the output node [7].

Figure 3 shows the structure of the BP neural network used in this study, which consists of an input layer with 21 nodes, a hidden layer with 16 nodes and an output layer with one node.

Table 3. Product attribute list used in training the proposed model

Product attributes	Unit	Level of information
Mass	kg	quantitative, speified
Ceramics	%mass	quantitative, speified
Fibers	%mass	quantitative, speified
Ferrous metals	%mass	quantitative, speified
Non-ferrous metals	%mass	quantitative, speified
Plastics	%mass	quantitative, speified
Paper/Cardboard	%mass	quantitative, speified
Chemicals	%mass	quantitative, speified
Wood	%mass	quantitative, speified
Other materials	%mass	quantitative, speified
Lifetime	hours	quantitative, speified
Use time	hours	quantitative, speified
Operation mode	dimentionless	qualitative, speified
Additional consumable	dimentionless	qualitative, binary
Energy source	dimentionless	qualitative, speified
Power consumption	Watt	quantitative, speified
Modularity	dimentionless	qualitative, binary
Serviceability	dimentionless	qualitative, binary
Disassembleability	dimentionless	qualitative, binary

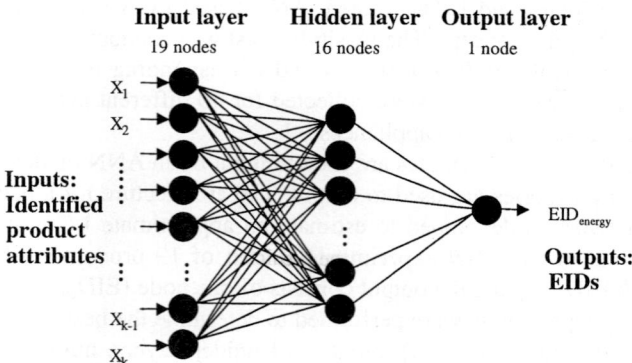

Fig. 3. Structure of the BP neural network to estimate approximate LCA of products

The predicted results of grouping products are shown in table 4 and figure 4. The approximate LCA using ANNs with the identified EIDs and product attributes gives good results except for heater. It is shown that classification of products provides reasonable LCA predictions of the grouping products. The absolute errors of LCA predicted by the ANN model ranged from 0.11 to 12 percent of the levels given by the actual LCA, so the results obtained by this ANN model seem to be satisfactory.

Table 4. The predicted results of grouping products by using ANN

Product	Actual LCA	Predicted LCA	Relative error (%)
Vacuum cleaner	5110	4686.84	4.23
Mini-Vacuum Cleaner	176	122.21	5.72
Radio	207	164.53	3.75
Heater	24800	39471.19	-12.02
Coffeemaker	3980	5097.76	-5.22
Washing Machine	54500	49682.29	-0.11
Refrigerator (small)	2686.19	3002.98	2.27
Refrigerator (large)	18777.79	20507.22	1.23
TV	24320.37	26807.69	-2.92
LCD TV	24813.73	24430.42	0.5
Average absolute error			3.79
Maximum absolute error			12.02

* Training sample size is 30, ** Test sample size is 10

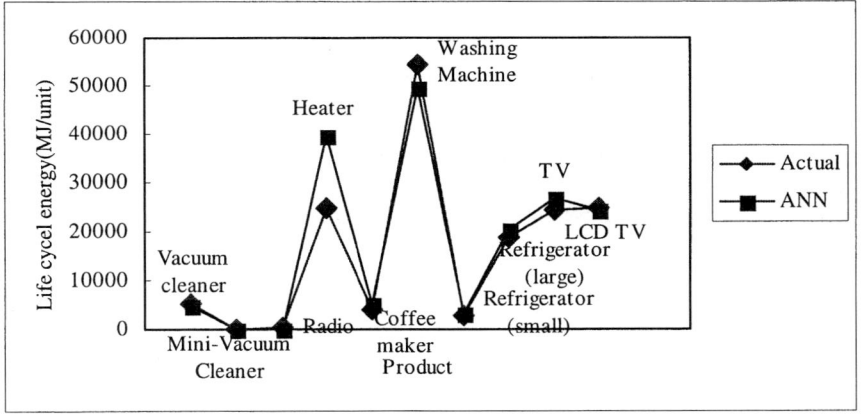

Fig. 4. Comparison of the predicted results of LCA

4 Conclusions

The lack of analytic LCA for early conceptual design stage motivated the development of this estimation model. This paper explored the possibility of an ANN based LCA model for the conceptual design phase by classifying products into groups according to their environmental and product characteristics. This paper presented an exploratory approach to develop a product classification system for this purpose. The proposed approach is to guide a systematic identification of environmentally based categories. The product classification ultimately identified was used to create classification schemes with the C4.5 decision tree algorithm. C4.5 decision tree algorithms generated automated classification systems with different structures and error estimations by varying algorithm parameters and product attributes.

The product attributes and EIDs were identified to predict the product's environmental impacts. Then an ANN based approach with product attributes as inputs and EIDs as outputs was developed to predict approximate LCA of grouping members and the predicted results seemed to be satisfactory.

The proposed approach does not replace a full LCA but designers can use this guideline to optimize their effort and guide their decisions at the conceptual design phase of environmentally conscious product design.

Future work is needed to select a suitable classification scheme to support specialized proposed LCA models.

References

1. Curran, M. A.: Environmental Life-Cycle Assessment. McGraw-hill (1996)
2. Goedkoop, M. et al.: The Eco-indicator 99: A Damage Oriented Method for Life Cycle Impact Assessment. Pre consultants B. V., Netherlands (1999)
3. SimaPro 4 User's Manual. The Netherlands: PRe Consultants BV (1999)
4. Eisenhard, J.: Product Descriptors for Early Product Development: an Interface between Environmental Expert and Designers. MS of Science in Mechanical Engineering, MIT (2000)
5. Sousa, I., Eisenhard, J., and Wallace, D.R.: Approximate Life-Cycle Assessment of Product Concepts using Learning Systems. Journal of Industrial Ecology, Vol. 4. No. 4. (2001) 61-81
6. Quinlan, J.R.: C4.5 Programs for Machine Learning. Morgan Kaufmann, San Mateo, CA (1993)
7. Haykin, S. and Simon, S.: Neural Networks: A Comprehensive Foundation, Prentice Hall, (1998)

New Heuristics for No-Wait Flowshop Scheduling with Precedence Constraints and Sequence Dependent Setup Time

Young Hae Lee[1] and Jung Woo Jung[2]

[1] Department of Industrial Engineering, Hanyang University,
Sa 1 Dong, Ansan, Gyeonggi-Do, 426-791, Korea
yhlee@hanyang.ac.kr

[2] Department of Industrial Engineering, Hanyang University,
Sa 1 Dong, Ansan, Gyeonggi-Do, 426-791, Korea
jungjw@scm.hanyang.ac.kr

Abstract. This paper addresses a method to obtain the best sequence for no-wait flowshop scheduling with precedence constraints and sequence dependent setup time. The system is made up of a set of machines of various types and there is no interruption between tasks in a job. The objective is to determine the job sequence for processing with minimum makespan. The sequencing problem with precedence constraints and sequence dependent setup times is equivalent to the traveling salesman problem. A mixed integer programming (MIP) model is presented to obtain the best schedule. It is known that the MIP model for no-wait flowshop scheduling is NP-hard. Heuristic algorithms to gain the best job sequence are developed. From the experiments, it is found that the proposed algorithm generates the best job sequence efficiently.

1 Introduction

In chemical processes, all tasks of a job have to be processed without the interruption between consecutive machines. Interruptions reduce the quality and increase the cost of the product. The manufacturing system for these processes is known as the no-wait flowshop [2]. These processes are found in petroleum refineries, chemical plants, and steel plants. The processes in this system seem more like separate entities than a series of connected operations. In this system, starting time for the first operation of a job directly affects the job schedule of the other machines involved in the process.

The objective of the no-wait flowshop problem is to determine the job sequence to minimize makespan. Rajendran [8] proposed a heuristic algorithm to minimize makespan in the m-machine no-wait flowshop. Lin and Cheng [6] proved that the no-wait flowshop scheduling problem is NP-hard, and they developed an algorithm to minimize makespan. Wang and Cheng [10] proposed a heuristic algorithm considering the available constraints in a two machine no-wait flowshop. Recently, Aldowaisan and Allahverdi [1] obtained the better solutions by integrating meta-heuristics such as simulated annealing (SA) and genetic algorithm (GA) with heuristic algorithms developed by Rajendran [8] and Gangadgaran and Rajendran [3].

In many chemical processes, precedence constraints differentiate one product from another even when the products consist of the same materials. Moreover, sequence dependent setup time can be had an effect to determine the job sequence to minimize makespan. Precedence constraints and sequence dependent setup time are important factors in deciding the optimal job sequence for the no-wait flowshop. However, there are few studies having considered precedence constraints and sequence dependent setup time. In this paper, the determination of the job sequence in the no-wait flowshop with precedence constraints and sequence dependent setup time is treated.

If more than 3 machines are involved, the no-wait flowshop scheduling problem is proved to be NP-hard by Röck [9]. Therefore, new heuristic algorithms are developed to solve this problem.

The remainder of this paper is structured as follows. In section 2, the mathematical model of this problem is presented. Section 3 addresses heuristic algorithms with precedence constraints and sequence dependent setup time. In section 4, the numerical example and computational performance of heuristics is presented. Section 5 concludes the paper.

2 Mathematical Model

Emmons and Mathur [2] asserted the no-wait flowshop scheduling problem is equivalent to the traveling salesman problem (TSP). Therefore, the decision of the job sequence to minimize makespan can be recognized as that of the path to minimize cost in the TSP. It is assumed that there is no interruption between the setup time and process time for a task in a job. According to the problem description, we develop a m-machine no-wait flowshop scheduling model with precedence constraints and sequence dependent setup time.

Indices
i, j : index for jobs
k : index for machines

Notations
n : number of jobs
m : number of machines in the flowshop
d_{ij} : increased makespan for the job j following the job i , $d_{ii}=0$ for all i
u_i : the order of processing job i in a round sequence
s_{ij}^k : setup time for the job j followed by the job i on the machine k
p_{ik} : processing time for the job i on the machine k
c_{ij} : 1, if job i has precedence over job j, 0, otherwise

Decision Variable
x_{ij} : 1, if job j is followed by job i, 0, otherwise

The complete MIP model for the m-machine no-wait flowshop scheduling model with precedence constraints and sequence dependent setup time can be summarized as follows:

$$Minimize \quad \sum_{i=1}^{n} \sum_{j=1}^{n} d_{ij} \cdot x_{ij}$$

subject to

$$\sum_{i=1}^{n} x_{ij} = 1, for \quad j = 1, 2, \ldots, n \qquad (1)$$

$$\sum_{j=1}^{n} x_{ij} = 1, for \quad i = 1, 2, \ldots, n \qquad (2)$$

$$u_i - u_j + n \cdot x_{ij} \leq n - 1, for \quad i = 2, 3, \ldots, n, j = 2, 3, \ldots, n, i \neq j \qquad (3)$$

$$d_{ij} = MAX_{k=1,\ldots,m} \left(\sum_{h=1}^{k} p_{ih} + s_{ij}^{k} + \sum_{h=k}^{m} p_{jh} \right) - \sum_{k=1}^{m} p_{ih}$$
$$for \quad i = 1, 2, \ldots, n, j = 1, 2, \ldots, n, i \neq j \qquad (4)$$

$$x_{ij} \leq c_{ij}, for \quad j = 1, 2, \ldots, n \qquad (5)$$

$$x_{ij} \in \{0, 1\} \qquad (6)$$

$$c_{ij} \in \{0, 1\} \qquad (7)$$

$$u_i, u_j \geq 0 \qquad (8)$$

The objective function determines the job sequence to minimize makespan. Equations (1)-(3) are constraints of the TSP. Equation (3) is the constraint for the Hamiltonian cycle. In equation (4), increased makespan for the job j following the job i is presented. Precedence constraints are ensured by an equation (5). Equations (6)-(8) are the restrictions on the decision variable and parameters.

3 Heuristic Algorithm

The no-wait flowshop scheduling problem was proven to be NP-hard by Röck [9]. Precedence constraints and sequence dependent setup time must be considered in this paper. Therefore, we apply heuristic algorithms to find the best solution. Becauseany studies on this topic cannot be found, we develop three heuristic algorithms for no-wait flowshop scheduling, referred to as HNFS-1, HNFS-2, and HNFS-3 in this section.

3.1 HNFS-1 Algorithm

The first heuristic algorithm based on SA for the no-wait flowshop scheduling problem is suggested. SA is the representative methodology for finding the best solution of NP-hard problems in various meta-heuristic algorithms. This methodology is used in various problems such as the project scheduling problem[5] and the lot sizing problem [4], and has been proven to be a good tool for solving complex problems. Particularly, SA is recognized as an effective heuristic algorithm for the flowshop scheduling problem[7]. However, there have been few studies about no-wait flowshop scheduling with precedence constraints and sequence dependent setup time using SA. Therefore, we develop the HNFS-1 algorithm using SA. In the HNFS-1 algorithm the initial solution is a randomly generated job sequence. The adjacent job sequence is generated by changing the positions of two randomly chosen jobs. If the job sequence whose makespan is not less than that of the best job sequence is accepted and updated, the temperature is decreased. This algorithm is executed until the number of generated adjacent job sequences is equal to the defined number. This algorithm can be represented as follows.

HNFS-1 Algorithm:

Variable
    σ : the randomly generated job sequence
    B : the best job sequence
Begin
    $B \leftarrow \sigma$;
    Repeat
        A: Generate a new sequence σ' by exchanging positions of two
           randomly chosen jobs in σ ;
        If σ' does not satisfy precedence constraints
           Then go to A ;
        If makespan of σ' is less than that of σ
           Then $\sigma \leftarrow \sigma'$;
           If makespan of σ is less than that of B
               Then $B \leftarrow \sigma$;
        Generate a random number ;
        Else If random number is less than the temperature
           Then
               Begin
                   $\sigma \leftarrow \sigma'$;
                 Decrease the temperature;
               End;
    Until the completeness condition is satisfied;
End.

3.2 HNFS-2 Algorithm

The second heuristic algorithm, HNFS-2, based on the previous heuristic algorithm for the no-wait flowshop scheduling problem. Rajendran[8] developed an algorithm for the no-wait flowshop scheduling problem, and his algorithm is known as a very effective algorithm. Since this algorithm cannot deal with precedence constraints and sequence dependent setup time, we develop a new algorithm based on Rajendran[8]'s algorithm. There are some important differences between this algorithm and Rajendran's algorithm. First, precedence constraints and sequence dependent setup time is newly considered in this algorithm, but they are not considered in Rajendran's algorithm. Second, there is a difference in insertion techniques. The first chosen job in π is inserted in all positions of the partial schedule σ in this algorithm, but in Rajendran's algorithm it is inserted in vth position of the partial schedule σ where $(h+1)/2 \leq v \leq h+1$, if h is the number of jobs in σ. Because of this difference in insertion technique, more candidate solutions can be searched in this algorithm than in Rajendran's algorithm. The following algorithm describes the algorithm to find an initial sequence in HNFS-2.

HNFS-2 Algorithm:

Variable
  σ : the null set
Begin
  Divide all jobs into A, B groups where
    $A = \{i | MP_i \geq (1+m)/2\}, B = \{i | MP_i < (1+m)/2\}$,
    $MP_i = (\sum_i^m k \cdot p_{ik} / \sum_i^m p_{ik})$
  Generate A' : arranging elements from A in increasing order of MP_i ;
  Generate B' : arranging elements from B in decreasing order of MP_i ;
  Get the initial sequence of $\pi = A'B'$;
  $v = 0$;
  Repeat
    $v = v + 1$;
    If the sequence in π does not satisfy the vth precedence constraint,
       Then, exchange positions of two elements restricted
         by the vth precedence constraint ;
  Until v is equal to the number of precedence constraints ;
  Remove the element in the first position in π, and add in σ ;
  Repeat
    Remove the element in the first position in π, and insert
      in the all available position in the partial schedule σ ;
    Choose the best sequence σ^* with minimum makespan ;
    If the chosen sequence σ^* does not satisfy precedence constraints,
       Then, exchange positions of two elements restricted
         by precedence constraints ;
    $\sigma \leftarrow \sigma^*$;
  Until π is empty set ;
End.

3.3 HNFS-3 Algorithm

Finally, the third heuristic algorithm based on the above mentioned heuristic algorithms, HNFS-1 and HNFS-2, for the no-wait flowshop scheduling problem is suggested. HNFS-1 is the algorithm based on SA which is a kind of metaheuristic algorithm, and HNFS-2 is a heuristic algorithm. If SA can be used to find best sequence with a good initial sequence, we can expect to obtain the better sequence. Therefore, the integrated algorithm, HNFS-3, is newly developed with integrating HNFS-1 and HNFS-2. This algorithm is the similar as that for HNFS-1, but the solution of HNFS-2 is used as the initial solution in HNFS-3. The integrated algorithm, HNFS-3, can be illustrated as follows:

HNFS-3 Algorithm:

Variable
    σ : the job sequence in the result of HNFS-2 algorithm
        for an initial sequence
    B : the best job sequence
Begin
    $B \leftarrow \sigma$;
    Repeat
        A: Generate a new sequence σ' by exchanging positions of two
            randomly chosen jobs in σ ;
        If σ' does not satisfy precedence constraints
            Then go to A ;
        If makespan of σ' is less than that of σ
            Then $\sigma \leftarrow \sigma'$;
        If makespan of σ is less than that of B
            Then $B \leftarrow \sigma$;
        Generate a random number ;
        Else If random number is less than the temperature
            Then
                Begin
                      $\sigma \leftarrow \sigma'$;
                      Decrease the temperature;
                End;
    Until the completeness condition is satisfied;
End.

4 Experiments and Evaluation

The performance of three heuristics is evaluated in this section. The processing time and sequence dependent setup time is obtained from a discrete uniform distribution with a range of [1,100]. The no-wait flowshop which has 30, 50, 70, and 90 jobs and 5, 7, and 9 machines with randomly generated 5 precedence constraints are considered to find the effectiveness of three algorithms with fixed number of precedence constraints. Moreover, randomly generated 5, 10, and 15 precedence constraints are considered for the no-wait flowshop scheduling problem which has 50 jobs-5 machines, 70 jobs-7 machines, and 90 jobs-9 machines.

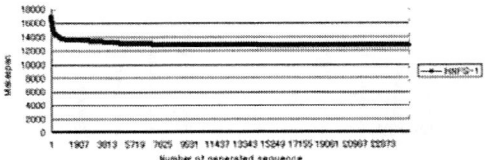

Fig. 1. Convergence of HNFS-1

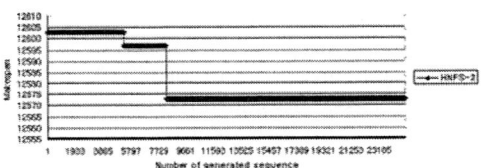

Fig. 2. Convergence of HNFS-3

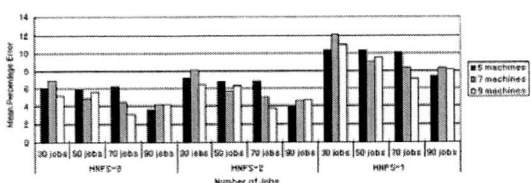

Fig. 3. Mean percentage error versus number of jobs (5 precedence constraints)

The initial temperature is 0.5 and reduced 30% if the solution whose makespan is more than that of the best solution is accepted. This algorithm is executed until the number of generated adjacent job sequences equals 25000.

Figures 1-2 show that it is necessary 25000 times observation to find convergence of heuristic algorithms, HNFS-1 and HNFS-3, for the biggest problem which has 90 jobs and 9 machines in this study. It is not necessary to consider the convergence of HNFS-2, since this does not include SA.This experiment is replicated 100 times for each problem. The performance of each algorithm is measured by mean percentage error, and standard deviation. The percentage error is defined as the percentage of the difference of makespan of between heuristic solution and best heuristic solution found in the experiment to makespan of best heuristic solution found in the experiment, which is also used in the work of Rajendran [8] and Aldowaisan and Allahverdi[1].

The results are depicted in Tables 1 and Figures 3-8. Table 1 shows average makespan and average runtime to get good sequence in replicated experiments. HNFS-2 and HNFS-3 can obtain the better sequence than HNFS-1 in shorter runtimes. HNFS-3 can find better sequence than HNFS-2, but it takes more

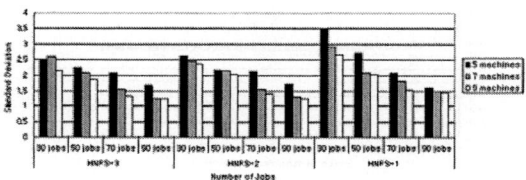

Fig. 4. Standard error versus number of jobs (5 precedence constraints)

runtime. However, the maximum difference is only few seconds for the worst case. In figures 3-4, mean percentage error of HNFS-3 is smaller than that of HNFS-1 and HNFS-2 for all cases. Standard deviation of HNFS-3 is smaller than that of HNFS-1 and HNFS-2. But, HNFS-2 outperforms HNFS-1. This means that the integration of SA with HNFS-1 is effective. In figures 5-6, smaller mean percentage error and standard deviation of HNFS-3 for changing number of precedence constraints is found. That is, the performance of HNFS-3 is better than that of HNFS-1 and HNFS-2. The above results show that HNFS-3 is a good algorithm for the no-wait flowshop scheduling with precedence constraints and sequence dependent setup time.

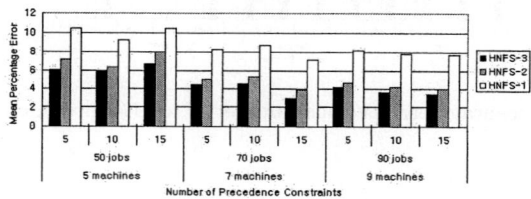

Fig. 5. Mean percentage error versus number of precedence constraints

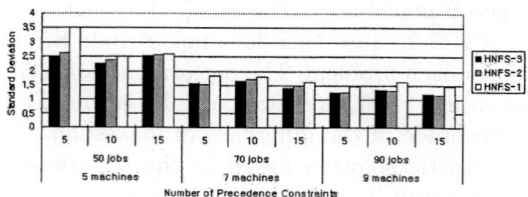

Fig. 6. Standard error versus number of precedence constraints

Table 1. Average makespan and average runtimes (ms)

Number of jobs	Number of machines	HNFS-1 Maksepan	HNFS-1 Runtimes	HNFS-2 Makespan	HNFS-2 Runtimes	HNFS-3 Makespan	HNFS-3 Runtimes
30	5	3873.01	217.55	3757.9	111.26	3720.42	135.34
	7	4360.04	241.09	4203.42	107.31	4159.28	140.08
	9	4781.86	224.68	4584.82	108.88	4534.94	145.72
50	5	6178.86	1925.53	5975.78	398.55	5931.01	508.22
	7	6921.99	1350.57	6706.07	406.93	6657.58	552.54
	9	7526.4	1387.4	7310.62	409.7	7256.55	592.02
70	5	8401.97	3581.7	8151.24	996.02	8106.26	1327.21
	7	9449.23	4046.58	9167.88	1058.31	9119.4	1542.05
	9	10244.88	4386.23	9926.15	979.21	9869.16	1586.25
90	5	10657.34	7334.81	10325.21	2030.19	10317.7	3016.9
	7	11979.9	7594.49	11575.86	2052.4	11528.4	2925.82
	9	12956.54	8121.23	12538.91	2009.28	12484.64	3177.28

5 Conclusion

The no-wait flowshop scheduling problem can be easily found in chemical processes or steel factories. In chemical processes, precedence constraints are an important consideration. Moreover, sequence dependent setup time should be considered in order to generate realistic schedule with minimum makespan. However this problem is NP-hard and any algorithms for this problem cannot be found. In this paper, we treated the no-wait flowshop scheduling problem by considering precedence constraints and sequence dependent setup time. First, the mathematical model aimed to provide the best job sequence with minimum makespan was developed. Three approaches, HNFS-1, HNFS-2 and HNFS-3, were proposed to obtain the best job sequence. The experimental results showed that HNFS-3 performed better than HNFS-1 and HNFS-2 in terms of the mean percentage error and standard deviation. Moreover, HNFS-3 can obtain the better job sequence faster than HNFS-1. Despite the advantages of the proposed algorithm, it should be improved. In chemical processes, there are many factors that affect the manufacturing processes. It is necessary to develop a new heuristic algorithm which considers all relevant characteristics of the products.

References

1. Aldowaisan, T., Allahverdi, A.: New Heuristics for No-wait Flowshops to Minimize Makespan. Comput. Oper. Res.30 (2003) 1219-1231
2. Emmons, H., Mathur, K.: Lot Sizing in a No-wait Flowshop. Oper. Res. Lett. 17 (1995) 159-164
3. Gangadharan, R., Rajendran, C.: Heuristic Algorithms for Scheduling in the No-wait Flow-shop. Int. J. Prod. Econ. 32 (1993) 285-290

4. Hung, Y.F., Chien, K.L.: A Multi-class Multi-level Capacitated Lot Sizing Model. J. Oper. Res. Soc. 51 (2000) 1309-1318
5. Lee, J.K., Kim, Y.D.: Search Heuristics for Resource Constrained Project Scheduling. J. Oper. Res. Soc. 47 (1996) 678-689
6. Lin, B.M.T., Cheng, T.C.E.: Batch Scheduling in the No-wait Two-machine Flowshop to Minimize the Makespan. Comput. Oper. Res. 28 (2001) 613-624
7. Ogbu, F.A., Smith, D.K.: The Application of the Simulated Annealing Algorithm to the Solution of the n/m/Cmax Flowshop Problem. Comput. Oper. Res. 17 (1990) 243-253
8. Rajendran, C.: A No-wait Flowshop Scheduling Heuristic to Minimize Makespan. J. Oper. Res. Soc. 45 (1994) 472-478
9. Röck, H.: The Three-machine No-wait Flow Shop is NP-complete. J. ACM. 31 (1984) 336-345
10. Wang, G., Cheng, T.C.E.: Heuristics for Two-machine No-wait Flowshop Scheduling with an Availability Constraint. Inform. Process. Lett. 80 (2001) 305-309

Efficient Dual Methods for Nonlinearly Constrained Networks*

Eugenio Mijangos

University of the Basque Country,
Department of Applied Mathematics and
Statistics and Operations Research,
P.O. Box 644, 48080 Bilbao, Spain
http://www.ehu.es/~mepmifee

Abstract. The minimization of nonlinearly constrained network flow problems can be performed by exploiting the efficiency of the network flow techniques. It lies in minimizing approximately a series of (augmented) Lagrangian functions including only the side constraints, subject to balance constraints in the nodes and capacity bounds. One of the drawbacks of the multiplier methods with quadratic penalty function when is applied to problems with inequality constraints is that the corresponding augmented Lagrangian function is not twice continuously differentiable even if the cost and constraint functions are. The author's purpose is to put forward two methods that overcome this difficulty: the exponential multiplier method and the ε-subgradient method, and to compare their efficiency with that of the quadratic multiplier method and that of the codes MINOS and LOQO. The results are encouraging.

1 Introduction

Consider the nonlinearly constrained network flow problem (**NCNFP**)

$$\underset{x}{\text{minimize}} \quad f(x) \tag{1}$$

$$\text{subject to} \quad x \in \mathcal{F} \tag{2}$$

$$\alpha \leq c(x) \leq \beta, \tag{3}$$

where:

– The set $\mathcal{F}$ is

$$\mathcal{F} = \{x \in \mathrm{I\!R}^n \mid Ax = b,\ 0 \leq x \leq \overline{x}\},$$

where A is a node-arc incidence $m \times n$-matrix, b is the production/demand m-vector, x are the flows on the arcs of the network represented by A, and $\overline{x}$ are the capacity bounds imposed on the flows of each arc.

---

* The research was partially supported by grant MCYT DPI 2002-03330.

- The side constraints (3) are defined by $c : \mathbb{R}^n \to \mathbb{R}^r$, such that $c = [c_1, \cdots, c_r]^t$, where $c_i(x)$ is linear or nonlinear and twice continuously differentiable on the feasible set $\mathcal{F}$ for all $i = 1, \cdots, r$.
- $f : \mathbb{R}^n \to \mathbb{R}$ is nonlinear and twice continuously differentiable on $\mathcal{F}$.

Throughout this work the gradient of f at x is defined as the column vector $\nabla f(x)$, and matrix $\nabla c(x) = [\nabla c_1(x), \cdots, \nabla c_r(x)]$ is the transpose of the *Jacobian* of c at x, though here, for convenience, it will simply be called the Jacobian.

Many nonlinear network flow problems (in addition to the balance constraints on the nodes and the capacity constraints on the arc flows) have nonlinear side constraints. These are termed nonlinearly constrained network flow problems, **NCNFP**. In recent works [9, 10, 11, 12], **NCNFP** has been solved using partial augmented Lagrangian methods with quadratic penalty function and superlinear-order multiplier estimates. They are the quadratic multiplier methods (QMM) cited in this paper. When $\alpha = \beta = 0$, these are based on the solution of a series of nonlinear network subproblems **(NNS)** of the form

$$\min_{x \in \mathcal{F}} \; l_\rho(x, \mu) \qquad (4)$$

where $\rho \in \mathbb{R}$, such that $\rho > 0$, and $\mu \in \mathbb{R}^r$ are fixed, and $l_\rho(x, \mu) = f(x) + \mu^t c(x) + \frac{1}{2}\rho c(x)^t c(x)$.

Thus, the only variables in **NNS** are the flows x on the arcs of the network represented by $\mathcal{F}$ (i.e., **NNS** is a pure nonlinear network flow problem), so the efficiency of network flow techniques can be exploited [4, 22].

A rough draft of this method of solving **NCNFP** is

1. Solve **NNS** for $\rho > 0$ and μ fixed. If the solution obtained by solving **NNS** is infeasible with respect to the constraints (3), go to step 2. Otherwise, the procedure ends.
2. Update the estimate μ of the Lagrange multipliers of the constraints (3) and (if necessary) the penalty coefficient ρ; then return to step 1.

Since Step 1 is efficiently solved, the key of this algorithm lies in Step 2 and especially in the updating of μ; see [15, 9, 10, 11, 12].

On the other hand, when $\alpha < \beta$ in (3) an approach that requires one multiplier per each two-sided inequality constraint is what uses the penalty function $p_j[c_j(x), \mu_j, \rho]$, which is not twice continuously differentiable; see [1, 14]. This gives rise to an augmented Lagrangian function that is continuously differentiable, but not twice continuously differentiable, even if the cost and constraint functions are. As a result, serious difficulties can arise when Newton-like methods are used to minimize the subproblem **NNS**. This motivates alternative twice differentiable augmented Lagrangian functions to handle inequality constraints. The author puts forward a way of overcoming this difficulty is to use the exponential multiplier method, as is suggested by Bertsekas in [2].

Another alternative approach to solve **NCNFP** is also considered, it is based on dual methods. In particular, we focus on the primal problem

$$\text{minimize} \quad f(x) \tag{5}$$
$$\text{subject to:} \quad x \in \mathcal{F} \tag{6}$$
$$c(x) \leq 0. \tag{7}$$

and its dual problem $\max_{\mu \geq 0} q(\mu)$, where $q(\mu) = \min_{x \in \mathcal{F}} \{f(x) + \mu^t c(x)\}$ is the dual function.

In this work to solve the dual problem we use an approximate subgradient method given that the dual function q is *approximately* computed [13].

The main aim of this work is to compare the efficiency of the exponential multiplier method (EMM) [2] and that of the approximate subgradient method (ASM) [13] with that of the quadratic multiplier method (QMM) [10, 11, 12], using as a reference the results of the codes MINOS [16] and LOQO [23]. Numerical experiments confirm the high efficiency and robustness of ASM and EMM.

This paper is organized as follows: Section 2 presents the exponential multiplier method; Section 3, the approximate subgradient method; Section 4, the solution to the nonlinearly constrained network flow problem; and Section 5 puts forward the numerical tests.

2 Exponential Multiplier Method

According to the previous section, a trouble of the quadratic multiplier method in the solution of **NCNFP** is that the penalty function $p(t, \mu, \rho)$ is not twice continuously differentiable [1]. A way of overcoming this trouble is by expressing the problem **NCNFP** in the form

$$\begin{aligned}
& \text{minimize} \quad f(x) \\
& \text{subject to:} \quad x \in \mathcal{F} \\
& \quad c_i(x) - \beta_i \leq 0, \quad i = 1, \ldots, r \\
& \quad -c_i(x) + \alpha_i \leq 0, \quad i = 1, \ldots, r
\end{aligned}$$

and introducing the exponential penalty function $\psi(t) = \exp(t) - 1$ as follows (see [2])

$$\mathcal{E}_\rho(x, \mu_+, \mu_-) = f(x) + \sum_{i=1}^r \frac{\mu_{i+}}{\rho} \psi[\rho\,(c_i(x) - \beta_i)] + \sum_{i=1}^r \frac{\mu_{i-}}{\rho} \psi[\rho\,(-c_i(x) + \alpha_i)], \tag{8}$$

where $\mu_{i+}, \mu_{i-} > 0$ for $i = 1, \ldots, r$, and each ρ is a positive penalty parameter.

Note that as Bertsekas showed in Chapter 4 of [2], for $\mu_{i+} > 0$ ($\mu_{i-} > 0$), the corresponding ith penalty term in (8)

▷ tends to ∞, as $\rho \to \infty$, for all infeasible x.
▷ tends to 0, as $\rho \to \infty$, for all feasible x.

Moreover, for fixed ρ, as $\mu_{i+}^k \to 0$ ($\mu_{i-}^k \to 0$), the penalty term goes to zero for all x, feasible or infeasible. This is contrary to what happens in QMM. Nguyen and Strodiot showed the convergence of the exponential multiplier method for nonconvex problems [18].

The application of this method to the solution of **NCNFP** is summarized in the following steps:

Step 1: At the kth iteration, for positive ρ, μ_+^k and μ_-^k, it is solved the nonlinear network subproblem

$$\underset{x \in \mathcal{F}}{\text{minimize}} \quad \mathcal{E}_\rho(x, \mu_+^k, \mu_-^k).$$

If the solution obtained x^k is feasible, the problem is already solved; otherwise,

Step 2: The multiplier estimates μ_+ and μ_- are updated according to

$$\mu_{i+}^{k+1} = (\mu_{i+}^k) \exp(\rho[c_i(x^k) - \beta_i])$$
$$\mu_{i-}^{k+1} = (\mu_{i-}^k) \exp(\rho[-c_i(x^k) + \alpha_i])$$

for $i = 1, \ldots, r$, and if necessary the penalty parameter ρ. Go to Step 1.

3 Subgradient Method

We express the dual problem of the primal problem (5-7) as

$$\text{maximize} \quad q(\mu) = \min_{x \in \mathcal{F}} l(x, \mu) = \min_{x \in \mathcal{F}} \{f(x) + \mu^t c(x)\} \tag{9}$$

$$\text{subject to:} \quad \mu \in \mathcal{M}, \tag{10}$$

where $\mathcal{M} = \{\mu \mid \mu \geq 0,\ q(\mu) > -\infty\}$. We assume throughout this section that:

▷ the constraint set $\mathcal{M}$ is closed and convex, and q is continuous on $\mathcal{M}$;
▷ for every $\mu \in \mathcal{M}$ some vector $x(\mu)$ that minimizes $l(x, \mu)$ over $x \in \mathcal{F}$ can be calculated, yielding a subgradient $c(x(\mu))$ of q at μ.

The subgradient method [6, 19, 21] consists of the iteration

$$\mu^{k+1} = [\mu^k + s_k c^k]^+, \tag{11}$$

where c^k is the subgradient $c(x(\mu^k))$, $[\cdot]^+$ denotes the projection on the closed convex set $\mathcal{M}$, and s_k is a positive scalar stepsize. The subgradient iteration looks like the gradient projection method, except that the subgradient c^k is used in place of the gradient, which may not exist. As is well known, unlike the gradient projection method, the new iteration may not improve the cost for all values of the stepsize s_k. However, what makes the subgradient method work is that for a sufficiently small stepsize s_k, the distance from the current iterate to the optimal solution set is reduced; see [21].

3.1 Approximate Subgradient Methods

When, as happens in this work, for a given $\mu \in \mathcal{M}$, the dual function value $q(\mu)$ is calculated by minimizing approximately $l(x, \mu)$ over $x \in \mathcal{F}$ (see (9)), the subgradient obtained (as well as the value of $q(\mu)$) will involve an error.

In order to analyze such methods, it is useful to introduce a notion of approximate subgradient [21, 2]. In particular, given a scalar $\varepsilon \geq 0$ and a vector $\overline{\mu}$ with $q(\overline{\mu}) > -\infty$, we say that c is an ε-*subgradient at* $\overline{\mu}$ if

$$q(\mu) \leq q(\overline{\mu}) + \varepsilon + c^t(\mu - \overline{\mu}), \qquad \forall \mu \in \mathbb{R}^r. \tag{12}$$

The set of all ε-subgradients at $\overline{\mu}$ is called the ε-*subdifferential at* $\overline{\mu}$ and is denoted by $\partial_\varepsilon q(\overline{\mu})$. Note that every subgradient at a given point is also an ε-subgradient for all $\varepsilon > 0$. Generally, however, an ε-subgradient need not be a subgradient, unless $\varepsilon = 0$.

An approximate subgradient method is defined by

$$\mu^{k+1} = [\mu^k + s_k c^k]^+, \tag{13}$$

where c^k is an ε_k-subgradient at μ^k and s_k a positive stepsize.

In our context, we minimize approximately $l(x, \mu^k)$ over $x \in \mathcal{F}$, thereby obtaining a vector $x^k \in \mathcal{F}$ with

$$l(x^k, \mu^k) \leq \inf_{x \in \mathcal{F}} l(x, \mu^k) + \varepsilon_k. \tag{14}$$

As is shown in [13], the corresponding constraint vector, $c(x^k)$, is an ε_k-subgradient at μ^k. If we denote $q_{\varepsilon_k}(\mu^k) = l(x^k, \mu^k)$, by definition of $q(\mu^k)$ and using (14) we have

$$q(\mu^k) \leq q_{\varepsilon_k}(\mu^k) \leq q(\mu^k) + \varepsilon_k \qquad \forall k. \tag{15}$$

3.2 Dynamic Stepsize Rule with Relaxation-Level Control

The dynamic stepsize rule consists in

$$s_k = \gamma_k \frac{q^* - q(\mu^k)}{\|c^k\|^2}, \tag{16}$$

where $q^* = \sup_{\mu \in \mathcal{M}} q(\mu)$, $c^k \in \partial q(\mu^k)$, $0 < \underline{\gamma} \leq \gamma_k \leq \overline{\gamma} < 2$. This stepsize was introduced by Poljak in [20] (see also Shor [21]).

Unfortunately, in most practical problems q^* and $q(\mu^k)$ are not known. The latter is approximated by $q_{\varepsilon_k}(\mu^k)$, which fulfils the inequalities (15). Regarding q^*, we may modify the dynamic stepsize (16) by replacing q^* with an estimate. This leads to the stepsize rule

$$s_k = \gamma_k \frac{q_{lev}^k - q_{\varepsilon_k}(\mu^k)}{\|c^k\|^2}, \qquad 0 < \underline{\gamma} \leq \gamma_k \leq \overline{\gamma} < 2, \qquad \forall k \geq 0 \tag{17}$$

where q_{lev}^k is an estimate of q^* and $c^k \in \partial_{\varepsilon_k} q(\mu^k)$ is bounded for $k = 0, 1, \ldots$

The dynamic stepsize rule with relaxation-level control is based on the algorithm given by Brännlund [3], whose convergence was shown by Goffin and Kiwiel in [7] when $\varepsilon_k = 0$ for all k. The following algorithm was fitted by Mijangos for the dual problem (9 -10) and $\varepsilon_k \to 0$ in [13].

Algorithm 1

Step 0 (*Initialization*): Select μ^0, $\delta_0 > 0$, and $B > 0$.
Set $\sigma_0 = 0$ and $q_{rec}^{-1} = \infty$.
Set $k = 0$, $l = 0$ and $k(l) = 0$, where $k(l)$ will denote the iteration k when the lth update of q_{lev}^k occurs. Then $k(l) = k$ will be set.

Step 1 (*Function evaluation*): Compute $q_{\varepsilon_k}(\mu^k)$ and $c^k \in \partial_{\varepsilon_k} q(\mu^k)$.
If $q_{\varepsilon_k}(\mu^k) > q_{rec}^{k-1}$, set $q_{rec}^k = q_{\varepsilon_k}(\mu^k)$.
Otherwise set $q_{rec}^k = q_{rec}^{k-1}$.

Step 2 (*Stopping rule*): If $\|c^k\| = 0$, terminate with $\mu^* = \mu^k$.

Step 3 (*Sufficient ascent detection*): If $q_{\varepsilon_k}(\mu^k) \geq q_{rec}^{k(l)} + \frac{1}{2}\delta_k$, set $k(l+1) = k$, $\sigma_k = 0$, $\delta_{l+1} = \delta_l$, $l := l+1$, and go to Step 5.

Step 4 (*Oscillation detection*): If $\sigma_k > B$, set $k(l+1) = k$, $\sigma_k = 0$, $\delta_{l+1} = \frac{1}{2}\delta_l$, $l := l+1$.

Step 5 (*Iterate update*): Set $q_{lev}^k = q_{rec}^{k(l)} + \delta_l$. Choose $\gamma \in [\underline{\gamma}, \overline{\gamma}]$ and compute μ^{k+1} by means of (13) with the stepsize s_k obtained by (17).

Step 6 (*Path long update*): Set $\sigma_{k+1} = \sigma_k + s_k\|c^k\|$, $k := k+1$, and go to Step 1.

Some remarks about this algorithm:

- q_{rec}^k keeps the record of the highest value attained by the iterates that are generated so far; i.e., $q_{rec}^k = \max_{0 \leq j \leq k} q_{\varepsilon_j}(\mu^j)$.
- The algorithm uses the same target level $q_{lev}^k = q_{rec}^{k(l)} + \delta_l$ for $k = k(l), k(l)+1, k(l)+2, \ldots, k(l+1) - 1$.
- The target level is updated only if sufficient ascent or oscillation is detected (in Step 3 and Step 4, respectively).
- It can be shown that the value σ_k is an upper bound on the length of the path travelled by iterates $\mu^{k(l)}, \ldots, \mu^k$ for $k < k(l+1)$.
- Whenever σ_k exceeds the given upper bound B on the path length, the parameter δ_i is decreased, which decreases the target level q_{lev}^k.

In [13] we analize the convergence of the ε-subgradient method (13) with the stepsize (17) for q_{lev} given by this algorithm.

3.3 Extension to Two-Sided Inequalities

The theory developed so far for the approximate subgradient methods when the side constraints are $c(x) \leq 0$ (see (5-7)) can be extended to handle two-sided inequality contraints of the form $\alpha_i \leq c_i(x) \leq \beta_i$, for $i = 1, \ldots, r$. A constraint of this type can be converted into the inequality constraints $c_i(x) - \beta_i \leq 0$ and $-c_i(x) + \alpha_i \leq 0$. In particular, if we replace each two-sided inequality constraint in the original problem **NCNFP** (see (1-3)) by two inequalities as above, then the dual function for **NCNFP** can be expressed as

$$q(\mu_+, \mu_-) = \min_{x \in \mathcal{F}} l(x, \mu_+, \mu_-),$$

where $\mu_+ = (\mu_{1+}, \ldots, \mu_{r+})$, $\mu_- = (\mu_{1-}, \ldots, \mu_{r-})$, and μ_{i+} and μ_{i-} are the multiplier estimates of the constraints $c_i(x) - \beta_i \leq 0$ and $-c_i(x) + \alpha_i \leq 0$. So we have

$$l(x, \mu_+, \mu_-) = f(x) + \sum_{i=1}^{r} \mu_{i+}(c_i(x) - \beta_i) + \sum_{i=1}^{r} \mu_{i-}(-c_i(x) + \alpha_i).$$

In addition, if x^k is an approximate solution of the minimization of the Lagrangian function over $x \in \mathcal{F}$ for (μ_+^k, μ_-^k), then we use the iteration

$$\mu_{i+}^{k+1} = \begin{cases} (\mu_{i+}^k + s^k(c_i^k - \beta_i))^+, & \text{if } c_i^k - \beta_i > 0 \\ 0, & \text{otherwise} \end{cases}$$

$$\mu_{i-}^{k+1} = \begin{cases} (\mu_{i-}^k + s^k(-c_i^k + \alpha_i))^+, & \text{if } -c_i^k + \alpha_i > 0 \\ 0, & \text{otherwise,} \end{cases}$$

where $c_i^k = c_i(x^k)$, and $(c_i^k - \beta_i)$, $(-c_i^k + \alpha_i)$ are the corresponding ε_k-subgradients.

4 Solution to NCNFP

An algorithm is given below for solving **NCNFP**. This algorithm uses the approximate subgradient method described in Section 3 for $c(x) \leq 0$ constraints. It can easily be extended to two-sided inequality constraints, as can be seen in Section 3.3.

The value of the dual function $q(\mu^k)$ is estimated by minimizing approximately $l(x, \mu^k)$ over $x \in \mathcal{F}$ (the set defined by the network constraints) so that the optimality tolerance, τ_x^k, becomes more rigorous as k increases; i.e., the minimization will be *asymptotically exact* [1]. In other words, we set $q_{\varepsilon_k}(\mu^k) = l(x^k, \mu^k)$, where x^k minimizes approximately the nonlinear network subproblem **NNS$_k$**

$$\min_{x \in \mathcal{F}} l(x, \mu^k)$$

in the sense that this minimization stops when we obtain a x^k such that

$$\|Z^t \nabla_x l(x^k, \mu^k)\| \leq \tau_x^k$$

where $\lim_{k \to \infty} \tau_x^k = 0$ and Z represents the reduction matrix whose columns form a base of the null subspace of the subspace generated by the rows of the matrix of active network constraints of this subproblem (see [17]).

Let $\bar{x}^k$ be the minimizer of this subproblem approximated by x^k. As is shown in [13], there exists a positive w, such that for $k = 1, 2, \ldots$ we have

$$l(x^k, \mu^k) \leq l(\bar{x}^k, \mu^k) + w\tau_x^k$$

If we set $\varepsilon_k = \omega\tau_x^k$, this inequality becomes (14) such that $\lim_{k\to\infty}\varepsilon_k = 0$. Consequently, we can denote $q_{\varepsilon_k} = l(x^k, \mu^k)$, which holds the inequality (15), and we may use the method described in Section 3.

Algorithm 2 (Approximate subgradient method for NCNFP)

Step 0 *Initialize.* Set $k = 1$, N_{max}, τ_x^1, ϵ_μ and τ_μ. Set $\mu^1 = 0$.

Step 1 *Compute* the dual function estimate, $q_{\varepsilon_k}(\mu^k)$, by solving **NNS$_k$**, so that if $\|Z^t\nabla_x l(x^k, \mu^k)\| \leq \tau_x^k$, then $x^k \in \mathcal{F}$ is an approximate solution, $q_{\varepsilon_k}(\mu^k) = l(x^k, \mu^k)$, and $c^k = c(x^k)$ is an ε_k-subgradient of q in μ^k.

Step 2 *Check the stopping rules* for μ^k.

T_1: Stop if $\displaystyle\max_{i=1,\ldots,r}\left\{\frac{(c_i^k)^+}{1+(c_i^k)^+}\right\} < \tau_\mu$, where $(c_i^k)^+ = \max\{0, c_i(x^k)\}$.

T_2: Stop if $\left|\dfrac{q^k - (q^{k-1} + q^{k-2} + q^{k-3})/3}{1+q^k}\right| < \epsilon_\mu$, where $q^n = q_{\varepsilon_n}(\mu^n)$.

T_3: Stop if $\dfrac{1}{4}\displaystyle\sum_{i=0}^{3}\|\mu^{k-i} - \mu^{k-i-1}\|_\infty < \epsilon_\mu$.

T_4: Stop if k reaches a prefixed value N_{max}.

If μ^k fulfils one of these tests, then it is optimal, and the algorithm stops. Without a duality gap, (x^k, μ^k) is a primal-dual solution.

Step 3 *Update* the estimate μ^k by means of the iteration

$$\mu_i^{k+1} = \begin{cases} \mu_i^k + s_k c_i^k, & \text{if } \mu_i^k + s_k c_i^k > 0 \\ 0, & \text{otherwise} \end{cases}$$

where s_k is computed using some stepsize rule. Go to Step 1.

In Step 0, for the checking of the stopping rules, $\tau_\mu = 10^{-5}$, $\epsilon_\mu = 10^{-5}$ and $N_{max} = 200$ have been taken. In addition, $\tau_x^0 = 10^{-1}$ by default.

Step 1 of this algorithm is carried out by the code PFNL, described in [14], which is based on the specific procedures for nonlinear networks flows [4, 22] and the active set procedure of Murtagh and Saunders [17] using a spanning tree as the basis matrix of the network constraints. Moreover,

$$\tau_x^{k+1} = \alpha\tau_x^k, \qquad \text{for a fixed } \alpha \in (0,1).$$

In this work, $\alpha = 10^{-1}$ by default. Note that in this case, $\varepsilon_k = \tau_x^k \omega = 10^{-k-1}\omega$.

In Step 2, alternative heuristic tests have been used for practical purposes. T_1 checks the feasibility of x^k, as if it is feasible the duality gap is zero, and then (x^k, μ^k) is a primal-dual solution for **NCNFP**. T_2 and T_3 mean that μ does not improve for the last N iterations. Note that $N = 4$.

To obtain s_k in Step 3, we have used the iteration (13) with the dynamic stepsize rule given by the level control Algorithm 1, in which B is replaced with $B_l = \max\{\overline{B}, B/l\}$ when an oscillation is detected, for $B = 10^{-3}\|x^1 - x^0\|$ and

Table 1. Comparison of the CPU-times in seconds

Problem	t_{MIN}	t_{LOQO}	t_{QMM}	t_{EMM}	t_{ASM}
D12e2	4.0	2.8	2.3	1.5	0.9
D13e2	5.4	40.6	4.3	1.6	1.0
D14e2	–	47.1	32.9	11.9	7.3
D12n1	178.7	1831.4	104.6	93.9	89.6
D13n1	294.3	1570.3	141.1	105.6	108.8
D14n1	–	2514.1	371.7	240.1	236.6
D21e2	32.3	170.5	5.0	4.1	3.6
D22e2	55.5	511.0	12.5	5.2	4.4
D23e2	1457.1	72.9	46.1	12.8	13.6
D31e1	88.1	–	4.7	3.2	3.8
D31e2	63.5	35.1	2.9	2.2	2.2

$\overline{B} = 0.01$. As can be seen, $\sum_{l=1}^{\infty} B_l = \infty$. In addition, we set $\gamma_k = 1$ for all k, $\delta = 10^{-5}|l(x_0, \mu_0)|$, $\delta_0 = 0.5\|(c^1)^+\|B$

The values given above have been heuristically chosen. The implementation in Fortran-77 of the previous algorithm, termed PFNRN05 (see [8]), was designed to solve large-scale nonlinear network flow problems with nonlinear side constraints.

5 Numerical Tests

The *artificial problems* used in these tests were created by means of the following DIMACS-random-network generators: Rmfgen and Gridgen; see [5]. These generators provide linear flow problems in networks without side constraints. The inequality nonlinear side constraints for the DIMACS networks were generated through the *Dirnl* random generator described in [14], which gives rise to the problems whose names start with **D**. The last two letters indicate the type of objective function that we have used for DIMACS problems: Namur functions, **n***, and EIO1 functions, **e***. More information about the features of these problems can be found in [15].

In Table 1 we have the CPU-times spent by PFNRN using the exponential multiplier method, t_{EMM}, the approximate subgradient method, t_{ASM}, and the quadratic multiplier method, $_{QMM}$ [12], are compared with those spent by MINOS [16], t_{MIN}, and LOQO [23], t_{LOQO}. The results of MINOS and LOQO have been obtained on the NEOS server with AMPL input. PFNRN with QMM, ASM, and EMM, has solved the problems on a Sun Sparc Ultra-250 work station under UNIX, which has a similar speed to that of the UFFIZI machine of the NEOS server.

The comparison of the CPU-times spent by these PFNRN's versions with MINOS and LOQO indicate clearly that PFNRN is more efficient than MINOS and LOQO for this kind of problems.

Likewise, Table 1 points out that PFNRN for both EMM and ASM has a similar efficiency, which is slightly better than for QMM. In addition, EMM has improved the times of QMM in all of these problems, but is highly dependent on the initial penalty parameter, which does not exist for ASM.

References

1. Bertsekas, D.P. Constrained Optimization and Lagrange Multiplier Methods. Academic Press, New York (1982)
2. Bertsekas, D.P. Nonlinear Programming. 2nd ed., Athena Scientific, Belmont, Massachusetts (1999)
3. Brännlund, U. On relaxation methods for nonsmooth convex optimization. Doctoral Thesis, Royal Institute of Technology, Stockholm, Sweden (1993)
4. Dembo, R.S. A primal truncated Newton algorithm with application to large-scale nonlinear network optimization. Mathematical Programming Studies, Vol. 31 (1987) 43–71
5. DIMACS. The first DIMACS international algorithm implementation challenge : The bench-mark experiments. Technical Report, DIMACS, New Brunswick, NJ, USA (1991)
6. Ermoliev, Yu. M. Methods for solving nonlinear extremal problems. Cybernetics, Vol. 2 (1966) 1–17
7. Goffin, J.L. and Kiwiel, K. Convergence of a simple subgradient level method. Mathematical Programming, Vol. 85 (1999) 207–211
8. Mijangos, E. PFNRN03 user's guide. Technical Report 97/05, Dept. of Statistics and Operations Research. Universitat Politècnica de Catalunya, 08028 Barcelona, Spain (1997) (downloadable from website http://www.ehu.es/~mepmifee/)
9. Mijangos, E. On superlinear multiplier update methods for partial augmented Lagrangian techniques. Qüestiio, Vol. 26 (2002) 141–171
10. Mijangos, E. An implementation of Newton-like methods on nonlinearly constrained networks. Computers and Operations Research, Vol. 32(2) (2004) 181–199
11. Mijangos, E. On the efficiency of multiplier methods for nonlinear network problems with nonlinear constraints. IMA Journal of Management Mathematics, Vol. 15(3) (2004) 211–226
12. Mijangos, E. An efficient method for nonlinearly constrained networks. European Journal of Operational Research, Vol. 161(3) (2005) 618–635
13. Mijangos, E. Approximate subgradient methods for nonlinearly constrained network flow problems. Submitted to Journal on Optimization Theory and Applications (2004)
14. Mijangos, E. and Nabona, N. The application of the multipliers method in nonlinear network flows with side constraints. Technical Report 96/10, Dept. of Statistics and Operations Research. Universitat Politècnica de Catalunya, 08028 Barcelona, Spain (1996) (downloadable from website http://www.ehu.es/~mepmifee/)
15. Mijangos, E. and Nabona, N. On the first-order estimation of multipliers from Kuhn-Tucker systems. Computers and Operations Research, Vol. 28 (2001) 243–270
16. Murtagh, B.A. and Saunders, M.A. MINOS 5.5. User's guide. Report SOL 83-20R, Department of Operations Research, Stanford University, Stanford, CA, USA (1998)

17. Murtagh, B.A. and Saunders, M.A. Large-scale linearly constrained optimization. Mathematical Programming, Vol. 14 (1978) 41–72
18. Nguyen, V.H. and Strodiot, J.J. On the Convergence Rate for a Penalty Function of Exponential Type. Journal of Optimization Theory and Applications, Vol. 27(4) (1979) 495–508
19. Poljak, B.T. A general method of solving extremum problems, Soviet Mathematics Doklady, Vol. 8 (1967) 593–597
20. Poljak, B.T. Minimization of unsmooth functionals, Z. Vyschisl. Mat. i Mat. Fiz., Vol. 9 (1969) 14–29
21. Shor, N.Z. Minimization methods for nondifferentiable functions, Springer-Verlag, Berlin (1985)
22. Toint, Ph.L. and Tuyttens, D. On large scale nonlinear network optimization. Mathematical Programming, Vol. 48 (1990) 125–159
23. Vanderbei, R.J. LOQO: An interior point code for quadratic programming. Optimization Methods and Software, Vol. 12 (1999) 451–484

A First-Order ε-Approximation Algorithm for Linear Programs and a Second-Order Implementation

Ana Maria A.C. Rocha[1], Edite M.G.P. Fernandes[1], and João L.C. Soares[2]

[1] Departamento de Produção e Sistemas, Universidade do Minho, Portugal
{arocha, emgpf}@dps.uminho.pt
[2] Departamento de Matemática, Universidade de Coimbra, Portugal
jsoares@mat.uc.pt

Abstract. This article presents an algorithm that finds an ε-feasible solution relatively to some constraints of a linear program. The algorithm is a first-order feasible directions method with constant stepsize that attempts to find the minimizer of an exponential penalty function. When embedded with bisection search, the algorithm allows for the approximated solution of linear programs. We present applications of this framework to set-partitioning problems and report some computational results with first-order and second-order implementations.

1 Introduction

Set-covering, -partitioning and -packing models arise in many applications like crew scheduling (trains, buses or airplanes), political districting, protection of microdata, information retrieval, etc. Typically these models are suboptimally solved by heuristics because an optimization framework (usually of the branch-and-price type) has to be rather specialized, if feasible at all. Moreover, a branch-and-price framework requires the solution of large linear programs at every node of the branch-and-price tree and these linear programs may take a long time and storage to be solved to optimality.

Our framework attempts to find reasonable approximate solutions to those models quickly and without too much storage, along the lines of Lagrangian relaxation. The approximation obtained may serve the purpose of speeding-up the optimal basis identification by simplex-type algorithms. We will be looking for an approximated solution of a linear program in the following form

$$\begin{aligned} z_* \equiv \min\ & cx \\ \text{s.t.}\ & Ax \geq b\ , \\ & x \in P\ , \end{aligned} \qquad (1)$$

where A is a real $m \times n$ matrix, b is a real m-dimensional vector and $P \subseteq \mathbb{R}^n$ is a compact set (possibly, a lattice) over which optimizing linear programs is considered "easy". For example, in a set-covering model, A is a matrix of zeros and ones, b is a vector of all-ones, and the set P is the lattice $\{0,1\}^n$, or the hypercube $[0,1]^n$ in the fractional version. If P includes a budget constraint then P becomes the feasible region of a $0-1$ knapsack problem or a fractional $0-1$ knapsack problem, respectively. We also assume that conv(P) is a polyhedron.

In recent years, several researchers have developed algorithms (see [3, 4, 2]) that seek to produce approximate solutions to linear programs of this sort, by constructing an exponential potential function which serves as a surrogate for feasibility and optimality.

We focus on obtaining a reasonable approximation to the optimal solution of (1) by an ε-feasible solution. We say that a point x is ε-feasible relatively to the constraints $Ax \geq b$ if

$$\lambda(x) \equiv \max_{i=1,\ldots,m} (b_i - a_i x) \leq \varepsilon , \qquad (2)$$

where a_i denotes the ith row of the matrix A. To achieve this, we will attempt to solve the following convex program

$$\Phi(\alpha, z) = \min \phi_\alpha(x) \equiv \sum_{i=1}^{m} \exp\left(\alpha\left(b_i - a_i x\right)\right) \qquad (3)$$
$$\text{s.t. } x \in P'(z) \equiv \text{conv}\left(P \cap \{x : cx \leq z\}\right)$$

where conv$(\cdot)$ denotes convex hull, for several adequate values of the parameters α and z. The scalar α is a penalty parameter and z is a guess for the value of z_*. To solve the nonlinear program (3) we propose a first-order feasible directions method with constant stepsize whose running time depends polynomially on $1/\varepsilon$ and the *width* of the set $P'(z)$ relatively to the constraints $Ax \geq b$. Significantly, the running time does not depend explicitly on n and, hence, it can be applied when n is exponentially large, assuming that, for a given row vector y, there exists a polynomial subroutine to optimize yAx over $P'(z)$.

This paper is organized as follows. In Sect. 2 we describe the ε-approximation algorithm used in this work. Then, in Sect. 3 we present the first-order algorithm with fixed stepsize. Our computational experience on set-partitioning problems is presented in Sect. 4 and Sect. 5 contains some conclusions and future work.

2 Main ε-Approximation Algorithm

If x is feasible in (1) then $\phi_\alpha(x) \leq m$, while if x is simply ε-feasible then $\phi_\alpha(x) \leq m\exp(\alpha\varepsilon)$. On the other hand, if x is not ε-feasible then $\phi_\alpha(x) > \exp(\alpha\varepsilon)$. We will choose α so that it may be possible to assert whether x is ε-feasible from the value of $\phi_\alpha(x)$, as formally stated in the next lemma.

Lemma 1. *If $\alpha \geq \ln((1+\varepsilon)m)/\varepsilon$ then,*

1. *if there is no ε-feasible solution relatively to '$Ax \geq b$' such that $x \in P'(z)$ then $\Phi(\alpha, z) > (1+\varepsilon)m$.*
2. *if $\phi_\alpha(x) \leq (1+\varepsilon)m$ then x is an ε-feasible solution relatively to '$Ax \geq b$'.*

Proof. If there is no ε-feasible solution relatively to '$Ax \geq b$' such that $x \in P'(z)$ then $\varepsilon < \varepsilon' \equiv \min\{\lambda(x)\colon x \in P'(z)\}$. Thus, $\Phi(\alpha, z) > \exp(\alpha\varepsilon') \geq (1+\varepsilon)m$. If $\phi_\alpha(x) \leq (1+\varepsilon)m$ then $\exp(\alpha(b_i - a_i\bar{x})) \leq (1+\varepsilon)m$, for any $i = 1, \ldots, m$. Equivalently, $b_i - a_i\bar{x} \leq \ln((1+\varepsilon)m)/\alpha \leq \varepsilon$, for any $i = 1, \ldots, m$. □

Keeping the value of α fixed, we will use bisection to search for the minimum value of z such that $P'(z) \cap \{x\colon Ax \geq b\}$ is nonempty, being driven by (3). The bisection search maintains an interval $[z_a, z_b]$ such that $P'(z_a) \cap \{x\colon Ax \geq b\}$ is empty, and there is some $x \in P'(z_b)$ that is ε-feasible relatively to '$Ax \geq b$'. The search is interrupted when $z_b - z_a$ is small enough to guarantee $z_b \leq z_*$. This does not imply any bound on how much z_b differs from z_*. It may be possible that z_b is much less than z_* though very unlikely for α large, as it is implicit from Proposition 1 below.

Proposition 1. *Let the sequence $\{\alpha_k\}$ be such that $\lim_k \alpha_k = +\infty$ and y^k be a vector defined componentwise by*

$$y_i^k = \alpha_k \exp\left(\alpha_k \left(b_i - a_i x^k\right)\right) \quad (i = 1, 2, \ldots, m) , \tag{4}$$

where x^k is optimal in $\Phi(\alpha_k, z_*)$. Then, for every accumulation point $(\bar{x}, \bar{y})$ of the sequence $\{(x^k, y^k)\}$, $\bar{x}$ is optimal for program (1).

Proof. Assume $\lim_{k \in K}(x^k, y^k) = (\bar{x}, \bar{y})$. Since P is closed, $\bar{x} \in P'(z_*)$. If we show that $A\bar{x} \geq b$ then $\bar{x}$ must be optimal in (1). Since $\lim_{k \in K} \alpha_k \exp\left(\alpha_k\left(b_i - a_i x^k\right)\right)$ exists then $\bar{y}_i = \lim_{k \in K} \alpha_k \exp\left(\alpha_k (b_i - a_i \bar{x})\right)$, from where we conclude that $a_i \bar{x} \leq b_i$, for every $i = 1, 2, \ldots, m$. □

If $[z_a^k, z_b^k]$ is the last bisection interval of the search for a given ε_k then, we may restart the bisection search with $[z_b^k, z_b]$ for some $\varepsilon_{k+1} < \varepsilon_k$, for example, $\varepsilon_{k+1} = \varepsilon_k/2$. The value of z_b denotes a proven upper bound on the value of z_*. Note that an ε_k-feasible solution is known at the left extreme of the new interval and furthermore it belongs to $P'(z)$, for any $z \in [z_b^k, z_b]$. Thus, it may serve as initial solution for the minimization of ϕ_α when $\varepsilon = \varepsilon_{k+1}$ and will not *overflow* the exponential function evaluations.

Each iteration of our main algorithm consists of a number of iterations of bisection search on the z value for a fixed value of the parameter ε. Then, ε is decreased, the bisection interval is updated and bisection search restarts. This process is terminated when ε is small enough.

From Lemma 1, if $\Phi(\varepsilon, z) \leq (1 + \varepsilon)m$ then there is $x \in P'(z)$ that is ε-feasible relatively to '$Ax \geq b$' (for example, the optimal solution); otherwise, there is no feasible solution x in (1) satisfying $cx \leq z$. The new interval $[z_a^{k+1}, z_b^{k+1}]$ is adequately updated and k is increased by one unit. Termination occurs when $z_b^k - z_a^k$ is small enough. The algorithm is formally described in Algorithm 1. Step 2 of the algorithm **Bisection search** involves applying an off-the-shelf convex optimization method to (3). Note that, we always have $x_a^k \in P'(z)$ which makes it a natural starting point for the corresponding optimization algorithm.

Algorithm 1. Bisection search $(\varepsilon, x_a, z_a, x_b, z_b)$
Given $0 < \varepsilon$, and $x_a \in P'(z_a)$, $x_b \in P'(z_b)$ such that

$$\Phi(\varepsilon, z_b) \leq \phi_\varepsilon(x_b) \leq (1+\varepsilon)m < \Phi(\varepsilon, z_a) \leq \phi_\varepsilon(x_a) . \tag{5}$$

Initialization: Set $(x_a^0, z_a^0, x_b^0, z_b^0) := (x_a, z_a, x_b, z_b)$ and $k := 0$.

Generic Iteration k:

Step 1: If $z_b^k - z_a^k = 1$ then set $(x_a, z_a, x_b, z_b) := (x_a^k, z_a^k, x_b^k, z_b^k)$ and STOP.
Step 2: Set $z := \lceil (z_a^k + z_b^k)/2 \rceil$ and obtain $\bar{x} \in P'(z)$ such that

$$\text{either} \quad \Phi(\varepsilon, z) \leq \phi_\varepsilon(\bar{x}) \leq (1+\varepsilon)m , \tag{6}$$

$$\text{or} \quad (1+\varepsilon)m < \Phi(\varepsilon, z) \leq \phi_\varepsilon(\bar{x}) . \tag{7}$$

Step 3: Set $(x_a^{k+1}, z_a^{k+1}, x_b^{k+1}, z_b^{k+1}) := \begin{cases} (x_a^k, z_a^k, \bar{x}, z) & \text{if (6) holds,} \\ (\bar{x}, z, x_b^k, z_b^k) & \text{if (7) holds.} \end{cases}$
Set $k := k+1$.

We may now present a formal description of the **Main** algorithm (see Algorithm 2). On entry: $z_a = \lceil \min\{cx - y(Ax - b) : x \in P\} \rceil$ is an integral lower bound on the value of z_*, for some fixed $y \geq 0$; $x_a \in P'(z_a)$ and Δ is a positive integer related to the initial amplitude of the starting intervals in each bisection search. Overall, we have the following convergence result. After a finite number of calls, the last interval $[z_b - 1, z_b]$ of the **Bisection search** routine is such that $z_b = \lceil \bar{z} \rceil$, where $\bar{z} = \min\{z : x \in P'(z), Ax \geq b\}$. In general $\bar{z} \leq z_*$, with equality if P is a polyhedron.

Algorithm 2. Main $(z_a, x_a, z_b, x_b, \Delta)$
Given $z_a \leq z_*$, $x_a \in P'(z_a)$ and a positive integer Δ.
Initialization:
  Choose $\varepsilon > 0$ so that x_a do not *overflow* $\phi_\varepsilon(x_a)$ and $\Phi(\varepsilon, z_a) > (1+\varepsilon)m$.
  While $\Phi(\varepsilon, z_a + \Delta) > (1+\varepsilon)m$ redefine x_a and set $z_a := z_a + \Delta$.
  Define x_b as the last solution found and set $z_b := z_a + \Delta$.
Generic Iteration:

Step 1: Call **Bisection search** $(\varepsilon, x_a, z_a, x_b, z_b)$.
Step 2: While $(\Phi(\varepsilon/2, z_b) \leq (1+\varepsilon/2)m$ and ε is not small enough)
          redefine x_b and set $\varepsilon := \varepsilon/2$.
      If (ε is small enough)
      Then   set x_b as the last solution found and STOP.
Step 3: Set $x_a := x_b$ and $z_a := z_b$.
      While $(\Phi(\varepsilon, z_b + \Delta) > (1+\varepsilon)m$) redefine x_a, z_a and set $z_b := z_b + \Delta$.
      Set x_b as the last solution found and set $z_b := z_b + \Delta$.
      Repeat *Step 1*.

3 A First-Order Algorithm with Fixed Stepsize

Our conceptual algorithm to solving (3) is a first-order iterative procedure with a fixed stepsize. The algorithm coincides with the algorithm Improve-Packing proposed in [5] but the stepsize and the stopping criterion are different. The direction of movement at a generic iterate $\bar{x} \in P'(z)$, that is not ε-feasible relatively to the constraints '$Ax \geq b$', is determined from solving the following linear program

$$\begin{aligned} \min\ & \phi_\varepsilon(\bar{x}) + \nabla\phi_\varepsilon(\bar{x})(x - \bar{x}) \\ \text{s.t.}\ & x \in P'(z) \ . \end{aligned} \qquad (8)$$

If $\hat{x}$ is optimal in (8) then we reset $\bar{x}$ to $\bar{x} + \hat{\sigma}(\hat{x} - \bar{x})$, for some fixed stepsize $\hat{\sigma} \in (0, 1]$, and proceed analogously to the next iteration. The conceptual algorithm is halted when $\phi_\varepsilon(\bar{x}) \leq (1+\varepsilon)m$ or a maximum number of iterations is reached. Of course, in practice other stopping criteria should be accounted for. For example, notice that (8) is equivalent to

$$\begin{aligned} \max\ & (\bar{y}A)\, x \\ \text{s.t.}\ & cx \leq z \ , \\ & x \in P \ , \end{aligned} \qquad (9)$$

for $\bar{y}$ defined componentwise by $\bar{y}_i = \alpha\exp(\alpha(b_i - a_i\bar{x}))$, for $i = 1, 2, \ldots, m$. This is essentially because $\nabla\phi_\varepsilon(\bar{x}) = -\bar{y}A$. Then, since $\bar{y} \geq 0$, the inequality '$\bar{y}Ax \geq \bar{y}b$' is valid for the polyhedron $\{x\colon Ax \geq b\}$. Thus, if at some point of the solution procedure of (3) we have that the optimal value of (9) is smaller than $\bar{y}b$ then we may immediately conclude that $P'(z) \cap \{x\colon Ax \geq b\}$ is empty.

Theorem 1 below presents one particular choice for the fixed stepsize $\hat{\sigma}$. It depends on the following quantity, introduced as the *width* of $P'(z)$ relatively to the constraints '$Ax \geq b$' in [5],

$$\rho \equiv \left\{ \begin{array}{l} \sup \|Ax - Ay\|_\infty \\ \text{s.t.}\ x, y \in P'(z) \end{array} \right\} = \left\{ \max_{i=1,2,\ldots,m} \left(\begin{array}{l} \sup |a_ix - a_iy| \\ \text{s.t.}\ x, y \in P'(z) \end{array} \right) \right\} \ . \qquad (10)$$

In [5], ρ is defined differently depending when whether the matrix A is such that $Ax \geq 0$, for every $x \in P'(z)$, or not. If yes, then the two definitions coincide with $\rho = \max_i \max_{x \in P'(z)} a_ix$. If not, then our definition of ρ is half of the ρ that is proposed in [5].

Theorem 1. *([6]) Assume that $z_* \leq z$, $\bar{x} \in P'(z)$ and it is not ε-feasible (relatively to '$Ax \geq b$'), for some $\varepsilon \in (0,1)$, $\hat{x} \in P'(z)$ is optimal for program (8), ρ is given by (10) and*

$$\alpha \geq \max\left(\frac{\ln(m(3+\varepsilon))}{\varepsilon}, \frac{1}{\rho \ln 2} \right) \ . \qquad (11)$$

Then, for $\hat{\sigma} = 1/(\alpha\rho)^2$ we have that

$$\phi_\varepsilon\left(\bar{x} + \hat{\sigma}(\hat{x} - \bar{x})\right) < \left(1 - \frac{1}{4(\alpha\rho)^2}\frac{1+\varepsilon}{3+\varepsilon}\right)\phi_\varepsilon(\bar{x}) \ . \qquad (12)$$

In summary, assuming that $z_* \leq z$, if the first k iterates are not ε-feasible then

$$\phi_\varepsilon(x^{k+1}) < \left(1 - \frac{1}{4(\alpha\rho)^2}\frac{1+\varepsilon}{3+\varepsilon}\right)^k \phi_\varepsilon(x^0) , \qquad (13)$$

where, we note, that the right hand side goes to zero as k goes to $+\infty$. The following corollary states a worst-case complexity bound on the number of iterations of the algorithm.

Corollary 1. *([6]) If α satisfies (11) and $\varepsilon \in (0,1)$ then our algorithm, using $\hat{\sigma} = 1/(\alpha\rho)^2$ and starting from $x^0 \in P'(z)$, terminates after*

$$\frac{\ln(m) + \ln(1+\varepsilon) - \ln\phi_\varepsilon(x^0)}{\ln\left(1 - \frac{1}{4(\alpha\rho)^2}\frac{1+\varepsilon}{3+\varepsilon}\right)} < 16\alpha^3\rho^2\lambda(x^0) \qquad (14)$$

iterations, with an $x \in P'(z)$ that is ε-feasible relatively to '$Ax \geq b$' or, otherwise, with the proof that there is no x feasible in (1) such that $cx \leq z$.

Note that the right hand side of (14) is $\mathcal{O}(\ln^3(m)\varepsilon^{-3}\rho^2\lambda(x^0))$. If $\lambda(x^0) = \mathcal{O}(\varepsilon)$ then only $\mathcal{O}(\ln^3(m)\varepsilon^{-2}\rho^2) = \tilde{\mathcal{O}}(\varepsilon^{-2}\rho^2)$ iterations are required. This complexity result is related to Karger and Plotkin's [4–Theorem 2.5] ($\tilde{\mathcal{O}}(\varepsilon^{-3}\rho)$) and Plotkin, Shmoys and Tardos's [5–Theorem 2.12] ($\tilde{\mathcal{O}}(\varepsilon^{-2}\rho\ln\varepsilon^{-1})$). The result of Karger and Plotkin is valid even if the budget constraint is included in the objective function of (3) without counting in the definition of ρ. We recall that a function $f(n)$ is said to be $\tilde{\mathcal{O}}(g(n))$ if there exists a constant c such that $f(n)\ln^c(n) \geq \mathcal{O}(g(n))$.

Given an initial point $\bar{x} \in P'(z)$, the Algorithm 3 contains a practical implementation of the first-order algorithm to solving (3).

Algorithm 3. Solve_subproblem $(\bar{x}, \text{flag})$
Given $\bar{x} \in P'(z)$ and a small positive tolerance δ.
Generic Iteration:

Step 1: If $(\phi_\varepsilon(\bar{x}) \leq (1+\varepsilon)m)$
  Then set flag := **CP1** and STOP;
  Else If (maximum number of iterations is reached)
    Then set flag := **CP4** and STOP.

Step 2: Let $\hat{x}$ be optimal for (9).
  If ($\phi_\varepsilon(\bar{x}) + \nabla\phi_\varepsilon(\bar{x})(\hat{x} - \bar{x}) > (1+\varepsilon)m$)
  Then set flag := **CP2** and STOP;
  Else If $(\phi_\varepsilon(\bar{x}) + \nabla\phi_\varepsilon(\bar{x})(\hat{x} - \bar{x}) > \Phi_\varepsilon(\bar{x}) - \delta))$
    Then set flag := **CP3** and STOP;
    Else   If $(\bar{y}A\hat{x} < \bar{y}b)$
      Then set flag := **CP5** and STOP.

Step 3: Set $\bar{x} := \bar{x} + \sigma_*(\hat{x} - \bar{x})$,
  where $\sigma_* \in \arg\min\{\phi_\varepsilon(\bar{x} + \sigma(\hat{x} - \bar{x})) : \sigma \in (0,1]\}$.
  Repeat *Step 1*.

On exit of this routine, $\bar{x}$ should be understood as optimal in (3) when flag $\neq$ **CP4**. When this is the case, output should be interpreted as follows: $\Phi(\varepsilon, z) \leq (1+\varepsilon)m \iff$ flag $\in \{\textbf{CP1}, \textbf{CP3}\}$.

Step 2 of the routine **Solve_subproblem** requires solving a program with a linear objective function. Interesting possibilities are: (1) $P = [0,1]^n$, in which case (9) can be solved by a greedy type algorithm; (2) $P = \{0,1\}^n$, in which case (9) is a *0-1 knapsack* problem and can be solved by the CPLEX MIP solver; (3) P is a generic polyhedron in which case (9) can be solved by the CPLEX solver.

In what follows we will expand on how the line search is performed assuming that we are looking for an ε-feasible solution relatively to an equality system $Ax = b$, as this is the case with the test problems studied. Step 3 of the algorithm **Solve_subproblem** aims at finding a minimizer σ_* of the function $g: [0,1] \to \mathbb{R}$ defined by $g(\sigma) \equiv \phi_\varepsilon(\bar{x} + \sigma d) \equiv \sum_{i=1}^m \phi_i(\bar{x} + \sigma d) \equiv 2 \sum_{i=1}^m \cosh(a_i \bar{x} - b_i + \sigma a_i d)$, where $d = \hat{x} - \bar{x}$ and cosh denotes de hyperbolic cosine function. Note that g is convex and $g'(0) < 0$. The numerical method of choice would be the unidimensional Newton's method for it is, in this case, globally convergent and possesses locally quadratic convergence.

However, since the functions ϕ_i are defined by exponential functions, Newton's method may require too many iterations to reach its region of quadratic convergence. A method like bisection search may be more adequate at early stages of the minimization process. Our initial bisection interval is implied by the following result.

Theorem 2. *([6]) For every $i \in \{1, 2, \ldots, m\}$, let U_i be an upper bound on the value of $\phi_i(\bar{x} + \sigma_* d)$. Then, for every i such that $a_i d \neq 0$, the following holds,*

$$\sigma_* \leq \frac{1}{\alpha a_i d} \ln \left(\frac{U_i + \text{sgn}(a_i d)\sqrt{U_i^2 - 4}}{2 \exp(\alpha(a_i \bar{x} - b_i))} \right). \tag{15}$$

where $\text{sgn}(\cdot)$ *denotes the sign function.*

The bound (15) is important for the search for σ_* not to overflow the exponential function evaluations.

Bienstock [2] proposed to use bisection search until a sufficiently small interval is found and only then start with Newton's method. We propose a slightly different procedure. We propose to try two Newton's steps departing from the left limit of the current bisection interval. By using these two Newton points we make a judgement of whether the region of quadratic convergence for Newton's method was achieved. If yes, we switch to the pure Newton's method. More precisely, if σ_{N1} is the first Newton point and σ_{N2} is the second Newton point then we consider that the region of quadratic convergence is achieved if

$$\left(c_1 g'(\sigma_{N2})\right)^2 \leq |g'(\sigma_{N1})| \text{ if } |g'(\sigma_{N2})| > 1) \text{ or } \left(|g'(\sigma_{N2})| \leq c_2 g'(\sigma_{N1})^2 \text{ if } |g'(\sigma_{N2})| \leq 1\right) \tag{16}$$

where c_1, c_2 are positive constants. Otherwise, the interval is updated by using the information on these points, the midpoint σ_{mid} is also tried and the bisection interval is again updated. The process restarts.

4 Computational Results with Set-Partitioning Problems

In this section we report computational results for the first-order method previously discussed and with a second-order nonlinear programming package. The computational tests were performed on a PC with a 2.66GHz Pentium IV microprocessor and 512Mb of memory running RedHat Linux 8.0. The algorithm was implemented in AMPL (Version 7.1) modeling language.

Here, we are looking for finding "good" ε-feasible solutions of fractional set-partitioning problems arising in airline crew scheduling. Some of these linear programs are extremely difficult to solve with traditional algorithms. Generically, fractional set-partitioning problems are of the form

$$\begin{aligned} \min \; & cx \\ \text{s.t.} \; & Ax = \mathbb{1} \\ & x \in [0,1]^n \; , \end{aligned} \qquad (17)$$

where A is a $m \times n$ matrix, with 0-1 coefficients, and $\mathbb{1}$ denotes a vector of ones. Our framework was tested on real-world set-partitioning problems obtained from the OR-Library (http://www.brunel.ac.uk/depts/ma/research/jeb/info.html).

Table 1. Set-partitioning problems

Name	m	n	z^{LP}	Vol It.	Vol Dual	Vol Primal	Vol Viol.	Vol Time
sppnw08	24	434	35894	356	35894.0	36188.0	0.01889	0.02
sppnw10	24	853	68271	501	68146.8	68510.8	0.01974	0.04
sppnw12	27	626	14118	447	14101.4	14222.6	0.01859	0.03
sppnw15	31	467	67743	483	67713.8	67407.0	0.01934	0.03
sppnw20	22	685	16626	408	16603.4	16645.2	0.01991	0.02
sppnw21	25	577	7380	358	7370.8	7387.1	0.01875	0.03
sppnw22	23	619	6942	340	6916.8	6917.4	0.01903	0.02
sppnw23	19	711	12317	357	12300.1	12412.8	0.01971	0.03
sppnw24	19	1366	5843	313	5827.8	5885.3	0.01924	0.04
sppnw25	20	1217	5852	305	5829.8	5851.3	0.01763	0.04
sppnw26	23	771	6743	304	6732.8	6742.4	0.01584	0.03
sppnw27	22	1355	9877.5	363	9870.0	9934.8	0.01933	0.05
sppnw28	18	1210	8169	342	8160.1	8167.5	0.01995	0.04
sppnw32	19	294	14570	430	14559.9	14559.6	0.01748	0.01
sppnw35	23	1709	7206	334	7194.0	7262.6	0.01633	0.06
sppnw36	20	1783	7260	631	7259.2	7225.3	0.01995	0.13
sppnw38	23	1220	5552	321	5540.5	5526.0	0.01951	0.04

Before a call to **Main** algorithm we apply the volume algorithm, developed by Barahona and Anbil [1], that is an extension of the subgradient algorithm, which produces approximate feasible primal and dual solutions to a linear program, much more quickly than solving it exactly. This algorithm approximately solves the Lagrangian relaxation of the problem (17), i.e.,

$$\max \min \{cx - y(Ax - 1\!\!1) : x \in [0,1]^n\}$$
$$\text{s.t.} \quad y \in \mathbb{R}^m$$

requiring a not too demanding number of iterations.

The main characteristics of the selected problems are described in Table 1.

The first four columns of this table identify the problem by its name, m, n, and z^{LP}, the known optimal value of the linear programming relaxation. The remaining columns are related to the volume algorithm [1], namely, the number of iterations up to finding a primal vector where each constraint is violated by at most 0.02 or the difference between the dual lower bound and the primal value is less than 1%. We also present the time required by the volume algorithm (in seconds).

From the (dual) lower bound derived from volume algorithm we obtain an integral lower bound z_a on the value of z_*. The initial x_a was set to all-zeros because often the primal solution obtained through the volume algorithm would not satisfy $cx \leq z_a$. Note that, since $c > 0$, $x_a \in P'(z_a)$. In our experiments we have chosen to set $\Delta = 1$.

We implemented the first-order feasible directions minimization algorithm, presented in Sect. 3, to solve the nonlinear problem (3). In Step 2 of the **Solve_subproblem** algorithm we used the software CPLEX (version 7.1) to get an optimal solution for the linear problem.

Table 2 is divided into four parts. The first part identifies the set-partitioning problem. Next, we have the initial value for z_a (input to **Main**), that is the lower bound found by volume algorithm.

Table 2. Results of solving (3) with first-order and second-order implementations

Name	Initial z_a	Final ε	First-order Max. Viol.	Out. It.	Time (sec.)	Second-order Max. Viol.	Out. It.	Time (sec.)
sppnw08	35894	6.10E-05	4.20E-07	1	469.86	1.28E-14	1	21.48
sppnw10	68147	0.01563	0.00269	4*	26404.26	9.15E-10	6	3546.41
sppnw12	14102	0.00781	0.00121	3*	17301.78	7.41E-14	4	134.80
sppnw15	67714	0.03125	0.00521	1*	4522.86	5.79E-09	5	124.87
sppnw20	16604	0.03125	0.00819	1*	5653.53	9.88E-09	4	275.65
sppnw21	7371	0.03125	0.00573	1*	4752.56	8.96E-10	4	110.24
sppnw22	6917	0.01563	0.00470	3*	7919.76	2.68E-10	5	230.26
sppnw23	12301	0.01563	0.00385	2*	9034.06	2.74E-11	5	15173.3
sppnw24	5828	0.01563	0.00296	3*	44584.35	1.38E-09	4	2107.83
sppnw25	5830	0.01563	0.00280	6*	59902.49	1.14E-09	5	1718.74
sppnw26	6733	0.01563	0.00308	3*	16722.60	1.91E-08	4	275.63
sppnw27	9871	0.00781	0.00175	3*	23535.21	2.18E-14	5	1588.36
sppnw28	8161	0.00781	0.00104	4*	23150.30	8.65E-15	4	1147.54
sppnw32	14560	0.01563	0.00296	3*	3059.28	8.72E-10	4	12.51
sppnw35	7194	0.01563	0.00393	2*	19234.21	1.92E-09	5	5420.31
sppnw36	7260	0.01563	0.00302	1*	32498.39	4.23E-13	1	3049.8
sppnw38	5541	0.01563	0.00457	2*	26496.43	1.43E-10	4	1677.83

In the third part, we present the results of the first-order implementation. For all these problems the initial value for ε is 1, because at the beginning of the algorithm the primal vector x_a is a vector of zeros and $\varepsilon = \|\mathbb{1} - Ax_a\|_\infty = 1$. We exhibit the final value for ε, the maximal violation obtained for the constraints, the number of outer iterations, which corresponds to the number of bisection calls, and the time (in seconds) required. For all problems, except sppnw08, no optimal $\bar{x}$ was found (flag=**CP4**), as the maximum number of iterations (set to 20000) was reached, as pointed out with the character $*$ in the table. Nevertheless, the maximal constraint violation obtained is reduced at least 60% (compare with Table 1).

Then we analyze the strategy of solving the nonlinear subproblem (3) using solver LOQO 6.0. LOQO is an implementation of a primal-dual interior point method for solving nonlinear constrained optimization problems. The fourth part of Table 2 summarizes the results. We remark that, for all the problems, the algorithm halted because the stopping criterion in the **Main** algorithm ($\varepsilon < 10^{-4}$) was achieved. The maximal violation of the constraints is now much smaller than the one obtained with the volume algorithm. A different ε reduction scheme ($\varepsilon := \varepsilon/10$) in the Step 2 of the **Main** algorithm was tried and, in general, the number of outer iterations decreased by one or two units and the time spent was smaller in 71% of the problems.

5 Conclusions

In this paper we propose an algorithm for finding an ε-feasible solution relatively to the constraints of a linear program. The first-order version of the algorithm attempts to minimize a linear approximation of an exponential penalty function using feasible directions and a constant stepsize. The second-order implementation uses the software LOQO to minimize the exponential function.

Our preliminary computational experiments show that the first-order method does not perform as expected when solving set-partitioning problems. In particular, the algorithm converges very slowly and does not reach an ε-approximate solution, for $\varepsilon < 10^{-4}$, for all tested problems but for sppnw08.

We may also conclude that the second-order implementation of the algorithm works quite well, specially for problems that are not large-scale, finding an ε-feasible solution in (1) satisfying $cx < z$, for a small and adequate ε value. Future developments will focus on improving convergence of the first-order algorithm.

References

1. Barahona, F., Anbil, R.: The volume algorithm: Producing primal solutions with a subgradient method. Math. Prog. **87** (2000) 385–399
2. Bienstock, D.: Potential Function Methods for approximately solving linear programming problems: theory and practice. Kluwer Academic Publishers (2002)
3. Grigoriadis, M.D., Khachiyan, L.G.:Fast approximation schemes for convex programs with many blocks and coupling constraints. SIAM J. Optim. **4** 1994 86–107

4. Karger, D., Plotkin, S.: Adding multiple cost constraints to combinatorial optimization problems, with applications to multicommodity flows. Proceedings of the 27th Annual ACM Symposium on Theory of Computing (1995) 18 – 25
5. Plotkin, S., Shmoys, D.B., Tardos, E.: Fast approximation algorithms for fractional packing and covering problems. Math. Oper. Res. **20** (1995) 495–504
6. Rocha, A.M.: Fast and stable algorithms based on the Lagrangian relaxation. PhD Thesis (in portuguese), Universidade do Minho, Portugal (2004) (forthcoming)

Inventory Allocation with Multi-echelon Service Level Considerations

Jenn-Rong Lin[1], Linda K. Nozick[2], and Mark A. Turnquist[2]

[1] Graduate Institute of Transportation and Logistics, National Chia-Yi University
300 University Rd. Chia-Yi, Taiwan 600, ROC
jrlin@mail.ncyu.edu.tw

[2] School of Civil and Environmental Engineering, Cornell University
Hollister Hall, Ithaca, NY 14853, USA
{lkn3, mat14}@cornell.edu

Abstract. This paper develops a METRIC-like method to optimize inventory allocations for low-demand items in a multi-echelon finished goods distribution system. The allocations are made by trading-off total inventory investment against service levels as measured by the stockout probabilities. We focus on a one-product three-echelon system and develop a solution procedure for such a system. We then apply that model and solution procedure to an illustrative example and compare the resulting inventory policies to optimal policies for a two-echelon system.

1 Introduction

The purpose of this paper is to formulate and analyze a multi-product multi-echelon inventory system for finished goods with Poisson demand at the final retail locations. The key decisions are the inventory levels of various products to be held at each stocking location. The design decisions are made with concern for both total inventory investment and service levels (measured by the average stockout probability at each echelon). The system structure and operation is assumed known and fixed. This specifies the distribution network, potential stocking locations, material flows, the average in-transit lead times on links, the average manufacturing lead time at the plants and the demand processes at the retail locations. The concerns in this model are the tactical decisions of how much of each product to stock at each location.

The general structure of the finished goods distribution system under study is represented in Figure 1. This structure represents the distribution of finished goods from production plants to final demand (represented by retail outlets). To take advantage of bulk shipment rates, distribution centers (DC's) are established near retail outlets. A network of DC's allows larger shipments from the plants to the DC's, potentially decreasing unit transportation costs as a result of the economies-of-scale in transportation. Such consolidation may reduce transportation rates significantly.

The finished goods distribution system may involve thousands of different products or stock keeping units (SKU's). Average demand rates for individual

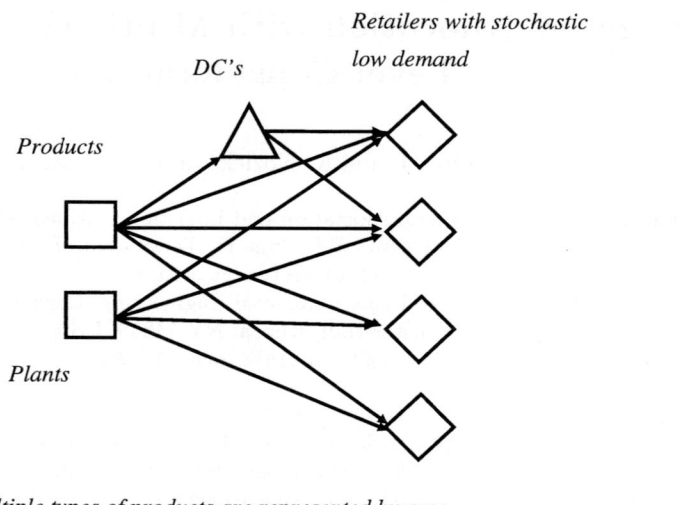

Fig. 1. Distribution network structure

products at each retail outlet are often relatively low due to the proliferation of products. We represent this demand using a Poisson model. Inventory control in such a system can take advantage of a multi-echelon structure. Muckstadt and Thomas [1] show that multi-echelon inventory systems may reduce the total inventory cost substantially when most products have low demand rates. However, Hausman and Erkip [2] demonstrate that the reduced inventory cost may not be significant enough to offset the organizational benefits and managerial simplicity of the single-echelon model. The debate motivates us to evaluate the benefit of stocking inventory at the DC's. From an inventory perspective, the existence of the DC's creates a potential opportunity for a third echelon of inventory, and raises the questions of whether or not the DC's should be stocking locations, and if so, for what products.

The purpose of this paper is to develop a multi-product multi-echelon inventory allocation model for a finished goods distribution system and to investigate the potential worth of stocking inventory at the DC's. We will first focus on allocation of a single product (made at a single plant) across three echelons, and then illustrate how to extend the model to include more products and multiple plants. The remainder of this paper is organized as follows. In section 2 we briefly review the literature of continuous review policies for multi-echelon inventory systems. In section 3 we present a mathematical programming formulation of a one-product three-echelon inventory allocation problem. In section 4 we develop a solution procedure for three-echelon systems. In section 5 we describe an illustrative example. In section 6 we discuss oppurtunities for future research.

2 Literature Review

Researchers have studied continuous-review policies for multi-echelon inventory systems in various ways, and there is an extensive literature on this topic. For review articles on multi-echelon inventory systems, we refer to Nahmias [3], Axsäter [4], and Diks et al. [5]. We briefly review some key works of particular relevance to this research. Sherbrooke [6] develops the well-known METRIC (Multi-Echelon Technique for Recoverable Item Control) model for determining the optimal allocation of repairable items in a two-echelon inventory system. The METRIC approximation has been used successfully in practice, especially in military applications involving thousands of items with low demand rates. Several applications in the Air Force Logistics Command have been reported by Demmy and Presutti [7]. Several extensions of METRIC considering multi-indenture items and more general batch ordering policies have been developed, and discussions of these models can be found in the review articles mentioned above. Sherbrooke [8] also discusses these extensions in detail.

METRIC-like models approximate the steady state distribution of inventory level for a stockage policy and use it to determine the average cost associated with the policy. Axäster [9] develops a more efficient and direct approach to find the minimum-cost inventory policy for an inventory system with one warehouse and N retailers.

Nozick and Turnquist [10] take a METRIC-like approach to formulate a one-product two-echelon inventory system. They consider two types of stockout situations and assume that the stockout penalties can be explicitly estimated and expressed as equivalent dollars. Their formulation is a nonlinear optimization model to determine the inventory levels that minimize the sum of the costs resulting from stockouts and inventory holding costs. We extend their model to a multi-product three-echelon system and develop a solution procedure to find the optimal inventory allocation.

3 Problem Definition

The inventory allocation problem can be summarized as follows. For each product determine how much to stock at each location so as to strike the appropriate balance between inventory holding costs and stock out costs, given the following parameters: locations of the plants, distribution centers and warehouses; the average transportation time between facilities; the average manufacturing lead times at the plants; and the demand processes at the retailers.

Let us first look at a one-plant, one-product, three-echelon system, which is a special case of the more general multi-plant, multi-product, three-echelon system. This particular type of three-echelon system consists of a group of retailers, a group of DC's, and a plant. Customer demands occur only at retailers. If the retailer has stock on hand, the order is satisfied. Otherwise, it is backordered. Once an order occurs at a retailer, it will request replenishment from the higher echelon - either the plant or the DC through which it receives its shipments.

Once a request occurs at a DC, it will request replenishment from the plant. There is no lateral supply among retailers and there is also no lateral supply among DC's.

We define the following subscripts, sets, decision variables, and input parameters.

Subscripts and sets:

- $c \in C$ indicate the distribution centers (DC's) including a dummy distribution center to represent products flows that go directly from a plant to a retailer.
- $d \in D$ indicate the retailers.
- $D(c)$ is the set of retailers $d \in D$ that receive products from distribution center c.

Decision variables:

- S_0 : established stock level at the plant.
- $\vec{S_c}$: vector of established stock levels at DC's, with elements S_c, the established stock level at DC c.
- $\vec{S_d}$: vector of established stock levels at retailers, with elements S_d, the established stock level at retailer d.
- $\rho_0(S_0)$: stockout probability at the plant (which is a function of S_0 and the average manufacturing lead time at the plant).
- $\rho_c(S_c, S_0)$: stockout probability at DC c (which is a function of S_c, S_0 and the average in-transit lead time from the plant to DC c).
- $\rho_d(S_d, S_c, S_0)$: stockout probability at retailer d (which is a function of S_d, S_c, S_0 and the average in-transit lead time from DC c to retailer d).

Input parameters:

- λ_d : is the average demand rate at retailer d.
- ψ : is the average manufacturing lead time at the plant.
- ζ_c : is the average in-transit lead time from the plant to DC c.
- ζ_{cd} : is the average in-transit lead time from DC c to retailer d.
- ζ_{0d} : the average in-transit lead time from plant to retailer d (if shipments are made direct, rather than through a DC).
- h : is the unit inventory holding cost per period.
- β_1 : is the penalty incurred per incident when the retailer is stocked out but inventory is available at the DC.
- β_2 : is the penalty incurred per incident when both the retailer and DC are stocked out but inventory is available at the plant. We assume $\beta_2 > \beta_1$.
- β_3 : is the penalty incurred per incident when there is a stockout at all three levels. We assume $\beta_3 > \beta_2 > \beta_1$.

The mathematical programming formulation of this one-product, three-echelon inventory model is then:

$$minC(\vec{S_d}, \vec{S_c}, S_0) = h \times (S_0 + \sum_{c \in C} S_c + \sum_{d \in D} S_d)$$
$$+ \beta_1 \times \sum_{c \in C}[(1 - \rho_c) \times \sum_{d \in D(c)} \lambda_d \rho_d] + \beta_2 \times (1 - \rho_0) \times \sum_{c \in c}(\rho_c \times \sum_{d \in D(c)} \lambda_d \rho_d)$$
$$+ \beta_3 \times \rho_0 \times \sum_{c \in C}(\rho_c \times \sum_{d \in D(c)} \lambda_d \rho_d) \tag{1}$$

such that

$$S_d \geq 0, \forall d \in D; S_c \geq 0, \forall c \in C; S_0 \geq 0 \tag{2}$$

The objective is to minimize inventory holding costs at the plant, DC's and retailers plus penalty costs for stockout. Since we assume that each facility replenishes inventory on a one-for-one basis and that the demand at each retailer follows a Poisson distribution with average demand λ_d, we can determine the accumulated demand for each product at the plant and each DC. The average replenishment lead times and the stockout probabilities for each product at each location are discussed in Feeney and Sherbrooke [11] and Sherbrooke [6]. For products flows that do not use a DC, the stockout probability at the dummy DC is 1.0. Equations (3) and (4) determine the average demand for each product at each DC and the plant. Equations (5) and (6) determine the average waiting time and average replenishment lead time for each product from the plants to DC's. Equations (7) and (8) determine the average waiting time and average replenishment lead time for each product from DC's to retailers. Equations (9)-(11) determine the stockout probabilities for each product at the plant, DC's and retailers.

$$\Lambda_c = \sum_{d \in D(c)} \lambda_d, \forall c \in C \tag{3}$$

$$\Lambda_0 = \sum_{c \in C} \Lambda_c = \sum_{d \in D} \lambda_d \tag{4}$$

$$W_0(S_0) = \frac{\sum_{x > S_0}(X - S_0) p(x|\Lambda_0 \psi)}{\Lambda_0} \tag{5}$$

$$\zeta'_c = \zeta_c + W_0(S_0) \tag{6}$$

$$W_c(S_c, S_0) = \frac{\sum_{x > S_c}(x - S_c) p(X|\Lambda_c \zeta'_c)}{\Lambda_c}, \forall c \in C \tag{7}$$

$$\zeta'_{cd} = \zeta_{cd} + W_c(S_c, S_0), \forall c \in C, \forall d \in D(c) \tag{8}$$

$$\rho_0(S_0) = \sum_{q=S_0+1}^{\infty} \frac{e^{-\Lambda_0\psi}(\Lambda_0\psi)^q}{q!} \tag{9}$$

$$\rho_c(S_c, S_0) = \sum_{q=S_c+1}^{\infty} \frac{e^{-\Lambda_c\zeta'_c}(\Lambda_c\zeta'_c)^q}{q!}, \forall c \in C \tag{10}$$

$$\rho_d(S_d, S_c, S_0) = \sum_{q=S_d+1}^{\infty} \frac{e^{-\lambda_d\zeta'_{cd}}(\lambda_d\zeta'_{cd})^q}{q!}, \forall c \in C, \forall d \in D(c) \tag{11}$$

This model is a nonlinear optimization whose only constraints are non-negativity restrictions on the inventory level at each location. Since the objective function is not jointly convex in $\vec{S_d}, \vec{S_c}, S_0$, some additional care must be taken to find the global optimum. In the next section, we discuss the solution procedure.

4 Solution Procedure

To solve a three-echelon inventory allocation problem, we first have to know how to solve a two-echelon inventory allocation problem. Consider a two-echelon inventory system where we do not stock any inventory at the DC's. This is a particular case of three-echelon inventory system, where the stockout probability at each DC is 1. This will simplify the objective function (1) as (12), denoted as $C(\vec{S_d}, S_0)$.

$$\begin{aligned}
C(\vec{S_d}, S_0) &= C(\vec{S_d}, \vec{S_c}, S_0 | \vec{S_c} = \vec{0}) \\
&= h \times (S_0 + \sum_{d \in D} S_d) + \beta_2 \times (1 - \rho_0) \times \sum_{d \in D} \lambda_d \rho_d + \beta_3 \times \rho_0 \times \sum_{d \in D} \lambda_d \rho_d \\
&= h \times S_0 + \sum_{d \in D} h \times S_d + [\beta_2 \times (1 - \rho_0) + \beta_3 \times \rho_0] \times \lambda_d \rho_d \\
&= h \times S_0 + \sum_{d \in D} (h \times S_d + \beta' \times \lambda_d \rho_d) \\
&= h \times S_0 + \sum_{d \in D} C(S_d | S_0)
\end{aligned} \tag{12}$$

The cost function of a two-echelon inventory system $C(\vec{S_d}, S_0)$ is still not jointly convex in both the inventory levels at retailers ($\vec{S_d}$) and the inventory level at the plant (S_0). However, for a given inventory level at the plant (S_0), it is separable for each retailer. Each retailer inventory cost function, denoted $C(S_d|S_0)$, is convex if the inventory level (S_d) is greater than the average lead time demand of the retailer. Based on incremental cost analysis, we observe that $C(S_d|S_0)$ is either a monotone increasing function in the inventory level held at the retailer (S_d) or it has two local minima. In the first case, setting the retailer inventory to zero is the global optimum. In the second case, one of the two local optima is when the inventory level at the retailer is zero, and the other lies in the convex region. Based on this observation, we develop an algorithm to efficiently find the optimal inventory level for the retailers.

1. Compare the value of the maximal reduction of stockout penalty $[\beta' \times \lambda_d \times p(x = \lfloor \lambda_d \zeta'_d \rfloor | \lambda_d \zeta'_d)]$ with the unit holding cost (h). If its value is less than h, return the optimal inventory level (S_d^*) equal to zero and terminate. Otherwise, continue on step 2.
2. Starting from the integer part of the lead time demand ($x = \lfloor \lambda_d \zeta'_d \rfloor$), find the smallest x such that the reduced stockout penalty by holding one more unit is less than the unit holding cost. (Denote this x as (S_d^*).)
3. Compare the total cost of stocking (S_d^*) at the retailer with the total cost of not stocking inventory at the retailer and choose the one with lower total cost.

Based on the solution procedure for the retailers, we can easily construct a systematic search method to find the optimal inventory allocation between plants and retailers. More specifically, the algorithm is as follows.

1. Find the lower and upper bounds of the optimal inventory level at the plant.
2. For each possible inventory level at the plant (S_o), find the optimal inventory level held at a retailer (S_d) for each retailer separately (applying the solution algorithm above).
3. Find the optimal combination of inventory levels at the retailers and the plant $(\overrightarrow{S_d^*}, S_o^*)$ that has the smallest total inventory cost $C(\overrightarrow{S_d^*}, S_o^*)$.

We apply Axsäter's idea [9] to find the lower bound and upper bound of the optimal inventory level at the plant. The approach is summarized as follows. First, find the lower bound and upper bounds of the optimal inventory level at retailers for the shortest and longest replenishment lead times, respectively. Then find the lower bound of the optimal level at the plant by optimizing the total cost of the whole system, for the upper bounds of the optimal inventory level at retailers. Similarly, we can find the upper bound of the optimal inventory level at the plant by optimizing the total cost of the whole system, for the lower bounds of the optimal inventory level at retailers. Once we know how to find the optimal inventory allocation of a two-echelon inventory system, finding the optimal inventory allocation of a three-echelon system is straightforward. The solution algorithm for a three-echelon system is as follows.

1. Find the lower and upper bounds of the optimal inventory level at the plant.
2. For each possible inventory level at the plant (S_0) within the allowable range, find the optimal inventory allocation for each two-echelon sub-system that consists of a DC with those retailers receiving products via it (applying the solution algorithm for two-echelon inventory systems).
3. Choose the optimal combination of inventory levels at the retailers, the DC's and the plant $(\overrightarrow{S_d^*}, \overrightarrow{S_c^*}, S_0^*)$ that minimizes the total inventory costs.

5 Illustrative Example

We demonstrate the computations of inventory allocation on a small example with 1 plant, 1 DC and 4 retailers, as illustrated in Figure 2. The material flows, the average in-transit lead times on links, the average manufacturing lead time, and the proportions of total average demand at the various retailers are also shown in Figure 2.

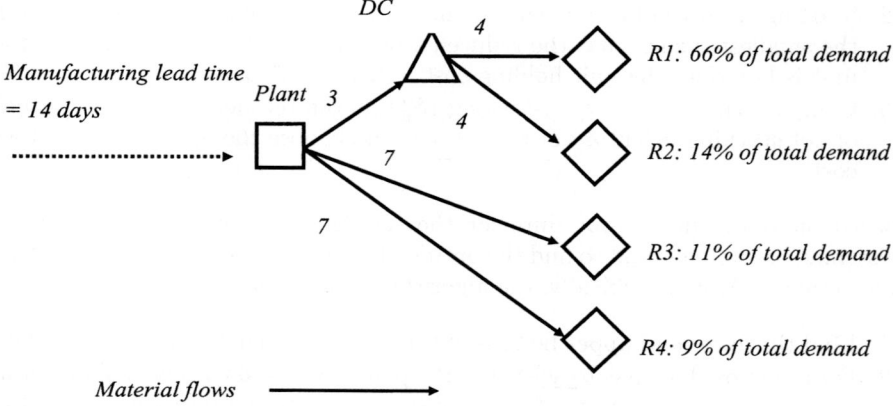

Fig. 2. Example network

Computations have been done for a range of products whose values vary from $18,000 to $48,000 per unit (the products are automobiles). For purposes of this example, we will assume that the annual inventory holding cost is 24%. The optimal inventory allocation depends on the unit holding costs and the values of the three stockout penalties. We assume that the three per unit stockout penalties are $200, $350 and $2,100 respectively. The resulting system-wide fill rate is about 86% (across all products and locations) in both the two-echelon and three-echelon solutions. Thus, the overall performance of the selected two-echelon and three-echelon solutions is similar, but the inventory allocations for individual products are quite different. The optimal inventory allocations for six sample products (automobile model configurations) are as shown in Table 1, and the average stockout probabilities at each location are shown in Table 2.

In general, more inventory is placed at the retailers with higher demands. For example, we stock more inventory at R1 than at R2. For those products with extremely low demand (e.g., ID's 88 and 224), we centralize inventory at the plant and do not stock inventory at the retailers at all. The optimal inventory allocation for the products with medium demand depends on the product value. We stock the low-value products (e.g., ID 376) at all locations. However we centralize inventory of high-value products (e.g., ID 36) at the plant and stock only a few units at the retailers with highest demand. The inventory

Table 1. Inventory policies for illustrative example

Product ID	Unit Price	Annual demand at plant	Two-echelon Inventory Level					Three-echelon Inventory Level					
			Plant	R1	R2	R3	R4	Plant	Dc	R1	R2	R3	R4
17	48000	608	31	12	0	0	0	24	7	8	2	3	3
36	48000	266	15	6	0	0	0	14	4	0	0	0	0
88	48000	30	2	0	0	0	0	2	0	0	0	0	0
224	38000	12	1	0	0	0	0	1	0	0	0	0	0
376	18000	106	6	4	1	1	1	5	2	3	1	0	1

Table 2. Stockout probabilities for illustrative example

Product ID	Two-echelon Stockout Probability(%)					Three-echelon Stockout Probability(%)					
	Plant	R1	R2	R3	R4	Plant	DC	R1	R2	R3	R4
17	7.3	9.7	100.0	100.0	100.0	47.0	27.9	10.1	26.0	17.4	12.6
36	9.6	13.7	100.0	100.0	100.0	15.3	13.0	100.0	100.0	100.0	100.0
88	29.2	100.0	100.0	100.0	100.0	29.2	100.0	100.0	100.0	100.0	100.0
224	42.9	100.0	100.0	100.0	100.0	42.9	100.0	100.0	100.0	100.0	100.0
376	18.4	4.6	33.5	27.0	23.6	30.5	35.0	11.4	22.0	28.5	24.9

levels of high-value products with larger relative demand (e.g., ID 17) show the difference between the two-echelon and three-echelon allocations. We centralize most inventory at the plant in the two-echelon system, but in the three-echelon system we stock inventory at all locations.

6 Conclusion

Future research is useful in at least the following directions. First, a useful extension of the model is to production systemsk where the same product is manufactured in more than one plant. Second, we do not consider emergency shipments when a stockout incident occurs. If the in-transit lead times are the bottleneck of the system, we may want to consider expedited transportation options. Third, we assume the system structure and material flows are determined by the transportation options. The in-transit lead time of indirect shipments may increase as a result of increased route length. On the other hand, it may decrease as a result of a more frequent shipping schedule. While re-engineering the whole distribution system, it would be useful to explicitly consider the tradeoff between inventory costs and transportation costs caused by the transportation options.

References

1. Muckstadt, J.A., Thomas, L.J. : Are Multi-Echelon Inventory Methods Worth Implementing in Systems with Low-Demand-Rate Items?. Management Science. **26** (1980) 483-494.
2. Hausman, W.H., Erkip, N.K. : Multi-echelon vs. Single-echelon Inventory Control Policies for Low-demand Items. Management Science. **40** (1994) 597-602.
3. Nahmis, S.: Managing Repairable Item Inventory System: A Review. in Multi-level Production/Inventory Control Systems: Theory and Practice, edited by Schwarz L. New York: North-Holland, (1981).
4. Axsater, S.: Continuous Review Policies for Multi-Level Inventory Systems with Stochastic Demand. in Logistic of Production and Inventory, Handbook in Operations Research and Management Science 4, edited by Graves, S.C., Rinnooy Kan, A.H.G., Zipkin, P.H., Amsterdam: Elsevier (North-Holland), (1993).
5. Diks, E.B. et al.: Multi-echelon Systems: A Service Measure Perspective. European Journal of Operational Research **95** (1996) 241-263.
6. Sherbrooke, C.C. Optimal Inventory Modeling of Systems: Multi-Echelon Techniques. New York: John Wiley, Sons, Inc. (1992).
7. Demmy, W.S., Pressuti, V.J., Multi-Echelon Inventory Theory in the Air Force Logistics Command. in Multi-level Production/Inventory Control Systems: Theory and Practice, edited by Schwarz L., New York: North-Holland, (1981).
8. Sherbrooke, C.C. METRIC: A Multi-echelon Technique for Recoverable Inventory Control. Operation Research **16** (1968) 122-141.
9. Axsater, S. Simple Solution Procedures for a Class of Two-Echelon Inventory Problems. Operations Research **38** (1990) 64-69.
10. Nozick, L.K. and M.A. Turnquist A Two-echelon Inventory Allocation and Distribution Center Location Analysis. Transportation Research Part E **37** (2001) 425-441.
11. Feeney, G.J. and C.C. Sherbrooke. The (S-1,S) Inventory Policy under Compound Poisson Demand. Management Science **12** (1966) 391-411.

A Queueing Model for Multi-product Production System

Ho Woo Lee[1] and Tae Hoon Kim[2]

[1] Dept. of Systems Management Engineering,
Sungkyunkwan University, Su Won, KOREA 440-746
hwlee@skku.edu
http://web.skku.edu/~or
[2] Wireless Communication Track, School of Engineering,
Information and communications University,
Tae Jon, KOREA 305-600
hoony78@lycos.co.kr

Abstract. In this paper, we propose a queueing model for a multi-product production system in which the manufacturing environments are changing over time between some predetermined states. We model this system by the discrete-time Markovian Arrival Stream (MAS) and derive the mean level of WIP (work-in-process) inventory. A numerical example is presented.

1 Introduction

Most conventional models on production systems have assumed that raw materials arrive according to renewal processes and processing times are independent of the arrival processes. But in most real production and manufacturing systems, this is far from the truth. In most real-world production systems, arrivals are correlated and service times are not independent of each other. This research was motivated by a manufacturing process in which

(a) there are multiple types of products,
(b) there are multiple machine conditions,
(c) decision on the acceptance of orders depend on the machine conditions,
(d) the processing times are dependent on the machine conditions.

As an example, consider a job-shop in which the condition of a machine is classified as good or bad. During the time the machine is in a bad condition, the shop accepts only type-1 orders. During the time the machine is in a good condition, both type-1 and type-2 orders are accepted. The machine conditions can be extended to management-level manufacturing environments. For example, when the company is in some bad managerial situation such as labor dispute or workers' strike, it can accept only particular types of orders.

Considering that time units are discrete in many manufacturing systems, this system can be modeled by a discrete-time queueing system with multiple

Markovian arrival streams (MAS). The multiple Markovian arrival stream can be regarded as an extension of the well-known single-stream MAP (Markovian Arrival process) to the multiple arrival streams with phase-dependent service times. For more detailed discussion of MAS, readers are advised to see Takine [11].

In this paper, we assume that the processing times (service times) of raw materials (customers) differ depending on the product types (classes) to which they belong. This is a special case of Takine [12] who considered a FCFS queueing system in which the service times of customers differ depending upon the phases of the underlying Markov chain (UMC) before and after arrival instances and customer classes. We note that we can specialize Takine's model to our manufacturing setting at the cost of phase-dependent service times. The reward is the easy computation of mean waiting times (and thereby mean WIP inventory level).

It is known that it is impossible to use the conventional matrix analytic method to treat the MAS queueing system. Also, the supplementary variable technique based on the joint Markovian process of the queue length and the remaining (or elapsed) service time are not applicable, either. Takine and Hasegawa [16] was the first successful work on the analytical treatment to analyze such queues. They showed simple formulas for workload in a MAP/G/1 queue with state-dependent service times. Later, Takine [12] succeeded in obtaining a purely algorithmic procedure to compute the stationary joint queue length distribution in the work-conserving FIFO multiple stream queue. He used the attained waiting time to characterize the joint queue length distribution as the distribution of the number of customers arriving during the attained waiting time. Then, he finally obtained an equation that relates the joint queue length distribution with the sojourn time distributions of customers from respective arrival streams. Readers are advised to see Takine [11-15] and Masuyama and Takine [6, 7, 8] for recent works on the queuing systems with (batch) multiple Markovian arrival streams.

It was Noh and Choi [9] who extended the work of Takine [12] to discrete-time settings for the first time. They applied their model to the analysis of multimedia conference system. They first derived the workload distribution, and then the distributions of the actual waiting time, the sojourn time of a customer of each class and the joint distribution of the numbers of customers of each class. They provided a recursion formula to compute the joint queue length distribution.

In this paper, we are interested in deriving the mean level of WIP inventory in the manufacturing system described above. Instead of using the complicated joint queue length distribution employed by Noh and Choi [9], we accomplish our goal by setting up the equations for virtual waiting time first and then directly deriving the mean waiting time from the virtual waiting time results. Then, we can use the celebrated Little's formula to derive the mean WIP inventory level. In this way, we do not need to go through the complex computational algorithm. This is possible because the processing times are not dependent on the arrival phase of the orders, but only on the order types.

2 Description of the Arrival Process

The continuous-time multiple Markovian arrival stream (MAS) was first introduced by Asmussen and Koole [1] and was well described in Takine [11, 12, 13]. The counting process associated with MAS is a marked MAP introduced by He [2] (He called this process MMAP (He and Alfa [3])). See also He and Neuts [4] for more discussion about MMAP.

Based on Takine [12], we describe the discrete-time marked MAP in the context of our manufacturing settings. There exist K product-types and m machine states. Let us consider a type-k order which arrives to the system when the machine states is in i and changes the machine state to j after its arrival. We call this order $E_{ij}^{(k)}$-type. $E_{ij}^{(k)}$-type orders represent the situations in which the current machine status needs to be upgraded in order to meet the product requirement. During a unit time period, an $E_{ij}^{(k)}$-type order exists with probability $D_{ij}^{(k)}$. The current machine state-i may change to state-j by the system owner with probability $D_{ij}^{(0)}$.

We note that $\sum_{j=1}^{m} D_{ij}^{(0)} + \sum_{k=1}^{K} \sum_{j=1}^{m} D_{ij}^{(k)} = 1$. Thus $\sum_{j=1}^{m} D_{ij}^{(0)} + \sum_{k=1}^{K} D_{ij}^{(k)}$ is the probability that the machine state changes to j at the start of the next period given that the machine state is in i at the start of the current period.

If we use $\boldsymbol{D}_0$ as a matrix of $D_{ij}^{(0)}$ and $\boldsymbol{D}_k$ as a matrix of $D_{ij}^{(k)}$, we have

$$\boldsymbol{D}_k = \begin{pmatrix} D_{(1,1)}^{(k)} & D_{(1,2)}^{(k)} & \cdots & D_{(1,m)}^{(k)} \\ \vdots & \vdots & \vdots & \vdots \\ D_{(m,1)}^{(k)} & D_{(m,2)}^{(k)} & \cdots & D_{(m,m)}^{(k)} \end{pmatrix}.$$

Let us define $\boldsymbol{D} = \sum_{k=1}^{K} \boldsymbol{D}_k$. Then, $(\boldsymbol{D}_0 + \boldsymbol{D})$ is a transition probability matrix of the UMC and we have $(\boldsymbol{D}_0 + \boldsymbol{D})\mathbf{e} = \mathbf{e}$ where $\mathbf{e}$ is the $(m \times 1)$ vectors of 1's. If we define $\boldsymbol{\pi} = (\pi_1, \pi_2, \ldots, \pi_m)$ as the stationary vector of the UMC, then, $\boldsymbol{\pi}$ satisfies $\boldsymbol{\pi} = \boldsymbol{\pi}(\boldsymbol{D}_0 + \boldsymbol{D})$, $\boldsymbol{\pi}\mathbf{e} = 1$.

Let us consider the class-k customer who arrives to the phase i of the UMC and changes the phase to j after its arrival. Let $S_{ij}^{(k)}$ be the service time random variable of this customer. If we define the service time distribution $h_{ij}^{(k)}(n)$ as $h_{ij}^{(k)}(n) = Pr[S_{ij}^{(k)} = n, (n \geq 1)]$, we can combine $D_{ij}^{(k)}$ and the service time as the product $D_{ij}^{(k)} h_{ij}^{(k)}(n)$, which leads to a matrix $\boldsymbol{D}_k(n)$ as follows;

$$\boldsymbol{D}_k(n) = \begin{pmatrix} D_{(1,1)}^{(k)} h_{(1,1)}^{(k)}(n) & D_{(1,2)}^{(k)} h_{(1,2)}^{(k)}(n) & \cdots & D_{(1,m)}^{(k)} h_{(1,m)}^{(k)}(n) \\ \vdots & \vdots & \vdots & \vdots \\ D_{(m,1)}^{(k)} h_{(m,1)}^{(k)}(n) & D_{(m,2)}^{(k)} h_{(m,2)}^{(k)}(n) & \cdots & D_{(m,m)}^{(k)} h_{(m,m)}^{(k)}(n) \end{pmatrix}.$$

Then, we can represent the MMAP as $\{\boldsymbol{D}_0, \boldsymbol{D}_1(n), \ldots, \boldsymbol{D}_K(n)\}$. Then, we have $\boldsymbol{D}_k = \sum_{n=1}^{\infty} \boldsymbol{D}_k(n) = \sum_{n=1}^{\infty} \boldsymbol{D}_k h^{(k)}(n)$ and $\boldsymbol{D}(n) = \sum_{k=1}^{K} \boldsymbol{D}_k(n)$. The MAS

(Markovian arrival streams) of Takine [11, 12] is the MMAP integrated with such service times.

The arrival rate $\lambda_{k,i,k}$ of $E_{ij}^{(k)}$-type orders becomes $\lambda_{k,i,j} = \pi_i D_{i,j}^{(k)}$. Then, the total arrival rate of type-k orders becomes $\lambda_k = \pi D_k \mathbf{e}$ and the total arrival rate λ of orders becomes $\lambda = \sum_{k=1}^{K} \lambda_k = \sum_{k=1}^{K} \pi D_k \mathbf{e}$.

3 Virtual Waiting Time

In this section, we derive the virtual waiting time distribution. This will be the basis of deriving the mean actual waiting time which will be discussed in the next section. From now on, we will also use the terms customer, service time and server in place of order, processing time and machine.

In continuous-time systems, if there are no server vacations, the virtual waiting time is equal to the unfinished work (workload) under FCFS. But, in discrete-time systems under late-arrival model, the unfinished work is the sum of the virtual waiting time and the amount of work that is brought in during a unit period (slot). Readers are referred to Takagi [10] for the relationship between the virtual waiting time and the workload under the late arrival model.

Let us define the following notations;

$U(n)$: unfinished work (workload) at the start of slot-n,
$V(n)$: waiting time of the virtual customer who arrives in slot-n,
$J(n)$: UMC phase at the start of slot-n,
$u_j(l,n) = Pr[U(n) = l, \ J(n) = j], \ (1 \leq j \leq m)$
$v_j(l,n) = Pr[V(n) = l, \ J(n) = j], \ (1 \leq j \leq m)$

If we define $\boldsymbol{u}(l)$ as an $(1 \times m)$ vector whose jth element is $u_j(l) = \lim_{n \to \infty} u_j(l,n)$. Then, it is not difficult to obtain the steady-state vector system equations as follows;

$$\boldsymbol{u}(0) = \boldsymbol{u}(0)\boldsymbol{D}_0 + \boldsymbol{u}(1)\boldsymbol{D}_0, \tag{3.1}$$

$$\boldsymbol{u}(l) = \boldsymbol{u}(0)\boldsymbol{D}(l) + \sum_{k=0}^{l} \boldsymbol{u}(l-n+1)\boldsymbol{D}(n) + \boldsymbol{u}(l+1)\boldsymbol{D}_0, \ (l \geq 0). \tag{3.2}$$

Let us define the vector generating function (GF) $\boldsymbol{u}^*(\omega) = \sum_{l=0}^{\infty} \boldsymbol{u}(l)\omega^l$. If we multiply (3.2) by ω^l, sum over $l = 1, 2, \ldots$ and use (3.1), we get

$$\boldsymbol{u}^*(\omega) = \boldsymbol{u}(0)(\omega - 1)[\omega \boldsymbol{I} - \boldsymbol{D}_0 - \boldsymbol{D}^*(\omega)]^{-1}[\boldsymbol{D}_0 + \boldsymbol{D}^*(\omega)]. \tag{3.3}$$

Let $v_j(l) = \lim_{n \to \infty} v_j(l,n)$ be the steady-state probability that the virtual waiting time is l and the UMC phase is j at the arrival instance of the virtual customer. Let $\boldsymbol{v}(l) = (v_1(l), \ldots, v_m(l))$ and $\boldsymbol{v}^*(\omega) = \sum_{n=0}^{\infty} \boldsymbol{v}(n)\omega^n$. Note that $[\boldsymbol{D}_0 + \boldsymbol{D}^*(\omega)]$ in (3.3) is the matrix GF of the workload that is brought in by the customer who arrives during a period (slot). Thus by dropping $[\boldsymbol{D}_0 + \boldsymbol{D}^*(\omega)]$ from (3.3), we get

$$\boldsymbol{v}^*(\omega)[\omega \boldsymbol{I} - \boldsymbol{D}_0 - \boldsymbol{D}^*(\omega)] = \boldsymbol{u}(0)(\omega - 1). \tag{3.4}$$

3.1 Evaluation of $u(0)$

We need to evaluate $u(0)$ contained in (3.4). Let τ be the first passage time the workload becomes 0. Define $P_{ij}(n)$ as

$$P_{ij}(n) = Pr\left[J(\tau) = j \mid U(0) = n, \; J(0) = i\right],$$

and $\boldsymbol{P}(n)$ as the $(m \times m)$ matrix of $P_{ij}(n)$. Then, conditioning on the events that occur during a period, we have

$$\boldsymbol{P} = \boldsymbol{P}(1) = \boldsymbol{D}_0 + \sum_{n=1}^{\infty} \boldsymbol{D}(n)\boldsymbol{P}(n),$$

which becomes, after using $\boldsymbol{P}(n+k) = \boldsymbol{P}(n)\boldsymbol{P}(k)$ (Takine and Hasegawa [16]),

$$\boldsymbol{P} = \boldsymbol{P}(1) = \boldsymbol{D}_0 + \sum_{n=1}^{\infty} \boldsymbol{D}(n)\boldsymbol{P}^n. \tag{3.5}$$

Note that, according to the similar reason given in Takine and Hasegawa [16] and Lucantoni et al. [5] for continuous time systems, $\boldsymbol{P}$ can be considered as a transition probability matrix of a Markov chain that is obtained by excising busy periods. If we let $\boldsymbol{\phi} = (\phi_1, \ldots, \phi_m)$ as the stationary vector of $\boldsymbol{P}$ that satisfies

$$\boldsymbol{\phi}\boldsymbol{P} = \boldsymbol{\phi}, \quad \boldsymbol{\phi}\boldsymbol{e} = 1, \tag{3.6}$$

then, ϕ_i is the probability that at the start of an arbitrary period during an idle period, the UMC phase is i. Then, we have

$$\boldsymbol{u}(0) = (1 - \rho)\boldsymbol{\phi}. \tag{3.7}$$

Let $w_{k,j}(l)$ be the probability that the waiting time of an arbitrary class-k customer is l and the UMC phase just after its arrival is j. Let $r_{k,j}(l)$ be the probability that the sojourn time (waiting time + service time) of an arbitrary class-k customer is l and the UMC phase just after its arrival is j. We define $\boldsymbol{w}_k(l)$ and $\boldsymbol{r}_k(l)$ as the vectors whose jth element is $w_{k,j}(l)$ and $r_{k,j}(l)$ respectively. Then, we have the following well-known relationship (Takine [12]);

$$\boldsymbol{w}_k(l) = \frac{\boldsymbol{v}(l)\boldsymbol{D}_k}{\lambda_k}, \quad (k = 1, 2, \ldots, K), \tag{3.8}$$

$$\boldsymbol{r}_k(l) = \frac{\sum_{n=0}^{l} \boldsymbol{v}(n)\boldsymbol{D}_k(l-n)}{\lambda_k}, \quad (k = 1, 2, \ldots, K). \tag{3.9}$$

From these, we get

$$\sum_{k=1}^{K} \lambda_k \boldsymbol{w}_k(l) = \sum_{k=1}^{K} \boldsymbol{v}(l)\boldsymbol{D}_k = \boldsymbol{v}(l)\boldsymbol{D}, \tag{3.10}$$

$$\sum_{k=1}^{K} \lambda_k \boldsymbol{r}_k(l) = \sum_{k=1}^{K} \sum_{n=0}^{l} \boldsymbol{v}(n)\boldsymbol{D}_k(l-n) = \sum_{n=0}^{l} \boldsymbol{v}(n)\boldsymbol{D}(l-n). \tag{3.11}$$

4 Direct Computation of Average WIP Inventory

In this section, we derive the average level of the WIP inventory by first deriving the mean waiting time directly from the virtual waiting time and then using the Little's law. First, we need to compute $\boldsymbol{u}(0) = (1-\rho)\boldsymbol{\phi}$ (see (3.7)). Noting that $\boldsymbol{\phi}$ is the stationary vector of $\boldsymbol{P}$ which is given by (see (3.5))

$$\boldsymbol{P} = \boldsymbol{P}(1) = \boldsymbol{D}_0 + \sum_{n=1}^{\infty} \boldsymbol{D}(n)\boldsymbol{P}^n,$$

$\boldsymbol{P}$ can be numerically obtained as the limit $\boldsymbol{P}_\infty$ from the following recursion

$$\boldsymbol{P}_{k+1} = \boldsymbol{D}_0 + \sum_{n=1}^{\infty} \boldsymbol{D}(n)(\boldsymbol{P}_k)^n, \qquad (4.1)$$

starting with $\boldsymbol{P}_0 = \boldsymbol{D}_0$ (Takine and Hasegawa [16]).

Without loss of generality, we assume two classes, i.e., $K = 2$. Cases of three or more classes are natural extensions. Since we assume that service times are iid within classes, we have

$$\boldsymbol{D}^*(\omega) = \boldsymbol{D}_1^*(\omega) + \boldsymbol{D}_2^*(\omega) = \boldsymbol{D}_1 S_1^*(\omega) + \boldsymbol{D}_2 S_2^*(\omega), \qquad (4.2)$$

where $S_k^*(\omega)$ is the GF of the service time of class-k customers.

If we define $\boldsymbol{D}_k^{(n)} = \frac{d^n}{d\omega^n}\boldsymbol{D}_k^*(\omega)\big|_{\omega=1}$ and $\boldsymbol{D}^{(n)} = \frac{d^n}{d\omega^n}\boldsymbol{D}^*(\omega)\big|_{\omega=1}$, we have

$$\boldsymbol{D}^{(1)} = \boldsymbol{D}_1^{(1)} + \boldsymbol{D}_2^{(1)} = \boldsymbol{D}_1 E(S_1) + \boldsymbol{D}_2 E(S_2), \qquad (4.3a)$$

$$\boldsymbol{D}^{(2)} = \boldsymbol{D}_1^{(2)} + \boldsymbol{D}_2^{(2)} = \boldsymbol{D}_1 E[S_1(S_1-1)] + \boldsymbol{D}_2 E[S_2(S_2-1)]. \qquad (4.3b)$$

Let us write (3.4) as follows;

$$\boldsymbol{v}^*(\omega)\omega - \boldsymbol{v}^*(\omega)\boldsymbol{D}_0 - \boldsymbol{v}^*(\omega)\boldsymbol{D}^*(\omega) = \boldsymbol{u}(0)(\omega - 1). \qquad (4.4)$$

Differentiating (4.4) with respect to ω, we get

$$\boldsymbol{u}(0) = \omega \frac{d}{d\omega}\boldsymbol{v}^*(\omega) + \boldsymbol{v}^*(\omega) - \frac{d}{d\omega}\boldsymbol{v}^*(\omega)\boldsymbol{D}_0 - \left[\frac{d}{d\omega}\boldsymbol{v}^*(\omega)\right]\boldsymbol{D}^*(\omega)$$
$$- \boldsymbol{v}^*(\omega)\frac{d}{d\omega}\boldsymbol{D}^*(\omega). \qquad (4.5)$$

Using $\omega = 1$ in (4.5) and adding $\boldsymbol{v}^{*(1)}\boldsymbol{e}\boldsymbol{\pi}$, we get

$$\boldsymbol{v}^{*(1)}(1)\left(\boldsymbol{I} - \boldsymbol{D}_0 - \boldsymbol{D} + \boldsymbol{e}\boldsymbol{\pi}\right) = \boldsymbol{v}^{*(1)}(1)\boldsymbol{e}\boldsymbol{\pi} + \left[\boldsymbol{u}(0) - \boldsymbol{e}\boldsymbol{\pi} + \boldsymbol{\pi}\boldsymbol{D}^{(1)}\right]. \qquad (4.6)$$

Multiplying by $(\boldsymbol{I} - \boldsymbol{D}_0 - \boldsymbol{D} + \boldsymbol{e}\boldsymbol{\pi})^{-1}$ and using $\boldsymbol{\pi}(\boldsymbol{I} - \boldsymbol{D}_0 - \boldsymbol{D} + \boldsymbol{e}\boldsymbol{\pi})^{-1} = \boldsymbol{\pi}$, we get

$$\boldsymbol{v}^{*(1)}(1) = \boldsymbol{v}^{*(1)}(1)\boldsymbol{e}\boldsymbol{\pi} + \left[\boldsymbol{u}(0) - \boldsymbol{\pi} + \boldsymbol{\pi}\boldsymbol{D}^{(1)}\right](\boldsymbol{I} - \boldsymbol{D}_0 - \boldsymbol{D} + \boldsymbol{e}\boldsymbol{\pi})^{-1}. \qquad (4.7)$$

Multiplying (4.7) by $D_1 e$ and using $\pi D_1 e = \lambda_1$ yields

$$v^{*(1)}(1)D_1 e = \lambda_1 v^{*(1)}(1)e + \left[u(0) - \pi + \pi D^{(1)}\right](I - D_0 - D + e\pi)^{-1} D_1 e. \tag{4.8a}$$

Multiplying (4.7) by $D_2 e$ and using $\pi D_2 e = \lambda_2$ yields

$$v^{*(1)}(1)D_2 e = \lambda_2 v^{*(1)}(1)e + \left[u(0) - \pi + \pi D^{(1)}\right](I - D_0 - D + e\pi)^{-1} D_2 e. \tag{4.8b}$$

Differentiating (4.5) with respect to ω, we have

$$0 = \omega \frac{d^2}{d\omega^2} v^*(\omega) + 2\frac{d}{d\omega} v^*(\omega) - \frac{d^2}{d\omega^2} v^*(\omega) D_0 - \left[\frac{d^2}{d\omega^2} v^*(\omega)\right] D^*(\omega)$$
$$- 2\left[\frac{d}{d\omega} v^*(\omega)\right]\left[\frac{d}{d\omega} D^*(\omega)\right] - \left[v^*(\omega) \frac{d^2}{d\omega^2} D^*(\omega)\right]. \tag{4.9}$$

Using $\omega = 1$ in (4.9), multiplying by e and using $(I - D_0 - D)e = 0$, we get

$$v^{*(1)}(1)e = v^{*(1)}(1)D^{(1)}e + \frac{1}{2}\pi D^{(2)}e. \tag{4.10}$$

Using (4.10) in (4.8a,b), we have

$$v^{*(1)}(1)D_1 e = \lambda_1 v^{*(1)}(1)D^{(1)}e + \frac{\lambda_1}{2}\pi D^{(2)}e$$
$$+ \left[u(0) - \pi + \pi D^{(1)}\right](I - D_0 - D + e\pi)^{-1} D_1 e, \tag{4.11a}$$

$$v^{*(1)}(1)D_2 e = \lambda_2 v^{*(1)}(1)D^{(1)}e + \frac{\lambda_2}{2}\pi D^{(2)}e$$
$$+ \left[u(0) - \pi + \pi D^{(1)}\right](I - D_0 - D + e\pi)^{-1} D_2 e. \tag{4.11b}$$

Using $D^{(1)} = D_1 E(S_1) + D_2 E(S_2)$, $\lambda_1 E(S_1) = \rho_1$ and $\lambda_2 E(S_2) = \rho_2$ in (4.11a,b), we get

$$(1 - \rho_1)v^{*(1)}(1)D_1 e = \lambda_1 E(S_2)v^{*(1)}(1)D_2 e + \frac{\lambda_1}{2}\pi D^{(2)}e \tag{4.12a}$$
$$+ \left[u(0) - \pi + \pi D^{(1)}\right](I - D_0 - D + e\pi)^{-1} D_1 e.$$

$$(1 - \rho_2)v^{*(1)}(1)D_2 e = \lambda_2 E(S_1)v^{*(1)}(1)D_1 e + \frac{\lambda_2}{2}\pi D^{(2)}e \tag{4.12b}$$
$$+ \left[u(0) - \pi + \pi D^{(1)}\right](I - D_0 - D + e\pi)^{-1} D_2 e.$$

From (3.8) and (3.9), the mean waiting time $W_{q(k)}$ of an arbitrary class-k customer can be obtained from

$$W_{q(k)} = v^{*(1)}(1)\frac{D_k}{\lambda_k}e. \tag{4.13a}$$

The mean waiting time W_q of an arbitrary customer can be obtained from

$$W_q = \sum_{k=1}^{2} \frac{\lambda_k}{\lambda} W_{q(k)} = v^{*(1)}(1) \frac{D}{\lambda} \mathbf{e}. \qquad (4.13b)$$

If we divide (4.12a,b) by $\lambda_1(1-\rho_1)$ and $\lambda_2(1-\rho_2)$ respectively and use (4.13a,b), we get

$$W_{q(1)} = \frac{\rho_2}{(1-\rho_1)} W_{q(2)} + \frac{1}{2(1-\rho_1)} \pi D^{(2)} \mathbf{e} \qquad (4.14a)$$
$$+ \frac{1}{(1-\rho_1)\lambda_1} \left[\mathbf{u}(0) - \boldsymbol{\pi} + \boldsymbol{\pi} D^{(1)} \right] (\mathbf{I} - D_0 - D + \mathbf{e}\boldsymbol{\pi})^{-1} D_1 \mathbf{e},$$

$$W_{q(2)} = \frac{\rho_1}{(1-\rho_2)} W_{q(1)} + \frac{1}{2(1-\rho_2)} \pi D^{(2)} \mathbf{e} \qquad (4.14b)$$
$$+ \frac{1}{(1-\rho_2)\lambda_2} \left[\mathbf{u}(0) - \boldsymbol{\pi} + \boldsymbol{\pi} D^{(1)} \right] (\mathbf{I} - D_0 - D + \mathbf{e}\boldsymbol{\pi})^{-1} D_2 \mathbf{e}.$$

$W_{q(1)}$ and $W_{q(2)}$ can be obtained by solving (4.14a,b) simultaneously. The mean waiting time of an arbitrary customer can be obtained from

$$W_q = \frac{\lambda_1}{\lambda} W_{q(1)} + \frac{\lambda_2}{\lambda} W_{q(2)} \qquad (4.15)$$

and the mean number of waiting customers can be obtained from the Little's law as follows;

$$L_{q(1)} = \lambda_1 W_{q(1)}, \quad L_{q(2)} = \lambda_2 W_{q(2)}, \qquad (4.16a)$$

$$L_q = L_{q(1)} + L_{q(2)} = \lambda W_q. \qquad (4.16b)$$

Finally the mean WIP inventory level L can be obtained from

$$L_{(1)} = L_{q(1)} + \rho_1, \quad L_{(2)} = L_{q(2)} + \rho_2, \qquad (4.17a)$$

$$L = L_{(1)} + L_{(2)} = L_q + \rho, \qquad (4.17b)$$

where $\rho_1 = \lambda_1 E(S_1)$, $\rho_2 = \lambda_2 E(S_2)$, and $\rho = \lambda E(S) = \rho_1 + \rho_2$.

5 Numerical Example

In this section, we present a numerical example and compute the mean waiting times and the mean queue lengths of each class under the following parameter matrices,

$$D_0 = \begin{pmatrix} 0.3 & 0.1 \\ 0.1 & 0.4 \end{pmatrix}, \quad D_1 = \begin{pmatrix} 0.1 & 0.1 \\ 0.2 & 0.1 \end{pmatrix}, \quad D_2 = \begin{pmatrix} 0.2 & 0.2 \\ 0.1 & 0.1 \end{pmatrix}.$$

Then, from $\boldsymbol{\pi}(D_0 + D_1 + D_2) = \boldsymbol{\pi}$, $\boldsymbol{\pi}\mathbf{e} = 1$, and $\lambda_k = \boldsymbol{\pi} D_k \mathbf{e}$, we get $\lambda_1 = 0.25$ and $\lambda_2 = 0.3$. We consider two cases:

(i) (Case 1) Service times follow geometric distribution with mean 1/0.65 and 1/0.55 for respective classes.
(ii) (Case 2) Service times follow geometric distribution with mean 1/0.7 and 1/0.6 for respective classes.

For (case 1), we get $\rho_1 = \lambda_1 E(S_1) = 0.3836$, $\rho_2 = \lambda_2 E(S_2) = 0.5455$, $\rho = \rho_1 + \rho_2 = 0.9301$ and for (case 2), we get $\rho_1 = 0.3571$, $\rho_2 = 0.5$, $\rho = 0.8571$.

Table 1. Performance measures

Measure	Case 1	Case 2
$E(S_1)$	1/0.65	1/0.7
$E(S_2)$	1/0.55	1/0.6
λ_1	0.25	0.25
λ_2	0.3	0.3
ρ_1	0.3846	0.3571
ρ_2	0.5455	0.5
ρ	0.9301	0.8571
$W_{q(1)}$	9.3849	3.4464
$W_{q(2)}$	9.4381	3.4941
W_q	9.4963	3.4724
$L_{(1)}$	2.7517	1.2188
$L_{(2)}$	3.4014	1.5482
L	6.1531	2.7670

Table 1 shows the various performance measures. To verify the accuracies of our performance measures, we compared our results with the simulation estimates. We used SIMSCRIPT II.5 for our simulations and running times were 5,000,000 unit times for all cases. The percentage errors were less than 0.9% for all cases which is not shown here.

Acknowledgement. This research was supported by KOSEF through Statistical Research Center for Complex Systems (SRCCS) at Seoul National University.

References

1. Asmussen, S. and Koole, G.: Marked point processes as limits of Markovian arrival streams. J. Appl. Prob. **30** (1993) 365–372
2. He, Q.M.: Queues with marked customer. Adv. Appl. Prob. **28** (1996) 567–587
3. He, Q.M. and Alfa, A.S.: The MMAP[K]/PH[K]/1 queues with a last-come -first-served preemptive service discipline. Queueing Systems **29** (1998) 269–291
4. He, Q.M. and Neuts, M.F.: Markov chains with marked transitions. Stochastic Process. Appl. **74** (1998) 37–52
5. Lucantoni, D.M., Meier-Hellstern, K.S. and Neuts, M.F.: A single server queue with server vacations and a class of non-renewal arrival processes. Adv. Appl. Prob. **22** (1990) 676–709

6. Masuyama, H. and Takine T.: Analysis of an infinite-server queue with batch Markovian arrival streams. Queueing Systems **42(3)** (2002) 269–296
7. Masuyama, H. and Takine, T.: Analysis and Computation of the Joint Queue Length Distribution in a FIFO Single-Server Queue with Multiple Batch Markovian Arrival Streams. Stochastic Models **19(3)** (2003) 349–381
8. Masuyama, H. and Takine, T.: Stationary Queue Length in a FIFO Single-Server Queue with Service Interruptions and Multiple Batch Markovian Arrival Streams. Journal of the Operations Research Society of Japan **46(3)** (2003) 319–341
9. Noh, S.H. and Choi, B.D.: Discrete time queues with Markovian arrival streams and state-dependent service times, IEICE Trans. Commun. **E86-B(6)** (2003) 1884–1892
10. Takagi, H.: Queueing Analysis: Volume 3. Discrete-Time Systems, North-Holland (1993)
11. Takine, T.: A recent progress in algorithmic analysis of FIFO queues with Markovian arrival streams. J. Korean Math. Soc. **38(4)** (2001) 807–842
12. Takine, T.: Queue length distribution in a FIFO single-server queue with multiple arrival streams having different service distributions. Queueing Systems **39** (2001) 349–375
13. Takine, T.: Distribution form of Little's law for FIFO queues with multiple arrival streams and its application to queue with vacations. Queueing Systems **37** (2001) 31–63
14. Takine, T.: Subexponential asymptotics of the waiting time distribution in a single-server queue with multiple Markovian arrival streams. Stochastic Models **17(4)** (2001) 429–448
15. Takine, T.: Matrix product-form solution for an LCFS-PR single-server queue with multiple arrival streams governed by a Markov chain. Queueing Systems **42(2)** (2002) 131-151
16. Takine, T and Hasegawa, T.: The workload in the MAP/G/1 queue with state-dependent services: its application to a queue with preemptive resume priority. Stochastic Models **10(1)** (1994) 183–204

Discretization Approach and Nonparametric Modeling for Long-Term HIV Dynamic Model

Jianwei Chen[1], Jin-Ting Zhang[2], and Hulin Wu[3]

[1] Department of Biostatistics and Computational Biology, University of Rochester,
601 Elmwood Avenue, Box 630, Rochester, New York 14642, USA
jchen@bst.rochester.edu
[2] Department of Statistics and Applied Probability, National University of Singapore,
Lower Kent Ridge Road 119260, Singapore
stazjt@nus.edu.sg
[3] Department of Biostatistics and Computational Biology, University of Rochester,
601 Elmwood Avenue, Box 630, Rochester, New York 14642, USA
hwu@bst.rochester.edu

Abstract. Modeling viral load response in long-term HIV dynamics is important for AIDS clinical study. It allows one to examine how a patient recuperates for an antiviral treatment. This paper investigates a discretization approach to the viral load based on a long-term HIV dynamic model. We propose a two-stage nonparametric procedure for estimating the viral load and the time-varying new productive virus. Simulation study shows that the proposed estimated methods are efficient and powerful for fitting long-term viral load measurements.

1 Introduction

HIV dynamic modeling is important in AIDS research. By modeling viral load trajectory after initiation of potent therapy, it has led to a new understanding of the pathogenesis of HIV infection and provides reasonable antiviral treatment strategies for treating AIDS patients. Many dynamic models have been proposed by AIDS researchers in the last decade for providing theoretical methods to fit the HIV dynamic models, see the review paper by Perelson and Nelson (1999) and the book by Nowak and May (2000). A well-known HIV viral dynamic model can be expressed as

$$\begin{array}{l} \frac{d}{dt}T(t) = \lambda - \rho T(t) - kT(t)V(t), \\ \frac{d}{dt}T^*(t) = kT(t)V(t) - \delta T^*(t), \\ \frac{d}{dt}V(t) = N\delta T^*(t) - cV(t), \end{array} \qquad (1)$$

where $T(t)$, $T^*(t)$ and $V(t)$ denote the numbers of target uninfected cells, infected cells, and virions at time t, respectively; λ represents the rate at which new T cells are generated from source within; ρ is the death rate per T cell; k is the infection rate of T cell infected by virus; δ is the rate of death for infected cells; c is the clearance rate of free virion and N is the number of virion produced from infected cell during its life-time. The main objective in HIV modeling is to estimate the parameters, to construct a predicted HIV dynamic model based on the observed AIDS clinical data and to guide treatment strategies for HIV-infected patients. Only the total viral load $V(t)$ is observable in an AIDS clinical study. The total viral load, which can be measured at different treatment time, is an important index for characterizing infectious situation of a patient.

Perelson et al. (1996) from physical experiments found that the concentration of uninfected CD4+ cell can usually be well approximated by a constant during a short time period (week). Specially, they investigated a short-term HIV viral dynamic model under the assumption of a constant target uninfected cells $T(t)$. Given certain initial conditions, a closed form solution of viral load for short-term HIV model (1) is given by

$$V(t) = V_0 e^{-ct} + \frac{cV_0}{c-\delta}\left\{\frac{c}{c-\delta}\left[e^{-\delta t} - e^{-ct}\right] - \delta t e^{-ct}\right\}, \qquad (2)$$

where $V_0 = V(0)$ denotes a baseline value of the viral load. Based on the short-term viral model (2), the parameters can be estimated by fitting a nonlinear parametric model as

$$\begin{aligned}Y(t) &= V(t) + e(t),\\ V(t) &= V_0 e^{-ct} + \tfrac{cV_0}{c-\delta}\left\{\tfrac{c}{c-\delta}\left[e^{-\delta t} - e^{-ct}\right] - \delta t e^{-ct}\right\},\end{aligned} \qquad (3)$$

where $Y(t)$ is observed viral load at time t and $e(t)$ is unobserved error with the mean zero and the variance $\sigma^2(t)$. Over the past several years, non-linear exponential-type parametric models have been used to fit short time clinical data from AIDS clinical trials, see Perelson et al. (1996, 1997), Wu, Ding and DeGruttola (1998), Grossman et al. (1999), Michele et al. (2004) and Han and Chaloner (2004), among others.

Although the model (3) is useful and convenient to describe short-term viral dynamics, it only can be used to characterize the viral load during the initial stage of therapy. For long-term AIDS treatment (months), the protease inhibitor may be imperfect and the concentration of target uninfected cells $T(t)$ is changing over time. Therefore, the short-term model (3) will not be applicable for long-term treatment. In general, most of immune response models such as the viral HIV dynamic model (1) are differential equation models without closed form solutions. The computation for the numerical solution will be very intensive if computation-based statistical method such as MCMC are applied for parameter estimation, see Mittler and Perelson (2001), Putter et al (2002), Huang et al (2003) and Michele et al (2004).

In this paper, we propose a general framework for fitting the HIV differential dynamic equation (1) using a direct discretization and a nonparametric smoothing technique. By using local polynomial kernel smoothing technique, a two stage estimation procedure is developed for estimating the viral road and the new productive virus. Simulation studies show that the proposed time-dependent discretization model can well capture the changes of the viral load data from the long-term viral load dynamics (1). Applications to the real data from AIDS clinical trial also demonstrate that the proposed approach can fit effectively the viral load measurements. The rest of paper is organized as follows. Section 2 introduces a time-dependent discretization model to approach the viral load HIV dynamic model. A two stage local estimation is proposed in Section 3. Simulation examples are presented in Section 4. Some concluding remarks are given in Section 5.

2   A Discretization Model for the Viral Load

From (1) and (3), a general viral differential dynamic model may be written as

$$Y(t) = V(t) + e(t),$$
$$\frac{dV(t)}{dt} = u(t) - cV(t), \qquad (4)$$
$$V(0) = V_0,$$

where $Y(t)$ is observed viral load at time t and $e(t)$ is unobserved error with the mean zero and the variance $\sigma^2(t)$. In the model (4), $V(t)$ is the true viral load at time t and $u(t) = N\delta T^*(t)$ denotes the new productive virus at time t while c indicates the clearance rate of free virion which can be estimated by the short-term model (3).

Suppose that we have the observed viral load sample $Y(t_i)$, $(i = 1, \cdots, n)$ from the model (4) at discrete time points: $t_1 < t_2 < \cdots < t_n$. The aim of this paper is to estimate the viral road $V(t)$ and the new productive virus $u(t)$ based on $Y(t_i)$, $(i = 1, \cdots, n)$, and then establish a predicable model for the changes of the viral load. In this section, we propose a direct discretization approximation to the differential equation model (4). Let

$$DV(t_i) = V(t_{i+1}) - V(t_i), \qquad \Delta t_i = t_{i+1} - t_i.$$

We use the simple Euler method to discretize the differential equation (4) as

$$DV(t_i) = [u(t_i) - cV(t_i)]\Delta t_i + e_2(t_i), \qquad i = 1, \cdots, n, \qquad (5)$$

where $e_2(t_i)$, $(i = 1, \cdots, n)$, are the discretization error with the mean zero and the variance $\sigma_2^2(t_i)$ which can be regarded as the conditional variance of $V(t_{i+1})$ given $V(t_i)$. Note that when the difference Δt_i are small, the equation $DV(t_i)/\Delta t_i = [u(t_i) - cV(t_i)]$ will be a good approximation of the model (4). We call the discretization model (5) as the time-dependent discretization model for the virus load response. Thus the model (4) can be discretized as

$$Y(t_i) = V(t_i) + e(t_i),$$
$$DV(t_i) = [u(t_i) - cV(t_i)]\Delta t_i + e_2(t_i), \qquad (6)$$
$$V(0) = V_0, \qquad i = 1, \cdots, n.$$

Now we turn to study estimation of $V(t)$ and $u(t)$. However, our data $Y(t_i)$ are measurements of $V(t_i)$ with error. One way to deal with the measurement error is to smooth $Y(t_i)$ to obtain the estimates of $V(t_i)$ using a smoothing technique such as local polynomial or spline method. Then plugging these estimates into the model (6) so that it becomes a standard nonparametric model. Some standard nonparametric regression method will be used to estimate the time-varying parameter $u(t)$. In next section, we propose a two stage local linear regression method to estimate the $V(t_i)$ and $u(t_i)$ $(i = 1, \cdots, n)$. It should be pointed out that the Euler discretization method is a first-order method. The higher order method such as Runge-Kutta may be used, but the associated model will be more complex and the computation will be more intensive.

3 Estimation

3.1 Two-Stage Local Linear Estimation

We employ the local polynomial modeling (Fan and Gijbels (1996)) to fit the model (6). The local polynomial kernel technique is a useful nonparametric method. It inherits many good sampling properties of the least-squares method. Here we develop a two-stage local estimation to manage the estimate problem of (6). The procedure consists of the following two stages. The first stage involves estimating $V(t)$ based on the first equation of the model (6). In the second stage, a local linear regression is fitted for estimating $u(t)$ using the estimated values $\hat{V}(t)$. We choose two different bandwidths in the two-stage estimation.

In the first stage, we aim to estimate $V(t)$. Assume that the function $V(t)$ possesses the 2th derivative. For each given time point t_0, we approximate $V(t)$ locally by

$$V(t) \approx \beta_0(t_0) + \beta_1(t_0)(t - t_0),$$

for t in a neighborhood of t_0. The estimator $\hat{V}(t_0) = \hat{\beta}_0(t_0)$ can be obtained by minimizing the following locally weighted least-squares function

$$\sum_{i=1}^{n} \{Y(t_i) - [\beta_0(t_0) + \beta_1(t_0)(t_i - t_0)]\}^2 K_{h_1}(t_i - t_0), \tag{7}$$

where $K_{h_1}(\cdot) = K(\cdot/h_1)/h_1$ with K being a kernel function and h_1 being a bandwidth. In the second stage, we use the local linear method to estimate the function $u(t)$ of (6). We assume that $u(t)$ has the second order derivative. Then $u(t)$ can be approximated around t_0 by

$$u(t) \approx \alpha_0(t_0) + \alpha_1(t_0)(t - t_0).$$

Thus, for a given initial estimator c, $u(t)$ can be estimated via minimizing locally weighted function.

$$\sum_{i=1}^{n-1} \left\{ \widehat{DV}(t_i) - [(\alpha_0(t_0) + \alpha_1(t_0))(t_i - t_0) - c\hat{V}(t_i)]\Delta t_i \right\}^2 K_{h_2}(t_i - t_0), \tag{8}$$

where $\widehat{DV}(t_i) = \hat{V}(t_{i+1}) - \hat{V}(t_i)$ $(i = 1, \cdots, n-1)$, and $K_{h_2}(.)$ is a kernel function and h_2 is a bandwidth. Let $(\hat{\alpha}_0, \hat{\alpha}_1)$ minimize the local weighted function (8). Then the local linear estimator of $u(t_0)$ is $\hat{u}(t_0) = \hat{\alpha}_0(t_0)$.

3.2 Bandwidth Selection

The selection of the bandwidth is important in any nonparametric techniques. There are several common methods including the plug-in method, preasmptotic data-driven method, the empirical bias bandwidth selection method, cross-validation and generalized cross-validation. In the paper, we have two bandwidths, h_1 and h_2, to select. We use a data-driven bandwidth selection method for the first bandwidth h_1 since it refers to local polynomial means estimation. The second bandwidth h_2 can be obtained using the generalized cross-validation that minimizes the following GCV score:

$$\text{GCV}(h_2) = \frac{\sum_{i=1}^{n-1}\{\widehat{DV}(t_i) - \hat{u}(t_i) - \hat{c}\hat{V}(t_i)\}^2}{\{1 - tr(\mathbf{S}_{h_2})/(n-1)\}^2}, \qquad (9)$$

where $\hat{V}(t_i)$ and $\widehat{DV}(t_i)$ have been estimated in the first stage local linear fit with the bandwidth h_1, and $\mathbf{S}_{h_2}$ is a smoother matrix.

4 Implementation

Here we describe an algorithm for estimating the constant death rate c, the viral road $V(t)$ and the new productive virus $u(t)$. Alternatively we may employ the short-term model (3) to estimate the constant rate c and apply the nonparametric methods developed in this paper to estimate $V(t)$ and $u(t)$. The procedure is summarized as follows:

Step 1. Estimate the constant death rate c of the viral load denoted by $\hat{c}$. Based on the analytical solution of the short term model (3), we fit a nonlinear regression model using the initial viral road data.

Step 2. Calculate the local linear estimator $\hat{V}(t_0)$ for $V(t_0)$ with a bandwidth h_1. In the same way, we also find the estimator $\hat{V}(t_i)$ for each given time point t_i ($i = 1, \cdots, n$).

Step 3. Calculate the local linear estimator $\hat{u}(t_0)$ with the second bandwidth h_2.

Step 4. Establish a predicted formulation of the viral load by using the estimated viral load $\hat{V}(t_i)$ and the estimated new productive virus $\hat{u}(t_i)$ in the time-dependent discretization model (6), i.e.

$$\hat{V}^*(t_{n+l}) = \begin{cases} \hat{V}(t_{n+l-1}) + \left[\hat{u}(t_{n+l-1}) - \hat{c}\hat{V}(t_{n+l-1})\right]\Delta t_{n+l}, & l \leq 1, \\ \hat{V}^*(t_{n+l-1}) + \left[\hat{u}(t_{n+l-1}) - \hat{c}\hat{V}^*(t_{n+l-1})\right]\Delta t_{n+l}, & l > 1. \end{cases}$$

5 Simulation

We conduct extensive Monte Carlo simulations to evaluate the finite properties of the proposed nonparametric estimators. The purposes are to examine whether the proposed approach can fit effectively the data from the HIV dynamic model with the long-term treatments. The observed viral load data are generated according to the HIV dynamic model (1) with errors. We simulate data $\{(t_i, Y(t_i)), i = 1, \cdots, n\}$ based on model:

$$Y(t_i) = V(t_i) + e(t_i), \qquad (10)$$

where the design time point $t_i = \text{day} \times t_i^*$ with t_i^* following a uniform distribution $U(0,1)$ and $V(t_i), i = 1, \cdots, n$, are numerical solutions of the viral differential dynamic model (1) with the known coefficients. The coefficient values used in our simulation studies are $\lambda = 7.0$, $p = 0.007$, $k = 20 \times 10^{-7}$, $\delta = 0.0999$, $N = 1000.0$, $c = 0.2$ and $V_0 = 10000.0$. This HIV dynamic model example had been studied by Xia (2003). The error term $e(t_i)$ follows an independent and identically normal distribution with mean 0 and variance $\sigma^2(t) = (1+t)\sigma^2$. 400 simulations are conducted with sample sizes $n = 100$ and 200.

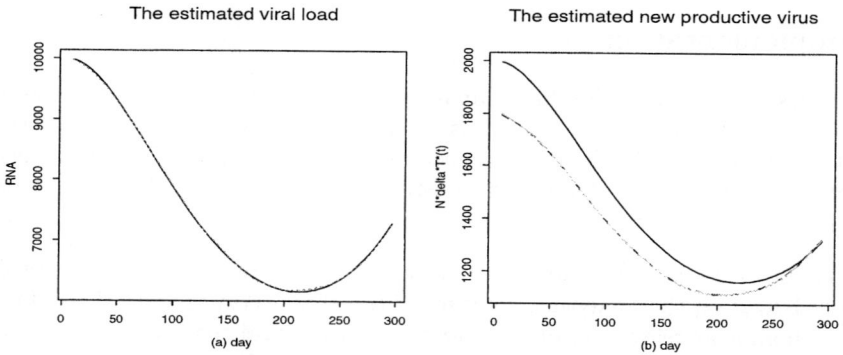

Fig. 1. Simulation results for the long-term HIV dynamics model with the error variance function $\sigma(t) = (1+t)^{1/2}100$. (a) The changes of the viral load as time (the solid line indicates the true values and the dashed line is the local estimator $\hat{V}(t)$); (b) The changes of the new productive virus with time. (the solid line indicates the true values and the dashed line is the local estimator $\hat{u}(t)$).

We apply the proposed local estimators $\hat{V}(t)$ for estimating the viral load $V(t)$ while the local estimator $\hat{u}(t)$ is used to estimate the new productive virus $u(t)$. The results are displayed in Figures 1 and 2. In the implementation, we employed the Epanechnikov kernel $K(t) = 0.75(1 - t^2)_+$. Figure 1 shows the estimated results for the simulation with the variance constant $\sigma = 100$. Panel (a) plots the changes of the estimated viral load over time while Panel (b) displays the changes of the estimated new productive virus as time. It can be seen that the estimators $\hat{V}(t)$ and $\hat{u}(t)$ (dashed lines) performs well compare with the corresponding true values $V(t)$ and $u(t)$ (solid lines), respectively. Although the variances of the measurement error are increased, Figure 2 shows that the estimated results with

$\sigma = 200$ continue to keep reasonable estimates. It indicates that the proposed discretization approach and local polynomial smoothing methods can capture effectively the data changes from the HIV dynamic model.

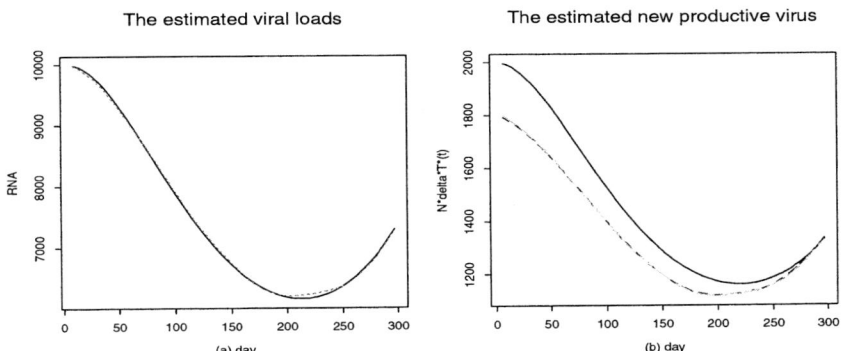

Fig. 2. Same caption as for Figure 1 but now for the error variance function $\sigma(t) = (1+t)^{1/2} 200$.

6 Discussion

We have developed a time-dependent discretization framework for modeling the viral load based on the HIV dynamics model. The local linear kernel techniques have been applied to estimate the viral load and the new productive virus. Simulation study shows that the proposed discretization model is efficient for fitting long-term viral load. Both the short-term or the long-term treatment viral dynamics can be well approached by the time dependent discretization model. In the longer version of this paper, we applied the methodology to a HIV dynamic data set from the AIDS clinical trials ACTG 387 study. The results also show that the proposed new approach can fit effectively the viral load measurement in the long-term treatment.

Based on the viral dynamic differential equation model and the Euler discretization, we have established a time-dependent discretization model for the long-term AIDS treatment. It should be pointed out that we only use a first order approximation in the Euler discretization. The higher order method may be good if we want to obtain more accurate approximation to the viral dynamic model. Stanton (1997) introduced a drift and diffusion estimator via higher-order approximation and kernel estimation. Fan and Zhang (2003) applied the Euler first order discretization to approach the diffusion model. They showed that higher-order method can reduce the numerical approximation error in asymptotic biases, but the variance escalates nearly exponentially with the order of approximation. Whether this phenomena is behind the deterministic viral dynamic model, further work will be needed. Although the discretization approach and nonparametric methods developed in this paper are based on the one variate differential equation model, these techniques can be extended to more complicated models. We will keep to work on this direction.

References

1. Aerts, M., Chaeskens, G.: Local polynomial estimation in multiparameter likelihood models. Journal of the American Statistical Association **92** (1997) 1536-1545.
2. Fan, J., Gijbels, I.: Local Polynomial Modeling and Its Applications. Chapman and Hall, London. (1996).
3. Fan, J., Yao, Q.: Efficient estimation of conditional variance functions in stochastic regression. Biometrika **85** (1998) 645-660.
4. Fan, J., Zhang, C.: A re-examination of Stanton's diffusion estimations with applications to financial model validation. Journal of American Statistical Association **98** (2003) 118–134.
5. Fitzgerald, A. P., DeGruttola, V. G., Vaida, F.: Modeling HIV viral rebound using non-linear mixed effects models. Statistics in Medicine **21** (2002) 2093–2108.
6. Han, C., Chaloner, K.: Bayesian experimental design for nonlinear mixed-effects models with application to HIV dynamics. Biometrics **60** (2004) 25–33.
7. Ho, D. D., Neumann, A. U., Perelson, A. S., Chen, W., Leonard, J. M., Markowitz, M.: Rapid turnover of plasma virions and CD4 lymphocytes in HIV-1 infection. Nature **374** (1995) 123–126.
8. Huang, Y., Rosenkranz, S. L., Wu, H.: Modeling HIV dynamics and antiviral responses with consideration of time-varying drug exposures, sensitivities and adherence. Mathematical Biosciences **184** (2003) 165–186.
9. Michele, D. M., Ruy, M. R., Martin, M., Ho, D. D., Perelson, A. S.: Modeling the long-term control of viral load in HIV-1 infected patients treated with antiretroviral therapy. Mathematical Biosciences **188** (2003) 47–62.
10. Nowak, M. A., May, R. M.: Virus dynamics: mathematical principles of immunology and virology. Oxford: Oxford University Press (2000).
11. Perelson, A. S., Nelson, P. W.: Mathematical analysis of HIV-1 dynamics *in vivo*. SIAM Review **41** (1999) 3–44.
12. Perelson, A. S., Neumann, A. U., Markowitz, M., Leonard, J.M., Ho, D. D.: HIV-1 dynamics *in vivo*: Virion clearance rate, infected cell life-span, and viral generation time. Science **271** (1996) 1582–1586.
13. Putter, H., Heisterkamp, S. H., Lange, J. M. A., De Wolf, F.: A Bayesian approach to parameter estimation in HIV dynamical models. Statistics in Medicine **21** (2002) 2199-2214.
14. Stanton, H.: A nonparametric models of term structure dynamics and the market price of interest rate risk. Journal of Finance **5** (1997) 1973–2002.
15. Verotta, D., and Schaedeli, F.: Non-linear dynamics models characterizing long-term virological data from AIDS clinical trials. Mathematical Biosciences **176** (2002) 163–183.
16. Wei, X. Ghosh, S. K., Taylor, M. E., Johonson, V. A., Emini, E.A., Hahn, B. H., Saag, M. S., Shaw, G. M.: Viral dynamics in human immunodeficiency virus type 1 infection. Nature **373** (1995) 11-7-122.
17. Wu, H., Ding, A.A., Gruttola, V. D.; Estimation of HIV dynamic parameters. Statistics in Medicine **17** (1998) 2463–2485.

18. Wu, H., Ding, A.A.: Population HIV-1 dynamics in vivo: Applicable models and inferential tool for virological data from AIDS clinical trials. Biometrics **55** (1999) 410-418.
19. Wu, H. and Zhang, J.T.: The study of long-term HIV dynamics using semi-parametric non-linear mixed-effects models. Statistics in Medicine **21** (2002) 3655–3675.
20. Wu, L.: A joint model for nonlinear mixed-effects models with censoring and covariates measured with error, with application to AIDS studies. Journal of the American Statistical Association **97** (2002) 955–963.

Performance Analysis and Optimization of an Improved Dynamic Movement-Based Location Update Scheme in Mobile Cellular Networks

Jang Hyun Baek[1]*, Jae Young Seo[1], and Douglas C. Sicker[2]

[1] Dept of Industrial and Information Systems Eng., Chonbuk Natl. Univ., Korea
[2] Dept of Computer Science, University of Colorado at Boulder, USA

Abstract. In this study, we propose an improved version of the dynamic movement-based location update scheme that stores all cells visited by the MS, and their neighboring cells, to reduce the location update cost of the dynamic movement-based location update scheme. We show that a significant reduction in the location update cost is achieved in our scheme. We also show that our scheme has optimal total cost when the threshold is rather small. Additionally, the MS requires a minimal amount of memory to successfully implement our scheme.

1 Introduction

In a mobile cellular network, location update is the process by which a mobile station (MS) notifies the network of its location, status, and other characteristics [1, 2]. The MS informs the network of its location and status so that the network can efficiently page the MS when a call arrives at the MS. Many location update schemes have been proposed with the aim of minimizing location update costs [1–7]. Important schemes include time-based [3, 4], movement-based [2, 3, 5, 6], and distance-based [2, 7] location update schemes. In this paper, we consider a movement-based location update scheme, and its improved versions, which are simple to implement and show good performance.

Due to the limited battery life and computational capacity in an MS, one desirable feature of a location update scheme is its easy implementation in the MS. A movement-based location update scheme [5] is one of the most easily implemented approaches. In this scheme, each MS maintains a counter, whose value is compared to a movement threshold D to trigger a location update. The MS performs a location update when the number of cells visited by the MS exceeds the movement threshold D. In this scheme, however, the MS increases its counter value even when the MS reenters the cell already visited, which triggers an unnecessary location update.

To overcome this deficiency, and to improve the performance of the movement-based update scheme, a dynamic location update scheme based on the number

---

* This work was supported by Korea Science & Engineering Foundation (R05-2004-000-11569-0).

of movements and the movement history was proposed [2]. We designate the proposed scheme in [2] as the DMBU (dynamic movement-based update) scheme for convenience. The basic idea in the DMBU scheme is to avoid the location update when the MS moves around the cell where the last location update was performed. This is achieved by utilizing simple movement history information.

In the DMBU scheme, each MS has three memory elements and one counter. Using them, the unnecessary location updates are not performed when the MS moves around the recently visited cell. Simulation results demonstrate that the DMBU scheme can achieve a significant cost reduction when compared to the movement-based scheme, especially when the call-to-mobility ratio (CMR) is low.

Even though the DMBU scheme shows a good performance, when the MS has sufficient memory to store all the cells visited by the MS and their neighbors, we can improve the performance of the DMBU scheme considerably. In this study, we propose an improved DMBU (iDMBU) scheme that reduces the location update cost of the DMBU scheme by maintaining a more complete history of visited cells and their neighbors. We also demonstrate that the iDMBU scheme continuously outperforms the DMBU scheme. For the sake of conciseness, we do not repeat the model description of [2] in detail, but rather concentrate on new results and comparing them with the DMBU scheme in [2].

2 Improved DMBU

In this section we describe an improved DMBU scheme, which results in less tracking cost than conventional DMBU scheme.

2.1 Dynamic Movement-Based Location Update Scheme [2]

Let us briefly describe the DMBU scheme. In the DMBU scheme, each MS has three memory elements (CURRENT, HISTORY, and UPDATE) and one counter. The CURRENT stores the identification (ID) of the cell where the MS is currently residing. The UPDATE stores the ID of the cell where the last location update was performed. The HISTORY stores the ID of the cell from which the MS moves into a new cell, which results in the counter value increment. The counter value is incremented, maintained, or reset at each cell movement according to the proposed location update algorithm. The counter value is incremented i) when the MS moves farther from the cell corresponding to the contents of the HISTORY and ii) when it moves out of the location update cell. Therefore, the contents of the HISTORY are updated in both these cases.

To determine whether an MS'ss movement increases the distance from the cell corresponding to the HISTORY, it is necessary to check that the cell is also one of the neighboring cells in the newly visited cell, or that it has been visited. For this process, each MS needs to have information on the IDs of the cell currently being visited, and its neighboring cells. In this paper, it is assumed that each BS broadcasts a beaconing signal that carries IDs for its own, and its neighboring cells, to the MS's within its coverage area [2]. For example, the BS in cell A broadcasts its own cell ID (A) and its six neighboring cell IDs. An MS moving into a new cell A uses this information to decide on a location update. Let NewCell_ID denote the ID of the newly moved-in cell. Let EnvironCell_DB denote a set of IDs for the newly moved-in cell and its neighboring cells. Notice here that the EnvironCell_DB includes the ID of the newly moved-in cell.

Using three memory elements, a counter and EnvironCell_DB, the unnecessary location updates are not performed when the MS moves around the recently visited cell. The pseudo-code for the DMBU scheme is as follows:

```
IF (cell boundary crossing)
    IF (NewCell_ID = UPDATE)
        counter = 0; HISTORY = NULL; CURRENT = NewCell_ID
    else IF (HISTORY is member of EnvironCell_DB)
            CURRENT = NewCell_ID
        else
            counter = counter + 1
            IF (counter < location update threshold)
                HISTORY = CURRENT; CURRENT = NewCell_ID
            else
                Location Update; Initialize
```

As described previously, an MS has three memory elements and a counter. When it moves into a new cell, it also temporarily maintains the newly moved-in cell and its six neighboring cells. Notice here that the memory size required in this scheme does not depend on the movement threshold value of D, which may be positive for memory size; however, it can prevent improved performance.

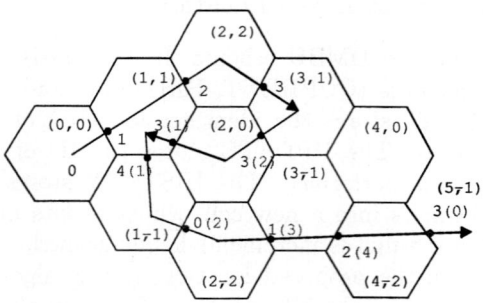

Fig. 1. An example path of an MS

2.2 Motivation

Let us consider the example in Figure 1. It is assumed D=5. In the figure, (x, y) indicates cell ID and $N_1(N_2)$ means that the counter value of the DMBU is N_1

and the counter value of the iDMBU is N_2 when entering the cell. When an MS reenters cell (1, 1) in the movement-based location update scheme, the counter value becomes 5, and a location update is performed.

In the DMBU scheme, when the MS reenters the cell (1, 1) it doesn't perform a location update because the counter value is maintained at 3 (the counter value of all the surrounding cells of cell (2, 2) stored in HISTORY). Finally, the MS updates its location when it visits cell (2, -2). In this way, the number of location updates in the DMBU scheme is always less than or equal to the number of location updates in the movement-based location update scheme.

However, the possibility to enhance performance still exists in the DMBU scheme. If we make an MS with more memory space and store more cells visited by the MS and neighboring cells after the last location update, as far as memory space permits, we can improve the performance of the DMBU scheme significantly. For example, consider the path where the MS reenters cell (1, 1) and visits cell (1, -1) through cell (5, -1), as shown in Figure 1.

If the MS stores more cells visited by the MS after the last location update and their neighboring cells, then when it reenters cell (1, 1) (that is in level-1 cells of the HISTORY) its counter value becomes 1. Further, it enters cell (2, 2) (that is in level-1 cells of the HISTORY), its counter value remains 1. Finally, the MS updates its location when it visits cell (5, -1). In this way, if we make the MS have more memory space, and store more cells visited by the MS and their neighboring cells, as far as memory space permits, we can improve the performance of the DMBU scheme significantly.

It is obvious that the iDMBU scheme needs more memory space than the DMBU scheme. However, our proposed scheme is sufficiently competitive because it is not economically difficult to extend memory space, considering the very low price of memory on the market. In addition, our scheme achieves optimal total cost when the threshold is rather small, and the MS requires small amount of memory to implement our scheme. This will be proved in section 4. In this study, we call our improved DMBU scheme an iDMBU scheme. Now, we will describe the iDMBU scheme in detail.

2.3 Improved Location-Based Location Update Scheme

The iDMBU scheme is basically similar to the DMBU scheme, except for the memory size, the rule of arrangement, and the rule of removal. Similar to the DMBU scheme, each MS maintains three memory elements (CURRENT, UPDATE, and HISTORY) and a counter. The CURRENT stores the ID of the cell where the MS is currently residing. The UPDATE stores the ID of the cell where the last location update was performed. In contrast, the HISTORY stores the ID of the cell visited by the MS, which is different from the iDMBU scheme. Figure 2 shows the data structure of the HISTORY when the threshold is D. The cells in the HISTORY are arranged in increasing order of levels, as shown in the figure. In the figure, n_i indicates number of cells estimated to be in ring "i", judging from up to date information, and level-i cells indicate the corresponding n_i cells.

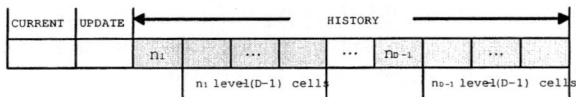

Fig. 2. Structure of memory element

In the hexagonal cell configuration, maximal n_i is 6i. Consequently, the maximal number of cells that can be stored in the HISTORY is $\sum_{i=1}^{D-1} 6i = 3D(D-1)$. For the sake of convenience, in this study it is assumed that the HISTORY can store the maximal number of cells.

The counter value is incremented, maintained, decremented, or reset at each cell movement according to the proposed location update algorithm. The counter value is incremented by one when the MS moves into a cell that is not in the HISTORY and the HISTORY is updated in this case. The counter value j is decremented by one when the MS moves into the level-(j-1) cell that is in the HISTORY. The counter value j is maintained when the MS moves into a level-j cell that is in the HISTORY. The counter value is reset when the MS updates its location.

Similar to the DMBU scheme, it is assumed that each BS broadcasts beaconing signals carrying IDs for its own, and its neighboring cells (*EnvironCell_DB*), to the MS's within its coverage area. An MS moving into a new cell uses this information to make a decision on location update. Using three memory elements, a counter, and *EnvironCell_DB*, the unnecessary location updates are not performed when the MS moves around the already visited cell. When the MS with counter value "k" enters cell j from cell i, the rules to store cell IDs in the HISTORY, and update the counter are as follows:

1. If cell j is not in the HISTORY, counter C increases by one.
   a) If C=D, a location update is triggered and the memory is initiated with cell j;
   b) Otherwise, cell i and its neighboring cells are added to the HISTORY.
2. If cell j is already in the HISTORY, two cases are possible.
   a) If it is a level-k cell, cell i and its neighboring cells are added to the HISTORY.
   b) If it is a level-(k-1) cell, the counter value k is decremented by one and cell i and its neighboring cells are added to the HISTORY.

Note that when a cell and its neighboring cells are added to the HISTORY each existing duplicated cell should be included in the lower level of the HISTORY. Note also that when the MS with counter value k enters cell j, it is sufficient to check in level-k and level-(k-1) cells to decide whether cell j is in the HISTORY and to determine what the counter value becomes.

3 Performance of the Improved Location-Based Location Update Scheme

We will show that the location update cost of the iDMBU scheme is less than or equal to that of the DMBU scheme. In other words, if we give the MS in the DMBU scheme more memory space and store all the cells visited by the MS and their neighboring cells as far as memory space permits, we can show that such a scheme improves the performance of the DMBU scheme considerably. However, in this study, we do not consider memory cost since it is a very low, one-time cost. As such it scarcely affects the superiority of the proposed scheme. Thus, we analyze and compare the location update cost of each scheme on radio channels, assuming that the memory cost is low enough that it can be ignored.

Let us first introduce some random variables to formulate the location update cost, and then compare the performances analytically.

M : number of entering cells between call arrivals
N_A: number of entering new cells that are not in the memory of the A scheme between call arrivals
D_A: cumulative sum of the decreased counter value of the A scheme between call arrivals, whenever the MS reenters the cell that is already in the memory.
L_A: N_A - D_A

Let us examine the above random variables for Figure 1. Table 1 represents values of random variables from cell (0, 0) through cell (5,-1). Figure 1 and Figure 3 represent corresponding stored cell IDs of the memory elements in the iDMBU scheme.

For these random variables, we can see that the following property is given:

Property: $N_{DMBU} \geq N_{iDMBU}, D_{DMBU}(= 0) \leq D_{iDMBU}, L_{DMBU} \geq L_{iDMBU}$
for every cell visited by the MS between call arrivals

We also can obtain following location update costs between call arrivals [6]:

$$C_U^{DMBU} = U \sum_{i=1}^{\infty} i \sum_{j=iD}^{\infty} \alpha(j) \sum_{k=iD}^{(i+1)D-1} Pr[L_{DMBU} = k|M = j]$$

$$C_U^{iDMBU} = U \sum_{i=1}^{\infty} i \sum_{j=iD}^{\infty} \alpha(j) \sum_{k=iD}^{(i+1)D-1} Pr[L_{iDMBU} = k|M = j]$$

where U is the unit location update cost required for one location update, and $\alpha(j)$ is the probability that an MS enters j cells between call arrivals. In the above, we obtain the value of $Pr[L|M]$ using computer simulations. We decided to use computer simulation because the complexity of the calculations made obtaining results rather difficult. Finally, we can obtain the following corollary:

Table 1. Memory elements and values of random variables for two schemes

Location (M)	DMBU scheme Memory			C	N	n [a]	iDMBU(2D) scheme Memory											C	N	D	L	n
0,0(0)	0,0		0,0	0	0	3	0,0	0,0										0	0	0	0	2
1,1(1)	1,1	0,0	0,0	1	1	3	1,1	0,0	5	0,2	1,-1	0,-2	-1,-1	-1,1				1	1	0	1	7
2,2(2)	2,2	1,1	0,0	2	2	3	2,2	0,0	6	0,2	1,1	1,-1	0,2	-1,-1	-1,1	2	1,3	2,0	2	2	0	2	10
3,1(3)	3,1	2,2	0,0	3	3	3	3,1	0,0	6	0,2	1,1	1,-1	0,2	-1,-1	-1,1	3	1,3	2,2	3	3	0	3	13
							2,0	2	2,4	3,3													
2,0(4)	2,0	2,2	0,0	3	3	3	2,0	0,0	6	0,2	1,1	1,-1	0,2	-1,-1	-1,1	3	1,3	2,2	2	3	1	2	15
							2,0	3	2,4	3,3	3,1	2	4,2	4,0									
1,1(5)	1,1	2,2	0,0	3	3	3	1,1	0,0	5	0,2	1,-1	0,-2	-1,-1	-1,1	3	1,3	2,2	2,0	1	3	2	1	17
							5	2,4	3,3	3,1	3,-1	2,-2	2	4,2	4,0								
1,-1(6)	1,-1	1,1	0,0	4	4	3	1,1	0,0	5	0,2	1,1	0,-2	-1,-1	-1,1	4	1,3	2,2	2,0	1	3	2	1	17
							2,-2	4	2,4	3,3	3,1	3,-1	2	4,2									
2,-2(7)	2,-2	N	2,-2	0	5	3	2,-2	0,0	6	0,2	1,1	1,-1	0,-2	-1,-1	-1,1	4	1,3	2,2	2	4	2	2	18
							2,0	1,-3	4	2,4	3,3	3,1	3,-1	2	4,2	4,0							
3,-1(8)	3,-1	2,-2	2,-2	1	6	3	3,-1	0,0	6	0,2	1,1	1,-1	0,-2	-1,-1	-1,1	4	1,3	2,2	3	5	2	3	19
							2,0	1,-3	5	2,4	3,3	3,1	3,-1	2,-4	2	4,2	4,0						
4,-2(9)	4,-2	3,-1	2,-2	2	7	3	4,-2	0,0	6	0,2	1,1	1,-1	0,-2	-1,-1	-1,1	4	1,3	2,2	4	6	2	4	20
							2,0	1,-3	6	2,4	3,3	3,1	3,-1	3,-3	2,-4	2	4,2	4,0					
4,-2(10)	5,-1	4,-2	2,-2	3	8	3	5,-1	5,-1										0	7	2	5	2

[a] n: number of stored cell IDs ($n = 2 + \sum_{i=1}^{D-1} n_i$ for iDMBU)

Corollary: For every cell visited by the MS between call arrivals, the number of location updates for the iDMBU scheme is less than or equal to that of the DMBU scheme. That is,

$$C_U^{iDMBU} = U \sum_{i=1}^{\infty} i \sum_{j=iD}^{\infty} \alpha(j) \sum_{k=iD}^{(i+1)D-1} Pr[L_{iDMBU} = k | M = j]$$

$$\leq C_U^{DMBU} = U \sum_{i=1}^{\infty} i \sum_{j=iD}^{\infty} \alpha(j) \sum_{k=iD}^{(i+1)D-1} Pr[L_{DMBU} = k | M = j]$$

Proof is omitted because it is trivial in accordance with the above property. Finally, we can obtain the following total signaling cost between call arrivals C_T for the location update and paging:

$$C_T = C_U + C_P = C_U + V[1 + 3D(D-1)]$$

where C_P is the paging cost between call arrivals, V is the cost of paging a cell, and [1+3D(D-1)] is the total number of cells in a location area. Note that it is assumed that all cells in the location area are paged simultaneously whenever an incoming call arrives as in most cellular systems.

4 Numerical Results

For illustrative purposes, a hexagonal cell configuration of [2,5] is assumed. The cell residence time and the incoming call arrivals are assumed to follow an exponential distribution with λ_m, and a Poisson process with λ_c, respectively. It is also assumed that U=10 and V=1, as in [2,5].

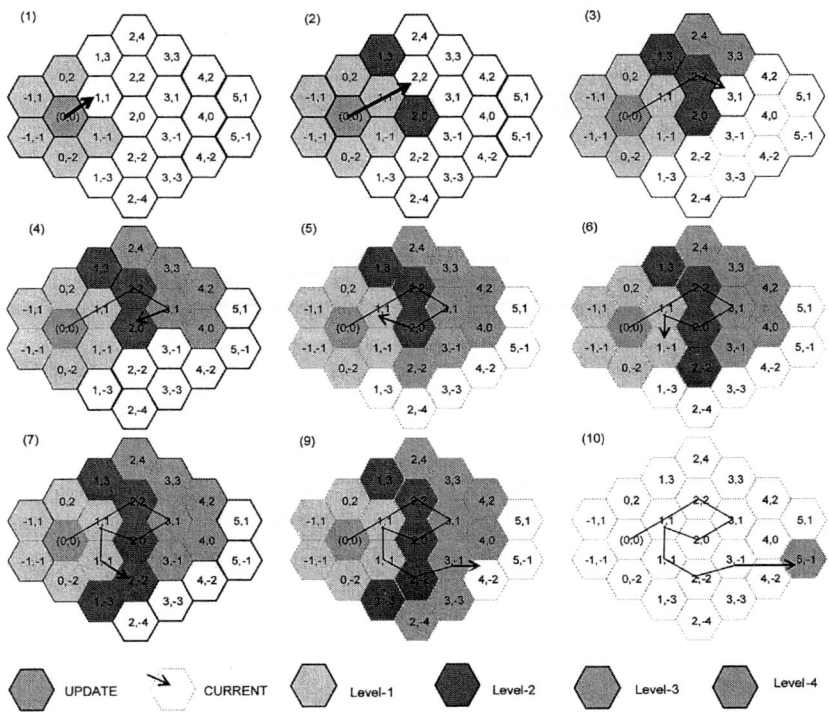

Fig. 3. Stored cell IDs of the memory elements in the iDMBU scheme

Figure 4 shows the location update cost of two schemes for different CMR values, with varying values of the movement threshold D. Note that CMR is defined as λ_c/λ_m and low CMR represents high mobility. In the figure, we can see that update cost of the iDMBU scheme is always less than that of the DMBU scheme. Furthermore, we can see that the iDMBU scheme achieves a significant cost reduction, compared to the DMBU, especially when the threshold is high. Figure 4 also shows that the update cost decreases as the threshold increases, but that the reduction ratio increases as the threshold increases.

Figure 5 shows the location update cost of the two schemes for varying values of CMR. In this figure, we can see that the update cost decreases as CMR increases, but that the reduction ratio (DMBU-iDMBU)/DMBU is almost constant as CMR increases. This indicates that the reduction ratio (DMBU-iDMBU)/DMBU is almost independent from the CMR value, as shown in Figure 4.

Figure 6 shows the location update cost, paging cost, and the total signaling cost for the two schemes with varied values of the movement threshold D, given as CMR=0.1. The paging cost of each scheme is equal because simultaneous paging is assumed. In the figure, we can see that both the schemes achieve optimal

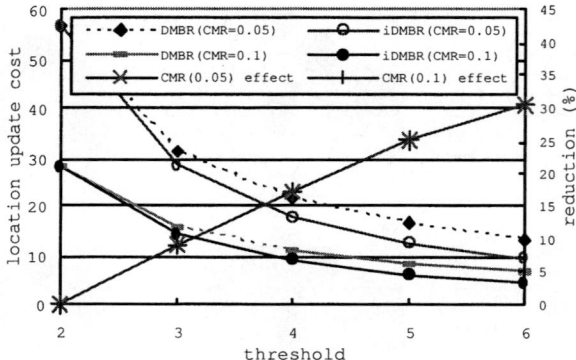

Fig. 4. Location update cost of the iDMBU scheme for the different CMR values

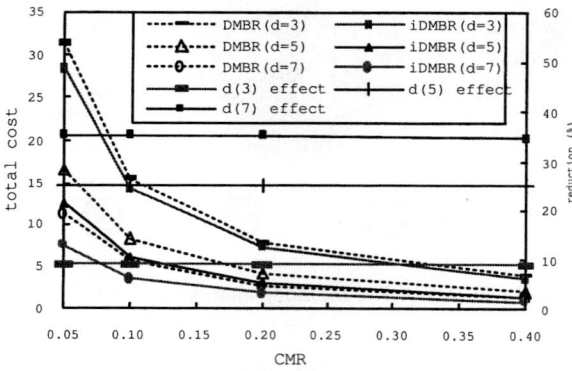

Fig. 5. Total cost for the two schemes

total cost when the threshold D=3, and then approximately 4.2% reduction of the total cost (around 9.2% reduction in registration cost) is achieved in the proposed scheme. It is expected that a more significant cost reduction will be achieved in the proposed scheme as the CMR decreases. For example, when CMR=0.05, both of the schemes have the optimal total cost when the threshold D=3, and around 5.7% reduction in total cost is achieved in the proposed scheme. We can also see that our scheme has the optimal total cost when the threshold is rather small, and when the MS requires a small amount of memory to implement our scheme.

5 Conclusion

An improved version of the dynamic movement-based location update scheme was presented. In our scheme, each MS can store more cells (those visited by

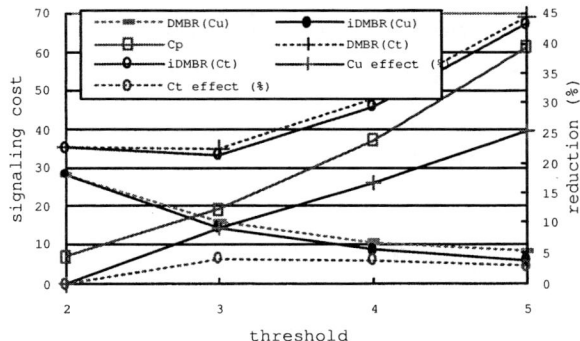

Fig. 6. Optimal toal cost for the two schemes

the MS and their neighboring cells) to reduce the location update cost of the dynamic movement-based location update scheme. We showed that a significant reduction in the location update cost is achieved in our scheme. We also showed that a significant reduction in the location update cost is achieved in our scheme. Finally, we showed that our scheme achieves the optimal total cost when the threshold is rather small, and when the MS needs a small amount of memory to implement the ELU scheme.

References

1. TIA/EIA/IS-95-B, MS-BS compatibility standard for dual-mode wideband spread spectrum cellular system (1999)
2. Park, J., Choi, J., Choi, M.: A Dynamic Location Update Scheme Based on the Number of Movements and the Movement History for Personal Communications Networks. IEICE Tr. on Communications, Vol. E85-B, No. 10 (2002) 2300-2310
3. Bar-Noy, A., Kessler, I., Sidi, M.: Mobile users: to update or not to update? Wireless Networks, Vol. 1, No. 2 (1995) 175-185
4. Lee, J. M., Kwon, B. S., Maeng, S. R.: Call Arrival History-Based Strategy: Adaptive Location Tracking in Personal Communication Networks. IEICE Tr. on Communications, Vol. E83-B, No. 10 (2000) 2376-2385
5. Akyildiz, I. F., Ho, J. S. M., Lin, Y. B.: Movement-based location update and selective paging for PCS networks. IEEE/ACM Tr. on Networking, Vol. 4, No. 4 (1996) 629-638
6. Baek J. H., Ryu, B. H.: An Improved Movement-Based Location Update and Selective Paging for PCS Networks. IEICE Tr. on Communications, Vol. E83-B, No. 7 (2000) 1509-1516
7. Baek J. H., Ryu, B. H.: Modeling and Analysis of Distance-Based Registration with Implicit Registration. ETRI Journal, Vol. 25, No. 6 (2003) 527-530

Capacitated Disassembly Scheduling: Minimizing the Number of Products Disassembled

Jun-Gyu Kim[1], Hyong-Bae Jeon[1], Hwa-Joong Kim[2]
Dong-Ho Lee[1], and Paul Xirouchakis[2]

[1] Department of Industrial Engineering, Hanyang University,
Sungdong-gu, Seoul 133-791, Korea
one@ihanyang.ac.kr, leman@hanyang.ac.kr, meonji7@freechal.com
[2] Institute of Production and Robotics (STI-IPR-LICP),
Swiss Federal Institute of Technology (EPFL), Lausanne CH-1015, Switzerland
{hwa-joong.kim, paul.xirouchakis}@epfl.ch

Abstract. Disassembly scheduling is the problem of determining the timing and quantity of disassembling used products to satisfy the demands of their parts or components over the planning horizon. This paper focuses on the case of single product type without parts commonality while the resource capacity restrictions are explicitly considered. The problem is formulated as an integer program for the objective of minimizing the number of products to be disassembled, and an optimal algorithm, after deriving the properties of the problem, is suggested. Computational experiments are done on randomly generated test problems, and the results are reported.

1 Introduction

Considerable attention has been given to various material and product recovery processes due to increasing legislation pressures to collect and upgrade used or end-of-life products in an environmentally conscious way. Disassembly, one of important material and product recovery processes, represents a way of obtaining constituent materials, parts and/or subassemblies from used or end-of-life products with necessary sorting operations. There have been a number of research works on various disassembly problems such as design for disassembly, disassembly process planning, disassembly scheduling, etc. For literature reviews on these problems, see Jovane et al. [2], Lee et al. [4], and O'Shea et al. [8].

This paper focuses on disassembly scheduling, which is the problem of determining the disassembly schedules of used or end-of-life products while satisfying the demand of their parts and components over the planning horizon. Disassembly scheduling, which corresponds to production planning in assembly systems, is one of the important operational problems in disassembly systems. In other words, from its solution, we can determine the quantity and timing of disassembly.

Most previous articles on disassembly scheduling are uncapacitated ones, i.e., the resource capacity constraints are not considered. Gupta and Taleb [1] consider the basic case, i.e., single product type without parts commonality, and suggest a simple algorithm without explicit objective function, and Lee and Xirouchakis [6] suggest a heuristic algorithm for the objective of minimizing the costs related with disassembly process. For the extended models with parts commonality and/or multiple product types, see Kim et al. [3], Taleb and Gupta [9], and Taleb et al. [10]. Recently, Lee et al. [5] present integer programming models for all the uncapacitated cases together with their performances. Unlike these, Lee et al. [7] consider the resource capacity constraints explicitly and suggest an integer programming model with various cost factors occurred in disassembly processes. Although the model can give optimal solutions, its application is limited only to the small sized problems. In fact, the computational results show that it is not adequate for practical sized problems due to its excessive computation times.

This paper focuses on the capacitated version of the problem for the basic case of single product type without parts commonality. As in the production planning problems in assembly systems, the resource capacity restrictions should be considered in the disassembly scheduling problems so that the resulting solutions are more applicable. The objective considered in this paper is to minimize the number of products to be disassembled. The problem is formulated as an integer program and an optimal solution algorithm, which is an extension of the existing algorithm without considering the capacity restrictions, is suggested in this paper. Computational experiments are done on randomly generated problems, and the test results are reported.

2 Problem Description

Before describing the problem, we first explain the disassembly product structure. In the disassembly structure, the root item represents the product itself and a leaf item represents any item that is not disassembled further. A child item represents any item that has a parent and a parent item is any item that has at least two child items. In the disassembly structure without parts commonality, each item can have at most one parent and hence there is no dependency among parts or components.

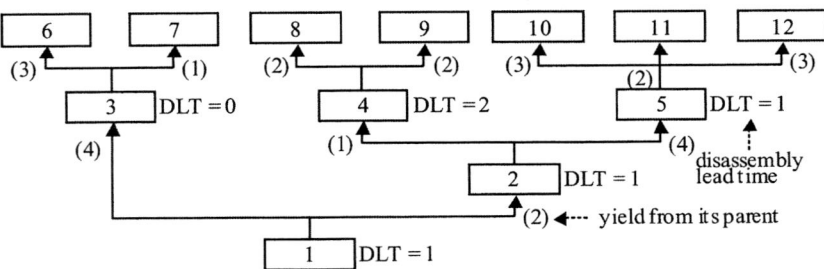

Fig. 1. Disassembly product structure: an example

Figure 1 shows an example of disassembly product structure, obtained from Gupta and Taleb [1]. Item 1 is the root item, and items 6 to 12 are leaf items. The number in parenthesis represents the yield of the item when its parent is disassembled, e.g., disassembly of one unit of item 5 results in three units of item 10, two units of item 11, and three units of item 12. Here, item 5 is called parent item, while items 10, 11 and 12 are called child items. Also, disassembly lead time (DLT) is the time required to disassemble a certain parent item.

Now, the capacitated disassembly scheduling problem considered in this paper can be described as follow: *for a given disassembly structure, the problem is to determine the quantity and timing of disassembling each parent item (including the root item) to meet the demands of leaf items over a planning horizon while satisfying the capacity restriction in each period of the planning horizon.* The capacity restriction in a period is the limit to assign the disassembly operations to that period. That is, there is an upper limit on the available time in each period of the planning horizon, and each disassembly operation assigned to a period consumes a portion of the available time of that period. The objective is to minimize the number of root items to be disassembled.

The disassembly product structure is assumed to be given from the corresponding disassembly process plan that specifies all disassembly operations and their processing times. It is assumed that there is no shortage of products. That is, the required number of products can be supplied whenever they are ordered. Other assumptions made in the problem are: (a) demands of leaf items are given and deterministic; (b) backlogging is not allowed and hence demand should be satisfied on time; (c) any defective parts resulting from disassembly are not considered; and (d) disassembly lead times with discrete time scale are given and deterministic.

The problem considered in this paper can be formulated as an integer program. Note that the integer program is a reversed form of the capacitated multilevel lot sizing problem that determines delivery times and quantities of parts and subassemblies in order to satisfy the demand of the product while satisfying the capacity restrictions. In the formulation, all items are numbered with integers $1, 2, \ldots i_l, \ldots I$. Here, i_l denotes the index for the first leaf item, and therefore the indices that are larger than or equal to i_l represent leaf items. The notations used are summarized below.

Parameters

g_i disassembly processing time of parent item i
C_t available capacity in period t
D_{it} demand requirement of leaf item i in period t
a_{ij} number of item j obtained by disassembling one unit of item i $(i < j)$
s_{it} external scheduled receipt of item i in period t

$\varphi(i)$ parent of item i
l_i disassembly lead time (DLT) of item i
I_{i0} initial inventory of item i

Decision variables

X_{it} amount of item i disassembled in period t
I_{it} inventory level of item i at the end of period t

The integer program is given below.

[**P**] Minimize $\sum_{t=1}^{T} X_{1t}$

subject to

$$I_{it} = I_{i,t-1} + s_{it} + a_{\varphi(i),i} \cdot X_{\varphi(i),t-l_{\varphi(i)}} - X_{it}$$
$$\text{for } i = 2, 3, \ldots i_l - 1 \text{ and } t = 1, 2, \ldots T \quad (1)$$

$$I_{it} = I_{i,t-1} + s_{it} + a_{\varphi(i),i} \cdot X_{\varphi(i),t-l_{\varphi(i)}} - D_{it}$$
$$\text{for } i = i_l, i_l + 1, \ldots I \text{ and } t = 1, 2, \ldots T \quad (2)$$

$$\sum_{i=1}^{i_l-1} g_i \cdot X_{it} \leq C_t \qquad \text{for } t = 1, 2, \ldots T \quad (3)$$

$$X_{it} \geq 0 \text{ and integer} \quad \text{for } i = 1, 2, \ldots i_l - 1 \text{ and } t = 1, 2, \ldots T \quad (4)$$

$$I_{it} \geq 0 \text{ and integer} \quad \text{for } i = 2, 3, \ldots I \text{ and } t = 1, 2, \ldots T \quad (5)$$

The objective function denotes minimizing the number of root items to be disassembled. Constraint (1) represents the inventory balance of each parent item. That is, at the end of each period, the inventory level of the parent item is what we had the period before, increased by the external scheduled receipt and the quantity obtained from disassembling its corresponding parent item, and decreased by the quantity of the item disassembled in that period. Here, the inventory balance constraint of the root item is omitted because it is not necessary to have surplus inventories of the root item. Also, the inventory balance of each leaf item is represented by constraint (2), which is different from (1) in that the demand requirement is used instead of the amount of items disassembled. Also, constraint (3) represents the capacity constraint in each period. That is, the total processing times of disassembly operations assigned to each period should be less than or equal to the given capacity. Finally, the constraints (4) and (5) represent the conditions on the decision variables. In particular, constraint (5) guarantees that backlogging is not permitted.

3 Optimal Solution Algorithm

This section presents the optimal solution algorithm which is an extension of the algorithm of Gupta and Taleb [1], to be called the GT algorithm hereafter. The optimal algorithm suggested in this paper consists of two main stages: obtaining an initial solution and modifying the initial solution.

Stage 1: Obtaining an Initial Solution

The initial solution is obtained using the GT algorithm that solves the uncapactited problem. The algorithm works as follows: at each level of the disassembly structure, the demands of parts or components are translated into an equivalent demand at the next level towards the root. In this fashion, the algorithm determines the disassembly schedule of the root item and its subassemblies, such that the demands of the parts or components are satisfied. See Gupta and Taleb [1] for more details.

The proposition given below characterizes the property of the solutions obtained by the GT algorithm. See Lee and Xirouchakis [6] for more details on its proof.

Proposition 1. *The solution obtained from the GT algorithm is the minimal latest disassembly schedule.*

In the above proposition, the minimum latest disassembly schedule implies that it satisfies the demands of leaf items as late as possible with the minimum amount of disassembly quantity. Therefore, we can see that the GT algorithm gives the optimal number of products disassembled for the uncapacitated problem.

Stage 2: Modifying the Initial Solution

In this stage, the optimal solution is obtained using the initial solution given from the first stage. Here, if the initial solution satisfies the capacity constraint, we can see that it is also optimal for the capacitated problem [P]. Otherwise, the initial solution should be modified to satisfy the capacity constraint while the objective value of the initial solution remains the same.

Before explaining the modification method, two properties are derived that lead to the optimal solutions.

Proposition 2. *The objective values of feasible solutions of problem [P] are greater than or equal to that obtained from the GT algorithm.*

Proof. From proposition 1, we can see that the solution obtained from the GT algorithm satisfies the demands of leaf items with the minimum amount of disassembly quantity, i.e., the minimal latest disassembly schedule. Therefore, no feasible solution of problem [P] can have less objective value than that of the GT algorithm. □

Proposition 2 implies that if the initial solution obtained from the first stage is modified repeatedly without increasing the initial objective value, we can obtain the optimal solution. In this paper, the modification is done with a backward move. Let X_{it}, for $i=1, 2, \ldots i_l - 1$ and $t = 1, 2, \ldots T$, be the solution obtained from the GT algorithm. Consider X_{iu} and X_{iv} for an arbitrary parent item i ($u < v$). Suppose that there exists remaining capacity in period u, defined as

$$\Delta C_u = C_u - \sum_{i=1}^{i_l-1} g_i \cdot X_{iu}. \tag{6}$$

Let $\Delta_{uv} = \min\{\lfloor \Delta C_u/g_i \rfloor, \min_{t=u,u+1,...v-1}(I_{it}), X_{iv}\}$, where $\lfloor \bullet \rfloor$ is the largest integer that is less than or equal to $\bullet$. Here, Δ_{uv} implies the maximum quantity that can be moved from period v to period u. Note that the move does not violate the capacity constraints since the amount is calculated based on the remaining capacity of period u. Then, if we move the amount Δ_{uv} from period v to period u, we obtain

$$X'_{iu} + X'_{iv} = (X_{iu} + \Delta_{uv}) + (X_{iv} - \Delta_{uv}) = X_{iu} + X_{iv},$$

where X'_{it} is the solution after the move. From this, we can derive the following relation for the parent of item i.

$$X'_{\varphi(i),u-l_{\varphi(i)}} + X'_{\varphi(i),v-l_{\varphi(i)}}$$
$$= (I_{iu} + X'_{iu} + s_{iu} - I_{i,u-1})/a_{\varphi(i),i} + (I_{iv} + X'_{iv} + s_{iv} - I_{i,v-1})/a_{\varphi(i),i}$$
$$= (I_{iu} + X_{iu} + s_{iu} - I_{i,u-1})/a_{\varphi(i),i} + (I_{iv} + X_{iv} + s_{iv} - I_{i,v-1})/a_{\varphi(i),i}$$
$$= X_{\varphi(i),u-l_{\varphi(i)}} + X_{\varphi(i),v-l_{\varphi(i)}}.$$

Therefore, for the root item, we obtain

$$X_{1,1} + X_{1,2} + \cdots + X'_{1,u-l_1} + \cdots + X'_{1,v-l_1} + \cdots + X_{1,T}$$
$$= X_{1,1} + X_{1,2} + \cdots + X_{1,u-l_1} + \cdots + X_{1,v-l_1} + \cdots + X_{1,T},$$

which implies that the move does not affect the objective value, i.e. number of products to be disassembled. Note that a forward move from period u to period v results in an infeasible solution since the disassembly schedule obtained from the GT algorithm is the latest one (Proposition 1).

Based on the two propositions and the backward move described above, we explain the method to obtain the optimal solution, i.e., the method to modify the initial solution into a feasible one with respect to the capacity constraint without increasing the initial objective value. The basic idea is to fill the capacity remained in each period as much as possible using the backward move(s) from a later period to that period. Here, the remaining capacities are considered from the first period to the last, and as described earlier, the amount of move of an arbitrary item i from period v to period u can be determined as

$$\Delta_{uv} = \min\left\{\left\lfloor \frac{\Delta C_u}{g_i} \right\rfloor, \min_{t=u,u+1,...v-1}(I_{it}), X_{iv}\right\} \tag{7}$$

where the first term implies the quantity that can be moved without violating the capacity constraint. Here, in the case of root item, the second term $\min_{t=u,u+1,...v-1}(I_{it})$ is not needed since there is no inventory for the root item.

The next proposition shows the feasibility condition of the backward move of non-root parent items. Note that no feasibility condition is needed for the root item since it has no inventory balance constraint.

Proposition 3. *For a given disassembly schedule X_{it}, the move of item k from period v to u $(u < v)$ is feasible if*

$$\Delta_{uv} \leq I_{kt} \quad \text{for } t = u, u+1, \ldots v-1.$$

Proof. The inventory level of item k, I'_{kt} for $t = u, u+1, \ldots v-1$, after the move of amount Δ_{uv} from period v to u can be calculated as follows. (The others remain the same.)

$$I'_{ku} = I_{k,u-1} + s_{ku} + a_{\varphi(k),k} \cdot X_{\varphi(k),u-l_k} - (X_{ku} + \Delta_{uv}) = I_{ku} - \Delta_{uv}$$

$$\begin{aligned}
I'_{kt} &= I'_{k,t-1} + s_{kt} + a_{\varphi(k),k} \cdot X_{\varphi(k),t-l_k} - X_{kt} \\
&= (I_{k,t-1} - \Delta_{uv}) + s_{kt} + a_{\varphi(k),k} \cdot X_{\varphi(k),t-l_k} - X_{kt} \\
&= (I_{k,t} + s_{kt} + a_{\varphi(k),k} \cdot X_{\varphi(k),t-l_k} - X_{kt}) - \Delta_{uv} \\
&= I_{kt} - \Delta_{uv} \quad \text{for } t = u+1, u+2, \ldots v-1
\end{aligned}$$

$$\begin{aligned}
I'_{kv} &= I'_{k,v-1} + s_{kv} + a_{\varphi(k),k} \cdot X_{\varphi(k),v-l_k} \\
&= (I_{k,v-1} - \Delta_{uv}) + s_{kv} + a_{\varphi(k),k} \cdot X_{\varphi(k),v-l_k} \\
&= I_{k,v-1} + s_{kv} + a_{\varphi(k),k} \cdot X_{\varphi(k),v-l_k} - \Delta_{uv} = I_{kv}
\end{aligned}$$

Then, the result follows from the condition $I'_{kt} \geq 0$ for $t = u, u+1, \ldots v-1$. □

Now, we explain the method to modify the initial solution without increasing the initial objective value. As stated earlier, remaining capacities are filled from the first period to the last. Consider an arbitrary period t. In this paper, the backward move is done from the items assigned to the next period $t+1$. Here, if there is remaining capacity after the moves for the items in period $t+1$, those in the next period $t+2$ (except for root item) are considered to be moved and this is done to the last period T until there is no remaining capacity in period t. According to (7), the amount of move in the case of root item is

$$\Delta_{1t} = \min\left\{ \left\lfloor \frac{\Delta C_t}{g_1} \right\rfloor, X_{1,t+1} \right\}, \tag{8}$$

where $\Delta C_t > 0$ and $X_{1,t+1} > 0$. Also, in the case of non-root parent items, it is needed to determine which item is moved first since there may be two or more non-root parent items assigned to period $t+1$ in the initial solution. To do this, we select the item i^* such that

$$i^* = \arg\min_{i \in R} \left\{ \frac{\Delta C_t}{g_i} - \left\lfloor \frac{\Delta C_t}{g_i} \right\rfloor \right\}, \tag{9}$$

where $R = \{2, 3, \ldots i_l - 1\}$, i.e., set of non-root parent items. That is, we select the item that results in the smallest remaining capacity after the backward move. The amount of move of the selected item i^* is

$$\Delta_{i^*t} = \min\left\{ \left\lfloor \frac{\Delta C_t}{g_{i^*}} \right\rfloor, I_{i^*t}, X_{i^*,t+1} \right\}, \tag{10}$$

which represents the maximum quantity of item i^* that can be moved from period $t+1$ to period t without violating the capacity constraint in period t.

The procedure of the second stage is summarized below. Note that the initial solution is given from the first stage.

Procedure. (*Modifying the initial solution*)

Step 1. Set $t = 1$

Step 2. Do the following steps:

1) Set $t_r = t + 1$ (for the root item) and $t_p = t + 1$ (for non-root items)
2) Calculate the remaining capacity in period t using (6).
3) If the disassembly quantity of the root item in period t is positive, calculate the movable quantity from period t_r to period t using (8), and modify the disassembly schedule of the root item. (The inventory levels of the root item and its child items are also modified.)
4) If there is remaining capacity in period t after the move of the root item, select a non-root parent item using (9) and calculate the movable quantity of the selected item from period t_p to period t using (10). Modify the disassembly schedule of the selected item. (The inventory levels of the item and its child items are also modified.) This step is repeatedly done until there is no remaining capacity in period t or all non-root parent items in period t_p are considered.
5) If there is remaining capacity in period t after the moves of all the non-root items in period t_p, set $t_p = t_p + 1$. If $t_p \leq T$, go to (4). Otherwise, set $t_r = t_r + 1$ and go to (3). If $t_r > T$ or there is no remaining capacity, go to Step 3.

Step 3. If $t < T$, set $t = t + 1$ and go to (1). Otherwise, stop and save the solution.

4 Computational Experiments

To show the performance of the algorithm suggested in this paper, computational tests were done on a number of randomly generated problems, and the results are given in this section. In this paper, the algorithm is compared with CPLEX 9.0, commercial integer programming software, and the performance measure used is the CPU seconds since both give the optimal solutions. The algorithm and the program to generate integer programs were coded in C and tests were done on a personal computer with a Pentium processor operating at 2.0 GHz clock speed.

Table 1. Test results on the suggested algorithm

Number Item	Number Period	Loose		Tight	
		Alg[1]	CPLEX[2]	Alg	CPLEX
20	10	0.001	0.068	0.000*	0.143
	20	0.000	0.256	0.000	0.751
	30	0.000	0.297	0.000	1.001
40	10	0.000	0.096	0.001	0.134
	20	0.000	0.522	0.001	0.776
	30	0.000	1.808	0.000	2.449
60	10	0.000	0.133	0.001	0.350
	20	0.000	1.488	0.000	1.921
	30	0.001	3.385	0.001	5.239
80	10	0.000	0.204	0.000	0.349
	20	0.000	2.431	0.001	3.162
	30	0.000	6.205	0.001	8.169
100	10	0.000	0.259	0.006	0.410
	20	0.002	3.181	0.000	4.046
	30	0.001	7.478	0.002	11.144

[1] algorithm suggested in this paper
[2] CPLEX 9.0
* average CPU second is less than 0.0005s

For the test, we generated 750 problems, i.e., 25 problems for each combination of two levels of capacity tightness (loose and tight), five levels of the number of items (20, 40, 60, 80, and 100) and three levels of the number of periods (10, 20, and 30). For each level of the number of items, 5 disassembly structures were randomly generated, in which the number of child items for each parent and its yield were generated from $DU(2,5)$ and $DU(1,3)$, respectively. Here, $DU(a,b)$ is the discrete uniform distribution with $[a,b]$. (The other data generation methods are omitted here because of the space limitation.)

The test results are summarized in Table 1 that shows the average CPU seconds of the suggested algorithm and CPLEX 9.0, respectively. Note that the algorithm suggested in this paper gave the optimal solutions for all the test problems. It can be seen from the table that the CPU seconds of the optimal algorithm are much smaller than those of CPLEX 9.0. In fact, its CPU seconds were within 0.001 seconds in average. This implies that the algorithm can be used to solve practical sized problems within very short computation time, and also can be used to solve the subproblems of more general problems such as those with general cost-based objectives.

5 Concluding Remarks

This paper considered the disassembly scheduling problem with resource capacity constraints for the objective of minimizing the number of products to be disassembled and suggested an optimal solution algorithm that extends the existing one for the uncapacitated problem. Computational experiments were done and the results showed that the algorithm suggested in this paper gave the optimal solutions within very short computation time.

This research can be extended in several ways. First, it is needed to consider the cost-based objectives, which make the problem much more difficult. Second, more general problems, such as those with parts commonality, multiple product types, etc., are worth to be considered.

Acknowledgements

The financial support the Korea Research Foundation project under grant number 2004-03-D00468 is gratefully acknowledged.

References

1. Gupta, S. M., Taleb, K. N.: Scheduling Disassembly, International Journal of Production Research, Vol. 32 (1994) 1857–1886
2. Jovane, F., Alting, L., Armoillotta, A., Eversheirm, W., Feldmann, K., Seliger, G., Roth, N.: A Key Issue in Product Life Cycle: Disassembly, Annals of the CIRP, Vol. 42 (1993) 651–658
3. Kim, H.-J., Lee, D.-H., Xirouchakis, P., Züst, R.: Disassembly Scheduling with Multiple Product Types, Annals of the CIRP, Vol. 52 (2003) 403–406
4. Lee, D.-H., Kang, J.-G., Xirouchakis, P.: Disassembly Planning and Scheduling: Review and Further Research, Proceedings of the Institution of Mechanical Engineers: Journal of Engineering Manufacture – Part B, Vol. 215 (2001) 695–710
5. Lee, D.-H., Kim, H.-J., Choi, G. Xirouchakis, P.: Disassembly Scheduling: Integer Programming Models, Proceedings of the Institution of Mechanical Engineers: Journal of Engineering Manufacture – Part B, Vol. 218 (2004) 1357–1372
6. Lee, D.-H., Xirouchakis, P.: A Two-Stage Heuristic for Disassembly Scheduling with Assembly Product Structure, Journal of the Operational Research Society, Vol. 55 (2004) 287–297
7. Lee, D.-H., Xirouchakis, P., Züst, R.: Disassembly Scheduling with Capacity Constraints, Annals of the CIRP, Vol. 51 (2002) 387–390
8. O'Shea, B., Grewal, S. S., Kaebernick, H.: State of the Art Literature Survey on Disassembly Planning, Concurrent Engineering: Research and Applications, Vol. 6 (1998) 345–357
9. Taleb, K. N., Gupta, S. M., Disassembly of Multiple Product Structures, Computers and Industrial Engineering, Vol. 32 (1997) 949–961
10. Taleb, K. N., Gupta, S. M., Brennan, L.: Disassembly of Complex Product Structures with Parts and Materials Commonality, Production Planning and Control, Vol. 8 (1997) 255–269

Ascent Phase Trajectory Optimization for a Hypersonic Vehicle Using Nonlinear Programming

H.M. Prasanna[1], D. Ghose[1], M.S. Bhat[1], C. Bhattacharyya[2], and J. Umakant[3]

[1] Department of Aerospace Engineering
[2] Department of Computer Science and Automation
[3] Aerodynamics Division, DRDL, Hyderabad, India
[1,2] Indian Institute of Science, Bangalore 560012, India
prasannahm@rediffmail.com, {dghose, msbdcl}@aero.iisc.ernet.in,
chiru@csa.iisc.ernet.in, umakantj@yahoo.com

Abstract. In this paper we present a nonlinear programming solution to one of the most challenging problems in trajectory optimization. Unlike most aerospace trajectory optimization problems the ascent phase of a hypersonic vehicle has to undergo large changes in altitude and associated aerodynamic conditions. As a result, its aerodynamic characteristics, as well as its propulsion parameters, undergo drastic changes. Further, the data available through wind tunnel tests are not always smooth. The challenge in solving such problems lies both in the preprocessing of the data as well as in the judicious use of optimization techniques. In this paper we advocate approximation of the infinite dimensional optimal control problem, derived from practical considerations of a hypersonic vehicle ascending from an altitude of 16 kms to an altitude of 32 kms with specified mach numbers, into a set of finite dimensional nonlinear programming problems. This finite dimensional approximation is shown to produce acceptable optimized results in terms of angle-of-attack control histories and state behaviour. A modification, that exploits the ultimate scheme of linear interpolation of the optimal discrete history, is proposed and is shown to produce accurate results with smaller number of grid points.

1 Introduction

Ascent phase trajectory optimization of hypersonic vehicles is one of the most challenging real-life optimization problems [1]. Numerical algorithms for these problems are difficult to develop due to the inherent nature of the problem [2]. Our paper is an effort in this direction.

Numerical solution of optimal control problems has two important components: optimization and numerical integration of differential equations [3, 4]. A naive approach to solve such problems is a "cut-and-paste" strategy that uses numerical integration and optimization packages in the form shown in Figure 1(a). A sophisticated

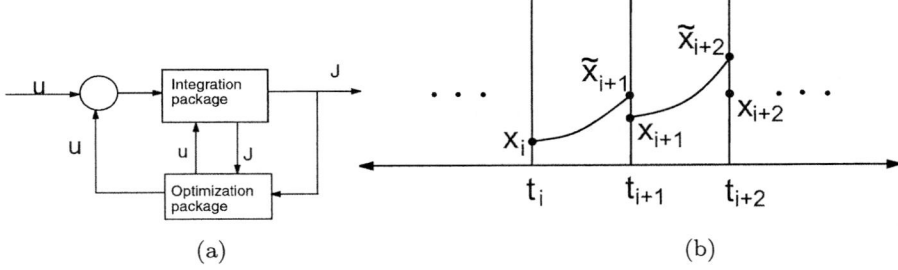

Fig. 1. (a) The naive cut-and-paste method (b) Direct transcription method

approach exploits the intimate relation that exists between the methods used for solving differential equations and the methods used for optimization [3]. In this paper we will describe such an approach based on nonlinear programming (NLP) and illustrate the method with an example of optimization of the trajectory for an experimental hypersonic vehicle during its ascent phase. The basic approach in solving a trajectory optimization problem using NLP is to represent an infinite dimensional optimal control problem by one or more simpler finite dimensional subproblems.

While using this technique, several practical issues arise that delineate these problems from textbook problems. In fact, taking care of these issues requires a sound knowledge of the physics behind the problem. Apart from the well understood issues of infeasible constraints, rank deficient constraints, constraint redundancy, and scaling, there are other issues of missing data in the aerodynamic and propulsion parameters as well as non-smoothness (which may be due to engineering errors or real) which may confuse standard spline interpolation methods. These problems require careful pre-processing of data before they can be used in the optimization process.

2 Methods for Solving Optimal Control Problems

2.1 The Optimal Control Problem

An optimal control problem may be formulated to find the control vector $u(t)$ to minimize the performance index [3, 5]

$$\phi[y(t_f), t_f] \tag{1}$$

evaluated at the final time t_f, which may be fixed or free. The dynamics of the system are given by the state equations and initial and final conditions,

$$\dot{y} = f[y(t), u(t), t], \ \chi[y(t_0), u(t_0), t_0] \equiv \chi_0 = 0, \ \psi[y(t_f), u(t_f), t_f] \equiv \psi_f = 0 \tag{2}$$

where, y is the state vector. Also, we have state and control constraints,

$$\xi_L \leq \xi[y(t), u(t), t] \leq \xi_U, \ y_L \leq y(t) \leq y_U, \ u_L \leq u(t) \leq u_U \tag{3}$$

2.2 Transcription Method

The optimal control problem is an infinite dimensional extension of a finite dimensional NLP problem. This infinite dimensional problem can be transcribed to a finite dimensional approximation. The transcription approach treats the state and control variable at discrete times as optimization variables. The time points are defined as, $t_I = t_1 < t_2 < \ldots < t_M = t_F$. These points are referred to as node, mesh, or grid points. The values of the state and control variables at these intermediate steps are treated as a set of NLP variables. The differential equation is thus replaced by a finite set of constraints. The transcription method decomposes the dynamic trajectory into several phases or intervals. If there are n number of intervals, the number of grid points are $n + 1$. This method uses the value of both state and control at each grid point as NLP (nonlinear programming) variables. We may include initial or final time as one of the NLP variables. The continuity in the states across the phase boundary is achieved by forming an NLP constraint such that, the propagated value of the state from previous phase matches with the value of the state variable at current phase. In the direct transcription method the propagation is done by using algebraic formulas derived from numerical integration schemes. This makes the propagation function approximate (with increasing levels of accuracy depending on the order of the integration scheme). However, as the interval size decreases the accuracy will increase. Let the propagation scheme be given by a function $P(.)$ so that the propagated state at t_{i+1} be defined as $\tilde{x}_{i+1}$. Then

$$P(x_i) = \tilde{x}_{i+1} \qquad (4)$$

where, x_i is the assumed state at time t_i. The constraint at t_{i+1} is,

$$x_{i+1} - \tilde{x}_{i+1} = x_{i+1} - P(x_i) = 0 \qquad (5)$$

Figure 1(b) explains the concept of NLP formulation. The choice of P is important. The control which is obtained from solving NLP problem is interpolated to make it continuous. The accuracy of discrete trajectory, obtained from solving the NLP problem, is assessed by using the interpolated control to integrate the state equations (ODEs). It is expected that as the number of intervals increases, the error between the simulation result and optimization will reduce.

2.3 Discretization Methods

Euler Method: This is a first order method and the constraints are given by,

$$0 = y_{k+1} - y_k - h_k f_k \equiv c(x) \qquad (6)$$

Trapezoidal Method: This is a second order method and the constraints are,

$$0 = y_{k+1} - y_k - (h_k/2)[f_k + f_{k+1}] \equiv c(x) \qquad (7)$$

Hermite-Simpson (Compressed) (HSC): This is of order four and the constraints are given by,

$$0 = y_{k+1} - y_k - (h_k/6)[f_{k+1} + 4f((y_{k+1} + y_k)/2 + h_k(f_k - f_{k+1})/8, \\ \bar{u}_{k+1}, t_k + h_k/2) + f_k] \equiv c(x) \qquad (8)$$

The NLP variables are given as $x^T = (t_k, y_1, u_1, \bar{u}_2, y_2, u_2, \bar{u}_3, \ldots, \bar{u}_M, y_M, u_M)$.

Hermite-Simpson (Separated) (HSS): The constraints are given by,

$$0 = \bar{y}_{k+1} - (1/2)(y_{k+1} + y_k) - (1/8)h_k(f_k - f_{k+1}) \tag{9}$$

$$0 = y_{k+1} - y_k - (1/6)h_k(f_{k+1} + 4\bar{f}_{k+1} + f_k) \tag{10}$$

Eqn. (9) defines the hermite interpolation for the state at the interval midpoint, while (10) enforces the Simpson quadrature over the interval. The NLP variables are, $x^T = (t_k, y_1, u_1, \bar{u}_2, y_2, u_2, \bar{y}_3, \bar{u}_3, \ldots, \bar{y}_M, \bar{u}_M, y_M, u_M)$.

In general, the duration of a phase may be variable, in which case, the set of NLP variables must be augmented to include the variable time(s) t_I and/or t_F. Also, we must alter the discretization such that the step size is $h_k = \tau_k(t_F - t_I) = \tau_k \Delta t$, where, $\Delta t = (t_F - t_I)$ with constants $0 < \tau_k < 1$ chosen so that the grid points are located at fixed fractions of the total phase duration.

3 Basic Models for Ascent Phase Trajectory Optimization of a Hypersonic Vehicle

The experimental hypersonic vehicle is air dropped from a carrier aircraft and boosted to Mach number 3.5 at an altitude of 16 km using solid rocket motors. The boosters are separated upon achieving this condition. Then the ascent phase commences and the final condition of Mach number 6.0 and altitude 32 km is achieved. The problem is to optimize the ascent phase trajectory so that the least fuel is expended (so that maximum remaining fuel can be utilized during the subsequent cruise phase). The constraints and other parameters of the problem are angle of attack $0° \leq \alpha \leq 8°$, maximum range during ascent phase ≤ 250 km, maximum time to reach the cruise conditions ≤ 250 s, maximum longitudinal acceleration during ascent phase $< 4g$, maximum lateral acceleration $< 5g$, weight of empty vehicle= 2000 kg (excluding fuel), fuel weight = 1600 kg.

The aerodynamic characteristics [6] of the vehicle are usually available in tabular form as values of the normal force coefficient (c_N) and axial force coefficient (c_A) which are functions of Mach number (M) and angle of attack (α). The lift and drag coefficients are computed using the relations $c_L = c_N \cos\alpha - c_A \sin\alpha$ and $c_D = c_N \sin\alpha + c_A \cos\alpha$. The resulting variation of lift and drag coefficients with mach number and alpha are shown in Figure 2.

Mass Flow Rate: The mass flow rate of air ($\dot{m}_{air}$) as function of Mach number and α is given in Table 1(a), for density corresponding to altitude 32.5 km. The mass flow rate of air at any altitude is, $\dot{m}_{air} = \rho v A$, where, A is nozzle area, v is free stream velocity at the nozzle, and ρ is density at given altitude. At 32.5 km, let the density be $\rho_{32.5}$, speed of sound be $S_{32.5}$, and the speed corresponding to a given mach numbers M is $V_{32.5} = M S_{32.5}$. We non-dimensionalize Table 1(a) and determine $\dot{m}_{air}$ as,

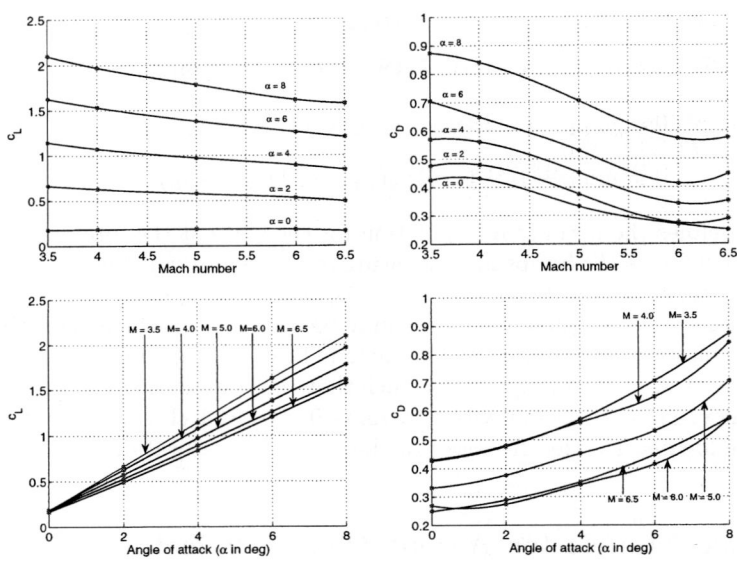

Fig. 2. Variation of lift and drag coefficients with α and Mach number

Table 1. (a) Mass flow rate of air (-kg/s) corresponding to altitude 32.5 km (b) Specific impulse (I_{sp} in seconds)

M/α	0°	2°	4°	6°	8°
3.5	6.95	7.24	8.30	8.94	9.60
4.0	8.94	9.69	10.67	11.67	12.69
5.0	12.79	14.71	16.63	18.61	20.32
6.0	17.74	21.20	24.73	26.87	29.19
6.5	21.51	25.17	29.61	33.18	35.26

(a)

	Altitude (km)						
M	12.5	15.0	20.0	25.0	30.0	35.0	40.0
3.0	1060	1054	1045	1025	1006	977	943
4.0	1060	1054	1045	1025	1006	977	943
5.0	970	964	952	931	910	879	847
6.0	855	849	838	817	799	775	750
7.0	724	719	713	698	687	669	644
8.0	596	594	593	581	570	542	499

(b)

$$\dot{m}_{nonD} = \dot{m}_{air}/(\rho_{32.5}V_{32.5}A) \qquad (11)$$

$$\underbrace{\dot{m}_{air}}_{at\ given\ altitude} = \dot{m}_{nonD}\underbrace{\rho v A}_{at\ given\ altitude} = \dot{m}_{air}/(\rho_{32.5}V_{32.5})\underbrace{\rho v}_{at\ given\ altitude} \qquad (12)$$

Specific Impulse: Specific impulse is a function of free stream Mach number and altitude (Table 1(b)). For Mach number between 3.0 to 4.0, specific impulse

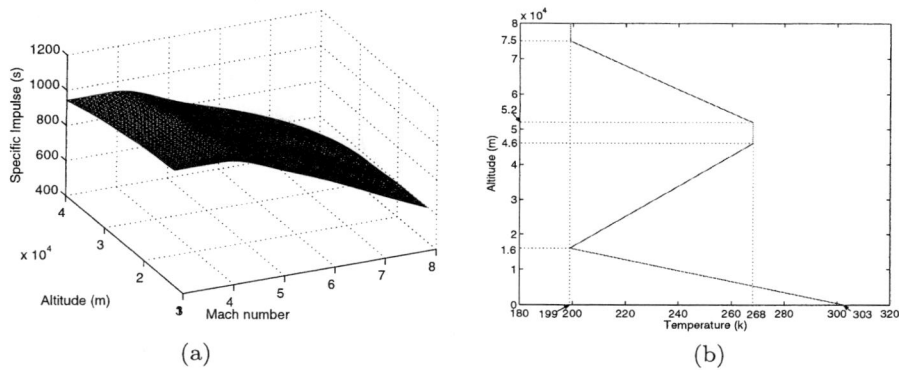

Fig. 3. (a) Specific impulse using spline interpolation (b) Temperature variation in Indian standard atmosphere

is constant. So linear interpolation is used for this region and spline interpolation for the rest of the data (Figure 3(a)).

Thrust: Thrust is calculated as $T = I_{sp}\dot{m}_f g$. Assuming equivalence ratio $\phi=1$ and stoichiometric ratio for fuel as 15, we have $\dot{m}_f = \dot{m}_{air}/15$, where, $\dot{m}_{air}$ at given altitude is obtained from the procedure described above.

Standard Atmosphere: The Indian standard atmosphere is assumed. The variation of temperature with altitude is shown in Figure 3(b). The atmospheric pressure at sea level is $P_0 = 100500$ (N/m^2) and specific gas constant is $R=287.05307$ (J/kgK).

4 Problem Formulation and Computational Results

During ascent phase the objective is to minimize fuel consumption (or maximize mass at final time). Thus, the optimization problem is formulated as,

$$\max \ m_f \tag{13}$$

$$\text{s.t.,} \quad \dot{h} = v \sin \gamma, \quad \dot{x} = v \cos \gamma, \quad \dot{m} = -T/(I_{sp} \ g) \tag{14}$$

$$\dot{v} = (1/m)[T \cos \alpha - D] - \mu \sin \gamma/(R_e + h)^2 \tag{15}$$

$$\dot{\gamma} = (1/(mv))[T \sin \alpha + L] + \cos \gamma[v/(R_e + h) - \mu/v(R_e + h)^2] \tag{16}$$

where, h is the altitude (m), v the velocity (m/sec), α the angle of attack (rad), γ the flight path angle (rad), m the mass (kg), μ the gravitational constant, g the gravitational acceleration, and R_e the radius of the earth. The trajectory related constraints

$$t_f \leq 250 \text{ s}, \quad x_f \leq 250000 \text{ m}, \quad 0° \leq \alpha \leq 8°, \quad a_{long} \leq 4g, \quad a_{lat} \leq 5g$$

are also imposed. where, a_{long} and a_{lat} are the longitudinal and lateral accelerations, respectively. defined as

$$a_{lat} = (1/m)[T\sin\alpha + L] + \cos\gamma[v^2/(R_e+h) - \mu/(R_e+h)^2] \quad (17)$$

$$a_{long} = (1/m)[T\cos\alpha - D] - \mu\sin\gamma/(R_e+h)^2 \quad (18)$$

The lift (L) and drag (D) forces are defined by,

$$D = (1/2)\rho Sv^2 c_D, \qquad L = (1/2)\rho Sv^2 c_L \quad (19)$$

where, S is the aerodynamic reference area of the vehicle, and ρ is the atmospheric density. The following constants complete the formulation of the problem:

$$h(0) = 16000.0 \text{ m}, \quad h(t_f) = 32000.0 \text{ m}, \quad v(0) = 989.77 \text{ m/s}$$
$$v(t_f) = 1847.0 \text{ m/s}, \quad \gamma(0) = 0.0698 \text{ rad}, \quad \gamma(t_f) = 0 \text{ rad}$$
$$m(0) = 3600.0 \text{ kg}, \quad x(0) = 0 \text{ m}, \quad \mu = 3.9853 \times 10^{14} \text{ m}^3/\text{s}^2$$
$$g = 9.8 \text{ m/s}^2, \quad R_e = 6.378388 \times 10^6 \text{ m}, \quad S = 1 \text{ m}^2$$

The optimization was started with 5 grid points initially and subsequently higher number of grid points (i.e., 9, 17, 33, 65) were considered. For 5 and 9 grid points,

Table 2. Results of ascent phase trajectory optimization

Grid points	method	h_f (m)	v_f (m/s)	γ_f (°)	m_f (kg)	$m_{f(\text{fuel})}$ (kg)	x_f ($\times 10^5$ m)	time (s)
5	NLP	32000	1847.0	0	2854.8	854.8	2.793	143.13
5	Step	38431	1937.1	0.3553	2764.3	764.3	2.3341	143.13
5	Linear	32305	1768.0	-3.4665	2907.4	907.4	2.1672	143.13
5	Hermite	33200	1759.8	-3.3375	2911.7	911.7	2.168	143.13
9	NLP	32000	1847	0	2855.4	855.4	2.1799	142.02
9	Step	34701	1888.4	-0.8626	2816.4	816.4	2.2623	142.02
9	Linear	31730	1824.2	-1.0466	2871.8	871.8	2.1743	142.02
9	Hermite	31236	1820.0	-0.9807	2875.6	875.6	2.1637	142.02
17	NLP	32000	1847	0	2855.2	855.2	2.1544	139.86
17	Step	32710	1872.2	-0.1342	2834.4	834.4	2.1863	139.86
17	Linear	31623	1845.2	0.0754	2857.3	857.3	2.1469	139.86
17	Hermite	31569	1844.8	0.0643	2857.7	857.7	2.1463	139.86
33	NLP	32000	1847	0	2855.2	855.2	2.1552	139.92
33	Step	32509	1861.0	-0.0954	2843.6	843.6	2.1747	139.92
33	Linear	31961	1846.9	-0.0100	2855.4	855.4	2.1548	139.92
33	Hermite	31958	1846.9	-0.0179	2855.4	855.4	2.155	139.92
65	NLP	32000	1847	0	2855.2	855.2	2.1552	139.93
65	Step	32267	1854.0	-0.0475	2849.5	849.5	2.1652	139.93
65	Linear	31994	1846.9	-0.0033	2855.3	855.3	2.1552	139.93
65	Hermite	31993	1846.9	-0.0054	2855.3	855.3	2.1552	139.93

the NLP constraints are formed using trapezoidal discretization and for higher number of grid points Hermite-Simpson discretization is used. Each variable is scaled by its upper bound to make the ranges, over which they vary, similar in their order of magnitude. This helps better numerical conditioning of the NLP. The optimization is carried out only at the grid points and the shape of the trajectory between the grid points is not known. To assess the accuracy of the result, we integrated (14) - (16) using control histories obtained from interpolation of the NLP generated discrete control. We used three interpolation schemes (step, linear, and hermite). The initial conditions for $2n-1$ grid points were obtained by linearly interpolating the optimized values from n grid points.

The results of optimization and integration (using different interpolation schemes for control) are given in Table 2. From Table 2, we observe that the error between the optimization and integration results are significantly large when number of grid points are few and the error reduces as the number of grid points increases. The integration using control generated by step interpolation has large error as compared to linear and hermite interpolations. The results for 33, and 65 grid points show very little improvement in the optimization results (see Figure 4). Figure 5 shows variation of altitude, velocity, flight path angle, mass, and range with time using linearly interpolated control for 65 grid points.

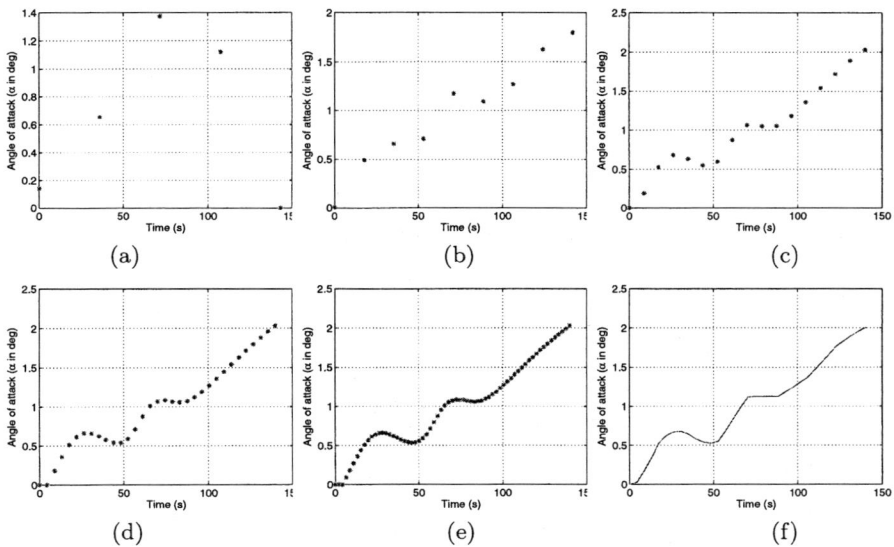

Fig. 4. (a)–(e) Angle of attack vs. time for 5, 9, 17, 33, and 65 grid points in the original NLP (f) Angle of attack vs. time for 65 grid points with modified method

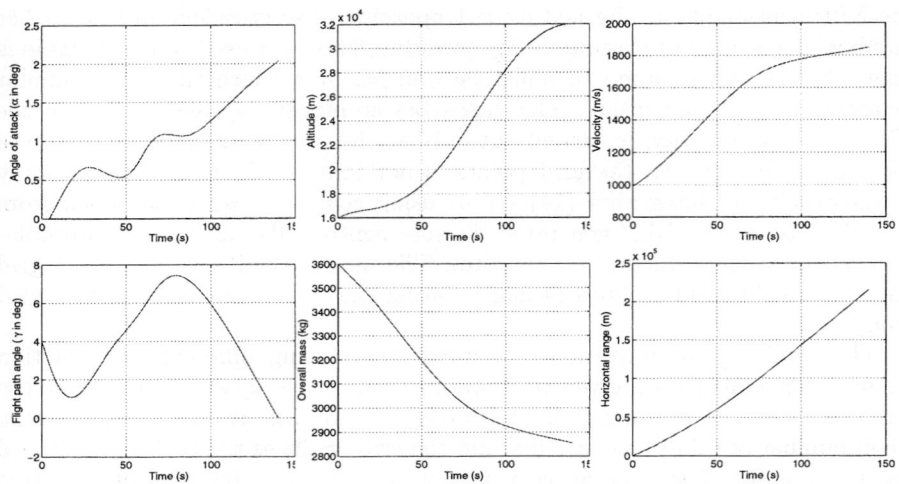

Fig. 5. State histories (for linearly interpolated control from 65 grid points)

5 Improving Computational Accuracy

Since the final step in the transcription method is implemented by linearly interpolating the control history between grid points, we use this knowledge by assuming a linear dynamics for control variable. So, we assume,

$$\dot{\alpha} = \beta \tag{20}$$

Table 3. Results of ascent phase trajectory optimization (with modified method)

Grid points	method	h_f (m)	v_f (m/s)	γ_f (°)	m_f (kg)	$m_{f(\text{fuel})}$ (kg)	x_f ($\times 10^5 m$)	time (s)
5	NLP	32000	1847	0	2854.9	854.9	2.2394	146.92
5	Linear	31517	1768.8	-3.7185	2909.0	909.0	2.2202	146.92
9	NLP	32000	1847	0	2855.4	855.4	2.1821	142.15
9	Linear	31723	1824.2	-1.0467	2871.7	871.7	2.1764	142.15
17	NLP	32000	1847	0	2855.2	855.2	2.1649	140.57
17	Linear	31908	1846	0.0610	2856.1	856.1	2.1615	140.57
33	NLP	32000	1847	0	2855.2	855.2	2.1654	140.62
33	Linear	31999	1847	0.0006	2855.2	855.2	2.1654	140.62
65	NLP	32000	1847	0	2855.2	855.2	2.1655	140.62
65	Linear	32000	1847	1.86e-005	2855.2	855.2	2.1654	140.62

where, β is now used as the control variable. Then, the equations of motion will include (20) in addition to (14) - (16). The set of NLP variables will consequently include an additional variable β at each grid point. With this modification, the results of optimization and integration are given in Table 3. Figure 4(f) shows variation of angle of attack against time for 65 grid points. Comparing with the α history in the earlier scheme we see a flat region between 70 and 90 seconds in the new scheme. The modified method also gives better accuracy. In fact it does so for just 33 grid points, whereas the original scheme produces similar accurate results at 65 grid points. However, the additional NLP variable and constraint increases the computational time slightly.

6   Conclusions

In this paper, we addressed the challenging application problem of ascent phase trajectory optimization of a hypersonic vehicle. It was shown that the linear interpolation scheme used to create the continuous control history for verification can itself be exploited while forming the dynamical equations to get faster convergence and higher accuracy.

Acknowledgements. This work was partially supported by a grant from DRDL.

References

1. P.F. Gath and A.J. Calise, Optimization of Launch Vehicle Ascent Trajectories with Path Constraints and Coast Arcs, *Journal of Guidance, Control, and Dynamics*, Vol. 24, 2001, pp. 296-304.
2. J.T. Betts, Survey of Numerical Methods for Trajectory Optimization, *Journal of Guidance, Control, and Dynamics*, Vol.21, 1998, pp. 193-207.
3. J.T. Betts, *Practical Methods for Optimal Control Using Nonlinear Programming*, The Society for Industrial and Applied Mathematics, Philadelphia, 2001.
4. E. Gill, Murray, and H. Wright, *Practical Optimization*, Academic Press, 1981.
5. A.E. Bryson, and Y.-C. Ho, *Applied Optimal Control*, John Wiley, New York, 1975.
6. J.D. Anderson, *Introduction to Flight*, McGraw-Hill, Singapore, 2000.

Estimating Parameters in Repairable Systems Under Accelerated Stress

Won Young Yun[1] and Eun Suk Kim[2]

[1] Department of Industrial Engineering, Pusan National University 30,
Changjeon-Dong, Kumjeong-Ku, Pusan 609-735, Korea
[2] Department of Industrial Engineering, Pusan National University 30,
Changjeon-Dong, Kumjeong-Ku, Pusan 609-735, Korea

Abstract. In this paper, the imperfect maintenance model and proportional intensity model are considered for the repairable systems under accelerated stress. The age reduction model (Malik's model) is assumed for imperfect maintenance. The stress acts multiplicatively on the baseline cumulative intensity. The log-likelihood function is derived and a maximizing procedure is proposed. In simulation studies, we investigate the accuracy and some properties of the estimation method.

1 Introduction

In the real world, all products and systems are unreliable in the sense that they degrade with age and/or usage ultimately. Product reliability contributes much to quality and competitiveness. Major decisions about reliability are based on life-test data. Analyses of data from such a normal(or accelerated) test yields necessary information on product life under design conditions. Accelerating testing saves much time and money.If the product is repairable, it can be possible to continue testing the product after failures. In many cases, failures seen at the system level are due to the failure of one component. For example, an electronic board's failure induced by a capacitor failure can be easily repaired and testing can be resumed. A repairable system is a system that, after failing to perform one or more of its functions satisfactorily, can be restored to satisfactory performance by any method other than replacement of the entire system. Malik (1979) and Brown, Mahoney and Sivazlian (BMS, 1983) proposed general approaches to model the maintenance effect in which maintenance reduces the age of the unit with respect to the rate of occurrence of failures. In this paper, Malik's model was considered. To evaluate testing under accelerated conditions, we used the Proportional Hazards Model (PHM), which is considered to be a tool to introduce the environmental factors to the analysis of lifetime data. Refer Kalbfleisch and Prentice (1980), Cox and Oakes (1984), Bagdonavicius and Nikulin (2002), and Nelson (1990) for further explanation. There are many kinds of environmental factors (stress) but in this paper, we consider a single and constant accelerating variable with m stress levels. The parameter, ρ, which is called the age (or intensity) reduction factor or the improvement factor, describes the maintenance effect in Malik's Model. If ρ goes to 0, the state of the maintained unit is almost as the same as that of pre-maintenance, and if ρ goes to 1, maintenance renews the unit. In this paper, we assume that the age reduction factor is constant. An estimation of the parameters of a repairable system under accelerated stress conditions is studied. The method of estimating parameters is maximum likelihood method using a Genetic Algorithm. Guida and Giorgio (1995) conducted a reliability analysis of accelerated life-time data from a repairable system. Guida and Giorgio (1995) analyzed the accelerated testing of a repairable item modeled by a nonhomogeneous Poisson process with covariates, analyzed a single accelerating variable with 2 stress levels, and derived closed-form maximum likelihood (ML) solutions. Guerin, Dumon and Lantieri (2004) studied accelerated life testing on repairable systems, defining two accelerated life models for repairable systems: the Arrhenius-exponential model and the Peck-Weibull model.

Notation

$h(t \mid s)$: intensity function under stress s
$H(t \mid s)$: cumulative intensity function under stress s
$g(a \cdot s)$: a positive-valued function $(= exp(a \cdot s))$
m : number of stress levels
n_i : total number of failures under i stress, $i = 1, 2, ... m$
N : total failure number $(= \sum_{i=1}^{m} n_i)$
$t_{i,j}$: j th failure-time under stress i, $i = 1, 2, ... m, j = 1, 2, ... n_i$
$z_{i,j}$: Inter-failure time between j th and j -1 th under stress i,
$\quad z_{i,j} = t_{i,j} - t_{i,j-1}, i = 1, 2, ... m, j = 1, 2, ... n_i$
$x_{i,j}$: virtual age just before repair under stress i,
$\quad x_{i,j} = 4 \sum_{k=1}^{j}(1-\rho) \cdot z_{i,k-1} + z_{i,j}, i = 1, 2, ... m, j = 1, 2, ... n_i$
$y_{i,j}$: virtual age right after repair under stress
$\quad y_{i,j} = \sum_{k=1}^{j}(1-\rho) \cdot z_{i,k}, i = 1, 2, ... m, j = 1, 2, ... n_i$
ρ : age reduction factor
t : operating time
s_i : normalized scale of stress level, $i = 1, 2, ... m$
α : scale parameter on normal condition, $\alpha \geq 0$
β : shape parameter, $\beta \geq 1$

Baseline pdf:

$$f(x \mid s) = \left[\frac{e^{a \cdot s}}{\alpha}\right]^{\beta} \cdot \beta \cdot x^{\beta-1} \cdot \exp\left[-\left(\frac{e^{a \cdot s} \cdot x}{\alpha}\right)^{\beta}\right]$$

Reliability function:

$$R(x \mid s) = \exp\left[-\left(\frac{e^{a \cdot s} \cdot x}{\alpha}\right)^{\beta}\right]$$

2 Model

Note that the IFT's $Z_1, Z_2, ...Z_n$ can be transformed to the failure times $T_1, T_2, ...T_n$. Since the transformed is one-to-one and the Jacobian of the transformation is 1, the joint pdf of the IFT's can be easily obtained from that of the failure times. When the $y_{i,j}$ is given, the conditional distribution is given by:

$$f\{x_{i,j}|y_{i,j-1}\} = \frac{f(x_{i,j})}{R(y_{i,j-1})}$$
$$= \left[\frac{e^{a \cdot s}}{\alpha}\right]^\beta \cdot \beta \cdot x_{i,j}^{\beta-1} \cdot \exp\left[-\left(\frac{e^{a \cdot s} \cdot x_{i,j}}{\alpha}\right)^\beta\right] \cdot \exp\left[\left(\frac{e^{a \cdot s} \cdot y_{i,j-1}}{\alpha}\right)^\beta\right]. \quad (1)$$

By substituting virtual age for $x_{i,j}$ and $y_{i,j}$, the conditional distribution is as follows:

$$f(x|s) = \left[\frac{e^{a \cdot s}}{\alpha}\right]^\beta \cdot \beta \cdot \left(\sum_{k=1}^{j}(1-\rho) \cdot z_{i,k-1} + z_{i,j}\right)^{\beta-1}$$
$$\cdot \exp\left[-\left(\frac{e^{a \cdot s} \cdot \left(\sum_{k=1}^{j}(1-\rho) \cdot z_{i,k-1} + z_{i,j}\right)}{\alpha}\right)^\beta\right] \cdot \exp\left[\left(\frac{e^{a \cdot s} \cdot \left(\sum_{k=1}^{j-1}(1-\rho) \cdot z_{i,k}\right)}{\alpha}\right)^\beta\right]. \quad (2)$$

The likelihood function is:

$$L(\alpha,\beta,\rho,a) = \prod_{i=1}^{m}\left\{\prod_{j=1}^{n_i}\left\{\left[\frac{e^{a \cdot s_i}}{\alpha}\right]^\beta \cdot \beta \cdot \left(\sum_{k=1}^{j}(1-\rho) \cdot z_{i,k-1} + z_{i,j}\right)^{\beta-1}\right.\right.$$
$$\left.\left.\cdot \exp\left[-\left(\frac{e^{a \cdot s_i} \cdot \left(\sum_{k=1}^{j}(1-\rho) \cdot z_{i,k-1}+z_{i,j}\right)}{\alpha}\right)^\beta\right]\right\} \cdot \prod_{j=1}^{n_i-1}\exp\left[\left(\frac{e^{a \cdot s_i} \cdot \left(\sum_{k=1}^{j-1}(1-\rho) \cdot z_{i,k}\right)}{\alpha}\right)^\beta\right]\right\}. \quad (3)$$

The log-likelihood function is:

$$\log L(\alpha,\beta,\rho,a) = \sum_{i=1}^{m} n_i \cdot \beta \cdot \ln\left(\frac{e^{a \cdot s_i}}{\alpha}\right) + N \cdot \ln\beta$$
$$+(\beta-1)\sum_{i=1}^{m}\sum_{j=1}^{n_i}\ln\left(\sum_{k=1}^{j}(1-\rho) \cdot z_{i,k-1}+z_{i,j}\right)$$
$$-\sum_{i=1}^{m}\sum_{j=1}^{n_i}\left[\frac{e^{a \cdot s_i} \cdot \left(\sum_{k=1}^{j}(1-\rho) \cdot z_{i,k-1}+z_{i,j}\right)}{\alpha}\right]^\beta + \sum_{i=1}^{m}\sum_{j=1}^{n_i-1}\left[\frac{e^{a \cdot s_i} \cdot \left(\sum_{k=1}^{j}(1-\rho) \cdot z_{i,k}\right)}{\alpha}\right]^\beta. \quad (4)$$

It is difficult to maximize Equation (4). A Genetic Algorithm is used to find the set of $(\hat{\alpha}, \hat{\beta}, \hat{\rho}, \hat{a})$ maximizing the log-likelihood function.

3 Numerical Examples

Simulations were carried out to investigate the accuracy of the estimation of model parameters. Failure times are generated for some specified parameter sets and we estimate model parameters using the proposed method with the generated failure data. Mean and mean squared error (MSE) are calculated to show the performance of the estimation. When the number of parameters to be estimated is large, it is difficult to find sets of values maximizing likelihood functions through classical search techniques. So, for finding the set of parameters maximizing the likelihood functions, a Genetic Algorithm (GA) is used. In GA, an individual is expressed vector of a fixed point figure. The method for encoding and decoding is very simple; some digits are allocated to one parameter in order of α, β, ρ, a i.e. (3156, 1578, 2560) is decoded to $\alpha = 3.156$, $\beta = 1.578$ and $\rho = 0.2560$. We tested the discrete recombination and selected section recombination. As a result of test, selected section recombination is better than discrete recombination. And we use the uniform mutation. We tested a variety of the rates of Crossover and Mutation. As a result of the test, 0.60(Crossover) and 0.35(Mutation)were selected, respectively.

3.1 Generate Failure Data

When $(k-1)$th failure data generated, kth failure time is given by:

$$U = F_n(x) = P(X_n < x_n | X_n > y_{n-1}) = \frac{F(X_n) - F(y_{n-1})}{R(y_{n-1})}$$

it is expressed as follows:

$$X_n = \frac{\alpha \cdot \left(\left(\frac{e^{a \cdot s} \cdot y_{n-1}}{\alpha}\right)^\beta - Log(1-U)\right)^{1/\beta}}{e^{a \cdot s}} \text{ and, } Z_n = X_n - y_{n-1}. \quad (5)$$

3.2 Computational Experiments

We investigate the effect of the number of failures for a unit. The input values of the parameters are set to $\alpha=2.0$, $\beta=1.5$, $\rho=0.3$, and $a=0.2$. Table 1 and Figure 1 show that as the number of failures for a unit increases, the estimates for α, β, ρ and a become close to true values. Second, the effect of the shape parameter of the intensity function is investigated. The shape parameter of the baseline intensity function is convex, linear, and concave when $\beta > 2$, $\beta = 2$, and $\beta < 2$, respectively. The input values of the parameters are set to $\alpha=2.0$, $\rho=0.3$, and $a=0.2$. The results are shown in Table 2 and Figure 2. It shows that the estimation for ρ becomes more accurate as the shape parameter increases. Third, the effect of the scale parameter is investigated. The input values of the parameters are set to $\beta=1.5$, $\rho=0.3$, and $a=0.2$. The results are shown in Table 3

Table 1. Effect of the number of failures for a unit

	$\alpha(2.0)$		$\beta(1.5)$		$\rho(0.3)$		$a(0.2)$	
n	Mean	MSE	Mean	MSE	Mean	MSE	Mean	MSE
30	2.397	0.557	1.345	0.702	0.376	0.945	0.449	0.358
50	1.974	0.225	1.484	0.552	0.328	0.636	0.246	0.222
100	2.090	0.196	1.505	0.112	0.326	0.520	0.222	0.233

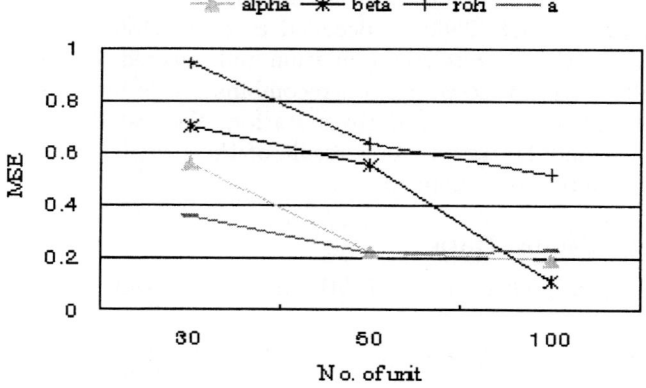

Fig. 1. Plot of MSE based on the number of failures

Table 2. Effect of the shape parameter

	$\alpha(2.0)$		β		$\rho(0.3)$		$a(0.2)$	
β	Mean	MSE	Mean	MSE	Mean	MSE	Mean	MSE
1.5	2.397	0.557	1.345	0.702	0.376	0.945	0.449	0.358
2.0	2.025	0.411	1.823	0.669	0.367	0.863	0.432	0.547
3.0	1.883	0.423	2.789	0.887	0.339	0.587	0.303	0.266

and Figure 3. It shows that the estimation for α, ρ becomes accurate as the scale parameter increases. Fourth, the effect of the age reduction factor is investigated. The input values of the parameters were set to α=2.0, β=1.5, and a=0.2. The results are shown in Table 4 and Figure 4. The degree of reduction factor is between 0 and 1 where the case of 0 corresponds to minimal repair and the case of 1 to perfect repair. The estimation accuracy for α depends on the value of reduction factor.

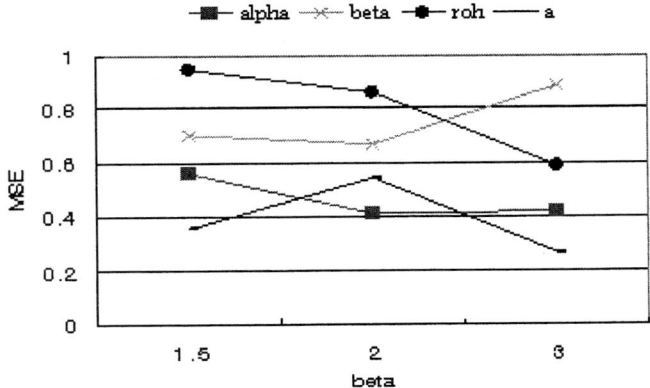

Fig. 2. Plots of MSE with different shape parameters

Table 3. Effect of the scale parameter

α	α		$\beta(1.5)$		$\rho(0.3)$		$a(0.2)$	
	Mean	MSE	Mean	MSE	Mean	MSE	Mean	MSE
2.0	2.397	0.557	1.345	0.702	0.376	0.945	0.449	0.358
3.0	2.840	0.401	1.423	0.080	0.316	0.402	0.317	0.111
5.0	4.413	0.791	1.341	0.071	0.267	0.210	0.315	0.141

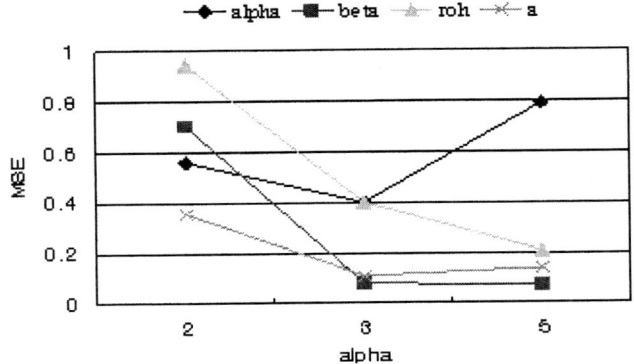

Fig. 3. Plots of MSE with different scale parameters

Table 4. Effect of the age reduction factor

	$\alpha(2.0)$		$\beta(1.5)$		ρ		$a(0.2)$	
ρ	Mean	MSE	Mean	MSE	Mean	MSE	Mean	MSE
0.0	1.914	0.151	1.364	0.062	0.095	0.034	0.358	0.160
0.2	1.838	0.352	1.230	0.075	0.232	0.332	0.332	0.142
0.5	1.939	0.717	1.331	0.076	0.484	0.185	0.361	0.148
0.8	2.059	0.760	1.354	0.069	0.721	0.157	0.376	0.172
1.0	2.387	0.613	1.303	0.103	0.962	0.166	0.270	0.119

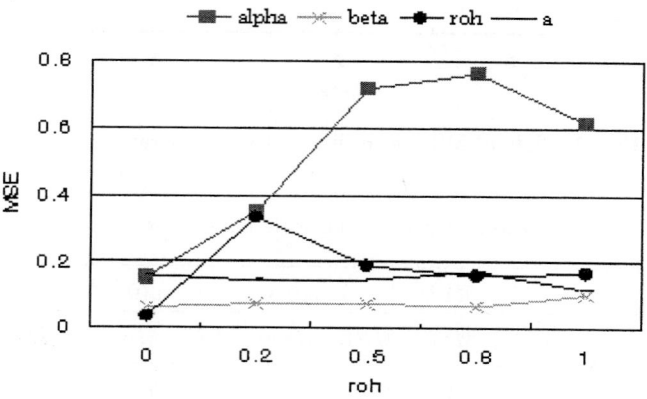

Fig. 4. Plots of MSE with different age reduction factors

4 Conclusion

The imperfect maintenance model and the proportional intensity model are considered to model a repairable system under accelerated stress. The maintenance effect is assumed to reduce the age of the system and the stress acts multiplicatively on the baseline cumulative intensity. Malik's model is used to determine the effect of the maintenance. The log-likelihood function is derived and an estimation procedure with a Genetic Algorithm is proposed. In simulation studies, we investigate the accuracy and some properties of the estimation methods. It is shown that the accuracy of the estimation is improved as the number of failure increases. Further work will be focused on the study of a variety of stress loadings and different models of age reduction factor.

Acknowledgement

This work has been supported by "Research Center for Future Logistics Information Technology" hosted by the Ministry of Education in Korea.

References

1. Bagdonavius, V. and Nikulin, M., *Accelerated Life Models*, Chapman & Hall, 2002.
2. Brown, J.F., Mahoney, J.F. and Sivazlian, B. D., "Hysteresis Repair in Discounted Replacement Problems", *IIE Transactions*, 15(1983), 156-165.
3. Cox, D.R. and Oakes, D., *Analysis of Survival Data*, Chapman & Hall, (1984).
4. Guerin, F., Dumon, B. and Lantieri, P., "Accelerated Life Testing on Repairable Systems", Proceedings of Reliability and Maintainability Symposium (2004), 340-345.
5. Guida, M. and Giorgio, M., "Reliability Analysis of Accelerated Life-Test Data from a Repairable System", *IEEE Transactions on Reliability*, 44(2)(1995), 337-346.
6. Kalbfleisch, J.D. and Prentice, R. L., *The Statistical Analysis of Failure Time Data*, John Wiley & Sons, Canada, (1980).
7. Malik, M.A.K., "Reliable Preventive Maintenance Scheduling", *AIIE Transactions*, (11)(1979), 221-228.
8. Nelson, W., *Accelerated Testing; Statistical Models, Test Plans and Data Analysis*, John Wiley & Sons, New York, (1990).

Optimization Model for Remanufacturing System at Strategic and Operational Level

Kibum Kim[1], Bongju Jeong[1], and Seung-Ju Jeong[2]

[1] Department of Industrial and Information Engineering, Yonsei University,
134 Shinchon-dong, Seodaemun-gu, Seoul, 120-749, Korea
{kibum, bongju}@yonsei.ac.kr
[2] The Korea Transport Institute,
2311 Daehwa-dong, Koyang-si, Gyeonggi-do, 411-701, Korea
sjj@koti.re.kr

Abstract. As many companies are becoming environmentally conscious, and as stringent environmental laws are being passed, remanufacturing has received growing attention. In this paper, we discuss the remanufacturing system where the manufacturer overhauls returned products and bringing back to 'as new' conditions. We propose two mathematical models for consecutive decision making at strategic and operational level in remanufacturing system. And sensitivity analyses are conducted on various parameters to gain insight into the proposed models by using a set of data reflecting a real business situation.

1 Introduction

Remanufacturing is an industrial process in which worn-out products are restored to like-new condition. Through a series of industrial process in a factory environment, a discarded product is completely disassembled. Usable parts are cleaned, refurbished, and put into inventory. Then the new products are reassembled from the old and, where necessary, new parts to produce a fully equivalent and sometimes superior in performance and expected lifetime to the original products (Lund, 1998). An example is Hewlett-Packard, who collects empty laser-printer cartridges from the customers for reuse (Jorjani et al., 2004).

Previous researches on remanufacturing can be classified into two categories, strategic level and operational level. At strategic level, the economics of remanufacturing is studied by Ferrer (1997). In his study, the feasibility of remanufacturing is investigated considering the value of recoverable parts and market specification. Salomon et al. (1996) develop a design of product return network. And Guide et al. (2001) find the optimal selling and acquisition process for remanufacturing product. However, most previous researches on remanufacturing focus on operational level, which are forecasting, production planning/control, inventory control/management and scheduling. A good overview on quantitative models for recovery production planning and inventory control is given in Fleischmann et al. (1997). Der Laan et al. (1997) extend the PUSH and PULL strategies to control a system in which all returned products are remanufactured and no planned disposal occur. Jayaraman et al. (2003) discuss reverse distribution, and propose a mathematical programming model and solution procedure for a reverse distribution problem. Their model builds upon the single-source plant location model developed by Pirkul et al. (1996). And Caruso et al. (1993) describe a solid waste management system, including collection, transportation, incineration, composting, recycling and disposal.

Although several researchers have suggested various remanufacturing models, few research are found for decision making problems at both strategic and operational levels. In this study, we propose some mathematical models for decision making at both strategic and operational levels in remanufacturing environment. The practical meaning of the proposed models is examined through computational experiments and analysis.

2 Mathematical Models

2.1 Problem Definition

Remanufacturing can be crucial to the survival of manufacturing companies because the permanent goodwill of the companies is at stake. But a lot of manufacturing companies are not willing to introduce the remanufacturing for many reasons, for example high installation cost for remanufacturing facilities and longer payback period than other alternative investment plans. Therefore the optimal investment model for remanufacturing facilities and operational model for remanufacturing execution are increasingly needed. In this paper, we propose such models for remanufacturing decision making problems. Two consistent models are developed for both strategic and operational decision making. The models are based on a comprehensive conceptual remanufacturing framework as shown in Fig.1.

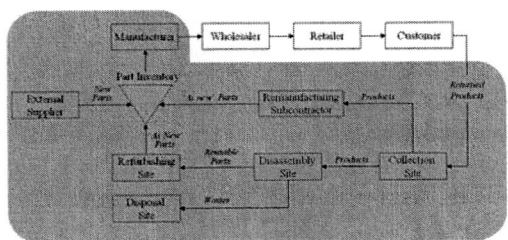

Fig. 1. Conceptual framework for remanufacturing system

Remanufacturing system starts with the returned products including end-of-life product from customers and they are collected to collection facility. Some of them are disassembled to remanufacturing and the rests, out of the bounds of remanufacturing capacity, are subcontracted. Disassembled parts are classified into reusable parts and non-reusable parts. The former goes to refurbishing site for repairing and cleaning, the latter dealing with wastes goes to disposal site to landfill or incinerate. After refurbishing process, 'as new' parts are stocked as part inventory together with new parts from external supplier and subcontractor. In this case, the manufacturing company has two alternatives for supplying parts: Either ordering the required parts to external suppliers or overhauling the returned products and bringing those back to 'as new' conditions.

Actually most manufacturing companies who are willing to introduce the proposed framework for remanufacturing are facing two major problems. One is how much money they need to invest for the capacity of remanufacturing facilities such as collection, disassembly, and refurbishing sites. The other is how many returned products should be thrown into the remanufacturing process for 'as new' condition and how many new parts need to be purchased from external suppliers. These problems are equivalent to two consecutive decision makings on facility investment and production planning which are formulated as two models as shown in Fig. 2. Model 1 is for a decision making at strategic level. The optimal investment scale for the capacity of remanufacturing facilities is determined in this model by trading off the internal and external constraints for remanufacturing such as governmental regulation, expected quantity of product return, and available investment scale for remanufacturing facilities. In Model 2, decision making at operational level is accomplished under the optimal capacity of remanufacturing facilities obtained from Model 1.

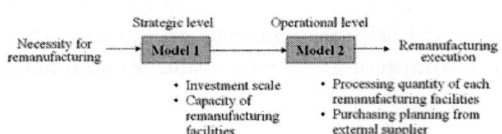

Fig. 2. Consecutive decision making model for remanufacturing

All the notations for variables and parameters used in the both models are summarized in Appendix A.

2.2 Models Description

Model 1. Remanufacturing Decision Making at Strategic Level

The objective of Model 1 is to determine the optimal investment scale for the capacity of each remanufacturing facilities, which are collection, disassembly, and refurbishing site, respectively. The Model 1 can be stated as follows:

Maximize

$$\sum_{i=1}^{I}(Inr_i \times CPAQr_i \times DifCost_i)-$$

$$\left[MCr \times \sum_{i=1}^{I} Inr_i \times CPAQr_i \right. \quad (1)$$

$$\left. + MCd \times \sum_{p=1}^{P} Ind_p \times CPAQd_p + MCc \times \sum_{p=1}^{P} Inc_p \times CPAQc_p \right]$$

Subject to

$$r_p^{gov} \times D_p^{avg} \leq Inc_p \times CPAQc_p \qquad , \forall p \quad (2)$$

$$r_p^{gov} \times D_p^{avg} \leq Ind_p \times CPAQd_p \qquad , \forall p \quad (3)$$

$$Ind_p \times CPAQd_p \leq Inc_p \times CPAQc_p \qquad , \forall p \quad (4)$$

$$Inr_i \times CPAQr_i \leq \sum_{p=1}^{P} Ind_p \times CPAQd_p \times BOM_{pi} \qquad , \forall i \quad (5)$$

$$\sum_{i=1}^{I} r_p^{gov} \times D_p^{avg} \times BOM_{pi} \leq Inr_i \times CPAQr_i \qquad , \forall i \quad (6)$$

$$(1-\lambda_i) \times \sum_{p=1}^{P} Ind_p \times CPAQd_p \times BOM_{pi} \leq Inr_i \times CPAQr_i \quad , \forall i \quad (7)$$

$$\sum_{i=1}^{I} Inr_i + \sum_{p=1}^{P} (Ind_p + Inc_p) \leq TOI \qquad (8)$$

$$Inr_i \geq 0 \qquad , \forall i \quad (9)$$

$$Ind_p, Inc_p \geq 0 \qquad , \forall p \quad (10)$$

The objective function (1) is to maximize the saving from the investment on remanufacturing facilities, which is the cost saving subtracted maintenance cost of each facilities. Constraints (2) and (3) represent that each capacity of collection and disassembly site should be larger than total amount of returned products by government regulation. Constraint (4) guarantees the capacity of disassembly site cannot exceed the capacity of collection site. Constraints (5) and (6) represent the upper bound and lower bound for the refurbishing capacity considering the number of parts from specific product (BOM_{pi}) to be disassembled. Constraint (7) gives the lower bound for refurbishing capacity considering the limit on the disposal ratio (λ_i). Constraint (8) ensures the sum of capacity investment for all remanufacturing facilities should be less than total available investment (TOI). Constraint set (9) and (10) check for the non-negativity of decision variables. The investment scale determined by Model 1 is converted to the capacity for each remanufacturing facility as follows:

$$Inc_p \times CPAQc_p = CSCp_p \qquad , \forall p \qquad (11)$$

$$Ind_p \times CPAQd_p = DCp_p \qquad , \forall p \qquad (12)$$

$$Inr_i \times CPAQr_i = RCp_i \qquad , \forall i \qquad (13)$$

Model 2. Remanufacturing Decision Making at Operational Level

The objective of Model 2 is to determine the optimal processing quantities for each remanufacturing facility by maximizing the total cost saving. The capacity information for remanufacturing facility obtained from Model 1 is used in this model. The Model 2 can be stated as follows:

Maximize

$$\sum_{t=1}^{T}\sum_{i=1}^{I}(PPC_i \times R_{it})-$$
$$\left[\sum_{t=1}^{T}\sum_{p=1}^{P}(CPIC_p CPI_{pt} + OUTC_p OUT_{pt} + DSUC_p DSU_{pt} + DSVC_p DP_{pt})+\right.$$
$$\sum_{t=1}^{T}\sum_{i=1}^{I}(DspC_i WPart_{it} + RSC_i RSU_{it} + RVC_i RPart_{it} + PIC_i PI_{it}+$$
$$\left. PPC_i PPart_{it}) + \sum_{t=1}^{T}\sum_{p=1}^{P}(1-DSU_{pt})DIDC + \sum_{t=1}^{T}\sum_{i=1}^{I}(1-RSU_{it})RIDC\right]$$

$$(14)$$

Subject to

$$DPart_{it} = \sum_{p=1}^{P} BOM_{pi} DP_{pt} \qquad , \forall i, t \qquad (15)$$

$$OPart_{it} = \sum_{p=1}^{P} BOM_{pi} OUT_{pt} \qquad , \forall i, t \qquad (16)$$

$$CP_{pt} + CPI_{p,t-1} = OUT_{pt} + DP_{pt} + CPI_{pt} \qquad , \forall p, t \qquad (17)$$

$$RPart_{it} + WPart_{it} = DPart_{it} \qquad , \forall i, t \qquad (18)$$

$$RPart_{it} + PPart_{it} + OPart_{it} + PI_{i,t-1} = R_{it} + PI_{it} \qquad , \forall i, t \qquad (19)$$

$$\sum_{i=1}^{I} VPart_i \, PI_{it} \leq PICp \qquad , \forall t \qquad (20)$$

$$OUT_{pt} + DP_{pt} \leq CSCp_p \qquad , \forall p, t \qquad (21)$$

$$DP_{pt} \leq DCp_p \qquad , \forall p, t \qquad (22)$$

$$RPart_{it} \leq RCp_i \qquad , \forall i, t \qquad (23)$$

$$RPart_{it} \leq M \cdot RSU_{it} \qquad , \forall i, t \qquad (24)$$

$$DP_{pt} \leq M \cdot DSU_{pt} \qquad , \forall p, t \qquad (25)$$

$$\sum_{t=1}^{T} WPart_{it} \leq \lambda_i \sum_{t=1}^{T} DPart_{it} \qquad , \forall i \qquad (26)$$

$$RSU_{it} \in \{0,1\} \qquad , \forall i, t \qquad (27)$$

$$DSU_{pt} \in \{0,1\} \qquad , \forall p, t \qquad (28)$$

$$OUT_{pt}, DP_{pt}, CPI_{pt} \geq 0 \qquad , \forall p, t \qquad (29)$$

$$RPart_{it}, PPart_{it}, OPart_{it}, WPart_{it}, PI_{it} \geq 0 \qquad , \forall i, t \qquad (30)$$

The objective function (14) is to maximize the sum of cost savings from remanufacturing process. It is measured by the gap of purchasing cost from external supplier and remanufacturing cost, which is composed of remanufactirng processing cost and idle cost at each remanufacturing facilities. Constraints (15) and (16) compute the number of parts by disassembling the products from collection site and remanufacturing subcontractor. Constraints (17)–(19) represent the balance equations for product and part inventory. Constraint (20) ensures the inventory quantities of part cannot exceed the predetermined capacity of part inventory ($PICp$). Constraint (21) ensures that the number of products from the collection site cannot exceed its capacity ($CSCp_p$). Constraints (22) and (23) guarantee the disassembly and refurbishing quantity cannot exceed the capacity of disassembly site (DCp_p) and refurbishing site (RCP_i), respectively. Constraints (24) and (25) are Big M constraints for setup at refurbishing and disassembly site. Constraint (26) shows that the manufacturing companies cannot exceed λ_i of disassembled parts to dispose, where λ_i is a predetermined value. Constraint set (27) and (28) check for binary variables and the last two constraint sets (29) and (30) check for the non-negativity of decision variables.

3 Insight into the Proposed Models

In this section, we present the insight into the proposed models. Data used in experiment is not real, but reflecting the real business situation. A set of data used in experiment includes 3 types of products, 5 types of parts, and 10 time periods. All input data is shown in Table 1, 2, and 3. We used an OPL Studio 3.0 for solving proposed models on a PC with Pentium IV 3.0 GHz.

For reflecting a real business situation, the gap between unit purchasing and remanufacturing cost ($DifCost_i$) is set to be 300, 200, 100, 150 and 300 for each part type. And the maintenance cost per unit capacity (MCc, MCd, MCr) is set to be 10, 70, and 100 for collection, disassembly, and remanufacturing site, respectively.

In order to gain some insight into the proposed models, sensitivity analyses are conducted on various parameters. At first, the behavior of cost savings in

Table 1. Returned product p at time t

	$t=1$	$t=2$	$t=3$	$t=4$	$t=5$	$t=6$	$t=7$	$t=8$	$t=9$	$t=10$
Product1	160	160	140	130	180	180	170	150	170	140
Product2	170	120	150	200	130	140	130	210	180	150
Product3	160	170	120	180	170	130	160	150	110	130

Table 2. Number of part i in each product p

	Part1	Part2	Part3	Part4	Part5
Product1	3	5	2	4	3
Product2	3	0	5	0	1
Product3	0	2	3	2	4

Table 3. Required quantity of part i at time t

	$t=1$	$t=2$	$t=3$	$t=4$	$t=5$	$t=6$	$t=7$	$t=8$	$t=9$	$t=10$
Part1	1470	1860	1530	1560	1710	1710	1800	1750	1800	1740
Part2	2970	3860	3010	3300	3630	3330	3720	3210	3520	3460
Part3	2510	3370	2310	2780	3390	2820	3120	2490	2800	3060
Part4	1480	1920	1700	1860	1620	1480	2000	1750	1890	1400
Part5	2010	2720	2050	2500	2550	2030	2680	2570	2490	2100

Model 1 is examined with respect to changes in total available investment (TOI). For the first analysis, the return ratio of product p by the government regulation is set to be 0.6, 0.7, and 0.5 for each types of product, and the upper bound for disposal rate of disassembled part is set to be 0.15 for all parts. The results are given in Fig. 3. We can see that the cost savings seem to be proportional to the available investments until 18,000 units investment that is equivalent to the sufficient capacity for remanufacturing all returned products, and after that the savings are likely to be constant. This means that there exists an optimal investment scale (in our case, 18,000 units investment) to maximize the cost saving from remanufacturing.

We also examined the cost savings according to the various return ratio (r_p^{gov}) by government as shown in Fig. 4. It is easily noticed that the cost saving remain constant until the return ratio reaches a specific point (in our case, 0.35), and then it decreases as the return ratio increases. This analysis shows that the company is always willing to invest for remanufacturing facility whenever it can obtain the maximum cost savings at the specific return ratio which is greater than the government one.

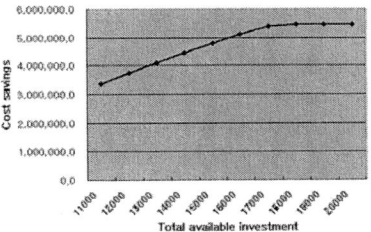

Fig. 3. Sensitivity analysis for total investment (TOI) vs. cost savings

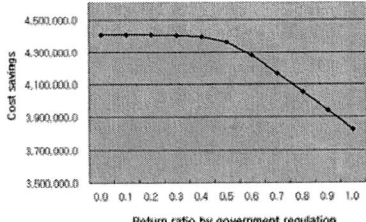

Fig. 4. Sensitivity analysis for the return ratio (r_p^{gov}) vs. cost savings

Figure 5, 6, and 7 show the proportion of capacity investments for processing each product or part in three different processing sites, according to the total available investments. From the figures, we can find that at each total investment level, the investment priority and amount for each product or part can be determined. For example, at collection site, if the total investment is less than 13,500, product 1 should be selected as the first priority for investment and also its capacity needs to be increased as the total investment increases.

Similarly, the sensitivity analyses for Model 2 are conducted for the capacity of remanufacturing facilities, which is the most important factor of Model 2. The results are shown in Fig. 8. For all graphs, the profits are maximized at the specific capacity level and from that level they remain constant. Due to difference of two objective functions, the level does not correspond to the result from Model 1 exactly. However, this analysis shows that we can obtain the maximum profit with the minimum capacity for each facility, which achieves the 'optimal remanufacturing'. In other words, there exists an optimal investment

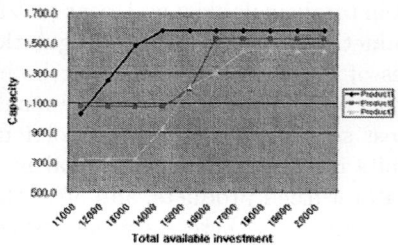
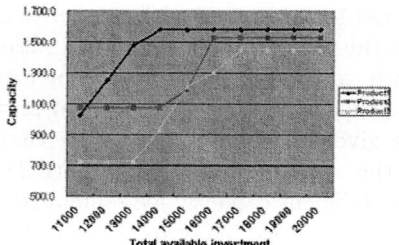

Fig. 5. Sensitivity analysis for total investment (TOI) vs. capacity for each product in collection site ($CSCp_p$)

Fig. 6. Sensitivity analysis for total investment (TOI) vs. capacity for each product in disassembly site (DCp_p)

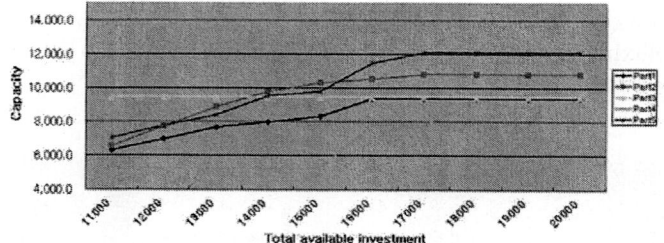

Fig. 7. Sensitivity analysis for total available investment (TOI) vs. capacity for each part in refurbishing site (RCP_i)

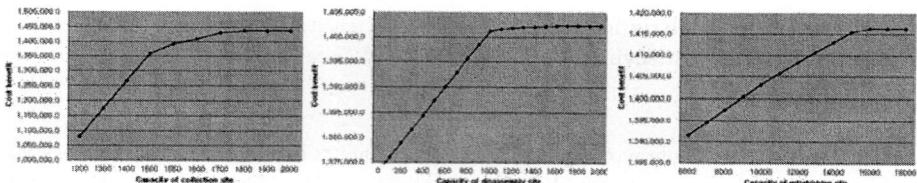

Fig. 8. Sensitivity analysis for the capacity of remanufacturing facilities($CSCp_p, DCp_p, RCp_i$) vs. cost saving

level for remanufacturing facilities, which is one of the most significant results of Model 1.

4 Conclusions

In this paper, we presented a framework of remanufacturing processes and proposed the mathematical models for decision-making at strategic and operational level. At strategic level, the model determines how much the company needs to invest for the remanufacturing facilities while maximizing the cost savings. Using this model, we can also find the investment priorities for facilitating each product or part at given total investment. The proposed model for operational level determines the processing quantity of each remanufacturing facility by maximizing the total profit. The post-optimal analysis for the operational model shows that there exists an optimal remanufacturing capacity for each facility at which the company achieves the maximal cost saving, which is compatible to the result of strategic model. This analysis indicates that the strategic model may be integrated into the operational model without loss of integrity.

As future research, we can delve into the integration model for all decision-making levels and need to develop a practical and efficient heuristics for solving the large-scale problems.

References

Caruso, C., Colorni, A., Paruccini, M.: The Regional Urban Solid Waste Management System: A Modeling Approach. Eur. J. Oper. Res. **70** (1993) 16–30

Der Laan, E., Salomon, M.: Production Planning and Inventory Control with Remanufacturing and Disposal. Eur. J. Oper. Res. **102** (1997) 264–278

Guide, V.D., Van Wassenhove, L.N.: Managing Product Returns for Remanufacturing. Prod. Oper. Manag. **10 2** (2001) 145–155

Ferrer, G.: The Economics of Personal Computer Remanufacturing. Resour. Conserv. Recy. **21** (1997) 79–108

Fleischmann, M., Bloemhof-Ruwarrd, J.M., Dekker, R., Van Der Lann, E., Van Nunen, J.A.E.E., Van Wassenhove, L.N.:Quantitative Models for Reverse Logistics: A Review. Eur. J. Oper. Res. **103** (1997) 1–17

Jayaraman, V., Patterson, R. A., Rolland, E.: The Design of Reverse Distribution Networks: Models and Solution Procedures. Eur. J. Oper. Res. **150** (2003) 128–149

Jorjani, S., Leu, J., Scott, C.: Model for the allocation of electronics components to reuse options. Int. J. Prod. Res. **42 6** (2004) 1131–1145

Lund, R.: Remanufacturing: United States Experience and Implications for Developing Nations. The World Bank, Washington, DC. (1998)

Pirkul, H., Jayaraman, V.: Production, Transportation, and Distribution Planning in a Multi-Commodity Tri-Echelon System. Transport. Sci. **30** (1996) 291–303

Salomon, M., Thierry, M., van Nunen, J., Van Wassenhove, L.: Distribution Logistics and Return Flow Optimization. Management Report Series. **254** (1996)

Appendix: Nomenclatures for The Proposed Models

Indices

p : Product index, $p = \{1, ..., P\}$
i : Part index, $i = \{1, ..., I\}$
t : Time period, $t = \{1, ..., T\}$, (T: Planning horizon)

Parameters

r_p^{gov}	: Product index	BOM_{pi}	: Number of part i from disassembling on unit of product
D_p^{avg}	: Average demand for product p in time t	$CSCp_p$	: Capacity of process in collection site for product p
$CPAQr_i$	: Capacity acquisition for part i by unit investment on refurbishing site	RCp_i	: Capacity of refurbishing site for part i
$CPAQd_p$	: Capacity acquisition for product p by unit investment on disassembly site	DCp_p	: Capacity of disassembly site for product p
$CPAQc_p$	: Capacity acquisition for product p by unit investment on collection site	$VPart_i$	: Volume occupied by one unit of part i
R_{it}	: Required quantity of part i at time t	$PICp$	: Inventory capacity of part inventory
CP_{pt}	: Collected quantity of product p at time t	λ_i	: Upper bound of disposal rate for disassembled part i

Cost Parameters

TOI	: Total available investment	$DSVC_p$	: Unit operation cost for disassembling collected product p
MCr	: Maintenance cost of Refurbishing site per unit capacity	$OUTC_p$	: Subcontracting cost for product p
MCd	: Maintenance cost of Disassembly site per unit capacity	RSC_i	: Set-up cost for refurbishing disassembled part i
MCc	: Maintenance cost of Collection site per unit capacity	RVC_i	: Unit operation cost of refurbishing disassembled part i
$DifCost_i$	: Gap of purchasing cost for new parts and unit remanufacturing cost	PIC_i	: Unit inventory holding cost of part i in part inventory
$CPICc_p$	: Unit inventory holding cost of collected product p in collection site	PPC_i	: Unit purchasing cost of part i from supplier at time t
$DspC_i$	: Disposal cost of disassembled part i	$DIDC$	: Idle cost of disassembly facility
$DSUC_p$	: Set-up cost for disassembling collected product p	$RIDC$	: Idle cost of refurbishing facility

Decision Variables

Inr_i	: Investment on refurbishing site for part i	$DPart_{it}$	: Number of disassembled part i at time t
Ind_p	: Investment on disassembly site for product p	PI_{it}	: Inventory level of part i at time t
Inc_p	: Investment on collection site for product p	$PPart_{it}$	: Number of purchased part i at time t
DP_{pt}	: Number of disassembled product p at time t	RSU_{it}	: 1, if refurbishing set-up for part i occurred at time t; 0, O/W
$RPart_{it}$	: Number of refurbished part i at time t	DSU_{pt}	: 1, if disassembled set-up for collected product p occurred at time t; 0, O/W
$WPart_{it}$	: Number of disposed part i at time t	OUT_{pt}	: Number of outsourcing product p at time t
CPI_{pt}	: Inventory level of colleted product p at time t	$OPart_{it}$	: Number of part i from subcontractor at time t

A Novel Procedure to Identify the Minimized Overlap Boundary of Two Groups by DEA Model*

Dong Shang Chang[1] and Yi Chun Kuo[2]

[1] Department of Business Administration, National Central University,
Chung-Li, Taiwan, Republic of China
changds@mgt.ncu.edu.tw

[2] Department of Business Administration, National Central University,
Chung-Li, Taiwan, Republic of China,
Department of International Trade, Vanung University,
Chung-Li, Taiwan, Republic of China
chun@msa.vnu.edu.tw

Abstract. In discriminant analysis (DA), the overlap among groups is a major source of misclassification. The identification of overlap boundary may provide additional information for the decision makers in risk management. Most effort of previous researches aim to improve the hit-ratio of correct classification, but few concentrates on the identification of overlap boundary. In this paper, a novel procedure which is based on data envelopment analysis (DEA) is proposed to identify the overlap boundary of two groups. The minimized overlap boundary can be obtained after taking linear transformation. An important merit of the proposed approach is to resolve the problem of calibrating overlap boundary with parametric approach in the case of small sample size.

1 Introduction

The classification problem of assigning observations into one of several groups plays a key role in decision making. Binary classification problem, in which the observations are restricted to one of two groups, has wide applicability in problems ranging from credit scoring, default prediction and direct marketing to applications in biology and medical domain [2].In discriminant analysis (DA), the overlap of two groups is a major source of misclassification. However, most effort of previous researches aim to improve the hit-ratio of correct classification, few concentrates on the identification of overlap boundary.While the overlap boundary can be identified, the decision maker will pay more attention to the new observation which is predicted to appear within the boundary. For example, in the case of loan evaluation, the manager may make the binary decision from

---

* This research is supported by the National Science Council of Republic of China under Contract NSC 93-2213-E-008-035.

the result of DA and obtain the additional information for risk control if the applicant is identified within the overlap boundary.

The statistical technique such as Fisher's linear discriminant function [9], Goal Programming (GP) approach initially proposed by Freed and Glover [11] and machine learning techniques such as neural network [17] are known as popular approaches to solve the discriminant problem, although these approaches can count the hit-ratio for the performance of classification, none of them are capable to identify the overlap boundary of two groups. Sueyoshi's DEA-DA [18] [19] [20] [21]propose a GP-based DA technique to identify the overlap which is obtained by minimizing the total deviation of misclassified observations, but it fails to calibrate the boundary of overlap. Moreover, in the case of small sample size, it is hard to fit the overlap boundary by parametric techniques.

In this paper, a novel procedure which is based on data envelopment analysis (DEA) is proposed to identify the overlap boundary of two groups. The basic idea is inspired that observations belonging to the same group should have the same production possibility set (PPS) and be dominated by the same frontier. If intersection exists between two PPSs, the observed overlap boundary can be obtained from the matching frontiers of two groups.

The remainder of this paper is organized as follows. Section 2 introduces the proposed procedure including overlap identifying and linear transformation. An example is examined in section 3.Conclusion is summarized in section 4.

2 Methodology

2.1 Overlap Identifying

DEA was introduced by Charnes et al. [6]. Consider n production units or Decision Making Units (DMUs) that are to be evaluated, all using the same m inputs to produces s different outputs. Let X_j be the input consumption vector from DMU j where $X_j = (x_{1j}, \cdots, x_{mj})^T$, and Y_j the output production vector where $Y_j = (y_{1j}, \cdots, y_{sj})^T$. The DEA input-oriented efficiency score ϕ' is given by

$$\phi' = \min_{\phi, \lambda} \phi$$
$$\text{s.t. } \phi X' \geq \sum_{j=1}^{n} \lambda_j X_j \qquad (1)$$
$$Y' \leq \sum_{j=1}^{n} \lambda_j Y_j$$

From the viewpoint of DEA, the definition of PPS with the pairs of input and output vectors (X_j, Y_j) of n DMUs is $P = \{(x,y)|x \geq \lambda X, y \leq \lambda Y, \lambda \geq 0\}$. In a simple two-dimensional case, by considering of discriminating variables, Z_1 and Z_2, as input or output variables, we can identify the frontier and the dominated PPS for each group respectively. If Z_1 and Z_2 are considered as output variables

($y_1 = Z_1, y_2 = Z_2$), and given one as the value of input ($x = 1$), it can identify the frontier and PPS as shown in Fig. 1 (a). Points C, D and E which use unit input to produce most output than other points are identified as benchmarks on the frontier. Similarly, Fig. 1 (b) shows another type of frontier by fixing the output value ($y = 1$) and treating Z_1 and Z_2 as input variables ($x_1 = Z_1, x_2 = Z_2$), point A, B and C are on the frontier by using least input than other points to produce one unit output.

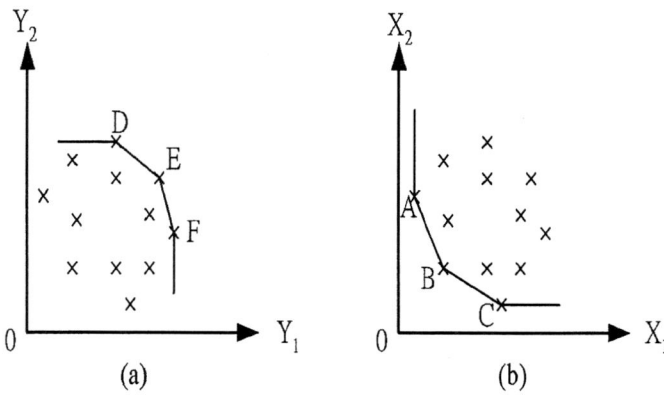

Fig. 1. Illustration of the frontiers identified by output-oriented DEA model in (a) and input-oriented DEA model in (b)

Two possible types of frontier for each group are shown in Fig. 2 (a) and (b). An adequate frontier for one group is defined that it can separate its own DMUs from another group. In Fig. 2 (a), the frontier G_1^I of group 1 (notate as G_1) is obtained by treating all the factors as input variables and fixing output value as one, which can exclude most DMUs of G_2 from G_1. Similarly, the frontier G_2^O can be generated for G_2 by regarding the factors as output variables and fixing input value as one. In contrast with preceding situation, G_1^O and G_2^I (shown in Fig. 2 (b)) are unadapted frontiers for two groups with all DMUs enveloped between them. The rule of determining a proper DEA model for each group will be discussed in Section 2.2.

The intersection of two PPS_s (notate as $PPS_1^1 \bigcap PPS_2^1$) determined by G_1^I and G_2^O might not catch the overlap in general (Such as Fig.3(a)(b)). With the convexity assumption on DEA model, DMUs locating in the margin of PPS will be identified as the benchmarks. If the data sets are coincident as the one depicted in Fig. 3 (c), the intersection of PPS_1^1 and PPS_2^1 may catch the overlap of two groups well. Coordinates rotated for data sets by the technique of linear transformation is further employed to deal this problem. (See Section 2.2)

The stratification model of DEA [22] can be applied to further identify the sequential layers of frontier for two groups. Two important reasons are that (i)it helps to detect whether outlier observation exists. By removing the first

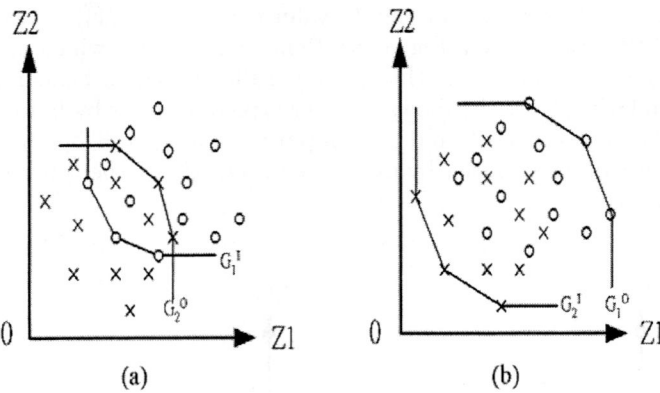

Fig. 2. Illustration of correct intersection identification in (a) and incorrect one in (b)

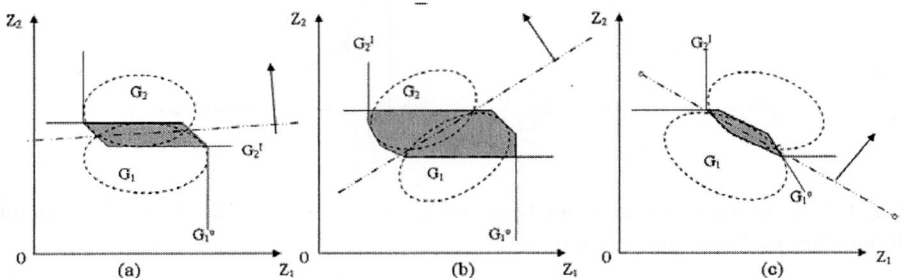

Fig. 3. Illustration of various intersections between two PPSs which are formed by different rotation of original data set in (a), (b) and (c)

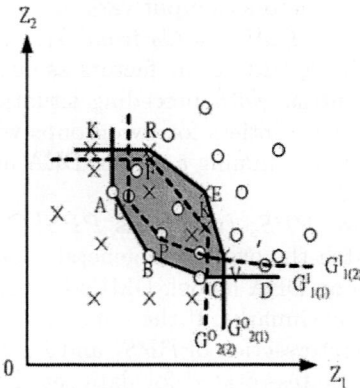

Fig. 4. Illustration of stratified boundaries of overlap

layer of frontier (such as $G^I_{1(1)}$ and $G^O_{2(1)}$ in Fig. 4), the DMUs within the set of $PPS^1_1 \cap PPS^1_2$ are further identified as each group's benchmarks to form the second layer of frontier (mark as $G^I_{1(2)}$ and $G^O_{2(2)}$). The second layer of frontiers can dominate a subset of PPS_1 and PPS_2 (here refer to as PPS^2_1 and PPS^2_2) separately and then reduce the intersection of PPS^2_1 and PPS^2_2 (notate as $PPS^2_1 \cap PPS^2_2$). If the number of observations shows sharp declines from $PPS^1_1 \cap PPS^1_2$ to $PPS^2_1 \cap PPS^2_2$, the first layer of frontier may be formed by the outlier data.(ii) Decision makers may pay more attention to the observations which lead to higher risk level of misclassification in a smaller range of overlap.

2.2 Linear Transformation

We illustrate the process of linear transformation by a two-dimensional coordinate system which can be extended to any high dimensional coordinate system easily.

Consider a discriminant problem with two factors. Give a $(n \times 2)$ matrix ($n = n_1 + n_2$, n_1 and n_2 are the number of observations from G_1 and G_2) to represent the factor value of n DMUs. Hence x_{nj} is the jth factor for the nth DMU. Three steps of linear transformation are described as below:

Step1: Determine the rotation angle and define the rotation matrix **T**.

In order to give the data set being rotated to a proper direction (say 45 degrees), we need to determine the rotation angle which is from the orthogonal unit vector of the classification line to the unit vector. The unit vector being defined as $\vec{u}_1 = \left[\frac{1}{\sqrt{2}}, \frac{1}{\sqrt{2}}\right]$ is 45 degrees from $\vec{u}_1$ to each Cartesian coordinate axes, the classification line can be solved by MSD (Minimizing the sum of individual deviations) [12] which is a well known linear discriminant technique and is formulated as followed:

$$\min \sum_{i=1}^{n_1+n_2} d_i$$

$$\text{s.t.} \sum_{j=1}^{2} z_{ij}w_j + d_i \geq c \quad for \quad i \in G_1 \qquad (2)$$

$$\sum_{j=1}^{2} z_{ij}w_j - d_i < c \quad for \quad i \in G_2$$

Where $\vec{W} = [w_1, w_2]$ is defined as orthogonal vector of the classification line equation $L: \sum z_{ij}w_j + d_i = c$, the rotation angle from $\vec{W}$ to $\vec{u}_1$ can make the data sets rotated to the direction as it shown in Fig. 3 (c). The angle θ can be computed by dot product of these two vectors, and it can be expressed as below

$$\theta = \cos^{-1}\left[\frac{\vec{W} \cdot \vec{u}_1}{|\vec{W}| \cdot |\vec{u}_1|}\right] \tag{3}$$

Where two Cartesian coordinate systems with a rotation angle θ can be expressed the rotation matrix **T** as below:

$$\mathbf{T} = \begin{bmatrix} \cos(\theta) & \sin(\theta) \\ -sin(\theta) & \cos(\theta) \end{bmatrix} \tag{4}$$

Step 2: Conduct linear transformation for coordinate rotation.

Assume that $\vec{Z}_k (k=1,2,\cdot,n)$ is the coordinate of the kth DMU, after linear transformation, the new coordinate of this DMU can be expressed as

$$\begin{bmatrix} z'_{k1} \\ z'_{k2} \end{bmatrix} = \begin{bmatrix} \cos(\theta) & \sin(\theta) \\ -sin(\theta) & \cos(\theta) \end{bmatrix} \cdot \begin{bmatrix} z_{k1} \\ z_{k2} \end{bmatrix} \tag{5}$$

$$\vec{Z}'_k = \mathbf{T} \cdot \vec{Z}_k \quad k=1,2,\cdots,n$$

Step 3: Apply adequate coordinate translation for the positive constraint of factor value in DEA model.

The new vector $\vec{Z}'_k$ may not satisfy the positive constraint on the factors of DMUs in DEA model, it is important to check whether factor values are all positive. If any one of the factors takes on negative value, the adequate positive constant value must be added to the factor value of all DMUs with the absolute value of the most negative value among these DMUs plus one. Therefore, the linear transformation and coordinate translation can be expressed as below:

$$\begin{bmatrix} z''_{k1} \\ z''_{k2} \end{bmatrix} = \begin{bmatrix} z'_{k1} \\ z'_{k2} \end{bmatrix} + \begin{bmatrix} d_1 \\ d_2 \end{bmatrix} \tag{6}$$

$$\vec{Z}''_k = \vec{Z}'_k + \vec{D} \quad k=1,2,\cdots,n$$

d_1 = the absolute value of the most negative value of x_{k1} plus one
d_2 = the absolute value of the most negative value of x_{k2}

After linear transformation, how to treat the factors correctly to obtain an adequate frontier for each group is determined by the rule: if $\sum z_{ij} w_j + d_i > c$ is true for the DMUs of one group with positive value of the first element of $\vec{W}$, after rotation from $\vec{W}$ to $\vec{u}_1$, the factors value of most DMUs will be larger than another group. In order to identify a lowest boundary for such group, all discriminating factos are treated as input variables to identify a frontier as G_1^I (shown in Fig. 3(a)). If $\sum z_{ij} w_j + d_{in} > c$ is true and the first parameter value of $\vec{W}$ is negative, or the first parameter value of $\vec{W}$ is positive but $\sum z_{ij} w_j + d_{in} < c$, the factors are regarded as output variables to form a upper bound of PPS like G_2^O.

3 Example

A two-dimensional case is chosen to illustrate the procedure with visualization of the identified overlap boundary. The data set contains 100 data which is collected to classify the breed of salmons [13], 50 from Canadian salmons and 50 from Alaskan ones. Two attributes are considered for classification, x_1 is the diameter of rings for the first-year freshwater growth and x_2 is marine growth. After linear transformation for data (the rotation angle θ is about 90 degrees computed by formula (3)), the overlap boundary of two groups is identified and shown in Fig. 6 (a). Because the overlap boundary obtained by applying raw data is different from the result by rotated data, in order to confirm that 90 degrees is optimal rotation angle in this case, the number of observations included in the overlap by different rotation angle is examined. Fig. 5 and Fig. 6 show that the 90 degrees rotation angle, which rotate the data set from the direction of -45 degrees to 45 degrees, can obtain the minimized overlap boundary which includes the lowest number of observations (See Fig.5).

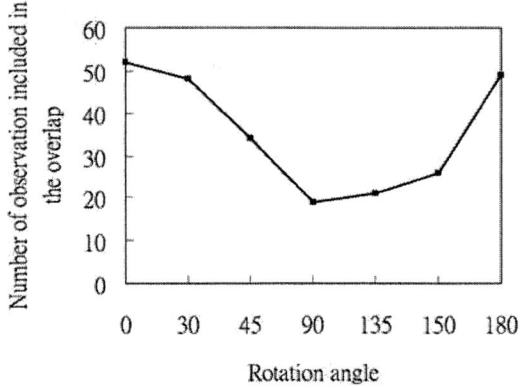

Fig. 5. Numbers of observations included in the overlap

4 Conclusion

In this paper, a minimized overlap boundary of two groups is identified by DEA without the need to specify the boundary function. The obtained overlap boundary provides additional information for decision makers to catch the observations within the overlap for risk management. It is also useful in other applications such as marketing strategies, the customers falling in the overlap can be viewed as a special market segment with similar characteristics. Therefore, the method of overlap boundary identification by DEA may be employed to explore the potential opportunities like data mining does.

The overlap boundary formed by the benchmarks of two groups can further be used as discriminant frontier. With the idea from the characteristic of PPS,

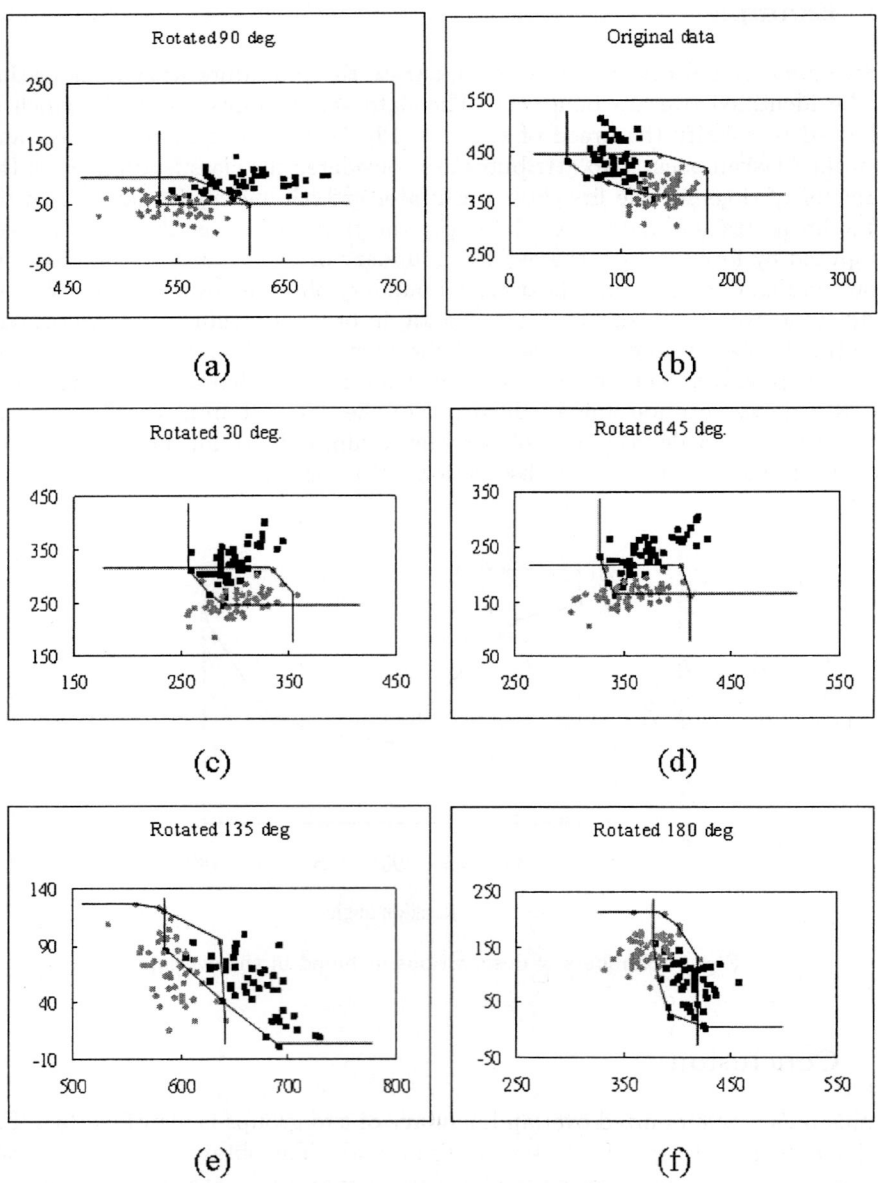

Fig. 6. Illustration of overlap boundaries identified by the data set with different rotation angle

it can establish the discriminant function by the frontier of groups rather than most of the existing approaches which minimize the total deviation of all the observations from their group mean. This clearly is an issue for future researches.

References

1. Bajgier, S. M., Hill, A.V.: An Experimental Comparison of Statistical and Linear Programming Approaches to the Discriminant Problem. Decision Sciences. 13 (1982) 604-618
2. Bhattacharyya S., Pendharkar P.C.: Inductive, Evolutionary, and Neural Computing Techniques for Discrimination: A Comparative Study. Decision Sciences. 29 (1998) 871-899
3. Billings, S.A., Lee, K.L.: Nonlinear Fisher Discriminant Analysis Using A Minimum Squared Error Cost Function and the Orthogonal Least Squares Algorithm. Neural Networks. 15 (2002) 263-270
4. Charnes A.: Data Envelopment Analysis: Theory, Methodology, and Application. Boston: Kluwer Academic Publishers (1994)
5. Charnes, A., Cooper, W.W.: Goal Programming and Multiple Objective Optimization, European Journal of Operational Research. 1 (1977) 39-54
6. Charnes, A., Cooper, W.W., Rhodes, E.: Measuring the Efficiency of Decision Making Units, European Journal of Operational Research. 2 (1978) 429-444
7. Cielen, A., Peeters, L., Vanhoof, K.: Bankruptcy Prediction Using a Data Envelopment Analysis, European Journal of Operational Research. 154 (2004) 526-532
8. Cooper W.W, Seiford L.M., Tone K.: Data Envelopment Analysis: A Comprehensive Text with Models, Applications, References, and DEA-Solver Software, Boston: Kluwer Academic Publishers (2000)
9. Fisher, R.A.: The Use of Multiple Measurements in Taxonomic Problems, Annals of Eugenics. 7 (1936) 179-188
10. Freed, N., Glover, F.: Simple But Powerful Goal Programming Models for Discriminant Problems, European Journal of Operational Research. 7 (1981) 44-66
11. Freed, N., Glover, F.: A Linear Programming Approach to the Discriminant Problem, Decision Sciences. 12 (1981) 68-74
12. Freed, N., Glover, F.: Evaluating Alternative Linear Programming Models to Solve the Two-Group Discriminant Problem. Decision Sciences. 17 (1986) 151-162
13. Johnson, R.A., Wichern, D.W.: Applied Multivariate Statistical Analysis. 3rd edn. Prentice-Hall,Englewood Cliffs New Jersey (1992) 519-521
14. Koehler, G.J., Erenguc, S.S.: Minimizing Misclassifications in Linear Discriminant Analysis. Decision Sciences. 21 (1990) 63-85
15. Lam, K.F., Moy, J.W.: An Experimental Comparison of Some Recently Developed Linear Programming Approaches to the Discriminant Problem. Journal of Computers and Operations Research. 24 (1997) 593-599
16. Lam, K.F., Moy, J.W.: A Piecewise Linear Programming Approach to the Two-Group Discriminant Problem-an Adaptation to Fisher's Linear Discriminant Function Model. European Journal of Operational Research. 145 (2003) 471-481
17. Rumelhart, D.E., Hinton, G. E., and Williams, R.J.: Learning Internal Representations by Error Propagation. In: Rumelhart D.E., McClelland J.L. (eds.): Parallel Distributed Processing: Exploration in The Microstructure of Cognition, Vol.1. Cambridge, MA: MIT Press (1986)
18. Sueyoshi, T.: DEA Discriminant Analysis in the View of Goal Programming. European Journal of Operational Research. 115 (1999) 564-582
19. Sueyoshi, T.: Extended DEA-Discriminant Analysis. European Journal of Operational Research. 131 (2001) 324-351
20. Sueyoshi, T.: Mixed Integer Programming Approach of Extended DEA-Discriminant Analysis. European Journal of Operational Research. 152 (2004)45-55

21. Sueyoshi, T., Kirihara, Y.: Efficiency Measurement and Strategic Classification of Japanese Banking Institutions. International Journal of Systems Science. 29 (1998) 1249-1263
22. Zhu, J.: Quantiative Models for Performance Evaluation and Benchmarking: Data Envelopment Analysis with Spreadsheets and DEA Excel Solver, Boston: Kluwer Academic Publishers, (2003)

Appendix

The formula (4) illustrate in the text is useful for two-dimensional case, but it is hard to extend to n dimension. For this reason, we rewrite formula (4) in another form. It is known that the rotation matrix **T** for two Cartesian coordinate systems which have the bases $[\vec{u}_1, \vec{u}_2]$ and $[\vec{u}'_1, \vec{u}'_2]$ respectively can be expressed as below:

$$\mathbf{T} = \begin{bmatrix} \vec{u}_1 \cdot \vec{u}'_1 & \vec{u}_1 \cdot \vec{u}'_2 \\ \vec{u}_2 \cdot \vec{u}'_1 & \vec{u}_2 \cdot \vec{u}'_2 \end{bmatrix}_{2 \times 2} \quad (I)$$

Four elements are included in above (2 × 2) matrix with each element being referred to as direction cosine. The direction cosine can be obtained by the dot product of different Cartesian coordinate bases. The bases are defined as:

$\vec{u}_1$ = unit vector of $[1, 1]$ = $[0.707, 0.707]$

The included angle between $\vec{u}_1$ and each Cartesian axes is 45 degrees.

$\vec{u}_2$ = null vector of $\vec{u}_1$ = $[-0.707, 0.707]$

$\vec{u}'_1$ = unit vector of $\vec{W} = \dfrac{[w_1, w_2]}{\sqrt{w_1^2 + w_2^2}}$

$\vec{u}'_2$ = null vector of $\vec{u}'_1$

The transformation matrix **T** obtained by formula (I) is equal to the matrix by formula (4). Formula (I) is more useful than formula (4) if it is applied to n-dimensional case. Assume $\vec{W} = [w_j]$, $j = 1, \cdots, n$ is generated by solving MSD, the transformation matrix **T** can be expressed as:

$$\mathbf{T} = \begin{bmatrix} \vec{u}_1 \cdot \vec{u}'_1 & \vec{u}_1 \cdot \vec{u}'_2 & \cdots & \vec{u}_1 \cdot \vec{u}'_n \\ \vec{u}_2 \cdot \vec{u}'_1 & \vec{u}_2 \cdot \vec{u}'_2 & \cdots & \vec{u}_2 \cdot \vec{u}'_n \\ \vdots & \vdots & \vdots & \vdots \\ \vec{u}_n \cdot \vec{u}'_1 & \vec{u}_n \cdot \vec{u}'_2 & \cdots & \vec{u}_n \cdot \vec{u}'_n \end{bmatrix}_{n \times n} \quad (II)$$

$\vec{u}_1$ = unit vector of $[1, 1, \cdots, 1]_{1 \times n} = \left[\dfrac{1}{\sqrt{n}}, \dfrac{1}{\sqrt{n}}, \cdots, \dfrac{1}{\sqrt{n}}\right]_{1 \times n}$

$\vec{u}_2$ = null vectors of $\vec{u}_1$

$\vec{u}'_1$ = unit vector of $\vec{W} = \dfrac{[w_1, w_2, \cdots, w_n]}{\sqrt{w_1^2 + w_2^2 + \cdots + w_n^2}}$

$\vec{u}'_2, \cdots, \vec{u}'_n$ = null vectors of $\vec{u}'_1$.

A Parallel Tabu Search Algorithm for Optimizing Multiobjective VLSI Placement

Mahmood R. Minhas and Sadiq M. Sait

College of Computer Sciences & Engineering,
King Fahd University of Petroleum & Minerals,
Dhahran-31261, Saudi Arabia
{minhas, sadiq}@ccse.kfupm.edu.sa

Abstract. In this paper, we present a parallel tabu search (TS) algorithm for efficient optimization of a constrained multiobjective VLSI standard cell placement problem. The primary purpose is to accelerate TS algorithm to reach near optimal placement solutions for large circuits. The proposed technique employs a candidate list partitioning strategy based on distribution of mutually disjoint set of moves among the slave processes. The implementation is carried out on a dedicated cluster of workstations. Experimental results using ISCAS-85/89 benchmark circuits illustrating quality and speedup trends are presented. A comparison of the obtained results is made with the results of a parallel genetic algorithm (GA) implementation.

1 Introduction and Related Work

General iterative heuristics such as tabu search and genetic algorithms (GAs) have been widely used to solve numerous hard problems [1]. This interest is attributed to their generality, ease of implementation, and ability to reach near optimal solutions by escaping from local minima. However, depending on size of a problem, such heuristics may have huge runtime requirements. This is also true for VLSI placement problem of modern industry-size circuits for which, iterative heuristics require huge run times to reach near optimal solutions [2, 3]. With rapidly increasing density of VLSI circuits, the run time dilemma of iterative techniques is aggravating and hence there is a need of accelerating their search process.

One way to adapt iterative techniques to large problems and to efficient traversing of large search space is their parallelization [4, 5]. An effective parallelization strategy must consider issues such as proper partitioning of the problem to facilitate uniform distribution of computationally intensive tasks. At the same time, it should be capable of thorough traversal of the complex search space. In the subsequent paragraphs, we present a brief review of some earlier efforts towards parallelization of TS.

A number of parallelization techniques have been reported in literature [6]. A taxonomy of parallel tabu search strategies was given by Crainic et. al [7]. In the most straightforward and widely adopted approach, k tabu search processes are spawned and run concurrently on k processors where each processor carries out independent search [6, 8]. Malek et al. suggested linking independent searches where each slave runs a copy of a serial TS but with different parameter settings [9]. After specified time intervals, all the slave processes are halted and the master process selects the overall best solution found so far and broadcasts it to all the slave processes. Each slave process then restarts its search process from the best solution it receives from the master.

Another proposed approach is to parallelize search process within an iteration of TS. In this approach, each process is given the task of exploring a subset of the neighborhood of the current solution. Here two approaches are followed: *synchronous* and *asynchronous*. In the synchronous approach, various processes are always working with the same solution but exploring different partitions of the neighborhood. The master process orchestrates the activities of the slave processes [8]. In the asynchronous approach, all processes are peer and are usually not working with the same current solution [6]. Both of these approaches require that the set of possible moves be partitioned among the available processors so that each processor can explore a distinct sub-region of the neighborhood.

Some other suggestions for efficient parallelization of TS include partitioning the search space, or partitioning the problem into smaller sub-problems with determining the best moves for each sub-problem, and then performing a compound move [6].

Attempts to solve some hard optimization problems have also been reported in literature. Parallel tabu search algorithms for the vehicle routing problem are presented in [8, 10]. A massively parallel implementation of tabu search for the Quadratic Assignment Problem (QAP) is reported in [11]. The parallel algorithm was implemented on a Connection Machine CM-2 (a massively parallel SIMD machine). A reduction in runtime per iteration was reported when compared to some other sequential and parallel implementations [11, 12].

In this paper, we present a parallel TS algorithm for efficient optimization of a hard optimization problem namely, constrained multiobjective VLSI standard cell placement. The rest of the paper is organized as follows: The next section briefly discusses the placement problem and the related cost functions. In Section 3, some implementation details of the proposed parallel tabu search algorithm are presented, followed by experimental results and comparisons in Section 4. Section 5 presents some concluding remarks.

2   The Placement Problem and Cost Functions

We address a multiobjective VLSI standard cell placement problem where the objectives are optimization of and total wirelength, power consumption, and timing performance (delay), whereas layout width is considered as a constraint. Cell placement is one of the intermediate steps in VLSI physical design and is a generalization of QAP.

The VLSI cell placement problem can be stated as follows [13]: Given a collection of cells or modules with ports (inputs, outputs, power and ground pins) on the boundaries, the dimensions of these cells (height, width, etc), and a collection

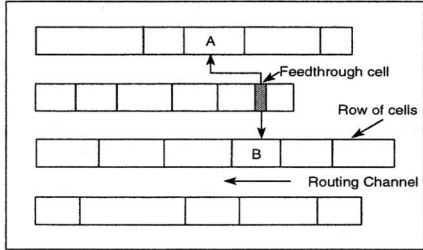

Fig. 1. A Typical Standard Cell Layout

of nets (which are sets of ports that are to be wired together), *placement* problem consists of finding suitable physical locations for each cell on the layout. By suitable we mean those locations that minimize given objective functions, subject to some constraints imposed by the designer, the implementation process, or layout strategy and the design style.

In this work, we deal with standard cell placement, where all the circuit cells are constrained to have the same height, while the width of the cell is variable and depends upon its complexity. A typical standard cell layout is shown in Figure 1. As can be seen in the figure, cells are arranged in rows with routing channels between the rows. Due to varying width of cells, row widths may be unequal depending on the type and number of cells placed in a row. An approximation would be to treat cells as points, but in order to have a more accurate estimate of wirelength, widths of cells are taken into account. Heights of routing channels are estimated using the vertical constraint graphs constructed during the channel routing phase. With this information, a fairly accurate estimate of power dissipation, delay and total wirelength can be obtained [13]. Next, we formulate the cost function used in our optimization process.

2.1 Cost Functions

Now we formulate cost functions for our three objectives and for the width constraint.

Wirelength Cost: Interconnect wirelength of each net in the circuit is estimated using steiner tree approximation. Total wirelength is computed by adding all these individual estimates:

$$Cost_{wire} = \sum_{i \in M} l_i \qquad (1)$$

where l_i is the wirelength estimation for net i and M denotes total number of nets in circuit.

Power Cost: Power consumption p_i of a net i in a circuit can be given as:

$$p_i \simeq C_i \cdot S_i \qquad (2)$$

where C_i is total capacitance of net i, and S_i is the switching probability of net i. Here, C_i depends on wirelength of net i, hence Equation 2 can be written as:

$$p_i \simeq l_i \cdot S_i \tag{3}$$

The cost function for total power consumption in the circuit can be given as:

$$Cost_{power} = \sum_{i \in M} p_i = \sum_{i \in M} (l_i \cdot S_i) \tag{4}$$

Delay Cost: The delay cost is determined by computing delay along the longest path in a circuit. The delay of any given path is computed as summation of the delays of all the nets belonging to the path and the switching delays of the cells driving these nets. Hence, the delay T_π of a path π consisting of nets $\{1, 2, ..., k\}$ is expressed as:

$$T_\pi = \sum_{i=1}^{k-1} (CD_i + ID_i) \tag{5}$$

where CD_i is the switching delay of the cell driving net i and ID_i is the interconnect delay of net i. The placement phase affects ID_i only because CD_i is technology dependent and is hence independent of the placement process. The cost function for delay of the circuit can be given as:

$$Cost_{delay} = max\{T_\pi\} \tag{6}$$

Width Cost: Width cost is given by the maximum of all the row widths in the layout. We have constrained layout width not to exceed a certain positive ratio α to the average row width w_{avg}, where w_{avg} is the minimum possible layout width obtained by dividing the total width of all the cells in the layout by the number of rows in the layout. We can express width constraint as below:

$$Width - w_{avg} \leq \alpha \times w_{avg} \tag{7}$$

2.2 Fuzzification of Multiobjectives

Since we are targeting to optimize three (possibly conflicting) objectives simultaneously, we need to formulate an aggregating cost function which can expresses the costs of the objectives in form of a single quantity. We resorted to using fuzzy in designing our aggregating cost function. Fuzzy logic allows to describe the objectives in terms of linguistic variables following which, the membership functions can be defined. The membership functions for *SMALL wirelength*, *LOW power consumption*, and *SHORT delay* are same as described in [3]. Finally, fuzzy rules are used to design the aggregate cost function. In this work, we have used the following fuzzy rule:

Rule 1: IF a solution has *SMALL wirelength* **AND** *LOW power consumption* **AND** *SHORT delay* **THEN** it is an *GOOD* solution.

The above rule is translated to *and-like* OWA fuzzy operator [14] and the membership $\mu(x)$ of a solution x in fuzzy set *GOOD solution* is obtained by:

$$\mu(x) = \begin{cases} \beta \cdot \min_{j=l,p,d}\{\mu_j(x)\} + (1-\beta) \cdot \frac{1}{3}\sum_{j=l,p,d} \mu_j(x); \\ \qquad \text{if } Width - w_{avg} \leq \alpha \cdot w_{avg}, \\ 0; \qquad \text{otherwise.} \end{cases} \quad (8)$$

Here $\mu_j(x)$ for $j = \{l, p, d\}$ are the membership values in the fuzzy sets *SMALL wirelength*, *LOW power consumption*, and *SHORT delay* respectively, whereas β is a constant in range $[0, 1]$. A placement solution that results in a higher value of $\mu(x)$ is considered a better solution.

3 Parallel Tabu Search Algorithm for Optimizing VLSI Placement

The parallel tabu search strategy adopted in this work was engineered after a careful performance analysis of our sequential TS implementation [3]. The analysis was performed using some profiling tools (like GNU profiler) to obtain insight into determining the time consuming operations of the code and the usage of resources. For the circuits experimented on, it was found that almost 60% to 80% of the total run time was spent on cost computation of the three objectives and their fuzzification. Furthermore, experiments with parameters revealed that for our hard optimization problem with conflicting multiobjectives, large sizes of candidate list (upto 120) were required to obtain high quality solutions. Since cost computation for all moves in the candidate list was the most time consuming operation, the proposed algorithm was designed with a view of partitioning this compute-intensive task. The proposed algorithm employs a candidate list partitioning strategy based on distribution of mutually disjoint set of moves among the slave processes. The pseudo code of the master and the slave processes in the proposed parallel TS are shown in Figures 2 and 3 respectively.

According to taxonomy given by Crainic et. al [7], our approach can be classified as a synchronous master-slave (one master and remaining slaves), 1-control (each process is responsible for its search), Rigid Synchronous (RS) (all processes are forced to establish communication and exchange information at specific points) and Single Point Single Strategy (SPSS) (all the processes start with the same initial solution and follow the same strategy).

In this implementation, there is an initialization step during which, the master process generates and sends an initial solution and a disjoint (non-overlapping) partial candidate list (PCL) to each slave process. A move in a PCL assigned to a slave in a particular iteration does not appear in PCLs assigned to other slaves. Each slave process searches its local neighborhood by trying each move in the partial candidate list on the initial solution and computes gains due to them. Then it sends the best move and its corresponding cost (or gain) to the

Algorithm. MasterProcess;
Begin
  (* S_0 is the initial solution. *)
  (* $BestS$ is the best solution. *)
  (* PCL is the Partial Candidate List. *)
  (* p is the number of slave processors. *)
  (* OBM is the Overall Best Move. *)
  Generate S_0 and p number of $PCLs$;
    Send S_0 and a PCL to each slave process;
    While $iteration\text{-}count < max\text{-}iterations$
      Receive best move and cost from each slave;
        Find OBM subject to tabu restrictions;
        Generate P number of $PCLs$;
      Send OBM and a PCL to each slave process;
      Update $BestS$; /*by applying OBM on $BestS$*/;
    EndWhile
  Return ($BestS$)
End. /*MasterProcess*/

Fig. 2. The master process in parallel TS

Algorithm. SlaveProcess;
Begin
  Receive S_0 and a PCL from the master process;
  $CurS = S_0$; (* Current Solution *)
  While $iteration\text{-}count < max\text{-}iterations$
    Try each move in PCL and compute cost;
    Send the best move and its cost to the master process;
    Receive OBM and a PCL from the master process;
    Update $CurS$ /* by applying OBM on $CurS$ */;
  EndWhile
End. /*SlaveProcess*/

Fig. 3. The slave process in parallel TS

master process. The master process selects the overall best move (OBM) among the moves it received from slave processes subject to tabu restrictions. Then in each subsequent iteration, the master process sends the overall best move and a new partial candidate list to each slave process. Each slave process now starts by performing the received overall best move so that all the slave processes start their iteration from the same solution. Each slave process searches its local neighborhood and sends the best move and its cost to the master process.

4 Experimental Results and Discussions

The experimental setup consists of the a homogeneous cluster of 8 machines (where 1 machine is always working as a master processor), x86 architecture, Pentium-4 of 2 GHz clock speed, and 256 MB of memory. These machines are connected by 100Mbit/s Ethernet switch. Operating system used in RedHat Linux 7.3 (kernel 2.4.7-10). The paradigm used for parallelization is MPI (Message Passing Interface). Specifically, MPICH (a portable implementation of MPI standard 1.1) is used in our implementation. In terms of GFlops measure, the maximum performance of the cluster, with NAS Parallel Benchmarks was found to be 1.6 GFlops, (using NAS's LU, Class A, for 8 processors). Using this same benchmark for a single processor, the individual performance of one machine was found out to be 0.3 GFlops. The maximum bandwidth that was achieved using PMB was 91.12 Mbits/sec, with an average latency of 68.69 μsec per message. ISCAS-85/89 circuits are used as performance benchmarks for evaluating the proposed parallel TS placement technique. These circuits are of various sizes in terms of number of cells and paths, and thus offer a variety of test cases.

For comparison purposes, we also implemented a parallel genetic algorithm (GA) which is a derivative of a standard distributed GA and follows the island model, with independently evolving sub-populations and periodic exchanges of solutions through migration [15, 16]. A pseudo-diversity approach is taken, wherein similar solutions are not permitted in the population at any time. This diversity serves to widen the search, while limiting the possibility of premature convergence in local minima solution space. The initial population is constructed at the master process and distributed among N slave processes which start running serial GA on their allocated population for a predefined number of iterations called the Migration Frequency (MF). Then each slave process sends MR (Migration Rate) number of its best solutions to the master process, which selects MR overall best solutions and broadcasts them to all slave processes. Each slave process absorbs the incoming best solutions into its population (if they are not already found) by replacing the weakest solutions. Each slave process then continues with the serial GA for another MF iterations. Standard PMX crossover is used to generate offsprings [1].

The quality of solution obtained and run time required using different number of processors for both TS and GA are tabulated in Table 1. For each circuit, the number of cells are given in the table. The '$\mu(s)$ TS' and '$\mu(s)$ GA' columns show the aggregate fuzzy membership of solution obtained by TS and by GA respectively, whereas 'p' denotes the number of processors used. It should be noted that run times shown are for achieving a certain fixed quality.

In case of large circuits, parallel GA was unable to find a reasonable quality solution even after running for a large amount of time. Even for smaller circuits, the solution quality obtained using TS is significantly superior to that obtained using GA, and also the speedup trend is very consistent for TS. On the other hand, parallel GA did not show such performance or trend.

The proposed parallel TS has shown a consistent trend in terms of speedup with increasing number of processors. Figure 4 shows a run-time as well as a speedup plot of parallel TS for some selected large circuits, and demonstrates an almost linear speedup. As can be seen, there is a consistent decreasing trend in run time. This trend is more pronounced for medium to large circuits than for smaller ones and reveals the good scalability of the proposed approach.

Table 1. Run times and solution quality $\mu(s)$ for achieving a target membership for serial and parallel TS/GA approaches. X indicates unreasonably high run time requirement

Circuit Name	Number of Cells	$\mu(s)$ TS	Time for Serial TS	Time for Parallel TS						$\mu(s)$ GA	Time for Serial GA	Time for Parallel GA		
				p=2	p=3	p=4	p=5	p=6	p=7			p=3	p=5	p=7
s386	172	0.688	52	28	20	17	16	15	14	0.504	15	9.9	5.7	6.7
s641	433	0.785	934	472	332	239	205	171	151	0.616	793	307	390	289
s832	310	0.644	74	40	33	23	22	20	19	0.479	128	43	37	39
s953	440	0.661	195	98	71	53	46	41	36	0.511	309	136	91	108
s1196	561	0.653	374	187	132	97	88	78	67	0.484	988	327	262	205
s1488	667	0.603	259	131	93	69	63	55	49	0.482	1883	677	435	418
s1494	661	0.601	268	137	96	72	65	57	51	0.496	1405	847	638	479
c3540	1753	0.665	2142	1146	703	547	440	370	344	-	X	X	X	X
s3330	1961	0.699	1186	590	451	313	245	210	184	-	X	X	X	X
c5378	2993	0.669	1850	914	601	467	371	312	264	-	X	X	X	X
s9234	5844	0.631	5571	2855	2006	1525	1272	1062	849	-	X	X	X	X

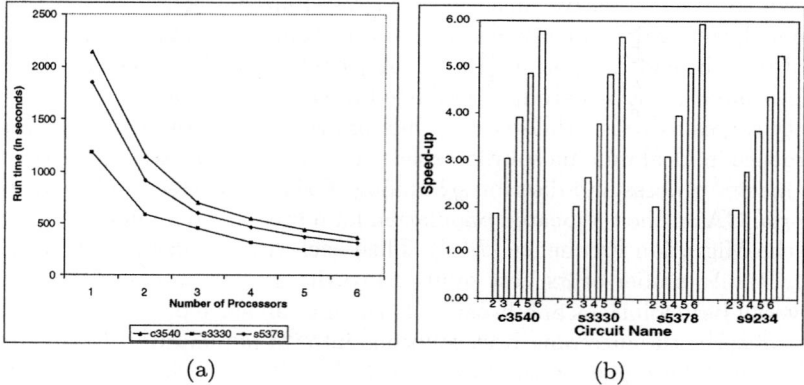

(a) (b)

Fig. 4. (a) Decrease in run times for selected large circuits with increasing number of processors for parallel TS. (b) Speedup obtained for selected large circuits for parallel TS

5 Conclusions

In this work, we presented a parallel tabu search strategy for accelerating the solution to a constrained multiobjective VLSI cell placement problem. The proposed strategy belongs to p-control, RS, MPSS class and was implemented on a dedicated cluster of workstations. A distributed parallel GA was also implemented for the comparison purposes. Experimental results on ISCAS-85/89 benchmarks exhibit that the proposed parallel TS shows an excellent trend in terms of speedup and requires far lesser run times as compared to serial TS for obtaining the same quality of placement solutions. When compared with results obtained by parallel GA, the proposed parallel TS clearly outperforms both in terms of solution quality as well as run time.

Acknowledgment

The authors would like to thank King Fahd University of Petroleum & Minerals, Dhahran, Saudi Arabia, for support under project code # COE/CELLPLACE/263.

References

1. Sadiq M. Sait and Habib Youssef. *Iterative Computer Algorithms and their Application to Engineering*. IEEE Computer Society Press, December 1999.
2. Sadiq M. Sait, Habib Youssef, Aiman El-Maleh, and Mahmood R. Minhas. Iterative Heuristics for Multiobjective VLSI Standard Cell Placement. *Proceedings of IJCNN'01, International Joint Conference on Neural Networks*, 3:2224–2229, July 2001.
3. Sadiq M. Sait, Mahmood R. Minhas, and Junaid A. Khan. Performance and Low-Power Driven VLSI Standard Cell Placement using Tabu Search. *Proceedings of the 2002 Congress on Evolutionary Computation*, 1:372–377, May 2002.
4. Prithviraj Banerjee. *Parallel Algorithms for VLSI Computer-Aided Design*. Prentice Hall International, 1994.
5. Van-Dat Cung, Simone L. Martins, Celso C. Riberio, and Catherine Roucairol. Strategies for the Parallel Implementation of metaheuristics. *Essays and Surveys in Metaheuristics*, pages 263–308, Kluwer 2001.
6. I. De Falco, R. Del Balio, E. Tarantino, and R. Vaccaro. Improving Search by Incorporating Evolution Principles in Parallel Tabu Search. In *Proc. of the first IEEE Conference on Evolutionary Computation- CEC'94*, volume 1, pages 823–828, June 1994.
7. T. G. Crainic, M. Toulouse, and M. Gendreau. Towards a Taxonomy of Parallel Tabu Search Heuristics. *INFORMS Journal of Computing*, 9(1):61–72, 1997.
8. Bruno-Laurent Garica, Jean-Yves Potvin, and Jean-Marc Rousseau. A Parallel Implementation of the Tabu Search Heuristic for Vehicle Routing Problems with Time Window Constraints. *Computers & Operations Research*, 21(9):1025–1033, November 1994.
9. M. Malek, M. Guruswamy, M. Pandya, and H. Owens. Serial and parallel Simulated Annealing and Tabu Search algorithms for the Traveling Salesman Problem. *Annals of Ops. Res.*, 21:59–84, 1989.
10. E. Taillard. Parallel Iterative Search Methods for the Vehicle Routing Problem. *Networks*, 23:661–673, 1993.
11. J. Chakrapani and J. Skorin-Kapov. Massively Parallel Tabu Search for the Quadratic Assignment Problem. *Annals of Operations Research*, 41:327–341, 1993.
12. E. Taillard. Robust Tabu Search for the Quadratic Assignment Problem. *Parallel Computing*, 17:443–455, 1991.
13. Sadiq M. Sait and Habib Youssef. *VLSI Physical Design Automation: Theory and Practice*. World Scientific, Singapore, 2001.
14. Ronald R. Yager. On Ordered Weighted Averaging Aggregation Operators in Multicriteria Decision Making. *IEEE Transaction on Systems, MAN, and Cybernetics*, 18(1), January 1988.
15. M. Toulouse, T. G. Crainic, and M. Gendreau. Issues in Designing Parallel and Distributed Search Algorithms for Discrete Optimization Problems. *Publication CRT-96-36, Centre de recherche sur les transports, Université de Montréal, Montréal, Canada*, 1996.
16. Erick Cant-Paz. A Survey of Parallel Genetic Algorithms. *Calculateurs Parallles, Reseaux et Systems Repartis*, 1998.

A Coupled Gradient Network Approach for the Multi-machine Earliness and Tardiness Scheduling Problem

Derya Eren Akyol and G. Mirac Bayhan

Department of Industrial Engineering, University of Dokuz Eylul,
35100 Bornova-Izmir, TURKEY
{derya.eren, mirac.bayhan}@deu.edu.tr

Abstract. This paper considers the earliness and tardiness problem of sequencing a set of independent jobs on non-identical multi-machines, and explores the use of artificial neural networks as a valid alternative to the traditional scheduling approaches. A coupled gradient network approach is employed to provide a shop scheduling analysis framework. The methodology is based on a penalty function approach used to construct the appropriate energy function and a gradient type network. The mathematical formulation of the problem is firstly presented and six coupled gradient networks are constructed to model the mixed nature of the problem. After the network architecture and the energy function were specified, the dynamics are defined by steepest gradient descent algorithm.

1 Introduction

Although in its earliest version the earliness and tardiness (ET) problem was called the minimum weighted absolute deviation problem, since about 1990 it has been commonly denominated the ET problem [1]. Simply, the problem is to schedule N jobs on a single machine to minimize the sum of the weighted differences between job completion times and due dates. Interestingly, objective of the ET problem fits perfectly to Just In Time (JIT) production control policy where an early or a late delivery of an order causes an increase in the production costs. In general, the early cost can be regarded as a holding cost for finished goods, deterioration of perishable goods and opportunity costs. The tardy cost can be considered as the backlogging cost that includes performance penalties, lost sales and lost goodwill. In other words, the total tardiness criterion measures the conformance to due dates but ignores the consequences of jobs completing early, and penalizes only the jobs that finish late. With the growing interest in JIT scheduling, jobs that complete before their due dates are also penalized.

In the literature, much effort has been devoted to earliness and tardiness scheduling models. For a detailed survey of the earlier applications see [2] and [3]. Although the single machine scheduling problems with earliness and tardiness penalties have been studied extensively, there are only a few published articles addressing the problem of scheduling jobs on multiple machines. [4], [5] and [6] are among the first researchers interested in the problem of scheduling N jobs on M identical parallel machines with the objective of minimizing the total penalty costs for earliness and tardiness. In [7], the problem of scheduling N jobs with sequence dependent setup times on identical parallel machines to minimize the sum of weighted earliness and tardiness is studied and a heuristic algorithm is proposed to solve it. For the same objective, two genetic algorithm approaches to the parallel machine scheduling problem including sequence dependent setup times are presented [8], and in [9] a compact mixed integer formulation for scheduling jobs on uniform parallel machines is proposed and different solution approaches to solve small sized problems are also presented. In another study [10], simulated annealing to the sequence dependent setup time scheduling problem is employed to obtain near optimal solutions. An exact procedure [11] is presented for scheduling N jobs on non identical machines to minimize the sum of earliness and tardiness penalties. The performance of two metaheuristic approaches proposed to solve the identical parallel machine scheduling problem with sequence dependent setup times are compared [12]. Following this work, a dynamic programming algorithm and two heuristics to minimize the total weighted earliness and tardiness for identical parallel machine scheduling problem are proposed [13].

The major distinction of our approach is that it utilizes the neural networks to tackle with the problem of scheduling a set of independent jobs including sequence dependent setup times, on non-identical multiple machines to minimize the total weighted earliness and tardiness. Thus, this study will be the first attempt to solve the early/tardy problems in the area of production scheduling using neural networks. It involves two stages, which may be followed sequentially: (1) Mathematical model presentation stage, (2) Gradient network design stage (a) architecture and (b) energy function construction.

2 Problem Formulation

This paper attempts to deal with the problem of job scheduling on non-identical multiple machines to minimize the total penalty costs for earliness and tardiness with sequence dependent setup times. The problem includes non-common due dates and non-uniform cost penalties. We are given N independent jobs $J = J_1, , J_N$ to be scheduled on M non-identical machines. Let X_i be the completion time of job i. Job i is early if $X_i < d_i$; it is tardy if $X_i > d_i$; and it is on time if $X_i = d_i$. The problem of scheduling N jobs with non common due dates, varying processing times and sequence dependent setup times on M machines is NP-hard. Therefore heuristic approaches have been developed to obtain good near optimal solutions.

In this study, we use the notation of $N/M/ET$ with sequence-dependent setup times to designate this problem. The notation for the problem is as follows:

β: a large number
d_i: due date for job i

e_i: penalty per unit earliness of job i
E_i: earliness of job i
P_{im}: processing time for job i using machine m
M: number of machines available
N: number of jobs to be scheduled
s_{ji}: setup time for job i when it immediately follows job j
s_{0i}: setup time for job i when it is the first in queue
t_i: penalty per unit tardiness of job i
T_i: tardiness of job i
X_i: completion time of job i
Y_{ijm}: 1 if job i precedes job j on machine m, 0 otherwise
Z_{im}: 1 if job i is processed on machine m, 0 otherwise

The multi-machine earliness-tardiness problem can be formulated by using the following mixed integer programming (MIP) model [14]. The formulation given below differs from the model given in [14] by the addition of the sixth constraint. Rather than writing the third inequality constraint for $i = 0,,N$, in our model it is written for $i = 1,,N$. The reason for making this modification in the model is correcting the model which was assigning two jobs at the same time to the first position.

Objective function:

$$min \sum_{i=1}^{N}(e_i E_i + t_i T_i)$$

Subject to:

$$X_i - T_i + E_i = d_i \quad i = 1, ..., N. \quad (1)$$

$$\sum_{m=1}^{M} Z_{im} = 1 \quad i = 1, ..., N. \quad (2)$$

$$\sum_{j=1}^{N} Y_{ijm} \leq Z_{im} \quad j \neq i, \quad i = 1, ..., N, \quad m = 1, ..., M. \quad (3)$$

$$\sum_{i=0}^{N} Y_{ijm} = Z_{jm} \quad i \neq j, \quad j = 1, ...N, \quad m = 1, ..., M. \quad (4)$$

$$X_i - X_j - \beta Y_{jim} \geq P_{im} + s_{ji} - \beta \quad j \neq i, \\ i = 1, ..., N, \quad j = 0, 1, ..., N, \quad m = 1, ..., M. \quad (5)$$

$$\sum_{i=1}^{N} Y_{0im} = 1 \quad \forall m = 1, ..., M. \quad (6)$$

All decision variables are non-negative and the binary variables Y_{ijm} and $Z_{im} \in 0, 1$. A dummy, always first and always present, *job 0* is introduced to simplify the writing of the constraints. Obviously, $X_0 = 0$. The objective is to minimize the sum of cost-weighted deviations in job completion times from the job due dates. The first constraint is a defining equation that measures the degree to which each job is tardy or early. The second constraint ensures that each job is processed on exactly one machine. The third and fourth constraints ensure that each job (but not the last scheduled job) must come immediately before, and each job (but not the first scheduled job) must come immediately after, only one other job. The fifth constraint ensures that the completion time of a job i is far enough after that of job j to include the processing time and setup time for job i, and the sixth constraint ensures the introduction of a dummy job 0 at the beginning of the sequence before all the real jobs on each machine.

3 Formulation of the Coupled Gradient Network for the Problem

In this section, the coupled gradient network, which is an extension of the original formulation given in [15] and [16] is used to represent the problem. Six coupled gradient networks that operate in parallel are constructed to obtain an optimal solution.

3.1 The Network Architecture

The coupled network will consist of six interacting recurrent networks: three networks called the E, T, X networks to represent real valued variables, and three networks called the Z, Y and YO networks to represent binary valued variables. That is, the variables $VE_1, VE_2, ..., VE_N; VT_1, VT_2, ..., VT_N; VX_1, VX_2, ..., VX_N$; and $VZ_{11}, VZ_{12}, ..., VZ_{NM}$ will be the node outputs (states) of the E network, T network, X network and the Z network, respectively. While the Y network's node outputs correspond to the variables $VY_{111}, VY_{112}, ..., VY_{NNM}$, node outputs of the YO network correspond to the variables $VY_{011}, VY_{012}, ..., VY_{0NM}$. Here, in order to make the program code easier, we use Y network to represent the Y variables for $i = 1, 2, ..., N$ and YO network to represent the dummy Y variables having the index $i = 0$.

The input to the ith node will be denoted by UE_i in the E network, by UT_i in the T network, by UX_i in the X network, by UZ_{im} in the Z network, by UY_{ijm} in the Y network, and by UY_{0jm} in the YO network. The dynamics of the coupled net will be defined in terms of these input variables.

For the nodes in the E, T and X network, the activation function will be linear. The activation function for the nodes of the Z, Y and YO network will take the usual sigmoidal form with slopes λ_Z, λ_Y and λ_{YO}.

3.2 The Energy function

Instead of using linear programming or the k-out-of-N rules to develop the energy function, we directly formulate the cost function according to the constraints term by term. The energy function for this network is constructed using a penalty function approach. That is the energy function E consists of the objective function $\sum_{i=1}^{N}(e_i E_i + t_i T_i)$ plus a penalty function to enforce the constraints. For our problem, the penalty function $P(E, T, X, Y, YO, Z)$ will include the following eight penalty terms : $P1, P2, P3, P4, P5, P6, P7$ and $P8$. Here, the first term $P1$ adds a positive penalty if the solution does not satisfy any of the equality constraints below.

$$X_i - T_i + E_i = d_i \quad i = 1, ..., N$$

In this case, $P1$ will take the following form, $P1 = \sum_{i=1}^{N}(X_i - T_i + E_i - d_i)^2$. Therefore, this term yields zero when these equality constraints are satisfied. The second term $P2$ adds a positive penalty if the solution does not satisfy any of the equality constraints $\sum_{m=1}^{M} Z_{im} = 1 \quad i = 1, ...N$
In accordance with this constraint, $P2$ will take the following form,

$$P2 = \sum_{i=1}^{N}(\sum_{m=1}^{M} Z_{im} - 1)^2$$

The third term $P3$ adds a positive penalty if the solution does not satisfy any of the inequality constraints

$$\sum_{j=1}^{N} Y_{ijm} \leq Z_{im} \quad j \neq i, \quad i = 1, ..., N, \quad m = 1, ..., M$$

$P3$ can be defined as; $P3 = \sum_{i=1}^{N} \sum_{m=1}^{M} \nu(\sum_{j=1, i \neq j}^{N} Y_{ijm} - Z_{im})$ Here, ν represents the penalty function, where $\upsilon(\varepsilon) = \varepsilon^2$ for all $\varepsilon > 0$ and $\upsilon(\varepsilon) = 0$ for all $\varepsilon \leq 0$.

The fourth term P4 adds a positive penalty if any of the equality constraints $\sum_{i=0}^{N} Y_{ijm} = Z_{jm} \quad i \neq j, \quad j = 1, ..., N, \quad m = 1, ..., M$ is violated.

P4 will take the following form; $P4 = \sum_{j=1}^{N} \sum_{m=1}^{M}(\sum_{i=0, i \neq j}^{N} Y_{ijm} - Z_{jm})^2$
The fifth term P5 adds a positive penalty if the solution does not satisfy any of the inequality constraints $X_i - X_j - \beta Y_{jim} \geq P_{im} + s_{ji} - \beta \quad j \neq i, \quad i = 1, ...N, \quad j = 0, 1, ..., N, \quad m = 1, ..., M$
P5 may be written as

$$\sum_{i=1, i \neq j}^{N} \sum_{j=0}^{N} \sum_{m=1}^{M} \nu(X_j - X_i + \beta(Y_{jim} - 1) + P_{im} + s_{ji})$$

To deal with introduction of a dummy *job* 0 at the beginning of the sequence before all the real jobs on each machine, the sixth term P6 can be defined as $P6 = \sum_{m=1}^{M}(\sum_{i=1}^{N} Y_{0im} - 1)^2$ which adds a positive penalty if any of the equality constraints $\sum_{i=1}^{N} Y_{0im} = 1 \quad \forall m = 1, ..., M$ is violated.

We require that Y_{ijm} and $Z_{im} \in 0, 1$. These constraints will be captured by the seventh and the eighth terms P7 and P8 which add a positive penalty if the binary constraints Y_{ijm} and $Z_{im} \in 0, 1$ are violated.

P7 may be written as: $\sum_{i=1}^{N} \sum_{m=1}^{M} Z_{im}(Z_{im} - 1)$ and P8 may be written as: $\sum_{i=0}^{N} \sum_{j=1}^{N} \sum_{m=1}^{M} Y_{ijm}(1 - Y_{ijm})$.

Correspondingly, the total penalty function $P(E, T, X, Y, YO, Z)$ with all constraints can be induced as follows:

$$P = \sum_{i=1}^{N}(X_i - T_i + E_i - d_i)^2 + \sum_{i=1}^{N}(\sum_{m=1}^{M} Z_{im} - 1)^2 + \sum_{i=1}^{N} \sum_{m=1}^{M} \nu(\sum_{j=1, i \neq j}^{N} Y_{ijm} - Z_{im}) +$$

$$\sum_{j=1}^{N} \sum_{m=1}^{M}(\sum_{i=0, i \neq j}^{N} Y_{ijm} - Z_{jm})^2 + \sum_{i=1}^{N} \sum_{j=0}^{N} \sum_{m=1}^{M} \nu(X_j - X_i + \beta(Y_{jim} - 1) + P_{im} + s_{ji}) +$$

$$\sum_{m=1}^{M}(\sum_{i=1}^{N} Y_{0im} - 1)^2 + \sum_{i=1}^{N} \sum_{m=1}^{M} Z_{im}(1 - Z_{im}) + \sum_{i=0}^{N} \sum_{j=1}^{N} \sum_{m=1}^{M} Y_{ijm}(1 - Y_{ijm})$$

So, the energy function for the coupled gradient network can be defined as

$$A \sum_{i=1}^{N}(e_i E_i + t_i T_i) + B \sum_{i=1}^{N}(X_i - T_i + E_i - d_i)^2 + C \sum_{i=1}^{N}(\sum_{m=1}^{M} Z_{im} - 1)^2 +$$

$$D \sum_{i=1}^{N} \sum_{m=1}^{M} \nu(\sum_{j=1, i \neq j}^{N} Y_{ijm} - Z_{im}) + E \sum_{j=1}^{N} \sum_{m=1}^{M}(\sum_{i=0, i \neq j}^{N} Y_{ijm} - Z_{jm})^2 +$$

$$F \sum_{i=1}^{N} \sum_{j=0}^{N} \sum_{m=1; i \neq j}^{M} \nu(X_j - X_i + \beta(Y_{jim} - 1) + P_{im} + s_{ji}) + G \sum_{m=1}^{M}(\sum_{i=1}^{N} Y_{0im} - 1)^2 +$$

$$H \sum_{i=1}^{N} \sum_{m=1}^{M} Z_{im}(1 - Z_{im}) + I \sum_{i=0}^{N} \sum_{j=1}^{N} \sum_{m=1}^{M} Y_{ijm}(1 - Y_{ijm}) \quad (7)$$

where A, B, C, D, E, F, G, H and I are the penalty coefficients. We may omit the seventh and the eighth penalty terms from the energy function because these constraints may be satisfied by using a sigmoidal type activation function for variables Z_{im} and Y_{ijm} in obtaining the output values. If we rewrite the energy function in terms of the output (state) variables, we may obtain

$$E(VE, VT, VX, VY, VZ) = A \sum_{i=1}^{N}(e_i VE_i + t_i VT_i) + B \sum_{i=1}^{N}(VX_i - VT_i + VE_i - d_i)^2 +$$

$$C \sum_{i=1}^{N}(\sum_{m=1}^{M} VZ_{im} - 1)^2 + D \sum_{i=1}^{N} \sum_{m=1}^{M} \nu(\sum_{j=1, i \neq j}^{N} VY_{ijm} - VZ_{im}) +$$

$$E \sum_{j=1}^{N} \sum_{m=1}^{M}(\sum_{i=0, i \neq j}^{N} VY_{ijm} - VZ_{jm})^2 + F \sum_{i=1}^{N} \sum_{j=0}^{N} \sum_{m=1; i \neq j}^{M} \nu(VX_j - VX_i + \beta(VY_{jim} - 1) +$$

$$P_{im} + s_{ji}) + G \sum_{m=1}^{M}(\sum_{i=1}^{N} VY_{0im} - 1)^2 \qquad (8)$$

3.3 The Dynamics

The dynamics for the coupled gradient network are obtained by gradient descent on the energy function. The equations of motion are obtained as follows

$$\frac{dUE_i}{dt} = -\eta_E \frac{\partial E}{\partial VE_i} = -\eta_E[Ae_i + 2B(VE_i - VX_i - VT_i - d_i)] \qquad (9)$$

$$\frac{dUT_i}{dt} = -\eta_T \frac{\partial E}{\partial VT_i} = -\eta_T[At_i + 2B(VT_i + d_i - VE_i - VX_i)] \qquad (10)$$

$$\frac{dUX_i}{dt} = -\eta_x \frac{\partial E}{\partial VX_i} = \qquad (11)$$

$$-\eta_X \left[\begin{array}{c} 2B(VX_i + VE_i - d_i - VT_i) + \\ F\sum_{j=0}^{N}\sum_{m=1}^{M} -\nu'(VX_j - VX_i + \beta(VY_{jim} - 1) + S_{ji} + P_{im}) + \\ F\sum_{i=1}^{N}\sum_{m=1}^{M} \nu'(VX_j - VX_i + \beta(VY_{jim} - 1) + S_{ji} + P_{im}) \end{array} \right]$$

$$\frac{dUZ_{im}}{dt} = -\eta_z \frac{\partial E}{\partial VZ_{im}} = \qquad (12)$$

$$-\eta_Z \left[\begin{array}{c} 2C(\sum_{m=1}^{M} VZ_{im} - 1) - D\nu'(\sum_{j=1,i\neq j}^{N} VY_{ijm} - VZ_{im}) \\ 2E(VZ_{im} - \sum_{j=0,i\neq j}^{N} VY_{jim}) \end{array} \right]$$

$$\frac{dUY_{ijm}}{dt} = -\eta_Y \frac{\partial E}{\partial VY_{ijm}} = \qquad (13)$$

$$-\eta_Y \left[\begin{array}{c} D\nu'(\sum_{j=1,i\neq j}^{N} VY_{ijm} - VZ_{im}) + 2E(\sum_{i=0,i\neq j}^{N} VY_{ijm} - VZ_{jm}) \\ +\beta\nu'((\beta(VY_{ijm} - 1) + VX_i - VX_j + S_{ij} + P_{jm})) \end{array} \right]$$

$$\frac{dUY_{0jm}}{dt} = -\eta_{Y0} \frac{\partial E}{\partial VY_{0jm}} \qquad (14)$$

$$= -\eta_{Y0} \left[\begin{array}{c} 2E[(VY_{0jm}) + \sum_{i=1,i\neq j}^{N} VY_{ijm} - VZ_{jm}] \\ +\beta\nu'[\beta(VY_{0jm} - 1) - VX_j + S_{0j} + P_{jm}] \end{array} \right]$$

where η_E, η_T, η_X, η_Z, η_Y and η_{Y0} are positive coefficients which will be used to scale the dynamics of the six networks and ν' is the derivative of the penalty function ν. $\nu'(\varepsilon) = 2\varepsilon$ for all $\varepsilon > 0$ and $\nu'(\varepsilon) = 0$ for all $\varepsilon \leq 0$ The states of the neurons are updated at iteration k as follows.

$$UE_i^k = UE_i^{k-1} - \eta_E \frac{\partial E}{\partial VE_i} \quad (15)$$

$$UT_i^k = UT_i^{k-1} - \eta_T \frac{\partial E}{\partial VT_i} \quad (16)$$

$$UX_i^k = UX_i^{k-1} - \eta_X \frac{\partial E}{\partial VX_i} \quad (17)$$

$$UY_{ijm}^k = UY_{ijm}^{k-1} - \eta_Y \frac{\partial E}{\partial VY_{ijm}} \quad (18)$$

$$UY_{0jm}^k = UY_{0jm}^{k-1} - \eta_{Y0} \frac{\partial E}{\partial VY_{0jm}} \quad (19)$$

$$UZ_{im}^k = UZ_{im}^{k-1} - \eta_Z \frac{\partial E}{\partial VZ_{im}} \quad (20)$$

Neuron outputs can be calculated by $V = f(U)$, where $f(.)$ is the activation function, U is the input and V is the output of the neuron.

3.4 Selection of the Parameters

In order to simulate the coupled gradient networks for our ET problem, some parameters should be determined by trial and error. These are the penalty coefficients A, B, C, D, E, F and G; the activation function slopes λ_Z, λ_Y and λ_{YO}; the step sizes $\eta_E, \eta_T, \eta_X, \eta_Z, \eta_Y, \eta_{YO}$, and initial conditions for the input and output variables. So, we should assign initial values for $UE_i^0, UT_i^0, UX_i^0, UZ_{im}^0, UY_{ijm}^0, VE_i^0, VT_i^0, VX_i^0, VZ_{im}^0, VY_{ijm}^0$ for all $i = 1, 2, ..., N$; $j = 1, 2, ..., N$; $m = 1, 2, ..., M$, also for UY_{0jm}^0 and VY_{0jm}^0 for all $j = 1, 2, ...N; m = 1, 2, ...M$.

Because there is no theoretically established method for choosing the values of the penalty coefficients for an arbitrary optimization problem, the appropriate values for these coefficients can be determined by empirically running simulations and observing the optimality and/or feasibility of the resulting equilibrium points of the system [17]. The network can be initialized to small random values, and from its initialized state synchronous updating of the network will then allow a minimum energy state to be attained. The binary constraints Y_{ijm} and $Z_{im} \in 0, 1$ can be satisfied by increasing the activation slopes. In order to ensure smooth convergence, the selection of step size must be done carefully.

The dynamics of the coupled gradient network will converge to local minima of the energy function E. Since the energy function includes seven terms, each of which is competing to be minimized, there are many local minima and a tradeoff exists be-tween the terms to be minimized. An infeasible solution may be obtained when at least one of the constraint penalty terms is non-zero. In this case, the objective function term will generally be quite small but the solution will not be feasible. Alternatively, even if all constraints are satisfied, a local minimum which causes a feasible solution but not good may be encountered. In order to satisfy each penalty term, its associated penalty parameter can be increased. But this causes an increase in other penalty terms and a tradeoff occurs. The optimal values of the penalty parameters should be found that result a feasible and a good solution which minimizes the objective function [18].

4 Conclusion

With the growing interest in JIT manufacturing, it is recognized that early and tardy jobs incur costs. Therefore, both earliness and tardiness minimization must be considered in the objective of a schedule. Although a large body of literature exists for solving single machine scheduling problems involving earliness and tardiness penalties, there are few papers aim to minimize the sum of weighted earliness and tardiness, and dealing with non-identical multi machine scheduling problems involving sequence dependent setup times and distinct due dates. To the best of our knowledge, there is no previously published article that tried to solve this NP-hard problem using neural networks. So, we believe that this attempt to solve the non-identical multi machine scheduling problem including sequence dependent setups will make a contribution to the literature.

In this paper, we tried to describe the general methodology of constructing the coupled network consisting of six recurrent networks, and explained how to formulate our constrained problem as an unconstrained minimization problem. Then, equations of motion are given. The steps of the solution procedure and parameter selection process are explained. The future directions for this study are to run simulations to find feasible and good solutions, to test the performance of the proposed method on different size of scheduling models and to compare the results with those of a standard linear programming (LP) solver. Other future directions are to improve the performance of the proposed method by including time varying penalty coefficients, in order to overcome the tradeoff problem.

References

1. Ahmed, M.U., Sundararaghavan, P.S.: Minimizing the weighted sum of late and early completion penalties in a single machine. IIE Transactions, 22 (1990) 288-290
2. Rachavachari, M.: Scheduling problems with non-regular penalty functions - a review. Opsearch. 25 (1988) 144-164
3. Baker, K.R.., Scudder, G.D.: Sequencing with earliness and tardiness penalties: a review. Operations Research. 38 (1990) 22-36
4. Arkin, E., Roundy, R.O.: Weighted-tardiness scheduling on parallel machines with proportional weights. Operations Research. 39 (1991) 64-81
5. De, P., Ghosh, J.B., Wells, C.E.: Due dates and early/tardy scheduling on identical parallel machines. Naval Research Logistics. 41 (1994) 17-32
6. Sundararaghavan, P., Ahmed, M.U.: Minimizing the sum of absolute lateness in single-machine and multimachine scheduling. Naval Research Logistics Quarterly. 31 (1984) 25-33
7. Zhu, Z., Heady, R.: Minimizing the Sum of Job Earliness and Tardiness in a Multimachine System. International Journal of Production Research. 36 (1998) 1619-1632

8. Sivrikaya-Serifoglu, F., Ulusoy, G.: Parallel machine scheduling with earliness and tardiness penalties. Computers & Operations Research. 26 (1999) 773-787
9. Balakrishan, N., Kanet, J.J., Sridharan, S'. V.: Early/tardy scheduling with sequence dependent setups on uniform parallel machines. Computers & Operations Research. 26 (1999) 127-141
10. Radhakrishnan, S., Ventura, J.A.: Simulated annealing for parallel machine scheduling with earliness-tardiness penalties and sequence dependent setup times. International Journal of Operational Research. 8 (2000) 2233-2252
11. Croce, F.D., Trubian, M.: Optimal idle time insertion in early-tardy parallel machines scheduling with precedence constraints. Production Planning & Control. 13 (2002) 133-142
12. Mendes, A.S., Mller, F.M., Frana, P.M., Moscato, P.: Comparing meta-heuristic approaches for parallel machine scheduling problems. Production Planning & Control. 13 (2002) 143-154
13. Sun, H., Wang, G.: Parallel machine earliness and tardiness scheduling with proportional weights. Computers & Operations Research. 30 (2003) 801-808
14. Zhu, Z., Heady, R.B.: Minimizing the sum of earliness/tardiness in multi-machine scheduling: a mixed integer programming appraoch. Computers & Industrial Engineering. 38 (2000) 297-305
15. Hopfield, J.: Neurons with graded response have collective computational properties like those of two-state neurons. Proceedings of the National Academy of Sciences of the USA. 81 (1984) 3088-3092
16. Hopfield, J., Tank, T.W.: Neural computation of decisions in optimization problems. Biological Cybernetics. 52 (1985) 141-152
17. Watta, P.B.: A coupled gradient network approach for static and temporal mixed-integer optimization. IEEE Transactions on Neural Networks, 7 (1996) 578-593
18. Smith, K.: Neural Networks for Combinatorial Optimization: A Review of More Than a Decade of Research. Informs Journal on Computing. 11 (1999) 15-34

An Analytic Model for Correlated Traffics in Computer-Communication Networks

Si-Yeong Lim and Sun Hur

Department of Industrial Engineering,
Hanyang Univ., Korea
hursun@hanyang.ac.kr

Abstract. It is well known that the traffic in computer-communication systems is autocorrelated and the correlation makes an great effect on the performances. So it is important to study the correlated arrival process to better estimate the performances at the system and Markov renewal process is considered here to model the autocorrelated arrival stream. We derive the expected number of packets at arbitrary epoch and expected delay time using supplementary variable method.

1 Introduction

Queueing theory is a good tool for measuring the performance of computer-communication networks. Researchers have considered Poisson process to be an proper representation for network traffic in real systems, which is based upon the independence assumption in the traffics and made the computation involved substantially simple. But as the long-range dependency(LRD) and self-similar properties seem to be adequate to model the traffic characteristics in many recent literatures, the independency assumptions in the traffic should be doubted[1]. As an example, it is known that the distribution of packet interarrivals clearly differs from exponential, especially in the local area and wide area network traffic[2]. In addition, the behavior of a time-dependent process showed statistically significant correlations across large time scales(LRD). And recent studies on networks empirically observed that aggregate packet flows were statistically self-similar in nature, i.e., the statistical properties of the aggregate network traffic remain the same over an extremely wide rage of time scales or over all time scales[1]. From that point of time, many researchers began to study how they could represent the real traffic of computer communication systems. But the description of the real traffic by means of LRD could not quantify the parameters definitively, but estimate only, and even worse, the estimated parameters are often known to produce conflicting results [1]. From this point of view, it is required to introduce other random processes which can represent the autocorrelation characteristics and are mathematically manageable as well. Some arrival processes such as MMPP(Markov modulated Poisson process) and Markovian Arrival process(MAP) appear to play an important role to model the real network traffic because they have Markovian property and correlations (MAP were introduced first by [3] and [4]). But the analyses of these processes often lead to matrix-geometric forms and it was difficult to derive the parameters needed from the real traffic. Another correlated process with Markovian property is Markov renewal process(MR). [5] showed that the traffics in a queue with instantaneous Bernoulli feedback, and Palm's overflow traffic are Markov renewal processes.

Markov renewal process(MRP) is a quite general random process which has many well known processes such as Poisson process, Markov process, and renewal process as special cases. It has correlation in itself so we can use MRP as the correlated arrival process to a queueing system. When we regard a queue of which the arrival stream is correlated as an independent GI/G/1 system, we cannot take the effect of correlations embedded in the traffics into consideration and so may seriously under-estimate the performance measures of the systems [6][7]. [6] obtained the expected number of customers, mean waiting time and sojourn time at arrival epoch for the model having MR arrival and exponential service. [8] obtained the distribution of the number of customers at arrival epoch and lag-r correlations of arrival intervals and showed the effect of correlation on the number of customers in the steady state. [9] studied the effect that the transition matrix has on the waiting time of nth customer as well as on the stationary waiting time. He extended the results of [7] by taking general service times, but only examined the relations from the stochastic ordering point of view. Besides, [10] studied the correlation's effect on the system performances using simulations. [11], [12] and [13] studied the queueing systems where interarrival times and service times are correlated.

In this study, we analyze the performances of a single server queueing system which has a general service time distribution and a Markov renewal process as its arrival process. We obtain the probability distribution and the expected number of customers at arbitrary time using supplementary variable method. And from this, using Little's theorem, we obtain the expected waiting time. These results could shed light on the better estimation of the performance measures of the computer communication networks whose internal traffic is non-renewal which is more common in the real systems.

2   Notations and System Equations

In this study, we made an correlated arrival process to clarify the effect of correlation between arrivals on the queueing performance. And we manifested the effect of correlation without the notion of LRD and self-similarity despite its simplicity. For brevity of the paper we do not provide the definition and characteristics of the MRP here. One can find the detailed explanation on the MRP in any textbooks, for example, see [14]. We assume that the number of states(or types) of the underlying Markov chain is two. Then the semi-Markov kernel, $A(x)$, of the Markov renewal arrival process can be written

$$A(x) = [A_{ij}] = \begin{bmatrix} p_{11}F_{11}(x) & p_{12}F_{12}(x) \\ p_{21}F_{21}(x) & p_{22}F_{22}(x) \end{bmatrix},$$

where $F_{ij}(x)$ is the distribution function of the inter-arrival times given that the previous arrival is of type i and the next is of type j. Especially, we assume that $p_{ij} = p$ if $i = j$ and $p_{ij} = 1 - p$ if $i \neq j$, and $F_{ij}(x) = F_j(x) = 1 - e^{-\lambda_j x}$ for simplicity. This can simplify greatly our derivation without loss of any generality because it still contain the correlated nature in itself. Even though the MRP given above is too simple to represent the real life telecommunication traffic, we adopt it to focus on the effect of correlation of the arrival process on the system performance. One can extend this form of MRP into which has more states and general type of distribution function by following similar steps described below.

The following notations are used throughout the paper.
$N(t)$: the number of customers at time t.
$z_b(t)$: the type of the most recent arrival before time t.
$z_n(t)$: the type of the next arrival after time t.
$S, s(x), S^*(\theta)$: random variable, probability density function and Laplace-transform of the service time, respectively.
$S_R(t)$: the remaining service time at time t.
$P_0(t; i, j) = \Pr(N(t) = 0, z_b(t) = i, z_n(t) = j)$.
$P_n(x, t; i, j)dx = \Pr(N(t) = n, S_R(t) \in (x, x+dx), z_b(t) = i, z_n(t) = j)$.
$P_0(; i, j) = \lim_{t \to \infty} P_0(t; i, j)$.
$P_n(x; i, j) = \lim_{t \to \infty} P_n(x, t; i, j)$.
$\overline{P}_n^*(\theta; i, j) = \int_0^\infty e^{-\theta x} P_n(x; i, j) dx$, the Laplace transform of $P_n(x; i, j)$.

By chasing the probability flows during dt, we can obtain the following equations:

$$P_0(t+dt; 1, 1) = (1 - \lambda_1 dt)P_0(t; 1, 1) + P_1(0, t; 1, 1)s(x)dt + o(dt), \quad (1)$$

$$P_1(x - dt, t + dt; 1, 1)dt = (1 - \lambda_1 dt)P_1(x, t; 1, 1)dt + P_2(0, t; 1, 1)dts(x)dt$$
$$+ p\lambda_1 dt P_0(; 1, 1)s(x)dt + p\lambda_1 dt P_0(; 2, 1)s(x)dt + o(dt) \quad (2)$$

Similar equations can be derived for $P_n(x - dt, t + dt; i, j)dt$, $n \geq 2 (i, j = 1, 2)$, and we omit them. Then, by takeing $t \to \infty$, we can have the following steady-state equilibrium system of equations:

$$\lambda_1 P_0(; 1, 1) = P_1(0; 1, 1), \quad (3)$$

$$\lambda_2 P_0(; 1, 2) = P_1(0; 1, 2), \quad (4)$$

$$\lambda_1 P_0(; 2, 1) = P_1(0; 2, 1), \quad (5)$$

$$\lambda_2 P_0(; 2, 2) = P_1(0; 2, 2), \quad (6)$$

$$-\frac{d}{dx} P_1(x; 1, 1) = -\lambda_1 P_1(x; 1, 1) + p\lambda_1 P_0(; 1, 1)s(x)$$
$$+ p\lambda_1 P_0(; 2, 1)s(x) + P_2(0; 1, 1)s(x), \quad (7)$$

$$-\frac{d}{dx}P_1(x;1,2) = -\lambda_2 P_2(x;1,2) + (1-p)\lambda_1 P_0(;1,1)s(x)$$
$$+ (1-p)\lambda_1 P_0(;2,1)s(x) + P_2(0;1,2)s(x), \qquad (8)$$

$$-\frac{d}{dx}P_1(x;2,1) = -\lambda_1 P_1(x;2,1) + (1-p)\lambda_2 P_0(;2,2)s(x)$$
$$+ (1-p)\lambda_2 P_0(;1,2)s(x) + P_2(0;2,1)s(x), \qquad (9)$$

$$-\frac{d}{dx}P_1(x;2,2) = -\lambda_2 P_2(x;2,2) + p\lambda_2 P_0(;1,1)s(x)$$
$$+ p\lambda_2 P_0(;2,1)s(x) + P_2(0;2,2)s(x), \qquad (10)$$

and for $n \geq 2$,

$$-\frac{d}{dx}P_n(x;1,1) = -\lambda_1 P_n(x;1,1) + p\lambda_1 P_{n-1}(x;1,1)$$
$$+ p\lambda_1 P_{n-1}(x;2,1) + P_{n+1}(0;1,1)s(x), \qquad (11)$$

$$-\frac{d}{dx}P_n(x;1,2) = -\lambda_2 P_n(x;1,2) + (1-p)\lambda_1 P_{n-1}(x;1,1)$$
$$+ (1-p)\lambda_1 P_{n-1}(x;2,1) + P_{n+1}(0;1,2)s(x), \qquad (12)$$

$$-\frac{d}{dx}P_n(x;2,1) = -\lambda_1 P_n(x;2,1) + (1-p)\lambda_2 P_{n-1}(x;2,2)$$
$$+ (1-p)\lambda_2 P_{n-1}(x;1,2) + P_{n+1}(0;2,1)s(x), \qquad (13)$$

$$-\frac{d}{dx}P_n(x;2,2) = -\lambda_2 P_n(x;2,2) + p\lambda_2 P_{n-1}(x;2,2)$$
$$+ p\lambda_2 P_{n-1}(x;2,1) + P_{n+1}(0;2,2)s(x) . \qquad (14)$$

We define the generating functions: $\overline{P}^*(\theta,z;i,j) = \sum_{n=1}^{\infty} \overline{P}_n^*(\theta;i,j)z^n$ and $\overline{P}(0,z;i,j) = \sum_{n=1}^{\infty} P_n(0;i,j)z^n$.

By taking Laplace transforms on the equations (3)-(14), we obtain:

$$(\theta - \lambda_1 + p\lambda_1 z)\overline{P}^*(\theta,z;1,1)$$
$$= (1 - \frac{S^*(\theta)}{z})\overline{P}(0,z;1,1) - p\lambda_1 z \overline{P}^*(\theta,z;2,1)$$
$$- \lambda_1 z S^*(\theta)P_0(;2,1) + \lambda_1 S^*(\theta)(1-zp)P_0(;1,1), \qquad (15)$$

$$(\theta - \lambda_1)\overline{P}^*(\theta,z;2,1)$$
$$= (1 - \frac{S^*(\theta)}{z})\overline{P}(0,z;2,1) - (1-p)\lambda_2 z(\overline{P}^*(\theta,z;1,2) + \overline{P}^*(\theta,z;2,2))$$
$$- (1-p)\lambda_2 z S^*(\theta)(P_0(;1,2) + P_0(;2,2)) + \lambda_1 S^*(\theta)P_0(;2,1), \qquad (16)$$

$$(\theta - \lambda_2)\overline{P}^*(\theta, z; 1, 2)$$
$$= (1 - \frac{S^*(\theta)}{z})\overline{P}(0, z; 1, 2) - (1-p)\lambda_1 z(\overline{P}^*(\theta, z; 1, 1) + \overline{P}^*(\theta, z; 2, 1))$$
$$- (1-p)\lambda_1 z S^*(\theta)(P_0(; 1, 1) + P_0(; 2, 1)) + \lambda_2 S^*(\theta) P_0(; 1, 2), \qquad (17)$$

$$(\theta - \lambda_2 + p\lambda_2 z)\overline{P}^*(\theta, z; 2, 2)$$
$$= (1 - \frac{s^*(\theta)}{z})\overline{P}(0, z; 2, 2) - p\lambda_2 z \overline{P}^*(\theta, z; 1, 2)$$
$$- \lambda_2 z S^*(\theta) P_0(; 1, 2) + \lambda_2 S^*(\theta)(1 - zp) P_0(; 2, 2) \ . \qquad (18)$$

Let us denote

$$\overline{P}^*(\theta, z) = \overline{P}^*(\theta, z; 1, 1) + \overline{P}^*(\theta, z; 2, 1),$$
$$\overline{Q}^*(\theta, z) = \overline{P}^*(\theta, z; 1, 2) + \overline{P}^*(\theta, z; 2, 2),$$
$$\overline{P}(0, z) = \overline{P}(0, z; 1, 1) + \overline{P}(0, z; 2, 1),$$
$$\overline{Q}(0, z) = \overline{P}(0, z; 1, 2) + \overline{P}(0, z; 2, 2),$$
$$P_0 = P_0(; 1, 1) + P_0(; 2, 1),$$
$$Q_0 = P_0(; 1, 2) + P_0(; 2, 2) \ .$$

Now, by means of the above notations, we obtain

$$(\theta - p\lambda_1 + p\lambda_1 z)\overline{P}^*(\theta, z)$$
$$= (1 - \frac{S^*(\theta)}{z})\overline{P}(0, z) + p\lambda_1 S^*(\theta)(1 - z) P_0$$
$$+ (1-p)(\lambda_1 \overline{P}^*(\theta, z) - \lambda_2 z \overline{Q}^*(\theta, z))$$
$$+ (1-p)S^*(\theta)(\lambda_1 P_0 - \lambda_2 z Q_0), \qquad (19)$$

$$(\theta - p\lambda_2 + p\lambda_2 z)\overline{Q}^*(\theta, z)$$
$$= (1 - \frac{S^*(\theta)}{z})\overline{Q}(0, z) + p\lambda_2 S^*(\theta)(1 - z) Q_0$$
$$+ (1-p)(\lambda_2 \overline{Q}^*(\theta, z) - \lambda_1 z \overline{P}^*(\theta, z))$$
$$+ (1-p)S^*(\theta)(\lambda_2 Q_0 - \lambda_1 z P_0) \ . \qquad (20)$$

By plugging $\theta = p\lambda_1 - p\lambda_1 z$ and $\theta = p\lambda_2 - p\lambda_2 z$ into equations (19) and (20) respectively, and letting $\theta = 0$, we can obtain the following equations (21), (22).

$$\overline{P}^*(0, z) = \frac{z(S^*(p\lambda_1 - p\lambda_1 z) - 1)\{p\lambda_1(1-z)P_0 + (1-p)\lambda_1 P_0 - (1-p)\lambda_2 z Q_0\}}{-p\lambda_1(1-z)(S^*(p\lambda_1 - p\lambda_1 z) - z)}$$
$$+ \frac{(z-1)(1-p)\{\lambda_1 \overline{P}^*(p\lambda_1 - p\lambda_1 z, z) - \lambda_2 z \overline{Q}^*(p\lambda_1 - p\lambda_1 z, z)\}}{-p\lambda_1(1-z)(S^*(p\lambda_1 - p\lambda_1 z) - z)}$$
$$- \frac{(z - S^*(p\lambda_1 - p\lambda_1 z))(1-p)\{\lambda_1 \overline{P}^*(0, z) - \lambda_2 z \overline{Q}^*(0, z)\}}{-p\lambda_1(1-z)(S^*(p\lambda_1 - p\lambda_1 z) - z)}, \qquad (21)$$

$$\overline{Q}^*(0,z) = \frac{z(S^*(p\lambda_2 - p\lambda_2 z) - 1)\{p\lambda_2(1-z)Q_0 + (1-p)\lambda_2 Q_0 - (1-p)\lambda_1 z P_0\}}{-p\lambda_2(1-z)(S^*(p\lambda_2 - p\lambda_2 z) - z)}$$
$$+ \frac{(z-1)(1-p)\{\lambda_2 \overline{Q}^*(p\lambda_2 - p\lambda_2 z, z) - \lambda_1 z \overline{P}^*(p\lambda_2 - p\lambda_2 z, z)\}}{-p\lambda_2(1-z)(S^*(p\lambda_2 - p\lambda_2 z) - z)}$$
$$- \frac{(z - S^*(p\lambda_2 - p\lambda_2 z))(1-p)\{\lambda_2 \overline{Q}^*(0,z) - \lambda_1 z \overline{P}^*(0,z)\}}{-p\lambda_2(1-z)(S^*(p\lambda_2 - p\lambda_2 z) - z)}. \quad (22)$$

It can be shown that the two-dimensional process $\{(N(t), z_n(t)); t \geq 0\}$ is reversible and so $\lambda_1 \overline{P}^*(0,z) = \lambda_2 \overline{Q}^*(0,z)$ and $\lambda_1 P_0 = \lambda_2 Q_0$. Using these relations, we rearrange the equations (21), (22).

$$\lambda_1 \overline{P}^*(0,z) = \frac{z\lambda_1 P_0(1 - S^*(p\lambda_1 - p\lambda_1 z))}{S^*(p\lambda_1 - p\lambda_1 z) - z}$$
$$+ \frac{(1-p)\{\lambda_1 \overline{P}^*(p\lambda_1 - p\lambda_1 z, z) - \lambda_2 z \overline{Q}^*(p\lambda_1 - p\lambda_1 z, z)\}}{S^*(p\lambda_1 - p\lambda_1 z) - z}, (23)$$

$$\lambda_2 \overline{Q}^*(0,z) = \frac{z\lambda_2 Q_0(1 - S^*(p\lambda_2 - p\lambda_2 z))}{S^*(p\lambda_2 - p\lambda_2 z) - z}$$
$$+ \frac{(1-p)\{\lambda_2 \overline{Q}^*(p\lambda_2 - p\lambda_2 z, z) - \lambda_1 z \overline{P}^*(p\lambda_2 - p\lambda_2 z, z)\}}{S^*(p\lambda_2 - p\lambda_2 z) - z}. (24)$$

3 Expected Number of Customers at Arbitrary Time

Using the boundary condition, $P_0 + Q_0 + \overline{P}^*(0,1) + \overline{Q}^*(0,1) = 1$, we obtain the values of variables as follows.

$$\overline{P}^*(0,1) = \frac{2\lambda_1 \lambda_2 E(S)}{\lambda_1 + \lambda_2} \frac{\lambda_2}{\lambda_1 + \lambda_2}, \quad (25)$$

$$\overline{Q}^*(0,1) = \frac{2\lambda_1 \lambda_2 E(S)}{\lambda_1 + \lambda_2} \frac{\lambda_1}{\lambda_1 + \lambda_2}, \quad (26)$$

$$P_0 = \frac{\lambda_1 + \lambda_2 - 2\lambda_1 \lambda_2 E(S)}{\lambda_1 + \lambda_2} \frac{\lambda_2}{\lambda_1 + \lambda_2}, \quad (27)$$

$$Q_0 = \frac{\lambda_1 + \lambda_2 - 2\lambda_1 \lambda_2 E(S)}{\lambda_1 + \lambda_2} \frac{\lambda_1}{\lambda_1 + \lambda_2}. \quad (28)$$

By differentiating the equations (23), (24) and plugging $z = 1$ in equations (19), (20), the expected number of customer in systems at arbitrary time is given by

$$E(N) = \frac{a^2 E(S^2)}{2(1-\rho)} + \rho + \frac{\rho}{2(1-\rho)} \frac{1}{1-p} \frac{(\lambda_1 - \lambda_2)^2}{(\lambda_1 + \lambda_2)^2}, \quad (29)$$

where $a = 2\lambda_1\lambda_2/(\lambda_1+\lambda_2)$, $\rho = aE(S)$. In [7], they provide the expected number of customers in system *at arrival epochs* and their result, however, contains an unknown value P_0, which is the probability that the arriving customer sees the system empty. In this study, we derived the expected number of customers *at arbitrary times* in the equation (30), which is more general with no unknown value. In addition, [7] and [9] showed that the system size grows infinity as p goes to 1 but failed to obtain an explicit relationship between the system performance and the parameters like p, λ_1, and λ_2, while we do in this paper.

By Little's theorem, the expected waiting time is given by

$$E(W_q) = \frac{aE(S^2)}{2(1-\rho)} + \frac{E(S)}{2(1-\rho)}\frac{1}{1-p}\frac{(\lambda_1-\lambda_2)^2}{(\lambda_1+\lambda_2)^2} \ . \tag{30}$$

From the above result, we can see the monotone relations between the transition probability, p, of states and the expected waiting time.

4 Experiment

We compare our results with simulation and the GI/G/1 system. [15] obtained the expected waiting time of GI/G/1 systems which is given by

$$E(W_q) = \frac{3(\lambda_1^2+\lambda_2^2)-2\lambda_1\lambda_2}{(\lambda_1+\lambda_2)4\lambda_1\lambda_2(1-\rho)} + \frac{a^2\text{Var}(S)+(1-\rho)^2}{2a(1-\rho)} \\ -\frac{\lambda_1^2+\lambda_2^2}{\lambda_1\lambda_2(\lambda_1+\lambda_2)} \ . \tag{31}$$

We use $\lambda_1 = 1, \lambda_2 = 1/7$ and exponential service time with mean 3(i.e., $\rho = 0.75$). The correlations of arrival process are increased(from -0.212 to 0.212) as the value of p becomes bigger.

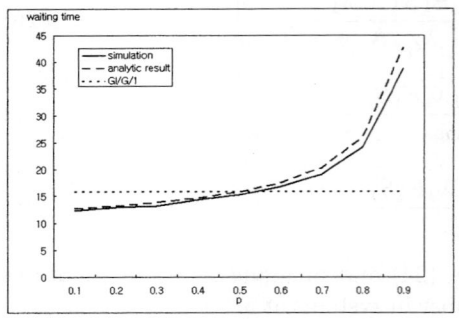
(a) Exponential service time

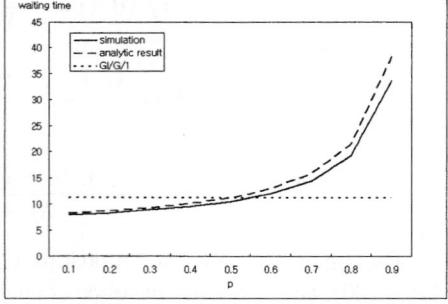
(b) Deterministic service time

Fig. 1. Comparison result : analytic result vs GI/G/1 and simulation

Fig(1). shows a very close agreement in the mean waiting times between our analytic calculation and the simulation. For values of p below 0.7 the errors are within 5%. The discrepancy becomes larger as p becomes bigger because of the computational overflows but the error still lies within 10% when $p = 0.9$. This validates our analytic formula derived in the equation (30). Also, it shows that if one regards a queue with correlated arrivals as if the arrivals are renewal then it might lead a serious under-estimation in the performance measures. As p goes to 1, which implies higher correlation, the under-estimation errors become drastically increased, as one can see in the Fig(1).

5 Conclusion

In this paper, we consider the queue with a Markov renewal arrival process and general service time distribution. Using supplementary variable method, we derive the probability distribution of the number of customers at arbitrary time and the performance measures like the expected number of customers and waiting time. We showed, by numerical experiments, that our analytic expression, which solves the incomplete results in the previous literatures, is valid. As pointed out in many previous papers, it might be quite misleading if one assumes the packet interarrival times are *iid* in the computer communication network and the performance measures could be highly underestimated if they are correlated, which is very common in the real system. Therefore, if one can model a traffic as an MRP then more accurate performance estimation could be possible. Extension to the more generalized MRP case than given in this paper should be investigated and the way to represent a real traffics as an MRP could be considered for further research.

References

[1] Karagiannis, T., Molle, M., Faloutsos, M.: Long-Range Dependence:Ten Years of Internet Traffic Modeling. IEEE Internet Computing (2004) 2–9
[2] Paxon, V., Floyd, S.: Wide Area Traffic:The Failure of Poisson Modeling. IEEE/ACM Transactions on Networking **3(3)** (1995) 226–244
[3] Saito, H.: The Departure Process of an N/G/1 queue. Performance Evaluation **11** (1990) 241–251
[4] Lucantoni, D.M.: New Results on the Single Server Queue with a Batch Markovian Arrivals. Commnu. Statist.- Stochastic Models **7(1)** (1991) 1–46
[5] Disney, R.L., Kiessler, P.C.: Traffic Processes in Queueing Networks - A Markov Renewal Approach. The Johns Hopkins University Press.(1987)
[6] Patuwo, B.E.: The Effect of Dependency in the Markov Renewal Arrival Process on the Various Performance Measures of Exponential Server Queue. Ph.D. thesis Virginia Polytechnic Institute and State University (1989)
[7] Szekli, R., Disney, R.L., Hur, S.: MR/GI/1 queues with positively correlated arrival stream. Journal of Applied Probability **31** (1994) 497–514
[8] Hur, S.: A note on varying the number of states in the arrival process of MR/GI/1 queue. Computers and OR **24** (1997) 1113–1118

[9] Bäuerle, N.: Monotonicity Results for MR/G/1 Queues. J.Appl.Prob. **34** (1997) 514–524
[10] Livny, M., Melamed, B., Tsiolis, A.K.: The Impact of Autocorrelation on Queueing Systems. Management Science **39(3)** (1993) 322-339
[11] Chao, X.: Monotone effect of dependency between interarrival and service time in a simple queueing system. Operations Research Letters **17** (1995) 47–51
[12] Boxma, O.J., Perry, D.: A queueing model with dependence between service and interarrival times. European Journal of Operational Research **128** (2001) 611–624
[13] Müller, A.: On the waiting times in queues with dependency between interarrival and service times. Operations Research Letters **26** (2000) 43–47
[14] Çinlar, E.: Introduction to Stochastic processes. Prentice-Hall Inc., Englewood Cliffs, N.J.(1975)
[15] Marshall, K.T.: Some inequalities in queueing. Opns.Res. **16** (1968) 651–665

Product Mix Decisions in the Process Industry[*]

Seung J. Noh[1] and Suk-Chul Rim[2]

[1] School of Business, Myongji University,
120-728 Seoul, Korea
sjnoh@mju.ac.kr
[2] School of Industrial and Information Systems Engineering,
Ajou University, 443-749 Suwon, Korea
scrim@ajou.ac.kr

Abstract. A resin manufacturer in Korea operates a large plant for synthetic resin products. The production process consists of two stages where a line in the first stage is a bottleneck of the whole process. Some low-profit products have consumed considerable amount of bottleneck capacity, and caused opportunity loss in profit generation. This is due to the traditional management policy that allows several marketing business units to plan their annual target sales individually and independently, without careful consideration of the effective use of the production capacity. Noting such marketing-production misalignment, we developed linear programming models to help determine desirable product mix while coping with market demands and production capacity.

1 Introduction

A resin manufacturer in Korea produces technology-intensive, value-added chemicals including styrene derivatives, engineering plastics, materials used in electronic products and special function composite. Although domestic and worldwide markets have been oversupplied for the past few years, the company has been prosperous in an extremely competitive environment. Production is mainly triggered by customer orders and sales volume is heavily dependent on the marketing capability of the company. Marketing division of the company consists of three domestic and one oversea marketing units. Traditionally, the four units individually plan their own annual target sales based on past sales and market forecasts. The planned volume of final products are then integrated and adjusted by the Strategic Business Unit under the consideration of yearly production capacity and target profit level set by top management. In the decision making process, the company has not used any optimization techniques in determining the target sales. For this reason, the company has experienced marketing-production misalignments we elaborate hereafter.

The plant operates a two-stage polymer (resin) production system where the products of stage 1 are used to produce final products in stage 2. Stage 1, the

---

[*] This research was supported by Myongji University and IITA of Korea.

polymerization process, is to produce nineteen grades of polymer powder using four production lines. All grades of polymer powder are then temporarily stored in several silos. Stage 2, the compounding process, is to mix one or more types of polymer powder with additives and color to produce over 2,600 different final products. Final products can be classified into 188 different groups according to the mix rate (recipe) of the polymer powder. To simplify our discussion we use the term final products to indicate 188 final product groups throughout the paper.

Line 1 and line 2 produce six grades of type A polymer powder and five grades of type B polymer powder, respectively. Both lines 3 and 4 produce eight grades of type C polymer powder. Polymer powder is compounded into final products according to the predetermined recipe. Production volume of the polymer powder therefore depends upon the volume of final products. It was observed that line 1 in the stage 1 is the bottleneck of the whole process. Thus, effective use of type A polymer powder produced in this bottleneck line is crucial to profit generation. A preliminary analysis indicated that some low-profit final products consume significant amount of bottleneck polymer powder. If the annual sales volume of such products is reduced to a certain extent, the saved amount of type A polymer powder grades can be used for high-profit products, thereby increasing the total profit.

In the company there had been debates among the middle level managers regarding product mix under the consideration of limited capacity and market orders. However, the company had maintained conventional product mix for years since none of the decision makers had any experience with operations research or mathematical modeling that helps quantify the effects of the changes in product mix on the profit level.

Throughout this paper, we use the term unit profit to indicate the difference between the market price and the direct material cost of a ton of final product. Unit profit varies from product to product. This term, sometimes called unit contribution, which is conventionally used in the process industry, is different from the marginal profit used in accounting, in a sense that it does not consider other cost factors such as labor cost and overheads. We also use the term total profit to indicate the sum of unit profits multiplied by annual sales volume of final products.

Given a unique recipe matrix, unequal unit profit for each final product, and limited production capacity, optimal product mix problem naturally arises [Assad 1992]. We developed three linear programming models to help figure out desirable product mix and assess additional capacity needed for the bottleneck line. The decision variable, X_j, denotes the annual production volume (in tons) of final product j. It was assumed that the demand of each final product is controllable within a certain range, which imposes lower and upper limits on X_j as constraints. The lower and upper limits imply the minimum and maximum amount to be supplied to the market, respectively. This assumption is practically reasonable since marketing capability of the company can, to a certain extent, control the annual sales of each final product.

2   The Models

Model 1: Maximizing Total Profit
In this model we tried to figure out desirable changes in product mix which would increase the total profit under the constraints of production capacity and current market demands (Appendix). The objective function we tried to maximize is the sum of the profits generated by 188 final products, and the constraint set includes the capacities of polymer powder production lines in the production stage 1. The constraint set is more complicated than the representation in the appendix because illustrating some of the details in the production process characteristics requires expanded mathematical representations, which we have omitted.

We also included in the constraint set the upper and lower bounds on the sales volume of each product by ±10% from the current level. This is because sharp changes in sales volume would not be possible in a short period of time. For example, time and cost are required to increase the sales of the high-profit products by expanding current market or opening up a new market. Likewise, decreasing the sales volume of the low-profit products would require the company to induce customers to gradually migrate from low-profit products to more value-added ones. Marketing business units in the company agreed that their current capability could adjust the sales of the products by approximately ±10% from the current level. Therefore, we employed this number in the constraints set.

We solved the model using spreadsheet user interface [Ragsdale 1995], and compared the results with the actual production volume of the year 2003. Results suggested that it would have been beneficial if 54 out of 188 final products had been produced and sold 10% less, while the rest of the final products had been produced and sold 10% more. These changes in sales volume would have enabled the company to earn additional U$ 3.3 million (approximately 3.7% of the total profit in 2003) even without expanding the capacity of the bottleneck line (Table 1.). Note that this additional profit could be created by merely 1.5% increase in total production volume. The amount of type B and type C polymers would have been also increased with the changes in product mix without violating the capacity constraints for lines 2, 3, and 4. In addition, sensitivity analysis [Mathur and Solow 1994] listed the final products in the order of their marginal contributions, which helps the marketing division identify the products that they should focus on.

Model 2: Minimizing the Use of Bottleneck Polymer Powder
In this model we tried to figure out desirable changes in product mix that would decrease the total consumption of type A polymer powder produced from bottleneck line 1, under the constraints of production capacity and total profit level required. One of the primary concerns of the production side is the profitable use of the bottleneck polymer powder for two reasons. First, type A polymer powder is more costly since they contain more expensive raw materials than other types. Second, line 1 producing type A polymer powder is a bottleneck, and investment

Table 1. Slight changes in product mix would help use the bottleneck line more profitably and increase the total profit by 3.3 million dollars

	As was in 2003	Model 1: max. total profit
Total Volume of Final Products	199,155.9 tons	202,077.4 tons
Utilization of Line 1	100 %	100 %
Type A Powder Used	55,229 tons	55,230 tons
Type B Powder Used	8,051 tons	8,388 tons
Type C Powder Used	99,891 tons	100,099 tons
Total Annual Profit	U$ 88.8 million	U$ 92.1 million

Table 2. Slight changes in product mix would also help reduce the usage of bottleneck polymer powder while maintaining the level of profit of the year 2003. The saved capacity, estimated at 3 million dollars, could be assigned to more profitable products

	As was in 2003	Model 2: min. bottleneck polymer powder
Total Volume of Final Products	199,155.9 tons	194,160.6 tons
Utilization of Line 1	100 %	96.2 %
Type A Powder Used	55,229 tons	53,119 tons
Type B Powder Used	8,051 tons	8,297 tons
Type C Powder Used	99,891 tons	96,347 tons
Total Annual Profit	U$ 88.8 million	U$ 88.8 million

in capacity expansion is too large. For these reasons, we decided to find ways to better use of the type A polymer powder by changing product mix. Our aim was to identify final products whose sales volume should be increased or decreased to a certain extent while maintaining the level of the total profit as that of the year 2003.

The objective function we tried to minimize is the sum of the type A polymer powder produced in line 1, and the constraint set includes the capacities of the lines 2 through 4 in the production stage 1. We also included in the constraint set the minimum total profit level. Again we assumed that the marketing capability could change the sales volume of each product by $\pm 10\%$ as in the model 1 (Appendix).

Results from model 2 suggested that, by a slight change in product mix, 3.8% of the current capacity of line 1 could have been saved, while maintaining the level of total profit as in 2003. To achieve this, 58 of 188 final products should have been produced and sold 10% less, while the rest of the products 10% more. We noticed that this result is very similar to those obtained from Model 1. The changes in product mix resulted in a slight decrease in the total volume of the final products. This implies that some capacity of the production stage 2 could also have been saved. The estimated worth of the saved capacity in line 1 and in stage 2 was approximately U$ 3.0 million. Furthermore, the company could have saved some capacity in the production of type C powder (Table 2.).

Model 3: Capacity Expansion of the Bottleneck Line
Our preliminary analysis indicated that capacity of the stage 2 far exceeds that of line 1 in stage 1. That is, if we increase the bottleneck capacity to a certain extent, we can fully utilize the capacity of stage 2 and, in turn, increase annual sales volume. In this model we tried to assess additional capacity of the bottleneck line required to balance the production stages 1 and 2. We tried to maximize the total profit while allowing infinite capacity for line 1 (Appendix). Then the solution could be used to assess how much of additional capacity for line 1 is required. We included in the constraint set the upper and lower bounds on the sales volume of each product by ±20% from the current level. This is because we found that just ±10% changes in final products were not enough to make stages 1 and 2 balanced.

Results from model 3 suggested that additional 15% of the capacity of line 1 is required to balance the stages 1 and 2, which would result in 18% increase in the total annual profit (Table 3). Production lines 2, 3, and 4 in stage 1 for type B and C powder remained unsaturated even in this case. Unfortunately, it turned out that adding additional capacity to line 1 is not economically justified since it would cost tens of millions of dollars. Moreover, 20% increase in the sales of high-profit products would not be possible in a short period of time. However, the results could be used in a further economic analysis to assess the amount of bottleneck polymer powder that may be purchased from outside vendors.

Table 3. The capacity of the bottleneck line should be increased by 15% to balance the production stages 1 and 2. Additional 16 million dollars in the total profit would be achievable. The rest of the lines in stage 1 turned out to have enough capacities as of now

	As was in 2003	Model 3: max. profit by balancing two production stages
Total Volume of Final Products	199,155.9 tons	230,170.5 tons
Utilization of Line 1	100 %	115 %
Type A Powder Used	55,229 tons	63,531.3 tons
Type B Powder Used	8,051 tons	9,638.9 tons
Type C Powder Used	99,891 tons	114,452.2 tons
Total Annual Profit	U$ 88.8 million	U$ 104.8 million

3 Using the Models in the Real World

When the models and their implications were presented to the middle level managers, their responses were twofold. Production personnel unanimously agreed with the recommendations that the models suggested. On the other hand, marketing side claimed that practically it would be very difficult to control the demands in such a make-to-order based market environment. Even so, it was clear that the recommendations suggested by the models could be highly valuable in the decision making process if we could address the decision makers' concerns successfully. The optimal solutions obtained from the models were considered as a starting point for reshaping the marketing strategy rather than a final solution of the product mix problem. We did not even anticipate that the results of the models could be successfully taken into actions. Fortunately, however, top management quickly decided to reset the target sales of the year 2004 right after we presented the models and recommendations. Since then, the company has been in the process of shifting its target market towards more profitable products. Technology division of the company is currently in the process of changing the recipe of final products for more efficient use of bottleneck powder to produce more value-added products. And some of the lines in stage 2 are being expanded. The models will then be continuously refined for annual sales planning to cope with such changes in recipe, production capacity, and the market.

4 Conclusions

We developed three LP models to suggest better product mix policies for a resin manufacturer with spreadsheet user interface that allowed what-if scenarios to be easily evaluated and presented. The first two models identify final products that should have been sold more or less than the current level in order to increase the total profit without any capacity expansion. The third model assessed additional capacity of the bottleneck process needed to balance the whole production process. The models we developed helped the decision makers in many ways. The results successfully assisted marketing people in reshaping the business strategies towards more profitable target market. Along with this, production side took steps towards refining the recipe for efficient use of polymer powder. Capacity expansion of several lines in stage 2 is also under consideration. Most of all, decision makers realized the power of optimization techniques in systematically describing the nature of the business and finding out appropriate solutions.

We know that mathematical models are valuable tools to support a wide range of managerial decision making. In practice, we learned that integrating the quantitative and simplified solutions into actual decision making process requires much more effort than needed in constructing and solving the models. Another valuable lesson taken from the implementation process of the models is the importance of the role of top management in various steps of the decision making process. Understanding the nature of the problems, setting up ultimate goals to pursue, checking and expediting the progress of work, and encouraging personnel to do their best in resolving problems are all must-do steps that the top management ought to take for successful implementation of the recommendations.

References

[1] Assad, A., Wasil, E., Lilien, G.: Excellence in Management Science in Practice. Prentice-Hall, Inc., New Jersey (1992)
[2] Mathur, K., Solow, D.: Management Science. Prentice-Hall, Inc., New Jersey (1994)
[3] Ragsdale, C.: Spreadsheet Modeling and Decision Analysis. Course Technology, Inc., Cambridge MA (1995)

Appendix: The Models

Notation

X_j :production volume of final product $j, j = 1, 2, \ldots, 188$ (decision variables, in tons)
b_j :unit profit of product j
c_i :annual production quantity of polymer powder $i, i = 1, 2, \ldots, 19$ (in tons)
a_{ij} :amount of polymer powder i compounded into one ton of product j (in tons)
k_i :hourly production quantity of polymer powder i (in tons)
T :total annual production time (in hours) available for polymer production lines
X_j^0 :production volume of product j in 2003 F : total production capacity of stage 2 for production of final products
P :current total profit

Model 1
Maximize
$$\sum b_j X_j \qquad (1)$$

Subject to

$$\sum_j a_{ij} X_j = c_i \qquad (2)$$

$$\sum_i \frac{c_i}{k_i} \leq T, \quad i = 1, \ldots, 6 \qquad (3)$$

$$\sum_i \frac{c_i}{k_i} \leq T, \quad i = 7, \ldots, 11 \qquad (4)$$

$$\sum_i \frac{c_i}{k_i} \leq 2T, \quad i = 12, \ldots, 19 \qquad (5)$$

$$0.9 X_j^0 \leq X_j \leq 1.1 X_j^0 \qquad (6)$$

The objective function in (1) is to maximize total annual profit. Constraint (3) represents the capacity limit of line 1 for the production of type A polymer powder. Likewise, constraints (4) and (5) represent the capacity limits on type B and type C polymer powder production, respectively. Note that the right-hand side of constraint (5) is doubled because type C powder can be produced in both of lines 3 and 4. Constraint (6) represents the marketing capability that can adjust the sales volume of each final product by $\pm 10\%$ from the level of the year 2003. Nonnegativity constraints for X_j's are unnecessary because of constraint (6).

Model 2
Minimize

$$\sum_i \sum_j a_{ij} X_j, \quad i = 1, \ldots, 6 \tag{7}$$

Subject to

$$\sum_j a_{ij} X_j = c_i, \quad i = 7, \ldots, 19 \tag{8}$$

$$\sum_i \frac{c_i}{k_i} \leq T, \quad i = 7, \ldots, 11 \tag{9}$$

$$\sum_i \frac{c_i}{k_i} \leq 2T, \quad i = 12, \ldots, 19 \tag{10}$$

$$\sum_j b_j X_j \geq P \tag{11}$$

$$0.9 X_j^0 \leq X_j \leq 1.1 X_j^0 \tag{12}$$

The objective function in (7) is to minimize the total volume of the type A polymer powder grades, 1 through 6, which are produced from the bottleneck line 1. Note that, in (8), we consider only the powder other than the six bottleneck powder. Constraints (9) represents the capacity limit of line 2 which produces type B powder, and (10) represents the capacity limits of lines 3 and 4 which produce type C powder, respectively. Constraint (11) is to secure at least the current level of total profit.

Model 3
Maximize

$$\sum b_j X_j \tag{13}$$

Subject to

$$\sum_j a_{ij} X_j = c_i \tag{14}$$

$$0.8X_j^0 \leq X_j \leq 1.2X_j^0 \qquad (15)$$

$$\sum_j X_j \leq F \qquad (16)$$

The objective function in (13) is to maximize total annual profit. All the capacity constraints for the lines in stage 1 were eliminated to allow infinite capacities so that additional capacities required could be measured. Changes in the volume of each product are allowed by ±20% from the current level in (15). Constraint (16) imposes the production capacity limit of the production stage 2.

On the Optimal Workloads Allocation of an FMS with Finite In-process Buffers

Soo-Tae Kwon

Department of Information Systems,
School of Information Technology and Engineering, Jeonju University

Abstract. This paper considers a workload allocation problem of a flexible manufacturing system composed of several parallel workstations each with both input and output buffers where two automated guided vehicles (AGVs) are used for input and output material handling. The problem is divided into 4 types according to the capacities of input and output buffers, and then analyzed to yield the highest throughput for the given FMS model. Some interesting properties are derived that are useful for characterizing optimal allocation of workloads, and some numerical results are presented.

Keywords: FMS, Queueing Network, Throughput, Workload.

1 Introduction

The term flexible manufacturing system (FMS) is used to describe a network of automated workstations linked by a common computer controlled material handling device to transport work pieces from one workstation to another. Unlike a transfer line where all work pieces follow a sequential route through the system, an FMS permits work pieces to visit workstations in any arbitrary sequence as desired. And, FMSs have been introduced in an effort to increase productivity by reducing inventory and increasing the utilization of machining centers simultaneously.

Because of high equipment costs involved in FMSs, there is a basic need to pay attention to the factory design phase during which the main decision on resources are made. Main topics in the operation and management of the FMS are well classified as follows:

1) Design problems considering of selection of part families, selection of FMS production system, selection of material handling system, selection of pallets and fixtures, etc., and
2) Operational problems consisting of planning, grouping, machine loading and scheduling.

The problem of workload allocation relates quite generally to issues in the management of FMSs, since there often exists considerable latitude in the allocation of work among groups of flexible machines. Determining the rate which products should be dispatched on the various routes is known as the routing mix problem and routes may be assigned to optimize the allocation of workload. Production mix can often be varied in the short run to affect a particular allocation of workload. Also, it can be seen that the decision regarding the allocation of servers to stations is closely linked the workload allocation problem.

Much research has concentrated on queueing network model analyses to evaluate the performance of FMSs, and concerned with mathematical models to address the optimization problems of complex systems such as routing optimization, server allocation, workload allocation, buffer allocation under the performance model. Vinod and Solberg (1985) have presented a methodology to design the optimal system configuration of FMSs and modelled as a closed queueing networks of multi-server queues. Dallery and Stecke(1990) have derived some properties useful for characterizing the optimal allocation servers and workload in single-class, multi-server closed queueing networks. Stecke and Morin(1985) have analyzed the optimality of balancing workloads to maximize the expected production in a single-server closed queueing network model of FMS. Calabrese(1992) has examined the general problem of workload allocation in an open Jackson network of multi-server queues. Ma and Matsui(2002) have discussed the performance evaluation of the flexible machining/assembly systems of a central server type, and considered a fixed, dynamic versus an ordered-entry routing rule. And, Ooijn and Bertrand(2003) have investigated the effect of a simple workload dependent arrival rate control policy on the throughput and WIP of a simple model of a job shop.

In the above-mentioned reference models, machines were assumed not to be blocked, that is, not to have any output capacity restriction. These days, the automated guided vehicle (AGV) is commonly used to increase potential flexibility. By the way, it may not be possible to carry immediately the finished parts from the machines which are subject to AGV's capacity restriction. The restriction can cause any operation blocking at the machines, so that it may be desirable to provide some storage space to reduce the impact of such blocking. Sung and Kwon(1994) have investigated a queueing network model for an FMS composed of several parallel workstations each with both limited input and output buffers where two AGVs are used for input and output material handling.

In this paper, as a workload allocation problem, the problem of determining the routing probability is considered to yield the highest throughput for the given FMS model (Sung and Kwon 1994). The problem is divided into 4 types according to the capacities of input and output buffers, and then analyzed. Some interesting properties are derived that are useful for characterizing optimal allocation of workloads, and some numerical results are presented.

2 The Performance Evaluation Model

The FMS model is identical to that in Sung and Kwon(1994). The network consists of a set of n workstations. Each workstation $i(i = 1, \cdots, n)$ has machine with both limited input and output buffers. The capacities of input and output buffers are limited up to IB_i and OB_i respectively, and the machines perform in an exponential service time distribution. All the workstations are linked to an automated storage and retrieval system (AS/RS) by AGVs which consist of AGV(I) and AGV(O). The capacity of the AS/RS is unlimited, and external arrivals at the AS/RS follow a Poisson process with rate λ.

The FCFS (first come first served) discipline is adopted here for the services of AGVs and machines. AGV(I) delivers the input parts from the AS/RS to each input buffer of workstations, and AGV(O) carries the finished parts away from each output buffer of workstations to the AS/RS, with corresponding exponential service time distributions. Specifically, AGV(I) distributes all parts from the AS/RS to the workstations according to the routing probabilities $\gamma_i (\sum_{i=1}^n \gamma_i = 1)$ which can be interpreted as the proportion of part dispatching from the AS/RS to workstation i.

Moreover, any part (material) can be blocked on arrival (delivery) at an input buffer which is already full with earlier-arrived parts. Such a blocked part will be recirculated instead of occupying the AGV(I) and waiting in front of the workstation (block-and-recirculate mechanism). Any finished part can also be blocked on arrival at an output buffer which is already full with earlier-finished parts. Such a blocked part will occupy the machine to remain blocked until a part departure occurs from the output buffer. During such a blocking time, the machine cannot render service to any other part that might be waiting at its input buffer (block-and-hold mechanism).

Sung and Kwon(1994) have developed an iterative algorithm to approximate system performance measures such as system throughput and machine utilization. The approximation procedure decomposes the queueing network into individual queues with revised arrival and service processes. These individual queues are then analyzed in isolation. The individual queues are grouped into two classes. The first class consists of input buffers and machines, and the second one consists of output buffers and AGV(O). The first and second classes are called the first-level queue and the second-level queue, respectively.

The following notations are used throughout this paper $(i = 1, \cdots, n)$:

λ	external arrival rate at AS/RS
λ_i	arrival rate at each input buffer i in the first-level queue
λ_i^*	arrival rate at each input buffer i in the second-level queue
μ	service rate of AGV
μ_i	service rate of machine i
$P(k_1, \cdots, k_n)$	probability that there are k_i units at each output buffer i in the second-level queue with infinite capacity.
$P(idle)$	probability that there is no unit in the second-level queue with infinite capacity
$\prod(k_1, \cdots, k_n)$	probability that there are k_i units at each output buffer i in the second-level queue with finite capacity.
$\prod(idle)$	probability that there is no unit in the second-level queue with finite capacity

The second-level queue is independently analyzed first to find the steady-state probability by using the theory of reversibility. The steady-state probability is derived as follows.

Lemma 1. *(refer to Sung and Kwon 1994, Theorem 2)*

The steady-state probability of the second-level queue is derived as

$$\prod(k_1, \cdots, k_n) = P(k_1, \cdots, k_n)/G$$

$$\prod(idle) = P(idle)/G \qquad (1)$$

where,

$$A = \{(k_1, \cdots, k_n) | 0 <= k_i <= OB_i, 1 <= i <= n\},$$
$$G = \sum_{(k_1, \cdots, k_n) \in A} P(k_1, \cdots, k_n) + P(idle),$$
$$P(k_1, \cdots, k_n) = (1-\rho) \cdot \rho^{(k_1 + \cdots + k_n + 1)} \cdot \frac{(k_1 + \cdots + k_n)!}{k_1! \cdots k_n!} \cdot q_1^{k_1} \cdots q_n^{k_n},$$
$$P(idle) = 1 - \rho,$$
$$\rho = \sum_{i=1}^{n} \lambda_i^* / \mu,$$
$$q_i = \lambda_i^* / \sum_{i=1}^{n} \lambda_i^*.$$

It is followed by finding the clearance service time accommodating all the possible blocking delays that a part might undergo due to the phenomenon of blocking. The clearance time is derived from the steady-state probability of second-level queue. Then, the first-level queues are analyzed by this expected clearance time in the approach of the M/M/1/K queueing model.

It requires, in advance, the knowledge of the effective input rate. When the capacity of the queue i is infinite, the effective input rates at these individual queues can immediately be calculated. However, when the capacity of the queue i is finite, the effective input rates at these queues can not be figured out immediately. These various situations are all put together to present the algorithm in an iterative procedure whereby the true value of the effective input rate at each finite first-level queue can successively and approximately be computed.

3 The Workload Allocation Problem

In this section, as a workload allocation problem, the problem of determining the routing probability is considered to yield the highest throughput for the given performance evaluation model. The problem can be divided into 4 types depending on the capacities of input and output buffers as follows;

Type 1 : infinite capacity of input buffers & infinite capacity of output buffers
Type 2 : infinite capacity of input buffers & finite capacity of output buffers
Type 3 : finite capacity of input buffers & infinite capacity of output buffers
Type 4 : finite capacity of input buffers & finite capacity of output buffers

It is unnecessary to consider the Type 1, because the system throughput is equal to the system arrival rate regardless of the workload allocation.

In case of Type 2, the system throughput is derived as follows :

$$f(\lambda_1^*, \cdots, \lambda_n^*, OB_1, \cdots, OB_n, \mu) = \mu \cdot (1 - \prod(idle)) = \mu \cdot (1 - \frac{1-\rho}{G}) \quad (2)$$

where

$$\rho = \frac{\sum_{i_1=1}^{n} \lambda_i^*}{\mu},$$

$$G = (1-\rho) + \sum_{k_1=0}^{OB_1} \cdots \sum_{k_n=0}^{OB_n} (1-\rho) \cdot \rho^{(k_1+\cdots+k_n+1)} \cdot \frac{(k_1+\cdots+k_n)!}{k_1! \cdots k_n!} \cdot q_1^{k_1} \cdots q_n^{k_n},$$

λ_i^* and OB_i denote the input rate and the capacity of output buffer i. And, the throughput function $f(\lambda_1^*, \cdots, \lambda_n^*, OB_1, \cdots, OB_n, \mu)$ is characterized as follows.

Property 1. In the second-level queue, the throughput is maximized at the balanced input rate.

Proof :

For simplification, the proof will be completed only for the case of 2 output buffers with one capacity ($n=2$, and $OB_1=OB_2=1$), respectively. The general case can easily be derived in the similar process.

By the definition of throughput and G,

$$f(\lambda_1^*, \lambda_2^*, OB_1, OB_2, \mu) = \mu \cdot (1 - \frac{(1-\rho)}{G})$$

$$= \mu[1 - \frac{(1-\rho)}{((1-\rho)(1+\rho+\rho^2+2\rho^3 q_1 q_2))}]$$

$$= \mu[1 - \frac{1}{(1+\rho+\rho^2+2\rho^3 q_1 q_2)}]$$

Since $q_1+q_2 = 1$ and $q_1 q_2$ is maximized at $q_1=q_2= 0.5$, $f(\lambda_1^*, \lambda_2^*, OB_1, OB_2, \mu)$ is maximized at the same input rate ($\lambda_1^* = \lambda_2^*$).
This completes the proof.

In case of Type 3, in order to maximize the system throughput according to the routing probability, it is necessary to consider the throughputs of the first-level queues. It is recalled that the first-level queue is composed of finite input buffer and a single machine. Since arrivals follow a Poisson process with rate λ at the AS/RS, arrivals at each first-level queue follow a Poisson process with rate $\lambda_i (= \lambda \cdot \gamma_i)$ due to the property of Poisson decomposition, and the service times of machines are exponentially distributed.

The throughput of the first-level queue is derived as follows :

$$f(\lambda_i, IB_i, \mu_i) = \frac{\lambda_i \cdot \mu_i(\lambda^{IB_i} - \mu^{IB_i})}{\lambda^{IB_i+1} - \mu^{IB_i+1}} \quad \text{if} \quad \lambda_i \neq \mu_i \quad (3)$$

$$\frac{\lambda_i \cdot IB_i}{IB_i + 1} \quad \text{if} \quad \lambda_i = \mu_i$$

where λ_i and μ_i denote the Poisson input rate and the exponential service rate, respectively. And, the throughput $f(\lambda_i, IB_i, \mu_i)$ is characterized as follows.

Lemma 2. *The throughput of the first-level queue is a monotonically increasing concave function of input rate, that is, for $0 < c < 1$.*

$$cf(\lambda_i, IB_i, \mu_i) < f(c \cdot \lambda_i, IB_i, \mu_i) < f(\lambda_i, IB_i, \mu_i) \qquad (4)$$

The system configuration is composed of n parallel workstations so that the system throughput is given in the sum of throughputs of the first-level queues. Thus, in order to determine the routing probabilities maximizing the system throughput, a mathematical formulation is derived as follows.

Problem P :

$$\begin{aligned} maximize \quad & \sum_{i=1}^{n} f(\lambda_i, IB_i, \mu_i) \\ subject\ to \quad & \sum_{i=1}^{n} \gamma_i = 1 \\ & 0 <= \gamma_i <= 1 \qquad for\ all\ \ i \end{aligned} \qquad (5)$$

The throughput function of the problem P is concave function of input rate by the result of Lemma 2. Since the objective function is concave and the constraints form a convex set of decision variable γ_i, the problem is the convex programming problem, which is solved by using the Lagrangean method. This leads to the following property to maximize the system throughput.

Property 2. The system throughput is maximized at the routing probabilities γ_i satisfying the relation $f'(\lambda_i, IB_i, \mu_i) = f'(\lambda_j, IB_j, \mu_j)$ for all i, j, where f' is the first derivative f with respect to γ_i.

Proof :

Let $x_i = \frac{\lambda \cdot \gamma_i}{\mu_i}$, then $f(\lambda_i, IB_i, \mu_i) = \mu \cdot (1 - \frac{1-x_i}{1-x_i^{IB_i+1}})$.

And, let $f(x_i) = 1 - \frac{1-x_i}{1-x_i^{IB_i+1}}$, then the problem P is transformed to problem PR.

Problem PR :

$$\begin{aligned} maximize \quad & \sum_{i=1}^{n} \mu_i \cdot f(x_i) \\ subject\ to \quad & \sum_{i=1}^{n} \mu_i \cdot x_i = \lambda \end{aligned} \qquad (6)$$

The formulation of the Lagrangean is then derived as

$$L(x_1, \cdots, x_n, \theta) = \sum_{i=1}^{n} \mu_i \cdot f(x_i) - \theta(\sum_{i=1}^{n} \mu_i \cdot x_i - \lambda) \qquad (7)$$

where θ is called Lagrange multiplier.

To obtain the solution value of x_i , we must solve the equations.

$$\begin{aligned} \frac{\partial L}{\partial x_i} &= \mu_i \cdot f'(x_i) - \theta\mu_i = 0 \qquad for\ all\ \ i \\ \frac{\partial L}{\partial \theta} &= \sum_{i=1}^{n} \mu_i \cdot x_i - \lambda = 0 \end{aligned} \qquad (8)$$

The solution of the equations yields
$$\theta = f'(x_i) = f'(x_j) \quad \text{for all} \quad i,j \tag{9}$$
This completes the proof.

The result of Property 2 implies that the routing probabilities should be determined in view of load balance. For example, if the workstations have the same input buffer capacities, the routing probabilities ($\gamma_i = \frac{\mu_i}{\sum_{i=1}^{n} \mu_i}$) maximize the system throughput.

In case of Type 4, the problem of determining the routing probability can be analyzed by using the results of Property 1 and 2, and the system throughput can be obtained by using the given performance evaluation model. In order to maximize the system throughput according to the routing probability, the given performance evaluation model is applied to the whole system with parameter set 1 ($\lambda = 1$, $\mu_1 = \mu_2 = 1$, $\mu = 1$, $IB_1 = IB_2 = 2$, $OB_1 = OB_2 = 1$), parameter set 2 ($\lambda = 1$, $\mu_1 = \mu_2 = 1$, $\mu = 1$, $IB_1 = IB_2 = 1$, $OB_1 = OB_2 = 2$), parameter set 3 ($\lambda = 1$, $\mu_1 = 1$, $\mu_2 = 0.5$, $\mu = 1$, $IB_1 = IB_2 = 2$, $OB_1 = OB_2 = 1$), where λ, μ_i, and μ denote the arrival rate at the AS/RS, the machine service rate, and the AGV service rate, respectively. The results of system throughput according to the routing probability are shown in Figure 1.

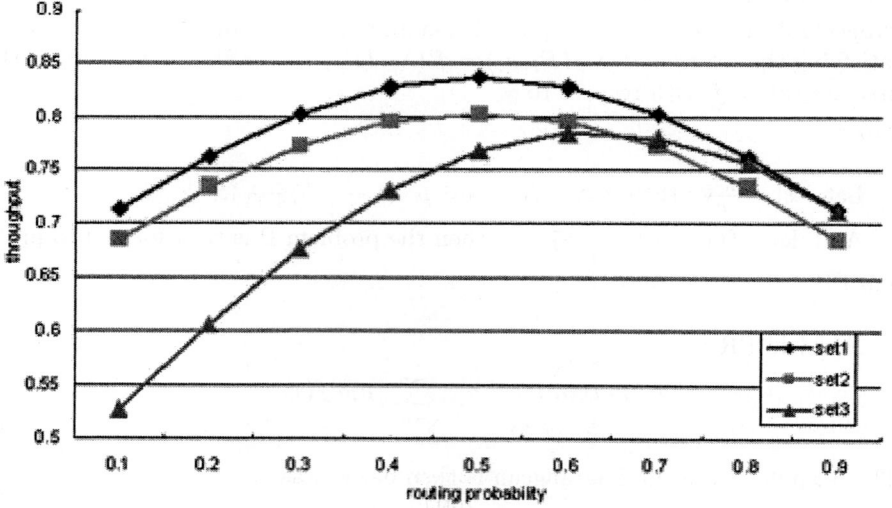

Fig. 1. Throughput according to the routing probability γ_1

The result of parameter set 1 and 2 represent that the throughput is maximized at the same routing probability($\gamma_1 = \gamma_2 = 0.5$) when the workstations have the same buffer capacity ($IB_1 = IB_2, OB_1 = OB_2$) and machine service rate ($\mu_1 = \mu_2$). And, in view of throughput, it is better to allocate the available spaces to input buffer than output buffer by comparison with results of parameter set 1 and 2. From the results of parameter set 3, in order to maximize the system throughput, the routing probability should be determined in proportion to the machine capability when the parameters are the same except the machine service rate.

4 Conclusions

In this paper, a design aspect of a flexible manufacturing system composed of several parallel workstations each with both input and output buffers where two AGVs are used for input and output material handling is considered. The optimal design decision is made on the allocation of workload (routing probability) under the given performance evaluation model.

Some interesting properties are derived that are useful for characterizing optimal allocation of workload. The computational results show that the system throughput is maximized as the routing probability is determined in view of load balance.

Further research is to consider the optimal buffer allocation problem and also to extend these concepts to general production systems.

References

1. Calabrese, J.M.: Optimal Workload Allocation in Open Networks of Multiserver Queues. Management Science, vol. 38, No. 12, (1992) 1792-1802.
2. Dallery, Y. and Stecke, K.E.: On the Optimal Allocation of Servers and Workloads in Closed Queueing Networks. Operation Research, Vol.38, No. 4, (1990) 694-703.
3. Dowdy, L.W., Eager, D.L., Gordon, K.D. and Saxton, L.V.: Throughput Concavity and Response Time Convexity. Information Processing Letters, **19**, (1984) 209-212.
4. Everett, H.: Generalized Lagrange Multiplier Method for Solving Problems of Optimum Allocation of Resources. Operations Research, **11**, (1963) 399-417.
5. Ma, J. and Matsui, M.: Performance evaluation of a flexible machining/assembly system and routing comparisons. International Journal of Production Research, **40**(7), (2002) 1713-1724.
6. Martin, G.E.: Optimal Design of Production Lines. International Journal of Production Research, **32**(5), (1994) 989-1000.
7. Ooijen, H.P.G and Bertrand, J.W.M.: The effects a simple arrival rate control policy on throughput and work-in-process in production systems with workload dependent processing rates. International Journal of Production Economics, **85**, (2003) 61-68.
8. Stecke, K.E. and Morin, T.L.: The Optimality of Balancing Workloads in Certain Types of Flexible Manufacturing Systems. European Journal of Operational Research, Vol. 20, (1985) 68-82.
9. Sung, C.S. and Kwon, S.T.: Performance Modelling of an FMS with Finite Input and Output Buffers. International Journal of Production Economics, **37**, (1994) 161-175.
10. Vinod, B. and Solberg, J.J.: The Optimal Design of Flexible Manufacturing Systems. International Journal of Production Research, **23**(6), (1985) 1141-1151.

NEOS Server Usage in Wastewater Treatment Cost Minimization

I.A.C.P. Espírito-Santo[1], Edite M.G.P. Fernandes[1],
M.M. Araújo[1], and E.C. Ferreira[2]

[1] Systems and Production Department, Minho University, Braga, Portugal
{iapinho, emgpf, mmaraujo}@dps.uminho.pt
[2] Centre of Biological Engineering, Minho University, Braga, Portugal
ecferreira@deb.uminho.pt

Abstract. This paper describes the optimal design and operation of an activated sludge system in wastewater treatment plants. The optimization problem is represented as a smooth programming problem with linear and nonlinear equality and inequality constraints, in which the objective is to minimize the total cost required to design and operate the activated sludge system under imposed effluent quality laws. We analyze four real world plants in the Trás-os-Montes region (Portugal) and report the numerical results obtained with the FILTER, IPOPT, SNOPT and LOQO optimizers.

1 Introduction

Wastewater treatment plants (WWTP's) are nowadays emerging everywhere as authorities concerned with environmental issues legislate tighter laws on water quality. The high costs associated with the plant installation and operation require a wise optimization of the process.

A typical WWTP is usually defined by a primary treatment, a secondary treatment and in some cases a tertiary treatment. The primary treatment is a physical process and aims to eliminate the gross solids and grease, so avoiding the blocking up of the secondary treatment. As its cost does not depend too much on the characteristics of the wastewater, we chose not to include it in the optimization procedure. The secondary treatment is a biological process and is the most important treatment in the plant because it eliminates the soluble pollutants. When the wastewater is very polluted and the secondary treatment does not provide the demanded quality, a tertiary treatment, usually a chemical process, can be included.

This paper is part of an ongoing research project in which we are engaged to optimize the design and the operation of WWTP's in terms of minimum total cost (investment and operation costs). The work herein presented focus solely on the secondary treatment, in particular on an activated sludge system, because this is the chosen secondary treatment to be used by the four plants that we propose to analyze - Alijó, Murça, Sabrosa de Aguiar and Sanfins do Douro -

that are located in Trás-os-Montes region. This is a poor country region in the north of Portugal that produces high quality wines and has, besides domestic effluents, significant effluent variations in terms of amount of pollution and flow, during the vintage season. The mentioned system consists of an aeration tank and a secondary settler. The influent enters the aeration tank where the biological reactions take place, in order to remove the dissolved carbonaceous matter and nitrogen. The sludge that leaves this tank enters in the secondary settler to remove the suspended solids. After this treatment, the treated final effluent leaves the settling tank and the thickened sludge is recycled to the aeration tank and part of it is wasted.

The aim of this paper is to determine the optimal design and operation of the four above mentioned WWTP's, guaranteeing the water quality with pollution levels lower than the maxima defined by portuguese laws.

The optimal design and operation consists of finding the optimal aeration tank volume, sedimentation area and depth of the secondary settler tank, the air flow needed, to name a few, which yield the lowest total cost of the system.

The mathematical modelling of the system results in a smooth nonconvex nonlinear constrained optimization problem that is to be solved by NEOS Server (http://www-neos.mcs.anl.gov/neos/) optimization tools.

To the best of our knowledge, apart the work done by Tyteca et al. [5], that uses simple models to describe the aeration tank and the secondary settler, no WWTP real optimization has been published until now. Previous published work on activated sludge systems using ASM type models [4], [11] and, either the ATV [1] or the double exponential model [12] for settling tanks, focus on obtaining the best combination of the state variables testing by simulation two or three alternative designs and choosing the one with the lowest cost [8], [9], [10], [13]. The simulation is carried out using WEST++ (http://www.hemmis.be), GPS-X (http://www.hydromantis.com) and DESASS [8].

This paper is organized as follows. In Sect. 2 we present a not too much technical description of the equations of the mathematical model. Section 3 is devoted to the listing of some optimization tools in the NEOS Server for use to the public. Section 4 reports on the numerical experiments done on four real world problems and Sect. 5 contains the conclusions.

2 Mathematical Modelling

The system under study consists of an aeration tank, where the biological reactions take place, and a secondary settler for the sedimentation of the sludge and clarification of the effluent. To describe the aeration tank we chose the activated sludge model n.1, described by Henze et al. [4], which considers both the elimination of the carbonaceous matter and the removal of the nitrogen compounds. This model is widely accepted by the scientific community, as it produces good predictive values by simulations. This means that all state variables keep their biological interpretation. The tank is considered a completely stirred tank reactor (CSTR) in steady state. For the settling tank the ATV design procedure

[1] is used, which is a very simple model but describes the settling process very well, besides considering also peak flow events.

The problem contains seven sets of constraints. The first set results from mass balances around the aeration tank using the Peterson matrix of the ASM1 model [4]. The generic equation for a mass balance around a certain system considering a CSTR is

$$\frac{Q}{V_a}(\xi_{in} - \xi) + r_\xi = \frac{d\xi}{dt},$$

where Q is the flow that enters the tank, V_a is the aeration tank volume, ξ and ξ_{in} are the concentrations of the component around which the mass balances are being made inside the reactor and on entry, respectively. It is convenient to refer that in a CSTR the concentration of a compound is the same at any point inside the reactor and at the effluent of that reactor. The reaction term for the compound in question, r_ξ, is obtained by the sum of the product of the stoichiometric coefficients, $\nu_{\xi j}$, with the expression of the process reaction rate, ρ_j, of the ASM1 Peterson matrix [4]

$$r_\xi = \sum_j \nu_{\xi j} \rho_j.$$

In steady state, the accumulation term given by $\frac{d\xi}{dt}$ is zero, because the concentration is constant in time. A WWTP in labor for a sufficiently long period of time without significant variations can be considered at steady state. As our purpose is to make cost predictions in a long term basis it is reasonable to do so. The ASM1 model involves 8 processes incorporating 13 different components, such as the substrate, the bacteria, dissolved oxygen, among others. For the sake of clearness, we include here the mass balance equation related to one of the components - the soluble substrate (S_S):

$$\frac{-\mu_H}{Y_H} \frac{S_S}{K_S + S_S} \left(\frac{S_O}{K_{OH} + S_O} + \eta_g \frac{K_{OH}}{K_{OH} + S_O} \frac{S_{NO}}{K_{NO} + S_{NO}} \right) X_{BH}$$

$$+ k_h \frac{X_{BH}}{K_X X_{BH} + X_S} \left(\frac{S_O}{K_{OH} + S_O} + \eta_h \frac{K_{OH}}{K_{OH} + S_O} \frac{S_{NO}}{K_{NO} + S_{NO}} \right) X_S$$

$$+ \frac{Q}{V_a}(S_{S_{in}} - S_S) = 0.$$

We denote all the soluble components by $S_?$ and the particulates by $X_?$. All the other symbols are stoichiometric or kinetic parameters for the wastewater considered. (See [4] for details on how to obtain all the other equations.)

The second group of constraints concern the secondary settler and are set using the ATV procedure design [1]. Traditionally the secondary settler is underestimated when compared with the aeration tank. However, it plays a crucial role in the activated sludge system. When the wastewater leaves the aeration tank, where the biological treatment took place, the treated water should be separated from the biological sludge, otherwise, the chemical oxygen demand would be higher than it is at the entry of the system. The most common way of achieving this purpose is by sedimentation in tanks. The optimization of the sedimentation area and depth must rely on the sludge characteristics, which in turn are related with the performance of the aeration tank. So, the operation of the biological reactor influences directly the performance of the settling tank and for that reason, one should never be considered without the other. The ATV design procedure contemplates the peak wet weather flow (PWWF) events, during which there is a reduction in the sludge concentration. To turn around this problem, a certain depth is allocated to support the fluctuation of solids during these events ($h_3 = \Delta X V_a \frac{DVSI}{480 A_s}$). This way a reduction in the sedimentation area (A_s) is allowed. A compaction zone ($h_4 = X_p \frac{DVSI}{1000}$), where the sludge is thickened in order to achieve the convenient concentration to return to the biological reactor, also has to be contemplated and depends only on the characteristics of the sludge. $DVSI$ is the diluted volumetric sludge index and ΔX is the variation of the sludge concentration inside the aeration tank in a PWWF event. A clear water zone (h_1) and a separation zone (h_2) should also be considered and are set empirically ($h_1 + h_2 = 1$, say). The depth of the settling tank, h, is the sum of these four zones.

The sedimentation area is still related to the peak flow, Q_p, by the expression

$$\frac{Q_p}{A_s} \leq 2400 \left(X_p DVSI\right)^{-1.34}.$$

The other important group of constraints are a set of linear equalities and define composite variables. In a real system, some state variables are, most of the time, not available for evaluation. Thus, readily measured composite variables are used instead. For example, the chemical oxygen demand (COD) is composed by soluble and particulate components, that are related by the equation

$$COD = S_I + S_S + X_I + X_S + X_{BH} + X_{BA} + X_P.$$

Similar equations can be defined for the volatile suspended solids (VSS), total suspended solids (TSS), biochemical oxygen demand (BOD), total nitrogen of Kjeldahl (TKN) and total nitrogen (N).

The system behavior, in terms of concentration and flows, may be predicted by balances. In order to achieve a consistent system, these balances must be done around the entire system and not only around each unitary process. They were done to the suspended matter, dissolved matter and flows and these correspond to the fourth group of constraints. The equations for particulate compounds (organic and inorganic) have the following form

$$(1+r)Q_{inf} X_{?_{ent}} = Q_{inf} X_{?_{inf}} + (1+r)Q_{inf} X_? - \frac{V_a X_?}{SRT X_{?_r}} (X_{?_r} - X_{?_{ef}})$$

$$- Q_{inf} X_{?_{ef}}$$

and for the solubles we have

$$(1+r)Q_{inf} S_{?_{in}} = Q_{inf} S_{?_{inf}} + r Q_{inf} S_{?_r}$$

where r is the recycle rate, SRT is the sludge retention time and $Q_?$ represents the volumetric flows. As to the subscripts, inf concerns the influent wastewater, ent the entry of the aeration tank, r the recycled sludge and ef the treated effluent.

It is also necessary to add some system variables definitions, in order to define the system correctly. In this group we include the sludge retention time, the recycle rate, hydraulic retention time (HRT), recycle rate in a PWWF event (r_p), recycle flow rate in a PWWF event (Q_{r_p}) and maximum overflow rate ($\frac{Q_\text{p}}{A_\text{s}}$):

$$SRT = \frac{V_\text{a} X}{Q_\text{w} X_\text{r}}$$

$$HRT = \frac{V_\text{a}}{Q}$$

$$r = \frac{Q_\text{r}}{Q_\text{inf}}$$

$$r_\text{p} = \frac{0.7\, TSS}{TSS_{\text{max}_\text{p}} - 0.7\, TSS}$$

$$Q_{r_\text{p}} = r_\text{p} Q_\text{p}$$

$$\frac{Q_\text{p}}{A_\text{s}} \leq 2 \ .$$

A fixed value for the relation between volatile and total suspended solids was considered

$$\frac{VSS}{TSS} = 0.7 \ .$$

All the variables are considered nonnegative, although more restricted bounds are imposed to some of them due to operational consistencies. For example, the dissolved oxygen has to be always greater or equal to 2 mg/L. These conditions define a set of simple bounds on the variables.

Finally, the quality of the effluent has to be imposed. The quality constraints are usually derived from law restrictions. The most used are related with limits in the COD, N and TSS at the effluent. In mathematical terms, these constraints are defined by portuguese laws as $COD_\text{ef} \leq 125$, $N_\text{ef} \leq 15$ and $TSS_\text{ef} \leq 35$.

The objective cost function used represents the total cost and includes both investment and operation costs. The operation cost is usually on annual basis, so it has to be updated to a present value using the adequate economic factors of conversion. Each term in the objective function is based on the basic model $C = aZ^b$ [5], where a and b are the parameters to be estimated, C is the cost and Z is the characteristic of the unitary process that most influences the cost. For example, for the investment cost of the aeration tank, the volume (V_a) and air flow (G_S) are considered. The parameters a and b are estimated by the least squares technique, using real data collected from a WWTP building company. The operation cost of the aeration tank considers the air flow, and the investment and operation costs of the secondary settler depend on the sedimentation area, A_s, and the depth, h. Summing up all these terms, we get the following objective cost function:

$$F_\mathrm{obj} = 174.2 V_\mathrm{a}^{1.07} + 12487 G_\mathrm{S}^{0.62} + 114.8 G_\mathrm{S} + 955.5 A_\mathrm{s}^{0.97} + 41.3 \left(A_\mathrm{s} h\right)^{1.07} \ . \quad (1)$$

3   NEOS Server Usage

NEOS Server provides the possibility to run problems on powerful machines in a user friendly manner through the internet.

Depending on the type of optimization problem, the user has a list of solvers to choose from. The choice of solver is also dictated by the language used to define the optimization problem. Our problem was coded in AMPL format (http://www.ampl.com/cm/cs/what/ampl/).

The solvers for smooth nonlinear constrained optimization problems with AMPL input format are the following: FILTER, IPOPT, LOQO, SNOPT, KNITRO, LANCELOT, MINOS, MOSEK and PENNON.

From the list, we excluded immediately the MOSEK optimizer as it does not work for nonconvex problems. KNITRO, LANCELOT, MINOS and PENNON were also excluded because the first converged only for some of the carried out runs and the others did not converge at all. The remaining four optimizers converged in all runs although not all to the same solution. A brief description of each one follows.

FILTER is a software developed by R. Fletcher and S. Leyffer that is based on a Filter-SQP algorithm and implements a Sequential Quadratic Programming trust region algorithm with a filter to promote global convergence [2]. The idea of a filter is motivated by the aim of avoiding the need to use penalty parameters as required by l_1 or augmented Lagrangian merit functions.

IPOPT is an optimizer developed by A. Wächter, L. T. Biegler, A. Raghunathan and Yi-Dong Lang. It implements a primal-dual interior point algorithm with a filter line search strategy. As a barrier method, the algorithm computes approximate solutions for a sequence of barrier problems (associated with the original problem) for a decreasing sequence of positive barrier parameters converging to zero. The barrier problems are solved using a filter line search algorithm. We refer to the Technical Report [7] for details.

LOQO (http://www.princeton.edu/~rvdb/loqo/) solver was developed by R. J. Vanderbei and H. Y. Benson, and it is based on an infeasible primal-dual interior point method with an l_2 penalty merit function to ensure progress toward feasibility and optimality [6].

SNOPT (http://www.sbsi−sol−optimize.com/asp/sol_product_snopt.htm) is a sequential quadratic programming method for large-scale optimization problems involving general linear and nonlinear constraints that uses an active-set approach [3].

We do not aim to analyze the performance of the solvers but rather to solve our design problem which, being a medium-scale problem, turns out to be a quite difficult one.

4 Computational Results

The problem of the optimal design and operation of the activated sludge system consists of finding the volume of the aeration tank, the air flow needed for the aeration tank, the sedimentation area, the secondary settler depth, the recycle rate, the effluent flow and concentration of total suspended solids, carbonaceous matter and total nitrogen in the treated water, to name a few, in such a way that, verifying the aeration tank balances as well as the system balances, satisfy the composite variables constraints, the secondary settler constraints, the system variables definition constraints, the quality constraints and the simple bounds on the variables, and minimize the cost function (1). Our formulated problem has 57 parameters, 82 variables and 64 constraints, where 28 are nonlinear equalities, 35 are linear equalities and there is only one nonlinear inequality. Seventy one variables are bounded below and eleven are bounded below and above. The chosen values for the stoichiometric, kinetic and operational parameters that appear in the mathematical formulation of the problem are the default values presented in the simulator GPS-X, and they are usually found in real activated sludge based plants for domestic effluents.

The collected data from the four analyzed small towns are listed in Table 1. These data consider the population equivalent, the influent flow, the peak flow, the influent COD, the influent TSS and define average conditions that are crucial for the dimensioning of the plant.

Table 1. Data collected from the four small towns

	\multicolumn{4}{c}{Location of the WWTP}			
	Alijó	Murça	Sabrosa	Sanfins
pop. eq.	6850	3850	2750	3100
influent flow (m^3/day)	1050	885	467.5	530
peak flow (m^3/h)	108	86.4	48.6	54
COD (Kg/m^3)	2000	1750	1250	1250
TSS (Kg/m^3)	750	660	610	610

Several experiences were done for the WWTP's under study, using the available NEOS Server solvers mentioned in Sect. 3, and considering different values of COD reduction in the preliminary treatment. This reduction typically varies from 40 to 70%. Table 2 presents the effect of the primary treatment efficiency on the cost of the activated sludge system for the WWTP from Alijó. The solver used was the FILTER. The values of the total cost are in millions of euros. In the table, we report the number of iterations needed by FILTER to converge to

Table 2. Comparison of the results in the WWTP from Alijó considering different COD reductions in the primary treatment

	COD reduction		
	40%	55%	70%
total cost	11.33	7.62	5.25
iterations	28	20	21
func. eval.	15	11	10
cons. eval.	34	20	21

Table 3. Results for the studied WWTPs for different solvers, considering 70% of COD reduction in the preliminary treatment

Solver		Location of the WWTP			
		Alijó	Murça	Sabrosa	Sanfins
FILTER	total cost	5.25	4.03	6.23	1.46
	iterations	21	20	19	40
	func. eval.	10	12	1	29
	cons. eval.	21	21	23	40
IPOPT	total cost	5.25	4.03	1.33	1.46
	iterations	53	50	68	63
	func. eval.	119	57	70	68
	cons. eval.	119	57	70	68
SNOPT 6.2	total cost	8.37	4.03	1.56	1.46
	iterations	346	529	853	665
	func. eval.	53	113	404	277
	cons. eval.	52	112	403	276
LOQO 6.06	total cost	8.36	5.91	1.56	1.70
	iterations	85	74	41	45
	func. eval.	85	74	41	45
	cons. eval.	85	74	41	45

the solution, the number of function evaluations and the number of constraints evaluations. As shown, the efficiency of a primary treatment is crucial because the higher is the achieved COD reduction, the lower is the investment and operation cost of the secondary treatment. We remark that the cost of the preliminary treatment is also related with its efficiency, although not as dramatic as the cost of the activated sludge system. Thus, for the remaining experiences we assume the most favorable situation, i.e., we assume that the preliminary treatment has an efficiency of 70%.

Table 3 reports on the minimum total cost (in millions of euros), of the four WWTP's under study, the number of iterations up to finding a solution and the number of function and constraints evaluations, using each one of the listed solvers. The solvers find the solutions using their default settings.

Table 4. Results of the optimal design and operation solution for the studied WWTPs considering 70% of COD reduction in the preliminary treatment

	Alijó	Murça	Sabrosa	Sanfins
$V_a(m^3)$	1673	1203	395	448
$A_s(m^2)$	217	173	97	108
$h(m)$	5.4	5.0	3.6	3.6
$G_S(m^3/\text{day STP})$	8707	6039	1147	1300
$COD_{ef}(g\,COD/m^3)$	98.8	99.6	125	125
$TSS_{ef}(g/m^3)$	35.0	35	35	35
$N_{ef}(g\,N/m^3)$	8.2	9.5	13.0	13.0

Some conclusions may be drawn. The solution found by each of the four solvers is not always the same. We also observe an overall advantage in using IPOPT optimizer as it converges to a solution with the lowest total cost in all plants.

Table 4 reports on optimal values of the aeration tank volume, sedimentation area, settler depth, air flow, chemical oxygen demand at the effluent, total suspended solids at the effluent and nitrogen at the effluent obtained by IPOPT optimizer for each plant. We remark that although the achieved values of TSS correspond to the imposed law limit, the same does not occur with COD and N.

The nitrogen that enters in the system is only the quantity requested to ensure the growth of the bacteria present in the biological sludge. This means that in this kind of populations the nitrogen levels are not considered pollutant. As it can be seen in Table 1, the nitrogen does not appear as an entering parameter. For that reason, the nitrogen at the effluent never reaches the limit imposed by law.

As to the COD we have a different situation. In the largest WWTP's (Alijó and Sanfins) the imposed law limit is not reached because to be able to achieve the TSS limit, the system is capable of removing more COD than the demanded. The opposite occurs in the other two plants. As they are very small, the minimum cost is achieved only when the COD reaches the limit.

5 Conclusions

In this paper we consider the optimal design and operation, in terms of minimum installation and operation cost, of an activated sludge system in WWTP's from the north of Portugal, based on portuguese real data and effluent quality law limits. Four real WWTP's were analyzed and the optimization of the problems was carried out running NEOS Server solvers (FILTER, IPOPT, LOQO and SNOPT).

From our numerical experiences, we may conclude that the efficiency of the primary treatment influences directly and in a very expressive way the resulting cost of the biological treatment. To have a more realistic idea of the best solution, the whole treatment plant should be considered as future developments.

Acknowledgement. The authors acknowledge the company Factor Ambiente (Braga, Portugal) for the data provided.

References

1. Ekama, G. A., Barnard, J. L., Günthert, F. W., Krebs, P., McCorquodale, J. A., Parker, D. S., Wahlberg, E. J.: Secondary Settling tanks: Theory, modelling, design and operation, Technical Report 6. IAWQ - international association on water quality (1978)
2. Fletcher, R., Leyffer, S., Toint, P. L.: On the Global Convergence of a Filter-SQP Algorithm, Technical Report NA/197 (2002)
3. Gill, P. E., Murray, W., Saunders, M. A.: SNOPT: An SQP algorithm for large-scale constrained optimization. SIAM J. Optim. **12** (2002) 979–1006
4. Henze, M., Grady Jr, C. P. L.,Marais, G. V. R., Matsuo, T.: Activated sludge model no 1, Technical Report 1. IAWPRC Task Group on Mathematical Modelling for design and operation of biological wastewater treatment (1986)
5. Tyteca, D., Smeers Y., Nyns, E. J.: Mathematical Modeling and Economic Optimization of Wastewater Treatment Plants. CRC Crit. Rev. in Environ. Control **8**(1) (1977) 1–89
6. Vanderbei, R. J., Shanno, D. F.: An Interior-Point Algorithm for Nonconvex Nonlinear Programming. Comp. Optim. and Appl. **13**(1997) 231–252
7. Wächter, A., Biegler, L. T.: On the Implementation of an Interior-Point Filter Line-Search Algorithm for Large-Scale Nonlinear Programming. Technical Report (http://www.research.ibm.com/people/a/andreasw/papers/ipopt.pdf) (2004)
8. Seco, A., Serralta, J., Ferrer, J.: Biological Nutrient Removal Model No.1 (BNRM1). Wat. Sci. Tech. **50**(6) (2004) 69–78
9. Otterpohl, R., Rolfs, T., Londong, J.: Optimizing operation of wastewater treatment plants by offline and online computer simulation. Wat. Sci. Tech. **30**(2) (1994) 165–174
10. Gillot, S., De Clercq, B., Defour, F., Gernaey, K., Vanrolleghem, P. A.: Optimization of Wastewater Treatment Plant Design and Operation using Simulation and Cost Analysis. In: Proceedings 72nd Annual WEF Conference and Exposition. New Orleans, USA (1999) 9–13
11. Henze, M., Gujer, W., Mino, T., Matsuo,T., Wentzel, M. C., Marais, G. V. R., Van Loosdrecht, M. C. M.: Activated Sludge Model No. 2d, (ASM2d). Wat. Sci. Tech. **39**(1) (1999) 165–182
12. Takács, I., Patry, G. G., Nolasco, D.: A Dynamic Model of the Clarification-Thickening Process. Wat. Res. **25**(10) (1991) 1263–1271
13. Copp, J. B., editor: The Cost Simulation Benchmark - Description and Simulator Manual. Office for Official Publications of the European Comunities (2002)

Branch and Price Algorithm for Content Allocation Problem in VOD Network

Jungman Hong[1] and Seungkil Lim[2]

[1] Consulting Division, LG CNS,
Seoul, 100-768, Korea
jmhong@lgcns.com
[2] Division of e-businessIT, Sungkyul University,
Anyang, 430-742, Korea
seungkil@sungkyul.edu

Abstract. This paper considers an optimal video file allocation problem in a video-on-demand (VOD) network. The objective of the problem is to find an optimal video file allocation strategy, giving both types of videos and the number of copies of each video file type to be carried at each distributed server and each local server in the VOD network, that minimizes the associated storage and transmission cost subject to each server capacity. The problem is formulated as a mixed integer programming problem and solved by a column generation approach. Computational results show that the associated column generation algorithm with three exploited valid inequalities applied and a branch-and-bound procedure can together solve practical size problems in reasonable time.

1 Introduction

Recent advances in optical fiber transmission technology and high speed switching technology, and development of huge storage systems are expected to allow communication networks to provide customers with interactive broadband multimedia services such as video on demand (VOD), interactive multimedia entertainment and distance learning. VOD service is the essential part of various multimedia applications in the sense that it allows geographically distributed users to interactively access video files supported from a network of VOD servers. To provide complete VOD services, the issue of how to allocate (assign) video files among servers should be resolved first [1]. A variety of different video file allocation strategies have been studied for various structures of VOD networks as seen in the literature. Hwang et al. [1] have considered a video file allocation heuristic algorithm for finding a feasible number of video copies to be stored at each server to satisfy stochastic demands, while Kwong et al. [2] have considered not an video file allocation problem but a time-dependent capacity planning (analyzing) problem for video servers. Barnett et al. [3] have compared a centralized approach with a distributed one for video file allocation in a non-hierarchical VOD network and concluded that the distributed approach may incur less storage cost and also offer considerable advantages in terms of smaller bandwidth

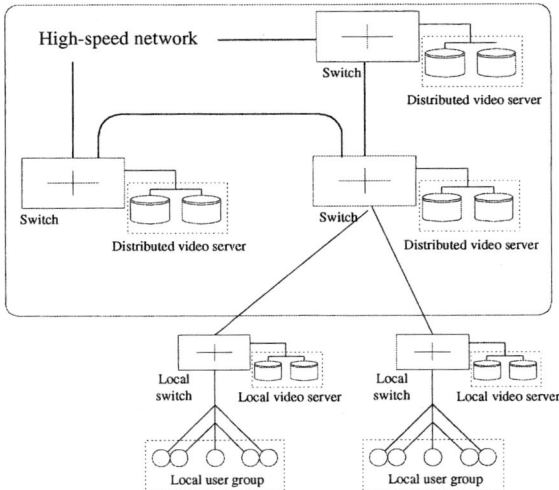

Fig. 1. Overall structure of the VOD network architecture

requirement and better service quality. Cidon et al.[4] have developed a distributed optimal location algorithm in a hierarchical network architecture for the distribution of multimedia content.

The two-level hierarchical VOD network to be considered in this paper is represented as a distributed network composed of a mesh type (general type) of high-speed network in the upper level and many logical star types of local access networks in the lower level. In the network, deterministic demands are considered. Moreover, it is assumed that any video files carried at the distributed video server can be shared among all the local user groups in the network, while the local video servers can serve their associated local user groups only. Figure 1 illustrates such a VOD network. In such a two-level network, most popular video files are commonly carried at local video servers so that most local user group's requests can be served by the local servers. However, if any video file requested by a local user group is not carried at a local video server and if the local video server is filled to his capacity with other video files, then the request for the video file will be forwarded to the distributed video server in the upper level where the requested video file is kept in storage. Since the number of users (customers) who are simultaneously served by a single VOD server is bounded above due to each VOD processing capacity, the capacity constraint is considered such that the number of video files to be carried at each single VOD server is limited.

In allocating such video files among the video servers in the proposed VOD network, this paper considers a cost function composed of storage cost (cost of storing at hard disk or memory) of video file copies at servers and transmission cost (charge for communication network service or penalty for any service delay) of video file copies on each associated distribution link in the network. Transmission cost can be minimized by storing as many video file copies as possible at the local video servers, while storage cost can be minimized by storing as many video copies as possible at the distributed video servers. Any video file copy carried at a distributed video server can be shared among all the local user groups and hence it may be better to carry many video file copies at such a distributed video server, since storage cost at the distributed video server can be less expensive due to its economy-of-scale effect. This implies that there is a trade-off relation between transmission cost and storage cost. Such complex interrelationship between the two major cost elements, being concerned with reviewing a huge number of possible video file allocations, may make it extremely hard to find the optimal VOD network design which does optimally trade off the two costs. Therefore, this paper wants to propose a column generation approach by exploiting some valid inequalities.

2 Problem Modeling

This section introduces some notation and definitions, and then presents an integer programming formulation of the proposed problem. For the problem formulation, it is assumed that an associated network layout $G = (N, E)$, a set of video files to be distributed and their demands from each user group, and a set of distributed video servers and their associated storage costs for each video file and their transmission costs on each transmission link are all given. Some notation and definitions are now introduced to derive the integer programming formulation.

E: set of undirected links each connecting two switches,
A: set of directed links each connecting two switches,
K: set of video file types,
L: set of local video servers
M: set of distributed video servers,
N: set of all video servers, $N = L \cup M$,
c_i: capacity of server i, represented by the maximum number of video file copies to be carried at server $i \in N$,
λ_{lk}: expected demand for video file type $k \in K$ from the local group of users to whom local video server $l \in L$ is supposed to provide his service,
μ_k: size of video file type $k \in K$,
$\alpha_{(ij)}$: transmission cost per unit size of video file on link $(i,j) \in A$,
β_i: storage cost per unit size of video file at server $i \in N$,
$x_{ik}=$ variable representing the number of copies of video file type $k \in K$ assigned to server $i \in N$

Undirected and directed links are distinctively denoted by $\{i,j\}$ and (i,j), respectively. Recalling that λ_{lk} corresponds to the expected number of copies of video file type k to satisfy the demand from local user group l, we assume, for simplicity, that the data of λ_{lk} and c_i are given to satisfy the relation $\lceil \lambda_{lk} \rceil < c_i$ for each $l \in L$, $k \in K$ and $i \in N$, where multiple copies of each video file are allowed to serve simultaneously, and where $\lceil z \rceil$ is the smallest integer value greater than or equal to real value z. The proposed problem can be transformed into a binary integer programming problem which will be treated as the master problem in this paper. Let $[a,b]^p$ $(p = 1, 2, \cdots, P)$ be the incident vectors of all the possible storage strategies of a single video file type, say k, where a_i^p denotes the number of video files (copies) of the single type that are allocated at server $i \in N$ in the corresponding column p, and the vector b represents the single type such that b_k^p has the value 1 if column p corresponds to video file type $k \in K$, and 0, otherwise. Each of the vectors represents a strategy for a single video file type to be assigned at all the servers. By using the incident vectors, the problem can be formulated into the following binary integer programming problem, called as Problem MP which will be treated as the master problem for the proposed problem.

Problem MP:

$$Z_{MP} = \min \sum_p \gamma^p y^p \quad (1)$$

s.t.

$$\sum_p a_i^p y^p \leq c_i, \quad \forall i \in N, \quad (2)$$

$$\sum_p b_k^p y^p = 1, \quad \forall k \in K, \quad (3)$$

$$y^p \in \{0, 1\}, \quad p = 1, 2, \cdots, P \quad (4)$$

In the objective function, γ^p denotes the total cost being composed of the storage cost and transmission cost of a video file type corresponding to column p. Constraints (2) ensure that each server has to satisfy its capacity restriction, and constraints (3) ensure that only one column should be selected for each video file type. Then, Problem MP is to find the optimal (cost minimal) column p with its corresponding video file type $k \in K$, where column p corresponds to one of the combinations of video file types and their allocations (copies) among all the servers. Problem MP is, however, not in a complete form in the sense that all the columns P can not explicitly be specified yet, since the parameters a_i^p and b_k^p are too many to define so as to represent all the possible (exponentially many) combinations of video file types and copies. Therefore, a column generation procedure will be exploited to solve the master problem as discussed in the next section. The linear programming relaxation of Problem MP, denoted by LPMP, can be derived by relaxing the variables y^p of constraints (4) to be nonnegative ones, and its objective function value is denoted by Z_{LPMP}. Problem LPMP has exponentially many y^p variables. However, it will be shown in the next section that the LP relaxation of Problem MP, LPMP, can be solved efficiently by the column generation method which has successfully been used in solving many communication network configuration problems[5, 6, 7, 8].

3 Solution Approach

3.1 Column Generation Problem

Let u_i and v_k be the nonnegative dual variables associated with constraints (2) and (3), respectively. Also, let u_i^* and v_k^* be the optimal values of the dual variables solved from any current Problem LPMP. The column generation problem for video file type $k \in K$, called Problem SP$(k)'$, can now be derived by use of the optimal values, v_i^* and v_k^*, of the dual variables [9] and path variables ϕ_{il}^k (representing the flow amount from distributed server i to local server l).

Problem SP$(k)'$:

$$\min \sum_{i \in N} (\mu_k \beta_i - u_i^*) x_{ik} + \mu_k \sum_{i \in M} \sum_{l \in L} p_{il} \phi_{il}^k - v_k^* \quad (5)$$

s.t.

$$\sum_{l \in L} \phi_{il}^k \leq x_{ik}, \quad \forall i \in M, \quad (6)$$

$$\sum_{i \in M} \phi_{il}^k + x_{lk} \geq \lambda_{lk}, \quad \forall l \in L, \quad (7)$$

$$x_{ik} \in \{0, 1, \cdots, c_i\}, \quad \forall i \in N, \qquad \phi_{il}^k \geq 0, \quad \forall i \in M, \ \forall l \in L, \quad (8)$$

where p_{il} denotes the length (transmission distance) of the shortest path from distributed server i to local user group (server) l for video file type k with link weight $\alpha_{(ij)}$, and path variable ϕ_{il}^k denotes the flow amount of video file type k from distributed server i to local server l to satisfy the demand of local user group l. Constraints (6) ensures that any flow amount of video file type k initiated from an upper-level node i should not be greater than the capacity of the type k at the node i. Similarly, constraints (7) ensures that the total flow-in amount and capacity of video file type k at a lower-level node l should be enough to satisfy the demand for the type k at the node l. If the resulting optimal objective function value of Problem SP$(k)'$ is less than zero, then the newly generated column can be added to the current (restricted) master problem. Otherwise, no column is generated with the video file type $k \in K$. If no more column is generated with all $k \in K$, then the current solution to Problem LPMP will become optimal. Moreover, if this optimal solution is integral, then that solution will become an optimal solution to Problem MP. A branch and cut algorithm is used to solve Problem SP$(k)'$ using some valid inequalities given in the next subsection.

3.2 Valid Inequalities

Some of the valid inequalities to be used in the cutting plane procedure for problem SP$(k)'$ are now exploited. We denote the convex hull of the set of solutions to constraints (6)-(8) by polytope $P_{SP(k)'}$. Then, some valid inequalities will be exploited for polytope $P_{SP(k)'}$. Note that polytope $P_{SP(k)'}$ is of full dimensional because the following $|N|+|M||L|+1$ affinely independent integer solutions are in polytope $P_{SP(k)'}$;

(1) $x_{lk} = \lceil \lambda_{lk} \rceil$, $l \in L$, $x_{ik} = 0$, $i \in M$, $\phi_{il}^k = 0$, $i \in M$, $l \in L$
(2) $x_{\bar{l}k} = \lceil \lambda_{\bar{l}k} \rceil + 1$, $x_{lk} = \lceil \lambda_{lk} \rceil$, $l \in L \setminus \{\bar{l}\}$, $x_{ik} = 0$, $i \in M$, $\phi_{il}^k = 0$, $i \in M$, $l \in L$
(3) $x_{jk} = 1$, $x_{lk} = \lceil \lambda_{lk} \rceil$, $l \in L$, $x_{ik} = 0$, $i \in M \setminus \{j\}$, $\phi_{il}^k = 0$, $i \in M$, $l \in L$
(4) $x_{jk} = 1$, $\phi_{j\bar{l}}^k = 1$, $x_{lk} = \lceil \lambda_{lk} \rceil$, $l \in L$, $x_{ik} = 0$, $i \in M \setminus \{j\}$, $\phi_{il}^k = 0$, $(i,l) \in \{(i,l) | i \in M, l \in L\} \setminus \{(j, \bar{l})\}$,

Now, we want to introduce the concept of a facet. For the facet introduction, let $\sum_{i \in N} \pi_i^1 x_{ik} + \sum_{i \in M} \sum_{l \in L} \pi_{il}^2 \phi_{il}^k \leq \pi_0$ be a valid inequality for a polytope. Then, it can be noted by referring to [9] that the valid inequality $\sum_{i \in N} \pi_i^1 x_{ik} + \sum_{i \in M} \sum_{l \in L} \pi_{il}^2 \phi_{il}^k \leq \pi_0$ defines a facet of the full dimensional polytope if and only if the dimension of polytope minus one affinely independent feasible integer solutions satisfy $\sum_{i \in N} \pi_i^1 x_{ik} + \sum_{i \in M} \sum_{l \in L} \pi_{il}^2 \phi_{il}^k = \pi_0$.

Now, consider both the sets $H \subset L$ and $Q \subset M$ which satisfy the following two conditions for some positive integers m_q, $q \in Q$;

(i) $\sum_{h \in H} \lambda_{hk} + \eta = \sum_{q \in Q} m_q$, for $0 < \eta < 1$
(ii) for $m_q < c_q$, for all $q \in Q$,

$$\sum_{q \in Q} \sum_{h \in H} \phi_{qh}^k + (1-\eta) \sum_{q \in Q} (m_q - x_{qk}) \leq \sum_{h \in H} \lambda_{hk} \quad (9)$$

Lemma 1. *Inequality (9) is a valid inequality for Problem $SP(k)'$.*

Proof. See Hong [10] for more details on its proof.

Now, as a special case of the valid inequality (9), consider the situation where the cardinality of Q is equal to one as, $|Q| = 1$. Then, the valid inequality (9) can be a facet of polytope $P_{SP(k)'}$ as proved in Theorem 1.

Theorem 1. *If set $H \subset L$ and $i \in M$ satisfy the condition $0 < \eta < 1$, where $\sum_{h \in H} \lambda_{hk} + \eta = m_i < c_i$, then the valid inequality*

$$\sum_{h \in H} \phi_{ih}^k + (1-\eta)(m_i - x_{ik}) \leq \sum_{h \in H} \lambda_{hk} \quad (10)$$

will be a facet of polytope $P_{SP(k)'}$

Proof. This can be proved easily by showing that there exist $|N|+|M||L|$ affinely independent integer solutions, which satisfy the valid inequality (10) at equality [9]. See Hong [10] for more details on its proof.

Similarly, two other valid inequalities can be derived as in Theorems 2 and 3. Proofs for these theorems are omitted since we can prove them similarly as Theorem 1. See Hong [10] for more details on the proofs.

Theorem 2. *If set $H \subset L$ satisfies the conditions*

(i) $\sum_{h \in H} \lceil \lambda_{hk} \rceil = \lceil \sum_{h \in H} \lambda_{hk} \rceil$, and
(ii) $\lceil \lambda_{hk} \rceil > \lambda_{hk}$, for all $h \in H$,

then the following inequality will be a facet of polytope $P_{SP(k)'}$;

$$\sum_{i \in M} x_{ik} + \sum_{h \in H} x_{hk} \geq \sum_{h \in H} \lceil \lambda_{hk} \rceil. \tag{11}$$

Theorem 3. *If the relation $\lceil \lambda_{l^*k} \rceil > \lambda_{l^*k}$ holds, then the valid inequality (12) will be a facet of polytope $P_{SP(k)'}$;*

$$\sum_{i \in M} x_{ik} + x_{l^*k} \geq \lceil \lambda_{l^*k} \rceil \tag{12}$$

3.3 Separation Procedure

Let $(x_{ik}^*, \phi_{il}^{k*})$ be a current fractional solution of the LP relaxation of Problem SP$(k)'$. Then in order to cut off the current fractional solution, we need to solve the associated separation problem which is to find if there exists a node $i \in M$, from the fractional solution indices, which violates the inequality (10) (that is, the cutting plane(10)) and also the associated set $H \subset L$ which satisfies the relations

$$0 < \eta = \lceil x_{ik}^* \rceil - \sum_{h \in H} \lambda_{hk} < 1, \quad \sum_{h \in H} \phi_{ih}^{k*} + (1-\eta)(\lceil x_{ik}^* \rceil - x_{ik}^*) > \sum_{h \in H} \lambda_{hk}.$$

In order to find if inequality (10) is violated at any $i \in M$, we shall use a greedy approach. The first step of our greedy algorithm (separation procedure) is to sort all nodes $i \in M$ in the decreasing order of values $\lceil x_{ik}^* \rceil - x_{ik}^*$. Then the algorithm will continue to check the nodes one by one in the sorted order until any node $i \in M$ satisfies the following conditions;

$$\sum_{h \in H_i} \phi_{ih}^{k*} + (1-\eta)(\lceil x_{ik}^* \rceil - x_{ik}^*) > \sum_{h \in H_i} \lambda_{hk},$$

where $H_i = \{h | \phi_{ih}^{k*} > 0, h \in L\}$. The checking process will stop when any node $i \in M$ satisfying the above conditions is found. Similarly, the separation procedure can be adapted for inequality (11). See Hong [10] for more details on the procedure. Note that inequality (12) that satisfies the condition $\lceil \lambda_{lk} \rceil > \lambda_{lk}$, $l \in L$ can be directly added to Problem SP$(k)'$ at the initial step of the algorithm.

3.4 Overall Solution Algorithm

This section proposes the overall solution algorithm. In the overall algorithm, the initially restricted LPMP starts with the columns corresponding to constraints

(3) and with some sufficiently large values pre-assigned to the objective coefficients γ^p so that the initial LPMP can become feasible. After solving the initial LPMP, the overall solution algorithm proceeds iteratively to generate the most attractive column (greatest contributing to the objective function) at each iteration by solving Problem $SP(k)'$ by use of the cutting plane procedure with all the above derived valid inequalities incorporated and the associated branch-and-bound procedure. Correspondingly, at each iteration, the overall solution algorithm solves Problem LPMP with such generated most attractive column added. This overall process is repeated until Problem LPMP is optimized (in other words, until no more attractive column is generated from Problem $SP(k)'$). If the optimal solution to Problem LPMP is fractional, then a simple branch-and-bound procedure will be used to determine an integer solution, by using the objective function value of Problem LPMP as the bound needed at each branching step of the procedure.

4 Computational Results

This section gives the computational results of numerical examples solved by the above mentioned overall solution procedure. In this experiment, the CPLEX callable library is used as an LP solver. All the numerical problems are solved on a Pentium PC by the proposed overall solution algorithm coded in C language. The performance of the proposed overall solution algorithm is evaluated for its efficiency and effectiveness with randomly generated networks. The distributed servers in the numerical problem graphs are randomly placed in a 5000-by-5000 square grid. Links connecting pairs of servers are placed at probability $P(i,j) = A \cdot exp\left(\frac{-d(i,j)}{B \cdot C}\right)$, where $d(i,j)$ represents the distance between servers i and j, and C represents the maximum possible distance between the two servers [11]. The parameters A and B have real values defined in the range $(0,1]$. It has been observed in [12] that this link placing method with appropriate values of parameters A and B can give networks that may resemble real-world networks. Accordingly, in our experimental test, A and B are set to values 0.05 and 0.4, respectively, and each distributed server has the minimum degree of 2 and the maximum degree of 4 so that removing any one link out of the network should not lead to separating the network. The local servers are connected with one (randomly selected) of the generated distributed servers, but the maximum number of local servers to be connected with a distributed server is, for making them evenly distributed, restricted to $\lceil |L|/|M| \rceil$. In this experiment, we assume that the video demand probabilities follow the Zipf's distribution [13]. The transmission cost per unit size of every video file type on link $(i,j) \in \bar{A}$, $\alpha_{(ij)}$, is generated in association with the Euclidean distance between the endpoints such as $\alpha_{(ij)} = d(i,j)/\alpha$, where $d(i,j)$ is the Euclidean distance and $\bar{A} = \{(i,j)|i \in M, j \in M\}$. The cost for transmission to local servers from their associated distributed server is generated for each link from the uniform distribution

Table 1. Computational results; Z_{LPMP} and Z_{MP} are the objective values of the problems $LPMP$ and MP, respectively. CPU Time[a] and CPU Time[b] are processing times at the root node and on the entire solution algorithm, respectively

$|M|$	$|L|$	Z_{LPMP}	CPU Time[a](s)	Z_{MP}	CPU Time[b](s)
5	20	639.8	1694.8	639.8	1695.0
	25	757.0	3171.8	757.0	3172.0
	30	863.2	6698.7	863.2	6698.8
	35	972.8	9492.8	972.8	9493.0
10	20	713.9	6593.1	729.7	6595.1
	25	810.4	11023.2	819.8	11025.4
	30	942.7	32039.0	964.7	32039.0
	35	1063.0	107721.1	1092.1	107829.1

Table 2. Computational results with varying video size

Range of μ_k	Z_{LPMP}	CPU Time[a](s)	Z_{MP}	CPU Time[b](s)
[0.8,1.2]	639.8	1694.8	639.8	1695.0
[1.8,2.2]	1222.9	1514.9	1249.7	1515.0
[2.8,3.2]	1837.8	1202.6	1877.6	1202.6
[3.8,4.2]	2453.5	1162.4	2506.5	1162.5
[4.8,5.2]	3076.4	1337.7	3142.6	1337.7

defined over the range [0.8,1.2]. The storage cost per unit size of every video file type at server $i \in N$, β_i, is generated from the uniform distributions defined over the range [0.8,1.2], for both the distributed servers and the local servers. The number of video file copies that can be carried at each of the distributed servers and the local servers are generated from the uniform distributions defined over the ranges [100,150] and [50,60], respectively. The size of every video file type is generated from the uniform distribution defined over the range [0.8,1.2]. The demand for each video file type is generated from the Zipf's distribution. The demand for video file type $k \in K$ from local user group $l \in L$ is generated as the associated random value multiplied by the Zipf's probability defined as $\lambda_{lk} = \Lambda_l * g/k^{1-\theta}$, where Λ_l is drawn for each local user group from the uniform distribution defined over the range [8, 10].

In this experiment, reasonable sizes of graphs are considered to reflect the real world as having 5 or 10 distributed servers ($|M|$ = 5 or 10), number of various local servers ranged from 20 to 35 ($|L|$=20, 25, 30, or 35) and number of various video file types being set to the value 200 ($|K|$=200). Table 1 summarizes the experimental results. As shown in the table, the proposed algorithm has found the optimal solutions in all the cases of the numerical test. Tables 2 shows the computational results of the various problem instances (here, $|M|$=5, $|L|$=20, $|K|$=200) with various video sizes which are incorporated in transmission cost. In all of the tables, the results are given in average value, derived from the results of 20 random problem instances at each instances.

5 Conclusion

This paper considers an optimal video file allocation problem in a VOD network which is in a two-level hierarchical topology with the higher level sub-network for distributed servers and the lower level sub-network for local servers. The problem is to find an optimal video allocation strategy which gives both the optimal types of videos and the optimal number of copies of each video file type to be carried at each server such that the total cost composed of storage and transmission costs should be minimized subject to capacity of each server. The problem is formulated as a mixed integer programming problem with parameters representing all the possible (exponentially many) combinations of video types and copies. The problem is then transformed into a binary integer programming problem, for which a column generation problem is exploited to solve the associated linear programming relaxation. Some strong valid inequalities are exploited for the column generation problem. Computational results show that the proposed solution algorithm works well.

References

1. Hwang, R., Sun, Y.: Optimal video placement for hierarchical video-on-demand systems. IEEE transactions on broadcasting. **44** (1998) 392-401
2. Kwong, Y.K., Cvetko, J.: Capacity requirements of video servers in broadcast television facilities. Smpte journal. **108** (1999) 477-480
3. Barnett, S.A., Anido, G.J.: A cost comparison of distributed and centralized approaches to video-on-demand. IEEE Journal on selected areas in communications. **14** (1996) 1173-1183
4. Cidon, I., Kutten, S., Soffer, R.: Optimal allocation of electronic content. Computer Networks. **40** (2002) 205-218
5. Lee, K., Park, K., Park, S.: Design of capacitated networks with tree configurations. Telecommunication systems. **6** (1996) 1-19
6. Park, K., Kang, S., Pakr, S.: An integer programming approach to the bandwidth packing problem. Management science. **42** (1996) 1277-1291
7. Parker, M., Ryan, J.: A column generation algorithm for bandwidth packing. IEEE journal on selected areas in communications. **2** (1994) 185-195
8. Sung, C.S., Hong, J.M.: Branch and price algorithm for a multicast routing problem. Journal of the operational research society. **50** (1999) 1168-1175
9. Nemhauser, G.L., Wolsey, L.A.: Integer and combinatorial optimization. John wiley and sons. (1988)
10. Hong, J.M.: Some Optimization Issues in Designing Multimedia Communication Networks. Ph. D. Dissertation, KAIST. (2000)
11. Waxman, B.: Routing of multipoint connections. IEEE journal on selected areas in communications. **6** (1988) 1617-1622
12. Parsa, M., Uha, Q., Garcia-Luna-Aceves, J.J.: An iterative algorithm for delay-constrained munimum-cost multicasting. IEEE/ACM transactions on networking. **6** (1998) 461-474
13. Chen, Y.S., Chong, P.: Mathematical modeling of empirical laws in computer application : a case study Computers and mathematics with applications. **24** (1992) 77-87

Regrouping Service Sites: A Genetic Approach Using a Voronoi Diagram

Jeong-Yeon Seo[1], Sang-Min Park[2], Seoung Soo Lee[3], and Deok-Soo Kim[1]

[1] Department of Industrial Engineering, Hanyang University,
17 Haengdang-Dong, Sungdong-Ku, Seoul, 133-791, South Korea
{jyseo, smpark}@voronoi.hanyang.ac.kr
dskim@hanyang.ac.kr
[2] Voronoi Diagram Research Center, Hanyang University,
17 Haengdang-Dong, Sungdong-Ku, Seoul, 133-791, South Korea
[3] CAESIT, Konkuk University,
1 Hwayang-Dong, Gwangjin-Ku, Seoul, 143-701, South Korea
sslee@konkuk.ac.kr

Abstract. In this paper, we consider the problem of regrouping service sites into a smaller number of service sites called *centers*. Each service site is represented as a point in the plane and has service demand. We aim to group the sites so that each group has balanced service demand and the sum of distances between sites and their corresponding center is minimized. By using Voronoi diagrams, we obtain topological information among the sites and based on this, we define a mutation operator of a genetic algorithm. The experimental results show improvements in running time as well as cost optimization. We also provide a variety of empirical results by changing the relative importance of the two criteria, which involve service demand and distances, respectively.

1 Introduction

Suppose there are n service sites spread over an area and each site has service demand. We would like to partition the area into k disjoint regions. More precisely, we want to select k service sites called *centers* each of which serves the corresponding region. Each selected center will replace a group of neighboring service sites. We aim to group the sites so that each group has balanced service demand and the sum of distances between sites and their corresponding center is minimized. We assume that every pair of sites are connected by an edge whose weight is the Euclidean distance between the two sites.

This kind of problems have been considered in numerous areas including operations research, pattern recognition, image processing and computer graphics. Grouping is usually called *clustering* or sometimes *decomposition* in terms of graphs or d-dimensional space. The problem of finding the optimal clustering is NP-hard in general[3] and approximate solutions have been proposed[4][12]. Similar problems can be found in facility location which is to locate a set of supply stations which optimizes a given objective function. The k-center problem and the k-median problem are well-known facility location problems. The k-center problem is to select k sites(centers) among the sites of a given set so that the maximum of the distances from each site to the closest center is minimized. In the k-median problem, the total sum of the distances should be minimized[1].

The objective of regrouping problem is to consolidate resources in general. The selected centers are expected to provide higher quality services and at the same time reduce the cost of services. For example, suppose that a company plans to restructure the system of branch offices for efficient operation. You can merge a certain number of offices into one in order to cope with the decrease of overall service demand. This solution can be applied to various service sites such as medical service centers, educational sites, warehouses, shops, etc.

Regrouping problem was addressed by Mansour et al.[2] whose service sites are given as a graph. Thus the connection between sites are defined by a given set of edges and their weights. The goal of the problem is to locate the centers so that the total travel distance between service sites and their corresponding centers within a region is minimized. At the same time, it should have balanced distribution of services over the different regions. They presented a two-phase method: first decompose the graph using a tuned hybrid genetic algorithm, and then find a suitable center for each region by a heuristic algorithm.

In this paper we present a genetic algorithm to solve the regrouping problem in a geometric setting. Thus we decompose the 2D plane whereas Mansour et al. decompose the graph. A Voronoi diagram is a useful structure which provides proximity information among points. It can be a good tool for optimizing the sum of distances. For balanced distribution of service demand, we adopt genetic approach. We suggest a hybrid genetic algorithm that is combined with Voronoi diagrams. To demonstrate the efficiency of our algorithm, we compare it with two other algorithms: One uses a pure genetic approach and the other uses Voronoi diagrams. The experimental results show improvements in running time as well as cost optimization. We also provide a variety of empirical results by changing the relative importance of the two criteria, which involve service demand and distances, respectively.

In the next section, we present the formal definition of the problem and describe the objective function. Then we give an overview of genetic algorithms. Section 3 gives the definition of the Voronoi diagram and explains how it works in our algorithm. In Section 4 & 5, we elaborate on our genetic algorithm and present experimental results.

2 Preliminary

In this section, we formally introduce the definition of regrouping problem. Next, we briefly describe the concept of genetic algorithm.

2.1 Problem Definition

Assume that n sites are distributed in 2D plane R. Each site v_i has x- and y-coordinates (x_i, y_i) and weight w_i. An edge e_{ij} between v_i and v_j also has the weight l_{ij} which is the Euclidean distance between (x_i, y_i) and (x_j, y_j). More formally, we are given the following undirected graph $G = <V, E>$ with vertex set $V = \{ v_i \mid i = 1, 2, \ldots, n \}$ and edge set $E = \{e_{ij} = (v_i, v_j) \mid i, j = 1, 2, \ldots, n \}$ where each $v_i = (x_i, y_i)$ is associated with w_i and each e_{ij} is associated with l_{ij}.

Goal: Partition the plane R into k disjoint regions $r_1, r_2, \ldots, r_k$ and select a center within each region so that the objective function O is minimized. We define the function O with regard to two criteria:

- The deviation of the total vertex weights in a region r_j is minimized.

$$WD_{r_j} = | \sum_{v_i \in r_j} w_i - \sum_{v_i \in V} \frac{w_i}{k} | \qquad (1)$$

- The sum of distances to the center within a region r_j is minimized.

$$L_{r_j} = \sum_{v_i \in r_j} l_{ic} \quad \text{where } v_c \text{ is the center of } r_j. \qquad (2)$$

The cost of the region r_j is the sum of the two terms and we use α as an experimental parameter.

$$RC_{r_j} = \alpha WD_{r_j} + (1-\alpha) L_{r_j} \qquad (3)$$

The objective function O to be minimized is given by the average of the region cost:

$$O = \frac{1}{k} \sum_{r_j \in R} RC_{r_j} \qquad (4)$$

To summarize, we want to find k centers of n sites so that the weights of vertices are distributed over the regions as evenly as possible and the sum of distances to the center in each region is minimized.

2.2 Overview of Genetic Algorithms

The motivation of the genetic algorithm is the biological paradigm of natural selection as first articulated by Holland[7]. A genetic algorithm starts with a population. This population consists of a set of possible solutions, called *chromosomes* or *individuals*. This population then evolves into different populations for several iterations. For each iteration, the evolution process proceeds as selection, crossover, and mutation in concert with the value of objective function. For more details on genetic algorithms, refer to [5] and [6].

The main operations of genetic algorithms are selection, crossover and mutation. A *selection* operation decides whether each individual of current population survives or not according to its fitness value. Then, two individuals of population are randomly selected and combined through a *crossover* operation. The resulting offspring individuals are modified by a *mutation* operation. Now, a new population is created and the evolution process is repeated until the terminal condition is met.

3 Why Voronoi Diagram?

Let $P = \{v_1, v_2, \cdots, v_n\}$ be a finite set of sites in the plane. A *Voronoi diagram* of P partitions the plane such that each region corresponds to one of the sites and all the points in one region are closer to the corresponding site than to any other site. More formally, a *Voronoi region* $V(v_i)$ of a site v_i is defined as

$$V(v_i) = \{p \mid d(p, v_i) < d(p, v_j) \text{ for any } j(\neq i)\}$$

where $d(p,q)$ is the Euclidean distance between p and q. Note that a Voronoi region is always a bounded convex polygon as shown in Figure 1. Voronoi diagrams have been extensively studied as a useful tool in various fields such as astronomy, biology, cartography, crystallography and marketing [10, 11]. The survey on Voronoi Diagrams can be found in [8, 9].

In the regrouping problem, we have two criteria to determine the quality of a solution. One is the balanced distribution of service demands and the other is the sum of distances to the centers. We Noting that a Voronoi diagram provides the most compact and concise representation of the proximity information in Euclidean space, we adopt this structure in our algorithm. In fact, the Voronoi diagram can give the solution that minimizes the sum of distances provided that a set of centers is known. The regrouping problem consists of two parts - decomposing the plane into k regions and selecting a center for each region. In our algorithm, we first select centers with regard to the distribution of the service demands. With this set of centers, we decompose the plane using the structure of Voronoi diagrams. These two steps are used as basic operations of our genetic algorithm to optimize the solution.

4 Regrouping Using Genetic Approach and Voronoi Diagram

In this section, we propose a genetic algorithm using a Voronoi diagram, which we will call VGA, for the problem of regrouping service sites. To verify the performance of VGA, we implement two more algorithms as well as VGA: A pure genetic algorithm(GA) and an algorithm that only uses Voronoi diagrams(VD). In the followings, we describe each of the three algorithms.

4.1 GA Algorithm

We first give GA algorithm which is loyal to the original concept of genetic algorithms. Each chromosome in our problem is represented by n-elements row [$C(v_1), C(v_2), C(v_3), ..., C(v_n)$]. The i-th site v_i takes a value $C(v_i)$ from $1, 2, ..., k$ (k is the number of regions) which represents a center whose region contains the site. In general, the initial population of possible solutions is randomly generated. We use roulette-wheel selection and the fitness value of the individuals to be minimized is the objective function O shown in Equation (4). The crossover operation used is the double-point crossover and the crossover rate is set to be 0.7. It is applied to a randomly-selected pair of individuals. The probability of mutation is 0.2. We perform a mutation operation by switching the regions of two randomly-selected individuals. The following gives the outline of GA algorithm.

```
Program GA
{ Input : graph G =< V, E > }
{ Output: subgraphs : R_1, R_2, ..., R_k }
Calculate the probability of each site :Pr(V);
Generate the initial population by randomly assigning sites
to subgraphs;
Evaluate fitness of each individual;
Repeat (150 times)
{
      Rank individuals;
      Give probability to each individual;
      Apply selection, crossover, and mutation;
      Evaluate fitness of individuals;
}
```

4.2 VD Algorithm

Instead of generating random values which form the individuals of the initial population of GA algorithm, we use a Voronoi diagram for the initial population of VD algorithm. We first assign the probability to each site according to its weight. Then, we choose k centers with high probability as the generators of Voronoi diagram and then construct the Voronoi diagram of the k centers. The sites that are contained a Voronoi region belong to the same group. The outline of generating the regrouping solution using a Voronoi Diagram is given as follows.

```
Program VD
{ Input : graph G =< V, E > }
{ Output: subgraphs : R_1, R_2, ..., R_k }
Assign probability to each site according to its weight
Choose k sites ( for each individual )
Compute VD of k sites
Classify v_i into one of k Voronoi regions for each i
Evaluate O
Consider the individual with the minimum O as the solution
```

4.3 VGA Algorithm

In VGA algorithm, the basic method is a genetic algorithm and at the generation of the initial population and the mutation, we employ the Voronoi diagram structure that gives topological information among the sites. First we start with the same initial population as VD algorithm. In order to classify the sites into one of k regions, we construct the Voronoi diagram for the k centers, which we call VD_k. To take advantage of the topology and geometry information of VD_k, we define a mutation operator which is applied only on the sites near the boundaries of VD_k regions. But it is not clear which sites are close to the boundaries. Thus we construct the Voronoi diagram of all the sites, $VD(V)$ and find sites whose neighboring site belongs to a different group. If such a site is selected as a mutation site, the site moves to one of the neighboring regions. Refer to Figure 1: site A mutates into Region 1 and site B arbitrarily chooses one of Region 1 and Region 2 and moves to it.

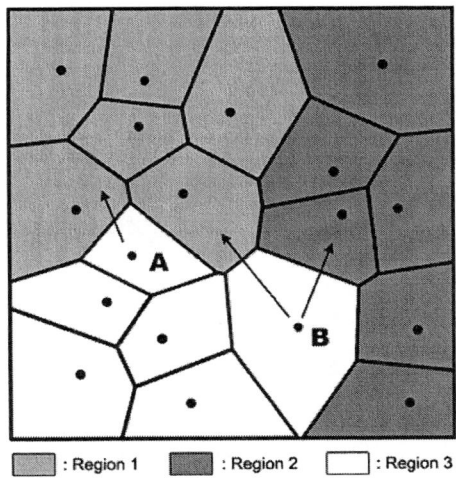

Fig. 1. Mutation on the boundaries of Voronoi regions

```
Program VGA
{ Input : graph G =< V,E > }
{ Output: subgraphs: R_1, R_2, ..., R_k }
Calculate the probability of each site : Pr(V);
Compute VD(V);
Choose k sites ( for each individual )
Compute Voronoi diagram of k sites
Classify v_i into one of k Voronoi regions for each i
Evaluate O for each individual
Repeat (50 times)
{
        For ( each individual )
        {
                Make a boundary site list for each region;
                Apply a mutation operation on the list;
        }
        Evaluate O for each individual
        Apply selection;
}
```

We set the probability of mutation to be 0.2 as in GA algorithm. With the resulting population the next iteration is executed. Crossover operations turned out to be unsuitable for this population because some individuals created by crossover may have either more than or less than k centers. We give the outline of VGA algorithm as above.

Table 1. Experimental results ($\beta = 1 - \alpha$)

α	β	Objective fn. Value of VD (running time(sec))			Objective fn. Value of VGA (running time(sec))			Objective fn. Value of GA (running time(sec))		
		MIN	MAX	AVG	MIN	MAX	AVG	MIN	MAX	AVG
0.1	0.9	87.198 (0.266)	104.974 (0.281)	98.702 (0.267)	88.375 (55.704)	101.548 (57.1)	95.329 (59.055)	163.087 (70.656)	194.515 (69.797)	180.222 (69.042)
0.2	0.8	81.456 (0.281)	106.378 (0.25)	96.931 (0.265)	77.442 (68.25)	100.785 (62.265)	90.201 (59.983)	148.497 (68.344)	182.755 (72)	167.310 (69.007)
0.3	0.7	76.661 (0.265)	104.826 (0.265)	91.720 (0.266)	65.621 (65.319)	89.758 (52.156)	80.137 (61.452)	139.013 (75.343)	172.492 (70.141)	155.423 (70.585)
0.4	0.6	72.561 (0.266)	102.119 (0.266)	86.431 (0.261)	61.206 (67.062)	77.337 (66.719)	71.927 (64.012)	125.942 (71.344)	149.409 (64.125)	138.218 (72.619)
0.5	0.5	65.781 (0.266)	103.804 (0.255)	84.032 (0.270)	56.115 (65.813)	68.903 (59.562)	63.737 (65.560)	106.154 (77.375)	126.329 (56.340)	120.302 (71.937)
0.6	0.4	54.399 (0.261)	86.665 (0.281)	74.740 (0.266)	48.465 (64.145)	58.820 (54.390)	55.849 (66.198)	91.967 (66.537)	112.300 (68.125)	102.213 (72.180)
0.7	0.3	54.243 (0.265)	75.661 (0.265)	63.017 (0.265)	35.715 (74.406)	54.143 (62.255)	44.476 (67.632)	77.819 (90.937)	95.665 (68.625)	84.187 (72.623)
0.8	0.2	40.177 (0.265)	81.830 (0.266)	56.467 (0.270)	27.626 (63.234)	38.240 (65.625)	33.011 (65.933)	60.581 (90.687)	72.513 (68.469)	64.362 (74.973)
0.9	0.1	44.413 (0.250)	68.840 (0.25)	57.978 (0.261)	17.440 (65.706)	27.593 (60.759)	22.312 (63.938)	34.412 (80.152)	50.986 (67.813)	43.921 (72.544)

5 Experimental Results

In our implementation, we choose the size of the population to be equal to $2 \times n \times k$. And x- and y- coordinates and weight of site are randomly generated between 10 and 100. The average, min and max values are computed over 10 problem instances. These instances are derived from graphs with $n = 50$ that are decomposed into 5 different regions ($k = 5$). Experiments were performed on a PC with Pentium IV 2.4 GHz CPU and 512 MB of RAM. The implementation was done using Visual C++.

Recall that the objective function is computed for j-th group as follows (refer to Equation (3)).

$$RC_{r_j} = \alpha W D_{r_j} + \beta L_{r_j} \quad \text{where} \quad \alpha + \beta = 1$$

Note that the first term is related to the service demand and the second term is related to the distances. To give the diversity of the simulation, we scale the value of α from 0 to 1 range in our experiment. Hence α is used to investigate how the results of three algorithms are influenced by the change of relative importance of the two criteria.

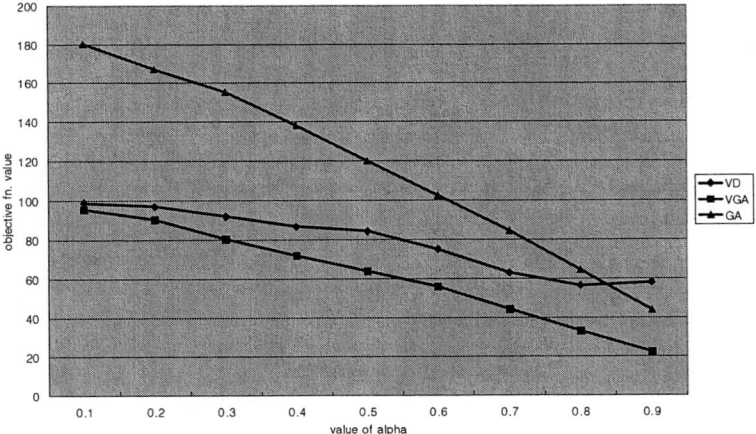

Fig. 2. Comparison of three algorithms

The results obtained in our experiments are shown in Table 1, which is also shown as a graph in Figure 2. The overall performance of VGA algorithm is superior to the other two. When α is small, VD algorithm is good because it is mainly focussed on minimizing the sum of distances. When α is close to 1, GA wins VD which makes little efforts to optimize the service demand term. VGA algorithm which uses both Voronoi diagrams and genetic approaches produces high quality solutions for the regrouping instances compared to GA algorithm (pure GA approach) and VD algorithm (pure geometric approach). The value of the objective function of the solutions of VGA algorithm is 47.5 % better than that of GA algorithm on the average. In terms of running time, the simple VD algorithm is the fastest. GA algorithm is able to obtain smaller values of the objective function than those in Table 1, but it takes much longer than VGA algorithm. We set the number of iteration for VGA algorithm and GA algorithm to be 50 and 150 respectively to complete both in a similar time. On that condition, VGA algorithm produces a better objective function value than GA algorithm. Therefore, the suggested VGA algorithm gives good results in terms of both quality of grouping as well as running time. To summarize these considerations, the followings can be said:

- VGA algorithm is distinguished in the overall performance.
- VD algorithm gives good solution when α is small because it is mainly focussed on minimizing the sum of distances.
- GA algorithm gives good solution when α is close to 1 because it makes efforts to optimize the service demand term compared to VD.
- Between two algorithms using genetic approach, VGA algorithm is much faster than GA algorithm when the same quality of the solution is required.

An example of the output of our VGA algorithm is shown in Figure 3. Each region of k groups is filled with different patterns. The center of each region is represented as a black point while the other sites are shown as white ones.

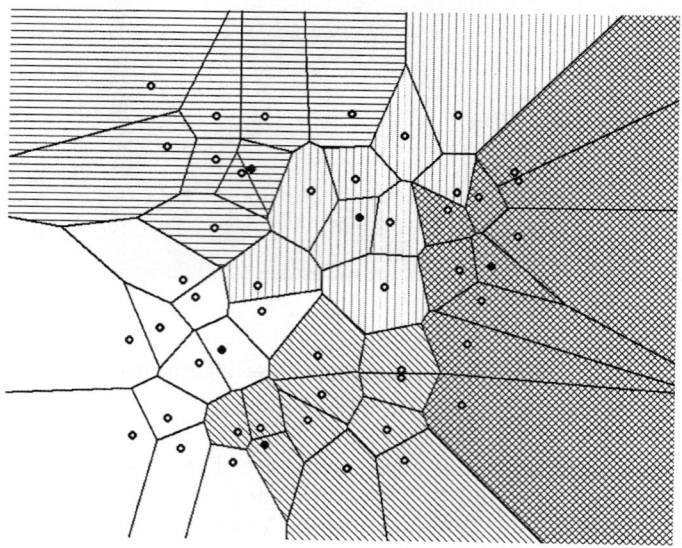

Fig. 3. VGA solution (objective function value = 65.37 , $\alpha = \beta = 0.5$)

6 Conclusion

We proposed a new genetic algorithm which produces good and efficient solutions for the regrouping problem. The idea of combining the genetic algorithm with the proximity information provided by Voronoi diagram has led to good solutions as shown in the empirical results. Possible extension of our work would be grouping problems with other kinds of quality measure such as minimizing the maximum radius, diameter or area of the group regions. We expect the Voronoi diagram to be a useful method for those variants as well. In addition, algorithms employing Voronoi diagrams provide the grouping whose elements are concentrated around their center rather than being spread around, which is significant in an aesthetic sense.

References

1. Arora, S., Raghavan, P., Rao, S.: Approximation schemes for Euclidean k-medians and related problems. In Proceedings of the 30th Annual ACM Symposium on Theory of Computing. ACM, New York (1998) 106-113
2. Mansour, N., Tabbara, H., Dana, T.: A genetic algorithm approach for regrouping service sites. Computers & Operations Research, Vol. 31. (2004) 1317-1333
3. Garey, M.R., Johnson, D.S.: Computers and intractability: a guide to the theory of NP-completeness. Freeman and Co., New York (1979)
4. Vazirani, V.: Approximation Algorithms. Springer, Berlin (2001)
5. Chambers, L.D.: Practical handbook of genetic algorithms. Vols. I and II. CRC Press, Boca Raton, FL (1995)
6. Goldberg, D.E.: Genetic algorithms in search, optimization and machine learning. Addison-Wesley, Reading, MA (1989)
7. Holland, J.: Adaptation In Natural and Artificial Systems. The University of Michigan Press, Ann Arbour (1975)
8. Okabe, A., Boots, B., Sugihara, K.: Spatial Tessellations Concetps and Applications of Voronoi Diagrams. John Wiley & Sons (1992)
9. Aurenhammer, F.: Voronoi diagrams - a survey of a fundamental geometric data structure. ACM Computing Surveys, Vol. 23. (1991) 345-405
10. Kim, D.S., Chung, Y.C., Kim, J.J., Kim, D., Yu, K.: Voronoi diagram as an analysis tool for spatial properties for ceramics. Journal of Ceramic Processing Research, Vol. 3, No. 3, PartII. (2002) 150-152
11. Seo, D.I., Moon, B.R.: Voronoi Quantized Crossover for Traveling Salesman Problem. In Genetic and Evolutionary Computation Conference. (2002) 544-552
12. Bui, T.N., Moon, B.R.: Genetic algorithm and Graph partitioning. IEEE Transactions on computers, Vol.45. No.7. (1996) 841-855

Profile Association Rule Mining Using Tests of Hypotheses Without Support Threshold

Kwang-Il Ahn and Jae-Yearn Kim

Industrial Engineering, Hanyang University, Seoul, Korea
jyk@hanyang.ac.kr

Abstract. Association rule mining has been a core research topic in data mining. Most of the past researches focused on discovering relationships among items in the transaction database. In addition, mining algorithms for discovering association rules need the support threshold to discover frequent itemsets. But the essence of association rule mining is to find very associated relationships among itemsets not to discover frequent itemsets. In this paper, we deal with mining the relationships among the customer profile information and the purchased items. We make the sample databases from the original database and use the tests of hypotheses on the interestingness of the rules from the sample data. Our approach can speed up mining process by storing the sample database into main memory and provide insights by presenting the rules of low support but high association.

1 Introduction

Since its introduction, association rule mining has become an important field in data mining. Mining association rules is useful for discovering relationships among data in large databases. Usually, it has been applied to sales transaction databases. Given a transaction database, an example of association rule might be that customers who buy bread tend to buy milk.

Association rule mining can be decomposed into two subproblems: finding frequent itemsets, i.e., finding the subsets of items that occur frequently in a large database, and extracting interesting rules of association between the frequent itemsets. The usual measures in association rule mining are support and confidence. The support of an itemset X means the percentage of the number of transactions containing X in total transactions. An itemset X is referred to as a frequent(or large) itemset if the support of X is no less than a user-defined minimum support(called *minsup*). An association rule has the form $X \rightarrow Y$, where both X and Y are sets of items, and $X \cap Y = \phi$. The confidence of a association rule is the conditional probability that a transaction contains Y, given that it contains X [1]. It is given as $confidence(X \rightarrow Y) = support(X \cup Y)/support(X)$.

Much research has focused on deriving efficient algorithms for finding frequent itemsets. Since the introduction of the Apriori algorithm [2], other algorithms have been developed such as DHP [9], Max-Miner [5], FP-tree [7], VIPER [10],

Common-Item Tree [8], Diffset-based algorithm [13], CT-ITL [11], and Modified-diffset method [4].

However, the above researches have been conducted for discovering the relationships among items in the transaction database. Recently, the issue of mining the relationships between items in the transaction database and attributes in the customer database was investigated in [12]. An example of such an association rules might be "customers whose majors are data-mining tend to buy the book 'Database'." This kind of rule is called 'profile association rule' [3] and useful for product recommendation. Here, we deal with mining association rules with an expression *attribute values* → *itemsets*.

Generally, association rule mining algorithms are based on support-based pruning. So high level of support threshold can ignore the very correlated items that do occur rarely. And low level of support threshold results in generating too many frequent itemsets. The essence of association rule mining is not to find frequent itemsets but to discover associated rules. In our approach, minimum support is not given. But instead, we use the tests of hypotheses with sampling data and present the very correlated relationships of customer attribute values and items. The rest of this paper is organized as follows: In Sec. 2, we define the problem of mining profile association rules and review the tests of hypotheses. In Sec. 3. the procedures of the proposed algorithm are described. In Sec. 4, the experimental results on synthetic databases are presented. Our conclusions are presented in Sec. 5.

2 Preliminaries

In this section, we provide the definitions of the terms that are used in this paper. We also present the hypotheses to be tested.

2.1 Profile Association Rules

A profile association rule is one in which the antecedent of the rule consists of customer profile information from the customer database and the consequent of the rule does customer behavior information from transaction database [3]. The goal of discovering profile association rules is to identify customer profiles for promoting sales of products [3].

Let $v_{i,j}$ be the associated value j at the customer attribute i and X an itemset. A profile association rule has the form *'conjunction of $v_{i,j}$ → X'*. The nature of customer attribute data is quantitative or categorical. We assume that customer profile data is given in categorical form. For quantitative data, we first convert it into categorical data by discretizing the data.

In order to evaluate whether a rule is interesting or not, we use 'improvement' [6] as a measure of interest. The improvement of a rule $v_{i,j} \to X$ is defined by Equation 1.

$$improvement(v_{i,j} \to X) = \frac{support(v_{i,j} \cup X)/support(v_{i,j})}{support(X)} \qquad (1)$$

Improvement means the influencing power of customer profile for promoting sales of items. In Equation 1, if the value of improvement measure is greater than 1, it means customers who has attribute $v_{i,j}$ buy an itemset X more frequently than others.

An interesting profile association rule is defined in Definition 1.

Definition 1. *Let C be a conjunction of attribute values $v_{i,j}$ and X an itemset. The profile association rule $C \to X$ is interesting if the improvement of the rule is greater than 1.*

Our study does not consider support threshold. So, some of interesting discovered rules may have low supports.

2.2 Tests of Hypotheses

Hypotheses tests are tests based on a sample of data to determine which of two different statements of nature is true. The two statements of nature are commonly called the null hypothesis and the alternative hypothesis. There are one-sided and two-sided alternative hypotheses. In this study, one-sided alternative hypotheses are appropriate since we are interested in making a decision whether the improvement of the rule is greater than the minimum threshold or not.

Suppose that the random variable M represents the improvement of an original data. In addition, we assume that both the mean μ of M and the variance σ^2 are unknown. The hypotheses would be Equation 2.

$$H_0 : \mu \leq 1$$
$$H_1 : \mu > 1 \qquad (2)$$

Assume that a random sample of size n, say $M_1, M_2, \ldots, M_n$ is available, and let $\overline{M}$ and S^2 be the sample mean and variance, respectively. The test procedure is based on the statistic in Equation 3.

$$t_0 = \frac{\overline{M} - 1}{S/\sqrt{n}} \qquad (3)$$

which follows the t-distribution with $n - 1$ degrees of freedom if the null hypothesis H_0 is true. We calculate the test statistic t_0 in Equation 3 and reject H_0 if $t_0 > t_{\alpha, n-1}$, where α is the significance level of the test.

3 Discovering Interesting Profile Association Rules

In this section, we describe our algorithm, which uses the sample database and the tests of hypotheses. In Sec. 3.1, we introduce the advantages of sampling and the vertical data format. Section 3.2 describes the types of rules to be mined in this study. In Sec. 3.3, all procedures of our algorithm are described.

3.1 Sample Database

A record in the transaction database contains the customer identifier and the purchased items. And a customer database consists of the customer identifier and his(or her) attributes with specific values. The sizes of the databases are very large. So, the scanning the databases in mining process is a time consuming task.

In order to overcome the limit, we discover rules from the sample data. By random sampling, we select n records from the transaction database and the customer database. We can make sample database which is able to be stored in main memory. This can reduce time of scanning databases.

For effective counting process, the sample database takes the form of the vertical data format in which column stores an item identifier(or an attribute value)and a list of transaction identifiers. A vertical data format is very effective to count support since only those columns that are relevant to the following scan of the mining process are accessed from disk.

3.2 Types of Rules

In this section, we describe the various types of profile association rules. The fist type of rules is 'Basic rule'. A basic rule is an interesting profile association rule which has one attribute value and one item. For example, $income = high \rightarrow car$ is a basic rule. Basic rules become the bases for discovering the other types of rules.

Second type of rules is 'Combined rule'. Let $v_{i,j}$ and $v_{i,k}$ be the associated values j and k, respectively, at the customer attribute i and X an itemset. A combined rule is $(v_{i,j} \ or \ v_{i,k}) \rightarrow X$. A combined rule has the following property:

Property 1. If both $v_{i,j} \rightarrow X$ and $v_{i,k} \rightarrow X$ are interesting profile association rules, the combined rule $(v_{i,j} \ or \ v_{i,k}) \rightarrow X$ is always interesting.

Proof. By Equation 1,

$$\frac{support(v_{i,j} \cup X)/support(v_{i,j})}{support(X)} > 1 \text{ and } \frac{support(v_{i,k} \cup X)/support(v_{i,k})}{support(X)} > 1$$

that is,
$$support(v_{i,j} \cup X) > support(v_{i,j})support(X) \text{ and}$$
$$support(v_{i,k} \cup X) > support(v_{i,k})support(X)$$

Adding the two inequalities results in
$$support(v_{i,j} \cup X) + support(v_{i,k} \cup X)$$
$$> support(v_{i,j})support(X) + support(v_{i,k})support(X)$$

Since $transactions \ with \ v_{i,j} \cap transactions \ with \ v_{i,k} = \phi$,
$$support((v_{i,j} \ or \ v_{i,k}) \cup X) > support(v_{i,j} \ or \ v_{i,k})support(X)$$

As a result, $\frac{support((v_{i,j} \ or \ v_{i,k}) \cup X)/support(v_{i,j} \ or \ v_{i,k})}{support(X)} > 1$

and it means $(v_{i,j} \ or \ v_{i,k}) \rightarrow X$ is interesting.

Third type of rules is 'Multi-dimensional rule'. A multi-dimensional rule means the one with multiple attributes or items. We generate multi-dimensional rules if there are basic rules whose consequent parts are the same, or whose conditional parts are the same. Let $v_{i,j} \rightarrow target1$, $v_{i',k} \rightarrow target1$ and $v_{i,j} \rightarrow target2$ be the basic rules. Multi-dimensional rules can be $(v_{i,j} \cup v_{i',k}) \rightarrow target1$ and $v_{i,j} \rightarrow target1 \cup target2$. In this study, the number of condition(or consequent) part of the rules is limited to 3 since long rules are generally difficult to be interpreted.

In addition to the above types, there are exceptional profile association rules which cannot be generated from basic rules. This kind of rules are interesting but difficult to analyze. In this study, we do not deal with this kind of rules. Our scope of this research is to find the rules to be applicable for promoting sales of products.

3.3   Algorithm

In this section, we describe our algorithm, which uses a sample database and the tests of hypotheses. The minimum support is not given. The steps for mining profile association rules are as follows:

Step 1. Sampling Data and Counting support

(a) Make the vertical sample database SDB.
(b) Count the supports of items and attribute values in the SDB. This is conducted as the SDB is made.

Step 2. Discovering Basic Rules

(a) Make the candidate basic rules with the type $v_{i,j} \rightarrow X$, where $v_{i,j}$ is the specific value j at the customer attribute i and X an item.
(b) Count the support of each $v_{i,j} \rightarrow X$ by scanning the sample database.
(c) Discover $(v_{i,j} \cup X)$ whose support is greater than $support(v_{i,j})support(X)$, which is necessary and sufficient condition to satisfy improvement threshold on the ground of Equation 1.

Step 3. Generating Combined Rules

(a) Select $(v_{i,j} \cup X)$ and $(v_{i,k} \cup X)$ for $j \neq k$ among the basic rules.
(b) Generate the combined rules $(v_{i,j} \text{ or } v_{i,k}) \rightarrow X$. This step does not need to scan SDB since the improvement of the combined rules are always greater than 1 and can be calculated with the information given in Step 1, 2.

Step 4. Generating Multi-Dimensional Rules

(a) Select $(v_{i,j} \cup X_m)$ for $m = 1, 2, 3, \ldots, n$ among the basic and combined rules.
(b) Make the candidate rules with the type $v_{i,j} \rightarrow Y$, where Y is the combination of X_m.
(c) Select $(v_{i,j} \cup X)$ for $i = 1, 2, 3, \ldots, n$ among the basic and combined rules.

(d) Make the candidate rules with the type $V \to X$, where V is the combination of $v_{i,j}$.
(e) Count the supports of each $v_{i,j} \to Y$ and $V \to X$ by scanning the sample database.
(f) Discover the multi-dimensional rules $v_{i,j} \to Y$ and $V \to X$, whose improvements are greater than 1.

Step 5. Collecting improvements of the discovered rules

(a) Make more sample databases like Step 1.
(b) Collect the information of improvements of each discovered rules.

Step 6. Tests of interestingness

(a) Calculate the sample mean and variance of improvement of each rules
(b) Test the interestingness of the rules by using Equations 2, 3.

4 Experimental Results

We tested the proposed algorithm with two synthetic databases. The synthetic databases mimic the transactions in a retailing environment. Table 1 shows the width and height of each database.

Table 1. Width and height of the databases

Databases	Records	Record width	Items	Attributes
T5I4D10kN35A3	10,000	5	35	3
T10I6D10kN55A3	10,000	10	55	3

The parameters in the databases are as follows: 'T' means the average length of transaction in the database, 'I' means the average size of the maximal potentially frequent itemsets, 'D' means the total number of transactions, 'N' means the number of items, and 'A' means the number of attributes.

We made the 30 sample databases with size 100 from the synthetic databases. Table 2 shows the number of the generated rules in the sample databases. Many multi-dimensional rules were generated, though the size of the sample database is small. We can notice that the basic rules with low supports were generated. The last column in Table 2 shows interesting basic rules with low support.

The proposed algorithm has the following advantages. First, this algorithm can reduce database scanning time by using the sample databases. Second, the profile association rules of low support but high correlation can be discovered. But, there is a limitation that many multi-dimensional rules, which should be tested, are generated, if there are many items.

Table 2. Number of rules

Databases	Basic($supp < 0.05$)	Combined	Multi-dimension	Interesting basic
T5I4D10kN35A3	55(12)	3	218	5
T10I6D10kN55A3	105(28)	9	465	15

5 Conclusion

In this paper, we have proposed a new algorithm for discovering profile association rules. We made the sample databases from the customer database and the transaction database. In the proposed algorithm, the sample databases were scanned instead of the original databases. This can save time of scanning databases. And the tests of hypotheses were conducted on the interestingness of the rules. The experimental results showed that the profile association rules of low support were discovered since the interestingness was high. This provides insights for promoting sales of items. As further researches, it will be interesting to discover the exceptional profile association rules.

References

1. Agrawal, R., Imielinski, T. and Swami A. N.: Mining association rules between sets of items in large databases. Proceedings of the ACM-SIGMOD International Conference on Management of Data (1993) 207-216
2. Agrawal, R., Srikant, R.: Fast algorithms for mining association rules in large databases. Proceedings of the 20th International Conference on Very Large Data Bases (1994) 487-499
3. Aggarwal, C. C., Sun, Z., Yu, P. S.: Online algorithms for finding profile association rules. Proceedings of International Conference on Information and Knowledge Management (1998) 86-95
4. Ahn, K. I., Kim, J. Y.: Efficient mining of frequent itemsets and a measure of interest for association rule mining. Journal of Information and Knowledge Management **3**(3) (2004) 245-257
5. Bayardo, R. J.: Efficiently mining long patterns from databases. Proceedings of the ACM- SIGMOD International Conference on Management of Data (1998) 85-93
6. Berry, M., Linoff, G.: Data mining techniques for marketing, sales, and customer support. John Wiley & Sons, Inc (1997)
7. Han, J., Pei, J., Yin, Y.: Mining frequent patterns without candidate generation. Proceedings of the ACM-SIGMOD International Conference on Management of Data (2000) 1-12
8. Kim, C. O., Ahn, K. I., Kim, S. J., Kim, J. Y.: An efficient tree structure method for mining association rules. Journal of the Korean Institute of Industrial Engineers **27**(1) (2001) 30-36
9. Park, J. S., Chen, M. S., Yu, P. S.: An effective hash based algorithm for mining association rules. Proceedings of the ACM-SIGMOD International Conference on Management of Data (1995) 175-186

10. Shenoy, P., Haritsa, J. R., Sudarshan, S.: Turbo-charging vertical mining of large databases. Proceedings of the ACM-SIGMOD International Conference on Management of Data (2000) 22-23
11. Sucahyo, Y. G., Gopalan, R. P.: CT-ITL: Efficient frequent item set mining using a compressed prefix tree with pattern growth. Proceedings of the Australasian Database Conference (2003) 95-104
12. Tsai, P. S. M., Chen, C. M.: Mining interesting association rules form customer databases and transaction databases. Information Systems **29** (2004) 685-696
13. Zaki, M. J., Gouda, K.: Fast vertical mining using diffsets. Proceedings of the 9th International Conference on Knowledge Discovery and Data Mining (2003) 326-335

The Capacitated max-k-cut Problem

Daya Ram Gaur[1] and Ramesh Krishnamurti[2]

[1] Department of Math & Computer Sc,
University of Lethbridge, Lethbridge, AB, T1K 4G9
gaur@cs.uleth.ca
http://www.cs.uleth.ca/ gaur
[2] School of Computing Science,
Simon Fraser University, Burnaby, BC, V5A 1S6
ramesh@cs.sfu.ca
http://www.cs.sfu.ca/ ramesh

Abstract. In this paper we study a capacitated version of the classical max-k-cut problem. Given a graph we are to partition the vertices into k equal-sized sets such that the number of edges across the sets is maximized. We present a deterministic approximation algorithm for this problem with performance ratio $(k-1)/k$. Our algorithm is based on local search, a technique that has been applied to a variety of combinatorial optimization problems.

1 Introduction

The classical max-k-cut problem is one of partitioning the set of vertices V in a graph into k partitions such that the number of edges across the partitions is maximum. Note that there is no constraint imposed on the size of each partition. In the capacitated version we study in this paper, we are to partition the vertices into k equal-sized sets such that number of edges across the sets is maximized. We let cmax-k-cut denote the capacitated version of the max-k-cut problem. We present a deterministic approximation algorithm for this problem which is based on local search, and show that the performance ratio is $(k-1)/k$.

2 Related Work

One of the first deterministic approximation algorithms for max-2-cut was based on the idea of local search. Starting with an arbitrary partition of size 2 the idea is to move vertices from one partition to the other until there is no improvement. The performance ratio of the local search algorithm is 1/2 [7]. A randomized algorithm for the max-2-cut problem with a performance ratio of 1/2 is due to Sahni and Gonzalez [8].

Until the seminal work of Goemans and Williamson [4] this was the best known result for max-2-cut. Goemans and Williamson [4] gave a randomized rounding approximation algorithm for max-2-cut with a performance ratio of

0.87856 based on the semi-definite programming relaxation. This algorithm was later derandomized by Mahajan and Ramesh [6]. Feige and Jerrum [3] generalize the results in [4] and show that the performance ratio for the max-k-cut problem is α_k, where α_k satisfy $\alpha_k > 1 - 1/k$, $\alpha_k \sim (2 \log k/)k^2$ and $\alpha_2 \geq 0.878567, \alpha_3 \geq .800217, \alpha_4 \geq 0.850304, \alpha_5 \geq 0.874243, \alpha_{10} \geq 0.926642, \alpha_{100} \geq 0.990625$. Goemans and Williamson [5] improved the results in [3] for the max-3-cut problem by giving an approximation algorithm with a performance ratio of 0.83601. The algorithm in [5] is based on complex semi-definite programming. An approximation algorithm based on local search (similar to the one outlined in Papadimitriou and Stieglitz [7]) for the max-k-cut problem has a performance ratio of $(k-1)/k$ (Problem 2.14 in [9]). However, no constraint is imposed on the size of each partition. Andersson [2] generalizes the results from [3] to obtain a $1-1/k+\Theta(1/k^3)$ approximation algorithm for the capacitated max-k-cut problem. Andersson [2] also noted that a simple randomized algorithm which partitions the vertices into k equal-sized sets has a performance ratio of $(k-1)/k$. For the case when all the partitions are of arbitrary sizes, Ageev and Sviridenko [1] describe a 1/2 approximation algorithm using a new technique for randomized rounding called pipage rounding. It should be noted that both the algorithms of Andersson [2] and Ageev and Sviridenko [1] are randomized in nature. In this paper, we provide a deterministic approximation algorithm with a performance ratio of $(k-1)/k$.

We describe below an adaptation of the local improvement heuristic described for this problem [7]. For the sake of exposition, we describe this adaptation for cmax-2-cut and cmax-3-cut, and show that the performance ratio for these problems is at least 1/2 and 2/3 respectively. We then provide a proof for the general case, and show that the performance ratio is $(k-1)/k$.

3 A 1/2-Approximation Algorithm for cmax-2-cut

In this section we address the problem of partitioning the set of n vertices into two equal-sized sets (of size $n/2$ each) such that the number of edges that cross the partition is maximum.

Consider the following greedy algorithm for solving cmax-2-cut: start with an arbitrary partition (L,R) such that L is of size $n/2$ and R is of size $n/2$. While there exists a pair $u \in L$ and $v \in R$ such that switching them increases the number of edges in the cut, perform the switch. We show that the cut discovered by the greedy algorithm is at least 1/2 the size of the optimal cut.

Given a vertex u and a set of vertices V, e_{uV} denotes the number of edges from u which are incident on vertices in set V. Let the optimal partition be $L_o = \{V_1 \cup V_3\}$ and $R_o = \{V_2 \cup V_4\}$, and the partition computed by the greedy algorithm be $L_g = \{V_1 \cup V_2\}$ and $R_g = \{V_3 \cup V_4\}$. Let e_{ij} be the number of edges from V_i to V_j. Note that $e_{ij} = e_{ji}$. We note that the number of edges in the optimal cut denoted OPT $= e_{12} + e_{23} + e_{14} + e_{34}$ and the number of edges in the greedy cut denoted ALG $= e_{13} + e_{23} + e_{14} + e_{24}$.

Theorem 1. The performance ratio of the greedy algorithm is at least 1/2.

Proof: Consider a pair of vertices u, v, where $u \in L_g = V_1 \cup V_2$ and $v \in R_g = V_3 \cup V_4$. As the greedy algorithm is locally optimal, swapping vertices u and v should not increase the number of edges crossing the cut. In the equation below, the right hand side (left hand side) represents the number of edges incident on u and v that cross the cut before (after) swapping.

$$e_{uV_1} + e_{uV_2} + e_{vV_3} + e_{vV_4} \leq e_{uV_3} + e_{uV_4} + e_{vV_1} + e_{vV_2} \tag{1}$$

Equation 1 is valid for four cases, corresponding to $u \in V_1$ and $v \in V_3$, $u \in V_1$ and $v \in V_4$, $u \in V_2$ and $v \in V_3$, $u \in V_2$ and $v \in V_4$.

We consider the case when $u \in V_1$ and $v \in V_3$ in detail below:
Summing up Equation 1 above over all $u \in V_1$,

$$2e_{11} + e_{12} + |V_1|(e_{vV_3} + e_{vV_4}) \leq e_{13} + e_{14} + |V_1|(e_{vV_1} + e_{vV_2}) \tag{2}$$

Summing up Equation 2 above over all $v \in V_3$,

$$|V_3|(2e_{11} + e_{12}) + |V_1|(2e_{33} + e_{34}) \leq |V_3|(e_{13} + e_{14}) + |V_1|(e_{13} + e_{23}) \tag{3}$$

Dropping the e_{jj} terms in the LHS in Equation 3 above,

$$|V_3|e_{12} + |V_1|e_{34} \leq |V_3|(e_{13} + e_{14}) + |V_1|(e_{13} + e_{23}) \tag{4}$$

Similarly, for the cases when $u \in V_1, v \in V_4$, $u \in V_2, v \in V_3$, and $u \in V_2, v \in V_4$, we get the following three equations respectively:

$$|V_4|e_{12} + |V_1|e_{34} \leq |V_4|(e_{13} + e_{14}) + |V_1|(e_{14} + e_{24}) \tag{5}$$
$$|V_3|e_{12} + |V_2|e_{34} \leq |V_3|(e_{23} + e_{24}) + |V_2|(e_{13} + e_{23}) \tag{6}$$
$$|V_4|e_{12} + |V_2|e_{34} \leq |V_4|(e_{23} + e_{24}) + |V_2|(e_{14} + e_{24}) \tag{7}$$

Adding Equations 2, 5, 6, and 7,

$$2(|V_3|+|V_4|)e_{12} + 2(|V_1|+|V_2|)e_{34} \leq (|V_1|+|V_2|+|V_3|+|V_4|)(e_{13}+e_{23}+e_{14}+e_{24}) \tag{8}$$

Since $|V_1| + |V_2| = n/2$ (and $|V_3| + |V_4| = n/2$). Equation 8 can now be rewritten as

$$2(n/2)e_{12} + 2(n/2)e_{34} \leq n(e_{13} + e_{23} + e_{14} + e_{24}) \tag{9}$$

Adding $n(e_{14} + e_{23})$ to both sides of Equation 9,

$$2(n/2)e_{12} + 2(n/2)e_{34} + n(e_{14} + e_{23}) \leq n(e_{13} + e_{23} + e_{14} + e_{24}) \\ + n(e_{14} + e_{23}) \leq 2n(e_{13} + e_{23} + e_{14} + e_{24}) \tag{10}$$

Equation 10 above can be rewritten as

$$e_{12} + e_{34} + e_{14} + e_{23} \leq 2(e_{13} + e_{23} + e_{14} + e_{24}) \tag{11}$$

Since the left hand side of Equation 11 is OPT and its right hand side is ALG, it follows that

$$\frac{ALG}{OPT} \geq \frac{1}{2} \tag{12}$$

□

4 A 2/3-Approximation Algorithm for cmax-3-cut

In this section we consider the problem of partitioning the vertices of a graph into three equal-sized sets of size $n/3$ each such that the number of edges across the sets is maximized. Once again we start with an arbitrary partition of the set into three subsets, each of size $n/3$. While there exists a pair of vertices such that interchanging them increases the edges in the cut, we perform the switch. Let the optimal partition be $\{V_1, V_4, V_7\}, \{V_2, V_5, V_8\}, \{V_3, V_6, V_9\}$ whereas the locally optimal partition discovered by the greedy algorithm is $\{V_1, V_2, V_3\}, \{V_4, V_5, V_6\}, \{V_7, V_8, V_9\}$ as shown in Figure 1. Once again e_{ij} is the number of vertices from set V_i to set V_j and $e_{ij} = e_{ji}$. We note that
$OPT = e_{12} + e_{13} + e_{15} + e_{16} + e_{18} + e_{19} + e_{23} + e_{24} + e_{26} + e_{27} + e_{29} + e_{34} + e_{35} + e_{37} + e_{38} + e_{45} + e_{48} + e_{46} + e_{49} + e_{56} + e_{57} + e_{59} + e_{67} + e_{68} + e_{78} + e_{79} + e_{89}$ and
$ALG = e_{14} + e_{15} + e_{16} + e_{17} + e_{18} + e_{19} + e_{24} + e_{25} + e_{26} + e_{27} + e_{28} + e_{29} + e_{34} + e_{35} + e_{36} + e_{37} + e_{38} + e_{39} + e_{47} + e_{48} + e_{49} + e_{57} + e_{58} + e_{59} + e_{67} + e_{68} + e_{69}$.

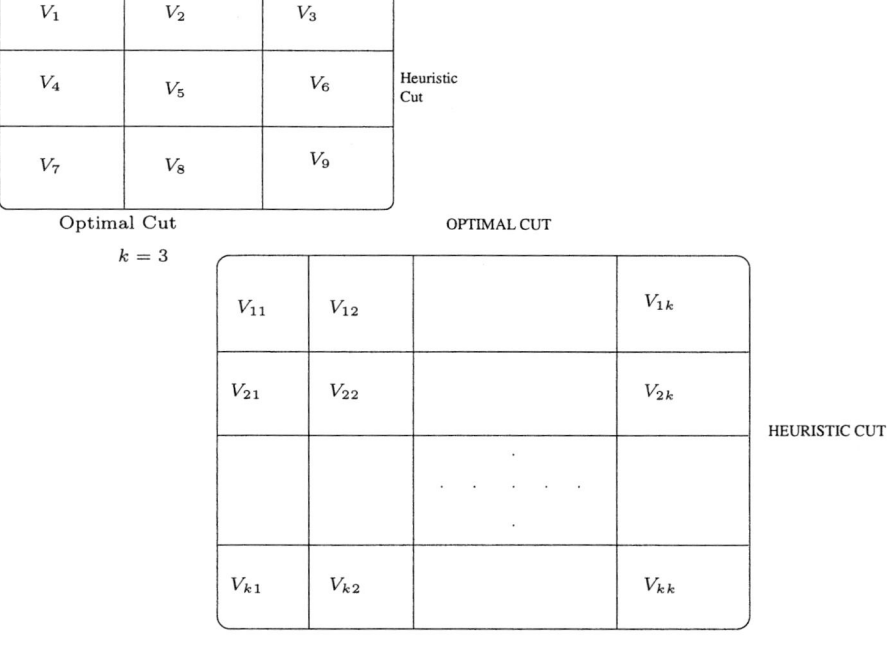

Fig. 1. Heuristic v/s Optimal cut

Theorem 2. The greedy algorithm has performance ratio 2/3.

Proof: As the greedy solution is locally optimal, for $a \in \{V_1, V_2, V_3\}$ and $b \in \{V_4, V_5, V_6\}$ the following holds:

$$e_{aV_1}+e_{aV_2}+e_{aV_3}+e_{bV_4}+e_{bV_5}+e_{bV_6} \le e_{aV_4}+e_{aV_5}+e_{aV_6}+e_{bV_1}+e_{bV_2}+e_{bV_3} \quad (13)$$

Summing up Equation 13 over $a \in V_1$ we get,

$$(2e_{11}+e_{12}+e_{13})+|V_1|(e_{bV_4}+e_{bV_5}+e_{bV_6}) \le (e_{14}+e_{15}+e_{16})+|V_1|(e_{bV_1}+e_{bV_2}+e_{bV_3}) \quad (14)$$

Summing up Equation 14 over $b \in \{V_4, V_5, V_6\}$ we get,

$$|V_4|(2e_{11}+e_{12}+e_{13})+|V_1|(2e_{44}+e_{45}+e_{46})$$
$$\le |V_4|(e_{14}+e_{15}+e_{16})+|V_1|(e_{14}+e_{24}+e_{34}) \quad (15)$$

$$|V_5|(2e_{11}+e_{12}+e_{13})+|V_1|(e_{45}+2e_{55}+e_{56})$$
$$\le |V_5|(e_{14}+e_{15}+e_{16})+|V_1|(e_{15}+e_{25}+e_{35}) \quad (16)$$

$$|V_6|(2e_{11}+e_{12}+e_{13})+|V_1|(e_{46}+e_{56}+2e_{66})$$
$$\le |V_6|(e_{14}+e_{15}+e_{16})+|V_1|(e_{16}+e_{26}+e_{36}) \quad (17)$$

Summing up Equation 13 over $a \in V_2$ we get,

$$(e_{12}+2e_{22}+e_{23})+|V_2|(e_{bV_4}+e_{bV_5}+e_{bV_6}) \le (e_{24}+e_{25}+e_{26})+|V_2|(e_{bV_1}+e_{bV_2}+e_{bV_3}) \quad (18)$$

Summing up Equation 18 over $b \in \{V_4, V_5, V_6\}$ we get,

$$|V_4|(e_{12}+2e_{22}+e_{23})+|V_2|(2e_{44}+e_{45}+e_{46})$$
$$\le |V_4|(e_{24}+e_{25}+e_{26})+|V_2|(e_{14}+e_{24}+e_{34}) \quad (19)$$

$$|V_5|(e_{12}+2e_{22}+e_{23})+|V_2|(e_{45}+2e_{55}+e_{56})$$
$$\le |V_5|(e_{24}+e_{25}+e_{26})+|V_2|(e_{15}+e_{25}+e_{35}) \quad (20)$$

$$|V_6|(e_{12}+2e_{22}+e_{23})+|V_2|(e_{46}+e_{56}+2e_{66})$$
$$\le |V_6|(e_{24}+e_{25}+e_{26})+|V_2|(e_{16}+e_{26}+e_{36}) \quad (21)$$

Summing up Equation 13 over $a \in V_3$ we get,

$$(e_{13}+e_{23}+2e_{33})+|V_3|(e_{bV_4}+e_{bV_5}+e_{bV_6}) \le (e_{34}+e_{35}+e_{36})+|V_2|(e_{bV_1}+e_{bV_2}+e_{bV_3}) \quad (22)$$

Summing up Equation 22 over $b \in \{V_4, V_5, V_6\}$ we get,

$$|V_4|(e_{13}+e_{23}+2e_{33})+|V_3|(2e_{44}+e_{45}+e_{46})$$
$$\le |V_4|(e_{34}+e_{35}+e_{36})+|V_3|(e_{14}+e_{24}+e_{34}) \quad (23)$$

$$|V_5|(e_{13}+e_{23}+2e_{33})+|V_3|(e_{45}+2e_{55}+e_{56})$$
$$\le |V_5|(e_{34}+e_{35}+e_{36})+|V_3|(e_{15}+e_{25}+e_{35}) \quad (24)$$

$$|V_6|(e_{13}+e_{23}+2e_{33})+|V_3|(e_{46}+e_{56}+2e_{66})$$
$$\le |V_6|(e_{34}+e_{35}+e_{36})+|V_3|(e_{16}+e_{26}+e_{36}) \quad (25)$$

Similarly, for $a \in \{V_1, V_2, V_3\}$ and $b \in \{V_7, V_8, V_9\}$ the following holds,

$$e_{aV_1} + e_{aV_2} + e_{aV_3} + e_{bV_7} + e_{bV_8} + e_{bV_9} \leq e_{aV_7} + e_{aV_8} + e_{aV_9} + e_{bV_1} + e_{bV_2} + e_{bV_3} \quad (26)$$

and summing up over all a, b we get the following nine equations.

$$|V_7|(2e_{11} + e_{12} + e_{13}) + |V_1|(2e_{77} + e_{78} + e_{79})$$
$$\leq |V_7|(e_{17} + e_{18} + e_{19}) + |V_1|(e_{17} + e_{27} + e_{37}) \quad (27)$$
$$|V_8|(2e_{11} + e_{12} + e_{13}) + |V_1|(e_{78} + 2e_{88} + e_{89})$$
$$\leq |V_8|(e_{17} + e_{18} + e_{19}) + |V_1|(e_{18} + e_{28} + e_{38}) \quad (28)$$
$$|V_9|(2e_{11} + e_{12} + e_{13}) + |V_1|(e_{79} + e_{89} + 2e_{99})$$
$$\leq |V_9|(e_{17} + e_{18} + e_{19}) + |V_1|(e_{19} + e_{29} + e_{39}) \quad (29)$$
$$|V_7|(e_{12} + 2e_{22} + e_{23}) + |V_2|(2e_{77} + e_{78} + e_{79})$$
$$\leq |V_7|(e_{27} + e_{28} + e_{29}) + |V_2|(e_{17} + e_{27} + e_{37}) \quad (30)$$
$$|V_8|(e_{12} + 2e_{22} + e_{23}) + |V_2|(e_{78} + 2e_{88} + e_{89})$$
$$\leq |V_8|(e_{27} + e_{28} + e_{29}) + |V_2|(e_{18} + e_{28} + e_{38}) \quad (31)$$
$$|V_9|(e_{12} + 2e_{22} + e_{23}) + |V_2|(e_{79} + e_{89} + 2e_{99})$$
$$\leq |V_9|(e_{27} + e_{28} + e_{29}) + |V_2|(e_{19} + e_{29} + e_{39}) \quad (32)$$
$$|V_7|(e_{13} + e_{23} + 2e_{33}) + |V_3|(2e_{77} + e_{78} + e_{79})$$
$$\leq |V_7|(e_{37} + e_{38} + e_{39}) + |V_3|(e_{17} + e_{27} + e_{37}) \quad (33)$$
$$|V_8|(e_{13} + e_{23} + 2e_{33}) + |V_3|(e_{78} + 2e_{88} + e_{89})$$
$$\leq |V_8|(e_{37} + e_{38} + e_{39}) + |V_3|(e_{18} + e_{28} + e_{38}) \quad (34)$$
$$|V_9|(e_{13} + e_{23} + 2e_{33}) + |V_3|(e_{79} + e_{89} + 2e_{99})$$
$$\leq |V_9|(e_{37} + e_{38} + e_{39}) + |V_3|(e_{19} + e_{29} + e_{39}) \quad (35)$$

Similarly, for $a \in \{V_4, V_5, V_6\}$ and $b \in \{V_7, V_8, V_9\}$ the following holds,

$$e_{aV_4} + e_{aV_5} + e_{aV_6} + e_{bV_7} + e_{bV_8} + e_{bV_9} \leq e_{aV_4} + e_{aV_5} + e_{aV_6} + e_{bV_1} + e_{bV_2} + e_{bV_3} \quad (36)$$

and summing up over all a, b we get the following nine equations.

$$|V_7|(2e_{44} + e_{45} + e_{46}) + |V_4|(2e_{77} + e_{78} + e_{79})$$
$$\leq |V_7|(e_{47} + e_{48} + e_{49}) + |V_4|(e_{47} + e_{57} + e_{67}) \quad (37)$$
$$|V_8|(2e_{44} + e_{45} + e_{46}) + |V_4|(e_{78} + 2e_{88} + e_{89})$$
$$\leq |V_8|(e_{47} + e_{48} + e_{49}) + |V_4|(e_{48} + e_{58} + e_{68}) \quad (38)$$
$$|V_9|(2e_{44} + e_{45} + e_{46}) + |V_4|(e_{79} + e_{89} + 2e_{99})$$
$$\leq |V_9|(e_{47} + e_{48} + e_{49}) + |V_4|(e_{49} + e_{59} + e_{69}) \quad (39)$$
$$|V_7|(e_{45} + 2e_{55} + e_{56}) + |V_5|(2e_{77} + e_{78} + e_{79})$$
$$\leq |V_7|(e_{57} + e_{58} + e_{59}) + |V_5|(e_{47} + e_{57} + e_{67}) \quad (40)$$

$$|V_8|(e_{45} + 2e_{55} + e_{56}) + |V_5|(e_{78} + 2e_{88} + e_{89})$$
$$\leq |V_8|(e_{57} + e_{58} + e_{59}) + |V_5|(e_{48} + e_{58} + e_{68}) \tag{41}$$
$$|V_9|(e_{45} + 2e_{55} + e_{56}) + |V_5|(e_{79} + e_{89} + 2e_{99})$$
$$\leq |V_9|(e_{57} + e_{58} + e_{59}) + |V_5|(e_{49} + e_{59} + e_{69}) \tag{42}$$
$$|V_7|(e_{46} + e_{56} + 2e_{66}) + |V_6|(2e_{77} + e_{78} + e_{79})$$
$$\leq |V_7|(e_{67} + e_{68} + e_{69}) + |V_6|(e_{47} + e_{57} + e_{67}) \tag{43}$$
$$|V_8|(e_{46} + e_{56} + 2e_{66}) + |V_6|(e_{78} + 2e_{88} + e_{89})$$
$$\leq |V_8|(e_{67} + e_{68} + e_{69}) + |V_6|(e_{48} + e_{58} + e_{68}) \tag{44}$$
$$|V_9|(e_{46} + e_{56} + 2e_{66}) + |V_6|(e_{79} + e_{89} + 2e_{99})$$
$$\leq |V_9|(e_{67} + e_{68} + e_{69}) + |V_6|(e_{49} + e_{59} + e_{69}) \tag{45}$$

Summing up equations 9, 10, 11, 13, 14, 15, 17, 18, 19, 21, ..., 29, 31, ..., 39, we get the following inequalities. For ease of exposition, we let $R_1 = |V_1| + |V_2| + |V_3|$, $R_2 = |V_4| + |V_5| + |V_6|$, and $R_3 = |V_7| + |V_8| + |V_9|$ denote the sum of the sets in Row 1, Row 2, and Row 3 respectively.

$$\begin{aligned} &(e_{12} + e_{13} + e_{23})(2(R_2 + R_3)) \\ &+ (e_{45} + e_{46} + e_{56})(2(R_1 + R_3)) \leq \\ &+ (e_{78} + e_{79} + e_{89})(2(R_1 + R_2)) \end{aligned} \quad \begin{aligned} &(e_{14} + e_{15} + e_{16} \\ &+ e_{24} + e_{25} + e_{26} \\ &+ e_{34} + e_{35} + e_{36})(R_1 + R_2) \\ &+ (e_{17} + e_{18} + e_{19} \\ &+ e_{27} + e_{28} + e_{29} \\ &+ e_{37} + e_{38} + e_{39})(R_1 + R_3) \\ &+ (e_{47} + e_{48} + e_{49} \\ &+ e_{57} + e_{58} + e_{59} \\ &+ e_{67} + e_{68} + e_{69})(R_2 + R_3) \end{aligned}$$

If all the partitions are of the same size $(n/3)$ then $R_1 = R_2 = R_3 = n/3$, and we get

$$4n/3(e_{12} + e_{13} + e_{23} + e_{45} + e_{46} + e_{56} + e_{78} + e_{79} + e_{89}) \leq (2n/3)ALG$$

Equivalently, $\quad (e_{12} + e_{13} + e_{23} + e_{45} + e_{46} + e_{56} + e_{78} + e_{79} + e_{89}) \leq ALG/2.$

Furthermore $e_{15} + e_{16} + e_{18} + e_{19} + e_{24} + e_{26} + e_{27} + e_{29} + e_{34} + e_{35} + e_{37} + e_{38} + e_{48} + e_{49} + e_{57} + e_{59} + e_{67} + e_{68} \leq ALG$. Adding this with the previous equation we get $OPT \leq 3/2ALG$. □

5 Generalization

In this section we generalize the results presented in the previous sections. The problem is to partition the vertices of a graph into k equal-sized sets (of size

n/k each) such that the number of edges across the sets is maximized. Once again we start with an arbitrary partition of the desired size. While there exists a pair of vertices such that interchanging them increases the edges in the cut, we perform the switch. Let the optimal partition and the locally optimal partition discovered by the greedy algorithm be as shown in Figure 1. The set of vertices in each row corresponds to a partition obtained by the greedy algorithm, and the set of vertices in each column corresponds to a partition obtained by the optimal. Thus, the first row, comprising vertex sets $V_{11}, V_{12}, \ldots, V_{1k}$, corresponds to a partition $V_{1x} = V_{11} \cup V_{12} \cup \ldots V_{1k}$ obtained by the greedy algorithm, and the first column comprising vertex sets $V_{x1} = V_{11}, V_{22}, \ldots, V_{k1}$, corresponds to a partition $V_{11} \cup V_{21} \cup \ldots V_{k1}$ obtained by the optimal. In general, V_{px} denotes the set of vertices in row p (and therefore in a greedy partition), and V_{xq} denotes the set of vertices in column q (and therefore in an optimal partition) in Figure 1. Finally, we let $e_{pq,rs}$ denote the edges between sets V_{pq} and V_{rs}. We now provide the proof for the general case.

Theorem 3. The performance ratio of the greedy algorithm is $(k-1)/k$.

Proof: Consider two vertices u, v, where $u \in V_{pq}$ and $v \in V_{rs}$, where $p \neq r$. Since $p \neq r$, the vertices u and v lie in different partitions of the greedy cut. The number of edges that cross the greedy cut currently (but would not if vertices u and v were swapped), is given by

$$e_{uV_{rx}} + e_{vV_{px}}$$

The number of edges that do not cross the greedy cut currently (but would if the vertices u and v were swapped), is given by

$$e_{uV_{px}} + e_{vV_{rx}}$$

When the cut derived by the algorithm is locally optimal,

$$e_{uV_{rx}} + e_{vV_{px}} \leq e_{uV_{px}} + e_{vV_{rx}}$$

Summing the LHS and RHS over all vertices $u \in V_{pq}$ and $v \in V_{rs}$,

$$|V_{pq}| \sum_{j=1, j \neq s}^{k} e_{rs,rj} + |V_{rs}| \sum_{j=1, j \neq q}^{k} e_{pq,pj} \leq |V_{pq}| \sum_{j=1}^{k} e_{rs,pj} + |V_{rs}| \sum_{j=1}^{k} e_{pq,rj}$$

The LHS corresponds to edges that cross the optimal cut (summed over all vertices $u \in V_{pq}, v \in V_{rs}$), but do not cross the greedy cut. The RHS corresponds to edges that cross the greedy cut (summed over all vertices $u \in V_{pq}, v \in V_{rs}$), some of which may cross the optimal cut.

For each term $e_{pq,pj}, j > q$ on the LHS, the coefficient is given by $\sum_{i=1, i \neq p}^{k} 2V_{ix}$. Since there are $2(k-1)$ partitions, each of which has n/k elements, each coefficient equals $2(n/k)(k-1)$. Similarly, for each term of the form $e_{pq,rs}, r > p$, the coefficient is given by $V_{px} + V_{rx}$. Since each of the these partitions has cardinality n/k, the coefficient for each term on the RHS equals $2(n/k)$.

From the above, it follows that

$$\sum_{p=1}^{k}\sum_{j=p+1}^{k}\sum_{q=1}^{k} e_{pq,pj} \leq \frac{1}{k-1}\sum_{p=1}^{k}\sum_{r=p+1}^{k}\sum_{q=1}^{k}\sum_{j=1}^{k} e_{pq,rj}$$

Note that the term $\sum_{p=1}^{k}\sum_{q=1}^{k}\sum_{r=p+1}^{k}\sum_{j=1}^{k} e_{pq,rj}$ on the RHS of the equation above is the size of the greedy cut (denoted by ALG). Adding this to both the LHS and RHS, we get

$$\sum_{p=1}^{k}\sum_{q=1}^{k}\sum_{r=p+1}^{k}\sum_{j=1}^{k} e_{pq,rj} + \sum_{p=1}^{k}\sum_{j=p+1}^{k} e_{pq,pj}$$
$$\leq \sum_{p=1}^{k}\sum_{q=1}^{k}\sum_{r=p+1}^{k}\sum_{j=1}^{k} e_{pq,rj} + \frac{1}{k-1}\sum_{p=1}^{k}\sum_{q=1}^{k}\sum_{r=p+1}^{k}\sum_{j=1}^{k} e_{pq,rj}$$

Since OPT is at most the LHS, we get $OPT \leq (1+\frac{1}{k-1})ALG$, which implies that $ALG \geq \frac{k-1}{k}OPT$. □

6 Conclusion

In this paper we studied a capacitated version of the max-k-cut problem. We showed that a fast approximation algorithm based on local search has a performance ratio of $(k-1)/k$. To the best of our knowledge this is the first deterministic approximation algorithm for the capacitated version of the max-k-cut. It should be noted that the problem has applications in stowage of containers in ports and ships [1, 2], parking vehicles in lots and other constrained versions of packing problems.

Acknowledgements. The authors would like to thank Rajeev Kohli for bringing the first two references to their attention.

References

1. M. Avriel, M. Penn, and N. Shpirer. Container ship stowage problem: complexity and connection to the coloring of circle graphs. *Discrete Appl. Math.*, 103(1-3):271–279, 2000.
2. M. Avriel, M. Penn, N. Shpirer, and S. Witteboon. Stowage planning for container ships to reduce the number of shifts. *Annals of Operations Research*, 76:55–71, 1998.
3. A. Frieze and M. Jerrum. Improved approximation algorithms for max k-cut and max bisection. *Algorithmica*, 18(1):67–81, 1997.
4. M. Goemans and D. Williamson. Improved approximation algorithms for maximum cut and satisfiability problems using semidefinite programming. *Journal of the ACM*, 42(6):1115–1145, 1995.
5. M. Goemans and D. Williamson. Approximation algorithms for max-3-cut and other problems via complex semidefinite programming. In *Proceedings of the thirty-third annual ACM symposium on Theory of computing*, pages 443–452. ACM Press, 2001.

6. S. Mahajan and H. Ramesh. Derandomizing approximation algorithms based on semidefinite programming. *SIAM Journal on Computing*, 5:1641–1663, 1999.
7. C. Papadimitriou. *Computational Complexity*. Addison Wesley, 1994.
8. S. Sahni and T. Gonzalez. P-complete approximation problems. *Journal of the ACM*, 23:555–565, 1976.
9. V. Vazirani. *Approximation Algorithms*. Springer, 2001.

A Cooperative Multi Colony Ant Optimization Based Approach to Efficiently Allocate Customers to Multiple Distribution Centers in a Supply Chain Network

Srinivas[1], Yogesh Dashora[2], Alok Kumar Choudhary[2], J.A. Harding[1], and M.K. Tiwari[2]

[1] Wolfson School of Mechanical and Manufacturing Engineering,
Loughborough University UK, LE11 3TU
{s.srinivas, j.a.harding}@lboro.ac.uk
[2] Department of Manufacturing Engineering, NIFFT, Ranchi-834005, India
mkt09@hotmail.com

Abstract. With the rapid change of world economy, firms need to deploy alternative methodologies to improve the responsiveness of supply chain. The present work aims to minimize the workload disparities among various distribution centres with an aim to minimize the total shipping cost. In general, this problem is characterized by its combinatorial nature and complex allocation criterion that makes its computationally intractable. In order to optimally/near optimally resolve the balanced allocation problem, an evolutionary Cooperative Multi Colony Ant Optimization (CMCAO) has been developed. This algorithm takes its governing traits from the traditional Ant Colony optimization (ACO). The proposed algorithm is marked by the cooperation among "sister ants" that makes it compatible to the problems pertaining to multiple dimensions. Robustness of the proposed algorithm is authenticated by comparing with GA based strategy and the efficiency of the algorithm is validated by ANOVA.

Keywords: Supply chain, Balanced allocation, Cooperative Multi Colony Ants, ANOVA.

1 Introduction

A supply chain network (SCN) is typically described as a network of suppliers, fabrication/assembly sites, distribution centers, ware houses, retail locations, and customers. One of the key issues in the immaculate linkage of supply chain members is related to the expansion of distribution centers. In this paper an at tempt has been made to protably expand the distribution centers by optimally allocating the customers to the distribution centers in order to make the system balanced. Several prominent researchers have contributed to body of the liter ature pertaining to location-allocation problem [1], [2]. The complex balanced allocation problem has attracted considerable attention in the recent years due to its crucial impact on the supply chain [6]. The intricacy of the problem can be assessed by the fact that the optimal allocation scheme has to be chosen among a number of feasible alternative combinations that increase exponentially with the increase of customers and distribution centers. Thus, the problem can be characterized as NP-hard and comes under the broad category of combinatorial optimization problems. The main motive behind the construction of supply chain network is its cost efficiency that takes into account the minimization of total logistic cost (shipping cost, inventory carrying cost, etc.) for each distributor. While the balancing of Supply Chain Network is being carried out, 'equitability' of supply chain network with respect to the product ows is also considered. Here, the word 'equitable' refers to allotment of comparative amount of work load to all distribution centers.

This paper proposes a new Cooperative Multi Colony Ant Optimization (CM CAO) algorithm to solve the complex balanced allocation problem that utilizes the search strategy based on the cooperation of information between the 'sister ants' of various colonies. The proposed approach is the modied version Ant Colony Optimization (ACO), which is a generic meta-heuristic that aims to obtain optimal/near optimal solutions for the complex problems that are NP hard in nature [3], [4], [5]. The ant algorithm takes its inspiration from the behavior of real ants that they could find the shortest path from their colony to the feeding zone. The proposed CMCAO algorithm utilizes the traits of ant algorithms to solve multidimensional problem sets. It outperforms other prevailing solution methodologies due to its appealing property of handling even complex problems of comparatively large data sets with greater ease. Rest of the paper has been arranged in the following sequence: Section 2 mathematically models the balanced allocation problem. The description of CMCAO algorithm along with its application over balanced allocation problem is covered under Section 3. The case study is discussed in the section 4. The paper is concluded in section 5.

2 Mathematical Modeling

In the balanced allocation problem, the shipping cost between distribution centers and the customer affects the profit of manufacturers. Hence, the minimization of the aggregate shipping cost of all distribution centers along with the fulfillment of balanced allocation criteria has been taken as the main objective function. The mathematical formulation represented by equation (1) and (2) depicts the objective functions and it had been strategically incorporated in the local and global search in the proposed CMCAO algorithm. The mathematical formulation of the problem can be given as:

$$Minimize \quad \{DI\} \qquad (1)$$

$$Minimize \sum_{i=1}^{N} \sum_{j=1}^{C_{nj}} SC_{ij} \times \Delta_{ij} \qquad (2)$$

$$\Delta_{ij} = \begin{cases} 1, \text{ if } j^{th} \text{ customer is allocated to } i^{th} \text{ distribution center,} \\ 0, \text{ otherwise} \end{cases} \qquad (3)$$

DI : Degree of Imbalance
SC_{ij} : Shipping Cost of jth customer from i^{th} distribution center.
C_{nj} : Number of customer allotted to distribution centre J.

3 Proposed Cooperative Multi Colony Ant Optimization (CMCAO) Algorithm

In this paper, a new Cooperative Multi Colony Ant Optimization algorithm has been proposed to deal with, and employed to solve the balanced allocation problem of a supply chain. The CMCAO algorithm is peculiar in the way it deals with multidimensional data sets. In the problems pertaining to such data sets, each dimension represents a definite collection of nodes with similar attributes. Each node has its characteristic dimensions and a different ant colony moves in each dimension with the mutual cooperation of 'sister' ants. Thus, The optimum path selection is based on mutual cooperation of sister ants.

3.1 The Algorithm

The proposed algorithm uses M colonies, with each colony comprised of N ants. Each of the ant in a colony has sister ants in the other colonies, which work together to get the best path. Initially, each ant is randomly placed on a node of its characteristic colony. Having placed all the ants on the nodes, they are moved randomly for a few cycles without any visibility or pheromone trail identifying capacity. This takes the algorithm out of initial stagnation and provides initial guidelines for the ants to start with. To do away with the repetition of the nodes in an ant's path, two tabu lists are associated with each ant. First list i.e. $Tabu1^k$ keeps the track of the nodes encountered in the path of ant type k, while the second list i.e. $Tabu2^k$ stores the information about the feasible nodes those are available for the ant type k to move. After completion of the tour, the pheromone is updated on each node which was initialized as zero. The pheromone updation is based on the following relation:

$$\Gamma_{i,j} = (1 - \partial)\Gamma_{i,j} + \sum_{k=1}^{n} \Delta\Gamma_{i,j}^{k} \qquad (4)$$

$$\Delta\Gamma_{i,j}^{k} = \begin{cases} \Theta \backslash W_k, & if\ edge(i,j)\ is\ chosen\ in\ the\ path\ of\ ant\ type\ k, \\ 0, & otherwise \end{cases} \qquad (5)$$

$\Gamma_{i,j}$: Pheromone intensity on edge(i,j)
∂ : Pheromone evaporation rate.
Θ : A predefined constant.
W_k : weightage of the tour completed by ant type k

N = Number of ants = Number of customers
M = Number of colonies = Number of distribution centers

Here, the weightage is the cumulative weightage of all sister ants of ant type k. In node transition phase, each ant is assigned its starting node randomly in its own characteristic dimension. Each ant chooses the next node probabilistically. For k^{th} ant type of mth dimension, placed on a node i, the probability to choose the next node j is given by

$$\Xi_{i,j}^k = \begin{cases} \frac{(\Gamma_{i,j})^\alpha (\nu_{i,j})^\beta}{\sum_{m \in tabu2^k}(\Gamma_{i,j})^\alpha (\nu_{i,j})^\beta} \\ 0, \quad otherwise \end{cases} \quad (6)$$

$\Xi_{i,j}^k$: probability of ant type k to choose edge (i,j)
ν_{ij} : visibility of the node j from node i
α : constant determining the dependency on the pheromone trail intensity.
β : constant determining the dependency on visibility of the node

After the completion of a tour, pheromone trail on each edge is updated using equation 4 and equation 5. The parameters utilized by the algorithm to come out of local optima are discussed in the following subsection.

Avoidance of Stagnation and Quick Convergence: Stagnation can be depicted as the situation where algorithm stops searching for new paths. If the selection of ∂ is not proper, the efficiency of probabilistic selection (based on the pheromone trail density and visibility) is hampered. This causes stagnation by making the search procedure as biased search procedure. Aforementioned problem is resolved by generating a random number 's', that is to be compared with the intensity of pheromone trails. For very high pheromone trail $\Gamma_{i,j}$,'s' will be very small and for very low pheromone trail the same 's' is very large. Depending on 's' if the trail is insignificant, ants overlook the path and randomly select the next node. Thus there is always a possibility to look for new paths and in turn stagnation is avoided. Another problem encountered is pertaining to the swift convergence of the algorithm. This problem is tackled using a parameter 'μ_o' which is defined as:

$$\mu_o = \frac{log_e(iter_n)}{log_e(max_i)}; \quad \mu_o \epsilon [0,1] \quad (7)$$

Where, $iter_n$ is the iteration number and max_i is maximum iteration. In initial cycles, μ_o remains close to zero and in the later phases μ_o approaches to a unit value. Another random number u is generated and is compared with μ_o. If μ is smaller, then the probabilistic approach is applied .But, if μ_o is small, then the factor s is used. Hence, the search for new paths is encouraged in the initial phases and the algorithm becomes more powerful in avoiding the entrapment in local optima.

Pseudo Code: The Pseudo code of the algorithm is described as follows:

1. *Initialization*
    - Calculate the visibility matrix.
    - Set the algorithm parameters (max_i, α, β, etc.)
    - Set number of colonies M to be used as per problem requirement. Mark a sister ant for each ant in every other colony.
    - Randomly place ants on each node. /* Ants are placed on nodes of their characteristic colonies */
    - Set iter = 0.
    - Let each ant complete the tour. /* Having moved through all the nodes allocated to ant confirms completion.*/
    - Set initial pheromone intensity by using equation (3) and equation (4)
    - Choose the next node to move randomly.
2. *Node Transition*
    - If iter ¿maxi go to 4.
    - Set k = 0; /* counter for ant type*/
    - **Label:** Generate random number $\mu (0 \leq \mu \leq 1)$.
    - If $\mu \leq \mu_o$ then proceed, otherwise choose the next node randomly.
    - Create Tabu1 listing the moved nodes and Tabu2 having probability of nodes that can be visited. Probability is calculated as per equation (6).
    - Mark the nodes of Tabu1 as closed for the ant type that has moved.
    - Choose the node with highest probability
    - Generate a random number $s(0 \leq s \leq 1)$
    - If, $s \geq \xi_{i,j}^k$ then choose the next node on random basis. Else, choose the node with highest probability.
    - Add the chosen node to Tabu1 for further reference.
    - k = k+ 1;
    - If (k < M), go to Label.
3. *Updating*
    - iter→ iter + 1.
    - If best-cost ¿ best-iter, then best-cost = best-iter .
    - Enter the best tour in array best-tour.
    - Update pheromone matrix.
    - Empty all tabu lists.
    - $\mu_o = \frac{\log_e(iter_n)}{\log_e(max_i)}$; go to 2
4. *Result*
    - Give the best-tour along with best-cost as output

3.2 Application of CMCAO Algorithm to the Balanced Allocation Problem

Let there be M distribution centers with N customer centers to be allocated. Hence, M colonies of ants are used i.e. each distribution center is marked by a different colony. Each distribution center is characterized by N number of nodes (customers) and each of the N customers has a chance to be allocated to any of the distribution center. Each colony has N ants to be moved. Each ant of a colony has a sister ant in every other colony with which it share its experiences of the encountered paths. Prior to the initiation of the algorithm, it is to be decided that how many customers (C_n) are to be allocated to each distribution center for the fulfillment of balanced allocation criterion.

$$C_n = \lceil \frac{N}{M} \rceil + \Omega \tag{8}$$

$$\Omega = \begin{cases} 1, \; if \; C_n \epsilon Dis \\ 0, \; otherwise \end{cases} \tag{9}$$

$$Dis = set \; of \; randomly \; chosen \; (N - (N \backslash M)) \; distributors. \tag{10}$$

Now, a colony 'c' is chosen to initiate the movement. Among the N ants of a colony an ant 'k' is chosen randomly and is set to move. When the ant has moved through the number of nodes (customers) allocated to it, the tour of ant type k of colony c is considered to be completed. Now, the remaining ants of the colony are one by one moved in the same fashion. Again, the next colony to move is chosen randomly from the remaining unmoved colonies and ants are set to move. Here, all those nodes that are marked as closed by the $Tabu1^k$ are also restricted for the ant type k. Now, the ants of next colony start moving on the nodes and aforementioned process is continued until all the colonies have completed their respective tours. When the tour of all ants in the initialization phase has been completed, the pheromone is laid on each edge of the path traversed by any of the ant. The quantity of laid pheromone is calculated with the help of equation 4 and equation 5. Now in node transition phase, the ants are distributed on the nodes in the same fashion as in initialization phase. In this phase, the next node is chosen probabilistically as per equation (6). After the completion of a cycle the pheromone on each node is updated keeping in view the objective of minimizing the deviation in the cost of distribution centers. Equation (4) and equation (5) governs basic pheromone updation, the weightage of the tour is defined by the degree of imbalance that is to be minimized. The degree of imbalance is defined as:

$$DI = \frac{SC^M_{max} - SC^M_{min}}{SC^M_{avg}} \tag{11}$$

SC^M_{max} : Maximum Shipping cost among all the distributors
SC^M_{min} : Minimum Shipping cost among all the distributors
SC^M_{avg} : Average Shipping cost

It is noteworthy that the degree of imbalance represents the weightage of the tour and hence, is placed in the denominator of expression that represents pheromone updation. The singularity (i.e. situation having zero DI), or extremely low DI may cause heavy imbalance in the quantity of pheromone to be laid. This in turn causes the algorithm to follow biased search and the problem of entrapment in the local minima is visualized. To do away with such situations, a critical minimum value of DI is set as DI_{min} and can be incorporated in equation 4 and 5 as:

$$\Delta \Gamma_{i,j}^k = \begin{cases} \Theta \backslash DI_k, & if\ edge(i,j)\ is\ chosen\ in\ the\ path\ of\ ant\ type\ k, \\ and\ DI_k > DI_{min} \\ \Theta \backslash DI_k + \Delta \overline{\Gamma}_{avg}, & if\ edge(i,j)\ is\ chosen\ in\ the\ path\ of\ ant\ type\ k, \\ and\ 0 < DI_k <= DImin \\ 0, & otherwise \end{cases} \quad (12)$$

$\Delta \overline{\Gamma}_{avg}$:average pheromone laid by all the ants with $DI > DI_{min}$

Thus, the path with extremely low/zero DI will be intensively favored and that would cause more and more ants to choose the path. A variable array 'best-tour' keeps the track of the best tour travelled during the search process, and variable 'best-cost' retains the value of corresponding cost of the tour obtained. As the number of iterations reaches the maximum allowable iterations, the algorithm stops and best-tour along with best-cost is given as the output.

4 Case Study

In order to test the robustness of the proposed CMCAO algorithm for the balanced allocation problem, the data set has been adopted from [6]. In this case study, the firm produces chain link fences and other related hardware items. These are distributed to a total of 21 customers across the country. Firm has future plans to move its current distribution center to a new location in order to avoid overlapped distribution and duplicated delivery efforts within the existing centers. It was found that seven sites are most appropriate for reallocations that are marked as $(DC_1, DC_2, ...DC_7)$. The shipping costs from various distribution centers to the customers can be obtained from [6]. In this illustration firm has DC_1, DC_2, and DC_3 as the potential distribution centers to which 21 customer centers are to be allocated while fulfilling the balanced allocation criteria. By applying the proposed CMCAO algorithm over the undertaken data set, the optimal/sub optimal allocation sequence obtained is given in Table 1. The convergence of result starts on very few iterations, thus the computational burden and CPU time have been considerably reduced. This convergence trend is validated by the graph shown in Figure 1, between the average shipping cost and number of iterations, for the data set. The average cost obtained for all the distribution centers accord with the costs obtained by the implementation of GA based strategy of Zhou [6]. When the problem size is comparatively smaller, both the strategies, i.e. GA and CMCAO, produce the optimum solution. When large and real size data sets with greater number of customers and distribution centers are considered, the proposed CMCAO algorithm outperforms the GA based application. The comparative results for the average cost, obtained by the

Table 1. Optimal/Sub optimal allocation sequence

Distributors	Customers allocated							Shipping Cost
D_1	C_1	C_2	C_{10}	C_{11}	C_{12}	C_{17}	C_{20}	22.40
D_2	C_6	C_7	C_8	C_9	C_{13}	C_{14}	C_{16}	22.40
D_3	C_3	C_4	C_5	C_{15}	C_{18}	C_{19}	C_{21}	22.40

Table 2. Comparative results for the average shipping cost

Number of Customers	30	40	50	60	70	80	90	100
Number of Distributors	3	4	5	6	7	8	9	10
Average Shipping Cost (GA based approach)	311.7	318.0	338.0	333.5	277.9	264.5	295.4	282.4
Average shipping cost (CMCAO approach)	310.3	315.7	334.56	332.3	275.8	263.9	280.4	281.2

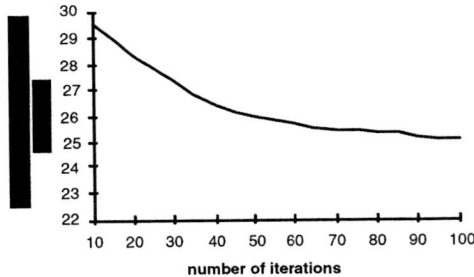

Fig. 1. Number of iterations vs average shipping cost

GA based strategy and the proposed CMCAO algorithm, have been displayed in Table 2. Here, the shipping costs are randomly and uniformly distributed in the range of [20, 50] To assess the applicability of the algorithm performance analysis for varying problem sizes and parameters related to the problem environment are performed. The relevant parameters are organized into four categories named as very small, small, medium and large data set, to have a better appraisal of the algorithm performance. The parameter values used for the testing purpose are summarized in Table 3.

A new parameter termed as 'Cost index' has been utilized to evaluate the performance of the algorithm. It can be mathematically defined as:

$$Costindex = \frac{Worstcase - BestCase}{Worstcase} \qquad (13)$$

Where, the Worst cost and Best cost described in the equation (14) are the shipping costs of ants with maximum and minimum cost, respectively, after

Table 3. Parameter values related to the data sets of problem

Classification	Number of Distributors	Number Of Customers
Very Small (VS)	3	30-40
	4	40-50
Small (S)	5	50-60
	6	60-70
Medium (M)	7	70-80
	8	80-90
Large (L)	9	90-100
	10	100-105

Table 4. Cost Index for various data sets

Number of Distributors	Number of customers	Cost Index
3	30	.01796
3	35	.01638
4	40	.02345
4	45	.01569
5	50	.01447
5	55	.01379
6	60	.01563
6	65	.01753
7	70	.02638
7	75	.01536
8	80	.01379
8	85	.03137
9	90	.03264
9	95	.01119
10	100	.01897
10	105	.02897

100 iterations. From the denition of Cost Index, it can be envisaged that the near optimal solution of the problem is achieved if its value is very small. The computational results for dierent categories of data sets are provided in Table 4. The variation of Cost Index with increasing number of iterations has been described in Figure 2

As the number of iterations increase, more and more ants tend to move on the optimal or near optimal paths that in turn decrease the Cost Index. It is evident from Figure 2 that Cost Index constantly decreases as the number of iterations increase and its very low value at the later stages provide the near optimal solution and establishes the efficacy of the proposed algorithm. Table 5 provides the average Cost Indices for different problem sizes. A two way ANOVA without replication was performed to assess the significance of the problem parameters. The results of the ANOVA test are provided in the Table 6. The results obtained by the ANOVA, prove the accuracy of the proposed algorithm to solve the

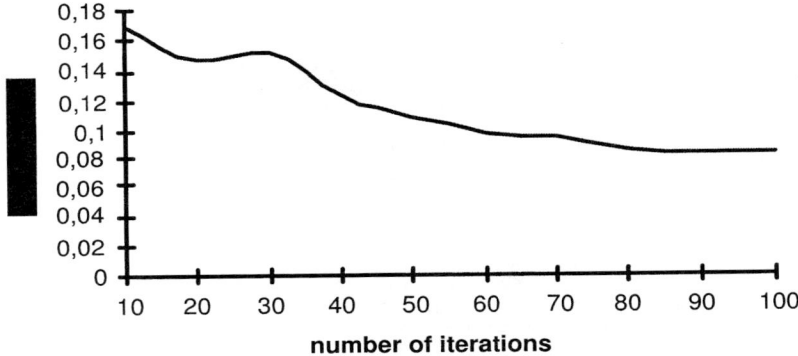

Fig. 2. Number of iterations vs cost index

Table 5. Average cost index for dierent problem size(L: Average (Cost Index) for the smaller number of customers in the respective categories. H: Average (Cost Index) for the larger number of customers in the respective categories)

	L	H	Average
VS	.01717	.01957	.01837
S	.01413	.01658	.016005
M	.02087	.02258	.021675
L	.02192	.02397	.022943

Table 6. Results of ANOVA analysis(α=0.005, SS: Sum of Squares; df: Degree of freedom; MS: Mean Square Error; F: F value; F crit: F critical value)

Source of Variation	SS	df	MS	F	P-value	F crit
Rows	0.007048	3	0.002349	395.8443	0.000215	9.276619
Columns	0.000927	1	0.000927	156.1443	0.001105	10.12796
Error	1.78E-05	3	5.93E-06			
Total	0.007992	7				

balanced allocation problem. F test is carried out at 99.5 percent confidence level which is highly significant. Thus, it statistically validates the robustness of the algorithm.

Exhaustive experiments over various algorithm parameters have been carried out and their effect on the performance of the algorithm has been carefully studied. The relative values of α and β are responsible for the dependence of the local search on pheromone trail intensity and the visibility. The algorithm has been tested by varying α and β in the range of [0.25 - 10] and [0.5 - 5] respectively. To infer the results obtained, a 3-D plot has been drawn between the values of α, β and the average shipping cost after 100 iterations, as is shown in Figure 3. It can be concluded from Figure 3 that the algorithm performs better with the

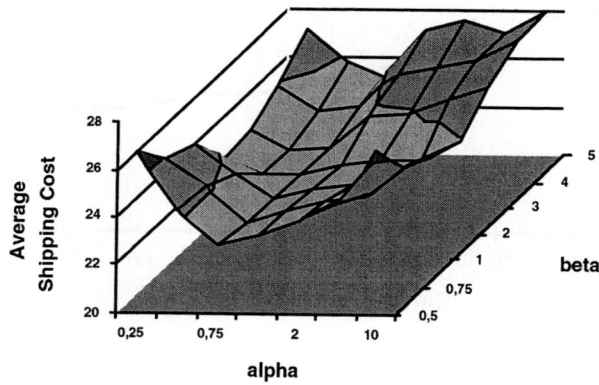

Fig. 3. Variation of average shipping cost vs variation in a and β

value of α around 1 and the value of β ranging between 2-3. The relatively low values of α and β and makes the algorithm more prone to be trapped in the local optima, while their higher values increases the computational cost. In nutshell, the aforementioned computational results not only authenticate the efficacy and supremacy of the proposed strategy but also provide a new dimension to the solution of complex allocation problems in real time.

5 Conclusion

In this paper, a new Cooperative Multi Colony Ant Optimization (CMCAO) has been proposed to solve the balanced allocation problem. The most alluring feature of the algorithm is its flexibility and simple structure that enables it to be tailored in such a fashion that optimal/sub optimal solutions for the problems of varied dimensions are achieved quite easily. The results of the exhaustive computer simulation of the algorithm shows its out performance over GA based strategy. Low value of cost index authenticates the optimality/sub optimality of the results. The ANOVA test statistically validates the efficacy and robustness of the algorithm. Also, based on these results, the use of CMCAO algorithm seems to be promising in supporting the notion of seamless and more efficient supply chain.

References

1. Aikens, C. H.: X Facility Location Models for distribution planning. European Journal of Operations Research,22, (1985) 263-279.
2. Current, J. R., Min, H., and Schilling, D. A.: Multiobjective analysis of location decisions. European Journal of Operational Research, 49,(1990) 295-307.

3. Dorigo, M., Maniezzo, V. and Colorni, A., The ant systems: optimization by a colony of cooperative agents . IEFE-Trans. Man, Machine and Cybernatics-Part B, 26(1),(1996) 29-41.
4. Maniezzo, V. and Colorni, A., and Dorigo, M.: Algodesk: an Experimental Comparison of Eight Evolutionary Heuristics Applied to the Quadratic Assignment Problem, European Journal of Operation Research, 181, (1995) 188-205
5. Kumar, R., Tiwari, M. K., Shankar, R.: Shedulingof Flexible Manufacturing Systems: an ant colony optimization approach. I. Mech. E., Part B., Journal of Engineering Manufacture.,217,(2003) 1443-1453
6. Zhou Gengui, Min Hokey and Gen Mitsuo: The balanced allocation of customers to multiple distribution centers in the supply chain network: a genetic algorithm approach, Computers and Industrial Engineering, 43, (2002) 251-261.

Experimentation System for Efficient Job Performing in Veterinary Medicine Area

Leszek Koszalka[1] and Piotr Skworcow[2]

[1] Chair of Systems and Computer Networks,
Wroclaw University of Technology, 50-370 Wroclaw, Poland
leszek.koszalka@pwr.wroc.pl
[2] Control Theory and Applications Centre, Coventry University,
Coventry CV15FB, the United Kingdom
p.skworcow@coventry.ac.uk

Abstract. In this paper we present the experimentation system with database for efficient job performing, using as an example an animal clinic. The structure of tasks and operations is based on the real procedures performed at an animal clinic. The assumed model implemented in the system is more sophisticated than a typical job-shop because of a multitude of various parameters, such as fatigue factor and the presence of uncertainty. The used heuristic priority algorithm enables us to analyze easily the impact of varied factors on the produced work-plan, and emphasize or switch off the impact of any considered factor. The system has been implemented in Matlab environment. In this work we present the opportunities of the proposed system on two examples. The first example of research is focused on choosing the optimal subset of performers for a given work. The second example concerns some work-rest policies, needed for evaluating the optimal work-rest model.

1 Introduction

The first job-shop scheduling problems with presence of the fatigue factor were considered by Eilon [1] in sixtieths of the former century. Since then, lot of algorithms were developed, as there are plenty of practical issues concerning scheduling problems, e.g. productions processes [2]. The majority of these algorithms are based on polynomial and pseudo-polynomial methods [3], neural networks [4], genetic and evolutionary approaches [5]. The problem discussed in this paper is more sophisticated than a typical job-shop problem [6] because not only a fatigue factor, but multi-skilled performers, random human-factor, objects dependent on processing times and other introduced factors are taken into consideration. For the considered model of performing we propose an heuristic priority job scheduling algorithm to minimizing the completion time. This algorithm uses dispatching rules similar to those described by Panwalkar and Iskander [7]. The algorithm is an element of the designed and implemented complex experimentation system. The flexible construction of the system gives opportunities for making various investigations. In this paper we focus on the work-rest policies and efficiency of performers. Bechtold and Thompson proposed in [8] an optimal work-rest policy for a work group, but they considered the model of process with no idle time caused by precedence constraints. Moreover, their work-rest policy assumed single break for all group at one time, thus their results can not be applied to the problem stated in this paper.

The paper is organized as follows: The problem statement and the proposed models of processes are described in Sect. 2. In Sect. 3 we present the proposed job scheduling algorithm. Section 4 contains description of the experimentation system, including the introduced measure for efficiency. The results of investigations are shown and discussed in Sect. 5. The final remarks appear in Sect. 6.

2 Problem Statement

We have a set of n performers $D = \{d_1, d_2, ...d_n\}$ and a set of m objects $P = \{p_1, p_2, ...p_m\}$. In this work we used an animal clinic as a background of a problem, thus performers and objects are meant as doctors and patients, respectively. The clinic has its own database, which contains w defined tasks $Z = \{t_1, t_2, ...t_w\}$ and r operations $O = \{o^1, o^2, ...o^r\}$. Each task is a sequence of operations (with precedence constraints) to be executed by the performers on the objects. The i-th task may be denoted as $t_i = \{o_{1,i}, o_{2,i}, ...o_{j,i}, ...o_{r_i,i},\}$ where $o_{j,i}$ is the j-th operation in the i-th task, $o_{j,i} \in O$ (e.g. $o_{1,2} = o^3$) and r_i is the total number of operations forming i-th task. The four groups of features (entry sets, parameters) are distinct.

I. The Operations Features: (i) standard time of performing stp (given in t.u. - time units, $stp \in N \backslash \{0\}$), (ii) operation difficulty $operd$. Obviously stp and $operd$ are independent, e.g. an interview with a patient can take more time (big stp) but it's less fatiguing than setting a broken thighbone (big $operd$).

II. The Objects Features: (i) tasks to be performed on objects (t_i - one per object), (ii) object difficulty $objd$, and (iii) external priority ep interpreted e.g. as a bribe paid for a patient to be served faster.

III. The Performers Features: (i) endurance en, (ii) recovery rate rr, (iii) current level of stamina st_T [0 - 100%], (iv) matrix S of levels of skills in operations and (v) the maximal allowed value of max_skill_val. Let k-th row of S is $S_k = \{s_{1,k}, s_{2,k}, ...s_{j,k}, ...s_{r,k}\}$, where $s_{j,k}$ is level of the k-th performer's skill in j-th operation, $0 \leq s_{j,k} \leq max\_skill\_val$.

IV. Additional Entry Parameters: (i) fatigue threshold ft - the time of continuous work, above which fatigue has an impact on performers (they start loosing stamina), (ii) minimum stamina st_{MIN} - the minimal level of stamina, below which we put a performer to rest, (iii) full recovery $full\_rec$, (iv) continuity weight w_cont which determines how important is performing every task with as few breaks as possible, (v) external priority weight w_ep, (vi) standard deviation $stddev$ of a normal distribution, and (vii) global recovery rate $global\_rr$.

The key point of the process of performing job (tasks on the given set of objects) is that each object needs to go through the series of operations (which forms a task corresponded to a given object) to be executed by the performers. This process in the real life is very complex - it is impossible to perfectly model the influence of behaviour of human-being and disturbances, therefore we cannot accurately predict a time of performing operation. To model this uncertainty we introduce a random "human-factor" denoted as *rand*.

For the purposes of this paper, it is assumed that: (i) each operation from any task can be executed by any performer that is able to do it, thus one object may be "served" by several performers, but is being executed by only one at the moment, (ii) each performer can work on several objects during the whole process, but only on a fixed one at a time, (iii) a performer may be idle at some moments, when his skills or precedence constraints don't allow him to work, (iv) the setup time [6] is neglected, (v) tasks are divisible (breaks between operations are permitted), but operations are not divisible.

The time of performing the operation j on the object (patient) i by the performer (doctor) k is expressed by (1).

$$T_{i,j,k} = \frac{stp_j \cdot objd_i}{s_{j,k} \cdot st_T^k} \cdot rand \qquad (1)$$

where st_T^k is a level of stamina of k-th doctor at the beginning of performing operation j, $rand$ is of normal distribution with mean equal to 1 and standard deviation $stddev$, we also set its lower limit at 0.75, thus the time of performing the operation can be shortened by random factor at most by 25%.

When a performer works, he loses his stamina (not permanently). The level of stamina has an impact on the performer's efficiency. Following [9], the lost of stamina by k-th doctor may be given by (2).

$$st_T^k = st_{T-1}^k - \frac{1.1^{(cw_k - ft)}}{100} \cdot \frac{operd}{en_k} \qquad (2)$$

where st_{T-1}^k represents stamina at previous t.u, cw_k is the time of continuous work (in t. u.) of k-th doctor. If $st^k < st_{MIN}$ then k-th doctor is resting until: $st^k > st_{MIN}$ for $full\_rec = 0$, or $st^k = 100\%$ for $full\_rec = 1$. Note, that $full\_rec$ takes into account only performers who are put to rest, not those that are idle. When a performer is put to rest or is idle (Fig.1), he recovers his stamina. We consider a linear recovering expressed by (3).

$$st_T^k = st_{T-1}^k + rr_k \cdot global\_rr \qquad (3)$$

In this work we have chosen the $finish\_time$ (the completion time C_{MAX} to the whole job) as the efficiency index (the shorter, the better). Because of uncertainty the whole performing process is non-deterministic i.e. identical entry sets of features can result in the different completion times.

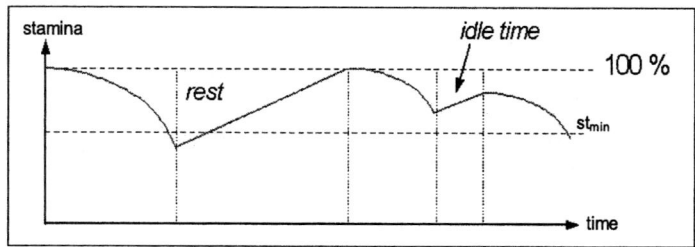

Fig. 1. An example of the loss and recovery of stamina for $full\_rec = 1$

3 Job Scheduling Algorithm

In this section we present our heuristic priority-algorithm. Even a standard $job-shop$ problem without any fatigue factor is a strongly non-polynomially hard (strongly NP) for $n \geq 2$ [10]. The problem stated here is much more sophisticated than a classic job-shop and we are unaware of any readily available optimal or sub-optimal algorithm that could solve it, even excepting a random factor.

In each iteration (at each distinct t.u.), we determine the two sets:

The Set of Available Performers. $AD = \{ad_1, ad_2, ...ad_{anT}\}$, where anT means a number of available performers at iteration T. Note, that these are idle performers, whose $st_T \geq st_{MIN}$.

The Set of Available Operations. $AO = \{ao^1, ao^2, ...ao^{arT}\}$, where arT means a number of operations not yet executed till time moment T. These are the initial operations or those, for which predecessors have already been completed.

Having AO and AD we calculate priorities $p_{k,j}, k = 1..anT, j = 1..arT$ for each possible pair of ao and ad. Next, we choose the following pairs with the highest priorities, until no more pairs can be assigned. The procedure of designating AO and AD, counting priorities and assigning operation-performer pairs is repeated for $T+1, T+2, ...$ until all the tasks are accomplished.

We assume, that priorities are calculated using (4).

$$p_{k,j} = \frac{w_s \cdot s_{j,k} + w\_ep \cdot ep_j + w_{st} \cdot st_T^k + w\_cont \cdot initial}{w_{nop} \cdot nop_j + w_{stp} \cdot stp_j} \quad (4)$$

where ep_j is external priority of the object on which ao^j is supposed to be performed, $initial = 0$ if ao^j is an initial operation or $initial = 1$, otherwise, nop_j is a number of items in the task to which operation j belongs. Weights in (4) are user-defined (w_ep, w_cont) or automatically fixed ($w_s, w_{st}, w_{nop}, w_{stp}$). We may wish to finish the tasks already started as soon as possible, as we deal with alive animals. Adjusting w_cont, we can raise or lower the priorities of operations of those tasks. In future all fixed weights will be adjusted through a learning process, at present, they all are equal to 1.

4 Experimentation System

The system consists of three modules: the simulator based on the model described in Sect. 2 and the algorithm presented in Sect. 3, the input (plan of experiment) - output (presentations of results) module and the database editor. The system (Fig. 2) was implemented in MatLab environment, because (i) algorithm contains plenty of matrix operations, for which MatLab is very efficient, (ii) it allows easy connection with database using SQL commands. Thanks to user-friendly graphic interface the user is able to define and modify sets D and P and other input parameters. The scale of input sets is limited only by the size of the screen, as the algorithm was developed for the sets of any size and structure. The programs make also possible to record input sets and to read them later. As the result of a single simulation we receive the work-plan for the doctors in a form of the Gantt's graph and information about the anticipated completion time.

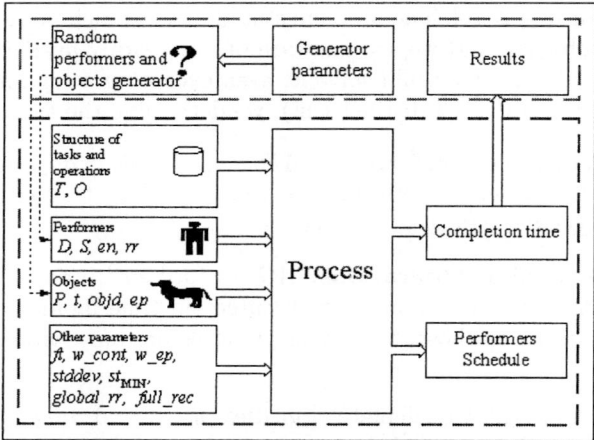

Fig. 2. The block-scheme of the experimentation system

The simulator can be used as a managing tool e.g. to find the best subset of a given set of all available performers for a known set of objects. The program generates all subsets of performers that are able to execute the required operations, and launches series (because of random factor) of simulations for each subset, as a result it displays mean completion times for all considered subsets.

The database editor application was built in Sybase environment [11] with using Power Builder 6.5. The application enables the user to delete/alter/create operations and structure of tasks. In Fig. 3 the conceptual data model of the database is presented.

Using the application as a complex experimentation system we define all the parameters, except entry sets D and P, as they are generated randomly. Thanks to randomly generated sets (in automatic manner) we can evaluate some global job performing policies, useful for most cases in real situations.

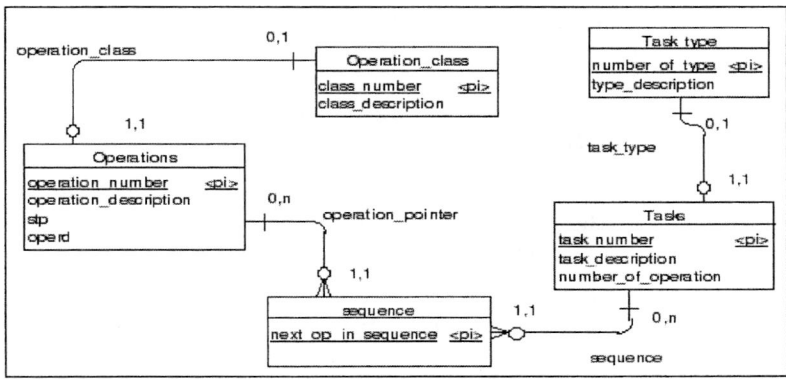

Fig. 3. Conceptual model of the database

The generator creates the sets: $\Delta = \{D_1, D_2, ...\}$ - set of sets D, and $\Pi = \{P_1, P_2, ...\}$ - set of sets P. The user can fix the following parameters: (i) *iteration* $\overline{\overline{\Delta}} = \overline{\overline{\Pi}}$ i.e. number of generated entry sets of performers and objects, (ii) *trials* a number of trials for each pair $<D_i, P_i>$, (iii) max_perf, and (iv) max_obj.

In order to generate sets P and D the following *procedure* was applied:

Step. 1 Generate $\forall_{D_i \in \Delta} \overline{\overline{D_i}} = random(1..max\_perf)$ and
$\forall_{P_i \in \Pi} \overline{\overline{P_i}} = random(1..max\_obj)$
Step. 2 Generate $\forall_{D_i \in \Delta} \forall_{1 \leq j \leq \overline{\overline{D_i}}}\ num_{i,j} = random(1..\overline{\overline{O}})$
Step. 3 Find $\forall_{D_i \in \Delta} \forall_{1 \leq j \leq \overline{\overline{D_i}}}\ V_{i,j} = \{val_1, ...val_k, ...\}$ where
$\forall_{D_i \in \Delta} \forall_{1 \leq j \leq \overline{\overline{D_i}}} \forall_{1 \geq k \geq num j}\ val_k = random(1...max\_skill\_val)$
Step 4. Let $\forall_{D_i \in \Delta} \forall_{1 \leq j \leq \overline{\overline{D_i}}}\ perm_{i,j}$ be random permutation $\{1, 2, ...\overline{\overline{O}}\}$
Step 5. $\forall_{D_i \in \Delta} \forall_{1 \leq j \leq \overline{\overline{D_i}}}$ the first $num_{i,j}$ items from the sequence $perm_{i,j}$ are the chosen operations performed by d_j, their skill values are given by $V_{i,j}$.
Step 6. Generate $\forall_{P_i \in \Delta} \forall_{1 \leq i \leq \overline{\overline{P_i}}}\ t_i = random(1...\overline{\overline{T}})$
Step 7. Check whether $\forall_{D_i \in \Delta, P_i \in \Pi}\ D_i$ is able to perform all the tasks from P_i, if not add skills to performer using $random(1...max\_skill\_val)$.
Step 8. Randomly generate other parameters: $en, rr, objd$ and ep.

In order to eliminate an impact of *rand* factor we repeat trials for the same initial conditions and then calculate mean of results (5).

$$mft_i^j = \frac{1}{trials} \cdot \sum_{l=1}^{trials} (C_{MAX_i^j})_l \qquad (5)$$

However, the mft_i^j for a given pair $<D_i, P_i>$ may be hundreds times greater than of another pair, thus we need to normalize the results along with (6).

$$\forall_{1\leq i\leq iteration}\forall_{1\leq j\leq res}\quad mft\_norm_i^j = mft_i^j \cdot \frac{g\_mean}{mean_i} \qquad (6)$$

$$mean_i = \frac{1}{res} \cdot \sum_{j=1}^{res} mft_i^j, \quad g\_mean = max_{1\leq i\leq iteration}(mean_i)$$

where res is the total number of elements in Δ. Finally, the efficiency index of i-th simulation eff_i (the lower, the better) is given by (7).

$$eff_i = \frac{1}{iteration} \cdot \sum_{j=1}^{iteration} mft\_norm_i^j \qquad (7)$$

Carrying out experiments for various values of the same parameter (e.g. ft or w_cont) and comparing efficiency indices eff_i of each simulation we can determine the global influence (regardless of entry sets) of this parameter on the completion time.

5 Investigations

Experiment 1. The objective was to find the subset consisting of three performers chosen from a given set of $n=5$ performers for job associated with the known set of objects such that the completion time (mean $Cmax$) be minimal.

The features of $r=28$ operations defined for the animal clinic were: standard time $stp=1$ for $j=2$-9,12,14,17,19,24,27,28, $stp=2$ for $j=1,11,13,16,18,20,22$, $stp=3$ for $j=10,15,21,23,25,26$, operation difficulties $operd=1$ for $j=1$-9,11,12,14-17,27, $operd=2$ for $j=10,19$-22,24,28, $operd=3$ for $j=13,25,26$, and $operd=4$ for $j=18,23$. There were distinguished $w=10$ tasks, their characteristics are given as below, where the indices j of operations which form tasks are in {} brackets and parameters $objd$ and ep are in () brackets: $t_1(5,5)=\{1,4,6,17,21,13\}$, $t_2(1,20) = \{1,4,16,17,19,11,24,22,13,12\}$, $t_3(3,10) = \{1,3,4,5,6,7,8,14,15,2\}$, $t_4(4,0)=\{1,6,4,16,17,19,11,22,20,12\}$, $t_5(3,0)=\{1,6,5,9,2,16,17,18,19,11,24,23,12\}$, $t_6(2,0) = \{1,16,6,5,9,25,26,12\}$, $t_7(1,0) = \{1,16,6,5,17,11,24,22,20,13,12\}$, $t_8(1,20)=\{1,3,6,5,7,8,10,14,15,2\}$, $t_9(2,5) = \{1,4,6,5,7,28,16,17,11,22,13,20,12\}$, $t_{10}(5,15)=\{1,6,5,8,14,10,9,28,12\}$.

Two sets of objects were considered: the first set consisted of $m=8$ objects (case 1) required performing associated tasks of $t_3, t_2, t_1, t_5, t_7, t_9, t_{10}, t_4$ and the second set consisted of $m=4$ objects (case 2) which required tasks of t_3, t_8, t_6, t_5. The performers had features en and rr as specified in brackets: $d_1(2,2)$, $d_2(3,1)$, $d_3(1,3)$, $d_4(3,3)$, $d_5(5,5)$. The matrix S of skills of performers is presented in Table 1. Other parameters values were: $w\_ep=0$, $ft=1$, $min\_stamina=10\%$, $std\_dev=0.1$, $full\_rec=0$, $w\_cont=0$, $global\_rr=2\%$. The number of simulations for each feasible subset containing three performers (team of doctors) was equal to 100.

Table 1. Matrix S – performer's skills in operations

j	1	2	3	4	5	6	7	8	9	10	11	12	13	14
d_1	2	3	1	0	1	0	3	0	3	0	2	0	1	0
d_2	0	2	0	3	0	1	0	1	0	2	1	3	0	2
d_3	2	1	0	0	3	0	0	2	0	0	1	0	1	1
d_4	0	1	2	1	2	1	2	1	2	2	2	0	3	3
d_5	1	1	0	2	2	0	0	1	1	0	0	2	2	0
j	15	16	17	18	19	20	21	22	23	24	25	26	27	28
d_1	2	0	2	0	3	0	3	0	3	0	1	0	2	1
d_2	2	3	0	2	0	1	0	1	2	1	0	1	0	1
d_3	2	2	2	2	1	3	3	1	1	1	2	2	1	0
d_4	2	2	1	1	1	0	0	0	0	3	2	1	1	1
d_5	0	2	2	0	0	2	0	0	0	2	2	0	0	0

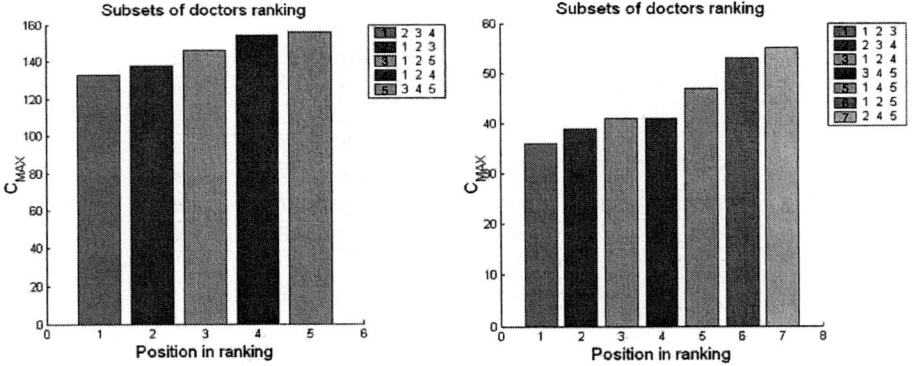

Fig. 4. Ranking of teams

The obtained results are presented in Fig. 4, on the left for case 1 and on the right for case 2. In case 1, the team {d_2, d_3, d_4} was the best. For this subset the estimated mean C_{MAX} was of of 133 t.u. In case 2, the best was the team {d_1, d_2, d_3} with mean C_{MAX} equal to 36 t.u.

Comparing these results with input data we may observe that (i) it would be expected that the performer d_5 with the greatest en and rr becomes a member of the best team but it isn't, quite the contrary the performer d_2 having the lowest rr is a member of 'winning' teams in both cases and d_5 is a member of teams located in third (case 1) and fourth (case 2) place, (ii) the performer d_4 with the average best skills (see Table 1) and relatively good features en and rr is also not a member of the best team in case 2.

These observations and others confirm that simulation and investigation are very useful while making decision about choosing a team of performers for job performing.

Experiment 2. The objective was to find the best values of *min_stamina* and *full_rec* for different *global_rr* and to make an attempt in determining a global work-rest policy for the model proposed.

The values of parameters were fixed as: $w\_cont = w\_ep = 0$, $trials = max\_perf = max\_obj = 10$, $ft = 1$, $stdev = 0.20$, $iteration = 500$. We made series of experiments for all possible combinations of *min_stamina* and *full_rec*. We compared *eff* for constant *full_rec* and various values of *min_stamina*. In Tab. 2 the results of research i.e. the three best values of *min_stamina* and *eff* corresponded to these values are presented. Note, that one can not compare these values of *eff*, which are not in the same row of the table, as they concern different sets of performers and objects, generated independently for each considered *global_rr*.

Table 2. Relationship between stamina and efficiency

	\multicolumn{6}{c}{$full\_rec = 0$}
	\multicolumn{6}{c}{*min_stamina* and corresponding *eff*}
global_rr	1th best	eff	2th best	eff	3th best	eff
0.05%	90%	14966	80%	15383	70%	15697
0.2%	90%	3209	80%	3315	70%	3426
0.5%	90%	2354	80%	2445	70%	2560
2%	90%	865	80%	890	70%	920
5%	90%	844	80%	861	70%	885
20%	90%	480	80%	481	70%	483
	\multicolumn{6}{c}{$full\_rec = 1$}
	\multicolumn{6}{c}{*min_stamina* and corresponding *eff*}
global_rr	1th best	eff	2th best	eff	3th best	eff
0.05%	60%	24192	70%	24577	50%	25052
0.2%	90%	5766	80%	6209	70%	6319
0.5%	90%	2248	80%	2365	70%	2451
2%	90%	519	80%	530	70%	540
5%	90%	839	80%	851	70%	872
20%	90%	495	80%	497	70%	502

It may be observed, that (i) when we want to use full recovery policy (which is easier to apply in reality causing fewer breaks during the work period), then globally the best *min_stamina* value is of 90% (except very slowly recovering performers), (ii) if *global_rr* = 20%, the differences between *min_stamina* = 70%, 80% and 90% are not remarkable, (iii) the lower is *global_rr*, there are more locally worst results for globally the best arrangement of parameters, therefore we need to be careful while planning a work-rest policy, especially for slowly recovering performers.

6 Final Remarks

The system presented in this paper seems to be very useful for investigations concerning efficient job performing policies. The proposed model of processes is based on the job-shop model, but is much more complex. At present, the database is applied to veterinary clinic. However, the flexible structure of the experimentation system allows the user to adapt the model to the required reality by altering the database. Adjusting the parameters, the user is able to estimate their impact on efficiency index, and to create some convenient job performing policies.

Future work leading to development of knowledge-based decision making system can be directed to (i) improvement of scheduling algorithm by adjusting the weights in (4) with using learning methods [12, 13], (ii) taking into consideration other efficiency measures, and (iii) various extensions of functions of experimentation system.

References

1. Eilon, S.: On mechanistic approach to fatigue and rest periods. International Journal of Production Research, **3** (1964) 327-332
2. Geyik, F.: A hybrid frameworkfor the multicriteria production schedulins. Proc. of International Conference of Responsive Manufacturing ICRM02 (2002) 782-788
3. Shmoys, D.B., Stein, C.: Improved approximation algorithms for shop scheduling problems. SIAM, **23** (1994) 617-632.
4. Sabuncuoglu, I., Gurgun, B.: A neural network model for scheduling problem. EJOR, **93** (1996) 288-299
5. Cheng, R., Gen, M.: Genetic algorithms and engineering design. New York (1997)
6. Jain, A. S., Meeran, S.: Deterministic job shop scheduling: past, present and future. International Journal of Production Research, **36** (1998) 1249-1272
7. Panwalkar, S.S., Iskander, W.: A survey of scheduling rules. Operations Research, **25** (1997)
8. Bechtold, S.E., Thompson, G.M.: Optimal scheduling of a flexible-duration rest period for a work group. Operations Research, **41** (1993) 1046-1054
9. Bechtold, S. E., Summers, D. E.: Optimal work-rest scheduling with exponential work-rate decay. Management Science, **34** (1988) 547-552
10. Gola, M., Kasprzak, A.: Exact and approximate algorithms for two-criteria topological design problem of WAN with budget and delay constraints. LNCS, **3045** (2004) 611-620
11. Pozniak-Koszalka, I.: Relational databases in Sybase environment: modelling, designing, applications. Wroclaw University of Technologi Press, (2004).
12. Sutton, R., Barto, A.: Reinforcement Learning. MIT Press, Cambridge, Bradford Book (1998)
13. Wozniak, M.: Proposition of boosting algorithm for probabilistic decision support system. LNCS, **3036**, (2004) 675-678

An Anti-collision Algorithm Using Two-Functioned Estimation for RFID Tags

Jia Zhai[1] and Gi-Nam Wang[2]

[1] Graduate Student, Industrial & Information Systems Engineering Department,
Ajou University, Suwon, Korea
zhaijiaws@vip.sina.com
[2] Industrial & Information Systems Engineering Department, Ajou University,
Suwon, Korea
gnwang@ajou.ac.kr

Abstract. Radio Frequency Identification (RFID) has recently played an important role in ubiquitous sensing technology. While more advanced applications have been equipped with RFID devices, sensing multiple passive tags simultaneously becomes especially important. In this paper, using complementarily two-functioned estimation, we propose an identification method based on the stochastic process. The underlying mathematical principles and parameters estimation models have also been well discussed. Numerical examples are given to verify the proposed two-functioned estimation identification method within a given expected accuracy-level. Key Words: RFID (Radio Frequency Identification), Anti-collision Algorithm, Two-functioned Estimation.

1 Introduction

The Radio Frequency Identification (RFID) technology is one of the key technologies in today's ubiquitous computing system. It is a fast, secure, and efficient identification procedure that influences a lot of different application domains. One application scene that we can easily imagine might at the check desk of a supermarket. Much more complicated applications could be employed in the areas such as auto-distribution production line and warehouse security moving in and out check. Furthermore, this emerging technology can be deployed under some extremely hazard circumstances where human can not reach an extreme high temperature production process in a pharmaceutical plant, and also under special conditions used in chip manufacturing plants which require a completely antiseptic condition. Compared with the existing bar code system, RFID can do everything bar codes can and much more, e.g., it does not require line-of-sight, it can be reprogrammed easily, it can be used in harsh environment, it can store more data, and it can read many data of tags simultaneously with high accuracy-level[9]. With all the superiorities mentioned above and the fact that RFID system will decrease to a relatively low price, we might anticipate RFID could be a good solution in object identification and tracking material status under a ubiquitous computing environment.

RFID system basically consists of three components: transceiver (reader), transponder (tag), and data management subsystem[2]. Transceiver, or reader, can both read data from and write data to a transponder, and it also works as a power supplier in the passive RFID systems. The transponder, or tag, is usually attached to an object to be identified and store data of the object. There are two types of RFID tags: the active one which generates power by itself usually a battery and the passive one which gets energy from the transceiver by radio frequency. The passive tag seems to be more attractive because of the price factors in that the preferable price for pervasive deployment of RFID tags is about $0.05[2]. The last but not least part is the subsystem usually a data management system that in charge of the huge information linking the physical world. It may contain both the application layer software and the middle layer software like Savant[3]. These three parts cooperate efficiently to implement all kinds of applications using RFID technology.

The ability to identify many tags simultaneously is crucial for more advanced applications such as to check out lots of items at the supermarket. It might take a lot of time to identify them one by one, which requires an efficient identification method checking large number of goods at one time without any delay. However, if multiple RFID tags are being read synchronously, the radio signals will interfere with each other, and may cancel each other's communication data. Widely used are two kinds of methods to solve this type of collision problem. One is based on a probabilistic method and the other is based on a deterministic method.

In this paper, we present a complementarily two-functioned estimation algorithm for identifying multiple RFID tags simultaneously. In the following sections, we will look over some related research works firstly. Then the mathematical basis of our research work especially some statistics conceptions are discussed. Based on the mathematical foundations, we propose two estimation functions for computing the number of tags in one transceivers reading zone. Also presented is an efficient our proposed algorithm, complementarily two-functioned estimation, to solve anti-collision problem with an illustrative examples. Conclusion and further studies are also discussed in final section.

2 Related Works

The previous works could be classified into two different approaches: one is based on deterministic model, and the other is probability model. The deterministic method has been discussed in some papers, like Marcel Jacomet's work[6]. His algorithm used a bit-wise arbitration model during the identification phase of the target device. In Weis's paper[4], he showed one deterministic algorithm based on the binary-tree walking scheme. Christian Floerkemeier's research gave some illustrations of the sources of error by some already existing methods which deal with the collision problem[7]. He also pointed out how to make a system more appropriate to handle this multiple identifying situation.

The probability approach has been performed specifically to solve collision problem of RFID tags, which is related to our works. In Sarma's research work[2], it gave a clear definition of tags collision and discussion of difficulties to solve the problem. Moreover, they pointed out probabilistic algorithms were mainly based on the Aloha scheme presenting the performance metrics of the anti-collision algorithm. There was one project which is very similar to our research work, the Vogt's paper[5]. He also gave an anti-collision algorithm based on the probabilistic method, and his algorithm only employed one simple estimation function to compute the tags number, which could be inaccurate from our research results. An attempt is given to improve the previous Vogt's work by presenting complementarily two-functioned estimation, which possibly the previous improper points could be corrected or improved by our approach.

3 Problem Definition and Mathematical Basis

3.1 Framed Aloha

In the probabilistic system, the reader can broadcast one reading command causing all the tags in the interrogation zone responding, and then all the tags send back the required data by radio wave. Such process is called one reading cycle, and corresponding time named one frame time. Then we divide one frame time into lots of small time slots, so that each tag can occupy one slot time to communicate with the reader without interfering with other tags. We use sn to represent the slot number of one frame in the rest of this paper. The time of one reading cycle depends on the sn value we choose, for easily programming in machine language, we will only use different number of 16, 32, 64, 128, and 256 as sn value.

In our Framed Aloha system, the tags occupy each slot randomly at each reading cycle ensuring after a certain amount of reading cycles, most of the tags can be read by the reader and achieve an expected reading accuracy-level such as 0.99. Figure 1 gives an illustration of one completed Slotted Aloha reading process for identifying all tags, the value of sn used here is 8. We use tn denoting tags number, so the value of tn is also 8 here. There are three occupation cases in each slot: by just one tag so that the data can be received by the reader correctly; by many tags, all the data of those tags will get lost; and by no tags. After four reading cycles, all the data can be read correctly by the reader.

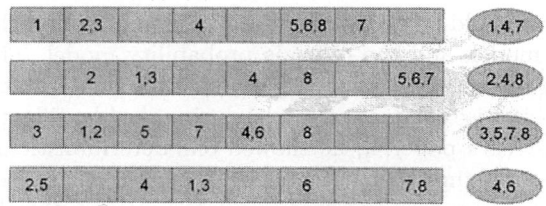

Fig. 1. An illustration of Framed Aloha reading process. After four reading cycles, all the data can be read correctly by the reader

3.2 Binomial Distribution

Given certain slot number sn and tags number tn, the number of tags in one slot, denoted by t, is a typical binomially distributed with parameters tn and 1/sn:

$$B_{tn,\frac{1}{sn}}(t) = \binom{tn}{t}(\frac{1}{sn})^t(1-\frac{1}{sn})^{tn-t} \quad (1)$$

Using (1), we can compute the probability that t numbers of tags are definitely occupied in one slot.

3.3 Occupation Probability

Identifying distribution of tags to certain number of slots is a kind of problem called Occupation Probability[8]. Based on (1), the expected value of the number of slots that occupied by t tags is given by:

$$expected\ value = sn\binom{tn}{t}(\frac{1}{sn})^t(1-\frac{1}{sn})^{tn-t} \quad (2)$$

This is a crucial equation because we will use it to implement the parameters estimation functions.

3.4 Markov Process

The whole reading process can be regarded as a Markov Process in that the number of newly identified tags of each reading cycle only depends on the previously known number of tags during last reading cycle. Hence we could use the transition matrix of Markov Process to compute a lower bound of the number of reading cycles, which is necessary to identify all tags with a given accuracy-level. How to compute the transition matrix and how to compute the necessary reading cycles were already well shown in Vogt's work[5]. In our research, focuses are given on estimating tn, sn, and reading cycle within a given accuracy-level. After we need to know sn and the accuracy-level, by given value of tn, we can easily compute the necessary reading cycles number. Therefore, how to decide the value of tn and sn is the key point and we will discuss latter.

4 Parameters Estimation

In a lot of identification applications, the exactly tags number is unknown at the beginning. We propose parameter estimation functions in order to estimate the tags number. For having done this, we assume that after each reading cycle, we can check the reading performance by three parameters: (C0, C1, Ck)[5]. C0: the number of slots that the slot is empty. C1: the number of slots that the slot is occupied by only one tag. Ck: the number of slots that the slot is collided by several tags: a collision occurs.

4.1 The First Estimation Function

Because most of the collision is just between two tags, the first estimation function is easily obtained by[5]:

$$tn = C1 + Ck * 2 \tag{3}$$

4.2 The Second Estimation Function

When the number of collision which is caused by many tags but not just two tags increased, the first function can not still work accurately enough. As a result, we need another function to compute the tags number more precisely. By the mathematical basis we discussed previously and (1), (2), we can compute the expected outcome (C0, C1, Ck) with already known tn and sn value. We run the code below in Matlab and make it a function:

$$(C0, C1, Ck) \leftarrow (sn, tn) \tag{4}$$

```
function [c0,c1,ck]=getc(sn,tn)
 c0=sn*((1-(1/sn))^tn);
 c1=tn*((1-(1/sn))^(tn-1));
 ck=sn-c0-c1;
End
```

Given typical sn, tn value, we can attain some useful numerical data of expected (C0, C1, Ck) shown in Fig. 2.

The Chebyshev's inequality indicated that the outcome of a random experiment involving a random variable is most likely somewhere near the expected outcome value. Therefore, the way to use these data is to compute more expected value of (C0, C1, Ck) by more possible tags number, e.g. tn= 1 to 400 step 1. Then make an expected outcome value table similar to the tables shown in Fig. 2. After each reading cycle, we could compare the experiment result value of (C0, C1, Ck) with the expected outcome value table. By choosing the value which has the smallest wrong-weight, we can attain the estimated sn, tn value finally.

Another method to use this function is to choose a suitable sn value at the beginning. Because there are five variables in (4), after getting a check result (C0, C1, Ck), we can easily compute the last variable tn in this function.

5 Two-Functioned Estimation Algorithm

We have already showed two estimation functions, and discussed the methods for using them. Consequently how to use them complementarily is a key concern for writing our algorithm. Because the second function is a relatively steady function, we might check the accuracy-level of the first function to evaluate its performance according to different value of sn and tn. The wrong-weight of the first estimation function can be obtained by (3), (4) and the expected outcome value table shown in Fig. 2.

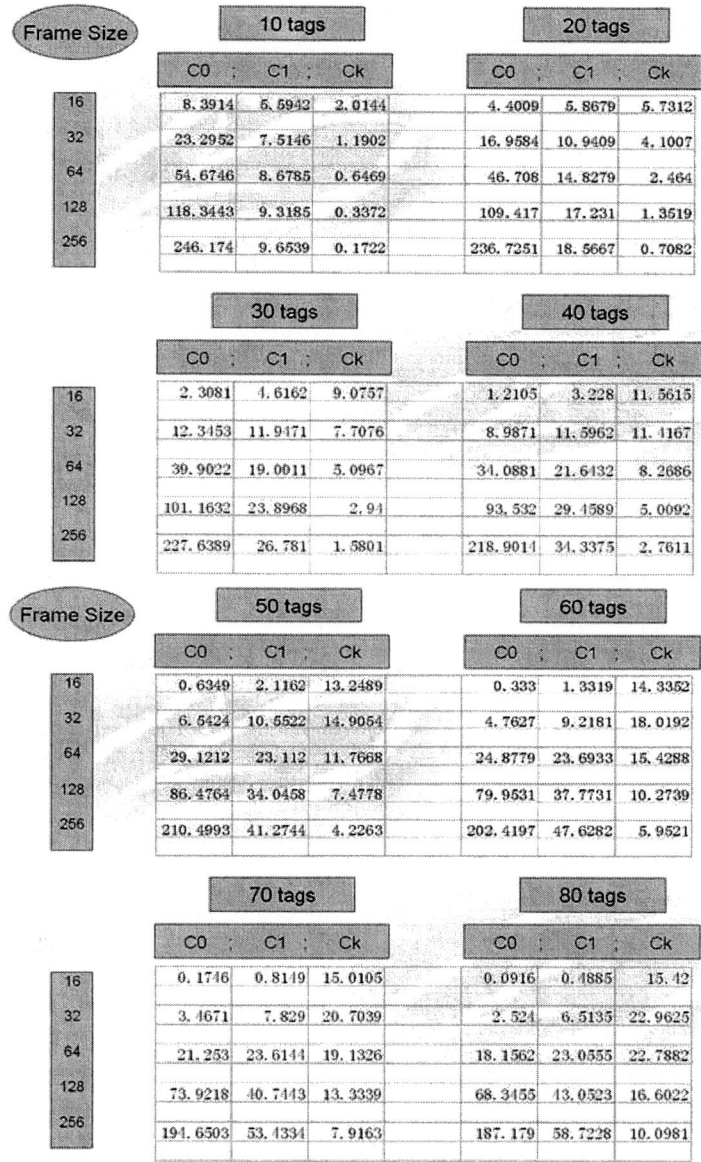

Fig. 2. Some typical expected value of (C0, C1, Ck) by given certain tn and sn

$$wrong\ weight = \frac{estimated\ tags\ number}{real\ tags\ number} \quad (5)$$

By the data shown in Fig. 3, we can clearly see the deficiency if we just use the first estimation function. As the accuracy-level is divided by 0.95, we choose the

	10	20	30	40	50	60	70	80
16	9.623	17.3303	22.7676	26.351	28.614	30.0023	30.8359	31.3285
32	9.895	19.1423	27.3623	34.4296	40.363	45.2565	49.2368	52.0899
64	9.9723	19.7559	29.1945	38.1804	46.6456	54.5509	61.8796	68.6319
128	9.9929	19.9348	29.7768	39.4773	49.0014	58.3209	67.4121	76.2567
256	9.9983	19.9831	29.9412	39.8597	49.727	59.5324	69.266	78.919

	10	20	30	40	50	60	70	80	90	100
16	0.9623	0.8665	0.7589	0.6588	0.5723	0.5	0.4405	0.3916	0.3513	0.3778
32	0.9895	0.9571	0.9121	0.8607	0.8073	0.7543	0.7034	0.7441	0.611	0.5701
64	0.9972	0.9878	0.9732	0.9545	0.9329	0.9092	0.884	0.8579	0.8313	0.8247
128	0.9993	0.9967	0.9926	0.9869	0.98	0.972	0.963	0.9532	0.9427	0.9315
256	0.9998	0.9992	0.998	0.9965	0.9945	0.9922	0.9895	0.9865	0.9832	0.9795

Fig. 3. The upper table shows the expected result employing the first estimation function, and the below table is the wrong-weight computed by (5). We divide the accuracy by 0.95

second function when the accuracy of the first function falls below 0.95. Moreover, we can obtain, when the value of sn is 256, the accuracy-level will below 0.95 after 170 tags which is not shown in the table.

After using the estimation functions rightly to compute the tags number, we might hope to adapt the sn value again by the newly tn value for better identifying performance. By the data shown in Fig. 2, we can compute the successful reading slot rate by:

$$the\ successful\ reading\ rate = \frac{the\ successful\ reading\ slot\ number}{slot\ number} \quad (6)$$

From Fig. 4, we can find out the highest successful rate is usually around 0.37 which is very close to the maximum throughput of Aloha protocol. If we look over the data by each column, we can figure out the best frame size for each tn value, e.g. when tn is 10 or 20, the best sn value are both 16. Based on more data computed and much more research works, as well as considering the reading time factor of different sn value, we give the table below for adapting sn value:

We will not consider the condition that the sn value is smaller than 16 or bigger than 256, because it makes no sense in practical usage.

At present, by the data and functions discussed above, we are going to propose our complementarily two-functioned estimation algorithm. The algorithm assumes that the tags set in the interrogation zone is static, which means after

	10	20	30	40	50	60	70	80
16	0.3496	0.3667	0.2885	0.2018	0.1323	0.0832	0.0509	0.0305
32	0.2348	0.3419	0.3733	0.3624	0.3298	0.2881	0.2447	0.2035
64	0.1356	0.2317	0.2969	0.3382	0.3611	0.3702	0.369	0.3602
128	0.0728	0.1346	0.1867	0.2301	0.266	0.2951	0.3183	0.3363
256	0.0377	0.0725	0.1046	0.1341	0.1612	0.186	0.2087	0.2294

Fig. 4. The successful reading rate of different sn and tn

Table 1. Best fit sn value of different tags number

sn	16	32	64	128	256
low	1	16	31	61	121
high	15	30	60	120	∞

a set of tags entering the reading field they must stay in that area until all tags have been identified within our expected accuracy-level. No tags are allowed to either enter or exit during the reading process. The state diagram of the reading process is shown in Fig. 5.

Firstly, we set the starting sn value by the middle one 64, and run a reading cycle to get the performance feedback (C0, C1, Ck). Then we use the first estimation function to compute tn value and check its accuracy-level by the data shown in Fig. 3. If it is a satisfied result, we use this tn value to choose a best-fit sn size and output the result. If it is not, we have to substitute the first estimation function by the second one and also resize sn by newly tn value.

After having done the above process, we have already got satisfied sn, tn value and could end the process. Yet for much better accuracy, we could take further consideration. For doing this, we might run the estimation process again after adapting best-fit sn value. However, most applications reluctantly waste another reading cycle time, so usually the process ceases after just running one reading cycle.

We use the function below and Table 1 to adapt the value of sn by a new tn value. Because sn value starts by 64, the function need checking twice whether it has to decrease or increase the value of sn in order to reach every possible value.

```
function [sn]=adaptsn(sn,tn)
  if (tn<low(sn)) then sn=sn/2
  if (tn<low(sn)) then sn=sn/2
  if (tn>high(sn)) then sn=sn*2
  if (tn>high(sn)) then sn=sn*2
end
```

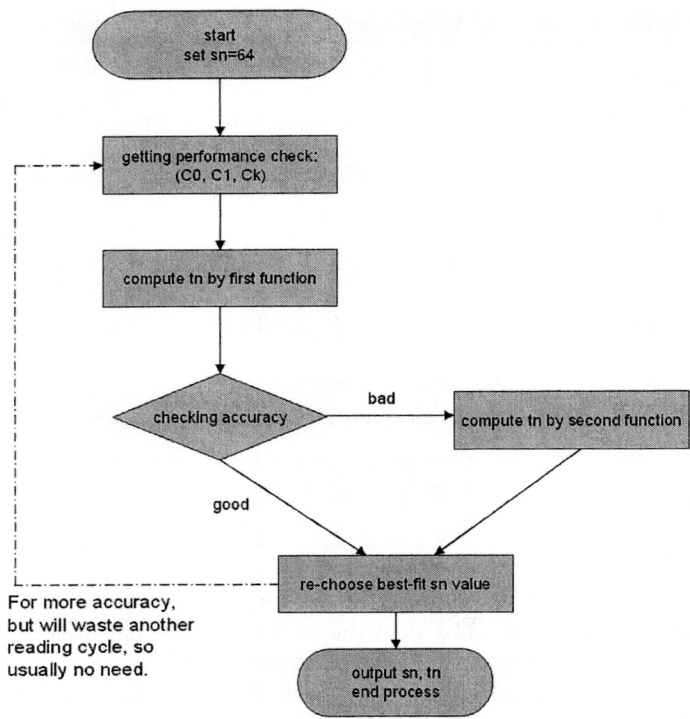

Fig. 5. The state diagram of complementarily two-functioned estimation algorithm

The output of the above complementarily two-functioned estimation algorithm is sn, tn value. By these two values we can easily compute the time of reading cycles within our desired accuracy-level by regarding the reading process as a Markov Process like what we have discussed previously.

6 Conclusion

We have demonstrated a complementarily two-functioned estimation algorithm for identifying multiple passive RFID tags synchronously. The mathematical issues required for this method has been well discussed in this paper. Furthermore, we presented two estimation functions for computing the unknown tags number and the best fit slots number which may give contribution to other anti-collision methods as well. Finally, we proposed an advanced algorithm for identifying passive tags. Illustrative numerical examples are given to verify that the proposed two-functioned estimation algorithm could be utilized efficiently. With the consolidated mathematical basis and the high precise level of our estimation method we proved either by formula or by data, we could make a conclusion that the method shown in our paper will definitely work well and provide high-level accuracy in real application. As a further research, application specific algorithms could be developed based on our research work, which could facilitate the pervasive deployment of RFID system.

References

1. Klaus Finkenzeller: RFID Handbook: Fundamentals and Applications in Contactless Smart Card and Identification. Second Edition. John Wiley & Sons. (2002) 206–211
2. Sanjay E. Sarma, Stephen A. Weis, and Daniel W. Engels: RFID Systems and Security and Privacy Implications. CHES 2002, LNCS **2523** (2003) 454–469
3. Amit Goyal: Technical Report: Savant Guide. Auto-ID Center MIT. (2003)
4. Stephen A. Weis, Sanjay E. Sarma, Ronald L. Rivest and Daniel W. Engels: Security and Privacy Aspects of Low-Cost Radio Frequency Identification Systems. Security in Pervasive Computing 2003, LNCS **2802** (2004) 201–212
5. Vogt, H.: Multiple object identification with passive RFID tags. Systems, Man and Cyber-netics, 2002 IEEE International Conference on, Volume: **3** Oct. (2002) 6–9
6. Marcel Jacomet, Adrian Ehrsam, and Urs Gehrig: Contactless Identification Device With Anticollision Algorithm. University of Applied Science Berne, Switzerland
7. Christian Floerkemeier and Matthias Lampe: Issues with RFID usage in ubiquitous comput-ing applications. PERVASIVE 2004, LNCS **3001** (2004) 188–193
8. Rajeev Motwani and Prabhakar Raghavan: Randomized Algorithms. Cambridge University Press. (1995)
9. Susy d'Hont: The Cutting Edge of RFID Technology and Applications for Manufacturing and Distribution. Texas Instrument TIRIS

A Proximal Solution for a Class of Extended Minimax Location Problem

Oscar Cornejo[1] and Cristian Michelot[2]

[1] Facultad de Ingeniería, Universidad Católica de la Ssma. Concepción,
Casilla 297 - Concepción - Chile
ocornejo@ucsc.cl
[2] Laboratoire Analyse Applique et Optimisation
Université de Bourgogne, Dijon-France
Christian.Michelot@u-bourgogne.fr

Abstract. We propose a proximal approach for solving a wide class of minimax location problems which in particular contains the round trip location problem. We show that a suitable reformulation of the problem allows to construct a Fenchel duality scheme the primal-dual optimality conditions of which can be solved by a proximal algorithm. This approach permits to solve problems for which distances are measured by mixed norms or gauges and to handle a large variety of convex constraints. Several numerical results are presented.

Keywords: Continuous location, minimax location, round-trip location problem, proximal method, Fenchel duality, partial inverse method.

1 Introduction

The aim of this paper is to propose a proximal approach for solving an important class of minimax continuous location problems which in particular contains the round trip location problem [1] and the weighted extended minimax location problem with set up costs [2]. The round trip location problem consists in finding the location of a new facility so that the maximum weighted round trip distance between the new facility and n pairs of existing facilities (or demand points) is minimized. We mean by round trip distance the total distance travelled starting from the new facility via a pair of existing facilities and going back to the new facility. As example, A.W.Chan and D.W. Hear consider the location of a delivery service. Customers have goods to be delivered from warehouses to retail stores and the objective is to minimize the maximum delivery time.

The extended minimax location problem considered by Drezner [2] is a generalization of the single facility minimax location problem. We want to locate two new facilities such that the maximum trip distance via n fixed existing facilities (or demand points) is minimized. Here we mean by trip distance the total distance travelled starting from the first new facility via a demand point and going back to the second new facility. As suggested application Drezner considers the problem of locating emergency hospital services. The total time for dispatching an ambulance and bringing the patient to the hospital consists of the travel time of the ambulance, some set up cost time and travelling back to the hospital. The problem is to shorten the response time for the farthest customer. Observing that placing the ambulance service on the hospital may not be optimal Drezner proposes to consider a priori different sites for the hospital and the ambulance station. Since the time the ambulance takes to get to the patient and the time it takes to bring the patient to the hospital have not the same importance Drezner also suggests different measures of distance for outward journey and journey back. That justifies the use of different norms or gauges.

In our model we are faced to new facilities the locations of which should be optimaly determined via a min-max criterion. It is well known that min-max criteria induce nondifferentiability in optimization problems. In the framework of continuous location analysis, this nondifferentiability is also due to the fact that a norm is never differentiable at the origin and to the (possible) use of polyhedral norms such as the ℓ_1 norm, the Tchebychev norm, etc. The nondifferentiability prevents the use of standard optimization methods and leads to solving problems by adapted procedures. Algorithms based on linear programming have already been developed for round trip problems involving the rectilinear norm [4]. Drezner proposed to solve unconstrained round trip location problems by a trajectory approach. Other procedures like those explained in [3] have also been studied. All these approaches are not completely satisfactory in particular because they often cannot be extended to problems involving mixed norm and/or non linear constraints. The aim of this paper is to investigate the interest of considering proximal procedures, like those developed in [5], and which exploit in depht the very special structure of minimax location problems.

A major technical difficulty which frequently arises when one wants to implement proximal algorithms is the computation of the proximal iteration. We will show that this difficulty can be overcome by working on an equivalent decomposable formulation of the problem. The idea is to incorporate the non linear constraints in the objective function via penalization terms, to split the objective function in independent terms and to handle by duality all the original linear constraints and the linear relations induced by the splitting. The role of the splitting is to decompose the objective function as a sum of independent terms in such a way the proximal mapping associated to the subdifferential of each term can be effectively and easily computed. This can be done by several tricks as duplication of some variables and alternative representation of certain convex functions. The original linear constraints as well as the linear relations induced by the splitting are conserved as constraints because they can be easily treated by Fenchel duality. Another fact which often militates against the use of proximal procedures is the slow rate of convergence. This second difficulty cannot be completely eliminated. However, as already observed in [5] proximal procedures seem to perform rather well on location problems and their efficiency can be significantly improved by a judicious scaling on the data. These procedures have also important advantages, as robustness and stability. Their use allows a great flexibility in using mixed norms and different types of (convex) constraints.

2 The General Formulation

We consider a general minimax problem

$$\begin{cases} \text{Minimize } \max_{1 \leq i \leq n} [\gamma_i(X - A_i) + c_i] \\ X \in \Omega_1 \times \Omega_2; \ X \in \Omega \cap L \end{cases}$$

where:
$X = (x^1, x^2)$ represents the locations of two new facilities to be placed,
$A_i = (a_i^1, a_i^2)$, $i = 1, 2, \ldots, n$, represents the locations of two fixed facilities,
γ_i is a gauge on $\mathbb{R}^2 \times \mathbb{R}^2$ of the form $\gamma_i(X) = \gamma_i^1(x^1) + \gamma_i^2(x^2)$ where γ_i^1 and γ_i^2 are two gauges on $\mathbb{R}^2$,
c_i is a nonnegative constant (a fixed cost) associated with $A_i = (a_i^1, a_i^2)$,
$\Omega^1 = \bigcap_{j=1}^{m_1} \Omega_j^1$ and $\Omega^2 = \bigcap_{j=1}^{m_2} \Omega_j^2$, are two sets of non linear constraints, expressed as intersection of closed convex sets,
Ω is a closed convex set representing (coupling) nonlinear constraints
$L \subset \mathbb{R}^2 \times \mathbb{R}^2$ is a linear subspace.

Taking $\Omega^1 = \Omega^2$, $\Omega = \mathbb{R}^2 \times \mathbb{R}^2$, $L := \{X; \ x^1 = x^2\}$, $\gamma_i^1(x^1) = w_i\gamma(x^1)$, $\gamma_i^2(x^2) = w_i\gamma(x^2)$ and $c_i = w_i\gamma(a_i^1 - a_i^2)$ with $w_i > 0$ and γ a norm on $\mathbb{R}^2$, we get a constrained version of the classical round trip location problem studied by Chan and Hearn [1]. Taking $a_i^1 = a_i^2$, $L = \mathbb{R}^2 \times \mathbb{R}^2$, we obtain a constrained version of the weighted minimax location problem introduced and studied by Drezner [2].

3 Problem Transformation

As explained in the introduction we will reformulate the problem with the aim of splitting the objective function as a sum of independent terms. Noting that our model is in fact a single minimax facility location problem involving superfacilities $A_i = (a_i^1, a_i^2) \in \mathbb{R}^4$ which gather two elementary facilities $a_i^1 \in \mathbb{R}^2$ and $a_i^1 \in \mathbb{R}^2$, the first idea which comes to our mind is to use the classical reformulation of minmax problems which consists in rewriting the problem as the one of finding a point (X, α) in the epigraph of the objective function and with lowest height α. Proceeding that way and duplicating the variables X and α with a number of copies equal to the number of fixed superfacilities A_i, one could easily obtain a decomposable reformulation of our original problem. Additional copies of X can be introduced to decompose nonlinear constraints, if necessary. However, and according to what is explained in the next section, our proximal approach leads to compute series of projections on the epigraphs of the norms used and thus would lead to serious technical difficulties due to the structure of these norms. Efficient routines for computing the projection on the epigraph of a norm can be developed when the structure of the unit ball is well known, as for Euclidean norms, ℓ_p norms or polyhedral norms for which the extreme points of their unit ball are accessible. Unfortunately, in a modelling phase one generally controls the choice of γ_i^1 and γ_i^2 whose structure can thus be assumed to be well known, but we do not control the choice of the γ_i's which are only known via their analytical expression $\gamma_i(X) = \gamma_i^1(x^1) + \gamma_i^2(x^2)$. This is the main reason why a more subtle reformulation based on the following observation is required. An equivalent formulation of a minimax problem of the form: minimize $\max_{1 \leq i \leq n} f_i^1(x) + f_i^2(x)$; $x \in \mathbb{R}^n$, is given by

$$\widehat{P} \begin{cases} \text{Minimize } \frac{1}{n}\sum_{i=1}^{n}(\alpha_i^1 + \alpha_i^2) \\ f_i^r(x) \leq \alpha_i^r, \; i=1,2,\ldots n; \; r=1,2 \\ \alpha_1^1 + \alpha_1^2 = \alpha_2^1 + \alpha_2^2 = \cdots = \alpha_n^1 + \alpha_n^2. \end{cases}$$

Now, it is not difficult to prove that problem P and problem $\widehat{P}$ are equivalent in the following sense:

- If $\bar{x}$ is optimal for P then any pair $(\bar{x}, \bar{\alpha})$ such that $f_i^1(\bar{x}) \leq \bar{\alpha}_i^1$, $f_i^2(\bar{x}) \leq \bar{\alpha}_i^2$ and $\bar{\alpha}_i^1 + \bar{\alpha}_i^2 = \bar{\xi}$ with $\bar{\xi} = \max_i (f_i^1(\bar{x}) + f_i^2(\bar{x}))$ is optimal for $\widehat{P}$.
- If $(\bar{x}, \bar{\alpha})$ is optimal for $\widehat{P}$ then $\bar{x}$ is optimal for P.

One can also observe that the non linear constraints of the transformed problem can be expressed in geometrical form as $(x, \alpha_i^1) \in \text{Epi } f_i^1$ and $(x, \alpha_i^2) \in \text{Epi } f_i^2$. Using this transformation one easily obtains the following equivalent version of our original minimax location problem

$$\begin{cases} \text{Minimize } \frac{1}{n}\sum_{i=1}^{n}(\alpha_i^1 + \alpha_i^2) \\ (x^r - a_i^r, \alpha_i^r - \frac{c_i}{2}) \in \text{Epi } \gamma_i^r, \; i=1,2,\ldots n; \; r=1,2 \\ x^r \in \Omega_j^r, \; j=1,2,\ldots,m_r; \; r=1,2 \\ \alpha_1^1 + \alpha_1^2 = \alpha_2^1 + \alpha_2^2 = \cdots = \alpha_n^1 + \alpha_n^2 \\ (x^1, x^2) \in \Omega \cap L. \end{cases}$$

This new formulation still contains nonlinear constraints which remain coupled. A classical trick to obtain a complete decomposable problem is to duplicate the variables x^1 and x^2 with the additional (linear) condition that the copies x_i^1 of x^1 (resp. x_i^2 of x^2) should be equal. Another trick which facilitates the construction of a Fenchel duality scheme is to conserve only the linear relations as constraints and to incorporate the nonlinear one in the objective function as penalization terms. This can be done by adding to the objective the indicator functions of the nonlinear constraint sets. The new formulation, has a completely separable convex objective function made up of independent terms and has only linear constraints. This reformulation of the original problem immediately induces a Fenchel dual which has a similar and symmetrical form (see [9]).

4 Duality Scheme and Optimality Conditions

The Fenchel dual consists in minimizing the conjugate of the objective function on the orthogonal complement of the primal linear constraints. Using the conjugate of a function $\varphi : \mathbb{R}^n \times \mathbb{R} \to \mathbb{R} \cup \{+\infty\}$ of the form: $\varphi(x, \alpha) := \frac{\alpha}{n} + \psi_{Epi\gamma}(x - a, \alpha - r)$ where γ is a norm, $n \in \mathbb{N}$ and $(a, r) \in \mathbb{R}^n \times \mathbb{R}$ we get as Fenchel dual problem, with the change of variables $\lambda_i = \frac{1}{n} - \beta_i^1 = \frac{1}{n} - \beta_i^2$

$$\begin{cases} \text{Minimize } \sum_{i=1}^{n} [\langle a_i^1, p_i^1 \rangle - \frac{c_i}{2}\lambda_i] + \sum_{i=1}^{n} [\langle a_i^2, p_i^2 \rangle - \frac{c_i}{2}\lambda_i] \\ \qquad + \sum_{j=1}^{m_1} \psi_{\Omega_j^1}^*(p_{n+j}^1) + \sum_{j=1}^{m_2} \psi_{\Omega_j^2}^*(p_{n+j}^2) + \psi_{\Omega}^*(p_0^1, p_0^2) \\ (\gamma_i^r)^0(p_i^r) \leq \lambda_i, \quad i = 1, 2, \ldots, n; \ r = 1, 2 \\ \sum_{i=1}^{n} \lambda_i = 1; \ (\sum_{i=0}^{n+m_1} p_i^1, \sum_{i=0}^{n+m_2} p_i^2) \in L^\perp \end{cases}$$

From this duality scheme we can directly deduce the (primal-dual) optimality conditions. These conditions are of two types, a first set of five conditions which are nonlinear and define a maximal monotone operator T and the remaining conditions which gather the (linear) primal and dual constraints and define two complementary linear subspaces respectively denoted by A and $A^\perp$. Note that the maximality and monotonicity of T directly follows from the fact that these nonlinear conditions are governed by normal cones hence by subdifferentials.

5 Resolution Method

5.1 The Proximal Solution Procedure

Solving the optimality conditions amounts to finding a pair $[(x, \alpha), (p, \beta)]$ in the graph of the maximal monotone operator T as defined previously and in the product $A \times A^\perp$ formed by the linear primal and dual constraints. This task can be done by the partial inverse method introduced by J. Spingarn [11]. Indeed, according to [11] $[(x, \alpha), (p, \beta)]$ satisfies the optimality conditions if and only if the sum $(x, \alpha) + (p, \beta)$ is a zero of the partial inverse T_A of T with respect to A (or equivalently of the partial inverse $T_{A^\perp}^{-1}$ of T^{-1} with respect to $A^\perp$). The partial inverse operator T_A being also maximal monotone (see [11]) one can apply the proximal point algorithm [10] to generate a zero z of T_A and subsequently project z onto A and $A^\perp$ to rediscover a primal solution (x, α) and a dual solution (p, β). This procedure has been called *Partial inverse Method* by Spingarn. The partial inverse T_A is not directly accessible but only given via its graph (see [11]). However, one can express without difficulty the proximal mapping associated with T_A in terms of T, A and $A^\perp$. Roughly speaking, the procedure consists in alternately projecting on the graph of T and on the product $A \times A^\perp$. Actually, the graph of T being non convex, the usual orthogonal projection is replaced by a proximal step which can be also viewed as a kind of modified projection. More precisely at iteration k the method generates two pairs $[(\tilde{x}^{(k)}, \tilde{\alpha}^{(k)}), (\tilde{p}^{(k)}, \tilde{\beta}^{(k)})]$ and $[(x^{(k+1)}, \alpha^{(k+1)}), (p^{(k+1)}, \beta^{(k+1)})]$. As a result of the first step which is called the proximal phase, the iterate $[(\tilde{x}^{(k)}, \tilde{\alpha}^{(k)}), (\tilde{p}^{(k)}, \tilde{\beta}^{(k)})]$ which, by construction, belongs to the graph of T but is not primal-dual feasible, is the unique solution of the particular system. So that we can start by computing $(\tilde{x}^{(k)}, \tilde{\alpha}^{(k)})$ via the proximal mapping associated with T and deduce the corresponding dual pair $(\tilde{p}^{(k)}, \tilde{\beta}^{(k)})$ e.g. for technical reasons, to compute $(\tilde{p}^{(k)}, \tilde{\beta}^{(k)})$ via the proximal mapping associated with T^{-1} and recover the primal variables. As a result of the second step called the projection phase and which in fact constitutes the updating rules, the variables $(x^{(k+1)}, \alpha^{(k+1)})$ which are, by construction, primal feasible are obtained by projecting $(\tilde{x}^{(k)}, \tilde{\alpha}^{(k)})$, onto A. In the same way the dual variables $(p^{(k+1)}, \beta^{(k+1)})$ are obtained by projecting $(\tilde{p}^{(k)}, \tilde{\beta}^{(k)}))$ onto $A^{\perp}$. As T is separable the proximal phase leads to independent proximal substeps which are simple and can be computed in parallel. A first series of substeps involves the proximal mapping associated to the subdifferential of the function φ introduced in the previous section. Thus these substeps lead to orthogonal projections onto the Epigraph of the norm γ. Such a projection is very simple in the Euclidean case. Special and efficient subroutines can also be developed for other usual norms, like ℓ_p norms in the plane, rectilinear norms, Tchebychev norms or more generally for polyhedral norms given by the extreme points of their unit balls. A second series of substeps involves the proximal map associated to the subdifferential of the indicator function of a constraint. This proximal map is thus an orthogonal projection on the constraint. For projecting on the constraints specific routines can be developed. However it is well known that in practice, the projection on a convex set is not an easy task except for sets having a simple structure like hyperplanes, halfspaces, subspaces given by linear relations or boxes with faces parallel to the axes. For polyhedra defined by inequalities or given by their extreme points various efficient algorithms have been developed (see e.g. [6]). For more complex structure one can try to decompose the constraints as intersection of sets with simple structure. As illustration any polyhedron described by a series of inequalities can be viewed as intersection of halfspaces. Our approach completely allows to exploit such a decomposition. The projection phase leads to very simple updating rules. According to the structure of A the projection on this subspace can be decomposed as a projection onto

$$A_x = \left\{ x;\ \begin{array}{l} x_0^1 = x_1^1 = \cdots = x_n^1 = x_{n+1}^1 = \cdots = x_{n+m_1}^1 \\ x_0^2 = x_1^2 = \cdots = x_n^2 = x_{n+1}^2 = \cdots = x_{n+m_2}^2 \\ (x_0^1, x_0^2) \in L \end{array} \right\}$$

and a projection onto $A_\alpha = \{\alpha;\ \alpha_1^1 + \alpha_1^2 = \alpha_2^1 + \alpha_2^2 = \cdots = \alpha_n^1 + \alpha_n^2\}$.

The orthogonal projection on A_α can be explicitly computed. Indeed, the linear subspace A_α can be described as $A_\alpha = \{\alpha;\ M\alpha = 0\}$ where M is an $(n-1) \times 2n$ matrix. Since the inverse of (MM^T) is explicitly given by $(MM^T)^{-1} = \frac{1}{2n}[nI_{n-1} - 1\!\!1_{n-1}]$ where I_n is the $n \times n$ identity matrix and $1\!\!1_n$ is the $n \times n$

matrix whose all elements are equal to 1, thus the projection matrix is equal to $P = I_{2n} - M^T(MM^T)^{-1}M$.

5.2 The Partial Inverse Algorithm

Initialization phase:

Choose arbitrarily $(x^{1(0)}, x^{2(0)})$, choose $p_i^{1(0)}$, $i = 0, 1, \ldots, n + m_1$ and $p_i^{2(0)}$, $i = 0, 1, \ldots, n+m_2$ such that $(\sum_{i=0}^{n+m_1} p_i^{1(0)}, \sum_{i=0}^{n+m_2} p_i^{2(0)}) \in L$, $(\alpha_i^{1(0)}, \alpha_i^{2(0)})$, $i = 1, 2, \ldots, n$ such that $\alpha_1^{1(0)} + \alpha_1^{2(0)} = \alpha_2^{1(0)} + \alpha_2^{2(0)} = \cdots = \alpha_n^{1(0)} + \alpha_n^{2(0)}$ and $\beta_i^{(0)}$, $i = 1, 2, \ldots, n$ such that $\sum_{i=1}^n \beta_i^{(0)} = 1$. Then, given the kth-iterates $(x^{1(k)}, x^{2(k)})$, $p_i^{1(k)}$, $i = 0, 1, \ldots, n + m_1$ and $p_i^{2(k)}$, $i = 0, 1, \ldots, n + m_2$, $(\alpha_i^{1(k)}, \alpha_i^{2(k)})$, $i = 1, 2, \ldots, n$ and $\beta_i^{(k)}$, $i = 1, 2, \ldots, n$, define the next iterates as follows.

Proximal phase:

1. for $i = 1, 2, \ldots, n$ and $r = 1, 2$, calculate
$$(\tilde{x}_i^{r(k)}, \tilde{\alpha}_i^{r(k)}) = (a_i^r, \tfrac{c_i}{2}) + \text{Proj}_{\text{Epi}\,\gamma_i^r}(x^{r(k)} + p_i^{r(k)}, \alpha_i^{r(k)} + \beta_i^{(k)})$$
2. for $j = 1, 2, \ldots, m_r$ and $r = 1, 2$, calculate
$$\tilde{x}_{n+j}^{r(k)} = \text{Proj}_{\Omega_j^r}(x^{r(k)} + p_{n+j}^{r(k)})$$
3. calculate $(\tilde{x}_0^{1(k)}, \tilde{x}_0^{2(k)}) = \text{Proj}_\Omega(x^{1(k)} + p_0^{1(k)}, x^{2(k)} + p_0^{2(k)})$

Projection phase:

1. calculate $(\bar{x}^1, \bar{x}^2) = (\frac{1}{n+m_1+1} \sum_{\ell=0}^{n+m_1} \tilde{x}_\ell^{1(k)}, \frac{1}{n+m_2+1} \sum_{\ell=0}^{n+m_2} \tilde{x}_\ell^{2(k)})$
2. calculate $(x^{1(k+1)}, x^{2(k+1)}) = \text{Proj}_L(\bar{x}^1, \bar{x}^2)$
3. calculate $(\bar{p}^1, \bar{p}^2) = (\frac{1}{n+m_1+1} \sum_{\ell=0}^{n+m_1} \tilde{p}_\ell^{1(k)}, \frac{1}{n+m_2+1} \sum_{\ell=0}^{n+m_2} \tilde{p}_\ell^{2(k)})$
4. calculate $(\hat{p}^1, \hat{p}^2) = \text{Proj}_L(\bar{p}^1, \bar{p}^2)$
5. for $i = 1, 2, \ldots, m_1$ and for $j = 1, 2, \ldots, m_2$, calculate
$$p_i^{1(k+1)} = \tilde{p}_i^{1(k)} - \hat{p}^1 \text{ and } p_j^{2(k+1)} = \tilde{p}_j^{2(k)} - \hat{p}^2$$
6. for $i = 1, 2, \ldots, n$, calculate
$$\alpha_i^{1(k+1)} = \tfrac{1}{2n}\sum_{\ell=1}^n (\tilde{\alpha}_\ell^{1(k)} + \tilde{\alpha}_\ell^{2(k)}) + \tfrac{1}{2}(\tilde{\alpha}_i^{1(k)} - \tilde{\alpha}_i^{2(k)})$$

and
$$\alpha_i^{2(k+1)} = \tfrac{1}{2n}\sum_{\ell=1}^n (\tilde{\alpha}_\ell^{1(k)} + \tilde{\alpha}_\ell^{2(k)}) + \tfrac{1}{2}(\tilde{\alpha}_i^{2(k)} - \tilde{\alpha}_i^{1(k)})$$

7. for $i = 1, 2, \ldots, n$, calculate
$$\beta_i^{(k+1)} = -\tfrac{1}{2n}\sum_{\ell=1}^n (\tilde{\beta}_\ell^{1(k)} + \tilde{\beta}_\ell^{2(k)}) + \tfrac{1}{2}(\tilde{\beta}_i^{1(k)} + \tilde{\beta}_i^{2(k)})$$

5.3 Numerical Results

We have implemented our algorithm and tested it with a set of artificial facilities to emphasize the behavior of the algorithm. The given algorithm has been implemented in FORTRAN using a PC Pentium IV. Single precision arithmetic was used throughout. The objective function to be minimized achieves the general form Minimize $\max_{1\leq i\leq n} [\gamma_i(X - A_i) + c_i]$. In order to improve the performance, a suitable transformation of the data was performed having the following transformations of the primal variables $(\tilde{x}, \tilde{\alpha}) = (\eta x, \tau \eta \alpha)$ and the dual variables $(\tilde{p}, \tilde{\beta}) = (\tau p, \beta)$ where $\eta > 0$ and $\tau > 0$ are suitable parameters. Thus we get the following objective function

$$\text{Minimize} \max_{1\leq i\leq n} \tau[\gamma_i(X - A_i/\eta) + c_i/\tau\eta]$$

We used the relative error as stopping rule, that is, $\|z^{k+1} - z^k\|/\|z^k\| < \epsilon$ and the value of ϵ was fixed to 10^{-7}. This stopping rule was defined on the product space, thus, we have $z^k = (x + p, \alpha + \lambda) = (x, \alpha) + (p, \lambda)$ as the current point, and z^{k+1} as the new point, which lead us to $z^{k+1} = (x^+ + p^+, \alpha^+ + \lambda^+)$. Some results in the particular case of the round-trip problem are presented below, where both the Euclidean and the L_1 norms were used. The tables that we show below summarize the performance of algorithm where we use the following notation:

NumFac : Number of fixed facilities n
INP : Initial point $x_0 = (x^{1(0)}, x^{2(0)})$
Nit : Number of iterations of algorithm
VarDual : Number of dual variables different zero
Par : Optimal value parameters (τ, η)
fp : Optimal value primal objective function
fd : Optimal value dual objective function

Initially the dual variables are set to zero. Without scaling on the data, convergence needs more iterations i.e. τ and η equal to one the algorithm was stopped when the maximum iteration count was reached ($ITMAX = 10000$). For this

Table 1. Norm L_2

Norm L_2	Example 1	Example 2	Example 3	Example 4	Example 5
NumFac	30	50	150	1500	5000
INP	(1,1)	(1,1)	(1,1)	(1,1)	(1, 1)
Nit	860	1364	3101	3789	6340
VarDual	2	2	2	2	2
Par(τ, η)	(2.5, 11)	(2.5, 7.4)	(5.0, 5.0)	(5.3, 6.0)	(7.5, 5.0)
fp	86.03088570	90.60901184	246.2697144	562.8785634	1786.1920468
fd	86.03109589	90.60993762	246.2467835	562.8755136	1786.1898775

Table 2. Norm L_1

Norm L_1	Example 1	Example 2	Example 3	Example 4	Example 5
NumFac	30	50	60	1500	5000
INP	(1,1)	(1,1)	(1,1)	(1, 1)	(1, 1)
Nit	636	1216	1188	2545	2879
VarDual	2	2	2	2	2
Par(τ,η)	(11, 11.5)	(8.8, 8.0)	(25.0, 9.0)	(18.0, 5.0)	(12.0, 19.0)
fp	95.80012124	9.579991516	1.191994285	0.878934525	2.118794501
fd	95.79968930	9.579930144	1.192009265	0.899856118	2.120956443

reason, we introduce a suitable change of scale. According to this, we modified the objective function assigning positive values for the parameters τ and η. Tables 1 and 2 show the values of τ and η that improve the algorithm performance regarding the number of iterations and CPU time. Another aspect of relevance is the strong relationship between the convergence speed of the algorithm and the scale factor being used. The value of the primal function is calculated using the dual variables in order to avoid the analytical determination of the norms that

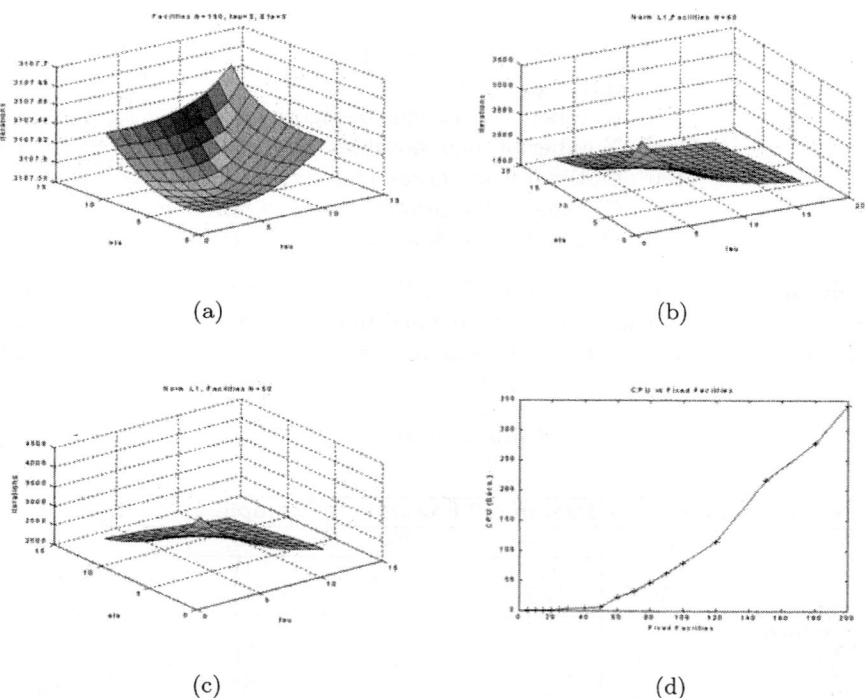

Fig. 1. Facilities Versus Parameters

define the objective function of the problem. The numerical results obtained so far show that when convergence is achieved, almost all the dual variables are zero. Our main interest are precisely those dual variables that do not converge to zero, since these will determine the active constrains at the optimum, as it can be seen in the above tables.

Figure 1, (a), (b) and (c) shows the relationship between the iteration number and the value of the parameters. Since these surfaces were obtained plotting empirical data, it would be of extremely useful to determine an analytical expression relating parameter values with convergence speed, which will lead to the problem of finding an optimal value for the set of parameters, achieving maximum convergence speed. Figure 1 (d) shows the relationship between CPU time and the number of facilities. It can be seen that for $n \geq 30$, the algorithm performance was very high, whereas for greater values of n, the convergence speed decreased notoriously.

In order to enlarge the performance evaluations of the above mentioned algorithm, we can consider a combination of different types of norms. This test is proposed as future work. Another important issue is the parallel computation schemes which have shown to be very efficient particularly in the resolution of numerical problems where the main task can be split up into smaller ones, which is the case of the problems that have been treated in this work.

References

1. CHAN, A.W., and HEARN, D.W. (1977) A rectilinear distance round-trip location problem, *Transportation Science* 11, 107–123.
2. DREZNER, Z. (1991) The weighted minimax location problem with set-up costs and extensions, *Recherche Opérationnelle/Operations Research* 25, 55–64.
3. FRENK, J.B.G., GROMICHO, J. & ZHANG, S. (1996) General models in min-max continuous location: theory and solution techniques. *Journal of Optimization Theory and Applications* 89, 39–63.
4. ICHIMORI, T. and NISHIDA, T. (1985) Note on a rectilinear distance round-trip location problem, *Transportation Science* 19, 84–91.
5. IDRISSI, H., LEFEBVRE, O. and MICHELOT, C. (1988) A primal-dual algorithm for a constrained Fermat-Weber problem involving mixed gauges, *RAIRO Operations Research* 22, 313-330.
6. MIFFLIN, R.(1974) A stable method for solving certain constrained least squares problems, *Mathematical Programming* 16, 141–158
7. PLASTRIA, F. (1995a) Continuous location problems. In Drezner, Z. (Ed.), *Facility Location: A Survey of Applications and Methods* (pp. 225–262). New York: Springer-Verlag.
8. PLASTRIA, F. (1995b) Fully geometric solutions to some planar minimax location problems. *Studies in Locational Analysis* 7, 171–183.
9. ROCKAFELLAR, R.T. (1970) *Convex Analysis*, Princeton, New Jersey: Princeton University Press.
10. ROCKAFELLAR, R.T. (1976) Monotone operators and the proximal point algorithm", *SIAM Journal on Control and Optimization* 14, 877-898.
11. SPINGARN, J.E. (1983) Partial inverse of a monotone operator, *Applied Mathematics and Optimization* 10, 247-265.

A Lagrangean Relaxation Approach for Capacitated Disassembly Scheduling

Hwa-Joong Kim[1], Dong-Ho Lee[2], and Paul Xirouchakis[1]

[1] Institute of Production and Robotics (STI-IPR-LICP),
Swiss Federal Institute of Technology (EPFL),
Lausanne CH-1015, Switzerland
{hwa-joong.kim, paul.xirouchakis}@epfl.ch
[2] Department of Industrial Engineering,
Hanyang University, Sungdong-gu, Seoul 133-791, Korea
leman@hanyang.ac.kr

Abstract. We consider the problem of determining the disassembly schedule (quantity and timing) of products in order to satisfy the demand of their parts or components over a finite planning horizon. This paper focuses on the capacitated version of the problem for the objective of minimizing the sum of setup, disassembly operation, and inventory holding costs. The problem is formulated as an integer program, and to solve the problem, a Lagrangean heuristic algorithm is developed after reformulating the integer program. To show the performance of the heuristic algorithm, computational experiments are done on randomly generated test problems, and the test results show that the algorithm suggested in this paper works well.

1 Introduction

Disassembly, one of the basic material and product recovery processes, represents a way of obtaining constituent materials, parts, subassemblies, or other groupings from used or end-of-life products with necessary sorting operations. Due to environmental and economic reasons, a number of manufacturing firms have been paying considerable attention to disassembly. Meanwhile, a number of research works have been done on various disassembly areas such as design for disassembly, disassembly process planning, disassembly scheduling, etc. For literature reviews on these problems, see Lee *et al.* [3].

This paper focuses on the problem of determining the quantity and timing of disassembling (used or end-of-life) products in order to satisfy the demand of their parts or components over a planning horizon, which is called disassembly scheduling in the literature. Most previous research articles on disassembly scheduling consider uncapacitated problems. Gupta and Taleb [1] suggest a reversed form of the material requirement planning (MRP) algorithm without an explicit objective function. Recently, Lee and Xirouchakis [5] suggest a heuristic algorithm for the objective of minimizing various costs related with the disassembly process. For more extended problems with parts commonality and/or multiple product types, see Kim *et al.* [2], Lee *et al.* [4], Neuendorf *et al.* [7], Taleb and Gupta [8], and Taleb *et al.* [9]. Unlike these, Lee *et al.* [6] consider the resource capacity constraints explicitly and suggest an integer programming model with various cost factors occurred in disassembly processes. Although the model can give optimal solutions, its application is limited only to the small sized problems. In fact, the computational results show that it is not adequate for practical sized problems due to its excessive computation times.

This paper focuses on the capacitated problem for the case of single product type without parts commonality. The objective is to minimize the sum of setup, disassembly operation, and inventory holding costs. This paper extends the research result of Lee *et al.* [6] with respect to two points. First, we consider the fixed setup costs incurred whenever disassembly operations are done over the planning horizon. Second, as pointed out in further research in Lee *et al.* [6], we suggest a Lagrangean heuristic algorithm that can give near optimal solutions up to large sized problems within a reasonable amount of computation time.

2 Problem Description

This section begins with explaining the disassembly product structure. Its root item represents the product to be disassembled and the leaf items are the parts or components to be demanded and not to be disassembled further. A child item denotes an item that has one parent and a parent item is an item that has more than one child item. Note that each child item has only one parent in the problem considered in this paper, i.e., the problem without parts commonality. Figure 1 shows an example of the disassembly product structure obtained from Gupta and Taleb [1]. The number in parenthesis represents the yield of the item when its parent is disassembled. Also, the disassembly lead time (DLT) of a parent item implies the total time required to receive the item after placing an order of disassembling the item.

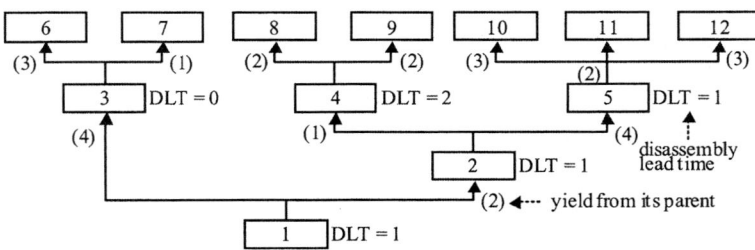

Fig. 1. Disassembly product structure: an example

The capacitated disassembly scheduling problem considered in this paper is defined as follows: *for a given disassembly structure, the problem is to determine the quantity and timing of disassembling each parent item to meet the demand* of leaf items over a planning horizon while satisfying the capacity restriction in each period of the planning horizon. The capacity restriction of a period is considered in the form of a limit to assign in a disassembly operations to that period. That is, there is an upper limit on the available time in each period of the planning horizon, and each disassembly operation assigned to a period consumes a portion of the available time of that period. The objective is to minimize the sum of setup, disassembly operation, and inventory holding costs. The setup cost implies the cost required for preparing the corresponding disassembly operation. It is assumed that the setup cost occurs in a period if any disassembly operation is performed in that period. The disassembly operation cost is the cost proportional to the labor or machine processing time required for performing the corresponding disassembly operation, and the inventory holding cost occurs when items are stored to satisfy future demand, and they are computed based on the end-of-period inventory.

Assumptions made in this problem are summarized as follows: (a) demand of leaf items is given and deterministic; (b) backlogging is not allowed and hence demands are satisfied on time; (c) parts/components are perfect in quality, i.e., no defective parts/components are considered; and (d) each disassembly operation is done in one and only one period and cannot be done over two or more periods.

To describe the problem mathematically, we present an integer programming model. In the model, without loss of generality, all items are numbered with integers 1, 2, ... i_l, ... N, where 1 and i_l denote the indices for the root item and the first leaf item, respectively. The notations used are summarized below.

Parameters

s_i setup cost of parent item i
p_i disassembly operation cost of parent item i
h_i inventory holding cost of item i
g_i disassembly operation time of parent item i
C_t capacity in period t
d_{it} demand of leaf item i in period t
a_{ij} number of units of item j obtained by disassembling one unit of item i ($i < j$)
I_{i0} initial inventory of item i
l_i disassembly lead lime of item i
$\varphi(i)$ parent of item i
M arbitrary large number

Decision variables

X_{it} amount of disassembly operations of item i in period t
Y_{it} = 1 if there is a setup for item i in period t, and 0 otherwise
I_{it} inventory level of item i at the end of period t

Now, the integer program is given below.

[**P1**] Minimize $\sum_{i=1}^{i_l-1}\sum_{t=1}^{T} s_i \cdot Y_{it} + \sum_{i=1}^{i_l-1}\sum_{t=1}^{T} p_i \cdot X_{it} + \sum_{i=2}^{N}\sum_{t=1}^{T} h_i \cdot I_{it}$

s.t. $I_{it} = I_{i,t-1} + a_{\varphi(i),i} \cdot X_{\varphi(i),t-l_{\varphi(i)}} - X_{it}$

$\qquad\qquad\qquad$ for $i = 2, 3, \ldots i_l - 1$ and $t = 1, 2, \ldots T$ $\quad(1)$

$I_{it} = I_{i,t-1} + a_{\varphi(i),i} \cdot X_{\varphi(i),t-l_{\varphi(i)}} - d_{it}$

$\qquad\qquad\qquad$ for $i = i_l, i_l + 1, \ldots N$ and $t = 1, 2, \ldots T$ $\quad(2)$

$\sum_{i=1}^{i_l-1} g_i \cdot X_{it} \leq C_t \qquad$ for $t = 1, 2, \ldots T$ $\qquad\qquad\qquad\qquad\quad(3)$

$X_{it} \leq M \cdot Y_{it} \qquad$ for $i = 1, 2, \ldots i_l - 1$ and $t = 1, 2, \ldots T$ $\quad(4)$

$Y_{it} \in \{0, 1\} \qquad\;\;$ for $i = 1, 2, \ldots i_l - 1$ and $t = 1, 2, \ldots T$ $\quad(5)$

$X_{it} \geq 0$ and integer $\quad$ for $i = 1, 2, \ldots i_l - 1$ and $t = 1, 2, \ldots T$ $\quad(6)$

$I_{it} \geq 0$ and integer $\quad$ for $i = 2, 3, \ldots N$ and $t = 1, 2, \ldots T$ $\qquad(7)$

The objective function denotes the sum of setup, disassembly operation, and inventory holding costs. Constraints (1) and (2) define the inventory level of non-root items at the end of each period, called the inventory flow conservation constraint. Note that no inventory flow conservation constraint is needed for the root item since its surplus-inventory of the root item results in unnecessary cost increase. Also, constraint (3) represents the capacity constraint in each period. That is, the total time required to perform the disassembly operations assigned to each period should be less than or equal to the given time capacity of that period. Constraint (4) guarantees that a setup cost in a period is incurred if any disassembly operation is performed at that period. Constraints (5), (6), and (7) represent the conditions on the decision variables.

3 Solution Algorithm

Before explaining the heuristic algorithm suggested in this paper, we reformulate the original problem [P1] as another integer program so that the Lagrangean relaxation technique can be applied more effectively. Then, the Lagrangean relaxation heuristic algorithm is presented, together with a method to find good feasible solutions while considering the trade-offs among the relevant costs.

3.1 Problem Reformulation

The first step is to replace the inventory variables of the original formulation [P1] using the following equation:

$$I_{it} = I_{i0} + \sum_{j=1}^{t} (a_{\varphi(i),i} \cdot X_{\varphi(i),j-l_{\varphi(i)}} - Q_{ij}) \qquad (8)$$

where $Q_{it} = X_{it}$ for $i = 2, 3, \ldots i_l - 1$ and $t = 1, 2, \ldots T$, and $Q_{it} = d_{it}$ for $i = i_l, i_l + 1, \ldots N$ and $t = 1, 2, \ldots T$. Also, using (7) and (8), and changing the indices, i.e., i and k are used instead of $\varphi(i)$ and i, the nonnegative constraint (7) can be changed into

$$\sum_{j=1}^{t} a_{ik} \cdot X_{ij} \geq \sum_{j=1}^{t} Q_{k,j+l_i} - I_{k0}$$
$$\text{for } i = 1, 2, \ldots i_l - 1, \ k \in H(i), \text{ and } t = 1, 2, \ldots T \quad (7')$$

Hereafter, the above constraint is called the demand constraint since it can used to represent the demand requirements in the reformulation.

In the second step, we add a new demand constraint that the disassembly quantity of a parent should be at least more than or equal to the demand of its leaf items. To explain this, let $L(i)$ and $P(i,j)$ denote the set of leaf items among successors of parent item i and the path from parent item i to leaf item j, respectively. For example, $L(2) = \{8, 9, 10, 11, 12\}$ and $P(2,12) = 2 \to 5 \to 12$ in Figure 1. Then, the new demand constraint for item i in period t can be represented as

$$\sum_{j=1}^{t} X_{ij} \geq \sum_{j=1}^{t} D_{ij}^{e} - A_{i0}^{e} + I_{i0} \qquad \text{for } e \in L(i) \quad (9)$$

where D_{it}^{e} and A_{i0}^{e}, associated with leaf item e, denote the transformed demand and the transformed initial inventory of item i in period t, respectively. More specifically,

$$D_{it}^{e} = \frac{D_{k,t+l_i}^{e}}{a_{ik}}, \quad D_{et}^{e} = d_{et} \qquad \text{for } k \in H(i) \cap P(i,e), \text{ and}$$

$$A_{i0}^{e} = \frac{A_{k0}^{e}}{a_{ik}} + I_{i0}, \quad A_{e0}^{e} = I_{e0} \qquad \text{for } k \in H(i) \cap P(i,e).$$

Then, the constraint (9) is further changed into

$$\sum_{j=1}^{t} X_{ij} \geq \max_{e \in L(i)} \left\{ \sum_{j=1}^{t} D_{ij}^{e} - A_{i0}^{e} + I_{i0} \right\}$$
$$\text{for } i = 1, 2, \ldots i_l - 1 \text{ and } t = 1, 2, \ldots T \quad (10)$$

since the demand requirement of each of the items in $L(i)$, i.e., leaf items among successors of parent item i, can be satisfied by disassembling item i to the maximum amount to satisfy the demand requirements of all the leaf items in $L(i)$.

Now, the integer program [P1] can be reformulated as follows.

[P2] Minimize $\sum_{i=1}^{i_l-1} \sum_{t=1}^{T} s_i \cdot Y_{it} + \sum_{i=1}^{i_l-1} \sum_{t=1}^{T} c_{it} \cdot X_{it}$

s.t. (3) – (6), (7'), and (10)

In the reformulation [P2], the new constraint (10) does not affect the optimal solution since all demand requirements can be satisfied with only constraint (7'). However, it is added to the reformulation because we design the solution algorithm to satisfy the demand requirements after the constraint (7') is relaxed. In fact, our algorithm is based on the relaxation of the constraints (7') and (3), and hence there is no way to satisfy demand requirements of leaf items in the relaxed problem without the constraint (10).

3.2 The Lagrangean Heuristic

As stated earlier, our Lagrangean heuristic is based on the relaxation of (7') and (3) in [P2]. First, the objective function of the relaxed problem becomes

$$\sum_{i=1}^{i_l-1} \sum_{t=1}^{T} s_i \cdot Y_{it} + \sum_{i=1}^{i_l-1} \sum_{t=1}^{T} c_{it} \cdot X_{it} + \sum_{t=1}^{T} \mu_t \cdot \left\{ \sum_{i=1}^{i_l-1} g_i \cdot X_{it} - C_t \right\}$$

$$- \sum_{i=1}^{i_l-1} \sum_{k \in H(i)} \sum_{t=1}^{T} \lambda_{ikt} \cdot \left\{ \sum_{j=1}^{t} (a_{ik} \cdot X_{ij} - Q_{k,j+l_i}) - I_{k0} \right\}$$

where λ_{ikt} and μ_t are the Lagrangean multipliers corresponding to (7') and (3), respectively. Then, the relaxed problem is summarized in the following:

[LR] $Z(\lambda, \mu) = \min \left\{ \sum_{i=1}^{i_l-1} \sum_{t=1}^{T} s_i \cdot Y_{it} + \sum_{i=1}^{i_l-1} \sum_{t=1}^{T} v_{it} \cdot X_{it} + F \right\}$

s.t. $\lambda_{ikt} \geq 0$ for $i = 1, 2, \ldots i_l - 1, k \in H(i)$, and $t = 1, 2, \ldots T$
$\mu_t \geq 0$ for $t = 1, 2, \ldots T$
and (4) − (6) and (10)

where λ and μ denote the vectors representing the Lagrangean multipliers,

$$v_{it} = c_{it} + \mu_t \cdot g_i - \sum_{k \in H(i), k \notin L(i)} \sum_{j=t}^{T} a_{ik} \cdot \lambda_{ikj} + \sum_{j=t-l_{\varphi(i)}}^{T} \lambda_{\varphi(i), ij} \text{ and}$$

$$F = \sum_{i=1}^{i_l-1} \sum_{t=1}^{T} \left\{ \sum_{k \in H(i) \cap L(i)} \lambda_{ikt} \cdot \sum_{j=1}^{t} d_{k,j+l_i} - \sum_{k \in H(i)} \lambda_{ikt} \cdot I_{k0} \right\} - \sum_{t=1}^{T} \mu_t \cdot C_t.$$

Here, the term F is a constant and can be removed without further consideration.

The relaxed problem [LR] can be decomposed into the following mutually independent subproblems $[SP_i]$, $i = 1, 2, \ldots i_l - 1$.

[SP_i] Minimize $\sum_{t=1}^{T} s_i \cdot Y_{it} + \sum_{t=1}^{T} v_{it} \cdot X_{it}$

s.t. $\sum_{j=1}^{t} X_{ij} \geq \max_{e \in L(i)} \left\{ \sum_{j=1}^{t} D_{ij}^e - A_{i0}^e + I_{i0} \right\}$ for $t = 1, 2, \ldots T$ (11)

$X_{it} \leq M \cdot Y_{it}$ for $t = 1, 2, \ldots T$

$Y_{it} \in \{0, 1\}$ for $t = 1, 2, \ldots T$

$X_{it} \geq 0$ and integer   for $t = 1, 2, \ldots T$

$\lambda_{ikt} \geq 0$ for $k \in H(i)$ and $t = 1, 2, \ldots T$

$\mu_t \geq 0$ for $t = 1, 2, \ldots T$

Each of the subproblems [SP$_i$], $i = 1, 2, \ldots i_l - 1$, is the single item lot-sizing problem that can be solved easily in a polynomial time using the algorithm suggested by Wagelmans et al. [10]. In fact, the above formulation [SP$_i$] is the same as that of Wagelmans et al. [10] except for the maximum term. Therefore, a lower bound can be obtained by solving subproblems [SP$_i$] for $i = 1, 2, \ldots i_l - 1$ and the best one can be obtained by solving the following Lagrangean dual problem:

$$Z(\boldsymbol{\lambda}^*, \boldsymbol{\mu}^*) = \max_{\boldsymbol{\lambda}, \boldsymbol{\mu}} (\boldsymbol{\lambda}, \boldsymbol{\mu})$$

The Lagrangean multipliers are updated using the well-known subgradient optimization algorithm. The subgradient optimization algorithm generates a sequence of Lagrangean multipliers using the following rule:

$$\lambda_{ikt}^{w+1} = \max\left[0, \lambda_{ikt}^w - \alpha_w \cdot \left\{\sum_{j=1}^{t}(a_{ik} \cdot X_{ij}^* - Q_{k,j+l_i}^*) + I_{k0}\right\}\right]$$

for $i = 1, 2, \ldots i_l - 1$, $k \in H(i)$, and $t = 1, 2, \ldots T$ (12)

$$\mu_t^{w+1} = \max\left[0, \mu_t^w + \beta_w \cdot \left\{\sum_{i=1}^{i_l - 1} g_i \cdot X_{ij}^* - C_t\right\}\right] \quad \text{for } t = 1, 2, \ldots T \quad (13)$$

where λ_{ikt}^{w+1} and μ_t^{w+1} denote the values of the multipliers updated at iteration w, and X_{it}^* denote the optimal solution of the relaxed problem at iteration w. Also, α_w and β_w denote the step sizes at iteration w, updated by

$$\alpha_w = \delta_w \cdot \frac{\bar{Z} - Z(\boldsymbol{\lambda}^*, \boldsymbol{\mu}^*)}{\sum_{i=1}^{i_l-1}\sum_{k \in H(i)}\sum_{t=1}^{T}\left\{\sum_{j=1}^{t}(a_{ik} \cdot X_{ij}^* - Q_{k,j+l_i}^*) + I_{k0}\right\}} \quad (14)$$

$$\beta_w = \delta_w \cdot \frac{\bar{Z} - Z(\boldsymbol{\lambda}^*, \boldsymbol{\mu}^*)}{\sum_{t=1}^{T}\left\{\sum_{i=1}^{i_l-1} g_i \cdot X_{it}^* - C_t\right\}} \quad (15)$$

where $\bar{Z}$ is the best upper bound, $\boldsymbol{\lambda}^w$ and $\boldsymbol{\mu}^w$ denote the vectors of the Lagrangean multipliers at iteration w, and δ_w, $0 \le \delta_w \le 2$, is a constant.

Now, we explain the Lagrangean heuristic that makes the solution of the relaxed problem [LR] feasible. To obtain a feasible one (upper bound) using the solution of the relaxed problem, first, we generate a solution feasible to demand constraints, and then, we generate a solution feasible to capacity constraints.

To generate a solution feasible to the demand constraint (7'), we solve another subproblem recursively from parent item $i_l - 1$ to the root item 1 using the algorithm suggested by Wagelmans et al. [10]. In the subproblem, the following constraint is used instead of (11) in [SP$_i$]:

$$\sum_{j=1}^{t} X_{ij} \ge \max_{k \in H(i)} \frac{\sum_{j=1}^{t} Q_{k,j+l_i} - I_{k0}}{a_{ik}} \quad \text{for } t = 1, 2, \ldots T$$

To generate a solution feasible to capacity constraints, the solution obtained from the first step is modified iteratively. This is done by moving the amount of the overloaded disassembly quantity in an earlier (backward move) or later period (forward move). More specifically, in the moves, the overloaded disassembly quantity of an item assigned to the selected period is moved to one period earlier (or later) while considering the demand constraints (7′) and the cost changes associated with the move. This is done for each of the items assigned to the selected period, and the best one that gives the minimum cost increase is selected.

Now, the Lagrangean heuristic suggested in this paper is summarized as follows. The algorithm is terminated after a predetermined number of iterations, i.e., when the iteration count (w) reaches a predetermined limit (W).

Procedure 1. (The Lagrangean heuristic algorithm)

Step 1. Let $w = 1$ and $\lambda_{ikt}^0 = 0$ and $\mu_t^0 = 0$ for all i, k, and t. Let the upper and lower bounds be a big number and 0, respectively. Calculate the transformed demands, initial inventories, and the marginal disassembly costs.

Step 2. Calculate the Lagrangean cost v_{it} for all i and t, and the constant value. With the Lagrangean cost, solve subproblem [SP$_i$], $i = 1, 2, \ldots i_l - 1$, using the algorithm suggested by Wagelmans et al. [10].

Step 3. Obtain a lower bound by computing the objective function value using the solution of [LR]. If the lower bound is improved, update the lower bound. Also, find an upper bound using the method explained earlier. If it is improved, update the upper bound and save the solution.

Step 4. Update Lagrangean multipliers using (12) and (13) with the step sizes, α_w and β_w determined by (14) and (15). Set $w = w + 1$. If $w > W$, stop and otherwise, go to Step 2.

4 Computational Experiments

To show the performance of the algorithm suggested in this paper, computational tests were done on randomly generated problems using the disassembly structure given in Figure 1. Two performance measures were used in the test: percentage deviation from the lower bound obtained by solving the Lagrangean dual problem; and percentage deviation from the optimal solution value obtained using CPLEX 8.1, commercial integer programming software. Here, due to the excessive computational burden, we set the time limit as 3600 seconds for CPLEX 8.1.

For the disassembly structure given in Figure 1, we generated 60 problems, i.e., 10 problems for each combination of two levels of capacity tightness (loose and tight) and three levels of the number of periods (10, 20, and 30). For each problem, disassembly operation costs, inventory holding costs and setup cost were generated from $DU(50, 100)$, $DU(5, 10)$, and $DU(500, 1000)$, respectively. Here, $DU(a, b)$ is the discrete uniform distribution with $[a, b]$. Capacity per period was

Table 1. Test results for the suggested algorithm

(a) Case of loose capacity

Problem	Number Period					
	Deviation			CPU		
	10	20	30	10	20	30
1	0.32[1](0.13[2])	0.18(0.07)	0.17(0.10)	0.63[3](0.25[4])	1.94(1.80)	3.87(10.08)
2	0.29(0.16)	0.30(0.17)	0.28(0.22)	0.64(0.26)	1.97(49.50)	3.91(228.15)
3	0.45(0.22)	0.19(0.13)	0.29(·)	0.65(0.42)	1.92(5.78)	3.93(·)
4	0.61(0.42)	0.16(0.06)	1.07(·)	0.61(0.45)	1.87(62.86)	4.14(·)
5	0.83(0.69)	1.09(0.87)	0.37(0.2)	0.65(0.29)	2.02(75.63)	3.95(3173.76)
6	0.31(0.06)	1.11(0.92)	0.48(·)	0.61(0.51)	1.99(66.37)	3.97(·)
7	0.38(0.19)	0.23(0.19)	0.17(0.10)	0.62(0.52)	1.92(1.12)	3.88(20.68)
8	0.08(0.00)	0.12(0.03)	0.61(·)	0.62(0.15)	1.89(17.80)	4.06(·)
9	0.05(0.00)	1.34(1.01)	0.14(0.10)	0.62(0.17)	2.06(46.33)	3.93(24.76)
10	2.07(0.12)	0.36(0.26)	0.15(·)	0.73(1.52)	1.94(12.63)	3.95(·)

[1] percentage deviation from a lower bound obtained from Lagrangean dual problem
[2] percentage deviation from an optimal solution obtained from CPLEX 8.1 (dots imply that optimal solutions were not obtained within 3600 seconds)
[3] CPU seconds of Lagrangean heuristic algorithm
[4] CPU seconds of CPLEX 8.1

(b) Case of tight capacity

Problem	Number Period					
	Deviation			CPU		
	10	20	30	10	20	30
1	0.35(0.07)	1.55(·)	0.57(·)	0.69(0.52)	2.59(·)	4.20(·)
2	2.21(1.65)	0.29(0.06)	1.11(·)	0.69(1.16)	1.98(273.92)	4.95(·)
3	1.01(0.81)	0.91(·)	0.68(·)	0.68(0.36)	2.09(·)	4.30(·)
4	1.09(0.49)	0.74(·)	0.92(·)	0.66(23.94)	2.49(·)	4.46(·)
5	0.59(0.16)	0.46(0.31)	1.22(·)	0.66(1.00)	2.16(71.49)	4.25(·)
6	0.55(0.35)	0.69(0.33)	0.12(0.05)	0.64(0.21)	2.38(1004.58)	4.04(404.83)
7	1.20(0.72)	0.35(0.11)	1.33(·)	0.75(1.42)	2.16(70.93)	4.34(·)
8	1.37(0.89)	3.39(·)	1.16(·)	0.80(0.95)	2.44(·)	4.30(·)
9	1.14(0.42)	1.48(1.13)	0.32(0.22)	0.75(5.84)	2.17(477.51)	4.15(112.6)
10	1.23(0.84)	0.85(0.31)	0.93(·)	0.81(0.84)	2.15(612.47)	4.41(·)

set to 400 and disassembly time was generated from $U(1, 4)$. Here, $U(a, b)$ is the continuous uniform distribution with $[a, b]$. (The other data generation methods are omitted here because of the space limitation.) Also, the Lagrangean heuristic requires specific values for several parameters. After a preliminary experiment, these were set as follows: the iteration limit W was set to 5000; and δ_w was set to 2 initially and halved if the lower bound has not been improved in 90 iterations.

The test results are summarized in Table 1. The percentage deviations from optimal solution values are not reported for some problem sets since we could not obtain the optimal solutions using CPLEX 8.1 within 3600 seconds. It can be seen from the table that the Lagrangean heuristic suggested in this paper gives near optimal solutions. That is, the percentage deviations from lower bounds and optimal solutions were less than 2% and 1%, respectively. Also, computation times for the Lagrangean heuristic were significantly shorter than those for CPLEX 8.1, i.e., less than 5 seconds were required while CPLEX for many problems required more than 3600 seconds. This implies that the Lagrangean heuristic suggested in this paper can be used to solve practical sized problems within a reasonable amount of computation times.

5 Concluding Remarks

In this paper, we considered the problem of determining the disassembly schedule of products to satisfy the demands of their parts or components over a finite planning horizon. The capacitated problem with single product type without parts commonality is considered for the objective of minimizing the sum of setup, disassembly operation, and inventory holding costs. The problem is solved using a Lagrangean relaxation approach in which the relaxed problem becomes the single item lot-sizing problem after decomposition. Test results on randomly generated problems showed that the heuristic can give near optimal solutions within very short computation time.

Acknowledgements

The financial supports from the Swiss National Science Foundation under contract number 2000-066640 and the Korea Research Foundation under grant number 2004-03-D00468 are gratefully acknowledged.

References

1. Gupta, S. M., Taleb, K. N.: Scheduling Disassembly, International Journal of Production Research, Vol. 32 (1994) 1857–1886
2. Kim, H.-J., Lee, D.-H., Xirouchakis, P., Züst, R.: Disassembly Scheduling with Multiple Product Types, Annals of the CIRP, Vol. 52 (2003) 403–406
3. Lee, D.-H., Kang, J.-G., Xirouchakis, P.: Disassembly Planning and Scheduling: Review and Further Research, Proceedings of the Institution of Mechanical Engineers: Journal of Engineering Manufacture – Part B, Vol. 215 (2001) 695–710
4. Lee, D.-H., Kim, H.-J., Choi, G., Xirouchakis, P.: Disassembly Scheduling: Integer Programming Models, Proceedings of the Institution of Mechanical Engineers: Journal of Engineering Manufacture – Part B, Vol. 218 (2004) 1357–1372
5. Lee, D.-H., Xirouchakis, P.: A Two-Stage Heuristic for Disassembly Scheduling with Assembly Product Structure, Journal of the Operational Research Society, Vol. 55 (2004) 287–297

6. Lee, D.-H., Xirouchakis, P., Züst, R.: Disassembly Scheduling with Capacity Constraints, Annals of the CIRP, Vol. 51 (2002) 387–390
7. Neuendorf, K.-P., Lee, D.-H., Kiritsis, D., Xirouchakis, P.: Disassembly Scheduling with Parts commonality using Petri-nets with Timestamps, Fundamenta Informaticae, Vol. 47 (2001) 295–306
8. Taleb, K. N., Gupta, S. M., Disassembly of Multiple Product Structures, Computers and Industrial Engineering, Vol. 32 (1997) 949–961
9. Taleb, K. N., Gupta, S. M., Brennan, L.: Disassembly of Complex Product Structures with Parts and Materials Commonality, Production Planning and Control, Vol. 8 (1997) 255–269
10. Wagelmans, A., Hoesel, S. V., Kolen, A.: Economic Lot Sizing: an $O(n \log n)$ Algorithm that Runs in Linear Time in the Wagner-Whitin Case, Operations Research, Vol. 40 (1992) 145–156

DNA-Based Algorithm for 0-1 Planning Problem

L. Wang, Z.P. Chen, and X.H. Jiang

Department of Computer and Information Science, Fujian University of Technology,
Fuzhou, Fujian, 350014, RP China
wanglei_hn@hn165.com, zp_chen@hnu.net, xh_jiang@csu.net

Abstract. Biochemical reaction theory based DNA computation is of much better performance in solving a class of intractable computational problems such as NP-complete problems, it is important to study the DNA computation. A novel algorithm based on DNA computation is proposed, which solves a special category of 0-1 planning problem by using the surface-based fluorescence labeling technique. The analysis show that our new algorithm is of significant advantages such as simple encoding, low cost and short operating time, etc.

1 Introduction

Along with the development of computers, some complex problems such as nonlinearity and NP-completeness emerged in the new field of engineering, which could not be solved by our current electronic computers, because of their slow operating speed and limited storage. So scientists are searching for other kinds of processors, which can break through the speed limitation of the traditional silicon microelectronic circuitry in parallel computation. Since the techniques of DNA computation can meet the needs of highly data-parallel computation and can be used to settle those above complex problems, so studying the DNA computing technologies is of great importance [1-3].

In recent years, some useful research has been done about the techniques of DNA computation. In 1994, *Alderman*[4] experimentally demonstrated that DNA molecules and common molecular biology techniques could be used to solve complicated combinational problems such as the famous Hamiltonian path problem. One year later, *Lipton*[5] generalized Alderman's work, and presented a DNA computing model, which can solve the satisfy-ability (SAT) problem. In 1996, *Frank*[6] successfully expressed the binary digits 0 and 1 by using DNA molecular, and on the basis of that, built a DNA-based model, which can complete simple additive operation. In 1999, *Bernard*[7] proposed an improved DNA-based model, which can achieve more complex additive operation than the Frank's model through separating the input strands from the operator strands. In the same year, *Oliver*[8] proposed a new DNA-based computing model, which can perform much more complicated operation such as the matrix multiplication. In 2000, *Sakamoto*[9] exploited the hairpin formation by single-stranded DNA molecules, and settled a 3-SAT problem by using molecule computation. In 1999, *Liu*[10] developed a novel surface based approach, which solved the SAT problem too. Two years later, *Wu*[11] improved *Liu*'s algorithm in 2001. 0-1 planning problems are very important problems in operational researches. In 2003, *Yin*[12] applied DNA computing to the 0-1 planning problems firstly, and solved a kind of special 0-1 problems by using fluorescence-labeling techniques.

On the basis of concepts such as index of constraint-equations group and determinative factor of constraint equations proposed in this paper, an novel DNA-based algorithm for the 0-1 planning problem is presented, which uses the surface-based fluorescence-labeling technique, and can obtain all of the feasible solutions to the special integer-planning problem easily. The analysis show that our new algorithm is of significant advantages such as simple encoding, low cost and short operating time, etc.

2 Model of DNA Computation

DNA (Deoxyribonucleic acid), is a molecule that usually exists as a right-handed double helix, is the hereditary material of most organisms. In double-stranded DNA (dsDNA), there exists only four kinds of nucleotide bases such as adenine (A), guanine (G), Cytosine (C) and thymine (T), which form coplanar base pairs by hydrogen bonding. The rules of base pairing are: A with T: the purine adenine (A) always pairs with the pyrimidine thymine (T), and C with G: the pyrimidine cytosine (C) always pairs with the purine guanine (G). This type of base pairing is called *complementary*. The actual base pairing structures are illustrated in Fig.1.

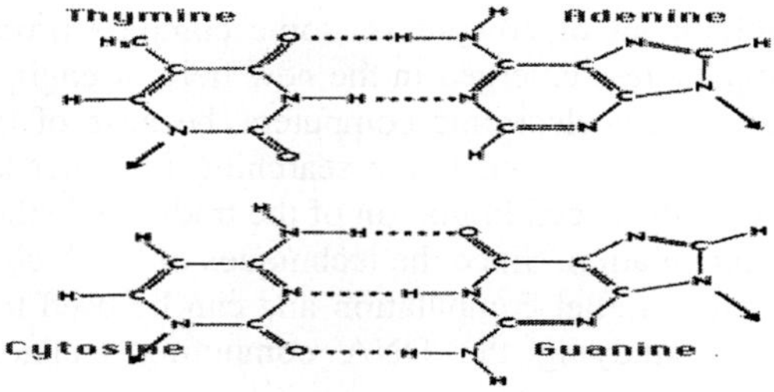

Fig. 1. Base pairing structures

The main idea of the DNA computation is to use the complementary base pairing rules and DNA's special double helical structure to encode information. After having encoded information into DNA molecules, all of the operating objects can be mapped into DNA strands, which will form all kinds of data pools after having reacted with the biological enzyme. By using those data pools, then we can obtain the final solutions to the target problems through high parallel biochemical reactions.

3 0-1 Planning Problem

The 0-1 planning is a very important problem in operational researches. In the 0-1 planning problem, the variable x_i can be 0 or 1 only. The common form of the 0-1 planning problem can be described as follows:

Definition 1: 0-1 planning problem:

$$Max(min)u = c_1x_1 + c_2x_2 + \ldots + c_nx_n \qquad (1)$$

$$\begin{cases} a_{11}x_1 + a_{12}x_2 + \ldots + a_{1n}x_n \leq (=, \geq) b_1 \\ a_{21}x_1 + a_{22}x_2 + \ldots + a_{2n}x_n \leq (=, \geq) b_2 \\ \ldots\ldots \\ a_{m1}x_1 + a_{m2}x_2 + \ldots + a_{mn}x_n \leq (=, \geq) b_m \\ x_1, x_2, \ldots, x_n = 0, 1 \end{cases} \qquad (2)$$

Where c_j, b_i, a_{ij} are any integer, $i=1,2,\ldots,m$, and $j=1,2,\ldots,n$.

In this paper, we have solved the following special category of 0-1 planning problem:

Definition 2: A special category of 0-1 planning problem:

$$Max(min)u = c_1x_1 + c_2x_2 + \ldots + c_nx_n \qquad (3)$$

$$\begin{cases} a_{11}x_1 + a_{12}x_2 + \ldots + a_{1n}x_n \leq (=, \geq) b_1 \\ a_{21}x_1 + a_{22}x_2 + \ldots + a_{2n}x_n \leq (=, \geq) b_2 \\ \ldots\ldots \\ a_{m1}x_1 + a_{m2}x_2 + \ldots + a_{mn}x_n \leq (=, \geq) b_m \\ x_1, x_2, \ldots, x_n = 0, 1 \end{cases} \qquad (4)$$

Where c_j and b_i are any integer, a_{ij} belongs to $\{-1, 0, 1\}$, as for $i=1,2,\ldots,m$, and $j=1,2,\ldots,n$.

In the above definition 2, the formula (3) is called *target function*, formula (4) is called *constraint-equations group*, and each equation in formula (4) is called a *constraint equation*. In any constraint equation in formula (4), if $b_i<0$, then after multiplying -1 to both sides of the constraint equation, we can suppose that $b_i \geq 0$, where $i=1, 2, \ldots, m$. Now, we present the corresponding algorithm for the special category of 0-1 planning problem as follows according to the definition 2:

Step 1: Generate all possible solutions, consisted of all possible combinations of 0 and 1, for the given special category of 0-1 planning problem.

Step 2: Delete the non-feasible solutions from the remnant possible solutions by using the constraint equations.

Step 3: Keep all of the remnant solutions.

Step 4: Repeat steps 2 and 3, then we can eliminate all of the non-feasible solutions, and obtain all of the feasible solutions to the given special category of 0-1 planning problem, after all of the constraint equations have been applied.

Step 5: Compute the target function's value for each feasible solution, and Compare those values, then we can obtain the optimal solutions to the given special category of 0-1 planning problem finally.

Before describing our new DNA-based algorithm for the given special category of 0-1 planning problem, some relative definitions will be given firstly.

Definition 3 (*Index of constraint equations group*): As for the constraint equations group given by formula (4), let $t_j = \max\limits_{1 \leq i \leq m} \{|a_{ij}|\}$, where $1 \leq j \leq n$. $T = \sum\limits_{j=1}^{n} t_j$ is called index of the given constraint equations group.

Definition 4 (*Negative index of constraint equation*): As for each constraint equation in formula (4) such as $a_{s1}x_1 + a_{s2}x_2 + \ldots + a_{sn}x_n \leq (=, \geq) b_s$, where $s \in [1, m]$. Let $a_{st_1}, a_{st_2}, \ldots, a_{st_l}$ are all coefficients that equal -1, $R = |\sum\limits_{j=1}^{l} a_{st_j}|$ is called negative index of the given constraint equation.

Definition 5 (*determinative factor of constraint equation*): As for each constraint equation in formula (4) such as $a_{s1}x_1 + a_{s2}x_2 + \ldots + a_{sn}x_n \leq (=, \geq) b_s$, where $s \in [1, m]$, let R is the negative index of the given constraint equation, and T is the index of the given constraint equations group of formula (2). $\Delta = T - R - b_s$ is called determinative factor of the given constraint equation.

Definition 6 (*Complement links of constraint equation*): Let n kinds of oligonucleotides denote the n variables $x_1, x_2, \ldots, x_n$, and other n kinds of oligonucleotides denote the n variables $\bar{x}_1, \bar{x}_2, \ldots, \bar{x}_n$. As for any given constraint equation $\delta_{s1}x_1 + \delta_{s2}x_2 + \ldots + \delta_{sn}x_n \leq (=, \geq) b_s$, $s \in [1, m]$, $\delta_{sj} \in \{-1, 1\}$, $j = 1, 2, \ldots, n$, we call $x''_1, x''_2, \ldots, x''_n$ as the *complement link* of the given constraint equation, where

$$x''_j = \begin{cases} x'_j, & \text{if } \delta_{sj} = 1 \\ \bar{x}'_j, & \text{if } \delta_{sj} = -1 \end{cases}$$

, the oligonucleotide corresponding to x_j represents $x_j = 1$, the oligonucleotide corresponding to $\bar{x}_j$ represents $x_j = 0$, and the oligonucleotide corresponding to x'_j is the complementary of the oligonucleotide corresponding to x_j, for any $j = 1, 2, \ldots, n$.

Theorem 1: As for any given constraint equation in formula (4) such as $\delta_{s1}x_1 + \delta_{s2}x_2 + \ldots + \delta_{sn}x_n \Theta b_s$, where $s \in [1, m]$, $\delta_{sj} \in \{-1, 1\}$, $j = 1, 2, \ldots, n$, $\Theta \in \{\leq, =, \geq\}$, and let Δ be its determinative factor, then after the complement link of the given constraint equation is added to surface, on which DNA strands with fluorescence labeling are fixed:

(1) If Θ is $\geq$, then the variables serial corresponding to a DNA strand satisfy the given constraint equation $\Leftrightarrow$ there are at most Δ kinds of colors left on the surface of the DNA strand when it is observed through the laser confocal microscope.

(2) If Θ is $=$, then the variables serial corresponding to a DNA strand satisfy the given constraint equation $\Leftrightarrow$ there are just Δ kinds of colors left on the surface of the DNA strand when it is observed through the laser confocal microscope.

(3) If Θ is $\leq$, then the variables serial corresponding to a DNA strand satisfy the given constraint equation $\Leftrightarrow$ there are at least Δ kinds of colors left on the surface of the DNA strand when it is observed through the laser confocal microscope.

Proof: Since the proofs of (1), (2) and (3) are similar, we give the detailed proof of (1) only.

Firstly, as for the given constraint equation: $\delta_{s1} x_1 + \delta_{s2} x_2 + ... + \delta_{sn} x_n \geq b_s$, we suppose that $\delta_{s1}, \delta_{s2},..., \delta_{sr}$ equal -1, and $\delta_{r+1 1}, \delta_{r+2 1},..., \delta_{n1}$ equal 1. Let $b = b_s$, and then the given constraint equation can be simplified as $-x_1 - x_2 -...- x_r + x_{r+1} + x_{r+2} +...+ x_n \geq b$. Let T be index of the constrain equations group described by formula (4), then it is obvious that $T \geq n$, and $\Delta = T - b - r$. Let $f_1, f_2,..., f_r, f_{r+1}, f_{r+2},..., f_n, f_{n+1}, f_{n+2},..., f_T$ be the surface-fixed DNA strand with fluorescence labeling.

($\Rightarrow$): Supposing that the number of 1 in $f_1, f_2,..., f_r$ is t, Since $f_1, f_2,..., f_r, f_{r+1}, f_{r+2},..., f_n, f_{n+1}, f_{n+2},..., f_T$ satisfy the given constraint equation $\Rightarrow -f_1 - f_2 -...- f_r + f_{r+1} + f_{r+2} +...+ f_n \geq b \Rightarrow f_{r+1} + f_{r+2} +...+ f_n \geq b+t$, so the number of 1 in $f_{r+1}, f_{r+2},..., f_n$ is at least $b+t$. After the complement link $\overline{x}'_1, \overline{x}'_2,..., \overline{x}'_r, x'_{r+1}, x'_{r+2},..., x'_n$ with fluorescence quencher corresponding to the given constraint equation is added to surface, since all of the oligonucleotides corresponding to $x'_{r+1}, x'_{r+2},..., x'_n$ are complements of the oligonucleotide corresponding to $1 \Rightarrow$ There are at least $b+t$ kinds of fluorescence dyes in $f_{r+1}, f_{r+2},..., f_n$ that will be extinguished after the complement link $x'_{r+1}, x'_{r+2},..., x'_n$ is added to surface $\Rightarrow$ There exist at most $T - r - b - t$ kinds of fluorescence dyes left in $f_{r+1}, f_{r+2},..., f_n, f_{n+1}, f_{n+2},..., f_T$. Since the number of 1 in $f_1, f_2,..., f_r$ is $t \Rightarrow$ The number of 0 in $f_1, f_2,..., f_r$ is $r-t$. Since all of the oligonucleotides corresponding to $\overline{x}'_1, \overline{x}'_2,..., \overline{x}'_r$ are complements of the oligonucleotide corresponding to $0 \Rightarrow$ There are at least $r-t$ kinds of fluorescence dyes in $f_1, f_2,..., f_r$ that will be extinguished after the complement link $\overline{x}'_1, \overline{x}'_2,..., \overline{x}'_r$ is added to surface $\Rightarrow$ There are t kinds of fluorescence dyes in $f_1, f_2,..., f_r \Rightarrow$ There are at most $T - r - b - t + t = T - b - r = \Delta$ kinds of fluorescence dyes in $f_1, f_2,..., f_r, f_{r+1}, f_{r+2},..., f_n, f_{n+1}, f_{n+2},..., f_T$.

($\Leftarrow$): Supposing that the number of 1 in $f_1, f_2,..., f_r$ is t, and the number of 1 in $f_{r+1}, f_{r+2},..., f_n$ is p. After the complement link $\overline{x}'_1, \overline{x}'_2,..., \overline{x}'_r, x'_{r+1}, x'_{r+2},..., x'_n$ with fluorescence quencher corresponding to the given constraint equation is added to surface, since the number of 1 in $f_{r+1}, f_{r+2},..., f_n$ is p, and all of the oligonucleotides corresponding to $x'_{r+1}, x'_{r+2},..., x'_n$ are complements of the oligonucleotide corresponding to $1 \Rightarrow$ There are at least p kinds of fluorescence dyes in $f_{r+1}, f_{r+2},..., f_n$ that will be extinguished after the complement link $x'_{r+1}, x'_{r+2},..., x'_n$ is added to surface. Since the number of 1 in $f_1, f_2,..., f_r$ is $t \Rightarrow$ The number of 0 in $f_1, f_2,..., f_r$ is $r-t$. Since all of the oligonucleotides corresponding to $\overline{x}'_1, \overline{x}'_2,..., \overline{x}'_r$ are complements of the oligonucleotide corresponding to $0 \Rightarrow$ There are at least $r-t$ kinds of fluorescence dyes in $f_1, f_2,..., f_r$ that will be extinguished after the complement link $\overline{x}'_1, \overline{x}'_2,..., \overline{x}'_r$ is added to surface. $\Rightarrow$ There are t kinds of fluorescence dyes in $f_1, f_2, \ldots, f_r$ $\Rightarrow$ There exist at most $T-p-r+t$ kinds of fluorescence dyes in f_1, $f_2, \ldots, f_r, f_{r+1}, f_{r+2}, \ldots, f_n, f_{n+1}, f_{n+2}, \ldots, f_T$. Since we have known that there are at most $\Delta = T - b - r$ kinds of fluorescence dyes in $f_1, f_2, \ldots, f_r, f_{r+1}, f_{r+2}, \ldots, f_n, f_{n+1}, f_{n+2}, \ldots, f_T \Rightarrow T - r - p + t \leq T - b - r \Rightarrow p \geq b + t \Rightarrow f_1, f_2, \ldots, f_r, f_{r+1}, f_{r+2}, \ldots, f_n$ satisfy the given constraint equation $\Rightarrow f_1, f_2, \ldots, f_r, f_{r+1}, f_{r+2}, \ldots, f_n, f_{n+1}, f_{n+2}, \ldots, f_T$ satisfy the given constraint equation.

4 DNA-Based Algorithm for 0-1 Planning Problem

As for a constraint equations group with n variables and m constraint equations, we present our innovative DNA-based algorithm for the special category 0-1 planning problem given by definition 2 as follows.

Step 1: Generate $4n$ kinds of different oligonucleotides, and divide them into 4 groups. Let the n kinds of different oligonucleotides in the first group denote the variables $x_1, x_2, \ldots, x_n$, the n kinds of different oligonucleotides in the second group denote the variables $\overline{x}_1, \overline{x}_2, \ldots, \overline{x}_n$, the n kinds of different oligonucleotides in the third group denote $x'_1, x'_2, \ldots, x'_n$ that are corresponding complements of the variables in the first group, the n kinds of different oligonucleotides in the fourth group denote $\overline{x}'_1, \overline{x}'_2, \ldots, \overline{x}'_n$ that are corresponding complements of the variables in the second group. Then construct DNA strands as follows by using the $2n$ kinds of different oligonucleotides in the first two groups:

  (1) Construct 2^n different combinations of the $2n$ kinds of different oligonucleotides, and each combination shall include the oligonucleotides corresponding to the n different variables. Then we will obtain 2^n kinds of different DNA strands.
  (2) Link n kinds of different fluorescence dyes to each DNA strand from top to down.
  (3) Fix the DNA strands on surface.

Step 2: Add the complement link with fluorescence quencher of each constraint equation to surface, after hybridization, observe the fluorescence quenching degree of the surface-fixed DNA strand through the laser confocal microscope, and determine that whether the DNA strand satisfies the given constraint equation according to the theorem 1.

Step 3: Heat the products of step 2 to unwrap the double-strands, and wash away all of the complement links.

Step 4: Repeat steps 2 and 3 for $m-1$ times. Then we can obtain all of the feasible solutions that satisfy the constraint equations group.

Step 5: Compute the target function's value for each feasible solution, and Compare those values, then we can obtain the optimal solutions to the given special category 0-1 planning problem finally.

5 Analysis

In this chapter, a simple example will be presented in detail to describe our new algorithm's computing processes.

$$Min\ u = 2x_1 + x_2 + 3y_1 - 2z_1 \qquad (5)$$

$$\begin{cases} x_1 + x_2 + y_1 - z_1 \geq 2 \\ x_1 + z_1 \leq 1 \\ y_1 + z_1 \leq 1 \\ x_1 - x_2 = 0 \end{cases} \qquad (6)$$

It is obvious that the index of formula (6) is 4, and the negative index of the first constrain equation is 1, so he first constraint equation's determinative factor $\Delta_1 = 4-2-1=1$. Similarly, it is easy to know that the second constrain equation's determinative factor $\Delta_2 = 3$, the third constrain equation's determinative factor $\Delta_3 = 3$, and the fourth constrain equation's determinative factor $\Delta_4 = 3$.

Step 1: According to the step1 proposed by chapter4, we construct 4*4 (=16) kinds of different oligonucleotides, and divide them into 4 groups. Let the 4 kinds of different oligonucleotides in the first group denote the variables x_1, x_2, y_1, z_1, the 4 kinds of different oligonucleotides in the second group denote the variables $\overline{x}_1, \overline{x}_2, \overline{y}_1, \overline{z}_1$, the 4 kinds of different oligonucleotides in the third group denote x'_1, x'_2, y'_1, z'_1, the 4 kinds of different oligonucleotides in the fourth group denote $\overline{x}'_1, \overline{x}'_2, \overline{y}'_1, \overline{z}'_1$. We select the following 16 kinds of different oligonucleotides illustrated in Fig.2 for our example.

x_1: AACCTGGT, x_2: ACCATAGC, y_1: AGAGTCTC, z_1: GATCATTA
$\overline{x}_1$: CCAAGTTG, $\overline{x}_2$: GTTGGGGT, $\overline{y}_1$: AGCTTGCA, $\overline{z}_1$: AGTCTATA
x'_1: TTGGACCA, x'_2: TGGTATCG, y'_1: TCTCAGAG, z'_1: CTAGTAAT
$\overline{x}'_1$: GGTTCAAC, $\overline{x}'_2$: CAACCCAA, $\overline{y}'_1$: TCGAACGT, $\overline{z}'_1$: TCAGATAT

Fig. 2. 16 kinds of different oligonucleotides

Then, construct 16 kinds of different DNA strands by using the 8 kinds of different oligonucleotides in the first two groups, and then, as illustrated in Fig.3, fix them on surface, and Link 4 kinds of different fluorescence dyes such as Red, Green, yellow and Blue to each DNA strand from top to down.

Step 2: According to the step2 proposed by chapter4, link the fluorescence quenchers of Blue, yellow, Green and Red to x'_1, x'_2, y'_1, z'_1 separately. Later, add the positive complement link x'_1, x'_2, y'_1, $\overline{z}'_1$ of the first equation in

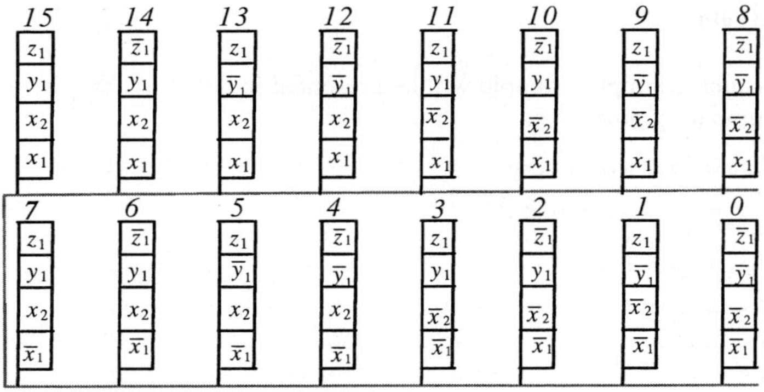

Fig. 3. Surface-fixed DNA strands with fluorescence dyes

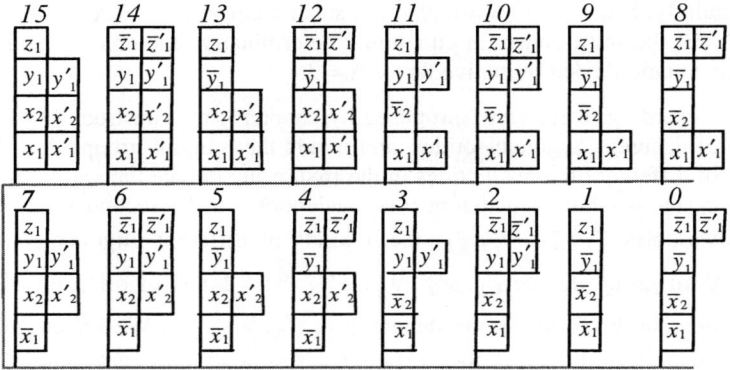

Fig. 4. graph obtained by adding complement link of the first constraint equation

formula (6) to surface. It is easy to know that the DNA strand with no more than $\Delta_1=1$ kinds of colors satisfy the first constraint equation according to theorem 1. So, as illustrated in Fig.4, the strands with No.15, No.14, No.12, No.10 and No.6 will satisfy the first constraint equation. Take photos to keep the results.

Step 3: According to the step3 proposed by chapter4, Heat the products to unwrap the double-strands, and wash away all of the complement links.

Step 4: According to the step4 proposed by chapter4, repeat steps 2 and 3.

(1) Add the complement link x'_1, z'_1 of the second equation in formula (6) to surface. It is easy to know that the DNA strand with no lesser more than $\Delta_2=3$ kinds of colors satisfy the second constraint equation according to theorem1. So, the strands with No.14, No.12, No.10 and No.6 will satisfy the second constraint equation in those DNA strands with No.15, No.14, No.12, No.10 and No.6. Take photos to keep the results (as for the limitation of paper, the graph is omitted).

(2) Add the complement link y'_1, z'_1 of the third equation in formula (6) to surface. It is easy to know that the DNA strand with no less than $\Delta_3=3$ kinds of colors satisfy the third constraint equation according to theorem1. So, the strands with No.14, No.12, No.10 and No.6 will satisfy the third constraint equation in those DNA strands with No.14, No.12, No.10 and No.6. Take photos to keep the results (as for the limitation of paper, the graph is omitted).

(3) Add the complement link x'_1, $\overline{x}'_2$ of the fourth equation in formula (6) to surface. It is easy to know that the DNA strand with just $\Delta_4=3$ kinds of colors satisfy the fourth constraint equation according to theorem1. So, the strands with No.14 and No.12 will satisfy the third constraint equation in those DNA strands with No.14, No.12, No.10 and No.6. Take photos to keep the results (as for the limitation of paper, the graph is omitted).

(4) Because the variable serial corresponding to DNA strands with No.14 and No.12 are 1,1,1,0 and 1,1,0,0 separately. So we can know that the feasible solutions to the given special category of 0-1 planning problem is $x_1=1$, $x_2=1$, $y_1=1$, $z_1=0$, and $x_1=1$, $x_2=1$, $y_1=0$, $z_1=0$.

Step 5: Apply $x_1=1$, $x_2=1$, $y_1=1$, $z_1=0$, and $x_1=1$, $x_2=1$, $y_1=0$, $z_1=0$ to the target function separately, it is obvious that the value of the target function is 6 and 3 separately. So, the optimal solution to the given special category of 0-1 planning problem is $x_1=1$, $x_2=1$, $y_1=0$, $z_1=0$, and the minimum value of the target function is 3.

6 Conclusions

DNA computation is of excellent advantages in solving the NP-complete problems because of its high parallel ability. On the basis of fluorescence labeling technology, a novel DNA-based algorithm for 0-1 planning problem is proposed in this paper, which solves a special category of 0-1 planning problem, and at the same time is of good characteristics such as simple encoding, low cost and short operating time, etc. Our next goal is to solve the general 0-1 planning problem and integer-planning problem by using DNA-based computing techniques.

References

1. Gao Lin, Xu Jin. DNA solution of vertex cover problem based on sticker model. Chinese Journal of Electronics.Vol.15, (2003) 1496-1500.
2. Bach, E. et. al. DNA Models and Algorithms for NP-Complete Problems. Proceedings of the 11th Annual Meeting on DNA Based Computers.Vol.44, (1999)151-161.
3. Xu Jin, Zhang Lei. DNA Computer Principle, Advances and Difficulties (1): Biological Computing System and Its Applications to Graph Theory. Chinese Journal of Computers. Vol.26, (2003) 1-11.
4. Frank G, Makiko F. Carter B. Making DNA add. Science. Vol.273, (1996) 220-223.
5. Yurke B, Mills Jr. A P. Cheng Siu Lai. DNA implementation of addition in which the input strands are separate from the operator strands. Bio-systems.Vol.52, (1999) 165-174.

6. Oliver J S. Computation with DNA: Matrix multiplication. DIAMACS series. Discrete Mathematics and Theoretical Computer Science. Vol.52, (1999) 165-171.
7. Alderman, L. M. Molecular computations to combinatorial problems. Science. Vol.266, (1994) 1021-1024
8. R Lipton. Using DNA to solve NP-Complete Problems. Science. Vol.268, (1995) 542-545.
9. Sakamoto et.al. Molecular computation by DNA hairpin formation. Science. Vol.288, (2000) 1223-1226.
10. Liu Q, Guo Z, Fei Z et al. A surface based approach to DNA computation. Discrete Mathmatics and Theoretical Computer Science. Vol.44, (1999) 123-132.
11. Wu Hao-Yang. An improved surface based method for DNA computation. Bio-systems. Vol.59, (2001) 1-5.
12. Yin Zhixiang, Zhang Fengyue, Xu Jin. 0-1 planning problem based on DNA computing. Chinese Journal of electronics and Information. Vol.15, (2003) 1-5.

Clustering for Image Retrieval via Improved Fuzzy-ART

Sang-Sung Park[1], Hun-Woo Yoo[2,*], Man-Hee Lee[1], Jae-Yeon Kim[3], and Dong-Sik Jang[1]

[1] Industrial Systems and Information Engineering, Korea University,
1, 5-ka, Anam-Dong, Sungbuk-Ku, Seoul 136-701, South Korea
{hanyul, vision, jang}@korea.ac.kr
[2] Center for Cognitive Science, Yonsei University,
134, Shinchon-Dong, Seodaemun-Ku, Seoul, 120-749, South Korea
paulyhw@yonsei.ac.kr
[3] Industrial Engineering, DongYang University,
1, Kyo Chon Dong, Young-ju, KyoungSangBukDo, 750-711, South Korea
jykim@phenix.dyu.ac.kr

Abstract. Clustering technique is essential for fast retrieval in large database. In this paper, new image clustering technique is proposed for content-based image retrieval. Fuzzy-ART mechanism maps high-dimensional input features into the output neuron. Joint HSV histogram and average entropy computed from gray-level co-occurrence matrices in the localized image region is employed as input feature elements. Original Fuzzy-ART suffers unnecessary increase of the number of output neurons when the noise input is presented. Our new Fuzzy-ART mechanism resolves the problem by differently updating the committed node and uncommitted node, and checking the vigilance test again. To show the validity of our algorithm, experiment results on image clustering performance and comparison with original Fuzzy-ART are presented in terms of recall rates.

1 Introduction

Images have always been an essential and effective medium for presenting visual data. With advances in today's computer technologies, it is not surprising that in many applications, much of the data is images.

In this paper, we deal with content-based image retrieval, which is a technique to retrieve images based on their visual properties such as color [1, 2], texture [3], and shape [4]. Systems [5, 6, 7] are well known for supporting this content-based image retrieval.

Fast retrieval in databases has been one of the active research areas. In that process, without any clustering schemes and adequate indexing structures, retrievals of similar images are time-consuming because the database system must compare the query image to each image in the database. This cost can be particularly prohibitive if the database images are very large and their features tend to have high-dimensionality. This high-dimensional indexing structure increases the retrieval time and memory space exponentially, as the member of feature dimension increases. Thus, frequently,

---

* Corresponding author.

it does not have any advantages against the simple sequential search. So, fast search algorithms, which can deal with high-dimensional feature data, are often an essential component of the image database. There have been a number of indexing data structures suggested to handle high-dimensional data: SS-tree[8], TV-tree[9], and X-tree[10].

In this paper, we present new image clustering technique that is useful for speedily finding the images from a large image database system. In this scheme, similar images are clustered based on the image feature and associated clustering algorithm. When the query is presented, similar images to the query are retrieved only from the most similar cluster to the query, thus full-database searches are not necessary.

We use improved Fuzzy-ART as clustering technique for narrowing the search space. The Fuzzy-ART [11] is a neurally-motivated unsupervised learning technique which has been used in many data analysis tasks. The Fuzzy-ART forms a nonlinear mapping of a high-dimensional input space to one cluster in artificial output neurons. This is possible by the fact that weight vectors in its neurons are trained and updated to the values that represent input pattern. Patterns that are mutually similar in respect to the given extracted feature are thus located same cluster.

However, the original Fuzzy-ART system yields undesirable results when the noise input is presented. The number of output clusters increase due to abrupt weight change caused by noise input. Our new Fuzzy-ART mechanism resolves the problem by differently updating the committed neuron and uncommitted neuron.

2 Image Features

The performance of the image clustering depends on the image features to describe the image content and adequate clustering mechanism. In this paper, color and texture information are used to represent image features. For color, joint HSV histogram extracted from local region is employed. For texture, entropies computed from local region are employed. These features extracted from each image in the database are used as input vector to the Fuzzy-ART network.

Color: For representing color, we used HSV (Hue, Saturation, Value) color model because this model is closely related to human visual perception. For color quantization, we uniformly quantized HSV space into 18 bins for hue (each bin consisting of a range of 20 degree), 3 bins for saturation and 3 bins for value for lower resolution.

In order to represent the local color histogram, we divided image into equal-sized 3×3 rectangular regions and extract HSV joint histogram that has quantized 162 bins for each region. And to obtain compact representation, we extract from each joint histogram the bin that has the maximum peak. Take hue h, saturation s, and value v associated to the bin as representing features in that rectangular region and normalize to be within the same range of [0,1]. Thus, each image has the $3 \times 3 \times 3 (= 27)$ dimensional color vector.

Texture: Most natural images include textures. Scenes containing pictures of wood, grass, etc. can be easily classified based on the texture rather than color or shape.

Therefore, it may be useful to extract texture features for image clustering. Like as color feature, we include a texture feature extracted from localized image region.

As a texture feature, we used the entropy extracted from the co-occurrence matrix[5]. Detailed feature extraction is performed as follows:

1. Conversion of color image to gray image
2. Dividing image into 3×3 rectangular regions as in color case.
3. Obtaining co-occurrence matrix for four (horizontal 0^0, vertical 90^0 and two diagonal 45^0 and 135^0) orientation in region and normalize entries of four matrices to [0, 1] by dividing each entry by total number of pixels.
4. Extracting average entropy value from four matrices.

$$e = \frac{-\sum_k \sum_i \sum_j p(i,j) \log(p(i,j))}{4}, \quad k = 1, 2, 3, 4 \quad (1)$$

5. Constructing texture feature vector by concatenating entropies over all rectangular regions.

Thus, each image has the $3 \times 3 (= 9)$ dimensional texture vector.

3 Artificial Neural Network

The ART network is an unsupervised vector classifier that accepts input vectors that are classified according to the stored pattern they most resemble [12]. It also provides for a scheme allowing adaptive expansion of the output layer of neurons until an adequate size is reached based on the number of classes, inherent in the observation. The ART network can adaptively create a new neuron corresponding to an input pattern if it is determined to be "sufficiently" different from existing clusters. In particular, ART network is designed to attempt to address the stability-plasticity dilemma; it provides a mechanism by which the network can learn new patterns without forgetting (or degrading) old knowledge.

In our research, we use a Fuzzy-ART model among ART networks for image clustering purpose. The reasons of a Fuzzy-ART use are: it is an unsupervised self-organizing network, it does not forget previously learned patterns, and in particular, it deals with real-valued data as an input. The first two reasons are adequate to index images into the database and the third reason is appropriate to deal with our real-valued color and texture features.

A Fuzzy-ART model consists of three fields such as input layer F_0, comparison layer F_1 and output layer F_2. Each field has M, 2M, and N nodes, respectively. In the layers F_1 and F_2, two kinds of weights such as bottom-up weights and top-down weights are connected each other. Input vector a is preprocessed by so-called complement coding in the layer F_0. Layer F_1 compares the similarity between the input vector and top-down weight vector and layer F_2 chooses the node with the maximum competitive signal of bottom-up weights when an input vector is presented. Fig. 1 shows the general architecture of Fuzzy-ART network.

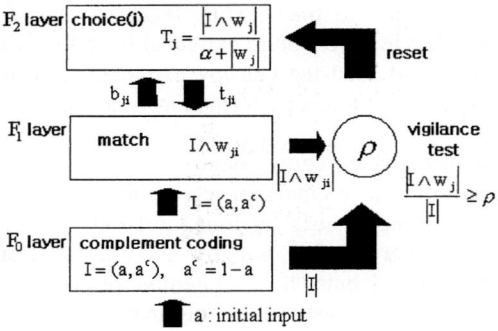

Fig. 1. The architecture of Fuzzy-ART

In the figure, representative vector W_J of Fuzzy-ART contains both the top-down weights and bottom-up weights. When the winning node J of F_2 layer is chosen, its top-down and bottom-up weights are updated in the same manner. Here, since both weights have same values, either weight can be considered to be representative pattern (prototype). Hence, fast learning is possible by updating only one of both weights.

$$t_J^{(new)} = \beta(I \wedge t_J^{(old)}) + (1-\beta)t_J^{(old)} \tag{2}$$

$$b_J^{(new)} = \beta(I \wedge b_J^{(old)}) + (1-\beta)b_J^{(old)} \tag{3}$$

where β is learning rate parameter, I is input vector, and T_J and B_J is top-down and bottom-up weight vectors associated with node J, respectively.

However, Above Fuzzy-ART inherently yields problems according to the range of learning parameter β. In the fast learning $\beta = 1$, prototype can be fast stable owing to update rule based on the fuzzy AND operator ($\wedge$) between input vector and weight vector. However, when the inputs with noises are presented, it can tend to recognize falsely, thus unnecessarily allocate new category in F_2. On the other hand, in the slow learning $0 < \beta < 1$, prototype has less impact from noise input, but it is slowly stable. Therefore, fast-commit and slow-recode scheme are necessary.

In this paper, we present new modified Fuzzy-ART algorithm in order to achieve fast stable with less impact from noise input. The proposed algorithm is based on that fast learning is applied to the committed node, whereas slow learning is applied to the uncommitted node.

4 Modified Fuzzy-ART

New Fuzzy-ART model employs different update rule on top-down weights and bottom-up weights. This scheme resolves unnecessary increase of output node, which is caused by abrupt weight changes when the noise input is presented. New algorithm is described as following steps:

Step 1: Initialization
Parameter Initialization:
choice parameter $\alpha \in (0, \infty)$, learning rate parameter $\beta \in [0, 1]$, vigilance parameter $\rho \in [0, 1]$, weight parameter $\eta \in [0, 1]$

Input vector Initialization:
If input vector $I = (a_1, a_2, ..., a_M)$ where each of its components is value ranged [0, 1] is presented to F_0, vector I that has 2M dimension is transformed and presented to F_1 such that

$$I = (a, a^c) = (a_1, a_2, ..., a_M, a_1^c, a_2^c, ..., a_M^c) \qquad (4)$$

where $a_k^c = 1 - a_k$.
This type of transformation is called complement coding.

Weight Initialization:
Weights associated to each node j ($j=1, 2, 3, ..., N$) in output layer F_2 are initialized.

$$b_{ij} = t_{ji} = 1, \quad (1 \leq i \leq 2M, \ 1 \leq j \leq N) \qquad (5)$$

where b_{ij} and t_{ji} is bottom-up weights and top-down weights connected between node i of the F_1 layer and node j of the F_2 layer, respectively. Here, it is assumed that all the nodes in F_2 are uncommitted (i.e., they did not experience weight updates).

Step 2: Category Choice
Upon presentation of an input I, a choice function T_j is computed for each category j in the F_2 layer

$$T_j(I) = \frac{|I \wedge b_j|}{\alpha + |b_j|} \qquad (6)$$

where the norm operator $|\cdot|$ is defined as $|x| = \sum_{i=1}^{2M} |x_i|$, the symbol $\wedge$ denotes the fuzzy AND operator, that is $x \wedge y = (\min(x_1, y_1), ..., \min(x_{2M}, y_{2M}))$, and α is a user-defined choice parameter, $\alpha \in (0, \infty)$. The category J for which the choice function is maximal, that is, $T_J = \max\{T_j : j = 1, 2, ..., N\}$ is chosen as a winning node. If two nodes have same maximal value T_J, node with small index j is chosen.

Step 3: Vigilance Test
In this step, similarity between top-down weight vector of the winning node J in Step 2 and input vector is compared based on vigilance parameter in order to check whether the input vector should be allocated to existing category or to new category.

Resonance:
Resonance occurs if the match function M_J meets the vigilance criteria ρ:

$$M_J = \frac{|I \wedge t_J|}{|I|} \geq \rho \qquad (7)$$

In this case, input I is allocated to the winning node J and top-down weight T_J is updated as described later.

Reset:
If the match function M_J does not meet the vigilance criteria, then mismatch reset occurs. The value of T_J in the current node is set to -1 for the duration of the current input presentation and another category is chosen in Step 2. This process is repeated until the chosen node J meets the vigilance test. If all F_2 nodes do not meet the test, new category is allocated.

Step 4: Learning (Weight Updates)
Learning takes place only in resonance case. The proposed weight update algorithm learns the pattern differently in the committed node and uncommitted node.

Uncommitted Node:
If all the nodes in F_2 layer does not meet vigilance test, a new category is allocated and receives current input. Thus, this node has no experience of weight update and does not impact on the existing category. In this case, weights update of new node is based on the existing learning rule (fast learning $\beta = 1$).

$$t_J^{(new)} = \beta(I \wedge t_J^{(old)}) + (1-\beta)t_J^{(old)} \tag{8}$$

$$b_J^{(new)} = \beta(I \wedge b_J^{(old)}) + (1-\beta)b_J^{(old)} \tag{9}$$

Committed Node:
Since committed node passed the vigilance test, it has experience of weight update before. In this case, learning takes place only at the chosen category. Proposed update rule performed top-down weights and bottom-up weights differently. Top-down weight vector t_J is updated based on the relative weight between current input vector and current top-down weight vector. Bottom-up weight vector b_J is updated based on fuzzy AND operator between updated top-down weight $t_J^{(new)}$ and current bottom-up weight vector.

$$t_J^{(new)} = \eta I + (1-\eta)t_J^{(old)} \tag{10}$$

$$b_J^{(new)} = t_J^{(new)} \wedge b_J^{(old)} \tag{11}$$

When the noise input is presented, equation (10) plays as a buffer against abrupt top-down weight change and equation (11) prevents bottom-up weight from tending to be unstable. In equations (10) and (11), only bottom-up weights vector always decreases to be stable. Thus we consider this vector as representative pattern (prototype). However, if the input vector always increases, the prototype b_J can be outside of chosen category.

Step 5: Re-vigilance Test and Re-weight Update
In order to deal with new allocation of unnecessary category, vigilance test is performed again by the following equation:

$$\frac{\left|b_J^{(new)} \wedge t_J^{(new)}\right|}{|I|} \geq \rho \tag{12}$$

If the (12) fails, since the b_J is outside of chosen category, we force the b_J to be allocated to chosen category by updating as follows:

$$b_J^{(new)} = t_J^{(new)} \tag{13}$$

The flow chart of the proposed algorithm is depicted in Fig. 2.

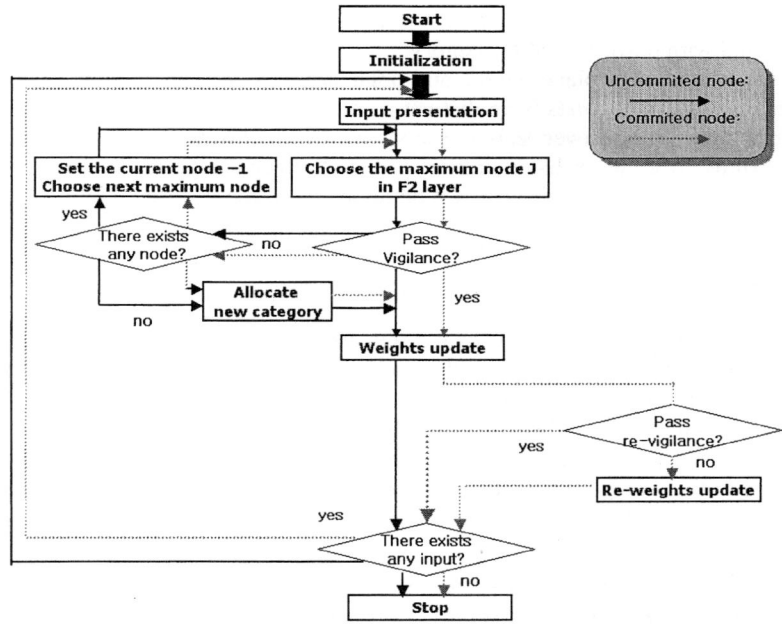

Fig. 2. The flow chart of proposed Fuzzy-ART algorithm

5 Experiments

We experimented on 200 images where most of them have dimensions of 192×128 pixels. The 200 images can be categorized such as airplane, eagle, elephant, horse, lion, polar bear, rose, zebra, tiger, valley, sunset, dolphin, flower, and bear.

We performed three experiments: 1) clustering results according to the change of the number of output nodes in F_2 layer, 2) clustering results according to the ordered presentation and the randomly presentation of input vectors, 3) clustering results using Fuzzy-ART and proposed modified Fuzzy-ART. For measure of clustering performance, we choose a recall rate according to each category as follows:

$$\text{clustering performance} = \frac{\text{the number of similar images in each cluster}}{\text{total number of similar images in database}} \quad (14)$$

5.1 Clustering Results According to the Change of the Number of Output Nodes

For this experiments, parameters are initialized as follows. $\alpha = 0.0001$, $\beta = 0.85$, $\eta = 0.85$, and $\rho = 0.91$. Learning is performed three times. In our experimental data, the network was stable after three learning epochs. Input vector is 72-dimensional since it has 27 for color and 9 for texture plus 36 for complement coding. Input vectors for learning is ordered list (that is input data are ordered based on the associated file names), Table 1 shows clustering results after three epochs. Here, row denotes the each category and number in parenthesis denotes the number of database images associated with corresponding category.

For the case of airplane, horse, and polar bear, recall rates is more than 0.9 since images in those categories have distinct color and texture information between object and background. However, zebra case shows the lowest recall rate because its color and texture information have similar to that of other categories such as rion, horse, and elephant. Zebra images does not be allocated to the corresponding category with up to 10 output nodes, but allocated with more than 11 nodes.

5.2 Clustering Results According to the Presentation Order of Input Vector

For this experiments, two kinds of indexes are used. First, images are ordered manually according to the file names. Thus, neighboring images are likely to belong to the same category. Then, color and texture features from each image are extracted and indexed into the database. Result in Table 1 is based on the ordered input list. Second, image list is mixed using random function. Thus, neighboring images are likely to belong to other categories. Color and texture features from each image are extracted and indexed into the database. Parameters are initialized similarly like as $\alpha = 0.0001$, $\beta = 0.85$, $\eta = 0.85$, and $\rho = 0.91$. Learning is performed three times. Table 2 shows comparison of clustering results after three epochs in the cases of random input. Compared with ordered input in Table 1, clustering results using random input order showed less recall rate as a whole.

5.3 Clustering Using Fuzzy-ART and Proposed Fuzzy-ART

For this experiments, we fixed output node n=12 and used ordered input. Initializing parameters are set to $\alpha = 0.0001$, $\beta = 0.85$, $\eta = 0.85$, and $\rho = 0.91$ in the same manner. The learning is performed three times. Fig. 3 shows comparison of clustering results using Fuzzy-ART and improved Fuzzy-ART. Existing Fuzzy-ART shows average recall rate 0.741, whereas improved Fuzzy-ART shows average recall 0.827.

Table 1. Clustering results for each category according to the different number of output nodes in F2 layer

	F_2=9	F_2=10	F_2=11	F_2=12	F_2=13	F_2=14	Average recall
Airplane(10)	10	10	10	10	10	10	1.000
Eagle(15)	12	12	12	11	11	11	0.767
Elephant(10)	7	8	7	7	7	7	0.717
Horse(20)	19	19	19	19	19	19	0.950
Lion(4)	3	3	3	3	3	3	0.750
Polar bear(19)	18	18	18	17	16	16	0.903
Rose(19)	15	13	12	14	14	14	0.719
Zebra(6)	0	0	5	5	5	5	0.555
Tiger(20)	14	15	17	18	17	17	0.817
Valley(20)	17	18	18	18	18	17	0.883
Sunset(15)	11	11	12	12	12	12	0.778
Dolphin(20)	12	12	14	13	13	13	0.642
Flower(10)	8	8	8	9	8	7	0.800
Bear(12)	10	10	10	10	10	10	0.833
Average recall	0.747	0.753	0.821	0.827	0.813	0.802	0.794

Table 2. Clustering results for each category using random input

	$F_2=9$	$F_2=10$	$F_2=11$	$F_2=12$	$F_2=13$	$F_2=14$	Average recall
Airplane(10)	9	9	10	10	10	10	0.967
Eagle(15)	12	11	12	11	11	11	0.756
Elephant(10)	7	7	7	7	7	7	0.707
Horse(20)	18	19	19	19	19	19	0.942
Lion(4)	2	3	3	3	3	3	0.703
Polar bear(19)	17	17	18	19	15	15	0.886
Rose(19)	14	13	12	13	14	14	0.702
Zebra(6)	0	1	5	5	5	5	0.583
Tiger(20)	14	14	14	15	15	15	0.725
Valley(20)	17	18	18	18	17	17	0.875
Sunset(15)	10	10	10	11	11	11	0.700
Dolphin(20)	12	13	13	13	13	13	0.642
Flower(10)	8	8	8	9	8	7	0.800
Bear(12)	8	8	9	9	9	9	0.722

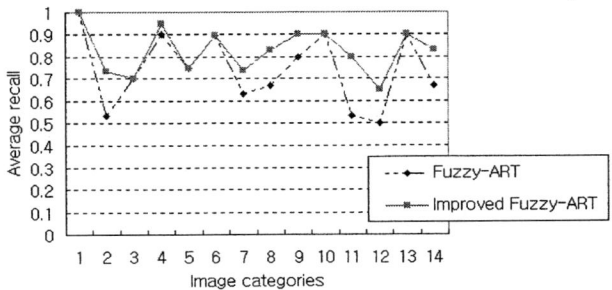

Fig. 3. Plot of average recall using Fuzzy-ART and proposed Fuzzy-ART

6 Conclusions

In this paper, an improved Fuzzy-ART network was used as clustering technique for content-based image retrieval. The proposed algorithm can reduce the retrieval time substantially since the images are examined only on the similar cluster rather than on the full-database. As input elements to the network, HSV joint histogram and average entropy computed gray-level co-occurrence matrices from localized image regions were used.

An original Fuzzy-ART model is sensitive to the noise inputs. Thus, we presented modified clustering network algorithm that was robust against the noises. The proposed algorithm updates committed node and uncommitted node differently. It prevents increase of unnecessary output categories by reducing abrupt weight changes.

References

1. Swain, M., Ballard, D.: Color indexing. International Journal of Computer Vision, Vol. 7, Num.1. (1991) 11-32

2. Smith, J.R., Chang, S.F.: Tools and techniques for color image retrieval. In Proc. SPIE: Storage and Retrieval for Image and Video Databases IV, Vol. 2670. (1996) 426-437
3. Manjunath, B.S., Ma, W.Y.: Texture features for browsing and retrieval of image data. Tech. Rep. CIPR TR-95-06 (1995)
4. Jain, A.K., Vailaya, A.: Shape-based retrieval: A case study with trademark image databases. Pattern Recognition, Vol. 31, Num. 9. (1998) 1369-1390
5. Flickner, M., Sawhney, H., Niblack, W., Ashley, J., Huang, Q., Dom, B., Gorkani, M., Hafner, J., Lee, D., Petkovic, D., Steele, D., Yanker, P.: Query by image content: The QBIC system. IEEE Computer, Vol. 28, Num. 9. (1995) 23-31
6. Smith, J.R., Chang, S.E.: VisualSEEK: A fully automated content-based image query system. in Proc. ACM Multimedia, (1996) 87-98
7. Carson, C., Belongie, S., Greenspan, H., Malick, J.: Blobworld: Image segmentation using expectation-maximization and its application to image querying. IEEE Trans on Pattern Analysis and Machine Intelligence, Vol. 24, Num. 8. (2002) 1026-1638
8. White, D.A., Jain, R.: Similarity indexing with the SS-tree. In Proc. 12th IEEE International Conference on Data Engineering, (1996) 516-523
9. Lin, K.I., Jagadish, H.V., Faloutsos, C.: The TV-tree: An index structure for high-dimensional data. VLDB Journal, Vol. 3, Num. 4. (1994) 517-549
10. Berchtold, S., Keim, D.A., Kriegel, H.P.: The X-tree: An index structure for high-dimensional data. In Proc. 22th Int. Conf. on Very Large Data Bases, (1996) 28-39
11. Carpenter, G.A., Grossberg, S., Rosen, D.B.: Fuzzy-ART: Fast stable learning and categorization of analog patterns by an adaptive resonance system. Neural Networks, Vol. 4. (1991) 759-771
12. Carpenter, G.A., Grossberg, S.: A massively parallel architecture for a self-organizing neural pattern recognition machine. Computer Vision, Graphics, and Image Processing, Vol. 37. (1987) 54-115

Mining Schemas in Semi-structured Data Using Fuzzy Decision Trees

Sun Wei and Liu Da-xin

College of Computer Science and Technology, HARBIN Engineering University,
HARBIN Heilongjiang Province, China
sun97611@sina.com.cn

Abstract. It is well known that World Wide Web has become a huge information resource. The semi-structured data appears in a wide range of applications, such as digital libraries, on-line documentations, electronic commerce. After we have obtained enough data from WWW, we then use data mining method to mine schema knowledge from the data. Therefore, it is very important for us to utilize schema information effectively. This paper proposes a method of schema mining based on fuzzy decision tree to get useful schema information on the web. This algorithm includes three stages, represented using Datalog, incremental clustering, determining using fuzzy decision tree. Using this algorithm, we can discover schema knowledge implicit in the semi-structured data. This knowledge can make users understand the information structure on the web more deeply and thoroughly. At the same time, it can also provide a kind of effective schema for the querying of web information. In the future, we will further the work on extract association rules using machine learning method and study the clustering method in semi-structured data knowledge discovery.

1 Introduction

It is well known that World Wide Web has become a huge information resource. Therefore, it is very important for us to utilize this kind of information effectively. However, the information on WWW can't be queried and manipulated in a general way. Although some sites may provide search engines, the queries are performed through keyword match operations, and query results are still in HTML format. The information still needs to be viewed on the corresponding web sites through browsers. Users are difficult to obtain the information structure and schema information of the whole web site. In fact, this kind of information does not have any predefined structure, and it is also called semi-structured data. It appears in a wide range of applications, such as digital libraries, on-line documentations, electronic commerce. After we have obtained enough data from WWW, we then use data mining method to mine schema knowledge from the data. Unfortunately, most of existing methods focus on the knowledge discovery from relational data, we must design a new method to deal with the hierarchy and irregularity of the semi-structured data.

Semi-structured data differ from structured data in traditional databases in that the data are not confined by a rigid schema which is defined in advance, and often exhibit irregularities. Furthermore, these data evolve rapidly under the WWW environments.

Typical models proposed for managing such data are labeled directed graphs[1,2,3], which are schemaless and self-describing, e.g. Object Exchange Model (OEM)[3] data graph. To cope with the irregularities of semi-structured data, regular path expressions are widely used in query languages for semi-structured systems [1,2]. However, the lack of schema poses great difficulties in query optimization. For example, to process a query embedded with regular path expressions, query evaluation inevitably involves an exhaustive search over the complete data graph.

Based on automata equivalence, DataGuide is proposed [4]. It is a concise and accurate graph schema for the data graph in terms of label paths. Its accuracy lies in that the schema and the data have the same sets of label paths, and its conciseness lies in that every label path in the schema is unique. However for large cyclic graphs, constructing a DataGuide may require exponential time and result in a DataGuide which is much larger than the original data graph. In general, it is expensive to construct and maintain an accurate graph schema for a data graph when the data are irregular and evolve rapidly. Furthermore, an accurate schema may be too large and hence impractical for query optimization. In [5], several heuristic-based strategies are proposed to build approximate DataGuide by merging "similar" portions of the DataGuide during construction. However these strategies are situation specific. [7] and [9] aim to type objects in the data set approximately by combining objects with similar type definitions, instead of constructing an approximate structural summary for the data graph. Their approximation methods are sensitive to the predetermined external parameters (such as the threshold θ, the number of clusters K, etc.); and it is hard to decide the optimal values for different data graphs in advance.

In this paper, we propose to construct an approximate graph schema by clustering objects with similar incoming and outgoing edge patterns using an incremental conceptual clustering method [8]. Our approach has the following unique features. No predetermined parameters are required to approximate the schema. It is cheap to construct and maintain the resultant schema and it is small in size. Moreover, we propose a query evaluation strategy for processing regular path expression queries with assistance of the schema.

The rest of the paper is organized as follows. The background knowledge is outlined in Sect. 2. Section 3 introduces the accurate graph schema, namely DataGuide. In Sect. 4, the algorithm to extract the approximate graph schema is given. Section 5 describes the query evaluation strategies and Sect. 6 presents the preliminary experimental results. Finally, Section 7 concludes the paper.

2 Related Work

Semi-structure data get schema form data, so one of the important aspects of schema research is the schema mining how to extract schema from data. The problem of schema mining is to automatically compute and extract schema from appointed instances of data without any predefined knowledge. If several schemas match the instances of data, the best one should be choose to express the instances. For the moment, the methods of extracting schema mainly have DataGuide, extracting rules based on Datalog, some clustering and classification methods, such as [6] proposed conceptual clustering method.

2.1 Graph Schema Extracting

DataGuide is emphasis on extracting graph schema from semi-structured data. DataGuide is the graph schema of Lore[2], which developed by Stanford as a semi-structured data management system. The system is based on the semi-structured data model OEM. DataGuide has two characteristics, accuracy and concise. Accuracy means that any path in data graph presents in DataGuide, and any path in DataGuide presents in data graph. Concise means that any path in DataGuide presents only once.

Building DataGuide for an appointed data source equivalences translating NFA to DFA. According as the method of automata, a data source may have several DataGuide. In Lore system, Strong DataGuide is chosen. Strong DataGuides guarantee that whenever one of nodes can be reached by multiple paths, these paths have identical target sets in the data source. Conversely, simple paths with the same target set in the database always lead to the same node in a strong DataGuide. The deterministic schema graph of a data source with cycles and node sharing may be of exponential size. For instance, assume a graph database as the one shown in Fig.1 (a). Its two a edges below the root need to be merged in the DataGuide because each path must only be represented once. There are two possibilities to do this, depicted in (b) and (c), respectively. In (b), the DataGuide is a weak one, and (c) is a strong DataGuide. The algorithm of strong DataGuide is simple, and retrieval data graph on deep-first to check all paths and nodes for edges and nodes of schema graph.

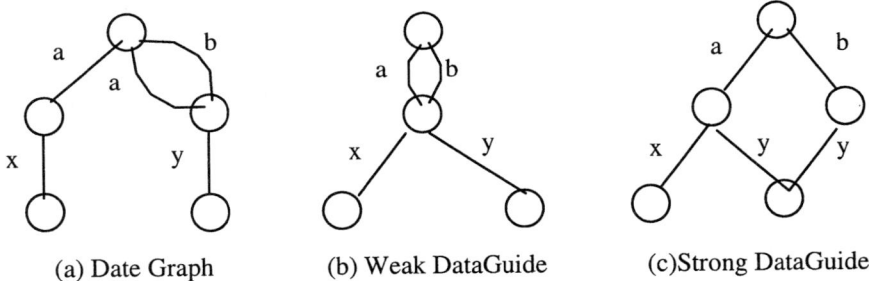

(a) Date Graph　　　(b) Weak DataGuide　　　(c)Strong DataGuide

Fig. 1. Weak DataGuide and Strong one

2.2 Datalog Rule Extracting

When schema of XML is described using Datalog, the extracting schema is how to get a set of datalog rules from semi-structured data automatically. There are some research on datalog, reference [10] proposes the method of extracting schema by typing the semi-structured data using a set of datalog rules. The process is as follow. First, a rule is defined for every complex object of data graph according the incoming and outgoing edge patterns of complex object. Second, classify set of knowledge and all potential types and compute the greatest fixpoint, so we get a type classification of all data object. Third, if the object sets of two types of the classification of data object is the same, the two type should be composed to reduce the number of types. At last, if the number of types is even great, it can be done that clustering the types by choosing proper methods to reduce the number more.

Through above process of data graph, the approximate types can be gained, which is small comparing to the perfect types and is easy to use in actual disposal.

2.3 Clustering and Classification

Some techniques of artificial intelligent are introduced to researches of semi-structured data, and the researchers propose several methods of extracting schema based on machine learning. They mainly adopt the methods of clustering or classification. The method of CG(Conceptual Graph) , proposed by the Chinese University of Hong Kong, introduces the idea of an incremental clustering of machine learning. CG is the approximate schema of semi-structured data, its extracting uses an incremental clustering method which is proposed in machine learning. It clusters the objects of data graph according to the similar degree of incoming and outgoing edge patterns of objects. The algorithm disposes an object of data graph once and joins it to an existing type or establishes a new type according to making the efficient function best. The schema graph got by above method may include the paths that not exist at data graph or duplicate paths, so it is not accuracy. The merit of it is to reduce the size of schema graph and to enhance practicability. The CG application on query optimization and its comparison to DataGuide are shown in reference [8].

3 Method Summary

There are several strategies of schema extraction that have been proposed. The main techniques have been introduced above. However these strategies have defects as follows: first, their approximation methods are sensitive to the predetermined external parameters and not define the approximation type; second, they handle easily extra edges but difficultly missing edges. Other, the schema graph of semi-structured data is not accurate.

The goal of this work is to be able to approximately type a large collection of semi-structured data efficiently. We are therefore led to making simplifying assumptions and introducing heuristics to be able to process this large collection in an effective way. In this section, we present the problem and main idea of solving in rather general terms. The various steps are detailed in the section 4.

First, the schema of semi-structured data is not precision and it is a character of semi-structured data. So the schema is not precision that we are mining from semi-structured data. We use fuzzy decision tree to decide the schema of the clusters of data. In the second, the K-cluster is widely used in mining schema and the number of clusters K requires to be appointed according to persons' experience, and it is difficult to determine the proper type. So we propose to construct approximate data set by clustering objects with similar incoming and outgoing edge patterns using an incremental clustering method. No predetermined parameters are required. The third and final, the defect of the mining schema includes two patterns of excess and deficit. In the case of relational and object data that are very regular and with the proper typing program, we obtain a perfect classification of the objects. In general, one should not expect this to happen. Suppose we have a program P that proposes a typing for a database D. We need a measure of how well P types D. A first measure is the number of ground facts in D that are not used to validate the type of any object. We call this measure the *excess* since it captures the number of facts that are in excess. Excess is rather easy to capture with datalog programs and the greatest fixpoint semantics. The deficit, some information that may be missing, is much less so. In our method, we use benefit function to cluster objects and the benefit function is consist of the conditional probability of an edge label in clustered type and the conditional probability of types including an edge label.

4 Mining Schema

In this section, we present a method of mining a perfect schema from semi-structured data. In the method, there are three stages that will be introduced in follow.

4.1 Stage 1: Represented by Datalog

The first key issue is the choice of a description language for types. Our typing is inspired by the typing found in object databases although it is more general since we allow objects to live in many incomparable classes, i.e., have multiple roles. This aspect is also a clear departure from previously proposed typings for semistructured data. We believe (and some early experiments support this belief) that multiple roles are essential when the data is fairly irregular. We define a typing in terms of a monadic datalog program. The intensional relations correspond to object classes and the rules describe the inner structure of objects in classes. The greatest fixpoint semantics of the program defines the class extents.

In this section, we present the model of semi-structured data and the types that are used in the present paper. We assume some familiarity with relational databases and more particularly with the datalog query language.

We model semistructured data in the style of [3] as a labeled directed graph. The nodes in the graph represent objects and the labels on the edges convey semantic information about the relationships between objects. The sink nodes (nodes without outgoing edges) in the graph represent *atomic objects* and have values associated with them. The other nodes represent *complex objects*. In a standard manner, we represent the graph using two (base) relations defined as follows:

link(FromObj, ToObj, Label): Relation *link* contains all the edge information. Precisely, link (o_1, o_2, l) corresponds to an edge labeled l from object o_1 to o_2. Note that there may be more that one edge from o_1 to o_2, but, in our model, for a particular l, there is at most one such edge labeled l.

atomic(Obj,Value): This relation contains all the value information. The fact atomic(o,v) corresponds to object o being atomic and having value v.

We also require that (1) each atomic object has exactly one value, i.e. Obj is a key in relation atomic, and (2) each atomic object has no outgoing edges, i.e., the first projections of *link* and *atomic* are disjoint.

In this paper, we consider that a typing is specified by a datalog program of a specific form (to be described shortly). The only two extensional relations (EDB's) of the typing program are *link* and *atomic*. The intensional relations (IDB's) are all monadic and correspond to the various types defined by the program.

For each complex objects O_k, assign a unique type predicate $type_k$. The rule for $type_k$ will contain $\overleftarrow{l}^i$ iff there is an edge labeled l from o_i to o_k, and $\overrightarrow{l}^i$ if there is an edge labeled l from o_k to o_i. And the rule for $type_k$ will contain $\overrightarrow{l}^o$ iff there is an edge labeled l from o_i to some atomic object.

4.2 Stage 2: Incremental Clustering

In most cases, the home type will have too many types to be useful as a summary of the data set. There will be many 'similar' types that intuitively can be clustered into one type thus dramatically reducing the size and complexity of the whole type. In this section, we outline how to cluster the types to reduce the number of types while keeping the defect low.

Let τ_i be a type which has presented at home types. We use $A(\tau_i)$ and $R(\tau_i)$ to denote the sets of labels associated with the outgoing edges and incoming edges of a type τ_i, respectively. A guiding principle in the clustering is to predict $A(\tau_i)$ and $R(\tau_i)$ as accurately as possible. For each label l in $A(\tau_i) \cup R(\tau_i)$, the predictability is reflected by the conditional probability $P(l|\tau_i)$. The higher is the probability, the more probable that a label l included in τ_i. Another principle, minimizing the size of the whole types is to minimize the number of appearances of each label l. This is equivalent to maximizing the predictiveness of l, which is reflected by the conditional probability $P(\tau_i|l)$. The higher is the predictiveness, the more probable that a type τ_i having an edge labeled with l. The expected tradeoff is defined as utility function $U(\tau_i)$ as follow.

$$U(\tau_i) = \frac{1}{|R(\tau_i) \cup A(\tau_i)|} \sum_{l \in R(\tau_i) \cup A(\tau_i)} P(l|\tau_i) \cdot P(\tau_i|l)$$

At the beginning, there is only one type corresponding to the root. Thereafter, for each time we visit a type τ_j from τ_i along the edge labeled with l, the type τ_j will be either an existing type or a newly created one, which results in the highest utility.

4.3 Sage 3: Mining Schema by Fuzzy Decision Tree

Through above two stages, we can get K-clusters which have low defects. In this section, we use fuzzy decision tree, a machine learning method, to determine a proper type in every cluster. For the sake of no precision of schema of semi-structured data, we get the fuzzy results of the schema, for every schema has its probability.

Definition 1. (Including-type)
Cluster K_i includes type set of $\tau_{i_1}, \tau_{i_2}, \ldots, \tau_{i_n}$. If type τ includes label l and label l is present in every type τ_{i_j} ($1 \leq j \leq n$), type τ is a including-type of K_i. Denoted by including-type(K_i) = τ.

Definition 2. (Included-type)
Cluster K_i includes type set of $\tau_{i_1}, \tau_{i_2}, \ldots, \tau_{i_n}$. If type τ includes label l and label l is present in not less than one of type τ_{i_j} ($1 \leq j \leq n$), type τ is a including-type of K_i. Denoted by included-type(K_i) = τ.

The proper type of Cluster K_i is between included-type(K_i) and including-type(K_i), so we use fuzzy decision tree to determine the proper type. We have two strategies as follows: adding labels from including-type (K_i) to proper type and reducing labels from included-type (K_i) to proper type. In the paper, we choose the first one. We defined firstly the comparability of a type for every instance in cluster K_i.

Definition 3. (comparability)
The comparability can be defined by a ratio, The change from an instance to type τ is
a× the number of adding labels + b× the number of reducing labels, a and b are power parameters. The ratio $= 1 - \dfrac{change(ins\tan ce_i)}{totaloflabel(ins\tan ce_i)}$.

For example, there is a cluster, including $\overleftarrow{\overrightarrow{a\ b}}$, $\overleftarrow{\overrightarrow{a\ b\ c}}$, $\overleftarrow{\overrightarrow{a\ b\ d}}$, $\overleftarrow{\overrightarrow{a\ c}}$, $\overleftarrow{\overrightarrow{a\ d}}$ five instances (several instances maybe the same type).

Decision trees have proven to be a valuable tool for description, classification and generalization of data. Today, a wealth of algorithms for the automatic construction of decision trees can be traced back to the ancestors ID3 [11] or CART. In many practical applications, the data used are inherently of imprecise and subjective nature. A popular approach to capture this vagueness of information is the use of fuzzy logic, going back to Zadeh [12]. The basic idea of fuzzy logic is to replace the "crisp" truth values 1 and 0 by a degree of truth in the interval[0,1]. In many respects, one can view classical logic as a special case of fuzzy logic, providing a more fine grained representation for imprecise human judgments. Consider, for instance, the question whether a jacket is fashionable. Here, a simple yes/no judgment looses information, or might even be infeasible. To combine the advantages of decision trees and fuzzy logic, the concept of a decision tree has been generalized to fuzzy logic, and there is a number of algorithms creating such trees, e.g. [13,14].

We apply fuzzy decision tree to determine the approximate type of example above. We use the strategy that adds labels to including-type of K_i. The added label is chosen that the function Gain is best. Function Gain is as follow:

$$Gain(K_i, a) = \frac{\sum comparality(ins\tan ce)}{Num(ins\tan ce)}$$

The probability of a type τ is the ratio of the number of instances that match the type τ in all the instances.

We can get the including-type in example 1, $\overleftarrow{a}$. We can add label b, c and d to the including-type. So we calculate the gain of label b, c and d.
$Gain(K_i, b) = 0.46$, $Gain(K_i, c) = 0.33$, and $Gain(K_i, d) = 0.33$.

we add label b to the type $\overleftarrow{a}$, show in Fig.2. There three instances accord with the type $\overleftarrow{a}\overrightarrow{b}$ in five instances so the positive of $\overleftarrow{a}\overrightarrow{b}$ is 0.6 and negative is 0.4. Thereafter, for each time we choose the label that have the best of gain function to add and calculate the positive and negative of the type.

In example 1, the positive of type $\overleftarrow{a}\overrightarrow{b}$ is 0.6, and the positive of type $\overleftarrow{a}\overrightarrow{b}\overrightarrow{c}$ is 0.2. So we should choose type $\overleftarrow{a}\overrightarrow{b}$ as the approximate type of cluster K_i.

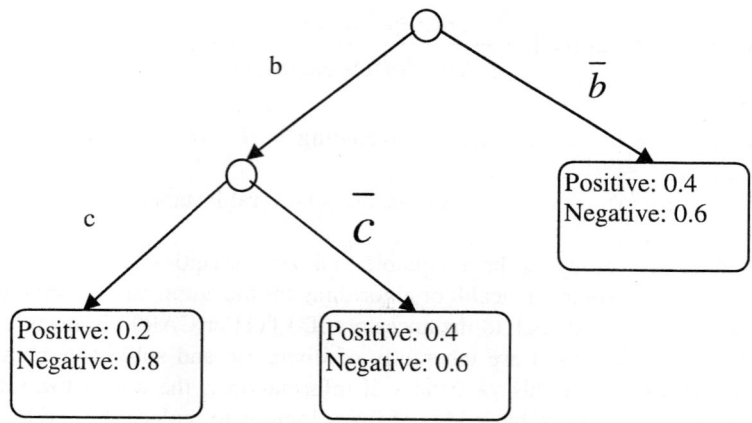

Fig. 2. A fuzzy decision tree in Example 1

5 Discussion

Semi-structured data are represented as labeled directed graphs and queried with regular path expressions. The accurate graph schema, DataGuide, which does not contain missing, duplicate or false paths in the schema, is not effective for optimization of a wide range of regular path expression queries due to the fact that its size can be much larger than the original data graph. In this paper, we proposed an approximate schema based on an incremental clustering method and fuzzy decision tree method. Our method showed that for different data graphs the sizes of approximate schemas were smaller than that of accurate schemas. Also, for regular path expression queries, query optimization with the approximate graph schema outperformed the accurate graph schema. In the near future, we plan to explore various approximate schemas and conduct an extensive performance study in order to evaluate their effectiveness.

References

1. S. Abiteboul, D. Quass, J. McHugh, J. Widom, and J. Wiener. The lorel query language for semi-structured data. *International Journal on Digital Libraries*, 1(1):68-88, 1997.
2. P. Bueman, S. Davidson, G. Hillebrand, and D. Suciu. A query language and optimization techniques for unstructured data. In *Proceedings of ACM SIGMOD International Conference on Management of Data*, 1996.
3. Y. Papakonstantinou, H. Garcia-Molina, and J. Widom. Object exchange across heterogeneous information sources. In *Proceedings of International Conference on Data Engineering*, 1995.
4. R. Goldman and J. Widom. Dataguides: Enabling query formulation and optimization in semistructured databases. In *Proceedings of the 23rd International Conference on Very Large Data Bases*, 1997.
5. R. Goldman and J. Widom. Approximate dataguides. Technical report, Stanford University, 1998.
6. Wang Qiuyue, Jeffrey Xu Yu, Huang Jinhui. Approximate Graph Schema Extraction for Semi-structured Data. In proc. Of EDBT 2000, Germany, March 2000.
7. S. Nestorov, S. Abiteboul, and R. Motwani. Inferring structure in semistructured data. In *Proceedings of the Workshop on Management of Semistructured Data*, 1997.
8. D. Fisher. Knowledge acquisition via incremental conceptual clustering. In J. Shavlik and T. Dietterich, editors, *Readings in Machine Learning*. Morgan Kaufmann Publishers, 1990.
9. S. Nestorov, S. Abiteboul, and R.Motwani. Extracting schema from semistructured data. In *Proceedings of ACM SIGMOD International Conference on Management of Data*, 1998.
10. Nestorov S, Abiteboul S, Motwani R. Extracting Schema from Semistructured Data. In Proc. of ACM SIGMOD Conf. On Management of Data. Seattle, WA,1998.
11. J. R. Quinlan. Induction on decision trees. *Machine Learning*, 1:81–106, 1986.
12. L. A. Zadeh. Fuzzy sets. *Information and Control*, 8:407–428, 1965.
13. Cezary Z. Janikow. Fuzzy decision trees: Issues and methods. *IEEE Transactions on Systems, Man, and Cybernetics*, 28(1):1–14, 1998.
14. L. Breiman, J.H. Friedman, R.A. Olshen, and C.J. Stone. *Classification and Regression Trees*. Wadsworth & Brooks Advanced Books and Software, Pacific Grove, CA, 1984.

Parallel Seismic Propagation Simulation in Anisotropic Media by Irregular Grids Finite Difference Method on PC Cluster

Weitao Sun[1,2], Jiwu Shu[2], and Weimin Zheng[2]

[1] ZHOU PEI-YUAN Center for Applied Mathematics, Tsinghua University,
Beijing 100084, China
[2] Department of Computer Science & Technology, Tsinghua University,
Beijing 100084, China
sunwt@tsinghua.edu.cn

Abstract. A 3D Finite Difference Method (FDM) with spatially irregular grids is developed to simulate the seismic propagation in anisotropic media. Staggered irregular grid finite difference operator with second-order time and spatial accuracy are used to approximate the velocity-stress elastic wave equations. The parallel codes are implemented with Message Passing Interface (MPI) library and c language. The 3D model with complex earth structure geometry is split into more flexible subdomains by the proposed irregular method. The spurious diffractions from "staircase" interfaces can be easily eliminated without grid densification and costs less computing time. Parallel simulation scheme is described by pseudo codes. The spatial parallelism on PC cluster makes it a promising method for geo-science numerical computing. Parallel computation shows that the message passing between different CPUs are composed of the subdomain boundary information and need a considerable communication. The excellent parallelism speedup can be achieved through reasonable subdomain division and fast network connection.

Keywords: Seismic simulation; Finite difference; Cluster computing.

1 Introduction

Seismic propagation simulation in complex media is widely used in oil/gas exploration, earthquake prediction and many other fields. The governing equation is a set of hyperbolic Partial Differential Equations (PDE) and is commonly modeled by numerical method, such as Finite Difference Method (FDM). The model size in seismic exploration is very large and can be several kilometers in each side. In the seismic inversion problems, the seismic propagation has to be simulated iteratively, which are extremely computationally intensive. The realistic-sized seismic propagation simulation in anisotropic media cost too much memory and time to run on a single-processor machine. Many efforts have been made to speed up the simulation, including algorithm improvement and parallel computing.

Early researches on FDM for elastic wave modeling in complex media include Alterman and Karal[1], Boore[2], Alford[3], Kelly[4], Virieux[5,6] et al. Levander[7] applied fourth-order approximation in space to the P-SV scheme. Dablain[8] gave high order finite difference (FD) scheme to save memory and reduce computational costs. Graves[9] presented a 3D fourth-order velocity-stress scheme with effective material parameters. Conventional FDM is performed on regular Cartesian grids and has a "staircase fashion" boundary at curved interface, which lead to spurious diffractions in media with complex geometry and heterogeneous elastic constants. In addition, variations of local physical parameters require fine regular grids of whole model field, which greatly increases computation costs. G.H. Shortley[10] firstly studied the FDM of Laplace equations with irregular grid. Jastram and Tessmer[11], Falk [12] et al developed irregular grid FDM on staggered grid. Tessmer[13], Hestholm and Ruud[14] simulated curved surface with deformed rectangular grid. Mufti[15] researched the problems for "staircase fashion" interfaces. Oprsal[16] et al studied irregular grid for heterogeneous media. Pitarka[17] developed a rectangular irregular grids FD scheme. Nordström[18] derived high order FD with deformed grids in curvilinear coordinate. Sun[19] studied the non-rectangular irregular grids FDM for elastic wave propagation in complex media.

Earthquake modeling with parallel FDM was studied by Olsen, Archuleta and Matarese[20]. Minkoff[21] gave a spatial parallelism scheme for elastic wave propagation in isotropic media. This paper presents a irregular grids parallel FDM scheme. Elastic wave equations are discretized on staggered irregular grids. Deformed non-rectangular grids are used at complex geometrical boundaries and in vicinities of material discontinuities. Non-physical diffractions from rough interfaces in regular grids and rectangular irregular grids FD can be greatly weaken without increasing the grid density. The spatial decomposition can be carried out according to the earth structure geometries. The parallel simulation codes are implemented by MPI and c language on PC clusters. The parallel computation of synthetic models shows that the parallel implementation has the same accuracy as the serial code. Reasonable subdomain division can achieve high parallelism speedup.

2 Formulation

The formulations described in abbreviated form here are derived and discussed in detail in [19]. More classical reference to elastic wave propagation can be found in [22]. The 3D elastic wave equation in a 3D anisotropic media is solved in this paper.

2.1 Governing Equations

Velocity-stress elastic wave equations in Cartesian coordinate are:

$$\rho \dot{V} = D \cdot T \ . \tag{1}$$

$$\dot{T} = C \cdot D^{T} \cdot V \ . \tag{2}$$

where $\rho(x)$ is the media density. The velocity vector $V(x,t)$, stress vector $T(x,t)$, differential operator matrix D and orthorhombic anisotropic elastic coefficient matrix C are defined as:

$$\boldsymbol{V}^{\mathrm{T}} = (v_x, v_y, v_z) \, . \tag{3}$$

$$\boldsymbol{T}^{\mathrm{T}} = (\tau_{xx}, \tau_{yy}, \tau_{zz}, \tau_{yz}, \tau_{xz}, \tau_{xy}) \, . \tag{4}$$

$$\boldsymbol{D} = \begin{bmatrix} \frac{\partial}{\partial x} & 0 & 0 & 0 & \frac{\partial}{\partial z} & \frac{\partial}{\partial y} \\ 0 & \frac{\partial}{\partial y} & 0 & \frac{\partial}{\partial z} & 0 & \frac{\partial}{\partial x} \\ 0 & 0 & \frac{\partial}{\partial z} & \frac{\partial}{\partial y} & \frac{\partial}{\partial x} & 0 \end{bmatrix} . \tag{5}$$

$$\boldsymbol{C} = \begin{bmatrix} c_{11} & c_{12} & c_{13} & 0 & 0 & 0 \\ & c_{22} & c_{23} & 0 & 0 & 0 \\ & & c_{33} & 0 & 0 & 0 \\ & & & c_{44} & 0 & 0 \\ & & & & c_{55} & 0 \\ \text{symmetric} & & & & & c_{66} \end{bmatrix} . \tag{6}$$

2.2 Finite Difference Formulas

The first-order partial differential operator is discretized on staggered non-rectangular irregular grids. Grid configurations in Cartesian x-z coordinate are shown in Fig. 1. The wave field values at the relevant nodes m, n and l can be extended as Taylor series, which consists of the spatial nodes intervals and the wave field gradients. The first-order spatial differential operator for 3D problems in x axis is obtained from the Taylor series.

$$\frac{\partial \phi}{\partial x} \cong \sum_{q=1}^{12} N_q^x(x_{m-1},\ldots,x_{l+2}; y_{m-1},\ldots,y_{l+2}; z_{m-1},\ldots,z_{l+2}) \cdot \phi_q \, . \tag{7}$$

Here N_q^x is a function related to coordinate values of the relevant spatial nodes. ϕ_q is the wave field value at the q th node. The subscript $q=1,\ldots,12$ indicates the spatial index $m-1,\ldots,m+2, n-1,\ldots,n+2, l-1,\ldots,l+2$ respectively. It is straightforward to get $\frac{\partial \phi}{\partial y}$ and $\frac{\partial \phi}{\partial z}$ in the same way.

$$\frac{\partial \phi}{\partial y} \cong \sum_{q=1}^{12} N_q^y(x_{m-1},\ldots,x_{l+2}; y_{m-1},\ldots,y_{l+2}; z_{m-1},\ldots,z_{l+2}) \cdot \phi_q \, . \tag{8}$$

$$\frac{\partial \phi}{\partial z} \cong \sum_{q=1}^{12} N_q^z(x_{m-1},\ldots,x_{l+2}; y_{m-1},\ldots,y_{l+2}; z_{m-1},\ldots,z_{l+2}) \cdot \phi_q \, . \tag{9}$$

The first derivative operator with respect to time at time $j\Delta t$ is as following.

$$\frac{\partial \phi^j}{\partial t} \cong \frac{\phi^{j+1} - \phi^j}{\Delta t}. \tag{10}$$

The detailed derivation and stability analysis can be found in [19].

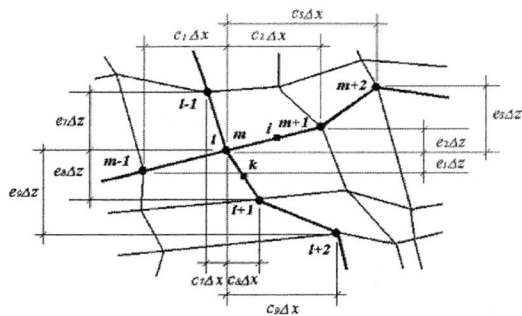

Fig. 1. The irregular grids FDM at *x-z* plane. Mid point of (*m*, *l*) and (*m*+1, *l*) is grid i. Mid point of (*m*, *l*) and (*m*, *l*+1) is grid *k*. The grid intervals are not equal to each other

2.3 Absorbing Boundary Condition

Higdon's (Higdon [23]) absorbing boundary condition is applied to the outside boundaries of the irregular grid to minimize the reflections. The absorbing boundary condition operator for elastic wave propagation problems is

$$B = \prod_{j=1}^{m} \left(c_j \frac{\partial}{\partial t} - \alpha \frac{\partial}{\partial x} \right) \tag{11}$$

The *j*th operator is perfectly absorbing for the P wave traveling at angles of incidence $\pm \cos^{-1} c_j$ and for the S wave traveling at angles of incidence $\pm \cos^{-1}(\frac{\beta}{\alpha} c_j)$. The absorbing boundary condition operator for spatial step Δx_k can be approximated as:

$$D_k(E_x, E_t^{-1}) = \prod_{j=1}^{m} \left[c_j \left(\frac{I - E_t^{-1}}{\Delta t} \right) ((1-a)I + aE_x) - \alpha \left(\frac{E_x - I}{\Delta x_k} \right) ((1-b)I + bE_t^{-1}) \right]. \tag{12}$$

Forward shift operators in *x* and *t* are:

$$E_x f_{m,n,k}^j = f_{m+1,n,k}^j \qquad E_t f_{m,n,k}^j = f_{m,n,k}^{j+1}. \tag{13}$$

Parameters *a* and *b* have different values for different FDM schemes.

3 Parallel Seismic Propagation Simulation

Realistic-sized seismic propagation simulation costs huge amount of memory and computation time, which is hard to run on single-processor computer. Shared memory multiple processor machines have been used to solve computationally intensive problems. The connection between the processors and memory is through some form of interconnection network. This kind of specially designed computer system is very expensive and only large universities and companies can afford it. The emergence of PC cluster provides a novel way for large-scale science and engineering computation. Each computer consists a processor and local memory and is connected through an interconnection network. Each processor sends messages to other processors for data communications. Although the bandwidth and latency speeds on clusters tend to run 1-3 orders of magnitude slower than on shared memory multiple processor machines, the lower costs of building clusters is still so attractive compared with the special designed machines.

3.1 Subdomain Decomposition

Message passing between multiple processors introduces additional time costs to the whole simulation problem. Spatial decomposition for explicit FDM scheme provides a way to minimize the communication and achieve a good speedup. In spatial parallelism, the physical problem domain is usually split into cubic-shaped subdomains in one, two or three dimensions. Each processor solves its own subdomain problem and communicates the boundary values with other processors. The flexible irregular girds FDM scheme in this paper makes it possible to divide physical problem into irregular shaped subdomains according to the geological structures. A 2D decomposition is illustrated in Fig. 2. A locally uniform media subdomain is produced by the irregular FDM scheme and is assigned to each processor. Such uniform subdomain problem can be solved with its own dispersion and stability condition in parallelism.

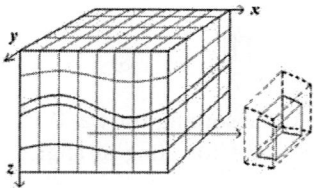

Fig. 2. 3D decomposition for spatial parallelism with irregular grid FDM. Single processor subdomain (with ghost cells shown as dashed lines) taken from global irregular grid domain is shown. The curved geological structure is considered in the spatial decomposition

Each processor calculates the finite difference solutions for its subdomain independently, except for the boundary value communication. Padded subdomains (ghost cells) of memory are created. The irregular grids FDM needs tow extra planes of memory to be created on each face of the subdomain. The ghost cells of the subdomains at physical model boundary have to be processed by the absorbing boundary contition. Otherwise, the processor just uses communicated boundary values in the ghost cells.

3.2 Pseudo Codes

The spatial parallelism codes are written in c language with MPI library. The model parameters such as size, velocities, elastic coefficients, source and receiver position are read from the parameter file by the master process. The master process verifies dispersion and stability conditions and broadcasts the model parameters to each slave process. After receiving the parameters, each process begins to create its temporary arrays and solve the subdomain problem. The subdomain boundary values are sent to neighboring processors in the Cartesian dimensions and the grid values of velocity and stress are updated after the message passing. The grid values on the physical model boundaries are processed by the absorbing boundary condition after the grid value updating. The master process collects wavefield snapshot and synthetic seismic records after certain time-step loop. Then the master process writes the synthetic seismogram to result files. The pseudo spatial parallelism codes are given as follows.

Pseudo codes of the spatial parallel seismic propagation simulation with MPI library in c language.

```
Establish the MPI environment;
  Master process read the dimension of dynamically cre-
ated array from parameter file;
  Master process broadcast the dynamic array dimension
to all processes;
  All processes create the dynamic array;
  Master process read computation parameters from
parameter file;
  Master process broadcast parameters to all processes;
  Each process begin to simulate:
    Check the FDM stability condition;
    Create temporary arrays;
    Convert the elastic coefficients for anisotropic
media;
    Begin the time-step loop:
      Compute the source amplitude;
      Communicate boundary values with neighboring
processors;
      Update grid velocity and stress;
      Process physical model boundary values with
absorbing boundary condition;
      Send grid velocity and stress to master process
if snapshot and synthetic records need to be write out;
    End the time-step loop;
    Free the temporary arrays;
  Free the dynamically created arrays;
  End the simulation;
Terminate the MPI environment.
```

4 Numerical Examples

Although a much larger real date example can be used to test the limits of the parallel algorithm, it is more important to illustrate code accuracy and efficiency by a relatively small typical example. An anisotropic media model with curved layer is used to valid parallel irregular grids FDM codes. The model size is 200×100×100 meters in x-y-z coordinate and the thin layer is located between the depth of z=30 meters and z=80 meters (Fig. 3 (a)). A total number of 100×50×50 grid points are used. The irregular grid configuration at x-z cross section is shown in Fig. 3 (b). The explosive Gauss source with 50Hz peak frequency is located at (25m, 50m, 20m). A receiver is located at (80m, 50m, 20m). Time step is 0.0001second and the total computation time is 0.2 second. The elastic parameters and densities are listed in Table 1.

The grid discretion is more flexible than regular grid scheme and can be make according to the model geometrical structures. The synthesis seismograms at the receiver are computed by serial code and spatially parallel code respectively. The parallel simulation program was run on a 1Gbps Ethernet connected PC cluster with a 2×4 processor grid. Each grid computer has two 2.4 GHz Xeon CPUs and 1Gbyte memory. The 3D model simulation costs 3719.68 seconds on single processor and 850.02 seconds on 2×4 processor grid. The parallelism efficiency for this problem on PC cluster approaches to 54.7%. Fig. 4 shows the seismograms simulated by parallel method and the serial method. The snapshot contours of velocity in x direction (Vx) at y=50 meters cross section are given in Fig. 5. The Vx snapshots at 0.12 seconds computed by serial and parallel codes are shown in Fig. 6. In Fig.5 and Fig.6 the wave field features computed by parallel codes are almost the same as by the serial codes at different time. However, the computation costs are greatly cut down.

Table 1. The elastic coefficients and densities of the 3D model with curved layers

Parameters	Layer I	Layer II	Layer III
ρ (kg/m^3)	2211	1000	2000
c_{11} c_{22} c_{33} (GPa)	12.0	2.25	18.0
c_{12} c_{13} c_{23} (GPa)	5.0	2.25	6.44
c_{44} c_{55} c_{66} (GPa)	3.5	0	5.78

All the figures show that the parallel simulation results are identical to the ones form the serial simulation code. The wave field values computed by the proposed method have the same accuracy no matter the serial or parallel implementation. But the message passing costs a lot of resources in spatially parallelism scheme. The balancing between message passing and wave field computation is a key factor for improving parallel speedup. Small subdomain division has a very high message-passing/computing ratio, which weakens the parallel speedup effect. Here a 25×50×25 block is assigned to each processor.

Although PC cluster has many attractive features, such as low price, high flexibility and high expandability, the interconnection network bandwidth and latency are still a bottleneck. Even if the 1Gbps Ethernet connected PC cluster, the typical latency is still at the 10^{-3} s level. Bandwidth and latency on distributed shared memory PC clusters tend to run 1-3 orders of magnitude slower than on the shared memory multiprocessor system. Cluster computing is hard to achieve very high parallel efficiency. The leading edge research and application of Fibre Channel network and Infiniband provide many promising directions for the future high performance computing.

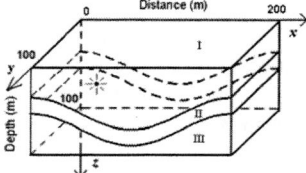 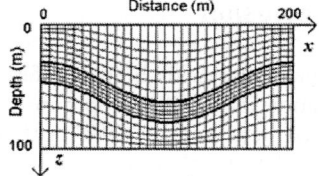

Fig. 3. (a) (left figure) 3D model with curved layer in Cartesian coordinate; (b) (right figure) Irregular grids discretization at x-z cross section

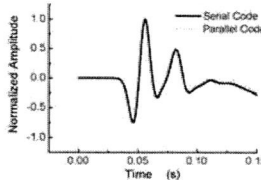

Fig. 4. Synthetic seismograms comparison for the parallel and serial seismic propagation simulation codes. The seismograms are synthesized at the receiver location (80m, 50m, 20m)

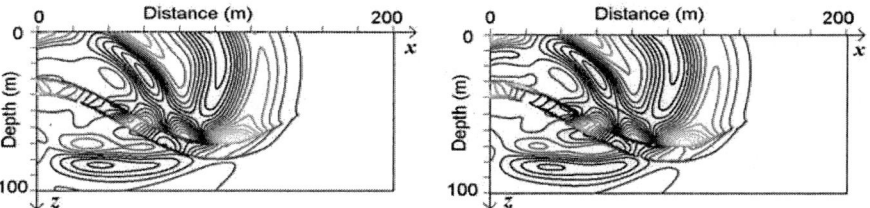

Fig. 5. Snapshot contours of velocity in x direction at the y=50 meters cross-section. The snapshot time is 0.066 seconds. (a) (left figure) serial code snapshot contour (b) (right figure) the parallel code snapshot contour

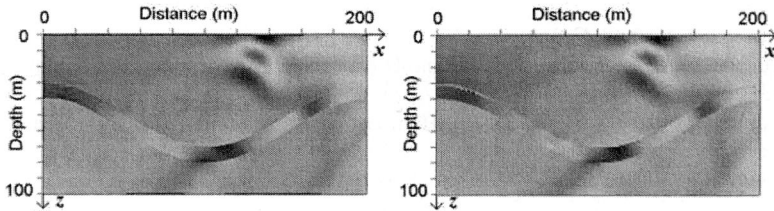

Fig. 6. Snapshot of velocity in x direction at the y=50 meters cross-section. The snapshot time is 0.12 seconds. (a) (left figure) serial code snapshot (b) (right figure) the parallel code snapshot

5 Conclusion

Non-rectangular grids Finite Difference Method is developed to simulate the seismic propagation in 3D anisotropic media model with complex earth structure geometry. Staggered grid scheme and spatial parallelism are used to provide computation speedup. The parallel codes are implemented in c language with Message Passing libraries. The physical model is divided into subdomains for each processor with more flexibility. Model geometries are fully considered during the spatial decomposition procedure by the proposed method. The seismic simulation program was run on PC cluster composed of Xeon CPUs. High bandwidth and low latency are provided by 1 Gbps Ethernet interconnections. In the numerical example, the message passing and computation are relatively balanced with a $25 \times 50 \times 25$ subdomain block size. A 54.7% parallel efficiency can be achieved on a 2×4 processor grid for this model.

Acknowledgement

The works described in this paper are supported by National Natural Science Foundation of China (Grant No.10402015, Grant No.60473101), China Postdoctoral Foundation (No.2004035309) and the National Key Basic Research and Development (973) Program of China (Grant No. 2004CB318205).

References

1. Alterman Z., Karal, F. C. Jr.: Propagation of Elastic Waves in Layered Media by Finite Difference Methods. Bull. Seism. Soc. Am., Vol. 58(1) (1968) 367-398
2. Boore, D. M.: Finite Difference Methods for Seismic Wave Propagation in Heterogeneous Materials. In: Bolt, B. A. (eds): Computational physics. Academic Press. Inc., (1972)
3. Alford, R. M., Kelly, K. R. and Boore, D. M.: Accuracy of Finite-Difference Modeling of the Acoustic Wave Equation. Geophysics, Vol. 39 (1974) 834-842
4. Kelly, K. R., Ward, R. W. and Treitel, S. et al.: Synthetic Seismograms, a Finite Difference Approach. Geophysics, Vol. 41(1976) 2-27
5. Virieux, J.: SH-Wave Propagation in Heterogeneous Media: Velocity-Stress Finite-Difference Method. Geophysics, Vol. 49(1984) 1933-1942
6. Virieux, J.: P-SV Wave Propagation in Heterogeneous Media: Velocity-Stress Finite-Difference Method. Geophysics, Vol. 51(1986) 889-901
7. Levander, A. R.: Fourth-Order Finite-Difference P-SV Seismograms. Geophysics, Vol. 53(1988) 1425-1436
8. Dablain, M. A.: The Application of High-Order Differencing to the Scalar Wave Equation. Geophysics, Vol. 51(1986) 54-66
9. Graves, R. W.: Simulating Seismic Wave Propagation in 3D Elastic Media Using Staggered-Grid Finite Differences. Bull. Seism. Soc. Am., Vol. 86(1996) 1091-1106
10. Shortley, G. H. and Weller, R.: Numerical Solution of Laplace's Equation. J. Appl. Phys., Vol. 9(1938) 334-348
11. Jastram, C. and Tessmer, E.: Elastic Modeling on a Grid with Vertically Varying Spacing. Geophys. Prosp., Vol. 42(1994) 357-370

12. Falk, J., Tessmer, E. and Gajevski, D.: Tube Wave Modeling by the Finite-Differences Method with Varying Grid Spacing. Pageoph, Vol. 14(1996) 77-93
13. Tessmer, E., Kosloff, D. and Behle, A.: Elastic Wave Propagation Simulation in the Presence Surface Topography. Geophys. J. Internat. , Vol.108(1992) 621-632
14. Hestholm, S. O. and Ruud, B. O.: 2-D Finite-Difference Elastic Wave Modeling Including Surface Topography. Geophys. Prosp., Vol. 42(1994) 371-390
15. Mufti, I. R.: Large-Scale Three-Dimensional Seismic Models and Their Interpretive Significance. Geophysics, Vol.55(9) (1990) 1166-1182
16. Ivo. O. and Jiri, Z.: Elastic Finite-Difference Method for Irregular Grids. Geophysics, Vol. 64(1) (1999) 240-250
17. Pitarka, A.: 3D Elastic Finite-Difference Modeling of Seismic Motion Using Staggered Grids With Nonuniform Spacing. Bull. Seism. Soc. Am. , Vol. 89 (1) (1999) 54-68
18. Nordström, J. and Carpenter, M. H.: High-Order Finite Difference Methods, Multidimensional Linear Problems and Curvilinear Coordinates. Journal of Computational Physics, Vol. 173 (2001) 149-174
19. Sun, W.T., Yang, H.Z.: Seismic Propagation Simulation in Complex Media with Not-Rectangular Irregular-Grid Finite-Difference. ACTA MECHANICA SINICA, Vol. 20(3) (2004) 299-306
20. Olsen, K., Archuleta, R. and Matarese, J.: Three-Dimensional Simulation of a Magnitude 7.75 Earthquake on the San Andreas Fault. Science, Vol. 270 (1995) 1628-1632
21. Minkoff, S.: Spatial Parallelism of a 3D Finite Difference Velocity-Stress Elastic Wave Propagation Code. SIAM J. Sci. Comput., Vol. 24(1) (2002) 1-19
22. Aki, K. and Richards, P.: Quantitative Seismology, Theory and Methods. W. H. Freeman, San Francisco, CA (1980)
23. Higdon, R L.: Absorbing Boundary Conditions for Difference Approximations to the Multidimensional Wave Equation. Math. Comp., Vol. 47 (1986) 437-459

The Web Replica Allocation and Topology Assignment Problem in Wide Area Networks: Algorithms and Computational Results

Marcin Markowski and Andrzej Kasprzak

Chair of Systems and Computer Networks,
Wroclaw University of Technology,
Wybrzeze Wyspianskiego 27, 50-370 Wroclaw, Poland
marcin.markowski@pwr.wroc.pl
andrzej.kasprzak@pwr.wroc.pl

Abstract. This paper studies the problem of designing wide area networks. In the paper the web replica allocation and topology assignment problem with budget constraint is considered. The goal is select web replica allocation at nodes, network topology, channel capacities and flow routes in order to minimize average delay per packet and web replica connecting cost at nodes subject to budget constraint. The problem is NP-complete. Then, the branch and bound method is used to construct the exact algorithm. Also an approximate algorithm is presented. Some computational results are reported. Based on computational experiments, several properties of the considered problem are formulated.

1 Introduction

The Wide Area Network (WAN) designing process incorporates such aspects as topology assignment, capacity and flow assignment and resource allocation. Because of huge number of users connected to the network, resources often have to be replicated. The optimal network configuration is very important – it allows minimize the cost of the network and guarantee the quality of service in the WAN. Proper resource replication allows to minimize the traffic rate in the network what ensures an efficiency and reliability of the wide area network. The flow demands in the network are changing in time because new users are joining the network and new resources (servers) are added. That is why the optimization of the network parameters (such as topology, capacity, resource allocation) must be performed regularly for the existing network. In the designing and optimization process different criteria of the network quality are taking into account. Very useful criteria are the investment cost of the network and the quality of service in the network. In the paper the exact and approximate algorithms for simultaneous assignment of web replica allocation, network topology, channels capacity and flow assignment are considered. We use the combined criterion composed of the quality of service in the network and of the investment cost represented by the connecting cost of replicas at nodes. As the quality of service we assume the total average delay per packet in the network given by the Kleinrock's formula [1]. The considered problem is formulated as follows:

given: user allocation at nodes, for every web server the set of nodes to which web server may be connected, number of replicas, budget of the network, traffic requirements user-user and user-server, set of potential channels and their possible capacities and costs (i.e. cost-capacity function)
minimize: linear combination of the total average delay per packet and the connecting costs of replicas at nodes
over: network topology, channel capacities, multicommodity flow (i.e. routing), web replica allocation
subject to: multicommodity flow constraints, channel capacity constraints, budget constraint, web replica allocation constraints.

We assume that channels' capacities can be chosen from discrete sequence defined by ITU-T (International Telecommunication Union – Telecommunication Sector) recommendations. Then, the formulated above web replica allocation and topology assignment problem is NP-complete [2,3].

In the literature there are some algorithms for wide area networks optimization. They concentrate on topology assignment [4,5], capacity and flow assignment [1,2,6] or resource allocation and replication [3,7,8]. Some algorithms solving topology assignment problem can be found in [4,5]. In the papers [7,8] there are some algorithms for optimal replica allocation in the network when the network topology is known. The considered problem is more general than the problems considered in the literature because it combines simultaneous topology assignment and replica allocation. In the literature such formulated problem has not been considered yet.

2 Problem Formulation

Consider a WAN consisting of n nodes and b potential channels which may be used to build the network. For each potential channel i there is the set $\overline{C}^i = \{c_1^i,...,c_{s(i)-1}^i\}$ of alternative values of capacities from which exactly one must be chosen if the i-th channel was chosen to build the WAN. Let d_j^i be the cost of leasing capacity c_j^i [\$/month]. Let $c_{s(i)}^i = 0$ for $i = 1,...,b$. Then $C^i = \overline{C}^i \cup \{c_{s(i)}^i\}$ be the set of alternative capacities from among which exactly one must be used to channel i. If the capacity $c_{s(i)}^i$ is chosen then the i-th channel is not used to build the wide area network.

Let x_j^i be the decision variable which is equal to one if the capacity c_j^i is assigned to channel i and x_j^i is equal to zero otherwise. Since exactly one capacity from the set C^i must be chosen for channel i, then the following condition must be satisfied:

$$\sum_{j=1}^{s(i)} x_j^i = 1 \text{ for } i = 1,...,b. \tag{1}$$

Let $W^i = \{x_1^i,...,x_{s(i)}^i\}$ be the set of variables x_j^i which correspond to the i-th channel. Let X_r' be the permutation of values of all variables x_j^i for which the condition (1) is satisfied, and let X_r be the set of variables, which are equal to one in X_r'.

Let K denotes the total number of servers, which must be allocated in WAN and let LK_k denotes the number of replicas of k-th web server. Let M_k be the set of nodes to which k-th web server (or replica of k-th server) may be connected, and let $e(k)$ be the number of all possible allocation for k-th server. Since only one replica of server may be allocated in one node then the following condition must be satisfied

$$LK_k \le e(k) \text{ for } k=1,...,K. \tag{2}$$

Let y_{kh} be the decision binary variable for k-th web server allocation. y_{kh} is equal to one if the replica of k-th web server is connected to node h, and equal to zero otherwise. Since LK_k replicas of k-th web server must be allocated in the network then the following condition must be satisfied

$$\sum_{h \in M_k} y_{kh} = LK_k \text{ for } k=1,...,K. \tag{3}$$

Let Y_r be the set of all variables y_{kh}, which are equal to one. The pair of sets (X_r, Y_r) is called a selection. Let $\Re$ be the family of all selections. X_r determines capacities of channels and Y_r determines the replicas allocation at nodes of WAN.

Let $T(X_r, Y_r)$ be the minimal average delay per packet in WAN in which values of channel capacities are given by X_r and traffic requirements are given by Y_r (depending on web replica allocation). $T(X_r, Y_r)$ can be obtained by solving a multi-commodity flow problem in the network [1]. Let $A(Y_r)$ be the cost of connecting all web replicas at nodes. Let $Q(X_r, Y_r)$ be linear combination of the total average delay per packet and the connecting costs of web servers at nodes

$$Q(X_r, Y_r) = \alpha T(X_r, Y_r) + \beta A(Y_r) \tag{4}$$

where α and β are the positive coefficients; $\alpha \in (0,1)$ and $\beta = 1-\alpha$.

Let B be the budget (maximal feasible leasing capacity cost of channels) of WAN. Then, the considered web replica allocation and topology assignment problem in WAN can be formulated as follows

$$\min_{(X_r, Y_r)} Q(X_r, Y_r) \tag{5}$$

subject to

$$(X_r, Y_r) \in \Re \tag{6}$$

$$d(X_r) = \sum_{x_j^i \in X_r} x_j^i d_j^i \le B \tag{7}$$

3 The Branch and Bound Algorithm

Assuming that $LK_k = 1$ for $k=1,...,K$ and $C^i = \overline{C}^i$ for $i=1,...,b$, the problem (5-7) is resolved itself into the "host allocation, capacity and flow assignment problem". Since the host allocation, capacity and flow assignment problem is NP-complete [3] then the problem (5-7) is also NP-complete as more general. Then, the branch and bound method can be used to construct the exact algorithm for solving the considered problem. The detailed description of the calculation scheme of the branch and bound method may be found in the paper [9].

Starting with the initial selection (X_1, Y_1) we generate a sequence of selections. The variables from the sets X_1 or Y_1 are called normal. The variables, which do not belong to X_1 or Y_1 are called reverse. A replacement of any normal variable by the reverse variable is called complementing. The generation of new selection involves the choice of the certain normal variable from X_1 or Y_1 for complementing; it is called the branching rules. The choice of the normal and reverse variables is based on local optimization criterion. For each selection (X_r, Y_r) we calculate lower bound to check the possibility of the generation of the selection (X_s, Y_s) with less value of a criterion Q than already found. If such (X_s, Y_s) does not exist, we abandon (X_r, Y_r) and all its possible successors and next backtrack to the predecessor from which (X_r, Y_r) was generated. If new selection (X_s, Y_s) is generated from (X_r, Y_r) by complementing the normal variable by the reverse variable, then we constantly fix the reverse variable. It means that this reverse variable cannot be complemented by any other in every possible successor of (X_s, Y_s). If we backtrack from (X_s, Y_s) to (X_r, Y_r) by reverse variable of certain normal variable from (X_r, Y_r) we momentarily fix this normal variable. So, for each (X_r, Y_r) we constantly fix a set F_r and momentarily fix the set F_r^t. The reverse variables in F_r are constantly fixed. Each momentarily fixed variable in F_r^t is the reverse variable abandoned during backtracking process. Variables which do not belong to F_r or F_r^t are called free in (X_r, Y_r). If we want backtrack from (X_1, Y_1) then the algorithm terminates.

3.1 Branching Rules

The purposes of branching rules is to find the normal variable from the selection (X_r, Y_r) for complementing and generating a successor (X_s, Y_s) of the selection (X_r, Y_r) with the least possible value of the criterion function (4). We can choose a variable x_j^i or a variable y_{kh}.

The choice criterion on variables x_j^i may be formulated in the same way like in the classical topology assignment problem, because changing the capacity of any channel does not change $A(Y_r)$. Then, in this case we can use the criterion given in the form of the following theorem [5].

Let γ be the total average packet rate from external sources, and let ε be the minimal feasible difference between the capacity and the flow in each channel.

Theorem 1. Let $(X_r, Y_r) \in \Re$. If the selection (X_s, Y_s) is obtained from the selection (X_r, Y_r) by complementing the variable $x_j^i \in X_r$ by the variable $x_l^i \in X_s$ then $Q(X_s, Y_s) \le Q(X_r, Y_r) - \Delta_{jl}^{ir}$, where

$$\Delta_{jl}^{ir} = \begin{cases} \dfrac{\alpha}{\gamma}\left(\dfrac{f_r^i}{c_j^i - f_r^i} - \dfrac{f_r^i}{c_l^i - f_r^i}\right) & \text{if } c_l^i \ge f_r^i \\ \dfrac{\alpha}{\gamma}\left(\dfrac{f_r^i}{c_j^i - f_r^i} - \dfrac{1}{\varepsilon}\left(\max_{x_z^i \in W^i - F_r} c_z^i\right)\right) & \text{otherwise} \end{cases} \quad (8)$$

and f_r^i is the flow in the i-th channel obtained by solving the multicommodity flow problem for network topology and channel capacities given by the selection X_r.

To formulate the choice criterion on variables y_{kh} we use the following theorem.

Theorem 2. Let $(X_r, Y_r) \in \Re$. If the selection (X_s, Y_s) is obtained from the selection (X_r, Y_r) by complementing the variable $y_{kh} \in X_r$ by the variable $y_{km} \in X_s$ then $Q(X_s, Y_s) \le Q(X_r, Y_r) - \delta_{hm}^k$, where

$$\delta_{hm}^k = \begin{cases} \dfrac{\alpha}{\gamma}\sum_{x_j^i \in X_r}\dfrac{f_r^i}{x_j^i c_j^i - f_r^i} - \dfrac{\alpha}{\gamma}\sum_{x_j^i \in X_r}\dfrac{\tilde{f}^i}{x_j^i c_j^i - \tilde{f}^i} + \beta(a_{kh} - a_{km}) \\ \qquad \text{if } \tilde{f}^i < x_j^i c_j^i \text{ for } x_j^i \in X_r \\ \infty \qquad \text{otherwise} \end{cases} \quad (9)$$

$\tilde{f}^i = f_r^i - f_{ik}' + f_{ik}''$, f_{ik}' and f_{ik}'' are parts of flow in i-th channel; f_{ik}' corresponds to the packets sent from all users to those replica of k-th web server which is allocated at node h (before complementing) and from this replica to all users. f_{ik}'' corresponds to the packets exchanged between all users and replica after reallocating replica from node h to node m. a_{kh} is the cost of connecting the server k at node h.

Let $E_r = (X_r \cup Y_r) - F_r$, and let G_r be the set of all reverse variables of normal variables, which belong to the set E_r. We want to choose a normal variable the complementing of which generates a successor with the possible least value of criterion function (4). We should choose such pairs $\{(y_{kh}, y_{km}): y_{kh} \in E_r, y_{km} \in G_r\}$ or $\{(x_j^i, x_l^i): x_j^i \in E_r, x_l^i \in G_r\}$ for which the value of criterion δ_{hm}^k or Δ_{jl}^{ir} is maximal.

3.2 Lower Bound

The lower bound LB_r of the criterion function (4) for every possible successor (X_s, Y_s) generated from the selection (X_r, Y_r) may be obtained by relaxing or omitting some constraints in the problem (5-7). To find the lower bound LB_r we reformulate the problem (5-7) in the following way:

- we assume that the variables x_j^i and y_{kh} are continuos variables,
- we omit the constraint (2),
- we approximate the discrete cost-capacity curves (given by the set C^i) with the lower linear envelope [1]. In this case, the constraint (7) may be relaxed by the constraint $\Sigma d^i c^i \leq B$, where $d^i = \min_{x_j^i \in W^i} (d_j^i / c_j^i)$ and c^i is the capacity of the channel i (continuos variable).

The solution of such reformulated problem gives the lower bound LB_r. To solve the above mentioned reformulated problem we can use the method proposed and described in the paper [3]. In the paper [3], this method has been used to find the lower bound for the host allocation (without web replicas), capacity and flow assignment problem.

4 Approximate Algorithm

The presented exact algorithm involves the initial selection $(X_1, Y_1) \in \Re$ for which the constraints (6) and (7) are satisfied [9]. Moreover, the initial selection should be the near-optimal solution of the problem (5-7). To find the initial selection the following approximate algorithm is proposed. Of course, this approximate algorithm may be also used to design of the WAN when the optimal solution is not necessary.

Step 1. For each k-th web server, choose LK_k nodes from the set M_k such that paths between these nodes and users of k-th web server are shortest in the network. Allocate the k-th web replicas at these nodes. Next, solve the classical topology assignment problem [5]. If this problem has no solution then the algorithm terminates – the problem (5-7) has no solution. Otherwise, perform $Q^* = Q$, where Q is value of criterion (4) obtained by solution the topology assignment problem.

Step 2. Choose the web replica location for which the value of δ_{hm}^k given by expression (9) is maximal. Change this location of web replica on location indicated by expression (9).

Step 3. Solve the topology assignment problem. If the obtained value Q is less than Q^* then perform $Q^* = Q$, and go to step 2. Otherwise, the algorithm terminates. The feasible solution was found. The network topology and web replicas allocation associated with the current Q^* is approximate solution of the problem (5-7).

5 Computational Results

The presented exact and approximate algorithms were implemented in C++ code. Extensive numerical experiments have been performed with these algorithms for many different network topologies and for many possible web replica locations. The experiments were conducted with two main purposes in mind: first, to test the computational efficiency of the algorithms and second, to examine the impact of various parameters on solutions to find properties of the problem (5-7) important from practical point of view.

In order to evaluate the computational efficiency of the exact algorithm many networks were considered. For each network the number of iteration of the algorithm was checked. These number of iterations were compared for all considered networks.

Let D_{max} be the maximal building cost of the network, and let D_{min} be the minimal building cost of the network; the problem (5-7) has no solution for $B < D_{min}$. To compare the results obtained for different wide area networks topologies and for different web replica locations we introduce the normalized budget $NB = ((B - D_{min})/(D_{max} - D_{min})) \cdot 100\%$.

Moreover, let $P^i(NB)$ be the number of iterations of the branch and bound algorithm to obtain the optimal value of Q for normalized budget equal to NB for i-th considered network topology. Let

$$P(u,v) = \frac{1}{Z}\sum_{i=1}^{Z}\left(\sum_{NB\in[u,v]}P^i(NB) \bigg/ \sum_{NB\in[1,100]}P^i(NB)\right)\cdot 100\%$$

be the arithmetic mean of the relative number of iterations for $NB\in [u,v]$ calculated for all considered network topologies and for different web replica locations, where Z is the number of considered wide area networks [4]. Fig. 1 shows the dependency of P on divisions [0%,10%), [10%,20%), ..., [90%,100%] of normalized budget NB. It follows from Fig. 1 that the exact algorithm is especially effective from computational point of view for $NB\in [40\%,100\%]$.

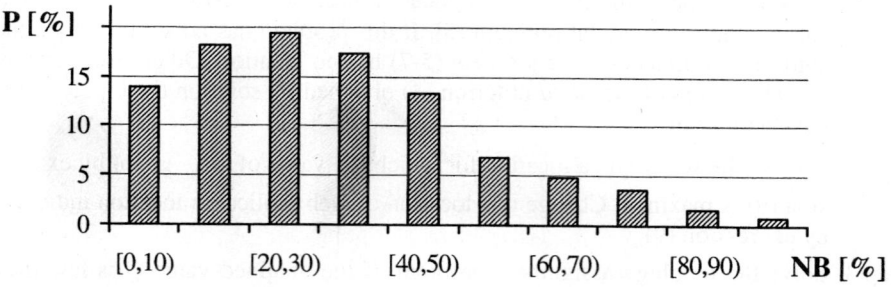

Fig. 1. The dependence of P on normalized budget NB

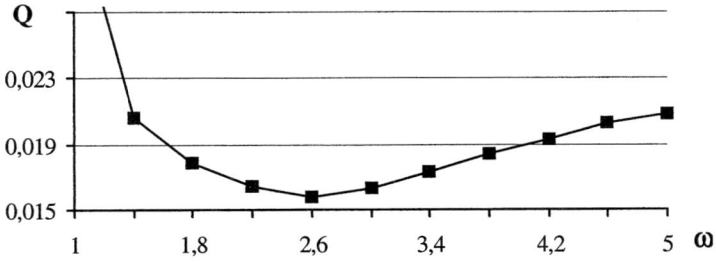

Fig. 2. Dependence Q on average number of web replicas ω

Let ω be the average number of web replicas in the wide area network.

$$\omega = \frac{1}{K} \sum_{k=1}^{K} LK_k$$

Moreover, let $Q(\omega)$ be the optimal value of criterion (4) obtained by solving the problem (5-7) for average number of web replicas equal to ω.

The dependence of the optimal value of the criterion Q on the average numbers of web replicas ω has been examined. Fig. 2 shows the typically dependence of the criterion Q on average number of web replicas ω. It follows from the numerical experiments and from the Fig. 2 that the function $Q(\omega)$ is convex. Moreover, there exists the minimum of the function $Q(\omega)$. The observations following from the computer experiments may be formulated in the form of the below conclusions.

Conclusion 1. The performance quality of the wide area networks depends on the average number of web replicas.

Conclusion 2. There exists such value of average number of web replicas $\bar{\omega}$ for which the function $Q(\omega)$ is minimal.

Let ω_{min} be the average number of replicas in the network, for which the value of $Q(\omega_{min})$ is minimal. Fig. 3 shows the typical dependence of ω_{min} on the numbers of nodes n of the wide area network.

Conclusion 3. The average number of web replicas ω_{min} in the WAN, for which the value of Q is minimal, depends on the number of nodes of the WAN. This dependence is linear.

Because the algorithm for the network topology and web replica allocation problem assume that the numbers of replicas for each web server are given then the properties described in conclusions 2 and 3 allow to simplify the designing process of the networks. In first step of the designing process we may calculate the number of replicas for each web server and in the next step we may solve the web replica allocation and network topology assignment problem using the presented exact algorithm or approximate algorithm.

The typical dependence of the optimal value of Q on budget B for different values of parameter α is presented in the Fig. 4. It follows from Fig. 4 that there exists such

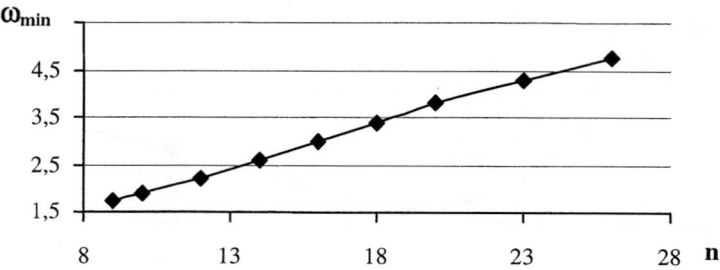

Fig. 3. The dependence of ω_{min} on the number of nodes n

budget B^*, that the problem (5-7) has the same solution for each B greater than or equal to B^*.

Conclusion 4. In the problem (5-7), for $B \geq B^*$ the constraint (7) may be substituted by constraint $d(X_r) \leq B^*$.

This observation shows that the influence of the building cost (budget) on the optimal solution of the problem (5-7) is very limited for greater values of budget B.

The distance between approximate and optimal solution has been also examined. Let Q^{app} be the solution obtained by approximate algorithm and Q^{opt} be the optimal value obtained by the exact algorithm for the problem (5-7). Let r denotes the distance between heuristic and optimal solutions: $r = |Q^{app} - Q^{opt}| / Q^{opt} * 100\%$. The value r shows how the results obtained using the approximate algorithm are worse than the optimal solution. Let

$$R[a,b] = \frac{\text{number of solutions for which } r \in [a,b]}{\text{number of all solutions}} * 100\%$$

denotes number of solutions obtained from approximate algorithm (in percentage) which are greater than optimal solutions more than $a\%$ and less than $b\%$. Fig. 5 shows the dependence R on divisions [0%–1%), [1%–5%), ..., [80%–100%).

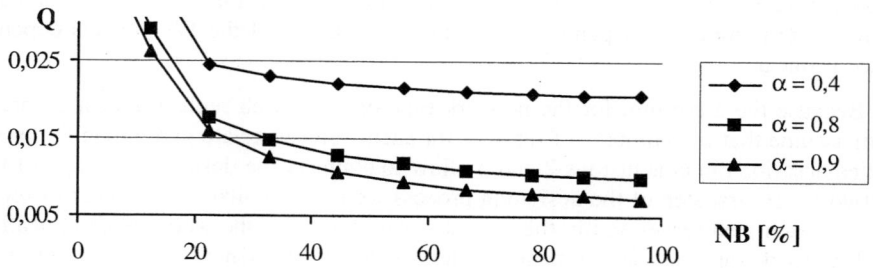

Fig. 4. Typical dependence of Q on B for different values of coefficient α

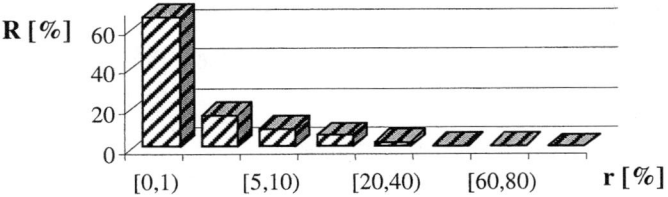

Fig. 5. Difference between approximate and optimal solutions

6 Conclusion

The exact and approximate algorithms for solving the web replica allocation and network topology assignment problem in WAN are presented. The considered problem is more general than the similar problems presented in the literature. It follows from computational experiments (Fig. 5) that about 60% approximate solutions differ from optimal solutions at most 1%. Moreover, we are of the opinion that the network property formulated as conclusions 2 and 3 is very important from practical point of view. This property says that there is the influence of the number of web replicas on performance quality of the WAN. This work was supported by a research project of The Polish State Committee for Scientific Research in 2005-2007.

References

1. Fratta, L., Gerla, M., Kleinrock, L.: The Flow Deviation Method: an Approach to Store-and-Forward Communication Network Design. Networks 3 (1973) 97-133
2. Kasprzak, A.: Topological Design of the Wide Area Networks. Wroclaw University of Technology Press, Wroclaw (2001)
3. Markowski M., Kasprzak A.: An Exact Algorithm for Host Allocation, Capacity and Flow Assignment Problem in WAN. in: Internet Technologies, Application and Societal Impact. Kluwer Academic Publishers, Boston (2002) 73-82
4. Kasprzak A.: Exact and Approximate Algorithms for Topological Design of Wide Area Networks with Non-simultaneous Single Commodity Flows. Lecture Notes in Computer Science 2660 (2003) 799-808
5. Gola M., Kasprzak A.: Exact and Approximate Algorithms for Two-Criteria Topological Design Problem of WAN with Budget and Delay Constraints. Lecture Notes in Computer Science 3045 (2004) 611-620
6. Walkowiak, K.: A New Method of Primary Routes Selection for Local Restoration. Lecture Notes in Computer Science 3042 (2004), 1024-1035
7. Seonho K., Miyoun Y., Yongtae S.: Placement Algorithm of Web Server Replicas. Lecture Notes in Computer Science 3043 (2004) 328-336
8. Qiu L., Padmanabhan V.N., Voelker G.M.: On the Placement of Web Server Replicas. Proceedings of 20th IEEE INFOCOM, Anchorage (2001) 1587–1596
9. Wolsey, L.A.: Integer Programming. Wiley-Interscience, New York (1998)

Optimal Walking Pattern Generation for a Quadruped Robot Using Genetic-Fuzzy Algorithm

Bo-Hee Lee[1], Jung-Shik Kong[2], and Jin-Geol Kim[3]

[1] School of Electrical Engineering, Semyung University, ShinWal-Dong, Chechon, Korea
bhlee@semyung.ac.kr
[2] Dept. of Automation Eng., Inha University, YongHyun-Dong, Nam-Gu, Inchon, Korea
tempus@dreamwiz.com
[3] School of Electrical Eng., Inha University, YongHyun-Dong, Nam-Gu, Inchon, Korea
john@inha.ac.kr

Abstract. In this paper, we described an optimal walking pattern generation using genetic-fuzzy algorithm that can assist walking robot avoid obstacles. In order to walk on an uneven terrain, a quadruped robot must recognize obstacles and take a trajectory that fits with the environment. In that respect, the robot should have two decision-making algorithms that will help its structural limitation. The first algorithm is to generate a body movement that can be related to the movement of the legs, and the other is to make legs' movements smooth in order to reduce jerks. The research presented in this paper, using genetic-fuzzy algorithm, suggests how to find an optimal path movement and smooth walk for quadruped robots. To realize such movement, a relationship between body path and legs trajectory was defined, and a rule based on genetic-fuzzy algorithm was made. From that rule, the optimal legs' trajectory could be determined and the body path generated. As a result, a quadruped robot could walk and avoid obstacles with smoothness.

1 Introduction

The concerns for the human-interactive robots have been increasing rapidly, especially in the field of entertainment. Robots bring some social pleasure to people and also assist disabled persons moving from indoors to outdoors in spite of the existence of uneven terrain and obstacles like stairway. In order to be able to operate in such environment, it is important to design the legs of the human-interactive robot in a way that they recognize the uneven terrain and obstacles. Besides, they should have autonomous capability based on AI technologies such as voice management, motor control microprocessor application technology, etc. Recently, these concepts have been implemented in robots such as ASIMO of Honda [1], AIBO of Sony [2, 3], HOAP-1 of Fujitsu [4], and these types of robots have walking capability, interact with human in the restricted fields, such as in indoor or laboratory environments.

For those kinds of robots, the important issues are how to recognize the obstacle at an unknown environment and how to realize the free walking. To recognize external environment, it is necessary to recognize obstacles and have the ability to transfer its movement to the legs [5, 6]. Path generation is developed continuously. However, for quadruped robots, since the relationship between leg trajectory and body path has not been clearly determined, the optimal leg trajectory still need to be defined. Therefore, in order to guarantee free walking ability of a quadruped, optimal body path generation and smooth leg motion generation were considered in this study.

This paper suggests a fuzzy-genetic algorithm, which implemented in a quadruped robot should make it possible for it to avoid obstacles smoothly and produce optimal leg trajectory simultaneously. To solve this problem, a fuzzy rule base to generate optimal gait was made by genetic algorithm. One step distance and angle are the fundamental elements of the rule base. When the quadruped robot receives a messages informing of obstacles, can find the optimal gait through the fuzzy algorithm. As a result, the optimal path guarantees the robot walking ability. Furthermore, it makes it possible for the robot to avoid obstacles and at the same time reduce jerks of joints. All the processes were simulated with a PC program.

2 A Quadruped Robot Configuration

A quadruped robot, called SERO-V with 14 joints and a main controller was designed. Each of the joints was actuated by an RC-motor and the main controller was an Atmega103 8-bit microprocessor from ATMEL. The main controller exchanged control data with a PC simulator in order to collect some recognition signals and data about each angle from the remote controller. As it is well known, the RC-motor is simply a position-controlling device based on pulse train, which can easily control many actuators simultaneously. Fig. 1 shows the coordinates of SERO-V.

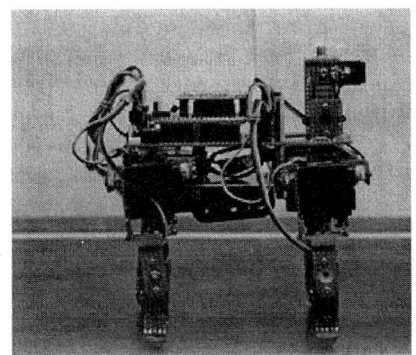
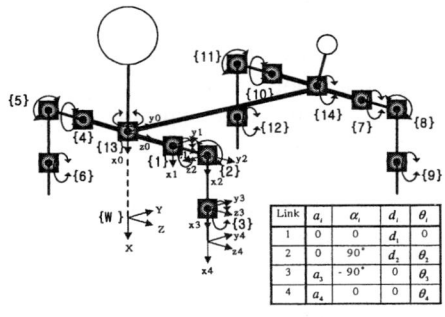

Fig. 1. General configuration and coordinates of SERO V

Since the robot had a symmetric leg structure, the kinematics analysis was done only with the front-left leg, and was latter easily replicated to the others. The inverse kinematics of the front-left leg was as follows. Equations (1), (2), and (3) express the inverse kinematics of front left leg and equation (4) shows the each parameter of equation (1). The parameters P_x, P_y, P_z, a_3, a_4, d_1 and d_2 are the position vectors of the end points and the link parameter θ_2 was the shoulder's pitch joint, θ_3 was the shoulder's yaw joint, and θ_4 was the knee joint. The inverse kinematics of the other legs was obtained by only changing the base coordinates. The basic leg trajectory was generated with the fifth order polynomial equation (5). The parameters were determined with the velocities and accelerations of via points.

$$\theta_2 = \sin^{-1}\left(\frac{-\beta \pm \sqrt{\beta^2 - \alpha\gamma}}{\alpha}\right) \quad (1)$$

$$\theta_3 = \sin^{-1}\left(\frac{2ka_3}{p_x^2 + p_y^2 - a_4^2 + a_3^2 + (p_z - d_1 - d_2)^2}\right) \quad (2)$$

$$\theta_4 = \cos^{-1}\left(\frac{p_x^2 + p_y^2 - a_4^2 + a_3^2 + (p_z - d_1 - d_2)^2}{2a_3 a_4} - \frac{a_3}{a_4}\right) \quad (3)$$

$$\alpha = p_x^2 + p_y^2 \quad \beta = a_4 p_y s_4 \quad \gamma = a_4^2 s_4^2 - p_x^2 \quad (4)$$

$$p(t) = a_0 + a_1 t + a_2 t^2 + a_3 t^3 + a_4 t^4 + a_5 t^5 \quad (5)$$

3 Walking Pattern Generation

3.1 Optimal Body Path Generation

To generate obstacles avoidance path, a robot has to recognize obstacles from external environments. Thus, a PSD sensor and a RC-motor were installed. The robot finds information about the obstacle by using the following method.

At first, a robot establishes a relationship between its body and the obstacle ahead. That relationship can be expressed with two parameters. One of them is the shortest distance from the robot to the obstacle, and the other is the shortest that will make the avoidance of the obstacle possible (Fig. 2).

After the robot determines the appropriate distance and angle necessary to avoid the obstacle, it searches the body projection angle that it will follow in its movement. In this situation, we use the fuzzy algorithm related with direction and angle values. As it is well known, in complex situations, the fuzzy algorithm can successfully find an optimal solution faster [7].

Table 1 represents the rule-base values for the fuzzified input angle related to information regarding the obstacle. In the table, D-4 to D+4 are the fuzzified values of the shortest obstacle avoiding width (w). The values from E-3 to E+3 are the fuzzified distances (d) between the robot and the obstacles.

We chose the 56 rule-bases, and their outputs represent the available body propulsion angles range from –90 to +90 degree. From this rule-base, the robot could search the optimal values for body direction without collision.

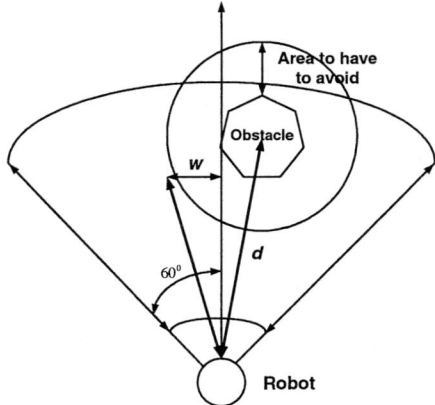

Fig. 2. Obstacle environments

Table 1. Rule-base for w and d

	E-3	E-2	E-1	E0	E+1	E+2	E+3
D-4	-9	-7	-6	-5	-4	-3	-3
D-3	-8	-6	-5	-4	-4	-3	-2
D-2	-7	-5	-4	-3	-3	-2	-2
D-1	-6	-3	-3	-2	-2	-2	-1
D0	0	0	0	0	0	0	0
D+1	5	3	3	2	2	2	1
D+2	6	5	4	3	3	2	2
D+3	7	6	5	4	4	3	2
D+4	9	7	6	5	4	3	3

3.2 Gait Generation

While generating obstacle avoidance path, the robot should satisfy smooth activity simultaneously. The process was divided into two parts. One was to make procedure for the fuzzy rule-base data by genetic algorithm, and the other was for the generation of an optimal leg's trajectory based on the via points information from rule-base.

3.2.1 Rule-Base

Generally, a genetic algorithm is used in searching for an optimum. It is not only efficient in searching optimal solution [8-10], but also dose so without solving any differential equation for the dynamic system. In this study, the genetic algorithm created new generation parameters from the regeneration, crossover and mutation processes (Table 2).

Table 2. Parameters of GA

Parameter	Value	Parameter	Value
Generation No.	250	Population No.	100
Crossover Rate	0.7	Mutation Rate	0.2
String No.	240		
Fitness function	$\dfrac{1}{\sum(\dot{\theta}_{i+1}-\dot{\theta}_i)^2 + \sum(\ddot{\theta}_{i+1}-\ddot{\theta}_i)^2}$		

3.2.2 Optimal Gait Generation

Since the via-points of the legs strongly influence the robot, their selection is an important element. If the robot walks on a pre-defined terrain, the use of predefined trajectory should be very effective. If the routine is not pre-determined, leg trajectory is no more valid. A fuzzy algorithm was suggested to deal with that situation. The fuzzy algorithm is an optimal solution algorithm that imitates human's decision-making pattern. It can find with case the optimal solution quickly [11]. Fig.3 describes all the processes of the proposed algorithm, which was classified into three blocks.

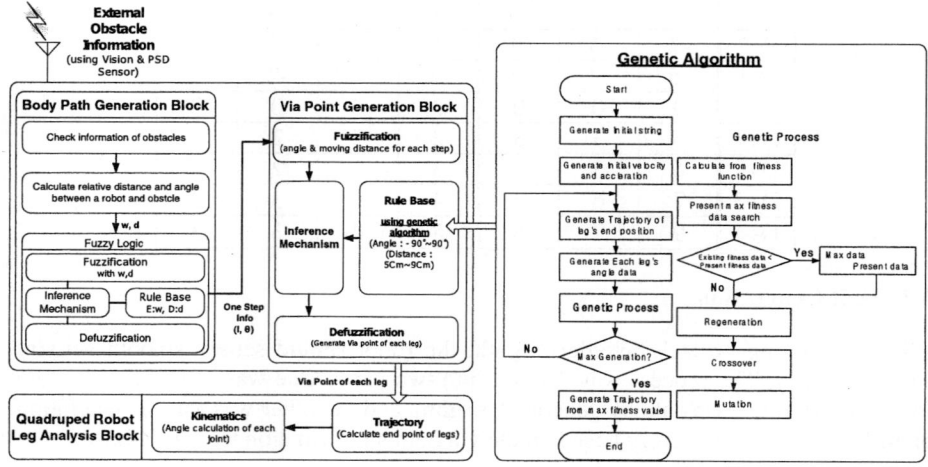

Fig. 3. Flowchart of proposed algorithm

One block was for the body path generation block related to section 3.1. It determined the optimal stride from information about the obstacles it received from sensors. Another block was the via-point generation block, and it generated optimal via-point of legs information from one-step information about the previous block. The last block was the leg analysis block, which made the real joint movement based on the via-point data using leg's trajectory and kinematics.

The sequences to get the via-point were the same. First, the robot collected the body propulsion data such as the relative distance and relative orientation between itself and the obstacles. Then, after the fuzzification, the inference engine produced the fuzzified velocity and acceleration for the via-point with the rule base made by genetic algorithm. The interference engine had 91 if-then-else-logics with the related 7 walking distances and 13 walking angles data using this rule base. After that, the information was converted into real velocity and acceleration at via points by defuzzification process using the center of area. Finally these data were converted into the motor angle through kinematics and trajectory generation algorithm.

3.2.3 Walking Pattern Generation

In making the walking pattern from the method mentioned above, unless the leg's motion was not restricted, it increased to infinitely. Therefore it was impossible to make the walking pattern. Fig. 4 depicts the leg flow in function of time. Because at least three legs of a quadruped robot have to remain on the surface to maintain a stable walking, the determination of the legs movement with respect to the period of time was very important issue. Fig. 4 illustrates the foot position according to walking phase.

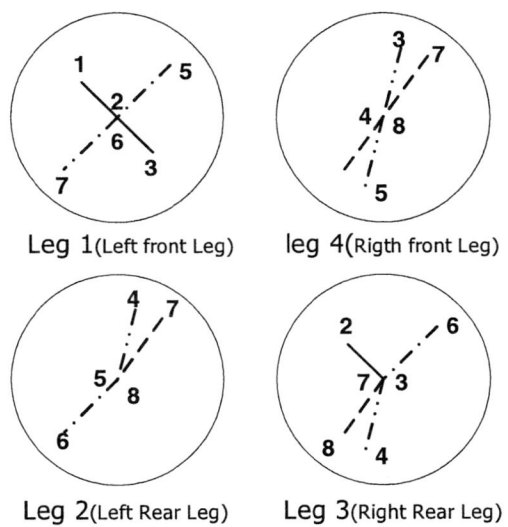

Fig. 4. Plain view of foot position according to walking phase

The number represents a plain view of the position of the sole of the foot with respect to time. Table 3 shows the foot position with respect to walking distance and angle determined by the fuzzy algorithm. Herein x, y, and z represents the moving directional vectors, as shown in Fig. 1. Using the one-step information, the robot could determine its own leg moving positions. The legs were designed to move just a half of body propulsion time. First the legs touched down on the surface, and then propelled the body to the other half. In this case, two legs propelled the body at the same time. One leg remained in swing phase, while the other legs remained on the surface to ensure stability and free motion. This walking procedure was made from phase 1 to phase 8, repeatedly.

Table 3. Foot positions w.r.t. direction and orientation of a robot body

	Leg1			Leg4		
	x	y	z	x	y	z
1	$l_1 \cos\theta_1$	0	$l_1 \sin\theta_1$	$-l_0 \cos\theta_0$	0	$-l_0 \sin\theta_0$
2	0	0	0	0	-0.05	0
3	$-l_1 \cos\theta_1$	0	$-l_1 \sin\theta_1$	$l_2 \cos\theta_2$	0	$l_2 \sin\theta_2$
4	0	-0.05	0	0	0	0
5	$l_3 \cos\theta_3$	0	$l_3 \sin\theta_3$	$-l_1 \cos\theta_1$	0	$-l_1 \sin\theta_1$
6	0	0	0	0	-0.05	0
7	$-l_3 \cos\theta_3$	0	$-l_3 \sin\theta_3$	$l_4 \cos\theta_4$	0	$l_4 \sin\theta_4$
8	0	-0.05	0	0	0	0
	Leg2			Leg3		
	x	y	z	x	y	z
1	0	0	0	0	-0.05	0
2	$-l_2 \cos\theta_2$	0	$-l_2 \sin\theta_2$	$l_1 \cos\theta_1$	0	$l_1 \sin\theta_1$
3	0	-0.05	0	0	0	0
4	$l_3 \cos\theta_3$	0	$l_3 \sin\theta_3$	$-l_2 \cos\theta_2$	0	$-l_2 \sin\theta_2$
5	0	0	0	0	-0.05	0
6	$-l_4 \cos\theta_4$	0	$-l_4 \sin\theta_4$	$l_3 \cos\theta_3$	0	$l_3 \sin\theta_3$
7	0	-0.05	0	0	0	0
8	$l_5 \cos\theta_5$	0	$l_5 \sin\theta_5$	$-l_4 \cos\theta_4$	0	$-l_4 \sin\theta_4$

Once the external data were successfully converted, the robot could determine the velocity and acceleration of each via point by fuzzy algorithm. That method was discussed in detail in section 3.2.2. After processing all the sequences, the robot could generate the walking patterns for a smooth movement.

4 Simulation Result

All the algorithms were made with Visual C++ in a PC environment. The first sequence of walking patterns collected information related to the obstacle, and calculated the adequate motion data (walking distance and angle data of the body), and then generated the legs trajectory as proposed. Both velocity and acceleration of via point were supported to be '0'. The Fig. 5 depicts a 3D motion snapshot. The surface was considered uneven and the three sensor interfaces located at (14cm, -2cm), (45cm, 4cm) and (65cm, 0cm) were considered as obstacles. Those obstacles were squares of 5cm, 8cm, and 2cm respectively. The total walking time lasted 11.6s, and the robot made 20 body propulsions.

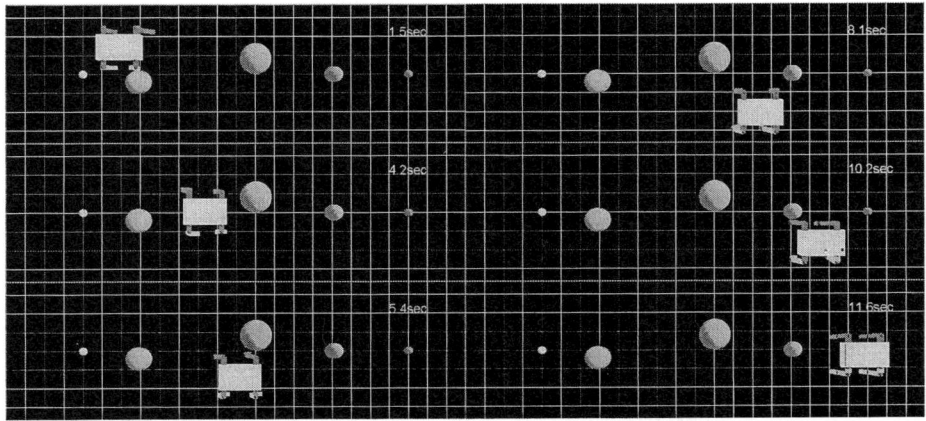

Fig. 5. Simulation snapshot of 3D walking sequence

Fig. 6 shows the full trajectory position comparison with and without proposed algorithm, and Fig. 7 shows the full step trajectory acceleration of each axis with and without the genetic-fuzzy algorithm. In Fig. 8, we can see the comparison of the angular acceleration obtained when using or not the algorithm. The deviation rates of the acceleration were reduced to 25% in applying the genetic-fuzzy algorithm.

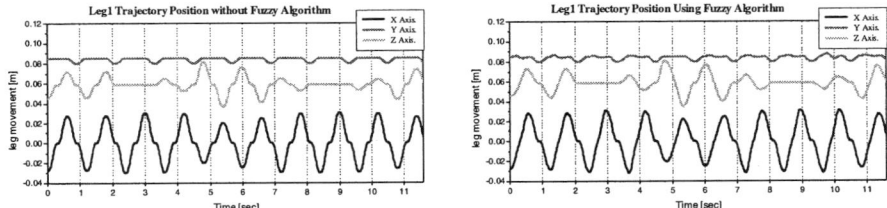

Fig. 6. Position comparison of leg1

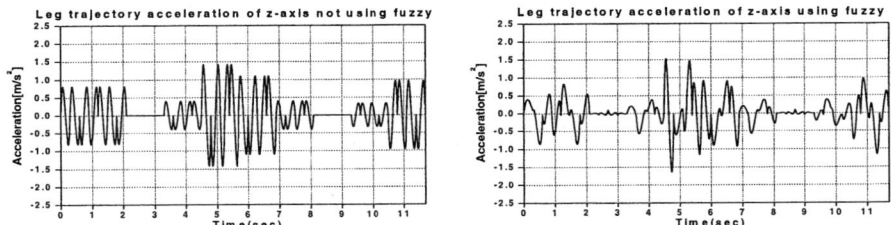

Fig. 7. Comparison of the acceleration of leg1

Investigating the improvement of the acceleration of joints with and without the algorithm, the sum of the acceleration of the shoulder's pitch joint (θ_2) was reduced

25%, and the shoulder's yaw joint (θ_3) was improved about 28%. The acceleration data of knee joint (θ_4) improved about 6%. Angular acceleration reduction implies the smooth motion of the robot. As a result, the robot joints were less influenced after applying the algorithm.

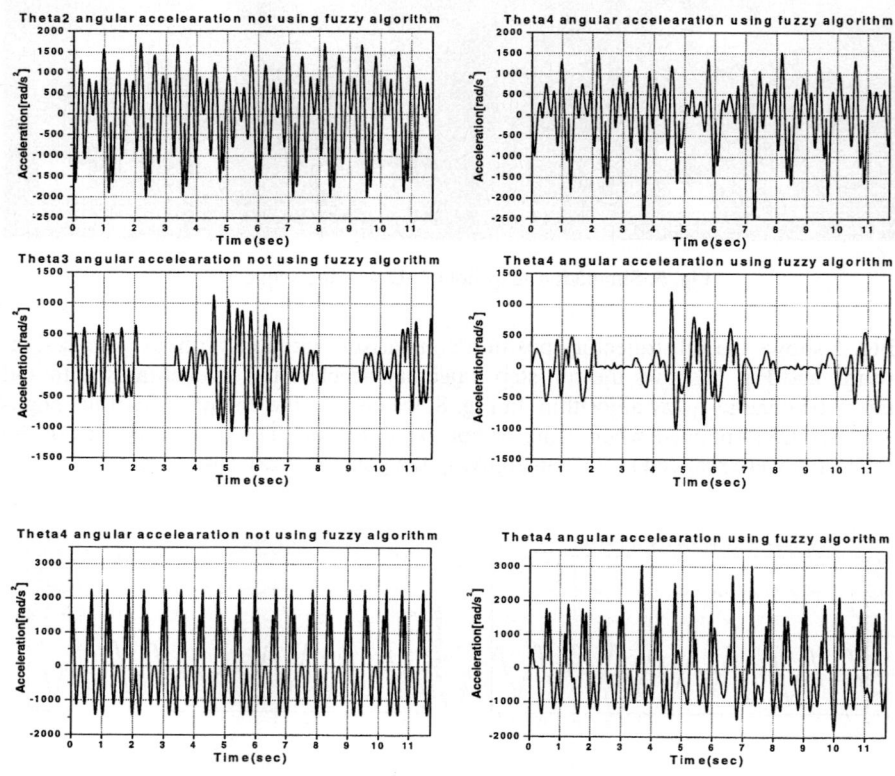

Fig. 8. Joint angular acceleration of leg 1

5 Conclusion

It is difficult to generate a trajectory that can foresee external environments for mobile robot. Especially, the body drive trajectory of a quadruped robot can't be directly related to its leg's trajectory with a deterministic function. Therefore a relationship between the two trajectories is required. The other difficulty in designing a robot is how to realize a smooth motion for its legs. Furthermore, it is essential for a robot to have a capability of stable walking. In this paper, the genetic-fuzzy algorithm was suggested to solve those problems. It is suggested that the genetic algorithm is employed for a leg's trajectory generation method based on rule-base, and the fuzzy algorithm is applied for the optimal movable path to avoid obstacles. From those algorithms, a quadruped robot could create an optimal walking path while considering the uneven terrain. In addition the optimal leg trajectory could be created after trajectory planning from this path.

The proposed method will be verified with the real robot SERO-V using the information of real sensors. As future works, a vision system and various sensors will be implemented to obtain actual data from external environment so as to incorporate human friendly behavior to robots.

Acknowledgement

This work was supported by Grant No. R01-2003-000-10364-0 from Korea Science & Engineering Foundation.

References

1. Y. Sakagami, R. Watanabe, C. Aoyarna, S. Matsunaga, N. Higaki, and K. Fujimura: The intelligent ASIMO: System overview and integration, IEEE/RSJ International Conference, Vol. 3 (2002) pp. 2478-2483
2. G. S. Hornby, S. Takamura, J. Yokono, O. Hanagata, T. Yamamoto and M. Fujita: Evolving Robust Gaits with AIBO, IEEE Robotics and Automation, Vol. 3 (2000) pp. 3040-3045
3. P. Buschka, A. Saffiotti and Z. Wasik: Fuzzy Landmark-Based Localization for a Legged Robot, IEEE/RSJ International Conference, Vol. 2 (2000) pp. 1205-1210
4. Shotaro Kamio, Hongwei Liu, Hideyuki Mitsuhasi and Hitoshi Iba: Researches on Ingeniously Behaving Agents, NASA/DoD Conference (2003) pp. 208-217
5. T. Shibata, T. Matsumoto, T. Kuwahara, M. Inaba and H. Inoue: Hyper Scooter: A Mobile Robot Sharing Visual Information with a Human, IEEE Intl. Conf. on Robotics and Automation, Vol. 1 (1995) pp. 1074-1079
6. S. Kagami, K. Okada, M. Kabasawa, Y. Matsumoto, A. Konno, M. Inaba and H. Inoue: A Vision-based Legged Robot as a Research Platform, IEEE/RSJ Intl. Conf. on Intelligent Robots and Systems (1998) pp. 235-240
7. Xuedong Chen, Keigo Watanabe, Kazuo Kiguchi and Kiyotaka Izumi: An ART-Based Fuzzy Controller for a Adaptive Navigation of a Quadruped Robot, IEEE/ASME Transactions on Mechatronics, Vol. 7, No. 3 (2002) pp.318-328
8. H.Zhung, J.Wu, and W.Hung: Optimal Planning of Robot Calibration Experiments by Genetic Algorithms, IEEE Conf. on Robotics and Automation, Vol. 2 (1996) pp. 981-986
9. K. Shimojima, N. Kubota, and T. Fukuda: Trajectory Planning of Reconfigurable Redundant Manipulator Using Virus- Evolutionary Genetic Algorithm, IEEE Conf. on Industrial Electronics, Control, and Instrumentation, Vol. 2 (1996) pp. 836-841
10. K. H. Shim, B. H. Lee, and J.G. Kim: Control Balancing Weight for IWR Robot by Genetic Algorithm, Proceeding of the 11th KACC (1996) pp. 1185-1188
11. Xuedong Chen, Keigo Watanabe, Kazuo Kiguchi, and Kiyotaka Izumi: An ART-Based Fuzzy Controller for a Adaptive Navigation of a Quadruped Robot, IEEE/ASME Transactions on Mechatronics, Vol. 7, No. 3 (2002) pp. 318-328

Modelling of Process of Electronic Signature with Petri Nets and (Max, Plus) Algebra

Ahmed Nait-Sidi-Moh and Maxime Wack

Laboratory of Systems and Transports,
University of Technology of Belfort-Montbeliard,
90010 Belfort Cedex, France
{ahmed.nait-sidi-moh, maxime.wack}@utbm.fr

Abstract. This article discusses the modelling and the evaluation of process of electronic signature (ES). According to a certain point of view, this process can be shown as a class of Dynamic Discrete Event Systems (DDES). It is in this framework that we study this class with using Petri Nets (PN) and the theory of linear systems in (max, +) algebra. We introduce these two formalisms with the aim to describe the graphical and analytical behaviours of studied process. The resolution of (max, +) model which describes the system enables us to evaluate the process performances in terms of occurrence dates of various events that compose it (authentication, hashcoding, signature, timestamping). To illustrate our methodology, we finish this article with a numerical example.

1 Introduction

The security of the information systems presents a primordial task because of a strong growth of the technology. Actually, the treatment technique of the information occupies a dominating place in the data-processing applications. With the growth of the electronic exchanges, the security becomes increasingly crucial. To improve the quality of service (QoS) of the electronic exchanges it is necessary to have the means of analysis and the appropriate methods. This QoS must be validated with specification of models that repose on formal properties like robustness, reliability ...

The information systems constitute a particular class of DDES whose dynamic is governed by various phenomenons as synchronisation, concurrency and parallelism. These systems result essentially from the human design, according to a certain point of view, their activities are due to asynchronous occurrences of discrete events. It is thus possible to model them via various adapted and dedicated tools.

The objective of this paper, which enters within the framework of the security and the legalization of data-processing communications, is to study processes evolving in a space of discrete states. Our contribution concerns, more precisely, the adaptation of the concepts and the theoretical results which formalized graphically by PN and mathematically by (max, +) algebra. This is in order to model and analyse the data-processing exchanges between the various actors of confidence in the process of electronic delivery of signature hashcoding, authentication, signature, timestamping. Our study is thus concretized by the development of models able to bring solutions to the problems of safety and improvement of service quality of ES. These models facilitate the structural and behavioural analysis of signature process.

2 Digital Signature and Message Signing Basics

Digital signature is the courant way of authenticating of an electronic data. It is the achievement of many researches on asymmetric key cryptography and hashcoding.

2.1 Asymmetric Key Cryptography Concepts

When a sender entity (a person, a server ...) needs to securely send a message to a receiver entity, it encrypts the message using the receiver's public key. This key is published so that any sender can make use of the receiver's public key to encrypt data. The encrypted message is then unintelligible and can not be decrypted without the corresponding private key. This key must be securely stored by the receiver that does not publish it. Only the receiver would be able to decrypt the encrypted message. Asymmetric key cryptography achieves privacy and confidentiality. The most widely used asymmetric key cryptography algorithms are RSA [15], triple-DES [9].

2.2 Hashcoding Overview

Hashcoding [6] aims at creating a fixed-length message digest from any arbitrary data stream. This digest is independent of the size of the source stream. Let's consider h(), a one-way hash function used to compute a digest on a given stream s. The most important property of this function prevents source stream reconstruction only if the computed digest is known. Although reconstruction the original stream s from a given digest d may theoretically be possible, it appears to be computationally unfeasible:

$(h(s)=d) \Rightarrow (p(h\text{-}1(d) = s) \rightarrow 0)$

Moreover, the probability p that two different streams s1 and s2 obtain identical digests with a given hashcoding algorithm ha reaches zero. The hash function is then said collision resistant:

$(s1 \neq s2) \Rightarrow (p(h(s1,ha) = h(s2,ha)) \rightarrow 0$

Many digest algorithm such as MD2 [5], MD4 [14] and RIPEMD [2] [11] have been developed and they are specifically designed to compute digital signatures.

2.3 Digital Signature

Digital signature is the analogue of the handwritten signature, it shares with this latter the following properties:

- the reader must be convinced that the document was signed by the signer;
- the signature can not be falsified;
- the signature is not reusable. It is a part of the signed document and con not be moved on another document;
- a signed document is inalterable. Once signed, one can not modify it.

Digital signature generation is nothing but the application of asymmetric key cryptography over streams hashcodes. Unlike data encryption, digital signature's purpose does not consist of data confidentiality but rather in providing [4]:

- *data integrity*: digital signature to detect source streams alteration, i.e. unauthorized data modification;
- *authentication*: as the signature key is (theoretically) owned by the signer only, it is impossible for anyone else to generate the sender's signature on a given data stream. The stream is authenticated by comparing the signature with the signer's corresponding verification key;
- *non-repudiation:* this authentication-based service is a proof of transaction effectiveness. The signature entity can not deny being the author of the signature because nobody else could possibly have created such a signature on a given data stream.

Digital signature is generally computed on hashcodes rather than directly using data source streams. Although digital signature makes it possible to authenticate the received data, it does not identify the entity that signed the data (the signer) from the receiver point of view. Thus, any irrefutable link exists between the signer and its signature key. Such identification is provided by electronic certificates.

2.4 Electronic Certificates

An electronic qualified certificate (i.e. certificate) is an electronic proof of identity (Fig.1.). It is designed to allow senders identification by signed messages receivers. Certificates trust depends on their issuers trust. Only Certificate Authorities (CAs) are accredited by governments to deliver electronic certificates.

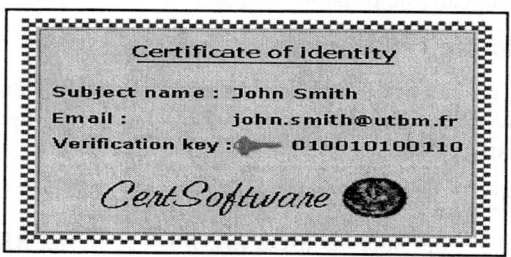

Fig. 1. Certificate description.

2.5 Various Types of Signature

The National Institute of Standards and Technology [10] enumerate a certain number of signatures currently used in handwritten form. In order to reproduce the traditional diagrams of signatures, the ES must provide an equivalent of all types of signature. In the framework of digital signature, we can distinguish several types of signature [13]:

- *Simple and single signature*
- It is the simplest signature which a single user signs a document (the document term is used here in the broad sense, and can thus indicate any numerical document (text, video, image, etc.)).
- *Simple multiple signature or Co-signature*
- This type of signature is presented when several users sign the same document.

- *Enveloping multiple signature or on-signature*
- This type is used when several individuals representing a given hierarchical structure sign a document. In this case, the individuals sign all the same document but the last signatures also sign the preceding signatures.
- *Against-signature*
- This type of signature is used when several users representing a hierarchical structure sign a document. In this case, the users sign the same document but the last signatures sign only the preceding signatures, and not the document.
- *Simple signature by parts of document*
- It is the case where several signatures appear on a document, but each signature applies to a different part of document.
- *Enveloping multiple signature with addition of information only*
- This type is used when a person wishes to add information to a document and sign them. Its signature covers all the anterior information and signatures.
- *Enveloping multiple signature with addition and suppression of information*
- This type is used when a signer can unilaterally modify a document which contains multiple signatures previous to his signature.

In what follows, we propose to model the process of one of these types of signatures. We choose then the process of simple and single signature. This one is composed of two processes: "signature process" where the signer signs a document, "verification process" where recipient checks the document authenticity and all signatures that it contains. In this paper, we will be interested in the signature process.

3 Simple and Single Signature

The act to sign a document can be regarded as a phenomenon of synchronization where resources (document and signer) must be available at the moment of signature. This can be seen like an appointment phenomenon. To sign a document, the signer must have a signature key locked and stored on a personal and confidential support (Cdcarte, smart card, USB key, etc.) on one hand (the safety in general is ensured by a secret code but whose means can be used according to the desired level of safety (identification retinal, digital, thermal, etc.)). On the other hand, the document to be signed must be "to hash" in order to obtain a single stamp which is clean for him. The signature of document is done when the private key of signer signs the stamp. One thus obtains a single signature on document. This signature is then subjected to one or several timestamping authorities who affix their signatures and preserve a trace of the transaction as well as the stamp of document. In this signature process, we will be interested in the dates of the operations to carry out on the documents from their creation to their signature and timestamping. To model this process, we use the adapted tools to the representation of systems whose dynamic is governed by the occurrence of events in a space of discrete state on one hand, and on the other hand, governed by various phenomena among which the synchronisation. We use then the Petri nets and (max, +) algebra.

3.1 Graphical and Mathematical Modelling

3.1.1 Graphical Modelling

3.1.1.1 Basic Petri Nets Definitions

In this section, we recall the basic Petri net tool that we will use in our modelling. A complete introduction can be found in [12]. A PN is a graph composed of two nodes: places and transitions. The oriented arcs connect certain places to certain transitions, or conversely. With each arc, we associate a weight (non negative integers). In a formal way, a PN is a 5-tuple PN = (P, T, A, W, M_0) where:

$P = \{P_1, ..., P_n\}$ is a finite set of places (represented with circles);
$T = \{T_1, ..., T_m\}$ is a finite set of transitions (represented with line segments);
$A \subseteq (P \times T) \cup (T \times P)$ is a finite set of arcs;
$W = A \rightarrow \{1, 2 ...\}$ is the weight function associated with arcs;
$M_0 = P \rightarrow \{0, 1, 2 ...\}$ is the initial making of graph.

An important class of PN that we will use in this paper is Timed Event Graph (TEG). In this class, each place has only one input and output transition. The aim of using TEG is to be able to describe, in a simple way, the behaviour of the process by mathematical and linear equations in (max, +) algebra [7].

Timed Event Graph Model

In order to comprehend the process working, we represent it by a graphical model with using a TEG tool. In this model, the transitions model the events (authenticity, hashcoding, signature, etc.) and their firings model the occurrence of these events. The places (resp. associated times) model the transit (reps. necessary times for transit) from one operation to another. In the figure 2, we represent the TEG model which models the two operations of signature and timestamping of a document.

In order to interpret the obtained graphical model with the mathematical equations in (max, +) algebra, we associate a variable to each model transition (see Fig. 3.). Thus, we associate with input transitions the input variables (denoted u_1 and u_2), with internal transitions the state variables ($x_1, x_2, x_3, x_4, x_5, x_6, x_7$), and finally we associate an output variable (denoted y) with the output transition. By introducing these new variables, and the associated temporisation with each place, we obtain the graphical model of Fig. 3.

3.1.2 Mathematical model

3.1.2.1 Some Elements of (Max, +) Algebra

We describe the considered process signature by a state representation in (max, +) algebra. First of all, we give some elements of this algebra [1] and [3].

We denote by IR_{max} the dioïd ($IR \cup -\infty, \oplus, \otimes$) where the operators "$\oplus$" and "$\otimes$" are respectively "max" and "the usual addition" ($\forall$ a, b in IR_{max}, a $\oplus$ b = max(a, b) and a $\otimes$ b = a + b). The neutral elements for the operators $\oplus$ and $\otimes$ are respectively $\varepsilon = -\infty$ and e = 0 ($\forall$ a $\in IR_{max}$, a $\oplus \varepsilon$ = a and a $\otimes$ e = a). Like other algebraic structures, the (max, +) algebra have a properties and characteristics, such that the associatively of

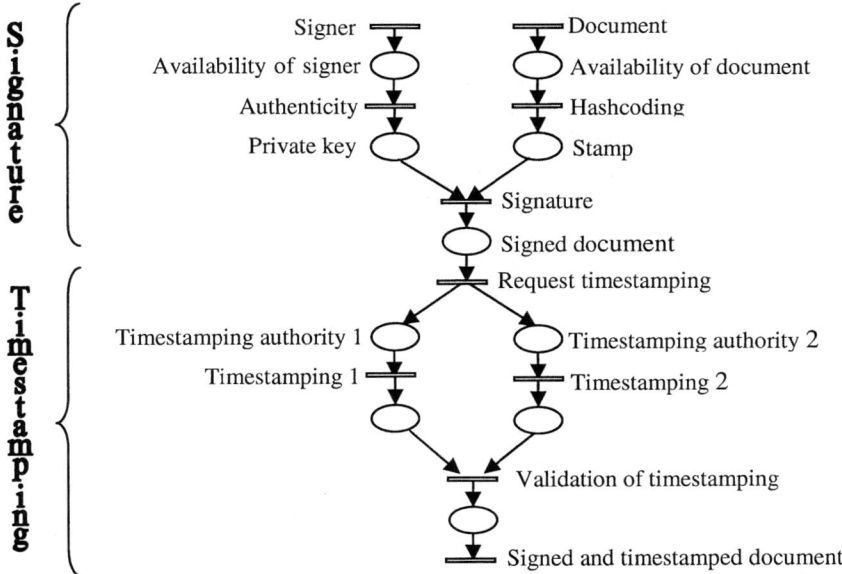

Fig. 2. Graphical model of simple and single signature process

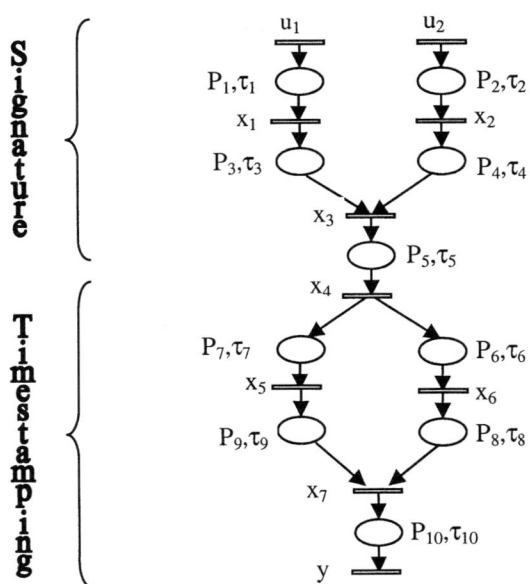

Fig. 3. TEG model of signature process: definition of variables and temporisations

addition, and the multiplication, the commutatively of addition, the distributivity of multiplication, existence of zero element (denoted ε). Let us consider a and b two elements in a dioïd D, the quantity a^*b is the smallest solution of the equation $x = a\,x \oplus b$, where the expression of a^* is given by : $a^* = e \oplus a \oplus a^2 \oplus \ldots$. In addition, each solution x satisfies $x = a^*x$ the greatest solution of $f(x) = a\,x \le b$ is : $x = b / a$. (the operator "/" (resp. "\") represent the subtraction in the right (resp. the left) in the (max, +) algebra. Finally, from a scalar dioïd D, let us consider $A \in D^{m \times n}$ and $B \in D^{m \times p}$, the sum and product of matrices are defined conventionally from the sum and product of scalars in D. In what follows, and without no risk of ambiguity, we omit "⊗" and let us write a ⊗ b like ab or a.b.

3.1.2.2. (Max, Plus) Linear Model
Let us present, in this study, only the associated equations with the signature part (see figure 3). In the same way, we can obtain all equations which represent the graphical model (the two parts: signature and timestamping).

The sate variable $x_i(k)$ (for $1 \le i \le 3$), called also "dater" is associated to each model transition x_i. It represents, the date of the kth firing of the transition x_i. In the same way, we define the input daters and the output daters $u_1(k)$, $u_2(k)$ and $y(k) = x_4(k)$ (here we consider only the system output that corresponds with the event "signed document"). By using all defined variables and the associated daters, we obtain the various equations that model the studied process. For example, we write the state equation associated with the transition x_3 :

$\forall\,k \ge 1$, $x_3(k) = \max(\tau_3 + x_1(k), \tau_4 + x_2(k))$: this signifies that the firing of the transition x_3 for the kth time (signature of kth document) occurs if the private key and the stamp are both available.

In the same way, we obtain the other equations of system. With replacing "max" by the operator "⊕" and "+" by "⊗", we describe all equations in the (max, +) algebra. For the given example, we have: $\forall\,k \ge 1$, $x_3(k) = \tau_3 \otimes x_1(k) \oplus \tau_4 \otimes x_2(k)$.

The (max, +) model that represents the "signature" part is then given by:

$\forall\,k \ge 1,$
$$\begin{cases} x_1(k) = \tau_1 \otimes u_1(k) \\ x_2(k) = \tau_2 \otimes u_2(k) \\ x_3(k) = \tau_3 \otimes x_1(k) \oplus \tau_4 \otimes x_2(k) \\ y(k) = x_3(k) \end{cases} \quad (1)$$

In order to put this system in the matrix form, we define the following vectors:
$X(k) = [x_1(k), x_2(k), x_3(k)]^T$: state vector;
$U(k) = [u_1(k), u_2(k)]^T$: input vector;
$Y(k) = y(k)$: output of system.

With using these vectors, we obtain the following matrix form:

$\forall\,k \ge 1,$
$$\begin{cases} X(k) = A\,X(k) \oplus B\,U(k) \\ Y(k) = C\,X(k) \end{cases} \quad (2)$$

A, B and C the characteristic matrices of the model and whose elements represent the data of system.

$$A = \begin{bmatrix} \varepsilon & \varepsilon & \varepsilon \\ \varepsilon & \varepsilon & \varepsilon \\ \tau_3 & \tau_4 & \varepsilon \end{bmatrix}, B = \begin{bmatrix} \tau_1 & \varepsilon \\ \varepsilon & \tau_2 \\ \varepsilon & \varepsilon \end{bmatrix} \text{ and } C = [\varepsilon \ \varepsilon \ e].$$

The first equation of the system (2) is an implicit equation. In order to solve it, and find all dates where various operations of process occur, we replace in the second member of (2), and in iterative way, $X(k)$ by its expression, thus we obtain:

$\forall\, k \geq 1$,

$$X(k) = A\,(A\,X(k) \oplus B\,U(k)) \oplus B\,U(k) = A^2\,X(k) \oplus A\,B\,U(k) \oplus B\,U(k)$$
$$= \ldots \tag{3}$$
$$= A^n\,X(k) \oplus A^{n-1}\,B\,U(k) \oplus A^{n-2}\,B\,U(k) \oplus \ldots \oplus A\,B\,U(k) \oplus B\,U(k)$$
$$= A^n\,X(k) \oplus (A^{n-1} \oplus A^{n-2} \oplus \ldots \oplus A \oplus E)\,B\,U(k)$$

Where n is the order of the matrix A (in our case n = 3). E is the identity matrix in dioid algebra (it composed of the element "e" on diagonal and "ε" elsewhere).

In the equation (3), it figures the expression of the quasi-inverse of the matrix A (denoted $A^* = A^{n-1} \oplus A^{n-2} \oplus \ldots \oplus A \oplus E$, and called Kleene star). As $A^n = \varepsilon$, (because the graphical model in figure 3 does not contain no strongly connected component), then the smallest solution of the first equation of system (2) is given by:

$\forall\, k \geq 1$, $\qquad\qquad X(k) = A^*\,B\,U(k) \tag{4}$

From (4), the expression of the output system is given by:

$\forall\, k \geq 1$, $\qquad\qquad Y(k) = C\,A^*\,B\,U(k) \tag{5}$

$$\text{With: } A^* = \begin{bmatrix} e & \varepsilon & \varepsilon \\ \varepsilon & e & \varepsilon \\ \tau_3 & \tau_4 & e \end{bmatrix}, A^*B = \begin{bmatrix} \tau_1 & \varepsilon \\ \varepsilon & \tau_2 \\ \tau_1\tau_3 & \tau_2\tau_4 \end{bmatrix} \text{ and } C\,A^*B = [\tau_1\tau_3 \ \ \tau_2\tau_4].$$

From the expressions (4) and (5), we deduce and evaluate the dates of authenticity and signature of document.

3.2 Example of Application

With the aim to include the process of the considered signature, we illustrate our method with an example of application for which we associate to each operation of process an execution time. On the table 1, we affect to each parameter τ_i ($1 \leq i \leq 5$) a numerical value. We note that the firing of each transition is immediate, and each validate transition is fired.

Table 1. Numerical values of the operations of signature process

Parameter τ_i	Signification	Numerical value (seconde)
τ_1	Availability of signer	60 s
τ_2	Availability of document	120 s
τ_3	creation of the private key	10 s
τ_4	creation of the stamp	15 s
τ_5	Signed document	0 s

According to these numerical values, the equations (4) and (5) that give the solutions of system (2) become:

$$\forall\, k \geq 1, \quad X(k) = \begin{bmatrix} x_1(k) \\ x_2(k) \\ x_3(k) \end{bmatrix} = \begin{bmatrix} 60 & \varepsilon \\ \varepsilon & 120 \\ 70 & 135 \end{bmatrix} \otimes \begin{bmatrix} u_1(k) \\ u_2(k) \end{bmatrix} \quad (6)$$

$$\forall\, k \geq 1, \quad Y(k) = [70 \quad 135] \otimes \begin{bmatrix} u_1(k) \\ u_2(k) \end{bmatrix} \quad (7)$$

We suppose that we want to sign a document (k^{th} document) by a signer (k^{th} signer). If we choose the system input such that $u_1(k) = u_2(k) = 0$ ($= e$); this corresponds to the dates of starting of signature process (creation of document and call of signer). From the equation (6), we obtain what follows:

Authenticity date $= x_1(k) = 60 \otimes e \oplus \varepsilon \otimes e = 60$ s;
Hashcoding $= x_2(k) = \varepsilon \otimes e \oplus 120 \otimes e = 120$ s;
Signature date $= x_3(k) = y(k) = 70 \otimes e \oplus 135 \otimes e = 135$ s.

In this example, we show that it is possible to evaluate in a formal way a signature process with (max, +) equations, by determining the various dates of occurrence of events that constitute it, like the authenticity, the timestamping, the signature ...

4 Conclusion

After describing the electronic signature as well as the various operations of its process starting from the creation of document to its signature, we treated a problematic related to the modelling and the analysis of process of electronic signature. For this objective, we proposed a formal approach based on adapted tools to the representation of the information systems which represent a class of dynamic discrete event systems. We use then Petri nets and (max, +) algebra. For these tools, we could present and evaluate the process of electronic signature. This demonstration of feasibility must enable us to develop a more complete model by completing it, and adding the whole of actors of the electronic signature.

In our next contribution, we will define a process control based on residuation theory in dioid algebra.

References

[1] Baccelli F., Cohen G., Olsder G. L. and Quadrat J. P., "Synchronisation and linearity : an algebra for Discrete Event Systems". Wiley, 1992.

[2] Dobbertin H., Bosselaers A., Preneel B., "RIPEMD-160, a strengthened version of RIPEMD", Fast Software Encryption, LNCS vol. 1039, D. Gollmann Ed., pp. 71-82, 1996.

[3] Gaubert S., "Théorie des systèmes linéaires dans les dioïdes", Thèse de doctorat, Ecole National Supérieure des mines de Paris. Juillet 1992.

[4] Kaeo M., "Designing Network Security", Macmillan Technical Publishing, USA, ISBN 1-57870-043-4, 1999.

[5] Kaliski Jr B. S., "RFC 1319: The MD2 Message-Digest Algorithm", RSA Laboratories, Janvier 1992.

[6] Menezes A. J., Van Oorschot P. C., Vanstone S. A., "Handbook of Applied Cryptography", CRC Press, USA, ISBN 0-8493-8523-7, Février 2001.

[7] Nait-Sidi-Moh A., "Contribution à la modélisation, à l'analyse et à la commande des systèmes de transport public par les réseaux de Petri et l'algèbre (Max, Plus)". Thèse de doctorat en Automatique et Informatique, Université de Technologie de Belfort-Montbéliard et Université de Franche-Comté, Décembre 2003.

[8] National Institute of Standards and Technology, "Secure Hash Standard (SHS)", Federal Information Processing Standards Publication, FIPS PUB 180-1, Avril 1995.

[9] National Institute of Standards and Technology, "Data Encryption Standard (DES)", Federal Information Processing Standards Publication, FIPS PUB 46-3, Octobre 1999.

[10] National Institute of Standards and Technology, "Digital Signature Standard (DSS)", Federal Information Processing Standards Publication, FIPS PUB 186-2, Janvier 2000.

[11] Preneel B., Bosselaers A., Dobbertin H., "The cryptographic hash function RIPEMD-160", CryptoBytes, vol. 3, No. 2, pp. 9-14, 1997.

[12] René David et Hassane Alla, "du grafcet aux réseaux de Petri". Série Automatique, hermes, Paris 1992.

[13] Rieupet D. , Wack M. , Cottin N. , Assossou D. , "Signature électronique multiple", Atelier Sécurité des Systèmes d'Information, XXIIème Congrès INFORSID, Mai 2004.

[14] Rivest R. L., "RFC 1320: The MD4 Message-Digest Algorithm", MIT Laboratory for Computer Science and RSA Data Security, Avril 1992.

[15] RSA Data Security Inc., "Public Key Cryptography Standards, PKCS 1-12", disponible en ligne à ftp://ftp.rsa.com/pub/pkcs, 1993.

Evolutionary Algorithm for Congestion Problem in Connection-Oriented Networks

Michał Przewoźniczek[1] and Krzysztof Walkowiak[2]

[1] Division of Computer Science, Faculty of Computer Science and Management,
Wroclaw University of Technology, Wybrzeze Wyspianskiego 27, 50-370 Wroclaw, Poland
`Michal.Przewozniczek@pwr.wroc.pl`
[2] Chair of Systems and Computer Networks, Faculty of Electronics,
Wroclaw University of Technology, Wybrzeze Wyspianskiego 27, 50-370 Wroclaw, Poland
`Krzysztof.Walkowiak@pwr.wroc.pl`

Abstract. The major objective of this paper is to deploy an effective evolutionary algorithm (EA) for the congestion problem in connection-oriented networks. The network flow is modeled as non-bifurcated multicommodity flow. The main novelty of this work is that the proposed evolutionary algorithm consists of two levels. The *high* level applies typical EA operators. The *low* level idea is based on the hierarchical algorithm idea. However, the presented approach is not a classical hierarchical algorithm. Therefore, we call the algorithm *quasi-hierarchical*. We present a brief description of the algorithm and results of simulations run over various networks.

1 Introduction

In this paper, we tackle the problem of flow assignment in computer networks by an innovative evolutionary algorithm. The considered optimization problem can be formulated as follows. Given a network, a set of origin-destination node pairs (commodities) and demand for the commodities, the congestion non-bifurcated problem (NBC) consists in routing the flow of commodities using only one path for each commodity. The objective is to maximize the minimum value of network link residual capacity (difference between link capacity and flow allocated on this link). Additionally, the capacity constraint must be satisfied, i.e. flow of arc (calculated as a sum of all commodities using that arc), cannot exceed capacity of the arc. We call a solution of the NBC feasible if the capacity constraint is satisfied. It must be noted that NBC problem is NP-complete [13]. Examples of popular network techniques applying the connection-oriented (c-o) flow are: Asynchronous Transfer Mode (ATM), MultiProtocol Label Switching (MPLS).

Our main focus is survivability of c-o networks. Survivability of computer networks is very important since loss of service means loss of revenues, especially in high-speed fiber networks. Therefore, network flows must be assigned to enable effective restoration when a failure occurs. Since, the residual capacity, which is not used in normal, failure-free situation is necessary for allocation of backup paths that restore broken connections, it is important to maximize the residual capacity of links. More information on survivable networks can be found in [6].

In this work we present an effective evolutionary algorithm that can be applied to NBC problem. The main novelty of our approach is that the proposed algorithm consists of two levels. The *high* level applies typical EA operators. The *low* level idea is based on the hierarchical algorithm idea. However, the presented method is not a classical hierarchical algorithm. Therefore, we call the algorithm *quasi-hierarchical*.

2 Congestion Problem

We model the network as a directed graph $G=(N,A)$ where N is a set of n nodes (vertices) representing network switches, A is a set of m arcs (directed edges). In this work we apply link-path formulation of the non-bifurcated multicommodity (m.c.) flow [6],[13]. It is obtained by providing for each commodity $i \in P$ a set of candidate routes $\Pi_i = \{\pi_i^k : k = 1,...,l_i\}$ connecting end nodes of commodity i. One commodity can use only one route π_i^k. Let x_i^k denote a 0/1 variable, which equals one if π_i^k is route for commodity i and is equal to 0 otherwise. Another binary variable a_{ij}^k indicates whether or not arc $j \in A$ belongs to π_i^k. Using this representation of m.c. flow the non-bifurcated congestion (NBC) problem is as follows

$$\max \; z \tag{1}$$

subject to

$$\sum_{\pi_i^k \in \Pi_i} x_i^k = 1 \quad \forall i \in P \tag{2}$$

$$x_i^k \in \{0,1\} \quad \forall i \in P; \pi_i^k \in \Pi_i \tag{3}$$

$$f_j = \sum_{i \in P} \sum_{\pi_i^k \in \Pi_i} a_{ij}^k x_i^k Q_i \tag{4}$$

$$f_a \le c_a \quad \forall a \in A \tag{5}$$

$$z \le c_a - f_a \quad \forall a \in A \tag{6}$$

The objective is to maximize the minimum arc residual capacity of network function (1). Since we consider the non-bifurcated multicommodity flow, condition (2) states that each commodity can use only one route. Constraint (3) ensures that decision variables x_i^k are binary ones. (4) is a definition of a link flow. Formula (5) is a capacity constraint. Finally, constraint (6) measures the residual capacity of each link. Variable z is the minimum value of the residual capacity over all arcs in the network. Therefore, the objective function can be also formulated as

$$\max z = \min_{a \in A} \; (c_a - f_a) \tag{7}$$

It should be noted that the non-bifurcated flow problem is also called unsplittable flow problem (UFP) [9]. Congestion problems formulated similarly to our formulation given by (1-6) can be found in the literature. The *maximum concurrent flow* problem consists in maximizing the throughput of the network, i.e. as large a common percentage of each demand as possible should be routed while not exceeding the capacity constraint. In the *minimum congestion* problem full demands must be routed in order to minimize the maximum load of an arc and satisfy the capacity constraint. Also the *relative congestion* defined as the maximum ratio over all arcs of flow f_a divided by the capacity c_a can be applied an objective function [1-3], [5], [9].

As mentioned above, our main focus in this work is on the network survivability of c-o networks. Therefore, in this work we apply formulation (1-6) of the congestion problem using non-bifurcated m.c. flow, which is more convenient for use in optimization of flows in survivable networks. However, some of algorithms developed for maximum concurrent flow and minimum congestion problems can be also applied to solve NBC problem.

Most of literature on problems related to NBC either concerns congestion problems of bifurcated m.c. flows, or some limited versions of UFP in the context of congestion. A comprehensive treatment of various algorithms developed for maximum concurrent flow and minimum congestion problems is given in [2]. Author focuses only on bifurcated m.c. flows. Chlamatac et al. proposed an algorithm for optimization of virtual paths in ATM networks that minimizes the load on networks links [3]. Since ATM is a connection-oriented technique, the network flow is modeled as non-bifurcated m.c. flow. The central idea behind the algorithm is to assign random weights to the links in a special way that assures, for any possible realization of the weights, that the minimum weight path between two nodes will be inevitably a minimum-hop path. Authors prove that for large networks the solution provided by the algorithm is within a small factor of the best possible solution.

There are many papers concentrating on developing constant-factor approximation algorithms and bounds for various versions of unsplittable flow problems (UFP). Some of them consider a limited version of UFP – the *single-source unsplittable flow problem*, which consists in allocating a set of commodities having the same source node to single paths [8]. Approximation algorithms for multi-source UFP are discussed in [9]. However, works concentrating on constant-factor approximation algorithms lack numerical experiments presenting performance of algorithms in real networks. Main focus is reduced to mathematical properties of algorithms, while the way how to use these algorithms for optimization of flows in computer networks is not discussed, what limits significance of such papers for real applications.

The NBC problem is an integer 0/1 problem with linear constraints. However, size of the problem is very large even for relatively small networks. A popular method to solve 0/1 problems is the branch and bound (B&B) approach. Such algorithms were applied to many problems related to NBC. Nevertheless, branch and bound algorithms are intractable for networks of medium and large size. Moreover, a large reduction of candidate routes is required to make the problem feasible for branch and bound algorithm [7], [13].

All references discussed above consider the static optimization of network flows. There are also related works that focus on dynamic optimization of network demands. In dynamic routing, requests arrive one-by-one and future demands are unknown. Shortest-widest path (SWP) algorithm that enables dynamic allocation of paths according to bottleneck metrics like residual capacity or link utilization is proposed in

[11]. Dynamic routing algorithms can be used also for static optimization of network flows by applying the same algorithm sequentially for all demands. Since all demands are known a priori in static optimization, ordering of processed demands can be changed to improve the results.

3 Algorithm Description

In this section we present the main characteristics and objectives of new evolutionary algorithm called HEFAN (**H**ierarchical **E**volutionary algorithm for **F**low **A**ssignment in **N**on-bifurcated multicommodity flow). Ideas of presented algorithm are based on the approach presented in [14]. From the general point of view the algorithm has only one objective: to find a feasible and the best in terms of fitness function, set of routes between the specified node pairs in the oriented network. A typical evolutionary algorithm can easily gain this objective, but only if we support a proper collection of routes for which the search will be done. Such a collection can not be neither too narrow nor too wide – too narrow collection will force us to search too small part of possible solutions space, which does not contain the best of them. Too wide set of routes will make the problem intractable due to the tremendous number of possibilities. These are the reasons why presented system is conceptually divided into three abstract parts: the routes proposals database, the high and the low level. The high level works as a typical EA. The low level is an intelligent searcher of new candidate routes, based on EA ideas, using some of its features but not being such an algorithm in fact.

The high level of the presented algorithm contains typical EA operators that can be found in any publications and books of this topic [4], [10], [12]. The low level idea is based on the hierarchical algorithm idea [15], even though it's existence does not make the presented algorithm hierarchical in fact, as well as it makes the whole algorithm not being a typical EA at all. This is the reason why the algorithm presented in this work was called by authors *quasi-hierarchical*. Because of lack of space in this publication only the main outline of presented algorithm will be shown.

First, we explain the high level of the algorithm. The representation of the considered problem is as follows. One chromosome is the set of route proposals for all commodities in the network. For instance, chromosome 237 means that first commodity uses the route number 2; second commodity uses route number 3 and etc.

Main operators of high level are introduced in the following. The selection method is based on a roulette wheel method. There are two crossing operators supported: linear and random. The decision which of them is going to be used is made randomly and the user sets proportions. In the linear crossing we randomly pick up a crossing breakpoint and create the offspring by copying the first x genes from first parent and the rest from second; second offspring creation process is analogous. The difference between random order crossing operator and traditional linear crossing is only that there is no crossing point in such operation, instead of that we randomly copy the genes from one of the parents to their offspring in a way that the offsprings are always opposite (if on n position first offspring has a gene from the second parent the second offspring has gene from first parent). Mutation operator is replacing randomly picked up gene (route between the two nodes) by a different proposition of such gene (different route between the same nodes) from the routes database.

The *brainstorm* operator is dealing, for a period of time and in special way, with every individual's fitness function value. Brainstorm isn't a typical operator, it is rather a population state operator. The turn on condition is fulfilled when mean value of population average fitness value of last n generations is not greater then the same average of the previous n generations increased by $x\%$. Consequently, if the following formula is satisfied the brainstorm is turned on

$$\sum_{i=c-n}^{c} F_{avg}(i)/n \leq \left(\sum_{i=c-2n}^{c-n} F_{avg}(i)/n\right)\left(1+\frac{x}{100}\right) \quad (8)$$

where c is the number of current generation, $F_{avg}(i)$ denotes the average value of fitness function for generation number i. The following function defines the value of brainstorm modifier $BM(i)$ for generation i

$$BM(i) = \begin{cases} 0 & \text{if turned off} \quad i < t \\ F_{avg}(i) & \text{if turned on} \quad t \leq i < t+y \\ \frac{(t+y+z)-i}{z} F_{avg}(i) & \text{if turning off} \quad t+y \leq i < t+y+z \end{cases} \quad (9)$$

where t is number of generation for which the brainstorm is turned on. The n, x, y, z and t values are defined by user. The brainstorm operator (population state) is the answer for a typical situation when the population starts to concentrate only around one local maximum area. In such population all individuals are very close to each other and their fitness is also very similar. That is why the whole average fitness starts to become constant. In HEFAN this situation triggers the brainstorm that helps to go out from this situation (so the population is diverse again). Brainstorm modifier is introduced to the fitness function for chromosome p in generation i

$$F(i,p) = z(p) + BM(i) \quad (10)$$

where $z(p)$ denotes value of congestion obtained for routes given in chromosome p.

In the low level of the algorithm the chromosome is represented by a list of network nodes identifiers that the route is going through. Node identifiers are the individual's genes, which means that the number of genes for every individual can be different and that its neighbors determine the value of a gene. For instance, chromosome 2456 denotes a route originating in node 2, going through nodes 4 and 5, and finally reaching the node 6.

The low level operators of HEFAN are as follows. There is no regular fitness function defined for low level. The selection procedure is based on the simple idea which is best described by sentence: "If we can not directly tell which routes are somehow better then the others then we have to believe that we will probably find better routes in better fitting sets of them (the better high level individuals)". In order to select two low level parents we select the high level one and we randomly pick up one of his genes which is in fact the low level individual after that we select another high level individual and take one of his genes that is representing the route between the same nodes as the first already chosen low level parent. In the crossing operator the high level individuals for which the low level parents are from, are rewritten into the new high level population. If the chance decides the low level crossing operator is not to be used, both high level individuals stay the same like they did before. If the low level crossing operator is used they will contain the modified version of routes that were randomly chosen to be the low level parents. The two selected low level individuals of the same type (two routes with the same source and destination node identifier) are crossed by random division of both individuals into two pieces. The first part from first parent is then glued with the second part of second parent giving us the first offspring while the second offspring is made in an analogical way. The two pieces does not have to fit to each other (in fact they will not in most of the cases) because of that we glue them using the randomly generated route between the glued nodes from the routes database.

The low level mutation operator is first checking the mutation chance for every high level individual. If the low mutation operator is to be used we randomly pick up the high level individual gene, which is low level individual and mutate it by replacing the randomly chosen part of it by the proper (having the same origin and destination nodes identifiers as the replaced part) route proposition from the routes database. It should be noted that the low level cross and mutation operators might produce the route that is containing empty loops, which are removed from it before it is shown to the routes database as a new route proposition.

The routes database contains the list of known routes. The new propositions generated during the system work are accepted if they do not exist in the database yet, they are rejected if they already exist. The database asked for the route proposition between the specified nodes pair randomly picks up one of known routes that have the proper origin and destination node identifiers. There must be at least one route proposition between the pair of nodes to start the algorithm.

4 Results

The algorithm was coded in C++. Performance of HEFAN was studied in detail on a family of networks consisting of a master network and a series of progressively sparser networks derived from the master network. In Fig. 1 we show the master network that consists of 36 nodes and 162 directed links, which is referred to as 162. All other tested network called 144, 128, 114, and 104 also have 36 nodes, however the number of links is 144, 128, 114, and 104, respectively. Capacity of network links is set to 5000 BU (bandwidth units). For each network we tested various demand matrices. Each demand matrix includes a full mesh of 1260 demands, i.e. there is a demand for each source-destination node pair and each demand has the same bandwidth requirement. The number of tested demands matrices is 7, 7, 12, 7, and 10, respectively for networks 162, 144, 128, 114, and 104.

The first step of numerical experiments was tuning of the algorithm. Due to experience with HEFAN, preliminary tests and previous research [14] we decided to apply the following values of algorithm's parameters. Probabilities of high level crossing, high level mutation, low level crossing, low level mutation are set to the following values {0.0/0.0/0.7/0.5}, {0.5/0.1/0.0/0.0}, {0.5/0.1/0.3/0.2},

{0.5/0.1/0.6/0.4}. All these four combinations are tested with and without brainstorm modifier. Other parameters are fixed to the same value for all experiments, i.e. number of generations - 1000, the population size - 1000, random crossing possibility – 0.75.

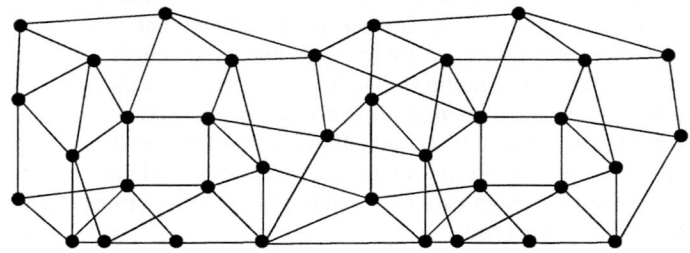

Fig. 1. Topology of network 162

In order to make performance evaluation easier, we introduce a concept of *competitive ration*. The competitive ration, which indicates how well an algorithm performs for a given set of parameters, is defined as the difference between result obtained for a particular simulation and the maximum value of congestion obtained for a considered network topology and traffic matrix. For instance, if for the experiment consisting of 8 simulations (8 different sets of parameters values) the maximum value of congestion is 2000 and the results obtained for one considered combination of parameters is 1500; the competitive ration of this set of parameters is calculated as follows: (2000-1500)/2000=0.25. The competitive ration indicates quality of obtained result of a given parameters compared to results obtained for other parameters. Low value of competitive ration means that the considered combination of parameters yields a result very close to the best obtained result. For presentation of aggregate results we apply the *aggregate competitive ration*, which is a sum of competitive rations over all considered experiments.

In Table 1 we report values of aggregate competitive ration obtained for all 8 tested combinations of algorithm's parameters. The best results for each network is typed bold. Analysis of results suggests that the best performance is provided by the combination 0.5/0.1/0.6/0.4 without brainstorm. It is interesting to note that the algorithm without brainstorm performs a little bit better than with brainstorm. When applied to another non-bifurcated problem with objective function that is a sum of all link metrics our algorithm performed much better with brainstorm operator [14]. The largest improvement was observed for highly congested network in which only a small subset of the solution space is feasible due to the capacity constraint. Thus, the algorithm can focus on a population that concentrates on a small local area, in which individuals are very similar to each other. A direct consequence of this is that the fitness function starts to become constant what starts the brainstorm operator to diversify the population. However, the congestion function considered in this work includes the capacity constraint. In other words, when maximizing the minimum arc residual capacity, the algorithm tries to select feasible (in terms of capacity constraint) combinations of routes. Otherwise, if the capacity constraint is not satisfied, the objective function is negative. Thus, the brainstorm modifier when applied to NBC problem isn't so important as in other non-bifurcated problems. Moreover, in some cases brainstorm can unnecessarily add to the population more diverse chromosomes what prolongs the algorithm's convergence. Also Fig. 2, in which we show convergence of the algorithm as a function of generation number for network 128, we can observe that the algorithm performs better without brainstorm.

Table 1. Aggregate competitive ration of various combinations of algorithm's parameters

Probabilities	Brainstorm	Networks				
		104	114	128	144	162
0.0/0.0/0.7/0.5	On	0.23	0.92	1.33	1.50	0.82
0.5/0.1/0.0/0.0	On	1.25	1.06	1.76	2.14	0.94
0.5/0.1/0.3/0.2	On	**0.02**	0.66	0.38	0.84	**0.00**
0.5/0.1/0.6/0.4	On	0.05	0.70	0.36	0.20	**0.00**
0.0/0.0/0.7/0.5	Off	0.32	0.85	1.25	1.21	0.94
0.5/0.1/0.0/0.0	Off	1.25	1.06	1.83	2.25	0.94
0.5/0.1/0.3/0.2	Off	**0.02**	0.59	0.98	0.11	**0.00**
0.5/0.1/0.6/0.4	Off	0.06	**0.04**	**0.75**	**0.09**	**0.00**

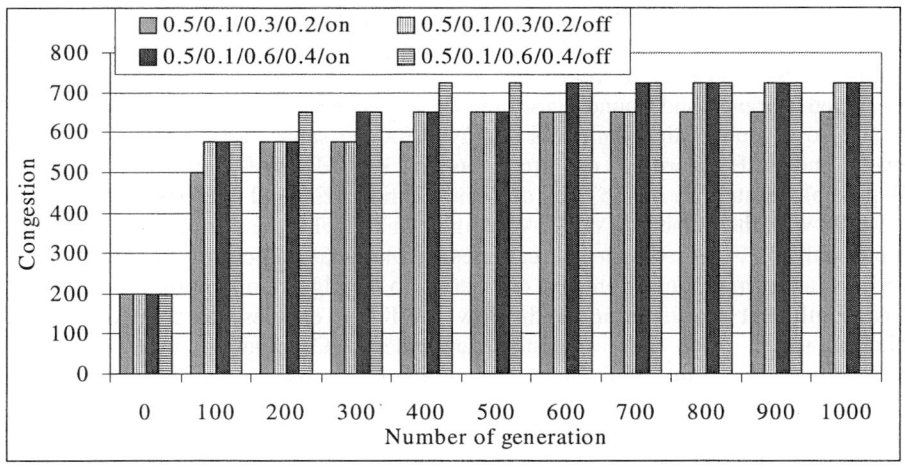

Fig. 2. Convergence of algorithm for various sets of parameters obtained for network 128

To evaluate quality of obtained results we test existing heuristics in the same simulation scenario. For reference we select the following algorithms developed of non-bifurcated versions of the congestion problem. We implement algorithm CFZ proposed by Chlamatac et al. [3]. CFZ minimizes the network load, however the capacity constraint is not taken into account, i.e. links have no limits on the load they can carry. Therefore, we propose a modification of CFZ algorithm (called CFZMod), in which only feasible (in terms of capacity constraint) paths are considered in Step 4 of CFZ. We also apply the shortest-widest path (SWP) algorithm [11]. In particular, we develop two versions of the algorithm using shortest-widest path (SWP) method. In SWPNorm algorithm paths are processed sequentially without any sorting. In SWPSort paths are sorted according to their residual capacity, starting from the path with smallest value of residual capacity. In Table 2 we present relative performance HEFAN and other heuristics.

Generally, HEFAN outperforms other algorithms. Only for network 162 SWP gives the same results as HEFAN. However, HEFAN requires much more time (about 1 hour on a PC with 2GHz processor and 512 MB of RAM) to find a solution while other heuristics need less than second.

Table 2. Aggregate competitive ration of EA and other heuristics

Algorithms	Networks				
	104	114	128	144	162
HEFAN	0.19	0.00	0.00	0.07	0.00
SWP	0.32	1.27	5.26	2.21	0.00
SWPSort	2.50	1.27	1.66	1.29	2.59
CFZMod	8.79	6.54	11.00	6.18	5.29

It is difficult to evaluate the goodness of our method because, as mentioned above, the only way to find optimal solution for 0/1 problems is branch and bound method, which can be used only to very small networks (about 10 nodes). One of authors developed a branch and bound algorithm for similar to NBC non-bifurcated problem [17]. Also a genetic algorithm was deployed for the same problem [16]. However, genetic and B&B algorithms can process only a limited set of candidate routes – not all possible routes are taken into account. Summarizing all 309 tests made for 8 networks with the number of nodes varying from 10 to 14, the genetic algorithm gave results only 0.7% worse than optimal ones. Evolutionary algorithm proposed in this work includes important extensions comparing to the algorithm proposed in [16]. The most significant modification is the low level that enables searching the whole solution space. Also new operators are added. Concluding, we expect that evolutionary algorithm HEFAN presented above is able to find results close to optimal ones.

5 Concluding Remarks and Further Work

The main idea of developed algorithm is based on the two-level structure. The low level enables expanding the search space to all possible routes in the network. Previous genetic algorithms developed for non-bifurcated flow problems usually process on a fixed set of candidate routes [16]. Also the branch-and-bound algorithm requires a big reduction of candidate routes if the solution is to be found in acceptable time. Therefore, low level is a very important modification from the perspective of computational efficiency. It should be underlined that the proposed algorithm can be applied for a wide range of optimization problems encountered in c-o networks using as objective functions network cost, delay and many other criteria. One of possible modifications of presented approach may be changing the current algorithm so that it can also design the network topology for a given demand and route sets.

Acknowledgements. This work was supported by a research project of the Polish State Committee for Scientific Research carried out in years 2005-2007.

References

1. Bienstock, D., Gunluk, O.: Computational experience with a difficult multicommodity flow problem. Math. Programming, 68 (1995), 213-238
2. Bienstock, D.: Potential function methods for approximately solving linear programming problems, Theory and Practice. Kluwer Academic Publishers (2002)
3. Chlamatac, I., Fargo, A., Zhang, T.:Optimizing the System of Virtual Paths. IEEE/ACM Trans. Networking, 6 (1994), 581-587
4. Davis. L.: Handbook of Genetic Algorithm. Van Nostrand Reinhold. New York (1996)
5. Duhamel, C., Vatinlen, B., Mahey, P., Chauvet F.: Minimizing congestion with a bounded number of paths, in Proceedings of ALGOTEL'03 (2003), 155-160
6. Grover, W.: Mesh-based Survivable Networks: Options and Strategies for Optical, MPLS, SONET and ATM Networking. Prentice Hall PTR, Upper Saddle River, New Jersey (2004)
7. Kasprzak, A.: Exact and Approximate Algorithms for Topological Design of Wide Area Networks with Non-simultaneous Single Commodity Flows. Lectures Notes In Computer Science, LNCS 2660, (2003) 799-808
8. Kleinberg, J.: Single-source unsplittable flow. In Proc. of FOCS '96, pp. 68-77, 1996.
9. Kolman, P., Scheideler, C.: Improved bounds for the unsplittable flow problem. In Proc. of the Symposium on Discrete Algorithms (2002), 184-193
10. Kwasnicka, H.: Genetic and Evolutionary Algorithms - an Overview. Division of Computer Science, Wroclaw University of Technology (1998)
11. Ma, Q., Steenkiste, P.: On Path Selection for Traffic with Bandwidth Guarantees. In Proc. of IEEE International Conference on Network Protocols (1997), 191-202
12. Michalewicz, Z.: Genetic Algorithms + Data Structures = Evolution Programs. 3rd edn. Springer-Verlag, Berlin Heidelberg New York (1996)
13. Pióro, M., Medhi, D.: Routing, Flow, and Capacity Design in Communication and Computer Networks. Morgan Kaufman Publishers (2004)
14. Przewozniczek. M.: Genetic algorithms in use of routes finding in computer connection oriented networks. Diploma work. Chair of Systems and Computer Networks, Wroclaw University of Technology (2003)
15. Radcliffe. N.. Surry. P.: Co-operation through Hierarchical Competition in Genetic Data Mining. Technical Report EPCC-TR94-09, Edinburgh Parallel Computing Center (1994)
16. Walkowiak, K.: Genetic approach to virtual paths assignment in survivable ATM networks. In Proc. of 7th International Conference on Soft Computing MENDEL (2001) 13-18
17. Walkowiak K.: A Branch and Bound Algorithm for Primary Routes Assignment in Survivable Connection Oriented Networks. Computational Optimization and Applications 2 (2004), 149-171.

Design and Development of File System for Storage Area Networks

Gyoung-Bae Kim[1], Myung-Joon Kim[2], and Hae-Young Bae[3]

[1] Dept. of Computer Education, Seowon University, 231 Mochung-dong Heungduk-gu,
Cheongju-shi Chungbuk, South Korea
gbkim@seowon.ac.kr
[2] Internet Server Group, Digital Home Research Division, ETRI,
161 Gajung-dong, Yusung-gu, Daejeon, South Korea
joonkim@etri.re.kr
[3] Dept. of Computer Science Engineering, Inha University, 250 Younghyun-dong,
Nam-gu, Incheon, South Korea
hybae@inha.ac.kr

Abstract. By merging network and channel interfaces, resulting interfaces allow multiple computers to physically share storage devices. A storage area network (SAN) is a high-speed special-purpose network (or subnetwork) that interconnects different kinds of data storage devices with associated data servers on behalf of a larger network of users. In SAN, computers service local file requests directly from shared storage devices. Direct device access eliminates the server machines as bottlenecks to performance and availability. Communication is unnecessary between computers, since each machine views the storage as being locally attached. SAN provides us to very large physical storage up to 64-bit address space, but traditional file systems can't adapt to the file system for SAN because they have the limitation of scalability.

In this paper, we present architectures and features of SANtopia that allows multiple machines to access and share disk and tape devices on a Fibre Channel or SCSI storage network in Linux system. It performs well as a local file system, as a traditional network file system running over IP environments, and as a high performance cluster file system running over storage area networks like Fibre Channel. SANtopia provides a key cluster enabling technology for Linux, helping to bring the availability, scalability, and load balancing benefits of clustering to Linux.

1 Introduction

In recent years, speed and responsiveness of computer systems is often determined by network and storage subsystem performance. Faster and more scalable networking interfaces like Fibre Channel (FC) and Gigabit Ethernet provide the scaffold from which higher performance implementations may be constructed, but new methods are required about how machines interact with network enabled storage devices [1].

Computer architectures have for many years struggled with the problem of fast and efficient transfer of data between many memory and external storage devices. Recently, it had been convenient to point out that disk devices were three or four orders of magnitude slower than main memory and that their rate of capacity increase, access time decrease, and bandwidth increase were below the corresponding rate for both processor and main memory.

Data communication has traditionally been divided into two quite distinct worlds, the network world and the channel world. The channel usually deals with well-structured, closed and fixed environments. Usually there is one host system and several attached devices, like a master-slave setup. Communication is usually restricted to occur only between the master and one of the attached devices. This means that control over the communication is simpler and more of the resources can be used for actual data transport. The bandwidth is often high and latency low on such buses. However, the architecture is inflexible and fixed. The basic SCSI bus with just one host computer is an example of such a channel environment [2].

In the network the picture is quite different. Here the environment is unstructured, open and behaves much more unpredictably. All devices can talk to any other device at any time. One consequence of this is that more control is needed to correctly handle connections, access permissions, route information and other aspects of correct behavior. The result is that Fibre Channel often are characterized by high throughput and low overhead, while networks tend to have lower throughput and high overhead.

This interface technology combines both network and storage features and provides an industry standard, high bandwidth, switched interconnection network between client and devices. Theses developments made possible us to designing new storage architectures, one that assumes disks are highly capable peer devices available directly in a network.

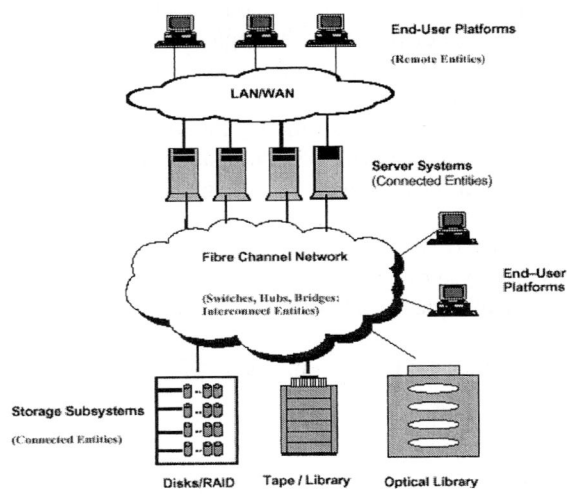

Fig. 1. SAN Environment

Traditional local and distributed file systems support a persistent name space by creating a mapping between blocks found on disk drives and a set of files, file names, and directories. These file systems view devices as local: devices are not shared, hence there is no need in the file system to enforce device sharing semantics. Instead, the focus is on aggressively caching and aggregating file system operations to improve performance by reducing the number of actual disk accesses required for each file system operation [3], [4].

New networking technologies allow multiple machines to share the same storage devices. File systems that allow these machines to simultaneously mount and access files on these shared devices are called shared file systems [5], [6]. Shared file systems provide a server-less alternative to traditional distributed file systems where the server is the focus of all data sharing. As shown in Figure 1, machines attach directly to devices across a storage area network (SAN)[7]. A SAN consist of a local area network that allows storage devices to be directly attached to the network.

In this paper we present a global shared file system called SANtopia[8],[9],[10],[11] that allows multiple machines to access and share disk and tape devices on a Fibre Channel or SCSI storage network in Linux system. It will perform well as a local file system, as a traditional network file system running over IP environments, and as a high performance cluster file system running over storage area networks like Fibre Channel. SANtopia provides a key cluster enabling technology for Linux, helping to bring the availability, scalability, and load balancing benefits of clustering to Linux.

Our goal is to develop a scalable, in number of clients and devices, capacity, connectivity server-less global file sharing system for the Fibre channel-based SAN. It exploits the speed and device scalability of SAN clusters, and provides the client scalability and network interoperability of SAN.

SANtopia views storage as a large storage pool-a collection of network attached storage devices logically grouped to provide node machines with a unified large storage space. Theses storage pools provided by logical volume manager (LVM) are not owned or controlled by any one machines but rather act as shared storage to all machines and devices on the network.

The rest of the paper is organized as follows: Section 2 describes and analyzes traditional file system. Section 3 describes SANtopia architecture and each component. In Section 4, we explain in detail logical volume manager of SANtopia that provides client to storage pools and software RAID. In Section 5, we present file manager of SANtopia for global shared file system. The file manager provides VFS interfaces, metadata managements, system calls, and recovery. In section 6, we describe buffer and lock manager of SANtopia. In SANtopia, we integrated lock manager and buffer manager to reduce communication overhead. In section 7, we evaluate SANtopia file system compare with traditional file system, such as EXT2 and GFS. We conclude the paper in Section 8.

2 Related Works

Most UNIX file system types have a similar general structure, although the exact details vary quite a bit. The central concepts are *superblock*, *inode*, *data block*, *directory block*, and *indirection block*. The superblock contains information about the file system as a whole, such as its size (the exact information here depends on the file system). An inode contains all information about a file, except its name. The name is stored in the directory, together with the number of the inode. A directory entry consists of a filename and the number of the inode which represents the file. The inode contains the numbers of several data blocks, which are used to store the data in the file. There is space only for a few data block numbers in the inode, however, and if more are needed, more space for pointers to the data blocks is allocated dynamically. These dynamically allocated blocks are indirect blocks; the name indicates that in order to find the data block, one has to find its number in the indirect block first.

Linux is a Unix-like operating system, which runs on PC-386 computers. It was implemented first as extension to the Minix operating system and its first versions included support for the Minix file system only. The Minix file system contains two serious limitations: block addresses are stored in 16 bit integers, thus the maximal file system size is restricted to 64 mega bytes, and directories contain fixed-size entries and the maximal file name is 14 characters. We have designed and implemented two new file systems that are included in the standard Linux kernel. These file systems, called "Extended File System" (Ext fs) and "Second Extended File System" (Ext2 fs) [12] raise the limitations and add new features. Every Linux file system implements a basic set of common concepts derivate from the Unix operating system files are represented by inodes, directories are simply files containing a list of entries and devices can be accessed by requesting I/O on special files.

Frangipani [13] is a new scalable distributed file system that manages a collection of disks on multiple machines as a single shared pool of storage. The machines are assumed to be under a common administration and to be able to communicate securely. One distinguish feature of Frangipani is that it has a very simple internal structure-a set of cooperating machines use a common store and synchronize access to that store with locks. This simple structure enables us to handle system recovery, reconfiguration, and load balancing with very little machinery.

Frangipani is layered on top of Petal[14], an easy-to-administer distributed storage system that provides virtual disks to its clients. Like a physical disk, a Petal virtual disk provides storage that can be read or written in blocks. Unlike physical disk, a virtual disk provides a sparse 2^{64} bytes address space, with physical storage allocated only on demand. Petal also provides efficient snapshots to support consistent backup. Frangipani inherits much of its scalability, fault tolerance, and easy administration from the underlying storage system, but careful design was required to extend these properties to the file system level.

XFS [15] is the next generation local file system for SGI's workstations and servers. It is a general purpose Unix file system that runs on workstations with 16 megabytes of memory and a single disk drive and also on large SMP network servers with gigabytes of memory and terabytes of disk capacity. In this paper we describe the XFS file system with a focus on the mechanisms it uses to manage large file systems on large computer systems. The most notable mechanism used by XFS to increase the scalability of the file system is the pervasive use of B+ trees [16]. B+ trees are used for tracking free extents in the file system rather than bitmaps. B+ trees are used to index directory entries rather than using linear lookup structures. B+ trees are used to manage file extent maps that overflow the number of direct pointers kept in the inodes. Finally, B+ trees are used to keep track of dynamically allocated inodes scattered throughout the file system.

GFS(Global File System) [17] is the industry's most advanced and mature scalable file system. Recognized as the de facto cluster file system on Linux, GFS is a highly stable solution for enterprise and technical computing applications requiring reliable access to data. GFS allows multiple servers on a Storage Area Network

(SAN) to have read and write access to a single file system on shared SAN devices, delivering the strength, safety and simplicity demanded by enterprise and technical computing environments.

3 Architecture of SANtopia

Figure 2 shows the architecture of SANtopia supports global file sharing in SAN environments. The SANtopia consists of four parts: global file manager, system manager, global buffer and lock manager, and logical volume manager.

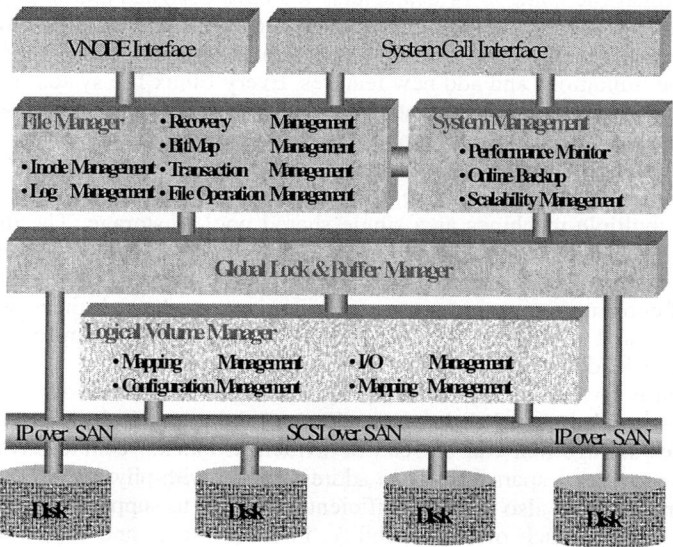

Fig. 2. Architecture of SANtopia

The users require the operation of SANtopia file system using VNODE operation and system call in VFS (Virtual File System).The global file manager of SANtopia processes metadata structures and its operation for functions of large volume file sharing. Each operation is defined as transaction that is a unit of file operation and recovery. The system manager provides performance monitoring, on-line backup, and the system configuration of SANtopia. Using system manager tools, system manager is able to attach or detach some nodes, and add and delete shared disks for logical volume. The global lock and buffer manager gives the function of global buffer sharing to minimize the disk I/O. Especially, to reduce communication overhead between local server and global server for coherency of buffer, we integrate the lock manager controls the file access and the buffer manager provides file sharing in SANtopia.

The logical volume manager provides file manager to a large logical volume (storage pool) is coalesced into a heterogeneous collection of shared storages. There are several function in LVM, 1) addition of physical disk into logical volume, 2) software

RAID, 3) mapping logical address into physical address, 4) transfer data page form disk into buffer. In the rest of this paper, we illustrate designs approaches each component of SANtopia.

4 Volume Manager

The Logical Volume Manager (LVM) supports the abstraction of a single unified logical storage address space for SANtopia client. The Logical Volume Manager is implemented in a device driver level on top of the Fibre Channel Drivers and the SCSI device. It also provides SANtopia File Manager to coalesce a heterogeneous collection of shared storage into a single logical volume. This manager translates from logical address of the file system to physical address space of each device.

A logical volume layout consists of private partition and public partition. The private partition, one of physical partition in physical disk, contains disk identifier, logical partition information, and information of logical volume for management of logical volume layer. The public partitions consist of sub-logical partitions. Each sub-logical partition has allocation bitmap, mapping information, and data blocks. To resize SANtopia file system the LVM The LVM is able to attach or detach sub-logical volumes.

The Redundant Array of Inexpensive Disks (RAID) is a popular way of improving the system I/O performance and reliability. There are different levels of disk array that cover the range of possibilities between: improving system I/O performance and increased reliability. The large file systems require the capacity of several disks, but most file systems must be created on a single disk. A hardware RAID device is one solution to this problem. There are many excellent benefit of hardware RAID, but it is an expensive solution because a logical large single disk is constructed many small disks in SAN environment. The SANtopia LVM provides clients with a high performance and reliable SAN storage capacity in software RAID. The LVM software RAID driver will implement RAID levels 0, 1, 0+1, 5, concatenation (linear).

The mapping manager of LVM provides file manager with the virtualization of physical storage. The LVM is able to move data between logical partitions for performance or reorganization disk data such as disk space is full and also to support snapshots. Snapshots provide an efficient way to backup an image of any file system built on an LVM. The LVM easily implements snapshot using the functions of mapping manager.

5 File Manager

5.1 Layout of Huge File System

In the traditional file system, it divides the allocation area into inodes, directories, inode bit maps, and data blocks. The number of inode bit map entries for each inode is allocated one entry per 4KB data blocks in the Linux file system. This causes serious problems since the inode table will be full, in spite of the fact that the file system has free space, when a huge number of huge small files exist in file system.

For supporting large file systems, SANtopia uses 64-bit address space. The layout of the SANtopia file system consists of a boot area, super block, allocation area, and bitmap area. The boot area stores the booting images for the operating system. The super block stores the information on the file system such as the total size of the file system, block size, bitmap size, and several traditional super block data.

Therefore, there is no division in the allocation area among inodes, directories, and data blocks. Since information on the allocation area manages the bitmap table using two bits, we assigned the value of bit map as follows: 00 is a free allocation blocks (extent), 01 is an inode, 10 is a directory entry, and 11 is a data extent. Using this method, there is no limitation in the number of inode table entries.

5.2 Dynamic Bitmap

The proposed mechanism divides disk between allocation area and bitmap area. Each bitmap has the information of allocation area. One bit of bitmap corresponds to one allocation group such as extent. For example, if block size is 1,024 bytes and allocation unit is 1 block and there are 1,000 bits for allocation group, then allocation group size is 1,000.

The bitmap size in file system decided by follows:

$$BTS = (DS - (SBT + BTS)) / (1 - (BLS - BHS))$$

BS : size of bitmap
DS : total disk size
BTS : size of booting area
SBS : size of super bock
BLS : size of block
BHS : size of bitmap header

Fig.3 shows the structure of dynamic bitmap. It consists of bitmap header and allocation bitmap area. The bitmap header has control information of dynamic bitmap such as inode type, total bitmap bit, and used bit count. In Fig.3, bitmap type is inode and 40 bits are used among 60 bits

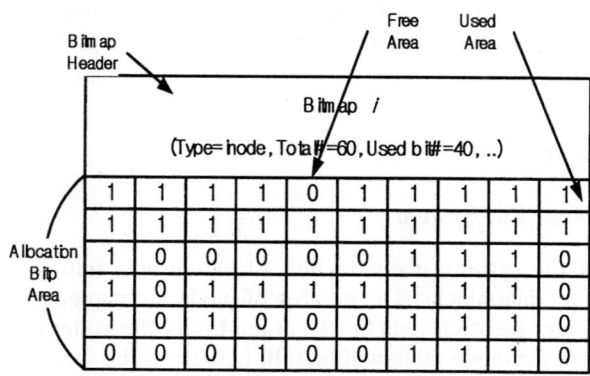

Fig. 3. Structure of the Dynamic bitmap

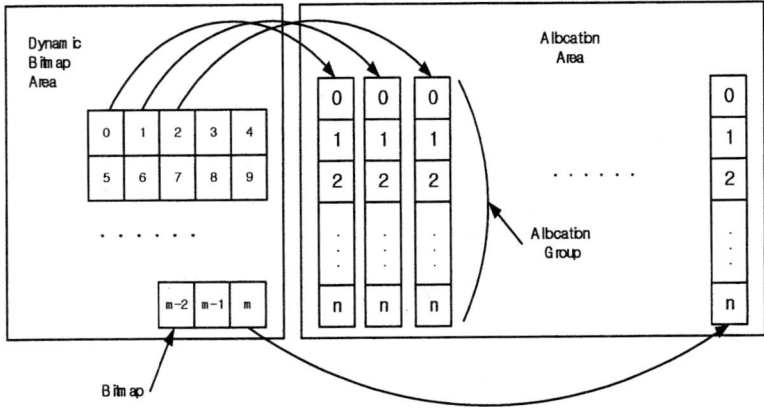

Fig. 4. Mapping dynamic bitmap and allocation area

Figure 4 describes the mapping dynamic bitmap and allocation area.

Each allocation area manages allocation group, which divides all disk allocation area. In Fig.4, the size of allocation group assigned in file system is $n+1$, and total number of bitmap is $m+1$. So, total size of file system is $(n+1) * (m+1)$.

5.3 Directory Management

It is one of the serious problems when we are adapted to traditional file systems in SAN environment that they don't perform well is large directories. Most traditional file systems, such as UNIX file system (UFS) and Linux file system (Ext2), were designed for small size file system. So, they have limitation makes number of files in directory. And also they store files in directory as unsorted linear list of directory entries. This is satisfactory for small directories, but it increases directory operations time for big ones. On average, we must search through half of the directory to find any one entry in the directory. In traditional directory structure, large directories can take up megabytes of disk a space, so this is costly not only in CUP time but also it takes executive time for I/O to disks. There are two alternatives are proposed for large directory. One approach is using B-tree for entries in directory such as xFS file system. The other approach is Extensible Hashing such as GFS.

The SANtopia file system is adapted to Extensible Hashing for directory structure. Extensible Hashing provides a way of storing a directory entry so that any on entry can be found very quickly. But basic Extensible Hashing method is not design for large directory, so we modified traditional Extendible Hashing data structure for extent-based large directory is called Extent-based Multi-Level Extensible Hashing (EBMLEH). Fig.5 shows our EBMLEH structure.

The EBMLEH uses a multi-bit hash of each filename. A subset of the bit is used a unique index in a hash table. The allocation unit of EBMLEH is the extent is decided in making file system. Each pointer of leaf hash table points to leaf extent block that

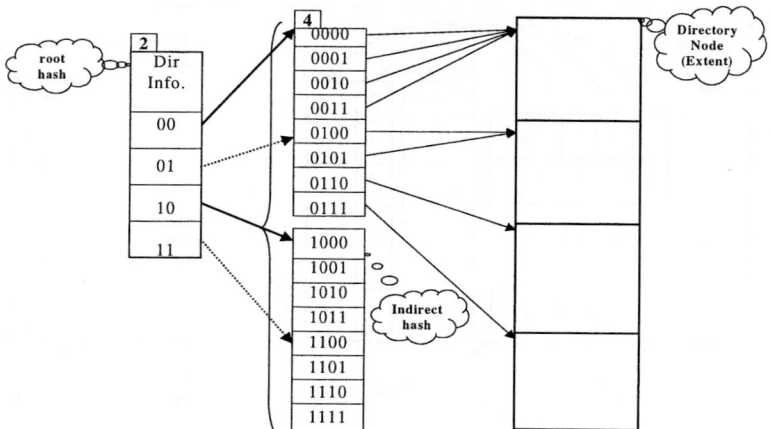

Fig. 5. Extent-Based Multi-Level EH for directory

contain the directory entries. The EBMLEH has multi-level hashing structures that contain a root node and indirect nodes that are set of hash values and block pointers in the hashing tables. Each node has indicator that represents the bit size of hash value to be compared in the hash table.

The small size directory only uses root hash node, the pointer of root hash node directly point to directory entries block. But, in large directory, the capacity of root node will be reached its limitation when directory entries are increased more and more. The EBMLEH uses indirect hash nodes for that case. In Fig. 5 example, if the capacity of root node is 4 and indicator of hash bits is 2 in root node, the directory entries already exceeded to its limitation, so the root node has 4 indirect hash nodes. The root node points to the indirect hash node and each indirect hash nodes point to real block stores directory entries. We need two steps to fine any particular directory entry in Fig. 5. The number of directory entries, is able to point directory blocks, is increased rapidly when the root node splits into indirect hash nodes in EBMLEH.

6 Buffer and Lock Manager

The performance of file system is very sensitive to the buffer management scheme. The SANtopia provides file system to local buffer and global buffer. To reduce the disk I/O, each node will need a local buffer for recently accessed data. On the other hands, global buffer has the total information of each local buffer.

The clients of SANtopia require at first the particular page to local buffer manager, if it not found the page in local buffer lists then the local buffer manager requires this page to global buffer manager. The global buffer manager searches the required page by local buffer manager for global buffer list and transfers the page.

To reduce the disk I/O, each node will need a local buffer list for recently accessed data. Buffer coherency needs to be maintained such that if any data page is updated by a node, the old copies of this page presents in the buffer of other nodes must be invalidated. The buffer invalidation scheme reduces the buffer hit probability in SAN environments. For non-shared system, a large buffer implies a higher buffer hit probability, and with sufficient buffer, we can archive any buffer hit probability. However, in global data sharing system, a large buffer also leads to a higher buffer invalidation rate.

In SANtopia, we integrate buffer and lock message. The lock manager controls data access. All users must be obtained the permission of some file access from the lock manager. When client requires the lock of file, the local lock manager transfers concurrently lock message pigged with local buffer list from local buffer to global buffer manager.

There are two types' hosts in SANtopia. The one is the host only has local buffer and lock manager and manages list of local resource. The other is the host not only has local buffer and lock manager but also has global buffer list and global lock list. Each local manager transfers the local information to global manager and gets the global information from global manager.

7 Performance Evaluation

To evaluate the performance of the SANtopia, we develop under Redhat 8.0. Each mode has HP machine with 750 MHz Dual CPU and 512MB main memory. For verifying our experiments, we make use of arbitrary file read/write system calls in hosts. Table 1. shows the parameters and values using performance evaluation.

Table 1. Performance Parameters

Parameter	Value
Total disk size	1 Tera Byte
Allocation sizes	4,096 Byte
Size of bitmap block	1,024 Byte
Size of bitmap header	24 Byte
# Of inode per bitmap	1,000
Size of unit inode	128 Byte
File size	10MB ~ 500MB
Block size	4KB
Transfer Size	1MB

Total disk size is 1 Tera byte, allocation sizes 4KB, and the size of bitmap block is 1,024B. In Fig. 6 and 7 shows the performance result of the proposed dynamic bitmap scheme. Fig. 6 show total number of files when file size growing. Our approach has more files than static bitmap until file size comes to 4KB.

In Fig. 7 our dynamic bitmap outperform than static bitmap in the disk usage outperform.

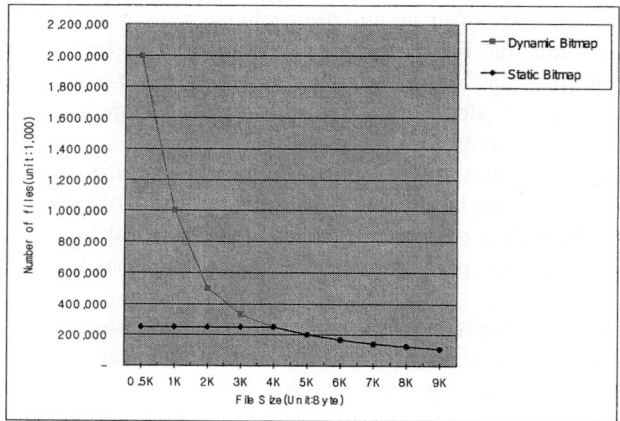

Fig. 6. Number of files according to file size

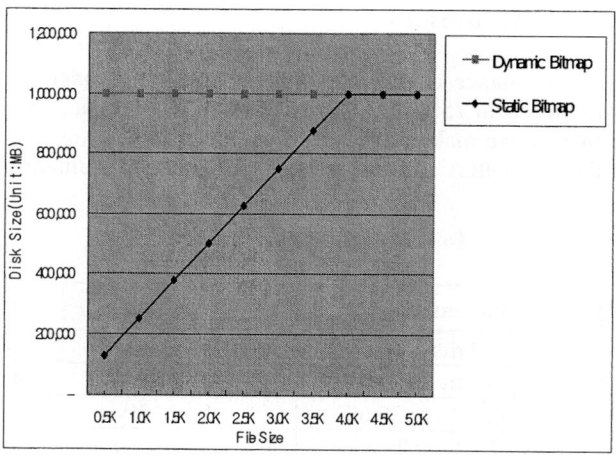

Fig. 7. Disk usage according file size

We evaluate the average response time for making a comparison between EXT2, GFS, and SANtopia. In Fig.8, the data size varies from 10MB to 500MB. The result time consists of computation time and blocks I/O time for metadata and file data. The graph increases as the data size grow in graceful manner. But, SANtopia provides the journaling faculty that it caused by more journaling I/O. In Fig. 8' below graph show an average response time between GFS and SANtopia. The semi-flat structure of SANtopia reduces the metadata I/O in terms of an inode extension.

In Fig. 9, we evaluate SANtopia's creation time for making a comparison between EXT2 and GFS. In Fig.9's upper graph, SANtopia creation time is superior to Ext2. This result show that our file system's extent-based allocation reduces an average creation time. As show in Fig. 9's below graph, the directory creation time of SANtopia and GFS show that the semi-flat structure of inode reduces the indirect block's I/O.

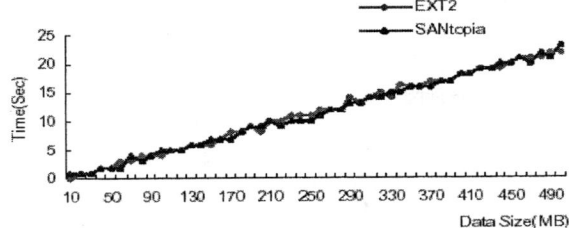
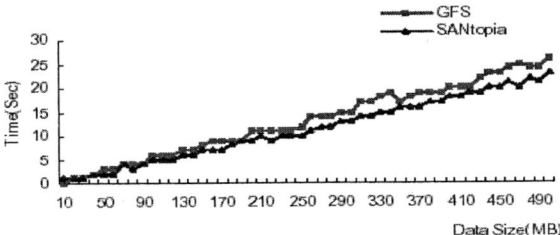

Fig. 8. Average response time

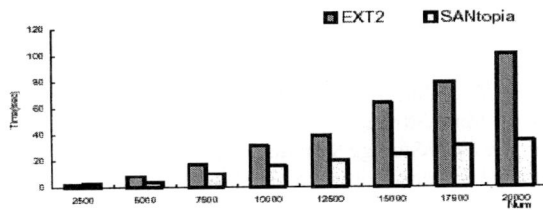
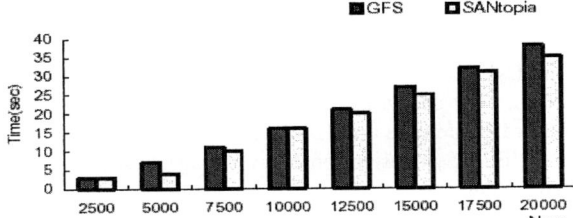

Fig. 9. Ext2/GFS/SANtopia directory creation time

8 Conclusion

In this paper, we present architectures and design features of shared disk file system, so-called SANtopia, that allows multiple machines to access and share disk and tape devices on a Fibre Channel or SCSI storage network in Linux system. It will perform well as a local file system, as a traditional network file system running over IP environments, and as a high performance cluster file system running over storage area networks. The SANtopia provides a key cluster enabling technology for Linux, helping to bring the availability, scalability, and load balancing benefits of clustering to Linux.

SANtopia views storage as a large storage pool-a collection of network attached storage devices logically grouped to provide node machines with a unified large storage space. Theses storage pools provided by logical volume manager (LVM) are not owned or controlled by any one machines but rather act as shared storage to all machines and devices on the network. And we evaluate the SANtopia file system compare with Ext2 and GFS. The results of performance evaluations show that SANtopia outperforms than EXT2 and GFS.

Acknowledgment

This research was supported by the MIC(Ministry of Information and Communication), Korea, under the ITRC(Information Technology Research Center) support program supervised by the IITA(Institute of Information Technology Assessment)

References

1. Kenneth W. et al., "A 64-bit, Shared Disk File System for Linux," The 7th NASA Goddard Conference on Mass Storage System and Technologies in cooperation with the 16th IEEE Symposium on Mass Storage Systems, pp.22-41, San Diego, USA, March 1999.
2. Erik Riedel, "Storage System: Not just a bunch of disks anymore," ACM QUEUE, pp.32-41, 2003.
3. L.McVoy and S.Kleiman, "Extent-like performance from a Unix file system," Proceedings of the 1991 Winter USEUNIX Conference, pp.33-34, Dallas, TX, June 1991.
4. Uresh Vahalia, Unix Internals: The New Frontiers, Prentice-Hall, 1996.
5. Matthew T. O'Keefe, "Standard file systems and fibre channel", The Sixth Goddard Conference on Mass Storage System and Technologies in cooperation with the Fifteen IEEE Symposium on Mass Storage Systems, pp.1-16, Colleage Park, Maryland, March 1998.
6. Steve Soltis et al., "The design and performance of a shared disk file system for IRIX", The Sixth Goddard Conference on Mass Storage System and Technologies in cooperation with the Fifteen IEEE Symposium on Mass Storage Systems, pp.41-66, Colleage Park, Maryland, March 1998.
7. K. Matthews. "Implementing a Shared File System on a HiPPi disk array," Fourteenth IEEE Symposium on Mass Storage Systems, pp.77-88, September 1995.
8. G.B.Kim et al, "Global File Sharing System for SAN," The 3rd International Conference on Advanced Communication Technology, pp.870-874, Feb. 2001.

9. G.B.Kim et al, "A dynamic bitmap for huge file system in SANs," Recent Advances in Computers, Computing and Communications, WSEAS press, pp.229-234, 2002.
10. Yong-Ju Lee and Gyoung-Bae Kim, " Design and Implementation of a Large-Scale Directory Structure in Linux System," 2nd International Conference on Computer and Information Science, KIPS, pp.386-390, 2002.
11. G.B.Kim et al, "Design and Implementation of File System for Storage Cluster," " The 5th International Conference on Advanced Communication Technology, pp.121-126, Feb. 2003.
12. R. Card, "Design and Implementation of the Second Extended File System," Proceeding of the first Dutch International Symposium on Linux, 1995.
13. Chandramohan A. et. al, "Frangipani: A Scalable Distributed File System," Proceedings of the 16th ACM Symposium on Operating Systems Principles, pp. 224-237, St. Malo, France, October 1997.
14. Edward K. Lee and Chandramohan A. Thekkath, "Petal: Distributed Virtual Disks," Proceedings of the 7th ACM International Conference on Architectural Support for Programming Languages and Operating Systems, pp. 84-92, Cambridge, Massachusetts, October 1996.
15. http://oss.sgi.com/projects/xfs/
16. Douglas Comer, "The Ubiquitous B-Tree," Computing Survey, 11(2), pp.121-137, 1979.
17. http://www.sistina.com/gfs/

Transaction Reordering for Epidemic Quorum in Replicated Databases

Huaizhong Lin[1,*], Zengwei Zheng[1,2], and Chun Chen[1]

[1] College of Computer Science, Zhejiang University, 310027 Hangzhou, China
{linhz, zhengzw, chenc}@zju.edu.cn
[2] City College, Zhejiang University, 310015 Hangzhou, China
zhengzw@zucc.edu.cn

Abstract. Epidemic model gives an efficient approach to transaction processing of replication systems in weakly connected environments. The approach has the advantages of high adaptation, support for low-bandwidth network, committing updates in an entirely decentralized control fashion, and so on. Optimistic voting protocol of epidemic model, which introduces condition and order vote in the voting process of transactions, takes an optimistic approach in conflict reconciliation. Though it brings down transaction abort rate dramatically when compared to other implementing protocols, optimistic voting protocol also rollbacks transactions unnecessarily. In this paper, transaction reordering technique is proposed in optimistic voting protocol to reduce the probability of transaction aborts. The protocol ensures one-copy serializability and does not lead to any deadlock situation. Experimental results suggest that the transaction reordering technique decreases abort rate and improves average response time of transactions markedly.

1 Introduction

Weakly connected environments, which are characterized by low bandwidth, excessive latency, instability of connection, and constant disconnection, are used more and more frequently [1]. Data replication is the common approach to improve system performance and availability. But due to the massive communication overhead in weakly connected environments, eager replication may bring about unacceptable number of failed or blocked transactions, and result in dramatic drop of system performance [2].

Epidemic model [3-7] gives another approach for managing replicated data. In an epidemic approach, sites perform update operations locally and communicate peer-to-peer in a lazy manner to propagate updates. Transactional consistency is achieved by decentralized conflict detect and reconciliation. Sites communicate in a way that maintains the causal order of updates and the communication can pass through one or more intermediate sites. Therefore, the epidemic model provides an environment that is tolerant of communication failures and doesn't require continuous connection

---

* Supported by the Natural Science Fundation of Zhejiang Province, China (Grant no. M603230) and the Research Fund for Doctoral Program of Higher Education from Ministry of Education, China (Grant no. 20020335020).

between sites. Epidemic model is suitable for transaction processing of replication systems in weakly connected environments.

Epidemic algorithms mimic the spread of a contagious disease [8]. Just as infected individuals pass on a virus to those with whom they come into contact. Epidemic algorithms have recently gained popularity as a potentially effective solution for information dissemination in large-scale systems, particularly peer-to peer systems deployed on Internet or ad hoc networks.

Several protocols have been proposed for implementing epidemic model, like ROWA (Read-One Write-All) protocol [3], quorum protocol [4], voting protocol [6,7], and optimistic voting protocol [9], etc. Optimistic voting protocol of epidemic model, which introduces condition and order vote in the voting process of transactions, takes an optimistic approach in conflict reconciliation. Condition vote postpones the final decisions on conflicting transactions and therefore improves the chances for transactions to get yes vote. Order vote prescribes the commit order of transactions that have read-write and write-write conflicts and thus eliminates transaction aborts due to these kinds of data conflicts. Though it brings down transaction abort rate dramatically when compared to other implementing protocols, like ROWA protocol, quorum protocol, and voting protocol, optimistic voting protocol also rollbacks transactions unnecessarily.

In this paper, we propose transaction reordering technique in optimistic voting protocol to further reduce the probability of transaction aborts. The transaction reordering technique works as follows. When a transaction T is tested for conflict at a given replication server, we not only test if T can be serialized after other transactions, but also find out if T can be serialized somewhere before other transactions. By this way, we can eliminate some write-read conflict. In the vote process we introduce advance vote which tells servers to commit the transaction before certain other transactions and when it is committed really after certain other transactions, some updates of the transaction should be omitted. The protocol ensures one-copy serializability of update transactions and does not lead to any deadlock situation. The transaction reordering technique decreases abort rate and improves average response time of transactions markedly when compared to other protocols.

The rest of the paper is organized as follows. In section 2, we develop the necessary background and introduce the optimistic voting protocol briefly. In section 3, we give a detail description of transaction reordering technique. In section 4, we perform the performance evaluation. We conclude the paper in section 5.

2 Optimistic Voting Protocol

2.1 Epidemic Model

We consider a distributed system consisting of n sites labeled $S_1, S_2, ..., S_n$ and data items replicated fully or partially at all sites. Epidemic model assumes a fail-stop model of site failures and an unreliable communication medium. Sites communicate each other through messages passing in a pair-wise manner. Messages can arrive in any order, take an unbounded amount of time to arrive, or may be lost entirely, however, messages will not arrive corrupted. For this reason, timeout is not used in the protocols to detect conflicts and deadlocks.

Epidemic model is based on the causal delivery of log records where each record corresponds to one transaction instead of one operation. An event model [4] is used to describe the system execution, $(E, \rightarrow)$, where E is a set of transaction events and $\rightarrow$ is the happened-before relation which is a partial order on all events in E. The partial order $\rightarrow$ satisfies the following two conditions:

(1) Events occurring at the same site are totally ordered;
(2) If e is a sending event and f is the corresponding receiving event, then e$\rightarrow$f.

Vector clocks are used to ensure the property that if two events are causally ordered, their effects should be applied in that order at all sites. Each site S_k keeps a two-dimension time-table, which corresponds to S_k's most recent knowledge of the vector clocks at all sites. Upon communication, S_k sends a message including its own time-table and all records that receiving site hasn't received. Then the receiving site processes the events according to causal order and incorporates the time-table in an atomic step to reflect the new information from S_k.

2.2 A Transaction's Life

Upon completion of operations, a read-only transaction can be committed locally whereas an update transaction pre-commits and becomes a candidate. The read set, write set, and the update values of the candidate are recorded in log. Then sites exchange their respective log records to detect global conflicts and propagate values written by the transaction. A candidate is voted on and is eventually either committed (if it wins a plurality of the total system votes) or aborted.

When a transaction pre-commits, it is attached with a global distinct timestamp denoted by (local_ts, site_index), which is composed of a local timestamp and a distinct site index.

Formally, we define a total order < on timestamps as follows. Suppose two timestamps of transactions $ts(T_1)=(local\_ts_1, site\_index_1)$ and $ts(T_2)=(local\_ts_2, site\_index_2)$, then $ts(T_1)<ts(T_2)$ if and only if:

(1) $local\_ts_1 < local\_ts_2$, or
(2) $local\_ts_1 = local\_ts_2$ and $site\_index_1 < site\_index_2$.

The information of local timestamp is piggybacked in the usual epidemic messages and a site adjusts its local timestamp as follows: when site A receives a message from site B, it advances its local timestamp to max{ $local\_ts_A$, the $local\_ts_B$ carried by message}.

2.3 Condition and Order Vote

We present the following definitions.

Definition 1. For each transaction T, we define two sets: the ReadSet(T), consisting of the data items that are read by T; the WriteSet(T), consisting of the data items that are written by T.

Definition 2. For each data item x in the database, there is a nonnegative integer called version number to be attached with it. The version number is increased whenever the replica is updated and reflects the order in which the data item is updated. ReadVN(T,x) denotes the version number of data item x that transaction T read. WriteVN(T,x) denotes the version number that is provided for data item x by the write operation of T. CurrVN(S_k,x) denotes the version number of data item x currently in site S_k.

Definition 3. Two transactions T_i and T_j are said to be conflict if they both contain an operation that accesses the same data item and at least one of them is a write operation, denoted by conflict(T_i,T_j). The conflicts can be classified into read-write, write-write, and write-read conflicts, which are denoted by rw_conflict(T_i,T_j), ww_conflict(T_i,T_j), and wr_conflict(T_i,T_j) respectively.

Suppose two conflicting transactions T_i and T_j are issued by two sites concurrently. To maintain serializability, previous epidemic protocols consider that there is only one transaction can be committed and each site can only cast yes vote to one transaction in election, for example T_i. In optimistic voting protocol, to increase the chances to get yes vote for transaction T_j, sites can cast condition vote on it (whereas it is cast no vote in quorum or voting protocols). The condition vote on T_j can be transformed to yes vote if T_i is aborted. The use of condition vote postpones the final vote decision on transactions.

Definition 4. When voting on transaction T, suppose C={T_1,...,T_p} is the set in which each transaction conflicts with T, the condition vote cond(C) means that it can be transformed to yes vote in case each transaction in C is aborted, otherwise to no vote.

The transform rules of condition vote are as follows:

(1) If $\exists T_i \in C$, T_i has been aborted, then cond(C) → cond(C-T_i);
(2) If $\forall T_i \in C$, T_i has been aborted, then cond(C) → yes;
(3) If $\exists T_i \in C$, T_i has been committed, then cond(C) → no.

For two transactions T_i and T_j that only have read-write and write-write conflicts, if the correct order can be preserved at all sites, e.g. T_i is committed before T_j, then the two conflicting transactions T_i and T_j can all be committed maintaining consistency. Order vote prescribes the commit order of these kinds of conflicting transactions. Additionally, it is easily observed that condition and order vote can coexist on one transaction T.

Definition 5. When voting on transaction T, suppose C={T_1,...,T_p} is the set in which each transaction has only read-write and write-write conflicts with T, the order vote order(C) means that it can be transformed to yes vote when all transactions in C have been committed or aborted at one site.

The transform rule of order vote is as follows:

If at one site, $\forall T_i \in C$, T_i has been committed or aborted, then order(C) → yes.

3 Transaction Reordering

Transaction reordering technique can be viewed as an extension to the classical certification introduced by Kung and Robinson [10]. Intuitively, Kung-Robinson's technique consists in constructing a serialization order, according to which committed transactions appear to have executed. Given that transactions { $T_k,...,T_n$ } have committed after T had started, T can be committed if

$$(WriteSet(T_k) \cup ... \cup WriteSet(T_n)) \cap ReadSet(T) = \emptyset$$

Kung-Robinson's technique aborts concurrent transactions whose executions were not lucky enough to be delivered in a favorable order. Our reordering technique is based on the observation that the transaction serialization order does not need to be the same as the transaction arrival order at sites. The idea is to dynamically build a serialization order, in such a way that less aborts are produced. By reordering a transaction to a position other than the current one, our protocol increases the probability of success.

Definition 6. When voting on transaction T, suppose {$T_1,...,T_n$} is the transaction list in a site based on serialization order. If $(ReadSet(T_k) \cup ... \cup ReadSet(T_n)) \cap WriteSet(T) = \emptyset$, then T can be inserted before T_k and the vote to T is the advance vote, which is noted as adv(C) (suppose C={T_i| T_i in{$T_k,...,T_n$} and T has write-write or read-write conflict with T_i}).

The advance vote means that T should be committed before {$T_k,...,T_n$}. In the actual execution the committing of $T_k,...$, and T_n may be earlier than T. So when T is committed, some updates of T should be omitted to keep the right serialization order.

The disorder of actual committing may bring two problems. One problem is that if a transaction T' starts between the committing of $T_k,..., T_n$ and the committing of T, it may see an inconsistent state. The problem is resolved in the following algorithm by aborting at least one transaction of T and T' to maintain one-copy serialization.

Another problem is the version number management. Each update on the data item increases the version number by 1. Omitted update also increase version number to preserve ultimate equality of version number at all sites. But the version number read by a transaction may not indicate the right transaction that really updates the data item because of disorder committing of transactions. So, it is difficult to detect if a value read by a transaction has been overwritten at a site. To resolve this problem, we attach each version number with the global timestamp that the update transaction has. By this mean, the write-read relation can be detected properly. For brevity, the modification of version number is not shown in the following algorithm.

Each site S_k maintains a list of transactions by serialization order. Let $list_k$ denote the transactions list in which the vote on each candidate by the site is not no vote. Each transaction in $list_k$ is in one state of committed, vote yes, vote condition, vote order, vote advance (vote condition, order, and advance can be combined). The committed transactions in the list head that the commit results have been known by all sites can be pruned from the list. When S_k receives a new candidate T_{new}, it votes on T_{new} according to the following algorithm in Fig. 1.

```
cond-set=∅; order-set=∅; adv-set=∅
if ∃x ∈ ReadSet(T_new), ReadVN(T_new,x)<CurrVN(S_k,x),
    it means that the value read by T_new has been overwritten, then vote no
for T_i in list_k
    if (wr_conflict(T_i, T_new)) then
        if (ts(T_i)<ts(T_new))
            cond_set=cond_set ∪ T_i
            continue
        else if (not wr_conflict(T_new, T_i ∪ ... ∪ T_n))
            adv_set={T |T in {T_i,...,T_n}, rw_conflict(T_new,T) or ww_conflict(T_new,T) is true }
            if (∃T ∈ adv_set, ts(T)<ts(T_new)) then vote no and break
            vote(cond_set,order_set,adv_set) and break
        else
            vote no and break
    else if(rw_conflict(T_i, T_new)) then
        if (ts(T_i)<ts(T_new)) then
            order_set=order_set ∪ T_i
        else
            vote no and break
    else if(ww_conflict(T_i, T_new)) then
        if (ts(T_i)<ts(T_new)) then
            order_set=order_set ∪ T_i
        else if(not wr_conflict(T_new, T_i ∪ ... ∪ T_n))
            adv_set={T |T in {T_i,...,T_n}, rw_conflict(T_new,T) or ww_conflict(T_new,T) is true }
            if (∃T ∈ adv_set, ts(T)<ts(T_new)) then vote no and break
            vote(cond_set,order_set,adv_set) and break
        else
            vote no and break
    end if
end for
if hasn't vote
    vote(cond-set, order-set, adv-set)
```

Fig. 1. Algorithm for optimistic voting protocol with reordering

3.1 An Example

We explain reordering technique with an example. Suppose three transactions T_1, T_2 and T_3 ($ts(T_1)<ts(T_2)<ts(T_3)$).

Because three transactions have data conflict with each other, only one transaction can be committed in voting protocol. Fig. 2 shows optimistic voting process, which avoid the abort of T_3 by use of order vote. As to optimistic voting with reordering (fig. 3), three transactions can all be committed. From the figures, we observe that T_2 in optimistic voting can be committed earlier than in voting protocol by use of order vote. For clarity, we omit some unimportant information exchanges in Fig. 2 and Fig. 3.

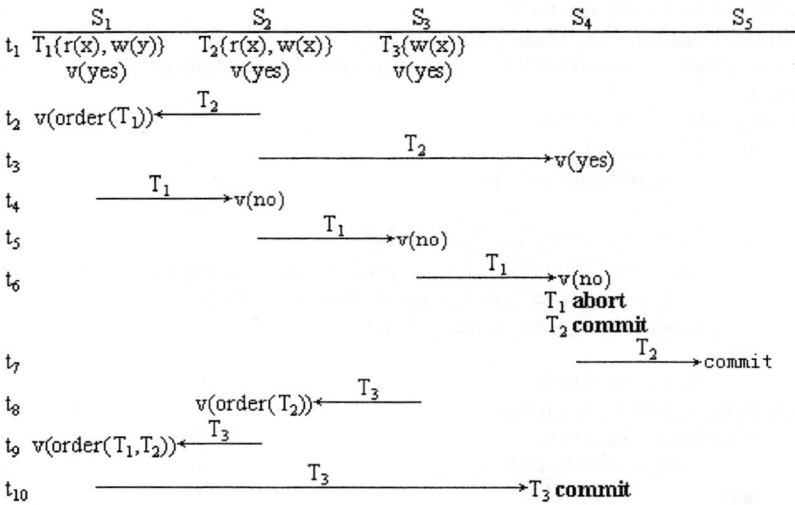

Fig. 2. Optimistic voting protocol

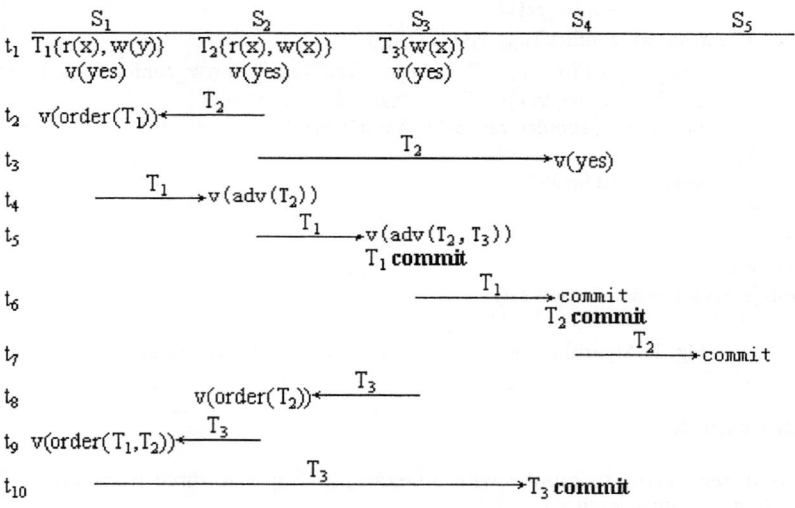

Fig. 3. Optimistic voting with reordering

3.2 Correctness

The correctness proof of optimistic voting can be found in [9]. Here, for brevity, we only give a brief explanation that the optimistic voting with reordering also preserves eventual consistency for update transactions. In the following discussion we assume a failure free execution.

We assume that every site provides local concurrent control and applies update transactions in a serial order locally. Let $\rightarrow_k$ be the total order relation on the local transaction set at S_k. The one-copy serializability is based on a global total order that is conflict compatible with all local serialization orders.

With the use of advance vote, transactions may be committed out of the serialization order. However, we can prove that this disorder will not affect the validity of committed transactions in optimistic voting with reordering. In other words, suppose $T_i \rightarrow_k T_j \rightarrow_k T_m$ in site S_k and there is write-read conflict between T_i (T_j) and T_m (we assume that T_m is launched at S_k). Because T_j may execute before T_i (it indicates that T_i must have vote adv(T_j)), T_m will probably only see the results of T_j. We now prove that at least one of T_i and T_m should be aborted to assure the validity of transactions.

As T_j may execute before T_i, from the algorithm, if T_i cannot get adv(T_j) at a site, it will only get no vote at this site. The epidemic communication model assures that the vote on T_m will after T_j at all sites. At each site, there are two cases:

1. If T_i arrived before T_m. because there is write-read conflict between T_i and T_m, T_m will get cond(T_i) at best;
2. If T_i arrived after T_m. It is obvious that T_i must arrive after T_j. Because T_j is before T_m by serialization order and there is write-read conflict between T_i and T_m, T_i cannot insert before T_j. So, T_i can only get no vote.

From above voting process, we can conclude that only one of T_i and T_m will get majority votes and be committed.

Another property of the algorithm is deadlock free. The condition, order, and advance votes of a transaction are dependent on other transactions. This dependency relation in the algorithm is one-way, i.e. a transaction with condition and order vote can only depend on transactions that have smaller global timestamp than it, likewise a transaction with advance vote can only insert before transactions that have larger global timestamp than it. The one-way dependency ensures that there are no cycles among transactions and therefore no global deadlocks. Therefore, the algorithm guarantees that a transaction will eventually arrive at the same election result among all sites and a site will vote for a transaction at most once.

4 Performance Evaluation

We perform experiments to show performance improvement attained by optimistic voting protocol and reordering technique. Additionally, we investigate the representative epidemic replication scheme from the literature, voting protocol [6] (quorum protocol [4] is similar to voting protocol). The evaluations are done at 10 desktops connected via a 10Mbps Ethernet network.

The simulation assumes that data items are fully replicated at all sites and tickets are uniformly distributed among sites. Each site generates transactions randomly according to a global transaction generation rate. Data items are accessed uniformly by transactions. Each site periodically initiates a synchronization session with a given synchronization interval by sending a pull request to another randomly selected site.

Since we focus on the transaction abort rate and commit delay of different protocols, we don't model any read-only transactions. Each transaction read 5-10 data items and write 5 data items which may be not in the read set with 5% chance. The main parameters and settings used in the experiments are summarized in Table 1.

Table 1. Experimental parameters

Parameters	Descriptions	Values
N	Site number	10
Sync. interval	Average synchronization interval	1-3s
Trans. rate	Average generation rate of update transactions	0.2-20/s
Data items	Total data item number	500

Fig.4 and Fig.5 illustrate the transaction abort rate of three protocols for various values of transaction generation rate and synchronization interval. From the figures, it is obvious that optimistic voting protocol with reordering outperforms the other two protocols. The reordering technique improves the performance by about 8% at best when compared to optimistic voting protocol, and about 18% to voting protocol.

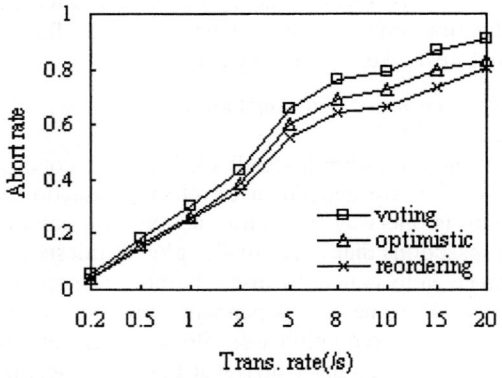

Fig. 4. Abort rate vs. transaction generation rate (Synchronization interval=1.0s)

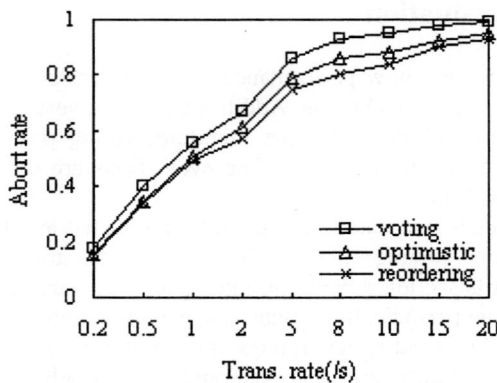

Fig. 5. Abort rate vs. transaction generation rate (Synchronization interval=3.0s)

In asynchronous transaction processing systems, the commit times at sites typically exhibit wide variability. We adopt the first and last commit delay to evaluate average response time of transactions. From the experiments we learn that reordering technique reduces commit delay slightly. But the differences are too small and we omit the figures in the paper. The slight improvement of transaction response time is come from the use of advance vote which is regarded as yes vote in the voting process and therefore it make a transaction get a plurality of votes much earlier.

5 Conclusion

Epidemic replication schemes are used extensively in transaction processing in weakly connected environments. Some continuously connected systems also use epidemic model to improve system efficiency. The reordering technique presented in this paper improves system performance in epidemic model and is of high practical values.

References

1. Huaizhong Lin and Chun Chen. An agent-based approach of transaction processing in mobile environments. In Proceedings of 6th International Conference for Young Computer Scientists. Hangzhou, China, 2001. 299-303.
2. J. Gray, P. Helland, P. O'Neil, and D. Shasha. The dangers of replication and a solution. In Proceedings of ACM SIGMOD International Conference on Management of Data. Montreal, Canada, 1996. 173-182.
3. D. Agrawal, A. El Abbadi, and R. Steinke. Epidemic algorithms in replicated databases. In Proceedings of 16th ACM SIGACT-SIGMOD-SIGART Symposium on Principles of Database Systems. Tucson, Arizona, 1997. 161-172.
4. J. Holliday, R. Steinke, D. Agrawal, and A. El Abbadi. Epidemic quorums for managing replicated data. In Proceedings of 19th IEEE International Performance, Computing, and Communications Conference. Phoenix, Arizona, 2000. 93-100.
5. K. Petersen, M. J. Spreitzer, D. B. Terry, M. M. Theimer, and A. J. Demers. Flexible update propagation for weakly consistent replication. In Proceedings of 16th ACM Symposium on Operating System Principles. St. Malo, France, 1997. 288-301.
6. U. Çetintemel, P. J. Keleher, and M. J. Franklin. Support for speculative update propagation and mobility in Deno. In Proceedings of 21st International Conference on Distributed Computing Systems. Phoenix, Arizona: IEEE Computer Society Press, 2001. 509-516.
7. P. J. Keleher. Decentralized replicated object protocols. In Proceedings of 18th Annual ACM Symposium on Principles of Distributed Computing. Atlanta, Georgia, 1999. 143-151.
8. Patrick T. Eugster, Rachid Guerraoui, Anne-Marie Kermarrec, and Laurent Massoulié. Epidemic Information Dissemination in Distributed Systems. IEEE Computer, 2004, (5): 60-67.
9. Huaizhong Lin and Chun Chen. Optimistic voting for managing replicated data. Journal of Computer Science and Technology, 2002, 17(6): 874-881.
10. H. T. Kung and J. T. Robinson. On optimistic methods for concurrency control. ACM Transactions on Database Systems, 1981, 6(2):213-226.

Automatic Boundary Tumor Segmentation of a Liver

Kyung-Sik Seo[1] and Tae-Woong Chung[2]

[1] MOMED Company, Chosun Univ. ICVC 5111,
375 Seosuk-dong, Dong-gu, Gwangju, Korea
momed@paran.com
[2] Dept. of Diagnostic Radiology,
Chonnam National University Hospital, Gwangju, Korea
twchung@jnu.ac.kr

Abstract. This paper proposes automatic boundary tumor segmentation for the computer aided liver diagnosis system. As pre-processing, the liver structure is first segmented using histogram transformation, multi-modal threshold, C-class maximum a posteriori decision, and binary morphological filtering. After binary transformation of the liver structure, the image based bounding box is created and convex deficiencies are segmented. Large convex deficiencies are selected by pixel area estimation and selected deficiencies are transformed to gray-level deficiencies. The boundary tumor is selected by estimating the variance of deficiencies. In order to test the proposed algorithm, 225 slices from nine patients were selected. Experimental results show that the proposed algorithm is very useful for diagnosis of the abnormal liver with the boundary tumor.

1 Introduction

Liver cancer, which is the fifth most common cancer, is more serious in areas of western and central Africa and eastern and southeastern Asia [1]. The average incidence of liver cancer in these areas is 20 per 100,000, and liver cancer is the third highest death cause from cancer [1]. In Korea, the incidence of liver cancer is quite high at 19% for males and 7% for females [2]. New cases of liver cancer in the Seoul area have an approximate rate per year of 34.1 for males and 11.5 for females per 100,000 people [2]. In order to improve the curability of liver cancer, early detection is critical. Liver cancer, like other cancers, manifests itself with abnormal cells, conglomerated growth, and tumor formation. If the hepatic tumor is detected early, treatment and curing of a patient may be easy, and human life can be prolonged.

Liver segmentation using CT images has been vigorously performed because CT is a very conventional and non-invasive technique. Bae et al [3] used priori information about liver morphology and image processing techniques such as gray-level thresholding, Gaussian smoothing, mathematical morphology techniques, and B-splines. Luomin et al [4] shows automatic liver segmentation technique for three-dimensional visualization of CT data. This segmentation includes a global histogram, morphologic operators, and the parametrically deformable contour model. Also, boundaries of the thresholded liver volume are modified section-by-section by exploiting information from adjacent sections. Volume-rendered images are then created. Hyunjin et al [5] built a probabilistic atlas of the brain and extended abdominal segmentation including the liver, kidneys, and spinal cord. The warping transform and mutual information are used as the similarity measure. Its segmentation by incorporating the atlas information into the Bayesian framework is improved over a standard unsupervised segmentation method. Gao et al [6] developed automatic liver segmentation using a global histogram, morphologic operations, and the parametrically deformable contour model. Saitoh et al. [7] performed automatic segmentation of liver region based on extracted blood vessels. Also, Seo et al. [8] proposed fully automatic liver segmentation based on the spine.

However, previous research has been concentrated on only liver segmentation and volume construction. In this paper, an automatic boundary tumor segmentation method of the liver using CT images is proposed. An automatic hepatic tumor segmentation method is presented in the following section. In the next section, experiments and results are described. Finally, the conclusion will be given in the last section.

2 Boundary Tumor Segmentation

In this section, an automatic boundary tumor segmentation of the liver is presented. A liver structure is first segmented from the CT image. The image-based bounding box is created and convex deficiencies are segmented. Then the boundary tumor is extracted by variance estimation.

2.1 Liver Segmentation

As pre-processing, a liver structure is segmented. The first important work to segment a boundary tumor is to segment a liver structure. The liver is extracted using histogram transformation [9], multi-modal threshold [10], and maximum a posteriori decision [11]. In order to eliminate other abdominal organs such as the heart and right kidney, binary morphological (BM) filtering is performed by dilation, erosion, closing, and filling [12, 13, 14]. Fig. 1(a) shows a sample CT image and Fig. 1(b) is the liver structure.

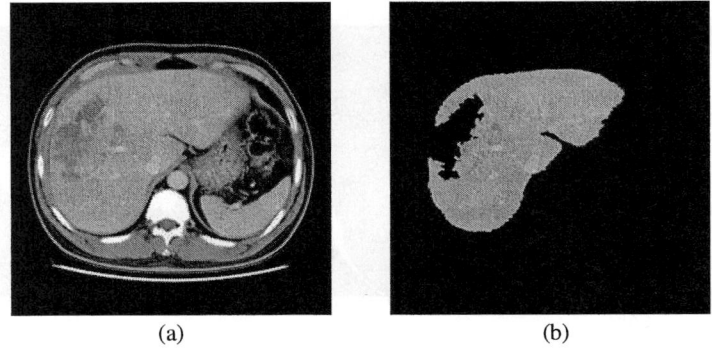

Fig. 1. Liver segmentation. (a) Sample CT image (b) Liver segmentation

2.2 Creation of Bounding Box

After extraction of the liver structure, the bounding box is created. The bounding box is the smallest rectangle containing the binary object. Let $I : Z^2 \rightarrow Z$ be the gray-level image of the live structure and (m, n) be a pixel location. Then, $I(m,n) \in Z$. Let B be the binary image of I as shown in Figure 2(a). The horizontal and vertical lines with the same height and width including the binary object are drawn. Then four coordinates of the image-based bounding box such as the top-left, top-right, bottom-left, and bottom-right point are calculated from meeting points of the horizontal and vertical lines. The image-based bounding box is created using these four coordinates as shown in Figure 2(b).

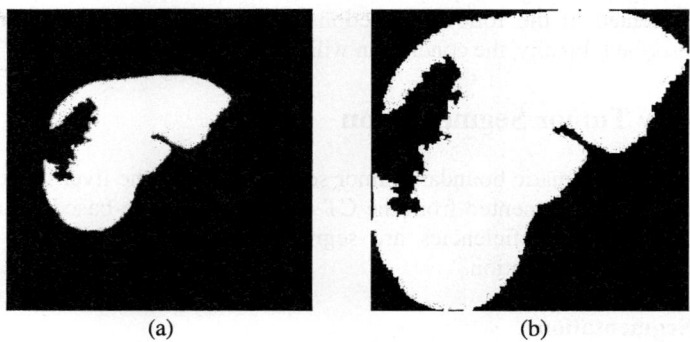

Fig. 2. Creation of the bounding box. (a) Binary transformation of the sample image. (b) Image-based bounding box

2.3 Segmentation of Convex Deficiencies

In order to create convex deficiencies of the binary object, a convex hull is built in the segmented bounding box. The convex hull is the smallest enclosed convex polygon containing the binary object as shown in Figure 3(a). Let B_C be the convex

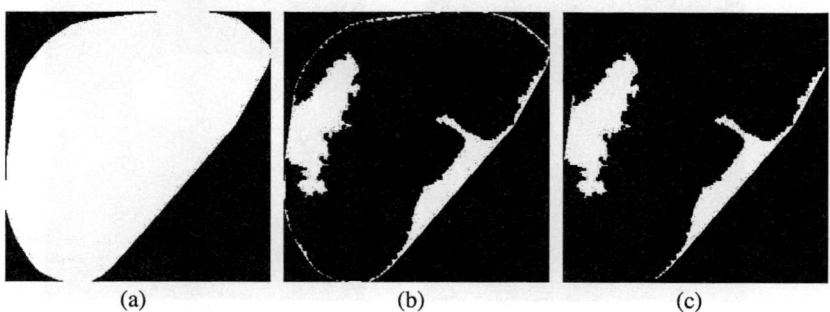

Fig. 3. Segmentation of convex deficiencies. (a) Convex hull. (b) Convex deficiencies. (c)Convex deficiencies after removal of small deficiencies

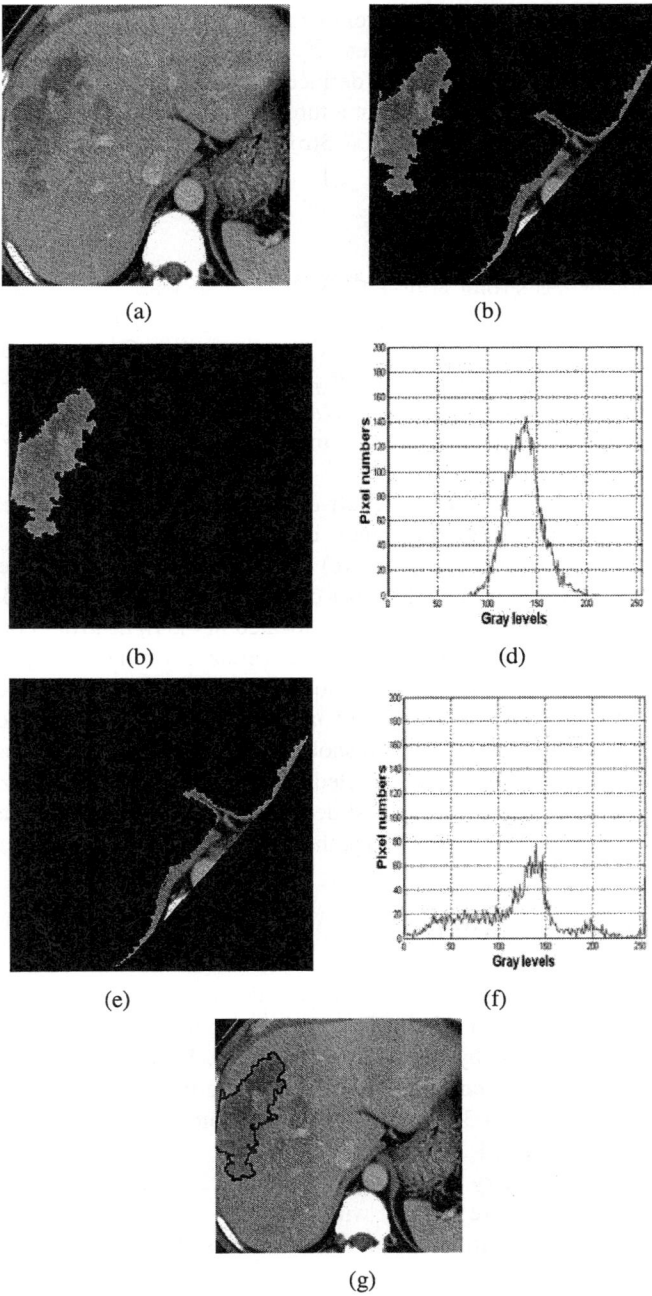

Fig. 4. Tumor extraction. (a) Bounding box of the CT image. (b) Gray-level convex deficie-ncies Tumor extraction. (c) First gray-level deficiency. (d) Histogram of the first gray-level deficiency. (e) Second gray-level deficiency. (f) Histogram of the second gray-level deficiency. (g) Boundary tumor in the CT image

hull and B_{cd} be the set of convex deficiencies. The set of convex deficiencies is extracted from the difference between B_C and B, that is, $B_{cd} = B_C - B_{ROI}$ as shown in Figure 3(b). Then the B_{cd} is defined as $B_{cd} = \{B_{cd\_1}, B_{cd\_2}, ..., B_{cd\_n}\}$.

As small convex deficiencies are not a tumor, theses small convex deficiencies are removed from B_{cd} as shown in Figure 3(c). Then the new set of convex hulls is defined as $B'_{cd} = \{B'_{cd\_1}, B'_{cd\_2}, ..., B'_{cd\_m}\}$.

2.4 Tumor Extraction

A boundary tumor has strong gray-level pixel homogeneity same as an inner tumor. The boundary tumor is segmented by estimating this pixel homogeneity. Let I_{ROI} be the segmented gray-level image from the CT image using four coordinates as shown in Figure 4(a). Convex deficiencies of B'_{cd} are transformed to the gray-level convex deficiencies defined as $I'_{cd} = \{(m,n) \mid B'_{cd}(m,n) \otimes I_{ROI}(m,n)\}$ where $B'_{cd}(m.n)$ is the binary image and $I_{ROI}(m,n)$ is the CT image. Figure 4(b) shows the transformed gray-level convex deficiencies.

Each deficiency is labeled from the left-upper corner to the right-lower corner. The histogram of each gray-level deficiency using this labeling order is generated to calculate statistical properties. Figure 4(c) and (d) show the first labeled deficiency and its histogram located in the left-upper corner. Also, .Figure 4(e) and (f) show the second labeled deficiency and its histogram located in the right-lower corner.

In order to select tumors from I'_{cd}, the variance of the histogram of I'_{cd} is estimated. As the tumor has stronger homogeneity than other convex hulls, the variance of tumors is relatively small. Let h'_{cd} be histograms of I'_{cd}. Then, I'_{cd} is the tumor if $\{\text{var}(h'_{cd}) < z\}$ where z is threshold value of variance. After estimating the variance of deficiency, the tumor is selected as shown in Figure 4(c). From comparing the variance of two deficiencies, the first deficiency as the tumor has smaller variance than the second deficiency. Figure 4(g) is the boundary tumor in the CT image.

3 Experiments and Analysis

CT images to be used in this research were provided by Chonnam National University Hospital in Kwangju, Korea. The CT scans were obtained by using a LightSpeed Qx/i, which was produced by GE Medical Systems. Scanning was performed with intravenous contrast enhancement. Also, the scanning parameters used a tube current of 230 mAs and 120 kVp, a 30 cm field of view, 5 mm collimation and a table speed of 15 mm/sec (pitch factor, 1:3).

In order to test boundary tumor segmentation, 225 slices of nine patients were selected. These slices were varied with the size and location of tumors. One radiologist in Chonnam National University Hospital took part in this research to diagnose abnormality of the liver boundary. Test data were evaluated by four basic possibilities such as true negative (TN), false positive (FP), false negative (FN), and true positive (TP). As the evaluation measure, sensitivity, specificity, and accuracy were calculated [15]. Table 1 shows the data of evaluated slices followed by slice numbers.

As the evaluation measure, sensitivity, specificity, and accuracy were calculated. As sensitivity represents the fraction of patients with disease who test positive, sensitivity is defined as

$$Sensitivity = \frac{TP}{TP + FN}. \tag{1}$$

As specificity represents the fraction of patients without disease who test negative, specificity is defined as

$$Specificity = \frac{TN}{TN + FP}. \tag{2}$$

Also, accuracy is defined as

$$Accuracy = \frac{TP + TN}{TP + TN + FP + FN}. \tag{3}$$

As the data measure, sensitivity is 0.7273, specificity is 0.9257, and accuracy is 0.8578. These results show the proposed method is very useful for diagnosis of the abnormal liver with the boundary tumor.

Table 1. Data of evaluated slices

Patients	Number of slices	Number of TN slices	Number of FP slices	Number of FN slices	Number of TP slices
Patient 01	22	20	1	0	1
Patient 02	27	10	0	7	10
Patient 03	22	11	0	3	8
Patient 04	28	10	0	1	17
Patient 05	29	19	0	3	7
Patient 06	26	14	2	1	9
Patient 07	23	19	2	1	1
Patient 08	26	14	6	5	1
Patient 09	22	20	0	0	2
Total number	225	137	11	21	56

4 Conclusions

In this paper, we proposed automatic boundary tumor segmentation for the computer aided liver diagnosis system. As pre-processing, the liver structure was segmented using histogram transformation, multi-modal threshold, C-class maximum a posteriori decision, and binary morphological filtering. After binary transformation of the liver structure, the image based bounding box was created and convex deficiencies were segmented. Then large convex deficiencies were transformed to gray-level deficiencies. The boundary tumor was selected by estimating the variance of deficiencies. In order to test the proposed algorithm, 225 slices from nine patients were selected. Experimental results show that the proposed algorithm is very useful for diagnosis of the abnormal liver with the boundary tumor. In order to evaluate the proposed method, 262 slices from 10 patients were selected. From the evaluation results, we had 0.7273 of sensitivity, 0.9257 of specificity, and 0.8578 of accuracy. These results show that the proposed method is very useful for diagnosis of the abnormal liver with the boundary tumor.

References

1. Parkin, D. M.: Global cancer statistics in the year 2000. Lancet Oncology, Vol. 2. (2001) 533-54
2. Lee H.: Liver cancer. The Korean Society of Gastroenterology, Seoul Korea (2001)
3. Bae, K. T., Giger, M. L., Chen, C. T., Kahn, Jr. C. E.: Automatic segmentation of liver structure in CT images. Med. Phys.,Vol. 20. (1993) 71-78
4. Gao, L., Heath, D. G., Kuszyk, B. S., Fishman, E. K.: Automatic liver segmentation technique for three-dimensional visualization of CT data. Radiology, Vol. 201. (1996) 359-364
5. Park, H., Bland, P. H., Meyer, C. R.: Construction of an abdominal probabilistic atlas and its application in segmentation. IEEE Trans. Med. Imag., Vol. 22. No. 4. (2003) 483-492
6. Tsai, D.: Automatic segmentation of liver structure in CT images using a neural network. IEICE Trans. Fundamentals, Vol. E77-A. No. 11, (1994) 1892-1895
7. Saitoh, T., Tamura, Y., Kaneko, T.: Automatic segmentation of liver region based on extracted blood vessels. System and Computers in Japan, Vol. 35. No. 5. (2004) 1-10.
8. Seo, K., Ludeman, L. C., Park S., Park, J.: Efficient liver segmentation based on the spine. LNCS, Vol. 3261. (2004) 400-409.
9. Orfanidis, S. J.: Introduction to signal processing. Prentice Hall, Upper Saddle River NJ (1996)
10. Schilling, R. J., Harris, S. L.: Applied numerical methods for engineers. Brooks/Cole Publishing Com., Pacific Grove CA (2000)
11. Ludeman, L. C.: Random processes: filtering, estimation, and detection. Wiley & Sons Inc., Hoboken NJ (2003)
12. Gonzalez, R. C., Woods, R. E.: Digital image processing. Prentice Hall, Upper Saddle River NJ (2002)
13. Shapiro, L. G., Stockman, G. C.: Computer vision. Prentice-Hall, Upper Saddle River NJ (2001)
14. Parker, J.R.: Algorithms for image processing and computer vision. Wiley Computer Publishing, New York (1997)
15. Rangayyan R.M.: Biomedical signal analysis. Wiley, New York NY (2002)

Fast Algorithms for l1 Norm/Mixed l1 and l2 Norms for Image Restoration*

Haoying Fu[1], Michael K. Ng[2], Mila Nikolova[3], Jesse Barlow[1], and Wai-ki Ching[2]

[1]Department of Computer Science and Enginnering,
Pennsylvania State University,
[2]Department of Mathematics, The University of Hong Kong
[3]Centre de Mathématiques et de Leurs Applications, France

Abstract. Image restoration problems are often solved by finding the minimizer of a suitable objective function. Usually this function consists of a data-fitting term and a regularization term. For the least squares solution, both the data-fitting and the regularization terms are in the $\ell 2$ norm. In this paper, we consider the least absolute deviation (LAD) solution and the least mixed norm (LMN) solution. For the LAD solution, both the data-fitting and the regularization terms are in the $\ell 1$ norm. For the LMN solution, the regularization term is in the $\ell 1$ norm but the data-fitting term is in the $\ell 2$ norm. The LAD and the LMN solutions are formulated as the solutions of a linear and a quadratic programming problems respectively, and solved by interior point methods. At each iteration of the interior point method, a structured linear system must be solved. The preconditioned conjugate gradient method with factorized sparse inverse preconditioners is employed to such structured inner systems. Experimental results are used to demonstrate the effectiveness of our approach. We also show the quality of the restored images using the minimization of $\ell 1$ norm/mixed $\ell 1$ and $\ell 2$ norms is better than that using $\ell 2$ norm approach.

Keywords: image restoration, least absolute deviation, least mixed norm, interior point method.

1 Introduction

The problem of image restoration is considered. The observed image is the convolution of a shift invariant blurring function with the true image plus some additive noise. Let **f**, **g** and **w** be the discretized original scene, observed scene and additive noise respectively. Let H be the blurring matrix of appropriate size

---

* The work has been supported by a grant from the PROCORE - France/Hong Kong Joint Research Scheme sponsored by the Research Grants Council of Hong Kong and the Consulate General of France in Hong Kong F-HK 30/04T, and RGC grant nos. HKU7130/02P, HKU7046/03P, HKU7035/04P, and 10205541.

built according to the discretized point spread function **h**. Then the discretized image formation process can be modeled as

$$\mathbf{g} = H\mathbf{f} + \mathbf{w}. \tag{1}$$

Assuming that the discretized scenes have $m \times n$ pixels, then **f**, **g** and **w** are vectors of length mn, and H is a matrix of $mn \times mn$.

It is well known that image restoration problems tend to be very ill-conditioned. Directly solving (1) will yield a solution that is extremely sensitive to noise, therefore regularization methods are needed to stabilize the solution. For example, the least squares solution with Tikhonov's regularization [12] takes the following form:

$$\min_{\mathbf{f}} \|\mathbf{g} - H\mathbf{f}\|_2^2 + \alpha \|R\mathbf{f}\|_2^2. \tag{2}$$

In this optimization problem, the second term is a regularization term that measures the "irregularity" of the solution. We call R the regularization operator and α the regularization parameter. Very often R is chosen to be the difference operator, e.g., the first-order or the second order finite difference operator.

The implicit assumption behind (2) is that the additive noise is white Gaussian and the prior distribution of **f** is also Gaussian. Under this assumption, the solution of (2) can be interpreted as the "maximum *a posteriori* (MAP) estimator" of the original scene, see [3] for instance. However, due to the presence of edges, the prior distribution of an image rarely satisfies the Gaussian assumption well. In many cases, the additive noise does not satisfy the Gaussian assumption either, for instance, the noise follows a Laplace distribution. In the literature, there has been a growing interest in using $\ell 1$ norm for parameter estimation [1, 3, 2, 4], and for image restoration [6, 9, 10]. The advantage of using the $\ell 1$ norm is that the solution is more robust than $\ell 2$ norm in statistical estimation problems. In particular, a small number of outliers usually do not change the solution much, see for instance [4, 9].

Since edges in an image lead to outliers in the regularization term, it is natural to consider using ℓ_1 norm for the regularization term:

$$\min_{\mathbf{f}} \|\mathbf{g} - H\mathbf{f}\|_2^2 + \alpha \|R\mathbf{f}\|_1. \tag{3}$$

We call the solution to (3) the Least Mixed Norm (LMN) solution.

If the additive noise does not satisfy the Gaussian assumption either, we consider using ℓ_1 norm for both the data-fitting and the regularization terms:

$$\min_{\mathbf{f}} \|\mathbf{g} - H\mathbf{f}\|_1 + \alpha \|R\mathbf{f}\|_1. \tag{4}$$

We call the solution to (4) the Least Absolute Deviation (LAD) solution.

The rest of this paper is organized as follows. In Section II, we formulate (4) and (3) as a linear or quadratic programming problems and solve them by interior point methods. The major work of each iteration of interior point methods is to solve a large linear system. In Section III, we propose to solve this system by preconditioned conjugate gradient method. Experimental results are presented in Section IV, and conclusions are made in Section V.

2 Linear Programming, Quadratic Programming, and Interior Point Methods

2.1 Computing the LAD Solution by Linear Programming

Usually, images satisfy $\mathbf{f} \geq 0$, that is, pixels have non-negative intensity values. This constraint is often omitted in restoration methods, mainly because of the numerical intricacies it entails. In this work, we provide the option to take this constraint into account.

Problem (4), with the non-negativity constraint imposed, can be formulated as a linear programming problem as follows. Little modification is needed if the non-negativity constraint is not desired.

Let $\mathbf{u} = H\mathbf{f} - \mathbf{g}$, let $\mathbf{v} = \alpha R\mathbf{f}$. We split $\mathbf{u}$ and $\mathbf{v}$ into their non-negative and non-positive parts. That is, $\mathbf{u} = \mathbf{u}^+ - \mathbf{u}^-$ and $\mathbf{v} = \mathbf{v}^+ - \mathbf{v}^-$, where $\mathbf{u}^+ = \max(\mathbf{u}, 0)$, $\mathbf{u}^- = \max(-\mathbf{u}, 0)$, $\mathbf{v}^+ = \max(\mathbf{v}, 0)$, and $\mathbf{v}^- = \max(-\mathbf{v}, 0)$. The problem can now be written as

$$\min_{\mathbf{f}, \mathbf{u}^+, \mathbf{u}^-, \mathbf{v}^+, \mathbf{v}^-} \mathbf{1}^T \mathbf{u}^+ + \mathbf{1}^T \mathbf{u}^- + \mathbf{1}^T \mathbf{v}^+ + \mathbf{1}^T \mathbf{v}^- \tag{5}$$

subject to

$$H\mathbf{f} - \mathbf{g} = \mathbf{u}^+ - \mathbf{u}^-$$
$$\alpha R\mathbf{f} = \mathbf{v}^+ - \mathbf{v}^-$$
$$\mathbf{u}^+, \mathbf{u}^-, \mathbf{v}^+, \mathbf{v}^-, \mathbf{f} \geq 0$$

Here **1** denotes the vector of all ones of appropriate size. This notation is used throughout this paper.

Clearly (5) is a linear programming problem in the standard form:

$$\min_{\mathbf{x}} \mathbf{c}^T \mathbf{x} \quad \text{subject to} \quad A\mathbf{x} = \mathbf{b}, \quad \mathbf{x} \geq 0, \tag{6}$$

where A, $\mathbf{b}$, $\mathbf{c}$ and $\mathbf{x}$ are defined as follows.

$$A = \begin{bmatrix} H & -I & I & 0 & 0 \\ \alpha R & 0 & 0 & -I & I \end{bmatrix}, \quad \mathbf{b} = \begin{bmatrix} \mathbf{g} \\ 0 \end{bmatrix},$$

$$\mathbf{x} = \begin{bmatrix} \mathbf{f} \\ \mathbf{u}^+ \\ \mathbf{u}^- \\ \mathbf{v}^+ \\ \mathbf{v}^- \end{bmatrix} \quad \text{and} \quad \mathbf{c} = \begin{bmatrix} 0 \\ 1 \\ 1 \\ 1 \\ 1 \end{bmatrix}.$$

The Lagrangian function for (6) is

$$\mathcal{L}(\mathbf{x}, \boldsymbol{\lambda}, \mathbf{s}) = \mathbf{c}^T \mathbf{x} - \boldsymbol{\lambda}^T (A\mathbf{x} - \mathbf{b}) - \mathbf{s}^T \mathbf{x}. \tag{7}$$

Here $\boldsymbol{\lambda}$ and $\mathbf{s}$ are the Lagrange multiplier vectors for the constraints $A\mathbf{x} = \mathbf{b}$ and $\mathbf{x} \geq 0$ respectively. For clarity, we partition $\boldsymbol{\lambda}$ as

$$\boldsymbol{\lambda} = \begin{bmatrix} \boldsymbol{\lambda}_u \\ \boldsymbol{\lambda}_v \end{bmatrix}, \tag{8}$$

where $\boldsymbol{\lambda}_u$ is the Lagrange multiplier vector for the constraint $H\mathbf{f} - \mathbf{u}^+ + \mathbf{u}^- = \mathbf{g}$, and $\boldsymbol{\lambda}_v$ is for $\alpha R - \mathbf{v}^+ + \mathbf{v}^- = 0$.

2.2 Interior Point Methods

Primal-dual interior point methods have become a common choice for solving large linear programming problems. We briefly outline our adapted interior point method below. A detailed description of interior point methods for linear programming can be found in [13–Chapter 1] or [11–Chapter 14].

The optimality condition for (6) is as follows.

$$F(\mathbf{x}, \boldsymbol{\lambda}, \mathbf{s}) = \begin{bmatrix} A^T\boldsymbol{\lambda} + \mathbf{s} - \mathbf{c} \\ A\mathbf{x} - \mathbf{b} \\ XS\mathbf{1} \end{bmatrix} = 0, \quad \mathbf{x} \geq 0, \quad \mathbf{s} \geq 0, \qquad (9)$$

where $X = \text{diag}(\mathbf{x})$ and $S = \text{diag}(\mathbf{s})$. Primal-dual interior point methods have their origin in Newton's method for the system of nonlinear equations (9). Newton's method starts with some initial guess for the solution, and calculates a search direction at each iteration by solving a linearized model of the original system. A detailed description of Newton's method for nonlinear systems can be found in [11–Chapter 11]. In the primal-dual interior point algorithm, the basic Newton step is modified such that the search directions are aimed at points on the *central path* $(\mathbf{x}_\tau, \boldsymbol{\lambda}_\tau, \mathbf{s}_\tau)$, defined as

$$F(\mathbf{x}_\tau, \boldsymbol{\lambda}_\tau, \mathbf{s}_\tau) = \begin{bmatrix} 0 \\ 0 \\ \tau\mathbf{1} \end{bmatrix}, \quad \mathbf{x}_\tau > 0, \quad \mathbf{s}_\tau > 0. \qquad (10)$$

Very often τ is written as $\sigma\mu$, where $\sigma \in [0, 1]$ is a *centering parameter*, and μ is the *duality measure* defined by

$$\mu = \frac{1}{n}\sum_{i=1}^{n} x_i s_i = \frac{\mathbf{x}^T\mathbf{s}}{n}. \qquad (11)$$

The step length at each iteration is chosen such that the new iterate is strictly positive, that is, $\mathbf{x} > 0$ and $\mathbf{s} > 0$.

The Newton search direction, $(\triangle\mathbf{x}, \triangle\boldsymbol{\lambda}, \triangle\mathbf{s})$, is computed by solving the following linear system:

$$\begin{bmatrix} 0 & A^T & I \\ A & 0 & 0 \\ S & 0 & X \end{bmatrix} \begin{bmatrix} \triangle\mathbf{x} \\ \triangle\boldsymbol{\lambda} \\ \triangle\mathbf{s} \end{bmatrix} = \begin{bmatrix} -\mathbf{r}_c \\ -\mathbf{r}_b \\ -\mathbf{r}_a \end{bmatrix}, \qquad (12)$$

where

$$\mathbf{r}_c = A^T\boldsymbol{\lambda} + \mathbf{s} - \mathbf{c}, \quad \mathbf{r}_b = A\mathbf{x} - \mathbf{b}, \quad \mathbf{r}_a = XS\mathbf{1} - \sigma\mu\mathbf{1}.$$

By eliminating $\triangle\mathbf{s}$ in (12), we obtain

$$\begin{bmatrix} -X^{-1}S & A^T \\ A & 0 \end{bmatrix} \begin{bmatrix} \triangle\mathbf{x} \\ \triangle\boldsymbol{\lambda} \end{bmatrix} = \begin{bmatrix} -\hat{\mathbf{r}}_c \\ -\mathbf{r}_b \end{bmatrix} \qquad (13)$$

where $\hat{\mathbf{r}}_c = \mathbf{r}_c - X^{-1}\mathbf{r}_a$. Let $D = S^{-1/2}X^{1/2}$, then (13) can be written as

$$\begin{bmatrix} -D^{-2} & A^T \\ A & 0 \end{bmatrix} \begin{bmatrix} \triangle \mathbf{x} \\ \triangle \lambda \end{bmatrix} = \begin{bmatrix} -\hat{\mathbf{r}}_c \\ -\mathbf{r}_b \end{bmatrix}. \qquad (14)$$

Note that since the components of $\mathbf{x}$ and $\mathbf{s}$ are strictly positive, all the diagonal elements of D are well defined. Here

$$D = \begin{bmatrix} D_1 & 0 & 0 & 0 & 0 \\ 0 & D_2 & 0 & 0 & 0 \\ 0 & 0 & D_3 & 0 & 0 \\ 0 & 0 & 0 & D_4 & 0 \\ 0 & 0 & 0 & 0 & D_5 \end{bmatrix} \quad \hat{\mathbf{r}}_c = \begin{bmatrix} \hat{\mathbf{r}}_{c1} \\ \hat{\mathbf{r}}_{c2} \\ \hat{\mathbf{r}}_{c3} \\ \hat{\mathbf{r}}_{c4} \\ \hat{\mathbf{r}}_{c5} \end{bmatrix}$$

$$\hat{\mathbf{r}}_b = \begin{bmatrix} \mathbf{r}_{b1} \\ \mathbf{r}_{b2} \end{bmatrix}.$$

By eliminating $\triangle \mathbf{u}^+$, $\triangle \mathbf{u}^-$, $\triangle \mathbf{v}^+$ and $\triangle \mathbf{v}^-$, and then $\triangle \lambda_u$ and $\triangle \lambda_v$, we obtain

$$\left[D_1^{-2} + H^T(D_2^2 + D_3^2)^{-1}H + \alpha^2 R^T(D_4^2 + D_5^2)^{-1}R \right] \triangle \mathbf{f} = \tilde{\mathbf{r}}_{c1}, \qquad (15)$$

where

$$\tilde{\mathbf{r}}_{c1} = \hat{\mathbf{r}}_{c1} - H^T(D_2^2 + D_3^2)^{-1}\hat{\mathbf{r}}_{b1} - \alpha R^T(D_4^2 + D_5^2)^{-1}\hat{\mathbf{r}}_{b2}.$$

Once (15) has been solved, the other unknowns in (12) can be easily recovered.

It is easy to check that the coefficient matrix in (15) is symmetric positive definite. However, this system gets ill conditioned as the iterates get close to the solution. In Section III, we employ a preconditioning method to solve how to precondition this system efficiently.

2.3 Computing the LMN Solution by Quadratic Programming

Problem (3), with or without the non-negative constraint, can be formulated as a quadratic programming problem in a similar way that we formulated (4) as a linear programming problem. For this problem, the linear system to be solved at each iteration of the interior point method can be reduced such that it has the form of

$$(D_1 + L^T D_2 L + 2H^T H)\triangle \mathbf{x} = \mathbf{r}, \qquad (16)$$

where D_1 and D_2 are diagonal matrices with positive diagonal elements. This system is also solved by the preconditioned conjugate gradient method.

3 Preconditioning the Inner Systems

Since the linear systems (15) and (16) are ill conditioned when the interior point iterations are close to the solution, preconditioners are needed to accelerate the convergence of the conjugate gradient iterations. Our experimental results indicate that the Factorized Banded Inverse Preconditioner (FBIP) [7] is very effective for these systems.

The approach is given as follows. Let B be a symmetric positive definite matrix, and its Cholesky factorization be $B = CC^T$. The idea of factorized sparse inverse preconditioner is to find the lower triangular matrix L with sparsity pattern $\mathcal{S}$ such that
$$\|I - CL\|_F$$
is minimized, where $\|\cdot\|_F$ denotes Frobenius norm. We say that $B \in \mathcal{R}^{m \times n}$ has lower bandwidth p if $b_{ij} = 0$ whenever $i > j+p$. The upper bandwidth of a matrix is defined similarly. Note that by this definition, a tridiagonal matrix has upper and lower bandwidth of 1.

Lin et al. [7] proved the following theorem:

Theorem 1. *Let T be a Hermitian Toeplitz matrix. Denote the k-th diagonal of T by t_k. Assume the diagonals of T satisfy*
$$|t_k| \le ce^{-\gamma|k|} \tag{17}$$
for some $c > 0$ and $\gamma > 0$, or
$$|t_k| \le c(|k|+1)^{-s} \tag{18}$$
for some $c > 0$ and $s > 3/2$. Then for any given $\epsilon > 0$, there exists $p' > 0$ such that for all $p > p'$,
$$\|L_p - C^{-1}\| \le \epsilon,$$
where L_p is the FBIP of T with the lower bandwidth p, and C is the Cholesky factor of T.

This theorem indicates that if a Toeplitz matrix T has certain off-diagonal decay property, then the FBIP will be a good approximation to T^{-1}. Note that if a Toeplitz matrix is banded, then both (17) and (18) are satisfied trivially.

Lin et al. [7] also considered Toeplitz-related systems of the form $I + T^T DT$, where D is a positive diagonal matrix. They showed that FBIPs are effective for these systems if T decays as stated in (17) or (18).

The matrices we are trying to precondition are the coefficient matrices in (15) and (16), that is, $B = D_1^{-2} + H^T(D_2^2 + D_3^2)^{-1}H + \alpha^2 R^T(D_4^2 + D_5^2)^{-1}R$ and $B = D_1 + \alpha^2 R^T D_2 R + 2H^T H$. While the focus of this research is on two-dimensional images, we have the following lemma for the one dimensional problem.

Lemma 1. *Let T be a Toeplitz matrix with its diagonals satisfying (17) or (18). Let D_1, D_2 and D_3 be diagonal matrices with positive diagonal entries. Let*
$$B = D_1 + T^T D_2 T + R^T D_3 R, \tag{19}$$
be a well-conditioned matrix, where R is the first-order or the second-order difference operator. Then for any given $\epsilon > 0$, there exists $p' > 0$ such that for all $p > p'$,
$$\|L_p - C^{-1}\| \le \epsilon,$$
where L_p is the FBIP of B with lower bandwidth p, and C is the Cholesky factor of B.

Lemma 1 indicates that the linear system in the one-dimensional LMN and LAD problems can be efficiently preconditioned by the FBIP.

For two-dimensional problems, we assume that the blurring matrix H has a block-level off-diagonal decay property, and each block also has off-diagonal decay property. This is true if the blurring function decays in spatial domain, or if the support of the blur is small. In this case, we can set the FBIP to be a triangular block banded matrix with each block being a banded matrix.

4 Experimental Results

We present experimental results on a high resolution image reconstruction problem [5, 8]. In this problem, the blurring matrix is only periodically shift invariant, that is, periodically BTTB. Here we employ factorized banded inverse preconditioners (FBIP) [7] for (15) and (16). Assuming the number of low resolution frames is small, which in general is true, then the blurring matrix satisfies the decay properties well. According to the theoretical results in [7], we expect fast convergence will be observed in the tests.

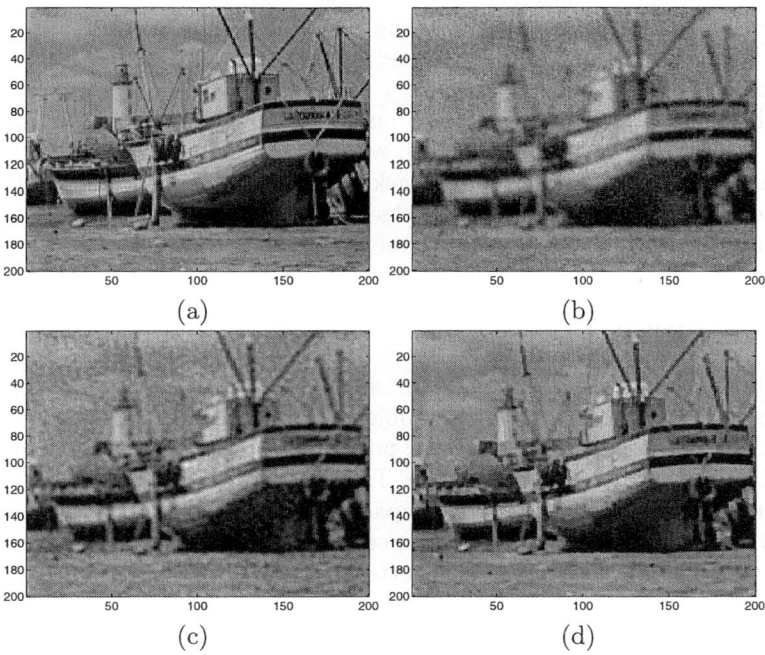

Fig. 1. (a) Original boat image, (b) Observed high resolution image (c) Least Squares restoration, PSNR=29.25 (d) the LAD solution, PSNR=35.26

The experiments are based on the boat image shown in Figure 1(a). This is a 200×200 grey level image, with grey levels varying from 0 to 1. A 4×4 sensor

Table 1. CPU times and number of conjugate gradient iterations needed at each interior point iteration, with or without the FBIP

PD Itn	No Pre		FBIP	
	No. iter	CPU time	No. iter	CPU time
10	286	262.68	25	41.27
12	415	392.57	32	50.99
14	657	621.32	43	64.03
16	1049	1007.38	63	89.88
18	1445	1349.26	99	130.66
20	>2000	>2000.00	163	212.41

array is used to retrieve sixteen 50×50 under-sampled frames. The displacements of the sensors are randomly perturbed around the ideal displacements. 50% of the low resolution pixels are further contaminated by a Gaussian white noise with a variance of 0.05. Figure 1(b) shows the observed high resolution image, that is, the image obtained by merging the low resolution frames. The first order difference operator is used as the regularization operator.

The least squares solution is shown in Figure 1(c). The optimal regularization parameter is used. Figure 1(d) shows the LAD solution, which has a much high PSNR, and exhibits significantly better visual quality than the least squares solution.

Table 1 shows the CPU time and number of conjugate gradient iterations needed at each interior point iteration, with or without the FBIP. The sparsity pattern of the FBIP is chosen to be block banded with banded blocks, the block bandwidth and bandwidth of the blocks are both chosen to be 3. We see that FBIP considerably accelerates the convergence rate.

5 Conclusion

In conclusion, we have proposed efficient algorithms for finding the LAD or LMN solution for the image restoration problem. We formulated these problems as solutions to smooth linear or quadratic programming problems, which are solved by primal-dual interior point methods. The linear system at each interior point iteration is first reduced to a more compact system and then solved by the PCG method. We show that the FBIP can speed up the convergence of the conjugate gradient method if a proper sparsity pattern is chosen.

References

1. S. Alliney. Digital filters as absolute norm regularizers. *IEEE transactions on signal processing*, 40:1548–1562, 1992.
2. S. Alliney. A property of the minimum vectors of a regularizing functional defined by means of absolute norm. *IEEE transactions on signal processing*, 45:913–917, 1997.

3. S. Alliney and S. Ruzinsky. An algorithm for the minimization of mixed ℓ1 and ℓ2 norms with application to bayesian estimation. *IEEE transactions on signal processing*, 42(3):618–627, March 1994.
4. P. Bloomfield and W. Steiger. *Least absolute deviations: theory, applications, and algorithms.* Boston, Birkhuser, 1983.
5. N. K. Bose and K. J. Boo. High-resolution image reconstruction with multisensors. *International Journal of Imaging Systems and Technology*, 9:294–304, 1998.
6. T. Chan and S. Esedoglu. Aspects of total variation regularized ℓ1 function approximation. Technical report, University of California at Los Angeles, CAM report, 2004.
7. F. Lin, M. Ng, and W. Ching. Factorized banded inverse preconditioners for matrices with Toeplitz structure. *SIAM Journal on Scientific Computing*, to appear.
8. M. K. Ng and A. M. Yip. A fast map algorithm for high-resolution image reconstruction with multisensors. *Multidimensional Systems and Signal Processing*, 12:143–164, 2001.
9. M. Nikolova. Minimizers of cost-functions involving nonsmooth data-fidelity terms: application to the processing of outliers. *SIAM Journal on Numerical Analysis*, 40:965–994, 2002.
10. M. Nikolova. A variational approach to remove outliers and impulse noise. *Journal of Mathematical Imaging and Vision*, 20:99–120, 2004.
11. J. Nocedal and S. Wright. *Numerical Optimization.* Springer, New York, 1999.
12. A. Tikhonov and V. Arsenin. *Solution of ill-posed problems.* Winston, Washington, DC, 1977.
13. S. Wright. *Primal-Dual Interior-Point Methods.* SIAM publications, Philadelphia, 1997.

Intelligent Semantic Information Retrieval in Medical Pattern Cognitive Analysis

Marek R. Ogiela[*], Ryszard Tadeusiewicz[*], and Lidia Ogiela[**]

[*] AGH University of Science and Technology,
Institute of Automatics
[**] Department of Company Management,
Al. Mickiewicza 30, PL-30-059 Kraków, Poland
{mogiela, rtad, logiela}@agh.edu.pl

Abstract. This paper will present a new approach to the interpretation of semantic information retrieved from complex X-ray images. The tasks of the analysis and the interpretation of cognitive meaning of selected medical diagnostic images are made possible owing to the application of graph image languages based on tree grammars. One of the main problems in the fast accessing and analysis of information collected in various medical examinations is the way to transform efficiently the visual information into a form enabling intelligent recognition and understanding of semantic meaning of selected patterns. Another problem in accessing for useful information in multimedia databases is the creation of a method of representation and indexing of important objects constituting the data contents. In the paper we describe some examples presenting ways of applying picture languages techniques in the creation of intelligent cognitive multimedia systems for selected classes of medical images showing especially wrist structures.

1 Introduction

One the main problems in accessing information collected in medical databases is the way to transform efficiently the visual information of patterns into a form enabling intelligent selection of cases obtained as an answer to queries directed at selected elements of contents of searched-for images. Therefore this paper will present the possibilities of application of picture languages and graph grammars used as the ordering factor indexing and supporting commitment and semantically-oriented search for visual information in multimedia medical databases.

It is worthwhile to emphasise the very essence of semantic analysis which allows for content-based grouping of images sometimes differing in form though conveying similar diagnostic information about the disease. Generally speaking, we have to consider a great variety of examined medical cases [5, 7].

In medical images an actual shape of anomalies or lesions can vary between the cases due to the fact that human organs vary between individuals, differing in shape, size and location while the forms and progress of pathological lesions (e.g. caused by neoplasm or chronic inflammation process) are unforeseeable [1, 7]. On the other hand, every type of disease leads to some characteristic changes in the shapes of visualised organs; therefore this type of information, obtained owing to the application of the method of structural pattern analysis, will constitute information label determining the image content. Techniques proposed in [7] allow the change of a pattern into its syntactic description in such a way that the automatically generated language formula transforms precisely the basic pattern content: the shape of the examined organ and its anomaly caused by disease. Those formalised, automatically generated descriptions of shapes of objects seen on a pattern placed or searched-for in a database, allow for separating the indexing process from the secondary formal features of the recorded patterns. Accordingly, the description is focused on the most important contents.

Graph grammar description algorithms as presented in this paper expand the traditional methods of computer-aided analysis through the interpretation of possibilities directed at tasks supporting medical diagnostics. Additionally, semantic information enables the use of such techniques in tasks of semantically-oriented search for some concrete disease cases in medical image data bases. In practice such tasks were difficult to implement, sometimes even impossible due to difficulties in creating indexing keys that describe image semantics [7]. Expanding analysis possibilities by meaning interpretation allows us to find an answer to questions concerning the medical meaning of the analysed image, semantic information specified by the features of the examined image and classification possibilities of disease units based on lesions on the image. The analysis of images conducted in this paper will go in the direction pointed out by the formulated questions. Its objective will be, in particular, to present and evaluate the possibilities of expansive graph grammar application for the recognition and intelligent meaning analysis of wrist and bones radiogrammes.

2 Linguistic Interpretation of Medical Images

The idea presented here and associated with creating indexing keys allows for an efficient search and categorization of both information specifying the type of medical imaging or the examined structure and meaningful semantic information specifying the looked-for object within the framework of one database. In a special case, apart from the main indexing key allowing the search or archiving of a specified type of medical images e.g. palm radiogrammes, spinal cord images etc., it is possible to create additional indexing labels specifying successive layers of semantic details of image contents. First of all, this information tells us about the progress of a disease detected with the use of the described grammars and semantic actions defined in them. The indexing information is also a description of the external morphology of the imaged organ. This type of description takes the form of a terminal symbol sequence introduced while grammars are defined for individual types of images and organs visible on them. The shape morphology described in this way requires a much smaller memory and computation input for the execution of the archiving operations and for searching for a given pattern. Finally, the lowest level of information useful for a detailed search are the types of recognised lesions and sequences of production numbers leading to the generation of a linguistic description of those lesions. Such productions are defined in the sequential and graph grammars introduced in papers [5]. Their sequences describing successive patterns of morphological lesions can constitute important information useful for a quick search for such irregularities on image data. Also graph grammars are applied for the task of description of important shape morphology features for the wrist structure [3]. The application of the description created in this way is analogous.

3 Wrist Radiograms Syntactic Description

Real examples of images showing pathological lesions in the form of wrist bone dislocation, and fusion have been shown on Figure 1. Such irregularities are to be detected and interpreted correctly by the syntactic image recognition method described in this paper [3].

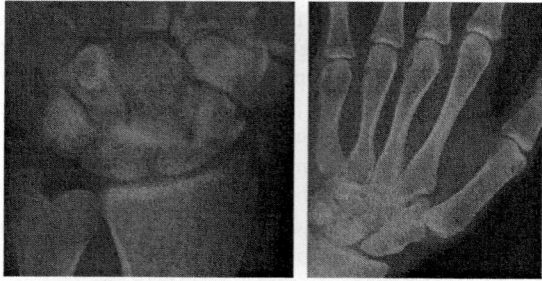

Fig. 1. A) Image showing lesions in the form of the lunate dislocation. B) Fusion of the scaphoid with distal row of carpal bones

For making the representation of the examined structures in the form of EDG graphs (graphs with directed peaks and with labelled peaks and edges) [6] it is necessary to define an appropriate linguistic formalism that is an appropriate graph grammar defining a language. The language is defined in such a way that one could describe using it, without any ambiguities, every image representing a spatial system composed of elements similar to the wrist bone system. In this way we create a tool describing all possible shapes and locations of wrist bones, both the correct and pathological ones. The linguistic formalism that we propose in this paper in order to execute the task of mirroring real medical image forms into graph formulas fit for computer processing, will be an expansive graph grammar [7]. After defining such a grammar, every X-ray image will be converted into a linguistic formula built in accordance with the rules of that grammar. The effective parsing of that formula conducted by the computer, compliant with the rules of the created grammar will lead to an automatic assessment of photograph contents. This will make it possible in particular to determine whether the built of a wrist falls within the norm or whether it has pathological deviations.

For the analysis of wrist radiogrammmes an xpansive graph grammar was defined.

$$G_{exp}=(N, \Sigma, \Gamma, P, S)$$

Non-terminal set of peak labels
   N= {ST, ULNA SCAPHOID, LUNATE, TRIQUETRUM, PISIFORM, TRAPEZIUM, TRAPEZOID, CAPITATE, HAMATE, M1, M2, M3, M4, M5}

Terminal set of peak labels Σ={r, u s, l, t, p, tm, tz, c, h, m1, m2, m3, m4, m5}
Γ - edge label set {s<p<q<r<s<t<u<w<x<y<z}
Start symbol S=ST
P – is a finite production set presented on Figure 2.

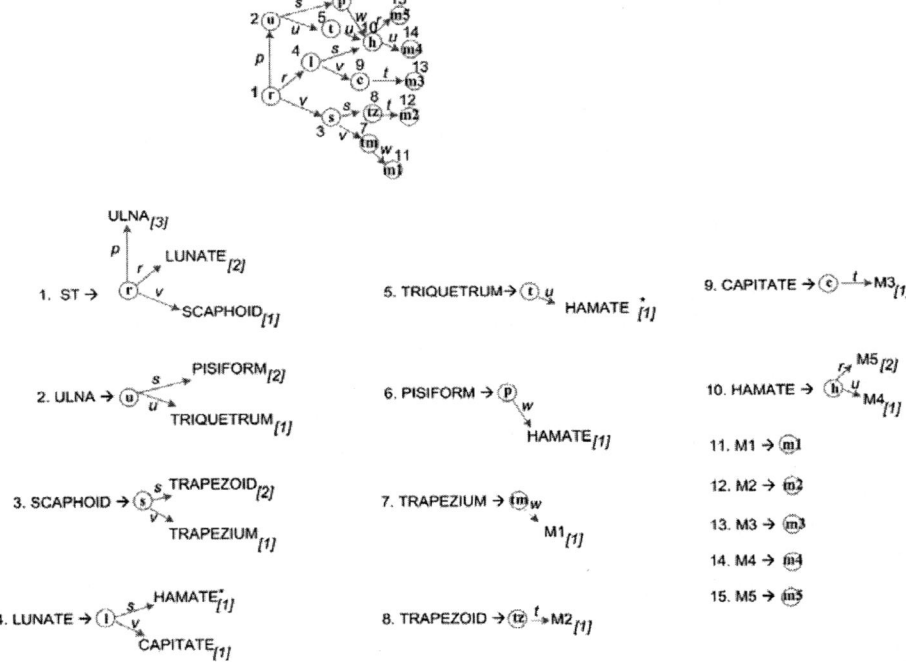

Fig. 2. Production set introducing a representation of the correct build and the number of bones in the wrist

4 Selected Results

As a result of cognitive multimedia analysis using linguistic approach it is possible to quite efficiently describe, mine and search pathogenesis information of the deformations viewed on selected x-ray images, what means the possibility of recognize and quick search some kind of diseases even on images absolutely not similar one to other.
   The methods prepared were aimed at building an automatic detection and semantic interpretation system for detected regularities as well as the diagnosed irregularities in

the carpus bones. It is worth notice, however, that the test data set used for defining the output rules in the presented grammar was composed of only about 30 radio-grammes. Despite such an insignificant representation and, owing to a great descriptive and generation power of the used graphs, it is possible to classify practically any number of analysed images. The initially assessed efficiency exceeds the 90% threshold. On the other hand, the appearance of difficult and ambiguous cases can be quickly considered by enriching grammar formalisms by new (so far not considered due to lack of empirical data) reasoning rules and by specifying the meaning interpretation or such new cases. This is also the direction that will be followed in the conducted research.

5 Conclusions

The semantic approach to indexing and searching of multimedia medical databases seems to be a more efficient and appropriate method than the traditional indexing methods [2, 4] due to the presence of distinguishable objects visible in the discussed medical patterns, which by virtue of their shape define pathological disease symptoms. A structural description of medical image contents becomes easier and more unambiguous than an analogous description applied to a different category of patterns, for examples scenes. Nevertheless, after defining an appropriate graph grammar, the methodology described here can be utilised to describe any patterns. It will enable therefore the creation of object-oriented semantic description of contents of those data and it will also constitute the key to their indexation and search. An additional advantage of structural description methods is a potential of additional analysis of the examined images in the course of archiving and defining the semantic meaning of lesions visible on them; this is performed by imitating a qualified professional's understanding of medical images.

Syntactic information, together with contour representation in the form of terminal symbols as well as production number sequences describing discovered pathologies, constitute the representation of patterns placed in the database or searched-for [7].
In the course of the analysis of selected images in each case we obtain a type of recognized symptom and a sequence of production numbers, which lead to grammar derivation of shape description of such lesions. Such sequences create the proper description of analyzed shapes and are stored in indexing record. In every case those spots are highlighted where any irregularities have been identified. For each of them there are detailed descriptions in the program representation relating to the previously listed information used as indexing keys.

The approach to the generation of structural-semantic representation of medical patterns in multimedia databases with the use of context-free and EDT graph grammars [7] presented in this study is an entirely new solution. Preliminary research reveals that such approach proves to be an extremely efficient and universal tool enabling visual data compression and unambiguous data representation. An important point is that the proposed indexation technique has been optimised with an aim to find diagnostic information, easily observed in the shape of organs visible on patterns and dependent on disease symptoms.

Acknowledgement

This work was supported by the AGH University of Science and Technology under Grant No. 10.10.120.39.

References

[1] Bankman, I. (ed.): Handbook of Medical Imaging: Processing and Analysis. Academic Press (2002)
[2] Berchtold S., Keim D. A., Kriegel H-P., Seidl T.: Indexing the solution space: a new technique for nearest neighbor search in high-dimensional space. IEEE Transactions on Knowledge & Data Engineering, 12(1) (2000) 45-57
[3] Burgener, F.A., Kormano, M.: Bone and Joint Disorders. Thieme, Stuttgart (1997)
[4] Martinez A. M., Serra J. R.: A new approach to object-related image retrieval. Journal of Visual Languages & Computing, 11(3) (2000) 345-363
[5] Ogiela, M.R., Tadeusiewicz, R.: Artificial Intelligence structural imaging techniques in visual pattern analysis and medical data understanding. Pattern Recognition, 36 (2003) 2441-2452
[6] Skomorowski M.: Parsing of random graphs for scene analysis. Machine GRAPHICS & VISION International Journal, 7 (1998) 313-323
[7] Tadeusiewicz, R, Ogiela, M.R.: Medical Image Understanding Technology. Springer, Berlin-Heidelberg (2004)

FSPN-Based Genetically Optimized Fuzzy Polynomial Neural Networks

Sung-Kwun Oh[1], Seok-Beom Roh[2], Daehee Park[2], and Yong-Kab Kim[2]

[1] Department of Electrical Engineering, The University of Suwon, San 2-2 Wau-ri,
Bongdam-eup, Hwaseong-si, Gyeonggi-do, 445-743, South Korea
ohsk@suwon.ac.kr
[2] Department of Electrical Electronic and Information Engineering, Wonkwang University,
344-2, Shinyong-Dong, Iksan, Chon-Buk, 570-749, South Korea

Abstract. In this paper, we introduce a new topology of Fuzzy Polynomial Neural Networks (FPNN) that is based on a genetically optimized multilayer perceptron with fuzzy set-based polynomial neurons (FSPNs) and discuss its comprehensive design methodology involving mechanisms of genetic optimization, especially genetic algorithms (GAs). The proposed FPNN gives rise to a structurally optimized structure and comes with a substantial level of flexibility in comparison to the one we encounter in conventional FPNNs. The structural optimization is realized via GAs whereas in case of the parametric optimization we proceed with a standard least square method-based learning. Through the consecutive process of such structural and parametric optimization, an optimized and flexible fuzzy neural network is generated in a dynamic fashion. The performance of the proposed gFPNN is quantified through experimentation that exploits standard data already used in fuzzy modeling. These results reveal superiority of the proposed networks over the existing fuzzy and neural models.

1 Introduction

Recently, a lot of attention has been directed towards advanced techniques of complex system modeling. While neural networks, fuzzy sets and evolutionary computing as the technologies of Computational Intelligence (CI) have expanded and enriched a field of modeling quite immensely, they have also gave rise to a number of new methodological issues and increased our awareness about tradeoffs one has to make in system modeling [1], [2], [3], [4]. The most successful approaches to hybridize fuzzy systems with learning and adaptation have been made in the realm of CI. Especially neural fuzzy systems and genetic fuzzy systems hybridize the approximate inference method of fuzzy systems with the learning capabilities of neural networks and evolutionary algorithms [5]. As one of the representative design approaches which are advanced tools, a family of fuzzy set-based polynomial neuron (FSPN)-based SOPNN(called "FPNN" as a new category of neuro-fuzzy networks)[6] were introduced to build predictive models for such highly nonlinear systems. The FPNN algorithm exhibits some tendency to produce overly complex networks as well as a repetitive computation load by the trial and error method and/or the repetitive parameter adjustment by designer like in case of the original GMDH algorithm. In this study, in addressing the above problems with the conventional SOPNN (especially, FPN-based SOPNN called "FPNN" [6], [9]) as well as the GMDH algorithm, we introduce a new genetic design approach; as a consequence we will be referring to these networks as GA-based FPNN (to be called "gFPNN"). The determination of the optimal values of the parameters available within an individual FSPN (viz. the number of input variables, the order of the polynomial, the number of membership functions, and a collection of the specific subset of input variables) leads to a structurally and parametrically optimized network.

2 The Architecture and Development of Fuzzy Polynomial Neural Networks (FPNN)

The FSPN consists of two basic functional modules. The first one, labeled by **F**, is a collection of fuzzy sets that form an interface between the input numeric variables and the processing part realized by the neuron. The second module (denoted here by **P**) is about the function – based nonlinear (polynomial) processing. This nonlinear processing involves some input variables. In other words, FSPN realizes a family of multiple-input single-output rules. Each rule reads in the form

$$\text{if } x_p \text{ is } A_k \text{ then } z \text{ is } P_{pk}(x_i, x_j, a_{pk})$$
$$\text{if } x_q \text{ is } B_k \text{ then } z \text{ is } P_{qk}(x_i, x_j, a_{qk})$$
(1)

where a_{lk} is a vector of the parameters of the conclusion part of the rule while $P_{lk}(x_i, x_j, a)$ denotes the regression polynomial forming the consequence part of the fuzzy rule which uses several types of high-order polynomials besides the constant function forming the simplest version of the consequence; refer to Table 1. The activation levels of the rules contribute to the output of the FSPN being computed as a weighted average of the individual condition parts (functional transformations) P_K (note that the index of the rule, namely "K" is a shorthand notation for the two indexes of fuzzy sets used in the rule (1), that is $K = (l, k)$).

$$z = \sum_{l=1}^{\text{total inputs}} \left(\sum_{k=1}^{\text{total\_rules related to input } l} \mu_{(l,k)} P_{(l,k)}(x_i, x_j, a_{(l,k)}) \middle/ \sum_{k=1}^{\text{total\_rules related to input } l} \mu_{(l,k)} \right)$$
$$= \sum_{l=1}^{\text{total inputs}} \left(\sum_{k=1}^{\text{total\_rules related to input } l} \tilde{\mu}_{(l,k)} P_{(l,k)}(x_i, x_j, a_{(l,k)}) \right)$$
(2)

$$\tilde{\mu}_{(l,k)} = \frac{\mu_{(l,k)}}{\sum_{k=1}^{\text{total\_rules related to input } l} \mu_{(l,k)}}$$
(3)

GAs is a stochastic search technique based on the principles of evolution, natural selection, and genetic recombination by simulating "survival of the fittest" in a population of potential solutions(individuals) to the problem at hand [7]. For the optimization of the FPNN model, GA uses the serial method of binary type, roulette-wheel used in the selection process, one-point crossover in the crossover operation, and a binary inversion (complementation) operation in the mutation operator. To retain the best individual and carry it over to the next generation, we use elitist strategy [8].

Table 1. Different forms of the regression polynomials standing in the consequence part of the fuzzy

No. of inputs Order of the polynomial	1	2	3
0 (Type 1)	Constant	Constant	Constant
1 (Type 2)	Linear	Bilinear	Trilinear
2 (Type 3)	Quadratic	Biquadratic-1	Triquadratic-1
2 (Type 4)		Biquadratic-2	Triquadratic-2

1: Basic type, 2: Modified type

3 The Algorithms and Design Procedure of Genetically Optimized FPNN

The framework of the design procedure of the Fuzzy Polynomial Neural Networks (FPNN) based on genetically optimized multi-layer perceptron architecture comprises the following steps.

[Step 1] *Determine system's input variables*
[Step 2] *Form training and testing data*
The input-output data set $(x_i, y_i)=(x_{1i}, x_{2i}, ..., x_{ni}, y_i)$, $i=1, 2, ..., N$ is divided into two parts, that is, a training and testing dataset.
[Step 3] *Decide initial information for constructing the FPNN structure*
[Step 4] *Decide FSPN structure using genetic design*
When it comes to the organization of the chromosome representing (mapping) the structure of the FPNN, we divide the chromosome to be used for genetic optimization into three sub-chromosomes. The 1st sub-chromosome contains the number of input variables, the 2nd sub-chromosome involves the order of the polynomial of the node, the 3rd sub-chromosome contains the number of membership functions(MFs), and the last sub-chromosome(remaining bits) includes input variables coming to the corresponding node (fuzzy set-based PN:FSPN). All these elements are optimized when running the GA.
[Step 5] *Carry out fuzzy inference and coefficient parameters estimation for fuzzy identification in the selected node (FSPN)*
Regression polynomials (polynomial and in the specific case, a constant value) standing in the conclusion part of fuzzy rules are given as different types of Type 1, 2, 3, or 4, see Table 1. In each fuzzy inference, we consider two types of membership functions, namely triangular and Gaussian-like membership functions. The consequence parameters are produced by the standard least squares method
[Step 6] *Select nodes (FSPNs) with the best predictive capability and construct their corresponding layer*

The generation process can be organized as the following sequence of steps

Sub-step 1) We set up initial genetic information necessary for generation of the FPNN architecture.
Sub-step 2) The nodes (FSPNs) are generated through the genetic design.
Sub-step 3) We calculate the fitness function. The fitness function reads as

$$F(\text{fitness function}) = 1/(1+EPI) \tag{4}$$

where *EPI* denotes the performance index for the testing data (or validation data).
Sub-step 4) To move on to the next generation, we carry out selection, crossover, and mutation operation using genetic initial information and the fitness values obtained via ***sub-step 3***.
Sub-step 5) We choose several FSPNs characterized by the best fitness values. Here, we use the pre-defined number *W* of FSPNs with better predictive capability that need to be preserved to assure an optimal operation at the next iteration of the FPNN algorithm. The outputs of the retained nodes (FSPNs) serve as inputs to the next layer of the network. There are two cases as to the number of the retained FSPNs, that is

(i) If $W^* < W$, then the number of the nodes retained for the next layer is equal to z.

Here, W^* denotes the number of the retained nodes in each layer that nodes with the duplicated fitness values are moved.
(ii) If $W^* \geq W$, then for the next layer, the number of the retained nodes is equal to *W*.

Sub-step 6) For the elitist strategy, we select the node that has the highest fitness value among the selected nodes (*W*).
Sub-step 7) We generate new populations of the next generation using operators of GAs obtained from ***Sub-step 4***. We use the elitist strategy. This sub-step carries out by repeating ***sub-step 2-6***.
Sub-step 8) We combine the nodes (*W* populations) obtained in the previous generation with the nodes (*W* populations) obtained in the current generation.
Sub-step 9) Until the last generation, this sub-step carries out by repeating ***sub-step 7-8***.

[Step 7] *Check the termination criterion*

As far as the performance index is concerned (that reflects a numeric accuracy of the layers), a termination is straightforward and comes in the form,

$$F_1 \leq F_* \tag{5}$$

Where, F_1 denotes a maximal fitness value occurring at the current layer whereas F_* stands for a maximal fitness value that occurred at the previous layer. In this study, we use a measure (performance index) that is the Root Mean Squared Error (RMSE).

$$E(PI \text{ or } EPI) = \sqrt{\frac{1}{N}\sum_{p=1}^{N}(y_p - \hat{y}_p)^2} \tag{6}$$

[Step 8] *Determine new input variables for the next layer*

If (5) has not been met, the model is expanded. The outputs of the preserved nodes $(z_{1i}, z_{2i}, \ldots, z_{Wi})$ serves as new inputs to the next layer $(x_{1j}, x_{2j}, \ldots, x_{Wj})(j=i+1)$. This is captured by the expression

$$x_{1j} = z_{1i}, x_{2j} = z_{2i}, \ldots, x_{wj} = z_{wi} \tag{7}$$

The FPNN algorithm is carried out by repeating steps 4-8.

4 Experimental Studies

The performance of the GA-based FPNN is illustrated with the aid of well-known and widely used dataset of the chaotic Mackey-Glass time series [10], [11], [12], [13], [14], [15], [16], [17].

The time series is generated by the chaotic Mackey-Glass differential delay equation comes in the form

$$\dot{x}(t) = \frac{0.2x(t-\tau)}{1+x^{10}(t-\tau)} - 0.1x(t) \tag{8}$$

To obtain the time series value at each integer point, we applied the fourth-order Runge-Kutta method to find the numerical solution to (8). From the Mackey-Glass time series $x(t)$, we extracted 1000 input-output data pairs in the following format:

$$[x(t-24), x(t-18), x(t-12), x(t-6), x(t); x(t+6)]$$

where, t=118 to 1117. The first 500 pairs were used as the training data set while the remaining 500 pairs formed the testing data set. To come up with a quantitative

Table 2. Summary of the parameters of the genetic optimization

	Parameters	1st layer	2nd to 5th layer
GA	Maximum generation	150	150
	Total population size	100	100
	Selected population size (W)	30	30
	Crossover rate	0.65	0.65
	Mutation rate	0.1	0.1
	String length	3+3+2+30	3+3+2+30
FPNN	Maximal no.(Max) of inputs to be selected	1≤l≤Max(2~5)	1≤l≤Max(2~5)
	Polynomial type (Type T) of the consequent part of fuzzy rules	1≤T≤4	1≤T≤4
	Consequent input type to be used for Type T (*)	Type T*	Type T
	Membership Function (MF) type	Triangular Gaussian	Triangular Gaussian
	No. of MFs per input	2≤M≤5	2≤M≤5

l, P, Max: integers, T* means that entire system inputs are used for the polynomial in the conclusion part of the rules.

evaluation of the network, we use the standard RMSE performance index as given by (6). Table 2 summarizes the list of parameters used in the genetic optimization of the network.

Table 3. Performance index of the network of each layer versus the increase of maximal number of inputs to be selected (in case of using Type T* at the 1st layer and triangular MF)

Max	1st layer						2nd layer				
	Node		T	M	PI	EPI	Node		T	M	PI	EPI
2	1		3	3	5	7.175e-4 7.015e-4	21		25	1	3	6.309e-4 6.034e-4
3	1	3	0	3	5	7.175e-4 7.015e-4	20	23	29	2	5	5.937e-4 5.595e-4
4	1	3	4	5	4	5	5.583e-4 5.614e-4	5	7	15	19	2	4	4.540e-4 4.556e-4
5	1	3	4	5	0	4	5	5.583e-4 5.614e-4	2	8	19	25	30	1	3	4.488e-4 4.415e-4

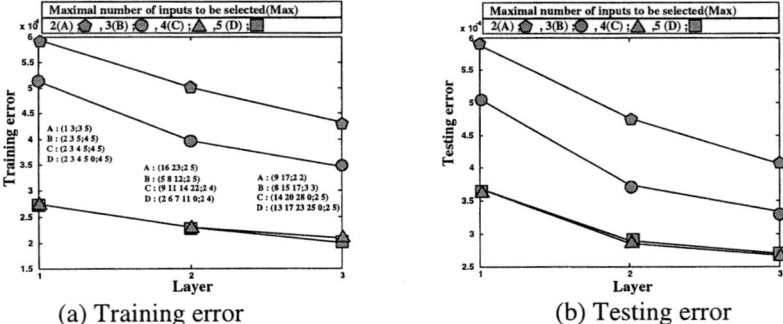

(a) Training error (b) Testing error

Fig. 1. Performance index of gFPNN with respect to the increase of number of layers (in case of using Type T* at the 1st layers and Gaussian-like MF)

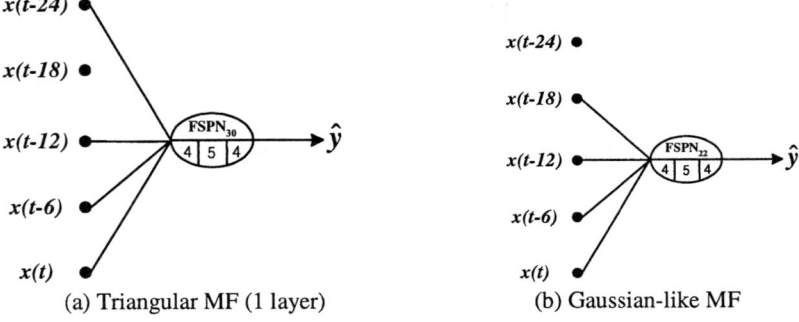

(a) Triangular MF (1 layer) (b) Gaussian-like MF

Fig. 2. Genetically optimized FPNN(gFPNN) architecture(Max=4 and Type T*)

Table 3 summarizes the performance of the 1st and the 2nd layer of the network when changing the maximal number of inputs to be selected; here Max was set up to 2 through 5.

Fig. 1 depicts the performance index of each layer of gFPNN according to the increase of maximal number of inputs to be selected. In Fig. 1, the premise part of A:(•;•)- D:(•;•) denotes the optimal node numbers at each layer of the network, namely those with the best predictive performance, and their consequent part stands for the polynomial order and the number of MFs. Fig. 2(a) illustrates the detailed optimal topologies of gFPNN for 1 layer when using Max=4 and triangular MF. And also Fig. 2(b) illustrates the detailed optimal topology of gFPNN for 1 layer in case of using Max= 4 and Gaussian-like MF. Table 4 gives a comparative summary of the network with other models.

Table 4. Comparative analysis of the performance of the network; considered are models reported in the literature

Model				PI	Performance index PI_s	EPI_s	NDEI[*]
Wang's model[10]				0.044 0.013 0.010			
Cascaded-correlation NN[11]							0.06
Backpropagation MLP[11]							0.02
6th-order polynomial[11]							0.04
ANFIS[12]					0.0016	0.0015	0.007
FNN model[13]					0.014	0.009	
Recurrent neural network[14]				0.0138			
SONN**[15]	Type I	Basic (5th layer)	Case 1		0.0011	0.0011	0.005
			Case 2		0.0027	0.0028	0.011
		Modified (5th layer)	Case 1		0.0012	0.0011	0.005
			Case 2		0.0038	0.0038	0.016
	Type II	Basic (5th layer)	Case 1		0.0003	0.0005	0.0016
			Case 2		0.0002	0.0004	0.0011
Proposed gFPNN	Type T*	Triangular (2nd layer)	Max=4		4.54e-4	4.55e-4	
		Gaussian (2nd layer)	Max=4		2.29e-4	2.85e-4	

[*]Non-dimensional error index (NDEI) as used in [16] is defined as the root mean square errors divided by the standard deviation of the target series. ** is called "conventional optimized FPNN".

5 Concluding Remarks

In this study, the GA-based design procedure of Fuzzy Polynomial Neural Networks (FPNN) along with their architectural considerations has been investigated. The design methodology comes as a hybrid structural optimization and parametric learning being viewed as two fundamental phases of the design process. The comprehensive experimental studies involving well-known datasets quantify a superb performance of the network in comparison to the existing fuzzy and neuro-fuzzy models. Most importantly, through the proposed framework of genetic optimization we can efficiently search for the optimal network architecture (structurally and parametrically optimized network) and this becomes crucial in improving the performance of the resulting model.

Acknowledgement. This work has been supported by KESRI(I-2004-0-074-0-00), which is funded by MOCIE(Ministry of commerce, industry and energy)

References

1. Cherkassky, V., Gehring, D., Mulier, F.: Comparison of adaptive methods for function estimation from samples. IEEE Trans. Neural Networks. **7** (1996) 969-984
2. Dickerson, J.A., Kosko, B.: Fuzzy function approximation with ellipsoidal rules. IEEE Trans. Syst., Man, Cybernetics. Part B. **26** (1996) 542-560
3. Sommer, V., Tobias, P., Kohl, D., Sundgren, H., Lundstrom, L.: Neural networks and abductive networks for chemical sensor signals: A case comparison. Sensors and Actuators B. **28** (1995) 217-222
4. Kleinsteuber, S., Sepehri, N.: A polynomial network modeling approach to a class of large-scale hydraulic systems. Computers Elect. Eng. **22** (1996) 151-168
5. Cordon, O., et al.: Ten years of genetic fuzzy systems: current framework and new trends. Fuzzy Sets and Systems. **141**(1) (2004) 5-31
6. Oh, S.K., Pedrycz, W.: Self-organizing Polynomial Neural Networks Based on Polynomial and Fuzzy Polynomial Neurons: Analysis and Design. Fuzzy Sets and Systems. **142**(2) (2003) 163-198
7. Michalewicz, Z.: Genetic Algorithms + Data Structures = Evolution Programs. 3rd edn. Springer-Verlag, Berlin Heidelberg New York (1996)
8. Jong, D., K. A.: Are Genetic Algorithms Function Optimizers?. Parallel Problem Solving from Nature 2, Manner, R. and Manderick, B. eds., North-Holland, Amsterdam (1992)
9. Oh, S.K., Pedrycz, W.: Fuzzy Polynomial Neuron-Based Self-Organizing Neural Networks. Int. J. of General Systems, **32** (2003) 237-250
10. Wang, L.X., Mendel, J.M..: Generating fuzzy rules from numerical data with applications. IEEE Trans. Systems, Man, Cybern. **22** (1992) 1414-1427
11. Crowder, R.S. III.: Predicting the Mackey-Glass time series with cascade-correlation learning. In: D. Touretzky, G. Hinton, and T. Sejnowski, editors, Proceedings of the 1990 Connectionist Models Summer School, Carnegic Mellon University, (1990) 117-123
12. Jang, J.S.R.: ANFIS: Adaptive-Network-Based Fuzzy Inference System. IEEE Trans. System, Man, and Cybern. **23** (1993) 665-685
13. Maguire, L.P., Roche, B., McGinnity, T.M., McDaid, L.J.: Predicting a chaotic time series using a fuzzy neural network. Information Sciences. **112** (1998) 125-136
14. James, C.L., Huang, T.Y.: Automatic structure and parameter training methods for modeling of mechanical systems by recurrent neural networks. Applied Mathematical Modeling. **23** (1999) 933-944
15. Oh, S.K., Pedrycz, W., Ahn, T.C.: Self-organizing neural networks with fuzzy polynomial neurons. Applied Soft Computing. **2** (2002) 1-10

16. Lapedes, A.S., Farber, R.: Non-linear Signal Processing Using Neural Networks: Prediction and System Modeling. Technical Report LA-UR-87-2662, Los Alamos National Laboratory, Los Alamos, New Mexico **87545** (1987)
17. Mackey, M.C., Glass, L.: Oscillation and chaos in physiological control systems. Science, **197** (1977) 287-289

Unsupervised Color Image Segmentation Using Mean Shift and Deterministic Annealing EM

Wanhyun Cho[1], Jonghyun Park[2], Myungeun Lee[3], and Soonyoung Park[3]

[1] Department of Statistics, Chonnam National University, Chonnam, South Korea
whcho@chonnam.ac.kr
[2] Institute for Robotics and Intelligent Systems, University of Southern California, Los Angeles, CA, 90089, USA
jonghyun@iris.usc.edu
[3] Department of Electronics Engineering, Mokpo National University, Chonnam, South Korea
melee, sypark@mokpo.ac.kr

Abstract. We present an unsupervised segmentation algorithm combining the mean shift procedure and deterministic annealing expectation maximization (DAEM) called MS-DAEM algorithm. We use the mean shift procedure to determine the number of components in a mixture model and to detect their modes of each mixture component. Next, we have adopted the Gaussian mixture model (GMM) to represent the probability distribution of color feature vectors. A DAEM formula is used to estimate the parameters of the GMM which represents the multi-colored objects statistically. The experimental results show that the mean shift part of the proposed MS-DAEM algorithm is efficient to determine the number of components and initial modes of each component in mixture models. And also it shows that the DAEM part provides a global optimal solution for the parameter estimation in a mixture model and the natural color images are segmented efficiently by using the GMM with components estimated by MS-DAEM algorithm.

1 Introduction

The segmentation of natural color image into an unknown number of distinct and homogeneous regions is a difficult problem and becomes a fundamental issue in low-level computer vision tasks. Given an image, feature vectors are extracted from local neighborhoods and mapped into the space spanned by their coordinates. Significant features in the image then correspond to high-density regions in this space. The finite mixture of multivariate probability distributions has been used as the statistical modeling of a continuous feature space. The widely often used assumption in modeling by using a finite mixture of distribution is that the number of components or clusters is small and known a priori and the individual components obey multivariate normal distributions. That is, the feature space can be modeled as a finite mixture of Gaussian distributions with a known number of components.

However, we cannot recognize the number of colors composing an observed real image before analyzing its image. So, we need the method that automatically estimate the number of mixture components. And also Gaussian mixture model is commonly used to represent the probability distribution of the feature vector in feature space. The EM algorithm is naturally used for estimating the parameters of Gaussian mixture model. But the estimates of parameters obtained by EM algorithm are strongly dependent upon their initial values and they are sometimes achieved by the local maximum of total log likelihood.

In this paper, to overcome this problem, we are going to consider an unsupervised segmentation algorithm combining the mean shift procedure with deterministic annealing EM, which will be referred as MS-DAEM algorithm. We show how to apply it for the estimation of components and parameters in a mixture model. We adopt the Gaussian mixture model to represent the probability distribution of the observed feature vector and perform the image segmentation using this model. And this paper demonstrates the performance of our segmentation algorithm from natural color scenes.

2 Mean Shift Procedure

2.1 Density Gradient Estimation

Let $\{\mathbf{X}_i\}, i=1,\cdots,n$ be the set of n data points in a d-dimensional Euclidean space. The multivariate kernel density estimate obtained with kernel $K(\mathbf{x})$ and window radius h, computed at point $\mathbf{x}$ is defined as:

$$\hat{f}_K(\mathbf{x}) = \frac{1}{nh^d}\sum_{i=1}^{n}K(\frac{\mathbf{x}-\mathbf{X}_i}{h}). \tag{1}$$

Here, we are interested only in a class of radially symmetric kernels satisfying

$$K(\mathbf{x}) = c_{K,d}k(\|\mathbf{x}\|^2),$$

in which case it suffices to define the function $k(x)$ called the profile of the kernel, only for $x \geq 0$ and $c_{K,d}$ is the normalized constant which makes $K(\mathbf{x})$ integrate to one.

The differentiation of the kernel allows one to define the estimate of the density gradient as the gradient of the kernel density estimate:

$$\nabla \hat{f}_K(\mathbf{x}) = \frac{1}{nh^d}\sum_{i=1}^{n}\nabla K(\frac{\mathbf{x}-\mathbf{X}_i}{h}) = \frac{2c_{K,d}}{nh^{d+2}}\sum_{i=1}^{n}(\mathbf{x}-\mathbf{X}_i)k'(\|\frac{\mathbf{x}-\mathbf{X}_i}{h}\|^2). \tag{2}$$

We define the derivative of the kernel profile as a new function

$$g(x) = -k'(x),$$

and assume that this exists for all $x \geq 0$, except for a finite set of points. Now, if we use a function for profile, the kernel is defined as

$$G(\mathbf{x}) = c_{G,d}g(\|\mathbf{x}\|^2),$$

where $c_{G,d}$ is the corresponding normalization constant. In this case, the kernel $K(\mathbf{x})$ is called the shadow of kernel $G(\mathbf{x})$.

If we use a function $g(x)$ in formula (2), then the gradient of the density estimator is written by

$$\nabla \hat{f}_K(\mathbf{x}) = \frac{2c_{K,d}}{nh^{d+2}} \sum_{i=1}^{n} (\mathbf{x} - \mathbf{X}_i) g(\|\frac{\mathbf{x} - \mathbf{X}_i}{h}\|^2)$$

$$= \frac{2c_{K,d}}{nh^{d+2}} \sum_{i=1}^{n} g\left(\|\frac{\mathbf{x}-\mathbf{X}_i}{h}\|^2\right) \left(\frac{\sum_{i=1}^{n} g(\|\frac{\mathbf{x}-\mathbf{X}_i}{h}\|^2)\mathbf{X}_i}{\sum_{i=1}^{n} g(\|\frac{\mathbf{x}-\mathbf{X}_i}{h}\|^2)} - \mathbf{x} \right) \quad (3)$$

Here, this is given as the product of two terms having special meaning. The first term in the expression (3) is proportional to the density estimate at $\mathbf{x}$ computed with the kernel $G(\mathbf{x})$

$$\hat{f}_G(\mathbf{x}) = \frac{1}{nh^d} \sum_{i=1}^{n} G(\frac{\mathbf{x}-\mathbf{X}_i}{h}) = \frac{c_{G,d}}{nh^d} \sum_{i=1}^{n} g(\|\frac{\mathbf{x}-\mathbf{X}_i}{h}\|^2) ,$$

and the second term is defined as the mean shift vector

$$\mathbf{m}_G(\mathbf{x}) = \left(\frac{\sum_{i=1}^{n} g\left(\left\|\frac{x-X_i}{h}\right\|^2\right) X_i}{\sum_{i=1}^{n} g\left(\left\|\frac{x-X_i}{h}\right\|^2\right)} - x \right). \quad (4)$$

This vector is the difference between the weight mean using the kernel $G(\mathbf{x})$ for weights and the center of the kernel. Then, we can rewrite the expression (3) as

$$\nabla \hat{f}_K(\mathbf{x}) = \frac{2c_{K,d}}{h^2 c_{G,d}} \hat{f}_G(\mathbf{x}) \mathbf{m}_G(\mathbf{x}),$$

which yields

$$\mathbf{m}_G(\mathbf{x}) = \frac{1}{2} h^2 c \frac{\nabla \hat{f}_K(\mathbf{x})}{\hat{f}_G(\mathbf{x})} . \quad (5)$$

The expression (5) shows the mean shift vector being proportional to the gradient of the density estimate at the point it is computed. As the vector points in the direction of maximum increase in density, it can define a path leading to a local density maximum which becomes a mode of density. It also exhibits a desirable adaptive behavior, with the mean shift step being large for low-density regions and decreases as a point $\mathbf{x}$ approaches a mode. Each data point thus becomes associated to a point of convergence, which represents a local mode of the density in the d-dimensional space.

2.2 Detection of Modes Based on Mean Shift procedure

Let us denote by $\{\mathbf{y}_1, \mathbf{y}_2, \cdots\}$ the sequence of successive locations of kernel $G(\mathbf{x})$, where these points are computed by the following formula

$$\mathbf{y}_j = \frac{\sum_{i=1}^{n} g\left(\|\frac{\mathbf{x}-\mathbf{X}_i}{h}\|^2\right)\mathbf{X}_i}{\sum_{i=1}^{n} g\left(\|\frac{\mathbf{x}-\mathbf{X}_i}{h}\|^2\right)} \quad j=1,2,\cdots \quad . \tag{6}$$

This is the weighted mean at $\mathbf{y}_j$ computed with kernel $G(\mathbf{x})$ and $\mathbf{y}_1$ is the center of the initial position of the kernel, $\mathbf{x}$. The corresponding sequence of density estimates computed with shadow kernel $K(\mathbf{x})$ is given by

$$\hat{f}_K(j) = \hat{f}_K(\mathbf{y}_j) \quad j=1,2,\cdots .$$

Here, if the kernel has a convex and monotonically decreasing profile, two sequence $\{\mathbf{y}_1, \mathbf{y}_2, \cdots\}$ and $\{\hat{f}_K(1), \hat{f}_K(2), \cdots\}$ converge and $\{\hat{f}_K(1), \hat{f}_K(2), \cdots\}$ is monotonically increasing. After that, let us denote by $\mathbf{y}_c$ and $\hat{f}_K^c$ the convergence points of their sequences respectively. Here, we can get two kinds of implications from the convergence result. First, the magnitude of the mean shift vector converges to zero. In fact, the j-th mean shift vector is given as

$$\mathbf{m}_G(\mathbf{y}_j) = \mathbf{y}_{j+1} - \mathbf{y}_j ,$$

and this is equal to zero at the limit point, $\mathbf{y}_c$. In other words, the gradient of the density estimate computed at $\mathbf{y}_c$ is zero. That is,

$$\nabla \hat{f}_K(\mathbf{y}_c) = 0.$$

Hence, $\mathbf{y}_c$ is a stationary point of density estimate, $\hat{f}_K(\mathbf{x})$. Second, since $\{\hat{f}_K(1), \hat{f}_K(2), \cdots\}$ is monotonically increasing, the trajectories of mean shift iterations are attracted by local maximum if they are unique stationary points. That is, once $\mathbf{y}_j$ gets sufficiently close to a mode of density estimate, it converges to mode.

The theoretical results obtained from the above implications suggest a practical algorithm for mode detection:

    Step1: Run the mean shift procedure to find the stationary points of density estimates.
    Step2: Prune these points by retaining only the local maximum.

This algorithm automatically determines the number and location of modes of estimated density function. We shall use the detected mode or cluster centers from the mean shift procedure to be manifestations of underlying components of the mixture model for our image segmentation task.

3 Segmentation of Color Image Using Gaussian Mixture Model

3.1 Modeling of Color Images with GMM and Parameter Estimation Using DAEM Algorithm

Normally, natural images consist of several objects and they have native color stimuli. Color model for representing their colors is commonly used with the RGB model. To use HIS (Hue, Intensity, Saturation) color model as a user-oriented color model in this paper, we first translate the RGB color space into the HIS color space. A color distribution is obtained by projecting the pixel values in the selected object into the color space. Here, we will employ a GMM to characterize the distribution of color feature vectors observed from each object consisting of a natural color image.

Suppose that a color image consists of a set of disjoint pixel labeled 1 to N, and that each pixel is assumed to belong to one of the K distinct regions. We let the groups $G_1, \cdots, G_K$ represent the K possible regions. Also we let $\mathbf{y}_i$ denote the finite dimensional feature vector observed from i th pixel ($i = 1, \cdots, N$). Then, we adopt the GMM for a distribution of each feature vector $\mathbf{y}_i$ as defined as the following model

$$p(\mathbf{y}; \Theta) = \sum_{k=1}^{K} \pi_k N(\boldsymbol{\mu}_k, \Sigma_k),$$

where $N(\boldsymbol{\mu}_k, \Sigma_k)$ denote a bivariate or trivariate normal distribution with mean vector $\boldsymbol{\mu}_k$ and a covariance matrix Σ_k, and Θ is the parameter vector used to characterize each object or region. Further, we let $\mathbf{Z}_1, \cdots, \mathbf{Z}_N$ denote the unobservable group indicator vectors, where the k th element z_{ik} of $\mathbf{Z}_i$ is taken to be one or zero according to the case in which the i th pixel does or does not belong to the k th group. Here, if parameter vector, π is denoted as the prior probability in which each pixels belongs to a particular group, then the probability function of $\mathbf{Z}$ is given as follows:

$$p(\mathbf{z}; \pi) = \prod_{k=1}^{K} \pi_k^{z_{ik}}.$$

Then the distribution of a color image is expressed by the joint distribution of a complete data vector, $\mathbf{x} = (\mathbf{y}, \mathbf{z})$, and the log likelihood function that can be formed on the basis of the complete data $\mathbf{x}$ if we adopt the GMM for an observed feature vector, is given by

$$\log L_C(\Theta | \mathbf{x}) = \log p(\mathbf{y} | \mathbf{z}; \Theta) + \log p(\mathbf{z}; \pi)$$
$$= \sum_{k=1}^{K} \sum_{i=1}^{N} Z_{ik} \log(N(\boldsymbol{\mu}_k, \Sigma_k)) + \sum_{k=1}^{K} \sum_{i=1}^{N} Z_{ik} \log(\pi_k) \quad (7)$$

where Θ is the vector containing the elements of Θ and π.

The problem of maximum likelihood estimation of Θ given the observed vector $\mathbf{y}$ can be solved by applying the EM algorithm proposed by Dempster et al. for the incomplete data. However the EM algorithm has two kinds of disadvantages. The first is hard to avoid unfavorable local maximum of the log-likelihood and the second is overfitting problem. Thus we have to think about the new method that is able to improve the EM algorithm. It is known as a DAEM algorithm. This is to use the principle of maximum entropy to estimate the parameter [8].

We consider the complete data log likelihood $\log L_c(\Theta | \mathbf{x})$ as a function of the hidden variable $\mathbf{z}$ for fixed parameter vector Θ, and define a cost function on the hidden variable space Ω_z as follows:

$$H(\mathbf{z}; \mathbf{y}, \Theta) = -\log L_c(\Theta | \mathbf{x}) \tag{8}$$

Then we need to minimize $E(H(\mathbf{z}; \mathbf{y}, \Theta))$ with respect to probability distribution $p(\mathbf{z}; \pi)$ over the distribution space subject to a constraint on the entropy. It yields a quantity, which is known as the generalized free energy in statistical physics. Introducing a Lagrange parameter β, we arrive at the following objective function:

$$\vartheta(P_z^{(t)}, \Theta) = E_{P_z^{(t)}}(H(\mathbf{z}; \mathbf{y}, \Theta)) + \beta \cdot E_{P_z^{(t)}}(\log P_z^{(t)}) \tag{9}$$

The solution of the minimization problem associated with the generalized free energy in $\vartheta(P_z^{(t)}, \Theta)$ with respect to probability distribution $p(\mathbf{z}; \pi)$ with a fixed parameter Θ is the following Gibbs distribution:

$$P_\beta(\mathbf{z} | \mathbf{y}, \Theta) = \frac{1}{\sum_{\mathbf{z}' \in \Omega_z} \exp(-\beta H(\mathbf{z}'))} \cdot \exp(-\beta H(\mathbf{z})) \tag{10}$$

Hence we can obtain a new posterior distribution, $p_\beta(\mathbf{z} | \mathbf{y}, \Theta)$ parameterized by β. Next, we should find the minimum of $\vartheta(P_z^{(t)}, \Theta)$ with respect to Θ with fixed posterior $p_\beta(\mathbf{z} | \mathbf{y}, \Theta)$. It means finding the estimates $\Theta^{(t)}$ that minimizes $\vartheta(P_z^{(t)}, \Theta)$. The generalized free energy, $\vartheta(P_z^{(t)}, \Theta)$ can be written by the following form:

$$\vartheta(P_z^{(t)}, \Theta) = Q_\beta(\Theta) + \beta \cdot E_{P_z^{(t)}}(\log P_z^{(t)})$$

Since the second term on the right hand side of the generalized free energy is independent of Θ, we should find the value of Θ minimizing the first term

$$Q_\beta(\Theta) = E_{P_z^{(t)}}(H(\mathbf{z}; \mathbf{y}, \Theta))$$

To achieve this purpose, we can add a new β-loop, which is called annealing loop, to the original EM-algorithm and replace the original posterior with the new posterior distribution, $p_\beta(\mathbf{z} | \mathbf{y}, \Theta)$ parameterized by β.

Finally, after finishing fully iteration, we can obtain the conditional expectation of the hidden variable, Z_{ik} given the observed feature data from E-step. This is given by

$$\tau_k^\beta(\mathbf{y}_i) = E(Z_{ik}) = \frac{(\hat{\pi}_k N(\mathbf{y}_i; \hat{\boldsymbol{\mu}}_k, \hat{\boldsymbol{\Sigma}}_k))^\beta}{\sum_{j=1}^{K}(\hat{\pi}_j N(\mathbf{y}_i; \hat{\boldsymbol{\mu}}_j, \hat{\boldsymbol{\Sigma}}_j))^\beta} \qquad (11)$$

And also we can obtain the estimators of mixing proportions, a component mean vector and covariance matrix from M-step. These are respectively given as

$$\hat{\pi}_k = \frac{1}{N}\sum_{i=1}^{N}\tau_k^\beta(\mathbf{y}_i), \; k=1,\cdots,K,$$

$$\hat{\boldsymbol{\mu}}_k = \frac{\sum_{i=1}^{N}\tau_k^\beta(\mathbf{y}_i)\mathbf{y}_i}{\sum_{i=1}^{N}\tau_k^\beta(\mathbf{y}_i)}, \; k=1,\cdots,K$$

and

$$\hat{\boldsymbol{\Sigma}}_k = \frac{\sum_{i=1}^{N}\tau_k^\beta(\mathbf{y}_i)(\mathbf{y}_i-\hat{\boldsymbol{\mu}}_k)(\mathbf{y}_i-\hat{\boldsymbol{\mu}}_k)^t}{\sum_{i=1}^{N}\tau_k^\beta(\mathbf{y}_i)}, \; k=1,\cdots,K \qquad (12)$$

3.2 Segmentation of Natural Color Image

Suppose that a natural image consists of a set of the K distinct objects or regions. We usually segment a natural image to assign each pixel to some regions or objects. To do this, we need a posterior probability of the i th pixel belonging to k th region. These probabilities have already been estimated by DAEM algorithm in last section. That is, given the observed data set, $S = \{\mathbf{y}_i\}_{i=1}^{N}$ and a knowledge of estimated parameter vector $\hat{\Theta}$, the posterior probability of the i th pixel belongs to k th region is given by the formula (12).

Next, we try to find what the component or region has the maximum value among the estimated posterior probabilities. This is define as

$$\hat{Z}_i = \arg\max_{1\leq k\leq K} \tau_k^\beta(\mathbf{y}_i), \; i=1,\cdots,N.$$

Then, we can segment a natural image by assigning each pixel to the region or object having the maximum a posterior probability.

4 Experimental Results

To demonstrate the performance of the proposed segmentation algorithm, we have conducted several experiments for the synthetic data and real images. Fig. 1 shows synthetic data sets consisting of 4000 samples generated from three different Gaussian mixture functions. We have first applied mean shift procedure to the data sets in Fig. 1 to find the number and location of clusters. In Fig. 2, we can see that each mode is correctly detected regardless of the cluster structures.

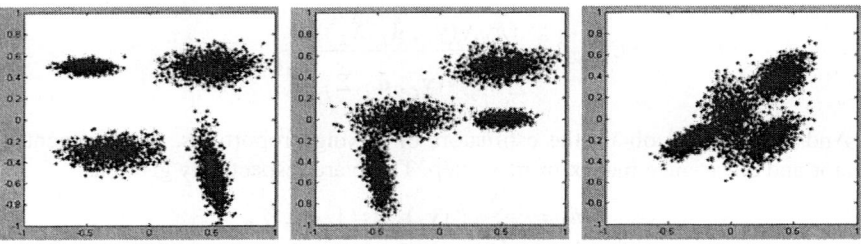

Fig. 1. Synthetic data representing three different Gaussian mixture functions

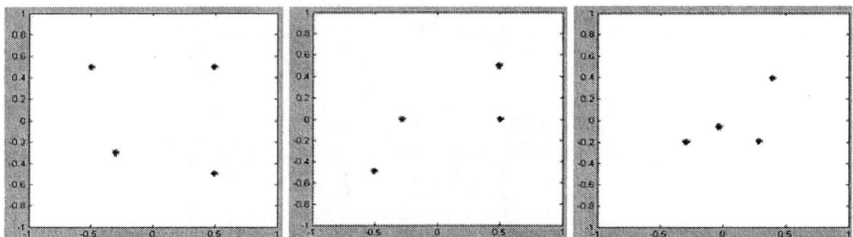

Fig. 2. Mode detection results of the mean shift procedure for the data set in Fig.1

To examine the influence of the bandwidth (window radius) parameter h, we have obtained trajectories by applying the mean shift procedure to the synthetic data set consisting of 6000 samples with multivariate Gaussian density functions in Fig. 3.

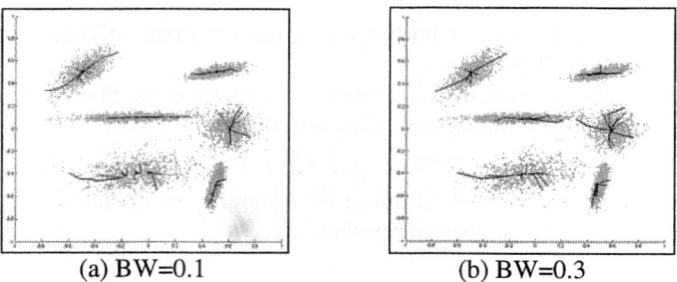

(a) BW=0.1   (b) BW=0.3

Fig. 3. Trajectory comparison of the mean shift procedures with two different bandwidths

Randomly selected 60 points were used as the initial window location and bandwidth of 0.1 and 0.3 were chosen for the experiment. As being expected, we can observe that the trajectories exhibit the path leading to a local density maximum from each initial location. Unfavorable effect, however, has occurred from the cluster with scattering samples. Number of modes increase when the small bandwidth is employed to the cluster with large covariance matrix. This is due to the unstable variations of local density estimated by a small bandwidth. This kind of artifacts can be eliminated by merging the closely located modes to the one corresponding to the highest density [7]. As the similar reason, the paths with bandwidth 0.3 follow much smoother trajectory toward the mode than the ones with bandwidth 0.1.

Now we have applied the proposed MS-DAEM algorithm to the color images "text" and "hand", in Fig. 4(a) and 5(a), respectively. The conversion of the RGB color image to HSI model is carried out and the hue and saturation components are only used as values of the feature vectors. The segmentation results using the MS-DAEM algorithm are shown in Fig. 4(b), and 5(b). We can see that the homogeneous objects are partitioned into the same region accurately and the fine structure is preserved.

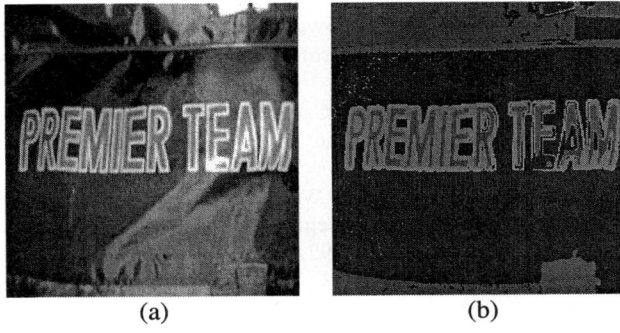

Fig. 4. Text image. (a) Original   (b) Segmentation result using MS-DAEM

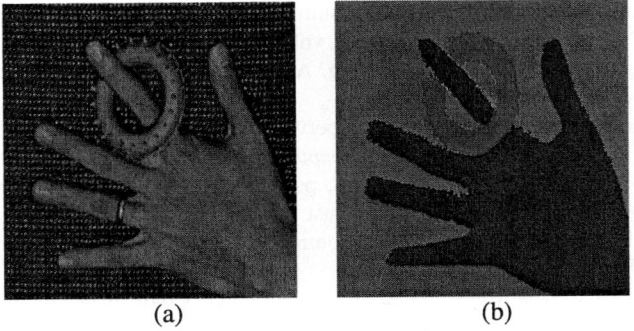

Fig. 5. Hand image. (a) Original   (b) Segmentation result using MS-DAEM

5 Conclusions

In this paper, we have proposed MS-DAEM algorithm combining the mean shift procedure and the deterministic annealing EM algorithm for unsupervised segmentation of natural color image. The mean shift procedure using a gradient of a kernel density estimates provides an unsupervised mode detection when the number of components is not known a priori. The DAEM algorithm is the estimation method of various parameters in GMM derived from the principle of maximum entropy to overcome the local maximum problem associated with the conventional EM algorithm.

We conclude from the experiments for the synthetic data and real images that the mean shift procedure has been proven to perform well in detecting the number of components or clusters in complicated feature spaces of many real images, and the DAEM algorithm is robust to initial conditions and it provides a global optimal solution for the ML estimators.

References

1. G.J. McLachlan, S.K.Nguyen, G.J.Galloway, and D.Wang, "Clustering of Magnetic Resonance Images", Technical Report, Department of Mathematics, University of Queensland, 1998.
2. H. Permuter, J. Francos, I. H. Jermyn, "Gaussian mixture models of texture and colour for image database retrieval," Proc. of the IEEE International Conference on Acoustics, Speech and Signal Processing, 2003.
3. Yves Delignon, Abdelwaheb Marzouki, Wojciech Pieczynski, "Estimation of Generalized Mixtures and Its application in Image Segmentation," IEEE Transactions on Image Processing, vol. 6, no. 10, pp. 1364-1375, 1997.
4. Dempster A. P., Laird N. M., Rubin D. B., "Maximum likelihood from incomplete data via the EM algorithm," Journal of Royal Statistical Society B, 39, 1-39, 1977.
5. T.Hofmann and J.M.Buhman, "Pairwise data clustering by deterministic annealing", IEEE Transactions on PAMI, vol. 19, no. 1, pp. 1-13, 1998.
6. N.Ueda and R.Nakano, "Deterministic annealing EM algorithm", Neural Networks, vol. 11, pp. 271-282, 1998.
7. D. Comaniciu, and P. Meer, "Mean Shift: A Robust Approach Toward Feature Space Analysis", IEEE Transactions on PAMI, vol. 24, no. 5, pp. 1-17, 2002.
8. Y. Cheng, "Mean Shift, Mode Seeking, And Clustering", IEEE Transactions on PAMI, vol. 17, no. 8, pp. 790-799, 1995.
9. A. H. Kam and W. J. Fitzgerald, "Unsupervised Multiscale Image Segmentation", In *Proc. 3rd International Conference on Computer Vision, Pattern Recognition and Image Processing (CVPRIP 2000)*, volume II, pages 54-57, Atlantic City, New Jersey, USA,
10. H. Wang and D. Suter, "A Novel Robust Method for Large Numbers of Gross Errors", Seventh International Conference on Control, Automation, Robotics And Vision, pp 326-331, 2002.
11. Jonghyun Park, Wanhyun Cho and Soonyoung Park, "Color Image Segmentation Using a Gaussian Mixture Model and a Mean field Annealing EM", IEICE Transactions on Information and Systems, Vol. E86-D, No 10, pp 2240-2248, 2003.

Identity-Based Key Agreement Protocols in a Multiple PKG Environment[*]

Hoonjung Lee[1], Donghyun Kim[1], Sangjin Kim[2], and Heekuck Oh[1]

[1] Department of Computer Science and Engineering,
Hanyang University, Republic of Korea
{leehj, kimdh, hkoh}@cse.hanyang.ac.kr
[2] School of Internet Media Engineering,
Korea University of Technology and Education, Republic of Korea
sangjin@kut.ac.kr

Abstract. To date, most identity-based key agreement protocols are based on a single PKG (Private Key Generator) environment. In 2002, Chen and Kudla proposed an identity-based key agreement protocol for a multiple PKG environment, where each PKG shares identical system parameters but possesses a distinct master key. However, it is more realistic to assume that each PKG uses different system parameters. In this paper, we propose a new two party key agreement protocol between users belonging to different PKGs that do not share system parameters. We also extend this protocol to a tripartite key agreement protocol. Our two party protocol requires the same amount of pairing computation as Smart's protocol for a single PKG environment and provides PKG forward secrecy. We show that the proposed key agreement protocols satisfy every security requirements of key agreement protocols.

Keywords: ID-based cryptosystem, bilinear map, key agreement protocol, multiple PKG.

1 Introduction

Key establishment protocol is a cryptographic primitive that is used to share a common secret key between entities. This secret key is normally used as a session key to construct a secure channel between the entities concerned. Key establishment protocol can be subdivided into key transport protocol and key agreement protocol. In key transport protocol, one of the participant creates the shared key and distributes it to others securely. On the other hand, in key agreement protocol, each entity computes the common secret key using the information contributed by all the entities involved. In key transport protocol, all the participants have to trust the entity responsible for creating the new shared key. Therefore, normally a trusted third party is used as a server that creates and distributes shared keys. However, in this setting, this entity becomes a good

---

[*] This work was supported by the Ministry of Information and Communication, Korea, under the university HNRC-ITRC program supervised by the IITA.

target for attack and the success of such attack is catastrophic. Furthermore, this entity may become a bottleneck point. In contrast, key agreement protocols does not suffer from these problems, since no single entity determines the shared key.

1.1 Related Work

In 1984, Shamir introduced the concept of identity-based public key cryptosystem where public keys of users' are derived from their own unique identity information such as an email address [1]. In identity-based cryptosystem, private keys of users' are issued by a trusted authority called the PKG. The private key of a user is generated using the master key of the system. Therefore, PKG can inherently decrypt any ciphertext or forge signatures of any users'. In 2001, Boneh and Franklin proposed a first practical identity-based encryption scheme based on the Weil pairing [2]. Since then, most researches on identity-based cryptosystem are based on this system.

Identity-based two-party key agreement protocol was first proposed by Smart in 2001 [3]. This protocol is based on Boneh and Franklin's work and requires two pairing computation to compute the session key. However, Smart's protocol does not satisfy PKG forward secrecy. In 2002, Chen and Kudla introduced three new key agreement protocols. One of them extended Smart's protocol to provide PKG forward secrecy property. Another one extended Smart's protocol to multiple PKG environment where users who acquired their private keys from different PKG can share a common key [4]. However, in their setting, every PKG shares the common system parameters, but possess distinct master key.

Research on identity-based tripartite key agreement protocol was initiated by Joux in 2000 [5]. In 2003, Al-Riyami et al. pointed out that Joux's method does not support message authentication between participants, and proposed a new tripartite key agreement protocol that cures the drawback of Joux's [6]. A first identity-based tripartite key agreement protocol was introduced by Zhang et al. in 2002 [7]. In 2003, Shim also proposed another identity-based tripartite key agreement protocol, which requires less computation than Zhang et al.'s [8]. In 2004, Cheng et al. proposed a new identity-based tripartite key agreement protocol which is different from Shim's and Zhang et al.'s system [9]. However, this protocol has a serious flaw that allows an adversary to acquire the private key of a user easily.

1.2 Our Contribution

In identity-based cryptosystems, users acquire their private key from the PKG. A single PKG may be responsible for issuing private keys to members of a small-scale organization, but it is unrealistic to assume that a single PKG will be responsible for issuing private keys to members of different organizations, let alone the entire nation or the entire world. Furthermore, it is also unrealistic to assume that different PKGs will share common system parameters and differ only in the master key as done by Chen and Kudla. Therefore, we must consider multiple PKG environment where all the PKGs use different system parameters.

To date, most of the identity-based key agreement protocols are based on a single PKG environment [3,4,5,6,7,8,9,10]. In order to extend these protocols to a setting where multiple PKGs exist, there should be some way to combine entities' contribution from different settings into a single common value. The most obvious solution to accomplish this is to combine the results of separate key agreement protocols executed in each PKG environment. However, since each entity only has its private key from its PKG, direct execution of existing protocol is infeasible. To ameliorate this situation, in this paper, we propose a new efficient ID-based two-party key agreement protocol for multiple PKG environment. We also extends this protocol to a tripartite version. Our two-party key agreement protocol is as efficiency as any previous ID-based two-party key agreement protocol that provides PKG forward secrecy.

2 Security Attributes of Key Agreement Protocol

The followings are the security requirements of key agreement protocols, some of which are specific to ID-based key agreement protocol.

Known-key security: Each run of the key agreement protocol should generate an unique and independent session key. An adversary must have non-negligible advantage on compromising future session keys, even though it compromised past session keys.

Forward secrecy: An adversary must have non-negligible advantage on compromising past session keys, even though it compromised long-term private keys of one or more participants. The notion of forward secrecy can be further extended to the following two types of secrecy.
- **Perfect forward secrecy:** The forward secrecy must be preserved even if long-term private keys of all the participants involved are compromised.
- **PKG forward secrecy:** The forward secrecy must be preserved even if the master key of the PKG is compromised.

Key-compromise resilience: An adversary must have non-negligible advantage on impersonating others to A, even if it has compromised A's private key.

Unknown key share resilience: An adversary must have non-negligible advantage on coercing others into sharing a key with other entities when it is actually sharing with a different entity.

Key control: An adversary must have non-negligible advantage on forcing the session key to be a preselected value.

The PKG forward secrecy is a stronger notion of perfect forward secrecy that applies only to ID-based protocols. In ID-based cryptosystems, if the master key is compromised, an adversary can compute all the participants' private keys. Therefore, If ID-based key agreement protocol satisfies PKG forward secrecy, then it also satisfies perfect forward secrecy too. However, the opposite is not true. We can also define perfect PKG forward secrecy in multiple PKG environment. This secrecy implies that forward secrecy is preserved even if all the master keys of PKGs are compromised.

3 Mathematical Background

3.1 Pairings

Bilinear pairings such as Weil pairing and Tate pairing reduces the discrete logarithm problem on elliptic curves to that in a finite field. Originally pairings were introduced as a tool that can be used to attack cryptosystems based on elliptic curves. Using pairings, decision Diffie-Hellman problem on elliptic curves can be easily solved. This is a well-known MOV (Menezes, Okamoto, Vanstone) reduction [11] and FR (Frey, Ruck) attack [12]. In recent years, bilinear pairings have found positive applications in cryptography to construct new cryptographic primitives. In 2000, Joux showed that the Weil pairing can be used to construct a simple tripartite Diffie-Hellman key agreement protocol [5]. Since then, most of identity-based primitives exploit pairing to achieve their goals.

From now on, we will use the following notations: 1) q means a large prime number, 2) $\mathbb{G}_1$ and $\mathbb{G}_2$ are two groups with the same order q, where $\mathbb{G}_1$ is an additive group on an elliptic curve, and $\mathbb{G}_2$ is a multiplicative group of a finite field, 3) P, Q, and R are random elements of $\mathbb{G}_1$, and 4) a, b, x, y, and z are random elements of $\mathbb{Z}_q^*$.

Definition 1 (Admissible Bilinear Map). *A map $e : G_1 \times G_2 \to G_2$ is an admissible bilinear map only if it satisfies the following properties:*

- **Bilinear**: Given P, Q, and R, the followings hold.
  - $e(P+Q,R) = e(P,R) \cdot e(Q,R)$
  - $e(P,Q+R) = e(P,Q) \cdot e(P,R)$

  This property also implies the followings:
  $e(aP,bQ) = e(P,bQ)^a = e(aP,Q)^b = e(P,Q)^{ab} = e(abP,Q) = e(p,abQ)$.
- **Non-degenerate**: If P and Q are not identity elements of $\mathbb{G}_1$, then $e(P,Q) \neq O$, where O is an identity element of $\mathbb{G}_2$.
- **Computable**: There exists an efficient algorithm to compute $e(P,Q)$ for all $P,Q \in \mathbb{G}_1$

3.2 Cryptographic Problems

Definition 2 (Discrete Logarithm Problem (DLP) in $\mathbb{G}_1$). *DLP is as follow: Given $\langle P, xP \rangle$, compute $x \in \mathbb{Z}_q$.*

Definition 3 (Computational Diffie-Hellman Problem (CDHP) in $\mathbb{G}_1$). *CDHP is as follow: Given $\langle P, xP, yP \rangle$, compute $xyP \in \mathbb{G}_1$.*

Definition 4 (Bilinear Diffie-Hellman Problem (BDHP) in $\mathbb{G}_1$ and $\mathbb{G}_2$). *BDHP is as follow: Given $\langle P, xP, yP, zP \rangle$, compute $e(P,P)^{xyz} \in \mathbb{G}_2$.*

Currently, solving DLP, CDHP, and BDHP is computationally infeasible. For more detail, refer to [2].

4 The Protocols

4.1 System Setup

Basically, the system setup phase is similar to that of Boneh and Franklin's work. However, in our system, there are total n different PKGs, which do not share common system parameters. Therefore, each PKG must configure its parameters as follows:

- Each PKG$_i$ chooses its basic system parameter: $\langle \mathbb{G}_1^{(i)}, \mathbb{G}_2^{(i)}, e^{(i)} \rangle$, where $\mathbb{G}_1^{(i)}$ is an additive group of order $q^{(i)}$, $\mathbb{G}_2^{(i)}$ is a multiplicative group of order $q^{(i)}$, and $e^{(i)}$ is admissible bilinear map between $\mathbb{G}_1^{(i)}$ and $\mathbb{G}_2^{(i)}$.
- PKG$_i$ chooses $P^{(i)}$, a random generator of $\mathbb{G}_1^{(i)}$. It also chooses cryptographic hash functions $H_1^{(i)} : \{0,1\}^* \to \mathbb{G}_1^{(i)}$ and $H_2^{(i)} : \mathbb{G}_2^{(i)} \to \{0,1\}^k$, where k is a length of partial session key.
- Finally, PKG$_i$ chooses its master key $s^{(i)} \in \mathbb{Z}_{q^{(i)}}^*$ randomly. It also computes its public key $P_{pub}^{(i)} = s^{(i)} P^{(i)}$.

After completing the system setup phase, each PKG publishes its public system parameters:

$$\langle \mathbb{G}_1^{(i)}, \mathbb{G}_2^{(i)}, P^{(i)}, P_{pub}^{(i)}, H_1^{(i)}, H_2^{(i)}, e^{(i)} \rangle.$$

4.2 Identity-Based Two-Party Key Agreement Protocol

In this section, we will introduce a new key agreement protocol between two entity A and B, each of which have acquired its private key from PKG$_1$ and PKG$_2$, respectively. We assume that A and B have chosen their ephemeral key $a \in \mathbb{Z}_{q_1}^*$ and $b \in \mathbb{Z}_{q_2}^*$ respectively. Then, the protocol runs as follows:

Message 1: $A \to B$: $T_{AB}^{(2)} = a^{(2)} P^{(2)}, W_A^{(1)} = a^{(1)} P_{pub}^{(1)}$

Message 2: $B \to A$: $T_{BA}^{(1)} = b^{(1)} P^{(1)}, W_B^{(2)} = b^{(2)} P_{pub}^{(2)}$

Protocol 1

After the messages are exchanged, each participant computes the two partial session keys as follows:

- A computes partial session keys, $K_{AB}^{(1)} = e^{(1)}(a^{(1)} S_A^{(1)}, T_{BA}^{(1)})$, and $K_{AB}^{(2)} = e^{(2)}(a^{(2)} Q_B^{(2)}, W_B^{(2)})$.
- B computes partial session keys, $K_{BA}^{(1)} = e^{(1)}(b^{(1)} Q_A^{(1)}, W_A^{(1)})$, and $K_{BA}^{(2)} = e^{(2)}(b^{(2)} S_B^{(2)}, T_{AB}^{(2)})$.

Now, each entity uses the two partial session key to compute the common session key. In detail, A computes the common session key $SK_{AB} = H(H_2^{(1)}(K_{AB}^{(1)}), H_2^{(2)}(K_{AB}^{(2)}))$, where H is a general hash function such as SHA-1. Similarly, B computes the session key $SK_{BA} = H(H_2^{(1)}(K_{BA}^{(1)}), H_2^{(2)}(K_{BA}^{(2)}))$. We can show that both participant have agreed on the same session key $SK = SK_{AB} = SK_{BA}$ by the followings:

$$\begin{aligned}
K_{AB}^{(1)} &= e^{(1)}(a^{(1)}S_A^{(1)}, T_{BA}^{(1)}) & K_{AB}^{(2)} &= e^{(2)}(a^{(2)}Q_B^{(2)}, W_B^{(2)}) \\
&= e^{(1)}(a^{(1)}s^{(1)}Q_A^{(1)}, b^{(1)}P^{(1)}) & &= e^{(2)}(a^{(2)}Q_B^{(2)}, b^{(2)}P_{pub}^{(2)}) \\
&= e^{(1)}(Q_A^{(1)}, P^{(1)})^{a^{(1)}s^{(1)}b^{(1)}} & &= e^{(2)}(a^{(2)}Q_B^{(2)}, b^{(2)}s^{(2)}P^{(2)}) \\
&= e^{(1)}(b^{(1)}Q_A^{(1)}, a^{(1)}s^{(1)}P^{(1)}) & &= e^{(2)}(Q_B^{(2)}, P^{(2)})^{a^{(2)}b^{(2)}s^{(2)}} \\
&= e^{(1)}(b^{(1)}Q_A^{(1)}, a^{(1)}P_{pub}^{(1)}) & &= e^{(2)}(b^{(2)}s^{(2)}Q_B^{(2)}, a^{(2)}P^{(2)}) \\
&= e^{(1)}(b^{(1)}Q_A^{(1)}, W_A^{(1)}) & &= e^{(2)}(b^{(2)}S_B^{(2)}, T_{AB}^{(2)}) \\
&= K_{BA}^{(1)}, & &= K_{BA}^{(2)}.
\end{aligned}$$

4.3 ID-Based Tripartite Key Agreement Protocol

In this subsection, we introduce a new tripartite key agreement protocol which is evolved from our two-party key agreement protocol. We assume that there are three participants A, B, and C, each of which have acquired its private key from PKG_1, PKG_2, and PKG_3, respectively. We also assume that A, B, and C have chosen their ephemeral key $a \in \mathbb{Z}_{q^{(1)}}^*$, $b \in \mathbb{Z}_{q^{(2)}}^*$, and $c \in \mathbb{Z}_{q^{(3)}}^*$, respectively.

The First Round. The protocol is divided into two discrete rounds. In the first round, each entity constructs separate secure channel between others. To achieve this goal, every entity exploits our two-party key agreement protocol with the others individually. First of all, each participant broadcast messages to the others as follows:

Message 1: $A \rightarrow B, C$: $T_{AB}^{(2)} = a^{(2)}P^{(2)}, T_{AC}^{(3)} = a^{(3)}P^{(3)}, W_A^{(1)} = a^{(1)}P_{pub}^{(1)}$
Message 2: $B \rightarrow C, A$: $T_{BA}^{(1)} = b^{(1)}P^{(1)}, T_{BC}^{(3)} = b^{(3)}P^{(3)}, W_B^{(2)} = b^{(2)}P_{pub}^{(2)}$
Message 3: $C \rightarrow A, B$: $T_{CA}^{(1)} = c^{(1)}P^{(1)}, T_{CB}^{(2)} = b^{(2)}P^{(2)}, W_C^{(3)} = c^{(3)}P_{pub}^{(3)}$

First Round of Protocol 2

After broadcasting, each entity computes partial session keys. In detail, A computes partial keys K_{AB}, which is used to construct secure channel between A and B, and K_{AC}, which is also used to construct secure channel between A and C.

$$K_{AB} = H(H_2^{(2)}(e^{(2)}(a^{(2)}Q_B^{(2)}, W_B^{(2)})), H_2^{(1)}(e^{(1)}(a^{(1)}S_A^{(1)}, T_{BA}^{(1)})))$$
$$K_{AC} = H(H_2^{(3)}(e^{(3)}(a^{(3)}Q_C^{(3)}, W_C^{(3)})), H_2^{(1)}(e^{(1)}(a^{(1)}S_A^{(1)}, T_{CA}^{(1)})))$$

Similarly, B also computes its partial session keys, which is used for building secure channel with A and C, respectively.

$$K_{BA} = H(H_2^{(1)}(e^{(1)}(b^{(1)}Q_A^{(1)}, W_A^{(1)})), H_2^{(2)}(e^{(2)}(b^{(2)}S_B^{(2)}, T_{AB}^{(2)})))$$
$$K_{BC} = H(H_2^{(3)}(e^{(3)}(b^{(3)}Q_C^{(3)}, W_C^{(3)})), H_2^{(2)}(e^{(2)}(b^{(2)}S_B^{(2)}, T_{CB}^{(2)})))$$

Finally, C computes its partial session keys.

$$K_{CA} = H(H_2^{(1)}(e^{(1)}(c^{(1)}Q_A^{(1)}, W_A^{(1)})), H_2^{(3)}(e^{(3)}(c^{(3)}S_C^{(3)}, T_{AC}^{(3)})))$$
$$K_{CB} = H(H_2^{(2)}(e^{(2)}(c^{(2)}Q_B^{(2)}, W_B^{(2)})), H_2^{(3)}(e^{(3)}(c^{(3)}S_C^{(3)}, T_{BC}^{(3)})))$$

At the end of this phase, each entity can computes pairwise keys with other entities, separately as being introduced in our two-party key agreement protocol.

The Second Round. In the second round, each entity exchanges some values with others in order to share common secret key. First of all, each entity broadcasts some values, each of which includes the ephemeral key of each entity's. Protocol runs as follows:

Message 4: $A \to B, C$: $\{a^{(1)}P^{(1)}\}_{K_{AB}}, \{a^{(1)}P^{(1)}\}_{K_{AC}}$
Message 5: $B \to C, A$: $\{b^{(2)}P^{(2)}\}_{K_{BA}}, \{b^{(2)}P^{(2)}\}_{K_{BC}}$
Message 6: $C \to A, B$: $\{c^{(3)}P^{(3)}\}_{K_{CA}}, \{c^{(3)}P^{(3)}\}_{K_{CB}}$

Second Round of Protocol 2

After broadcasting, A computes a common session key SK as follows:

$$K_{ABC}^{(1)} = e^{(1)}(b^{(1)}P^{(1)}, c^{(1)}P^{(1)})^{a^{(1)}} = e^{(1)}(P^{(1)}, P^{(1)})^{a^{(1)}b^{(1)}c^{(1)}}$$
$$K_{ABC}^{(2)} = e^{(2)}(b^{(2)}P^{(2)}, c^{(2)}P^{(2)})^{a^{(2)}} = e^{(2)}(P^{(2)}, P^{(2)})^{a^{(2)}b^{(2)}c^{(2)}}$$
$$K_{ABC}^{(3)} = e^{(3)}(b^{(3)}P^{(3)}, c^{(3)}P^{(3)})^{a^{(3)}} = e^{(3)}(P^{(3)}, P^{(3)})^{a^{(3)}b^{(3)}c^{(3)}}$$

$$SK = H(H_2^{(1)}(K_{ABC}^{(1)}), H_2^{(2)}(K_{ABC}^{(2)}), H_2^{(3)}(K_{ABC}^{(3)}))$$

B also computes a common session key SK as follows:

$$K_{ABC}^{(1)} = e^{(1)}(aP^{(1)}, cP^{(1)})^{b} = e^{(1)}(P^{(1)}, P^{(1)})^{abc}$$
$$K_{ABC}^{(2)} = e^{(2)}(aP^{(2)}, cP^{(2)})^{b} = e^{(2)}(P^{(2)}, P^{(2)})^{abc}$$
$$K_{ABC}^{(3)} = e^{(3)}(aP^{(3)}, cP^{(3)})^{b} = e^{(3)}(P^{(3)}, P^{(3)})^{abc}$$

$$SK = H(H_2^{(1)}(K_{ABC}^{(1)}), H_2^{(2)}(K_{ABC}^{(2)}), H_2^{(3)}(K_{ABC}^{(3)}))$$

Finally, C compute a common session key SK as follows:

$$K_{ABC}^{(1)} = e^{(1)}(aP^{(1)}, bP^{(1)})^{c} = e^{(1)}(P^{(1)}, P^{(1)})^{abc}$$
$$K_{ABC}^{(2)} = e^{(2)}(aP^{(2)}, bP^{(2)})^{c} = e^{(2)}(P^{(2)}, P^{(2)})^{abc}$$
$$K_{ABC}^{(3)} = e^{(3)}(aP^{(3)}, bP^{(3)})^{c} = e^{(3)}(P^{(3)}, P^{(3)})^{abc}$$

$$SK = H(H_2^{(1)}(K_{ABC}^{(1)}), H_2^{(2)}(K_{ABC}^{(2)}), H_2^{(3)}(K_{ABC}^{(3)}))$$

5 Analysis

In this section, we give the security and efficiency analysis of our proposed protocols. We first heuristically argue that our protocols satisfy the security requirements of the key agreement protocol. We then discuss the efficiency of our protocols by comparing the number of pairing computation required with other protocols.

5.1 Security Analysis

We only discuss the security of two-party key agreement protocol. If an adversary can obtain the session key of two-party agreement protocol, the adversary can decrypt the ciphertext exchanged during message 4 to 6 of tripartite protocol. However, if it is computationally infeasible for the adversary to obtain any one of the ephemeral key a, b, and c, it is also infeasible to obtain $K_{ABC}^{(i)}$ due to the difficulty of BDHP. It is computationally infeasible to obtain the ephemeral key from the publicly available information due to the difficulty of DLP.

1. **Man-in-the-middle-attack:** This kind of attack can be foiled if the origin authentication of values exchanged can be provided. Although, origin authentication is not provided, the way the final session key is computed prevents this kind of attack. If an attacker intercepts the two messages and sends the following to A:

$$T_{CA}^{(1)} = cP^{(1)}, W_C^{(2)} = cP_{pub}^2,$$

   the computed partial key will be as follows:

$$K_{AC}^{(1)} = e^{(1)}(aS_A^{(1)}, T_{CA}^{(1)}) = e^{(1)}(Q_A^{(1)}, P^{(1)})^{acs^{(1)}},$$

$$K_{AC}^{(2)} = e^{(2)}(aQ_B^{(2)}, W_C^{(2)}) = e^{(2)}(Q_B^{(2)}, P^{(2)})^{acs^{(2)}}.$$

   Although, the attacker can compute $K_{AC}^{(1)}$, it is infeasible for the attacker to compute $K_{AC}^{(2)}$ without acquiring ephemeral key a of A, or the master key s^2 of PKG_2.

2. **Known-key security:** In our protocol, ephemeral keys such as a, b, and c are used to construct the session key. As a result, each run of the protocol creates unique session key which is independent to past or future session keys. Therefore, compromise of past session keys do not affect the security of future session keys.

3. **PKG forward secrecy:** To satisfy PKG forward secrecy, the compromise of master key of PKG_1 and PKG_2 must not affected the security of past session keys. The past session key can be computed if the corresponding partial session keys $K_{AB}^{(1)}$ and $K_{BA}^{(2)}$ can be computed. If we assume that adversary knows $s^{(1)}$, $K_{AB}^{(1)} = e^{(1)}(Q_A^{(1)}, P^{(1)})^{as^{(1)}b}$ can be computed if the adversary can compute $e^{(1)}(aQ_A^{(1)}, bP^{(1)})$, $e^{(1)}(bQ_A^{(1)}, aP^{(1)})$, $e^{(1)}(Q_A^{(1)}, abP^{(1)})$, or $e^{(1)}(abQ_A^{(1)}, P^{(1)})$. Without acquiring ephemeral key a or b, it is infeasible to obtain any of these values from the publicly available information. The same argument also applies to $K_{BA}^{(2)}$.

4. **Key-compromise resilience:** Since both parties private key is needed to compute the session key, the compromise of A's private key does not help the adversary to impersonate others to A.

5. **Unknown key-share resilience:** Since a party always uses the other party's authenticated public key as one of the input used to compute the session key, an adversary cannot deceive a party into falsely believing the identity of the other party in concern.

6. **Key control:** Since each party contributes a fresh ephemeral key as one of the input used to compute the session key, one of the party cannot force the session key to be some preselected value.

Table 1. Comparison of pairing computation in two-party key agreement protocol

protocol	In single PKG environment		In multiple PKG environment	
	pairing(each)	pairing(all)	pairing(each)	pairing(all)
Chen and Kudla's protocol	2	4	2*	4
Our protocol			2**	4

*: each PKGs shares the common system parameters but distinct master keys.
**: each PKGs uses different system parameters.

Table 2. Comparison of pairing computation in tripartite key agreement protocol

protocol	In single PKG environment		In multiple PKG environment	
	pairing(each)	pairing(all)	pairing(each)	pairing(all)
Cheng et al's protocol	5	15	15	45
Our protocol 2			7	21

5.2 Efficiency Analysis

In this subsection, we compare the number of pairing computation performed by each user with other protocols. Since pairing computation out weigh other computations, the protocol that requires less pairing computation can be considered as a more efficient protocol.

In table 1, we compare our two party protocol with Chen and Kudla's protocol that provides PKG forward secrecy. We can see that the efficiency of our protocol is equal to Chen and Kudla's even though in our setting each PKGs uses different system parameters.

6 Conclusion

In this paper, we have proposed a new efficient ID-based two-party key agreement protocol for multiple PKG environment. We have also extended this protocol to a tripartite version. We have showed that our proposed protocols satisfy all the security requirements including the PKG forward secrecy. The efficiency of two-party key agreement protocol is equal to previous ID-based protocols for single PKG environment. The security of our proposed key agreement protocols are based on the difficulty of DLP, CDHP, and BDHP on an elliptic curve.

References

1. Shamir, A.: Identity-based Cryptosystems and Signature Schemes. In Advances in Cryptology, Crypto 1984. Lecture Notes in Computer Science, Vol. 196. Springer-Verlag (1985) 47–53

2. Boneh, D., Franklin, M.: Identity-based Encryption from Weil pairing. In Advances in Cryptology, Crypto 2001. Lecture Notes in Computer Science, Vol. 2139. Springer-Verlag (2001) 213–229
3. Smart, N.: An Identity-based Authenticated Key Agreement Protocol Based on Weil Pairing. In Electronic Letters, Vol. 38. IEEE (2002) 630–632
4. Chen, L., Kudla C.: Identity-based Authenticated Key Agreement Protocols from Pairings. In Proceedings of the 16th IEEE Computer Security Foundations Workshop. IEEE Computer Society Press (2003) 219–233
5. Joux, A.: A One Round Protocol for Tripartite Diffie-Hellman. In Proceedings of Algorithmic Number Theory Symposium, ANTS-IV. Lecture Notes in Computer Science, Vol. 1838. Spinger-Verlag (2000) 385–394
6. Al-Riyami, S., Patterson, K.: Tripartite Authenticated Key Agreement Protocols from Pairings. In Proceedings of IMA Conference on Cryptography and Coding. Lecture Notes in Computer Science, Vol. 2898. Spinger-Verlag (2003) 332–359
7. Zhang, F., Liu, S., Kim, K.: ID-Based One Round Authenticated Tripartite Key Agreement Protocols with Pairings. Crypology ePrint Archive, Report 2002/122
8. Shim, K.: Efficient One Round Tripartite Authenticated Key Agreement Protocol Based on Weil Pairing. In Electronic Letters, Vol. 39, IEEE (2003) 208–209
9. Cheng, Z., Vasiu, L., Comley, R.: Pairing-Based One-Round Tripartite Key Agreement Protocols. Cryptology ePrint Archive, Report 2004/079
10. Chen, L., Kudla, C.: Identity-based Authenticated Key Agreement Protocols from Pairings. Cryptology ePrint Archive, Report 2002/184
11. Menezes, A., Okamoto, T., Vanstone, S.: Reducing Elliptic Curve Logarithms to Logarithms in a Finite Field. Transaction of Information Theory, Vol. 39. IEEE (1993) 1639–1646
12. Frey, G., Ruck, H.: A Remark Concerning m-divisibility and The Discrete Logarithm in The Divisor Class Group of Curves. Mathematics of Computation, Vol. 62. (1994) 865–874

Evolutionally Optimized Fuzzy Neural Networks Based on Evolutionary Fuzzy Granulation

Sung-Kwun Oh[1], Byoung-Jun Park[2], Witold Pedrycz[3], and Hyun-Ki Kim[1]

[1] Department of Electrical Engineering, The University of Suwon, San 2-2 Wau-ri,
Bongdam-eup, Hwaseong-si, Gyeonggi-do, 445-743, South Korea
ohsk@suwon.ac.kr
[2] Department of Electrical Electronic and Information Engineering, Wonkwang University,
South Korea
[3] Department of Electrical and Computer Engineering, University of Alberta, Edmonton,
AB T6G 2G6, Canada
and Systems Research Institute, Polish Academy of Sciences, Warsaw, Poland

Abstract. In this paper, new architectures and comprehensive design methodologies of Genetic Algorithms (GAs) based Evolutionally optimized Fuzzy Neural Networks (EoFNN) are introduced and the dynamic search-based GAs is introduced to lead to rapidly optimal convergence over a limited region or a boundary condition. The proposed EoFNN is based on the Fuzzy Neural Networks (FNN) with the extended structure of fuzzy rules being formed within the networks. In the consequence part of the fuzzy rules, three different forms of the regression polynomials such as constant, linear and modified quadratic are taken into consideration. The structure and parameters of the EoFNN are optimized by the dynamic search-based GAs.

1 Introduction

Lately, CI computing technique becomes hot issue of IT (Information technology) and abilities of that interest. The omnipresent tendency is the one that exploits techniques of CI [1] by embracing neurocomputing imitating neural structure of a human [2], fuzzy modeling using linguistic knowledge and experiences of experts [3], [4], and genetic optimization based on the natural law [5,6]. Especially the two of the most successful approaches have been the hybridization attempts made in the framework of CI. Neuro-fuzzy systems are one of them [7], [8]. A different approach to hybridization leads to genetic fuzzy systems [6], [9].

In this paper, new architectures and comprehensive design methodologies of Genetic Algorithms (GAs) based Evolutionally optimized Fuzzy Neural Networks (EoFNN) are introduced for effective analysis and solution of nonlinear problem and complex systems. The proposed EoFNN is based on the Fuzzy Neural Networks (FNN). In the consequence part of the fuzzy rules, three different forms of the regression polynomials such as constant, linear and modified quadratic are taken into consideration. The polynomial of a fuzzy rule results from that we look for a fuzzy subspace (a fuzzy rule) influencing the better output of a model, and then raise the order of polynomial of the fuzzy rule (subspace). Contrary to the former, we make a simplified form for the representation of a fuzzy subspace lowering of the performance of a model. This methodology can effectively reduce the number of parameters and improve the performance of a model. GAs being a global optimization technique determines optimal parameters in a vast search space. But it cannot effectively avoid a large amount of time-consuming iteration because GAs finds optimal parameters by using a given space (region). To alleviate the problems, the dynamic search-based GAs is introduced to lead to rapidly optimal convergence over a limited region or a boundary condition. To assess the performance of the proposed model, we exploit a well-known numerical example.

2 Polynomial Fuzzy Inference Architecture of FNN

The overall network of conventional rule based fuzzy neural networks (FNN [8], [9]) consists of fuzzy rules as shown in (1) and (2). The networks are classified into the two main categories according to the type of fuzzy inference, namely, the simplified and linear fuzzy inference. Two different fuzzy inference methods lead us to the topologies visualized in Fig. 1. The learning of the FNN is realized by adjusting connection weights w_i or w_{ki} of the nodes and as such it follows a standard Back-Propagation (BP) algorithm.

$$R^i : If \ x_1 \ is \ A_{1i} \ and \ \cdots \ x_k \ is \ A_{ki} \ then \ Cy_i = w_i \tag{1}$$

$$R^i : If \ x_1 \ is \ A_{1i} \ and \ \cdots \ x_k \ is \ A_{ki} \ then \ Cy_i = w_{0i} + w_{1i} \cdot x_1 + \cdots + w_{ki} \cdot x_k \tag{2}$$

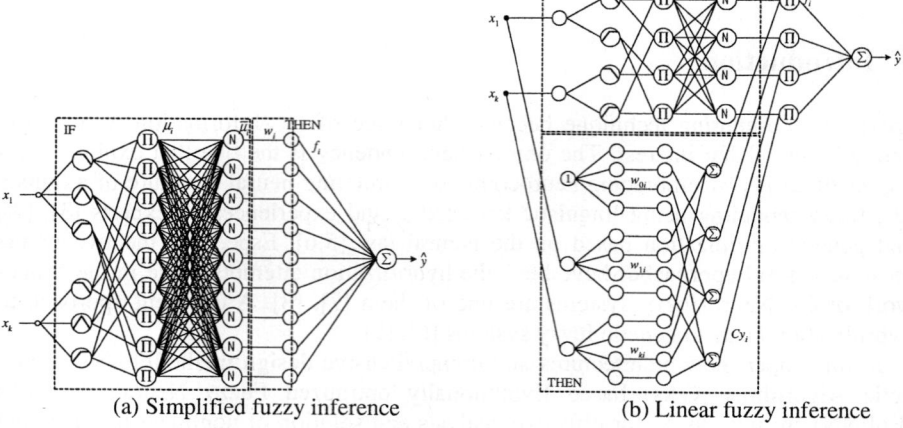

(a) Simplified fuzzy inference   (b) Linear fuzzy inference

Fig. 1. Architecture of conventional FNN

In this paper, we propose the polynomial fuzzy inference based FNN (pFNN). The proposed pFNNs are obtained from the integration and extension of conventional FNNs. The topology of the proposed pFNN is show in Fig. 2 and consists of the aggregate of fuzzy rules such as (3). The consequence part of pFNN consists of summation of a constant term, a linear sum of input variables and a linear sum of combination of two variables for the entire input variables. The polynomial structure of the consequence part in pFNN emerges into the networks with connection weights shown in Fig. 2. This network structure of the modified quadratic polynomial involves simplified (Type 0), linear (Type 1), and polynomial (Type 2) fuzzy inferences. And fuzzy inference structure of pFNN can be defined by the selection of Type (order of a polynomial).

$$R^i : \text{If } x_1 \text{ is } A_{1i} \text{ and } \cdots x_k \text{ is } A_{ki}$$
$$\text{then } Cy_i = w_{0i} + w_{1i} \cdot x_1 + \cdots + w_{ki} \cdot x_k + w_{k+1i} \cdot x_1 \cdot x_2 + \cdots + w_{k+ji} \cdot x_k \cdot x_k \quad (3)$$

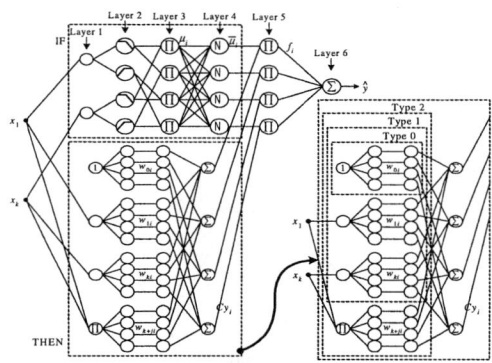

Fig. 2. Topology of the polynomial fuzzy inference based FNN

[Layer 1] Input layer
[Layer 2] Computing activation degrees of linguistic labels
[Layer 3] Computing firing strength of premise rules
[Layer 4] Normalization of a degree activation (firing) of the rule
[Layer 5] Multiplying a normalized activation degree of the rule by connection weight

$$f_i = \bar{\mu}_i \times Cy_i \text{ where, } \begin{cases} \text{Type 0}: Cy_i = w_{0i} \\ \text{Type 1}: Cy_i = w_{0i} + w_{1i} \cdot x_1 + \cdots + w_{ki} \cdot x_k \\ \text{Type 2}: Cy_i = w_{0i} + w_{1i} \cdot x_1 + \cdots + w_{k+1i} \cdot x_1 \cdot x_2 + \cdots \end{cases} \quad (4)$$

[Layer 6] Computing output of pFNN

$$\hat{y} = \sum_{i=1}^{n} f_i = \sum_{i=1}^{n} \bar{\mu}_i \cdot Cy_i = \sum_{i=1}^{n} \frac{\mu_i \cdot Cy_i}{\sum_{i=1}^{n} \mu_i} \quad (5)$$

The learning of the proposed pFNN is realized by adjusting connection weights w, which organize the consequence networks of pFNN in Fig. 2. The standard Back-propagation (BP) algorithm is utilized as the learning method in this study. The complete update formulas combining the momentum components are (6)~(8).

i) Type 0 :

$$\Delta w_{0i}(t+1) = 2\cdot\eta\cdot(y_p - \hat{y}_p)\cdot\overline{\mu}_i + \alpha\cdot\Delta w_{0i}(t) \tag{6}$$

ii) Type 1 :

$$\begin{cases} \Delta w_{0i}(t+1) = 2\cdot\eta\cdot(y_p - \hat{y}_p)\cdot\overline{\mu}_i + \alpha\cdot\Delta w_{0i}(t) \\ \Delta w_{ki}(t+1) = 2\cdot\eta\cdot(y_p - \hat{y}_p)\cdot\overline{\mu}_i\cdot x_k + \alpha\cdot\Delta w_{ki}(t) \end{cases} \tag{7}$$

iii) Type 2 :

$$\begin{cases} \Delta w_{0i}(t+1) = 2\cdot\eta\cdot(y_p - \hat{y}_p)\cdot\overline{\mu}_i + \alpha\cdot\Delta w_{0i}(t) \\ \Delta w_{ki}(t+1) = 2\cdot\eta\cdot(y_p - \hat{y}_p)\cdot\overline{\mu}_i\cdot x_k + \alpha\cdot\Delta w_{ki}(t) \\ \Delta w_{k+ji}(t+1) = 2\cdot\eta\cdot(y_p - \hat{y}_p)\cdot\overline{\mu}_i\cdot x_k\cdot x_k + \alpha\cdot\Delta w_{k+ji}(t) \end{cases} \tag{8}$$

Where, $\Delta w_{ki}(t) = w_{ki}(t) - w_{ki}(t-1)$. η and α are constrained to the unit interval.

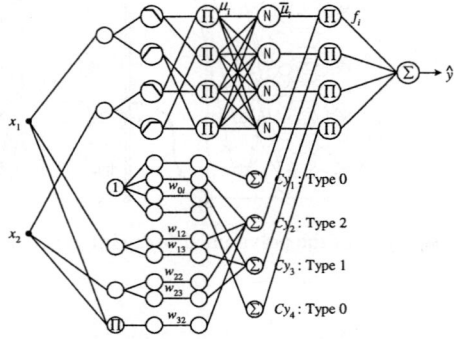

Fig. 3. pFNN architecture; polynomials of consequence have different orders for each fuzzy rules

$$\begin{aligned} R^1 &: If\ x_1\ is\ A_{11}\ and\ x_2\ is\ A_{21}\ then\ Cy_1 = w_{01} \\ R^2 &: If\ x_1\ is\ A_{11}\ and\ x_2\ is\ A_{22}\ then\ Cy_2 = w_{02} + w_{12}\cdot x_1 + w_{22}\cdot x_2 + w_{32}\cdot x_1\cdot x_2 \\ R^3 &: If\ x_1\ is\ A_{12}\ and\ x_2\ is\ A_{21}\ then\ Cy_3 = w_{03} + w_{13}\cdot x_1 + w_{23}\cdot x_2 \\ R^4 &: If\ x_1\ is\ A_{12}\ and\ x_2\ is\ A_{22}\ then\ Cy_4 = w_{04} \end{aligned} \tag{9}$$

The proposed pFNN can be designed to adapt a characteristic of a given system, also, that has the faculty of making a simple structure out of a complicated model for a nonlinear system, because the pFNN comprises consequence structure with various orders (Types) for fuzzy rules as shown in Fig. 3 and (9).

3 Evolutionarily Optimized Fuzzy Neural Networks

In this chapter, we introduce new architectures and comprehensive design methodologies of genetic algorithms (GAs [5], [6]) based evolutionarily optimized Fuzzy neural networks (EoFNN). For the evolutionally optimized architecture, the dynamic search-based GAs is proposed, and also the efficient methodology of chromosomes application of GAs for the identification of architecture and parameters of EoFNN is discussed.

To search a global solution in process of optimization using GAs is stagnated by various causes. One factor of them is to suitably give a definition of the search space for a global solution. Generally, search space (range) is defined for a given system and string length (the number of bit) is set for the defined range. Seeking for the optimal solution is based these. If the search range is defined as a large scale, then the number of bit increases in size for the given space. This cause the request of many computing time, tardy search of solutions, etc, so, GAs show a drop in performance. If a small-scale for the solution space is given, then the string length is sorted. This can ease the computing-burden, but debase the quality (detailed drawing) of solutions. Therefore, in order to improve these problems, we introduce the dynamic search-based GAs. This methodology discovers an optimal solution through adjusting search range. Adjustment of a range is based on the moving distance of a basis solution. A basis solution is previously determined for sufficiently large space. The procedure of adjusting space for each step in the dynamic search-based GAs is shown in Fig. 4.

In order to generate the proposed EoFNN, the dynamic search based GAs is used in the optimization problems of structures and parameters. From the point of fuzzy rules, these divide into the structure and parameters of the premise part, and that of consequence part. The structure issues in the premise of fuzzy rules deal with how to use of input variables (space) influencing outputs of model. The selection of input variables and the division of space are closely related to generation of fuzzy rules that determine the structure of FNN, and govern the performance of a model. Moreover, a number of input variable and a number of space divisions induce some fatal problems such as the increase of the number of fuzzy rules and the time required. Therefore, the relevant selection of input variables and the appropriate division of space are required. The structure of the consequence part of fuzzy rules is related to how represents a fuzzy subspace. Universally, the conventional methods offer uniform types to each subspace. However, it forms a complicated structure and debases the output quality of a model, because it does not consider the correlation of input variables and reflect a feature of fuzzy subspace. In this study, we apply the various forms in expressions of a fuzzy subspace. The form is selected according to an influence of a fuzzy subspace for an output criterion and provides users with the necessary information of a subspace for a system analysis.

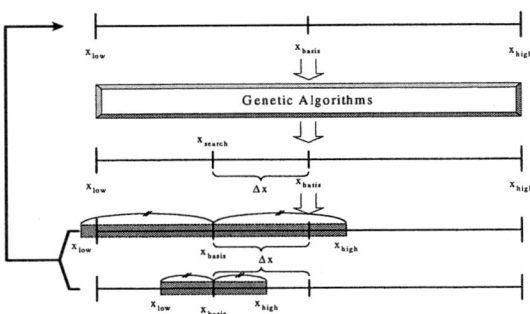

Fig. 4. Dynamic search based Gas

4 Experimental Studies

In this experiment, we use three-input nonlinear function as in [3], [4]. This dataset was analyzed using Sugeno's method [3]. We consider 40 pairs of the original input-output data. The performance index (PI) is defined by (10). 20 out of 40 pairs of input-output data are used as learning set and the remaining part serves as a testing set.

$$E(PI \text{ or } EPI) = \frac{1}{n}\sum_{p=1}^{n}\frac{|y_p - \hat{y}_p|}{y_p} \times 100(\%) \tag{10}$$

Table 1 summarizes the results of the EoFNN architectures. This table includes the tuning methodologies using dynamic search based GAs. Ⓐ case includes two auto-tuning processes, namely, structure and parameter tuning processes. In first process, structure of a given model is tuned, that is input variables of premise and consequence, membership function, and order of polynomial are set. And then, parameters of the identified structure are tuned in second process. Ⓑ case includes structure and parameter tuning processes, however, two tuning processes is not done separately but done at the same time. That is, input variables of premise and consequence, parameters of membership function, and order of polynomial are tuned. Ⓐ$_k$ and Ⓑ$_k$ add the condition of being restricted in the number of inputs of a model to Ⓐ and Ⓑ.

Table 1. Performance index of EoFNN02 for the nonlinear function

Case	Premise Inputs	MFs	Para.	Consequence Inputs	Order	PI	E_PI
Ⓐ	GAs (x_1,x_2,x_3)	2×2×2	Min-Max	GAs	Gas	1.948	4.401
	Tuned	2×2×2	GAs	Tuned	Tuned	0.423	0.990
Ⓑ	GAs (x_1,x_2,x_3)	2×2×2	GAs	GAs	Gas	0.241	0.357
Ⓐ$_2$	≤2 GAs	2×2	Min-Max	GAs	Gas	2.068	5.164
	Tuned (x_2,x_3)	2×2	GAs	Tuned	Tuned	0.232	1.013
Ⓑ$_2$	≤2 GAs (x_2,x_3)	2×2	GAs	GAs	Gas	0.224	0.643

$$\begin{aligned}
&R^1 : \text{If } x_1 \text{ is } A_{11} \text{ and } x_2 \text{ is } A_{21} \text{ and } x_3 \text{ is } A_{31} \text{ then } Cy_1 = w_{01} + w_{31}x_3 \\
&R^2 : \text{If } x_1 \text{ is } A_{11} \text{ and } x_2 \text{ is } A_{21} \text{ and } x_3 \text{ is } A_{32} \text{ then } Cy_2 = w_{02} + w_{22}x_2 + w_{32}x_3 \\
&R^3 : \text{If } x_1 \text{ is } A_{11} \text{ and } x_2 \text{ is } A_{22} \text{ and } x_3 \text{ is } A_{31} \text{ then } Cy_3 = w_{03} + w_{13}x_1 + w_{23}x_2 + w_{33}x_3 \\
&R^4 : \text{If } x_1 \text{ is } A_{11} \text{ and } x_2 \text{ is } A_{22} \text{ and } x_3 \text{ is } A_{32} \text{ then } Cy_4 = w_{04} + w_{24}x_2 \\
&R^5 : \text{If } x_1 \text{ is } A_{12} \text{ and } x_2 \text{ is } A_{21} \text{ and } x_3 \text{ is } A_{31} \text{ then } Cy_5 = w_{05} + w_{15}x_1 + w_{25}x_2 + w_{35}x_3 \\
&R^6 : \text{If } x_1 \text{ is } A_{12} \text{ and } x_2 \text{ is } A_{21} \text{ and } x_3 \text{ is } A_{32} \text{ then } Cy_6 = w_{06} + w_{16}x_1 \\
&R^7 : \text{If } x_1 \text{ is } A_{12} \text{ and } x_2 \text{ is } A_{22} \text{ and } x_3 \text{ is } A_{31} \text{ then } Cy_7 = w_{07} + w_{27}x_2 + w_{37}x_3 \\
&R^8 : \text{If } x_1 \text{ is } A_{12} \text{ and } x_2 \text{ is } A_{22} \text{ and } x_3 \text{ is } A_{32} \text{ then } Cy_8 = w_{08} + w_{28}x_2
\end{aligned} \tag{11}$$

$$\begin{aligned}
&\text{then } Cy_1 = w_{01} + w_{11}x_1 \\
&\text{then } Cy_2 = w_{02} \\
\cdots \quad &\text{then } Cy_3 = w_{03} + w_{13}x_1 \\
&\text{then } Cy_4 = w_{04} \\
&\text{then } Cy_5 = w_{05} + w_{15}x_1 + w_{25}x_2 + w_{35}x_3 \\
\cdots \quad &\text{then } Cy_6 = w_{06} + w_{26}x_2 + w_{36}x_3 + w_{66}x_2x_3 \\
&\text{then } Cy_7 = w_{07} + w_{17}x_1 + w_{37}x_3 \\
&\text{then } Cy_8 = w_{08} + w_{38}x_3
\end{aligned} \qquad (12)$$

EoFNN designed by Ⓐ method consists of 8(=2^3) fuzzy rules such as (11). Ⓑ carries out the tuning process of structure and parameters at the same time unlike Ⓐ. EoFNN designed by Ⓑ method consists of 8(=2^3) fuzzy rules such as (12). The premise structures of the fuzzy rule are the same that of Ⓐ. The output characteristic of the architecture obtained by means of Ⓑ is better than that of Ⓐ. As shown in (11) and (12), the consequence structure of fuzzy rules includes 22 parameters in (11) and 19 parameters in (12), respectively. Therefore, EoFNN generated by Ⓑ is preferred as an optimal architecture for the output performance and simplicity of a model. The preferred EoFNN results from Ⓑ$_2$. This architecture consists of 2 inputs, x_1 and x_3, in premise part, and 4 fuzzy rules such as (13). Let compare the EoFNN shown in (12) with the EoFNN shown in (13). The generalization ability of (13) is less than that of (12), however, EoFNN by (13) is constructed by 4 fuzzy rules with 2 inputs in the premise and 18 parameters of the consequence part, namely this architecture is simpler than EoFNN by (12).

$$\begin{aligned}
R^1 &: \text{If } x_2 \text{ is } A_{21} \text{ and } x_3 \text{ is } A_{31} \text{ then } Cy_1 = w_{01} + w_{11}x_1 + w_{31}x_3 + w_{51}x_1x_3 \\
R^2 &: \text{If } x_2 \text{ is } A_{21} \text{ and } x_3 \text{ is } A_{32} \text{ then } Cy_2 = w_{02} + w_{12}x_1 + w_{22}x_2 + w_{32}x_3 + w_{42}x_1x_2 + w_{52}x_1x_3 + w_{62}x_2x_3 \\
R^3 &: \text{If } x_2 \text{ is } A_{22} \text{ and } x_3 \text{ is } A_{31} \text{ then } Cy_3 = w_{03} + w_{13}x_1 + w_{23}x_2 + w_{33}x_3 \\
R^4 &: \text{If } x_2 \text{ is } A_{22} \text{ and } x_3 \text{ is } A_{32} \text{ then } Cy_4 = w_{04} + w_{14}x_1 + w_{34}x_3
\end{aligned} \qquad (13)$$

Fig. 5 shows topologies of EoFNN designed by (13) with 2 inputs. In order to solve a given nonlinear problem, these architectures are designed and generated in flexibility that can cope with an environment (condition).

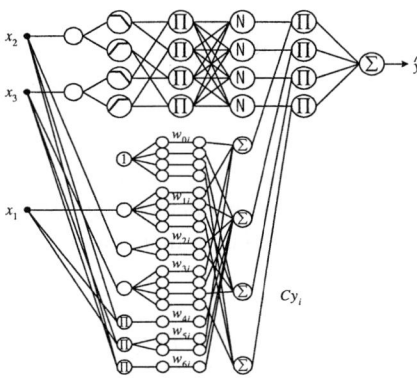

Fig. 5. Topology of EoFNN02 with 2 inputs for the nonlinear function

Table 2 covers a comparative analysis including several previous models. Sugeno's model I and II were fuzzy models based on linear fuzzy inference method while Shin-ichi's models formed fuzzy rules by using learning method of neural networks. The study of literature [9] is based on fuzzy-neural networks using HCM clustering and evolutionary fuzzy granulation. Multi-FNN consists of 3 FNN structures. The proposed EoFNNs come with higher accuracy and improved prediction capabilities.

Table 2. Comparison of performance with other modeling methods

Model		PI	E_PI	No. of rules
Linear model [4]		12.7	11.1	
GMDH [4,10]		4.7	5.7	
Sugeno's [3,4]	Fuzzy I	1.5	2.1	3
	Fuzzy II	1.1	3.6	4
Shin-ichi's [7]	FNN Type 1	0.84	1.22	$8(2^3)$
	FNN Type 2	0.73	1.28	$4(2^2)$
	FNN Type 3	0.63	1.25	$8(2^3)$
FNN [9]	Simplified	2.865	3.206	9(3+3+3)
	Linear	2.670	3.063	9(3+3+3)
Multi-FNN [9]	Simplified	0.865	0.956	9(3+3+3)
	Linear	0.174	0.689	9(3+3+3)
Proposed model (EoFNN)		0.241	0.357	8(2×2×2)
		0.224	0.643	4(2×2)

5 Concluding Remarks

In this paper, new architectures and comprehensive design methodologies of Evolutionally optimized Fuzzy Neural Networks (EoFNN) has discussed for effective analysis and solution of nonlinear problem and complex systems. Also, the dynamic search-based GAs has introduced to lead to rapidly optimal convergence over a limited region or a boundary condition. The proposed EoFNN is based on the Fuzzy Neural Networks (FNN) with the extended structure of the fuzzy rules. The structure and parameters of the proposed EoFNN are optimized by GAs. The proposed EoFNN can be designed to adapt a characteristic of a given system, also, that has the faculty of making a simple structure out of a complicated model for a nonlinear system. This methodology can effectively reduce the number of parameters and improve the performance of a model. The proposed EoFNN can be efficiently carried out both at the structural as well as parametric level for overall optimization by utilizing the separate (Ⓐ) or consecutive (Ⓑ) tuning technology. From the results, Ⓑ methodologies simultaneously tuning both structure and parameters, reduce parameters of consequence part, and offer the output performance better than the Ⓐ. Namely, Ⓑ method is effective in identifying a model than Ⓐ.

Acknowledgement. This work has been supported by KESRI(I-2004-0-074-0-00), which is funded by MOCIE(Ministry of commerce, industry and energy).

References

1. Pedrycz, W., Peters, J.F.: Computational Intelligence and Software Engineering. World Scientific. Singapore (1998)
2. Chan, L.W., Fallside, F.: An Adaptive Training Algorithm for Back Propagation Networks. Computer Speech and Language. **2** (1987) 205-218
3. Kang, G., Sugeno, M.: Fuzzy Modeling. Transactions of the Society of Instrument and Control Engineers. **23**(6) (1987) 106-108
4. Park, M.Y., Choi, H.S.: Fuzzy Control System. Daeyoungsa, Korea (1990) 143-158
5. Goldberg, D.E.: Genetic Algorithms in search, Optimization & Machine Learning. Addison-wesley. (1989)
6. Cordon, O. et al.: Ten years of genetic fuzzy systems: current framework and new trends. Fuzzy Sets and Systems. **141**(1) (2004) 5-31
7. Horikawa, S. I., Furuhashi, T., Uchigawa, Y.: On Fuzzy Modeling Using Fuzzy Neural Networks with the Back Propagation Algorithm. IEEE Transactions on Neural Networks. **3**(5) (1992) 801-806
8. Park, H.S., Oh, S.K.: Rule-based Fuzzy-Neural Networks Using the Identification Algorithm of GA Hybrid Scheme. International Journal of Control, Automation, and Systems. **1**(1) (2003) 101- 110
9. Park, H.S., Oh, S.K.: Multi-FNN Identification Based on HCM Clustering and Evolutionary Fuzzy Granulation. International Journal of Control, Automation and Systems. **1**(2) (2003) 194-202
10. Kondo, T.: Revised GMDH algorithm estimating degree of the complete polynomial. Transactions of the Society of Instrument and Control Engineers. **22**(9) (1986) 928-934

Multi-stage Detailed Placement Algorithm for Large-Scale Mixed-Mode Layout Design[*]

Lijuan Luo[1], Qiang Zhou[2], Xianlong Hong[2], and Hanbin Zhou[1]

[1] Dept. of Computer Science and Technology, Tsinghua Univ., Beijing, China
{luolj03, zhouhb}@mails.tsinghua.edu.cn
[2] Dept. of Computer Science and Technology, Tsinghua Univ., Beijing, China
{zhouqiang, hxl-dcs}@mail.tsinghua.edu.cn

Abstract. Quadratic/Analytical placement methods have been widely used in the latest IC design process. This kind of placement needs a powerful detailed placement method for large-scale mixed macro and standard cell placement. An efficient and effective multi-stage iterative detailed placement (MSIP) algorithm is proposed in this paper. It combines a better initial placement based on combinatorial optimization method with a deterministic local search for post optimization. Various strategies are used for saving computation time. Experimental results show that it can get an average of 22% wire length improvement comparing to PAFLO [6] in comparable runtime.

1 Introduction

Quadratic placement (e.g. [17]-[21]) is widely accepted in modern VLSI physical design. This kind of placements is carried out in two successive stages: global placement and detailed placement. Global placement roughly spreads out cells over the chip area by alternating global optimization (i.e. solving a quadratic programming problem) and cell shifting steps. Which method is adopted in cell shifting step is the departure point of various quadratic placements, but no matter it is partitioning strategy (e.g. [20][21]) or force-directed strategy(e.g.[17][19]) or grid warping(e.g.[18]), cells are inevitably overlapped after global placement and this problem is especially serious with greater chip scale, since the granularity of grid structure in global placement can not be too fine for complexity. So detailed placement solution – namely how to assign cells to cell rows, eliminate overlaps and further improve placement quality – is influential to both the final placement quality and the total running time for most modern placement tools.

With developing into SOC era, mixed-mode placement comes up, where the chip is divided into many discrete layout areas by macros which may be as large as 10,000 times of standard cells [5]. In this case most of existing detailed placement algorithm is inapplicable and since small difference in detailed placement strategy can result in remarkable movement of cells, detailed placement algorithm is receiving a great challenge.

---

[*] This work was supported partly by Hi-Tech Research and Development (863) Program of China (2002AA1Z1460), and National Science Foundation of China (90407005 and 6076014).

Existing detailed placement algorithms can be classified into constructive techniques and iterative improvement techniques.

Constructive algorithms include efficient algorithm from combinatorial optimization [9][22] and mathematical programming methods, such as the detailed placement of RITUAL [3]. They are unsatisfactory either for low quality owing to lack of post optimization("post optimization" in [22] only tries to minimize cell movement from global position but fails to search for optimum position in terms of half-perimeter wire-length) or for the excessive CPU time.

Iterative improvement algorithms are completed though two successive stages: First, initial placement, namely transferring global placement to a physical legal placement without overlaps; then, further optimizing wire-length by iteration. This kind of algorithms generally present two problems: First, the initial solution has two opposite extremities: (1) too complex and time-consuming such as the annealing-based method in NRG [11]; (2) too simple such as the partition-and-mapping method in FAME[4] and PAFLO[6] which not only lengthens the post optimization process but also inevitably effects final placement quality [10]. Second, in the process of post optimization, stochastic perturbation strategies such as swapping heuristic[1], optimal interleaving [14], enumeration[2][4][6], intra-column swaps[23] and so on always make the post optimization process time-consuming, we can see this by comparison in section 5.

A new iterative improvement detailed placement algorithm – MSIP for mixed-mode IC design is proposed in this paper. The input is global placement result with macros fixed without overlaps and standard cells placed in global optimal positions which may be physically illegal [7] [13] [16]. The output is a legal placement with all standard cells placed in rows without overlaps. The optimization objective is half-perimeter wire-length minimization, however we think that the proposed method can be further extended to other objectives as well.

Contribution of MSIP lies in: First, at initial stage, the task is divided into three sub-tasks and each sub-task is formulated as a combinatorial optimization problem (e.g. the problem of minimum-cost flow and the problem of LPP) and solved optimally; meanwhile thanks to the divide-and-conquer strategy and the dynamic selection strategy a good legal initial placement is got at the minimum cost of CPU time. Second, a deterministic local search method is adopted in the post optimization procedure so that not only greatly improves final placement quality but also achieves significant time reduction. As a whole, the new algorithm optimally solves the detailed placement problem in every sub-stage. It's verified to be effective and efficient.

Organization of this paper is as follows: Section 2 generalizes the fundamental thinking of the whole algorithm. Section 3 and Section 4 concretely introduce the two stages of MSIP. Experimental results are shown in Section 5 and some concluding remarks make up the final section.

2 Overview of the Algorithm

As other iterative methods, our new detailed placement algorithm is divided into two main stages: initial placement and post optimization.

During initial placement, the input is the global placement result which is, in mathematical sense, "optimal" with respect to squared wire-length. Thus when balancing cell distribution, minimum-cost-flow (MCF) method is adopted to minimize deviation from the global optimal position, namely analogizing moving distance by cost and moving cell area by flow; and later Optimal Single-Row Placement (OSRP) is used to resolve overlaps while maintaining the order of cells within a row. However, with the size of placement problem becomes larger and larger, the traditional flat placement strategies are no longer suitable for complexity reason. Therefore, a hierarchical technique is proposed to greatly decrease computational time without sacrificing quality.

Following the initial placement, it comes to the post optimization step. In contrast to the statement that post-optimization procedure is not worth the computation time [9], our experiments show that an effective post-optimization procedure can greatly optimize the final half-perimeter wire-length. That is because although squared wire-length in global placement is a good prediction of half-perimeter wire-length, it is still different from the latter and great improvement potential can still be developed in post-optimization.

3 Initial Placement

This is the first stage of detailed placement which inherits the global placement result and mainly completes two tasks: balancing local distribution and resolving overlaps.

In mixed-mode placement, rows are divided into zones by macros (namely a zone is a maximal part of a row which is not intersected by macros) and the criterion of balancing local distribution is to ensure that every zone can accommodate all cells within it. Here, a hierarchical strategy is adopted in order to minimize complexity. In detail, the chip is divided into $m \times n$ equivalent square areas (usually as high as several rows but narrower than the width of rows) called regions and zones are further divided into sub-zones by region boundaries.

Thus, initial placement process is implemented through three phases: balancing the regions – the higher hierarchy (Section 3.1), balancing the sub-zones of every region in turn – the lower hierarchy (Section 3.2), merging sub-zones and resolving overlaps in zones – restoring to flat strategy (Section 3.3).

3.1 Balancing the Regions

The task of this phase is to ensure that every region can not only hold cells in it but also have some free space in order to further balance sub-zones within the region. So define a max density, and if the occupy ratio of one region is higher than that density it is treated as overloaded. Region capacity equals to the total area of the region minus the area occupied by macros.

In order to minimize cell movement from global optimal position, it's natural to solve a minimum-cost flow problem, where vertices correspond to the regions and edges join neighbor regions (namely adjacent regions). The cost of an edge is defined as the distance between the two region centers. Of course, the overloaded regions are the sources, while regions with free capacity are possible sinks. Refer to [9] for details.

The calculated flow is realized by actually moving cells from an overloaded region to a neighbor region. Since there are much more cells in the so called region here than in [9], we adopt a greedy but more efficient method [10] to choose cells to move between regions instead of the dynamic programming method of [9].

3.2 Balancing the Sub-zones

The task of this phase is to eliminate capacity violation in sub-zones within every region in turn. Basically the same framework as 3.1 is adopted. Note that since the distribution of sub-zones is irregular caused by the existence of macros, two sub-zones are called neighbors as long as they are subsequent in the same cell row or they are in adjacent rows with overlap in x direction.

Unlike [22] which iterates the balancing procedure until no zone contains more circuits than fit into it, balancing procedure here only operates once and a little overload is still permissible. The reason is that test-suites of [22] is especially sparse (the total size of the standard circuits divided by the total unblocked area of the chip averages 33.96%) and can be easily balanced in a few iterations, but that is not the case for most of other circuits.

3.3 Resolving Overlaps in Zones

Since we can't ensure the absolute capacity balance of sub-zones only by the above network flow method, a greedy method [10] is adopted to further balance zones if necessary. After it's guaranteed that no zone is overloaded, begin to resolve cell overlaps each zone subsequently. We adopt the algorithm of Optimal Single-Row Placement (OSRP) [15] and make two points of improvement [10]. Besides, experimental results show that if the free space of a zone is little, wire length reduction of the above optimal approach is very limited compared with a simple implementation and usually not worth the computation time. So for simplicity a valve-density is given. If the zone density is lower than that valve, adopt the optimal approach, and otherwise adopt a simple mapping method like [6]. This dynamic selection strategy contributes a lot to the reduction of CPU time.

4 Post Optimization

This phase inherits initial legal placement and improves current placement iteratively by local adjustment. Every iteration can be generalized as follows: for each cell, find the corresponding optimal position scope and adjust cell position to approach the scope if such adjustment can decrease total wire length. Iteration continues until no more obvious improvement can be obtained. Since every new configuration is generated according to the deterministic optimal scope, optimal placement can be achieved much more quickly than other stochastic perturbation strategy. The method for determining optimal position scope of every cell and the strategy of generating new configuration according to this optimal position scope are respectively stated in Section4.1 and Section4.2.

4.1 Determination of Optimal Position Scope

Given a cell c, the problem is to find an optimal position scope for c; if c is placed within this scope, total half perimeter wire length of the nets connected to c is minimized. Now only consider the scope in x direction, and it's the same to y.

As shown in Fig.1, suppose there are n nets connected to c. Then for each net, get its leftmost and rightmost x coordinates excluding c. We totally record $2 \times n$ positions in a monotonously increasing array x[0]...x[2n-1]. Then the span from x[n-1] to x[n] is the optimal position of c in x direction. Similar method is adopted in [12] for finding optimal slots for cells. But here cells can move successively in x direction, so we turn to the concept of derivative for proof.

Proof: To simplify notations, we use the coordinates of cell centers instead of pins. Since the size of cell is small compared with the placement area, this approximation will not cause much error. Denote the total wire length by $f(x)$ and the wire length of a single net by $f_i(x) (1 \leq i \leq n)$. Here x is the coordinate of cell c.

$$f(x) = \sum_{i=1}^{n} f_i(x) \tag{1}$$

Derivative of $f_i(x)$ is determined as follows:

$$\frac{df_i(x)}{dx} = p_i(x) - q_i(x)$$

$$p_i(x) = \begin{cases} 1 & x \geq r_i \\ 0 & x < r_i \end{cases}, \quad q_i(x) = \begin{cases} 1 & x < l_i \\ 0 & x \geq l_i \end{cases} \tag{2}$$

l_i (r_i) is leftmost(rightmost) x coordinate of the net excluding cell c. Thus

$$\frac{df(x)}{dx} = \sum_{i=1}^{n} \frac{df_i(x)}{dx} = \sum_{i=1}^{n} p_i(x) - \sum_{i=1}^{n} q_i(x) \tag{3}$$

Define $p(x) = \sum_{i=1}^{n} p_i(x)$, which equals to the number of r_i to the left of x (including x).

Define $q(x) = \sum_{i=1}^{n} q_i(x)$, which equals to the number of l_i to the right of x, then

$$\frac{df(x)}{dx} \begin{cases} <0 & \text{when } x < x[n-1] \quad (\text{because } p(x) < q(x)) \\ =0 & \text{when } x[n-1] \leq x < x[n] \quad (\text{because } p(x) = q(x)) \\ >0 & \text{when } x \geq x[n] \quad (\text{because } p(x) > q(x)) \end{cases} \tag{4}$$

4.2 New State Generator

Fig.2 shows our algorithm for generating new placement configuration according to the optimal position scope of a cell. The so called sinking zone in Fig.2 means one zone which intersects with the optimal scope and has the largest free space.

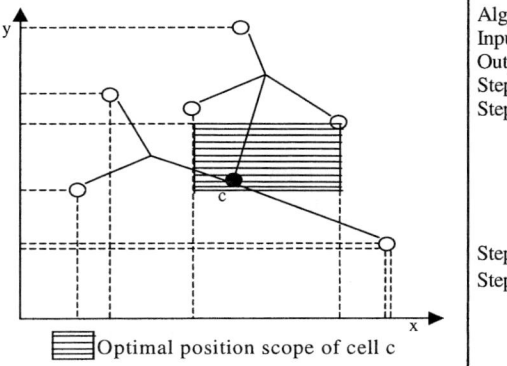

```
Algorithm: newStateGenerate(c,z)
Input: cell c, original zone z where c lies
Output: new placement state
Step 1: determine optimal position scope R for c
Step 2: IF z intersects with R
            move c to optimal position within z
        ELSE Select a sinking zone t
            IF t can accommodate c
                move c to optimal position in t
            ELSE move c to optimal position in z
Step 3: calculate moving cost △W
Step 4: IF △W<0
            eliminate overlaps caused by Step 2
        ELSE
            restore c to original position
```

Fig. 1. Determine Optimal Position   **Fig. 2.** New state generator

Moving cost is calculated in the following manner:

$$\triangle W = \triangle L + \triangle C \tag{5}$$

Here $\triangle L$ is the changed length for those nets connected to the cell which is moved to the optimal scope (obviously $\triangle L$ is non-positive). $\triangle C$ is the changed length for those nets connected to the cells shifted to avoid overlapping.

4.3 Analysis of Post-optimization Result

In order to testify the efficiency of this deterministic post-optimization procedure, results of the optimization process for one benchmark are listed in table 1.The table shows that significant wire-length improvement is obtained in the first iteration (about 9.7%). This strongly proves what we have previously stated, namely half-perimeter wire-length differs from squared wire-length and great potential for wire-length improvement still exists in detailed placement.

On the other hand, wire-length improvement is less and less obvious in later iteration, which can also be observed from the decrease of moving cells. This phenomenon is also reasonable for an iterative method, since in the process of optimization, cells are gradually placed to the optimum positions and no more space is reserved for improvement.

Experiments show that generally three times of iteration are suitable for good trade-off of placement quality and CPU time.

Table 1. Post_optimization of ibm01

Phase		WL	Moving cells	
			within zone	cross zones
Initial placement		4.21e6		
optimization	Iteration 1	3.80e6	2952	3024
	Iteration 2	3.68e6	2486	711
	Iteration 3	3.67e6	1612	230

5 Experimental Results

The whole algorithm of MSIP was implemented in C++ on a 750MHz/8GB memory Sun workstation running on Solaris. Tested circuits are downloaded from ISPD02 benchmarks [8] in LEF/DEF format. All testcases are large-scale, mixed macro and standard cell placement benchmarks, except for circuit ibm05. The macros are all hard blocks with fixed aspect ratios and pin locations.

Given the same output of global placement [24], we apply both MSIP and PAFLO [6], an improved version of the detailed placement (the author speaks of final placement) algorithm adopted in [7]. Table 2 shows the results. We can see that significant wire length reductions (an average of 22%) are obtained in comparable time for the implementation of PAFLO is so simple and inefficient with partition-and-mapping method as initial placement and enumeration as post optimization strategy.

Table 2. Placement Result Comparison with PAFLO

circuit	PAFLO		MSIP		
	WL	time(s)	WL	WL _impr	time(s)
ibm01	4.18e6	174	3.67e6	12.20%	99
ibm02	9.69e6	380	8.28e6	14.55%	211
ibm03	1.86e7	518	1.29e7	30.65%	259
ibm04	2.29e7	815	1.22e7	46.72%	289
ibm05	1.62e7	518	1.39e7	14.20%	315
ibm06	1.61e7	771	1.27e7	21.12%	347
ibm07	2.36e7	902	1.94e7	17.80%	640
ibm08	2.39e7	1029	2.04e7	14.64%	852
ibm09	2.96e7	1157	2.21e7	25.34%	806
ibm10	1.10e8	2398	6.06e7	44.91%	2931
ibm11	3.96e7	1562	2.97e7	25.00%	1341
ibm12	6.92e7	1710	5.74e7	17.05%	2524
ibm13	5.63e7	2142	3.94e7	30.02%	1960
ibm14	6.83e7	2710	6.01e7	12.01%	4209
ibm15	9.50e7	4194	7.86e7	17.26%	4586
ibm16	1.11e8	3971	9.24e7	21.70%	6532
ibm17	1.29e8	4176	1.06e8	17.83%	7473
ibm18	8.10e7	4002	6.87e7	15.19%	4672
avg.				22.12%	

Table 3. Result Comparison with Combining Algorithm

circuits	Comb Algorithm		MSIP		
	WL	time(s)	WL	WL_impr	time(s)
ibm01	4.04e6	247	3.67e6	9.16%	99
ibm05	1.53e7	535	1.39e7	9.15%	315
ibm09	2.46e7	1493	2.21e7	10.16%	806
ibm13	4.39e7	2965	3.94e7	10.25%	1960
ibm18	7.48e7	6224	6.87e7	8.16%	4672
avg.				9.40%	

Since the initial detailed placement of PAFLO is rather simple compared with ours, to prove the efficiency of our post-optimization, we also list results of the method combining our initial solution with PAFLO's post-optimization in Table 3. It's clear by comparison that the suggested post-optimization algorithm in this paper has preferable CPU time besides wire length improvement (average of 9.4%) compared with the enumeration and swapping method of PAFLO.

For further showing the contribution of our new detailed placement to whole placement quality, table 4 lists the overlap degree after global placement, which is computed according to the equation given in [22] and wire-length after global placement as well as wire-length after our new detailed placement. Averagely, wire-length increases -9.79% in the process of MSIP, much better than the average of 0.82% in [22] (where average overlap after global placement is equal to 1.477).

Admittedly, the extent of wire-length increase and the final wire-length are also related to the particular global placement method and in order to show the efficiency of our algorithm accompanied with other global placement method, we also apply our algorithm to the global placement result of FastPlace[17] on some benchmarks downloaded from ISPD04[25]. These benchmarks are all standard cells (since FastPlace[17] can not handle macro blocks) and we name them st_IbmXX to distinguish with ibmXX benchmarks in mixed-mode format. Table 5 shows that wire-length after MSIP averagely increases -1.09%, which is also much better than the average of 0.82% in [22]. We also list the result of FastPlace(namely using the detailed placement of FastPlace after global placement of FastPlace) given in [17] in the last column of table , and comparison shows that when employing MSIP as detailed placement, wire-length of FastPlace can be further improved.

Table 4. Comparison wire-length before and after MSIP

circuit	Overlap	WL_GP	WL_MSIP	WL_increase
ibm01	2.00471	3.94e6	3.67e6	-6.85%
ibm02	6.17317	9.30e6	8.28e6	-10.97%
ibm03	2.01259	1.31e7	1.29e7	-1.53%
ibm04	2.05413	1.34e7	1.22e7	-8.96%
ibm05	2.00323	1.60e7	1.27e7	-20.63%
Avg.				-9.79%

Table 5. Wire-length result when combining MSIP with FastPlace Global Placement

circuits	overlap	WL_GP	WL_MSIP	WL_inc	WL_FP
st_Ibm01	1.77188	1.95e6	1.92e6	-1.54%	1.91e6
st_Ibm02	1.77655	4.09e6	3.97e6	-2.93%	4.02e6
st_Ibm03	1.78554	5.26e6	5.29e6	0.57%	5.45e6
st_Ibm04	1.82647	6.40e6	6.36e6	-0.63%	6.63e6
st_Ibm05	1.77677	1.08e7	1.07e6	-0.93%	1.10e7
Avg.	1.78744			-1.09%	

6 Concluding Remarks

This paper proposes an effective multi-stage detailed placement algorithm combining a good initial placement with a novel post optimization procedure in mixed-mode IC design. The initial solution uses network-flow method to satisfy row capacity constraint from a global view, and then refers to the thought of LPP to optimally resolve cell overlaps. Moreover, owing to the divide-and-conquer strategy and dynamic selection strategy the problem complexity is minimized. At the stage of post optimization, a deterministic method is presented to create new configuration so that optimal placement result can be achieved very fast. Promising experimental results are got by the new algorithm.

A natural progression of this work is to study more in-deep relationship between the initial detailed placement and the post optimization, so that placement quality can be further improved.

References

1. A.B. Kahng, P.Tucker, A.Zelikovsky, "Optimization of linear placements for wirelength minimization with free sites," in Proc. Of the Asia and South Pacific Design Automation Conference, 1999, pp.241-244.
2. A. E. Caldewll, A. B. Kahng and I. L. Markov, "Optimal partitioners and end-case placers for standard-cell layout," in Proc. ISPD'99, pp. 90-94, 1999.
3. A. Srinivasan, K. Chaudhary, and E. S. Kuh, "RITUAL: a performance driven placement algorithm," in IEEE Trans. on Circuits and Systems. II, vol. 39, no. 11, pp. 825-840, Nov. 1992.
4. B. Yao, W. T. Hou, X. L. Hong and Y. C. Cai, " FAME: A Fast Detailed Placement Algorithm for Standard_Cell Layout Based on Mixed Mincut and Enumeration," in Proc. Int. Conf. On CAD/CG', pp. 616-621, Shanghai, 1999.
5. Chin-Chinh Chang, Jason Cong and Xin Yuan, "Multi-level placement for large-scale mixed-size IC designs", Proc. Asia South Pacific Design Automation Conference, pp 325-330, January 2003.
6. Hanbin Zhou, Weimin Wu, Xianlong Hong, " PAFLO: a fast standard-cell detailed placement algorithm," in Proceeding of ICCAS 2002, pp.1401-1405.
7. Hong Yu, Xianlong Hong, Yici Cai, "MMP: a novel placement algorithm for combined macro block and standard cell layout design," in Proceeding of ASP-DAC, pp.271-276, 2000.
8. http://vlsicad.eecs.umich.edu/BK/ISPD02bench.

9. Jens Vygen, "Algorithms for detailed placement of standard cells," in Proc. of the Conference Design Automation and Test in Europe (DATE' 1998), IEEE 1998, pp.321-324.
10. Lijuan Luo, Qiang Zhou, Xianlong Hong, Hanbin Zhou, "Effective algorithm for optimal initial solution in mixed-mode detailed placement," ASICON2003,pp.170-173.
11. Majid Sarrafzadeh, Maogang Wang, "NRG: global and detailed placement," in Proc. Of IEEE Intl. Conference on Computer Aided Design, pp. 532-537,IEEE Computer Society Press, 1997.
12. S. Goto, "An Effective Algorithm for the Two-Dimensional Placement Problem in Electrical Circuit Layout", IEEE Trans. on Circuits Syst., vol. CAS-28, No. 1, January 1981.
13. S.N. Adya and I. L. Markov, "Consistent placement of macro-block using floorplanning and standard-cell placement," in Proc. Int. Symp. On Physical Design, pp. 12-17,2002
14. S.-W. Hur and J. Lillis, "Mongrel: hybrid techniques for standard cell placement". In International Conference on Computer-Aided Design, pages 165-170. IEEE, 2000.
15. U.Brenner, J.Vygen, "Faster optimal single-row placement with fixed ordering," in Proceedings of the conference on Design automation and test in Europe, pp.117-121, March 2000-March 2000, Paris,France.
16. Wu Weimin, Hong Xianlong, Cai Yici, Yang Changqi, Gu Jun, "A mixed mode placement algorithm for combined design of macro blocks and standard cells," in Proceeding of ASICON 2001, 2001.
17. N. Viswanathan and C. CN Chu. Fastplace: Efficient analytical placement using cell shifting, iterative local refinement and a hybrid net model. In ACM/IEEE International Symposium on Physical Design, pages 26-33, Phoenix, AZ, 2004.
18. Zhong Xiu, James D. Ma, Suzanne M.Fowler, Rob A. Rutenbar, " Large-Scale Placement by Grid-Warping", Proc. Design Automation Conference, pp 351-356, June 2004.
19. H.Eisenmann and F.M.Johannes,"Generic Global Placement and Floorplanning",in Proc. of the 35th Design Automation Conference,pp.269-274,1998
20. KLeinhans, J.M. Sigl, G., Johannes, F.M. and Antreich, K. "Gordian : VLSI Placement by Quadratic Programming and Slicing Optimization", IEEE Trans. CAD, 10,3(Mar. 1991),356-365.
21. Sigl,G., Doll, K., and Johannes, F.M. Analytical Placement : A Linear or a Quadratic Objective Function? Design Automation Conf., 1991,427-432.
22. Ulrich Brenner, Anna Pauli, Jens Vygen,"Almost Optimum Placement Legalization by Minimum Cost Flow and Dynamic Programming" in Proc. ISPD'04, pp. 2-8, 2004.
23. K. Doll, F.M. Johannes,K.J. Antreich,"Iterative placement improvement by network flow methods," Proc. IEEE Trans. CAD, vol.13,no.10 Oct. 1994.
24. Changqi Yang, Xianlong Hong, Hannah Honghua Yang, Qiang Zhou, Yici Cai, Yongqiang Lu. Recursively Combine Floorplan and Q-Place in Mixed Mode Placement Based on Circuit's Variety of Block Configuration. Proceeding of IEEE International Symposium on Circuits and Systems, 2004 Vancouver, Canada: Vol. 5, 81-84.
25. http://www.public.iastate.edu/~nataraj/ISPD04_Bench.html

Adaptive Mesh Smoothing for Feature Preservation

Weishi Li, Li Ping Goh, Terence Hung, and Shuhong Xu

Institute of High Performance Computing
1 Science Park Road, #01-01 The Capricorn, Science Park II, Singapore 117528
(liws, gohlp, terence, xush)@ihpc.a-star.edu.sg

Abstract. A simple algorithm is presented in this paper to preserve the feature of the mesh while the mesh is smoothed. In this algorithm, the bilateral filter is modified to incorporate local first-order properties of the mesh to enhance the effectiveness of the filter in preserving features. The smoothing process is error-bounded to avoid over-smoothing the mesh. Several examples are given to demonstrate the effectiveness of this algorithm in preserving the feature while removing noise from the mesh.

1 Introduction

Even with high-fidelity scanners, noise is inevitable in the acquired data, and therefore the data should be pre-processed to remove the noise for visualisation, manufacturing, and many other digital geometry applications that rely on local differential properties of the surface [1]. As most scanners are capable of producing triangular meshes, and the connectivity information implicitly defines the surface topology and serves as a means of fast access to the neighbouring samples, meshes, instead of raw data points, are smoothed to remove the noise.

Typically, mesh smoothing methods are based on image filtering techniques. Taubin [2] extended signal processing to mesh smoothing based on the definition of Laplacian operator on meshes. Subsequently, geometric diffusion algorithm for meshes was introduced by Desbrun *et al* [3], and locally adaptive Wiener filtering to meshes was introduced by Peng *et al* [4]. Although these algorithms are efficient in smoothing meshes, they are all isotropic, and therefore features of the mesh are blurred while the mesh is smoothed.

Feature-preserving mesh smoothing methods are mostly inspired by anisotropic diffusion [5, 6, 7, 8]. Clarenz *et al* [6] formulated a discrete anisotropic diffusion for meshes. Taubin [8] proposed diffusion-type smoothing on the normal field, and the mesh is smoothed via evolving the surface to match the smoothed normal. Although these methods are superior to the isotropic methods in preserving the feature, they are time consuming [9].

Recently, Fleishman *et al.* [10] adopted bilateral filtering, a non-linear filter derived from Gaussian blur, to remove noise from the mesh (referred to as *DenoisePoint* in this paper). This method is simple to implement, fast in computation and effective in preserving the feature. However, the parameters of the filters are interactively assigned and uniform for all vertices. In practice, it is not straightforward to find good parameters for a given mesh. In addition, filters with uniform parameters are not suitable to smooth irregularly sampled meshes, such as the mesh obtained from multiresolution processing [11]. A similar mesh smoothing algorithm (referred to as *RobustEstimation* in this paper), also for regularly sampled meshes, was presented by Jones *et al* [9].

The existing algorithms, except for *RobustEstimation,* smooth the mesh iteratively. The number of iteration required for a particular mesh is determined by the user according to his/her visual judgement, and over-smoothing is prevalent. Ohtake *et al* [12] proposed to control the displacement distance in each of the iterations and place no control on the accumulated smoothing error, thus the possibility of over-smoothing is reduced, but still unavoidable.

In this paper, a simple mesh smoothing algorithm is developed based on bilateral filtering. To improve the performance of bilateral filter in preserving sharp features of the mesh, local first-order properties, i.e. normals of the mesh at the vertices, are incorporated into the filter. We propose to determine particular filter parameters for each vertex according to the local geometric information of the vertex to smooth the irregularly sampled mesh. We also develop a simple scheme to control the accumulated smoothing error to avoid over-smoothing the mesh.

The organisation of this paper is as follows. A brief introduction to bilateral filter is given in Section 2. In Section 3, the smoothing algorithm is presented. Examples are shown in Section 4 to demonstrate the effectiveness of the algorithm, followed by a conclusion section at the end of the paper.

2 Bilateral Filtering

Bilateral filtering was developed by Smith and Brady [13] as an alternative to anisotropic diffusion (see also Tomasi and Manduchi [14]) in image processing. It is a non-linear filter where the output is a weighted average of the input. It starts with standard Gaussian filtering with a spatial kernel f. However, the weight of a pixel also depends on a function g in the intensity domain, which decreases the weight of pixels with large intensity differences. The output of the bilateral filter for a pixel s is then:

$$J_s = \frac{1}{k(s)} \sum_{p \in \Omega} f(p-s) g(I_p - I_s) I_p \quad (1)$$

where Ω is the whole image, $k(s)$ is a normalisation term

$$k(s) = \sum_{p \in \Omega} f(p-s) g(I_p - I_s)$$

In practice, they use a Gaussian for f in the spatial domain, and a Gaussian for g in the intensity domain. Therefore, the value at a pixel s is influenced mainly by pixels that are spatially close and have a similar intensity.

Despite its simplicity, this bilateral filtering successfully competes with other image smoothing algorithm [13, 14, 15, 16]. Its effectiviness to meshes has also been demonstrated by Fleishman *et al* [10] and Jones *et al* [9].

3 Mesh Smoothing

A triangular mesh is denoted as a pair *(P,K)*, where *p* is a set of vertex positions $P = \{\mathbf{p}_i \in \mathbf{R}^3 \mid i = 0,...,n\}$, and *K* is a simplicial complex which contains all the topological information, i.e. adjacency information. The complex *K* is a set of subset of $\{0,...,n\}$. The subsets are called simplices and come in three types: the 0-simplices $\{i\} \in K$ are called vertices *V*, the 1-simplices $(i,j) \in K$ are called edges *E*, and the 2-simplices $(i,j,k) \in K$ are called faces *F*. In the following sections, we also refer to the vertex position as vertex. The correct meaning can always be deduced from the context. The 1-ring neighbours of vertex *i* are the set $N_1(i)$ of vertex *j* connected to *i* by an edge (i,j), that is $N_1(i) = \{j \mid (i,j) \in E\}$, and the 2-ring neighbours of a vertex *i* are set

$$N_2(i) = \bigcup_{k \in N_1(i)} N_1(k) \setminus \{i\} \quad (2)$$

The set difference $A \setminus B$ is defined by $A \setminus B = \{x \mid x \in A \text{ and } x \notin B\}$ [17].

Mesh smoothing is to estimate a new position for each vertex, i.e. determine a vector $\Delta \mathbf{p}_i$ and distance d_i for each vertex *i=0,...n*, to update the vertex position $\mathbf{p}_i$ as follows:

$$\hat{\mathbf{p}}_i = \mathbf{p}_i + d_i \Delta \mathbf{p}_i \quad (3)$$

to obtain a smoother mesh.

The displacement of the vertex position along its tangent plane will cause vertex-shift, and further results in an increase in the irregularity of the mesh [11], we move vertex positions along the normal direction, i.e. $\Delta \mathbf{p}_i = \mathbf{n}_i$. The normal at a vertex is the weighted average of the normals to the triangles in the 1-ring neighbourhood of the vertex. Then, the remaining problem is to determine displacement distance d_i for the vertex.

3.1 Displacement Distance

To preserve the feature of the mesh, the displacement distance of each vertex must reflect the local frequency of the mesh at the vertex. Otherwise, if high frequencies are considered as low ones or low frequencies are considered as high ones, the filtering will be distorted and the result is either blurring of the feature or inadequate smoothing. In Laplacian smoothing of mesh, this is called as frequency confusion [3].

Fleishman et al [10] proposed to use the heights of the incident vertices over the tangent plane at a vertex to reflect the frequency information and preserve the feature. In Fig. 1(a), points $\mathbf{p}_0$, $\mathbf{p}_1$, $\mathbf{p}_2$, $\mathbf{p}_3$ are 1-ring neighbours of $\mathbf{p}$. Points $\mathbf{p}'_0$ and $\mathbf{p}'_2$ are the projection of $\mathbf{p}_0$ and $\mathbf{p}_2$ on the tangent plane at vertex $\mathbf{p}$, and $\|\mathbf{p}'_0 - \mathbf{p}_0\| \neq \|\mathbf{p}'_2 - \mathbf{p}_2\|$. Therefore, $\mathbf{p}_0$ and $\mathbf{p}_2$ will contribute differently to the displacement of $\mathbf{p}$ and the feature is preserved during the smoothing process of *DenoisePoint*. However, the heights of the incident vertices over the tangent plane of a vertex can not reveal the frequency information completely. If $\|\mathbf{p}-\mathbf{p}_0\|=\|\mathbf{p}-\mathbf{p}_2\|$ and $\mathbf{n}_0 \neq \mathbf{n}_2$, as depicted in Fig. 1(b), $\mathbf{p}_0$ and $\mathbf{p}_2$ should also contribute differently to the displacement of $\mathbf{p}$. In other words, the first-order properties of the incident vertices should also contribute to the displacement of the vertex.

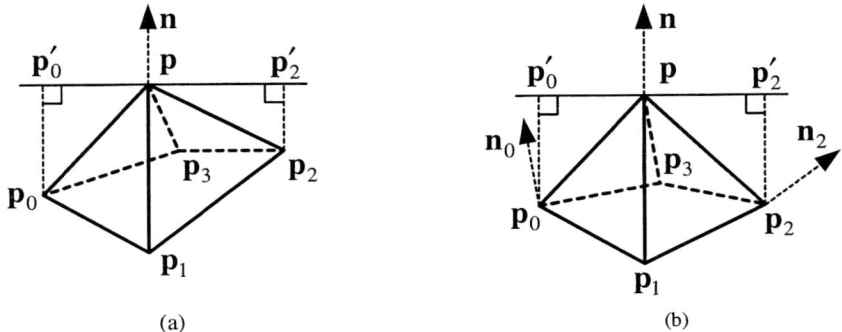

Fig. 1. Frequency confusion: (a) Fleishman et al's solution, (b) counterexample of Fleishman et al's solution

In order to incorporate the normal to calculate the displacement distance, we present the following function.

$$d_i = \frac{1}{k(i)} \sum_{j \in N(i)} f(l_j) g(h_j) v(\hat{n}_j) h_j \tag{4}$$

where $k(i) = \sum_{j \in N(i)} f(l_j) g(h_j)$, $l_j = \|\mathbf{p}_i - \mathbf{p}_j\|$, $h_j = (\mathbf{p}_i - \mathbf{p}_j, \mathbf{n}_i)$, and $v(\hat{n}_j) = e^{(-\hat{n}_j^2/2\sigma_v^2)}$ with $\hat{n}_j = \sqrt{1-(\mathbf{n}_i,\mathbf{n}_j)^2}$.

Standard Gaussian filters with parameters σ_f and σ_g, respectively, are used for f and g. Same as *DenoisePoint*, the spatial weight f depends on the distance between the neighbours and the central vertex $\mathbf{p}_i$, and the height h_j of the neighbours over the tangent plane T_i to the mesh at $\mathbf{p}_i$ is analogous with the grey level of a pixel of an image. The improvement is that a new coefficient $v(\hat{n}_j)$ is introduced to incorporate the normal of the mesh to preserve the feature, especially the sharp feature. For two incident vertices $j_0, j_1 \in N(i)$ and $h_{j_0} = h_{j_1}$, if $\angle(n_i, n_{j_0}) < \angle(n_i, n_{j_1})$, then vertex j_0 will contribute more to the displacement of vertex i according to Eqn. (4). Thus, the features of the mesh are preserved. Parameter σ_v also influences the performance of the algorithm in preserving features, i.e. smaller σ_v will result in better feature-preserving performance.

Parameters σ_f and σ_g are set for each vertex as follows:

$$\sigma_f = \sqrt{\sum_{j \in N_1(i)} l_j^2 / (n-1)} \quad (5)$$

$$\sigma_g = \sqrt{\sum_{j \in N_1(i)} h_j^2 / (n-1)} \quad (6)$$

As σ_f and σ_g are set according to the local shape information of each vertex, the smoothing operation is adaptive to the vertex. For the irregularly sampled mesh, this kind of adaptive method is better than the uniform methods used in *DenoisePoint* and *RobustEstimation*, and will benefit the removal of noise and at the same time preserving the feature.

Distance d_i can be determined by bilateral filtering over the 1-ring or 2-ring neighbours. The choice of 1-ring versus 2-ring neighbours depends to a large extent on the sampling frequency of the mesh. If the sampling frequency is very high and the mesh is very dense, 1-ring neighbours are not sufficient for the removing of noise, and 2-ring neighbours are suggested.

3.2 Error Control

It is important to control smoothing error in applications. If the error is not controlled, the mesh may be over-smoothed and the features are blurred due to excessive smoothing iterations. In most existing algorithms, the position of each vertex is modified along different directions in different iterations, only the error in a single iteration is controlled [12].

In this section, a simple scheme is developed to control the accumulated smoothing error. The error bound consisted of an upper limit and a lower limit. In mesh smoothing, all vertices of the mesh are assigned upper limit t_i^+ and lower limit t_i^- for $i = 0,...n$. For the original mesh, $t_0^+ = -t_0^- = ... = t_i^+ = -t_i^- = ... = t_n^+ = -t_n^-$. In mesh smoothing, several smoothing iterations may be needed to obtain the acceptable mesh. If each vertex is moved along the same particular direction during the smoothing iterations, instead of different direction in different iteration as most algorithms, the smoothing error can be dealt with more directly. In this paper, each vertex is displaced along the normal of the original vertex. The error-bounded smoothing algorithm is summarised as follows.

Mesh smoothing algorithm

1. For vertex i with error bound t_i^- and t_i^+,
2. Compute distance d_i using Eqn. (4).
3. If $t_i^- \le d_i \le t_i^+$, $\hat{t}_i^+ = t_i^+ - d_i$, $\hat{t}_i^- = t_i^- - d_i$.
4. Otherwise, if $d_i < t_i^-$, $d_i = t_i^-$, $\hat{t}_i^- = 0$, $\hat{t}_i^+ = t_i^+ - d_i$.
5. Otherwise, $d_i = t_i^+$, $\hat{t}_i^- = t_i^- - d_i$, $\hat{t}_i^+ = 0$.
6. Update the vertex position $\hat{\mathbf{p}}_i$ using Eqn. (3).

The advantage of this scheme is that the accumulated error is bounded without calculation of the distance between the smoothed mesh and the original mesh. This will considerably reduce the complexity of the mesh smoothing operation for large-scale datasets. This scheme also tends to constrain the shrinkage of the mesh.

4 Implementations and Discussions

We have implemented the algorithm described in the previous section and compared our results to the results of *DenoisePoint* and *RobustEstimation*. All meshes are rendered with flat shading to show faceting. The number of smoothing iterations for our algorithm is the same as that for *DenoisePoint* in each example.

Five examples are shown in this section. The first example is the Fandisk model (see Fig. 2). As *RobustEstimation* performs not very well in preserving sharp features [10], we compare the result of our algorithm with that of *DenoisePoint*. It is obvious the sharp edges of the result of *DenoisePoint* algorithm are blurred (refer to Fig. 2(b)) while the edges are preserved in the result of our algorithm (refer to Fig. 2(c)).

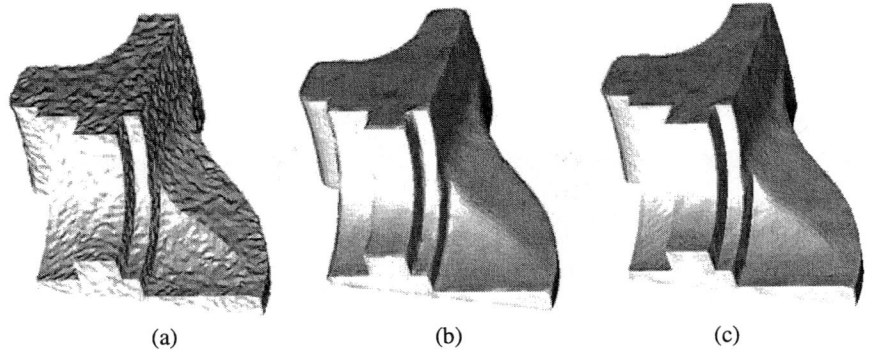

Fig. 2. The Fandisk model: (a) the original model, (b) the model smoothed by DenoisePoint, and (c) the model smoothed by our algorithm

The second example is an irregularly sampled ball-like mesh (see Fig. 3). The maximum and minimum radiuses of the results are shown in Table 1. The variance of the radius of our result is less than that of *DenoisePoint*, which means our result is

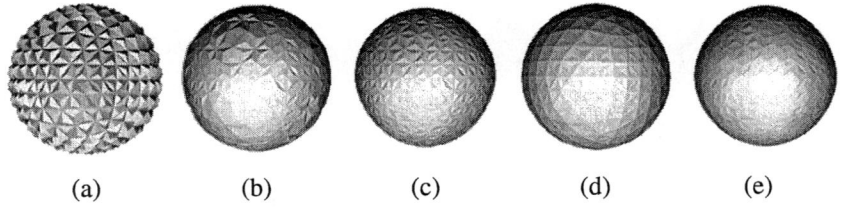

Fig. 3. The ball-like model: (a) the original model, (b) and (c) the model smoothed by DenoisePoint for one iteration and five iterations, respectively, and (d) and (e) the model smoothed by our algorithm for one iteration and five iterations, respectively

Table 1. The maximum and minimum radius of the results

	DenoisePoint			Our algorithm		
	Max. Radius	Min. Radius	Diff.	Max. Radius	Min. Radius	Diff.
1 iteration	1.0245	0.9695	0.0550	1.0109	0.9879	0.0230
5 iterations	0.9627	0.9155	0.0472	0.9743	0.9458	0.0285

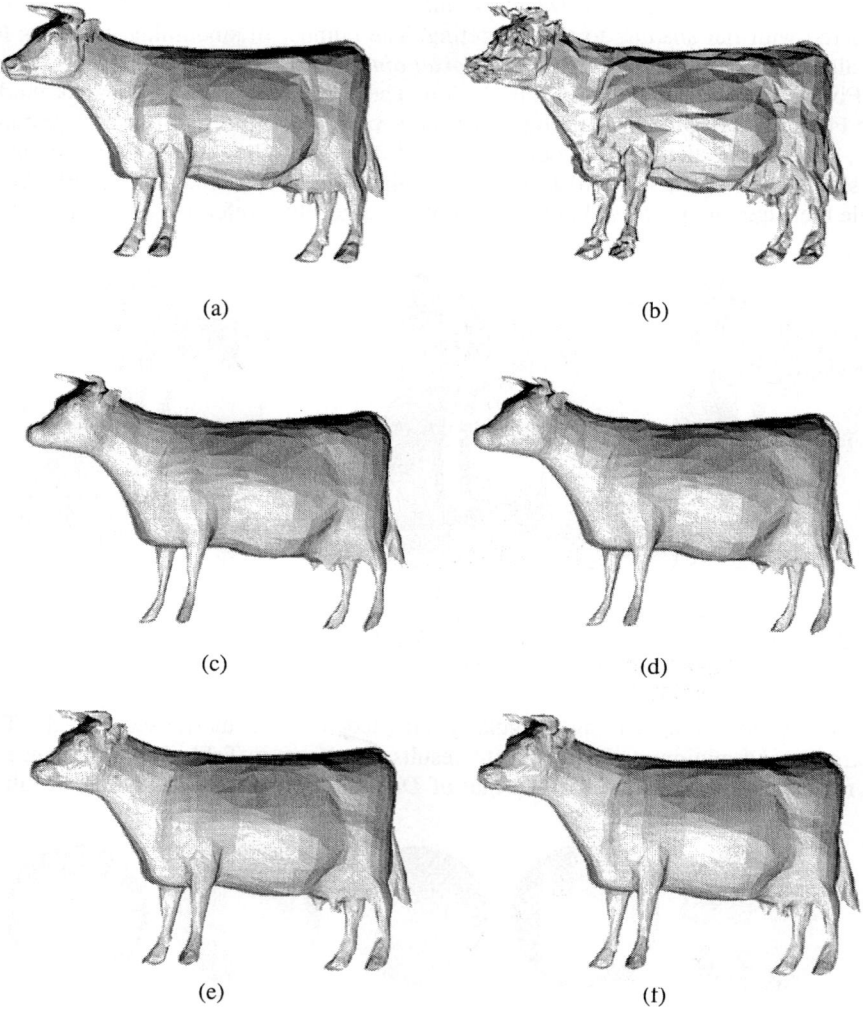

Fig. 4. The cow model: (a) the original model, (b) the disturbed model, (c) and (d) the model smoothed by DenoisePoint for one iteration and five iterations, respectively, and (e) and (f) the model smoothed by our algorithm for one iteration and five iterations, respectively

smoother than that of *DenoisePoint*. The visual effect of our algorithm is also obviously better. Another advantage of our algorithm is that the shrinkage is less than the result of *DenoisePoint*. This example also demonstrates that *DenoisePoint* algorithm with uniform parameters does not work well in removing noise from isotropic models if they are not regularly sampled.

The third example is a cow model (see Fig. 4). We disturb it with noise. Our algorithm also performs better than *DenoisePoint* algorithm, especially in the areas of noses, eyes and legs (refer to Fig. 4(c), (e)). As the smoothing error is bounded in our algorithm, over-smoothing is prevented. So, after five smoothing iterations, the features are still preserved as illustrated in Fig. 4(f). In comparison, the noses and eyes are almost smoothed out in Fig. 4(d).

The above two examples shown in Fig. 3 and Fig. 4 are highly irregularly sampled meshes. Our algorithm performs much better in terms of smoothing irregularly samples meshes. This is because the parameters of the filter are adaptively set to each vertex in our algorithm.

The fifth example is the head of a dragon model (see Fig. 5). The results of *DenoisePoint* and our algorithm (refer to Fig. 5(b), (d)) are quite similar. As for the result of *RobustEstimation* (refer to Fig. 5(c)), the beard around the lower jaw is blurred, and the noise is not removed in several regions.

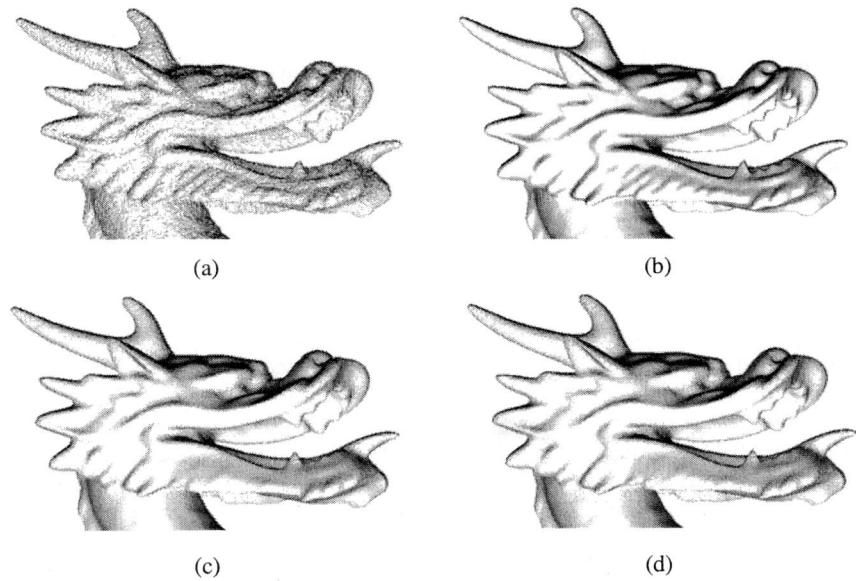

Fig. 5. The dragon head model: (a) the original model, (b) the model smoothed by DenoisePoint, (c) the model smoothed by RobustEstimation, and (d) the model smoothed by our algorithm

The mesh in the above example shown in Fig. 5 is very dense and nearly regularly sampled. For this kind of meshes, we find, if there is no sharp features in the meshes, the results of our algorithm are comparable with that of *DenoisePoint* if the filter parameters of *DenoisePoint* are carefully selected.

Other than effectiviness, our algorithm also has advantage over the other two algorithms in parameter setting. For our algorithm, only one parameter σ_v is needed to be set by the user while two parameters are needed for the other two algorithms. It is also very easy to set parameter σ_v in our algorithm. If there are sharp features in the mesh and they are to be preserved, value smaller than 0.5 (according to the sharpness of the features, sharper the features are, smaller the value should be) could be assigned to parameter σ_v. Otherwise, value greater than or equal to 0.5 could be assigned to parameter σ_v. As we have tested, for most meshes that have no sharp features, parameter σ_v can be set to be 0.5.

5 Conclusions

We have developed a feature-preserving mesh smoothing technique. We presented a scheme to incorporate the local first-order properties of the mesh at the vertices into the bilateral filter to preserve the sharp features. The incident vertices are projected to the tangent plane of the central vertex, and the updated position of the central vertex is computed as the modified bilateral filtering of the 1-ring or 2-ring neighbours treated as height field. The parameters are automatically assigned particularly for each vertex. We have demonstrated our algorithm on several models, and compared our results with that results of *DenoisePoint* algorithm presented by Fleishman et al and *RobustEstimation* algorithm presented by Jones et al. Our algorithm performs better than the other two algorithms, especially in preserving sharp features and smoothing irregularly sampled meshes.

References

1. Meyer M, Desbrun M, Schröder P, Barr AH. Discrete differential-geometry operators for triangulated 2-manifolds. In: Proceeding of Visualization and Mathematics 2002.
2. Taubin G. A signal processing approach to fair surface design. Computer Graphcis 1995; 351-358.
3. Desbrun M, Meyer M, Schröder P, Barr AH. Implicit fairing of irregular meshes using diffusion and curvature flow. Computer Graphics 1999, 317-324.
4. Peng J, Strela V, Zorin D. A simple algorithm for surface denoising. In: IEEE Visualization 2001; 107-112.
5. Bajaj C, Xu G. Anisotropic diffusion on surfaces and functions on surfaces. ACM Trans. Graph. 2003; 22(1): 4-32.
6. Clarenz U, Diewald U, Rumpf. M. Anisotropic geometric diffusion in surface Processing. In: IEEE Visualization 2000, 397-405.
7. Tasdizen T, Whitaker R, Burchard P, Osher S. Geometric surface smoothing via anisotropic diffusion of normals. In: IEEE Visualization 2002; 125-132.
8. Taubin G. Linear anisotropic mesh filtering. Tech. Rep. IBM Research Report RC 2213.
9. Jones TR, Durand F, Desbrun M. Non-iterative, feature-preserving mesh smoothing. ACM Trans. Graph. 2003; 22(3): 943-949.

10. Fleishman S, Drori I, Cohen-Or D. Bilateral mesh denoising. ACM Trans. Graph. 2003; 22(3): 950-953.
11. Guskov I, Sweldens W, Schröder P. Multiresolution signal processing for meshes. Computer Graphics 1999; 325-334.
12. Ohtake Y, Belyaev A, Bogaevski I. Mesh regularization and adaptive smoothing. Computer-Aided Design 2001; 33: 789-800.
13. Smith S, Brady J. SUSAN-a new approach to low level image processing. IJCV 1997; 23: 45-78.
14. Tomasi C, Manduchi R. Bilateral filtering for gray and color images. In: Proc. IEEE Int. Conf. on Computer Vision 1998; 836-846.
15. Durand F, Dorsey J. Fast bilateral filtering for the display of high-dynamic-range images. ACM Trans. Graph. 21, 3, 257-266.
16. Elad M. On the bilateral filter and ways to improve it. IEEE Trans. On Image Processing 2002; 11(10): 1141-1151.
17. Harris JW, Stocker H. Handbook of mathematics and computational science. New York: Springer-Verlag, 1998.

A Fuzzy Grouping-Based Load Balancing for Distributed Object Computing Systems*

Hyo Cheol Ahn and Hee Yong Youn

School of Information and Communications Engineering,
Sungkyunkwan University, 440-746, Suwon, Korea
`ahhyo7942@hotmail.com`, `youn@ece.skku.ac.kr`

Abstract. In the distributed object computing systems a set of server objects are made available over the network for computations on behalf of remote clients. In a typical distributed system setting the existing load balancing algorithms usually do not consider the global system state. In this paper we propose a new approach for improving the performance of distributed system using fuzzy grouping-based load balancing. It utilizes membership graphs in terms of the amount of CPU time and memory used for inferencing the service priority. Extensive computer simulation reveals that the proposed approach allows consistently higher performance than other approaches in terms of response time and throughput for various number of servers and tasks.

Keywords: Distributed object systems, fuzzy grouping, load balancing, membership graph, service priority.

1 Introduction

Currently, a number of technologies exist for the development of distributed and parallel computing applications. Various commercial software products are employing distributed object-based approaches such as CORBA (common object request broker architecture), DCOM, and Java RMI (remote method invocation) [3]. In the distributed object-based approach, an application is constructed as a group of interacting objects. These objects are distributed over multiple computers which interact with each other through well predefined protocols. It is common that there exist multiple objects (possibly on heterogeneous platforms) providing the same service in a distributed object computing system. This is mainly to achieve high availability and scalability.

A tremendous increase of the interest has been witnessed for the past few years in object-oriented client/server applications and, in general, in distributed object systems. Availability of powerful, low-cost computers and high-bandwidth communication links, together with recent developments in the area of object-oriented languages and systems, boosted performance of object-oriented distributed object systems,

---

* This research was supported by the Ubiquitous Autonomic Computing and Network Project, 21st Century Frontier R&D Program in Korea and the Brain Korea 21 Project in 2004. Corresponding author: Hee Yong Youn.

where a set of server objects are made available over the network for computations on behalf of remote clients. In a typical distributed system setting, requests for services arrive at different servers in a random fashion. This causes a situation of non-uniform load to occur across the system servers. Load imbalance is observed by the existence of servers that are highly loaded while others are lightly loaded or even idle. Such situations are harmful to the system performance in terms of mean response time of services and resource utilization.

The requirements in load balancing and the environments are quite different from those assumed with classical load balancing algorithms. Most importantly, the existing load balancing algorithms used in the distributed object computing systems are usually based on simple techniques such as round-robin or random algorithm, which may not give rise to an optimized performance because they do not consider the information of the global system state. An effective load balancing scheme requires knowledge of the global system state (e.g., the workload distribution). However, in a distributed computing system, the global state is swiftly and dynamically changing, and it is very difficult to accurately model the system using an analytical approach. This uncertainty in the global state has been a primary issue in the design of efficient distributed computing systems. Increasing the frequency of information exchange between the nodes is not necessarily a practical solution because messaging overheads may adversely affect the efficiency of the system. Moreover, the overheads of load balancing mechanisms can be highly detrimental to the performance of the system under heavy system load condition.

Therefore, we propose a new distributed load balancing mechanism that characterizes uncertainty in the decision making process and reduces messaging overheads. The proposed mechanism employs a fuzzy grouping-based load balancing approach to model the state variables that cause uncertainty in the global state. However, the degree of uncertainty in the global state is not a tangible object because the actual global state cannot be measured. To overcome this problem, we find the sources that cause uncertainty in the global state. They include the system load condition (the amount of CPU time and memory used) in individual servers. However, each server cannot precisely measure the system load condition. To deal with these problems, we use linguistic variables [7] that represent the system load condition, and the servers are divided into groups to reduce message overheads [10]. Respective group has a load monitor. The load monitor checks the amount of CPU time and memory used for individual server in the group. Then we make fuzzy inference rules that map the qualitative representation of the sources of uncertainty to that of the thresholds. Finally, we infer quantitative thresholds by an inference mechanism of fuzzy control. There have been some attempts for load balancing using fuzzy-based approaches. Thus, we implement and evaluate the proposed approach of fuzzy-grouping, and compare it with an existing scheme. Extensive computer simulation reveals that the proposed approach allows consistently higher performance than other approaches in terms of response time and throughput for various number of servers and tasks.

The rest of the paper is organized as follows. Section 2 briefly reviews the fuzzy set theory, load balancing, and grouping used in this paper. Section 3 proposes a new fuzzy-grouping-based load balancing approach. Section 4 evaluates the performance of the proposed approach. Finally, we conclude the paper in Section 5.

2 Related Work

2.1 Fuzzy Set Theory

This subsection briefly introduces the notion of fuzzy sets, linguistic variables, and possibility distributions, and fuzzy if-then rules [7, 8].

- *Fuzzy sets*: sets with smooth grades
- *Linguistic variables*: variables whose values are both qualitatively and quantitatively described by a fuzzy set
- *Possibility distributions*: constraints on the value of a linguistic variable imposed by assigning it a fuzzy set
- *Fuzzy if-then rules*: a knowledge representation scheme describing a functional mapping or a logic formula that generalizes an implication in two-valued logic

A fuzzy set is defined as a class of objects with a continuum of grades of membership. A linguistic variable represents a variable whose values are words or sentences in a natural or artificial language. For example, if system load is regarded as a linguistic variable, it may have the following terms as its values: light, moderate, and heavy. If 'light' is interpreted as a load below about one task, 'moderate' as a load between about 3 tasks and 5 tasks, and 'heavy' as a load above about 7 tasks, these terms may be described as fuzzy sets as shown in Figure 1, where µ represents the grade of membership [6].

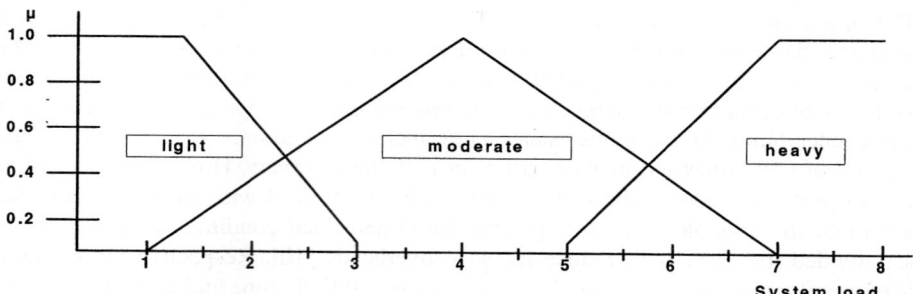

Fig. 1. The membership graph of system load

When a fuzzy set is assigned to a variable whose precise value is unknown, the fuzzy set serves as a constraint on the degree of ease for the variable to take a certain value. This degree of ease is called the 'possibility degree'. Thus, a possibility distribution of a variable is a function that maps elements in the variable's universe of discourse to their possibility degree. Finally, a fuzzy if-then rule is a scheme for representing knowledge and association that is inexact and imprecise in nature.

The if-part of a fuzzy rule is called the rule's antecedent, and the then-part of a rule is called its consequent. Reasoning using fuzzy if-then rules has three major features. First, it enables a rule that partially matches the input data to make an inference. Second, it typically infers the possibility distribution of an output variable from the possibility distribution of an input variable. Third, the system combines the inferred conclusions from all the rules to form an overall conclusion. Also, many fuzzy rule-based systems that need to produce a precise output use a defuzzification process to convert the inferred possibility distribution of an output variable to a representative precise value. Therefore, most applications of fuzzy logic use fuzzy if-then rules.

2.2 Load Balancing Approaches

Typical load balancing approaches for distributed object systems are round-robin, random, bid [1], and phase [2]. Recently, new load balancing schemes adopt java-space, request redirection, or fuzzy logic-based request redirection approach [3]. Request redirection approach is based on dynamic redirection of client requests from overloaded machines to underloaded machines. Redirection begins when the load of the overloaded machine exceeds a certain threshold value. The request redirection algorithm can be divided into two parts: (1) static scheduling based on round-robin ordering, and (2) request redirection. In round-robin ordering, the global scheduler schedules client requests to remote server objects in a fixed order. The global scheduler does not take server states into account when distributing the requests. On the other hand, request redirection is a common practice in balancing the load of a Web server system in which HTTP requests are redirected from overloaded servers to underloaded servers.

The basic components of the system include a global scheduler, a set of local schedulers and load monitors, and a group of server and client objects. The global scheduler communicates with the local schedulers residing in the server machines. Based on the current server state informed by a load monitor, a local scheduler decides whether to execute a remote call locally or redirect it to another server. The load monitor counts the number of concurrent calls being sent to the server object. It then compares the count with the threshold value. If the load condition changes, it sends a multicast message to notify other local schedulers. Each local scheduler starts a specific thread to listen to the load messages arriving at the multicast address from other servers.

In fuzzy logic-based request redirection approach, a fuzzy logic controller is incorporated into the request redirection system as a core component. The role of the fuzzy logic controller is to manage the decisions on the request redirection of each server. Instead of manually setting the threshold value for each server, the fuzzy logic mechanism allows the servers to make a decision based on the server ranking assigned by the fuzzy logic controller [3]. Thus, the higher the service rank, the more the server redirects the request. In other words, a server with lower service rank is likely to receive extra requests. Each server refers to its own rank and other servers' ranks to decide where it should redirect a remote call. As the name implies, fuzzy logic controller has to make use of fuzzy information to perform logic control. Several linguistic variables, server load, mean deviation of server load, and server rank are used in the fuzzy logic algorithm.

2.3 Forming Groups

In order to reduce the communication overhead, nodes at close distance may form a group [9]. For instance, all or some of the nodes in the same local area network

(LAN) may form a group. Nodes may join or leave their groups dynamically. A node may join a group by sending a message to its group's DR (designated representative). The DR communicates with the DRs in other peer groups. For a node leaving a group, it has to move its entire load to some other nodes. This task is implemented as follows:

1) The leaving node sends a "LEAVE" message to its DR about its leave.
2) The DR finds the nodes suitable for load transfer.
3) The leaving node transfers its loads in its queue without accepting loads anymore. If its loads are not completely transferred after a certain time period, then the remaining jobs in the queue are dropped. The time period is a variable and, for instance, set to at least twice the maximum allowance of waiting time for making a load transfer request.

3 The Proposed Approach

In this section we propose a fuzzy grouping-based load balancing approach, and then describe the fuzzy inference engine which is a part of the fuzzy load balancing system. Thereafter, the interactions between the fuzzy inference engine and other components are discussed in detail.

3.1 Fuzzy-Based Load Balancing

Fuzzy logic control-based load balancer attempts to capture intuition in the form of IF-THEN rules, and conclusions are drawn from the rules [4]. System parameters based on both intuitive and expert knowledge can be modeled as linguistic variables and their corresponding membership functions can be designed. Thus, nonlinear system with great complexity and uncertainty can be effectively controlled by fuzzy rules without complex, uncertain, and error-prone mathematical models [4]. The proposed fuzzy load balancer uses an algorithm for efficient load balancing. This algorithm of fuzzy rule-based inference consists of three basic steps and an additional optional step.

1) Fuzzy Matching: Calculate the degree to which the input data match the condition of the fuzzy rules.
2) Inference: Derive the rule's conclusion based on the degree of matching.
3) Combination: Combine the conclusion inferred by all fuzzy rules into a final conclusion.
4) (Optional) Defuzzification: For applications that need a crisp output (e.g., in control systems), an additional step is used to convert a fuzzy conclusion into a crisp one.

We propose a membership graph where the amount of CPU time and memory used in individual servers are input data and service priority is output data as shown in Figure 2(a). The definition of fuzzy set based on the amount of CPU time used is as follows: {low (L), less than moderate (LMO), moderate (MO), high (H), very high (VH)}. The fuzzy set definition based on the amount of memory used is as follows:

{small (s), less than medium (LME), medium (ME), large (L)}. Figure 2(b) shows the membership graph for the amount of memory used. We measure the amount of CPU time and memory used using the method, *main_cpu()* and *GlobalMemoryStatus()*, respectively, implemented by C++ language. The values are measured in each server. We use service priority (SP) to classify services into seven categories as shown in Figure 2(c). The fuzzy set of SP is as follows: {very low (VL), low (L), less than medium (LME), medium (ME), more than medium (MME), high (H), very high (VH)}. The load balancer sends the request from a client to a server of lower service priority. A set of inference rules is defined as shown in Figure 2(d) after defining the fuzzy variables above.

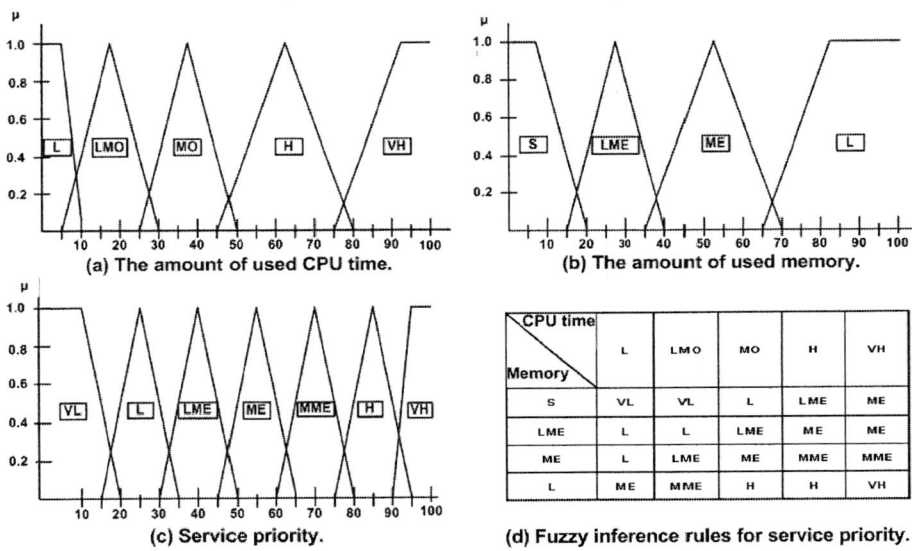

Fig. 2. The membership graphs and fuzzy rules for service priority

By applying the fuzzy inference rules in Figure 2(d), a decision can be made based on both antecedents. For example, if the amount of CPU time used is LMO and the amount of memory used is S, SP is VL. Having these fuzzy inference rules and membership graphs, the fuzzification and defuzzification process can be carried out as follows. First, the input values of the amount of CPU time and memory used are mapped to their respective membership degree values on their membership graphs. The degree values are compared and the minimum of the two is then projected onto the membership function of their consequence graph. The output graph, usually in the shape of a trapezium [4], represents the output of one inference rule. After the output graph is generated, defuzzification of the fuzzy output into a crisp or numeric value can be carried out. This defuzzification is a mean of maximum (MOM) and center of area (COA) [4]. We use the COA to defuzzify the output.

3.2 Load Monitor

For load balancing that requires information on the actual system state, load monitor is used. The job of load monitor is to collect load information and pass it to the point where it is needed for load balancing. Fuzzy load balancer depends on the information delivered by load monitor.

As shown in Figure 3, the proposed fuzzy grouping-based load balancing approach includes two servers and a load monitor in each group. The load monitor measures load information (the amount of CPU time and memory used) of two servers every two seconds and sends it to the fuzzy load balancer in the form of an asynchronous distributed event. The fuzzy load balancer then makes a decision whether it forwards the client's request to an appropriate server or ignores the load information. Note that, instead of adopting polling policy in which the load balancer periodically queries the server load condition from the load monitor, pushing policy is employed in which the load monitor pushes measured load value to the fuzzy load balancer.

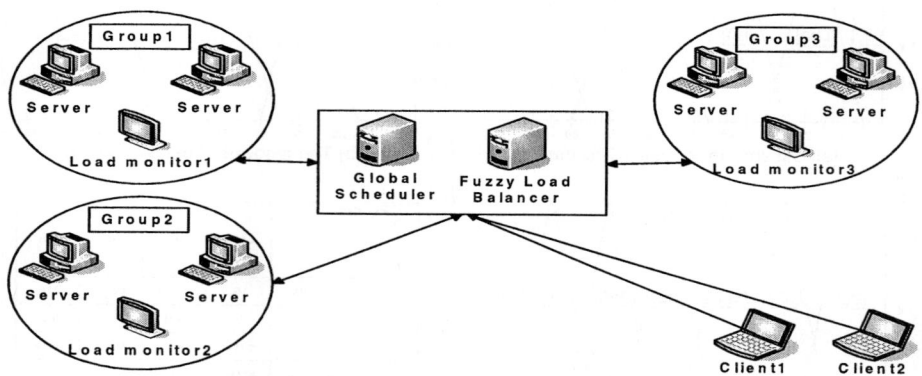

Fig. 3. The structure of the proposed fuzzy grouping-based load balancing system

3.3 Grouping

As shown in Figure 3, the proposed approach consists of three groups. In order to form a group, the load monitor in each group measures not only the information of the load of the servers but also the performances. A group is formed according to the performance index of CPU and memory. The performance index of CPU is based on SPEC (Standard Performance Evaluation Corporation) [5], while the performance index of memory is based on the memory size of participating servers. Figure 4 shows that the servers are ordered according to the rank. This rank was decided by the sum of the performance index of CPU and memory. If the number of active servers are six, for example, grouping is performed as follows:

The number of groups = The number of active servers (6) / size of a group (2)
= Rank number + 3
Grouping: [{1, 4}, {2, 5}, {3, 6}]

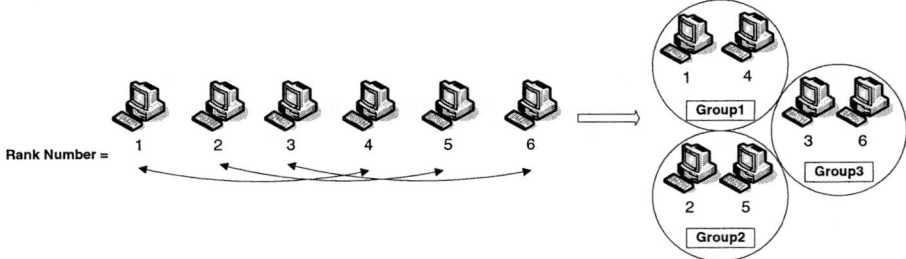

Fig. 4. The grouping operation

Grouping is needed if the following event occurs:

1) New server addition: The global scheduler moves a server to the group of highest load condition.
2) Server failure: The load of it are reassigned to other servers according to their load condition.
3) New group addition: The global scheduler assigns services to the groups according to their updated service priority.

The proposed fuzzy grouping-based load balancing approach reduces messaging overhead using the grouping approach.

3.4 The Operation Flow

The relationship between the components of the proposed fuzzy load balancing is shown in Figure 5. It can be briefly summarized as follows:

1) A client sends a request to the service.
2) The load monitor sends the information on the amount of CPU time and memory used by the fuzzy load balancer for fuzzy inference.
3) When new information of the amount on CPU time and memory used arrives, the fuzzy load balancer deduces the service priority of it.
4) The fuzzy load balancer sends it to the global scheduler.
5) The global scheduler sends a request to the server according to the service priority.

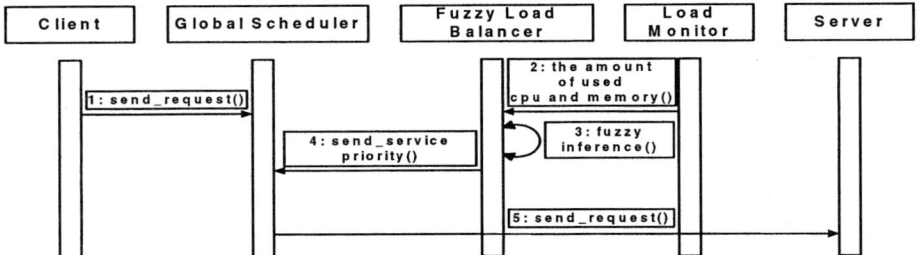

Fig. 5. The diagram of the flow of fuzzy load balancing operation

4 Performance Evaluation

We compare the performance of round-robin, fuzzy, and the proposed fuzzy group-based load balancing scheme by measuring the response time and throughput through computer simulation. The task arrival pattern at each server is modeled as a Poisson distribution with $\lambda = 6$, and it is assumed that all servers have the same arrival rate. Also, the service time for each request and the amount of CPU time and memory used are uniformly distributed. We have performed an extensive simulation with different combinations of numbers of servers and tasks, which vary form 2 to 12 and 100 to 3000, respectively.

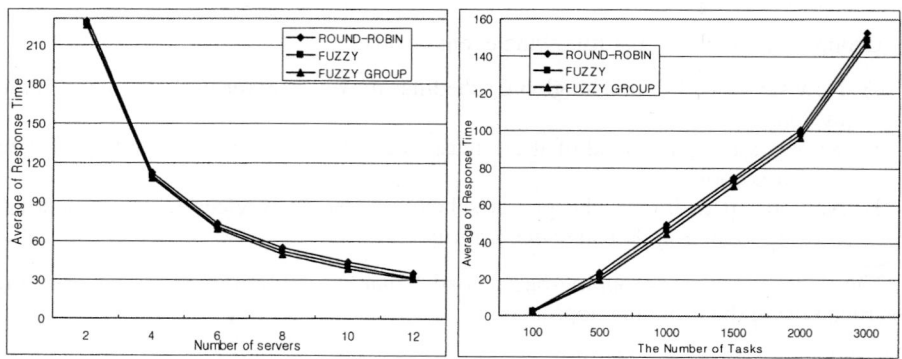

Fig. 6. Average response time with varying number of servers and tasks

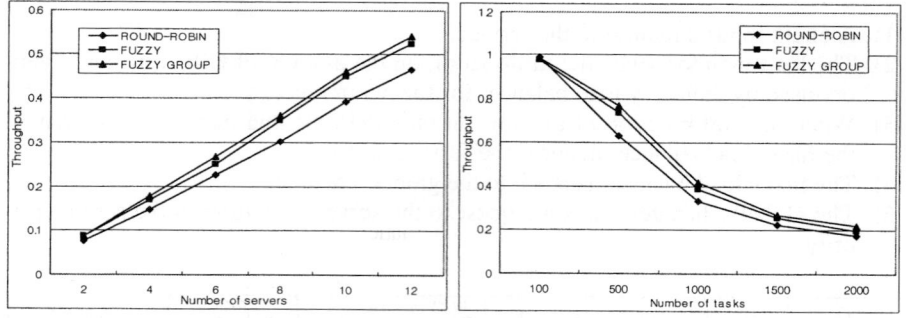

Fig. 7. Average throughput with varying number of servers and tasks

The average response times of the three load balancing approaches as the number of servers and tasks vary are shown in Figure 6. It illustrates that the proposed fuzzy group-based approach consistently outperforms the other approaches for different number of servers and tasks. Figure 7 shows how the throughput varies for each load balancing approach. In the simulation the throughput with varying number of servers is for 1500 tasks in each server, while the throughput with varying number of tasks is for 6 servers in each task, respectively. As can be seen from Figure 7, throughput increases as the number of servers increases. On the other hand, throughput decreases as the number of tasks increases. Again, throughput of the proposed fuzzy group-based load balancing is the highest among the three. This is because it assigns more tasks to the servers with relatively low loads using the proposed grouping concept. The simulation reveals that uncertainty in global state should be reflected in the design of load balancing mechanisms for better performance.

5 Conclusion

In this paper we have proposed a load balancing approach which effectively overcomes the load imbalance problem in the distributed object systems using fuzzy grouping. This approach dynamically selects a set of participating servers and distributes the load based on the total load of the tasks arriving in a given time interval. We have compared three typical load balancing approaches; round-robin, fuzzy-based, and the proposed fuzzy grouping-based load balancing approach. The experimental results show that the proposed approach is more efficient and flexible than the other two approaches in terms of response time and throughput for various number of servers and tasks. Future work will be adaptation of the proposed approach to an agent platform and CORBA.

References

[1] R. Smith: The contract net protocol: High-level communication and control in a distributed problem solver. IEEE Trans. Computer. Vol. 29. (1980) 1104-1113
[2] T.L. Casavant, J.G. Kuhl: Effects of response and stability on scheduling in distributed computing systems. IEEE Trans. Software Eng, Vol. 14. (1988) 1578-1587
[3] L.-S. Cheung,Y.-K. Kwok.: On Load Balancing Approaches for Distributed Object Computing Systems. The Journal of Supercomputing. Vol. 27. (2004) 149-175
[4] B. Kosko, Neural Networks and Fuzzy Systems: A Dynamical Systems Approach to Machine Intelligence. Prentice Hall, New Jersey (1992)
[5] The Standard Performance Evaluation Corporation, http://www.spec.org/
[6] P. Chulhye, G. Kuhl: A fuzzy-based distributed load balancing algorithm for large distributed systems. Proceedings of the 2nd International Symposium on Autonomous Decentralized Systems (1995) 266–273
[7] L.A. Zadeh: Fuzzy logic. IEEE Computer (1988) 83-93
[8] El-Abd, Aly. E: Load Balancing in Distributed Computing Systems Using Fuzzy Expert Systems. Proceedings of the International conference (2002) 141-144
[9] Yang Jia, Jizhou sun, Zunce Wei: Load balance in a new group communication system for the WAN. Electrical and Computer Engineering, IEEE CCECE 2003 (2003) 931-934
[10] Zhuang. Y. C, Ce-Kuen Shieh, Tyng-Yue Liang: A group-based load balance scheme for software distributed shared memory systems. First IEEE/ACM International Symposium (2001) 371-378

DSP-Based ADI-PML Formulations for Truncating Linear Debye and Lorentz Dispersive FDTD Domains

Omar Ramadan

Department of Computer Engineering, Eastern Mediterranean University,
Gazimagusa, Mersin 10, Turkey
omar.ramadan@emu.edu.tr

Abstract. An efficient and unconditional stable formulations of the perfectly matched layer (PML) are presented for truncating linear Debye and Lorentz dispersive Finite Difference Time Domain (FDTD) grids. The formulations incorporate the Digital Signal Processing algorithms developed for digital filters into the Alternating Direction Implicit FDTD method. Numerical examples are included to validated the proposed formulations.

1 Introduction

In the last decade, the explicit Finite Difference Time Domain (FDTD) method has been widely used for solving Maxwell's curl equations [1]. This method, however, is conditionally stable, i.e., the maximum time step is limited by the Courant-Friedrichs-Lewy (CFL) stability condition. Since the CFL is determined by the smallest cell size in the domain, the FDTD analysis of fine geometric structures requires large number of time iterations. Hence, the elimination of the CFL stability limit is one of the latest challenges in the FDTD research. Recently, the unconditional stable Alternating Direction Implicit FDTD (ADI-FDTD) method [2], has been used as an alternative tool to the FDTD algorithm for solving electromagnetic problems. This method removes the CFL stability condition required by the conventional FDTD scheme.

To model open region problem efficiently using the FDTD or the ADI-FDTD methods, suitable absorbing boundary conditions (ABCs) are needed to truncate the computational domain. Berengers Perfectly Matched Layer (PML) [3] has been shown to be one of the most efficient FDTD ABCs. In [4], the convolutional theorem has been incorporated with the ADI-FDTD formulations of the PML to truncate lossless FDTD domains. Very recently, different unconditionally stable PML formulations, based on the stretched coordinate PML [5] or the anisotropic PML [6], have been introduced for truncating lossy and dispersive FDTD domains [7], [8]. In these formulations, the Auxiliary Differential Equation (ADE) technique has been used to model linear Debye dispersive media.

In this paper, the Digital Signal Processing (DSP) algorithms developed for digital filters [9] are combined with the ADI-FDTD scheme to truncate linear

Debye and Lorentz FDTD domains. The proposed method has the advantage of simplicity as it allows direct time discretizations of Maxwell's equations in the PML region. Numerical examples are included to show the validity of the proposed formulations.

The paper is organized as follows. In section 2, the formulations of the proposed algorithm is presented. Section 3 includes the results of several numerical tests which show the effectiveness of the proposed method. Finally, a summary and conclusions are included in section 4.

2 Formulation

Using the PML formulations of [5], the field equations for an electromagnetic wave propagating in a linear and electrically dispersive one-dimensional domain can be written as

$$j\omega\widehat{\varepsilon}_r(\omega)\widetilde{E}_z = c\frac{1}{S_x}\frac{\partial}{\partial x}\widetilde{H}_y \tag{1}$$

$$j\omega\widetilde{H}_y = c\frac{1}{S_x}\frac{\partial}{\partial x}\widetilde{E}_z \tag{2}$$

where $\widetilde{E}_z$, and $\widetilde{H}_y$ are the Fourier transform of the corresponding fields, $\widehat{\varepsilon}_r(\omega)$ is the complex relative permittivity of the medium, c is the speed of light, and S_x is the PML stretched coordinate variable [5] defined as

$$S_x = 1 + \frac{\sigma_x}{j\omega\varepsilon_o} \tag{3}$$

where σ_x is the PML conductivity profile along the x-direction. For a lossy and linear dispersive media, $\widehat{\varepsilon}_r(\omega)$ can be expressed as

$$\widehat{\varepsilon}_r(\omega) = \varepsilon_\infty + \frac{\sigma}{j\omega\varepsilon_o} + \chi(\omega) \tag{4}$$

where ε_∞ is the infinite frequency permittivity, σ is the conductivity, and $\chi(\omega)$ is the electric susceptibility of the medium. Substituting (4) into (1), the following can be obtained

$$j\omega\varepsilon_\infty\widetilde{E}_z + j\omega\widetilde{Q}_z + \frac{\sigma}{\varepsilon_o}\widetilde{E}_z = c\frac{1}{S_x}\frac{\partial}{\partial x}\widetilde{H}_y \tag{5}$$

where $\widetilde{Q}_z$ is given by the relation

$$\widetilde{Q}_z = \chi(\omega)\widetilde{E}_z \tag{6}$$

Using the relation $j\omega \rightarrow \partial/\partial t$, (5) can be written in the time domain as

$$\varepsilon_\infty\frac{\partial}{\partial t}E_z + \frac{\partial}{\partial t}Q_z + \frac{\sigma}{\varepsilon_o}E_z = c\mathbf{S}_x(t) * \frac{\partial}{\partial x}H_y \tag{7}$$

where $\mathbf{S}_x(t)$ is the inverse Fourier transform of S_x^{-1} and $*$ represents the convolution operation. The convolutional term on the right hand side of (7) can be computed efficiently by using the Z-transform [10]. Hence, using the Bilinear transformation relation [9]:

$$j\omega \Rightarrow \frac{4}{\Delta_t}\frac{1-\mathcal{Z}^{-1/2}}{1+\mathcal{Z}^{-1/2}} \qquad (8)$$

S_x^{-1} can be written in the $\mathcal{Z}$-domain as

$$\mathbf{S}_x(\mathcal{Z}) = \frac{b_0 + b_1 \mathcal{Z}^{-1/2}}{1 + a_1 \mathcal{Z}^{-1/2}} = B_0 + \frac{B_1}{1 + a_1 \mathcal{Z}^{-1/2}} \qquad (9)$$

where Δ_t is the time step, $B_0 = b_1/a_1$, $B_1 = b_0 - b_1/a_1$ and

$$a_1 = -\frac{1-p}{1+p}, \quad b_0 = \frac{1}{1+p}, \text{ and } b_1 = -\frac{1}{1+p} \qquad (10)$$

with

$$p = \frac{\Delta_t \sigma_x}{4\varepsilon_o} \qquad (11)$$

It must be noted that the superscript of $\mathcal{Z}$ in (8) is based on a half-step to match the ADI updating scheme which based on splitting each time step into two sub-steps, i.e., $n \to n+1/2 \to n+1$ [2]. Taking the $\mathcal{Z}$-transform of the right hand side of (7), the following can be obtained

$$\mathcal{Z}\left[c\mathbf{S}_x(t) * \frac{\partial}{\partial x}H_y\right] = c\mathbf{S}_x(\mathcal{Z})\frac{\partial}{\partial x}H_y = cB_0\frac{\partial}{\partial x}H_y + f_{zx} \qquad (12)$$

where f_{zx} is given by

$$f_{zx} = \frac{cB_1}{1 + a_1 \mathcal{Z}^{-1/2}}\frac{\partial}{\partial x}H_y = -a_1 \mathcal{Z}^{-1/2} f_{zx} + cB_1 \frac{\partial}{\partial x}H_y \qquad (13)$$

Transforming (13) to the discrete time domain using the $\mathcal{Z}$-transform relation $\mathcal{Z}^{-m}G(\mathcal{Z}) \to G^{n-m}$ [9], discretizing the space derivative of H_y and discretizing the left hand side of (7) at the first ADI-FDTD sub-iteration at the $n+1/2$ time step by using the conventional FDTD algorithm [1], the following can be obtained:

$$Q_{z_i}^{n+1/2} + p^+ E_{z_i}^{n+1/2} = Q_{z_i}^n + p^- E_{z_i}^n + \zeta B_{0_i}(H_{y_{i+1/2}}^{n+1/2} - H_{y_{i-1/2}}^{n+1/2}) + f_{zx_i}^{n+1/2} \qquad (14)$$

where Δ is the space cell size, $\zeta = c\Delta_t/2\Delta$, $p^\pm = \varepsilon_\infty \pm \sigma\Delta_t/4\varepsilon_0$, and $f_{zx_i}^{n+1/2}$ is given by

$$f_{zx_i}^{n+1/2} = -a_{1_i} f_{zx_i}^n + \zeta B_{1_i}\left(H_{y_{i+1/2}}^{n+1/2} - H_{y_{i-1/2}}^{n+1/2}\right) \qquad (15)$$

In the same manner, (2) can be written in the first ADI sub-iteration as

$$H_{y_{i+1/2}}^{n+1/2} = H_{y_{i+1/2}}^n + \zeta B_{0_{i+1/2}}\left(E_{z_{i+1}}^{n+1/2} - E_{z_i}^{n+1/2}\right) + g_{zx_{i+1/2}}^{n+1/2} \qquad (16)$$

where $g_{zx_{i+1/2}}^{n+1/2}$ is given by

$$g_{zx_{i+1/2}}^{n+1/2} = -a_{1_{i+1/2}} g_{zx_{i+1/2}}^{n} + \zeta B_{1_{i+1/2}} \left(E_{z_{i+1}}^{n+1/2} - E_{z_i}^{n+1/2} \right) \quad (17)$$

For a linear dispersive media, $\chi(\omega)$ can be written in the s-domain as

$$\chi(s) = \frac{\sum_{m=0}^{N} \widetilde{d}_m s^m}{\sum_{m=0}^{N} \widetilde{e}_m s^m} \quad (18)$$

where $\widetilde{d}_m$ and $\widetilde{e}_m$, $(m = 0, 1, \cdots, N)$, are the coefficients of the rational polynomials and N is the maximum order of the dispersive media. Using (8), (18) can be written in the Z-domain as

$$\chi(\mathcal{Z}) = \frac{\sum_{m=0}^{N} d_m \mathcal{Z}^{(1-m)/2}}{\sum_{m=0}^{N} e_m \mathcal{Z}^{(1-m)/2}} \quad (19)$$

where the coefficients d_m and e_m, $(m = 0, 1, \cdots, N)$ are related to $\widetilde{d}_m$ and $\widetilde{e}_m$, and the time step Δ_t. Using (19), and taking the Z-transform of (6), Q_z can be written the discrete time domain as

$$Q_{z_i}^{n+1/2} = \frac{d_0}{e_0} E_{z_i}^{n+1/2} + \frac{1}{e_0} \sum_{m=1}^{N} \left(d_m E_{z_i}^{n+(1-m)/2} - e_m Q_{z_i}^{n+(1-m)/2} \right) \quad (20)$$

Substituting (20) into (14), the following can be obtained:

$$E_{z_i}^{n+1/2} = C_0 E_{z_i}^{n} + C_1 Q_{z_i}^{n} + C_1 \left[\zeta B_{0_i} \left(H_{y_{i+1/2}}^{n+1/2} - H_{y_{i-1/2}}^{n+1/2} \right) + f_{zx_i}^{n+1/2} \right] + C_2 \Psi_{e_i}^{n} \quad (21)$$

where

$$\Psi_{e_i}^{n} = \sum_{m=1}^{N} \left(d_m E_{z_i}^{n+(1-m)/2} - e_m Q_{z_i}^{n+(1-m)/2} \right) \quad (22)$$

and

$$C_0 = \frac{p^-}{q}, \quad C_1 = \frac{1}{q}, \text{ and } C_2 = -\frac{1}{qe_0} \quad (23)$$

with

$$q = p^+ + \frac{d_0}{e_0} \quad (24)$$

It is clear from (21) that $E_{z_i}^{n+1/2}$ can not be updated directly as it depends on $H_{y_{i+1/2}}^{n+1/2}$. Substituting (15), (16) and (17) into (21), the following can be obtained

$$-A_i^- E_{z_{i-1}}^{n+1/2} + A_i E_{z_i}^{n+1/2} - A_i^+ E_{z_{i+1}}^{n+1/2} = C_0 E_{z_i}^{n} - a_{1_i} C_1 f_{zx_i}^{n} + r_i \zeta C_1$$
$$\times \left[\left(H_{y_{i+1/2}}^{n} - H_{y_{i-1/2}}^{n} \right) - \left(a_{1_{i+1/2}} g_{yx_{i+1/2}}^{n} - a_{1_{i-1/2}} g_{yx_{i-1/2}}^{n} \right) \right]$$
$$+ C_2 \Psi_{e_i}^{n} + C_1 Q_{z_i}^{n} \quad (25)$$

where

$$r_i = B_{0_i} + B_{1_i}, \ \Lambda = \zeta^2 C_1 r_i, \ A_i^{\pm} = \Lambda r_{i\pm 1/2}, \text{ and } A_i = 1 + A_i^+ + A_i^- \qquad (26)$$

It should be pointed out that (25) forms a tri-diagonal matrix which can be solved easily [2]. After computing $E_z^{n+1/2}$, $Q_z^{n+1/2}$, $g_{zx}^{n+1/2}$, $H_y^{n+1/2}$, and $f_{zx}^{n+1/2}$ can be updated explicitly from (20), (17), (16) and (15), respectively. Similar formulations can be obtained in the second ADI sub-iteration at the $n+1$ time step. Furthermore, the above formulations can be applied in the non-PML regions by setting only the coefficients a_1, b_0, and b_1 defined in (10) to unity. Finally, it must be noted that the coefficients used in (19) depend on the susceptibility of the computational domain. In this paper, the susceptibility of the medium is assumed to have a first and second order frequency dispersion, i.e., Debye and Lorentz type media.

2.1 Linear Debye Dispersion

The frequency dependent susceptibility of a linear Debye dispersion is given by

$$\chi(\omega) = \frac{\Delta \varepsilon}{1 + j\omega t_0} \qquad (27)$$

where $\Delta \varepsilon = \varepsilon_s - \varepsilon_\infty$, where ε_s is the static permittivity and t_0 is the relaxation time of the medium. In this case, the coefficients of (19) can be written

$$d_0 = \Delta \varepsilon p, \ d_1 = \Delta \varepsilon p,$$
$$e_0 = 1 + p, \text{ and } e_1 = -(1 - p) \qquad (28)$$

with

$$p = \frac{\Delta_t}{4t_0} \qquad (29)$$

2.2 Linear Lorentz Dispersion

The susceptibility of a linear Lorentz dispersion is given by the following frequency dependent function

$$\chi(\omega) = \frac{\Delta \varepsilon \omega_0^2}{\omega_0^2 + j2\delta\omega - \omega^2} \qquad (30)$$

where ω_0 is the resonance radial frequency, and δ is the damping constant. In this case, the coefficients of (19) can be written

$$d_0 = \Delta \varepsilon p^2, \ d_1 = 2\Delta \varepsilon p^2, \ d_2 = \Delta \varepsilon p^2,$$
$$e_0 = 1 + 2q + p^2, \ e_1 = -2(1 - p^2), \text{ and } e_2 = 1 - 2q + p^2 \qquad (31)$$

with

$$p = \frac{\omega_0 \Delta_t}{4}, \text{ and } q = \frac{\delta \Delta_t}{4} \qquad (32)$$

2.3 Stability and Dispersion Analysis

The analytical stability and dispersion analysis of the proposed scheme can be studied by following the technique introduced in [2]. Due to the limit of the space in this paper, the details of the theoretical proof are not shown here. However, an experiment was performed to numerically show the proposed scheme is unconditionally stable as described in the following section.

3 Simulation Study

To validate the proposed formulations, two numerical tests are included. In the first test, a z−polarized Gaussian pulse was used to excite a one-dimensional isotropic, homogeneous, lossy and electrically dispersive Debye medium which have the following parameters: $\varepsilon_\infty = 50$, $\varepsilon_s = 160$, $t_0 = 5.88 ns$, and $\sigma = 0.62 S/m$. This model was used to approximate the muscle tissue in the frequency range 100-915MHz. The excitation pulse was located at the center of the computational domain. The computational domain extends in the x−direction with the size of 100Δ, where $\Delta = 01.mm$, and terminated by eight additional PML layers with the parameters of PML[8, 2, 0.001%], as defined in [3]. In this test, the simulation was carried out for the first 32768 time steps.

Figure 1 shows the time response of the E_z field recorded 20 cells away from the excitation point as obtained with the proposed formulations for different values of CFL number (CFLN) defined as CFLN=$\Delta t/\Delta t_{max}^{FDTD}$ where Δt_{max}^{FDTD} is the maximum stability limit of the conventional FDTD algorithm. For the sake

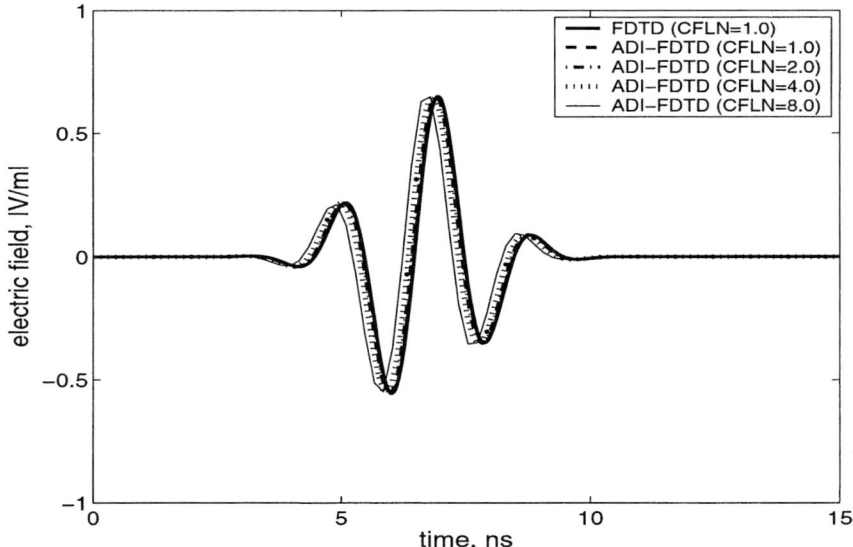

Fig. 1. Electric field as observed 20 cells from the excitation point

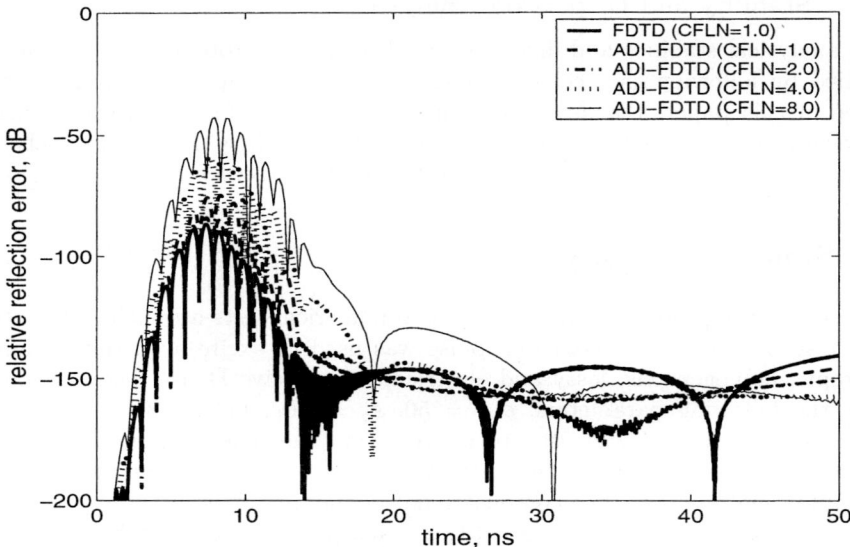

Fig. 2. Relative reflection error as a function of time as observed one space cell from the PML/ FDTD computational domain interface for Debye dispersive case

of comparison, the results obtained using the conventional PML implementation of (1)–(2) [10] for CFLN=1.0 is also shown in Fig. 1. As can be seen from Fig. 1, the proposed formulations remain stable beyond the stability limit. However, it must be noted that the accuracy of the proposed formulations decreases with the increase of the CFLN value. This is due to the increase of the numerical dispersion error of the ADI-FDTD method [4]. To measure the PML reflection errors a reference FDTD solution is needed. In this study, a larger domain with the size of $1000\Delta x$ and truncated by additional 32 PML layers with the parameters of PML[32, 4, 0.001%] was used. Figure 2 shows the relative reflection error as observed one cell away from the PML/ computational domain interface for the conventional PML formulations [10] and for proposed formulations with different values of CFLN. The relative reflection error was computed as

$$\mathcal{R}_{dB} = 20\log_{10}\frac{|E_z^R(t) - E_z^T(t)|}{\max{[|E_z^R(t)|]}} \qquad (33)$$

where $E_z^T(t)$ is the field computed using the test domain, and $E_z^R(t)$ is the reference field computed using the larger domain. The reference field was computed for each CFLN value in order to isolate the PML reflection errors from the grid dispersion error of the ADI-FDTD scheme [4]. Similar to the results reported in [4], it is apparent from Fig. 1 that the relative reflection error of the proposed formulations increases as the CFLN value increases. Figure 3 shows the frequency spectrum of both the PML reflection coefficient and the excitation

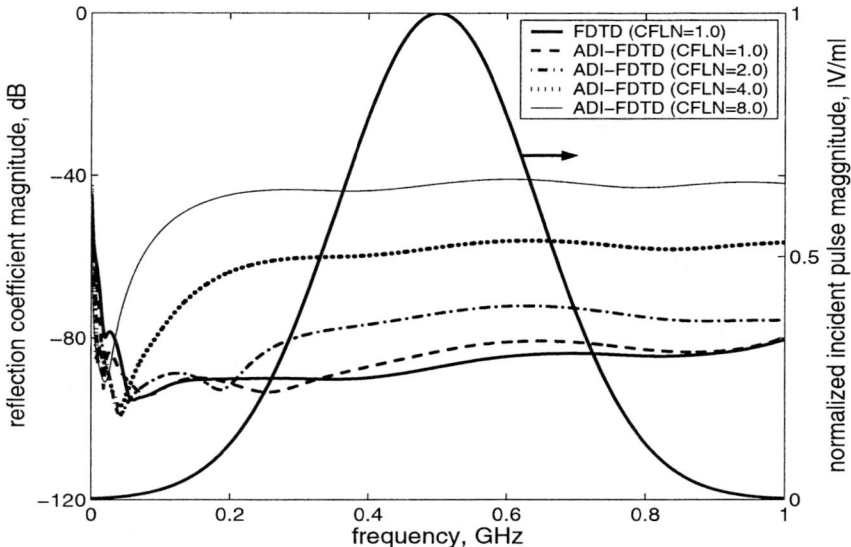

Fig. 3. Spectrum of the reflection coefficient magnitude as observed one space cell from the PML/ FDTD computational domain interface for Debye dispersive case

source used in this test. The reflection coefficient was computed one cell away from the PML/computational domain interface as

$$\mathcal{R}_{dB}(f) = 20 \log_{10} \left| \frac{\mathcal{F}\left\{ E_z^R(t) - E_z^T(t) \right\}}{\mathcal{F}\left\{ E_z^R(t) \right\}} \right| \tag{34}$$

where $\mathcal{F}\{.\}$ is the Fourier transform operation. As can be seen from Fig. 3, the reflection coefficient of the proposed scheme increases with the CFLN value. From the previous results, it can be noted that the time step of the proposed formulations is not restricted by the CFL stability limit, but by the required accuracy level.

In the second test, the FDTD domain was assumed to be a linear Lorentz media with parameters: $\varepsilon_\infty = 1$, $\varepsilon_s = 2.25$, $\omega_0 = 4 \times 10^{16} rad/s$, and $\delta = 0.28 \times 10^{16} s^{-1}$. The space cell size was chosen as $\Delta = 1 \times 10^{-10}$ and the computational domain was terminated by eight additional PML layers with the parameters of PML[8, 2, 0.001%]. Figure 4 shows the frequency spectrum of both the PML reflection coefficient and the excitation source used in this test. Similar to the results reported a single above, the reflection coefficient of the proposed scheme increases with the CFLN value. Also, it must be noted there is a sharp increase in the reflection error near the resonance frequency. However, this is not important because this frequency range is strongly attenuated for the chosen dielectric parameters.

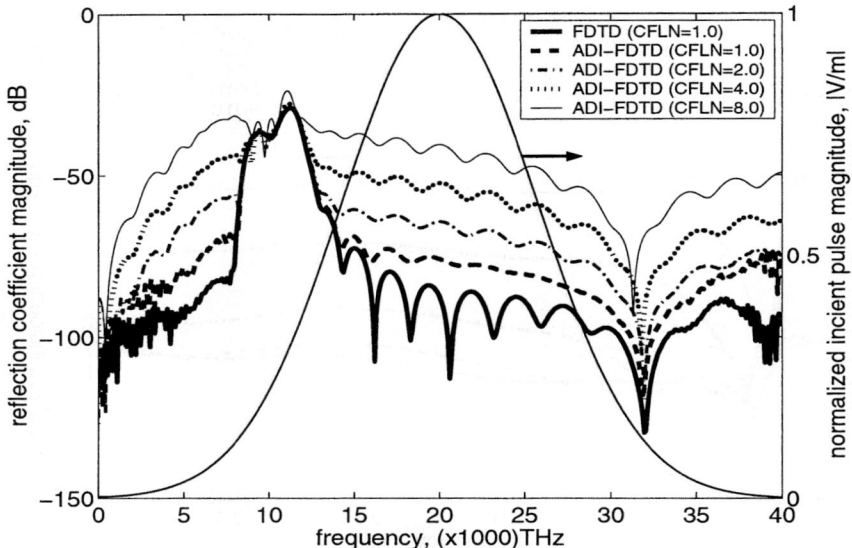

Fig. 4. Spectrum of the reflection coefficient magnitude as observed one space cell from the PML/ FDTD computational domain interface for Lorentz dispersive case

4 Conclusion

In this paper, the DSP algorithms developed for digital filters are incorporated with the ADI-PML formulations to truncate linear Debye and Lorentz dispersive FDTD domains. Numerical examples carried out in one dimension show that the formulations are unconditionally stable, i.e., the time step can be increased beyond the stability limit. This will allow decreasing the simulation time, but at the cost of the accuracy due to the increase of the numerical dispersion error of the ADI-FDTD method. Hence, the time step of the proposed formulations is not restricted by the CFL stability limit, but by the required accuracy level. In a similar manner, the proposed formulations can be extended to the two and three dimensional cases.

References

1. Taflove, A.: Computational electrodynamics: The Finite-Difference Time-Domain Method. Artech House, Boston, London (1995).
2. Namiki, T.: A new FDTD algorithm based on alternating-direction implicit. IEEE Transactions on Microwave Theory and Techniques **47** (1999) 2003-2007
3. Berenger, J.-P.: A perfectly matched layer for the absorption of electromagnetic waves. Journal of Computational Physics **114** (1994) 185-200
4. Gedney, S.D., Liu, G., Roden, J.A., and Zhu, A.: Perfectly matched layer media with CFS for an unconditionally stable ADI-FDTD method. IEEE Transactions on Antennas and Propagation **49** (2001) 1554-1559

5. Chew W.C., and Weedon W.H.: A 3-D perfectly matched medium from modified Maxwell's equations with stretched coordinates. Microwave and Optical Technology Letters **7** (1994) 599-604
6. Gedney S.D.: An anisotropic perfectly matched layer absorbing medium for the truncation of FDTD lattices. IEEE Transactions on Antennas and Propagation **44** (1996) 1630-1639
7. Ramadan, O.: Unconditionally stable ADI-FDTD implementation of PML for frequency dispersive debye media. Electronics Letters **40** (2004) 230-232
8. Ramadan, O.: Unconditionally stable ADI-FDTD formulations of the APML for frequency-dependent media. IEEE Microwave and Wireless Components Letters **14** (2004) 537-539
9. Proakis, J.G., and Manolakis, D.G.: Digital signal processing: principles, algorithms and applications. Prentice Hall, International Editions, third edition (1995)
10. Ramadan, O., Oztoprak, A.Y.: DSP techniques for implementation of perfectly matched layer for truncating FDTD domains. Electronics Letters **38**, (2002) 211-212

Mobile Agent Based Adaptive Scheduling Mechanism in Peer to Peer Grid Computing

SungJin Choi[1], MaengSoon Baik[1], ChongSun Hwang[1],
JoonMin Gil[2], and HeonChang Yu[3]

[1] Dept. of Computer Science and Engineering, Korea University,
5-1 Anam-dong, Sungbuk-gu, Seoul 136-701, Republic of Korea
{lotieye, msbak, hwang}@disys.korea.ac.kr
[2] Korea Institute of Science and Technology Information (KISTI)
jmgil@kisti.re.kr yuhc@comedu.korea.ac.kr
[3] Dept. of Computer Science Education, Korea University
yuhc@comedu.korea.ac.kr

Abstract. In a peer to peer grid computing environment, volunteers are exposed to failures such as crash and link failures. In addition, since volunteers can dynamically join and leave executions and they are not dedicated only to a peer to peer grid computing, the executions of volunteers are stopped or suspended more frequently than in a grid computing environment. These failures result in the delay and blocking of the executions of tasks and even partial or entire loss of the executions. In addition, these failures make it difficult for a volunteer server to schedule tasks and manage the allocated tasks as well as volunteers. Existing peer to peer grid computing systems, however, do not deal with these failures in scheduling mechanisms. Moreover, since existing scheduling mechanisms are performed only by a volunteer server in a centralized way, there is a high overhead.

To solve these problems, we propose a mobile agent based adaptive scheduling mechanism (MAASM). We implemented MAASM in Korea@Home and ODDUGI mobile agent system. The MAASM reduces the overhead of volunteer server by using mobile agents in scheduling procedure in a distributed way. In addition, it tolerates the various failures(especially, *volunteer autonomy failures*) which frequently occur in a peer to peer grid computing environment. Consequently, MAASM guarantees reliable and continuous executions in spite of the failures, so it decreases total execution time.

1 Introduction

A grid computing system is a platform that provides the access to various computing resources owned by institutions by making virtual organization[5]. On the other hand, a peer to peer grid computing system is a platform that achieves a high throughput computing by harvesting a number of idle desktop computers owned by individuals(which is called volunteers) on the edge of Internet using peer to peer computing technologies [1, 2, 3, 4, 5, 6, 7, 8]. The peer to peer grid computing systems usually support embarrassingly parallel applications which consist of a lot of instances of the same computation with each own data. The applications are usually involved with scientific problems which need large amounts of processing capacity over long periods of time. In recent years, there has been a rapidly growing interest in peer to peer grid computing systems because of the success of the most popular examples such as SETI@Home [1], distributed.net [2].

A peer to peer grid computing environment mainly consists of clients, volunteers and volunteer servers. A client is a parallel job submitter. A volunteer is a resource provider which donates its computing resources. A volunteer server is a central manager which controls submitted jobs and volunteers. A client submits a parallel job to a volunteer server. The job is divided into sub-jobs which have each own input data. The sub-job is called a *task* in this paper. A task consists of a parallel code and data. The volunteer server allocates the tasks to volunteers by using scheduling mechanisms. Each volunteer executes its task during idle time while continuously requesting data to the volunteer server. When each volunteer finishes the task, it returns the result of the task to the volunteer server. Finally, the volunteer server returns the final result of the job back to the client.

Since peer to peer grid computing is based on desktop computers at the edge of Internet, volunteers are exposed to link and crash failures. In addition, volunteers are voluntary participants, so they can freely join and leave in the middle of the executions without any constraints. Thus, a *public execution*(i.e. the execution of a task as a volunteer) is stopped arbitrarily. Moreover, volunteers are not totally dedicated only to peer to peer grid computing, so public executions get temporarily suspended by a *private execution* (i.e. the execution of a private job as a personal user). In this paper, we regard these situations as *volunteer autonomy failures* because they lead to the delay and blocking of the execution of tasks and even partial or entire loss of the executions. The volunteer autonomy failures occur more frequently than in a standard grid environment because a peer to peer grid computing system is based on dynamic desktop computers.

Existing peer to peer grid computing systems, however, do not deal with volunteer autonomy failures in scheduling mechanisms. In addition, the volunteer autonomy failures make it difficult for a volunteer server to schedule tasks and manage the allocated tasks as well as volunteers. As a result, the overhead occurs. Moreover, in existing peer to peer grid computing systems, since the scheduling and management mechanisms are performed only by the volunteer server in a centralized way. In other words, the volunteer server schedules the tasks and manages all volunteers through direct connection with each volunteer. As a result, the overhead is increased more and more. Therefore, we should consider volunteer autonomy failures to guarantee reliable and continuous executions. In addition, we should handle scheduling and fault tolerance procedures in a distributed way.

In this paper, we propose a mobile agent based adaptive scheduling mechanism (MAASM). We make use of mobile agents which partially take over scheduling and fault tolerance procedures from the volunteer server in order to reduce the overhead. That is, the scheduling and fault tolerance mechanisms are implemented as a mobile agent. The mobile agents are distributed to volunteer groups which are constructed by volunteer properties such as *volunteering service time* and *volunteer availability*. Then, they process the scheduling and fault tolerance procedures in a distributed way without the direct control of their volunteer server. We implemented the MAASM in Korea@Home [13] and ODDUGI mobile agent system [14]. The MAASM reduces the overhead of volunteer server by using mobile agents in a distributed way. In addition, it can tolerate the volunteer autonomy failures which frequently occur in a peer to peer grid computing environment. Consequently, it guarantees reliable and continuous executions in spite of failures, so it decreases total execution time.

The rest of the paper is structured as follows. Section 2 describes why mobile agent is used. Section 3 describes MAASM in details. Section 4 presents the implementation issues and experimental results. Section 5 reviews related work which has been studied in this area. Section 6 concludes the paper.

2   Why Mobile Agent?

In existing peer to peer grid computing system, a volunteer server suffers from high overhead. The volunteer server maintains properties of volunteers such as CPU, memory, OS, location, and so on. According to the properties, the volunteer server has responsibility for scheduling in a centralized way. In addition, the volunteer server performs fault tolerant mechanism if volunteers fails. Since the scheduling mechanism is performed only by the volunteer server, various scheduling mechanisms are not performed at a time according to volunteer properties. To solve these problems, we make use of mobile agent technology.

There are some advantages in using mobile agent in a peer to peer grid computing environment.

1) *Several scheduling mechanisms can be performed at a time according to the properties of volunteers.* For example, various scheduling mechanisms can be implemented as mobile agents. If there are various volunteer groups, the best appropriate scheduling mobile agent is assigned to the volunteer group according to the property of volunteer group. Existing peer to peer grid computing system, however, cannot apply various scheduling mechanisms at a time because one scheduling mechanism is performed only by a volunteer server.

2) *A mobile agent can decrease the overhead of volunteer server by performing scheduling and fault tolerance procedures in a decentralized way.* The scheduling mobile agents are distributed to volunteer groups. After that, they perform each scheduling and fault tolerance procedures in each volunteer group. Therefore, the volunteer server does not undergo the overhead any more.

3) *A mobile agent can adapt to a dynamical peer to peer grid computing environment.* In a peer to peer grid computing environment, volunteers can join and leave at any time. The scheduling mobile agent can tolerate the volunteer autonomy failures by using migration and replication functionalities which the mobile agent itself provides.

3 Mobile Agent Based Adaptive Scheduling Mechanism

In this section, we describe the mobile agent based adaptive scheduling mechanism (MAASM) in detail. The MAASM is categorized into four phases : 1) *Constructing volunteer groups*, 2) *Allocating scheduling mobile agents to volunteer groups*, 3) *Distributing task mobile agents to group members*, 4) *Handling failures*.

3.1 Constructing Volunteer Groups

Classifying Volunteers. When classifying volunteers, their CPU and memory capacities are important factors. The most important factors, however, are location, volunteering time, and volunteer availability because a peer to peer grid computing systems are based on dynamic desktop computers. In a peer to peer grid computing environment, the capacities of desktop computers are similar, while the volunteering time and availability are very various [10]. Therefore, the computation time is more affected by the latter factors. In this paper, we classify the volunteers according to location, volunteering time, and volunteer availability. The volunteering time and volunteer availability are defined as follows.

Definition 1 (Volunteering Time). *Volunteering time (Υ) is the period when a volunteer is supposed to donates its resources.*

$$\Upsilon = \Upsilon_R + \Upsilon_S$$

Here, the *reserved volunteering time*(Υ_R) is reserved time when a volunteer provides its computing resources. Volunteers mostly perform public execution during Υ_R, rarely private execution. On the other hand, the *selfish volunteering time* (Υ_S) is unexpected volunteering time. Thus, volunteers usually perform private execution during the Υ_S, sometimes public execution.

Definition 2 (Volunteer Availability). *Volunteer availability (α_v) is the probability that a volunteer will be operational correctly and be able to deliver the volunteer services during volunteering time Υ*

$$\alpha_v = \frac{MVT}{MVT + MTTVAF}$$

Here, MVT means "mean volunteering time" and $MTTVAF$ means "mean time to volunteer autonomy failures". The α_v reflects the degree of volunteer autonomy failures, while traditional availability in distributed systems mainly is related with crash failure. Although the MVT usually reflects both Υ_R and Υ_S, we consider only Υ_R in this paper because most tasks are executed during Υ_R.

Volunteers are categorized into region volunteers or home volunteers according to their location. *Home volunteers* are referred to as resource donators at home. *Region volunteers* are a set of resource donators who generally are affiliated with organizations such as university, institution, and so on. Region volunteers are connected with LAN or Intranet, while home volunteers are connected with WAN or Internet.

Volunteers are categorized into four classes (i.e. A, B, C, D classes) according to Υ and α_v like Fig. 1 (a).

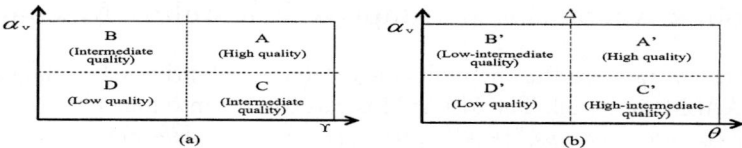

Fig. 1. The classification of volunteers and volunteer groups

Classifying and Making Volunteer Groups. A volunteer server selects volunteers as volunteer group members according to the properties of volunteers such as location, volunteer availability, and volunteering service time.

Volunteer service time is defined as follows.

Definition 3 (Volunteering Service Time). *Volunteering service time (Θ) is the expected service time when a volunteer participates in public execution during Υ*

$$\Theta = \Upsilon \times \alpha_v$$

In scheduling, Θ is more appropriate than Υ because Θ represents the time when a volunteer actually executes each task in the presence of Λ. Therefore, volunteer groups are constructed according to Θ, not Υ.

If volunteer groups are constructed on the basis of the location, region volunteers belong to the same group and home volunteers are formed into the same group in order to reduce communication cost between members.

If both α_v and θ are considered in grouping volunteers, volunteer groups are categorized into four classes like Fig. 1 (b). Here, Δ is the expected computation time of a task. The class A' and C' mainly execute the tasks, while the class B' and D' are used as replication or migration places when A' and C' are suffering from failures.

Fig. 2 shows the algorithm of volunteer group construction.

```
// To classify the registered volunteers(V) into home or region volunteers
ClassifyVolunteersByLocation(V);
// To classify the home and region volunteers into A, B, C, D classes, respectively
ClassifyVolunteers(V);
// To construct volunteer groups
if (V_i.Θ ≥ Δ) then     // V_i : one of the classified volunteers
    if (V_i ∈ V_A||V_i ∈ V_B) then    // V_A : A class, V_B: B class
        V_i → VG_A';    // → : assign, VG_A' : A' class
    else    // V_i ∈ V_C||V_i ∈ V_D, here, V_C : C class, V_D : D class
        V_i → VG_C';    // VG_C' : C' class
else    // V_i.Θ < Δ
    if (V_i ∈ V_A||V_i ∈ V_B) then
        V_i → VG_B';    // VG_B' : B' class
    else
        V_i → VG_D';    // VG_D' : D' class
```

Fig. 2. Algorithm of volunteer group construction

```
// DVS : deputy volunteer set
// CDVS : candidate deputy volunteer set
// TDVS : temporal deputy volunteer set
// CDVS ⊂ VG_{A'}
TDVS = OrderedBy(CDVS.α_v);
DVS = OrderedBy(TDVS.(Θ+HC+NB));  // HC : harddisk capacity, NB : network bandwidth
DV = PopBestDV(DVS) // Pop the best deputy volunteer from DVS
```

Fig. 3. Algorithm of deputy volunteers selection

3.2 Allocating Scheduling Mobile Agents to Scheduling Groups

After constructing volunteer groups, a volunteer server allocates scheduling mobile agents(S-MA) to volunteer groups. However, it is not practical to allocate S-MAs directly to the volunteer groups in a scheduling procedure because some volunteer groups are not perfect for finishing the tasks reliably. Therefore, we need making new scheduling groups by combining the volunteer groups each other : A'D' & C'B', A'B' & C'D', and A'C' & B'D'. In this paper, we consider the first combination in scheduling because B' volunteer group compensates for C' volunteer group with regard to volunteer availability.

The S-MA is executed at a deputy volunteer. The deputy volunteer is selected by the algorithm like Fig. 3.

3.3 Distributing Task Mobile Agents to Group Members

After the S-MAs are allocated to scheduling groups, each S-MA distributes task mobile agents(T-MA) which consist of parallel code and data to the members of scheduling group. According to the type of scheduling groups, the S-MAs perform different scheduling and fault tolerance mechanisms other than existing peer to peer grid computing systems. Fault tolerance mechanisms are described in the next subsection.

The S-MA of A'D' scheduling group performs scheduling algorithm as follows.

1) *Order A' volunteer group by $α_v$ and then by $Θ$.* 2) *Distribute T-MA to members of the A' volunteer group.* 3) *If T-MA fails, replicate the failed task new volunteers selected in D' volunteer group.*

The S-MA of C'B' scheduling group performs scheduling algorithm as follows.

1) *Order C' and B' volunteer groups by $α_v$ and then by $Θ$.* 2) *Distributes T-MA to members of C' volunteer group.* 3) *If T-MA fails, replicate or migrate the failed task to new volunteers selected in the ordered B' volunteer group.*

Tasks are distributed to firstly A'D' scheduling group and then C'B' one. In addition, the tasks are firstly distributed to the volunteers which have high $α_v$ and long $Θ$. In the scheduling algorithm, the task is not allocated to B' and D' volunteer groups, because they have no time enough to finish the task reliably. Therefore, they are usually used to assist the main volunteer groups(i.e. A' or C'). Especially, in the C'B' scheduling group, B' volunteer group compensates for C' volunteer group with regard to availability. In other words, if a volunteer in C' volunteer group suffers from volunteer autonomy failures, volunteers in B' volunteer group are used to tolerate the failures.

3.4 Handling Failures

We describe how the scheduling mobile agent and task mobile agent works in the presence of failures in this subsection.

Failure Model. We assume that volunteers suffer from crash, link, and volunteer autonomy failures. Especially, volunteer autonomy failures occur much more often than crash and link failures in a peer to peer grid computing environment.

The peer to peer grid computing system respects the autonomy of volunteers. In other words, volunteers can leave arbitrarily in the middle of public execution, and volunteers are allowed to execute private execution at any time while interrupting the public execution. The former is referred to as *volunteer volatility failure*(Φ), the latter *volunteer interference failure*(Ψ). We call these failures as *volunteer autonomy failures* (Λ) because the public execution is stopped and suspended. Φ is related to the completion of public execution. Ψ is related to the continuity of public execution.

Φ and Ψ are different from crash failure in that the operating system is alive in the presence of Φ and Ψ, while the system shuts down in the presence of crash failure [11, 12]. Φ is different from crash failure in that Φ occurs by the will of volunteer [11, 12]. Ψ is different from Φ in that a peer to peer grid computing system is alive in the presence of Ψ, while it is not operating in case of Φ.

The S-MA sends alive message to the volunteer server. Similarly, the T-MA sends alive message to the S-MA. The T-MAs in D' volunteer group do not send alive message in order to reduce the management overhead of D' volunteer group. The volunteer server detects the crash failure of S-MA by using timeout. Similarly, the S-MA detects the crash failure of T-MA. Volunteers can detect volunteer autonomy failures by oneself because its operating system does not shut down.

Failure of S-MA. S-MAs rarely suffer from the volunteer autonomy failures because they are executed at the deputy volunteers which are selected among A' volunteer groups. The S-MA stores information such as scheduling group lists, scheduling table, and task results in a stable storage. If S-MA fails, the information is sent to new deputy volunteer.

Fig. 4 shows the fault tolerant algorithm. In the algorithm, in case of volunteer interference failure, the S-MA does not take any actions because it can perform scheduling in that the peer to peer grid computing system is alive.

Failure of T-MA. T-MAs suffer from volunteer autonomy failures more often than S-MA, because they have low availability relatively. The T-MA checkpoints the execution state at the rate of $\frac{1}{V.\alpha_v}$ in order to reduce the overhead.

Fig. 5 shows the fault tolerant algorithm. In the algorithm, there is no fault tolerant mechanism for D' volunteer group in the presence of failures during the execution in order to reduce management overhead. Since A' volunteer group has enough availability and volunteering service time, it can finish tasks without the help of D' volunteer group. Therefore, the D' volunteer group executes tasks for test such as calculating availability or volunteering time.

```
[ If Volunteer server(VS) detects the crash failure of S-MA ]
V_m = SelectDeputyVolunteer(A');
SendS-MA'(V_m);  // send S-MA'(the latest checkpointed S-MA) to V_m
[ If Φ occurs ]
// At the S-MA side
S-MA' = Checkpoint(S-MA);
Send VolatilityFailure message to VS;
Send S-MA' to VS;
// At the volunteer server side
if (VS is informed of the volunteer volatility failure) then
    if (VS does not received the Rejoin message within the interval) then
        V_m = SelectDeputyVolunteer(A');
        SendS-MA'(V_m);
```

Fig. 4. Fault tolerant algorithm in presence of failures of S-MA

```
[ If S-MA detects the crash failure of T-MA ]
if (T-MA ∈ A' volunteer group) then
    V_m = SelectNewVolunteer(A');
else if (T-MA ∈ C' volunteer group) then
    V_m = SelectNewVolunteer(C');
else if (T-MA ∈ B' volunteer group) then
    V_m = SelectNewVolunteer(B');
SendT-MA'(V_m);  // send T-MA'(the latest checkpointed T-MA) to V_m
[ If Φ occurs ]
// At the T-MA side
T-MA' = Checkpoint(T-MA);
if (T-MA ∈ A' || T-MA ∈ C' || T-MA ∈ B') then
    Send VolatilityFailure message to S-MA;
    Send T-MA' to S-MA;
// At the S-MA side
if (S-MA is informed of the volatility failure) then
    if (S-MA does not receive the Rejoin message within the interval) then
        if (T-MA ∈ A') then
            V_m = SelectNewVolunteer(D');
        else if (T-MA ∈ C' || T-MA ∈ B') then
            V_m = SelectNewVolunteer(B');
        SendT-MA'(V_m);
[ If Ψ occurs ]
// At the T-MA side
Checkpoint(T-MA);
if (T-MA is not restarted within the interval) then
    if (T-MA ∈ A' || T-MA ∈ C' || T-MA ∈ B') then
        Send InterferenceFailure message to S-MA;
    if (T-MA receives the candidate volunteer V_m) then
        ShadowMigrateT-MA(V_m);
// At the S-MA side
if (S-MA receives InterferenceFailure message) then
    if (T-MA ∈ A' volunteer group) then
        V_m = SelectNewVolunteer(D');
    else if (T-MA ∈ C' || T-MA ∈ B') then
        V_m = SelectNewVolunteer(B');
    SendCandidateVolunteer(V_m);
```

Fig. 5. Fault tolerant algorithm in presence of failures of T-MA

4 Implementation and Evaluation

4.1 Implementation

We implemented our scheduling mechanism in the "Korea@Home" [13] and "ODDUGI" mobile agent system [14]. Korea@Home project aims to harness a massive computing power by harvesting the number of desktop computers over the network. In addition, the ODDUGI which was developed by our laboratory is a mobile agent system that supports reliable, secure and fault tolerant execution of mobile agents. Fig. 6 shows the volunteer execution screen shot in Korea@Home and ODDUGI mobile agent system.

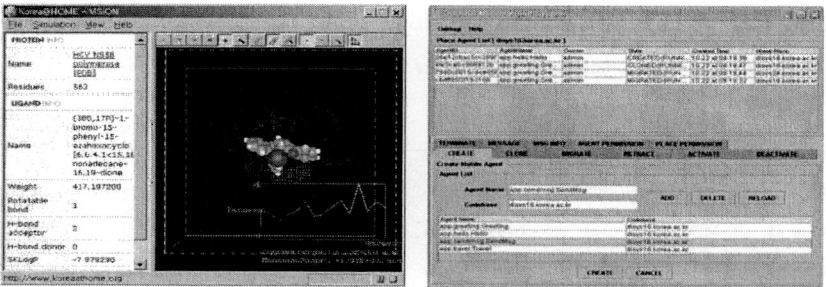

Fig. 6. Screen shot of Korea@Home and ODDUGI mobile agent system

4.2 Preliminary Experimental Results

Table 1 shows the experimental environment. A few volunteers are connected with 10MB/sec and most volunteers with 100MB/sec. The 210 volunteers have various volunteer availability like Table 1. Here, λ represents the volunteer autonomy failures rate of the system. τ represents $MTTVAF$. We evaluate the 210 volunteers according to $MTTVAF$, that is, 0.05, 0.1 and 0.3, respectively. The number of tasks is 2,000. The execution time of a task is 20 minutes. The volunteering time of each volunteer is 3 hours.

Fig. 7 shows the number of tasks according to volunteer availability. (a) and (c) show the total number of tasks which 30 volunteers which have different λ complete, while (b) and (d) show the success rate as well as the sum of tasks completed at all volunteers. In other words, Fig. 7 shows how volunteer availability affects the peer to peer grid computing. As the volunteer availability is lower, the number of completed tasks is decreased.

In Fig. 7, (c) and (d) show the results applied to our MAASM. On the other hand, (a) and (b) show the results performed in a centralized way. We found that the total number of tasks in (c) and (d) is larger than that in (a) and (b).

Table 1. Experimental Environment

λ(/ hour)	α_v ($\tau = 0.05$ hour)	α_v ($\tau = 0.1$ hour)	α_v ($\tau = 0.3$ hour)	# of volunteers
0.5	0.97	0.95	0.87	30
1	0.95	0.91	0.77	30
2	0.91	0.83	0.63	30
4	0.83	0.71	0.45	30
6	0.77	0.63	0.36	30
12	0.63	0.45	0.22	30
20	0.5	0.3	0.14	30
	mean=0.73	mean=0.64	mean=0.49	sum=210

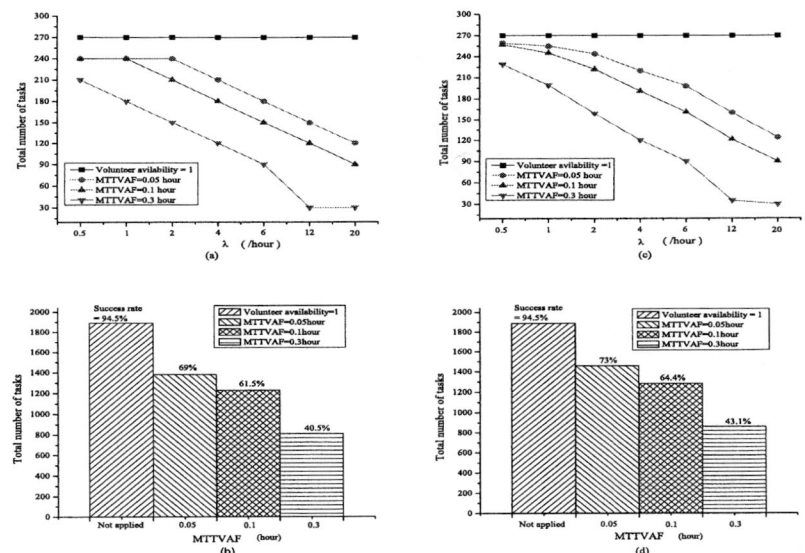

Fig. 7. Total number of tasks according to volunteer availability

The MAASM can complete about 3 ~ 4% more tasks because it reduces the overhead occurred in a centralized way. As a result, the MAASM can reduce the total execution time.

5 Related Work

AgentTeamwork [9] proposed a mobile agent based PC Grid middleware. Agent-Teamwork makes use of mobile agents for resource search or job migration in the presence of crash failure. AgentTeamwork is similar to our system in terms of using mobile agents. However, our system provides mobile agent based scheduling mechanism which reflects the volunteer autonomy failures differently from Agent-

Teamwork. Furthermore, our system provides volunteer groups based scheduling algorithm according to volunteering service time and volunteer availability.

In SETI@Home [1] project, the central server has a role of scheduling and management of volunteers, so it has high overhead. A volunteer takes a checkpoint periodically for the purpose of tolerating crash failure. However, SETI@Home does not handle volunteer autonomy failures. Furthermore, SETI@Home does not consider fault tolerant scheduling mechanism which reflects the volunteer availability and volunteer autonomy failures.

Bayanihan [6] and Javelin [7] provided fault tolerant scheduling mechanisms, that is, eager scheduling in the presence of crash failure. However, they do not deal with the volunteer autonomy failures. They reduce the overhead of central server by using the multiple servers, while our system makes use of mobile agents.

6 Conclusion and Future Work

We proposed a mobile agent based adaptive scheduling mechanism (MAASM) in a peer to peer grid computing environment. We implemented our adaptive scheduling mechanism in Korea@Home and ODDUGI mobile agent system. We showed that the MAASM has higher success rate of tasks because it considers volunteer availability and volunteer autonomy failures in a scheduling mechanism. In addition, the MAASM can reduce the overhead of volunteer server by using scheduling mobile agents according to volunteer groups in a distributed way. In the future, we will study and implement computational replication strategies according to volunteer autonomy failures, volunteer availability and volunteering service time.

Acknowledgment

This work was supported by the Korea Institute of Science and Technology Information.

References

1. SETI@home, "http://setiathome.ssl.berkeley.edu"
2. Distributed.net, "http://distributed.net"
3. D. S. Milojicic, V. Kalogeraki, R. Lukose, K. Nagaraja, J. Pruyne, B. Richard, S. Rollins, and Z. Xu, "Peer-to-Peer Computing", HP Laboratories Palo Alto HPL-2002-57, March 2002.
4. Ian Foster and Adriana Iamnitchi, "On Death, Taxes, and the Convergence of Peer-to-Peer and Grid Computing", 2nd International Workshop on Peer-to-Peer Systems (IPTPS'03), February 2003.
5. F. Berman, G. C. Fox, and A. J. G. Hey, "Grid Computing : Making the Global Infrastructure a Reality", Wiley, 2003

6. L. F. G. Sarmenta, S. Hirano. "Bayanihan: Building and Studying volunteer computing Systems Using Java", Future Generation Computer Systems Special Issue on Metacomputing, Vol. 15, No. 5/6., 1999.
7. M. O. Neary, S. P. Brydon, P. Kmiec, S. Rollins, and P. Cappello, "Javelin++: Scalability Issues in Global Computing", Concurrency: Parctice and Experience, pp. 727-735, December 2000.
8. G. Fedak, C. Germain, V. Neri, and F. Cappello, "XtremWeb: A Generic Global Computing System", CCGrid'01 workshop on Global Computing on Personal Devices, pp. 582 -587, May 2001.
9. M. Fukuda, Y. Tanaka, N. Suzuki and L. F. Bic, "A Mobile-Agent-Based PC Grid", Autonomic Computing Workshop AMS'03, pp. 142-150, June 2003.
10. D. Kondo, M. Taufer, J. Karanicolas, C. L. Brooks, H. Casanova and A. Chien, "Characterizing and Evaluating Desktop Grids: An Empirical Study", IPDPS'04, April 2004.
11. P. Jalote, "Fault Tolerance in Distributed Systems", Prentice-Hall, 1994
12. A. S. Tanenbaum and M. V. Steen, "Distributed Systems: Principles and Paradigms", Prentice Hall, 2002.
13. Korea@Home, http://www.koreaathome.org/eng/
14. ODDUGI mobile agent system, http://oddugi.korea.ac.kr/
15. M. Baik, S. Choi, C. Hwang, J. Gil, H. Yu, "Adaptive Group Computation Approach in the Peer-to-peer Grid Computing Systems", Concurrency and Computation: Practice and Experience, 2005

Comparison of Global Optimization Methods for Drag Reduction in the Automotive Industry

Laurent Dumas[1], Vincent Herbert, and Frédérique Muyl[2]

[1] Laboratoire Jacques-Louis Lions,
Université Pierre et Marie Curie,
75252 Paris Cedex 05, France
dumas@ann.jussieu.fr,
[2] PSA Peugeot Citroën
2 route de Gisy, 78943 Vélizy Villacoublay, France
{vincent.herbert, frederique.muyl}@mpsa.com

Abstract. Various global optimization methods are compared in order to find the best strategy to solve realistic drag reduction problems in the automotive industry. All the methods consist in improving classical genetic algorithms, either by coupling them with a deterministic descent method or by incorporating a fast but approximated evaluation process. The efficiency of these methods (called HM and AGA respectively) is shown and compared, first on analytical test functions, then on a drag reduction problem where the computational time of a GA is reduced by a factor up to 7.

1 Introduction

The topic of drag reduction in the automotive industry has been extensively studied since many years because of its great importance in terms of fuel consumption reduction. However, a computational and automatic approach of this problem has been unreachable until recently because of its difficulty due to the main two reasons: the complexity of the cost function to minimize and the computation time of each evaluation. The first attempt in this direction has been presented, up to our knowledge, by the present authors in [1]: this article was describing the drag minimization of a simplified 3d car shape with a global optimization method that coupled a genetic algorithm and a second order BFGS method. The interest of such hybrid method had been clearly shown on analytic cases where the convergence speed up was spectacular compared to a classical genetic algorithm. Unfortunately, the improvement for the industrial case was not so important because of the lack of accuracy of the gradient computation and so forth of the hessian approximation. The present article is intended to go further in this direction by performing a large comparison of different strategies to enhance the convergence of genetic algorithms, either with a hybrid method or with an approximated evaluation process. All these global optimization methods, described in paragraph 2, are compared on two analytic cases in paragraph 3 and on an industrial problem of car drag reduction in paragraph 4.

2 Global Optimization Methods

There exists many methods for minimizing a cost function J defined from a set $\mathcal{O} \subset \mathbb{R}^n$ to $\mathbb{R}_+$. Among them, the class of genetic algorithms (GA), which main principles are recalled in the next subsection, has the major advantage to seek for a global minimum. Unfortunately, this method is very time consuming because of the large number of cost function evaluations that are needed. All the hybrid optimization methods presented in subsection 2.2 greatly reduce this time cost by coupling a GA with a deterministic descent method. Another way to speed up the convergence of a GA is described in subsection 2.3 and consists in doing fast but approximated evaluations during the optimization process. In this last case, an improvement of an existing approximation model is proposed.

2.1 Genetic Algorithms (GA)

Genetic algorithms are global optimization methods directly inspired from the Darwinian theory of evolution of species ([2]). They consist in following the evolution of a certain number N_p of possible solutions, $(x_i)_{1 \leq i \leq N_p} \in \mathcal{O}^{N_p}$, also called population. To each element (or individual) x_i of the population is affected a fitness value inversely proportional to $J(x_i)$ in case of a minimization problem. The population is regenerated N_g times by using three stochastic principles called selection, crossover and mutation, that mimic the biological law of the 'survival of the fittest'. These principles are applied in the following way: at each generation, $\frac{N_p}{2}$ couples are selected by using a roulette wheel process with respective parts based on the fitness rank of each individual in the population. To each selected couple, the crossover and mutation principles are then successively applied with a respective probability p_c and p_m. The crossover of two elements consists in creating two new elements by doing a barycentric combination of them with random and independent coefficients in each coordinate. The mutation principle consists in replacing a member of the population by a new randomly chosen in its neighborhood. A one-elitism principle is added in order to be sure to keep in the population the best element of the previous generation. Thus, the algorithm writes as:

- Randomly choose the initial population $P_1 = \{x_{i,1} \in \mathcal{O}, 1 \leq i \leq N_p\}$
- $n_g = 1$. Repeat until $n_g = N_g$
  - Evaluate $\{J(x_{i,n_g}), 1 \leq i \leq N_p\}$ and $m = \min\{J(x_{i,n_g}), 1 \leq i \leq N_p\}$
  - 1-elitism: if $n_g \geq 2$ & $J(X_{n_g-1}) < m$ then $x_{i,n_g} = X_{n_g-1}$ for a random i
  - Affect a fitness value to each element. Call X_{n_g} the best element
  - for k from 1 to $\frac{N_g}{2}$
    - Selection of $(x_{\alpha,n_g}, x_{\beta,n_g})$ with respect to the fitness value
    - with probability p_c: replace $(x_{\alpha,n_g}, x_{\beta,n_g})$ by $(x'_{\alpha,n_g}, x'_{\beta,n_g})$ by crossover
    - with probability p_m: replace $(x'_{\alpha,n_g}, x'_{\beta,n_g})$ by $(x''_{\alpha,n_g}, x''_{\beta,n_g})$ by mutation
  - end for
- $n_g = n_g + 1$. Generate the new population P_{n_g}

2.2 Hybrid Methods (HM)

The principle of hybrid optimization has been introduced in the previous decade ([3],[4],[5]) in order to improve the convergence speed of an evolutionary algorithm, such as a genetic algorithm, in the case of computationally expensive optimization problems under limited computational budget. The general idea is to couple a GA with a deterministic descent method which will explore more rapidly the local minima of the objective function. More recently, this idea has been used in conjunction with computationally cheap surrogate approximation models (see the next paragraph for more details). In the present approach, the local search is only performed starting from the best current element X_{n_g} at a generation n_g after a stagnation in the GA process has been observed during $N_{stag} \in \mathbb{N}$ generations. Moreover, this procedure is done with exact evaluations of the cost function. Thus, the algorithm of the previous subsection is modified in the following way:

- (Affect a fitness value to each element. Call X_{n_g} the best element)
- if $X_{n_g-N_{stag}-1} \neq X_{n_g}$ and $X_{n_g-N_{stag}} = ... = X_{n_g}$ then apply a descent method starting from X_{n_g}
- (for k from 1 to $\frac{N_g}{2}$)

Such hybrid method is called $\text{HM}_{N_{stag}}$. Note in particular that HM_0 consists to apply a descent method to the current best element of the population after each improvement in the GA process. In [1], a hybrid method of type HM_3 has been used.

The descent method can be of any type, first or second order (as in [1] when the BFGS algorithm has been used). In the present article, a first order gradient method with a backtracking line search strategy has been selected in order to limit the influence of the approximation eventually done in the gradient evaluation. A maximal number of five iterations in the descent method has also been fixed.

2.3 Genetic Algorithms with Approximated Evaluations (AGA)

Another idea to speed up the GA convergence when the computation time of the cost function $x \mapsto J(x)$ is high, is to take benefit of the large and growing data base of exact evaluations by making fast and approximated evaluations $x \mapsto \tilde{J}(x)$ leading to what is called surrogate or meta-models (see [6] and [7] for an overview). This general idea has also been used recently in conjunction with a hybrid process: in [8], a strategy for coupling an evolutionary algorithm with local search and quadratic response surface methods is proposed whereas a parallel hybrid evolutionary algorithm framework that leverages surrogate models for solving computationally expensive design problems with general constraints is presented in [9] and further extended. In the present work, the surrogate model is developed independently of the hybrid process and consists to perform exact evaluations only for all the best fitted elements of the population (in the sense of $\tilde{J}$) and for one randomly chosen element. The new algorithm, called AGA is thus deduced from the algorithm of section 2.1 and writes as:

- ($n_g = 1$. Repeat until $n_g = N_g$)
- if $n_g = 1$ then evaluate $\{J(x_{i,n_g}), 1 \leq i \leq N_p\}$
- elseif $n_g \geq 2$
-     for i from 1 to N_p
-         Evaluate $\tilde{J}(x_{i,n_g})$.
-         if $\tilde{J}(x_{i,n_g}) < J(X_{n_g-1})$ then evaluate $J(x_{i,n_g})$
-     end for
-     for a random i: evaluate $J(x_{i,n_g})$
- end elseif
- (1-elitism)

The interpolation method chosen here comes from the field of neural networks and is called RBF (Radial Basis Function) interpolation ([10]). Suppose that the function J is known on N points $\{T_i, 1 \leq i \leq N\}$, the idea is to approximate J at a new point x by making a linear combination of radial functions of the type:

$$\tilde{J}(x) = \sum_{i=1}^{n_c} \psi_i \Phi(||x - \hat{T}_i||)$$

where:

- $\{\hat{T}_i, 1 \leq i \leq n_c\} \subset \{T_i, 1 \leq i \leq N\}$ is the set of the $n_c \leq N$ nearest points to x for the euclidian norm $||.||$, on which an exact evaluation of J is known.
- Φ is a radial basis function chosen in the following set:

$$\Phi_1(u) = \exp(-\frac{u^2}{r^2})$$

$$\Phi_2(u) = \sqrt{u^2 + r^2}$$

$$\Phi_3(u) = \frac{1}{\sqrt{u^2 + r^2}}$$

$$\Phi_4(u) = \exp(-\frac{u}{r})$$

for which the parameter $r > 0$ is called the attenuation parameter.

The scalar coefficients $(\psi_i)_{1 \leq i \leq n_c}$ are obtained by solving the least square problem of size $N \times n_c$:

$$\text{minimize} \quad err(x) = \sum_{i=1}^{N} (J(T_i) - \tilde{J}(T_i))^2 + \lambda \sum_{j=1}^{n_c} \psi_j^2$$

where $\lambda > 0$ is called the regularization parameter.

In order to attenuate or even remove the dependancy of this model to its attached parameters, a secondary global optimization procedure (namely a classical GA) has been over-added in order to determine for each x, the best values (with respect to $err(x)$) of the parameters n_c, $r \in [0.01, 10]$, $\lambda \in [0, 10]$ and $\Phi \in \{\Phi_1, \Phi_2, \Phi_3, \Phi_4\}$. As this new step introduces a second level of global optimization, it is only reserved to cases where the time evaluation of $x \mapsto J(x)$ is many orders of magnitude higher than the time evaluation of $x \mapsto \tilde{J}(x)$, as in a car drag reduction problem.

3 Comparison of Global Optimization Methods on Two Analytic Cases

Before applying them on a real drag reduction problem, all the previous global optimization algorithms have been tested and compared on two analytic test functions. These functions have been constructed in order to exhibit two behaviors that are supposed to be representative of a large number of realistic optimization problems. The first one is a Rastrigin type function with 3 parameters, directly inspired from the original one with 2 arguments:

$$Rast(x) = \sum_{i=1}^{3} \left(x_i^2 - \cos(18x_i) \right) + 3$$

defined on $\mathcal{O} = [-2, 2]^3$ for which there exists many local minima (more than a hundred) and only a global minimum located at $x_m = (0, 0, 0)$ equal to 0.

The second test function is a modification of the classical Griewank function with 30 parameters:

$$Griew(x) = \sum_{i=1}^{30} \frac{x_i^2}{50} - \prod_{i=1}^{30} \cos(\frac{x_i}{\sqrt{i}})$$

defined on $\mathcal{O} = [-10, 10]^{30}$ which has also a unique global minima at the origin but only few local minima. In order to achieve a quasi-certain convergence with a simple genetic algorithm, the population number and the maximal generations number are respectively fixed to $(N_p, N_g) = (30, 300)$ for the Rastrigin function and $(N_p, N_g) = (100, 160)$ for the Griewank function. The crossover and mutation probability have been set to their best observed value in this case, that is $p_c = 0.3$ and $p_m = 0.9$ (see [11] for more details).

Figure 1 displayed below gives an example of convergence history for the Rastrigin and Griewank function respectively, with six different global optimization methods that have been previously presented in paragraph 2 (GA, HM_0, HM_2, HM_3, HM_4 and AGA). Note that each gradient computation is counted as $2n$ function evaluations where n is the dimension number of the search domain (3 or 30 here) and that each approximated evaluation of the function in the AGA algorithm is not counted as an evaluation. Due to the large number of curves displayed, the authors must apologize for the poor visibility on a black and white copy.

On this figure, it can be seen that each method that has been constructed gives better results than a simple GA on two different aspects: the first one is the computational time which is assumed to be directly related to the evaluation number of the function to minimize and the second one is the accuracy level that is reached at the end of the computation.

In order to give more quantitative results, a statistical study based on a set of 100 independent optimization processes has been realized. Table 1 gives the approximated average gain compared to a classical genetic algorithm (ie the evaluation number reduction rate for a given convergence level) that has been observed for three global optimization methods. Note that the results obtained

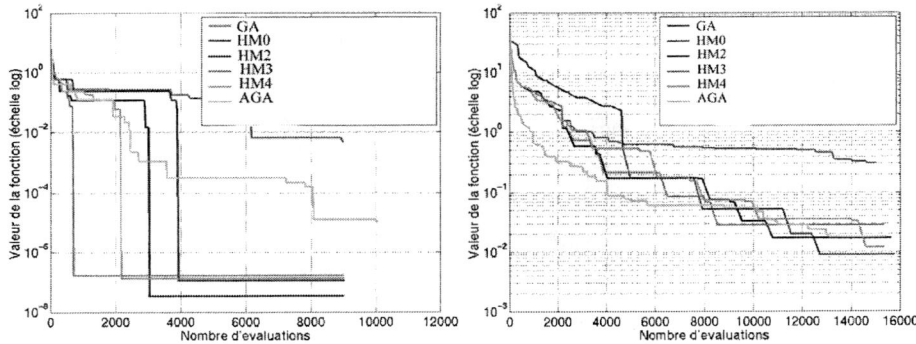

Fig. 1. Example of convergence history for the Rastrigin function (left) and the Griewank function (right) with 6 different global optimization methods

Table 1. Evaluation number reduction rate for HM_0, HM_3 and AGA compared to a simple GA for each test function

	Rastrigin function	Griewank function
HM_0	2	2
HM_3	10	4
AGA	4	10

for HM_1, HM_2 and HM_4 have not been displayed as they were not as good as those of HM_0 and HM_3.

This table confirms the interest of all the global optimization methods previously presented, but on a various degree. For example, it can be observed that HM_3 give better results than HM_0, which could not have been easily forecast before these tests. Thus, HM_3 and AGA seem to be the most promising ones in terms of convergence time reduction. One one hand, HM_3 appears to be more efficient than AGA on the Rastrigin function as the latter exhibits many local minima that can be rapidly tracked by the local search process. On the other hand, as the number of parameters increases, the AGA method performs better than the HM_3 method as it does not need any gradient computation which becomes very costly by finite differencies.

In view of their promising results, these global optimization methods are now used and compared in the next paragraph in the context of a car drag reduction problem.

4 Comparison of Global Optimization Methods for Car Drag Reduction

A classical 3D car drag reduction problem ([12]), already investigated in [1], has been extensively studied with the global optimization methods presented in

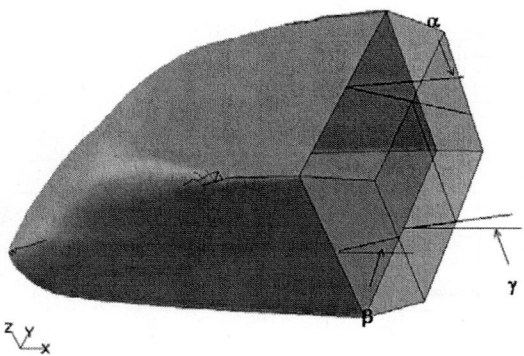

Fig. 2. 3D car shape parametrized by its three rear angles α, β and γ

paragraph 2. It consists in minimizing the drag coefficient (also called C_d) of a simplified car shape with respect to the three geometrical angles defining its rear shape (see Figure 2): the back light angle (called α), the boat-tail angle (β) and the ramp angle (γ).

The drag coefficient to be minimized is defined by the following expression:

$$C_d \equiv C_d(\alpha, \beta, \gamma) = \frac{F_x}{\frac{1}{2}\rho V_\infty^2 S}$$

where ρ is the mass density, S the front surface, V_∞ the freestream velocity and F_x the longitudinal component of the aerodynamic force exerted on the car. The latter is obtained after a 3D turbulent Navier Stokes computation around the car. This computation, very costly and sensitive to the car geometry, explains the major difficulty of such optimization problem.

In order to improve the first optimization attempt presented in [1], the search domain has been reduced: here, $(\alpha, \beta, \gamma) \in [15, 25] \times [5, 15] \times [15, 25]$ (degrees) and the aerodynamic computation is done with a finer grid, namely with a 6 million cells mesh. In this context, one C_d evaluation, done with a commercial CFD code, takes 14 hours CPU time on a single processor machine.

Three different types of global optimization methods have been compared on this problem. The first and reference type is a classical GA with a population number N_p equal to 20 and $(p_c, p_m) = (0.3, 0.9)$. The second type consists of hybrid methods, HM_0 and HM_3, as described in paragraph 2.2 with gradient evaluations computed by centered finite differences. The third type is a GA with fast and approximated evaluations (called AGA) with or without a secondary optimization of the interpolation parameters n_c, λ, r and Φ: see paragraph 2.3.

The convergence history of all these optimization methods for the present drag reduction problem is depicted in Figure 3. In order to achieve a reasonable computational time, parallel evaluations on a cluster of workstations have been done.

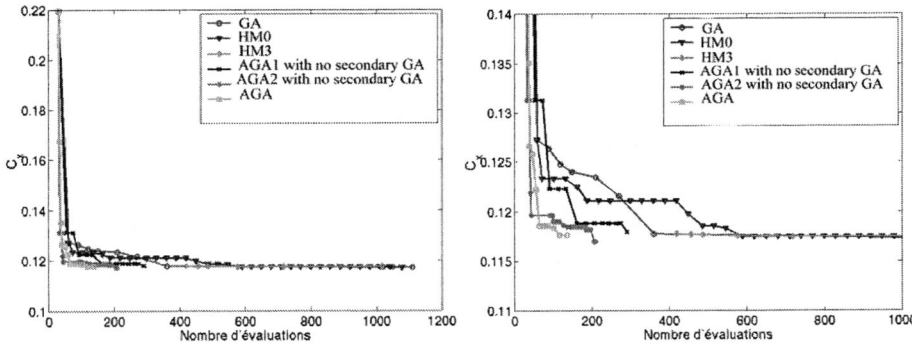

Fig. 3. Convergence history for drag reduction of a 3d car shape with six different global optimization methods (right figure: zoom of left figure)

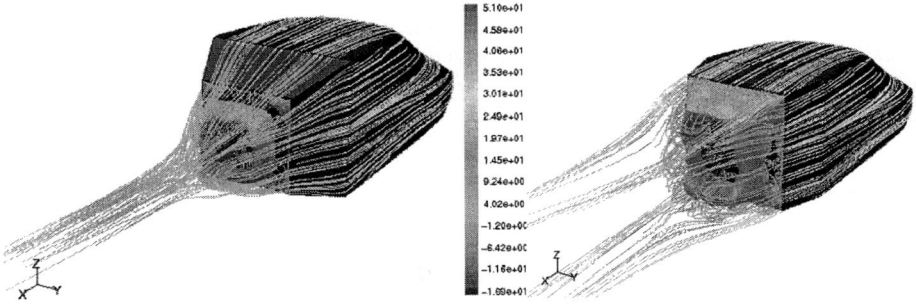

Fig. 4. Iso pressure streamlines coloured by the longitudinal speed for the lowest drag (left) and a high drag (right) car shape

This figure show in particular that all the methods have nearly reached the same drag value, around 0.117, starting with an almost double value, but with a different number of cost function evaluations. More precisely, the AGA algorithm has permitted to reduce this number by a factor 7 compared to a classical GA. Note that the AGA methods with a fixed set of parameters (called AGA1 and AGA2) exhibit a lower gain, which seems to justify the interest of a secondary optimization of the approximation parameters. On the other hand, the hybrid method HM_0 and HM_3 don't exhibit a significant improvement compared to the GA, likely because of the lack of accuracy in the gradient computation.

The optimal angles obtained by all the global optimization methods are nearly equal to $(\alpha, \beta, \gamma) = (17.7, 10, 18.4)$. These values have been experimentally confirmed to be associated with the lowest drag value that can be reached. The associated computational aerodynamic wake flow is depicted in Figure 4 and compared to an example of a high drag shape ($Cd = 0.22$).

It can be seen in particular that the optimized shape exhibits a narrow and regular recirculation volume behind the vehicle as predicted by many authors.

Note that such aerodynamic interpretation must be done on a 3D level and not from a longitudinal projection because of the real 'three dimensionality' of the flow.

5 Conclusion

In this article, two types of global optimization methods have been compared with the classical genetic algorithm method (GA). The first type, called hybrid methods (HM), consists in including a local search process for some well chosen individuals during the GA evolution. The second type, called AGA, incorporates a fast but approximated evaluation process for some individuals. For all the tested problems, the needed number of cost function evaluations to achieve global convergence has been largely decreased, between a factor 2 and 10 compared to a classical GA. For instance, in a classical car drag optimization problem, the AGA method with a new surrogate model has permitted to obtain the lowest drag car shape 7 times quicker than any other existing method. Such promising method is now ready to be applied to reduce the aerodynamic drag coefficient of more and more realistic car shapes with a larger number of parameters.

References

1. Muyl F., Dumas L. and Herbert V.: Hybrid method for aerodynamic shape optimization in automotive industry. Computers and Fluids, **33** (2004) 849-858
2. Goldberg D.E.: Genetic Algorithms in Search, Optimization, and Machine Learning. Addison-Wesley (1989)
3. Poloni C.: Hybrid GA for multi objective aerodynamic shape optimization. Genetic algorithms in engineering and computer science. John Wiley and Sons, **33** (1995) 397-415
4. Renders J.M. and Flasse S.P.: Hybrid methods using genetic algorithms for global optimization. IEEE Transactions on systems, man and cybernetics, **26** (1996) 243-258
5. Vicini A. and Quagliarella D.: Airfoil and wing design through hybrid optimization strategies. AIAA paper (1998).
6. Ong Y.S., Nair P.B., Keane A.J. and Wong K. W.: Surrogate-Assisted Evolutionary Optimization Frameworks for High-Fidelity Engineering Design Problems. Knowledge Incorporation in Evolutionary Computation. Studies in Fuzziness and Soft Computing Series, Springer Verlag (2004) 307-331
7. Jin Y.: A survey on fitness approximation in evolutionary computation. Journal of Soft Computing, **9** (2005) 3-12
8. Jin Y., Olhofer M. and Sendhoff B.: A framework for evolutionary optimization with approximate fitness functions. IEEE Transactions on Evolutionary Computation, **6** (2002) 481-494
9. Ong Y.S., Nair P.B. and Keane A.J., Evolutionary Optimization of Computationally Expensive Problems via Surrogate Modeling. AIAA Journal, **41** (2003) 687-696

10. Giannakoglou K.C.: Acceleration of GA using neural networks, theoretical background. GA for optimization in aeronautics and turbomachinery, VKI Lecture Series, (2000)
11. Muyl F.: Méthode d'optimisation hybrides: application à l'optimisation de formes arodynamiques automobiles. Phd thesis Université Paris 6 (2003)
12. Sagi C.J., Han T., Hammond D.C.: Optimization of bluff body for minimum drag in ground proximity. AIAA paper (1992)

Multiple Intervals Versus Smoothing of Boundaries in the Discretization of Performance Indicators Used for Diagnosis in Cellular Networks

Raquel Barco[1], Pedro Lázaro[1], Luis Díez[1], and Volker Wille[2]

[1] Departamento de Ingeniería de Comunicaciones. Universidad de Málaga,
Campus Universitario de Teatinos, 29071 Málaga. Spain
{rbm, plazaro, diez}@ic.uma.es
[2] Nokia Networks, Performance Services, Ermine Business Park,
Huntingdon, Cambridge PE29 6YJ, UK
Volker.Ville@nokia.com

Abstract. Most real-world applications of diagnosis involve continuous-valued attributes, which are normally discretized before the existing classification algorithms are applied. The discretization may be based on data or on human expertise. In cellular networks the number of classified examples is very limited. Thus, the diagnosis experts should specify the boundaries of the intervals for each discretized symptom. The large number of values makes it difficult to specify precise parameters. Even if boundaries are obtained from classified examples, due to the limited number of cases, the obtained values are not very accurate. In this paper two techniques to improve the performance of diagnosis systems based on Bayesian Networks are compared. Some empirical results are presented for diagnosis in a GSM network. The first method, Smooth Bayesian Networks, is shown to be more robust to imprecise setting of boundaries. The second method, Multiple Uniform Intervals, is superior if accurately defined boundaries are available.

1 Introduction

During the last years the mobile telecommunication industry has undergone extraordinary changes brought about by the introduction of new technologies and market forces. As a consequence, the operation of the radio network is becoming increasingly complex. The only viable option for operators to reduce operational costs is to increase the level of automation in the work process. In recent years operators have shown an increasing interest to automate troubleshooting in the radio access network. Troubleshooting consists of detecting problems (e.g. cells with a high number of dropped calls), identifying the cause (e.g. interference) and solving the problem (e.g. improving the frequency plan). The most difficult task is the diagnosis, which is currently a manual process accomplished by experts in the radio network. These experts are personnel dedicated to daily analysis of main performance indicators and alarms of the cells, aiming at isolating the cause of the problems. *Bayesian Networks* (BN) [1] are the technique that has been adopted for the automated fault diagnosis in cellular networks. BNs have been extensively used for diagnosis applications in many domains, such as diagnosis of medical diseases. Normally, the continuous variables are discretized before being used as input evidence for a discrete BN. Thus, numerous algorithms have been proposed for the discretization of continuous attributes [2, 3, 4]. Many of them are based on large databases of classified examples. Nevertheless, in domains such as cellular networks it would be very costly to obtain labelled cases and, therefore, the boundaries of the intervals have to be elicited by experts. Experience shows that when the number of symptoms is large, the task of defining the parameters of the model (boundaries for the intervals and probabilities) is very demanding and experts feel quite reluctant to specify more than two intervals per symptom. However, the higher the number of intervals, the better the discrete symptoms approximate the continuous variables. Furthermore, the huge amount of parameters to be defined leads to imprecise definition of the parameters.

In this paper, two techniques to improve the modelling of the continuous symptoms without increasing the number of parameters required from the experts are compared. The first method, called *Smooth Bayesian Networks* (SBN), was presented in [5]. In SBNs, continuous symptoms are discretized into intervals whose boundaries (*thresholds*) are smoothed. In this paper we propose a second technique to improve the diagnosis accuracy, *Multiple Uniform Intervals* (MUI), which increases the number of intervals based on uniform distributions of the continuous symptoms. Both systems have been used in experiments to diagnose the cause of excessive dropped calls in GSM cells. The performance comparison is addressed attending to two main criteria: classification error and sensitivity to imprecise definition of boundaries.

This paper is organised as follows. Section 2 reviews previous work in the area under study. Section 3 introduces the basis of the diagnosis of problems in the radio access network of cellular systems and revises the concept of Bayesian Networks. The two methods proposed to increase the diagnosis accuracy, SBNs and MUIs, are described in Section 4. Section 5 describes the experiments and the achieved results. Finally, some conclusions and future work are outlined in Section 6.

2 Related Work

Automated diagnosis based on Bayesian Networks has been successfully applied in many fields, such as diagnosis of diseases in medicine, of problems with printers, of problems in communication networks, etc. However, up to our knowledge, no references can be found on automated diagnosis in the radio access network of cellular systems. The research in cellular networks has been focused on fault detection [6] and on alarm correlation [7]. Although alarm correlation can be considered a first step in the diagnosis of faults, normally alarms do not provide conclusive information to identify the cause of the problem, especially if the possible causes are not only faults in pieces of equipment but other non-hardware related issues. For example, faults such as interference or lack of coverage are difficult to identify if performance indicators are not considered.

In [8] a system for automated diagnosis of faults in cellular systems which took into account both performance indicators and alarms was proposed. The reasoning mechanism consisted of a *Naive Bayesian Classifier* [9], which can be represented as a BN following the structure of a *Naive Bayes Model*. In [10] an alternative system based on the knowledge of troubleshooting experts instead of data was proposed. The knowledge acquisition was eased by means of a tool that automatically created the BN from the experts' answers to simple questions [11]. The continuous symptoms were discretized and the experts had to elicit the probabilities and the cut boundaries for the discretized intervals. The main problems arose due to the fact that the continuous symptoms were discretized into a reduced number of states and, therefore, a very small change in the value of a symptom could lead to a big change in the probability of the real cause. This reasoning mechanism is far from the actual way of thinking of a human expert, which is "smoother" in its conclusions.

Several authors have tried to incorporate this continuity of human reasoning into Bayesian Networks. In [12, 13] an integration of Bayesian Networks and fuzzy logic is proposed. [14] describes a methodology to estimate the probability density function of a continuous variable from the probabilities of the discrete variables in a BN. In [5] Smooth Bayesian Networks were presented, which can be understood as a method to add uncertainty about the state of the symptoms, especially in the proximity of the boundaries. In that way, the transitions between states are smoother than the steep boundaries of traditional BNs.

3 Automated Diagnosis Systems and Bayesian Networks

3.1 Diagnosis in the Radio Access Network of Cellular Systems

A *problem* is a situation occurring in a cell which has an influence on the service offered by the cell. Operators use different methods to identify the problematic cells. The dropped call rate (DCR) is a good indicator about the quality of the network, being normally around 1 or 2% in mature networks. Once the cells with problems are isolated, a diagnosis of the cause of the problems should be done separately for each problematic cell. A *cause* or *fault* is the defective behaviour of some logical or physical component in the cell that provokes failures and generates a high DCR, e.g. interference, hardware fault, etc. A *symptom* is a performance indicator whose value can be a manifestation of a fault, e.g. the number of handovers due to interference.

The performance of the network can be measured by means of multiple symptoms. The most important ones, called *Key Performance Indicators* (KPI), are collected daily by the *Network Management System* (NMS) with the help of counters situated at different points of the network. Besides, the NMS provides thousands of alarms from network elements, which may help to identify the cause of the problem. Consequently, when a fault is causing problems in a cell the value of some performance indicators change from their nominal values and some alarms may also be triggered. Therefore, the aim of the diagnosis system is to identify the cause of a problem based on the values of some symptoms.

3.2 Bayesian Networks

A Bayesian Network is a pair (D, P) that allows efficient representation of a joint probability distribution over a set of random variables $U = \{X_1, ..., X_n\}$. D is a *directed acyclic graph* (DAG), whose vertices correspond to the random variables $X_1, ..., X_n$ and whose edges represent direct dependencies between the variables. The second component, P, is a set of conditional probability functions, one for each variable, $p(X_i|\pi_i)$, where π_i is the parent set of X_i in U.

The set P defines a unique joint probability distribution over U given by

$$P(U) = \prod_{i=1}^{n} p(X_i|\pi_i) \qquad (1)$$

An important BN structure is the Simple Bayes Model (SBM) or Naive Bayes Model. In [8] the proposed method to perform the diagnosis was a naive Bayesian classifier, which may be represented as a SBM. The difference with the diagnosis system proposed in [10] is that the former worked with continuous probability functions whose parameters were obtained from a database of classified examples, whereas the latter used a SBM with discrete variables whose parameters were assessed by experts in the radio access network. In both cases, the states of the parent node were the possible causes of dropped calls, whereas the children were the symptoms.

4 Discretization of Continuous Performance Indicators

This paper is focused on systems whose continuous symptoms have been discretized. The boundaries for the intervals are normally assessed by troubleshooting experts, although alternatively they can be calculated from data in case enough labelled cases are available. In order to simplify the knowledge acquisition process, symptoms are discretized in only two intervals. In the following sections we present two methods to improve the diagnosis performance without requiring more information than that supplied for symptoms with two states. The first technique, Smooth Bayesian Networks, maintains the two states of the original BN, but it smooths the transition between the two intervals. The second algorithm, Multiple Uniform Intervals, designs a discrete BN with 3 states per symptom, based on the 2-state BN.

4.1 Two States: Smooth Bayesian Networks (SBN)

In traditional BNs a variable is exactly in one of its states. Smooth Bayesian Networks [5] consider that there is not certainty about which of the mutually exclusive states is taken by a given variable, specially when the value of the continuous symptom is near the threshold. This type of evidence is called *likelihood evidence* [1, 15]. As a consequence, in SBNs the shape of the posterior probabilities of the causes given the symptoms is smooth in comparison with the steep shape of the probabilities in conventional BNs.

Let c_S be a a continuous symptom and S its corresponding discretized variable. The value of the virtual evidence depends on the *belief mapping function*, $f_i^S(s)$, which models the belief in the value of c_S being s given that S is in the state s_i. For example, let's consider the continuous symptom $c\_S =$ "Percentage of handovers due to interference" and let's assume that the discrete symptom S has two states: low / high. The probability density function (pdf) $f_i^S(20\%) = P(c\_S = 20\%|S = low)$ represents the pdf of the symptom c_S evaluated at 20%, knowing that the symptom is low.

There is a trade-off in selecting the adequate belief mapping functions: on the one hand, achieving high classification accuracy, and, on the other hand, simplifying the knowledge acquisition process. In [5] three types of belief mapping functions were proposed: rectangular, trapezoidal and rect-gaussian. A SBN with rectangular functions is equivalent to a conventional discrete BN. The trapezoidal and rect-gaussian functions depend on the threshold T and on a parameter p, which is related to the width of the transition between states and has been named *degree of smoothness*.

4.2 Three States: Multiple Uniform Intervals (MUI)

A straightforward approach to improve the diagnosis performance is increasing the number of intervals in which the continuous symptoms are discretized. The drawback is that the knowledge acquisition is complicated because the number of parameters highly increases. The algorithm proposed in this section, which will be called Multiple Uniform Intervals, increases the number of intervals while maintaining the amount of information required from experts when only two intervals are used.

Let c_S be a continuous symptom, which has been discretized into two intervals and let S be the resulting discrete symptom. The parameters elicited by the experts are the probability of the first state given each cause, p_{1k}, and the boundary between the two states, T.

$$p_{1k} = P(S = s_1 | C_k) \, ; \quad S = s_1 \quad \text{if } c\_S \leq T, \quad \text{else} \quad S = s_2 \quad (2)$$

The algorithm consists of creating a third interval centered in T, whose width p is a parameter equivalent to the degree of smoothness characteristic of SBNs. In order to set the probability of the new state, it is assumed that the continuous symptom follows a uniform distribution. Thus, the probability of the second state is

$$p_{2k}^* = \frac{p}{b-a} \quad (3)$$

where a and b are the lower and upper limits, respectively, of the continuous symptom.

The probabilities of the other two states have to be updated to

$$p^*_{1k} = p_{1k} \cdot \left(1 - \frac{p}{b-a}\right); \; p^*_{3k} = (1 - p_{1k}) \cdot \left(1 - \frac{p}{b-a}\right) \quad (4)$$

This algorithm may easily be extended to multiple intervals.
For the SBM it can be demonstrated that the posterior probability of the cause C_i given the symptoms is

$$P(C_i|S_1,...,S_N) = P(C_i) \cdot \frac{\prod_{j<>x} P(S_j|C_i)}{\sum_k P(C_k) \prod_{j<>x} P(S_j|C_k)} \quad (5)$$

where S_x are the symptoms whose state is the second one.

5 Empirical Study

5.1 Reference Cases

A *case* is a set of values for some symptoms together with the actual cause. Contrary to other domains such as medical diagnosis, in which there are large knowledge databases, in the cellular network domain these reference cases are not normally available. The historical databases in the NMS of the network contain a large number of values describing the performance of cells with neither an indication of whether there was a problem in the cell nor the cause of such problem.

With the purpose of obtaining the required reference cases, a three month trial in a real GSM network was carried out. Everyday some problematic cells were identified and their faults were diagnosed by troubleshooting experts. The data from the network were combined with human expertise to obtain 500 classified cases, whose causes were considered the gold standard when comparing different diagnosis systems. A detailed explanation of the algorithm to generate the reference cases can be found in [5].

5.2 Experimental Set-Up

The first model created was a discrete Simple Bayes Model composed of 7 causes and 24 discrete symptoms. The probability tables were elicited by GSM experts with the help of a knowledge acquisition tool. The symptoms in the model were originally continuous, but they were discretized in two intervals using a modification of the discretization algorithm presented in [3]. Some SBNs with two types of belief mapping functions (trapezoidal and rect-gausian) were also designed, based on the previous discrete BN. Finally, a system composed of discrete symptoms with 3 states was built following the uniform approach proposed in section 4.2.

The classification criterion was diagnosing the cause with the highest probability (zero-one loss). The evaluated performance measures have been the classification error, the true diagnosis and the rank order. The classification error is the percentage of cases incorrectly classified. The true diagnosis is the average over all the cases of the probability of the real cause assessed by the diagnosis system. Finally, the rank order is the average over all the cases of the order of the real cause in the rank of the diagnosed probabilities.

The second aspect to consider was the sensitivity of the results to changes in the parameters of the mapping functions. In [5] a method to empirically analyze the sensitivity of BNs to imprecise definition of thresholds was proposed. It was based on previous studies about the sensitivity of BNs to imprecise probabilities [16, 17, 18]. The method was based on adding random noise to the thresholds and examining the effects on the diagnostic performance.

5.3 Results

The objective of the first experiment was to study the posterior probability of the causes depending on the cut-boundaries. Figure 1a presents a case whose real fault cause was interference in the downlink. This case was diagnosed by a 2-state BN, a SBN with rect-gaussian functions and a 3-state BN. The parameter p was set to 6% (symptoms in the model are measured as percentages). A hardware fault was erroneously diagnosed by the 2-state BN, whereas both the SBN and the 3-state BN achieved a correct diagnostic. Figure 1a depicts the probability of the two causes vs. the value of the threshold for a quality related symptom.

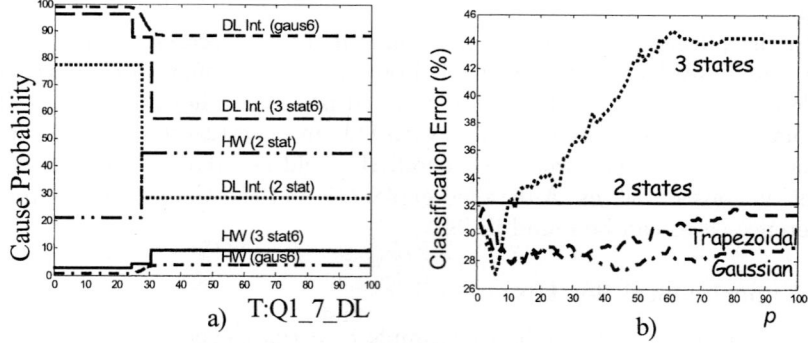

Fig. 1. a) Cause probabilities (DL interference and hardware) depending on the threshold of the symptom Q1_7_DL, b)Classification error vs. degree of smoothness

The next step was to compare the performance of a 2-state BN, a 3-state BN and SBNs with different belief mapping functions. The performance measures over the 500 reference cases are shown in Table 1. It can be observed that the lowest error is achieved with the 3-state BN for $p=6\%$.

A conclusion from Table 1 is that the performance depends on the parameter p. This is the reason why the following study was aimed at measuring the influence of that parameter. Figure 1b shows the classification error obtained

Table 1. Performance measures for diagnosis models

Model	Classification Error (%)	True diagnosis		Rank order	
		Mean	St.Dev.	Mean	St.Dev.
Discrete BN	32.2	65.0	37.3	1.50	0.93
SBN, Trap6	30.4	67.1	36.4	1.49	0.95
SBN, Trap10	28.4	66.8	36.2	1.48	0.98
SBN, Gaus6	28.4	67.2	35.7	1.47	0.96
SBN, Gaus10	28	66.4	35.7	1.48	0.99
3 st., p=6	27	66.3	35.6	1.40	0.81
3 st., p=10	32.2	63.8	35.6	1.51	0.96

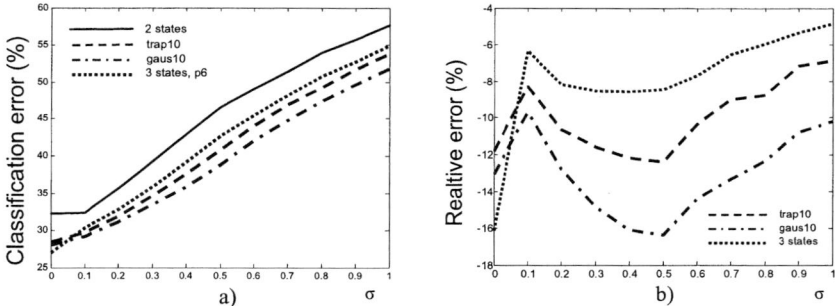

Fig. 2. a) Error for increasing levels of noise; b) Error relative to the 2-state BN

for the 500 reference cases with the different diagnosis systems depending on the value of p. For the 2-state BN the error is constant because the parameter p does not exist in conventional BNs. For the SBNs, it can be observed that the performance is improved with respect to the discrete BN, regardless of the value of p. In the case of the 3-state BN, it can be appreciated that, except for values of p lower than 10, the results are worst than for the 2-state network. Therefore, building a BN with 3 states based on the 2-state BN following the uniform approach may improve the diagnosis performance in comparison to the discrete BN and the SBNs, but only for low values of p.

The second type of experiments was related to the sensitivity of the system performance to an imprecise definition of the thresholds. Noise following a log-odd normal distribution generated independently for each of the 500 reference cases was added to the thresholds. The level of noise was changed from $\sigma = 0$ to 1 in steps of 0.1, generating 200 networks independently for each level of noise. The experiment was repeated for a 2-state BN, two SBNs with trapezoidal and rect-gaussian functions both with $p = 10$, and a 3-state BN with $p = 6$. Therefore, the total number of networks used in our experiments was 4 million, comprised of 10 levels of noise x 200 runs x 500 cases x 4 networks, plus 4 networks without noise.

Figure 2a shows the classification error vs. the standard deviation of the noise for the different systems. Figure 2b presents the error relative to the 2-state BN for different levels of imprecision. For no noise ($\sigma = 0$) the results are the same as the ones exposed in Table 1, i.e. the best accuracy is obtained with a 3-state BN. On the other hand, except for low values of σ, the error of SBNs is lower than the one obtained with the 3-state BN. These statements are more noticeable for high levels of imprecision in the thresholds. Therefore, SBNs are more robust to imprecise definition of thresholds than discrete BNs, regardless of whether the number of states is two or three, which is especially important in domains where the number of classified examples is scarce or non-existent and the variance of the symptoms is large.

5.4 Conclusions and Future Work

In this paper we have compared two methods to overcome the difficulties in the construction of a model for diagnosing faults in the radio access part of cellular networks. Smooth Bayesian Networks, were already presented in [5]. In this paper we have also proposed a novel method, Multiple Uniform Intervals, to increase the number of states of the discretized symptoms. The main advantages of MUI are both its simplicity and the fact that it does not increase the amount of information provided by the domain experts in comparison to that required for a 2-state BN.

Some models have been built based on data from real GSM networks and cases classified by diagnosis experts. The 3-state BN (MUI) has shown to achieve the best results when the width of the middle interval is small. When the thresholds are precisely defined, the 3-state BN has presented less error than the other methods. However, as the level of imprecision increases SBNs show better performance. Regardless of the imprecision in the thresholds, both methods have shown to be superior to traditional BNs with 2 states.

It is believed that the benefits of using SBNs or MUIs would be higher if the thresholds in the model were specified by experts instead of being calculated from data, because the imprecision is normally worst in the former case. Repeating the experiments using thresholds specified by experts is a task planned to be addressed in the near future.

The BNs used in the experiments have been based on a Simple Bayes Model. Additional research would be required in order to confirm if the conclusions can be generalized to other BN structures.

Acknowledgements

This work has been partially supported by the Spanish Ministry of Science and Technology under project TIC2003-07827

References

1. Pearl, J.: Probabilistic Reasoning in Intelligent Systems, Morgan Kaufmann, San Francisco, 1988.
2. Dougherty, K., Kohavi, R., Sahami, M.: Supervised and unsupervised discretization of continuous features, in: Proc.12th Intern. Conference on Machine Learning, Tahoe City, CA (Morgan Kaufmann, San Mateo, CA), 1995, pp. 194-202.
3. Fayyad, U.M., Irani, K.B.: Multi-interval discretization of continuous valued attributes for classification learning, in: Proc.13th Intern.Joint Conf. on Artificial Intelligence, Morgan Kaufmann, 1993, pp. 1022-1027.
4. Yang, Y., Webb, G.I.: A Comparative Study of Discretization Methods for Naive-Bayes Classifiers, in: Proceddings of PKAW 2002, Tokyo, Japan, 2002, pp. 159-173.
5. Barco, R., Díez, L., Wille, V.: Smoothing boundaries of states in Bayesian Networks: Application to diagnosis in cellular systems, submitted to Artificial Intelligence, Elsevier, Feb.2005.
6. Laiho, J., Kylvj, M., Hglund, A.: Utilisation of Advanced Analysis Methods in UMTS Networks, in: Proc.IEEE 55th Vehicular Techn. Conf., 2002, pp. 726-730.
7. Jakobson, G., Weissman, M.D.: Alarm Correlation, IEEE Network, Vol.7 (6) (1993), 52-59.
8. Barco, R., Wille, V., Díez, L.: System for automatic diagnosis in a cellular network based on performance indicators, European Transactions on Telecommunications, Vol.15, Wiley, 2005.
9. Domingos, P., Pazzani, P.: On the optimality of the Simple Bayesian Classifier under Zero-One Loss, Machine Learning, 29 (1997) 103-130.
10. Barco, R., et al.: Automated troubleshooting of mobile networks using Bayesian Networks, in: Proc.IASTED Intern.Conf. on Communication Systems and Networks, Acta Press, 2002, pp. 105-110.
11. Barco, R., et al.: Automated troubleshooting of a mobile communication network using Bayesian networks, in: Proc.4th IEEE Intern.Workshop on Mobile and Wireless Communications Network, 2002, pp. 606 - 610.
12. Pan, H., McMichael, D.: Fuzzy causal probabilistic networks - a new ideal and practical inference engine, in: Proc.1st International Conference on Multisource-Multisensor Information Fusion, 1998.
13. Pan, H.: Fuzzy Bayesian Networks - A General Formalism for Representation, Inference and Learning with Hybrid Bayesian Networks, International Journal of Pattern Recognition and Artificial Intelligence, 14 (7) (2000) 941-963.
14. Yang, C.C.: Fuzzy Bayesian Inference, in: Proc.IEEE International Conference on Systems, Man and Cybernetics, Orlando, FL, 1997, pp.2707-2712.
15. Valtorta, M., Vomlel, J.: Soft Evidential Update for Probabilistic Multiagent Systems, International Journal of Approximate Reasoning, 29 (1) (2002) 71-106.
16. Pradhan, M., et al.: The sensitivity of belief networks to imprecise probabilities: An experimental investigation, Artificial Intelligence 85 (1-2) (1996) 363-397.
17. Kipersztok, O., Wang, H.: Another Look at Sensitivity of Bayesian Networks to Imprecise Probabilities, in: Proc.8th International Workshop on Artificial Intelligence and Statistics, 2001.
18. Henrion, M., et al.: Why is diagnosis using belief networks insensitive to imprecision in probabilities?, in: Proc.12th Conf. on Uncertainty in Artificial Intelligence, Morgan Kaufmann, 1996, pp. 307-314.

Visual Interactive Clustering and Querying of Spatio-Temporal Data

Olga Sourina and Dongquan Liu

School of Electrical & Electronic Engineering,
Nanyang Technological University, Block S2,
Nanyang Avenue, Singapore 639798
eosourina@ntu.edu.sg
http://www.ntu.edu.sg/home/eosourina

Abstract. Visualization techniques increase the user involvement in the interactive process of data mining and querying of spatio-temporal data. This paper describes a novel geometric approach to clustering and querying of spatio-temporal data. We propose the uniform geometric model based on function representation of solids to cluster and query time-dependent data. Clustering and querying are integrated with visualization techniques in one GUI. First, visual clustering with blobby model allows the user to see the result of clustering on the screen for different time points and/or time intervals and set the appropriate parameters interactively. After that, the user gets the data of clusters for the chosen time frames. Then, the user can visually query the cluster/clusters he/she is interested in with geometric primitive solids which currently are cubes, spheres/ellipsoids, cylinders, etc. Geometric operations of union, intersection and/or subtraction can be performed over the geometric primitive solids to get the final query shape. The user visually clusters spatio-temporal data and queries the clusters with geometric shapes through graphics interface accessing dynamically 3D projections of multidimensional points from database, warehouses or files. With the uniform geometric model of the clustering and querying of spatio-temporal data, 3D visualization tools can be naturally incorporated in one system to allow the user to visualize and query clusters changing over time.

1 Introduction

Clustering and querying of spatial data are classical problems in databases, and warehousing. Clustering algorithms can be applied for similarity search, customer segmentation, pattern recognition, trend analysis, etc. Numbers of algorithms for clustering multidimensional data have been proposed in the last few years, e.g. partitioning method such as k-means [1] and k-medoids [2], hierarchical method such as CURE [3], density-based method such as DBSCAN [4] and DENCLUE [5], etc. However, analysis of spatio-temporal data including clustering has received less attention. To be able to run on spatio-temporal data spatial clustering algorithms need temporal extensions [6, 7].

It becomes more and more important for the modern clustering systems to give the user an easy understanding of both the data set and the results [8]. Visualization offers the user an intuitive way of analysis that can help to discover data patterns and structures. Data visualization techniques [9, 10] when incorporated with clustering algorithms could improve interpretability and usability of the data and clustering process. Visualization techniques could be used not only for the interpretation of the results but also for the interpretation of the whole process of clustering in order to help the user to come up with hypothesis and to set the values of parameters. For instance, the user could select the projection directions [11] for high dimension data set. To incorporate visualization techniques, the existing clustering algorithms use the result of clustering algorithm as the input for visualization system [10]. The drawback of such approach is that it can be costly and inefficient. The better solution is to combine two processes together, which means to use the same model in clustering and visualization.

On the other hand, development of query methods and graphical user interfaces is a new trend in data mining [12]. Querying of time-dependent data is a classical problem in temporal databases and warehousing. The goal of works in this area is to propose data representation model and query model able to handle time-dependent geometries including those changing continuously that describe moving objects [13, 14]. Spatio-temporal predicates are introduced to query time-dependent data [14].

In work [15], we proposed and fully described geometric query model with implicit functions. Then, in work [16], we proposed a solid-based clustering method. In this paper, we extend the uniform geometric model to handling time-dependent data. We proposed to apply solid-base clustering algorithm to data changing over time. The extended uniform geometric model allows integrate solid-based visual clustering and visual querying of time-dependent data in one GUI. Implicit function is used in spatio-temporal predicate implementation. This allows us to pose complex shape queries changing over the time. Our extended model allows us to integrate clustering, querying and visualization of spatio-temporal data. Spatio-temporal query languages are not discussed in this paper.

The paper is organized as follows. Section 2 introduces the model defined with implicit functions that is used for interactive visual clustering of time-dependent data and describes similarity of the model with the model of density-based methods. Querying of time-dependent data based on the geometric query model is described in Section 3. Implementation of the system and examples of visual clustering and querying are discussed in Section 4. In Section 5, conclusion and future work are considered.

2 Solid-Based Clustering of Spatio-Temporal Data

The implicit modeling techniques are relatively new. This approach has become more sophisticated, generating new interest in computer graphics and related fields [17]. It uses implicit function instead of parametric function or explicit function as its mathematical foundation. In work [16], we proposed to define a cluster as a solid reconstructed on the points with the implicit functions. In this paper, we define cluster as a solid existing at time point. The solid not only describes the granular property of the cluster but also describes its boundary. With this definition, a new object could be easily identified to which cluster it belongs.

Let **P** be a set of multidimensional points **P**={[$p_1, p_2, ..., p_n$, t]} = {[**P**, t]} in n dimensional Euclidean space E^n, and t is time in the n-dimensional Euclidean space E^n. Then, a solid reconstructed on the points can be described with function-based representation as follows:

$f(\mathbf{X},\mathbf{P}) \geq 0$, where **X** belongs to E^n. The function can be defined by procedure. Such function defines closed n-dimensional geometric solid in E^n space under the following conditions:

$f(\mathbf{X}, \mathbf{P})>0$ for the points inside the solid,
$f(\mathbf{X}, \mathbf{P})=0$ for the points on the solid boundary,
$f(\mathbf{X}, \mathbf{P})<0$ for the points outside the object,
where $\mathbf{X}= \{x_1,x_2,...,x_n\}$ is a position vector of the point in E^n.

The zero set of these functions provides surfaces, and the values that are greater or equal to zero define multidimensional geometric solid objects. We consider each cluster as a different solid per time point. For querying we use function-based representation of the query solid as well. For clustering, we implement solid-based subdivision algorithm for each time frame. In solid-based subdivision algorithm that was proposed and described for spatial data in work [18], we build the solid by computing the density function of points in the whole field, which means adding up all the influence functions of points inside the field. The sum is a field density function that consists of all influence functions of the points. The field density function can make a complete description of the whole data space. Parabolic function, square wave function and Gaussian function are some examples of basic influence function in density-based algorithms. The functions used in blobby, meta-balls or soft objects in Computer Graphics can also serve as the basic influence function for better efficiency of our method.

In this work, we use blobby model that is similar to the model used in density-based clustering methods. Blobby model was first accomplished by Blinn, and now the term blobby always includes other related models and is not limited only to the original model. The blobby primitive is described as follows:

$$f(r)=a*e^{-(r/b)}$$

where r is the distance from the center point of a primitive, a is the height of the function and b is related to the standard deviation which is Gaussian. At any point of the surface, the isosurface "potential" is equal to the sum of all the primitives' contributions using the following function:

$$F(X,P)=\sum_{i=1}^{N} f(r)-T \geq 0$$

where N is the total number of blobby primitives and T is the threshold constant that determines the value of the isosurface.

In blobby model, the distance from the center point of the primitives describes the range that the primitive can influence on, the summary function adds up the influences of all primitives and the threshold determines the level of the isosurface built. In our method, the same blobby model is used in the subdivision algorithm. By connecting all points with the same potential value (i.e. the threshold value), we can build a smooth implicit surface around the cluster. If there are many clusters in the data set, we will get many implicit surfaces with the same potential value. And each of the implicit surfaces will wrap the cluster. In our model, implicit surfaces serve as the boundaries of clusters. By substituting one point's coordinates into the blobby function and compare the result with T, we can easily know whether the point is inside or outside of the implicit surface. Thus, the solid-based subdivision algorithm is implemented as follows. We connect each point of each time point dataset with other points and sort the points into the clusters checking the values along the line connecting two points.

As we mentioned before, the formulae of blobby model is similar to the density-based clustering model of DBSCAN, OPTICS and DENCLUE. In density-based methods, Gaussian function is used as follows:

$$f^r(\vec{x}) = e^{-\frac{d(\vec{x},\vec{r})^2}{2\sigma^2}}$$

, where $d(\vec{x},\vec{r})$ is a distance between two points.

We implemented visual clustering with blobby functions. The blobby formula has additional parameter a which can make cluster shape "thinner". We have to set interactively three parameters for our model a – scale factor, b – exponential factor, and T – threshold value.

We have to note that the solid-based subdivision algorithm can also act as a stand alone clustering algorithm even without being integrated with visualization techniques into the system.

3 Querying of Spatio-Temporal Data

After we visualize clusters per time point or/and time interval using interactively set parameters, we can query the clusters with geometric objects. In this section, we describe our geometric query model consisting from geometric objects generally changing over time and operations. The proposed model is an extension of the model that was introduced first in work [15]. As it was shown there, geometric interpretation of relational algebra selection operation can be phrased as follows: "find out the points that belong to the solid." In our model, the query solid can be a complex geometric solid. The complex query solid can be created with union, intersection, and other operations over primitive solids that are generally hyperhalfspaces, hypercuboids, hyperellipsoids, etc. Selection operation of relational algebra can be found in geometry as point/solid classification predicate. In this paper, we extend the model with time dimension. Then, the point/solid classification predicate can be described as follows. Let P be a point in Euclidean space E^n and t is time, G_1 be a query solid described with implicit function f_1 defined with time-dependent parameters and location changing over time, bG_1 be a boundary of G_1 and iG_1 be an interior of G_1. Then a point/solid predicate is described with the implicit function representation of the geometric object G_l by a 3-valued predicate:

$$S_3(P, G_1) = \begin{cases} 0, \text{if } f_1(x_1, x_2, \ldots, x_n, t) < 0 & P \notin G_1 \\ 1, \text{if } f_1(x_1, x_2, \ldots, x_n, t) = 0 & P \in bG_1 \\ 2, \text{if } f_1(x_1, x_2, \ldots, x_n, t) > 0 & P \in iG_1 \end{cases}$$

In our model, query solid can have time-dependent parameters and/or coordinates that can be defined analytically or by procedure. Thus, the geometric query model consists of the following geometric objects:

- n-dimensional points P = $\{[x_1, x_2 \ldots \ldots x_n, t]\}$ where t is time;
- time-dependent 3-dimensional primitive geometric objects for the construction of a query solid using geometric operations.

The following is an implicit function representation of the primitive time-dependent 3-dimensional geometric solids that could be used for construction of geometric criteria:

Halfspace:
G_1: $f_1(\mathbf{X}, t) = f_1(x_1, x_2, x_3, t) = (x_i - a[t]) \geq 0$
Where a is some real number ($a \in R$)

Sphere:
G_1: $f_1(\mathbf{X}, t) = r[t]^2 - (x_1 - x_{0,1}[t])^2 - (x_2 - x_{0,2}[t])^2 - (x_3 - x_{0,3}[t])^2 \geq 0$
Where $x_{0,1}, x_{0,2}, x_{0,3} \in R$

Ellipsoid:
G_1: $f_1(\mathbf{X}, t) = 1 - ((x_1 - x_{0,1}[t])/a_1[t])^2 - ((x_2 - x_{0,2}[t])/a_2[t])^2 - ((x_3 - x_{0,3}[t])/a_3[t])^2 \geq 0$
where $x_{0,1}, x_{0,2}, x_{0,3} \in R$ and $a_1, a_2, a_3 \in R$.

Cone:
G_1: $f_1(\mathbf{X}, t) = ((x_1 - x_{0,1}[t])/a_1[t])^2 - ((x_2 - x_{0,2}[t])/a_2[t])^2 - ((x_3 - x_{0,3}[t])/a_3[t])^2 \geq 0$
where $x_{0,1}, x_{0,2}, x_{0,3} \in R$ and $a_1, a_2, a_3 \in R$.

Cylinder:
G_1: $f_1(\mathbf{X}, t) = ((x_1 - x_{0,1}[t])/a_1[t])^2 - ((x_2 - x_{0,2}[t])/a_2[t])^2 \geq 0$
Where $x_{0,1}, x_{0,2} \in R$ and $a_1, a_2 \in R$.

By further declaring that our model is open to any type of objects that can be defined implicitly with some functions $f(x_1, x_2, x_3, t) \geq 0$, we could avoid the problem of a minimum set of primitives and to change this set depending on the application problem to be solved.

Geometric operations are applied to primitive geometric objects to obtain complex geometric shapes at each time point. The analytical definition of set-theoretic operations is realized in the form proposed by Ricci [19], where operations over implicit functions are considered. Affine transformations (translation, rotation and scaling) are also used to increase an expressive power of the proposed geometric model. Geometric operations include set-theoretic union, intersection, difference and orthographic projection.

Mathematically,

Union: $G_3 = G_1 \cup G_2$ of two objects $G_1 \subset E^n$ and $G_2 \subset E^n$ with the descriptive functions f_1 and f_2 will be defined as $f_3 = f_1 \vee f_2 = \max(f_1, f_2) \geq 0$, where $G_3 \subset E^n$. *Intersection:*

$G_3=G_1 \cap G_2$ of two objects $G_1 \subset E^n$ and $G_2 \subset E^n$ with the descriptive functions f_1 and f_2 will be defined as $f_3=f_1 \wedge f_2=\min(f_1,f_2) \geq 0$, where $G_3 \subset E^n$. **Complement:** $G_2=\neg G_1$ of object $G_1 \subset E^n$ with the descriptive functions f_1 will be defined as $f_2=-f_1 \geq 0$. **Difference:** $G_3=G_1 \setminus G_2$ between objects $G_1 \subset E^n$ and $G_2 \subset E^n$ with descriptive functions f_1 and f_2 will be defined as $f_3=f_1 \wedge (-f_2)=\min(f_1,-f_2) \geq 0$, where $G_3 \subset E^n$. **Translation:** $G_2=T(G_1)$ of object $G_1 \subset E^k$ with descriptive functions f_1 by $a_1, a_2,...,a_n$ will be defined as $f_1(x_1-a_1, x_2-a_2, ..., x_n-a_n) \geq 0$. **Rotation:** $G_2=R(G_1)$ of object $G_1 \subset E^k$ with descriptive functions f_1 of angle α about some axis will be defined as $f_1(x'_1, x'_2,...,x'_n) \geq 0$ where $[x'_1\ x'_2 ...\ x'_n\ 1]=R^{-1}[x_1\ x_2...\ x_n\ 1]$ and R^{-1} is an inverse matrix of rotation. **Scaling:** $G_2=S(G_1)$ of object $G_1 \subset E^k$ with descriptive functions f_1 in $s_1, s_2,..., s_n$ times will be defined as $f_1(x_1/s_1, x_2/s_2, ..., x_n/s_n) \geq 0$.

Thus, with this model we can query clusters changing over time with the final query shape.

4 Visual Clustering and Querying

Our system for visual interactive 3D clustering and querying of spatio-temporal data is based on the uniform geometric model with implicit functions. Visual clustering allows the user to set interactively the appropriate parameters for clustering data changing over time. We developed the graphical user interface based on the geometric algebra. We use the geometric concepts to cluster and query spatio-temporal data and apply visualization techniques to interpret the clustering process and querying. The points mapped from the database and the clustering process both is visualized. To get initial clusters on 3-dimensional points clouds changing over time default parameters are used. After initial visual clustering, parameters can be changed at any time point. Cluster can be queried with a complex geometric query solid with union, intersection or other operations over primitive geometric solids. These primitive query solids currently are cuboids (box), cylinders, cones, and ellipsoid. We also can fit interactively any chosen cluster with the wrapping solid and keep the cluster implicit formulae in the database. In addition, any point belonging to the visualized solid can be located and identified in the database, data warehouse or file.

Let us consider examples of the visual 3-dimensional clustering of data changing over time. First, 3D projections of multidimensional points from database or file are visualized as clouds of points and can be viewed at any time point and/or interval. Then, the points are clustered visually with blobby functions and subdivision algorithm and can be viewed at any time point and/or interval as well. In Fig. 1, the visual clustering of spatio-temporal data is shown. In Fig. 2, an example of querying of time-dependent data is shown. First, 3-dimensional projection of multidimensional points is mapped from the database and visualized as point clouds. A blobby solid can be reconstructed at each time point to show shape of point clouds changing over time. Then, a solid query is posed and time interval is chosen. Here, a query solid is a cylinder that does not change its parameters and location over time interval. The result of the query is time-dependent data and is visualized as set of snapshots or as file with animation.

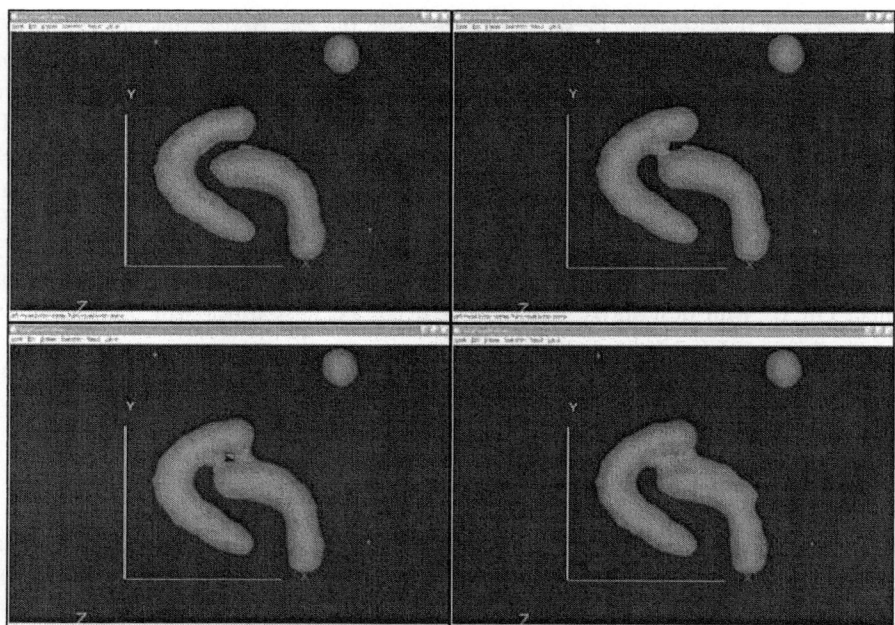

Fig. 1. 3D visual clustering of spatio-temporal data

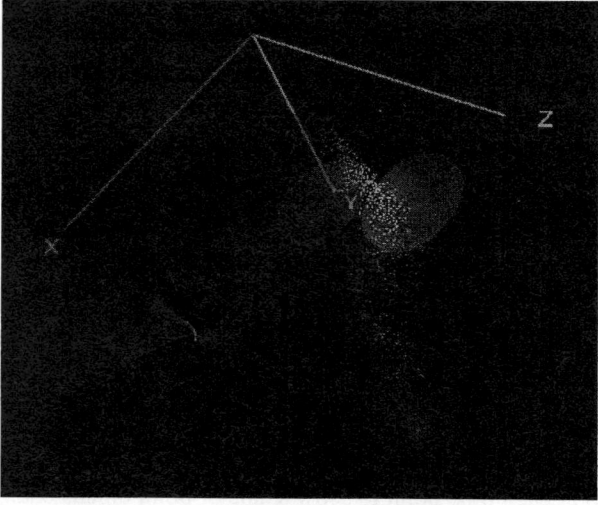

Fig. 2. Querying of the cluster with cylinder shape

With the proposed query model, the user could specify a query solid for each time point defining time-dependent primitive solids parameters and location analytically or by procedure. Currently, with the implemented GUI, the user constructs the query shape that does not change over time interval.

Geometric objects can be drawn opaque or transparent. We employ visualization techniques and advanced computer graphics algorithms for the implementation of the user interface.

In Fig. 3 wrapping with ellipsoids and union of ellipsoids is shown. The system is implemented with the software Visualization Toolkit (VTK) where visualization is implemented with marching cube algorithm [20].

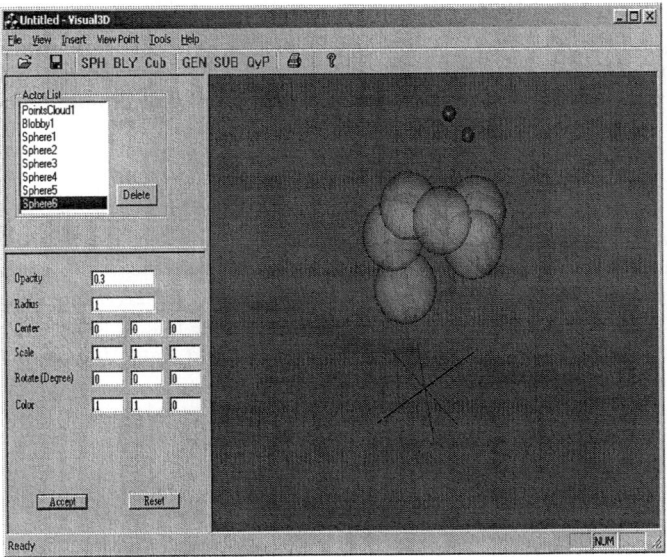

Fig. 3. Wrapping the cluster with union of ellipsoids

5   Conclusion and Future Work

In this paper, we introduced geometric approach to clustering and querying of time-dependent data in spatio-temporal databases, and data warehouses. We have presented visual interactive 3-dimensional method of clustering and querying spatio-temporal data. Visualization, clustering and querying are integrated in the system prototype. We conclude that our interactive method has a great potential for interactive data clustering and querying of data changing over time. The nature of our geometric model has the advantage of easy integration with visualization techniques. Thus, the user can be involved into the clustering and querying process in order to make more efficient and intuitive decisions on the data changing over time. The work completed by now mainly has focused on the testing of the method on small datasets. We are planning to improve and test our algorithms on large data changing over time and to design the system for large spatio-temporal databases and warehouses. In future, we also are planning to continue our research in the interactive geometric clustering looking for the optimal parameters. We are going to make further improvements on the algorithms and visualization techniques to make the clustering and querying process more efficient and intuitive to the user. The human vision is the most experienced in the interpretation of realistic representations. The application of advanced computer graphics algorithms and visualization techniques for graphical data mining languages and representation of the data mining results could help to explore the data through more intuitive interface employing even modern VR tools.

References

1. J. Hartigan, and M. Wong, "A K-means Clustering Algorithm", *Applied Statistics, 28,* 1979, pp. 100-108.
2. L. Kaufman, and P. Rousseeuw, *Finding Groups in Data: A Introduction to Cluster Analysis.* New York, John Wiley and Sons, 1990.
3. Sudipto Guha, R. Rastogi, K. Shim, *CURE: A Clustering Algorithm for large databases.* Technical report, Bell Laboratories, Murray Hill, 1997.
4. M. Ester, Hans-Peter Kriegel, Jorg Sander, Xiaowei. Xu, "A Density-Based Algorithm for Discovering Clusters in Large Spatial Databases with Noise", *Proc of KDD-1996*, 1996.
5. A. Hinneburg, D.A. Keim, An Efficient Approach to Clustering in Large Multimedia Databases with Noise, *American Association for Artificial Intelligence,* 1998.
6. U.M. Fayyad, G. Piatetsky-Shapiro, P. Smyyh, From data mining to knowledge discovery: An Overview, In *Advances in Knowledge Discovery and Data Mining*, Cambridge, MA: MIT Press, 1996, pp.1-34.
7. N.J. Miller, and J. Han, Geographic data mining and knowledge discovery: An Overview, In Geographic Data Mining and Knowledge Discovery, London, New York: Taylor & Fransis, 2001, pp. 3-32.
8. D.A. Keim, "Information Visualization and Visual Data Mining", *IEEE Transactions on Visualization and Computer Graphics*, vol. 8, 1, 2002, pp. 1-8.
9. T.C. Sprenger, M.H. Gross, A. Eggenberger, M. Kaufmann, A Framework for Physically-Based Information Visualization, *in Proceedings of Eurographics Workshop on Visualization '97*, Boulogne sur Mer, France, April 28-30, 1997, pp. 77-86.
10. T. C. Sprenger, R. Brunella, M. H. Gross. "H-BLOB: A Hierarchical Visual Clustering Method Using Implicit Surfaces," Department of Computer Science, Swiss Federal Institute of Technology (ETH), Zurich, Switzerland
11. A. Hinneburg, D.A. Keim, and M. Wawryniuk, "HD-Eye: Visual Mining of High-Dimensional Data", *IEEE Computer Graphics and Applications,* September/October 1999, pp. 22-31.
12. J. Han, M. Kamber, *Data Mining Concepts and Techniques.* San Francisco, CA: Morgan Kaufmann Publishers, 2000.
13. R. H. Güting et al, A Foundation for Representing and Querying Moving Objects. ACM Transaction on Database Systems, Vol. 25, No. 1, March 2000, pp. 1-42.
14. M. Erwig, M. Schneider, Developments in Spatio-Temporal Query Languages, *IEEE Int. Workshop on Spatio-Temporal Data Models and Languages (STDML),* 1999, pp. 441-449.
15. O. Sourina, S.H. Boey, Geometric Query Types for Data Retrieval in Relational Databases, *Data & Knowledge Engineering,* Elsevier Science B.V., Vol. 27, 2, 1998, pp. 207 – 229

16. O. Sourina, and L. Dongquan, "Geometric approach to clustering and querying in databases and warehouses", in *Proc. of Cyberworlds 2003*, Singapore, Dec. 2003, pp. 326-333.
17. Bloomenthal J., *An Introduction to Implicit Surfaces*, Morgan-Kaufmann, 1997.
18. O. Sourina, and D. Liu, Visual interactive 3-dimensional clustering with implicit functions, In *Proc. of IEEE CIS 2004*, Dec. 2004.
19. Ricci A., A constructive geometry for computer graphics, *The Computer Journal*, Vol. 16, 2, 1973, pp. 157-160.
20. Schroeder W., Martin K., Loresen B., *The Visualization Toolkit*, Prentice Hall, 1998.

Breakdown-Free ML(k)BiCGStab Algorithm for Non-Hermitian Linear Systems

Kentaro Moriya[1] and Takashi Nodera[2]

[1] Aoyama Gakuin University, 5-10-1 Fuchinobe,
Sagamihara, Kanagawa 229-8558, Japan
[2] Keio University, 3-14-1 Hiyoshi,
Kohoku, Yokohama 223-8522, Japan

Abstract. ML(k)BiCGStab algorithm stabilizes BiCGStab algorithm by using k pseudo Krylov subspaces. However, if k is too large, the computation cost often becomes expensive, and the performance of BiCGStab algorithm is not always improved. In this paper, a new variant of ML(k)BiCGStab algorithm is proposed. In the proposed scheme, k is varied and pseudo Krylov subspaces are recomputed when the Lanczos breakdown occurs. Numerical experiments are reported which indicate that the proposed scheme performs better than the original ML(k)BiCGStab algorithm and BiCGStab algorithm.

1 Introduction

We now discuss the large and sparse linear system

$$Ax = b, \quad A \in \mathbf{R}^{n \times n}, \quad x,\ b \in \mathbf{R}^n, \qquad (1)$$

where the coefficient matrix A is regular and non-Hermitian. It is known that the linear system (1) are often solved by non-stationary iterative method.

Recently Yeung et al. [4] has proposed ML(k)BiCGStab algorithm for solving the linear system (1). This algorithm uses k Lanczos starting vectors $q_0, q_1, \ldots, q_{k-1}$ and combines BiCGStab algorithm [3] with k pseudo Krylov subspaces:

$$K(q_0,\ A^T),\ K(q_1,\ A^T),\ \ldots,\ K(q_{k-1},\ A^T).$$

The number of pseudo Krylov subspaces, to say k, is usually determined by users. Therefore, the orthogonal cost occasionally becomes so expensive as k is increased. In order to reduce the computation cost for orthogonalization, we propose a new scheme of varying k when the Lanczos breakdown occurs. The proposed scheme increases k and reproduces k pseudo Krylov subspaces so that the Lanczos breakdown is less likely to occur. The proposed scheme is also implemented on the MIMD parallel computer, Compaq Beowulf. It's also with the original BiCGStab algorithm and ML(k)BiCGStab algorithm. We show that the proposed scheme performs more effectively than these original algorithms.

This paper is organized as follows. We introduce the original BiCGStab algorithm and ML(k)BiCGStab algorithm in Section 2 and Section 3, respectively. We discuss the Lanczos breakdown and propose the adaptive determination of k based on the Lanczos breakdown in Section 4. In Section 5 some numerical examples are reported and the concluding remarks are given in Section 6.

2   BiCGStab Algorithm

BiCGStab algorithm [3] accelerates the convergence of the residual norm by computing the product of the residual vector of BiCG algorithm [1] and an appropriate 1-step MR polynomial. The l-th residual vector of BiCG algorithm and degree one MR polynomial are defined as $\tilde{r}_l$ and

$$P_l(A) = I - \alpha_l A, \tag{2}$$

respectively. The l-th residual vector of BiCGStab algorithm can be described as

$$r_l = P_l(A) P_{l-1}(A) P_{l-2}(A) \ldots P_0(A) \tilde{r}_l, \tag{3}$$

where α_l is determined so that the residual norm $\|r_l\|_2$ is always minimized. The drawback of BiCGStab algorithm is that the Lanczos breakdown often occurs when the pivot becomes near zero or exact zero. The more details of the Lanczos breakdown are given in Subsection 4.1.

3   ML(k)BiCGStab Algorithm

The residual vector of ML(k)BiCG algorithm is orthogonalized to all of k pseudo Krylov subspaces

$$K(q_0,\ A^T),\ K(q_1,\ A^T),\ \ldots,\ K(q_{k-1},\ A^T).$$

We now define the vectors p_{ik+j} as

$$p_{ik+j} = (A^T)^i q_j, \quad j = 0, 1, \ldots, k-1, \quad i = 0, 1, \ldots, \tag{4}$$

where q_j are starting Lanczos vectors and are orthogonalized to each other. The l-th residual vector ML(k)BiCG algorithm is implemented so that the orthogonal condition

$$r_l \perp \mathrm{Span}\{p_0,\ p_1,\ \ldots, p_l\} \tag{5}$$

is satisfied, where $l = ik + j$. The implementation of ML(k)BiCG algorithm is shown in Figure 1. Three vectors x_l, $\tilde{r}_l$ and g_l are the approximate solution, the residual vector and the direction vector, respectively. The residual vector of ML(k)BiCGStab algorithm can be obtained by computing the product of the residual vector of ML(k)BiCG algorithm and a degree one MR polynomial

```
                      ML(k)BiCG algorithm

  choose $x_0$ and $q_0$
  Compute $q_1$, ..., $q_{k-1}$ by Arnoldi process
  $\tilde{r}_0 = b - Ax_0$,   $p_0 = q_0$,   $g_0 = \tilde{r}_0$,   $l = 1$
  while "$\|\tilde{r}_l\|_2 / \|\tilde{r}_0\|_2 \geq \epsilon$" do
      $\alpha_l = (p_{l-1}, \tilde{r}_{l-1}) / (p_{l-1}, Ag_{l-1})$
      $x_l = x_{l-1} + \alpha_l g_{l-1}$
      $\tilde{r}_l = \tilde{r}_{l-1} - \alpha_l Ag_{l-1}$
      for $i = \max(l - k, 0)$ to $l - 1$ do
          $z = \tilde{r}_l + \sum_{j=\max(l-k,\,0)}^{i-1} \left[\beta_j^{(l)} g_j\right]$,   $\beta_i^{(l)} = -(p_i, Az) / (p_i, Ag_i)$
      enddo
      $g_l = \tilde{r}_l + \sum_{j=\max(l-k,\,0)}^{l-1} \left[\beta_j^{(l)} g_j\right]$
      Compute $p_l$, using the formula (4)
      $l = l + 1$
  enddo
```

Fig. 1. ML(k)BiCG algorithm

(2). The main advantage of ML(k)BiCGStab algorithm is distributing the risk of the Lanczos breakdown to k pseudo Krylov subspaces. If there is only one pseudo Krylov subspace and k iterations are done, the residual vector has to be orthogonalized to a same pseudo Krylov subspace k times. Moreover, as the iterations are increased, the dimension of only one pseudo Krylov subspace is higher and the pseudo Krylov subspace is more likely to be ill-conditioned. On the other hand, if there is k pseudo Krylov subspaces and k iterations are done, the residual vector can be orthogonalized to each pseudo Krylov subspace only once. Then the dimensions of k pseudo Krylov subspaces are not so high as the former case. Therefore, in the latter case, the Lanczos breakdown is less likely to occur than the former case. Especially in case of $k = 1$, the residual vector of ML(k)BiCGStab algorithm is orthogonalized to only one pseudo Krylov subspace $K(q_0, A^T)$. Therefore, ML(1)BiCGStab algorithm is equivalent to BiCGStab algorithm mathematically. For the more details of ML(k)BiCG algorithm and ML(k)BiCGStab algorithm, see Yeung et al. [4].

4 Varying k

In this section, we define the Lanczos breakdown. Moreover, we also propose the new scheme that varies k and reproduces the pseudo Krylov subspaces when the Lanczos breakdown occurs.

4.1 The Criterion of Testing the Lanczos Breakdown

We now assume that there are k pseudo Krylov subspaces

$$K(q_j,\ A^T) = \{q_j,\ A^T q_j,\ (A^T)^2 q_j, \ldots\ldots\},\qquad j = 0,\ 1,\ \ldots, k-1.$$

Based on Van der Vorst [3], we define the pivot of the residual vector and q_j as

$$\psi_j = \frac{|(r_l, q_j)|}{(\|r_l\|_2 \cdot \|q_j\|_2)}. \quad (6)$$

Especially in case of $k = 1$, since only one pseudo Krylov subspace $K(q_0, A^T)$ is used, ML(k)BiCGStab algorithm is just the same as BiCGStab algorithm.

In BiCGStab algorithm or ML(k)BiCGStab algorithm, it becomes difficult to orthogonalize the residual vector to the pseudo Krylov subspace $K(q_j, A^T)$ when the pivot ψ_j is near zero or exact zero. These phenomena are usually called the Lanczos breakdown. When at least one of ψ_j in the equation (6) becomes near zero or exact zero, the residual vector of BiCGStab algorithm or ML(k)BiCGStab algorithm is no longer orthogonalized to the pseudo Krylov subspace $K(q_j, A^T)$. Therefore, we consider that the Lanczos breakdown occurs when the following condition

$$\psi_j < \varepsilon, \quad \exists j = 0, 1, \ldots, k-1 \quad (7)$$

is satisfied. It is usually taken $\varepsilon = \sqrt{\xi}$, where ξ is the relative machine precision. For instance, if 64 bits are used for double precision arithmetic, a suitable value of ξ is about 1.0×10^{-16}. Therefore, in this case it's suitable to set $\varepsilon = 1.0 \times 10^{-8}$. Especially in case of $\psi_j = 0$, it is called the serious breakdown. However, this case seldom occurs.

4.2 Reproducing the Pseudo Krylov Subspaces

We now propose a new variant of ML(k)BiCGStab algorithm. The proposed scheme varies k based on the Lanczos breakdown. Then it also recomputes the Lanczos starting vectors q_j and reproduces the pseudo Krylov subspaces. We combine ML(k)BiCGStab algorithm with the following three steps.

Step (i) If the condition (7) is satisfied, the current residual vector is normalized and is set to q_0.

Step (ii) If $k \leq k_{\max}$ is satisfied, k is increased by one. Otherwise, if $k = k_{\max}$, k is returned to 1.

Step (iii) If $k \geq 2$, the remaining Lanczos starting vectors $q_1, q_2, \ldots, q_{k-1}$ are recomputed, using the Arnoldi process.

We also notice that $k_{\max}$ is the maximum value of k. In the Step (i), when at least one of these pivots is too small, the new Lanczos starting vector q_0 is obtained by normalizing the current residual vector. By carrying out the Step (i), the pivot ψ_j is able to be set to 1.0, the maximum value of the pivot ψ_j. In the Step (ii), the algorithm is stabilized by increasing k. However, if k has already reached $k_{\max}$, k gets back to 1 and the Lanczos starting vectors $q_0, q_1, \ldots, q_{k-1}$ are reproduced. This means that even if $k_{\max}$ Lanczos starting vectors have already been used, there is still the possibility that the Lanczos breakdown occurs. In this

$$\underline{\text{ML}(\le k_{\max})\text{BiCGStab algorithm}}$$

choose x_0, q_0
$r_0 = b - Ax_0$, $\quad g_0 = r_0$, $\quad l = 0$, $\quad k = 1$
while "$\|r_l\|_2 / \|r_0\|_2 \ge \epsilon$" do
$\quad w_l = Ag_l$, $\quad c_l = (q_0, w_l)$, $\quad \alpha_{l+1} = (q_0, r_l)/c_l$
$\quad u_{l+1} = r_l - \alpha_{l+1} w_l$
$\quad \rho_{l+1} = -(u_{l+1}, Au_{l+1}) / \|Au_{l+1}\|_2^2$
$\quad x_{l+1} = x_l - \rho_{l+1} u_{l+1} + \alpha_{l+1} g_l$
$\quad r_{l+1} = \rho_{l+1} Au_{l+1} + u_{l+1}$
$\quad \hat{l} = l + 1$
$\quad$for $i = 1$ to k do
$\quad\quad z_d = u_{l+i}$, $\quad z_g = r_{l+i}$, $\quad z_w = 0$
$\quad\quad$for $s = i$ to $k - 1$ do
$\quad\quad\quad \beta_{l-k+s} = -\left(q_{s+1}, z_d\right)/c_{l-k+s}$
$\quad\quad\quad z_d = z_d + \beta_{l-k+s} d_{l-k+s}$
$\quad\quad\quad z_g = z_g + \beta_{l-k+s} g_{l-k+s}$
$\quad\quad\quad z_w = z_w + \beta_{l-k+s} w_{l-k+s}$
$\quad\quad$enddo
$\quad\quad \beta_l = -\frac{(q_0, r_{l+1}) + \rho_{l+1}(q_0, z_w)}{\rho_{l+1} c_l}$
$\quad\quad z_g = z_g + \beta_l g_l$
$\quad\quad z_w = \rho_{l+1}(z_w + \beta_l w_l)$
$\quad\quad z_d = r_{l+1} + z_w$
$\quad\quad$for $s = 1$ to $i - 1$ do
$\quad\quad\quad \beta_{l+s} = -\left(q_{s+1}, z_d\right)/c_{l+s}$
$\quad\quad\quad z_d = z_d + \beta_{l+s} d_{l+s}$
$\quad\quad\quad z_g = z_g + \beta_{l+s} g_{l+s}$
$\quad\quad$enddo
$\quad\quad d_{l+i} = z_d - u_{l+i}$, $\quad g_{l+i} = z_g + z_w$
$\quad\quad$if $i < k$ then
$\quad\quad\quad c_{l+i} = \left(q_{i+1}, d_{l+i}\right)$
$\quad\quad\quad \alpha_{l+i+1} = \left(q_{i+1}, u_{l+i}\right)/c_{l+i}$
$\quad\quad\quad u_{l+i+1} = u_{l+i} - \alpha_{l+i+1} d_{l+i}$
$\quad\quad\quad x_{l+i+1} = x_{l+i} + \rho_{l+1} \alpha_{l+i+1} g_{l+i}$
$\quad\quad\quad w_{l+i} = Ag_{l+i}$
$\quad\quad\quad r_{l+i+1} = r_{l+i} - \rho_{l+1} \alpha_{l+i+1} w_{l+i}$
$\quad\quad\quad \hat{l} = \hat{l} + 1$
$\quad\quad$endif
$\quad$enddo
$\quad l = \hat{l}$
$\quad \langle\!\langle$ **beginning of new process** $\rangle\!\rangle$
$\quad$for $j = 0$ to $k - 1$ do
$\quad\quad \psi_j = |\left(r_l, q_j\right)|/(\|r_l\|_2 \cdot \|q_j\|_2)$
$\quad\quad$if $\psi_j < \varepsilon$ then
$\quad\quad\quad q_0 = r_l / \|r_l\|_2$
$\quad\quad\quad$if $k = k_{\max}$ then
$\quad\quad\quad\quad k = 1$
$\quad\quad\quad$else
$\quad\quad\quad\quad k = k + 1$
$\quad\quad\quad\quad$compute $q_1, \ldots, q_{k-1}$ by Arnoldi method
$\quad\quad\quad$endif
$\quad\quad\quad$**break** current "for loop"
$\quad\quad$endif
$\quad$enddo
$\quad \langle\!\langle$ **end of new process** $\rangle\!\rangle$
enddo

Fig. 2. ML($\le k_{\max}$)BiCGStab algorithm

case, we consider that it is necessary to recompute the Lanczos starting vectors and reproduce the pseudo Krylov subspaces. In the Step (iii), using the vector q_0 obtained from the Step (i), the remaining Lanczos starting vectors $q_1, \ldots, q_{k-1}$ are recomputed. However, since only one pseudo Krylov subspace $K(q_0, A^T)$ is used in case of $k = 1$, the Step (iii) does not have to be carried out. Even if the number of pseudo Krylov subspaces has already reached $k_{\max}$, recomputing the Lanczos starting vectors $q_0, q_1, \ldots, q_{k-1}$ may be carried out in order to avoid the Lanczos breakdown. We call ML(k)BiCGStab algorithm with the proposed scheme "ML($\leq k_{\max}$)BiCGStab algorithm". Figure 2 illustrates the sequences of ML($\leq k_{\max}$)BiCGStab algorithm. This Figure is given based on Yeung et al. [4]. The terms shown between "⟨⟨ **beginning of new process** ⟩⟩" and "⟨⟨ **end of new process** ⟩⟩" corresponds to the proposed scheme.

5 The Numerical Examples

All of the numerical examples are implemented on the MIMD parallel machine, COMPAQ Beowulf. The system is composed by the following components.

- **Cell processor:** Alpha 600MHz × 16
- **Local memory;** 1.0GB
- **Floating point arithmetic:** double precision (64bit)
- **Programing language:** C
- **Communication library:** MPI

5.1 The Performance of the Proposed Scheme

We compare the computation time of ML($\leq k_{\max}$)BiCGStab algorithm with BiCGStab algorithm and ML(k)BiCGStab algorithm. The following conditions are applied to all of the numerical examples.

The initial guess of the linear system: $x_0 = 0$
The stopping criterion: $\|r_l\|_2 / \|b\|_2 < 1.0 \times 10^{-12}$
ε, **the threshold of the condition (7):** 1.0×10^{-8}

The parameters, k and $k_{\max}$ are set to 1, 2, 4 and 8. As mentioned in Section 3, ML(1)BiCGStab algorithm is not compared because BiCGStab algorithm is the same as ML(1)BiCGStab algorithm. However, ML(≤ 1)BiCGStab algorithm is not equivalent to BiCGStab algorithm or ML(1)BiCGStab algorithm because at least Step (i) has to be executed in ML(≤ 1)BiCGStab algorithm. Therefore, ML(≤ 1)BiCGStab algorithm is tested in the numerical examples. We consider that one iteration step is carried out when the order of Krylov subspace $K(r_0, A)$ is increased by 1. No preconditioners are used for any algorithms.

[**Example 1**]. We study the boundary value problem of 3-dimensional linear partial differential equation [2] on the square region $\Omega = [0, 1]^3$,

$$a_1 u_{xx} + a_2 u_{yy} + a_3 u_{zz} + 10.0(a_4 u_x + a_5 u_y + a_6 u_z) + a_7 u = f \text{ on } \Omega$$

Table 1. The numerical results in Example 1, (time: computation time (sec), iter: iterations)

Algorithm	time	iter
BiCGStab	16.0	338
ML(2)BiCGStab	17.0	392
ML(4)BiCGStab	19.0	416
ML(8)BiCGStab	30.0	480
ML(≤ 1)BiCGStab	12.0	279
ML(≤ 2)BiCGStab	14.0	336
ML(≤ 4)BiCGStab	15.0	333
ML(≤ 8)BiCGStab	21.0	440

Table 2. Example 1: The occurrence of the Lanczos breakdown of ML($\leq k_{\max}$)BiCGStab algorithm

$k_{\max}$	ψ_0	ψ_1	ψ_2	ψ_3	ψ_4	ψ_5	ψ_6	ψ_7
1	6	–	–	–	–	–	–	–
2	4	0	–	–	–	–	–	–
4	3	1	0	0	–	–	–	–
8	1	1	0	0	1	0	0	0

where

$$a_1 = 2 + \sin(2\pi x)\cos(2\pi y)\cos(2\pi z),$$
$$a_2 = 2 + \cos(2\pi x)\sin(2\pi y)\cos(2\pi z),$$
$$a_3 = 2 + \cos(2\pi x)\cos(2\pi y)\sin(2\pi z),$$
$$a_4 = \sin(4\pi x), \ a_5 = \sin(4\pi y), \ a_6 = \sin(4\pi z),$$
$$a_7 = \sin(2\pi x)\sin(2\pi y)\sin(2\pi z).$$

The right-hand side f and the boundary conditions are determined so that the exact solution is

$$u(x, y, z) = \sin(2\pi x)\cos(2\pi y)\sin(2\pi z).$$

We discretize this equation by using the finite differential method with 80^3 grid points. This yields the linear system of order of 512,000. In Table 1, we show the computation time and iterations required for satisfying the stopping criterion. In most of the cases, ML($\leq k_{\max}$)BiCGStab algorithm performs better than BiCGStab algorithm and ML(k)BiCGStab algorithm. Especially, the computation time of ML(≤ 8)BiCGStab algorithm is decreased by 30% in compared with ML(8)BiCGStab algorithm. Table 2 gives the occurrence that the pivots of ML($\leq k_{\max}$)BiCGStab algorithm satisfy the condition (7). In any cases, the Lanczos breakdown of ψ_0 is more likely to occur than the other pivots $\psi_1, \ldots, \psi_{k-1}$. It's also shown that the Lanczos breakdown is less likely to occur as k is increased.

[Example 2]. We study the boundary value problem of the 3-dimensional nonlinear partial differential equations [2] on the square region $\Omega = [0, 1]^3$

$$u_{xx} + u_{yy} + u_{zz} + 0.1(uu_x + vu_y + wu_z) + u = f_1 \text{ on } \Omega$$
$$v_{xx} + v_{yy} + v_{zz} + 0.1(uv_x + vv_y + wv_z) + v = f_2 \text{ on } \Omega$$
$$w_{xx} + w_{yy} + w_{zz} + 0.1(uw_x + vw_y + ww_z) + w = f_3 \text{ on } \Omega.$$

The right-hand side functions f_1, f_2, f_3 and the boundary conditions are defined so that the exact solutions are as follows:

$$u = \sin(\pi x)\cos(\pi y)\cos(\pi z), \quad v = \cos(\pi x)\sin(\pi y)\cos(\pi z),$$
$$w = \cos(\pi x)\cos(\pi y)\sin(\pi z).$$

We let h denote the width between two grid points and set $h = 1/81$. Therefore, as for $i, j, k = 1, 2, \ldots 80$, the nonlinear system of order of 1,536,000

$$q(g_{i,j,k}) = a_{1,i,j,k} g_{i,j,k-1} + a_{2,i,j,k} g_{i,j-1,k} + a_{3,i,j,k} g_{i-1,j,k} + (h^2 - 6)g_{i,j,k}$$
$$+ a_{4,i,j,k} g_{i,j,k+1} + a_{5,i,j,k} g_{i,j+1,k} + a_{6,i,j,k} g_{i+1,j,k} - h^2 f_{i,j,k} = 0 \quad (8)$$

is generated, where

$$a_{1,i,j,k} = 1 - \frac{0.1h}{2} w_{i,j,k}, \quad a_{2,i,j,k} = 1 - \frac{0.1h}{2} v_{i,j,k}, \quad a_{3,i,j,k} = 1 - \frac{0.1h}{2} u_{i,j,k},$$
$$a_{4,i,j,k} = 1 + \frac{0.1h}{2} w_{i,j,k}, \quad a_{5,i,j,k} = 1 + \frac{0.1h}{2} v_{i,j,k}, \quad a_{6,i,j,k} = 1 + \frac{0.1h}{2} u_{i,j,k},$$
$$g_{i,j,k} = [u_{i,j,k}, v_{i,j,k}, w_{i,j,k}]^T, \quad f_{i,j,k} = [f_{1,i,j,k}, f_{2,i,j,k}, f_{3,i,j,k}]^T.$$

Moreover, $u_{i,j,k}$, $v_{i,j,k}$, $w_{i,j,k}$, $f_{1,i,j,k}$, $f_{2,i,j,k}$, $f_{3,i,j,k}$ are the values of the functions u, v, w, f_1, f_2, f_3 on the grid point (ih, jh, kh), respectively. For the nonlinear system (8), the approximate solution $g_{i,j,k}$ is obtained, using Newton method. We let s_l and $q_l(s_l)$ denote l-th approximate solution and residual vector of Newton method, respectively and define as follows:

$$s_l = [g_{l,1,1,1}^T, g_{l,2,1,1}^T, \ldots, g_{l,80,1,1}^T, g_{l,1,2,1}^T, g_{l,2,2,1}^T, \ldots, \ldots, g_{l,80,80,80}^T]^T$$
$$q_l(s_l) = [q(g_{l,1,1,1})^T, q(g_{l,2,1,1})^T, \ldots, q(g_{l,80,1,1})^T,$$
$$q(g_{l,1,2,1})^T, q(g_{l,2,2,1})^T, \ldots, \ldots, q(g_{80,80,80})^T]^T.$$

s_l and $q_l(s_l)$ are the vector of order of 1,536,000 and the Newton recurrence can be described as

$$s_{l+1} = s_l - J_l^{-1}(s_l) q_l(s_l), \quad (9)$$

where $J_l(s_l)$ is the Jacobi matrix of the vector $q_l(s_l)$. The initial guess s_0 is computed by using the linear Lagrange polynomial which connects the grid points $(ih, jh, 0)$ and $(ih, jh, 1)$ as for $\forall i, j$. In order to do one step of Newton method, the computation of $J_l^{-1}(s_l) q_l(s_l)$ has to be carried out once. Therefore the linear system

$$J_l(s_l) x = q_l(s_l) \quad (10)$$

are solved by BiCGStab algorithm, ML(k)BiCGStab algorithm and ML($\leq k_{\max}$) BiCGStab algorithm. The stopping criterion of the Newton formula (9) is set to

$$\|q_l(s_l)\|_2 / \|q_0(s_0)\|_2 < 1.0 \times 10^{-12} \quad (11)$$

Table 3. The numerical results for Example 2, (time: the computation time for the linear system (10) (sec), iter: iterations for the linear system (10))

algorithm	The Newton iteration						total time
	1st step		2nd step		3rd step		
	time	iter	time	iter	time	iter	
BiCGStab	34.0	249	39.0	268	40.0	278	113.0
ML(2)BiCGStab	35.0	262	37.0	282	44.0	334	106.0
ML(4)BiCGStab	43.0	292	47.0	320	60.0	404	150.0
ML(8)BiCGStab	71.0	360	89.0	448	88.0	440	248.0
ML(≤ 1)BiCGStab	27.0	188	36.0	242	43.0	292	106.0
ML(≤ 2)BiCGStab	31.0	211	35.0	240	39.0	269	105.0
ML(≤ 4)BiCGStab	30.0	209	39.0	264	46.0	315	115.0
ML(≤ 8)BiCGStab	37.0	241	44.0	289	47.0	314	128.0

Table 4. Example 2: The occurrence of the Lanczos breakdown of ML($\leq k_{\max}$)B-iCGStab algorithm (1st iteration of Newton recurrence (9))

$k_{\max}$	ψ_0	ψ_1	ψ_2	ψ_3	ψ_4	ψ_5	ψ_6	ψ_7
1	8	–	–	–	–	–	–	–
2	6	0	–	–	–	–	–	–
4	4	1	0	0	–	–	–	–
8	2	1	1	0	1	0	0	0

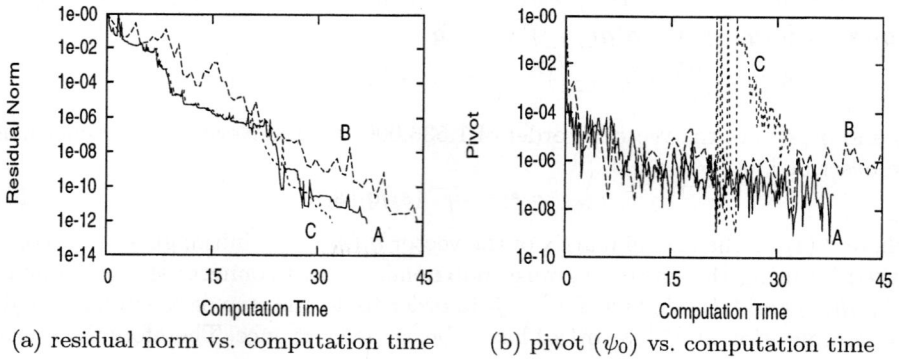

(a) residual norm vs. computation time (b) pivot (ψ_0) vs. computation time

Fig. 3. Example 2: The behavior of two values vs. computation time (1st iteration step of Newton method), A: BiCGStab algorithm, B: ML(4)BiCGStab algorithm, C: ML(≤ 4)BiCGStab algorithm

About the numerical results, the number of iterations of the Newton recurrence (9) is required three times for satisfying the stopping criterion (11).

Therefore, as for three steps of Newton method, the computation time and iterations for solving the linear system (10) are shown in Table 3. The notation "total time" in Table 3 means the total computation time for solving the linear system (10) three times. Roughly speaking, the total computation time of ML($\leq k_{\max}$)BiCGStab algorithm is less expensive than the two algorithms. Especially, in the best case, the computation time of ML(≤ 8)BiCGStab algorithm is required only 50% of ML(8)BiCGStab algorithm.

We now focus on the 1st iteration of the Newton recurrence (9). We analyze BiCGStab algorithm, ML(4)BiCGStab algorithm and ML(≤ 4)BiCGStab algorithm for solving the linear system (10). The behaviors of the residual norm and the pivot ψ_0 are shown in Figure 3. Figure 3.(a) shows that the convergences of ML(≤ 4)BiCGStab algorithm and BiCGStab algorithm are almost the same about 15 seconds. However, going past about 15 second, the convergence of ML(≤ 4)BiCGStab algorithm is more accelerated than BiCGStab algorithm. the residual norm of ML(≤ 4)BiCGStab algorithm converges about 10% faster than BiCGStab algorithm. On the other hand, ML(4)BiCGStab algorithm converges more slowly than the two other algorithms. According to Figure 3.(b), the pivot ψ_0 of BiCGStab algorithm becomes less than 1.0×10^{-8} and finally never recover from the Lanczos breakdown. About ML(4)BiCGStab algorithm, the pivot ψ_0 is a little bit larger than BiCGStab algorithm. However, its value always lies in near 1.0×10^{-8}. On the other hand, the pivot ψ_0 of ML(≤ 4)BiCGStab algorithm get back to 1.0 whenever the Lanczos breakdown occurs. Moreover, it never falls down less than 1.0×10^{-8} after about 25 seconds. Therefore it is shown that ML(≤ 4)BiCGStab algorithm is more stabilized than the two original algorithms. The occurrence of the Lanczos breakdown is also shown in Table 4. Just like Example 1, Table 4 illustrates that the breakdown of ψ_0 is more occurred than the other pivots $\psi_1, \psi_2, \ldots, \psi_{k-1}$. About the pivot ψ_0 of ML(≤ 4)BiCGStab algorithm, the condition (7) is satisfied eight times. In Example 2, it also shown that the Lanczos breakdown is less likely to occur as the number of the pseudo Krylov subspaces is increased.

6 The Concluding Remarks

In this paper, we have proposed ML($\leq k_{\max}$)BiCGStab algorithm, the new variant of ML(k)BiCGStab algorithm. As mentioned in the Example 1 and 2, ML($\leq k_{\max}$)BiCGStab algorithm performs better than BiCGStab algorithm and ML(k)BiCGStab algorithm. Moreover, from Figure 3.(b) in Example 2, the pivots of BiCGStab algorithm and ML(k)BiCGStab algorithm become smaller as the computation time is passed. On the other hand, ML($\leq k_{\max}$)BiCGStab algorithm avoids the Lanczos breakdown and the pivot ψ_0 returns to 1.0 several times. Therefore, the main conclusion is that ML($\leq k_{\max}$)BiCGStab algorithm can be alternative to the original BiCGStab algorithm and ML(k)BiCGStab algorithm.

References

1. Fletcher, R.: Conjugate Gradient Methods for Indefinite Systems, *Lecture Notes in Math.*, No. 506, pp. 73–89, (1976).
2. Schönauer, W.: Scientific Computing on Vector Computers, North Holland, (1987).
3. Van der Vorst, H. A.: Bi-CGSTAB: A Fast and Smoothly Converging Variant of Bi-CG for the Solution of Non-Symmetric Linear Systems, *SIAM J. Sci. Stat. Comput.*, Vol. 13, pp. 631–644, (1992).
4. Yeung, M. and Chan, T, F.: ML(k)BiCGStab: A BiCGSTAB Variant based on Multiple Lanczos Vectors, *SIAM J. Sci. Comput.*, Vol. 21, No, 4, pp. 1263–1290, (1999).

On Algorithm for Efficiently Combining Two Independent Measures in Routing Paths*

Moonseong Kim[1], Young-Cheol Bang[2], and Hyunseung Choo[1]

[1] School of Information and Communication Engineering,
Sungkyunkwan University, 440-746, Suwon, Korea +82-31-290-7145
{moonseong, choo}@ece.skku.ac.kr
[2] Department of Computer Engineering,
Korea Polytechnic University,
429-793, Gyeonggi-Do, Korea +82-31-496-8292
ybang@kpu.ac.kr

Abstract. This paper investigates the routing efficiency problem with quality of service (QoS). A solution to this problem is needed to provide real-time communication service to connection-oriented applications, such as video and voice transmissions. We propose a new weight parameter by efficiently combining two independent measures, the cost and the delay. The weight ω plays on important role in combining the two measures. If the ω approaches 0, then the path delay is low. Otherwise the path cost is low. Therefore if we decide an ω, we then find the efficient routing path. A case study shows various routing paths for each ω. We also use simulations to show the variety of paths for each ω. When network users have various QoS requirements, the proposed weight parameter is very informative.

1 Introduction

The advanced multimedia technology with high-speed networks generates a bunch of real-time applications. Since high end-services such as videoconferencing, demand based services (Video, Music, and News on Demand), Internet broadcasting, etc. are popularized, the significance of real-time transmission has grown rapidly to support Quality of Service (QoS). As network users using high-bandwidth applications increase, efficient usage of the networks is required for the better utilization of the network resources. Also, it requires that networks offer various kinds of qualities of services, in terms of time and cost, to satisfy users' demands. It essentially becomes an optimization problem.

For real-time applications, the path delay should be acceptable while its cost should be as low as possible to meet users demands. This is called the Delay Constrained Least Cost (DCLC) path problem, which is NP-hard [4]. There are

---

* This work was supported in parts by Brain Korea 21 and the Ministry of Information and Communication in Republic of Korea. Dr. H. Choo is the corresponding author.

a lot of heuristics to solve it; CBF [13], DCUR [8], ELS [5] and so on. But the CBF has exponential running time and the DCUR frequently takes both the least delay (LD) path and the least cost (LC) path from a current node to a destination node. The LD path cost is relatively more expensive than the LC path cost, and moreover, the LC path delay is relatively higher than the LD path delay. As we have seen, the DCLC desires to find the path that takes into account both cost and delay. Even though there exists a delay sacrifice for a lower cost, the delay should be constrained by the delay bound. The negotiation between cost and delay is important. Hence, we introduce the new parameter that simultaneously regulates both the cost and the delay. Our proposed parameter can adjust them by weight $\omega \in [0,1]$.

The rest of paper is organized as follows. In Section 2, the network model and the interval estimation are explained, Section 3 presents details of the new parameter and illustrates with example. Then we analyze and evaluate the performance of the proposed parameter by simulation in Section 4. Section 5 concludes this paper.

2 Preliminaries

2.1 Network Model

We consider that a computer network is represented by a directed graph $G = (V, E)$ with n nodes and l links or arcs, where V is a set of nodes and E is a set of links, respectively. Each link $e = (i, j) \in E$ is associated with two parameters, namely link cost $c(e) > 0$ and link delay $d(e) > 0$. The delay of a link, $d(e)$, is the sum of the perceived queueing delay, transmission delay, and propagation delay. We define a path as sequence of links such that $(u, i) \to (i, j) \to \ldots \to (k, v)$, belongs to E. Let an ordering set $P(u,v) = \{(u,i), (i,j), \ldots, (k,v)\}$ denote the path from node u to node v. If all nodes u, i, j, $\ldots$, k, v in $P(u,v)$ are distinct, then we say that it is a simple directed path. For a given source node $s \in V$ and a destination node $d \in V$, $(2^{s \to d}, \infty)$ is the set of all possible paths from s to d. We define the length of the path $P(u,v)$, denoted by $n(P(u,v))$, as a number of links in $P(u,v)$. The path cost of P is given by $\phi_C(P) = \sum_{e \in P} c(e)$ and the path delay of P is given by $\phi_D(P) = \sum_{e \in P} d(e)$. $(2^{s \to d}, \Delta)$ is the set of paths from s to d for which the end-to-end delay is bounded by Δ. Therefore $(2^{s \to d}, \Delta) \subseteq (2^{s \to d}, \infty)$. The DCLC problem is to find the path that satisfies $min\{ \phi_C(P_k) \mid P_k \in (2^{s \to d}, \Delta), \forall k \in \Lambda \}$, where Λ is an index set.

2.2 Statistic Interval Estimation

An interval estimate of a parameter θ is an interval (θ_1, θ_2), the endpoints of which are functions $\theta_1 = g_1(X)$ and $\theta_2 = g_2(X)$ of the observation vector X. The corresponding random interval (θ_1, θ_2) is the interval estimator of θ. We shall say that (θ_1, θ_2) is a γ confidence interval of θ if $Prob\{\theta_1 < \theta < \theta_2\} = \gamma$. The constant γ is the confidence coefficient of the estimate and the difference $\alpha = 1 - \gamma$ is the confidence level. Thus γ is a subjective measure of our confidence that the unknown θ is in the interval (θ_1, θ_2) [7]. The $100(1-\alpha)\%$ confidence interval for the sample mean $\bar{X}$ of the X can be described by $(\bar{X} - z_{\alpha/2}\frac{S}{\sqrt{n}}, \bar{X} + z_{\alpha/2}\frac{S}{\sqrt{n}})$ when unknown variance and S is sample variance. If we would like to have the 95% confidence interval, then the solution of the following equation is $z_{\alpha/2} = 1.96$ as the percentile which means:

$$2\int_0^{z_{\alpha/2}} \frac{1}{\sqrt{2\pi}} e^{-\frac{x^2}{2}} dx = 0.95 .\quad (1)$$

3 Proposed Weight Parameter

3.1 New Parameter for the Negotiation Between Cost and Delay

In this paper, we assume that the co-domain of cost function is equal to the co-domain of delay function as a matter of convenience. We compute two paths P_{LD} and P_{LC} from s to d in a given network G. Since only link-delays are considered to compute $P_{LD}(s,d)$, $\phi_C(P_{LD})$ is always greater than or equal to $\phi_C(P_{LC})$. If the path cost, $\phi_C(P_{LD})$, is decreased by $100\big(1 - \frac{\phi_C(P_{LC})}{\phi_C(P_{LD})}\big)\%$, $\phi_C(P_{LD})$ is obviously equal to $\phi_C(P_{LC})$. Let $\bar{C} = \frac{\phi_C(P_{LD})}{n(P_{LD})}$ be the average of link costs $c(e)$ along P_{LD} with $e \in P_{LD}$. To decrease $100\big(1 - \frac{\phi_C(P_{LC})}{\phi_C(P_{LD})}\big)\%$ for $\phi_C(P_{LD})$, we consider the confidence interval $2 \times 100\big(1 - \frac{\phi_C(P_{LC})}{\phi_C(P_{LD})}\big)\%$ and should calculate its percentile. Because the normal density function is symmetric to the mean, if the value that has to be decreased is greater than or equal to 50% then we interpreted as a 99.9% confidence interval.

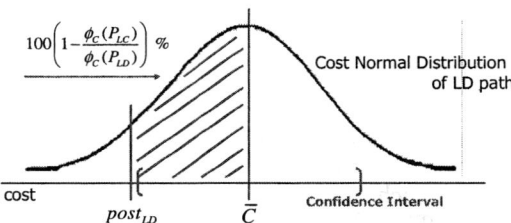

Fig. 1. $post_{LD}$

As shown in Fig. 1, $post_{LD}$ is the detection point to change the path cost from s to d. So, it is essential to find the percentile $z_{\alpha/2}^{LD}$. In order to obtain it, the cumulative distribution function (CDF) is employed. Ideally, the PDF is a discrete function but we assume that the PDF is a continuous function in convenience through out this paper. Let the CDF be $F(x) = \int_{-\infty}^x \frac{1}{\sqrt{2\pi}} e^{-\frac{y^2}{2}} dy$. Then, the percentile, $z_{\alpha/2}^{LD}$, is a solution of the following equation:

Table 1. The percentile

\multicolumn{10}{|l|}{$\eta^x = \lfloor 100 (1 - \frac{\phi_C(P_{\sim x})}{\phi_C(P_x)}) \rfloor$ %, where $LD =\sim LC$.}

$z^x_{\alpha/2} = 3.29$ if $\eta^x \geq 50$.

η^x	$z^x_{\alpha/2}$	η^x	$z^x_{\alpha/2}$	η^x	$z^x_{\alpha/2}$	η^x	$z^x_{\alpha/2}$	η^x	$z^x_{\alpha/2}$
49	2.33	48	2.05	47	1.88	46	1.75	45	1.65
44	1.56	43	1.48	42	1.41	41	1.34	40	1.28
39	1.23	38	1.18	37	1.13	36	1.08	35	1.04
34	0.99	33	0.95	32	0.92	31	0.88	30	0.84
29	0.81	28	0.77	27	0.74	26	0.71	25	0.67
24	0.64	23	0.61	22	0.58	21	0.55	20	0.52
19	0.50	18	0.47	17	0.44	16	0.41	15	0.39
14	0.36	13	0.33	12	0.31	11	0.28	10	0.25
9	0.23	8	0.20	7	0.18	6	0.15	5	0.13
4	0.1	3	0.08	2	0.05	1	0.03	0	0.00

$$F(z^{LD}_{\alpha/2}) - \frac{1}{2} = 1 - \frac{\phi_C(P_{LC})}{\phi_C(P_{LD})} \qquad (2)$$

which means

$$z^{LD}_{\alpha/2} = F^{-1}(\frac{3}{2} - \frac{\phi_C(P_{LC})}{\phi_C(P_{LD})}) \quad \text{if} \quad \lfloor 100(1 - \frac{\phi_C(P_{LC})}{\phi_C(P_{LD})}) \rfloor\% < 50\% \ . \qquad (3)$$

Table 1 shows the percentile calculated by Mathematica.
After calculating the percentile, we compute $post_{LD}$:

$$post_{LD} = \bar{C} - z^d_{\alpha/2} \frac{S_{LD}}{\sqrt{n(P_{LD})}} \qquad (4)$$

where S_{LD} is the sample standard deviation,

$$S_{LD} = \sqrt{\frac{1}{n(P_{LD}) - 1} \sum_{e \in P_{LD}} (c(e) - \bar{C})^2} \ . \qquad (5)$$

If $n(P_{LD}) = 1$, then $S_{LD} = 0$. The new cost value of each link is as follow:

$$Cfct(e, \omega) = max\{ 1, 1 + (c(e) - post_{LD}) \frac{\omega}{0.5} \}, \text{ where } 0 \leq \omega \leq 1 \qquad (6)$$

Meanwhile, $P_{LC}(s,d)$ is computed by only taking the link-cost into account. So, $\phi_D(P_{LC})$ is always greater than or equal to $\phi_D(P_{LD})$. If $\phi_D(P_{LC})$ is decreased by $100(1 - \frac{\phi_D(P_{LD})}{\phi_D(P_{LC})})\%$, then the decreased value is to be $\phi_D(P_{LD})$. Since the new delay value of each link can be derived by the same manner used in the case of P_{LD}:

$$Dfct(e, \omega) = max\{ 1, 1 + (d(e) - post_{LC}) \frac{1 - \omega}{0.5} \} \qquad (7)$$

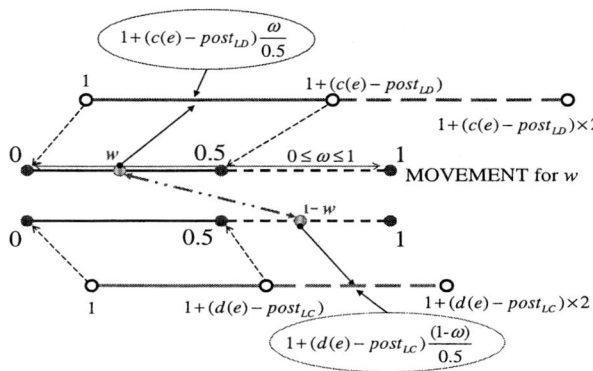

Fig. 2. The movement for an ω

Once the $Cfct(e,\omega)$ and the $Dfct(e,\omega)$ are computed, we calculate the new value $Cfct(e,\omega) \times Dfct(e,\omega)$ for each link in G. Because the best reasonable greedy method could be satisfied both the new cost value and the new delay value. Thus, links with low value of $Cfct(e,\omega) \times Dfct(e,\omega)$ should be selected. We will now get down to the center of this new parameter. The weight ω goes deep into the heart of this parameter. Here is the Fig. 2 which shows the role of ω. If ω is nearly 0, then link values are concentrated on the low delay. Nearly 1 contrasts with nearly 0. One of the notable features of ω is the regulation of path cost and delay. We use the Dijkstra's shortest path algorithm with the new weight parameter.

3.2 A Case Study

The following steps explain a process for obtaining new parameter.

Steps to calculate the New Parameter

1. Compute two paths P_{LD} and P_{LC} for the source and a destination.
2. Compute $\bar{C} = \frac{\phi_C(P_{LD})}{n(P_{LD})}$ and $\bar{D} = \frac{\phi_D(P_{LC})}{n(P_{LC})}$
3. Compute $F^{-1}\big(\frac{3}{2} - \frac{\phi_C(P_{LC})}{\phi_C(P_{LD})}\big)$ and $F^{-1}\big(\frac{3}{2} - \frac{\phi_D(P_{LD})}{\phi_D(P_{LC})}\big)$ i.e., $z_{\alpha/2}^{LD}$ and $z_{\alpha/2}^{LC}$. The function F is Gaussian distribution function.
4. Compute $post_{LD} = \bar{C} - z_{\alpha/2}^{LD} \frac{S_{LD}}{\sqrt{n(P_{LD})}}$ and $post_{LC} = \bar{D} - z_{\alpha/2}^{LC} \frac{S_{LC}}{\sqrt{n(P_{LC})}}$ $S_{(\cdot)}$ is a standard deviation.
5. Compute $Cfct(e,\omega) = max\{\,1,\, 1+(c(e)-post_{LD})\frac{\omega}{0.5}\,\}$ and $Dfct(e,\omega) = max\{\,1,\, 1+(d(e)-post_{LC})\frac{1-\omega}{0.5}\,\}$
6. We obtain the new value, $Cfct(e,\omega) \times Dfct(e,\omega)$, for each link in G.

Fig. 3 and 4 are good illustrative examples of the new weight parameter. Fig. 3 shows a given network topology G. Link costs and link delays are shown to each link as a pair $(cost, delay)$. To construct a path from the source node v_0

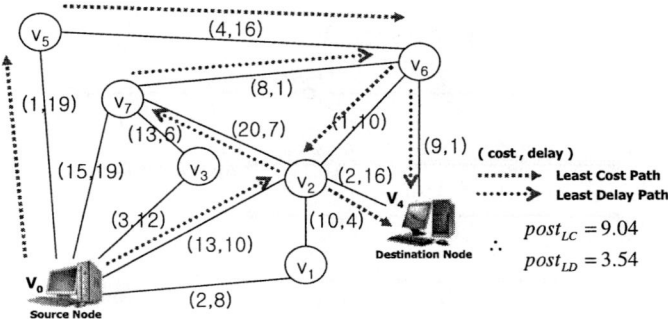

Fig. 3. A given network G, least cost path P_{LC} and least delay path P_{LD}

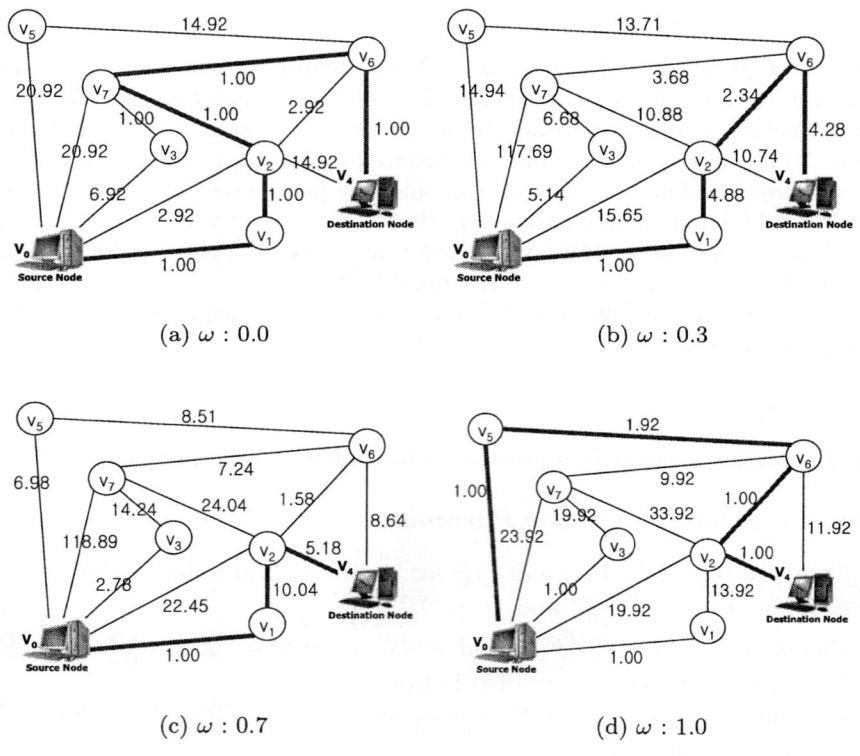

Fig. 4. The variety of paths for each ω

to the destination node v_4, we consider either link cost or link delay. The paths selected as P_{LC} and P_{LD} are shown in Fig. 3. Fig. 4 shows the paths computed by the new parameter for each weight ω. The new weight parameter is obtained as follows:

$$\bar{C} = \frac{13 + 20 + 8 + 9}{4} = 12.5 \tag{8}$$

$$S_{LD} = \sqrt{\frac{(13-\bar{C})^2 + (20-\bar{C})^2 + (8-\bar{C})^2 + (9-\bar{C})^2}{4-1}} = \sqrt{29.67} \tag{9}$$

$$\lfloor 100(1 - \frac{\phi_C(P_{LC})}{\phi_C(P_{LD})}) \rfloor\% = \lfloor 100(1 - \frac{8}{50}) \rfloor\% = 84\%. \tag{10}$$

$$\therefore z^{LD}_{\alpha/2} \approx 3.29 \tag{11}$$

$$post_{LD} = 12.5 - 3.29 \times \frac{\sqrt{29.67}}{\sqrt{4}} = 3.54 \tag{12}$$

$$Cfct(e,\omega) = max\{ 1, 1 + (c(e) - 3.54)\frac{\omega}{0.5} \}, \forall e \in E \tag{13}$$

$$\bar{D} = \frac{19 + 16 + 10 + 16}{4} = 15.25 \tag{14}$$

$$S_{LC} = \sqrt{\frac{(19-\bar{D})^2 + (16-\bar{D})^2 + (10-\bar{D})^2 + (16-\bar{D})^2}{4-1}} = \sqrt{14.25} \tag{15}$$

$$\lfloor 100(1 - \frac{\phi_D(P_{LD})}{\phi_D(P_{LC})}) \rfloor\% = \lfloor 100(1 - \frac{19}{61}) \rfloor\% = \lfloor 68.85 \rfloor\% = 69\%. \tag{16}$$

$$\therefore z^{LC}_{\alpha/2} \approx 3.29 \tag{17}$$

$$post_{LC} = 15.25 - 3.29 \times \frac{\sqrt{14.25}}{\sqrt{4}} = 9.04 \tag{18}$$

$$Dfct(e,\omega) = max\{ 1, 1 + (d(e) - 9.04)\frac{1-\omega}{0.5} \}, \forall e \in E \tag{19}$$

$$\therefore Cfct(e,\omega) \times Dfct(e,\omega), \forall e \in E. \tag{20}$$

In Fig. 4(b), we calculate $Cfct((v_1,v_2), 0.3) = max\{ 1, 1+(10-3.54)\frac{0.3}{0.5}\} = 4.88$ and $Dfct((v_1,v_2), 0.3) = max\{ 1, 1+(4-9.04)\frac{1-0.3}{0.5}\} = 1$ at link (v_1,v_2) with $\omega = 0.3$. By the same manner, we obtain all new values in the given network. Fig. 4 shows the paths constructed by the new parameter for each weight ω.

As indicated in Table 2, the path cost order is $\phi_C(P_{LC}) \leq \phi_C(P_{\omega:1.0}) \leq \phi_C(P_{\omega:0.7}) \leq \phi_C(P_{\omega:0.3}) \leq \phi_C(P_{\omega:0.0}) \leq \phi_C(P_{LD})$ and the path delay order is $\phi_D(P_{LD}) \leq \phi_D(P_{\omega:0.0}) \leq \phi_D(P_{\omega:0.3}) \leq \phi_D(P_{\omega:0.7}) \leq \phi_D(P_{\omega:1.0}) \leq \phi_D(P_{LC})$. Therefore, our method is quite likely a performance of a kth shortest path algorithm that has the high time complexity.

4 Performance Evaluation

4.1 Random Network Topology for the Simulation

Random graphs are the acknowledged model for different kinds of networks, communication networks in particular. There are many algorithms and programs, but the speed is usually the main goal, not the statistical properties. In the last decade the problem was discussed, for example, by B.M. Waxman (1993) [12],

Table 2. The comparison with example results

P_{LD}		$P_{\omega:0.0}$		$P_{\omega:0.3}$	
$\phi_C(P_{LD})$	$\phi_D(P_{LD})$	$\phi_C(P_{\omega:0.0})$	$\phi_D(P_{\omega:0.0})$	$\phi_C(P_{\omega:0.3})$	$\phi_D(P_{\omega:0.3})$
50	19	49	21	22	23
$P_{\omega:0.7}$		$P_{\omega:1.0}$		P_{LC}	
$\phi_C(P_{\omega:0.7})$	$\phi_D(P_{\omega:0.7})$	$\phi_C(P_{\omega:1.0})$	$\phi_D(P_{\omega:1.0})$	$\phi_C(P_{LC})$	$\phi_D(P_{LC})$
14	28	8	61	8	61

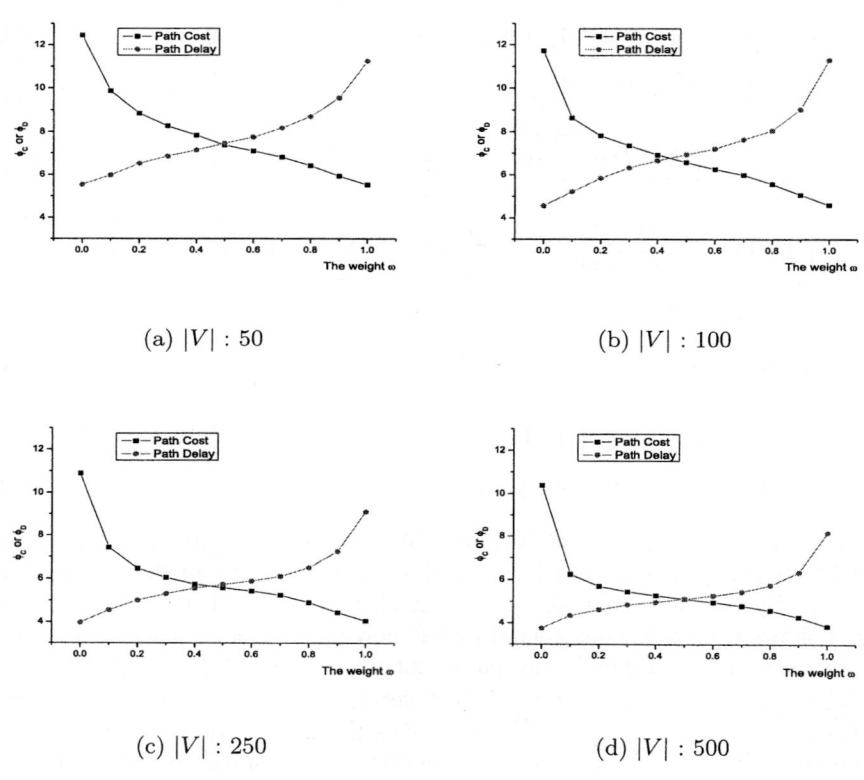

(a) $|V|$: 50

(b) $|V|$: 100

(c) $|V|$: 250

(d) $|V|$: 500

Fig. 5. Performance comparison for each P_e : 0.3 and $|V|$

M. Doar (1993, 1996) [2, 3], C.-K. Toh (1993) [11], E.W. Zegura, K.L. Calvert, and S. Bhattacharjee (1996) [14], K.L. Calvert, M. Doar, and M. Doar (1997) [1], R. Kumar, P. Raghavan, S. Rajagopalan, D. Sivakumar, A. Tomkins, and E. Upfal (2000) [6]. They have presented fast algorithms that allow generate random graph with different properties, similar to real communication networks, in particular. But none of them have discussed the stochastic properties of generated random graphs. A.S. Rodionov and H. Choo [9, 10] have formulated two major demands to the generators of random graph: attainability of all graphs with required properties and the uniformity of their distribution. If the second demand is sometimes difficult to prove theoretically, it is possible to check the distribution statistically. The random graph is similar to real networks. The method uses parameters n - the number of nodes in networks, and P_e - the probability of edge existence between any node pair.

4.2 Simulation Results

Comparing the path costs and the path delays for the new weight parameter are described here, respectively. The proposed weight parameter is implemented in C^{++}. The 10 different network environments are generated for each size of given 50, 100, 250, and 500 nodes. A source and a destination node are randomly selected in the network topology. We simulate 100 times (total $10 \times 100 = 1000$) for each network topology and P_e=0.3. Fig. 5 shows the simulation results for our method. The average path cost is decreasing as ω approaches a value of 1. Similarly, the average path delay is increasing as ω approaches a value of 1. Therefore, ω plays on important role in combining the two independent measures, cost and delay. If a delay bound is given, then we may find the path that is appropriate for the path cost. Since the new parameter simultaneously takes into account both the cost and the delay, it seems reasonable to use the new weight parameter.

5 Conclusion

This paper investigated the efficiency routing problem in point-to-point connection-oriented networks with a QoS. We formulated the new weight parameter that simultaneously took into account both the cost and the delay. The cost of least delay path is relatively more expensive than the cost of least cost path, and moreover, the delay of least cost path is relatively higher than the delay of least delay path. The weight ω plays on important role in combining the two the measures. If the ω is nearly 0, then the path delay is low. Otherwise the path cost is low. Thus, the efficient routing path can be determined once the ω is selected. It seems reasonable to use the new weight parameter. When network users have various QoS requirements, the proposed weight parameter proves to be very informative.

References

1. K.L. Calvert, M. Doar, and M. Doar, "Modelling Internet Topology," IEEE Communications Magazine, pp. 160-163, June 1997.
2. M. Doar, Multicast in the ATM environment. PhD thesis, Cambridge Univ., Computer Lab., September 1993.
3. M. Doar, "A Better Mode for Generating Test Networks," IEEE Proc. GLOBECOM'96, pp. 86-93, 1996.

4. M. Garey and D. Johnson, Computers and intractability: A Guide to the Theory of NP-Completeness, New York: Freeman, 1979.
5. M. Kim, Y.-C. Bang, and H. Choo, "Estimated Link Selection for DCLC Problem," IEEE ICC 2004, vol. 4, pp. 1937-1941, June 2004.
6. R. Kumar, P. Raghavan, S. Rajagopalan, D Sivakumar, A. Tomkins, and E Upfal, "Stochastic models for the Web graph," Proc. 41st Annual Symposium on Foundations of Computer Science, pp. 57-65, 2000.
7. A. Papoulis and S. U. Pillai, Probability, Random Variables, and Stochastic Processes, 4th ed. McGraw-Hill, 2002.
8. D.S. Reeves and H.F. Salama, "A distributed algorithm for delay-constrained unicast routing," IEEE/ACM Transactions on Networking, vol. 8, pp. 239-250, April 2000.
9. A.S. Rodionov and H. Choo, "On Generating Random Network Structures: Trees," Springer-Verlag Lecture Notes in Computer Science, vol. 2658, pp. 879-887, June 2003.
10. A.S. Rodionov and H. Choo, "On Generating Random Network Structures: Connected Graphs," Springer-Verlag Lecture Notes in Computer Science, vol. 3090, pp. 483-491, September 2004.
11. C.-K. Toh, "Performance Evaluation of Crossover Switch Discovery Algorithms for Wireless ATM LANs," IEEE Proc. INFOCOM'96, pp. 1380-1387, 1993.
12. B.M. Waxman, "Routing of Multipoint Connections," IEEE JSAC, vol. 9, pp. 1617-1622, 1993.
13. R. Widyono, "The Design and Evaluation of Routing Algorithms for Real-Time Channels," International Computer Science Institute, Univ. of California at Berkeley, Tech. Rep. ICSI TR-94-024, June 1994.
14. E.W. Zegura, K.L. Calvert, and S. Bhattacharjee, "How to model an Internetwork," Proc. INFOVCOM'96, pp. 594-602, 1996.

Real Time Hand Tracking Based on Active Contour Model

Jae Sik Chang[1], Eun Yi Kim[2], KeeChul Jung[3], and Hang Joon Kim[1]

[1] Dept. of Computer Engineering, Kyungpook National Univ., Daegu, South Korea
{jschang, hjkim}@ailab.knu.ac.kr
[2] Scool of Internet and Multimedia, NITRI*, Konkuk Univ., Seoul, South Korea
eykim@konkuk.ac.kr
[3] School of Media, College of Information Science, Soongsil University
kcjung@ssu.ac.kr

Abstract. This paper presents active contours based method for hand tracking using color information. The main problem in active contours based approach is that results are very sensitive to location of the initial curve. Initial curve far form the object induces more heavy computational cost, low accuracy of results, as well as missing the object that has a large movement. Therefore, this paper presents a hand tracking method using a mean shift algorithm and active contours. The proposed method consists of two steps: hand localization and hand extraction. In the first step, the hand location is estimated using mean shift. And the second step, at the location, evolves the initial curve using an active contour model. To assess the effectiveness of the proposed method, it is applied to real image sequences which include moving hand.

1 Introduction

Vision based gesture recognition is an important technology for perceptual human-computer interaction, and has received more and more attention in recent years [1]. Hand tracking is an essential step for gesture recognition, where location or shape of the hand must be known before recognition.

Recently, active contour models are successfully used for object boundary detection and tracking because of their ability to effectively descript curve and elastic property. So, they have been applied to many applications such as non-rigid object (hand, pedestrian and etc.) detection and tracking, shape warping system and so on [2, 3, 4].

In the tracking approaches based on active contour models, the object tracking problem is considered as a curve evolution problem, i.e., initial curve is evolved until it matches the object boundary of interest [2, 3]. The curve evolution based approaches have been used due to their following advantages: 1) saving computation time, and 2) avoiding local optima. Generally, the curve evolutions are computed in narrow band around the current curve. This small computation area induces low computation cost. And starting point of evolution (initial curve) near the global optimum (object boundary) guarantees practically the convergence to global optimum.

---

* Next-Generation Innovative Technology Research Institute.

However the advantages are very sensitive to conditions of the initial curve such as location, scale and shape. Among these conditions, location of the initial curve has a high effect on the results. Initial curve far from the object induces more heavy computational cost, low accuracy of results, as well as missing the object that has a large movement.

Accordingly, this paper proposes a method for hand tracking using mean shift algorithm and active contours. The method consists of two steps: hand localization and hand extraction. In the first step, the hand location is estimated using mean shift. And the second step, at the location, evolves the initial curve using an active contour model. The proposed method not only develops the advantage of the curve evolution based approaches but also adds the robustness to large amount of motion of the object. Additionally, we use skin color information as hand feature, which is represented by a 2D-Gaussian model. The use of skin color information endows the proposed method for robustness to noise.

The remainder of the paper is organized as follows. Chapter 2 illustrates how to localize the hand using mean shift algorithm and active contours based hand detection method is shown in chapter 3. Experimental results are presented in chapter 4. Finally, chapter 5 concludes the paper.

2 Hand Localization

2.1 Mean Shift Algorithm

The mean shift algorithm is a nonparametric technique that climbs the gradient of a probability distribution to find the nearest dominant mode (peak) [5, 6]. The algorithm has recently been adopted as an efficient technique for object tracking [6, 7].

The algorithm simply replacing the search window location (the centroid) with a object probability distribution $\{P(I_{ij}|\alpha_o)\}_{i,j=1,...,IW,IH}$ (IW: image width, IH: image height) which represent the probability of a pixel (i,j) in the image being part of object, where α_o is its parameters and I is a photometric variable. The search window location is simply computed as follows [5, 6, 7]:

$$x = M_{10}/M_{00} \quad \text{and} \quad y = M_{01}/M_{00}, \tag{1}$$

where M_{ab} is the $(a+b)th$ moment as defined by

$$M_{ab}(W) = \sum_{i,j \in W} i^a j^b P(I_{ij} | \alpha_o).$$

The object location is obtained by successive computations of the search window location (x,y).

2.2 Hand Localization Using Mean Shift

The mean shift algorithm for hand localization is as follows:

1. Set up initial location of search window W in the current frame with final location in the previous frame, and repeat Steps 2 to 3 until terminal condition is satisfied.

2. Generate a hand probability distribution within W.
3. Estimate the search window location using Eq. (1).
4. Output the window location as the object location.

If the variation of the window location is smaller than a threshold value, then the terminal condition is satisfied.

In the mean shift algorithm, instead of calculating the hand probability distribution over the whole image, the distribution calculation can be restricted to a smaller image region within the search window. This results in significant computational savings when the hand does not dominate the image [5].

3 Hand Extraction

3.1 Active Contours Based on Region Competition

Zhu and Yuille proposed a hybrid approach to image segmentation, called region competition [8]. Their basic functional is as follows:

$$E[\Gamma,\{\alpha_i\}] = \sum_{i=1}^{M}\left\{\frac{\mu}{2}\int_{R_i} ds - \log P(\{I_s : s \in R_i\}|\alpha_i) + \lambda\right\} \quad (2)$$

where Γ is the boundary in the image, $P(\cdot)$ is a specific distribution for region R_i, α_i is its parameters, M is the number of the regions, s is a site of image coordinate system, and μ and λ are two constants.

To minimize the energy E, steepest descent can be done with respect to boundary Γ. For any point $\vec{v}$. On the boundary Γ we obtain:

$$\frac{d\vec{v}}{dt} = -\frac{\delta E[\Gamma,\{\alpha_i\}]}{\delta \vec{v}} \quad (3)$$

where the right-hand side is (minus) the functional derivative of the energy E.

Taking the functional derivative yields the motion equation for point $\vec{v}$:

$$\frac{d\vec{v}}{dt} = \sum_{k \in Q_{(\vec{v})}}\left\{-\frac{\mu}{2}k_{k(\vec{v})}\vec{n}_{k(\vec{v})} + \log P(I_{(\vec{v})}|\alpha_k)\vec{n}_{k(\vec{v})}\right\} \quad (4)$$

where $Q_{(\vec{v})} = \{k \mid \vec{v} \text{ lies on } \Gamma_k\}$, i.e., the summation is done over those regions R_k for which $\vec{v}$ is on Γ_k. $k_{k(\vec{v})}$ is the curvature of Γ_k at point $\vec{v}$ and $\vec{n}_{k(\vec{v})}$ is the unit normal to Γ_k at point $\vec{v}$.

Region competition contains many of the desirable properties of region growing and active contours. Indeed we can derive many aspects of these models as special cases of region competition [8, 9]. Active contours can be a special case in which there are two regions (object region R_o and background region R_b) and a common boundary Γ as shown in follows:

$$\frac{d\vec{v}}{dt} = -\mu k_{o(\vec{v})}\vec{n}_{o(\vec{v})} + \left(\log P(I_{(\vec{v})}|\alpha_o) - \log P(I_{(\vec{v})}|\alpha_b)\right)\vec{n}_{o(\vec{v})} \quad (5)$$

3.2 Level Set Implementation

The active contour evolution was implemented using the level set technique. We represent curve Γ implicitly by the zero level set of function $u : \mathcal{R}^2 \to \mathcal{R}$, with the region inside Γ corresponding to $u > 0$. Accordingly, Eq. (5) can be rewritten by the following equation, which is a level set evolution equation [2, 3]:

$$\frac{du(s)}{dt} = -\mu k_s \|\nabla u\| + \left(\log P(I_s \mid \alpha_o) - \log P(I_s \mid \alpha_b)\right)\|\nabla u\|, \qquad (6)$$

where

$$k = \frac{u_{xx} y_y^2 - 2u_y u_x u_{xy} + u_{yy} u_x^2}{(u_x^2 + u_y^2)^{3/2}}.$$

3.3 Hand Extraction Using Active Contours

The aim of the hand extraction is to find closed curve that separates the image into hand and background regions. The hand to be tracked is assumed to be characterized by skin color which has a 2-D Gaussian distribution $P(I_s \mid \alpha_o)$ in chromatic color space. Unlike in the hand region, the background is difficult to be characterized a simple probability distribution. The distribution is not clustered in a small area of a feature space due to their variety. However, it is spread out across the whole space uniformly for a variety of background regions. From that, we can assume that the photometric variable of background is uniformly distributed in the space. Thus, the distribution $P(I_s \mid \alpha_b)$ can be proportional to a constant value.

Active contour model based hand boundary extraction algorithm is as follows:

1. Set up initial level values u, and repeat Steps 2 to 3 until terminal condition is satisfied.
2. Update level values using Eq. (6) within narrow band around curve, zero level set.
3. Reconstruct the evolved curve, zero level set.
4. Output the final evolved curve as the object boundary.

To set up the initial level values, we use a Euclidian distance mapping technique. Euclidian distance between each pixel of the image and initial curve is assigned to the pixel as a level value. In general active contours, the search area for optimal boundary curve is restricted to the narrow band around curve. This not only save computational cost but also avoid the local optima when the initial curve is near the hand boundary. However it makes the evolving curve miss the boundary when the curve is far from the hand.

After updating the level values, the approximated final propagated curve, the zero level set, is reconstructed. Curve reconstruction is accomplished by determining the zero crossing grid location in the level set function. The terminal condition is satisfied when the difference of the number of pixel inside contour Γ is less than a threshold value chosen manually.

4 Experimental Results

This paper presents a method for tracking hand which has a distribution over color. This section focuses on evaluating the proposed method. In order to assess the effectiveness of the proposed method, it was compared with those obtained using the active contours for distribution tracking proposed by Freedman et al. [2].

Freedman's method finds the region such that the sample distribution of the interior of the region most closely matches the model distribution using active contours. For matching distribution, the method examined Kullback-Leibler distance and Bhattacharyya measure. In this experiment, we only have tested the former.

4.1 Evaluation Function

To quantitatively evaluate the performance of the two methods, The Chamfer distance was used. This distance has been many used as matching measure between shapes [10]. To calculate the distance, ground truths are manually extracted from images to construct accurate boundaries of each hand. Then, the distances between the ground truth and the hand boundaries extracted by the respective method are calculated.
The Chamfer distance is the average over one shape of distance to the closet point on the other and defined as

$$C(\Gamma_H, \Gamma_G) = \frac{1}{3}\sqrt{\frac{1}{n}\sum_{i=1}^{n} v_i^2} \qquad (7)$$

where Γ_H and Γ_G are hand boundary detected by the proposed method and manually, respectively. In Eq. (7), v_i are the distance values from each point on Γ_H to the closet point on Γ_G and n is the number of points in the curve. The distance values v_i were described in [10].

4.2 Hand Tracking Results

For photometric variable which describe the hands, we use skin-color information which is represented by a 2D-Gaussian model. In the RGB space, color representation includes both color and brightness. Therefore, RGB is not necessarily the best color representation for detecting pixels with skin color. Brightness can be removed by dividing the three components of a color pixel (R, G, B) according to intensity. This space is known as chromatic color, where intensity is a normalized color vector with two components (r, g). The skin-color model is obtained from 200 sample images. Means and covariance matrix of the skin color model are as follows:

$$m = (\bar{r}, \bar{g}) = (117.588, 79.064),$$

$$\Sigma = \begin{bmatrix} \sigma_r^2 & \rho_{X,Y}\sigma_g\sigma_r \\ \rho_{X,Y}\sigma_r\sigma_g & \sigma_g^2 \end{bmatrix} = \begin{bmatrix} 24.132 & -10.085 \\ -10.085 & 8.748 \end{bmatrix}.$$

The hand tracking result in real image sequence is shown in Fig. 1. In the first frame, an initial curve was manually selected around the object, and then the curve was evolved using only active contours. The proposed method is successful in tracking through the entire 80-frame sequence. Freedman's method also succeeds in the hand tracking in the sequence, because the sequence has high capture rate and hand has not a large movement. However Freedman's method takes lager time to track the hand than the proposed method as shown in Table 1. The Chamfer distances of the two methods are shown in Fig. 2. In the case of the proposed method, hand localization using mean shift is considered as the first iteration. The distance in the proposed method decreases more dramatically and the method satisfies the stopping criteria after less iteration than Freedman's method.

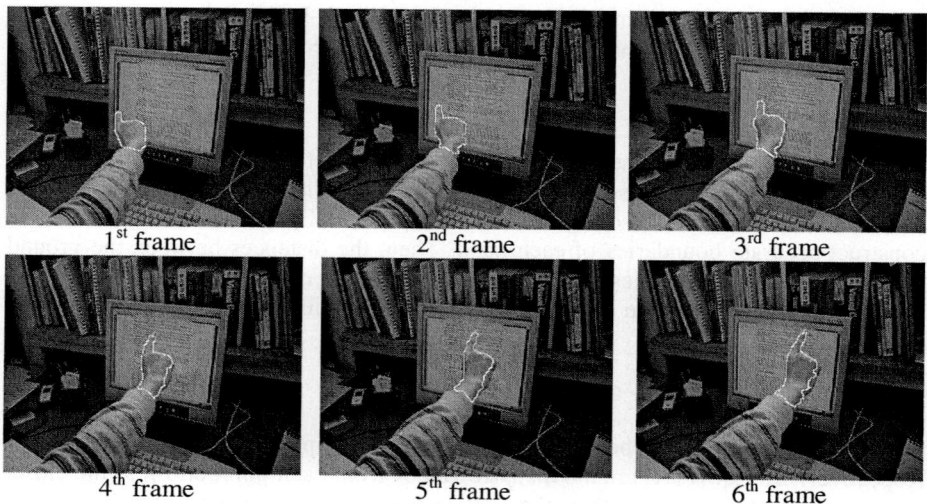

Fig. 1. Tracking hand with the proposed method

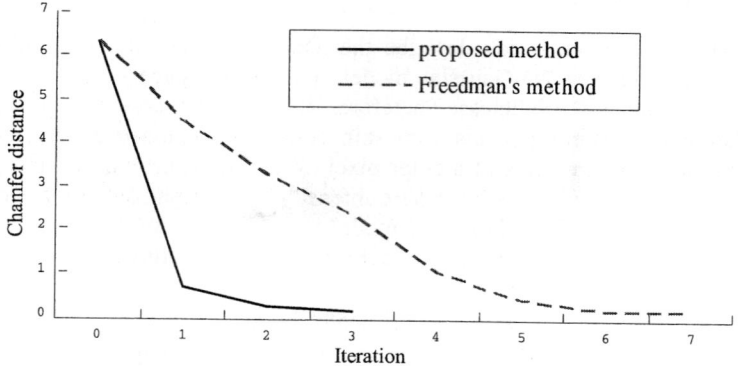

Fig. 2. Comparison of two methods in term of the Chamfer distance

One of the problems of almost active contours is that the search areas for optima are limited to the narrow band around curve. Because of it, the active contours have difficulties to track objects that have large amount of motion. The other side, in the proposed method, the initial curve is moved near the global optimum before curve evolution. Accordingly, the method is more effective to track the hand that have large amount of motion. Fig.3 and 4 show the tracking results, in an image sequence which includes hand that have large amount of motion, extracted by proposed and Freedman's methods, respectively. As shown in Fig.3 and 4, the proposed method track hand boundary accurately, on the contrary, the Freedman's method fails to track the hand.

Table 1. Time taken for tracking in an image sequence (sec.)

	1st frame	2nd frame	3rd frame	4th frame	5th frame	6th frame
Feedman's method	0.192000	0.360000	0.359000	0.453000	0.188000	0.438000
proposed method	0.192000	0.188000	0.187000	0.218000	0.156000	0.188000

Fig. 3. Tracking hand in an image sequence which includes a large amount of motion by proposed method

Fig. 4. Tracking hand in an image sequence which includes a large amount of motion by Freedman's method

5 Conclusions

In this paper, we have proposed an active contour model based hand tracking with mean shift algorithm. In the approaches based on active contour models, the object tracking problem is considered as a curve flow problem and their results are very sensitive to condition of initial contour. Bad initial condition induces a heavy computational cost, low accuracy of results, and missing the object that has a large movement. Accordingly, the proposed method consisted of two steps: hand localization and hand extraction. The first step finds the hand location using a mean shift algorithm. And at the location, the initial curve is evolved using an active contour model to find object boundary. The experimental results shown demonstrate that the proposed method yields accurate tracking results despite low computational cost.

Acknowledgement

This work was supported by grant No. F01-2004-000-10402-0 from the International Cooperative Research Program of the Korea Science & Engineering Foundation.

References

1. Shan, C., Wei, Y., Tan, T., Ojardias, F.: Real Time Hand Tracking by Combining Particle Filtering and Mean Shift. Proceeding of the Sixth IEEE International Conference on Automatic Face and Gesture Recognition. (2004) 669-674
2. Freedman, D., Zhang, T.: Active Contours for Tracking Distributions. IEEE Transactions on Image Processing. Vol. 13, No. 4 (2004) 518-526
3. Chan, T. F., Vese, L. A.: Active Contours Without Edges. IEEE Transactions on Image Processing. Vol. 10, No. 2 (2001) 266-277
4. Gastaud, M., Barlaud, M., Aubert, G.: Combining Shape Prior and Statistical Features for Active Contour Segmentation. IEEE Transactions on Circuits and Systems for Video Technology. Vol. 14. No. 5 (2004) 726-734
5. Kim, K. I., Jung, K., Kim, J. H.:Texture-Based Approach for Text Detection in Image Using Support Vector Machines and Continuously Adaptive Mean Shift Algorithm. IEEE Transactions on Pattern Analysis and Machine Intelligence. Vol. 25, No. 12 (2003) 1631-1639
6. Bradski, G. R.: Computer Vision Face Tracking For Use in a Perceptual User Interface. Intel Technology Journal 2nd quarter (1998) 1-15
7. Jaffre, G., Crouzil, A.: Non-rigid Object Localization From Color Model Using Mean Shift. In Proceedings of the International Conference on Image Processing, Vol. 3 (2003) 317-319
8. Zhu, S. C., Yuille, A.: Region Competition: Unifying Snakes, Region Growing, and Bayes/MDL for Multiband Image Segmentation. IEEE Transactions on Pattern Analysis and Machine Intelligence. Vol. 18, No 9 (1996) 884-900
9. Mansouri, A.: Region Tracking via Level Set PDEs without Motion Computation. IEEE Transactions on Pattern Analysis and Machine Intelligence. Vol. 24, No. 7 (2002) 947-961
10. Borgefors, G.: Hierarchical Chamfer Matching: A Parametric Edge Matching Algorithm. IEEE Transactions on Pattern Analysis and Machine Intelligence. Vol. 10. No. 11 (1998) 849-865

Hardware Accelerator for Vector Quantization by Using Pruned Look-Up Table

Pi-Chung Wang[1], Chun-Liang Lee[1], Hung-Yi Chang[2], and Tung-Shou Chen[3]

[1] Telecommunication Laboratories, Chunghwa Telecom Co., Ltd.
7F, No. 11 Lane 74 Hsin-Yi Rd. Sec. 4, Taipei, Taiwan 106, R.O.C.
{abu, chlilee}@cht.com.tw
[2] Department of Information Management,
I-Shou University, Kaohsiung, Taiwan 840, R.O.C.
leorean@isu.edu.tw
[3] Institute of Computer Science and Information Technology,
National Taichung Institute of Technology, Taichung, Taiwan 404, R.O.C.
rcchens@ntit.edu.tw

Abstract. Vector quantization (VQ) is an elementary technique for image compression. However, searching for the nearest codeword in a codebook is time-consuming. The existing schemes focus on software-based implementation to reduce the computation. However, such schemes also incur extra computation and limit the improvement. In this paper, we propose a hardware-based scheme *"Pruned Look-Up Table"* (PLUT) which could prune possible codewords. The scheme is based on the observation that the minimum one-dimensional distance between the tested vector and its matched codeword is usually small. The observation inspires us to select likely codewords by the one-dimensional distance, which is represented by bitmaps. With the bitmaps containing the positional information to represent the geometric relation within codewords, the hardware implementation can succinctly reduce the required computation of VQ. Simulation results demonstrate that the proposed scheme can eliminate more than 75% computation with an extra storage of 128 Kbytes.

1 Introduction

VQ is an important technique for image compression, and has been proven to be simple and efficient [1]. VQ can be defined as a mapping from k-dimensional Euclidean space into a finite subset C of R^k. The set C is known as the *codebook* and $C = \{c_i | i = 1, 2, \ldots, N\}$, where c_i is a *codeword* and N is the codebook size. To compress an image, VQ comprises two functions: an encoder and a decoder. The VQ encoder first divides the image into $N_w \times N_h$ blocks (or vectors). Let the block size be k ($k = w \times h$), then each block is a k-dimensional vector. VQ selects an appropriate codeword $c_q = [c_{q(0)}, c_{q(1)}, \ldots, c_{q(k-1)}]$ for each image vector $x = [x_{(0)}, x_{(1)}, \ldots, x_{(k-1)}]$ such that the distance between x and c_q is the smallest, where c_q is the closest codeword of x and $c_{q(j)}$ denotes the jth-dimensional value of the codeword c_q. The distortion between the image vector x and each codeword c_i is measured by the *squared Euclidean distance*, i.e.,

$$d(x, c_i) = \|x - c_i\|^2 = \sum_{j=0}^{k-1}[x_{(j)} - c_{i(j)}]^2. \tag{1}$$

After the selection of the closest codeword, VQ replaces the vector x by the *index* q of c_q. The VQ decoder has the same codebook as that of the encoder. For each index, VQ decoder can easily fetch its corresponding codeword, and piece them together into the decoded image.

The codebook search is the major bottleneck in VQ. From equation (1), the calculation of the squared Euclidean distance needs k subtractions and k multiplications to derive k $[x_{(j)} - c_{i(j)}]^2$s. Since the multiplication is a complex operation, it increases the total computational complexity of equation (1). Therefore, speeding up the calculation of the squared Euclidean distance is a major hurdle. Furthermore, an efficient hardware implementation is also attractive to reduce the VQ computation.

Many methods have been proposed to shorten VQ encoding time [2,3,4,5,6]. These schemes emphasize computation speed, table storage and image quality. The existing schemes focus on software-based implementation to reduce the computation. However, such schemes also incur extra computation and limit the improvement. Moreover, these schemes did not utilize the geometrical information implied in the codewords. In this work, we propose an adaptive scheme *"Pruned Look-Up Table"* (PLUT) which selects the computed codewords. The new scheme uses bitmaps to represent the geometric relation within codewords. Accordingly, the search procedure could refer the information to sift unlikely codewords easily. Since the lookup procedure is simple enough, the proposed scheme is suitable for hardware implementation. With the bitmaps containing the positional information to represent the geometric relation within codewords, the hardware implementation can succinctly reduce the required computation of VQ. Simulation results demonstrate the effectiveness. The rest of this paper is organized as follows. The proposed scheme and implementation are presented in Section 2. Section 3 addresses the performance evaluation. Section 4 concludes the work.

2   PLUT Method

To compress an image through VQ, the codebook must be generated first. The codebook is gathered through approach, like the Lindo-Buzo-Gray (LBG) algorithm [7], based on one or multiple images. The quality of the compressed images ties to whether the codebook is well trained, i.e., the squared Euclidean distance between the tested vector and the matched codeword in the adopted codebook is small. Thus, a well trained codebook could improve the compression quality. As implied in the equation of squared Euclidean distance calculation, a well-trained codebook can lead to the implication that the one-dimensional distance, $|x_{(j)} - c_{M(j)}|$ where $0 \leq j \leq k-1$, between the tested vector x and the matched codeword c_M should be relatively small.

To further verify our assumption, the distribution of the smallest one-dimensional distance $\min_{j=0}^{k-1}|x_{(j)} - c_{M(j)}|$ between the tested vectors and their matched codewords is presented in Fig. 1. The codebook is trained according to the image "Lena", then the six images are compressed by full search VQ. The quality of the images are estimated by the *peak signal-to-noise ratio* (PSNR), which is defined as PSNR=$10 \cdot log_{10}(255^2/MSE)$ dB. Here the *mean-square error* (MSE) is defined as MSE= $(1/m)^2 \sum_{i=0}^{m-1} \sum_{j=0}^{m-1} [\alpha_{(i,j)} - \beta_{(i,j)}]^2$ for an $m \times m$ image, where $\alpha_{(i,j)}$ and $\beta_{(i,j)}$ denote the original and quantized gray level of pixel (i, j) in the image, respectively. A larger PSNR value has been proven to have preserved the original image quality better. For the compressed images with better quality, including "Lena" and "Zelda", most of their smallest one-dimensional distances are less than 8. Furthermore, 99% smallest one-dimensional distances are less than 4. However, the ratio is reduced to 93% $\sim$ 97% for the other images since their quality of compression is also decreased.

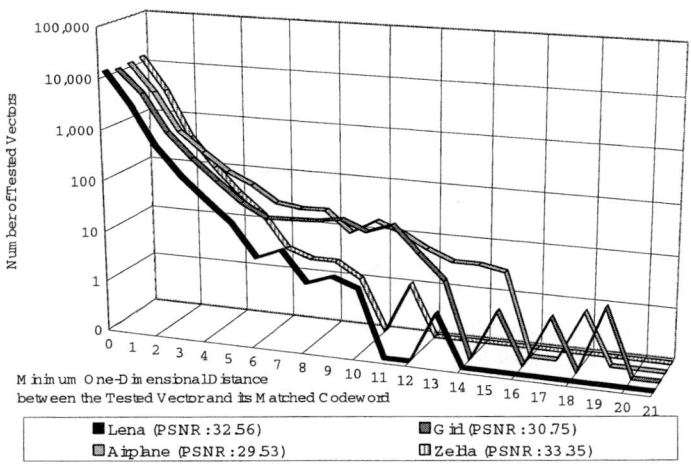

Fig. 1. The Distribution of the Smallest One-dimensional Distance for Different Images

We use a two-dimensional VQ as an example. There are two codewords, C_1 (3, 1) and C_2 (2, 2). To calculate the nearest codeword for the tested vector, V_1 (1, 2), the squared Euclidean distances to C_1 and C_2 are 4 and 2, respectively. Hence C_2 is chosen as the result. Also, C_2 is the nearest codeword for V_2 at (2,3).

Since the smallest one-dimensional distance between the tested vector and the selected codeword is small with a well-trained codebook, the property can be utilized to fasten VQ computation. Our idea is to represent the positional information by bitmaps and refer the bitmaps to select likely codewords. For each

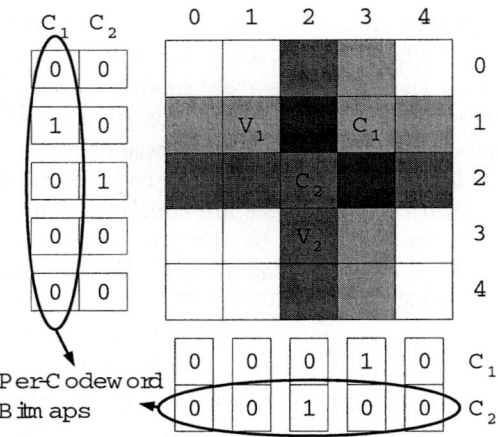

Fig. 2. Two-dimensional Per-Codeword Bitmaps ($R = 0$)

codeword i, we adopt k per-codeword bitmaps to record their positional information for each dimension. Each bitmap consists of m bits to correspond every position. The b_{th} bit in the per-codeword bitmap for dimension j of codeword i is set to one if b is within a certain range of $c_{i(j)}$, say R. The per-codeword bitmaps for the previous example are shown in Fig. 2. The range R is equal to zero. For the first tested vector V_1, it is within the designated range of C_1 in the vertical dimension, and only C_1 is considered for vector quantizing. Similarly, V_2 is within the range of C_2 in the horizontal dimension. Thus C_2 is selected as the closest codeword for V_2 directly.

Although the scheme could sift likely codewords easily, it is not totally accurate. In Fig. 2, C_1 is presumed as the closest codeword for V_1. However, C_2 is the one with the smallest Euclidean distance to V_1, and *false match* is caused. In addition, two kinds of bricks would cause problems: unoccupied bricks (e.g. bricks at (0,0) or (1,3)) and repeatedly occupied ones (e.g. bricks at (2,1) or (3,2)). If the tested vectors locate in the unoccupied bricks, they are not assigned to any codeword, i.e. every codeword must be computed to decide the closest one, and there is no speedup. For the vectors locating in the repeatedly occupied bricks, the codewords whose range occupies the vectors would be calculated for the Euclidean distance, thus the speedup is lessened.

To less the problem, a wider range could be adopted, as shown in Fig. 3 where the renewed bitmaps for $R = 1$ are presented. With the new range, most bricks are occupied by at least one codeword's square. However, the conjunct bricks are also increased due to the larger occupied region.

A suitable range is thus important to the performance of the proposed scheme since a wider range will increase the number of candidates while a narrow range might result in a null set. In our experiments, various ranges are investigated to evaluate the performance and the image quality. Next, the construction/lookup procedure of the searchable data structure is introduced.

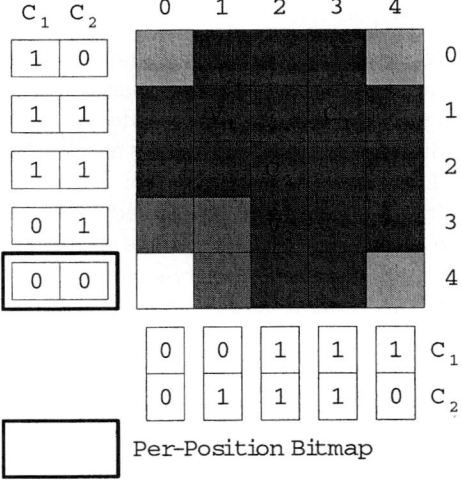

Fig. 3. Two-dimensional Per-Codeword Bitmaps ($R = 1$)

2.1 The Construction of the Searchable Data Structure - Positional Bitmaps

Although the per-codeword bitmaps could present the positional information, they are not searchable. This is because accessing bitmaps for each codeword is inefficient. To utilize the bitmaps based on the proposed concept, the per-position bitmaps are generated from the per-codeword bitmaps. In Fig. 3, we also illustrate the relationships between the per-position bitmaps and the per-codeword bitmaps.

The per-position bitmap for position p at dimension j is defined as $B_{j,p}^R$, where D is the preset range. The i_{th} bit is defined as $B_{j,p}^R(i)$ which is set to one if $p - R \leq c_{i(j)} \leq p + R$. The pseudo code is given in Fig. 4. For each range R, the required storage is $m \times N$ per dimension. With a typical 16-dimensional codebook with 256 entries and 256 gray levels, the occupied memory is 128 Kbytes.

```
Bitmap-Filling Algorithm
For each dimension j, ∀j ∈ {0, k − 1} BEGIN
    For each position p, ∀p ∈ {0, m − 1} BEGIN
        For each codeword i, ∀i ∈ {0, N − 1} BEGIN
            If p − R ≤ c_{i(j)} ≤ p + R, B_{j,p}^R(i) = 1.
            Otherwise, B_{j,p}^R(i) = 0.
        END
    END
END
```

Fig. 4. Bitmap-Filling Algorithm

2.2 The Lookup Procedure

The **PLUT** scheme combines bitmap pruning and **TLUT** to achieve fast processing. For a tested vector, the j_{th} value x_j is used to access the bitmap B_{j,x_j}^R. Each set bit indicates that the corresponding codeword is within a range R from the tested vector at dimension j. Accordingly, the Euclidean distance is calculated by accessing TLUT. The pseudo code for lookup procedure is listed in Fig. 5. First, the multiple bitmaps are performed **OR** operations to derive the representative bitmap D^R. To check whether the i_{th} bit in D^R is set, we further perform **AND** operation with D^R and a pre-generated bitmap with only i_{th} bit set $(00\ldots010\ldots0)$. If the value is larger than zero, then codeword i is one of the candidate.

Vector Quantization by PLUT Algorithm
For each vector x **BEGIN**
    Fetch the B_{j,x_j}^R, where $j \in dim$.
    $D^R = \bigcup_{j \in \{0, k-1\}} B_{j,x_j}^R$.
    For each set bit $D^R(i)$ **BEGIN**
        Calculate Euclidean distance $d(x, c_i)$ where
        $d(x, c_i) = \sum_{j=0}^{k-1} TLUT_1[|x_{(j)}, c_{i(j)}|]$.
        If $d(x, c_i) \leq min\_distance$ **BEGIN**
            $min\_distance\_id = i$
            $min\_distance = d(x, c_i)$
        END
    END
    $min\_distance\_id$ is the quantized index for x.
END

Fig. 5. Vector Quantization by PLUT Algorithm

We use the previous example in Fig. 2 to explain the procedure, where $R = 0$. For the tested vector V_1 "11", the second per-position bitmap "00" at x-axis and second one "10" at y-axis are fetched. The representative bitmap "10" is derived by performing **OR** to these two bitmaps. Consequently, the representative bitmap is performed **AND** operation with "10" to indicate that the first codeword is one of the candidate and the computation for the squared Euclidean distance between V_1 and C_1 is thus carried out. Next, the representative bitmap is performed **AND** operation with "01" again. Since no set bit is found in the resulted bitmap, the calculation for the squared Euclidean distance between V_1 and C_2 is omitted.

2.3 Hardware Implementation

The hardware implementation is preferable for the PLUT scheme. This is because PLUT requires memory bus with N-bit wide (typically $N = 256$). Even in the modern software platform, the memory bus is less than 128 bits. In Fig.

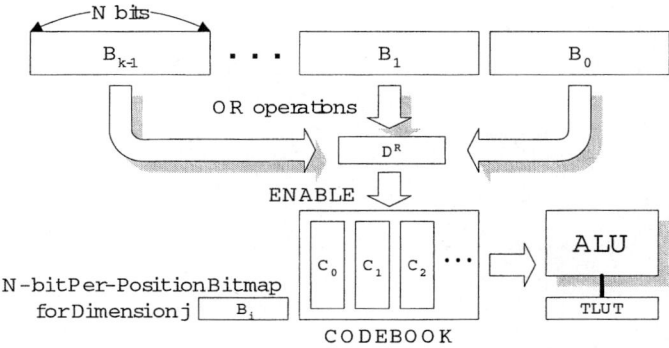

Fig. 6. Hardware Implementation of PLUT Scheme

6, we present a conceptual model for the hardware implementation. This implementation includes K independent RAM modules for per-position bitmaps. Bitmap of each dimension is located in a storage. To perform the search, the per-position bitmaps are fetched from RAM modules simultaneously and performed **OR** operation. Then, the resulted bitmap D^R enables the codewords in the candidate for calculating the Euclidean distance in ALU. Notably, this architecture is suitable for parallelized hardware or pipelining.

3 Performance Evaluation

We have conducted several simulations to show the efficiency of PLUT. All images used in these experiments were 512 × 512 monochrome still images, with each pixel of these images containing 256 gray levels. These images were then divided into 4 × 4 pixel blocks. Each block was a 16-dimensional vector. We used image "Lena" as our training set to generate codebook C. In the previous literature [1,2], the quality of an image compression method was usually estimated by the following five criteria: compression ratio, image quality, execution time, extra memory size, and the number of mathematical operations. All of our experimental images had the same compression ratio, hence only the latter four criteria are listed to evaluate the performance of the proposed scheme. The quality of the images are estimated by the PSNR, which is addressed in Section 2. The extra memory denotes the storage needed for executing PLUT scheme. As for the mathematical operations, the number of the calculated codewords is also considered since the operations for each codeword are identical. In addition, the compression time is evaluated based on software implementation since the performance of hardware implementation can be illustrated from the number of calculated codewords.

The decompressed images based on the PLUT scheme with different ranges are shown in Fig. 7. Basically, the image quality of PLUT is improved gradually as the range increases, such as the PSNR value for $R = 0$ is worse than that

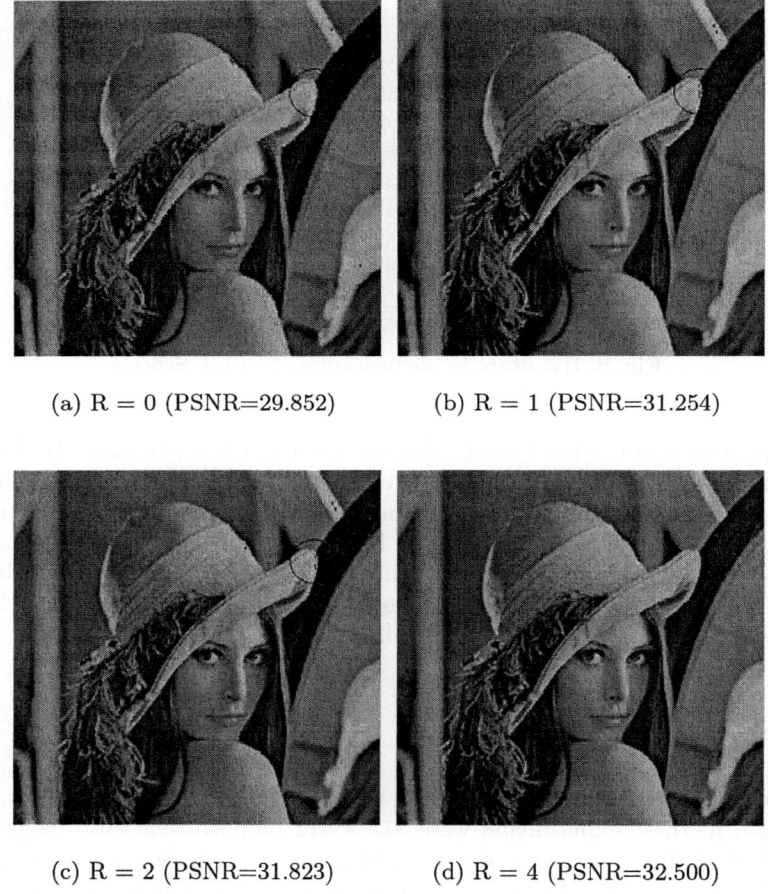

Fig. 7. The Decompressed Lena Images of PLUT Scheme

for $R = 1$ and $R = 2$. However, the quality of some area shows different trend, as shown in the circles of Fig. 7(a),7(b),7(c). This is mainly because for several blocks, there is no candidate derived by PLUT with $R = 0$, thus full search is executed for these blocks. As the range increases to 1 or 2, some codewords are selected for calculation of Euclidean distance. Nevertheless, the codewords cannot yield better precision than full search. The occurrence of such faults ties to the quality of the used codebook. Also, these faults can be alleviated by adopting larger range or enabling full search as the squared Euclidean distance is larger than a certain value. As shown in Fig. 7(d), the image quality is almost identical to VQ and TLUT while PLUT range is enlarged to 4.

The performance of the software-based implementation is illustrated in Table 1. The experiments were performed on an IBM PC with a 500-MHz Pentium CPU. VQ indicates the vector quantization without any speedup. The ranges for PLUT vary from 0 to 8. With a smaller range, the image quality is degraded since the occurrence of false matches is increased. Nevertheless, the calculated codewords are reduced by the per-position bitmaps, the execution time is lessened as well. Full search requires no extra storage while TLUT needs 256 bytes. For PLUT scheme, the extra storage is 128 Kbytes for bitmap and 256 bytes for TLUT. If the hardware implementation is considered, the bitwise operation cab be parallelized to further shorten the vector quantizing time.

Table 1. The Performance of PLUT with Different Ranges

Lena	Full Search	TLUT	PLUT Scheme				
			R=0	R=1	R=2	R=4	R=8
PSNR	32.56	32.56	29.85	31.25	31.82	32.50	32.55
Time (sec.)	1.30	1.09	0.23	0.42	0.53	0.67	0.83
Codewords	256	256	19	44	59	78	99
Storage (byte)	0	256	128K (PLUT) + 256 (TLUT)				

Table 2 illustrates the performance of PLUT based on different images. For the images with better compression quality in full search, PLUT generates more candidates since the codewords are usually close to the compressed blocks. While the range is enlarged to 4, PLUT can derived compressed images with comparable quality to full search while requiring only half execution time.

Table 2. The Performance of PLUT based on Different Images (N=256)

Images	Lena			Girl			Airplane			Zelda		
Metrics	Codewords	Time	PSNR	Codewords	Time	PSNR	Codewords	Time	PSNR	Codewords	Time	PSNR
Full Search	256	1.30	32.56	256	1.30	30.75	256	1.30	29.53	256	1.30	33.35
TLUT	256	1.09	32.56	256	1.11	30.75	256	1.11	29.53	256	1.09	33.35
PLUT,R=0	19	0.23	29.85	17	0.21	29.08	14	0.18	27.57	20	0.24	31.98
PLUT,R=1	44	0.42	31.25	40	0.39	30.14	32	0.33	28.86	44	0.42	33.06
PLUT,R=2	58	0.53	31.82	54	0.50	30.35	44	0.41	29.15	59	0.54	33.25
PLUT,R=4	78	0.67	32.50	72	0.64	30.45	58	0.52	29.35	78	0.67	33.32

In summary, with $R = 2$, the proposed scheme can reduce more than 50% computation without losing image quality. If a hardware implementation is adopted, 25% computation can be further eliminated since only a fourth of codewords are calculated for squared Euclidean distance. Therefore, only a fourth of computation is required.

4 Conclusion

In this study, we present a new novel algorithm "PLUT" for codebook search in VQ. The new scheme is based on the observation that the minimal one-dimensional distance between the tested vector and the matched codeword is usually small. To represent the geometrical information, PLUT adopts bitwise data structure, which is simple and storage efficient. By setting a given range, the PLUT can sift out unfeasible codewords easily, hence it is suitable for hardware implementation. A conceptual hardware implementation is also revealed. Since the performance of PLUT ties to the quality of codebook, PLUT is suitable for high-quality image compression. The performance evaluation further demonstrates that 75% computation can be reduced with an extra 128 Kbytes storage.

References

1. Gersho, A., Gray, R. M.: Vector Quantization and Signal Compression. Boston, MA: Kluwer (1992).
2. Chen, T. S., Chang, C. C.: An Efficient Computation of Euclidean Distances Using Approximated Look-Up Table. IEEE Trans. Circuits Syst. Video Technol., Vol. 7 (2000) 594-599.
3. Davidson, G. A., Cappello, P. R., Gersho A., Systolic architectures for vector quantization, IEEE Trans. Acoust. Speech, Signal Processing, Vol. 36 (1988) 1651-1664.
4. Park, H., Prasana, V. K.: Modular VLSI architectures for real-time full-search-based vector quantization. IEEE Trans. Circuits Syst. Video Technol., Vol. 3 (1993) 309-317.
5. Ramamoorthy, P. A., Potu, B., Tran, T.: Bit-serial VLSI implementation nof vector quantizer for real-time image coding. IEEE Trans. Circuits Syst., Vol. 36 (1989) 1281-1290.
6. Rizvi, S. A., Nasrabadi, N. M.: An efficient euclidean distance computation for quantization using a truncated look-up table. IEEE Trans. Circuits Syst. Video Technol., Vol. 5 (1995) 370-371.
7. Linde, Y., Buzo, A., Gray, R. M.: An algorithm for vector quantizer design. IEEE Trans. Communications, Vol. 28 (1980) 84-95.
8. Chang, H. Y., Wang, P. C., Chen, R. C., Hu, S. C.: Performance Improvement of Vector Quantization by Using Threshold. Lecture Notes in Computer Science, Vol. 3333 (2004) 647-654.

Optimizations of Data Distribution Localities in Cluster Grid Environments

Ching-Hsien Hsu[1*], Shih-Chang Chen[1], Chao-Tung Yang[2], and Kuan-Ching Li[3]

[1] Department of Computer Science and Information Engineering,
Chung Hua University, Hsinchu 300 Taiwan
chh@chu.edu.tw
[2] Department of Computer Science and Information Engineering,
Tunghai University, Taichung 40704 Taiwan
ctyang@mail.thu.edu.tw
[3] Department of Computer Science and Information Management,
Providence University, Taichung 43301 Taiwan
kuancli@pu.edu.tw

Abstract. The advent of widely interconnected computing resources introduces the technologies of ubiquitous computing, peer to peer computing and grid computing. In this paper, we present an efficient data distribution scheme for optimizing data localities of SPMD data parallel programs on cluster grid, a typical computational grid environment consists of several clusters located in multiple campuses that distributed globally over the Internet. Because of the Internet infrastructure of cluster grid, the communication overhead becomes as key factor to the performance of parallel applications. Effectiveness of the proposed distribution mechanism is to reduce inter-cluster communication overheads and to speed the execution of data parallel programs in the underlying distributed cluster grid. The theoretical analysis and experimental results show improvement of communication costs and scalable of the proposed techniques on different hierarchical cluster grids.

1 Introduction

One of the virtues of high performance computing is to integrate massive computing resources for accomplishing large-scaled computation problems. The common characteristic of these problems is enormous data to be processed. In this aspect, clusters have been employed as a platform for a number of such applications including supercomputing, commercial applications and grand challenge problems. The use of cluster of computers as a platform for high-performance and high-availability computing is mainly due to their cost-effective nature. As the growth of Internet technologies, the computational grids become widely accepted paradigm for solving these applications.

---
[*] The correspondence address

Computing grid system [8] integrates geographically distributed computing resources to establish a virtual and high expandable parallel machine; cluster grid is a typical paradigm in which each cluster is geographically located in different campus and is connected by software of computational grids through the Internet. In cluster grid, computers might exchange data through network to other computers to run job completion. This consequently incurs two kinds of communication between grid nodes in a cluster grid. If the two grid nodes are geographically belong to different clusters, the messaging should be accomplished through the Internet. We refer this kind of data transmission as *external communication*. If the two grid nodes are geographically in the same space domain, the communications take place within a cluster; we refer this kind of data transmission as *interior communication*. Intuitively, the external communication is usually with higher communication latency than that of the interior communication sine the data should be routed through numbers of layer-3 routers or higher-level network devices over the Internet. Therefore, to efficiently execute parallel programs on cluster grid, it is extremely critical to avoid large amount of external communications.

This paper presents an extended processor reordering technique for minimizing external communications of data parallel program on cluster grid. We employ the problem of data alignments and realignments in data parallel programming languages to examine the effective of the proposed data to logical processor mapping technique. As researches discovered that many parallel applications require different access patterns to meet parallelism and data locality during program execution. This will involve a series of data transfers such as array redistribution. For example, a 2D-FFT pipeline involves communicating images with the same distribution repeatedly from one task to another. Consequently, the computing nodes might decompose local data set into sub-blocks uniformly and remapped these data blocks to designate processor group. From this phenomenon, we propose a processor-reordering scheme to reduce the volume of external communications of data parallel programs in cluster grid. The key idea is that of distributing data to grid/cluster nodes according to a mapping function at data distribution phase initially instead of in numerical-ascending order. We also evaluate the impact of the proposed techniques. The theoretical analysis and experiments results of the processor-reordering technique on mapping data to logical grid nodes show improvement of volume of external communications and conduce to better performance of data alignment in different cluster grid topologies.

This paper is organized as follows. Section 2 briefly surveys the related works. In section 3, we define the distribution localization problems as preliminaries; then we describe some terminologies regarding the communication model of data distribution in cluster grid. Section 4 describes the extended processor-reordering technique for distribution localizations. Section 5 discusses theoretical analysis of performance and experimental results on real computing grid environment. Finally, conclusions and future work are given in section 6.

2 Related Work

PC clusters have been widely used for solving grand challenge applications due to their good price-performance nature. With the growth of Internet technologies, the computational grids [5] become newly accepted paradigm for solving these applications. As the number of clusters increases within an enterprise and globally, there is the need for a software architecture that can integrate these resources into larger grid of clusters. Therefore, the goal of effectively utilizing the power of geographically distributed computing resources has been the subject of many research projects like Globus [7, 9] and Condor [10]. Frey *et al.* [10] also presented an agent-based resource management system that allowed users to control global resources. The system is combined with Condor and Globus, gave powerful job management capabilities is called Condor-G.

Researches on computing grid have been broadly discussed on different aspects, such as security, fault tolerance, resource management [10, 2, 4], job scheduling [18, 19, 20], and communication optimizations [21, 6]. From the issue of communication optimizations, Dawson *et al.* [6] and Zhu *et al.* [21] addressed the problems of optimizations of user-level communication patterns in local space domain for cluster-based parallel computing. Plaat *et al.* analyzed the behavior of different applications on wide-area multi-clusters [17, 3]. Similar researches were studied in the past years over traditional supercomputing architectures [13, 14]. Guo *et al.* [12] eliminated node contention in communication step and reduced communication steps with schedule table. Y. W. Lim *et al.* [16] presented an efficient algorithm for block-cyclic data realignments. A processor mapping technique presented by Kalns and Ni [15] can minimize the total amount of communicating data. Namely, the mapping technique minimizes the size of data that need to be transmitted between two algorithm phases. Lee *et al.* [11] proposed similar method to reduce data communication cost by reordering the logical processors' id. They proposed four algorithms for logical processor reordering. They also compared the four reordering algorithms under various conditions of communication patterns.

There are significant improvements of the above researches for parallel applications on distributed memory multi-computers. However, most techniques applicable only for applications running on local space domain, like single cluster or parallel machine. For a global grid of clusters, these techniques become inapplicable due to various factors of Internet hierarchical and its communication latency. In this following discussion, our emphasis is on minimizing the communication costs for data parallel programs on cluster grid and on enhancing data distribution localities.

3 Data Distribution over Clusters

Data parallel programming model has become a widely accepted paradigm for parallel programming on distributed memory multicomputers. To efficiently execute a parallel program, appropriate data distribution is critical for balancing the computational load. A typical function to decompose the data equally can be accomplished via the BLOCK distribution directive in data parallel programs.

Many previous studies have shown that the data reference patterns of some parallel applications might be changed dynamically. As they evolve, a good mapping of data to logical processors must change adaptively in order to ensure good data locality and reduce inter-processor communication during program running. For example, a global array could be equally allocated to a set of processors initially in BLOCK distribution manner. As the algorithm goes into another phase that requires to access fine-grain sub-block data patterns, processors might divide their own local data set into sub-blocks locally and then exchange these sub-blocks with corresponding processors. Figure 1 shows an example of this scenario. In the initial distribution, the global array is evenly decomposed into nine data sets and distributed over processors that are selected from three clusters. In the target distribution, each node divides its local data into three sub-blocks evenly and distributes them to the same processor set in a similar manner. Because these data blocks might be required and located in different processors during runtime, efficient communications of inter-processors or inter-clusters become the major subject in term of performance for these applications. Our following emphasis is on how to reduce the inter-cluster communications of data parallel programs when performing dynamic data realignment on cluster grid.

Initial Distribution								
Cluster-1			Cluster-2			Cluster-3		
P_0	P_1	P_2	P_3	P_4	P_5	P_6	P_7	P_8
A	B	C	D	E	F	G	H	I
Target Distribution								
Cluster1	Cluster2	Cluster3	Cluster1	Cluster2	Cluster3	Cluster1	Cluster2	Cluster3
P_0 P_1 P_2	P_3 P_4 P_5	P_6 P_7 P_8	P_0 P_1 P_2	P_3 P_4 P_5	P_6 P_7 P_8	P_0 P_1 P_2	P_3 P_4 P_5	P_6 P_7 P_8
a_1 a_2 a_3	b_1 b_2 b_3	c_1 c_2 c_3	d_1 d_2 d_3	e_1 e_2 e_3	f_1 f_2 f_3	g_1 g_2 g_3	h_1 h_2 h_3	i_1 i_2 i_3

Fig. 1. Data distributions over cluster grid

We first formulate the discussing problem in order to facilitate the explication of the proposed approach. Given a global array and processors' grid, the global array is distributed over processors in BLOCK manner at the initiation of program execution. Processors are requested to partition their local data block into K (*partition factor*) equally sub-blocks and distribute them over corresponding processors in next computational phase. Due to intricate assemblage of cluster grid, this paper also assumes that each cluster provides the same number of nodes. According to this assumption, we use C to denote the number of clusters in the grid; n to represent the number of processors provided by a cluster; and P to be the total number of processors in the cluster grid.

Now, we derive the cost model for evaluating the communication costs in cluster grid and for demonstrating performance analysis in the following sections. Since cluster grid is composed of heterogeneous cluster systems, the overheads of interior communication in different clusters might different and therefore should be identified individually. Let T_i represents the time of two processors both in *Cluster-i* to transmit per unit data; I_i is the total number of interior communications within cluster i; for external communication between cluster i and cluster j, T_{ij} is used to represent the time of processor p in cluster i and processor q in cluster j to transmit per unit data; similarly, the total number of external communications between cluster i and cluster j is denoted by E_{ij}. According to these declarations, we can have equation $T_{comm} = \sum_{i=1}^{C} I_i \times T_i + \sum_{i,j=1, i \neq j}^{C} (E_{ij} \times T_{ij})$. This equation explicitly defines the communication costs of a parallel program running on a cluster grid. However, there are various factors might cause unstable communication delay over internet; it is difficult to estimate accurate costs. As the need of a criterion for performance modeling, integrating the interior and external communications among all clusters into points is an alternative mechanism to get legitimate evaluation. Therefore, we totted up the number of these two terms as $|I| = \sum_{i=1}^{C} I_i$, the number of interior communications, and $|E| = \sum_{i,j=1, i \neq j}^{C} E_{ij}$, the number of external communications for the following discussion.

4 Optimization for Localities of Data Distribution

4.1 Motivating Example

We use the example in Figure 1 to motivate the proposed optimization technique. In order to accomplish the target distribution, processors do the same operation as processor P_0 divides its data block A into a_1, a_2, and a_3; it then distributes these three sub-blocks to processors P_0, P_1 and P_2, respectively. Because processors P_0, P_1 and P_2 belong to the same cluster with P_0; therefore, these three communications are interior. However, the same situation on processor P_1 generates three external communications. Because processor P_1 divides its local data block B into b_1, b_2, and b_3. It then distributes these three sub-blocks to processors P_3, P_4 and P_5, respectively. As processor P_1 belongs to *Cluster* 1 and processors P_3, P_4 and P_5 belong to *Cluster* 2. Therefore, this results three external communications. Figure 2 summarizes all messaging patterns of this example into communication table. We noted that messages $\{a_1, a_2, a_3\}$, $\{e_1, e_2, e_3\}$ and $\{i_1, i_2, i_3\}$ are interior communications ($|I| = 9$); all the others are external communications ($|E| = 18$).

DP\SP	P_0	P_1	P_2	P_3	P_4	P_5	P_6	P_7	P_8
P_0	a_1	a_2	a_3						
P_1				b_1	b_2	b_3			
P_2							c_1	c_2	c_3
P_3	d_1	d_2	d_3						
P_4				e_1	e_2	e_3			
P_5							f_1	f_2	f_3
P_6	g_1	g_2	g_3						
P_7				h_1	h_2	h_3			
P_8							i_1	i_2	i_3
	Cluster-1			Cluster-2			Cluster-3		

Fig. 2. Communication table of data distribution over cluster grid

4.2 Algorithm

The proposed localization optimization of data distribution was achieved by a processor reordering approach. The main idea of this technique is to employ the concept of changing data to logical processor mapping; and expect to translate remote data exchange into local or group message passing. Such techniques were used in several previous researches to minimize data transmission time of runtime array redistribution. In cluster grid, the similar concept can be applied. In order to localize the communication, we need to derive a mapping function produces sequence of logical processors for grouping communications into local cluster. Figure 3 shows the concept of our processor reordering technique. A reordering agent is used to accomplish this process. The source data is partitioned and distributed to processors into initial distributions ($ID(P_X)$) according to the processor sequence derived from reordering agent, where x is the processor id. To accomplish the target distribution ($TD(P_{X'})$), the initial data is divided into sub-blocks and remapped to processors according to the new processors id X' that is also derived from the reordering agent. Given distribution factor K and cluster grid with C clusters, for the case of $K=n$, the reordering agent is functioned by the following mapping function.

$$F(X) = X' = \lfloor X/C \rfloor + (X \bmod C) * K$$

For general cases, i.e., $K \neq n$, the reordering agent is functioned by the following *processor reordering algorithm* as shown in Figure 4.

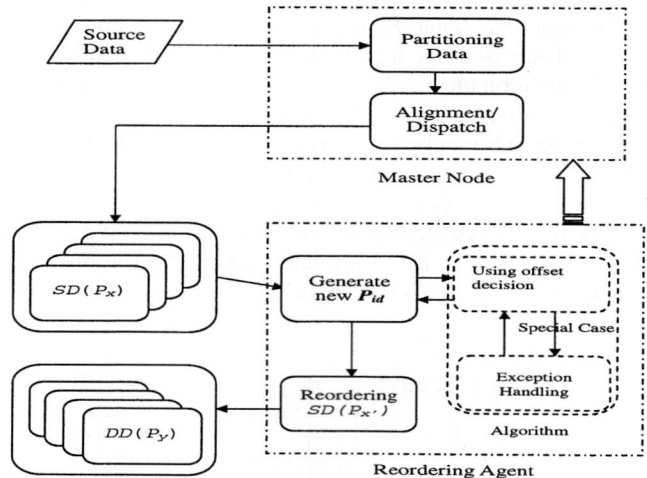

Fig. 3. The flow of data to logical processor mapping using processor reordering algorithm

Figure 5 shows the communication table of the same example after applying the above reordering scheme. The source data is distributed according to the reordered sequence of processors' id, i.e., <P_0, P_3, P_6, P_1, P_4, P_7, P_2, P_5, P_8> which is computed by mapping function. In the target distribution, processor P_0 distributes three sub-blocks to processors P_0, P_1 and P_2 in the same cluster. Similarly, processor P_3 sends three sub-blocks to processors P_3, P_4 and P_5 that are in the same cluster with P_3; and processor P_6 sends e_1, e_2 and e_3 to processors P_6, P_7 and P_8 that causes three interior communications. All other processors generate three interior communications too.

```
Algorithm Processor_Reordering (C, K, n, SD, DD, P)
1.   offset = ⌊K/n⌋
2.   For each source processor P_i
3.   {
4.      start position: d_i = (partition factor (K) + processor rank i) mod P
5.      // d_i is the index of corresponding destination processor for the first
6.      // (assume ℜ) of sub-blocks of source processor P_i
7.      According to offset distance, determine the appropriate target
8.      cluster for at most n consecutive sub-blocks leading by ℜ.
9.      target cluster: m = (d_i + offset) / n
10.     Arm = Select_New_Rank() // select a new id for P_i
11.     If (ARm < n ) {
12.        F(X) = ARm
13.     } else
14.        F(X) = X'= find_next_cluster();
15.  }
end_of_Processor_Reordering
```

Fig. 4. Processor reordering algorithm

DP\SP	P_0	P_1	P_2	P_3	P_4	P_5	P_6	P_7	P_8
P_0	a_1	a_2	a_3						
P_3				b_1	b_2	b_3			
P_6							c_1	c_2	c_3
P_1	d_1	d_2	d_3						
P_4				e_1	e_2	e_3			
P_7							f_1	f_2	f_3
P_2	g_1	g_2	g_3						
P_5				h_1	h_2	h_3			
P_8							i_1	i_2	i_3
	Cluster-1			Cluster-2			Cluster-3		

Fig. 5. Communication table of data distribution over cluster grid with processor reordering

There is no external communication incurred in this example. Therefore, we have $|I| = 27$ and $|E| = 0$.

5 Performance Evaluation

5.1 Theoretical Analysis

This section presents the theoretical value of processor reordering technique in different hierarchy of cluster grid. For different number of clusters (C) and partition factors (K), the amount of interior communications is computed and shown in Figure 6.

For the grid consists of four clusters ($C=4$), the values of K vary from 4 to 10 ($K \geq C$). The results in Figure (a) show that the processor reordering technique provides more interior communications than the method without processor reordering. Figure 6(b) gives the number of interior communications for both methods when $n \neq K$. Note that Figure 6 reports the theoretical results which will not be affected by internet traffic.

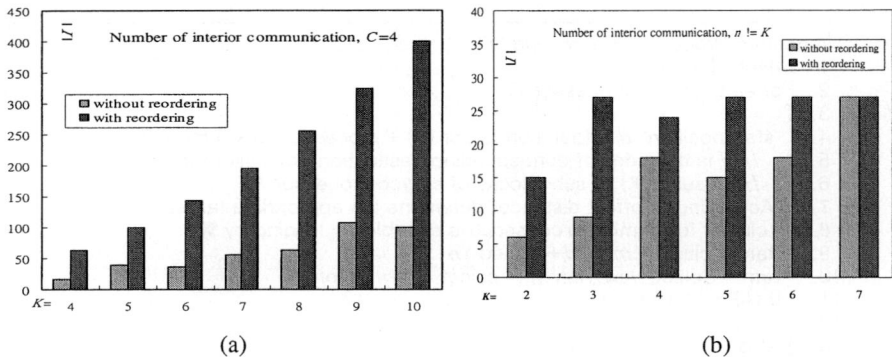

Fig. 6. The number of interior communications (a) $C=4$ and $K = n$ (b) $C = n = 3$

5.2 Simulation Results

To evaluate the performance of the proposed technique, we have implemented the processor reordering method and tested the realignment programs on *Taiwan UniGrid* in which eight universities' clusters are geographically internet-connected. Each owns different number of computing nodes. The programs were written in the single program multiple data (*SPMD*) programming paradigm with C+MPI codes.

Figure 7 shows the execution time of the methods with and without processor reordering to perform data realignment when $C=3$ and $K=3$. Figure 7(a) gives the result of 1MB test data that without file system access (I/O). The result for 10MB test data that is accessed via file system (I/O) is given in Figure 7(b). Different combinations of clusters denoted as *NTI*, *NTC*, *NTD*, etc. were tested. The composition of these labels is summarized in Table 1.

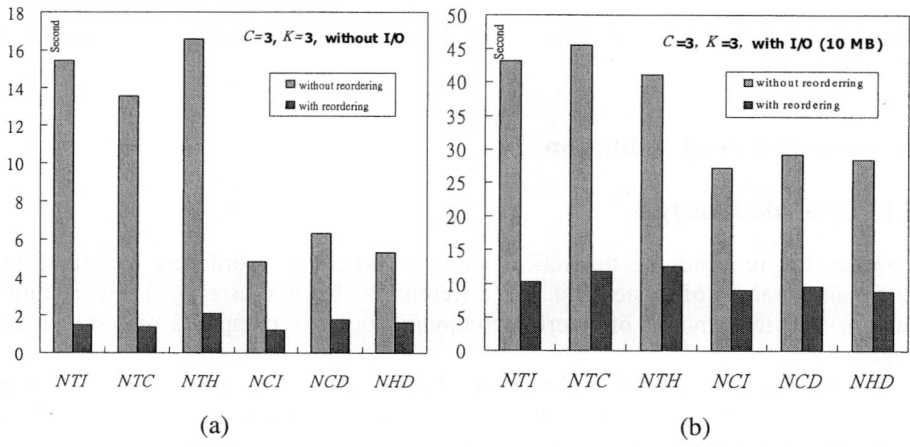

Fig. 7. Execution time of different methods to perform data realignments on cluster grid when $C = K = 3$

In this experiment, method with processor reordering technique outperforms the method that without processor reordering. Compare to the results given in Figure 6, this experiment matches the theoretical predictions. It also satisfying reflects the efficiency of the processor reordering technique.

Table 1. Labels of different cluster grid

Label	Cluster-1	Cluster-2	Cluster-3	Label	Cluster-1	Cluster-2	Cluster-3
NTI	NCHC	NTHU	IIS	NCI	NCHC	CHU	IIS
NTC	NCHC	NTHU	CHU	NCD	NCHC	CHU	NDHU
NTH	NCHC	NTHU	THU	NHD	NCHC	THU	NDHU
NCDI	NCHC	CHU	NDHU	IIS			

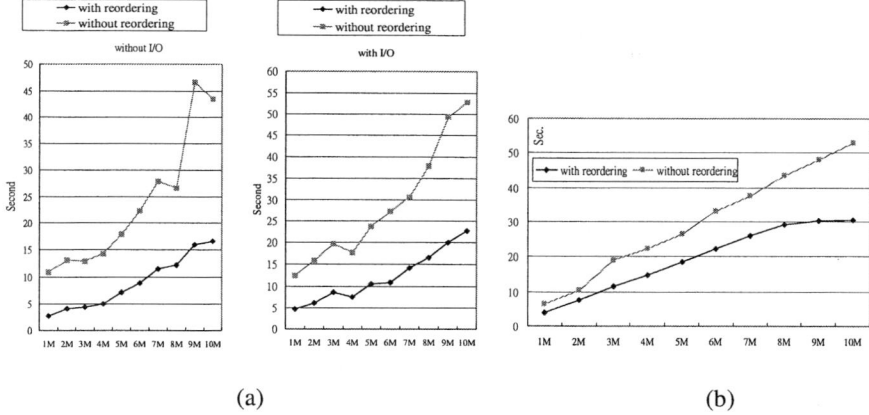

Fig. 8. Execution time of both methods on different data sets. (a) $C = K = n = 4$ using mapping function (b) $C = n = 3$ and $K = 5$ using reordering algorithm

Figures 8(a) and 8(b) show the results of applying mapping function and reordering algorithm, respectively. The test-bed for case $C = K = n$ in Figure 8(a) is on four clusters (*NCDI*). Figure 8 (b) reports the performance for generalized reordering technique on cluster grid. The experimental results show that processor reordering techniques provide significant improvement.

6 Conclusions

In this paper, we have presented an efficient data distribution scheme for optimizing data localities of SPMD data parallel programs on cluster grid. The theoretical analysis and experimental results of the distribution schemes on mapping data to logical grid

grid nodes show improvement of communication costs and scalable on different hierarchical cluster grids.

There is some research issues remained in this paper. The current work of our study restricts the number of computing nodes provided by different clusters to be identical. In the future, we plan to demonstrate a generalized method for solving non-identical applications. Besides, the issues of different grid topologies and analysis of network communication latency are also interesting and will be investigated.

References

1. Taiwan UniGrid, http://unigrid.nchc.org.tw
2. O. Beaumont, A. Legrand and Y. Robert, "Optimal algorithms for scheduling divisible workloads on heterogeneous systems," *Proceedings of the 12th IEEE Heterogeneous Computing Workshop,* 2003.
3. Henri E. Bal, Aske Plaat, Mirjam G. Bakker, Peter Dozy, and Rutger F.H. Hofman, "Optimizing Parallel Applications for Wide-Area Clusters," *Proceedings of the 12th International Parallel Processing Symposium IPPS'98,* pp 784-790, 1998.
4. M. Faerman, A. Birnbaum, H. Casanova and F. Berman, "Resource Allocation for Steerable Parallel Parameter Searches," *Proceedings of GRID'02,* 2002.
5. J. Blythe, E. Deelman, Y. Gil, C. Kesselman, A. Agarwal, G. Mehta and K. Vahi, "The role of planning in grid computing," *Proceedings of ICAPS'03,* 2003.
6. J. Dawson and P. Strazdins, "Optimizing User-Level Communication Patterns on the Fujitsu AP3000," *Proceedings of the 1st IEEE International Workshop on Cluster Computing,* pp. 105-111, 1999.
7. I. Foster, "Building an open Grid," *Proceedings of the second IEEE international symposium on Network Computing and Applications,* 2003.
8. I. Foster and C. Kessclman, "The Grid: Blueprint for a New Computing Infrastructure," Morgan Kaufmann, ISBN 1-55860-475-8, 1999.
9. I. Foster and C. Kessclman, "Globus: A metacomputing infrastructure toolkit," *Intl. J. Supercomputer Applications,* vol. 11, no. 2, pp. 115-128, 1997.
10. James Frey, Todd Tannenbaum, M. Livny, I. Foster and S. Tuccke, "Condor-G: A Computation Management Agent for Multi-Institutional Grids," *Journal of Cluster Computing,* vol. 5, pp. 237 – 246, 2002.
11. Saeri Lee, Hyun-Gyoo Yook, Mi-Soon Koo and Myong-Soon Park, "Processor reordering algorithms toward efficient GEN_BLOCK redistribution," *Proceedings of the 2001 ACM symposium on Applied computing,* 2001.
12. M. Guo and I. Nakata, "A Framework for Efficient Data Redistribution on Distributed Memory Multicomputers," *The Journal of Supercomputing,* vol.20, no.3, pp. 243-265, 2001.
13. Florin Isaila and Walter F. Tichy, "Mapping Functions and Data Redistribution for Parallel Files," *Proceedings of IPDPS 2002 Workshop on Parallel and Distributed Scientific and Engineering Computing with Applications, Fort Lauderdale,* April 2002.
14. Jens Koonp and Eduard Mehofer, "Distribution assignment placement: Effective optimization of redistribution costs," *IEEE TPDS,* vol. 13, no. 6, June 2002.
15. E. T. Kalns and L. M. Ni, "Processor mapping techniques toward efficient data redistribution," *IEEE TPDS,* vol. 6, no. 12, pp. 1234-1247, 1995.
16. Y. W. Lim, P. B. Bhat and V. K. Parsanna, "Efficient algorithm for block-cyclic redistribution of arrays," *Algorithmica,* vol. 24, no. 3-4, pp. 298-330, 1999.

17. Aske Plaat, Henri E. Bal, and Rutger F.H. Hofman, "Sensitivity of Parallel Applications to Large Differences in Bandwidth and Latency in Two-Layer Interconnects," *Proceedings of the 5th IEEE High Performance Computer Architecture HPCA'99*, pp. 244-253, 1999.
18. Xiao Qin and Hong Jiang, "Dynamic, Reliability-driven Scheduling of Parallel Real-time Jobs in Heterogeneous Systems," *Proceedings of the 30th ICPP*, Valencia, Spain, 2001.
19. S. Ranaweera and Dharma P. Agrawal, "Scheduling of Periodic Time Critical Applications for Pipelined Execution on Heterogeneous Systems," *Proceedings of the 30th ICPP*, Valencia, Spain, 2001.
20. D.P. Spooner, S.A. Jarvis, J. Caoy, S. Saini and G.R. Nudd, "Local Grid Scheduling Techniques using Performance Prediction," *IEE Proc. Computers and Digital Techniques*, 150(2): 87-96, 2003.
21. Ming Zhu, Wentong Cai and Bu-Sung Lee, "Key Message Algorithm: A Communication Optimization Algorithm in Cluster-Based Parallel Computing," *Proceedings of the 1st IEEE International Workshop on Cluster Computing*, 1999.

Abuse-Free Item Exchange

Hao Wang[1], Heqing Guo[1], Jianfei Yin[1], Qi He[2], Manshan Lin[1], and Jun Zhang[2]

[1] School of Computer Science & Engineering, South China University of Technology,
Guangzhou, China 510640
{iswanghao, yjhhome, lmshill}@hotmail.com
guozhou@scut.edu.cn,
[2] Computer Engineering School, Nanyang Technological University, Singapore 639798
{qihe0001, jzhang}@ntu.edu.sg

Abstract. Electronic exchange is widely used in e-commerce systems. This paper mainly discusses abuse-freeness in general item exchange protocol for two parties. Based on convertible signature scheme and adapted designated verifier proof, an efficient item exchange protocol is proposed to satisfy many interesting requirements including fairness, timeliness and *strong abuse-freeness*.

1 Introduction

The goal of fair protocols is to guarantee fairness of web-based electronic exchange in applications like e-commerce and e-government. In other words, they should assure that no party could falsely deny involvement in the exchange or having sent/received the specific item[1]. Assuming Alice wants to exchange an item with another item belonging to Bob, the protocol must assure that wherever the protocol ends, both of them either get the expected item (and non-repudiation evidences) or nothing. To take one step further, the protocol should be abuse-free, which means before the exchange ends, either party cannot prove to outside party that s/he can control the exchange outcome: success or aborted. A Trusted Third Party (TTP) is involved as Pagnia and Garner [10] have proved that no definite fairness can be achieved without a TTP.

Fairness issue has been studied in different scenarios: fair exchange [1][8], contract signing [5], payment[13][14], non-repudiable message transmission [15], and so on. But these protocols are inter-transformable, e.g., a fair exchange protocol can be easily transformed to be a contract signing protocol.

In 1996, Asokan et al.[1] and Zhou et al.[15] proposed optimistic approach and presents fair protocols with offline TTP, in which TTP intervenes only when an error occurs (network error or malicious party's cheating). But the recovered messages are different from those produced by the sender or the recipient, which make the protocols suffer from bad publicity and weak fairness, as the recovered messages may lose some functionalities of the original ones. *Invisible TTP* is first introduced by Micali [9] to solve this problem. The TTP can generate exactly the same evidences as the sender or the recipient. In this way, judging the outcome evidences and received items cannot decide whether the TTP has been involved, so that the recovery is done in a transparent way.

---

[1] Item can be a signature, electronic goods, payment, email, message, and so on.

Using *convertible signatures* (CS) is the recently focused approach to realize transparent recovery. It means to firstly send a partial committed signature that can be converted into a full signature (that is a normal signature) by both the TTP and the signer. Boyd and Foo [2] have proposed a fair payment protocol using the RSA-based convertible signatures scheme proposed by Gennaro et al. [6] (GKR signature scheme). This protocol generates standard RSA signature as final evidence. But it is not efficient and practical enough because it involves an interactive verification process. Under the asynchronous network condition, less interaction the better.

Abuse-freeness, as a new requirement of fair protocols, is first mentioned by Boyd and Foo [2], and formally presented by Garay et al. [5]. In their definition, abuse-freeness means that before the protocol ends, no party can prove to an outside party that he can choosing whether to complete or to abort the transaction. But as sequential analysis work by Chadha et al. [3][4] said, in any fair, optimistic, timely protocol, an optimistic party yields an advantage to his/her opponent. That means that the opponent has both a strategy to complete and to abort the exchange. So we have to change our direction: to prove a stronger version of abuse-freeness, that is, before the protocol ends, no party can prove to an outside party that his/her opponent is participating in the exchange.

In this paper, we propose a protocol which allows exchange of two items and their evidences of origin/receipt. It guarantees fairness, timeliness and especially, strong abuse-freeness. And it only contains 4 steps, which is the minimum in this scenario. As in [2], we continue to use the GKR scheme in order to generate standard RSA signatures as final evidences. To assure abuse-freeness, we use the designated verifier proofs presented by Jakobsson et al. [7] and strengthened by [12].

In Section 2, we present several requirements for fair exchange. In Section 3, we present the modified convertible signature scheme and the general item exchange protocol, which is analyzed in Section 4. Section 5 gives some concluding remarks.

2 Requirements for Fair Exchange

In [13][14], we have studied on the requirements of fair payment. For continuation, we use similar definitions only with major change on abuse-freeness.

Definition 1. Effectiveness
A fair protocol is *effective* if (independently of the communication channels quality) there exists a successful execution of the protocol.

Definition 2. Fairness
A fair protocol is *fair* if (the communication channels quality being fixed) when the protocol run ends, both exchangers either get their expected item (and non-repudiation evidences) or nothing useful.

Definition 3. Timeliness
A fair protocol is *timely* if (the communication channels quality being fixed) the protocol can be completed in a finite amount of time while preserving fairness for both exchangers.

Former definitions on timeliness do *not* bring into account the highly time-sensitive items such as electronic events tickets or airline tickets. After the scheduled time, the item will be outdated and of no use, so there should be some mechanism to guarantee this kind of exchange to have a high success probability while assuring fairness. Now available solution is to separate the item receiving over network and delivering it to user by utilizing trusted hardware on user's machine [11].

Definition 4. Non-repudiability

A fair protocol is *non-repudiable* if when the exchange succeeds, either exchanger cannot deny (partially or totally) his/her participation.

Definition 5. Abuse-Freeness

A fair protocol is *abuse-free* if before the protocol ends, no party is able to prove to an outside party that s/he has the power to terminate (abort) or successfully complete the protocol.

Definition 6. Strong Abuse-freeness

A fair protocol is *strongly abuse-free* if before the protocol ends, no party is able to prove to an outside party that his/her opponent is participating in the protocol.

Clearly, we can see strong abuse-freeness implies abuse-freeness as if a party cannot prove that the other is participating, he cannot prove even he actually can control the outcome of the protocol.

3 A General Item Exchange Protocol

Alice and Bob needs to exchange two items' partial and final evidences of origin/receipt, that is, 8 evidence (**EOO$_A$, EOR$_A$, EOO$_B$, EOR$_B$, NRO$_A$, NRR$_A$, NRO$_B$, NRR$_B$**) needs to be exchanged. So there are several steps in the main protocol and the recovery protocol that have multiple purposes. We modify the GKR convertible signature scheme with designated verifier proof method.

Let n be the Alice's RSA modulus. n is a strong prime and it satisfies $n=pq$ where $p=2p'+1$ and $q=2q'+1$ (p,q,p',q' are primes). Her public key is the pair (e,n) and private key is d. To make the signature convertible, d is multiplicative divided in d_1 and d_2, satisfying $d_1 d_2 e = 1 \mod \phi(n)$. d_1 (chosen by the TTP) is the secret key shared between Alice and TTP, it will be used to convert the partial signature to a final one.

To describe the protocol, we use following notation.

- $X \rightarrow Y$: transmission from entity X to Y
- $h()$: a collision resistant one-way hash function
- $E_k()/D_k()$: a symmetric-key encryption/decryption function under key k
- $E_X()/D_X()$: a public-key encryption/decryption function under pk_X
- $S_X()$: ordinary signature function of X
- k: the key used to cipher goods
- pk_X/sk_X: public/secret key of X
- $PS_X()$: partial signature function of X
- $FS_X()$: the final signature function of entity X
- $item_X$: the item X wants to send
- $descr_X$: a description of $item_X$, detailed enough to identify the item
- k_X: the session key X uses to cipher $item_X$
- pk_X: public key of X
- sk_X: secret key of X
- $cipher_X = E_k(item_X)$: the cipher of $item$ under k_X

- l: a label that in conjunction with (A,B) uniquely identifies a protocol run
- f: a flag indicating the purpose of a message

Also, we must clearly state assumptions before describing the protocol:

Communication Network. We assume the communication channel between Alice and Bob is unreliable and channels between exchangers (Alice/Bob) and TTP are resilient. Messages in a resilient channel can be delayed but will eventually arrive. On the contrary, messages in unreliable network may be lost.

Cryptographic Tools. Encryption tools including symmetric encryption, asymmetric encryption and normal signature is secure. In addition, the adopted signature scheme is message recovery.

Honest TTP. The TTP should send a valid and honest reply to every request. Honest means that when the TTP is involved, if a recover decision is made, Alice gets the payment and Bob gets the goods; if a abort decision is made, Alice and Bob get the abort confirmation and they cannot recover the exchange in any future time.

3.1 Registration Protocol

The registration protocol between the registering party (Alice/Bob) and TTP needs to be run only once. And the resulting common parameters can be used for any number of exchanges. Alice requests for key registration by sending her public key pair (e, n) to the TTP. TTP checks the validity of n (by checking its certificate, the checking is denoted by $check\_pk()$), if passes, it sends d_1 to Alice (for security, d_1 should be encrypted some way). Then Alice chooses a reference message ω and computes $PS(\omega) = \omega^{d_2}$ and send them to TTP. After TTP checks (using the function denoted by $check\,\omega()$) whether

$$\omega \equiv PS(\omega)^{d,e} \pmod{n}$$

If it holds, he will send a certificate $cert_A = S_{TTP}(A, e, n, \omega, PS(\omega))$ to Alice.

Registration Protocol
$A \rightarrow TTP$: f_{Reg}, TTP, pk_X
    TTP: **if not** $check\_pk()$ **then** stop
$TTP \rightarrow A$: f_{Share}, A, $E_A(d_1)$
$A \rightarrow TTP$: f_{Ref}, $\omega, PS(\omega)$
    TTP: **if not** $check\,\omega()$ **then** stop
$TTP \rightarrow A$: f_{cert}, A, $cert_A$

With the certificate, Bob can be convinced that TTP can convert the partial signatures once they are signed by the same d_2 as $PS(\omega)$. Bob also need to involve such a registration protocol to get his own certificate $cert_B$. Note that they may send the same reference message to the TTP, which won't affect the security of the verification protocol.

3.2 Main Protocol

The item to be sent is divided into two parts: the cipher and the key. The main protocol contains 4 steps. In this scheme, the partial signature is defined as

$$PS(m) = m^{d_2} \pmod{n}$$

and it is converted to be the final signature using

$$FS(m) = PS(m)^{d_1} \pmod{n}$$

It works because

$$FS(m)^e \equiv PS(m)^{d_1 e} \equiv m^{d_1 d_2 e} \equiv m \pmod{n}$$

holds.

Following we focus our attention on our non-interactive verification protocol. We assume that Alice knows $PS_B(\omega)$ and Bob knows $PS_A(\omega)$.

Generating Proofs. X selects $\alpha, \beta, u \in Z_q$ and calculates

$$\begin{cases} s = \omega^\alpha PS_Y(\omega)^\beta \bmod n \\ \Omega = \omega^u \bmod n \\ M = m^u \bmod n \\ v = h(s, \Omega, M) \\ r = u + d_x(v + \alpha) \bmod q \end{cases}$$

The proof of the $PS_X(m)$, denoted by $pf(PS_X(m))$, is ($\alpha, \beta, \Omega, M, r$).

Verifying Proofs. When Y gets the $PS_X(m)$ and $pf(PS_X(m))$, s/he will calculate

$$\begin{cases} s = \omega^\alpha PS_Y(\omega)^\beta \bmod n \\ v = h(s, \Omega, M) \end{cases}$$

and verifies

$$\begin{cases} \Omega PS_X(\omega)^{h+\alpha} = \omega^r \bmod n \\ M PS_X(m)^{h+\alpha} = m^r \bmod n \end{cases}$$

Simulating Transcripts. Y can simulate correct transcripts by selecting $t < n, \gamma < n, \eta < n$ and calculate

$$\begin{cases} s = \omega^\gamma \bmod n \\ \Omega = \omega^t PS_X(\omega)^{-\eta} \bmod n \\ M = m^t PS_X(m)^{-\eta} \bmod n \\ v = h(s, \Omega, M) \\ \mu = \eta - h \bmod q \\ r = (\gamma - \mu)d_y^{-1} \bmod q \end{cases}$$

So Y cannot convince any outside party of the validity of the partial signature of Alice.

We note these verifying operations as the predicate

$verify(pf(PS_x(m)), m, PS_x(m), \omega, PS_x(\omega))$. If the verification fails, it returns **false**.

In our protocol, we denote the content to be signed by the item sender as $a_X=($ f_{NROX}, Y, l, $h(k)$, $cipher$, $E_{TTP}(k))$, then the $\mathbf{EOO}_X = PS_X(a_X)$ plus pf $(PS_X(a_X))$ and $\mathbf{NRO}_X = FS_X(a_X)$. Similarly, let $b_X=($ f_{NRRX}, X, $l)$, then $\mathbf{EOR}_X = PS_Y(b_X)$ plus pf $(PS_Y(b_X))$ and $\mathbf{NRR}_X = FS_Y(b_X)$.

Main Protocol

$A \rightarrow B$: f_{EOOA}, f_{EORB}, B, l, $h(k_A)$, $cipher_A$, $E_{TTP}(k_A)$, $\mathbf{EOO}_A$, $\mathbf{EOR}_B$
 B: **if not** $verify(\mathbf{EOO}_A)$ **or not** $verify(\mathbf{EOR}_B)$ **then** B stop
$B \rightarrow A$: f_{EORA}, f_{EOOB}, A, l, $\mathbf{EOR}_A$, $h(k_B)$, $cipher_B$, $E_{TTP}(k_B)$, $\mathbf{EOO}_B$
 A: **if** times out **or not** ($verify(\mathbf{EOR}_A)$ **and** $verify(\mathbf{EOO}_B)$) **then** abort
$A \rightarrow B$: f_{NROA}, f_{NRRB}, B, l, k_A, $\mathbf{NRO}_A$, $\mathbf{NRR}_B$
 B: **if** times out **then** recover$[X:=B,Y:=A]$
$B \rightarrow A$: f_{NRRA}, f_{NROB}, A, l, $\mathbf{NRR}_A$, k_B, $\mathbf{NRO}_B$
 A: **if** times out **then** recover$[X:=A,Y:=B]$

When one party gets the partial evidences, s/he needs to verify the partial signature. If the verification of $\mathbf{EOO}_A$ fails, Bob can simply quit the exchange without any risks. But if times out or the verification in step 2 or step 3 fails, Alice and Bob need respectively to run the abort protocol to prevent later recovery by the other party. If Alice and Bob time out respectively in step 4 and step 5, they can run the recover protocol to complete the exchange.

3.3 Recover Protocol and Abort Protocol

Recover protocol is executed when an error happens, one party needs TTP's help to decrypt the key k and generate the final evidences for him/her. One party submits an abort request using abort protocol, preventing the other party may recover in a future time which s/he will not wait. And $\mathbf{Rec}_X = S_X(f_{RecX}, Y, l)$ is the recover request; $\mathbf{Abort} = S_X(f_{Abort}, TTP, l)$ is the abort request; $\mathbf{Con}_a = S_{TTP}(f_{cona}, A, B, l)$ is the abort confirmation.

Recover Protocol

$X \rightarrow TTP$: f_{RecX}, Y, l, $h(cipher_A)$, $h(k_A)$, $E_{TTP}(k_A)$, $h(cipher_B)$, $h(k_B)$, $E_{TTP}(k_B)$, $\mathbf{Rec}_X$, $\mathbf{EOR}_A$, $\mathbf{EOO}_A$, $\mathbf{EOR}_B$, $\mathbf{EOO}_B$
 TTP: **if** $h(k_A) \neq h(D_{TTP}(E_{TTP}(k_A)))$ **or** $h(k_B) \neq h(D_{TTP}(E_{TTP}(k_B)))$ **or** $aborted()$ **or** $recovered()$ **then** stop
    else $recovered$=true
$TTP \rightarrow A$: f_A, A, l, k_B, $\mathbf{NRR}_A$, $\mathbf{NRO}_B$
$TTP \rightarrow B$: f_B, B, l, k_A, $\mathbf{NRR}_B$, $\mathbf{NRO}_A$

Abort Protocol

$X \rightarrow TTP$: f_{Abort}, l, Y, abort
 TTP: **if** $aborted()$ **or** $recovered()$ **then** stop
    else $aborted$=true
$TTP \rightarrow A$: f_{Cona}, A, B, l, $\mathbf{Con}_a$
$TTP \rightarrow B$: f_{Cona}, A, B, l, $\mathbf{Con}_a$
---

4 Discussions

Following is the analysis with respect to requirement definitions in section 2. As the proof of effectiveness, fairness, timeliness and non-repudiation requirement is similar with those in [13], so we omit here for space reason.

Claim 1. *Assuming the channel between Alice and Bob is unreliable, the protocol satisfies the effectiveness requirement.*

Claim 2. *Assuming the channels between the TTP and exchangers (Alice and Bob) are resilient and the TTP is honest, the protocol satisfies the fairness requirement.*

Claim 3. *Assuming the channels between the TTP and exchangers (Alice and Bob) are resilient, the protocol satisfies timeliness requirement.*

Claim 4. *Assuming the channels between the TTP and exchangers (Alice and Bob) are resilient, and the adopted convertible signature scheme is secure, the protocol satisfies non-repudiation requirement.*

Claim 5. *Assuming the channels between the TTP and exchangers (Alice and Bob) are resilient and the adopted convertible signature scheme and designated verifier proof are secure, the protocol guarantees strong abuse-freeness.*

Proof: Assume Alice is the honest one while Bob is trying to prove to gain advantage. During the main protocol, before Alice sends $\mathbf{NRO}_A$, $\mathbf{NRR}_B$, Bob only gets partial evidence $\mathbf{EOO}_A$, $\mathbf{EOR}_B$. And as the evidences' proofs are designated to him and no other can be convinced that the evidences are actually generated by Alice, so at this stage, Bob cannot prove to outside party that Alice is in the exchange. After Alice sends the final evidences, even Bob doesn't send back his final evidences, Alice still can attain them by requesting recover to the TTP and the exchange is completed in success and Bob gets no advantage. In all, before getting the final evidences, no party can prove to outside party that his/her opponent is participating in the protocol.

5 Conclusions

With convertible RSA signature scheme and adapted designated verifier proof, we present a carefully built item exchange protocol satisfying many interesting requirements including fairness, timeliness and strong abuse-freeness.

When we consider different instantiations of item in many application scenarios, we can achieve fair payment with electronic services, abuse-free contract-signing, and etc. Our future work will be focused on further verification of our item exchange protocol based on the theoretical game model and try to prove completeness.

References

1. N. Asokan, M. Schunter, and M. Waidner. Optimistic protocols for fair exchange. In *Proceedings of the fourh ACM Conference on Computer and Communications Security*, ACM press, 1997.
2. C. Boyd, E. Foo. Off-line Fair Payment Protocols using Convertible Signatures. In *Advances in Cryptology---ASIA CRYPT'98*, SpringerVerlag, 1998.

3. R. Chadha and M. Kanovich and A. Scedrov. Inductive methods and contract-signing protocols. In *Proceedings of 8-th ACM confererence on Computer and Communications Security(CCS-8)*. ACM Press, 2001.
4. R. Chadha, J. Mitchell, A. Scedrov and V. Shmatikov. Contract signing, optimism and advantage. In *Proceedings of CONCUR 2003, LNCS 2761*, Springer-Verlag, 2003.
5. J. Garay, M. Jakobsson, and P. MacKenzie. Abuse-free optimistic contract signing. In *Advances in Cryptology - CRYPTO '99*, SpringerVerlag, 1999.
6. R. Gennaro, H. Krawczyk, and T. Rabin. RSA-based undeniable signatures. In *Advances in Cryptology --- CRYPTO '97, LNCS 1296*. Springer Verlag, 1997.
7. M. Jakobsson, K. Sako, R. Impagliazzo. Designated verifier proofs and their applications. In *Eurocrypt'96, LNCS 1070*, Springer Verlag, 1996.
8. O. Markowitch and S. Saeednia. Optimistic fair-exchange with transparent signature recovery. In *Proceedings of 5th International Conference, Financial Cryptography 2001*, Springer-Verlag, 2001.
9. S. Micali. Certified e-mail with invisible post offices. Available from author: an invited presentation at the RSA'97 conference, 1997.
10. H. Pagnia and F. C. Gartner. On the impossibility of fair exchange without a trusted third party. *Tech. Rep. TUD-BS-1999-02 (March)*, Darmstadt University of Technology, 1999.
11. H. Pagnia, H. Vogt, F.C. Gärtner, and U.G. Wilhelm. Solving Fair Exchange with Mobile Agents. *LNCS 1882*, Springer, 2000.
12. S. Saeednia, S. Kremer and O. Markowitch, An efficient strong designated verifier scheme. In *Proceedings of 6th International Conference on Information Security and Cryptology (ICISC 2003), LNCS*, Springer-Verlag, 2003.
13. H. Wang and H. Guo. Fair Payment Protocols for E-Commerce. In *Proceedings of Fourth IFIP Conference on e-Commerce, e-Business, and e-Government (I3E'04). Building the E-Society: E-Commerce, E-Business and E-Government*, Kluwer academic publishers, 2004.
14. H. Wang, H. Guo and Manshan Lin. New Fair Payment Protocols. In *Proceedings of 1st International Conference on E-business and Telecommunication Networks (ICETE'04)*. INSTICC press, 2004.
15. J. Zhou and D. Gollmann. An Efficient Non-repudiation Protocol. In *Proceedings of 1997 IEEE Computer Security Foundations Workshop (CSFW 10)*, 1997.

Transcoding Pattern Generation for Adaptation of Digital Items Containing Multiple Media Streams in Ubiquitous Environment

Maria Hong[1], DaeHyuck Park[2], YoungHwan Lim[2], YoungSong Mun[3], and Seongjin Ahn[4]

[1] Digital Media Engineering, Anyang University, Anyang, Kyonggi-Do, Korea
`maria@anyang.ac.kr`
[2] Department of Media, Soongsil University, Seoul, Korea
`hotdigi@ssu.ac.kr , yhlim@computing.ssu.ac.kr`
[3] Department of computer Science, Soongsil University, Seoul, Korea
`mnu@computing.ssu.ac.kr`
[4] Department of Computer Education, Sungkyunkwan University, Seoul, Korea
`sjahn@comedu.skku.ac.kr`

Abstract. Digital item adaptation (DIA) is one of main parts in MPEG21. The goal of the DIA is to achieve interoperable transparent access to multimedia contents by shielding users from network and terminal installation, management and implementation issues. In general, a digital item consists of multiple discrete or continuous media and adaptation process is to be applied to one of its multiple media. Therefore in the process of adapting the given digital item, the transcoding order of its media has great impact on the adaptation performance because the adaptation process may stop at the point of satisfying a adaptation QoS(Quality of Services). In order to reduce the transcoding time which result in reducing the initial delay time, we propose a pattern based transcoding model for DIA. Because the performance of the model is dependent on the pattern, we also suggest a EPOB based pattern generation method. Finally a sample pattern was suggested based on the experimental results of the method.

1 Introduction

Digital item adaptation (DIA) is one of main parts in MPEG21. The goal of the DIA is to achieve interoperable transparent access to multimedia contents by shielding users from network and terminal installation, management and implementation issues. As suggested by the MPEG21 group, the combination of resource adaptation and descriptor adaptation produces newly adapted Digital Item. In general, a digital item consists of multiple discrete or continuous media and adaptation process is to be applied to one of its multiple media. Therefore in the process of adapting the given digital item, the transcoding order of its media has great impact on the adaptation performance because the adaptation process may stop at the point of satisfying a adaptation QoS(Quality of Services). In this paper, firstly, digital items in ubiquitous environment are classified into 3 types depending on its consisting media. We would like to discuss a method of generating appropriate transcoding patterns for each type of digital item.

1.1 Classification of Digital Items

The amount of data, a digital item requires, depends on the types of media comprising the digital item. Each media comprising the digital item can be divided into discrete media (image and text) and continuous media (audio and video). The total required amount of data changes according to the percentage of each media. These digital items may be classified into three different kinds of cases:

Case 1: where the amount of discrete media exceeds that of continuous media(PT1);
Case 2: where there is approximately equal amount of each media(PT2);
Case 3: continuous media exceeds discrete media(PT3).

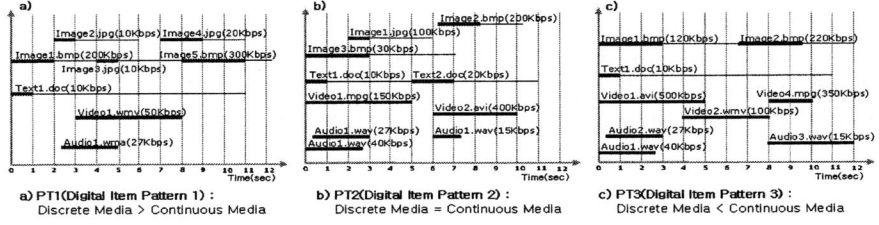

Fig. 1. Classification of Digital Item by Media Constituent

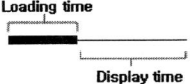

Fig. 2. Notation of Discrete Media

Of the different kinds of media, discrete media, once loaded, does not use bandwidth during continuous display. Thus, in Fig. 1, discrete media cannot be shown as it is in Fig. 2. In terms of discrete media, loading time is the period of time in which all the images are received, while display time is the period of time in which the images are shown in the digital item. Bandwidth is used only during loading time. In contrast, continuous media keeps using bandwidth during the display time. Fig. 2 shows how much bandwidth the continuous media portion of a digital item requires.

1.2 Problems and Survey

There are several critical problems in displaying the above-mentioned digital item through ubiquitous environment. In order to access digital items in a server via ubiquitous devices, the contents should be adopted according to the system environment device characteristics and user preferences. In ubiquitous environment, those devices dependent adaptation requirements are not statically determined but unpredictable until all the factors of the adaptation requirements such as the user, the terminal, network path, user preference get fixed. Therefore an application specific adaptation mechanism can not be applied to a general digital item adaptation engine.

To find a resolution to playing digital items in real time, there have been continuous studies regarding the prefetch techniques [4],[5],[6],[7],[8]. The prefetch technique is a basic concept that uses common streaming services including a wired Internet environment. But, for a wireless environment, an appropriate optimization is needed for the prefetch time according to the characteristics of each wireless network. Also, this technique has limitations when using mobile terminals because of their low memory capacity. To relieve this problem, transcoding techniques were introduced as [9],[10],[11],[12].Basically, digital contents contains a huge amount of multimedia data such as images, video, sound etc., so mobile terminals are not adequate enough to connect to the Internet directly due to limitations of bandwidths, transaction capacities and memory sizes when compared to desktop computers. To alleviate these constraints and deliver stable multimedia data to mobile terminals, an adaptation of digital item to the point that prefetched play is possible is required between the multimedia server and mobile terminal. However, transcoding all the streams in digital items without any priority would take too much time, and this would result in an increase in the early delay time that occurred before playing on mobile terminals.

1.3 Research Directions

There are many different policies in selecting one stream in a digital item to apply a adaptation process as follow;

Table 1. Steam Selection Scheduling Policy

Policy 1: The Stream Requiring the Highest bps, First
Policy 2: The Stream Requiring the Least bps, First
Policy 3: The Stream Requiring the Highest Transcoding Ratio, First
Policy 4: The Stream Requiring the Longest Transcoding Time, First
Policy 5: The Stream Across Maximum number of Segments, First
Policy 6: The Stream Having the Maximum File Size, First

For a given digital item, the appropriate policy may be different depending on its constituents. In the following, we propose a transcoding pattern based model of DIA(digital item adaptation) and suggest a method of generating an appropriate transcoding policy for each type of digital items.

2 A Pattern Based Transcoding Model for DIA (Digital Item Adaptation)

In general, when a multimedia digital item get adapted for a ubiquitous terminal by using transcoders, the amount of data tends to be reduced significantly [10][11]. In other words, if every stream of the digital item under goes transcoding, the bandwidth required will fall within the range of the network bandwidth, which will enable play on the terminal. However, it takes a significant amount of time to transcode every stream, which increases initial delay time before display. Thus, this paper proposes a way of generating transcoding patterns which is used to guide the order of streams in transcoding process, which will minimize transcoding time for a multimedia contents composed of diverse streams. A digital item, which has been transformed to match the speed of the mobile communication network and thus, has minimized initial delay time during the display, can be delivered and played normally as following procedure. In order not to exceed network bandwidth ($N(t)$), streams should be selected in an order that will minimize transcoding time. In other words, the initial delay time of playing a digital item can be reduced by first determining the criteria by which streams will be prioritized for transcoding, and then scheduling them according to their priority rankings. Thus, the process shown in Table 2 must be followed in compliance with the QoS of the user's mobile terminal.

Table 2. Presentation Procedure on Mobile Terminals

① Check playability of a given digital item, and if yes, done(start to deliver)
② Else apply a transcoding process, which changes unplayable digital item into playable item on mobile terminals, check the playability, and if yes, start to deliver the adapted digital item
③ Else do negotiation of adaptation QoS. If negotiable then repeat else declare "the digital item is unplayable".

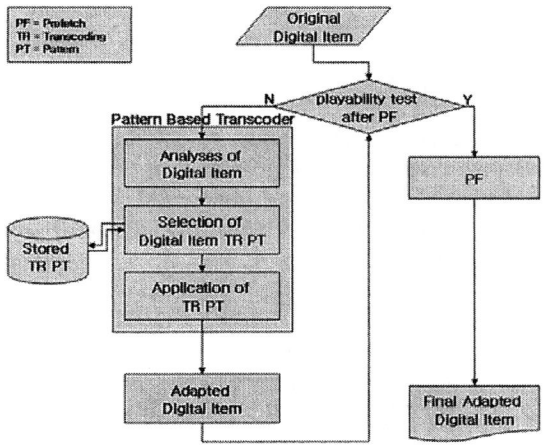

Fig. 3. A Pattern Based Transcoding Model

The procedure may be depicted as the Fig. 3, in which the pattern based transcoding process is more emphasized. The key idea of this model is that it has the transcode scheduling patterns and simply select and apply an appropriate transcoding pattern to the digital item whenever available. Then the model may reduce significantly transcoding time by eliminating the on line and time consuming scheduling process. Another advantage is that its transcoding patterns may be generated in off-line mode. That means we may spend time sufficiently to generate optimized transcoding patterns, whose performance is dependent on the transcoding elements, called transcoders. In this section, we will briefly discuss the components, "Playbility with prefetch test" and "Pattern Based Transcoder". In the following chapter, we will described the pattern generation method in great detail.

2.1 Playability Test

It is necessary to find out whether a digital item is playable on mobile terminals. Each multimedia data stream of the resource in a digital item should be calculated, which comes out as an input value for the required bandwidth per second, and then it should be determined whether the requirements are satisfied when compared to the network bandwidth. Therefore, playing segments should be divided for identification. Using this playing segment, it can be determined whether a digital item is playable on mobile terminals or not. If $\sum_{i=0}^{n} S_i \leq N(t)$ is satisfied (playing segments indicate S and network bandwidth indicates $N(t)$), then the item is said to be playable to a mobile terminal. If not, the adaptation process should be applied.

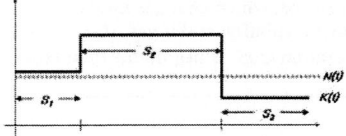

Fig. 4. Playing Segment

2.2 Playability with Prefetch

In most cases, digital items without prefetch are not playable due to the limitation of ubiquitous network bandwidth. So we need a method of deciding the playability after prefetch the digital item. To apply the prefetch and transcode technique to digital item, which is determined as unplayable, the concept of **EPOB**(End Point Over Bandwidth) will be introduced.

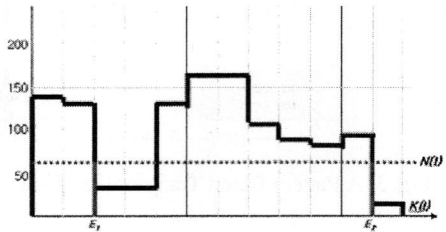

Fig. 5. EPOB(End Point of Over Bandwidth)

The EPOB can be explained as a point t, when the required amount of data of S requires the amount of data below the network bandwidth after requiring data over the network bandwidth. This can be defined as follows:

Definition 1. End Point of Over Bandwidth

X : Digital item, $N(t)$: Network bandwidth, $K(t)$: Required bandwidth of playing segment

$$\lim_{t \to E-0} K(t) > N(t) \text{ and, } \lim_{t \to E+0} K(t) \leq N(t) \quad (1)$$

Time t, which satisfies formula (1), is called the EPOB, and is also written as **E**.

The reason to find **E** is that it is used as a critical unit for prefetch and transcode. It is obvious that every unplayable digital item has at least one **E**. By applying the prefetch to the **X** with **E**, it is possible to overcome limited memory size and overhead of mobile terminals in comparison with the technique of prefetching the entire **X**.

The algorithm, which determines playability by using the **E**-based prefetch technique, is as follows:

Algorithm 1. E -based prefetch technique

① **Divide a digital item into segments based on E.**
② **Calculate the amount of prefetch after comparing the required bandwidth from zero to E_i and network bandwidth until E_i.**
③ **Compare the amount of prefetch calculated in ② and the mobile buffer size.**
④ **If the buffer size in ③ is bigger than the prefetching data size, the presentation is determined as playable and prefetch is applied according to the prefetch policy.**
⑤ **If the buffer size is smaller than the prefetching data size, the digital item is determined as unplayable and is transferred to the transcoding phase.**

If there are more than two E_s in a digital item, prefetch to E_1 is applied to determine playability. If the result is affirmative, extension of the area to E_2 is applied and determine playability.

2.3 Pattern Based Transcoder

Use For our model, we assume an optimized transcoding patterns for each type of digital item are already stored in a Pattern Database. The Pattern Data consists of a pair type of digital item and scheduling policy as (type, policy). The process of transcoding the digital item is very simple as follow;

1) For a given digital item, analyze and decide its type PT
2) Select an appropriate scheduling policy SP, for the given PT, from the Pattern DB
3) Apply transcoding process, to the digital item, based on the scheduling policy SP

The performance of this part is dependent on the scheduling policy stored in the transcoding scheduling pattern DB.
Now we will describe the method of generating the patterns in the following chapter.

3 EPOB Based Transcoding Pattern Generation

The presentation has now been determined as unplayable through the playability test in chapter 2. Now another analysis method must be applied to test whether transcoding would make the presentation playable, and, if so, to find out how to applying the transcoding.

3.1 Transcode

Transcode is a way of transferring quality of service that is related to multimedia streams, such as data format, color depth, frame rate, and size, etc.

Definition 2. Definition of Transcode

Src : Source resource,   Dest : Destination resource
When   QoS(Dest data) = TR {QoS (Src data)},
if QoS (Dest data) ≠ Qos (Src data) then TR is called transcode

As can be seen in Definition 2, in cases when it is necessary to exchange data between different kinds of terminals, the destination terminal may be unable to handle the source data. To solve this problem, a certain type of transformation function, which is here called "*TR*", or transcode, can be utilized. We assume that the model we propose a set of unit transcoders whose functions are limited to one transcoding function. The performance of each transcoder is implementation dependent. In deciding the transcoding policy, we don't have to apply the transcoder itself. The only information we should have is their attributes. Transcode has two major attributes: transfer rate per second (rate of data transfer); and transaction rate per second. The attribute of transfer rate is ratio of source and destination data after transcode for a second and the transaction rate is the time to transcode the source for a second. A sample table showing the relationship between transaction rate and transfer rate, considering each stream's transcoding characteristics, is shown in Table 3. We assume that our model has the transcoder DB which has the attributes described above.

Table 3. Table of Sample Transcodes

Transcode(TR)	Trasnfer Rate (TR.tf)	Transaction Rate (TR.ta)
BMP $\xrightarrow{bmpTRf_{jpg}}$ JPG	About 33: 1	7Mbits /sec
BMP(640X480) $\xrightarrow{640x480TRs_{64x48}}$ BMP(64X48)	About 100: 1	100Mbits/sec
AVI $\xrightarrow{aviTRf_{mpeg4}}$ MPEG4	About 100: 1	3.5Mbits/sec
MPEG2 $\xrightarrow{mpeg2TRf_{mpeg4}}$ MPEG4	About 10: 1	1.8Mbits/sec
AVI(640X480) $\xrightarrow{640x480TRs_{64x48}}$ AVI(64X48)	About 100: 1	100Mbits/sec
WAV $\xrightarrow{wavTRf_{Adpcm}}$ ADPCM	About 30: 1	400Kbits/sec

3.2   EPOB-Based Transcode Scheduling Method

The following algorithm will check whether Formula 1 in Chapter 2 and the time for transcoding are both satisfied during each transcoding level. The transcoding time of the level, as in Formula 1, is the total transcoding time calculated by the given scheduling -- that is, the initial delay time. Stream selection policy used in Algorithm 2 is as follows.

Algorithm 2. Stream Selection Algorithm for EPOB-Based Transcode Scheduling

Level 1: In compliance with the given selection policy, choose stream S for transcoding.
Level 2: Convert stream S, selected at level 1, into S' by applying an appropriate transcoders
  in the transcoder DB.
Level 3: Replace S with S' of the original presentation and compose a new presentation.
Level 4: In the digital item, consider the left side of Formula 1 as X and calculate

$$X = TR[\int_0^{EPOBi} S(t)dt].$$

Level 5: If $X \leq \int_0^{EPOBi} N(t)dt$, selection is complete.

Level 6: If the result does not satisfy level 5, start from level 1 again.

Thus, if Algorithm 2 and Table 1's policies are applied to PT3, the following example can be composed.

Table 4. An Example of Algorithm 2

Stream :
Image1.bmp(34kbps) + Image2.bmp(44 kbps) + Image3.bmp(54bps) + Image4.jpg(14 kbps) + Video1.avi(100 kbps) + Video2.mpeg(68 kbps) + *Video3.avi(103 kbps)* + Video4.mpg(72 kbps) + Audio1.wav(27 kbps) + Audio2.wav(21 kbps)
TR : Video 3.avi(103kbps) → TRf,TRs[Video3'.mpeg4(0.515kbps)]
EPOB : S(t) 865.06> N(t) 616
TR Time = 0.007msec

3.3 Analysis of the Experiment Results

The aims of the experiment are as follows: 1) understand the composition of the streams in the given digital items; 2) divide them according to types as in chapter 1; 3) estimate transcoding time of their streams; and 4) calculate the scheduling policy to reduce initial delay time. Fig.8. shows the stream selection and Fig.6 shows the transcoding time of the presentation shown in Fig.7. The results are as follows: when the bandwidth is low (14kbps, 24kbps), selection policies 5 and 6 were less effective in terms of selecting policies and in terms of transcoding as well. When the network bandwidth was 56kbps, selection policy 1 by far needed the fewest numbers in transcoding, and less transcoding time. When the network bandwidth was above 144k, overall numbers of stream selection were almost the same. But selection policies 1 and 3 showed the least time in transcoding, while Policy 2 kept transcoding many streams without interruption and needed the most time in transcoding.

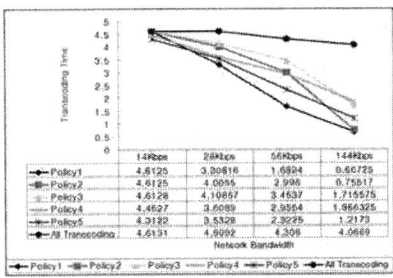

Fig. 6. Selection Policy Transcoding Time Results

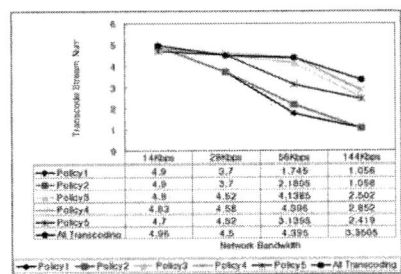

Fig. 7. Selection Policy Transcoding Stream Number Results

Type 1		Selection Policy 1	Selection Policy 2	Selection Policy 3	Selection Policy 4	Selection Policy 5
14Kbps (PCS)	Selection Order	V1, I0, I3, I7, I9, I1, I6, I2, I8, I4, A1	A1, I4, I8, I2, I1, I6, I0, I3, I7, I9, V1	V1, I0, I1, I2, I3, I4, I6, I7, I8, I9, A1	A1, V1, I0, I1, I2, I3, I4, I6, I7, I8, I9	A1, V1, I0, I3, I7, I9
	Units	11 out of 11	11 out of 11	11 out of 11	11 out of 11	6 out of 11
	TR Time	27.5646sec	27.5646sec	27.5646sec	27.5646sec	27.5538sec
28Kbps (Cell)	Selection Order	V1, I0, I3, I7, I9, I1, I6, I2, I8, I4, A1	A1, I4, I8, I2, I1, I6, I0, I3, I7, I9, V1	V1, I0, I1, I2, I3, I4, I6, I7, I8, I9, A1	A1, V1, I0, I1, I2, I3	A1, V1, I0, I3
	Units	11 out of 11	11 out of 11	11 out of 11	6 out of 11	4 out of 11
	TR Time	27.5646sec	27.5646sec	27.5646sec	27.5075sec	27.5026sec
56Kbps (CDMA)	Selection Order	I0, I1, I2, V1, I7, I8, I9	A1, I0, I1, I2, I4, I6, I3, V1, I7, I8, I9	I0, I1, I2, I3, A1, V1, I7, I8, I9,	A1, I0, I1, I2, V1, I7, I8, I9	A1, I0, I1, I2, V1, I7, I8, I9
	Units	7 out of 11	11 out of 11	9 out of 11	8 out of 11	8 out of 11
	TR Time	1.93504sec	27.5646sec	27.5606sec	27.535sec	27.535sec
144Kbps (CDMA 2000 IMT2000)	Selection Order	I0, I1, V1	A1, I0, I1, I4, I2, I6, I3, V1	I0, I1, V1	A1, I0, I1, V1	A1, I0, I1, V1
	Units	3 out of 11	8 out of 11	3 out of 11	4 out of 11	4 out of 11
	TR Time	1.87968sec	27.5114sec	1.87968sec	27.4797sec	27.4797sec

Fig. 8. Transcoding Time of Type 1 According to Each Selection Policy

3.4 Transcode Scheduling Pattern

Based on the experimental result, we can find a transcoding scheduling pattern as in the Table 5. Due to the limited experiments, the number of types is simple.

Table 5. A Pattern of Transcoding Schedule

Digital Item Type	The Best Policy Transcoding
PT1(Digital Item Pattern 1) : (Discrete Media > Continuous Media)	Policy 6 : The Stream Having the Maximum File Size, First
PT1(Digital Item Pattern 2) (Discrete Media = Continuous Media)	Policy 1 : The Stream Requiring the Highest bps, First
PT1(Digital Item Pattern 3) (Discrete Media < Continuous Media)	Policy 1 : The Stream Having the Maximum File Size, First

However, we are assure that further accumulation of experiments may provided with more sophisticated pattern.

4 Conclusion

This paper proposes a model of transcoding streams of a digital item according to selection policies. The aim is to minimize initial delay time, which occurs when playing multimedia presentations in mobile terminals. To generate a transcoding pattern, digital items were divided into different types and 6 scheduling policies are applied in various network bandwidths. As a result, we are able to choose an appropriate scheduling policy a policy for all the type of digital items. Now we have a complete transcoding model by having the pattern in the transcoding pattern DB. For future works, we hope to refine the type of digital items in more detail and to do experiments to find more appropriate patterns.

Acknowledgments

This paper was supported by grant No. R01-2004-000-10618-0 from the Basic Research Program of the Korea Science & Engineering Foundation.

References

1. Young-hwan Yim, Sun-hye Lee, and Myung-soo Yim, "Study on real time presentations of multimedia mail on the internet," Journal of Korea Information Processing Society, vol. 6, no. 4 (1999): 877-889.
2. In-Ho Lin and Bih-Hwang Lee, "Synchronization Model and Resource Scheduling for Distributed Multimedia Presentation System," IEICE TRANS. INF. & SYST., vol. 83-D, no. 4 (April 2000).
3. Dae-won Park, Maria Hong, Kyu-jung Kim, and Young-hwan Yim, "Study on sending image streams to mobile phones without additional software," Journal of Korea Information Processing Society, vol. 3, no. 3 (2001): 55-66.
4. Javed I. Khan and Qingping Tao, "Prefetch Scheduling for Composite Hypermedia," IEEE International Conference on Communication, vol. 3 (2001): 768-773.
5. Frank H.P. Fitzek, "A prefetching protocol for continuous media streaming in wireless environments," IEEE Journal on Selected Areas in Communications, vol. 19, no. 10 (October 2001): 2015-2028.
6. Rita Cucchiara, Massimo Piccardi, and Andrea Prati, "Temporal analysis of cache prefetching strategies for multimedia applications," IEEE International Conference on Performance, Computing, and Communications (2001): 311 –318.
7. Taeil Jeong, JeaWook Ham, and Sungfo Kim, "A Pre-scheduling Mechanism for Multimedia Presentation Synchronization," Proceedings from IEEE International Conference on Multimedia Computing and Systems (June 1997): 379-386.
8. Won-hee Choi, "Buffer prefetch method for playing multimedia in mobile terminals" (Master's thesis, Soongshil University, 2002), 1-25.
9. Niklas Bjork and Charilaos Christopoulos, "Trans-coder Architectures for Video Coding," IEEE Transactions on Consumer Electronics, vol. 44, no. 1 (1998).
10. Eui-Sun Kang, "Guided Search Method Research to convert MPEG2 P frame into H.263 P frame under compressed condition" (Master's thesis, Soongshil University, 2001), 1-15.
11. Jo-Won Lee, "Study on improving transcoder efficiency in reducing image size in compressed segments" (Ph.D. diss., Soongshil University, 2001), 1-18.
12. Sung-Mi Jeon, "Converting path creation methods to play multimedia with different service quality between vertical sections" (Ph.D. diss., Soongshil University, 2003), 2-31.

Identity-Based Aggregate and Verifiably Encrypted Signatures from Bilinear Pairing

Xiangguo Cheng, Jingmei Liu, and Xinmei Wang

State Key Laboratory of Integrated Services Network,
Xidian University, Xi'an 710071, P. R. China
{chengxiangguo, jmliu, xmwang}@xidian.edu.cn

Abstract. Aggregate signatures are digital signatures that allow n players to sign n different messages and all these signatures can be aggregated into a single signature. This single signature enables the verifier to determine whether the n players have signed the n original messages. Verifiably encrypted signatures are used when Alice wants to sign a message for Bob but does not want Bob to possess her signature on the message until a later date. In this paper, we first propose an identity (ID)-based signature scheme from bilinear pairing and show that such a scheme can be used to generate an ID-based aggregate signature. Then, combining this ID-based signature with the short signature given by Boneh, Lynn and Shacham, we come up with an ID-based verifiably encrypted signature. Due to the nice properties of the bilinear pairing, the proposed signatures are simple, efficient and have short signature size.

1 Introduction

The concept of ID-based public key cryptography, first introduced by Shamir [1], allows a user to use his identity information such as name, Email address or telephone number as his public key. It is a good alternative for certificate-based public key cryptography. Ever since Boneh and Franklin gave a practical ID-based encryption scheme from Weil pairing [2] in 2001, several ID-based signatures and short signatures from bilinear pairing have been proposed [3-7].

Aggregate signatures, recently proposed by Boneh et al. [8], are digital signatures that allow n members of a given group of potential signers to sign n different messages and all these signatures can be aggregated into a single signature. This single signature will convince the verifier that the n players did indeed sign the n original messages. Aggregate signatures are useful for reducing the size of certificate verification chains (by aggregating all signatures in the chain) and for reducing message size in secure routing protocols such as SBGP. It is also useful in other special area where the signatures on many different messages generated by many different users need to be compressed.

Verifiably encrypted signatures are used in online contract signing [9, 10] to provide fair exchange. When Alice wants to sign a message for Bob but does not want him to possess her signature on the message immediately. Alice can achieve this by encrypting her signature using the public key of a trusted third party (Adjudicator), and sending the result to Bob along with a proof that she has given him a valid encryption of her signature. Bob can verify that Alice has signed the message but cannot deduce any information about her signature. At a later stage, Bob can obtain the signature either from Alice or resort to the adjudicator who can reveal Alice's signature. Previous constructions of such scheme [9, 11] require zero knowledge proofs to verify an encrypted signature. Boneh et al. [8] gave a verifiably encrypted signature scheme as an application of their aggregate signature. Zhang et al. [12] also gave a verifiably encrypted signature scheme based on their short signature scheme from bilinear pairing [7]. Both signatures are short and can be validated efficiently.

To our best knowledge, no ID-based aggregate and verifiably encrypted signatures schemes have been found so far.

In this paper, we first propose an ID-based signature scheme. It is in fact a variant of the ID-based signature scheme given by Yi [4]. We show that this scheme can be used to construct an ID-based aggregate signature. Then combining the above ID-based signature and the short signature due to Boneh, Lynn, and Shacham [6], we propose an ID-based verifiably encrypted signature scheme. All these signatures are based on the bilinear pairing. Like all other pairing-based signatures, they are simple, efficient and have short signature size.

The remaining sections are organized as follows. In the next section we will give a brief introduction to bilinear pairing and review a short signature scheme. Section 3 presents an ID-based signature scheme and analyzes its security. We propose an ID-based aggregate signature and give its security analysis in Section 4. Section 5 introduces an ID-based verifiably encrypted signature scheme and analyzes its security. Conclusion is drawn in the last section.

2 Preliminaries

We first briefly introduce some mathematical theory related to the following schemes.

2.1 Bilinear Pairing

Let G_1 be a cyclic additive group generated by P, whose order is a prime q, and G_2 a cyclic multiplicative group of the same order q. A bilinear pairing $e: G_1 \times G_1 \to G_2$ is a map with the following properties:

(1) **Bilinear**: $e(aR, bQ) = e(R, Q)^{ab}$ for all $R, Q \in G_1$ and $a, b \in \mathbb{Z}_q^*$.
(2) **Non-degenerate**: $e(P, P)$ is a generator of G_2.
(3) **Computable**: There is an efficient algorithm to compute $e(R, Q)$ for all $R, Q \in G_1$.

2.2 GDH Group

Assuming that the discrete logarithm problem (DLP) in G_1 and G_2 is hard. We consider the following two problems in G_1.

(1) *Computational Diffie-Hellman Problem (CDHP)*: Given (P, aP, bP), compute abP.
(2) *Decision Diffie-Hellman Problem (DDHP)*: Given (P, aP, bP, cP) decide whether $c = ab \pmod{q}$.

The hardness of *CDHP* in G_1 depends on the hardness assumption of *DLP* in G_1 [2, 6]. But *DDHP* is easy in G_1 since $c = ab(\bmod q) \Leftrightarrow e(aP, bP) = e(P, cP)$.

A group G is called a Gap Diffie-Hellman (*GDH*) group if *CDHP* is hard but *DDHP* is easy in G. From the bilinear pairing, we can obtain the *GDH* groups. Such groups can be found on supersingular elliptic curves or hyperelliptic curves over the finite fields, and the bilinear pairings can be derived from Weil or Tate pairing [2].

Schemes in this paper can work on any *GDH* group. Throughout this paper, we define the system parameters in all schemes as follows: G_1, G_2, e, q, P are described as above. Define two cryptographic hash functions: $H_1 : \{0,1\}^* \times G_1^* \rightarrow \mathbb{Z}_q^*$ and $H_2 : \{0,1\}^* \rightarrow G_1^*$. All these system parameters are denoted as $Params = \{G_1, G_2, e, q, P, H_1, H_2\}$.

2.3 BLS Short Signature Scheme

A signature scheme consists of four algorithms: a system parameters generation algorithm *ParamGen*, a key generation algorithm *KeyGen*, a signature generation algorithm *Sign* and a signature verification algorithm *Verify*.

We recall a short signature scheme given by Boneh, Lynn, and Shacham [6]:

(1) *ParamGen*: Given a security parameter k, it outputs system parameters *Params*.
(2) *KeyGen*: The signer chooses a random number $x \in \mathbb{Z}_q^*$ and computes $X = xP$. The private-public key pair of the signer is (x, X).
(3) *Sign*: Given a message m, the signer computes $Y = H_2(m)$ and $\sigma = xY$. The signature on message m is σ.
(4) *Verify*: Given a signature σ on m, the verifier computes $e(P, \sigma)$ and $e(X, Y)$. He accepts the signature if $e(P, \sigma) = e(X, Y)$.

Theorem 1. [6] The above short signature is secure against existential forgery under adaptively chosen message attack in the random oracle model with the assumption that *CDHP* in G_1 is hard.

3 ID-Based Signature and Its Security

To construct the ID-based aggregate and verifiably encrypted signatures, we first propose an ID-based signature. It is in fact a variant of the scheme given by Yi [4].

3.1 ID-Based Signature Scheme

The proposed ID-based signature is described as follows:

(1) *ParamGen*: Sharing the same *Params* with the BLS short signatures.
(2) *KeyGen*: Given an identity *ID*, the key generate center (*KGC*) picks a random number $s \in \mathbb{Z}_q^*$ and computes $P_{pub} = sP$, $Q_{ID} = H_2(ID)$ and $D_{ID} = sQ_{ID}$. s and P_{pub} are the master key and public key of the system, respectively. The private-public key pair corresponding to *ID* is (D_{ID}, Q_{ID}).

(3) *Sign*: Given a message m, the signer chooses a random number $r \in \mathbb{Z}_q^*$ and computes $R = rP$, $h = H_1(m,R)$ and $S = rP_{pub} + hD_{ID}$. The signature on message m under ID is $\sigma = (R,S)$.

(4) *Verify*: Given a signature σ, the verifier computes $Q_{ID} = H_2(ID)$, $h = H_1(m,R)$, $T = R + hQ_{ID}$. He accepts the signature if $e(P,S) = e(P_{pub},T)$.

3.2 Security Analysis

Cha and Cheon gave a security notion of an ID-based signature scheme: *Security against existential forgery on adaptively chosen message and ID attack* [3]. We refer the readers to [3] for details.

Theorem 2. The proposed scheme is secure against *existential forgery on adaptively chosen message and ID attack* in the random oracle model if *CDHP* in G_1 is hard.

Proof: Using the similar method given in [3], we can obtain a result: If there is a polynomial time algorithm $\mathcal{A}_0$ for an adaptively chosen message and *ID* attack to our scheme, then there exists an algorithm $\mathcal{A}_1$ with the same advantage for adaptively chosen message and given *ID* attack.

In the following we will show that the scheme is secure against existential forgery on adaptively chosen message and given *ID* attack if *CDHP* in G_1 is hard.

We assume that the given identity is ID, the corresponding public-private key pair is (Q_{ID}, D_{ID}). According to the Forking Lemma in [13], if there exists an efficient algorithm $\mathcal{A}_1$ for an adaptively chosen message and given ID attack to our scheme, then there exists an efficient algorithm $\mathcal{B}_0$ which can produce two valid signatures (M,R,h_1,S_1) and (M,R,h_2,S_2) such that $h_1 \neq h_2$. Based on $\mathcal{B}_0$, an algorithm $\mathcal{B}_1$, which is as efficient as $\mathcal{B}_0$, can be constructed as follows. Let inputs to algorithm $\mathcal{B}_1$ be P, $P_{pub} = sP$ and $Q_{ID} = tP$ for some $t \in \mathbb{Z}_m^*$. $\mathcal{B}_1$ picks a message M and runs $\mathcal{B}_0$ to obtain two forgeries (M,R,h_1,S_1) and (M,R,h_2,S_2) such that $h_1 \neq h_2$, and $e(P,S_1) = e(P_{pub}, R + h_1 Q_{ID})$, $e(P,S_2) = e(P_{pub}, R + h_2 Q_{ID})$. That is $e(P,(S_1 - S_2) - (h_1 - h_2)D_{ID}) = 1$. Since e is non-degenerate, we have $(S_1 - S_2) - (h_1 - h_2)D_{ID} = \mathcal{O}$ and $D_{ID} = (h_1 - h_2)^{-1}(S_1 - S_2)$. It means that algorithm $\mathcal{B}_1$ can solve an instance of *CDHP* in G_1 since $D_{ID} = sQ_{ID} = stP$.

There is no efficient algorithm for an adaptively chosen message and given ID attack to our scheme since *CDHP* in G_1 is hard. Therefore, our scheme is secure against *existential forgery under adaptively chosen message and ID attack*.

4 ID-Based Aggregate Signature and Its Security

Based on the ID-based signature proposed in Section 3, we can construct an ID-based aggregate signature.

4.1 ID-Based Aggregate Signature Scheme

The proposed ID-based aggregate signature consists of 6 algorithms: *ParamGen*, *KeyGen*, *Sign* and *Verify* are the same as that in the ordinary ID-based signature, the signature aggregation algorithm *AggSign* and the aggregate signature verification algorithm *AggVerify* provide the aggregation capability.

(1) *ParamGen*: Sharing the same *Params* with the original ID-based signature.
(2) *KeyGen*: Let $P_1, P_2, \cdots, P_n$ denote all the players to join the signing. The identity of P_i is denoted as ID_i, the corresponding private-public key pair is $(D_{ID}^{(i)}, Q_{ID}^{(i)})$.
(3) *Sign*: Given n different messages $m_1, m_2, \cdots, m_n$, without lose of generality, we assume that P_i signs message m_i. He randomly picks a number $r_i \in \mathbb{Z}_q^*$, computes and broadcasts $R_i = r_i P$. Let $R = \sum_{i=1}^{n} R_i$, $h_i = H_1(m_i, R)$ and $\sigma_i = (R_i, S_i)$, where $S_i = r_i P_{pub} + h_i D_{ID}^{(i)}$. The signature on m_i given by P_i is σ_i.
(4) *Verify*: Anyone can be designated to aggregate all these single signatures. The designated player (*DP*) first verifies the validity of each single signature. Having received all the single signatures, *DP* computes $R = \sum_{i=1}^{n} R_i$, $h_i = H_1(m_i, R)$ and $T_i = R_i + h_i Q_{ID}^{(i)}$. He accepts the signature if $e(P, S_i) = e(P_{pub}, T_i)$.
(5) *AggSign*: We assume that the single signatures are all valid. *DP* computes $S = \sum_{i=1}^{n} S_i$. The aggregate signature on n different messages $m_1, m_2, \cdots, m_n$ given by n players $P_1, P_2, \cdots, P_n$ is $\sigma = (R, S)$.
(6) *AggVerify*: After receiving $\sigma = (R, S)$, the verifier computes $h_i = H_1(m_i, R)$ and $T = R + \sum_{i=1}^{n} h_i Q_{ID}^{(i)}$. He accepts the aggregate signature if $e(P, S) = e(P_{pub}, T)$.

Correctness of the aggregate signature:

$$e(P, S) = e(P, \sum_{i=1}^{n} S_i) = \prod_{i=1}^{n} e(P, S_i) = \prod_{i=1}^{n} e(P_{pub}, T_i) = e(P_{pub}, \sum_{i=1}^{n} T_i)$$
$$= e(P_{pub}, \sum_{i=1}^{n} (R_i + h_i Q_{ID}^{(i)})) = e(P_{pub}, (R + \sum_{i=1}^{n} h_i Q_{ID}^{(i)})) = e(P_{pub}, T).$$

4.2 Security Analysis

We allow an adversary $\mathcal{A}$ to corrupt all but one honest signer P_n while analyzing the security of our aggregate signature.

Theorem 3. The proposed aggregate signature is secure against existential forgery under chosen message and ID attack in the random oracle model.

Proof: Let $\mathcal{A}$ be a polynomial time adversary for the proposed aggregate signature scheme. We will construct an adversary $\mathcal{B}$ for the underlying signature scheme with the same advantage as $\mathcal{A}$.

$\mathcal{B}$ has the public key $Q_{ID}^{(n)}$ of P_n and access to the random hash oracle and the signing oracle. First $\mathcal{B}$ gives $\mathcal{A}$ the public key $Q_{ID}^{(n)}$. Then $\mathcal{A}$ outputs the set of other $n-1$ private-public key pairs $(D_{ID}^{(1)}, Q_{ID}^{(1)}), (D_{ID}^{(2)}, Q_{ID}^{(2)}), \cdots, (D_{ID}^{(n-1)}, Q_{ID}^{(n-1)})$. Whenever $\mathcal{A}$ asks P_n to join an aggregate signature generation protocol on some messages m_1, $m_2, \cdots, m_n$, $\mathcal{B}$ forwards the query to its signing oracle and returns the reply back to $\mathcal{A}$. At some point, $\mathcal{A}$ outputs an attempted forgery $\sigma = (R, S)$ of some messages m_1, $m_2, \cdots, m_n$. Then $\mathcal{B}$ computes $h_i = H_1(m_i, R)(i = 1, 2, \cdots, n-1)$. He can easily generate the single signatures $\sigma_i = (R_i, S_i)$ on message m_i for $i = 1, 2, \cdots, n-1$ since he knows the private-public key pairs $(D_{ID}^{(1)}, Q_{ID}^{(1)}), (D_{ID}^{(2)}, Q_{ID}^{(2)}), \cdots, (D_{ID}^{(n-1)}, Q_{ID}^{(n-1)})$. Then, he computes $R_n = R - \sum_{i=1}^{n-1} R_i$, $S_n = S - \sum_{i=1}^{n-1} S_i$ and obtains a forgery $\sigma_n = (R_n, S_n)$ on message m_n given by player P_n. It is easy to see that $\mathcal{B}$ will be able to succeed in forgery whenever $\mathcal{A}$ is successful. Theorem 2 has proved the security of the underlying signature. Therefore, the proposed aggregate signature is also secure.

5 ID-Based Verifiably Encrypted Signature and Its Security

Combining the above ID-based signature and the **BLS** short signature, we are able to construct an ID-based verifiably encrypted signature.

5.1 ID-Based Verifiably Encrypted Signature

A verifiably encrypted signature scheme consists of three entities: *signer, verifier and adjudicator*. There are eight algorithms: Five, *ParamGen, KeyGen, Sign, Verify* and *AdjKeyGen* are analogous to those in ordinary signature scheme. The others, *VerSign, VerVerify* and *Adjudicate*, provide the verifiably encrypted signature capability.

(1) *ParamGen*: Sharing the same *Params* with the original ID-based signature.
(2) *KeyGen*: It is analogous to that in the original ID-based signature.
(3) *AdjKeyGen*: It is the same as that in the **BLS** short signature. The adjudicator's private-public key pair is (x, X).
(4) *Sign*: It is analogous to that in the original ID-based signature.
(5) *Verify*: It is the same as that in the original ID-based signature.
(6) *VerSign*: Given a message m, the signer runs the *Sign* algorithm and obtains the signature $\sigma = (R, S)$ on the message m under ID. Then he randomly picks $r \in \mathbb{Z}_q^*$ computes $\eta = rP$ and $\mu = rX$, where X is the public key of the adjudicator. Let $\upsilon = S + \mu$. The triple $\omega = (R, \upsilon, \eta)$ is the verifiably encrypted signature of message m under ID and the adjudicator's public key X.
(7) *VerVerify*: Given a signature $\omega = (R, \upsilon, \eta)$, the verifier computes $Q_{ID} = H_2(ID)$, $h = H_1(m, R)$ and $T = R + hQ_{ID}$. He accepts the signature if $e(P, \upsilon) = e(P_{pub}, T)e(\eta, X)$.

(8) *Adjudicate*: Given a verifiably encrypted signature $\omega=(R,\upsilon,\eta)$, the adjudicator first checks that the signature is valid, then computes $\tau=x\eta$ and $S'=\upsilon-\tau$. $\sigma'=(R,S')$ is the signature of message m under ID.

5.2 Security Analysis

To analyze the security of the verifiably encrypted signature, the security properties of *validity*, *unforgeability* and *opacity* of the scheme should be considered.

(1) **Validity:** $VerVerify(m, VerSign(m))$ and $Verify(m, Adjudicate(VerSign(m)))$ hold for all properly generated user's key pairs and adjudicator's key pairs.
(2) **Unforgeability:** It is difficult to forge a valid verifiably encrypted signature.
(3) **Opacity:** Given a verifiably encrypted signature, it is difficult to extract an ordinary signature on the same message under the same identity ID.

Theorem 4. The proposed signature has the property of validity.

Proof: If $\omega=(R,\upsilon,\eta)$ is a valid verifiably encrypted signature of message m under ID and X, then
$$e(P,\upsilon)=e(P,S+\mu)=e(P,S)\cdot e(P,\mu)=e(P_{pub},T)\cdot e(P,rX)=e(P_{pub},T)e(\eta,X).$$
This means $VerVerify(m,VerSign(m))$ holds. Moreover, for a valid verifiably encrypted signature $\omega=(R,\upsilon,\eta)$, the extracted signature $\sigma'=(R,S')$ satisfies
$$e(P,S')=e(P,\upsilon-\tau)=e(P,S+\mu-\tau)=e(P,S+rX-x\eta)=e(P,S)=e(P_{pub},T).$$
Therefore the output $\sigma'=(R,S')$ of *Adjudicate* is a valid signature of message m under ID. This means $Verify(m, Adjudicate(VerSign(m)))$ holds.

Theorem 5. The proposed signature has the property of unforgeability.

Proof: Given a forger algorithm $\mathcal{A}$ for the ID-based verifiably encrypted signature scheme, we construct a forge algorithm $\mathcal{B}$ for the underlying ID-based signature. $\mathcal{B}$ simulates the challenger and interacts with $\mathcal{A}$ as follows:

(1) *Setup*: $\mathcal{B}$ generates a private and public keys pair (x,X), which serves as the adjudicator's keys.
(2) *Hash Queries*: $\mathcal{A}$ requests a hash on some message m. $\mathcal{B}$ makes a query on m to its own hash oracle and gives the value back to $\mathcal{A}$.
(3) *ID Queries*: $\mathcal{A}$ requests the private and public keys corresponding to some identity ID. $\mathcal{B}$ makes a query on ID and gives the keys to $\mathcal{A}$.
(4) *VerSign Queries*: $\mathcal{A}$ requests a signature for some message and identity pair (m,ID). $\mathcal{B}$ queries its signing oracle for (m,ID), obtaining $\sigma=(R,S)$. It then randomly chooses a number $r\in\mathbb{Z}_q^*$ computes $\eta=rP$, $\mu=rX$ and $\upsilon=S+\mu$, returns to $\mathcal{A}$ the triple $\omega=(R,\upsilon,\eta)$.
(5) *Adjudicate Queries*: $\mathcal{A}$ requests adjudication for $\omega=(R,\upsilon,\eta)$. $\mathcal{B}$ checks that the signature is valid, then computes $\tau=x\eta$, $S'=\omega-\tau$ and returns $\sigma'=(R,S')$.
(6) *Output*: $\mathcal{A}$ outputs a forge $\omega^*=(R^*,\upsilon^*,\eta^*)$, a verifiably encrypted signature on a message m^* under ID^* and adjudicator's public key X. If $\mathcal{A}$ is successful, $\mathcal{B}$

computes $\tau^* = x\eta^*$ and $S^* = \omega^* - \tau^*$. Then $\sigma^* = (R^*, S^*)$ is a valid signature on message m^* under ID^*.

We note that $\mathcal{B}$ succeeds in forging a signature with the same probability of $\mathcal{A}$. Theorem 1 has shown that the underlying signature is unforgeable. As a result, our scheme is also unforgeable.

Theorem 6. The proposed signature has the property of opacity.

Proof: Suppose given a verifiably encrypted signature $\omega = (R, \upsilon, \eta)$ on a message m under identity ID and adjudicator's public key X, an adversary $\mathcal{A}$ wants to compute the signature $\sigma = (R, S)$ on the message m under ID. $\mathcal{A}$ either directly forges the signature or extract a signature $\sigma' = (R, S')$ from $\omega = (R, \upsilon, \eta)$ such that $e(P, S') = e(P_{pub}, T)$. We note that it is impossible for the adversary $\mathcal{A}$ to directly forge a signature since Theorem 2 has shown that the underlying signature is secure against existential forgery. In the following, we will show that it is also impossible to extract a valid signature σ' from the verifiably encrypted signature $\omega = (R, \upsilon, \eta)$. From the generation of the verifiably encrypted signature $\omega = (R, \upsilon, \eta)$, we know that $\upsilon = S + \mu$, where $\sigma = (R, S)$ is an ID-based signature and $\mu = rX$ is in fact a **BLS** short signature on the adjudicator's public key X (X can be viewed as a hash value of some message) under the public key η ($r \in \mathbb{Z}_q^*$ can be viewed as the private key corresponding to η). If the adversary $\mathcal{A}$ can extract a signature $\sigma' = (R, S')$ from the signature $\omega = (R, \upsilon, \eta)$ such that $e(P, S') = e(P_{pub}, T)$. Since $\omega = (R, \upsilon, \eta)$ satisfies $e(P, \upsilon) = e(P_{pub}, T) \cdot e(\eta, X)$, it can be easily derived that $\upsilon - S'$ satisfies $e(P, \upsilon - S') = e(\eta, X)$. This indicates that $\upsilon - S'$ is a **BLS** short signature of X under the public key η, which means that $\mathcal{A}$ has forged a short signature. Theorem 1 has shown that the short signature is existential unforgeable. Therefore, it is impossible to extract a valid signature from the verifiably encrypted signature and the scheme has the property of opacity.

6 Conclusion

In this paper, we first proposed an ID-based signature. Based on this signature, we constructed an ID-based aggregate signature. Combining this ID-based signature with the **BLS** short signature, we presented an ID-based verifiably encrypted signature. All these signatures are based on bilinear pairing. Just like all other pairing based cryptosystems, they are simple, efficient and have short signature size.

References

1. Shamir, A.: Identity-Based Cryptosystems and Signature Schemes. In: Advance in cryptology-Crypto'84, Lecture Notes in Computer Science, Vol. 196, Springer-Verlag, Berlin Heidelberg New York (1987), 47-53.

2. Boneh, D., Franklin, M.: Identity Based Encryption from the Weil Pairing. In: Advance in cryptology-Crypto'01, Lecture Notes in Computer Science, Vol. 2139, Springer-Verlag, Berlin Heidelberg New York (2001), 213-229.
3. Cha, J.C., Cheon, J.H.: An Identity-Based Signature from Gap Diffie-Hellman Groups. In: Advance in Public Key Cryptography-PKC 2003, Lecture Notes in Computer Science, Vol. 2139, Springer-Verlag, Berlin Heidelberg New York (2003), 18-30.
4. Yi, X.: An Identity-Based Signature Scheme from the Weil Pairing. IEEE Communications Letters, Vol. 7(2), 2003, 76-78.
5. Hess, F.: Efficient Identity Based Signature Schemes Based on Pairings. In: Proceeding of Select Areas in Cryptography, SAC 2002, Springer-Verlag (2003), 310-324.
6. Boneh, D., Lynn, B., Shacham, H.: Short Signatures from the Weil Pairing. In: Advance in cryptology-Asiacrypt'01, Lecture Notes in Computer Science, Vol. 2248, Springer-Verlag, Berlin Heidelberg New York (2001), 514-532.
7. Zhang, F., Safavi, R., Susilo, W.: An Efficient Signature Scheme from Bilinear Pairings and Its Applications. In: Advance in Public Key Cryptography-PKC 2004, Lecture Notes in Computer Science, Vol. 2947, Springer-Verlag, Berlin Heidelberg New York (2004), 227-290.
8. Boneh, D., Gentry, C., Lynn, B., Shacham, H.: Aggregate and Verifiably Encrypted Signatures from Bilinear Maps. In: Advance in cryptology-Eurocrypt'03, Lecture Notes in Computer Science, Vol. 2656, Springer-Verlag, Berlin Heidelberg New York (2003), 272-293.
9. Asokan, N., Shoup, V., Waidner, M.: Optimistic Fair Exchange of Digital Signatures. IEEE J. Selected Areas in Comm., Vol. 18(4), 2000, 593-610.
10. Bao, F., Deng, R., Mao, W.: Efficient and Practical Fair Exchange Protocols with Offline TTP. In: Proceedings of IEEE Symposium on Security and Privacy, 1998, 77-85.
11. Poupard, G., Stern, J.: Fair Encryption of RSA Keys. In: Advance in cryptology-Eurocrypt'00, Lecture Notes in Computer Science, Vol. 1807, Springer-Verlag, Berlin Heidelberg New York (2000), 172-189.
12. Zhang, F., Safavi, R., Susilo, W.: Efficient Verifiably Encrypted Signature and Partially Blind Signature from Bilinear Pairings. In: Indocrypt'03, Lecture Notes in Computer Science, Vol. 2904, Springer-Verlag, Berlin Heidelberg New York (2003), 191-204.
13. Pointcheval, D., Stern, J.: Security Arguments for Digital Signatures and Blind Signatures. J. Cryptology, Vol. 13(3), 2000, 361–396.

Element-Size Independent Analysis of Elasto-Plastic Damage Behaviors of Framed Structures

Yutaka Toi and Jeoung-Gwen Lee

Institute of Industrial Science, University of Tokyo,
Komaba 4-6-1, Meguro-ku, Tokyo 153-8505, Japan
toi@iis.u-tokyo.ac.jp

Abstract. The adaptively shifted integration (ASI) technique and continuum damage mechanics are applied to the nonlinear finite element analysis of framed structures modeled by linear Timoshenko beam elements. A new form of evolution equation of damage, which is a function of plastic relative rotational angles, is introduced in order to remove the mesh-dependence caused by the strain-dependence of damage. The elasto-plastic damage behavior of framed structures including yielding, damage initiation and growth can be accurately and efficiently predicted by the combination of the ASI technique and the new damage evolution equation. Some numerical studies are carried out in order to show the validity, especially the mesh-independence of the proposed computational method.

1 Introduction

The occurrence and growth of a number of microscopic defects such as microcracks and microvoids in materials cause reduction of the stiffness, strength and toughness as well as the remaining life of materials. Continuum damage mechanics (abbreviated to CDM) is the theory that can take into account the effects of such microscopic defects on the mechanical properties of solids in the framework of continuum mechanics. CDM has been applied to the finite element analysis of various damage and failure problems of structural members in many literatures [1-5]. The so-called local approach to fracture based on damage mechanics and the finite element method can consistently model the mechanical behaviors from the initiation and evolution of damage through the propagation of macrocracks, however, it is pointed out as a problem that the calculated results considerably depend upon the assumed finite element mesh [4].

The damage analysis of framed structures based on CDM has been studied by many researchers [6-16]. Krajcinovic [6] defined the isotropic damage variable (the damage modulus) related to the fracture stress and used it to calculate the ultimate moment carrying capacities of concrete beams. Chandrakanth and Pandey [7] carried out the elasto-plastic damage analysis of Timoshenko layered beams. Cipollina, Lopez-Inojosa and Florez-Lopez [8], Florez-Lopez [9], Thomson, Bendito and Florez-Lopez [11], Perdomo, Ramirez and Florez-Lopez [12], Marante and Florez-Lopez [15] presented the formulation for the damage analysis of RC frames by the lumped dissipation model and implemented it in the commercial finite element program.

Florez-Lopez [10] gave a unified formulation for the damage analysis of steel and RC frame members. Inglessis, Gomez, Quintero and Florez-Lopez [13], Inglessis, Medina, Lopez, Febres and Florez-Lopez [14], Febres, Inglessis and Florez-Lopez [16] conducted the analysis of steel frames considering damage and local buckling in tubular members. However, no discussion has been made for the mesh-dependence of the finite element solutions for the damage problem of framed structures in the existing literatures [6-16].

The linear Timoshenko beam element is generally used in the finite element analysis of framed structures considering the effect of shear deformation [17]. Toi [18] derived the relation between the location of a numerical integration point and the position of occurrence of a plastic hinge in the element, considering the equivalence condition for the strain energy approximations of the finite element and the computational discontinuum mechanics model composed of rigid bars and connection springs. The computational method identified as the adaptively shifted integration technique [19] (abbreviated to the ASI technique) was developed, based on this equivalence condition. The ASI technique, in which the plastic hinge can be formed at the exact position by adaptively shifting the position of a numerical integration point, gives accurate elasto-plastic solutions even by the modeling with the minimum number of elements. The ASI technique has been applied to the static and dynamic plastic collapse analysis of framed structures [19-22], through which the validity of the method has been demonstrated with respect to the computational efficiency and accuracy.

In the present study, a new computational method is formulated for the elasto-plastic damage analysis of framed structures, based on the ASI technique for the linear Timoshenko beam element and the concept of CDM. The non-layered approach, in which the stress-strain relation is expressed in terms of the resultant stresses and the corresponding generalized strains, is employed in order to reduce the computing time for the large-scale framed structures. A new form of damage evolution equation, which is expressed in terms of plastic relative rotational angles instead of plastic curvature changes, is proposed in order to remove the mesh-dependence of solutions in the damage analysis. The present method is applicable to the collapse analysis of framed structures including elasto-plasticity, damage initiation, its evolution and fracture. Numerical studies for simple frames are conducted to show accuracy, efficiency and the mesh-independence of the proposed method.

2 Formulation for Elasto-Plastic Damage Analysis

In the first subsection of the present section, the ASI technique is described for the elasto-plastic damage analysis using linear Timoshenko beam elements based on the non-layered approach. The ASI technique is expected to provide high efficiency and accuracy of the finite element solutions for the collapse analysis of framed structures. The elasto-plastic constitutive equation considering damage is formulated in the second subsection, in which the tangential stress-strain matrix is derived for the elasto-plastic damage analysis. The bending moments, the axial force and the torsional moment are used as resultant stresses in the formulation. The corresponding generalized strains are respectively the curvature changes, the average axial strain and the torsional angle. The damage evolution equation is discussed in the third subsection.

A new form of damage evolution equation, which is expressed in terms of plastic relative rotational angles instead of plastic curvature changes, is proposed in order to remove the mesh-dependence of finite element solutions in the elasto-plastic damage analysis of framed structures.

2.1 ASI Technique

One of the authors Toi considered an equivalence condition for the strain energy approximations of the linear Timoshenko beam element (the upper figure in Fig. 1) and the computational discontinuum mechanics model which is composed of rigid bars connected with two types of springs resisting relative rotational and transverse displacement respectively (lower figure in Fig. 1) [18]. The strain energy approximation of the linear Timoshenko beam element is a function of the location of a numerical integration point s_1, while the strain energy function of the discontinuum mechanics model depends upon the position of the connection springs r_1. As a result, the following relation was obtained as the equivalence condition for both discrete models:

$$s_1 = -r_1 \tag{1}$$

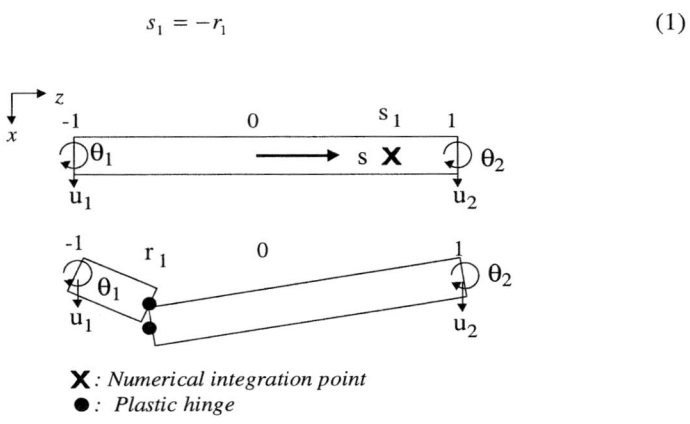

X: Numerical integration point
●: Plastic hinge

Fig. 1. Linear Timosenko beam element and its physical equivalent

When the equivalence condition given by eq. (1) is satisfied, the linear Timoshenko beam element and the computational discontinuum mechanics model are completely the same. The concept of plastic hinges can be easily, explicitly and accurately taken into account by reducing the rotational spring constant in the latter physical model. Therefore, it is clear that a plastic hinge can be formed an arbitrary position in the linear Timoshenko beam element by adaptively shifting the numerical integration point according to eq. (1). When the integration point is located at the right edge in the element, a plastic hinge is formed at the left edge and vice versa. This case is actually important when plastic hinges are formed at member joints or concentratedly loaded points, since they cannot be formed at exact positions when the numerical integration point is located at the central point in each element as is usually done. The details of the ASI technique are as follows.

The numerical integration point is located at the center of an element ($s_1 = 0$) while the element is entirely elastic. The incremental stiffness equation for the element is then given by

$$\mathbf{k}\,d\mathbf{u} = d\mathbf{f} \tag{2a}$$

where

$$\mathbf{k} = L_{elm}\,\mathbf{B}(0)^t\,\mathbf{D}_e(0)\mathbf{B}(0) \tag{2b}$$

In equations (2a) and (2b), the following notations are used: $\mathbf{k}$; the elastic stiffness matrix, $d\mathbf{u}$; the nodal displacement increment vector, $d\mathbf{f}$; the nodal external force increment vector, L_{elm} ; the element length, $\mathbf{B}(s_1)$; the strain-displacement matrix, $\mathbf{D}_e(r_1)$; the elastic stress-strain matrix. The generalized strain increment vector is calculated as

$$d\boldsymbol{\varepsilon}(0) = \mathbf{B}(0)\,d\mathbf{u} \tag{3}$$

The resultant stress increment vector is evaluated as

$$d\mathbf{R}(0) = \mathbf{D}_e(0)\,d\boldsymbol{\varepsilon}(0) \tag{4}$$

The distribution of resultant stresses in the elastically deformed element is determined by the following form of equation [12]:

$$d\mathbf{R}(s) = \mathbf{T}(s)\,d\mathbf{R}(0) \tag{5}$$

where $\mathbf{T}(s)$ is the interpolation function matrix given in [19]. The location of the cross-section in the element which reaches a fully plastic state at first can be determined by comparing the calculated distribution of resultant stresses with the assumed yield function.

$$f[\mathbf{R}(r_1)] = \max_{-1 \le s \le 1}\{f[\mathbf{R}(s)]\} \tag{6}$$

Immediately after the occurrence of the fully plastic section, the numerical integration point is shifted to the new point ($s_1 = -r_1$) according to equation (1) so as to form a plastic hinge exactly at the position of the fully plastic section. For instance, if a fully plastic section occurs at the right edge in the element ($r_1 = 1$), the numerical integration point is shifted to the left edge of the element ($s_1 = -1$) and vice versa. The incremental stiffness equation at the following incremental step is then given by

$$\mathbf{k}\,d\mathbf{u} = d\mathbf{f} \tag{7a}$$

where

$$\mathbf{k} = L_{elm}\,\mathbf{B}(-r_1)^T\,\mathbf{D}_{epd}(r_1)\,\mathbf{B}(-r_1) \tag{7b}$$

In equation (7b), $\mathbf{D}_{epd}(r_1)$ is the stress-strain matrix for elasto-plastic deformation considering damage. The generalized strain increment vector is calculated as

$$d\boldsymbol{\varepsilon}(r_1) = \mathbf{B}(-r_1)\,d\mathbf{u} \tag{8}$$

The resultant stress increment vector is evaluated as

$$d\mathbf{R}(r_1) = \mathbf{D}_{epd}(r_1) d\varepsilon(r_1) \tag{9}$$

The numerical integration point returns to the center when the unloading occurs, and it is shifted again after reyielding.

2.2 Elasto-Plastic Damage Constitutive Equation

The elasto-plastic damage constitutive equation is formulated for the incremental analysis of framed structures in the present subsection, based on the previous study for the elasto-plastic analysis of framed structures by the non-layered approach [23] and the concept of CDM [2].

The dissipation potential of the system is the sum of the plastic potential and the damage potential, which is given by the following equation:

$$F = F_P(\overline{\mathbf{R}}, R; D) + F_D(Y; r, D) \tag{10}$$

where F_P is the plastic potential for the evolution of plastic strains that is a function of the effective resultant stress ($\overline{\mathbf{R}}$), the isotropic hardening stress variable (R) and the scalar damage variable (D). F_D is the damage potential for the evolution of damage that is a function of the strain energy density release rate (Y), the strain of isotropic hardening (r) and the scalar damage variable (D).

The damage increment is obtained by the following equation:

$$dD = d\lambda (\partial F / \partial Y) = d\lambda (\partial F_D / \partial Y) \tag{11}$$

where $d\lambda$ is a proportional coefficient. The concrete form of this equation is discussed in the next subsection.

The yield function is assumed as follows:

$$f = \overline{\sigma}_{eq} - R - \sigma_0 = 0 \tag{12}$$

where the equivalent effective stress $\overline{\sigma}_{eq}$ is given as follows:

$$\overline{\sigma}_{eq}^2 = (\overline{R}_1/Z_{x0})^2 + (\overline{R}_2/Z_{y0})^2 + (\overline{R}_3/A)^2 + (\overline{R}_4/W_p)^2 \tag{13}$$

in which R_1, R_2, R_3 and R_4 are the two components of bending moments, the axial force and the torsional moment respectively. A is the cross-sectional area. Z_{x0}, Z_{y0} and W_p are the plastic sectional factors [23]. Each effective resultant stress component is given by the following equation:

$$\overline{R}_i = R_i/(1-D) \quad (i = 1, \cdots, 4) \tag{14}$$

The following equation is assumed to hold on the yield surface considering damage:

$$df = (\partial f / \partial \overline{\mathbf{R}}) d\overline{\mathbf{R}} + (\partial f / \partial R) dR + (\partial f / \partial D) dD = 0 \tag{15}$$

Using the yield function of equation (10) as the plastic potential F_p, the generalized plastic strain increment ($d\varepsilon_p$) and the strain increment of isotropic hardening (dr) are given by the following equations:

$$d\varepsilon_p = d\lambda(\partial F/\partial \overline{R}) = d\lambda(\partial F_p/\partial \overline{R}) \tag{16}$$

$$dr = -(d\lambda/A)(\partial F/\partial R) = -(d\lambda/A)(\partial F_p/\partial R) \tag{17}$$

where $d\lambda$ is a proportional coefficient.

The total strain increment in the plastic state is the sum of the elastic strain increment and the plastic strain increment. As a result, the following equation is obtained:

$$d\overline{R} = \mathbf{C}\,d\varepsilon_e = \mathbf{C}(d\varepsilon - d\varepsilon_p) = \mathbf{C}\,d\varepsilon - \mathbf{C}\,d\lambda(\partial F_p/\partial \overline{R}) \tag{18}$$

where $\mathbf{C}$ is the resultant stress-generalized strain matrix. $d\varepsilon_e$ and $d\varepsilon_p$ are the generalized elastic and plastic strain increment respectively.

The plastic hardening parameter and its increment are assumed as follows:

$$R = Kr^n \tag{19}$$

$$dR = nKr^{n-1}dr = Hdr = H\,d\lambda/A \tag{20}$$

where K and n are the material constants.

Substituting equations (11), (12), (16) (18) and (20) into equation (15), the proportional coefficient $d\lambda$ is calculated as follows:

$$d\lambda = \left(\frac{\partial F_p}{\partial \overline{R}}\right)^T \mathbf{C}\,d\varepsilon \left/ \left[\frac{H}{A} + \left(\frac{\partial F_p}{\partial \overline{R}}\right)^T \mathbf{C}\frac{\partial F_p}{\partial \overline{R}} + \frac{\overline{\sigma}_{eq}}{1-D}\frac{\partial F_D}{\partial Y}\right]\right. \tag{21}$$

Substituting equation (21) into (18), the following incremental relation between effective resultant stresses and generalized strains can be obtained:

$$d\overline{R} = \overline{\mathbf{C}}\,d\varepsilon = \mathbf{C}\left[1 - \frac{\dfrac{\partial F_p}{\partial \overline{R}}\left(\dfrac{\partial F_p}{\partial \overline{R}}\right)^T \mathbf{C}}{\dfrac{H}{A} + \left(\dfrac{\partial F_p}{\partial \overline{R}}\right)^T \mathbf{C}\dfrac{\partial F_p}{\partial \overline{R}} + \dfrac{\overline{\sigma}_{eq}}{1-D}\dfrac{\partial F_D}{\partial Y}}\right]d\varepsilon \tag{22}$$

where $\overline{\mathbf{C}}$ is the tangential, elasto-plastic damage stiffness matrix.

The incremental relation between resultant stresses and generalized strains is given by the following equation:

$$d\mathbf{R} = (1-D)d\overline{\mathbf{R}} - \overline{\mathbf{R}}dD = \mathbf{D}_{epd}\,d\varepsilon \tag{23}$$

$$= \left[(1-D)\mathbf{C} - \left\{(1-D)\mathbf{C}\frac{\partial F_p}{\partial \mathbf{R}} + \overline{\mathbf{R}}\frac{\partial F_d}{\partial Y}\right\} \frac{\left(\frac{\partial F_p}{\partial \overline{\mathbf{R}}}\right)^T \mathbf{C}}{\frac{H}{A} + \left(\frac{\partial F_p}{\partial \overline{\mathbf{R}}}\right)^T \mathbf{C}\frac{\partial F_p}{\partial \overline{\mathbf{R}}} + \frac{\overline{\sigma}_{eq}}{1-D}\frac{\partial F_D}{\partial Y}}\right] d\varepsilon$$

where $\mathbf{D}_{epd}$ is the tangential, elasto-plastic damage matrix relating resultant stress increments with generalized strain increments to be used in equations (7b) and (9).

2.3 Damage Evolution Equation

The following damage evolution equation given by Lemaitre [2] is used as equation (11) in the preceding subsection:

$$dD = (Y/S)^s dp \quad \text{when} \quad p \geq p_D \tag{24a}$$

where

$$dp = dr/(1-D) \tag{24b}$$

S and s in eq. (24a) are material constants. p and p_D are the accumulated equivalent generalized plastic strain and its critical value for the initiation of damage. The equivalent generalized plastic strain increment $dp(\kappa)$ is given as follows:

$$dp(\kappa) \propto \frac{\overline{R}_1}{Z_{x0}^2} EI_x d\kappa_x + \frac{\overline{R}_2}{Z_{y0}^2} EI_y d\kappa_y + \frac{\overline{R}_3}{A^2} EAd\varepsilon + \frac{\overline{R}_4}{W_p^2} GAd\theta_z' \tag{25}$$

where $d\kappa_x$ and $d\kappa_y$ are the curvature change increments. $d\varepsilon$ and $d\theta_z'$ are average axial strain and the torsional rate respectively. The notation $dp(\kappa)$ indicates that the equivalent generalized plastic strain increment dp is a function of curvature changes here. The strain energy release rate Y is given as follows:

$$Y = \sigma_{eq}^2 / 2E(1-D)^2 \tag{26}$$

where E is the Young's modulus. The time-independent damage that evolves with an increase of the equivalent stress and the equivalent plastic strain is assumed in the present analysis.

The generalized strain increment in the elasto-plastic behavior is the sum of the elastic component and the plastic component. Therefore the relation with the nodal displacement increment is expressed by the following equation:

$$d\varepsilon_e + d\varepsilon_p = \mathbf{B}(d\mathbf{u}_e + d\mathbf{u}_p) \tag{27}$$

The relation between the curvature change increment and the nodal rotational angle increment for the linear Timoshenko beam element is given by the following equation:

$$dK_e + dK_p = d\theta_e/L_{elm} + d\theta_p/L_{elm} = (d\theta_{e2} - d\theta_{e1})/L_{elm} + (d\theta_{p2} - d\theta_{p1})/L_{elm} \qquad (28)$$

where L_{elm} is the element length. The subscripts 1 and 2 indicate nodes at both edges of the element. The plastic relative rotational angle $d\theta_p$ at the plastic hinge can be accurately calculated by the application of the ASI technique, not depending on the element length. This can be proved by the fact that the plastic collapse load of framed structures calculated by the ASI technique coincides with the exact solution given by the theoretical plastic analysis [24], independent of the number of elements [19]. On the other hand, the calculated curvature changes are mesh-dependent.

The damage evolution calculated by equations (24), (25) and (26) extremely depends on the element length (L_{elm}) since the damage evolution equation is expressed in terms of the curvature change increments as shown in equation (25). Then, equation (25) is replaced with the following equation:

$$dp(\theta) \propto \frac{\overline{R}_1}{Z_{x0}^2} EI_x \frac{d\theta_x}{L_{eff,x}} + \frac{\overline{R}_2}{Z_{y0}^2} EI_y \frac{d\theta_y}{L_{eff,y}} + \frac{\overline{R}_3}{A^2} EAd\varepsilon + \frac{\overline{R}_4}{W_p^2} GAd\theta_z \qquad (29)$$

where

$$d\theta_{px} = d\theta_{px2} - d\theta_{px1}, \quad d\theta_{py} = d\theta_{py2} - d\theta_{py1} \qquad (30)$$

where $L_{eff,x}$ and $L_{eff,y}$ are the effective element length dependent on the shape and dimension of the cross-section and the material property, which are the parameters relating the curvature change increments with the plastic relative rotational angles. The plastic relative rotational angle is an important parameter in the plastic analysis of framed structures [24], in which the plastic collapse load and the residual strength of plastic hinges are calculated and discussed by using this parameter. The effective element length as well as the other material constants concerning damage should be determined in the experiments containing bending tests of frame members. However, the tentative values are used in numerical examples in the next chapter. It is expected that the use of $dp(\theta)$ in equation (29) instead of $dp(\kappa)$ in equation (25) will remove the mesh-dependence of the finite element solutions for the elasto-plastic damage analysis of framed structures.

3 Numerical Example

Due to space limitation, only one example is illustrated in the present section. Figure 2 shows the analyzed space frame as well as the results calculated by the finite element method using the ASI technique based on the damage evolution equation expressed in terms of the equivalent plastic strain increment $dp(\theta)$ given in equation (29). As shown in the figure, the mesh-dependence has almost been removed and the highest computational efficiency and accuracy have been achieved by the combined use of the ASI technique and the new damage evolution equation expressed in terms of the plastic relative rotational angles. It should be noted that the minimum number of linear Timoshenko beam elements for the subdivision of each member is two, because only one plastic hinge can be formed in the element. One-element modeling per member is possible, when cubic elements based on Bernoulli-Euler hypothesis are used [19].

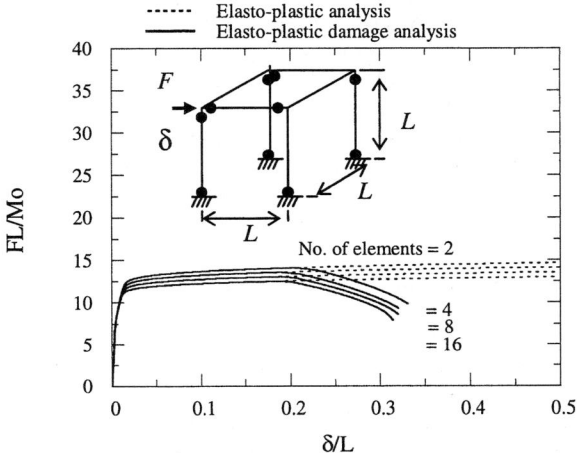

Fig. 2. Load-displacement curves for a space frame by the ASI technique using $dp(\theta)$

4 Concluding Remarks

A new finite element formulation for the elasto-plastic damage analysis of framed structures has been proposed by the combined use of the ASI technique for linear Timoshenko beam elements and the new damage evolution equation expressed in terms of plastic relative rotational angles. It has been confirmed through some numerical studies that the present method is almost mesh-independent and two-element idealization per member is enough for practical purpose. The present computational method can analyze the collapse behavior of large-scale framed structures considering elasto-plasticity, damage and fracture with the highest computational efficiency and accuracy. The tentative value was used as the effective element length. However, it should be determined by experiments, which will be conducted in near future.

References

1. Kachanov, L. M.: Introduction to Continuum Damage Mechanics. Martinus Nijhoff Publishers, (1986)
2. Lemaitre, J.: A Course on Damage Mechanics (Second Edition). Springer, (1996)
3. Krajcinovic, D.: Damage Mechanics. Elsevier, (1996)
4. Skrzypek, J., Ganczarski, A.: Modeling of Material Damage and Failure of Structures: Theory and Applications. Springer, (1999)
5. Kattan, P.I., Voyiadjis, G.Z.: Damage Mechanics with Finite Elements. Springer, (2001)

6. Krajcinovic, D.: Distributed Damage Theory of Beams in Pure Bending. Journal of Applied Mechanics, Transactions of ASME, 46 (1979) 592-596
7. Chandrakanth, S., Pandey, P.C.: Damage Coupled Elasto-Plastic Finite Element Analysis of a Timoshenko Layered Beam. Computers and Structures, 69 (1988) 411-420
8. Cipollina, A., Lopez-Inojosa, A., Florez-Lopez, J.: A Simplified Damage Mechanics Approach to Nonlinear Analysis of Frames. Computers and Structrures, 54 (1995) 1113-1126
9. Florez-Lopez, J.: Simplified Model of Unilateral Damage for RC Frames. J. of Structural Engineering-ASCE, 121(12) (1995) 1765-1772
10. Florez-Lopez, J.: Frame Analysis and Continuum Damage Mechanics. European Journal of Mechanics A/Solids, 17(2) (1998) 269-283
11. Thomson, E., Bendito, A., Florez-Lopez, J.: Simplified Model of Low Cycle Fatigue for RC Frames. Journal of Structural Engineering-ASCE, 124(9) (1998) 1082-1085
12. Perdomo, M.E., Ramirez, A., Florez-Lopez, J.: Simulation of Damage in RC Frames with Variable Axial Forces. Earthquake Engineering & Structural Dynamics, 28(3) (1999) 311-328
13. Inglessis, P., Gomez, G., Quintero, G., Florez-Lopez, J.: Model of Damage for Steel Frame Members. Engineering Structures, 21(10) (1999) 954-964
14. Inglessis, P., Medina, S., Lopez, A., Febres, R., Florez-Lopez, J.: Modeling of Local Buckling in Tubular Steel Frames by Using Plastic Hinges with Damage. Steel & Composite Structures, 2(1) (2002) 21-34
15. Marante, M.E., Florez-Lopez, J.: Three-Dimensional Analysis of Reinforced Concrete Frames Based on Lumped Damage Mechanics. International Journal of Solids and Structures, 40(19) (2003) 5109-5123
16. Febres, R., Inglessis, P., Florez-Lopez, J.: Modeling of Local Buckling in Tubular Steel Frames Subjected to Cyclic Loading. Computers & Structures, 81(22-23) (2003) 2237-2247
17. Bathe, K. J.: Finite Element Procedures, Prentice Hall, (1996)
18. Toi, Y.: Shifted Integration Technique in One-Dimensional Plastic Collapse Analysis Using Linear and Cubic Finite Elements. International Journal for Numerical Methods in Engineering, 31 (1991) 1537-1552
19. Toi, Y., Isobe, D.: Adaptively Shifted Integration Technique for Finite Element Collapse Analysis of Framed Structures. International Journal for Numerical Methods in Engineering, 36 (1993) 2323-2339
20. Toi, Y., Isobe, D.: Finite Element Analysis of Quasi-Static and Dynamic Collapse Behaviors of Framed Structures by the Adaptively Shifted Integration Technique. Computers and Structures. 58 (1996) 947-955
21. Toi, Y., Lee, J.G.: Finite Element Crash Analysis of Framed Structures by the Adaptively Shifted Integration Technique. JSME International Journal, Series A, 43(3) (2000) 242-251
22. Isobe, D., Toi, Y.: Analysis of Structurally Discontinuous Reinforced Concrete Building Frames Using the ASI Technique. Computers and Structures, 76(4) (2000) 242-251
23. Toi, Y., Yang, H.J.: Finite Element Crush Analysis of Framed Structures. Computers and Structures, 41(1) (1991) 137-149
24. Hodge, P.G.Jr.: Plastic Analysis of Structures, McGraw-Hill, (1959)

On the Rila-Mitchell Security Protocols for Biometrics-Based Cardholder Authentication in Smartcards

Raphael C.-W. Phan[1] and Bok-Min Goi[2,*]

[1] Information Security Research (iSECURES) Lab,
Swinburne Sarawak Institute of Technology,
93576 Kuching, Malaysia
rphan@swinburne.edu.my

[2] Multimedia University, 63100 Cyberjaya, Malaysia
bmgoi@mmu.edu.my

Abstract. We consider the security of the Rila-Mitchell security protocols recently proposed for biometrics-based smartcard systems. We first present a man-in-the-middle (MITM) attack on one of these protocols and hence show that it fails to achieve mutual authentication between the smartcard and smartcard reader. In particular, a hostile smartcard can trick the reader into believing that it is a legitimate card and vice versa. We also discuss security cautions that if not handled carefully would lead to attacks. We further suggest countermeasures to strengthen the protocols against our attacks, as well as to guard against the cautions highlighted. Our emphasis here is that seemingly secure protocols when implemented with poor choices of parameters would lead to attacks.

Keywords: Smartcards, biometrics, cardholder authentication, attacks.

1 Introduction

A protocol [3] is a set of rules that define how communication is to be done between two or more parties. In a common networked environment where the communication channel is open to eavesdropping and modifications, security is a critical issue. In this context, *security protocols* are cryptographic protocols that allow communicating parties to perform mutual authentication, key exchange or both. In [7], Rila and Mitchell proposed several security protocols intended for use with biometrics-based smartcard systems [6]. In this paper, we attack one of the protocols and show that it is insecure against man-in-the-middle (MITM) attacks, contrary to the designers' claims [7]. We also discuss security cautions, namely how poor choices of security parameters would lead to attacks.

---

* The second author acknowledges the Malaysia IRPA grant (04-99-01-00003-EAR).

1.1 Standard Security Criteria

We describe standard security criteria expected of any security protocol:

Criterion 1: Mutual Authentication [7]. A smartcard reader must be assured that the smartcard inserted is a legitimate one, and vice versa.

Criterion 2: Resistance to Man-in-the-Middle (MITM) attacks [8]. An MITM attack is where an attacker places himself between two legitimate parties and can impersonate one or both of them. A security protocol should achieve this criterion else it entirely fails to achieve its standard objective of providing authentication between legitimate parties.

Criterion 3: Standard Collision Occurrence [8]. A collision in an n-bit value should only occur with the negligible probability of 2^{-n}.

In this paper, the above are the security criteria of interest to us since we will be showing in the ensuing sections situations where the Rila-Mitchell security protocols will fail to achieve them. The interested reader is further referred to [7] for details of other standard security criteria for protocols.

1.2 The Adversarial Model

The adversarial model used in our paper follows directly from the one considered by the designers themselves, Rila and Mitchell in [7]. They assumed that active attackers are allowed, namely those able to not only eavesdrop on communicated messages but also modify them to their liking. They also assumed that though an attacker can insert a hostile smartcard into a legitimate smartcard reader and also use a hostile smartcard reader to read legitimate smartcards, they claimed that such instances would be unsuccessful since their protocols are supposed to detect such violations. We later show in Section 3 that Protocol 3 does not fulfill this.

As is common with any security protocol, the following are assumed: An adversary could be an insider, i.e., a legitimate party in the network and who can initiate protocol sessions, introduce new messages, receive protocol messages from other parties intended for itself, etc. Note further that encryption only provides confidentiality but not integrity, meaning that though an attacker does not know the secret key used for encrypting any message parts, he could still replay previously valid encrypted parts.

We review in Section 2 the security protocols of [7]. In Section 3, we present our MITM attack. In Section 4, we discuss security cautions for the protocols and how these may cause attacks. We also suggest countermeasures to strengthen the protocols. We conclude in Section 5.

2 Rila-Mitchell Security Protocols

The notations used throughout this paper as follows:
Rila and Mitchell proposed in Section 3 of [7] two similar protocols that they admitted were secure only against passive attacks [3, 9], i.e. an attacker is unable

C	The smartcard
R	The smartcard reader
N_A	The nonce generated by entity A (which may be C or R)
$BioData$	The captured fingerprint (biometric) image
EF	The extracted features from $BioData$
$\|$	Concatenation
$m_K(\cdot)$	A MAC function keyed by secret key, K
$LSB_i(x)$	The i least significant bits (rightmost) bits of x
$MSB_i(x)$	The i most significant bits (leftmost) bits of x
$x << y$	Cyclic shift (rotate) of x left by y bits
$x >> y$	Cyclic shift (rotate) of x right by y bits

to modify existing messages or create new messages. We remark that such an assumption is very impractical by today's standards because the communication link between the smartcard and the reader is commonly accessible by the public. This is true due since a smartcard could be used in various situations, and smartcard readers owned by diverse individuals. Ensuring that an attacker can only mount passive replay attacks is hence not feasible at all.

Therefore, henceforth we concentrate on Rila and Mitchell's suggestion in Section 4 of their paper [7], that their two protocols can be secured against active attacks by replacing the internal hash function, h with a message authentication code (MAC), m_K. We strongly feel that the MAC variants of the protocols are more practical than their hash function counterparts. In addition to these two, Rila and Mitchell also proposed a protocol to allow the reader to verify that the card inserted is a legitimate one. For lack of better names, we denote the three protocols in Section 4 of [7] as Protocols 1, 2 and 3.

Protocol 1 (Using Nonces and BioData).

Message 1: $R \to C$ N_R
Message 2: $C \to R$ $N_C \| BioData \| m_K(N_C \| N_R \| BioData)$
Message 3: $R \to C$ $EF \| m_K(N_R \| N_C \| EF)$

In the first step, the smartcard reader, R generates a random number, N_R and sends it to the smartcard, C. Then C using its built-in fingerprint sensor captures the fingerprint image, $BioData$, and generates a random number, N_C, and sends both these along with a MACed value of $N_C \| N_R \| BioData$ as message 2 to R. Next, R re-computes the MAC and verifies that it is correct. It then extracts the features, EF of $BioData$ and uses this to form message 3, along with a MAC of $N_R \| N_C \| EF$ to C. Then C re-computes the MAC and verifies that it is correct.

Protocol 2 (Using BioData as a Nonce).

Message 1: $R \to C$ N_R
Message 2: $C \to R$ $BioData \| m_K(BioData \| NR)$
Message 3: $R \to C$ $EF \| m_K(EF \| m_K(BioData \| N_R))$

Protocol 2 is very similar to protocol 1 except instead of generating its own random number, C uses the captured fingerprint image, $BioData$ as a random number. Rila and Mitchell note that this relies on the assumption that two different measurements of the same biometric feature of the same person are very likely to be different [7]. Further, to assure that the smartcard has not been inserted into a hostile card reader and vice versa, Rila and Mitchell proposed a separate authentication protocol, as follows:

Protocol 3 (Using Nonces only, without BioData).

Message 1: $R \to C$ N_R
Message 2: $C \to R$ $N_C \| m_K(N_C \| N_R)$
Message 3: $R \to C$ $m_K(N_R \| N_C)$

R generates a random number N_R, and sends it as message 1 to C. Then C generates a random number, N_C and sends this along with a MACed value of $N_C \| N_R$ as message 2 to R. Next, R re-computes this MAC and verifies its correctness. It generates a MAC of $N_R \| N_C$ which is sent as message 3 to C. Finally, C re-computes this MAC and verifies its correctness.

3   A Man-in-the-Middle (MITM) Attack on Protocol 3

We present a man-in-the-middle (MITM) attack on Rila and Mitchell's Protocol 3, showing that a smartcard reader can be bluffed by an inserted hostile smartcard into thinking it is legitimate, and vice versa. This disproves their claim in [7] that with this protocol the card reader can verify that the card inserted is a legitimate one.

An attacker places himself between a valid card, C and a valid reader, R. He puts C into a hostile cloned reader, R', and inserts a hostile smartcard, C' into R.

$\alpha.1 : R \to C'$ N_R
$\beta.1 : R' \to C$ N_R
$\beta.2 : C \to R'$ $N_C \| m_K(N_C \| N_R)$
$\alpha.2 : C' \to R$ $N_C \| m_K(N_C \| N_R)$
$\alpha.3 : R \to C'$ $m_K(N_R \| N_C)$
$\beta.3 : R' \to C$ $m_K(N_R \| N_C)$

Once C' is inserted into R, R generates a random number, N_R and issues it as message $\alpha.1$. This is captured by C' who immediately forwards it to R'. R' replays $\alpha.1$ as message $\beta.1$ to the valid card, C, which returns the message $\beta.2$. This is captured by R' and forwarded to C' which replays it as message $\alpha.2$ to R. R responds with message $\alpha.3$ to C', thereby the hostile card, C' is fully authenticated to the legitimate reader, R. C' forwards this message to R', which replays it as message $\beta.3$ to C, and the hostile reader, R' is authenticated to the legitimate card, C.

This MITM attack resembles the Grand Chessmaster problem [8] and Mafia fraud [2] that can be applied on identification schemes. One may argue that this is a passive attack and does not really interfere in any way since the protocol would appear to be the same whether the attacker is present or not. However, the essence of this attack is that both the legitimate card and reader need not even be present at the same place, but what suffices is that the MITM attack leads them to believe the other party is present. The hostile card and reader would suffice to be in the stead of their legitimate counterparts. This is a failure of mutual authentication between the legitimate card and reader, which should both be present in one place for successful mutual authentication. Thus, Protocol 3 fails to achieve criteria 1 and 2 outlined in Section 1.1.

4 Further Security Cautions and Countermeasures

We discuss further cautions on practically deploying the Rila-Mitchell security protocols. In particular, we show that when specifications are not made explicit, the resultant poor choices of such specifications during implementations may cause the protocols to fail criterion 3 of standard collision resistance, further leading to attacks that cause a failure of mutual authentication (criterion 1).

4.1 Collisions and Attacks on Protocol 1

We first present two attacks on Protocols 1 in this subsection, while attacks on Protocols 2 and 3 will be described in the next subsection.

Collision Attack 1. Let N_R be the random number generated by R in a previous protocol session, and n denotes its size in bits. Further, denote N'_R as the random number generated by R in the current session. Then for the case when the following two conditions are met:

$$N'_R = LSB_{n-r}(N_R), \qquad (1)$$

$$N'_C = N_C \| MSB_r(N_R) \qquad (2)$$

for $r \in \{0, 1, ..., n-1\}$, then the same MAC value and hence a collision would be obtained. This collision is formalized as:

$$m_K(N_C \| N_R \| BioData) = m_K(N'_C \| N'_R \| BioData). \qquad (3)$$

There are n possible cases for the above generalized collision phenomenon. Let m be size of $N_C \| N_R$ in bits. Then, the probability that the collision in (3) occurs is increased from $\frac{1}{2^m}$ to $\frac{n}{2^m}$. This is clearly an undesirable property since a securely used m-bit value should only have collisions with probability $\frac{1}{2^m}$. Under such cases, Protocol 1 would fail criterion 3.

We describe how this can be exploited in an attack. Let α be the previous run of the protocol, and β the current run. The attack proceeds:

$\alpha.1 : R \to C \qquad N_R$
$\alpha.2 : C \to R \qquad N_C \| BioData \| m_K(N_C \| N_R \| BioData)$
$\alpha.3 : R \to C \qquad EF \| m_K(N_R \| N_C \| EF)$
$\beta.1 : R \to I_C \qquad N'_R$
$\beta.2 : I_C \to R \qquad N'_C \| BioData \| m_K(N_C \| N_R \| BioData)$
$\beta.3 : R \to I_C \qquad EF \| m_K(N'_R \| N'_C \| EF)$

An attacker, I has listened in on a previous protocol run, α and hence has captured all the messages in that run. Now, he inserts a hostile smartcard into the reader, R and so initiates a new protocol run, β which starts with the reader, R generating and sending a random number, N'_R to the card, I_C. The attacker's card checks N'_R to see if it satisfies (4). If so, it chooses its own random number, N'_C to satisfy condition (5), and also replays the previously captured fingerprint image, $BioData$ as well as the previously captured MAC, $m_K(N_C\|N_R\|BioData)$ in order to form the message $\beta.2$. When the reader, R receives this message, it would re-compute the MAC, $m_K(N'_C\|N'_R\|BioData)$ and indeed this will be the same value as the received MAC in message $\beta.2$. It therefore accepts the hostile smartcard as valid and fully authenticated. Protocol 1 therefore fails in such circumstances to achieve criterion 1 of mutual authentication (see Section 1.1).

Collision Attack 2. Consider now the case when:

$$MSB_r(N_R) = MSB_r(BioData), \qquad (4)$$

$$N'_R = N_R << r, \qquad (5)$$

$$N'_C = N_C \| MSB_r(NR). \qquad (6)$$

Then:

$$BioData' = LSB_{b-r}(BioData) \qquad (7)$$

where b denotes the size of $BioData$ in bits. Let m be the size of $N_C\|N_R$, then since there are n cases of the above generalized collisions, the probability of such collisions occurring is $\frac{n}{2^{m+2r}}$ instead of the expected $\frac{1}{2^{m+2r}}$ as in the case of an $(m+2r)$-bit value in equations (4) to (7).

We can do better than that and discard the restriction in (4). When:

$$N'_R = N_R >> r, \qquad (8)$$

$$N_C = N'_C \| LSB_r(N_R) \qquad (9)$$

are met, then:

$$BioData' = LSB_r(N_R)\|BioData. \qquad (10)$$

The above generalization occurs with probability $\frac{n}{2^m}$ instead of $\frac{1}{2^m}$.

By exploiting either of the above generalizations which occur with a resultant probability of $\frac{n}{2^{m+2r}} + \frac{n}{2^m} \approx \frac{n}{2^{m-1}}$, our attack then proceeds similarly as Attack 1. The steps in the initial four messages are the same. Then, prior to constructing the message $\beta.2$, the hostile smartcard checks N'_R to see if it satisfies (5) or (8). If so, it chooses its own random number, N'_C to satisfy condition (6) or (9) respectively, and also chooses the new fingerprint image, $BioData'$ as according to the condition (7) or (10) respectively. To complete message $\beta.2$, it replays the previously captured MAC, $m_K(N_C\|N_R\|BioData)$. When the reader, R receives this message, it would re-compute the MAC, $m_K(N'_C\|N'_R\|BioData')$ and indeed this will be the same value as the received MAC in message $\beta.2$. It therefore accepts the hostile smartcard as valid and fully authenticated. Protocol 1 therefore fails in such cases to provide authentication and allows a hostile smartcard to pass off with a fake *BioData*.

4.2 Collision and Attack on Protocol 2

Protocol 2 relies on the assumption that the fingerprint image, *BioData* captured from the same person is random enough. However, the authors admitted that the difference between every two fingerprint captures would be small. We remark however that although *BioData* on its own may be unique enough, the concatenation of *BioData* and N_R may not be, i.e. if:

$$BioData' = MSB_r(BioData), \tag{11}$$
$$N'_R = LSB_{b-r}(BioData)\|N_R, \tag{12}$$

then the same MAC and hence a collision would result! This is given as:

$$m_K(BioData\|N_R) = m_K(BioData'\|N'_R). \tag{13}$$

There are b possible such cases of collisions so the probability of this is $\frac{b}{2^m}$ instead of an expected $\frac{1}{2^m}$. Another scenario for this is when:

$$BioData' = BioData\|MSB_r(N_R), \tag{14}$$
$$N'_R = LSB_{b-r}(N_R), \tag{15}$$

then again a collision as in (13) results. There are similarly b possible such cases of collisions and hence the same probability of occurrence. The resultant probability for the two collision scenarios above is $\frac{b}{2^{m-1}}$. Protocol 2 therefore fails to achieve criterion 3. Our attack follows:

$$\alpha.1 : R \to C \quad N_R$$
$$\alpha.2 : C \to R \quad BioData\|m_K(BioData\|N_R)$$
$$\alpha.3 : R \to C \quad EF\|m_K(EF\|m_K(BioData\|N_R))$$
$$\beta.1 : R \to I_C \quad N'_R$$
$$\beta.2 : I_C \to R \quad BioData'\|m_K(BioData\|N_R)$$
$$\beta.3 : R \to I_C \quad EF'|m_K(EF'\|m_K(BioData'\|N'_R))$$

The steps in the first 4 messages are similar to the attacks in Section 3.1. Prior to constructing the message $\beta.2$, the hostile smartcard checks N'_R to see if it satisfies (12) or (15). If so, it chooses its fingerprint image, *BioData'* to satisfy condition (11) or (14) respectively, and replays the previously captured MAC, $m_K(BioData\|N_R)$ in order to completely form the message $\beta.2$. When the reader, R receives this message, it would re-compute the MAC, $m_K(BioData'\|N'_R)$ and indeed this will be the same value as the received MAC in message $\beta.2$. It therefore accepts the hostile smartcard as valid and fully authenticated. In this case, Protocol 2 therefore fails to achieve criterion 1 of mutual authentication, but instead allows a hostile smartcard to pass off with a fake *BioData'*.

4.3 Collision and Attack on Protocol 3

Protocol 3 is claimed in [7] to assure the smartcard that it has not been inserted into a hostile card reader, and vice versa. However, we have disproved this claim by mounting an MITM attack in Section 3. Here, we will further show how collisions occurring in Protocol 3 would allow for an additional attack to be mounted on this protocol.

Collision Attack. The attacker, who inserts a hostile smartcard into a valid reader, waits until the collision occurs:

$$N'_R = N_C. \tag{16}$$

He then chooses:

$$N'_C = N_R. \tag{17}$$

This allows him to replay previously captured MACs, as follows:

$\alpha.1 : R \to C \quad N_R$
$\alpha.2 : C \to R \quad N_C \| m_K(N_C \| N_R)$
$\alpha.3 : R \to C \quad m_K(N_R \| N_C)$
$\beta.1 : R \to I_C \quad N'_R$
$\beta.2 : I_C \to R \quad N'_C \| m_K(N_R \| N_C)$
$\beta.3 : R \to I_C \quad m_K(N'_R \| N'_C)$

Here, a valid previous protocol run, α, whose messages are captured by the attacker. He then monitors every new message $\beta.1$ until the collision in (16) occurs, upon which he immediately chooses N'_C to satisfy condition (17). This he uses together with a replay of $m_K(N_R \| N_C)$ to form message $\beta.2$ to R. This MAC will be accepted as valid by R, who then returns with message $\beta.3$.

Further Cautions. We conclude by stating two other cautions regarding Protocol 3. Firstly, the case when $N_R = N_C$, then the MAC in message 2 can be replayed as message 3! Secondly, note that any attacker can use C as an oracle to generate $m_K(N_C \| x)$, where x can be any bit sequence sent to C as message 1! To do so, one would merely need to intercept the original message 1 from R to C, and replace it with x. Such an exploitation is desirable in some cases to mount knownz− or chosen−plaintext attacks [3] which are applicable to almost all cryptographic primitives such as block ciphers, stream ciphers, hash functions or MACs.

4.4 Countermeasures

The concerns we raised on Protocols 1 and 2 are due to their designers not fixing the length of the random numbers, N_R and N_C but left it as a flexibility of the protocol implementer. We stress that such inexplicitness can result in subtle attacks on security protocols [1, 4, 5, 10].

We also recommend to encrypt and hence keep confidential the sensitive information such as *BioData* and *EF* rather than transmitting them in the clear! This prevents them from being misused not only in the current system but elsewhere as such information would suffice to identify an individual in most situations. This improvement also makes the protocols more resistant to attacks of the sort that we have presented.

Figure 2 in the original Rila and Mitchell paper [7] also shows that the Yes/No decision signal to the Application component of the smartcard system is accessible externally and not confined within the card. This therefore implies that it will always be possible for an attacker to replay a Yes signal to the Application component regardless of whether an attacker is attacking the rest of the system. We would recommend that for better security, this Yes/No signal as well as the application component should be within the tamper-proof card, and not as otherwise indicated in Figure 2 of [7]. However, such a requirement poses additional implementation restrictions, especially when the application is access-control based, for instance to control access to some premises.

5 Concluding Remarks

Our attack on Rila and Mitchell's Protocol 3 shows that it fails to achieve the claim of allowing the reader to verify that an inserted card is legitimate and vice versa. Our cautions in Section 4 further serve as a general reminder to protocol designers and implementers that all underlying assumptions and potential security shortcomings should be made explicitly clear in the protocol specification, as has also been shown and reminded in [1]. We have also suggested some countermeasures to strengthen the protocols against such problems. However, our suggestions are not entirely exhaustive, hence further analysis needs to be conducted on them.

It would be interesting to consider how to secure the entire smartcard system to encompass the reader as well, since current systems only assure that the card is tamper-resistant while the reader is left vulnerable to tampering, and the communication line between the smartcard and the reader can be eavesdropped easily.

References

1. M. Abadi. Explicit Communication Revisited: Two New Attacks on Authentication Protocols. *IEEE Transactions on Software Engineering*, vol. 23, no. 3, pp. 185-186, 1997.
2. Y. Desmedt, C. Goutier, S. Bengio. Special Uses and Abuses of the Fiat-Shamir Passport Protocol. In *Proceedings of Crypto '87*, LNCS, vol. 293, Springer-Verlag, pp. 21-39, 1988.
3. N. Ferguson, B. Schneier. *Practical Cryptography*. Wiley Publishing, Indiana, 2003.
4. ISO/IEC. *Information Technology - Security Techniques (Entity Authentication Mechanisms Part 2: Entity authentication using symmetric techniques*, 1993.
5. G. Lowe. An attack on the Needham-Schroeder public-key protocol. *Information Processing Letters*, vol. 56, pp. 131-133, 1995.

6. L. Rila, C.J. Mitchell. Security Analysis of Smartcard to Card Reader Communications for Biometric Cardholder Authentication. *5th Smart Card Research and Advanced Application Conference (CARDIS '02)*, USENIX, pp. 19-28, 2002.
7. L. Rila, C.J. Mitchell. Security Protocols for Biometrics-Based Cardholder Authentication in Smartcards. *Applied Cryptography and Network Security (ACNS '03)*, LNCS, vol. 2846, Springer-Verlag, pp. 254-264, 2003.
8. B. Schneier. *Applied Cryptography: Protocols, Algorithms, and Source Code in C*, 2nd edn, John Wiley & Sons, New York, 1996.
9. D.R. Stinson. *Cryptography: Theory and Practice*, 2nd edn, Chapman & Hall/CRC, Florida, 2002.
10. P. Syverson. A Taxonomy of Replay Attacks. *7th IEEE Computer Security Foundations Workshop*, pp. 131-136, 1994.

On-line Fabric-Defects Detection Based on Wavelet Analysis

Sungshin Kim[1], Hyeon Bae[1], Seong-Pyo Cheon[1], and Kwang-Baek Kim[2]

[1] School of Electrical and Computer Engineering,
Pusan National University,
Jangjeon-dong, Geumjeong-gu,
609-735 Busan, Korea
{sskim, baehyeon, buzz74}@pusan.ac.kr
http://icsl.ee.pusan.ac.kr
[2] Department of Computer Engineering, Silla University, Korea
gbkim@silla.ac.kr

Abstract. This paper introduces a vision-based on-line fabric inspection methodology for woven textile fabrics. The current procedure for the determination of fabric defects in the textile industry is performed by humans in the off-line stage. The proposed inspection system consists of hardware and software components. The hardware components consist of CCD array camera, a frame grabber, and appropriate illumination. The software routines capitalize on vertical and horizontal scanning algorithms to reduce the 2-D image into a stream of 1-D data. Next, wavelet transform is used to extract features that are characteristic of a particular defect. The signal-to-noise ratio (SNR) calculation based on the results of the wavelet transform is performed to measure any defects. Defect detection is carried out by employing SNR and scanning methods. Learning routines are called upon to optimize the wavelet coefficients. Test results from different types of defect and different styles of fabric demonstrate the effectiveness of the proposed inspection system.

1 Introduction

The textile industries, as with any industry today, desire to produce the highest quality goods to meet customer demands and to reduce the costs associated with off-quality in the shortest amount of time [1], [2]. Currently, much of fabric inspection is done manually after a significant amount of fabric is produced, removed from the weaving machine, batched into large rolls (1,000-2,000 yards or more) and then sent to an inspection frame. Only about 70% of defects are detected in off-line inspection, even with the most highly trained inspectors. Off-quality sections in the rolls must be re-rolled to remove them.

An automated defect detection and identification system enhances product quality. It also provides a robust method to detect weaving defects. Higher production speeds make the timely detection of fabric defects more important than ever. Newer weaving technologies tend to include larger roll sizes, and this translates into a greater potential for off-quality production before inspection. Computer vision systems are free of some of the limitations of humans while offering the potential for robust defect detection with few false alarms. The advantage of the on-line inspection system is not only defect detection and identification, but also quality improvement by a feedback control loop to adjust setpoints.

In this paper we introduce a computer-vision-based automatic inspection system that can be effectively used to detect and identify faults in various kinds of fabrics. The four major textile defects and the defects of the different styles are illustrated in Figures 1(a) and (b), respectively. Automatic inspection of textile fabrics can be achieved by employing feature analysis algorithms. The feature extraction and identification problem is in fact a problem of classifying features into different categories [3], [4]. It may be viewed as a process of mapping from the feature space to the decision space [5]. Product characterization is an important application area of these algorithms.

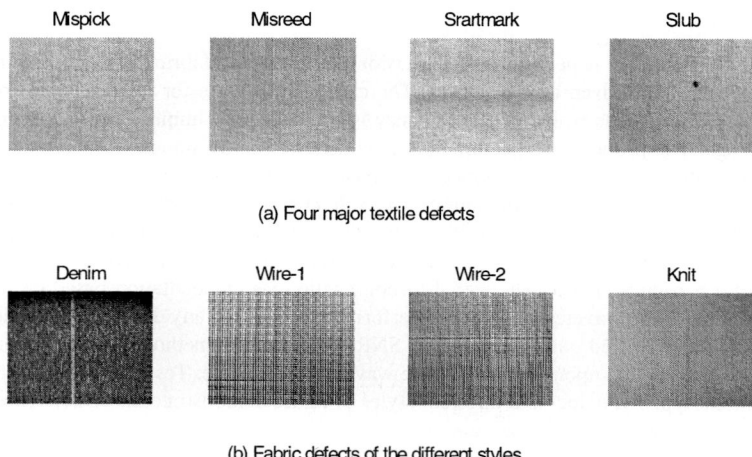

Fig. 1. (a) Four major textile defects (b) Fabric defects of the different styles

This paper is organized as follows. In Section II, preprocessing is presented. Wavelet transform is introduced in Section III. A defect detection process based on the signal-to-noise ratio is presented in Section IV. Experimental results to illustrate the robustness of the proposed approach are shown in Section V. Finally, some conclusions are drawn.

2 Preprocessing

The proposed inspection system architecture consists of preprocessing, feature extraction and decision support systems, as shown in Figure 2.

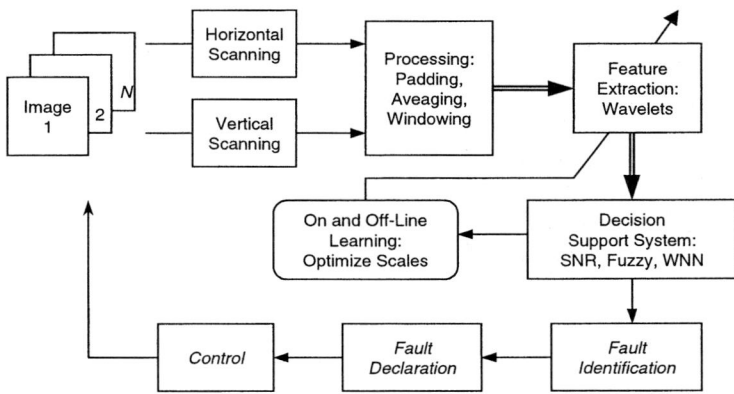

Fig. 2. The inspection system architecture

2.1 Vertical and Horizontal Scan

Preprocessing should actually work together with feature extraction, or at least with the feature extractor's objective. So, the objective of this procedure is to reduce the search space in order to ease the operation of feature extraction, and to enhance the real signal to improve the quality of feature extraction. Since the time requirement is critical in an on-line inspection system, reduction of search space is the first priority in scanning.

A projection method is proposed in consideration of the uniqueness of fabric defects. It is not difficult to perceive that defects in clothes sit mostly either horizontally or vertically. As a matter fact, this is determined by the way in which fabrics are made. Since only the line-type of information is important for fabrics, there is no need to tackle time-consuming 2-D identification unless defect details are required to be identified to almost an extreme degree. Thus in this study, we projected 2-D images horizontally and vertically into two 1-D signals. Vertical projection is called horizontal scan and produces horizontal signals. This type of projection is illustrated in Figure 3. By projection, more specifically we mean that we average all pixel values along a specific direction and use this average value (a point) to represent all of the pixels (a line). Mathematically, we can express this kind of projection as below:

$$horizontal\ signal: P_h(i) = \sum_{j=1}^{n} A(i,j)/n$$
$$vertical\ signal:\quad P_v(i) = \sum_{i=1}^{m} A(i,j)/m \tag{1}$$

where $A(i, j)$ is the scanned image matrix and $i = 1,...,m,\ j = 1,...,n$. These 1-D procesing approach as a preprocessing will prepare for the feature extractor.

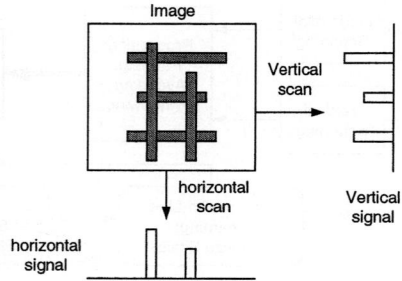

Fig. 3. Projection of 2-D image into two 1-D signals

2.2 End Artifact

For convolution (or transform) with wavelets, we need to pad at the start and finish ends of the 1-D signal. To avoid an edge distortion artifact the data must merge smoothly with an appropriate padding. We have tried (1) padding a constant value; (2) padding an average value; (3) padding a mirrored part of a signal; and (4) padding a repeated part of a signal. We now propose to pad a rotated part of a signal. An example of the rotational padding is shown in Figure 4. The benefits of this method are the following:

- a part of signal not a man-made one is used;
- no high frequency area is created at the ends;
- a defect very close to the ends is amplified;
- the center of the wavelet is aligned with the start of the signal.

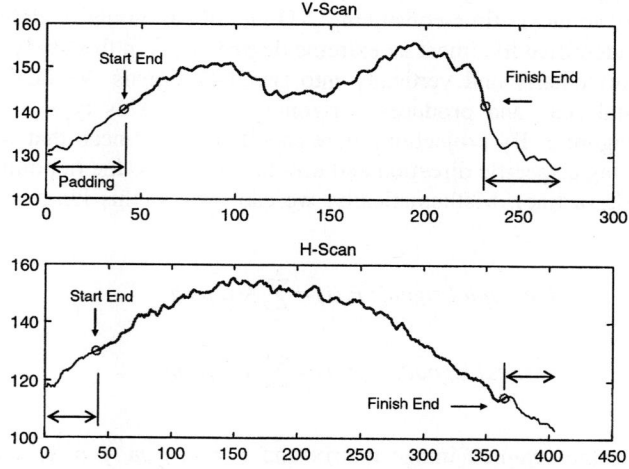

Fig. 4. Rotating padding method applied to V-, H-scanned signals

3 Feature Extraction

Automatic inspection of textile fabrics can be achieved by employing feature analysis algorithms. The feature extraction and identification problem is in fact a problem of classifying features into different categories. It may be viewed as a mapping from the feature space to the decision space.

The overall block diagram of the proposed method is shown in Figure 5. The algorithm consists of wavelet analysis, optimization of the wavelets' coefficients, and the signal-to noise ratio (SNR). These components for feature extraction, detection, and identification are discussed in the following sections.

3.1 Wavelet Analysis for Defect Detection

Wavelet transform (WT) has been widely described in [6], [7] and consists of the convolution product of a function with an analyzing wavelet. The input signal $x(t)$ is in the form of a stream of 1-D data. This data undergoes preprocessing in order to reduce its noise content and increase its usability.

WT, with different wavelet functions to extract features from the signal, is shown in Figure 5. WT provides an alternative to classical short time Fourier transform (STFT) and Gabor Transform [8] for non-stationary signal analysis. The basic difference is that, in contrast to STFT, which uses a single analysis window, WT employs short windows at high frequencies and long windows at low frequencies [9]. Basis functions, called wavelets, constitute the underlying element of wavelet analysis. They are obtained from a single prototype wavelet via compression and shifting operations. The prototype is often called the mother wavelet. The notion of scale is introduced as an alternative to frequency, leading to the so-called time-scale representation.

Let $x(t) \in L^2(R)$ be the signal to be analyzed. Let $\alpha, \beta \in R$, where α is a scaling factor and β is a translation in time. A family of signals, called wavelets, is chosen, $\{\psi_{\alpha,\beta}\} \in L^2(R)$, for different values of α and β, given by

$$\Psi_{\alpha,\beta}(t) \equiv |\alpha|^{-\frac{1}{2}} \psi\left(\frac{t-\beta}{\alpha}\right) \qquad \forall \alpha, \beta \in R$$
$$\int_{-\infty}^{\infty} \psi(t)dt = 0 \qquad (2)$$

where $\psi(t)$ is called the mother wavelet.

The coefficients of WT, for some α and β, are defined as the inner products in $L^2(R)$ of $x(t)$ and $\psi_{\alpha,\beta}(t)$, as

$$c_{\alpha,\beta} = <x, \psi_{\alpha,\beta}> = \int_{-\infty}^{\infty} x(t)\psi_{\alpha,\beta}(t)dt. \qquad (3)$$

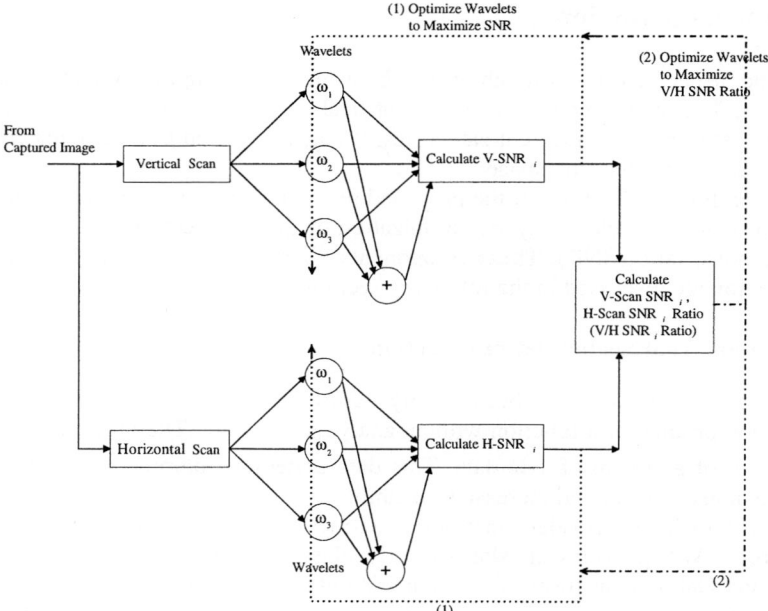

Fig. 5. Feature extraction and optimization using wavelet coefficients

For a discrete case, the wavelet coefficients are obtained as:

$$c_{\alpha,\beta} = \sum_{j=0}^{N} x(j)\psi_{\alpha,\beta}(j) \qquad (4)$$

where N is the number of samples for which $\psi_{\alpha,\beta}(t) \neq 0$.

By choosing a wavelet which is the second derivative of a smoothing function, the wavelet coefficients become proportional to the second derivative of the smoothed signal. The Mexican hat transform involves the wavelet

$$\psi(x;\alpha,\beta) = \left[1 - \left(\frac{x-\beta}{\alpha}\right)^2\right] \exp\left[\frac{-(x-\beta)^2}{2\alpha^2}\right]. \qquad (5)$$

The transformation applied to the wavelet coefficients at the different scales includes segmentation, generation of windows, and multiresolution recombination using a coarse-to-fine approach. We use three wavelets of different scale in Equation (5). The wavelet coefficients are optimized based on the SNR defined in the next section. As an optimization method, the simplex method [10] for nonlinear programming problems is employed. The objective function is defined as

$$J = \text{maximize}\left[\min_i\left(\max_j(SNR_j^i)\right)\right] \qquad (5)$$

where i is the number of sampled images and j is the results of 1-D signals. The optimized three wavelets with scale factors α =0.29, 0.58, and 0.78, are shown in Figure 6.

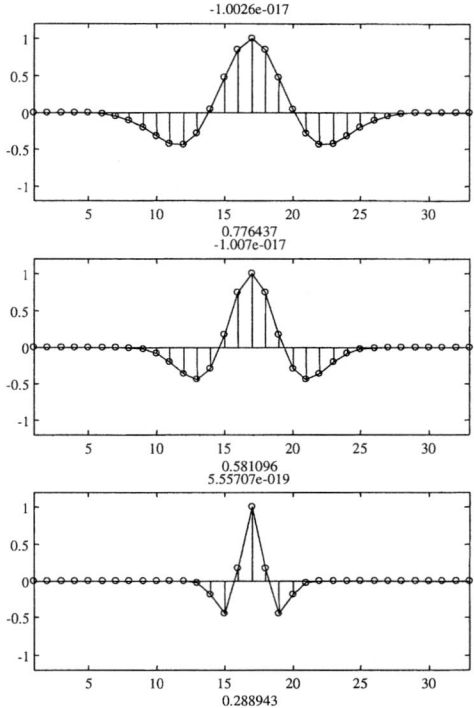

Fig. 6. The optimized wavelets

3.2 Signal-to-Noise Ratio

The maximum and the average of the waveform can be used to calculate the SNR. The SNR is then max(waveform)/average(waveform). This method sometimes gave us good results. But in some other cases, this method did not work well. The reason is that the average of the waveform includes the signal (or feature). Thus, the SNR becomes smaller for large signals.

The new method is proposed in order to separate the signal from the noise. A window is applied to the waveform to pick up the signal. The SNR is calculated as F(signal)/F(noise) where F is a function that could be Max, Area, or Energy (power).

The energy consideration allows us to use more information in the signal. Hence, the resulting gap between the SNR with signal and the SNR with no signal could be relatively greater than that resulted from the first method. This wider SNR gap eases thresholding and thus increases detectability. The SNR used in this paper is as follows:

$$SNR = \frac{\max_i(|s_i|)}{(\sum_{i=1}^{m}|s_i| - \sum_{j=1}^{n}|p_j|)/(m-n)}, \quad p_j \in s_i \qquad (6)$$

$$p_j : s_{k-(n-1)/2} \leq p_j \leq s_{k+(n-1)/2}$$

$$p_{\max} = \max_i(|s_i|) \quad \text{at} \quad i = k.$$

4 Experimental Results

The captured images were tested with the optimized three wavelets that helped to detect low-, medium-, and high-frequency defects. The images included different defects. Also we tested the different styles of fabric. The analyzed results show that defects were detected according to the different defect types and styles of fabric. For example, the horizontal defect in Figure 7 was detected from the vertical scan in the *high*-frequency SNR, 16.56. The vertical defect in Figure 8 was detected from the horizontal scan in the *medium*-frequency SNR, 7.62. The slub defect in Figure 9 was detected from the vertical scan in the *high*-frequency SNR, 9.86. These results show the robustness of the proposed method. The decision support system for identification was performed using these vertical and horizontal SNR information.

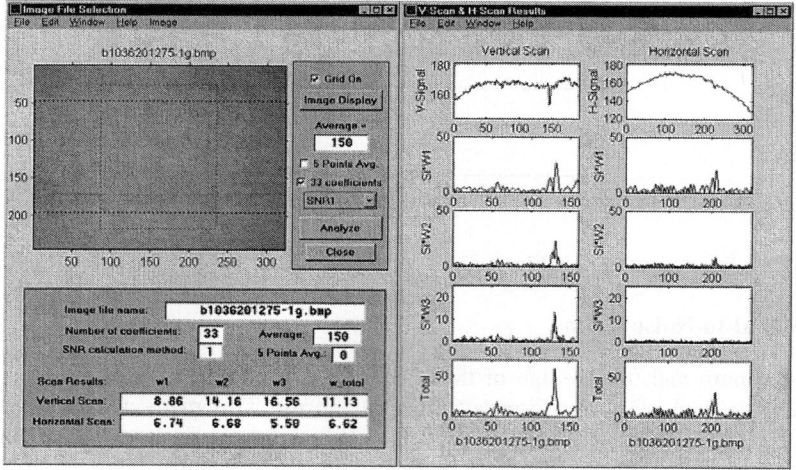

Fig. 7. Analysis results: Broken Pick

5 Conclusions

This paper introduces a vision-based on-line fabric inspection methodology for woven textile fabrics. Due to the inherent periodicity, variability, and noise of textile fabrics, it is not easy to use the traditional frequency techniques to perform an adequate analysis. The proposed inspection system consists of the following modules: capturing images, vertical and horizontal scanning algorithms, wavelet transform to extract features, SNR calculation, and defect declaration routines. The optimization process attempts to choose the best wavelet scales for a given mother wavelet. The test results from different types of defect and different styles of fabric demonstrate the effectiveness of the proposed inspection system.

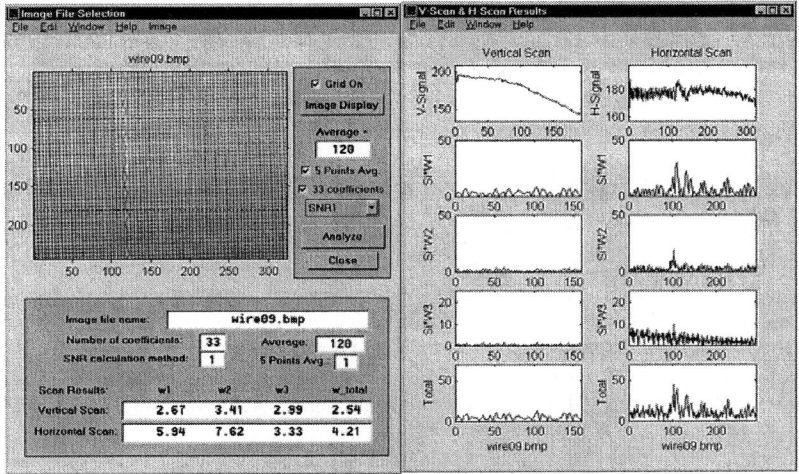

Fig. 8. Analysis results: Warp End

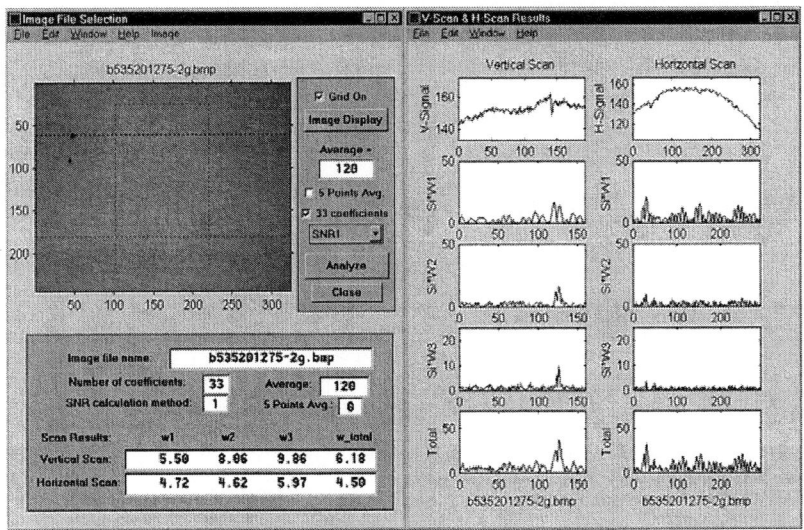

Fig. 9. Analysis results: Slub

Acknowledgement. This work was supported by "Research Center for Logistics Information Technology (LIT)" hosted by the Ministry of Education & Human Resources Development in Korea.

References

1. Kim, S., Vachtsevanos, G., and Dorrity, J.L.: An intelligent approach to integration of textile processes. ASME '96 Intl. Congress & Exposition, Atlanta, GA, (1996) 73-79
2. Vachtsevanos, G., Dorrity, J.L., Kumar, A., Kim, S.: Advanced application of statistical and fuzzy control to textile processes. IEEE Trans. on Industry Applications. 30 (1994) 510-516
3. Jasper, W.J., Garnier, S.J.,Potlapalli, H.: Texture characterization and defect detection using adaptive wavelets. Optical Engineering. 35 (1996) 3140-3149
4. Brad R. and Brad R.: Quality Assurance by Automated Defect Detection of Textile Fabrics. Proc. of XI-th Int. Symp. SINTES, vol. II. Craiova Romania (2003) 487-491
5. Chan C.H. and Pang G.: Fabric defect detection by Fourier analysis. IEEE Trans. Ind. Application. 36 (2000) 1267-1276
6. Mallat, S.: Wavelets for a vision. Proceedings of the IEEE. 84 (1996) 604-614
7. Strang, G., Nguyen, T.: Wavelets and filter banks. Wellesley-Cambridge Press (1996)
8. Masters, T.: Signal and Image Processing with Neural Networks. John Wiley and Sons, Inc. (1994)
9. Rioul, O., Vetterli, M.: Wavelet and Signal Processing. IEEE Signal Processing Magazine. 8 (1991) 14-38
10. Rao, S.S.: Optimization-Theory and Applications. New York, John Wiley Sons (1984) 292-300

Application of Time-Series Data Mining for Fault Diagnosis of Induction Motors

Hyeon Bae[1], Sungshin Kim[1], Yon Tae Kim[1], and Sang-Hyuk Lee[1]

[1] School of Electrical and Computer Engineering, Pusan National University,
30 Jangjeon-dong, Geumjeong-gu, 609-735 Busan, Korea
{baehyeon, sskim, dream0561, leehyuk}@pusan.ac.kr
http://icsl.ee.pusan.ac.kr

Abstract. The motor is the workhorse of industries. The issues of preventive and condition-based maintenance, online monitoring, system fault detection, diagnosis, and prognosis are of increasing importance. This paper introduces a technique to detect faults in induction motors. Stator currents are measured by current meters and stored by time domain. The time domain is not suitable for representing current signals, so the frequency domain is used to display the signals. Fourier transform is used to convert the signals onto frequency domain. After the signals have been converted, the features of the signals are extracted by the signal processing methods like the wavelet analysis, spectrum analysis, and other methods. The discovered features are entered to a pattern classification model such as a neural network model, a polynomial neural network, a fuzzy inference model, or other models. This paper describes the results of detecting fault using Fourier and wavelet analysis.

1 Introduction

The most popular way of converting electrical energy to mechanical energy is an induction motor. This motor plays an important role in modern industrial plants. The risk of motor failure can be remarkably reduced if normal service conditions can be arranged in advance. In other words, one may avoid very costly expensive downtime by replacing or repairing motors if warning signs of impending failure can be headed. In recent years, fault diagnosis has become a challenging topic for many electric machine researchers. The major faults of electrical machines can be broadly classified as follows [1]:

- Broken rotor bar or cracked rotor end-rings
- Static and dynamic air-gap irregularities
- Bent shaft (akin to dynamic eccentricity)
- Bearing and gearbox failure

The diagnostic methods to identify the faults listed above may involve several different types of fields of science and technology [1], [2]. Several methods used to detect faults in induction motors are as follows:

- Electromagnetic field monitoring
- Temperature measurements

- Infrared recognition
- Radio frequency (RF) emissions monitoring
- Noise and vibration monitoring
- Motor current signature analysis (MCSA)
- AI and NN based techniques

Although the Fourier transform is an effective method and widely used in signal processing, the transformed signal may lose some time domain information. The limitation of the Fourier transform in analyzing non-stationary signals leads to the introduction of time-frequency or time scale signal processing tools, assuming the independence of each frequency channel when the original signal is decomposed. This assumption may be considered as a limitation of this approach.

Wavelet transform is a method for time varying or non-stationary signal analysis, and uses a new description of spectral decomposition via the scaling concept. Wavelet theory provides a unified framework for a number of techniques, which have been developed for various signal processing applications. One of its feature is multi-resolution signal analysis with a vigorous function of both time and frequency localization. Mallat's pyramidal algorithm based on convolutions with quadratic mirror filters is a fast method similar to FFT for signal decomposition of the original signal in an orthonormal wavelet basis or as a decomposition of the signal in a set of independent frequency bands. The independence is due to the orthogonality of the wavelet function [3].

2 Fault Detection of Induction Motor

Many types of signals have been studied for the fault detection of induction motors. However, each technique has advantages and disadvantages with respect to the various types of faults. Table 1 shows classifiable and unclassifiable faults corresponding to technique. As shown in Table 1, the MCSA technology is the best detection method among those compared [4].

Table 1. Comparison of detection technologies

Method	Faults it can detect				
	Insulation	Stator Winding	Rotor Winding	Rotor eccentricity	Bear damage
Vibration	×	×	○	○	○
MCSA	×	○	○	○	○
Axial flux	×	○	○	○	×
Lubricating oil debris	×	×	×	×	○
Cooling gas	○	○	○	×	×
Partial discharge	○	×	×	×	×

2.1 Bearing Faults

Though almost 4~50% of all motor failures are bearing related, very little has been reported in the literature regarding bearing related fault detection techniques. Bearing faults might manifest themselves as rotor asymmetry faults from the category of eccentricity related faults [5]. Harmonic components introduced by bearing failures in the line current spectrum are given by [4]:

$$f_{bg} = f \pm kf_b \tag{1}$$

$$f_b = Zf_r / d(1 - \frac{d^2}{D^2}\cos^2\alpha) \tag{2}$$

where, Z is the number of balls in the bearing, D is the diameter of the pitch circle, and α is the contact angle in radians.

Artificial intelligence or neural networks have been researched to detect bearing related faults on line. Also, adaptive, statistical time frequency methods have been studied to locate bearing faults.

2.2 Rotor Faults

Rotor failures now account for 5-10% of total induction motor failures. Broken rotor bars give rise to a sequence of side-bands given by:

$$f_b = (1 \pm 2ks)f \qquad k = 1, 2, 3, \ldots \tag{3}$$

where f is the supply frequency and s is the slip. Frequency domain analysis and parameter estimation techniques have been widely used to detect these types of faults.

In practice, current side bands may exist even when the machine is healthy [6]. Also rotor asymmetry, resulting from rotor ellipticity, misalignment of the shaft with the cage, magnetic anisotropy, and other problems shows up with the same frequency components as the broken bars [7].

2.3 Eccentricity Related Faults

This fault is the result of an unequal air-gap between the stator and rotor. It is called static air-gap eccentricity when the position of the minimal radial air-gap length is fixed in space. In dynamic eccentricity, the center of rotor is not at the center of rotation, so the minimum air-gap changes as the rotor turns. This maybe caused by a bent rotor shaft, bearing wear or misalignment, mechanical resonance at critical speed, or other conditions. In practice an air-gap eccentricity of up to 10% is permissible.

Using MCSA the equation describing the frequency components of interest is:

$$f\left[(kR \pm n_d)\frac{(1-s)}{p} \pm v\right] \tag{4}$$

where $n_d=0$ (in case of static eccentricity), and $n_d=1, 2, 3, \ldots$ (in case if dynamic eccentricity). f is the fundamental supply frequency, R is the number of rotor slots, s is slip, p is the number of pole pairs, k is any integer and υ is the order of the stator time harmonics.

Even though it is obvious, it bears stating, sometimes different faults produce nearly the same frequency components or behave like healthy machine, which make the diagnosis impossible. This is another reason new techniques must be developed. Figures 1 shows the system structure.

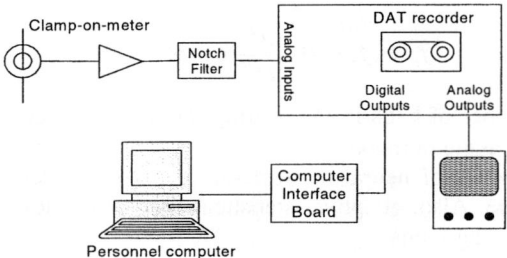

Fig. 1. System structure for data collection and detection

3 Wavelet Transformation

A wavelet is a function ψ belonging to $L^2(R)$ with a zero average. It is normalized and centered in the in the neighborhood of $t=0$. A family of time-frequency atoms is obtained by scaling ψ by a^j and translating is by b:

$$\psi_{a,b} = |a|^{\frac{-j}{2}} \psi\left(\frac{t-b}{a^j}\right) \tag{5}$$

These atoms also remain normalized. The wavelet transform of f belonging to $L^2(R)$ at the time b and scale a^j is:

$$Wf(b, a^j) = \langle f, \varphi_{b,a^j} \rangle = \int_{-\infty}^{+\infty} f(t) \frac{1}{\sqrt{a^j}} \varphi^*(\frac{t-b}{a^j}) dt \tag{6}$$

A real wavelet transform is complete and maintains an energy conservation as long as the wavelet satisfies a weak admissibility condition which is:

$$\int_0^{+\infty} \frac{|\Psi(w)|^2}{|w|} dw = \int_{-\infty}^0 \frac{|\Psi(w)|^2}{|w|} dw = C_\psi < +\infty \tag{7}$$

When $Wf(b, a^j)$ is known only for $a<a_0$, to recover f we need a complement of information corresponding to $Wf(b, a^j)$ for $a<a_0$. This is obtained by introducing a *scaling function* ϕ that is an aggregation of wavelets at scales larger than 1. $\hat{\psi}(w)$ and $\hat{\varphi}(w)$ are Fourier transforms of $\psi(t)$ and $\varphi(t)$ respectively. $\psi(t)$ is a band pass filter, and $\varphi(t)$ is a low-pass filter. Taking positive frequency into account $\hat{\varphi}(w)$ has infor-

mation in [0, π] and $\hat{\psi}(w)$ in [π, 2π]. Therefore they both have complete signal information without any redundancy. Decomposing the signal in [0, π] using Mallat's algorithm gives:

$$h(n) = \langle 2^{-j}\varphi(2^{-1}t)\varphi(t-n)\rangle \qquad j = 0,1,2,\ldots \qquad (8)$$
$$g(n) = \langle 2^{-j}\psi(2^{-1}t)\varphi(t-n)\rangle$$

Wavelet decomposition does not involve the signal in [π, 2π]. In order to decompose the signal in whole frequency bands, wavelet packets can be used for this purpose. A single act of decomposition results in 2^l frequency bands each with the same bandwidth. That is:

$$\left[\frac{(i-1)f_n}{2}, \frac{if_n}{2}\right] \qquad i = 1,2,\ldots,2^j \qquad (9)$$

where, f_n is the Nyquist Frequency, in the ith frequency band.

A wavelet can decompose an original signal that is non-stationary or stationary into independent frequency bands with multi-resolution.

4 Experimental Results

4.1 Current Signals and Data Preprocessing

The motor ratings applied in this paper depend on electrical conditions. The rated voltage, speed, and horsepower are 220V, 3450RPM, and 0.5HP, respectively. Also implemented motor specifications include the number of slots, the number of poles, slip, and other factors. The specifications for used motors are 34 slots, 4 poles, and 24 rotor bars. Figure 2 shows the experiment equipment.

Fig. 2. The data acquisition equipment

The current signals are measured under fixed conditions that consider the sensitivity of the measuring signals that is, the output of the current probes. The sensitivity of each channel is 10mV/A, 100mV/A, and 100mV/A, respectively. The specification of the measured input current signal under this condition consists of 16,384 sampling numbers, 3*kHz* maximum frequency, and 2.1333 measuring time. Therefore the sampling time is 2.1333 over 16,384. Fault types used in this study are broken rotor, faulty bearing, bowed rotor, unbalance, and static and dynamic eccentricity.

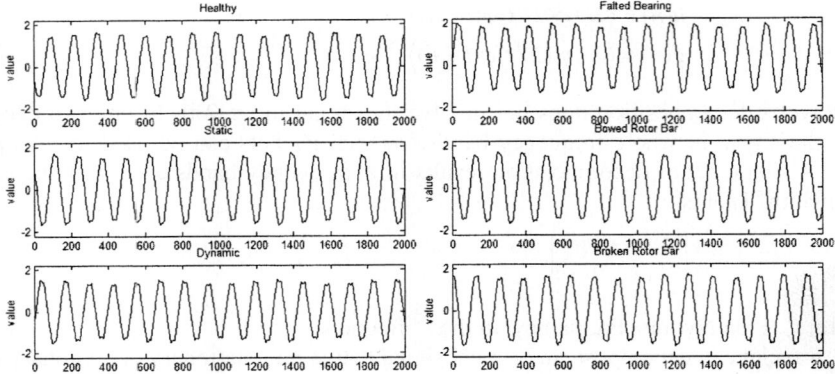

Fig. 3. Current signals measured by equipped motors

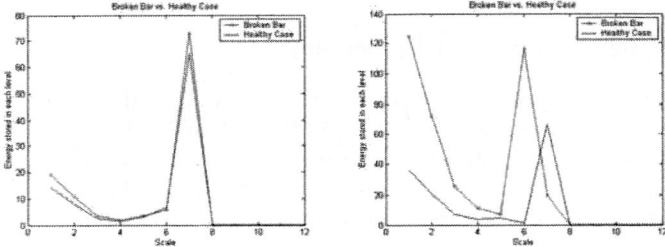

Fig. 4. The wavelet result of a unsynchronized signal

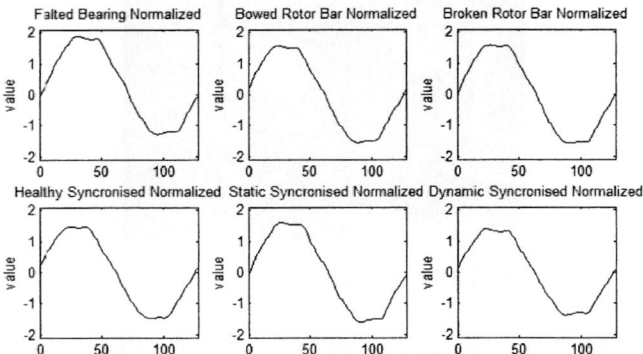

Fig. 5. Synchronized and normalized signals

When wavelet decomposition is used to detect faults in induction motors, the unsynchronized current phase problem should have great influence on the detection results. Figure 3 shows original current signals that are not synchronized with the origin point. If the target signals are not synchronized with each other, unexpected results will appear in wavelet decomposition as shown in Fig. 4. Therefore, the signals have to be re-sampled with the starting origin to phase zero. In this study, 64 of 128 cycles of the signals had to be re-sampled. The average value divided by one cycle signal is used to reduce the noise of the original signals. Figure 5 shows the synchronized and normalized signals used in data preprocessing.

4.2 Feature Extraction Using Wavelet Analysis

In this paper, wavelet analysis was used to detect the faults of induction motors as shown in Fig. 6. Wavelet transform is suitable for time series data mining for extracting the features of faulty motors because the wavelet transform does not lose time information while transforming. In this paper, the specific scale of wavelet decomposition is considered in fault detection, as shown in the figure. Because this detail scale contains most of the information for several faults, the 6th decomposition scale value was analyzed for fault detection. Trial-and-error resulted in the total decomposition of 12 scales. The Coif-let wavelet function was selected based on experimental experience as providing better performance to perform the decomposition. If the mother wavelet function is changed, the performance will also change.

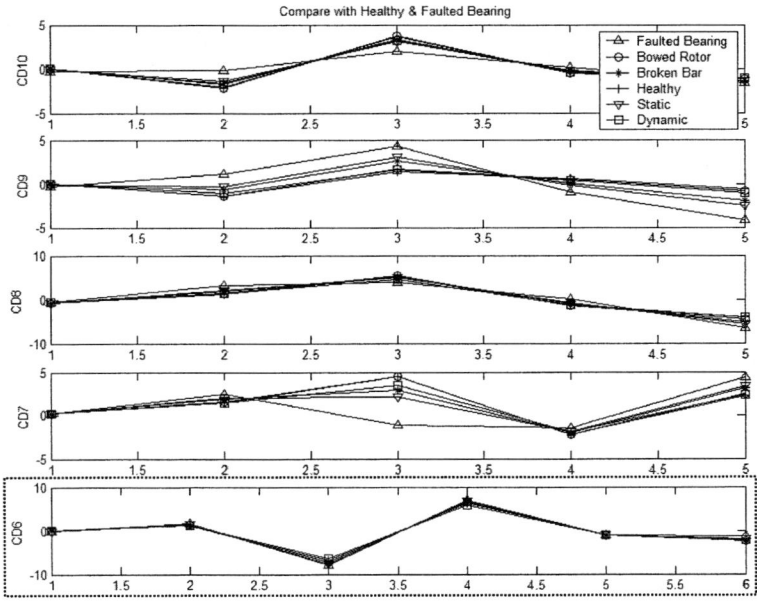

Fig. 6. Details and approximate signals for full loads in wavelet decomposition

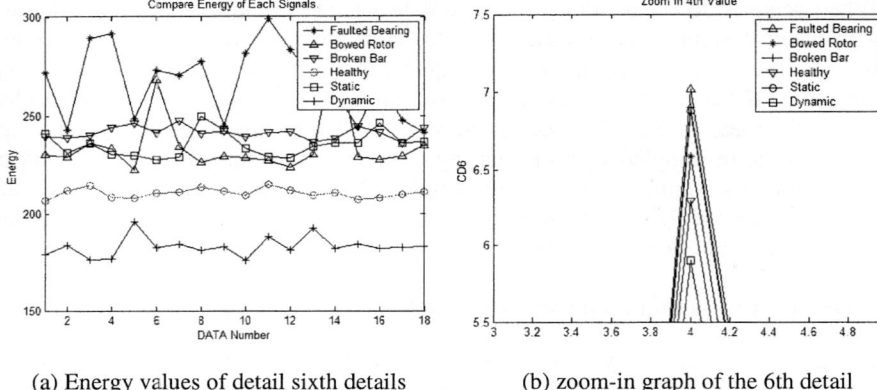

(a) Energy values of detail sixth details (b) zoom-in graph of the 6th detail

Fig. 7. Energy values of detail scale 6 for fault detection and zoom-in graph around the 4th values of detail scale 6

Table 2. The results of feature extraction

Con.	Faulted Bearing		Broken Rotor Bar		Healthy		Static Eccentricity		Dynamic Eccentricity	
	Gradient	Peak	Gradient	Peak	Gradient	Peak	Gradient	Peak	Gradient	Peak
1	14.88	7.0172	14.49	6.8991	13.04	6.2908	14.51	6.8780	12.29	5.9028
2	14.50	6.8379	14.09	6.7671	13.40	6.4371	14.01	6.6925	11.79	5.7362
3	14.84	7.0021	13.93	6.7198	13.42	6.4537	14.36	6.8110	12.21	5.8725
4	14.56	6.8790	14.09	6.7804	13.27	6.3826	14.36	6.7807	11.69	5.6959
5	14.63	6.9143	13.85	6.6965	13.36	6.3996	14.10	6.6862	12.68	6.1149
6	15.29	7.2088	13.90	6.7061	13.06	6.3142	14.05	6.6599	12.48	5.9872
7	14.80	6.9674	13.99	6.7557	13.20	6.3737	14.10	6.6882	12.64	6.0213
8	14.87	7.0154	13.95	6.7262	13.14	6.3528	14.82	7.0270	12.37	5.9286
9	14.46	6.8168	14.29	6.8534	13.44	6.4484	14.59	6.9077	11.96	5.7996
10	14.70	6.9265	14.04	6.7558	13.51	6.4572	13.82	6.6394	11.98	5.8011
11	14.80	6.9789	13.64	6.5992	13.13	6.3512	14.08	6.6932	11.63	5.6883
12	14.71	6.9384	13.91	6.7101	13.32	6.4012	14.08	6.6860	11.90	5.7664
13	14.67	6.9236	14.06	6.7501	13.42	6.4213	14.28	6.7739	12.39	5.9980
14	15.01	7.0807	14.08	6.7694	13.40	6.4247	14.34	6.7925	12.18	5.8778
15	14.52	6.8524	13.98	6.7455	13.26	6.3643	14.34	6.7925	12.32	5.9288
16	14.72	6.9335	14.25	6.8368	13.25	6.3611	14.70	6.9562	12.55	6.0009
17	14.50	6.8386	14.22	6.7934	13.42	6.4254	13.71	6.6383	12.56	6.0024
18	14.45	6.8124	13.91	6.7165	12.82	6.2286	14.39	6.8116	11.83	5.7474
19	14.78	6.9692	14.12	6.7988	13.17	6.3510	13.90	6.6699	12.13	5.8449
20	14.80	6.9924	13.99	6.7362	13.39	6.4127	14.28	6.7652	12.14	5.8456
AVRG	14.72418	6.94526	14.03978	6.75580	13.27085	6.38259	14.24050	6.76749	12.18707	5.87802
Order	6		4		2		5		1	
STD	0.20525	0.09667	0.18330	0.06394	0.17154	0.05856	0.28803	0.10682	0.31817	0.11973
CORR	-0.20594	-0.15736	0.13360	0.04461	1.00000	1.00000	-0.25017	-0.19993	0.07075	0.02187
COVAR	-0.00689	-0.00085	0.00399	0.00016	0.02795	0.00326	-0.01174	-0.00121	0.00367	0.00015

The energy values of decomposed wavelets has been used to detect broken bar faults in past studies but this approach is not useful for detecting several types of faults. The energy values were only used to distinguish properly functioning machines from faulty ones. The energy values are calculated by the squared sum of Wavelet decomposition detail 6 of 18 data. The energy value of properly functioning machines is much different from those of faulty ones, as shown in Fig. 7 (a). The collected features are a gradient value between the 3rd and 4th values and a peak value of the 4th of the 6th as shown in Fig. 7 (b). These features classified the faults best.

Figure 8 represents the results of wavelet analysis and Table 2 shows the results of feature extraction using the Wavelet analysis. The decomposition results show the specific features among the faults used in this study. Most faults can be detected by wavelet analysis except for broken rotor bars and static eccentricity as shown in the figure. This is because the original signals of those two faults have similar shapes. Broken rotor bars and static eccentricity could not be classified by Wavelet analysis.

To solve this problem, another approach was required to detect both faults and Fourier transform was applied to detect broken rotor bars. This analysis was suitable for detecting broken rotor bar, because this analysis shows that broken rotor bars have a specific side band around supply frequency, as shown in Fig. 9. Broken rotor bars can be detected by analyzing the Fourier transform as shown in the figure. From the analysis result, it is possible to detect the some faults using the Fourier and wavelet analysis. Performance of fault detection and diagnosis can be improved using wavelet analysis and Fourier analysis together.

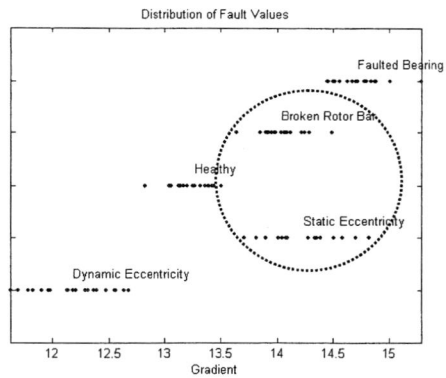

Fig. 8. The distribution of wavelet features of faults

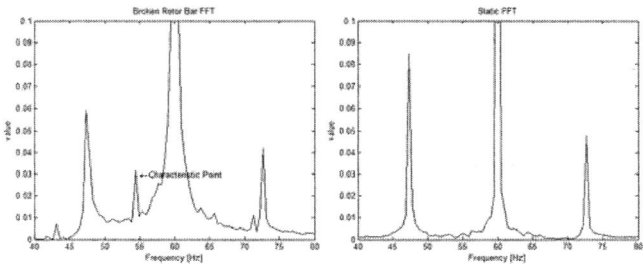

Fig. 9. Comparison between two cases using FFT

5 Conclusion

This paper described the time-series data mining that extracts features from time-series data such as supplied current signals of target motors. Fourier and wavelet transform are the typical techniques for the time-series data mining.

The wavelet analysis method can detect most of the faults of induction motors. The exception is a mechanical fault like bearing bowed fault. The gradient and peak values of the 6th detail result of 12 scales of wavelet decomposition were used.

The features of broken rotor bar and static eccentricity yield similar results in the wavelet analysis, but were different in Fourier analysis. Therefore the use of both types of analysis together can distinguish the faults. In future work, classification models such as neural networks should be used to automatically classify faults.

Acknowledgement

This work was supported by "Research Center for Logistics Information Technology (LIT)" hosted by the Ministry of Education & Human Resources Development in Korea.

References

1. P. Vas, Parameter Estimation, Condition Monitoring, and Diagnosis of Electrical Machines, Clarendron Press, Oxford, 1993.
2. G. B. Kliman and J. Stein, "Induction motor fault detection via passive current monitoring," International Conference in Electrical Machines, Cambridge, MA, pp. 13-17, August 1990.
3. K. Abbaszadeh, J. Milimonfared, M. Haji, and H. A. Toliyat, "Broken Bar Detection In Iduction Motor via Wavelet Transformation," IECON'01: The 27th Annual Conference of the IEEE Industrial Electronics Society, pp. 95-99, 2001.
4. Y. E. Zhongming and W. U. Bin, "A Review on Induction Motor Online Fault Diagnosis," The Third International Power Electronics and Motion Control Conference (PIEMC 2000), vol. 3, pp. 1353-1358, Aug. 15-18, 2000.
5. Masoud Haji and Hamid A. Toliyat, "Patern Recognition-A Technique for Induction Machines Rotor Fault Detection Eccentricity and Broken Bar Fault," Conference Record of the 2001 IEEE Industry Applications Conference, vol. 3, pp. 1572-1578, 30 Sept.-4 Oct. 2001.
6. S. Nandi, H. A. Toliyat, "Condition Monitoring and Fault Diagnosis of Electrical Machines – A Review," IEEE Industry Applications Conference, vol. 1, pp. 197-204, 1999.
7. B. Yazici, G. B. Kliman, "An Adaptive Statistical Time-Frequency Method for Detection of Broken Bars and Bearing Faults in Motors Using Stator Current," IEEE Trans. On Industry Appl., vol. 35, no. 2, pp. 442-452, March/April 1999.

Distortion Measure for Binary Document Image Using Distance and Stroke

Guiyue Jin[1] and Ki Dong Lee[2]

[1] Doctorial student, School of Electrical Eng. & Computer Science,
Yeungnam Univ., 214-1, Dae-Dong, Kyungsan, Kyungbuk, 712-749, Korea
jinguiyue@yumail.ac.kr
[2] Associate Professor, School of Electrical Eng. & Computer Science,
Yeungnam Univ., 214-1, Dae-Dong, Kyungsan, Kyungbuk, 712-749, Korea
kdrhee@yu.ac.kr

Abstract. This paper proposes a new objective distortion measure for binary document images. This measure is based on flipped pixel's immediate neighbors. And it considers the distance between the flipped pixel and neighbor pixels and stroke direction of the neighbor pixels. The simulation and experiments results show that the proposed distortion measure for binary document images matches well to subjective evaluation by human visual perception.

1 Introduction

Distortion measures, which give a numerical measure of picture quality, play an important role in many fields of image processing, watermarkning and data hiding. Digital document image has been receiving more and more attention at present. Digital document images are essentially binary images. In some binary document image applications, visual distortion may be present, and it is very important to measure such distortion for performance evaluation. There are two ways to measure visual distortion [1]. One way is subjective measure, and the other is objective measure. Subjective measure is costly, but it is important, since a human is the ultimate viewer. On the other hand, objective measure is repeatable and easier to implement, although such a measure do not always agree with the subjective one. There are some objective measures giving simple mathematical deviations between the original and the reproduced image. The most widely used such measure is the peak signal-to-noise ratio (PSNR). Although such measures quantify the error in mathematical terms, they do not necessarily reflect the observer's visual perception of the errors.

Human visual perception of document images is different from that of natural images, which are usually continuous multiple-frequency images. Document images are essentially binary, and there are only two levels, black and white. Moreover, documents are mostly consisting of characters, which are more impressive symbols rather than physical objects in natural images. Hence, the HVS models built for natural images may not be well suited for document images. So the perception of distortion in document images is also different from that in natural images. In a particular language, such as English, people know very well what a certain alphabetic character should look like. Hence, distortion in document image could be more obtrusive than distortion in natural images. And the distortion measures proposed for natural images are not often applicable to binary document images.

Several methods for measuring distortion for document images have been proposed in literature. Wu et al [2] measure the visual distortion in data hiding through the change in smoothness and connectivity caused by flipping a pixel. However, the analysis involved is too complex. DRDM method [3] uses a weight matrix to measure the distortion, but this method does not consider the flipped pixel and the stroke direction of character.

In this paper, we propose a new distortion measure algorithm for binary document images. This paper is organized as follows. In section 2, the proposed algorithm is introduced. Section 3 shows the simulation and experiment results. Finally, this paper is concluded.

2 Proposed Distortion Measure Scheme

For a binary document image, we can find that the distance between pixels and pixels plays an important role in their mutual interference perceived by human eyes. As discussed above, readers are so familiar with alphabetic characters that even a single-pixel distortion can be perceived easily. Therefore, the main factors are whether the distortion is in a viewer's focus and the stroke direction of focusing character. The distortion of one pixel is more visible when it is in the field of view of the pixel in focus and when the pixel is in the horizontal or vertical stroke direction than in the diagonal or anti-diagonal stroke direction. Further, from a magnified viewing, each pixel is essentially a black or white square. Therefore, the flipped pixel is more influencing to horizontal or vertical neighbor pixel than to diagonal or anti-diagonal neighbor pixel since the diagonal or anti-diagonal pixel is farther than the horizontal or vertical neighbor pixel. Base on these observations, we proposed an objective distortion measure for binary document images. We name it the distance and stroke distortion measure (DSDM) scheme

2.1 The Weight Matrix

First, this scheme measures the distortion using a weighed matrix with each of its weights determined by the reciprocal of a distance measured from the center pixel. The weight matrix W_m is of size m×m, m =2×n +1, n=1, 2, 3, 4, ..., The center element of this matrix is at (i_c, j_c), $i_c = j_c = (m+1)/2$. $W_m(i, j), 1 \leq i, j \leq m$, is defined as follows:

$$W_m(i, j) = \begin{cases} 1/2, & for\ i = i_c\ and\ j = j_c \\ \dfrac{1}{\sqrt{(i-i_c)^2 + (j-j_c)^2}}, & otherwise. \end{cases} \quad (1)$$

This matrix is normalized to form the normalized weight matrix W_{Nm}.

$$W_{Nm}(i, j) = \frac{W_m(i, j)}{\sum_{i=1}^{m} \sum_{j=1}^{m} W_m(i, j)} \qquad (2)$$

2.2 The Influence Coefficient

See the Fig.1, if the central pixel in image a1 or image b1 is flipped, we can feel different visual distortion level in image a2 and image b2, but the same weight matrix value can be got in image a2 and image b2, so only using the weight matrix, we can get the same distortion value and draw the wrong conclusion.

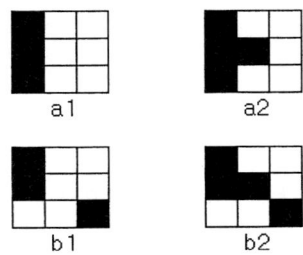

Fig. 1. Image a and Image b having the same weight matrix

It takes human visual perception that the eye is most sensitive to vertical and horizontal lines and edges in an image and is least sensitive to lines and edges with a 45_degree orientation [4], so the line finding operator [5] can be applied to a binary text document image, and line or edge directions are obtained. One possibility is a convolution mask of size 5 × 5, we show only four orientations of the line finding convolution mask of this size in Fig.2. They represent vertical, horizontal, diagonal and anti-diagonal directions.

$$h1 = \begin{bmatrix} 0 & 0 & 0 & 0 & 0 \\ 0 & -1 & 2 & -1 & 0 \\ 0 & -1 & 2 & -1 & 0 \\ 0 & -1 & 2 & -1 & 0 \\ 0 & 0 & 0 & 0 & 0 \end{bmatrix} \qquad h3 = \begin{bmatrix} 0 & 0 & 0 & 0 & 0 \\ -1 & 2 & -1 & 0 & 0 \\ 0 & -1 & 2 & -1 & 0 \\ 0 & 0 & -1 & 2 & -1 \\ 0 & 0 & 0 & 0 & 0 \end{bmatrix}$$

$$h2 = \begin{bmatrix} 0 & 0 & 0 & 0 & 0 \\ 0 & -1 & -1 & -1 & 0 \\ 0 & 2 & 2 & 2 & 0 \\ 0 & -1 & -1 & -1 & 0 \\ 0 & 0 & 0 & 0 & 0 \end{bmatrix} \qquad h4 = \begin{bmatrix} 0 & 0 & 0 & 0 & 0 \\ 0 & 0 & -1 & 2 & -1 \\ 0 & -1 & 2 & -1 & 0 \\ -1 & 2 & -1 & 0 & 0 \\ 0 & 0 & 0 & 0 & 0 \end{bmatrix}$$

Fig. 2. Convolution masks

If line or edge obtained from line finding computation around the flipped pixel is horizontal or vertical direction, big influence coefficient is given to the computation of distortion. If line or edge obtained from line finding computation around the flipped pixel is diagonal or anti-diagonal direction, the small influence coefficient is given to that of distortion.

2.3 The Distortion Evaluation

Then compute the visual distortion, suppose that there are N flipped (from black pixel to white pixel or from white pixel to black pixel) pixels in the output image g(x, y), For the kth flipped pixel in the image g(x, y), the resulted distortion is calculated from an m*m block B_k in original image f(x, y) that is centered at $(x, y)_k$, The distortion measured is given by

$$DSD_k = \sum_{i,j} \alpha_k [D(i,j) \times W_{Nm}(i,j)] \quad (3)$$

α_k is the influence coefficient

Where the (i, j)th element of the difference matrix is given by

$$D_k(i,j) = |B_k(i,j) - g[(x,y)_k]| \quad (4)$$

For the possibly flipped pixels near the image edge, where m×m neighborhood maybe does not exist, it is possible to expand the rest of the m×m neighborhood with the same value as the flipped pixel, which is equivalent to just ignoring the rest of the neighbors.

2.4 The Valid Areas

Assume there is only one flipped pixel in an m×m block B_k, then we consider all flipped pixel in the image, the image distortion is given as

$$DSD = \frac{\sum_{k=1}^{N} DSD_k}{NUBN} \quad (5)$$

Where NUBN is to estimate the valid area in the image and it is defined as the number of non-uniform (not all white pixels) 8×8 blocks in f(x, y). However, if the flipped pixels aggregate some areas rather than distribute in equilibrium, different visual distortion occurs. So we modify the formula (5) to:

$$DSD_{eff} = \frac{\sum_{k=1}^{N} DSD_k}{NUBN_{eff}} \quad (6)$$

In this formula, Where $NUBN_{eff}$ is to estimate the valid area in the image and it is defined as the number of non-uniform (not all white pixels) 8×8 blocks including the flipped pixel in g(x, y).

This DSDM scheme is an efficient and simple method to measure visual distortion for binary document images. This result can be demonstrated by the simulation and experiment results presented below.

3 Simulation and Experiment Results

We carried out the simulation and experiments to test how well the distortion measure proposed is matched with human visual perception.

3.1 Simulation Results and Analysis

The simulations are carried out to test the distortion measure. We designed the image converted from Microsoft Word to Photo shop with a resolution of 72 dots per inch (dpi). The image is of size 198×109. First, we calculate the DRD_k [3] for all pixels, then the pixels are grouped according to the DRD_k, that is, the pixels having the same DRD_k are classified to the same group, distortion generator is designed to do random flipping in different groups. Every time 51 pixels are randomly chosen for flipping. 500 test images are generated for the simulation. And we divided test images into three groups according to the DSD value calculated. One set of the test images generated is shown in Fig.2.

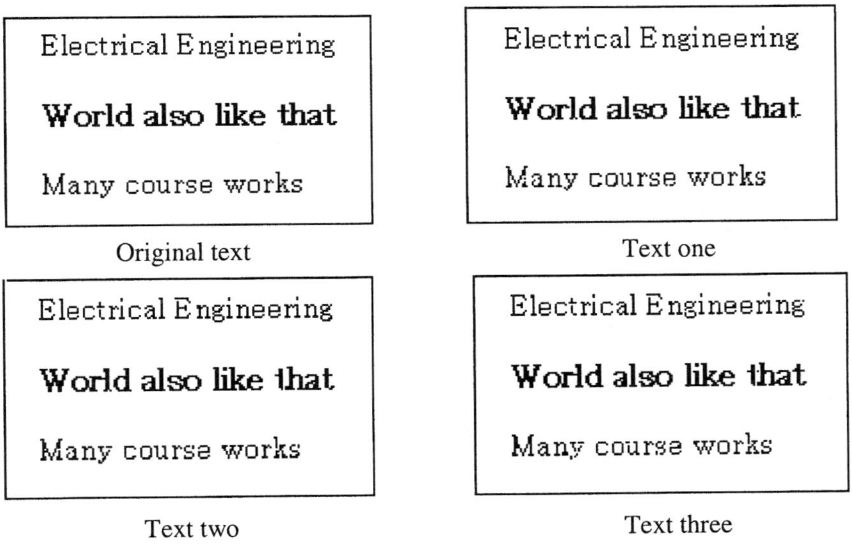

Fig. 3. Original image and Three reproduced images having the same DRD

Table 1. Compared measure data in Fig.3

	Text one	Text two	Text three
PSNR	26.2652	26.2652	26.2652
DRD	0.277813	0.277813	0.277813
DSD	0.222251	0.270759	0.333376

From the Fig.3, we can feel the different distortion in the eye. And in the table 1, there are different DSD distortion values and we can differentiate the distortion, but using DRDM, we can not decide the visual distortion difference. So DSDM better agrees with the subject measure.

Then let's see the effect of flipped pixels distribution. See the Fig.4, we can feel different distortion. But in the table 2, these two images have the same DRD value and DSD value, and we can not differentiate the visual distortion. But they have different DSD_{eff} values and can correctly evaluate the visual distortion.

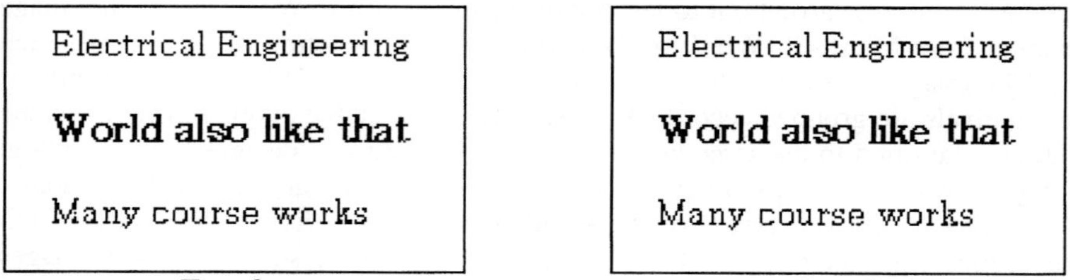 Electrical Engineering **World also like that** Many course works	Electrical Engineering **World also like that** Many course works
Text four	Text five

Fig. 4. Different flipped pixel distribution having the same DSD

Table 2. Compare with measure data in Fig.4

	Distortion Test one	Distortion Test two
PSNR	26.2652	26.2652
DRD	0.277813	0.277813
DSD	0.222251	0.222251
DSD_{eff}	0.594847	0.546617

3.2 Experiment Results and Analysis

We printed the Fig. 3 and Fig. 4 using a HP LaserJet 5000 printer. And we asked the observers to rank the visual distortion level to Fig.3 and Fig.4 according to the visual distortion that the observers perceive when they views the images at a suitable distance under normal lighting conditions. Text one, text two and text three are group1 and the other two are group2. Group1 has three distortion levels (high, middle, low) and group2 has two distortion levels (high, low). We collected the ranking data from 50 observers and got the statistics data as table 3.

Table 3. The statistics data in subjective measure

		Distortion level 1	Distortion level 2	Distortion level 3
Group 1	Text one	0%	0%	100%
	Text two	10%	90%	0%
	Text three	90%	10%	0%
Group 2	Text four	92%	8%	
	Text five	8%	92%	

In the table 3, for Group 1, Distortion level 1 denotes the highest distortion, Distortion level 2 denotes middle distortion and Distortion level 3 denotes the lowest distortion. We can see 100% people consider that Text one distorts least, and 90% people consider that Text three distorts most. For Group 2, Distortion level 1 denotes higher distortion and Distortion level 2 denotes lower distortion. 92% people consider that Text four distorts more than Text five. Compared with the simulation results, we can get almost the same evaluation although objective measure does not totally agree with the subjective measure.

4 Conclusion

In this paper, we proposed a new objective distortion measure for binary document image. This measure is based on DRDM method, but it can evaluate more accurately. It considers the flipped pixel influence, stroke direction near the flipped pixel and flipped pixels distribution. And simulation and experiment results show that it matched with human visual perception very well and it is very simple, effective and it can be implemented on-line. This measure can be used in a wide range of applications involving visual distortion in digital binary document images, such as watermarking and data hiding.

References

1. Yun, Q. S. and Huifang, S.: Image and Video Compression for Multimedia Engineering: Fundamental, Algorithm, and Standards. Boca Raton, FL: CRC, (1999).
2. Min, W. and Bede, L.: Data hiding in digital binary image. Proc. IEEE Int. Conference on Multimedia and Expo, Vol. 1, New York City, July 31-Aug. 2 (2000) 393-396.
3. Haiping, L., Jian, W., Alex C. K. and Yun Q. S.: An objective distortion measure for binary document images based on human visual perception. Proceeding International Conference Pattern Recognition, vol. 4, Quebec, Canada, Aug. (2002) 239-242.
4. Taylor, M. M.: Visual Discrimination and Orientation. Journal of the Optical Society of America A, (1963) 763-765.
5. Milan, S., Vaclav, H. and Roger, B.: Image Processing, Analysis, and Machine Vision. Thomson-Engineering, the second edition (1998).

Region and Shape Prior Based Geodesic Active Contour and Application in Cardiac Valve Segmentation

Yanfeng Shang[1], Xin Yang[1], Ming Zhu[2], Biao Jin[2], and Ming Liu[2]

[1] Institute of Image Processing & Pattern Recognition, Shanghai Jiaotong University,
Shanghai 200030, P.R. China
{aysyf, yangxin}@sjtu.edu.cn
[2] Xinhua Hospital, Attached to Shanghai Second Medical University,
Shanghai 200092, P.R. China
zhuming58@vip.sina.com, {king1969, lesserniuniu}@hotmail.com

Abstract. Geodesic active contour is a useful image segmentation method. But it may fail to segment objects disturbed by complex noises. Prior knowledge on certain object is a powerful guidance in image segmentation. We represent region and shape prior of certain object in a form of speed field and incorporate it into Geodesic Active Contours. Region prior constrains the zero level set evolving in certain region and shape prior pulls the curve to the ideal contour. Applications in a large quantity of cardiac valve echocardiographic sequences have shown that the algorithm is a more accurate and efficient image segmentation method.

1 Introduction

The algorithm presented in this paper was originally developed for a project of reconstruction of mitral valve leaflet from 3D ultrasound images sequences, which leads to a better understanding of cardiac valve mechanics and the mechanisms of valve failure. And Segmenting valve efficiently could also aid surgery in diagnosis and analysis of cardiac valve disease.

Among medical imaging techniques, ultrasound is particularly attractive because of its real time, noninvasiveness and relatively low cost. In clinical practice, segmentation of ultrasound images still relies on manual or semi-automatic outlines produced by expert physicians, especially when it comes to the object as complex as cardiac valve. And the large quantities of data of 3D volume sequences make a manual procedure impossible. Efficient processing of the information contained in ultrasound image calls for automatic object segmentation techniques.

Geodesic active contour is a powerful image segmentation model, but it may fail in echocardiographic sequences. For the inherent noise, intensity similarity between object and background, and complex movements, make it difficult to segment the cardiac valve accurately. We could make full use of the valve prior, and segment it under the guidance of prior knowledge to improve accuracy and reduce the manual intervention. We can segment the valve automatically guided by the following prior:

1) The valve moves in a relatively fixed region between neighbor slices.
2) The valve has a relatively stationary shape at a certain position.

We developed an algorithm based on geodesic active contour, and represented the prior as a speed field. The speed field drives the zero level converging on the ideal contour and then the valve of heart is segmented. Section 2 of this paper gives an overview of some of the existing prior based object segmentation methods. The proposed algorithm is described and formulated in Section 3. The application results are presented in Section 4 and conclusion follows in Section 5.

2 Review

When segmenting or locating an object, prior information about the subject can be a significant help. Prior based image segmentation which incorporates the prior information of certain object into the segmenting process makes the ambiguous, noise-stained or occluded contour clear and the final result becomes more robust, accurate and efficient. Snake and level set are the two models which are usually used to incorporate prior knowledge to segment certain object.

There were some applications of prior based image segmentation under snake framework in the past several years. The snake methodology define an energy function over a curve as the sum of an internal and external energy of the curve, and evolves the curve to minimize the energy[1]. Prior could be incorporated into the function as an energy item freely. D. Cremers[2,3] established a prior PDM(Point distribute Model) of certain object, then calculated the contour's post Bayesian probability and made it an Energy item to control the evolving process. I. Mikic[4] modified the internal energy to preserve the thickness of the cardiac valve leaflet. But in snake framework, the object is described as a point serial which makes it hardly to deal with topology changing. The goal of valve segmentation is to study the mechanism of movement, which makes the cardiac wall that cooperates with the valve as important as valve. And the valve leaflet may be divided into several isolate parts in certain slices. All these mean that the topology of contour changes when the valve moves. So it's too hard for snake model to be applied to our project. And what is more, snake model needs an initial outline close to the contour which makes an automatic segmentation process impossible.

Level set based segmentation embeds an initial curve as the zero level set of a higher dimensional surface, and evolves the surface so that the zero level set converges on the boundary of the object to be segmented. Since it evolves in a higher dimension, it can deal with topology changing naturally. But it is difficult to incorporate prior knowledge into the evolution process. M.E. Leventon[5] calculated the difference between the zero level and the statistical prior contours, and embedded it in the evolution process. Y.M. Chen[6] made it be applicable to linear transform. They all got a more robust result than traditional method. But calculating the difference between zero level and prior statistical contour at each evolution step is time-consuming, and a scalar of contour similar metric can't guide the evolution directly. In the following, we present a new algorithm which represents the region and shape prior knowledge in a form of speed field. The speed vector could drive the zero level set directly to the ideal cardiac valve contour.

3 Prior Based Geodesic Active Contour

Given a region or shape prior, the information can be folded into the segmentation process. This section represents the region and shape prior as speed fields and embeds them to the level set evolution equation to pull the surface to the ideal contour, which could reduce the manual procedure in great degree.

Our approach is based on geodesic active contour [7, 8],

$$\frac{\partial \phi}{\partial t} = u(x)(k+v_0)|\nabla \phi| + \nabla u \cdot \nabla \phi \quad (1)$$

Where: $\quad u(x) = -|\nabla G_\sigma * I|$

where v_0 is an image-dependent balloon force added to force the contour to flow outward. In this level set framework, the surface φ evolves at every point perpendicular to the level sets as a function of the curvature at that point and the image gradient.

New speed items are added to the evolution equation of geodesic active contour. The new prior knowledge item forms a total force with internal and external force of the original image and drive the zero level set converge to the ideal contour. There are two levels of the additional prior force, one is the lower level of region prior constrain which make the zero level evolve in certain region, the other is the higher level of shape prior constrain which makes the final contour converges to the prior shape of object. Then we get a new equation:

$$\frac{\partial \phi}{\partial t} = u(x)(k+v_0)|\nabla \phi| + \nabla u \cdot \nabla \phi + \sum F_i \cdot \nabla \phi \quad (2)$$

where F_i is the region, shape or the other prior knowledge forces. The speed force is more directly and efficient than the similarity metric of contour. But what is difficult is how to transform the prior knowledge into a speed field. The algorithm is presented in details in following.

3.1 Region Prior Based Geodesic Active Contours

The movement of cardiac valve is very complex. It goes with beating of heart, turns around the valve root, and makes great distortion by itself. But the whole valve moves in a relatively stationary region which is just in the ventricle. When it comes to 3D echocardiographic sequence, the valve could share the same region in different slice of the same time. So does it in different time at the same sample position. Then we can segment several images under the guidance of the same region. When the zero level set is limited to evolve in the fix region, the segmenting process will be more robust and efficient. The whole 3D echocardiographic sequence can be segmented guided by several prior regions.

Consider a prior region Ω within which the valve moves, there is a region function $J(x,y)$:

$$J(x,y) = \begin{cases} 1 & (x,y) \in \Omega \\ 0 & (x,y) \notin \Omega \end{cases} \quad (3)$$

A speed field is created outside the prior region. The force of the field is zero inside the prior region and direct to the prior region outside it. The power of the force has close relation to the distance from the point to the prior region. So the speed field has a potential to drive zero level set to the prior region. And the prior force would get a balance with the inflating force nearby the boundary of the region. When segmenting the valve, appropriate contour could be got at the root of valve lest the zero level set evolve to the whole cardiac wall. The distance from point X to prior region Ω is defined as:

$$\gamma(X) = \begin{cases} d(X) & X \notin \Omega \\ 0 & X \in \Omega \end{cases} \quad (4)$$

$$d(X) = \min(|X - X_I|) \quad X_I \in \Omega \quad (5)$$

Then the speed field of the prior region is:

$$F_{region}(X) = [f_r(d) + c_1] \frac{\nabla \gamma}{|\nabla \gamma|} \quad (6)$$

where c_1 equal or a little less than v_0; $f_r()$ makes the prior force almost c_1 nearby Ω and rise to c_1+c_2 far away from Ω. We take:

$$f_r(d) = c_2 \left(1 - \exp\left(-\frac{d^2}{\sigma^2}\right)\right) \quad (7)$$

Then a speed field is gotten which could drive the zero level set to Ω.

The final region prior based geodesic active contour equation is:

$$\frac{\partial \phi}{\partial t} = u(x)(k + v_0)|\nabla \phi| + \nabla u \cdot \nabla \phi + [f_r(d) + c_1] \frac{\nabla \gamma}{|\nabla \gamma|} \cdot \nabla \phi \quad (8)$$

We could get a final result which includes the cardiac valve, the root of valve and the raised cardiac wall which cooperates with the valve by some post-processing. Erode the prior region Ω and get Ω', fill the segmentation result with foreground color outside Ω', make an intensity reversion, and then the final result is got.

Pre-processing is very important for noises which are inevitable in the ultrasound image. Modified curvature diffusion equation (MCDE)[9,10] is adopted to filter the original image which could preserve the edge and smooth the noise. The equation is given as:

$$u_t = |\nabla u|\nabla \cdot c(|\nabla u|)\frac{\nabla u}{|\nabla u|} \qquad (9)$$

There is an example of cardiac valve segmented by region prior based geodesic active contour in Fig. 1.

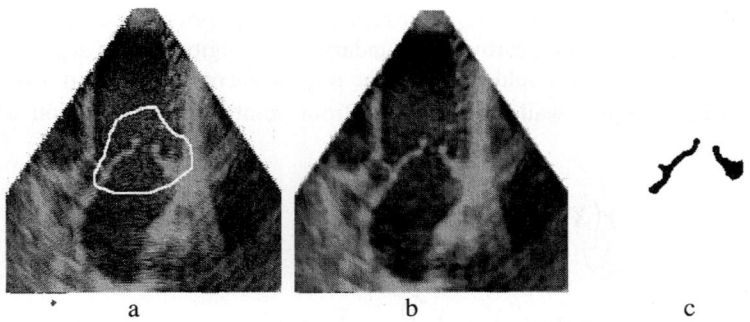

      a   b   c

Fig. 1. Cardiac valve segmented by region prior based geodesic active contour. (a) An image from echocardiographic sequence and prior valve region. (b) Result of preprocessing. (c) The final segmenting result

3.2 Shape Prior Based Geodesic Active Contour

Because of the intrinsic noise of the echocardiographic image and the blur caused by movement of the cardiac valve, it is unavoidable that there would be some segmentation errors. To get a more accurate contour, we need make full use of the prior shape of heart valve in segmenting process.

A speed field is set nearby the prior shape which directs to the nearest point of shape. The force F_{shape} pulls all the points nearby to the prior shape. Consider a prior contour C, we define the distance from point X to contour C as ε:

$$\varepsilon(X) = d(X) = \min(|X - X_I|) \quad X_I \in C \qquad (10)$$

Then the speed field produced by prior contour is:

$$F_{shape}(X) = f_s(d)\frac{\nabla \varepsilon}{|\nabla \varepsilon|} \qquad (11)$$

F_{shape} directs to the nearest shape point, the magnitude of F_{shape} is controlled by $f_s()$.

Attribute of the prior shape force is very important. There are two kinds of force. One is like the elastic force which would be more powerful far away from prior shape, the other is just like the force in electric field, the closer to the shape, the more powerful it would be. It is hoped that F_{shape} could only take effect in the field nearby the prior contour and leave it alone far away from the prior contour. And the closer, the more powerful. So the second is taken. It's supposed that the farthest neighborhood distance be δ:

$$f_s(d) = \begin{cases} k(\delta - d) & d \leq \delta \\ 0 & d > \delta \end{cases} \quad (12)$$

The final shape prior base geodesic active contour equation is gotten:

$$\frac{\partial \phi}{\partial t} = u(x)(k + v_0)|\nabla \phi| + \nabla u \cdot \nabla \phi + f_s(d)\frac{\nabla \varepsilon}{|\nabla \varepsilon|} \cdot \nabla \phi \quad (13)$$

The algorithm is demonstrated on synthetic image. We find that we could get a better result guided by prior shape field (Fig. 1). We set $\delta = 15$ in segmenting the circle.

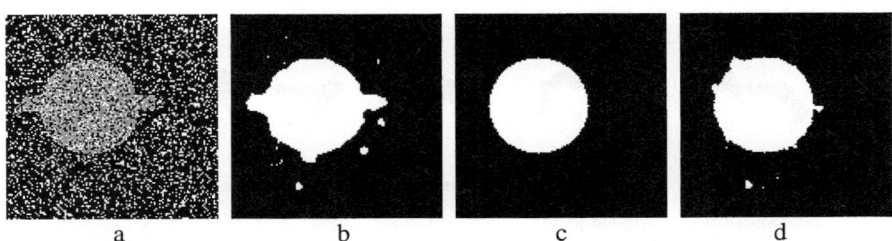

Fig. 2. Circle segmented by shape prior based geodesic active contour. (a)Circle stained by a bar and salt & pepper noise. (b) Circle segmented by Geodesic active contour. (c) Prior circle shape. (d) Circle Segmented by shape prior based geodesic active contour

4 Application and Results

We put the algorithm into practice and efficiently segmented heart valve leaflets from ten 3D echocardiographic sequences, each covering one complete cardiac cycle. The 3D sequences were recorded using the Philips Sonos 750 TTO probe which scanned object rotationally. There are 13-17 frames per cardiac cycle and the angular slice spacing was 3 degrees resulting in 60 images slices in each frames. Therefore, there are about one thousand images in each 3D sequence. The resolution of image was 240*256.

It was obviously too tedious to segment all the images manually by traditional method. Our methods dealt with all the images without too much manual intervention. And it is neither sensitive to the initial zero level set nor prior region as long as it is just between the valve and cardiac wall. Given several prior regions, our method could segment the whole image sequence automatically without too much parameter choosing or adjusting. For the images of the same scanning time could share the same prior region, and so does the images of neighboring times. Some results of region prior based Geodesic active contour are shown below (Fig. 3). The images in Fig. 3 are at the same scanning position of different scanning time from a 3D valve sequence. To facilitate display, we cut the valve region out.

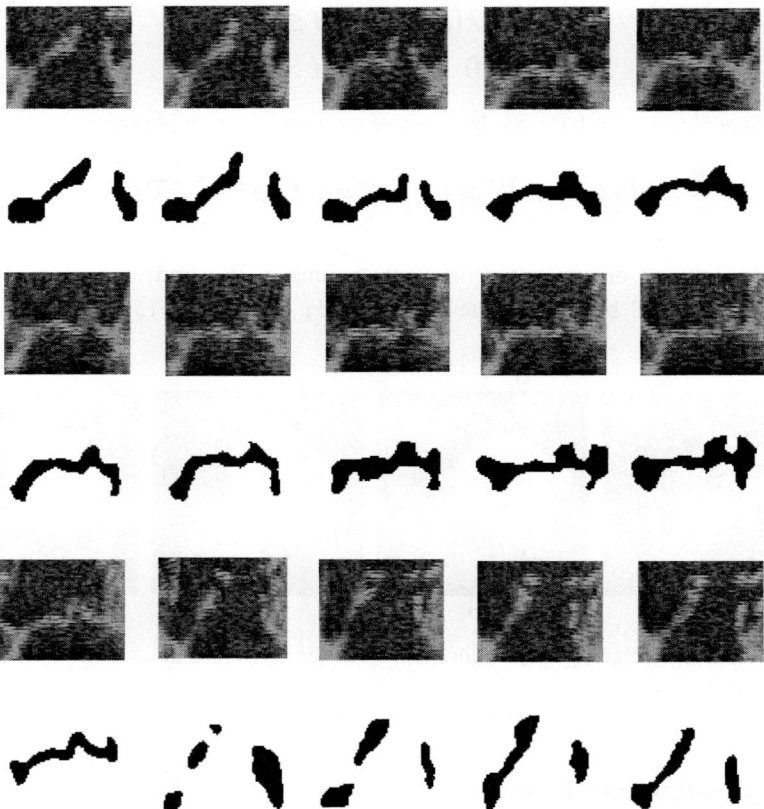

Fig. 3. Valves segmented by region prior based geodesic active contour

Most of segmentation results above could satisfy the needs of 3D reconstruction and diagnosis. Because of contamination of noise and movement blur, there will be inevitable errors in some contours, such as the eighth the twelfth, and the thirteenth contour in Figure 3. We could segment this kind of noise-disturbed images guided by a prior shape. The prior shape could either come from an output of neighbor slices, or from a manual outline by expert physicians. In manual outlining process, it is unnecessary to draw out the whole shape, but the part of stained edge is just enough. We show some results segmented by shape prior based geodesic active contour in Fig. 4. Almost all segmentation error could be restored under the guidance of the shape prior.

Overlapping the contour slices of the same scanning time at different scanning position, a 3D valve could be gotten (Figure 5). These contours are segmented guided by a rectangle prior region. A few of them are resegmented guided by prior shape outlined by physicians. It shows that our segmenting algorithms could get approving valve contours. To facility static display, only 9 of 60 slices were shown.

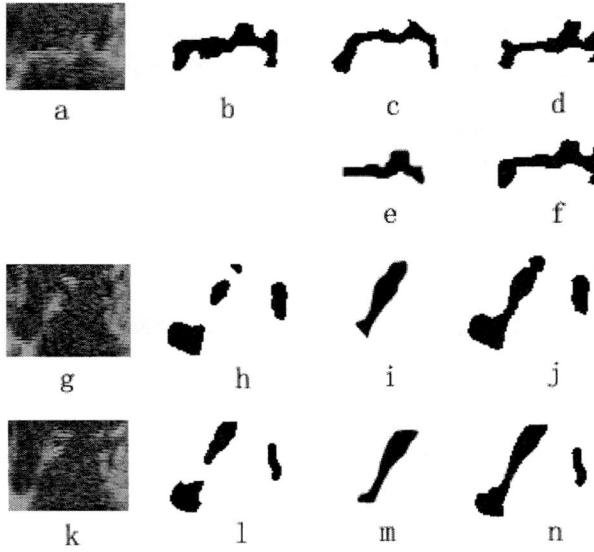

Fig. 4. Cardiac valve segmented by shape prior based Geodesic active contour. (a), (g), (k) The initial echocardiographic image. (b), (h), (l) Result guided by region prior. (c) Segmentation result of neighbor slice. (d) Result guided by shape prior of c. (e), (i), (m) Manual outline. (f), (j), (n) Result guided by shape prior

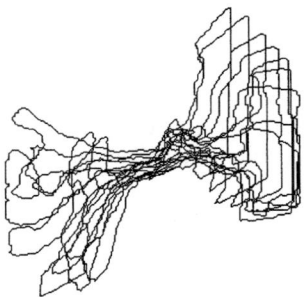

Fig. 5. Cardiac valve reconstructed by rotationally scanned slices which is segmented by region and shape prior based Geodesic Active Contour

5 Conclusions

The inherent noise, blur and the large quantity of data of echocardiographic sequences make it difficult to segment the valve structures efficiently, which holds back the clinical application. We present a new algorithm to incorporate prior knowledge into geodesic active contour. The priors are expressed as a speed field which directly draws the zero level set to the ideal contour. Region prior limits the zero level set evolving in certain region and shape prior draws the curve to the ideal contour. The actual application on 3D echocardiographic sequences shows that the algorithm segments the valve structure accurately and reduces the manual procedure greatly.

Prior based image segmentation is an active subject at present. We could express more prior knowledge as speed field and embed them in image segmentation process in future work. Guided by prior information, the image segmentation will be more accurate and efficient.

Acknowledgements

This work was partially supported by National Basic Research Program of China (No.2003CB716104) and Shanghai Science and Technology Development Foundation (034119820).

References

1. Kass, M., Witkin, A., and Terzopoulos, D.: Snakes: Active contour models. International Journal of Computer Vision, Vol. 1(1988) 321-331.
2. Daniel, C., Florian, T., Joachim, W. and Christoph, S.: Diffusion Snakes: Introducing Statistical Shape Knowledge into the Mumford-Shah Function. International Journal of Computer Vision, Vol. 50(2002) 295-313.
3. Daniel, C., Timo, K. and Christoph, S.: Shape Statistics in Kernel Space for Variational Image Segmentation. Pattern Recognition, Vol. 36(2003) 1929-1943.
4. Ivana, M., Slawomir, K. and James D.T.: Segmentation and Tracking in Echocardiographic sequences: Active Contour Guided by Optical Flow Estimates. IEEE Trans. on Medical Imaging, Vol. 17(1998) 274-284.
5. Michael, E.L., Grimson, W.E.L. and Olivier, F.: Statistical Shape Influence in Geodesic Active Contours. Computer Vision and Image Understanding, Vol. 1(2000) 316-323
6. Chen, Y., Hemant, D., Tagare, S.T. etc.: Using Prior Shapes in Geometric Active Contours in a Variational Framework. International Journal of Computer Vision, Vol. 50(2002) 315-328.
7. Caselles, V., Kimmel, R. and Sapiro. G.: Geodesic Active Contours. International Journal of Computer Vision, Vol. 22(1997) 61-79.
8. Kichenassamy, A., Kumar, A., Olver, P. etc.: Gradient Flows and Geometric Active Contour Models. In IEEE Int'l Conf. Comp. Vision, (1995) 810-815.
9. Pietro, P. and Jalhandra, M.: Scale-space and Edge Detection Using Anisotropic Diffusion. IEEE Trans. on Pattern Analysis Machine Intelligence, Vol. 12(1990) 629-639.
10. Whitaker, R. and Xue, X.: Variable-Conductance, Level-Set Curvature for Image Processing, ICIP, (2001) 142-145.

Interactive Fluid Animation Using Particle Dynamics Simulation and Pre-integrated Volume Rendering

Jeongjin Lee[1], Helen Hong[2], and Yeong Gil Shin[1]

[1] School of Electrical Engineering and Computer Science, Seoul National University,
{jjlee, yshin}@cglab.snu.ac.kr
[2] School of Electrical Engineering and Computer Science BK21: Information Technology,
Seoul National University, San 56-1 Shinlim-dong Kwanak-gu, Seoul 151-742, Korea
hlhong@cse.snu.ac.kr

Abstract. This paper proposes an interactive fluid animation using an enhanced particle dynamics simulation method with pre-integrated volume rendering. The particle dynamics simulation of fluid flow can be conducted in real-time using the Lennard-Jones model. The computational efficiency is achieved using small number of particles for representing a significant volume. To get a high-quality rendering image with small data, we use the pre-integrated volume rendering technique. Experimental results show that the proposed method can simulate and render the fluid motion at interactive speed with acceptable visual quality.

1 Introduction

An active area of research in computer graphics is the animation of natural fluid phenomena. Recently, the demand for interactive fluid animation has increased for 3D computer games and virtual reality applications. However, it is very difficult to animate natural fluid phenomena at interactive speed, because their motions are so complex and irregular that intensive simulation and rendering time is needed.

In the previous research, only off-line fluid animation methods have been reported [1-4]. In general, fluid animation is carried out by physical simulation immediately followed by visual rendering. For the physical simulation of fluids, the most frequently used practices are the numerical simulation of isolated fluid particles using particle dynamics equations and the continuum analysis of flow via the Navier-Stokes equation. Miller et al. [5] proposed a spring model among particles to represent viscous fluid flow. Terzopoulos et al. [6] introduced molecular dynamics to consider interactions between particles. In these approaches, when the number of particles increases significantly, the number of related links between particles exponentially increases. Therefore, it takes too much time for realistic fluid simulation due to the large number of particles for describing complex fluid motions. Stam [1] proposed a precise and stable method to solve the Navier-Stokes equations for any time step. Foster [3] applied a 3D incompressible Navier-Stokes equation. Above methods using the Navier-Stokes equations yield a realistic fluid motion when properly conditioned, but still need huge calculations of complex equations. The second limitation is the time complexity of visual rendering. Global illumination has been widely used for natural fluid animation. Jensen et al. [7] proposed a photon mapping method currently used in many applications. Global illumination is generally successful in rendering premium-quality images, but too slow to be used in interactive applications.

In this paper, we propose a novel technique for interactive fluid animation on a general PC. For rapid analysis of the motion of fluids, we use a modified form of particle dynamic equations. The fluid interaction was approximated by the attractive and repulsive forces between adjacent particles using the Lennard-Jones model to emulate fluid viscosity. To get a high quality rendering image with a smaller volume data, we use a pre-integrated volume rendering method [8]. Various fluid effects can be animated by real-time parameter controls. Experimental results show that our method generates fluid animation at interactive speed with acceptable image quality.

The organization of the paper is as follows. In Section 2, we discuss the particle dynamics simulation of our method, and describe how the simulation data are rendered. In Section 3, experimental results show various examples of fluid animation. This paper is concluded with brief discussions of the results in Section 4.

2 Interactive Fluid Animation

The proposed fluid animation method is composed of three steps in Fig. 1. In the first step, the simulation and rendering parameters are determined in order to reflect the characteristics of the type of fluid. Major simulation parameters to represent physical characteristics of the fluid are density, number of particles and the radius of a particle. Major rendering parameter to define visual properties of the fluid is an opacity transfer function which controls the color and opacity of each particle. In the second step, we set the boundary and initial conditions to define the initial distribution of fluid fields. For boundary conditions, boundary faces such as walls are chosen. For initial conditions, the injection position and velocity of the fluids are initialized. In the third step, interactive fluid animation is performed by the particle dynamics simulation followed by the pre-integrated volume rendering. The user can change parameters interactively to achieve various effects.

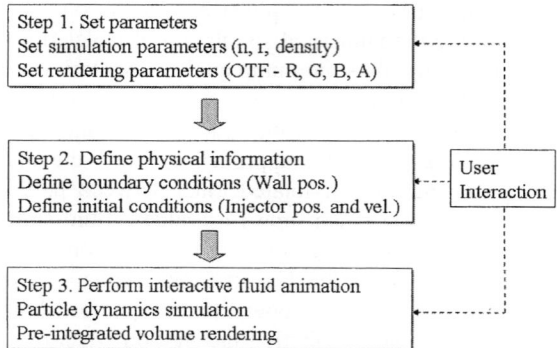

Fig. 1. The overview of interactive fluid animation

2.1 Particle Dynamics Simulation

Two approaches, particle dynamics and continuum dynamics, have been widely used for fluid simulation. The continuum dynamics approach is not suitable for interactive applications due to its high time complexity of calculating the Navier-Stokes equation [1-4]. In our approach, a simple particle dynamics approach is chosen since it is much faster than a continuum dynamics approach based on the Navier-Stokes equation.

In particle dynamics, a spherical particle is assumed to be the basic element that makes an object such as for solid, liquid and gas, and used for calculating interactions between particles. For N spherically symmetric particles, the total inter-particle potential energy $E(\mathbf{r}^N)$ is the sum of isolated pair interactions according to pair-wise addition.

$$E(\mathbf{r}^N) = \sum_{i=1}^{N} \sum_{j=1}^{N} u(r_{ij}), \ i \neq j , \qquad (1)$$

where $\mathbf{r}^N$ is the set of vectors that locate centers of mass, i.e. $\mathbf{r}^N = \{\mathbf{r}_1, \mathbf{r}_2, \mathbf{r}_3, ..., \mathbf{r}_N\}$ and r_{ij} is the scalar distance between particles i and j.

The elementary potential energy $u(r_{ij})$ is taken from the Lennard-Jones (LJ) potential model [9-10]. For two particles i and j separated by a distance r_{ij}, the potential energy $u(r_{ij})$ between the both can be defined as

$$u(r_{ij}) = 4\varepsilon \left(\left(\frac{\sigma}{r_{ij}} \right)^{12} - \left(\frac{\sigma}{r_{ij}} \right)^{6} \right) . \qquad (2)$$

The force field f_{ij} created by two particles i and j can be given as

$$f_{ij} = -\frac{du(r_{ij})}{dr_{ij}} = \left(\frac{48\varepsilon}{\sigma^2} \right) \left(\left(\frac{\sigma}{r_{ij}} \right)^{14} - \frac{1}{2} \left(\frac{\sigma}{r_{ij}} \right)^{8} \right) r_{ij} . \qquad (3)$$

Since the inter-particle potential forces are conservative within a given potential field, the overall potential force $F_{i,p}$ acting on particle i is related to the potential by

$$F_{i,p} = -\frac{\partial E(\mathbf{r}^N)}{\partial \mathbf{r}_i} = m_i \ddot{\mathbf{r}}_i , \qquad (4)$$

where m_i is the mass of particle i. Fig. 2 illustrates the Lennard-Jones potential and the force extended over a modest range of pair separations. The critical distance at which the positive sign of the inter-particle force becomes negative can be considered as the particle radius.

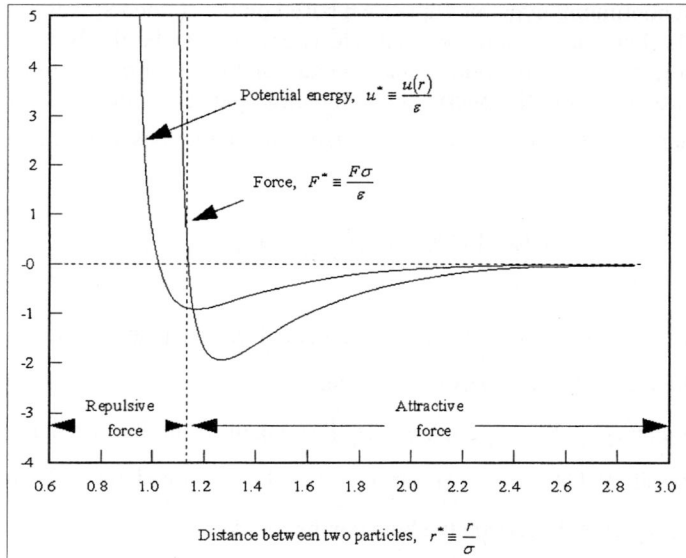

Fig. 2. The Lennard-Jones potential and the force

The friction force $F_{i,f}$ on particle i can be given as

$$F_{i,f} = -\zeta \dot{\mathbf{r}}_i, \tag{5}$$

where ζ is the friction coefficient. The governing equation of the force balance on particle i can hence be written as

$$-\frac{\partial E(\mathbf{r}^N)}{\partial \mathbf{r}_i} + \zeta \dot{\mathbf{r}}_i = 0. \tag{6}$$

Eq. (6) can be used to calculate the interaction between particles by assuming them as slightly deformable soft spheres. The soft sphere collision method is useful for treating interactions between particles with the separation distance within a moderate multiple of the critical distance. For the collision of particles between adjacent particles that can be assumed as hard spheres, the velocity of particle i and j after the collision can be given as follows.

$$u_i = \frac{1}{m_i + m_j}\left[(m_i - em_j)u_i + m_j(1+e)u_j\right]$$

$$u_j = \frac{1}{m_i + m_j}\left[(m_i - em_j)u_j + m_i(1+e)u_i\right]$$

$$v_i = \frac{1}{m_i + m_j}\left[(m_i - em_j)v_i + m_j(1+e)v_j\right]$$

$$v_j = \frac{1}{m_i + m_j}\left[(m_i - em_j)v_j + m_i(1+e)v_i\right] \quad (7)$$

$$w_i = \frac{1}{m_i + m_j}\left[(m_i - em_j)w_i + m_j(1+e)w_j\right]$$

$$w_j = \frac{1}{m_i + m_j}\left[(m_i - em_j)w_j + m_i(1+e)w_i\right],$$

where e is the restitution coefficient.

Fluids have both particle and continuum characteristics. A particle model is used to emphasize particle characteristics, whereas a continuum model is used to emphasize continuum characteristics. Neither model alone is sufficient since real fluids exhibit both characteristics. Our method overcomes this difficulty by controlling simulation parameters – the radius and number of particles. When values of these parameters are small, the particle characteristics are emphasized, as in a splashing effect. When values of these parameters are large, the continuum characteristics are emphasized, as in a flowing effect. We can interactively control the mixture ratio of particle and continuum characteristics with two parameters for desired fluid effects.

2.2 Fast Visual Rendering of Simulation Data

A photon mapping method [7] accomplishes the global illumination effect using ray tracing, which is excessively slow for interactive applications. For interactive rendering without the loss of image quality, we use a pre-integrated volume rendering on graphics hardware. Rendering of simulation data is accomplished in the following three steps.

In the first step, we transform the simulation data into volume data. We divide the 3D space, in which particles of simulation data are stored, into regular cells having unit length d. The density of each cell is determined by the volume fraction value as Eq. (8). These density values are used for volume rendering.

$$\text{Volume fraction} = \frac{n \times \frac{4}{3}\pi r^3}{d^3}, \quad (8)$$

where n is the number of particles in the cell and r is the radius of particles.

In the second step, we visualize volume data using a pre-integrated volume rendering technique. The color and opacity between two neighboring slices of volume data are pre-integrated and stored in the graphics hardware texture memory for acceleration [11]. Using the pre-integrated volume rendering technique accelerated by graphics hardware we can get the high quality image with a smaller volume data at the interactive rate.

In the third step, we can control the color of fluids by controlling the opacity transfer function, which assigns different colors and opacity values according to the value of volume fraction. Various visual effects can be generated by interactively modifying the opacity transfer function.

3 Experimental Results

All of the implementations and tests have been performed on an Intel Pentium IV PC containing 2.4 GHz CPU and 1.0 GB of main memory with GeForce FX 5800 graphics hardware. Fig. 4 and 5 show the animation of water flowing from a bottle to a cup. The opacity transfer function of this animation is given in Fig. 3. Red, green, blue color and opacity are assigned in each control point V1, V2, V3 and V4. As shown in Fig. 4, when the number of particles (n) is 2000 and the radius of particles (r) is 0.002 [m], particle characteristics of the water are emphasized. This animation occurs in 7 ~ 8 fps at 730 x 520 resolution. Fig. 5 shows that when the number of particles (n) is 10000 and the radius of particles (r) is 0.003 [m], continuum characteristics of the water are emphasized. This animation occurs in 3 ~ 4 fps at 730 x 520 resolution.

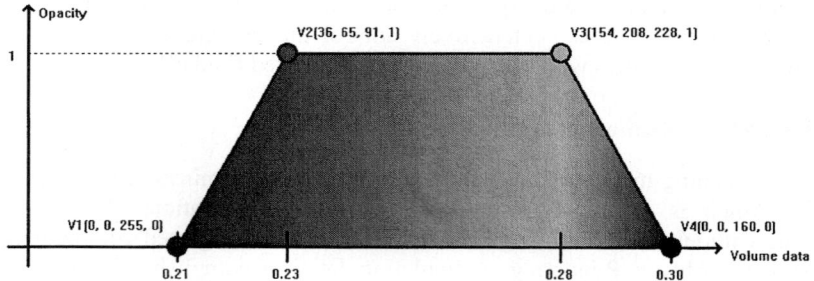

Fig. 3. Opacity transfer function for water animation

Fire and smoke is animated in about 10 fps, as shown in Fig. 7. A realistic color of fire and rather blurred boundary colors of fire can be achieved. If we change the color at V1 in Fig. 6 to white ((R, G, B) = (255, 255, 255)), we can add the effect of smoke around boundaries of fire, as shown in Fig. 8. Because boundary regions have low value of volume data, we assign white in this region.

Interactive Fluid Animation Using Particle Dynamics Simulation 1117

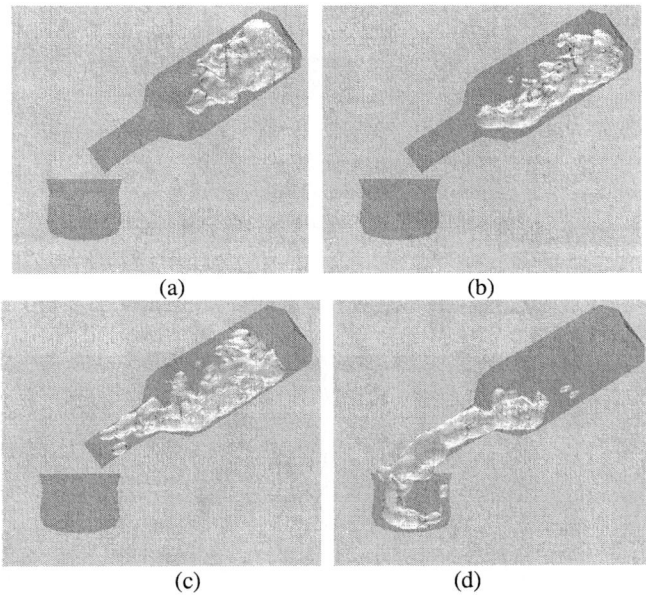

Fig. 4. The animation of pouring water from a bottle into a cup (n = 2000, r = 0.002 [m]) (a) t = 0.12 [s] (b) t = 0.18 [s] (c) t = 0.33 [s] (d) t = 1.02 [s]

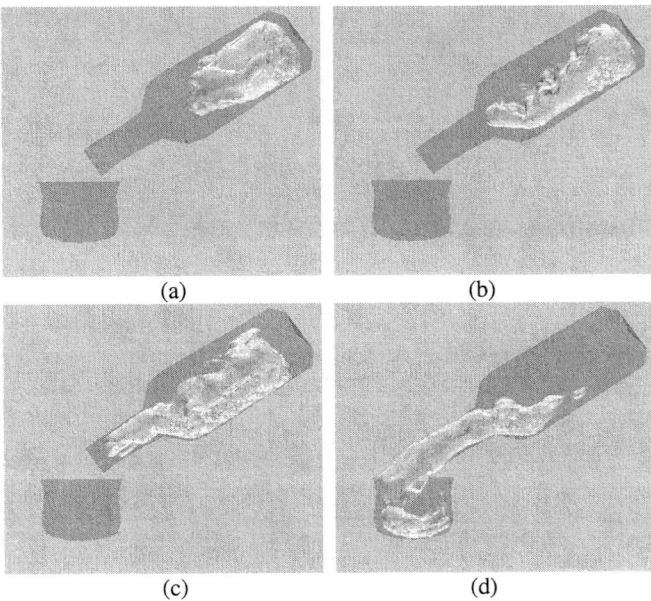

Fig. 5. The animation of pouring water from a bottle into a cup (n = 10000, r = 0.003 [m]) (a) t = 0.12 [s] (b) t = 0.18 [s] (c) t = 0.33 [s] (d) t = 1.02 [s]

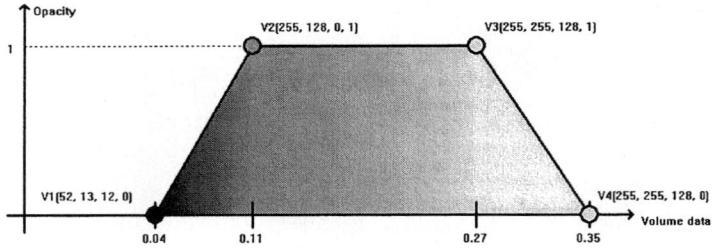

Fig. 6. Opacity transfer function for fire animation

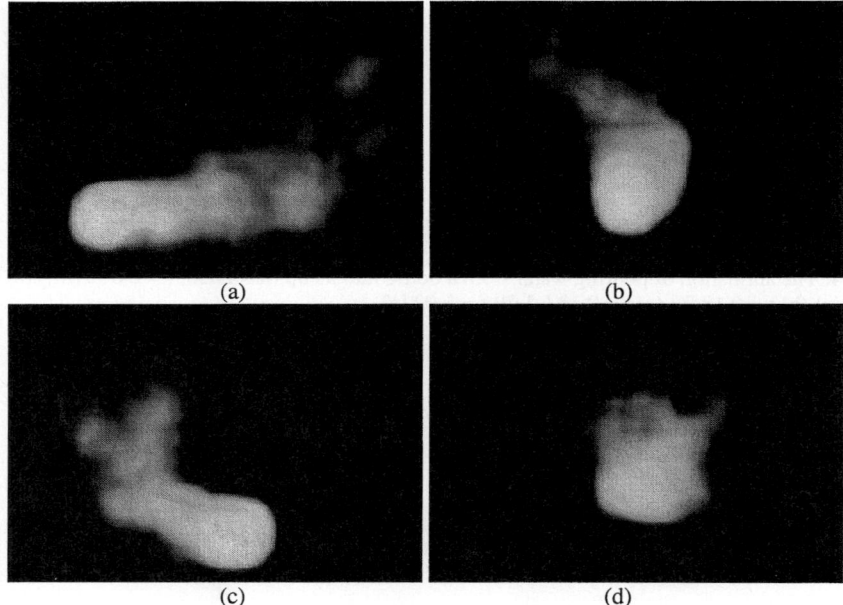

Fig. 7. The animation of fire

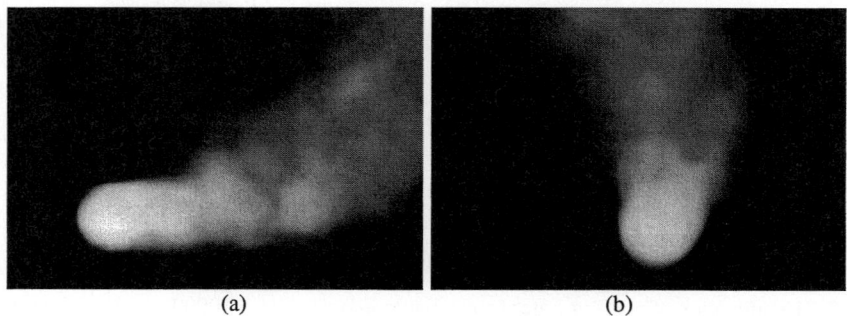

Fig. 8. The animation of fire with smoke

4 Conclusion

This paper presents a novel technique of fluid animation, which integrates particle dynamics simulation and pre-integrated volume rendering. The particle dynamics simulation can be conducted in real-time using the Lennard-Jones model. Furthermore, pre-integrated volume rendering allowed us to avoid the unnaturalness of images usually obtained with particle dynamics. In the animation of water, both the particle and continuum characteristics of the water can be realistically displayed by manipulating simulation parameters. For the animation of fire and smoke, various effects can be generated by manipulating rendering parameters. Experimental results show that our method can simulate and render the fluid motion at interactive speed with acceptable visual quality.

References

1. Stam, J., Stable Fluids, ACM SIGGRAPH (1999) 121-128
2. Foster, N., Practical Animation of Liquids, ACM SIGGRAPH (2001) 23-30
3. Foster, N., Realistic Animation of Liquids, Graphical Models and Image Processing (1996) 204-212
4. Enright, D., Animation and Rendering of Complex Water Surfaces, ACM SIGGRAPH (2002) 734-744
5. Miller, Gavin S. P., Pearce, A., Globular Dynamics: A Connected Particle System for Animating Viscous Fluids, Computers and Graphics Vol. 13, No. 3 (1989) 305-309
6. Terzopoulos, Platt, Fleischer, From Goop to Glop: Melting Deformable Models, Graphics Interface (1989)
7. Jensen, H.W., Christensen, P.H., Efficient Simulation of Light Transport in Scenes with Participating Media using Photon Maps, ACM SIGGRAPH (1998) 311-320
8. Engel, K., Kraus, M., Ertl, T., High-Quality Pre-Integrated Volume Rendering Using Hardware-Accelerated Pixel Shading, Siggraph/Eurographics Workshop on Graphics Hardware (2001) 9-16
9. Rapaport, D.C., The Art of Molecular Dynamics Simulation, Cambridge University Press (1995)
10. McQuarrie, D.A., Statistical Mechanics, Harper and Row (1976)
11. Rottger., S., Kraus, M., Ertl, T., Hardware-accelerated Volume and Isosurface Rendering, In Proc. Of Visualizastion (2000) 109-116

Performance of Linear Algebra Code: Intel Xeon EM64T and ItaniumII Case Examples

Terry Moreland and Chih Jeng Kenneth Tan*

OptimaNumerics Ltd,
Belfast, United Kingdom
{tmoreland, cjtan}@OptimaNumerics.com

Abstract. The canonical form of many large-scale scientific and technical computing problems are often linear algebra problems. As such, routines such as matrix solvers find their use in a wide range of applications. The performance of matrix solvers are often critical in determining the performance of the application programs. This paper investigates the performance of common linear algebra routines on the current architectures of interest to supercomputing users, namely the Intel Xeon EM64T and ItaniumII, with examples from OptimaNumerics Libraries. Performance issues and myths are also shown and diffused in this paper.

Keywords: Linear solver, Systems of linear algebraic equations, Architecture specific tuning, Diagonalizer, SVD, Eigenvalues, Eigensolvers.

1 Introduction

A large number of scientific and technical computing problems can be reduced to three classes of "basic" problems: linear algebra, Fourier transform, and randomized simulation. Of these three classes, linear algebra is often the dominant class of problems solved in many supercomputing centers.

Routines such as matrix solvers and diagonalizers find their use in a wide range of applications: from computational chemistry to quantum physics to structural analysis. The performance of linear algebra code are often critical in determining the performance of the application programs. Over the years, a de facto standard, Linear Algebra Package (LAPACK) [5], has arisen, for handling common linear algebra operations in a portable manner.

Today, LAPACK routines are widely accepted and used as standard linear algebra routines. LAPACK implementations are available from high performance computing software vendors, OptimaNumerics included. Also, hardware manufacturers often provide versions of LAPACK that are intended to be specifically tuned for the particular architecture, along with the lower level Basic Linear Algebra Subroutines (BLAS) [1].

---

* Also at: School of Computer Science, The Queen's University of Belfast, UK.

Here we will investigate the performance of common linear algebra routines on the current architectures of interest to supercomputing users, namely Intel Xeon EM64T and ItaniumII. The problems and myths of performance are also shown and diffused in this paper. This is a work in progress, following on previously reported results in [6] and [7].

2 Common Linear Algebra Operations

The common linear algebra operations that scientific and technical computing applications rely upon are: Cholesky and LU factorizations and solvers, least squares solver, eigensolver, singular value decomposition and QR factorization. A brief sketch of these operations are given below.

2.1 Cholesky Solver

Positive definite systems of linear equations are among the most important classes of special $Ax = b$ problems. A matrix $A \in \mathbb{R}^{n \times n}$ is positive definite if $x^T A x > 0$ for all nonzero $x \in \mathbb{R}^n$. If A is symmetric and nonsingular, then the factorization $A = LDL^T$, where L is a lower triangular matrix and D is a diagonal matrix, exists and is stable to compute. In addition, all elements of D are positive. Since, the elements of D, d_k are positive, then the matrix $G = L \operatorname{diag}(\sqrt{d_1}, \ldots, \sqrt{d_n})$ is a real lower triangular matrix with positive diagonal elements. Therefore, the factorization can be rewritten as $A = GG^T$, and is known as the Cholesky factorization. Following from this:

$$Ax = b \tag{1}$$
$$(GG^T) x = b \tag{2}$$
$$G(G^T x) = b. \tag{3}$$

Letting $G^T x = y$, then $Gy = b$. Thus, once the Cholesky factorization is computed, we can easily solve the $Ax = b$ problem by solving the triangular systems $Gy = b$ and $G^T x = y$.

2.2 LU Solver

Given a matrix $A \in \mathbb{R}^{n \times n}$, there exist a factorization $A = LU$ where $L, U \in \mathbb{R}^{n \times n}$. L is a unit lower triangular matrix and U is an upper triangular matrix. Therefore, the system of linear equations $Ax = b$ can be solved as follows:

$$Ax = b \tag{4}$$
$$LUx = b \tag{5}$$

Letting $Ux = y$, therefore $Ly = b$ and $Ux = y$. The solution can then be obtained by back substitution.

2.3 Generalized Least Squares Solver

For the problem of solution linear systems of equations $Ax = b$ where $A \in \mathbb{R}^{m \times n}$ and $m > n$, a unique solution x does not exist as the system is overdetermined. In these cases it is possible to find a best-fit solution that minimizes $\|Ax - b\|_p$. p is commonly chosen to be 2, producing a least squares solution, as the function $\frac{\|Ax - b\|_2^2}{2}$ is differentiable in x and because the L_2-norm is preserved under orthogonal transformations.

2.4 QR Factorization

For the matrix $A \in \mathbb{R}^{m \times n}$ there exists a factorization $A = QR$, where $Q \in \mathbb{R}^{m \times m}$ is an orthogonal matrix and $R \in \mathbb{R}^{m \times n}$ is a upper trapezodial matrix.

2.5 Eigensolver

Eigenvalue and eigenvector computation is another core linear algebra problem. For $A \in \mathbb{R}^{n \times n}$ the n roots of the characteristic polynomial

$$p(\lambda) = \det(\lambda I - A) \tag{6}$$

are the eigenvalues of A. For each $\lambda \in \lambda(A)$, there exist

$$Ax = \lambda x \tag{7}$$

where the vectors x are known as eigenvectors.

2.6 Singular Value Decomposition

Singular Value Decomposition is among the most important classes of diagonalization problems. For a real $m \times n$ matrix A there exists orthogonal matrices U and V such that:

$$U = [u_1, u_2, \ldots u_m] \in \mathbb{R}^{n x m} \tag{8}$$
$$V = [v_1, v_2, \ldots v_m] \in \mathbb{R}^{n x n} \tag{9}$$
$$U^T A V = D \tag{10}$$

where $D = \text{diag}(\sigma_1, \sigma_2, \ldots \sigma_p) \in \mathbb{R}^{m x n}$ and $p = \min(m, n)$.

3 What's Wrong with Linear Algebra Algorithms on Contemporary Hardware?

While many traditional linear algebra algorithms may be very attractive in terms of theoretical performance, they perform poorly in real-world applications, when implemented on today's computers. This is partly due to the fact that operation counts used in formulating the theoretical order of complexity do not take into account data loading latencies or data locality. While in theory, it is possible to load data instantaneously, it is not achievable in practice on real hardware. If the theory does not reflect what will be observed in practice, then the value of the theory would be questionable.

An inherent problem in the traditional algorithms is that they do not lend themselves easily to make use of hardware facilities such as instruction prediction, data prefetching, and the various cache levels of the CPU. Mathematicians and computer scientists who pay attention to the systems' memory hierarchy, Translation Look-aside Buffers, or CPU architecture while formulating new algorithms remain as minorities. Significant proportion of mathematicians deriving numerical algorithms also fail to pay attention to the capabilities and functionalities of the compilers. If an algorithm is designed void of reality of computing hardware, then the practical validity for algorithm would be questionable.

These deficiencies lead to algorithms which appear to be efficient in theory on paper, but perform poorly when implemented and executed on a real machine.

3.1 Myths of Linear Algebra Packages

Over the years, computer hardware manufacturers, independent software vendors and end users have standardized to use BLAS and LAPACK for common linear algebra computations.

Implementations of BLAS could be obtained for virtually every computer platform of relevance to technical and scientific computing today. Many hardware manufacturers have their own implementations of BLAS as well. The three levels of BLAS, provide routines for vector-vector operations (BLAS Level 1), vector-matrix operations (BLAS Level 2), and matrix-matrix operations (BLAS Level 3). The operations performed do not go much beyond operations such as matrix multiplication and rank-k operations.

It is possible to achieve BLAS performance close to peak chip performance on a variety of hardware today [3, 2, 4]. But it is often misunderstood that if BLAS performance is close to peak chip performance, then naturally, the performance of LAPACK, which is build on top of BLAS, would deliver similar levels of high performance. This is nothing more than a myth.

It is also a myth that all implementations of BLAS and LAPACK are created equal. Some implementations deliver better performance than others.

The benchmarks that we have performed at OptimaNumerics show that while BLAS performance close to peak chip performance is attainable, LAPACK performance is either lacking, or worse, well below peak chip performance. The benchmarks we used may be made available for any interested parties to use.

4 Performance Tests

We conducted performance tests to benchmark the performance of the Cholesky solver, QR Factorization, SVD, and Eigensolvers using using routines from the OptimaNumerics Libraries on Intel Xeon EM64T and ItaniumII architectures. The benchmarks were conducted with no other load on the machines.

4.1 Intel Xeon EM64T

On the Intel Xeon EM64T (Nocona, IA-32E) architecture, the benchmarks were conducted on machines with Xeon EM64T CPUs running at 3GHz. There were 2 CPUs in the machine, but only 1 CPU was used. The compilers used were Intel Fortran Compiler version 8.1, and Intel C++ Compiler version 8.1. The matrices used were generated uniformly distributed random matrices. The memory available on the machine was 4GB of SDRAM. Each CPU has 12kB instruction L1 cache, 20kB data L1 cache, and 1024kB L2 on-chip cache.

4.2 Intel ItaniumII

On the Intel ItaniumII (IA-64) architecture, the benchmarks were conducted on machines with ItaniumII CPUs running at 900MHz. There were 8 CPUs in the machine, but only 1 CPU was used. Intel Fortran Compiler version 8.1, and Intel C++ Compiler version 8.1 were used. The matrices used were generated uniformly distributed random matrices. The memory available on the machine was 15GB of SDRAM. The CPUs have 16kB instruction L1 cache, 16kB data L1 cache, 256kB L2 on-chip cache, and 1.5MB L3 on-chip cache.

4.3 OptimaNumerics Libraries: OptimaNumerics Linear Algebra Module

The OptimaNumerics Linear Algebra Module is part of OptimaNumerics Libraries. OptimaNumerics Linear Algebra Module provides a complete LAPACK implementation. The routines incorporated in OptimaNumerics Linear Algebra Module features algorithms which makes efficient use of the CPU and memory available. In addition to exploiting the hardware features in the CPU, the algorithms take into account the memory architecture and processor architecture on the machine as well.

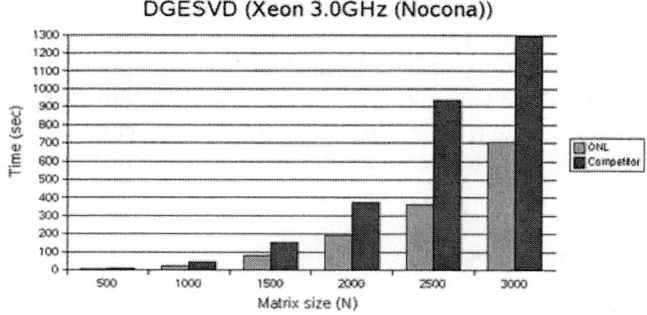

Fig. 1. Performance of SVD routine from the OptimaNumerics Libraries compared to the closest competitor on Intel Xeon EM64T (Nocona) CPU

4.4 Benchmark Results

The results of the double precision benchmarks conducted are shown in Figures 1, 2, 3, 4, 5, 6, 7, 8, 9, 10, 11, 12, and 13.

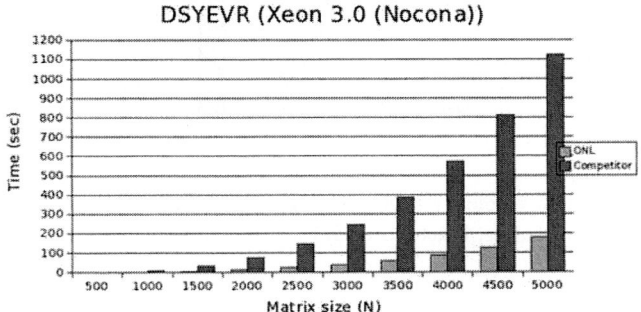

Fig. 2. Performance of symmetric eigensolver from the OptimaNumerics Libraries compared to the closest competitor on Intel Xeon EM64T (Nocona) CPU

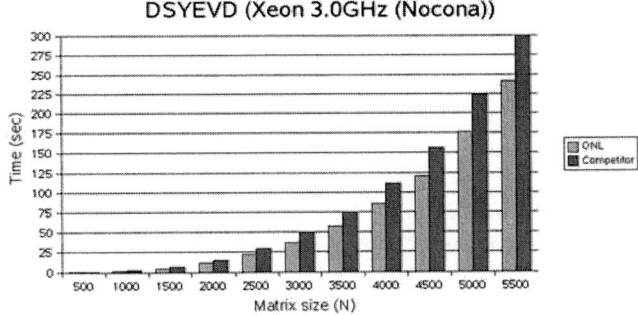

Fig. 3. Performance of symmetric eigensolver from the OptimaNumerics Libraries compared to the closest competitor on Intel Xeon EM64T (Nocona) CPU

5 Discussions and Conclusion

As seen in the performance graphs, it is evident that LAPACK routines provided by the hardware manufacturers – the closest competitors – are under-performing, compared to OptimaNumerics Libraries.

It is to be noted that OptimaNumerics Libraries routines are implemented in high level languages – C and Fortran – rather than in assembly language. The code base is 100% portable.

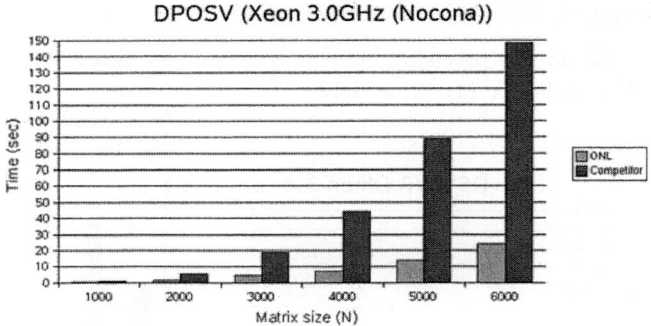

Fig. 4. Performance of Cholesky solver from the OptimaNumerics Libraries compared to the closest competitor on Intel Xeon EM64T (Nocona) CPU

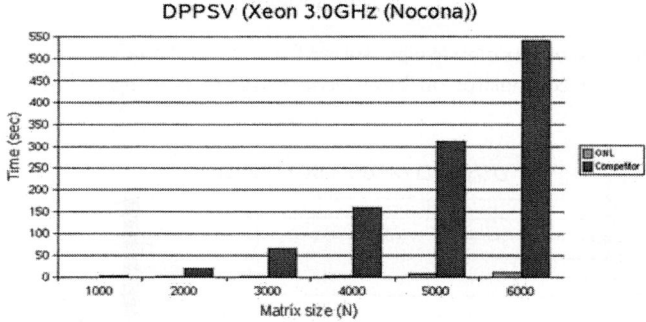

Fig. 5. Performance of Cholesky solver from the OptimaNumerics Libraries compared to the closest competitor on Intel Xeon EM64T (Nocona) CPU

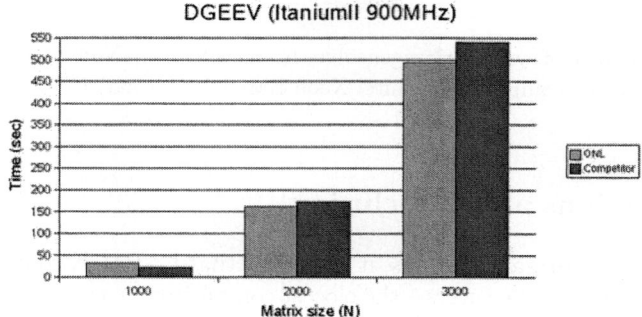

Fig. 6. Performance of generalized eigensolver from the OptimaNumerics Libraries compared to the closest competitor on Intel ItaniumII CPU

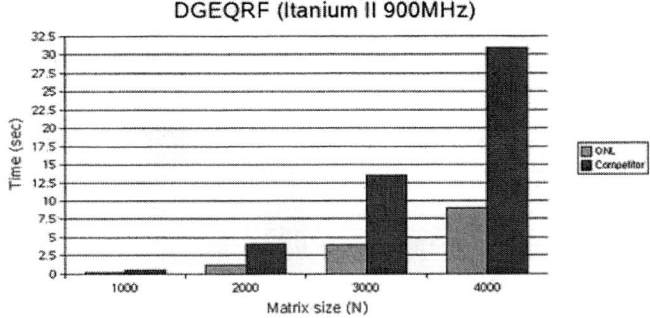

Fig. 7. Performance of QR factorization routine from the OptimaNumerics Libraries compared to the closest competitor on Intel ItaniumII CPU

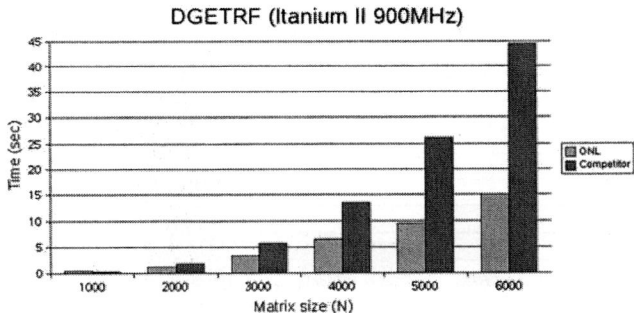

Fig. 8. Performance of LU factorization routine from the OptimaNumerics Libraries compared to the closest competitor on Intel ItaniumII CPU

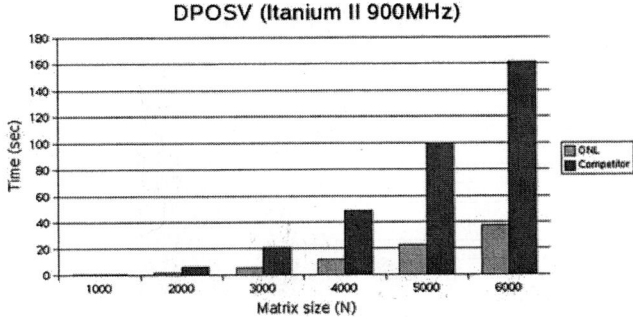

Fig. 9. Performance of Cholesky solver from the OptimaNumerics Libraries compared to the closest competitor on Intel ItaniumII CPU

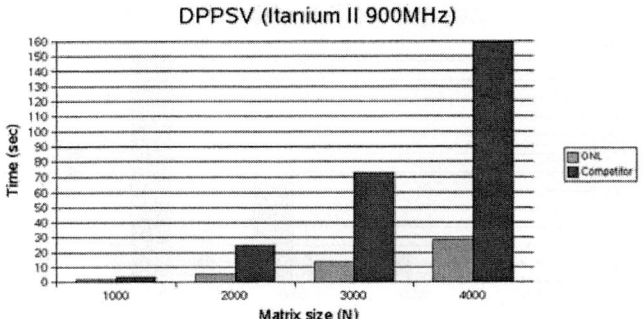

Fig. 10. Performance of Cholesky solver from the OptimaNumerics Libraries compared to the closest competitor on Intel ItaniumII CPU

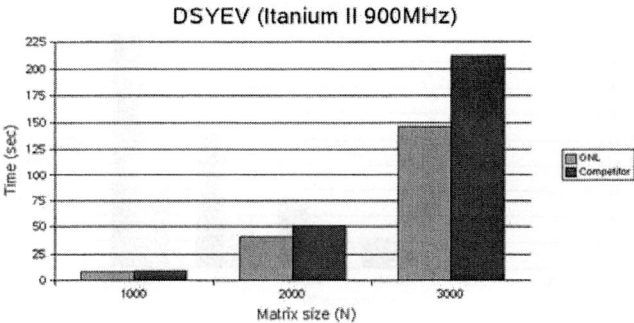

Fig. 11. Performance of eigensolver from the OptimaNumerics Libraries compared to the closest competitor on Intel ItaniumII CPU

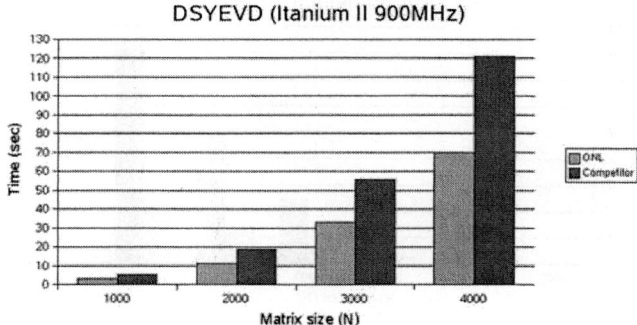

Fig. 12. Performance of eigensolver from the OptimaNumerics Libraries compared to the closest competitor on Intel ItaniumII CPU

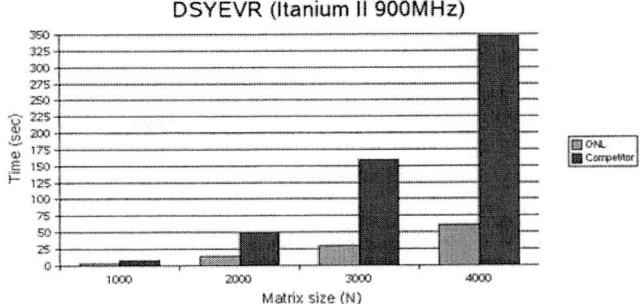

Fig. 13. Performance of eigensolver from the OptimaNumerics Libraries compared to the closest competitor on Intel ItaniumII CPU

We can therefore draw the following conclusions:

1. High performance can be achieved for a CPU-intensive computation problem, with code written in C and Fortran, using highly efficient, novel algorithms.
2. It is possible to achieve performance significantly higher than that attainable using hardware manufacturers' libraries.
3. While BLAS level code may be efficient, an implementation of LAPACK layered above the efficient BLAS is not guaranteed to be similarly efficient.

Also, complexity measures based on operation counts need re-thinking as their present state does not reflect the reality and therefore cannot be relied upon to draw judgements on efficiencies of algorithms.

As shown in [7], efficiencies of scientific computing has great financial implications. In addition, one cannot assume that since Moore's Law states that performance doubles every 18 months, one can simply keep buying new hardware to achieve better performance. For example, since on the Intel Xeon EM64T CPU, the OptimaNumerics Libraries Cholesky solver is almost 50 times faster than the closest competitor, it will be more than 8.5 years (102 months) before one can achieve the same level of performance, assuming Moore's Law holds!

References

1. Basic Linear Algebra Subroutines (BLAS). http://www.netlib.org/blas/.
2. BIENTINESI, P., GUNNELS, J. A., GUSTAVSON, F. G., HENRY, G. M., MYERS, M. E., QUINTANA-ORTI, E. S., AND VAN DE GEIJN, R. A. The Science of Programming High-Performance Linear Algebra Libraries. In *Proceedings of Performance Optimization for High-Level Languages and Libraries (POHLL-02)* (2002), Association for Computing Machinery.
3. GOTO, K., AND VAN DE GEIJN, R. On Reducing TLB Misses in Matrix Multiplication. Tech. Rep. TR-2002-55, University of Texas at Austin, 2003. FLAME Working Note 9.

4. GUNNELS, J. A., HENRY, G. M., AND VAN DE GEIJN, R. A. A Family of High-Performance Matrix Algorithms. In *Computational Science – 2001, Part I* (2001), V. N. Alexandrov, J. J. Dongarra, B. A. Juliano, R. S. Renner, and C. J. K. Tan, Eds., vol. 2073 of *Lecture Notes in Computer Science*, Springer-Verlag, pp. 51 – 60.
5. Linear Algbra Package (LAPACK). http://www.netlib.org/lapack/.
6. TAN, C. J. K. Performance Evaluation of Matrix Solvers on Compaq Alpha and Intel Itanium Processors. In *Proceedings of the 2002 International Conference on Parallel and Distributed Processing Techniques and Applications (PDPTA 2002)* (2002), H. R. Arabnia, M. L. Gavrilova, C. J. K. Tan, and et al., Eds., CSREA.
7. TAN, C. J. K., HAGAN, D., AND DIXON, M. A Performance Comparison of Matrix Solvers on Compaq Alpha, Intel Itanium, and Intel Itanium II Processors. In *Computational Science and Its Applications: ICCSA 2003* (2003), V. Kumar, M. L. Gavrilova, C. J. K. Tan, and P. L'Ecuyer, Eds., vol. 2667 of *Lecture Notes in Computer Science*, Springer-Verlag, pp. 818 – 827.

Dataset Filtering Based Association Rule Updating in Small-Sized Temporal Databases

Jason J. Jung[1] and Geun-Sik Jo[1]

Intelligent E-Commerce Systems Laboratory,
School of Computer Science and Engineering, Inha University,
253 Yonghyun-dong, Incheon, Korea 402-751
j2jung@intelligent.pe.kr, gsjo@inha.ac.kr

Abstract. Association rule mining can uncover the most frequent patterns from large datasets. This algorithm such as *Apriori*, however, is time-consuming task. In this paper we examine the issue of maintaining association rules from newly streaming dataset in temporal databases. More importantly, we have focused on the temporal databases of which storage are restricted to relatively small sized. In order to deal with this problem, temporal constraints estimated by linear regression is applied to dataset filtering, which is a repeated task deleting records conflicted with these constraints. For conducting experiments, we simulated datasets made by synthetic data generator.

1 Introduction

Since association rule mining algorithms were introduced in [1], there have been many studies focusing on how to find frequent patterns from a given itemset such as market basket analysis. Traditionally, *Apriori* algorithm [2] and FP-Growth [3] have been the most well-known methods. These algorithms, however, have considered only static datasets. It means that the streaming datasets like online transactional logs are difficult to be driven by generic *Apriori*-like algorithms. In fact, many applications on the Web and real world have focused on mining sequential patterns from data streams. For example, on-line newspaper article recommendation, web proxy server for prefetching content, and user preference extraction for supporting adaptive web browsing can be told as the domains relevant to analyzing data streams from many clients.

Several studies thereby have been proposed for maintaining the set of mined association rules. FUP (Fast UPdate) is an incremental updating technique based on *Apriori* and DHP (Direct Hashing and Pruning) [5]. After a set of new transactions are piled up, FUP finds out new large itemsets from a new dataset and compared them with old ones based on heuristics, in order to determine which operation should be executed like removing losers, generating candidate sets, and finding winners. Furthermore, FUP_2 was more generalized algorithm of FUP, as handling other maintenance problems [6]. In [4], DELI (Difference Estimation for Large Itemsets) was proposed as a way of estimating the difference between the association rules in a database before and after they are updated. DELI is used as an indicator for whether the FUP_2 should be applied to the database to accurately find out new association rules. However, these algorithms are highly time consuming, because they are basically composed of the repetition of the same tasks such as scanning dataset, counting itemsets, and measuring their supports in order to generate the candidate set finding out the large itemsets iteratively.

In this paper, we have been focusing on how to update association rules in a temporal database of which size is relatively small. Each association rule extracted from the given transaction dataset can be qualified as the rule measures like support and confidence. We assume that the temporal transitions of these rule measures are very important information including the trend of the streaming dataset of temporal databases. With the sequence of the rule measures of each association rule, we simply exploit linear regression method to predict the rule measure values of association rules at the next step. Based on this prediction, therefore, a certain part of dataset related to each association rule can be determined whether it is filtered or not.

In the following section, we describe and design the problem in small-sized temporal databases. In section 3 and 4, we address how to establish the set of temporal constraints by using linear regression method and how to continuously update the association rules by filtering datasets. In section 5, experimental results will be shown, and then, we discuss several issues related to dataset filtering. Finally, in section 6, we draw conclusions and introduce our future work.

2 Problem Description

In this paper, we assume that the new dataset db is added to old dataset DB, as shown in Fig. 1.

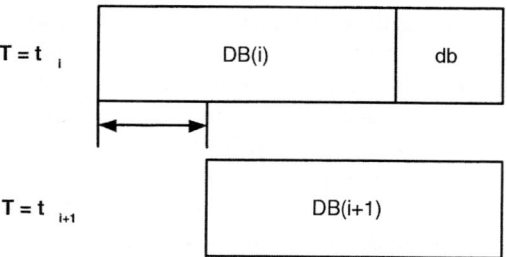

Fig. 1. Dataset filtering for updating temporal database

The total size of temporal database TDB is $size(TDB) = size(DB) + size(db) = N$, which is relatively small. In general temporal databases, there is a buffer for temporally storing the new dataset. The size of this buffer $size(db)$ depends on the rate of temporal sensitivity $R_{TS} = size(db)/size(DB)$. During merging these two datasets, heuristic-based approaches such as sliding windows are needed to remove some parts of itemsets of DB and db. We suppose that updating association rules is the same as mining association rules from databases merged new dataset. A set of literals, called *items* is represented as X. Let DB (and db) be a set of transactions, where each transaction T_i is the i-th itemset $\{x_1, x_2, \ldots, x_\alpha, \ldots, x_n\}_i$. By using simple data mining algorithm, we obtain the set of association rules discovered from DB, denoted by R_{DB}. Each element in this set is represented as a predicate form (e.g., $x_i \Rightarrow x_j$) with rule measurements like support and confidence. For example, a rule "diaper $\Rightarrow$ beer" should be attached "[support=10%, confidence=80%]". Therefore, we note the two main issues concentrated for updating association rules from streaming dataset in this paper, as follows.

- **Dataset filtering.** Conceptually, a data mining problem is dealt with simple generate-and-test algorithm, which is the evaluation of all possible combinations among items in a given database. Dataset filtering can be regarded as pruning the search space for improving the performance of these tasks.
- **Gathering information while updating.** The sequential patterns of rule measurements can be applied to predict the corresponding association rules at the next step. While updating frequent itemsets from streaming new dataset, the constraints may be tighter or looser, according to the predicted rule measurements.

Users generally define the minimum support (or minimum confidence) as the threshold value for mining frequent patterns, according to their own preferences. In this paper, the minimum support, which is regarded as an explicit constraint for detecting noisy patterns such as outliers, should be determined at the very low level, because of the association rules of which supports is expectable to be increased. Another constraint can be generated by the mathematical prediction for dataset filtering.

3 Temporal Constraints by Linear Regression

With the repetition of updating temporal database with new dataset, we can simply compute the rule measurement of discovered frequent patterns (or association rules). Once a sequence of these data is obtained, we can predict next step by using various regression schemes.

3.1 Prediction by Linear Regression

A classic statistical problem is to try to determine the relationship between two random variables T and RM [11]. Linear regression attempts to explain this relationship with a straight line fit to the data. The linear regression model postulates that

$$RM = a + b \times T + e \qquad (1)$$

where the "residual" e is a random variable with mean zero. The coefficients a and b are determined by the condition that the sum of the square residuals is as small as possible. Given s data points of the form $(t_1, rm_1), (t_2, rm_2), \ldots,$ (t_s, rm_s), then the slope coefficients b can be computed with

$$b = \frac{\sum_{i=1}^{s}(t_i - \bar{t})(rm_i - r\bar{m})}{\sum_{i=1}^{s}(t_i - \bar{t})^2} \quad (2)$$

where $\bar{t}$ and $r\bar{m}$ are the averages of $t_1, t_2, \ldots, t_s$ and $rm_1, rm_2, \ldots, rm_s$, respectively. The coefficient a can be simply calculated by $a = r\bar{m} - b \times r\bar{m}$.

3.2 Temporal Constraint Generation with Normalization

Now, we have to consider how to generate temporal constraints with normalization. We can obtain the supports of frequent patterns at each updating time, with the equation $Supp(fp) = freq(fp)/N$. Then, as shown in Fig. 2, by using the sequences of their supports at t_{i-1} and t_i, the next supports at t_{i+1} can be estimated.

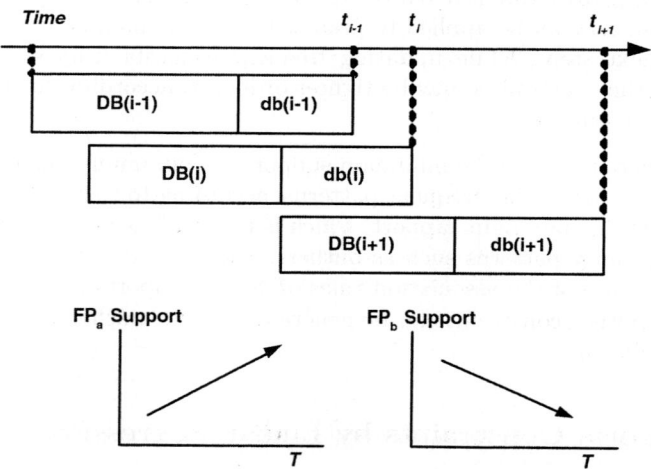

Fig. 2. Temporal constraints by linear regression

Based on the estimated support of a certain frequent pattern $S\tilde{u}pp(fp_k)^{t_{i+1}}$, above all, the constraints for $freq(fp_k)^{t_{i+1}}$ can be approximately derived by

$$freq(fp_k)^{t_{i+1}} \approx size(\tilde{DB}^{t_{i+1}}) \times S\tilde{u}pp(fp_k)^{t_{i+1}} \quad (3)$$

where $size(\tilde{DB})$ is the estimated size of DB. Depended on the tightness of constraints, the number of filtered dataset becomes different. For example, if the constraints are very tight, their propagation can cause DB to be influenced and removed many part of datasets. More importantly, we have to consider the opposite case, which means the constraints are very loose. Thereby, the normalization process is needed, by using the constraint

$$\sum_{k=1}^{K} \left[\tilde{Supp}(fp_k)^{t_{i+1}} \times size(DB + db) \right] \leq \frac{size(db)}{R_{TS}} \quad (4)$$

where K is the total number of frequent patterns of which supports are larger than the minimum support. As previously mentioned, the R_{TS} is defined by users and means the rate of temporal sensitivity controlling the size of db. The small R_{TS} can make it more frequently updating association rules.

4 Maintaining Association Rules Based on Dataset Filtering

In order to increase the performance of updating association rules from temporal databases, we have tried to exploit the temporal constraints predicted by linear regression. We want to note two kinds of constraints, which are user-defined and statistically generated constraints as follows.

- **User-defined constraints.** This should be, in advance, explicitly configurated. The minimum support for filtering rare associations and the R_{TS} for controlling the frequency of updating are classified into this category. Additionally, the taxonomies and hierarchical relationship between items can be used as constraints.
- **Estimated constraints by regression.** These constraints are inductively generated during on-line mining tasks. For this kind of constraints, we need to establish several functions such as Eq. (3) and (4).

A bunch of constraints are organized in a form of graph. Then, this constraint graph is applied to filter a particular part of dataset conflicted with constraints. The constraint graph should be configured through the user's requests represented as predefined operators. This graph therefore can keep adjusting itself to on-line streaming dataset. As the sliding window is shifted (in Fig. 2) by newly inserted data, consistency checking should be applied to test their satisfiability.

4.1 Consistency Checking by Constraint Graphs

We have described how to organize constraint graph by users. Basically, in order to reduce search space of a given problem, consistency checking can be conducted. It finds out the redundant parts that we should not scan any more. We are focusing on node-consistency (*NC*) and arc-consistency (*AC*). *NC* checking is based on unary constraints involved with a particular item x_i. Algorithm *NC* presents the pseudo code for node-consistency achievement:

Algorithm NC
Input:
Time Window, $TW = [T_0, \ldots, T_N]$;
Old Dataset, DB; New Dataset, db;
Set of Frequent Patterns Discovered from DB, FP_{DB};
Constraint Graph, CG;
Procedure:
begin
  $i \leftarrow N$;
  while $i \geq (N - size(db))$ **and** $T_i \in TW$ **do**
  begin
    for each $x_j \in T_i$ **do**
      $Update(x_j)$;
      if (**not** $Satisfies(x_j, CG_1(x_j))$) **then**
        $Prune(x_j)$
    $i \leftarrow i - 1$;
  end
  $Prune(<listofconflicteditems>, DB)$;
end.

AC checking is based on binary constraints involved with a pair of item x_i and x_j. The AC achievement algorithm is shown below:

Algorithm AC
Input:
Time Window, $TW = [T_0, \ldots, T_N]$;
Old Dataset, DB; New Dataset, db;
Set of Frequent Patterns Discovered from DB, FP_{DB};
Constraint Graph, CG;
Procedure:
begin
  $k \leftarrow N$;
  while $k \geq (N - size(db))$ **and** $T_k \in TW$ **do**
  begin
    for each $x_i \in T_k$ **do**
      $NC(x_i)$;
      if (**not** $Satisfies(x_i, CG_2(x_i))$) **then**
        $Prune(x_i)$
    $k \leftarrow k - 1$;
  end
  $Prune(<listofconflicteditems>, DB)$;
end.

In these codes, the function $Update(x_j)$ represents the aggregation operations related to input nodes, such as counting. The function $Satisfies(x_j, CG_1(x_j))$ evaluates input node x_j with unary constraints involved in the corresponding node. More importantly, the function *Prune* removes the transactions conflicted with from old dataset DB. For example, let the minimum support of an item x_i be $\theta_{Sup}(x_i)$. During checking NC of new dataset db, transactions including x_i can be pruned, as shown in the following equation:

$$count(x_i, db) \geq \theta_{Sup}(x_i) \times (size(DB) + size(db)) - Supp(x_i, DB) \times size(DB) \quad (5)$$

where *count* is the function for measuring the frequency of input parameter by counting the itemset including an item in a given dataset. After NC of a certain item, we can retrieve binary constraints by function CG_2.

5 Experimental Results and Discussion

In order to conduct our experiments, we used sequential datasets generated using the generator from the IBM Almaden Quest research group[1]. These synthetic datasets contain three fields, which are a customer index (CustID), a transaction index (TransID), and a set of items (Item). Three different temporal databases we designed to be capable of storing ascii-formatted data are limited to 50 KBytes (TDB50KB), 100 KBytes (TDB100KB), and 200 KBytes (TDB200KB). With respect to the size of databases, the datasets were segmented and streamed into databases. For mining frequent patterns, we simply employed *Apriori* algorithm.

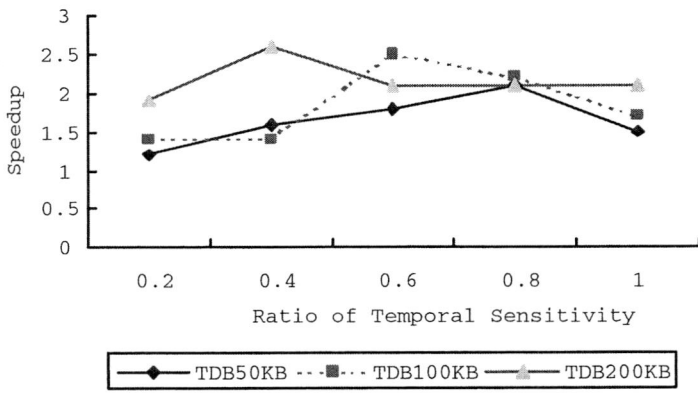

Fig. 3. Evaluating the effect of coefficient R_{TS}

First experiment found out the relationship between the coefficient R_{TS} and the speedup of updating association rules. As shown in Fig. 3, we were able to

---
[1] Website: http://www.almaden.ibm.com/software/quest/Resources/index.shtml

discover that $R_{TS} = 0.8$ is the optimal value. With this optimal R_{TS}, we compared the amount of dataset filtered by estimated constraints and the frequency of updating.

Table 1. Evaluating the performance of updating

Database	TDB50KB	TDB100KB	TDB200KB
Filtered dataset	16.3%	21.1%	23.7%
Number of updating	283	210	165

As shown in Table. 1, TDB200KB was the most powerful database to maintain the streaming dataset.

6   Conclusion and Future Work

In this paper, we have considered the problem of analyzing the streaming data for efficiently updating association rules. We have proposed consistency checking scheme based on user-defined constraints like minimum supports, as filtering redundant part of data. Moreover, regression based information gathering while updating has been proposed to adaptively control the tightness of constraints of given problems. While scanning datasets for finding frequent large itemsets, constraints can be adaptive to new datasets. As a matter of fact, due to the difficulties of the description of constraints, users have to be supported, as constraint information is notified. To do this, we need to define some problem-depended functions for retrieving new information from transaction data. During shopping, as an example, a group of customers under the similar circumstance (e.g., preferences and economical condition) have almost the same behavioral patterns such as the number of items, the total price of all items, and the quality of items in a basket.

As a future work, we are expecting context-awareness computing for mobile devices of which computation power and storage capacity are relatively small. Furthermore, we need the additional research for applying not only *NC* and *AC*, but also path consistency (*PC*) checking.

Acknowledgement. This work was supported by INHA UNIVERSITY Research Grant. (INHA-2005)

References

1. Agrawal, R., Imielinski, T., Swami, A.: Mining association rules between sets of items in large databases. In Proc. of the ACM SIGMOD Conference on Management of Data (1993) 207–216
2. Agrawal, R., Srikant, R.: Fast Algorithms for Mining Association Rules. In Proc. of the 20th VLDB Conference (1994)

3. Han, J., Pei, J.: Mining Frequent Patterns by Pattern-Growth: Methodology and Implications. ACM SIGKDD Explorations (2000) 31–36
4. Lee, S.D., Cheung, D.W.: Maintenance of Discovered Association Rules: When to Update? In Proc. of ACM SIGMOD Workshop on Data Mining and Knowledge Discovery (DMKD) (1997)
5. Cheung, D.W., Han, J., Ng, V.T., Wong, C.Y.: Maintenance of Discovered Rules in Large Databases: An Incremental Updating Technique. In Proc. of Int. Conf. on Data Engineering (1996) 106–114
6. Cheung, D.W., Lee, S.D., Kao, B.: A General Incremental Technique for Maintaining Discovered Association Rules. In Proc. of Int. Conf. on Database Systems for Advanced Applications (DASFAA) (1997) 185–194
7. Zheng, Q., Xu, K., Ma, S.: When to Update the Sequential Patterns of Stream Data? In: Whang, K.-Y., Jeon, J., Shim, K., Srivastava, J. (eds.): Advances in Knowledge Discovery and Data Mining. Lecture Notes in Artificial Intelligence, Vol. 2637. Springer-Verlag (2003) 545–550
8. Hidber, C.: Online Association Rule Mining. In Proc. of the ACM SIGMOD Conference on Management of Data (1999) 145–156
9. Pudi, V., Haritsa, J.: How Good are Association-rule Mining Algorithm? In Proc. of the 18th Int. Conf. on Data Engineering (2002)
10. Wojciechowski, M., Zakrzewicz, M.: Dataset Filtering Techniques in Constraint-Based Frequent Pattern Mining. In: Hand, D.J., Adams, N.M., Bolton, R.J. (eds.): Pattern Detection and Discovery. Lecture Notes in Computer Science, Vol. 2447 Springer-Verlag (2002) 77–91
11. Papoulis, A., Pillai, S.U.: Probability, Random Variables and Stochastic Processes. 4th edn. McGraw-Hill (2002)

A Comparison of Model Selection Methods for Multi-class Support Vector Machines*

Huaqing Li, Feihu Qi, and Shaoyu Wang

Department of Computer Science and Engineering,
Shanghai Jiao Tong University, Shanghai 200030, P.R. China
waking_lee@cs.sjtu.edu.cn

Abstract. Model selection plays a key role in the performance of support vector machines (SVMs). At present, nearly all researches are based on binary classification and focus on how to estimate the generalization performance of SVMs effectively and efficiently. For problems with more than two classes, where a classifier is typically constructed by combining several binary SVMs [8], most researchers simply select all binary SVM models simultaneously in one hyper-parameter space. Though this *all-in-one* method works well, there is another choice – the *one-in-one* method where each binary SVM model is selected independently and separately. In this paper, we compare the two methods for multi-class SVMs with the *one-against-one* strategy [8]. Their properties are discussed and their performance is analyzed based on experimental results.

1 Introduction

Support vector machine (SVM) was originally designed for binary classification problems. It has powerful learning ability and good generalization ability. When dealing with problems involving more than two classes, a multi-class SVM is typically constructed by combining several binary SVMs [8].

Generally, SVM works as follows for binary classification [1]: First the training examples are mapped, through a mapping function Φ, into a high (even infinite) dimensional feature space $\mathcal{H}$. Then the optimal separating hyperplane in $\mathcal{H}$ is searched for to separate examples of different classes as possible, while maximizing the distance from either class to the hyperplane. In implementation, the use of kernel functions avoids the explicit use of mapping functions and makes SVM a practical tool. However, as different kernel functions lead to different SVMs with usually quite different performance, it turns to be very important, yet very hard, to select appropriately the type and parameter(s) of the kernel function for a given problem.

At present, nearly all model selection researches are based on binary classification and focus on how to estimate the generalization performance of SVMs

---

* This work is supported by the National Natural Science Foundation of China under grant No. 60072029 and No.60271033.

effectively and efficiently [6, 2, 5, 4]. For multi-class SVMs, most researchers take it for granted to select one set of hyper-parameters for all binary SVMs involved. Though this *all-in-one* method works well, it has some disadvantages. The most obvious one is that it is not flexible and all binary SVM models have to be re-selected when new classes enter the training set or existing classes get out of the training set. Such re-selection can be very burdensome for real life problems where the change of classes is common.

In this paper, we investigate an alternative model selection method, the *one-in-one* method, for multi-class SVMs. The new method conducts model selection for each binary SVM involved independently and separately. Thus it is flexible and can deal efficiently with the change of classes in the training set. Such a virtue can be of great value to some real life problems. The rest of the paper is organized as follows: In Section 2, we briefly review the basic theory of SVMs. Section 3 describes the two model selection methods for multi-class SVMs and compares them theoretically. Experimental results and corresponding analysis are presented in Section 4. Finally, Section 5 concludes the paper.

2 SVMs for Pattern Classification

Given a set of linearly separable training examples $\{x_i, y_i\}$, $i = 1, 2, \ldots, l$, where $x_i \in R^n$ is the i-th training vector and $y_i \in \{-1, 1\}$ is the corresponding target label. $y_i = 1$ denotes that x_i is in the first class and $y_i = -1$ denotes that x_i is in the second class. An SVM searches for the optimal separating hyperplane which separates the largest possible fraction of examples of the same class on the same side. This can be formulated as follows:

$$\min \ \frac{1}{2}||w||^2 \ , \tag{1}$$

$$\text{s.t.} \quad y_i(w \bullet x_i + b) - 1 \geq 0 \quad \forall i \ ,$$

where w is the normal to the hyperplane, b is the threshold, $||\cdot||$ is the Euclidean norm, • stands for dot product. Introducing Lagrangian multipliers α_i, we obtain

$$\min \ L_P = \frac{1}{2}||w||^2 - \sum_{i=1}^{l} \alpha_i y_i (x_i \bullet w + b) + \sum_{i=1}^{l} \alpha_i \ . \tag{2}$$

Since this is a convex quadratic programming problem, we can equally solve the Wolfe dual

$$\max \ L_D = \sum_{i=1}^{l} \alpha_i - \frac{1}{2} \sum_{i=1}^{l} \sum_{j=1}^{l} \alpha_i \alpha_j y_i y_j (x_i \bullet x_j) \ , \tag{3}$$

subject to

$$\sum_{i=1}^{l} \alpha_i y_i = 0 \ ,$$

$$\alpha_i \geq 0 \quad \forall i ,$$

with the solution

$$\mathbf{w} = \sum_{i=1}^{l} \alpha_i y_i \mathbf{x_i} . \tag{4}$$

For a test example x, the classification is then,

$$f(\mathbf{x}) = sign(\mathbf{w} \bullet \mathbf{x} + b) . \tag{5}$$

The above linearly separable case results in an SVM with hard margin, i.e. no training errors occur. If the training set is nonlinearly separable, we can first map, through a mapping function Φ, the original inputs into a high (even infinite) dimensional feature space $\mathcal{H}$ wherein the mapped examples are linearly separable. Then we search for the optimal separating hyperplane in $\mathcal{H}$. The corresponding formulas of (3) is

$$\max \quad L_D = \sum_{i=1}^{l} \alpha_i - \frac{1}{2} \sum_{i=1}^{l} \sum_{j=1}^{l} \alpha_i \alpha_j y_i y_j (\Phi(\mathbf{x_i}) \bullet \Phi(\mathbf{x_j})) . \tag{6}$$

And the solution becomes

$$\mathbf{w} = \sum_{i=1}^{l} \alpha_i y_i \Phi(\mathbf{x_i}) . \tag{7}$$

The corresponding classification rule is

$$\begin{aligned} f(\mathbf{x}) &= sign(\mathbf{w} \bullet \Phi(\mathbf{x}) + b) \\ &= sign(\sum_{i=1}^{l} \alpha_i y_i (\Phi(\mathbf{x_i}) \bullet \Phi(\mathbf{x})) + b) . \end{aligned} \tag{8}$$

As the only operation between mapped examples is dot product, kernel functions can be employed to avoid the explicit use of the mapping function Φ via $K(\mathbf{x_i}, \mathbf{x_j}) = \Phi(\mathbf{x_i}) \bullet \Phi(\mathbf{x_j})$. The most popular kernels include the radius basis function (RBF) kernel, the polynomial kernel and the sigmoid kernel.

RBF Kernel : $\quad K(\mathbf{x_i}, \mathbf{x_j}) = \exp(-\sigma ||\mathbf{x_i} - \mathbf{x_j}||^2)$,
Polynomial Kernel : $\quad K(\mathbf{x_i}, \mathbf{x_j}) = (\sigma + \gamma(\mathbf{x_i} \bullet \mathbf{x_j}))^\delta$,
Sigmoid Kernel : $\quad K(\mathbf{x_i}, \mathbf{x_j}) = \tanh(\sigma(\mathbf{x_i} \bullet \mathbf{x_j}) - \delta)$.

If the training set is inseparable, slack variables ξ_i have to be introduced. Then the constraints of (1) are modified as

$$y_i(\mathbf{w} \bullet \mathbf{x_i} + b) - 1 + \xi_i \geq 0 \quad \forall i . \tag{9}$$

Two objectives exist under such cases. One is the so called L1 soft margin formula:

$$\min \quad \frac{1}{2}||\mathbf{w}||^2 + C \sum_{i=1}^{l} \xi_i , \tag{10}$$

where C is the penalty parameter. The other is the so called L2 soft margin formula:

$$\min \ \frac{1}{2}||\mathbf{w}||^2 + \frac{C}{2}\sum_{i=1}^{l}\xi_i^{\ 2}\ . \tag{11}$$

The two formulas mainly differ in that (11) can be treated as hard margin cases through some transformation while (10) can not. However, (11) is more easily affected by outliers than (10). Hence we employ (10) in this paper. The kernel employed is the RBF kernel. Thereby two parameters need to be tuned by model selection algorithms; the penalty parameter C and the kernel parameter σ.

2.1 Extending SVMs to Multi-class Problems

As SVM was dedicated to binary classification, two popular methods have been proposed to apply it to multi-class problems. Suppose we are dealing with a k-class problem. One method can be used is *one-against-rest* [8], which trains one SVM for each class to distinguish it from all the other classes. Thus k binary SVMs need to be trained. The other scheme is *one-against-one* [7], which trains $\frac{k(k-1)}{2}$ binary SVMs, each of which discriminate two of the k classes. In this paper, the latter method is employed.

3 Model Selection for Multi-class SVMs

When selecting models for multi-class SVMs, two issues should be considered; classifier's generalization performance estimation and the selection method. Recently the first issue attracted lots of researches, while the second one is less considered by most researchers.

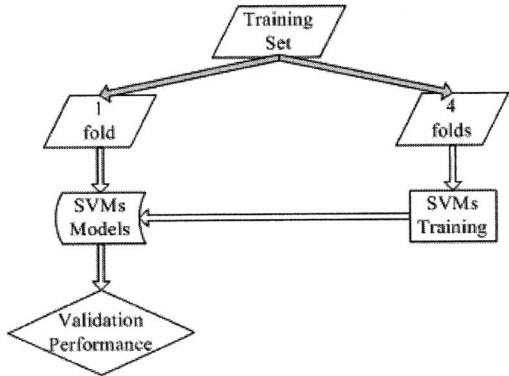

Fig. 1. Procedure of the 5-fold cross validation algorithm

3.1 Generalization Performance Estimation

Several algorithms exist to estimate SVMs generalization performance. In this paper, we employ the most popular and robust one – the cross validation (CV) algorithm [3,4]. The CV algorithm first divides the original training set into several subsets of nearly the same size. Then each subset is sequentially used as the validation set while the others are used as the training set. Finally SVMs performance on all validation sets is summed to form the cross validation rate. The procedure of the 5-fold CV algorithm is demonstrated in Fig. 1.

Generally the CV algorithm employs an exhaustive grid-search strategy in some predefined parameter ranges. In [2], Chung et al. pointed out that trying exponentially growing sequences of C and σ is a practical method to identify good parameters for SVMs with the RBF kernel. However, a standard grid-search is very computational expensive when dealing with even moderate problems.

In [4], Staelin proposed a coarse-to-fine search strategy for the CV algorithm based on ideas from design of experiments. Experimental results showed that it is robust and works effectively and efficiently. The strategy can be briefly described as follows: Start the search with a very coarse grid covering the whole search space and iteratively refine both the grid resolution and search boundaries, keeping the number of samples roughly constant at each iteration. In this paper, a similar search strategy like this is employed for the CV algorithm.

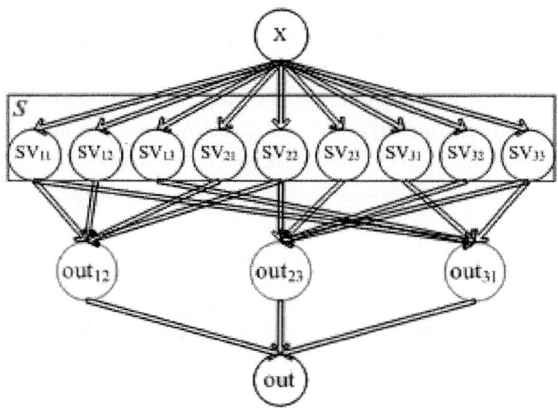

Fig. 2. Multi-class SVM obtained by the *all-in-one* method

3.2 The *All-in-One* Method

The *all-in-one* method is employed in most literature where *all-in-one* means *all binary SVMs in one hyper-parameter space*. With this method, all binary SVM models are selected simultaneously, and the hyper-parameters are chosen on the basis of the predicted performance of the global multi-class SVM in discriminating all classes. Thus the resulted binary SVMs live in one same hyper-parameter space. Take for example we are dealing with a three-class problem.

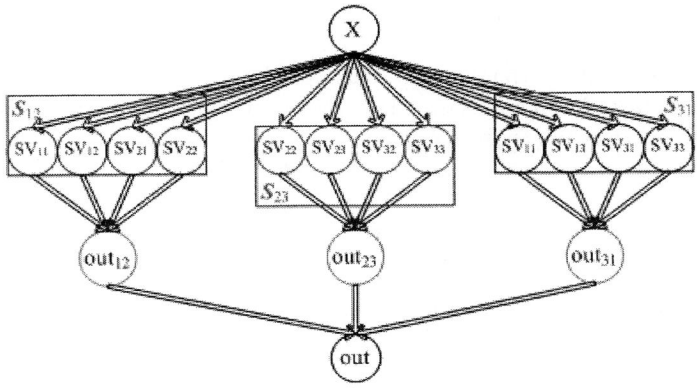

Fig. 3. Multi-class SVM obtained by the *one-in-one* method

Three binary SVMs need to be trained; SVM_{12}, SVM_{23} and SVM_{31}, where SVM_{ij} separates class i from class j. We make the following assumptions: The chosen hyper-parameter space is $\mathcal{S}$; SVM_{12} takes SV_{11}, SV_{12}, SV_{21}, and SV_{22} as its support vectors; SVM_{23} takes SV_{22}, SV_{23}, SV_{32}, and SV_{33} as its support vectors; SVM_{31} takes SV_{31}, SV_{33}, SV_{11}, and SV_{13} as its support vectors. Where SV_{ij} is the j-th support vector from class i. Then the multi-class SVM achieved can be illustrated in Fig. 2 in a neural network form.

3.3 The *One-in-One* Method

In this paper, we investigate another selection method called *one-in-one*, where *one-in-one* indicates that *one binary SVM in one hyper-parameter space*. With this method, each binary SVM model is selected independently and separately on the basis of its predicted performance in discriminating the two classes involved. Thus the resulted binary SVMs live in different hyper-parameter spaces. Consider the example described in Section 3.2. Assume the support vectors of each binary SVM are unchanged, but SVM_{ij} lives in its specific hyper-parameter space marked as $\mathcal{S}_{ij}$. The resulted multi-class SVM can be illustrated in Fig. 3 in a neural network form.

3.4 Comparison of the Two Methods

From Fig. 3 we can see that the structure of the obtained multi-class SVM by the *one-in-one* method is very adaptive. Since each binary SVM is separately tuned, the change of classes (new classes emerge or existing classes disapear) can be easily treated. On the contrary, when class change happens, the *all-in-one* method has to re-select all binary SVM models as shown in Fig. 2. This can be very burdensome for some real life problems where the change of classes is not unusual.

However when employing the *one-in-one* method, the binary SVM models are chosen on the basis of their predicted performance in classifying the two classes involved, rather than on the basis of the predicted performance of the global multi-class classifier. Hence the obtained binary SVMs may have greater tendency to overfit than those obtained by the *all-in-one* method. This may have a bad influence on the performance of the obtained global multi-class SVM.

3.5 Choose a Winner from Several Equi-Performance Spaces

Empirically the grid search procedure often results in several hyper-parameter spaces with the same best validation performance. Thus some rule is needed to pick out one space as the winner. In this paper, we choose the one with a smallest C for the *all-in-one* method as done in [9]. As to the *one-in-one* method, besides the smallest C strategy, the largest margin strategy is considered – i.e. the space in which a binary SVM has the largest margin is chosen as the winner.

4 Experiments

Experiments are carried out to compare the performance of the two model selection methods on several problems from the Statlog Collection [11] and the UCI Repository of machine learning databases [10]. Problem statistics are listed in Table 1. Note that in our investigation, we use the scaled version of these problems as done in [8]. For problems without testing data, we divide the original training set into two parts; $\frac{2}{3}$ of the data are used for training and the rest $\frac{1}{3}$ data are used for testing.

Table 1. Statistics of the problems used

statistics	iris	wine	glass	vowel	vehicle	dna	satimage
#training data	150	178	214	528	846	2000	4435
#testing data	0	0	0	0	0	1186	2000
#class	3	3	6	11	4	3	6
#attributes	4	13	13	10	18	180	36

Model selection is done in the $\log_2$-space of both parameters. The parameter ranges are $\log_2 C \in \{-5, -4, \ldots, 15\}$ and $\log_2 \sigma \in \{-15, -14, \ldots, 3\}$. Totally five iterations are performed for the 5-fold CV algorithm. At each iteration five points uniformly distributed in the latest range are examined. LIBSVM [3] is employed for SVMs training and testing. The experimental results are shown in Table 2. Where *one-in-one*^{C} employs the smallest C strategy to deal with equi-performance hyper-parameter spaces, and *one-in-one*^{M} employs the largest margin strategy. The methods are compared by their classification rate on the testing data.

Table 2. Performance comparison of the two selection methods on several problems

methods	iris	wine	glass	vowel	vehicle	dna	satimage
all-in-one	100%	96.23%	62.5%	62.34%	85.71%	95.35%	91.2%
one-in-one^C	100%	96.23%	60.25%	59.84%	85.65%	93.14%	91.23%
one-in-one^M	100%	96.23%	60.25%	61.14%	85.48%	93.97%	91.18%

From Table 2, we can see that on all problems the *all-in-one* method performs best except *satimage*. On most problems, the three algorithms has very comparative performance. However, on problems *glass, vowel, dna*, the two *one-in-one* algorithms have poorer performance. Observing Table 1, we find that, compared with others, these three problems have more classes and/or much more attributes. Thereby their training sets may be not representative enough. This coincides with the analysis in Section 3.4 that binary SVMs obtained by the *one-in-one* method tend to overfit the training sets. These results, from another point of view, imply that when there are enough representative training data, the *one-in-one* method works well.

It is interesting to note that the two *one-in-one* algorithms have very comparative performance on all problems except *vowel*. This indicates that, although simple, the smallest C strategy is effective. We own this to the fact that a smaller C usually leads to SVMs with larger margins.

5 Conclusion

In this paper, we study model selection for multi-class SVMs with the *one-against-one* strategy. Two methods are investigated; the *all-in-one* method and the *one-in-one* method. The former chooses all binary SVM models simultaneously in one hyper-parameter space, while the latter chooses models for each binary SVM independently and separately. Both methods have advantages and disadvantages. The *all-in-one* method is robust but not adaptive to changes of training classes. On the contrary, the *one-in-one* method is adaptive to changes of training classes but not that robust. Hence the decision of which method to be used must be made according to the specific characteristics of a given problem. At present, we are investigating the two methods on much larger problems. Corresponding results will be reported in the near future.

References

1. Burges, C.J.: A Tutorial on Support Vector Machines for Pattern Recognition. Data Mining and Knowledge Discovery. **2** (1998) 121–267
2. Chung, K.-M., Kao, W.-C., Sun, T., Wang, L.-L., Lin, C.-J.: Radius Margin Bounds for Support Vector Machines with the RBF Kernel. Neural Computation. **11** (2003) 2643–2681

3. Chang, C.-C., Lin, C.-J.: LIBSVM: A Library for Support Vector Machines. (2002) Online at http://www.csie.ntu.edu.tw/~cjlin/papers/libsvm.pdf
4. Staelin, C.: Parameter Selection for Support Vector Machines. (2003) Online at http://www.hpl.hp.com/techreports/2002/HPL-2002-354R1.pdf
5. Li, H.-Q., Wang, S.-Y., Qi, F.-H: Minimal Enclosing Sphere Estimation and Its Application to SVMs Model Selection. In: Yin,F.L., Wang, J., Guo, C.G. (eds.): Advances in Neural Networks – ISNN 2004. (2004) 487-493
6. Chapelle, O., Vapnik, V., Bousquet, O., Mukherjee, S: Choosing Multiple Parameters for Support Vector Machines. Machine Learning. **46** (2002) 131-159
7. Kressel, U.: Pairwise Classification and Support Vector Machines. In: Schölkopf, B., Burges, C., Smola, A. (eds.): Advances in Kernel Methods: Support Vector Learning. MIT Press (1999) 255-268
8. Hsu, C.-W., Lin, C.-J.: A comparison of methods for multi-class support vector machines. IEEE Trans. on Neural Networks. **13** (2002) 415-425
9. Wu, T.-F., Lin, C.-J., Weng, R.C.: Probability Estimates for Multi-Class Classification by Pairwise Coupling. Journal of Machine Learning Research. **5** (2004) 975-1005
10. Blake, C.L., Merz, C.J.: UCI Repository of Machine Learning Databases. (1998) Online at http://www.ics.uci.edu/$\sim$mlearn/MLRepository.html
11. Michie, D., Spiegelhalter, D.J., Taylor, C.C.: Machine Learning, Neural and Statistical Classification. Ellis Horwood, London. (1994) Data available at ftp://ftp.ncc.up.pt/pub/statlog

Fuzzy Category and Fuzzy Interest for Web User Understanding*

SiHun Lee[†], Jee-Hyong Lee[†,**], Keon-Myung Lee[‡], and Hee Yong Youn[†]

[†] School of Information and Communication Eng.,
Sungkyunkwan University, Korea
c1soju@skku.edu, jhlee@ece.skku.ac.kr, youn@ece.skku.ac.kr
[‡] School of Electric and Computer Eng.,
Chungbuk National University, Korea
kmlee@cbnu.ac.kr

Abstract. Web usage mining is a research field for searching potentially useful and valuable information from web log file. Web log file is a simple list of pages that users refer. Therefore, it is not easy to analyze user's current interest field from web log file. This paper presents web usage mining method for finding users' current interest based on Fuzzy category. We consider not only how many times a user visits pages but also when he visits. We describe a user's current interest with a fuzzy interest degree to categories. Based on fuzzy categories and fuzzy interest degrees, we also propose a method for understanding web users. For this, we define the category vector space. We also present experiment results which shows how our method helps to understand web users.

1 Introduction

Data mining can be defined as searching high-capacity database for useful but unknown information that cannot be drawn by simple queries [1, 5, 7, 8]. Web mining is a searching for useful patterns in data stored in web site or web usage data. Usually, web mining includes web structure mining, web contents mining and web usage mining [4, 6].

Web usage mining is a research field for searching potentially useful and valuable information from web log file or web usage data. One of the most interesting information to find out through web usage mining is web users' interest fields and the models of users who have similar interest fields. Web log file is usually used for web usage mining. Web log file is a simple list of pages that users visited. So, it is not easy to find out which contents a user has interest in. For example, a user visited page A, B and C many times : page A is about football, page B baseball and page C basketball. Then what we can obtain from page

---

* This research was supported by Ubiquitous Computing Technology Research Institute(04A2-B2-32) funded by the Korean Ministry of Science and Technology.
** Corresponding author.

visit analysis may be very simple : "The user much visited page A, B and C than the others." However, we may want to have much useful information such as that the user has interest in sports rather than he has interest in page A, B and C. So, we need to consider the contents of the page he visited in order to understand users and users' interest and preference.

Most of existing methods mainly use the page visit count to obtain users' interest [2,3]. However, since users' interests may change as time goes on, we need to also consider the page visit time. If a user has visited a page many times, then it is concluded that he has interest in that page. Since a user's interest is changing, simply considering visit counts may not enough. For example, a user visited page A 100 times last month and this month page B 100 times. Then is it reasonable to conclude that he has the same interest in page A as in page B? Even though the visit counts of both pages are the same, page B is recently visited than page A so we may conclude that he has more interest in page B than A. Thus, we need to consider the visit time as well as the visit count.

The purpose of our research is proposing a method analyzing web log files to obtain better user understanding. For this we define fuzzy categories to classify the contents of web pages and fuzzy interests in categories reflecting not only visit counts but also visit times. That is, our method mines web user interest fields using fuzzy categories of web contents.

We also propose a method for understanding web page visit counts and time users based on fuzzy interests in the category vector space. We define the category vector space as a space whose axes are the categories given by a web administrator. We map users' fuzzy interests onto the category vector space and analyze them in the space to find users who have similar interest and create user models.

In section 2, we define fuzzy interests as well as fuzzy categories. The category vector space is described in section 3 and experiments for fuzzy interests and user analysis in the category vector space are presented in section 4. Finally we conclude in section 5.

2   User's Fuzzy Interest

2.1   Fuzzy Category

In order to find a user's interest fields from the web pages the user visited, we first have to know what contents the pages contain. For describing the contents of pages, we introduce fuzzy categories. A topic in a page may belong to a single category or several categories. For example, sports shoes may belong to the sport category as well as the shoes category, thus so is the contents of a web page for sports shoes. We use the degree to which a page belongs to a category. For example, a page P_1 for sports shoes may belongs to the sports category with a degree of 0.3 and to the shoes with 0.7. We will represent those as $\mu_{sports}(P_1) = 0.3$ and $\mu_{shoes}(P_1) = 0.7$.

Since the contents of a page are fuzzily categorized, we call it fuzzy category. Before mining users' interest fields in a web site, we should have the category

Table 1. Membership degrees to categories

	C_1	C_2	C_3	C_4	C_5
P_1	0.1	0	0	0.3	0.6
P_2	0.4	0	0	0.1	0.5
P_3	0.1	0	0.3	0.1	0.5
P_4	0	0	0.4	0.6	0
P_5	0	0	1	0	0
P_6	0	0	0	0.2	0.8
P_7	0.8	0.2	0	0	0
P_8	0	0	0	0.1	0.9

degree (or membership degree) of each web page to each category. A web administrator may choose categories of interests and assign membership degrees to each page according to its contents. Table 1 is an example of fuzzy categories of a web-site with 8 pages from P_1 to P_8. The administrator chooses five categories from C_1 to C_5. In this web site, page P_1 contains topics which belong to C_1 with a degree of 0.1, C_4 with 0.3 and C_5 with 0.6.

2.2 Fuzzy Interest

For mining a user's interest fields or categories, we have to look at the pages he visited. If a user has an interest in a certain field, he may frequently visit the page containing it. We have to investigate what contents the visited pages include. From the investigation, we have to infer user's interest field. Instead of choosing a single field as user's interest, we evaluate user's interest degrees to the fields. If a user visits web pages containing a certain field, his interest degree of the field will be high.

One of important factors in obtaining user's interest field is a time factor. User's interest changes as time goes on. However, most existing web mining methods do not consider factors of visit time. We reflect the time when a user visits pages to read user's interest fields. Before defining fuzzy interest, we define the category counter as follows:

$$Count(C) = \sum_{t=1}^{T} \mu_{T_t}(C).$$

It represents the number of pages including contents of category C. T is the number of transactions a user has made. $\mu_{T_t}(C)$, which represents how many pages included in the tth transaction belongs to category C, is defined as follows:

$$\mu_{T_t}(C) = \frac{\sum_{p \in T_t} \mu_C(p)}{\text{number of pages included in } T_t}.$$

$Interest(C)$ is a user's degree of interest in category C. It is defined as follows:

$$Interest(C) = \frac{Count(C)}{\sum_{t=1}^{T} t - \sum_{t=1}^{T}\{(t \times \mu_{T_t}(C)) + T\}}.$$

Table 2. Transaction 1

	C_1	C_2	C_3	C_4	C_5
P_1	0.1	0	0	0.3	0.6
P_6	0	0	0	0.2	0.8
P_8	0	0	0	0.1	0.9
$\mu_{C_i}(T_1)$	0.03	0	0	0.20	0.77

It assigns a higher degree to the categories included in recently visited pages. Since $Interest(C)$ is between 0 and 1, we call it fuzzy interests.

For example, a user visited a web site of Table 1 twice. As the first visit he made transactions $T_1 = P_1, P_6, P_8$ and at the second $T_2 = P_2, P_3, P_6$. That is, the user visited pages P_1, P_6 and P_8 at the first visit, and P_2, P_3 and P_6 at the second. Tables 2 and 3 show the fuzzy category of the pages in each transaction and the degrees that the transactions include each category, i.e. $\mu_{C_i}(T_j)$ for $i = 1, 2, \ldots, 5$ and $j = 1, 2$.

Table 3. Transaction 2

	C_1	C_2	C_3	C_4	C_5
P_2	0.4	0	0	0.1	0.5
P_3	0.1	0	0.3	0.1	0.5
P_6	0	0	0	0.2	0.8
$\mu_{C_i}(T_2)$	0.17	0	0.10	0.13	0.60

Then, user's degree of interest in C_5 can be evaluated as follows:

$$Count(C_5) = \sum_{t=1}^{2} \mu_{T_t}(C_5) = 0.77 + 0.60 = 1.37$$

$$Interest(C_5) = \frac{Count(C_5)}{\sum_{t=1}^{T} t - \sum_{t=1}^{T} \{(t \times \mu_{T_t}(C_5)) + T\}}$$

$$= \frac{1.37}{3 - \{1 \times 0.77 + 2 \times 0.60\} + 2} = 0.452$$

We may say that the user has interest in C_5 with a degree of 0.452.

2.3 Attributes of Fuzzy Interests

User's interest degree may change as time goes on, but it may have a tendency in changes. We have investigated it and identify four basic attributes of changes through time.

1. If a user does not refer pages including a field, he does not have interest in that field.

2. If a user refers only pages including only a field, he has the most interest in that field.
3. The more a user visits a page, the more interest he has in the field of the page.
4. Even if a user equally visits two pages, he has more interest in the topic of the pages visited more recently than the other.

Attribute 1 and 2 are the boundary condition, attribute 3 is the monotonicity and attribute 4 is the recentness. Our definition of the interest degree also satisfies the above attributes.

Followings are the attributes rewritten in the context of our definitions:

1. $\forall T_n, \mu_{T_n}(C) = 0 \rightarrow Interest(C) = 0$.
2. $\forall T_n, \mu_{T_n}(C) = 1 \rightarrow Interest(C) = 1$.
3. If a user makes only a transaction T including $C_1, C_2, \ldots$ and C_c, $Interest(C_i)$ increases as the number of transactions increases for $i = 1, 2, \ldots, c$.
4. If a user makes n transaction T_1s first and n transaction T_2s next, where T_1 includes C_1, T_1 includes C_2 and $\mu_{T_1}(C_1) = \mu_{T_2}(C_2)$ $Interest(C_1) < Interest(C_2)$.

Proofs are following:

Attribute 1: It is clear by definition.
Attribute 2: It is clear by definition.
Attribute 3: Since a user always makes the same transaction, $Count(C)$ is independent of the number of transactions. Thus for $m < n$,
$(\sum_{t=1}^{n} t - \sum_{t=1}^{n}((t \times \mu_{T_t}(C)) + n\ ) - (\sum_{t=1}^{m} t - \sum_{t=1}^{m}((t \times \mu_{T_t}(C)) + m\ ) > 0$. Therefore, $Interest(C)$ for m transactions is smaller than $Interest(C)$ for n.
Attribute 4: Since $\mu_{T_1}(C_1) = \mu_{T_1}(C_2)$, $Count(C_1) = Count(C_2)$. Thus, $\sum_{t=1}^{T} t \times \mu_{T_2}(C_2) > \sum_{t=1}^{T} t \times \mu_{T_1}(C_1)$. Therefore, $Interest(C_1) < Interest(C_2)$.

Through those analysis, we can know that the fuzzy interest also satisfies the basic attributes which user's interest may have.

3 Category Vector Space

We have described a method to find out user's interest degree. To provide better services to users in a web site it needs to understand users and user groups with similar interests. For web user analysis, we suppose that categories are conceptually independent from each other. That is, the interest degree of a category cannot be inferred from other categories. For example, web pages are fuzzily categorized into 3 categories: C_1, C_2 and C_3. We assume that we cannot infer a user's interest degree of C_3 from his interest degree of C_1 or C_2. This is analogue to a vector space whose axes C_1, C_2 and C_3 are. Thus we can create a vector space whose axes are C_1, C_2 and C_3. We call this space the category vector space.

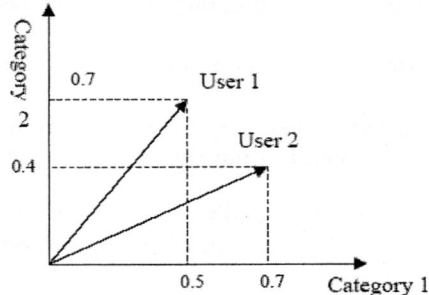

Fig. 1. Two-category space

All users' fuzzy interest degrees can be mapped on points in the category vector space. For example, there are category C_1 and C_2, and User 1's fuzzy interest degree is (0.5, 0.7) and User 2's fuzzy interest degree is (0.7, 0.4). Then, we make a space where C_1 and C_2 are axes. The interest degrees of each user can mapped onto a point each in the space as shown in Figure 1.

If two users' interests are similar, those will be located near from each other in the category vector space. Thus, the distance between two points can represent the similarity of two users' interests.

4 Experiments

4.1 Fuzzy Interest Analysis

Our fuzzy interest is compared with the method using only visit counts. We compare with the following count-based method:

$$Interest(Page) = \frac{\text{Number of transactions that contain the } Page}{\text{Total number of transactions}}.$$

It is used in [6] for finding association rules. For experiments, we assume that we have a web log for 60 days. We use 30-day time window, that is, we use recent 30 days web log data for interest analysis. Through the experiments, we verify how our method reflects time factors. We performed two experiments: simple and realistic. We assume that there are two pages A and B each of which contains only one topic. Figure 2 shows the comparison of the simple case. A user's interest is changing from page A to B at time 30. He visits only page A from time 0 to 30 and only B after 30. Figure 2(a) shows the number of visits, (b) the interest degrees by the counter-based method and (c) our fuzzy degrees. Since the counter-based method considers only the visit count and does not the

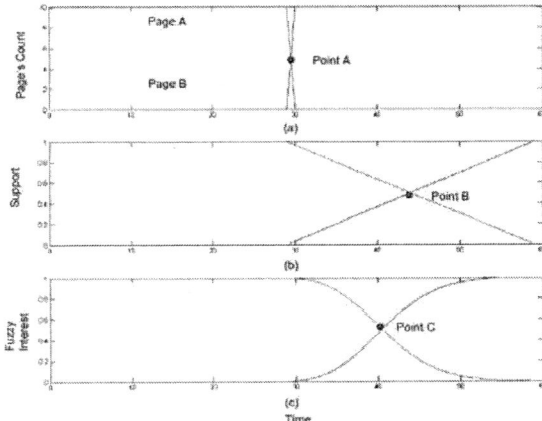

Fig. 2. (a)visit count (b)count-based interest(c)fuzzy interest for a simple case

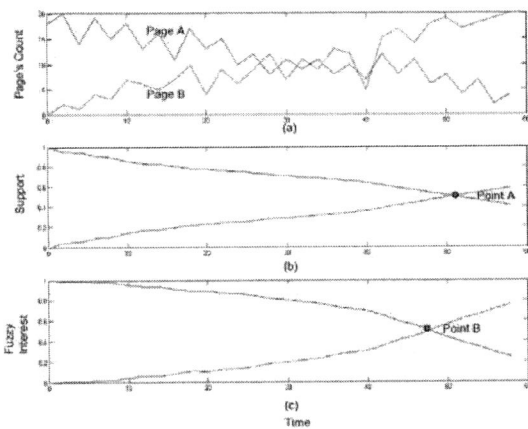

Fig. 3. (a)visit count (b)count-based interest(c)fuzzy interest for a realistic case

visit time. Its user's interest slowly reflects the user's interest change. Since our method gives more weight on the recent visits, it reflects the interest change more properly than the count-based method.

Figure 3 shows (a) the visit counts, (b) the interest degrees by the count-based method and (c) the fuzzy degree. The second experiment is for the realistic case. In the real life, a user's interest or visit count may be changing in a complex pattern. The user has more interest in page A at time 0 and more interest in page B at last but the change pattern is complex. Figure 3 is the result of this case. We may note that our method properly reflect the user's interest change.

4.2 User Analysis in the Category Vector Space

In this section, we apply our method for user analysis of a realistic web site. We build a simple web site by referring to a real internet shopping mall. The web site has nine pages of which contains the information of shoes and bags. We let seven users visit the web site. We make a log for their visits and analyze using our method.

Table 4. Fuzzy category degrees of the sample pages

	Shoes	Sports	Brand name	Bags
Page 1	0.7	0.3	0.8	0
Page 2	0.4	0.6	0.8	0
Page 3	0.7	0.3	0.2	0
Page 4	0.3	0.7	0.3	0
Page 5	1.0	0	0.5	0
Page 6	0	0.7	0.7	0.3
Page 7	0	0.7	0.4	0.3
Page 8	0	0	0.9	1
Page 9	0	0	0.1	1

For describing web contents, we choose four categories: shoes, bags, sports and brand name. Table 4 shows the category degrees of the nine pages to each category. For example, goods in Page 1 are related to shoes with a degree of 0.7, to shoes with 0.0, to sports with 0.3 and to brand name with 0.8. Table 5 shows users' visit ratio of each page. In this experiment, we will show how users' interests may be analyzed, so we present only the visit ratio. A user's visit ratio of a page is the ratio of the number of transactions including the page to the number of total transactions. Thus a high ratio means a high visit count.

Table 5. Users' visit ratio

	Page1	Page2	Page3	Page4	Page5	Page6	Page7	Page8	Page9
User A	0.52	0.56	0.08	0.48	0.2	0.66	0.3	0.74	0.12
User B	0.68	0.8	0.06	0.08	0.06	0.36	0.1	0.84	0.74
User C	0.28	0.22	0.18	0.28	0.8	0.2	0.2	0.1	0.06
User D	0.58	0.64	0.38	0.36	0.52	0.5	0.52	0.12	0.16
User E	0.8	0.18	0.64	0.24	0.66	0.04	0.12	0.52	0.42
User F	0.12	0.1	0.14	0.1	0.16	0.62	0.48	0.8	0.96
User G	0.4	0.42	0.5	0.4	0.78	0.84	0.02	0.42	0.38

As we see from Table 5, it is very difficult to analyze the users: which field who has interest in, who has a similar interest field to whose, what tendency

Table 6. Users' fuzzy interest degrees

	Shoes	Sports	Brand name	Bags
User A	0.11	0.21	0.54	0.14
User B	0.1	0.13	0.52	0.25
User C	0.6	0.08	0.3	0.03
User D	0.25	0.26	0.42	0.06
User E	0.37	0.1	0.4	0.13
User F	0.03	0.1	0.33	0.55
User G	0.25	0.2	0.43	0.12

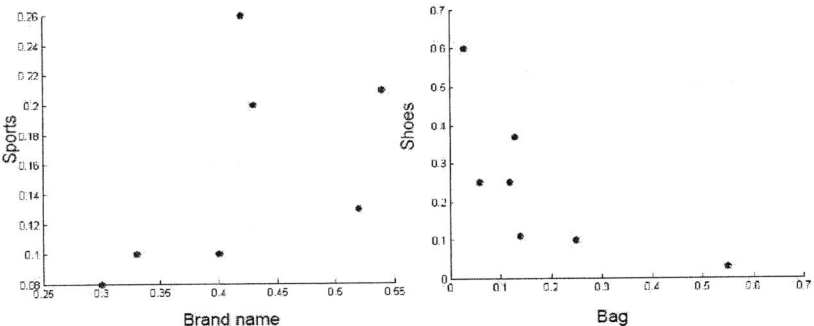

Fig. 4. Fuzzy interest in category space

the users visiting this site have, etc.. For example, can we say that User A has the highest interest in bags just because User A's visit ratio of Page 8 is the highest and Page 8 contains information on bags? It is not easy. However, if we apply our fuzzy interest and the category vector space, we can easily answer to those questions. Table 6 shows each user's fuzzy interest to the categories in Table 4. For example, User A and B have interest in brand names, User C in shoes, etc.. If we map the fuzzy interest degrees onto the category vector space, we can get more interesting information. Who has the most similar interest filed to User A's? If we evaluate the distance from User A's fuzzy interest, i.e., (0.11, 0.21, 0.54, 0.14) to the others' and then select the closest one to User A, we can find. User B has the most similar interest field to User A. Also we can analyze the users' tendency. If we use only sports and band name as axes, plot the fuzzy interests on the two axis space and cluster them, we get the tendency in the viewpoint of sports and brand name. Figure 4(a) shows the results: Three users locates at the right-upper corner and other three users at the left-lower corner, so we may conclude that the users visiting the test web site have interest in brand names if they have interest in sports. Figure 4(b) shows the analysis of users in the viewpoint of shoes and bags. We can find a tendency of the users that they do not have interest in both shoes and bags because there is no user who locates at the right-upper corner. From the experiments, we know that our method using fuzzy interest and the category vector space is very useful for user understanding and analysis.

5   Conclusion

This research is for web user understanding using fuzzy category and fuzzy interest. We defined fuzzy category, and presented a method to find out fuzzy interest degrees reflecting the time factor of transactions. We identified the attributes of interest changes through time, and proved that our fuzzy interest satisfied those. Also we investigated how fuzzy interest reflected the time factor through the experiment and compared with the existing method. For the analysis of web users, we defined the category vector space whose axes are the categories given by a web administrator. Users' fuzzy interests were mapped onto the space and analyzed. We applied it to a realistic web site log file and drew interesting results. We could easily analyze a web log file with fuzzy categories, fuzzy interests and the category vector space. Now, we continue our research by using real web usage data to find out valuable information from a real web site.

References

1. R. Agrawal, R. Srikant, "Fast Algorithms for Mining Association Rules, Proc. of VLDB Conference," pp.487-499, 1994.
2. H. Yi, Y.C. Chen, L.P. Chen, "Enabling Personalized Recommendation on the Web Based on User Interests and Behaviors," Proc. of 11th International Workshop, IEEE, pp.1066-1077, 2001.
3. A. Gyenesei, "A Fuzzy Approach for Mining Quantitative Association Rules," TUCS Technical Reports, no. 336, 2000.
4. J.S. Jang, S.H. Jun, K.W. Oh, "Fuzzy Web Usage Mining for User Modeling," International Journal of Fuzzy Logic and Intelligent Systems, vol. 2, no. 3, pp.204-209, 2002.
5. R. Cooley, B. Mobasher, J. Srivastava, "Data Preparation for Mining World Wide Web Browsing Patterns," Journal of Knowledge and Information System, vol. 1, no. 1, pp.8-19, 1999.
6. R. Cooley, B. Mobasher, J. Srivastava, "Web mining : Information and Pattern Discovery on the World Wide Web," Proc. of the 9th IEEE International Conf. on Tools with Artificial Intelligence, pp.61-62, 1997.
7. M. Spiliopoulou, "Web Usage Mining for Web Site Evaluation," Communications of the ACM, 43, pp.127-134, 2000.
8. B. Mohasher, R. Cooley, J. Srivastava, "Automatic personalization based on Web usage mining," Communications of the ACM, vol. 43, pp.142-152, 2000.

Automatic License Plate Recognition System Based on Color Image Processing

Xifan Shi[1], Weizhong Zhao[2], and Yonghang Shen[2]

[1] College of Computer Science and Technology, Zhejiang University,
[2] Department of Physics, College of Science, Zhejiang University,
310027, Hangzhou, Zhejiang, China
zjufan@hotmail.com, physyh@zju.edu.cn

Abstract. A License plate recognition (LPR) system can be divided into the following steps: preprocessing, plate region extraction, plate region thresholding, character segmentation, character recognition and post-processing. For step 2, a combination of color and shape information of plate is used and a satisfactory extraction result is achieved. For step 3, first channel is selected, then threshold is computed and finally the region is thresholded. For step 4, the character is segmented along vertical, horizontal direction and some tentative optimizations are applied. For step 5, minimum Euclidean distance based template matching is used. And for those confusing characters such as '8' & 'B' and '0' & 'D', a special processing is necessary. And for the final step, validity is checked by machine and manual. The experiment performed by program based on aforementioned algorithms indicates that our LPR system based on color image processing is quite quick and accurate.

1 Introduction

The automatic identification of vehicles has been in considerable demand especially with the sharp increase in the vehicle related crimes and traffic jams. It can also play a crucial role in security zone access control, automatic toll road collection and intelligent traffic management system. Since the plate can identify a car uniquely, it is of great interest in recent decade in using computer vision technology to recognize a car and several results have been achieved [2-14].

A typical LPR system can be divided into the following modules: preprocessing (including image enhancement and restoration), plate region extraction, plate region thresholding, character segmentation, character recognition and post-processing (validity checking). The first two modules, which only concern the shape and back/fore ground color of a plate and irrespective of character set in a plate, are the front end of the system. Module 4 and 5, on the contrary, are related to character set in a plate and regardless of the shape and back/fore ground color of a plate, so they are the back end of the system. Module 3, however, should take the shape and back/fore ground color of a plate as well as character set in a plate into consideration. Therefore, it is hard to say which end it can be categorized into.

To develop an automatic recognition system of a car plate, a stable recognition of a plate region is of vital importance. Techniques such as edge extraction [1][6], Hough transformation [7] and morphological operations [8] have been applied. An edge-based approach is normally simple and fast. However, it is too sensitive to the unwanted edges, which may happen to appear in the front of a car. Therefore, this method cannot be used independently. Using HT is very sensitive to deformation of a plate boundary and needs much memory. Though using gray value shows better performance, it still has difficulties recognizing a car image if the image has many similar parts of gray values to a plate region, such as a radiator region [11][12]. Morphology has been known to be strong to noise signals, but it is rarely used in real time systems because of its slow operation. So in recent years, color image processing technology [4][5] is employed to overcome these disadvantages. First, all of the plate region candidates are found by histogram. After that, each one is verified by comparing its WHR (Width to Height Ratio), foreground and background color with current plate standard and eliminated if it is definitely not of plate region. And finally, for each survivor, an attempt to read plate information is made by invoking the back end.

In the back end, first channel is selected and the plate region is thresholded in the selected channel. And then, each character is extracted by histogram and some optimizations such as the merge of unconnected character (i.e. Chuan, or 川), the removal of space mark, frame and pin, the correction of top and bottom coordinates in y direction and tilt correction are done during this phase. Next, each character is recognized by using minimum Euclidean distance based template matching since it's more noise tolerant than structural analysis based method [2][3]. And for those confusing characters, '8' & 'B' and '0' & 'D', for instance, a special processing is necessary to improve the accuracy. Finally, validity checking is performed against vehicle related crimes.

2 Plate Region Extraction

In principle, image should first be preprocessed, namely, enhanced and restored. But the experiment shows that it doesn't deserve its relatively heavy computational cost, so this step is skipped.

The basic idea of extraction of a plate region is that the color combination of a plate (background) and character (foreground) is unique and this combination occurs almost only in a plate region [14]. The correctness of this assumption is proved by the success of plate region extraction.

Altogether there are 4 kinds of plates in China mainland. They are yellow background and black characters plate for oversize vehicle, blue background and white characters plate for light-duty vehicle, white background and black or red characters plate for police or military vehicle, black background and white characters plate for vehicle of embassy, consulate and foreigners. At first, RGB model is used to classify all the pixels into the following 6 categories: blue, white, yellow, black, red and other, but unfortunately it fails because of the wide RGB value difference under different illumination. So HLS model is introduced, and this time the desired result is achieved, but it is too slow, namely, it takes PIII 1G roughly 1 second to processing a 1024X768 photo. Clearly, the bottleneck is the conversion from RGB value to HLS value while the key to its success is insensitivity under different illumination. Naturally, an ideal algorithm must retain this insensitivity under different illumination while eliminating the conversion between the two color models. Hence, the pixels are classified into 13 categories instead of 6 according to variance of illumination in the RGB domain. They are dark blue, blue, light blue, dark yellow, yellow, light yellow, dark black, black, gray black, gray white, white, light white and other. Here, red is not take into account because this color appears only once in the center or right part of the police or military vehicle plates whose dominant character color is black. Thus, it is enough to identify the plate by checking the black pixels. The speed is increased to 0.5 second per photo while the correct extraction rate remains the same to HLS. But, that's not enough. Actually, the dot and line interlace scan method is used and the time cost is reduced to 1/4 of the non-interlaced one. After the plate is extracted, the region is verified by its shape, i.e. WHR. In China mainland, there are three WHR values, which are 3.8 for police or military vehicle plates, 2.0 for rear edition of oversize vehicle plate and 3.6 for others. Because 3.6 and 3.8 is too close, they are merged into one. So if the WHR of the extracted plate is sufficiently close to 3.7 or 2.0, the verification is passed.

According to Amdahl's law, frequent case should be favored over the infrequent case. In China mainland, the most common plate is white characters with blue background. Therefore, plate is first tried to be recognized as a white blue pair, then as a black yellow pair, next as a white black pair and finally as a black white pair.

Taking a white blue pair for example, this process can be illustrated as follows.

Fig. 1. Extraction of a plate region in vertical

As shown in Figure 1, the whole image is scanned and only the number of dark blue pixels exceeds the given threshold, say 1000, so it can be deduced that it is a dark blue background plate. Thereby, the plate region in vertical direction is identified by thresholding the histogram of dark blue pixels.

It is evident that the only candidate is the middle one (For the top, the number of lines where number of dark blue pixels exceeds the threshold is too small and thus omitted. If two adjacent plate regions are sufficiently close, then they are merged into one.). In addition, owing to the favor of frequent case and the fact that the plate region is generally occurred in the lower part of an image, the scan is done from bottom to top and hence the middle one is first found. The extracted one is in Figure 2. Similarly, by thresholding in horizontal direction, the plate region is obtained, as illustrated in Figure 3.

Fig. 2. Extraction of a plate region in horizontal

Fig. 3. The extracted plate region

To confirm the extraction, the shape or terminologically WHR is examined [2][3]. Here, it is 310/85=3.65, sufficiently close to 3.7, so the verification is passed.

3 Character Segmentation and Recognition

3.1 Thresholding

The thresholding procedure should introduce as little noise as possible, since subsequent steps may be seriously affected by a poor thresholding algorithm. Also, because the lighting conditions vary widely over a plate, locally adaptive thresholding is required. Empirical methods are devised and they succeed in thresholding the plate region.

There are a variety of threshold algorithms, but the experiments show that "simple is the best", if considering both speed and accuracy, so bimodal histogram segmentation [13] is introduced. As Figure 4 shows, if the pixels of objects form one of its peaks, while pixels of the background form another peak, then the histogram is called bimodal. It is the case provided that an image consists of objects of approximately the same gray level that differs from the gray level of the background. Fortunately, this condition is satisfied, for the color of characters, or, the object, is almost the same and the color of the background of the plate region is also almost the same, which makes this simple segmentation algorithm works. Since there are three (R, G and B) channels in an image, the channel is selected by the largest standard deviation of the three. Larger standard deviation means longer distance between the two peaks while longer distance between the two peaks means the clearer division between background and object and less sensitive to the noise introduced by thresholding. In the case of the plate region in Figure 3, the standard deviations in red, green, blue channels are 74.57, 72.51, 59.98, respectively, so the red channel is selected for thresholding. This is reasonable, because the background is blue and the object is white, which has blue component and naturally, standard deviation in the blue channel must be the smallest.

Without loss of generality, it is assumed that the object is white and the background is black before thresholding (If not, the color is reversed and this process is only needed for black yellow pair and black white pair). It can be proved that after thresholding, the number of white pixel is 68%~85% of the plate region. Suppose V is the value making 85% of the plate become white and U is the average value of the remaining. Then threshold value is U minus DetalV, which is from 5 to 10. Correct thresholding is accomplished by this rule of thumb.

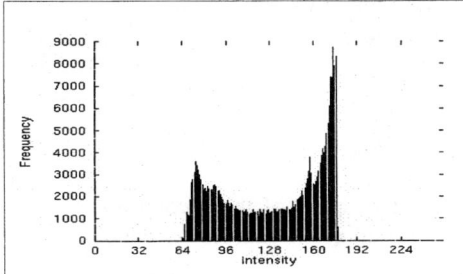

Fig. 4. Bimodal histogram

3.2 Segmentation

First, according to its WHR, the plate is classified as either double line or single line. The threshold is 1/10 and 1/6 of the width of the plate for the former and the latter, respectively. Then the line whose number of black pixels exceeds the threshold is selected, and if two adjacent selected regions are sufficiently close, then they are merged into one. Next, the WHR of each segmented region is verified, if it is too large, it is discarded as frame. This process is shown in Figure 5.

Similar process (including threshold acquisition, selection, merge and discard) can be done in horizontal direction, as illustrated in Figure 6 and 7. The characters are segmented, but the performance is not quite satisfactory, and therefore some optimizations are carried out during this stage.

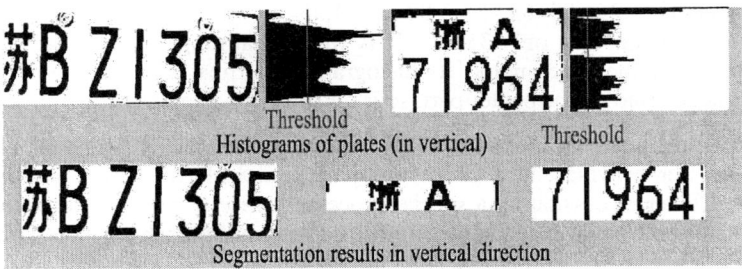

Fig. 5. Segmentation in vertical direction

Fig. 6. Segmentation in horizontal direction

苏BZ1305

Fig. 7. Segmented characters

Removal of Space Mark. The space of the second and the third character is much larger than that of any other adjacent characters, which can be formalized into the following rule of thumb:

The ratio of the largest space to the second largest space between the adjacent characters is 1.25~1.45.

Fig. 8. Mis-segmented characters (due to space mark)

This rule is helpful in removing the space mark, as illustrated in Figure 8. After segmentation, 8 characters are found including the space mark, the third character. The largest space is 55 while the second largest space is 53. The ratio is 55 / 53 = 1.04, not within the range of 1.25~1.45. It is suspicious of the existence of space mark. If it is indeed the case, 55 must be the second largest space and the largest space is from 69 (1.25X55=68.75) to 80 (1.45X55=79.75). By addition of the adjacent number, it is obvious that only 35+36=71 is within that range. Hence, the third character is probably the space mark. Its histogram in vertical direction shows that the pixel is concentrated on the center part, so it must the space mark and removed.

Merge of Unconnected Character. The first one on the plate of China mainland, the abbreviation for province, is a Chinese character and all characters are connected except for Chuan, necessitating a special process. A case in point is shown in Figure 9. The character Chuan is segmented into its three strokes, which must be merged. The largest space is 148 while the second largest space is 113. The ratio is 148 / 113 = 1.31 within the range of 1.25~1.45. So the fourth character should be the second character, which means the first character is a union of the first three characters. Merge is done in right to left order until the WHR of the merged character is within

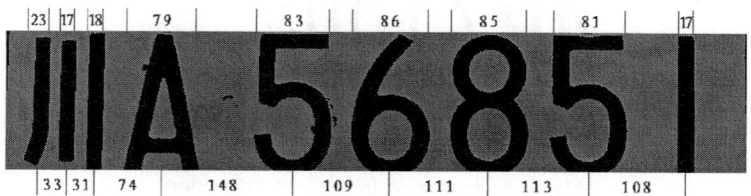

Fig. 9. Mis-segmented Chinese character (due to disconnectivity)

normal range. In this case, only by merge the first three characters can the WHR be satisfied, which leads to the correct merge of the unconnected character Chuan.

Correction of Top and Bottom. Coordinates. Because plate may be tilted, the top and bottom coordinates are probably not correct (see in Figure 7). This process is required and the coordinates of each character are rectified by utilizing its histogram in vertical direction. The correction result of the plate in Figure 7 is shown in Figure 10.

Fig. 10. Correction result

Removal of Frame. In Figure 10, there's some noise in the lower right part of the last character, which comes from the frame. But the last character is a digit, which is connected. This property makes the removal of frame possible and the comparison is shown in Figure 11.

Fig. 11. Removal of frame

Fig. 12. Degraded top coordinate correction due to pin

Removal of Pin. Because of the pin, on some occasions, the results of the correction of the top and bottom coordinates are degraded, rather than upgraded, as illustrated in Figure 12. But owing to the linearity of the top/bottom coordinates, the top/bottom coordinate of each character must between the top/bottom coordinates of their left and right neighbor. In Figure 12, it is beyond doubt that the top coordinate of the second is large than that of the first and third, so the top coordinate is substituted by the average of that of the first and third and thus the pin is successfully removed. Pin on the sixth can be removed in the same way and the result is shown in Figure 13.

沪AJ2824

Fig. 13. Pin removal

Tilt Correction. For every segmented character, there must be a top pixel whose y value is the biggest. The x and y coordinate of top pixel of character i is x_i and y_i, respectively. Owing to the linearity of the top coordinates, the relationship between x and y can be expressed as the following formula:

$$y = a + bx.$$

By minimizing

$$Q = \sum_{i=1}^{N} (y_i - a - bx_i)^2,$$

We obtain:

$$\bar{x} = \frac{1}{N}\sum_{i=1}^{N} x_i, \quad \bar{y} = \frac{1}{N}\sum_{i=1}^{N} y_i,$$

$$L_{xx} = \sum_{i=1}^{N} x_i^2 - \frac{1}{N}\left(\sum_{i=1}^{N} x_i\right)^2, \quad L_{yy} = \sum_{i=1}^{N} y_i^2 - \frac{1}{N}\left(\sum_{i=1}^{N} y_i\right)^2, \quad L_{xy} = \sum_{i=1}^{N} x_i y_i - \frac{1}{N}\left(\sum_{i=1}^{N} x_i\right)\left(\sum_{i=1}^{N} y_i\right).$$

The coefficient a, b and fitting coefficient γ are as follows:

$$b = \frac{L_{xy}}{L_{xx}}, \quad a = \bar{y} - b\bar{x}, \quad \gamma = \frac{L_{xy}}{\sqrt{L_{xx}L_{yy}}} = b\sqrt{\frac{L_{xx}}{L_{yy}}}.$$

And the top tilting degree is $\arctan b$. By the same token, the bottom tilting degree can be calculated. If the top tilting degree and the bottom tilting degree are all positive or negative, the plate is deemed to be tilted. The tilting degree is the average of top tilting degree and bottom tilting degree weighed by top fitting coefficient and bottom fitting coefficient respectively. In the case of Figure 10, the top tilting degree and bottom tilting degree is −2.46 and −1.82, respectively. The fitting coefficients are both −1.00. So its tilting degree is:

$$\frac{(-2.46)\times(-1.00)+(-1.82)\times(-1.00)}{(-1.00-1.00)} = -2.14.$$

It's more than 2, so rotation is needed.

Fig. 14. Plate after tilt correction

The rotation is performed by the following formula:

$$g(x', y') = f(x'\cos\theta + y'\sin\theta + x_0, -x'\sin\theta + y'\cos\vartheta + y_0) = f(x, y),$$
$$x = x'\cos\theta + y'\sin\theta + x_0,$$
$$y = -x'\sin\theta + y'\cos\vartheta + y_0.$$

where x' and y' are the coordinates of the new and x and y are those of the old, θ the rotation degree and (x_0, y_0) the rotation center. But in most cases x_0 or y_0 is not integer, so linear interpolation is employed and the result is shown in Fig. 14 and 15.

Fig. 15. Segmented characters of Figure 14 (before other optimizations)

3.3 Character Recognition

If the WHR of the character is less than 1/3, it is tried to be recognized as '1'.

For '1' candidates, if its pixel fill rate is more than 0.6, it is recognized as '1', otherwise discarded.

For other characters, first, its size is normalized to 32X64. Then, minimum Euclidean distance based template matching is used to recognize each character [2][3]. And for those confusing characters such as '8' & 'B' and '0' & 'D', a special processing is necessary. Pixels do differ in the left top triangle and in the left bottom triangle of these 4 characters. This property endows us the opportunity to distinguish '8' from 'B' or '0' from 'D' by checking these two triangles. Also, in China mainland, the second character is alphabetic, the third and fourth character is alphanumeric and the last three is numeric, this can constrain the matching in the alphanumeric template set and eliminate the unnecessary incorrect recognition from letter to digit or vice versa.

3.4 Validity Checking

Validity is checked by machine and manual. For machine, the plate is searched in database to see whether it indeed exists. If the matched record does exist and is retrieved, the color of background and foreground of the plate is compared to those in it. If either of the former conditions fails, the vehicle will be stopped. And if the plate is on the blacklist, say, wanted by the police, it should be detained either. For manual, the type (oversize or light-duty), the brand (Benz or BMW) and the color of the current car body are compared to the information in the database. Again, if it fails, the vehicle will be held.

4 Conclusion

The experiment performed by program based on aforesaid algorithms indicates that our LPR system based on color image processing is quite quick and accurate. Even on a PIII 1G PC, 90% of the photos under various illuminations are read correctly within 0.3s.

In this article, the automatic Chinese LPR system based on color image processing is proposed. The using of color image processing instead of grayscale, the further division from 6 colors to 13 colors to gain the robustness under various illuminations and the selection of channel are the major breakthroughs. And there are also some empirical rules, such as the computation of threshold value, the optimizations during character segmentation and the special processing to distinguish '8' from 'B' or '0' from 'D'. The end justifies the means. Last but not least, the validity check is performed. It is of absolute necessity to be introduced into a practical LPR system.

References

1. D.H. Ballard, Computer vision. Prentice-Hall Inc., (1991)
2. Ahmed, M. J., Sarfraz, M., Zidouri, A., and Alkhatib, W. G., License Plate Recognition System, The Proceedings of The 10th IEEE International Conference On Electronics, Circuits And Systems (ICECS2003), Sharjah, United Arab Emirates (UAE).
3. Sarfraz, M., Ahmed, M., and Ghazi, S. A. (2003), Saudi Arabian License Plate Recognition System, The Proceedings of IEEE International Conference on Geoemetric Modeling and Graphics-GMAG2003, London, UK, IEEE Computer Society Press.
4. Shyang-Lih Chang, Li-Shien Chen, Yun-Chung Chung, Sei-Wan Chen, Automatic license plate recognition, IEEE Transactions on Intelligent Transportation Systems, Vol.: 5, Issue: 1, (2004) 42-53
5. Guangzhi Cao, Jianqian Chen, Jingping Jiang, An adaptive approach to vehicle license plate localization, IECON '03. The 29th Annual Conference of the IEEE Industrial Electronics Society, Vol.: 2 (2003) 1786-1791
6. K. Lanayama, Y. Fujikawa, K. Fujimoto, M. Horino, Development of Vehicle License Number Recognition System Using Real-time, Image Processing and its Application to Travel-Time Measurement. Proceedings of the 41st IEEE Vehicular Technology Conference (1991) 798-804
7. K. M. Kim, B. J. Lee, K. Lyou, G. T. Park. The automatic Recognition of the Plate of Vehicle Using the Correlation Coefficient and Hough Transform, Journal of Control, Automation and Systems Engineering, Vol.3, No.5, (1997) 511-519
8. M. Shridhar, J. W. Miller, G. Houle, L. Bijnagte, Recognition of License Plate Images: Issues and Perspectives. Proceedings of International Conference on Document Analysis and Recognition, (1999) 17-20
9. Sunghoon Kim, Daechul Kim, Younbok Ryu, Gyeonghwan Kim, A Robust License-plate Extraction Method under Complex Image Conditions, Proceedings. 16th International Conference on Pattern Recognition, (ICPR'02) Vol. 3 (2002) 216-219
10. H. J. Choi, A Study on the Extraction and Recognition of a Car Number Plate by Image Processing, Journal of Korea Institute of Telematics and Electronics(KITE) (1987) Vol. 24 No. 2, 309-315
11. B. T. Cheon et al, The Extraction of a Number Plate from a Moving car, Proc. of First Workshop on Character Recognition (1993) 133-136
12. H. S. Chong and H. J. Cho, Locating Car License Plate Using Subregion Features, Journal of the KISS (1994) Vol. 21 No. 6, 1149-1159
13. Prewitt, J.M.S. and Mendelsohn, M.L. The analysis of cell images, in Ann. N.Y. Acad. Sci, (1966) 1035-1053
14. E. R. Lee, P. K. Kim, H. J. Kim, Automatic Recognition of a Car License Plate Using Color Image Processing, IEEE International Conference on Image Processing, Vol. 2, (1994) 301-305

Exploiting Locality Characteristics for Reducing Signaling Load in Hierarchical Mobile IPv6 Networks*

Ki-Sik Kong, Sung-Ju Roh, and Chong-Sun Hwang

Dept. of Computer Science and Engineering, Korea Univ.
1, 5-Ga, Anam-Dong, Sungbuk-Gu, Seoul 136-701, Republic of Korea
{kskong, sjroh, hwang}@disys.korea.ac.kr

Abstract. Hierarchical Mobile IPv6 (HMIPv6) aims to reduce the number of the binding update messages in the backbone networks, and also improve handoff performance. However, this does not imply any change to the periodic binding refresh message to the home agent and the correspondent node, and now a mobile node (MN) additionally should send it to the mobility anchor point (MAP). In addition, the MAP should encapsulate and forward incoming packets to the MN. These facts indicate that the reduction of the number of the binding update messages in the backbone networks can be achieved at the expense of increase of the signaling load within a MAP domain. On the other hand, it is observed that an MN may habitually stay for a relatively long time or spend much time on connecting to the Internet in a specific cell (hereafter, *home cell*) covering its home, office or laboratory, etc. Thus, when we consider the preceding facts and observation in HMIPv6 networks, HMIPv6 may not be particularly favorable during a home cell residence time in terms of the signaling load; In this case, it may be preferable that the MN uses Mobile IPv6 (MIPv6), not HMIPv6. In this paper, therefore, we presents a new efficient mobility management scheme to enable an MN to selectively switch its mobility management scheme according to whether it is currently in its home cell or not in HMIPv6 networks, which can reduce the signaling load while maintaining the same level of handoff latency as HMIPv6. The numerical results indicate that compared with HMIPv6, the proposed scheme has apparent potential to reduce the signaling load in HMIPv6 networks.

1 Introduction

The tremendous growth of wireless technology and the popularization of laptop/notebook computers have prompted research into mobility support in networking protocols. Although Mobile IPv6 (MIPv6) [1] is one of the dominating protocols that provide mobility support in IPv6 networks, it is not scalable; For

---

* This research was supported by University IT Research Center Project.

example, if the home agent (HA) or correspondent nodes (CNs) are far from the MN even if the MN moves across the adjacent subnet, the binding update (BU) messages may travel across several IP networks. In addition, as the number of the MNs increases in the networks, the number of the BU messages[1] also increases proportionally and this phenomenon may result in significant signaling and processing load through the networks. In order to overcome these drawbacks, Hierarchical MIPv6 (HMIPv6) [2,3] has been proposed. HMIPv6 introduces a new entity, the mobility anchor point (MAP) which works as a proxy for the HA in a foreign network. When an MN moves into a network covered by a new MAP, it is assigned two new care-of-addresses (CoAs): a regional CoA on the MAP's subnet (RCoA) and an on-link address (LCoA). If an MN changes its LCoA within a MAP domain, it only needs to register the new address with the MAP. In contrast, the RCoA registered with the HA and CN does not change.

HMIPv6 has been designed to reduce the number of the BU messages in the backbone networks and also improve handoff performance by reducing handoff latency. However, this does not imply any change to the periodic binding refresh (BR) message to the HA and the CN, and now an MN additionally should send it to the mobility anchor point (MAP). In addition, the MAP should encapsulate and forward incoming packets directly to the MN. These facts indicate that the reduction of the number of the BU messages in the backbone networks can be achieved at the expense of increase of the signaling load within a MAP domain. On the other hand, it is observed that an MN may habitually stay for a relative long time or spend much time on connecting to Internet in a specific cell[2] (i.e., *home cell*) than in the rest of the cells (hereafter, *ordinary cells*) within a MAP domain. Thus, when we consider the preceding facts and observation in HMIPv6 networks, HMIPv6 may not be particularly favorable during a home cell residence time in terms of the signaling load; In this case, it may be preferable that the MN uses Mobile IPv6 (MIPv6), not HMIPv6, even if the MN is HMIPv6-aware. In a large-scale wireless/mobile network, localized mobility management scheme (e.g., HMIPv6) will be widely used. Therefore, in such an environments, especially when HMIPv6 is applied to a large-scale wireless/mobile network, the efforts toward reducing the signaling load in the networks should be more emphasized because a huge number of the MNs will be serviced by the MAP.

This paper presents an efficient mobility management scheme to reduce the signaling load in HMIPv6 networks, while maintaining the same level of handoff latency as HMIPv6, which enables an MN to selectively switch its mobility management scheme according to whether it is currently in its home cell or not. The remainder of this paper is organized as follows. Section 2 briefly mentions the

---

[1] The BU message may also imply the periodic binding refresh (BR) message, which is generated by an MN whenever the binding lifetime is close to expiration.

[2] In this paper, we assume that HMIPv6 is applied to a large-scale wireless/mobile network, and that the coverage area of the mobile network is partitioned into *cells*.

background and motivation of this paper. In Sect.3, we introduce our proposed scheme for HMIPv6 networks, called HHMIPv6. In Sect.4, we conduct the analysis of signaling load between HMIPv6 and HHMIPv6. Numerical results will be given in Sect.5. Conclusions and future work will be given in Sect.6.

2 Background and Motivation

There have been a lot of recent researches for efficient mobility management exploiting the MN's mobility/traffic pattern in wireless mobile networks.

In [4], the authors proposed an optimal update strategy which determines whether or not a mobile terminal (MT) should update in each location area (LA), and minimizes the average location management cost derived from an MT-specific mobility model and call generation pattern. In [5], they proposed a simple, yet efficient location management scheme to reduce the paging cost. While an MT is residing in an LA with a cell called anchor-cell, where the MT usually stays for a significant period, an intra-LA location update is performed whenever the MT changes its location between the anchor-cell and the rest of cells in the LA. For an incoming call, either the anchor cell or the rest of cells in the LA is paged to locate the MT. Thus, the paging cost is greatly reduced, especially when the called MT is located in its anchor-cell.

In [6], the authors made the key observation that while the potential set of sources for the MN may be large, the set of sources that a given MN communicates most frequently with is very small. Based on this observation, they developed the concept of a working set of nodes for the MN. In addition, they proposed an adaptive location management scheme that enables an MN to dynamically determine its working set and trade-off routing and update costs in order to reduce the total cost. They also pointed out that most Internet users tend to have a relatively unchanging mobility behavior, which is closely related to the environment of the Internet user [7]. In [8], the authors performed a comparative analysis of MIPv6 and HMIPv6. In this paper, they investigated the effects of various parameters such as the average speed of an MN, its packet arrival rate, and the binding lifetime. Their results demonstrated that in terms of the signaling bandwidth consumption, the signaling load generated by an MN during its average domain residence time in HMIPv6, gets larger than that in MIPv6 as the average speed of an MN gets lower (i.e., an MN's average subnet/domain residence time gets longer) and its packet arrival rate gets higher.

As shown in these literatures, we can see that a lot of studies exploit the MN's mobility/traffic pattern for efficient mobility management. Therefore, once these characteristics are maintained in the MN's history, mobility management may become not only easier but also more efficient. Inspired by the idea in [4,5] and the observation and facts in [6,7,8], we propose a simple, yet efficient history-based mobility management scheme for HMIPv6 networks.

3 Exploiting Locality Characteristics for Reducing Signaling Load in HMIPv6 Networks

It is observed that an MN may stay for a significant period or spend much time on connecting to the Internet in a specific cell than in the rest of the cells within a MAP domain. For example, an MN may mainly stay or connect to the Internet in the cell covering its home, office or laboratory than in the rest of the cells within a MAP domain. We refer to a cell within a MAP domain where the MN stays for a considerable time and spends much time on connecting to the Internet as *home cell*, and refer to a MAP domain containing a home cell as *home domain*, respectively. In addition, we refer to the rest of the cells other than home cell within a home domain as *ordinary cells*. Based on the preceding observation, we propose an efficient history-based auxiliary mobility management scheme for HMIPv6 networks, called HHMIPv6.

The motivation of HHMIPv6 is to exploit the MN's mobility/traffic locality characteristics in order to reduce the signaling load in HMIPv6 networks. In HHMIPv6, we assume that each MN keeps its own home cell addresses, which are obtained based on its mobility/traffic history information. To estimate the home cell for a particular MN, its mobility/traffic pattern throughout the days or weeks are observed over a long period of time.[3] In HHMIPv6, a new flag, *O flag* is added in the option field of the BU message to indicate whether an MN is in its home cell or not.

Figure 1 shows the new BU message format in HHMIPv6. When an MN enters its home cell, the *O flag* is set to 1. Otherwise, the *O flag* is unset. The operation of HHMIPv6 is exactly the same as that of HMIPv6 except either when an MN enters/leaves its home cell or while it stays in its home cell. The binding update procedures in HHMIPv6 may be slightly different in terms of the home cell crossing. In other words, there are four possible different cases according to the following movement types in terms of the home cell crossing.

- Case 1: ordinary cell ⇒ home cell ⇒ ordinary cell
- Case 2: ordinary cell ⇒ home cell ⇒ outside cell[4]
- Case 3: outside cell ⇒ home cell ⇒ ordinary cell
- Case 4: outside cell ⇒ home cell ⇒ outside cell

Due to the space limitation, from now on, we will mainly describe HHMIPv6 for the Case 1 only (For more details, refer to the binding update procedure in HHMIPv6 shown in Fig.2. All the cases from Case 1 through Case 4 can be described in Fig.2. Note that, in Fig.2, when an MN moves from the outside cell

---

[3] The various ways to obtain a good estimate of the home cell for each MN from these information is an important issue by itself and is beyond the scope of this paper.

[4] For the description of the movement types, we refer to the cell within another MAP domain other than home domain as *outside cell*.

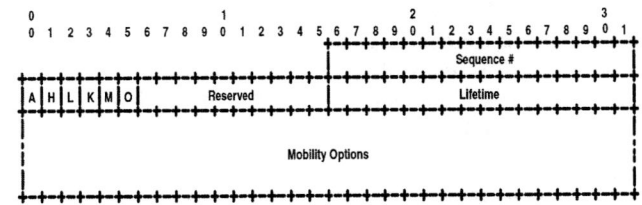

Fig. 1. The new BU message format in HHMIPv6

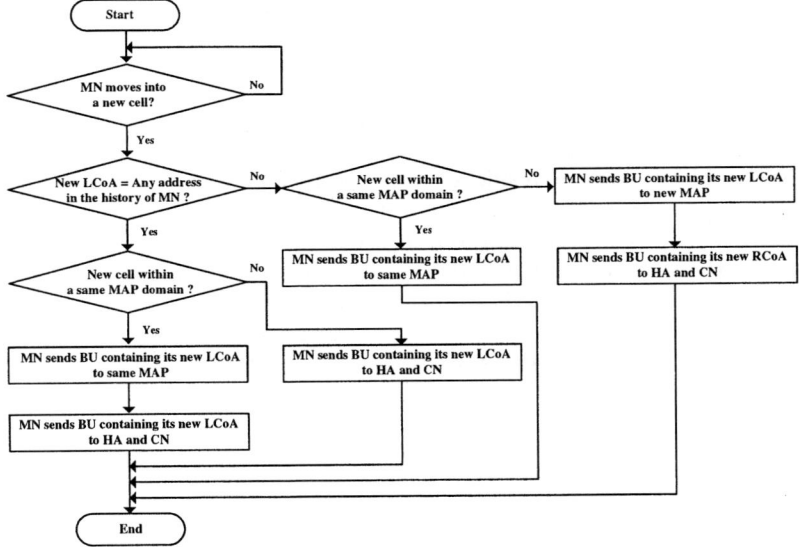

Fig. 2. The operation of binding update procedure in HHMIPv6

into its home cell located in the border of the home domain[5], it does not need to send the BU message to the new MAP. Instead, it sends it to the new MAP, for the first time, right after it moves from its home cell to the ordinary cell).

The binding update procedure in HHMIPv6 is described as follows. Whenever an MN enters a new cell, it checks its home cell addresses to see if there is any address equal to new LCoA. If the same address is found, an MN decides that it just moved into its home cell. Once an MN knows that it just entered its home cell, it now behaves as if it operates just like in MIPv6, not in HHMIPv6, until it leaves it. That is, after sending the BU message to the MAP just like in HMIPv6, an MN switches its mobility management scheme to MIPv6 by

---

[5] In this case, just like the BU procedure by the inter-MAP movement for a smooth inter-MAP handover in HMIPv6 [2], the MN may send a BU message to the previous MAP requesting it to forward packets addressed to the MN's home cell. This will allow the MN to continue to receive packets while updating the HA and the CNs.

additionally sending the BU messages containing its LCoA to the HA and the CN. Note that the reason for sending the BU message to the MAP is to maintain the same level of handoff latency as HMIPv6. However, neither an MN nor the MAP sends the periodic BR message or the BR request message to each other since they know from the BU message with *O flag* set to 1 that an MN will use MIPv6 during its home cell residence time. On the other hand, when an MN leaves its home cell, it switches its mobility management scheme back to HMIPv6 by additionally sending the BU message containing its RCoA to the HA and the CN after sending the BU message to the MAP.

4   Analysis of Signaling Load

In this section, the total signaling loads (i.e., the sum of the signaling load incurred by the binding update (BU), the binding refresh (BR), the binding acknowledgement (BAck) and the packet tunneling (PT)) generated by an MN during its MAP domain residence time in HMIPv6 and HHMIPv6 are analyzed. For the simplicity, we assume that an MN crosses K cells during its MAP domain residence time. In addition, in order to capture deterministic or quasi-deterministic MN's movement pattern, we assume that an MN has a home cell within a MAP domain, and that during a MAP domain residence time, it enters its home cell after crossing n ($0 < n < K-1$) ordinary cells, stays once there, and then leaves it.[6] For the analysis, the following notations are used.

- C_Y^X: operation Y cost during an MN's MAP domain residence time in X scheme
- C_{Y-Z}^X: operation Y cost for Z in X scheme
- d_Z: average number of hops between an MN's default access router and Z
- T_Z: binding lifetime for the MN at Z

Table 1. Values and meanings for the subscripts

Parameter	Description	Values [Meaning]
X	Mobility Management Scheme	Hmip [HMIPv6], HHmip [HHMIPv6]
Y	Operation Type	BU [binding update], BR [binding refresh], PT [packet tunneling], Total [BU+BR+PT]
Z	Network Entity	ha [HA], cn [CN], map [MAP]

The values that subscripts shown in the above notations can take are summarized in Table 1. In addition, the parameters for the analysis are shown in Table 2.

---

[6] As already mentioned in Sect.3, in case an MN moves from the outside cell into its home cell located in the border of the home domain, it does not need to send the BU message to the new MAP. Thus, the signaling bandwidth generated by BU/BAck message can be saved more than that under this scenario.

Table 2. Parameters for the performance analysis

Parameter	Description
p^{hc}	Average packet arrival rates for an MN during its home cell residence time
p_i^{oc}	Average packet arrival rates for an MN during the i-th visited ordinary cell residence time
q	Probability [A single packet is routed directly to the MN without being intercepted by the HA]
K	The number of cells that an MN has crossed during its MAP domain residence time
N	Average number of the bindings for the CNs maintained in the MN's binding update list during its MAP domain residence time
S_{bu}	Signaling bandwidth consumption generated by a BU/BAck message
S_{pt}	Signaling bandwidth consumption generated by tunneling per packet

4.1 Total Signaling Load in HMIPv6

In HMIPv6, when an MN first enters a MAP domain, it needs to register with the MAP and the HA. Then, when an MN moves into a new cell within the same MAP domain, it registers with the MAP only. Therefore, according to the BU procedure in HMIPv6, C_{BU}^{Hmip} can be expressed as follows.

$$C_{BU}^{Hmip} = \frac{(2Kd_{map} + 2d_{ha} + Nd_{cn}) \times S_{bu}}{t_{total}} \qquad (1)$$

Note that the HA and the MAP must return a BAck message to the MN, but the CN does not need to return a BAck message to the MN. In the above equation, t_{total} means an MN's MAP domain residence time ($t_{total} = t^{hc} + \sum_{i=1}^{K-1} t_i^{oc}$), and t^{hc} and t_i^{oc} are an MN's cell residence times in its home cell and i-th visited ordinary cell, respectively. On the other hand, C_{BR}^{Hmip} can be expressed as

$$C_{BR}^{Hmip} = \frac{C_{BR-ha}^{Hmip} + C_{BR-map}^{Hmip} + C_{BR-cn}^{Hmip}}{t_{total}} \qquad (2)$$

where C_{BR-ha}^{Hmip}, C_{BR-map}^{Hmip} and C_{BR-cn}^{Hmip} are as follows.

$$C_{BR-ha}^{Hmip} = 2d_{ha} \times S_{bu} \times \lfloor \frac{t_{total}}{T_{ha}} \rfloor \qquad (3)$$

$$C_{BR-map}^{Hmip} = 2d_{map} \times S_{bu} \times (\sum_{i=1}^{K-1} \lfloor \frac{t_i^{oc}}{T_{map}} \rfloor + \lfloor \frac{t^{hc}}{T_{map}} \rfloor) \qquad (4)$$

$$C_{BR-cn}^{Hmip} = Nd_{cn} \times S_{bu} \times \lfloor \frac{t_{total}}{T_{cn}} \rfloor \qquad (5)$$

On the other hand, C_{PT}^{Hmip} can be derived as follows.

$$C_{PT}^{Hmip} = \frac{\{qD_{dir}^{Hmip} + (1-q)D_{indir}^{Hmip}\} \times (p^{hc}t^{hc} + \sum_{i=1}^{K-1} p_i^{oc}t_i^{oc})}{t_{total}} \quad (6)$$

where D_{dir}^{Hmip} and D_{indir}^{Hmip} are the packet tunneling cost generated by a direct packet delivery (not intercepted by the HA), and the packet tunneling cost generated by delivering a packet routed indirectly via the HA, while an MN stays in a cell (regardless of either home cell or ordinary cell) in HMIPv6, respectively. Therefore, these costs can be expressed as follows.

$$D_{dir}^{Hmip} = d_{map} \times S_{pt} \quad (7)$$
$$D_{indir}^{Hmip} = d_{ha} \times S_{pt} \quad (8)$$

Finally, using Eq. (1), (2) and (6), C_{Total}^{Hmip} can be expressed as follows.

$$C_{Total}^{Hmip} = C_{BU}^{Hmip} + C_{BR}^{Hmip} + C_{PT}^{Hmip} \quad (9)$$

4.2 Total Signaling Load in HHMIPv6

Similar to Eq.(1), C_{Bu}^{HHmip} can be expressed as follows.

$$C_{BU}^{HHmip} = \frac{\{2Kd_{map} + 3(2d_{ha} + Nd_{cn})\} \times S_{bu}}{t_{total}} \quad (10)$$

Note that the cost of $(2d_{ha} + Nd_{cn}) \times S_{bu}$ is additionally generated twice when an MN enters its home cell and when an MN leaves it, respectively. On the other hand, C_{BR}^{HHmip} can be expressed as

$$C_{BR}^{HHmip} = \frac{C_{BR-ha}^{HHmip} + C_{BR-map}^{HHmip} + C_{BR-cn}^{HHmip}}{t_{total}} \quad (11)$$

where C_{BR-ha}^{HHmip}, C_{BR-map}^{HHmip} and C_{BR-cn}^{HHmip} are as follows.

$$C_{BR-ha}^{HHmip} = 2d_{ha} \times S_{bu} \times (\lfloor \frac{\sum_{i=1}^{n} t_i^{oc}}{T_{ha}} \rfloor + \lfloor \frac{t^{hc}}{T_{ha}} \rfloor + \lfloor \frac{\sum_{i=n+1}^{K-1} t_i^{oc}}{T_{ha}} \rfloor) \quad (12)$$

$$C_{BR-map}^{HHmip} = 2d_{map} \times S_{bu} \times \sum_{i=1}^{K-1} \lfloor \frac{t_i^{oc}}{T_{map}} \rfloor \quad (13)$$

$$C_{BR-cn}^{HHmip} = Nd_{cn} \times S_{bu} \times (\lfloor \frac{\sum_{i=1}^{n} t_i^{oc}}{T_{cn}} \rfloor + \lfloor \frac{t^{hc}}{T_{cn}} \rfloor + \lfloor \frac{\sum_{i=n+1}^{K-1} t_i^{oc}}{T_{cn}} \rfloor) \quad (14)$$

On the other hand, C_{PT}^{HHmip} can be expressed as follows.

$$C_{PT}^{HHmip} = \frac{\bar{C}_{hc}^{HHmip} + \bar{C}_{oc}^{HHmip}}{t_{total}} \quad (15)$$

where $\bar{C}_{hc}^{HHmip}$ and $\bar{C}_{oc}^{HHmip}$ are the packet tunneling costs in HHMIPv6 while an MN stays in its home cell and the ordinary cells within a MAP domain, respectively. Therefore, these costs are as follows.

$$\bar{C}_{hc}^{HHmip} = \{q\bar{D}_{dir}^{HHmip} + (1-q)\bar{D}_{indir}^{HHmip}\} \times p^{hc}t^{hc} \qquad (16)$$

$$\bar{C}_{oc}^{HHmip} = \{q\hat{D}_{dir}^{HHmip} + (1-q)\hat{D}_{indir}^{HHmip}\} \times \sum_{i=1}^{K-1} p_i^{oc}t_i^{oc} \qquad (17)$$

where $\bar{D}_{dir}^{HHmip}$ and $\bar{D}_{indir}^{HHmip}$ are the packet tunneling cost generated by a direct packet delivery, and the packet tunneling cost generated by delivering a packet routed indirectly via the HA, while an MN stays in its home cell in HHMIPv6, respectively. Similarly, $\hat{D}_{dir}^{HHmip}$ and $\hat{D}_{indir}^{HHmip}$ are the packet tunneling cost generated by a direct packet delivery, and the packet tunneling cost generated by delivering a packet routed indirectly via the HA, while an MN stays in the i-th ordinary cell in HHMIPv6, respectively. Therefore, $\bar{D}_{dir}^{HHmip}$, $\bar{D}_{indir}^{HHmip}$, $\hat{D}_{dir}^{HHmip}$ and $\hat{D}_{indir}^{HHmip}$ can be expressed as follows.

$$\bar{D}_{dir}^{HHmip} = 0 \qquad (18)$$

$$\bar{D}_{indir}^{HHmip} = d_{ha} \times S_{pt} \qquad (19)$$

$$\hat{D}_{dir}^{HHmip} = D_{dir}^{Hmip} = d_{map} \times S_{pt} \qquad (20)$$

$$\hat{D}_{indir}^{HHmip} = D_{indir}^{Hmip} = d_{ha} \times S_{pt} \qquad (21)$$

Finally, using Eq. (10), (11) and (15), C_{Total}^{HHmip} can be expressed as follows.

$$C_{Total}^{HHmip} = C_{BU}^{HHmip} + C_{BR}^{HHmip} + C_{PT}^{HHmip} \qquad (22)$$

5 Numerical Results

The total signaling loads in both HMIPv6 and HHMIPv6, generated by an MN during its MAP domain residence time, are evaluated by investigating the relative signaling load in HHMIPv6. The *relative signaling load in HHMIPv6* is defined as the ratio of the signaling load in HHMIPv6 to that in HMIPv6. For the analysis, we set d_{map}, d_{ha}, d_{cn}, q and n to be 2, 8, 6, 0.7 and 5, respectively [9]. T_{map}, T_{ha} and T_{cn} are set to be 5 mins [10]. And, t_i^{oc} and p_i^{oc} are assumed to be uniformly distributed with $U[5, 60]$ mins and $U[0.01, 0.5]$ kilopkt/hr, respectively. Also, the size of a BU/BAck message is assumed to be equal to the size of an IPv6 header (40 bytes) plus the size of a binding update extension header (28 bytes), so 68 bytes ($=S_{bu}$) [2,3]. In addition, the additional signaling bandwidth consumption generated by tunneling per packet is equal to the size of IPv6 header, so 40 bytes ($=S_{pt}$). Figure 3 indicates the relative signaling load in HHMIPv6 for K=7 and 15. In both figures, for almost all the conditions except when both t^{hc} and p^{hc} are very small, the relative signaling load in HHMIPv6 gets smaller as t^{hc} and p^{hc} get larger. This is due to the results caused by both no periodic BR/BAck messages between the MAP and the MN, and no packet tunneling from the MAP to the MN during its home cell residence time.

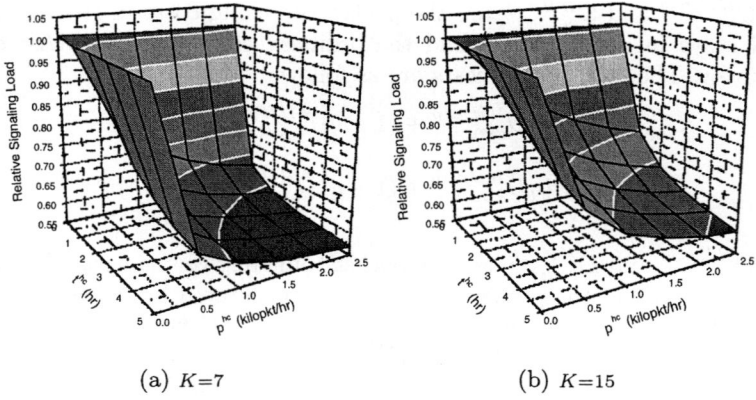

(a) $K=7$ (b) $K=15$

Fig. 3. Relative signaling load in HHMIPv6

6 Conclusions and Future Works

The reduction of the signaling load associated with IP mobility management is one of the significant challenges to IP mobility support protocols [11]. By exploiting the observation that the MN may have mobility/traffic locality characteristics in several specific cells (i.e., home cell), we proposed a simple, yet efficient history-based mobility management scheme for HMIPv6 networks, called HH-MIPv6. HHMIPv6 enables an MN to selectively switch its mobility management scheme according to whether it is currently in its home cell or not in HMIPv6 networks. The numerical results indicated that the signaling load in HHMIPv6 gets smaller than that in HMIPv6 as an MN's home cell residence time or the packet arrival rate in its home cell gets larger. Our future research subjects include the design of more flexible MN's history management policy and the analytical modelling using Marchov Chain model for performance evaluation.

References

1. D. Johnson and C. Perkins, "Mobility Support in IPv6," RFC 3775, Jun. 2004.
2. H.Soliman, C.Castelluccia, K.Malki, and L.Bellier, "Hierarchical Mobile IPv6 Mobility Management (HMIPv6)," draft-ietf-mipshop-hmipv6-03.txt, Oct. 2004.
3. C. Castelluccia, "HMIPv6: A Hierarchical Mobile IPv6 Proposal," ACM Mobile Computing and Communications Review, vol.4, no.1, pp.48-59, Jan. 2000.
4. S. K. Sen, A. Bhattacharya and S. K. Das, "A Selective Location Update Strategy for PCS Users," ACM/Baltzer Wireless Networks, pp.313-326, 1999.
5. Mao, Z, "An Intra-LA Location Update Strategy for Reducing Paging Cost," IEEE Communications Letters, vol.6, no.8, pp.334-336, Aug. 2002.
6. S. Rajagopalan and B.R. Badrinath, "An Adaptive Location Management Strategy for Mobile IP," Proc. Mobicom'95, Nov. 1995.
7. T.Imielinski, B.R.Badrinath, "Wireless Computing: Challenges in Data Management", Comm. of the ACM, pp.19-28, Oct. 1994.

8. K.-S. Kong, S.-J. Roh and C.-S.Hwang, "A Comparative Analytical Study on the Performance of IP Mobility Protocols: Mobile IPv6 and Hierarchical Mobile IPv6," MoMM 2004, pp.437-446, Sept. 2004.
9. S. Pack and Y. Choi, "A Study on Performance of Hierarchical Mobile IPv6 in IP-Based Cellular Networks," IEICE Trans. Commun., vol.E87-B, no.3 pp.462-469, Mar. 2004.
10. R. Ramjee, K. Varadhan, L. Salgarelli, S. Thuel, W. Yuan, T. Porta, "HAWAII: A Domain-Based Approach for Supporting Mobility in Wide-Area Wireless Networks," IEEE/ACM Trans. Networking, vol.10, no.3, pp.396-410, Jun. 2002.
11. I.F. Akyildiz, J. Xie, and S. Mohanty, "A Survey of Mobility Management in Next-Generation All-IP-Based Wireless Systems," IEEE Wireless Commun., pp.16-28, Aug. 2004.

Parallel Feature-Preserving Mesh Smoothing

Xiangmin Jiao and Phillip J. Alexander

Computational Science and Engineering,
University of Illinois, Urbana, IL 61801, USA
{jiao, palexand}@cse.uiuc.edu

Abstract. We present a parallel approach for optimizing surface meshes by redistributing vertices on a feature-aware higher-order reconstruction of a triangulated surface. Our method is based on a novel extension of the fundamental quadric, called the *medial quadric*. This quadric helps solve some basic geometric problems, including detection of ridges and corners, computation of one-sided normals along ridges, and construction of higher-order approximations of triangulated surfaces. Our new techniques are easy to parallelize and hence are particularly beneficial for large-scale applications.

Keywords: Computational geometry; feature detection; mesh smoothing; quadric.

1 Introduction

In this paper, we devise new techniques for estimating normals and identifying geometric features for triangulated surfaces, and apply them to redistribute vertices of surface meshes on parallel machines. Mesh smoothing is an important problem in many computational applications [4]. It is frequently used as a post-processing step in mesh generation, and is critical in numerical simulations with deforming geometry. Compared to two-dimensional meshes, surface meshes are particularly difficult to optimize, because curved shapes and sharp features of geometric models must be preserved, frequently without the availability of the underlying CAD models. Therefore, feature-aware higher-order approximations must be constructed by analyzing discrete surfaces. In large-scale scientific simulations, the problem is even more challenging, because meshes are partitioned and distributed across multiple processes on a parallel machine, making it difficult to apply some traditional analytic and reconstruction techniques.

To achieve our objectives, we first develop new techniques to estimate surface normals, especially one-sided normals along ridges, and devise a new vertex-based scheme for detecting ridges and corners of a triangulated surface. Our techniques are based on an extension of the well-known fundamental quadric [2, 9] and tensor voting [12]. These previous techniques provide insights into the local geometry of a discrete surface, but suffer from ambiguities such as undetermined signs of the estimated normals and indistinction between near cusps and smooth surfaces. Our extension, called the *medial quadric*, implicitly uses a local coordinate frame with origin approximately on the medial axis to resolve these ambiguities. Utilizing the results of the medial quadric, we then present a feature-aware higher-order reconstruction of a surface triangulation, and integrate them to deliver a parallel method for surface mesh smoothing.

2 Estimation of Vertex Normals

Surface mesh smoothing, like many other geometric problems, requires accurate estimation of vertex normals. We present a novel concept called the *medial quadric*, which connects two seemingly unrelated classes of normal estimations (namely, *weighted averaging* [13] and *tensor voting* [12]) and subsequently develop a new estimation method.

Weighted Averaging. A commonly used approach for estimating vertex normals is to average the (potentially weighted) normals of the faces incident on a vertex. There is no consensus on the best choice of weights [15]. The simplest weighting scheme is to use unit weight for every face [5]. Other popular schemes include area-weighted average [13] and angle-weighted average [17]. Another scheme was recently derived to recover the exact normal if the mesh is a tessellation of a sphere [11]. Empirically, these weighting schemes produce nearly identical results for well-shaped smooth surfaces. For well-shaped surfaces with singularities, angle-weighted averaging tends to deliver balanced weights and hence better results, but it is sensitive to perturbation for surfaces with obtuse (especially nearly 180°) triangles. We therefore propose a *guarded angle-weighting scheme* to take the smaller of the edge angle at a vertex and its complement as the weight for each face, i.e., $w_i = \min\{\theta_i, \pi - \theta_i\}$, where θ_i is the edge angle in the ith incident face.

Quadric and Tensor Voting. Another class of estimation is obtained through eigen-decomposition of a quadric. Let γ be a plane containing a point $p \in \mathbb{R}^3$ with unit normal vector $\hat{n}$. The *offset* of γ from the origin is $\delta = -p^T \hat{n}$. The signed distance of γ from any point $x \in \mathbb{R}^3$ is then

$$d(x, \gamma) = (x - p)^T \hat{n} = x^T \hat{n} + \delta. \tag{1}$$

Given a collection of planes $\{\gamma_i\}$ (in particular, the tangent planes of the triangles incident on a vertex), let $\hat{n}_i$ denote their unit outward normals, δ_i their offsets, and w_i their associated *positive* weights. The weighted sum of squared distances to γ_i from x is

$$Q(x) = \sum_i w_i d^2(x, \gamma_i) = x^T A x + 2 b^T x + c, \tag{2}$$

where $A = \sum_i w_i \hat{n}_i \hat{n}_i^T$, $b = \sum_i w_i \delta_i \hat{n}_i$, and $c = \sum_i w_i \delta_i^2$. The metric Q is the well-known *fundamental quadric* [2,9], which is minimized in $\mathbb{R}^3$ at the solution of the 3×3 linear system

$$A x = -b. \tag{3}$$

In general, A is symmetric and positive semi-definite, and its eigenvalues are all real and nonnegative. For an introduction to eigenvalue problems, see textbooks such as [6].

Let λ_i be the eigenvalues of A such that $\lambda_1 \geq \lambda_2 \geq \lambda_3$, and $\hat{e}_i$ their corresponding orthonormal eigenvectors. Based on the spectrum theorem, A can be decomposed into

$$A = \sum_{i=1}^{3} \lambda_i \hat{e}_i \hat{e}_i^T. \tag{4}$$

If the neighborhood of a vertex v is smooth, then $\hat{e}_2$ and $\hat{e}_3$ are approximately the principal directions at v, and $\hat{e}_1$ approximates the normal direction [9, 12]. This approximation scheme is referred to as *tensor voting* [12] or *normal voting* [14] in the literature. However, it has the following limitations:

- the direction of $\hat{e}_1$ is also sensitive to weights, similar to weighted averaging
- the sign of $\hat{e}_1$ is undetermined and may point inward or outward
- if the vertex v is on a sharp ridge with dihedral angle $> \pi/2$, then $\hat{e}_2$ instead of $\hat{e}_1$ provides a more meaningful approximation to the normal
- if the ridge nearly forms a right angle, then none of the eigenvectors provides a meaningful approximation to the normal

Another popular approach, which is the dual of tensor voting, is to take the eigenvector corresponding to the smallest eigenvalue of the matrix $\sum_i w_i \hat{t}_i \hat{t}_i^T$, where $\hat{t}_i$ is a tangent vector of the ith face [16]. This approach has limitations similar to tensor voting.

Medial Quadric. To overcome the above limitations, suppose there is a point o such that all faces have the same *negative* offset δ to o. As the quadric is scale- and position-independent, without loss of generality, we take $\delta = -1$ and place the origin of the coordinate frame at o. Because o is on the *medial axis* of the surface, we refer to this quadric as the *medial quadric*. This quadric is minimized by the solution of (3) with

$$b = -\sum_i w_i \hat{n}_i. \qquad (5)$$

The unit vector of $-b$ (i.e., the right-hand side of (3)) is the *weighted-average* outward normal. The solution x is the position vector from o to v, and its unit vector $\hat{x}$ delivers another approximation to the outward normal at v. Unlike the weighted-average normals or eigenvectors, however, $\hat{x}$ is independent of the weights given that the point o exists (because if it exists, o is uniquely defined regardless of the weights).

Another geometric interpretation of the medial quadric is that the origin o is at the intersection of the planes that are parallel to the γ_i with a normal distance -1. When such an intersection does not exist exactly, the origin would then be at the intersection in the least squares sense. When the planes γ_i are nearly parallel to each other, this intersection and in turn $\hat{x}$ are sensitive to perturbation. Numerically, this sensitivity corresponds to a large condition number of A. To solve for x robustly, we constrain it within the primary space of A, i.e. the space spanned by the eigenvectors corresponding to relatively large eigenvalues. Let d be the dimension of the primary space, and V be a $3 \times d$ matrix, whose ith column is $\hat{e}_i$. The solution of Eq. (3) then reduces to finding a vector $s \in \mathbb{R}^d$, such that Q is minimized at $x = Vs$. The vector s is the solution to

$$\left(V^T A V\right) s = -V^T b, \qquad (6)$$

which is an $d \times d$ linear system. The condition number of Eq. (6) is then λ_1/λ_d. The solution to (6) is $s_i = -\hat{e}_i^T b/\lambda_i$, and hence

$$x = \sum_{i=1}^d s_i \hat{e}_i = \sum_{i=1}^d -\hat{e}_i^T b \hat{e}_i/\lambda_i. \qquad (7)$$

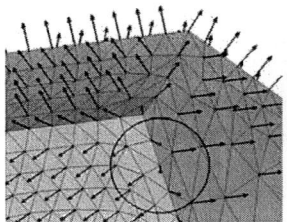 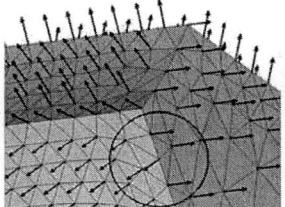 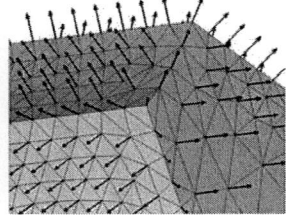

Fig. 1. Demonstration of effect of imbalance of weights. From left to right, arrows indicate estimated normals from averaging, tensor voting, and medial quadric, all weighted by area

In particular, when $d = 1$ (i.e., for smooth points), x is along the direction of $\hat{e}_1$ where the sign is determined by the weighted average. For $d \geq 2$ (such as at a ridge or corner), x is then a linear combination of the eigenvectors and delivers an approximation *mean normal*, which is insensitive to weights. Fig. 1 compares the normal estimations from weighted-averaging, tensor voting, and medial quadric. Only the medial quadric delivers consistent weight-insensitive approximation along features.

One-Sided Normals. The medial quadric is particularly useful in estimating one-sided normals along ridges. In particular, given a face σ, let $\hat{n}_\sigma$ be its face normal and t_σ its average tangent pointing from v to its opposite edge center. Given that the weights are balanced between the two sides of the ridge, the one-sided normal on the side of σ at v is

$$\hat{n}_+ = \left(\sqrt{\lambda_I}\hat{x} + \text{sign}\left(\hat{n}_\sigma^T \hat{y}\right)\sqrt{\lambda_J}\hat{y}\right)/\sqrt{\lambda_1 + \lambda_2}, \qquad (8)$$

where $I = \text{argmax}_i\{|s_i| \mid 1 \leq i \leq 2\}$, $J = 3 - I$, and $\hat{y}$ is the binormal $\hat{x} \times \hat{e}_3$. To arrive at the coefficients, assuming the total weight $w \equiv \sum w_i$ is balanced between the two sides at a ridge vertex, then λ_1 and λ_2 with dihedral angle θ are $w\max\{\sin^2(\theta/2), \cos^2(\theta/2)\}$ and $w\min\{\sin^2(\theta/2), \cos^2(\theta/2)\}$, respectively. Therefore, the guarded angle-weighted scheme tends to produce reasonable estimates except next to obtuse triangles. For meshes with obtuse triangles, a more accurate estimation can be obtained at additional cost, by constructing and solving one-sided quadrics for each ridge vertex, i.e., $A_+ x_+ = b_+$, where A_+ and b_+ are constructed using the faces of which $\text{sign}(\hat{n}_i^T \hat{y}) > 0$.

3 Vertex-Based Geometric Features

Extracting features (or singularities) from a discretized surface is an important subject for geometric applications. Most feature detection schemes are *edge-based*, in that they first identify ridge edges and then classify vertices based on edge classification; see e.g. [1, 10]. To facilitate feature detection for a partitioned surface mesh on a parallel machine, we present a vertex-based detection scheme, which extends the approach of Medioni et al. [12] with inspirations from the medial quadric.

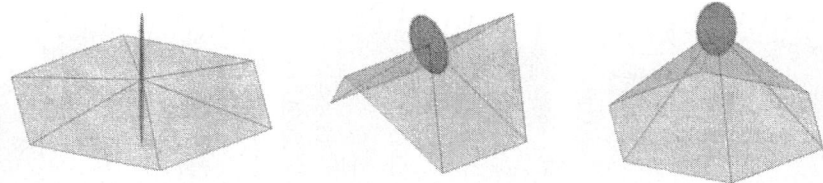

Fig. 2. Differing behavior of eigenvalues at smooth, ridge, and corner points. Eigenvalues and eigenvectors are depicted by ellipsoids, whose axes are aligned along eigenvectors, with semi-axes proportional to their corresponding eigenvalues

Vertex Classification. The relative sizes of eigenvalues of the matrix $\boldsymbol{A}$ of Eq. (3) are closely related to the local smoothness at a vertex v, as illustrated in Fig. 2. More formally, $\boldsymbol{A}$ can be expressed as the linear combination of

$$\boldsymbol{A} = (\lambda_1 - \lambda_2)\boldsymbol{E}_1 + (\lambda_2 - \lambda_3)\boldsymbol{E}_2 + \lambda_3\boldsymbol{E}_3, \qquad (9)$$

where $\boldsymbol{E}_d \equiv \sum_{i=1}^{d} \hat{e}_i \hat{e}_i^T$ are the *saliency tensors* for surface, curve (ridge), and point (corner) for $d = 1$, 2, and 3, respectively [12]. The relative sizes of these components were used in [12] and [14] to classify features. Similar to such approaches, we define

$$r = \lambda_3 / \max\{\lambda_1 - \lambda_2, \lambda_2 - \lambda_3\}$$

as the *corner saliency* and consider a vertex as a corner if s is larger than a threshold β. A tiny (nearly zero) β would classify all vertices as corners and a huge (nearly infinity) β would classify no corners. Given a user-specified ridge-angle threshold ψ, as a rule of thumb, we take $\beta \approx \cot \psi$. When r is small, unlike previous methods, we consider λ_3 and its corresponding eigenvector as noise in the model, and hence classify a ridge by comparing λ_2/λ_1 against a threshold α. This approach leads to a more reliable classification for ridges in practice than previous methods. For a ridge with dihedral angle $\theta \leq \pi/2$, the eigenvalues satisfy $\lambda_2/\lambda_1 \approx \tan^2(\theta/2)$, and therefore we set $\alpha = \tan^2(\psi/2)$.

Because matrix $\boldsymbol{A}$ is independent of the signs of normals, the proceeding approach may falsely classify a sharp ridge (e.g., a near cusp) as a smooth surface and classify a sharp corner as a ridge. To resolve this issue, we observe that acute angles are accompanied by the reversal of the order of $\left|\boldsymbol{x}^T \hat{e}_i\right| = \left|s_i \boldsymbol{b}^T\right|/\lambda_i$ (c.f. Eq. (7)), and this order is nearly independent of the weights. Therefore, we introduce a safeguard $g_i = \left|\hat{\boldsymbol{b}}^T \hat{e}_i\right| / \min\{\varepsilon \lambda_1, \lambda_i\}$, where ε (say 10^{-7}) avoids potential division by zero. In summary, a vertex is classified as follows:

1. if $\mathrm{argmax}_i\{g_i\} = 3$ or $\lambda_3 \geq \beta \max\{\lambda_1 - \lambda_2, \lambda_2 - \lambda_3\}$, then v is a corner
2. otherwise, if $\mathrm{argmax}_i\{g_i\} = 2$ or $\lambda_2 \geq \alpha \lambda_1$, then v is on a ridge
3. otherwise, v is at a smooth point

To demonstrate the robustness of this new method, Fig. 3 highlights the ridge vertices detected by our approach with $\psi = 20°$ and by the normal-voting scheme [14] with

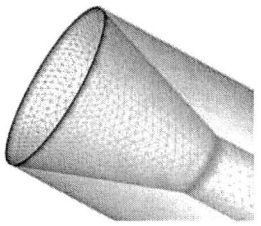 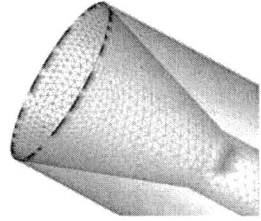 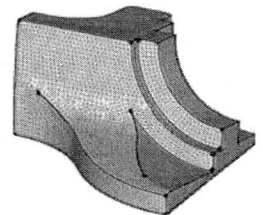

Fig. 3. Comparison of ridge detection with our new method (left) and normal voting (right)

Fig. 4. Features on fandisk detected by our method

$(\lambda_1 - \lambda_2)/(\lambda_0 - \lambda_1) \geq \cos\psi / \sin^2(\psi/2)$ for $\psi = 19°$ (because all vertices would be classified to be smooth with $\psi \geq 20°$). Our scheme is clearly more reliable and less sensitive to perturbation.

Edge Classification. If vertex v is a ridge vertex, then the eigenvector $\hat{e}_3$ is approximately tangential to the ridge, and its incident ridge edges are nearly parallel to $\hat{e}_3$. In addition, the other vertex of either of its incident ridge edges is most likely also a ridge or corner vertex. Therefore, we identify ridge edges as follows. Let $\hat{t}_\tau$ denote the unit tangent of an edge τ incident on v pointing away from v. For each ridge vertex, compute the largest (positive) and the smallest (negative) values of $s_\tau \equiv m_\tau \hat{e}_3^T \hat{t}_\tau$, where m_τ is the number of incident ridge or corner vertices of τ. An incident edge is on the ridge if its associated s_σ has either of the two extreme values and $|s_\tau| \geq 2\cos\psi$, i.e., τ is nearly parallel to $\hat{e}_3$. After classifying all ridge edges, a ridge vertex is upgraded to a corner if it is incident on more than two ridge edges or $|s_\tau| < 2\cos\psi$ for either of its extreme values of s_τ. Fig. 4 shows the corners and ridges after these adjustments for the fandisk model, where even weak features were detected accurately by our method.

4 Feature-Preserving PN Triangles

Utilizing normal estimation and feature detection, we now develop a feature-aware higher-order approximation of a surface triangulation suitable for parallelization. In particular, we employ curved point-normal triangles, or *PN triangles* [18], which provide a simple and convenient way to construct a piecewise cubic approximation of a surface with triangular Bézier patches from vertex coordinates and normals. The resulting approximation is C^1 continuous at vertices and C^0 continuous everywhere else, and a separate quadratic approximation of normals recovers continuity of normals. PN triangles are constructed triangle-by-triangle, without using additional neighbor information, and therefore make a good candidate for distributed meshes.

Summary of PN Triangles. The key component of PN triangles is to determine the seven non-vertex control points for each triangle (two along each edge and one inside the face) from the vertices. The construction first linearly interpolates the control points, so that edge points uniformly subdivide the edges and the face point is at the centroid.

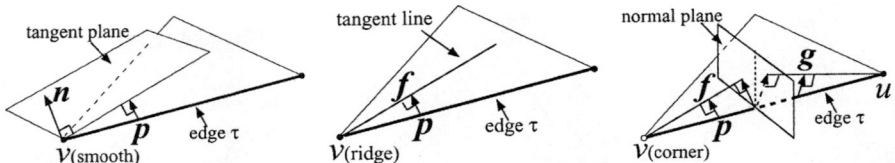

Fig. 5. Construction of control points along edges for feature-preserving PN triangles at smooth, ridge, and corner vertices, respectively

Each control point p on edge τ is then moved by a displacement f, as we will describe shortly. The centroid is moved by a displacement equal to 1.5 times the average displacement of the six edge points, so that quadratic polynomial patches can be reconstructed exactly [3]. To construct a continuous normal field, the normal direction at the midpoint of an edge τ is set to the mirror image of the average normal of the vertices of τ against the normal plane. For details, readers are referred to [18]. This simple construction delivers good results for smooth surfaces, but some amendment is needed at the presence of sharp features.

Feature Treatments. To deliver a systematic treatment at sharp ridges and corners for the geometric construction of PN triangles, we leverage the results of our medial quadric. Given an edge τ, let p be a control point on τ, vertex v the end-point of τ closer to p, and vector v its coordinates. Suppose τ and its end-points have been classified by feature detection. As illustrated in Fig. 5, we evaluate the displacement f at p as follows:

1. If v is at a smooth point, then project p onto the tangent plane at v, i.e., $f = (v - p)^T \hat{n}\hat{n}$ (c.f. Fig. 5(left)), where $\hat{n} = \hat{e}_1$.
2. If both v and τ are on a ridge (c.f. Fig. 5(middle)), then project p onto the tangent line of the ridge at v, i.e., $f = (v - p) - (v - p)^T \hat{e}_3 \hat{e}_3$.
3. If v is on a ridge but τ is not, then project p onto the one-sided tangent plane at v (c.f. Fig. 5(left)), i.e., $f = (v - p)^T \hat{n}_+ \hat{n}_+$, where $\hat{n}_+$ is the one-sided normal.
4. If both vertices of τ are corners, then consider the edge as straight and take $f = 0$.
5. If v is at a corner and the other vertex u of τ is not (c.f. Fig. 5(right)), then compute f as the displacement g of u mirrored against the normal plane of τ, i.e., $f = g - 2g^T tt$, where t is the tangent of τ.

The first two cases are equivalent to the projections in [18], but Cases 3 through 5 are introduced here to avoid large errors near features. In Case 5, the mirroring operation allows higher-order reconstruction at a corner.

5 Parallel Surface Mesh Smoothing

We now leverage the above techniques to address the problem of parallel surface mesh smoothing, i.e., to achieve better mesh quality by redistributing smooth or ridge vertices while preserving the shape and features of a surface.

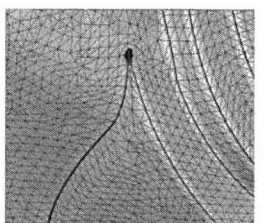 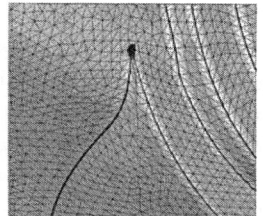 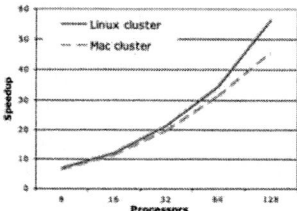

Fig. 6. Before and after smoothing of fandisk model   **Fig. 7.** Parallel performance

Smoothing Algorithm. Given a smooth or ridge vertex v, we project its incident faces (i.e., its *star*) onto its tangent space, i.e., the space spanned by $\hat{e}_2$ and $\hat{e}_3$ if v is at a smooth point, or $\hat{e}_3$ if v is on a ridge. Let T be a rectangular matrix, whose column vectors form the orthonormal vector base of the tangent space. We perform a coordinate transformation locally so that v becomes the origin. The projection of a neighbor vertex (denoted by u_i) of v onto the tangent space is $T^T(u_i - v)$. We compute the *center* of the star of v in the tangent space, where the definition of center depends on the specific metric in use, but is typically a weighted sum of the vertices in its star [4], i.e.,

$$d = \left(\sum_i w_i T^T(u_i - v)\right) / \sum_i w_i, \qquad (10)$$

where the w_i are metric-dependent weights. If v is at a smooth vertex, we set the weights to the sum of the edge angles at v in the incident faces of edge vu_i. If v is a ridge vertex, then we set the weights for its neighbor ridge vertices along the ridge to 1 and those for its neighbor smooth vertices to 0.

After obtaining d, in general we move v for a fraction of d, say ad. To avoid foldover of triangles, we choose a to be ≤ 0.5 and to be small enough so that the barycentric coordinates of ad corresponding to the other two vertices in each triangle are no greater than $c/3$ for some $c \in (0,1)$ (say $c = 0.5$). To utilize the higher-order constructions, in particular the PN triangles, we locate the triangle σ that contains the new point $p = v + ad$ and then map p onto the Bézier patch constructed by feature-preserving PN triangles. Because the new positions of smooth vertices may depend on those of ridge vertices but not vice versa, we first perform vertex redistribution within ridges and then redistribute smooth vertices using the new locations of ridge vertices. A simple Jacobi iteration is adopted within either redistribution stage. When performing smoothing for multiple iterations, we interpolate the normals using the quadric reconstruction for better efficiency. Fig. 6 shows the fandisk model (c.f. Fig. 4) near the junction before and after smoothing, where the dark curves indicate the detected features. The shapes of the triangles were improved noticeably without distorting the features.

Parallel Implementation. In large-scale computational applications, mesh smoothing, including feature detection and surface projection, must be performed on a mesh that is partitioned and distributed across multiple processors on a parallel machine. The techniques and algorithms presented above are all easily parallelized, as we have algorithmically localized their calculations, of which the most noteworthy are classification of feature vertices and calculation of one-sided normals. These operations are difficult to compute for vertices along partition boundaries using traditional methods, unless a process has access to the remote faces, vertices, and feature edges next to its boundary vertices.

Our algorithms do require a few communication steps, all of which are reduction operations on the vertices shared by more than one process along partition boundaries. These include the summation operations in the construction of A and b for the medial quadric and in the numerator and denominator in Eq. (10) for vertex redistribution. In addition, classification of ridge edges requires reduction to the maximum and minimum values of s_σ for shared vertices. To broadcast the displacements of each shared vertex in its containing PN triangle, we first zero out the displacements for shared vertices without a local containing triangle on each process, and then reduce to the values of the largest magnitude for each component.

Fig. 7 shows the scalability results of our straightforward parallel implementation for a fixed problem with a total number of 30768 vertices and 59236 triangles. The experiments were conducted on a Linux cluster with dual 1GHz Pentium III processors per node and Myrinet interconnection, and on a faster Mac cluster with dual 2GHz G5 processors per node and also Myrinet interconnection, both located at the University of Illinois. Our method delivers nearly linear scalability for this modest size problem up to 128 processors, and better scalability was achieved on the Linux cluster due to higher ratio of bandwidth to processing power. Better scalability is expected for larger problems and further optimization of the implementation.

6 Conclusion

We have developed a parallel method for surface mesh smoothing based on a new concept called the medial quadric. This quadric facilitates the solution of a number of geometric primitives, including a reliable vertex-based scheme for feature detection, which is easier to parallelize than edge-based schemes, and accurate one-sided normal estimation. These primitives are then used to construct feature-aware higher-order approximation for surface triangulations based on PN triangles. We presented some preliminary but promising experimental results of our surface mesh smoothing algorithm. We are currently conducting more tests, especially for distributed meshes on parallel machines, and integrating our methods into large-scale scientific simulations at the Center for Simulations of Advanced Rockets at University of Illinois [7, 8]. Future directions include more extensive experimental study of our algorithms, detailed comparison against other methods, systematic analysis of normal estimation and feature detection schemes, and extension to estimation of curvatures.

Acknowledgments

Supported by the U.S. Department of Energy through the University of California under subcontract B523819, and in part by NSF and DARPA under CARGO grant #0310446. We thank Prof. John Hart for helpful discussions and references on curved PN triangles,

Prof. Michael Garland for discussions on quadrics, and Prof. Herbert Edelsbrunner for suggestions on enhancing the rigorousness of the paper.

References

1. T. Baker. Identification and preservation of surface features. In *13th Int. Meshing Roundtable*, pages 299–310, 2004.
2. H. Edelsbrunner. *Geometry and Topology for Mesh Generation*. Cambridge University Press, 2001.
3. G. Farin. Smooth interpolation to scattered 3D data. In R. E. Barnhill and W. Boehm, editors, *Surfaces in Computer-Aided Geometric Design*, pages 43–63, 1983.
4. P. J. Frey. *Mesh Generation: Application to finite elements*. Hermes, 2000.
5. H. Gouraud. Continuous shading of curved surfaces. *IEEE Trans. Computers*, 20:623–629, 1971.
6. M. T. Heath. *Scientific Computing: An Introductory Survey*. McGraw–Hill, New York, 2nd edition, 2002.
7. M. T. Heath and W. A. Dick. Virtual prototyping of solid propellant rockets. *Comput. Sci. & Engr.*, 2:21–32, 2000.
8. M. T. Heath and X. Jiao. Parallel simulation of multicomponent systems. In *6th Int. Conf. on High Performance Computing for Computational Science*, Valencia, Spain, 2004.
9. P. S. Heckbert and M. Garland. Optimal triangulation and quadric-based surface simplification. *Comput. Geom.*, pages 49–65, 1999.
10. X. Jiao and M. T. Heath. Feature detection for surface meshes. In *Proc. of 8th Int. Conf. on Numerical Grid Generation in Computational Field Simulations*, pages 705–714, 2002.
11. N. Max. Weights for computing vertex normals from facet normals. *J. Graphics Tools*, 4:1–6, 1999.
12. G. Medioni, M.-S. Lee, and C.-K. Tang. *A computational framework for segmentation and grouping*. Elsevier, 2000.
13. D. Meek and D. Walton. On surface normal and Gaussian curvature approximations of given data sampled from a smooth surface. *Comput. Aid. Geom. Des.*, 17:521–543, 2000.
14. D. L. Page, A. F. Koschan, Y. Sun, J. K. Paik, and M. A. Abidi. Robust crease detection and curvature estimation of piecewise smooth surfaces from triangle mesh approximations using normal voting. In *Proc. Intl. Conf. on Computer Vision*, volume 1, pages 162–167, 2001.
15. S. Petitjean. A survey of methods for recovering quadrics in triangle meshes. *ACM Comput. Surv.*, 34:211–262, 2002.
16. G. Taubin. Estimating the tensor of curvature of a surface from a polyhedral approximation. In *Proc. of Int. Conf. on Computer Vision*, pages 902–907, 1995.
17. G. Thürmer and C. A. Wüthrich. Computing vertex normals from polygonal facets. *J. Graphics Tools*, 3:43–46, 1998.
18. A. Vlachos, J. Peters, C. Boyd, and J. L. Mitchell. Curved PN triangles. In *Proc. of 2001 Symposium on Interactive 3D graphics*, pages 159–166, 2001.

On Multiparametric Sensitivity Analysis in Minimum Cost Network Flow Problem

Sanjeet Singh[a], Pankaj Gupta[b,*], and Davinder Bhatia[c]

[a] Scientific Analysis Group, Defence R & D Organization,
Ministry of Defence, Metcalfe House, Delhi–110054, India
[b] Department of Mathematics, Deen Dayal Upadhyaya College
(University of Delhi), Shivaji Marg, Karampura, New Delhi–110015, India
[c] Department of Operational Research, Faculty of Mathematical Sciences,
University of Delhi, Delhi–110007, India

Abstract. In this paper, we study multiparametric sensitivity analysis for minimum cost network flow problem using linear programming approach. We discuss supply/demand, arc capacity and cost sensitivity analysis using the concept of maximum volume region within the tolerance region. An extension of multiparametric sensitivity analysis to multicommodity minimum cost network flow problem is also presented. Numerical examples are given to illustrate the results.

Keywords: Minimum cost network flow; multicommodity minimum cost network flow; multiparametric sensitivity analysis; tolerance approach; parametric analysis.

1 Introduction

The minimum cost flow problem is the most fundamental of all network flow problems defined as: To determine a least cost shipment of a commodity through a network in order to satisfy demands at certain nodes from available supplies at other nodes. This problem has number of familiar applications: the distribution of a product from manufacturing plants to warehouses ; or from warehouses to retailers ; the flow of raw material and intermediate goods through the various machine stations in a production line ; the routing of automobiles through an urban street network ; and the routing of calls through the telephone system besides a number of indirect applications; see Ahuja et. al. [1] for the state of the art.

In practice one would like to know 'how the output of a model varies as a function of variation in the input data and the model parameters' using sensitivity analysis. The purpose of studying sensitivity analysis for minimum cost flow problem is to determine changes in the optimal solution resulting from changes in the data (supply/demand vector or the capacity or the cost of any arc). There are

---

[*] Corresponding Author: Flat No-01, Kamayani Kunj, Plot No-69, I.P.Extension, Delhi-110092, India. Email:pankaj_gupta15@yahoo.com

two different ways of performing sensitivity analysis in network flow problems: (1) using simplex-based methods from linear programming, and (2) using combinatorial methods. Each method has its advantage. For example, although combinatorial methods obtain better worst-case time bounds for performing sensitivity analysis, simplex-based methods might be more efficient in practice; see Ahuja et. al. [1] for the state of the art. Using linear programming approach, Ravi and Wendell [3] applied Tolerance approach [6,7,8] to discuss sensitivity analysis for network linear program. In general, main focus of sensitivity analysis is on simultaneous and independent perturbations of the parameters. Besides this, all the parameters are required to be analyzed at their independent levels of sensitivity. Wang and Huang [5] proposed the concept of maximum volume in the tolerance region for the sensitivity analysis of a multiparametric single objective LPP to allow the parameters to be investigated at their independent levels of sensitivity.

In this paper, we specialize the approach of Wang and Huang [5] to minimum cost network flow problem. We study supply/demand, arc capacity and cost multiparametric sensitivity analysis by investigating each parameter at its independent level of sensitivity. An extension of multiparametric sensitivity analysis to multicommodity minimum cost network flow problem is also presented. Numerical examples are given to illustrate the results developed in the paper.

2   Problem Formulation and Sensitivity Model

Let $G = (N, A)$ be a directed network defined by a set N of n nodes and a set A of m directed arcs. Each arc $(i,j) \in A$ has an associated cost c_{ij} that denotes the cost per unit flow on that arc. It is assumed that flow cost varies linearly with the amount of flow. We also associate with each arc $(i,j) \in A$, a capacity u_{ij} that denotes the maximum amount that can flow on the arc. We associate with each node $i \in N$, an integer $b(i)$, which represent its supply or demand. If $b(i) > 0$, node i is a supply node ; if $b(i) < 0$, node i is a demand node and if $b(i) = 0$, node i is a transshipment node. The decision variables in minimum cost flow problem are arc flows and we represent the flow on an arc $(i,j) \in A$ by x_{ij}. The mathematical model of the minimum cost flow problem can be stated as follows:

(MCF) Minimize $z(x) = \sum_{(i,j) \in A} c_{ij} x_{ij}$

subject to $\sum_{\{j:(i,j) \in A\}} x_{ij} - \sum_{\{j:(j,i) \in A\}} x_{ji} = b(i)$ for all $i \in N$,   (1)

$0 \leq x_{ij} \leq u_{ij}$ for all $(i,j) \in A$.   (2)

It is assumed that the supply/demand at the various nodes satisfy the condition $\sum_{i \in N} b(i) = 0$. Constraints of the form (1) are called mass balance constraints and the constraints of the form (2) are called the bundle constraint.

Definition 2.1 ([1]). A tree T is a spanning tree of a directed network G if T is a spanning subgraph of G.

Definition 2.2 ([1]). For any feasible solution, x, an arc (i,j) is a free arc if $0 < x_{ij} < u_{ij}$ and the arc (i,j) is a restricted arc if $x_{ij} = 0$ or $x_{ij} = u_{ij}$.

Definition 2.3 ([1]). A feasible solution x and an associated spanning tree of the network is a spanning tree solution if every non-tree arc is a restricted arc.

A spanning tree solution partitions the arc set A into three subsets (i) T, the arcs in the spanning tree; (ii) L, the non-tree arcs whose flow is restricted to zero (iii) U, the non-tree arcs whose flow is restricted in value to the arc's flow capacities. We refer to the triple (T, L, U) as a spanning tree structure.

Since the minimum cost flow problem is a linear programming problem, we can use linear programming optimality conditions to characterize optimal solutions to the problem (MCF). As the linear programming problem formulation (MCF) has one bundle constraint for every arc (i,j) of the network and one mass balance constraint for each node, the dual linear program has two types of dual variables: a price w_{ij} on each arc (i,j) and a node potential $\pi(i)$ for each node i. Using these dual variables, we define the reduced cost c_{ij}^{π} of arc (i,j) as follows :

$$c_{ij}^{\pi} = c_{ij} + w_{ij} - \pi(i) + \pi(j).$$

The minimum cost flow optimality conditions for the problem (MCF) are stated as under:

Let $x = (x_{ij})$ be the spanning tree solution associated with the spanning tree structure (T, L, U). A spanning tree structure (T, L, U) is an optimal spanning tree structure if it is feasible and for some choice of node potentials π, the arc reduced costs c_{ij}^{π} satisfy the following conditions:

(a) $c_{ij}^{\pi} = 0$ for all $(i,j) \in T$
(b) $c_{ij}^{\pi} \geq 0$ for all $(i,j) \in L$
(c) $c_{ij}^{\pi} \leq 0$ for all $(i,j) \in U$

The purpose of sensitivity analysis in the problem (MCF) is to determine changes in the optimal solution resulting from changes in the data i.e., changes in the supply/demand vector or the capacity or cost of any arc.

To address multiparametric perturbations in the cost c_{ij} of the arc (i,j), capacity u_{ij} of the arc (i,j) and in the supply/demand $b(i)$ at node i; we consider the following perturbed model of the problem (MCF):

(PMCF) Minimize $z'(x) = \sum_{(i,j) \in A} (c_{ij} + \Delta c_{ij}) x_{ij}$

subject to $\sum_{\{j:(i,j) \in A\}} x_{ij} - \sum_{\{j:(j,i) \in A\}} x_{ji} = b(i) + \Delta b(i)$ for all $i \in N$,

$0 \leq x_{ij} \leq u_{ij} + \Delta u_{ij}$ for all $(i,j) \in A$, where $\Delta c_{ij} = \sum_{h=1}^{H} \alpha_{ijh} t_h$, $\Delta b(i) = \sum_{h=1}^{H} \beta_{ih} t_h$, $\Delta u_{ij} = \sum_{h=1}^{H} \gamma_{ijh} t_h$ are the multi-parametric perturbations defined by the perturbation parameter $t = (t_1, t_2, \ldots, t_H)^T$. Here H is the total number of parameters.

Let S be a general notation for a critical region. In the following propositions, we construct critical regions for simultaneous and independent perturbations with respect to c_{ij}, u_{ij} and $b(i)$.

Proposition 2.1. When cost c_{ij}, arc capacity u_{ij} of a non-tree arc (i,j) and supply/demand $b(i)$ at node i are perturbed simultaneously and independently, the critical region S of the problem (PMCF) is given by

$$S = \left\{ t = (t_1, t_2, \cdots, t_H)^T \Big| \hat{c}_{ij}^\pi + \sum_{h=1}^{H} \alpha_{ijh} t_h \geq 0 \ for \ (i,j) \in L; \right.$$

$$\sum_{i \in N} \sum_{h=1}^{H} \beta_{ih} t_h = 0, \ \sum_{h=1}^{H} \beta_{ih} t_h \leq \delta,$$

$$\left. \sum_{i \in N} \sum_{h=1}^{H} \gamma_{jjh} t_h = 0, \ \sum_{h=1}^{H} \gamma_{ijh} t_h \leq \delta' \right\}.$$

or

$$S = \left\{ t = (t_1, t_2, \cdots, t_H)^T \Big| \hat{c}_{ij}^\pi + \sum_{h=1}^{H} \alpha_{ijh} t_h \leq 0 \ for \ (i,j) \in U; \right.$$

$$\sum_{i \in N} \sum_{h=1}^{H} \beta_{ih} t_h = 0, \ \sum_{h=1}^{H} \beta_{ih} t_h \leq \delta,$$

$$\left. \sum_{i \in N} \sum_{h=1}^{H} \gamma_{jjh} t_h = 0, \ \sum_{h=1}^{H} \gamma_{ijh} t_h \leq \delta' \right\}.$$

Proof. Let $x = (x_{ij})$ denote an optimal solution of the minimum cost flow problem. Let (T, L, U) denote a corresponding spanning tree structure and π be some set of node potentials.

Cost Sensitivity Analysis. In this case, changing the cost of the non-tree arc (i,j) does not change the node potentials of the current spanning tree structure. For the current optimal solution to remain optimal, modified reduced set $\hat{c}_{ij}^\pi$ of the non-tree arc (i,j) must satisfy the optimality condition (b) or (c) whichever is appropriate, i.e.,

$\hat{c}_{ij}^\pi \geq 0 \ for \ (i,j) \in L$

$\Longrightarrow$

$(c_{ij} + \Delta c_{ij}) + w_{ij} - \pi(i) + \pi(j) \geq 0$

$\Longrightarrow$

$\hat{c}_{ij}^\pi + \sum_{h=1}^{H} \alpha_{ijh} t_h \geq 0 \ for \ (i,j) \in L.$

or

$$\hat{c}_{ij}^{\pi} \leq 0 \ for \ (i,j) \in U$$
$$\Longrightarrow$$
$$(c_{ij} + \Delta c_{ij}) + w_{ij} - \pi(i) + \pi(j) \leq 0$$
$$\Longrightarrow$$
$$\hat{c}_{ij}^{\pi} + \sum_{h=1}^{H} \alpha_{ijh} t_h \leq 0 \ for \ (i,j) \in U.$$

Supply/Demand Sensitivity Analysis: Suppose that supply/demand of a node i is changed to $b(i) + \Delta b(i)$ then in order to maintain the feasibility of the problem (MCF), we must satisfy the condition $\sum_{i \in N} b(i) = 0$, therefore supply/demand of another node l must be changed to $b(l) - \Delta b(i)$. The mass balance constraints of the problem (MCF) require that we must transport $\Delta b(i)$ units of flow from node i to node l. Let P be the unique tree path from node i to node l in the optimal spanning tree structure.

Let $\overline{P}$ and $\underline{P}$ respectively, denote the sets of arcs in P that are along and opposite to the direction of the path. The maximum flow change δ_{ij} on an arc $(i,j) \in P$ that preserves the flow bounds is

$$\delta_{ij} = \begin{cases} u_{ij} - x_{ij} & if \quad (i,j) \in \overline{P} \\ x_{ij} & if \quad (i,j) \in \underline{P} \end{cases}$$

Let $\delta = \min\{\delta_{ij} : (i,j) \in P\}$. Therefore, to maintain the feasibility of the current solution, $\Delta b(i)$ must be less that δ i.e., $\sum_{h=1}^{H} \beta_{ih} t_h \leq \delta$. Also the net flow from the node i should sum to zero i.e., $\sum_{i \in N} \sum_{h=1}^{H} \beta_{ih} t_h = 0$.

Arc Capacity Sensitivity Analysis: When the capacity of an arc (i,j) increases by Δu_{ij} units, the current optimal solution remains feasible; to determine whether this solution remains optimal, we check the optimality condition (a). If the non-tree arc (i,j) is at its lower bound, increasing its capacity u_{ij} by Δu_{ij} does not affect the optimality condition for this arc. However, If the non-tree arc (i,j) is at its upper bound and its capacity increases by Δu_{ij} units, this creates an excess of Δu_{ij} units at node j and deficit of Δu_{ij} at node i. To achieve feasibility, we must send Δu_{ij} units from node j to node i. This objective can be achieved using supply/demand sensitivity analysis as follows:

Let P_1 be unique tree path from node j to node i in the optimal spanning tree structure. Let $\overline{P}_1$ and $\underline{P}_1$ respectively denote the sets of arcs in P_1 that are along and opposite to the direction to the path. The maximum flow change δ'_{ij} on an arc $(i,j) \in P_1$ that preserves the flow bounds is

$$\delta'_{ij} = \begin{cases} u_{ij} - x_{ij} & if \quad (i,j) \in \overline{P}_1 \\ x_{ij} & if \quad (i,j) \in \underline{P}_1 \end{cases}$$

Let $\delta' = \min\{\delta'_{ij} : (i,j) \in P_1\}$. Therefore to maintain the feasibility of the current solution, Δu_{ij} must to less than δ' i.e., $\sum_{h=1}^{H} \gamma_{ijh} t_h \leq \delta'$. Also the net flow from the node i and j should sum to zero i.e., $\sum_{i \in N} \sum_{h=1}^{H} \gamma_{ijh} t_h = 0$. ■

Proposition 2.2. When cost c_{ij}, arc capacity u_{ij} of a tree arc (i,j) and supply/demand $b(i)$ at node i are perturbed simultaneously and independently, the critical region S_1 of the problem (PMCF) is given by

$$S_1 = \left\{ t = (t_1, t_2, \cdots, t_H)^T \,\middle|\, c_{ij}^\pi + \sum_{h=1}^{H} \alpha_{ijh} t_h \geq 0 \text{ for } (i,j) \in L \cup [D(j), \underline{D(j)}], \right.$$

$$c_{ij}^\pi + \sum_{h=1}^{H} \alpha_{ijh} t_h \leq 0 \text{ for } (i,j) \in U \cup [D(j), \underline{D(j)}]\,;\, \sum_{i \in N} \sum_{h=1}^{H} \beta_{ih} t_h = 0,\, \sum_{h=1}^{H} \beta_{ijh} t_h \leq \delta,$$

$$\left. \sum_{i \in N} \sum_{h=1}^{H} \gamma_{jjh} t_h = 0,\, \sum_{h=1}^{H} \gamma_{ijh} t_h \leq \delta' \right\}.$$

Proof. Cost Sensitivity Analysis: In this case, changing the cost of arc (i,j) changes some node potentials. If arc (i,j) is an upward pointing arc in the current spanning tree, potentials of all the nodes in $D(i)$ changes by Δc_{ij} and if (i,j) is a downward pointing arc, potentials of all the nodes $D(j)$ changes by $\sum_{h=1}^{H} \alpha_{ijh} t_h$. Now these changes in node potentials modify the reduced costs of those non-tree arcs that belong to the cut $[D(j), \underline{D(j)}]$. Therefore the current spanning tree structure remains optimal if the following conditions are satisfied;

$$c_{ij}^\pi + \sum_{h=1}^{H} \alpha_{ijh} t_h \geq 0 \text{ for } arcs\ (i,j) \in L \cup [D(j), \underline{D(j)}]$$

$$c_{ij}^\pi + \sum_{h=1}^{H} \alpha_{ijh} t_h \leq 0 \text{ for } arcs\ (i,j) \in U \cup [D(j), \underline{D(j)}].$$

Supply/demand and arc capacity sensitivity analysis can be carried out in the same manner as in Proposition 2.1. ■

In multiparametric sensitivity analysis, to investigate the parameters, a maximum volume region (MVR) is defined, which is bounded by symmetrically rectangular parallelepiped. The MVR is characterized by a maximization problem. This approach is a significant improvement over the earlier approaches to sensitivity analysis in the problem (MCF) because it handles the perturbation parameters with greater flexibility by allowing them to be analyzed at their independent levels of sensitivity.

Since the critical region is a polyhedral set, there exists $L = [\ell_{ij}] \in R^{I \times H}$, $d = \{d_i\} \in R^I$, $I, H \in N$, where I and H are the number of constraints and variables of S, respectively, such that $S = \{t = (t_1, t_2, ..., t_H)^T \mid Lt \leq d\}$.

Remark 2.1. It follows from Propostions 2.1 and 2.2 that $t = 0$ belongs to S, and thus we have $d \geq 0$.

Definition 2.4 ([5]). The (MVR) B_S of a polyhedral set $S=\{t=(t_1,t_2,\ldots,t_H)^T \mid Lt \leq d\} = \{t = (t_1, t_2, \ldots, t_H)^T \mid \sum_{j=1}^{H} l_{ij} t_j \leq d_i, i = 1, 2, \ldots, I\}$, where $d_i \geq 0$ for $i = 1, 2, \ldots, I$ and $\sum_{i=1}^{I} |l_{ij}| > 0$ for $j = 1, 2, \ldots, H$, is $B_S = \{t = (t_1, t_2, \ldots, t_H)^T \mid |t_j| \leq k_j^*, j = 1, 2, \ldots, H\}$.

The volume of B_S is $\text{Vol}(B_S) = 2^H k_1^* \cdot k_2^* \cdot \ldots \cdot k_H^*$.

Here $k^* = (k_1^*, k_2^*, \ldots, k_H^*)^T$ is uniquely determined with the following two cases:

(i) If $d_i > 0$ for $i = 1, 2, \ldots, I$, then k^* is the unique optimal solution of the problem (P1), where $|L|$ is obtained by changing the negative elements of matrix L to be positive

$$(P1) \quad \text{Max} \prod k_j$$
$$\text{subject to } |L|k \leq d$$
$$k \geq 0.$$

(ii) If $d_i = 0$ for some i, let $I^\circ = \{i | d_i = 0, i = 1, 2, \ldots, I\} \neq \phi$ and $I^+ = \{i | d_i > 0, i = 1, 2, \ldots, I\}$ then we have
   (a) If $I^+ = \phi$ then $k^* = 0$ is the unique optimal solution
   (b) If $I^+ \neq \phi$ then let $\Omega = \bigcup_{i \in I^\circ} \{j \mid l_{ij} \neq 0, j = 1, 2, \ldots, H\}$ be the index set of focal parameters that appear in some constraints with right-hand-side $d_i = 0$. Then $k_j^* = 0$ for all j belonging to Ω. The others, $k_j^*, j \notin \Omega$, can be uniquely determined as follows: After deleting all variables t_j, $j \in \Omega$ and constraints with right-hand-side $d_i = 0$ from the system of constraints S, let the remaining subsystem be in the form of (3) with $d_i' > 0$ for all index i as below:

$$S' = \{t' = [t_j]^T, j \notin \Omega | L't' \leq d'\} \tag{3}$$

then $k^{*\prime}$ (i.e., $k_j^*, j \notin \Omega$) can be uniquely determined by solving the following problem (P2)

$$(P2) \quad \text{Max} \prod_{j \notin \Omega} k_j$$
$$\text{subject to } |L'|k' \leq d'$$
$$k' \geq 0.$$

Multiparametric sensitivity analysis in the problem (MCF) can now be performed as follows: Obtain the critical region S by considering perturbations in the cost coefficients, arc capacities and supply/demand at a given node. The MVR of the critical region is obtained by solving the problem (P1)/(P2). The problem (P1)/(P2) can be solved by existing techniques such as Dynamic Programming. The detailed algorithm can be found in Wang and Huang [4]. Software GINO [2] can also be used to solve the nonlinear programming problem (P1)/(P2).

Numerical Example. To illustrate the results of multiparametric sensitivity analysis, we consider a minimum cost flow problem with the data given in the figure (a) and its optimal solution in figure (b).

The minimal spanning tree structure for the current optimal flow is given in the following figure.

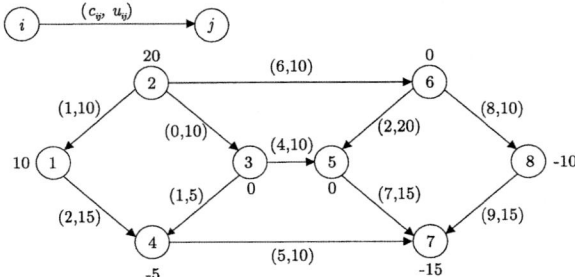

Fig. a.

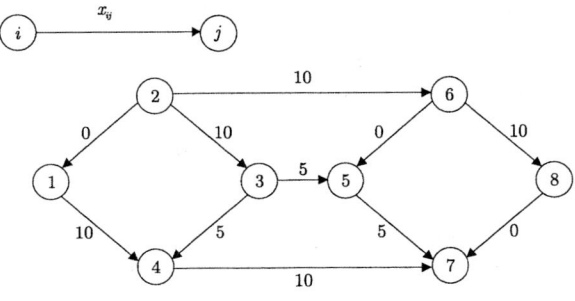

Fig. b.

The following are the node potentials computed from the minimal spanning tree in figure c.
$$\pi = (0, -1, -1, -2, -5, -7, 12, -21)$$

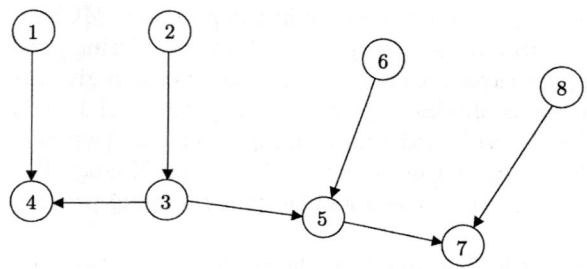

Fig. c.

Using these node potentials we obtain the following reduced costs:

$$c^\pi_{14} = 0 \ \ c^\pi_{21} = 2 \ \ c^\pi_{32} = 0 \ \ c^\pi_{35} = 0$$
$$c^\pi_{41} = 0 \ \ c^\pi_{43} = 0 \ \ c^\pi_{53} = 0 \ \ c^\pi_{56} = 0$$
$$c^\pi_{57} = 0 \ \ c^\pi_{62} = 0 \ \ c^\pi_{47} = 5 \ \ c^\pi_{75} = 0$$
$$c^\pi_{78} = 0 \ \ c^\pi_{86} = 6.$$

Also, $T = \{(1,4),(2,3),(3,4),(3,5),(5,6),(5,7),(7,8)\}$
$L = \{(2,1)\}$, $U = \{(2,6),(4,7),(6,8)\}$

Now we consider the following perturbations:

$\Delta c_{21} = 3t_1 - 4t_2 + 2t_3, \ \Delta c_{47} = 2t_1 + t_2 + 3t_3 + 2t_4$
$\Delta b(1) = 3t_1 + t_2 + 3t_3 + 2t_4, \ \Delta b(4) = -3t_1 - t_2 - 3t_3 - 2t_4$
$\Delta u_{37} = 3t_1 - 4t_2 + t_3$

Thus, $\delta = u_{14} - x_{14} = 5, \ \delta' = 5$.
Therefore critical region for the problem PMCF is given by

$S = \{t = (t_1,t_2,t_3,t_4)^T | 2 + 3t_1 - 4t_2 + 2t_3 \geq 0, \ 5 + 2t_1 + t_2 + 3t_3 + 2t_4 \leq 0, 3t_1 + t_2 + 3t_3 + 2t_4 \leq 5, \ 3t_1 - 4t_2 + t_3 \leq 5\}$.

The MVR (B_S) is obtained by solving the following maximization problem:

$\text{Max} V(k) = k_1.k_2.k_3.k_4,$
$\quad 3k_1 + 4k_2 + 2k_3 \leq 2, \qquad 2k_1 + k_2 + 3k_3 + 2k_4 \leq 5,$
$\quad 3k_1 + k_2 + 3k_3 + 2k_4 \leq 5, \ 3k_1 + 4k_2 + k_3 \leq 5,$
$\quad k_1, k_2, k_3, k_4 \geq 0.$

The optimal solution of the above problem is $k^* = (0.2161, 0.1904, 0.2951, 1.638)$.
MVR $(B_S) = \{t = (t_1,t_2,t_3,t_4)^T ||t_1| \leq 0.2161, |t_2| \leq 0.1904, |t_3| \leq 0.2951, |t_4| \leq 1.638\}$.
$\text{Vol}(B_S) = 2^4(0.2161).(0.1904).(0.2951).(1.638) = 0.3182$.

3 Extension

Multicommodity network flow models provide optimization problems whose solution gives the best routing of a set of k different types of flows (the commodities) through the arcs of a network. This kind of problems arise naturally when modelling applications in, e.g., routing, telecommunications networks and transportation networks.

In this section, we will deal with fairly general formulation, where arcs are considered to have a capacity for each commodity, and a mutual capacity for all the commodities. The general multicomodity flow problem (MMCF) can be formulated as follows:

$$\text{(MMCF)} \quad \text{Min} \sum_{k=1}^{p} c^k x^k$$

$$\text{subject to} \quad A_k x^k = b^k \quad \forall \ k; 1 \leq k \leq p,$$

$$\sum_{k=1}^{p} x_{ij}^k \leq u_{ij} \quad \forall \ (i,j) \in A,$$

$$x_{ij}^k \geq 0, \quad \forall \ k; 1 \leq k \leq p, \ \forall \ (i,j) \in A.$$

Here x_{ij}^k is the flow of commodity k on arc (i,j) with unit cost c_{ij}^k, u_{ij} is the mutual capacity of arc(i,j), A_k is the node-arc incidence matrix for commodity k, b^k is the vector of supplies/demands for commodity i at the nodes of the network. We shall assume that the flow variables x_{ij}^k have no individual flow bounds; that is, each $u_{ij}^k = +\infty$.

Remark 3.1. For a given directed graph with m number of nodes, n number of arcs and k number of commodities, the multicommodity flow problem is a linear programming problem with $km + n$ constraints and $(k+1)n$ variables. In some real-world models, k can be very large e.g., $k = n^2$. For instance, in many telecommunication problems we have a commodity for the flow of data/voice to be sent between each pair of nodes of the network. Thus, the resulting linear programming problem can be huge even for graphs of moderate size.

Let $\pi^k(i)$, the node potential for commodity k on node i, and w_{ij}, which is the arc price, be the dual variables corresponding to the arc $(i,j) \in A, 1 \leq k \leq p$. Using these dual variables, we define the reduced cost as follows:

$$\bar{c}_{ij}^{\pi,k} = c_{ij}^k + w_{ij} - \pi^k(i) + \pi^k(j), \quad \forall \ (i,j) \in A, \ 1 \leq k \leq p.$$

The complementary slackness optimality conditions for the problem (MMCF) are stated as under:

The commodity flows x_{ij}^k are optimal in problem (MMCF) with each $u_{ij}^k = \infty$ if and only if they are feasible and for some choice of (nonnegative) arc prices w_{ij} and node potentials $\pi^k(i)$ (unrestricted in sign), the reduced costs and arc flows satisfy the following conditions:

(a) $w_{ij}\left(\sum_{k=1}^{p} x_{ij}^k - u_{ij}\right) = 0$, $\forall\ (i,j) \in A$.

(b) $(c_{ij}^k + w_{ij} - \pi^k(i) + \pi^k(j))x_{ij}^k = 0$, $\forall\ (i,j) \in A$.

(c) $\bar{c}_{ij}^{\pi,k} \geq 0$, $\forall\ (i,j) \in A$.

To address perturbations in the cost vector c^k, we consider the following perturbed model of the problem (MMCF):

$$\text{(PMMCF) Minimize } \sum_{k=1}^{p}(c^k + \Delta c^k)x^k$$

$$\text{subject to } \sum_{k=1}^{p} x_{ij}^k \leq u_{ij}\ \forall\ (i,j) \in A,$$

$$A_k x^k = b^k\ \forall\ k, 1 \leq k \leq p,$$

$$x_{ij} \geq 0\ \forall\ k, 1 \leq k \leq p,\ \forall\ (i,j) \in A,$$

where $\Delta c_{ij}^k = \sum_{h=1}^{H} \alpha_{ijh} t_h$ are the multiparametric perturbations defined by the perturbation parameter $t = (t_1, t_2, \ldots, t_H)^T$. Here H is the total number of parameters.

Proposition 3.1. When the cost vector c^k is perturbed simultaneously and independently, the critical region for the problem (PMMCF) is given by

$$S = \left\{t = (t_1, t_2, \ldots, t_H)^T \mid \bar{c}_{ij}^k + \sum_{h=1}^{H} \alpha_{ijh} t_h \geq 0, \forall\ (i,j) \in A, 1 \leq k \leq \gamma\right\}.$$

Remark 3.2. To consider multiparametric sensitivity analysis corresponding to perturbations in the arc capacities and supply/demand vector in the problem (MMCF), we can consider the dual problem to the (MMCF) and apply the results of cost sensitivity analysis.

Numerical Example. To illustrate the results, we consider the following multi-commodity minimum cost flow (MMCF) problem, corresponding to the network in Figure d.

The optimal solution of the above problem (MMCF) is given by

$$x_{32}^2 = x_{25}^2 = x_{56}^2 = 2, x_{12}^1 = x_{25}^1 = x_{54}^1 = 3$$
$$x_{14}^1 = 2, \bar{C}^1 = [8, 10, 6, 2], \bar{C}^2 = [2, 6, 6, 8]$$

Consider the following perturbations in cost vector c^K, $1 \leq k \leq 2$ and in the arc capacity u_{25}:

$$\Delta c^1 = [2t_1 - t_2 + 3t_3, 0, t_1 + t_2 - 2t_3, 0]$$
$$\Delta c^2 = [0, 4t_1, 2t_2 + 3t_3, 3t_1 - t_2 - t_3, 0]$$

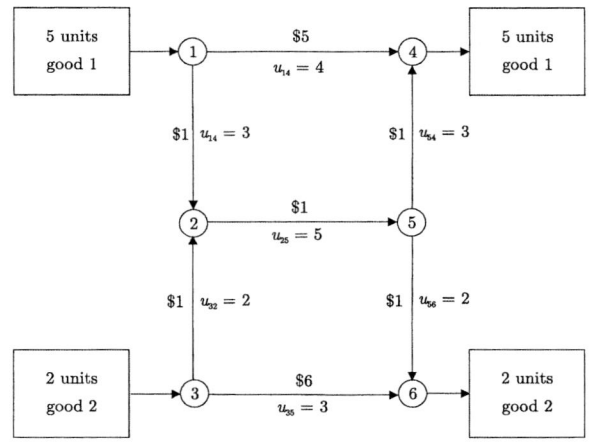

Fig. d. Multicomodity minimum cost flow network

Now, we construct critical region S, where the index set of optimal basic variables remains optimal.

$$S = \{t = (t_1, t_2, t_3)^T \mid 8 + 2t_1 - t_2 + 3t_3 \geq 0, 6 + t_1 t_2 - 2t_3 \geq 0,$$
$$6 + 4t_1 + 2t_2 + 3t_3 \geq 0, 6 + 3t_1 - t_2 - t_3 \geq 0\}.$$

The MVR of S is obtained by solving the following maximization problem:

$$\text{Max } V(k) = k_1 \cdot k_2 \cdot k_3$$
$$\text{subject to } 2k_1 + k_2 + 3k_3 \leq 8, k_1 + k_2 + 2k_3 \leq 6$$
$$4k_1 + 2k_2 + 3k_3 \leq 6, 3k_1 + k_2 + k_3 \leq 6$$
$$k_1, k_2, k_3 \geq 0.$$

The optimal solution of the above problem is $k^* = (0.5, 1.0, 0.667)$.

MVR $(B_S) = \{t = (t_1, t_2, t_3)^T \mid |t_1| \leq 0.5, |t_2| \leq 1.0, |t_3| \leq 0.667\}$.
Vol$(B_S) = 2^3 (0.5).(1.0).(0.667) = 2.668$.

Acknowledgements. The authors wish to express their deep gratitude to Professor R.N. Kaul (Retd.), Department of Mathematics, University of Delhi, Delhi and Professor M.C. Puri, Department of Mathematics, I.I.T., Delhi for their inspiration throughout the preparation of this paper.

References

[1] Ahuja, R.K., Magnanti, T.L. and Orlin, J.B.: *Network Flows: Theory, Algorithms and Applications*, Prentice Hall, Englewood Cliffs, New Jersey (1980).

[2] Liebman, J., Ladson, L., Scharge, L. and Waren, A.: *Modeling and optimization with GINO*, The Scientific Press, San Francisco, CA (1986).
[3] Ravi, N. and Wendell, R.E.: *The tolerance approach to sensitivity analysis in network linear programming*, Networks **18** (1988) 159–171.
[4] Wang, H.F. and Huang, C.S. : *The maximal tolerance analysis on the constraint matrix in linear programming*, Journal of the Chinese Institute of Engineers **15(5)**(1992) 507–517.
[5] Wang, H.F. and Huang, C.S.: *Multi-parametric analysis of the maximum tolerance in a linear programming problem*, European Journal of Operational Research **67** (1993) 75–87.
[6] Wendell, R.E.: *A preview of a tolerance approach to sensitivity analysis in linear programming*, Discrete Mathematics **38** (1982) 121–124.
[7] Wendell, R.E.: *Using bounds on the data in linear programming: The tolerance approach to sensitivity analysis*, Mathematical Programming **29**(1984) 304–322.
[8] Wendell, R.E.: *The tolerance approach to sensitivity analysis in linear programming*, Management Science **31** (1985) 564–578.

Mining Patterns of Mobile Users Through Mobile Devices and the Musics They Listen

John Goh and David Taniar

School of Business Systems,
Monash University Vic 3800 Australia
{Jen.Goh, David.Taniar}@infotech.monash.edu

Abstract. Mobile data mining [8-11] is about the analysis of data generated by mobile activities, in search for useful patterns in order to support different types of decision making requirement. Mobile devices are loaded with features such as the capability to listen to radio from a mobile phone. Mobile users who listen to radios on their mobile phones are a source of data generated from mobile activities. The location dependent data [9] and the song they listen to can be combined and analysed in order to better understand the behaviour of mobile users. This paper shows how this can be done by using taste template, which categorises a behaivoural type in order to match mobile users into one of these categories. Conclusion from this research project confirms a new way to learning behaviour of mobile users.

1 Introduction

Mobile data mining [8-11, 18] is an emerging field of research that focuses on analysing the data generated by mobile users. The data generated by mobile users include: communication history among mobile users, physical location visit history of mobile users, and the activities done by mobile users using their mobile devices, such as surfing a particular type of website. This research project develops two ways to analyze the listening behavior of mobile users through their radio or mp3 players on their mobile phone in order to predict their psychological nature for the purpose of supporting other activities that requires psychological knowledge of the mobile users. One such example will be the sending of relevant and useful marketing materials in contrast to irrelevant and annoying marketing materials may be viewed as spam.

As mobile devices are getting more and more popular with the penetration rate of mobile devices in many developing and developed countries increasing every year [1], many data are generated from these mobile activities. From these mobile activities, comes an interesting opportunity to search for useful knowledge by means of analysis of these data. The analysis of these data have the potential to support different decision making requirements from different fields, especially marketing, whereby mobile data mining enables the marketers to better understand their potential consumers.

As mobile users changes channel or changes the musis he or she is listening to, it can be due to personal preference such as whether the song is suitable for what he or she is currently doing. The change of channel is the source of selection, whereby the change of channel tells the data mining machine whether the mobile user likes or dislikes the particular type of song. By implementing a way to separate the likes and dislikes and match the mobile user into a particular group of behaviour, it allows better prediction of their preference over a certain product to another.

Mobile devices these days are incorporated with multiple functions. One trend that is getting popular is the incorporation of radio function for listening to public radios. Some of the mobile phones have even television function incorporated. Another trend of the mobile phone technology is the incorporation of functionalities from personal digital assistant (PDA) in order for the device to work as both mobile phone and personal digital assistant at the same time. The functionalities incorporated from personal digital assistant includes the date planning functions, the ability to surf the internet with WiFi speed, the ability to incorporate large amount of software and functionalities such as encyclopedia and global positioning system by using secure digital (SD) cards easily into the card slots.

2 Background

Music is defined as "a brief composition adapted for singing" in dictionary.com. Music has existed from many thousands of years since ancient times, and all natives seem to have developed their own style of music. Music is a way for human being to express their emotions, from happy ones to sad ones. There are music created in the history of mankind for all kinds of ocassions: celebration of a nation (national anthemn), expressing a happy moment (love songs), expressing a sad moment (sad love songs), and even teaching kids how to learn a language (the famous ABC song).

It can be established that music is an important part of all human being's life [19]. With the technology of radio receiving capability installed in newer versions of mobile phones, mobile users can listen to different songs when they are on their move. The current technology allows the access to current location of mobile user and the music that they are currently listening to.

Music can be separated into different genres [19], such as ancient music, band music, blues, children's music, choral music, country music, dance music, electronic music, film music, folk music, jazz, opera, orchestral music, rap / hip hop, reggae / ska, religious music, rock / pop / alternative, seasonal music, wedding music [19]. In each different genre, the words in the music express the emotions and the message of the composer. A person who likes to listen to the music indicates a degree of empathy to the subject of the music, that is, the emotions conveyed. Music can also be separated technically by means of their rhtym, and loudness, which can be translated into engineering terms of frequency of sound and amplitude.

Mobility has the advantage in the data mining field [8]. Based on the fact that mobile devices follows the owner everywhere they go [9], and the owner uses the mobile device to interact with many other contacts, there are two major sources of dataset which are volumous. First, the source dataset that describes the physical locations which the mobile device has been through. Second, the source dataset that describes the list of contacts that the mobile device have contacted [9]. The reason that these source datasets are volumous is that activities are performed at every minute and datasets that describes these activities are continuously formed. Therefore, when dealing with source datasets drawn from mobile environment, care must be taken to ensure either a proper strategy is in place to manage the volumous data or that the dataset is reduced to a degree that is less volumous but still acceptably accurate for mobile data mining purposes.

Data mining is a wide research field with the main objective of finding patterns from a source of raw dataset, so that knowledge or prepositions can be verified. In the area of analyzing the market basket that are the transactions generated by a customer checking out from a grocery shop, association rules [2, 14, 15, 17] and sequential patterns [3] has been developed. Source data comes in different sizes, such as time series [4, 12, 13], geographical [5, 15] based datasets. When it comes to mobility, another related field is location dependency [6, 16] which concerns the context in a particular location itself. Data mining further extends itself to mine web log datasets [7].

3 Proposed Method: Mobile Data Mining by Music Analysis

The objective of music analysis in mobile data mining is to use the listening of music activities from mobile users in order to find out which types of songs the mobile users are favoured to, and which ones are not favoured. This can be done by analysing how the mobile users' changes their channel or music and the durations the mobile users spent on different pieces of songs. Longer duration means that the mobile users have empathy to a particular song, and lesser duration indicates that the mobile users dislikes the song, or is searching for more interesting songs.

The process of mobile data mining by music analysis can be separated into the following steps:

1. Categorise musics into different personalities.
2. Identify the songs that the mobile users like and dislikes.

Step 1: Categorise Musics into Different Personalities

In order to categorise musics into different personalities, it can be done by using qualitative or quantitative methods. Qualitative method is to use a panel of human beings from different backgrounds, and match different songs into a set of predefined personalities, such as idealist, traditionalist, rationalist, and hedonist. The figure below shows a sample type of personality chart.

Table 1. A Personality Chart

Idealist	Traditionalist
Conscientious	Leisurely
Sensitive	Serious
Vigilant	Self-Sacrificing
Dramatic	Devoted
Rationalist	**Hedonist**
Aggressive	Self-Confident
Idiosyncratic	Adventurous
Inventive	Mercurial
Solitary	Artistic

In order to speed the process up, the song can be passed into a system that analyses it's frequency and amplitude that represents whether the song is a fast song or the song is an active song such as rock songs that is louder than other songs, in order to determine the type of song and place them into different personality types. This process, however have inaccuracies traded off with performance. The algorithm for classifying is shown below. The program code is predefined with the following personality definitions: *idealist.frequency* = 100, idealist.amplitude = 100, traditionalist.frequency = 200, traditionalist.amplitude = 200, rationalist.frequency = 300, rationalist.amplotide = 300, hedonist.frequency = 400, hedonist.amplotude = 400.

Fig. 1. Algorithm to Classify Musics

```
Function Classify Music (Music) {
   Music Info = Music.Play();
   Case Info.Frequency, Info.Amplitude {
      Frequency <= 100 && Amplitude <=100 Then
         Music.Personality = Idealist;
      Frequency <= 200 && Amplitude <= 200 Then
         Music.Personality = Traditionalist;
      Frequency <= 300 && Amplitude <= 300 Then
Music.Personality = Rationalist;
      Frequency <= 400 && Amplitude <= 400 Then
         Music.Personality = Hedonist;
   }
   Return Music.Personality;
}
```

Step 2: Identify Songs Mobile Users Likes and Dislikes

In this step, both the songs mobile users' likes and dislikes are important to be identified. The likes builds stronger confidence into a particular personality, and the dislikes removes the inaccurate confidence into a particular personality by reducing the proabability of the mobile user being matched to the inaccurate personality type. In a music listening process, a mobile user can potentially act the following:

1. Listen to the whole song.
2. Skip the song because the mobile user dislikes the song.
3. Skip the song because the mobile user wishes to check out other songs.
4. Bored with the song half way through listening.
5. First time listening to the song.

Only when the mobile user have listened the majority part of the song, only then the mobile user is considered showing empathy to the song. This can be handled by using the following variables, namely *song_start, song_end, song_duration, listen_start, listen_stop, listen_duration*. The following table describes the variables.

Table 2. Variables to Determine Degree of Empathy

Variable Name	Meaning
song_start	The time which the song starts to play. It is time 0.
song_end	The time which the song ends completely. It is the duration of the song, equals to song_duration.
song_duration	The total duration of the song from start to end. Calculated by song_end − song_start = song_end − 0 = song_end.
listen_start	It is the time which the mobile user starts the listening process. If listening starts from the beginning, listen_start = 0. If listening starts half way through, say time 10, then listen_start = 10.
listen_stop	It is the time which the mobile user stops listening the song, and can be a full stop or a switch to another song. If the whole song is listened, then listen_stop = song_duration. If the listening stops 5 seconds after the song starts, then listen_stop = 5.
listen_duration	It is the total duration which the mobile user have spent listening to a particular song. It is calculated by listen_stop − listen_start.

In order to ensure that only the songs that mobile users have shown a sufficient degree of empathy are selected as "likes", separating itself from the "dislikes" an algorithm is required. The algorithm below shows filters out the "likes" and "dislikes" by using the above variables, which looks at the degree of empathy shown by a mobile user to a particular song.

```
Likes_Or_Dislikes (Music) {
  Return Likes If {
    Music.listen_duration >= 0.8(Music.song_duration);
    Music.listen_stop <= Music.song_duration - 5;
  }
  Return Dislikes If {
    Music.listen_duration <= 0.8(Music.song_duration);
    Music.listen_stop <= Music.song_start + 5;
  }
}
```

Fig. 2. Algorithm to Measure Degree of Empathy

After the determination of degree of empathy a mobile user has on a particular song by the above algorithm, it is recorded and stored to a repository, such as a database or a data warehouse. The repository will contain results in the following manner.

```
User {1, 2, 3, …, n}
  Likes:
    Song {1, 2, 3, … , n};
  Dislikes:
    Song {1, 2, 3, …, n};
```

Fig. 3. Format of Results in Repository

The personality match repository format is shown in the following diagram, and is matched against the result repository from above diagram in order to predict the personality of the mobile users.

```
Personality.Idealist:
  Song {1, 2, 3, …, n};
Personality.Traditionalist:
  Song {1, 2, 3, …, n};
Personality.Rationalist:
  Song {1, 2, 3, …, n};
Personality.Hedonist:
  Song {1, 2, 3, …, n};
Personality.Unknown_To_Be_Classified:
  Song {1, 2, 3, …, n};
```

Fig. 4. Format of Personality Definition in Repository

The end result of this mobile data mining method can be summarised in the following diagram. The songs that mobile users have listened to are checked with the

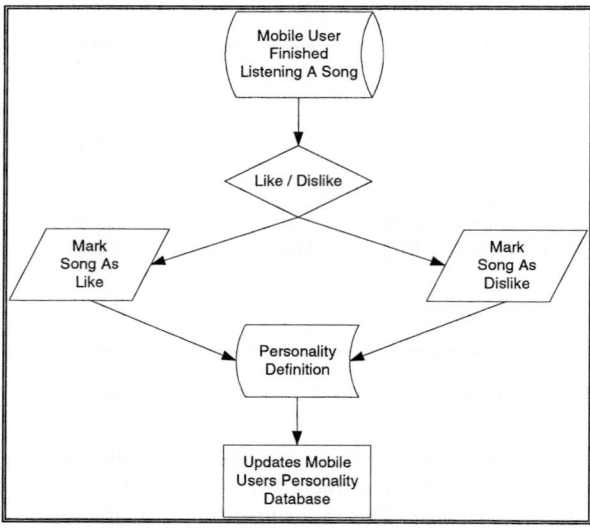

Fig. 5. Summary for Mobile Data Mining by Music Analysis

degree of empathy, and selected as likes or dislikes. The personality and song matching is done before hand by using either qualitative measures by human being or quantitative measurs by measuring the frequency and amplitude. The final outcome is to match every single mobile user into one of the personality category.

4 Performance Evaluation

Performance evaluation is done on Pentium IV machine with 256MB of RAM and 10GB of hard disk space. Different songs from different genre will be gathered in appropriate format and analysed using Matlab. The objective of performance evaluation is to compare the qualitative definition of song personality done by human being and the quantitative definition of song personality done by computers using frequency and amplitude. Four songs are selected and their personality is predetermined by using a personal that is neutral to this project. The songs are then fed into Matlab for frequency and amplitude analysis and determine their personality matching as if they were fed for technical personality selection. The chart below shows the result whereby samples are taken from each song and their amplitude are graphed in order to show both the amplitude and frequency. The chart illustrates that each song have different level of amplitude and frequency.

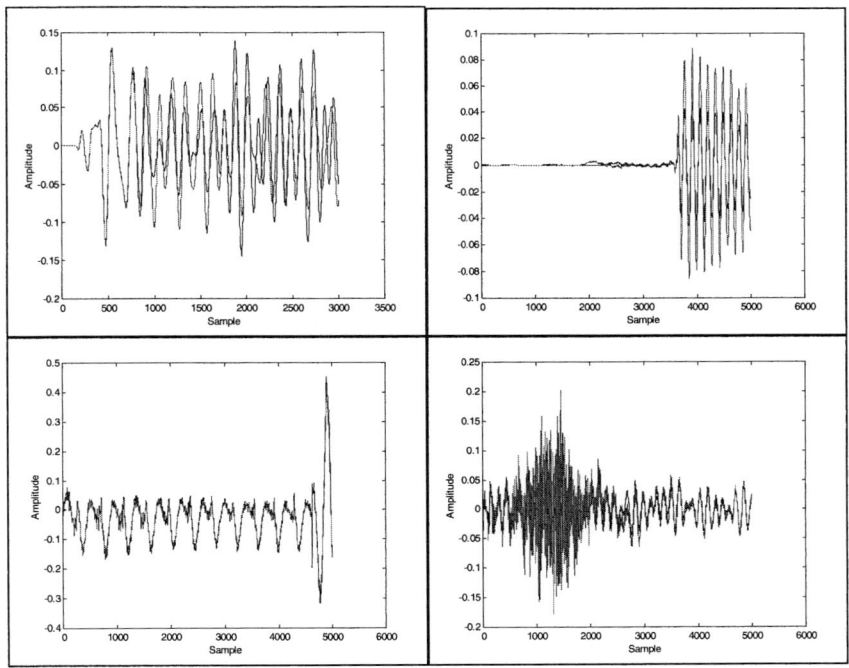

Fig. 6. Amplitude Analysis of Song 1, 2, 3 and 4

Table 3. Analysis of Performance Charts

Song	Amplitude	Frequency	Technical Result	Human Result	Match? (YES/NO)
Song 1	High	Medium	H	I	NO
Song 2	Medium	Medium	T	T	NO
Song 3	Low	Low	I	R	NO
Song 4	Medium	High	R	H	NO

(I = Idealist, T = Traditionalist, R = Rationalist, H = Hedonist)

The analysis above tells whether does technical result is good enough compared to human result. After the comparison between the result generated by the technical analysis and the result generated by the human being who actually listened to the song, the results from computer and human appear to differ completely. There is no single matching of result. This can only be caused by two ways: imperfection of the technical algorithm or the personality bias of the human being. Nevertheless, the point is that, the effort for using human result would be more cost effective in viewing of the charting of song titles, and the ability for users to vote their favourate songs. Therefore, although technical analysis can be a possibility, it is more realistic to use human analysis.

5 Conclusion and Future Work

This project has confirmed that music analysis approach can be used for mobile data mining. In addition to this, it has also been found that using algorithms in order to determine the personality type of a person who listens to a particular song based on technical parameter will not work successfully in real life implementation due to the imperfection of the algorithms. Things such as feelings are unable to be identified accurately and parameterised by using numbers and symbols. Therefore, it is essential the using a wide base of users to determine a song's personality type when performing real life implementation of such systems. Future works is to focus on issues such as privacy of listening behaviour data, and better clasiffication methods to classify different songs into different personality sets.

References

1. Bloomberg.com.2004-11-03. Nokia's Ollila Says Global Mobile-Phone Users Reach 1.7 Billion.
   http:// quote.bloomberg.com/apps/news?pid=71000001&refer=europe&sid=aatIQhKrl1, 2004
2. R. Agrawal and R. Srikant. Fast Algorithms for Mining Association Rules. *In Proc. 20th Int. Conf. Very Large Data Bases,* pp. 487-499, 1994.
3. R. Agrawal and R. Srikant. Mining Sequential Patterns. *In Proc. 11th Int. Conf. on Data Engineering,* pp. 3-14, 1995.

4. D. Barbar´a, P. Chen, and Z. Nazeri. Self-Similar Mining of Time Association Rules. *PAKDD 2004,* pp. 86-95, 2004.
5. I. Claude, J.-L. Daire, and G. Sebag. Fetal Brain MRI: Segmentation and Biometric Analysis of the Posterior Fossa. *IEEE Transactions on Biomedical Engineering*, vol. 51, no. 4, pp. 617-626, 2004.
6. P. Dourish. What We Talk About When We Talk About Context. *ACM Journal of Pervasive and Ubiquitous Computing*, vol. 8, no. 1, pp. 19-30, 2004.
7. M. Eirinaki and M. Vazirgaiannis. Web Mining for Web Personalization. *ACM Transactions on Internet Technology*, vol. 3, no. 1, pp. 1-27, 2003.
8. J. Goh and D. Taniar. Mining Frequency Pattern from Mobile Users. In Proc. *Knowledge-Based Intelligent Information & Eng. Sys.*, 2004. (To Appear)
9. J. Goh and D. Taniar. Mobile Data Mining by Location Dependencies. In Proc. *5th Int. Conf. on Intelligent Data Engineering and Automated Learning*, 2004. (To Appear)
10. J. Goh and D. Taniar. Mining Physical Parallel Pattern from Mobile Users. In Proc. *Int. Conf. on Embedded and Ubiquitous Computing*, 2004. (To Appear)
11. J. Goh and D. Taniar. An Efficient Mobile Data Mining Model. In Proc., 2004. (Submitted)
12. J. Han, G. Dong, and Y. Yin. Efficient Mining of Partial Periodic Patterns in Time Series Database. *In Proc. of Int. Conf. on Data Engineering,* pp. 106-115, 1999.
13. J. Han, W. Gong, and Y. Yin. Mining Segment-Wise Periodic Patterns in Time Related Databases. *In Proc. 4th Int. Conf. on Knowledge Discovery and Data Mining*, vol. no. pp. 214-218, 1998.
14. J. Han, J. Pei, and Y. Yin. Mining Frequent Patterns without Candidate Generation. *In Proc. Int. Conf. SIGMOD,* pp. 1-12, 2000.
15. K. Koperski and J. Han. Discovery of Spatial Association Rules in Geographical Information Databases. *4th Int. Symp. on Advances in Spatial Databases,* pp. 47-66, 1995.
16. D. L. Lee, J. Xu, B. Zheng, and W.-C. Lee. Data management in location-dependent information services. *Pervasive Computing, IEEE*, vol. 1, no. 3, pp. 65-72, 2002.
17. J. Li, B. Tang, and N. Cercone. Applying Association Rules for Interesting Recommendations Using Rule Templates. *PAKDD 2004,* pp. 166-170, 2004.
18. E.-P. Lim, Y. Wang, K.-L. Ong, and et al. In Search of Knowledge About Mobile Users. *ERCIM News*, vol. 1, no. 54, pp. 10, 2003.
19. J. C. Mowen. The 3M model of motivation and personality: theory and empirical applications to consumer behavior. Kluwer Academic 2000.

Scheduling the Interactions of Multiple Parallel Jobs and Sequential Jobs on a Non-dedicated Cluster

Adel Ben Mnaouer

School of Computer Engineering,
Center of MultiMedia and Network Technology,
Nanyang Technological University,
Block N4, #02a-32, Nanyang Avenue, Singapore 639798
Tel: (65) 6790-6253 Fax: (65) 6792-6559
adelm@pmail.ntu.edu.sg

Abstract. This paper presents a performance evaluation of the interactions between local sequential processes running on behalf of interactive applications and parallel processes running as part of parallel applications on a nondedicated distributed computing environment. To control the interactions between the two types of processes, we propose to constrain the scheduling of local interactive processes by a measure of the maximum response time (MRT) expected by the workstation (WS) user. We propose a mathematical model of the scheduling problem based on the usage of the MRT measure. In addition, we provide a scheduling scheme that within the MRT cycle computes the time quanta needed to satisfy the requirements of both local interactive processes and the parallel task processes present in the system. Analytical and simulation results have shown the effectiveness of the proposed scheduling scheme in allowing the parallel tasks to ensure a minimum speedup even in heavy load situations and to maximize the speedup adaptively depending on the load conditions.

1 Introduction

A cluster of processing units can be either dedicated or non-dedicated, with regards to the execution of parallel application. Dedicated clusters of processing units are useful for running parallel applications since the computing power is all available for them without being slowed down by any sequential processes (except OS processes) [4]. However, dedicated clusters, may get underused if there aren't enough parallel applications that will keep such systems busy.

In a non-dedicated cluster environment, the processing unit (e.g., a workstation) may belong to particular owners, who have agreed to share the computing resources of their workstations. Parallel applications running on such systems may get their parallel tasks allocated to specific workstations (WSs) and subjected to contentions (for the CPU and other resources) with the local owners' Interactive Processes (IPs).

A path of research in cluster computing has focused on finding means to exploit the idle CPU cycles in such nondedicated environments to run parallel programs. The main objective was on exploiting workstations' idle cycles for executing very large parallel jobs in presence of workstation owner's interactive jobs while minimizing the side effects of different workload interactions [1], [3], [7], [9].

The main consideration in this type of research is whether owner processes are given equal or higher precedence over parallel tasks running on the owner's WS.

Our major objective in this research, is to provide insights into how lightly loaded nondedicated homogeneous clusters can be utilized to provide a sort of free environment for executing large parallel jobs, while preserving the response time of interactive processes.

Thus, we propose a mathematical model for the problem of scheduling local sequential processes and parallel processes based on the usage of the The response time is assumed delimited by a given upper bound limit, called the Max-response time (MRT) measure. This bound may be obtained by conducting empirical studies that determine the maximum time a WS user can wait till he/she gets a response back from the system before feeling that the system is unusually slow. We consider a system with multiple parallel jobs executed on the distributed computing environment. Then, we propose a scheduling algorithm that within the MRT cycle computes time quanta to satisfy both local interactive processes and the parallel processes present in the system.

The scheduling algorithm is then modeled using a timed Colored Petri Net (CPN) that is simulated to compute speedup measures for different settings of IP arrival rates, execution times and number of nodes in the cluster.

This work represents an extension of the single parallel task system proposed by the author earlier [6] where analytical and simulation results showed the effectiveness of the proposed scheme.

1.1 Relate Work

Lots of studies on the effective mixing and scheduling of sequential and parallel workloads on a non-dedicate cluster have been reported in the literature.

In [8], the performance of parallel and sequential jobs in a heterogeneous nondedicated environment have been studied. Their model was based on determining the power weight of each workstation in the cluster and preserving a proper portion of it for the parallel task running on the workstation. The performance of parallel jobs is guaranteed whether workstations are lightly or heavily loaded by local jobs.

Lieuteneger and Xian studied job interactions assuming that user jobs have priority over parallel jobs [12]. One single task of a Parallel job was assumed running and it is granted access to the CPU to finish one unit of work whenever an owner process has finished, and it is preempted when a new owner process is started. The authors showed that the high variance of the workstation's owner processes and a small ratio of the parallel job to the interactive job demand significantly degrades the performance of the parallel job. This approach avoids process migration but does not guarantee a good performance of parallel jobs.

Our approach is more conservative than the above ones with regards to the response time of interactive processes, while ensuring dynamic and adaptive harvesting of available CPU cycles depending on the presence and number of IPs in the system. In addition, our approach ensures a proportional partitioning of the available CPU cycles among the parallel tasks according to their priorities.

The remainder of this paper is organized as follows: section II gives a formulation of the problem. Section III describes the analytical model. Section IV summarizes the analytical and simulation environments and main results. Conclusions and future extensions are given section V.

2 Problem Formulation

Our target is to design a new processor allocation scheme to achieve a reasonable speedup for parallel jobs, while preserving a maximum bound of response time for the interactive sequential processes, subject to the number of IPs in the system.

In order to achieve a good performance for both types of jobs, an efficient scheduling technique is needed to handle the interactions between them. We propose a local scheduling technique where each workstation schedules its processes independently.

Due to the interactive nature of the workstations owners' jobs, the response time of those jobs should be minimized. Thus, our scheme is based on using the MRT measure (explained above) as the CPU scheduling cycle. This cycle is divided into two major quanta. One reserved for the available interactive processes and the remaining is allocated to the parallel tasks (that are running as part of several tasks of the parallel jobs executed on the cluster). We make following assumptions:

- There are m homogeneous workstations.
- Each parallel job can be partitioned in upto m tasks.
- Parallel jobs' tasks are perfectly balanced and there is no communication / synchronization between the tasks except in the final phase.
- random number of interactive processes is generated in each workstation.
- Workstation owners are in continuous cycle of thinking and using the workstations.

3 The Analytical Model

The analytical model used to describe the scheduling of both types of jobs is based on partitioning the CPU cycle into two quanta q_1 and q_2, where q_1 is allocated to the interactive Processes and q_2 is divided proportionally among all parallel tasks present on the WS based on their respective priorities.

3.1 The Model

To ensure an acceptable, minimal performance for parallel jobs the number of IPs per cycle should be limited to:

$$0 \leq N_{ip}^i \leq \lfloor(\tau - (q_{pp} * \sum_{j=1}^{NPJ} Pr^j))/q_{min}\rfloor \qquad (1)$$

Thus, the maximum number of interactive processes per cycle should leave at least one unit of work q_{pp} for the execution of each parallel task at each cycle. This implies that Max_{ip}^i in a cycle i should be set as:

$$Max_{ip}^i = \lfloor(\tau - (q_{pp} * \sum_{j=1}^{NPJ} Pr^j))/q_{min}\rfloor \qquad (2)$$

Hence, the quantum reserved for a parallel tasks q_2 is bounded by $\sum_{j=1}^{NPJ} Pr^j * q_{pp}$ and τ. It is equal to $\sum_{j=1}^{NPJ} Pr^j * q_{pp}$ in a heavily loaded cycle when only one unit of each parallel task is executed. It is equal to τ when there are no IPs in the cycle ($N_{ip}^i = 0$), that is:

$$\sum_{j=1}^{NPJ} Pr^j * q_{pp} \leq q_2 \leq \tau \qquad (3)$$

If the above condition is not met, that is, if the number of IPs becomes big, then the proposed scheme becomes not applicable and migration should be considered. The processing time J_x of any parallel task x allocated to a computing node is equal to the total parallel job execution time demand L_x (where x is in $[1..NPJ]$) divided by the number of homogeneous workstations in the cluster m and is given by:

$$J_x = L_x/m \qquad (4)$$

Note that the above requirement can be easily relaxed, and still have the approach applicable. However, the overall speedup may differ depending on the granularity level considered.

Using the desirable quantum for an interactive process q_{dip} and the number of IPs in a cycle i, N_{ip}^i, the quantum for IPs q_1^i can be computed as

$$q_1^i = q_{dip} * N_{ip}^i \qquad (5)$$

After computing q_1^i, we check whether it allows the parallel tasks to execute their minimum requirement, i.e., $q_{pp} * \sum_{j=1}^{NPJ} Pr^j$. If not, then the quantum reserved for IPs is reset to the minimum quantum q_{min}, as follows:

$$if(q_1^i > (\tau - (q_{pp} * \sum_{j=1}^{NPJ} Pr^j))) \text{ then } q_{ip} = q_{dip}$$
$$\text{else } q_{ip} = q_{min} \quad (6)$$

Here, we assume for simplification purposes that q_{pp} is set in a way that allows any parallel task to execute at least one unit of work.

Then, given the quantum for an interactive process q_{ip} and the number of IPs in a cycle i, N_{ip}^i, the quantum for all IPs q_1^i can be computed as

$$q_1^i = q_{ip} * N_{ip}^i \quad (7)$$

Once the quantum for IPs q_1^i is calculated, the quantum for parallel tasks q_2^i can be computed as

$$q_2^i = \tau - q_1^i \quad (8)$$

And subsequently the quantum of each parallel task x at a cycle i can be obtained using the parallel jobs priorities as

$$q_2^{i,x} = q_2^i * \frac{Pr^x}{\sum_{j=1}^{NPJ} Pr^j} \quad (9)$$

Therefore, the total time needed by a parallel task x during its lifetime is given by

$$\sum_{i=1}^{k} q_2^{i,x} = J_x \quad (10)$$

Where k is the last cycle in which the parallel task completes its execution.

Then, the time required for running a parallel task x up to completion can be approximated as

$$T_p^x = \sum_{i=1}^{k}(q_1^i + q_2^i) = \sum_{i=1}^{k} q_1^i + \sum_{i=1}^{k} q_2^i$$
$$= \sum_{i=1}^{k} q_1^i + J_x + \sum_{i=1}^{k} \sum_{j=1}^{NPJ-1} q_2^{i,j} \quad (11)$$

where $j \neq x$

It follows that, the time needed for running the parallel task x is approximately bounded by:

$$J_x + \sum_{i=1}^{k} \sum_{j=1}^{NPJ-1} q_2^{i,j} + \beta \leq T_p^x \leq \tau * (\frac{J_x}{q_{pp} * \frac{Pr^x}{\sum_{j=1}^{NPJ} Pr^j}}) + \alpha \quad (12)$$

where α is the total context-switching overhead in presence of IPs and β is the one considered when no IPs are present in the system during the lifetime of the parallel task x.

The lower bound is reached when there are no IPs in the system during the lifetime of the parallel task, while the upper bound is reached when only one unit of parallel task q_{pp} is executed per cycle.

Given the mean execution time (MET) and the mean inter-arrival time of IPs (MIIT), the numbers of IPs per cycle can be estimated as:

$$EN_{ip} = \lceil MET/q_{ip} \rceil * \lceil \tau/MIIT \rceil \qquad (13)$$

where $\lceil MET/q_{ip} \rceil$ represents the number of cycles needed by the interactive process, and $\lceil \tau/MIIT \rceil$ represents the IPs arrival rate per cycle τ.

After computing the expected number of IPs, EN_{ip}^i, it is compared to Max_{ip}^i. If it is found to be greater than Max_{ip}^i, it is set equal to Max_{ip}^i, in order to ensure a feasible solution for the parallel task (Alternatively, a migration can be considered at this point), that is:

$$if(EN_{ip}^i > Max_{ip}^i) \ then \ EN_{ip}^i = Max_{ip}^i \qquad (14)$$

The above constraint can be a practical one in case specific contracts of workstation usage should be negotiated through special agreements, with the WS owners like the "social contract" scheme proposed in [2]. For instance the value of τ can also be negotiated if there are specials incentives for the WS owners to tolerate a bigger delay.

Consequently, the value of EN_{ip}^i is used to determine the proper quantum to be allocated to the IPs that would be either q_{min} or q_{dip}, as follows:

$$if(EN_{ip}^i * q_{ip}) > \tau \ then \ q_{ip} = q_{min} \ else \ q_{ip} = q_{dip} \qquad (15)$$

The total number of cycles needed by a parallel task x can be estimated by:

$$NC^x = \lceil \frac{J_x}{[(\tau - q_{ip} * EN_{ip}) * \frac{Pr^x}{\sum_{j=1}^{NPJ} Pr^j}]} \rceil \qquad (16)$$

Where $(\tau - q_{ip} * EN_{ip})$ represents q_2, i.e., the time quantum allocated to the parallel tasks per cycle. Whereas $Pr^x / \sum_{j=1}^{NPJ} Pr^j$ represents a normalization factor needed for allocating the parallel task quantum to the parallel tasks proportionally with respect to their priorities.

In case of a single parallel task x present, equation (14) is reduced to

$$NC = \lceil \frac{J_x}{(\tau - q_{ip} * EN_{ip})} \rceil \qquad (17)$$

The total context-switching overhead during the lifetime of a parallel task x can be computed as:

$$\alpha^x = (EN_{ip} + NPJ - 1) * NC^x * CS \qquad (18)$$

Where $(EN_{ip} + NPJ)$ represents the total number of processes in a cycle, and $(EN_{ip} + NPJ - 1)$ represents the total number of preemptions.

For a single parallel job x running on the cluster, it becomes:

$$\alpha = EN_{ip} * NC * CS \qquad (19)$$

After computing the number of cycles NC^x needed by a parallel task x and the total context-switching overhead α^x, the total time needed for running the parallel task in the cluster, T_p^x, can be computed by multiplying NC^x by the cycle size τ and adding the total context-switching overhead as:

$$T_p^x = \tau * NC^x + \alpha^x \qquad (20)$$

Finally the speedup, σ can be computed by taking the ratio of the total time needed to execute the parallel job on a single dedicated sequential workstation to the time needed to execute the same job on the cluster as:

$$\sigma^x = T_s^x / T_p^x \qquad (21)$$

We propose next, a procedure $findSpeedup$ that allows to find the speedup of a particular parallel job when more than one job is running in the cluster.

Procedure: $findSpeedup$

1. Initialize total number of cycles executed (CE) to zero.
2. Compute the number of cycles of each parallel job currently in the cluster using (13), (15) and (16).
3. Select the job that requires the minimum number of cycles (assume job A with cycles of Y)
4. Using this number of cycles Y, estimate the execution time of the parallel job (ET) using equations (16), (18) and (20).
5. Get the remaining time for each parallel job as $RT = J - ET$.
6. Update the execution times of the parallel job ($J = RT$).
7. Add these cycles to the total number of cycles executed $CE = CE + Y$.
8. Compute the Speedup for job A.
9. Exclude job A from the cluster and update the total of priorities.
10. Repeat step 2 to step 10 until the speedups of the all the jobs in the cluster are computed.

Next we propose a scheduling algorithm that is used to allocate the proper time quanta for both IPs and the parallel task on a single workstation.

The algorithm is called for execution to compute time quanta whenever a parallel task is encountered. It uses the information about the number of interactive and parallel processes present in the ready queue.

3.2 The Scheduling Algorithm (alloc_quanta)

```
while (a parallel task is in the Ready Queue) do
{
PT = J                       // Set parallel task time to J
Cycle = 0                    // Initialize cycle count
while (PT <> 0) do           // Loop while the parallel
                             // task is not completed
{   cycle = cycle + 1        // Increment cycle count
    get(N^i_ip)              // Get the number of IPs in
                             // the system
```
$q_1^i = N_{ip}^i * q_{dip}$ // Compute the time quantum for IPs
                             // Check whether the time quantum
                             // allocated for IPs does not allow
                             // at least one unit of work
                             // for the parallel task
// if it doesn't allow then reset the time quantum for IPs to
// q_{min} instead of q_{ip}

if $(q_1^i > (\tau - q_{pp} * \sum_{j=1}^{NPJ} Pr^j))$ then
{ $q_1^i = N_{ip}^i * q_{min}$
// Check q_1^i again. If still doesn't leave time for the parallel
// tasks then quit (No feasible solution, consider migration)

if $(q_1^i > (\tau - q_{pp} * \sum_{j=1}^{NPJ} Pr^j))$ then exit
}
// Compute the time quantum for the parallel task
// If parallel task's time quantum is less than
// its remaining time then subtract that quantum
// from the total execution time
$q_2^{i,x} = (\tau - q_1^i) * \frac{Pr^x}{\sum_{j=1}^{NPJ} Pr^j}$

if $(q_2^i < PT)$ then $PT = PT - q_2^i$
else $PT = 0$ // If remaining time is less than the
                     // allocated quantum, then set PT to zero
} // end inner-while
} // end outer-while

4 Description of the Simulation Environment

The above scheduling algorithm was modeled using a Colored Petri Net (CPN) [10] and simulated using the Design/CPN tool [11]. CPNs represent a class of High-level Petri nets that provides powerful tools for the specification, modeling and simulation of large-scale complex systems.

The detailed description of the CPN models can be found in [5]. Interested readers may consult [13] for details and literature about Colored Petri Nets.

In the performance analysis, various scenarios for the parallel jobs were considered (combinations of mixed priorities with mixed task sizes), and the performance of each parallel job is evaluated based on the analytical model and the corresponding CPN based simulation. Each parallel job is assigned a priority and proportionally to this priority the scheduler allocates to this job, the proper portion from the parallel jobs' quantum. All the simulations are performed using the same settings of case 1.

We have evaluated the effect of the cluster size on the speedup of the parallel jobs. The results illustrated first, the agreement between analytical and simulation studies. They also showed how the proposed scheme could allow a parallel job to adapt to the load conditions of the WS and strive to achieve a minimum speed up that commensurate with its priority and maximizes its speedup adaptively.

The analytical and simulation studies were omitted for lack of space.

5 Conclusion and Future Works

This study presented an analytical model for the scheduling and regulation of CPU cycle stealing by parallel tasks in a non-dedicated cluster environment. The proposed scheme is used to control the interactions between local IPs and processes running on behalf of parallel Jobs executed on the cluster. It strives to conciliate the two conflicting requirements of higher speedup for parallel jobs and low response times for interactive local processes. In fact, this proposed scheme can be used as a base for brokered processing power usage in a nondedicated distributed processing environment.

The scheme proposes a measure of maximum response time (MRT) expected by the interactive user of the workstation as a scheduling cycle to compute time quanta that will satisfy the conflicting requirements of both local IPs present in the system and parallel task processes. The MRT measure can either be set according to some empirical studies (investigating interactive user tolerance level) or can be a negotiated factor, on which will depend some billing policies.

We have proposed a mathematical formulation of the problem together with a scheduling scheme that was effectively modeled and simulated using a Colored Petri Net (CPN).

Analytical and simulation results showed agreement and revealed the effectiveness of the proposed scheduling scheme in allowing the parallel tasks to ensure a minimum speedup even in heavy loaded situations and to maximize the speedup adaptively depending on the load conditions.

References

1. A. Acharya, G. Edjlali, and J. Saltz, *The Utility of Exploiting Idle Workstations for Parallel Computation*, Proc. of ACM SIGMETRICS Intl. Conf. on Measurement and Modeling of Computer Systems, June 15-18, Seattle, pp. 225-236, 1997.

2. R. Arpaci, A. Dusseau, A. Vahdat, L. Liu, T. Anderson and D. Patterson, *The Interaction of Parallel and Sequential Workloads on a Network of Workstations*, Proc. of the 1995 ACM SIGMETRICS Intl. Conf. on Measurement and Modeling of Computer Systems, pp. 267-278, May 1995.
3. A. Baratloo, M. Karaul, H. Karl, and Z. M. Kedem, *An Infrastructure for Network Computing with Java Applets*, Proc. of the ACM Workshop on Java for High-Performance Computing, Feb. 1998.
4. R. Buyya, *High Performance Cluster Computing: Architictures and System; and Programming and Application*, Vol 1, and Vol 2, Prentice Hall, 1999.
5. A. B. Mnaouer and B. Al-Riyami: *Colored Petri Nets Based Modeling and Simulation of Mixed Workload Interaction in a Nondedicated Cluster*, proc. of the 7th Intl. Conf. on High Performance Computing and Grid in Asia, HPC Asia 2004, pp.294-303, July 20-22, 2004, Tokyo, Japan.
6. A. B. Mnaouer and B. Al-Riyami: *Effective Scheduling of Local Interactive and Parallel Processes in a Non-Dedicated Cluster Environment*, Accepted in the JPDC Journal (to appear).
7. L. Vanhelsuwe, *Create your own Supercomputer with Java*, JavaWorld 2(1), 1997.
8. X. Du and X. Zhang, *Coordinating Parallel Processes on Network of Workstations*, Journal of Parallel and Distributed Computing, 46, 2,pp. 125-135, 1997.
9. D. Finkel, C. E. Wills, M. J. Chiaraldi, K. Amorin, A. Covati, and M. Lee, *An Applet based Anonymous Distributed Computing System*, In Internet Research: Electronic Networking Applications and Policy, Vol. 11, No. 1, 2001.
10. K. Jensen, *Coloured Petri Nets: Basic Concepts, Analysis Methods and Practical Use*, Vol.1 and Vol.2, Monographs in Theoretical Computer Science, Springer-Verlag, 1992, 1994.
11. K. Jensen, S. Christensen, P. Huber, and M. Holla, *Design/CPN: A reference Manual*, Computer Science Dept, University of Aarhus, Denmark, 1996.
12. S. Leutenegger and X. Sun, *Limitations of Cycle Stealing for Parallel Processing on a Network of Homogeneous Workstations*, JPDC, 43, 2, pp. 169-178, June 1997.
13. A comprehensive site about Colored Petri Nets at *http://www.daimi.au.dk/ cpn/*, online as per June 1st 2003.

Feature-Correlation Based Multi-view Detection

Kuo Zhang, Jie Tang, JuanZi Li, and KeHong Wang

Knowledge Engineering Lab, Department of Computer Science, Tsinghua University,
Beijing 100084, P.R.China
{zkuo99, j-tang02}@mails.tsinghua.edu.cn,
{ljz, wkh}@keg.cs.tsinghua.edu.cn

Abstract. A view validation algorithm has been shown to predict whether or not the views are sufficiently compatible for solving a particular learning task. But it only works when a natural split of features exists. If the split does not exist, it will fail to manufacture a feature split to build the best views. In this paper, we present a general algorithm **CCFP** (Correlation and Compatibility based Feature Partitioner) to automate multi-view detection. CCFP first labels the large amount of unlabeled examples using single view algorithm, then calculates the conditional SU (Symmetric Uncertainty) between every pair of features and the IG (Information Gain) of each feature given the examples labeled previously by single view algorithm with high-confidence predictions. According to the estimated values of SU and IG, all the features will be partitioned into two *views* that are low correlated, compatible and sufficient enough. The experiment results show that multi-view learner with views generated by CCFP outperforms learner with views generated by other means clearly.

1 Introduction

Because in many machine learning settings, unlabeled examples are significantly easier to come by than labeled ones [1], recently, there has been interest in multi-view learning algorithms, such as Co-EM and Co-training, that utilize unlabeled data to improve classification performance. When learning from both labeled and unlabeled examples, multi-view algorithms explicitly leveraging natural multiple views, i.e., several disjoint subsets of features, may outperform single-view algorithms that do not. For instance, one can classify web pages based on two views, one of which is the words occurring in the pages, and the other is the words occurring in the hyperlinks pointing to the pages. The Co-EM algorithm [2][3] combines multi-view learning framework with the semi-supervised EM approach. The semi-supervised EM (Expectation Maximization) algorithm [4] is probably the most prominent single-view approach to learning from labeled and unlabeled data [3], in which an initial classifier h is learned from a small set of labeled examples with Naive Bayes as underlying learner first, and then repeatedly performs a two-step procedure: (1) use h to probabilistically label all the unlabeled examples; (2) learn a new *maximum a posteriori* (MAP) hypothesis h from the examples labeled in step (1). Co-EM algorithm runs EM in each view first and, before each new EM iteration, inter-changes the probabilistic labels generated in each view.

Nevertheless, multi-view algorithms do not outperform single view algorithms all the time especially when the following assumptions are violated [5]:

- Each of the views is sufficient to learn the concept of interest. For instance, one can sufficiently learn the web page classifier *either* on the words in the pages *or* on the words in the hyperlinks pointing to them when given enough labeled data.
- The views are low correlated enough given an example's label.
- The views are compatible enough.

A view validation algorithm has been shown to predict weather or not the views are sufficiently compatible for solving a particular task [5]. But there are three limitations in their view validation algorithm. First, because the view validation algorithm needs lots of learning tasks as training examples, it can be applied only to problems in which the same views are used to solve a large number of different learning tasks. Second, it only provides validation on existing views; when a natural split of features does not exist, it can not detect the best views. Third, it does not take account of feature correlation.

In this paper, we present an algorithm, called **CCFP** (Correlation and Compatibility based Feature Partitioner), to automate multi-view detection. In our framework, the large amount of unlabeled examples are labeled using single view algorithm first, then our algorithm calculates the conditional SU (Symmetric Uncertainty) between every pair of features and the IG (Information Gain) of each feature given the examples labeled previously by single view algorithm with high-confidence predictions. According to the estimated values of SU and IG, all the features will be partitioned into two *views* that are low correlated, compatible and sufficient enough. Experiment results show that our multi-view detection algorithm outperforms other means of feature split.

The paper is organized as follows: in section 2 we introduce terminologies and background related. In section 3, our approach of feature correlation based multi-view detection is presented detailedly. Section 4 describes the experimental results and analysis. Before concluded with discussions, the related works are given in section 5.

2 Terminologies and Background

2.1 Multi-view Setting

The multi-view setting applies to learning tasks that have a natural or artificial way to partition their features into subsets, each of which is called a *view* [6]. In such tasks, each example is described by different sets of features. For instance, in a domain with two views V_1 and V_2, any example x can be presented as a triple $[x_1, x_2, l]$, where x_1 and x_2 are descriptions of x in V_1 and V_2, and l is its label [5].

2.2 View Compatibility and Correlation

It has proved that using two views to bootstrap each other, a target concept can be learned from a small set of labeled examples and a large set of unlabeled examples given that the views are uncorrelated and compatible enough [6]. The *uncorrelated* means that for any example $[x_1, x_2, l]$, x_1 and x_2 are independent given l. The intuition behind this is that when views are independent, the training sets derived from each other are selected randomly.

Definition 1. Views V_1 and V_2 are uncorrelated when $x_1 \square V_1$, $x_2 \square V_2 : p(x_1, x_2|l) = p(x_1|l)p(x_2|l)$ [2].

The *compatible* means that the target concepts in the two views label all the examples with the same value. Otherwise the classifiers may get worse and worse when training each other for several times.

Definition 2. Views V_1 and V_2 are compatible with target concept $t : x \rightarrow y$ when there are hypotheses $h_1 : V_1 \rightarrow \{-1, +1\}$ and $h_2 : V_2 \rightarrow \{-1, +1\}$ such that, for all $x = (x_1, x_2), f_1(x_1) = f_2(x_2) = t(x)$ [2].

2.3 Semi-supervised Algorithm

One way to reduce the amount of labeled examples required is to use algorithms that combine labeled and unlabeled examples when learning [7]. Algorithms learn from both labeled and unlabeled examples are called semi-supervised algorithms. Semi-supervised learning can significantly reduce the amount of labeled data when remaining high accuracy. EM and Co-training are two famous semi-supervised algorithms, and both of them have proved effective on reducing amount of labeled data [6][7][8].

3 Multi-view Detection

3.1 Overview of the Multi-view Detection Algorithm

In some real world problems, because of lacking of natural feature split, it is necessary to partition the features according to their correlation and compatibility. Traditional methods use a strategy of random feature partition to generate multiple views, and prove useful in several specific domains. However, it is not always true since the views generated by random partition can't always be sufficiently uncorrelated and compatible. In this section, we introduce a *view detection* algorithm CCFP: for a given problem, our algorithm partitions the features into two views, which are uncorrelated, compatible and sufficient enough. Multi-view algorithms can perform effectively on the views generated by our algorithm.

In order to calculate the correlation between features given class labels, it is required that large amount of examples should be labeled first. In our approach, only the examples labeled with high confidence prediction will be used. Figure 1 shows the pseudo-code of CCFP. From the 2nd line to the 3rd line the conditional Symmetric Uncertainty (a measure to feature correlation) between every pair of features are calculated. The 4th line and the 5th line compute the Information Gain of each feature given class labels. In the 6th line, features are partitioned into two subsets (F_1, F_2) according to the correlation between features and the information gain of features. Features with high correlation are preferred to be divided into the same view. And both feature sets should include informative features, therefore, the views can be guaranteed sufficient and compatible enough. In the 7th line, the two views (F_1, F_2) are tested weather or not sufficient and compatible enough. If they are not, our algorithm adjusts the *weight* for IG, and re-partitions the feature set.

```
Algorithm GetViews(E_l, E_u, F)
Objective: compute the best two views
Input:
  E_l: the small set of labeled examples.
  E_u: the large set of unlabeled examples.
  F:   the set of features involved in the learning
task.
Output:
  Feature sets (views): F_1 and F_2.
Method:
1. Using EM to label examples in E_u predicted with
high confidence. Get a labeled example set E_ul
2. forall pairs of features p = [f_1, f_2], f_1, f_2 ∈ F
3.     SU[p] = GetSymmetricUncertainty(f_1, f_2, E_ul)
4. forall feature f ∈ F
5.     IG[f] = GetInformationGain(f, E_ul)
6. F_1, F_2 = PartitionFeatures(F, SU, IG, weight)
7. while TestCompatibility(F_1, F_2, E_l, E_u) is false
8.     Adjust weight for IG
9.     F_1, F_2 = PartitionFeatures(F, SU, IG, weight)
10. return F_1, F_2
```

Fig. 1. Overview of **CCFP** Algorithm

3.2 Information Gain and Symmetric Uncertainty

In our approach, we use a correlation measure based on the information-theoretical concept of *entropy*, a measure of the uncertainty of a random variable. The entropy of a random variable X is defined as follows:

$$H(X) = -\sum_{i} P(x_i) \log_2(P(x_i)) \qquad (1)$$

The entropy of X after giving another random variable is called *conditional entropy*, which is defined as:

$$H(X|Y) = -\sum_{j} P(y_j) \sum_{i} P(x_i|y_j) \log_2(P(x_i|y_j)) \qquad (2)$$

Where $p(x_i)$ is the prior probability for random variable $X = x_i$, and $p(x_i|y_j)$ is the conditional probability of x_i given the variable $Y = y_j$. The amount by which the entropy of variable X decreases reflects additional information about X provided by Y and is called *Information Gain* [9][10]. As proved in [9], information gain is symmetrical for two random variables X and Y. So a feature is more informative for the classification problem if it has larger information gain given class labels:

$$IG(f|C) = H(f) - H(f|C) \qquad (3)$$

Where f denotes a feature and C denotes class label.

However, information gain and entropy is biased in favor of features with more values. In order to ensure that values have the same effect, we choose *Symmetric Uncertainty* [11] which is normalized to range [0, 1]. The conditional symmetric uncertainty between two features (f_1, f_2) given class labels is defined as:

$$SU(f_1, f_2 \mid C) = 2 \left[\frac{H(f_1 \mid C) - H(f_1 \mid f_2, C)}{H(f_1 \mid C) + H(f_2 \mid C)} \right] \quad (4)$$

This measure needs all of the features to be nominal to normalize its values to range [0, 1]. In our approach, continuous features are discretized properly in advance to apply this measure method.

3.3 Feature Partitioning

As for feature partitioning, we here mainly address three problems: (1) how to generate two views low correlated enough; (2) how to ensure that the views are compatible; and (3) how to ensure that the views are sufficient. We develop an algorithm to deal with these questions. To simplify the following description, we assume that the number of views needed to detect is two.

The input of this step are a feature set $F = \{f_1, f_2, \ldots, f_n\}$, where n is the number of features, the conditional symmetric uncertainty $SU(f_i, f_j|C)$ between every pair of features, where $1 \leq i, j \leq n$, and the information gains $IG(f_k|C)$, where $1 \leq k \leq n$.

We first choose two features (f_a, f_b) with the lowest conditional symmetric uncertainty as initial centers of the two target feature sets.

$$(f_a, f_b) = \arg\min_{1 \leq i, j \leq n}(SU(f_i, f_j \mid C)) \quad (5)$$

Once feature set centers have been selected, let the first feature set or *view* be $F_1 = \{f_a\}$, and the second be $F_2 = \{f_b\}$. Let feature set F_{left} be the collection of features left, i.e. $F_{left} = \{f \mid f \in F, f \notin F_1, f \notin F_2\}$.

Then each view F_m ($m = 1$ or 2) picks a new feature f_{new} from F_{left} which is both correlated to F_m and informative to add to F_m. For each view, f_{new} will be selected according to:

$$f_{new} = \arg\max_{f_i \in F_{left}} \left(\frac{\sum_{f_j \in F_m} SU(f_i, f_j \mid C)}{|F_m|} + IG(f_i \mid C) * w \right) \quad (m=1, 2) \quad (6)$$

Where w is used to adjust the weight between correlation and compatibility. Each time we let $F_{left} = F_{left} - \{f_{new}\}$, $F_m = F_m + \{f_{new}\}$. CCFP iterates this pickup until F_{left} is empty. Then feature set F is partitioned into two subsets F_1 and F_2.

One view of F_1 and F_2 got from the partitioning algorithm may be too weak to sufficiently learn the target concept. In this case, they are not compatible and the weak view may hurt the overview performance while training them each other in multi-view learning.

To cope with this problem, CCFP learns two classifiers from both views using single-view EM algorithm and labels all the examples with the two classifiers. Then we use the percentage of examples labeled identically by both of the views to determine whether the views are compatible and sufficient or not. If the percentage is smaller than a threshold θ, the feature set F will be re-partitioned with a bigger weight w for information gain until the percentage is bigger than θ. Now we just determine this threshold empirically.

4 Empirical Results

The objective of this section is to evaluate CCFP in terms of accuracy with various numbers of labeled examples.

4.1 Dataset and Experiment Setup

Dataset. CCFP is a domain independent multi-view detection algorithm. We apply it to text classification domain here on the well-known 20-newsgroups dataset [12]. Each newsgroup consists of 1000 articles. We select ten pairs of newsgroups from the dataset to establish ten binary-class problems. For every problem, each class is partitioned into five folders each of which consists of 200 articles. When tokenizing these articles, words on a standard stoplist are removed. And stemming is performed with WordNet [13], using the base form of each word. And the number of words in each document is normalized to avoid bias. *Error Rate* is used as our metric to evaluate the classifiers:

$$Error\ Rate = \frac{N_e}{N_t}$$

Where N_e is the number of examples miss classified, and N_t is the total number of examples in testing set.

Experiment Setup. We compare three algorithms: Naive Bayes, EM and Co-EM. On each of the ten pairs of newsgroups, all of these three algorithms run five times. For Naive Bayes algorithm, one of the folders is used as test set, and the other four folders are used as training set every time. For EM and Co-EM, one of the folders is used for test, with n (n=2,4,8,16,32,64,128) labeled examples and (800-n) unlabeled examples for training every time. The n labeled examples are chosen randomly from the other four folders. In order to study the influence of view compatibility and correlation to the performance of multi-view learners, when testing Co-EM, we use three different feature splits: (1) B-W (Best and Worst) splits the best half of features (with high information gain) into a best view, and the worst half of features into a worst view. This can be regarded as the worst possibility of random feature split. (2) BG (Both Good) splits the features into two views which are comparative (both have features with high and low information gain). This can be regarded as the best possibility of random feature split. (3) CCFP splits features using our approach.

4.2 Experiment Results and Discussion

As shown in table 1, when the number of labeled examples is fixed to 2 or 4, Co-EM with CCFP obtains the lowest error rates. These empirical results deserve several comments.

Table 1. Classfication error rates on the newsgroups dataset. Each error rate is the average of 50 runs (ten pairs of newsgroups and each for five runs)

	NaiveBayes	EM	Co-EM		
			B-W	BG	CCFP
Labeled	800	2	2	2	2
Unlabeled	0	798	798	798	798
Error Rate	3.1 ± 0.2%	29.66 ± 11.9%	31.05 ± 3.9%	13.66 ± 5.5%	10.35 ± 4.2%
Labeled	—	4	4	4	4
Unlabeled	—	796	796	796	796
Error Rate	—	14.21 ± 5.6%	28.95 ± 2.2%	6.25 ± 1.3%	3.82 ± 0.5%

First, views generated by CCFP outperform that generated by BG. Intuitively, the power comes from that CCFP splits features into two views which are sufficient, compatible and low correlated while the views generated by BG are only sufficient and compatible but not low correlated. For instance, when a view labels an example as *class1* by mistake, the second view is more likely to predict the same example as *class2* given it is low correlated with the first one. Then the first view may be corrected. Otherwise, if the other view is high correlated, it is more likely for the second view to predict the example as the same mistake class, too.

Second, even though they has the best view which consists of the best half of all the features, the views generated by B-W perform very badly due to their incompatibility. The intuition behind this is that the bad view is not sufficient, so that it may give wrong information to the good view.

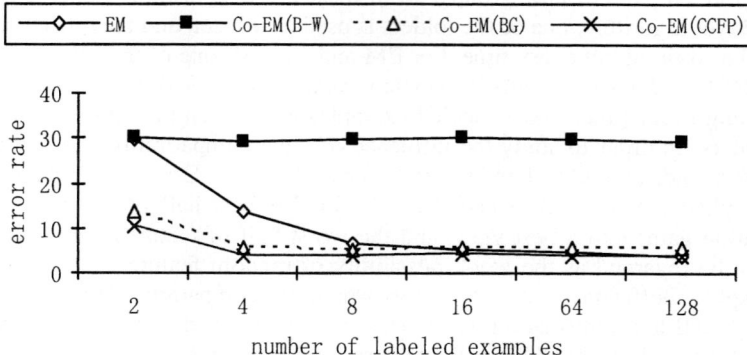

Fig. 2. Error rates for different number of labeled examples. Each error rate is the average of 50 runs

We also examine the influence of the number of labeled examples on classification accuracy shown in figure 2. Each of the four lines displays the learning curves obtained on the problems with various numbers of labeled examples. When the number of labeled examples is small, multi-view learner clearly outperforms single-view learner. Because the views with low correlation correct each other in every iteration until they are converged. We can conclude that multi-view learner can achieve higher accuracy with very few labeled examples. But, multi-view learner's accuracy does not increase as much as single learner when the number of labeled examples increases. We conjecture that each of the subsets of features may have a capability limitation, so they are not as sufficient as the full set of features.

5 Related Work

Blum and Mitchell [6] provided the first formalization of learning in multi-view framework. They show that PAC-like guarantees on learning with labeled and unlabeled data hold under the assumptions that each view is sufficient for classification and the views are weak correlated. Blum and Mitchell also introduced the Co-training, which is the first general-purpose, multi-view algorithm. Nigam and Ghani [3] use Expectation Maximization for multi-view learning. Co-EM can be seen as the closest implementation of the theoretical framework proposed in [6]. Ghani [8] uses Error-Correcting Output Codes to allow Co-training and Co-EM to scale up well to problems with a large number of classes. Muslea et al. [14] extend the Co-EM to incorporate active learning and show its robust behavior on a large amount of problems. All of these works are based on existing views or the views generated by splitting features randomly.

Muslea et al. [5] developed an adaptive view validation algorithm, which is the first meta-learning approach to deciding whether or not multi-view learning is appropriate for new, unseen tasks. The view validation algorithm uses several solved problem instances to train a classifier that discriminates between instances for which the views are sufficiently/insufficiently compatible for multi-view learning. But, it can not automatically partition the domain's features into views that are adequate for multi-view learning.

6 Conclusion

The results described in this paper lead us to believe that multi-view detection approach is indeed useful for learning with few labeled examples. When there is no natural feature split, it is necessary to partition the features according to their correlation and compatibility. The views generated based on symmetric uncertainty and information gain by our approach are low correlated, compatible and sufficient. We have shown that multi-view learner with views generated by our approach outperforms learner with views generated by other means. When the amount of labeled examples is small, it also outperforms single-view learner clearly.

There are still several limitations in our approach, which we plan to cope with in the near future work. First, views generated by our approach always have the same number of features, so some correlated features may be partitioned into the same view. Second, once the views have been generated, how to assign the weight of each view properly? Third, we assign the value of θ mentioned in section 3.3 manually now, so how to assign the value automatically?

References

1. D. Yarowsky. Unsupervised word sense disambiguation rivaling supervised methods. *Proceedings of the 33rd Annual Meeting of the Association for Computational Linguistics*, pages 189-196, 1995.
2. Ulf Brefeld, Tobias Scheffer. Co-EM support vector learning. *Proceedings of the 21st international conference on Machine learning*, 2004.
3. Nigam, K., Ghani, R. Analyzing the effectiveness and applicability of co-training. *Proceedings of Information and Knowledge Management*, 2000.
4. Dempster, A., Laird, N., & Rubin, D. Maximum likelihood from incomplete data via the EM algorithm. *Journal of the Royal Statistical Society B, 39*, 1977.
5. Muslea, Ion, Steven Minton, & Craig Knoblock. Adaptive view validation: A first step towards automatic view detection. *In The 19th International Conference on Machine Learning (ICML-2002)*, pages 443–450, 2002.
6. Blum, A., & Mitchell, T. Combining labeled and unlabeled data with co-training. *Proceedings of the Conference on Computational Learning Theory*, pages 92–100, 1998.
7. Ghani, Rayid. Combining labeled and unlabeled data for text classification with a large number of categories. *Proceedings of IEEE Conference on Data Mining.* 2001.
8. Ghani, Rayid. Combining labeled and unlabeled data for multiclass text classification. *Proceedings of the 19th International Conference on Machine Learning (ICML-2002)*, pages 187–194. 2002.
9. Lei Yu, Huan Liu. Feature Selection for High-Dimensional Data: A Fast Correlation-Based Filter Solution. *Proceedings of the 19th International Conference on Machine Learning (ICML-2003)*, pages 856-863, 2003.
10. Quinlan, J. *C4.5: Programs for machine learning.* Morgan Kaufmann. (1993).
11. Press, W. H., Flannery, B. P., Teukolsky, S. A., & Vetterling, W. T. *Numerical recipes in C.* Cambridge University Press, Cambridge, 1988.
12. T. Joachims. A probabilistic analysis of the Rocchio algorithm with TFIDF for text categorization. *In Proceedings of ICML '97*, 1997.
13. Miller. G. WordNet: An online lexical database. *International Journal of Lexicography.* MIT Press, 1990.
14. Muslea, Ion, Steven Minton, & Craig Knoblock. Active + Semi-supervised Learning = Robust Multi-view Learning. *In The 19th International Conference on Machine Learning (ICML-2002)*, pages 435–442, 2002.

BEST: Buffer-Driven Efficient Streaming Protocol

Sunhun Lee, Jungmin Lee, Kwangsue Chung, WoongChul Choi, and Seung Hyong Rhee

School of Electronics Engineering,
Kwangwoon University, Korea
{sunlee, jmlee}@adams.kw.ac.kr,
{kchung, wchoi, srhee}@daisy.kw.ac.kr

Abstract. The standard streaming protocol is mostly based on UDP with no end-to-end congestion control. For this reason, wide usage of multimedia applications in Internet might lead to congested networks. To avoid such a situation, the congestion controlled streaming protocols have been developed. However, by considering only the stability aspect of network, some works ignore the characteristics of multimedia streaming applications. Moreover, most of previous works have no consideration on the characteristics of wide spreading high-speed networks. In this paper, in order to overcome limitations of the previous streaming protocols, we propose a new streaming protocol called "BEST(Buffer-driven Efficient STreaming)". The BEST protocol takes a hybrid viewpoint that considers both user viewpoint and network viewpoint. Therefore, the BEST protocol improves the network stability by reducing the packet loss and it also provides the smoothed playback by preventing buffer underflow or overflow. The BEST protocol is designed to consider high-speed networks with longer delay. Through the simulation, it is proved that the BEST protocol has a better performance than previous works in high-speed network environments.

1 Introduction

The Internet has recently been experiencing an explosive growth in the use of audio and video streaming applications. Such applications are delay-sensitive, semi-reliable and rate-based. Thus they require isochronous processing and Quality-of-Service(QoS) from the end-to-end viewpoint. However, today's Internet does not attempt to guarantee an upper bound on end-to-end delay and a lower bound on available bandwidth. As a result, most of multimedia applications use UDP(User Datagram Protocol) that has no congestion control mechanism. However, the emergence of non-congestion-controlled real-time multimedia applications threatens unfairness to competing TCP(Transmission Control Protocol) traffic and possible congestion collapse [1, 2].

Studies on the congestion controlled streaming protocol has been increasingly done since the 1990s [2, 3, 4, 5, 6, 7, 8, 9, 10, 11, 12, 13]. These works attempt to guarantee the network stability and fairness with competing TCP traffic. However, by considering only the stability aspect of network, these works ignore the characteristics of multimedia streaming applications. Moreover, most of the previous works have no consideration on the longer-delay characteristics in high-speed networks.

In this paper, we propose a new streaming protocol called BEST(Buffer-driven Efficient STreaming). The BEST protocol takes a hybrid viewpoint that considers both user viewpoint and network viewpoint. Therefore it maintains the network stability as previous works and achieves the smoothed playback by controlling the sending rate and the video quality. Our protocol is also designed to consider the longer-delay characteristics of high-speed networks for a better performance.

The rest of this paper is organized as follows. In Section 2, we review some of the related works and in Section 3, we present various algorithms of the BEST protocol. Detailed description of our simulation results are presented in Section 4. Finally, Section 5 concludes the paper and discusses some of our future work.

2 Related Work

While data applications such as Web and FTP(File Transfer Protocol) are mostly based on TCP, multimedia applications will be based on UDP due to its real-time characteristics. However, UDP does not support congestion control mechanism. For this reason, UDP-based streaming protocols cause the unfairness with competing TCP traffic and the starvation of congestion controlled TCP traffic which reduces its bandwidth share during overload situation. Several congestion controlled streaming protocols have been developed recently.

Previous works can be classified into user viewpoint and network viewpoint. With the network viewpoint, the network stability and the fairness with competing TCP traffic are seeking. With the user viewpoint, they pursue the smoothed playback in a receiver side.

Previous works with the network viewpoint can be further categorized into three different approaches. The first approach mimics TCP's AIMD(Additive Increase Multiplicative Decrease) algorithm to achieve TCP-friendliness. Streaming protocols in this category were discussed in [2, 3, 4, 5, 6, 7]. The second approach adjusts its sending rate according to Padhye's TCP throughput equation. Streaming protocols in this category were presented in [8, 9, 10]. The third approach uses bandwidth estimation scheme that was employed in the TCP Westwood. This approach was presented in [11, 12].

A typical example of the network viewpoint protocol is the TFRC(TCP-Friendly Rate Control) for unicast streaming applications which uses the TCP throughput equation [14]. Padhye et al. presents an analytical model for the available bandwidth share(T) of TCP connection with S as the segment size, p as the packet loss rate, t_{RTT} as the RTT(Round Trip Time), t_{RTO} as the RTO(Retransmission Time-Out). The average bandwidth share of TCP depends mainly on t_{RTO} and p as shown in (1):

$$T = \frac{S}{t_{RTT}\sqrt{\frac{2p}{3}} + t_{RTO}(3\sqrt{\frac{3p}{8}})p(1+32p^2)} \quad (1)$$

The TFRC is responsive to network congestion over longer time period and changes the sending rate in a slowly responsive manner. Therefore, it is considered to be compatible to TCP flows and to reduce packet loss occurrence. But it has some disadvantages which are oscillations of sending rate and possibility of buffer underflow. Moreover, according to recent researches, the TCP throughput equation is influenced by the throughput model employed, the variability of loss events, and the correlation structure of the loss process. Also, revealed itself not to be friendly toward TCP since it mimics only the long term behaviour of TCP [15].

Unlike the network viewpoint protocol discussed above, the buffer-driven scheme is a streaming protocol with the user viewpoint. This scheme scales video quality and schedules data of transmission based on receiver buffer occupancy and sender buffer occupancy. Therefore, the buffer-driven scheme provides smoothed playback by preventing buffer underflow or overflow. More in details, during underload periods, the video quality is increased, which results in higher information rate, whereas during overload periods, the video quality is decreased. However, the buffer-driven scheme is inherited the unstability and unfairness from the UDP-based streaming protocols because it has no concern about network situation [13].

3 BEST Protocol Mechanism

3.1 Overall Architecture

The previous streaming protocols have only one viewpoint. With the user viewpoint, they provide the smoothed playback by preventing buffer underflow or overflow. With the network viewpoint, they improve the network stability. Moreover, both have no consideration on the longer-delay characteristics of high-speed networks. We propose the BEST(Buffer-driven Efficient STreaming) protocol that has both user and network viewpoint in high-speed network environments. It controls the sending rate and the video quality based on the buffer occupancy and the network state.

Figure 1 shows the overall architecture of the BEST protocol. Basically, the BEST protocol relies on the RTCP(Realtime Transport Control Protocol) feedback messages. The RTCP, periodically reports the information on network state to the sender [16]. The BEST protocol determines the network state as congestion_state or stable_state according to this information.

After deciding the network state, the BEST protocol estimates the receiver buffer occupancy in a sender side. Then it controls the sending rate or the video quality on the basis of the estimated receiver buffer state. If the estimated receiver buffer state is very high or low, then it controls the video quality. Otherwise, it controls the sending rate according to the network state.

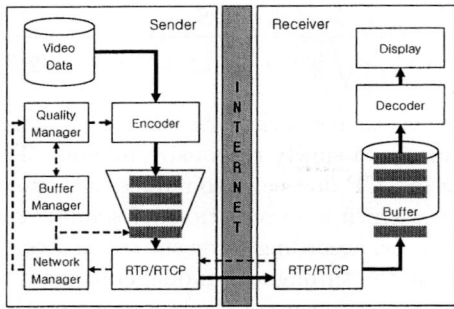

Fig. 1. Overall architecture

3.2 Decision on Network State

The BEST protocol decides network state based on the packet loss rate that can be calculated from the RR(Receiver Report) of RTCP. The packet loss rate is estimated in a receiver side by counting the gaps in the sequence numbers included in RTP(Realtime Transport Protocol) header of the data packets. It is calculated on the proportion of the number of received packets to the number of sent packets during the RTCP interval [16].

$$\text{Packet Loss Rate} = 1 - \frac{\text{Number of Packets Received}}{\text{Number of Packets Sent}} \quad (2)$$

Figure 2 shows the decision rule for network state. If the packet loss rate is higher than link error rate, then the network state is decided to be congestion_state. In this case, packet losses were caused by a congestion. Otherwise, the network state is decided to be stable_state.

If (Packet loss rate > Link error rate)
: Network state = congestion_state
Else (Packet loss rate < Link error rate)
: Network state = stable_state

Fig. 2. Decision rule for network state

3.3 Estimation on Receiver Buffer State

With user viewpoint, the BEST protocol controls the sending rate and the video quality by estimating the receiver's buffer state. To prevent buffer underflow or overflow, it estimates the receiver buffer state.

Figure 3 shows a buffer state in a receiver side with B_T as the total buffer size, $B_C(t)$ as the current buffer occupancy, $R_I(t)$ as the input data rate in a receiver. If we ignore the network transmission delay, $R_I(t)$ is equal to the outgoing rate in a sender. $R_C(t)$ is the data consuming rate in a receiver, as same as encoding

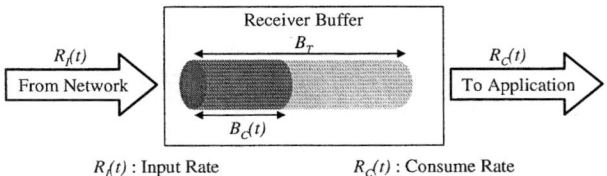

Fig. 3. Buffer state in a receiver side

rate of video streams. The sender can estimate the receiver buffer state(B_E) as follows:

$$B_E(t+1) = B_C(t) + \{R_I(t) - R_C(t)\} \times RTT \quad (3)$$

$$B_C(t) \approx B_E(t) \quad (4)$$

$$R_I(t) = \frac{\text{Number of Packets} \times PacketSize}{\text{RTCP Period}} \quad (5)$$

$$R_C(t) = \text{Encoding Rate at Time (t)} \quad (6)$$

After updating the next sending rate and video quality, $B_C(t)$ is re-calculated on the basis of the new sending rate and video quality. In (3), the BEST protocol predicts one-RTT-early the receiver buffer state. This scheme efficiently prevents buffer underflow or overflow with no additional overhead in a high-speed networks with longer delay.

3.4 Sending Rate and Video Quality Control

The BEST protocol firstly determines the network state as congestion_state or stable_state based on the packet loss rate. And then it estimates the receiver buffer state. Based on the network state and the buffer state, it controls the sending rate or the video quality.

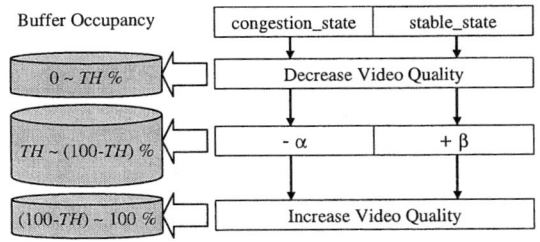

Fig. 4. Control scheme for BEST protocol

Figure 4 shows the scheme to controlling the sending rate and the video quality. For the congestion_state, the sender has to reduce the sending rate for the network stability. But, if the receiver buffer state is estimated as $(0 \sim TH\%)$, we decrease the video quality for preventing buffer underflow. For $\{(100\text{-}TH) \sim 100\%\}$ of buffer occupancy, the video quality is increased to prevent the buffer overflow. However, for the buffer state $\{TH \sim (100\text{-}TH)\%\}$, it simply reduces the sending rate by α for the network stability. Parameter α, the data reduction rate, is supposed to be less than differences among various levels of video qualities. In this paper, α is set to 1Mbps because we use five different video qualities(20, 16, 12, 8, 4Mbps) and the difference among them is 4Mbps.

For $(0 \sim TH\%)$ of the buffer occupancy in the stable_state, the sender decreases the video quality. However, for $\{(100\text{-}TH) \sim 100\%\}$, it increases the video quality. In the case of $\{TH \sim (100\text{-}TH)\%\}$, the BEST protocol increases the sending rate by β for competing with other traffics. Parameter β, the data increasing rate, is calculated as follows:

$$\beta = \alpha \times \frac{\text{Number of Packets}/sec}{\text{Number of Packets/Congestion Period}} \qquad (7)$$

Equation (7) for β is formulated to compensate the decrease of sending rate by α in the congestion period.

Depending on the buffer occupancy, we decide whether the video quality or the sending rate should be controlled. A threshold, TH, is a very important parameter in deciding the buffer occupancy level. For a large threshold(TH) value, the BEST protocol depends mainly on the buffer state instead of the network state. On the other hand, for a small threshold, the network state has more impact on the BEST protocol. Based on our experiments on the TH value, 25% is the optimal value for fairly reacting both the buffer state and the network state. Figure 5 shows the pseudo code of the BEST protocol algorithm.

```
If (Packet loss rate > Link error rate):         // congestion_state
    If (Buffer occupancy < TH)
        Decrease video quality
    Else if (TH < Buffer occupancy < (100 - TH))
        Sending rate -= α
    Else (Buffer occupancy > (100 - TH))
        Increase video quality
Else (Packet loss rate < Link error rate):       // stable_state
    If (Buffer occupancy < TH)
        Decrease video quality
    Else if (TH < Buffer occupancy < (100 - TH))
        Sending rate += β
    Else (Buffer occupancy > (100 - TH))
        Increase video quality
```

Fig. 5. BEST protocol algorithm

4 Simulation and Evaluation

In this Section, we present our simulation results. Using the ns2 simulator, the performance of the BEST(Buffer-driven Efficient STreaming) protocol has been measured, compared with the TFRC and the buffer-driven scheme [17]. To emulate the competing network conditions, background TCP traffic is introduced.

4.1 Simulation Environment

Figure 6 shows the topology for our simulations. Five different video qualities are used, such as 20Mbps, 16Mbps, 12Mbps, 8Mbps, and 4Mbps. 10Mbytes buffer space is allocated for the receiver buffer. We assume that the initial buffer occupancy is 50%.

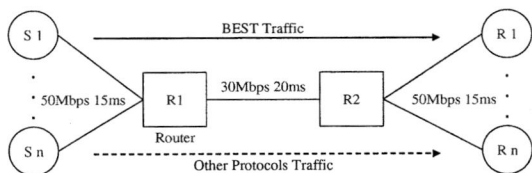

Fig. 6. Simulation environment

4.2 Performance Evaluation

To evaluate our BEST protocol, throughput, video quality, buffer occupancy, and packet loss are measured. Figure 7 (a) shows that our BEST protocol dynamically controls the video quality and the sending rate. On the beginning, the sender transmits the 20Mbps video stream. At about 16, 19 and 23 second, the sender increases the video quality in order to prevent buffer overflow. At about 15, 17, 18 and 21 second, the sender decreases the video quality in order to prevent buffer underflow.

Because the BEST protocol controls the sending rate or the video quality based on receiver buffer state, the estimation accuracy of the buffer state is very important. Figure 7 (b) compares the estimated buffer state in a sender side and the actual one in a receiver side. It shows that the estimated buffer state and the actual one are very close. Figure 7 (c) shows the packet losses of the BEST protocol, compared with the that of RTP. Unlike RTP, the BEST protocol can reduce packet losses by controlling the sending rate based on network state. Approximately, the BEST protocol reduces 50% of packet losses, compared with RTP.

In this paper, we compare the performance of the BEST protocol with the previous streaming protocols, the TFRC and the buffer-driven scheme. Figure 8 compares the BEST protocol and the previous streaming protocols in terms of the packet loss and the buffer occupancy. In Fig. 8 (a), the packet losses for each protocols are depicted. Because the buffer-driven scheme only controls the video quality based on the current buffer state without any consideration on the

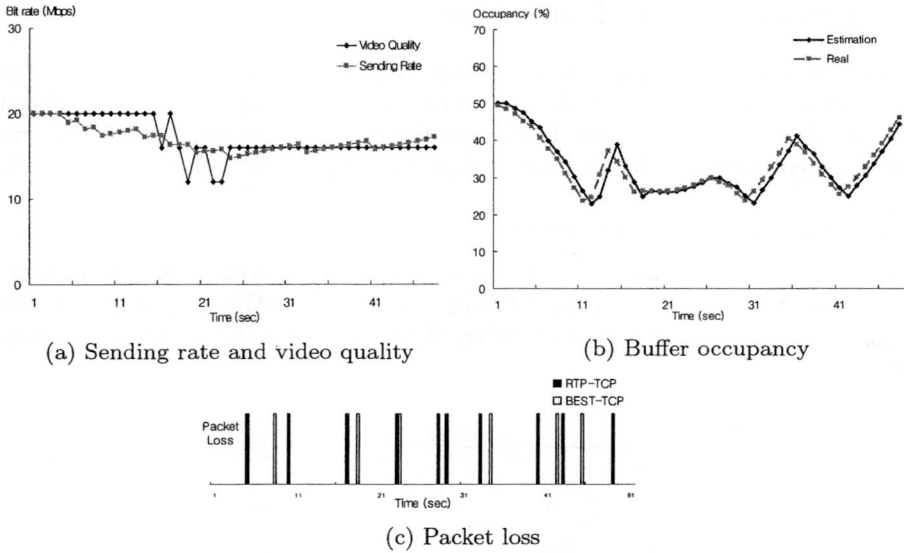

Fig. 7. Performance of the BEST protocol

network state, it suffers more packet losses than our BEST protocol. However, the number of packet loss for the BEST and the TFRC are about the same. From this result, it is shown that the BEST protocol has the approximately same performance with the TFRC in the network stability aspect. Figure 8 (b) shows the buffer occupancy changes. The TFRC experiences a serious buffer underflow, because it has no control on the video quality upon the buffer occupancy state. But, the BEST protocol and the buffer-driven scheme are successfully preventing the buffer underflow or overflow.

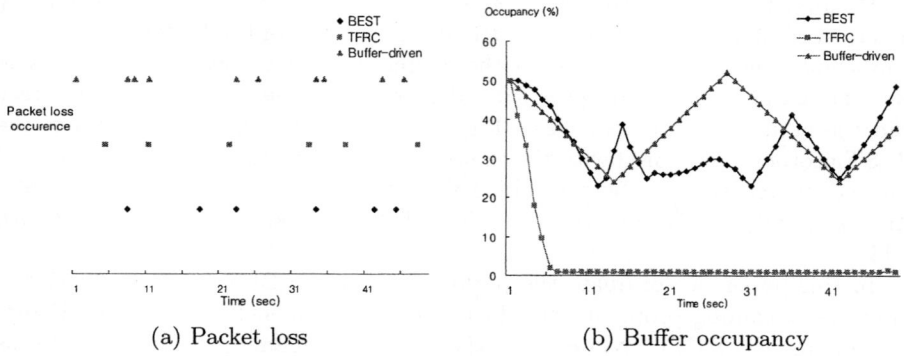

Fig. 8. Comparison between BEST and previous protocols

5 Conclusion

Most of streaming protocols are based on UDP with no end-to-end congestion control. For this reason, wide usage of multimedia applications in Internet might lead to congested networks. To avoid such a situation, several congestion controlled streaming protocols were proposed recently. However, by considering only the stability aspect of network, most of previous works ignore the characteristics of multimedia streaming applications.

In this paper, in order to overcome limitations of the previous streaming protocols, we propose a new streaming protocol, the BEST(Buffer-driven Efficient STreaming). Our protocol takes a hybrid viewpoint that considers both user viewpoint and network viewpoint. It controls the sending rate or the video quality on the basis of the buffer occupancy and the network state. Therefore, the BEST protocol improves the network stability by reducing the packet loss and it also provides the smoothed playback by preventing buffer underflow or overflow. Simulation results have shown that our BEST protocol has a better performance than previous approaches.

In the future, we will further enhance the BEST protocol for TCP-friendliness.

Acknowledgement

This research was supported by the MIC(Ministry of Information and Communication), Korea, under the ITRC(Information Technology Research Center) support program supervised by the IITA(Institute of Information Technology Assessment)

References

1. S. Floyd and F. Kevin: outer mechanisms to support end-to-end congestion control. Technical Report, LBL-Berkeley. (1997)
2. S. Cen, C. Pu, and J. Walpole: Flow and congestion control for internet streaming applications. Multimedia Computing and Networking. (1998)
3. R. Rejaie, M. Handley, and D. Estrin: RAP: An end-to-end rate based congestion control mechanism for real-time streams in the Internet. IEEE INFOCOMM. (1999)
4. B. Mukherjee and T. Brecht: Time-lined TCP for the TCP-friendly delivery of streaming media. International Conference on Network Protocols. (2000)
5. S. Na and J. Ahn: TCP-like flow control algorithm for real-time applications. IEEE International Conference. (2000)
6. I. Rhee, V. Ozdemir, and Y. Yi: TEAR: TCP emulation at receivers - flow control for multimedia streaming. Technical Report, NCSU. (2000)
7. D. Bansal, and H. Balakrishnan: Binomial Congestion Control Algorithms. IEEE INFOCOMM. (2001)
8. J. Padhye, J. Kurose, D. Towsley, and R. Koodli: A model based TCP-friendly rate control protocol. NOSSDAV. (1999)
9. Q. Zhang, Y. Zhang, and W. Zhu: Resource allocation for multimedia streaming over the Internet. IEEE Transactions on Multimedia. (2001)

10. B. Song, K. Chung, and Y. Shin: SRTP: TCP-friendly congestion control for multimedia streaming. 16th International Conference on Information Networking. (2002)
11. N. Aboobaker, D. Chanady, M. Gerla, and M. Sanadidi: Streaming media congestion control using bandwidth estimation. IFIP/IEEE Internation Conference on Management of Multimedia Networks and Services. (2002)
12. A. Balk, D. Maggiorini, M. Gerla, and M. Sanadidi: Adaptive MPEG-4 video streaming with bandwidth estimation. QoS-IP. (2003)
13. D. Ye, X. Wang, Z. Zhang, and Q. Wu: A buffer-driven approach to adaptively stream stored video over Internet. High Speed Networks and Multimedia Communications 5th International Conference. (2002)
14. J. Padhye, V. Firoiu, D. Towsley, and J. Kurpose: Modeling TCP throughput: A simple model and its empirical validation. ACM SIGCOMM. (1998)
15. L. Grieco and S. Mascolo: Adaptive rate control for streaming flows over the Internet. ACM Multimedia Systems Journal, Vol. 9. (2004)
16. H. Schulzrinne, S. Casner, R. Frederick, and V. Jacobson: RTP: A transport protocol for real-time applications. IETF, RFC 1889. (1996)
17. UCB LBNL VINT: Network Simulator ns (Version 2). http://www-mash.cs.berkeley.edu/ns/

A New Neuro-Dominance Rule for Single Machine Tardiness Problem

Tarık Çakar

Sakarya University Engineering Faculty,
Department of Industrial Engineering,
54187 Adapazarı – Turkey

Abstract. We present a neuro-dominance rule for single machine total weighted tardiness problem. To obtain the neuro-dominance rule (NDR), backpropagation artificial neural network (BPANN) has been trained using 5000 data and also tested using 5000 another data. The proposed neuro-dominance rule provides a sufficient condition for local optimality. It has been proved that if any sequence violates the neuro-dominance rule then violating jobs are switched according to the total weighted tardiness criterion. The proposed neuro-dominance rule is compared to a number of competing heuristics and meta heuristics for a set of randomly generated problems. Our computational results indicate that the neuro-dominance rule dominates the heuristics and meta heuristics in all runs. Therefore, the neuro-dominance rule can improve the upper and lower bounding schemes.

1 Introduction

Because of the fact that companies should struggle to survive in a strongly competitive commercial environment, a big emphasis is required to be placed on the coordination of the priorities of the companies through the functional fields. Jensen at al. [1] has stated the importance of a customer depending on a variety of factors; (i) the length of relationship of companies with the costumer, (ii) how frequently they provide business to the company, (iii) how much of the capacity of the companies they fill with orders and the potential of a costumer to provide orders in the future. Avoiding delay penalties and meeting due dates are the most important aims of scheduling in many applications. Tardy delivery costs, for example, costumer bad will, lost future sales and rush shipping costs, change significantly over orders and costumers, and in a job priority the implied strategic weight should be reflected. An extensive majority of the job shop scheduling literature is full with rules, which do not consider information about costumer importance or job tardiness penalty. Thus, the strategic priorities of the companies need the information related to costumer importance be incorporated into its shop floor control decisions. Furthermore, it may not be sufficient to measure the performance of shop floor by employing unweighted performance measurements alone that treat each job in the shop in the same importance in the presence of job tardiness penalties. In this paper, a novel superior rule for the single machine total weighted tardiness problem with penalties depending on job has been presented and implemented in lower and upper bounding schemes. Lawyer [2], presented total weighted tardiness problem as $1 \parallel \sum w_i T_i$ is strongly and gives pseudo polynomial algorithm for the total tardiness problem, $\sum w_i T_i$. For weighted and unweighted tardiness problems, a few various solution methods were presented in [3][4][8]. Emmons's study [4] is based on deriving several superior rules that limits the search for an optimum solution to the $1\|\sum T_i$ problem. The rules in stated study are used in both branch and bound (B&B) and dynamic programming algorithms (Fisher [5] and Potts and Van Wassenhove [6,7]). These results were extended to the weighted tardiness problem by Rinnooy Kan et al. [8]. Chambers et al. [3] presented a study based on improving novel heuristic superior rules and a flexible decomposition heuristic. Abdul-Razaq et al. [9] has tested the exact approaches, which are used in solving the weighted tardiness problem, and Emmons's superior rules were used in their paper to form a precedence graph to find upper and lower bounds. They demonstrated that the most promising lower bound both in time consumption and quality is the linear lower bound method by Potts and Van Wassenhove [6], obtained from Lagrangian slacking of the machine capacity constraints. Hoogeveen et al. [10] presented a study based on reformulating the problem using slack variables and demonstrated that better Lagrangian lower bounds could be used. Swarc has presented a study [11] based on proving the existence of a special ordering for the single machine earliness- tardiness (E/T) problem with job independent penalties where the arrangement of two adjacent jobs in an optimal schedule depends on their start time. A two-stage decomposition mechanism to $1 \parallel \sum w_i T_i$ problem when tardiness penalties are proportional to the processing time is presented by Szwarc and Liu [12]. As stated above the importance of a costumer depends on different factors, however it is important for manufacturing in the reflection of these priorities in the decisions of scheduling. For this reason, a new superior rule for the most general case of the total weighted tardiness problem is presented by us. Our proposed rule extends and also covers the Emmon's results and generalizations of Rinnooy Kan et al. under consideration of the time dependent orderings between each pair of jobs. Miscellaneous heuristics and dispatching rules were proposed, because of the fact that the implicit enumerative algorithms may need important computer resources both in terms of memory and computation times. With specified due dates and delay penalties for the weighted tardiness problem, Vepsalainen and Morton [13] has developed and tested efficient dispatching rules. An adequate condition for local optimality is provided by their proposed superior rule, and it generated schedules, which cannot be developed by adjacent job interchanges. In this paper, a trained BPANN to show how the proposed superior rule can be used to develop a sequence given by a dispatching rule. We also gave the proof of that if any sequence disturbs the proposed superior rule, and then switching the disturbing jobs either lowers the total weighted tardiness or leaves it unchanged. Because of the comprehensive computational requirements, according to the literature the weighted tardiness problem is NP-hard and the lower bounds do not have practical applications. Akturk and Yildirim[16] proposed more practical application about weighted tardiness problem and computing lower bound. A study of Potts and Van Wassenhove based on the linear lower bound is rather a weak lower bound, however the most promising one was presented by Abdul-Razaq et al.. His study is in contradiction with the conjecture about this subject that one should limit the search tree as much as possible with using the sharpest possible bounds. The linear lower bound computations are based on an initial sequence. In this paper, a solution that has a better upper bound value, which is near to optimal solution, is presented. Out solution also improves the lower bound value obtained from the linear lower bound method. Sabuncuoglu and Gurgun [14] proposed a new neural network approach to solve the single machine mean tardiness scheduling problem and the minimum makespan job shop scheduling problem. The proposed network by Sabuncuoglu and Gurgun combines the characteristics of neural networks and algorithmic approaches. Recently, Akturk and Ozdemir [15] proposed a new dominance rule for $1 \mid r_j \mid \sum w_i T_i$ problem that can be used in reducing the number of alternatives in any exact approach. Akturk and Yildirim used a interchange function, $\Delta_{ij}(t)$, is used to specify the new dominance properties, which gives the cost of interchanging adjacent jobs i and j whose processing starts at time t. Akturk and Yildirim found three breakpoints using the cost functions and obtained a number of rules by using the breakpoints.

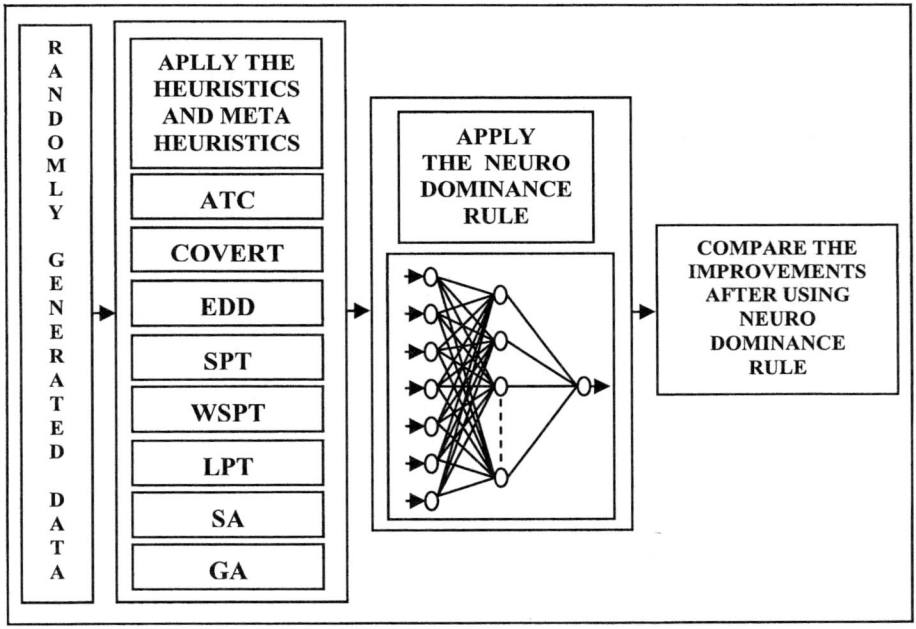

Fig. 1. Steps of the study from obtaining randomly data to comparison of the results

In this study, instead of extracting rules finding break point using cost functions, an artificial neural network was trained using sufficient number of data different from Aktürk and Yıldırım. When the necessary inputs were given, according to the total weighted tardiness problem criterion, it is decided that which job will come firstly among the adjacent jobs. This paper organized as follows; In the section 2, used parameters, modeling of the problem and how the proposed NDR works are discussed. In the section 3, used lower and upper bound schemes are explained. In the section 4, all of computational results and analysis are reported.

2 Problem Definition

The single machine problem may be explained as follows. Each job, which is numbered from 1 to n, should be processed with no interruption on a single machine, which can use only one job at a time. All of the jobs will be available to be processed at time "0". If a job is presented with i, it has parameters as p_i, d_i, and w_i, which refer to an integer processing time, a due date and a positive weights, respectively. The problem can be defined as finding a schedule S, which minimizes $f(S) = \sum_{i=1}^{N} w_i T_i$ function. The dominance rule may be introduced by considering schedules, where Q_1 and Q_2 are two disjoint subsequences the rest n-2 jobs, $S_1 = Q_1 ij Q_2$ and $S_2 = Q_1 ji Q_2$. $t = \sum_{k \in Q_1} p_k$ is the completion time of Q_1.

In this study, it is decided which job will be done firstly among two adjacent jobs according to the total wieighted tardiness criterion using a trained BPANN. The first job is taken as i and the second one is taken as j in these two adjacent jobs without taking care of due date or processing time. The used neural network has 7 inputs and 1 output, and there are 30 neurons in the hidden layer. The starting time of job i (T), the processing time of job i (p_i) due date of job i (d_i), the weight of job i (w_i), the processing time of job j(p_j), the due date of job j (d_j), the weight of job j (w_j) were given as inputs to the BPANN. "0"and "1" values were used to determine the precedence of the jobs. If output value of the BPANN is "0" then i should precede j (i→j). If output value of the BPANN is "1" then j should precede i (j→i). Structure of the used BPANN can be seen in Figure 2. The parameters related to the training and test of neural network were given in Table 1. It can be seen that how the NDR works in figure 3.

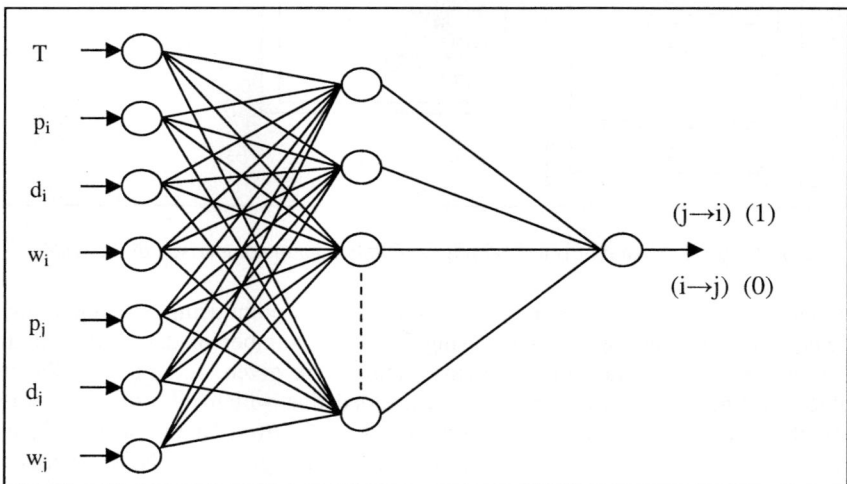

Fig. 2. Structure of the used BPANN. There are 7 input and 1 output

Table 1. Training and test parameters of the used BPANN

Sample size in training set	5000
Learned sample in training set	5000
Number of test data to test trained network	5000
Achievement rate of the test data (%)	%100
Activation function	Sigmoidal
Iteration number	4.000.000
Learning rate	0.35
Momentum rate	0.75

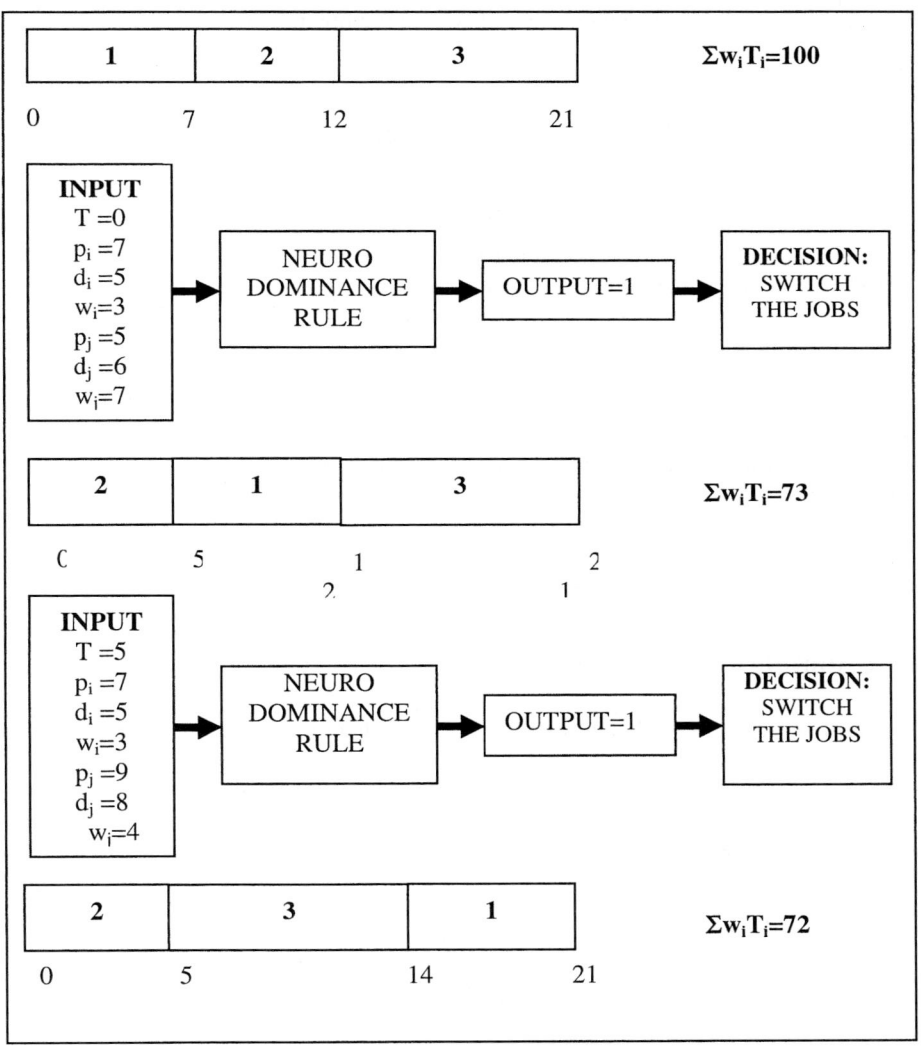

Fig. 3. An example: How the proposed neuro-dominance rule works

3 Linear Lower Bound

Potts and Wan Wassenhove [6] have originally obtained the linear lower bound based on using the Lagrangian Relaxation approach with subproblems, which are total weighted completion time problems. Abdul-Razaq and his co-workers have presented additional derivation of it based on reducing the total weighted tardiness criterion to a linear function, i.e. total weighted completion time problem. For the job i, i = 1 to n, $w_i \geq v_i \geq 0$ and C_i is the completion time of job i, we have

$$w_i T_i = w_i \max\{C_i - d_i, 0\} \geq v_i \max\{C_i - d_i, 0\} \geq v_i (C_i - d_i) \quad (1)$$

Suppose that $v=(v_1, \ldots, v_n)$ is a vector of linear weights, i.e. weights for the linear function $C_i - d_i$, chosen so that $0 \leq v_i \leq w_i$. If so a lower bound can be expressed by given linear function below:

$$\text{LB}_{\text{lin}}(v) = \sum_{i=1}^{n} v_i (C_i - d_i) \leq \sum_{i=1}^{n} w_i \max\{C_i - d_i, 0\} \quad (2)$$

This situation shows that the total weighted completion time problem solution gets a lower bound on the total weighted tardiness problem. For any given v value, the optimal solution of the total weighted completion problem may be realized by the WSPT rule in which the jobs are sequenced in non-increasing order of v_i/v_p. An initial sequence is needed in the determination of the job completion time C_i to obtain the linear lower bound. Afterwards, v, refers to the vector of linear weights, is chosen to maximize $\text{LB}_{\text{Lin}}(v)$ with the condition of that $v_i \leq w_i$ for each job i. In Abdul-Razzaq's study, several lower bounding approaches have been compared and according to the their computational results the linear lower bound is found superior to others, which were given in the literature, because of the its quick computability and low memory requirement. In this paper, the impact of an initial sequence on the linear lower bound value will be tested and tried to present having a better, i.e. near optimal, upper bound value will improve the lower bound value. This linear bound scheme also was used by Akturk and Yildirim.

4 Computational Results

In this study, each lower bounding scheme was tested on a set of randomly generated problems. We have tested the lower bounding scheme on problems with 50, 70 and 100 jobs, which were generated as: for each job i, p_i, and w_i were generated from two uniform distributions, [1, 10] and [1, 100] to create low or high variation, respectively. Here as stated early, p_i and w_i refers to an integer processing time and an integer weight, respectively. The proportional range of due dates (RDD) and average tardiness factor (TFF) were selected from the set {0.1, 0.3, 0.5, 0.7, 0.9}. d_i, an integer due date from the distribution [P(1-TF-RDD/2), P(1-TF+RDD/2)] was produced for each job i, here, P refers to total processing time, $\sum_{i=1}^{n} p_i$. As summarized in Table 2, we considered and evaluated 300 example sets and took 100 replications for each combination resulting among 30.000 randomly generated runs.

Table 2. Experimental design

Factors	Distribution range
Number of jobs	50, 70, 100
Processing time range	[1-10], [1-100]
Weight range	[1-10], [1-100]
RDD	0.1, 0.3, 0.5, 0.7, 0.9
TF	0.1, 0.3, 0.5, 0.7, 0.9

To find an initial sequence for the linear lower bound, a number of heuristics were selected and their priority indexes were given as a summary in Table 3. The WSPT, EDD, LPT and SPT can be given as examples of static dispatching rules, where as ATC and COVERT are dynamics ones. Vepsalainen and Morton [13] have mentioned in their paper as: the ATC rule is superior to other sequencing heuristics and they defined it close to the optimal for the $\sum w_i T_i$ problem.

Table 3. Priority Rules

RULE	RANK AND PRIORITY INDEX
COVERT	$\max\left[\dfrac{w_i}{p_i}\max\left(0,1-\dfrac{\max(0,d_i-t-p_i)}{kp_i}\right)\right]$
ATC	$\max\left[\dfrac{w_i}{p_i}\exp\left(-\dfrac{\max(0,d_i-t-p_i)}{k\overline{p}}\right)\right]$
WSPT	$\max\left(\dfrac{w_i}{p_i}\right)$
EDD	$\min(d_i)$
SPT	$\min(p_i)$
LPT	$\max(p_i)$

In addition to heuristics, two different meta heuristic, simulated annealing (SA) and genetic algorithms (GA), were used in this study. The parameters and operators used in SA to generate new solution were given. In this study, two different operator have been used to generate new negihbourhood solution. Operators are swap and inverse operator. Total weighted tardiness was taken as a fitness funtion. In SA, the best value, obtained from heuristics, was taken as a starting solution

Swap operator		Inverse operator	
Old solution	New solution	Old solution	New solution
1984563**7**2	19**7**45638**2**	19**8456**372	19**3654**872

SA has some weak points such as long running time and difficulty in selecting cooling parameter when the problem size becomes larger. A geometric ratio was used in SA as $T_{k+1} = \alpha T_k$, where T_k and T_{k+1} are the temperature values for k and k+1 steps, respectively. Geometric ratio is used more commonly in practice. In this study, the initial temperature was taken 10000 and 0.95 was used for cooling ratio (α).

In this study, when preparing initial populations in genetic algorithm, for any given problem, the solutions obtained from COVERT, ATC, EDD, WSPT, LPT; and SA methods, were also used. Others were randomly generated. Total weighted tardiness was taken as a fitness function. The parameters used in genetic algorithm were as given below.

Population size : 100   Crossover rate : 100%
Max generation : 200   Mutation rate   : 0.05

Linear Order Crossover (LOX) method has been applied to each chromosome independently. LOX works as follows:

1. Select the sublist from chromosomes randomly ;
chromosome #1 : 123**456**789
chromosome #2 : 645**713**298

2. Remove the sublist 2 from chromosome #1;
chromosome #1 : h2h456h89
chromosome #1 : 245hhh689

3. Remove the sublist 1 from chromosome #2;
chromosome #2 : hhh713298
chromosome #2 : 713hhh298

4. Insert sublist into holes to form offspring;
offspring #1 : 245713689
offspring #2 : 713456298

Mutation operator works as follows :
Select the randomly a chromosome and select the randomly two gene and swap the genes:

Selected genes   : 3**7**6541**2**98   Mutation   : 326541798

If any sequence violates the dominance rule, then the proposed algorithm either lowers the weighted tardiness or leaves it unchanged. Firstly, to find an initial sequence we used one of the dispatching rules, afterwards the algorithm was applied to get the sequence indicated as Heuristic+NDR. The average lower bound value was calculated for each heuristic before and after implementing the algorithm along with the average improvement ($impr$) and this situation is summarized in Table 4. ATC, COVERT, and WSPT seem to execute better than other heuristics in the literature when the dominance rule is applied to get the local optimal sequence. But, SA and GA meta heuristics perform better than the other heuristics. Each heuristic and meta heuristic over 10,000 runs for 50, 70 and 100 job states were tested by us and given in Table 5. As stated above, (>) denotes number of runs in which sequence gotten from Heuristic+NDR gives a higher linear lower bound value than the sequence gotten from the heuristic, where as (=) denotes number of runs in which Heuristic+NDR executes as well as heuristic, and (<) denotes number of runs in which Heuristic+NDR executes worse. For instance, the combination of EDD+NDR executed 4596 times better (>) than EDD rule. According to the large t-test values on the average improvement, the proposed dominance rule provides an important improvement on all rules and the amount of improvement is noteworthy at 99.5% confidence level for all heuristics.

Table 4. Computational results for n=70

Heuristics and Meta Heuristics	UPPER BOUND			LINEAR LOWER BOUND		
	Before	After (+NDR)	$\overline{impr}$ (%)	Before	After (+NDR)	$\overline{impr}$ (%)
COVERT	126487	124343	4.02	985654	992123	0.61
ATC	118467	117896	6.81	104356	104397	0.09
EDD	263255	126912	40.73	27615	95811	108.21
WSPT	151349	132845	48.96	101378	104689	1.12
SPT	229604	214657	19.27	17898	22330	7.46
LPT	538936	162927	83.09	20435	82341	152.78
SA	117945	116992	3.04	104423	104452	0.08
GA	116762	115952	2.97	104102	104153	0.08

Table 5. Comparison of the linear lower bound

	n=50				n=70				n=100			
	>	=	<	t-test	>	=	<	t-test	>	=	<	t-test
COVERT	437	9501	62	3.86	653	8879	468	4.35	983	8879	138	4.28
ATC	2500	7484	16	24.32	2567	7379	54	25.42	2578	7297	125	25.46
EDD	4596	5182	222	32.76	4444	5214	342	30.45	4351	5376	273	31.02
WSPT	2389	7604	7	21.45	2422	7572	6	23.42	2372	7612	16	20.88
SPT	4548	5181	271	25.44	3853	5876	271	23.68	3969	5789	242	29.66
LPT	4764	5012	224	33.43	4743	5012	245	33.59	4642	5169	189	33.12
SA+NDR	392	9596	12	3.45	426	9553	21	3.82	331	9642	27	3.66
GA+NDR	291	9706	3	3.39	372	9623	5	3.47	224	9772	4	3.52

5 Conclusion

In this study, we have developed a neuro-dominance rule for $1 \| \sum w_i T_i$ problem. A BPANN has been used obtaining the proposed neuro-dominance rule. Inputs of the trained BPANN are starting date of the first job (T), processing times (p_i and p_j), due dates (d_i and d_j) and weights of the jobs (w_i and w_j). Output of the BPANN is a decision of which job should precede. The proposed neuro-dominance rule provides a sufficient condition for local optimality. Therefore, a sequence obtained by the proposed neuro-dominance rule cannot be improved by adjacent job interchanges. Computational results over 30,000 randomly generated problems indicate that the amount of improvement is significant. For the future research, single machine total weighted tardiness problem with unequal release dates can be modeled by using artificial neural networks.

References

1. Jensen, J.B., Philipoom, P.R., Malhotra, M.K., Evaluation of scheduling Rule s with commensurate customer priorities in job shops, Journal of operation management,1995, 13, 213-228.
2. Lawler, E. L., A "Pseudopolynomial" algorithm for sequencing job to minimize total tardiness, Annals of Discrete Mathematics, 1997, 1, 331-342.
3. Chambers, R. J., Carraway, R.L., Lowe T.J. and Morin T.L., Dominance and decomposition heuristics for single machine scheduling, Operation Research,1991, 39, 639-647.
4. Emmons, H., One machine sequencing to minimize certain functions of job tardiness. Operation Research, 1969, 17, 701-715.
5. Fisher M.L., A dual algorithm for the one-machine scheduling problem, Mathematical Programming, 1976, 11, 229-251.
6. Potts, C.N. and Van Wassenhove, L.N., A Branch and bound algorithmfor total weighted tardiness problem, Operation Research, 1985, 33, 363-377.
7. Potts, C.N. and Van Wassenhove, L.N., Dynamic programming and decomposition approaches for the single machine total tardiness problem, European Journal of Operation Research, 1987, 32, 405-414.
8. Rinnooy Kan, A.H.G., Lageweg B.J. and Lenstra, J.K., Minimizing total costs in one machine scheduling, Operations Research, 1975, 23, 908-927.
9. Abdul-Razaq, T.S., Potts, C. N. And Van Wassenhove, L.N., A survey of algorithms for the single machine total weighted tardiness scheduling problem, Discrete Applied Mathematics, 1990, 26, 235-253.
10. Hoogeveen, J.A. and Van de Velde, S.L., Stronger Lagrangian bounds by use of slack variables: applications to machine scheduling problems. Mathematical Programming, 195, 70, 173-190.
11. Szwarc,W., Adjacent orderings in single machine scheduling with earliness and tardiness penalties, Naval research Logistics, 1993, 1993, 40, 229-243.
12. Szwarc,W., and Liu, J.J., Weighted Tardines single machine scheduling with proportional weights, Management Science, 1993, 39, 626-632.
13. Vepsalainen, A.P.J., and Morton, T.E., Priority rules for job shops with weighted tardiness cost, management Science, 1987, 33, 1035-1047.
14. Sabuncuoglu, I. And Gurgun, B., A neural network model for scheduling problems, European Journal of Operational research, 1996, 93(2), 288-299.
15. Akturk, M.S., Ozdemir, D., A new dominance rule to minimize total weighted tardiness with unequal release date, European Journal of Operational research, 2001, 135, 394-412.
16. Akturk, M.S., Yidirim, M.B., A new lower bounding scheme for the total weighted tardiness problem, Computers and Operational Research, 1998, 25(4), 265-278.

Sinogram Denoising of Cryo-Electron Microscopy Images

Taneli Mielikäinen[1] and Janne Ravantti[2]

[1] HIIT Basic Research Unit, Department of Computer Science,
University of Helsinki, Finland
`tmielika@cs.Helsinki.FI`
[2] Institute of Biotechnology and Faculty of Biosciences,
University of Helsinki, Finland
`ravantti@cs.Helsinki.FI`

Abstract. Cryo-electron microscopy has recently been recognized as a useful alternative to obtain three-dimensional density maps of macromolecular complexes, especially when crystallography and NMR techniques fail. The three-dimensional model is constructed from large collections of cryo-electron microscopy images of identical particles in random (and unknown) orientations.

The major problem with cryo-electron microscopy is that the images are very noisy as the signal-to-noise ratio can be below one. Thus, standard filtering techniques are not directly applicable. Traditionally, the problem of immense noise in the cryo-electron microscopy images has been tackled by clustering the images and computing the class averages. However, then one has to assume that the particles have only few preferred orientations. In this paper we propose a sound method for denoising cryo-electron microscopy images using their Radon transforms. The method assumes only that the images are from identical particles but nothing is assumed about the orientations of the particles. Our preliminary experiments show that the method can be used to improve the image quality even when the signal-to-noise ratio is very low.

1 Introduction

Structural biology studies how biological systems are built. Especially, determining three-dimensional electron density maps of macromolecular complexes, such as proteins or viruses, is one of the most important tasks in structural biology [1, 2].

Standard techniques to obtain three-dimensional density maps of such particles (at atomic resolution) are by X-ray diffraction (crystallography) and by nuclear magnetic resonance (NMR) studies. However, X-ray diffraction requires that the particles can form three-dimensional crystals and the applicability of NMR is limited to relatively small particles [3]. There are many well-studied viruses that do not seem to crystallize and are too large for NMR techniques.

A more flexible way to reconstruct density maps is offered by cryo-electron microscopy [2, 4]. Currently, the resolution of the cryo-electron microscopy reconstruction is not quite as high as resolutions obtainable by crystallography or NMR but it is improving steadily. Reconstruction of density maps by cryo-electron microscopy consists of the following subtasks:

Specimen preparation. A thin layer of water containing a large number of identical particles of interest is rapidly plunged into liquid ethane to freeze the specimen very quickly. Quick cooling prevents water from forming regular structures [2]. Moreover, the particles get frozen in random orientations in the iced specimen.

Electron microscopy. The electron microscope produces a projection of the iced specimen. This projection is called a *micrograph*. Unfortunately the electron beam of the microscope rapidly destroys the specimen so getting (multiple) accurate pictures from it is not possible.

Particle picking. Individual projections of particles are extracted from the micrograph. There are efficient methods to picking the projections, see e.g. [5, 6].

Orientation search. The orientations (i.e., the projection directions for each extracted particle) for the projections are determined. There are several (heuristic) approaches for finding the orientations, see e.g. [7, 8, 9, 10, 11, 12].

Reconstruction. If the orientations for the projections are known then quite standard tomography techniques can be applied to reconstruct the three-dimensional electron density map from the projections [2].

A main difficulty in cryo-electron microscopy is the very low signal-to-noise-ratio of the images. Thus, standard filtering techniques are not directly applicable since they filter also too much of the signal. Currently the high noise level is reduced by clustering the images to few clusters and computing class averages of the images [2]. Unfortunately, by doing that it is implicitly assumed that most of the particles are in few preferred orientations. Also, already relatively small variations in orientations make the class averages blurred.

In this paper we propose an alternative denoising method that does not assume anything about the projection directions. The method is based on some special properties of sinograms obtained from the (two-dimensional) projections of the same (three-dimensional) density map. The experiments show that it can be used to denoise even very noisy projections.

This paper is organized as follows. In Section 2 we define some central concepts of this work, in Section 3 we describe our denoising approach and in Section 4 we present some preliminary denoising experiments on simulated data. Section 5 is a short conclusion.

2 Projections and Sinograms

A *density map* is a mapping $D : \mathbb{R}^3 \to \mathbb{R}$ with a compact support. An *orientation* o is a rotation of the three-dimensional space and it can be described, e.g., by a three-dimensional rotation matrix. A *projection p* of a three-dimensional density map D to orientation o is the integral

$$p(x,y) = \int_{-\infty}^{\infty} D\left(R_o\, [x,y,z]^T\right) dz$$

where R_o is a rotation matrix, i.e., the mass of D is projected on a plane passing through the origin and determined by the orientation o. Some examples of projections from a density map used in our experiments (presented in Section 4) are shown in Figure 1. The brightness of a pixel is proportional to its mass.

The (three-dimensional) *Radon transform* $\hat{D}$ of the density map D is the mapping

$$\hat{D}(x,o) = \int_{-\infty}^{\infty}\int_{-\infty}^{\infty} D\left(R_o\,[x,y,z]^T\right) dy dz,$$

i.e., $\hat{D}$ consists of all one-dimensional projections of D. Similarly, the (two-dimensional) Radon transform of the projection p is the mapping

$$\hat{p}(x,\alpha) = \int_{\infty}^{\infty} p\left(R_\alpha\,[x,y]^T\right) dy$$

where R_α a two-dimensional rotation matrix with rotation angle α.

Any two projections p_i and p_j from the same density map have a desirable property: their Radon transforms $\hat{p}_i$ and $\hat{p}_j$ have one (one-dimensional) common projection, a *common line*, i.e., there are α_i and α_j such that $\hat{p}_i(x,\alpha_i) = \hat{p}_j(x,\alpha_j)$ for all x. This fact is known as the *Common Line Theorem* and it forms the central part of several orientation search techniques [9, 10, 13, 14, 15, 16, 17]. Unfortunately, the robust behavior of the orientation search methods requires that the the images are not too noisy. The recent results on the computational complexity of orientation search indicate that there is not much hope to a robust determination of orientations without considerable reduction of noise [18]. The standard approach to noise reduction in cryo-electron microscopy is to cluster the images and to compute the class averages [2]. However, this solution implicitly assumes that majority of imaging directions cluster well. In Section 3 we show how noise levels of cryo-electron microscopy images can be reduced without assuming the clusterability of the imaging directions.

In practice, a discrete versions of two-dimensional Radon transforms, called *sinograms*, are used instead of the continuous Radon transforms. A sinogram of an $m \times m$ image is an $l \times m$ matrix of rational numbers such that each row $i = 1,\ldots,l$ in the sinogram corresponds to one one-dimensional projection of the image to direction determined by the angle $(i-1)\pi/l$. Some examples of sinograms are shown in Figure 1. There are several efficient methods for computing sinograms [19, 20, 21].

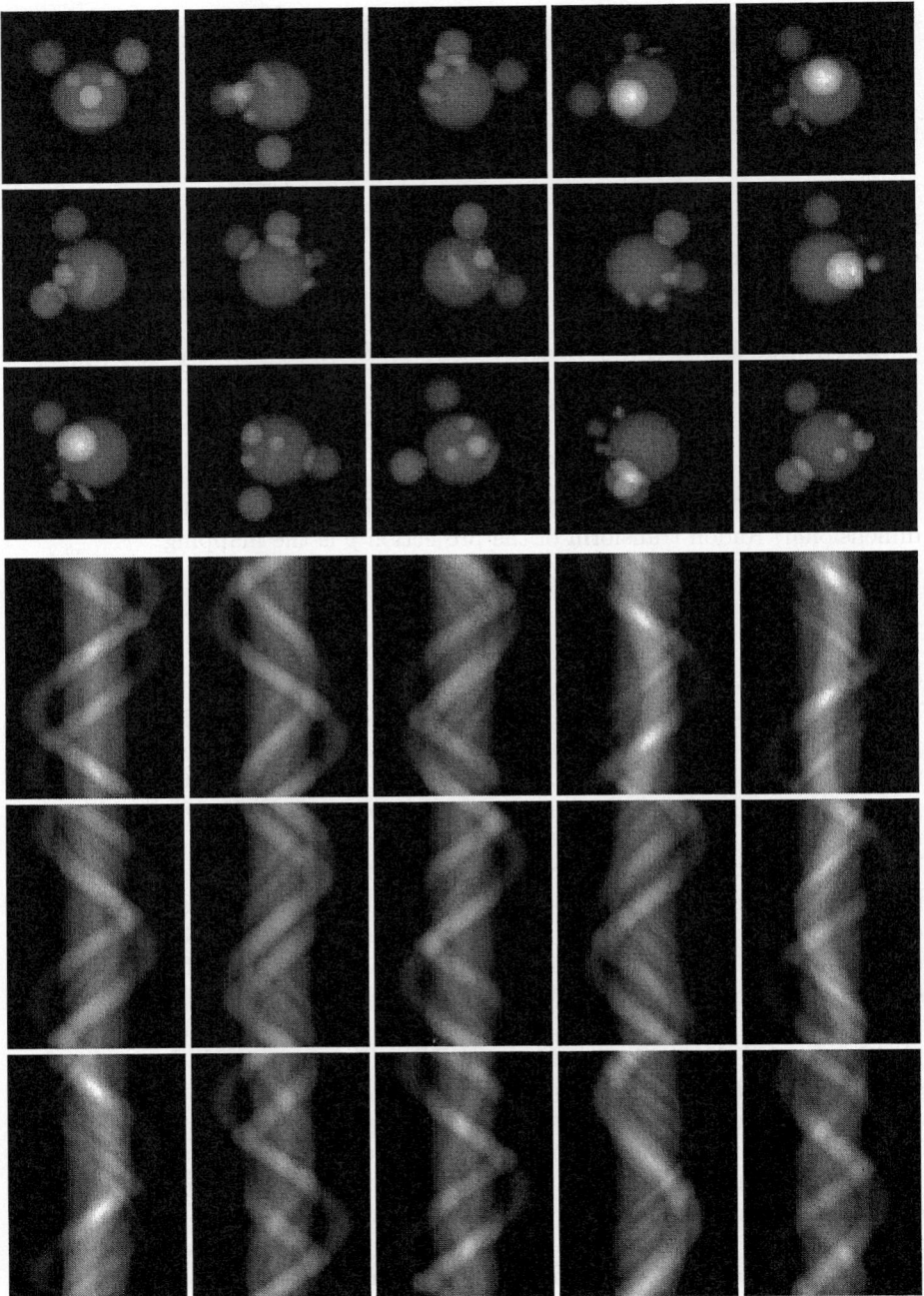

Fig. 1. Noiseless projections (top) of the density map from random directions and their sinograms (bottom)

3 Sinogram Denoising

According to the Common Line Theorem, for each two projections p_i and p_j there are angles α_i and α_j such that $\hat{p}_i(x, \alpha_i) = \hat{p}_j(x, \alpha_j)$ for all x. Due to discretization errors and high level of noise in cryo-electron microscopy images, the images of identical macromolecular complexes satisfy the theorem only approximately: the sinogram rows k_i and k_j corresponding to the true common line between projections $\hat{p}_i$ and $\hat{p}_j$ should be quite similar in the sinograms s_i of p_i and s_j of p_j if the noise level is not too high. Thus, the most similar sinogram rows in the sinograms s_i and s_j are good candidates to determine the common line between the projections p_i and p_j.

The standard approach to reduce the noise in cryo-electron microscopy images is to compute averages of several images. The same idea of computing averages can be exploited also without assuming the clusterability of the projection directions since there is always a common line between each two sinograms s_i and s_j of projections p_i and p_j of the same density map D. Thus, if k_i and k_j are the indices of the sinograms s_i and s_j corresponding to the common line between p_i and p_j, the sinogram rows $s_i[k_i]$ and $s_j[k_j]$ can be replaced by $(s_i[k_i] + s_j[k_j])/2$.

In general, when there are n projections from the same particle, a sinogram row $s_i[k_i]$ can be replaced by $\sum_{j=1}^{n} s_j[k_j]/n$ where the sinogram rows $s_j[k_j]$ correspond to the common lines with $s_i[k_i]$.

The method is described more precisely by Algorithm 1. The algorithm inputs n sinograms $s_1, \ldots, s_n$ of size $l \times m$ and outputs their denoised versions $t_1, \ldots, t_n$.

Algorithm 1. Sinogram denoising algorithm

1: **function** DENOISE-SINOGRAMS$(s_1, \ldots, s_n)$
2:    **for** $i = 1, \ldots, n$ **do**
3:        $t_i \leftarrow s_i$; $c_i \leftarrow 1$
4:    **end for**
5:    **for** $i = 1, \ldots, n-1$ **do**
6:        **for** $j = i+1, \ldots, n$ **do**
7:            Find the most similar sinogram lines $s_i[k_i]$ and $s_j[k_j]$ from s_i and s_j.
8:            $t_i[k_i] \leftarrow t_i[k_i] + s_j[k_j]$; $c_i[k_i] \leftarrow c_i[k_i] + 1$
9:            $t_j[k_j] \leftarrow t_j[k_j] + s_i[k_i]$; $c_j[k_j] \leftarrow c_j[k_j] + 1$
10:       **end for**
11:   **end for**
12:   **for** $i = 1, \ldots, n$ **do**
13:       **for** $k = 1, \ldots, l$ **do**
14:           $t_i[k] \leftarrow t_i[k]/c_i[k]$
15:       **end for**
16:   **end for**
17:   **return** $t_1, \ldots, t_n$
18: **end function**

Clearly, the effectiveness of this method depends on how well we are able to detect the (almost) true common lines between the sinograms of the projections.

In the next section we experiment the method using correlation as the measure of the similarity between the rows.

4 Experiments

We tested the effectiveness of the proposed denoising scheme with an artificial density map of size $64 \times 64 \times 64$ voxels. The density map consists of balls with different weights, radii and locations. We projected the density map to 6300 random directions, computed sinograms of these projections, added to the projections Gaussian noise with different variances, and computed the sinograms of noiseless and noisy projections (see Figure 1 and Figures 2–5, respectively) with angular step-size of 4 degrees. Furthermore, we denoised the sinograms of the noisy projections using the proposed denoising method. The similarity measure used between the sinogram rows was the correlation coefficient.

The results are shown in Figures 2, 3, 4, and 5. Denoising seems to work reasonably well, at least visually. Twilight zone of the straightforward implementation of the method is between signal-to-noise ratios 0.5 and 0.3. Note that the low contrast of the sinograms in Figure 5 is due to the normalization of intensity values.

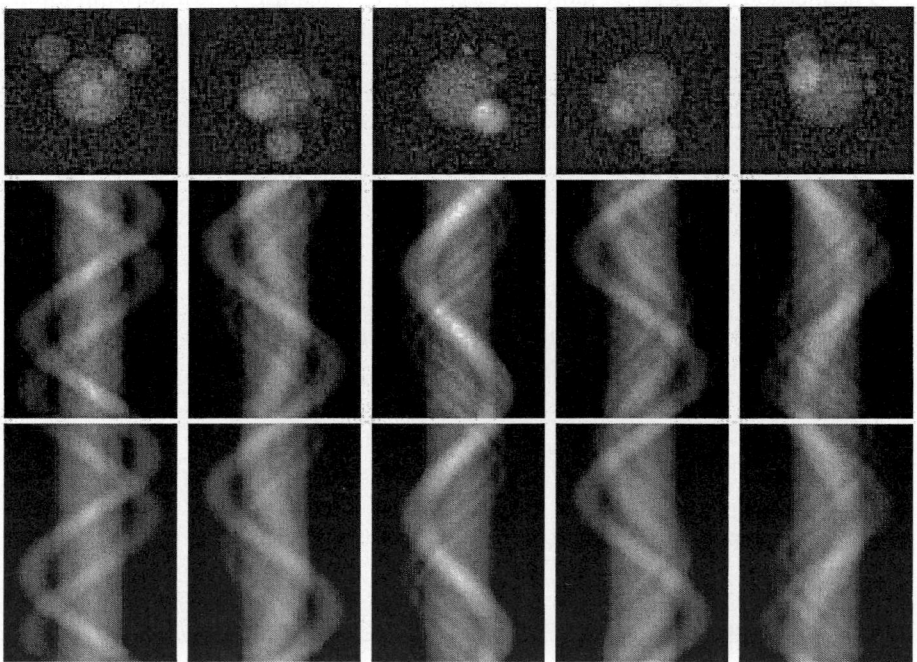

Fig. 2. Projections with signal-to-noise ratio 2 (top), noisy sinograms (middle) and denoised sinograms (bottom)

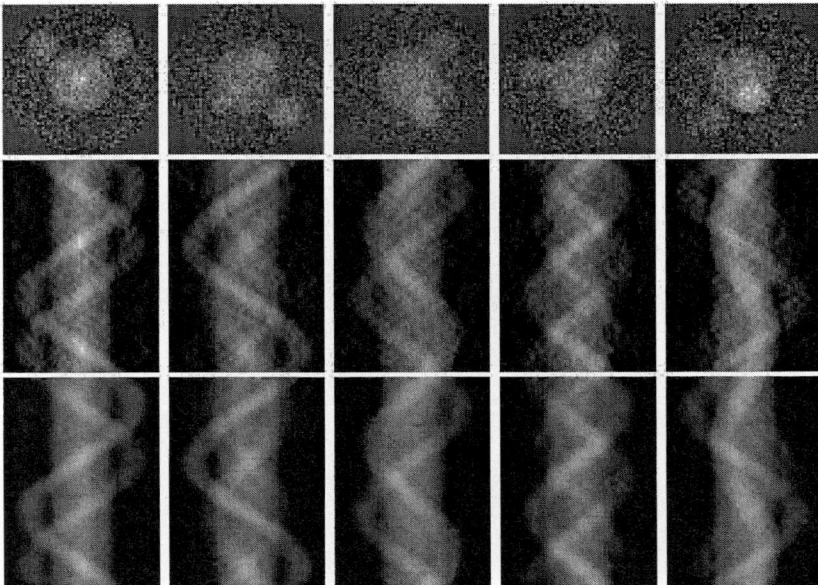

Fig. 3. Projections with signal-to-noise ratio 1 (top), noisy sinograms (middle) and denoised sinograms (bottom)

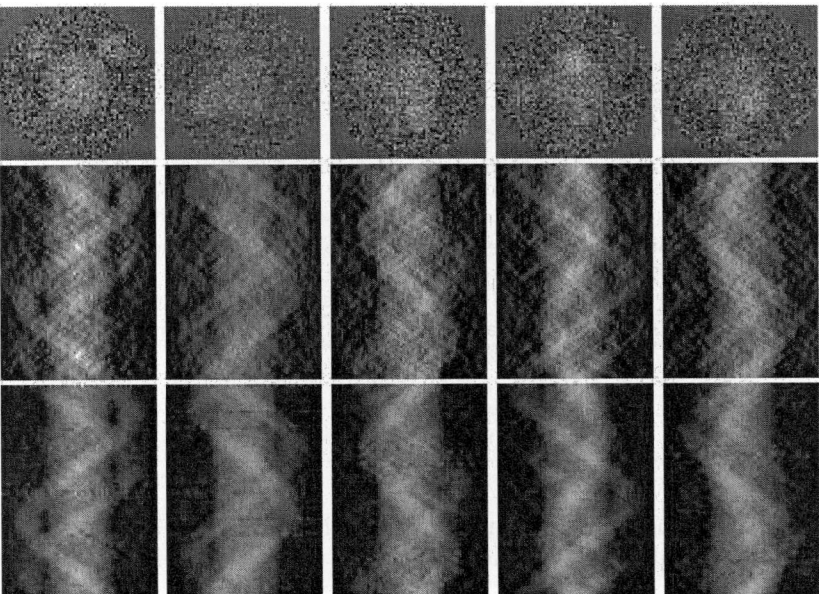

Fig. 4. Projections with signal-to-noise ratio 0.5 (top), noisy sinograms (middle) and denoised sinograms (bottom)

Fig. 5. Projections with signal-to-noise ratio 0.3 (top), noisy sinograms (middle) and denoised sinograms (bottom)

To get more quantitative information about our denoising method, we computed correlation coefficients between noiseless, noisy and denoised sinograms with different signal-to-noise ratios. The results are shown in Table 1.

Table 1. Correlations between noiseless, noisy and denoised sinograms

signal-to-noise ratio	noiseless vs. noisy		noiseless vs. denoised		noisy vs. denoised	
	mean	variance	mean	variance	mean	variance
2	0.9962	1.293e-07	0.9997	5.303e-09	0.9970	7.805e-08
1	0.9851	2.225e-06	0.9971	5.264e-07	0.9910	7.813e-07
0.5	0.9437	4.318e-05	0.9800	3.665e-05	0.9747	1.341e-05
0.3	0.8846	6.772e-04	0.8512	1.450e-02	0.8630	1.123e-02

Clearly, the sinogram denoising improved the correlation coefficients also in this experiment, although the method did not show improvement with respect to average of the correlation coefficient with signal-to-noise ratio 0.3. Note that although all correlations are quite high, the correlations are order of magnitude closer to the maximum correlation (i.e., correlation 1) for all signal-to-noise ratios except the signal-to-noise ratio 0.3.

Overall, the results are very promising, especially as the implementation used in the experiments is a straightforward implementation of Algorithm 1; for example, no additional filtering nor more sophisticated estimation of sinogram row similarities were used.

5 Conclusions

In this paper, we proposed a novel denoising method for cryo-electron microscopy images that, unlike the previously known approaches, does not assume anything about imaging directions of the images.

The described approach is based on denoising the sinograms of the cryo-electron microscopy images and thus it is directly applicable within several orientation search methods that determine the orientations using the pairwise similarities between sinogram rows [9, 10, 13, 14, 15, 16, 17]. We showed experimentally that even a straightforward implementation of the denoising approach is able to reduce noise even when the signal-to-noise ratios are very low.

Although the denoising method seems very promising, there is still plenty of room for improvements. The effectiveness of the method relies on reasonably robust common line detection which could clearly be improved. For example, as the common lines fix the relative orientations of the cryo-electron microscopy images, these constraints could be used to expose false common lines. Furthermore, the possible information about symmetry could be used to improve the signal-to-noise ratio. As future work we plan to study the improvements suggested above and other heuristics to further facilitate the effectiveness of the sinogram denoising approach proposed in this paper. Also, more sophisticated methods to evaluate the success of the denoising shall be investigated. The quality of the reconstruction of the density map depends also on the reconstruction method. Thus, the suitability of different reconstruction algorithms to be used in conjunction with the sinogram denoising is also of our interest.

Acknowledgements

We wish to thank Dennis Bamford, Teemu Kivioja and Esko Ukkonen for helpful discussions on sinograms and cryo-electron microscopy.

References

1. Baker, T.S., Olson, N.H., Fuller, S.D.: Adding the third dimension to virus life cycles: Three-dimensional reconstruction of icosahedral. Microbiology and Molecular Biology Reviews **63** (1999) 862–922

2. Frank, J.: Three-Dimensional Electron Microscopy of Macromolecular Assemblies. Academic Press (1996)
3. Carazo, J.M., Sorzano, C.O., Rietzel, E., Schröder, R., Marabini, R.: Discrete tomography in electron microscopy. In Herman, G.T., Kuba, A., eds.: Discrete Tomography: Foundations, Algorithms, and Applications. Applied and Numerical Harmonic Analysis. Birkhäuser (1999) 405–416
4. Crowther, R., DeRosier, D., Klug, A.: The reconstruction of a three-dimensional structure from projections and its application to electron microscopy. Proceedings of the Royal Society of London A **317** (1970) 319–340
5. Kivioja, T., Ravantti, J., Verkhovsky, A., Ukkonen, E., Bamford, D.: Local average intensity-based method for identifying spherical particles in electron micrographs. Journal of Structural Biology **131** (2000) 126–134
6. Nicholson, W.V., Glaeser, R.M.: Review: Automatic particle detection in electron microscopy. Journal of Structural Biology **133** (2001) 90–101
7. Baker, T.S., Cheng, R.H.: A model-based approach for determining orientations of biological macromolecules imaged by cryoelectron microscopy. Journal of Structural Biology **116** (1996) 120–130
8. Doerschuk, P.C., Johnson, J.E.: *Ab initio* reconstruction and experimental design for cryo electron microscopy. IEEE Transactions on Information Theory **46** (2000) 1714–1729
9. Fuller, S.D., Butcher, S.J., Cheng, R.H., Baker, T.S.: Three-dimensional reconstruction of icosahedral particles – the uncommon line. Journal of Structural Biology **116** (1996) 48–55
10. van Heel, M.: Angular reconstitution: a posteriori assignment of projection directions for 3D reconstruction. Ultramicroscopy **21** (1987) 11–124
11. Ji, Y., Marinescu, D.C., Chang, W., Baker, T.S.: Orientation refinement of virus structures with unknown symmetry. In: Proceedings of the International Parallel and Distributed Processing Symposium. IEEE Computer Society (2003) 49–56
12. Lanczycki, C.J., Johnson, C.A., Trus, B.L., Conway, J.F., Steven, A.C., Martino, R.L.: Parallel computing strategies for determining viral capsid structure by cryo-electron microscopy. IEEE Computational Science & Engineering **5** (1998) 76–91
13. Bellon, P.L., Cantele, F., Lanzavecchia, S.: Correspondence analysis of sinogram lines. Sinogram trajectories in factor space replace raw images in the orientation of projections of macromolecular assemblies. Ultramicroscopy **87** (2001) 187–197
14. Bellon, P.L., Lanzavecchia, S., Scatturin, V.: A two exposures technique of electron tomography from projections with random orientation and a *quasi*-Boolean angular reconstitution. Ultramicroscopy **72** (1998) 177–186
15. Lauren, P.D., Nandhakumar, N.: Estimating the viewing parameters of random, noisy projections of asymmetric objects for tomographic reconstruction. IEEE Transactions on Pattern Analysis and Machine Intelligence **19** (1997) 417–430
16. Penczek, P.A., Zhu, J., Frank, J.: A common-lines based method for determining orientations for $N > 3$ particle projections simultaneously. Ultramicroscopy **63** (1996) 205–218
17. Thuman-Commike, P.A., Chiu, W.: Improved common line-based icosahedral particle image orientation estimation algorithms. Ultramicroscopy **68** (1997) 231–255
18. Mielikäinen, T., Ravantti, J., Ukkonen, E.: The computational complexity of orientation search in cryo-electron microscopy. In Bubak, M., van Albada, G.D., Sloot, P.M.A., Dongarra, J.J., eds.: Computational Science – ICCS 2004. Volume 3036 of Lecture Notes in Computer Science. Springer-Verlag (2004)

19. Brady, M.L.: A fast discrete approximation algorithm for the Radon transform. SIAM Journal on Computing **27** (1998) 107–119
20. Brandt, A., Dym, J.: Fast calculation of multiple line integrals. SIAM Journal on Computing **20** (1999) 1417–1429
21. Lanzavecchia, S., Tosoni, L., Bellon, P.L.: Fast sinogram computation and the sinogram-based alignment of images. Cabios **12** (1996) 531–537

Study of a Cluster-Based Parallel System Through Analytical Modeling and Simulation

Bahman Javadi, Siavash Khorsandi, and Mohammad K. Akbari

Department of Computer Eng. and Information Technology,
Amirkabir University of Technology, Hafez Ave.,Tehran, Iran
{javadi, siavash, akbari}@ce.aut.ac.ir

Abstract. In this paper we present a new analytical model for cluster-based parallel systems based on Multi-Chain Open Queuing Network. The proposed model is general enough in terms of node design and clustering which can be extended to any parallel system. To have a good estimation of the model parameters, a benchmark suite has been applied to a real system and parameters calibration is performed based on actual measurements. In this study the hypercube topology was chosen for the interconnection network of the system. We present an analytical modeling and validate it by simulation. The QNAT is used for analytical modeling and the OMNet++ simulator has been used to carry out the simulations. Comparison of simulation and analytical results verifies validity of the model. System workload in simulations is generated using both analytical and experimental distributions and a close match with analytical results in both cases is observed.

1 Introduction

Nowadays, due to a great demand for parallel processing systems in many high performance applications, there is a subtle need for tools and techniques to properly evaluate and analyze the performance of these systems. Such a performance study may be used to: select the best architecture platform for an application domain, predict the performance of an application on a large scale configuration of an existing architecture, select the best algorithm for solving the problem on a given hardware platform and identify application and architectural bottlenecks in a parallel system to suggest application restructuring and architectural enhancements [1].

In this paper we intend to propose a model to evaluate a parallel processing system based on commodity clustering. Each node in this system is modeled with a Queuing Network Model (QNM). The model proposed here is scalable and can be used in any similar system. We estimate parameters of this model using data extracted from benchmark behaviors on a real parallel system. In other words, measurement technique is used for obtaining the model parameters. This model is solved analytically using QNAT [2]. We also developed a software model for our simulation studies which is performed with OMNeT++ [3]. The results obtained through these studies are used to validate the model.

There are few prior researches and papers on analytical modeling of cluster computing systems [4], [5]. In [4], a distributed shared memory system is considered which is different from our message passing cluster system. In [5] analytic modeling and simulation are used to evaluate network servers' performance implemented on clusters of workstations. More specially, in order to determine the potential performance benefits of locality-conscious network servers, they developed a simple queuing network to model locality-conscious and oblivious servers.

The rest of the paper is organized as follows. We describe the system under study, the proposed system model and the techniques used to estimate system parameters in Section 2. In Section 3, we present the analytical results, description of the system model and scenarios used in simulations. Numerical results are covered in Section 4 and they are also compared with analytical results. Finally, Section 5 summarizes our findings and concludes the paper.

2 System Model

In this part, we introduce a model that has been designed to evaluate the Beowulf cluster systems. This model is a general class of cluster systems which can represent various configurations. For instance, we choose the Hypercube topology in which there exists 2^d nodes where d is the dimension of cube and each node has d-connection to the adjacent nodes. Since there is a network switch to connect the dispatcher to all nodes, therefore each node has a dedicated network interface for connection to the switch. The network switch is only used for the dispatch operations. In the following sections, we introduce the system under study and present the model of a node in the system in detail as well.

2.1 The System Under Study

The system used in this study, is called *AkuCluster*. This is a Beowulf-class [6] system that is made in Amirkabir University of Technology for academic studies. This system has 32 Pentium III Processing Elements with Red Hat Linux operating system and Fast Ethernet communication network in a flexible interconnection network. Flexible topology gives the capability to choose the best topology for various applications. By changing the topology, proper routing tables will automatically be assigned to the nodes. Although, the routing tables are static but routing algorithm is designed to balance the network load.

There are two phase for executing a parallel program on this system. First phase, *the Dispatch Phase*, each jobs distribute to the each node. Then, the *Processing Phase* will be started, and nodes process the jobs and communicate to each other up to end of execution. That is, the *Dispatcher* node performs at dispatch phase and *don't has any role in the processing phase*. After that execution complete the program results gather in the dispatcher node to represent to user. *It's obvious we don't have a master/slave environment for parallel processing.* There is cluster management software which can be use at two execution phases. This machine provides the opportunity to test all of our ideas.

2.2 Our Modeling

Our model is designed on the basis of Queuing Network Model. Fig. 1 depicts a primitive structure of our model for $d=2$ which is a Multi-Chain Open Queuing

Network with two classes of customers. One class is for jobs (class number one) and another class is for data (class number two). As we know the dispatcher node distributes workload to all nodes and collects the results. So, in this model the dispatcher generates jobs and dispatches to each node. These jobs enter the Network Interface (NI) queue and then CPU queue for the processing. Due to probable data dependency of parallel jobs during processing, they may need to communicate with each other. Therefore, the process which is exited from CPU queue has a probability to be sent to other nodes or to the dispatcher when is a completed job. P is the probability when it's sent to other nodes and $1-P$ when goes back to the dispatcher. In this case, the model will have a loop for the chain of class number one (Fig. 1). In the each iteration of the loop, a part of job proceeds and communicates with other nodes through NI queue. When class number one processes reach the output the NI queues (queues 5 and 6) they get duplicated, one copy goes toward feedback path to CPU queue and another proceeds to neighbor nodes.

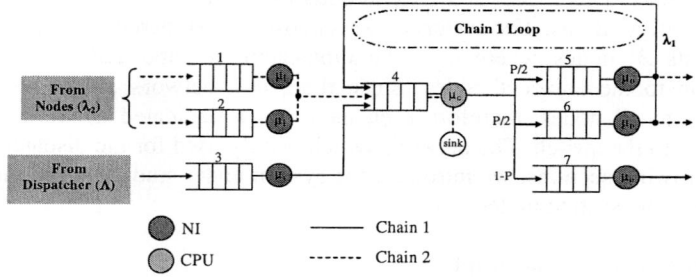

Fig. 1. Model for a Node in the Cluster System ($d=2$)

In other words, processes of class number one (jobs) after getting service from CPU queue are sent to a neighbor node with data format and P probability. Also, there is one additional path to CPU queue (feedback path), for continuing the process of jobs along with data which has come from neighbor nodes. It has been shown that in the real system, average input workload is nearly same as output workload and our model considers this equality. In the mean time, the processes of class number two (data) are consumed in the CPU queue (sink point).

In our model, we didn't employ any disk systems. Because, disks are originally designed for initial/temporary data storage, in addition applying disk systems for virtual memories diminishes the system performance dramatically. In the model, the dispatcher node is simply modeled as a generator and sinker. It generates jobs and sinks completed jobs. The network switch model is a delayed bi-directional buffer, which redirects received packets to their destinations.

2.3 System Model Parameters Estimation

To have estimations for the parameters in the model, we must estimate service rate for NI queues, CPU queue and routing probability P. For this, we applied measurement techniques on a real system using a benchmark suite included NAS Parallel

Benchmarks (NPB) [7] and High Performance Linpack (HPL) [8] benchmarks. NPB benchmarks are compiled for A and B classes and HPL is compiled for problem size of equal to 5k and 10k, so the benchmark suite has 16 programs. We ran these benchmarks on our real system (AkuCluster) and monitored status of the programs on one node and the dispatcher node. Measurement tool, which was used for monitoring of network traffic and execution time was ProcMeter-3.0 [9]. This tool can be run as background process and log the status of a node in specific intervals

Fig. 2 shows input/output network traffic, as average byte was transmitted/received on instance node of the real system. This figure shows that average network input and output traffic to/from a node in this system is nearly equal. This fact confirms our model assumption.

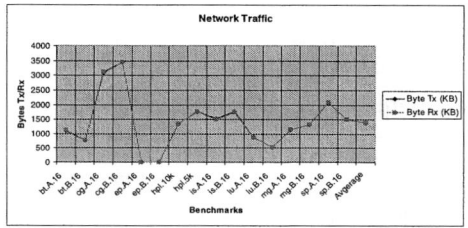

Fig. 2. Average Input/Output Network Traffic on a node

To compute the service time of NI queue, the average packet size of each NI has to be divided by the line bit rate. Since the system uses Fast Ethernet standard, so the typical bit rate of line should be 100Mbps. To have an accurate result, we measured real bit rate of line with ping-pong test and variable packet size. As a result the maximum bit rate was measured for 88Mbps with packet size above 2Mbytes. To measure the packet size distribution, the IPTraf-2.7.0 network monitoring tool [10] was used and its measurements show that the average size of sending and receiving packets of all benchmarks is about 733.1848396 bytes. Since average input network traffic is nearly equal to output network traffic, therefore the average bit rate of sending/receiving packets should be half of the total line bit rate. Thus,

$$\text{Average Packet Tx/Rx Time} = \frac{733.1848396 \ Bytes}{88 \times \frac{1}{2} \ Mbps} = 0.000133306 \ sec ,$$

and service rate of NI queue can be computed as follows:

$$\mu_o = \mu_i = \frac{1}{0.000133306} = 7501.518994$$

The CPU queue service time is the inter-arrival time between two consecutive packet transmissions. Average packet transmission rate for each benchmark can be obtained from ProcMeter output log file which is depicted in Fig. 3. Inter-arrival time is reverse of this rate, but this computed values don't give the exact value of CPU service time, because the whole CPU time isn't dedicated for execution of benchmarks. Thus, the exact CPU queue service time is average of numbers which is computed as follows:

CPU queue service time =
Average Percentage of CPU usage × Packet Transmission Inter-arrival time
$$= 0.035114634 \text{ sec}$$

and it's service rate,

$$\mu_c = \frac{1}{0.035114634} = 28.4781549$$

This rate is much smaller than service rate of NI queue, which is completely anticipated.

At last, we have to compute the routing probability P. This value can be calculated by division of the transmission packet rate to neighbor nodes to the total transmission packet rate. As it was mentioned above, we can write:

$$P = \frac{\text{Transmission Packets Rate to Neighbor Nodes}}{\text{Total Transmission Packet Rate}}$$

$$P = \frac{1382.017959}{1545.900312} = 0.893989055$$

Now, all the system parameters are ready and we can analyze the model.

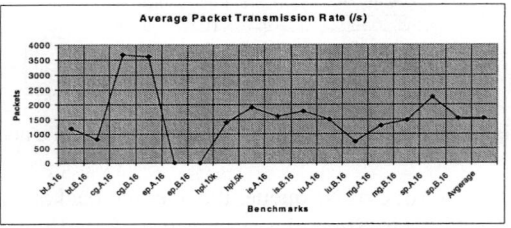

Fig. 3. Average Packet Transmission Rate

3 Analytical Solution and Simulation Scenario

3.1 Analytical Solution

In the previous sections, we presented the node model and its parameter estimation. We will now give an analytical solution for this model. Based on Jackson queuing network theory [11], we apply a product form solution where each node is analyzed in isolation. We assume the following assumptions in our analysis:

1. All queues in the model are M/M/1.
2. The service time of Network Interface input, output and CPU queues have exponential distribution with parameters μ_i, μ_o and μ_c respectively.
3. Input rate of chain one and two are λ_1 and λ_2 respectively. Also, generation rate of the dispatcher is Λ.

In Fig. 1 you can see the model with mentioned assumptions. So we can obtain following equations:

$$\lambda_1 = \lambda_2$$
$$\lambda_1 = \frac{\Lambda}{1-P} \quad (1)$$

Study of a Cluster-Based Parallel System Through Analytical Modeling and Simulation   1267

To analysis this model, the utilization of CPU queue (ρ) was chosen as a metric. Now, the utilization equation can be given as follows:

$$\rho = \lambda_2 \cdot \frac{1}{\mu_c} + (\Lambda + \lambda_1) \cdot \frac{1}{\mu_c} \qquad (2)$$

and with equations (1), we can obtain:

$$\rho = \lambda_2 \cdot \frac{1}{\mu_c} + \lambda_1 \left[(1-P) \cdot \frac{1}{\mu_c} + \frac{1}{\mu_c} \right]$$

$$\Rightarrow \rho = \lambda_1 \cdot (3 \cdot \frac{1}{\mu_c} - P \cdot \frac{1}{\mu_c}) \qquad (3)$$

However, after assigning a value for utilization of CPU queue, λ_1, λ_2 and Λ can be calculated from equations (3) and (1) respectively. In this regard, Table 1 depicts the above rates based on three given values of ρ with the assumption of $\mu c=0.03511$ and $P=0.894$. Then the QNAT (Queuing Network Analysis Tool) [2] was used to analyze our model using the values of Table 1 and also obtain the metrics of each queue.

Table 1. Utilization of CPU queue and Input Rates of the Model

ρ	$\lambda_1=\lambda_2$	Λ
0.3	4.1	0.43
0.6	8.11	0.89
0.9	12.17	1.29

The first output metric is the average delay in each chain of the model. These delays can be obtained as follows:

$$T_{chain1} = T_{queue3} + Visit\ Count_{queue4} \cdot (T_{queue4})$$
$$+ Visit\ Count_{queue5\ (6)} \cdot (T_{queue5(6)}) + T_{queue7} \qquad (4)$$

$$T_{chain2} = T_{queue1(2)} + T_{queue4}$$

In addition, we choose input rate of CPU queue (queue No. 4) and its utilization as other items of output metrics. The results of analytical model analysis are listed in Table 2. Comparing the values of CPU queue utilizations in this table and Table 1, shows a minor difference between them which is related correctly to the implementation of QNAT analysis formula [2]. It is obvious that visit count of queue No.5/No.6 in the chain one is one unit less than visit count of CPU queue in the same chain. All visit counts in the chain two are uniquely one.

Table 2. Analytical Modeling Results

ρ	λ_{cpu}	Avg. delay of chain 1	Avg. delay of chain 2	Visit count (CPU queue)
0.286378	8.1566	0.464958354	0.049265017	9.434
0.569597	16.2232	0.7703785	0.081640034	9.434
0.85457	24.3398	2.278388644	0.241488051	9.434

3.2 Simulation Scenario

After analytical solution, the OMNeT++ (Objective Modular Network Testbed in C++) tool was used to simulate the system model. OMNeT++ tool is an object-oriented modular discrete event simulator which accepts the description of model topology in NED language. The NED language supports modular description of a network.

For example, to model the hypercube interconnection network, we provide a topology template in this regard, which gets the number of dimensions (d) to generate a hypercube network of $2d$ nodes. Also, the dispatcher node and the switch model are included in the topology template. Since each component in this template is a separate module, so it is very flexible for any change. The node's model requires $2d+3$ queues, of which $2d+2$ queues are used for NI queues and one queue for the CPU queue. As mentioned before, the dispatcher node posses a message generator/sinker, so it can be modeled simply with a fixed structure.

Table 3. The Parameters of the System Model

NI queue Service Time(sec)	0.00013
CPU queue Service Time (sec)	0.03511
Routing Probability (P)	0.894
Switch Delay (sec)	0.00001

4 Numerical Results

To verify the system model, we compared simulation and analytical results. For this, we did run two different simulations. In the first simulation runs, we assume that all queues in the system have inter-arrival times with exponential distribution, and the probability distribution of the service time, also exponential with mean $1/\mu$ second. In the mean time, to have an accurate simulation, we need the network switch delay. Fineberg et al. in [12] have measured the delay of Fast Ethernet switches about 10μs.

Now, all the model parameters can be listed as in Table 3. Simulations are done for each value of Λ and this parameter is obtained from analytical modeling. In other words, the dispatcher node generates a specific workload for the system, with this the simulation continues until no job remains unprocessed. Finally, the completed processes are captured by sink in the dispatcher node.

We run each simulation for 20,000 jobs which are generated by dispatcher. Our experiment has shown that no considerable changes in the output values would be observed if the number of jobs is increased. These jobs have exponential distribution with parameter Λ. Output results from this simulation for three values of dispatcher generation rate are shown in Table 4. The outcome of comparison of analytical modeling results (Table 2) with simulation results proved the validation of our model. The difference between average delay of chains in the simulation runs and analytical modeling results is due to the switch delay.

Table 4. Simulation Results with Exponential Distribution Workload

Λ	ρ	λ_{cpu}	Avg. delay of chain 1	Avg. delay of chain 2	Visit count (CPU queue)
0.43	0.27	7.8	0.60	0.09	9.41
0.89	0.54	15.24	0.92	0.12	9.43
1.29	0.81	22.9	2.32	0.25	9.46

In the second simulation, we write two experimental distribution functions for NI and CPU queues. This is done in two steps. First, we generate "unif (0, 1)" random numbers, i.e., random numbers that are uniformly distributed in the range 0 through 1. Next, we transform these into random numbers having the desired distribution. For the transformation step, let Y be a unif (0, 1) random variable, then

$$X = G^{-1}(Y) \rightarrow Y = G(X) \tag{5}$$

That is, the distribution of X is $G(x)$. The $G(x)$ is a mapping function which is resulted from measured data of section 2.3, and is used to provide the distribution functions for NI and CPU queues distinctively. We applied these functions, and rerun the first simulation to provide the results which are shown in Table 5. By comparing the results of two simulations, we come to the point that average delay of two chains in the second run has been increased and in the same time the CPU queue utilization is increased too, which is completely rational. It should be noted that the utilization rates of the second simulation runs are closer to the analytical modeling results.

Table 5. Simulation Results with Experimental Distribution Workload

Λ	ρ	λ_{cpu}	Avg. delay of chain 1	Avg. delay of chain 2	Visit count (CPU queue)
0.43	0.28	7.73	0.66	0.25	9.41
0.89	0.57	15.54	1.10	0.34	9.41
1.29	0.83	23.03	2.54	0.57	9.43

To have a concrete proof for the validity of our model, we extend our experiment by defining one more parameter, called CPU queue length Probability Distribution Function (PDF). In this regard, we calculated CPU queue length PDF for three values of utilization (ρ). For each value of the utilization, we provided PDFs from two previous simulations and the analytical formula in [11]. It has been shown that the queue length PDF of a simple M/M/1 queue is *geometric* with parameter ρ, and can be written as follows:

$$P(n) = \rho^n \cdot (1-\rho) \tag{6}$$

The values of ρ were obtained from Table 2 of analytical solution. The results of this experience are shown in the Fig. 4 for the CPU queue utilizations of 0.85. As it can be seen, these figures have similar geometric shapes and this is a strong proof for the validity of our model. To justify the extra delay times of each chain in the simulation runs, one has to consider that the probability of CPU queue length is more than correspondence probability in analytical model curve, especially for low quantity of queue lengths.

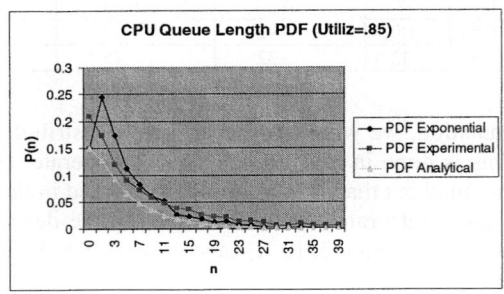

Fig. 4. CPU queue length Probability Distribution Function ($\rho=0.85$)

5 Conclusions

In this paper we developed a new model based on the Queuing Networks to evaluate the cluster-based parallel systems. We run a benchmark suite on a real system to measure the required data. These data were used for estimation of model parameters. We used QNAT to get the analytical solution of the model and then OMNet++ for simulation. This gives a good flexibility for any modification and future expandability. Verification of the model is done by comparing analytical results with two different simulation runs. A close match can be observed between analysis and simulations. Therefore, we can use the same approach for other parallel configurations and performance studies.

Acknowledgments. We would like to thank Dr.A.T.Haghighat and M.Kalantari for remarks and discussions and A. Jalalzadeh for his comments and help to prepare this paper.

References

1. Lei Hu and Ian Gorton. Performance Evaluation for Parallel Systems: A Survey, in *Technical Report No. UNSW-CSE-TR-9707, Department of Computer Systems, School of Computer Science and engineering, University of NSW*, Oct. 1997.
2. D. Manjunath, D.M. Bhaskar, Hema Tahilrmani, Sanjay K.Bose, M.N. Umesh. QNAT: A Graphical Tool for the Analysis of Queuing Networks, *IEEE TENCON'98 International Conference*, New Delhi, India, Dec. 1998.
3. Nicky van Foreest. Simulation Queuing Networks with OMNet++, in *Tutorial of OMNnet++ Simulator*, Department of Telecommunications, Budapest University of Technology and Economics, Apr. 2002.
4. Xing Du, Xiaodong Zhang, Zhichun Zhu. Memory Hierarchy Consideration for Cost-Effective Cluster Computing, *IEEE Transaction on Computers*, Vol.49, No. 9, Sep. 2000.

5. E. V. Carrera and Ricardo Bianchini. Analytical and Experimental Evaluation of Cluster-Based Network Servers, in *Technical Report 718*, University of Rochester, Aug. 1999.
6. T. Sterling. Beowulf Cluster Computing with Linux, *MIT Press*, Cambridge, MA, 2002.
7. Rob F. Van Der Wijngaart. NAS Parallel Benchmarks, Version 2.4, in *NAS Technical Report NAS-02-007*, Oct. 2002.
8. Petitet, R. C. Whaley, J. Dongarra, A. Cleary. HPL- A Portable Implementation of the High-Performance Linpack Benchmark for Distributed-Memory Computers, Innovative Computing Laboratory, Department of Computer Science, University of Tennessee, Sep. 2000.
9. ProcMeter System Monitor Tool, URL=http://www.gendanken.demon.co.uk/procmeter3/
10. IPTraf Network Monitoring Software, URL=http://cebu.mozcom.com/riker/iptraf/
11. D. Bertsekas, R. Gallager. Data Networks, *Prentice Hall*, New Jersey, 1992.
12. S. A. Fineberg and K. T. Pedretti. Analysis of 100Mb/s Ethernet for Whitney Commodity Computing Testbed, in *NAS Technical Report NAS-97-025*, Oct. 1997.

Robust Parallel Job Scheduling Infrastructure for Service-Oriented Grid Computing Systems

J.H. Abawajy

School of Information Technology,
Deakin University,
Geelong, VIC., Australia

Abstract. Recent trends in grid computing development is moving towards a service-oriented architecture. With the momentum gaining for the service-oriented grid computing systems, the issue of deploying support for integrated scheduling and fault-tolerant approaches becomes paramount importance. To this end, we propose a scalable framework that loosely couples the dynamic job scheduling approach with the hybrid replications approach to schedule jobs efficiently while at the same time providing fault-tolerance. The novelty of the proposed framework is that it uses passive replication approach under high system load and active replication approach under low system loads. The switch between these two replication methods is also done dynamically and transparently.

1 Introduction

Grid computing [6] has emerged as a global platform for coordinated sharing of services (i.e., compute, communication, storage, distributed data, resources, applications, and processes). Recent trends in grid computing development is moving towards a service-oriented architecture as exemplified by the Open Grid Services Architecture (OGSA) [7]. Such platforms connect service providers and consumers of services and data, while shielding them from details of the underlying infrastructure. As a result, service-oriented grid computing is attracting increasing attention from the grid computing research community. Although Grid computing systems can potentially furnish enormous amounts of computational and storage resources to solve large-scale problems, grid computing systems are highly susceptible to a variety of failures including node failure, inter-connection network failure, scheduling middleware failure, and application failure. Due to these vulnerableness, achieving large-scale computing in a seamless manner on grid computing systems introduces not only the problem of efficient utilization and satisfactory response time but also the problem of fault-tolerance.

There are numerous grid scheduling policies (e.g., [15]). Unfortunately, fault-tolerance have not been factored into the design and development of most existing scheduling strategies. Research coverage of fault tolerant scheduling is limited as the primary goal for nearly all scheduling algorithms developed so far has been high performance by exploiting as much parallelism as possible. One of the reasons for this is that achieving integrated scheduling and fault-tolerance goal is a difficult proposition as the job scheduling and fault-tolerance are difficult problems to solve in their own right. However, with the momentum gaining for the service-oriented grid computing systems, the issue of deploying support for integrated scheduling and fault-tolerant approaches becomes paramount importance [2], [19]. Moreover, as grids are increasingly used for applications requiring high levels of performance and reliability, the ability to tolerate failures while effectively exploiting the variably sized pools of grid computing resources in an scalable and transparent manner must be an integral part of grid computing systems [9],[21], [8]. To this end, we propose a fault-tolerant dynamic scheduling policy that loosely couples dynamic job scheduling with job replication scheme such that jobs are efficiently and reliably executed. The novelty of the proposed algorithm is that it uses passive replication approach under high system load and active replication approach under low system loads. The switch between these two replication methods is also done dynamically and transparently.

The rest of the paper is organized as follows. In Section 2, a formal definition of the fault-tolerant scheduling problem is given. This section also establishes the fact that, to a large extent, the problem considered in this paper has not been fully addressed in the literature. Section 3 presents the proposed fault-tolerant scheduling policy. Finally, the conclusion and future directions are presented in Section 4.

2 Problem Statement and Related Work

Fault-tolerance is a major issue in Grid computing. This is because as the system increases both in size and complexity, the possibility of a component (e.g., a node, link, scheduler) failure also increases. The primary sources of failures are the system software (e.g., kernel panics, device driver) bugs, hardware failures (e.g., memory errors), link outage, reboots due to resource exhaustion (e.g., file descriptors) and so forth. Thus, the ability to tolerate failures while effectively exploiting the Grid computing resources in a scalable and transparent manner must be an integral part of Grid computing infrastructure. However, fault-tolerance has received the least attention in Grid computing literature as access to remote resources was the main motivation for building Grid computing, and it remains the primary goal today. Currently, potential users have to spend a significant amount of time and effort in order to use Grid computing, which can become a serious obstacle to its adoption and use. In this section, we formulate the problem and discuss some related works.

2.1 Problem Statement

The fault-tolerant scheduling problem (FTSP) addressed in this paper can be formally stated as shown in 1. In this paper, we assume that the system components may fail and can be eventually recovered from failure. Also, we assume that both hardware and software failures obey the fail-stop [16] failure mode. As in [11], we assume that faults can occur on-line at any point in time and the total number of faulty processors in a given cluster may never exceed a known

---

Given: A set of n jobs, $\mathbf{J}=\{J_1, ..., J_n\}$, where each job, J_i, arrives in a stochastic manner into a system composed of m independent clusters, $\mathbf{S}=\{C_1,...,C_m\}$.

1. Each job, J_i, can be decomposed into t tasks, $\mathbf{T}=\{T_1,...,T_t\}$. Each task T_i executes sequential code and is fully preemptable.
2. Each cluster, C_j, is composed of $\mathbf{P}$ shareable (i.e., community-based) processors. Each processor may fail with probability f, $0 \leq f \leq 1$, and be repaired independently.

FTSP: Our goal is to design an on-line scheduling policy such that:

1. applications are efficiently and reliably executed to their logical termination;
2. mean response time is minimized; and
3. the scheduler has no knowledge of: (1) the service time of the jobs or the tasks; (2) the job arrival times; (3) how many processors each job needs until the job actually arrives; (4) and the set of processors available for scheduling the jobs.

Fig. 1. Fault-tolerant grid scheduling problem

fraction. We also assume that node failures are independent from each other [21]. In addition, we assume that every cluster scheduler in the system is reachable from any other cluster scheduler unless there is a failure in the network or the node housing the cluster scheduler. A scheme to deal with node, scheduler and link failures is discussed in [3].

2.2 Related Work

A variety of successful Grid infrastructures that focuses on simplifying access and usage of Grid computing has been developed over the past few years (e.g., [10]). These infrastructures have allowed a great deal of Grid applications, tools, and systems development[9]. Recently, interest in making Grid computing systems fault tolerant has been receiving attention [3],[9],[21], [20]. For example, several fault detection service architecture have been developed for grid computing systems (e.g., [4], [17], [18]).

Fault-tolerance in the context of Grid computing can be generally divided into three main categories:

1. application−level fault tolerance − deals with reliability techniques incorporated within the application software;
2. system−level fault tolerance − involves reliability techniques incorporated within the system hardware such as a workstation and network; and
3. middleware−level fault tolerance − deals with the reliability of grid middlewares;

While both application-level fault-tolerance and system-level fault-tolerance have received some attention in Grid computing, middleware-level fault-tolerance has not [3]. Middleware-level services in Grid environments encompasses Grid brokers (i.e., schedulers), information services components (e.g., resource discovery service), security components and so forth. These components hide the complexity of resource discovery, management and scheduling from the end-user. However, these middleware services can fail for a number of reasons. For example, when the node they are running on fails or the capacity of the system is exhausted. Also, when middleware services are interoperating with other systems, this can lead to failures. Currently, system administrators and users manually handle middleware failures. Unless significant support is provided from a Grid resource management system, the necessity of manually performing this task would significantly slow down the utilization and hence proliferation of Grid systems.

In this paper, we will focus on the scheduling middleware. Although job scheduling and fault-tolerance are active areas of research in cluster computing environments, these two areas have largely been and continue to be developed independent of one another each focusing on a different aspects of computing. Research in scheduling has focused on efficiency by exploiting as much parallelism as possible while assuming that the resources are 100% reliable [1],[15]. Also, existing solutions for grid computing systems, to a large extent, are based on requiring static and dynamic application and system resource information, and performance prediction models. This kind of information is not always available and is often difficult to obtain. Moreover, most of the conventional grid-based systems use a static scheduling model (e.g., LSF [22]).

Similarly, checkpoint-recovery [11] and job replication [20] techniques are popular fault-tolerance approaches on distributed systems. However, as noted in [14], these fault-tolerant approaches typically ignore the issue of processor allocation. This can lead to a significant degradation in response time of the applications [14] and to counter this effect an efficient job scheduling policy is required.

Some studies have addressed scheduling and fault-tolerance jointly, but under some unrealistic assumptions such as requiring a particular programming paradigm [13], a membership services [12], and ample available processors for replica scheduling [20]. The problem with some of these approaches such as the state machine approach (which has embodied in toolkits like ISIS [5]) is that they make heavy demands, do not scale well and have proven hard to apply to networked environments [12]. Also, most of these approaches require a specialized and complex software layer that must be installed at each computation node (e.g., [12]). In addition, these systems apply to one cluster and one job situation. We have not found any literature for multi-programmed and multi-cluster environments. Also, all these systems use static scheduling policy whereas we focus here on the dynamic fault-tolerant scheduling approach.

3 Fault-Tolerant Scheduling Policies

In this section, the proposed fault-tolerant scheduling policy is discussed. The proposed policy is called *Dynamic Fault-Tolerant Scheduling (DFTS)* policy and it is a fault-tolerant version of the *Adaptive Hierarchical Scheduling (AHS)* policy

[1] augmented with a scheme that automatically replicates jobs and tasks over several sites and processors, keep track of the number of replicas, instantiate them on-demand and delete the replicas when the primary copies of the jobs and tasks successfully complete execution. In DFTS, the core system architecture is designed around **L**-levels of virtual hierarchy, which we refer to as a *cluster tree*, as shown in Figure 2. At the top of the cluster tree, there is a *system scheduler* while at the leaf level there is a *local scheduler (LS)* for each node. In between the *system scheduler* and the *local schedulers*, there exists a hierarchy of *cluster schedulers (CS)*.

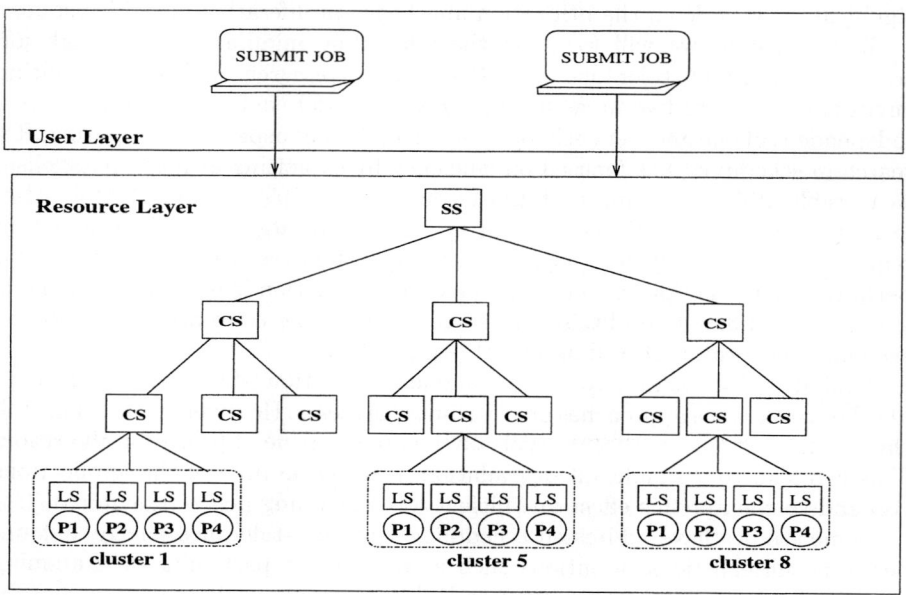

Fig. 2. An example of cluster tree (SS: System scheduler; CS: Cluster scheduler; LS: Local scheduler; and P_i: workstation i)

We refer to all processors reachable from a given node in the cluster tree as its *partition-reach*. We associate a parameter called *base load level* with each node in the cluster tree. For non-leaf nodes, the base load level is set to zero. For all the leaf-level nodes, the base load level is the same as the multiprogramming level (MPL) of the node. The MPL parameter of a node controls the maximum number of tasks that can concurrently execute at any given time. Since the processors may have different processing speed, the MPL of processor P_i is determined as follows:

$$MPL(P_i) = \left\lceil \frac{speed(P_i) \times \text{Base MPL}}{speed(P_{slow})} \right\rceil \quad (1)$$

where $speed(P_{slow})$ is the speed of the slowest workstation in the system.

The DFTS policy has two main components namely; *Fault Management* component and *Job and Task Scheduling* component. Without lose of generality, we assume that all incoming jobs are submitted to the system scheduler where they are placed in the *job wait queue* until a placement decision is made. As in [20], the user specifies if fault-tolerance is required and the number of desired replicas t the time of job submission. We now describe these two components in detail.

3.1 Failure Management

The policy maintains some state information for failure and recovery detections in Application Status Table (AST). Also, a fail-over strategy is used when a link or a node failure is detected. A detailed discussion of the fail-over strategy is given in [2], [3]. In this section, we present the replica creation, placement and monitoring components of the failure management subsystem.

Job Replication
The replica creation and placement ensures that a job and its constituent task are stored in a number of locations in the cluster tree. Jobs are replicated over clusters while tasks are replicated over processors. Specifically, When a job with fault-tolerance requirement arrives into the system, DFTS undertakes the following steps:

1. create a replica of the job;
2. keep the replica and send the original job to a child that is alive and reachable; and
3. update the application status table (AST) to reflect where the job replicas are located. This process recursively follows down the cluster tree until we reach the lowest level cluster scheduler (LCS) at which point the replica placement process terminates.

Replica Management
The DFTS monitors applications at job-level (between non-leaf nodes and their parents) and at task-level (between leaf nodes and their parents). A monitoring message exchanged between a parent and a leaf-level node is called a *report* while that between non-leaf nodes is called a *summary*. A report message contains status information of a particular task running on a particular node and sent every $REPORT$-$INTERVAL$ time units. In contrast, the *summary* message contains a collection of many reports and sent every $SUMMARY$-$INTERVAL$ time periods such that $REPORT$-$INTERVAL < SUMMARY$-$INTERVAL$.

When a processor completes execution of a task, the report message contains a $FINISH$ message. In this case, the receiving scheduler deletes the corresponding replica and informs the backup scheduler to do the same. When the last replica of a given job is deleted, the job is declared as successfully completed. In this case, the cluster scheduler immediately sends a summary message that contains the $COMPLETED$ message to the parent scheduler, which deletes the copy of the job and forward the same message to its parent. This process continues recursively until all replicas of the job are deleted.

Replica Management
After each assignment, the children periodically inform their parents the health of the computations as discussed above. If the parent does not receive any such message from a particular child in a given amount of time, then the parent suspects that the child has failed. In this case, it notes this fact in the AST and sends a request for report message to the child. If a reply from the child has not been received within a specific time frame, the child is declared dead. The replica of a job is then scheduled on a helath node.

3.2 Job and Task Scheduling

Self-scheduling
The DFTS policy is demand-driven where nodes in the system look for work when their load is below a given threshold. Specifically, whenever the *current load level* of a non-root node in the cluster tree falls below its *base load level*, the node sends a *Request for Computation (RFC)* message asking for $\mathbf{T_{req}}$ units of computation to its parent, where $\mathbf{T_{req}}$ is computed as follows:

$$T_{req} = \text{base load level} - \text{current load}. \tag{2}$$

After sending RFC message to its parent, the node updates its *base load level* to ensure that it can have only one outstanding RFC at any given time.

Job and Task Transfer
When a parent receives a RFC, if it has no job to send to the child, the new RFC is bocklogged and processed when work becomes available. Otherwise, the RFC recursively ascends the cluster tree until the RFC reaches either the *system scheduler* or a node that has unassigned jobs. In the later case, a set of jobs/tasks are transfered down the hierarchy along the path the RFC has traveled. This amount is determined dynamically during parent and child negotiations and the number of unscheduled jobs. First, we determine an ideal number of jobs/tasks that can possible be sent to a child scheduler as follows:

$$T_{target} = \lceil T_r \times \text{number of tasks queued} \rceil \tag{3}$$

where T_r is the *transfer factor* and is computed as follows:

$$T_r = \frac{\text{partition-reach of the child node}}{\text{partition-reach of the parent node}} \tag{4}$$

Once the target number of jobs is determined, the algorithm then considers the size of the RFC from the child as a hint to adjust the number of jobs that will actually be transferred down one level to the child as follows:

$$T_{target} = \begin{cases} min(T_{req}, \text{queue length}) & \text{if } T_{req} > T_{target} \\ min(T_{req}, \Phi_{child}) & \text{Otherwise.} \end{cases} \tag{5}$$

where Φ_{child} is the child node partition-reach.

Job and Task Selection
Finally, the algorithm selects jobs that have their replicas within the partition reach of the requesting schedulers. If there are no such jobs, then jobs belonging to the left sibling of the requesting node is searched. If this fails the jobs of the right sibling of the requesting node are selected. This process continues until the exact number of jobs to be sent to the requesting node is reached. The motivation for this job selection scheme is that we minimize replica management overheads (e.g., the replica instantiation latency) in case the original job fails. We also reduce the job transfer latency as we have to only send control messages to the child scheduler if the replica is already located there. Finally, it reduces the time that a child scheduler waits for the jobs to arrive, which increases system utilization.

After dispatching the jobs to a child, the parent informs the backup scheduler about the assignment and then updates the application status table (AST) to reflect the new assignment.

4 Conclusion and Future Directions

In this paper, we presented a scalable framework that loosely couples the dynamic job scheduling approach with the hybrid (i.e., passive and active replications) approach to schedule jobs efficiently while at the same time providing fault-tolerance. The main advantage of the proposed approach is that fail-soft behaviour (i.e., graceful degradation) is achieved in a user-transparent manner. Furthermore, being a dynamic algorithm estimations of execution or communication times are not required. An important characteristic of our algorithm is that it makes use of some local knowledge like faulty/intact or busy/idle states of nodes and about the execution location of jobs.

We are currently conducting extensive experiments using simulations. Our preliminary results show that the proposed approach performs quite well under various failure scenarios. The results show that the proposed policy performed at low cost in order to support fault-tolerance. Thus the results we obtained encourage us to continue our research in this direction. In the proposed fault-tolerant distributed framework, the latency of detecting the errors might be affected by message traffic in the communication network. To address this problem, we intend to develop an on-line mechanism to dynamically measure the round-trip time of the underlying network and calculate the error latency accordingly. We configured the system with two replicas, but this default value may change adaptively depending on the reliability of the cluster-computing environment. This will be a subject to be addressed in the future.

References

1. Jemal H. Abawajy and Sivarama P. Dandamudi. Parallel job scheduling on multi-cluster computing systems. In *Proceedings of IEEE International Conference on Cluster Computing (CLUSTER'03)*, pages 11–21, 2003.

2. Jemal H. Abawajy and Sivarama P. Dandamudi. A reconfigurable multi-layered grid scheduling infrastructure. In Hamid R. Arabnia and Youngsong Mun, editors, *Proceedings of the International Conference on Parallel and Distributed Processing Techniques and Applications, PDPTA '03, June 23 - 26, 2003, Las Vegas, Nevada, USA, Volume 1*, pages 138–144. CSREA Press, 2003.
3. Jemal H. Abawajy and Sivarama P. Dandamudi. Fault-tolerant grid resource management infrastructure. *Journal of Neural, Parallel and Scientific Computations*, 12:208–220, 2004.
4. J.H. Abawajy. Fault detection service architecture for grid computing systems. In *Lecture Notes in Computer Science*, volume 3044/2004, pages 107 – 115. Springer-Verlg, 2004.
5. Kenneth P. Birman. The process group approach to reliable distributed computing. Technical report, Department of Computer Science, Cornell University, Jul 1991.
6. Ian Foster. The grid: A new infrastructure for 21st century science. *Physics Today*, 55(2):42–47, 2002.
7. Ian T. Foster, Carl Kesselman, and Steven Tuecke. The anatomy of the grid - enabling scalable virtual organizations. *CoRR*, cs.AR/0103025, 2001.
8. Jrn Gehring and Achim Streit. Robust resource management for metacomputers. In *HPDC '00: Proceedings of the Ninth IEEE International Symposium on High Performance Distributed Computing (HPDC'00)*, page 105. IEEE Computer Society, 2000.
9. Soonwook Hwang and Carl Kesselman. Gridworkflow: A flexible failure handling framework for the grid. In *12th International Symposium on High-Performance Distributed Computing (HPDC-12 2003), 22-24 June 2003, Seattle, WA, USA*, pages 126–137. IEEE Computer Society, 2003.
10. I. Foster and C. Kesselman. Globus: A Toolkit-Based Grid Architecture. In *The Grid: Blueprint for a Future Computing Infrastructure*, pages 259–278. MORGAN-KAUFMANN, 1998.
11. Leon Juan, Fisher Allan L., and Steenkiste Peter. Fail-safe PVM: A Portable Package for Distributed Programming with Transparent Recovery. Technical report, CMU, Department of Computer Science, Feb 1993.
12. K. Marzullo L. Alvisi. Waft: Support for fault-tolerance in wide-area object oriented systems. In *Proceedings of ISW'98*, pages 5–10, 1998.
13. A. Nguyen-Tuong, A. S. Grimshaw, and J. F. Karprovich. Fault-tolerance via replication in coarse grain data-flow. Technical Report CS-95-38, Department of Computer Science, University of Virginia, 1995.
14. James S. Plank and Wael R. Elwasif. Experimental assessment of workstation failures and their impact on checkpointing systems. In *Symposium on FTC'98*, pages 48–57, 1998.
15. Anuraag S., Alok S., and Avinash S. A scheduling model for grid computing systems. In *Proceedings of Grid'01*, pages 111–123. IEEE Computer Society, 2001.
16. Fred B. Schneider. Byzantine generals in action: Implementing failstop processors. *ACM Transactions on Computer Systems*, 2(2):145–154, 1984.
17. P. Stelling, I. Foster, C. Kesselman, and G. von Laszewski. C.Lee. A fault detection service for wide area distributed computations. In *Proc. 7th Symposium on High Performance Computing*, pages 268–278, 1998.
18. Brian Tierney, Brian Crowley, Dan Gunter, Mason Holding, Jason Lee, and Mary Thompson. A monitoring sensor management system for grid environments. In *HPDC*, pages 97–104, 2000.

19. Namyoon W., Soonho C., Hyungsoo J., and Park Y. & Park H. Jungwhan M., Heon Y. Y. Mpich-gf: Providing fault tolerance on grid environments. In *Proceedings of 3rd IEEE/ACM International Symposium on Cluster Computing and the Grid*, 2003.
20. J. B. Weissman. Fault-tolerant wide area parallel computation. In *Proceedings of IDDPS'2000 Workshops*, pages 1214–1225, 2000.
21. Jon B. Weissman. Fault tolerant computing on the grid: What are my options? In *HPDC '99: Proceedings of the The Eighth IEEE International Symposium on High Performance Distributed Computing*, page 26. IEEE Computer Society, 1999.
22. Ming Q. Xu. Effective metacomputing using LSF multicluster. In *CCGRID '01: Proceedings of the 1st International Symposium on Cluster Computing and the Grid*, pages 100 – 106. IEEE Computer Society, 2001.

SLA Management in a Service Oriented Architecture

James Padgett, Mohammed Haji, and Karim Djemame

School of Computing,
University of Leeds,
Leeds LS2 9JT, UK

Abstract. This paper presents a Service Level Agreement (SLA) management architecture for the Grid. SLAs are an essential component in building Grid systems where commitments and assurances are specified, implemented and monitored. Targeting CPU type resources, we show how a SLA manager is able to interface with a broker designed for user applications that require resources on demand. The broker uses a novel three-phase commit protocol which provides the means to secure resources that meet the application's requirements through SLAs. Experiments are carried out on a Grid testbed to show how a SLA for a compute service is specified. Experimental results show that the broker provides performance enhancement in terms of the time taken from submission of application requirements until a job begins execution.

1 Introduction

The Grid [1] offers scientists and engineering communities high performance computational resources in a seamless virtual organisation. In a Grid environment, users and resource providers often belonging to multiple management domains are brought together. Users must be given some form of commitments and assurances on top of the allocated resources (this is sometimes referred to as Quality of Service), and it is the resources provider responsability to deal with erroneous conditions, fail over policies etc. A key goal of Grid computing is to deliver the commitments and assurances on top of the allocated resources which include, for example, availability of resources (compute resources, storage etc), security and network performance (latency, throughput) [7].

Commitments and assurances are implemented through the use of Service Level Agreements (SLA), which determine the *contract* between the user and the Grid Service provider. A SLA is defined as an explicit statement of expectations and obligations that exist in a business relationship between the user and the Grid Service provider. A formalised representation of commitments in the form of SLA documents is required, so that information collection and SLA evaluation may be automated. At any given point in time many SLAs may exist, and each SLA in turn may have a number of objectives to be fulfilled. In the context of a Grid application consolidation of management information is required when resources are spread across geographically distributed domains. SLAs may be distributed, and their validation depends on local measurements. With this is mind, the paper presents a SLA management architecture for automated SLA negotiation, monitoring and policing mechanisms. The current Open Grid Services Architecture (OGSA) [6] specification is moving towards a Web service derived technology, following the convergence of standards from the Grid and Web Services communities, most notably through the Web Services Resource Framework [3]. SLA management is defined as a high level service supporting SLAs within the Grid. Thus, a Grid user accessing Grid services on demand and with quality of service agreements enabling commitments to be fulfilled is a primary requirement. The SLA Manager negotiates a SLA for the rights to execute a Grid service, management of SLA involves monitoring achieved using performance measurement data obtained from a set of Grid monitoring tools. Policing is performed using violation data obtained through automated monitoring of the SLA against real-time performance measurement data. SLA policing mechanisms must enforce changes to the execution to meet SLA guarantees.

The aims of this paper are: 1) to present a Grid SLA management architecture and show how a SLA can be specified in a service oriented architecture; 2) taking a SLA for a compute service as a motivation, discuss how the interaction between the SLA manager and a resource broker is performed in order to guarantee the allocation of compute resources, and 3) to present the design and implementation of a resource broker which provides the means to negotiate and acquire resources that meet the user's requirements through SLAs. The resource broker incorporates a three-phase commit protocol that provides services to ensure decisions are made with up-to-date information, resources are differentiated and the nominated resources are secured before jobs are submitted.

The structure of the paper is as follows: in section 2, the Grid SLA management architecture to formalise QoS requirements between users and resource providers is presented. Section 3 describes the interaction between the SLA manager and different resource brokers. Section 4 provides an overview of the architecture and design of a resource broker within the Service Negotiation and Acquisition Protocol (SNAP) framework. In section 5 a SLA specification is presented using a compute service as an example. Section 6 presents some experiments involving the performance evaluation of the resource broker, and discusses the experimental results obtained on a Grid testbed. Related work is described in section 7, followed by a conclusion and discussion on future work.

2 Service Level Agreement Management Architecture

SLAs and their management are an important way to implement and formalise QoS requirements between users and Grid service providers. SLA management can be classified as a high-level service which needs interfaces to the factory, registration and discovery service for finding resources based on user's QoS requirements. Figure 1 shows the proposed SLA manager architecture. Once the SLA Manager has instantiated an agreement and the execution is running, the

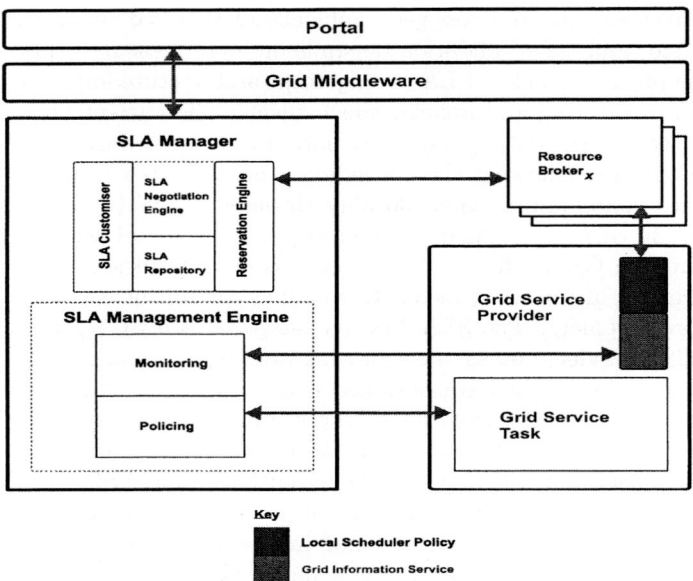

Fig. 1. SLA Manager Architecture

user can interact with the underlying Grid service. Interaction between the SLA interface and the Grid execution interface is maintained throughout the agreement life cycle to enforce the SLA guarantees. Service level negotiation takes place through the SLA Manager Factory interface using SNAP [4], which provides negotiation and acquisition of resources. The requirements will be formally captured in a number of *Service Level Objectives* (SLO). The type of reservations made can be resource based or service based and will be executed through the Reservation Engine. Once an agreement has been reached the SLA is formalised by a SLA Manager Grid service instance. This Grid service has a number of service data elements which hold metadata relating to the negotiated SLA. The SLA Manager Grid service instance has a service interface through which the SLA is enforced. All the functionality contained within the SLA manager is accessed through the service interface. Once an agreement is signed it is stored in the SLA repository where it can be called whenever validation is needed. The SLA Customiser has the ability to make changes to the signed SLA after the agreement has been signed. This could include changes to the state of the SLA to record violations. Based on the service requirements of the signed SLA, the reservation engine will attempt to acquire the resources needed to guarantee them. Its responsibility is to make reservations and control the reservation status. Once a SLA has been agreed and reservation of the selected resources has occurred, the managed Grid service can begin execution. The SLA Manager contains an SLA management engine which is tasked with automated monitoring of the metrics needed to enforce the guarantees in the Service Level Objectives. It uses an external Grid Monitoring Service [17] to select the Grid monitoring tools which will be needed to monitor the SLA. Guarantees that can be offered fall into three categories: performance, usage, and reliability. For performance guarantees considered in this work, the service requirements are specified in the SLOs. The *Service Level Indicators* (SLI) specify the level at which the SLOs have to be maintained. Guaranteeing service performance is important in maintaining perceived QoS during Grid service execution. SLI specification vary, and can be expressed as a min/max threshold or as a distribution. The SLA policing engine will adapt the execution of the managed Grid service, either in response to a violation or to prevent violation from occurring. The method used to adapt the Grid service execution is set down in the SLA and based on the policies of local resources. Adaptation has the potential to significantly improve the performance of applications bound with SLAs. An adaptive application can change its behaviour depending on available resources, optimising itself to its dynamic environment.

3 SLA Manager and Resource Broker Interaction

The SLA manager is designed to automate SLA life cycle from negotiation to termination. A SLA is specified as an XML document. Strict rules governing what can be specified in the SLA must be provided in the form of a schema document. Presenting the SLA in XML format is necessary in a service oriented architecture such as the one proposed here. It allows the SLA to traverse the Grid with the users execution as it moves between resources. The SLA manager can thus be installed on all Grid resources as part of the middleware as a high level Grid Service. It can parse the XML SLA documents and instantly configure the SLA Management system for the users execution (see section 5).

The SLA Manager's reservation engine is able to interface with a number of different resource brokers providing reservations for different types of Grid resources. All such interactions take place between the Grid service interfaces of the SLA Manager and the resource brokers, as shown in Figure 1. In addition, such an implementation allows access to standard Grid service functionality [2]. The example used in this work is a SNAP-based resource broker which provides reservations for compute resources. The SLA Manager can contact this resource broker to provision a SLA based on the task requirements. This is similar to the Task Service Level Agreement (TSLA) defined within the SNAP framework. This represents an agreement which specifies the desired performance of the task [4], where the task represents a job submission, execution or Grid service invokation. A modular approach to SLA formation is thus provided. The SLA manager enters into an agreement with the resource broker service which provides a reservation guarantee, but it is the individual brokers which form the reservation agreements with the local Grid resources.

4 SNAP-Based Resource Broker

The resource broker is designed to insulate the user from Grid middleware, enabling transparent submission of jobs to the Grid. The resource broker takes

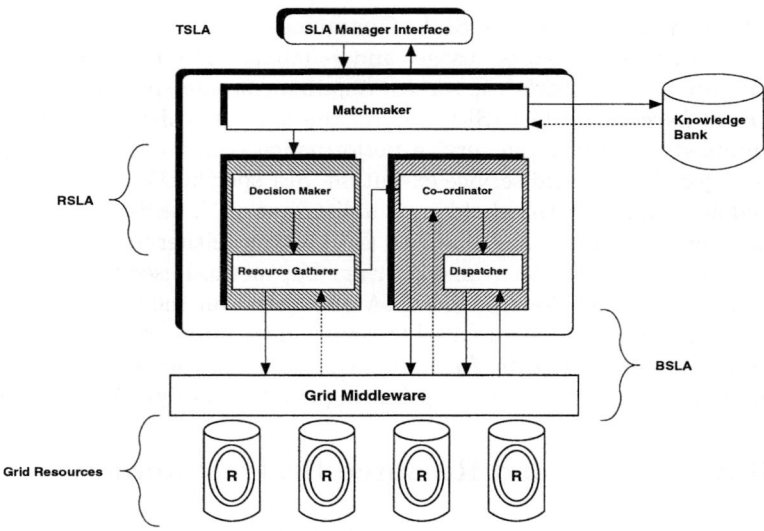

Fig. 2. Resource Broker Architecture

the user's requirements as specified in the SLA (e.g. number of CPUs, operating system) and job description and contacts resources that may support these requirements to gather information on their current state (e.g. current load). A decision is made as to which resource(s) will be used to run the job and this is followed by a negotiation with these resources. This negotiation is based on the framework provided by SNAP [4], whereby guarantees are obtained that the user's requirements will be fulfilled by the use of an SLA. Three types of SLA are used within the SNAP framework: 1) Task Service Level Agreement (TSLA) in which a clear objective specification of the task and its resource requirements is given; 2) Resource Service Level Agreement (RSLA), which involves the negotiation for the right to consume a resource. An RSLA characterises a resource in terms of its capabilities, without necessarily specifying what the resource will be used for, and 3) Binding Service Level Agreement (BSLA) associates the task with the resources.

The broker's architecture is shown in Figure 2. A Matchmaker uses the parsed user requirements to contact a *Knowledge Bank* (KB). The latter is a repository that stores static information on all resources. The broker can access this information on behalf of the user for each resource he/she is entitled to use. The information stored in the KB as attributes include the number of CPUs, the operating system, memory, storage capacity and past behaviour performance of a resource. The KB stores a history profile of past performance of resources, to enable the broker to differentiate and categorise resources into different levels. An analogy to the KB is a telephone directory where information stored directs to a particular service that caters for the users' needs. Further it could also be used to store economic accounting details. Details on the Decision Maker, the Resource Gatherer, the Co-ordinator and the Dispatcher are found in [8].

Once the resources are secured the final procedure (binding) is executed by the Dispatcher by submitting the task and binding it to the resources.

Two versions of the broker are developed: 1) Broker with reservation: this broker adopts a mechanism that secures resources for utilisation through the means of immediate reservation, and 2) Three-phase Commit broker: the motivation here is that the Grid is dynamic and not centralised with a signal administrator such as some traditional systems. Thus at a single point in time there might be several users competing for the same resources without each others knowledge of their existence or interest. Because the resource status is constantly changing, as a consequence the process of contacting the information provider could lead into an infinite oscillation between the broker and the information provider without a successful job submission. This is why a three-phase commit protocol is adopted that acknowledges this fact. As a result it sets up *probes* during the information gathering process to ensure rapid update of any status change of a resource. Details on the three-phase commit protocol are found in [8].

Globus MDS (Monitoring and Directory Service) [5] is used in the broker's architecture to gather dynamic information. However the information service provision offered by the default Globus installation has been extended. Specifically, the MDS is deriving information from the local resource manager, Sun Grid Engine (SGE) [16].

5 Example: SLA Specification of a Compute Service

An example SLA for a Compute service is specified in Table 1. It gives indication of the components which make up the SLA generated by the SLA Manager. Figure 3 gives an indication of the components which make up the TSLA generated by the SLA manager as well as the resulting XML representation of a job submission example based on the task requirements.

The SLO's represent a qualitative guarantee such as CPU, RAM or HDD SLA. They comprise a set of SLI parameters which represent quantitative data describing the level of the SLO guarantee, such as CPU_COUNT or RAM_COUNT. The SLI values may take a number of forms, two which will

Table 1. SLA Specification of a Compute Service

Component	Observation
Purpose	Run a Grid job with guarantees to ensure user's requirements are met
Parties	The user, the resource broker, the compute resources
Scope	Compute service
Service Level Objective	Ensure availability of a certain number of resources that satisfy user's requirements for the duration of the job execution
SLO Attributes	CPU count, CPU type, CPU speed, Operating system and version
SLIs	For each SLO attribute, its value is a service level indicator
Administration	The SLA's objectives are met through resource brokering and adaptation (Section 4)

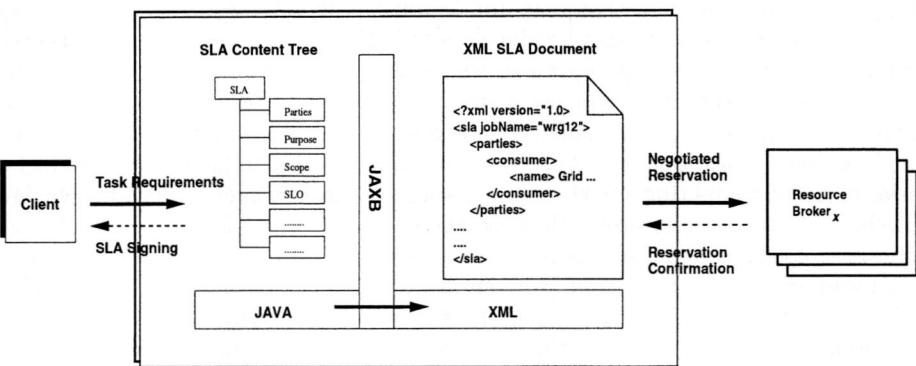

Fig. 3. Compute Service SLA Specification and Interaction with Resource Broker

be used are (1) a parameter distribution where the value of the SLI must be in a range or (2) a list where the parameter must equal a definite value.

The SLA document is XML based, whereas the SLA is represented internally by a content tree made up of Java objects. The SLA Manager supports a number of service guarantees through differentiated classes of service; best effort, best effort with adaptation, reservation, reservation with adaptation.

6 Experimental Results

The experiments's objective is a comparison of the performance of a SNAP broker that incorporates the three-phase commit protocol and one that only uses reservation as a means to secure the nominated resources once an SLA has been specified. This is in terms of the time interval between submission (to the broker) of user requirements and the job beginning execution. The experiments are performed on a Grid test-bed consisting of 10 machines running Linux 2.4 with kernel 2.4.20-18.8. Globus 2.4 is installed on all machines and a Grid Resource Information Service (GRIS) is associated with each of them. Communication occurs with a fast (100 Mbps) Local Area Network. The SLO is to ensure the availability of 3 resources prior to job execution. Each of the two brokers is considered in turn. Each is given access to only 3 of the resources and on each of these, another job is running for a fixed duration. The experiments performed are based on this scenario. In the experiment, the additional job that is submitted so that the resources are unavailable has a fixed duration of 30 seconds. This job is submitted immediately before the broker is executed. The information provider response time (i.e. the GRIS response time) is then varied between 10 and 90 seconds to reflect real Grid systems. On the Grid testbed the information provider response time is fairly stable at 8 seconds for the GRIS on each machine but the response time was varied by adding a synthetic delay into the code. The time taken between the broker beginning execution and the broker becoming aware that the resources are free is then recorded, in addition to the time taken before the job begins execution. Further the number of contacts made to the information provided is also recorded.

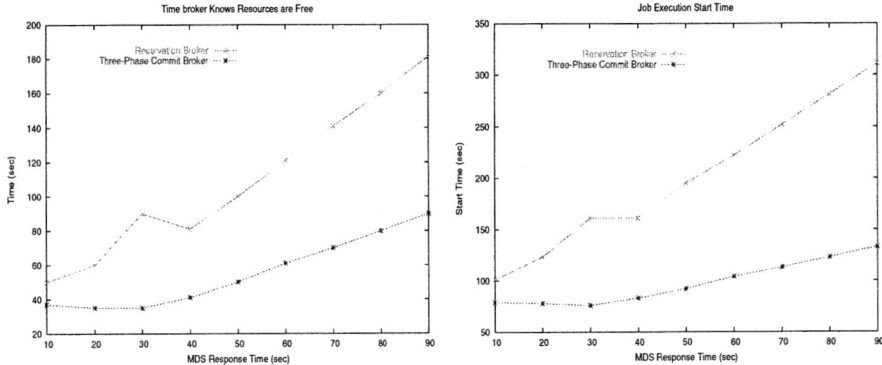

Fig. 4. (1) Time taken for broker to determine resources are free, as a function of MDS response time; (2) Time taken for job to begin execution

The results of the experiments are shown in Figure 4: 1) time taken between the broker beginning execution and the broker becoming aware that the resources are free, as a function of the information provider response time, and 2) time taken between the broker beginning execution and the jobbeginning execution. Clearly, the three-phase commit SNAP broker becomes aware that the status of the resources has changed, much faster than the broker with reservation only, as a consequence of the use of probes. Consequently, the job begins execution sooner when the three-phase commit SNAP broker is used. Usually, the longer the response time, the longer it takes before either broker is aware that the resources are free. For the three-phase commit SNAP broker, there is only one contact with the information provider. Hence increasing the response time has little effect until it exceeds the 30 second period for which the resources are taken. For the broker with reservation, the effect is apparent even for faster response times. However, note that the time taken before this broker becomes aware of the change in resource status is shorter when the response time is 40 seconds than when the response time is 30 seconds. This is due to the fact that when the response time is 30 seconds, the broker with reservation needs to contact the information provider four times before it is aware of the change in status, while if the response time is 40 seconds, only three contacts are required. These experiments demonstrate that the three-phase commit protocol ensures that the broker has access to fast updates on the status of the resources enabling a performance enhancement.

7 Related Work

There have been a number of attempts at defining an SLA management architecture for both Web and Grid services. Architectures from Hewlett-Packard Laboratories [12] and IBM T.J. Watson Research Center [10] concentrate on service level agreements within commercial Grids. The service level agreement language used is that presented by Ludwig et al in [11]. The Global Grid Forum have defined a draft of OGSI-Agreement [2] which is an agreement-based Grid service management specification designed to support Grid service management during the lifetime of a service instance. Two other important works are related to automated SLA monitoring for Web services [9] and analysis of service level agreements for Web services [13].

Resource management and scheduling in the Grid is receiving increased attention [14]. However, to the best of our knowledge, there has been no work on the specific problems addressed here, namely implementation of SNAP-based resource brokering and protocols for securing resources before job submission. An overview of some current Grid scheduling efforts is found in [15]. Overall the brokers found in the literature have not shown ways to secure resources before job submission, which is essential for SLAs in the dynamic environment of the Grid.

8 Conclusion

SLA management is important in formalising QoS implementation within Grid services. In this paper, an Grid SLA management architecture is presented. Making simplifying assumptions in the prototype developed, this work considers CPU type resources, and the interaction between the SLA manager and a resource broker that is designed to guarantee the resource allocation is discussed . Experimental results on a Grid testbed show that the resource broker is a viable contender for use in future Grid implementations. Future work in SLA management will involve monitoring and policing which are achieved using performance measurement data obtained from a set of monitoring tools. Observed data needs to be validated against commitments specified in SLAs. As the Grid integrates heterogeneous resources with varying quality and availability, this places importance on its ability to monitor the state of these resources and adapt to changes in their availability over time. Although this work within the field of SLA management targets CPU type resources, research is under way to exploit other types of resources such as data storage and network bandwidth.

References

1. F. Berman, G.C. Cox, and A.J.G. Hey, editors. *Grid Computing - Making the Global Infrastructure a Reality.* Wiley, 2003.
2. K. Czajkowski, A. Dan, J. Rofrano, S. Tuecke, and M. Xu. Agreement-based Grid Service Management (OGSI-Agreement) Version 0. Global Grid Forum, Aug 2003.

3. K. Czajkowski, D. Ferguson, I. Foster, J. Frey, S. Graham, T. Maguire, D. Snelling, and S. Tuecke. From Open Grid Services Infrastructure to WS-Resource Framework: Refactoring and Extension, Feb 2004. http://www.globus.org/wsrf/specs/ogsi_to_wsrf_1.0.pdf.
4. K. Czajkowski, I. Foster, C. Kesselman, V. Sander, and S. Tuecke. SNAP: a Protocol for Negotiating Service Level Agreements and Coordinating Resource Management in Distributed Systems. In *Proceedings of JSSPP'2002*, Edinburgh, Scotland, Jul 2002.
5. S. Fitzgerald, I. Foster, C. Kesselman, G. von Laszewski, W. Smith, and S. Tuecke. A Directory Service for Configuring High-Performance Distributed Computations. In *Proceedings of the 6th IEEE Symposium on High-Performance Distributed Computing (HPDC 6)*, pages 365–376, Portland, OR, Aug 1997.
6. I. Foster, C. Kesselman, J.M. Nick, and S. Tuecke. The Physiology of the Grid, chapter 8, pages 217–249. In Berman et al. [1], 2003.
7. I. Foster, C. Kesselman, and S. Tuecke. The Anatomy of the Grid, chapter 6, pages 171–197. In Berman et al. [1], 2003.
8. M. Haji, P. Dew, K. Djemame, and I. Gourlay. A SNAP-based Community Resource Broker using a Three-Phase Commit Protocol. Proceedings of the International Parallel and Distributed Processing Symposium (IPDPS'2004), Santa Fe, New Mexico, Apr 2004.
9. L. Jin, V. Machiraju, and A. Sahai. Analysis on Service Level Agreement of Web Services. Technical Report HPL-2002-180, HP Labs., 2002.
10. A. Leff, J.T. Rayfield, and D.M. Dias. Service-Level Agreements and Commercial Grids. *IEEE Internet Computing*, 7(4):44–50, Aug 2003.
11. H. Ludwig, A. Keller, A. Dan, and R. King. A Service Level Agreement Language for Dynamic Electronic Services. In *Proceedings of the 4th International Workshop on Advanced Issues of E-commerce and Web-based Information Systems (WECWIS'2002)*, Newport Beach, CA., Jun 2002. IEEE Computer Society.
12. A. Sahai, S. Graupner, V. Machiraju, and A. van Moorsel. Specifying and Monitoring Guarentees in Commercial Grids through SLA. In *Proceedings of the 3rd IEEE/ACM International Symposium on Cluster Computing and the Grid (CC-GRID'2003)*, Tokyo, Japan, May 2003. IEEE Computer Society.
13. A. Sahai, V. Machiraju, M. Sayal, L.J. Jin, and F. Casati. Automated SLA Monitoring for Web Services. Technical Report HPL-2002-191, HP Labs., 2002.
14. Scheduling and Resource Management Working Group. Global Grid Forum, 2003. https://forge.gridforum.org/projects/srm.
15. J.M. Schopf. A General Architecture for Scheduling on the Grid. Technical Report ANL/MCS-P1000-10002, Argonne National Laboratory, 2002.
16. Sun Microsystems. Sun Grid Engine. http://www.sun.com/software/gridware.
17. B. Tierney, R. Aydt, D. Gunter, W. Smith, M. Swany, V. Taylor, and R. Wolski. *A Grid Monitoring Architecture*. GGF Performance Working Group, Jan 2002.

Attacks on Port Knocking Authentication Mechanism

Antonio Izquierdo Manzanares, Joaquín Torres Márquez,
Juan M. Estevez-Tapiador, and Julio César Hernández Castro

Universidad Carlos III de Madrid, Avda de la Universidad 30,
28911 Leganés (Madrid), Spain
{aizquier, jtmarque, jestevez, jcesar}@inf.uc3m.es

Abstract. Research in authentication mechanisms has led to the design and development of new schemes. The security provided by these procedures must be reviewed and analyzed before they can be widely used. In this paper, we analyze some weaknesses of the port knocking authentication method that makes it vulnerable to many attacks. We will present the NAT-Knocking attack, in which an unauthorized user can gain access to the protected server just by being in the same network than an authorized user. We will also discuss the DoS-Knocking attack, which could lead to service disruptions due to attackers "knocking" on many ports of the protected server. Finally, we will review further implementation issues.

1 Introduction

Authentication has been an issue in communications security, as it is the mechanism that allows principals to identify each other involving some kind of operations prior to communication with each other. As we see, authentication is the "access point" to the communication, so it is a potential target for an attacker. This means the whole authentication process must be secured using protocols and mechanisms that allow each principal to verify each other's identity. Traditionally it has been said that authorization should rely on something the principal knows (such as a password or a pass phrase [10]), something the principal is (such as biometric values [1]), and/or something the principal has (e.g. a smart card [1]).

Port knocking [4] authorization mechanism relies on something the principal knows. In this case, the password is not a character sequence but a port sequence: The server we are willing to communicate keeps all of its network ports closed and these ports must be "knocked" in the correct sequence in order for the server to open the desired communication port. The procedure for "knocking" a port consists of sending a packet to that port, so the server will notice a connection attempt against a closed port and log it.

The main advantage of port knocking is that, as knocked ports are closed during the authentication procedure, it should not be possible to identify whether a server is using port knocking to authenticate the clients [6], resulting in a stealthy mechanism. However, this stealth property has been attacked in [5] as well as the "security through obscurity" that seems to cover the whole port knocking proposal.

In this paper we describe several further security issues of port knocking and we will discuss whether port knocking really provides some advantages over traditional methods.

1.1 Port Knocking

Port knocking, as described in [4] is an authentication method that provides network level access to services using a password composed of several ports. The server keeps all of its ports closed (e.g. using a firewall), and the clients transmit the port-sequence making connections to such ports. This is called "knocking" a port.

The server logs all these connection attempts, and an external process parses the logs and looks for correct sequences. When this process finds a valid sequence it updates the connection policies allowing communications from the client to the desired service. In order to achieve this the client's IP and the port to open must be encoded in the port sequence.

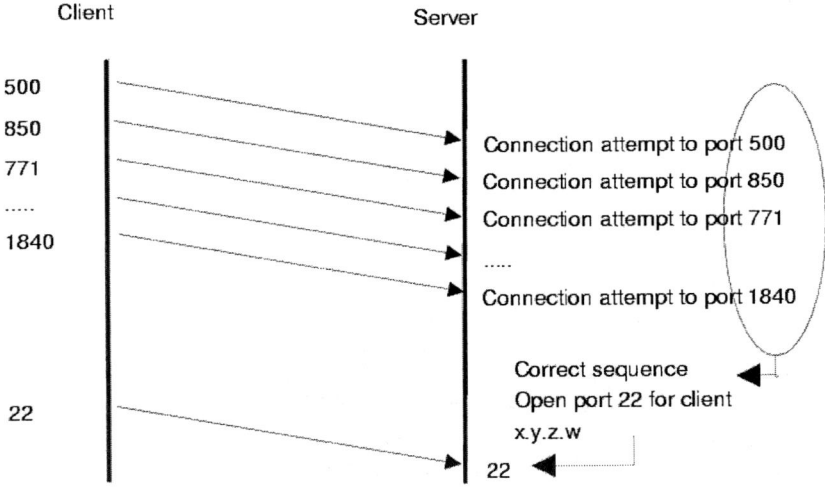

Fig. 1. Illustration of the port knocking authentication mechanism

Figure 1 shows an example of the whole process where the client computes the sequence to use a service (SSH in this example) based on its IP address (*500, 850, 771,..., 1840*). After that, the client starts making connections to those ports, in order to generate entries in the server's log. A process in the server is parsing that log, and when it detects a valid sequence it computes the service port

(port 22 in this example) and the client's address, and modifies the connection policies in order to open the requested port for that client. Consequently, from the client's point of view, the ports it has to "knock" on can be calculated as:

$$p_1 = f_1\left(port\_to\_open, client\_address, \ldots\right), \tag{1}$$

$$p_2 = f_2\left(port\_to\_open, client\_address, \ldots\right), \tag{2}$$

etc..., where p_1 is the first port used in the knocking sequence (500 in our example), p_2 is the second one, etc... From the server's point of view, all the parameters the port knocking server has to take into account when modifying the firewall policies can be expressed as:

$$opened\_port = f_3\left(p_1, p_2, \ldots, p_n\right), \tag{3}$$

$$IP\_allowed = f_4\left(p_1, p_2, \ldots, p_n\right), \tag{4}$$

and so on.

2 NAT-Knocking: The Problem of Sharing Network Addresses

As stated previously, port knocking is based on the idea of opening ports in the firewall just to clients that have provided the correct password through the appropriate "knocks" in the server. To identify the clients, port knocking makes the sequence dependant on the client's network address (as stated in equation 4). However, this raises the question of what might happen if two clients shared the same address, e. g.: if Network Address Translation (NAT, [7]) is used. As shown in Figure 2, when packets from inside the NAT exit the private network they all share the same source address (the public address for the NAT), and packets cannot be used to identify the source behind the NAT without accessing the router's NAT tables.

We can see in Figure 2 how two hosts A and B share the same public IP address, so they have to use NAT to access other networks. Packet 1 (created by A) is translated into packet 2, where the source address and port have changed to the public address and a free port in the NAT device. The same process is done with packet 3 from B. As we can see, from the outside of the NAT it is impossible to determine which packet comes from which source using the source address.

As port knocking relies in the network address to open required ports to trusted clients, it cannot differentiate among all the clients behind the same NAT device. As a result, when a client in a network that uses NAT identifies itself to the port knocking server and gets access to the service port, this server has given access to every device that uses the same NAT (and, consequently, shares the same public network address).

Furthermore, when the trusted user is "knocking" on the firewall there is no need for a potential attacker to watch the authentication sequence, she just needs to wait until the service port is open to have access to it without knowing the authentication sequence. An example can be seen in Table 1: Client 172.16.8.102 is the authorized user that authenticates using port knocking with server 163.117.149.93 requesting the opening of port 80. We can see all the knocks sent to the server (*frames 1 to 24*) and how after that the client is able to connect to port 80 and proceeds with normal communication (*frames 25 to 30*). However, client 172.16.8.102 is using NAT, and another client in the same

Table 1. Network capture of the NAT-Knocking Attack

No.	Source	Destination	Protocol	Info
1	172.16.8.102	163.117.149.93	TCP	32987 → 7682 [SYN]
2	172.16.8.102	163.117.149.93	TCP	32988 → 7697 [SYN]
3	172.16.8.102	163.117.149.93	TCP	32989 → 7810 [SYN]
4	172.16.8.102	163.117.149.93	TCP	32990 → 7800 [SYN]
5	172.16.8.102	163.117.149.93	TCP	32991 → 7811 [SYN]
6	172.16.8.102	163.117.149.93	TCP	32992 → 7809 [SYN]
7	172.16.8.102	163.117.149.93	TCP	32993 → 7673 [SYN]
8	172.16.8.102	163.117.149.93	TCP	32994 → 7686 [SYN]
9	172.16.8.102	163.117.149.93	TCP	32995 → 7603 [SYN]
10	172.16.8.102	163.117.149.93	TCP	32996 → 7682 [SYN]
11	172.16.8.102	163.117.149.93	TCP	32997 → 7602 [SYN]
12	172.16.8.102	163.117.149.93	TCP	32998 → 7887 [SYN]
13	172.16.8.102	163.117.149.93	TCP	32999 → 7699 [SYN]
14	172.16.8.102	163.117.149.93	TCP	33000 → 7808 [SYN]
15	172.16.8.102	163.117.149.93	TCP	33001 → 7629 [SYN]
16	172.16.8.102	163.117.149.93	TCP	33002 → 7602 [SYN]
17	172.16.8.102	163.117.149.93	TCP	33003 → 7686 [SYN]
18	172.16.8.102	163.117.149.93	TCP	33004 → 7663 [SYN]
19	172.16.8.102	163.117.149.93	TCP	33005 → 7655 [SYN]
20	172.16.8.102	163.117.149.93	TCP	33006 → 7692 [SYN]
21	172.16.8.102	163.117.149.93	TCP	33007 → 7992 [SYN]
22	172.16.8.102	163.117.149.93	TCP	33008 → 7839 [SYN]
23	172.16.8.102	163.117.149.93	TCP	33009 → 7637 [SYN]
24	172.16.8.102	163.117.149.93	TCP	33010 → 7990 [SYN]
25	172.16.8.102	163.117.149.93	TCP	33011 → 80 [SYN]
26	163.117.149.93	172.16.8.102	TCP	80 → 33011 [SYN, ACK]
27	172.16.8.102	163.117.149.93	TCP	33011 → 80 [ACK]
28	172.16.8.102	163.117.149.93	HTTP	GET /info.php HTTP/1.1
29	163.117.149.93	172.16.8.102	TCP	80 → 33011 [ACK]
30	163.117.149.93	172.16.8.102	HTTP	HTTP/1.1 200 OK
31	**172.16.8.100**	163.117.149.93	TCP	1196 → 80 [SYN]
32	163.117.149.93	**172.16.8.100**	TCP	80 → 1196 [SYN,ACK]
33	**172.16.8.100**	163.117.149.93	TCP	1196 → 80 [ACK]
34	**172.16.8.100**	163.117.149.93	HTTP	GET /info.php HTTP/1.1
35	163.117.149.93	**172.16.8.100**	TCP	80 → 1196 [ACK]

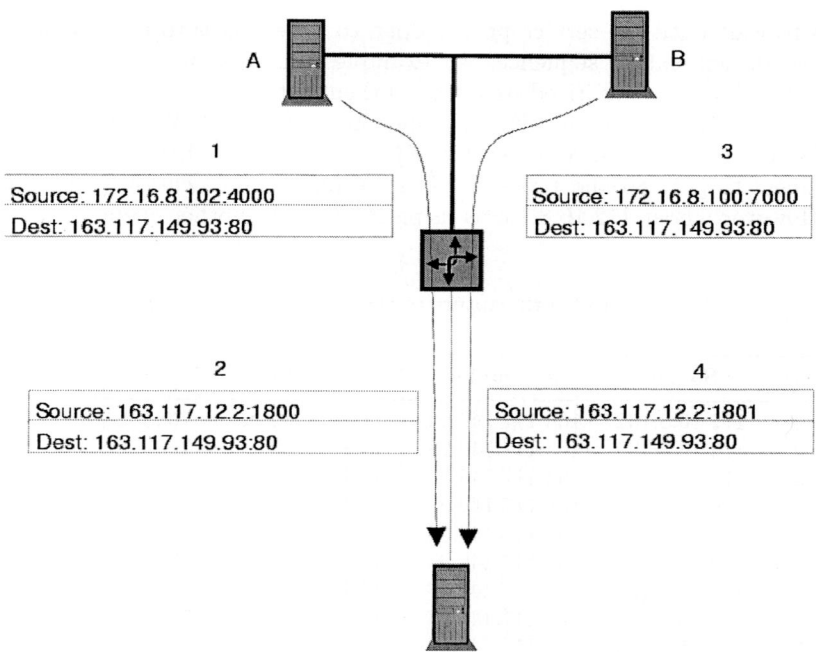

Fig. 2. Sample NAT configuration, with two hosts A and B sharing the same network address

network (*172.16.8.100*) can connect to the service opened by the authorized user without performing the knock sequence (*frames 31 to 35*).

3 DoS-Knocking Attack: Handling Large Amounts of Work

In order to work properly, port knocking server must control the connection attempts made against it, so it can look for authentication patterns and open ports when requested by authorized users. This implies that a process must parse the logs in real time to automatically detect the sequences.

When analyzing the way port knocking fulfills this duty we realize that the parser process must use a buffer for each different client that makes a connection against a closed port, so the process is able to track each authentication sequence.[1] If an attacker manages to send forged packets with random source network addresses (the same way some worms propagate [3]), the parser process

---

[1] Even if we select a reduced range of port to watch instead of considering the whole range of 65.535 ports as "knocking-available" we still need a large amount of them to provide the system with a minimum security level.

Table 2. Status of the port knocking server as reported by vmstat during the proposed DoS-Knocking attack, with time interval between measures being 5 seconds

PROCS		MEMORY				CPU		
R	B	SWPD	FREE	BUF	CACHE	US	SY	ID
1	0	0	7536	8920	119524	0	0	100
2	0	0	7536	8920	119524	2	0	98
2	0	0	7536	8920	119524	0	1	99
1	0	0	7536	8920	119524	0	1	99
1	0	0	3784	8924	119524	0	0	100
5	**0**	**0**	**3976**	**8924**	**117472**	**12**	**47**	**42**
6	**0**	**0**	**3264**	**8916**	**116648**	**23**	**77**	**0**
5	**0**	**0**	**3628**	**8920**	**112636**	**58**	**42**	**0**
4	**0**	**0**	**3708**	**8920**	**111668**	**81**	**19**	**0**
3	**0**	**0**	**3816**	**8920**	**110696**	**85**	**15**	**0**
5	**0**	**0**	**3960**	**8920**	**109720**	**92**	**8**	**0**
4	**0**	**0**	**3052**	**8920**	**109732**	**88**	**12**	**0**
5	**0**	**0**	**3576**	**8928**	**107916**	**50**	**50**	**0**
4	**0**	**0**	**3832**	**8928**	**106092**	**41**	**59**	**0**
3	**0**	**0**	**3908**	**8924**	**102180**	**89**	**11**	**0**
3	**0**	**0**	**3900**	**8928**	**102180**	**83**	**17**	**0**
4	**0**	**0**	**3900**	**8928**	**102180**	**89**	**11**	**0**
1	0	0	3900	8928	102180	78	13	9
1	0	0	3896	8928	102180	0	0	100
1	0	0	3896	8928	102180	0	0	100
2	**0**	**0**	**7688**	**8932**	**102180**	**1**	**0**	**99**
1	**0**	**0**	**7688**	**8932**	**102180**	**1**	**1**	**98**

would have to create a buffer for each one of the addresses, making this process to consume high amounts of memory. In the Appendix is included the source code to generate an attack consisting of sending packets with random fake source addresses to random ports on the target.

Another two considerations about DoS attacks on port knocking are about parameter encryption and the parser performance. [4] recommended parameter encryption so that integrity and confidentiality can be achieved, and [9] proposed a solution based on one-time-passwords to avoid replay attacks. However, when using encryption the server must decrypt the contents of the authorization sequence for all the clients that try to authenticate using port knocking. This can lead to high CPU loads, as shown in Table 2, where a Pentium-class computer (166 MHz and 32 MB of RAM) could overload a Pentium III-class server (500 MHz and 192 MB of RAM). Rows in bold are the status when the attack (60.000 packets with different forged source address aimed at random ports) took place. Last two bold rows of the table show status when the attack had finished, but the server was still processing the fake knocks. Furthermore, we should consider whether having a process parsing logs in real time is something our system can handle, or maybe that will heavily decrease its performance.

In order to prevent this kind of attacks it could be necessary to use port knocking just for the range of addresses that would use it and, if possible, to have separate logs for those knocks that come from those addresses and those that don't, so we reduce the work of the parser. Additionally, we should consider using a fast encryption algorithm that doesn't require a lot of CPU cycles to decrypt the contents of the port sequence, such as TEA [8].

4 Other Implementation Issues

Apart from the already discussed flaws in port knocking, a real implementation presents many problems that need to be faced, the first of all being the port range being used to knock. Although [5] criticized the shortage of ports being used from the cryptographic point of view, we can also assure that a small range to perform the knocks makes it possible for an attacker to discover (by capturing several authentication sequences) the range of valid ports and perform brute-force attacks. On the other hand, if the implementation had chosen a large range of valid ports, it would be cryptographically more securer, but it would be easier to perform a DoS-Knocking attack, as most of the ports are valid and the parser process would have to create buffers for each knock.

A proposed solution for this implementation issue would be to choose well-balanced groups of large ranges that take into account our needs of confidentiality and performance. If our main concern is performance, we should choose a handful of large ranges (for example, 10 ranges of 1000 ports) so the ranges are not small enough to permit identification with few data captures, and at the same time they are not large enough to favor DoS-Knocking attacks. On the other hand, if we must provide our system with maximum security we should consider using all available ports in order to maximize the key to our network (as [5] details).

Anyways, traffic analysis is always possible, giving an attacker the clue about the authentication mechanism being used. If an attacker can gather traffic data that include several authentications, she may realize that prior to any connection to a protected port there are a number of connection attempts against closed ports (and depending on the number of authentications captured, it could be possible to identify the valid ports ranges being used). Therefore, we have a situation that revokes the main advantage of port knocking: being an stealth method of authenticating. [5] proposes using UDP instead of TCP in order to make the authentication really stealth. However, given an attacker with enough data collected from the network, the amount of UDP traffic to closed ports would be clearly higher than expected and with no apparent reason, giving the attacker an idea of what it is going on.

5 Conclusions

The analysis of port knocking authentication method has revealed both some design flaws and implementation problems that could provide access to unauthorized users and/or cause a DoS attack on the port knocking server. We have described the NAT-Knocking attack and stated that the problem of access control based on network address when NAT is used cannot be solved as long as network address is the only identifier used. Regarding this, port knocking does not improve the security offered by other network-level devices and mechanisms, such as firewalls.

We have also described the DoS-Knocking attack and we have proven that a low-end computer is able to take a much more powerful server to the limits of its computing capabilities. We have seen that encryption and the clients-tracking mechanism tend to exhaust both memory and CPU, so it is possible that under peaks of work legitimate users could perform a DoS attack without pretending to.

Finally, partly due to the implementation and partly due to the design, we have shown how traffic analysis is possible for an attacker with enough data captured from the network, so she could have information about the authentication procedure. This fact attacks directly the basis of port knocking: being an stealth authentication scheme.

References

1. Anderson, R.: Security Engineering: A Guide to Building Dependable Distributed. J. Wiley & Sons (2001)
2. Claerhout, B.: http://cs.ecs.baylor.edu/~donahoo/NIUNet/hacking/hijack/
3. Knowles, D., Perriot, F., Szor, P.: W32.Blaster.Worm Report. Symantec Security Response (2003)
4. Krzywinski, M.: Port Knocking: Network Authentication Across Closed Ports. SysAdmin Magazine 12 (2003) 12-17
5. Narayanan, A.: A critique of Port Knocking NewsForge, August 8 (2004) http://software.newsforge.com/software/04/08/02/1954253.shtml
6. Schneier, B.: Port Knocking. Crypto-Gram Newsletter, March 15 (2004) http://www.schneier.com/crypto-gram-0403.html#5
7. Srisuresh, P., Egevang, K.: Traditional IP Network Address Translator (Traditional NAT). RFC 3022 (2001)
8. Wheeler, D., Needham, R.: TEA, a Tiny Encryption Algorithm. Fast Software Encryption 1994: 363-366
9. Worth, D.: CÖK - Cryptographic One-Time Knocking. BlackHat 2004.
10. Yan, J., Blackwell, A., Anderson, R., Grant, A.: The Memorability and Security of Passwords. Some Empirical Results. Technical Report No. 500, Computer Laboratory, University of Cambridge (2000)

Appendix: DoS-Knocking Sample Code

Code used to perform the DoS-Knocking attack.

```c
#include            "forgeit.h" #define INTERFACE       "eth0" #define INTERFACE_PREFIX 14

char SOURCE[16],DEST[16]; int SOURCE_P,DEST_P;

int main(int argc, char *argv[]) {
    int i, quantity, starting_port, range, fd_send;
    if(argc != 5) {
        printf("\tusage: %s ip_dst quantity init_port port_qty\n",
                argv[0]);
        exit(0);
    }

    DEV_PREFIX = INTERFACE_PREFIX;
    memset(SOURCE,0,16);
    memset(DEST,0,16);
    srand(time(NULL));
    strncpy(DEST,argv[1],15);
    quantity = atoi(argv[2]);
    starting_port = atoi(argv[3]);
    range = atoi(argv[4]);
    fd_send = open_sending();

    for (i=0;i<quantity;i++)
    {
        snprintf(SOURCE,15,"%d.%d.%d.%d"
                    ,1+(int)(254.0* rand()(RAND_MAX+1.0))
                    ,1+(int)(254.0*rand() / (RAND_MAX+1.0))
                    ,1+(int)(254.0*rand() / (RAND_MAX+1.0))
                    ,1+(int)(254.0*rand() / (RAND_MAX+1.0)));

        SOURCE_P=1024+(int)(60000.0*rand() / (RAND_MAX+1.0));
        DEST_P=starting_port+(int)(((float)range)*rand() /
                    (RAND_MAX+1.0));
        transmit_TCP (fd_send, SOURCE, SOURCE_P,DEST, DEST_P,SYN);
    }

    return 0;
}
```

(Example based on Brecht Claerhout's `ipspoof` and `sniper-rst` code [2], modified to improve packet-sending performance)

Marketing on Internet Communications Security for Online Bank Transactions

José M. Sierra[1], Julio C. Hernández[1], Eva Ponce[2], and Jaime Manera[3]

[1] Universidad Carlos III de Madrid
[2] Universidad Politécnica de Madrid
[3] Universidad Rey Juan Carlos
sierra@inf.uc3m.es

Abstract. Everyday more companies extend the use of Internet apart from informational uses. The progression to more sophisticated Internet usage will need that customers trust on the security of the online transactions and on the security mechanisms involved in this protection. For that reason, moreover the common marketing techniques, some special marketing on Internet security will be needed. In fact, customers already are not satisfied with existing protections in Internet e-commerce services and the future of this potential global marketplace will depend on security.

1 Introduction

In certain sense, the economics of the world are driven by developments in Information and Communications Technologies (ICT). In the last years, the exponential growth of Internet users has shown the e-commerce potential but at the same time some problems have also arisen. The continuous success of hacker's attacks and the dissemination of this type of news in the media have contributed to the customers distrust. Companies are aware about how these issues can affect to the expansion of e-commerce companies and they are disposed to assume a more important role in the protection of their systems and concerning customers about the security measures applied.

The reality appears to be that market forces will continue to fuel the rapid expansion of e-commerce, regardless of concerns over security. Additionally, it is difficult to encourage private sector investment in security solutions in the absence of a clear relationship between security and profitability.

Some of the solutions address to the standardization of Internet security and the introduction of legal measures that force companies to apply enough protection and to reinforce the punishment of electronic crime. Among the first option many standards have rise in recent years, ISO-IEC 17799 is the most accepted internationally; and for the legal measures, most of countries have developed specific legislation for the protection of e-commerce customers and companies, i.e. the European Union created the EU directive 2000/31/CE which established the protection needed for e-commerce companies and customers.

Organizations need to take responsibility for their own protection and understand that they are facing an increasingly dangerous e-commerce environment requiring them to have knowledge of the weaknesses of the systems that the firm is dependent on. The marketing of commercially viable e-commerce initiatives must promote customers trust on security and watch over how they perceive the security measures applied.

2 Online Economic Transactions Problems

The online banking objective is not to replace traditional bank offices, but it is to reduce the number of physical offices (with the consistent save of money for the banks) and to supply a more flexible service to their customers. Forecasts address that in a close future Internet banking could represents the 60% of the business [8]. In countries such as the United Kingdom traditional bank offices has been reduced in 3% in recent years. Companies like Deutsche Bank and ABN-Amor have announced the reorganization of their retail banking, reducing their employees and offices and encouraging their Internet e-banking services [10]. Gerlach [4] reported that more than 500 conventional banks in the USA currently offered customers online access to their accounts. In fact, major banks, such as Bank of America and Wells-Fargo have offered a variety of services such are: CDs, credit cards, funds transfer and loans; through their web sites.

The incredible growth of the Internet is changing the way corporations conduct business with consumers. Many new virtual banks[1] have entered the baking industry, providing customers with financial services over the Internet [9]. Since theses Internet-only banks usually have no branch offices, they can substantially reduce operating and fixed cost by replacing employees and physical facilities with information technology. These cost savings have helped Internet-based banks offer better and profitable services than traditional banks [4].

But although Internet banks have focused their attention on improving their banking service quality, these kind of banks are fastened to important problems, such are:

- The way to ensure that customer is who say they are. For example in the first step of their relationship, opening an account.
- Although a public network infrastructure is used, the protection of the communication must ensure the privacy of all contents transmitted.
- The technology involved, the Internet Communications Technology (ICT) is far of being understood by their customers, so the protections measures established by banks cannot be completely assimilated by their users.

Rose [13] evaluated the service quality of 23 USA Internet Banks, including 12 Internet-only banks, in terms of seven service categories:

1. Opening an account
2. Deposits and withdrawals

---

[1] For example, NetBank opened its virtual doors in February 1996 with a business model completely different to the landscape of the banking industry. http://www.netbank.com/about.htm

3. Rates and fees
4. Navigation and case of use
5. Bill paying
6. Security
7. Customer service

She found that most of the sampled banks showed an unsatisfactory level of service quality and argued that: " online banking today is often a maddening frustrating affair that can cause as many problems as it solves".

Our work is focus on the security issues of e-banking. Our aim is to determine the key issues of Internet security marketing that would facilitate that customers trust on e-banking. However, there are important security problems in the Internet technology and these problems can be inherited by the e-banking.

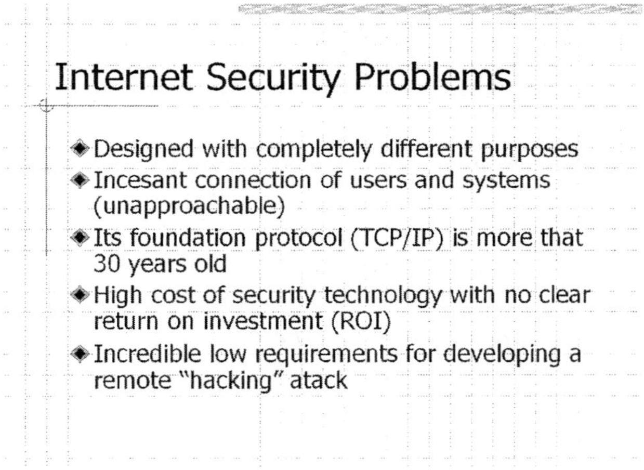

Internet Security Problems

- Designed with completely different purposes
- Incesant connection of users and systems (unapproachable)
- Its foundation protocol (TCP/IP) is more that 30 years old
- High cost of security technology with no clear return on investment (ROI)
- Incredible low requirements for developing a remote "hacking" atack

Fig. 1. Internet Security Problems

The social impact of the Internet security flaws or Internet hacking attacks is very important. The causes that provoke the broad dissemination of this type of news are difficult to determine, but it is a fact that the media use to emphasize their impact and this does not contribute in customers trust on Internet technology.

In this way, and although insecurity is an intrinsic feature of the Internet, is also true that the use of security measures can reduce the risk under assumable levels. The use of digital signature technology for the authentication and non-repudiation of entities solves one to the most important security problems. Ciphering can ensure the privacy of communication and finally the integrity can be obtained by the digital signature of the contents or by the use of other cryptographic algorithms. The standard "de facto" for the protection of online banking is the SSL Protocol (Secure Socket Layer) that supplies the authentication, integrity and confidentiality services for the contents exchanged by those types of connections.

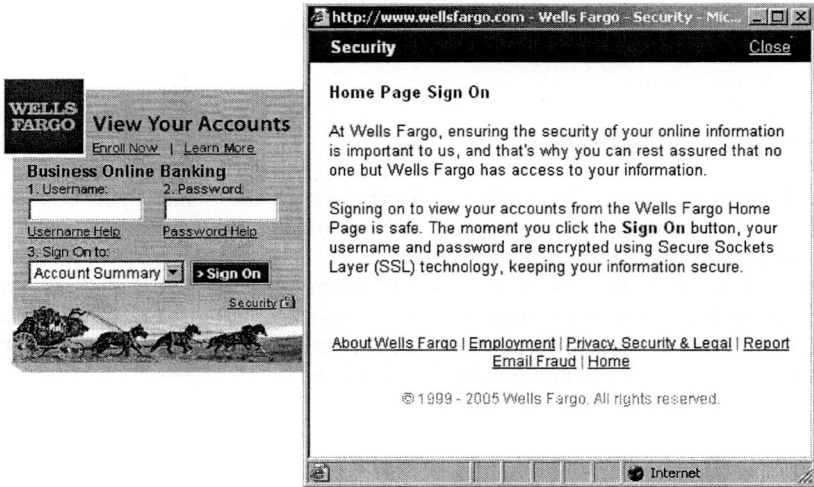

Fig. 2. Wells Fargo online, security statement based on SSL Protocol

However there are other Internet Security Problems that, for the moment, cannot be technologically solved. The clearest example are the Denial of Service attacks, or more properly, the Distributed Denial of Service attacks (DDoS) which in several occasions has disturbed the normal development of e-commerce companies.

3 Users Perception of Online Security

We are conscious that several factors have to be considered analyzing the perception of users about online bank transactions. Many authors have tried to identify the dimensions traditional banking service. In 1995, Johnston [7] examined the banking industry and found 18 service quality attributes, among them: integrity, reliability and security; where clearly joined to the protection of bank transactions.

For online banking, relatively little empirical research has addressed the issue of the key dimensions of the Internet banking service. We would like to remark Jayawardhena and Foley [6] in 2000 that suggested that the features of Internet banking web sites, such as: speed (to download), content, design, interactivity, navigation and security; which are critical to enhancing customer satisfaction.

More recently, Liao and Cheung [9] publish an empirical study of the consumer attitudes toward the usefulness and willingness to use Internet e-retail banking in Singapore. Their work showed that expectations of accuracy, security, network speed, user-friendliness, user involvement and convenience were the most important quality attributes underlying perceived usefulness.

The profile of the users shows that their knowledge on security issues is vague [12]. Even those who usually use online banking do not understand the security measures applied to protect electronic transactions, neither know the basic properties of digital information. However, it is also necessary to indicate, that this situation is not different in the case of traditional banking.

Although banks, in real world, are usually associated to high level of security, it is also true that real world banks use certain technologies that are widely consider insecure. An example are common electronic cards (Magnetic Stripe Cards); the security of these devices has been broken in numerous occasions, but however public still use them and not many consumers reject its use in favor of the known "smart cards". So, something happens because consumers are more concerned about flaws of their security and privacy in the cyber world. The information and communications technologies are a variable world where the fight between the security flaws and the protection mechanisms is very evident.

Banks are in an unsurpassable situation to encourage consumers to trust not only in their physical world but also in their online services. There is no doubt that they must achieve an important effort to explain to their customers the protection measures supplied. Nowadays, not many people believe that insecurity suspicion could hold up the e-commerce expansion, but we think that the marketing on Internet security is a duty for every single company, which is offering online services.

4 How Can Be Managed the Marketing on ICT Security

The Internet is evolving into a new medium for communicating and interacting with customers and thus will affect the roles that marketing play. Security will be a big issue for the companies that wish to offer their services to Internet using customers around the world. On the one hand, they must protect their information system to possible attacks or flaws; many security technologies can be used (firewalls, Intrusion Detection Systems, Virtual Private Networks, SSL communications and so on). But on the other hand, these companies must encourage customers to use a technology, which in certain circumstances can be insecure. Our work tries to establish how to manage the marketing on Internet communications security in order to transmit to the customers trust on the services supplied.

Trust is a very complex and fuzzy concept, which is very difficult to deal with. Under the same protection of certain service, two people can vary significantly on how both trust on it. Sometimes, if the customer knows much information about the ICT then it will be easy to trust on the security measures applied, but in other situation this wide knowledge can complicate the trust on the same service.

For this reason, our work tries to establish common patters that mainly trigger the trust on the security measures applied. Although more dimensions can be identified, we have selected only two of them, trying just to give a guide to the designing of marketing strategies.

According to the *Knowledge on ICT of the Customer*, we recognize that the marketing strategy could differ between a marketing based on quantitative description and marketing based on a qualitative description (fig. 3). Only if the customer's knowledge on ICT is high a qualitative description of the security techniques can be used. In this way, the qualitative perspective should change to a quantitative when customer's knowledge decreases.

However, it is also possible that if the customer knowledge is very modest, no marketing based on description of the security measures can be useful. In these situations, we propose other marketing strategy, where we intent to gain the customer's

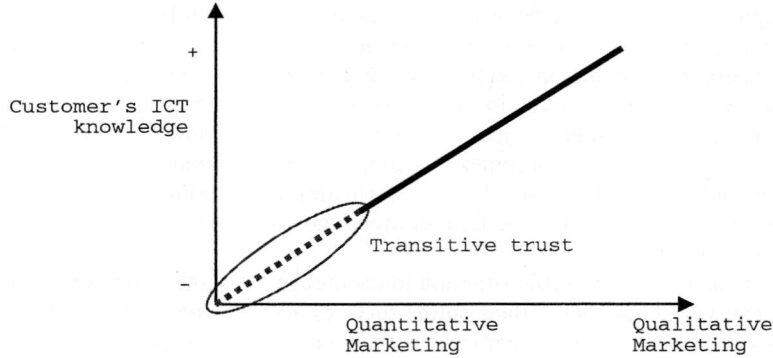

Fig. 3. Customer's ICT knowledge

trust in an indirect mode. We call transitive trust when other organization, which the customer trusts on, certifies the security level supplied by an online company. This is the case of well-known Verisign, or could be the case of a certification made by the government, etc. Customer's trust on this third party is transmitted to certain online service by the exhibition of the certification.

What is shown in figure 3, is that when the customer's knowledge on ICT is too low, the only way of gaining his trust is by the use of other entity, which allows to transmit customer's trust to our service.

The second dimension we have identified is the Service Potential Risk. It is reasonable that customers demand more information when the service used is subjected to more risk. In this way, when the potential risk is too high it will be practical to increase the marketing intensity, supplying to the customer the security information that confirms that always is under control. By the contrary, if the potential risk is very low, not much security information will be needed and the marketing intensity can be reduced (fig. 4).

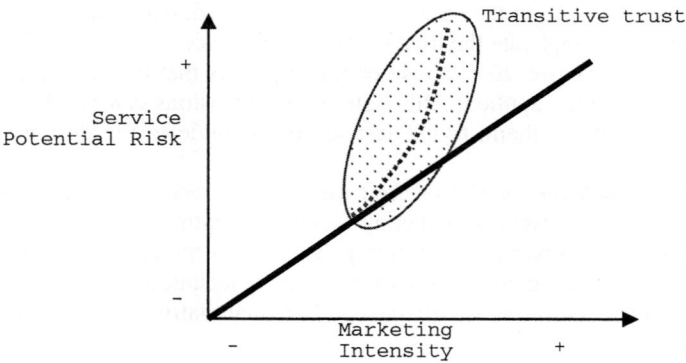

Fig. 4. Service Potential Risk

Nevertheless, if marketing intensity is increased too much, the customer can get the perception of a very heavy service. So, trying to avoid elevated rates of marketing intensity, transitive trust can be used to reduce the marketing effort by the use of other third party.

5 Conclusions

The expansion of e-commerce will require widespread faith in the ability of the technological systems that carry electronic transactions to function in a flawless mode. However, surveys all indicate that both customers and commercials users are in doubt concerning the security of the basic systems. The problem is no easy to solve, because the technological complexity of the cryptographic techniques and the ignorance of certain sectors of society do not facilitate trust of customers.

Online banks are everyday more worried about the security image of their services. In the following years, banks will give important investments on security for the improvement of their information systems and their online services. Furthermore, banks should attack new approaches to prompt customers to trust on it.

Our work tries to improve the success of marketing techniques on the online environment. We outline two dimensions to take into account in the designing of Internet security marketing. The first dimension identified is the Customer Knowledge on ICT and the second one is the Service Potential Risk, in both cases different marketing strategies must be applied. Finally our work explains how the transitive trust can modify the two dimensions commented.

References

[1] Arnott, D. C. and Bridgewater S. "Internet, interaction and implications for marketing". Marketing Intelligence & Planning, vol. 20, n. 2, pp. 86-95. 2002
[2] Barczak, G., Ellen, P. and Pilling B. "Developing Typologies of Consumer Motives for Use of Technologically Based Banking Services" Journal of Business Research, vol. 38 pp. 131-139. 1997.
[3] Forcht, K.A. "Doing business on the Internet: marketing and security aspects". Information Management & Computer Security. vol. 4, n.4. 1996
[4] Gerlach, D. "Put the money where your mouse is" PC World, March, pp. 191-199. 2000
[5] Heinen, J. "Internet marketing practices". Information Management & Computer Security. vol. 4, n.5. 1996
[6] Jayawardhena,C and Foley, P. "Changes in the banking sector – the case of Internet backing in the UK". Internet Research: Electronic Networking applications and Policy. Vol. 10, n. 1, pp. 19-30. 2000
[7] Johnston, R. "The Determinants of service quality: satisfiers and dissatisfies" International Journal of Service Industry Management, vol. 6, n.1, pp.19-30. 1995
[8] Jun, M. And Cai, S. "The key determinants of Internet banking service quality: a content analisys". International Journal of Bank Marketing, vol. 19, n.7. 2001
[9] Liao, Z. and Cheung, M. T. "Internet-based e-banking and consumer attitudes: an empirical study". Information and Management, vol. 39, pp. 283-295. 2002

[10] McCrohan, K. F. "Facing the threats to electronic commerce". Journal of Business and Industrial marketing, vol. 18, n. 2, pp. 133.145. 2003
[11] Ponce E. et al. "Security Consequences of Messaging Hubs in Many-to-Many E-procurement Solutions". ICCSA 2004.
[12] Roboff, G. and Charles, C. "Privacy of Financial Information in Cyberspace". Journal of Retail Banking Services; Autumn; vol. 20, n. 3, 1998
[13] Rose, S. "The truth about online banking". Money, vol. 29, n. 4, pp. 114-122. 2000

A Formal Analysis of Fairness and Non-repudiation in the RSA-CEGD Protocol

Almudena Alcaide, Juan M. Estévez-Tapiador,
Antonio Izquierdo, and José M. Sierra

Department of Computer Science, Carlos III University of Madrid,
Avda. Universidad 30, 28911, Leganés, Madrid (Spain)
{aalcaide, jestevez, aizquier, sierra}@inf.uc3m.es

Abstract. Recently, Nenadić et al. (2004) proposed the RSA-CEGD protocol for certified delivery of e-goods. This is a relatively complex scheme based on verifiable and recoverable encrypted signatures (VRES) to guarantee properties such as strong fairness and non-repudiation, among others. In this paper, we illustrate how an extended logic of beliefs can be helpful to analyze in a formal manner these security properties. This approach requires the previous definition of some novel constructions to deal with evidences exchanged by parties during the protocol execution. The study performed within this framework reveals the lack of non-repudiation in RSA-CEGD and points out some other weaknesses.

1 Introduction

Interest in protocols for fair exchange of information with non-repudiation stems from its importance in many applications where disputes among parties can occur. Assurance of these properties enables the deployment of a wide range of applications, such as certified e-mail or business transactions through communication networks. As a result, fair non-repudiation has experienced an explosion of proposals in recent years (see [9] for an excellent survey).

Nevertheless, fairness and non-repudiation have not been so extensively studied as other classic issues, such as confidentiality or authentication. Previous experience in these contexts has shown that designing security protocols is an error-prone task, and that formal analysis can aid in detecting weaknesses. However, many of the protocols proposed to achieve fair non-repudiation have not been subject to a formal security analysis. In the field of security protocols, formal methods have mainly focused on the analysis of authentication and key-establishment protocols. These techniques, however, cannot be directly applied to reasoning about properties such as fairness or non-repudiation without a previous formalization of what the protocol goals are. Furthermore, many assumptions usually done for other protocols are no longer valid in these environments. For instance, most of the analysis models developed consider the existence of an attacker with different capabilities, e.g. control over the communication channels, though the protocol parties are assumed to trust each other and behave according to the protocol rules. Nevertheless, in fair exchange or non-repudiation there is no adversary per se [15], such as the classic Dolev-Yao intruder model [4]. These protocols, on the contrary, are designed for scenarios wherein participants may misbehave with the aim of cheating the other party, thus obtaining her own profit.

The tools developed to verify fair non-repudiation are scarce when compared to those available for classic security requirements. To the best of our knowledge, the first attempt to reasoning about this kind of protocols in a formal manner was due to Kailar [6, 7]. He used a BAN-like logic wherein the central construction is the predicate CanProve, as in "A CanProve B says X". Properties such as non-repudiation can be easily formalized using this predicate. A similar approach was introduced by Zhou and Gollman in [17]. In this work, however, authors opted for the use of the SVO logic instead of BAN. An alternative was presented by Schneider in [14], in which a verification of a non-repudiation protocol is carried out by using CSP (*Communicating Sequential Processes*). This is an abstract language aimed at modeling the communication patterns of concurrent systems that interact through message passing.

Even though evidences of non-repudiation can be provided by means of classic cryptographic techniques, fairness is much more difficult to achieve and also to guarantee. Recent works due to Kremer and Raskin have focused on the consideration of exchange protocols as a game wherein messages exchanged by participants can be seen as their strategies [10, 11].

Despite efforts as those mentioned above, it still remains difficult to assure formally that a protocol guarantees fair non-repudiation. Consider, as an illustrative example, a protocol proposed in 1996 by Zhou and Gollman [16] that was verified and proved correct using three different methods [2, 14, 17]. Surprisingly, in 2002 Gürgens and Rudolph demonstrated the absence of fair non-repudiation in that protocol under reasonable assumptions [5]. In this case, possible attacks were detected after an analysis performed with a different formalism that considered scenarios not checked before.

Verifications of security properties must be carefully treated. A formal analysis does not establish security, but only with respect to a model which, in turn, is based on assumptions. Many of these assumptions may be violated in scenarios not considered by the protocol designers. As a result, assumptions are usually the inherent weaknesses, exactly in the sense pointed out by Denning: "the way to crack a system is to step outside the box, to break the rules under which the protocol was proved secure" [3]. Anyway, the formalization effort required to perform a verification serves to highlight assumptions and can clarify some flaws in the protocol design, thus pointing out possible vias of exploitation. In this paper, we show how an extended logic of beliefs can be useful to reasoning about properties such as fairness and non-repudiation. Our aim with this approach is not to prove security, but to: 1) identify critical steps in the protocol evolution; 2) analyze the evidences supporting security properties that each party possesses; and 3) reasoning about scenarios in which such evidences might be misused with success. As a proof-of-concept, we illustrate this methodology by presenting a partial analysis of the RSA-CEGD protocol [12] –an RSA-based scheme for certified e-goods delivery. Due to space restrictions, in this work we focus exclusively in the strong fairness and non-repudiation properties, showing that the latter is not satisfied.

The rest of this paper is organized as follows. For completeness, Section 2 presents a brief overview of the RSA-CEGD protocol. Section 3 introduces a formalization of the two features of this protocol considered in this work: strong fairness and non-repudiation. The core of the analysis is exposed in Section 4, while Section 5 discusses in detail the key points of our approach. Finally, Section 6 concludes the paper.

2 Overview of the RSA-CEGD Protocol

Throughout this paper, we will use the same notation introduced by the authors in the original paper [12]. The RSA-CEGD is an optimistic fair exchange protocol composed of two sub-protocols, as shown in Fig. 1. As usual, the exchange sub-protocol is used to carry out the exchange between parties without any TTP's involvement. In case the process fails to complete successfully, a recovery protocol can be invoked to handle this situation.

The notion of verifiable and recoverable encrypted signature (VRES) underlies at the core of the RSA-CEGD protocol. A VRES is basically an encrypted signature, which acts as a *receipt* from the receiver's point of view, with two main properties. First, it can be *verified*: the receiver is assured that the VRES contains the expected signature without obtaining any valuable information about the signature itself during the verification process. And second, the receiver is assured that the original signature can be *recovered* with the assistance of a designated TTP in case the original sender refuses to do it.

Due to these two properties, the VRES becomes an interesting cryptographic primitive upon which fairness can be provided. The RSA-CEGD protocol relies on this element within the general scheme we sketch in what follows:

1. A ciphers the message with an encryption key and sends it to B.
2. B generates the VRES of his signature and sends it back to A.
3. Upon successful verification of the VRES, A is assured that it is secure for her to send the decryption key to B, so he can access the message.
4. Finally, B sends his original signature to A as a receipt. In case he refuses, a TTP can recover the signature from the VRES, thus restoring fairness.

The RSA-CEGD protocol makes use of a novel VRES method based on the RSA system, hence its name. The idea stems from the so-called *theory of cross-decryption* [13], which establishes that an RSA encrypted text can be decrypted by using two different keys if both pairs of secret/public keys are appropriately chosen. Party B is enforced to use a key of this kind to encrypt the VRES, while the TTP retains the other. This way, if subsequently B refuses to provide A with his signature, the TTP is able to recover it from the VRES.

The exchange sub-protocol
E1: $P_a \to P_b : E_{k_a}(D_a), CertD_a, x_a, E_{sk_a}(h_a)$
E2: $P_b \to P_a : (x_b, xx_b, y_b), s_b, C_{bt}$
E3: $P_a \to P_b : r_a$
E4: $P_b \to P_a : r_b$

The recovery sub-protocol
R1: $P_a \to P_t : C_{bt}, y_b, s_b, y_a, r_a$
R2: $P_t \to P_a : r_b$
R3: $P_t \to P_b : r_a$

Definition of the protocol's items

$x_a = (r_a \times k_a) \mod n_a$: encryption of P_a's key k_a with random number r_a

$CertD_a = (desc_a, hd_a, h_a, ek_a, sign_{CA})$: certificate for D_a issued by the CA, where
$\quad desc_a =$ description (content summary) of D_a
$\quad hd_a = h(E_{k_a}(D_a))$: hash value of the encryption of D_a with the key k_a
$\quad h_a = h(D_a)$: hash value of D_a
$\quad ek_a = E_{pk_a}(k_a)$: encryption of the key k_a with P_a's public key, pk_a

$E_{sk_a}(hd_a)$: P_a's RSA signature on D_a serving as a proof of origin of D_a

r_a : random prime generated by P_a for the encryption of key k_a

$y_a = E_{pk_a}(r_a)$: RSA encryption of number r_a with key pk_a

r_b : random prime generated by P_b for the generation of the VRES (y_b, x_b, xx_b)

$rec_b = (h_a)^{d_b} \mod n_b$: P_b's receipt for P_a's e-goods D_a, i.e. P_b's RSA signature on D_a

(y_b, x_b, xx_b): P_b's VRES, where
$y_b = r_b^{e_b} \mod (n_b \times n_{bt})$: encryption of r_b with P_b's public key. Also recoverable by P_t
$x_b = (r_b \times (h_a)^{d_b}) \mod n_b = (r_b \times rec_b) \mod n_b$: encryption of rec_b with r_b
$xx_b = (r_b \times E_{sk_{bt}}(h(y_b))) \mod n_{bt}$: control number that confirms the correct use of r_b

$C_{bt} = (pk_{bt}, w_{bt}, s_{bt})$: P_b's RSA public-key certificate issued by P_t
$\quad pk_{bt} = (e_{bt}, n_{bt})$: public RSA key related to C_{bt}, with $e_{bt} = e_b$
$\quad sk_{bt} = (d_{bt}, n_{bt})$: private RSA key related to C_{bt}
$\quad w_{bt} = (h(sk_t, pk_{bt})^{-1} \times d_{bt}) \mod n_{bt}$
$\quad s_{bt} = E_{sk_t}(h(pk_{bt}, w_{bt}))$: P_t's signature on $h(pk_{bt}, w_{bt})$.

$s_b = E_{sk_b}(h(C_{bt}, y_b, y_a, P_a))$: P_b's recovery authorization token

Fig. 1. The RSA-CEGD protocol

3 Formalization of Protocol Objectives

The protocol RSA-CEGD presented in [12] is defined to satisfy a list of six different objectives. In this paper, we will only focus on the formal analysis of three of them, the ones related to strong fairness and non-repudiation. The specific definitions for each of those three goals are quoted literally from [12]. Moreover, our analysis focuses on the items considered by the protocol authors as proof of fairness and non-repudiation. In the following section, such objectives and evidences will be formalized using classic BAN notation.

3.1 Strong Fairness and Correctness

Strong fairness is defined by the protocol authors as follows: *"If and only if the sender has obtained the receiver's receipt or can obtain it with the assistance of a STTP, then the receiver has obtained the sender's e-goods or can obtain them with the assistance of the STTP"*. Consequently, the protocol is fair and correct if and only if the following formulae are verified:

- If P_b has got the goods, then P_a has the authorization token and the item $\langle rec_b \rangle_{r_b}$:

$$\frac{P_b \triangleleft D_a}{P_a \models s_b, \quad P_a \triangleleft \langle rec_b \rangle_{r_b}} \tag{1}$$

- If P_a has rec_b, then P_b has the goods:

$$\frac{P_a \triangleleft rec_b}{P_b \triangleleft D_a} \tag{2}$$

3.2 Non-repudiation of Origin of the E-goods (NRO)

As stated when defining the objectives for the RSA-CEGD, non-repudiation of origin is preserved when: *"The recipient is provided with a proof that the sender is indeed the originator of the e-goods"*. The protocol establishes the item $\{hash(D_a)\}_{sk_a}$ as the only proof of origin. To preserve NRO, P_b must be sure that P_a is the real sender of such a message, and P_a must also ensure that P_b cannot have access to such a message if P_a has not sent it. This is, NRO is preserved when these two formulae are satisfied:

$$\frac{P_b \triangleleft \{hash(D_a)\}_{sk_a}}{P_b \models P_a \models \{hash(D_a)\}_{sk_a}} \tag{3}$$

$$\frac{P_b \triangleleft \{hash(D_a)\}_{sk_a}}{P_a \models \{hash(D_a)\}_{sk_a}} \tag{4}$$

Regarding formula (3), P_b knows that the originator of the e-goods is P_a. However, P_b knows that the sender is indeed P_a, *only* because the channel between P_a and P_b is considered to be authenticated and confidential. Therefore, prior to the current run of the protocol P_a and P_b have been authenticated as the parties involved in the communication, and P_b can be sure that the sender of the signature that he receives is P_a. The formula $\{hash(D_a)\}_{sk_a}$ itself does not prove that P_a is the sender as well as the originator. Furthermore –formula (4)–, P_a cannot know whether P_b can obtain the proof of origin from a different source. The formula $\{hash(D_a)\}_{sk_a}$ may not belong to the current execution. P_b could make P_a responsible for a item that P_a never sent to him. The validation process will show how NRO cannot be verified.

3.3 Non-repudiation of Receipt of the E-goods (NRR)

RSA-CEGD authors consider the property of non-repudiation of receipt to be preserved when *"The sender of the e-goods is provided with a proof that the intended recipient has indeed received the e-goods"*. In this case, P_b's signature over the hash value of the e-goods is admitted as irrevocable proof that P_b received the e-goods. Therefore, NRR is preserved if the following formula is satisfied:

$$\frac{P_a \triangleleft \{hash(D_a)\}_{sk_b}}{P_a \models P_b \triangleleft D_a} \quad (5)$$

Here, the item $\{hash(D_a)\}_{sk_b}$ does link the entity P_b and the e-goods D_a. On the other hand, $CertD_a$ will join together the e-goods D_a and the originator entity P_a.

4 Security Analysis

The entire validation process is described in detail in [1]. In what follows, only the necessary steps for analyzing the objectives are shown.

4.1 Analysis of Strong Fairness and Correctness

As stated in Section 3.1, the protocol is fair and correct if and only if formulae (1) and (2) are verified. In the former case, the most important steps of its analysis are provided below:

1. Message   E1: $P_a \rightarrow P_b$: $E_{k_a}(D_a), CertD_a, x_a, E_{sk_a}(h_a)$
2. Items $\{D_a\}_{K_a}, CertD_a, \langle K_a \rangle_{r_a}, \{h_a\}_{sk_a}$
3. Message   E2: $P_b \rightarrow P_a$: $(x_b, xx_b, y_b), s_b, C_{bt}$
4. Conclusions $P_a \triangleleft \langle rec_b \rangle_{r_b}$, where $rec_b = \{h_a\}_{sk_b}$
   $P_a \models s_b$, where $s_b = \{hash(C_{bt}, y_b, y_a, P_a)\}_{sk_b}$
5. Message   E3: $P_a \rightarrow P_b$: r_a
6. Conclusions $P_b \triangleleft D_a$

The sequential order in which the exchange protocol is executed gives us the implication: $P_b \triangleleft D_a \Rightarrow P_a \models s_b, P_a \triangleleft \{rec_b\}_{r_b}$. Otherwise, P_t's honesty and *resilient communication channels* between P_b and P_t would ensure the property when the recovery protocol is invoked.

The analysis concerning formula (2) is completely analogous. In a similar way, the sequential order in which the exchange protocol is executed gives us the implication: $P_a \triangleleft rec_b \Rightarrow P_b \triangleleft D_a$. Again, P_t's honesty and *resilient communication channels* between P_a and P_t would ensure the property in case the recovery protocol has to be invoked.

Finally, note that due to the absence of any timeout routine as part of the protocol, a deadlock conflict could arise easily. For example, a network failure can prevent P_a from receiving message E2, while at the same time P_b —unaware of the failure— keeps awaiting to receive message E3.

4.2 Analysis of NRO

According to the definition provided in Section 3.2, NRO is verified if formulae (3) and (4) hold. In the former case, the most important steps of its analysis are provided in what follows:

1. Message   E1: $P_a \rightarrow P_b$: $E_{k_a}(D_a), CertD_a, x_a, E_{sk_a}(h_a)$
2. Items $\mathbf{P_b} \models \mathbf{P_a} \mathrel{\vert\!\sim} \mathbf{h_a}$
3. Conclusions $\mathbf{P_b} \triangleleft \mathbf{h_a}$, as h_a is part of $CertD_a$
4. Items $\mathbf{P_b} \models \stackrel{+\mathbf{pk_a}}{\mapsto} \mathbf{P_a} \Leftrightarrow \mathbf{P_b} \triangleleft \mathbf{h_a}$
5. Conclusions P_b verifies $h_a = hash(D_a)$ from $CertD_a$
                                        Therefore: $\mathbf{P_b} \triangleleft \{\mathbf{h_a}\}_{\mathbf{sk_a}}$
6. Conclusions   So P_b knows that P_a is the originator:
                                        $\mathbf{P_b} \models \mathbf{P_a} \mathrel{\vert\!\sim} \{\mathbf{h_a}\}_{\mathbf{sk_a}}$
7. Conclusions   Because the channel between P_a and P_b is considered to be authenticated and confidential, P_b knows that P_a is the sender. So P_a must *still believe* in such a message:
                                        $\mathbf{P_b} \models \mathbf{P_a} \models \{\mathbf{hash(D_a)}\}_{\mathbf{sk_a}}$

Therefore, the formula $\{hash(D_a)\}_{sk_a}$ itself does not prove that P_a is the sender as well as the originator. The item $\{hash(D_a)\}_{sk_a}$ could have been used before, i.e. in another instance of the same protocol. An entity, different from P_a, could have been part of a genuine run of the protocol and could now be reusing such a formula with malicious purposes (replay attack).

Furthermore, neither the implication (4) can be verified with the items provided by each party during the protocol execution, nor confidentiality of communications can assists in the validation process of implication (4). P_a cannot know whether P_b can obtain a proof of origin from a different source. In fact, the formula $\{hash(D_a)\}_{sk_a}$ may not belong to the current protocol execution. P_b could be reusing the item $\{hash(D_a)\}_{sk_a}$ obtained from an entity which participated with P_a in a previous run of the protocol. In such a case, P_b was not meant to be the receiver of item $\{hash(D_a)\}_{sk_a}$, even though it is possible that P_b could make P_a responsible for having sent a proof of origin that P_a never sent to P_b. As a result, NRO is not preserved.

4.3 Analysis of NRR

According to the definition given in Section 3.3, satisfying the double implication expressed by formula (5) will verify NRR. In this case, the sequential order in which the exchange protocol is executed gives us the implication: $P_a \triangleleft \{h_a\}_{sk_b} \Rightarrow P_b \triangleleft D_a$. Otherwise, *resilient communication channels* between P_b and P_t would ensure the property when the recovery protocol is invoked.

Finally, note that the communications channels between P_t and entities P_a and P_b, coupled with P_t's honesty, play a crucial role with regard to NRR and fairness.

5 Discussion

5.1 On the Analysis Process

The strong fairness and non-repudiation protocol RSA-CEGD has been formally analyzed using a BAN-like logic. The entire validation and analysis processes are described in [1]. BAN and related logics were never designed to validate security properties such as fairness or non-repudiation, but authentication and key-exchange protocols. However, the modifications and new rules added to BAN to pursue our goals have been minimal. Only new concepts as public keys certificates or VRES signatures have been described using new rules in order to formalize their definitions and behaviors.

A new operator has also been introduced to define the double implication "if and only if" between formulae. The double implication has been represented as a double line between formulae within the same rule.

5.2 Double Implication in Logic Rules

Double implication in the extended rules is introduced to gain completeness as well as correctness of the validation process. Fig. 2(a) depicts a general stage of the protocol validation process. For each stage, a set of hypotheses H_A is defined for a given entity A. Within that set H_A, A carries out some verifications and computations. If the verifications are negative then A will abort the current run of the protocol. Otherwise, if the verifications are positive, A can use the extended set of rules to infer the set $Results_A$ with a double implication, i.e: "if and only if".

The *correctness* of such a stage will be verified if A can obtain the expected set of objectives from the set H_A and the verifications carried out in it; but also, if A can prove that no other type of verifications or computations could result in contradicting the objectives. In this regard, the process ensures A that if results contradicting the set $Results_A$ can be derived from a different set of computations, then $Verifications_A$ would not have been satisfied, so A would have aborted the execution of the protocol. This situation is graphically explained in Fig. 2(c). Note that the set of hypotheses H_A is considered to be well defined and determined by the protocol. Only secret keys, public keys and certificates are part of the initial set H_A. For each stage, the formulae contained within the message sent and the set of results from previous stages are added to H_A.

The *completeness* of the validation process comes from the double implication. How can A be sure that, given the hypotheses in set H_A, a different set of verifications and computations, not related to the ones determined by the protocol, could not reach to the same results as in set $Results_A$?. This could be the case when an entity is trying to cheat on another or when an attacker is trying to carry out an attack. The situation is shown in the diagram depicted in Fig. 2(b). In this situation, the double implication would mean that $Verifications'_A \Rightarrow Verifications_A$, or equivalently $Verifications'_A \subseteq Verifications_A$. This way, any different set of verifications and computations should be part –or imply– those determined by the protocol.

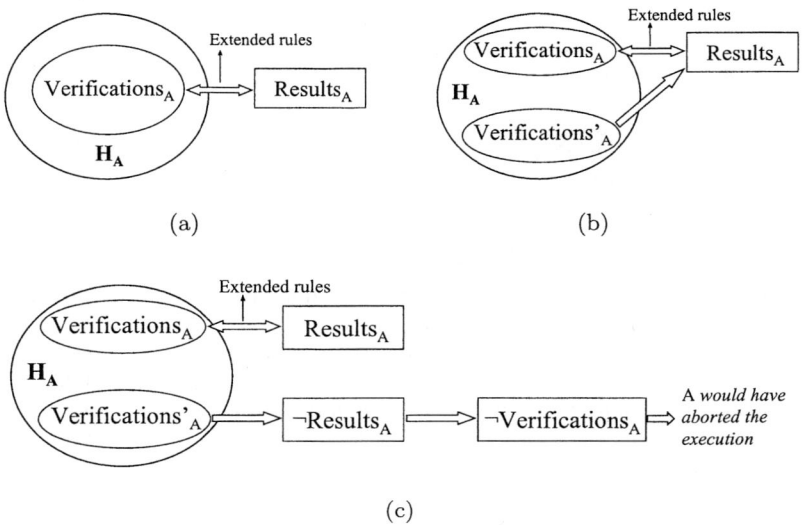

Fig. 2. Scheme of the analysis process

5.3 Reasoning About Evidences

Suppose entities P_a and P_b, and a message X being sent from P_a to P_b. The concept of freshness, represented in BAN as $\sharp(X)$, is the only operator to promote a belief such $P_a \mid\!\sim X$ onto $P_a \models X$. This is, "P_a once said X" onto "P_a still believes on X". We have seen in Section 4.2 how the formula $P_a \mid\!\sim X$ is not enough to prove origin of message X. Any given proof of origin should somehow link the sender, the originator of X and X itself with the undergoing protocol execution.

Something very similar applies to non-repudiation of receipt. A proof of receipt should link together the receiver and the message received with the undergoing protocol execution. In fact, these requirements are well understood in the design of non-repudiation protocols (useful guidelines can be found in [8, 9]).

6 Conclusions

In the approach outlined in this paper, we have attempted to reasoning formally about some security properties of a complex protocol, such as it is the case of RSA-CEGD. Formalization provides us with an useful framework for identifying weaknesses that can get easily unnoticed during an informal study. Moreover, the very process of analysis serves frequently not only to point out possible attacks, but also aids to understand how the protocol can be improved.

References

1. A. Alcaide and J. M. Estévez. "Formal Analysis of the RSA-CEGD protocol". *Technical Report*, January 2005.
2. G. Bella and L. Paulson. "Mechanical Proofs about a Non-repudiation Protocol". *Proc. 14th Intl. Conf. Theorem Proving in Higher Order Logic*. LNCS, pp.91–104. Springer-Verlag, 2001.
3. D. E. Denning. "The Limits of Formal Security Models". National Computer Systems Security Award Aceptance Speech. October, 1999. Available online at: http://www.cs.georgetown.edu/~denning/infosec/award.html.
4. D. Dolev and A. C. Yao. "On the Security of Public Key Protocols", *IEEE Trans. Inf. Theory*, IT-29(12):198–208, 1983.
5. S. Gürgens and C. Rudolph. "Security Analysis of (Un-) Fair Non-repudiation Protocols", *FASec 2002*, LNCS 2629, pp. 97–114. Springer-Verlag, 2002.
6. R. Kailar. "Reasoning about accountability in protocols for electronic commerce". *Proc. IEEE Symp. Security and Privacy*, pp. 236–250. IEEE Computer Security Press, 1995.
7. R. Kailar. "Accountability in electronic commerce protocols". *IEEE Trans. Software Engineering*, 5(22):313–328, 1996.
8. P. Louridas. "Some guidelines for non-repudiation protocols". *Computer Communication Review*, 30(5):29–38. October, 2000. ACM Press.
9. S. Kremer, O. Markowitch, and J. Zhou. "An intensive survey of fair non-repudiation protocols". *Computer Comunications*, 25(17):1606–1621. Elsevier Science B.V., 2002.
10. S. Kremer and J. F. Raskin. "A game approach to the verification of exchange protocols - application to non-repudiation protocols". *Workshop on Issues in the Theory of Security (WITS'00)*, July 2000.
11. S. Kremer and J. F. Raskin. "A Game-Based Verification of Non-Repudiation and Fair Exchange Protocols". *Journal of Computer Security*, 11(13):399–429, 2003.
12. A. Nenadić, N. Zhang, S. Barton. "A Security Protocol for Certified E-goods Delivery", *Proc. IEEE Int. Conf. Information Technology, Coding, and Computing (ITCC'04)*, Las Vegas, NV, USA, IEEE Computer Society, 2004, pp. 22–28.
13. I. Ray and I. Ray. "An Optimistic Fair Exchange E-commerce Protocol with Automated Dispute Resolution". *Proc. Int. Conf. E-Commerce and Web Technologies, EC-Web 2000*. LNCS 1875, pp. 84–93. Springer-Verlag, 2000.
14. S. Schneider. "Formal Analysis of a Non-repudiation Protocol". *IEEE Computer Security Foundations Workshop*. IEEE Computer Society Press, 1998.
15. P. Syverson and I. Cervestato. "The Logic of Authentication Protocols", R. Forcardi and R. Gorrieri, Eds.: *Foundations of Security Analysis and Design (FOSAD'00)*, Tutorial Lectures, LNCS 2171, pp. 63–136. Springer-Verlag, 2000.
16. J. Zhou and D. Gollman. "A fair non-repudiation protocol". *Proc. 1996 Symp. on Research in Security and Privacy*, pp. 55–61. Oakland, CA, USA. IEEE Computer Society Press, 1996.
17. J. Zhou and D. Gollman. "Towards verification of non-repudiation protocols". *Proc. 1998 Intl. Refinement Workshop and Formal Methods Pacific*, pp. 370–380. 1998.

Distribution Data Security System Based on Web Based Active Database

Sang-Yule Choi*, Myong-Chul Shin**, Nam-Young Hur**, Jong-Boo Kim*,
Tai-hoon Kim***, and Jae-Sang Cha****

* Dept. of Electronic Engineering , Induk Institute of Technology,
San 76 Wolgye-dong, Seoul, Korea
** School of Electrical and Computer Engineering, Sungkyunkwan University,
Suwon 440-746, Korea
*** 3San 7, Geoyeou-dong, Songpa-gu, Seoul, Korea
**** Dept. of Information and Communication Engineering,
SeoKyeong University, Seoul, Korea

Abstract. The electric utility has the responsibility to provide a good quality of electricity to their customers. Therefore, they have introduced DDSS(Distribution Data Security System) to automate the power distribution data security. DDSS engineers need a set of state-of-the-art applications, eg. managing distribution system in active manner and gaining economic benefits from a flexible DDSS architectural design. The existing DDSS functionally could not handle these needs. It has to be managed by operators whenever feeder faults data are detected. Therefore, it may be possible for propagating the feeder overloading area, if operator makes a mistake. And it utilizes closed architecture, therefore it is hard to meet the system migration and future enhancement requirements. This paper represents web based, platform-independent, flexible DDSS architectural design and active database application. The recently advanced internet technologies are fully utilized in the new DDSS architecture. Therefore, it can meet the system migration and future enhancement requirements. And, by using active database, DDSS can minimize feeder overloadings area in distribution system without intervening of operator, therefore, minimizing feeder overloadings area can be free from the mistake of operator.

1 Introduction

The electrical utility has the responsibility to provide a good quality of electricity to their customers. Therefore, the DDSS (Distribution Data Security System) is introduced to control and operate complex power distribution system in an economical and reliable fashion. It includes important functions such as data security, feeder automation, load control and telemetering. In korea, the electrical utility company is now facing deregulation and privatization. Several privatized distribution companies will be appear in few years. And they will require new way of thinking and new

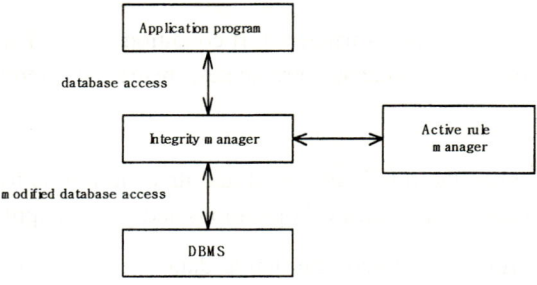

Fig. 1. Layered and built-in architecture

2.2 Active Rule Manager

In ECA rules, one event may triggers several defined rules, and some actions of triggered rules may trigger another rules. Therefore, forward-chaining rules can be fired by one event.

Active rule manager controls the firing of rules. It interfaces with integrity manager, which is used to check if integrity constraints are satisfied or not. It also responsible for aggregating composite primitive events that are signaled from integrity manager, and determine which rules are to be fired. When integrity manager sends signals, rules are fired.

2.3 Data Requirements for DDSS Database

Data requirements for DDSS database are substation, transformer, feeder, feeder section, switch, and information. And attributes of these requirements have to be easily applied to feeder reconfiguration program.

After feeder reconfiguration is executed, resulted data are restored to information table.

- SUBSTATION

Substation number(psn), transformer number(tsn)

- TRANSFORMER

Transformer number(tsn), substation number (psn), transformer capacity(tac), feeder number1 (fsn1), Feeder number2(fsn2), feeder number3 (fsn3), transformer loadings(premva), open/close status(close : 1, open : 0)

Transformer is connected to three feeders, and attribute transformer loadings value is sum of all feeder loadings of feeder number1, 2, 3 .

- FEEDER

Feeder number(fsn), connected transformer number (tsn), feeder loadings (fsnmva),Start feeder section number (fsnsec), close/open status (close:1, open:0).

- FEEDER SECTION

Feeder section number (fsnsec), switch number (ssn), feeder capacity 1(fnc), feeder capacity 2 (fanc), Resistance(rr), reactance(xx), feeder number(fsn), fault flag

(normal :0 ,fault :1), start point(fsnfbsn), end point(fsntbsn), section loadings(ssnmva), voltage(vv), minimum voltage(min_vv)
  Where, voltage (vv) = VL—Vdrop
  VL : Voltage of appropriate switch
  VL : Voltage drop of feeder section
  Fnc is 80 [%] of feeder capacity and fanc is 100 [%] of feeder capacity.
  Feeder section loadings and close/open status are the same value of switch because, by using active rule, they are updated by the same quality of the appropriate switch which close(or open) feeder section.
  ● SWITCH
  Switch number(ssn), feeder section number (fsnsec), loadings(ssnmva), current(aa), voltage (vv), close/open status (close:1, open:1)
  It is assumed that detected loadings, current, voltage values are updated by data acquisition system that is located in real distribution systems.
  ● INFORMATION
  Identification (rsn), total losses (ploss), the amount of loss change (dploss)

2.4 Conceptual Design for DDSS Databases

In this paper, conceptual design is represented by entity-relationship diagram, and relationship between entities is as follows:
  ● SUBSTATION :TRANSFORMER
  One substation can have several transformers, therefore, a relationship type is "have" and 1 : N relationship between the two entity types SUBSTATION and TRANSFORMER.
  ● TRANSFORMER : FEEDER
  One transformer can distribute power among several feeders, therefore, a relationship type is "distribute " and 1 : N relationship between the two entity types TRANSFORMER : FEEDER
  ● FEEDER : FEEDER SECTION
  One feeder is consisted of feeder sections, therefore, a relationship type is "consists-of" and 1 : N relationship between the two entity types FEEDER : FEEDER SECTION
  ● FEEDER : INFORMATION
  Information contains resulted data obtained from feeder reconfiguration program. And feeder reconfiguration program can be executed several times, therefore, a relationship type is "execute" and 1 : N relationship between the two entity types FEEDER : INFORMATION
  ● FEEDER SECTION : SWITCH
  one switch can close(or open) a appropriate feeder section, therefore, a relationship type is " close_open" and 1 : 1 relationship between the two entity types FEEDER SECTION : SWITCH

The above entity-relationship is displayed by means of the graphical notation known as ER diagram in fig.2.

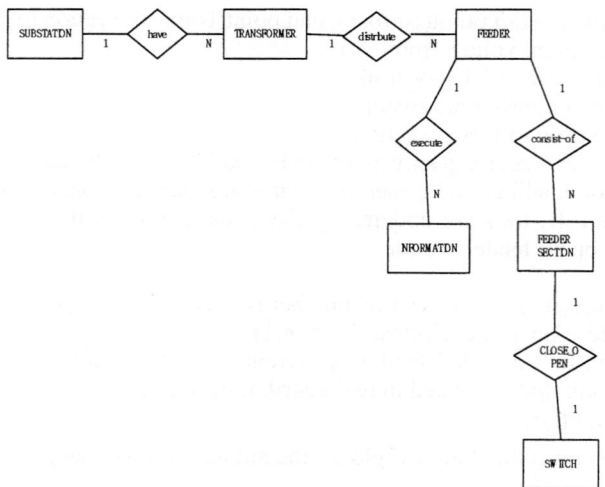

Fig. 2. ERD of DDSS Database

2.5 Rules Definition for DDSS Database

Data acquisition system which is located in distribution networks detects currents, section loadings, voltage and the status of close/open in real time, and the present attribute values of switch is updated by detected ones. DDSS active database monitors the present state of distribution networks by using updated values: If updated values trigger some events and condition are true, then they imply that distribution networks is in overloaded state. Therefore action for feeder reconfiguration is executed to relive feeder overloadings and minimize line losses. Switches to be open(or closed) are selected after feeder reconfiguration is executed.

Feeder reconfiguration can fire other rules which update attribute close/open status value of selected switchs and send close/open signal to appropriate intelligent switches installed in distribution networks.

Due to maintaining radial distribution networks structure, updated value of switch can trigger other rules that update close/open status value of another switch(not a selected one from feeder reconfiguration). And updated attribute value of switch can trigger rules which updates the close/open status value of appropriate feeder section that is connected to updated switch, because close/open status of feeder section is dependent on attribute values of appropriate switch object. Distribution networks can be operated with minimum losses, and radial structure can be maintained without intervening of operator with the definition of active rules which have update propagation characteristics.

The definition of active rules for DDSS are as follows:

Rule 1) If initially attribute close/open status value of switch is changed from open to close then attribute close/open status value of appropriate feeder section which is connected to the switch is changed from open to close

Rule R1 for switch
Event : update to st
Condition : updated(switch(S)), NEW S.st=close
Action: update feeder section.st=close where feeder section.ssn = switch.ssn

Rule 2) If initially attribute close/open status value of switch object is changed from close to open then close/open status value of appropriate feeder section object, which is connected to the switch is changed from close to open.

Rule R2 for switch
Event : update to st
Condition : updated(switch(S)), NEW S.st=open
Action: update feeder section.st=open where feeder section.ssn = switch.ssn

Rule 3) If attribute loadings value of switch object is updated then section loadings value of feeder section is updated by the same value of loadings of switch.

Rule R3 for switch
Event : update to ssnmva
Condition : True
Action : update feeder section.ssnmva = switch.ssnmva where feeder section.ssn = switch.ssn

Rule 4) If updated section loadings value of feeder section is exceeded by 80[%] of feeder capacity then feeder reconfiguration program is executed to minimize line losses and relive overloadings.

Rule R4 for feeder section
Event : update to ssnmva
Condition : updated(feeder section(FD)),
NEW FD.ssnmva > FD.fnc
Action : reconfiguration()

Rule 5) If reconfiguration() is executed and switch to be closed(or opened) is selected, then attribute close/open status of selected switch is updated by close(or open).

Rule R5 for reconfiguration()
Event : reconfiguration()
Conditon : TRUE
Action : (update switch.st=open where switch.ssn
= result of open reconfiguration())
&&(update switch.st=close where switch.ssn=
result of close reconfiguration())

where, result of open reconfiguration() : a selected switch to be opened resulting from feeder reconfiguration.
result of close reconfiguration() : a selected switch to be closed resulting from feeder reconfiguration.

Rule 6) If attribute voltage value of switch is updated then the attribute voltage value of appropriate feeder section is updated

> Rule R6 for switch
> Event : update vv
> Condition : updated(switch(S)), TRUE
> Action : update feeder section.vv = switch.vv—Vdrop
> where, feeder section.ssn = switch.ssn
> Where, Vdrop : Voltage drop of feeder section

Rule 7) If updated voltage value of feeder section is lower than minimum voltage value of feeder section then feeder reconfiguration program is executed to minimize line losses and relive overloadings.

> Condition : updated(feeder section(FD)),
> NEW FD.vv > FD.min_vv
> Action : reconfiguration()

Rule 8) If initially attribute close/open status value of switch is changed from open to close then Active database sends signal to intelligent switch to be closed

> Rule R8 for switch
> Event : update to st
> Condition : updated(switch(S)), NEW S.st=close
> Action: signal to intelligent switch to be closed

Rule 9) If initially attribute close/open status value of switch is changed from close to open then active database sends signal to appropriate intelligent switch to be opened

> Rule R9 for switch
> Event : update to st
> Condition : updated(switch(S)), NEW S.st=open
> Action: signal to intelligent switch to be opened

Rule 10) If feeder reconfiguration executed then information is updated by the resulted data form feeder reconfiguration.

> Rule R10 for reconfiguration
> Event : reconfiguration()
> Condition : TRUE
> Action :(update information.rsn = history) && (update information.ploss= result loss of reconfiguraion())&&(update information.dploss = result of change_of_loss of reconfiguration())
> Where, history = reconfiguration times
> result loss of reconfiguraion(): total loss after feeder reconfiguration
> result of change_of_loss of reconfiguration() : the amount of loss change resulting from feeder reconfiguration;

2.6 Active Rule Manager for DDSS Active Database

The execution of a rule's action may trigger another rule, whose action may trigger other rule's event. Active rule manager coordinates active rule interaction and the execution of active rules during transaction execution by interfacing constraint manager. In this paper, rule interaction is represented by means of Triggering Graph[2].

DEFINITION : Let R be an arbitrary active rule set. The triggering Graph is a directed graph {V,E}, where each node vi ∈ V corresponds to a rule ri ∈ R, A directed arc <jrj , rk > ∈ E means that the action of rule rj generates events which trigger rule rk . Fig 3 represents rule interaction using Triggering Graph.

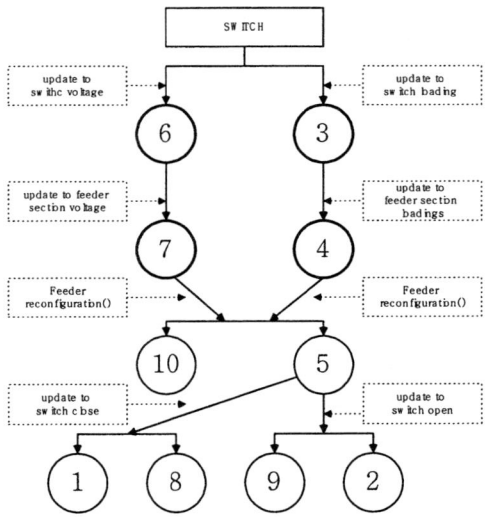

Fig. 3. Triggering graph for Feeder automation

In fig.3, when intelligent switch loadings of distribution networks is changed because of the increasing of customer's consuming, data acquisition system detects the amount of loadings change and sends updating signals switches of DDSs database. Due to the updating signals, rule R3 is triggered to update section loadings of appropriate feeder section that is connected to the switch. If this updated value exceeds 80[%] of feeder capacity then R4 is triggered to minimize line losses and to relive feeder overloadings by using feeder reconfiguration program. R4's action trigger R5 and R10 simultaneously. Due to R10's action, data resulted from feeder reconfiguration program is stored in information object. R5's action updates attribute close/open status value of switch selected from feeder reconfiguration program and triggers R1, R2, R8 and R9 simultaneously. Because of R1, R2, R8 and R9, close/open status of feeder section is updated by the same status of appropriate switch and close/open status of Intelligent switch located in distribution networks is changed by the same status of appropriate switch.

3 Web Based DDSS Active Database System Architecture

A flexible open DDSS system architecture has to satisfy the two-fold requirements[3] It is built to vendor neutral standards(easy in the use of software package).

It will provide the ability to enhance an existing DDSS without relaying one vendor(easy in the continuing development and maintenance of the software package) These requirements can be met by using the internet as the operating environment. Using internet technology for DDSS architecture will gain following benefits
First, it support cross-platform architecture:

In a standardized internet browser environment with HTML and TCP/IP protocols, users will continue using the platforms with which they are most familiar without conscious of different hardware platform.

Second, it follows open system standard :

By following Structured Query Language(SQL), HTML, HTTP, FTP, TCP/IP, and PPP, data exchange and system expansion are easily done with minimum efforts[4].

The proposed new web based DDSS architecture make uses of the Java 2 Enterprise Edition architecture which is a Web-based multitier architecture, as presented in fig. 4.

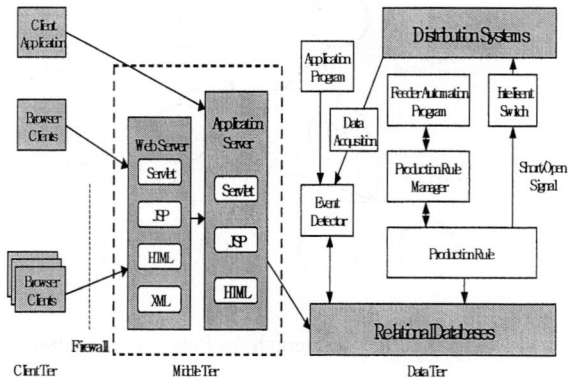

Fig. 4. Web based DDSS active database system architecture

The architecture can be expressed by subdividing two parts in fig.4 The one is web based architecture which support open system and the other is active database architecture which support DDSS feeder automation

For web based architecture, Its structure can be divided into three tiers[3]

Client tier : It provides user interface
Middle tier: It is subdivided into the Web server and the application server.
Data tier: It handles the information storage

In the middle tier, application server is transferring request from the web into appropriate functions in the system and also provide for interfacing different kinds of database. Web server acts as the gateway for Web-based clients to access the database.

Power distribution engineers and operators sent HTTP requests to Web server through the industrial standard web browser. If requested web page containing database SQL command, the database interprets SQL commands and returns matched data in the database back to Web server. Matched data is formulated as a web page and displayed web page in the client window[4].
For active database architecture shown in the right-side of fig.4.
Its structure can be expressed specifically by dividing into three parts

- Event detector :
It checks telemetered data and sends signal to production rule manager if integrity constraints are violated.
- Production rule manager
It accept signal from event detector and determines which rules to be fired.
- Production rule
It allows specification of data manipulation operations that are executed automatically whenever certain events occur or conditions are satisfied.

Data acquisition system located in distribution network detects switch's close/open status, current, voltage and loadings, and it sends a detected data to central DAS database system.

In DDSS database system, event detector accepts detected data sending from data acquisition system and sends data modification operation to database. If integrity constraints is violated because of data modification then event detector send signal to production rule manager. By using production rule manger, production rules are fired and resulted data is updated in database. After updating values in database, event detector checks if updating values satisfies integrity constraints or not. If integrity constraints are violated then event detector sends signals to production rule manager. And, when Rule R10 and Rule R5 are triggered, close/open signals are send to appropriate intelligent switch located in distribution networks.

4 Conclusion

This paper presents a new Web-based active database architectural design to the Distribution data security system applications. This architecture includes web-based architecture and active database parts. For web based parts, the authors utilize the Java 2 Enterprise Edition architecture for open system, which will easy in the continuing development and maintenance of the software package. Therefore, it is easy to meet open-access competitive market environment. For active database architecture, the author design production rule and production rule manager for DDSS feeder automation By utilizing proposed rules, distribution network can be operated reliably with minimum operators intervening

Acknowledgment

This work has been supported by 2004-0726-000, which is funded by Gyeonggi Province, Korea.

References

1. J. Widom, S.Ceri, "Active Database Systems : Triggers and rules for Advanced Database Processing", MORGAN KAUFMANN PUBLISHERS"
2. E. Baralis, S. Ceri and S.Paraboshi, "Compile-Time and Runtime Analysis of Active Behavior", IEEE Trans, On Knowledge and Data Engineering, Vol. 10, No.3, pp 353 – 370, 1998
3. S.Chen, F.Y. LU, "Web-Based Simulations of Power Systems", IEEE Computer Application in Power, January, 2002, pp35-40
4. J.T. Ma, T. Liu, L.F. Wu, "New Energy Management System Architectural Design and Intranet/Internet Applications to Power Systems" Conference Proc. Power industry computer application conference, 1995, pp 207- 212
5. E.Baralis, A.Bianco, "Performance Evaluation of Rule Execution Semantics in Active Database" Tech. Rep. DAI.EB. 96.1, Aug. 1996
6. IEEE Task Force on Power System Control Center Database, "Critical Issues Affecting Power System Control Center Database", IEEE Trans. On Power System, vol, 11, no.2 , May, 1996
7. G..S. Martire, D.J.H. Nuttall, "Open Systems and Database", IEEE Trans, On Power System, Vol. 8, NO. 2, May. 1993.

Data Protection Based on Physical Separation: Concepts and Application Scenarios

Stefan Lindskog[1], Karl-Johan Grinnemo[2], and Anna Brunstrom[1]

[1] Department of Computer Science,
Karlstad University, SE-651 88 Karlstad, Sweden
{Stefan.Lindskog, Anna.Brunstrom}@kau.se
[2] TietoEnator AB,
Lagergrens gata 2, SE-651 15 Karlstad, Sweden
Karl-Johan.Grinnemo@tietoenator.com

Abstract. Data protection is an increasingly important issue in today's communication networks. Traditional solutions for protecting data when transferred over a network are almost exclusively based on cryptography. As a complement, we propose the use of multiple physically separate paths to accomplish data protection. A general concept for providing physical separation of data streams together with a threat model is presented. The main target is delay-sensitive applications such as telephony signaling, live TV, and radio broadcasts that require only lightweight security. The threat considered is malicious interception of network transfers through so-called eavesdropping attacks. Application scenarios and techniques to provide physically separate paths are discussed.

1 Introduction

In the last few years, we have experienced a steadily growing interest in using the Internet as a vehicle for e-banking, e-commerce, virtual company networks, telephony, live TV, and other applications requiring secure communication. To this end, network security has become pivotal for the future of the Internet and Internet-based solutions.

Currently, network security is almost exclusively accomplished through encryption. For example, e-banking and e-commerce typically take place over Secure Sockets Layer (SSL) [9] or Transport Layer Security (TLS) [6] connections. However, although encryption gives adequate protection, it may lead to severely degraded network performance in terms of latency and throughput. Specifically, Apostolopoulos et al. [2] demonstrated a throughput reduction of more than 90 % when TLS (using RC4 and MD5) was used to access Netscape and Apache Web servers as compared to no encryption at all. Furthermore, Burke et al. [3] showed that applications running on high-end microprocessors are not even likely to saturate a T3 (approximately 45 Mbps) line. To this end, various selective encryption schemes [12, 19, 23, 24] that produce less overhead compared to ordinary encryption schemes, such as RC4, DES, and AES, have recently been proposed. The basic idea of selective encryption is to offer lightweight security by encrypting only a subset of the data. Such schemes are intended to be used when the computational overhead produced by encryption and/or decryption must be reduced and a less stringent security level is acceptable.

Using selective encryption is, however, only one way to accomplish lightweight security. Furthermore, as pointed out by Rushby and Randell [22], the basis for security is separation. Thus another, in a sense more straightforward, way to obtain lightweight security is to physically separate the data to be protected, i.e., to partition and send the data along different routes. Physical separation has previously been used in other contexts. For example, Deswarte et al. [5] proposed the use of physical separation to accomplish intrusion tolerance in a distributed file archiving system, i.e., to make the file archiving system resilient against single-point attacks. However, to our knowledge, physical separation has not been used to obtain transmission security in networks. To this end, this paper suggests using physical separation as a complement to encryption for delay-sensitive network applications that require only lightweight security. It is our belief that, similar to selective encryption, physical separation could be an alternative for applications such as telephony signaling, live TV, and other multimedia applications. In other words, it could be an alternative for applications where latency and throughput requirements outweigh the importance of absolute protection against malicious (eavesdropping) attacks. According to Pfleeger and Pfleeger [18], providing adequate protection is a key security principle [1]. The principle implies that data should be protected to a degree that is consistent with their value.

The remainder of the paper is organized as follows. Section 2 presents our idea of providing data protection through physical separation. The threat model is discussed in Section 3. Section 4 focuses on application scenarios and on how to provide multiple physically separate paths in reality. Finally, Section 5 concludes the paper with some final remarks and a few words on future work.

2   Data Protection Through Physical Separation

Security is typically implemented through one or more security services. In particular, a combination of protective and detective services is often used. Defense diversity is achieved by combining two or more security services. Diversity of defense is a general security principle used to enhance security [4]. Firewalls are used for example to block suspicious network traffic to and from internal networks, and intrusion detection systems (IDSs) are used as a complement to firewalls to detect insider and outsider intrusion attempts as well as successful intrusions. However, neither firewalls nor IDSs are suitable security tools for protecting data that are transferred over an insecure network, such as the Internet. Instead, some form of data protection service is needed. The type of service necessary depends on what is to be protected.

---

[1] The same principle is in [17] referred to as the *principle of timeliness*.

Data protection services are used to achieve data confidentiality, data integrity, data authenticity, and/or non-repudiation [25]. A data confidentiality service ensures that transmitted data are accessible only for reading by authorized parties, while a data integrity service ensures that only authorized parties are allowed to modify transmitted data. A data authenticity service ensures that the origin and/or the source of data are correctly identified. Finally, a non-repudiation service ensures that neither the sender nor the receiver of a message can deny the transmission.

Data protection for network transfers has traditionally been implemented exclusively through cryptographic separation, and various cryptographic systems are widely used today. The major disadvantage of cryptographic separation, however, is that it requires adequate computational resources (at least) at the endpoints. For this reason, we propose the use of physical separation as a complement to existing cryptography-based data protection services for applications with lightweight security requirements. Cryptography and physical separation may also be combined to achieve protection diversity.

The idea of physical separation is to simultaneously send messages belonging to the same data stream on multiple physical paths. Protection is thus provided by the geographical fragmentation and scattering of data. Figure 1 shows two communicating hosts, A and B. In this case, host A is the sender and host B the receiver. Both hosts are multi-homed and equipped with two physical network interfaces, A_1 and A_2, and B_1 and B_2, respectively.

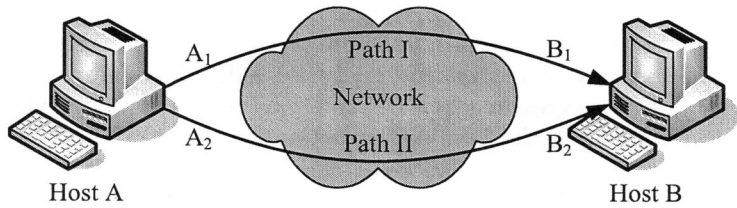

Fig. 1. Physical separation of two communication paths

Two distinct paths can be obtained, Path I and Path II, by using the different network interfaces, which is illustrated in Fig. 1. These two paths are then used simultaneously during transmission of data. Specifically, when connecting a multi-homed host to the network, each network interface is bound to a separate IP address. An application running on a multi-homed host can thus select the network interface on which to send data. Different strategies for distributing data on these paths can be used depending on the application. An even distribution of data on the different paths is suitable in some situations, while, in other situations, it is not. For example, when multimedia data are transferred, the amount of traffic sent on the different paths may need to be adjusted with respect to path characteristics such as bandwidth, delay, jitter, etc. The distribution may also be affected by security considerations. Imagine, for example, a map with a secret target marked with an "X". Neither the map nor the mark is sensitive by itself but together they are. Increased security may thus be achieved by simply sending the map and the "X" mark on two physically separate paths.

3 Threat Model

The specific threat considered in this paper is malicious interception of network traffic through so-called eavesdropping attacks. Eavesdropping attacks are a serious threat to the confidentiality (or secrecy) of data transferred in a network. The person conducting such an attack is referred to as an eavesdropper. In this paper, our hypothesis is that an eavesdropper needs to acquire access to all or at least most of the data sent over the different paths to successfully perform an eavesdropping attack. This means that an attacker must identify all used paths, gain access to the traffic, and finally decode the data.

An eavesdropping attack on a particular victim is most easily performed near the victim's physical network connection. If the wire that connects the victim's computer is accessible, a protocol analyzer can be used to intercept all traffic that passes through the wire. Another option is to use a so-called network sniffer. When a network sniffer is used in a broadcast network, such as a non-switched Ethernet network or an IEEE 802.11 wireless network, the network interface card (NIC) on the computer executing the sniffer is configured in promiscuous mode. Promiscuous mode implies that all traffic sent over the broadcast network is intercepted by the NIC and forwarded to the sniffer for further processing.

Interception of traffic to and/or from a particular user is much more complicated if the point for eavesdropping is many hops away from the end nodes. In the core network, traffic from a large set of users will pass. Thus, the data processing necessary to filter out the relevant traffic will require a great deal of computational resources. In addition, traffic to and/or from the victim may be routed on different paths from session to session and can also be re-routed within a session due to network failures etc. Furthermore, if proxies or network address translators (NATs) are used, it might not be evident who is actually communicating.

When data on multiple communication paths are fragmented and scattered, eavesdropping attacks become much more difficult. The highest degree of protection through physical separation is achieved when data are routed on fully distinct communication paths all the way from the sender to the receiver. However, in many IP-based network architectures, fully distinct paths can not easily be guaranteed. Still, since the endpoints are the most susceptible parts to eavesdropping attacks, we believe that it is satisfactory in many real situations to guarantee physical separation of the traffic closest to the endpoints.

4 Application Scenarios

It is evident that physical separation as a mean to achieve security is more applicable in some situations than others. In the following, we give some examples of application scenarios in which physical separation may be used to provide lightweight data protection and outline how physical separation can be realized in these scenarios.

4.1 SS7 over IP

Logically, the Public Switched Telephone Network (PSTN) comprises two networks: one call-traffic network for the transmission of speech, data, fax etc., and one signaling network for the transfer of control or signaling information. Specifically, the signaling network enables transfer of control information in between nodes in connection with: traffic control procedures such as call set-up, supervision, and release; services such as toll-free (800/888) and toll (900) wireline services, and roaming in wireless cellular networks; and, finally, network management procedures such as blocking and de-blocking of trunks.

The predominant signaling system used in today's PSTN is Signaling System #7 (SS7) and, with VoIP emerging as a competitive alternative to the PSTN, it has become increasingly important to enable interworking between SS7 and IP. To this end, the IETF Signaling Transport (SIGTRAN) working group has defined a framework architecture, the SIGTRAN architecture [15], that describes how to transport SS7 signaling information over an IP-based network. The SIGTRAN architecture defines a set of adaptation protocols that insulate the upper layers of the SS7 protocol stack from IP and thus make it possible to run SS7, essentially unaltered, over an IP-based network. Furthermore, the SIGTRAN architecture defines a new transport protocol, the Stream Control Transmission Protocol (SCTP) [26].

SCTP is the common denominator for all SIGTRAN networks and, although SCTP is now considered a general transport protocol on a par with TCP and UDP, it is still true that SCTP evolved from the SIGTRAN working group and the requirements of SS7 signaling. Similar to TCP, SCTP is a connection-oriented, reliable transport protocol offering ordered (sequential) delivery of user messages. However, contrary to TCP, SCTP is message oriented and provides support for multi-streaming and multi-homing. Multi-streaming enables SCTP to send separate transactions as independent units and thereby avoid so-called head-of-line blocking, which happens when a lost message in one transaction effectively blocks one or several other transactions.

Support for multi-homing was included in SCTP to provide quick failure detection and recovery, which is particularly important for SS7 signaling. Network-level timers running in today's SS7 networks monitor message delivery performance and may generate alarms that trigger message re-routing if they expire as a result of extraordinary delays. Thus, this network monitoring functionality must extend to VoIP networks to maintain end-to-end, carrier-grade telecommunications quality.

Figure 2 illustrates how the SCTP multi-homing support works. SEP-A and SEP-B are two dual-homed signaling endpoints. Each combination of source and destination IP addresses, e.g., (P_1,P_2), (A_1,A_2), (P_1,A_2) etc., constitutes a network path. One network path is always selected as the primary path, and (P_1,P_2) functions as the primary path in Fig. 2. Provided the primary path is available, all packets are sent on this path. The remaining paths function only

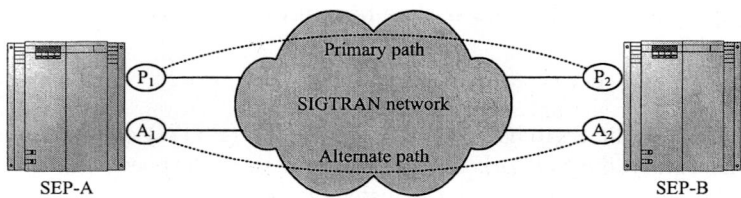

Fig. 2. Multi-homing support in SCTP

as backup or alternate paths. However, if the primary path becomes unavailable, one of the alternate paths takes over as the primary path. Thus the SCTP multi-homing support does not include load-balancing.

Although multi-homing in SCTP and SIGTRAN is exclusively intended for failure recovery, we believe that multi-homing could also be used to accomplish physical separation, and thus enhance security, for delay-sensitive signaling traffic. Specifically, we consider physical separation through the use of SCTP multi-homing a viable complement to IPSec [11], the security mechanism recommended for SIGTRAN by IETF—especially, since IPSec has been shown [13] to decrease the throughput of small packets, such as signaling packets, by approximately 60 % as compared to unencrypted transfer.

Since path switching in SCTP normally only takes place during failure recovery, physical separation must rely on some other mechanism than the SCTP failover mechanism. One possible solution would be to employ the SCTP changeover procedure, which provides for explicit changes of primary paths. However, the SCTP changeover procedure is primarily intended for infrequent path switching, and it might thus be necessary to improve its performance in order for it to be useful for physical separation. Another solution would be to consider one of the SCTP implementations that support load sharing, e.g., LS-SCTP [8] and RivuS [20].

To actually accomplish physical separation, it must be guaranteed that the primary and alternate paths are kept distinct from each other or, at least, only overlap in backbone routers and/or links. While this is not possible in the general case, with paths crossing several autonomous systems, it is indeed possible in controlled, dedicated SIGTRAN networks. As a matter of fact, since distinct paths also improve path redundancy and failure recovery, several telecommunication operators already plan to build their SIGTRAN networks with distinct primary and alternate paths between signaling endpoints.

Two techniques that have been proposed to establish distinct paths are redundant networks and MultiProtocol Label Switching (MPLS) [21]. Redundant networks are fairly straightforward and simply entail having the signaling endpoints connected to two or more distinct networks. However, to our knowledge this technique has not yet been deployed in any real SIGTRAN network. Thus it remains to be shown that the costs of building several separate networks are actually compensated for in terms of reliability, availability, and faster failure recovery.

While it is too early to completely rule out the idea of using redundant networks, MPLS is currently the most compelling technique for establishing distinct paths. An MPLS network or Internet consists of a set of nodes, called Label Switched Routers (LSRs), which are capable of switching and routing packets on the basis of a label that has been appended to each packet. Labels define a flow of packets between two endpoints or, in the case of multicast, between a source endpoint and a multicast group of destination endpoints. For each distinct flow, called a Forwarding Equivalence Class (FEC), a specific path through the network of LSRs is defined. Thus MPLS is a connection-oriented technology. Associated with each FEC is a traffic characterization that defines the QoS requirements for that flow. The LSRs do not need to examine or process the IP header, but rather simply forward each packet on the basis of its label value. The forwarding process is therefore simpler than with a traditional IP router. Furthermore, it is envisioned that label-switched paths could be determined, and perhaps modeled, with traffic engineering tools that reside on network management workstations.

4.2 Application-Controlled Multi-homing

In the previous section, multi-homing was handled at the transport layer. Another alternative is to directly control multi-homing at the application layer, or possibly in a supporting middleware. This could be suitable for multimedia applications, for example, or more generally for applications where the transmitted data essentially comprise several independent, or almost independent, parallel data streams.

The main advantage of controlling multi-homing directly in the application, as compared to controlling it from lower layers such as the transport layer, is that it gives greater control over how data are transmitted. It also has the advantage of not being dependent on a specific transport protocol. Furthermore, from a security perspective, the use of parallel connections adds some level of extra protection. An eavesdropper that has managed to collect all data over all paths still needs to identify the transport layer connections that carry data that belong to the application of interest. Furthermore, once the connections have been identified, it still remains for the eavesdropper to appropriately reassemble the data.

Application-controlled multi-homing is, however, not without problems. For one thing, it entails that the reassembly of packets from different paths must be taken care of by the application itself. Second, the problem of enforcing distinct paths remains, provided of course that the paths have to be completely separated in the first place. Physical separation at the endpoints can be achieved by simply sending traffic over several network interfaces, possibly using different network technologies and/or media. For example, some parts of the traffic could go over wireline paths while other parts are sent via wireless access lines. If greater control of the physical paths is required, MPLS [21] could be used in the same way as discussed in Section 4.1, and thus provide explicit label-switched paths between endpoints. However, application-controlled multi-homing also opens the way for application specific routing strategies through the use of so-called overlay networks [16].

An overlay network can be viewed as a logical network implemented on top of a real physical network. Each node in the overlay also exists in the underlying physical network. However from the viewpoint of the overlay, the nodes are not only capable of routing packets on the basis of their destination address, but are also able to process and forward packets in application-specific ways. Thus, an overlay network could assist a multi-homed endpoint in enforcing distinct paths. For example, a technique similar to the one proposed for Resilient Overlay Networks (RONs) [1] could be used. In a way similar to a RON, nodes could reside in a variety of routing domains and cooperate with each other to forward data. Since routing domains or Autonomous Systems (ASs) rarely share interior links, flows routed through different ASs are likely, but not guaranteed, to be forwarded on distinct paths.

Another technique for enforcing distinct paths when using overlay networks may be to employ a routing underlay network as suggested by Nakao et al. [14]. They propose a new architectural element, a routing underlay, that sits between an overlay network and the underlying IP network. The overlay network queries the routing underlay when it makes routing decisions. The routing underlay in turn extracts and aggregates topology information from the underlying IP network and answers the queries of the overlay network. Specifically, one of the services that could be offered to the overlay by the underlay network is to return a distinct path between an originating and destination endpoint. In fact, this is one of three services that are actually proposed in [14].

Finally, it can be mentioned that recent developments in routing technologies make possible new alternatives, apart from MPLS and overlay routing, to obtain distinct paths even though the paths cross several ASs. Examples of this include route controllers, such as the Peer Director [7], which assists BGP in routing traffic through multi-homed routing domains (domains connected to several ISPs), and the BANANAS framework [10], which permits source-based multipath inter-domain routing.

5 Conclusions

In this paper, we propose the use of physical separation as a complement to encryption for delay-sensitive applications that require only lightweight security. A threat model is presented. Further some application scenarios are highlighted, and alternative solutions for providing separate paths in these scenarios are discussed.

Since establishing physically distinct paths is a non-trivial task in IP networks, a large part of the description of application scenarios addresses this issue. Transport layer controlled multi-homing using SCTP in conjunction with either redundant networks or MPLS are suggested as feasible solutions. MPLS could also be used together with application-controlled multi-homing. In this context, application-specific routing through the use of overlay networks is another possible solution.

As a next step we intend to make a feasibility study of the use of SCTP for transportation of SS7 signaling traffic on physically separate paths. The purpose in particular is to evaluate the complexity in terms of reassembly of data and to study the consequences of simultaneously sending data on multiple paths with regard to the SCTP congestion control mechanisms.

Acknowledgments

This research is supported in part by grants from the Knowledge Foundation of Sweden and from the CMIT research platform at Karlstad University in Sweden. The second author would also like to thank TietoEnator AB for their support of his research.

References

1. D. Andersen, H. Balakrishnan, Frans Kaashoek, and Robert Morris. Resilient overlay networks. In *Proceedings of the 18th ACM Symposium on Operating System Principles (SOSP 2001)*, pages 131–145, Chateau Lake Louise, Canada, October 2001.
2. G. Apostolopoulos, V. Peris, and D. Saha. Transport layer security: How much does it really cost? In *Proceedings of the Conference on Computer Communications (IEEE INFOCOM)*, volume 2, pages 717–725, New York, New York, USA, March 1999.
3. J. Burke, J. McDonald, and T. Austin. Architectural support for fast symmetric cryptography. *ACM SIGOPS Operating Systems Review*, 34(5):178–189, December 2000.
4. D. B. Chapman and E. D. Zwicky. *Building Internet Firewalls*. O'Reilly & Associates, 1995.
5. Y. Deswarte, L. Blain, J. C. Fabre, and J. M. Pons. Security. In D. Powell, editor, *Delta-4: A Generic Architecture for Dependable Distributed Computing*, chapter 13, pages 329–339. Springer-Verlag, 1991.
6. T. Dierks and C. Allen. RFC 2246: The TLS protocol version 1.0, January 1999.
7. Radware: Peer Director. http://www.radware.com/content/products/pd, January 2, 2005.
8. A. A. El Al, T. Saadawi, and L. Myung. LS-SCTP: A bandwidth aggregation technique for stream control transmission protocol. *Computer Communications*, 27(10):1012–1024, June 2004.
9. A. Frier, P. Karlton, and P. Kocher. The SSL 3.0 protocol. Netscape Communication Corporation, November 1996.
10. H. Tahilramani Kaur, S. Kalyanaraman, A. Weiss, S. Kanwar, and A. Gandhi. BANANAS: An evolutionary framework for explicit and multipath routing in the Internet. In *Proceedings of the ACM SIGCOMM Workshop on Future Directions in Network Architecture (FDNA 2003)*, pages 277–288, Karlsruhe, Germany, 2003.
11. S. Kent and R. Atkinson. RFC 2401: Security architecture for the Internet protocol, November 1998.

12. S. Lindskog, J. Strandbergh, M. Hackman, and E. Jonsson. A content-independent scalable encryption model. In *Proceedings of the 2004 International Conference on Computational Science and its Applications (ICCSA'04), part I*, pages 821–830, Assisi, Italy, May 14–17, 2004.
13. S. Miltchev, S. Ioannidis, and A. D. Keromytis. A study of the relative costs of network security protocols. In *Proceedings of the FREENIX Track: 2002 USENIX Annual Technical Conference*, pages 41–48, Monterey, California, USA, June 2002.
14. A. Nakao, L. Peterson, and A. Bavier. A routing underlay for overlay networks. In *Proceedings of the ACM SIGCOMM 2003)*, pages 11–18, Karlsruhe, Germany, August 2003.
15. L. Ong, I. Rytina, M. Garcia, H. Schwarzbauer, L. Coene, H. Lin, I. Juhasz, M. Holdrege, and C. Sharp. RFC 2719: Framework architecture for signaling transport, October 1999.
16. L. Peterson, T. Anderson, D. Culler, and T. Roscoe. A blueprint for introducing disruptive technology into the Internet. In *Proceedings of the First ACM Workshop on Hot Topics in Networking (HotNets 2002)*, Princeton, New Jersey, USA, October 2002.
17. C. P. Pfleeger. *Security in Computing*. Prentice-Hall, 2nd edition, 1997.
18. C. P. Pfleeger and S. Lawrence Pfleeger. *Security in Computing*. Prentice-Hall, 3rd edition, 2003.
19. M. Podesser, H. P. Schmidt, and A. Uhl. Selective bitplane encryption for secure transmission of image data in mobile environments. In *Proceedings of the 5th IEEE Nordic Signal Processing Symposium (NORSIG'02)*, Tromsø/Trondheim, Norway, October 2002.
20. RivuS project homepage. http://sourceforge.net/projects/rivus/, January 2, 2005.
21. E. Rosen, A. Viswanathan, and R. Callon. RFC 3031: Multiprotocol label switching architecture, January 2001.
22. J. M. Rushby and B. Randell. A distributed secure system. In *Proceedings of the 1983 IEEE Symposium on Security and Privacy*, pages 127–135, Oakland, California, USA, April 1983.
23. A. Servetti and J. C. De Martin. Perception-based selective encryption of G.729 speech. In *Proceedings of the 2002 IEEE Internatinal Conference on Acoustics, Speech, and Signal Processing*, volume 1, pages 621–624, Orlando, Florida, USA, May 2002.
24. G. A. Spanos and T. B. Maples. Performance study of a selective encryption scheme for security of networked, real-time video. In *Proceedings of the 4th International Conference on Computer Communications and Networks (ICCCN'95)*, pages 72–78, Las Vegas, Nevada, USA, September 1995.
25. W. Stallings. *Cryptography and Network Security: Priniples and Practice*. Prentice-Hall, 2nd edition, 1998.
26. R. R. Stewart, Q. Xie, K. Morneault, C. Sharp, H. J. Schwarzbauer, T. Taylor, I. Rytina, M. Kalla, L. Zhang, and V. Paxson. RFC 2960: Stream control transmission protocol, October 2000.

Some Results on a Class of Optimization Spaces

K.C. Sivakumar[1] and J. Mercy Swarna[2]

[1] Department of Mathematics,
Indian Institute of Technology Madras,
Chennai 600 036, India
[2] MIT, Anna University, Chennai 600 044, India

Abstract. Let X be a partially ordered real Banach space, $a, b \in X$ with $a \leq b$. Let ϕ be a bounded linear functional on X. We call X a Ben-Israel-Charnes space (or a B-C space, for short) if the linear program: Maximize $\langle \phi, x \rangle$ subject to $a \leq x \leq b$ has an optimal solution. Such problems have been shown to be important in solving a class of problems known as Interval Linear Programs. B-C spaces were introduced by the first author in his doctoral dissertation. In this paper we identify new classes of Banach spaces that are B-C spaces. We also present sufficient conditions under which answers are in the affirmative for the following questions:

1 When is a closed subspace of a B-C space, a B-C space?
2 Is the range of a bounded linear map from a Banach space into a B-C space, a B-C space?

Keywords: Interval Linear Programs; Partially Ordered Banach spaces; Ben-Israel-Charnes Spaces.

AMS Subject Classification: Primary: 90C48, 90C05; Secondary: 47N10.

1 Introduction

Let X and Y be real Banach spaces with Y partially ordered (Definition 2.1 below), $A : X \to Y$ be a linear map and $a, b \in Y$ with $a \leq b$. Let ϕ be a linear functional on X. A class of linear programs of considerable study known as interval linear programs denoted by $ILP(a, b, \phi, A)$ are problems of the form:

$$\text{Maximize } \langle \phi, x \rangle$$

$$\text{subject to } a \leq Ax \leq b.$$

Ben-Israel and Charnes were the first to investigate ILP's [1]. They considered the case $X = \mathbb{R}^n$ and $Y = \mathbb{R}^m$ where they assumed that the matrix A is of full row-rank. Explicit optimal solutions for such problems were given in terms of the generalized inverses of A. These were later extended in [8 & 9] to the case when the restriction on A was removed. Kulkarni and Sivakumar [2, 3 & 4] investigated interval linear programs in the infinite dimensional setting and showed how some of the results in the finite dimensional case can be extended. This work is also reported in [6].

We now motivate the concept of a Ben-Isreal-Charnes space. Consider the linear equation $Ax = u$, Equivalently, $x = Tu + y$, $y \in N(A)$, where T is a $\{1\}$- inverse of A. Suppose that $\langle \phi, y \rangle = 0$, $\forall\, y \in N(A)$. Then $ILP(a, b, \phi, A)$ is equivalent to the problem $ILP(a, b, \psi, I)$:

$$\text{Maximize } \langle \psi, u \rangle$$

$$\text{subject to } a \leq u \leq b,$$

where $\langle \psi, u \rangle = \langle T^*\phi, u \rangle$, where T^* is the adjoint map of T. It is well-known that, under certain assumptions $ILP(a, b, \psi, I)$ is explicitly solvable [2, 3 & 6]. This in turn can be used to obtain explicit optimal solutions to $ILP(a, b, \phi, A)$. A partially ordered real Banach space where $ILP(a, b, \psi, I)$ has an optimal solution $\forall a, b \in X$, $a \leq b$ and $\phi \in X'$ is called a Ben-Israel- Charnes space in [6], in recognition of the work of these authors on finite interval linear programs. It must be mentioned that a class of production-planning and input-output problems have been shown to have explicit solutions in the setting of B-C spaces. This work appears in [4] and is the first instance of the use of the nomenclature B-C spaces in the literature.

In this paper we identify new classes of Banach spaces that are B-C spaces. We also address the following questions:

1. When is a closed subspace of a B-C space, a B-C space?
2. Is the range of a bounded linear map from a Banach space into a B-C space, a B-C space?

We present sufficient conditions under which the answers are in the affirmative. We organize the paper as follows. In section 2 we present the preliminary concepts and examples that will be used throughout. This also helps to fix the notation. Section 3 presents the idea of a Ben-Israel-Charnes space, where numerous examples are given as illustrations. The main results are Theorem 3.2, Theorem 3.5, Theorem 3.7, Theorem 3.15 and Theorem 3.19.

2 Preliminaries

This section serves to fix the notation in this paper and to review the notions of cones and lattices. The treatment is as found in [5].

Definition 2.1. *Let X be a real vector space. Then X is called a partially ordered vector space if X has a partially order $' \leq '$ defined on it satisfying the following: For $x, y \in X$ with $x \leq y$, we have $x + u \leq y + u$ for all $u \in X$ and $\alpha x \leq \alpha y$ for all $\alpha \geq 0$.*

Definition 2.2. *Let X be a partially ordered real vector space. Then the subset $\mathcal{C} := \{x \in X : x \geq \mathbf{0}\}$ is called the positive cone of X. $\mathcal{C}$ is called a pointed cone if the partial order ' $\leq$' is also antisymmetric, viz., $x \leq y$ and $y \leq x$ imply $x = y$ for every $x, y \in X$.*

Definition 2.3. *A vector space is said to be a partially ordered Banach space if it is a partially ordered vector space and also a Banach space with respect to some norm.*

Example 2.4. *$\mathbb{R}^n$ the Euclidean space is a partially ordered real Banach space with $\mathbb{R}_+^n := \{x = (x_1, x_2, \cdots, x_n) \in \mathbb{R}^n : x_i \geq 0 \forall i = 1, 2, \cdots, n\}$ as a pointed positive cone.*

Example 2.5. *Let μ be a σ-finite positive measure on a σ-algebra in a nonempty set Y and for $1 \leq p < \infty$, let $X = L^p(Y, \mu)$ denote the space of (equivalent classes of) measurable p-integrable functions on Y. Then X is a partially ordered Banach space with the pointed positive cone $\mathcal{C} := \{f \in X : f \geq 0 \, a.e.(\mu)\}$. In particular $l^p, 1 \leq p < \infty$ is a partially ordered Banach space with the pointed positive cone $\mathcal{P} := \{x \in l^p : x_i \geq 0 \forall i\}$.*

We next review the concept of vector lattices. Here the partial order is assumed to satisfy the antisymmetry condition also (Definition 2.2).

Definition 2.6. *A real vector lattice V is a partially ordered real vector space such that $sup\{x, y\}$ and $inf\{x, y\}$ exist for all $x, y \in V$.*

Definition 2.7. *Let V ba a vector lattice. A norm $\|.\|$ on V is called lattice norm if $|x| \leq |y| \Rightarrow \| x \| \leq \| y \|$ for all $x, y \in V$. Here $|x| := x^+ + x^-$, where $x^+ := sup\{x, 0\}$ and $x^- := sup\{-x, 0\}$. A normed lattice $(V, \|.\|)$ is called a Banach lattice if the norm is complete.*

Example 2.8. *The vector lattice of example ?? is a Banach lattice under the norm $\|f\| := sup \{ |f(b)| : b \in B\}$.*

Example 2.9. *$C([0, 1])$, the space of continuous real functions on $[0, 1]$ with the usual order and supremum norm is a Banach lattice.*

Definition 2.10. *A nonempty subset D of a vector lattice V is called directed if $\forall (x, y) \in D \times D$ there exists $z \in D$ such that $x \leq z$ and $y \leq z$. Let D be directed. Let $D_x := \{z \in D : x \leq z\}$. Then the family $\{D_x : x \in D\}$ is called the section filter of D. Let D be a directed subset of V such that $x_0 := sup \, D$ exists. Then the filter in V generated by D is the family of order intervals $\{ [x, x_0] : x \in D\}$ such that $\bigcap_{x \in D}[x, x_0] = \{x_0\}$. Here $[x, x_0] := \{y \in V : x \leq y \leq x_0\}$. A filter F on V is said to order-converge to $x \in V$ if F contains a family of order intervals with intersection $\{x\}$. A normed vector lattice is said to have an order continuous norm if every order convergent filter in V converges with respect to the norm on V.*

Example 2.11. *For $1 \leq p < \infty$, $L^p(\mu)$ are Banach lattices that have order continuous norms. c_0 the Banach lattice of real convergent sequences converging to zero has order continuous norm. It is well known that with the supremum norm $C([0, 1])$ is not order continuous. See [5] for more details.*

Definition 2.12. Let X_1 be a real Banach space and X_2 be a partially ordered real Banach space with $\mathcal{P}_2$ as the positive cone. Let $A : X_1 \to X_2$ be linear, ϕ be a bounded linear functional on X_1 and $a, b \in X_2$ with $a \leq b$. A vector $x^* \in X_1$ is said to be feasible for the problem $ILP(a, b, \phi, A)$ if $a \leq Ax^* \leq b$. The problem $ILP(a, b, \phi, A)$ is said to be feasible if there exists a feasible vector for it. A feasible vector x^* is said to be optimal if $\langle \phi, x^* - x \rangle \geq 0$ for every feasible vector x. The problem $ILP(a, b, \phi, A)$ is said to be bounded if $\sup \{\langle \phi, x \rangle : a \leq Ax \leq b\} < \infty$.

3 Ben-Israel-Charnes Spaces

Definition 3.1. Let X be a partially ordered real Banach space. Let I denote the identity map on X. We say that X is a Ben-Israel-Charnes space or a B-C space, in short if $ILP(a, b, \phi, I)$ has an optimal solution for all $a, b \in X$ with $a \leq b$ and for all bounded linear functionals ϕ on X.

First we identify a class of Banach spaces that are B-C spaces.

Theorem 3.2. Let X be a partially ordered real Banach space such that every interval $[a, b] := \{x \in X : a \leq x \leq b\}$ is weakly compact. Then X is a B-C space.

Proof. Since $[a, b]$ is weakly compact, for any bounded linear functional ϕ on X, $\sup \{\langle \phi, x \rangle : a \leq x \leq b\} < \infty$ and there exists x^0 such that $a \leq x^0 \leq b$ with $\langle \phi, x^0 \rangle \geq \langle \phi, x \rangle \forall x \in [a, b]$. Hence X is a B-C space. □

Remark 3.3. The question of whether the converse of Theorem 3.2 holds, remains open. It follows from the above theorem that c_0, the Banach lattice of all real convergent sequences converging to zero is a B-C space. (Also see example 6, pp 92, [5]).

Corollary 3.4. Let X be a Banach lattice with order continuous norm. Then X is a B-C space.

It can be shown that the Banach lattice $C([0, 1])$ of all real-valued continuous functions on $[0, 1]$ is not a B-C space. It may be observed that the supremum norm on $C([0, 1])$ is not order continuous.

Theorem 3.5. *(Example 4.1.6, [6])* Let μ be a σ-finite positive measure on a σ-algebra $\mathcal{M}$ in a nonempty set Y. Then $L^p(Y, \mu), 1 \leq p < \infty$ is a B-C space.

Remark 3.6. It follows from Theorem 3.5 that $\mathbb{R}^n$, the n-dimensional real Euclidean space is a B-C space. Also l^p, the space of real p-summable sequences for $1 \leq p < \infty$ is a B-C space. This follows by taking $Y = \mathbb{N}$ and μ to be the counting measure on $\mathbb{N}$.

Theorem 3.7. *(Lemma 4,[3]) Let H be a partially ordered real Hilbert space with $\mathcal{P}$ as the positive cone such that there exists an orthonormal basis $\{u^\alpha : \alpha \in J\}$, J an index set, of H with $u^\alpha \in \mathcal{P}$ $\forall \alpha \in J$. Then H is a B-C space.*

Remark 3.8. *Theorem 3.5, Remark 3.6 and Theorem 3.7 are all concerned with classes of Banach spaces that are B-C spaces. Even though the validity of these follow from Theorem 3.2, constructive proofs are available, (See for instance [3], [6]) whereas the proof of Theorem 3.2 is existential. (See also Remark 3.10.)*

Theorem 3.9. *The cartesian product of a finite number of B-C spaces is a B-C space with component wise operations.*

Remark 3.10. *We next consider the vector space $H = S\mathbb{R}^{n \times n}$ of real $n \times n$ symmetric matrices with the cone of positive semi-definite matrices. Identifying $S\mathbb{R}^{n \times n}$ with $\mathbb{R}^{n \times (n+1)/2}$ it follows that $S\mathbb{R}^{n \times n}$ is a Banach lattice with order continuous norm. Hence by Corollary 3.4, it follows that any ILP posed over $S\mathbb{R}^{n \times n}$ has an optimal solution. Thus $S\mathbb{R}^{n \times n}$ is a B-C space. We refer the reader to [7] for a proof without the language of B-C spaces but which uses the theory of cones in finite dimensional spaces.*
We next give an example of a cone different from the usual cone in $\mathbb{R}^n$ with respect to which it is a B-C space.

Example 3.11. *Let $\mathcal{P} = \{x = (x_1, x_2, x_3) \in \mathbb{R}^3 : x_1 \geq 0, x_3 \geq 0\}$ and $\{2x_1x_3 \geq x_2^2\}$. Note that $\mathcal{P}$ is an infinitely generated cone and that it does not satisfy the conditions of Theorem 3.7. However, since closed intervals are compact, it follows that $\mathbb{R}^3$ with the positive cone $\mathcal{P}$ is a B-C space.*

The next is an example of a cone in the real Euclidean space giving rise to Banach space that is not a *B-C* space.

Example 3.12. *Consider the lexicographic ordering of $\mathbb{R}^2$ defined by the cone $\mathcal{P} = \{x \in \mathbb{R}^2 :\}$Either $x_1 \geq 0$ or both $x_1 = 0$ and $\{x_2 \geq 0\}$. We show that $\mathbb{R}^2$ with this cone is not a B-C space. Let $a = (1,0), b = (2,0)$ and $c = (0,1)$. Consider $u^n = (1, n), n \in \mathbb{N}$. It can be verified that $a \leq u^n \leq b$ and $c^T u^n = n$. Thus the problem $ILP(a, b, c, I)$ is unbounded.*

We now address the first question in the introduction: when is a closed subspace of a *B-C* space, a *B-C* space? We present a sufficient condition under which we have a positive answer. We need the following definitions.

Definition 3.13. *Let X_1 and X_2 be partially ordered vector spaces with positive cones $\mathcal{P}_1$ and $\mathcal{P}_2$, respectively. A linear map $A : X_1 \to X_2$ is called positive if $A\mathcal{P}_1 \subseteq \mathcal{P}_2$.*

Definition 3.14. *An operator T on a Hilbert space H is called a projection on a subspace X if $T^2 = T$ and $R(T) = X$. A projection T is called orthogonal if $T^* = T$.*

Theorem 3.15. *Let Y be a B-C space. Let X be a closed subspace of Y such that there exists a positive projection on X. Then X is a B-C space.*

Proof. Let T be a positive projection on X. Then $R(T) = X$. It can be shown that $R(T^*) = X'$, where X' denotes the space of all bounded linear functionals on X. Now, consider the problem

$$\text{Maximize } \langle \phi, x \rangle$$
$$\text{subject to } a \leq x \leq b$$
$$x \in X,$$

for $a, b, c \in X$ with $a \leq b$. (Note that $a \leq b$ in X means that $b - a \in X \cap \mathcal{P}$). The subproblem

$$\text{Maximize } \langle \psi, x \rangle$$
$$\text{subject to } a \leq x \leq b,$$

where ψ is an extension of ϕ to Y, has an optimal solution $x^0 \in Y$. Let $Tx^0 = y^0$. We show that y^0 is optimal for the original problem. Since $a, b \in X = R(T)$, we have $a = Ta \leq Tx^0 = y^0 \leq Tb = b$. Thus y^0 is feasible for the original problem. Let x be such that $a \leq x \leq b$, $x \in X$. Then

$$\langle \phi, y^0 \rangle = \langle \phi, Tx^0 \rangle = \langle T^*\phi, x^0 \rangle = \langle \phi, x^0 \rangle \geq \langle \phi, x \rangle,$$

as $\phi \in X' = R(T^*)$. Since $\langle \psi, x \rangle = \langle \phi, x \rangle \forall x \in X$, this completes the proof. □

The next example demonstrates that the converse of Theorem 3.15 is not true.

Example 3.16. *Let ℓ^2 be endowed with the partial order $\mathcal{P} = \{x \in \ell^2 : x_1 \geq 0, x_3 \geq 0, 2x_1 x_3 \geq x_2^2\}$ and $\{x_i \geq 0, i \geq 4\}$. Then $(\ell^2, \mathcal{P})$ is a B-C space. Let X be the closed subspace of ℓ^2 defined by $X := \{x \in \ell^2 : x_1 = 0\}$. Let $\mathcal{Q} := X \cap \mathcal{P}$. Then $\mathcal{Q} = \{x \in \ell^2 : x_1 = x_2 = 0\}$, $\{x_i \geq 0, i \geq 3\}$. Then $(X, \mathcal{Q})$ is a B-C space. Let T be a projection (not necessarily orthogonal) of ℓ^2 on X. Then it follows that $T(x) = (0, x_1, x_2, \cdots)$, the right-shift operator on ℓ^2. If $x^0 = (1, -1, 1, 0, 0, \cdots)$ then $x^0 \in \mathcal{P}$ but T is not positive.*

We next give a sufficient condition to show that the second question in the introduction has an affirmative answer. We need the following definition.

Definition 3.17. *Let $A : X_1 \to X_2$ be linear. A linear map $Y : X_2 \to X_1$ is called a left-inverse of A if $YA = I$, the identity map on X_1.*

Example 3.18. *Let $A : \ell^2 \to \ell^2$ be the right sift operator, i.e., $A(x) = (0, x_1, x_2, \cdots)$ and let $Y = A^*$. Then Y is the left shift operator, i.e., $Y(x) = (x_2, x_3, \cdots)$. Note that A and Y are positive (with the usual order on ℓ^2) and that $YA = I$.*

Theorem 3.19. *Let X_1 be a B-C space and X_2 be a partially ordered real Banach space, with positive cones $\mathcal{P}_1$ and $\mathcal{P}_2$, respectively. Let $A \in BL(X_1, X_2)$ be positive, $R(A)$ be closed and $Y \in BL(X_2, X_1)$ be a positive left-inverse of A. Then $R(A)$ is a B-C space.*

Proof. Let $y^1, y^2 \in R(A)$ with $y^1 \leq y^2$ and $\phi \in R(A)'$, where $R(A)'$ denotes the space of all bounded linear functionals on $R(A)$. We show that $ILP(y^1, y^2, \phi, I)$ has an optimal solution. Consider $ILP(y^1, y^2, \phi, I)$:

$$\text{Maximize } \langle \phi, x \rangle$$
$$\text{subject to } y^1 \leq y \leq y^2, y \in R(A).$$

Let $x^1 = Yy^1$ and $x^2 = Yy^2$. Also $y^1 \leq y \leq y^2, y \in R(A) \Leftrightarrow x^1 \leq x \leq x^2$, $x = Yy \in X_1$. Let ψ be a bounded linear extension of ϕ to X_2. Then $ILP(y^1, y^2, \phi, I)$ is equivalent to:

$$\text{Maximize } \langle A^*\psi, x \rangle$$
$$\text{subject to } x^1 \leq x \leq x^2,$$

which has an optimal solution, as X_1 is a B-C space. □

Remark 3.20. *In addition to the assumptions of Theorem 3.19, let X_2 be a B-C space. Since AY is a positive projection on $R(A)$, the conclusion that $R(A)$ is a B-C space then follows from Theorem 3.15.*

It can be shown through examples that each of the conditions of Theorem 3.19 is indispensable. The next example shows that the converse of Theorem 3.19 is not true.

Example 3.21. *Let $X_1 = \mathbb{R}^2$ with the usual cone $\mathcal{P}_1$. Then X_1 is a B-C space. Consider $X_2 = \mathbb{R}^3$ with the cone $\mathcal{P}_2 = \{x \in \mathbb{R}^3 : x_1 \geq 0, x_3 \geq 0\}$ and $\{2x_1x_3 \geq x_2^2\}$. Then X_2 is a B-C space. Let $A : \mathbb{R}^2 \to \mathbb{R}^3$ be defined by $A = \begin{pmatrix} 1 & 1 \\ 0 & -1 \\ 1 & 1 \end{pmatrix}$ and let $Y = \begin{pmatrix} a & 1 & 1-a \\ b & -1 & -b \end{pmatrix}$ where $a, b \in \mathbb{R}$. Then A is positive, $YA = I$, (Y is the most general left-inverse of A). $R(A)$ is a B-C space with the cone $Q = \{x \in \mathbb{R}^3 : x_1 = x_3 \geq 0\}$ and $\{2x_1^2 \geq x_2^2\}$. But Y is not a positive operator. Thus there is no positive left-inverse for A.*

Remark 3.22. *In the light of Theorem 3.19 and Example 3.12 it follows that there cannot exist a nonnegative matrix having a nonnegative inverse from $\mathbb{R}^2$ with the lexicographic order into $\mathbb{R}^2$ with the usual cone. The reasoning is as follows. Let $u = (1,1)$ and $v = (1,-1)$, then u and v both belong to the lexicographic cone in $\mathbb{R}^2$. If A is such that Au and Av are nonnegative then the first two columns of A are identical. Hence A is not invertible.*

References

1. Ben-Israel, A. and Charnes, A., An explicit solution of a special class of linear programming problems, Operations Research, 16, 1166-1175, 1968.
2. Kulkarni, S. H. and Sivakumar, K. C., Applications of generalized inverses to interval linear programs in Hilbert spaces, Numer. Funct. Anal. and Optimiz., 16(7&8), 965-973, 1995.
3. Kulkarni, S. H. and Sivakumar, K. C., Explicit solutions of a special class of linear programming problems in Banach spaces, Acta. Sci. Math. (Szeged), 62, 457-465, 1996.
4. Kulkarni, S.H. and Sivakumar, K. C., Explicit Solutions of a special class of linear economic models, Ind. J. Pure Appl. Math., 26(3), 217-223, 1995.
5. Schaefer, H. H., Banach lattices and positive operators, Springer-verlag, 1970.
6. Sivakumar, K. C., Interval linear programs in infinite dimensional spaces, Ph.D. Dissertation, Indian Institute of Technology, Madras, 1994.
7. Wolkowicz, H., Explicit solutions for interval semidefinite linear programs, Lin. Alg. Appl., 236, 95-104, 1996.
8. Zlobec, S. and Ben-Israel, A., On explicit solutions of interval linear programs, Israel J. Math., 8, 12-22, 1970.
9. Zlobec, S. and Ben-Israel, A., Explicit solutions of interval linear programs, Operations Research, 21, 390-393, 1973.

Author Index

Abawajy, J.H.   III-60, IV-1272
Ahiska, S. Sebnem   IV-301
Åhlander, Krister   I-657
Ahmad, Uzair   II-1045
Ahn, Beumjun   IV-448
Ahn, Byeong Seok   IV-360
Ahn, Chang-Beom   I-166
Ahn, EunYoung   I-1122
Ahn, Hyo Cheol   IV-916
Ahn, Jaewoo   I-223
Ahn, Joung Chul   II-741
Ahn, Kwang-Il   IV-662
Ahn, Seongjin   I-137, I-242, I-398, II-676, II-696, II-848, IV-1036
Ahn, Yonghak   II-732
Akbari, Mohammad K.   IV-1262
Akyol, Derya Eren   IV-596
Alcaide, Almudena   III-729, IV-1309
Alexander, Phillip J.   IV-1180
Ali, A.   II-1045
Alkassar, Ammar   II-634
Aloisio, Giovanni   III-1
Alonso, Olga Marroquin   II-1156
Altas, Irfan   III-463
An, Sunshin   I-261
Anikeenko, A.V.   I-816
Anton, François   I-669, I-683
Aquino, Adélia J.A.   I-1004
Araújo, Madalena M.   IV-632
Aranda, Gabriela N.   I-1064
Arteconi, Leonardo   I-1093
Aylett, Ruth   IV-30

Bacak, Goksen   III-522
Baciu, George   I-737
Bae, Hae-Young   IV-812
Bae, Hanseok   I-388
Bae, Hyeon   IV-1075, IV-1085
Bae, Hyerim   III-1259
Bae, Ihn Han   II-169
Bae, Jongho   IV-232
Bae, Kyoung Yul   I-204
Baek, Dong-Hyun   IV-222
Baek, Jang Hyun   IV-528

Baek, Jang-Mi   III-964
Baek, Jun-Geol   IV-148
Baek, Sunkyoung   I-37
Baig, Meerja Humayun   I-806
Baik, MaengSoon   III-89, IV-936
Baker, Robert G.V.   III-143
Bang, Young-Cheol   IV-989
Bannai, Hideo   III-349
Bao, Hujun   III-215
Bao, XiaoMing   II-1167
Barbatti, Mario   I-1004
Barco, Raquel   IV-958
Barlow, Jesse   IV-843
Baumgartner, Robert   II-988
Bayhan, G. Mirac   IV-596
Bekker, Henk   IV-397
Bénédet, Vincent   I-838
Bernholdt, D.E.   III-29
Bernholt, Thorsten   I-697
Bertazzon, Stefania   III-118, III-152
Bhat, M.S.   IV-548
Bhatia, Davinder   IV-1190
Bhattacharyya, Chiranjib   IV-548
Bierbaum, Aron   III-1119
Borruso, Giuseppe   III-126
Bozer, Yavuz A.   IV-437
Braad, Eelco P.   IV-397
Brucker, Peter   IV-182
Brunstrom, Anna   IV-1331
Burns, John   II-1254
Burrage, Kevin   II-1245
Byun, Sang-Yong   III-788
Byun, Yung-Cheol   III-788

Caballero-Gil, Pino   III-719
Cafaro, Massimo   III-1
Cai, Guoyin   III-173
Çakar, Tarık   IV-1241
Caminati, Walther   I-1046
Castro, Julio César Hernández   IV-1292
Catanzani, Riccardo   I-921
Cattani, Carlo   III-604
Cechich, Alejandra   I-1064
Cha, ByungRae   II-254

Cha, Jae-Sang   II-332, II-341, II-373, II-411, II-429, II-449, IV-1319
Cha, Jeon-Hee   I-11
Cha, Jung-Eun   III-896
Cha, Kyungup   III-1269
Chae, Jongwoo   II-147
Chae, Kijoon   I-591
Chae, Oksam   II-732, IV-20
Chae, Soo-young   II-458
Chan, Choong Wah   II-657
Chanchio K.   III-29
Chang, Chun Young   I-1204
Chang, Dong Shang   IV-577
Chang, Elizabeth   II-1125
Chang, Hangbae   IV-128
Chang, Hung-Yi   IV-1007
Chang, Jae Sik   IV-999
Chang, Jae-Woo   I-77
Chang, Ok-Bae   III-758, III-836, III-878, III-945
Chang, Pei-Chann   IV-172, IV-417
Chang, Soo Ho   I-46
Chau, Rowena   II-956
Che, Ming   III-284
Che, Yinghui   III-225
Chen, M.L.   III-29
Chen, Chia-Ho   IV-417
Chen, Chun   IV-826
Chen, J.C.   IV-333
Chen, Jianwei   IV-519
Chen, Jinwen   II-1217
Chen, Ling   III-338
Chen, Shih-Chang   IV-1017
Chen, Taiyi   I-967
Chen, Tse-Shih   I-19
Chen, Tung-Shou   IV-1007
Chen, Weidong   I-865
Chen, Wen   II-806
Chen, Yefang   IV-1
Chen, Yun-Shiow   IV-172
Chen, Zhiping P.   IV-733
Cheng, Xiangguo   IV-1046
Cheon, Seong-Pyo   IV-1075
Chernetsov, Nikita   III-133
Chi, Jeong Hee   II-977
Ching, Wai-Ki   IV-342, IV-843
Cho, Byung Rae   IV-212
Cho, Cheol-Hyung   I-707, III-993
Cho, Chiwoon   III-1297
Cho, Dongyoung   I-232

Cho, Eun-Sook   III-778, III-868
Cho, Hyeon Seob   II-832
Cho, Kyung Dal   II-474
Cho, Miyoung   I-37
Cho, Nam Wook   III-1297
Cho, Seokhyang   I-498
Cho, Sok-Pal   II-781, II-896
Cho, Sung-Keun   IV-48
Cho, SungHo   I-204
Cho, Wanhyun   IV-867
Cho, Yongju   III-1289
Cho, Yongsun   I-1
Cho, Yongyun   II-1008
Cho, Yookun   II-353
Cho, You-Ze   I-378
Cho, Youngsong   I-707, I-716, III-993
Cho, YoungTak   IV-20
Choi, In Seon   II-889
Choi, DeaWoo   IV-103
Choi, Deokjai   I-195
Choi, Dong-seong   IV-62
Choi, Eunmi   II-187, III-858
Choi, Gyunghyun   IV-261
Choi, Hee-Chul   III-938
Choi, Honzong   III-1289
Choi, Hyang-Chang   II-82
Choi, Hyung Jo   II-1207
Choi, Ilhoon   III-1229
Choi, Jaemin   II-567
Choi, Jaeyoung   II-1008, II-1018, IV-10
Choi, Jong Hwa   IV-86
Choi, Jong-In   III-1148
Choi, Jonghyoun   I-271
Choi, Kee-Hyun   III-99
Choi, Kun Myon   I-448
Choi, Mi-Sook   III-778
Choi, Sang-soo   II-458
Choi, Sang-Yule   II-341, II-429, IV-1319
Choi, Sung-ja   II-215
Choi, SungJin   III-89, IV-936
Choi, Wonwoo   I-137
Choi, WoongChul   IV-1231
Choi, Yeon-Sung   II-71
Choi, YoungSik   I-186
Chong, Kil To   II-1207, II-1293
Chong, Kiwon   I-1
Choo, Hyunseung   I-291, I-448, I-468, I-529, I-540, IV-989
Choudhary, Alok Kumar   IV-680
Chow, K.P.   III-651

Chow, Sherman S.M.   III-651
Choy, Yoon-Chu   I-847
Chua, Eng-Huat   II-1167
Chun, Junchul   I-1135
Chun, Kilsoo   II-381
Chun, Kwang Ho   II-749
Chun, Kwang-ho   II-723
Chung, Chin Hyun   I-638, I-1213
Chung, Hyun-Sook   III-788
Chung, Jinwook   I-137
Chung, Kwangsue   IV-1231
Chung, Min Young   I-348, I-448, I-529
Chung, Mokdong   II-147
Chung, Tae-sun   IV-72
Chung, Tae-Woong   IV-836
Chung, Tai-Myung   I-146, I-468
Chung, YoonJung   II-92, II-274
Chung, Youn-Ky   III-769
Chunyan, Yu   I-875, I-974
Cornejo, Oscar   IV-712
Corradini, Flavio   II-1264
Costantini, Alessandro   I-1046
Cotrina, Josep   II-527, II-624
Couloigner, Isabelle   III-181
Croce, Federico Della   IV-202
Cruz-Neira, Carolina   III-1070, III-1119
Cui, Kebin   I-214
Cui, Shi   II-657

Das, Amitabha   I-994
Das, Sandip   I-827
Dashora, Yogesh   IV-680
Datta, Amitava   I-87, II-686, III-206
Da-xin, Liu   IV-753
Debels, Dieter   IV-378
de Frutos Escrig, David   II-1156
Deris, M. Mat   III-60
Dévai, Frank   I-726
Dew, Robert   III-49
Díez, Luis   IV-958
Dillon, Tharam S.   II-914, II-1125
Ding, Jintai   II-595
Ding, Yongsheng   III-69
Djemame, Karim   IV-1282
Doboga, Flavia   III-563
Dol'nikov, Vladimir   III-628
Dongyi, Ye   I-875, I-974
Du, Tianbao   I-1040
Duan, Pu   II-657

Dumas, Laurent   IV-948
Duong, Doan Dai   II-1066

Emiris, Ioannis   I-683
Enzi, Christian   II-988
Eong, Gu-Beom   II-42
Epicoco, Italo   III-1
Ercan, M. Fikret   III-445
Ergenc, Tanil   III-463
Escoffier, Bruno   IV-192, IV-202
Espírito-Santo, Isabel A.C.P.   IV-632
Estévez-Tapiador, Juan M.   IV-1292, IV-1309
Eun, He-Jue   II-10

Faudot, Dominique   I-838
Feng, Jieqing   III-1023
Fernandes, Edite M.G.P   IV-488, IV-632
Fernandez, Marcel   II-527, II-624
Ferreira, Eugenio C.   IV-632
Fiore, Sandro   III-1
Fong, Simon   II-1106
For, Wei-Khing   II-1167
Frank, A.O.   II-1018
Froeklich, Johannes   I-905, I-938
Fu, Haoying   IV-843
Fúster-Sabater, Amparo   III-719

Gaglio, Salvatore   III-39
Gálvez, Akemi   III-472, III-482, III-502
Gao, Chaohui   I-1040
Gao, Lei   III-69
Garcia, Ernesto   I-1083
Gardner, William   II-1125
Gatani, Luca   III-39
Gaur, Daya Ram   IV-670
Gavrilova, M.L.   I-816
Gavrilova, Marina L.   I-748
Geist A.   III-29
Gerardo, Bobby   II-71, II-205
Gervasi, Osvaldo   I-905, I-921, I-938
Ghelmez, Mihaela   III-563
Ghinea, G.   II-1018
Ghose, Debasish   IV-548
Gil, JoonMin   IV-936
Gimenez, Xavi   I-1083
Goetschalckx, Marc   IV-322
Goff, Raal   I-87
Goh, Dion Hoe-Lian   II-1177
Goh, John   IV-1203

Goh, Li Ping IV-906
Goi, Bok-Min I-488, IV-1065
Gold, Christopher M. I-737
Goldengorin, Boris IV-397
Goscinski, Andrzej III-49
Goswami, Partha P. I-827
Gower, Jason II-595
Grinnemo, Karl-Johan IV-1331
Großschädl, Johann II-665
Grząślewicz, Ryszard II-517
Gu, Mi Sug II-966
Guan, Xiucui IV-161
Guan, Yanning III-173
Guo, Heqing III-691, IV-1028
Guo, X.C. IV-1040
Gupta, Pankaj IV-1190

Ha, JaeCheol II-245
Haji, Mohammed IV-1282
Han, Chang Hee IV-222, IV-360
Han, In-sung II-904
Han, Joohyun II-1008
Han, Jung-Soo III-748, III-886
Han, Kyuho I-261
Han, SangHoon I-1122
Han, Young-Ju I-146
Harding, Jenny A. IV-680
Hartling, Patrick III-1070, III-1119
He, Ping III-338
He, Qi III-691, IV-1028
He, Yuanjun III-1099
Hedgecock, Ian M. I-1054
Heng, Swee-Huay II-603
Henze, Nicola II-988
Heo, Hoon IV-20
Herbert, Vincent IV-948
Hernández, Julio C. IV-1301
Herrlich, Marc II-988
Herzog, Marcus II-988
Higuchi, Tomoyuki III-381, III-389
Hirose, Osamu III-349
Hong, Changho IV-138
Hong, Choong Seon I-195, I-339
Hong, Chun Pyo I-508
Hong, Helen IV-1111
Hong, In-Sik III-964
Hong, Jung-Hun II-1
Hong, Jungman IV-642
Hong, Kicheon I-1154
Hong, Kiwon I-195

Hong, Maria I-242, IV-1036
Hong, Seok Hoo II-1076
Hong, Xianlong IV-896
Hou, Jia II-749
Hsieh, Min-Chi II-1055
Hsieh, Ying-Jiun IV-437
Hsu, Ching-Hsien IV-1017
Hu, Bingcheng II-1274
Hu, Guofei I-758
Hu, Xiaohua III-374
Hu, Xiaoyan III-235
Hu, Yifeng I-985
Hu, Yincui III-173
Hua, Wei III-215
Huang, Zhong III-374
Huettmann, Falk III-133, III-152
Huh, Eui-Nam I-311, I-628, I-1144
Hui, Lucas C.K. III-651
Hung, Terence I-769, IV-906
Hur, Nam-Young II-341, IV-1319
Hur, Sun II-714, IV-606
Huynh, Trong Thua I-339
Hwang, Chong-Sun III-89, IV-936, IV-1169
Hwang, Gi Yean II-749
Hwang, Ha Jin II-304, III-798
Hwang, Hyun-Suk II-127
Hwang, Jae-Jeong II-205
Hwang, Jeong Hee II-925, II-966
Hwang, Jun I-1170, I-1204
Hwang, Seok-Hyung II-1
Hwang, Suk-Hyung III-827, III-938
Hwang, Sun-Myung II-21, III-846
Hwang, Yoo Mi I-1129
Hwang, Young Ju I-619
Hwang, Yumi I-1129
Hyun, Chang-Moon III-927
Hyun, Chung Chin I-1177
Hyuncheol, Kim II-676

Iglesias, Andrés III-502, III-472, III-482, III-492, III-547, III-1157
Im, Chae-Tae I-368
Im, Dong-Ju II-420, II-474
Imoto, Seiya III-349, III-389
In, Hoh Peter II-274
Inceoglu, Mustafa Murat III-538, IV-56
Iordache, Dan III-614
Ipanaqué, Ruben III-492
Iqbal, Mahrin II-1045

Iqbal, Mudeem   II-1045
Izquierdo, Antonio   III-729, IV-1309

Jackson, Steven Glenn   III-512
Jahwan, Koo   II-696
Jalili-Kharaajoo, Mahdi   I-1030
Jang, Dong-Sik   IV-743
Jang, Injoo   II-102, II-111
Jang, Jongsu   I-609
Jang, Sehoon   I-569
Jang, Sung Man   II-754
Jansen, A.P.J.   I-1020
Javadi, Bahman   IV-1262
Jeon, Hoseong   I-529
Jeon, Hyong-Bae   IV-538
Jeon, Nam Joo   IV-86
Jeong, Bongju   IV-566
Jeong, Chang Sung   I-601
Jeong, Eun-Hee   II-322, II-585
Jeong, Eunjoo   I-118
Jeong, Gu-Beom   II-42
Jeong, Hwa-Young   I-928
Jeong, In-Jae   IV-222, IV-312
Jeong, JaeYong   II-353
Jeong, Jong-Youl   I-311
Jeong, Jongpil   I-291
Jeong, Kugsang   I-195
Jeong, KwangChul   I-540
Jeong, Seung-Ju   IV-566
Ji, Joon-Yong   III-1139
Ji, Junfeng   III-1167
Ji, Yong Gu   III-1249
Jia, Zhaoqing   III-10
Jiang, Chaojun   I-1040
Jiang, Xinhua H.   IV-733
Jiao, Xiangmin   IV-1180
Jin, Biao   IV-1102
Jin, Bo   III-299
Jin, Guiyue   IV-1095
Jin, Jing   III-416
Jin, YoungTaek   III-846
Jin, Zhou   III-435
Jo, Geun-Sik   IV-1131
Jo, Hea Suk   I-519
Joo, Inhak   II-1136
Joung, Bong Jo   I-1196, I-1213
Ju, Hak Soo   II-381
Ju, Jaeyoung   III-1259
Jun, Woochun   IV-48
Jung, Changho   II-537

Jung, Ho-Sung   II-332
Jung, Hoe Sang   III-1177
Jung, Hye-Jung   III-739
Jung, Jason J.   IV-1131
Jung, Jin Chul   I-252
Jung, Jung Woo   IV-467
Jung, KeeChul   IV-999
Jung, Kwang Hoon   I-1177
Jung, SM.   II-1028

Kang, Euisun   I-242
Kang, HeeJo   II-420, II-483
Kang, Kyung Hwan   IV-350
Kang, Kyung-Woo   I-29
Kang, MunSu   I-186
Kang, Oh-Hyung   II-195, II-284, II-295
Kang, Seo-Il   II-177
Kang, Suk-Hoon   I-320
Kang, Yeon-hee   II-215
Kang, Yu-Kyung   III-938
Karsak, E. Ertugrul   IV-301
Kasprzak, Andrzej   IV-772
Kemp, Ray   II-1187
Khachoyan, Avet A.   IV-1012
Khorsandi, Siavash   IV-1262
Kiani, Saad Liaquat   II-1096
Kim, B.S.   II-1028
Kim, Byunggi   I-118
Kim, Byung Wan   III-1306
Kim, Chang Han   II-647
Kim, Chang Hoon   I-508
Kim, Chang Ouk   IV-148
Kim, Chang-Hun   III-1080, III-1129, III-1139, III-1148
Kim, Chang-Min   I-176, III-817, IV-38
Kim, Chang-Soo   II-127
Kim, Chul-Hong   III-896
Kim, Chulyeon   IV-261
Kim, Dae Hee   II-1284
Kim, Dae Sung   I-1111
Kim, Dae Youb   II-381
Kim, Daegeun   II-1035
Kim, Deok-Soo   I-707, I-716, III-993, III-1060, IV-652
Kim, D.K.   II-1028
Kim, Do-Hyeon   I-378
Kim, Do-Hyung   II-401
Kim, Dong-Soon   III-938
Kim, Donghyun   IV-877
Kim, Dongkeun   I-857

Kim, Dongkyun   I-388
Kim, Dongsoo   III-1249
Kim, Donguk   I-716, III-993
Kim, Dounguk   I-707
Kim, Eun Ju   I-127
Kim, Eun Suk   IV-558
Kim, Eun Yi   IV-999
Kim, Eunah   I-591
Kim, Gi-Hong   II-771
Kim, Gui-Jung   III-748, III-886
Kim, Guk-Boh   II-42
Kim, Gye-Young   I-11
Kim, Gyoung-Bae   IV-812
Kim, Hae Geun   II-295
Kim, Hae-Sun   II-157
Kim, Haeng-Kon   II-1, II-52, II-62, II-137, III-769, III-906, III-916
Kim, Hak-Keun   I-847
Kim, Hang Joon   IV-999
Kim, Hee Sook   II-483, II-798
Kim, Hong-Gee   III-827
Kim, Hong-jin   II-781, II-896
Kim, HongSoo   III-89
Kim, Hoontae   III-1249
Kim, Howon   II-1146
Kim, Hwa-Joong   IV-538, IV-722
Kim, Hwankoo   II-245
Kim, HyoungJoong   IV-269
Kim, Hyun Cheol   I-281
Kim, Hyun-Ah   I-427, III-426, IV-38
Kim, Hyun-Ki   IV-887
Kim, Hyuncheol   I-137, II-676
Kim, Hyung Jin   II-789, II-880
Kim, InJung   II-92, II-274
Kim, Jae-Gon   IV-280, IV-322
Kim, Jae-Sung   II-401
Kim, Jae-Yearn   IV-662
Kim, Jae-Yeon   IV-743
Kim, Jang-Sub   I-348
Kim, Jee-In   I-886
Kim, Jeom-Goo   II-762
Kim, Jeong Ah   III-846
Kim, Jeong Kee   II-714
Kim, Jin Ok   I-638, I-1187
Kim, Jin Soo   I-638, I-1187
Kim, Jin-Geol   IV-782
Kim, Jin-Mook   II-904
Kim, Jin-Sung   II-31, II-567
Kim, Jong-Boo   II-341, IV-1319
Kim, Jong Hwa   III-1033

Kim, Jong-Nam   I-67
Kim, Jongsung   II-567
Kim, Jong-Woo   I-1177, II-127
Kim, Ju-Yeon   II-127
Kim, Jun-Gyu   IV-538
Kim, Jung-Min   III-788
Kim, Jungchul   I-1154
Kim, Juwan   I-857
Kim, Kap Sik   III-798
Kim, Kibum   IV-566
Kim, KiJoo   I-186
Kim, Kwan-Joong   I-118
Kim, Kwang-Baek   IV-1075
Kim, Kwang-Hoon   I-176, III-817, IV-38
Kim, Kwang-Ki   III-806
Kim, Kyung-kyu   IV-128
Kim, Mihui   I-591
Kim, Mijeong   II-1136
Kim, Minsoo   II-225, II-1136, III-1249, III-1259
Kim, Misun   I-550, I-559
Kim, Miyoung   I-550, I-559
Kim, Moonseong   IV-989
Kim, Myoung Soo   III-916
Kim, Myuhng-Joo   I-156
Kim, Myung Ho   I-223
Kim, Myung Won   I-127
Kim, Myung-Joon   IV-812
Kim, Nam Chul   I-1111
Kim, Pankoo   I-37
Kim, Sang Ho   II-977, IV-79
Kim, Sang-Bok   I-628
Kim, Sangjin   IV-877
Kim, Sangkyun   III-1229, III-1239, IV-122
Kim, Seungjoo   I-498, II-1146
Kim, Soo Dong   I-46, I-57
Kim, Soo-Kyun   III-1080, III-1129, III-1139
Kim, Soung Won   III-916
Kim, S.R.   II-1028
Kim, Sung Jin   II-1076
Kim, Sung Jo   III-79
Kim, Sung Ki   I-252
Kim, Sung-il   IV-62
Kim, Sung-Ryul   I-359
Kim, Sungshin   IV-1075, IV-1085
Kim, Tae Hoon   IV-509
Kim, Tae Joong   III-1279
Kim, Tae-Eun   II-474

Kim, Taeho   IV-280
Kim, Taewan   II-863
Kim, Tai-Hoon   II-341, II-429, II-468, II-491, IV-1319
Kim, Ungmo   II-936
Kim, Won-sik   IV-62
Kim, Wooju   III-1289, IV-103
Kim, Y.H.   III-1089
Kim, Yon Tae   IV-1085
Kim, Yong-Kah   IV-858
Kim, Yong-Soo   I-320, I-1162
Kim, Yong-Sung   II-10, II-31, III-954
Kim, Yongtae   II-647
Kim, Young Jin   IV-212, IV-232
Kim, Young-Chan   I-1170
Kim, Young-Chul   IV-10
Kim, Young-Shin   I-311
Kim, Young-Tak   II-157
Kim, Youngchul   I-107
Ko, Eun-Jung   III-945
Ko, Hoon   II-442
Ko, Jaeseon   II-205
Ko, S.L.   III-1089
Koh, Jae Young   II-741
Komijan, Alireza Rashidi   IV-388
Kong, Jung-Shik   IV-782
Kong, Ki-Sik   II-1225, IV-1169
Koo, Jahwan   II-696, II-848
Koo, Yun-Mo   III-1187
Koszalka, Leszek   IV-692
Kravchenko, Svetlana A.   IV-182
Kriesell, Matthias   II-988
Krishnamurti, Ramesh   IV-670
Kuo, Yi Chun   IV-577
Kurosawa, Kaoru   II-603
Kutyłowski, Jarosław   II-517
Kutyłowski, Mirosław   II-517
Kwag, Sujin   I-418
Kwak, Byeong Heui   IV-48
Kwak, Kyungsup   II-373, II-429
Kwak, NoYoon   I-1122
Kwon, Dong-Hee   I-368
Kwon, Gihwon   III-973
Kwon, Hyuck Moo   IV-212, IV-232
Kwon, Jungkyu   II-147
Kwon, Ki-Ryong   II-557
Kwon, Ki-Ryoung   III-1209
Kwon, Oh Hyun   II-137
Kwon, Soo-Tae   IV-624
Kwon, Soonhak   I-508

Kwon, Taekyoung   I-577, I-584
Kwon, Yong-Moo   I-913

La, Hyun Jung   I-46
Lægreid, Astrid   III-327
Laganà, Antonio   I-905, I-921, I-938, I-1046, I-1083, I-1093
Lago, Noelia Faginas   I-1083
Lai, Edison   II-1106
Lai, K.K.   IV-250
Lamarque, Loïc   I-838
Lázaro, Pedro   IV-958
Ledoux, Hugo   I-737
Lee, Bo-Hee   IV-782
Lee, Bong-Hwan   I-320
Lee, Byoungcheon   II-245
Lee, Byung Ki   IV-350
Lee, Byung-Gook   III-1209
Lee, Byung-Kwan   II-322, II-585
Lee, Chang-Mog   III-758
Lee, Chong Hyun   II-373, II-411, II-429, II-449
Lee, Chun-Liang   IV-1007
Lee, Dong Chun   II-714, II-741, II-762, II-889, II-896
Lee, Dong Hoon   I-619, II-381
Lee, Dong-Ho   IV-538, IV-722
Lee, DongWoo   I-232
Lee, Eun-Ser   II-363, II-483
Lee, Eung Jae   II-998
Lee, Eung Young   III-1279
Lee, Eunkyu   II-1136
Lee, Eunseok   I-291
Lee, Gang-soo   II-215, II-458
Lee, Geuk   II-754
Lee, Gi-Sung   II-839
Lee, Hakjoo   III-1269
Lee, Ho Woo   IV-509
Lee, Hong Joo   III-1239, IV-113, IV-122
Lee, Hoonjung   IV-877
Lee, Hyewon K.   I-97, I-118
Lee, Hyoung-Gon   III-1219
Lee, Hyun Chan   III-993
Lee, HyunChan   III-1060
Lee, Hyung-Hyo   II-82
Lee, Hyung-Woo   II-391, II-401, IV-62
Lee, Im-Yeong   II-117, II-177
Lee, Insup   I-156
Lee, Jae-deuk   II-420
Lee, Jaeho   III-1060

Lee, Jae-Wan   II-71, II-205, II-474
Lee, Jee-Hyong   IV-1149
Lee, Jeongheon   IV-20
Lee, Jeongjin   IV-1111
Lee, Jeoung-Gwen   IV-1055
Lee, Ji-Hyen   III-878
Lee, Ji-Hyun   III-836
Lee, Jongchan   I-107
Lee, Jong chan   II-781
Lee, Jong Hee   II-856
Lee, Jong-Hyouk   I-146, I-468
Lee, Joon-Jae   III-1209
Lee, Joong-Jae   I-11
Lee, Joungho   II-111
Lee, Ju-Il   I-427
Lee, Jun   I-886
Lee, Jun-Won   III-426
Lee, Jung   III-1080, III-1129, III-1139
Lee, Jung-Bae   III-938
Lee, Jung-Hoon   I-176
Lee, Jungmin   IV-1231
Lee, Jungwoo   IV-96
Lee, Kang-Won   I-378
Lee, Keon-Myung   IV-1149
Lee, Keun Kwang   II-474, II-420
Lee, Keun Wang   II-798, II-832, II-856
Lee, Key Seo   I-1213
Lee, Ki Dong   IV-1095
Lee, Ki-Kwang   IV-427
Lee, Kwang Hyoung   II-798
Lee, Kwangsoo   II-537
Lee, Kyunghye   I-408
Lee, Malrey   II-71, II-363, II-420, II-474, II-483
Lee, Man-Hee   IV-743
Lee, Mi-Kyung   II-31
Lee, Min Koo   IV-212, IV-232
Lee, Moon Ho   II-749
Lee, Mun-Kyu   II-314
Lee, Myung-jin   IV-62
Lee, Myungeun   IV-867
Lee, Myungho   IV-72
Lee, NamHoon   II-274
Lee, Pill-Woo   I-1144
Lee, S.Y.   II-1045
Lee, Sang Ho   II-1076
Lee, Sang Hyo   I-1213
Lee, Sangsun   I-418
Lee, Sang Won   III-1279
Lee, Sang-Hyuk   IV-1085

Lee, Sang-Young   II-762, III-945
Lee, Sangjin   II-537, II-567
Lee, SangKeun   II-1225
Lee, Sangsoo   II-816
Lee, Se-Yul   I-320, I-1162
Lee, SeongHoon   I-232
Lee, Seoung Soo   III-1033, IV-652
Lee, Seung-Yeon   I-628, II-332
Lee, Seung-Yong   II-225
Lee, Seung-youn   II-468, II-491, II-499
Lee, SiHun   IV-1149
Lee, SooBeom   II-789, II-880
Lee, Su Mi   I-619
Lee, Suk-Hwan   II-557
Lee, Sungchang   I-540
Lee, Sunghwan   IV-96
Lee, SungKyu   IV-103
Lee, Sungyoung   II-1096, II-1106, II-1115
Lee, Sunhun   IV-1231
Lee, Suwon   II-420, II-474
Lee, Tae Dong   I-601
Lee, Tae-Jin   I-448
Lee, Taek   II-274
Lee, TaiSik   II-880
Lee, Tong-Yee   III-1043, III-1050
Lee, Wonchan   I-1154
Lee, Woojin   I-1
Lee, Woongjae   I-1162, I-1196
Lee, Yi-Shiun   III-309
Lee, Yong-Koo   II-1115
Lee, Yonghwan   II-187, III-858
Lee, Yongjae   II-863
Lee, Young Hae   IV-467
Lee, Young Hoon   IV-350
Lee, Young Keun   II-420, II-474
Lee, YoungGyo   II-92
Lee, YoungKyun   II-880
Lee, Yue-Shi   II-1055
Lee, Yung-Hyeon   II-762
Leem, Choon Seong   III-1269, III-1289, III-1306, IV-79, IV-86, IV-113
Lei, Feiyu   II-806
Leon, V. Jorge   IV-312
Leung, Stephen C.H.   IV-250
Lezzi, Daniele   III-1
Li, Huaqing   IV-1140
Li, Jin-Tao   II-547
Li, JuanZi   IV-1222
Li, Kuan-Ching   IV-1017

Li, Li   III-190
Li, Minglu   III-10
Li, Peng   III-292
Li, Sheng   III-1167
Li, Tsai-Yen   I-957
Li, Weishi   I-769, IV-906
Li, Xiao-Li   III-318
Li, Xiaotu   III-416
Li, Xiaowei   III-266
Li, Yanda   II-1217
Li, Yun   III-374
Li, Zhanhuai   I-214
Li, Zhuowei   I-994
Liao, Mao-Yung   I-957
Liang, Xiaohui   III-225
Liang, Y.C.   I-1040
Lim, Cheol-Su   III-1080, III-1148
Lim, Ee-Peng   II-1177
Lim, Heui Seok   I-1129
Lim, Hyung-Jin   I-146
Lim, In-Taek   I-438
Lim, Jongin   II-381, II-537, II-567, II-647
Lim, Jong In   I-619
Lim, Jongtae   IV-138
Lim, Myoung-seob   II-723
Lim, Seungkil   IV-642
Lim, Si-Yeong   IV-606
Lim, Soon-Bum   I-847
Lim, YoungHwan   I-242, IV-1036
Lim, Younghwan   I-398, II-676, II-848
Lin, Huaizhong   IV-826
Lin, Jenn-Rong   IV-499
Lin, Manshan   III-691, IV-1028
Lin, Ping-Hsien   III-1050
Lindskog, Stefan   IV-1331
Lischka, Hans   I-1004
Liu, Bin   II-508
Liu, Dongquan   IV-968
Liu, Fenlin   II-508
Liu, Jiming   II-1274
Liu, Jingmei   IV-1046
Liu, Joseph K.   II-614
Liu, Ming   IV-1102
Liu, Mingzhe   II-1187
Liu, Xuehui   III-1167
Liu, Yue   III-266
Lopez, Javier   III-681
Lu, Chung-Dar   III-299
Lu, Dongming   I-865, I-985

Lu, Jiahui   I-1040
Lu, Xiaolin   III-256
Luengo, Francisco   III-1157
Luo, Lijuan   IV-896
Luo, Xiangyang   II-508
Luo, Ying   III-173
Luo, Yingwei   I-301, II-822

Ma, Fanyuan   II-1086
Ma, Liang   III-292
Ma, Lizhuang   I-776
Mackay, Troy D.   III-143
Mał afiejski, Michal   I-647
Manera, Jaime   IV-1301
Mani, Venkataraman   IV-269
Manzanares, Antonio Izquierdo   IV-1292
Mao, Zhihong   I-776
Maris, Assimo   I-1046
Markowski, Marcin   IV-772
Márquez, Joaquín Torres   IV-1292
Martoyan, Gagik A.   I-1012
Medvedev, N.N.   I-816
Meng, Qingfan   I-1040
Merelli, Emanuela   II-1264
Miao, Lanfang   I-758
Miao, Yongwei   III-1023
Michelot, Christian   IV-712
Mielikäinen, Taneli   IV-1251
Mijangos, Eugenio   IV-477
Million, D.L.   III-29
Min, Byoung Joon   I-252
Min, Byoung-Muk   II-896
Min, Dugki   II-187, III-858
Min, Hyun Gi   I-57
Min, Jihong   I-1154
Min, Kyongpil   I-1135
Min, Seung-hyun   II-723
Min, Sung-Hwan   IV-458
Minasyan, Seyran H.   I-1012
Minghui, Wu   I-875, I-974
Minhas, Mahmood R.   IV-587
Mirto, Maria   III-1
Miyano, Satoru   III-349
Mnaouer, Adel Ben   IV-1212
Mo, Jianzhong   I-967
Mocavero, Silvia   III-1
Moon, Hyeonjoon   I-584
Moon, Kiyoung   I-609

Morarescu, Cristian   III-556, III-563
Moreland, Terry   IV-1120
Morillo, Pedro   III-1119
Moriya, Kentaro   IV-978
Mourrain, Bernard   I-683
Mun, Ki-Young   I-311
Mun, Young-Song   I-97, I-118, I-242,
   I-271, I-398, I-408, I-459, I-550, I-559,
   I-569, I-628, II-676, II-848, IV-1036
Murat, Cécile   IV-202
Muyl, Frédérique   IV-948

Nait-Sidi-Moh, Ahmed   IV-792
Nakamura, Yasuaki   III-1013
Nam, Junghyun   I-498
Nam, Kichun   I-1129
Nam, Kyung-Won   I-1170
Nandy, Subhas C.   I-827
Nariai, Naoki   III-349
Nasir, Uzma   II-1045
Nassis, Vicky   II-914
Ng, Michael Kwok   IV-843
Ng, See-Kiong   II-1167, III-318
Nicolay, Thomas   II-634
Nie, Weifang   III-284, III-292, III-416
Nikolova, Mila   IV-843
Ninulescu, Valerică   III-635, III-643
Nodera, Takashi   IV-978
Noël, Alfred G.   III-512
Noh, Angela Song-Ie   I-1144
Noh, Bong-Nam   II-82, II-225
Noh, Hye-Min   III-836, III-878, III-945
Noh, Seung J.   IV-615
Nozick, Linda K.   IV-499
Nugraheni, Cecilia E.   III-453

Offermans, W.K.   I-1020
Ogiela, Lidia   IV-852
Ogiela, Marek R.   IV-852
Oh, Am-Sok   II-322, 585
Oh, Heekuck   IV-877
Oh, Nam-Ho   II-401
Oh, Sei-Chang   II-816
Oh, Seoung-Jun   I-166
Oh, Sun-Jin   II-169
Oh, Sung-Kwun   IV-858, IV-887
Ok, MinHwan   II-1035
Olmes, Zhanna   I-448
Omar M.   III-60

Ong, Eng Teo   I-769
Onyeahialam, Anthonia   III-152

Padgett, James   IV-1282
Páez, Antonio   III-162
Paik, Juryon   II-936
Pan, Hailang   I-896
Pan, Xuezeng   I-329, II-704
Pan, Yi   III-338
Pan, Yunhe   I-865
Pan, Zhigeng   II-946, III-190, III-245
Pandey, R.B   II-1197
Pang, Mingyong   III-245
Park, Myong-soon   II-1035
Park, Bongjoo   II-245
Park, Byoung-Jun   IV-887
Park, Byungchul   I-468
Park, Chan Yong   II-1284
Park, Chankwon   III-1219
Park, Cheol-Min   I-1170
Park, Choon-Sik   II-225
Park, Daehee   IV-858
Park, Dea-Woo   II-235
Park, DaeHyuck   IV-1036
Park, DongGook   II-245
Park, Eung-Ki   II-225
Park, Gyung-Leen   I-478
Park, Hayoung   II-442
Park, Hee Jun   IV-122
Park, Hee-Dong   I-378
Park, Hee-Un   II-117
Park, Heejun   III-1316
Park, Jesang   I-418
Park, Jin-Woo   I-913, III-1219
Park, Jonghyun   IV-867
Park, Joon Young   III-993, III-1060
Park, Joowon   II-789
Park, KwangJin   II-1225
Park, Kyeongmo   II-264
Park, KyungWoo   II-254
Park, Mi-Og   II-235
Park, Namje   I-609, II-1146
Park, Sachoun   III-973
Park, Sang-Min   IV-652
Park, Sang-Sung   IV-743
Park, Sangjoon   I-107, I-118
Park, Seon Hee   II-1284
Park, Seoung Kyu   III-1306
Park, Si Hyung   III-1033
Park, Soonyoung   IV-867

Park, Sung Hee   II-1284
Park, Sung-gi   II-127
Park, Sung-Ho   I-11
Park, Sungjun   I-886
Park, Sung-Seok   II-127
Park, Woojin   I-261
Park, Yongsu   II-353
Park, Youngho   II-647
Park, Yunsun   IV-148
Parlos, A.G   II-1293
Paschos, Vangelis Th.   IV-192, IV-202
Pedrycz, Witold   IV-887
Pei, Bingzhen   III-10
Peng, Jiming   IV-290
Peng, Qunsheng   I-758, III-1023
Penubarthi, Chaitanya   I-156
Pérez, María S.   III-109
Phan, Raphael C.-W.   I-488, III-661, IV-1065
Piattini, Mario   I-1064
Pietkiewicz, Wojciech   II-517
Ping, Lingdi   I-329, II-704
Pirani, Fernando   I-1046
Pirrone, Nicola   I-1054
Podoleanu, Adrian   III-556
Ponce, Eva   IV-1301
Porschen, Stefan   I-796
Prasanna, H.M.   IV-548
Przewoźniczek, Michał   IV-802
Pusca, Stefan   III-563, III-569, III-614

Qi, Feihu   IV-1140
Qing, Sihan   III-711
Qu, Na   III-225

Rahayu, Wenny   II-914, II-925
Rajugan, R.,   II-914, II-1125
Ramadan, Omar   IV-926
Rasheed, Faraz   II-1115
Ravantti, Janne   IV-1251
Raza, Syed Arshad   I-806
Re, Giuseppe Lo   III-39
Ren, Lifeng   IV-30
Rhee, Seung Hyong   IV-1231
Rhee, Seung-Hyun   III-1259
Rhew, Sung Yul   I-57
Riaz, Maria   II-1096
Ribagorda, Arturo   III-729
Riganelli, Antonio   I-905, I-921, I-938
Rim, Suk-Chul   IV-615

Rob, Seok-Beom   IV-858
Rocha, Ana Maria A.C.   IV-488
Roh, Sung-Ju   IV-1169
Rohe, Markus   II-634
Roman, Rodrigo   III-681
Rosi, Marzio   I-1101
Ruskin, Heather J.   II-1254
Ryou, Hwang-bin   II-904
Ryu, Keun Ho   II-925, II-977
Ryu, Han-Kyu   I-378
Ryu, Joonghyun   III-993
Ryu, Joung Woo   I-127
Ryu, Keun Ho   II-966
Ryu, Keun Hos   II-998
Ryu, Seonggeun   I-398
Ryu, Yeonseung   IV-72

Sadjadi, Seyed Jafar   IV-388
Sætre, Rune   III-327
Sait, Sadiq M.   IV-587
Salzer, Reiner   I-938
Sánchez, Alberto   III-109
Sarfraz, Muhammad   I-806
Saxena, Amitabh   III-672
Seo, Dae-Hee   II-117
Seo, Jae Young   IV-528
Seo, Jae-Hyun   II-82, II-225, II-254
Seo, Jeong-Yeon   IV-652
Seo, Jung-Taek   II-225
Seo, Kwang-Kyu   IV-448, IV-458
Seo, Kyung-Sik   IV-836
Seo, Young-Jun   I-928
Seong, Myoung-ho   II-723
Seongjin, Ahn   II-676, II-696
Serif, T.   II-1018
Shang, Yanfeng   IV-1102
Shao, Min-Hua   III-701
Shehzad, Anjum   II-1096, II-1106
Shen, Lianguan G.   III-1003
Shen, Yonghang   IV-1159
Sheng, Yu   I-985
Shi, Lie   III-190
Shi, Lei   I-896
Shi, Xifan   IV-1159
Shim, Bo-Yeon   III-806
Shim, Donghee   I-232
Shim, Young-Chul   I-427, III-426
Shimizu, Mayumi   III-1013
Shin, Byeong-Seok   III-1177, III-1187
Shin, Chungsoo   I-459

Shin, Dong-Ryeol   I-348, III-99
Shin, Ho-Jin   III-99
Shin, Ho-Jun   III-806
Shin, Hyo Young   II-741
Shin, Hyoun Gyu   IV-86
Shin, Hyun-Ho   II-157
Shin, In-Hye   I-478
Shin, Kitae   III-1219
Shin, Myong-Chul   II-332, II-341, II-499, IV-1319
Shin, Seong-Yoon   II-195, II-284
Shin, Yeong Gil   IV-1111
Shin, Yongtae   II-442
Shu, Jiwu   IV-762
Shumilina, Anna   I-1075
Sicker, Douglas C.   IV-528
Siddiqi, Mohammad Umar   III-661
Sierra, José M.   IV-1301, IV-1309
Sim, Jeong Seop   II-1284
Sim, Terence   III-1197
Simeonidis, Minas   III-569
Sinclair, Brett   III-49
Singh, Sanjeet   IV-1190
Sivakumar, K.C.   IV-1341
Skworcow, Piotr   IV-692
Smith, Kate A.   II-956
So, Yeon-hee   IV-62
Soares, João L.C.   IV-488
Soh, Ben   III-672
Soh, Jin   II-754
Sohn, Bangyong   II-442
Sohn, Chae-Bong   I-166
Sohn, Hong-Gyoo   II-771
Sohn, Sungwon   I-609
Sokolov, B.V.   IV-407
Solimannejad, Mohammad   I-1004
Son, Bongsoo   II-789, II-816, II-863
Song, Hoseong   III-1259
Song, Hui   II-1086
Song, Il-Yeol   III-402
Song, MoonBae   II-1225
Song, Teuk-Seob   I-847
Song, Yeong-Sun   II-771
Song, Young-Jae   I-928, III-886
Soriano, Miguel   II-527, II-624
Sourin, Alexei   III-983
Sourina, Olga   IV-968
Srinivas   IV-680
Steigedal, Tonje Stroemmen   III-327
Sterian, Andreea   III-585, III-643

Sterian, Andreea-Rodica   III-635
Sterian, Rodica   III-592, III-598
Strelkov, Nikolay   III-621, III-628
Su, Hua   II-1293
Suh, Young-Joo   I-368
Sun, Dong Guk   III-79
Sun, Jizhou   III-284, III-292, III-416, III-435
Sun, Weitao   IV-762
Sung, Jaechul   II-567
Sung, Ji-Yeon   I-1144
Suresh, Sundaram   IV-269
Swarna, J. Mercy   IV-1341

Ta, Duong Nguyen Binh   I-947
Tadeusiewicz, Ryszard   IV-852
Tae, Kang Soo   I-478
Tai, Allen H.   IV-342
Tamada, Yoshinori   III-349
Tan, Chew Lim   III-1197
Tan, Kenneth Chih Jeng   IV-1120
Tan, Soon-Heng   III-318
Tan, Wuzheng   I-776
Tang, Jiakui   III-173
Tang, Jie   IV-1222
Tang, Sheng   II-547
Tang, Yuchun   III-299
Taniar, David   IV-1203
Tao, Pai-Cheng   I-957
Tavadyan, Levon A.   I-1012
Techapichetvanich, Kesaraporn   III-206
Teillaud, Monique   I-683
Teng, Lirong   I-1040
Thuy, Le Thi Thu   II-1066
Tian, Haishan   III-1099
Tian, Tianhai   II-1245
Tillich, Stefan   II-665
Ting, Ching-Jung   IV-417
Tiwari, Manoj Kumar   IV-680
Toi, Yutaka   IV-1055
Toma, Alexandru   III-556, III-569
Toma, Cristian   III-556, III-592, III-598
Toma, Ghiocel   III-563, III-569, III-576, III-585, III-614
Toma, Theodora   III-556, III-569
Tomaschewski, Kai   II-988
Torres, Joaquin   III-729
Trunfio, Giuseppe A.   I-1054
Turnquist, Mark A.   IV-499
Tveit, Amund   III-327

Author Index

Ufuktepe, Ünal   III-522, III-529
Umakant, J.   IV-548
Urbina, Ruben T.   III-547

Vanhoucke, Mario   IV-378
Velardo, Fernando Rosa   II-1156
Varella, E.   I-938
Vita, Marco   II-1264
Vizcaíno, Aurora   I-1064

Wack, Maxime   IV-792
Wahala, Kristiina   I-938
Walkowiak, Krzysztof   IV-802
Wan, Zheng   I-329, II-704
Wang, Chen   II-1086
Wang, Chengfeng   I-748
Wang, Chuanpeng   III-225
Wang, Gi-Nam   IV-702
Wang, Guilin   III-701, III-711
Wang, Hao   III-691, IV-1028
Wang, Hei-Chia   III-309
Wang, Hui-Mei   IV-172
Wang, Jianqin   III-173
Wang, Jiening   III-284
Wang, K.J.   IV-333
Wang, Lei   IV-733
Wang, Pi-Chung   IV-1007
Wang, Ruili   II-1187
Wang, Shaoyu   IV-1140
Wang, Shu   IV-1
Wang, S.M.   IV-333
Wang, Weinong   II-806
Wang, Xiaolin   II-822
Wang, Xinmei   IV-1046
Wang, Xiuhui   III-215
Wang, Yanguang   III-173
Wang, Yongtian   III-266
Weber, Irene   III-299
Wee, H.M.   IV-333
Wee, Hyun-Wook   III-938, IV-333
Wei, Sun   IV-753
Weng, Dongdong   III-266
Wenjun, Wang   I-301
Wille, Volker   IV-958
Wirt, Kai   II-577
Won, Chung In   I-707
Won, Dongho   I-498, I-609, II-92, II-1146
Won, Hyung Jun   III-1259
Won, Jae-Kang   III-817

Wong, Duncan S.   II-614
Woo, Gyun   I-29
Woo, Seon-Mi   III-954
Woo, Sinam   I-261
Wu, C.G.   I-1040
Wu, Chaolin   III-173
Wu, Enhua   III-1167
Wu, Hulin   IV-519
Wu, Yong   III-1099
Wu, Yue   IV-250
Wu, Zhiping   II-595

Xia, Yu   IV-290
Xiaolin, Wang   I-301
Xinpeng, Lin   I-301
Xiong, Guomin   II-822
Xirouchakis, Paul   IV-538, IV-722
Xu, Bing   II-946
Xu, Dan   III-274
Xu, Guilin   IV-30
Xu, Jie   I-758
Xu, Qing   III-292
Xu, Shuhong   I-769, IV-906
Xu, Xiaohua   III-338
Xu, Zhuoqun   II-822
Xue, Yong   III-173

Yamaguchi, Rui   III-381
Yamamoto, Osami   I-786
Yamashita, Satoru   III-381
Yan, Chung-Ren   III-1043
Yan, Dayuan   III-266
Yan, Hong   III-357
Yang, Byounghak   IV-241
Yang, Chao-Tung   IV-1017
Yang, Ching-Nung   I-19
Yang, Dong Jin   II-647
Yang, Hae-Sool   II-1, II-52, III-739, III-827, III-938
Yang, Hongwei   II-946
Yang, Jie   III-416
Yang, KwonWoo   III-89
Yang, Tao   III-266
Yang, X.S.   III-1109
Yang, Xin   I-896, IV-1102
Yang, Yoo-Kil   III-1129
Yantır, Ahmet   III-529
Yao, Xin   II-1217
Yates, Paul   I-938
Yazici, Ali   III-463

Ye, Dingfeng   II-595
Ye, Lu   III-190, IV-30
Yeh, Chung-Hsing   II-956
Yen, Show-Jane   II-1055
Yi, Yong-Hoon   II-82
Yim, Wha Young   I-1213
Yin, Jianfei   III-691, IV-1028
Yin, Ming   II-1177
Yingwei, Luo   I-301
Yiu, S.M.   III-651
Yoo, Cheol-Jung   III-758, III-836, III-878, III-945
Yoo, Chun-Sik   II-31, III-954
Yoo, Hun-Woo   IV-458, IV-743
Yoo, Hyeong Seon   II-102, II-111
Yoo, Jin Ah   II-889
Yoo, Seung Hwan   I-252
Yoo, Seung-Jae   II-870
Yoo, Sun K.   II-1028
Yoon, Chang-Dae   II-332, II-373, II-429
Yoon, Mi-sun   IV-62
Yoon, Yeo Bong   III-19
Yoshida, Ryo   III-389
You, L.H.   III-197
You, Jinyuan   III-10
You, Peng-Sheng   IV-368
Youn, Chan-Hyun   I-320
Youn, Hee Yong   I-519, II-936, III-19, IV-916, IV-1149
Youn, Hyunsang   I-291
Youn, Ju-In   II-10
Younghwan, Lim   II-676
Youngsong, Mun   II-676
Yu, Eun Jung   III-1306
Yu, HeonChang   III-89, IV-936
Yu, Jiangying   III-225
Yu, Sang-Jun   I-166
Yuan, Qingshu   I-865, I-985
Yun, HY.   II-1028
Yun, Sung-Hyun   II-391, II-401, IV-62
Yun, Won Young   IV-558
Yunhe, Pan   I-875, I-974
Yusupov, R.M.   IV-407

Zantidis, Dimitri   III-672
Zaychik, E.M.   IV-407
Zhai, Jia   IV-702
Zhai, Qi   III-284, III-435
Zhang, Changshui   II-1217
Zhang, Fuyan   III-245
Zhang, Jian J.   III-197, III-1003, III-1109
Zhang, Jianzhong   IV-161
Zhang, Jiawan   III-292, III-416, III-435
Zhang, Jin-Ting   IV-519
Zhang, Jun   III-691, IV-1028
Zhang, Kuo   IV-1222
Zhang, Mingmin   II-946, III-190, III-245
Zhang, Mingming   IV-1, IV-30
Zhang, Qiaoping   III-181
Zhang, Qiong   I-967
Zhang, Shen   II-686
Zhang, Ya-Ping   III-274
Zhang, Yan-Qing   III-299
Zhang, Yi   III-435
Zhang, Yong-Dong   II-547
Zhang, Yu   III-1197
Zhang, Yang   I-214
Zhang, Yuanliang   II-1207
Zhao, Jane   II-1235
Zhao, Qinping   III-235
Zhao, Weizhong   IV-1159
Zhao, Yang   III-274
Zhao, Yiming   IV-1
Zheng, Jin Jin   III-1003
Zheng, Weimin   IV-762
Zheng, Zengwei   IV-826
Zhong, Shaobo   III-173
Zhou, Hanbin   IV-896
Zhou, Hong Jun   III-1003
Zhou, Jianying   III-681, III-701
Zhou, Qiang   IV-896
Zhou, Suiping   I-947
Zhou, Xiaohua   III-402
Zhu, Jiejie   IV-30
Zhu, Ming   IV-1102
Zhuoqun, Xu   I-301
Żyliński, Paweł   I-647

Lecture Notes in Computer Science

For information about Vols. 1–3392
please contact your bookseller or Springer

Vol. 3525: A.E. Abdallah, C.B. Jones, J.W. Sanders (Eds.), Communicating Sequential Processes. XIV, 321 pages. 2005.

Vol. 3517: H.S. Baird, D.P. Lopresti (Eds.), Human Interactive Proofs. IX, 143 pages. 2005.

Vol. 3510: T. Braun, G. Carle, Y. Koucheryavy, V. Tsaoussidis (Eds.), Wired/Wireless Internet Communications. XIV, 366 pages. 2005.

Vol. 3508: P. Bresciani, P. Giorgini, B. Henderson-Sellers, G. Low, M. Winikoff (Eds.), Agent-Oriented Information Systems II. X, 227 pages. 2005. (Subseries LNAI).

Vol. 3503: S.E. Nikoletseas (Ed.), Experimental and Efficient Algorithms. XV, 624 pages. 2005.

Vol. 3501: B. Kégl, G. Lapalme (Eds.), Advances in Artificial Intelligence. XV, 458 pages. 2005. (Subseries LNAI).

Vol. 3500: S. Miyano, J. Mesirov, S. Kasif, S. Istrail, P. Pevzner, M. Waterman (Eds.), Research in Computational Molecular Biology. XVII, 632 pages. 2005. (Subseries LNBI).

Vol. 3498: J. Wang, X. Liao, Z. Yi (Eds.), Advances in Neural Networks – ISNN 2005, Part III. L, 1077 pages. 2005.

Vol. 3497: J. Wang, X. Liao, Z. Yi (Eds.), Advances in Neural Networks – ISNN 2005, Part II. L, 947 pages. 2005.

Vol. 3496: J. Wang, X. Liao, Z. Yi (Eds.), Advances in Neural Networks – ISNN 2005, Part II. L, 1055 pages. 2005.

Vol. 3495: P. Kantor, G. Muresan, F. Roberts, D.D. Zeng, F.-Y. Wang, H. Chen, R.C. Merkle (Eds.), Intelligence and Security Informatics. XVIII, 674 pages. 2005.

Vol. 3494: R. Cramer (Ed.), Advances in Cryptology – EUROCRYPT 2005. XIV, 576 pages. 2005.

Vol. 3492: P. Blache, E. Stabler, J. Busquets, R. Moot (Eds.), Logical Aspects of Computational Linguistics. X, 363 pages. 2005. (Subseries LNAI).

Vol. 3489: G.T. Heineman, J.A. Stafford, H.W. Schmidt, K. Wallnau, C. Szyperski, I. Crnkovic (Eds.), Component-Based Software Engineering. XI, 358 pages. 2005.

Vol. 3488: M.-S. Hacid, N.V. Murray, Z.W. Raś, S. Tsumoto (Eds.), Foundations of Intelligent Systems. XIII, 700 pages. 2005. (Subseries LNAI).

Vol. 3483: O. Gervasi, M.L. Gavrilova, V. Kumar, A. Laganà, H.P. Lee, Y. Mun, D. Taniar, C.J.K. Tan (Eds.), Computational Science and Its Applications – ICCSA 2005, Part IV. LXV, 1362 pages. 2005.

Vol. 3482: O. Gervasi, M.L. Gavrilova, V. Kumar, A. Laganà, H.P. Lee, Y. Mun, D. Taniar, C.J.K. Tan (Eds.), Computational Science and Its Applications – ICCSA 2005, Part III. LXV, 1340 pages. 2005.

Vol. 3481: O. Gervasi, M.L. Gavrilova, V. Kumar, A. Laganà, H.P. Lee, Y. Mun, D. Taniar, C.J.K. Tan (Eds.), Computational Science and Its Applications – ICCSA 2005, Part II. LV, 1316 pages. 2005.

Vol. 3480: O. Gervasi, M.L. Gavrilova, V. Kumar, A. Laganà, H.P. Lee, Y. Mun, D. Taniar, C.J.K. Tan (Eds.), Computational Science and Its Applications – ICCSA 2005, Part I. LXV, 1234 pages. 2005.

Vol. 3479: T. Strang, C. Linnhoff-Popien (Eds.), Location- and Context-Awareness. XII, 378 pages. 2005.

Vol. 3477: P. Herrmann, V. Issarny (Eds.), Trust Management. XII, 426 pages. 2005.

Vol. 3475: N. Guelfi (Ed.), Rapid Integration of Software Engineering Techniques. X, 145 pages. 2005.

Vol. 3468: H.W. Gellersen, R. Want, A. Schmidt (Eds.), Pervasive Computing. XIII, 347 pages. 2005.

Vol. 3467: J. Giesl (Ed.), Term Rewriting and Applications. XIII, 517 pages. 2005.

Vol. 3465: M. Bernardo, A. Bogliolo (Eds.), Formal Methods for Mobile Computing. VII, 271 pages. 2005.

Vol. 3463: M. Dal Cin, M. Kaâniche, A. Pataricza (Eds.), Dependable Computing - EDCC 2005. XVI, 472 pages. 2005.

Vol. 3462: R. Boutaba, K. Almeroth, R. Puigjaner, S. Shen, J.P. Black (Eds.), NETWORKING 2005. XXX, 1483 pages. 2005.

Vol. 3461: P. Urzyczyn (Ed.), Typed Lambda Calculi and Applications. XI, 433 pages. 2005.

Vol. 3460: Ö. Babaoglu, M. Jelasity, A. Montresor, C. Fetzer, S. Leonardi, A. van Moorsel, M. van Steen (Eds.), Self-star Properties in Complex Information Systems. IX, 447 pages. 2005.

Vol. 3459: R. Kimmel, N.A. Sochen, J. Weickert (Eds.), Scale Space and PDE Methods in Computer Vision. XI, 634 pages. 2005.

Vol. 3456: H. Rust, Operational Semantics for Timed Systems. XII, 223 pages. 2005.

Vol. 3455: H. Treharne, S. King, M. Henson, S. Schneider (Eds.), ZB 2005: Formal Specification and Development in Z and B. XV, 493 pages. 2005.

Vol. 3454: J.-M. Jacquet, G.P. Picco (Eds.), Coordination Models and Languages. X, 299 pages. 2005.

Vol. 3453: L. Zhou, B.C. Ooi, X. Meng (Eds.), Database Systems for Advanced Applications. XXVII, 929 pages. 2005.

Vol. 3452: F. Baader, A. Voronkov (Eds.), Logic for Programming, Artificial Intelligence, and Reasoning. XI, 562 pages. 2005. (Subseries LNAI).

Vol. 3450: D. Hutter, M. Ullmann (Eds.), Security in Pervasive Computing. XI, 239 pages. 2005.

Vol. 3449: F. Rothlauf, J. Branke, S. Cagnoni, D.W. Corne, R. Drechsler, Y. Jin, P. Machado, E. Marchiori, J. Romero, G.D. Smith, G. Squillero (Eds.), Applications of Evolutionary Computing. XX, 631 pages. 2005.

Vol. 3448: G.R. Raidl, J. Gottlieb (Eds.), Evolutionary Computation in Combinatorial Optimization. XI, 271 pages. 2005.

Vol. 3447: M. Keijzer, A. Tettamanzi, P. Collet, J.v. Hemert, M. Tomassini (Eds.), Genetic Programming. XIII, 382 pages. 2005.

Vol. 3444: M. Sagiv (Ed.), Programming Languages and Systems. XIII, 439 pages. 2005.

Vol. 3443: R. Bodik (Ed.), Compiler Construction. XI, 305 pages. 2005.

Vol. 3442: M. Cerioli (Ed.), Fundamental Approaches to Software Engineering. XIII, 373 pages. 2005.

Vol. 3441: V. Sassone (Ed.), Foundations of Software Science and Computational Structures. XVIII, 521 pages. 2005.

Vol. 3440: N. Halbwachs, L.D. Zuck (Eds.), Tools and Algorithms for the Construction and Analysis of Systems. XVII, 588 pages. 2005.

Vol. 3439: R.H. Deng, F. Bao, H. Pang, J. Zhou (Eds.), Information Security Practice and Experience. XII, 424 pages. 2005.

Vol. 3437: T. Gschwind, C. Mascolo (Eds.), Software Engineering and Middleware. X, 245 pages. 2005.

Vol. 3436: B. Bouyssounouse, J. Sifakis (Eds.), Embedded Systems Design. XV, 492 pages. 2005.

Vol. 3434: L. Brun, M. Vento (Eds.), Graph-Based Representations in Pattern Recognition. XII, 384 pages. 2005.

Vol. 3433: S. Bhalla (Ed.), Databases in Networked Information Systems. VII, 319 pages. 2005.

Vol. 3432: M. Beigl, P. Lukowicz (Eds.), Systems Aspects in Organic and Pervasive Computing - ARCS 2005. X, 265 pages. 2005.

Vol. 3431: C. Dovrolis (Ed.), Passive and Active Network Measurement. XII, 374 pages. 2005.

Vol. 3429: E. Andres, G. Damiand, P. Lienhardt (Eds.), Discrete Geometry for Computer Imagery. X, 428 pages. 2005.

Vol. 3427: G. Kotsis, O. Spaniol (Eds.), Wireless Systems and Mobility in Next Generation Internet. VIII, 249 pages. 2005.

Vol. 3423: J.L. Fiadeiro, P.D. Mosses, F. Orejas (Eds.), Recent Trends in Algebraic Development Techniques. VIII, 271 pages. 2005.

Vol. 3422: R.T. Mittermeir (Ed.), From Computer Literacy to Informatics Fundamentals. X, 203 pages. 2005.

Vol. 3421: P. Lorenz, P. Dini (Eds.), Networking - ICN 2005, Part II. XXXV, 1153 pages. 2005.

Vol. 3420: P. Lorenz, P. Dini (Eds.), Networking - ICN 2005, Part I. XXXV, 933 pages. 2005.

Vol. 3419: B. Faltings, A. Petcu, F. Fages, F. Rossi (Eds.), Constraint Satisfaction and Constraint Logic Programming. X, 217 pages. 2005. (Subseries LNAI).

Vol. 3418: U. Brandes, T. Erlebach (Eds.), Network Analysis. XII, 471 pages. 2005.

Vol. 3416: M. Böhlen, J. Gamper, W. Polasek, M.A. Wimmer (Eds.), E-Government: Towards Electronic Democracy. XIII, 311 pages. 2005. (Subseries LNAI).

Vol. 3415: P. Davidsson, B. Logan, K. Takadama (Eds.), Multi-Agent and Multi-Agent-Based Simulation. X, 265 pages. 2005. (Subseries LNAI).

Vol. 3414: M. Morari, L. Thiele (Eds.), Hybrid Systems: Computation and Control. XII, 684 pages. 2005.

Vol. 3412: X. Franch, D. Port (Eds.), COTS-Based Software Systems. XVI, 312 pages. 2005.

Vol. 3411: S.H. Myaeng, M. Zhou, K.-F. Wong, H.-J. Zhang (Eds.), Information Retrieval Technology. XIII, 337 pages. 2005.

Vol. 3410: C.A. Coello Coello, A. Hernández Aguirre, E. Zitzler (Eds.), Evolutionary Multi-Criterion Optimization. XVI, 912 pages. 2005.

Vol. 3409: N. Guelfi, G. Reggio, A. Romanovsky (Eds.), Scientific Engineering of Distributed Java Applications. X, 127 pages. 2005.

Vol. 3408: D.E. Losada, J.M. Fernández-Luna (Eds.), Advances in Information Retrieval. XVII, 572 pages. 2005.

Vol. 3407: Z. Liu, K. Araki (Eds.), Theoretical Aspects of Computing - ICTAC 2004. XIV, 562 pages. 2005.

Vol. 3406: A. Gelbukh (Ed.), Computational Linguistics and Intelligent Text Processing. XVII, 829 pages. 2005.

Vol. 3404: V. Diekert, B. Durand (Eds.), STACS 2005. XVI, 706 pages. 2005.

Vol. 3403: B. Ganter, R. Godin (Eds.), Formal Concept Analysis. XI, 419 pages. 2005. (Subseries LNAI).

Vol. 3402: M. Daydé, J.J. Dongarra, V. Hernández, J.M.L.M. Palma (Eds.), High Performance Computing for Computational Science - VECPAR 2004. XI, 732 pages. 2005.

Vol. 3401: Z. Li, L.G. Vulkov, J. Waśniewski (Eds.), Numerical Analysis and Its Applications. XIII, 630 pages. 2005.

Vol. 3400: J.F. Peters, A. Skowron (Eds.), Transactions on Rough Sets III. IX, 461 pages. 2005.

Vol. 3399: Y. Zhang, K. Tanaka, J.X. Yu, S. Wang, M. Li (Eds.), Web Technologies Research and Development - APWeb 2005. XXII, 1082 pages. 2005.

Vol. 3398: D.-K. Baik (Ed.), Systems Modeling and Simulation: Theory and Applications. XIV, 733 pages. 2005. (Subseries LNAI).

Vol. 3397: T.G. Kim (Ed.), Artificial Intelligence and Simulation. XV, 711 pages. 2005. (Subseries LNAI).

Vol. 3396: R.M. van Eijk, M.-P. Huget, F. Dignum (Eds.), Agent Communication. X, 261 pages. 2005. (Subseries LNAI).

Vol. 3395: J. Grabowski, B. Nielsen (Eds.), Formal Approaches to Software Testing. X, 225 pages. 2005.

Vol. 3394: D. Kudenko, D. Kazakov, E. Alonso (Eds.), Adaptive Agents and Multi-Agent Systems II. VIII, 313 pages. 2005. (Subseries LNAI).

Vol. 3393: H.-J. Kreowski, U. Montanari, F. Orejas, G. Rozenberg, G. Taentzer (Eds.), Formal Methods in Software and Systems Modeling. XXVII, 413 pages. 2005.

Lecture Notes in Computer Science 3483

Commenced Publication in 1973
Founding and Former Series Editors:
Gerhard Goos, Juris Hartmanis, and Jan van Leeuwen

Editorial Board

David Hutchison
  Lancaster University, UK
Takeo Kanade
  Carnegie Mellon University, Pittsburgh, PA, USA
Josef Kittler
  University of Surrey, Guildford, UK
Jon M. Kleinberg
  Cornell University, Ithaca, NY, USA
Friedemann Mattern
  ETH Zurich, Switzerland
John C. Mitchell
  Stanford University, CA, USA
Moni Naor
  Weizmann Institute of Science, Rehovot, Israel
Oscar Nierstrasz
  University of Bern, Switzerland
C. Pandu Rangan
  Indian Institute of Technology, Madras, India
Bernhard Steffen
  University of Dortmund, Germany
Madhu Sudan
  Massachusetts Institute of Technology, MA, USA
Demetri Terzopoulos
  New York University, NY, USA
Doug Tygar
  University of California, Berkeley, CA, USA
Moshe Y. Vardi
  Rice University, Houston, TX, USA
Gerhard Weikum
  Max-Planck Institute of Computer Science, Saarbruecken, Germany